PRINCIPLES OF ANATOMY AND PHYSIOLOGY

NINTH EDITION

Gerard J. Tortora
Bergen Community College

Sandra Reynolds Grabowski
Purdue University

JOHN WILEY & SONS, INC.

New York • Chichester • Weinheim • Brisbane • Singapore • Toronto

Managing Editor: Wendy Earl
Production Supervisor: Sharon Montooth
Text and Cover Designer: Kathleen Cunningham
Layout Artist: Andrew Ogus
Copyeditor: Alan Titche
Art Coordinator: Claudia Durrell
Photo Editor: Mira Schachne
Indexer: Katherine Pitcoff

Illustration and photo credits follow the Glossary.

About The Cover
The cover was created by Kevin Somerville, a well-known contemporary medical illustrator who also has contributed many new drawings to the seventh, eighth, and ninth editions. Somerville was inspired by Leonardo DaVinci's so-called Great Lady Anatomy, drawn about 1508. Though DaVinci's drawing was beautiful, we now know that it was anatomically incorrect. By melding an aged appearance with contemporary science, Somerville reminds us of a long, highly valued tradition in medical illustration: the commitment to both beauty and accuracy in the study of human anatomy and physiology.

This book was typeset by GTS Graphics and printed and bound by Von Hoffmann Press, Inc. The cover was printed by Von Hoffman Press, Inc.

The paper in this book was manufactured by a mill whose forest management programs include sustained yield harvesting of its timberlands. Sustained yield harvesting principles ensure that the number of trees cut each year does not exceed the amount of new growth.

This book is printed on acid-free paper. ⊚

Library of Congress Cataloging-in-Publication Data
Tortora, Gerard J.
    Principles of anatomy and physiology / Gerard J. Tortora, Sandra
Reynolds Grabowski. – – 9th ed.
        p.   cm.
    ISBN 0-471-36692-7
    1. Human physiology.   2. Human anatomy.   I. Grabowski, Sandra
Reynolds.   II. Title.
    QP34.5.T67   2000   99-31201
    612--dc21   CIP

Printed in the United States of America.
10 9 8 7 6 5 4 3 2

ABOUT THE AUTHORS

Gerard J. Tortora is Professor of Biology at Bergen Community College in Paramus, New Jersey, where he teaches human anatomy and physiology as well as microbiology. The author of several best-selling science textbooks and laboratory manuals, Jerry is devoted first and foremost to his students and their aspirations. In recognition of this commitment, he was named Distinguished Faculty Scholar at Bergen Community College. In 1996 Jerry received a National Institute for Staff and Organizational Development (NISOD) excellence award from the University of Texas, and he was selected to represent Bergen Community College in a campaign to increase awareness of the contributions of community colleges to higher education.

Jerry received his bachelor's degree in biology from Fairleigh Dickinson University and his master's degree in science education from Montclair State College. He is a member of many professional organizations, such as the Human Anatomy and Physiology Society (HAPS), the American Society of Microbiology (ASM), the American Association for the Advancement of Science (AAAS), the National Education Association (NEA), and the Metropolitan Association of College and University Biologists (MACUB).

To my wife, Melanie, and to my children, Christopher, Anthony, and Andrew, who make it all worthwhile. — G.J.T.

Sandra Reynolds Grabowski is an instructor in the Department of Biological Sciences at Purdue University in West Lafayette, Indiana. For more than 20 years she has taught human anatomy and physiology to students in a wide range of academic programs. In 1992 students selected her as one of the top ten teachers in the School of Science at Purdue.

Sandy received her BS in biology and her PhD in neurophysiology from Purdue. She is an active member of the Human Anatomy and Physiology Society (HAPS), served as editor of *HAPS News* from 1990 through 1992, and was elected to serve a 3-year term as President Elect, President, and Past President from 1992 to 1995. In addition, she is a member of the American Association for the Advancement of Science (AAAS), the Association for Women in Science (AWIS), the National Science Teachers Association (NSTA), and the Society for College Science Teachers (SCST).

To my students, who continually inspire improvements in this work by the suggestions they make and the questions they ask. — S.R.G.

PREFACE

The study of human anatomy and physiology is an inquiry into the fascinating complexity of one's body. An anatomy and physiology course is a gateway to a rewarding career in a host of health-related fields, Massotherapy, and athletic training, and it provides a foundation for advanced scientific studies. Remaining true to the heritage of its predecessors, we wrote the ninth edition of *Principles of Anatomy and Physiology* to meet the exacting requirements of introductory anatomy and physiology courses. As educators, we offer this edition fully confident that it will advance the personal and professional aspirations of all its readers.

CORE VALUES

Over the years, our students have taught us to value simplicity, directness, and the power of clear illustrations. The noteworthy success of previous editions of this book, and its transformation into a standard-bearer in education, is a testament to the fact that many of our colleagues value these qualities too. Chapter after chapter, we built our core teaching values into *Principles of Anatomy and Physiology* by providing:

- clear, compelling, and up-to-date discussions of anatomy and physiology
- expertly executed and generously sized art
- classroom-tested pedagogy
- outstanding student study support.

Although this book offers a host of useful features that will appeal to a variety of learning styles, experience has taught us that features alone will not help students learn unless every discussion is easy to understand. By any measure, we trust that students will find the ninth edition of *Principles of Anatomy and Physiology* to be easier than ever to read.

CORE PRINCIPLES

Human anatomy and physiology is a formidable body of knowledge to present in an introductory course, and mastering the subject can require a great deal of memory work on a student's part. We have woven several interrelated strands of information throughout the ninth edition to fulfill the following three aims: helping readers develop a working knowledge of the subject, enabling them to use anatomical and physiological terminology confidently, and highlighting the practical application of anatomical and physiological concepts to students' career choices or in their everyday lives.

Introducing Homeostasis The dynamic physiological constancy known as homeostasis is the cardinal theme in *Principles of Anatomy and Physiology*. We immediately introduce this unifying concept in Chapter 1 and describe how various feedback mechanisms work to maintain physiological processes within the narrow range that is compatible with life. Homeostatic mechanisms are discussed throughout the book, and homeostatic processes are clarified and reinforced through our well-received series of homeostasis feedback illustrations. Moreover, we believe students can better understand normal physiological processes by examining situations in which those processes are undermined by diseases or disorders. The "Disorders: Homeostatic Imbalances" sections at the end of most chapters include concise discussions of major diseases and disorders that illustrate departures from normal homeostasis.

Laying a Foundation of Basic Science From the outset, students need to understand how individual structures relate to the composition of the entire body. For this reason, we introduce anatomical nomenclature such as regional names, directional terms, and the planes and sections that enable students to precisely describe the relationship of one body structure to another. We believe, too, that students learn more readily when they fully appreciate the chemical and cellular basis of anatomy and physiology. Accordingly, we extensively clarified and updated the chemistry and cellular biology coverage in the ninth edition to reflect current thinking on these subjects. Beautiful new three-dimensional illustrations of cellular structures and chemical processes complement this coverage.

Correlating Structure and Function Years of student feedback have convinced us that readers learn anatomy and physiology more readily when they remain mindful of the relationship between structure and function. The challenge, of course, is to correlate structure and function without overwhelming readers with extraneous details about either. As a writing team—an anatomist and a physiologist—our very different specializations offer practical advantages in fine-tuning the balance between anatomy and physiology. Throughout three editions, we have collaborated to clarify and refine discussions of the relationship between structure and function in each chapter.

Building a Professional Vocabulary Students—even the best ones—generally find it difficult at first to read and pronounce anatomical and physiological terms. Moreover, as educators we are sympathetic to the needs of the growing ranks of college students who speak English as a second language. For these reasons, we have striven since the book's inception to ensure that it has a strong and helpful vocabulary component.

ORGANIZATION

The book follows the same unit and topic sequence as its eight earlier editions. It is divided into five principal sections: Unit 1, "Organization of the Human Body," provides an understanding of the structural and functional levels of the body, from molecules to organ systems. Unit 2, "Principles of Support and Movement," analyzes the anatomy and physiology of the skeletal system, articulations, and the muscular system. Unit 3, "Control Systems of the Human Body," emphasizes the importance of neural communication in the immediate maintenance of homeostasis, the role of sensory receptors in providing information about the internal and external environment, and the significance of hormones in maintaining long-term homeostasis. Unit 4, "Maintenance of the Human Body," explains how body systems function to maintain homeostasis on a moment-to-moment basis through the processes of circulation, respiration, digestion, cellular metabolism, urinary functions, and buffer systems. Unit 5, "Continuity," covers the anatomy and physiology of the reproductive systems, development, and the basic concepts of genetics and inheritance.

SPECIAL TOPICS

Developmental Anatomy We often tell our students that they can better appreciate the "logic" of human anatomy by becoming aware of how various structures developed in the first place. As in previous editions, illustrated discussions of developmental anatomy are found near the conclusion of most body system chapters. Placing this coverage at the end of chapters enables students to master the anatomical termi-

nology they need before attempting to learn about embryonic and fetal structures. The fetus icon designates the start of each discussion.

Aging Students need to be reminded from time to time that anatomy and physiology is not static. As the body ages, its structure and related functions subtly change. Moreover, aging is a professionally relevant topic for the majority of this book's readers, who will go on to careers in health-related fields in which the average age of the client population is steadily advancing. For these reasons, many of the body system chapters explore age-related changes in anatomy and physiology at the end of most chapters.

Exercise Physical exercise can produce favorable changes in some anatomical structures and enhance many physiological functions, most notably those associated with the muscular, skeletal and cardiovascular systems. This information is especially relevant to readers embarking on careers in physical education, sports training, and dance. Hence, key chapters include brief discussions of exercise, which are signaled by a distinctive "running shoe" icon.

IMPROVED COVERAGE

Every chapter in the ninth edition of *Principles of Anatomy and Physiology* incorporates a host of improvements to both the text and the art, many suggested by reviewers, educators, and students. In addition, most chapters offer new Clinical Applications. Complete discussions of the improvements we made to each chapter are available in the *Professor's Resource Manual* to the ninth edition. Here are just some of the more noteworthy changes:

Chapter 3: The Cellular Level of Organization This foundational chapter was updated from front to back to reflect the latest thinking in cell biology. Students will benefit in particular from improved coverage of the plasma membrane and membrane transport mechanisms. Virtually every figure has been redrawn in a new, vibrant, three-dimensional style.

Chapter 4: Tissues The much-imitated tissue tables in this chapter were redesigned to improve student comprehension of histology. Now, every photomicrograph is accompanied by an interpretive illustration. We have added an orientation diagram next to each photomicrograph that indicates a location where a particular tissue type occurs in the body.

Chapter 9: Joints We clarified and strengthened this chapter by reorganizing the coverage according to the structural properties of joints and by offering more comprehensive coverage of movements at synovial joints. Students will benefit from new illustrated Exhibits that focus on the components and mechanics of four major joints: shoulder, elbow, hip, and knee. The improved coverage is further accentuated by clear and consistently stylized new range-of-motion photographs.

Chapter 10: Muscle Tissue   We reorganized and rewrote this chapter to provide clearer, simpler, and up-to-date explanations of the mechanics and physiology of muscle tissue. New, generously sized, three-dimensional illustrations of muscle tissue and its components complement our enhancements to this chapter's coverage.

Chapter 11: The Muscular System   The hallmark muscle Exhibits now feature a narrative tour of each muscle group. Clinical Applications are easier to spot, and a new feature, Relating Muscles to Movements, encourages readers to group related muscles according to their common actions. Innervation information now is provided in the discussion, a change that makes each Exhibit's table simpler and easier to study.

Chapter 16: The Special Senses   New three-dimensional drawings of major sensory organs by noted science illustrator Tomo Narashima superbly enhance the content and utility of this chapter. The chapter also offers enhanced coverage of extrinsic eye muscles and refraction abnormalities.

Chapter 21: Blood Vessels   In this chapter's Exhibits we provide a more consistent level of detail when discussing the arterial and venous circulation, especially in the limbs. Meticulously revised illustrations more clearly reveal the relevant blood vessel anatomy, and extended flow diagrams visually reinforce learning of relevant anatomy and blood vessel pathways.

HALLMARK FEATURES OF THE NINTH EDITION

The ninth edition of *Principles of Anatomy and Physiology* builds on the legacy of thoughtful and clearly designed features that distinguished its predecessors. Some regular features are totally redesigned for this edition. For example, the tables are easier to read and use and are now labeled as **Tables.** The term "Exhibit" is now reserved for a new, self-contained feature (described next). Some features are completely new. Here are the most noteworthy changes and additions:

More Helpful Exhibits   Students of anatomy and physiology need extra help learning the numerous structures that constitute certain body systems—most notably skeletal muscles, articulations, blood vessels, and nerves. As in previous editions, the chapters that present these topics are organized around **Exhibits,** each of which consists of a newly

expanded overview, a tabular summary of the relevant anatomy, and an associated suite of illustrations or photographs. Each Exhibit is prefaced by an Objective and closes with a Review Activity. We trust you will agree that our newly revised Exhibits, along with their spacious new design, make them ideal study vehicles for learning anatomically complex body systems.

Updated Clinical Focus   A perennial favorite among students, the numerous **Clinical Applications** in every chapter explore the clinical, professional, or everyday relevance of a particular anatomical structure or its related function. Some applications are new to this edition, and all have been reviewed for accuracy and relevance by a consulting panel of nurses. Each application directly follows the discussion to which it relates. Additionally, the **Homeostatic Imbalances** discussions found at the conclusion of most chapters have been updated and simplified.

New or Revised Study Tools   In response to students' and instructors' requests, we've added focused **Student Objectives** to the beginning of major sections of reading throughout each chapter. Complementing this change, new **Review Activities** appear at strategic intervals within chapters to give students the chance to validate their understanding as they read. As always, readers will benefit from the popular end-of-chapter **Study Outline** that is page referenced to the chapter discussions.

New study tools include the end-of-chapter **Self-Quiz Questions** written in a variety of styles that are calculated to appeal to readers' different testing preferences. Another new feature, **Critical Thinking Questions,** asks readers to apply concepts to real-life situations. The style of these questions ought to make students smile on occasion as well as think! Answers to the Self-Quiz and Critical Thinking Questions are located in Appendix D. Finally, the new **Student Companion CD-ROM** located inside the back cover provides a host of new interactive learning activities and testing opportunities.

Study Tools for Mastering Vocabulary   Every chapter **highlights key terms in boldface** type. We include **pronunciation guides** when major, or especially hard-to-pronounce, structures and functions are introduced in the discussions, Tables, or Exhibits. **Word roots** citing the Greek or Latin derivations of anatomical terms are offered as an additional aid. As a further service to readers, we provide a list of **Medical Terminology** at the conclusion of most chapters, and a comprehensive **Glossary** at the back of the book. The basic building blocks of medical terminology—**Combining Forms, Word Roots, Prefixes, and Suffixes**—are listed inside the back cover. Finally, the new Student Companion CD-ROM includes an audio pronunciation dictionary.

ENHANCEMENTS TO THE ILLUSTRATION PROGRAM

Teaching human anatomy and physiology is both a visual and a descriptive enterprise. Countless students have now benefited from the detailed and generously proportioned bone and muscle art that characterizes our book. Continuing in this tradition, you will find exciting new three-dimensional illustrations gracing the pages of Chapter 3, The Cellular Level of Organization, and Chapter 16, The Special Senses. Beyond this, over twenty-five percent of the art is new to this edition, and nearly every figure has been revised or improved in some way. Here are some of the new, as well as the tried-and-true, features of our illustration program:

New Histology-Based Art   As part of our plan of continuous improvement, many of the anatomical illustrations based on histological preparations have been replaced in this edition. See, for example, the beautifully rendered illustration of the layers of the gastrointestinal tract on page 820 or the new illustration of the interior of the eye on page 519.

Revised Feedback Loop Illustrations   As in editions past, this popular series of illustrations (see page 8) captures and clarifies the body's dynamic counterbalancing act in maintaining homeostasis. We subtly revised the feedback loops in the ninth edition to visually accentuate the roles that receptors, control centers, and effectors play in modifying a controlled physiological condition.

Helpful Orientation Diagrams   Students sometimes need help figuring out the plane of view of anatomy illustrations —descriptions alone do not always suffice. Every major anatomy illustration is accompanied by an orientation diagram that depicts and explains the perspective of the view represented in the figure.

Key Concept Statements   This art-related feature summarizes an idea that is discussed in the text and demonstrated in a figure. Each Key Concept Statement is positioned adjacent to its figure and is denoted by a distinctive "key" icon.

Revised Figure Questions   This highly applauded feature asks readers to synthesize verbal and visual information, think critically, or draw conclusions about what they see in a figure. Each Figure Question appears adjacent to its illustration and is highlighted in this edition by the distinctive "Q" icon. Answers are located at the back of each chapter.

Functions Banners   This feature succinctly lists the functions of an anatomical structure or body system depicted within a figure (see page 915). This juxtaposition of text and art further reinforces the connection between structure and function.

Cadaver Photos   As before, we provide an assortment of large, clear cadaver photos at strategic points in many chapters. What is more, many anatomy illustrations are keyed to the large cadaver photos available in *A Photographic Atlas of the Human Body*, by Gerard J. Tortora. This new atlas is available through John Wiley & Sons, Inc.

COMPLETE TEACHING AND LEARNING PACKAGE

Continuing the tradition of providing a complete teaching and learning package, the ninth edition of *Principles of Human Anatomy and Physiology* is available with a host of carefully planned supplementary materials that will help you and your students to attain the maximal benefit from our textbook. Complete descriptions of the supplements in the following list can be found on the inside front cover. Please contact your Wiley sales representative for additional information about any of these resources.

For Instructors:

- **Professor's Resource Manual** by Jerri K. Lindsey (0–471–37466–0)
- **Electronic Lecture Guide** (0–471–37461–X)
- **Printed Test Bank** by Pamela Langley (0–471–37469–5)
- **Microtest Computerized Test Bank** (Mac: 0–471–37464–2; Win: 0–471–37465–2)
- **Instructor's Presentation CD-ROM** (0–471–37459–8)
- **Full-color Overhead Transparencies** (0–471–37463–6)

For Students:

- **Student Companion CD-ROM**—free, at back of book
- **Learning Guide** by Kathleen Schmidt Prezbindowski (0–471–37467–9)
- **Illustrated Notebook** (0–471–37468-7)
- **Tortora/Grabowski Web Site** at http://www.wiley.com\college\bio\tortora
- **Photographic Atlas of the Human Body** by Gerard Tortora (0–471–37487–3)
- **A Brief Atlas of the Human Skeleton**—accompanies the text

Like each of our students, this book has a life of its own. The structure, content, and production values of *Principles of Anatomy and Physiology* are shaped as much by its relationship with educators and readers as by the vision that gave birth to the book nine editions ago. Today you, our readers, are the "heart" of this book. We invite you to continue the tradition of sending in your suggestions to us so that we can include them in the tenth edition.

Gerard J. Tortora
Department of Science and Health, S229
Bergen Community College
400 Paramus Road
Paramus, NJ 07652

Sandra Reynolds Grabowski
Department of Biological Sciences
1392 Lilly Hall of Life Sciences
Purdue University
West Lafayette, IN 47907-1392
email: SGrabows@bilbo.bio.purdue.edu

ACKNOWLEDGMENTS

For the ninth edition of *Principles of Anatomy and Physiology,* we have enjoyed the opportunity of collaborating with a group of dedicated and talented professionals. Accordingly, we would like to recognize and thank the members of our book team.

First, thanks go out to two contributors whose efforts enrich this edition. The challenging new end-of-chapter Self-Quiz Questions were written by Jerri K. Lindsey. Joan Barber contributed the new Critical Thinking Questions. We trust that readers will appreciate Joan's spirited and gently humorous approach to encouraging critical thinking. We greatly appreciate the work of Martha DePecol Sanner, Thomas Lancraft, Temma al-Mukhtar, Izak Paul, and James White in the development of the new Student Companion CD-ROM which enhances the integrated learning package of this text. We are grateful that Pamela Langly continues to lend her expert skills in developing test questions to our team. The value of her work cannot be underestimated. And a special thank you to Kathleen Prezbindowski, author of the Learning Guide, who has worked with us for many editions. The high quality of her work ensures student success.

The assistance from the editorial and production staff has been equally valuable. Kay Ueno, Senior Project Editor, worked very closely with us to guide the project to completion. Her directed comments, editorial wisdom, and continuous support and encouragement have been very much appreciated. Mark Wales, Senior Developmental Editor, guided our efforts to maintain a focus on pedagogy that directs and reinforces student learning. His insights, careful attention to detail, and years of experience have helped maintain the level of excellence of our textbook. Nicki Richesin, Editorial Assistant, facilitated the flow of manuscript and communications between us and other team members, which proved very useful to the timely completion of this book.

Wendy Earl, Managing Editor of Wendy Earl Productions, assumed the responsibility for producing *Principles of Anatomy and Physiology.* Her experience and professionalism are her trademarks and have meant so much to the book. Sharon Montooth, Production Editor, expertly coordinated the evolution of manuscript and galleys and then pages with efficiency and accuracy.

Claudia Durrell, Art Coordinator, has worked with us for many years, and her attention to design and knowledge of our art preferences have made our job so much easier and have enhanced the graphic elements of this book in ways we could never imagine. Mira Schachne, another long-time associate, was the Photo Coordinator. If the photo exists, Mira will find it—or pose for it herself! Her tenacity and sense of humor have been her signature for all the years we have known her.

Alan Titche, Copyeditor, helped in many ways to make the manuscript consistent in presentation and style and helped us to polish our transitions. Martha Ghent, Proofreader, alerted us to any needed corrections prior to publication. Katherine Pitcoff, our Indexer, assembled the comprehensive new index that will ably serve the needs of students and instructors alike.

The illustrations and photographs have always been a signature feature of *Principles of Anatomy and Physiology.* Respected scientific and medical illustrators Tomo Narashima, Steve Oh, and Wendy Hiller Gee join our team of outstanding artists: Leonard Dank, Sharon Ellis, Jean Jackson, Lauren Keswick, Lynn O'Kelley, Hilda Muinos, Nadine Sokol, Kevin Somerville, and Beth Willert. Mark Nielsen of the University of Utah provided many of the cadaver photos that appear in this edition. Jared Schneidman Design and Imagineering deserve special thanks for their computer graphics finesse.

We are particularly pleased by the challenging new interactive *Student Companion CD-ROM.* Lauren Fogel, Media Producer, and the talented design team at Red Hill Studios have created a CD-ROM that is outstanding in pedagogic sophistication and visual appeal.

We are extremely grateful to our colleagues who reviewed the manuscript and offered insightful suggestions for improvement. In addition, we were priviledged to have the advice of various consultants on anatomy, art, pathophysiology, and vocabulary issues. The contributions of all these people, who generously provided their time and expertise to help us maintain the book's accuracy and clarity, are acknowledged in the list that follows.

CONTRIBUTORS

Self-Quiz Questions

Jerri K. Lindsey, Tarrant County Junior College, Northeast

Critical Thinking Questions

Joan Barber, Delaware Technical and Community College

Student Companion CD-ROM Interactive Activities

Martha DePecol Sanner, Middlesex Community Technical College

Thomas Lancraft, Saint Petersburg Junior College

Student Companion CD-ROM Quiz Questions

Temma al-Mukhtar, San Diego Mesa College

Izak Paul, Mount Royal College

James White, Prairie State Community College

CONSULTANTS

Anatomy Art Consultant

Rose Leigh Vines, California State University, Sacramento

Art Advisory Panel

Steven Amdur, Nassau Community College

Robert Bauman, Jr., Amarillo College

Dan Porter, Amarillo College

Vernon Wiersema, Houston Community College

Clinical and Pathophysiology Consultants

Irene Aguilar, San Antonio College

Rebecca Bonugli, San Antonio College

Joan Parker Frizzell, LaSalle School of Nursing

Susan Gauthier, Temple University

Christine Vandenhouten, Bellin College of Nursing

Kathleen Zellner, Bellin College of Nursing

Vocabulary Consultant

Frances Fulton, Essex Community College

REVIEWERS

Steven Amdur, Nassau Community College

David Asai, Purdue University

Frank Baker, Golden West College

Robert Bauman, Amarillo College

Doris Benfer, Delaware County Community College

Charles Biggers, University of Memphis

Richard Blaney, Brevard Community College

Patrick Bufi, Northwest Institute of Acupuncture and Oriental Medicine

Ed Carroll, Marquette University

Margaret Chad, Saskatchewan Institute of Applied Sciences and Technology

David Dawkins, Redlands Community College

David Deutsch, Kent State University

Ellen Du Pre, Indian River Community College

Charles Egerton, Mississippi Gulf Coast Community College

Thomas Fahey, California State University, Chico

Harold Falls, Southwest Missouri State University

Eugene Fenster, Longview Community College

Lorraine Findlay, Nassau Community College

Mym Fowler, Tidewater Community College

Candice Francis, Palomar College

Ralph Fregosi, University of Arizona

Kathleen A. French, University of California, San Diego

Howard Fuld, Bronx Community College

Phyllis Gee, University of Manitoba

Gene Giggleman, Parker College of Chiropractic Medicine

Barry Golden, Yavapai Community College

Kimberly Hammond, University of California, Riverside

Linda Hardy, Saskatchewan Institute of Applied Sciences and Technology

Henry Hermo, Bronx Community College

Rosemary Hubbard, Marymount University

Paul A. Iaizzo, University of Minnesota, Minneapolis

James Janik, Miami University, Middletown

Paul Jarrell, Pasadena City College

Paul Kelly, Salem State College

Suzanne Kempke, Armstrong State Atlantic University

Jamie King, Craven Community College

Robert King, Northwest Community College

William Kleinelp, Jr., Middlesex County College

Gary Klinger, Nassau Community College

Edward Kolk, Nassau Community College

John J. Lepri, University of North Carolina, Greensboro

Steven Lewis, Penn Valley Community College

Kevin Lien, Portland Community College

Jerri K. Lindsey, Tarrant County Junior College

Joanne Lopez, Foothill College

Linda MacGregor, Bucks County Community College

Paul Malven, Purdue University

Patricia Mansfield, Santa Ana College

John Martin, Clark College

Randall McKee, University of Wisconsin

Katherine Mechlin, Wright State University

William Montgomery, Charles County Community College

C. Aubrey Morris, Pensacola Junior College

Theodore Namm, University of Massachusetts, Lowell

Judy Nesmith, University of Michigan, Dearborn

James Parker, North Georgia State University

Lisa Parks, Georgia State University

Robert Pollack, Nassau Community College

Jackie Reynolds, Richland Community College

Ronald Rodd, University of Evansville

Barbara Rundell, College of DuPage

Walter Saviuk, Daytona Beach Community College

Frank Schwartz, Ohio College of Massotherapy

Norm Scott, University of Waterloo

Marilyn Shopper, Johnson Community College

Wayne Simpson, University of Missouri, Kansas City

Janice Yoder Smith, Tarrant County Junior College
Timothy Stabler, Indiana University, Northwest Campus
James Stockand, Emory University
Frank Sullivan, Front Range Community College
Jenna Sullivan, University of Arizona
Steven Swoap, Williams College
Stuart Thompson, Stanford University
Karen Timberlake, Los Angeles Valley College
Steven Trautwein, Southwest Missouri State University
Jeffrey L. Travis, University of Albany-SUNY
Ann Vernon, Maryville University of St. Louis
Margaret Voss, Syracuse University
Richard L. Walker, University of Calgary
Marc Walters, Portland Community College
Joe Wheeler, North Lake College
James White, Prairie State Community College
Dorothy Winegar, Nassau Community College
Clarence Wolfe, North Virginia Community College
Mark Womble, Youngstown State University

FOCUS GROUP PARTICIPANTS
Human Anatomy and Physiology Society Meeting, 1998

Paul Barenberg, South Plains College
Robert Bauman, Jr., Amarillo College
Alease S. Bruce, University of Massachusetts, Lowell

Linda Burroughs, Rider University
Jay Dee Druecker, Chadron State College
Gary L. Johnson, Madison Area Technical College
Karen LaFleur, Greenville Technical College
Thomas Lancraft, Saint Petersburg Junior College
Pamela Langley, New Hampshire Technical Institute
Augustine Okocha, Milwaukee Area Technical College
David L. Parker, Northern Virginia Community College
Martha DePecol Sanner, Middlesex Community Technical College
Becky Tilton, South Plains College
Vernon Wiersema, Houston Community College, S.W.
Jackie Wright, South Plains College
Jim Young, South Plains College

Nassau County Community College, 1998

Steven Amdur
Lorraine Findlay
Gary Klinger
Edward Kolk
Lois Lucca
Walter Mondschein
Robert Pollack
A. J. Smeriglio
George Vossinas
Dorothy Winegar

TO THE STUDENT

SIX STEPS TO SUCCESSFUL LEARNING!

A Visual Guide to Principles of Anatomy and Physiology

Your book has a variety of special features to help you learn anatomy and physiology successfully and confidently. Each feature has been developed and refined based on feedback from the students and instructors who used previous editions. Here is a student tested, step by step method for mastering the study of anatomy and physiology.

1

ANTICIPATE

A preview reading of the **Study Outline** at the end of the chapter will help you recognize the important concepts you are about to explore. Use this balanced synopsis to acquaint yourself with the topics of the chapter.

> **STUDY OUTLINE**
>
> **PECTORAL (SHOULDER) GIRDLE (p. 218)**
> 1. Each pectoral (shoulder) girdle consists of a clavicle and scapula.
> 2. Each pectoral girdle attaches an upper limb to the axial skeleton.
>
> **UPPER LIMB (p. 221)**
> 1. There are 60 bones in the two upper limbs.
> 2. The bones of each upper limb include the humerus, the ulna, the radius, the carpals, the metacarpals, and the phalanges.
>
> **PELVIC (HIP) GIRDLE (p. 224)**
> 1. The pelvic (hip) girdle consists of two hip bones.
>
> 2. The female pelvis is adapted for pregnancy and childbirth. Gender-related differences in pelvic structure are listed and illustrated in Table 8.1.
>
> **COMPARISON OF PECTORAL AND PELVIC GIRDLES (p. 231)**
> 1. The pectoral girdle does not directly articulate with the vertebral column; the pelvic girdle does.
> 2. The glenoid fossae of the scapulae are shallow and maximize movement; the acetabula of the hip bones are deep and allow less movement.

2

SCAN

Examine the figures before reading a chapter. Study *all* the parts of a figure and read the labels carefully to familiarize yourself with the anatomical or physiological concepts it portrays. Start by reading the **legend,** which explains what the figure is about. Next, study the **key concept statement,** which reveals a basic idea portrayed in the figure. The **orientation diagram** located alongside some anatomy figures shows you the viewing angle of the featured image or where the structure is found in the body. Finally, consider the **figure question,** which asks you to draw conclusions about the figure. (Answers are at the end of the chapter.)

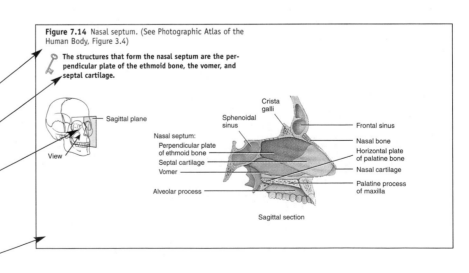

Figure 7.14 Nasal septum. (See Photographic Atlas of the Human Body, Figure 3.4)

🔑 The structures that form the nasal septum are the perpendicular plate of the ethmoid bone, the vomer, and septal cartilage.

Sagittal plane

View

Nasal septum:
Perpendicular plate of ethmoid bone
Septal cartilage
Vomer

Alveolar process

Crista galli

Sphenoidal sinus

Frontal sinus

Nasal bone

Horizontal plate of palatine bone

Nasal cartilage

Palatine process of maxilla

Sagittal section

3

READ

Start reading the chapter. Notice the following aids that are built in to help you learn:

ANATOMY AND PHYSIOLOGY DEFINED

OBJECTIVE

• *Define anatomy and physiology, and name several subdisciplines of these sciences.*

Two branches of science—anatomy and physiology—provide the foundation for understanding the body's parts and functions. **Anatomy** (a-NAT-ō-mē; *ana-* = up; *-tomy* = process of cutting) is the science of body structures and the relationships among structures. Anatomy was first studied by **dissection** (dis-SEK-shun; *dis-* = apart; *-section* = act of cutting), the careful cutting apart of body structures to study their relationships. Nowadays, many imaging techniques also contribute to the advancement of anatomical knowledge. We will compare some common imaging techniques

---

Student objectives reveal what you should learn from each major section.

A term in **boldface** is your signal that a word and the ideas associated with it are important to learn.

Pronunciation guides appear after the most important or hard-to-pronounce terms.

Some terms are followed by **word roots,** which can help you remember a term by telling you about its original meaning or how it is put together.

If you are unsure about the meaning of anatomy and physiology terms, check the **glossary** at the back of the book.

Figure 3.10 Facilitated diffusion of glucose across a plasma membrane. The transporter (GluT) binds to glucose in the extracellular fluid and releases it into the cytosol.

🔑 **Facilitated diffusion across a membrane requires a transporter but does not use ATP.**

☐ Extracellular fluid   ▨ Plasma membrane   ☐ Cytosol

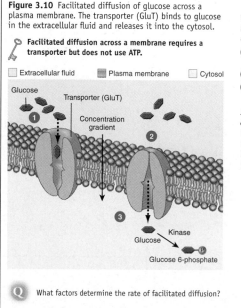

Ⓠ What factors determine the rate of facilitated diffusion?

Solutes that move across plasma membranes by facilitated diffusion include glucose, urea, fructose, galactose, and some vitamins. Glucose enters many body cells by facilitated diffusion, as follows (Figure 3.10):

❶ Glucose binds to a glucose transporter protein called GluT at the extracellular surface of the membrane.

❷ GluT undergoes a conformational change.

❸ GluT releases glucose on the other side of the membrane.

After glucose enters a cell by facilitated diffusion, an enzyme called a kinase attaches a phosphate group to produce a different molecule—namely, glucose 6-phosphate. This reaction keeps the intracellular concentration of glucose very low, so that the glucose concentration gradient always favors facilitated diffusion of glucose into, not out of, cells.

In some places, you will find **keyed narratives** that correlate the steps of a complex process described in the reading with an illustration of that same process.

You will encounter brief **clinical applications.** They will acquaint you with the clinical, professional, or personal significance of the preceding discussion.

CLINICAL APPLICATION
Sprain and Strain

A **sprain** is the forcible wrenching or twisting of a joint that stretches or tears its ligaments but does not dislocate the bones. It occurs when the ligaments are stressed beyond their normal capacity. Sprains also may damage surrounding blood vessels, muscles, tendons, or nerves. Severe sprains may be so painful that the joint cannot be moved. There is considerable swelling, which results from hemorrhage of ruptured blood vessels. The ankle joint is most often sprained; the lower back is another frequent location for

Concluding most chapters, you will find **Homeostatic Imbalances,** which are discussions of diseases and disorders and how these disrupt normal homeostatic processes, and **Medical Terminology,** a listing of related terms that supplements terms defined in the chapter.

DISORDERS: HOMEOSTATIC IMBALANCES

RHEUMATISM AND ARTHRITIS

Rheumatism (ROO-ma-tizm) is any of a variety of painful disorders of the supporting structures of the body—bones, ligaments, tendons, or muscles. **Arthritis** is a form of rheumatism in which the joints have become inflamed. Inflammation, pain, and stiffness may also involve adjacent muscles. Arthritis afflicts about 40 million people in the United States.

Three important classes of arthritis are (1) diffuse connective tissue diseases, such as rheumatoid arthritis, (2) degenerative joint disease, such as osteoarthritis, and (3) metabolic and endocrine diseases with associated arthritis, such as gouty arthritis.

Rheumatoid Arthritis

Rheumatoid arthritis (RA) is an autoimmune disease

irritation of the joints, and wear and abrasion. It is commonly known as "wear-and-tear" arthritis and is the leading cause of disability in older individuals.

Osteoarthritis is a progressive disorder of synovial joints, particularly weight-bearing joints. It is characterized by the deterioration of articular cartilage and by the formation of new bone in the subchondral areas and at the margins of the joint. The cartilage

MEDICAL TERMINOLOGY

Osteoarthritis (os'-tē-ō-ar-THRĪ-tis; *arthr* = joint) The degeneration of articular cartilage such that the bony ends touch; the resulting friction of bone against bone worsens the condition. Usually associated with the elderly.

Osteogenic sarcoma (os'-tē-Ō-JEN-ik sar-KŌ-ma; *sarcoma* = connective tissue tumor) Bone cancer that primarily affects

4

MASTER THE EXHIBITS

Some body systems contain scores of structures, to be memorized, which makes the study of anatomy and physiology particularly challenging. To help, chapters that present complex anatomy have special **Exhibits.** Each Exhibit is a self-contained learning exercise that consists of a narrative tour, a Table summarizing the structures to learn, and a group of illustrations. Spend as much time as you need mastering one Exhibit before moving to the next.

Exhibit 11.4 *Muscles that Move the Tongue—Extrinsic Muscles (Figure 11.7)*

OBJECTIVE

• Describe the origin, insertion, action, and innervation of the extrinsic muscles of the tongue.

The tongue is a highly mobile structure that is vital to digestive functions such as mastication, perception of taste, and deglutition (swallowing). It is also important in speech. The tongue's mobility is greatly aided by its suspension from the mandible, styloid process of the temporal bone, and hyoid bone.

The tongue is divided into lateral halves by a median fibrous septum. The septum extends throughout the length of the tongue and is attached inferiorly to the hyoid bone. Muscles of the tongue are of two principal types: extrinsic and intrinsic. **Extrinsic muscles** originate outside the tongue and insert into it. They move the entire tongue in various directions, such as anteriorly, posteriorly, and laterally. **Intrinsic muscles** originate and insert within the tongue. These muscles alter the shape of tongue rather than moving the entire tongue. The extrinsic and intrinsic muscles of the tongue insert into both lateral halves of the tongue.

When you study the extrinsic muscles of the tongue, you will notice that all of the names end in *glossus*, meaning tongue. You will also notice that the actions of the muscles are obvious, considering the positions of the mandible, styloid process, hyoid bone, and soft palate, which serve as origins for these muscles. For example, the **genioglossus** (originates on the mandible) pulls the tongue downward and forward, the **styloglossus** (originates on the styloid process) pulls the tongue upward and backward, the **hyoglossus** (originates on the hyoid bone) pulls the tongue downward and flattens it, and the **palatoglossus** (originates on the soft palate) raises the back portion of the tongue.

INNERVATION

All muscles are innervated by the hypoglossal (XII) cranial nerve, except the palatoglossus, which is innervated by the pharyngeal plexus, which contains axons from both the vagus (X) and the accessory (XI) cranial nerves.

CLINICAL APPLICATION
Intubation During Anesthesia

When general anesthesia is administered during surgery, a total relaxation of the genioglossus muscle results. This causes the tongue to fall posteriorly, which may obstruct the airway to the lungs. To avoid this, the mandible is either manually thrust forward and held in place, or a tube is inserted from the lips through the laryngopharynx (inferior portion of the throat) into the trachea (endotracheal intubation). ■

RELATING MUSCLES TO MOVEMENTS

Arrange the muscles in this exhibit according to the following actions on the tongue: (1) depression, (2) elevation, (3) protraction, and (4) retraction. The same muscle may be mentioned more than once.

When your physician says, "Open your mouth, stick out your tongue, and say ahh," so she can examine the inside of your mouth for possible signs of infection, which muscles do you contract?

MUSCLE	ORIGIN	INSERTION	ACTION
Genioglossus (jē'-nē-ō-GLOS-us; *geneion* = chin; *glossus* = tongue)	Mandible.	Undersurface of tongue and hyoid bone.	Depresses tongue and thrusts it anteriorly (protraction).
Styloglossus (stī'-lō-GLOS-us; *stylo* = stake or pole; styloid process of temporal bone)	Styloid process of temporal bone.	Side and undersurface of tongue.	Elevates tongue and draws it posteriorly (retraction).
Palatoglossus (pal'-a-tō-GLOS-us; *palato* = palate)	Anterior surface of soft palate.	Side of tongue.	Elevates posterior portion of tongue and draws soft palate down on tongue.
Hyoglossus (hī'-ō-GLOS-us)	Greater horn and body of hyoid bone.	Side of tongue.	Depresses tongue and draws down its sides.

Exhibit 11.4 *Muscles that Move the Tongue—Extrinsic Muscles (continued)*

Figure 11.7 Muscles that move the tongue.

The extrinsic and intrinsic muscles of the tongue are arranged in both lateral halves of the tongue.

Superior constrictor
Styloid process of temporal bone
Mastoid process of temporal bone
Digastric (posterior belly)
Middle constrictor
Styloid
Stylopharyngeus
HYOGLOSSUS
Hyoid bone
Inferior constrictor
Thyroid cartilage of larynx

STYLOGLOSSUS
PALATOGLOSSUS
Palatine tonsil
Hard palate (cut)
Tongue
GENIOGLOSSUS
Mandible (cut)
GENIOHYOID
Mylohyoid
Intermediate tendon of digastric
Fibrous loop for intermediate tendon of digastric
Thyrohyoid membrane (connects hyoid bone to larynx)

Right side deep view

Q What are the functions of the tongue?

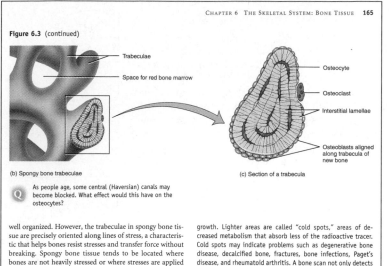

Figure 6.3 (continued)

Trabeculae

Space for red bone marrow

(b) Spongy bone trabeculae

Osteocyte

Osteoclast

Interstitial lamellae

Osteoblasts aligned along trabecula of new bone

(c) Section of a trabecula

Q As people age, some central (Haversian) canals may become blocked. What effect would this have on the osteocytes?

well organized. However, the trabeculae in spongy bone tissue are precisely oriented along lines of stress, a characteristic that helps bones resist stresses and transfer force without breaking. Spongy bone tissue tends to be located where bones are not heavily stressed or where stresses are applied from many directions.

Spongy bone tissue is different from compact bone tissue in two respects. First, spongy bone tissue is light, which reduces the overall weight of a bone so that it moves more readily when pulled by a skeletal muscle. Second, the trabeculae of spongy bone tissue support and protect the red bone marrow. The spongy bone tissue in the hip bones, ribs, breastbone, backbones, and the ends of long bones is the only site of red bone marrow and, thus, of hemopoiesis in adults.

growth. Lighter areas are called "cold spots," areas of decreased metabolism that absorb less of the radioactive tracer. Cold spots may indicate problems such as degenerative bone disease, decalcified bone, fractures, bone infections, Paget's disease, and rheumatoid arthritis. A bone scan not only detects abnormalities 3–6 months sooner than standard x-ray procedures, it exposes the patient to less radiation. ■

1. Why is bone considered a connective tissue?
2. Describe the four types of cells in bone tissue.
3. What is the composition of the matrix of bone tissue?
4. Distinguish between spongy and compact bone tissue in terms of its microscopic appearance, location, and function.

5

VALIDATE

As you read, answer the numbered **review activities,** which validate your understanding of the student objectives.

When you are finished, take the **self-quiz,** which poses questions in a variety of testing styles. (Answers are in Appendix D.)

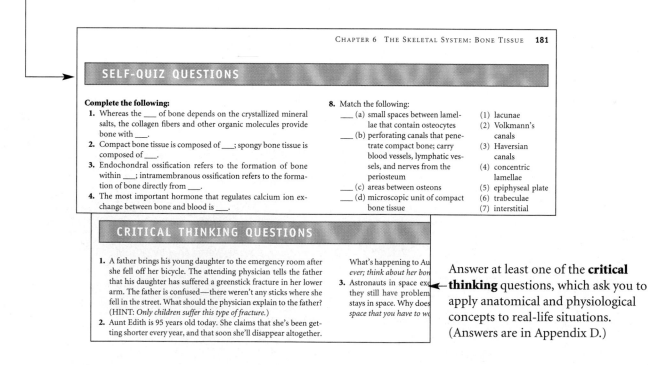

SELF-QUIZ QUESTIONS

Complete the following:

1. Whereas the ___ of bone depends on the crystallized mineral salts, the collagen fibers and other organic molecules provide bone with ___.
2. Compact bone tissue is composed of ___; spongy bone tissue is composed of ___.
3. Endochondral ossification refers to the formation of bone within ___; intramembranous ossification refers to the formation of bone directly from ___.
4. The most important hormone that regulates calcium ion exchange between bone and blood is ___.

8. Match the following:
___ (a) small spaces between lamellae that contain osteocytes
___ (b) perforating canals that penetrate compact bone; carry blood vessels, lymphatic vessels, and nerves from the periosteum
___ (c) areas between osteons
___ (d) microscopic unit of compact bone tissue

(1) lacunae
(2) Volkmann's canals
(3) Haversian canals
(4) concentric lamellae
(5) epiphyseal plate
(6) trabeculae
(7) interstitial

CRITICAL THINKING QUESTIONS

1. A father brings his young daughter to the emergency room after she fell off her bicycle. The attending physician tells the father that his daughter has suffered a greenstick fracture in her lower arm. The father is confused—there weren't any sticks where she fell in the street. What should the physician explain to the father? (HINT: *Only children suffer this type of fracture.*)
2. Aunt Edith is 95 years old today. She claims that she's been getting shorter every year, and that soon she'll disappear altogether.

What's happening to Au
ever; think about her bo
3. Astronauts in space exc
they still have problem
stays in space. Why does
space that you have to w

Answer at least one of the **critical thinking** questions, which ask you to apply anatomical and physiological concepts to real-life situations. (Answers are in Appendix D.)

6

GET INTERACTIVE!

CD-ROM

Use the **Student Companion CD-ROM** located inside the back cover. This engaging study tool offers a range of interactive activities and assessment tools that you can access to add to your knowledge or validate your understanding of the reading. Here's what you'll find:

Activities

Intriguing **feedback loop exercises** let you explore how the homeostasis of body systems is maintained under normal conditions, and how it is disrupted by diseases or disorders. Twelve **art exercises** allow you to investigate complex anatomical structures or to trace complex physiological processes step by step. In addition, **anatomy drag-and-drop exercises** in every chapter will sharpen your understanding of anatomical structures.

Testing

Validate your understanding by taking the 30-question **multiple choice quiz** available for each chapter. Each quiz scores automatically and provides you with immediate feedback about your answers. You can simulate a mid-term or final exam, too, by using the **test feature.** Select the specific chapters you want to be tested on and the number of questions you want.

Special Feature

Use the **pronunciation glossary** to review definitions and to hear the actual pronunciation of terms.

Learning Guide

This helpful study companion has chapter summaries, topic outlines and objectives, terminology exercises, and study questions cross-referenced to the book. Each chapter includes an extra Self-Quiz. Ask for the *Learning Guide to Principles of Anatomy and Physiology* (ISBN: 0–471–37467–9) at your college bookstore!

Web Site

For additional study help or to explore topics in more depth, please visit our web site at **http:\\www.wiley.com\college\bio\tortora**

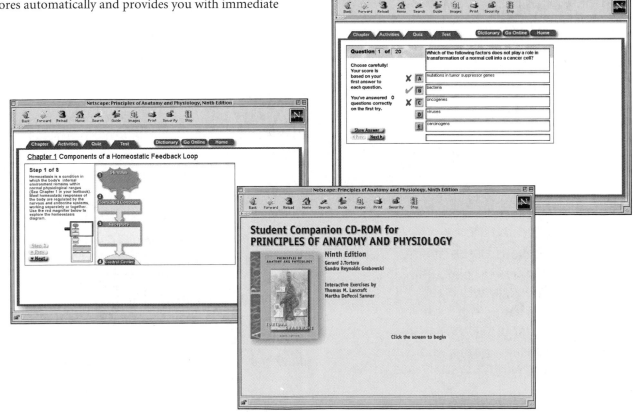

CONTENTS IN BRIEF

CONTENTS

UNIT 3
CONTROL SYSTEMS
OF THE HUMAN BODY

CHAPTER 12
NERVOUS TISSUE 378

CLINICAL APPLICATIONS

UNIT 4
MAINTENANCE OF THE HUMAN BODY

CHAPTER 19
THE CARDIOVASCULAR SYSTEM: THE BLOOD 610

CHAPTER 22
THE LYMPHATIC SYSTEM,
NONSPECIFIC RESISTANCE TO
DISEASE, AND IMMUNITY 738

UNIT 5
CONTINUITY

CLINICAL APPLICATIONS

CHAPTER 29
DEVELOPMENT AND
INHERITANCE 1022

AN INTRODUCTION TO THE HUMAN BODY

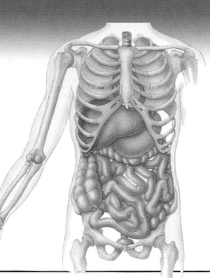

You are beginning a fascinating exploration of your own body. We will start by introducing anatomy and physiology as scientific disciplines, by considering how living things are organized, and by revealing properties that all living things share. Next, you will discover how the body regulates its own internal environment. This ceaseless process, called homeostasis, is a major theme in every chapter of this book. Finally, you will be introduced to a basic vocabulary that will help you speak about the body in a way that is understood by scientists and health-care professionals alike.

ANATOMY AND PHYSIOLOGY DEFINED

OBJECTIVE
• *Define anatomy and physiology, and name several subdisciplines of these sciences.*

Two branches of science—anatomy and physiology—provide the foundation for understanding the body's parts and functions. **Anatomy** (a-NAT-ō-mē; *ana-* = up; *-tomy* = process of cutting) is the science of body structures and the relationships among structures. Anatomy was first studied by **dissection** (dis-SEK-shun; *dis-* = apart; *-section* = act of cutting), the careful cutting apart of body structures to study their relationships. Nowadays, many imaging techniques also contribute to the advancement of anatomical knowledge. We will compare some common imaging techniques

later in this chapter. Whereas anatomy deals with structures of the body, **physiology** (fiz′-ē-OL-ō-jē; *physi-* = nature; *-ology* = study of) is the science of body functions—that is, how the body parts work. Table 1.1 describes several subdisciplines of anatomy and physiology.

Your body's structure, and how it works, is quite similar to other people's of the same age and gender. Accordingly, studying human anatomy and physiology usually involves learning about the "generic" structures and processes that are found in most adults. You are unique, however. You inherited from your parents a collection of traits that determine your individual physical appearance and potential. The study of how traits are passed from parents to their offspring is called **genetics,** the science of heredity. From time to time we will point out subtle differences in body structure and function, all the result of normal genetic variations.

You will learn about the human body by studying its anatomy and physiology together. Always keep in mind that the structure of a part of the body is adapted for performing certain functions. For example, the bones of the skull are tightly joined to form a rigid case that protects the brain. The bones of the fingers, by contrast, are more loosely joined to allow various types of movements. The teeth have different shapes for biting, tearing, and grinding food. The air sacs in the lungs are so thin that they easily permit the movement of oxygen into the blood for use by body cells and the movement of carbon dioxide out of the blood to be exhaled.

Table 1.1 Selected Subdisciplines of Anatomy and Physiology

SUBDISCIPLINES OF ANATOMY	STUDY OF	SUBDISCIPLINES OF PHYSIOLOGY	STUDY OF
Embryology (em'-brē-OL-ō-jē; *embry-* = embryo; *-ology* = study of)	Structures that emerge from the time of the fertilized egg through the eighth week in utero.	**Cell physiology**	Functions of cells.
Developmental anatomy	Structures that emerge from the time of the fertilized egg to the adult form.	**Neurophysiology** (NOO-rō-fiz-ē-ol'-ō-jē; *neuro-* = nerve)	Functional properties of nerve cells.
Cytology (sī-TOL-ō-jē; *cyt-* = cell)	Chemical and microscopic structure of cells.	**Endocrinology** (en'-dō-kri-NOL-ō-jē; *endo-* = within; *-crin* = secretion)	Hormones (chemical regulators in the blood) and how they control body functions.
Histology (hiss'-TOL-ō-jē; *hist-* = tissue)	Microscopic structure of tissues.	**Cardiovascular physiology** (kar-dē-ō-VAS-kyoo-lar; *cardi-* = heart; *-vascular* = blood vessels)	Functions of the heart and blood vessels.
Surface anatomy	Anatomical landmarks on the surface of the body through visualization and palpation.	**Immunology** (im'-yoo-NOL-ō-jē; *immun-* = not susceptible)	How the body defends itself against disease-causing agents.
Gross anatomy	Structures that can be examined without using a microscope.	**Respiratory physiology** (RES-pir-a-to'-rē; *respira-* = to breathe)	Functions of the air passageways and lungs.
Systemic anatomy	Structure of specific systems of the body such as the nervous or respiratory systems.	**Renal physiology** (RĒ-nal; *ren-* = kidney)	Functions of the kidneys.
Regional anatomy	Specific regions of the body such as the head or chest.	**Systemic physiology**	Functions of specific organ systems.
Radiographic anatomy (rā'-dē-ō-GRAF-ik; *radio-* = ray; *-graphic* = to write)	Body structures that can be visualized with x-rays.	**Exercise physiology**	Changes in cell and organ functions as a result of muscular activity.
Pathological anatomy (path'-ō-LOJ-i-kal; *path-* = disease)	Structural changes (from gross to microscopic) associated with disease.	**Pathophysiology** (PATH-ō-fiz-ē-ol'-ō-jē)	Functional changes associated with disease and aging.

CLINICAL APPLICATION
Palpation, Auscultation, and Percussion

Three noninvasive techniques are commonly used by health-care professionals and students of anatomy and physiology to assess certain aspects of body structure and function. In **palpation** (pal-PĀY-shun; *palpa-* = gently touching) the examiner feels body surfaces with the hands. An example is palpating a blood vessel (called an artery) to find the pulse and measure the heart rate. In **auscultation** (aus-cul-TĀY-shun; *ausculta-* = listening) the examiner listens to body sounds to evaluate the functioning of certain organs, often using a stethoscope to amplify the sounds. An example is auscultation of the lungs during breathing to check for crackling sounds associated with abnormal fluid accumulation in the lungs. In **percussion** (pur-KUSH-un; *percus-* = beat through) the examiner taps on the body surface with the fingertips and listens to the resulting echo. For example, percussion may reveal the abnormal presence of fluid in the lungs or air in the in-

testines. It is also used to reveal the size, consistency, and position of an underlying structure. ■

1. What body function might a respiratory therapist strive to improve?
2. Give an example of how the structure of a part of the body is related to its function.

LEVELS OF BODY ORGANIZATION
OBJECTIVES
• *Describe the levels of structural organization that make up the human body.*
• *Define the eleven systems of the human body, the organs present in each, and their general functions.*

The levels of organization of a language—letters of the alphabet, words, sentences, paragraphs, and so on—provide

Figure 1.1 Levels of structural organization in the human body.

🔑 **The levels of structural organization are chemical, cellular, tissue, organ, system, and organismal.**

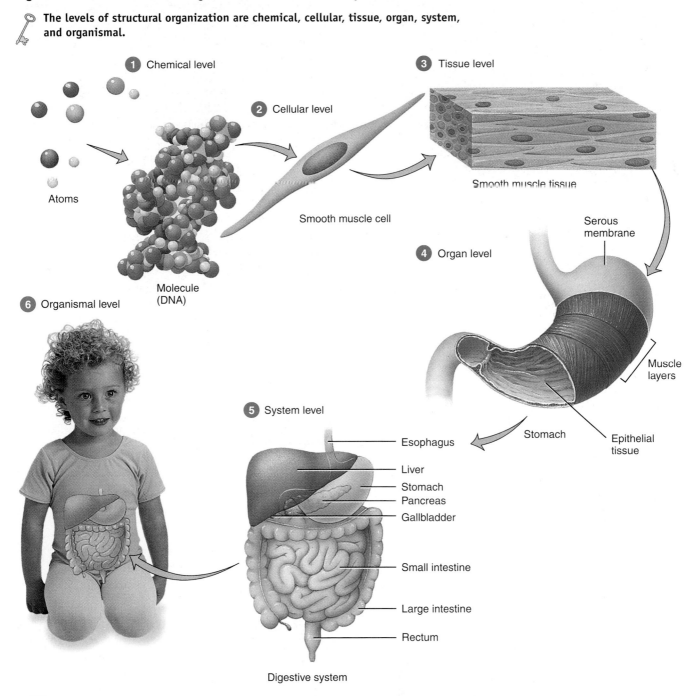

1 Chemical level

2 Cellular level

3 Tissue level

Atoms

Molecule (DNA)

Smooth muscle cell

Smooth muscle tissue

6 Organismal level

4 Organ level

Serous membrane

Muscle layers

Epithelial tissue

Stomach

5 System level

Esophagus
Liver
Stomach
Pancreas
Gallbladder

Small intestine

Large intestine

Rectum

Digestive system

❓ Which level of structural organization is composed of two or more different types of tissues that work together to perform a specific function?

a useful comparison to the levels of organization of the human body. Your exploration of the human body will extend from some of the smallest body structures and their functions to the largest structure—an entire person. From the smallest to the largest size of their components, six levels of organization are relevant to understanding anatomy and physiology: the chemical, cellular, tissue, organ, system, and organismal levels of organization (Figure 1.1).

1 The *chemical level* includes **atoms,** the smallest units of matter that participate in chemical reactions, and **molecules,** two or more atoms joined together. Certain atoms, such as carbon (C), hydrogen (H), oxygen (O), nitrogen (N), and calcium (Ca), are essential for maintaining life. Familiar examples of molecules found in the body are deoxyribonucleic acid (DNA), the genetic material passed from one generation to the next; the protein hemoglobin, which carries oxygen in the blood; and glucose, commonly known as blood sugar. Chapters 2 and 25 focus on the chemical level of organization.

2 Molecules, in turn, combine to form structures at the next level of organization—the *cellular level.* **Cells** are the basic structural and functional units of an organism and are the smallest living units in the human body. Among the many kinds of cells in your body are muscle cells, nerve cells, and blood cells. Shown in Figure 1.1 is a smooth muscle cell, one of three different types of muscle cells in your body. The focus of Chapter 3 is the cellular level of organization.

3 The next level of structural organization is the *tissue level.* **Tissues** are groups of cells and the materials surrounding them that work together to perform a particular function. There are just four basic types of tissue in your body: *epithelial tissue, connective tissue, muscle tissue,* and *nervous tissue,* which are described in Chapter 4. Smooth muscle tissue consists of tightly packed smooth muscle cells.

4 When different kinds of tissues are joined together, they form the next level of organization, the *organ level.* **Organs** are structures that are composed of two or more different types of tissues; they have specific functions and usually have recognizable shapes. Examples of organs are the stomach, heart, liver, lungs, and brain. Figure 1.1 shows how several tissues make up the stomach. The outer covering is a *serous membrane,* a layer of epithelial tissue and connective tissue that protects the stomach and other organs and reduces friction when the stomach moves and rubs against other organs. Underneath are the *muscle tissue layers,* which contract to churn and mix food and push it on to the next digestive organ, the small intestine. The innermost lining is an *epithelial tissue layer,* which contributes fluid and chemicals that aid digestive processes in the stomach.

5 The next level of structural organization in the body is the *system level.* A **system** consists of related organs that have a common function. An example is the digestive system, which breaks down and absorbs food. Its organs include the mouth, salivary glands, pharynx (throat), esophagus (food tube), stomach, small intestine, large intestine, rectum, liver, gallbladder, and pancreas. Sometimes an organ is part of more than one system. The pancreas, for example, is part of both the digestive system and the hormone-producing endocrine system.

6 The largest organizational level is the *organismal level.* An **organism** is any living individual. All the parts of the human body functioning with one another constitute the total organism—one living person.

In the chapters that follow you will study the anatomy and physiology of the major body systems. Table 1.2 introduces the components and functions of these systems in the order they are discussed in the book. You will also discover that all body systems influence one another. As an example, consider how just two body systems—the integumentary and skeletal systems—cooperate. The integumentary system, which includes the skin, hair, and nails, protects all other body systems, including the skeletal system, which is composed of all the bones and joints of the body. The skin serves as a barrier between the outside environment and internal tissues and organs. It also participates in the production of vitamin D, which is needed for proper deposition of calcium and other minerals into bone. The skeletal system, in turn, provides support for the integumentary system, serves as a reservoir for calcium—storing it in times of plenty and releasing it for other tissues in times of need—and generates cells that help the skin resist invasion by disease-causing organisms. As you study each of the body systems in more detail, you will discover how they work together to maintain health, provide protection from disease, and allow for reproduction of the human species.

1. Define the following terms: atom, molecule, cell, tissue, organ, system, and organism.
2. At what levels of organization does an exercise physiologist study the human body? (HINT: *Refer to Table 1.1.*)
3. Referring to Table 1.2, which organ systems help eliminate wastes?

CHARACTERISTICS OF THE LIVING HUMAN ORGANISM

OBJECTIVES
• *Define the important life processes of the human body.*
• *Define homeostasis and explain its relationship to interstitial fluid.*

Basic Life Processes

Organisms carry on certain processes that distinguish them from nonliving things. Following are the six most important life processes of the human body:

Metabolism (me-TAB-ō-lizm) is the sum of all the chemical processes that occur in the body. It includes breaking down large, complex molecules into smaller, simpler ones and building the body's structural and functional components. For example, proteins in food are split into amino acids, which are the building blocks of proteins. Amino acids can be used to build new proteins that make up body structures, for example, muscles and bones. Metabolism involves

Table 1.2 Eleven Systems of the Human Body

SYSTEM	COMPONENTS	FUNCTIONS	REFERENCE FIGURE
Integumentary system	Skin and structures derived from it, such as hair, nails, and sweat and oil glands.	Protects body; helps regulate body temperature; eliminates some wastes; helps produce vitamin D; detects sensations, such as pain, touch, hot, and cold.	Figure 5.1 on page 141.
Skeletal system	Bones and joints of the body and their associated cartilages.	Supports and protects body; aids body movements; houses cells that give rise to blood cells; stores minerals and lipids (fats).	Figure 7.1 on page 184.
Muscular system	Muscles composed of skeletal muscle tissue, so-named because it is usually attached to bones.	Produces body movements, such as walking; stabilizes body position (posture), generates heat.	Figure 11.3 on page 309.
Nervous system	Brain, spinal cord, nerves, and special sense organs, such as eye and ear.	Uses action potentials (nerve impulses) to regulate body activities; detects changes in the body's internal and external environment, interprets the changes, and responds by causing muscular contractions or glandular secretions.	Figures 13.2 and 14.1 on pages 415 and 447.
Endocrine system	Hormone-producing cells and glands, such as pituitary gland, thyroid gland, and pancreas.	Regulates body activities by releasing hormones, which are chemical messengers transported in blood from an endocrine gland to a target organ.	Figure 18.1 on page 567.
Cardiovascular system	Blood, heart, and blood vessels.	Heart pumps blood through blood vessels; blood carries oxygen and nutrients to cells and carbon dioxide and wastes away from cells and helps regulate acidity, temperature, and water content of body fluids; blood components help defend against disease and mend damaged blood vessels.	Figures 20.1, 21.18, and 21.24 on pages 637, 693, and 711.
Lymphatic and immune systems	Lymphatic fluid and vessels; also includes structures or organs that contain large numbers of white blood cells called lymphocytes, such as spleen, thymus gland, lymph nodes, and tonsils.	Returns proteins and fluid to blood; carries lipids from gastrointestinal tract to blood; are sites of maturation and proliferation of lymphocytes that protect against disease-causing organisms.	Figure 22.1 on page 739.
Respiratory system	Lungs and the airways leading into and out of them.	Transfers oxygen from inhaled air to blood and carbon dioxide from blood to exhaled air; helps regulate acidity of body fluids; air flowing out of lungs through vocal cords produces sounds.	Figure 23.1 on page 776.
Digestive system	Organs of gastrointestinal tract, a long tube that includes the mouth, esophagus, stomach, intestines, and anus; also includes accessory organs that assist in digestive processes, such as the salivary glands, liver, gallbladder, and pancreas.	Achieves physical and chemical breakdown of food; absorbs nutrients; eliminates solid wastes.	Figure 24.1 on page 819.
Urinary system	Kidneys, ureters, urinary bladder, and urethra.	Produces, stores, and eliminates urine; eliminates wastes and regulates volume and chemical composition of blood; maintains body's mineral balance; helps regulate red blood cell production.	Figure 26.1 on page 915.
Reproductive system	Gonads (testes or ovaries) and associated organs: uterine tubes, uterus, and vagina in females and epididymis, ductus deferens, and penis in males.	Gonads produce gametes (sperm or ova) that unite to form a new organism and release hormones that regulate reproduction and other body processes; associated organs transport and store gametes.	Figures 28.1 and 28.11 on pages 975 and 986.

using oxygen provided by the respiratory system and nutrients broken down in the digestive system to provide the chemical energy to power cellular activities.

Responsiveness is the body's ability to detect and respond to changes in its internal or external environment. Different cells in the body detect different sorts of changes and respond in characteristic ways. Nerve cells respond by generating electrical signals, known as nerve impulses. Muscle cells respond by contracting, which generates force to move body parts. Endocrine cells in the pancreas respond to an elevated blood glucose level by secreting the hormone insulin. Other body cells respond to insulin by taking up glucose, which lowers the blood glucose level to normal.

Movement includes motion of the whole body, individual organs, single cells, and even tiny structures inside cells. For example, the coordinated action of several leg muscles moves your whole body from one place to another when you walk or run. After you eat a meal that contains fats, your gallbladder contracts and squirts bile into the gastrointestinal tract to aid in the digestion of fats. When a body tissue is damaged or infected, certain white blood cells move from the blood into the tissue to help clean up and repair the area. And inside individual cells, various cell parts move from one position to another to carry out their functions.

Growth is an increase in body size that results from an increase in the size of existing cells, the number of cells, or both. In addition, a tissue sometimes increases in size because the amount of material found between cells increases. In growing bone, for example, mineral deposits accumulate around the bone cells, causing the bone to enlarge in length and width.

Each kind of cell in the body has a specialized form and function. **Differentiation** is the process a cell undergoes to develop from an unspecialized to a specialized state. Specialized cells differ in structure and function from the ancestor cells that gave rise to them. For example, red blood cells and several types of white blood cells differentiate from the same unspecialized ancestor cells in red bone marrow. Such ancestor cells, which can divide and give rise to progeny that undergo differentiation, are known as **stem cells.** Also, through differentiation, a fertilized egg develops into an embryo, and then into a fetus, an infant, a child, and finally an adult.

Reproduction refers either to the formation of new cells for growth, repair, or replacement, or to the production of a new individual. Some types of body cells, such as epithelial cells, reproduce continuously throughout a lifetime. Others, such as nerve and muscle cells, lose the ability to divide and proliferate and thus cannot be replaced if they are destroyed. Through formation of sperm and ova, life continues from one generation to the next.

Although not all of these processes are occurring in cells throughout the body all of the time, when they cease to occur properly the result is death of cells and then death of the human organism. Clinically, death in the human body is indicated by loss of the heartbeat, absence of spontaneous breathing, and loss of brain functions.

CLINICAL APPLICATION
Autopsy

An **autopsy** (AW-top-sē; *auto-* = self; *-opsy* = to see) is a postmortem (after death) examination of the body and dissection of its internal organs to confirm or determine the cause of death. An autopsy can uncover the existence of diseases not detected during life, determine the extent of injuries, and explain how those injuries may have contributed to a person's death. It also may support the accuracy of diagnostic tests, establish the beneficial and adverse effects of drugs, reveal the effects of environmental influences on the body, provide more information about a disease, assist in the accumulation of statistical data, and educate healthcare students. Moreover, an autopsy can reveal conditions that may affect offspring or siblings (such as congenital heart defects). An autopsy may be legally required, such as in the course of a criminal investigation, or to resolve disputes between beneficiaries and insurance companies about the cause of death. ■

Homeostasis

The French physiologist Claude Bernard (1813–1878) first proposed that the cells of many-celled organisms flourish because they live in the relative constancy of *"le milieu interieur"*—the internal environment—despite continual changes in the organisms' external environment. The American physiologist Walter B. Cannon (1871–1945) coined the term *homeostasis* to describe this dynamic constancy. **Homeostasis** (hō′-mē-ō-STĀ-sis; *homeo-* = sameness; *-stasis* = standing still) is the condition of equilibrium in the body's internal environment produced by the ceaseless interplay of all the body's regulatory processes. Homeostasis is a dynamic condition—in response to changing conditions, the body's equilibrium point can change over a narrow range that is compatible with maintaining life. For example, the level of glucose in the blood normally neither falls below 70 mg of glucose per 100 mL of blood nor rises above 110 mg/100 mL.* Each body structure, from the cellular level to the systemic level, contributes in some way to keeping the internal environment within normal limits.

Body Fluids

An important aspect of homeostasis is maintaining the volume and composition of **body fluids,** which are dilute, watery solutions found inside cells as well as surrounding them. The fluid within cells is called **intracellular fluid** (*intra-* = inside), abbreviated **ICF.** The fluid outside body cells is called **extracellular fluid** (*extra-* = outside), abbreviated **ECF.** Dissolved in the water of ICF and ECF are substances needed to maintain life, such as oxygen, nutrients, proteins, and a variety of electrically charged chemical particles called *ions.* The ECF that fills the narrow spaces between cells of tis-

---

*Appendix A describes metric measurements.

sues is known as **interstitial fluid** (in′-ter-STISH-al; *inter-* = between). The ECF within blood vessels is termed **plasma.**

As Bernard predicted, the proper functioning of body cells depends on precise regulation of the composition of their surrounding fluid. Because interstitial fluid surrounds all body cells, it is often called the body's *internal environment.* The composition of interstitial fluid changes as substances move back and forth between it and plasma. Such exchange of materials occurs across the thin walls of the smallest blood vessels in the body, the *blood capillaries.* This movement in both directions across capillary walls provides needed materials, such as glucose, oxygen, ions, and so on, to tissue cells and removes wastes, such as carbon dioxide, from interstitial fluid.

1. Which life process in the human body sustains all the others?
2. Describe the locations of intracellular fluid, extracellular fluid, interstitial fluid, and plasma.
3. Why is interstitial fluid called the internal environment of the body?

CONTROL OF HOMEOSTASIS

OBJECTIVES

• *Describe the components of a feedback system.*

• *Contrast the operation of negative and positive feedback systems.*

• *Explain why homeostatic imbalances cause disorders.*

Homeostasis in the human body is continually being disturbed. The disruption may come from the external (outside the body) environment in the form of physical insults such as intense heat or lack of oxygen. Other disruptions originate in the internal (within the body) environment—for example, a blood glucose level that is too low. Homeostatic imbalances may also occur due to psychological stresses in our social environment—the demands of work and school, for example. In most cases the disruption of homeostasis is mild and temporary, and the responses of body cells quickly restore balance in the internal environment. In other cases the disruption of homeostasis may be intense and prolonged, as in poisoning, overexposure to temperature extremes, severe infection, or death of one's spouse. Under these circumstances, regulation of homeostasis may fail.

Fortunately, the body has many regulating systems that bring the internal environment back into balance. Most often, the nervous system and the endocrine system, working together or independently, provide the corrective measures needed when homeostasis is disrupted. The nervous system regulates homeostasis by detecting deviations from the balanced state and then sending messages in the form of *nerve impulses* to organs that counteract the deviation. The endocrine system—a group of glands that secrete molecules called *hormones* into the blood—also regulates homeostasis.

Whereas nerve impulses typically cause rapid changes, hormones usually work more slowly. Both means of regulation, however, work toward the same end—maintaining homeostasis—and, as you will see, both work mainly by negative feedback systems.

Feedback Systems

The body can regulate its internal environment through a multitude of feedback systems. A **feedback system** is a cycle of events in which the status of a body condition is continually monitored, evaluated, changed, remonitored, reevaluated, and so on. Each monitored variable, such as body temperature, blood pressure, or blood glucose level, is termed a *controlled condition.* Any disruption that changes a controlled condition is called a *stimulus.* Three basic components make up a feedback system—a receptor, a control center, and an effector (Figure 1.2).

• A **receptor** is a body structure that monitors changes in a controlled condition and sends input in the form of nerve impulses or chemical signals to a control center. For example, nerve endings in the skin that sense temperature are one of the hundreds of different kinds of receptors in the body.

• A **control center** in the body sets the range of values within which a controlled condition should be maintained, evaluates the input it receives from receptors, and generates output commands when they are needed. Output from the control center can occur in several forms: nerve impulses, hormones, or other chemical signals.

• An **effector** is a body structure that receives output from the control center and produces a *response* or effect that changes the controlled condition. Nearly every organ or tissue in the body can behave as an effector. For example, when your body temperature drops sharply, your brain (control center) sends nerve impulses to your skeletal muscles (effectors) that cause you to shiver, which generates heat and raises your temperature.

A group of receptors and effectors communicating with their control center forms a feedback system that can regulate a controlled condition in the body's internal environment. In a feedback system, or *feedback loop,* the response of the system "feeds back" to change the controlled condition in some way. Feedback systems can produce either negative feedback or positive feedback. If the response reverses the original stimulus, as in the body temperature regulation example, the system is operating by *negative feedback.* If the response enhances or intensifies the original stimulus, the system is operating by *positive feedback.*

Negative Feedback Systems

A **negative feedback system** reverses a change in a controlled condition. First, a stimulus disrupts homeostasis by altering the controlled condition. The receptors that are part of the feedback system detect the change and send input to a

Figure 1.2 Operation of a feedback system. The dashed return arrow symbolizes negative feedback.

🔑 **The three basic components of a feedback system are receptors, a control center, and effectors.**

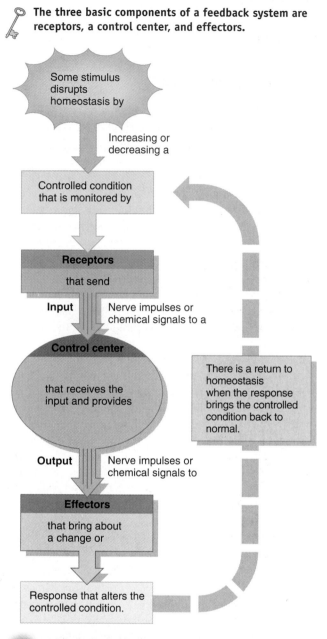

Q What is the basic difference between negative and positive feedback systems?

control center. The control center evaluates the input and, if necessary, issues output commands to an effector. The effector produces a physiological response that is able to return the controlled condition to its normal state.

Consider one negative feedback system that helps regulate blood pressure. Blood pressure (BP) is the force exerted by blood as it presses against the walls of blood vessels. When the heart beats faster or harder, BP increases. If some internal or external stimulus causes blood pressure (controlled condition) to rise, the following sequence of events occurs (Figure 1.3). The higher pressure is detected by *baroreceptors,* which are pressure-sensitive nerve cells (the receptors) located in the walls of certain blood vessels. The baroreceptors send nerve impulses (input) to the brain (control center), which interprets the impulses and responds by sending nerve impulses (output) to the heart (the effector). Heart rate decreases, which causes blood pressure to decrease (response). This sequence of events returns the controlled condition—blood pressure—to normal, and homeostasis is restored. Notice that the activity of the effector produces a result—a drop in BP—that opposes the stimulus—an increase in BP. Thus this is a negative feedback system.

Positive Feedback Systems

A **positive feedback system** tends to strengthen or reinforce a change in one of the body's controlled conditions. A positive feedback system operates similarly to a negative feedback system, except for the way the response affects the controlled condition. A stimulus alters a controlled condition, which is monitored by receptors that send input to a control center. The control center issues commands to an effector, but this time the effector produces a physiological response that *reinforces* the initial change in the controlled condition. The action of a positive feedback system continues until it is interrupted by some mechanism outside the system.

Normal childbirth provides a good example of a positive feedback system (Figure 1.4). The first contractions of labor (stimulus) push part of the baby into the cervix, the lowest part of the uterus, which opens into the vagina. Stretch-sensitive nerve cells (receptors) monitor the amount of stretching of the cervix (controlled condition). As stretching increases, they send more nerve impulses (input) to the brain (control center), which in turn releases the hormone oxytocin (output) into the blood. Oxytocin causes muscles in the uterine walls (effector) to contract even more forcefully. The contractions push the baby farther down the uterus, which stretches the cervix even more. The cycle of stretching, hormone release, and ever stronger contractions is interrupted only by the birth of the baby. Then, stretching of the cervix ceases and oxytocin is no longer released.

Figure 1.3 Homeostatic regulation of blood pressure by a negative feedback system. Note that the response is fed back into the system, and the system continues to lower blood pressure until there is a return to normal blood pressure (homeostasis).

🔑 **If the response reverses the stimulus, a system is operating by negative feedback.**

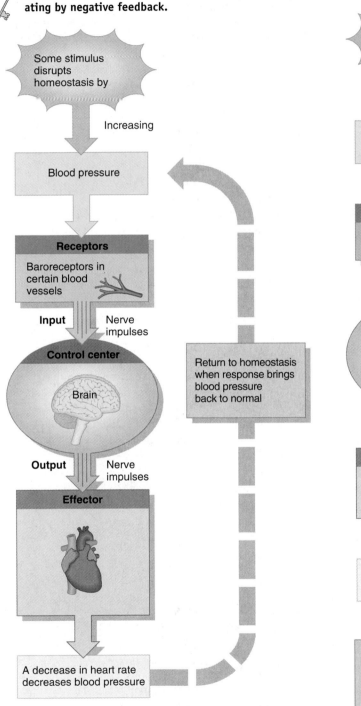

Figure 1.4 Positive feedback control of labor contractions during birth of a baby. The solid return arrow symbolizes positive feedback.

🔑 **If the response enhances or intensifies the stimulus, a system is operating by positive feedback.**

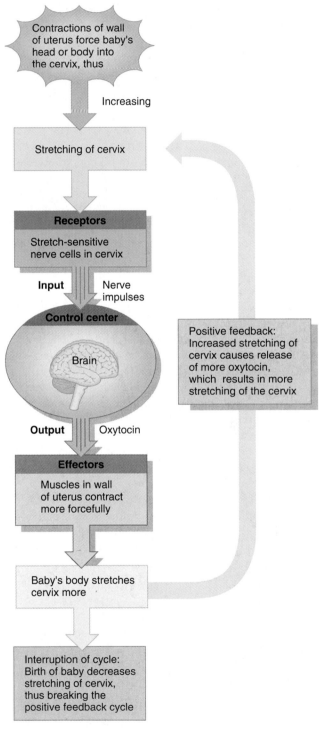

Ⓠ What would happen to heart rate if some stimulus caused blood pressure to decrease? Would this occur by way of positive or negative feedback?

Ⓠ Why do positive feedback systems that are part of a normal physiological response include a termination mechanism?

These examples suggest some important differences between positive and negative feedback systems. Because a positive feedback system continually reinforces a change in a controlled condition, it must be shut off by some event outside the system. If the action of a positive feedback system isn't stopped, it can "run away" and produce life-threatening conditions in the body. The action of a negative feedback system, on the other hand, slows and then stops as the controlled condition returns to its initial state. In general, positive feedback systems tend to reinforce conditions that don't happen very often, whereas negative feedback systems tend to regulate conditions in the body that are held fairly stable over long periods.

Homeostatic Imbalances

As long as all of the body's controlled conditions remain within certain narrow limits, body cells function efficiently, homeostasis is maintained, and the body stays healthy. Should one or more components of the body lose their ability to contribute to homeostasis, however, the normal equilibrium among body processes may be disturbed. If the homeostatic imbalance is moderate, a disorder or disease may occur; if it is severe, death may result.

A **disorder** is any derangement or abnormality of function. **Disease** is a more specific term for an illness characterized by a recognizable set of signs and symptoms. A *local disease* affects one part or a limited region of the body; a *systemic disease* affects either the entire body or several parts of it. Diseases alter body structures and functions in characteristic ways. A person with a disease may experience **symptoms,** which are subjective changes in body functions that are not apparent to an observer—for example, headache or nausea. Objective changes that a clinician can observe and measure, on the other hand, are called **signs.** Signs of disease can be either anatomical, such as swelling or a rash, or physiological, such as fever or paralysis.

The science that deals with why, when, and where diseases occur and how they are transmitted among individuals in a community is known as **epidemiology** (ep′-i-dē-mē-OL-ō-jē; *epi-* = upon; *-demi* = people). The science that deals with the effects and uses of drugs in the treatment of disease is called **pharmacology** (far′-ma-KOL-ō-jē; *pharmac-* = drug).

CLINICAL APPLICATION
Diagnosis of Disease

Diagnosis (dī′-ag-NŌ-sis; *dia-* = through; *-gnosis* = knowledge) is the science and skill of distinguishing one disorder or disease from another. A diagnosis is made on the basis of the patient's signs and symptoms, his or her medical history, a physical exam, and laboratory tests. Taking a *medical history* consists of collecting information about events that might be related to a patient's illness, including the chief complaint, history of present illness, past medical problems, family medical problems, social history, and review of symptoms. A *physical examination* is an orderly evaluation of the body and its functions. This process includes inspection (looking at or into the body using various instruments), palpation, auscultation, percussion, measuring vital signs (temperature, pulse, respiratory rate, and blood pressure), and sometimes laboratory tests. ∎

1. What types of disturbances can act as stimuli that initiate a feedback system?
2. How are negative and positive feedback systems similar? How are they different?
3. Contrast and give examples of symptoms and signs of a disease.
4. What is a diagnosis?

BASIC ANATOMICAL TERMINOLOGY
OBJECTIVES
• *Describe the orientation of the body in the anatomical position.*
• *Relate the common names to the corresponding anatomical descriptive terms for various regions of the human body.*
• *Define the anatomical planes and sections and the directional terms used to describe the human body.*

Scientists and health-care professionals use a common language of special terms when referring to body structures and their functions. The language of anatomy and physiology has precisely defined meanings that allow us to communicate without using unneeded or ambiguous words. For example, is it correct to say, "The wrist is above the fingers"? This might be true if your arms are at your sides. But if you hold your hands up above your head, your fingers would be above your wrists. To prevent this kind of confusion, anatomists developed a standard anatomical position and a special vocabulary for relating body parts to one another.

Body Positions

In the study of anatomy, descriptions of any region or part of the human body assume that the body is in a specific stance called the **anatomical position.** In the anatomical position, the subject stands erect facing the observer, with the head level and the eyes facing directly forward. The feet are flat on the floor and directed forward, and the arms are at the sides with the palms turned forward (Figure 1.5). Once the body is in the anatomical position, it is easier to visualize and understand how it is organized into various regions.

In the anatomical position, the body is upright. Two terms describe a reclining body. If the body is lying face down, it is in the **prone** position. If the body is lying face up, it is in the **supine** position.

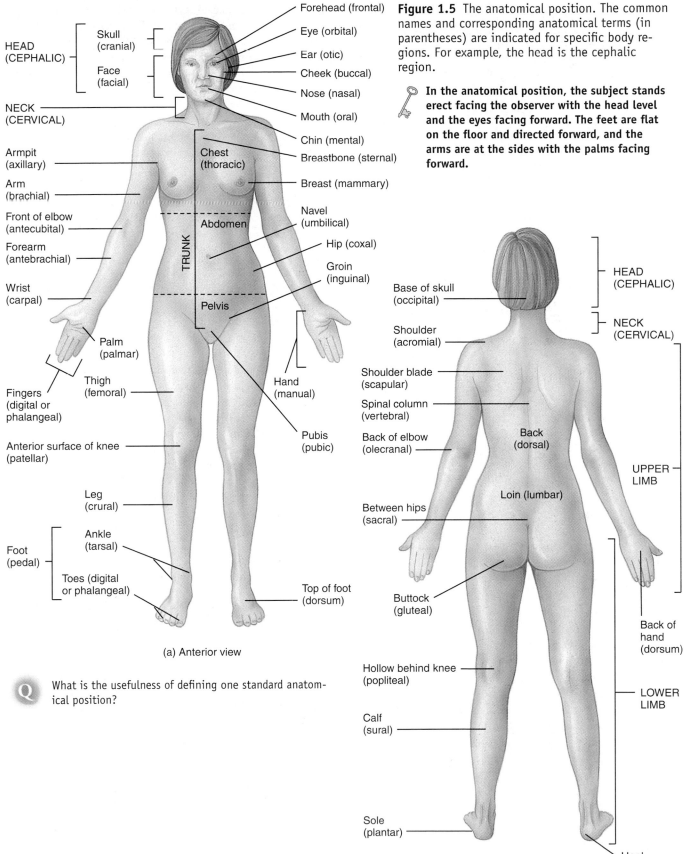

HEAD (CEPHALIC)
Skull (cranial)
Face (facial)

NECK (CERVICAL)

Armpit (axillary)

Arm (brachial)

Front of elbow (antecubital)

Forearm (antebrachial)

Wrist (carpal)

Palm (palmar)

Fingers (digital or phalangeal)

Thigh (femoral)

Anterior surface of knee (patellar)

Leg (crural)

Foot (pedal)
Ankle (tarsal)
Toes (digital or phalangeal)

TRUNK

Chest (thoracic)

Abdomen

Pelvis

Forehead (frontal)
Eye (orbital)
Ear (otic)
Cheek (buccal)
Nose (nasal)
Mouth (oral)
Chin (mental)
Breastbone (sternal)
Breast (mammary)
Navel (umbilical)
Hip (coxal)
Groin (inguinal)

Hand (manual)

Pubis (pubic)

Top of foot (dorsum)

(a) Anterior view

Figure 1.5 The anatomical position. The common names and corresponding anatomical terms (in parentheses) are indicated for specific body regions. For example, the head is the cephalic region.

🔑 **In the anatomical position, the subject stands erect facing the observer with the head level and the eyes facing forward. The feet are flat on the floor and directed forward, and the arms are at the sides with the palms facing forward.**

Base of skull (occipital)

Shoulder (acromial)

Shoulder blade (scapular)

Spinal column (vertebral)

Back of elbow (olecranal)

Between hips (sacral)

Buttock (gluteal)

Hollow behind knee (popliteal)

Calf (sural)

Sole (plantar)

HEAD (CEPHALIC)

NECK (CERVICAL)

Back (dorsal)

Loin (lumbar)

UPPER LIMB

Back of hand (dorsum)

LOWER LIMB

Heel (calcaneal)

(b) Posterior view

❓ What is the usefulness of defining one standard anatomical position?

Figure 1.6 Planes of the human body.

Frontal, transverse, sagittal, and oblique planes divide the body in specific ways.

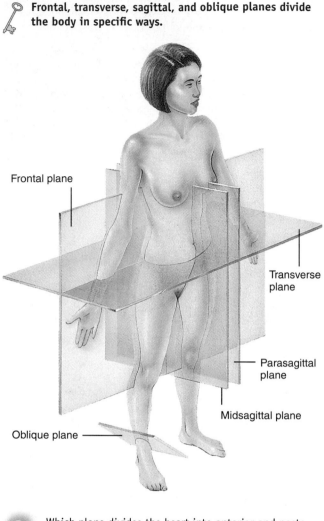

Frontal plane

Transverse plane

Parasagittal plane

Midsagittal plane

Oblique plane

Q Which plane divides the heart into anterior and posterior portions?

Regional Names

The human body is divided into several major regions that can be identified externally. The principal regions are the head, neck, trunk, upper limbs, and lower limbs (see Figure 1.5). The *head* consists of the skull and face. Whereas the skull encloses and protects the brain, the face is the anterior (front) portion of the head that includes the eyes, nose, mouth, forehead, cheeks, and chin. The *neck* supports the head and attaches it to the trunk. The *trunk* consists of the chest, abdomen, and pelvis. Each *upper limb* is attached to the trunk and consists of the shoulder, armpit, arm (portion of the limb from the shoulder to the elbow), forearm (portion of the limb from the elbow to the wrist), wrist, and hand. Each *lower limb* is also attached to the trunk and consists of the buttock, thigh (portion of the limb from the buttock to the knee), leg (portion of the limb from the knee to

the ankle), ankle, and foot. The *groin* is the area on the anterior surface of the body marked by a crease on each side, where the trunk attaches to the thighs.

Figure 1.5 shows the common names of major parts of the body. The corresponding anatomical descriptive form (adjective) for each part appears in parentheses next to the common name. For example, if you receive a tetanus shot in your *buttock*, it is a *gluteal* injection. Why does the descriptive form of a body part look different from its common name? The reason is that the descriptive form is based on a Greek or Latin word or "root" for the same part or area. The Latin word for armpit is *axilla* (ak-SIL-a), for example, and thus one of the nerves passing within the armpit is called the axillary nerve. You will learn more about the word roots of anatomical and physiological terms as you read this book.

Planes and Sections

You will also study parts of the body relative to planes—imaginary flat surfaces that pass through the body parts (Figure 1.6). A **sagittal plane** (SAJ-i-tal; *sagitt-* = arrow) is a vertical plane that divides the body or an organ into right and left sides. More specifically, when such a plane passes through the midline of the body or an organ and divides it into *equal* right and left sides, it is called a **midsagittal plane** or a **median plane.** If the sagittal plane does not pass through the midline but instead divides the body or an organ into *unequal* right and left sides, it is called a **parasagittal plane** (*para-* = near). A **frontal** or **coronal plane** (kō-RŌ-nal; *corona* = crown) divides the body or an organ into anterior (front) and posterior (back) portions. A **transverse plane** divides the body or an organ into superior (upper) and inferior (lower) portions. A transverse plane may also be termed a cross-sectional or horizontal plane. Sagittal, frontal, and transverse planes are all at right angles to one another. An **oblique plane,** by contrast, passes through the body or an organ at an angle between the transverse plane and either a sagittal or frontal plane.

When you study a body region, you often view it in **section,** meaning that you look at only one flat surface of the three-dimensional structure. It is important to know the plane of the section so you can understand the anatomical relationship of one part to another. Figure 1.7 indicates how three different sections—a *transverse section,* a *frontal section,* and a *midsagittal section*—provide different views of the brain.

1. Describe the anatomical position and explain why it is used.
2. Locate each region on your own body, and then identify it by its common name and the corresponding anatomical descriptive form.
3. What are the four types of planes that may be passed through the body? Explain how each divides the body.

Figure 1.7 Planes and sections through different parts of the brain. The diagrams (left) show the planes, and the photographs (right) show the resulting sections. (Note: The arrows in the diagrams indicate the direction from which each section is viewed. This aid is used throughout the book to indicate viewing perspective.)

🔑 **Planes divide the body in various ways to produce sections.**

(a) Transverse plane

View

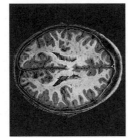

Transverse section

(b) Frontal plane

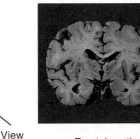

View

Frontal section

(c) Midsagittal plane

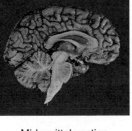

View

Midsagittal section

Q Which plane divides the brain into unequal right and left portions?

Directional Terms

To locate various body structures, anatomists use specific **directional terms,** words that describe the position of one body part relative to another. Several directional terms can be grouped in pairs that have opposite meanings—for example, anterior (front) and posterior (back). Exhibit 1.1 presents the principal directional terms.

Figure 1.8 Body cavities. The dashed lines in (a) and (b) indicate the border between the abdominal and pelvic cavities. (See Tortora, *A Photographic Atlas of the Human Body,* Figures 6.5, 6.11, and 11.12.)

🔑 **The two principal cavities are the dorsal and ventral body cavities.**

☐ DORSAL BODY CAVITY

▨ VENTRAL BODY CAVITY

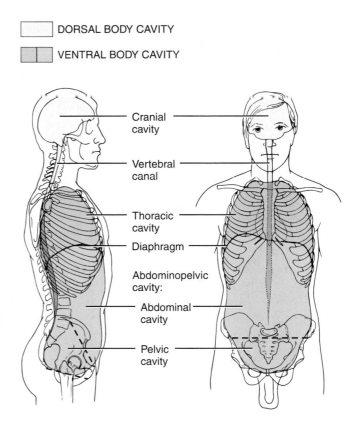

- Cranial cavity
- Vertebral canal
- Thoracic cavity
- Diaphragm
- Abdominopelvic cavity:
  - Abdominal cavity
  - Pelvic cavity

(a) Right lateral view   (b) Anterior view

Q In which cavities are the following organs located: urinary bladder, stomach, heart, pancreas, small intestine, lungs, internal female reproductive organs, thymus gland, spleen, liver? Use the following symbols for your response: T = thoracic cavity, A = abdominal cavity, or P = pelvic cavity.

Body Cavities

OBJECTIVE

• *Describe the principal body cavities, the organs they contain, and their associated linings.*

Body cavities are spaces within the body that help protect, separate, and support internal organs. Bones, muscles, and ligaments separate the various body cavities from one another. The two principal cavities are the dorsal and ventral body cavities (Figure 1.8).

Text continues on page 16

Exhibit 1.1	*Directional Terms Used to Describe the Human Body (Figure 1.9)*

Most of the directional terms used to describe the human body can be grouped into pairs that have opposite meanings. For example, **superior** means toward the upper part of the body, whereas **inferior** means toward the lower part of the body. Moreover, it is important to understand that directional terms have relative meanings; they only make sense when used to describe the position of one structure relative to another. For example, your knee is supe-

rior to your ankle, even though both are located in the inferior half of the body. Study the directional terms in the following table and the example of how each is used. As you read each example, look at Figure 1.8 to see the location of the structures mentioned.

Use each of the directional terms described in Exhibit 1.1 in a sentence. (HINT: Refer to the structures in Figure 1.9.)

DIRECTIONAL TERM	DEFINITION	EXAMPLE OF USE
Superior (soo'-PEER-ē-or) (**cephalic** or **cranial**)	Toward the head, or the upper part of a structure.	The heart is superior to the liver.
Inferior (in'-FEER-ē-or) (**caudal**)	Away from the head, or the lower part of a structure.	The stomach is inferior to the lungs.
Anterior (an-TEER-ē-or) (**ventral**)*	Nearer to or at the front of the body.	The sternum (breastbone) is anterior to the heart.
Posterior (pos-TEER-ē-or) (**dorsal**)*	Nearer to or at the back of the body.	The esophagus (food tube) is posterior to the trachea (windpipe).
Medial (MĒ-dē-al)	Nearer to the midline† or midsagittal plane.	The ulna is on the medial side of the forearm.
Lateral (LAT-er-al)	Farther from the midline or midsagittal plane.	The lungs are lateral to the heart.
Intermediate (in'-ter-MĒ-dē-at)	Between two structures.	The transverse colon is intermediate to the ascending and descending colons.
Ipsilateral (ip-si-LAT-er-al)	On the same side of the body as another structure.	The gallbladder and ascending colon are ipsilateral.
Contralateral (CON-tra-lat-er-al)	On the opposite side of the body from another structure.	The ascending and descending colons are contralateral.
Proximal (PROK-si-mal)	Nearer to the attachment of a limb to the trunk; nearer to the point of origin.	The humerus is proximal to the radius.
Distal (DIS-tal)	Farther from the attachment of a limb to the trunk; farther from the point of origin.	The phalanges are distal to the carpals.
Superficial (soo'-per-FISH-al)	Toward or on the surface of the body.	The ribs are superficial to the lungs.
Deep (DĒP)	Away from the surface of the body.	The ribs are deep to the skin of the chest.

*Ventral refers to the belly side, whereas dorsal refers to the back side. In four-legged animals anterior = cephalic (toward the head), ventral = inferior, posterior = caudal (toward the tail), and dorsal = superior.
†The midline is an imaginary vertical line that divides the body into equal right and left sides.

Exhibit 1.1   (continued)

Figure 1.9 Directional terms.

Directional terms precisely locate various parts of the body relative to one another.

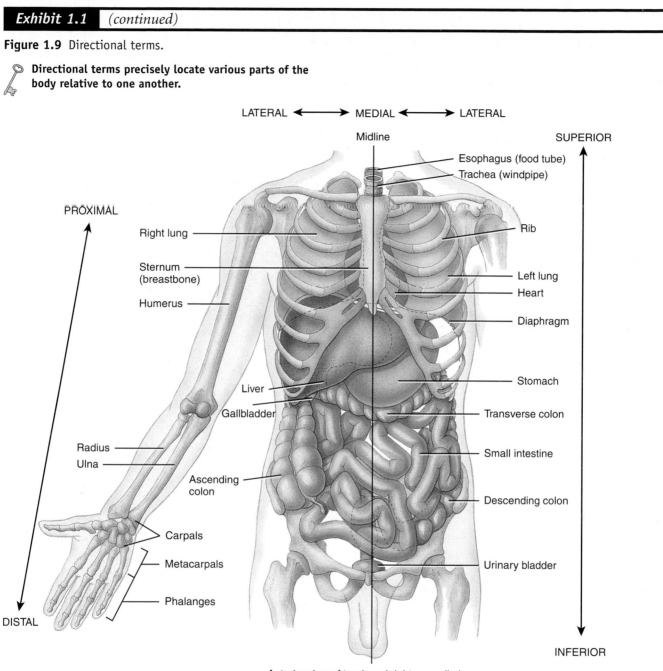

Anterior view of trunk and right upper limb

Q   Is the radius proximal to the humerus? Is the esophagus anterior to the trachea? Are the ribs superficial to the lungs? Is the urinary bladder medial to the ascending colon? Is the sternum lateral to the descending colon?

Figure 1.10 The thoracic cavity. The dashed lines indicate the borders of the mediastinum. Notice that the pericardial cavity surrounds the heart, and that the pleural cavities surround the lungs. (See Tortora, *A Photographic Atlas of the Human Body,* Figure 6.6.)

🔑 **The mediastinum is medial to the lungs; it extends from the sternum to the vertebral column and from the neck to the diaphragm.**

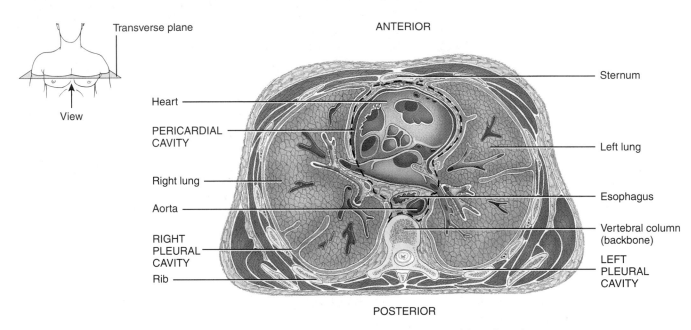

(a) Inferior view of transverse section of thoracic cavity

Dorsal Body Cavity

The **dorsal body cavity** is located near the dorsal (posterior) surface of the body and has two subdivisions, the cranial cavity and the vertebral canal. The **cranial cavity** is formed by the cranial bones and contains the brain. The **vertebral (spinal) canal** is formed by the bones of the vertebral column (backbone) and contains the spinal cord. Three layers of protective tissue, called **meninges** (me-NIN-jēz), line the dorsal body cavity.

Ventral Body Cavity

The other principal body cavity—the **ventral body cavity**—is located on the ventral (anterior) aspect of the body. The ventral body cavity also has two main subdivisions, the thoracic and the abdominopelvic cavities. The **diaphragm** (DĪ-a-fram; = partition or wall), the large dome-shaped muscle that powers lung expansion during breathing, forms the floor of the thoracic cavity and the roof of the abdominopelvic cavity. The organs inside the ventral body cavity are termed the **viscera** (VIS-er-a).

The superior portion of the ventral body cavity is the **thoracic cavity** (thor-AS-ik; *thorac-* = chest) or chest cavity

(Figure 1.10). The thoracic cavity is encircled by the ribs, the muscles of the chest, the sternum (breastbone), and the thoracic portion of the vertebral column or backbone. Within the thoracic cavity are three smaller cavities: the **pericardial cavity** (per′-i-KAR-dē-al; *peri-* = around; *-cardial* = heart), a fluid-filled space that surrounds the heart, and two **pleural cavities** (PLOOR-al; *pleur-* = rib or side). Each pleural cavity surrounds one lung and contains a small amount of fluid. The central portion of the thoracic cavity is called the **mediastinum** (mē′-dē-as-TĪ-num; *media-* = middle; *-stinum* = partition). It is located between the pleural cavities and extends from the sternum to the vertebral column, and from the neck to the diaphragm (see Figure 1.10b). The mediastinum contains all thoracic viscera except the lungs themselves. Among the structures in the mediastinum are the heart, esophagus, trachea, thymus gland, and several large blood vessels.

The inferior portion of the ventral body cavity is the **abdominopelvic cavity** (ab-dom′-i-nō-PEL-vik; see Figure 1.8), which extends from the diaphragm to the groin and is encircled by the abdominal wall and the bones and muscles of the pelvis. As the name suggests, the abdominopelvic cavity is divided into two portions, even though no wall separates

Figure 1.10 (continued)

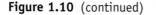

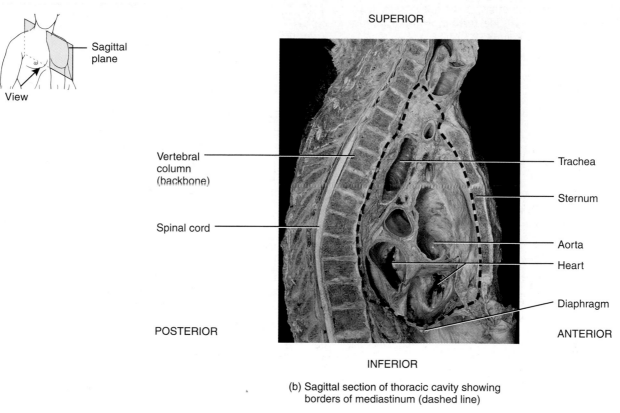

(b) Sagittal section of thoracic cavity showing borders of mediastinum (dashed line)

> **Q** Which of the following structures are contained in the mediastinum: right lung, heart, esophagus, spinal cord, aorta, rib, left pleural cavity?

them (Figure 1.11). The superior portion, the **abdominal cavity** (*abdomin-* = belly), contains the stomach, spleen, liver, gallbladder, small intestine, and most of the large intestine. The inferior portion, the **pelvic cavity** (*pelv-* = basin), contains the urinary bladder, portions of the large intestine, and internal organs of the reproductive system.

Thoracic and Abdominal Cavity Membranes

A thin, **slippery serous membrane** covers the viscera within the thoracic and abdominal cavities and also lines the walls of the thorax and abdomen. The parts of a serous membrane are (1) the *parietal layer* (pa-RĪ-e-tal), which lines the walls of the cavities, and (2) the *visceral layer,* which covers and adheres to the viscera within the cavities. Serous fluid between the two layers reduces friction, allowing the viscera to slide somewhat during movements—for example, when the lungs expand and deflate when a person is breathing.

The serous membrane of the pleural cavities is called the **pleura** (PLOO-ra). The *visceral pleura* clings to the surface of the lungs, whereas the *parietal pleura* lines the chest wall. In between is the pleural cavity, filled with a small volume of serous fluid. The serous membrane of the pericardial cavity is the **pericardium.** The *visceral pericardium* covers the surface of the heart, whereas the *parietal pericardium* lines the chest wall. Between them is the pericardial cavity. The **peritoneum** (per-i-tō-NĒ-um) is the serous membrane of the abdominal cavity. The *visceral peritoneum* covers the abdominal viscera, whereas the *parietal peritoneum* lines the abdominal wall. Between them is the peritoneal cavity.

A summary of body cavities and their membranes is presented in Table 1.3.

1. What landmarks separate the various body cavities from one another?
2. What is the mediastinum?

Figure 1.11 The abdominopelvic cavity. The horizontal black line approximates the point of separation of the abdominal and pelvic cavities. (See Tortora, *A Photographic Atlas of the Human Body,* Figure 12.2)

🔑 **The abdominopelvic cavity extends from the diaphragm to the groin.**

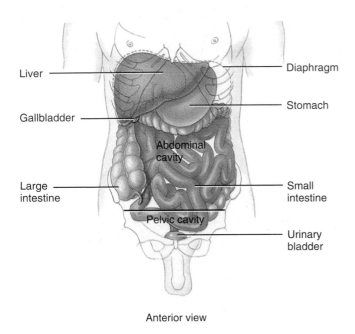

Anterior view

Ⓠ To which body systems do the organs shown here within the abdominal and pelvic cavities belong? (HINT: *Refer to Table 1.2 on page 5.*)

Table 1.3 Summary of Body Cavities and Their Membranes

CAVITY	COMMENTS
Dorsal	
Cranial	Formed by cranial bones and contains brain.
Vertebral	Formed by vertebral column and contains spinal cord.
Ventral	
Thoracic	Superior portion of ventral body cavity; contains pleural and pericardial cavities and the mediastinum.
Pleural	Each surrounds a lung; the serous membrane of the pleural cavities is the pleura.
Pericardial	Surrounds the heart; the serous membrane of the pericardial cavity is the pericardium.
Mediastinum	Central portion of thoracic cavity between the pleural cavities; extends from sternum to vertebral column and from neck to diaphragm; contains heart, thymus gland, esophagus, trachea, and several large blood vessels.
Abdominopelvic	Inferior portion of ventral body cavity; subdivided into abdominal and pelvic cavities.
Abdominal	Contains stomach, spleen, liver, gallbladder, pancreas, small intestine, and most of large intestine; the serous membrane of the abdominal cavity is the peritoneum.
Pelvic	Contains urinary bladder, portions of large intestine, and internal organs of reproduction.

Abdominopelvic Regions and Quadrants

OBJECTIVE

• *Name and describe the nine abdominopelvic regions and the four abdominopelvic quadrants.*

To describe the location of the many abdominal and pelvic organs more easily, anatomists and clinicians use two methods of dividing the abdominopelvic cavity into smaller compartments. In the first method, two transverse and two vertical lines, aligned like a tic-tac-toe grid, partition this cavity into nine **abdominopelvic regions** (Figure 1.12a). The *subcostal* (top transverse) *line* is drawn just inferior to the rib cage, across the inferior portion of the stomach; the *transtubercular* (bottom transverse) *line* is drawn just inferior to the tops of the hipbones. The left and right *midclavicular* (two vertical) *lines* are drawn through the midpoints of the clavicles (collar bones), just medial to the nipples. The four lines divide the abdominopelvic cavity into a larger middle section and smaller left and right sections. The names of the nine abdominopelvic regions are right hypochondriac, epigastric, left hypochondriac, right lumbar, umbilical, left lumbar, right inguinal (iliac), hypogastric (pubic), and left inguinal (iliac).

The second method is simpler and divides the abdominopelvic cavity into **quadrants** (KWOD-rantz; *quad*- = one-fourth), as shown in Figure 1.12b. In this method, a transverse plane and a midsagittal plane are passed through the **umbilicus** (um-bi-LĪ-kus; umbilic- = navel) or *belly button.* The names of the abdominopelvic quadrants are right upper quadrant (RUQ), left upper quadrant (LUQ), right lower quadrant (RLQ), and left lower quadrant (LLQ). Whereas the nine-region division is more widely used for anatomical studies, quadrants are more commonly used by clinicians for describing the site of abdominopelvic pain, tumor, or other abnormality.

1. Locate the nine abdominopelvic regions and the four abdominopelvic quadrants on yourself, and list some of the organs found in each.

Figure 1.12 Regions and quadrants of the abdominopelvic cavity.

🔑 The nine-region designation is used for anatomical studies, whereas the quadrant designation is used to locate the site of pain, tumor, or some other abnormality.

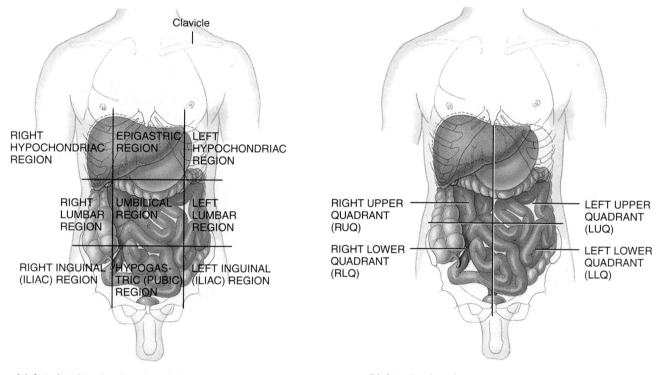

(a) Anterior view showing nine abdominopelvic regions

(b) Anterior view showing abdominopelvic quadrants

Q In which abdominopelvic *region* is each of the following found: most of the liver, transverse colon, urinary bladder, spleen? In which abdominopelvic *quadrant* would pain from appendicitis (inflammation of the appendix) be felt?

MEDICAL IMAGING

OBJECTIVE

• *Describe the principles and importance of medical imaging procedures in the evaluation of organ functions and the diagnosis of disease.*

Various kinds of **medical imaging** procedures allow visualization of structures inside our bodies and are increasingly helpful for precise diagnosis of a wide range of anatomical and physiological disorders. The grandparent of all medical imaging techniques is conventional radiography, in medical use since the late 1940s. The newer imaging technologies not only contribute to diagnosis of disease, but they also are advancing our understanding of normal physiology. Table 1.4 describes some commonly used medical imaging techniques. Other imaging methods, such as cardiac catheterization, will be discussed in other chapters.

1. Which forms of medical imaging use radiation? Which do not?
2. Which imaging modality best reveals the physiology of a structure?
3. Which imaging modality would best show a broken bone?

Table 1.4 Common Medical Imaging Procedures

Radiography

Procedure: A single barrage of x-rays passes through the body, producing an image of interior structures on x-ray–sensitive film. The resulting two-dimensional image is a *radiograph* (RĀ-dē-ō-graf'), commonly called an *x-ray*.

Comments: Produces clear images of bony or dense structures, which appear bright, but poor images of soft tissues or organs, which appear hazy or dark.

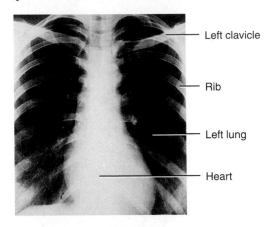

Left clavicle

Rib

Left lung

Heart

Anterior view of thorax

Computed tomography (CT)

[formerly called computerized axial tomography (CAT) scanning]

Procedure: Computer-assisted radiography in which an x-ray beam traces an arc at multiple angles around a section of the body. The resulting transverse section of the body, called a *CT scan,* is reproduced on a video monitor.

Comments: Visualizes soft tissues and organs with much more detail than conventional radiographs. Differing tissue densities show up as various shades of gray. Multiple scans can be assembled to build three-dimensional views of structures.

ANTERIOR

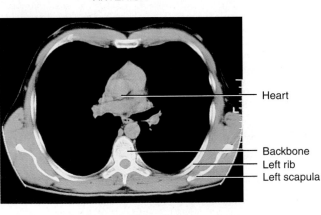

Heart

Backbone
Left rib
Left scapula

POSTERIOR

Inferior view of transverse section of the thorax

Note: In keeping with radiographic convention the section is viewed from the inferior aspect. Thus the left side of the body appears on the right side of the photograph.

Digital subtraction angiography (DSA)

Procedure: A computer compares radiographs of a body region before and after a dye is injected into blood vessels. Tissues around the blood vessels are erased (digitally subtracted) from the second image. The result, shown on a monitor, is an unobstructed view of the blood vessels.

Comments: Used primarily to study blood vessels in the brain and heart.

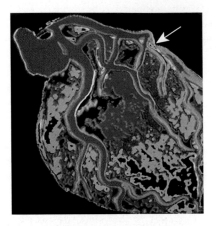

Blood vessels (red) surrounding heart
(arrow indicates narrowed vessel)

Sonography

Procedure: High-frequency sound waves produced by a handheld wand reflect off body tissues and are detected by the same instrument. The image, which may be still or moving, is called a **sonogram** (SŌ-nō-gram) and is reproduced on a video monitor.

Comments: Safe, noninvasive, painless, and uses no dyes. Most commonly used to visualize the fetus during pregnancy. Also used to observe the size, location, and actions of organs and blood flow through blood vessels.

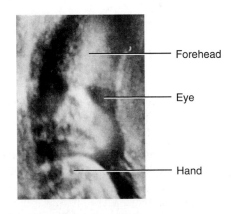

Forehead

Eye

Hand

Courtesy of Andrew Joseph Tortora
and Damaris Soler

Table 1.4 (continued)

Magnetic resonance imaging (MRI)

Procedure: The body is exposed to a high-energy magnetic field, which causes protons (small positive particles within atoms, such as hydrogen) in body fluids and tissues to arrange themselves in relation to the field. Then a pulse of radiowaves "reads" these ion patterns, and a color-coded image is assembled on a video monitor. The resulting image is a two- or three-dimensional blueprint of cellular chemistry.

Comments: Relatively safe, but can't be used on patients with metal in their bodies. Shows fine details for soft tissues but not for bones. Most useful for differentiating between normal and abnormal tissues. Used to detect tumors and artery-clogging fatty plaques, reveal brain abnormalities, and measure blood flow.

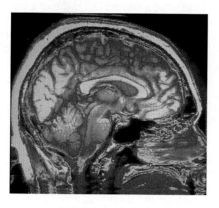

Sagittal section of brain

Positron emission tomography (PET)

Procedure: A substance that emits positrons (positively charged particles) is injected into the body, where it is taken up by tissues. The collision of positrons with negatively charged electrons in body tissues produces gamma rays (similar to x-rays) that are detected by gamma cameras positioned around the subject. A computer receives signals from the gamma cameras and constructs a *PET scan* image, displayed in color on a video monitor. The PET scan shows where the injected substance is being used in the body.

Comments: Used to study the physiology of body structures, such as metabolism in the brain or heart.

ANTERIOR

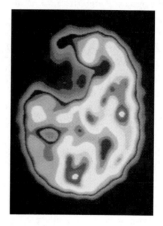

POSTERIOR

Transverse section showing blood flow through brain (darkened area at upper left indicates where a stroke has occurred)

STUDY OUTLINE

ANATOMY AND PHYSIOLOGY DEFINED (p. 1)

1. Anatomy is the science of body structures and the relationships among structures; physiology is the science of body functions.
2. Dissection is the careful cutting apart of body structures to study their relationships.
3. Some subdisciplines of anatomy are embryology, developmental anatomy, cytology, histology, surface anatomy, gross anatomy, systemic anatomy, regional anatomy, radiographic anatomy, and pathological anatomy (see Table 1.1 on page 2).
4. Some subdisciplines of physiology are cell physiology, neurophysiology, endocrinology, cardiovascular physiology, immunology, respiratory physiology, renal physiology, exercise physiology, and pathophysiology (see Table 1.1 on page 2).
5. Genetics is the science of heredity, the transmission of traits from parents to offspring.

LEVELS OF BODY ORGANIZATION (p. 2)

1. The human body consists of six levels of structural organization: chemical, cellular, tissue, organ, system, and organismal levels.
2. Cells are the basic structural and functional living units of an organism and the smallest living units in the human body.
3. Tissues are groups of cells and the materials surrounding them that work together to perform a particular function.
4. Organs are composed of two or more different types of tissues; they have specific functions and usually have recognizable shapes.
5. Systems consist of related organs that have a common function.
6. An organism is any living individual.
7. Table 1.2 on page 5 introduces the eleven systems of the human organism: the integumentary, skeletal, muscular, nervous, endocrine, cardiovascular, lymphatic and immune, respiratory, digestive, urinary, and reproductive systems.

CHARACTERISTICS OF THE LIVING HUMAN ORGANISM (p. 4)

1. All organisms carry on certain processes that distinguish them from nonliving things.
2. The most important life processes of the human body are metabolism, responsiveness, movement, growth, differentiation, and reproduction.
3. Homeostasis is a condition of equilibrium in the body's internal environment produced by the interplay of all the body's regulatory processes.
4. Body fluids are dilute, watery solutions. Intracellular fluid (ICF) is inside cells, and extracellular fluid (ECF) is outside cells. Interstitial fluid is the ECF that fills spaces between tissue cells, whereas plasma is the ECF within blood vessels.
5. Because it surrounds all body cells, interstitial fluid is called the body's internal environment.

CONTROL OF HOMEOSTASIS (p. 7)

1. Disruptions of homeostasis come from external and internal stimuli and psychological stresses.
2. When disruption of homeostasis is mild and temporary, responses of body cells quickly restore balance in the internal environment. If disruption is extreme, regulation of homeostasis may fail.
3. Most often, homeostasis is regulated by the nervous and endocrine systems acting together or separately. The nervous system detects body changes and sends nerve impulses to counteract the stress. The endocrine system regulates homeostasis by secreting hormones.
4. The components of a feedback system are (1) receptors that monitor changes in a controlled condition and send input to (2) a control center that sets the value at which a controlled condition should be maintained, evaluates the input it receives from receptors, and generates output commands when they are needed, and (3) effectors that receive output from the control center and produce a response (effect) that alters the controlled condition.
5. If a response reverses the original stimulus, the system is operating by negative feedback. If a response enhances the original stimulus, the system is operating by positive feedback.
6. One example of negative feedback is a system that helps to regulate blood pressure. If a stimulus causes blood pressure (controlled condition) to rise, baroreceptors (pressure-sensitive nerve cells, the receptors) in blood vessels send impulses (input) to the brain (control center). The brain sends impulses (output) to the heart (effector). As a result, heart rate decreases (response) and blood pressure drops back to normal (restoration of homeostasis).
7. One example of positive feedback occurs during the birth of a baby. When labor begins, the cervix of the uterus is stretched (stimulus), and stretch-sensitive nerve cells in the cervix (receptors) send nerve impulses (input) to the brain (control center). The brain responds by releasing oxytocin (output), which stimulates the uterus (effector) to contract more forcefully (response). Movement of the baby further stretches the cervix, more oxytocin is released, and even more forceful contractions occur. The cycle is broken with the birth of the baby.
8. Disruptions of homeostasis—homeostatic imbalances—can lead to disorders, diseases, and even death.
9. Disorder is a general term for any derangement or abnormality of function. A disease is an illness with a definite set of signs and symptoms.
10. Symptoms are subjective changes in body functions that are not apparent to an observer, whereas signs are objective changes that can be observed and measured.

BASIC ANATOMICAL TERMINOLOGY (p. 10)

Body Positions (p. 10)

1. Descriptions of any region of the body assume the body is in the anatomical position, in which the subject stands erect facing the observer, with the head level and the eyes facing directly forward. The feet are flat on the floor and directed forward, and the arms are at the sides, with the palms turned forward.
2. A body lying face down is prone, whereas a body lying face up is supine.

Regional Names (p. 12)

1. Regional names are terms given to specific regions of the body. The principal regions are the head, neck, trunk, upper limbs, and lower limbs.
2. Within the regions, specific body parts have common names and are specified by corresponding anatomical terms. Examples are chest (thoracic), nose (nasal), and wrist (carpal).

Planes and Sections (p. 12)

1. Planes are imaginary flat surfaces that are used to divide the body or organs into definite areas. A midsagittal plane divides the body or an organ into equal right and left sides; a parasagittal plane divides the body or an organ into unequal right and left sides; a frontal plane divides the body or an organ into anterior and posterior portions; a transverse plane divides the body or an organ into superior and inferior portions; and an oblique plane passes through the body or an organ at an angle between a transverse plane and either a midsagittal, parasagittal, or frontal plane.
2. Sections are flat surfaces resulting from cuts through body structures. They are named according to the plane on which the cut is made and include transverse, frontal, and sagittal sections.

Directional Terms (p. 12)

1. Directional terms indicate the relationship of one part of the body to another.
2. Exhibit 1.1 on page 14 summarizes commonly used directional terms.

Body Cavities (p. 13)

1. Spaces in the body that help protect, separate, and support internal organs are called body cavities.
2. The dorsal and ventral cavities are the two principal body cavities.
3. The dorsal cavity is subdivided into the cranial cavity, which contains the brain, and the vertebral canal, which contains the spinal cord. The meninges are protective tissues that line the dorsal cavity.

4. The ventral body cavity is subdivided by the diaphragm into a superior thoracic cavity and an inferior abdominopelvic cavity. The viscera are organs within the ventral body cavity. A serous membrane lines the wall of the cavity and adheres to the viscera.
5. The thoracic cavity is subdivided into three smaller cavities: a pericardial cavity, which contains the heart, and two pleural cavities, which contain the lungs.
6. The central portion of the thoracic cavity is the mediastinum. It is located between the pleural cavities and extends from the sternum to the vertebral column and from the neck to the diaphragm. It contains all thoracic viscera except the lungs.
7. The abdominopelvic cavity is divided into a superior abdominal and an inferior pelvic cavity.
8. Viscera of the abdominal cavity include the stomach, spleen, liver, gallbladder, pancreas, small intestine, and most of the large intestine.
9. Viscera of the pelvic cavity include the urinary bladder, portions of the large intestine, and internal organs of the reproductive system.
10. Serous membranes line the walls of the thoracic and abdominal cavities and cover the organs within them. They include the pleura, associated with the lungs; the pericardium, associated with the heart; and the peritoneum, associated with the abdominal cavity.
11. Table 1.3 on page 18 summarizes body cavities and their membranes.

Abdominopelvic Regions and Quadrants (p. 18)

1. To describe the location of organs more easily, the abdominopelvic cavity is divided into nine abdominopelvic regions: right hypochondriac, epigastric, left hypochondriac, right lumbar, umbilical, left lumbar, right inguinal (iliac), hypogastric (pubic), and left inguinal (iliac).
2. To locate the site of an abdominopelvic abnormality in clinical studies, the abdominopelvic cavity is divided into quadrants: right upper quadrant (RUQ), left upper quadrant (LUQ), right lower quadrant (RLQ), and left lower quadrant (LLQ).

MEDICAL IMAGING (p. 19)

1. Medical imaging techniques allow visualization of internal structures to diagnose abnormal anatomy and deviations from normal physiology.
2. Table 1.4 on page 20 summarizes several medical imaging techniques.

SELF-QUIZ QUESTIONS

Complete the following:

1. The basic structural and functional unit of the human organism is the ___.
2. The four basic types of tissue in the body are ___, ___, ___, and ___.
3. The phase of metabolism that involves the breaking down of large, complex molecules into smaller, simpler ones is called ___.
4. The condition in which the body's internal environment remains within certain physiological limits is termed ___.
5. Match the following:

___ (1) nervous system	(a) regulates body activities through chemicals transported in blood to various target organs of the body
___ (2) endocrine system	(b) regulates the volume and chemical composition of blood
___ (3) urinary system	(c) powers movements of the body and stabilizes body position
___ (4) cardiovascular system	(d) supports and protects the body, provides internal framework
___ (5) muscular system	(e) transports oxygen and nutrients to cells, protects against disease, carries wastes away from cells
___ (6) respiratory system	(f) regulates body activities through action potentials, receives sensory information, interprets the information, and responds to the information
___ (7) digestive system	(g) carries out the physical and chemical breakdown of food and absorption of nutrients
___ (8) skeletal system	(h) supplies oxygen, eliminates carbon dioxide

True or false:

6. In a negative feedback system, the response enhances or intensifies the original stimulus.
7. The ventral body cavity contains the heart, lungs, and abdominal viscera.

Choose the best answer to the following questions:

8. Which of the following statements are true of a feedback system? (1) A feedback system consists of a control center, a receptor, and an effector. (2) The control center receives input from the effector. (3) The receptor monitors environmental changes that affect the body. (4) The effector produces a response to some stress. (5) A feedback system involves a cycle of events in which information about the status of a condition is monitored and fed back to a central control region. (a) 1 and 2, (b) 1, 2, and 3, (c) 1, 3, and 5, (d) 1, 3, 4, and 5, (e) 2 and 4

9. Which of the following statements are true of a serous membrane? (1) It lines body cavities that do not open directly to the exterior. (2) It lines body cavities that open directly to the exterior. (3) It is a double-layered membrane. (4) Examples include the pleura, pericardium, and peritoneum. (5) It lines the cavity only and does not cover the organs within the cavity. (a) 1, 3, 4, and 5, (b) 1, 3, and 4, (c) 1, 3 and 5, (d) 1 and 3, (e) 1 and 5

10. In which abdominopelvic region is the appendix found? (a) right iliac, (b) right lumbar, (c) hypogastric (pubic), (d) umbilical, (e) epigastric.

11. A vertical plane that divides the body or an organ into right and left sides is termed a (an) (a) frontal plane, (b) sagittal plane, (c) transverse plane, (d) oblique plane, (e) coronal plane.

12. The two systems that regulate homeostatic responses of the body are the (a) nervous and cardiovascular systems, (b) respiratory and cardiovascular systems, (c) endocrine and cardiovascular systems, (d) cardiovascular and urinary systems, (e) endocrine and nervous systems.

13. The membrane associated with the heart is the (a) pleura, (b) peritoneum, (c) pericardium, (d) synovial membrane, (e) cutaneous membrane.

14. Match the following common and anatomical terms:

___ (1) axillary	(a) skull
___ (2) inguinal	(b) eye
___ (3) cervical	(c) cheek
___ (4) cranial	(d) armpit
___ (5) brachial	(e) arm
___ (6) orbital	(f) groin
___ (7) gluteal	(g) buttock
___ (8) buccal	(h) neck

15. Match each of the following directional terms to its definition:

___ (1) at the front of the body	(a) superior
___ (2) closer to the trunk	(b) inferior
___ (3) toward the upper part of a structure	(c) anterior (ventral)
___ (4) nearer to the midline of the body	(d) posterior (dorsal)
___ (5) farther from the midline of the body	(e) medial
___ (6) at the back of the body	(f) lateral
___ (7) farther from the trunk	(g) proximal
___ (8) toward the lower part of a structure	(h) distal

CRITICAL THINKING QUESTIONS

1. On her first anatomy and physiology exam, Heather defined homeostasis as "the condition in which the body approaches room temperature and stays there." Do you agree with Heather's definition? (HINT: *Does an oral thermometer measure down to 72°F or 25°C?*)

2. Elena complained of numbness and tingling sensations in both of her hands. Her physician recommended bilateral splints for carpal tunnel syndrome. Where would she wear these splints? (HINT: *The splints somewhat restrict movement of the hand.*)

3. Eighty-year-old Harold fell off a ladder and may have broken his arm. The emergency room physician ordered some x-rays, but Harold said that he "wants one of those fancy, new x-rays—an MRI. And while you're taking my picture," he continued, "check out that pacemaker near my heart, too." The physician refused to order an MRI. Why? (HINT: *Pacemakers contain metal.*)

ANSWERS TO FIGURE QUESTIONS

1.1 Organs have two or more different types of tissues that work together to perform a specific function.

1.2 The basic difference between negative and positive feedback systems is that in negative feedback systems, the response reverses the original stimulus, whereas in positive feedback systems, the response enhances the original stimulus.

1.3 When something causes blood pressure to decrease, then heart rate increases due to operation of this negative feedback system.

1.4 Because positive feedback systems continually intensify or reinforce the original stimulus, a termination mechanism is needed to end the response.

1.5 Having one standard anatomical position allows directional terms to be clear, and any body part can be described in relation to any other part.

1.6 The frontal plane divides the heart into anterior and posterior portions.

1.7 The parasagittal plane divides the brain into unequal right and left portions.

1.8 P, A, T, A, A, T, P, T, A, A.

1.9 No, No, Yes, Yes, No.

1.10 Structures in the mediastinum are the heart, esophagus, trachea, thymus gland, and several large blood vessels.

1.11 The illustrated abdominal cavity organs all belong to the digestive system (liver, gallbladder, stomach, appendix, small intestine, and most of the large intestine). Illustrated pelvic cavity organs belong to the urinary system (the urinary bladder) and the digestive system (part of the large intestine).

1.12 The liver is mostly in the epigastric region; the transverse colon is in the umbilical region; the urinary bladder is in the hypogastric region; the spleen is in the left hypochondriac region. The pain associated with appendicitis would be felt in the right lower quadrant (RLQ).

www.awl.com/bc

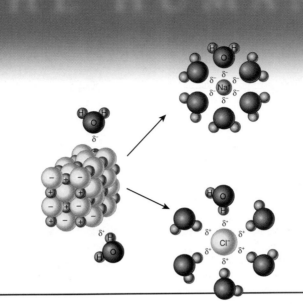

You learned in Chapter 1 that the chemical level of organization consists of atoms and molecules, which combine in the body to form structures and systems of astonishing size and complexity. In this chapter we will consider how atoms bond together in the body to form molecules. You will also discover how atoms and molecules release or store energy in processes known as chemical reactions, and how nearly every chemical reaction in the body is made possible by the body's water content. Finally, you will learn about five families of molecules whose unique properties are harnessed to assemble the body's structures or to power the processes that characterize life.

Chemistry (KEM-is-trē) is the science of the structure and interactions of matter. All living and nonliving things consist of **matter,** which is anything that occupies space and has mass. The amount of matter in any object is its **mass.** On Earth, *weight* is the force of gravity acting on matter. Things weigh less when they are farther from Earth because the pull of gravity is weaker. In outer space, weight is close to zero but mass remains the same as it is on Earth.

HOW MATTER IS ORGANIZED

OBJECTIVES

• *Identify the principal chemical elements of the human body.*

• *Describe the structures of atoms, ions, molecules, free radicals, and compounds.*

Chemical Elements

All forms of matter—both living and nonliving—are made up of a limited number of building blocks called **chemical elements.** Each element is a substance that cannot be split into a simpler substance by ordinary chemical means. Scientists now recognize 112 different elements; 92 occur naturally on Earth, and the rest are produced from the natural elements using devices such as particle accelerators or nuclear reactors. Elements are designated by **chemical symbols,** which are the first one or two letters of the element's name in English, Latin, or another language. Examples of elements and their chemical symbols include hydrogen (H), carbon (C), oxygen (O), nitrogen (N), calcium (Ca), sodium (Na; *natrium* = sodium), potassium (K; *kalium* = potassium), iron (Fe; *ferrum* = iron), and phosphorus (P).*

Twenty-six of the 92 naturally occurring elements normally are present in your body. Of these, just four elements—oxygen, carbon, hydrogen, and nitrogen—constitute about 96% of the body's mass. Nine others—calcium, phosphorus, potassium, sulfur (S), sodium, chlorine (Cl), magnesium (Mg), iodine (I), and iron—contribute an additional 3.9% of the body's mass. Table 2.1 lists these 13 elements that make up most of the body and provides an overview of their importance. An additional 13 elements,

---

*The periodic table of elements, which lists all of the known chemical elements, can be found in Appendix B.

called *trace elements* because they are present in tiny concentrations, account for the remaining 0.1% of the body's mass. The trace elements are aluminum (Al), boron (B), chromium (Cr), cobalt (Co), copper (Cu), fluorine (F), manganese (Mn), molybdenum (Mo), selenium (Se), silicon (Si), tin (Sn), vanadium (V), and zinc (Zn). Whereas some trace elements are known to have important functions in the body, the functions of others are unknown.

Structure of Atoms

Each element is made up of **atoms,** the smallest units of matter that retain the properties and characteristics of an element. Atoms are extremely small. Two hundred thousand of the largest atoms would fit on the period at the end of this sentence. Hydrogen atoms, the smallest atoms, have a diameter less than 0.1 nanometer (0.1×10^{-9} m = 0.0000000001 m), and the largest atoms are only five times larger.

Atoms consist of three major types of **subatomic particles:** protons, neutrons, and electrons (Figure 2.1). The dense central core of an atom is its **nucleus,** which contains positively charged particles called **protons (p^+)** and uncharged (neutral) particles called **neutrons (n^0).** The tiny negatively charged **electrons (e^-)** move about in a large space surrounding the nucleus. They do not follow a fixed path or orbit but instead form a negatively charged "cloud" that envelopes the nucleus (see Figure 2.1a). Even though their exact positions cannot be predicted, specific groups of electrons are most likely to be found within certain regions around the nucleus. These regions are called **electron shells,** which are drawn as circles about the nucleus even though their shapes are not all spherical. Each electron shell can hold only a limited number of electrons; the electron shell model best conveys this aspect of atomic structure (see Figure 2.1b). For instance, the electron shell nearest the nucleus never holds more than two electrons. This is the first shell. The second shell holds a maximum of eight electrons, whereas the third can hold up to 18 electrons. For example, notice in Figure 2.2 that sodium (Na) contains two electrons in the first shell, eight electrons in the second shell, and one electron in the third shell. Higher shells (there are as many as seven) can contain many more electrons. The most massive element present in the human body is iodine, which has a total of 53 electrons: 2 in the first shell, 8 in the second shell, 18 in the third shell, 18 in the fourth shell, and 7 in the fifth shell.

The number of electrons in an atom of an element always equals the number of protons. Because each electron carries one negative charge, the negatively charged electrons and the positively charged protons balance each other. Thus each atom is electrically neutral; its total charge is zero.

Atomic Number and Mass Number

The *number of protons* in the nucleus of an atom, designated by the atom's **atomic number,** distinguishes the atoms

Table 2.1 Main Chemical Elements in the Body

CHEMICAL ELEMENT (SYMBOL)	% OF TOTAL BODY MASS	SIGNIFICANCE
Oxygen (O)	65.0	Part of water and many organic (carbon-containing molecules); used to generate ATP, a molecule used by cells to temporarily store chemical energy.
Carbon (C)	18.5	Forms backbone chains and rings of all organic molecules—carbohydrates, lipids (fats), proteins, and nucleic acids (DNA and RNA).
Hydrogen (H)	9.5	Constituent of water and most organic molecules; ionized form (H^+) makes body fluids more acidic.
Nitrogen (N)	3.2	Component of all proteins and nucleic acids.
Calcium (Ca)	1.5	Contributes to hardness of bones and teeth; ionized form (Ca^{2+}) needed for blood clotting, release of hormones, contraction of muscle, and many other processes.
Phosphorus (P)	1.0	Component of nucleic acids and ATP; required for normal bone and tooth structure.
Potassium (K)	0.4	Ionized form (K^+) is the most plentiful cation (positively charged particle) in intracellular fluid; needed for nerve and muscle impulses.
Sulfur (S)	0.3	Component of some vitamins and many proteins.
Sodium (Na)	0.2	Ionized form (Na^+) is the most plentiful cation in extracellular fluid; essential for maintaining water balance; needed for nerve and muscle impulses.
Chlorine (Cl)	0.2	Ionized form (Cl^-) is the most plentiful anion (negatively charged particle) in extracellular fluid; essential for maintaining water balance.
Magnesium (Mg)	0.1	Ionized form (Mg^{2+}) needed for action of many enzymes, molecules that increase the rate of chemical reactions in organisms.
Iodine (I)	0.1	Part of thyroid hormones, which regulate metabolism.
Iron (Fe)	0.1	Ionized forms (Fe^{2+} and Fe^{3+}) are part of hemoglobin (oxygen-carrying protein in blood) and some enzymes.

of one element from those of another. Figure 2.2 shows that atoms of different elements have different atomic numbers because they have different numbers of protons. For example, oxygen has an atomic number of 8 because its nucleus

Figure 2.1 Two representations of the structure of an atom. Electrons move about the nucleus, which contains neutrons and protons. (a) In the electron cloud model of an atom, the shading represents the chance of finding an electron in regions outside the nucleus. (b) In the electron shell model, filled circles represent individual electrons, which are grouped into concentric circles according to the shells they occupy. Both models depict a carbon atom, with six protons, six neutrons, and six electrons.

An atom is the smallest unit of matter that retains the properties and characteristics of its element.

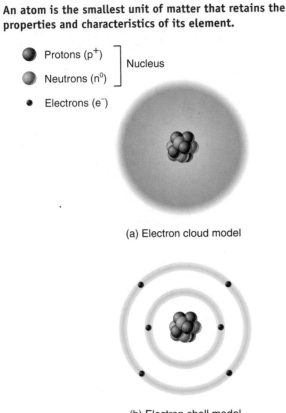

(a) Electron cloud model

(b) Electron shell model

Q How are the electrons of carbon distributed between the first and second electron shells?

has eight protons, whereas sodium has an atomic number of 11 because its nucleus has 11 protons.

The **mass number** of an atom is the sum of its protons and neutrons. For sodium, which has 11 protons and 12 neutrons, the mass number is 23 (see Figure 2.2). Although all atoms of one element have the same number of protons, they may have different numbers of neutrons and thus different mass numbers. Atoms of an element that have different mass numbers are called **isotopes.** In a sample of oxygen, for example, most atoms have eight neutrons, but a few have nine or ten, even though all have eight protons and eight electrons. Most isotopes are stable, which means that their nuclear structure does not change over time. The stable isotopes of oxygen are designated ^{16}O, ^{17}O, and ^{18}O (or O-16, O-17, and O-18). The numbers indicate the mass number

(total number of protons and neutrons) of each isotope. As you will discover, the chemical properties of an atom are a function of its electrons. Although the isotopes of an element have different numbers of neutrons, they have the same number of electrons and thus identical chemical properties.

Certain isotopes called **radioactive isotopes** are unstable; their nuclei decay into a simpler, more stable configuration. Examples are H-3, C-14, O-15, and O-19. As they decay, these atoms emit radiation—either subatomic particles or packets of energy—and in the process often transform into a different element. For example, the radioactive isotope of carbon, C-14, decays to N-14. The decay of a radioisotope may be fast, occurring in a fraction of a second, or slow, taking millions or even billions of years. Each radioactive isotope has a characteristic **half-life,** which is the time required for half of the radioactive atoms in a sample of that isotope to decay into a more stable form. The half-life of C-14 is 5600 years, whereas the half-life of I-131 is 8 days.

CLINICAL APPLICATION
Harmful and Beneficial Effects of Radiation

Radioactive isotopes can produce both harmful and helpful effects. They can pose a serious threat to the human body because their radiations can break apart molecules, which may produce tissue damage as well as cause various cancers. Although the decay of naturally occurring radioactive isotopes typically releases just a small amount of radiation into the environment, localized accumulations can occur. For instance, radon-222, a colorless and odorless gas, is a naturally occurring radioactive breakdown product of uranium that may seep out of the soil and accumulate in buildings. Radon exposure greatly increases the risk of lung cancer in smokers and causes many cases of lung cancer in nonsmokers. On the other hand, certain radioactive isotopes are used in medical imaging procedures and to treat cancer by killing cancerous cells. ■

Atomic Mass

The standard unit for measuring the mass of atoms and their subatomic particles is a **dalton,** also known as an *atomic mass unit (amu)*. A neutron has a mass of 1.008 daltons, and a proton has a mass of 1.007 daltons. The mass of an electron, however, is 0.0005 dalton, almost 2000 times smaller than the mass of a neutron or proton. The **atomic mass** (also called the *atomic weight*) of an element is the average mass of all its naturally occurring isotopes; it reflects the relative abundance of isotopes with different mass numbers. The atomic mass of chlorine, for example, is 35.45 daltons. About 76% of all chlorine atoms have 18 neutrons (mass number = 35), whereas 24% have 20 neutrons (mass number = 37). Typically, the atomic mass of an element is close to the mass number of its most abundant isotope.

The mass of a single atom is slightly less than the sum of the masses of its neutrons, protons, and electrons because

Figure 2.2 Atomic structures of several stable atoms.

🔑 **The atoms of different elements have different atomic numbers because they have different numbers of protons.**

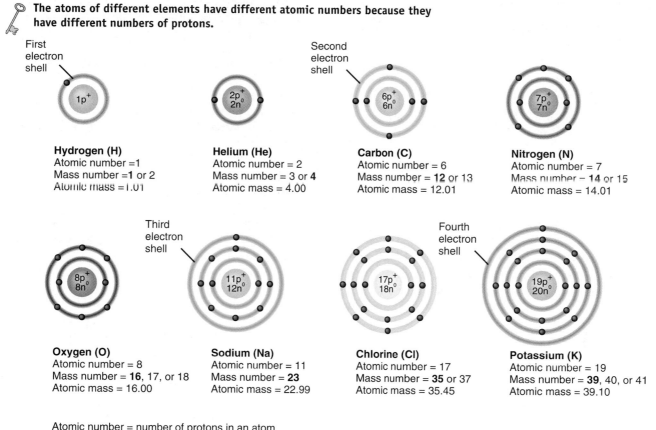

First electron shell

Hydrogen (H)
Atomic number =1
Mass number =**1** or 2
Atomic mass =1.01

Helium (He)
Atomic number = 2
Mass number = 3 or **4**
Atomic mass = 4.00

Second electron shell

Carbon (C)
Atomic number = 6
Mass number = **12** or 13
Atomic mass = 12.01

Nitrogen (N)
Atomic number = 7
Mass number – **14** or 15
Atomic mass = 14.01

Third electron shell

Oxygen (O)
Atomic number = 8
Mass number = **16**, 17, or 18
Atomic mass = 16.00

Sodium (Na)
Atomic number = 11
Mass number = **23**
Atomic mass = 22.99

Fourth electron shell

Chlorine (Cl)
Atomic number = 17
Mass number = **35** or 37
Atomic mass = 35.45

Potassium (K)
Atomic number = 19
Mass number = **39**, 40, or 41
Atomic mass = 39.10

Atomic number = number of protons in an atom
Mass number = number of protons and neutrons in an atom (boldface indicates most common isotope)
Atomic mass = average mass of all stable atoms of a given element

Ⓠ   Which four of these elements are present most abundantly in living organisms?

some mass (less than 1%) was lost when the atom's components came together to form an atom. This explains why an element's atomic mass can be slightly less than the mass number of its smallest stable isotope. For example, the atomic mass of sodium is less than 23.

Ions, Molecules, Free Radicals, and Compounds

As we discussed, atoms of the same element have the same number of protons. In addition, the atoms of each element have a characteristic way of losing, gaining, or sharing their electrons when interacting with other atoms. The way that electrons behave enables atoms in the body to exist in electrically charged forms called ions, or to join with each other into the complex combinations called molecules. If an atom either *gives up* or *gains* electrons, it becomes an **ion**— an atom that has a positive or negative charge because it has unequal numbers of protons and electrons. This process of giving up or gaining electrons is called *ionization*. An ion of

an atom is symbolized by writing its chemical symbol followed by the number of its positive (+) or negative (−) charges. For example, Ca^{2+} stands for a calcium ion that has two positive charges because it has lost two electrons.

In contrast, when two or more atoms *share* electrons, the resulting combination is called a **molecule** (MOL-e-kyool). A molecule may consist of two atoms of the same kind—for example, an oxygen molecule (Figure 2.3a). A *molecular formula* indicates the elements and the number of atoms of each element that make up a molecule. The molecular formula for a molecule of oxygen is O_2. The subscript 2 indicates that there are two atoms in the molecule. Two or more different kinds of atoms may also form a molecule, as in a water molecule: H_2O. Here one atom of oxygen shares electrons with two atoms of hydrogen.

A **free radical** is an electrically charged atom or group of atoms with an unpaired electron in its outermost shell. A common example is superoxide, which is formed by the addition of an electron to an oxygen molecule (Figure 2.3b).

Figure 2.3 Atomic structures of an oxygen molecule and a superoxide free radical.

 A free radical has an unpaired electron in its outermost electron shell.

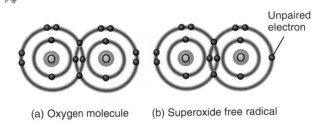

Unpaired electron

(a) Oxygen molecule   (b) Superoxide free radical

Q What substances in the body can inactivate oxygen-derived free radicals?

Having an unpaired electron makes a free radical unstable, highly reactive, and destructive to nearby molecules. Free radicals become stable by either giving up their unpaired electron to, or taking on an electron from, another molecule. In so doing, free radicals may break apart important body molecules.

A **compound** is a substance that can be broken down into two or more different elements by ordinary chemical means. Thus a compound always contains atoms of two or more different elements. Most of the atoms in the body are joined into compounds. Water (H_2O) and sodium chloride (NaCl), common table salt, are compounds. A molecule of oxygen (O_2), however, is not a compound because it consists of atoms of only one element.

 CLINICAL APPLICATION
Free Radicals and Their Effects on Health

Free radicals are produced by the absorption of energy from sources such as ultraviolet radiation or x-rays, by oxidation reactions (discussed shortly) that occur during normal metabolic processes, and by metabolic reactions that involve harmful substances such as carbon tetrachloride, a solvent used in dry cleaning. Among the many disorders and diseases linked to oxygen-derived free radicals are cancer, atherosclerosis, Alzheimer disease, emphysema, diabetes mellitus, cataracts, macular degeneration, rheumatoid arthritis, and deterioration associated with aging. Some evidence suggests that consuming more *antioxidants*—substances that inactivate oxygen-derived free radicals—can slow the pace of damage due to free radicals. Important dietary antioxidants include vitamins E and C, selenium, and beta-carotene. ■

1. List the names and chemical symbols of the 13 most abundant chemical elements in the human body.
2. Compare the meanings of atomic number, mass number, and atomic mass.
3. What are isotopes and radioisotopes of chemical elements?

CHEMICAL BONDS

OBJECTIVES

- *Describe how valence electrons form chemical bonds.*
- *Distinguish among ionic, covalent, and hydrogen bonds.*

Molecules and compounds are the products of **chemical bonds,** which act like a powerful "glue" that holds atoms close together. The likelihood that an atom will form a chemical bond with another atom depends on the number of electrons in its outermost shell, also called the **valence shell.** An atom with a valence shell holding eight electrons is *chemically stable,* which means it is unlikely to form chemical bonds with other atoms. Neon, for example, has eight electrons in its valence shell, and for this reason it does not easily bond with other atoms. Hydrogen and helium are exceptions; their valence shells only hold two electrons. Because helium has two valence electrons (see Figure 2.2), it too is stable and seldom bonds with other atoms.

The atoms of most biologically important elements do not have eight electrons in their valence shells. Under the right conditions, two or more atoms can interact in ways that produce a chemically stable arrangement of eight valence electrons for each atom. This chemical principle, called the **octet rule** (*octet* = set of eight), helps explain why atoms interact in preferential ways. That is to say, one atom is more likely to interact with another atom if doing so will leave both with eight valence electrons. For this to happen, an atom either empties its partially filled valence shell, fills it with donated electrons, or shares electrons with other atoms. The way that valence electrons are distributed determines what kind of chemical bond results. We will consider three kinds of chemical bonds: ionic bonds, covalent bonds, and hydrogen bonds.

Ionic Bonds

When an atom loses or gains a valence electron, ions are formed. Positively and negatively charged ions are attracted to one another—opposites attract. When this force of attraction holds ions having opposite charges together, it is called an **ionic bond.** Consider sodium and chlorine atoms to see how this happens. Sodium has one valence electron (Figure 2.4a). If sodium *loses* this electron, it is left with the eight electrons in its second shell, a complete octet. When this happens, however, the total number of protons (11) exceeds the number of electrons (10). As a result, the sodium atom becomes a **cation,** or positively charged ion. A sodium ion has a charge of 1+ and is written Na^+. On the other hand, chlorine has seven valence electrons (Figure 2.4b). If chlorine *gains* an electron from a neighboring atom, it will have a complete octet in its third electron shell. When this happens, the total number of electrons (18) exceeds the number of protons (17), and the chlorine atom becomes an **anion,** a negatively charged ion. The ionic form of chlorine is called a chloride ion. It has a charge of 1− and is written Cl^-. When an atom of sodium donates its sole valence elec-

Table 2.2 Common Ions in the Body

CATIONS		ANIONS	
NAME	**SYMBOL**	**NAME**	**SYMBOL**
Hydrogen ion	H^+	Fluoride ion	F^-
Sodium ion	Na^+	Chloride ion	Cl^-
Potassium ion	K^+	Iodide ion	I^-
Ammonium ion	NH_4^+	Hydroxide ion	OH^-
Hydronium ion	H_3O^+	Nitrate ion	NO_3^-
Magnesium ion	Mg^{2+}	Bicarbonate ion	HCO_3^-
Calcium ion	Ca^{2+}	Oxide ion	O^{2-}
Iron (II) ion	Fe^{2+}	Sulfide ion	S^{2-}
Iron (III) ion	Fe^{3+}	Phophate ion	PO_4^{3-}

tron to an atom of chlorine, the resulting positive and negative charges pull both ions tightly together into an ionic bond (Figure 2.4c). The resulting compound is sodium chloride, written NaCl.

In general, ionic compounds exist as solids, with an orderly, repeating arrangement of the ions, as in a crystal of NaCl (Figure 2.4d). A crystal of NaCl may be large or small—the total number of ions can vary—but the ratio of Na^+ to Cl^- is always 1:1. In the body, ionic bonds are found mainly in teeth and bones, where they give great strength to the tissue. Most other ions in the body are dissolved in body fluids. An ionic compound that dissociates into positive and negative ions in solution is called an **electrolyte** (e-LEK-trō-līt) because the solution can conduct an electric current. (The chemistry and importance of electrolytes are discussed in detail in Chapter 27.)

When a single positively charged ion of sodium and a single negatively charged ion of chlorine form an ionic bond, the net charge of the resulting compound, NaCl, is zero. The same is true for every compound with ionic bonds—the net charge of the compound is zero. In your study of the human body, however, you will discover many compounds that are positively or negatively charged. For example, ammonium (NH_4^+) and hydroxide (OH^-) are very common ionic compounds found in the body. The reason these compounds can carry a charge is that they are the result of covalent bonds (discussed next), not ionic bonds. Table 2.2 lists the names and symbols of the most common ions and ionic compounds in the body.

Covalent Bonds

When a **covalent bond** forms, neither of the combining atoms loses or gains electrons. Instead, the atoms form a molecule by *sharing* one, two, or three pairs of their valence electrons. The greater the number of electron pairs shared between two atoms, the stronger the covalent bond. The electrons in each pair are said to be shared because both

Figure 2.4 Ions and ionic bond formation. (a) A sodium atom can have a complete octet of electrons in its outermost shell by losing one electron. (b) A chlorine atom can have a complete octet by gaining one electron. (c) An ionic bond may form between oppositely charged ions. (d) In a crystal of NaCl, each Na^+ is surrounded by six Cl^-. In (a), (b), and (c), the electron that is lost or accepted is colored red.

🔑 **An ionic bond is the force of attraction that holds together oppositely charged ions.**

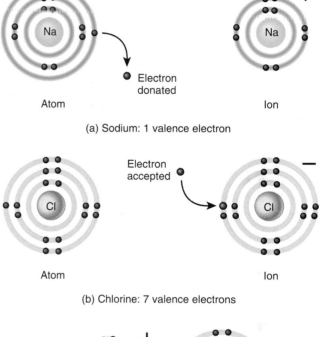

(a) Sodium: 1 valence electron

(b) Chlorine: 7 valence electrons

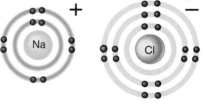

(c) Ionic bond in sodium chloride (NaCl)

(d) Packing of ions in a crystal of sodium chloride

Ⓠ What are cations and anions?

Figure 2.5 Covalent bond formation. The red electrons are shared *equally*. When writing the structural formula of a covalently bonded molecule, each pair of shared electrons is denoted by a straight line between the chemical symbols for two atoms. In molecular formulas, the number of atoms in each molecule is noted by subscripts.

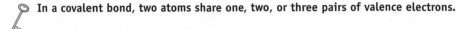 **In a covalent bond, two atoms share one, two, or three pairs of valence electrons.**

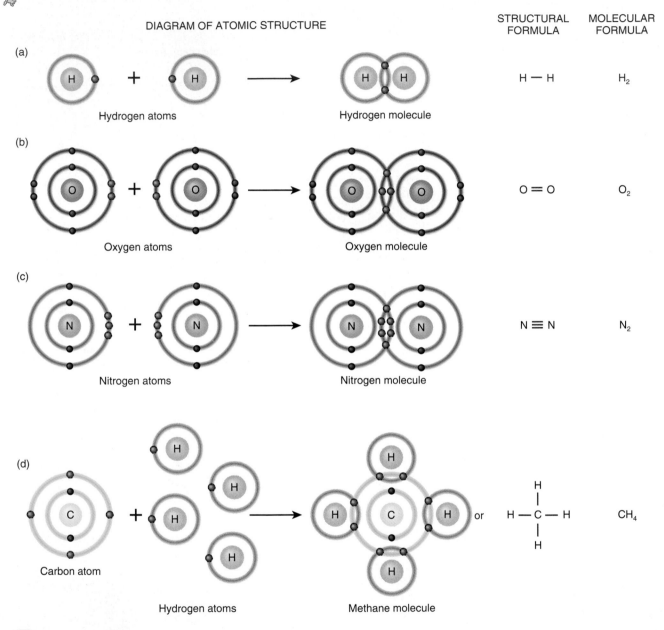

What is the principal difference between an ionic bond and a covalent bond?

spend most of their time in the region between the nuclei of the two atoms. Unlike ionic bonds, covalent bonds can form between atoms of the same element as well as between atoms of different elements. Covalent bonds are the most common chemical bonds in the body, and the compounds that result from them form most of the body's structures.

It is easiest to understand the nature of covalent bonds by considering those that form between atoms of the same

element. As noted, covalently bonded atoms may share up to three electron pairs. A **single covalent bond** results when two atoms share one electron pair. For example, a molecule of hydrogen forms when two hydrogen atoms share their single valence electrons (Figure 2.5a), which allows both atoms to have a full valence shell at least part of the time. A **double covalent bond** results when two atoms share two

Figure 2.6 Polar covalent bonds between oxygen and hydrogen atoms in a water molecule. The red electrons are shared *unequally*. Because the oxygen nucleus attracts the shared electrons more strongly, the oxygen end of a water molecule has a partial negative charge, written δ^-, and the hydrogen ends have a partial positive charge, written δ^+.

🔑 **A polar covalent bond occurs when one atomic nucleus attracts the shared electrons more strongly than does the nucleus of another atom in the molecule.**

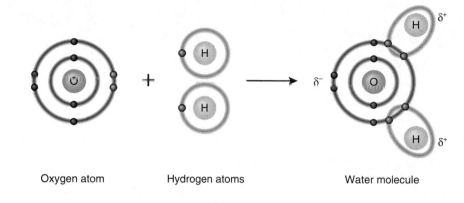

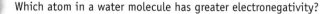

| Oxygen atom | Hydrogen atoms | Water molecule |

Q Which atom in a water molecule has greater electronegativity?

pairs of electrons, as is the case in an oxygen molecule (Figure 2.5b). A **triple covalent bond** occurs when two atoms share three pairs of electrons, as in a molecule of nitrogen (Figure 2.5c). Notice in the *structural formulas* for covalently bonded molecules in Figure 2.5 that the number of lines between the chemical symbols for two atoms indicates whether the bond is a single (—), double (=), or triple (≡) covalent bond.

The same principles of covalent bonding that apply to atoms of the same element also apply to covalent bonds between atoms of different elements. Methane (CH_4), a gas, contains covalent bonds formed between the atoms of two different elements (Figure 2.5d). The valence shell of the carbon atom can hold eight electrons but has only four of its own. The single electron shell of a hydrogen atom can hold two electrons, but each hydrogen atom has only one of its own. A methane molecule is the product of four separate single covalent bonds; each hydrogen atom shares one pair of electrons with the carbon atom.

In some covalent bonds, two atoms share the electrons equally—one atom does not attract the shared electrons more strongly than the other atom. This type of bond is a **nonpolar covalent bond.** The bonds between two identical atoms are always nonpolar covalent bonds (see Figure 2.5a–c). Another example of a nonpolar covalent bond is the single covalent bond that forms between carbon and each atom of hydrogen in a methane molecule (see Figure 2.5d).

In a **polar covalent bond,** the sharing of electrons between two atoms is unequal—one atom attracts the shared electrons more strongly than the other. When polar covalent bonds form, the resulting molecule has a partial negative charge near the atom that attracts electrons more strongly; this atom is said to have greater **electronegativity.** At least one other atom in the molecule then will have a partial posi-

Figure 2.7 Hydrogen bonding among water molecules. Each water molecule forms hydrogen bonds, indicated by dotted lines, with three to four neighboring water molecules.

🔑 **Hydrogen bonds occur because hydrogen atoms in one water molecule are attracted to the partial negative charge of the oxygen atom in another water molecule.**

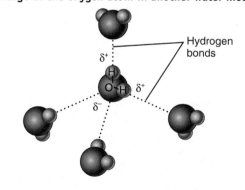

Q Why would you expect ammonia (NH_3) to form hydrogen bonds with water molecules?

tive charge. The partial charges are indicated by a lowercase Greek delta with a minus or plus sign: δ^- and δ^+. The most important example of a polar covalent bond in living systems is the bond between oxygen and hydrogen in a molecule of water (Figure 2.6). Later in the chapter we will see how polar covalent bonds allow water to dissolve many molecules that are important to life.

Hydrogen Bonds

The polar covalent bonds that form between hydrogen atoms and other atoms can give rise to **hydrogen bonds,** a third type of chemical bond (Figure 2.7). The polar covalent bond causes the hydrogen atom to have a partial positive charge (δ^+) that attracts the partial negative charge (δ^-) of neighboring electronegative atoms, most often oxygen or nitrogen in living organisms. Hydrogen bonds are weak—only about 5% as strong as covalent bonds. Consequently, they cannot bind atoms into molecules. However, they do establish links between molecules—for example, between water molecules or between various parts of a large molecule, such as a protein or nucleic acid (both discussed later in this chapter). Even though single hydrogen bonds are weak, very large molecules may contain hundreds of these bonds. Acting collectively, hydrogen bonds provide considerable strength and stability and help determine the three-dimensional shape of large molecules. As you will see later in this chapter, a large molecule's shape determines how it works in the body.

1. Which electron shell is the valence shell of an atom, and what is its significance?
2. How does valence relate to the octet rule?
3. What information is conveyed when you write the molecular or structural formula for a molecule?

CHEMICAL REACTIONS

OBJECTIVES
• *Define a chemical reaction.*
• *Describe the various forms of energy.*

A **chemical reaction** occurs when new bonds form or old bonds break between atoms. Chemical reactions are the foundation of all life processes, and as we have seen, the interactions of valence electrons are the basis of all chemical reactions. Consider how hydrogen and oxygen molecules react to form water molecules (Figure 2.8). The starting substances—H_2 and O_2—are known as the **reactants,** and the ending substances—two molecules of H_2O—are the **products.** The arrow in the figure indicates the direction in which the reaction proceeds. In a chemical reaction, the total mass of the reactants equals the total mass of the products, a relationship known as the **law of conservation of mass.** Because of this law, the number of atoms of each element is the same before and after the reaction. However, because the atoms are rearranged, the reactants and products have different chemical properties. Through thousands of different reactions, body structures are built and body functions are carried out. The term **metabolism** refers to all the chemical reactions occurring in the body.

Forms of Energy and Chemical Reactions

Each chemical reaction involves energy changes. **Energy** (*en-* = in; *-ergy* = work) is the capacity to do work. Two principal forms of energy are **potential energy,** energy stored by matter due to its position, and **kinetic energy,** the

Figure 2.8 The chemical reaction between two hydrogen molecules (H_2) and one oxygen molecule (O_2) to form two molecules of water (H_2O). Note that the reaction occurs by breaking old bonds and making new bonds.

🔑 **The number of atoms of each element is the same before and after a chemical reaction.**

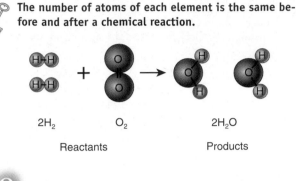

2H$_2$	O$_2$	2H$_2$O
Reactants		Products

Q Why does this reaction require two molecules of H_2?

energy associated with matter in motion. For example, the energy stored in a battery, in water behind a dam, or in a person poised to jump down some steps is potential energy. When the battery is used to run a clock, or when the gates of the dam are opened and the falling water turns a generator, or when the person jumps, potential energy is converted into kinetic energy. **Chemical energy** is a form of potential energy that is stored in the bonds of compounds and molecules. In your body, chemical energy in the foods you eat is eventually converted into various forms of kinetic energy, such as mechanical energy used to walk and talk, and heat energy used to maintain body temperature. The total amount of energy present at the beginning and end of a chemical reaction is the same—energy can be neither created nor destroyed, although it may be converted from one form to another. This principle is known as the **law of conservation of energy.**

Energy Transfer in Chemical Reactions

OBJECTIVES
• *Compare the basic differences between exergonic and endergonic chemical reactions.*
• *Describe the role of activation energy and enzymes.*

In chemical reactions, breaking old bonds requires energy, and forming new bonds releases energy. Because most chemical reactions involve both breaking bonds in the reactants and forming bonds in the products, the *overall reaction* may either release energy or absorb energy. **Exergonic reactions** (*ex-* = out) release more energy than they absorb. In an exergonic reaction, the energy released when new bonds form is *greater* than the energy needed to break apart old bonds, so surplus energy is released as the reaction occurs (Figure 2.9a). At the conclusion of an exergonic reaction, the products have *less* potential energy than the reactants. On

Figure 2.9 Energy transfer during exergonic and endergonic reactions.

🔑 **Exergonic reactions release energy, whereas endergonic reactions require energy.**

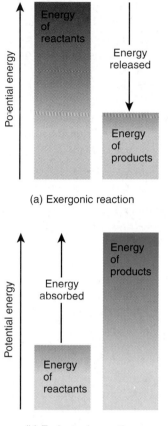

(a) Exergonic reaction

(b) Endergonic reaction

Ⓠ Which type of reaction creates products that have more chemical potential energy than the reactants?

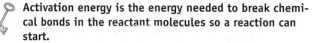

Figure 2.10 Activation energy.

🔑 **Activation energy is the energy needed to break chemical bonds in the reactant molecules so a reaction can start.**

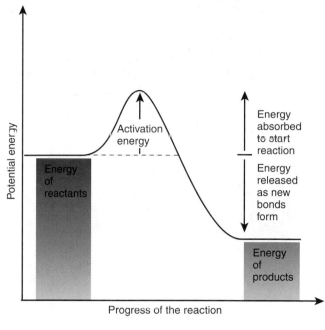

Progress of the reaction

Ⓠ Why is the reaction illustrated here exergonic?

the other hand, **endergonic reactions** (*end-* = within) absorb more energy than they release. In an endergonic reaction, the energy released when new bonds form is *less* than the energy needed to break apart old bonds, so energy must be absorbed for the reaction to take place (Figure 2.9b). At the conclusion of an endergonic reaction, the products have *more* potential energy than the reactants.

A key feature of the body's metabolism is the coupling of exergonic reactions and endergonic reactions, so that energy released from an exergonic reaction is used to drive an endergonic one. In general, exergonic reactions occur as nutrients, such as glucose, are broken down. Some of the energy released is temporarily stored in a special molecule called *adenosine triphosphate (ATP),* which will be discussed more fully later in this chapter. If a molecule of glucose is completely broken down, the chemical energy that was stored in its bonds can be used to produce as many as 38

molecules of ATP. The energy transferred to the ATP molecules is later used to drive endergonic reactions that lead to the building of body structures, such as muscles and bones. The energy in ATP is also used to do mechanical work, such as occurs during the contraction of muscle or the movement of substances into or out of cells.

Activation Energy

Because particles of matter such as atoms, ions, and molecules have kinetic energy, they are continuously moving and colliding with one another. A sufficiently forceful collision can disrupt the movement of valence electrons, causing an existing chemical bond to break or a new one to form. The collision energy needed to break chemical bonds in the reactants is called the **activation energy** (Figure 2.10). This is the initial energy "investment" that is needed to start a reaction. The reactants must absorb enough energy from their surroundings so that their chemical bonds become unstable and their valence electrons can interact to form new combinations. Then, as new bonds form, energy is released to the surroundings. Two factors influence the chance that a collision will occur and cause a chemical reaction:

1. *Concentration.* The more particles of matter present in a confined space, the greater the chance that they will collide. The concentration of particles increases when more

Figure 2.11 Comparison of energy needed for a chemical reaction to proceed with a catalyst (green curve) and without a catalyst (red curve).

🔑 **Catalysts speed up chemical reactions by lowering the activation energy.**

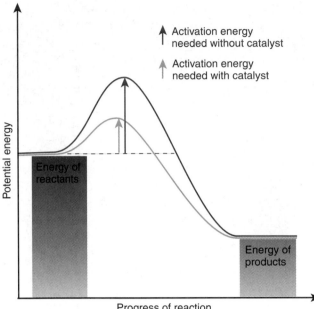

↑ Activation energy needed without catalyst

↑ Activation energy needed with catalyst

Energy of reactants

Potential energy

Energy of products

Progress of reaction

Ⓠ Does a catalyst change the potential energies of the products and reactants?

are added to a given space or when the pressure on the space increases, which forces the particles closer together so that they collide more often.

2. *Temperature.* As temperature rises, particles of matter move more rapidly. The result is that up to a point, the higher the temperature of matter, the more forcefully particles will collide, and the greater the chance that a collision will produce a reaction.

Catalysts

As we have seen, chemical reactions occur when chemical bonds break or form after atoms, ions, or molecules collide with one another. Normal temperature and pressure within the body are too low for most chemical reactions to occur rapidly enough to maintain life. Raising the temperature and the number of reacting particles of matter in the body could increase the frequency of collisions and thus increase the rate of chemical reactions, but doing so could also damage or kill the body's cells.

Substances called catalysts solve this problem. **Catalysts** are chemical compounds that speed up chemical reactions by lowering the activation energy needed for a reaction to occur (Figure 2.11). A catalyst does not alter the difference in

potential energy between the reactants and the products; it only lowers the amount of energy needed to get the reaction started.

For chemical reactions to occur, some particles of matter—especially large molecules—must not only collide with sufficient force, but they must "hit" one another at precise spots. A catalyst works by helping to orient the colliding particles of matter so that they touch at the spots that make the reaction happen. Although the action of a catalyst helps to speed up a chemical reaction, the catalyst itself is unchanged at the end of the reaction. Thus a single catalyst molecule can be used repeatedly to enable many chemical reactions to occur. The most important catalysts in the body are enzymes, which we will discuss later in this chapter.

Types of Chemical Reactions
OBJECTIVE
• *Describe synthesis, decomposition, exchange, reversible, and oxidation-reduction reactions.*

After a chemical reaction takes place, the atoms of the reactants are rearranged to yield products with new chemical properties. In this section we will look at the types of chemical reactions common to all living cells. Once you have learned them, you will be able to understand the chemical reactions discussed later in the book.

Synthesis Reactions—Anabolism

When two or more atoms, ions, or molecules combine to form new and larger molecules, the process is a **synthesis reaction.** The word *synthesis* means "to put together." Synthesis reactions can be expressed as follows:

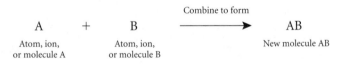

		Combine to form	
A	+	B ⟶	AB
Atom, ion, or molecule A		Atom, ion, or molecule B	New molecule AB

An example of a synthesis reaction is:

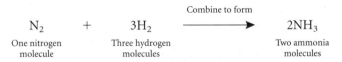

		Combine to form	
N_2	+	$3H_2$ ⟶	$2NH_3$
One nitrogen molecule		Three hydrogen molecules	Two ammonia molecules

All the synthesis reactions that occur in your body are collectively referred to as **anabolism** (a-NAB-ō-lizm). Overall, anabolic reactions are usually endergonic because they absorb more energy than they release. Combining simple molecules such as amino acids (discussed shortly) to form large molecules such as proteins is an example of anabolism.

Decomposition Reactions—Catabolism

In a **decomposition reaction** large molecules are split up into smaller atoms, ions, or molecules. A decomposition reaction is expressed as follows:

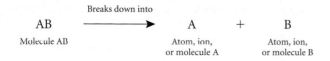

For example, under the proper conditions, methane can decompose into carbon and hydrogen molecules:

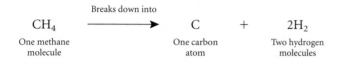

All decomposition reactions that occur in your body are collectively referred to as **catabolism** (ka-TAB-ō-lizm). Overall, catabolic reactions are usually exergonic because they release more energy than they absorb. The series of reactions that break down glucose to pyruvic acid, with the net production of two molecules of ATP, are important catabolic reactions in the body. These reactions will be discussed in Chapter 25.

Exchange Reactions

Many reactions in the body are **exchange reactions;** they consist of both synthesis and decomposition reactions. One type of exchange reaction works like this:

$$AB + CD \longrightarrow AD + BC$$

The bonds between A and B and between C and D break (decomposition), and new bonds then form (synthesis) between A and D and between B and C. An example of an exchange reaction is:

$$HCl + NaHCO_3 \longrightarrow H_2CO_3 + NaCl$$
Hydrochloric acid · Sodium bicarbonate · Carbonic acid · Sodium chloride

Notice that the ions in both compounds have "switched partners": The hydrogen ion (H^+) from HCl has combined with the bicarbonate ion (HCO_3^-) from $NaHCO_3$, and the sodium ion (Na^+) has combined with the chloride ion (Cl^-).

Reversible Reactions

A given chemical reaction may proceed in only one direction, from reactants to products, as previously indicated by the single arrows, or it may be reversible. In a **reversible reaction,** the products can revert to the original reactants. A reversible reaction is indicated by two half arrows pointing in opposite directions:

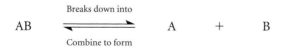

Some reactions are reversible only under special conditions:

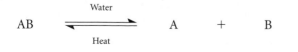

In that case, whatever is written above or below the arrows indicates the condition needed for the reaction to occur. In this case, AB breaks down into A and B only when water is added, and A and B react to produce AB only when heat is applied.

Oxidation-Reduction Reactions

Oxidation is the *loss of electrons* from a molecule, and it results in a decrease in the potential energy of the molecule. The process is called oxidation because the final acceptor of the lost electrons often is oxygen. Within cells, oxidation reactions often involve simultaneously removing a *hydrogen ion* (H^+, a hydrogen nucleus with no electrons) and a *hydride ion* (H^-, a hydrogen nucleus with two electrons) from a molecule. This reaction is equivalent to removing two hydrogen atoms ($H^+ + H^- = 2H$). An example of an oxidation reaction is the conversion of lactic acid, a waste produced by exercising muscles, into pyruvic acid, a molecule the body can use for ATP production. After the oxidation reaction is complete, two electrons have been lost together with the hydride ion:

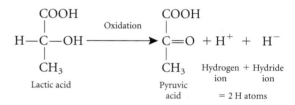

Reduction is the opposite of oxidation. It is the *gain of electrons* by a molecule, and it results in an increase in the potential energy of the molecule. An example of reduction is the conversion of pyruvic acid into lactic acid:

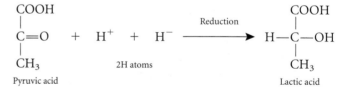

Within a cell, oxidation and reduction reactions are always coupled—whenever one substance is oxidized, another is almost simultaneously reduced. Such coupled reactions are referred to as **oxidation-reduction (redox) reactions.**

1. What is the relationship between reactants and products in a chemical reaction?
2. Compare potential energy and kinetic energy, and describe the law of conservation of energy.
3. Define activation energy. How do catalysts affect it?
4. In which type of chemical reaction does a molecule gain electrons from a hydride ion?
5. How are anabolism and catabolism related to synthesis and decomposition reactions, respectively?

INORGANIC COMPOUNDS AND SOLUTIONS

OBJECTIVES

- *Describe the properties of inorganic acids, bases, salts, and water.*

- *Distinguish among solutions, colloids, and suspensions.*

Most of the chemicals in your body exist in the form of compounds. Biologists and chemists divide these compounds into two principal classes: inorganic compounds and organic compounds. In general, **inorganic compounds** lack carbon and are structurally simple. They include water and many salts, acids, and bases. Inorganic compounds may have either ionic or covalent bonds. **Organic compounds,** on the other hand, always contain carbon, usually contain hydrogen, and always have covalent bonds. A few carbon-containing compounds—for instance, carbon dioxide (CO_2) and bicarbonate ion (HCO_3^-)—are classified as inorganic.

Inorganic Acids, Bases, and Salts

When inorganic acids, bases, or salts dissolve in water, they **dissociate** (dis′-sō-sē-ĀT); that is, they separate into ions and become surrounded by water molecules. An **acid** (Figure 2.12a) is a substance that dissociates into one or more **hydrogen ions (H^+)** and one or more anions. Because H^+ is a single proton with one positive charge, an acid also is a **proton donor.** A **base,** by contrast (Figure 2.12b), dissociates into one or more **hydroxide ions (OH^-)** and one or more cations. Because hydroxide ions have a strong attraction for protons, a base is a **proton acceptor.**

A **salt,** when dissolved in water, dissociates into cations and anions, neither of which is H^+ or OH^- (Figure 2.12c). In the body, salts are electrolytes that are important for carrying electrical currents (ions flowing from one place to another), especially in nerve and muscle tissue. The ions of salts also provide many essential chemical elements in intracellular and extracellular fluids such as blood, lymph, and the interstitial fluid of tissues.

Acids and bases react with one another to form salts. For example, the reaction of hydrochloric acid (HCl) and potassium hydroxide (KOH), a base, produces the salt potassium chloride (KCl) and water (H_2O). This exchange reaction can be written as follows:

$$HCl + KOH \longrightarrow H^+ + Cl^- + K^+ + OH^- \longrightarrow KCl + H_2O$$

Acid   Base   Dissociated ions   Salt   Water

Solutions, Colloids, and Suspensions

A **mixture** is a combination of elements or compounds that are physically blended together but not bound by chemical bonds. For example, the air you are breathing is a mixture of gases, mainly nitrogen, oxygen, and argon. Three common liquid mixtures are solutions, colloids, and suspensions.

Figure 2.12 Dissociation of inorganic acids, bases, and salts.

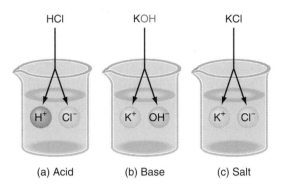

Dissociation is the separation of inorganic acids, bases, and salts into ions in a solution.

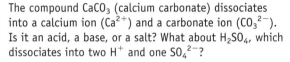

(a) Acid   (b) Base   (c) Salt

Q The compound $CaCO_3$ (calcium carbonate) dissociates into a calcium ion (Ca^{2+}) and a carbonate ion (CO_3^{2-}). Is it an acid, a base, or a salt? What about H_2SO_4, which dissociates into two H^+ and one SO_4^{2-}?

In a **solution,** a substance called the **solvent** dissolves another substance called the **solute.** Usually there is more solvent than solute in a solution. For example, your sweat is a dilute solution of water (the solvent) plus small amounts of salts (the solutes). Once mixed together, solutes remain evenly dispersed among the solvent molecules. Because the solute particles in a solution are very small, a solution looks clear and transparent.

A **colloid** differs from a solution mainly on the basis of the size of its particles. The solute particles in a colloid are large enough to scatter light, just as water droplets in fog scatter light from a car's headlight beams. For this reason, colloids usually look translucent or opaque. Milk is an example of a liquid that is both a colloid and a solution: The large milk proteins make it a colloid, whereas calcium salts, milk sugar (lactose), and other small particles are in solution. In both solutions and colloids, the solutes do not settle out and accumulate on the bottom of the container.

In a **suspension,** by contrast, the suspended material may mix with the liquid or suspending medium for some time, but eventually it will settle out. Blood is an example of a suspension. When freshly drawn from the body, blood has an even, reddish color. After it sits for a while in a test tube, the top layer appears pale yellow (see Figure 19.1a on page 612). This is the liquid portion of blood, called plasma, which is both a solution of ions and other small solutes and a colloid due to the presence of plasma proteins. Red blood cells that have settled out of the suspension drift to the bottom of the tube, making the lower layer look red.

The **concentration** of a solution may be expressed in several ways. Two common ways are by **percent,** which gives

Table 2.3 Percent and Molarity

DEFINITION	EXAMPLE
Percent (mass per volume) Number of grams of a substance per 100 milliliters (mL) of solution.	To make a 10% NaCl solution, take 10 gm of NaCl and add enough water to make a total of 100 mL of solution.
Molarity = moles (mol) per liter A 1 molar (1M) solution = 1 mole of a solute in 1 liter of solution.	To make a 1 molar (1M) solution of NaCl, dissolve 1 mole of NaCl (58.44 gm) in enough water to make a total of 1 liter of solution.

the relative mass of a solute found in a given volume of solution, and in units of **moles per liter (mol/liter),** which relate to the total number of molecules in a given volume of solution. A **mole** is the amount (in grams) of any substance that has a mass equal to the combined atomic masses of all its atoms. For example, one mole of the element chlorine (atomic mass = 35.45) is 35.45 grams and one mole of the salt sodium chloride (NaCl) is 58.44 grams (22.99 for Na + 35.45 for Cl). Just as a dozen always means 12 of something, a mole of anything has the same number of particles: 6.023×10^{23}. This huge number is called *Avogadro's number.* Thus measurements of substances that are stated in moles tell us about the numbers of atoms, ions, or molecules present. This is important when chemical reactions are occurring because each reaction requires a set number of atoms of specific elements. Table 2.3 describes these ways of expressing concentration.

Water

Water is the most important and abundant inorganic compound in all living systems. In fact, water is the medium in which nearly all of the body's chemical reactions occur. Water has many properties that explain why it is such an indispensable compound for life. The most important property of water is its polarity—the uneven sharing of valence electrons that confers a partial negative charge near the one oxygen atom and two partial positive charges near the two hydrogen atoms in a water molecule (see Figure 2.6). This property alone makes water an excellent solvent for other ionic or polar substances, gives water molecules cohesion (the tendency to stick together), and allows water to moderate temperature changes.

Water as a Solvent

Alchemists in medieval times tried to find a universal solvent, a substance that would dissolve all other materials. They found nothing that worked as well as water. Although it is the most versatile solvent known, it is not a "universal solvent." If it were, no container could hold it; the container would dissolve!

The versatility of water as a solvent for ionized or polar substances is due to its polar covalent bonds and its "bent"

Figure 2.13 How polar water molecules dissolve salts and polar substances. When a crystal of sodium chloride is placed in water, the slightly negative oxygen end (red) of water molecules is attracted to the positive sodium ions (Na$^+$), and the slightly positive hydrogen portions (gray) of water molecules are attracted to the negative chloride ions (Cl$^-$). Adapted from Karen Timberlake, *Chemistry* 6e, F8.5, p255 (Menlo Park, CA; Addison Wesley Longman, 1999). ©1999 Addison Wesley Longman, Inc.

🔑 **Water is a versatile solvent because its polar covalent bonds, in which electrons are shared unequally, create positive and negative regions.**

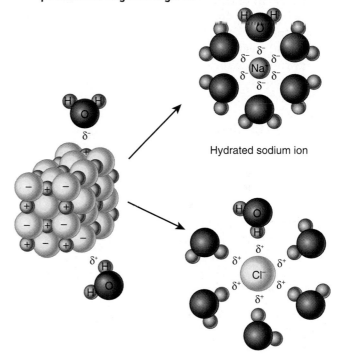

Hydrated sodium ion

Hydrated chloride ion

❓ Table sugar (sucrose) easily dissolves in water but is not an electrolyte. Is it likely that all the covalent bonds between atoms in table sugar are nonpolar bonds?

shape, which allows each water molecule to interact with four or more neighboring ions or molecules. Solutes that contain polar covalent bonds are **hydrophilic,** or "water loving," which means they dissolve easily in water. Molecules that contain mainly nonpolar covalent bonds, on the other hand, are **hydrophobic,** or "water fearing." They are not very water soluble.

To understand the dissolving capability of water, consider what happens when a crystal of a salt such as sodium chloride (NaCl) is placed in water (Figure 2.13). The sodium and chloride ions at the surface of the salt crystal are exposed to the polar water molecules. The electronegative oxygen portion of water molecules is attracted to the sodium ions (Na$^+$), and the electropositive hydrogen portions of water molecules are attracted to the chloride ions (Cl$^-$). Soon, water molecules surround and separate some Na$^+$ and Cl$^-$

from each other, breaking the ionic bonds that held NaCl together. In this way, water molecules dissociate the ions of the salt. As a salt dissociates in solution, each of its ions becomes surrounded by several water molecules. This sphere of water molecules lessens the chance that ions with opposite charges will come together and reform an ionic bond. In solution, some of the organic molecules in the body also exist as ions and acquire hydration spheres just like Na^+ and Cl^-. Other organic molecules are hydrophilic because they contain polar covalent bonds that give rise to partial negative and positive charges in different regions of the molecule.

The ability of water to form solutions and suspensions is essential to health and survival. Because water can dissolve or suspend so many different substances, it is an ideal medium for metabolic reactions. Water enables dissolved reactants to collide to form products. Water also dissolves waste products, which allows them to be flushed out of the body in the urine.

Water in Chemical Reactions

Water not only serves as the medium for most chemical reactions in the body, it also participates as a reactant or product in certain reactions. During digestion, for example, water can be added to large nutrient molecules to break them down into smaller molecules. This kind of decomposition reaction is called **hydrolysis** (hī-DROL-i-sis; *hydro-* = water; *-lysis* = to loosen), and it is necessary if the body is to use the energy in nutrients. On the other hand, when two smaller molecules join to form a larger molecule in a **dehydration synthesis reaction** (*de-* = from, down, or out; *hydra-* = water), a water molecule is removed. As you will see later in the chapter, such reactions occur during synthesis of proteins and other large molecules.

High Heat Capacity of Water

In comparison to most substances, water can absorb or release a relatively large amount of heat with only a modest change in its own temperature. For this reason, water is said to have a high *heat capacity*. The reason for this property is the large number of hydrogen bonds in water. As water absorbs heat energy, some of the energy is used to break hydrogen bonds. Less energy is then left over to increase the motion of water molecules and thus increase the water's temperature. The high heat capacity of water is the reason it is used in automobile radiators; it cools the engine by absorbing heat without its temperature rising to an unacceptably high level. The large amount of water in the body has a similar effect: It lessens the impact of environmental temperature changes, helping to maintain the homeostasis of body temperature.

Water also requires a large amount of heat to change from a liquid to a gas; its *heat of vaporization* is high. As water evaporates from the surface of the skin, it removes a large quantity of heat, providing an important cooling mechanism.

Cohesion of Water Molecules

The hydrogen bonds that link neighboring water molecules (see Figure 2.7) give water considerable *cohesion,* the tendency of like particles to stay together. The cohesion of water molecules creates a very high **surface tension,** a measure of the difficulty of stretching or breaking the surface of a liquid. At a boundary between water and air, water's surface tension is very high because the water molecules are much more attracted to one another than they are attracted to molecules in the air. The influence of water's surface tension on the body can be seen in the way it increases the work required for breathing. The air sacs of the lungs are coated by a thin film of watery fluid, so each inhalation must be forceful enough to overcome the opposing effect of surface tension as the air sacs stretch and enlarge when taking in air.

Water as a Lubricant

Water is a major part of mucus and other lubricating fluids. Lubrication is especially necessary in the chest and abdomen, where internal organs touch and slide over one another. It is also needed at joints, where bones, ligaments, and tendons rub against one another. In the gastrointestinal tract, water in mucus moistens foods, which aids their smooth passage.

Acid–Base Balance: The Concept of pH

OBJECTIVES
• *Define pH.*
• *Explain the role of buffer systems in homeostasis.*

To ensure homeostasis, intracellular and extracellular fluids must contain almost balanced quantities of acids and bases. The more hydrogen ions (H^+) dissolved in a solution, the more acidic the solution; conversely, the more hydroxide ions (OH^-), the more basic (alkaline) the solution. The chemical reactions that take place in the body are very sensitive to even small changes in the acidity or alkalinity of the body fluids in which they occur. Any departure from the narrow limits of normal H^+ and OH^- concentrations greatly disrupts body functions.

A solution's acidity or alkalinity is expressed on the **pH scale,** which runs from 0 to 14 (Figure 2.14). This scale is based on the concentration of H^+ in moles per liter. A pH of 7 means that a solution contains one ten-millionth (0.0000001) of a mole of hydrogen ions per liter. The number 0.0000001 is written as 1×10^{-7} in scientific notation, which indicates that the number is 1 with the decimal point moved seven places to the left. To convert this value to pH, the negative exponent (-7) is changed to a positive number (7). A solution with a H^+ concentration of 0.0001 (10^{-4}) moles per liter has a pH of 4; a solution with a H^+ concentration of 0.000000001 (10^{-9}) moles per liter has a pH of 9; and so on.

Figure 2.14 The pH scale. A pH below 7 indicates an acidic solution—more H^+ than OH^-. The lower the numerical value of the pH, the more acidic is the solution because the H^+ concentration becomes progressively greater. A pH above 7 indicates a basic (alkaline) solution; that is, there are more OH^- than H^+. The higher the pH, the more basic the solution.

🔑 **At pH 7 (neutrality), the concentrations of H^+ and OH^- are equal (10^{-7} mol/liter).**

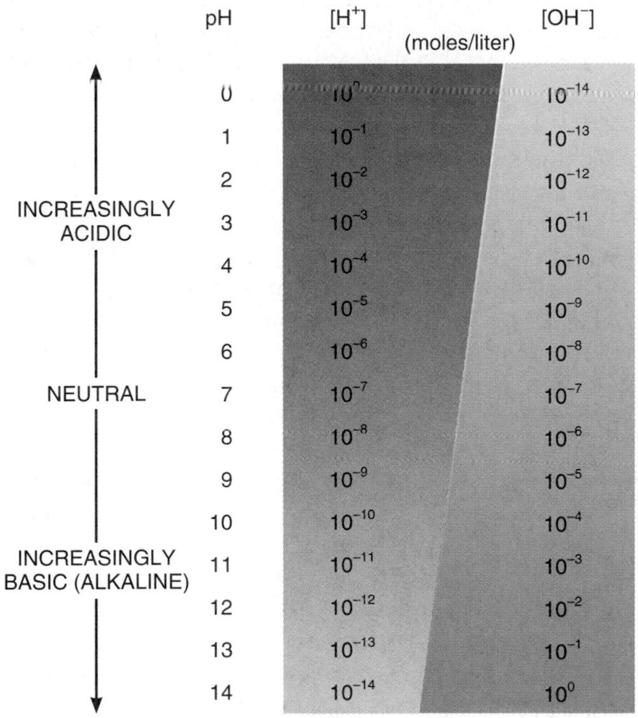

pH	[H⁺]	[OH⁻]
		(moles/liter)
0	10^0	10^{-14}
1	10^{-1}	10^{-13}
2	10^{-2}	10^{-12}
3	10^{-3}	10^{-11}
4	10^{-4}	10^{-10}
5	10^{-5}	10^{-9}
6	10^{-6}	10^{-8}
7	10^{-7}	10^{-7}
8	10^{-8}	10^{-6}
9	10^{-9}	10^{-5}
10	10^{-10}	10^{-4}
11	10^{-11}	10^{-3}
12	10^{-12}	10^{-2}
13	10^{-13}	10^{-1}
14	10^{-14}	10^0

INCREASINGLY ACIDIC — NEUTRAL — INCREASINGLY BASIC (ALKALINE)

Q What is the concentration of H^+ and OH^- at pH 6? Which pH is more acidic, 6.82 or 6.91? Which pH is closer to neutral, 8.41 or 5.59?

Table 2.4 pH Values of Selected Substances

SUBSTANCE	pH VALUE
• Gastric juice (found in the stomach)	1.2–3.0
Lemon juice	2.3
Vinegar	3.0
Carbonated soft drink	3.0–3.5
Orange juice	3.5
• Vaginal fluid	3.5–4.5
Tomato juice	4.2
Coffee	5.0
• Urine	4.6–8.0
• Saliva	6.35–6.85
Milk	6.8
Distilled (pure) water	7.0
• Blood	7.35–7.45
• Semen (fluid containing sperm)	7.20–7.60
• Cerebrospinal fluid (fluid associated with nervous system)	7.4
• Pancreatic juice (digestive juice of the pancreas)	7.1–8.2
• Bile (liver secretion that aids fat digestion)	7.6–8.6
Milk of magnesia	10.5
Lye	14.0

• Denotes substances in the human body.

The midpoint of the pH scale is 7, where the concentrations of H^+ and OH^- are equal. A substance with a pH of 7, such as distilled (pure) water, is considered to be neutral. A solution that has more H^+ than OH^- is an **acidic solution** and has a pH below 7. A solution that has more OH^- than H^+ is a **basic (alkaline) solution** and has a pH above 7. It is important to understand that a change of one whole number on the pH scale represents a *tenfold* change from the previous concentration: A pH of 1 denotes 10 times more H^+ than a pH of 2, whereas a pH of 3 indicates 10 times fewer H^+ than a pH of 2 and 100 times fewer H^+ than a pH of 1.

Maintaining pH: Buffer Systems

Although the pH of body fluids may differ, the normal limits for the various fluids are generally quite narrow. Table 2.4 compares the pH values for certain body fluids with those of some common substances. Homeostatic mecha-

nisms maintain the pH of blood between 7.35 and 7.45, slightly more basic than pure water. Saliva is slightly acidic, and semen is slightly basic. Because the kidneys help remove excess acid from the body, urine can be quite acidic. Even though strong acids and bases are continually taken into and formed by the body, the pH of fluids inside and outside cells remains almost constant. One important reason is the presence of **buffer systems,** which function to convert strong acids or bases into weak acids or bases. Strong acids (or bases) ionize easily and contribute many H^+ or (OH^-) to a solution. Therefore, they can change pH drastically, which can disrupt the body's metabolism. Weak acids (or bases) do not ionize as much and contribute fewer H^+ (or OH^-). Hence, they have less effect on the pH. The chemical compounds that can convert strong acids or bases into weak ones are called **buffers.**

One important buffer system in the body is the **carbonic acid–bicarbonate buffer system.** Because carbonic acid (H_2CO_3) can act as a weak acid, and the bicarbonate ion (HCO_3^-) can act as a weak base, this buffer system can compensate for either an excess or a shortage of H^+. For example, if there is an excess of H^+ (an acidic condition), HCO_3^- can function as a weak base and remove the excess H^+, as follows:

$$H^+ \ + \ HCO_3^- \longrightarrow H_2CO_3 \longrightarrow H_2O \ + \ CO_2$$

Hydrogen ion   Bicarbonate ion   Carbonic acid   Water   Carbon dioxide

On the other hand, if there is a shortage of H^+ (an alkaline condition), H_2CO_3 can function as a weak acid and provide needed H^+ as follows:

$$H_2CO_3 \longrightarrow H^+ \ + \ HCO_3^-$$

Carbonic acid (weak acid)   Hydrogen ion   Bicarbonate ion

Buffers and their roles in maintaining acid–base balance are described in more detail in Chapter 27.

1. How do inorganic compounds differ from organic compounds?
2. Describe two ways to express the concentration of a solution.
3. Describe the properties and functions of water in the body.
4. What is pH? Why does the body maintain a relatively constant pH?
5. What are the components of a buffer system? How is buffering an example of homeostasis?

ORGANIC COMPOUNDS

OBJECTIVES

• *Describe the functional groups of organic molecules.*
• *Identify the building blocks and functions of carbohydrates, lipids, proteins, enzymes, deoxyribonucleic acid (DNA), ribonucleic acid (RNA), and adenosine triphosphate (ATP).*

Water makes up 55–60% of an adult's total body mass; all other inorganic compounds add only an additional 1–2%. These relatively simple compounds, whose molecules have only a few atoms, cannot be used by cells to perform complicated biological functions. Many organic molecules, on the other hand, are relatively large and have unique characteristics that allow them to carry out complex functions. Important categories of organic compounds, which make up roughly 40% of body mass, include carbohydrates, lipids, proteins, nucleic acids, and adenosine triphosphate (ATP).

Carbon and Its Functional Groups

Carbon has several properties that make it particularly useful to living organisms. For one thing, it can form bonds with one to thousands of other carbon atoms to produce large molecules that can have many different shapes. This means that the body can build many different organic compounds, each of which has a unique structure and function.

Moreover, the large size of most carbon-containing molecules and the fact that some do not dissolve easily in water make them useful materials for building body structures.

Organic compounds are held together mostly or entirely by covalent bonds. Carbon has four electrons in its outer (valence) shell. It can bond covalently with a variety of atoms, including other carbon atoms, to form rings and straight or branched chains. Other elements that most often bond with carbon in organic compounds are hydrogen (one bond), oxygen (two bonds), and nitrogen (three bonds). Sulfur (two bonds) and phosphorus (five bonds) occur less often in organic compounds. Additional elements are present, but only in a few organic compounds.

The chain of carbon atoms in an organic molecule is called the **carbon skeleton.** Most of these carbons are bonded to hydrogen atoms. Attached to the carbon skeleton are distinctive **functional groups,** in which other elements form bonds with carbon and hydrogen atoms. Each type of functional group has a specific arrangement of atoms that confers characteristic chemical properties upon organic molecules. Table 2.5 lists the most common functional groups of organic molecules and describes some of their properties. Because organic molecules are often big, there are shorthand methods for representing their structural formulas. Figure 2.15 shows two ways to indicate the structure of the sugar glucose, a molecule with a ring-shaped carbon skeleton that has several hydroxyl groups attached.

Small organic molecules can combine into very large molecules called **macromolecules** (*macro-* = large). Macromolecules are usually **polymers** (*poly-* = many; *-mers* = parts), large molecules formed by covalent bonding of many identical or similar small building-block molecules called **monomers** (*mono-* = one). When two monomers join together, the reaction usually involves dehydration synthesis —the elimination of a hydrogen atom from one monomer and a hydroxyl group from the other. Macromolecules such as carbohydrates, lipids, proteins, and nucleic acids are assembled in the cell via dehydration synthesis. It is also possible for macromolecules to be broken down into monomers via hydrolysis.

Molecules that have the same molecular formula but different structures are called **isomers** (Ī-so-merz; *iso-* = the same). For example, the molecular formulas for the sugars glucose and fructose are both $C_6H_{12}O_6$. The individual atoms, however, are positioned differently along the carbon skeleton.

Carbohydrates

Carbohydrates include sugars, starches, glycogen, and cellulose. Even though they are a large and diverse group of organic compounds and have several functions, they represent only 2–3% of your total body mass. Plants store carbohydrate as starch and use the carbohydrate cellulose to build the cell wall. Cellulose is the most plentiful organic substance on Earth. Although humans eat cellulose, they cannot digest

Table 2.5 Major Functional Groups

NAME AND STRUCTURAL FORMULA*	OCCURRENCE AND SIGNIFICANCE
Hydroxyl R—O—H	*Alcohols* contain an —OH group, which is polar and hydrophilic due to its electronegative O atom. Molecules with many —OH groups dissolve easily in water.
Sulfhydryl R—S—H	*Thiols* have an —SH group, which is polar and hydrophilic due to its electronegative S atom. Certain amino acids, the building blocks of proteins, contain —SH groups, which help stabilize the shape of proteins.
Carbonyl $\overset{\text{O}}{\overset{\|}{\text{R—C—R}}}$ or $\overset{\text{O}}{\overset{\|}{\text{R—C—H}}}$	*Ketones* contain a carbonyl group within the carbon skeleton. The carbonyl group is polar and hydrophilic due to its electronegative O atom. *Aldehydes* have a carbonyl group at the end of the carbon skeleton.
Carboxyl $\overset{\text{O}}{\overset{\|}{\text{R—C—OH}}}$ or $\overset{\text{O}}{\overset{\|}{\text{R—C—O}^-}}$	*Carboxylic acids* contain a carboxyl group at the end of the carbon skeleton. All amino acids have a —COOH group at one end. The negatively charged form predominates at the pH of body cells and is hydrophilic.

NAME AND STRUCTURAL FORMULA*	OCCURRENCE AND SIGNIFICANCE
Ester $\overset{\text{O}}{\overset{\|}{\text{R—C—O—R}}}$	*Esters* predominate in dietary fats and oils and also occur in our body fat. Aspirin is an ester of salicylic acid, a pain-relieving molecule found in bark of the willow tree.
Phosphate $\text{R—O—}\overset{\overset{\text{O}}{\|}}{\underset{\underset{\text{O}^-}{\|}}{\text{P}}}\text{—O}^-$	*Phosphates* contain a phosphate group (PO_4), which is very hydrophilic due to the dual negative charges. An important example is adenosine triphosphate (ATP), which transfers chemical energy between organic molecules during chemical reactions.
Amino $\text{R—N}\overset{\diagup \text{H}}{\diagdown \text{H}}$ or $\text{R—}\overset{+}{\text{N}}\overset{\diagup \text{H}}{\underset{\diagdown \text{H}}{\text{—H}}}$	*Amines* have an —NH_2 group, which can act as a base and pick up a hydrogen ion, giving the amino group a positive charge. At the pH of body fluids, most amino groups have a charge of 1+. All amino acids have an amino group at one end.

*The letter R represents the carbon skeleton of the molecule.

---

it. It does, however, create bulk, which helps to move food and wastes along the gastrointestinal tract. In animals, the principal function of carbohydrates is to provide a readily available source of chemical energy to generate the ATP that drives metabolic reactions. Only a few carbohydrates form structural units in animals. One example is deoxyribose, a type of sugar that is a building block of deoxyribonucleic acid (DNA), the molecule that carries hereditary information. Other carbohydrates function as energy reserves. The prime storage carbohydrate in animals is glycogen, which is stored in the liver and skeletal muscles.

Carbon, hydrogen, and oxygen are the elements found in carbohydrates. The ratio of hydrogen to oxygen atoms is usually 2:1, the same as in water. Although there are exceptions, the general rule for carbohydrates is one carbon atom for each water molecule—the reason they are called carbohydrates, which means "watered carbon." Carbohydrates are divided into three major groups based on their size: monosaccharides, disaccharides, and polysaccharides (Table 2.6).

Figure 2.15 Alternate ways to write the structural formula for glucose.

In standard shorthand, carbon atoms are understood to be at locations where two bond lines intersect, and single hydrogen atoms are not indicated.

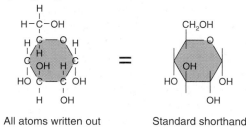

All atoms written out   Standard shorthand

Q How many hydroxyl groups does a molecule of glucose have? How many carbon atoms are part of glucose's carbon skeleton ring?

Table 2.6 Major Carbohydrate Groups

TYPE OF CARBOHYDRATE	EXAMPLES	TYPE OF CARBOHYDRATE	EXAMPLES
Monosaccharides	Glucose (blood sugar)	Polysaccharides	Glycogen (stored form of carbohydrate in animals)
	Fructose (found in fruits)		Starch (stored form of carbohydrate in plants and main carbohydrate in food)
	Galactose		Cellulose (part of cell walls in plants; not digested by humans but aids movement of food through intestines)
	Deoxyribose (in DNA)		
	Ribose (in RNA)		
Disaccharides	Sucrose (table sugar) = glucose + fructose		
	Lactose (milk sugar) = glucose + galactose		
	Maltose = glucose + glucose		

Figure 2.16 Structural and molecular formulas for the monosaccharides glucose and fructose and the disaccharide sucrose. In dehydration synthesis (read from left to right), two smaller molecules, glucose and fructose, are joined to form a larger molecule of sucrose. Note the loss of a water molecule. In hydrolysis (read from right to left), the addition of a water molecule to the larger sucrose molecule breaks the disaccharide into two smaller molecules, glucose and fructose.

🔑 **Monosaccharides are the monomeric building blocks of carbohydrates.**

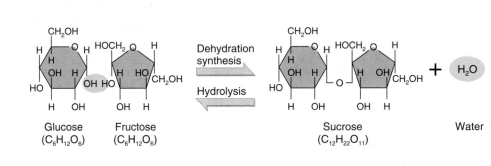

Q Is dehydration synthesis an anabolic or a catabolic reaction? How many carbons can you count in fructose? In sucrose?

Monosaccharides and Disaccharides: The Simple Sugars

Monosaccharides and disaccharides are known as **simple sugars.** The building blocks (monomers) of carbohydrates, **monosaccharides** (mon'-ō-SAK-a-rīds; *sacchar-* = sugar), contain from three to seven carbon atoms. The number of carbon atoms in the molecule is indicated by the prefix. For example, monosaccharides with three carbons are called trioses (*tri-* = three). There are also tetroses (four-carbon sugars), pentoses (five-carbon sugars), hexoses (six-carbon sugars), and heptoses (seven-carbon sugars). Glucose, a hexose, is the main energy-supplying molecule of the body.

Two monosaccharide molecules can combine by dehydration synthesis to form one **disaccharide** (dī-SAK-a-rīd; *di-* = two) molecule and a molecule of water. For example, molecules of the monosaccharides glucose and fructose combine to form a molecule of the disaccharide sucrose (table sugar), as shown in Figure 2.16. Although glucose and fructose have the same molecular formula, they are different monosaccharides because the relative positions of the oxygen and carbon atoms are different. Notice that the formula for sucrose is $C_{12}H_{22}O_{11}$, not $C_{12}H_{24}O_{12}$, because a molecule of water is split out as the two monosaccharides are joined.

Disaccharides can also be split into smaller, simpler molecules by hydrolysis. A molecule of sucrose, for example, may be hydrolyzed into its components, glucose and fructose, by the addition of water. Figure 2.16 also illustrates this reaction.

Polysaccharides

The third major group of carbohydrates, the **polysaccharides** (pol'-ē-SAK-a-rīds), contains tens or hundreds of monosaccharides joined through dehydration synthesis reactions. The principal polysaccharide in the human body is glycogen, which is composed of glucose units linked to each other. Glycogen is stored in the liver and skeletal muscles.

Like disaccharides, polysaccharides can be broken down into monosaccharides through hydrolysis reactions. For example, when the blood glucose level falls, liver cells have the ability to break down glycogen into glucose and release it into the blood. In this way, glucose is made available to body cells, where it can be broken down to synthesize ATP. Unlike simple sugars such as fructose and sucrose, however, polysaccharides usually are not soluble in water and do not taste sweet.

Lipids

A second important group of organic compounds is **lipids** (*lip-* = fat). Lipids make up 18–25% of body mass in lean adults. Like carbohydrates, lipids contain carbon, hydrogen, and oxygen. Unlike carbohydrates, they do not have a 2:1 ratio of hydrogen to oxygen. The proportion of electronegative oxygen atoms in lipids is usually smaller than in carbohydrates, so there are fewer polar covalent bonds. As a result, most lipids are insoluble in polar solvents such as water; they are *hydrophobic*. Nonpolar solvents such as chloroform and ether, however, readily dissolve lipids. Because they are hydrophobic, only the smallest lipids (some fatty acids) are dissolved in the watery blood. To become more soluble in blood plasma, other lipids form complexes with hydrophilic proteins; the resulting lipid and protein particles are **lipoproteins.**

The diverse lipid family includes triglycerides (fats and oils), phospholipids (lipids that contain phosphorus), steroids (lipids that contain rings of carbon atoms), eicosanoids (20-carbon lipids), and a variety of other lipids, including fatty acids, fat-soluble vitamins (vitamins A, D, E, and K), and lipoproteins. Table 2.7 summarizes the various types of lipids and highlights their roles in the human body.

Triglycerides

The most plentiful lipids in your body and in your diet are the **triglycerides** (trī-GLI-cer-īdes). At room temperature, triglycerides may be either solids (fats) or liquids (oils), and they are the body's most highly concentrated form of chemical energy. Triglycerides provide more than twice as much energy per gram as either carbohydrates or proteins. Our capacity to store triglycerides in adipose (fat) tissue is for all practical purposes, unlimited. Excess dietary carbohydrates, proteins, fats, and oils all have the same fate: They are deposited in adipose tissue as triglycerides.

A triglyceride consists of two types of building blocks: a single glycerol molecule and three fatty acid molecules. A three-carbon **glycerol** molecule forms the backbone of a triglyceride (Figure 2.17). Three **fatty acids** are attached by dehydration synthesis reactions, one to each carbon of the glycerol backbone. The chemical bond formed where each water molecule is removed is called an *ester linkage*. The reverse reaction, hydrolysis, breaks down a single molecule of a triglyceride into three fatty acids and glycerol.

Table 2.7 Types of Lipids in the Body

TYPE OF LIPID	FUNCTIONS
Triglycerides *(fats and oils)*	Protection, insulation, energy storage.
Phospholipids	Major lipid component of cell membranes.
Steroids	
Cholesterol	Minor component of all animal cell membranes; precursor of bile salts, vitamin D, and steroid hormones.
Bile salts	Needed for absorption of dietary lipids.
Vitamin D	Helps regulate calcium level in the body; needed for bone growth and repair.
Adrenocortical hormones	Help regulate metabolism, resistance to stress, and salt and water balance.
Sex hormones	Stimulate reproductive functions and sexual characteristics.
Eicosanoids	Have diverse effects on blood clotting, inflammation, immunity, stomach acid secretion, airway diameter, lipid breakdown, and smooth muscle contraction.
Other Lipids	
Fatty acids	Catabolized to generate adenosine triphosphate (ATP) or used to synthesize triglycerides and phospholipids.
Carotenes	Needed for synthesis of vitamin A, which is used to make visual pigments in the eyes.
Vitamin E	Promotes wound healing, prevents tissue scarring, contributes to the normal structure and function of the nervous system, and functions as an antioxidant.
Vitamin K	Required for synthesis of blood-clotting proteins.
Lipoproteins	Transport lipids in the blood, carry triglycerides and cholesterol to tissues, and remove excess cholesterol from the blood.

In later chapters we will refer to saturated, monounsaturated, and polyunsaturated fats. **Saturated fats** are triglycerides that contain only *single covalent bonds* between fatty acid carbon atoms. They do not contain any double bonds between fatty acid carbon atoms, and thus each carbon atom is *saturated with hydrogen atoms* (see, for example, palmitic acid and stearic acid in Figure 2.17c). Triglycerides with many saturated fatty acids tend to be solid at room temperature and occur mostly in animal tissues. They also occur in a few plant products, such as cocoa butter, palm oil, and coconut oil.

Monounsaturated fats contain fatty acids with *one* double covalent bond between two fatty acid carbon atoms, and thus they are not completely saturated with hydrogen atoms (see, for example, oleic acid in Figure 2.17c). Olive oil and peanut oil are rich in triglycerides with monounsaturated fatty acids.

Figure 2.17 The formation of a triglyceride from a glycerol and three fatty acid molecules. Each time a glycerol (a) and a fatty acid (b) are joined in dehydration synthesis, a molecule of water is lost. An ester linkage joins the glycerol to each of the three molecules of fatty acids, which vary in length and the number and location of double bonds between carbon atoms (C = C). Shown here (c) is a triglyceride molecule that contains two saturated fatty acids and a monounsaturated fatty acid. The kink (bend) in the oleic acid occurs at the double bond.

One glycerol and three fatty acids are the building blocks of triglycerides.

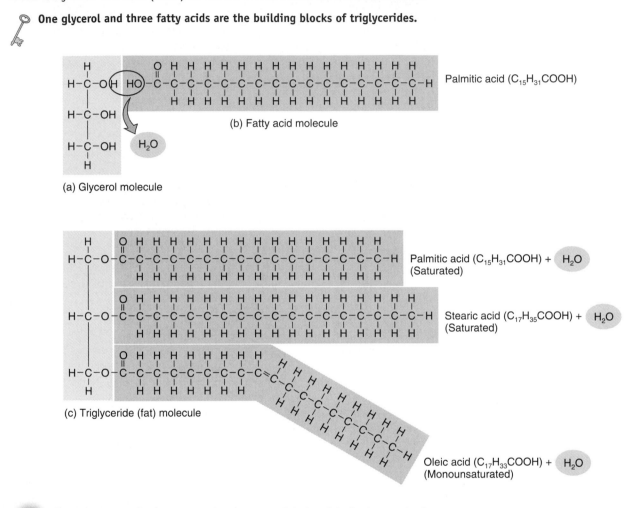

(a) Glycerol molecule

(b) Fatty acid molecule

Palmitic acid ($C_{15}H_{31}COOH$)

(c) Triglyceride (fat) molecule

Palmitic acid ($C_{15}H_{31}COOH$) + H_2O
(Saturated)

Stearic acid ($C_{17}H_{35}COOH$) + H_2O
(Saturated)

Oleic acid ($C_{17}H_{33}COOH$) + H_2O
(Monounsaturated)

Q Does the oxygen in the water molecule removed during dehydration synthesis come from the glycerol or from a fatty acid?

Polyunsaturated fats contain *more than one double covalent bond* between fatty acid carbon atoms. An example is linoleic acid. Corn oil, safflower oil, sunflower oil, cottonseed oil, sesame oil, and soybean oil contain a high percentage of polyunsaturated fatty acids.

Phospholipids

Like triglycerides, **phospholipids** have a glycerol backbone and two fatty acid chains attached to the first two carbons. In the third position, however, a phosphate group (PO_4^{3-}) links a small charged group that usually contains nitrogen (N) to the backbone (Figure 2.18). This portion of the molecule (the "head") is polar and can form hydrogen

bonds with water molecules. The two fatty acids (the "tails"), on the other hand, are nonpolar and can interact only with other lipids. Molecules that have both polar and nonpolar portions are said to be **amphipathic** (am-fi-PATH-ic; *amphi-* = on both sides; *-pathic* = feeling). Phospholipids, which have this unusual property, line up tails-to-tails in a double row to make up much of the membrane that surrounds each cell (Figure 2.18c).

Steroids

Steroids have four rings of carbon atoms, designated A, B, C, and D (Figure 2.19). Their structure differs considerably from that of the triglycerides. Other steroids may be

Figure 2.18 Phospholipids. (a) In the synthesis of phospholipids, two fatty acids attach to the first two carbons of the glycerol backbone. A phosphate group links a small charged group to the third carbon in glycerol. In (b), the circle represents the polar head region, and the two wavy lines represent the two nonpolar tails. Double bonds in the fatty acid hydrocarbon chain often form kinks in the tail.

🔑 **Phospholipids are amphipathic molecules, having both polar and nonpolar regions.**

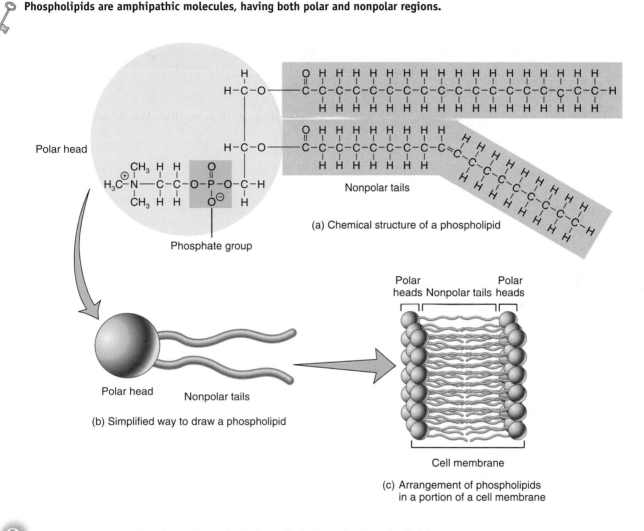

(a) Chemical structure of a phospholipid

(b) Simplified way to draw a phospholipid

(c) Arrangement of phospholipids in a portion of a cell membrane

Q Which portion of a phospholipid is hydrophilic, and which portion is hydrophobic?

synthesized from cholesterol (Figure 2.19a), which has a large nonpolar region consisting of the four rings and a hydrocarbon tail. In the body, the commonly encountered steroids, such as cholesterol, sex hormones, cortisol, bile salts, and vitamin D, are further classified as **sterols** because they have a hydroxyl (alcohol) group (—OH) attached to one or more of the rings. These polar hydroxyl groups make sterols weakly amphipathic. Although cholesterol can contribute to the buildup of fatty plaques in atherosclerosis, it also serves important functions—as a component of animal cell membranes and as the starting material for the synthesis of other steroids (see Table 2.7).

Eicosanoids and Other Lipids

Eicosanoids (ī-KŌ-sa-noids; *eikosan-* = twenty) are lipids derived from a 20-carbon fatty acid called arachidonic acid. The two principal subclasses of eicosanoids are the **prostaglandins** (pros′-ta-GLAN-dins) and **leukotrienes** (loo′-kō-TRĪ-ēns). Prostaglandins are involved in a wide variety of body functions. They modify responses to hormones, contribute to the inflammatory response (Chapter 22), prevent stomach ulcers, dilate (enlarge) airways to the lungs, regulate body temperature, and influence formation of blood clots, to name just a few effects. Leukotrienes participate in allergic and inflammatory responses.

Figure 2.19 Steroids. All steroids have four rings of carbon atoms (labeled A–D).

🔑 **Cholesterol is the starting material for synthesis of other steroids in the body.**

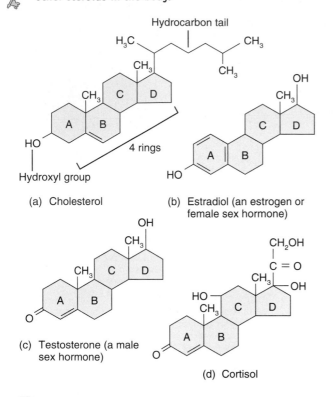

(a) Cholesterol

(b) Estradiol (an estrogen or female sex hormone)

(c) Testosterone (a male sex hormone)

(d) Cortisol

Q How is the structure of estradiol different from that of testosterone?

Body lipids also include fatty acids (which can undergo either hydrolysis to provide ATP or dehydration synthesis to build triglycerides and phospholipids); fat-soluble vitamins such as beta-carotenes (the yellow-orange pigments in egg yolk, carrots, and tomatoes that are converted to vitamin A); vitamins D, E, and K; and lipoproteins.

Proteins

A third principal group of organic compounds is **proteins,** which are much more complex in structure and have a larger range of functions than carbohydrates or lipids. A normal, lean adult body is 12–18% protein. Some proteins have a structural role—they are cellular building materials. Other proteins have physiological roles. For example, proteins in the form of enzymes speed up most essential biochemical reactions; other proteins provide the machinery for muscle contraction. Antibodies are proteins that defend against invading microbes. Some hormones that regulate homeostasis also are proteins. Table 2.8 describes several functions of proteins.

Table 2.8 Functions of Proteins

TYPE OF PROTEINS	FUNCTIONS
Structural	Form structural framework of various parts of the body.
	Examples: collagen in bone and other connective tissues, and keratin in skin, hair, and fingernails.
Regulatory	Function as hormones, regulate various physiological processes, control growth and development, and mediate responses of nervous system.
	Examples: insulin, which regulates blood glucose level, and substance P, which mediates sensation of pain in the nervous system.
Contractile	Allow shortening of muscle cells, which produces movement.
	Examples: myosin and actin.
Immunological	Aid responses that protect body against foreign substances and invading pathogens.
	Examples: antibodies and interleukins.
Transport	Carry vital substances throughout body.
	Example: hemoglobin, which transports most oxygen and some carbon dioxide in the blood.
Catalytic	Act as enzymes that regulate biochemical reactions.
	Examples: salivary amylase, lipase, and lactase.

Amino Acids and Polypeptides

Proteins always contain carbon, hydrogen, oxygen, and nitrogen; some also contain sulfur. Just as monosaccharides are the building blocks of polysaccharides, **amino acids** (a-MĒ-nō) are the building blocks (monomers) of proteins. Each of the 20 different amino acids has three important functional groups attached to a central carbon atom (Figure 2.20a): (1) an amino group (—NH$_2$), (2) a carboxyl group (—COOH), and (3) a side chain (R group). At the normal pH of body fluids, both the amino group and the carboxyl group are ionized (Figure 2.20b). The distinctive side chain gives each amino acid its individual chemical identity (Figure 2.20c).

Synthesis of a protein takes place in stepwise fashion—one amino acid is joined to a second, a third is then added to the first two, and so on. The covalent bond joining each pair of amino acids is called a **peptide bond.** It always forms between the carboxyl group (—COOH) of one amino acid and the amino group (—NH$_2$) of another. At the site of formation of a peptide bond, a molecule of water is removed (Figure 2.21). Thus this is a dehydration synthesis reaction. Breaking a peptide bond, as occurs during digestion of dietary proteins, is a hydrolysis reaction (see Figure 2.21).

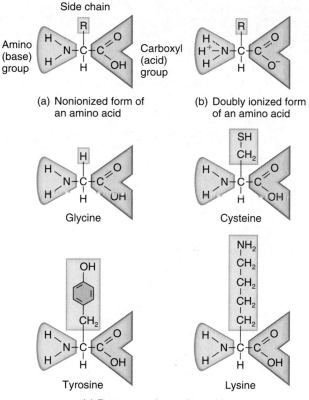

(a) Nonionized form of an amino acid

(b) Doubly ionized form of an amino acid

Glycine

Cysteine

Tyrosine

Lysine

(c) Representative amino acids

Figure 2.20 Amino acids. (a) In keeping with their name, amino acids have an amino group (shaded blue) and a carboxyl (acid) group (shaded red). The side chain (R group) is different in each amino acid. (b) At pH close to 7, both the amino group and the carboxyl group are ionized. (c) Glycine is the simplest amino acid; the side chain is a single H atom. Cysteine is one of two amino acids that contain sulfur (S). The side chain in tyrosine contains a six-carbon ring. Lysine has a second amino group at the end of the side chain.

🔑 **Body proteins contain 20 different amino acids, each of which has a unique side chain.**

Q   In an amino acid, what is the minimum number of carbon atoms? Of nitrogen atoms?

When two amino acids combine, a **dipeptide** results. Adding another amino acid to a dipeptide produces a **tripeptide.** Further additions of amino acids result in the formation of a chainlike **peptide** (4–10 amino acids) or **polypeptide** (10–2000 or more amino acids). Small proteins contain as few as 50 amino acids. In addition, a protein may consist of only one polypeptide chain or several that are folded together.

A great variety of proteins is possible because each variation in the number or sequence of amino acids can produce a different protein. The situation is similar to using an alphabet of 20 letters to form words: Each different amino acid is like a letter, and their various combinations (peptides, polypeptides, or proteins) are like the words.

Levels of Structural Organization in Proteins

Proteins exhibit four levels of structural organization. The **primary structure** is the unique sequence of amino acids that are linked by covalent peptide bonds to form a

Figure 2.21 Formation of a peptide bond between two amino acids during dehydration synthesis (read from left to right). In this example, glycine is joined to alanine (forming a dipeptide). Breaking a peptide bond occurs via hydrolysis (read from right to left).

🔑 **Amino acids are the monomeric building blocks of proteins.**

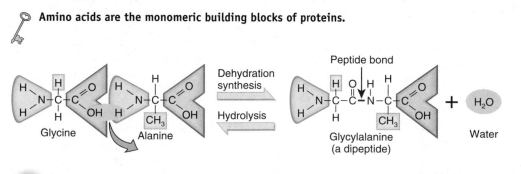

Glycine   Alanine

Dehydration synthesis

Hydrolysis

Peptide bond

Glycylalanine (a dipeptide)

Water

Q   What type of reaction takes place during catabolism of proteins?

Figure 2.22 Levels of structural organization in proteins. (a) The primary structure is the sequence of amino acids in the polypeptide. (b) Common secondary structures include alpha helixes and pleated sheets. For simplicity, the amino acid side groups are not shown here. (c) The tertiary structure is the overall folding pattern that produces a distinctive, three-dimensional shape. (d) The quaternary structure in a protein is the arrangement of two or more polypeptide chains relative to one another. Adapted from Neil Campbell, Jane Reece, and Larry Mitchell *Biology* 5e, F5.24, p75 (Menlo Park, CA; Addison Wesley Longman, 1999). ©1999 Addison Wesley Longman, Inc.

🔑 **The unique shape of each protein permits it to carry out specific functions.**

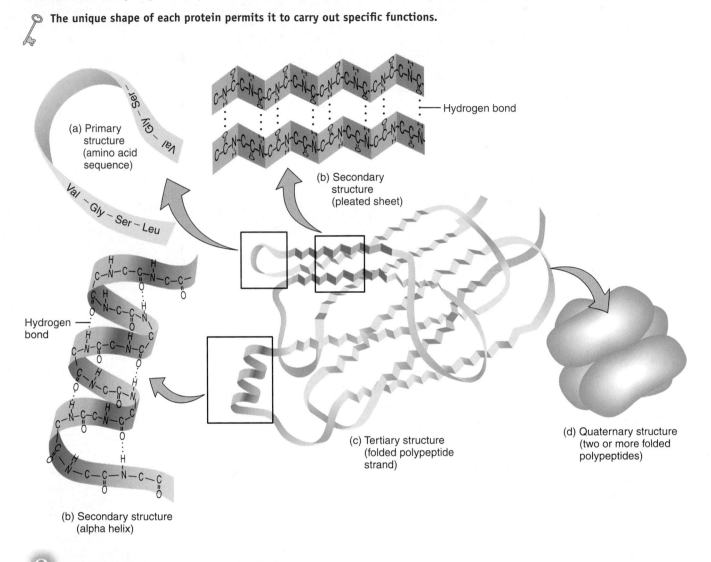

(a) Primary structure (amino acid sequence)

(b) Secondary structure (pleated sheet)

Hydrogen bond

(c) Tertiary structure (folded polypeptide strand)

(d) Quaternary structure (two or more folded polypeptides)

(b) Secondary structure (alpha helix)

Q Do all proteins have a quaternary structure?

polypeptide strand (Figure 2.22a). A protein's primary structure is genetically determined. Any changes in a protein's amino acid sequence can have serious consequences for the body. In sickle-cell disease, for example, a nonpolar amino acid (valine) replaces a polar amino acid (glutamate) at just two locations in the oxygen-carrying protein hemoglobin. This change of amino acids diminishes hemoglobin's water solubility. As a result, the altered hemoglobin tends to form crystals inside red blood cells, producing deformed, sickle-shaped cells that cannot properly squeeze through narrow blood vessels.

The **secondary structure** of a protein is the repeated twisting or folding of neighboring amino acids in the polypeptide chain (Figure 2.22b). Two common secondary

structures are a clockwise spiral called an alpha helix and pleated sheets. The secondary structure of a protein is stabilized by hydrogen bonds, which are found at regular intervals along the polypeptide backbone.

The **tertiary structure** (TUR-shē-er′-ē) refers to the three-dimensional shape of a polypeptide chain. The tertiary folding pattern may allow amino acids at opposite ends of the chain to be close neighbors (Figure 2.22c). Each protein has a unique tertiary structure that determines how it will function. Because most of the proteins in our body exist in watery surroundings, the folding process places most amino acids that have hydrophobic side chains in the central core of the protein, away from the protein's surface. Figure 2.23 shows the types of bonds that stabilize a protein's tertiary structure. The strongest but least common bonds are S—S covalent bonds, called *disulfide bridges.* These bonds form between the sulfhydryl groups of two monomers of the amino acid cysteine. Numerous weak bonds—hydrogen bonds, ionic bonds, and hydrophobic interactions—also determine the folding pattern.

When a protein contains more than one polypeptide chain, the arrangement of the individual polypeptide chains relative to one another is the **quaternary structure** (KWA-ter-ner′-ē; Figure 2.22d). The bonds that hold polypeptide chains together are basically the same as those that maintain the tertiary structure.

Proteins vary tremendously in structure. Different proteins have different architectures and different three-dimensional shapes. This variation in structure and shape is directly related to their diverse functions. Homeostatic mechanisms maintain the temperature and chemical composition of body fluids, which allow body proteins to keep their proper three-dimensional shapes. When a cell makes a protein, the polypeptide chain folds into a certain shape. One reason for folding of the polypeptide is that some of its parts are attracted to water (hydrophilic), and other parts are repelled by it (hydrophobic). Often, helper molecules known as chaperones aid the folding process.

In practically every case, the function of a protein depends on its ability to recognize and bind to some other molecule, much as a key fits in a lock. Thus a hormone binds to a specific protein on a cell whose function it will alter, and an antibody protein binds to a foreign substance (antigen) that has invaded the body. A protein's unique shape permits it to interact with other molecules to carry out specific functions.

If a protein encounters an environment in which temperature, pH, or electrolyte concentration is altered, it may unravel and lose its characteristic shape (secondary, tertiary, and quaternary structure). This process is called **denaturation.** Denatured proteins are no longer functional. A common example of permanent denaturation is seen in frying an egg. In a raw egg the soluble egg white protein (albumin) is a clear, viscous fluid. When heat is applied to the egg, however, the protein changes shape (denatures), becomes insoluble, and turns white.

Enzymes

In living cells, most catalysts are protein molecules called **enzymes** (EN-zīms). Some enzymes consist of two parts—a protein portion, called the **apoenzyme** (ā′-pō-EN-zīm), and a nonprotein portion, called a **cofactor.** The cofactor may be a metal ion (such as iron, magnesium, zinc, or calcium) or an organic molecule called a *coenzyme.* Coenzymes often are derived from vitamins. Together, the apoenzyme and cofactor form a **holoenzyme,** or whole enzyme. The names of enzymes usually end in the suffix *-ase.* All enzymes can be grouped according to the types of chemical reactions they catalyze. For example, *oxidases* add oxygen, *kinases* add phosphate, *dehydrogenases* remove hydrogen, *ATPases* split ATP, *anhydrases* remove water, *proteases* break down proteins, and *lipases* break down lipids.

Figure 2.23 Types of bonds that stabilize the tertiary structure of proteins. Adapted from Neil Campbell, Jane Reece, and Larry Mitchell *Biology* 5e, F5.22, p74 (Menlo Park, CA; Addison Wesley Longman, 1999). ©1999 Addison Wesley Longman, Inc.

The disulfide bridge is a strong covalent bond, whereas the hydrogen bond, hydrophobic interaction, and ionic bond are weak bonds.

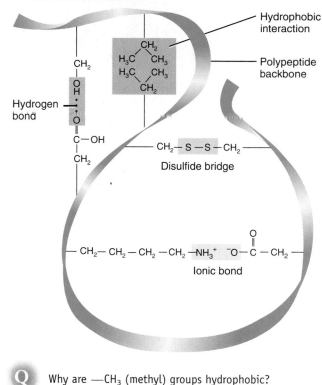

Q Why are —CH$_3$ (methyl) groups hydrophobic?

Enzymes catalyze selected reactions, doing so with great efficiency and with many built-in controls. Enzymes have three important properties:

1. *Enzymes are highly specific.* Each particular enzyme binds only to specific **substrates**—the reactant molecules on which the enzyme acts. In some cases, the portion of the enzyme that catalyzes the reaction, called the **active site,** is thought to "fit" the substrate like a key fits in a lock. In other cases the active site changes its shape to fit snugly around the substrate once the substrate enters the active site. This is known as an *induced fit.* Of the more than 1000 known enzymes in your body, each has a characteristic three-dimensional shape with a specific surface configuration, which allows it to recognize and bind to certain substrates.

Not only is an enzyme matched to a particular substrate, it also catalyzes a specific reaction. From among the large number of diverse molecules in a cell, an enzyme must recognize the correct substrate and then take it apart or merge it with another substrate to form one or more specific products.

2. *Enzymes are very efficient.* Under optimal conditions, enzymes can catalyze reactions at rates that are from 100 million to 10 billion times more rapid than those of similar reactions occurring without enzymes. The number of

Figure 2.24 How an enzyme works.

🔑 **An enzyme speeds up a chemical reaction without being altered or consumed.**

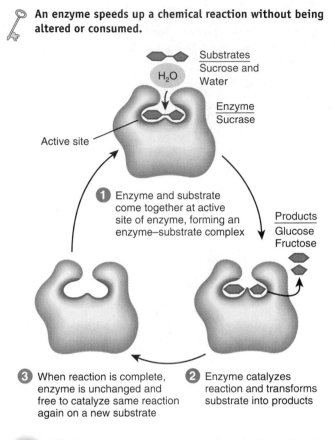

Substrates
Sucrose and
Water

H_2O

Enzyme
Sucrase

Active site

1 Enzyme and substrate come together at active site of enzyme, forming an enzyme–substrate complex

Products
Glucose
Fructose

3 When reaction is complete, enzyme is unchanged and free to catalyze same reaction again on a new substrate

2 Enzyme catalyzes reaction and transforms substrate into products

Q Why is it that sucrase cannot catalyze the formation of sucrose from glucose and fructose?

substrate molecules that a single enzyme molecule can convert to product molecules in one second is generally between 1 and 10,000 and can be as high as 600,000.

3. *Enzymes are subject to a variety of cellular controls.* Their rate of synthesis and their concentration at any given time are under the control of a cell's genes. Substances within the cell may either enhance or inhibit the activity of a given enzyme. Many enzymes have both active and inactive forms in cells; the rate at which the inactive form becomes active or vice versa is determined by the chemical environment inside the cell.

Enzymes lower the activation energy of a chemical reaction by decreasing the "randomness" of the collisions between molecules. They also help bring the substrates together in the proper orientation so that the reaction can occur. Figure 2.24 depicts the way an enzyme is thought to work:

1 The substrates make contact with the active site on the surface of the enzyme molecule, forming a temporary intermediate compound called the **enzyme–substrate complex.** In this reaction the substrates are the disaccharide sucrose and a molecule of water.

2 The substrate molecules are transformed by either the rearrangement of existing atoms, the breakdown of the substrate molecule, or the combination of several substrate molecules into the products of the reaction. Here the products are two monosaccharides: glucose and fructose.

3 After the reaction is completed and the reaction products move away from the enzyme, the unchanged enzyme is free to attach to other substrate molecules.

Sometimes a single enzyme may catalyze a reversible reaction in either direction, depending on the relative abundances of the substrates and products. For example, the enzyme *carbonic anhydrase* catalyzes the following reversible reaction:

$$CO_2 \quad + \quad H_2O \quad \underset{\text{Carbonic anhydrase}}{\rightleftharpoons} \quad H_2CO_3$$

Carbon
dioxide   Water   Carbonic acid

During exercise, when more CO_2 is produced and released into the blood, the reaction flows to the right, increasing the amount of carbonic acid in the blood. Then as you exhale CO_2, its level in the blood falls and the reaction flows to the left, converting carbonic acid to CO_2 and H_2O.

 CLINICAL APPLICATION
Galactosemia

Galactosemia (ga-lak-tō-SĒ-mē-a; *galactos-* = milk; *-emia* = in the blood) is an inherited disorder in which the monosaccharide galactose accumulates in the blood because it cannot be converted to glucose due to the absence of a required enzyme. An infant with the disorder will fail to thrive within a week after birth due to anorexia (loss of appetite), vomiting, and diarrhea. Because the main source of galactose in the infant's diet is milk sugar (lactose), treatment of galactosemia consists of eliminating milk from the diet. If treatment is delayed, the infant will remain physically small and may become mentally retarded. ■

Nucleic Acids: Deoxyribonucleic Acid (DNA) and Ribonucleic Acid (RNA)

Nucleic acids (noo-KLĒ-ic), so named because they were first discovered in the nuclei of cells, are huge organic molecules that contain carbon, hydrogen, oxygen, nitrogen, and phosphorus. Nucleic acids are of two varieties. The first, **deoxyribonucleic acid (DNA)** (dē-ok′-sē-rī-bō-noo-KLĒ-ik), forms the inherited genetic material inside each cell. Each **gene** is a segment of a DNA molecule. Our genes determine which traits we inherit, and by controlling protein synthesis they regulate most of the activities that take place in our cells throughout our lifetime. When a cell divides, its hereditary information passes on to the next generation of cells. **Ribonucleic acid (RNA),** the second type of nucleic acid, relays instructions from the genes to guide each cell's assembly of amino acids into proteins.

Figure 2.25 DNA molecule. (a) A nucleotide consists of a base, a pentose sugar, and a phosphate group. (b) The paired bases project toward the center of the double helix. The structure is stabilized by hydrogen bonds (dotted lines) between each base pair. There are two hydrogen bonds between adenine and thymine and three between cytosine and guanine.

🔑 **Nucleotides are the monomeric building blocks of nucleic acids.**

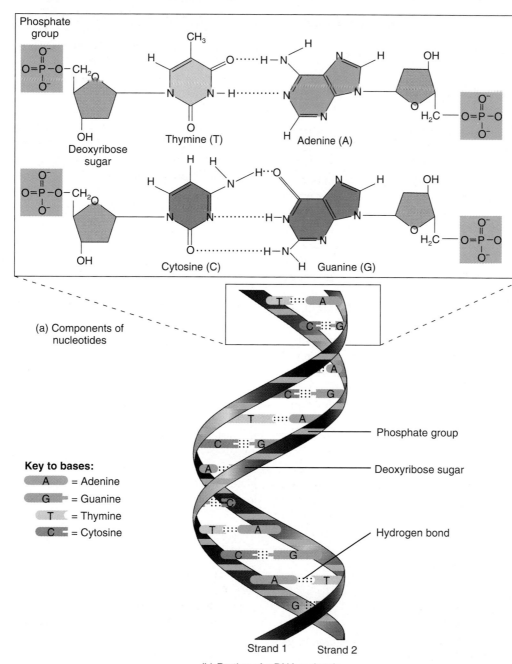

(a) Components of nucleotides

Key to bases:

A = Adenine
G = Guanine
T = Thymine
C = Cytosine

Phosphate group
Deoxyribose sugar
Hydrogen bond

Strand 1   Strand 2

(b) Portion of a DNA molecule

Q Which bases always pair with one another?

The monomers of nucleic acids are called **nucleotides.** A nucleic acid is a chain composed of repeating nucleotide units. Each nucleotide of DNA consists of three parts (Figure 2.25a):

1. *Nitrogenous base.* DNA contains four different **nitrogenous bases,** molecules that contain atoms of C, H, O, and N. In DNA the four bases are adenine (A), thymine (T),

cytosine (C), and guanine (G). Whereas adenine and guanine are larger, double-ring bases called **purines** (PYOO-rēns), thymine and cytosine are smaller, single-ring bases called **pyrimidines** (pī-RIM-i-dēns). The nucleotides are named according to the base that is present. For instance, a nucleotide containing thymine is called a thymine nucleotide, one containing adenine is called an adenine nucleotide, and so on.

2. *Pentose sugar.* A five-carbon sugar called **deoxyribose** attaches to each base in DNA.

3. *Phosphate group.* Alternating phosphate groups (PO_4^{3-}) and pentoses form the "backbone" of a DNA strand; the bases project inward from the backbone chain (Figure 2.25b).

In 1953, F. H. C. Crick of Great Britain and J. D. Watson, a young American scientist, published a brief paper describing how these three components might be arranged in DNA. Their insights into data gathered by others led them to construct a model so elegant and simple that the scientific world immediately knew it was correct! In the Watson–Crick **double helix** model, DNA resembles a spiral ladder (see Figure 2.25b). Two strands of alternating phosphate groups and deoxyribose sugars form the uprights of the ladder; paired bases, held together by hydrogen bonds, form the rungs. Adenine always pairs with thymine, and cytosine always pairs with guanine. If you know the sequence of bases in one strand of DNA, you can predict the sequence on the complementary (second) strand. Each time DNA is copied, as when an organism adds cells, the two strands unwind. Each strand serves as the template or mold on which to construct a new second strand. As you will see later, any change that occurs in the base sequence of a DNA strand is called a *mutation.* Some mutations can result in the death of a cell, cause cancer, or produce genetic defects in future generations.

RNA, the second variety of nucleic acid, differs from DNA in several respects. In humans, RNA is single-stranded. The sugar in the RNA nucleotide is the pentose **ribose,** and RNA contains the pyrimidine base uracil (U) instead of thymine. Cells contain three different kinds of RNA: messenger RNA, ribosomal RNA, and transfer RNA. Each has a specific role to perform in carrying out the instructions coded in DNA (described on page 87).

CLINICAL APPLICATION
DNA Fingerprinting

A technique called DNA fingerprinting is used in research and in courts of law to ascertain whether a person's DNA matches the DNA obtained from samples or pieces of legal evidence such as blood stains or hairs. In each person, certain DNA segments contain base sequences that are repeated several times. Both the number of repeat copies in one region and the number of regions subject to repeat are different from one person to another. DNA fingerprinting can be done with minute quantities of DNA—for example, from a single strand of hair, a drop of semen, or a spot of blood. It also can be used to identify a crime victim or a child's biological parents and even to determine whether two people have a common ancestor. ■

Adenosine Triphosphate

Adenosine triphosphate (a-DEN-ō-sēn) or **ATP** is the "energy currency" of living systems (Figure 2.26). It functions to store temporarily and then transfer the energy liberated in exergonic catabolic reactions to cellular activities that require energy (endergonic reactions). Among these cellular activities are muscular contractions, movement of chromosomes during cell division, movement of structures within cells, transport of substances across cell membranes, and synthesis of larger molecules from smaller ones. Structurally, ATP consists of three phosphate groups attached to an adenosine unit composed of adenine and the five-carbon sugar ribose.

When the terminal phosphate group (PO_4^{3-}), symbolized by Ⓟ in the following discussion, is hydrolyzed by the addition of a water molecule, the overall reaction liberates energy. This energy is used by the cell to power its activities. The enzyme that catalyzes the hydrolysis of ATP is called *ATPase.* Removal of the terminal phosphate group leaves a molecule called **adenosine diphosphate** (**ADP**). This reaction may be represented as follows:

ATP + H₂O $\xrightarrow{ATPase}$ ADP + Ⓟ + E

Adenosine triphosphate   Water   Adenosine diphosphate   Phosphate group   Energy

The energy supplied by the catabolism of ATP into ADP is constantly being used by the cell. As the supply of ATP at any given time is limited, a mechanism exists to replenish it: The enzyme *ATP synthase* catalyzes the addition of a phosphate group to ADP. The reaction may be represented as follows:

ADP + Ⓟ + E $\xrightarrow{ATP\ synthase}$ ATP + H₂O

Adenosine diphosphate   Phosphate group   Energy   Adenosine triphosphate   Water

As you can see from this reaction, energy is required to produce ATP. The energy needed to attach a phosphate group to ADP is supplied mainly by the catabolism of glucose in a process called cellular respiration. Cellular respiration has two phases, anaerobic and aerobic:

1. *Anaerobic phase.* In the absence of oxygen, glucose is partially broken down by a series of catabolic reactions into pyruvic acid. Each glucose that is converted into a pyruvic acid molecule yields two molecules of ATP.

2. *Aerobic phase.* In the presence of oxygen, glucose is completely broken down into carbon dioxide and water.

Figure 2.26 Structures of ATP and ADP. The two phosphate bonds that can be used to transfer energy are indicated by "squiggles" (~). Energy transfer typically involves hydrolysis of the terminal phosphate bond of ATP.

🔑 **ATP transfers chemical energy to power cellular activities.**

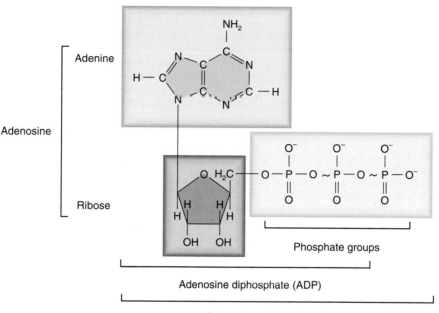

Q What are some cellular activities that depend on energy supplied by ATP?

These reactions generate heat and a large number of ATP molecules: 36–38 ATP from each glucose.

The details of cellular respiration will be discussed in Chapters 10 and 25.

1. How are carbohydrates classified?
2. Compare dehydration synthesis and hydrolysis.
3. Explain the importance to the body of the following lipids: triglycerides, phospholipids, steroids, lipoproteins, and eicosanoids.

4. Distinguish among saturated, monounsaturated, and polyunsaturated fats.
5. Define a protein. What is a peptide bond?
6. Discuss the classification of proteins on the basis of their level of structural organization.
7. How do DNA and RNA differ?
8. In the reaction catalyzed by ATP synthase, what are the substrates and products? Is this an exergonic or endergonic reaction?

HOW MATTER IS ORGANIZED (p. 26)

1. All forms of matter are composed of chemical elements.
2. Oxygen, carbon, hydrogen, and nitrogen make up about 96% of body mass.
3. Each element is made up of small units called atoms.
4. Atoms consist of a nucleus, which contains protons and neutrons, plus electrons that move about the nucleus in regions called electron shells.
5. The number of protons (the atomic number) distinguishes the atoms of one element from those of another element.
6. The mass number of an atom is the sum of its protons and neutrons.
7. Different atoms of an element that have the same number of protons but different numbers of neutrons are called isotopes. Radioactive isotopes are unstable and decay.
8. The atomic mass of an element is the average mass of all naturally occurring isotopes of that element.
9. An atom that *gives up* or *gains* electrons becomes an ion—an atom that has a positive or negative charge because it has unequal numbers of protons and electrons. Positively charged ions are cations; negatively charged ions are anions.
10. If two atoms share electrons, a molecule is formed. Compounds contain atoms of two or more elements.
11. A free radical is an electrically charged atom or group of atoms with an unpaired electron in its outermost shell. A common example is superoxide, which is formed by the addition of an electron to an oxygen molecule.

CHEMICAL BONDS (p. 30)

1. Atoms are held together by forces of attraction called chemical bonds. These bonds result from gaining, losing, or sharing electrons in the valence shell.
2. Most atoms become stable when they have an octet of eight electrons in their valence (outer) electron shell.
3. When the force of attraction between ions of opposite charge holds them together, an ionic bond has formed.
4. In a covalent bond, atoms share pairs of valence electrons. Covalent bonds may be single, double, or triple and either nonpolar or polar.
5. An atom of hydrogen that forms a polar covalent bond with an oxygen atom or a nitrogen atom may also form a weaker bond, called a hydrogen bond, with an electronegative atom. The polar covalent bond causes the hydrogen atom to have a partial positive charge (δ^+) that attracts the partial negative charge (δ^-) of neighboring electronegative atoms, often oxygen or nitrogen.

CHEMICAL REACTIONS (p. 34)

1. When atoms combine with or break apart from other atoms, a chemical reaction occurs. The starting substances are the reactants, and the ending ones are the products.
2. Energy, the capacity to do work, is of two principal kinds: potential (stored) energy and kinetic energy (energy of motion).
3. Endergonic reactions require energy, whereas exergonic reactions release energy. ATP couples endergonic and exergonic reactions.
4. The initial energy investment needed to start a reaction is the activation energy. Reactions are more likely when the concentrations and the temperatures of the reacting particles are higher.
5. Catalysts accelerate chemical reactions by lowering the activation energy. Most catalysts in living organisms are protein molecules called enzymes.
6. Synthesis reactions involve the combination of reactants to produce larger molecules. The reactions are anabolic and usually endergonic.
7. In decomposition reactions, a substance is broken down into smaller molecules. The reactions are catabolic and usually exergonic.
8. Small organic molecules are joined together to form larger molecules by a process called dehydration synthesis in which a molecule of water is removed. In the reverse process, called hydrolysis, large molecules are broken down into smaller ones by adding water.
9. Exchange reactions involve the replacement of one atom or atoms by another atom or atoms.
10. In reversible reactions, end products can revert to the original reactants.
11. In an oxidation-reduction (redox) reaction, one compound is oxidized (electrons are lost) while another compound is reduced (electrons are gained).

INORGANIC COMPOUNDS AND SOLUTIONS (p. 38)

1. Inorganic compounds usually are small and usually lack carbon. Organic substances always contain carbon, usually contain hydrogen, and always have covalent bonds.
2. Inorganic acids, bases, and salts dissociate into ions in water. An acid ionizes into anions and hydrogen ions (H^+); a base ionizes into cations and hydroxide ions (OH^-). A salt ionizes into neither H^+ nor OH^-.
3. Mixtures are combinations of elements or compounds that are physically blended together but are not bound by chemical bonds. Solutions, colloids, and suspensions are mixtures with different properties.
4. Two ways to express the concentration of a solution are percent and moles per liter. A mole (abbreviated mol) is the amount (in grams) of any substance that has a mass equal to the combined atomic mass of all its atoms.
5. Water is the most abundant substance in the body. It is an excellent solvent and suspending medium, participates in some metabolic reactions, and serves as a lubricant. Because of its many hydrogen bonds, water molecules are cohesive, which causes a high surface tension, and water has a high capacity for absorbing heat and a high heat of vaporization.
6. The pH of body fluids must remain fairly constant for the body to maintain homeostasis. On the pH scale, 7 represents neutrality. Values below 7 indicate acidic solutions, and values above 7 indicate alkaline solutions.
7. Buffer systems usually consist of a weak acid and a weak base. They react with excess H^+ and excess OH^-, which helps maintain pH homeostasis.
8. One important buffer system is the carbonic-acid–bicarbonate buffer system. The bicarbonate ion (HCO_3^-) acts as a weak base, and carbonic acid (H_2CO_3) acts as a weak acid.

ORGANIC COMPOUNDS (p. 42)

1. Carbon, with its four valence electrons, bonds covalently with other carbon atoms to form large molecules of many different shapes. Attached to the carbon skeletons of organic molecules are functional groups that confer distinctive chemical properties.

2. Carbohydrates provide most of the chemical energy needed to generate ATP. They may be monosaccharides, disaccharides, or polysaccharides.

3. Lipids are a diverse group of compounds that include triglycerides (fats and oils), phospholipids, steroids, and eicosanoids. Triglycerides protect, insulate, provide energy, and are stored. Phospholipids are important cell membrane components. Eicosanoids (prostaglandins and leukotrienes) modify hormone responses, contribute to inflammation, dilate airways, and regulate body temperature.

4. Proteins are constructed from amino acids. They give structure to the body, regulate processes, provide protection, help muscles contract, transport substances, and serve as enzymes. Levels of structural organization among proteins are primary, secondary, tertiary, and quaternary.

5. Deoxyribonucleic acid (DNA) and ribonucleic acid (RNA) are nucleic acids consisting of nitrogenous bases, five-carbon (pentose) sugars, and phosphate groups. DNA is a double helix and is the primary chemical in genes. RNA differs in structure and chemical composition from DNA and takes part in protein synthesis reactions.

6. Adenosine triphosphate (ATP) is the principal energy-transferring molecule in living systems. When it transfers energy to an endergonic reaction, it is decomposed to adenosine diphosphate (ADP) and Ⓟ. ATP is synthesized from ADP and Ⓟ using the energy supplied by various decomposition reactions, particularly those of glucose.

SELF-QUIZ QUESTIONS

Complete the following:

1. Atoms consist of ___, ___, and ___.
2. To be chemically stable, the atoms of all elements except hydrogen and helium must hold ___ electrons in the valence shell.
3. The chemical bond that most commonly serves as a link between molecules is the ___ bond.
4. The factors that influence the chance that a collision will occur and cause a chemical reaction are ___ and ___.
5. Match the following:

   ___ (a) acid
   ___ (b) base
   ___ (c) buffer
   ___ (d) enzyme
   ___ (e) pH
   ___ (f) ATP
   ___ (g) salt
   ___ (h) water

   (1) a polar covalent molecule that serves as a solvent, has a high heat capacity, creates a high surface tension, and serves as a lubricant
   (2) a substance that dissociates into one or more hydrogen ions and one or more anions
   (3) a substance that dissociates into cations and anions, neither of which is a hydrogen ion or a hydroxyl ion
   (4) a proton acceptor
   (5) a measure of hydrogen ion concentration
   (6) a chemical compound that can convert strong acids and bases into weak ones
   (7) a catalyst for chemical reactions that is specific, efficient, and under cellular control
   (8) a compound that functions to temporarily store and then transfer energy liberated in exergonic reactions to cellular activities that require energy

True or false:

6. Within a cell, when one substance is oxidized, another one may be almost simultaneously reduced.
7. Ionic bonds occur when atoms lose, gain, and share electrons in their valence shell.

Choose the best answer to the following questions:

8. Which of the following statements are true of covalent bonds? (1) The atoms form a molecule by sharing one, two, or three pairs of their valence electrons. (2) Covalent bonds can form only between atoms of different elements. (3) The greater the number of electron pairs shared, the stronger the covalent bond. (4) Covalent bonds are the least common chemical bonds in the body. (5) Covalent bonds can be either polar or nonpolar.
   (a) 1, 2, and 3, (b) 1, 3, and 5, (c) 2, 4, and 5, (d) 1, 2, and 5, (e) 1, 3, and 4

9. Organic compounds that function primarily to provide a readily available source of chemical energy to generate the ATP that drives metabolic reactions are (a) lipids, (b) nucleic acids, (c) water, (d) carbohydrates, (e) proteins.

10. The body's most highly concentrated form of chemical energy is (a) a triglyceride, (b) a lipoprotein, (c) glycogen, (d) a steroid, (e) an eicosanoid.

11. The organic compounds that serve in structural, regulatory, contractile, immunological, transport, and catalytic capacities are (a) lipids, (b) carbohydrates, (c) proteins, (d) nucleic acids, (e) buffers.

12. Which compounds form the inherited genetic material inside each cell and relay instructions from the genes to guide protein synthesis, respectively? (a) RNA and DNA, (b) ATP and DNA, (c) ATP and RNA, (d) DNA and ATP, (e) DNA and RNA.

13. Which of the following statements regarding the ATP molecule are true? (1) ATP is the energy currency for the cell. (2) The energy supplied by the hydrolysis of ATP is constantly being used by the cell. (3) Energy is required to produce ATP. (4) The production of ATP involves both aerobic and anaerobic phases. (5) The process of producing energy in the form of ATP is termed the law of conservation of energy.
   (a) 1, 2, 3, and 4, (b) 1, 2, 3, and 5, (c) 2, 4, and 5, (d) 1, 2, and 4, (e) 3, 4, and 5

14. Match the following:

___ (a) chemical element

___ (b) matter

___ (c) atom

___ (d) compound

___ (e) molecule

___ (f) isotope

___ (g) free radical

___ (h) ion

(1) an electrically charged atom or group of atoms with an unpaired electron in its outermost shell

(2) a combination resulting when two or more atoms share electrons

(3) the building block of all matter

(4) anything that occupies space and has mass

(5) an atom of an element that differs in the number of neutrons, and therefore mass number, from other atoms of the same element

(6) the smallest unit of matter that retains the properties and characteristics of an element

(7) an atom that has given up or gained electrons

(8) a substance that can be broken down into two or more different elements by ordinary chemical means

15. Match the following reactions with the term that describes them:

___ (a) $H_2 + Cl_2 \rightarrow 2HCl$

___ (b) $3NaOH + H_3PO_4 \rightarrow Na_3PO_4 + 3H_2O$

___ (c) $CaCO_3 + CO_2 + H_2O \rightarrow Ca(HCO_3)_2$

___ (d) $HNO_3 \rightarrow H^+ + NO_3^-$

___ (e) $NH_3 + H_2O \rightleftharpoons NH_4^+ + OH^-$

(1) synthesis reaction

(2) exchange reaction

(3) oxidation-reduction reaction

(4) decomposition reaction

(5) reversible reaction

CRITICAL THINKING QUESTIONS

1. When Selma joined the "Be Kind to Invertebrates Society" (their slogan is "Have you hugged a bug today?"), she decided to eat a diet free of sugars and fatty acids. How likely is it that Selma will stick to her diet? (HINT: *Selma's diet makes even bread and water look good.*)

2. Ming read an article in the *Daily News* describing a DNA fingerprinting technique being used as evidence in a criminal trial. The article explained that DNA is formed from 20 unique com-

binations of amino acids that are different in every individual. What should Ming's letter to the editor say? (HINT: *What do the initials "DNA" stand for?*)

3. When asked the pH of blood, math major Sean guessed "the same as water." Kareem told him that he was way off! Explain in terms that Sean will understand. (HINT: *Express the pH of blood mathematically.*)

ANSWERS TO FIGURE QUESTIONS

2.1 In carbon, the first shell contains two electrons and the second shell contains four electrons.

2.2 The four most plentiful elements in living organisms are oxygen, carbon, hydrogen, and nitrogen.

2.3 Antioxidants such as vitamins C and E can inactivate oxygen free radicals.

2.4 A cation is a positively charged ion; an anion is a negatively charged ion.

2.5 An ionic bond involves the *loss* and *gain* of electrons; a covalent bond involves the *sharing* of pairs of electrons.

2.6 The oxygen atom in a water molecule has greater electronegativity than the hydrogen atoms.

2.7 The N atom in ammonia is electronegative; it attracts electrons more strongly than do the H atoms, so the nitrogen end of ammonia acquires a slight negative charge. H atoms in water molecules (or in other ammonia molecules) can form hydrogen bonds with the nitrogen. Likewise, O atoms in water molecules can form hydrogen bonds with H atoms in ammonia molecules.

2.8 The law of conservation of mass stipulates that the number of hydrogen atoms in the reactants equals the number in the products—in this case, four hydrogen atoms total. Put another way, two molecules of H_2 are needed so that the number of H atoms and O atoms in the reactants is the same as the number of H atoms and O atoms in the products.

2.9 In an endergonic reaction, the products have more potential energy than do the reactants.

2.10 This reaction is exergonic because the reactants have more potential energy than do the products.

2.11 No. A catalyst does not change the potential energies of the products and reactants; it only lowers the activation energy needed to get the reaction going.

2.12 $CaCO_3$ is a salt, and H_2SO_4 is an acid.

2.13 Because sugar easily dissolves in a polar solvent (water), you can correctly predict that it has several polar covalent bonds.

2.14 $[H^+] = 10^{-6}$ mol/liter; $[OH^-] = 10^{-8}$ mol/liter. A pH of 6.82 is more acidic than a pH of 6.91. Both pH = 8.41 and pH = 5.59 are 1.41 pH units from neutral (pH = 7).

2.15 Glucose has five —OH groups and five carbon atoms in its ring.

2.16 Dehydration synthesis is an anabolic reaction. There are 6 carbons in fructose and 12 in sucrose.

2.17 The oxygen in the water molecule comes from a fatty acid.

2.18 The polar head region is hydrophilic, and the nonpolar tail region is hydrophobic.

2.19 The only differences are the number of double bonds in and the types of functional groups attached to ring A.

2.20 An amino acid has a minimum of two carbon atoms and one nitrogen atom (see the structure of glycine in Figure 2.20c).

2.21 Hydrolysis occurs during catabolism of proteins.

2.22 Proteins consisting of a single polypeptide chain do not have a quaternary structure.

2.23 Methyl groups (—CH_3) are hydrophobic because the bonds between carbon and hydrogen atoms are nonpolar covalent bonds; thus neither atom is electronegative.

2.24 Sucrase has specificity for the sucrose molecule and thus would not "recognize" glucose and fructose.

2.25 Thymine always pairs with adenine, and cytosine always pairs with guanine.

2.26 A few cellular activities that depend on energy supplied by ATP are muscular contractions, movement of chromosomes, transportation of substances across cell membranes, and synthesis (anabolic) reactions.

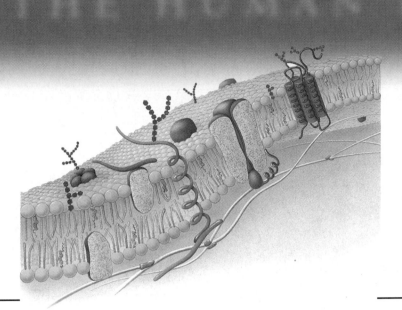

A **cell** is the basic, living, structural and functional unit of the body. Cells are composed of characteristic parts, the work of which is coordinated in a way that enables each type of cell to fulfill a unique biochemical or structural role. **Cytology** (sī-TOL-ō-jē; *cyt-* = cell; *-ology* = study of) is the scientific study of cellular structure, whereas **cell physiology** is the study of cellular function. As you study the various parts of a cell and their relationships to each other, you will learn that cell structure and function are interdependent and inseparable. Cells perform myriad chemical reactions to create life processes. They do so by compartmentalization, the isolation of specific kinds of chemical reactions within specialized structures inside the cell. The isolated reactions are coordinated with one another to maintain life in a cell, tissue, organ, system, and organism.

In very general terms, cells perform several basic kinds of work. For example, they regulate the inflow and outflow of materials to ensure that optimum conditions for life processes prevail inside the cell. Cells also use their genetic information (DNA) to guide the synthesis of most of their components and direct most of their chemical activities. Among these activities are the generation of ATP from the breakdown of nutrients, synthesis of molecules, transportation of molecules within and between cells, waste removal, and movement of parts of cells or even entire cells.

A GENERALIZED VIEW OF THE CELL

OBJECTIVE

• *Name and describe the three principal parts of a cell.*

Figure 3.1 is a generalized view of a cell that is a composite of many different cells in the body. While most cells have many of the features shown in this diagram, no one cell has all of them. For ease of study, we can divide a cell into three principal parts: the plasma membrane, cytoplasm, and nucleus.

• The **plasma membrane** forms the cell's sturdy outer surface, separating the cell's internal environment from the environment outside the cell. It is a selective barrier that orchestrates the flow of materials into and out of a cell to establish and maintain the appropriate environment for normal cellular activities. The plasma membrane also plays a key role in communication both among cells and between cells and their external environment.

• The **cytoplasm** (SĪ-tō-plazm; *-plasm* = formed or molded) is all the cellular contents between the plasma membrane and the nucleus. This compartment can be divided into two components: cytosol and organelles. **Cytosol** (SĪ-tō-sol) is the fluid portion of cytoplasm that consists mostly of water plus dissolved solutes and sus-

Figure 3.1 Generalized view of a body cell.

The cell is the basic, living, structural and functional unit of the body.

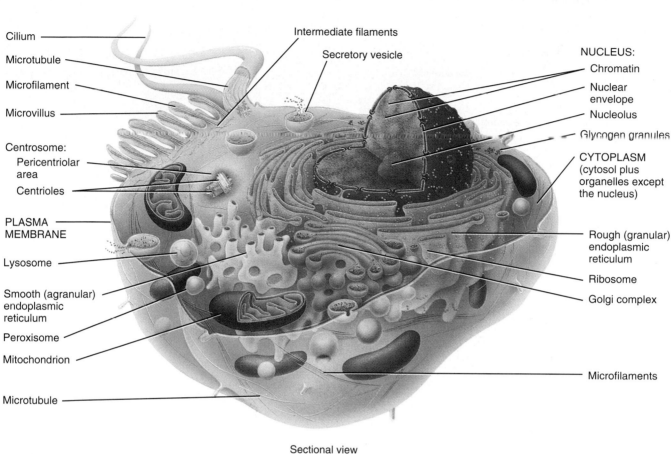

Cilium

Microtubule

Microfilament

Microvillus

Centrosome:
  Pericentriolar area
  Centrioles

PLASMA MEMBRANE

Lysosome

Smooth (agranular) endoplasmic reticulum

Peroxisome

Mitochondrion

Microtubule

Intermediate filaments

Secretory vesicle

NUCLEUS:
  Chromatin
  Nuclear envelope
  Nucleolus

Glycogen granules

CYTOPLASM (cytosol plus organelles except the nucleus)

Rough (granular) endoplasmic reticulum

Ribosome

Golgi complex

Microfilaments

Sectional view

Q What are the three principal parts of a cell?

pended particles. **Organelles** (or-ga-NELZ; = little organs) are highly organized subcellular structures, each having a characteristic shape and specific functions. Examples are the cytoskeleton, ribosomes, endoplasmic reticulum, Golgi complex, lysosomes, peroxisomes, and mitochondria.

• The **nucleus** (NOO-klē-us; = nut kernel), although technically an organelle, is considered separately because of its numerous and diverse functions. Within the nucleus are genes, which control cellular structure and most cellular activities.

THE PLASMA MEMBRANE

OBJECTIVE

• *Describe the structure and functions of the plasma membrane.*

The **plasma membrane** is a flexible yet sturdy barrier that surrounds and contains the cytoplasm of a cell. The *fluid mosaic model* describes its structure: The molecular arrangement of the plasma membrane resembles an ever-moving sea of lipids that contains a "mosaic" of many different proteins (Figure 3.2). The proteins may float freely like icebergs, be moored at specific locations, or be moved through the lipid sea. The lipids act as a barrier to the entry or exit of

Figure 3.2 Plasma membrane depicting the fluid mosaic arrangement of lipids and proteins.

 Membranes are fluid structures because the lipids and many of the proteins are free to rotate and move sideways in their own half of the bilayer.

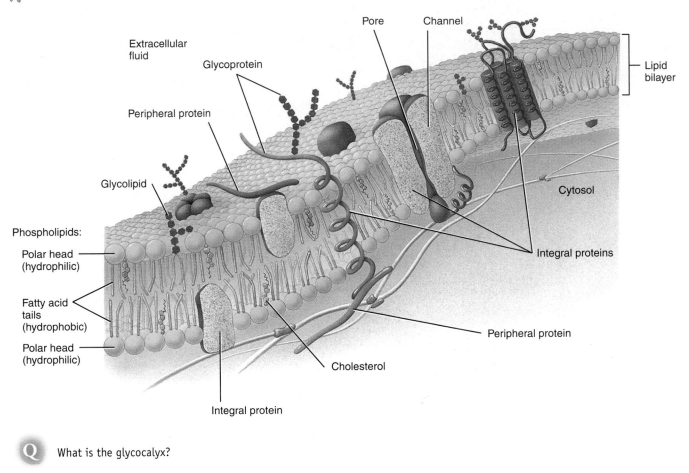

Q What is the glycocalyx?

charged or polar substances while some of the proteins in the plasma membrane act as "gatekeepers" that regulate the traffic of polar molecules and ions into and out of the cell. The plasma membranes of typical body cells are about a 50:50 mix by weight of proteins and lipids that are held together by noncovalent forces such as hydrogen bonds. But because membrane proteins are larger and more massive than membrane lipids, there are about 50 lipid molecules for each protein molecule.

The Lipid Bilayer

The basic framework of the plasma membrane is the **lipid bilayer,** two back-to-back layers made up of three types of lipid molecules—phospholipids, cholesterol, and glycolipids (see Figure 3.2). About 75% of the membrane lipids are **phospholipids,** lipids that contain phosphate groups. Present in smaller amounts are **cholesterol,** a steroid with an attached —OH (alcohol) group, and various **glycolipids,** lipids that have sugary carbohydrate groups attached.

The bilayer arrangement occurs because the lipids are **amphipathic** (am-fē-PATH-ik) molecules, which means that they have both polar and nonpolar parts. In phospholipids (see Figure 2.18 on page 47), the polar part is the phosphate-containing "head," which is hydrophilic (*hydro-* = water; *-philic* = loving). The nonpolar parts are the two long fatty acid "tails," which are hydrophobic (*-phobic* = fearing) hydrocarbon chains. Because "like seeks like," the phospholipid molecules orient themselves in the bilayer with their polar heads facing outward. In this way, the polar heads face a watery fluid on either side—cytosol on the inside and extracellular fluid on the outside. The hydrophobic fatty acid tails in each half of the bilayer point toward the center of the membrane, forming a nonpolar, hydrophobic region in the membrane's interior.

Glycolipids account for about 5% of membrane lipids. Their carbohydrate groups form a polar "head," whereas their fatty acid "tails" are nonpolar. Glycolipids appear only in the membrane layer that faces the extracellular fluid, which is one reason the bilayer is asymmetric, or different, on its two sides. The remaining 20% of plasma membrane lipids are weakly amphipathic cholesterol molecules (see Figure 2.19 on page 48), which are interspersed among the other lipids in both layers of the membrane. The tiny —OH group is the only polar region of cholesterol, and it forms hydrogen bonds with the polar heads of phospholipids and glycolipids. The stiff steroid rings and hydrocarbon tail of cholesterol are nonpolar; they fit among the fatty acid tails of the phospholipids and glycolipids. The plasma membrane typically has more cholesterol than do membranes of organelles.

Arrangement of Membrane Proteins

Membrane proteins are divided into two categories—integral and peripheral—according to whether or not they are embedded in the membrane (see Figure 3.2). **Integral proteins** extend into or through the lipid bilayer among the fatty acid tails and can be removed only by methods that disrupt membrane structure. **Peripheral proteins,** on the other hand, associate with membrane lipids or integral proteins at the inner or outer surface of the membrane and can be stripped away from the membrane by methods that do not disrupt membrane integrity. Like membrane lipids, integral membrane proteins are amphipathic. Their hydrophilic regions protrude into either the aqueous extracellular fluid or the cytosol, and their hydrophobic regions extend among the fatty acid tails. Integral proteins that stretch across the entire bilayer and project on both sides of the membrane are termed *transmembrane proteins.* Often, the portion of a protein that spans the nonpolar portion of the membrane consists of an alpha helix that is formed by amino acids having mainly nonpolar, hydrophobic side chains. In some cases, the protein's amino acid chain loops back and forth across the membrane several times; hydrophobic alpha helix segments that span the membrane alternate with polar, hydrophilic portions that protrude into the aqueous fluid on either side. Although many of the integral proteins can float laterally in the lipid bilayer, each individual protein has a specific orientation with respect to the "inside" and "outside" faces of the membrane, thus contributing to the membrane's asymmetry.

Many integral proteins are **glycoproteins,** proteins with sugary carbohydrate groups attached to the ends that protrude into the extracellular fluid. The carbohydrates are short, straight, or branched chains composed of 2–60 monosaccharides. The carbohydrate portions of glycolipids and glycoproteins form an extensive sugary coat called the **glycocalyx** (glī′-kō-KĀL-iks), which has several important functions. The composition of the glycocalyx acts like a molecular "signature" that enables cells to recognize one another. For example, a white blood cell's ability to detect a "foreign" glycocalyx is one basis of the immune response that helps the body to destroy invading organisms. In addition, the glycocalyx enables cells to adhere to one another in some tissues, and it protects the cell from being digested by enzymes in the extracellular fluid. The hydrophilic properties of the glycocalyx attract a film of fluid to the surface of many cells, an action that both makes red blood cells slippery as they flow through narrow blood vessels and protects cells that line the airways and the gastrointestinal tract from drying out.

Functions of Membrane Proteins

Generally, the types of lipids in cellular membranes vary only slightly from one membrane to another. Still, the membranes of different cells and various intracellular organelles feature remarkably different assortments of proteins, which determine many of the functions that plasma membranes can perform. Figure 3.3 shows several of the functions of plasma membrane proteins. Some transmembrane proteins are **channels** that have a *pore* or hole through which a specific substance, such as potassium ions (K^+), can flow into or out of the cell. Most channels are *selective;* they allow only a single type of ion to pass through. Other membrane proteins act as transporters. **Transporters** bind a polar substance at one side of the membrane, and, by changing their shape, move it to the other side of the membrane, where it is released. Integral proteins called **receptors** serve as cellular recognition sites. These proteins recognize and bind a specific molecule, such as the hormone insulin or a nutrient such as glucose, that is important for some cellular function. A specific molecule that binds to a receptor is called a *ligand* (LĪ-gand; *liga-* = tied) of that receptor. Some integral and peripheral proteins are **enzymes.** Membrane glycoproteins and glycolipids often are **cell-identity markers.** They may enable a cell to recognize other cells of the same kind during tissue formation or to recognize and respond to potentially dangerous foreign cells. The ABO blood type markers are one example of cell identity markers. When you receive a blood transfusion, the blood type must be compatible with your own. Integral and peripheral proteins may serve as **linkers,** which anchor proteins in the plasma membranes of neighboring cells to one another or to filaments inside and outside the cell.

Membrane Fluidity

Membranes are fluid structures, rather like cooking oil, because most of the membrane lipids and many of the membrane proteins easily rotate and move sideways in their own half of the bilayer. Neighboring lipid molecules exchange places about 10 million times per second and may wander completely around a cell in a few minutes! Membrane fluidity depends on both the number of double bonds in the fatty acid tails of the lipids that make up the bilayer and the amount of cholesterol present. Each double bond puts a "kink" in the fatty acid tail (see Figure 2.18 on page 47), which prevents lipid molecules from packing tightly in the membrane. Because of the fluidity of membrane lipids, the

Figure 3.3 Functions of membrane proteins.

🔑 **Membrane proteins largely reflect the functions a cell can perform.**

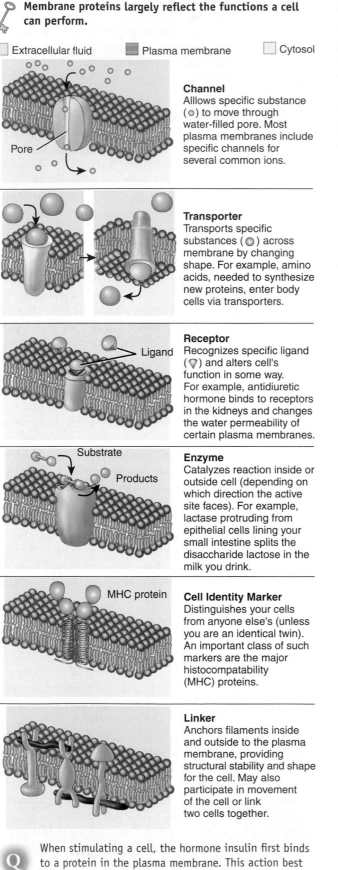

☐ Extracellular fluid   ▨ Plasma membrane   ☐ Cytosol

Channel
Alllows specific substance (○) to move through water-filled pore. Most plasma membranes include specific channels for several common ions.

Transporter
Transports specific substances (○) across membrane by changing shape. For example, amino acids, needed to synthesize new proteins, enter body cells via transporters.

Receptor
Recognizes specific ligand (▽) and alters cell's function in some way. For example, antidiuretic hormone binds to receptors in the kidneys and changes the water permeability of certain plasma membranes.

Enzyme
Catalyzes reaction inside or outside cell (depending on which direction the active site faces). For example, lactase protruding from epithelial cells lining your small intestine splits the disaccharide lactose in the milk you drink.

Cell Identity Marker
Distinguishes your cells from anyone else's (unless you are an identical twin). An important class of such markers are the major histocompatability (MHC) proteins.

Linker
Anchors filaments inside and outside to the plasma membrane, providing structural stability and shape for the cell. May also participate in movement of the cell or link two cells together.

Q When stimulating a cell, the hormone insulin first binds to a protein in the plasma membrane. This action best represents which membrane protein function?

lipid bilayer self-seals if it is torn or punctured. When a needle is pushed through a plasma membrane and pulled out, the puncture site seals spontaneously, and the cell does not burst. This property of membranes allows scientists to fertilize an ovum by injecting a sperm cell through a tiny syringe or to replace a cell's nucleus in a cloning experiment, such as was done to create the cloned sheep named Dolly.

Despite the great mobility of membrane lipids and proteins in their own half of the bilayer, they seldom flip-flop from one half of the bilayer to the other. This is because the hydrophilic parts pass through the hydrophobic core of the membrane only with great difficulty. Because flip-flopping is rare, the halves of the membrane bilayer remain asymmetric—the external half contains some lipids and proteins that are structurally and functionally different from those in the internal half.

Because of the way it forms hydrogen bonds with neighboring phospholipid and glycolipid heads and fills the space between bent fatty acid tails, cholesterol makes the lipid bilayer stronger but less fluid at normal body temperature. Interestingly, at low temperatures cholesterol has the opposite effect—it increases membrane fluidity, which is important in cold-blooded animals (but not in warm-blooded humans). When our plasma membranes accumulate an excessive amount of cholesterol, they become stiffer and less flexible. One consequence of atherosclerosis, commonly called "hardening of the arteries," is accumulation of cholesterol in the plasma membranes of cells that line blood vessels. The added cholesterol stiffens the plasma membranes, thus contributing to the decreased flexibility of blood vessels that characterizes this disease.

Membrane Permeability

A membrane is *permeable* to things that can pass through it and *impermeable* to those that cannot. Although plasma membranes are not completely permeable to any substance, they do permit some substances to pass more readily than others. This property of membranes is called **selective permeability.**

The lipid bilayer portion of the membrane is permeable to most nonpolar, uncharged molecules, such as oxygen, carbon dioxide, and steroids, but is impermeable to ions and charged or polar molecules, such as glucose. It is also permeable to water, an unexpected property given that water is a polar molecule. Even artificially made, pure phospholipid membranes are permeable to water, which is thought to pass through the lipid bilayer in the following way: As the fatty acid tails of membrane phospholipids and glycolipids randomly move about, small gaps briefly appear in the hydrophobic environment of the membrane's interior. Water molecules are small enough to move from one gap to another until they have crossed the membrane.

Transmembrane proteins that act as channels and transporters increase the plasma membrane's permeability to a variety of small- and medium-sized polar and charged substances (including ions) that cannot cross the lipid bilayer. These proteins are very selective—each one helps only a specific molecule or ion to cross the membrane. Macromolecules, such as proteins, are unable to pass through the plasma membrane except by vesicular transport (discussed later in this chapter).

Gradients Across the Plasma Membrane

Because the plasma membrane of a cell is selectively permeable, it admits some substances into the cytosol while excluding others. This property of the plasma membrane makes it possible for a living cell to maintain either larger or smaller concentrations of selected substances in the cytosol than in the extracellular fluid. The difference in the concentration of a chemical between one side of the plasma membrane and the other is called a *concentration gradient.* Many ions and molecules are more concentrated in either the cytosol or the extracellular fluid. For example, oxygen molecules and sodium ions (Na^+) are more concentrated in the extracellular fluid than in the cytosol, whereas the opposite is true for carbon dioxide molecules and potassium ions (K^+) (Figure 3.4a). The plasma membrane also creates a difference in the distribution of positively and negatively charged ions between one side of the plasma membrane and the other; this is an *electrical gradient,* also called the *membrane potential.* Typically, the inner surface of the membrane is more negatively charged and the outer surface is more positively charged (Figure 3.4b).

As you will see shortly, maintaining the concentration and electrical gradients are important to the life of the cell because they help move substances across the plasma membrane. In many cases a substance will move across a plasma membrane *down its concentration gradient.* That is to say, a substance will move "downhill," from where it is more concentrated to where it is less concentrated. Similarly, a positively charged substance will tend to move toward a negatively charged area, and a negatively charged substance will tend to move toward a positively charged area. Because ions are influenced by both a *concentration* gradient and an *electrical* gradient, the combined effect is termed an ion's **electrochemical gradient.**

1. Discuss the chemical forces that govern the arrangement of membrane lipids into a bilayer. In what respect is the lipid bilayer of the plasma membrane a barrier?
2. Defend this statement: The proteins present in a plasma membrane determine the functions that a membrane can perform.
3. Describe a functional advantage of membrane fluidity. What effects does cholesterol have on membrane fluidity?
4. Why are membranes said to have selective permeability?
5. Describe the two components of an electrochemical gradient.

Figure 3.4 Gradients across the plasma membrane. (a) In these concentration gradients, sodium ions and oxygen molecules are more concentrated in the extracellular fluid, whereas potassium ions and carbon dioxide are more concentrated in cytosol. (b) Because the inner surface of the plasma membrane of most cells is negative relative to the outer surface, an electrical gradient exists across the membrane.

Each substance tends to move across a membrane down its electrochemical gradient.

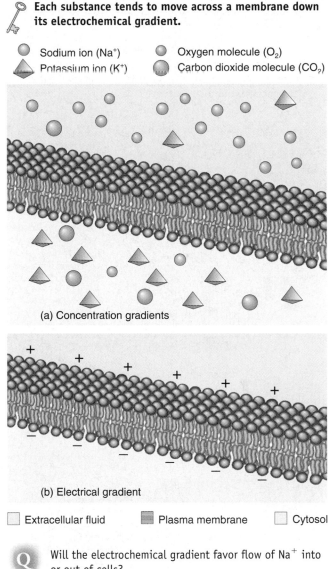

○ Sodium ion (Na^+)   ○ Oxygen molecule (O_2)
△ Potassium ion (K^+)   ◓ Carbon dioxide molecule (CO_2)

(a) Concentration gradients

(b) Electrical gradient

☐ Extracellular fluid   ▨ Plasma membrane   ☐ Cytosol

Q Will the electrochemical gradient favor flow of Na^+ into or out of cells?

TRANSPORT ACROSS THE PLASMA MEMBRANE

OBJECTIVE

• *Describe the processes that transport substances across the plasma membrane.*

Transport of materials across the plasma membrane is essential to the life of a cell. Certain substances must move into the cell to support metabolic reactions; others must be moved out because they have been produced by the cell for export or are cellular waste products. As you have seen, some

Figure 3.5 Processes for transport of materials across the plasma membrane. (a) In passive transport, a substance moves down its concentration gradient across the membrane, whereas in active transport, cellular energy is used to drive a substance against its concentration gradient. (b) In mediated transport, an integral membrane protein assists a substance to cross the membrane. Uniporters move a single substance across the membrane, whereas symporters move two substances across the membrane in the same direction and antiporters move two substances across the membrane in opposite directions.

🔑 **In passive transport, a substance moves down its concentration gradient, whereas in active transport, cellular energy is used to drive the substance "uphill," against its concentration gradient.**

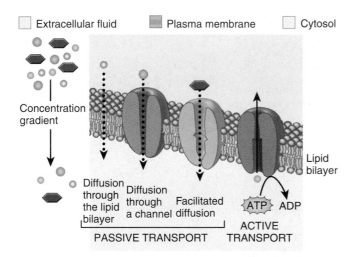

(a) Passive and active transport

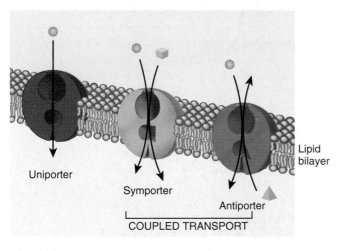

(b) Types of transporters in mediated transport

Ⓠ Which transport processes are mediated?

substances can cross the lipid bilayer portion of the plasma membrane, whereas others use channels or transporters formed by membrane proteins.

Substances cross cellular membranes via transport processes that are classified according to two criteria: (1) whether they are mediated or nonmediated and (2) whether they are active or passive. In *mediated transport*, materials move through the membrane with the assistance of a transporter protein, whereas in *nonmediated transport* no transporter protein is involved. In *passive transport*, a substance moves down its concentration gradient across the membrane using only its own kinetic energy of motion, whereas in *active transport* cellular energy, usually in the form of ATP, is used to drive the substance "uphill" against its concentration gradient.

Living cells use three passive transport processes (Figure 3.5a). Two of these are nonmediated: diffusion through the lipid bilayer and diffusion through a channel. One passive process, facilitated diffusion, is mediated. Among active transport processes, all of which are mediated, we can distinguish several types of transporters (Figure 3.5b). *Uniporters* move only a single substance across the membrane. In coupled transport, two substances move across the membrane together. Coupled transporters are either *symporters,* which move two substances in the same direction, or *antiporters,* which move two substances in opposite directions across the membrane.

Another way that materials may enter or leave cells is by processes that involve the formation of membrane-surrounded vesicles. In **vesicular transport** (see Figures 3.13–3.15), tiny vesicles either detach from the plasma membrane while bringing materials into the cell, a process called *endocytosis,* or merge with the plasma membrane to release materials from the cell, a process called *exocytosis.* Large particles, such as whole bacteria and red blood cells, and macromolecules, such as polysaccharides and proteins, may enter and leave cells by vesicular transport.

Principles of Diffusion

To learn why materials diffuse across membranes, you need to understand how diffusion operates in a solution. **Diffusion** is the random mixing of particles that occurs in a solution as a result of the particles' kinetic energy. Both the *solutes,* the dissolved substances, and the *solvent,* the liquid that does the dissolving, undergo diffusion. If a particular molecule is present in high concentration in one area of a solution and in low concentration in another area, more molecules diffuse from the region of high concentration to the region of low concentration than diffuse in the opposite direction. The difference in diffusion in the two directions is the overall or **net diffusion.** Substances undergoing net diffusion move from a high concentration to a low concentration and are said to move *downhill* or *down their concentration gradient.* An equilibrium is reached when the molecules are evenly distributed throughout the solution. At equilibrium, the molecules continue to move about randomly due to their kinetic energy, but there is no further change in their concentration.

An example of diffusion occurs if you place a crystal of dye in a water-filled container (Figure 3.6). Just next to the dye, the color is intense because the dye concentration is greatest there. At increasing distances, the color is lighter and lighter because the dye concentration is less. Some time later, at equilibrium, the solution of water and dye has a uniform color. The dye molecules have undergone net diffusion, down their concentration gradient, until they are evenly mixed in solution.

In the example of dye diffusion, no membrane was involved. Substances may also diffuse through a membrane, if the membrane is permeable to them. Several factors influence the diffusion rate of substances across plasma membranes:

1. *Steepness of the concentration gradient.* The greater the difference in concentration between the two sides of the membrane, the higher the rate of diffusion. When charged particles are diffusing, it is the steepness of both the concentration and electrical gradients—the electrochemical gradient—that determines diffusion rate across the membrane.
2. *Temperature.* The higher the temperature, the faster the rate of diffusion. In a person who has a fever, all of the body's diffusion processes occur more rapidly.
3. *Size or mass of the diffusing substance.* The larger the size or mass of the diffusing particle, the slower its diffusion rate. Molecules with smaller molecular weights diffuse more rapidly than larger ones.
4. *Surface area.* The larger the membrane surface area available for diffusion, the faster the diffusion rate. For example, the cells of the lungs have a large surface area available for diffusion of oxygen from the air into the blood. Some diseases, such as emphysema, reduce the surface area, which slows the rate of diffusion and makes breathing more difficult.
5. *Diffusion distance.* The greater the distance over which diffusion must occur, the longer it will take. Diffusion across a plasma membrane takes only a fraction of a second because the membrane is so thin.

We will begin to explore membrane transport by seeing how the solvent water crosses membranes and then continue by looking at the various categories of solutes.

Osmosis

Osmosis (oz-MŌ-sis) is the net movement of a solvent through a selectively permeable membrane. In living systems, the solvent is water, which moves by osmosis across plasma membranes from an area of *higher water concentration* to an area of *lower water concentration*. Another way to understand this idea is to consider the solute concentration: In osmosis, water moves through a selectively permeable

Figure 3.6 Diffusion. A crystal of dye placed in a cylinder of water dissolves (beginning) and there is net diffusion from the region of higher dye concentration to regions of lower dye concentration (intermediate). At equilibrium, dye concentration is uniform throughout.

🔑 **At equilibrium, net diffusion stops but random movement continues.**

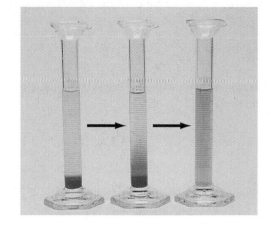

Beginning   Intermediate   Equilibrium

❓ How would having a fever affect body processes that involve diffusion?

membrane from an area of *lower solute concentration* to an area of *higher solute concentration*. Water molecules penetrate plasma membranes in two ways—via diffusion through the lipid bilayer, as previously mentioned, and through **aquaporins,** transmembrane proteins that function as water channels.

Osmosis occurs only when a membrane is permeable to water but not to certain solutes. A simple experiment can demonstrate osmosis. Consider a U-shaped tube in which a selectively permeable membrane separates the tube's left and right arms (Figure 3.7a). A volume of pure water is poured into the left arm, and the same volume of a solution containing a solute to which the membrane is impermeable is poured into the right arm (Figure 3.7a). Because the water concentration is higher on the left and lower on the right, net movement of water molecules—osmosis—occurs from left to right, as water moves down its concentration gradient. At the same time, the membrane prevents diffusion of the solute from the right arm into the left arm. As a result, the volume of water in the left arm decreases, and the volume of solution in the right arm increases (Figure 3.7b).

You might think that osmosis would continue until no water remained on the left side, but this is *not* what happens. In this experiment, the higher the column of solution in the right arm becomes, the more pressure it exerts on its side of the membrane. Pressure exerted in this way by a liquid is known as *hydrostatic pressure* and it forces water molecules

Figure 3.7 Principle of osmosis. Water molecules move through the selectively permeable membrane; the solute molecules in the right arm cannot pass through the membrane. (a) As the experiment starts, water molecules move from the left arm into the right arm, down the water concentration gradient. (b) After some time, the volume of water in the left arm has decreased and the volume of solution in the right arm has increased. At equilibrium, net osmosis has stopped: Hydrostatic pressure forces just as many water molecules to move from right to left as osmotic pressure forces water molecules to move from left to right. (c) If pressure is applied to the solution in the right arm, the starting conditions can be restored. This pressure, which stops osmosis, is equal to the osmotic pressure. Adapted from Martini, *Fundamentals of Anatomy and Physiology* 4e, F3.7, p75, (Upper Saddle River, NJ: Prentice Hall, 1998). © 1998 Prentice Hall, Inc.

 Osmosis is the movement of water molecules through a selectively permeable membrane.

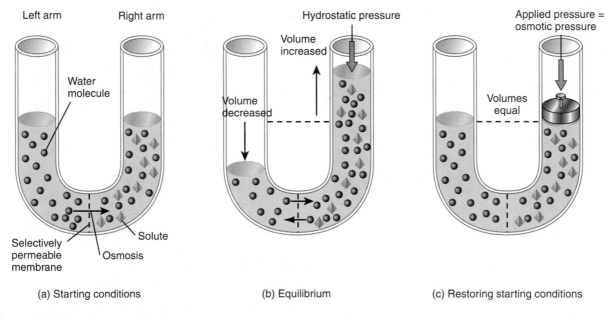

| (a) Starting conditions | (b) Equilibrium | (c) Restoring starting conditions |

Q Will the fluid level in the right arm rise until the water concentrations are the same in both arms?

to move back into the left arm. Equilibrium is reached when just as many water molecules move from right to left due to the hydrostatic pressure as move from left to right due to osmosis (Figure 3.7b). A solution containing solute particles that cannot cross the membrane exerts a force, called the **osmotic pressure.** The osmotic pressure of a solution is proportional to the concentration of the solute particles that cannot cross the membrane—the higher the solute concentration, the higher the solution's osmotic pressure. Consider what would happen if a piston were used to apply more pressure to the fluid in the right arm. With sufficient pressure, the volume of fluid in each arm could be restored to the starting volume, and the concentration of solute in the right arm would be the same as it was initially (Figure 3.7c). The amount of pressure needed to restore the starting condition matches the osmotic pressure. So another way to think about

osmotic pressure is that it is the pressure needed to stop the movement of water into a solution containing solutes when the two are separated by a membrane permeable only to the water. Notice that the osmotic pressure of a solution does not produce the movement of water during osmosis. Rather it is the pressure that would *prevent* such water movement.

The **tonicity** (*tonic* = tension) of a solution affects the volume and shape of body cells by causing osmosis of water into or out of them. Normally, the osmotic pressure of the cytosol is the same as the osmotic pressure of the interstitial fluid outside cells. Because the osmotic pressure on both sides of the plasma membrane (which is selectively permeable) is the same, cell volume remains relatively constant. Any solution in which a cell—for example, a red blood cell (RBC)—maintains its normal shape and volume is called an **isotonic**

Figure 3.8 Tonicity and its effects on red blood cells (RBCs). One example of an isotonic solution for RBCs is 0.9% NaCl.

🔑 **Cells placed in an isotonic solution stay the same shape because there is no net water movement into or out of the cell.**

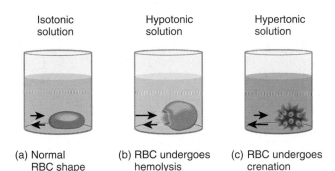

Isotonic solution	Hypotonic solution	Hypertonic solution
(a) Normal RBC shape	(b) RBC undergoes hemolysis	(c) RBC undergoes crenation

Q Will a 2% solution of NaCl cause hemolysis or crenation of RBCs?

solution (*iso-* = same) (Figure 3.8a). This is a solution in which the concentrations of solutes that cannot cross the plasma membrane are the *same* on both sides of the membrane. Under ordinary circumstances, a 0.9% NaCl (sodium chloride or table salt) solution, called a *normal saline solution,* is isotonic for RBCs. Whereas the RBC plasma membrane permits the water to move back and forth, it behaves as though it is impermeable to Na^+ and Cl^-, the solutes. Any Na^+ or Cl^- ions that enter the cell through channels or transporters are immediately moved back out by active transport or other means. These ions thus behave as though they cannot penetrate the membrane. When RBCs are bathed in 0.9% NaCl, water molecules enter and exit at the same rate, allowing the RBCs to keep their normal shape and volume.

A different situation results if RBCs are placed in a solution that has a *lower* concentration of solutes than the cytosol inside the RBCs. This is called a **hypotonic solution** (*hypo-* = less than) (Figure 3.8b). In this state, water molecules enter the cells faster than they leave, causing the RBCs to swell and eventually to burst. The rupture of RBCs in this manner is called **hemolysis** (hē-MOL-i-sis). Pure water is strongly hypotonic.

A **hypertonic solution** (*hyper-* = greater than) has a *higher* concentration of solutes than does the cytosol inside RBCs (Figure 3.8c). One example of a hypertonic solution is a 2% NaCl solution. In such a solution, water molecules move out of the cells faster than they enter, causing the cells to shrink. The shrinkage of RBCs in this manner is called **crenation** (kri-NĀ-shun). RBCs and other body cells may be damaged or destroyed if exposed to hypertonic or hypotonic solutions. For this reason, most intravenous (IV) solutions, which are infused into the blood of patients, are isotonic.

Diffusion Through the Lipid Bilayer

Nonpolar, hydrophobic molecules diffuse through the lipid bilayer of the plasma membrane, into and out of cells. Such molecules include oxygen, carbon dioxide, and nitrogen gases; fatty acids, steroids, and fat-soluble vitamins (A, E, D, and K); small alcohols; and ammonia. Because the plasma membrane is quite permeable to all these substances, they do not contribute to the osmotic pressure or tonicity of body fluids. Diffusion through the lipid bilayer is important in the movement of oxygen and carbon dioxide between blood and body cells, and between blood and air within the lungs during breathing. It also is the route for absorption of some nutrients and excretion of some wastes by body cells, activities that contribute to homeostasis.

Diffusion Through Membrane Channels

Most membrane channels are *ion channels;* they allow passage of small, inorganic ions, which are too hydrophilic to penetrate the nonpolar interior of the plasma membrane. Each ion can diffuse across the membrane only at sites containing channels that allow that specific ion to pass. In typical plasma membranes, the most numerous ion channels are selective for K^+ (potassium ions) or Cl^- (chloride ions); fewer channels are available for Na^+ (sodium ions) or Ca^{2+} (calcium ions). Diffusion of solutes through channels generally is slower than diffusion through the lipid bilayer because a smaller portion of the membrane's total surface area is occupied by channels. Still, diffusion through channels is a very fast process: More than a million potassium ions can flow through a K^+ channel in one second!

Although many ion channels are open all the time, others may be gated—a portion of the channel protein acts as a "gate," changing shape in one way to open the pore and in another way to close it (Figure 3.9). Some gates randomly alternate between the open and closed positions, whereas others are regulated by chemical or electrical changes inside and outside the cell. When the gates are open, ions diffuse into or out of cells, down their electrochemical gradients.

The plasma membranes of different types of cells may have fewer or more ion channels and thus display different permeabilities to various ions. The total number of channels that exist for a specific solute also may be regulated by signals from inside or outside the cell.

Facilitated Diffusion

Several solutes that are too polar or highly charged to diffuse through the lipid bilayer and too big to diffuse through membrane channels can cross the plasma membrane by **facilitated diffusion.** This process originally was envisioned as one in which a "carrier" protein bound to a solute and diffused with it across the membrane. The name is unfortunate, however, because the process is not truly diffusion. Instead, a solute binds to a specific transporter on one side of the membrane and is released on the other side after the transporter undergoes a conformational change.

Figure 3.9 Diffusion of K$^+$ through a gated membrane channel. Most membranes have channels that are selective for potassium ions (K$^+$), sodium ions (Na$^+$), calcium ions (Ca^{2+}), and chloride ions (Cl$^-$).

🔑 **Channels are integral membrane proteins that allow specific small, inorganic ions to pass through the membrane by diffusion.**

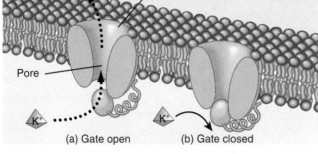

☐ Extracellular fluid   ▨ Plasma membrane   ☐ Cytosol

(a) Gate open   (b) Gate closed

Q How does a gated channel work?

Figure 3.10 Facilitated diffusion of glucose across a plasma membrane. The transporter (GluT) binds to glucose in the extracellular fluid and releases it into the cytosol.

🔑 **Facilitated diffusion across a membrane requires a transporter but does not use ATP.**

☐ Extracellular fluid   ▨ Plasma membrane   ☐ Cytosol

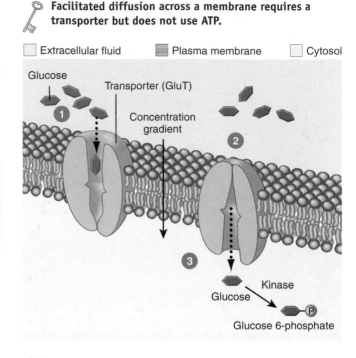

Q What factors determine the rate of facilitated diffusion?

For this reason, some researchers advocate changing the name to facilitated *transport*. Like diffusion, nevertheless, the net result is movement down a concentration gradient from a region of higher concentration to a region of lower concentration. This happens because the solute binds more often to the transporter on the side of the membrane where it is more concentrated before being moved to the side where it is less concentrated. Once the concentration is the same on both sides of the membrane, solute molecules bind to the transporter on the cytosolic side and move out to the extracellular fluid as rapidly as they bind to the transporter on the extracellular side and move into the cytosol. Thus the rate of facilitated diffusion is determined by the steepness of the concentration gradient between the two sides of the membrane. The number of facilitated diffusion transporters available in a plasma membrane places an upper limit, called the *transport maximum,* on the rate at which facilitated diffusion can occur. Once all the transporters are occupied, the transport maximum is reached, and a further increase in the concentration gradient does not produce faster facilitated diffusion. Thus the process of facilitated diffusion exhibits *saturation.*

Solutes that move across plasma membranes by facilitated diffusion include glucose, urea, fructose, galactose, and some vitamins. Glucose enters many body cells by facilitated diffusion, as follows (Figure 3.10):

1 Glucose binds to a glucose transporter protein called GluT at the extracellular surface of the membrane.

2 GluT undergoes a conformational change.

3 GluT releases glucose on the other side of the membrane.

After glucose enters a cell by facilitated diffusion, an enzyme called a kinase attaches a phosphate group to produce a dif-

ferent molecule—namely, glucose 6-phosphate. This reaction keeps the intracellular concentration of glucose very low, so that the glucose concentration gradient always favors facilitated diffusion of glucose into, not out of, cells.

The selective permeability of the plasma membrane is often regulated to achieve homeostasis. For example, the hormone insulin promotes the insertion of many copies of a specific type of glucose transporter into the plasma membranes of certain cells. Thus the effect of insulin is to increase the transport maximum for facilitated diffusion of glucose. With more transporters available, body cells can pick up glucose from the blood more rapidly.

Active Transport

Some polar or charged solutes that must enter or leave body cells cannot cross the plasma membrane through any form of passive transport because they would need to move "uphill," *against* their concentration gradient. Such solutes can cross the membrane by **active transport.** Active transport is a mediated, energy-requiring process in which transporter proteins move solutes across the membrane against a concentration gradient. Two sources of energy are used to drive active transport: (1) Energy obtained from hydrolysis of ATP is the source in *primary active transport,* and (2) energy stored in an ionic concentration gradient is the source in *secondary active transport.* Like facilitated diffusion, active transport processes exhibit a transport maximum and satu-

Figure 3.11 The sodium pump (Na$^+$/K$^+$ ATPase). Sodium ions (Na$^+$) are expelled and potassium ions (K$^+$) are imported into the cell. The pump will not work unless Na$^+$ and ATP are present in the cytosol and K$^+$ is present in the extracellular fluid.

🔑 **The sodium pump maintains a low intracellular concentration of sodium ions.**

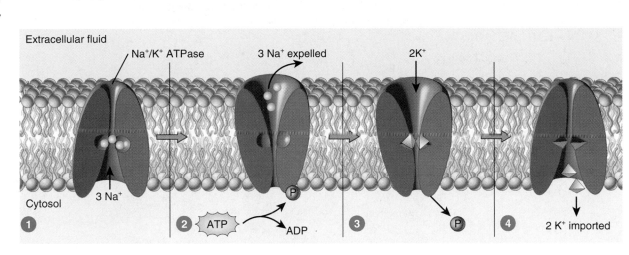

Q What is the role of ATP in the operation of this pump?

ration. Solutes actively transported across the plasma membrane include several ions, such as Na$^+$, K$^+$, H$^+$, Ca^{2+}, I$^-$, and Cl$^-$; amino acids; and monosaccharides. (Recall that some of these substances also cross the membrane by diffusion or facilitated diffusion when the proper channels and transporters are present.)

Primary Active Transport

In **primary active transport,** energy derived from hydrolysis of ATP changes the shape of a transporter protein, which "pumps" a substance across a plasma membrane against its concentration gradient. Indeed, transporter proteins that carry out primary active transport often are called **pumps.** A typical body cell expends about 40% of the ATP it generates on primary active transport. Drugs that turn off ATP production—for example, the poison cyanide—are lethal because they shut down active transport in cells throughout the body.

The most prevalent primary active transport mechanism expels sodium ions (Na$^+$) from cells and brings potassium ions (K$^+$) in. Because of the specific ions it moves, this antiporter is called the **Na$^+$/K$^+$ pump** or, more simply, the **sodium pump.** All cells have thousands of sodium pumps in their plasma membranes. Because a portion of the Na$^+$/K$^+$ pump acts as an enzyme that hydrolyzes ATP (an ATPase), another name for this pump is **Na$^+$/K$^+$ ATPase.** The sodium pump maintains a low concentration of sodium ions in the cytosol by pumping them into the extracellular fluid against the Na$^+$ concentration gradient. Concurrently, the sodium pump moves potassium ions into cells against the K$^+$ concentration gradient. Because K$^+$ and Na$^+$ slowly leak back across the plasma membrane down their electrochemi-

cal gradients—through passive transport or secondary active transport—the sodium pumps must operate continually to maintain a low concentration of sodium ions and a high concentration of potassium ions in the cytosol. Figure 3.11 depicts our current understanding of the sodium pump.

1️⃣ Three sodium ions (Na$^+$) in the cytosol bind to the pump protein.

2️⃣ Na$^+$ binding triggers the hydrolysis of ATP into ADP, a reaction that also attaches a phosphate group (Ⓟ) to the pump protein. This changes the shape of the pump protein such that the three Na$^+$ are released into the extracellular fluid. Now the shape of the pump protein favors binding of potassium ions in the extracellular fluid.

3️⃣ The binding of potassium ions (K$^+$) causes the pump protein to release the phosphate group, which again causes the shape of the pump protein to change.

4️⃣ As the pump protein reverts to its original shape, two K$^+$ are released into the cytosol. At this point, the pump is again ready to bind Na$^+$, and the cycle repeats.

Besides keeping the intracellular level of Na$^+$ low and K$^+$ high, continual operation of the sodium pumps also is important for maintaining normal cell volume. Recall that the osmotic pressure of a solution is proportional to the concentration of its solute particles that cannot penetrate the membrane. Because sodium ions that diffuse into a cell or enter through secondary active transport are immediately pumped out, it is as if they never entered. In effect, sodium ions behave as if they cannot penetrate the membrane. Thus sodium ions are an important contributor to the osmotic pressure of the extracellular fluid. A similar condition holds for potassium ions, and they contribute half of the osmotic pressure of the cytosol. By helping to maintain normal

Figure 3.12 Secondary active transport mechanisms. (a) Antiporters carry two substances across the membrane in opposite directions. (b) Symporters carry two substances across the membrane in the same direction.

> Secondary active transport mechanisms use the energy stored in an ionic concentration gradient (here, for Na$^+$). Because the gradient is maintained by primary active transport pumps that hydrolyze ATP, secondary active transport consumes ATP indirectly.

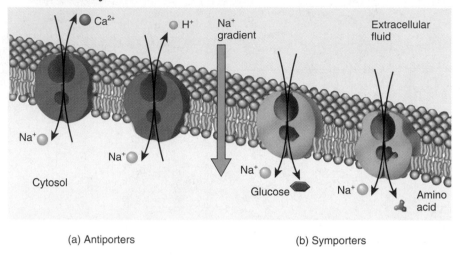

(a) Antiporters   (b) Symporters

Q What is the main difference between secondary and primary active transport mechanisms?

osmotic pressure on each side of the plasma membrane, the sodium pumps ensure that cells neither shrink nor swell due to osmotic movement of water out of or into cells by osmosis.

Secondary Active Transport

In **secondary active transport,** the energy stored in a sodium ion or hydrogen ion concentration gradient is used to drive other substances across the membrane against their own concentration gradients. Because a Na$^+$ or H$^+$ gradient is established by primary active transport, secondary active transport *indirectly* uses energy obtained from the hydrolysis of ATP.

The sodium pump maintains a steep concentration gradient of Na$^+$ across the plasma membrane. As a result the sodium ions have stored or potential energy, just as water behind a dam has potential energy. Accordingly, if there is a route for Na$^+$ to leak back in, some of the stored energy can be converted to kinetic energy (energy of motion) and used to transport other substances *against their concentration gradients.* In essence, secondary active transport proteins harness the energy in the Na$^+$ concentration gradient by providing easy routes for Na$^+$ to leak into cells. The protein acts as a coupled transporter by simultaneously binding to Na$^+$ and another substance and then changing its shape so that both substances cross the membrane at the same time.

Plasma membranes contain several antiporters and symporters that are powered by the Na$^+$ gradient (Figure 3.12a). In most body cells, the concentration of calcium ions (Ca^{2+}) is low in the cytosol because Na$^+$/Ca^{2+} antiporters let sodium ions enter the cell while ejecting calcium ions.

Likewise, Na$^+$/H$^+$ antiporters help regulate the cytosol's pH (H$^+$ concentration) by using the Na$^+$ gradient to expel excess H$^+$. Dietary glucose and amino acids, on the other hand, are absorbed into cells that line the small intestine by Na$^+$-glucose and Na$^+$-amino acid symporters (Figure 3.12b). In each case, sodium ions are moving down their concentration gradient while the coupled solutes move "uphill," against their concentration gradients. Keep in mind that all these symporters and antiporters are able to do their job because the sodium pumps maintain a low concentration of Na$^+$ in the cytosol.

CLINICAL APPLICATION
Digitalis

Because it strengthens the heartbeat, digitalis often is given to patients with heart failure, a condition of weakened pumping action by the heart. Digitalis exerts its effect by slowing the sodium pump, which lets more Na$^+$ accumulate inside heart muscle cells. The result is a smaller Na$^+$ concentration gradient across the membrane, which causes the Na$^+$/Ca^{2+} antiporters to slow down. As a result, more Ca^{2+} remains inside heart muscle cells. The slight increase in the level of calcium ions in the cytosol of heart muscle cells increases the force of their contractions and thus strengthens the heartbeat. As this example illustrates, the balance between the concentrations of Na$^+$ and Ca^{2+} in the cytosol and extracellular fluid is crucial to the normal functioning of muscle cells. ■

Vesicular Transport

A **vesicle** is a small, spherical membranous sac formed by budding off from an existing membrane. The vesicular membrane is selectively permeable, and the vesicle contains a small quantity of fluid plus dissolved or suspended particles. Vesicles transport substances from one structure to another within cells, take in substances from extracellular fluid, and release substances into extracellular fluid. Later in the chapter we will discuss vesicular transport between structures within a cell once you have a better understanding of the anatomy of a cell. At this point, we will concentrate on the two main types of vesicular transport between a cell and its extracellular fluid: (1) **endocytosis** (*endo-* = within), in which materials move into a cell in a vesicle formed from the plasma membrane and (2) **exocytosis** (*exo-* = out), in which materials move out of a cell by the fusion of vesicles with the plasma membrane. Both endocytosis and exocytosis require energy supplied by ATP.

Endocytosis

In endocytosis, the material brought into the cell is surrounded by an area of the plasma membrane, which buds off inside the cell to form a vesicle containing the ingested material. Here we consider three types of endocytosis: receptor-mediated endocytosis, pinocytosis, and phagocytosis.

Receptor-mediated Endocytosis This highly selective type of endocytosis enables a cell to take up particular ligands. A vesicle forms after a receptor protein in the plasma membrane recognizes and binds to a specific ligand in the extracellular fluid. By receptor-mediated endocytosis, cells take up transferrin (an iron-transporting protein in the blood), some vitamins, low density lipoproteins (LDLs), antibodies, certain hormones, and specific other macromolecules or particles. Receptor-mediated endocytosis occurs as follows (Figure 3.13):

1 *Binding.* On the extracellular side of the plasma membrane, a ligand binds to a specific receptor to form a receptor-ligand complex. The receptors are integral membrane proteins that are concentrated in specific regions of the plasma membrane, called *clathrin-coated pits*, because the membrane is lined on its cytoplasmic side by a peripheral protein called *clathrin*. The interaction between clathrin and the receptor-ligand complex causes the membrane to invaginate (fold inward).

2 *Vesicle formation.* The edges of the membrane around the clathrin-coated pit fuse and a portion of the membrane pinches off. The resulting vesicle, known as a *clathrin-coated vesicle,* contains the receptor-ligand complexes.

3 *Uncoating.* Almost immediately after it is formed, the clathrin-coated vesicle loses its clathrin coat to become an *uncoated vesicle.*

4 *Fusion with early endosome.* Several uncoated vesicles fuse with an *early endosome,* a vesicle that serves as a

Figure 3.13 Endocytosis: receptor-mediated endocytosis.

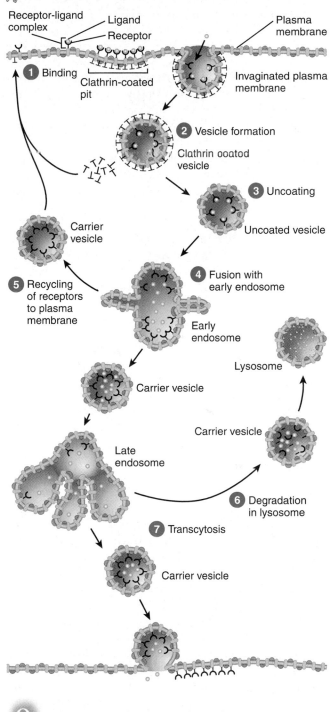

Receptor-mediated endocytosis imports materials that are needed by cells.

 What are several examples of ligands?

sorting compartment. Within an early endosome, ligands separate from their receptors. Ultimately, the contents of an early endosome are either recycled to the plasma membrane or transported to a lysosome, an organelle where degradation occurs.

5 *Recycling of receptors.* Many receptors within the early endosome enter a *carrier vesicle,* which buds from the early endosome and returns the receptors to the plasma membrane. Clathrin also is returned to the inner surface of the plasma membrane.

6 *Degradation in lysosomes.* Ligands (and their receptors) destined for degradation in lysosomes are transported from the early endosome, by one or more carrier vesicles, to a *late endosome.* The late endosome forms other carrier vesicles that fuse with lysosomes. Within lysosomes are digestive enzymes that break down a variety of large molecules such as proteins, polysaccharides, lipids, and nucleic acids.

7 *Transcytosis.* In some cases, ligands and receptors undergo **transcytosis,** a process whereby they are transported across the cell inside carrier vesicles that undergo exocytosis on the opposite side. As the carrier vesicles fuse with the plasma membrane, the vesicular contents are released into the extracellular fluid. Transcytosis occurs most often across the endothelial cells that line blood vessels and is a means for materials to move between blood plasma and interstitial fluid.

CLINICAL APPLICATION
Viruses and Receptor-mediated Endocytosis

Although receptor-mediated endocytosis normally imports needed materials, some viruses take advantage of this mechanism to enter and infect body cells. For example, the human immunodeficiency virus (HIV), which causes acquired immunodeficiency syndrome (AIDS), enter cells by attaching to a receptor called CD4 in the plasma membrane. This receptor is present at the surface of some body cells, such as white blood cells called helper T cells. After binding to CD4, HIV enters the cell via receptor-mediated endocytosis. ■

Phagocytosis **Phagocytosis** (fag′-ō-sī-TŌ-sis; *phago-* = to eat) is a form of endocytosis in which large solid particles, such as whole bacteria or viruses, are taken in by the cell (Figure 3.14). Phagocytosis begins as the particle binds to a plasma membrane receptor, causing the cell to extend projections of its plasma membrane and cytoplasm called **pseudopods** (SOO-dō-pods; *pseudo-* = false; *-pods* = feet). Pseudopods surround the particle outside the cell, and the membrane fuses to form a vesicle called a *phagosome,* which enters the cytoplasm. The phagosome fuses with one or more lysosomes, and the ingested material is broken down by lysosomal enzymes. In most cases, any undigested materials in the phagosome, now called a *residual body,* are released back into the extracellular fluid by exocytosis.

Phagocytosis occurs only in certain body cells, termed **phagocytes,** which are a class of cells that engulf and destroy bacteria and other foreign substances. Phagocytes include certain types of white blood cells and macrophages, which are present in most body tissues. The process of phagocytosis is a vital defense mechanism that helps protect the body from disease.

Figure 3.14 Endocytosis: phagocytosis.

🔑 **Phagocytosis is a vital defense mechanism that helps protect the body from disease.**

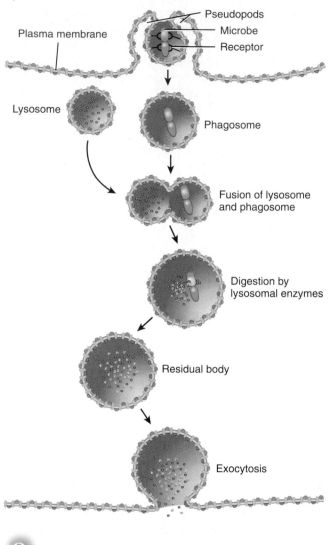

Plasma membrane — Pseudopods / Microbe / Receptor

Lysosome

Phagosome

Fusion of lysosome and phagosome

Digestion by lysosomal enzymes

Residual body

Exocytosis

Q What triggers pseudopod formation?

Pinocytosis **Pinocytosis** (pi-nō-sī-TŌ-sis; *pino-* = to drink) is a form of endocytosis that involves the *nonselective* uptake of tiny droplets of extracellular fluid (Figure 3.15). No receptor proteins are involved; all solutes dissolved in the extracellular fluid are brought into the cell. In pinocytosis, the plasma membrane folds inward, forming a *pinocyte vesicle* containing a droplet of extracellular fluid. The pinocytic vesicle detaches or "pinches off" from the plasma membrane and enters the cytosol. Within the cell, the pinocytic vesicle fuses with one or more lysosomes, where enzymes degrade the engulfed materials. As in phagocytosis, undigested materials accumulate in a residual body. In most cases the residual body fuses with the plasma membrane and expels its contents from the cell via exocytosis. Most body cells engage in pinocytosis.

Figure 3.15 Endocytosis: pinocytosis.

 Pinocytosis is the nonselective uptake of tiny droplets of extracellular fluid.

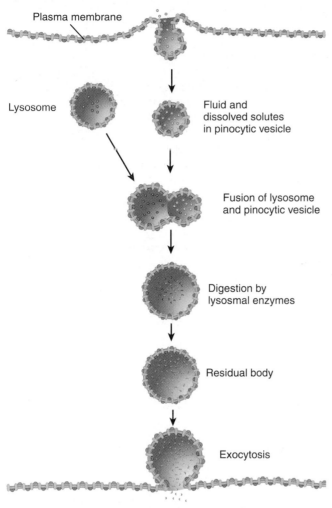

Plasma membrane

Lysosome

Fluid and dissolved solutes in pinocytic vesicle

Fusion of lysosome and pinocytic vesicle

Digestion by lysosmal enzymes

Residual body

Exocytosis

Q How do receptor-mediated endocytosis and phagocytosis differ from pinocytosis?

Exocytosis

As depicted in Figures 3.13 through 3.15, **exocytosis** involves the movement of materials out of a cell in vesicles that fuse with the plasma membrane. The ejected material may be either a waste product or a useful secretory product. All cells carry out exocytosis, but it is especially important in two types of cells: nerve cells, which release substances called *neurotransmitters,* and secretory cells, which secrete digestive enzymes or hormones. During exocytosis in secretory cells, membrane-enclosed vesicles called *secretory vesicles* form inside the cell, fuse with the plasma membrane, and release their contents into the extracellular fluid.

Portions of the plasma membrane that are lost through endocytosis are recovered or recycled by exocytosis. The balance between endocytosis and exocytosis keeps the surface area of the plasma membrane relatively constant. Membrane exchange is quite extensive in certain cells. Secretory cells of

the pancreas, for example, recycle an amount of plasma membrane equal to the cell's entire surface area every 90 minutes.

Table 3.1 summarizes the processes by which materials are transported into and out of cells.

1. Describe the key difference between passive and active transport.
2. Define uniporter, symporter, and antiporter.
3. What factors increase or decrease the rate of diffusion?
4. What is osmotic pressure?
5. Which substances diffuse directly through the lipid bilayer?
6. How does diffusion through membrane channels compare to facilitated diffusion?
7. What is the difference between primary and secondary active transport?
8. Define vesicular transport and distinguish among the three types.

CYTOPLASM

OBJECTIVE

• *Describe the structure and function of cytoplasm, cytosol, and organelles.*

Cytoplasm can be divided into two components: (1) cytosol and (2) organelles.

Cytosol

The **cytosol** is the fluid portion of the cytoplasm that surrounds organelles (see Figure 3.1), constituting about 55% of total cell volume. Although it varies in composition and consistency from one part of a cell to another, cytosol is 75–90% water plus various dissolved and suspended components. Among these are various ions, glucose, amino acids, fatty acids, proteins, lipids, ATP, and waste products. Also present are various organic molecules that aggregate into masses and are stored. These molecules may appear and disappear at various times in the life of a cell. Examples include *lipid droplets* that contain triglycerides and clusters of glycogen molecules called *glycogen granules* (see Figure 3.1).

The cytosol is the site of many chemical reactions required for a cell's existence. For example, enzymes in cytosol catalyze numerous chemical reactions. As a result of these reactions, energy is produced and captured to drive cellular activities. In addition, these reactions provide the building blocks for maintaining cell structure, function, and growth.

Organelles

As noted previously, **organelles** are specialized structures that have characteristic shapes and that perform specific functions in cellular growth, maintenance, and reproduction. Despite the many chemical reactions occurring in a cell at the same time, there is little interference between one type of reaction and another because they occur in different organelles. Each type of organelle has its own characteristic set of enzymes that carry out specific reactions, and each is a functional compartment where specific physiological

Table 3.1 Transport of Materials Into and Out of Cells

TRANSPORT PROCESS	DESCRIPTION	SUBSTANCES TRANSPORTED
OSMOSIS	Movement of water molecules across a selectively permeable membrane from an area of higher concentration of water to an area of lower concentration of water.	Solvent: water in living systems.
DIFFUSION	Random mixing of molecules or ions due to their kinetic energy. In net diffusion, a substance moves down a concentration gradient until it reaches equilibrium.	
Diffusion through the lipid bilayer	Passive diffusion of a substance through the lipid bilayer of the plasma membrane.	Nonpolar, hydrophobic solutes: oxygen, carbon dioxide, and nitrogen; fatty acids, steroids, and fat-soluble vitamins; glycerol, small alcohols; ammonia.
Diffusion through membrane channels	Passive diffusion of a substance down its electrochemical gradient through channels that cross a lipid bilayer; some channels are open all the time whereas others are gated.	Small inorganic solutes, mainly ions: K^+, Cl^-, Na^+, and Ca^{2+}.
FACILITATED DIFFUSION	Passive mediated transport of a substance down its concentration gradient via transmembrane proteins that act as transporters; maximum diffusion rate is limited by number of available transporters.	Polar or charged solutes: glucose, fructose, galactose, urea, and some vitamins.
ACTIVE TRANSPORT	Mediated transport in which cell expends energy to move a substance across the membrane against its concentration gradient through transmembrane proteins that act as transporters; maximum transport rate is limited by number of available transporters.	Polar or charged solutes.
Primary active transport	Transport of a substance across the membrane against its concentration gradient by pumps, transmembrane proteins that use energy supplied by ATP.	Na^+, K^+, Ca^{2+}, H^+, I^-. Cl^-, and other ions.
Secondary active transport	Coupled transport of two substances across the membrane using energy supplied by a Na^+ or H^+ concentration gradient maintained by primary active transport pumps. Antiporters move Na^+ (or H^+) and another substance in opposite directions across the membrane; symporters move Na^+ (or H^+) and another substance in the same direction across the membrane.	Antiport: Ca^{2+}, H^+ out of cells; symport: glucose, amino acids into cells.
VESICULAR TRANSPORT	Movement of substances into or out of a cell in vesicles that bud from the plasma membrane; requires energy supplied by ATP.	
Endocytosis	Movement of substances into a cell in vesicles.	
Receptor-mediated endocytosis	Ligand-receptor complexes trigger infolding of a clathrin-coated pit that forms a vesicle containing ligands.	Ligands: transferrin, low-density lipoproteins (LDLs); some vitamins, certain hormones, and antibodies.
Phagocytosis	"Cell eating"; movement of a solid particle into a cell after pseudopods engulf it to form a phagosome.	Bacteria, viruses, and aged or dead cells.
Pinocytosis	"Cell drinking"; movement of extracellular fluid into a cell by infolding of plasma membrane, forming pinocytic vesicle.	Solutes in extracellular fluid.
Exocytosis	Movement of substances out of a cell in secretory vesicles that fuse with the plasma membrane and release their contents into the extracellular fluid.	Neurotransmitters, hormones, and digestive enzymes.

processes take place. Moreover, organelles often cooperate with each other to maintain homeostasis.

The numbers and types of organelles vary in different cells, depending on the function of the cell. *Nonmembranous organelles* lack membranes and are in direct contact with the cytosol. Examples are the cytoskeleton, centrosomes, cilia, flagella, and ribosomes. On the other hand, *membranous or-*

ganelles are surrounded by one or two lipid bilayer membranes (similar to the plasma membrane) that separate their interiors from the cytosol. Examples of membranous organelles are the endoplasmic reticulum, Golgi complex, mitochondria, lysosomes, and peroxisomes. Although the nucleus is a membranous organelle, it is discussed separately because of its special importance in directing the life of a cell.

Figure 3.16 Cytoskeleton.

🔑 **The cytoskeleton is a network of three kinds of protein filaments that extend throughout the cytoplasm: microfilaments, intermediate filaments, and microtubules.**

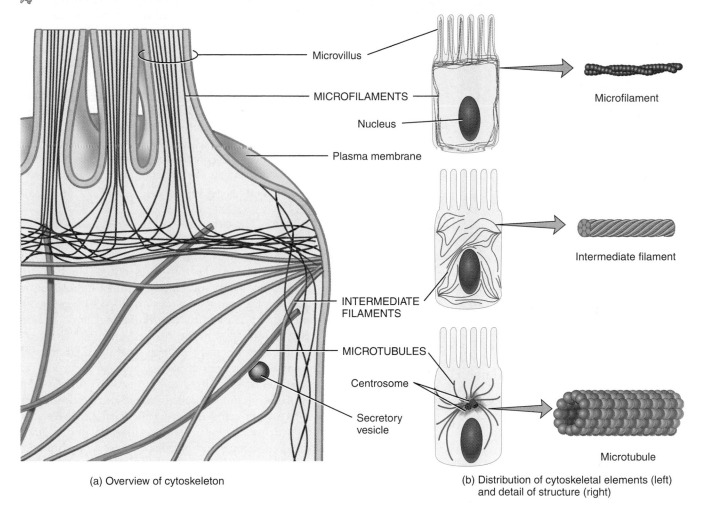

Microvillus

MICROFILAMENTS

Nucleus

Plasma membrane

Microfilament

Intermediate filament

INTERMEDIATE FILAMENTS

MICROTUBULES

Centrosome

Secretory vesicle

Microtubule

(a) Overview of cytoskeleton

(b) Distribution of cytoskeletal elements (left) and detail of structure (right)

Q Which cytoskeletal component helps form the structure of centrioles, cilia, and flagella?

The Cytoskeleton

The **cytoskeleton** is a network of several different kinds of protein filaments that extend throughout the cytosol (Figure 3.16). The cytoskeleton provides a structural framework for the cell, serving as a scaffold that helps to determine a cell's shape and organizes the cellular contents. The cytoskeleton is also responsible for cell movements, including the internal transport of organelles and some chemicals, the movement of chromosomes during cell division, and the movement of whole cells such as phagocytes. Although its name implies rigidity, the cytoskeleton is continually reorganizing as a cell moves and changes shape, as, for example, during cell division. The cytoskeleton is composed of three types of protein filaments. In the order of their increasing diameter, these structures are microfilaments, intermediate filaments, and microtubules.

Microfilaments These are the thinnest elements of the cytoskeleton. Most microfilaments are composed of the protein *actin* and are concentrated at the periphery of a cell (see Figure 3.16a and b). Microfilaments have two general functions: movement and mechanical support. With respect to movement, microfilaments are involved in muscle contraction, cell division, and cell locomotion, such as occurs during the migration of embryonic cells during development, the invasion of tissues by white blood cells to fight infection, or the migration of skin cells during wound healing.

Microfilaments provide much of the mechanical support that is responsible for the basic strength and shapes of cells. They anchor the cytoskeleton to integral proteins in the plasma membrane. Microfilaments also provide mechanical support for cell extensions called **microvilli** (*micro-* = small; *-villi* = tufts of hair), which are nonmotile, microscopic fin-

Figure 3.17 Centrosome.

🔑 **The pericentriolar area of a centrosome organizes the mitotic spindle during cell division, whereas the centrioles help form or regenerate flagella and cilia.**

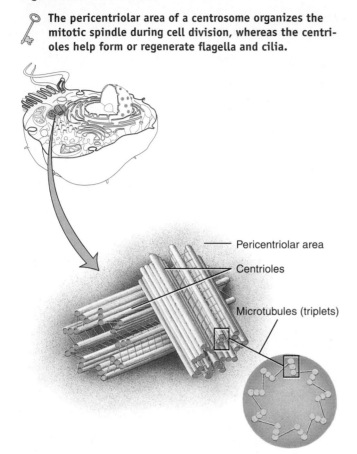

Pericentriolar area

Centrioles

Microtubules (triplets)

(a) Details of a centrosome (b) 9 + 0 array of centriole

Transverse section of centriole

Pericentriolar area

Longitudinal section of centriole

TEM 76,000x

(c) Centrioles

FUNCTIONS
1. Pericentriolar area serves as center for organizing microtubules in nondividing cells and for forming the mitotic spindle during cell division.
2. Centrioles play a role in formation and regeneration of cilia and flagella.

 If you observed that a cell did not have a centrosome, what could you predict about its capacity for cell division?

gerlike projections of the plasma membrane. Within a microvillus is a core of parallel microfilaments that supports it and attaches it to the cytoskeleton (see Figure 3.16a and b). Microvilli are abundant on the surfaces of cells involved in absorption, such as the epithelial cells that line the small intestine. Some microfilaments extend beyond the plasma membrane and help cells attach to one another or to extracellular material.

Intermediate Filaments As their name suggests, intermediate filaments are thicker than microfilaments but thinner than microtubules (see Figure 3.16a and b). Several different proteins can compose intermediate filaments, which are exceptionally strong. They are found in parts of cells subject to mechanical stress and also help anchor organelles such as the nucleus (see Figure 3.16a and b).

Microtubules The largest of the cytoskeletal components, microtubules are long, unbranched hollow tubes composed mainly of a protein called *tubulin*. An organelle called the centrosome (discussed shortly) serves as the initiation site for the assembly of microtubules. The microtubules grow outward from the centrosome toward the periphery of the cell (see Figure 3.16a and b). Microtubules help determine cell shape and function in the intracellular transport of organelles, such as secretory vesicles, and the migration of chromosomes during cell division. They also participate in the movement of specialized cell projections, such as cilia and flagella. *Motor proteins* called *kinesins* and *dyneins* are responsible for powering the movements in which microtubules participate. These proteins act like miniature engines that propel substances and organelles along a microtubule, much as a train engine pushes or pulls cars along a railroad track.

Centrosome

The **centrosome,** located near the nucleus, consists of two components: the pericentriolar area and centrioles (Figure 3.17a). The *pericentriolar area* is a region of the cytosol composed of a dense network of small protein fibers. This area is the organizing center for the mitotic spindle, which plays a critical role in cell division, and for microtubule formation in nondividing cells. Within the pericentriolar area is a pair of cylindrical structures called *centrioles,* each of which is composed of nine clusters of three microtubules (triplets) arranged in a circular pattern, an arrangement called a *9+0 array* (Figure 3.17b). The 9 refers to the nine clusters of microtubules, and the 0 refers to the absence of microtubules in the center. The long axis of one centriole is at a right angle to the long axis of the other (Figure 3.17c). Centrioles play a role in the formation or regeneration of cilia and flagella.

Cilia and Flagella

Microtubules are the dominant structural and functional components of cilia and flagella, both of which are motile projections of the cell surface. In the human body, cells that are firmly anchored in place use cilia to move fluids across their surfaces; motile cells, such as sperm cells, use flagella to propel themselves through a liquid medium.

Cilia (SIL-ē-a = eyelashes) are numerous, short, hairlike projections that extend from the surface of the cell (see Figure 3.1). Each cilium (the singular form) contains a core of microtubules surrounded by plasma membrane (Figure 3.18a). The microtubules are arranged such that one pair in the center is surrounded by nine clusters of two microtubules (doublets), an arrangement called a *9 + 2 array*. Each cilium is anchored to a *basal body* just below the surface of the plasma membrane. A basal body is similar in structure to a centriole. In fact, basal bodies and centrioles are considered to be two different functional manifestations of the same structure. The function of a basal body is to initiate the assembly of cilia and flagella. A cilium displays an oar-like pattern of beating in which it is relatively stiff during the power stroke, but relatively flexible during the recovery stroke (Figure 3.18b). The coordinated movement of numerous cilia on the surface of a cell ensures the steady movement of fluid along the cell's surface. Many cells of the respiratory tract, for example, have hundreds of cilia that help sweep foreign particles trapped in mucus away from the lungs. Their movement is paralyzed by nicotine in cigarette smoke. For this reason, smokers cough often to remove foreign particles from their airways. Cells that line the uterine (Fallopian) tubes also have cilia that sweep ova toward the uterus.

Flagella (fla-JEL-a; singular is *flagellum* = whip) are similar in structure to cilia but are much longer. Flagella usually move an entire cell. A flagellum generates forward motion along its axis by rapidly wiggling in a wavelike pattern (Figure 3.18c). The only example of a flagellum in the human body is a sperm cell's tail, which propels the sperm toward its rendezvous with an ovum.

Ribosomes

Ribosomes (RĪ-bō-sōms) are sites of protein synthesis. These tiny organelles are packages of **ribosomal RNA (rRNA)** and many ribosomal proteins. Ribosomes are so named because of their high content of *ribo*nucleic acid. Structurally, a ribosome consists of two subunits, one about half the size of the other (Figure 3.19). The two subunits are made separately in the nucleolus, a spherical body inside the nucleus. Once produced, they exit the nucleus and join together in the cytosol.

Some ribosomes, called *free ribosomes*, are unattached to any structure in the cytoplasm. Primarily, free ribosomes synthesize proteins used *inside* the cell. Other ribosomes, called *membrane-bound ribosomes*, attach to the nuclear membrane and to an extensively folded membrane called the endoplasmic reticulum. These ribosomes synthesize proteins destined for insertion in the plasma membrane or for export from the cell. Ribosomes are also located within mitochondria, where they synthesize mitochondrial proteins. Sometimes 10–20 ribosomes join together in a stringlike arrangement called a *polyribosome*.

Figure 3.18 Cilia and flagella.

🔑 **Whereas centrioles have a 9 + 0 array of microtubules, cilia and flagella have a 9 + 2 array.**

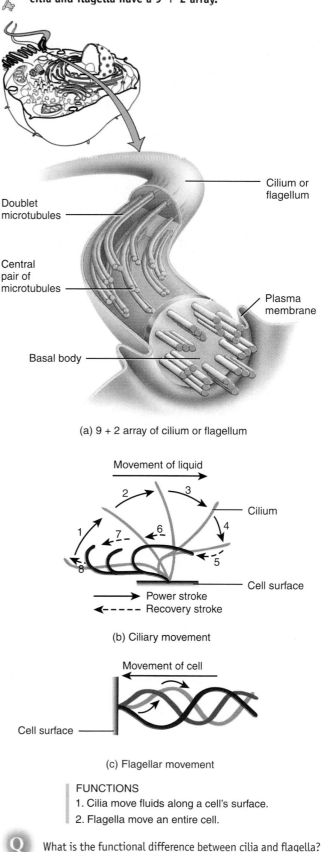

Doublet microtubules

Central pair of microtubules

Basal body

Cilium or flagellum

Plasma membrane

(a) 9 + 2 array of cilium or flagellum

Movement of liquid

Cilium

Cell surface

→ Power stroke
⇠ - - - - Recovery stroke

(b) Ciliary movement

Movement of cell

Cell surface

(c) Flagellar movement

FUNCTIONS
1. Cilia move fluids along a cell's surface.
2. Flagella move an entire cell.

Q What is the functional difference between cilia and flagella?

Figure 3.19 Ribosomes.

 Ribosomes are the sites of protein synthesis.

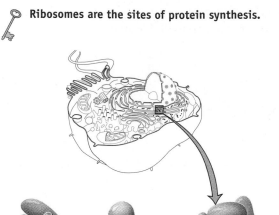

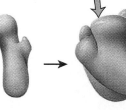

Large subunit   Small subunit   Complete functional ribosome

Details of ribosomal subunits

FUNCTIONS
1. Free ribosomes: synthesize proteins used within the cell.
2. Membrane-bound ribosomes: synthesize proteins destined for insertion in plasma membrane or export from cell.

Q Where are subunits of ribosomes synthesized and assembled?

Endoplasmic Reticulum

The **endoplasmic reticulum** (en'-dō-PLAS-mik re-TIK-yoo-lum; -*plasmic* = cytoplasm; *reticulum* = network) or **ER** is a network of membranes that form flattened sacs or tubules called **cisterns** (SIS-terns = cavity) (Figure 3.20). The ER extends from the membrane around the nucleus (nuclear envelope), to which it is connected, throughout the cytoplasm. The ER is so extensive that it constitutes more than half of the membranous surfaces within the cytoplasm of most cells.

Cells contain two distinct, but interrelated, forms of ER that differ in structure and function. The membrane of **rough ER** is continuous with the nuclear membrane and usually unfolds into a series of flattened sacs. The outer surface of rough ER is studded with ribosomes, the sites of protein synthesis. Proteins synthesized by ribosomes attached to rough ER enter cisterns within the ER for processing and sorting. In some cases, enzymes within the cisterns attach the proteins to carbohydrates to form glycoproteins. In other

Figure 3.20 Endoplasmic reticulum.

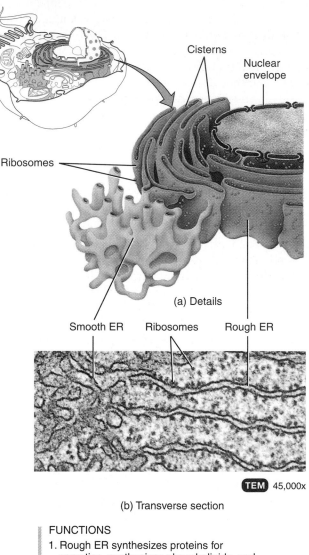 **The endoplasmic reticulum is a network of membrane-enclosed sacs or tubules called cisterns that extend throughout the cytoplasm and connect to the nuclear envelope.**

Cisterns

Nuclear envelope

Ribosomes

(a) Details

Smooth ER   Ribosomes   Rough ER

TEM 45,000x

(b) Transverse section

FUNCTIONS
1. Rough ER synthesizes proteins for secretion, synthesizes phospholipids, and forms new membranes for cellular structures.
2. Smooth ER synthesizes phospholipids, fats, and steroids.

Q What are the structural and functional differences between rough and smooth ER?

cases, enzymes attach the proteins to phospholipids, also synthesized by rough ER. These molecules may be incorporated into organelle membranes, or the plasma membrane. Thus rough ER is a factory for synthesizing secretory proteins and membrane molecules.

Figure 3.21 Golgi complex.

🔑 **The opposite faces of a Golgi complex differ in size, shape, content, enzymatic activity, and vesicles present.**

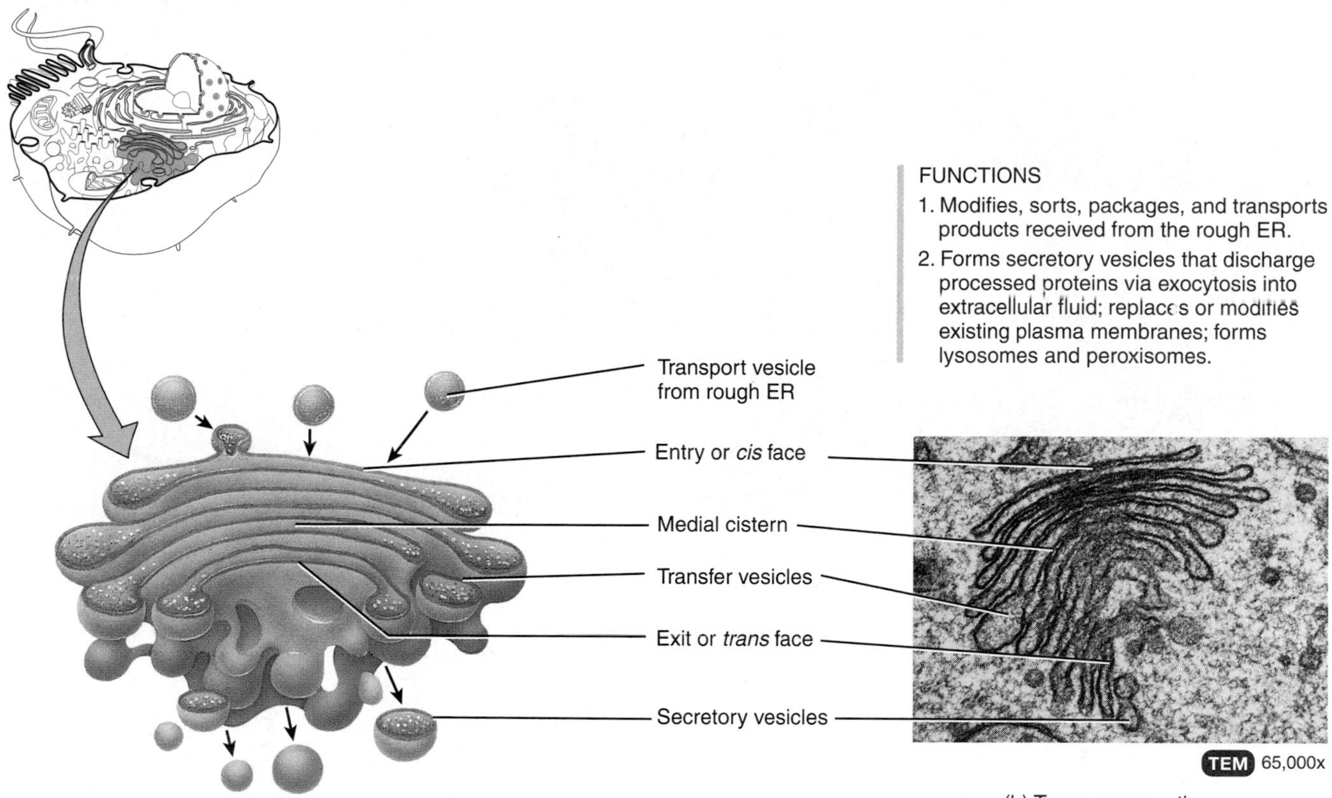

FUNCTIONS

1. Modifies, sorts, packages, and transports products received from the rough ER.
2. Forms secretory vesicles that discharge processed proteins via exocytosis into extracellular fluid; replaces or modifies existing plasma membranes; forms lysosomes and peroxisomes.

Transport vesicle from rough ER

Entry or *cis* face

Medial cistern

Transfer vesicles

Exit or *trans* face

Secretory vesicles

TEM 65,000x

(b) Transverse section

(a) Details

Q How do the entry and exit faces differ in function?

Smooth ER extends from the rough ER to form a network of membrane tubules (see Figure 3.20). Unlike rough ER, smooth ER does not have ribosomes on the outer surface of its membrane. However, smooth ER contains unique enzymes that make it functionally more diverse than rough ER. Although it does not synthesize proteins, smooth ER does synthesize phospholipids, as does rough ER. Smooth ER also synthesizes fats and steroids, such as estrogens and testosterone. In liver cells, enzymes of the smooth ER help release glucose into the bloodstream and inactivate or detoxify drugs and other potentially harmful substances, for example, alcohol. In muscle cells, calcium ions released from the sarcoplasmic reticulum, a form of smooth ER, trigger the contraction process.

Golgi Complex

Most of the proteins synthesized by ribosomes attached to rough ER are ultimately transported to other regions of the cell. The first step in the transport pathway is through an organelle called the **Golgi complex** (GOL-jē). It consists of 3–20 flattened membranous sacs with bulging edges, called **cisterns,** that resemble a stack of pita bread (Figure 3.21). The cisterns are often curved, giving the Golgi complex a cuplike shape. Most cells have only one Golgi complex, although some may have several. The Golgi complex is more extensive in cells that secrete proteins into the extracellular fluid. This fact is a clue to the organelle's role in the cell.

The cisterns at the opposite ends of a Golgi complex differ from each other in size, shape, content, enzymatic activity, and vesicles present (described shortly). The convex **entry** or *cis* **face** is composed of cisterns that face the rough ER. The concave **exit** or *trans* **face** is composed of cisterns that face the plasma membrane. Cisterns between the entry and exit faces are called **medial cisterns.**

Figure 3.22 Packaging of synthesized proteins by the Golgi complex.

🔑 **All proteins exported from the cell are processed in the Golgi complex.**

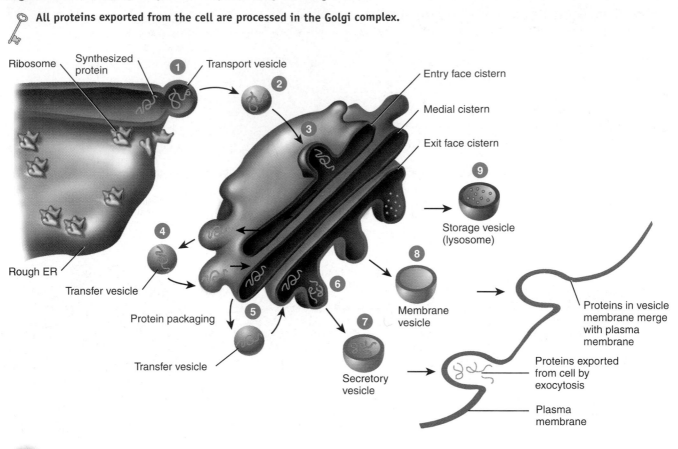

Q   What are the three general destinations for proteins that leave the Golgi complex?

The entry face, medial cisterns, and exit face of the Golgi complex contain different enzymes that permit each to modify, sort, and package proteins for transport to different destinations. For example, the entry face receives and modifies proteins produced by the rough ER. The medial cisterns add sugars to proteins and lipids to form glycoproteins and glycolipids, and they add proteins to lipids to form lipoproteins. The exit face modifies the molecules further and then sorts and packages them for transport to their destinations.

Proteins arriving at, passing through, and exiting the Golgi complex do so through a series of exchanges between vesicles (Figure 3.22):

1 Proteins synthesized by ribosomes on the rough ER are surrounded by a portion of the ER membrane, which eventually buds from the membrane surface to form a **transport vesicle.**

2 The transport vesicle moves toward the entry face of the Golgi complex.

3 The transport vesicle fuses with the entry face of the Golgi complex, releasing proteins into the cistern.

4 The proteins are modified and move from the entry face into one or more medial cisterns via **transfer vesicles** that bud from the cisterns' edges. Enzymes in the medial cisterns modify the proteins to form glycoproteins, glycolipids, and lipoproteins.

5 The products of the medial cisterns move via transfer vesicles into the interior of the exit face.

6 Within the exit face, the products are further modified and are sorted and packaged.

7 Some of the processed proteins leave the exit face in **secretory vesicles,** which deliver the proteins to the plasma membrane, where they are discharged by exocytosis into the extracellular fluid.

8 Other processed proteins leave the exit face in vesicles that deliver their contents to the plasma membrane for incorporation into the membrane. In doing so, the Golgi complex adds new segments of plasma membrane as existing segments are lost and modifies the number and distribution of membrane molecules.

9 Finally, some processed proteins leave the exit face in vesicles that are called **storage vesicles.** The major storage vesicle is a lysosome, whose structure and functions are discussed shortly.

CLINICAL APPLICATION
Cystic Fibrosis

Sometimes a disorder results from faulty routing of a molecule. Such is the case in **cystic fibrosis**, a deadly inherited disease that affects several body systems. The defective protein produced by the mutated cystic fibrosis gene fails to reach the plasma membrane, where it should be inserted to help pump chloride ions (Cl⁻) out of certain cells. Evidently, the would-be pump protein becomes stuck in the endoplasmic reticulum or Golgi complex and never reaches its correct destination. The result is an imbalance in the transport of fluid and ions across the plasma membrane that causes the buildup of thick mucus outside certain types of cells. The accumulated mucus clogs the airways in the lungs, causing breathing difficulty, and prevents proper secretion of digestive enzymes by the pancreas, causing digestive disturbances. ■

Lysosomes

Lysosomes (LĪ-sō-sōms; *lyso-* = dissolving; *-somes* = bodies) are membrane-enclosed vesicles that form in the Golgi complex (Figure 3.23). Inside are as many as 40 kinds of powerful digestive or hydrolytic enzymes that are capable of breaking down a wide variety of molecules. Lysosomal enzymes work best at acidic pHs, and the unique lysosomal membrane includes active transport pumps that move hydrogen ions (H⁺) into the lysosomes. Thus the lysosomal interior has a pH of 5, which is 100 times more acidic than the cytosolic pH of 7. The lysosomal membrane also allows the final products of digestion, such as sugars and amino acids, to be transported into the cytosol.

Lysosomal enzymes also recycle the cell's own structures. A lysosome can engulf another organelle, digest it, and return the digested components to the cytosol for reuse. In this way, old organelles are continually replaced. The process by which worn-out organelles are digested is called **autophagy** (aw-TOF-a-jē; *auto-* = self; *-phagy* = eating). A human liver cell, for example, recycles about half its cytoplasmic contents every week. In autophagy, the organelle to be digested is enclosed by a membrane derived from the ER to create a vesicle called an **autophagosome;** the vesicle then fuses with a lysosome. Lysosomal enzymes may also destroy their own cell, a process known as **autolysis** (aw-TOL-i-sis). Autolysis occurs in some pathological conditions and also is responsible for the tissue deterioration that occurs immediately after death.

Although most of the digestive processes involving lysosomal enzymes occur within a cell, there are some instances in which the enzymes operate in extracellular digestion. One example is the release of lysosomal enzymes during fertilization. The head of a sperm cell releases from its lysosomes enzymes that help the sperm cell penetrate the surface of the ovum.

Figure 3.23 Lysosomes.

Lysosomes arise in the Golgi complex and store several kinds of powerful digestive enzymes.

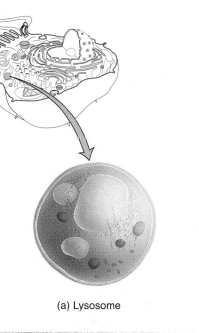

(a) Lysosome

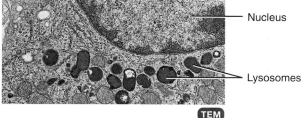

Nucleus

Lysosomes

TEM

(b) Several lysosomes in the cytoplasm

FUNCTIONS
1. Digests substances that enter a cell via endocytosis.
2. Autophagy—digestion of worn-out organelles.
3. Autolysis—digestion of entire cell.
4. Extracellular digestion.

 What is the name of the process by which worn-out organelles are digested by lysosomes?

CLINICAL APPLICATION
Tay-Sachs Disease

Some disorders are caused by faulty lysosomes. For example, **Tay-Sachs disease**, which most often affects children of eastern European Ashkenazi descent, is an inherited condition characterized by the absence of a single lysosomal enzyme. The missing enzyme normally breaks down a membrane glycolipid called ganglioside G$_{M2}$ that is especially

Figure 3.24 Mitochondria.

 Within mitochondria, chemical reactions called cellular respiration generate ATP.

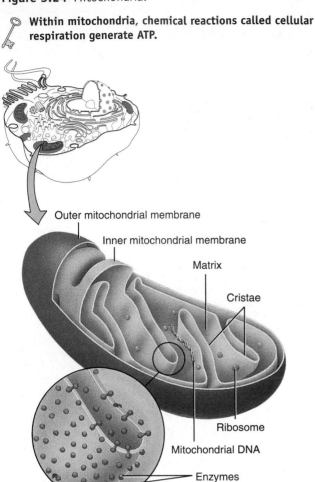

Outer mitochondrial membrane
Inner mitochondrial membrane
Matrix
Cristae
Ribosome
Mitochondrial DNA
Enzymes

(a) Mitochrondrial interior

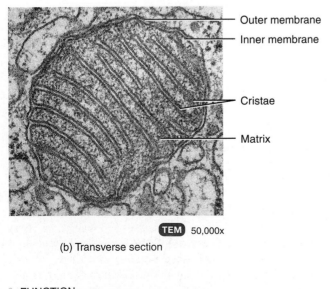

Outer membrane
Inner membrane
Cristae
Matrix

TEM 50,000x

(b) Transverse section

FUNCTION
Generates ATP through reactions of aerobic cellular respiration.

Q How do the cristae of a mitochondrion contribute to its ATP-producing function?

prevalent in nerve cells. As ganglioside G_{M2} accumulates, the nerve cells function less efficiently. Children with Tay-Sachs disease typically experience seizures and muscle rigidity. They gradually become blind, demented, and uncoordinated and usually die before the age of 5. The Tay-Sachs gene has been identified, and chromosome testing now can detect whether an adult is a carrier of the defective gene. ■

Peroxisomes

Another group of organelles similar in structure to lysosomes, but smaller, are called **peroxisomes** (pe-ROKS-i-sōms; *peroxi* = peroxide; *somes* = bodies; see Figure 3.1). Although peroxisomes were once thought to form by budding off the ER, it now is generally agreed that they form by the division of preexisting peroxisomes.

Peroxisomes contain one or more enzymes that can oxidize (remove hydrogen atoms from) various organic substances. For example, substances such as amino acids and fatty acids are oxidized in peroxisomes as part of normal metabolism. In addition, enzymes in peroxisomes oxidize toxic substances, such as alcohol. A byproduct of the oxidation reactions is hydrogen peroxide (H_2O_2), a potentially toxic compound. However, peroxisomes also contain an enzyme called *catalase*, which decomposes H_2O_2. Because the generation and degradation of H_2O_2 occurs within the same organelle, peroxisomes protect other parts of the cell from the toxic effects of H_2O_2.

Mitochondria

Because of their function in generating ATP, **mitochondria** (mī-tō-KON-drē-a; *mito-* = thread; *-chondria* = granules) are the "powerhouses" of the cell. A cell may have as few as a hundred or as many as several thousand mitochondria (singular is mitochondrion). Physiologically active cells, such as those found in the muscles, liver, and kidneys, have a large number of mitochondria because they use ATP at a high rate. Mitochondria are usually located in a cell where the energy need is greatest, such as between the contractile proteins in muscle cells. A mitochondrion is bounded by two membranes, each of which is similar in structure to the plasma membrane (Figure 3.24). The *outer mitochondrial membrane* is smooth, but the *inner mitochondrial membrane* is arranged in a series of folds called **cristae** (KRIS-tē; = ridges). The central fluid-filled cavity of a mitochondrion, enclosed by the inner membrane and cristae, is the **matrix.** The elaborate folds of the cristae provide an enormous surface area for the chemical reactions that are part of the aerobic phase of *cellular respiration.* These reactions produce most of a cell's ATP. Enzymes that catalyze cellular respiration are located in the matrix and on the cristae.

Like peroxisomes, mitochondria self-replicate, a process that occurs during times of increased cellular energy demand or before cell division. Each mitochondrion has multiple identical copies of a circular DNA molecule that contains 37 genes. The mitochondrial genes along with the genes in the cell's nucleus control the production of proteins that build mitochondrial components. Because ribosomes are also

Figure 3.25 Nucleus.

🔑 The nucleus contains most of a cell's genes, which are located on chromosomes.

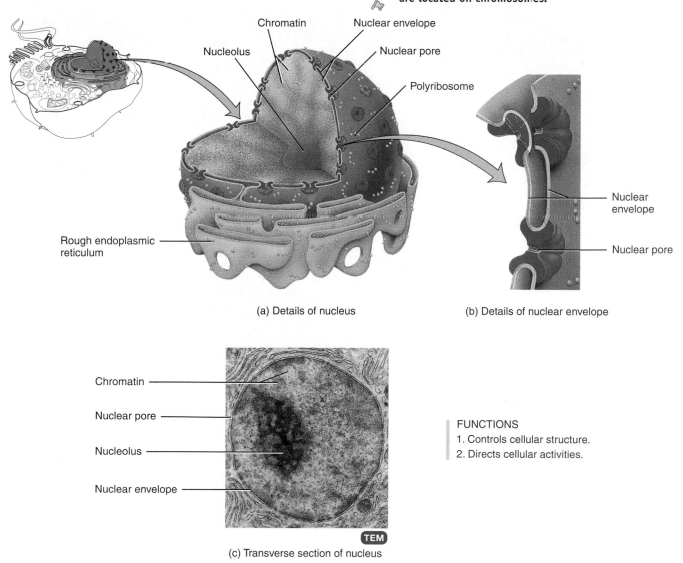

(a) Details of nucleus

(b) Details of nuclear envelope

(c) Transverse section of nucleus

FUNCTIONS
1. Controls cellular structure.
2. Directs cellular activities.

 What are the functions of nuclear genes?

present in the mitochondrial matrix, some protein synthesis occurs inside mitochondria.

Although the nucleus of each somatic cell contains genes from both your mother and father, mitochondrial genes usually are inherited only from your mother. The head of a sperm (the part that penetrates and fertilizes an ovum) normally lacks most organelles, such as mitochondria, ribosomes, endoplasmic reticulum, and Golgi complexes.

1. Distinguish between cytosol, cytoplasm, and organelles.
2. What are the components and functions of the cytoskeleton?
3. Briefly describe the structure and functions of centrosomes, cilia and flagella, ribosomes, endoplasmic reticulum, Golgi complex, lysosomes, peroxisomes, and mitochondria.

NUCLEUS

OBJECTIVE

• *Describe the structure and function of the nucleus.*

The **nucleus** is a spherical or oval-shaped structure that is usually the most prominent feature of a cell (Figure 3.25). Most body cells have a single nucleus, although some, such as mature red blood cells, have none. In contrast, skeletal muscle cells and a few other cells have several nuclei. A double membrane called the **nuclear envelope** separates the nucleus from the cytoplasm. Both layers of the nuclear envelope are lipid bilayers similar to the plasma membrane. The outer membrane of the nuclear envelope is continuous with rough ER and resembles it in structure. The nuclear envelope is perforated by numerous channels called **nuclear pores** (see Figure 3.25c). Each pore consists of a circular

Figure 3.26 Packing of DNA into a chromosome in a dividing cell. When packing is complete, two identical DNA and their histones form a pair of sister chromatids, held together by a centromere.

🔑 **A chromosome is a highly coiled and folded DNA molecule that is combined with protein molecules.**

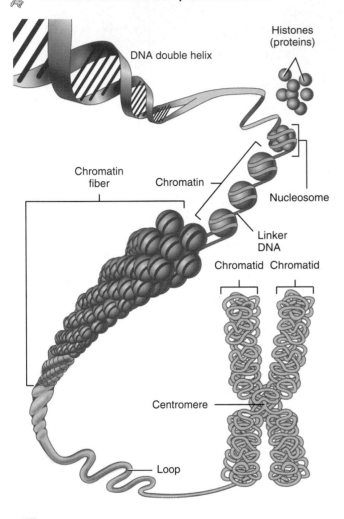

Q What are the components of a nucleosome?

arrangement of proteins that surrounds a large central channel, which is about 10 times larger than the pore of a channel protein in the plasma membrane.

Nuclear pores control the movement of substances between the nucleus and the cytoplasm. Small molecules and ions diffuse passively through open aqueous channels in the pores. Most large molecules, however, such as RNAs and proteins, cannot pass through the nuclear pores by diffusion. Instead, their passage involves an active transport process in which the molecules are recognized and selectively transported in one direction only. As large molecules are recognized and move through the regulated channels, the nuclear pores open up to accommodate them. This permits the selective transport of proteins from the cytosol into the nucleus, and of RNAs from the nucleus into the cytosol.

One or more spherical bodies called **nucleoli** (noo′KLĒ-ō-lī) are present inside the nucleus. Nucleoli (singular is nucleolus) are clusters of protein, DNA, and RNA that are not enclosed by a membrane. They are the sites of the assembly of ribosomes, which play a key role in protein synthesis. Nucleoli are quite prominent in cells that synthesize large amounts of protein, such as muscle and liver cells. Nucleoli disperse and disappear during cell division and reorganize once new cells are formed.

Within the nucleus are most of the cell's hereditary units, called **genes,** which control cellular structure and direct most cellular activities. Genes are arranged in single file along **chromosomes** (*chromo-* = colored). Human somatic (body) cells have 46 chromosomes, 23 inherited from each parent. Each chromosome is a long molecule of DNA that is coiled together with several proteins (Figure 3.26). In a cell that is not dividing, the 46 chromosomes appear as a diffuse, granular mass, which is called **chromatin.** Electron micrographs reveal that chromatin has a "beads-on-a-string" structure. Each "bead" is a **nucleosome** and consists of double-stranded DNA wrapped twice around a core of eight proteins called **histones,** which help organize the coiling and folding of DNA. The "string" between the "beads" is **linker DNA,** which holds adjacent nucleosomes together. Another histone promotes coiling of nucleosomes into a larger diameter **chromatin fiber,** which then folds into large loops. In cells that are not dividing, this is how DNA is packed. Just before cell division takes place, however, the DNA replicates (duplicates) and the loops condense even more, forming a pair of **sister chromatids.** Chromatids are easily seen as rod-shaped structures when viewed through a light microscope.

The main parts of a cell and their functions are summarized in Table 3.2.

| **1.** Describe how DNA is packed in the nucleus.

PROTEIN SYNTHESIS

OBJECTIVE

• *Describe the sequence of events involved in protein synthesis.*

Although cells synthesize many chemicals to maintain homeostasis, much of the cellular machinery is devoted to synthesizing large numbers of diverse proteins. The proteins, in turn, determine the physical and chemical characteristics of cells and, therefore, of organisms. Some proteins are used to assemble cellular structures such as the plasma membrane, the cytoskeleton, and other organelles. Other proteins serve as hormones, antibodies, and contractile elements in muscle tissue. Still others are enzymes that regulate the rates of the numerous chemical reactions that occur in cells.

The instructions for making proteins are found mainly in the DNA within the nucleus. To synthesize proteins, the information encoded in a region of DNA is first *transcribed* (copied) to produce a specific molecule of RNA (ribonucleic

Table 3.2 Cell Parts and Their Functions

PART	STRUCTURE	FUNCTIONS
PLASMA MEMBRANE	Fluid-mosaic lipid bilayer consisting of phospholipids, cholesterol, and glycolipids studded with proteins; surrounds cytoplasm.	Protects cellular contents; makes contact with other cells; contains channels, transporters, receptors, enzymes, cell-identity markers, and linker proteins; mediates the entry and exit of substances.
CYTOPLASM	Cellular contents between plasma membrane and nucleus—cytosol and organelles. Cytosol is composed of water, solutes, suspended particles, lipid droplets, and glycogen granules. Organelles are specialized membranous or nonmembranous structures with characteristic shapes and specific functions.	
Cytoskeleton	Network composed of three protein filaments: micro filaments, intermediate filaments, and microtubules.	Maintains shape and general organization of cellular contents; is responsible for cell movements.
Centrosome	Pericentriolar area plus paired centrioles (9 + 0 array of microtubules).	Pericentriolar area is organizing center for microtubules and mitotic spindle; centrioles form and regenerate cilia and flagella.
Cilia and flagella	Motile cell surface projections composed of 9 + 2 array of microtubules and a basal body.	Cilia move fluids over a cell's surface; flagella move an entire cell.
Ribosome	Composed of two subunits containing ribosomal RNA and ribosome proteins; may be free in cytosol or attached to rough ER.	Protein synthesis.
Endoplasmic reticulum (ER)	Membranous network of flattened sacs or tubules called cisterns. Rough ER is covered by ribosomes and is attached to nuclear membrane; smooth ER lacks ribosomes.	Rough ER synthesizes secretory proteins, phospholipids, and membranes; smooth ER secretes phospholipids, fats, and steroids.
Golgi complex	3–20 flattened membranous sacs called cisterns; structurally and functionally divided into entry face, medial cisterns, and exit face.	Entry face accepts proteins from rough ER; medial cisterns form glycoproteins, glycolipids, and lipoproteins; exit face stores, packages, and exports medial cistern products.
Lysosome	Vesicle formed from Golgi complex; contains digestive enzymes.	Fuses with and digests contents of late endosomes, pinocytic vesicles, and phagosomes; digests worn-out organelles (autophagy), entire cells (autolysis), and extracellular materials.
Peroxisome	Vesicle containing oxidative enzymes.	Detoxifies harmful substances.
Mitochondrion	Consists of outer and inner membranes, cristae, and matrix.	Site of aerobic cellular respiration reactions that produce most of a cell's ATP.
NUCLEUS	Consists of nuclear envelope with pores, nucleoli, and chromatin (or chromosomes).	Contains genes, which control cellular structure and most cellular activities.

acid) (Figure 3.27). Then, the information contained in RNA is *translated* into a corresponding sequence of amino acids that forms a protein molecule.

DNA has the potential to code for the production of many thousands of different proteins. Recall from Chapter 2 that only 20 amino acids are the building blocks of all proteins. For cells to synthesize proteins, ribosomes must join the appropriate amino acids in a sequence specified by a segment of RNA, which was produced according to the instructions coded on a corresponding segment of DNA.

Information is stored in DNA (and in RNA as well) in sets of three nitrogenous bases, or nucleotides. A sequence of three such nucleotides in DNA is called a **base triplet.** Each base triplet is transcribed (copied) as a complementary sequence of three RNA nucleotides, called a **codon.** A given

codon specifies one amino acid. The **genetic code** is the set of rules that relate the base triplet sequence of DNA to the corresponding codons of RNA and the amino acids they specify.

The first step in protein synthesis is transcription, which we examine next.

Transcription

During **transcription,** which occurs inside the nucleus, the genetic information represented by the sequence of base triplets in DNA serves as a template for copying the information into a complementary sequence of codons in a strand of RNA. Three kinds of RNA originate from the DNA template: **messenger RNA (mRNA),** which directs the synthesis of a

Figure 3.27 Overview of transcription and translation.

🔑 Whereas transcription is a nuclear event, translation is a cytoplasmic event.

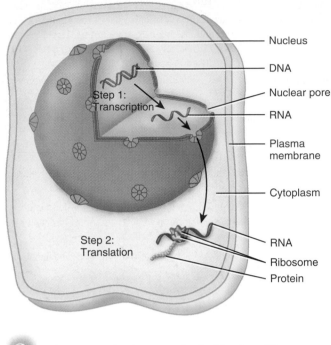

Nucleus
DNA
Nuclear pore
RNA
Plasma membrane
Cytoplasm
Step 1: Transcription
Step 2: Translation
RNA
Ribosome
Protein

Q Why are proteins important in the life of a cell?

protein; **ribosomal RNA (rRNA),** which joins with ribosomal proteins to make ribosomes; and **transfer RNA (tRNA),** which binds to an amino acid and holds it in place on a ribosome until it is incorporated into a protein during translation. There are more than 20 different types of tRNA, but each one can bind to only one of the 20 different amino acids. Thus, each gene in a segment of DNA is transcribed into a particular mRNA, rRNA, or tRNA.

Transcription of DNA is catalyzed by the enzyme *RNA polymerase.* However, the enzyme must be instructed where to start the transcription process and where to end it. The segment of DNA where transcription begins is a special nucleotide sequence called a **promoter,** which is located near the beginning of a gene (Figure 3.28a). This is where the RNA polymerase attaches to DNA. During transcription, bases pair in a complementary manner: The bases cytosine (C), guanine (G), and thymine (T) in the DNA template dictate guanine, cytosine, and adenine (A), respectively, in the RNA strand (Figure 3.28b). However, adenine in the DNA template dictates uracil (U), not thymine, in RNA:

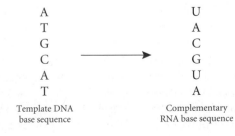

A	U
T	A
G	C
C	G
A	U
T	A

Template DNA base sequence → Complementary RNA base sequence

Figure 3.28 Transcription.

🔑 During transcription, the genetic information in DNA is copied to RNA.

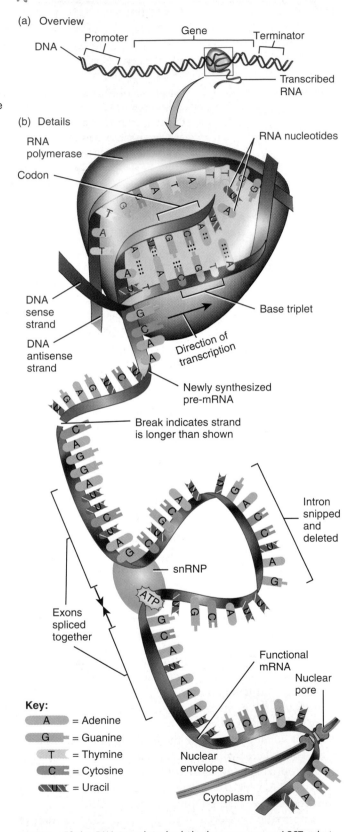

(a) Overview

DNA
Promoter
Gene
Terminator
Transcribed RNA

(b) Details

RNA polymerase
Codon
RNA nucleotides
DNA sense strand
DNA antisense strand
Direction of transcription
Base triplet
Newly synthesized pre-mRNA
Break indicates strand is longer than shown
snRNP
ATP
Intron snipped and deleted
Exons spliced together
Functional mRNA
Nuclear pore
Nuclear envelope
Cytoplasm

Key:
A = Adenine
G = Guanine
T = Thymine
C = Cytosine
U = Uracil

Q If the DNA template had the base sequence AGCT, what would be the mRNA base sequence? What enzyme catalyzes transcription of DNA?

Only one of the two DNA strands serves as a template for RNA synthesis. This strand is referred to as the **sense strand.** The other strand (the one not transcribed) is called the **antisense strand.**

Transcription of the DNA sense strand ends at another special nucleotide sequence called a **terminator,** which specifies the end of the gene (see Figure 3.28a). When RNA polymerase reaches the terminator, the enzyme detaches from the transcribed RNA molecule and the DNA sense strand.

Within a gene are regions called **introns** that do *not* code for parts of proteins, located between regions called **exons** that *do* code for segments of a protein. Immediately after transcription, mRNA includes information from both introns and exons and is called **pre-mRNA.** The introns are removed from pre-mRNA by **small nuclear ribonucleoproteins** (snRNPs, pronounced "snurps"; see Figure 3.28b), enzymes that cut out the introns and splice together the exons. The result is a functional mRNA molecule that passes through a pore in the nuclear envelope to reach the cytoplasm, where translation takes place.

Translation

Translation is the process whereby the nucleotide sequence in an mRNA molecule specifies the amino acid sequence of a protein. This process is accomplished by ribosomes in the cytoplasm. The small subunit of a ribosome has a *binding site* for mRNA; the large subunit has two binding sites for tRNA molecules (Figure 3.29). The first tRNA binding site is the *P site,* where the first tRNA molecule bearing its specific amino acid attaches to mRNA. The second is the *A site,* which holds the next tRNA molecule bearing its amino acid. Translation occurs in the following sequence (Figure 3.30):

1 An mRNA molecule binds to the small ribosomal subunit at the mRNA binding site. A special tRNA, called *initiator tRNA,* binds to the start codon (AUG) on mRNA, where translation begins.

2 The large ribosomal subunit attaches to the small subunit, creating a functional ribosome. The initiator tRNA fits into the P site on the ribosome. One end of a tRNA carries a specific amino acid, and the opposite end consists of a triplet of nucleotides called an **anticodon.** By base pairing, the tRNA anticodon attaches to the complementary mRNA codon. For example, if the mRNA codon is AUG, then a tRNA with the anticodon UAC would attach to it.

3 The anticodon of another tRNA with its amino acid attaches to the complementary mRNA codon at the A site of the ribosome.

4 The growing polypeptide separates from the tRNA at the P site and forms a peptide bond with the amino acid carried by the tRNA at the A site. An enzyme in the large ribosomal subunit catalyzes the formation of a peptide bond.

Figure 3.29 Ribosomes and translation.

Ribosomes have a binding site for mRNA, and a P site and an A site for tRNA.

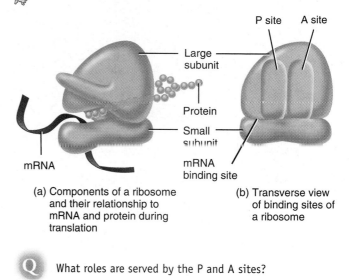

(a) Components of a ribosome and their relationship to mRNA and protein during translation

(b) Transverse view of binding sites of a ribosome

Q What roles are served by the P and A sites?

5 Following peptide bond formation, the tRNA at the P site detaches from the ribosome, and the ribosome shifts the mRNA strand by one codon. The tRNA in the A site bearing the newly forming protein shifts into the P site, allowing another tRNA with its amino acid to bind to a newly exposed codon at the A site. Steps 4 and 5 repeat again and again as the protein progressively lengthens.

6 Protein synthesis stops when the ribosome reaches a stop codon at the A site, at which time the completed protein detaches from the final tRNA. Then the tRNA vacates the A site, and the ribosome splits into its constituent subunits.

Protein synthesis progresses at a rate of about 15 amino acids per second. As the ribosome moves along the mRNA and before it completes synthesis of the whole protein, another ribosome may attach behind it and begin translation of the same mRNA strand. In this way, several ribosomes form a polyribosome that may be attached to the same mRNA. The simultaneous movement of several ribosomes along the same mRNA molecule permits the translation of one mRNA into several identical proteins in a short time.

CLINICAL APPLICATION
Recombinant DNA

Beginning in 1973, scientists developed techniques for inserting genes from other organisms into a variety of host cells. This manipulation causes the host organism to produce proteins it normally does not synthesize. Organisms so altered are called **recombinants,** and their DNA—a combination of DNA from different sources—is called **recombinant**

Figure 3.30 Protein elongation and termination of protein synthesis during translation.

🔑 **During protein synthesis the ribosomal subunits join, but they separate when the process is complete.**

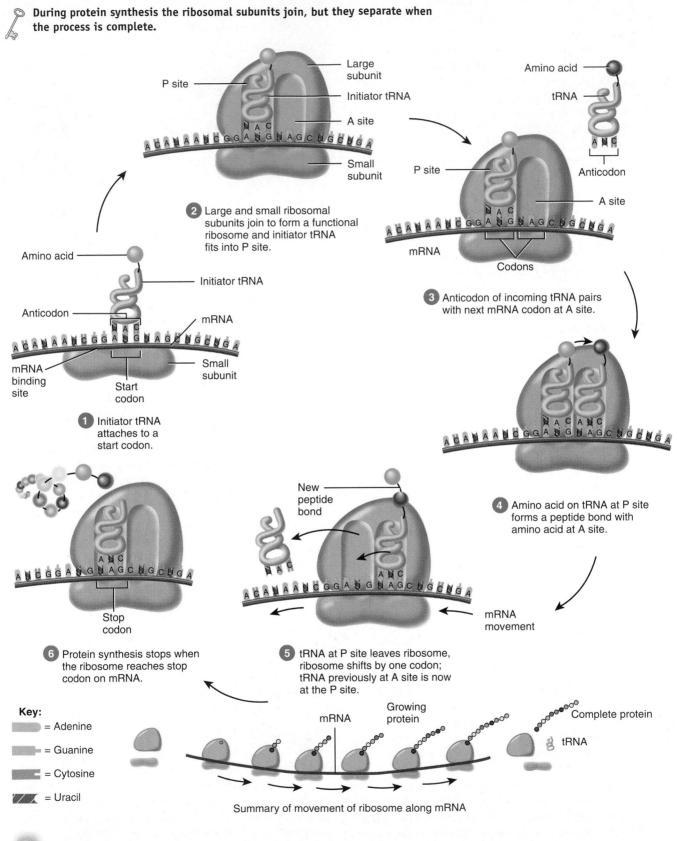

1 Initiator tRNA attaches to a start codon.

2 Large and small ribosomal subunits join to form a functional ribosome and initiator tRNA fits into P site.

3 Anticodon of incoming tRNA pairs with next mRNA codon at A site.

4 Amino acid on tRNA at P site forms a peptide bond with amino acid at A site.

5 tRNA at P site leaves ribosome, ribosome shifts by one codon; tRNA previously at A site is now at the P site.

6 Protein synthesis stops when the ribosome reaches stop codon on mRNA.

Key:

= Adenine

= Guanine

= Cytosine

= Uracil

Summary of movement of ribosome along mRNA

Q What is the function of a stop codon?

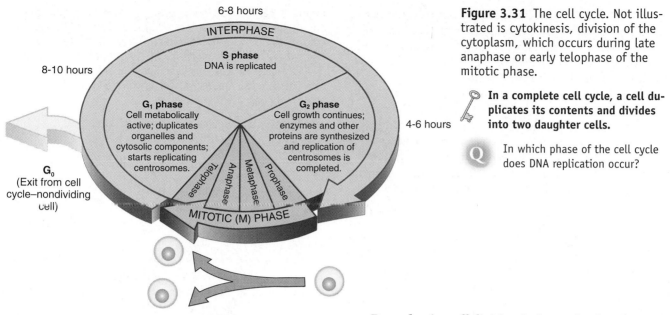

Figure 3.31 The cell cycle. Not illustrated is cytokinesis, division of the cytoplasm, which occurs during late anaphase or early telophase of the mitotic phase.

In a complete cell cycle, a cell duplicates its contents and divides into two daughter cells.

Q In which phase of the cell cycle does DNA replication occur?

DNA. When recombinant DNA functions properly, the host will synthesize the protein specified by the new gene it has acquired. The new technology that has arisen from manipulating genetic material is called **genetic engineering.**

The practical applications of recombinant DNA technology are enormous. Strains of recombinant bacteria are now producing large quantities of many important therapeutic substances. Just a few of these are *human growth hormone* (hGH), required for growth during childhood and important in adults' metabolism; *insulin,* a hormone that helps regulate blood glucose level and is used by diabetics; *interferon* (IFN), an antiviral (and possibly anticancer) substance; and *erythropoietin,* a hormone that stimulates production of red blood cells. ■

1. Distinguish between transcription and translation.
2. Summarize the steps in transcription and translation.

NORMAL CELL DIVISION

OBJECTIVE

• *Discuss the stages, events, and significance of somatic cell division.*

Most cellular activities maintain the life of the cell on a day-to-day basis. However, as somatic cells become damaged, diseased, or worn out, they are replaced by **cell division,** the process whereby cells reproduce themselves. We distinguish between two kinds of cell division: somatic (body) cell division and reproductive cell division.

In **somatic cell division,** a cell undergoes a nuclear division called **mitosis** and a cytoplasmic division called **cytokinesis** to produce two identical **daughter cells.** Each daughter cell has the same number and kind of chromosomes as the original cell. Somatic cell division replaces dead or injured cells and adds new ones for tissue growth.

Reproductive cell division is the mechanism that produces gametes—sperm and ova—the cells needed to form the next generation of sexually reproducing organisms. This process consists of a special two-step division called **meiosis,** in which the number of chromosomes in the nucleus is reduced by half. Meiosis is described in Chapter 28; here we focus on the division of somatic cells.

The Cell Cycle in Somatic Cells

The **cell cycle** is an orderly sequence of events by which a cell duplicates its contents and divides in two. Human cells, except for gametes, contain 23 pairs of chromosomes. The two chromosomes that belong to each pair—one contributed by the mother and one contributed by the father—are called **homologous chromosomes.** They have similar genes, which are usually arranged in the same order. When a cell reproduces, it must replicate (duplicate) all its chromosomes so that its genes may be passed on to the next generation of cells. The cell cycle consists of two major periods: interphase, when a cell is not dividing, and the mitotic (M) phase, when a cell is dividing (Figure 3.31).

Interphase

During **interphase** the cell replicates its DNA. It also manufactures additional organelles and cytosolic components in anticipation of cell division. Interphase is a state of high metabolic activity, and during this time the cell does most of its growing.

Interphase consists of three phases: G_1, S, and G_2 (Figure 3.31). The S stands for synthesis of DNA. Because the G-phases are periods when there is no activity related to DNA duplication, they are thought of as gaps or interruptions in DNA duplication. The **G_1 phase** is the interval between the mitotic phase and the S phase. During G_1, the cell is metabolically active; it duplicates its organelles and cytosolic components but not its DNA. Replication of centrosomes begins here but does not end until G_2; virtually all the cellular activities described in this chapter happen during G_1. For a typical body cell with a total cell cycle time of 24 hours, G_1 might last about 8–10 hours. The duration of this phase is quite variable, however, lasting from minutes, to hours, to years for different types of cells. The **S phase** is the interval between G_1 and G_2 and lasts about 6–8 hours. Its duration is also quite variable for different types of cells. During the S phase, DNA replication occurs, ensuring that the two daughter cells formed from cell division will have identical genetic material. The **G_2 phase** is the interval between the S phase and the mitotic phase. It lasts about 4–6 hours. During G_2, cell growth continues, enzymes and other proteins are synthesized in preparation for cell division, and the replication of centrosomes is completed. Cells that remain in G_1 for a very long time, perhaps destined never to divide again, are said to be in the **G_0 state.** For example, most nerve cells are in this state. Once a cell enters the S phase, however, it is committed to go through cell division.

When DNA replicates during the S phase, its helical structure partially uncoils, and the two strands separate at the points where hydrogen bonds connect base pairs (Figure 3.32). Each exposed base then picks up a complementary base (with its associated sugar and phosphate group). This uncoiling and complementary base pairing continues until each of the two original DNA strands is joined with a newly formed complementary DNA strand. The original DNA molecule has become two identical DNA molecules.

A microscopic view of a cell during interphase shows a clearly defined nuclear envelope, nucleolus, and chromatin (Figure 3.33a). The absence of visible chromosomes is another physical characteristic of interphase. Once a cell completes its activities during the G_1, S, and G_2 phases of interphase, the mitotic phase begins.

Mitotic Phase

The **mitotic (M) phase** of the cell cycle consists of nuclear division, or mitosis, and cytoplasmic division, or cytokinesis. The events that occur during this phase are plainly visible under a microscope because chromatin condenses into chromosomes.

Nuclear Division: Mitosis **Mitosis** is the distribution of the two sets of chromosomes, one set into each of two separate nuclei. The process results in the *exact* partitioning of genetic information. For convenience, biologists divide the process into four stages: prophase, metaphase, anaphase, and telophase. Mitosis is a continuous process, however, with one stage merging imperceptibly into the next.

- **Prophase.** During early prophase, the chromatin fibers condense and shorten (Figure 3.33b). The condensation process may prevent entangling of the long DNA strands as they move during mitosis. Because DNA replication

Figure 3.32 Replication of DNA. The two strands of the double helix separate by breaking the hydrogen bonds (shown as dotted lines) between nucleotides. New, complementary nucleotides attach at the proper sites, and a new strand of DNA is synthesized alongside each of the original strands. Arrows indicate hydrogen bonds forming again between pairs of bases.

 Replication doubles the amount of DNA.

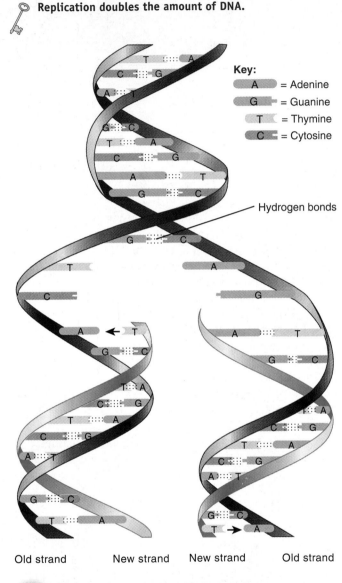

Key:
A = Adenine
G = Guanine
T = Thymine
C = Cytosine

Hydrogen bonds

Old strand   New strand   New strand   Old strand

Q Why is it crucial that DNA replication occurs prior to cytokinesis in somatic cell division?

took place during the S phase of interphase, each prophase chromosome contains a pair of identical double-stranded chromatids. Each chromatid pair is held together by a constricted region called a **centromere** that is required for the proper segregation of chromosomes. Attached to the outside of each centromere is a protein complex known as the **kinetochore** (ki-NET-ō-kor), whose function is described shortly.

Figure 3.33 Cell division: mitosis and cytokinesis. Begin the sequence at (a) at the top of the figure and read clockwise until you complete the process. Although sister chromatids are identical, they are differentiated here by color (red and yellow) to more easily follow their movements.

🔑 **In somatic cell division, a single diploid cell divides to produce two identical diploid daughter cells.**

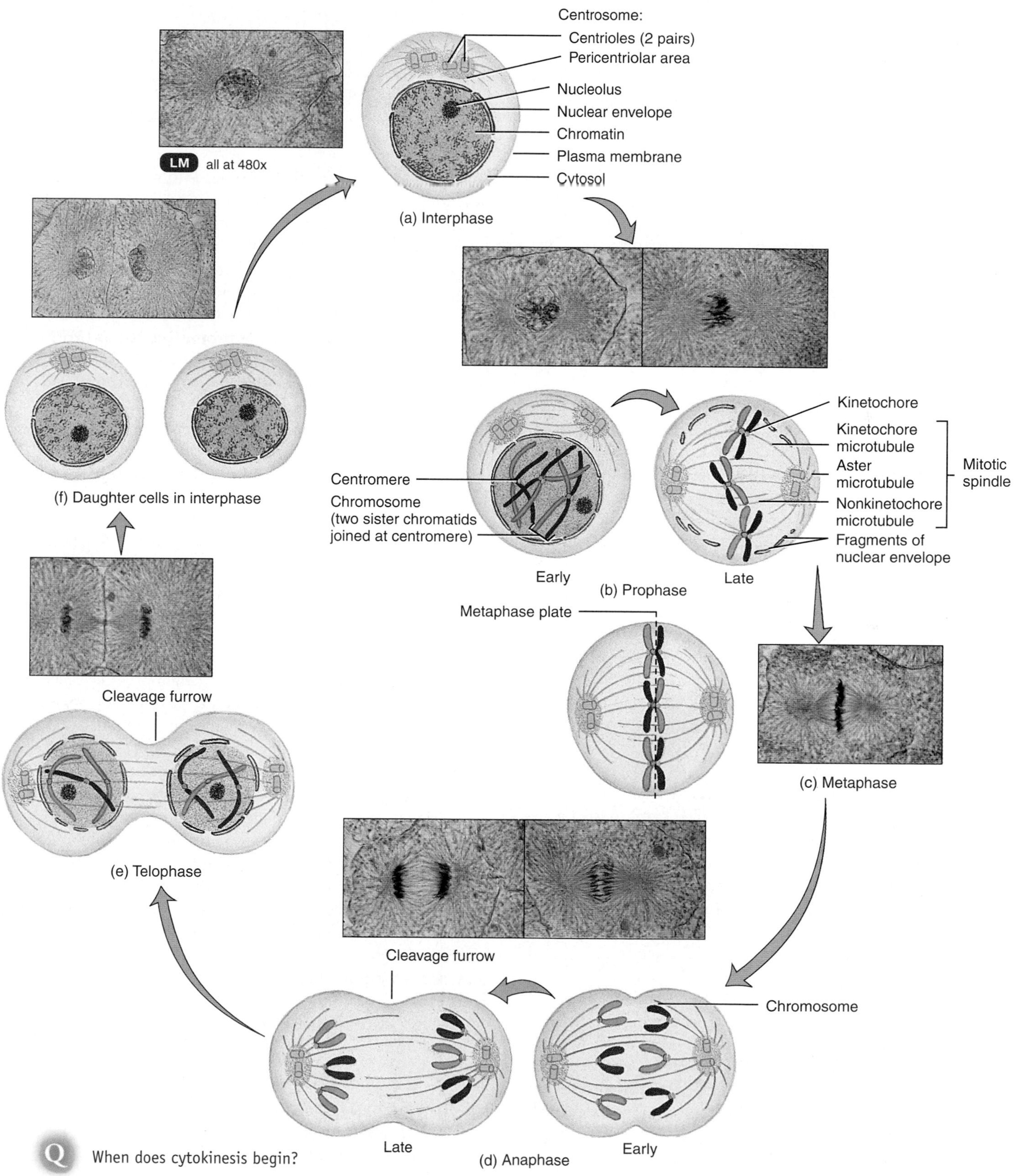

LM all at 480x

Centrosome:
Centrioles (2 pairs)
Pericentriolar area
Nucleolus
Nuclear envelope
Chromatin
Plasma membrane
Cytosol

(a) Interphase

Kinetochore
Kinetochore microtubule
Aster microtubule
Nonkinetochore microtubule
Fragments of nuclear envelope
Mitotic spindle

Centromere
Chromosome (two sister chromatids joined at centromere)

Early   Late
(b) Prophase

Metaphase plate

(c) Metaphase

(f) Daughter cells in interphase

Cleavage furrow

(e) Telophase

Cleavage furrow

Chromosome

Late   Early
(d) Anaphase

Q When does cytokinesis begin?

Later in prophase, the nucleolus disappears (see Figure 3.33b), and the nuclear envelope breaks down. In addition, each centrosome moves to an opposite pole (end) of the cell. As they do so, the pericentriolar area of the centrosomes starts to form the **mitotic spindle,** a football-shaped assembly of microtubules. The lengthening of microtubules between centrosomes pushes the centrosomes to the poles of the cell so that the spindle extends from pole to pole. As the mitotic spindle continues to develop, three types of microtubules form from the pericentriolar area of the centrosomes : (1) **nonkinetochore microtubules** extend between the two centrosomes but do not bind to kinetochores; (2) **kinetochore microtubules** extend inward and attach to kinetochores; and (3) **aster microtubules** radiate outward from the mitotic spindle. The spindle is responsible for the separation of chromatids to opposite poles of the cell.

- **Metaphase.** During metaphase, the kinetochore microtubules align the centromeres of the chromatid pairs at the exact center of the mitotic spindle (Figure 3.33c). This midpoint region is called the **metaphase plate** or **equatorial plane region.**

- **Anaphase.** During anaphase, the centromeres split, separating sister chromatids. This enables one chromatid of each pair to move toward opposite poles of the cell (Figure 3.33d). Once separated, the sister chromatids are called **daughter chromosomes.** As the chromosomes are pulled by the kinetochore microtubules during anaphase, they become V-shaped as the centromeres lead the way and seem to drag the trailing parts of the chromosomes toward the pole.

- **Telophase.** The final stage of mitosis, telophase, begins after chromosomal movement stops (Figure 3.33e). The identical sets of chromosomes now at opposite poles of the cell uncoil and revert to threadlike chromatin. A nuclear envelope forms around each chromatin mass, nucleoli reappear in the daughter nuclei, and eventually the mitotic spindle breaks up.

Cytoplasmic Division: Cytokinesis Division of a parent cell's cytoplasm and organelles is called **cytokinesis** (sī′-tō-ki-NĒ-sis; *cyto-* = cell; *-kinesis* = motion). This process begins in late anaphase or early telophase with formation of a **cleavage furrow,** a slight indentation of the plasma membrane, usually midway between the centrosomes, that extends around the center of the cell (see Figure 3.33d and e). The cleavage furrow is produced by the action of actin microfilaments that lie just inside the plasma membrane. The microfilaments form a *contractile ring* that pulls the plasma membrane progressively inward, constricting the center of the cell like a belt around the waist, and ultimately pinching it in two. The plane of the cleavage furrow is always perpendicular to the mitotic spindle, thereby ensuring that the two sets of chromosomes will be segregated into separate daughter cells. When cytokinesis is complete, interphase begins (Figure 3.33f).

Considering the cell cycle in its entirety, the sequence of events is

G_1 phase → S phase → G_2 phase → mitosis → cytokinesis

The events of the somatic cell cycle are summarized in Table 3.3.

Control of Cell Destiny

OBJECTIVE
- *Discuss the signals that induce cell division, and describe apoptosis.*

A cell has three possible destinies—to remain alive and functioning without dividing, to grow and divide, or to die. Homeostasis is maintained when there is a balance between cell proliferation and cell death. The signals that tell a cell when to exist in the G_0 phase, when to divide, and when to die have been the subjects of intense and fruitful research during the past decade. A key signal that induces cell division is a substance called **maturation promoting factor (MPF).** One component of MPF is a group of enzymes called **cdc2 proteins,** so named because they participate in the cell division cycle (cdc). Another component of MPF is a protein called **cyclin,** so named because its level rises and falls during the cell cycle. Cyclin builds up in the cell during interphase and activates cdc2 proteins and thus MPF. As a result, the cell undergoes mitosis.

Cellular death also is regulated. Throughout the lifetime of an organism, certain cells undergo an orderly, genetically programmed death, a process called **apoptosis** (ap-ō-TŌ-sis; = a falling off). In apoptosis, a triggering agent from either outside or inside the cell causes "cell-suicide" genes to produce enzymes that damage the cell in several ways, including disrupting its cytoskeleton and nucleus. As a result, the cell shrinks and pulls away from neighboring cells. The DNA within the nucleus fragments, and the cytoplasm shrinks, although the plasma membrane remains intact. Phagocytes in the vicinity then ingest the dying cell. Apoptosis, a normal type of cell death, contrasts with **necrosis** (ne-KRŌ-sis; = death), a pathological type of cell death that results from tissue injury. In necrosis, many adjacent cells swell, burst, and spill their cytoplasm into the interstitial fluid. The cellular debris usually stimulates an inflammatory response by the immune system that does not occur in apoptosis. Apoptosis is especially useful because it removes unneeded cells during development before birth. It continues to occur after birth to regulate the number of cells in a tissue and eliminate potentially dangerous cells such as cancer cells.

CLINICAL APPLICATION
Tumor-suppressor Genes

Abnormalities in genes that regulate the cell cycle or apoptosis are associated with many diseases. For example, some cancers are caused by damage to genes called **tu-**

mor-suppressor genes, which produce proteins that normally inhibit cell division. Loss or alteration of a tumor-suppressor gene called *p53* on chromosome 17 is the most common genetic change leading to a wide variety of tumors, including breast and colon cancers. The normal p53 protein arrests cells in the G_1 phase, which prevents cell division, and also is needed to initiate apoptosis. ■

1. Distinguish between the somatic and reproductive types of cell division. Why is each important?
2. Define interphase. When does DNA replicate itself?
3. Describe the principal events of each stage of the mitotic phase.
4. How is cell destiny controlled?

CELLS AND AGING

OBJECTIVE

• *Explain the relationship of aging to cellular processes.*

Aging is a normal process accompanied by a progressive alteration of the body's homeostatic adaptive responses. It produces observable changes in structure and function and increases vulnerability to environmental stress and disease. The specialized branch of medicine that deals with the medical problems and care of elderly persons is called **geriatrics** (jer′-ē-AT-riks; *ger-* = old age; *-iatrics* = medicine).

Although many millions of new cells normally are produced each minute, several kinds of cells in the body—heart cells, skeletal muscle cells, and nerve cells—do not divide because they are arrested permanently in the G_0 phase. Experiments have shown that many other cell types have only a limited capability to divide. Cells grown outside the body divide only a certain number of times and then stop. These observations suggest that cessation of mitosis is a normal, genetically programmed event. According to this view, "aging genes" are part of the genetic blueprint at birth, and they turn on at preprogrammed times, slowing down or halting processes vital to life.

Glucose, the most abundant sugar in the body, plays a role in the aging process. Glucose is haphazardly added to proteins inside and outside cells, forming irreversible cross-links between adjacent protein molecules. With advancing age, more cross-links form, which contributes to the stiffening and loss of elasticity that occur in aging tissues.

Free radicals produce oxidative damage in lipids, proteins, or nucleic acids by "stealing" an electron to accompany their unpaired electrons. Some effects are wrinkled skin, stiff joints, and hardened arteries. Normal cellular metabolism—for example, aerobic cellular respiration in mitochondria—produces some free radicals. Others are present in air pollution, radiation, and certain foods we eat. Naturally occurring enzymes in peroxisomes and in the cytosol normally dispose of free radicals. Certain dietary substances, such as vitamin E, vitamin C, beta-carotene, and selenium, are antioxidants that inhibit free radical formation.

Table 3.3 Events of the Somatic Cell Cycle

PHASE	ACTIVITY
INTERPHASE	Cell is between divisions; chromosomes not visible under light microscope.
G_1 phase	Metabolically active cell duplicates organelles and cytosolic components; centrosome replication begins.
S phase	DNA replicates.
G_2 phase	Cell growth, enzyme and protein synthesis continues; centrosome replication completed.
MITOTIC PHASE	Parent cell produces daughter cells with identical chromosomes; chromosomes visible under light microscope.
Mitosis	Nuclear division; distribution of two sets of chromosomes into separate nuclei.
Prophase	Chromatin fibers condense into paired chromatids; nucleolus and nuclear envelope disappear; centrosomes move to opposite poles of cell.
Metaphase	Centromeres of chromatid pairs line up at metaphase plate.
Anaphase	Centromeres divide; identical sets of chromosomes move to opposite poles of cell.
Telophase	Nuclear envelopes and nucleoli reappear; chromosomes resume chromatin form; mitotic spindle breaks up.
Cytokinesis	Cytoplasmic division; contractile ring forms cleavage furrow around center of cell, dividing cytoplasm into separate and equal portions.

Whereas some theories of aging explain the process at the cellular level, others concentrate on regulatory mechanisms operating within the entire organism. For example, the immune system may start to attack the body's own cells. This *autoimmune response* might be caused by changes in cell-identity markers at the surface of cells that cause antibodies to attach to and mark the cell for destruction. As changes in the proteins on the plasma membrane of cells increase, the autoimmune response intensifies, producing the well-known signs of aging.

1. What is aging?
2. What are some of the cellular changes that occur with aging?

CELLULAR DIVERSITY

To this point, we have investigated the *intracellular* complementarity of a cell's structure and function. Most (but not all) cells have a full complement of organelles, each of which controls a specific set of physiologic processes within the cell. However, not all cells in the body look the same; nor do they perform identical functional roles in the body.

Figure 3.34 Diverse shapes and sizes of human cells. The relative difference in size between the smallest and largest cells is actually much greater than shown here.

 The nearly 100 trillion cells in an average adult can be classified into about 200 different cell types.

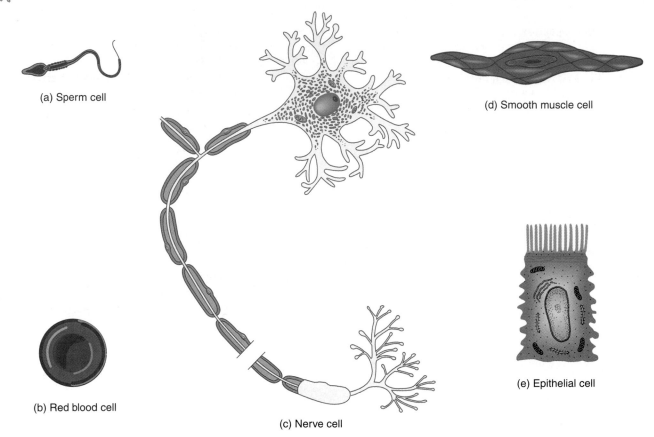

(a) Sperm cell

(d) Smooth muscle cell

(b) Red blood cell

(c) Nerve cell

(e) Epithelial cell

Q Why are sperm the only body cells that have a flagellum?

The body of an average human adult is composed of nearly 100 trillion cells, but all of these cells can be classified into about 200 different cell types. Cells vary considerably in size. High-powered microscopes are needed to see the smallest cells of the body. The largest cell, a single ovum, is barely visible to the unaided eye. The sizes of cells are measured in units called *micrometers.* One micrometer (μm) is equal to 1 one-millionth of a meter, or 10^{-6}m (1/25,000 of an inch). Whereas a red blood cell has a diameter of 8 μm, an ovum has a diameter of about 140 μm.

The shapes of cells also vary considerably (Figure 3.34). They may be round, oval, flat, cuboidal, columnar, elon-gated, star-shaped, cylindrical, or disc-shaped. A cell's shape is related to its function in the body. For example, a sperm cell has a long whiplike tail (flagellum) that it uses for loco-motion. The disc-shape of a red blood cell gives it a large sur-face area that enhances its ability to pass oxygen to other cells. Some cells contain microvilli, which greatly increase the cell's surface area. Microvilli are common in the epithelial cells that line the small intestine, where they enhance the absorption of digested food. Nerve cells have long extensions that permit them to transmit nerve impulses over great distances.

Cellular diversity is the stepping stone to more complex levels of biological organization such as tissues and organs.

CANCER

Cancer is a group of diseases characterized by uncontrolled cell proliferation. When cells in a part of the body divide without control, the excess tissue that develops is called a **tumor** or **neoplasm** (NĒ-ō-plazm; *neo-* = new). The study of tumors is called **oncology** (on-KOL-ō-jē; *onco-* = swelling or mass). Tumors may be cancerous and sometimes fatal, or they may be harmless. A cancerous neoplasm is called a **malignant tumor** or **malignancy.** One property of a malignant tumor is its ability to undergo **metastasis** (me-TAS-ta-sis), the spread of cancerous cells to other parts of the body. A **benign tumor** is a noncancerous growth. Although benign tumors do not metastasize, they may be surgically removed if they interfere with a normal body function or become disfiguring.

Types of Cancer

The name of the cancer is derived from the type of tissue in which it develops. Most human cancers are **carcinomas** (kar-sin-ŌM-az; *carcin-* = cancer; *-omas* = tumors), malignant tumors that arise from epithelial cells. **Melanomas** (mel-an-ŌM-az; *melan-* = black), for example, are cancerous growths of melanocytes, skin epithelial cells that produce the pigment melanin. **Sarcoma** (sar-KŌ-ma; *sarc-* = flesh) is a general term for any cancer arising from muscle cells or connective tissues. For example, **osteogenic sarcoma** (*osteo-* = bone; *-genic* = origin), the most frequent type of childhood cancer, destroys normal bone tissue. **Leukemia** (loo-KĒ-mē-a; *leuk-* = white; *-emia* = blood) is a cancer of blood-forming organs characterized by rapid growth of abnormal leukocytes (white blood cells). **Lymphoma** (lim-FŌ-ma) is a malignant disease of lymphatic tissue—for example, of lymph nodes.

Growth and Spread of Cancer

Cells of malignant tumors duplicate rapidly and continuously. As malignant cells invade surrounding tissues, they trigger **angiogenesis,** the growth of new networks of blood vessels. Proteins that trigger angiogenesis in tumors are called **tumor angiogenesis factors (TAFs).** As the cancer grows, it begins to compete with normal tissues for space and nutrients. Eventually, the normal tissue decreases in size and dies. Some malignant cells may detach from the initial (primary) tumor and invade a body cavity or enter the blood or lymph, then circulate to and invade other body tissues, establishing secondary tumors. Malignant cells resist the antitumor defenses of the body. The pain associated with cancer develops when the tumor presses on nerves or blocks a passageway in an organ so that secretions build up pressure.

Causes of Cancer

Several factors may trigger a normal cell to lose control and become cancerous. One cause is environmental agents: substances in the air we breathe, the water we drink, and the food we eat. A chemical agent or radiation that produces cancer is called a **carcinogen** (car-SIN-ō-jen). Carcinogens induce **mutations,** permanent structural changes in the DNA base sequence of a gene. The World Health Organization estimates that carcinogens may be associated with 60–90% of all human cancers. Examples of carcinogens are hydrocarbons found in cigarette tar, radon gas from the earth, and ultraviolet (UV) radiation in sunlight.

Viruses are a second cause of cancer. These agents are tiny packages of nucleic acids, either DNA or RNA, that are capable of infecting cells and converting them to virus-producers. Although the link between viruses and cancer is strongly established for a variety of animal cancers, the relation in human cancers is less clear, for the number of people infected with these viruses is much larger than the number who develop cancer. The evidence suggests that chronic viral infections are associated with up to one-fifth of all cancers.

Intensive research efforts are now directed toward studying cancer-causing genes, or **oncogenes** (ON-kō-jēnz). These genes, when inappropriately activated, have the ability to transform a normal cell into a cancerous cell. Oncogenes develop from normal genes that regulate growth and development, called **proto-oncogenes.** These genes may undergo some change that either causes them to produce an abnormal product or disrupts their control so that they are expressed inappropriately, making their products in excessive amounts or at the wrong time. Scientists believe that some oncogenes cause excessive production of growth factors, chemicals that stimulate cell growth. Other oncogenes may cause changes in a surface receptor, causing it to send signals as though it were being activated by a growth factor. As a result, the growth pattern of the cell becomes abnormal.

Proto-oncogenes in every cell carry out normal cellular functions until a malignant change occurs. It appears that some proto-oncogenes are activated to oncogenes by mutations in which the DNA of the proto-oncogene is altered. Other proto-oncogenes are activated by a rearrangement of the chromosomes so that segments of DNA are exchanged. Rearrangement activates proto-oncogenes by placing them near genes that enhance their activity. Viruses are believed to cause cancer by inserting their own oncogenes or proto-oncogenes into the host cell's DNA.

Carcinogenesis: A Multistep Process

Carcinogenesis (kar′-si-nō-JEN-e-sis), the process by which cancer develops, is a multistep process in which as many as ten distinct mutations may have to accumulate in a cell before it becomes cancerous. The progression of genetic changes leading to cancer is best understood for colon (colorectal) cancer. Such cancers, as well as lung and breast cancer, take years or decades to develop. In colon cancer, the tumor begins as an area of increased cell proliferation that results from one mutation. This growth then progresses to abnormal, but noncancerous, growths called adenomas. After two or three additional mutations, a mutation of *p53* (a tumor-suppressor gene) occurs and a carcinoma develops. The fact that so many mutations are needed for a cancer to develop indicates that cell growth is normally controlled with many sets of checks and balances.

Treatment of Cancer

Many cancers are removed surgically. However, when cancer is widely distributed throughout the body or exists in organs such as the brain whose functioning would be greatly harmed by surgery, chemotherapy and radiation therapy may be used instead. Chemotherapy involves administering drugs that poison cancerous cells.

Radiation therapy destroys the chromosomes of cancerous cells, thus preventing them from dividing. Because cancerous cells divide rapidly, they are more vulnerable to the destructive effects of chemotherapy and radiation therapy than are normal cells. Nevertheless, both chemotherapy and radiation therapy can kill or disrupt the function of some normal cells as well.

Treating cancer is difficult because it is not a single disease and because all the cells in a single tumor population rarely behave in the same way. Although most cancers are thought to derive from a single abnormal cell, by the time a tumor reaches a clinically detectable size, the cancer may contain a diverse population of abnor-

mal cells. For example, some cancerous cells metastasize readily, and others do not; some are sensitive to chemotherapy drugs and some are drug resistant. Because of differences in drug resistance, a single chemotherapeutic agent may destroy susceptible cells but permit resistant cells to proliferate.

Another stumbling block encountered by blood-borne anticancer drugs is the physical barrier developed by solid tumors, such as those that arise in the breasts, lungs, colon, and other organs. The high-pressure areas deep within such tumors collapse blood vessels in the tumor. This makes it difficult, if not impossible, for blood-borne anticancer agents to penetrate the tumor.

MEDICAL TERMINOLOGY

Anaplasia (an′-a-PLĀ-zē-a; *an* = not; *plassein* = to shape) The loss of tissue differentiation and function that is characteristic of most malignancies.

Atrophy (AT-rō-fē; *a* = without; *-trophy* = nourishment) A decrease in the size of cells, with a subsequent decrease in the size of the affected tissue or organ; wasting away.

Dysplasia (dis-PLĀ-zē-a; *dys-* = abnormal; *plassein* = to shape) Alteration in the size, shape, and organization of cells due to chronic irritation or inflammation; may progress to neoplasia (tumor formation, usually malignant) or revert to normal if the irritation is removed.

Hyperplasia (hī-per-PLĀ-zē-a; *hyper-* = over) Increase in the number of cells of a tissue due to an increase in the frequency of cell division.

Hypertrophy (hī-PER-trō-fē) Increase in the size of cells without cell division.

Metaplasia (met′-a-PLĀ-zē-a; *meta-* = change) The transformation of one type of cell into another.

Progeny (PROJ-e-nē; *pro-* = forward; *-geny* = production) Offspring or descendants.

STUDY OUTLINE

INTRODUCTION (p. 60)
1. A cell is the basic, living, structural and functional unit of the body.
2. Cytology is the scientific study of cellular structure. Cell physiology is the study of cellular function.

A GENERALIZED VIEW OF THE CELL (p. 60)
1. Figure 3.1 shows a generalized view of a cell that is a composite of many different cells in the body.
2. The principal parts of a cell are the plasma membrane; the cytoplasm, which consists of cytosol and organelles; and the nucleus.

THE PLASMA MEMBRANE (p. 61)
1. The plasma membrane surrounds and contains the cytoplasm of a cell.
2. The membrane is composed of a 50:50 mix by weight of proteins and lipids that are held together by noncovalent forces.
3. According to the fluid mosaic model, the membrane is a mosaic of proteins floating like icebergs in a bilayer sea of lipids.
4. The lipid bilayer consists of two back-to-back layers of phospholipids, cholesterol, and glycolipids. The bilayer arrangement occurs because the lipids are amphipathic, having both polar and nonpolar parts.
5. Integral proteins extend into or through the lipid bilayer, whereas peripheral proteins associate with membrane lipids or integral proteins at the inner or outer surface of the membrane.

6. Many integral proteins are glycoproteins that have sugar groups attached to the ends that face the extracellular fluid. Together with glycolipids, the glycoproteins form a glycocalyx on the extracellular surface of cells.
7. Membrane proteins have a variety of functions. Channels and transporters are membrane proteins that help specific solutes across the membrane; receptors serve as cellular recognition sites; and linkers anchor proteins in the plasma membranes to filaments inside and outside the cell. Some membrane proteins are enzymes and others are cell-identity markers.
8. Membrane fluidity is greater when there are more double bonds in the fatty acid tails of the lipids that make up the bilayer. Cholesterol makes the lipid bilayer stronger but less fluid at normal body temperature. Because of its fluidity, the lipid bilayer self-seals when torn or punctured.
9. The membrane's selective permeability permits some substances to pass more readily than others. The lipid bilayer is permeable to water and to most, nonpolar, uncharged molecules but impermeable to ions and charged or polar molecules other than water. Channels and transporters increase the plasma membrane's permeability to small- and medium-sized polar and charged substances, including ions, that cannot cross the lipid bilayer.
10. The selective permeability of the plasma membrane supports the existence of concentration gradients, differences in the concentrations of chemicals between one side of the membrane and the other.

TRANSPORT ACROSS THE PLASMA MEMBRANE (p. 65)

1. In mediated transport, the movement of a substance across a membrane is assisted by the binding of the substance to a transporter protein. Uniporters move a single substance across the membrane, whereas coupled transporters move two substances across the membrane. Symporters move two substances in the same direction, whereas antiporters move two substances in opposite directions across the membrane.

2. In passive transport, a substance moves down its concentration gradient across the membrane using its own kinetic energy of motion. In active transport, cellular energy is used to drive the substance "uphill" against its concentration gradient.

3. In vesicular transport, tiny vesicles either detach from or merge with the plasma membrane to move materials across the membrane, into or out of a cell.

4. Net diffusion is the net movement of molecules or ions from an area of higher concentration to an area of lower concentration until an equilibrium is reached.

5. The rate of diffusion across a plasma membrane is affected by the steepness of the concentration gradient, temperature, size or mass of the diffusing substance, surface area available for diffusion, and the distance over which diffusion must occur.

6. Osmosis is the net movement of water through a selectively permeable membrane from an area of higher water concentration to an area of lower water concentration.

7. In an isotonic solution, red blood cells maintain their normal shape; in a hypotonic solution, they undergo hemolysis; in a hypertonic solution, they undergo crenation.

8. Nonpolar, hydrophobic molecules, such as oxygen, carbon dioxide, nitrogen, steroids, fat-soluble vitamins (A, E, D, and K), small alcohols, and ammonia diffuse through the lipid bilayer of the plasma membrane.

9. Ion channels selective for K^+, Cl^-, Na^+, and Ca^{2+} allow diffusion across the plasma membrane of these small, inorganic ions, which are too hydrophilic to penetrate the membrane's nonpolar interior.

10. In facilitated diffusion, a solute such as glucose binds to a specific transporter on one side of the membrane and is released on the other side after the transporter undergoes a conformational change. Because a limited number of transporters are available, facilitated diffusion exhibits a transport maximum and saturation.

11. Substances can cross the membrane "uphill"—against their concentration gradient—by active transport. Actively transported substances includes several ions, such as Na^+, K^+, H^+, Ca^{2+}, I^-, Cl^-; amino acids; and monosaccharides.

12. Two sources of energy are used to drive active transport: Energy obtained from hydrolysis of ATP is the source in primary active transport, and energy stored in a Na^+ or H^+ concentration gradient is the source in secondary active transport.

13. The most prevalent primary active transport pump is the sodium pump, also known as Na^+/K^+ ATPase.

14. Secondary active transport mechanisms include both symporters and antiporters that are powered by either a Na^+ or H^+ concentration gradient.

15. Vesicular transport includes both endocytosis and exocytosis.

16. Receptor-mediated endocytosis is the selective uptake of large molecules and particles (ligands) that bind to specific receptors in membrane areas called clathrin-coated pits.

17. Phagocytosis is the ingestion of solid particles. It is an important process used by some white blood cells to destroy bacteria that enter the body.

18. Pinocytosis is the ingestion of extracellular fluid. In this process, the fluid becomes surrounded by a pinocytic vesicle.

19. Exocytosis involves movement of secretory or waste products out of a cell by fusion of vesicles with the plasma membrane.

CYTOPLASM (p. 75)

1. Cytoplasm is all the cellular contents between the plasma membrane and the nucleus.

2. It consists of cytosol and organelles.

Cytosol (p. 75)

1. Cytosol is the fluid portion of cytoplasm.

2. It contains mostly water, plus ions, glucose, amino acids, fatty acids, proteins, lipids, ATP, and waste products.

3. Cytosol is the site of many chemical reactions required for a cell's existence.

Organelles (p. 75)

1. Organelles are specialized structures with characteristic shapes that have specific functions.

2. Some organelles, such as the cytoskeleton, centrosome, cilia, flagella, and ribosomes, are nonmembranous (don't contain membranes).

3. Other organelles, such as the endoplasmic reticulum, Golgi complex, mitochondria, lysosomes, and peroxisomes, are membranous (contain one or more phospholipid membranes).

4. The cytoskeleton is a network of several kinds of protein filaments that extend throughout the cytoplasm.

5. Components of the cytoskeleton are microfilaments, intermediate filaments, and microtubules.

6. The cytoskeleton provides a structural framework for the cell and is responsible for cell movements.

7. The centrosome consists of a pericentriolar area and centrioles.

8. The pericentriolar area organizes microtubules in nondividing cells and the miotic spindle in dividing cells.

9. Centrioles function in the formation or regeneration of cilia and flagella.

10. Cilia and flagella are motile projections of the cell surface.

11. Cilia move fluid along the cell surface.

12. Flagella move an entire cell.

13. Ribosomes, composed of ribosomal RNA and ribosomal proteins, consist of two subunits made in the nucleus.

14. Free ribosomes are not attached to any cytoplasmic structure; membrane-bound ribosomes are attached to endoplasmic reticulum.

15. Ribosomes are sites of protein synthesis.

16. Endoplasmic reticulum (ER) is a network of membranes that form flattened sacs or tubules called cisterns; it extends from the nuclear membrane throughout the cytoplasm.

17. Rough ER is covered by ribosomes, and its primary function is protein synthesis. It also forms glycoproteins and attaches proteins to phospholipids, which it also synthesizes.

18. Smooth ER lacks ribosomes. It synthesizes phospholipids, fats, and steroids; releases glucose from the liver into the bloodstream; inactivates or detoxifies drugs and other potentially harmful substances; and releases calcium ions that trigger contraction in muscle cells.

19. The Golgi complex consists of flattened sacs called cisterns that differ in size, shape, content, enzymatic activity, and vesicles present. Entry face cisterns face the rough ER, exit face cisterns face the plasma membrane, and medial cisterns are between the two.
20. The Golgi complex receives synthesized products from the rough ER and modifies, sorts, packages, and transports them within vesicles to different destinations.
21. Some processed proteins leave the cell in secretory vesicles, some are incorporated into the plasma membrane, and some enter lysosomes.
22. Lysosomes are membrane-enclosed vesicles that contain digestive enzymes.
23. Late endosomes, phagosomes, and pinocytic vesicles deliver materials to lysosomes for degradation.
24. Lysosomes function in digestion of worn-out organelles (autophagy), digestion of a host cell (autolysis), and extracellular digestion.
25. Peroxisomes are similar to lysosomes, but smaller.
26. Peroxisomes oxidize various organic substances such as amino acids, fatty acids, and toxic substances and, in the process, produce hydrogen peroxide.
27. The hydrogen peroxide produced from oxidation is degraded by an enzyme in peroxisomes called catalase.
28. Mitochondria consist of a smooth outer membrane, an inner membrane containing cristae, and a fluid-filled cavity called the matrix. They are called "powerhouses" of the cell because they produce ATP.

NUCLEUS (p. 85)

1. The nucleus consists of a double nuclear envelope; nuclear pores, which control the movement of substances between the nucleus and cytoplasm; nucleoli, which produce ribosomes; and genes arranged on chromosomes.
2. Genes control cellular structure and most cellular functions.

PROTEIN SYNTHESIS (p. 86)

1. Most of the cellular machinery is devoted to synthesizing proteins.
2. Cells make proteins by transcribing and translating the genetic information encoded in DNA.
3. The genetic code refers to the set of rules that relate the base triplet sequences of DNA to the corresponding codons of RNA and the amino acids they encode.
4. In transcription, the genetic information in the sequence of base triplets in DNA serves as a template for copying the information into a complementary sequence of codons in messenger RNA.
5. Transcription begins on DNA in a region called a promoter.
6. Regions of DNA that code for protein synthesis are called exons; those that do not are called introns.
7. Newly synthesized pre-mRNA is modified before leaving the nucleus.
8. Translation is the process by which the nucleotide sequence of mRNA specifies the amino acid sequence of a protein.
9. In translation, mRNA binds to a ribosome, specific amino acids attach to tRNA, and anticodons of tRNA bind to codons of mRNA, thus bringing specific amino acids into position on a growing protein.
10. Translation begins at the start codon and terminates at the stop codon.

NORMAL CELL DIVISION (p. 91)

1. Cell division is the process by which cells reproduce themselves. It consists of nuclear division (mitosis or meiosis) and cytoplasmic division (cytokinesis).
2. Cell division that results in an increase in the number of body cells is called somatic cell division and involves a nuclear division called mitosis plus cytokinesis.
3. Cell division that results in the production of sperm and ova is called reproductive cell division and consists of a nuclear division called meiosis plus cytokinesis.

The Cell Cycle in Somatic Cells (p. 91)

1. The cell cycle is an orderly sequence of events in which a cell duplicates its contents and divides in two. It consists of interphase and a mitotic phase.
2. Before the mitotic phase, the DNA molecules, or chromosomes, replicate themselves so that identical chromosomes can be passed on to the next generation of cells.
3. A cell that is between divisions and is carrying on every life process except division is said to be in interphase, which consists of three phases: G_1, S, and G_2.
4. During the G_1 phase, the cell duplicates its organelles and cytosolic components; during the S phase, DNA replication occurs; during the G_2 phase, enzymes and other proteins are synthesized and centrosome replication is complete.
5. Mitosis is the replication of and distribution of two sets of chromosomes into separate and equal nuclei; it consists of prophase, metaphase, anaphase, and telophase.
6. Cytokinesis usually begins in late anaphase and ends in telophase.
7. A cleavage furrow forms at the cell's metaphase plate and progresses inward, cutting through the cell to form two separate portions of cytoplasm.

Control of Cell Destiny (p. 94)

1. A cell can either remain alive and functioning without dividing, grow and divide, or die.
2. Maturation promoting factor (MPF) induces cell division (both mitosis and meiosis). MPF consists of cyclin, which builds up during interphase, and cdc2 proteins, which are inactivated by cyclin.
3. Apoptosis is programmed cell death, a normal type of cell death. It first occurs during embryological development and continues for the lifetime of an organism.
4. Certain genes regulate both cell division and apoptosis. Abnormalities in these genes are associated with a wide variety of diseases and disorders.

CELLS AND AGING (p. 95)

1. Aging is a normal process accompanied by progressive alteration of the body's homeostatic adaptive responses.
2. Many theories of aging have been proposed, including genetically programmed cessation of cell division, the buildup of free radicals, and an intensified autoimmune response.

CELLULAR DIVERSITY (p. 95)

1. The almost 200 cell types in the body vary considerably in size and shape.
2. The sizes of cells are measured in micrometers. One micrometer equals 10^{-6} m (1/25,000 of an inch).
3. A cell's shape is related to its function.

SELF-QUIZ QUESTIONS

Complete the following:

1. The three principal parts of the cell are the ___, ___, and ___.
2. ___ refers to programmed cell death, whereas ___ refers to cell death resulting from tissue injury.
3. The fluid portion of the cytoplasm is the ___.
4. The site of ribosome formation is the ___.
5. Match the following:
   ___ (a) mitosis
   ___ (b) meiosis
   ___ (c) prophase
   ___ (d) metaphase
   ___ (e) anaphase
   ___ (f) telophase
   ___ (g) cytokinesis
   ___ (h) interphase

   (1) cytoplasmic division
   (2) somatic cell division resulting in identical daughter cells
   (3) reproductive cell division that reduces the number of chromosomes by half
   (4) stage of cell growth and metabolism
   (5) stage when chromatin fibers condense and shorten to form chromosomes
   (6) stage when centromeres split and sister chromatids move to opposite poles of the cell
   (7) stage when centromeres of chromatid pairs line up at the center of the mitotic spindle
   (8) stage when chromosomes uncoil and revert to chromatin

True or false:

6. Typically, the inner surface of the cell membrane is more positively charged and the outer surface is more negatively charged.
7. The sodium pump is an example of primary active transport.

Choose the best answer to the following questions:

8. The basic structural unit of the plasma membrane is the (a) lipid bilayer, (b) integral protein, (c) cholesterol molecule, (d) peripheral protein, (e) glycoprotein-glycolipid complex.
9. Integral proteins can function in the cell membrane in all the following ways *except* (a) as a channel, (b) as a transporter, (c) as a receptor, (d) as an exocytosis vesicle, (e) as a cell-identity marker.
10. Which of the following factors influence the diffusion rate of substances through a plasma membrane? (1) concentration gradient, (2) diffusion distance, (3) surface area, (4) size of diffusing substance, (5) temperature.
    (a) 1, 2, 3, and 5, (b) 1, 2, 3, 4, and 5, (c) 2, 3, 4, and 5, (d) 1, 2, and 5, (e) 2, 3, and 5.
11. A solution in which a cell maintains its normal shape and volume is called a (an) (a) hypertonic solution, (b) hypotonic solution, (c) isotonic solution, (d) solute solution, (e) solvent solution.
12. Which of the following statements regarding the nucleus are true? (1) Nucleoli within the nucleus are the sites of ribosome synthesis. (2) Most of the cell's hereditary units, called genes, are located within the nucleus. (3) The nuclear membrane is a solid, impermeable membrane. (4) Protein synthesis occurs within the nucleus. (5) In nondividing cells, DNA is found in the nucleus in the form of chromatin.
    (a) 1, 2, and 3, (b) 1, 2, and 4, (c) 1, 2, and 5, (d) 2, 4, and 5, (e) 2, 3, and 4.

13. Match the following:
    ___ (a) codon
    ___ (b) RNA polymerase
    ___ (c) intron
    ___ (d) exon
    ___ (e) transcription
    ___ (f) translation
    ___ (g) messenger RNA
    ___ (h) transfer RNA
    ___ (i) ribosomal RNA
    ___ (j) snRNP

    (1) DNA region that does not code for synthesis of a part of a protein
    (2) DNA region that codes for synthesis of a part of a protein
    (3) enzyme that removes all introns and joins remaining exons
    (4) a transcribed sequence of three RNA nucleotides
    (5) the copying of the DNA message onto messenger RNA
    (6) joins with ribosomal proteins to make ribosomes
    (7) binds to amino acids and holds them in place on a ribosome to be incorporated into a protein
    (8) directs synthesis of a protein
    (9) reading of messenger RNA for protein synthesis
    (10) enzyme that catalyzes transcription of DNA

14. Match the following:
    ___ (a) cytoskeleton
    ___ (b) centrosome
    ___ (c) ribosomes
    ___ (d) rough ER
    ___ (e) smooth ER
    ___ (f) Golgi complex
    ___ (g) lysosomes
    ___ (h) peroxisomes
    ___ (i) mitochondria

    (1) membrane-enclosed vesicles formed in the Golgi complex that contain strongly hydrolytic enzymes
    (2) network of protein filaments that extend through the cytoplasm
    (3) sites of protein synthesis
    (4) membrane-enclosed vesicles that contain enzymes that use molecular oxygen to oxidize various organic substances
    (5) a factory for synthesizing secretory proteins and membranes
    (6) functions in synthesizing phospholipids, secreting fats and steroids, helping liver cells release glucose into the bloodstream, and detoxification
    (7) function in ATP generation
    (8) modifies, sorts, packages, and transports products synthesized from the rough ER
    (9) organizing center for microtubule formation in nondividing cells and the formation or regeneration of cilia and flagella

15. Match the following:

___ (a) diffusion
___ (b) osmosis
___ (c) facilitated diffusion
___ (d) primary active transport
___ (e) secondary active transport

___ (f) vesicular transport
___ (g) phagocytosis
___ (h) pinocytosis
___ (i) exocytosis
___ (j) receptor-mediated endocytosis

(1) passive transport in which a solute binds to a specific transporter on one side of the membrane and is released on the other side

(2) movement of materials out of the cell by fusing of vesicles with the plasma membrane

(3) the random mixing of particles in a solution due to the kinetic energy of the particles

(4) transport of substances either into or out of the cell by means of a small, spherical membranous sac formed by budding off from existing membranes

(5) uses energy derived from hydrolysis of ATP to change the shape of a transporter protein, which "pumps" a substance across a cellular membrane against its concentration gradient

(6) indirectly uses energy obtained from the breakdown of ATP; involves symporters and antiporters

(7) type of endocytosis that involves the nonselective uptake of tiny droplets of extracellular fluid

(8) type of endocytosis in which large solid particles are taken in

(9) movement of water from an area of higher to an area of lower water concentration through a selectively permeable membrane

(10) process that allows a cell to take in large amounts of a substance without taking in a correspondingly large volume of extracellular fluid

CRITICAL THINKING QUESTIONS

1. In old detective novels, a dead body is sometimes called "a stiff." But the corpse doesn't remain stiff forever. After a time, it begins to soften. What causes this change in the tissues? (HINT: *Worn out organelles suffer a similar fate.*)

2. Mucin is a protein present in saliva and other secretions. When mixed with water, it becomes the slippery substance known as mucus. Trace the route taken by mucin through the cell, from its synthesis to its secretion, listing all the organelles and processes involved. (HINT: *All proteins are produced initially by the same organelle.*)

3. If you were trying to locate the DNA sequence that coded for an mRNA, would you look at a gene's introns or exons, at the sense strand or the antisense strand? (HINT: *The mRNA is complementary to specific parts of the DNA.*)

ANSWERS TO FIGURE QUESTIONS

3.1 Plasma membrane, cytoplasm, and nucleus.

3.2 The glycocalyx is the sugary coat on the extracellular surface of the plasma membrane that is composed of the carbohydrate portions of membrane glycolipids and glycoproteins.

3.3 The membrane protein that binds to insulin is a receptor.

3.4 Both the concentration gradient and the electrical gradient will favor flow of Na^+ into cells.

3.5 Facilitated diffusion and active transport are mediated transport processes.

3.6 Because fever involves an increase in body temperature, the rates of all diffusion processes would increase.

3.7 No, the water concentrations can never be the same in the two arms because the left arm contains pure water and the right arm contains a solution that is less than 100% water.

3.8 A 2% solution of NaCl will cause crenation of RBCs because it is hypertonic.

3.9 A gated channel opens or closes in response to chemical or electrical changes inside or outside the cell. When the gate is open, ions can diffuse through the channel.

3.10 The rate of facilitated diffusion depends on the concentration gradient and the number of available transporters.

3.11 ATP phosphorylates (adds a phosphate group to) the pump protein, which changes the pump's three-dimensional shape.

3.12 In secondary active transport, hydrolysis of ATP is used only indirectly to drive the activity of symporter or antiporter proteins, whereas this reaction directly powers the pump protein in primary active transport.

3.13 Cholesterol, iron, vitamins, and hormones are examples of ligands.

3.14 The binding of particles to a plasma membrane receptor triggers pseudopod formation.

3.15 Receptor-mediated endocytosis and phagocytosis involve receptor proteins, pinocytosis does not.

3.16 Microtubules.

3.17 A cell without a centrosome probably would not be able to undergo cell division.

3.18 Cilia move fluids across cell surfaces, whereas flagella move an entire cell.

3.19 Large and small ribosomal subunits are synthesized in the nucleolus in the nucleus and then join together in the cytoplasm.

3.20 Rough ER has attached ribosomes, whereas smooth ER does not. Rough ER synthesizes proteins that will be exported from the cell; smooth ER is associated with lipid synthesis and other metabolic reactions.

3.21 The entry face receives and modifies proteins from rough ER while the exit face modifies, sorts, packages, and transports molecules.

3.22 Some proteins are discharged from the cell by exocytosis, some are incorporated into the plasma membrane, and some occupy storage vesicles that become lysosomes.

3.23 Digestion of worn-out organelles by lysosomes is called autophagy.

3.24 Mitochondrial cristae increase the surface area available for chemical reactions and contain some of the enzymes needed for ATP production.

3.25 They control cellular structure and direct most cellular activities.

3.26 A nucleosome is a double-stranded DNA wrapped twice around a core of eight histones (proteins).

3.27 Proteins determine the physical and chemical characteristics of cells.

3.28 The DNA base sequence AGCT would be transcribed into the RNA base sequence UCGA. RNA polymerase catalyzes transcription of DNA.

3.29 The P site holds the tRNA attached to the growing protein. The A site holds the tRNA carrying the next amino acid to be added to the growing protein.

3.30 When a ribosome encounters a stop codon at the A site, the completed protein detaches from the final tRNA.

3.31 Chromosomes replicate during the S phase.

3.32 DNA replication occurs before cytokinesis so that each of the new daughter cells will have a complete set of genes.

3.33 Cytokinesis usually starts in late anaphase or early telophase.

3.34 Sperm, which use the flagella for locomotion, are the only body cells required to move considerable distances.

4

THE TISSUE LEVEL OF ORGANIZATION

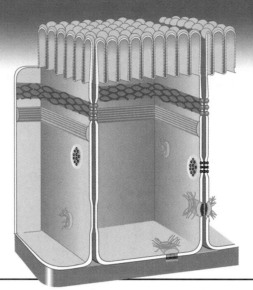

A cell is a complex collection of compartments, each of which carries out a host of biochemical reactions that make life possible. However, a cell seldom functions as an isolated unit in the body. Instead, cells usually work together in groups called tissues. A **tissue** is a group of similar cells that usually has a common embryonic origin and functions together to carry out specialized activities. The structure and properties of a specific tissue are influenced by factors such as the nature of the extracellular material that surrounds the tissue cells and the connections between the cells that compose the tissue. Tissues may be hard, semisolid, or even liquid in their consistency, a range exemplified by bone, fat, and blood. In addition, tissues vary tremendously with respect to the kinds of cells present, how the cells are arranged, and the types of fibers present, if any. **Histology** (hiss′-TOL-ō-jē; *hist-* = tissue; *-ology* = study of) is the science that deals with the study of tissues. A **pathologist** (pa-THOL-ō-gist; *patho-* = disease) is a scientist who specializes in laboratory studies of cells and tissues to help physicians make accurate diagnoses. One of the principal functions of a pathologist is to examine tissues for any changes that might indicate disease.

TYPES OF TISSUES AND THEIR ORIGINS

OBJECTIVE

• *Describe the characteristics of the four basic types of tissue that make up the human body.*

Body tissues can be classified into four basic types according to their function and structure:

1. **Epithelial tissue** covers body surfaces and lines hollow organs, body cavities, and ducts. It also forms glands.
2. **Connective tissue** protects and supports the body and its organs. Various types of connective tissue bind organs together, store energy reserves as fat, and help provide immunity to disease-causing organisms.
3. **Muscle tissue** generates the physical force needed to make body structures move.
4. **Nervous tissue** detects changes in a variety of conditions inside and outside the body and responds by generating nerve impulses. The nervous tissue in the brain helps to maintain homeostasis.

Epithelial tissue and connective tissue, except for bone tissue and blood, are discussed in detail in this chapter. The general features of bone tissue and blood will be introduced here, but their detailed discussion is presented in Chapters 6 and 19, respectively. Similarly, the structure and function of muscle tissue and nervous tissue are examined in detail in Chapters 10 and 12, respectively.

All tissues of the body develop from three **primary germ layers,** the first tissues that form in a human embryo: **ectoderm, endoderm,** and **mesoderm.** Epithelial tissues develop from all three primary germ layers. All connective tissues and most muscle tissues derive from mesoderm. Nervous tissue develops from ectoderm. (Table 29.1 on page 1029 provides a list of structures derived from the primary germ layers.)

Normally, most cells within a tissue remain anchored to other cells, to basement membranes (described shortly), and to connective tissues. A few cells, such as phagocytes, move freely through the body, searching for invaders. Before birth, however, many cells migrate extensively as part of the growth and development process.

CLINICAL APPLICATION
Biopsy

A **biopsy** (BĪ-op-sē; *bio-* = life, *-opsy* = to view) is the removal of a sample of living tissue for microscopic examination. This procedure is used to help diagnose many disorders, especially cancer, and to discover the cause of unexplained infections and inflammations. Both normal and potentially diseased tissues are removed for purposes of comparison. Once the tissue sample is removed, either surgically or through a needle and syringe, it may be preserved, stained to highlight special properties, or cut into thin sections for microscopic observation. Sometimes a biopsy is conducted while a patient is anesthetized during surgery to help a physician determine the most appropriate treatment. For example, if a biopsy of breast tissue reveals malignant cells, the surgeon can proceed with the most appropriate procedure immediately. ■

1. Define a tissue.
2. What are the four basic types of human tissue?

CELL JUNCTIONS
OBJECTIVE
• *Describe the structure and functions of the five main kinds of cell junctions.*

Most epithelial cells and some muscle and nerve cells are tightly joined into functional units. **Cell junctions** are contact points between the plasma membranes of tissue cells. Depending on their structure, cell junctions may serve one of three functions. Some cell junctions form fluid-tight seals between cells, like a "zip-lock" at the top of a sandwich bag; other cell junctions anchor cells together or to extracellular material; and others act as channels that allow ions and molecules to pass from cell to cell within a tissue. Here we consider the five most important cell junctions (Figure 4.1):

• **Tight junctions** connect the cells of tissues that line the surfaces of organs and body cavities (Figure 4.1a). At a tight junction, the outer surfaces of adjacent plasma membranes are fused together by a weblike strip of proteins. Tight junctions prevent the passage of substances between cells, and they are common between cells of epithelial tissues that line the stomach, intestines, and urinary bladder. Tight junctions prevent the contents of these organs from leaking into the blood or surrounding tissues, a situation that could be life threatening if it were to occur.

• **Adherens junctions** (ad-HER-ens) are made of *plaque*, a dense layer of proteins on the inside of the plasma membrane (Figure 4.1b). Microfilaments extend from the plaque into the cell's cytoplasm. Transmembrane glycoproteins (integral membrane proteins) anchored in the plaque of one cell cross the space between the membranes and connect with transmembrane proteins of the adjacent cell to attach the cells. In epithelial cells, adherens junctions often form extensive bands called *adhesion belts* that encircle the cell. The extensive lateral attachments provided by adhesion belts help epithelial surfaces to resist separation.

• **Desmosomes** (DEZ-mō-sōms; *desmo-* = band) are also composed of plaque and are linked by transmembrane glycoproteins that extend across the gap between adjacent cell membranes (Figure 4.1c). Thus they help attach cells to each other. Unlike adherens junctions, intermediate filaments extend from the plaque of desmosomes on one side of the cell across the cytoplasm to desmosomes on the opposite side of the cell. This structural arrangement contributes to the overall stability of the cells and tissue. Desmosomes are numerous among the cells that make up the epidermis, the most superficial layer of the skin, and between cardiac muscle cells of the heart.

• **Hemidesmosomes** (*hemi-* = half) look like half a desmosome (Figure 4.1d). They connect cells to extracellular material such as the basement membrane (discussed shortly). This structural arrangement anchors one kind of tissue to another in the body.

• **Gap junctions** are so named because the outer layers of adjacent plasma membranes converge but leave a tiny intercellular gap between them (Figure 4.1e). The gap is bridged by transmembrane protein channels called *connexons* that form minute fluid-filled tunnels. Ions and small molecules can diffuse through connexons from the cytosol of one cell to another. Gap junctions allow the cells in a tissue to communicate. In a developing embryo, some of the chemical and electrical signals that regulate growth and cell differentiation travel via gap junctions. Moreover, gap junctions enable nerve or muscle impulses to spread rapidly between cells, a process that is crucial for the normal operation of some parts of the nervous system and for the contraction of muscle in the heart and gastrointestinal tract.

1. Define a cell junction.
2. Which cell junctions are found in epithelial tissue?

EPITHELIAL TISSUE
OBJECTIVE
• *Describe the general features of epithelial tissue and the structure, location, and function for the different types of epithelium.*

Epithelial tissue (ep-i-THĒ-lē-al), or **epithelium** (plural is *epithelia*) consists of cells arranged in continuous sheets, in either single or multiple layers. The cells are

Figure 4.1 Cell junctions. Adapted from Lewis Kleinsmith and Valerie Kish, *Principles of Cell and Molecular Biology* 2e, F6.44, p237; F6.46, p238; F6.47, p239; F6.50, p241 (New York: HarperCollins, 1995). ©1995 HarperCollins College Publishers. By permission of Addison Wesley Longman.

Most epithelial cells and some muscle and nerve cells contain cell junctions.

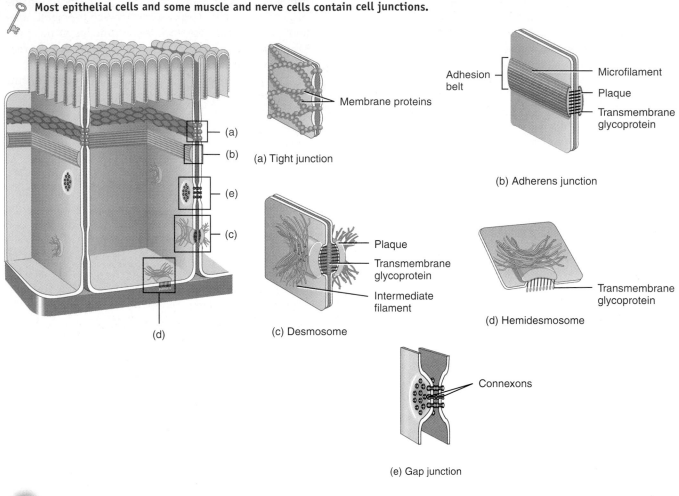

(a) Tight junction

(b) Adherens junction

(c) Desmosome

(d) Hemidesmosome

(e) Gap junction

Q Which type of cell junction functions in communication between adjacent cells?

densely packed, held tightly together by numerous cell junctions, and there is little extracellular space between adjacent plasma membranes (Figure 4.2). The **apical surface** of epithelial cells may be exposed to a body cavity, the lumen of an internal organ, or the exterior of the body. The **basal surface** adheres to an adjacent connective tissue.

The attachment between the basal surface and the connective tissue is a thin extracellular layer called the **basement membrane,** which commonly consists of two layers, the basal lamina and reticular lamina. The *basal lamina* (*lamina* = thin layer) contains collagen fibers (discussed shortly) and other proteins. The *reticular lamina* contains structural proteins such as reticular fibers and fibronectin. In addition to supporting epithelial tissue, the basement membrane serves as a filter in the kidneys and guides cells as they migrate during growth and tissue repair.

Epithelial tissue is **avascular** (*a-* = without; *-vascular* = vessel); that is, it lacks its own blood supply. The blood vessels that bring in nutrients and remove wastes are located in adjacent connective tissue. The exchange of these substances between connective tissue and epithelium occurs by diffu-

sion. Although epithelial tissue is avascular, it has a nerve supply.

Because epithelial tissue forms boundaries between the body's organs, or between the body and the external environment, it is repeatedly subject to physical breakdown and injury. But because it has a very high rate of cell division, epithelial tissue constantly renews and repairs itself by sloughing off dead or injured cells and replacing them with new ones. Epithelial tissue plays many different roles in the body, the most important of which include protection, filtration, secretion, absorption, and excretion. In addition, epithelial tissue combines with nervous tissue to form special organs for smell, hearing, vision, and touch.

Epithelial tissue may be divided into two types. **Covering and lining epithelium** forms the epidermis of the skin and the outer covering of some internal organs. It also forms the inner lining of blood vessels, ducts, and body cavities, and the interior of the respiratory, digestive, urinary, and reproductive systems. **Glandular epithelium** constitutes the secreting portion of glands, such as the thyroid gland, adrenal glands, and sweat glands.

Figure 4.2 Surfaces of epithelial cells and the structure and location of the basement membrane.

 The basement membrane is found between epithelium and connective tissue.

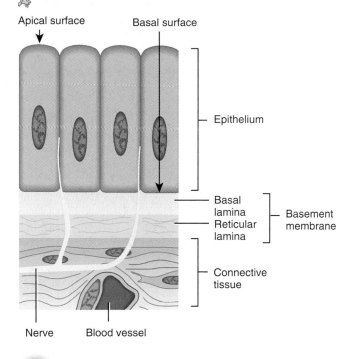

Apical surface

Basal surface

Epithelium

Basal lamina

Reticular lamina

Basement membrane

Connective tissue

Nerve   Blood vessel

Q What are the functions of the basement membrane?

Covering and Lining Epithelium

The types of covering and lining epithelial tissue are classified according to two characteristics: the arrangement of cells into layers and the shapes of the cells.

1. Arrangement of layers. Covering and lining epithelium tends to be arranged in one or more layers depending on the function the epithelium performs:

a. *Simple epithelium* is a single layer of cells that functions in diffusion, osmosis, filtration, *secretion* (the production and release of substances such as mucus, sweat, or enzymes), and *absorption* (the intake of fluids or other substances by cells).

b. *Stratified epithelium* consists of two or more layers of cells that protect underlying tissues in locations where there is considerable wear and tear.

c. *Pseudostratified epithelium* contains only a single layer of cells, but it appears to have multiple layers because cell nuclei lie at different levels and not all cells reach the apical surface. Cells that do extend to the apical surface are either ciliated or secrete mucus.

2. Cell shapes.

a. *Squamous* cells (SKWĀ-mus = flat) are thin and arranged like floor tiles, which allows the rapid transport of substances through them.

b. *Cuboidal* cells are shaped like cubes or hexagons. They function in either secretion or absorption.

c. *Columnar* cells are tall and cylindrical and protect underlying tissues. They may have cilia and may be specialized for secretion and absorption.

d. *Transitional* cells change shape, from columnar to flat and back, as body parts stretch, expand, or move.

Combining the arrangements of layers and cell shapes provides the following classification scheme of covering and lining epithelium:

   I. Simple epithelium
     A. Simple squamous epithelium
     B. Simple cuboidal epithelium
     C. Simple columnar epithelium
  II. Stratified epithelium
     A. Stratified squamous epithelium[*]
     B. Stratified cuboidal epithelium[*]
     C. Stratified columnar epithelium[*]
     D. Transitional epithelium
 III. Pseudostratified columnar epithelium

Each of these covering and lining epithelial tissues is described in the following sections and illustrated in Table 4.1. The tables throughout this chapter contain figures consisting of a photomicrograph, a corresponding diagram, and an inset that identifies a principal location of the tissue in the body. Along with the illustrations are descriptions, locations, and functions of the tissues.

Simple Epithelium

Simple Squamous Epithelium This tissue consists of a single layer of flat cells that resembles a tiled floor when viewed from the apical surface (Table 4.1A). The nucleus of each cell is oval or spherical and centrally located. Simple squamous epithelium is found in parts of the body where filtration (kidneys) or diffusion (lungs) are priority processes. It is not found in body areas that are subject to wear and tear.

The simple squamous epithelium that lines the heart, blood vessels, and lymphatic vessels is known as **endothelium** (*endo-* = within; *-thelium* = covering); the type that forms the epithelial layer of serous membranes is called **mesothelium** (*meso-* = middle). Endothelium and mesothelium are both derived from the embryonic mesoderm.

Simple Cuboidal Epithelium The cuboidal shape of the cells in this tissue (Table 4.1B) is obvious only when the tissue is sectioned and viewed from the side, as when a slice has been made through the epithelial tissue layer. Cell nuclei are usually round and centrally located. Simple cuboidal epithelium performs the functions of secretion and absorption.

---

[*]This classification is based on the shape of the cells at the apical surface.

Text continues on page 115

Table 4.1 Epithelial Tissues

COVERING AND LINING EPITHELIUM

A. Simple squamous
   epithelium

Description: Single layer of flat cells; centrally located nucleus.
Location: Lines heart, blood vessels, lymphatic vessels, air sacs of lungs, glomerular (Bowman's) capsule of kidneys, and inner surface of the tympanic membrane (eardrum); forms epithelial layer of serous membranes.
Function: Filtration, diffusion, osmosis, and secretion in serous membranes.

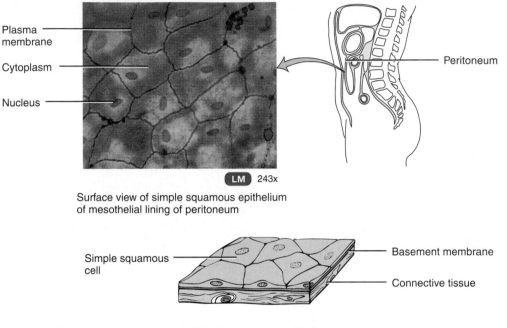

Plasma membrane

Cytoplasm

Nucleus

Peritoneum

LM 243x

Surface view of simple squamous epithelium
of mesothelial lining of peritoneum

Simple squamous cell

Basement membrane

Connective tissue

Simple squamous epithelium

B. Simple cuboidal
   epithelium

Description: Single layer of cube-shaped cells; centrally located nucleus.
Location: Covers surface of ovary, lines anterior surface of capsule of the lens of the eye, forms the pigmented epithelium at the back of the eye, lines kidney tubules and smaller ducts of many glands, and makes up the secreting portion of some glands such as the thyroid gland.
Function: Secretion and absorption.

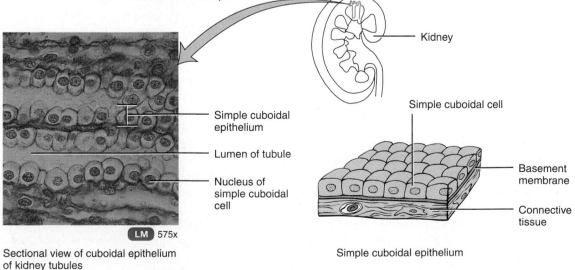

Kidney

Simple cuboidal epithelium

Lumen of tubule

Nucleus of simple cuboidal cell

LM 575x

Sectional view of cuboidal epithelium
of kidney tubules

Simple cuboidal cell

Basement membrane

Connective tissue

Simple cuboidal epithelium

Table 4.1 (continued)

COVERING AND LINING EPITHELIUM

C. Nonciliated simple columnar epithelium

Description: Single layer of nonciliated rectangular cells; nucleus at base of cell; contains goblet cells and cells with microvilli in some locations.
Location: Lines the gastrointestinal tract from the stomach to the anus, ducts of many glands, and gallbladder.
Function: Secretion and absorption.

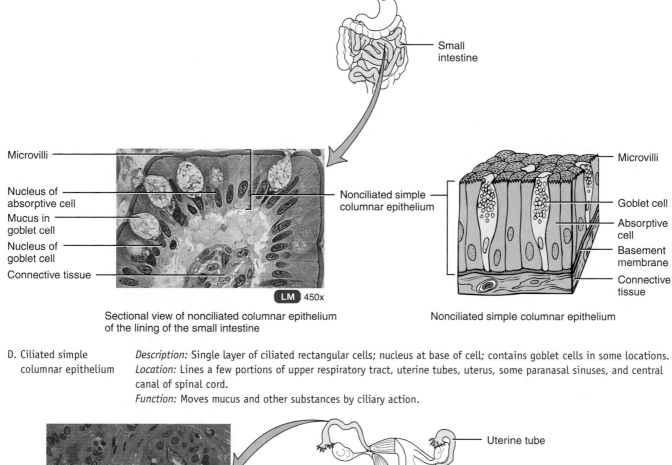

Sectional view of nonciliated columnar epithelium of the lining of the small intestine

Nonciliated simple columnar epithelium

D. Ciliated simple columnar epithelium

Description: Single layer of ciliated rectangular cells; nucleus at base of cell; contains goblet cells in some locations.
Location: Lines a few portions of upper respiratory tract, uterine tubes, uterus, some paranasal sinuses, and central canal of spinal cord.
Function: Moves mucus and other substances by ciliary action.

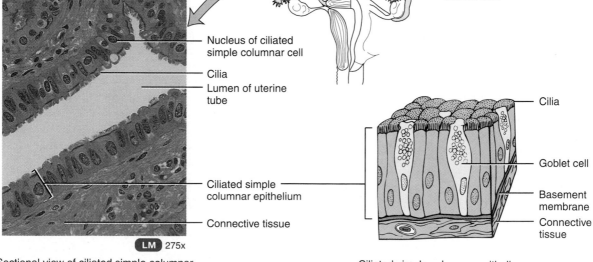

Sectional view of ciliated simple columnar epithelium of uterine tube

Ciliated simple columnar epithelium

Table 4.1 Epithelial Tissues (continued)

COVERING AND LINING EPITHELIUM

E. Stratified squamous epithelium

Description: Several layers of cells; cuboidal to columnar shape in deep layers; squamous cells in superficial layers; basal cells replace surface cells as they are lost.

Location: Keratinized variety forms superficial layer of skin; nonkeratinized variety lines wet surfaces, such as lining of the mouth, esophagus, part of epiglottis, and vagina, and covers the tongue.

Function: Protection.

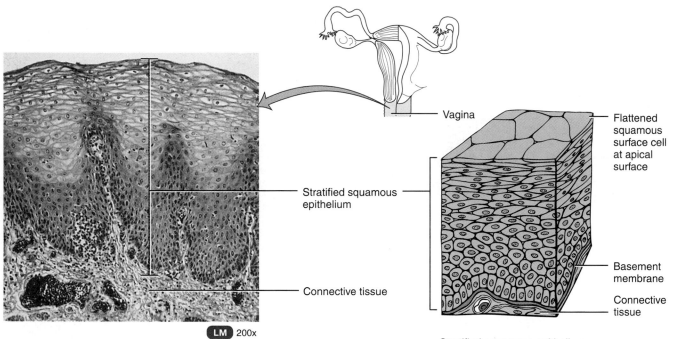

Vagina

Stratified squamous epithelium

Connective tissue

Flattened squamous surface cell at apical surface

Basement membrane

Connective tissue

LM 200x

Sectional view of stratified squamous epithelium of vagina

Stratified squamous epithelium

Table 4.1 (continued)

COVERING AND LINING EPITHELIUM

F. Stratified cuboidal epithelium

Description: Two or more layers of cells in which the cells at the apical surface are cube-shaped.
Location: Ducts of adult sweat glands and part of male urethra.
Function: Protection and limited secretion and absorbtion.

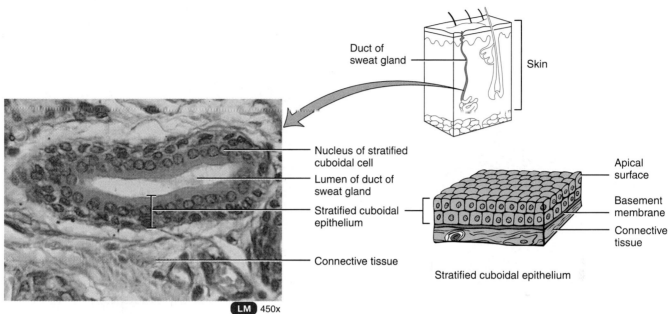

Sectional view of stratified cuboidal epithelium of the duct of a sweat gland

Stratified cuboidal epithelium

G. Stratified columnar epithelium

Description: Several layers of polyhedral cells; columnar cells are only in the apical layer.
Location: Lines part of urethra, large excretory ducts of some glands, small areas in anal mucous membrane, and a part of the conjunctiva of the eye.
Function: Protection and secretion.

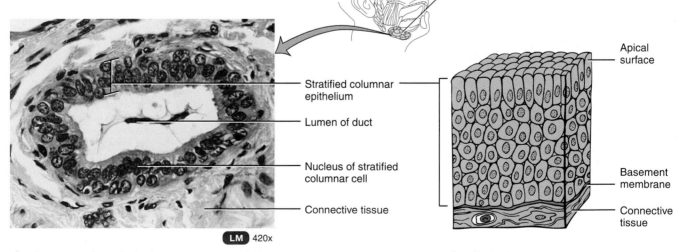

Sectional view of stratified columnar epithelium of the duct of a submandibular salivary gland

Stratified columnar epithelium

Table 4.1 Epithelial Tissues (continued)

COVERING AND LINING EPITHELIUM

H. Transitional
    epitheliumm

Description: Appearance is variable (transitional); shape of cells at apical surface ranges from squamous (when stretched) to cuboidal (when relaxed).
Location: Lines urinary bladder and portions of ureters and urethra.
Function: Permits distention.

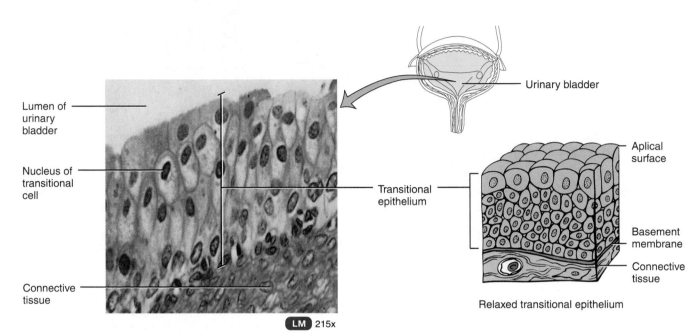

Sectional view of transitional epithelium of urinary bladder in relaxed state

Relaxed transitional epithelium

Table 4.1 (continued)

COVERING AND LINING EPITHELIUM

I. Pseudostratified columnar epithelium

Description: Not a true stratified tissue; nuclei of cells are at different levels; all cells are attached to basement membrane, but not all reach the apical surface.

Location: Pseudostratified ciliated columnar epithelium lines the airways of most of upper respiratory tract; pseudo-stratified nonciliated columnar epithelium lines larger ducts of many glands, epididymis, and part of male urethra.

Function: Secretion and movement of mucus by ciliary action.

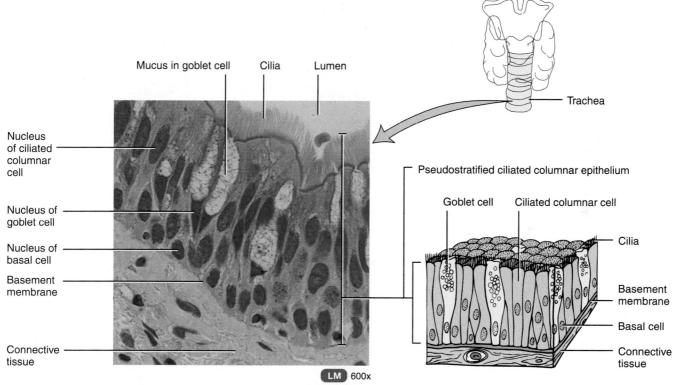

Sectional view of pseudostratified columnar epithelium of trachea

Pseudostratified columnar epithelium

Table 4.1 Epithelial Tissues (continued)

GLANDULAR EPITHELIUM

J. Endocrine glands
Description: Secretory products (hormones) diffuse into blood after passing through interstitial fluid.
Location: Examples include pituitary gland at base of brain, pineal gland in brain, thyroid and parathyroid glands near larynx (voice box), adrenal glands superior to kidneys, pancreas near stomach, ovaries in pelvic cavity, testes in scrotum, and thymus gland in thoracic cavity.
Function: Produce hormones that regulate various body activities.

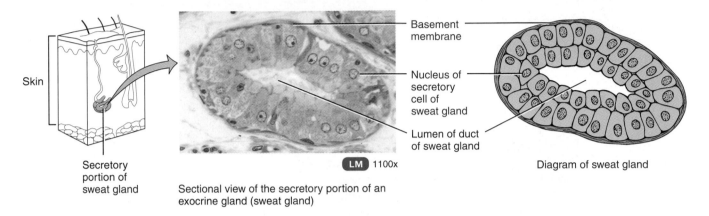

Skin

Secretory portion of sweat gland

Basement membrane

Nucleus of secretory cell of sweat gland

Lumen of duct of sweat gland

LM 1100x

Sectional view of the secretory portion of an exocrine gland (sweat gland)

Diagram of sweat gland

K. Exocrine glands
Description: Secretory products released into ducts.
Location: Sweat, oil, ear wax, and mammary glands of the skin; digestive glands such as salivary glands, which secrete into mouth cavity, and pancreas, which secretes into the small intestine.
Function: Produce mucus, perspiration, oil, ear wax, milk, saliva, or digestive enzymes.

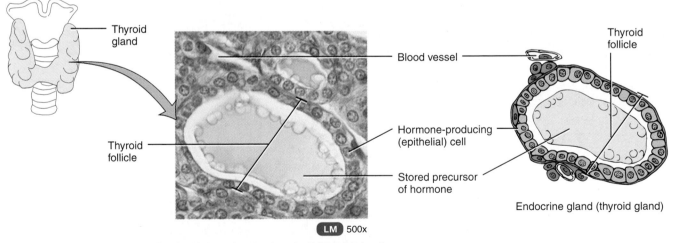

Thyroid gland

Thyroid follicle

Blood vessel

Hormone-producing (epithelial) cell

Stored precursor of hormone

Thyroid follicle

LM 500x

Sectional view of endocrine gland (thyroid gland)

Endocrine gland (thyroid gland)

Simple Columnar Epithelium When viewed from the side, the cells appear rectangular with oval nuclei near the base of the cells. Simple columnar epithelium exists in two forms: nonciliated simple columnar epithelium and ciliated simple columnar epithelium.

Nonciliated simple columnar epithelium contains microvilli and goblet cells (Table 4.1C). **Microvilli** are microscopic fingerlike cytoplasmic projections that increase the surface area of the plasma membrane (see Figure 3.2 on page 62). Their presence increases the rate of absorption by the cell. **Goblet cells** are modified columnar cells that secrete mucus, a slightly sticky fluid. Before it is released, mucus accumulates in the upper portion of the cell, causing that area to bulge out. The whole cell then resembles a goblet or wine glass. Secreted mucus serves as a lubricant for the linings of the digestive, respiratory, reproductive, and most of the urinary tracts. Mucus also helps to trap dust entering the respiratory tract, and it prevents destruction of the stomach lining by digestive enzymes.

Ciliated simple columnar epithelium (Table 4.1D) contains cells with cilia. In a few parts of the upper respiratory tract, ciliated columnar cells are interspersed with goblet cells. Mucus secreted by the goblet cells forms a film over the respiratory surface that traps inhaled foreign particles. The cilia wave in unison and move the mucus, and any foreign particles, toward the throat, where it can be coughed up and swallowed or spit out. Cilia also help to move an ovum through the uterine tubes into the uterus.

Stratified Epithelium

In contrast to simple epithelium, stratified epithelium has at least two layers of cells. Thus, it is more durable and can better protect underlying tissues. Some cells of stratified epithelia also produce secretions. The name of the specific kind of stratified epithelium depends on the shape of the cells at the apical surface.

Stratified Squamous Epithelium

Cells in the superficial layers of this type of epithelium are flat, whereas in the deep layers, cells vary in shape from cuboidal to columnar (Table 4.1E). The basal (deepest) cells continually undergo cell division. As new cells grow, the cells of the basal layer are pushed upward toward the apical surface. As they move farther from the deep layer and from their blood supply in the underlying connective tissue, they become dehydrated, shrunken, and harder. At the surface, the cells lose their cell junctions and are sloughed off, but they are replaced as new cells continually emerge from the basal cells.

Stratified squamous epithelium exists in both keratinized and nonkeratinized forms. In *keratinized stratified squamous epithelium,* a tough layer of keratin is deposited in the surface cells. **Keratin** is a protein that is resistant to friction and helps repel bacteria. *Nonkeratinized stratified squamous epithelium* does not contain keratin and remains moist.

CLINICAL APPLICATION
Papanicolaou Smears

A **Papanicolaou smear** (pa-pa-NI-kō-lō) or **Pap smear** involves collecting and microscopically examining cells that have sloughed off the surface of a tissue. A very common type of Pap smear involves examining the cells present in the secretions of the cervix and vagina. This type of Pap smear is performed mainly to detect early changes in the cells of the female reproductive system that may indicate cancer or a precancerous condition. Annual Pap smears are recommended for women over age 18 (earlier if sexually active) as part of a routine pelvic exam. ■

Stratified Cuboidal Epithelium This fairly rare type of epithelium sometimes consists of more than two layers of cells (Table 4.1F). Its function is mainly protective but also functions in limited secretion and absorption.

Stratified Columnar Epithelium Like stratified cuboidal epithelium, this type of tissue also is uncommon. Usually the basal layer or layers consist of shortened, irregularly polyhedral cells; only the apical cells are columnar in form (Table 4.1G). This type of epithelium functions in protection and secretion.

Transitional Epithelium This kind of epithelium is variable in appearance, depending on whether the organ it lines is relaxed or distended (stretched). In its relaxed state (Table 4.1H), transitional epithelium looks similar to stratified cuboidal epithelium, except that the apical cells tend to be large and rounded. This allows the tissue to be distended without the outer cells separating from one another. As the cells are stretched, they become squamous in shape, giving the appearance of stratified squamous epithelium. Because of its elasticity, transitional epithelium lines hollow structures that are subjected to expansion from within, such as the urinary bladder. Its function is to help prevent rupture of these organs.

Pseudostratified Columnar Epithelium

The third category of covering and lining epithelium is called pseudostratified columnar epithelium (Table 4.1I). The nuclei of the cells are at various depths. Even though all the cells are attached to the basement membrane in a single layer, some cells do not reach the apical surface. When viewed from the side, these features give the false impression of a multilayered tissue—thus the name *pseudo*stratified epithelium (*pseudo-* = false). In *pseudostratified ciliated columnar epithelium,* the cells that reach the apical surface either secrete mucus (goblet cells) or bear cilia that sweep away mucus and trapped foreign particles for eventual elimination from the body. *Pseudostratified nonciliated columnar epithelium* contains no cilia or goblet cells.

Figure 4.3 Multicellular exocrine glands. Gold represents the secretory portion; blue represents the duct.

🔑 **Structural classification of multicellular exocrine glands is based on the shape of the secreting portion and the branching pattern of the duct.**

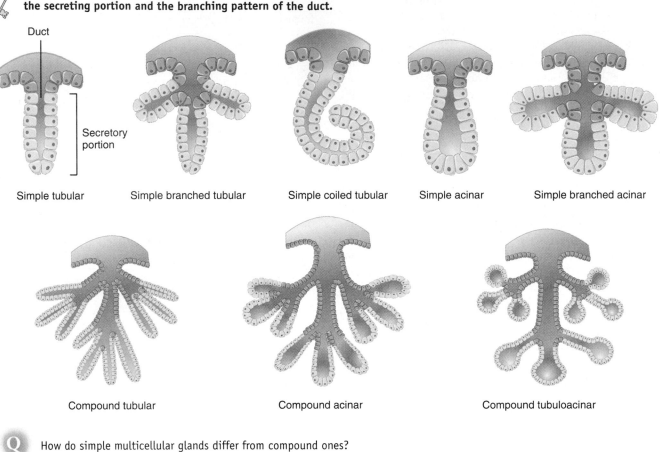

Simple tubular   Simple branched tubular   Simple coiled tubular   Simple acinar   Simple branched acinar

Compound tubular   Compound acinar   Compound tubuloacinar

Q How do simple multicellular glands differ from compound ones?

Glandular Epithelium

The function of glandular epithelium is secretion, which is accomplished by glandular cells that often lie in clusters deep to the covering and lining epithelium. A **gland** may consist of one cell or a group of highly specialized epithelial cells that secrete substances into ducts, onto a surface, or into the blood. The production of such substances always requires active work by the cells and results in an expenditure of energy. All glands of the body are classified as either endocrine or exocrine.

The secretions of **endocrine glands** (Table 4.1J) enter the extracellular fluid and then diffuse directly into the bloodstream without flowing through a duct. These secretions, called *hormones,* regulate many metabolic and physiological activities to maintain homeostasis. The pituitary, thyroid, and adrenal glands are examples of endocrine glands. Endocrine glands will be described in detail in Chapter 18.

Exocrine glands (*exo-* = outside; *-crine* = secretion; Table 4.1K) secrete their products into ducts (tubes) that empty at the surface of covering and lining epithelium or directly onto a free surface. The product of an exocrine gland may be released at the skin surface or into the lumen (cavity) of a hollow organ. The secretions of exocrine glands include mucus, sweat, oil, wax, and digestive enzymes. Examples of exocrine glands are sweat glands, which produce perspiration to help lower body temperature, and salivary glands, which secrete mucus and digestive enzymes.

Structural Classification of Exocrine Glands

Exocrine glands are classified into unicellular and multicellular types. **Unicellular glands** are single-celled. An important unicellular exocrine gland is the goblet cell. Although goblet cells do not contain ducts, they are classified as unicellular mucus-secreting exocrine glands. Most glands are **multicellular glands,** composed of cells that form a distinctive microscopic structure or macroscopic organ. Examples are sweat, oil, and salivary glands.

Multicellular glands are categorized according to two criteria: whether their glands are branched or unbranched, and the shape of their secretory portions (Figure 4.3). If the duct of the gland does not branch, then it is a **simple gland;** if the duct branches, it is a **compound gland.** Glands with tubular secretory portions are **tubular glands;** those with flasklike secretory portions are **acinar glands** (AS-i-nar; *acin-* = berry). **Tubuloacinar glands** have both tubular and flasklike secretory portions.

Combinations of the degree of duct branching and the shape of the secretory portion are the criteria for the following structural classification scheme for multicellular exocrine glands:

I. Simple gland: single, nonbranched duct.
   A. Simple tubular. Secretory portion is straight and tubular. Example: large intestinal glands.
   B. Simple branched tubular. Secretory portion is branched and tubular. Example: gastric glands.
   C. Simple coiled tubular. Secretory portion is coiled. Example: sweat glands.
   D. Simple acinar. Secretory portion is flasklike. Example: glands of penile urethra.
   E. Simple branched acinar. Secretory portion is branched and flasklike. Example: sebaceous (oil) glands.
II. Compound gland: branched duct.
   A. Compound tubular. Secretory portion is tubular. Example: bulbourethral (Cowper's) glands.
   B. Compound acinar. Secretory portion is flasklike. Example: mammary glands.
   C. Compound tubuloacinar. Secretory portion is both tubular and flasklike. Example: acinar glands of the pancreas.

Functional Classification of Exocrine Glands

The functional classification of exocrine glands is based on whether a secretion is released by itself from a cell or whether the secretion consists of entire or partial glandular cells themselves. **Merocrine glands** (MER-ō-krin; *mero-* = a part) simply form the secretory product and release it from the cell. The secretory product is formed by ribosomes attached to rough ER; processed, sorted, and packaged by the Golgi complex; and released from the cell in secretory vesicles via exocytosis (Figure 4.4a). Most exocrine glands of the body produce merocrine secretions. Examples of merocrine glands are the salivary glands and pancreas. **Apocrine glands** (AP-ō-krin; *apo-* = from) accumulate their secretory product at the apical surface of the secreting cell. That portion of the cell pinches off from the rest of the cell to form the secretion (Figure 4.4b). The remaining part of the cell repairs itself and repeats the process. Electron micrographic studies have called into question whether humans have apocrine glands. What were once thought to be apocrine glands—for example, mammary glands—are probably merocrine glands. The cells of **holocrine glands** (HŌ-lō-krin; *holo-* = entire) accumulate a secretory product in their cytosol. As the secretory cell matures, it dies and becomes the

Figure 4.4 Functional classification of multicellular exocrine glands.

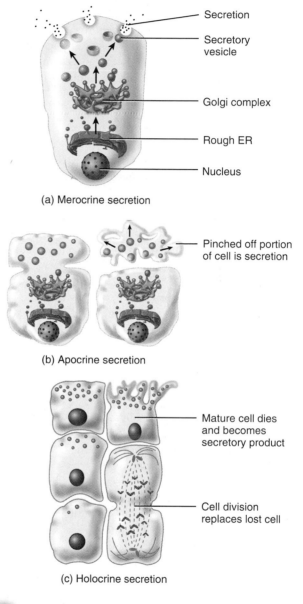

The functional classification of exocrine glands is based on whether a secretion is a product of a cell or consists of an entire or a partial glandular cell.

(a) Merocrine secretion

— Secretion
— Secretory vesicle
— Golgi complex
— Rough ER
— Nucleus

(b) Apocrine secretion

— Pinched off portion of cell is secretion

(c) Holocrine secretion

— Mature cell dies and becomes secretory product
— Cell division replaces lost cell

 What class of glands are sebaceous (oil) glands? Salivary glands?

secretory product (Figure 4.4c). The sloughed cell is replaced by a new cell. One example of a holocrine gland is a sebaceous (oil) gland of the skin.

1. Describe the various layering arrangements and cell shapes of epithelium.
2. What characteristics are common to all epithelial tissues?

3. Describe how the structure of the following kinds of epithelium is related to their functions: simple squamous, simple cuboidal, simple columnar (nonciliated and ciliated), stratified squamous (keratinized and nonkeratinized), stratified cuboidal, stratified columnar, transitional, and ciliated and nonciliated pseudostratified columnar.
4. Define endothelium and mesothelium.
5. What is a gland? Distinguish between endocrine glands and exocrine glands, and name and give examples of the three functional classes of exocrine glands.

CONNECTIVE TISSUE

OBJECTIVE

• *Describe the general features of connective tissue and the structure, location, and function of the various types of connective tissues.*

Connective tissue is the most abundant and widely distributed tissue in the body. In its various forms, connective tissue has a variety of functions: It binds together, supports, and strengthens other body tissues; protects and insulates internal organs; compartmentalizes structures such as skeletal muscles; is the major transport system within the body (blood, a fluid connective tissue); and is the major site of stored energy reserves (adipose, or fat, tissue).

General Features of Connective Tissue

Connective tissue consists of two basic elements: cells and a matrix. A **matrix** generally consists of fibers and a ground substance, the component of a connective tissue that occupies the space between the cells and the fibers. The matrix, which tends to prevent tissue cells from touching one another, may be fluid, semifluid, gelatinous, fibrous, or calcified. It is usually secreted by the connective tissue cells and determines the tissue's qualities. In blood, the matrix (which is not secreted by blood cells) is liquid. In cartilage, it is firm but pliable. In bone, it is hard and not pliable.

In contrast to epithelia, connective tissues do not usually occur on free surfaces, such as the covering or lining of internal organs, the lining of a body cavity, or the external surface of the body. However, joint cavities are lined by a type of connective tissue called areolar connective tissue. Also, unlike epithelia, connective tissues usually are highly vascular; that is, they have a rich blood supply. Exceptions include cartilage, which is avascular, and tendons, which have a scanty blood supply. Except for cartilage, connective tissues, like epithelia, have a nerve supply.

Next we examine in detail the components of connective tissue.

Components of Connective Tissue

Connective Tissue Cells

The cells in connective tissue are derived from mesodermal embryonic cells called mesenchymal cells. Each major type of connective tissue contains an immature class of cells whose name ends in -*blast*, which means "to bud or sprout." These immature cells are called *fibroblasts* in loose and dense connective tissue, *chondroblasts* in cartilage, and *osteoblasts* in bone. Blast cells retain the capacity for cell division and secrete the matrix that is characteristic of the tissue. In cartilage and bone, once the matrix is produced, the fibroblasts differentiate into mature cells whose names end in -*cyte*, such as chondrocytes and osteocytes. Mature cells have reduced capacity for cell division and matrix formation and are mostly involved in maintaining the matrix.

The types of cells present in various connective tissues depend on the type of tissue and include the following (Figure 4.5):

1. **Fibroblasts** (FĪ-brō-blasts; *fibro-* = fibers) are large flat, spindle-shaped cells with branching processes. They are present in all connective tissues, and they usually are the most numerous connective tissue cells. Fibroblasts migrate through the connective tissue, secreting the fibers and ground substance of the matrix.
2. **Macrophages** (MAK-rō-fā-jez; *macro-* = large; *-phages* = eaters), or *histiocytes,* develop from monocytes, a type of white blood cell. Macrophages have an irregular shape with short branching projections and are capable of engulfing bacteria and cellular debris by phagocytosis. Some are **fixed macrophages,** which means they reside in a particular tissue—for example, alveolar macrophages in the lungs or spleen macrophages in the spleen. Others are **wandering macrophages,** which roam the tissues and gather at sites of infection or inflammation.
3. **Plasma cells** are small and either round or irregular in shape. They develop from a type of white blood cell called a B lymphocyte. Plasma cells secrete *antibodies,* proteins that attack or neutralize foreign substances in the body. Thus, plasma cells are an important part of the body's immune system. Although they are found in many places in the body, most plasma cells reside in connective tissues, especially in the gastrointestinal tract and the mammary glands.
4. **Mast cells** are abundant alongside the blood vessels that supply connective tissue. They produce *histamine,* a chemical that dilates small blood vessels as part of the body's reaction to injury or infection.
5. **Adipocytes,** also called fat cells, are connective tissue cells that store triglycerides (fats). They are found below the skin and around organs such as the heart and kidneys.

Figure 4.5 Representative cells and fibers present in connective tissues.

 Fibroblasts are usually the most numerous connective tissue cells.

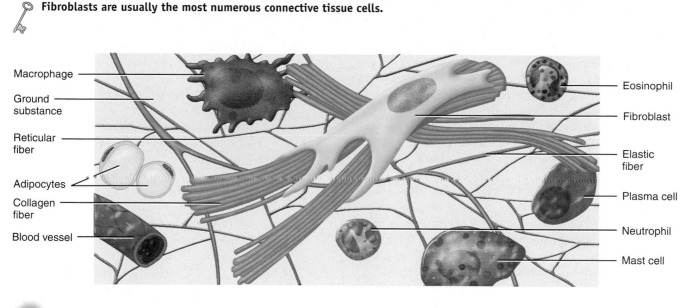

Ⓠ What is the function of fibroblasts?

---

6. **White blood cells** are not found in significant numbers in normal connective tissue. However, in response to certain conditions they migrate from blood into connective tissues, where their numbers increase significantly. For example, *neutrophils* increase in number at sites of infection, and *eosinophils* increase in number in response to allergic conditions and parasitic invasions.

Connective Tissue Matrix

Each type of connective tissue has unique properties, due to accumulation of specific matrix materials between the cells. Matrix derives its properties from a fluid, gel, or solid *ground substance* that contains protein *fibers.*

Ground Substance As previously mentioned, the ground substance is the component of a connective tissue between the cells and fibers. The ground substance supports cells, binds them together, and provides a medium through which substances are exchanged between the blood and cells. Until recently, it was thought to function mainly as an inert scaffolding that supports tissues. Now it is clear, however, that the ground substance plays an active role in how tissues develop, migrate, proliferate, and change shape, and in how they carry out their metabolic functions.

Ground substance contains an assortment of large molecules, many of which are complex combinations of polysaccharides and proteins. For example, **hyaluronic acid** (hī-a-loo-RON-ic) is a viscous, slippery substance that binds cells together, lubricates joints, and helps maintain the shape of

the eyeballs. It also appears to play a role in helping phagocytes migrate through connective tissue during development and wound repair. White blood cells, sperm cells, and some bacteria produce *hyaluronidase,* an enzyme that breaks apart hyaluronic acid and causes the ground substance of connective tissue to become watery. The ability to produce hyaluronidase enables white blood cells to move through connective tissues, and sperm cells to penetrate the ovum during fertilization. It also accounts for how bacteria spread through connective tissues. **Chondroitin sulfate** (kon-DROY-tin) is a jellylike substance that provides support and adhesiveness in cartilage, bone, skin, and blood vessels. The skin, tendons, blood vessels, and heart valves contain **dermatan sulfate,** whereas bone, cartilage, and the cornea of the eye contain **keratan sulfate.** Also present in the ground substance are **adhesion proteins,** which are responsible for linking components of the ground substance to each other and to the surfaces of cells. The principal adhesion protein of connective tissue is *fibronectin,* which binds to both collagen fibers (discussed shortly) and ground substance and crosslinks them together. It also attaches cells to the ground substance.

Fibers Fibers in the matrix strengthen and support connective tissues. Three types of fibers are embedded in the matrix between the cells: collagen fibers, elastic fibers, and reticular fibers.

Collagen fibers (*colla* = glue), of which there are at least five different types, are very strong and resist pulling forces, but they are not stiff, which promotes tissue flexibility. These fibers often occur in bundles lying parallel to one another (see Figure 4.5). The bundle arrangement affords great strength. Chemically, collagen fibers consist of the protein *collagen.* This is the most abundant protein in your body, representing about 25% of total protein. Collagen fibers are found in most types of connective tissues, especially bone, cartilage, tendons, and ligaments.

Elastic fibers, which are smaller in diameter than collagen fibers, branch and join together to form a network within a tissue. An elastic fiber consists of molecules of a protein called *elastin* surrounded by a glycoprotein named *fibrillin,* which is essential to the stability of an elastic fiber. Because of their unique molecular structure, elastic fibers are strong but can be stretched up to 150% of their relaxed length without breaking. Equally important, elastic fibers have the ability to return to their original shape after being stretched, a property called *elasticity.* Elastic fibers are plentiful in skin, blood vessel walls, and lung tissue.

Reticular fibers (*reticul-* = net), consisting of *collagen* and a coating of glycoprotein, provide support in the walls of blood vessels and form a network around fat cells, nerve fibers, and skeletal and smooth muscle cells. Produced by fibroblasts, they are much thinner than collagen fibers and form branching networks. Like collagen fibers, reticular fibers provide support and strength and also form the **stroma** (= bed or covering) or supporting framework of many soft organs, such as the spleen and lymph nodes. These fibers also help form the basement membrane.

CLINICAL APPLICATION
Marfan Syndrome

Marfan syndrome (MAR-fan) is an inherited disorder caused by a defective fibrillin gene. The result is abnormal development of elastic fibers. Tissues rich in elastic fibers are malformed or weakened. Structures affected most seriously are the covering layer of bones (periosteum), the ligament that suspends the lens of the eye, and the walls of the large arteries. People with Marfan syndrome tend to be tall and have disproportionately long arms, legs, fingers, and toes. A common symptom is blurred vision caused by displacement of the lens of the eye. The most life-threatening complication of Marfan syndrome is weakening of the aorta (the main artery that emerges from the heart), which can suddenly burst. ■

Classification of Connective Tissues

Classifying connective tissues can be challenging because of the diversity of cells and matrix present, and the differences in their relative proportions. Thus, the grouping of connective tissues into categories is not always clear-cut. We will classify them as follows:

I. Embryonic connective tissue
  A. Mesenchyme
  B. Mucous connective tissue
II. Mature connective tissue
  A. Loose connective tissue
    1. Areolar connective tissue
    2. Adipose tissue
    3. Reticular connective tissue
  B. Dense connective tissue
    1. Dense regular connective tissue
    2. Dense irregular connective tissue
    3. Elastic connective tissue
  C. Cartilage
    1. Hyaline cartilage
    2. Fibrocartilage
    3. Elastic cartilage
  D. Bone tissue
  E. Blood tissue
  F. Lymph

Note that our classification scheme has two major subclasses of connective tissue: embryonic and mature. **Embryonic connective tissue** is present primarily in the *embryo,* the developing human from fertilization through the first two months of pregnancy, and in the *fetus,* the developing human from the third month of pregnancy to birth.

One example of embryonic connective tissue found almost exclusively in the embryo is **mesenchyme** (MEZ-en-kīm), the tissue from which all other connective tissues eventually arise (Table 4.2A). Mesenchyme is composed of irregularly shaped cells, a semifluid ground substance, and delicate reticular fibers. Another kind of embryonic tissue is **mucous connective tissue (Wharton's jelly),** found primarily in the umbilical cord of the fetus. Mucous connective tissue is a form of mesenchyme that contains widely scattered fibroblasts, a more viscous jellylike ground substance, and collagen fibers (Table 4.2B).

The second major subclass of connective tissue, **mature connective tissue,** is present in the newborn and has cells produced from mesenchyme. Mature connective tissue is of several types, which we explore next.

Types of Mature Connective Tissue

The six types of mature connective tissue are (1) loose connective tissue, (2) dense connective tissue, (3) cartilage, (4) bone tissue, (5) blood tissue, and (6) lymph. We now examine each in detail.

Loose Connective Tissue

In **loose connective tissue** the fibers are loosely woven, and many cells are present. The types of loose connective tissue are areolar connective tissue, adipose tissue, and reticular connective tissue.

Table 4.2 Embryonic Connective Tissue

A. Mesenchyme

Description: Consists of irregularly shaped mesenchymal cells embedded in a semifluid ground substance that contains reticular fibers.

Location: Under skin and along developing bones of embryo; some mesenchymal cells are found in adult connective tissue, especially along blood vessels.

Function: Forms all other kinds of connective tissue.

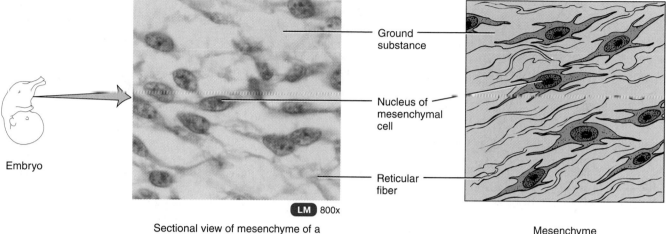

Embryo

Ground substance

Nucleus of mesenchymal cell

Reticular fiber

LM 800x

Sectional view of mesenchyme of a developing embryo

Mesenchyme

B. Mucous Connective Tissue

Description: Consists of widely scattered fibroblasts embedded in a viscous, jellylike ground substance that contains fine collagen fibers.

Location: Umbilical cord of fetus.

Function: Support.

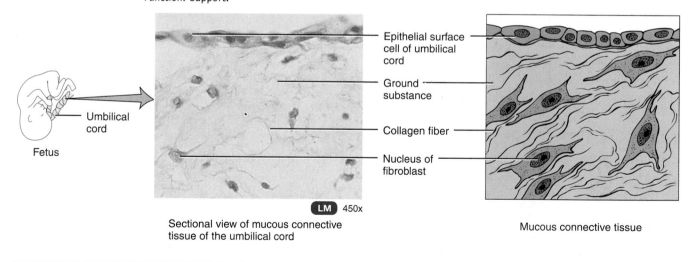

Umbilical cord

Fetus

Epithelial surface cell of umbilical cord

Ground substance

Collagen fiber

Nucleus of fibroblast

LM 450x

Sectional view of mucous connective tissue of the umbilical cord

Mucous connective tissue

Areolar Connective Tissue One of the most widely distributed connective tissues in the body is **areolar connective tissue** (a-RĒ-ō-lar; *areol-* = a small space). It contains several kinds of cells, including fibroblasts, macrophages, plasma cells, mast cells, adipocytes, and a few white blood cells (Table 4.3A). All three types of fibers—collagen, elastic, and reticular—are arranged randomly throughout the tissue. The ground substance contains hyaluronic acid, chondroitin sulfate, dermatan sulfate, and keratan sulfate. Combined with adipose tissue, areolar connective tissue forms the *subcutaneous layer*—the layer of tissue that attaches the skin to underlying tissues and organs.

Adipose Tissue Adipose tissue is a loose connective tissue in which the cells, called **adipocytes** (*adipo-* = fat) are specialized for storage of triglycerides (fats) (Table 4.3B). Adipocytes are derived from fibroblasts. Because the cell fills

Text continues on page 127

Table 4.3 Mature Connective Tissue

LOOSE CONNECTIVE TISSUE

A. Areolar connective tissue

Description: Consists of fibers (collagen, elastic, and reticular) and several kinds of cells (fibroblasts, macrophages, plasma cells, adipocytes, and mast cells) embedded in a semifluid ground substance.
Location: Subcutaneous layer of skin; papillary (superficial) region of dermis of skin; lamina propria of mucous membranes; and around blood vessels, nerves, and body organs.
Function: Strength, elasticity, and support.

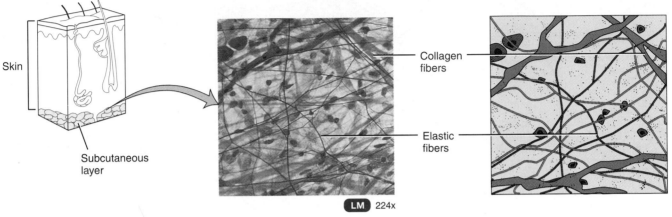

Sectional view of subcutaneous areolar connective tissue

Areolar connective tissue

B. Adipose tissue

Description: Consists of adipocytes, cells that are specialized to store triglycerides (fats) in a large central area in their cytoplasm; nuclei are peripherally located.
Location: Subcutaneous layer deep to skin, around heart and kidneys, yellow marrow of long bones, and padding around joints and behind eyeball in eye socket.
Function: Reduces heat loss through skin, serves as an energy reserve, supports, and protects. In newborns, brown fat generates considerable heat that helps maintain proper body temperature.

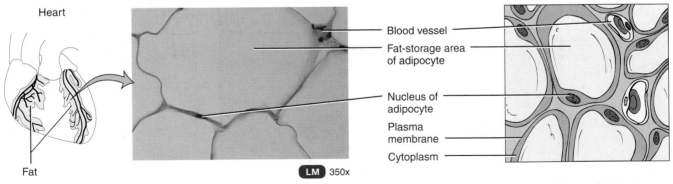

Sectional view of adipose tissue showing adipocytes of white fat

Adipose tissue

Table 4.3 (continued)

LOOSE CONNECTIVE TISSUE

C. Reticular connective
   tissue

Description: Consists of a network of interlacing reticular fibers and reticular cells.
Location: Stroma (supporting framework) of liver, spleen, lymph nodes; portion of bone marrow that gives rise to blood cells; reticular lamina of the basement membrane; and around blood vessels and muscle.
Function: Forms stroma of organs; binds together smooth muscle tissue cells.

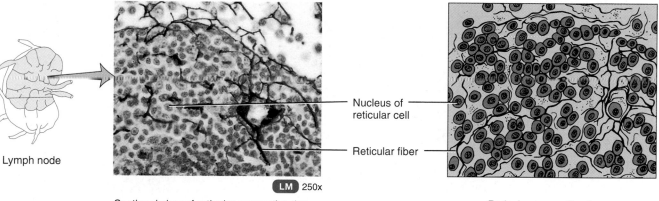

Lymph node

Nucleus of reticular cell

Reticular fiber

LM 250x

Sectional view of reticular connective tissue of a lymph node

Reticular connective tissue

DENSE CONNECTIVE TISSUE

D. Dense regular
   connective tissue

Description: Matrix looks shiny white; consists of predominantly collagen fibers arranged in bundles; fibroblasts present in rows between bundles.
Location: Forms tendons (attach muscle to bone), most ligaments (attach bone to bone), and aponeuroses (sheetlike tendons that attach muscle to muscle or muscle to bone).
Function: Provides strong attachment between various structures.

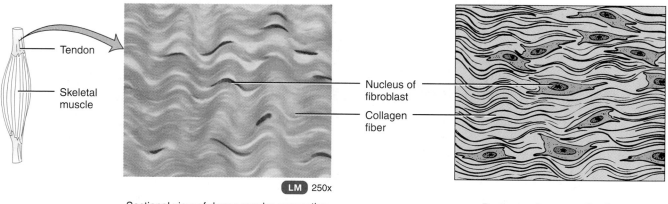

Tendon

Skeletal muscle

Nucleus of fibroblast

Collagen fiber

LM 250x

Sectional view of dense regular connective tissue of a tendon

Dense regular connective tissue

Table 4.3 Mature Connective Tissue (continued)

DENSE CONNECTIVE TISSUE

E. Dense irregular connective tissue

Description: Consists predominantly of collagen fibers, randomly arranged, and a few fibroblasts.

Location: Fasciae (tissue beneath skin and around muscles and other organs), reticular (deeper) region of dermis of skin, periosteum of bone, perichondrium of cartilage, joint capsules, membrane capsules around various organs (kidneys, liver, testes, lymph nodes), pericardium of the heart, and heart valves.

Function: Provides strength.

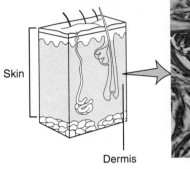

Skin

Dermis

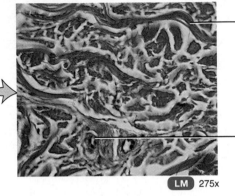

LM 275x

Sectional view of dense irregular connective tissue of reticular region of dermis

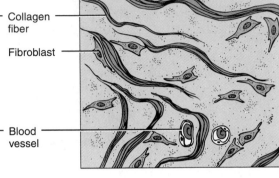

Collagen fiber

Fibroblast

Blood vessel

Dense irregular connective tissue

F. Elastic connective tissue

Description: Consists of predominantly freely branching elastic fibers; fibroblasts present in spaces between fibers.

Location: Lung tissue, walls of elastic arteries, trachea, bronchial tubes, true vocal cords, suspensory ligament of penis, and ligaments between vertebrae.

Function: Allows stretching of various organs.

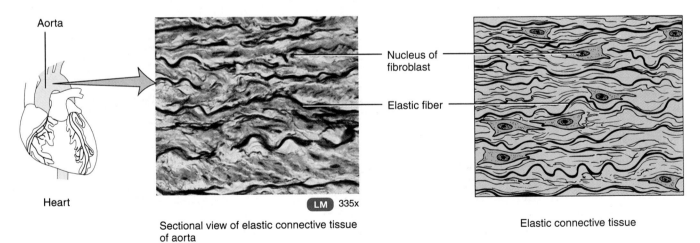

Aorta

Heart

LM 335x

Sectional view of elastic connective tissue of aorta

Nucleus of fibroblast

Elastic fiber

Elastic connective tissue

Table 4.3 (continued)

CARTILAGE

G. Hyaline cartilage

Description: Consists of a bluish-white, shiny ground substance with fine collagen fibers; contains numerous chondrocytes; most abundant type of cartilage.
Location: Ends of long bones, anterior ends of ribs, nose, parts of larynx, trachea, bronchi, bronchial tubes, and embryonic skeleton.
Function: Provides smooth surfaces for movement at joints, as well as flexibility and support.

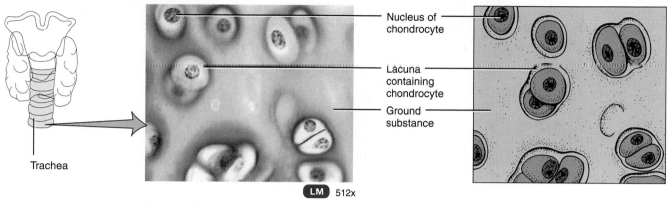

Trachea

Nucleus of chondrocyte

Lacuna containing chondrocyte

Ground substance

LM 512x

Sectional view of hyaline cartilage of trachea

Hyaline cartilage

H. Fibrocartilage

Description: Consists of chondrocytes scattered among bundles of collagen fibers within the matrix.
Location: Pubic symphysis (point where hipbones join anteriorly), intervertebral discs (discs between vertebrae), menisci (cartilage pads) of knee, and portions of tendons that insert into cartilage.
Function: Support and fusion.

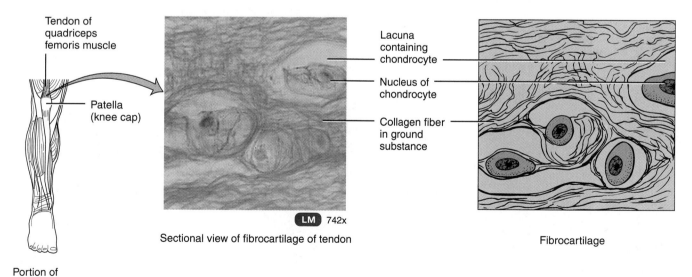

Tendon of quadriceps femoris muscle

Patella (knee cap)

Lacuna containing chondrocyte

Nucleus of chondrocyte

Collagen fiber in ground substance

LM 742x

Sectional view of fibrocartilage of tendon

Fibrocartilage

Portion of right lower limb

Table 4.3 Mature Connective Tissue (continued)

CARTILAGE

I. Elastic cartilage

Description: Consists of chondrocytes located in a threadlike network of elastic fibers within the matrix.
Location: Lid on top of larynx (epiglottis), external ear (auricle), and auditory (Eustachian) tubes.
Function: Gives support and maintains shape.

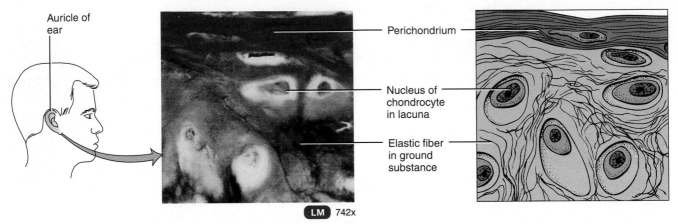

Auricle of ear

Perichondrium

Nucleus of chondrocyte in lacuna

Elastic fiber in ground substance

LM 742x

Sectional view of elastic cartilage of auricle of ear

Elastic cartilage

BONE TISSUE

J. Compact bone

Description: Compact bone tissue consists of osteons (Haversian systems) that contain lamellae, lacunae, osteocytes, canaliculi, and central (Haversian) canals. By contrast, spongy bone tissue (see Figure 6.3 on page 164) consists of thin columns called trabeculae; spaces between trabeculae are filled with red bone marrow.
Location: Both compact and spongy bone tissue make up the various parts of bones of the body.
Function: Support, protection, storage; houses blood-forming tissue; serves as levers that act together with muscle tissue to enable movement.

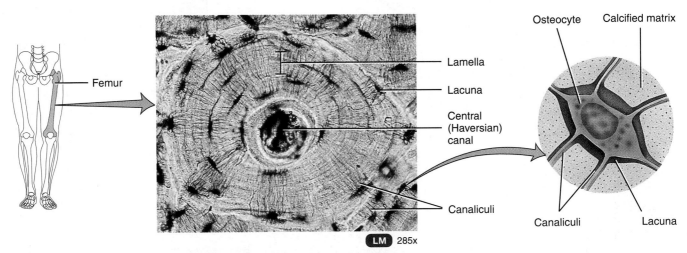

Femur

Osteocyte   Calcified matrix

Lamella

Lacuna

Central (Haversian) canal

Canaliculi

Canaliculi   Lacuna

LM 285x

Sectional view of an osteon (Haversian system) of femur (thigh bone)

Table 4.3 (continued)

BLOOD TISSUE

K. Blood

Description: Consists of plasma and formed elements. The formed elements are red blood cells (erythrocytes), white blood cells (leukocytes), and platelets.
Location: Within blood vessels (arteries, arterioles, capillaries, venules, and veins) and within the chambers of the heart.
Function: Red blood cells transport oxygen and carbon dioxide; white blood cells carry on phagocytosis and are involved in allergic reactions and immune system responses; platelets are essential for the clotting of blood.

Plasma

White blood cell
(leukocyte)

Red blood cell
(erythrocyte)

Platelet

LM 800x

Blood smear

Red blood cells

White blood cells

Platelets

Blood in blood vessels

up with a single, large triglyceride droplet, the cytoplasm and nucleus are pushed to the periphery of the cell. Adipose tissue is found wherever areolar connective tissue is located. Adipose tissue is a good insulator and can therefore reduce heat loss through the skin. It is a major energy reserve and generally supports and protects various organs.

Most of the fat in adults is **white fat,** the type just described. Another type, called **brown fat,** obtains its color from a very rich blood supply and numerous mitochondria, which contain colored pigments that participate in aerobic cellular respiration. Although brown fat is widespread in the fetus and infant, in adults only small amounts are present. Brown fat generates considerable heat and probably helps to maintain body temperature in the newborn. The heat generated by the many mitochondria is carried away to other body tissues by the extensive blood supply.

 CLINICAL APPLICATION
Liposuction

A surgical procedure, called **liposuction** (*lip-* = fat) or **suction lipectomy** (*-ectomy* = cut out), involves suctioning out small amounts of fat from various areas of the body. The technique can be used as a body-contouring procedure in a number of areas in which fat accumulates, such as the thighs, buttocks, arms, breasts, and abdomen. Post-surgical

complications that may develop include fat emboli (clots), infection, fluid depletion, injury to internal structures, and severe postoperative pain. ■

Reticular Connective Tissue   Reticular connective tissue consists of fine interlacing reticular fibers and reticular cells, cells that are connected to each other and form a network (Table 4.3C). Reticular connective tissue forms the stroma (supporting framework) of certain organs and helps bind together smooth muscle cells.

Dense Connective Tissue

Dense connective tissue contains more numerous, thicker, and denser fibers but considerably fewer cells than loose connective tissue. There are three types: dense regular connective tissue, dense irregular connective tissue, and elastic connective tissue.

Dense Regular Connective Tissue   In this tissue, bundles of collagen fibers are *regularly* arranged in parallel patterns that confer great strength (Table 4.3D). The tissue structure withstands pulling along the axis of the fibers. Fibroblasts, which produce the fibers and ground substance, appear in rows between the fibers. The tissue is silvery white and tough, yet somewhat pliable. Examples are tendons and most ligaments.

Dense Irregular Connective Tissue This tissue contains collagen fibers that are usually *irregularly* arranged (Table 4.3E), and it is found in parts of the body where pulling forces are exerted in various directions. The tissue usually occurs in sheets, such as in the dermis of the skin, which underlies the epidermis. Heart valves, the perichondrium (the membrane surrounding cartilage), and the periosteum (the membrane surrounding bone) are dense irregular connective tissues, although they have a fairly orderly arrangement of collagen fibers.

Elastic Connective Tissue Freely branching elastic fibers predominate in elastic connective tissue (Table 4.3F), giving the unstained tissue a yellowish color. Fibroblasts are present in the spaces between the fibers. Elastic connective tissue is quite strong and can recoil to its original shape after being stretched. Elasticity is important to the normal functioning of lung tissue, which recoils as you exhale, and elastic arteries, whose recoil between heart beats helps maintain blood flow.

Cartilage

Cartilage consists of a dense network of collagen fibers and elastic fibers firmly embedded in chondroitin sulfate, a rubbery component of the ground substance. Cartilage can endure considerably more stress than loose and dense connective tissues. Whereas the strength of cartilage is due to its collagen fibers, its resilience (ability to assume its original shape after deformation) is due to chondroitin sulfate.

The cells of mature cartilage, called **chondrocytes** (KON-drō-sīts; *chondro-* = cartilage), occur singly or in groups within spaces called **lacunae** (la-KOO-nē; = little lakes) in the matrix. (The singular is lacuna.) The surface of most cartilage is surrounded by a membrane of dense irregular connective tissue called the **perichondrium** (per′-i-KON-drē-um; *peri-* = around). Unlike other connective tissues, cartilage has no blood vessels or nerves, except in the perichondrium. Three kinds of cartilage are recognized: hyaline cartilage, fibrocartilage, and elastic cartilage.

Hyaline Cartilage This type of cartilage contains a resilient gel as its ground substance and appears in the body as a bluish-white, shiny substance. The fine collagen fibers are not visible with ordinary staining techniques, and prominent chondrocytes are found in lacunae (Table 4.3G). Most hyaline cartilage is surrounded by a perichondrium. The exceptions are the articular cartilage in joints and at the epiphyseal plates, the regions where bones lengthen as a person grows. Hyaline cartilage is the most abundant cartilage in the body. It affords flexibility and support and, at joints, reduces friction and absorbs shock. Hyaline cartilage is the weakest of the three types of cartilage.

Fibrocartilage Chondrocytes are scattered among clearly visible bundles of collagen fibers within the matrix of this type of cartilage (Table 4.3H). Fibrocartilage lacks a perichondrium. This tissue combines strength and rigidity and is the strongest of the three types of cartilage. One location for fibrocartilage is the discs between backbones.

Elastic Cartilage In this tissue, chondrocytes are located within a threadlike network of elastic fibers within the matrix (Table 4.3I). A perichondrium is present. Elastic cartilage provides strength and elasticity and maintains the shape of certain structures, such as the external ear.

Growth and Repair of Cartilage Cartilage grows slowly; metabolically, it is a relatively inactive tissue. When injured or inflamed, cartilage repair proceeds slowly, in large part because cartilage is avascular. Substances needed for repair and blood cells that participate in tissue repair must diffuse or migrate from the perichondrium into the cartilage. The growth of cartilage follows two basic patterns: interstitial growth and appositional growth.

In **interstitial growth,** the cartilage increases rapidly in size due to the division of existing chondrocytes and the continuous deposition of increasing amounts of matrix by the chondrocytes. These events cause the cartilage to expand from within—thus the term "*inter*stitial." This growth pattern occurs while the cartilage is young and pliable—that is, during childhood and adolescence.

In **appositional growth,** activity of cells in the inner chondrogenic layer of the perichondrium leads to growth. The deeper cells of the perichondrium, the fibroblasts, divide; some differentiate into chondroblasts. As differentiation continues, the chondroblasts surround themselves with matrix and become chondrocytes. As a result, matrix accumulates beneath the perichondrium on the surface of the cartilage, causing it to grow in width. Appositional growth starts later than interstitial growth and continues through adolescence.

Bone Tissue

Cartilage, joints, and bone make up the skeletal system. Bones are organs composed of several different connective tissues, including **bone** or **osseous tissue** (OS-ē-us), the periosteum, red and yellow bone marrow, and the endosteum (a membrane that lines a space within bone that stores yellow bone marrow). Bone tissue is classified as either compact or spongy, depending on how the matrix and cells are organized.

The basic unit of compact bone is an **osteon** or **Haversian system** (Table 4.3J). Each osteon is composed of four parts:

- The **lamellae** (la-MEL-lē; = little plates) are concentric rings of matrix that consist of mineral salts (mostly calcium and phosphates), which give bone its hardness, and collagen fibers, which give bone its strength.

- **Lacunae** (la-COO-nē) are small spaces between lamellae that contain mature bone cells called **osteocytes.**
- Projecting from the lacunae are **canaliculi** (CAN-a-lik-ū-lī; = little canals), networks of minute canals containing the processes of osteocytes. Canaliculi provide routes for nutrients to reach osteocytes and for wastes to leave them.
- A **central (Haversian) canal** contains blood vessels and nerves.

Spongy bone has no osteons; instead, it consists of columns of bone called **trabeculae** (tra-BEK-ū-lē; = little beams), which contain lamellae, osteocytes, lacunae, and canaliculi. Spaces between lamellae are filled with red bone marrow. Chapter 6 presents details of bone tissue histology.

The skeletal system has several functions. It supports soft tissues, protects delicate structures, and works with skeletal muscles to generate movement. Bone stores calcium and phosphorus; houses red bone marrow, which produces blood cells; and houses yellow bone marrow, which stores triglycerides as an energy source.

Blood Tissue

Blood tissue (or simply blood) is a connective tissue with a liquid matrix called **plasma,** a straw-colored fluid that consists mostly of water with a wide variety of dissolved substances—nutrients, wastes, enzymes, hormones, respiratory gases, and ions (Table 4.3K). Suspended in the plasma are formed elements—cells and cell fragments, including red blood cells (erythrocytes), white blood cells (leukocytes), and platelets. **Red blood cells** transport oxygen to body cells and remove carbon dioxide from them. **White blood cells** are involved in phagocytosis, immunity, and allergic reactions. **Platelets** participate in blood clotting. The details of blood are considered in Chapter 19.

Lymph

Lymph is interstitial fluid that flows in lymphatic vessels. It is a connective tissue that consists of a clear fluid similar to plasma but with much less protein. In the fluid are various cells and chemicals, whose composition varies from one part of the body to another. For example, lymph leaving lymph nodes includes many lymphocytes, a type of white blood cell, whereas lymph from the small intestine has a high content of newly absorbed dietary lipids. The details of lymph are considered in Chapter 22.

1. List the ways in which connective tissue differs from epithelium.
2. Describe the cells, ground substance, and fibers that make up connective tissue.
3. How are connective tissues classified? List the various types.
4. How are embryonic connective tissue and mature connective tissue distinguished?

5. Describe how the structure of the following mature connective tissues are related to their functions: areolar connective tissue, adipose tissue, reticular connective tissue, dense regular connective tissue, dense irregular connective tissue, elastic connective tissue, hyaline cartilage, fibrocartilage, elastic cartilage, bone tissue, and blood tissue.
6. Distinguish between interstitial and appositional growth of cartilage.

MEMBRANES

OBJECTIVE
• *Define a membrane, and describe the classification of membranes.*

Membranes are flat sheets of pliable tissue that cover or line a part of the body. The combination of an epithelial layer and an underlying connective tissue layer constitutes an **epithelial membrane.** The principal epithelial membranes of the body are mucous membranes, serous membranes, and the cutaneous membrane, or skin. (Skin is discussed in detail in Chapter 5, and will not be discussed here.) Another kind of membrane, a **synovial membrane,** lines joints and contains connective tissue but no epithelium.

Epithelial Membranes

Mucous Membranes

A **mucous membrane** or **mucosa** lines a body cavity that opens directly to the exterior. Mucous membranes line the entire digestive, respiratory, and reproductive systems, and much of the urinary system. They consist of both a lining layer of epithelium and an underlying layer of connective tissue.

The epithelial layer of a mucous membrane is an important feature of the body's defense mechanisms because it is a barrier that microbes and other pathogens have difficulty penetrating. Usually, tight junctions connect the cells, so materials cannot leak in between them. Goblet cells and other cells of the epithelial layer of a mucous membrane secrete mucus, and this slippery fluid prevents the cavities from drying out. It also traps particles in the respiratory passageways and lubricates food as it moves through the gastrointestinal tract. In addition, the epithelial layer secretes some of the enzymes needed for digestion and is the site of food and fluid absorption in the gastrointestinal tract. The epithelia of mucous membranes vary greatly in different parts of the body. For example, the epithelium of the mucous membrane of the small intestine is simple columnar (see Table 4.1C), whereas the epithelium of the large airways to the lungs is pseudostratified ciliated columnar (see Table 4.1I).

The connective tissue layer of a mucous membrane is areolar connective tissue and is called the **lamina propria** (LAM-i-na PRŌ-prē-a). The lamina propria is so named because it belongs to the mucous membrane (*propria* = one's own). The lamina propria supports the epithelium, binds it to the underlying structures, and allows some flexibility of the membrane. It also holds blood vessels in place and protects underlying muscles from abrasion or puncture. Oxygen and nutrients diffuse from the lamina propria to the epithelium covering it whereas carbon dioxide and wastes diffuse in the opposite direction.

Serous Membranes

A **serous membrane** (*serous* = watery) or **serosa** lines a body cavity that does not open directly to the exterior, and it covers the organs that lie within the cavity. Serous membranes consist of areolar connective tissue covered by mesothelium (simple squamous epithelium) and are composed of two layers. The portion of a serous membrane attached to the cavity wall is called the **parietal layer** (pa-RĪ-e-tal; *paries* = wall); the portion that covers and attaches to the organs inside these cavities is the **visceral layer** (*viscer-* = body organ). The mesothelium of a serous membrane secretes **serous fluid,** a watery lubricating fluid that allows organs to glide easily over one another or to slide against the walls of cavities.

The serous membrane lining the thoracic cavity and covering the lungs is the **pleura.** The serous membrane lining the heart cavity and covering the heart is the **pericardium.** The serous membrane lining the abdominal cavity and covering the abdominal organs is the **peritoneum.**

Synovial Membranes

Synovial membranes (sin-Ō-vē-al) line the cavities of the freely movable joints. (*Syn* = together, referring here to a place where bones come together.) Like serous membranes, synovial membranes line structures that do not open to the exterior. Unlike mucous, serous, and cutaneous membranes, they do not contain epithelium and are therefore not epithelial membranes. Synovial membranes are composed of areolar connective tissue with elastic fibers and varying numbers of adipocytes. Synovial membranes secrete **synovial fluid,** which lubricates and nourishes the cartilage covering the bones at movable joints; these membranes are called *articular synovial membranes.* Other synovial membranes line cushioning sacs, called *bursae,* and *tendon sheaths* in the hands and feet, thus easing the movement of muscle tendons.

1. Define the following kinds of membranes: mucous, serous, cutaneous, and synovial.
2. Where is each type of membrane located in the body? What are their functions?

MUSCLE TISSUE

OBJECTIVE

• *Describe the general features of muscle tissue, and contrast the structure, location, and mode of control of the three types of muscle tissue.*

Muscle tissue consists of fibers (cells) that are beautifully constructed to generate force. As a result of this characteristic, muscle tissue produces motion, maintains posture, and generates heat. Based on its location and certain structural and functional characteristics, muscle tissue is classified into three types: skeletal, cardiac, and smooth (Table 4.4).

Skeletal muscle tissue is named for its location—usually attached to the bones of the skeleton (Table 4.4A). It is also *striated;* the fibers contain alternating light and dark bands called *striations* that are perpendicular to the long axes of the fibers and visible under a light microscope. Skeletal muscle is *voluntary* because it can be made to contract or relax by conscious control. A single skeletal muscle fiber is very long, roughly cylindrical in shape, and has many nuclei located at the periphery of the cell. Within a whole muscle, the individual fibers are parallel to each other.

Cardiac muscle tissue forms the bulk of the wall of the heart (Table 4.4B). Like skeletal muscle, it is striated. However, unlike skeletal muscle tissue, it is *involuntary;* its contraction is not consciously controlled. Cardiac muscle fibers are branched and usually have only one centrally located nucleus, although an occasional cell has two nuclei. Cardiac muscle fibers attach end to end to each other by transverse thickenings of the plasma membrane called **intercalated discs** (*intercalat-* = to insert between), which contain both desmosomes and gap junctions. Intercalated discs are unique to cardiac muscle. The desmosomes strengthen the tissue and hold the fibers together during their vigorous contractions. The gap junctions provide a route for quick conduction of muscle action potentials throughout the heart.

Smooth muscle tissue is located in the walls of hollow internal structures such as blood vessels, airways to the lungs, the stomach, intestines, gallbladder, and urinary bladder (Table 4.4C). Its contraction helps constrict or narrow the lumen of blood vessels, physically break down and move food along the gastrointestinal tract, move fluids through the body, and eliminate wastes. Smooth muscle fibers are usually *involuntary,* and they are *nonstriated* (hence the term *smooth*). A smooth muscle fiber is small and spindle-shaped (thickest in the middle, with each end tapering) and contains a single, centrally located nucleus. Many individual fibers in some smooth muscle tissue—for example, in the wall of the intestines—are connected by gap junctions, which enable them to contract in synchrony. In other locations such as the iris of the eye, smooth muscle fibers contract individually, like skeletal muscle fibers, because gap junctions are absent. Chapter 10 provides a detailed discussion of muscle tissue.

1. Which muscles are striated and which are smooth?
2. Which muscles have gap junctions?

Table 4.4 Muscle Tissue

A. Skeletal muscle tissue *Description:* Long, cylindrical, striated fibers with many peripherally located nuclei; voluntary control.
Location: Usually attached to bones by tendons.
Function: Motion, posture, heat production.

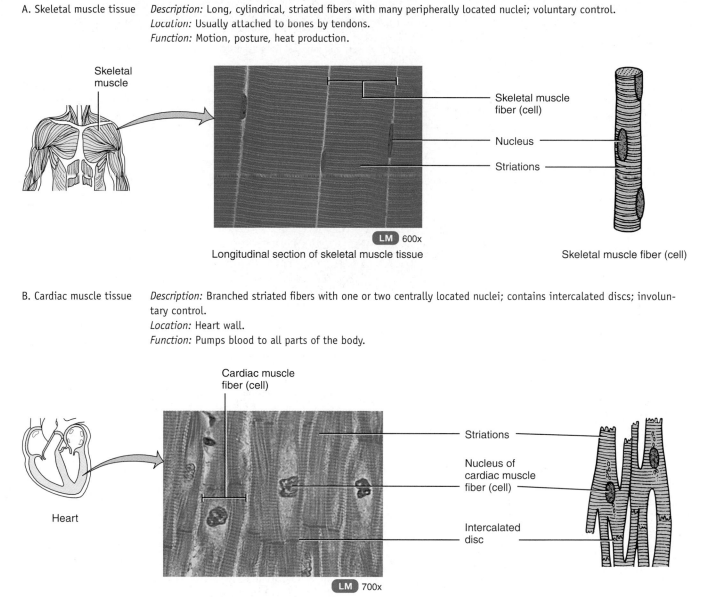

Skeletal muscle

Skeletal muscle fiber (cell)

Nucleus

Striations

LM 600x

Longitudinal section of skeletal muscle tissue

Skeletal muscle fiber (cell)

B. Cardiac muscle tissue *Description:* Branched striated fibers with one or two centrally located nuclei; contains intercalated discs; involuntary control.
Location: Heart wall.
Function: Pumps blood to all parts of the body.

Cardiac muscle fiber (cell)

Heart

Striations

Nucleus of cardiac muscle fiber (cell)

Intercalated disc

LM 700x

Longitudinal section of cardiac muscle tissue

Cardiac muscle fibers

Table 4.4 Muscle Tissue (continued)

C. Smooth muscle tissue   *Description:* Spindle-shaped, nonstriated fibers with one centrally located nucleus; involuntary control.
Location: Walls of hollow internal structures such as blood vessels, airways to the lungs, stomach, intestines, gall-bladder, urinary bladder, and uterus.
Function: Motion (constriction of blood vessels and airways, propulsion of foods through gastrointestinal tract, contraction of urinary bladder and gallbladder).

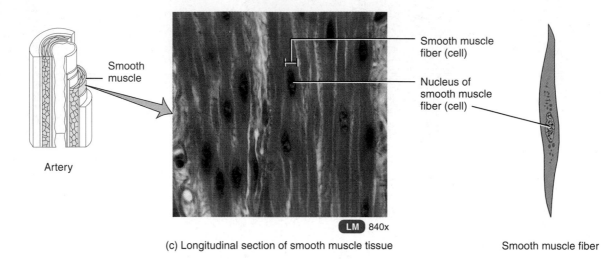

Artery

LM 840x

(c) Longitudinal section of smooth muscle tissue

Smooth muscle fiber

NERVOUS TISSUE

OBJECTIVE

• *Describe the structural features and functions of nervous tissue.*

Despite the awesome complexity of the nervous system, it consists of only two principal kinds of cells: neurons and neuroglia. **Neurons,** or nerve cells, are sensitive to various stimuli. They convert stimuli into nerve impulses (action potentials) and conduct these impulses to other neurons, to muscle tissue, or to glands. Most neurons consist of three basic portions: a cell body and two kinds of cell processes—dendrites and axons (Table 4.5). The **cell body** contains the nucleus and other organelles. **Dendrites** (*dendr-* = tree) are tapering, highly branched, and usually short cell processes. They are the major receiving or input portion of a neuron. The **axon** (*axo-* = axis) of a neuron is a single, thin, cylindrical process that may be very long. It is the output portion of a neuron, conducting nerve impulses toward another neuron or to some other tissue.

Even though **neuroglia** (noo-RŌG-lē-a; *-glia* = glue) do not generate or conduct nerve impulses, these cells do have many important functions (see Table 12.1 on page 386). The detailed structure and function of neurons and neuroglia are considered in Chapter 12.

CLINICAL APPLICATION
Tissue Engineering

In recent years a new technology called **tissue engineering** has emerged. Using a wide range of materials, scientists are learning to grow new tissues in the laboratory to replace damaged tissues in the body. Tissue engineers have already developed laboratory-grown versions of skin and cartilage. In the procedure, scaffolding beds of biodegradable synthetic materials or collagen are used as substrates that permit body cells such as skin cells or cartilage cells to be cultured. As the cells divide and assemble, the scaffolding degrades, and the new, permanent tissue is then implanted in the patient. Other structures being developed by tissue engineers include bones, tendons, heart valves, bone marrow, and intestines. Work is also underway to develop insulin-producing cells (pancreas) for diabetics, dopamine-producing cells (brain) for Parkinson's disease patients, and even entire livers and kidneys. ■

1. Distinguish between neurons and neuroglia.
2. What are the functions of the dendrites, cell body, and axon of a neuron?

TISSUE REPAIR: RESTORING HOMEOSTASIS

OBJECTIVE

• *Describe the role of tissue repair in restoring homeostasis.*

Table 4.5 Nervous Tissue

Description: Consists of neurons (nerve cells) and neuroglia. Neurons consist of a cell body and processes extending from the cell body multiple dendrites and a single axon. Neuroglia do not generate or conduct nerve impulses but have other important functions.

Location: Nervous system.

Function: Exhibits sensitivity to various types of stimuli, converts stimuli into nerve impulses, and conducts nerve impulses to other neurons, muscle fibers, or glands.

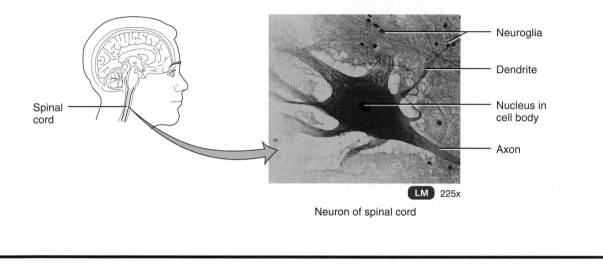

Neuron of spinal cord

Tissue repair is the process in which worn-out, damaged, or dead cells are replaced. New cells originate by cell division from either the **stroma,** the supporting connective tissue, or the **parenchyma,** cells that constitute the tissue's or organ's functioning part. In adults, each of the four basic tissue types (epithelial, connective, muscle, and nervous) has a different capacity for replenishing parenchymal cells lost by damage, disease, or other processes.

Epithelial cells, which endure considerable wear and tear (and even injury) in some locations, have a continuous capacity for renewal. In some cases, immature, undifferentiated cells called **stem cells** divide to replace lost or damaged cells. For example, stem cells reside in protected locations in the epithelia of the skin and gastrointestinal tract to replenish cells sloughed from the apical surface, and stem cells in red bone marrow continually provide new red and white blood cells and platelets. In other cases, mature, differentiated cells can undergo cell division; examples are hepatocytes (liver cells) and endothelial cells in blood vessels.

Some connective tissues also have a continuous capacity for renewal. One example is bone, which has an ample blood supply. Other connective tissues, such as cartilage or fibrous connective tissue, can replenish cells less readily in part because of a smaller blood supply.

Muscle tissue has a relatively poor capacity for renewal of lost cells. Existing cardiac muscle fibers cannot divide to form new fibers. Even though skeletal muscle tissue contains stem cells called *satellite cells,* they do not divide rapidly

enough to replace extensively damaged fibers. Smooth muscle fibers proliferate to some extent, but they do so much more slowly than the cells of epithelial or connective tissues.

Nervous tissue has the poorest capacity for renewal. Although experiments have recently revealed the presence of some stem cells in the brain, they normally do not undergo mitosis to replace damaged neurons. Discovering why this is so is a major goal of researchers seeking ways to encourage repair of damaged or diseased nervous tissue.

The Repair Process

The restoration of an injured tissue or organ to normal structure and function depends entirely on whether parenchymal or stromal cells are active in the repair process.

The cardinal factor in tissue repair is the capacity of parenchymal tissue to regenerate. This capacity, in turn, depends on the ability of the parenchymal cells to replicate quickly. If parenchymal cells accomplish the repair, tissue **regeneration** is possible, and a near-perfect reconstruction of the injured tissue may occur. However, if fibroblasts of the stroma are active in the repair, the replacement tissue will be a new connective tissue. The fibroblasts synthesize collagen and other matrix materials that aggregate to form scar tissue, a process known as **fibrosis.** Because scar tissue is not specialized to perform the functions of the parenchymal tissue, the original function of the tissue or organ is impaired.

When tissue damage is extensive, as in large, open

wounds, both the connective tissue stroma and the parenchymal cells are active in repair. This repair involves the rapid cell division of many fibroblasts and the manufacture of new collagen fibers to provide structural strength. Blood capillaries also sprout new buds to supply the healing tissue the materials it needs. All these processes create an actively growing connective tissue called **granulation tissue.** This new tissue forms across a wound or surgical incision to provide a framework (stroma) that supports the epithelial cells that migrate into the open area and fill it. The newly formed granulation tissue also secretes a fluid that kills bacteria.

CLINICAL APPLICATION
Adhesions

Scar tissue can form **adhesions,** abnormal joining of tissues. Adhesions commonly form in the abdomen around a site of previous inflammation such as an inflamed appendix, and they can develop after surgery. Although adhesions do not always cause problems, they can decrease tissue flexibility, cause obstruction (such as in the intestine), and make a subsequent operation more difficult. An *adhesiotomy,* the surgical release of adhesions, may be required. ■

Conditions Affecting Repair

Three factors affect tissue repair: nutrition, blood circulation, and age. Nutrition is vital because the healing process places a great demand on the body's store of nutrients. Adequate protein in the diet is important because most of the structural components of a tissue are proteins. Several vitamins also play a direct role in wound healing and tissue repair. For example, vitamin C directly affects the normal production and maintenance of matrix materials, especially collagen. Vitamin C also strengthens and promotes the formation of new blood vessels. With vitamin C deficiency, even superficial wounds fail to heal, and the walls of the blood vessels become fragile and are easily ruptured.

In tissue repair, proper blood circulation is essential to transport oxygen, nutrients, antibodies, and many defensive cells to the injured site. The blood also plays an important role in the removal of tissue fluid, bacteria, foreign bodies, and debris, elements that would otherwise interfere with healing.

AGING AND TISSUES

Generally, tissues heal faster and leave less obvious scars in the young than in the aged. In fact, surgery performed on fetuses leaves no scars. The younger body is generally in a better nutritional state, its tissues have a better blood supply, and its cells have a higher metabolic rate. Thus, cells can synthesize needed materials and divide more quickly. The extracellular components of tissues also change with age. Collagen fibers, responsible for the strength of tendons, increase in number and change in quality with aging. These changes in the collagen of arterial walls are as much responsible for their loss of extensibility as are the deposits associated with atherosclerosis, the deposition of fatty materials in arterial walls. Elastin, another extracellular component, is responsible for the elasticity of blood vessels and skin. It thickens, fragments, and acquires a greater affinity for calcium with age—changes that may also be associated with the development of atherosclerosis.

1. What is meant by tissue repair?
2. Distinguish between stromal and parenchymal repair.
3. What is the importance of granulation tissue?
4. What changes occur in epithelial and connective tissues with aging?

DISORDERS: HOMEOSTATIC IMBALANCES

SJÖGREN'S SYNDROME

Sjögren's syndrome (SHŌ-grenz) is a common autoimmune disorder that causes inflammation and destruction of exocrine glands, especially the lacrimal (tear) glands and salivary glands. An **autoimmune disease** is one in which antibodies produced by the immune system fail to distinguish what is foreign from what is self and attack the body's own tissues. Sjögren's syndrome is characterized by dryness of the mucous membranes in the eyes and mouth and salivary gland enlargement. Systemic effects include arthritis, difficulty swallowing, pancreatitis (inflammation of the pancreas), pleuritis (inflammation of the pleurae of the lungs), and migraine. The disorder affects more females than males by a ratio of 9:1. About 20% of older adults experience some signs of Sjögren's. Individuals with the condition have a greatly increased risk of developing malignant lymphoma. Treatment is supportive, including using artificial tears to moisten the eyes, sipping fluids, chewing sugarless gum, and using a saliva substitute to moisten the mouth.

SYSTEMIC LUPUS ERYTHEMATOSUS

Systemic lupus erythematosus (er-ē-the-ma-TŌ-sus), **SLE,** or simply lupus, is a chronic inflammatory disease of connective tissue occurring mostly in nonwhite women during their childbearing years. It is an autoimmune disease that can cause tissue damage in every body system. The disease, which can range from a mild condition in most patients to a rapidly fatal disease, is marked by periods of exacerbation and remission. The prevalence of SLE is about 1 in 2000 persons, with females more likely to be afflicted than males by a ratio of 8:1 or 9:1.

Although the cause of SLE is not known, genetic, environmental, and hormonal factors are implicated. The genetic component is suggested by studies of twins and family history. Environmental factors include viruses, bacteria, chemicals, drugs, exposure to excessive sunlight, and emotional stress. With regard to hormones, sex hormones, such as estrogens, may trigger SLE.

Signs and symptoms of SLE include painful joints, low-grade fever, fatigue, mouth ulcers, weight loss, enlarged lymph nodes and spleen, sensitivity to sunlight, rapid loss of large amounts of scalp hair, and anorexia. A distinguishing feature of lupus is an eruption across the bridge of the nose and cheeks called a "butterfly rash." Other skin lesions may occur, including blistering and ulceration. The erosive nature of some SLE skin lesions was thought to resemble the damage inflicted by the bite of a wolf—thus, the term *lupus* (= wolf). The most serious complications of the disease involve inflammation of the kidneys, liver, spleen, lungs, heart, brain, and gastrointestinal tract. Because there is no cure for SLE, treatment is supportive, including anti-inflammatory drugs, such as aspirin, and immunosuppressive drugs.

MEDICAL TERMINOLOGY

Atrophy (AT-rō-fē; *a-* = without; *-trophy* = nourishment) A decrease in the size of cells, with a subsequent decrease in the size of the affected tissue or organ.

Hypertrophy (hī-PER-trō-fē; *hyper-* = above or excessive) Increase in the size of a tissue because its cells enlarge without undergoing cell division.

Tissue rejection An immune response of the body directed at foreign proteins in a transplanted tissue or organ; immunosuppressive drugs, such as cyclosporine, have largely overcome tissue rejection in heart-, kidney-, and liver-transplant patients.

Tissue transplantation The replacement of a diseased or injured tissue or organ. The most successful transplants involve use of a person's own tissues or those from an identical twin; less-successful transplants involve less-compatible tissues from a donor.

Xenotransplantation (zen′ō-trans-plan-TĀ-shun; *xeno-* = strange, foreign) The replacement of a diseased or injured tissue or organ with cells or tissues from an animal. Only a few cases of successful xenotransplantation exist to date.

STUDY OUTLINE

TYPES OF TISSUES AND THEIR ORIGINS (p. 104)

1. A tissue is a group of similar cells that usually has a similar embryological origin and is specialized for a particular function.
2. The various tissues of the body are classified into four basic types: epithelial, connective, muscle, and nervous.
3. All tissues of the body develop from three primary germ layers, the first tissues that form in a human embryo: ectoderm, endoderm, and mesoderm.

CELL JUNCTIONS (p. 105)

1. Cell junctions are points of contact between adjacent plasma membranes.
2. Tight junctions form fluid-tight seals between cells; adherens junctions, desmosomes, and hemidesmosomes fasten cells to one another or the basement membrane; and gap junctions permit electrical or chemical signals to pass between cells.

EPITHELIAL TISSUE (p. 105)

1. The subtypes of epithelia include covering and lining epithelia and glandular epithelia.
2. An epithelium consists mostly of cells with little extracellular material between adjacent plasma membranes. It is arranged in sheets and attached to a basement membrane. Although it is avascular, it has a nerve supply. Epithelia are derived from all three primary germ layers and have a high capacity for renewal.

Covering and Lining Epithelium (p. 107)

1. Epithelial layers are simple (one layer), stratified (several layers), or pseudostratified (one layer that appears to be several);

the cell shapes may be squamous (flat), cuboidal (cubelike), columnar (rectangular), or transitional (variable).
2. Simple squamous epithelium consists of a single layer of flat cells (Table 4.1A). It is found in parts of the body where filtration or diffusion are priority processes. One type, endothelium, lines the heart and blood vessels. Another type, mesothelium, forms the serous membranes that line the thoracic and abdominopelvic cavities and cover the organs within them.
3. Simple cuboidal epithelium consists of a single layer of cube-shaped cells that function in secretion and absorption (Table 4.1B). It is found covering the ovaries, in the kidneys and eyes, and lining some glandular ducts.
4. Nonciliated simple columnar epithelium consists of a single layer of nonciliated rectangular cells (Table 4.1C). It lines most of the gastrointestinal tract. Specialized cells containing microvilli perform absorption. Goblet cells secrete mucus.
5. Ciliated simple columnar epithelium consists of a single layer of ciliated rectangular cells (Table 4.1D). It is found in a few portions of the upper respiratory tract, where it moves foreign particles trapped in mucus out of the respiratory tract.
6. Stratified squamous epithelium consists of several layers of cells; cells at the apical surface are flat (Table 4.1E). It is protective. A nonkeratinized variety lines the mouth; a keratinized variety forms the epidermis, the most superficial layer of the skin.
7. Stratified cuboidal epithelium consists of several layers of cells; cells at the apical surface are cube-shaped (Table 4.1F). It is found in adult sweat glands and a portion of the male urethra.

8. Stratified columnar epithelium consists of several layers of cells; cells at the apical surface are rectangular (Table 4.1G). It is found in a portion of the male urethra and large excretory ducts of some glands.

9. Transitional epithelium consists of several layers of cells whose appearance varies with the degree of stretching (Table 4.1H). It lines the urinary bladder.

10. Pseudostratified columnar epithelium has only one layer but gives the appearance of many (Table 4.1I). A ciliated variety contains goblet cells and lines most of the upper respiratory tract; a nonciliated variety has no goblet cells and lines ducts of many glands, the epididymis, and part of the male urethra.

Glandular Epithelium (p. 116)

1. A gland is a single cell or a group of epithelial cells adapted for secretion.

2. Endocrine glands secrete hormones into the blood (Table 4.1J).

3. Exocrine glands (mucous, sweat, oil, and digestive glands) secrete into ducts or directly onto a free surface (Table 4.1K).

4. Structural classification of exocrine glands includes unicellular and multicellular glands.

5. Functional classification of exocrine glands includes holocrine, apocrine, and merocrine glands.

CONNECTIVE TISSUE (p. 118)

1. Connective tissue is the most abundant body tissue.

2. Connective tissue consists of cells and a matrix of ground substance and fibers; it has abundant matrix with relatively few cells. It does not usually occur on free surfaces, has a nerve supply (except for cartilage), and is highly vascular (except for cartilage, tendons, and ligaments).

Components of Connective Tissue (p. 118)

1. Cells in connective tissue are derived from mesenchymal cells.

2. Cell types include fibroblasts (secrete matrix), macrophages (perform phagocytosis), plasma cells (secrete antibodies), mast cells (produce histamine), adipocytes (store fat), and white blood cells (migrate from blood in response to infections).

3. The ground substance and fibers make up the matrix.

4. The ground substance supports and binds cells together, provides a medium for the exchange of materials, and is active in influencing cell functions.

5. Substances found in the ground substance include hyaluronic acid, chondroitin sulfate, dermatan sulfate, keratan sulfate, and adhesion proteins.

6. The fibers in the matrix provide strength and support and are of three types: (a) collagen fibers (composed of collagen) are found in large amounts in bone, tendons, and ligaments; (b) elastic fibers (composed of elastin, fibrillin, and other glycoproteins) are found in skin, blood vessel walls, and lungs; and (c) reticular fibers (composed of collagen and glycoprotein) are found around fat cells, nerve fibers, and skeletal and smooth muscle cells.

Classification of Connective Tissue (p. 120)

1. The two major subclasses of connective tissue are embryonic connective tissue and mature connective tissue.

2. Mesenchyme forms all other connective tissues (Table 4.2A).

3. Mucous connective tissue is found in the umbilical cord of the fetus, where it gives support (Table 4.2B).

Types of Mature Connective Tissue (p. 120)

1. Mature connective tissue is connective tissue that differentiates from mesenchyme, and is present in the newborn. It is subdivided into several kinds: loose or dense connective tissue, cartilage, bone tissue, and blood tissue.

2. Loose connective tissue includes areolar connective tissue, adipose tissue, and reticular connective tissue.

3. Areolar connective tissue consists of the three types of fibers, several cells, and a semifluid ground substance (Table 4.3A). It is found in the subcutaneous layer, in mucous membranes, and around blood vessels, nerves, and body organs.

4. Adipose tissue consists of adipocytes, which store triglycerides (Table 4.3B). It is found in the subcutaneous layer, around organs, and in the yellow bone marrow of long bones.

5. Reticular connective tissue consists of reticular fibers and reticular cells and is found in the liver, spleen, and lymph nodes (Table 4.3C).

6. Dense connective tissue includes dense regular connective tissue, dense irregular connective tissue, and elastic connective tissue.

7. Dense regular connective tissue consists of parallel bundles of collagen fibers and fibroblasts (Table 4.3D). It forms tendons, most ligaments, and aponeuroses.

8. Dense irregular connective tissue consists of usually randomly arranged collagen fibers and a few fibroblasts (Table 4.3E). It is found in fasciae, the dermis of skin, and membrane capsules around organs.

9. Elastic connective tissue consists of branching elastic fibers and fibroblasts (Table 4.3F). It is found in the walls of large arteries, lungs, trachea, and bronchial tubes.

10. Cartilage contains chondrocytes and has a rubbery matrix (chondroitin sulfate) containing collagen and elastic fibers.

11. Hyaline cartilage is found in the embryonic skeleton, at the ends of bones, in the nose, and in respiratory structures (Table 4.3G). It is flexible, allows movement, and provides support.

12. Fibrocartilage is found in the pubic symphysis, intervertebral discs, and menisci (cartilage pads) of the knee joint (Table 4.3H).

13. Elastic cartilage maintains the shape of organs such as the epiglottis of the larynx, auditory (Eustachian) tubes, and external ear (Table 4.3I).

14. Cartilage enlarges by interstitial growth (from within) and appositional growth (from without).

15. Bone or osseous tissue consists of mineral salts and collagen fibers that contribute to the hardness of bone, and cells called osteocytes that are located in lacunae (Table 4.3J). It supports, protects, helps provide movement, stores minerals, and houses blood-forming tissue.

16. Blood tissue consists of plasma and formed elements—red blood cells, white blood cells, and platelets (Table 4.3K). Its cells function to transport oxygen and carbon dioxide, carry on phagocytosis, participate in allergic reactions, provide immunity, and bring about blood clotting.

17. Lymph flows in lymphatic vessels and is a clear fluid whose composition varies from one part of the body to another.

MEMBRANES (p. 129)

1. An epithelial membrane consists of an epithelial layer overlying a connective tissue layer. Examples are mucous, serous, and cutaneous membranes.
2. Mucous membranes line cavities that open to the exterior, such as the gastrointestinal tract.
3. Serous membranes line closed cavities (pleura, pericardium, peritoneum) and cover the organs in the cavities. These membranes consist of parietal and visceral layers.
4. The cutaneous membrane is the skin.
5. Synovial membranes line joint cavities, bursae, and tendon sheaths and consist of areolar connective tissue instead of epithelium.

MUSCLE TISSUE (p. 130)

1. Muscle tissue consists of fibers that are specialized for contraction. It provides motion, maintenance of posture, and heat production.
2. Skeletal muscle tissue is attached to bones and is striated, and its action is voluntary (Table 4.4A).
3. Cardiac muscle tissue forms most of the heart wall and is striated, and its action is involuntary (Table 4.4B).
4. Smooth muscle tissue is found in the walls of hollow internal structures (blood vessels and viscera) and is nonstriated, and its action is involuntary (Table 4.4C).

NERVOUS TISSUE (p. 132)

1. The nervous system is composed of neurons (nerve cells) and neuroglia (protective and supporting cells) (Table 4.5).
2. Neurons are sensitive to stimuli, convert stimuli into nerve impulses, and conduct nerve impulses.
3. Most neurons consist of a cell body and two types of processes called dendrites and axons.

TISSUE REPAIR: RESTORING HOMEOSTASIS (p. 132)

1. Tissue repair is the replacement of worn-out, damaged, or dead cells by healthy ones.
2. Stem cells may divide to replace lost or damaged cells.
3. If the injury is superficial, tissue repair involves parenchymal regeneration; if damage is extensive, granulation tissue is involved.
4. Good nutrition and blood circulation are important to tissue repair. Various vitamins and adequate dietary protein are needed.

AGING AND TISSUES (p. 134)

1. Tissues heal faster and leave less obvious scars in the young than in the aged; surgery performed on fetuses leaves no scars.
2. The extracellular components of tissues, such as collagen and elastic fibers, also change with age.

SELF-QUIZ QUESTIONS

1. Match the following:
   ___ (a) prevents organ contents from leaking into the blood or surrounding tissues
   ___ (b) forms adhesion belts that help epithelial surfaces resist separation
   ___ (c) makes tissues stable by linking the cytoskeletons of cells together
   ___ (d) connects cells to extracellular material such as the basement membrane; anchors one kind of tissue to another
   ___ (e) allows cells in a tissue to communicate; enable nerve or muscle impulses to spread rapidly between cells

   (1) gap junction
   (2) tight junction
   (3) desmosome
   (4) hemidesmosome
   (5) adherens junction

True or false:

2. The three primary germ layers are the endoderm, ectoderm, and mesoderm.
3. Nervous tissue consists of neurons, neuroglia, and fibroblasts.

Choose the best answer to the following questions:

4. The tissue type that can detect changes in the internal and external environments and respond to the changes is (a) nervous tissue, (b) muscle tissue, (c) connective tissue, (d) epithelium

5. Which of the following statements are true of epithelium? (1) The cells are arranged in continuous single- or multiple-layer sheets. (2) The attachment between the basal layer and the connective tissue is called the basement membrane. (3) This tissue has an extensive blood supply. (4) This tissue has a high rate of cell division. (5) This tissue functions in protection, secretion, absorption, and excretion.
   (a) 1, 2, 3, and 4, (b) 2, 3, 4, and 5, (c) 1, 2, 4, and 5, (d) 1, 3, and 5, (e) 2, 4, and 5
6. The type of exocrine gland that forms its secretory product and simply releases it from the cell is the (a) apocrine gland, (b) merocrine gland, (c) holocrine gland, (d) endocrine gland, (e) apical gland.
7. Connective tissue cells that secrete antibodies are termed (a) macrophages, (b) mast cells, (c) fibroblasts, (d) adipocytes, (e) plasma cells.
8. The membrane lining a body cavity that opens directly to the exterior is a (a) serous membrane, (b) mucous membrane, (c) synovial membrane, (d) plasma membrane, (e) basement membrane.
9. The muscle tissue that forms the bulk of the wall of the heart is (a) skeletal muscle, (b) smooth muscle, (c) involuntary, nonstriated muscle, (d) cardiac muscle, (e) striated, voluntary muscle.

10. Match the following:

___ (a) the tissue from which all other connective tissues eventually arise

___ (b) tissue found primarily in the fetal umbilical cord

___ (c) tissue consisting of several kinds of cells, containing all three fiber types, and found in the subcutaneous layer

___ (d) tissue specialized for triglyceride storage

___ (e) tissue that contains reticular fibers and reticular cells and forms the stroma of certain organs such as the spleen

___ (f) tissue with irregularly arranged collagen fibers found in the dermis of the skin

___ (g) tissue found in the lungs that is strong and can recoil back to its original shape after being stretched

___ (h) tissue that affords flexibility at joints and reduces joint friction

___ (i) tissue that provides strength and rigidity and is the strongest of the three types of cartilage

___ (j) avascular connective tissue that comes in three types

___ (k) tissue that forms the internal framework of the body and works with skeletal muscle to generate movement

___ (l) connective tissue with formed elements suspended in a liquid matrix called plasma

(1) blood
(2) fibrocartilage
(3) mesenchyme
(4) cartilage
(5) mucous connective tissue
(6) hyaline cartilage
(7) dense irregular connective tissue
(8) areolar connective tissue
(9) reticular connective tissue
(10) bone
(11) elastic connective tissue
(12) adipose tissue

11. Match the following:

___ (a) a single layer of flat cells found in the body where filtration (kidney) or diffusion (lungs) are priority processes

___ (b) lines the heart, blood vessels, and lymphatic vessels

___ (c) cube-shaped cells found in the kidney that function in secretion and absorption

___ (d) cells containing microvilli and goblet cells found in linings of the digestive, reproductive and urinary tracts

___ (e) cells found in the superficial part of skin that can resist friction and repel bacteria

___ (f) cells found in the urinary bladder that can change shape (stretch or relax)

___ (g) cells that are all attached to the basement membrane, although some do not reach to the apical surface; those cells that reach the apical surface secrete mucus or bear cilia

___ (h) a fairly rare type of epithelium that has a mainly protective function

(1) pseudostratified ciliated columnar epithelium
(2) endothelium
(3) transitional epithelium
(4) simple squamous epithelium
(5) simple cuboidal epithelium
(6) nonciliated simple columnar epithelium
(7) stratified cuboidal epithelium
(8) keratinized stratified squamous epithelium

Complete the following:

12. A group of cells that usually have a common origin and a specialized function is called a ___.

13. Secretions of ___ glands enter the extracellular fluid, and then diffuse into the bloodstream.

14. The three constituents of connective tissue are ___, ___, and ___.

15. Tissue repair accomplished by parenchymal cells is termed ___; tissue repair accomplished by fibroblasts is termed ___ and results in scar tissue.

CRITICAL THINKING QUESTIONS

1. Imagine that you live 50 years in the future, and you can custom-design a human to suit the environment. Your assignment is to customize the tissue makeup for life on a large planet with strong gravity; a cold, dry climate; and a thin atmosphere. What adaptations would you incorporate into the structure and/or amount of your tissues, and why? (HINT: *Think of the adaptations seen in people and animals that live in the Arctic versus those that live near the equator.*)

2. The neighborhood kids are walking around with straight pins and sewing needles stuck into their fingertips. There is no visible bleeding. What type of tissue have they pierced? What if they stuck a needle straight into the fingertip? (HINT: *If you pierced an earlobe, it would bleed.*)

3. Kendra was reading her *Principles of Anatomy and Physiology* text at the college cafeteria's all-you-can-eat salad bar. She was having trouble visualizing tissues, until her roommate plopped down a bowl of gelatin salad. It was pale pink with grapes, shredded carrots, and coconut. As the salad quivered in the bowl, Kendra suddenly understood connective tissue. Relate the structure of the salad to the structure of connective tissue. (HINT: *The fruit is suspended in the jellylike gelatin.*)

ANSWERS TO FIGURE QUESTIONS

4.1 Gap junctions allow the spread of electrical and chemical signals between adjacent cells.

4.2 The basement membrane provides physical support for epithelium, serves as a filter in the kidneys, and guides cell migration during development and tissue repair.

4.3 Simple multicellular exocrine glands have a nonbranched duct; compound multicellular exocrine glands have a branched duct.

4.4 Sebaceous (oil) glands are holocrine glands, and salivary glands are merocrine glands.

4.5 Fibroblasts secrete the fibers and ground substance of the matrix.

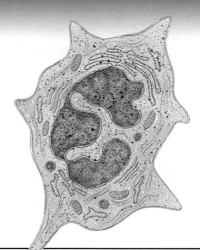

Skin and its accessory structures—hair and nails, along with various glands, muscles, and nerves—make up the **integumentary system** (in-teg-yoo-MEN-tar-ē; (*inte-* = whole; *-gument* = body covering). The integumentary system guards the body's physical and biochemical integrity, maintains a constant body temperature, and provides sensory information about the surrounding environment. This chapter discusses the body's most visible system, one that often defines our very self-image.

STRUCTURE OF THE SKIN

OBJECTIVE

• *Describe the various layers of the epidermis and dermis, and the cells that compose them.*

The **skin** consists of different tissues that are joined to perform specific functions. It is the largest organ of the body in surface area and weight. In adults, the skin covers an area of about 2 square meters (22 square feet) and weighs 4.5−5 kg (10−11 lb), about 16% of total body weight. It ranges in thickness from 0.5 mm on the eyelids to 4.0 mm on the heels. However, over most of the body it is 1−2 mm thick. **Dermatology** (der′-ma-TOL-ō-jē; *dermat-* = skin; *-ology* = study of) is the branch of medicine that specializes in diagnosing and treating skin disorders.

Structurally, the skin consists of two principal parts (Figure 5.1). The superficial, thinner portion, which is composed of *epithelial tissue,* is the **epidermis** (ep′-i-DERM-is; *epi-* = above). The deeper, thicker, *connective tissue* part is

the **dermis.** Deep to the dermis, and not part of the skin, is the **subcutaneous (subQ) layer.** Also called the **hypodermis** (*hypo-* = below), this layer consists of areolar and adipose tissues. Fibers that extend from the dermis anchor the skin to the subcutaneous layer, which, in turn, attaches to underlying tissues and organs. The subcutaneous layer serves as a storage depot for fat and contains large blood vessels that supply the skin. This region (and sometimes the dermis) also contains nerve endings called *lamellated (Pacinian) corpuscles* (pa-SIN-ē-an) that are sensitive to pressure (see Figure 5.1).

Epidermis

The **epidermis** is keratinized stratified squamous epithelium. It contains four principal types of cells: keratinocytes, melanocytes, Langerhans cells, and Merkel cells (Figure 5.2). About 90% of epidermal cells are **keratinocytes** (ker-a-TIN-ō-sīts; *keratino-* = hornlike; *-cytes* = cells) which produce the protein **keratin** (Figure 5.2a). Recall from Chapter 4 that keratin is a tough, fibrous protein that helps protect the skin and underlying tissues from heat, microbes, and chemicals. Keratinocytes also produce lamellar granules, which release a waterproofing sealant.

About 8% of the epidermal cells are **melanocytes** (MEL-a-nō-sīts; *melano-* = black), which produce the pigment melanin (Figure 5.2b). Their long, slender projections extend between the keratinocytes and transfer melanin granules to them. **Melanin** is a brown-black pigment that contributes to skin color and absorbs damaging ultraviolet (UV)

Figure 5.1 Components of the integumentary system. The skin consists of a thin, superficial epidermis and a deep, thicker dermis. Deep to the skin is the subcutaneous layer, which attaches the dermis to underlying organs and tissues.

🔑 The integumentary system includes the skin and its accessory structures—hair, nails, and glands—along with associated muscles and nerves.

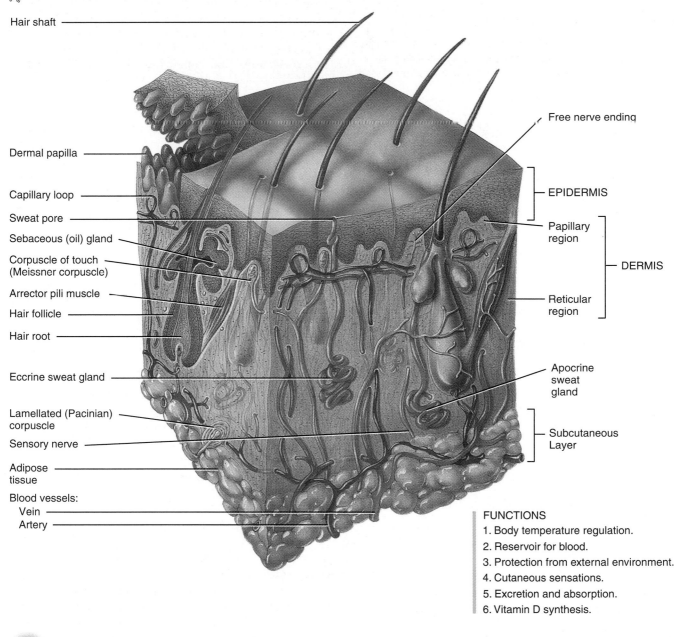

Hair shaft

Dermal papilla

Capillary loop

Sweat pore

Sebaceous (oil) gland

Corpuscle of touch (Meissner corpuscle)

Arrector pili muscle

Hair follicle

Hair root

Eccrine sweat gland

Lamellated (Pacinian) corpuscle

Sensory nerve

Adipose tissue

Blood vessels:
Vein
Artery

Free nerve ending

EPIDERMIS

Papillary region

DERMIS

Reticular region

Apocrine sweat gland

Subcutaneous Layer

FUNCTIONS
1. Body temperature regulation.
2. Reservoir for blood.
3. Protection from external environment.
4. Cutaneous sensations.
5. Excretion and absorption.
6. Vitamin D synthesis.

Q What kind of tissues make up the epidermis and the dermis?

light. Once inside keratinocytes, the melanin granules cluster to form a protective veil over the nucleus, on the side toward the skin surface. In this way they shield the nuclear DNA from UV light.

Langerhans cells (LANG-er-hans) arise from red bone marrow and migrate to the epidermis (Figure 5.2c), where

they constitute a small proportion of the epidermal cells. They participate in immune responses mounted against microbes that invade the skin, and are easily damaged by UV light.

Merkel cells are the least numerous of the epidermal cells. They are located in the deepest layer of the epidermis,

Figure 5.2 Types of cells in the epidermis. Besides keratinocytes, the epidermis contains melanocytes, which produce the pigment melanin; Langerhans cells, which participate in immune responses; and Merkel cells, which function in the sensation of touch. Adapted from Ira Telford and Charles Bridgman, *Introduction to Functional Histology* 2e, p84, p261, p262 (New York: HarperCollins, 1995). ©1995 HarperCollins College Publishers. By permission of Addison Wesley Longman.

🔑 **Most of the epidermis consists of keratinocytes, which produce the protein keratin (protects underlying tissues) and lamellar granules (contains a waterproof sealant).**

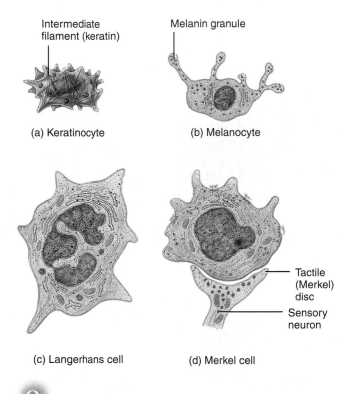

Intermediate filament (keratin)

Melanin granule

(a) Keratinocyte

(b) Melanocyte

(c) Langerhans cell

(d) Merkel cell

Tactile (Merkel) disc

Sensory neuron

Q What is the function of melanin?

where they contact the flattened process of a sensory neuron (nerve cell), a structure called a **tactile (Merkel) disc** (Figure 5.2d). Merkel cells and tactile discs function in the sensation of touch.

Several distinct layers of cells form the epidermis (Figure 5.3). In most regions of the body the epidermis has four strata or layers—stratum basale, stratum spinosum, stratum granulosum, and a thin stratum corneum. Where exposure to friction is greatest, such as in the fingertips, palms, and soles, the epidermis has five layers—stratum basale, stratum spinosum, stratum granulosum, stratum lucidum, and a thick stratum corneum.

Stratum Basale

The deepest layer of the epidermis is the **stratum basale** (*basal-* = base), composed of a single row of cuboidal or columnar keratinocytes, some of which are *stem cells* that undergo cell division to continually produce new keratinocytes. The nuclei of keratinocytes in the stratum basale are large, and their cytoplasm contains many ribosomes, a small Golgi complex, a few mitochondria, and some rough endoplasmic reticulum. The cytoskeleton within cells of the stratum basale includes intermediate filaments composed of keratin. These filaments attach to desmosomes, which bind cells of the stratum basale to each other and to the cells of the adjacent stratum spinosum. Keratin filaments also attach to hemidesmosomes, which bind the keratinocytes to the basement membrane between the epidermis and the dermis. Keratin also protects deeper layers from injury. Melanocytes, Langerhans cells, and Merkel cells with their associated tactile discs are scattered among the keratinocytes of the basal layer. The stratum basale is sometimes referred to as the **stratum germinativum** (jer′-mi-na-TĒ-vum; *germ-* = sprout) to indicate its role in forming new cells.

CLINICAL APPLICATION
Skin Grafts

New skin cannot regenerate if an injury destroys the stratum basale and its stem cells. Skin wounds of this magnitude require skin grafts in order to heal. A **skin graft** involves covering the wound with a patch of healthy skin taken from a donor site. To avoid tissue rejection, the transplanted skin is usually taken from the same individual (*autograft*) or an identical twin (*isograft*). If skin damage is so extensive that an autograft would cause harm, a self-donation procedure called *autologous skin transplantation* (aw-TOL-ō-gus) may be used. In this procedure, performed most often for severely burned patients, small amounts of an individual's epidermis are removed, and the keratinocytes are cultured in the laboratory to produce thin sheets of skin. The new skin is transplanted back to the patient so that it covers the burn wound and generates a permanent skin. ∎

Stratum Spinosum

Superficial to the stratum basale is the **stratum spinosum** (*spinos-* = thornlike), where 8–10 layers of polyhedral (many-sided) keratinocytes fit closely together. Cells in the more superficial portions of this layer become somewhat flattened. These keratinocytes have the same organelles as cells of the stratum basale, and some cells in this layer may retain their ability to undergo cell division. When cells of the stratum spinosum are prepared for microscopic examination, they shrink and pull apart such that they appear to be covered with thornlike spines (see Figure 5.2), although they are rounded and larger in living tissue. At each spinelike projection, bundles of intermediate filaments of the cytoskeleton insert into desmosomes, which tightly join the cells to one another. This arrangement provides both strength and flexibility to the skin. Projections of both Langerhans cells and melanocytes also appear in this stratum.

Figure 5.3 Layers of the epidermis.

🔑 **The epidermis consists of keratinized stratified squamous epithelium.**

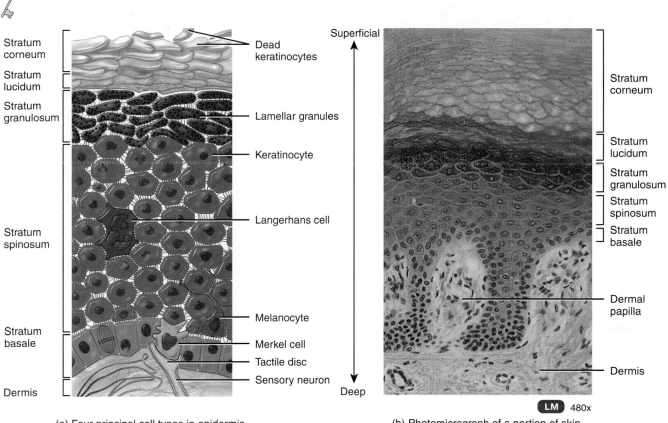

(a) Four principal cell types in epidermis

(b) Photomicrograph of a portion of skin

Q Which epidermal layer includes stem cells that continually undergo cell division?

Stratum Granulosum

At the middle of the epidermis, the **stratum granulosum** (*granulos-* = little grains) consists of three to five layers of flattened keratinocytes that are undergoing apoptosis. The nuclei and other organelles of these cells begin to degenerate, and intermediate filaments become more apparent. A distinctive feature of cells in this layer is the presence of darkly staining granules of a protein called **keratohyalin** (ker′-a-tō-HĪ-a-lin), which organizes intermediate filaments into even thicker bundles. Also present in the keratinocytes are membrane-enclosed **lamellar granules,** which release a lipid-rich secretion that fills the spaces between cells of the stratum granulosum and between more superficial cells of the epidermis. The lipid-rich secretion functions as a water-repellent sealant that retards loss of body fluids and entry of foreign materials. As their nuclei break down, the cells can no longer carry on vital metabolic reactions, and they die. Thus, the stratum granulosum marks the transition between the deeper, metabolically active strata and the dead cells of the more superficial strata.

Stratum Lucidum

The **stratum lucidum** (*lucid-* = clear) is present only in the skin of the fingertips, palms, and soles. It consists of three to five layers of clear, flat, dead keratinocytes that contain densely packed intermediate filaments and thickened plasma membranes.

Stratum Corneum

The **stratum corneum** (*corne-* = hard or hooflike) consists of 25–30 layers of dead, flat keratinocytes. The interior of the cells contains mostly densely packed intermediate filaments and keratohyalin. Between the cells are lipids from lamellar granules that help to make this layer water-repellent. These cells are continuously shed and replaced by cells from the deeper strata. The stratum corneum serves as an effective water-repellent barrier and also protects against injury and microbes. Constant exposure of skin to friction stimulates the formation of a *callus,* an abnormal thickening of the epidermis.

Table 5.1 Comparison of Epidermal Strata

STRATUM	DESCRIPTION
Basale	Deepest layer, composed of a single row of cuboidal or columnar keratinocytes that contain intermediate filaments composed of keratin; stem cells undergo cell division to produce new keratinocytes; melanocytes, Langerhans cells, and Merkel cells associated with tactile discs are scattered among the keratinocytes.
Spinosum	Eight to ten rows of polyhedral-shaped keratinocytes; includes projections of melanocytes and Langerhans cells.
Granulosum	Three to five rows of flattened keratinocytes, in which organelles are beginning to degenerate; cells contain the protein keratohyalin, which organizes intermediate filaments into thick bundles, and lamellar granules, which release a lipid-rich, water-repellent secretion.
Lucidum	Present only in skin of fingertips, palms, and soles; consists of three to five rows of clear, flat, dead keratinocytes with densely packed intermediate filaments.
Corneum	Twenty-five to 30 rows of dead, flat keratinocytes that contain densely packed intermediate filaments, keratohyalin, and lamellar granules.

Keratinization and Growth of the Epidermis

Newly formed cells in the stratum basale undergo a developmental process called **keratinization** as they are pushed to the surface. As the cells move from one epidermal layer to the next, they accumulate more and more keratin. Then they undergo apoptosis—the nucleus fragments, other organelles disappear, and the cells die. Eventually the keratinized cells slough off and are replaced by underlying cells that, in turn, become keratinized. The whole process by which cells form in the stratum basale, rise to the surface, become keratinized, and slough off takes about four weeks in average epidermis of 0.1 mm thickness. The rate of cell division in the stratum basale increases when the outer layers of the epidermis are stripped away, as occurs in abrasions and burns. The mechanisms that regulate this remarkable growth are not well understood, but hormonelike proteins such as **epidermal growth factor (EGF)** play a role.

Table 5.1 presents a summary of the distinctive features of the epidermal strata.

CLINICAL APPLICATION
Psoriasis

Psoriasis is a common and chronic skin disorder in which keratinocytes divide and move more quickly than normal from the stratum basale to the stratum corneum. They are shed prematurely in as little as 7–10 days. The immature keratinocytes make an abnormal keratin, which forms flaky, silvery scales at the skin surface, most often on the knees, elbows, and scalp (dandruff). Effective treatments—various topical ointments and UV phototherapy—suppress cell division, decrease the rate of cell growth, or inhibit keratinization. ■

Dermis

The second, deeper part of the skin, the **dermis,** is composed mainly of connective tissue containing collagen and elastic fibers. The few cells present in the dermis include fibroblasts, macrophages, and some adipocytes. Blood vessels, nerves, glands, and hair follicles are embedded in dermal tissue. Based on its tissue structure, the dermis can be divided into both a superficial papillary region and a deeper reticular region.

The **papillary region** is the superficial portion of the dermis, about one-fifth of the thickness of the total layer (see Figure 5.1). It consists of areolar connective tissue containing fine elastic fibers. Its surface area is greatly increased by small, fingerlike projections called **dermal papillae** (pa-PIL-ē; = nipples). These nipple-shaped structures indent the epidermis and contain loops of capillaries. Some dermal papillae also contain tactile receptors called **corpuscles of touch** or **Meissner corpuscles,** nerve endings that are sensitive to touch. Also present in the dermal papillae are **free nerve endings,** dendrites that lack any apparent structural specialization. Individual free nerve endings initiate signals that eventually are felt as sensations of warmth, coolness, pain, tickling, and itching.

The deeper portion of the dermis is called the **reticular region** (*reticul-* = netlike). It consists of dense, irregular connective tissue containing bundles of collagen and some coarse elastic fibers. The bundles of collagen fibers in the reticular region interlace in a netlike manner. Spaces between the fibers are occupied by a few adipose cells, hair follicles, nerves, sebaceous (oil) glands, and sudoriferous (sweat) glands.

The combination of collagen and elastic fibers in the reticular region provides the skin with strength, *extensibility* (ability to stretch), and *elasticity* (ability to return to original shape after stretching). The extensibility of skin can readily be seen in pregnancy and obesity. Small tears that occur in the dermis caused by extreme stretching produce *striae* (STRĪ-ē; *striae* = streaks), or stretch marks, which are visible as red or silvery white streaks on the skin surface.

The surface of the palms, fingers, soles, and toes is marked by a series of ridges and grooves. They appear either as straight lines or as a pattern of loops and whorls, as on the tips of the digits. These **epidermal ridges** develop during the third and fourth fetal months as the epidermis conforms to the contours of the underlying dermal papillae of the papillary region. The ridges function to increase the grip of the hand or foot by increasing friction. Because the ducts of sweat glands open on the tops of the epidermal ridges as sweat pores, the sweat and ridges form fingerprints (or footprints) when a smooth object is touched. The ridge pattern is genetically determined and is unique for each individual. Normally, the ridge pattern does not change during life, except to enlarge, and thus can serve as the basis for identification through fingerprints or footprints. A comparison of the structural features of the papillary and reticular regions of the dermis is presented in Table 5.2.

The Structural Basis of Skin Color

OBJECTIVE
• *Explain the basis for skin color.*

Melanin, carotene, and hemoglobin are three pigments that give skin a wide variety of colors. The amount of **melanin,** which is located mostly in the epidermis, causes the skin's color to vary from pale yellow to tan to black. Melanocytes are most plentiful in the mucous membranes, penis, nipples of the breasts and the area just around the nipples (areolae), face, and limbs. Because the *number* of melanocytes is about the same in all races, differences in skin color are due mainly to the *amount of pigment* the melanocytes produce and disperse to keratinocytes. In some people, melanin tends to accumulate in patches called *freckles.* As one grows older, *liver (age) spots* may develop. These are flat skin patches that look like freckles and range in color from light brown to black. Like freckles, liver spots are accumulations of melanin.

Melanocytes synthesize melanin from the amino acid *tyrosine* in the presence of an enzyme called *tyrosinase.* Synthesis occurs in an organelle called a **melanosome.** Exposure to UV light increases the enzymatic activity within melanosomes and leads to increased melanin production. Both the amount and darkness of melanin increase, which gives the skin a tanned appearance and further protects the body against UV radiation. Thus, within limits, melanin serves a protective function. As you will see, however, repeatedly exposing the skin to UV light may cause skin cancer. A tan is lost when the melanin-containing keratinocytes are shed from the stratum corneum.

Carotene (KAR-ō-tēn; *carot-* = carrot), a yellow-orange pigment, is the precursor of vitamin A, which is used to synthesize pigments needed for vision. Carotene is found in the stratum corneum and fatty areas of the dermis and subcutaneous layer.

When little melanin or carotene are present, the epidermis appears translucent. Thus, the skin of white people appears pink to red, depending on the amount and oxygen content of the blood moving through capillaries in the dermis. The red color is due to **hemoglobin,** the oxygen-carrying pigment in red blood cells.

Albinism (AL-bin-izm; *albin-* = white) is the inherited inability of an individual to produce melanin. Most albinos (al-BĪ-nōs), people affected by albinism, have melanocytes that are unable to synthesize tyrosinase. Melanin is missing from their hair, eyes, and skin. In another condition, called **vitiligo** (vit-i-LĪ-gō), the partial or complete loss of melanocytes from patches of skin produces irregular white spots.

Table 5.2 Comparison of Papillary and Reticular Regions of the Dermis

REGION	DESCRIPTION
Papillary	The superficial portion of the dermis (about one-fifth); consists of areolar connective tissue with elastic fibers; contains dermal papillae that house capillaries, corpuscles of touch, and free nerve endings.
Reticular	The deeper portion of the dermis (about four-fifths); consists of dense, irregular connective tissue with bundles of collagen and some coarse elastic fibers. Spaces between fibers contain some adipose cells, hair follicles, nerves, sebaceous glands, and sudoriferous glands.

CLINICAL APPLICATION
Skin Color as a Diagnostic Clue

The color of skin and mucous membranes can provide clues for diagnosing certain conditions. When blood is not picking up an adequate amount of oxygen in the lungs, such as in someone who has stopped breathing, the mucous membranes, nail beds, and skin appear bluish or **cyanotic** (sī-an-OT-ic; *cyan-* = blue). This occurs because hemoglobin that is depleted of oxygen looks deep, purplish blue. **Jaundice** (JON-dis; *jaund-* = yellow) is due to a buildup of the yellow pigment bilirubin in the blood. This condition gives a yellowish appearance to the whites of the eyes and the skin. Jaundice usually indicates liver disease. **Erythema** (er-e-THĒ-ma; *eryth-* = red), redness of the skin, is caused by engorgement of capillaries in the dermis with blood due to skin injury, exposure to heat, infection, inflammation, or allergic reactions. All skin color changes are observed most readily in people with lighter-colored skin and may be more difficult to discern in people with darker skin. ■

1. What is the integumentary system?
2. How does the subcutaneous layer relate to the skin?
3. Compare the structure of epidermis and dermis.
4. List the distinctive features of the epidermal layers from deepest to most superficial.
5. Compare the composition of the papillary and reticular regions of the dermis.
6. How are epidermal ridges formed?
7. Explain the factors that produce skin color.

ACCESSORY STRUCTURES OF THE SKIN

OBJECTIVE
• *Compare the structure, distribution, and functions of hair, skin glands, and nails.*

Accessory structures of the skin—hair, glands, and nails—develop from the embryonic epidermis. They have a host of important functions: Hair and nails protect the body, and sweat glands help regulate body temperature.

Hair

Hairs, or *pili* (PI-lē), are present on most skin surfaces except the palms, palmar surfaces of the digits and soles, and plantar surfaces of the digits. In adults, hair usually is most heavily distributed across the scalp, over the brows of the eyes, and around the external genitalia. Genetic and hormonal influences largely determine the thickness and pattern of distribution of hair.

Anatomy of a Hair

Each hair is composed of columns of dead, keratinized cells bonded together by extracellular proteins. The **shaft** is the superficial portion of the hair, most of which projects from the surface of the skin (Figure 5.4a). The shaft of straight hair is round in transverse section, whereas that of curly hair is oval. The **root** is the portion of the hair deep to the shaft that penetrates into the dermis, and sometimes into the subcutaneous layer. The shaft and root both consist of three concentric layers (Figure 5.4c, d). The inner *medulla* is composed of two or three rows of polyhedral-shaped cells containing pigment granules and air spaces. The middle *cortex* forms the major part of the shaft and consists of elongated cells that contain pigment granules in dark hair but mostly air in gray or white hair. The *cuticle* of the hair, the outermost layer, consists of a single layer of thin, flat cells that are the most heavily keratinized. Cuticle cells are arranged like shingles on the side of a house, with their free edges pointing toward the end of the hair (Figure 5.4b).

Surrounding the root of the hair is the **hair follicle,** which is made up of an external root sheath and an internal root sheath (see Figure 5.4c, d). The *external root sheath* is a downward continuation of the epidermis. Near the surface, it contains all the epidermal layers. At the base of the hair follicle, the external root sheath contains only the stratum basale. The *internal root sheath* forms a cellular tubular sheath between the external root sheath and the hair.

The base of each hair follicle is an onion-shaped structure, the **bulb** (see Figure 5.4c). This structure houses a nipple-shaped indentation, the **papilla of the hair,** which contains areolar connective tissue. The papilla of the hair contains many blood vessels that nourish the growing hair follicle. The bulb also contains a germinal layer of cells called the **matrix.** The matrix cells arise from the stratum basale, the site of cell division. Hence, matrix cells are responsible for the growth of existing hairs, and they produce new hairs when old hairs are shed. This replacement process occurs within the same follicle. Matrix cells also give rise to the cells of the internal root sheath.

Sebaceous (oil) glands (discussed shortly) and a bundle of smooth muscle cells are also associated with hairs (see Figure 5.4a). The smooth muscle is called **arrector pili** (*arrect-* = to raise). It extends from the superficial dermis of the skin to the side of the hair follicle. In its normal position, hair emerges at an angle to the surface of the skin. Under physiologic or emotional stress, such as cold or fright, autonomic nerve endings stimulate the arrector pili muscles to contract, which pulls the hair shafts perpendicular to the skin surface. This action causes "goose bumps" or "gooseflesh" because the skin around the shaft forms slight elevations.

Surrounding each hair follicle are dendrites of neurons, called **hair root plexuses,** that are sensitive to touch (see Figure 5.4a). The hair root plexuses generate nerve impulses if their hair shaft is moved.

Hair Growth

Each hair follicle goes through a *growth cycle,* which consists of a growth stage and a resting stage. During the **growth stage,** cells of the matrix differentiate, keratinize, and die. This process forms the root sheath and hair shaft. As new cells are added at the base of the hair root, the hair grows longer. In time, the growth of the hair stops and the **resting stage** begins. After the resting stage, a new growth cycle begins. The old hair root falls out or is pushed out of the hair follicle, and a new hair begins to grow in its place. Scalp hair grows for about 2–6 years and rests for about 3 months. At any time, about 85% of scalp hairs are in the growth stage.

Normal hair loss in an adult scalp is about 70–100 hairs per day. Both the rate of growth and the replacement cycle can be altered by illness, diet, high fever, surgery, blood loss, severe emotional stress, and gender. Rapid weight-loss diets that severely restrict calories or protein increase hair loss. An increase in the rate of hair shedding can also occur for 3–4 months after childbirth, with certain drugs, and after radiation therapies for cancer.

Hair Color

The color of hair is due primarily to the amount and type of melanin in its keratinized cells. Melanin is synthesized by melanocytes scattered in the matrix of the bulb and passes into cells of the cortex and medulla (see Figure 5.4c). Dark-colored hair contains mostly true melanin, whereas blond and red hair contain variants of melanin in which there is iron and more sulfur. Graying hair occurs because of a progressive decline in tyrosinase, whereas white hair results from accumulation of air bubbles in the medullary shaft.

Functions of Hair

Although the protection it offers is limited, hair on the head guards the scalp from injury and the sun's rays. It also decreases heat loss from the scalp. Eyebrows and eyelashes protect the eyes from foreign particles, as does hair in the nostrils and in the external ear canal. Touch receptors associated with hair follicles (hair root plexuses) are activated whenever a hair is even slightly moved. Thus, hairs function in sensing light touch.

Skin Glands

Several kinds of exocrine glands are associated with the skin: sebaceous (oil) glands, sudoriferous (sweat) glands, ceruminous glands, and mammary glands. Mammary

Figure 5.4 Hair.

Hairs are growths of epidermis composed of dead, keratinized cells.

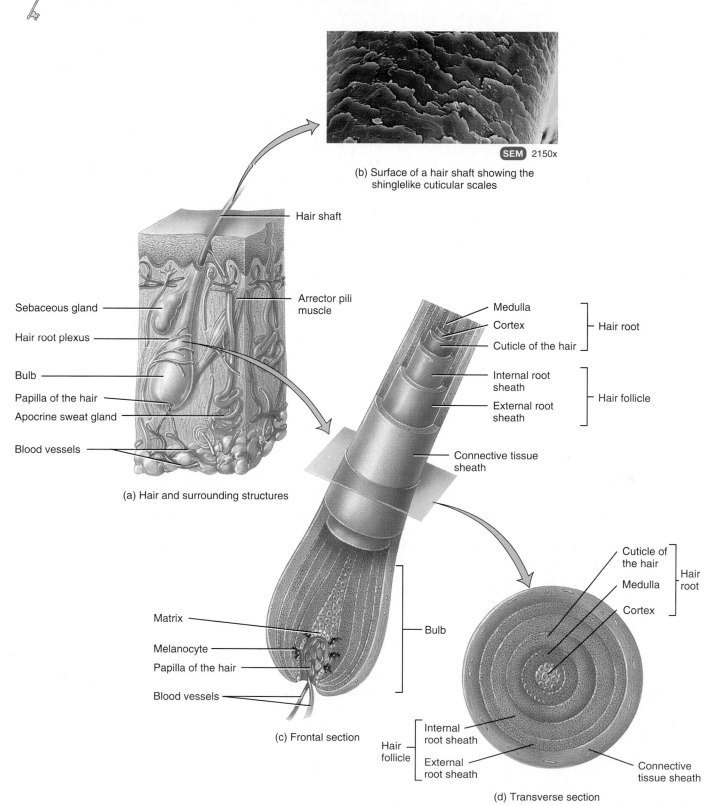

SEM 2150x

(b) Surface of a hair shaft showing the shinglelike cuticular scales

Hair shaft

Sebaceous gland

Hair root plexus

Bulb

Papilla of the hair

Apocrine sweat gland

Blood vessels

Arrector pili muscle

(a) Hair and surrounding structures

Medulla
Cortex — Hair root
Cuticle of the hair

Internal root sheath
External root sheath — Hair follicle

Connective tissue sheath

Matrix

Melanocyte

Papilla of the hair

Blood vessels

Bulb

(c) Frontal section

Hair follicle — Internal root sheath / External root sheath

Cuticle of the hair
Medulla — Hair root
Cortex

Connective tissue sheath

(d) Transverse section

Q Why does it hurt when you pluck a hair out but not when you have a haircut?

glands, which are specialized sudoriferous glands that secrete milk, are discussed in Chapter 28 along with the female reproductive system.

Sebaceous Glands

Sebaceous glands (se-BĀ-shus; *sebace-* = greasy) or **oil glands** are simple, branched acinar glands. With few exceptions, they are connected to hair follicles (see Figures 5.1 and 5.4a). The secreting portion of a sebaceous gland lies in the dermis and usually opens into the neck of the hair follicle. In other locations, such as the lips, glans penis, labia minora, and tarsal glands of the eyelids, sebaceous glands open directly onto the surface of the skin. Absent in the palms and soles, sebaceous glands vary in size and shape in the skin over other regions of the body. For example, they are small in most areas of the trunk and limbs, but large in the skin of the breasts, face, neck, and upper chest.

Sebaceous glands secrete an oily substance called **sebum** (SĒ-bum), which is a mixture of fats, cholesterol, proteins, inorganic salts, and pheromones. Sebum coats the surface of hairs and helps keep them from drying and becoming brittle. Sebum also prevents excessive evaporation of water from the skin, keeps the skin soft and pliable, and inhibits the growth of certain bacteria.

CLINICAL APPLICATION
Acne

Acne is an inflammation of sebaceous glands that usually begins at puberty, when the sebaceous glands grow in size and increase their production of sebum. Although testosterone, a male sex hormone, appears to play the greatest role in stimulating sebaceous glands, other steroid hormones from the ovaries and adrenal glands stimulate sebaceous secretions in females. Acne occurs predominantly in sebaceous follicles that have been colonized by bacteria, some of which thrive in the lipid-rich sebum. The infection may cause a cyst or sac of connective tissue cells to form, which can destroy and displace epidermal cells. This condition, called **cystic acne,** can permanently scar the epidermis. ■

Sudoriferous Glands

Three to four million **sweat glands,** or **sudoriferous glands** (soo'-dor-IF-er-us; *sudori-* = sweat; *-ferous* = bearing), release their secretions by exocytosis and empty them onto the skin surface through pores or into hair follicles. They are divided into two main types, eccrine and apocrine, based on their structure, location, and type of secretion.

Eccrine sweat glands (*eccrine* = secreting outwardly) are simple, coiled tubular glands and are much more common than apocrine sweat glands. They are distributed throughout the skin, except for the margins of the lips, nail beds of the fingers and toes, glans penis, glans clitoris, labia minora, and eardrums. Eccrine sweat glands are most numerous in the skin of the forehead, palms, and soles; their density can be as high as 450 per square centimeter (3000 per square inch) in the palms. The secretory portion of eccrine sweat glands is located mostly in the deep dermis (sometimes in the upper subcutaneous layer). The excretory duct projects through the dermis and epidermis and ends as a pore at the surface of the epidermis (see Figure 5.1).

The sweat produced by eccrine sweat glands (about 600 mL per day) consists of water, ions (mostly Na^+ and Cl^-), urea, uric acid, ammonia, amino acids, glucose, and lactic acid. The main function of eccrine gland sweat is to help regulate body temperature through evaporation. As sweat evaporates, large quantities of heat energy leave the body surface. Sweat, or perspiration, usually occurs first on the forehead and scalp, extends to the face and the rest of the body, and occurs last on the palms and soles. Under conditions of emotional stress, however, the palms, soles, and axillae are the first surfaces to sweat. Eccrine sweat also plays a small role in eliminating wastes such as urea, uric acid, and ammonia. Sweat that evaporates from the skin before it is perceived as moisture is referred to as **insensible perspiration.** Sweat that is excreted in larger amounts and is perceived as moisture on the skin is called **sensible perspiration.**

Apocrine sweat glands are also simple, coiled tubular glands. They are found mainly in the skin of the axilla (armpit), groin, areolae (pigmented areas around the nipples) of the breasts, and bearded regions of the face in adult males. These glands were once thought to release their secretions in an apocrine manner—by pinching off a portion of the cell. We now know, however, that their secretion is released by exocytosis, which is characteristic of the release of secretions by merocrine glands (see Chapter 4). Nevertheless, the term *apocrine* is still used. The secretory portion of these sweat glands is located mostly in the subcutaneous layer, and the excretory duct opens into hair follicles (see Figure 5.1). Their secretory product is slightly viscous compared to eccrine secretions and contains the same components as eccrine sweat plus lipids and proteins. In women, cells of apocrine sweat glands enlarge about the time of ovulation and shrink during menstruation. Whereas eccrine sweat glands start to function soon after birth, apocrine sweat glands do not begin to function until puberty. Apocrine sweat glands are stimulated during emotional stress and sexual excitement; these secretions are commonly known as a "cold sweat."

Table 5.3 presents a comparison of eccrine and apocrine sweat glands.

Ceruminous Glands

Modified sweat glands in the external ear, called **ceruminous glands** (se-ROO-mi-nus; *cer-* = wax), produce a waxy secretion. The secretory portions of ceruminous glands lie in the subcutaneous layer, deep to sebaceous glands. Their excretory ducts open either directly onto the surface of the external auditory canal (ear canal) or into ducts of sebaceous glands. The combined secretion of the ceruminous and sebaceous glands is called **cerumen,** or earwax. Cerumen, together with hairs in the external auditory canal, provides a sticky barrier that prevents the entrance of foreign bodies.

CLINICAL APPLICATION
Impacted Cerumen

Some people produce an abnormally large amount of cerumen in the external auditory canal. The cerumen may accumulate until it becomes impacted (firmly wedged), which prevents sound waves from reaching the eardrum. The treatment for impacted cerumen is usually periodic ear irrigation or removal of wax with a blunt instrument by trained medical personnel. The use of cotton-tipped swabs or sharp objects is not recommended for this purpose because they may push the cerumen farther into the external auditory canal and damage the eardrum. ■

Nails

Nails are plates of tightly packed, hard, keratinized epidermal cells. The cells form a clear, solid covering over the dorsal surfaces of the distal portions of the digits. Each nail consists of a nail body, a free edge, and a nail root (Figure 5.5). The **nail body** is the portion of the nail that is visible, the **free edge** is the part that may extend past the distal end of the digit, and the **nail root** is the portion that is buried in a fold of skin. Most of the nail body appears pink because of blood flowing through underlying capillaries. The free edge is white because there are no underlying capillaries. The whitish, crescent-shaped area of the proximal end of the nail body is called the **lunula** (LOO-nyoo-la; = little moon). It appears whitish because the vascular tissue underneath does not show through due to the thickened stratum basale in the area. Beneath the free edge is a thickened region of stratum corneum called the **hyponychium** (hī-'pō-NIK-ē-um; *hypo-* = below; *-onych* = nail), which secures the nail to the fingertip. The **eponychium** (ep'-ō-NIK-ē-um; *ep-* = above) or **cuticle** is a narrow band of epidermis that extends from and adheres to the margin (lateral border) of the nail wall. It occupies the proximal border of the nail and consists of stratum corneum.

The epithelium deep to the nail root is known as the **nail matrix,** the cells of which divide mitotically to produce growth. Nail growth occurs by the transformation of superficial cells of the matrix into nail cells. In the process, the harder outer layer is pushed forward over the stratum basale. The growth rate of nails is determined by the rate at which matrix cells divide, which is influenced by factors such as a person's age, health, and nutritional status. Nail growth also varies according to the season, the time of day, and environmental temperature. The average growth in the length of fingernails is about 1 mm (0.04 in.) per week. The growth rate

Table 5.3 Comparison of Eccrine and Apocrine Sweat Glands

FEATURE	ECCRINE SWEAT GLANDS	APPOCRINE SWEAT GLANDS
Distribution	Throughout skin, except for margins of lips, nail beds, glans penis and clitoris, labia minora, and eardrums.	Skin of the axilla, groin, areolae, and bearded regions of the face.
Location of secretory portion	Mostly in deep dermis.	Mostly in subcutaneous layer.
Termination of excretory duct	Surface of epidermis.	Hair follicle.
Secretion	Less viscous; consists of water, ions (Na$^+$, Cl$^-$), urea, uric acid, ammonia, amino acids, glucose, lactic acid.	More viscous; consists of the same components as eccrine sweat glands plus lipids and proteins.
Functions	Regulation of body temperature and waste removal.	Stimulated during emotional stress and sexual excitement.
Onset of function	Soon after birth.	Puberty.

is somewhat slower in toenails. The longer the digit, the faster the nail grows.

Functionally, nails help us grasp and manipulate small objects in various ways, provide protection against trauma to the ends of the digits, and allow us to scratch various parts of the body.

1. Describe the structure of a hair. What produces "goose bumps"?
2. Contrast the locations and functions of sebaceous (oil) glands, sudoriferous (sweat) glands, and ceruminous glands.
3. Describe the principal parts of a nail.

TYPES OF SKIN

OBJECTIVE

• *Compare structural and functional differences in thin and thick skin.*

Although the skin over the entire body is similar in structure, there are quite a few local variations related to thickness of the epidermis, strength, flexibility, degree of keratinization, distribution and type of hair, density and types of glands, pigmentation, vascularity (blood supply), and innervation (nerve supply). On the basis of certain structural and functional properties, we recognize two major types of skin: thin (hairy) and thick (hairless).

Figure 5.5 Nails. Shown is a fingernail.

Nail cells arise by transformation of superficial cells of the nail matrix into nail cells.

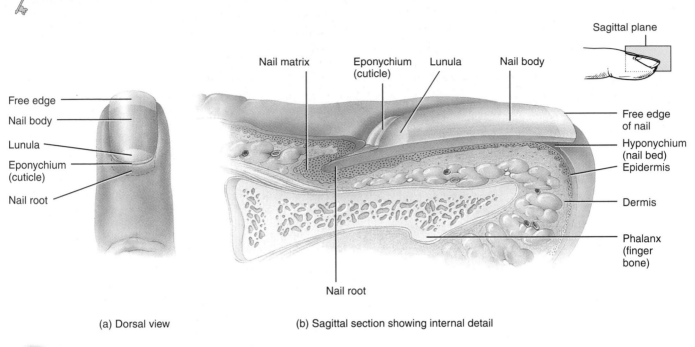

(a) Dorsal view   (b) Sagittal section showing internal detail

Q   Why are nails so hard?

Thin skin covers all parts of the body except for the palms, palmar surfaces of the digits, and the soles. Its epidermis is thin, just 0.10–0.15 mm. A distinct stratum lucidum is lacking, and the strata spinosum and corneum are relatively thin. Thin skin has lower, broader, and fewer dermal papillae, and thus lacks epidermal ridges. Additionally, even though thin skin has hair follicles, arrector pili muscles, and sebaceous (oil) glands, it has fewer sweat glands than thick skin. Finally, thin skin has a sparser distribution of sensory receptors than thick skin.

Thick skin covers the palms, palmar surfaces of the digits, and soles. Its epidermis is relatively thick, 0.6–4.5 mm, and features a distinct stratum lucidum as well as thicker strata spinosum and corneum. The dermal papillae of thick skin are higher, narrower, and more numerous than those in thin skin, and thus thick skin has epidermal ridges. Thick skin lacks hair follicles, arrector pili muscles, and sebaceous glands, and has more sweat glands than thin skin. Sensory receptors are more densely clustered in thick skin.

Table 5.4 presents a summary of the features of thin and thick skin.

FUNCTIONS OF THE SKIN

OBJECTIVE

• *Describe how the skin contributes to thermoregulation, protection, sensation, excretion and absorption, and synthesis of vitamin D.*

The skin helps to regulate body temperature, serves as a water-repellent and protective barrier between the external environment and internal tissues, contains sensory nerve endings, excretes a small amount of salts and several organic compounds and has some capacity to absorb substances, and helps to synthesize the active form of vitamin D. We next briefly examine each of these functions.

Thermoregulation

The skin contributes to *thermoregulation,* the homeostatic regulation of body temperature, in two ways: by liberating sweat at its surface and by adjusting the flow of blood in the dermis. In response to high environmental temperature or heat produced by exercise, the evaporation of sweat from the skin surface helps lower body temperature. In response to low environmental temperature, production of sweat is decreased, which helps conserve heat. During moderate exercise, the flow of blood through the skin increases, which increases the amount of heat radiated from the body.

The dermis houses an extensive network of blood vessels that carry 8–10% of the total blood flow in a resting adult. For this reason, the skin acts as a *blood reservoir.* During very strenuous exercise, however, skin blood vessels constrict (narrow) somewhat, diverting more blood to contracting skeletal muscles and the heart. Because of this shunting of blood away from the skin, however, less heat is lost from the skin, and body temperature tends to rise.

Table 5.4 Comparison of Thin and Thick Skin

FEATURE	THIN SKIN	THICK SKIN
Distribution	All parts of the body except palms, palmar surface of digits, and soles.	Palms, palmar surface of digits, and soles.
Epidermal thickness	0.10–0.15 mm.	0.6–4.5 mm.
Epidermal strata	Stratum lucidum essentially lacking; thinner strata spinosum and corneum.	Thick strata lucidum, spinosum, and corneum.
Epidermal ridges	Lacking due to poorly developed and fewer dermal papillae.	Present due to well-developed and more numerous dermal papillae.
Hair follicles and arrector pili muscles	Present.	Absent.
Sebaceous glands	Present.	Absent.
Sudoriferous glands	Fewer.	More numerous.
Sensory receptors	Sparser.	Denser.

Protection

The skin covers the body and provides protective physical, chemical, and biological barriers. Physically, the skin protects underlying tissues from abrasion, and the tightly interlocked keratinocytes resist invasion by microbes on the skin surface. Lipids released by lamellar granules retard evaporation of water from the skin surface, thus protecting the body from dehydration, and they also retard entry of water across the skin surface when we take a shower or go swimming. The oily sebum from the sebaceous glands also protects skin and hairs from drying out and contains bactericidal chemicals that kill surface bacteria. The pigment melanin provides some protection against the damaging effects of UV light. Protective functions that are biological in nature are carried out by epidermal Langerhans cells, which alert the immune system to the presence of potentially harmful microbial invaders, and by macrophages in the dermis, which phagocytize bacteria and viruses that manage to penetrate the skin surface.

Cutaneous Sensations

Cutaneous sensations are those that arise in the skin. These include tactile sensations—touch, pressure, vibration, and tickling—as well as thermal sensations such as warmth and coolness. Another cutaneous sensation, pain, usually is an indication of impending or actual tissue damage. Some of the wide variety of abundantly distributed nerve endings and receptors in the skin include the tactile discs of the epidermis, the corpuscles of touch in the dermis, and hair root plexuses around each hair follicle. Chapter 15 provides more details on the topic of cutaneous sensations.

Excretion and Absorption

The skin plays minor roles in *excretion*, the elimination of substances from the body, and *absorption*, the passage of materials from the external environment into body cells. Despite the almost waterproof nature of the stratum corneum, about 400 mL of water evaporates through it daily. A sedentary person loses an additional 200 mL per day of water as sweat, whereas a physically active person loses much more. Besides removing water and heat (by evaporation), sweat also is the vehicle for excretion of small amounts of salts, carbon dioxide, and two organic molecules that result from the breakdown of proteins—ammonia and urea.

The absorption of water-soluble substances through the skin is negligible, but certain lipid-soluble materials do penetrate the skin. These include fat-soluble vitamins (A, D, E, and K) and oxygen and carbon dioxide gases. Toxic materials that can be absorbed through the skin include organic solvents such as acetone (in some nail polish removers) and carbon tetrachloride (dry-cleaning fluid); salts of heavy metals such as lead, mercury, and arsenic; and the toxins in poison ivy and poison oak.

Synthesis of Vitamin D

What is commonly called vitamin D is actually a group of closely related compounds. Synthesis of vitamin D requires activation of a precursor molecule in the skin by UV rays in sunlight. Enzymes in the liver and kidneys then modify the activated molecule, finally producing *calcitriol*, the most active form of vitamin D. Calcitriol aids in the absorption of calcium in foods from the gastrointestinal tract into the blood.

CLINICAL APPLICATION
Transdermal Drug Administration

Most drugs are either absorbed into the body through the digestive system or injected into subcutaneous tissue or muscle. An alternative route, **transdermal drug administration,** enables a drug contained within an adhesive skin patch to pass across the epidermis and into the blood vessels of the dermis. The drug is released at a controlled rate over one to several days. Because the major barrier to penetration of most drugs is the stratum corneum, transdermal absorption is most rapid in regions of the skin where this layer is thin, such as the scrotum, face, and scalp. A growing number of drugs are available for transdermal administration. These drugs include nitroglycerin, for prevention of angina pectoris (chest pain associated with heart disease); scopolamine, for motion sickness; estradiol, used for estrogen-replacement therapy during menopause; and nicotine, used to help people stop smoking. ■

1. In what two ways does the skin help regulate body temperature?
2. In what ways does the skin serve as a barrier?
3. What sensations arise from stimulation of neurons in the skin?
4. What types of molecules can penetrate the stratum corneum?

MAINTAINING HOMEOSTASIS: SKIN WOUND HEALING

OBJECTIVE
• *Explain how epidermal wounds and deep wounds heal.*

When the integrity or continuity of the skin is damaged, a sequence of processes begins that repairs the skin to its normal (or near-normal) structure and function. Two kinds of wound-healing processes can occur, depending on the depth of the injury. Epidermal wound healing occurs following wounds that affect only the epidermis; deep wound healing occurs following wounds that penetrate the dermis or subcutaneous layer.

Epidermal Wound Healing

Even though the central portion of an epidermal wound may extend to the dermis, the edges of the wound usually involve only slight damage to superficial epidermal cells. Common types of epidermal wounds include abrasions, in which a portion of skin has been scraped away, and minor burns.

In response to an epidermal injury, basal cells of the epidermis surrounding the wound break contact with the basement membrane. The cells then enlarge and migrate across the wound (Figure 5.6a). The cells appear to migrate as a sheet until advancing cells from opposite sides of the wound meet. When epidermal cells encounter each other, they stop migrating due to a cellular response called **contact inhibition.** Migration of the epidermal cells stops completely when each is finally in contact with other epidermal cells on all sides.

While some basal epidermal cells migrate, epidermal growth factor, a hormone, stimulates basal stem cells to divide and replace the ones that have moved into the wound. Replacement of migrating basal cells continues until the wound is resurfaced (Figure 5.6b). Following this, the relocated cells divide to build new strata, thus thickening the new epidermis.

Deep Wound Healing

Deep wound healing occurs when an injury extends to the dermis and subcutaneous layer. Because multiple tissue layers must be repaired, the healing process is more complex than in epidermal wound healing. In addition, scar tissue is formed, and the healed tissue loses some of its normal function. Deep wound healing occurs in four phases: an inflammatory phase, a migratory phase, a proliferative phase, and a maturation phase.

During the **inflammatory phase,** a blood clot forms in the wound and loosely unites the wound edges (Figure 5.7a). This phase of deep wound healing involves **inflammation,** a vascular and cellular response that helps eliminate microbes, foreign material, and dying tissue in preparation for repair. The vasodilation and increased permeability of blood vessels associated with inflammation enhance delivery of white blood cells called neutrophils; of monocytes, which develop into macrophages that phagocytize microbes; and of mesenchymal cells, which develop into fibroblasts (Figure 5.7a).

The three phases that follow do the work of repairing the wound. In the **migratory phase,** the clot becomes a scab, and epithelial cells migrate beneath the scab to bridge the wound. Fibroblasts migrate along fibrin threads and begin synthesizing scar tissue (collagen fibers and glycoproteins), and damaged blood vessels begin to regrow. During this phase, the tissue filling the wound is called **granulation tissue.** The **proliferative phase** is characterized by extensive growth of epithelial cells beneath the scab, deposition by fibroblasts of collagen fibers in random patterns, and continued growth of blood vessels. Finally, during the **maturation phase,** the scab sloughs off once the epidermis has been restored to normal thickness. Collagen fibers become more organized, fibroblasts decrease in number, and blood vessels are restored to normal (Figure 5.7b).

As noted in Chapter 4, the process of scar tissue formation is called **fibrosis.** Sometimes, so much scar tissue is formed during deep wound healing that a raised scar results—that is, one that is elevated above the normal epidermal surface. If such a scar remains within the boundaries of the original wound, it is a **hypertrophic scar;** if it extends beyond the boundaries of the original wound into normal surrounding tissues, it is a **keloid scar.** Scar tissue differs from normal skin in that its collagen fibers are more densely arranged. Scar tissue also has fewer blood vessels and might not contain the same number of hairs, skin glands, or sensory structures as undamaged skin. Because of the arrangement of collagen fibers and the scarcity of blood vessels, scars are usually lighter in color than normal skin.

1. Why doesn't epidermal wound healing result in scar formation?

DEVELOPMENTAL ANATOMY OF THE INTEGUMENTARY SYSTEM

OBJECTIVE
• *Describe the development of the epidermis, its accessory structures, and the dermis.*

Figure 5.6 Epidermal wound healing.

🔑 **In an epidermal wound, the injury does not extend into the dermis.**

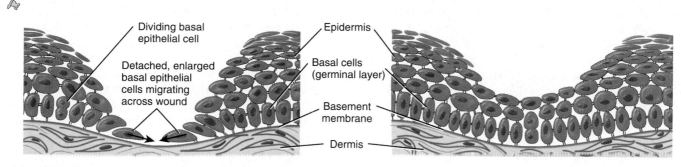

(a) Division of basal epithelial cells and migration across wound

(b) Thickening of epidermis

Ⓠ Would you expect an epidermal wound to bleed? Why or why not?

Figure 5.7 Deep wound healing. The initial inflammatory phase (a) is followed by a migratory phase, a proliferative phase, and finally a (b) maturation phase.

🔑 **In a deep wound, the injury extends deep to the epidermis.**

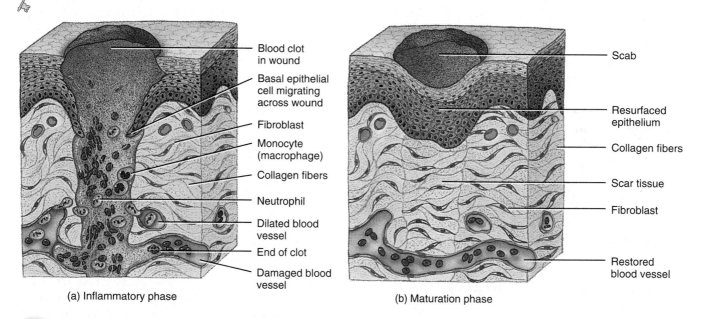

(a) Inflammatory phase

(b) Maturation phase

Ⓠ Why is the arrival of phagocytic white blood cells (neutrophils and macrophages) at the scene of a wound helpful?

At the end of many chapters throughout this book, we will discuss the developmental anatomy of body systems. Because the principal features of embryonic development are not described in detail until Chapter 29, it is necessary here to review a few terms and introduce others so that you can follow the development of organ systems.

As part of the early development of a fertilized egg, a portion of the developing embryo differentiates into three layers of tissue called **primary germ layers.** On the basis of position, the three primary germ layers are referred to as **ectoderm** (*ecto-* = outside), **mesoderm** (*meso-* = middle), and **endoderm** (*endo-* = within). They are the embryonic tissues from which all tissues and organs of the body will eventually develop (see Table 29.1 on page 1029).

The *epidermis* is derived from the **ectoderm.** At the beginning of the eighth week after fertilization the ectoderm consists of simple cuboidal epithelium. These cells become flattened and are known as the **periderm.** By the fourth month all layers of the epidermis are formed and each layer assumes its characteristic structure.

The *dermis* is derived from **mesodermal cells** in a zone beneath the ectoderm. There they undergo a process that changes them into the connective tissues that begin to form the dermis at about 11 weeks.

Nails are developed at about 10 weeks. Initially they consist of a thick layer of epithelium called the **primary nail field.** The nail itself is keratinized epithelium and grows distally from its base. It is not until the ninth month that the nails actually reach the tips of the digits.

Hair follicles develop between the ninth and twelfth weeks as downgrowths of the stratum basale of the epidermis into the deeper dermis. The downgrowths soon differentiate into the bulb and the papilla of the hair, beginnings of the epithelial portions of sebaceous glands, and other structures associated with hair follicles. By the fifth or sixth month, the follicles produce **lanugo** (delicate fetal hair), first on the head and then on other parts of the body. The lanugo is usually shed prior to birth.

The epithelial (secretory) portions of *sebaceous glands* develop from the sides of the hair follicles at about 16 weeks and remain connected to the follicles.

The epithelial portions of *sudoriferous glands* are also derived from downgrowths of the stratum basale of the epidermis into the dermis. They appear at about 20 weeks on the palms and soles and a little later in other regions. The connective tissue and blood vessels associated with the glands develop from **mesoderm.**

1. Which structures develop as downgrowths of the stratum basale?

AGING AND THE INTEGUMENTARY SYSTEM

OBJECTIVE

• *Describe the effects of aging on the integumentary system.*

At about the sixth month of fetal development, secretions from sebaceous glands mix with sloughed off epidermal cells and hairs to form a fatty substance called **vernix caseosa** (VER-niks KĀ-sē-ō-sa; *vernix* = varnish, *caseosa* = cheese). This substance covers and protects the skin of the fetus from the constant exposure to the amniotic fluid in which it is bathed. In addition, the vernix caseosa facilitates

the birth of the fetus because of its slippery nature. For most infants and children, relatively few problems are encountered with the skin as it ages. With the arrival of adolescence, however, some teens develop acne.

The pronounced effects of skin aging do not become noticeable until people reach their late forties. Most of the age-related changes occur in the dermis. Collagen fibers in the dermis begin to decrease in number, stiffen, break apart, and disorganize into a shapeless, matted tangle. Elastic fibers lose some of their elasticity, thicken into clumps, and fray, an effect that is greatly accelerated in the skin of smokers. Fibroblasts, which produce both collagen and elastic fibers, decrease in number. As a result, the skin forms the characteristic crevices and furrows known as wrinkles.

With further aging, Langerhans cells dwindle in number and macrophages become less-efficient phagocytes, thus decreasing the skin's immune responsiveness. Moreover, decreased size of sebaceous glands leads to dry and broken skin that is more susceptible to infection. Production of sweat diminishes, which probably contributes to the increased incidence of heat stroke in the elderly. There is a decrease in the number of functioning melanocytes, resulting in gray hair and atypical skin pigmentation. An increase in the size of some melanocytes produces pigmented blotching (liver spots). Walls of blood vessels in the dermis become thicker and less permeable, and subcutaneous fat is lost. For this reason, aged skin (especially the dermis) is thinner than young skin, and the migration of cells from the basal layer to the epidermal surface slows considerably. With the onset of old age, skin heals poorly and becomes more susceptible to pathological conditions such as skin cancer, itching, and pressure sores.

Growth of nails and hair slows during the second and third decades of life. The nails also may become more brittle with age, often due to dehydration or repeated use of cuticle remover or nail polish.

CLINICAL APPLICATION
Photodamage

Although basking in the warmth of the sun may feel good, it is not a healthy practice. Both the longer wavelength ultraviolet A (UVA) rays and the shorter wavelength UVB rays cause **photodamage** of the skin. Light-skinned and dark-skinned individuals alike experience the effect of acute overexposure to UVB rays—a sunburn. Even if sunburn does not occur, the UVB rays can damage the DNA in epidermal cells, which can produce genetic mutations that cause skin cancer. As UVA rays penetrate to the dermis, they produce oxygen free radicals that disrupt collagen and elastic fibers in the extracellular matrix. This is the main reason for the severe wrinkling that occurs in those who spend a great deal of time in the sun without protection. ■

DISORDERS: HOMEOSTATIC IMBALANCES

SKIN CANCER

Excessive exposure to the sun has caused virtually all of the 1 million cases of **skin cancer** diagnosed in the United States in the year 2000. There are three common forms of skin cancer. **Basal cell carcinomas** account for about 78% of all skin cancers. The tumors arise from cells in the stratum basale of the epidermis and rarely metastasize. **Squamous cell carcinomas,** which account for about 20% of all skin cancers, arise from squamous cells of the epidermis, and they have a variable tendency to metastasize. Most arise from preexisting lesions of damaged tissue on sun-exposed skin. Basal and squamous cell carcinomas are together known as *nonmelanoma skin cancer* and are 50% more common in males than in females. **Malignant melanomas** arise from melanocytes and account for about 2% of all skin cancers. They are the most prevalent life-threatening cancer in young women. The American Academy of Dermatology estimates that the lifetime risk of developing melanoma is currently 1 in 75, double the risk only 15 years ago. In part, this increase is due to depletion of the ozone layer, which absorbs some UV light high in the atmosphere. But the main reason for the increase is that more people are spending more time in the sun. Malignant melanomas metastasize rapidly and can kill a person within months of diagnosis.

The key to successful treatment of malignant melanoma is early detection. The early warning signs of malignant melanoma are identified by the acronym ABCD. *A* is for *asymmetry;* malignant melanomas tend to lack symmetry. *B* is for *border;* malignant melanomas have notched, indented, scalloped, or indistinct borders. *C* is for *color;* malignant melanomas have uneven coloration and may contain several colors. *D* is for *diameter;* ordinary moles tend to be smaller than 6 mm (0.25 in.), about the size of a pencil eraser. Once a malignant melanoma has the characteristics of A, B, and C, it is usually larger than 6 mm.

Among the risk factors for skin cancer are the following:

1. *Skin type.* Individuals with light-colored skin who never tan but always burn are at high risk.
2. *Sun exposure.* People who live in areas with many days of sunlight per year and at high altitudes (where ultraviolet light is more intense) have a higher risk of developing skin cancer. Likewise, people who engage in outdoor occupations and those who have suffered three or more severe sunburns have a higher risk.
3. *Family history.* Skin cancer rates are higher in some families than in others.
4. *Age.* Older people are more prone to skin cancer owing to longer total exposure to sunlight.
5. *Immunological status.* Individuals who are immunosuppressed have a higher incidence of skin cancer.

BURNS

A **burn** is tissue damage caused by excessive heat, electricity, radioactivity, or corrosive chemicals that destroy (denature) the proteins in the skin cells. Burns destroy some of the skin's important contributions to homeostasis—protection against microbial invasion and desiccation, and thermoregulation.

Burns are graded according to their severity. A *first-degree burn* involves only the epidermis. It is characterized by mild pain and erythema (redness) but no blisters. Skin functions remain intact. The pain and damage caused by a first-degree burn may be lessened by immediately flushing it with cold water. Generally, a first-degree burn will heal in about 3–6 days and may be accompanied by flaking or peeling. One example of a first-degree burn is a mild sunburn.

A *second-degree burn* destroys a portion of the epidermis and possibly parts of the dermis. Some skin functions are lost. In a second-degree burn, redness, blister formation, edema, and pain result. (Blister formation is caused by separation of the epidermis from the dermis due to the accumulation of tissue fluid between the layers.) Associated structures, such as hair follicles, sebaceous glands, and sweat glands, usually are not injured. If there is no infection, second-degree burns heal without skin grafting in about 3–4 weeks, but scarring may result. First- and second-degree burns are collectively referred to as *partial-thickness burns.*

A *third-degree burn* or *full-thickness burn* destroys a portion of the epidermis, the underlying dermis, and associated structures. Skin functions are lost. Such burns vary in appearance from marble-white to mahogany colored to charred, dry wounds. There is marked edema, and the burned region is numb because sensory nerve endings have been destroyed. Regeneration occurs slowly, and much granulation tissue forms before being covered by epithelium. Skin grafting may be required to promote healing and to minimize scarring.

The injury to the skin tissues directly in contact with the damaging agent is the *local effect* of a burn. Generally, however, the *systemic effects* of a major burn are a greater threat to life. The systemic effects of a burn may include (1) a large loss of water, plasma, and plasma proteins, which causes shock; (2) bacterial infection; (3) reduced circulation of blood; (4) decreased production of urine; and (5) diminished immune responses.

The seriousness of a burn is determined by its depth and extent of area involved, as well as the person's age and general health. According to the American Burn Association's classification of burn injury, a major burn includes third-degree burns over 10% of body surface area; or second-degree burns over 25% of body surface area; or any third-degree burns on the face, hands, feet, or perineum (per-i-NĒ-um, which includes the anal and urogenital regions).When the burn area exceeds 70%, more than half the victims die. A quick means for estimating the surface area affected by a burn in an adult is the **rule of nines:**

1. Count 9% if the anterior and posterior surfaces of the head and neck are affected.
2. Count 9% for the anterior and posterior surfaces of each upper limb (total of 18% for both upper limbs).
3. Count four times nine or 36% for the anterior and posterior surfaces of the trunk, including the buttocks.
4. Count 9% for the anterior and 9% for the posterior surfaces of each lower limb as far up as the buttocks (total of 36% for both lower limbs).
5. Count 1% for the perineum.

A more accurate way to estimate the amount of surface area affected by a burn is the **Lund–Browder method,** which estimates the extent by comparing the areas affected to the percentage of total surface area for body parts (Table 5.5). Because most of the proportions of the body change with growth, some percentages vary for different ages.

Table 5.5 Lund-Browder Burn Assessment Chart

AREA	AGE (YEARS)					AREA	AGE (YEARS)				
	0–1	1–4	5–9	10–15	ADULT		0–1	1–4	5–9	10–15	ADULT
Head	19	17	13	10	7	Left forearm	3	3	3	3	3
Neck	2	2	2	2	2	Right hand	$2\frac{1}{2}$	$2\frac{1}{2}$	$2\frac{1}{2}$	$2\frac{1}{2}$	$2\frac{1}{2}$
Anterior trunk	13	13	13	13	13	Left hand	$2\frac{1}{2}$	$2\frac{1}{2}$	$2\frac{1}{2}$	$2\frac{1}{2}$	$2\frac{1}{2}$
Posterior trunk	13	13	13	13	13	Right thigh	$5\frac{1}{2}$	$6\frac{1}{2}$	$8\frac{1}{2}$	$8\frac{1}{2}$	$9\frac{1}{2}$
Right buttock	$2\frac{1}{2}$	$2\frac{1}{2}$	$2\frac{1}{2}$	$2\frac{1}{2}$	$2\frac{1}{2}$	Left thigh	$5\frac{1}{2}$	$6\frac{1}{2}$	$8\frac{1}{2}$	$8\frac{1}{2}$	$9\frac{1}{2}$
Left buttock	$2\frac{1}{2}$	$2\frac{1}{2}$	$2\frac{1}{2}$	$2\frac{1}{2}$	$2\frac{1}{2}$	Right leg	5	5	$5\frac{1}{2}$	6	7
Genitalia	1	1	1	1	1	Left leg	5	5	$5\frac{1}{2}$	6	7
Right arm	4	4	4	4	4	Right foot	$3\frac{1}{2}$	$3\frac{1}{2}$	$3\frac{1}{2}$	$3\frac{1}{2}$	$3\frac{1}{2}$
Left arm	4	4	4	4	4	Left foot	$3\frac{1}{2}$	$3\frac{1}{2}$	$3\frac{1}{2}$	$3\frac{1}{2}$	$3\frac{1}{2}$
Right forearm	3	3	3	3	3						

PRESSURE SORES

Pressure sores, also known as *decubitus ulcers* (dē-KYOO-bi-tus) or *bedsores,* are caused by a constant deficiency of blood flow to tissues. Typically the affected tissue overlies a bony projection that has been subjected to prolonged pressure against an object such as a bed, cast, or splint. If the pressure is relieved in a few hours, redness occurs but no lasting tissue damage results. Blistering of the affected area may indicate superficial damage, whereas a reddish-blue discoloration may indicate deep tissue damage. Prolonged pressure results in tissue ulceration. Small breaks in the epidermis become infected, and the sensitive subcutaneous layer and deeper tissues are damaged. Eventually, the tissue dies. Pressure sores are seen most often in patients who are bedridden. With proper care, pressure sores are preventable.

MEDICAL TERMINOLOGY

Alopecia (al′-ō-PĒ-shē-a) Partial or complete lack of hair; may result from aging, endocrine disorders, chemotherapy for cancer, or skin disease.

Cold sore A lesion, usually in oral mucous membrane, caused by Type 1 herpes simplex virus (HSV) transmitted by oral or respiratory routes. The virus remains dormant until triggered by factors such as ultraviolet light, hormonal changes, and emotional stress. Also called a *fever blister.*

Contact dermatitis (der-ma-TĪ-tis; *dermat-* = skin; *-itis* = inflammation of) Inflammation of the skin characterized by redness, itching, and swelling and caused by exposure of the skin to chemicals that bring about an allergic reaction, such as poison ivy toxin.

Corn A painful conical thickening of the stratum corneum of the epidermis found principally over toe joints and between the toes, often caused by friction or pressure. Corns may be hard or soft, depending on their location. Hard corns are usually found over toe joints, and soft corns are usually found between the fourth and fifth toes.

Hemangioma (he-man′-jē-Ō-ma; *hem-* = blood; *-angi-* = blood vessel; *-oma* = tumor) Localized tumor of the skin and subcutaneous layer that results from an abnormal increase in blood vessels. One type is a **portwine stain,** a flat, pink, red, or purple lesion present at birth, usually at the nape of the neck.

Hives Condition of the skin marked by reddened elevated patches that are often itchy. Most commonly caused by infections, physical trauma, medications, emotional stress, food additives, and certain food allergies. Also called *urticaria* (yoor-ti-KAR-ē-a).

Impetigo (im′-pe-TĪ-gō) Superficial skin infection caused by *Staphylococcus* bacteria; most common in children.

Intradermal (in-tra-DER-mal; *intra-* = within) Within the skin. Also called *intracutaneous.*

Laceration (las-er-Ā-shun; *lacer-* = torn) An irregular tear of the skin.

Nevus (NĒ-vus) A round, flat, or raised area of pigmented skin that may be present at birth or may develop later. Varies in color from yellow-brown to black. Also called a *mole* or *birthmark.*

Pruritus (proo-RĪ-tus; *pruri-* = to itch) Itching, one of the most common dermatological disorders. It may be caused by skin disorders (infections), systemic disorders (cancer, kidney failure), psychogenic factors (emotional stress), or allergic reactions.

Topical In reference to a medication, applied to the skin surface rather than ingested or injected.

Wart Mass produced by uncontrolled growth of epithelial skin cells; caused by a papilloma virus. Most warts are noncancerous.

STUDY OUTLINE

STRUCTURE OF THE SKIN (p. 140)

1. The integumentary system consists of the skin and its accessory structures—hair, nails, glands, muscles, and nerves.
2. The skin is the largest organ of the body in surface area and weight. The principal parts of the skin are the epidermis (superficial) and dermis (deep).
3. The subcutaneous layer (hypodermis) is deep to the dermis and not part of the skin. It anchors the dermis to underlying tissues and organs, and it contains lamellated (Pacinian) corpuscles.
4. The types of cells in the epidermis are keratinocytes, melanocytes, Langerhans cells, and Merkel cells.
5. The epidermal layers, from deep to superficial, are the stratum basale, stratum spinosum, stratum granulosum, stratum lucidum (in thick skin only), and stratum corneum. Stem cells in the stratum basale undergo continuous cell division, producing keratinocytes for the other layers.
6. The dermis consists of papillary and reticular regions. The papillary region is composed of areolar connective tissue containing fine elastic fibers, dermal papillae, and Meissner corpuscles. The reticular region is composed of dense irregular connective tissue containing interlaced collagen and coarse elastic fibers, adipose tissue, hair follicles, nerves, sebaceous (oil) glands, and ducts of sudoriferous (sweat) glands.
7. Epidermal ridges provide the basis for fingerprints and footprints.
8. The color of skin is due to melanin, carotene, and hemoglobin.

ACCESSORY STRUCTURES OF THE SKIN (p. 145)

1. Accessory structures of the skin—hair, skin glands, and nails—develop from the embryonic epidermis.
2. A hair consists of a shaft, most of which is superficial to the surface, a root that penetrates the dermis and sometimes the subcutaneous layer, and a hair follicle.
3. Associated with each hair follicle is a sebaceous (oil) gland, an arrector pili muscle, and a hair root plexus.
4. New hairs develop from division of matrix cells in the bulb; hair replacement and growth occur in a cyclic pattern consisting of alternating growth and resting stages.
5. Hairs offer a limited amount of protection—from the sun, heat loss, and entry of foreign particles into the eyes, nose, and ears. They also function in sensing light touch.
6. Sebaceous (oil) glands are usually connected to hair follicles; they are absent in the palms and soles. Sebaceous glands produce sebum, which moistens hairs and waterproofs the skin. Clogged sebaceous glands may produce acne.
7. There are two types of sudoriferous (sweat) glands: eccrine and apocrine. Eccrine sweat glands have an extensive distribution; their ducts terminate at pores at the surface of the epidermis. Apocrine sweat glands are limited to the skin of the axilla, groin, and areolae; their ducts open into hair follicles. They begin functioning at puberty and are stimulated during emotional stress and sexual excitement. Mammary glands are specialized sudoriferous glands that secrete milk.
8. Ceruminous glands are modified sudoriferous glands that secrete cerumen. They are found in the external auditory canal (ear canal).
9. Nails are hard, keratinized epidermal cells over the dorsal surfaces of the distal portions of the digits.
10. The principal parts of a nail are the nail body, free edge, nail root, lunula, eponychium, and matrix. Cell division of the matrix cells produces new nails.

TYPES OF SKIN (p. 149)

1. Thin skin covers all parts of the body except for the palms, palmar surfaces of the digits, and the soles.
2. Thick skin covers the palms, palmar surfaces of the digits, and soles.

FUNCTIONS OF THE SKIN (p. 150)

1. Skin functions include body temperature regulation, protection, sensation, excretion and absorption, and synthesis of vitamin D.
2. The skin participates in thermoregulation by liberating sweat at its surface and by adjusting the flow of blood in the dermis.
3. The skin provides physical, chemical, and biological barriers that help protect the body.
4. Cutaneous sensations include tactile sensations, thermal sensations, and pain.

MAINTAINING HOMEOSTASIS: SKIN WOUND HEALING (p. 152)

1. In an epidermal wound, the central portion of the wound usually extends down to the dermis, whereas the wound edges involve only superficial damage to the epidermal cells.
2. Epidermal wounds are repaired by enlargement and migration of basal cells, contact inhibition, and division of migrating and stationary basal cells.
3. During the inflammatory phase of deep wound healing, a blood clot unites the wound edges, epithelial cells migrate across the wound, vasodilation and increased permeability of blood vessels enhance delivery of phagocytes, and mesenchymal cells develop into fibroblasts.
4. During the migratory phase, fibroblasts migrate along fibrin threads and begin synthesizing collagen fibers and glycoproteins.
5. During the proliferative phase there is extensive growth of epithelial cells.
6. During the maturation phase, the scab sloughs off, the epidermis is restored to normal thickness, collagen fibers become more organized, fibroblasts begin to disappear, and blood vessels are restored to normal.

DEVELOPMENTAL ANATOMY OF THE INTEGUMENTARY SYSTEM (p. 152)

1. The epidermis develops from the embryonic ectoderm, and the accessory structures of the skin (hair, nails, and skin glands) are epidermal derivatives.
2. The dermis is derived from mesodermal cells.

AGING AND THE INTEGUMENTARY SYSTEM (p. 154)

1. Most effects of aging begin to occur when people reach their late forties.
2. Among the effects of aging are wrinkling, loss of subcutaneous fat, atrophy of sebaceous glands, and decrease in the number of melanocytes and Langerhans cells.

Complete the following:

1. The two principal parts of the skin are the ___ and the ___.
2. The combination of collagen and elastic fibers in the reticular region of the dermis provides the skin with ___, ___, and ___.
3. The two regions of the dermis are the ___ and ___.
4. The pigments that give skin a wide variety of color are ___, ___, and ___.

True or false:

5. The color of hair is due primarily to melanin.
6. Modified sweat glands in the ear are called apocrine sweat glands.
7. Match the following:

   ___(a) liberation of sweat at the surface and adjustment of blood flow in the dermis
   ___(b) provision of a chemical barrier, resistance to microbial invasion, and protection of underlying tissues from abrasion
   ___(c) input of touch, pressure, pain, and heat and cold information
   ___(d) elimination of usually unneeded substances and passage of materials from the external environment into body cells
   ___(e) production of calcitriol

   (1) protection
   (2) sensation
   (3) thermoregulation
   (4) vitamin D synthesis
   (5) excretion/absorption

Choose the best answer to the following questions:

8. The substance that prevents excessive evaporation of water from the skin, keeps the skin soft and pliable, and inhibits the growth of bacteria is (a) sebum, (b) sweat, (c) cerumen, (d) carotene, (e) melatonin.
9. Epidermal ridges (a) indicate the predominant direction of underlying collagen fiber bundles, (b) increase the grip of the hand or foot, (c) provide the pigments of skin color, (d) synthesize vitamin D in the epidermis of the skin.
10. Which of the following are true for hair? (1) provides protection, (2) increases efficiency of heat loss, (3) matrix cells produce growth, (4) movement of shaft activates hair root plexus, (5) prevents dehydration.
    (a) 1, 2, and 3, (b) 1, 2, and 4, (c) 1, 2, 3, and 5, (d) 1, 3, 4, and 5, (e) 3, 4, and 5
11. Which of the following statements are true? (1) Nails are tightly packed, hard, keratinized cells of the epidermis that form a clear, solid covering over the dorsal surface of the terminal end of digits. (2) The free edge of the nail is white due to absence of capillaries. (3) Nails help us grasp and manipulate small objects. (4) Nails protect the ends of digits from trauma. (5) Nail color is due to a combination of melanin and carotene.
    (a) 1, 2, and 3, (b) 1, 3, and 4, (c) 1, 2, 3, and 4, (d) 2, 3, and 4, (e) 1, 3, and 5
12. The deepest layer of epidermis, composed of a single layer of keratinocytes that are mitotic, is the (a) stratum basale, (b) stratum granulosum, (c) stratum spinosum, (d) stratum lucidum, (e) stratum corneum.
13. Which of the following statements are true? (1) A first-degree burn involves only the surface epidermis. (2) In a second-degree burn, no skin function is lost. (3) First- and second-degree burns are collectively referred to as partial thickness burns. (4) A third-degree burn destroys the epidermis, dermis, and the epidermal derivatives. (5) When burns exceed 20% of body surface area, more than half the victims die.
    (a) 1, 2, and 3, (b) 2, 3, and 4, (c) 3, 4, and 5, (d) 1, 2, and 4, (e) 1, 3, and 4
14. Match the following:

    ___(a) produce the protein that helps protect the skin and underlying tissues from light, heat, microbes, and many chemicals
    ___(b) produce the pigment that contributes to skin color and absorbs ultraviolet light
    ___(c) cells that arise from red bone marrow, migrate to the epidermis, and participate in immune responses
    ___(d) cells thought to function in the sensation of touch
    ___(e) an abnormal thickening of the epidermis
    ___(f) release a lipid-rich secretion that functions as a water-repellent sealant in the stratum granulosum
    ___(g) pressure-sensitive cells found mostly in the subcutaneous layer
    ___(h) a fatty substance that covers and protects the skin of the fetus from the constant exposure to amniotic fluid

    (1) Merkel cells
    (2) callus
    (3) keratinocytes
    (4) Langerhans cells
    (5) melanocytes
    (6) lamellar granules
    (7) lamellated (Pacinian) corpuscles
    (8) vernix caseosa

15. Match the following:

____ (a) a skin disorder in which keratinocytes divide and move more quickly to the skin surface, are shed prematurely, and make an abnormal keratin

____ (b) an inherited inability to produce melanin

____ (c) a condition in which a partial or complete loss of melanocytes from patches of skin produce irregular white spots

____ (d) a yellowed appearance of the whites of the eyes and of light-colored skin, usually indicating liver disease

____ (e) a bluish appearance of mucous membranes, nail beds, and light-colored skin due to oxygen depletion

____ (f) redness of the skin caused by engorgement of capillaries in the dermis with blood

____ (g) partial or complete lack of hair

____ (h) an inflammation of sebaceous glands

____ (i) a precancerous skin lesion induced by sunlight

____ (j) the most common form of skin cancer

____ (k) a skin sore caused by long-standing, pressure-induced deficiency of blood flow to tissues overlying a bony projection

____ (l) the process of scar-tissue formation

____ (m) a vascular and cellular response that serves to remove microbes, foreign material, and dying tissue in preparation for healing

____ (n) a rapidly metastasizing, potentially fatal form of skin cancer

(1) fibrosis
(2) erythema
(3) vitiligo
(4) cyanosis
(5) psoriasis
(6) alopecia
(7) jaundice
(8) acne
(9) malignant melanoma
(10) solar keratosis
(11) decubitus ulcer
(12) basal cell carcinoma
(13) albinism
(14) inflammation

CRITICAL THINKING QUESTIONS

1. The amount of dust that collects in a house with an assortment of dogs, cats, and people is truly amazing. A lot of these dust particles had a previous "life" as part of the home's living occupants. Where did the dust originate on the human body? (HINT: *Dogs, cats, and humans all shed in their own way.*)

2. Kiko is fastidious about personal hygiene. She's convinced that body secretions on the skin are unsanitary and wants to have all her exocrine glands removed. Is this wise? (HINT: *There are ap-*

proximately 3–4 million sudoriferous glands alone.)

3. Your nephew has been learning about cells in science class, and now he refuses to take a bath. He asks, "If all cells have a semipermeable membrane, and the skin is made out of cells, then won't I swell up and pop if I take a bath?" Explain the relevant anatomy to your nephew before he starts attracting flies. (HINT: *Semipermeable membranes function in living cells.*)

ANSWERS TO FIGURE QUESTIONS

5.1 The epidermis is epithelial tissue, whereas the dermis is connective tissue.

5.2 Melanin protects the DNA of the nucleus of keratinocytes from damage from UV light.

5.3 The stratum basale is the layer of the epidermis that contains stem cells.

5.4 Plucking a hair stimulates hair root plexuses in the dermis, some of which are sensitive to pain. Because the cells of a hair shaft are already dead and the hair shaft lacks nerves, cutting

hair is not painful.

5.5 Nails are hard because they are composed of tightly packed, hard, keratinized epidermal cells.

5.6 Epidermal wounds do not bleed because there are no blood vessels in the epidermis.

5.7 In addition to phagocytizing microbes, neutrophils and macrophages can help clean up cellular debris that results from the wound.

6 THE SKELETAL SYSTEM: BONE TISSUE

Despite its simple appearance, bone is a complex and dynamic living tissue. Bone tissue engages in a continuous process of dynamic remodeling—building new bone and breaking down old bone. Bone is made up of several different tissues working together: bone or osseous tissue, cartilage, dense connective tissues, epithelium, various blood-forming tissues, adipose tissue, and nervous tissue. For this reason, each individual bone is an organ. The entire framework of bones and their cartilages together constitute the **skeletal system.** This chapter will survey the various components of bones so you can understand how bones form, how they age, and how exercise affects the density and strength of bones. The study of bone structure and the treatment of bone disorders is called **osteology** (os-tē-OL-ō-jē; *oste-* = bone; *-ology* = study of).

FUNCTIONS OF THE SKELETAL SYSTEM

OBJECTIVE
• *Discuss the functions of the skeletal system.*

Bone tissue and the skeletal system perform several basic functions:

1. *Support.* Bones serve as the structural framework for the body by supporting soft tissues and providing attachment points for the tendons of most skeletal muscles.

2. *Protection.* Bones protect many internal organs from injury. For example, cranial bones protect the brain, vertebrae protect the spinal cord, and the rib cage protects the heart and lungs.

3. *Assistance in movement.* When skeletal muscles contract, they pull on bones to produce movement.

4. *Mineral homeostasis.* Bone tissue stores several minerals, especially calcium and phosphorus, which contribute to the strength of the bone. Bone can release minerals into the bloodstream to maintain critical mineral balances and to distribute minerals to other organs.

5. *Blood cell production.* Within certain parts of bones a connective tissue called **red bone marrow** produces red blood cells, white blood cells, and platelets by a process called **hemopoiesis** (hēm-ō-poy-Ē-sis; *hemo-* = blood; *-poiesis* = to make). Red bone marrow, one of two types of bone marrow, consists of developing blood cells within a network of reticular fibers. Also present are adipocytes, macrophages, and fibroblasts.

6. *Triglyceride storage.* In the newborn, all bone marrow is red and is involved in hemopoiesis. However, with increasing age, blood cell production decreases, and most of the bone marrow changes from red to yellow. **Yellow bone marrow** consists primarily of adipocytes and a few scattered blood cells.

1. What kinds of tissues make up the skeletal system?
2. How do red and yellow bone marrow differ in composition, location, and function?

Figure 6.1 Parts of a long bone. The spongy bone tissue of the epiphysis and metaphysis contains red bone marrow, whereas the medullary cavity of the diaphysis contains yellow bone marrow (in adults).

🔑 **A long bone is covered by articular cartilage at its proximal and distal epiphyses and by periosteum around the diaphysis.**

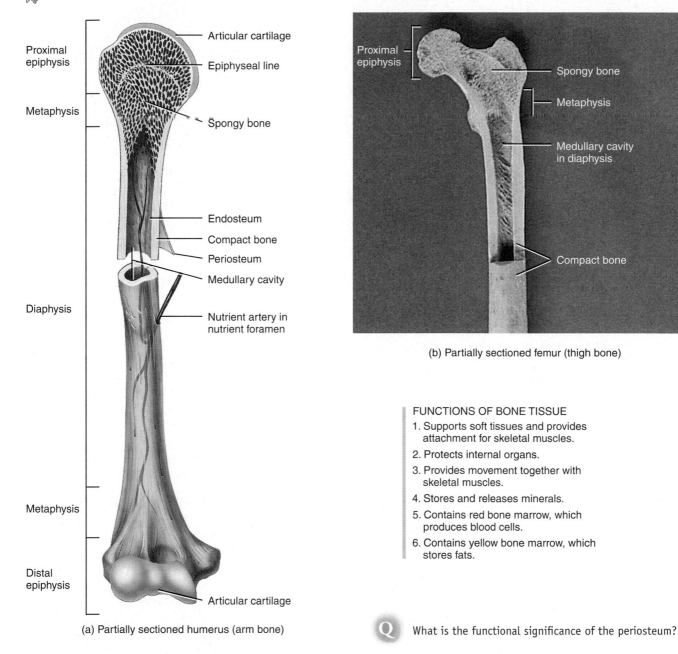

(a) Partially sectioned humerus (arm bone)

(b) Partially sectioned femur (thigh bone)

FUNCTIONS OF BONE TISSUE
1. Supports soft tissues and provides attachment for skeletal muscles.
2. Protects internal organs.
3. Provides movement together with skeletal muscles.
4. Stores and releases minerals.
5. Contains red bone marrow, which produces blood cells.
6. Contains yellow bone marrow, which stores fats.

Q What is the functional significance of the periosteum?

STRUCTURE OF BONE
OBJECTIVE
• *Describe the components of a long bone.*

The structure of bone may be analyzed by considering the parts of a long bone such as the humerus (the arm bone) or the femur (the thigh bone) (Figure 6.1). A long bone is one that has greater length than width. A typical long bone consists of the following parts:

1. The **diaphysis** (dī-AF-i-sis; *dia-* = through; *-physis* = growing) is the bone's shaft or body—the long, cylindrical, main portion of the bone.
2. The **epiphyses** (e-PIF-i-sēz; *epi-* = over) are the distal and proximal ends of the bone. (The singular is epiphysis.)

Figure 6.2 Types of cells in bone tissue.

 Osteogenic cells undergo cell division and develop into osteoblasts, which secrete bone matrix.

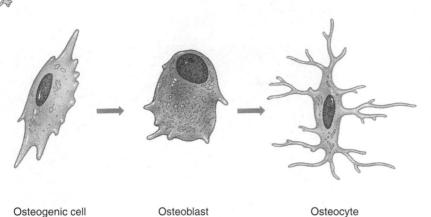 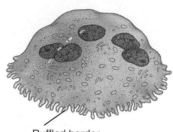

Ruffled border

Osteogenic cell
(develops into an
osteoblast)

Osteoblast
(forms bone
tissue)

Osteocyte
(maintains
bone tissue)

Osteoclast
(functions in resorption, the
destruction of bone matrix)

Q Why is bone resorption important?

3. The **metaphyses** (me-TAF-i-sēz; *meta-* = between) are the regions in a mature bone where the diaphysis joins the epiphyses. (The singular is metaphysis.) In a growing bone, the metaphyses are regions that include the epiphyseal plate, the point at which cartilage is replaced by bone. The epiphyseal plate is a layer of hyaline cartilage that allows the diaphysis of the bone to grow in length, but not in width.

4. The **articular cartilage** is a thin layer of hyaline cartilage covering the epiphysis where the bone forms an articulation (joint) with another bone. Articular cartilage reduces friction and absorbs shock at freely movable joints.

5. The **periosteum** (per′-ē-OS-tē-um; *peri-* = around) is a tough sheath of dense irregular connective tissue that surrounds the bone surface wherever it is not covered by articular cartilage. The periosteum contains bone forming cells that enable bone to grow in diameter or thickness, but not in length. It also protects the bone, assists in fracture repair, helps nourish bone tissue, and serves as an attachment point for ligaments and tendons.

6. The **medullary cavity** (MED-yoo-lar′-ē; *medulla-* = marrow, pith) or **marrow cavity** is the space within the diaphysis that contains fatty yellow bone marrow.

7. The **endosteum** (end-OS-tē-um; *endo-* = within) is a membrane that contains bone forming cells and lines the medullary cavity.

1. Diagram the parts of a long bone, and list the functions of each part.

HISTOLOGY OF BONE TISSUE
OBJECTIVE
• *Describe the histological features of bone tissue.*

Like other connective tissues, **bone,** or **osseous tissue** (OS-ē-us), contains an abundant matrix of intercellular materials that surround widely separated cells. In bone, the matrix is about 25% water, 25% protein fibers, and 50% crystallized mineral salts. There are four types of cells in bone tissue: osteogenic cells, osteoblasts, osteocytes, and osteoclasts (Figure 6.2).

1. **Osteogenic cells** (os′-tē-ō-JEN-ik; *-genic* = producing) are unspecialized stem cells derived from mesenchyme, the tissue from which all connective tissues are formed. They are the only bone cells to undergo cell division; the resulting daughter cells develop into osteoblasts. Osteogenic cells are found along the inner portion of the periosteum, in the endosteum, and in the canals within bone that contain blood vessels.

2. **Osteoblasts** (OS-tē-ō-blasts′; *-blasts* = buds or sprouts) are bone-building cells. They synthesize and secrete collagen fibers and other organic components needed to build the matrix of bone tissue, and they initiate calcification (described shortly). (Note: *Blasts* in bone or any other connective tissue secrete matrix.)

3. **Osteocytes** (OS-tē-ō-sīts′; *-cyte* = cell) are mature bone cells that are the principal cells of bone tissue. Osteocytes are derived from osteoblasts that have become entrapped

in matrix secretions. However, osteocytes no longer secrete matrix materials. Instead, they maintain the daily cellular activities of bone tissue, such as the exchange of nutrients and wastes with the blood. (Note: *Cytes* in bone or any other tissue maintain the tissue.)

4. **Osteoclasts** (OS-tē-ō-clasts′; *-clast* = break) are huge cells derived from the fusion of as many as 50 monocytes (a type of white blood cell) that are concentrated in the endosteum. On the side of the cell that faces the bone surface, the osteoclast's plasma membrane is deeply folded into a *ruffled border.* Here the cell releases powerful lysosomal enzymes and acids that digest the protein and mineral components of the underlying bone. This destruction of bone matrix is part of the normal development, growth, maintenance, and repair of bone. (Note: *Clasts* in bone destroy matrix.)

The matrix of bone, unlike that of other connective tissues, contains abundant inorganic mineral salts, primarily *hydroxyapatite* (calcium phosphate) and some calcium carbonate. In addition, bone matrix includes small amounts of magnesium hydroxide, fluoride, and sulfate. As these mineral salts are deposited in the framework formed by the collagen fibers of the matrix, they crystallize and the tissue hardens. This process of **calcification** or **mineralization** is initiated by osteoblasts.

Although a bone's *hardness* depends on the crystallized inorganic mineral salts, a bone's *flexibility* depends on its collagen fibers. Like reinforcing metal rods in concrete, collagen fibers and other organic molecules provide *tensile strength,* which is resistance to being stretched or torn apart. If the mineral salts in a bone are dissolved by soaking the bone in vinegar, for example, the bone becomes rubbery and flexible.

It was once thought that calcification simply occurred when enough mineral salts were present to form crystals. Now, however, we know that the process occurs only in the presence of collagen fibers. Mineral salts begin to crystallize in the microscopic spaces between collagen fibers. After the spaces are filled, mineral crystals accumulate around the collagen fibers. The combination of crystallized salts and collagen fibers is responsible for the hardness that is characteristic of bone.

Bone is not completely solid but has many small spaces between its hard components. Some spaces provide channels for blood vessels that supply bone cells with nutrients. Other spaces are storage areas for red bone marrow. Depending on the size and distribution of the spaces, the regions of a bone may be categorized as compact or spongy (see Figure 6.1). Overall, about 80% of the skeleton is compact bone and 20% is spongy bone.

Compact Bone Tissue

Compact bone tissue contains few spaces between its hard components. It forms the external layer of all bones and makes up the bulk of the diaphyses of long bones. Compact bone tissue provides protection and support and resists the stresses produced by weight and movement.

Compact bone tissue is arranged in units called **osteons** or **Haversian systems** (Figure 6.3a). Blood vessels, lymphatic vessels, and nerves from the periosteum penetrate the compact bone through **perforating (Volkmann's) canals.** The vessels and nerves of the perforating canals connect with those of the medullary cavity, periosteum, and **central (Haversian) canals.** The central canals run longitudinally through the bone. Around the canals are **concentric lamellae** (la-MEL-ē)—rings of hard, calcified matrix. Between the lamellae are small spaces called **lacunae** (la-KOO-nē; = little lakes), which contain osteocytes. Radiating in all directions from the lacunae are tiny **canaliculi** (kan′-a-LIK-yoo-lī; = small channels), which are filled with extracellular fluid. Inside the canaliculi are slender fingerlike processes of osteocytes (see inset at top left of Figure 6.3a). The canaliculi connect lacunae with one another and with the central canals. This intricate branching network provides many routes for blood-borne nutrients and oxygen to diffuse through the fluid to the osteocytes and for wastes to diffuse back to the blood vessels.

Osteons in compact bone tissue are aligned in the same direction along lines of stress. In the shaft, for example, they are parallel to the long axis of the bone. As a result, the shaft of a long bone resists bending or fracturing even when considerable force is applied from either end. Compact bone tissue tends to be thickest in those parts of a bone where stresses are applied in relatively few directions. The lines of stress in a bone are not static. They change as a person learns to walk and in response to repeated strenuous physical activity, such as occurs when a person undertakes weight training. The lines of stress in a bone also can change as a result of fractures or physical deformity. Thus, the organization of osteons changes over time in response to the physical demands placed on the skeleton.

The areas between osteons contain **interstitial lamellae,** which also have lacunae with osteocytes and canaliculi. Interstitial lamellae are fragments of older osteons that have been partially destroyed during bone rebuilding or growth. Lamellae that encircle the bone just beneath the periosteum are called **outer circumferential lamellae;** those that encircle the medullary cavity are called **inner circumferential lamellae.**

Spongy Bone Tissue

In contrast to compact bone tissue, **spongy bone tissue** does not contain true osteons (Figure 6.3b, c). It consists of lamellae that are arranged in an irregular lattice of thin columns of bone called **trabeculae** (tra-BEK-yoo-lē; = little beams). The macroscopic spaces between the trabeculae of some bones are filled with red bone marrow, which produces blood cells. Within each trabecula (singular form) are osteocytes that lie in lacunae. Radiating from the lacunae are

Figure 6.3 Histology of compact and spongy bone. (a) Sections through the diaphysis of a long bone, from the surrounding periosteum on the left, to compact bone in the middle, to spongy bone and the medullary cavity on the right. The inset at the upper left shows an osteocyte in a lacuna. (b and c) Details of spongy bone. A photomicrograph of compact bone tissue is provided in Table 4.3 on page 126.

🔑 **Osteocytes lie in lacunae arranged in concentric circles around a central canal in compact bone, and in lacunae arranged irregularly in trabeculae of spongy bone.**

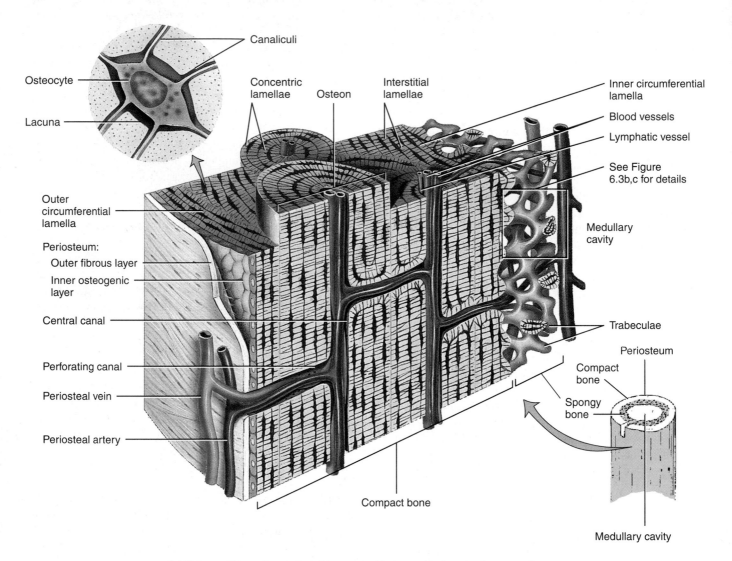

(a) Osteons (Haversian systems) in compact bone and trabeculae in spongy bone

canaliculi. Osteocytes in the trabeculae receive nourishment directly from the blood circulating through the medullary cavities.

Spongy bone tissue makes up most of the bone tissue of short, flat, and irregularly shaped bones; most of the epiphy-

ses of long bones; and a narrow rim around the medullary cavity of the diaphysis of long bones.

At first glance, the structure of the osteons of compact bone tissue may appear to be highly organized, whereas the trabeculae of spongy bone tissue may appear to be far less

Figure 6.3 (continued)

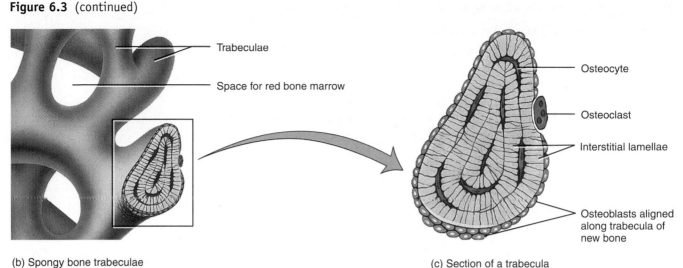

Trabeculae

Space for red bone marrow

Osteocyte

Osteoclast

Interstitial lamellae

Osteoblasts aligned along trabecula of new bone

(b) Spongy bone trabeculae

(c) Section of a trabecula

Q As people age, some central (Haversian) canals may become blocked. What effect would this have on the osteocytes?

well organized. However, the trabeculae in spongy bone tissue are precisely oriented along lines of stress, a characteristic that helps bones resist stresses and transfer force without breaking. Spongy bone tissue tends to be located where bones are not heavily stressed or where stresses are applied from many directions.

Spongy bone tissue is different from compact bone tissue in two respects. First, spongy bone tissue is light, which reduces the overall weight of a bone so that it moves more readily when pulled by a skeletal muscle. Second, the trabeculae of spongy bone tissue support and protect the red bone marrow. The spongy bone tissue in the hip bones, ribs, breastbone, backbones, and the ends of long bones is the only site of red bone marrow and, thus, of hemopoiesis in adults.

CLINICAL APPLICATION
Bone Scan

A **bone scan** is a diagnostic procedure that takes advantage of the fact that bone is living tissue. A small amount of a radioactive tracer compound that is readily absorbed by bone is injected intravenously. The degree of uptake of the tracer is related to the amount of blood flow to the bone. A scanning device measures the radiation emitted from the bones, and the information is translated into a photograph or diagram that can be read like an x-ray. Normal bone tissue is identified by a consistent gray color throughout because of its uniform uptake of the radioactive tracer. Darker or lighter areas, however, may indicate bone abnormalities. Darker areas are called "hot spots," areas of increased metabolism that absorb more of the radioactive tracer. Hot spots may indicate bone cancer, abnormal healing of fractures, or abnormal bone

growth. Lighter areas are called "cold spots," areas of decreased metabolism that absorb less of the radioactive tracer. Cold spots may indicate problems such as degenerative bone disease, decalcified bone, fractures, bone infections, Paget's disease, and rheumatoid arthritis. A bone scan not only detects abnormalities 3–6 months sooner than standard x-ray procedures, it exposes the patient to less radiation. ■

1. Why is bone considered a connective tissue?
2. Describe the four types of cells in bone tissue.
3. What is the composition of the matrix of bone tissue?
4. Distinguish between spongy and compact bone tissue in terms of its microscopic appearance, location, and function.

BLOOD AND NERVE SUPPLY OF BONE
OBJECTIVE
• *Describe the blood and nerve supply of bone.*

Bone is richly supplied with blood. Blood vessels, which are especially abundant in portions of bone containing red bone marrow, pass into bones from the periosteum. We will consider the blood supply to a long bone such as the mature tibia (shin bone) shown in Figure 6.4.

Periosteal arteries accompanied by nerves enter the diaphysis through numerous perforating (Volkmann's) canals and supply the periosteum and outer part of the compact bone (see Figure 6.3a). Near the center of the diaphysis, a large **nutrient artery** passes obliquely through the compact bone, through a hole called the **nutrient foramen** (see Figure 6.1a). On entering the medullary cavity, the nutrient artery divides into proximal and distal branches that supply

Figure 6.4 Blood supply of a mature long bone, the tibia (shinbone).

 Bone is richly supplied with blood vessels.

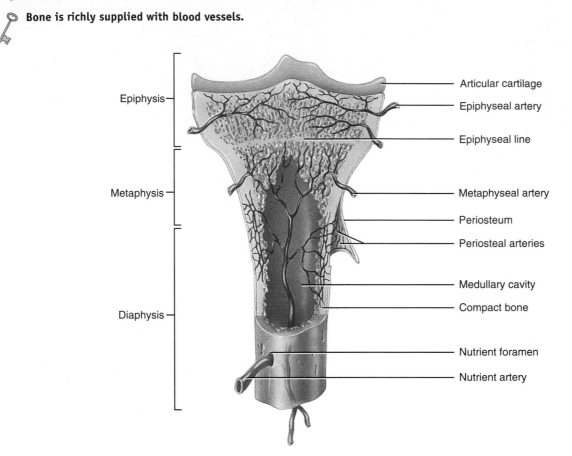

Q Where do periosteal arteries enter bone tissue?

both the inner part of compact bone tissue of the diaphysis and the spongy bone tissue and red marrow as far as the epiphyseal plates (or lines). Some bones, like the tibia, have only one nutrient artery; others like the femur (thigh bone) have several. The ends of long bones are supplied by the metaphyseal and epiphyseal arteries, which arise from arteries that supply the associated joint. The **metaphyseal arteries** enter the metaphyses of a long bone and, together with the nutrient artery, supply the red bone marrow and bone tissue of the metaphyses. The **epiphyseal arteries** enter the epiphyses of a long bone and supply the red bone marrow and bone tissue of the epiphyses.

Veins that carry blood away from long bones are evident in three places: (1) One or two **nutrient veins** accompany the nutrient artery in the diaphysis; (2) numerous **epiphyseal veins** and **metaphyseal veins** exit with their respective arteries in the epiphyses; and (3) many small **periosteal veins** exit with their respective veins in the periosteum.

Nerves accompany the blood vessels that supply bones. The periosteum is rich in sensory nerves, some of which carry pain sensations. These nerves are especially sensitive to

tearing or tension, which explains the severe pain resulting from a fracture or a bone tumor.

1. Explain the location and roles of the nutrient arteries, nutrient foramina, epiphyseal arteries, and periosteal arteries.

BONE FORMATION
OBJECTIVE
• *Describe the steps involved in intramembranous and endochondral ossification.*

The process by which bone forms is called **ossification** (os'-i-fi-KĀ-shun; *ossi-* = bone; *-fication* = making) or **osteogenesis.** The "skeleton" of a human embryo is composed of fibrous connective tissue membranes formed by condensed embryonic connective tissue (mesenchyme) or pieces of hyaline cartilage that resemble the shape of bones. These embryonic tissues provide the template for subsequent ossification, which begins during the sixth or seventh week of life and follows one of two patterns. These two kinds of ossification do not lead to differences in the structure of mature bones; they are simply different methods for bone formation.

- **Intramembranous ossification** (in′-tra-MEM-bra-nus; *intra-* = within; *-membran* = membrane) is the formation of bone directly on or within fibrous connective tissue membranes formed by condensed mesenchymal cells. Such bones form *directly* from mesenchyme without first going through a cartilage stage.
- **Endochondral ossification** (en′-dō-KON-dral; *-chondr* = cartilage) is the formation of bone within hyaline cartilage. In this ossification process, mesenchymal cells are transformed into chondroblasts, which initially produce a hyaline cartilage "model" of the bone. Subsequently, osteoblasts gradually replace the cartilage with bone.

Intramembranous Ossification

Of the two processes of bone formation, intramembranous ossification involves the fewest steps. The flat bones of the skull and mandible (lower jawbone) are formed in this way. The process occurs as follows (Figure 6.5):

1 At the site where the bone will develop, mesenchymal cells in fibrous connective tissue membranes condense (cluster) and differentiate, first into osteogenic cells and then into osteoblasts. The site of such a cluster is called a **center of ossification.** Osteoblasts secrete the organic matrix of bone until they are completely surrounded by it.

2 The secretion of matrix stops. The osteoblasts become osteocytes, which lie in lacunae and extend their narrow cytoplasmic processes into canaliculi that radiate in all directions. Within a few days, calcium and other mineral salts are deposited, and the matrix hardens or calcifies. Thus, calcification is just one aspect of ossification.

3 The bone matrix develops into *trabeculae,* which fuse with one another to create spongy bone. Blood vessels grow into the spaces between the trabeculae and the mesenchyme along the surface of the newly formed bone. Connective tissue that is associated with the blood vessels in the trabeculae differentiates into red bone marrow.

4 On the outside of the bone, the mesenchyme condenses and develops into the periosteum. Eventually, most superficial layers of the spongy bone are replaced by compact bone, but spongy bone remains at the center. Much of the newly formed bone is remodeled (destroyed and reformed), a process that slowly transforms the bone into its adult size and shape.

Endochondral Ossification

The replacement of cartilage by bone is called endochondral ossification. Although most bones of the body are

Figure 6.5 Intramembranous ossification. Refer to this figure as you read the corresponding numbered paragraphs in the text.

🔑 **Intramembranous ossification involves the formation of bone directly on or within fibrous connective tissue membranes formed by clusters of mesenchymal cells.**

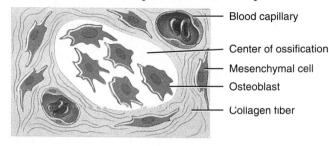

Blood capillary
Center of ossification
Mesenchymal cell
Osteoblast
Collagen fiber

1 Development of center of ossification

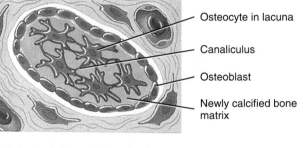

Osteocyte in lacuna
Canaliculus
Osteoblast
Newly calcified bone matrix

2 Osteocytes deposit mineral salts (calcification)

Mesenchyme condenses
Blood vessel
Trabeculae
Osteoblast

3 Formation of trabeculae

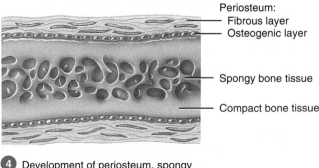

Periosteum:
— Fibrous layer
— Osteogenic layer
Spongy bone tissue
Compact bone tissue

4 Development of periosteum, spongy bone, and compact bone tissue

Q Which bones of the body develop by intramembranous ossification?

formed this way, the process is best observed in a long bone. It proceeds as follows (Figure 6.6):

1 *Development of the cartilage model.* At the site where bone is going to form, mesenchymal cells crowd together in the shape of the future bone. The mesenchymal cells differentiate into chondroblasts that produce a cartilage matrix; hence, the model consists of hyaline cartilage. In addition, a membrane called the **perichondrium** (per-i-KON-drē-um) develops around the cartilage model.

2 *Growth of the cartilage model.* Once chondroblasts become deeply buried in the cartilage matrix, they are called chondrocytes. The cartilage model grows in length by continual cell division of chondrocytes accompanied by further secretion of the cartilage matrix. This process results in an increase in length called **interstitial growth** (in′-ter-STISH-al), which means growth from within. In contrast, growth of the cartilage in thickness is due mainly to the addition of more matrix to the periphery of the model by new chondroblasts that develop from the perichondrium. This growth pattern in which matrix is deposited on the cartilage surface is called **appositional growth** (a-pō-ZISH-a-nal).

As the cartilage model continues to grow, chondrocytes in its midregion hypertrophy (increase in size), probably because they accumulate glycogen for ATP production and produce enzymes to catalyze additional chemical reactions. Some hypertrophied cells burst and release their contents, which changes the pH of the matrix. This change in pH triggers calcification. Other chondrocytes within the calcifying cartilage die because nutrients can no longer diffuse quickly enough through the matrix. As chondrocytes die, lacunae form and eventually merge into small cavities.

3 *Development of the primary ossification center.* A nutrient artery penetrates the perichondrium and the calcifying cartilage model through a nutrient foramen in the midregion of the cartilage model, stimulating osteogenic cells in the perichondrium to differentiate into osteoblasts. The cells secrete beneath the perichondrium a thin shell of compact bone called the **periosteal bone collar.** Once the perichondrium starts to form bone, it is known as the **periosteum.** Near the middle of the model, periosteal capillaries grow into the disintegrating calcified cartilage. These vessels and the associated osteoblasts, osteoclasts, and red bone marrow cells are known as the **periosteal bud.** Upon growing into the cartilage model, the capillaries induce growth of a **primary ossification center,** a region where bone tissue will replace most of the cartilage. Osteoblasts then begin to deposit bone matrix over the remnants of calcified cartilage, forming spongy bone trabeculae. As the ossification center enlarges toward the ends of the bone, osteoclasts break down the newly formed spongy bone trabeculae. This activity leaves a cavity, the medullary

(marrow) cavity, in the core of the model. The cavity then fills with red bone marrow. Primary ossification proceeds *inward* from the external surface of the bone.

4 *Development of secondary ossification centers.* The diaphysis (shaft), which was once a solid mass of hyaline cartilage, is replaced by compact bone, the core of which contains a red bone marrow–filled medullary cavity. When blood vessels enter the epiphyses, **secondary ossification centers** develop, usually around the time of birth. Bone formation is similar to that in primary ossification centers. One difference, however, is that spongy bone remains in the interior of the epiphyses (no medullary cavities are formed there). Secondary ossification proceeds *outward* from the center of the epiphysis toward the outer surface of the bone.

5 *Formation of articular cartilage and epiphyseal plate.* The hyaline cartilage that covers the epiphyses becomes the articular cartilage. Prior to adulthood, hyaline cartilage remains between the diaphysis and epiphysis as the *epiphyseal plate,* which is responsible for the lengthwise growth of long bones.

1. Outline the major events of intramembranous and endochondral ossification, and explain the main differences between the processes.

BONE GROWTH

OBJECTIVE
• *Describe how bone grows in length and thickness, and how nutrients and hormones regulate bone growth.*

During childhood, bones throughout the body grow in thickness by appositional growth, and long bones lengthen by the addition of bone material on the diaphyseal side of the epiphyseal plate. Bones stop growing in length at about age 25, although they may continue to thicken.

Growth in Length

To understand how a bone grows in length, you need to know some of the details of the structure of the epiphyseal plate (Figure 6.7). The **epiphyseal plate** (ep′-i-FIZ-ē-al) is a layer of hyaline cartilage in the metaphysis of a growing bone that consists of four zones (see Figure 6.7b).

1. *Zone of resting cartilage.* This layer is nearest the epiphysis and consists of small, scattered chondrocytes. The cells do not function in bone growth (thus the term *resting*). Instead, they anchor the epiphyseal plate to the bone of the epiphysis.

2. *Zone of proliferating cartilage.* Slightly larger chondrocytes in this zone are arranged like stacks of coins. The chondrocytes divide to replace those that die at the diaphyseal side of the epiphyseal plate.

3. *Zone of hypertrophic cartilage.* In this layer, the chondrocytes are even larger and remain arranged in columns.

Figure 6.6 Endochondral ossification.

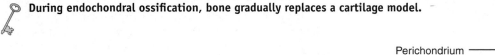

During endochondral ossification, bone gradually replaces a cartilage model.

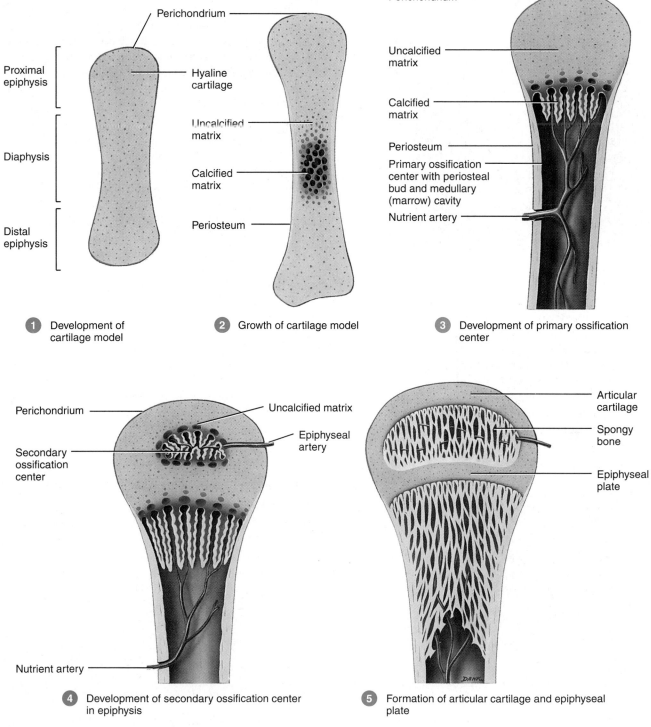

1 Development of cartilage model

2 Growth of cartilage model

3 Development of primary ossification center

4 Development of secondary ossification center in epiphysis

5 Formation of articular cartilage and epiphyseal plate

Q If radiographs of an 18-year-old basketball star show clear epiphyseal plates but no epiphyseal lines, is she likely to grow taller?

Figure 6.7 The epiphyseal plate is a layer of hyaline cartilage in the metaphysis of a growing bone. The epiphyseal plate appears as a dark band between whiter calcified areas in the radiograph.

🔑 **The epiphyseal plate allows the diaphysis of a bone to increase in length.**

(a) Radiograph showing the epiphyseal plate of the tibia of a 3-year-old

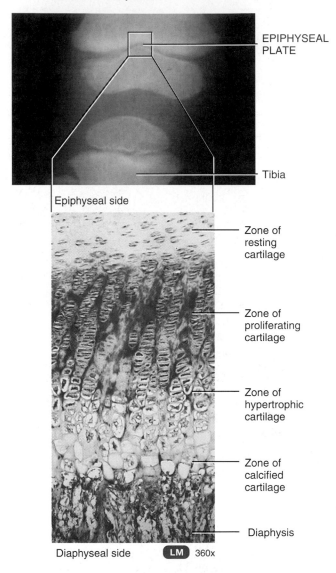

EPIPHYSEAL PLATE

Tibia

Epiphyseal side

Zone of resting cartilage

Zone of proliferating cartilage

Zone of hypertrophic cartilage

Zone of calcified cartilage

Diaphysis

Diaphyseal side   **LM** 360x

(b) Histology of the epiphyseal plate

Q What accounts for the lengthwise growth of the diaphysis?

The lengthening of the diaphysis is the result of cell divisions in the zone of proliferating cartilage and maturation of the cells in the zone of hypertrophic cartilage.

4. *Zone of calcified cartilage.* The final zone of the epiphyseal plate is only a few cells thick and consists mostly of dead chrondrocytes because the matrix around them has calcified. The calcified cartilage is dissolved by osteoclasts, and the area is invaded by osteoblasts and capillaries from the diaphysis. The osteoblasts lay down bone matrix, replacing the calcified cartilage. As a result, the diaphyseal border of the epiphyseal plate is firmly cemented to the bone of the diaphysis.

The activity of the epiphyseal plate is the only way that the diaphysis can increase in length. As a bone grows, chondrocytes proliferate on the epiphyseal side of the plate. New chondrocytes cover older ones, which are then destroyed by the process of calcification. Thus, the cartilage is replaced by bone on the diaphyseal side of the plate. In this way the thickness of the epiphyseal plate remains relatively constant, but the bone on the diaphyseal side increases in length.

Between the ages of 18 and 25, the epiphyseal plates close; that is, the epiphyseal cartilage cells stop dividing, and bone replaces the cartilage. The epiphyseal plate fades, leaving a bony feature called the *epiphyseal line.* The appearance of the epiphyseal line signifies that the bone has stopped growing in length. The clavicle is the last bone to stop growing. If a bone fracture damages the epiphyseal plate, the fractured bone may be shorter than normal once adult stature is reached. This is because damage to cartilage, which is avascular, accelerates closure of the epiphyseal plate, and growth in length of the bone is inhibited.

Growth in Thickness

Unlike cartilage, which can thicken by both interstitial and appositional growth, bone can grow in thickness or diameter only by appositional growth (Figure 6.8):

1 At the bone surface, periosteal cells differentiate into osteoblasts, which secrete the collagen fibers and other organic molecules that form bone matrix. The osteoblasts become surrounded by matrix and develop into osteocytes. This process forms bone ridges on either side of a periosteal blood vessel. The ridges slowly enlarge and create a groove for the periosteal blood vessel.

2 Eventually, the ridges fold together and fuse, and the groove becomes a tunnel that encloses the blood vessel. The former periosteum now becomes the endosteum that lines the tunnel.

3 Bone deposition by osteoblasts from the endosteum forms new concentric lamellae. The formation of additional concentric lamellae proceeds inward toward the periosteal blood vessel. In this way, the tunnel fills in, and a new osteon is created.

4 As an osteon is forming, osteoblasts under the periosteum deposit new outer circumferential lamellae, further increasing the thickness of the bone. As additional

Figure 6.8 Bone growth in diameter: appositional growth.

 Whereas cartilage can grow by both interstitial and appositional growth, bone can grow in diameter only by appositional growth.

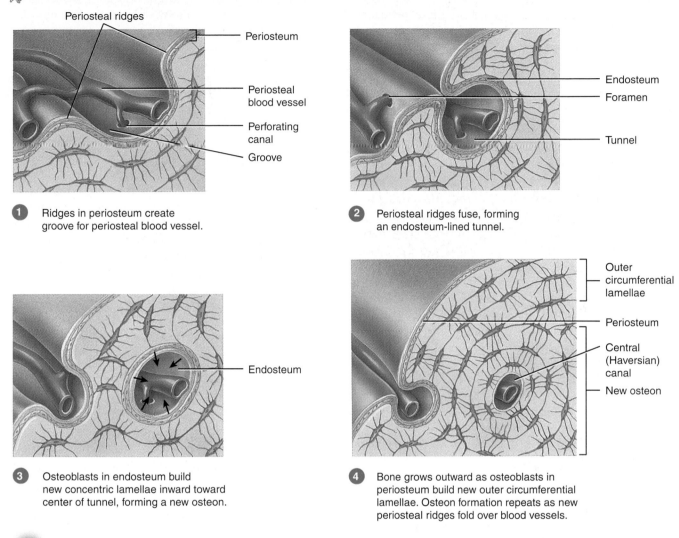

1 Ridges in periosteum create groove for periosteal blood vessel.

2 Periosteal ridges fuse, forming an endosteum-lined tunnel.

3 Osteoblasts in endosteum build new concentric lamellae inward toward center of tunnel, forming a new osteon.

4 Bone grows outward as osteoblasts in periosteum build new outer circumferential lamellae. Osteon formation repeats as new periosteal ridges fold over blood vessels.

Q How does the medullary cavity enlarge during growth in diameter?

periosteal blood vessels become enclosed as in Step 1, the growth process continues.

Recall that as new bone tissue is being deposited on the outer surface of bone, the bone tissue lining the medullary cavity is destroyed by osteoclasts in the endosteum. In this way the medullary cavity enlarges as the bone increases in diameter.

Factors Affecting Bone Growth

Growth and maintenance of bone depend on adequate dietary intake of minerals and vitamins, as well as on sufficient levels of several hormones. Considerable amounts of calcium and phosphorus are needed while bones are growing, as are smaller amounts of fluoride, magnesium, iron, and manganese. Vitamin C is needed for synthesis of the main bone protein—collagen—and also for differentiation

of osteoblasts into osteocytes. Vitamins K and B_{12} also are needed for protein synthesis, whereas vitamin A stimulates activity of osteoblasts.

During childhood, the most important hormones that stimulate bone growth are the insulinlike growth factors (IGFs), which are produced by bone tissue and also by the liver. The IGFs promote cell division at the epiphyseal plate and in the periosteum and enhance synthesis of the proteins needed to build new bone. IGF production, in turn, is stimulated by human growth hormone (hGH), which is produced by the anterior pituitary gland. Thyroid hormones (T_3 and T_4), from the thyroid gland, and insulin, from the pancreas, also are needed for normal bone growth.

At puberty, the ovaries in females and the testes in males start to secrete hormones known as sex steroids—estrogens from the ovaries and androgens from the testes. Although females have much higher levels of estrogens and males have higher levels of androgens, females also have low levels of androgens, and males have low levels of estrogens. This is because the adrenal glands of both sexes produce androgens, and other tissues, such as adipose tissue, can convert androgens to estrogens. Initially, the sex steroids cause the sudden growth or "growth spurt" that occurs during the teenage years. Estrogens also promote changes in the skeleton that are typical of females—for example, a wider pelvis. Ultimately, the sex steroids, especially estrogens in both sexes, shut down growth at the epiphyseal plates. At this point, elongation of the bones ceases. Lengthwise growth of bones typically is completed earlier in females than in males due to the higher levels of estrogens in females.

CLINICAL APPLICATION
Hormonal Abnormalities that Affect Height

Excessive or deficient secretion of hormones that normally govern bone growth causes a person to be abnormally tall or short. During childhood, oversecretion of hGH produces giantism, in which a person becomes much taller and heavier than normal, whereas undersecretion of hGH or thyroid hormones produces short stature. Because estrogens terminate growth at the epiphyseal plates, both men and women who lack either estrogens or receptors for estrogens on cells grow taller than normal. ■

1. Describe the zones of the epiphyseal plate and their functions.
2. Explain how bone grows in thickness or diameter.
3. What is the significance of the epiphyseal line?
4. Why are nutrients and hormones important in bone growth?

BONES AND HOMEOSTASIS

OBJECTIVES

• *Describe the processes involved in bone remodeling.*
• *Describe the sequence of events in repair of a fracture.*

Bone Remodeling

Even after bones have reached their adult shapes and sizes, they continue to be renewed. **Bone remodeling** is an ongoing process whereby osteoclasts first carve out small tunnels in old bone tissue and then osteoblasts rebuild it anew. A full cycle of bone remodeling may take as little as 2–3 months or last much longer, depending on where in the skeleton remodeling is occurring. For example, the distal portion of the thigh bone (femur) is fully remodeled about every four months. Remodeling normally serves two purposes: It renews bone tissue before deterioration sets in, and it redistributes bone matrix along lines of mechanical stress. Remodeling also is the way that injured bone heals.

The destruction of matrix by osteoclasts is called **bone resorption.** In this process, an osteoclast attaches tightly to

the bone surface at the endosteum or periosteum and forms a leak-proof seal at the edges of its ruffled border (see Figure 6.2). Then it releases protein-digesting lysosomal enzymes and several acids into the sealed pocket. The enzymes digest collagen fibers and other organic substances while the acids dissolve the bone minerals. Working together, several osteoclasts carve out a small tunnel in the old bone. The degraded bone proteins and matrix minerals, mainly calcium and phosphorus, enter an osteoclast by endocytosis, cross the cell in vesicles, and undergo exocytosis on the side opposite the ruffled border. Now in the interstitial fluid, the products of bone resorption diffuse into nearby blood capillaries. Once a small area of bone has been resorbed, osteoclasts depart and osteoblasts move in to rebuild the bone in that area.

To achieve homeostasis in bone, the bone-resorbing actions of osteoclasts must balance the bone-making actions of osteoblasts. If osteoblasts form too much new bone tissue, the bones become abnormally thick and heavy. If bone tissue becomes overly calcified, thick bumps called *spurs* may appear and may interfere with movement at joints. Alternatively, a loss of too much calcium or inadequate formation of new tissue weakens bone tissue.

The control of bone growth and bone remodeling is complex and not completely understood. Several circulating hormones as well as locally produced substances are involved. Sex steroids slow resorption of old bone and promote deposition of new bone. One way that estrogens slow resorption is by promoting apoptosis of osteoclasts. Other hormones that govern remodeling include parathyroid hormone and activated vitamin D, both of which also influence blood calcium level (see Figure 6.11).

Fracture and Repair of Bone

A **fracture** is any break in a bone. Fractures are given names that reflect their severity or the shape or position of the fracture line—or even the physician who first described them. Among the common kinds of fractures are the following (Figure 6.9):

• **Open (compound) fracture:** The broken ends of the bone protrude through the skin (Figure 6.9a). Conversely, a **closed (simple) fracture** does not break the skin.
• **Comminuted fracture** (KOM-i-nyoo-ted; *com-* = together; *-minuted* = crumbled): The bone splinters at the site of impact, and smaller bone fragments lie between the two main fragments (Figure 6.9b).
• **Greenstick fracture:** A partial fracture in which one side of the bone is broken and the other side bends; occurs only in children, whose bones are not yet fully ossified (Figure 6.9c).
• **Impacted fracture:** One end of the fractured bone is forcefully driven into the interior of the other (Figure 6.9d).
• **Pott's fracture:** A fracture of the distal end of the lateral leg bone (fibula), with serious injury of the distal tibial articulation (Figure 6.9e).

- **Colles' fracture** (KOL-ez): A fracture of the distal end of the lateral forearm bone (radius) in which the distal fragment is displaced posteriorly (Figure 6.9f).

    In some cases a bone may fracture without visibly breaking. For example, a **stress fracture** is a series of microscopic fissures in bone that forms without any evidence of injury to other tissues. In healthy adults, stress fractures result from repeated, strenuous activities such as running, jumping, or aerobic dancing. Stress fractures also result from disease processes that disrupt normal bone mineralization, such as osteoporosis (discussed later). About 25% of stress fractures involve the tibia (shin bone).

    The following steps occur in the repair of a bone fracture (Figure 6.10):

    1 *Formation of fracture hematoma.* As a result of the fracture, blood vessels crossing the fracture line are broken. These include vessels in the periosteum, osteons (Haversian systems), medullary cavity, and perforating canals. As blood leaks from the torn ends of the vessels, it forms a clot around the site of the fracture. This clot, called a **fracture hematoma** (hē'-ma-TŌ-ma; *hemat-* = blood; *-oma* = tumor), usually forms 6–8 hours after the injury. Because the circulation of blood stops when the fracture hematoma forms, bone cells at the fracture site die. The hematoma serves as a focus for the influx of cells that follows. Swelling and inflammation occurs in response to dead bone cells, producing additional cellular debris. Blood capillaries grow into the blood clot, and phagocytes (neutrophils and macrophages) and osteoclasts begin to remove the dead or damaged tissue in and around the fracture hematoma. This stage may last up to several weeks.

    2 *Fibrocartilaginous callus formation.* The infiltration of new blood capillaries into the fracture hematoma helps organize it into granulation tissue (see page 134), now called a **procallus.** Next, fibroblasts from the periosteum and osteogenic cells from the periosteum, endosteum, and red bone marrow invade the procallus. The fibroblasts produce collagen fibers, which help connect the broken ends of the bones. Phagocytes continue to remove cellular debris. Osteogenic cells develop into chondroblasts in areas of avascular healthy bone tissue and begin to produce fibrocartilage. Eventually, the procallus is transformed into a **fibrocartilaginous callus,** a mass of repair tissue that bridges the broken ends of the bones. The stage of the fibrocartilaginous callus lasts about 3 weeks.

    3 *Bony callus formation.* In areas closer to well-vascularized healthy bone tissue, osteogenic cells develop into osteoblasts, which begin to produce spongy bone trabeculae. The trabeculae join living and dead portions of the original bone fragments. In time, the fibrocartilage is converted to spongy bone, and the callus is then referred to as a **bony callus.** The stage of the bony callus lasts about 3–4 months.

Figure 6.9 Types of bone fractures.

🔑 **A fracture is any break in a bone.**

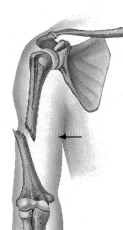
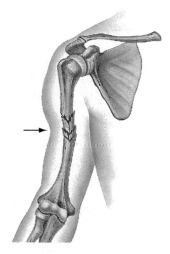

(a) Open fracture   (b) Comminuted fracture

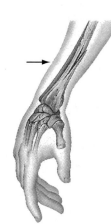
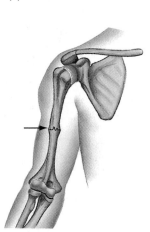

(c) Greenstick fracture   (d) Impacted fracture

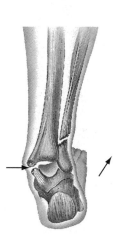
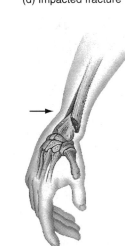

(e) Pott's fracture   (f) Colles' fracture

Q   What is a fibrocartilaginous callus?

Figure 6.10 Steps involved in repair of a bone fracture. From Priscilla LeMone and Karen
M. Burke, *Medical-Surgical Nursing,* p1560 (Menlo Park, CA: Benjamin/Cummings, 1996). ©1996 The
Benjamin/Cummings Publishing Company.

Bone heals more rapidly than cartilage because its blood supply is more plentiful.

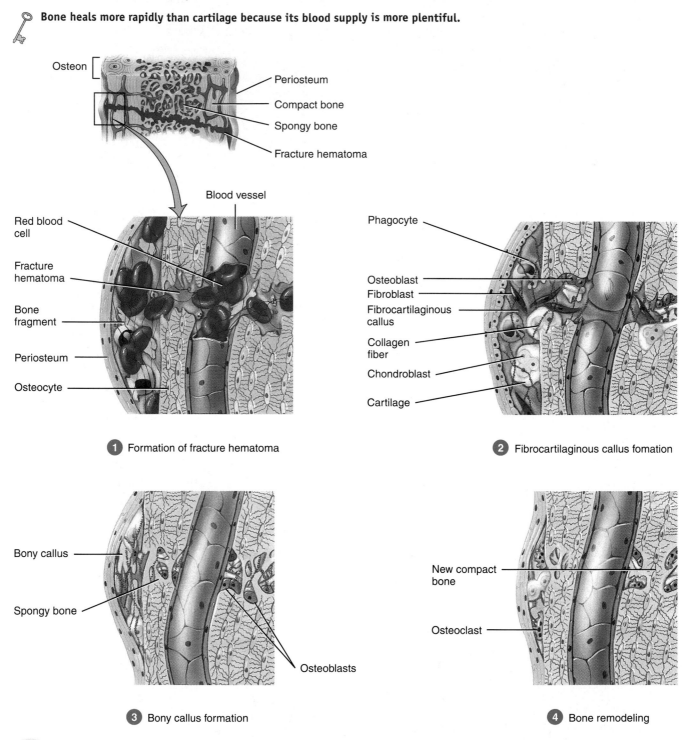

1 Formation of fracture hematoma

2 Fibrocartilaginous callus fomation

3 Bony callus formation

4 Bone remodeling

Q Why does it sometimes take months for a fracture to heal?

4 *Bone remodeling.* The final phase of fracture repair is **bone remodeling** of the callus. Dead portions of the original fragments of broken bone are gradually resorbed by osteoclasts. Compact bone replaces spongy bone around the periphery of the fracture. Sometimes, the repair process is so thorough that the fracture line is undetectable, even in a radiograph (x-ray). However, a thickened area on the surface of the bone may remain as evidence of a healed fracture.

Although bone has a generous blood supply, healing sometimes takes months. The calcium and phosphorus needed to strengthen and harden new bone are deposited only gradually, and bone cells generally grow and reproduce slowly. Moreover, the blood supply in a fractured bone sometimes is temporarily disrupted, which helps to explain the difficulty in the healing of severely fractured bones.

CLINICAL APPLICATION
Treatments for Fractures

Treatments for fractures vary according to the person's age, the type of fracture, and the bone involved. The ultimate goals of fracture treatment are the anatomic realignment of the bone fragments, immobilization to maintain realignment, and restoration of function. For bones to unite, the fractured ends must be brought into alignment, a process called **reduction,** and they must be immobilized during the time required for the fracture to heal. In **closed reduction,** the fractured ends of a bone are brought into alignment by manual manipulation, and the skin remains intact. In **open reduction,** the fractured ends of a bone are brought into alignment by a surgical procedure in which internal fixation devices such as screws, plates, pins, rods, and wires are used. Following reduction, a fractured bone may be kept immobilized by a cast, sling, splint, elastic bandage, external fixation device, or a combination of these devices. ■

Bone's Role in Calcium Homeostasis
OBJECTIVE

• *Describe the role of bone in calcium homeostasis.*

Bone is the body's major calcium reservoir, storing 99% of total body calcium. The level of calcium in the blood can be regulated by controlling the rate of calcium resorption from bone into blood and that of calcium deposition from blood into bone. Most functions of nerve cells depend on having just the right level of calcium ions (Ca^{2+}). Also, many enzymes require Ca^{2+} as a cofactor (an additional substance needed for an enzymatic reaction to occur), and blood clotting requires Ca^{2+}. For this reason, the blood plasma level of calcium ions is very closely regulated between 9 and 11 mg/100 mL. Even small changes in Ca^{2+} concentration outside this range may prove fatal—the heart may stop (cardiac arrest) if the concentration goes too high, or breathing may cease (respiratory arrest) if the level falls too low. The role of bone in calcium homeostasis is to "buffer" the blood calcium level, releasing Ca^{2+} into blood plasma when the level decreases, and taking Ca^{2+} back when the level rises. Hormones regulate these exchanges.

The most important hormone that regulates Ca^{2+} exchange between bone and blood is **parathyroid hormone (PTH),** secreted by the parathyroid glands (see Figure 18.13 on page 586). PTH secretion is linked to several negative feedback systems that adjust blood Ca^{2+} concentration. When a stimulus causes the blood Ca^{2+} level to fall, parathyroid gland cells (receptors) detect this change (Figure 6.11). The control center is the gene for PTH within the nucleus of a parathyroid gland cell. One input signal to the control center is an increased level of cyclic AMP (adenosine monophosphate) in the cytosol. Cyclic AMP accelerates reactions that "turn on" the PTH gene, PTH synthesis speeds up, and more PTH is released into the blood (output). PTH increases the number and activity of osteoclasts (effectors), which step up the pace of bone resorption. The resulting release of Ca^{2+} (and phosphate ions) from bone into blood plasma increases Ca^{2+} level and returns the blood Ca^{2+} level to normal.

Another hormone makes a potential contribution to the homeostasis of blood Ca^{2+} by its influence on bone. When blood Ca^{2+} rises above normal, **calcitonin (CT)** is secreted by *parafollicular cells* in the thyroid gland. CT inhibits activity of osteoclasts, speeds blood Ca^{2+} uptake by bone, and accelerates Ca^{2+} deposition into bones. The net result is that CT promotes bone formation and decreases blood Ca^{2+} level. Despite these effects, the role of CT in normal calcium homeostasis is uncertain because it can be completely absent or present in excess without causing clinical symptoms.

1. Define remodeling, and describe the roles of osteoblasts and osteoclasts in the process.
2. Define a fracture and outline the four steps involved in fracture repair.
3. How do hormones act on bone to regulate calcium homeostasis?

EXERCISE AND BONE TISSUE
OBJECTIVE

• *Describe how exercise and mechanical stress affect bone tissue.*

Within limits, bone has the ability to alter its strength in response to changes in mechanical stress. When placed under stress, bone tissue adapts by becoming stronger through increased deposition of mineral salts and production of collagen fibers. Another effect of stress is to increase the production of calcitonin, a hormone that inhibits bone resorption. Without mechanical stress, bone does not remodel normally because resorption outstrips bone formation. Removal of mechanical stress weakens bone through demineralization (loss of bone minerals) and collagen fiber reduction.

Figure 6.11 Negative feedback system for the regulation of blood calcium (Ca²⁺) concentration. PTH = parathyroid hormone.

> **Release of calcium from bone matrix and retention of calcium by the kidneys are the two main ways that blood calcium level can be increased.**

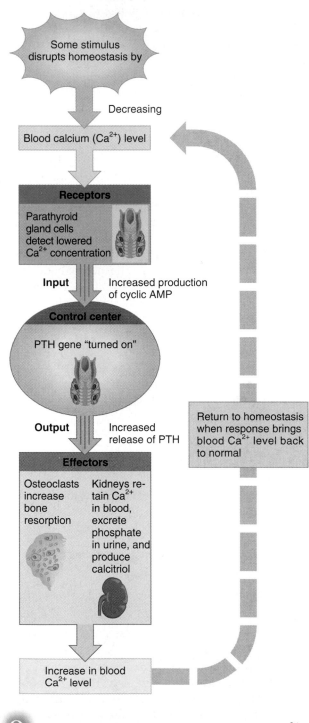

Q What body functions depend on proper levels of Ca²⁺?

The main mechanical stresses on bone result from the contraction of skeletal muscles and the pull of gravity. If a person is bedridden or has a fractured bone in a cast, the strength of the unstressed bones diminishes. Astronauts subjected to the microgravity of space also lose bone mass. In both cases, bone loss can be dramatic—as much as 1% per week. Bones of athletes, which are repetitively and highly stressed, become notably thicker than those of nonathletes. Weight-bearing activities, such as walking or moderate weight lifting, help build and retain bone mass. In recent years it has become increasingly clear that adolescents and young adults should engage in regular weight-bearing exercise prior to the closure of the epiphyseal plates. This helps to build total bone mass prior to its inevitable reduction with aging. Even elderly people can strengthen their bones by engaging in weight-bearing exercise.

1. Explain the types of mechanical stresses that may be used to strengthen bone tissue.

DEVELOPMENTAL ANATOMY OF THE SKELETAL SYSTEM

OBJECTIVE

• *Describe the development of the skeletal system and the limbs.*

As noted earlier, both intramembranous and endochondral ossification begin when *mesenchymal cells,* connective tissue cells derived from mesoderm, migrate into the area where bone formation will occur. In some skeletal structures, mesenchymal cells develop into chondroblasts that form cartilage. In other skeletal structures, mesenchymal cells develop into osteoblasts that form *bone tissue* by intramembranous or endochondral ossification.

Discussion of the development of the skeletal system provides an excellent opportunity to trace the development of the limbs. The limbs make their appearance about the fifth week as small elevations at the sides of the trunk called **limb buds** (Figure 6.12a). They consist of masses of general **mesoderm** covered by **ectoderm.** At this point, a mesenchymal skeleton exists in the limbs; some of the masses of mesoderm surrounding the developing bones will become the skeletal muscles of the limbs.

By the sixth week, the limb buds develop a constriction around the middle portion. The constriction produces distal segments of the upper buds called **hand plates** and distal segments of the lower buds called **foot plates** (Figure 6.12b). These plates represent the beginnings of the hands and feet, respectively. At this stage of limb development, a cartilaginous skeleton is present. By the seventh week (Figure 6.12c), the arm, forearm, and hand are evident in the upper limb bud, and the thigh, leg, and foot appear in the lower limb bud. Endochondral ossification has begun. By the eighth week (Figure 6.12d), as the shoulder, elbow, and wrist areas become apparent, the upper limb bud is appropriately called

Figure 6.12 Features of a developing human embryo during weeks 5 through 8 of development. The liver and heart prominences are surface bulges resulting from the underlying growth of these organs. The embryo receives oxygen and nutrients from its mother through blood vessels in the umbilical cord.

🗝️ **After the limb buds develop, endochondral ossification of the limb bones begins during the seventh embryonic week.**

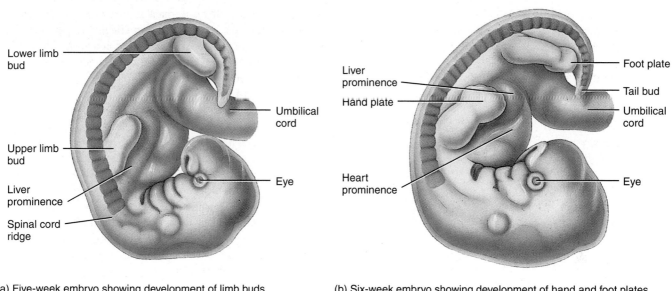

(a) Five-week embryo showing development of limb buds

(b) Six-week embryo showing development of hand and foot plates

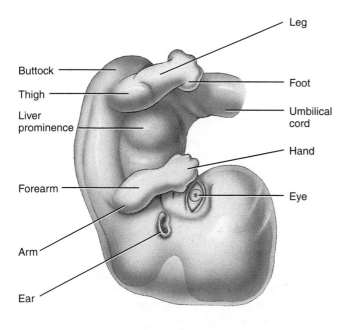

(c) Seven-week embryo showing development of arm, forearm, and hand in upper limb bud and thigh, leg, and foot in lower limb bud

(d) Eight-week embryo in which limb buds have developed into upper and lower limbs

Q Which of the three basic embryonic tissues—ectoderm, mesoderm, and endoderm—gives rise to the skeletal system?

the upper limb, and the lower limb bud is now the lower limb.

The **notochord** is a flexible rod of mesoderm that defines the midline of the embryo and also gives it some rigidity. The notochord lies in a position where the future vertebral column will develop. As the vertebrae form, the notochord becomes surrounded by the developing vertebral bodies. The notochord eventually disappears, except for remnants that persist as the *nucleus pulposus* of the intervertebral discs (see Figure 7.16 on page 202).

1. Describe when and how the limbs develop.

AGING AND BONE TISSUE

OBJECTIVE
• *Describe the effects of aging on bone tissue.*

From birth through adolescence, more bone tissue is produced than is lost during bone remodeling. In young adults the rates of bone deposition and resorption are about the same. As the level of sex steroids diminishes during middle age, especially in women after menopause, a decrease in bone mass occurs because bone resorption outpaces bone deposition. In old age, loss of bone through resorption occurs more rapidly than bone gain. Because women's bones generally are smaller and less massive than men's bones to begin with, loss of bone mass in old age typically has a greater adverse effect in women.

There are two principal effects of aging on bone tissue: loss of bone mass and brittleness. The first effect results from the loss of calcium and other minerals from bone matrix (demineralization). This loss usually begins after age 30 in females, accelerates greatly around age 45 as levels of estrogens decrease, and continues until as much as 30% of the calcium in bones is lost by age 70. Once bone loss begins in females, about 8% of bone mass is lost every ten years. In males, calcium loss typically does not begin until after age 60, and about 3% of bone mass is lost every ten years. The loss of calcium from bones is one of the problems in osteoporosis (described below).

The second principal effect of aging on the skeletal system, brittleness, results from a lowered rate of protein synthesis, which diminishes the organic portion of bone matrix, mainly collagen fibers, that gives bone its tensile strength. As a consequence, inorganic minerals gradually constitute a greater proportion of the bone matrix. The loss of tensile strength causes the bones to become very brittle and susceptible to fracture. In some elderly people collagen fiber synthesis slows, in part due to diminished production of human growth hormone. In addition to increasing the susceptibility to fractures, loss of bone mass also leads to deformity, pain, stiffness, loss of height, and loss of teeth.

1. What is demineralization, and how does it affect the functioning of bone?
2. What changes occur in the organic portion of bone matrix with aging?

DISORDERS: HOMEOSTATIC IMBALANCES

OSTEOPOROSIS

Osteoporosis (os′-tē-ō-pō-RŌ-sis; *por-* = passageway; *-osis* = condition) is literally a condition of porous bones (Figure 6.13). The basic problem is that bone resorption outpaces bone deposition. In large part this is due to depletion of calcium from the body—more calcium is lost in urine, feces, and sweat than is absorbed from the diet. Bone mass becomes so depleted that bones fracture, often spontaneously, under the mechanical stresses of everyday living. For example, a hip fracture might result from simply sitting down too quickly. In the United States, osteoporosis causes more than a million fractures a year, mainly in the hip, wrist, and vertebrae. Osteoporosis afflicts the entire skeletal system. In addition to fractures, osteoporosis causes shrinkage of vertebrae, height loss, hunched backs, and bone pain.

Thirty million people in the United States suffer from osteoporosis. The disorder primarily affects middle-aged and elderly people, 80% of them women. Older women suffer from osteoporosis more often than men for two reasons: Women's bones are less massive than men's bones, and production of estrogens in women

declines dramatically at menopause, whereas production of the main androgen, testosterone, in older men wanes gradually and only slightly. Besides gender, risk factors for developing osteoporosis include a family history of the disease, European or Asian ancestry, thin or small body build, an inactive lifestyle, cigarette smoking, a diet low in calcium and vitamin D, more than two alcoholic drinks a day, and the use of certain medications.

In postmenopausal women, treatment of osteoporosis may include estrogen replacement therapy (ERT; low doses of estrogens) or hormone replacement therapy (HRT; a combination of estrogens and progesterone, another sex steroid). Although such treatments help combat osteoporosis, they increase a woman's risk of breast cancer. The drug Raloxifene (Evista) mimics the beneficial effects of estrogens on bone without increasing the risk of breast cancer. A nonhormone drug called Alendronate (Fosamax) blocks resorption of bone by osteoclasts and is also used to treat osteoporosis.

Perhaps more important than treatment is prevention. Adequate calcium intake and weight-bearing exercise in her early years may be more beneficial to a woman than drugs and calcium supplements when she is older.

Figure 6.13 Comparison of spongy bone tissue from (a) a normal young adult and (b) a person with osteoporosis. Notice the weakened trabeculae in (b). Compact bone tissue is similarly affected by osteoporosis.

 In osteoporosis, bone resorption outpaces bone formation, so bone mass decreases.

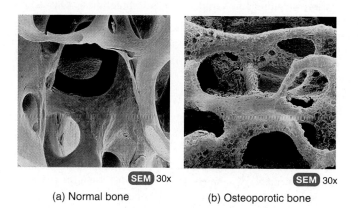

(a) Normal bone   (b) Osteoporotic bone

Q If you wanted to develop a drug to lessen the effects of osteoporosis, would you look for a chemical that inhibits the activity of osteoblasts, or that of osteoclasts?

RICKETS AND OSTEOMALACIA

Rickets and **osteomalacia** (os′-tē-ō-ma-LĀ-she-ah; -*malacia* = softness) are disorders in which bones fail to calcify. Although the organic matrix is still produced, calcium salts are not deposited, and the bones become "soft" or rubbery and easily deformed. Rickets affects the growing bones of children. Because new bone formed at the epiphyseal plates fails to ossify, bowed legs and deformities of the skull, rib cage, and pelvis are common results. In osteomalacia, sometimes called "adult rickets," new bone formed during remodeling fails to calcify. This disorder causes varying degrees of pain and tenderness in bones, especially in the hip and leg. Bone fractures also result from minor trauma.

MEDICAL TERMINOLOGY

Osteoarthritis (os′-tē-ō-ar-THRĪ-tis; *arthr* = joint) The degeneration of articular cartilage such that the bony ends touch; the resulting friction of bone against bone worsens the condition. Usually associated with the elderly.

Osteogenic sarcoma (os′-tē-Ō-JEN-ik sar-KŌ-ma; *sarcoma* = connective tissue tumor) Bone cancer that primarily affects osteoblasts and occurs most often in teenagers during their growth spurt; the most common sites are the metaphyses of the thighbone (femur), shinbone (tibia), and arm bone (humerus). Metastases occur most often in lungs; treatment consists of multidrug chemotherapy and removal of the malignant growth, or amputation of the limb.

Osteomyelitis (os′-tē-ō-mī-el-Ī-tis) An infection of bone characterized by high fever, sweating, chills, pain, and nausea, pus formation, edema, and warmth over the affected bone and rigid overlying muscles. It is often caused by bacteria, usually *Staphylococcus aureus*. The bacteria may reach the bone from outside the body (through open fractures, penetrating wounds, or orthopedic surgical procedures); from other sites of infection in the body (abscessed teeth, burn infections, urinary tract infections, or upper respiratory infections) via the blood; and from adjacent soft tissue infections (as occurs in diabetes mellitus).

Osteopenia (os′-tē-ō-PĒ-nē-a; *penia* = poverty) Reduced bone mass due to a decrease in the rate of bone synthesis to a level insufficient to compensate for normal bone resorption; any decrease in bone mass below normal. An example is osteoporosis.

INTRODUCTION (p. 160)

1. Bone is made up of several different tissues: bone or osseous tissue, cartilage, dense connective tissues, epithelium, various blood-forming tissues, adipose tissue, and nervous tissue.
2. The entire framework of bones and their cartilages constitute the skeletal system.

FUNCTIONS OF THE SKELETAL SYSTEM (p. 160)

1. The skeletal system functions in support, protection, movement, mineral storage and release, blood cell production, and triglyceride storage.

STRUCTURE OF BONE (p. 161)

1. Parts of a typical long bone are the diaphysis (shaft), proximal and distal epiphyses (ends), metaphyses, articular cartilage, periosteum, medullary (marrow) cavity, and endosteum.

HISTOLOGY OF BONE TISSUE (p. 162)

1. Bone tissue consists of widely separated cells surrounded by large amounts of matrix.
2. The four principal types of cells in bone tissue are osteogenic cells, osteoblasts, osteocytes, and osteoclasts.
3. The matrix of bone contains abundant mineral salts (mostly hydroxyapatite) and collagen fibers.
4. Compact bone tissue consists of osteons (Haversian systems) with little space between them.
5. Compact bone tissue lies over spongy bone tissue in the epiphyses and makes up most of the bone tissue of the diaphysis. Functionally, compact bone tissue protects, supports, and resists stress.
6. Spongy bone tissue does not contain osteons. It consists of trabeculae surrounding many red bone marrow–filled spaces.
7. Spongy bone tissue forms most of the structure of short, flat, and irregular bones, and the epiphyses of long bones. Functionally, spongy bone tissue trabeculae offer resistance along lines of stress, support and protect red bone marrow, and make bones lighter for easier movement.

BLOOD AND NERVE SUPPLY OF BONE (p. 165)

1. Long bones are supplied by periosteal, nutrient, and epiphyseal arteries; veins accompany the arteries.
2. Nerves accompany blood vessels in bone; the periosteum is rich in sensory neurons.

BONE FORMATION (p. 166)

1. Bone forms by a process called ossification (osteogenesis), which begins when mesenchymal cells become transformed into osteogenic cells. These undergo cell division and give rise to cells that differentiate into osteoblasts and osteoclasts.
2. Ossification begins during the sixth or seventh week of embryonic life. The two types of ossification, intramembranous and endochondral, involve the replacement of a preexisting connective tissue with bone.
3. Intramembranous ossification occurs within fibrous connective tissue membranes.
4. Endochondral ossification occurs within a hyaline cartilage model. The primary ossification center of a long bone is in the diaphysis. Cartilage degenerates, leaving cavities that merge to form the medullary cavity. Osteoblasts lay down bone. Next, ossification occurs in the epiphyses, where bone replaces cartilage, except for the epiphyseal plate.

BONE GROWTH (p. 169)

1. The epiphyseal plate consists of four zones: zones of resting cartilage, proliferating cartilage, hypertrophic cartilage, and calcified cartilage.
2. Because of the activity of the epiphyseal plate, the diaphysis of a bone increases in length.
3. Bone grows in thickness or diameter due to the addition of new bone tissue by periosteal osteoblasts around the outer surface of the bone (appositional growth).
4. Dietary minerals (especially calcium and phosphorus) and vitamins (C, K, and B_{12}) are needed for bone growth and maintenance. Insulinlike growth factors, human growth hormone, thyroid hormones, insulin, estrogens, and androgens stimulate bone growth. The sex steroids, especially estrogens, also shut down growth at the epiphyseal plates.

BONES AND HOMEOSTASIS (p. 172)

1. Remodeling is an ongoing process whereby osteoclasts carve out small tunnels in old bone tissue and then osteoblasts rebuild it anew.
2. In bone resorption, osteoclasts release enzymes and acids that degrade collagen fibers and dissolve mineral salts.
3. Sex steroids slow resorption of old bone and promote new bone deposition. Estrogens slow resorption by promoting apoptosis of osteoclasts.
4. A fracture is any break in a bone.
5. Fracture repair involves formation of a fracture hematoma, of fibrocartilaginous callus, of bony callus, and bone remodeling.
6. Types of fractures include closed (simple), open (compound), comminuted, greenstick, impacted, stress, Pott's, and Colles'.
7. Bone is the major reservoir for calcium in the body.
8. Parathyroid hormone (PTH) secreted by the parathyroid gland increases blood Ca^{2+} level, whereas calcitonin (CT) from the thyroid gland has the potential to decrease blood Ca^{2+} level.

EXERCISE AND BONE TISSUE (p. 175)

1. Mechanical stress increases bone strength by increasing deposition of mineral salts and production of collagen fibers.
2. Removal of mechanical stress weakens bone through demineralization and collagen fiber reduction.

DEVELOPMENTAL ANATOMY OF THE SKELETAL SYSTEM (p. 176)

1. Bone forms from mesoderm by intramembranous or endochondral ossification.
2. Limbs develop from limb buds, which consist of mesoderm and ectoderm.

AGING AND BONE TISSUE (p. 178)

1. The principal effect of aging is a loss of calcium from bones, which may result in osteoporosis.
2. Another effect is a decreased production of matrix proteins (mostly collagen fibers), which makes bones more brittle and thus more susceptible to fracture.

SELF-QUIZ QUESTIONS

Complete the following:

1. Whereas the ___ of bone depends on the crystallized mineral salts, the collagen fibers and other organic molecules provide bone with ___.
2. Compact bone tissue is composed of ___; spongy bone tissue is composed of ___.
3. Endochondral ossification refers to the formation of bone within ___; intramembranous ossification refers to the formation of bone directly from ___.
4. The most important hormone that regulates calcium ion exchange between bone and blood is ___.
5. Match the following:
   ___ (a) space within the shaft of the bone that contains either red or yellow bone marrow, depending on the age of the individual
   ___ (b) triglyceride storage tissue
   ___ (c) hemopoietic tissue
   ___ (d) thin layer of hyaline cartilage covering the ends of bones where they form a joint
   ___ (e) distal and proximal ends of bones
   ___ (f) the long, cylindrical main portion of the bone; the shaft
   ___ (g) in a growing bone, the region that contains the growth plate
   ___ (h) the tough membrane that surrounds the bone surface wherever cartilage is not present
   ___ (i) the lining of the inside of the bone
   ___ (j) a remnant of the active epiphyseal plate; a sign that the bone has stopped growing in length

   (1) articular cartilage
   (2) endosteum
   (3) medullary cavity
   (4) diaphysis
   (5) epiphyses
   (6) metaphysis
   (7) periosteum
   (8) red bone marrow
   (9) yellow bone marrow
   (10) epiphyseal line

True or false:

6. The activity of the epiphyseal plate is the only mechanism by which the diaphysis can increase in length.
7. Within limits, bone has the ability to alter its strength in response to mechanical stress; therefore, when placed under stress, bone becomes stronger through decreased deposition of mineral salts and production of collagen fibers.

8. Match the following:
   ___ (a) small spaces between lamellae that contain osteocytes
   ___ (b) perforating canals that penetrate compact bone; carry blood vessels, lymphatic vessels, and nerves from the periosteum
   ___ (c) areas between osteons
   ___ (d) microscopic unit of compact bone tissue
   ___ (e) tiny canals filled with extracellular fluid; connect lacunae to each other and to the central canal
   ___ (f) canals that run longitudinally through the bone and connect blood vessels and nerves to the osteocytes
   ___ (g) irregular lattice work of thin columns of bone found in spongy bone tissue
   ___ (h) rings of hard calcified matrix found around the central canals
   ___ (i) an opening in the shaft of the bone allowing an artery to pass into the bone
   ___ (j) a layer of hyaline cartilage in the area between the shaft and end of a growing bone

   (1) lacunae
   (2) Volkmann's canals
   (3) Haversian canals
   (4) concentric lamellae
   (5) epiphyseal plate
   (6) trabeculae
   (7) interstitial lamellae
   (8) canaliculi
   (9) nutrient foramen
   (10) osteon

Choose the best answer to the following questions:

9. Which of the following are functions of bone tissue and the skeletal system? (1) support, (2) excretion, (3) assistance in movement, (4) mineral homeostasis, (5) blood cell production.
   (a) 1, 2, and 3, (b) 2, 3, and 4, (c) 3, 4, and 5, (d) 1, 3, 4, and 5, (e) 2, 3, 4, and 5
10. Which of the following statements are true? (1) Osteogenic cells develop directly into osteoclasts. (2) Osteogenic cells are unspecialized stem cells. (3) Osteoblasts form bone. (4) Osteocytes are the principal cells of bone tissue. (5) Osteoclasts function in bone resorption. (6) Osteoblasts maintain daily cellular activities of bone tissue.
    (a) 1, 2, 4, 5, and 6, (b) 2, 3, 4, and 5, (c) 1, 3, 5, and 6, (d) 2, 4, and 6, (e) 1, 2, 4, 5, and 6
11. Which of the following statements is true? (a) The matrix of bone consists of inorganic mineral salts, primarily hydroxyapatite. (b) Spongy bone tissue makes up most of the bone tissue of the long bones of the limbs. (c) Yellow bone marrow is found only in embryonic bone. (d) Once bone forms, it is a dead tissue that exists for the life of the individual. (e) Red bone marrow is found in compact bone tissue only.

12. Which of the following statements are true? (1) Osteogenic cells are present in the periosteum. (2) The epiphyseal plate shapes the articular surfaces. (3) Bone can grow in thickness or diameter only by appositional growth. (4) Hormones play no role in bone growth; they function only in adult bone maintenance. (5) The epiphyseal line is formed prior to the formation of the epiphyseal plate.
(a) 1, 3, and 5, (b) 2, 4, and 5, (c) 1, 2, and 3, (d) 2, 4, and 5, (e) 1, 2, 3, and 5

13. Which of the following statements are true? (1) Weight-bearing activities help build and retain bone mass. (2) Removal of mechanical stress weakens bone. (3) In the absence of mechanical stress, resorption outstrips bone formation. (4) Mechanical stress decreases the production of calcitonin, a hormone that inhibits bone resorption. (5) The main mechanical stresses on bone are those resulting from the contraction of skeletal muscle and the pull of gravity.
(a) 1, 2, 3, and 5, (b) 2, 3, 4, and 5, (c) 1, 3, and 5, (d) 2, 4, and 5, (e) 2, 3, and 4

14. Which of the following is *not* required for normal bone growth, bone remodeling in the adult, and repair of fractured bone?
(a) insulinlike growth factors, (b) keratin, (c) vitamin C, (d) parathyroid hormone, (e) weight-bearing exercise.

15. Match the following:
___ (a) bone cancer primarily affecting osteoblasts in the bones of teenagers
___ (b) a condition of porous bones characterized by decreased bone mass and increased susceptibility to fractures
___ (c) splintered bone, with smaller fragments lying between main fragments
___ (d) a broken bone that does not break through the skin
___ (e) a partial break in a bone in which one side of the bone is broken and the other side bends
___ (f) a broken bone that protrudes through the skin
___ (g) microscopic bone breaks resulting from inability to withstand repeated stressful impact
___ (h) a degeneration of articular cartilage allowing the bony ends to touch; worsens due to friction between the bones
___ (i) condition characterized by failure of new bone formed by remodeling to calcify in adults
___ (j) an infection of bone

(1) closed (simple) fracture
(2) open (compound) fracture
(3) greenstick fracture
(4) stress fracture
(5) comminuted fracture
(6) osteoporosis
(7) osteomalacia
(8) osteomyelitis
(9) osteoarthritis
(10) osteogenic sarcoma

CRITICAL THINKING QUESTIONS

1. A father brings his young daughter to the emergency room after she fell off her bicycle. The attending physician tells the father that his daughter has suffered a greenstick fracture in her lower arm. The father is confused—there weren't any sticks where she fell in the street. What should the physician explain to the father? (HINT: *Only children suffer this type of fracture.*)

2. Aunt Edith is 95 years old today. She claims that she's been getting shorter every year, and that soon she'll disappear altogether. What's happening to Aunt Edith? (HINT: *Her mind is as sharp as ever; think about her bones.*)

3. Astronauts in space exercise as part of their daily routine, yet they still have problems with bone weakness after prolonged stays in space. Why does this happen? (HINT: *What is missing in space that you have to work against on Earth?*)

ANSWERS TO FIGURE QUESTIONS

6.1 The periosteum is essential for growth in bone diameter, bone repair, and bone nutrition. It also serves as a point of attachment for ligaments and tendons.

6.2 Bone resorption is necessary for the development, growth, maintenance, and repair of bone.

6.3 Because the central (Haversian) canals are the main blood supply to the osteocytes of an osteon, their blockage would lead to death of the osteocytes.

6.4 Periosteal arteries enter bone tissue through perforations (Volkmann's canals).

6.5 Flat bones of the skull and mandible (lower jawbone) develop by intramembranous ossification.

6.6 Yes, she probably will grow taller. Epiphyseal lines are indications of growth zones that have ceased to function. The absence of epiphyseal lines indicates that the bone is still lengthening.

6.7 The lengthwise growth of the diaphysis is caused by cell divisions in the zone of proliferating cartilage and maturation of the cells in the zone of hypertrophic cartilage.

6.8 The medullary cavity enlarges by activity of the osteoclasts in the endosteum.

6.9 A fibrocartilaginous callus is a mass of fibrocartilaginous repair tissue that bridges the ends of broken bones.

6.10 Healing of bone fractures can take months because calcium and phosphorus deposition is a slow process, and bone cells generally grow and reproduce slowly.

6.11 Heartbeat, respiration, nerve cell functioning, enzyme functioning, and blood clotting all depend on proper levels of calcium.

6.12 The skeletal system develops from mesoderm.

6.13 Inhibiting the activity of osteoclasts might lessen the effects of osteoporosis.

THE SKELETAL SYSTEM: THE AXIAL SKELETON

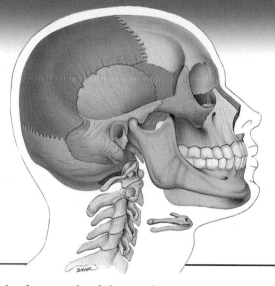

Because the skeletal system forms the framework of the body, a familiarity with the names, shapes, and positions of individual bones will help you locate other organs. For example, the radial artery, the site where the pulse is usually taken, is named for its proximity to the radius, the lateral bone of the forearm. The frontal lobe of the brain lies deep to the frontal (forehead) bone. The tibialis anterior muscle lies along the anterior surface of the tibia (shin bone). The ulnar nerve is named for its proximity to the ulna, the medial bone of the forearm.

Movements such as throwing a ball, biking, and walking require an interaction between bones and muscles. To understand how muscles produce different movements, you will learn where the muscles attach on individual bones and the types of joints acted on by the contracting muscles. The bones, muscles, and joints together form an integrated system called the **musculoskeletal system.**

DIVISIONS OF THE SKELETAL SYSTEM

OBJECTIVE

• *Describe how the skeleton is divided into axial and appendicular divisions.*

The adult human skeleton consists of 206 named bones, most of which are paired on the right and left sides of the body. Bones are grouped in two principal divisions: the 80 bones of the **axial skeleton** and the 126 bones of the **appendicular skeleton** (*appendic-* = to hang onto). Figure 7.1

shows how both divisions join to form the complete skeleton. The longitudinal **axis,** or center, of the human body is a vertical line that runs through the body's center of gravity, extending down through the head to the space between the feet. The axial skeleton consists of the bones arranged along the axis: skull bones, auditory ossicles (ear bones), hyoid bone (see Figure 7.4), ribs, breastbone, and bones of the backbone. The appendicular skeleton consists of the bones of the **upper** and **lower limbs (extremities),** plus the bones forming the **girdles** that connect the limbs to the axial skeleton. Functionally, the auditory ossicles in the middle ear are not part of either the axial or appendicular skeleton, but they are grouped with the axial skeleton for convenience. The auditory ossicles vibrate in response to sound waves that strike the eardrum and have a key role in the mechanism of hearing (see Chapter 16). Table 7.1 presents the standard grouping of the 80 bones of the axial skeleton and the 126 bones of the appendicular skeleton.

We will organize our study of the skeletal system around the two divisions of the skeleton, with emphasis on how the many bones of the body are interrelated. In this chapter we focus on the axial skeleton, looking first at the skull and then at the bones of the vertebral column (backbone) and the chest. In Chapter 8 we explore the appendicular skeleton, examining in turn the bones of the pectoral (shoulder) girdle and upper limbs, and then the pelvic (hip) girdle and the lower limbs. But before we examine the skull we direct our attention to some rather general characteristics of bones.

1. List the bones that make up the axial and appendicular divisions of the skeleton.

Figure 7.1 Divisions of the skeletal system. The axial skeleton is indicated in blue, the appendicular skeleton in yellow. (See Tortora, *A Photographic Atlas of the Human Body,* Figure 3.1)

Which of the following structures are part of the axial skeleton, and which are part of the appendicular skeleton? Skull, clavicle, vertebral column, shoulder girdle, humerus, pelvic girdle, and femur.

🔑 **The adult human skeleton consists of 206 bones grouped into axial and appendicular divisions.**

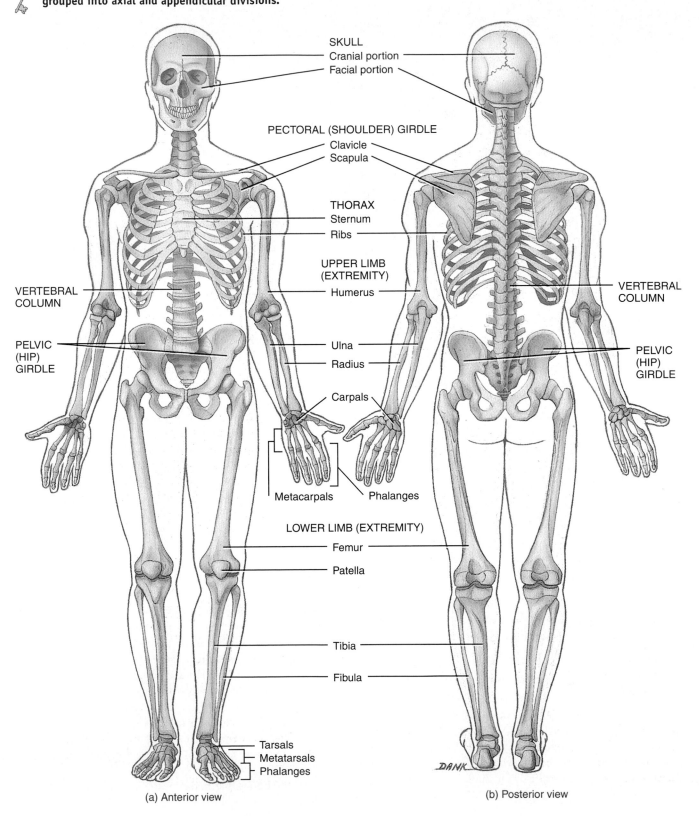

(a) Anterior view

(b) Posterior view

TYPES OF BONES

OBJECTIVE

• *Classify bones on the basis of shape and location.*

Almost all the bones of the body may be classified into five principal types on the basis of shape: long, short, flat, irregular, and sesamoid (Figure 7.2). **Long bones** have greater length than width and consist of a shaft and a variable number of extremities (ends). They are usually somewhat curved for strength. A slightly curved bone absorbs the stress of the body's weight at several different points, so that it is evenly distributed. If such bones were straight, the weight of the body would be unevenly distributed, and the bone would more easily fracture. Long bones consist mostly of compact bone tissue in their diaphyses but contain considerable amounts of spongy bone tissue in their epiphyses. Long bones include those in the thigh (femur), leg (tibia and fibula), arm (humerus), forearm (ulna and radius), and fingers and toes (phalanges).

Short bones are somewhat cube-shaped, nearly equal in length and width. They are composed of spongy bone tissue except at the surface, where there is a thin layer of compact bone tissue. Examples of short bones are the wrist or carpal bones (except for the pisiform, which is a sesamoid bone) and the ankle or tarsal bones (except for the calcaneus, which is an irregular bone).

Flat bones are generally thin and composed of two nearly parallel plates of compact bone tissue enclosing a layer of spongy bone tissue. Flat bones afford considerable protection and provide extensive areas for muscle attachment. Flat bones include the cranial bones, which protect the brain; the breastbone (sternum) and ribs, which protect organs in the thorax; and the shoulder blades (scapulae).

Irregular bones have complex shapes and cannot be grouped into any of the previous categories. They vary in the amount of spongy and compact bone present. Such bones include the vertebrae of the backbone and some facial bones.

Sesamoid bones (= like a sesame seed) develop in certain tendons where there is considerable friction, tension, and physical stress, such as the palms and soles. They may vary in number from person to person, are not always completely ossified, and typically measure only a few millimeters in diameter. A notable exception is the two patellae (kneecaps), which are large sesamoid bones that are normally present in everyone. Functionally, sesamoid bones protect tendons from excessive wear and tear, and they often change the direction of pull of a tendon, which improves the mechanical advantage at a joint.

An additional type of bone is not included in this classification by shape, but instead is classified by location. **Sutural bones** (SOO-chur-al; *sutur-* = seam) are small bones located within joints, called sutures, between certain cranial bones (see Figure 7.6). Their number varies greatly from person to person.

1. Give examples of long, short, flat, and irregular bones.

Table 7.1 The Bones of the Adult Skeletal System

DIVISION OF THE SKELETON	STRUCTURE	NUMBER OF BONES
Axial Skeleton		
	Skull	
	Cranium	8
	Face	14
	Hyoid	1
	Auditory ossicles	6
	Vertebral column	26
	Thorax	
	Sternum	1
	Ribs	24
	Subtotal =	80
Appendicular Skeleton		
	Pectoral (shoulder) girdles	
	Clavicle	2
	Scapula	2
	Upper limbs (extremities)	
	Humerus	2
	Ulna	2
	Radius	2
	Carpals	16
	Metacarpals	10
	Phalanges	28
	Pelvic (hip) girdle	
	Hip, pelvic, or coxal bone	2
	Lower limbs (extremities)	
	Femur	2
	Fibula	2
	Tibia	2
	Patella	2
	Tarsals	14
	Metatarsals	10
	Phalanges	28
	Subtotal =	126
	Total =	206

BONE SURFACE MARKINGS

OBJECTIVE

• *Describe the principal surface markings on bones and the functions of each.*

Bones have characteristic **surface markings,** structural features adapted for specific functions. There are two major types of surface markings: (1) *depressions and openings,*

Figure 7.2 Types of bones based on shape. The bones are not drawn to scale.

The shapes of bones largely determine their functions.

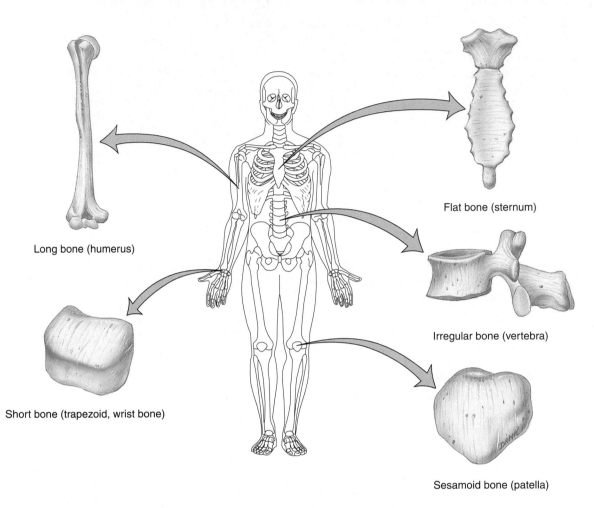

Long bone (humerus)

Flat bone (sternum)

Short bone (trapezoid, wrist bone)

Irregular bone (vertebra)

Sesamoid bone (patella)

Q Which type of bone primarily protects and provides a large surface area for muscle attachment?

which form joints or allow the passage of soft tissues (such as blood vessels and nerves), and (2) *processes,* which are projections or outgrowths that either help form joints or serve as attachment points for connective tissue (such as ligaments and tendons). Table 7.2 describes the various surface markings and provides examples of each.

1. List and describe 15 bone surface markings, and give an example of each. Check your list against Table 7.2.

SKULL

OBJECTIVE
• *Name the cranial and facial bones and indicate the number of each.*

The **skull,** which contains 22 bones, rests on the superior end of the vertebral column. It includes two sets of bones: cranial bones and facial bones. The **cranial bones**

(*crani-* = brain case) form the cranial cavity, which encloses and protects the brain. The eight cranial bones are the frontal bone, two parietal bones, two temporal bones, the occipital bone, the sphenoid bone, and the ethmoid bone. Fourteen **facial bones** form the face: two nasal bones, two maxillae (or maxillas), two zygomatic bones, the mandible, two lacrimal bones, two palatine bones, two inferior nasal conchae, and the vomer. Figures 7.3 through 7.8 illustrate these bones from different viewing directions.

General Features
Besides forming the large cranial cavity, the skull also forms several smaller cavities, including the nasal cavity and orbits (eye sockets), which open to the exterior. Certain skull bones also contain cavities that are lined with mucous membranes and are called paranasal sinuses. The sinuses open into the nasal cavity. Also within the skull are small cavities that house the structures that are involved in hearing and equilibrium.

Table 7.2 Bone Surface Markings

MARKING	DESCRIPTION	EXAMPLE
Depressions and Openings: Sites allowing the passage of soft tissue (nerves, blood vessels, ligaments, tendons) or formation of joints		
Fissure (FISH-ur)	Narrow slit between adjacent parts of bones through which blood vessels or nerves pass.	Superior orbital fissure of the sphenoid bone (Figure 7.13).
Foramen (fō-RĀ-men)	Opening (*foramen* = hole) through which blood vessels, nerves, or ligaments pass.	Optic foramen of the sphenoid bone (Figure 7.13).
Fossa (FOS-a)	Shallow depression (*fossa* = trench).	Coronoid fossa of the humerus (Figure 8.5d).
Sulcus (SUL-kus)	Furrow (*sulcus* = groove) along bone surface that accommodates a blood vessel, nerve, or tendon.	Intertubercular sulcus of the humerus (Figure 8.5d).
Meatus (mē-Ā-tus)	Tubelike opening (*meatus* = passageway).	External auditory meatus of the temporal bone (Figure 7.4).
Processes: Projections or outgrowths on bone that form joints or attachment points for connective tissue, such as ligaments and tendons.		
Processes that form joints:		
Condyle (KON-dīl)	Large, round protuberance (*condylus* = knuckle) at the end of a bone.	Lateral condyle of the femur (Figure 8.12a).
Facet	Smooth flat articular surface.	Superior articular facet of a vertebra (Figure 7.18d).
Head	Rounded articular projection supported on the neck (constricted portion) of a bone.	Head of the femur (Figure 8.12a).
Processes that form attachment points for connective tissue:		
Crest	Prominent ridge or elongated projection.	Iliac crest of the hip bone (Figure 8.9b).
Epicondyle	Projection above (*epi-* = above) a condyle.	Medial epicondyle of the femur (Figure 8.12a).
Line	Long, narrow ridge or border (less prominent than a crest).	Linea aspera of the femur (Figure 8.12b).
Spinous process	Sharp, slender projection.	Spinous process of a vertebra (Figure 7.17).
Trochanter (trō-KAN-ter)	Very large projection.	Greater trochanter of the femur (Figure 8.12b).
Tubercle (TOO-ber-kul)	Small, rounded projection (*tuber-* = knob).	Greater tubercle of the humerus (Figure 8.5a).
Tuberosity	Large, rounded, usually roughened projection.	Ischial tuberosity of the hip bone (Figure 8.9b).

Other than the auditory ossicles within the temporal bones, the mandible is the only movable bone of the skull. Most of the skull bones are held together by immovable joints called sutures, which are especially noticeable on the outer surface of the skull.

The skull has numerous surface markings, such as foramina and fissures through which blood vessels and nerves pass. You will learn the names of important skull bone surface markings when the various bones are described.

In addition to protecting the brain, the cranial bones also have other functions. Their inner surfaces attach to membranes (meninges) that stabilize the positions of the brain, blood vessels, and nerves. The outer surfaces of cranial bones provide large areas of attachment for muscles that move various parts of the head. The bones also provide attachment for some muscles that are involved in producing various facial expressions. Besides forming the framework of the face, the facial bones protect and provide support for the entrances to the digestive and respiratory systems. Together, the cranial and facial bones protect and support the delicate special sense organs for vision, taste, smell, hearing, and equilibrium.

Later in this chapter we will study in detail some of the special features of the skull, such as sutures, foramina, paranasal sinuses, the orbits (eye sockets), fontanels, and the nasal septum. These special features are best understood once you know the names, parts, and locations of the cranial and facial bones.

Cranial Bones

Frontal Bone

The **frontal bone** forms the forehead (the anterior part of the cranium), the roofs of the orbits, and most of the anterior part of the cranial floor (Figure 7.3). Soon after birth the left and right sides of the frontal bone are united by the *frontal suture*, which usually disappears by age 6. If it persists, it is referred to as the *metopic suture*.

Figure 7.3 Skull. (See Tortora, *A Photographic Atlas of the Human Body,* Figure 3.2)

The skull consists of two sets of bones: cranial and facial.

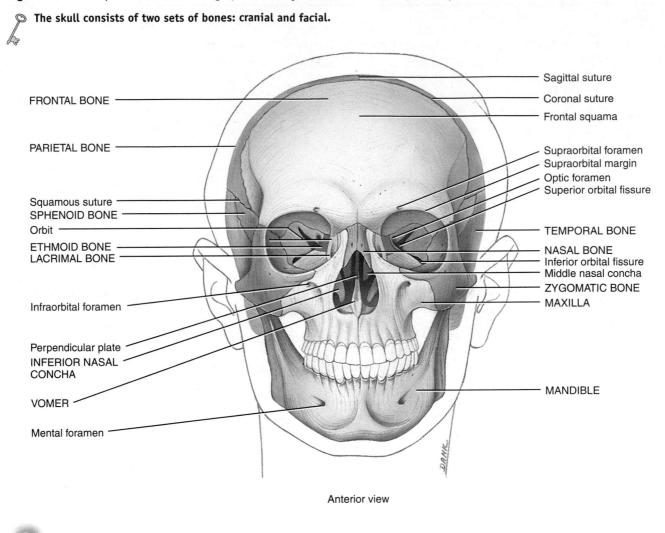

FRONTAL BONE

PARIETAL BONE

Squamous suture
SPHENOID BONE
Orbit
ETHMOID BONE
LACRIMAL BONE

Infraorbital foramen

Perpendicular plate
INFERIOR NASAL
CONCHA

VOMER

Mental foramen

Sagittal suture
Coronal suture
Frontal squama

Supraorbital foramen
Supraorbital margin
Optic foramen
Superior orbital fissure

TEMPORAL BONE

NASAL BONE
Inferior orbital fissure
Middle nasal concha
ZYGOMATIC BONE
MAXILLA

MANDIBLE

Anterior view

Q   Which of the bones shown here are cranial bones?

If you examine the anterior view of the skull in Figure 7.3, you will note the *frontal squama,* a scalelike plate of bone that forms the forehead. It gradually slopes inferiorly from the coronal suture, then angles abruptly and becomes almost vertical. Superior to the orbits the frontal bone thickens, forming the *supraorbital margin* (*supra-* = above; *-orbital* = wheel rut). From this margin the frontal bone extends posteriorly to form the roof of the orbit and part of the floor of the cranial cavity. Within the supraorbital margin, slightly medial to its midpoint, is a hole called the *supraorbital foramen.* As foramina (plural form) associated with cranial bones are discussed, refer to Table 7.4 on page 199 to note which structures pass through them. The *frontal sinuses* lie deep to the frontal squama. Among other functions, paranasal sinuses act as sound chambers that give the voice its resonance.

CLINICAL APPLICATION
Black Eyes

Just superior to the supraorbital margin is a sharp ridge. A blow to the ridge often fractures the bone or lacerates the skin over it, resulting in bleeding. Bruising of the skin over the ridge causes tissue fluid and blood to accumulate in the surrounding connective tissue. The resulting swelling and discoloration is called a **black eye.** ■

Parietal Bones

The two **parietal bones** (pa-RĪ-e-tal; *pariet-* = wall) form the greater portion of the sides and roof of the cranial cavity (Figure 7.4). The internal surfaces of the parietal bones contain many protrusions and depressions that accommodate the blood vessels supplying the dura mater, the superficial membrane (meninx) covering the brain.

Figure 7.4 Skull. Although the hyoid bone is not part of the skull, it is included in the illustration for reference. (See Tortora, *A Photographic Atlas of the Human Body,* Figure 3.3)

🔑 **The zygomatic arch is formed by the zygomatic process of the temporal bone and the temporal process of the zygomatic bone.**

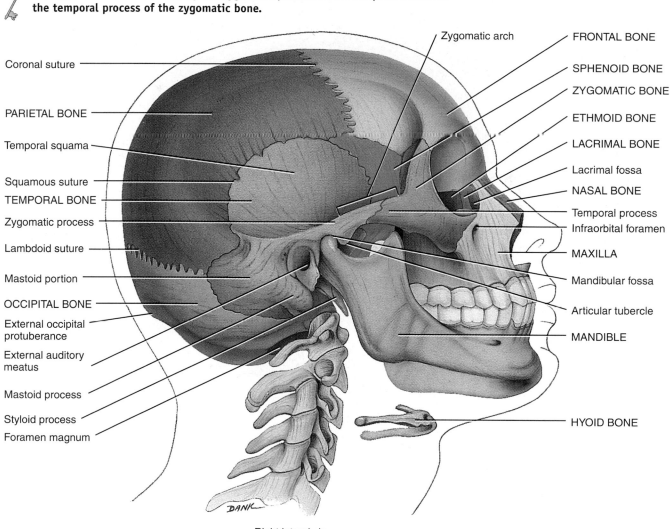

Right lateral view

Q What are the major bones on either side of the squamous suture, the lambdoid suture, and the coronal suture?

Temporal Bones

The two **temporal bones** (*tempor-* = temple) form the inferior lateral aspects of the cranium and part of the cranial floor. In the lateral view of the skull (see Figure 7.4), note the *temporal squama,* the thin, flat portion of the temporal bone that forms the anterior and superior part of the temple. Projecting from the inferior portion of the temporal squama is the *zygomatic process,* which articulates (forms a joint) with the temporal process of the zygomatic (cheek) bone. Together, the zygomatic process of the temporal bone and the temporal process of the zygomatic bone form the *zygomatic arch.*

On the inferior posterior surface of the zygomatic process of the temporal bone is a socket called the *mandibular fossa.* Anterior to the mandibular fossa is a rounded elevation, the *articular tubercle* (see Figure 7.4). The mandibular fossa and articular tubercle articulate with the mandible (lower jawbone) to form the *temporomandibular joint (TMJ).*

Located posteriorly on the temporal bone is the *mastoid portion* (*mastoid* = breast-shaped). See Figure 7.4. The mastoid portion is located posterior and inferior to the *external auditory meatus* (*meatus* = passageway), or ear canal, which directs sound waves into the ear. In the adult, this portion of

Figure 7.5 Skull. (See Tortora, *A Photographic Atlas of the Human Body*, Figure 3.4)

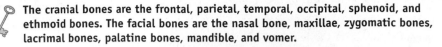

🔑 **The cranial bones are the frontal, parietal, temporal, occipital, sphenoid, and ethmoid bones. The facial bones are the nasal bone, maxillae, zygomatic bones, lacrimal bones, palatine bones, mandible, and vomer.**

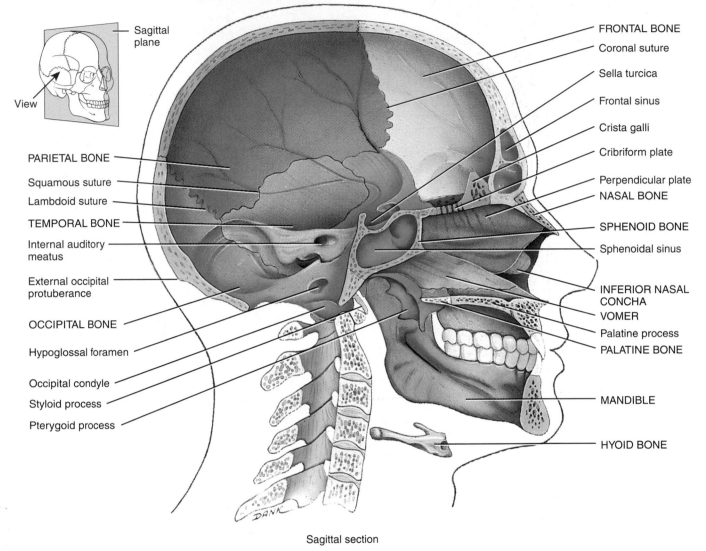

Sagittal plane

View

FRONTAL BONE
Coronal suture
Sella turcica
Frontal sinus
Crista galli
Cribriform plate
Perpendicular plate
NASAL BONE
SPHENOID BONE
Sphenoidal sinus
INFERIOR NASAL CONCHA
VOMER
Palatine process
PALATINE BONE
MANDIBLE
HYOID BONE

PARIETAL BONE
Squamous suture
Lambdoid suture
TEMPORAL BONE
Internal auditory meatus
External occipital protuberance
OCCIPITAL BONE
Hypoglossal foramen
Occipital condyle
Styloid process
Pterygoid process

Sagittal section

Q With which bones does the temporal bone articulate?

the bone contains several *mastoid "air cells."* These tiny air-filled compartments are separated from the brain by thin bony partitions. If **mastoiditis** (inflammation of the mastoid air cells) occurs, the infection may spread to the middle ear and to the brain.

The *mastoid process* is a rounded projection of the mastoid portion of the temporal bone posterior to the external auditory meatus. It serves as a point of attachment for several neck muscles. The *internal auditory meatus* (Figure 7.5) is the opening through which the facial (VII) and vestibulocochlear (VIII) cranial nerves pass. The *styloid process* (*styl-* = stake or pole) projects inferiorly from the inferior surface of the temporal bone and serves as a point of attachment for

muscles and ligaments of the tongue and neck (see Figure 7.4). Between the styloid process and the mastoid process is the *stylomastoid foramen* (see Figure 7.7).

At the floor of the cranial cavity (see Figure 7.8a) is the *petrous portion* of the temporal bone. This portion is triangular and located at the base of the skull between the sphenoid and occipital bones. The petrous portion houses the internal ear and the middle ear, structures involved in hearing and equilibrium. It also contains the *carotid foramen*, through which the carotid artery passes (see Figure 7.7). Posterior to the carotid foramen and anterior to the occipital bone is the *jugular foramen*, a passageway for the jugular vein.

Figure 7.6 Skull. The sutures are exaggerated for emphasis. (See Tortora, *A Photographic Atlas of the Human Body,* Figure 3.5)

🔑 **The occipital bone forms most of the posterior and inferior portions of the cranium.**

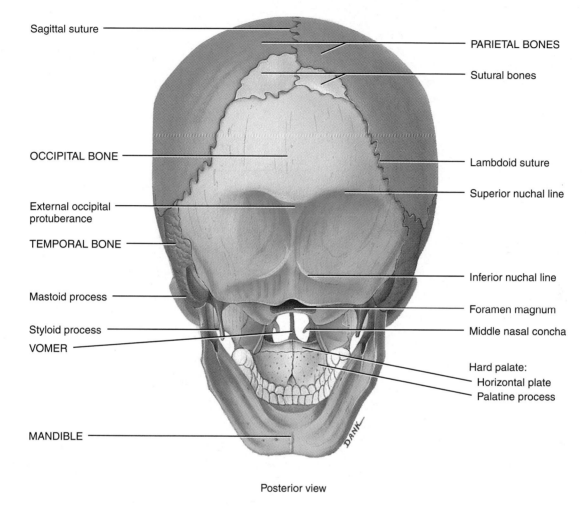

Posterior view

Ⓠ Which bones form the posterior, lateral portion of the cranium?

Occipital Bone

The **occipital bone** (ok-SIP-i-tal; *occipit-* = back of head) forms the posterior part and most of the base of the cranium (Figure 7.6; also see Figure 7.4). The *foramen magnum* (= large hole) is in the inferior part of the bone. Within this foramen, the medulla oblongata (inferior part of the brain) connects with the spinal cord. The vertebral and spinal arteries also pass through this foramen. The *occipital condyles* are oval processes with convex surfaces, one on either side of the foramen magnum (Figure 7.7). They articulate with depressions on the first cervical vertebra to form the *atlanto-occipital joint.* Superior to each occipital condyle on the inferior surface of the skull is the *hypoglossal foramen* (*hypo-* = under; *-glossal* = tongue) (see Figure 7.5).

The *external occipital protuberance* is a prominent midline projection on the posterior surface of the bone just superior to the foramen magnum. You may be able to feel this structure as a definite bump on the back of your head, just above your neck. (See Figure 7.4.) A large fibrous, elastic ligament, the *ligamentum nuchae* (*nucha-* = nape of neck), which helps support the head, extends from the external occipital protuberance to the seventh cervical vertebra. Extending laterally from the protuberance are two curved lines, the *superior nuchal lines,* and below these are two *inferior nuchal lines,* which are areas of muscle attachment (see Figure 7.7). It is possible to view the parts of the occipital bone, as well as surrounding structures, in the inferior view of the skull in Figure 7.7.

Figure 7.7 Skull. The mandible (lower jawbone) has been removed. (See Tortora, *A Photographic Atlas of the Human Body,* Figure 3.7)

🔑 **The occipital condyles of the occipital bone articulate with the first cervical vertebra to form the atlanto-occipital joints.**

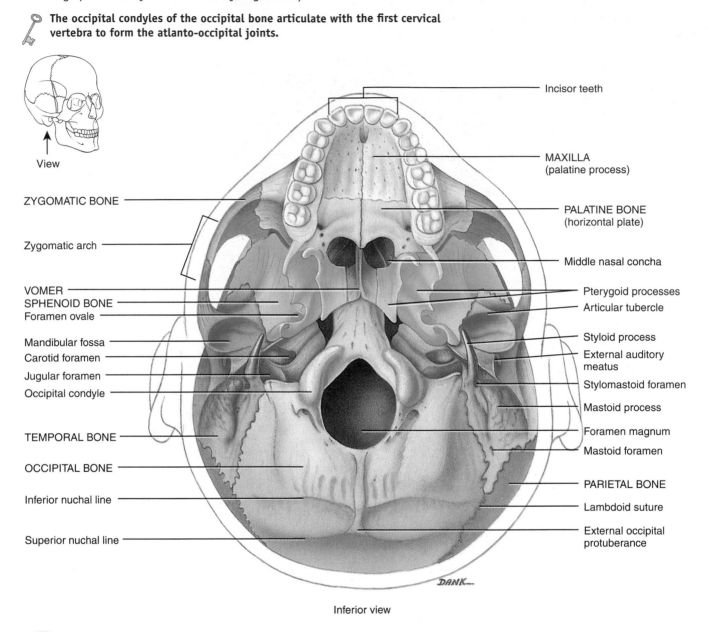

Inferior view

Q What parts of the nervous system join together within the foramen magnum?

Sphenoid Bone

The **sphenoid bone** (SFĒ-noyd; = wedge-shaped) lies at the middle part of the base of the skull (Figures 7.7 and 7.8). This bone is called the keystone of the cranial floor because it articulates with all the other cranial bones, holding them together. Viewing the floor of the cranium superiorly (Figure 7.8a), note the sphenoid articulations: anteriorly with the frontal bone, laterally with the temporal bones, and

posteriorly with the occipital bone. It lies posterior and slightly superior to the nasal cavity and forms part of the floor, sidewalls, and rear wall of the orbit (see Figure 7.13).

The shape of the sphenoid resembles a bat with outstretched wings (see Figure 7.8b). The *body* of the sphenoid is the cubelike medial portion between the ethmoid and occipital bones. It contains the *sphenoidal sinuses,* which drain into the nasal cavity (see Figure 7.11). On the superior surface of the body of the sphenoid is a depression called the

Figure 7.8 Sphenoid bone. (See Tortora, *A Photographic Atlas of the Human Body,* Figures 3.8 and 3.9)

The sphenoid bone is called the keystone of the cranial floor because it articulates with all other cranial bones, holding them together.

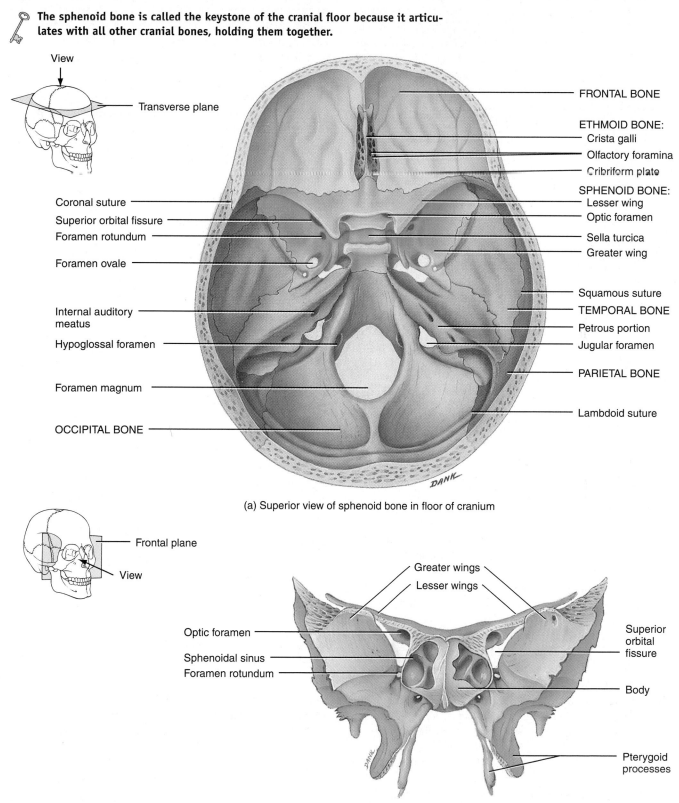

(a) Superior view of sphenoid bone in floor of cranium

(b) Anterior view of sphenoid bone

Q Starting at the crista galli of the ethmoid bone and going in a clockwise direction, what are the names of the bones that articulate with the sphenoid bone?

sella turcica (SEL-a TUR-si-ka; *sella* = saddle; *turcica* = Turkish), which cradles the pituitary gland.

The *greater wings* of the sphenoid project laterally from the body, forming the anterolateral floor of the cranium. The greater wings also form part of the lateral wall of the skull just anterior to the temporal bone and can be viewed externally. The *lesser wings,* which are smaller than the greater wings, form a ridge of bone anterior and superior to the greater wings. They form part of the floor of the cranium and the posterior part of the orbit of the eye.

Between the body and lesser wing just anterior to the sella turcica is the *optic foramen* (*optic* = eye). Lateral to the body between the greater and lesser wings is a somewhat triangular slit called the *superior orbital fissure.* This fissure may also be seen in the anterior view of the orbit in Figure 7.13.

In Figures 7.7 and 7.8b you can see the *pterygoid processes* (TER-i-goyd; = winglike) by looking at the inferior part of the sphenoid bone. These structures project inferiorly from the points where the body and greater wings unite and form the lateral posterior region of the nasal cavity. Some of the muscles that move the mandible attach to the pterygoid processes. At the base of the lateral pterygoid process in the greater wing is the *foramen ovale* (= oval). Another foramen associated with the sphenoid bone is the *foramen rotundum* (= round) located at the junction of the anterior and medial parts of the sphenoid bone.

Ethmoid Bone

The **ethmoid bone** (ETH-moid; = like a sieve) is a light, spongelike bone located on the midline in the anterior part of the cranial floor medial to the orbits (Figure 7.9). It is anterior to the sphenoid and posterior to the nasal bones. The ethmoid bone forms (1) part of the anterior portion of the cranial floor; (2) the medial wall of the orbits; (3) the superior portions of the nasal septum, a partition that divides the nasal cavity into right and left sides; and (4) most of the superior sidewalls of the nasal cavity. The ethmoid is a major superior supporting structure of the nasal cavity.

The *lateral masses* of the ethmoid bone compose most of the wall between the nasal cavity and the orbits. They contain 3 to 18 air spaces, or "cells," that give this bone a sievelike appearance. The ethmoidal cells together form the *ethmoidal sinuses* (see Figure 7.11). The *perpendicular plate* forms the superior portion of the nasal septum (see Figure 7.9). The *cribriform plate* (*cribri-* = sieve) lies in the anterior floor of the cranium and forms the roof of the nasal cavity. The cribriform plate contains the *olfactory foramina* (*olfact-* = smell). Projecting superiorly from the cribriform plate is a sharp triangular process called the *crista galli* (*crista* = crest; *galli* = cock). This structure serves as a point of attachment for the membranes that cover the brain.

The lateral masses contain two thin, scroll-shaped projections lateral to the nasal septum. These are called the *superior nasal concha* (KONG-ka; = shell) or *turbinate* and the *middle nasal concha (turbinate).* A third pair of conchae (plural form), the inferior nasal conchae, are separate bones

(discussed shortly). The conchae cause inhaled air to swirl; the result is that many inhaled particles strike and become trapped in the mucus that lines the nasal passageways. The action of the conchae helps cleanse inhaled air before it passes into the rest of the respiratory tract. Air striking the mucous lining of the conchae is also warmed and moistened.

Facial Bones

The shape of the face changes dramatically during the first two years after birth. The brain and cranial bones expand, the teeth form and erupt, and the paranasal sinuses increase in size. Growth of the face ceases at about 16 years of age.

Nasal Bones

The paired **nasal bones** meet at the midline (see Figure 7.3) and form part of the bridge of the nose. The major structural portion of the nose consists of cartilage.

Maxillae

The paired **maxillae** (mak-SIL-ē-; = jawbones) unite to form the upper jawbone and articulate directly with every bone of the face except the mandible, or lower jawbone (see Figures 7.4 and 7.7). They form part of the floors of the orbits, part of the lateral walls and floor of the nasal cavity, and most of the hard palate. The hard palate is a bony partition formed by the palatine processes of the maxillae and horizontal plates of the palatine bones that forms the roof of the mouth.

Each maxilla (singular form) contains a large *maxillary sinus* that empties into the nasal cavity (see Figure 7.11). The *alveolar process* (al-VĒ-ō-lar; *alveol-* = small cavity) is an arch that contains the *alveoli* (sockets) for the maxillary (upper) teeth. The *palatine process* is a horizontal projection of the maxilla that forms the anterior three-quarters of the hard palate. The uniting and fusion of the maxillary bones is normally completed before birth.

The *infraorbital foramen* (*infra-* = below), which can be seen in the anterior view of the skull in Figure 7.3, is an opening in the maxilla inferior to the orbit. A final structure associated with the maxilla and sphenoid bone is the *inferior orbital fissure.* It is located between the greater wing of the sphenoid and the maxilla (see Figure 7.13).

CLINICAL APPLICATION
Cleft Palate and Cleft Lip

Usually the palatine processes of the maxillary bones unite during weeks 10 to 12 of embryonic development. Failure to do so can result in a condition called **cleft palate.** The condition may also involve incomplete fusion of the horizontal plates of the palatine bones (see Figures 7.6 and 7.7). Another form of this condition, called **cleft lip,** involves a split in the upper lip. Cleft lip and cleft palate often occur together. Depending on the extent and position of the cleft,

Figure 7.9 Ethmoid bone. (See Tortora, *A Photographic Atlas of the Human Body*, Figure 3.10)

 The ethmoid bone forms part of the anterior portion of the cranial floor, the medial wall of the orbits, the superior portions of the nasal septum, and most of the sidewalls of the nasal cavity.

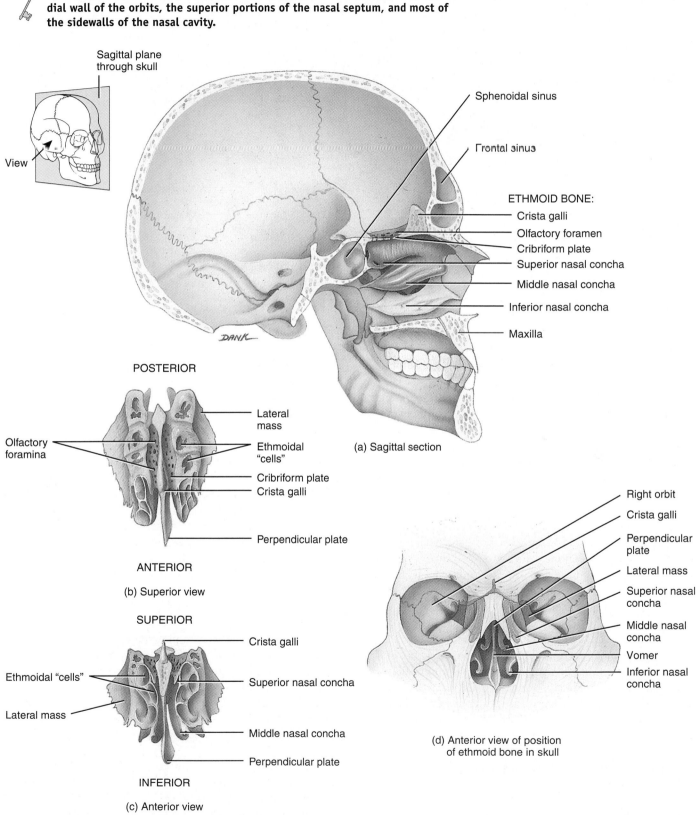

(a) Sagittal section

(b) Superior view

(c) Anterior view

(d) Anterior view of position of ethmoid bone in skull

 What part of the ethmoid bone forms the superior part of the nasal septum? The medial walls of the orbits?

Figure 7.10 Mandible.

🔑 **The mandible is the largest and strongest facial bone.**

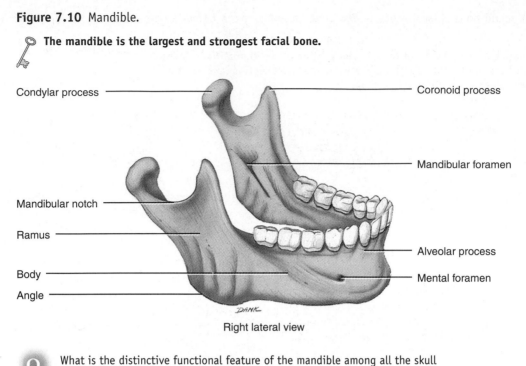

Right lateral view

Ⓠ What is the distinctive functional feature of the mandible among all the skull bones?

speech and swallowing may be affected. In addition, children with cleft palate tend to have many ear infections that can lead to hearing loss. Facial and oral surgeons recommend closure of cleft lip during the first few weeks following birth, and surgical results are excellent. Repair of cleft palate typically is done between 12 and 18 months of age, ideally before the child begins to talk. Speech therapy may be needed, because the palate is important for pronouncing consonants, and orthodontic therapy may be needed to align the teeth. Again, results are usually excellent. ∎

Zygomatic Bones

The two **zygomatic bones** (*zygo-* = yokelike), commonly called cheekbones, form the prominences of the cheeks and part of the lateral wall and floor of each orbit (see Figure 7.13). They articulate with the frontal, maxilla, sphenoid, and temporal bones.

Lacrimal Bones

The paired **lacrimal bones** (LAK-ri-mal; *lacrim-* = teardrops) are thin and roughly resemble a fingernail in size and shape (see Figures 7.3 and 7.4). They are the smallest bones of the face. These bones are posterior and lateral to the nasal bones and form a part of the medial wall of each orbit. The lacrimal bones each contain a *lacrimal fossa,* a vertical tunnel formed with the maxilla, that houses the lacrimal sac, a structure that gathers tears and passes them into the nasal cavity (see Figure 7.13).

Palatine Bones

The two **palatine bones** (PAL-a-tīn) are L-shaped (see Figure 7.7). They form the posterior portion of the hard

palate, part of the floor and lateral wall of the nasal cavity, and a small portion of the floors of the orbits. The posterior portion of the hard palate, which separates the nasal cavity from the oral cavity, is formed by the *horizontal plates* of the palatine bones (see Figures 7.6 and 7.7).

Inferior Nasal Conchae

The two **inferior nasal conchae** or **turbinates** are scroll-like bones that form a part of the inferior lateral wall of the nasal cavity and project into the nasal cavity inferior to the superior and middle nasal conchae of the ethmoid bone (see Figures 7.3 and 7.9a). They serve the same function as the superior and middle nasal conchae of the ethmoid bone; that is, they help swirl and filter air before it passes into the lungs. The inferior nasal conchae are separate bones and not part of the ethmoid.

Vomer

The **vomer** (VŌ-mer; = plowshare) is a roughly triangular bone on the floor of the nasal cavity that articulates superiorly with the perpendicular plate of the ethmoid bone and inferiorly with both the maxillae and palatine bones along the midline (see Figures 7.3 and 7.7). It is one of the components of the nasal septum, a partition that divides the nasal cavity into right and left sides.

Mandible

The **mandible** (*mand-* = to chew), or lower jawbone, is the largest, strongest facial bone (Figure 7.10). It is the only movable skull bone (other than the auditory ossicles). In the lateral view, you can see that the mandible consists of a curved, horizontal portion, the *body,* and two perpendicular portions, the *rami* (RĀ-mī; = branches). The *angle* of the mandible is the area where each ramus (singular form) meets the body. Each ramus has a posterior *condylar process* (KON-di-lar) that articulates with the mandibular fossa and articular tubercle of the temporal bone (see Figure 7.4) to form the **temporomandibular joint (TMJ).** It also has an anterior *coronoid process* (KOR-ō-noyd) to which the temporalis muscle attaches. The depression between the coronoid and condylar processes is called the *mandibular notch.* The *alveolar process* is an arch containing the *alveoli* (sockets) for the mandibular (lower) teeth.

The *mental foramen* (*ment-* = chin) is approximately inferior to the second premolar tooth. It is near this foramen that dentists reach the mental nerve when injecting anesthetics. Another foramen associated with the mandible is the *mandibular foramen* on the medial surface of each ramus, another site often used by dentists to inject anesthetics. The mandibular foramen is the beginning of the *mandibular canal,* which runs obliquely in the ramus and anteriorly to the body. Through the canal pass the inferior alveolar nerves and blood vessels, which are distributed to the mandibular teeth.

CLINICAL APPLICATION
Temporomandibular Joint Syndrome

One problem associated with the temporomandibular joint (TMJ) is **TMJ syndrome.** It is characterized by dull pain around the ear, tenderness of the jaw muscles, a clicking or popping noise when opening or closing the mouth, limited or abnormal opening of the mouth, headache, tooth sensitivity, and abnormal wearing of the teeth. TMJ syndrome can be caused by improperly aligned teeth, grinding or clenching the teeth, trauma to the head and neck, or arthritis. Treatment may involve applying moist heat or ice, eating a soft diet, taking pain relievers such as aspirin, muscle retraining, adjusting or reshaping the teeth, orthodontic treatment, or surgery. ■

Special Features of the Skull

OBJECTIVE

• *Describe the following special features of the skull: sutures, paranasal sinuses, fontanels, foramina, orbits, and nasal septum.*

Now that we have described the locations and parts of the cranial and facial bones, we will examine some of the special features of the skull, starting with sutures.

Sutures

A **suture** (SOO-chur; = seam) is an immovable joint that is found only between skull bones and that holds most skull bones together. Several types of sutures can be distinguished according to how the margins of the bones unite. In some sutures the margins of the bones are fairly smooth. In other sutures the margins overlap. In still other sutures the margins interlock in a jigsaw fashion. This latter arrange-

ment provides sutures added strength and decreases their chance of fracturing.

The names of many sutures reflect the bones they unite. For example, the frontozygomatic suture is between the frontal bone and the zygomatic bone. Similarly, the sphenoparietal suture is between the sphenoid bone and the parietal bone. In other cases, however, the names of sutures are not so obvious. Of the many sutures found in the skull, we will identify only four prominent ones:

1. The **coronal suture** (kō-RŌ-nal; *coron-* = crown) unites the frontal bone and both parietal bones (see Figure 7.4).
2. The **sagittal suture** (SAJ-i-tal; *sagitt-* = arrow) unites the two parietal bones on the superior midline of the skull (see Figure 7.6). The sagittal suture is so named because in the infant, before the bones of the skull are firmly united, the suture and the fontanels (soft spots) associated with it somewhat resemble an arrow.
3. The **lambdoid suture** (LAM-doyd) unites both parietal bones and occipital bones. This suture is so named because of its resemblance to the Greek letter lambda (Λ), as can be seen in Figure 7.6. Sutural bones may be found within the sagittal and lambdoid sutures.
4. The **squamous sutures** (SKWĀ-mus; *squam-* = flat) unite the parietal and temporal bones on the lateral aspects of the skull (see Figure 7.4).

Paranasal Sinuses

Even though they are not cranial or facial bones, this is an appropriate place to discuss the **paranasal sinuses** (*para-* = beside). These are paired cavities in certain cranial and facial bones near the nasal cavity and are clearly evident in a sagittal section of the skull (Figure 7.11). The paranasal sinuses are lined with mucous membranes that are continuous with the lining of the nasal cavity. Skull bones enclosing the paranasal sinuses are the frontal, sphenoid, ethmoid, and maxillary. Besides producing mucus, the paranasal sinuses serve as resonating chambers for sound as we speak or sing.

CLINICAL APPLICATION
Sinusitis

Secretions produced by the mucous membranes of the paranasal sinuses drain into the nasal cavity. An inflammation of the membranes due to an allergic reaction or infection is called **sinusitis.** If the membranes swell enough to block drainage into the nasal cavity, fluid pressure builds up in the paranasal sinuses, and a sinus headache results. Chronic sinusitis may also be caused by a severely deviated nasal septum or nasal polyps, growths that can be removed surgically. ■

Fontanels

The skeleton of a newly formed embryo consists of cartilage and fibrous connective tissue membrane structures shaped like bones. Gradually, ossification occurs—bone replaces the cartilage and fibrous connective tissue mem-

Figure 7.11 Paranasal sinuses. (See Tortora, *A Photographic Atlas of the Human Body,* Figure 3.4)

🔑 **Paranasal sinuses are mucous-membrane–lined spaces in the frontal, sphenoid, ethmoid, and maxillary bones that connect to the nasal cavity.**

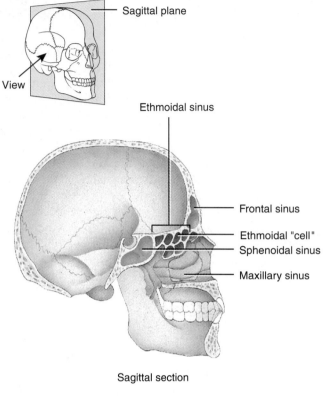

Sagittal plane

View

Ethmoidal sinus

Frontal sinus

Ethmoidal "cell"

Sphenoidal sinus

Maxillary sinus

Sagittal section

Ⓠ What are the functions of the paranasal sinuses?

Figure 7.12 Fontanels at birth. (See Tortora, *A Photographic Atlas of the Human Body,* Figure 3.12)

🔑 **Fontanels are membrane-filled spaces between cranial bones that are present at birth.**

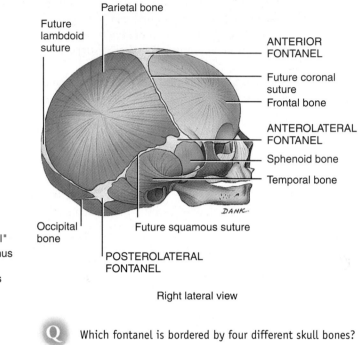

Parietal bone

Future lambdoid suture

ANTERIOR FONTANEL

Future coronal suture

Frontal bone

ANTEROLATERAL FONTANEL

Sphenoid bone

Temporal bone

DANK

Occipital bone

Future squamous suture

POSTEROLATERAL FONTANEL

Right lateral view

Ⓠ Which fontanel is bordered by four different skull bones?

branes. At birth, membrane-filled spaces called **fontanels** (fon-ta-NELZ; = little fountains) are found between cranial bones (Figure 7.12). Commonly called "soft spots," fontanels are areas of fibrous connective tissue membranes that will eventually be replaced with bone by intramembranous ossification and become sutures. Functionally, the fontanels enable the fetal skull to modify its size and shape as it passes through the birth canal and permit rapid growth of the brain during infancy. The amount of closure in fontanels helps a physician to gauge the degree of brain development. In addition, the anterior fontanel serves as a landmark for withdrawal of blood for analysis from the superior sagittal sinus (a large vein on the midline surface of the brain). Although an infant may have many fontanels at birth, the form and location of several are fairly constant (Table 7.3).

Foramina

Many **foramina** associated with the skull were mentioned in the descriptions of the cranial and facial bones they penetrate. As preparation for studying other systems of the body, especially the nervous and cardiovascular systems,

these foramina and the structures passing through them are listed in Table 7.4. For your convenience and for future reference, the foramina are listed alphabetically.

Orbits

Each **orbit** (eye socket) is a pyramid-shaped structure that contains the eyeball and associated structures. It is formed by seven bones of the skull (Figure 7.13) and has four regions that converge posteriorly to form an *apex* (posterior end):

1. The *roof* of the orbit consists of parts of the frontal and sphenoid bones.
2. The *lateral wall* of the orbit is formed by portions of the zygomatic and sphenoid bones.
3. The *floor* of the orbit is formed by parts of the maxilla, zygomatic, and palatine bones.
4. The *medial wall* of the orbit is formed by portions of the maxilla, lacrimal, ethmoid, and sphenoid bones.

Associated with the orbit are five openings:

- Optic foramen at the junction of the roof and medial wall.
- Superior orbital fissure at the superior lateral angle of the apex.
- Inferior orbital fissure at the junction of the lateral wall and floor.

Table 7.3 Fontanels (See also Figure 7.12)

FONTANEL		LOCATION	DESCRIPTION
Anterior (frontal)	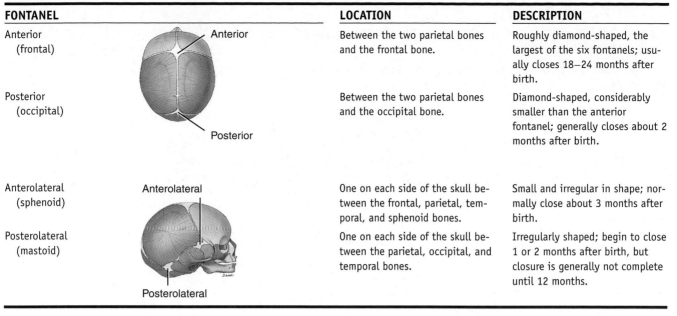	Between the two parietal bones and the frontal bone.	Roughly diamond-shaped, the largest of the six fontanels; usually closes 18–24 months after birth.
Posterior (occipital)		Between the two parietal bones and the occipital bone.	Diamond-shaped, considerably smaller than the anterior fontanel; generally closes about 2 months after birth.
Anterolateral (sphenoid)		One on each side of the skull between the frontal, parietal, temporal, and sphenoid bones.	Small and irregular in shape; normally close about 3 months after birth.
Posterolateral (mastoid)		One on each side of the skull between the parietal, occipital, and temporal bones.	Irregularly shaped; begin to close 1 or 2 months after birth, but closure is generally not complete until 12 months.

Table 7.4 Principal Foramina of the Skull

FORAMEN	LOCATION	STRUCTURES PASSING THROUGH
Carotid (relating to carotid artery in neck)	Petrous portion of temporal bone (Figure 7.7).	Internal carotid artery and sympathetic nerves for eyes.
Hypoglossal (*hypo-* = under; *glossus* = tongue)	Superior to base of occipital condyles (Figure 7.8).	Cranial nerve XII (hypoglossal) and branch of ascending pharyngeal artery.
Infraorbital (*infra-* = below)	Inferior to orbit in maxilla (Figure 7.13).	Infraorbital nerve and blood vessels and a branch of the maxillary division of cranial nerve V (trigeminal).
Jugular (*jugul-* = the throat)	Posterior to carotid canal between petrous portion of temporal bone and occipital bone (Figure 7.8).	Internal jugular vein, cranial nerves IX (glossopharyngeal), X (vagus), and XI (accessory).
Magnum (= large)	Occipital bone (Figure 7.7).	Medulla oblongata and its membranes (meninges), cranial nerve XI (accessory), and vertebral and spinal arteries.
Mandibular (*mand-* = to chew)	Medial surface of ramus of mandible (Figure 7.10).	Inferior alveolar nerve and blood vessels.
Mastoid (= breast-shaped)	Posterior border of mastoid process of temporal bone (Figure 7.7).	Emissary vein to transverse sinus and branch of occipital artery to dura mater.
Mental (*ment-* = chin)	Inferior to second premolar tooth in mandible (Figure 7.10).	Mental nerve and vessels.
Olfactory (*olfact-* = to smell)	Cribriform plate of ethmoid bone (Figure 7.8).	Cranial nerve I (olfactory).
Optic (= eye)	Between superior and inferior portions of small wing of sphenoid bone (Figure 7.13).	Cranial nerve II (optic) and ophthalmic artery.
Ovale (= oval)	Greater wing of sphenoid bone (Figure 7.8).	Mandibular branch of cranial nerve V (trigeminal).
Rotundum (= round)	Junction of anterior and medial parts of sphenoid bone (Figure 7.8).	Maxillary branch of cranial nerve V (trigeminal).
Stylomastoid (*stylo-* = stake or pole)	Between styloid and mastoid processes of temporal bone (Figure 7.7).	Cranial nerve VII (facial) and stylomastoid artery.
Supraorbital (*supra-* = above)	Supraorbital margin of orbit in frontal bone (Figure 7.13).	Supraorbital nerve and artery.

Figure 7.13 Details of the orbit (eye socket). (See Tortora, *A Photographic Atlas of the Human Body*, Figure 3.11)

 The orbit is a pyramid-shaped structure that contains the eyeball and associated structures.

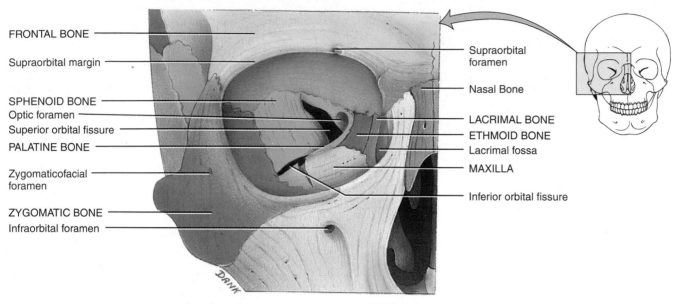

FRONTAL BONE

Supraorbital margin

SPHENOID BONE
Optic foramen
Superior orbital fissure
PALATINE BONE

Zygomaticofacial
foramen

ZYGOMATIC BONE
Infraorbital foramen

Supraorbital
foramen

Nasal Bone

LACRIMAL BONE
ETHMOID BONE
Lacrimal fossa

MAXILLA

Inferior orbital fissure

Anterior view showing the bones of the right orbit

Q   Which seven bones form the orbit?

Figure 7.14 Nasal septum. (See Tortora, *A Photographic Atlas of the Human Body*, Figure 3.4)

 The structures that form the nasal septum are the perpendicular plate of the ethmoid bone, the vomer, and septal cartilage.

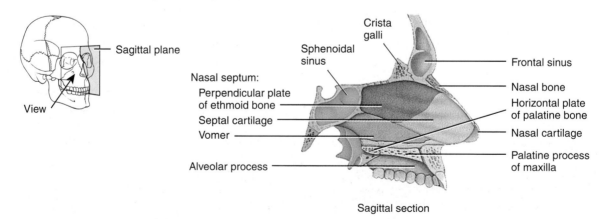

Sagittal plane

View

Nasal septum:
  Perpendicular plate
  of ethmoid bone
  Septal cartilage
  Vomer

Alveolar process

Sphenoidal
sinus

Crista
galli

Frontal sinus

Nasal bone

Horizontal plate
of palatine bone

Nasal cartilage

Palatine process
of maxilla

Sagittal section

Q   What is the function of the nasal septum?

- Supraorbital foramen on the medial side of the supraorbital margin of the frontal bone.
- Lacrimal fossa in the lacrimal bone.

Nasal Septum

The inside of the nose, called the nasal cavity, is divided into right and left sides by a vertical partition called the **nasal septum.** The septum is formed by the vomer, septal cartilage, and the perpendicular plate of the ethmoid bone (Figure 7.14). The anterior border of the vomer articulates with the septal cartilage, which is hyaline cartilage, to form the anterior portion of the septum. The superior border of the vomer articulates with the perpendicular plate of the ethmoid bone to form the remainder of the nasal septum.

CLINICAL APPLICATION
Deviated Nasal Septum

A **deviated nasal septum** is one that is deflected laterally from the midline of the nose. The deviation usually occurs at the junction of bone with the septal cartilage. Septal deviations may occur as a result of a developmental abnormality or trauma. If the deviation is severe, it may entirely block the nasal passageway. Even a partial blockage may lead to infection. If inflammation occurs, it may cause nasal congestion, blockage of the paranasal sinus openings, chronic sinusitis, headache, and nosebleeds. The condition usually can be corrected surgically. ■

HYOID BONE
OBJECTIVE
• *Describe the relationship of the hyoid bone to the skull.*

The single **hyoid bone** (= U-shaped) is a unique component of the axial skeleton because it does not articulate with any other bone (see Figure 7.4). Rather, it is suspended from the styloid processes of the temporal bones by ligaments and muscles. Located in the anterior neck between the mandible and larynx, the hyoid supports the tongue, providing attachment sites for some tongue muscles and for muscles of the neck and pharynx. The hyoid consists of a horizontal *body* and paired projections called the *lesser horns* and the *greater horns* (Figure 7.15). Muscles and ligaments attach to these paired projections.

The hyoid bone, as well as cartilages of the larynx and trachea, are often fractured during strangulation. As a result, they are carefully examined at autopsy when strangulation is suspected.

1. List the cranial and facial bones, and describe the general features of the skull.
2. Define the following: suture, paranasal sinus, fontanel, foramen.
3. What bones constitute the orbit?

Figure 7.15 Hyoid bone. (See Tortora, *A Photographic Atlas of the Human Body,* Figure 3.13)

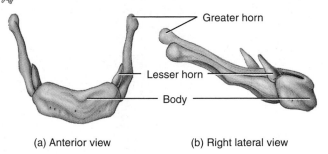

🔑 The hyoid bone supports the tongue, providing attachment sites for muscles of the tongue, neck, and pharynx.

Greater horn

Lesser horn

Body

(a) Anterior view　　　(b) Right lateral view

Q In what way is the hyoid bone different from all the other bones of the axial skeleton?

4. What structures make up the nasal septum?
5. What are the functions of the hyoid bone?

VERTEBRAL COLUMN
OBJECTIVE
• *Identify the regions and normal curves of the vertebral column.*

The **vertebral column,** also called the *spine* or *backbone,* together with the sternum and ribs, forms the skeleton of the trunk of the body. Whereas the vertebral column consists of bone and connective tissue, the spinal cord consists of nervous tissue. In effect, the vertebral column is a strong, flexible rod that both bends anteriorly, posteriorly, and laterally and rotates. It encloses and protects the spinal cord, supports the head, and serves as a point of attachment for the ribs, pelvic girdle, and muscles of the back.

The vertebral column makes up about two-fifths of the total height of the body and is composed of a series of bones called **vertebrae** (VER-te-brē; singular is *vertebra*). The length of the column is about 71 cm (28 in.) in an average adult male and about 61 cm (24 in.) in an average adult female. Between vertebrae are openings called *intervertebral foramina.* The spinal nerves that connect the spinal cord to various parts of the body pass through these openings.

The adult vertebral column is divided into five regions that contain 26 bones distributed as follows (Figure 7.16a):

- The cervical region (*cervic-* = neck) contains seven cervical vertebrae in the neck.
- The thoracic region (*thorac-* = chest) contains 12 thoracic vertebrae that lie posterior to the thoracic cavity.
- The lumbar region (*lumb-* = loin) contains five lumbar vertebrae that support the lower back.

Figure 7.16 Vertebral column. The numbers in parentheses in (a) indicate the number of vertebrae in each region. In (d), the relative size of the disc has been enlarged for emphasis. A "window" has been cut in the annulus fibrosus so that the nucleus pulposus can be seen. (See Tortora, *A Photographic Atlas of the Human Body*, Figure 3.15)

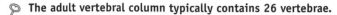

The adult vertebral column typically contains 26 vertebrae.

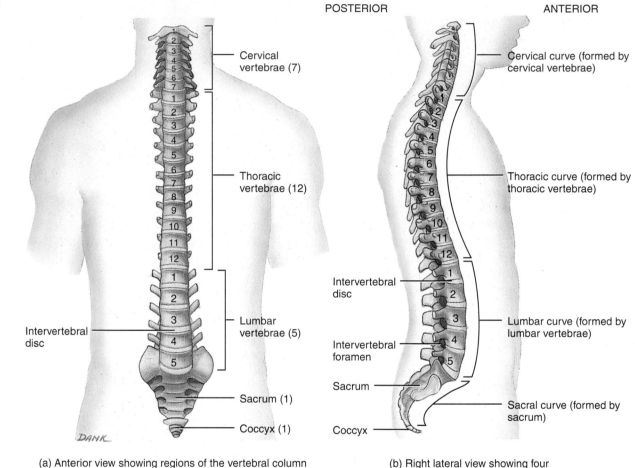

(a) Anterior view showing regions of the vertebral column

(b) Right lateral view showing four normal curves

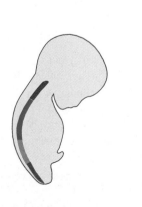

Single curve in fetus   Four curves in adult

(c) Fetal and adult curves

Normal intervertebral disc

Compressed intervertebral disc in a weight-bearing situation

(d) Intervertebral disc

Q Which curves are concave (relative to the anterior side of the body)?

- The sacral region contains the *sacrum* (SĀ-krum; = sacred bone), one bone consisting of five fused sacral vertebrae.
- The coccygeal region (kok-SIJ-ē-al) contains one bone (sometimes two bones) called the *coccyx* (KOK-siks; = resembling the bill of a cuckoo), which consists of usually four fused coccygeal vertebrae.

Thus, before the sacral and coccygeal vertebrae fuse, the total number of vertebrae is 33. Whereas the cervical, thoracic, and lumbar vertebrae are movable, the sacrum and coccyx are not. We will discuss each of these regions in greater detail shortly.

Intervertebral Discs

Between the bodies of adjacent vertebrae from the second cervical vertebra to the sacrum are **intervertebral discs** (Figure 7.16d). Each disc has an outer fibrous ring consisting of fibrocartilage called the *annulus fibrosus* (*annulus* = ring-like) and an inner soft, pulpy, highly elastic substance called the *nucleus pulposus* (*pulposus* = pulplike). The discs form strong joints, permit various movements of the vertebral column, and absorb vertical shock. Under compression, they flatten, broaden, and bulge from their intervertebral spaces. Superior to the sacrum, the intervertebral discs constitute about one-fourth the length of the vertebral column.

Normal Curves of the Vertebral Column

When viewed from the side, the vertebral column shows four **normal curves** (Figure 7.16b). The *cervical* and *lumbar curves* are anteriorly convex (bulging out), whereas the *thoracic* and *sacral curves* are anteriorly concave (cupping in). The curves of the vertebral column are important because they increase its strength, help maintain balance in the upright position, absorb shocks during walking, and help protect the column from fracture.

In the fetus, there is only a single anteriorly concave curve (Figure 7.16c). At approximately the third month after birth, when an infant begins to hold its head erect, the cervical curve develops. Later, when the child sits up, stands, and walks, the lumbar curve develops. The thoracic and sacral curves are called *primary curves* because they form first during fetal development. The cervical and lumbar curves are referred to as *secondary curves* because they begin to form later, several months after birth. All curves are fully developed by age 10. Later in the chapter we consider the causes of several types of abnormal curves—scoliosis, kyphosis, and lordosis (p. 214).

Parts of a Typical Vertebra

Even though vertebrae in different regions of the spinal column vary in size, shape, and detail, they are similar enough that we can discuss the structures (and the functions) of a typical vertebra (Figure 7.17). Vertebrae typically consist of a body, a vertebral arch, and several processes.

Body

The *body* is the thick, disc-shaped anterior portion that is the weight-bearing part of a vertebra. Its superior and inferior surfaces are roughened for the attachment of cartilaginous intervertebral discs. The anterior and lateral surfaces contain nutrient foramina for blood vessels.

Vertebral Arch

The *vertebral arch* extends posteriorly from the body of the vertebra. It and the body of the vertebra surround the spinal cord. The vertebral arch is formed by two short, thick processes, the *pedicles* (PED-i-kuls; = little feet), which project posteriorly from the body to unite with the laminae. The *laminae* (LAM-i-nē; = thin layers) are the flat parts that join to form the posterior portion of the vertebral arch. The *vertebral foramen* lies between the vertebral arch and body and contains the spinal cord, fat, areolar connective tissue, and blood vessels. Collectively, the vertebral foramina of all vertebrae form the vertebral (spinal) canal, which is the inferior portion of the dorsal body cavity. The pedicles exhibit superior and inferior notches called *vertebral notches*. When the vertebral notches are stacked on top of one another, there is an opening between adjoining vertebrae on each side of the column. Each opening, called an *intervertebral foramen*, permits the passage of a single spinal nerve.

Processes

Seven *processes* arise from the vertebral arch. At the point where a lamina and pedicle join, a *transverse process* extends laterally on each side. A single *spinous process* (*spine*) projects posteriorly from the junction of the laminae. These three processes serve as points of attachment for muscles. The remaining four processes form joints with other vertebrae above or below. The two *superior articular processes* of a vertebra articulate with the two inferior articular processes of the vertebra immediately superior to them. The two *inferior articular processes* of a vertebra articulate with the two superior articular processes of the vertebra immediately inferior to them. The articulating surfaces of the articular processes are referred to as *facets* (= little faces). The articulations formed between the bodies and articular facets of successive vertebrae are called *intervertebral joints*.

Regions of the Vertebral Column

We turn now to the five regions of the vertebral column, beginning superiorly and moving inferiorly. Note that vertebrae in each region are numbered in sequence, from superior to inferior.

Figure 7.17 Structure of a typical vertebra, as illustrated by a thoracic vertebra. In (b), only one spinal nerve has been included, and it has been extended beyond the intervertebral foramen for clarity. The sympathetic chain is part of the autonomic nervous system (see Figure 17.2). (See Tortora, *A Photographic Atlas of the Human Body,* Figure 3.16)

🔑 **A vertebra consists of a body, a vertebral arch, and several processes.**

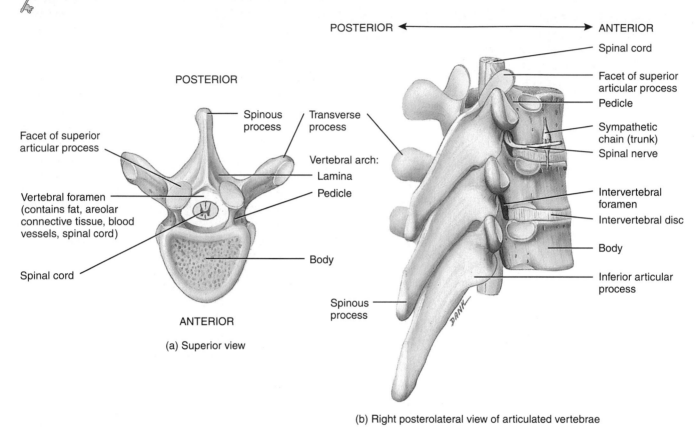

(a) Superior view

(b) Right posterolateral view of articulated vertebrae

Q What are the functions of the vertebral and intervertebral foramina?

Cervical Region

The bodies of **cervical vertebrae** (C1–C7) are smaller than those of thoracic vertebrae (Figure 7.18a). The vertebral arches, however, are larger. All cervical vertebrae have three foramina: one vertebral foramen and two transverse foramina (Figure 7.18b). The vertebral foramina of cervical vertebrae are the largest in the spinal column because they house the cervical enlargement of the spinal cord. Each cervical transverse process contains a *transverse foramen* through which the vertebral artery and its accompanying vein and nerve fibers pass. The spinous processes of C2 through C6 are often *bifid*—that is, split into two parts (Figure 7.18c, d).

The first two cervical vertebrae differ considerably from the others. Like the mythological Atlas, who supported the world on his shoulders, the first cervical vertebra (C1), the **atlas,** supports the head (Figure 7.18a, b). The atlas is a ring of bone with *anterior* and *posterior arches* and large *lateral masses.* It lacks a body and a spinous process. The superior

surfaces of the lateral masses, called *superior articular facets,* are concave and articulate with the occipital condyles of the occipital bone to form the *atlanto-occipital joints.* These articulations permit the movement seen when moving the head to signify yes. The inferior surfaces of the lateral masses, the *inferior articular facets,* articulate with the second cervical vertebra. The transverse processes and transverse foramina of the atlas are quite large.

The second cervical vertebra (C2), the **axis** (see Figure 7.18a, c), does have a body. A peglike process called the *dens* (= tooth) projects superiorly through the anterior portion of the vertebral foramen of the atlas. The dens makes a pivot on which the atlas and head rotate, as in moving the head to signify no. This arrangement permits side-to-side rotation of the head. The articulation formed between the anterior arch of the atlas and dens of the axis, and between their articular facets, is called the *atlanto-axial joint.* In some instances of

Figure 7.18 Cervical vertebrae. (See Tortora, *A Photographic Atlas of the Human Body,* Figure 3.17)

🔑 **The cervical vertebrae are found in the neck region.**

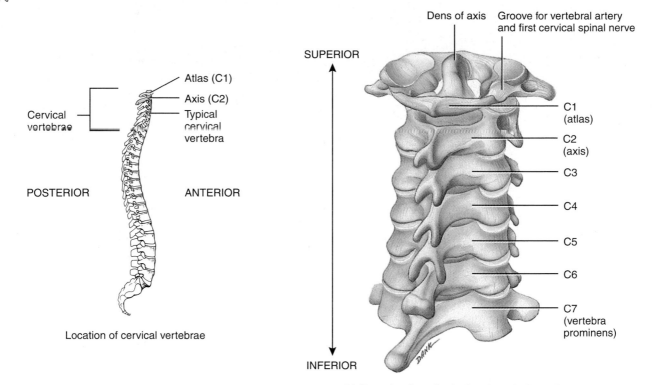

Location of cervical vertebrae

(a) Posterior view of articulated cervical vertebrae

figure continues

trauma, the dens of the axis may be driven into the medulla oblongata of the brain. When whiplash injuries result in death, this type of injury is the usual cause.

The third through sixth cervical vertebra (C3–C6), represented by the vertebra in Figure 7.18d, correspond to the structural pattern of the typical cervical vertebra previously described. The seventh cervical vertebra (C7), called the *vertebra prominens,* is somewhat different (see Figure 7.18a). It has a single large spinous process that may be seen and felt at the base of the neck.

Thoracic Region

Thoracic vertebrae (T1–T12; Figure 7.19) are considerably larger and stronger than cervical vertebrae. In addition, the spinous processes on T1 and T2 are long, laterally flattened, and directed inferiorly. In contrast, the spinous processes on T11 and T12 are shorter, broader, and directed more posteriorly. Compared to cervical vertebrae, thoracic vertebrae also have longer and larger transverse processes.

The best distinguishing feature of thoracic vertebrae is that they articulate with the ribs. The articulating surfaces of the vertebrae are called *facets* and *demifacets* (half-facets).

Except for T11 and T12, the transverse processes have facets for articulating with the *tubercles* of the ribs. The bodies of thoracic vertebrae also have facets or demifacets for articulation with the *heads* of the ribs. The articulations between the thoracic vertebrae and ribs are called *vertebrocostal joints.* As you can see in Figure 7.19a, T1 has a superior facet and an inferior demifacet, one on each side of the vertebral body. T2–T8 have a superior and inferior demifacet, one on each side of the vertebral body. T9 has a superior demifacet on each side of the vertebral body, and T10–T12 have a superior facet on each side of the vertebral body. Movements of the thoracic region are limited by thin intervertebral discs and by the attachment of the ribs to the sternum.

Lumbar Region

The **lumbar vertebrae** (L1–L5) are the largest and strongest in the vertebral column (Figure 7.20) because the amount of body weight supported by the vertebrae increases toward the inferior end of the backbone. Their various projections are short and thick. The superior articular processes are directed medially instead of superiorly and the inferior

Text continues on page 209

Figure 7.18 (continued)

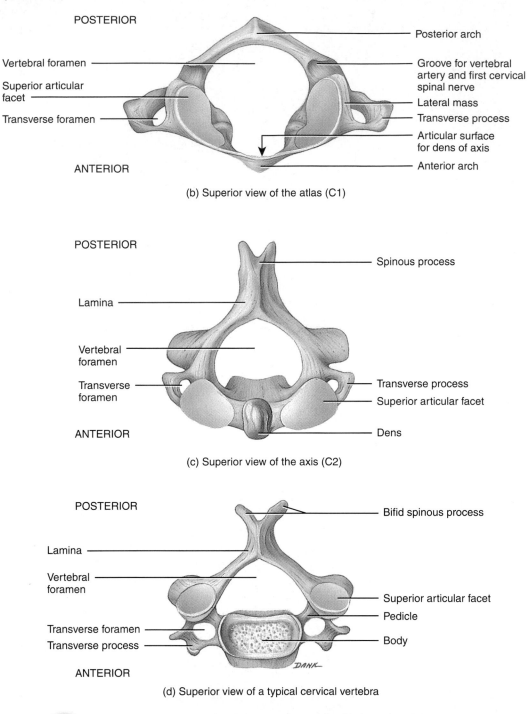

(b) Superior view of the atlas (C1)

(c) Superior view of the axis (C2)

(d) Superior view of a typical cervical vertebra

Q Which bones permit the movement of the head to signify no?

Figure 7.19 Thoracic vertebrae. (See Tortora, *A Photographic Atlas of the Human Body,* Figure 3.16)

🔑 **The thoracic vertebrae are found in the chest region and articulate with the ribs.**

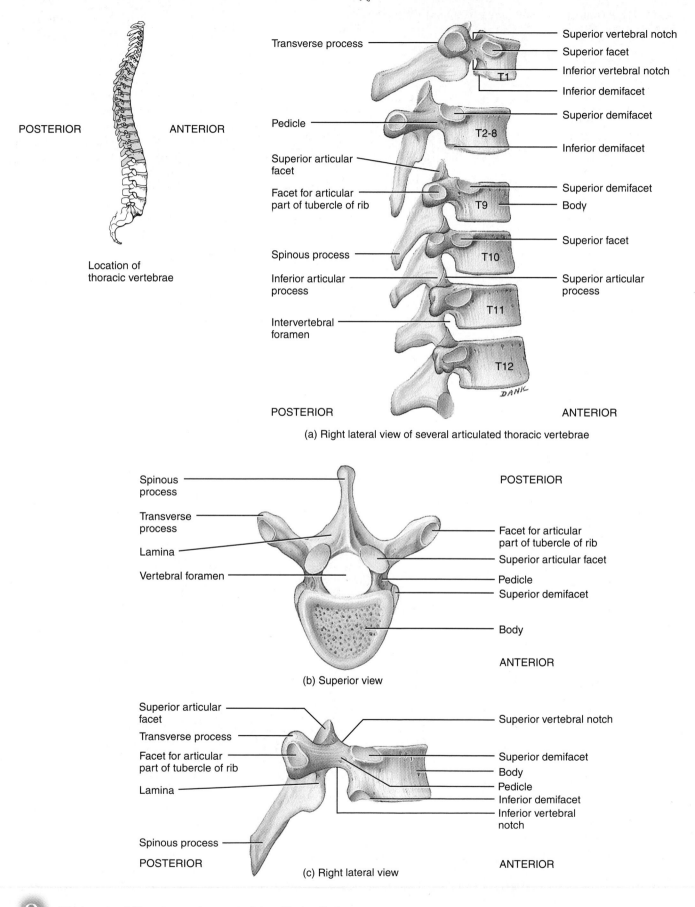

POSTERIOR   ANTERIOR

Location of thoracic vertebrae

Transverse process

Superior vertebral notch
Superior facet
Inferior vertebral notch
Inferior demifacet

T1

Pedicle

Superior demifacet

T2-8

Inferior demifacet

Superior articular facet

Facet for articular part of tubercle of rib

Superior demifacet

T9   Body

Spinous process

Superior facet

T10

Inferior articular process

Superior articular process

T11

Intervertebral foramen

T12

DANK

POSTERIOR   ANTERIOR

(a) Right lateral view of several articulated thoracic vertebrae

Spinous process

POSTERIOR

Transverse process

Lamina

Vertebral foramen

Facet for articular part of tubercle of rib
Superior articular facet
Pedicle
Superior demifacet

Body

ANTERIOR

(b) Superior view

Superior articular facet

Superior vertebral notch

Transverse process

Facet for articular part of tubercle of rib

Superior demifacet
Body
Pedicle
Inferior demifacet
Inferior vertebral notch

Lamina

Spinous process

POSTERIOR   ANTERIOR

(c) Right lateral view

Ⓠ Which parts of thoracic vertebrae articulate with the ribs?

Figure 7.20 Lumbar vertebrae. (See Tortora, *A Photographic Atlas of the Human Body,* Figure 3.18)

🗝 **Lumbar vertebrae are found in the lower back.**

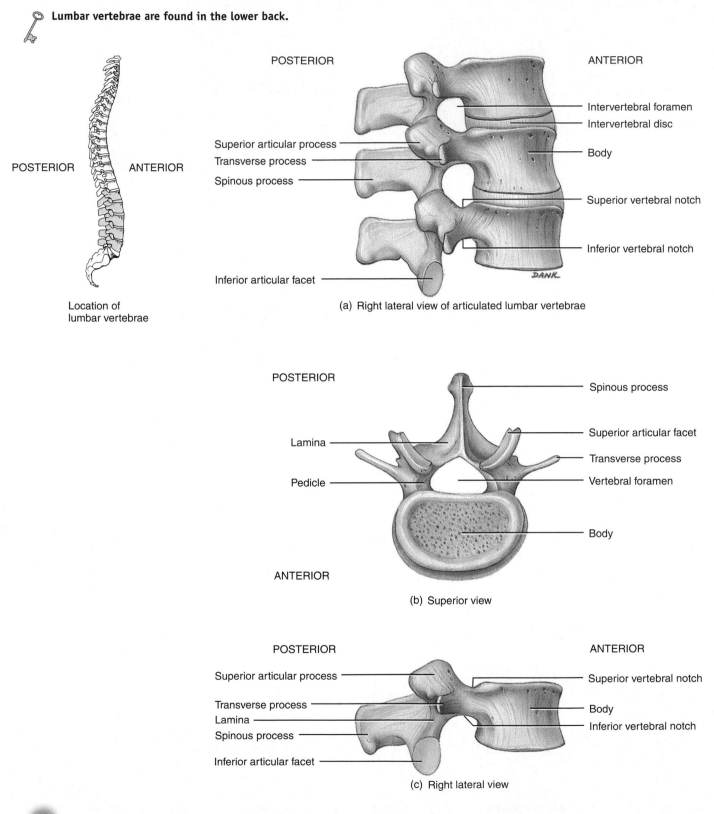

Location of lumbar vertebrae

POSTERIOR · ANTERIOR

POSTERIOR · ANTERIOR

Superior articular process
Transverse process
Spinous process

Inferior articular facet

Intervertebral foramen
Intervertebral disc

Body

Superior vertebral notch

Inferior vertebral notch

DANK

(a) Right lateral view of articulated lumbar vertebrae

POSTERIOR

Lamina
Pedicle

ANTERIOR

Spinous process

Superior articular facet
Transverse process
Vertebral foramen

Body

(b) Superior view

POSTERIOR · ANTERIOR

Superior articular process
Transverse process
Lamina
Spinous process
Inferior articular facet

Superior vertebral notch
Body
Inferior vertebral notch

(c) Right lateral view

Q Why are the lumbar vertebrae the largest and strongest in the vertebral column?

Table 7.5 Comparison of Principal Structural Features of Cervical, Thoracic, and Lumbar Vertebrae

CHARACTERISTIC	CERVICAL	THORACIC	LUMBAR
Overall structure	See Figure 7.18d, p. 206	See Figure 7.19b, p. 207	See Figure 7.20b, p. 208
Size	Small	Larger	Largest
Foramina	One vertebral and two transverse	One vertebral	One vertebral
Spinous processes	Slender and often bifid (C2–C6)	Long and fairly thick (most project inferiorly)	Short and blunt (project posteriorly rather than inferiorly)
Transverse processes	Small	Fairly large	Large and blunt
Articular facets for ribs	Absent	Present	Absent
Direction of articular facets			
Superior	Posterosuperior	Posterolateral	Medial
Inferior	Anteroinferior	Anteromedial	Lateral
Size of intervertebral discs	Thick relative to size of vertebral bodies	Thin relative to vertebral bodies	Massive

articular processes are directed laterally instead of inferiorly. The spinous processes are quadrilateral in shape, are thick and broad, and project nearly straight posteriorly. The spinous processes are well-adapted for the attachment of the large back muscles.

A summary of the major structural differences among cervical, thoracic, and lumbar vertebrae is presented in Table 7.5.

Sacrum

The **sacrum** is a triangular bone formed by the union of five sacral vertebrae, indicated in Figure 7.21a as S1–S5. The sacral vertebrae begin to fuse in individuals between 16 and 18 years of age, a process usually completed by age 30. The sacrum serves as a strong foundation for the pelvic girdle. It is positioned at the posterior portion of the pelvic cavity medial to the two hip bones. The female sacrum is shorter, wider, and more curved between S2 and S3 than the male sacrum (see Table 8.1).

The concave anterior side of the sacrum faces the pelvic cavity. It is smooth and contains four *transverse lines (ridges)* that mark the joining of the sacral vertebral bodies (see Figure 7.21a). At the ends of these lines are four pairs of *anterior sacral foramina.* The lateral portion of the superior surface contains a smooth surface called the *sacral ala* (= wing), which is formed by the fused transverse processes of the first sacral vertebra (S1).

The convex, posterior surface of the sacrum contains a *median sacral crest,* which is the fused spinous processes of the upper sacral vertebrae; a *lateral sacral crest,* which is the fused transverse processes of the sacral vertebrae; and four pairs of *posterior sacral foramina* (Figure 7.21b). These foramina communicate with the anterior sacral foramina through which nerves and blood vessels pass. The *sacral canal* is a continuation of the vertebral canal. The laminae of

the fifth sacral vertebra, and sometimes the fourth, fail to meet. This leaves an inferior entrance to the vertebral canal called the *sacral hiatus* (hī-Ā-tus; = opening). On either side of the sacral hiatus are the *sacral cornua,* the inferior articular processes of the fifth sacral vertebra. They are connected by ligaments to the coccyx.

The narrow inferior portion of the sacrum is known as the *apex.* The broad superior portion of the sacrum is called the *base.* The anteriorly projecting border of the base, called the *sacral promontory* (PROM-on-tō′-rē), is one of the points used for measurements of the pelvis. On both lateral surfaces the sacrum has a large *auricular surface* that articulates with the ilium of each hip bone to form the *sacroiliac joint.* Posterior to the auricular surface is a roughened surface, the *sacral tuberosity,* that contains depressions for the attachment of ligaments. The sacral tuberosity is another surface of the sacrum that unites with the hip bones to form the sacroiliac joints. The *superior articular processes* of the sacrum articulate with the fifth lumbar vertebra, and the base of the sacrum articulates with the body of the fifth lumbar vertebra, to form the *lumbosacral joint.*

Coccyx

The **coccyx** is also triangular in shape and is formed by the fusion of usually four coccygeal vertebrae, indicated in Figure 7.21 as Co1–Co4. The coccygeal vertebrae fuse when a person is between 20 and 30 years of age. The dorsal surface of the body of the coccyx contains two long *coccygeal cornua* that are connected by ligaments to the sacral cornua. The coccygeal cornua are the pedicles and superior articular processes of the first coccygeal vertebra. On the lateral surfaces of the coccyx are a series of *transverse processes,* the first pair being the largest. The coccyx articulates superiorly with the apex of the sacrum. In females, the coccyx points inferiorly; in males, it points anteriorly (see Table 8.1).

Figure 7.21 Sacrum and coccyx. (See Tortora, *A Photographic Atlas of the Human Body,* Figure 3.19)

🔑 **The sacrum is formed by the union of five sacral vertebrae, and the coccyx is formed by the union of usually four coccygeal vertebrae.**

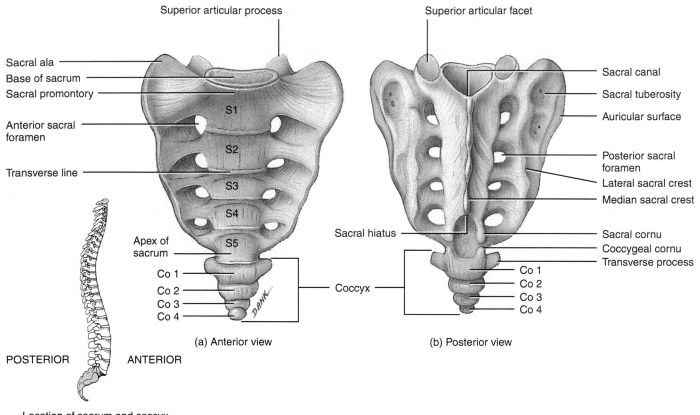

Superior articular process

Superior articular facet

Sacral ala
Base of sacrum
Sacral promontory

Anterior sacral foramen

Transverse line

S1
S2
S3
S4
S5

Apex of sacrum

Co 1
Co 2
Co 3
Co 4

Sacral canal
Sacral tuberosity
Auricular surface

Posterior sacral foramen
Lateral sacral crest
Median sacral crest

Sacral hiatus

Sacral cornu
Coccygeal cornu
Transverse process

Coccyx

Co 1
Co 2
Co 3
Co 4

Coccyx

(a) Anterior view

(b) Posterior view

POSTERIOR   ANTERIOR

Location of sacrum and coccyx

Q How many foramina pierce the sacrum, and what is their function?

CLINICAL APPLICATION
Caudal Anesthesia

Anesthetic agents that act on the sacral and coccygeal nerves are sometimes injected through the sacral hiatus, a procedure called **caudal anesthesia** or **epidural block.** The procedure is used most often to relieve pain during labor and to provide anesthesia to the perianal area. Because the sacral hiatus is between the sacral cornua, the cornua are important bony landmarks for locating the hiatus. Anesthetic agents may also be injected through the posterior sacral foramina. ∎

1. What are the functions of the vertebral column?
2. When do the secondary vertebral curves develop?
3. What are the principal distinguishing characteristics of the bones of the various regions of the vertebral column?

THORAX

OBJECTIVE
• *Identify the bones of the thorax.*

The term **thorax** refers to the entire chest. The skeletal portion of the thorax is a bony cage formed by the sternum, costal cartilages, ribs, and the bodies of the thoracic vertebrae (Figure 7.22). The thoracic cage is narrower at its superior end and broader at its inferior end. It is flattened from front to back. The thoracic cage encloses and protects the organs in the thoracic and superior abdominal cavities. It also provides support for the bones of the shoulder girdle and upper limbs.

Sternum

The **sternum,** or breastbone, is a flat, narrow bone measuring about 15 cm (6 in.) in length. It is located on the

Figure 7.22 Skeleton of the thorax. (See Tortora, *A Photographic Atlas of the Human Body,* Figure 3.20)

 The bones of the thorax enclose and protect organs in the thoracic cavity and in the superior abdominal cavity.

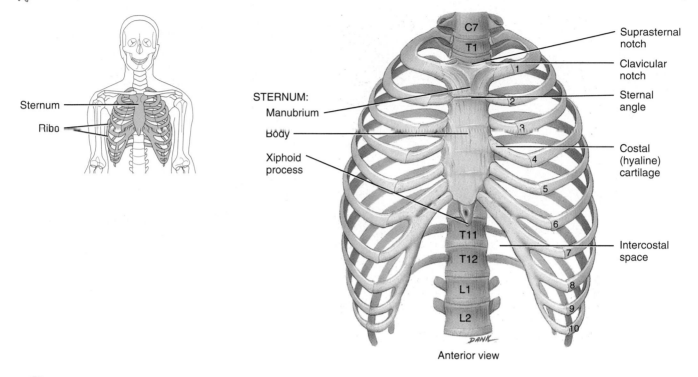

Anterior view

Q Which ribs are true ribs? Which are false ribs?

anterior midline of the thoracic wall. During thoracic surgery, it may be split midsagittally to allow surgeons access to structures in the thoracic cavity such as the thymus gland, heart, and great vessels of the heart.

The sternum consists of three portions (see Figure 7.22): the *manubrium* (ma-NOO-brē-um; = handle-like), the superior portion; the *body,* the middle and largest portion; and the *xiphoid process* (ZĪ-foyd; = sword-shaped), the inferior, smallest portion. The junction of the manubrium and body forms the *sternal angle.* The manubrium has a depression on its superior surface called the *suprasternal notch.* Lateral to the suprasternal notch are *clavicular notches* that articulate with the medial ends of the clavicles to form the *sternoclavicular joints.* The manubrium also articulates with the costal cartilages of the first and second ribs to form the *sternocostal joints.*

The body of the sternum articulates directly or indirectly with the costal cartilages of the second through tenth ribs. The xiphoid process consists of hyaline cartilage during infancy and childhood and does not ossify completely until a person is about age 40. It has no ribs attached to it but provides attachment for some abdominal muscles. Incorrect positioning of the hands of a rescuer during cardiopulmonary

resuscitation (CPR) may result in a fracture of the xiphoid process, driving it into internal organs.

Ribs

Twelve pairs of **ribs** give structural support to the sides of the thoracic cavity (see Figure 7.22). The ribs increase in length from the first through seventh, then decrease in length to the twelfth rib. Each articulates posteriorly with its corresponding thoracic vertebra.

The first through seventh pairs of ribs have a direct anterior attachment to the sternum by a strip of hyaline cartilage called *costal cartilage (cost- =* rib). These ribs are called *true (vertebrosternal) ribs.* The remaining five pairs of ribs are termed *false ribs* because their costal cartilages either attach indirectly to the sternum or do not attach to the sternum at all. The cartilages of the eighth, ninth, and tenth pairs of ribs attach to each other and then to the cartilages of the seventh pair of ribs. These false ribs are called *vertebrochondral ribs.* The eleventh and twelfth pairs of ribs are false ribs designated as *floating (vertebral) ribs* because the costal cartilage at their anterior ends does not attach to the sternum at all. These ribs attach only posteriorly to the thoracic vertebrae.

Figure 7.23 The structure of ribs. Each rib has a head, a neck, and a body. The facets and the articular part of the tubercle are where the rib articulates with a vertebra. (See Tortora, *A Photographic Atlas of the Human Body,* Figure 3.21)

🔑 **Each rib articulates posteriorly with its corresponding thoracic vertebra.**

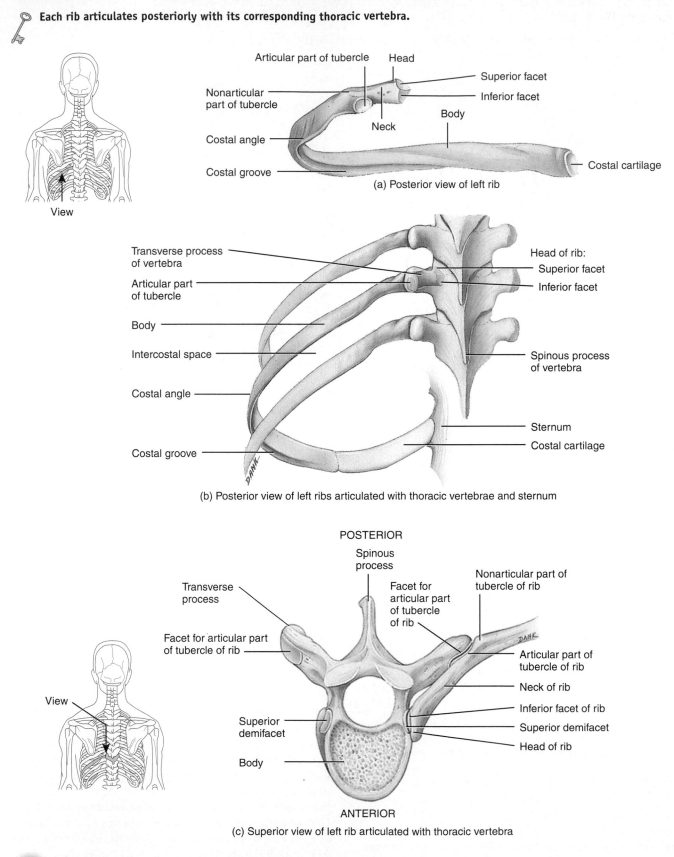

(a) Posterior view of left rib

(b) Posterior view of left ribs articulated with thoracic vertebrae and sternum

(c) Superior view of left rib articulated with thoracic vertebra

Q How does a rib articulate with a thoracic vertebra?

Figure 7.23a shows the parts of a typical (third through ninth) rib. The *head* is a projection at the posterior end of the rib. It consists of one or two *facets* that articulate with facets on the bodies of adjacent thoracic vertebrae to form *vertebrocostal joints.* The *neck* is a constricted portion just lateral to the head. A knoblike structure on the posterior surface where the neck joins the body is called a *tubercle* (TOO-ber-kul). It consists of a *nonarticular part* that affords attachment to the ligament of the tubercle, and an *articular part* that articulates with the facet of a transverse process of the inferior of the two vertebrae to which the head of the rib is connected. These articulations also form vertebrocostal joints. The *body (shaft)* is the main part of the rib. A short distance beyond the tubercle, an abrupt change in the curvature of the shaft occurs. This point is called the *costal angle.* The inner surface of the rib has a *costal groove* that protects blood vessels and a small nerve.

In summary, the posterior portion of the rib is connected to a thoracic vertebra by its head and the articular part of a tubercle. The facet of the head fits into a facet on the body of one vertebra or into the demifacets of two adjoining vertebrae. The articular part of the tubercle articulates with the facet of the transverse process of the vertebra.

Spaces between ribs, called *intercostal spaces,* are occupied by intercostal muscles, blood vessels, and nerves. Surgical access to the lungs or other structures in the thoracic cavity is commonly obtained through an intercostal space. Special rib retractors are used to create a wide separation between ribs. The costal cartilages are sufficiently elastic in younger individuals to permit considerable bending without breaking.

CLINICAL APPLICATION
Rib Fractures

Rib fractures are the most common chest injuries, and they usually result from direct blows, most often from impact with a steering wheel, falls, and crushing injuries to the chest. Ribs tend to break at the point where the greatest force is applied, but they may also break at their weakest point—the site of greatest curvature, which is just anterior to the costal angle. In some cases, fractured ribs may puncture the heart, great vessels of the heart, lungs, trachea, bronchi, esophagus, spleen, liver, and kidneys. Rib fractures are usually quite painful. ■

1. What bones form the skeleton of the thorax?
2. What are the functions of the bones of the thorax?
3. How are ribs classified on the basis of their attachment to the sternum?

DISORDERS: HOMEOSTATIC IMBALANCES

HERNIATED (SLIPPED) DISC

In their function as shock absorbers, intervertebral discs are constantly being compressed. If the anterior and posterior ligaments of the discs become injured or weakened, the pressure developed in the nucleus pulposus may be great enough to rupture the surrounding fibrocartilage (annulus fibrosus). If this occurs, the nucleus pulposus may herniate (protrude) posteriorly or into one of the adjacent vertebral bodies. This condition is called a **herniated (slipped) disc.** It occurs most often in the lumbar region because it bears much of the weight of the body, and it is the region of the most flexing and bending.

Most often the nucleus pulposus slips posteriorly toward the spinal cord and spinal nerves (Figure 7.24). This movement exerts pressure on the spinal nerves, causing acute pain. If the roots of the sciatic nerve, which passes from the spinal cord to the foot, are compressed, the pain radiates down the posterior thigh, through the calf, and occasionally into the foot. If pressure is exerted on the spinal cord itself, some of its neurons may be destroyed. A person with a herniated disc may undergo a *laminectomy,* a procedure in which parts of the laminae of the vertebra and intervertebral disc are removed to relieve pressure on the nerves.

Figure 7.24 Herniated (slipped) disc.

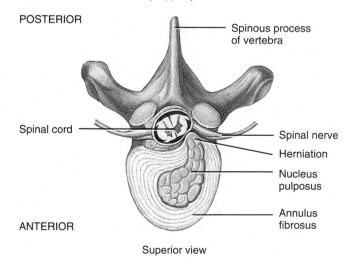

Superior view

ABNORMAL CURVES OF THE VERTEBRAL COLUMN

Various conditions may exaggerate the normal curves of the vertebral column, or the column may acquire a lateral bend, resulting in **abnormal curves** of the vertebral column.

Scoliosis (skō-lē-Ō-sis; *scolio-* = crooked) is a lateral bending of the vertebral column, usually in the thoracic region. This is the most common of the abnormal curves. It may result from congenitally (present at birth) malformed vertebrae, chronic sciatica, paralysis of muscles on one side of the vertebral column, poor posture, or one leg being shorter than the other.

Kyphosis (kī-Fō-sis; *kyphos-* = hump) is an exaggeration of the thoracic curve of the vertebral column. In tuberculosis of the spine, vertebral bodies may partially collapse, causing an acute angular bending of the vertebral column. In the elderly, degeneration of the intervertebral discs leads to kyphosis. Kyphosis may also be caused by rickets and poor posture. It is also common in females with advanced osteoporosis. The term *round-shouldered* is an expression for mild kyphosis.

Lordosis (lor-DŌ-sis; *lord-* = bent backward), sometimes called *swayback,* is an exaggeration of the lumbar curve of the vertebral column. It may result from increased weight of the abdomen as in pregnancy or extreme obesity, poor posture, rickets, or tuberculosis of the spine.

SPINA BIFIDA

Spina bifida (SPĪ-na BIF-i-da) is a congenital defect of the vertebral column in which laminae fail to unite at the midline. In serious cases, protrusion of the membranes (meninges) around the spinal cord or the spinal cord itself produces perilous problems, such as partial or complete paralysis, partial or complete loss of urinary bladder control, and the absence of reflexes. An increased risk of spina bifida is associated with low levels of folic acid, one of the B vitamins, during pregnancy. Spina bifida may be diagnosed prenatally by a test of the mother's blood for a substance, called alpha-fetoprotein, produced by the fetus; by sonography; or by amniocentesis (withdrawal of amniotic fluid for analysis).

STUDY OUTLINE

INTRODUCTION (p. 183)

1. Bones not only protect and afford movement; they also serve as landmarks for locating parts of other body systems.
2. The musculoskeletal system is composed of the bones, joints, and muscles working together.

DIVISIONS OF THE SKELETAL SYSTEM (p. 183)

1. The axial skeleton consists of bones arranged along the longitudinal axis. The parts of the axial skeleton are the skull, auditory ossicles (ear bones), hyoid bone, vertebral column, sternum, and ribs.
2. The appendicular skeleton consists of the bones of the girdles and the upper and lower limbs (extremities). The parts of the appendicular skeleton are the pectoral (shoulder) girdles, bones of the upper limbs, pelvic (hip) girdles, and bones of the lower limbs.

TYPES OF BONES (p. 185)

1. On the basis of shape, bones are classified as long, short, flat, irregular, or sesamoid. Sesamoid bones develop in tendons or ligaments.
2. Sutural bones are found within the sutures of certain cranial bones.

BONE SURFACE MARKINGS (p. 185)

1. Surface markings are structural features visible on the surfaces of bones.
2. Each marking—whether a depression, an opening, or a process—is structured for a specific function, such as joint formation, muscle attachment, or passage of nerves and blood vessels (see Table 7.2).

SKULL (p. 186)

1. The skull consists of cranial bones and facial bones. It is composed of 22 bones.
2. The eight cranial bones include the frontal, parietal (2), temporal (2), occipital, sphenoid, and ethmoid.
3. The 14 facial bones are the nasal (2), maxillae (2), zygomatic (2), lacrimal (2), palatine (2), inferior nasal conchae (2), vomer, and mandible.
4. Sutures are immovable joints that connect most bones of the skull. Examples are coronal, sagittal, lambdoid, and squamous sutures.
5. Paranasal sinuses are cavities in bones of the skull that communicate with the nasal cavity. They are lined by mucous membranes. The cranial bones enclosing paranasal sinuses are the frontal, sphenoid, ethmoid, and maxillae.
6. Fontanels are fibrous, connective tissue membrane–filled spaces between the cranial bones of fetuses and infants. The major fontanels are the anterior, posterior, anterolaterals, and posterolaterals (see Table 7.3). They fill in with bone and become sutures.
7. The foramina of the skull bones provide passages for nerves and blood vessels; see Table 7.4.
8. Each of the orbits (eye sockets) is formed by seven bones of the skull.
9. The nasal septum consists of the vomer, perpendicular plate of the ethmoid, and septal cartilage. It divides the nasal cavity into left and right sides.

HYOID BONE (p. 201)

1. The hyoid bone is a U-shaped bone that does not articulate with any other bone.
2. It supports the tongue and provides attachment for some tongue muscles and for some muscles of the pharynx and neck.

VERTEBRAL COLUMN (p. 201)

1. The vertebral column, sternum, and ribs constitute the skeleton of the body's trunk.
2. The 26 bones of the adult vertebral column are the cervical vertebrae (7), the thoracic vertebrae (12), the lumbar vertebrae (5), the sacrum (5 fused vertebrae), and the coccyx (usually 4 fused vertebrae).
3. The vertebral column contains normal curves (cervical, thoracic, lumbar, and sacral) that give strength, support, and balance.

4. The vertebrae are similar in structure, each usually consisting of a body, vertebral arch, and seven processes. Vertebrae in the different regions of the column vary in size, shape, and detail.

THORAX (p. 210)

1. The thoracic skeleton consists of the sternum, ribs and costal cartilages, and thoracic vertebrae.
2. The thoracic cage protects vital organs in the chest area and upper abdomen.

SELF-QUIZ QUESTIONS

True or false:

1. The axial skeleton consists of the bones that lie along the axis: skull bones, ear bones, hyoid bone, ribs, breastbone, and pelvic and pectoral girdles.
2. The first cervical vertebra is the axis, and the second is the atlas.

Compete the following:

3. The sella turcica of the sphenoid bone houses the ___.
4. Membrane-filled spaces between cranial bones that enable the fetal skull to modify its size and shape for passage through the birth canal are called ___.
5. The bone of the axial skeleton that does not articulate with any other bones is the ___ bone.
6. ___ separate adjacent vertebrae, from the second cervical vertebra to the sacrum.
7. Match the following:

   ___(a) supraorbital foramen
   ___(b) temporomandibular joint
   ___(c) external auditory meatus
   ___(d) foramen magnum
   ___(e) optic foramen
   ___(f) cribriform plate
   ___(g) palatine process
   ___(h) turbinates
   ___(i) ramus, body, and condylar process
   ___(j) transverse foramen, bifid processes
   ___(k) dens
   ___(l) tailbone
   ___(m) costal cartilages
   ___(n) xiphoid process
   ___(o) provides nerve supply to entire upper limb

   (1) temporal bone
   (2) sphenoid bone
   (3) inferior nasal conchae
   (4) cervical vertebrae
   (5) ethmoid bone
   (6) articulation of mandibular fossa and articular tubercle of the temporal bone to the mandible
   (7) occipital bone
   (8) frontal bone
   (9) maxillae
   (10) mandible
   (11) axis
   (12) coccyx
   (13) sternum
   (14) brachial plexus
   (15) ribs

8. Match the following (the same answer may be used more than once):

   ___ (a) bones that have greater length than width and consist of a shaft and a variable number of extremities
   ___ (b) cube-shaped bones that are nearly equal in length and width
   ___ (c) bones that develop in certain tendons where there is considerable friction, tension, and physical stress
   ___ (d) small bones located within joints between certain cranial bones
   ___ (e) thin bones composed of two nearly parallel plates of compact bone enclosing a layer of spongy bone
   ___ (f) bones with complex shapes, including the vertebrae and calcaneus
   ___ (g) includes patella and pisiform bone
   ___ (h) bones that provide considerable protection and extensive areas for muscle attachment
   ___ (i) include femur, tibia, fibula, humerus, ulna, and radius
   ___ (j) include cranial bones, sternum, and ribs
   ___ (k) include almost all of the carpal and tarsal bones

   (1) irregular bones
   (2) long bones
   (3) short bones
   (4) flat bones
   (5) sesamoid bones
   (6) sutural bones

Choose the best answer to the following questions:

9. Which of the following are true for the cranial and/or facial bones? (1) Their inner surfaces attach to membranes that stabilize positions of the brain, blood vessels, and nerves. (2) Their outer surfaces provide areas of attachment for muscles that move various parts of the head. (3) They protect and provide support for the entrances to the digestive, respiratory, and integumentary systems. (4) They provide attachment for the muscles of facial expression. (5) They protect and support the delicate special sense organs.
   (a) 1, 2, 4, and 5, (b) 2, 3, 4, and 5, (c) 1, 3, 4, and 5, (d) 1, 2, 3, and 5, (e) 1, 2, 3, 4, and 5

10. In which of the following bones are paranasal sinuses *not* found? (a) frontal bone, (b) sphenoid bone, (c) lacrimal bones, (d) ethmoid bone, (e) maxillae.

11. Which of the following are functions of the vertebral curves? (1) protection of the heart and lungs, (2) increase in strength, (3) maintenance of balance in the upright position, (4) absorption of shock during walking, (5) protection of vertebrae against fracture.
(a) 1, 2, 3, and 4, (b) 2, 3, 4, and 5, (c) 1, 3, 4, and 5, (d) 1, 2, 3, and 5, (e) 1, 2, 3, 4, and 5

12. An exaggeration of the thoracic curve of the vertebral column is called (a) scoliosis, (b) kyphosis, (c) lordosis, (d) spina bifida, (e) herniation.

13. The suture that unites the two parietal bones is the (a) coronal suture, (b) lambdoid suture, (c) squamous suture, (d) sagittal suture, (e) frontal suture.

14. Which of the following statements are true? (1) The thoracic cage encloses and protects the organs in the thoracic cavity and upper abdominal cavity. (2) The thorax provides support for the bones of the shoulder girdles and upper limbs. (3) The thorax is formed by the sternum, costal cartilages, ribs, and bodies of the thoracic vertebrae. (4) The thoracic ribs include true ribs, false ribs, and floating ribs. (5) The sternum consists of the manubrium, body, and xiphoid process.
(a) 1, 2, and 3, (b) 2, 3, and 4, (c) 1, 2, 3, and 5, (d) 2, 3, 4, and 5, (e) 1, 2, 3, 4, and 5

15. Match the following:

___(a) forms the forehead

___(b) form the inferior lateral aspects of the cranium and part of the cranial floor, contain zygomatic process and mastoid process

___(c) forms part of the anterior portion of the cranial floor, medial wall of the orbits, superior portions of nasal septum, most of the sidewalls of the nasal cavity; is a major supporting structure of the nasal cavity

___(d) form the prominence of the cheek and part of the lateral wall and floor of each orbit

___(e) the largest, strongest facial bone; is the only movable skull bone

___(f) a roughly triangular bone on the floor of the nasal cavity; one of the components of the nasal septum

___(g) form greater portion of the sides and roof of the cranial cavity

___(h) forms the posterior part and most of the base of the cranium; contains the foramen magnum

___(i) called the keystone of the cranial floor; contains the sella turcica, optic foramen, and pterygoid processes

___(j) form the bridge of the nose

___(k) the smallest bones of the face; contain a vertical groove that houses a structure that gathers tears and passes them into the nasal cavity

___(l) unite to form the upper jawbone and articulate with every bone of the face except the lower jawbone

___(m) form the posterior part of the hard palate, part of the floor and lateral wall of the nasal cavity, and a small portion of the floors of the orbits

___(n) scroll-like bones that form a part of the lateral walls of the nasal cavity; functions in the turbulent circulation and filtration of air

(1) frontal bone
(2) parietal bones
(3) temporal bones
(4) occipital bone
(5) sphenoid bone
(6) ethmoid bone
(7) nasal bones
(8) maxillae
(9) zygomatic bones
(10) lacrimal bones
(11) palatine bones
(12) vomer
(13) mandible
(14) inferior nasal conchae

CRITICAL THINKING QUESTIONS

1. While investigating her new baby brother, 4-year-old Latisha found a soft spot on the baby's skull and announced that the baby needed to go back because "it's not finished yet." Explain the presence of soft spots in the infant's skull. (HINT: *An infant should be "soft in the head," even if an adult should not.*)

2. Thirty-five-year-old Barbara (old enough to know better) was bouncing down the stairs in her stocking feet when she slipped and landed hard on her buttocks. She felt a pain sharp enough to bring tears to her eyes, and she needed to sit very gingerly on the

way to and at the emergency room. The attending physician said that she had broken a bone but that she wouldn't be given a cast. What bone do you think she broke? (HINT: *Your buttocks usually keep this bone from being banged when you sit.*)

3. The advertisement reads "New Posture Perfect Mattress! Keeps spine perfectly straight—just like when you were born! A straight spine equals a great sleep!" Would you buy a mattress from this company? Explain. (HINT: *Take a sideways view of the vertebral column.*)

ANSWERS TO FIGURE QUESTIONS

7.1 Axial skeleton: skull, vertebral column. Appendicular skeleton: clavicle, shoulder girdle, humerus, pelvic girdle, femur.

7.2 Flat bones protect and provide a large surface area for muscle attachment.

7.3 The frontal, parietal, sphenoid, ethmoid, and temporal bones are cranial bones.

7.4 Squamous suture: parietal and temporal bones. Lambdoid suture: parietal and occipital bones. Coronal suture: parietal and frontal bones.

7.5 The temporal bone articulates with the parietal, sphenoid, zygomatic, and occipital bones.

7.6 The parietal bones form the posterior, lateral portion of the cranium.

7.7 The medulla oblongata of the brain connects with the spinal cord in the foramen magnum.

7.8 Crista galli of ethmoid bone, frontal, parietal, temporal, occipital, temporal, parietal, frontal, crista galli of ethmoid bone.

7.9 The perpendicular plate of the ethmoid bone forms the superior part of the nasal septum, and the lateral masses compose most of the medial walls of the orbits.

7.10 The mandible is the only movable skull bone, other than the auditory ossicles.

7.11 The paranasal sinuses produce mucus and serve as resonating chambers for vocalization.

7.12 The anterolateral fontanel is bordered by four bones.

7.13 Bones forming the orbit are the frontal, sphenoid, zygomatic, maxilla, lacrimal, ethmoid, and palatine.

7.14 The nasal septum divides the nasal cavity into right and left sides.

7.15 The hyoid bone does not articulate with any other bone.

7.16 The thoracic and sacral curves are concave.

7.17 The vertebral foramina enclose the spinal cord, whereas the intervertebral foramina provide spaces for spinal nerves to exit the vertebral column.

7.18 The atlas moving on the axis permits movement of the head to signify "no."

7.19 The facets and demifacets on the body of the thoracic vertebrae articulate with the facets on the head of the ribs, and the facets on the transverse processes of these vertebrae articulate with the tubercle of the ribs.

7.20 The lumbar vertebrae are stout because the amount of body weight supported by vertebrae increases toward the inferior end of the vertebral column.

7.21 There are four pairs of sacral foramina, for a total of eight. Each anterior sacral foramen joins a posterior sacral foramen at the intervertebral foramen. Nerves and blood vessels pass through these tunnels in the bone.

7.22 True ribs: pairs 1–7; false ribs: pairs 8–12.

7.23 The facet on the head of a rib fits into a facet on the body of a vertebra, and the articular part of the tubercle of a rib articulates with the facet of the transverse process of a vertebra.

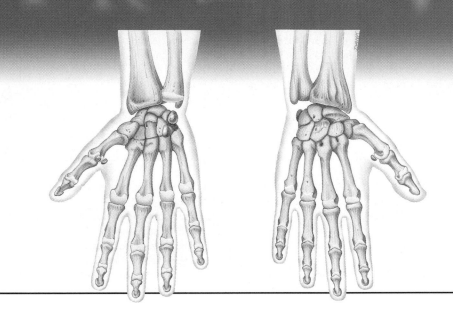

The two main divisions of the skeletal system are the axial skeleton and the appendicular skeleton. Whereas the axial skeleton serves mainly to protect and support internal organs, the appendicular skeleton, the focus of this chapter, mainly facilitates movement. The appendicular skeleton includes bones that make up the upper and lower limbs as well as bones, called girdles, that attach the limbs to the axial skeleton. As you study this chapter you will see how the bones of the appendicular skeleton act with each other and with skeletal muscles to bring about a variety of movements.

PECTORAL (SHOULDER) GIRDLE

OBJECTIVE
• *Identify the bones of the pectoral (shoulder) girdle and their principal markings.*

The **pectoral** (PEK-tō-ral) or **shoulder girdles** attach the bones of the upper limbs to the axial skeleton (Figure 8.1). Each of the two pectoral girdles consists of a clavicle and a scapula. The *clavicle* is the anterior component; it articulates with the *sternum* at the *sternoclavicular joint.* The posterior component, the *scapula*, articulates with the clavicle at the *acromioclavicular joint* and with the humerus at the *glenohumeral (shoulder) joint.* The pectoral girdles do not articulate with the vertebral column and are held in position instead by complex muscle attachments.

Clavicle

Each slender, S-shaped **clavicle** (KLAV-i-kul; = key) or *collarbone* lies horizontally in the superior and anterior part of the thorax superior to the first rib (Figure 8.2). The medial half of the clavicle is convex anteriorly, whereas the lateral half is concave anteriorly. The medial end of the clavicle, the *sternal extremity,* is rounded and articulates with the sternum to form the *sternoclavicular joint.* The broad, flat, lateral end, the *acromial extremity* (a-KRŌ-mē-al), articulates with the acromion of the scapula. This joint is called the *acromioclavicular joint.* (See Figure 8.1.) The *conoid tubercle* (KŌ-noyd; = conelike) on the inferior surface of the lateral end of the bone is a point of attachment for the conoid ligament. The *costal tuberosity* on the inferior surface of the medial end is a point of attachment for the costoclavicular ligament.

CLINICAL APPLICATION
Fractured Clavicle

The clavicle transmits mechanical force from the upper limb to the trunk. If the force transmitted to the clavicle is excessive, as in falling on one's outstretched arm, a **fractured clavicle** may result. The clavicle is one of the most frequently broken bones in the body. Because the junction of the clavicle's two curves is its weakest point, the clavicular midregion is the most frequent fracture site. ■

Scapula

Each **scapula** (SCAP-yoo-la), or *shoulder blade,* is a large, triangular, flat bone situated in the superior part of the posterior thorax between the levels of the second and seventh ribs (Figure 8.3). The medial borders of the scapulae (plural) lie about 5 cm (2 inches) from the vertebral column.

Figure 8.1 Right pectoral (shoulder) girdle.

🔑 **The clavicle is the anterior component of the pectoral girdle, and the scapula is the posterior component.**

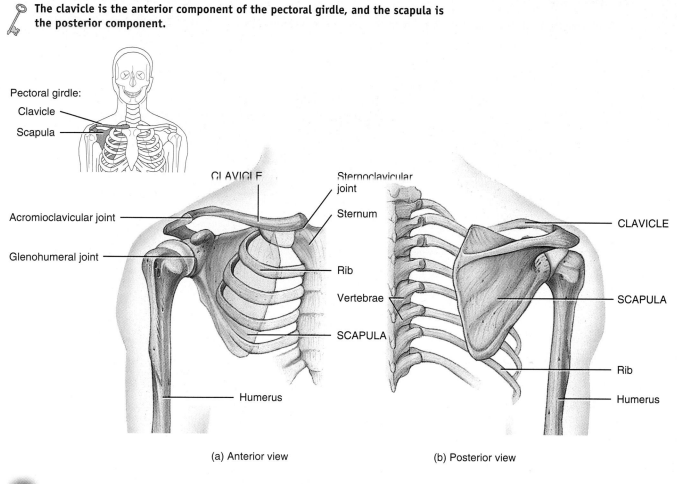

(a) Anterior view

(b) Posterior view

Q What is the function of the pectoral girdles?

Figure 8.2 Right clavicle.

🔑 **The clavicle articulates medially with the sternum and laterally with the acromion of the scapula.**

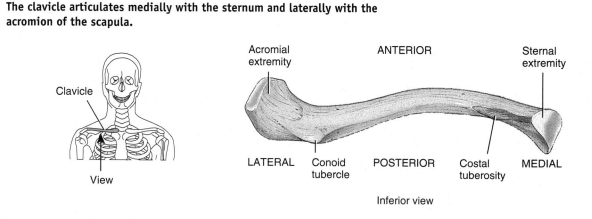

Inferior view

Q Which part of the clavicle is its weakest point?

Figure 8.3 Right scapula (shoulder blade). (See Tortora, *A Photographic Atlas of the Human Body,* Figure 3.22.)

The glenoid cavity of the scapula articulates with the head of the humerus to form the shoulder joint.

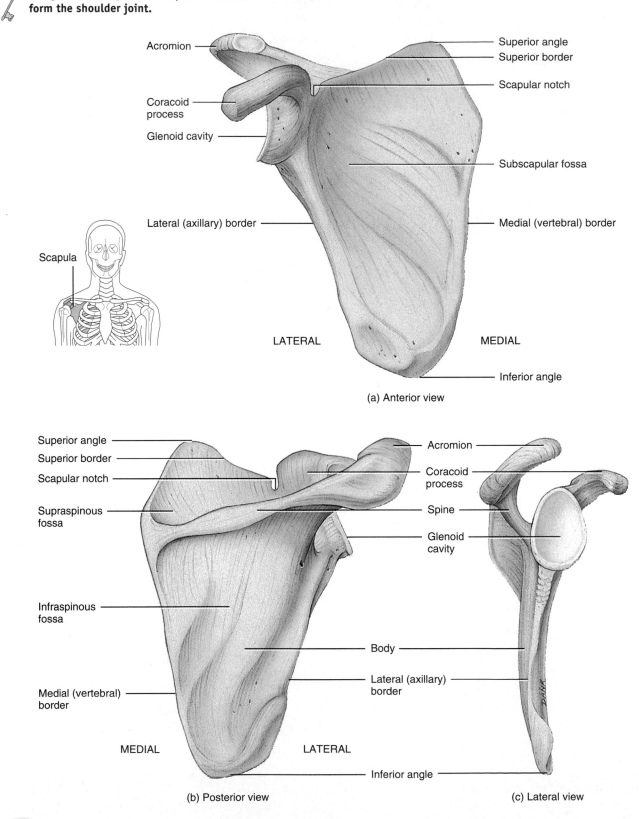

(a) Anterior view

(b) Posterior view

(c) Lateral view

Q Which part of the scapula forms the high point of the shoulder?

A sharp ridge, the *spine*, runs diagonally across the posterior surface of the flattened, triangular *body* of the scapula. The lateral end of the spine projects as a flattened, expanded process called the *acromion* (a-KRŌ-mē-on; *acrom-* = topmost), easily felt as the high point of the shoulder. Tailors measure the length of the upper limb from the acromion. The acromion articulates with the acromial extremity of the clavicle to form the *acromioclavicular joint*. Inferior to the acromion is a shallow depression called the *glenoid cavity*. This cavity accepts the head of the humerus (arm bone) to form the *glenohumeral joint* (see Figure 8.1).

The thin edge of the bone near the vertebral column is the *medial (vertebral) border*. The thick edge closer to the arm is the *lateral (axillary) border*. The medial and lateral borders join at the *inferior angle*. The superior edge of the scapula, called the *superior border,* joins the vertebral border at the *superior angle*. The *scapular notch* is a prominent indentation along the superior border through which the suprascapular nerve passes.

At the lateral end of the superior border of the scapula is a projection of the anterior surface called the *coracoid process* (KOR-a-koyd; = like a crow's beak), to which the tendons of muscles attach. Above and below the spine are two fossae: the *supraspinous fossa* (soo-pra-SPĪ-nus) and the *infraspinous fossa* (in-fra-SPĪ-nus), respectively. Both serve as surfaces of attachment for the tendons of shoulder muscles called the supraspinatus and infraspinatus muscles. On the anterior surface is a slightly hollowed-out area called the *subscapular fossa,* also a surface of attachment for the tendons of shoulder muscles.

1. Which bones or parts of bones of the pectoral girdle form the sternoclavicular, acromioclavicular, and glenohumeral joints?

UPPER LIMB (EXTREMITY)

OBJECTIVE

• *Identify the bones of the upper limb and their principal markings.*

The **upper limbs (extremities)** consist of 60 bones. Each upper limb includes the humerus of the arm; the ulna and radius of the forearm; and the carpals of the carpus (wrist), the metacarpals of the metacarpus (palm), and the phalanges (bones of the digits) of the hand (Figure 8.4).

Humerus

The **humerus** (HYOO-mer-us), or arm bone, is the longest and largest bone of the upper limb (Figure 8.5). It articulates proximally with the scapula and distally at the elbow with both the ulna and the radius.

The proximal end of the humerus features a rounded *head* that articulates with the glenoid cavity of the scapula to form the *glenohumeral joint.* Distal to the head is the

Figure 8.4 Right upper limb.

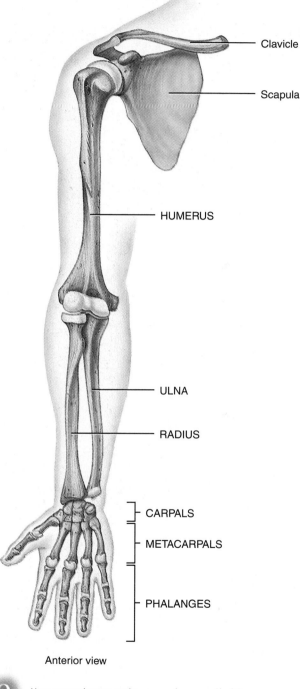

The upper limb consists of a humerus, an ulna, a radius, carpals, metacarpals, and phalanges.

— Clavicle

— Scapula

— HUMERUS

— ULNA

— RADIUS

CARPALS

METACARPALS

PHALANGES

Anterior view

Q   How many bones make up each upper limb?

anatomical neck, the site of the epiphyseal line, which is visible as an oblique groove. The *greater tubercle* is a lateral projection distal to the anatomical neck. It is the most laterally

Figure 8.5 Right humerus in relation to the scapula, ulna, and radius. (See Tortora, *A Photographic Atlas of the Human Body,* Figure 3.23.)

The humerus is the longest and largest bone of the upper limb.

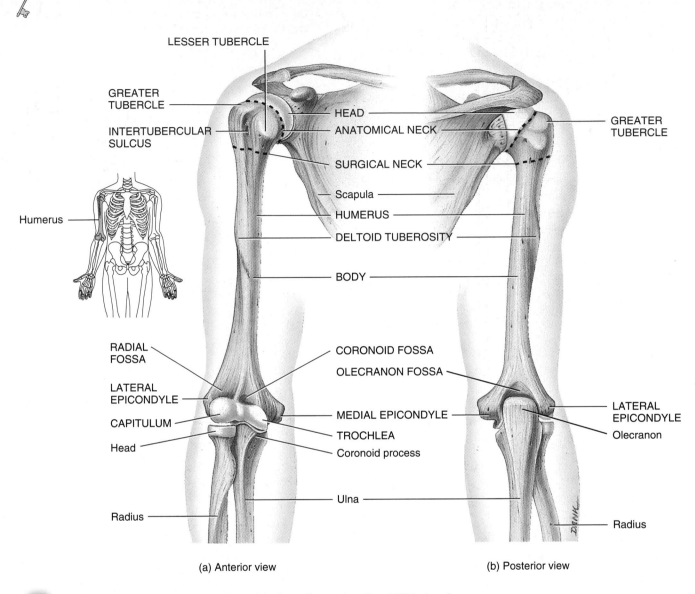

LESSER TUBERCLE

GREATER TUBERCLE

INTERTUBERCULAR SULCUS

Humerus

HEAD

ANATOMICAL NECK

SURGICAL NECK

Scapula

HUMERUS

DELTOID TUBEROSITY

BODY

GREATER TUBERCLE

RADIAL FOSSA

LATERAL EPICONDYLE

CAPITULUM

Head

Radius

CORONOID FOSSA

OLECRANON FOSSA

MEDIAL EPICONDYLE

TROCHLEA

Coronoid process

Ulna

LATERAL EPICONDYLE

Olecranon

Radius

(a) Anterior view

(b) Posterior view

Q Which parts of the humerus articulate with the radius at the elbow? With the ulna at the elbow?

palpable bony landmark of the shoulder region. The *lesser tubercle* projects anteriorly. Between both tubercles runs an *intertubercular sulcus.* The *surgical neck* is a constriction in the humerus just distal to the tubercles, where the head tapers to the shaft; it is so named because fractures often occur here.

The *body (shaft)* of the humerus is roughly cylindrical at its proximal end, but it gradually becomes triangular until it is flattened and broad at its distal end. Laterally, at the middle portion of the shaft, there is a roughened, V-shaped area called the *deltoid tuberosity.* This area serves as a point of attachment for the tendons of the deltoid muscle.

Several prominent features are evident at the distal end of the humerus. The *capitulum* (ka-PIT-yoo-lum; *capit-* = head) is a rounded knob on the lateral aspect of the bone that articulates with the head of the radius. The *radial fossa* is an

anterior depression that receives the head of the radius when the forearm is flexed (bent). The *trochlea* (TRŌK-lē-a), located medial to the capitulum, is a spool-shaped surface that articulates with the ulna. The *coronoid fossa* (KOR-o-noyd; = crown-shaped) is an anterior depression that receives the coronoid process of the ulna when the forearm is flexed. The *olecranon fossa* (ō-LEK-ra-non; = elbow) is a posterior depression that receives the olecranon of the ulna when the forearm is extended (straightened). The *medial epicondyle* and *lateral epicondyle* are rough projections on either side of the distal end to which the tendons of most muscles of the forearm are attached. The ulnar nerve lies on the posterior surface of the medial epicondyle and may easily be palpated by rolling a finger over the skin surface above the medial epicondyle.

Figure 8.6 Right ulna and radius in relation to the humerus and carpals. (See Tortora, *A Photographic Atlas of the Human Body,* Figure 3.24.)

In the forearm, the longer ulna is on the medial side, whereas the shorter radius is on the lateral side.

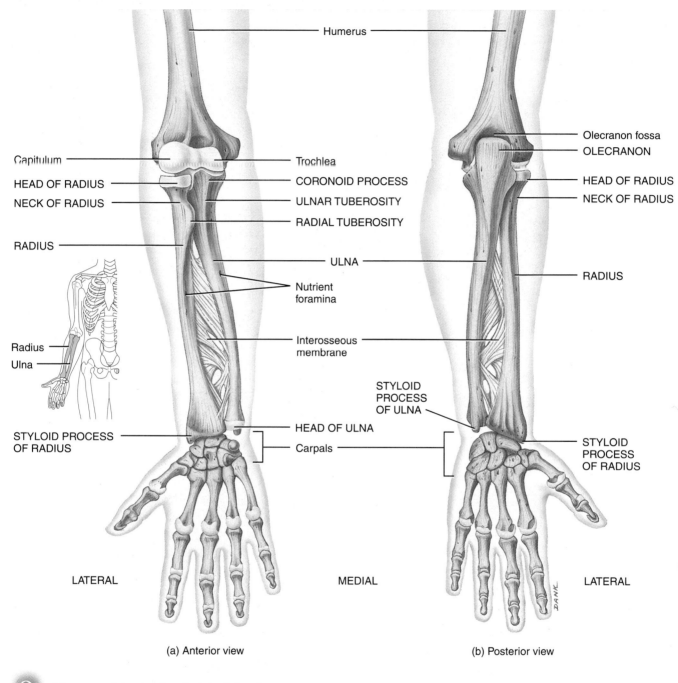

(a) Anterior view   (b) Posterior view

Q What part of the ulna is called the "elbow"?

Ulna and Radius

The **ulna** is located on the medial aspect (the little-finger side) of the forearm and is longer than the radius (Figure 8.6). It is connected to the radius by a broad, flat, fibrous connective tissue called the **interosseous membrane** (in-ter-OS-ē-us; *inter-* = between, *osse-* = bone). This membrane also provides a site of attachment for some tendons of

deep skeletal muscles of the forearm. At the proximal end of the ulna (Figure 8.6b) is the *olecranon* or *olecranon process,* which forms the prominence of the elbow. The *coronoid process* (Figure 8.6a) is an anterior projection that, together with the olecranon, receives the trochlea of the humerus. The *trochlear notch* is a large curved area between the olecranon and coronoid process that forms part of the elbow joint

Figure 8.7 Articulations formed by the ulna and radius. (a) Elbow joint. (b) Joint surfaces at proximal end of the ulna. (c) Joint surfaces at distal ends of radius and ulna.

The elbow joint is formed by two articulations: (1) the trochlear notch of the ulna with the trochlea of the humerus and (2) the head of the radius with the capitulum of the humerus.

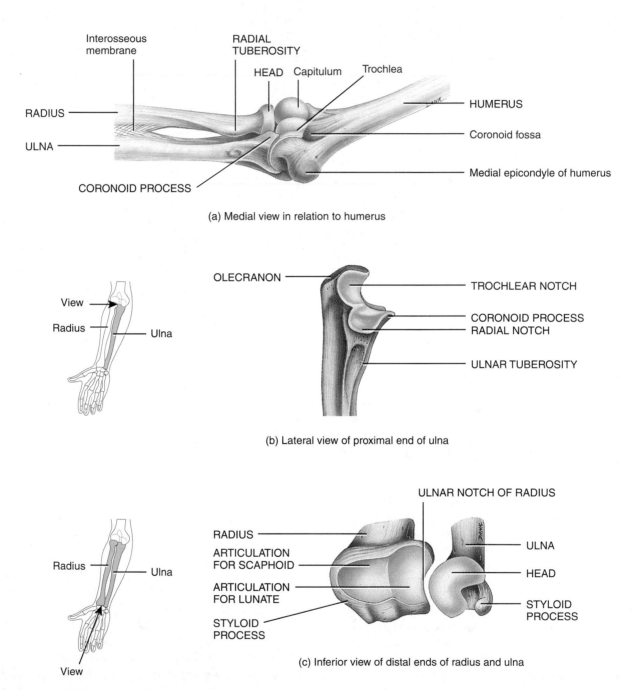

(a) Medial view in relation to humerus

(b) Lateral view of proximal end of ulna

(c) Inferior view of distal ends of radius and ulna

 How many joints are formed between the radius and ulna? What are they called?

(see Figure 8.7b). Just inferior to the coronoid process is the *ulnar tuberosity.* The distal end of the ulna consists of a *head* that is separated from the wrist by a fibrocartilage disc. A *styloid process* (*stylo-* = stake or pole) is on the posterior side of the distal end.

The **radius** is located on the lateral aspect (thumb side) of the forearm (see Figure 8.6). The proximal end of the radius has a disc-shaped *head* that articulates with the capitulum of the humerus and the radial notch of the ulna. Inferior to the head is the constricted *neck.* A roughened area inferior to the neck on the medial side, called the *radial tuberosity,* is a point of attachment for the tendons of the biceps brachii muscle. The shaft of the radius widens distally to form a *styloid process* on the lateral side.

The *elbow joint* is where the ulna and radius articulate with the humerus. This occurs in two places: where the head of the radius articulates with the capitulum of the humerus (Figure 8.7a), and where the trochlea of the humerus is received by the trochlear notch of the ulna (Figure 8.7b).

The ulna and radius also articulate with one another in two places. Proximally, the head of the radius articulates with the ulna's *radial notch,* a depression that is lateral and inferior to the trochlear notch (Figure 8.7b). This articulation is the *proximal radioulnar joint.* Distally, the head of the ulna articulates with the *ulnar notch* of the radius (Figure 8.7c). This articulation is the *distal radioulnar joint.* Finally, the distal end of the radius articulates with three bones of the wrist—the lunate, the scaphoid, and the triquetrum—to form the *radiocarpal (wrist) joint.*

Carpals, Metacarpals, and Phalanges

The **carpus** (wrist) is the proximal region of the hand and consists of eight small bones, the **carpals,** joined to one another by ligaments (Figure 8.8). Articulations between carpal bones are called *intercarpal joints.* The carpals are arranged in two transverse rows of four bones each. Their names reflect their shapes. The carpals in the proximal row, from lateral to medial, are the **scaphoid** (*scaph-* = boat), **lunate** (*lun-* = moon), **triquetrum** (*tri-* = three; *-quetrum* = corners), and **pisiform** (*pisi-* = pea-shaped). The carpals in the distal row, from lateral to medial, are the **trapezium** (*trapez-* = table), **trapezoid, capitate,** and **hamate** (*ham-* = hooked). The capitate is the largest carpal bone; its rounded projection, the head, articulates with the lunate. The hamate is named for a large hook-shaped projection on its anterior surface. In about 70% of carpal fractures, only the scaphoid is broken. This is because the force of a fall on an outstretched hand is transmitted from the capitate through the scaphoid to the radius.

The concave space formed by the pisiform and hamate (on the ulnar side), and the scaphoid and trapezium (on the radial side), plus the *flexor retinaculum* (deep fascia) is the **carpal tunnel.** The long flexor tendons of the digits and thumb and the median nerve pass through the carpal tunnel. Narrowing of the carpal tunnel may give rise to a condition called carpal tunnel syndrome (described on page 350).

The **metacarpus** (*meta-* = beyond), or palm, is the intermediate region of the hand and consists of five bones called **metacarpals.** Each metacarpal bone consists of a proximal *base,* an intermediate *shaft,* and a distal *head* (Figure 8.8b). The metacarpal bones are numbered I to V (or 1–5), starting with the one proximal to the thumb. The bases articulate with the distal row of carpal bones to form the *carpometacarpal joints.* The heads articulate with the proximal phalanges to form the *metacarpophalangeal joints.* The heads of the metacarpals are commonly called "knuckles" and are readily visible in a clenched fist.

The **phalanges** (fa-LAN-jēz, *phalan-* = a battle line), or bones of the digits, make up the distal region of the hand. There are 14 phalanges in the five digits of each hand and, like the metacarpals, the digits are numbered I to V (or 1–5), beginning with the thumb. A single bone of a digit is referred to as a **phalanx** (FĀ-lanks). Each phalanx consists of a proximal *base,* an intermediate *shaft,* and a distal *head.* There are two phalanges in the thumb (*pollex*), and three phalanges in each of the other four digits. In order from the thumb, these other four digits are commonly referred to as the index finger, middle finger, ring finger, and little finger. The first row of phalanges, the *proximal row,* articulates with the metacarpal bones and second row of phalanges. The second row of phalanges, the *middle row,* articulates with the proximal row and the third row. The third row of phalanges, the *distal row,* articulates with the middle row. The thumb has no middle phalanx. Joints between phalanges are called *interphalangeal joints.*

1. Name the bones that form the upper limb, from proximal to distal.
2. Describe the joints the bones of the upper limb form with one another.

PELVIC (HIP) GIRDLE

OBJECTIVE

• *Identify the bones of the pelvic girdle and their principal markings.*

The **pelvic (hip) girdle** consists of the two **hip bones,** also called **coxal bones** (KOK-sal; *cox-* = hip) (Figure 8.9). The hip bones are united to each other anteriorly at a joint called the **pubic symphysis** (PYOO-bik SIM-fi-sis). They unite posteriorly with the sacrum at the *sacroiliac joints.* The complete ring composed of the hip bones, pubic symphysis, and sacrum forms a deep, basinlike structure called the **bony pelvis** (*pelv-* = basin). Functionally, the bony pelvis provides a strong and stable support for the vertebral column and pelvic viscera. The pelvic girdle of the bony pelvis also accepts the bones of the lower limbs, connecting them to the axial skeleton.

Each of the two hip bones of a newborn consists of three bones separated by cartilage: a superior *ilium,* an inferior

Figure 8.8 Right wrist and hand in relation to the ulna and radius. (See Tortora, *A Photographic Atlas of the Human Body,* Figure 3.25)

 The skeleton of the hand consists of the proximal carpals, the intermediate metacarpals, and the distal phalanges.

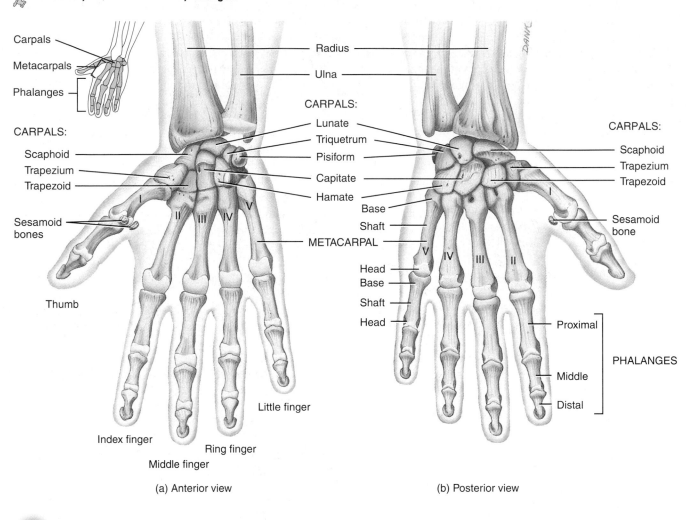

(a) Anterior view

(b) Posterior view

Q Which is the most frequently fractured wrist bone?

and anterior *pubis,* and an inferior and posterior *ischium* (IS-kē-um). Eventually, the three separate bones fuse together (Figure 8.10a). Although the hip bones function as single bones, anatomists commonly discuss them as if they still consisted of three bones.

Ilium

The **ilium** (= flank) is the largest of the three components of the hip bone (Figure 8.10b, c). It is divided into a superior *ala* (= wing) and an inferior *body,* which enters into the formation of the *acetabulum,* the socket for the head of the femur. Its superior border, the *iliac crest,* ends anteriorly in a blunt *anterior superior iliac spine.* Below this spine is the *anterior inferior iliac spine.* Posteriorly, the iliac crest ends in

a sharp *posterior superior iliac spine.* Below this spine is the *posterior inferior iliac spine.* The spines serve as points of attachment for the tendons of the muscles of the trunk, hip, and thighs. Below the posterior inferior iliac spine is the *greater sciatic notch* (sī-AT-ik), through which the sciatic nerve, the longest nerve in the body, passes. The medial surface of the ilium contains the *iliac fossa,* a concavity where the tendon of the iliacus muscle attaches. Posterior to this fossa are the *iliac tuberosity,* a point of attachment for the sacroiliac ligament, and the *auricular surface (auricular =* ear-shaped), which articulates with the sacrum to form the *sacroiliac joint* (see Figure 8.9). Projecting anteriorly and superiorly from the auricular surface is a ridge called the *arcu-ate line* (AR-kyoo-āt; *arc-* = bow). The other conspicuous

Figure 8.9 Bony pelvis. Shown here is the female bony pelvis. (See Tortora, *A Photographic Atlas of the Human Body,* Figure 3.27)

🔑 **The hip bones are united anteriorly at the pubic symphysis and posteriorly at the sacrum to form the bony pelvis.**

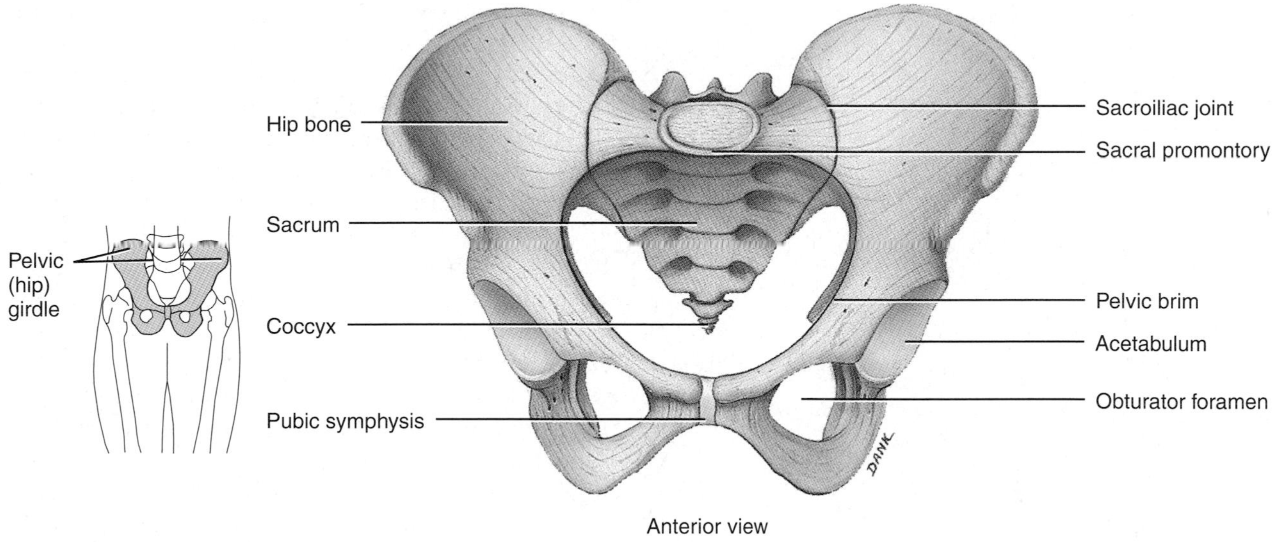

Anterior view

Q What are the functions of the bony pelvis?

markings of the ilium are three arched lines on its lateral surface called the *posterior gluteal line* (*glut-* = buttock), the *anterior gluteal line,* and the *inferior gluteal line.* The tendons of the gluteal muscles attach to the ilium between these lines.

Ischium

The **ischium** (*ischi-* = hip) is the inferior, posterior portion of the hip bone (see Figure 8.10b, c). It is composed of a superior *body* and inferior *ramus* (*ram-* = branch), which joins the pubis. The ischium contains the prominent *ischial spine,* a *lesser sciatic notch* below the spine, and a rough and thickened *ischial tuberosity.* This prominent tuberosity may hurt someone's thigh when you sit on their lap. Together, the ramus and the pubis surround the *obturator foramen* (OB-too-rā-ter; *obtur-* = closed up), the largest foramen in the skeleton. The foramen is so named because, even though blood vessels and nerves pass through it, it is nearly completely closed by the fibrous *obturator membrane.*

Pubis

The **pubis** or **os pubis,** meaning pubic bone, is the anterior and inferior part of the hip bone (Figure 8.10b, c). It consists of a *superior ramus,* an *inferior ramus,* and a *body* between the rami (plural) that contributes to the formation of the pubic symphysis. The anterior border of the body is known as the *pubic crest,* and at its lateral end is a projection, the *pubic tubercle.* This tubercle is the beginning of a raised

line, the *iliopectineal line* (il-ē-ō-pek-TIN-ē-al), which extends superiorly and laterally along the superior ramus to merge with the arcuate line of the ilium. These lines, as you will see shortly, are important landmarks for distinguishing the superior and inferior portions of the bony pelvis.

The pubic symphysis is the joint between the two hip bones (see Figure 8.9). It consists of a disc of fibrocartilage. The pubic arch is formed by the convergence of the inferior rami of the two pubes (plural) (see Table 8.1). The *acetabulum* (as-e-TAB-yoo-lum; = vinegar cup) is the deep fossa formed by the ilium, ischium, and pubis. It is the socket that accepts the rounded head of the femur. Together, the acetabulum and the femoral head form the *hip (coxal) joint.* On the inferior portion of the acetabulum is a deep indentation, the *acetabular notch.* It forms a foramen through which nutrient vessels and nerves enter the joint, and it serves as a point of attachment for ligaments of the femur (for example, the ligament of the head of the femur).

True and False Pelves

The bony pelvis is divided into superior and inferior portions by a boundary called the *pelvic brim* (Figure 8.11a). You can trace the pelvic brim by following the landmarks around parts of the hip bones to form the outline of an oblique plane. Beginning posteriorly at the *sacral promontory* of the sacrum, trace laterally and inferiorly along the *arcuate lines* of the ilium of the hip bones. Continue inferiorly along

Figure 8.10 Right hip bone. The lines of fusion of the ilium, ischium, and pubis depicted in (a) are not always visible in an adult. (See Tortora, *A Photographic Atlas of the Human Body,* Figure 3.26.)

🔑 **The acetabulum is the socket formed where the three parts of the hip bone converge.**

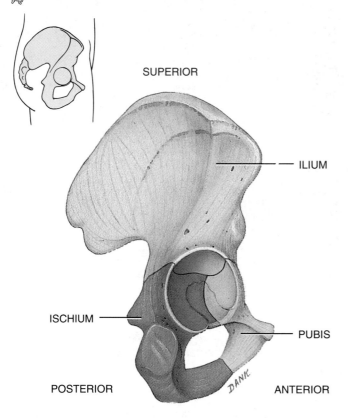

the *iliopectineal lines* of the pubis. Finally, trace anteriorly to the superior portion of the *pubic symphysis.* Together, these points form an oblique plane that is higher in the back than in the front. The circumference of this plane is the pelvic brim.

The portion of the bony pelvis above the pelvic brim is the **false (greater) pelvis** (see Figure 8.11b). It is bordered by the lumbar vertebrae posteriorly, the upper portions of the hip bones laterally, and the abdominal wall anteriorly. The space enclosed by the false pelvis is part of the abdomen; it does not contain pelvic organs, except for the urinary bladder (when it is full) and the uterus during pregnancy.

The portion of the bony pelvis below the pelvic brim is the **true (lesser) pelvis** (Figure 8.11b). It is bounded by the sacrum and coccyx posteriorly, inferior portions of the ilium and ischium laterally, and the pubic bones anteriorly. The true pelvis surrounds the pelvic cavity (see Figure 1.8 on page 13.). The superior opening of the true pelvis is the pelvic brim, also called the *pelvic inlet;* the inferior opening of the true pelvis is the *pelvic outlet.* The *pelvic axis* is an imaginary line that curves through the true pelvis and joins the central points of the planes of the pelvic inlet and outlet. During childbirth the pelvic axis is the route taken by the baby's head as it descends through the pelvis.

| **1.** Distinguish between the true and false pelves.

Q Which part of the hip bone articulates with the femur? With the sacrum?

(a) Lateral view showing portions of hip bone

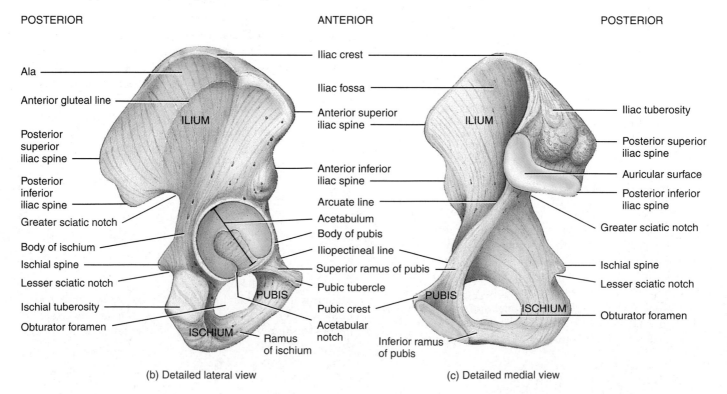

(b) Detailed lateral view

(c) Detailed medial view

Figure 8.11 True and false pelves. Shown here is the female pelvis. For simplicity, the landmarks of the pelvic brim are shown only on the left side of the body, and the outline of the pelvic brim is shown only on the right side. The entire pelvic brim is shown in Figure 8.9.

🔑 **The true and false pelves are separated by the pelvic brim.**

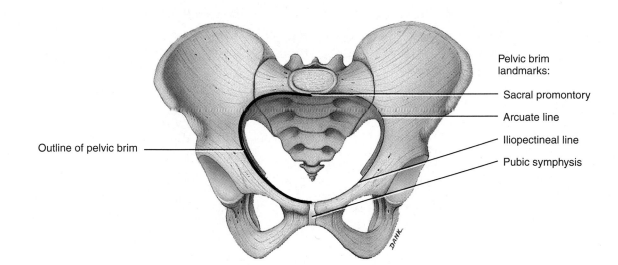

Pelvic brim landmarks:

Sacral promontory

Arcuate line

Iliopectineal line

Pubic symphysis

Outline of pelvic brim

(a) Anterior view of borders of pelvic brim

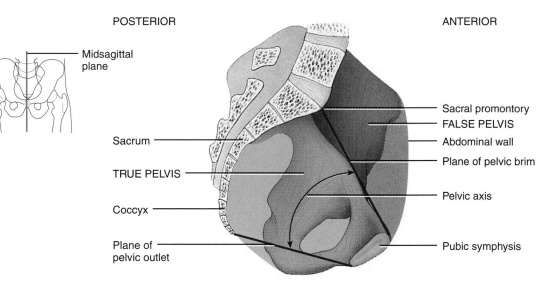

POSTERIOR

ANTERIOR

Midsagittal plane

Sacrum

TRUE PELVIS

Coccyx

Plane of pelvic outlet

Sacral promontory
FALSE PELVIS
Abdominal wall
Plane of pelvic brim
Pelvic axis
Pubic symphysis

(b) Midsagittal section indicating locations of true and false pelves

Q What is the significance of the pelvic axis?

Table 8.1 Comparison of Female and Male Pelves

POINT OF COMPARISON	FEMALE	MALE
General structure	Light and thin.	Heavy and thick.
False (greater) pelvis	Shallow.	Deep.
Pelvic brim (inlet)	Larger and more oval.	Smaller and heart-shaped.
Acetabulum	Small and faces anteriorly.	Large and faces laterally.
Obturator foramen	Oval.	Round.
Pubic arch	Greater than 90° angle.	Less than 90° angle.

Pubic arch (greater than 90°)   Pubic arch (less than 90°)

Anterior views

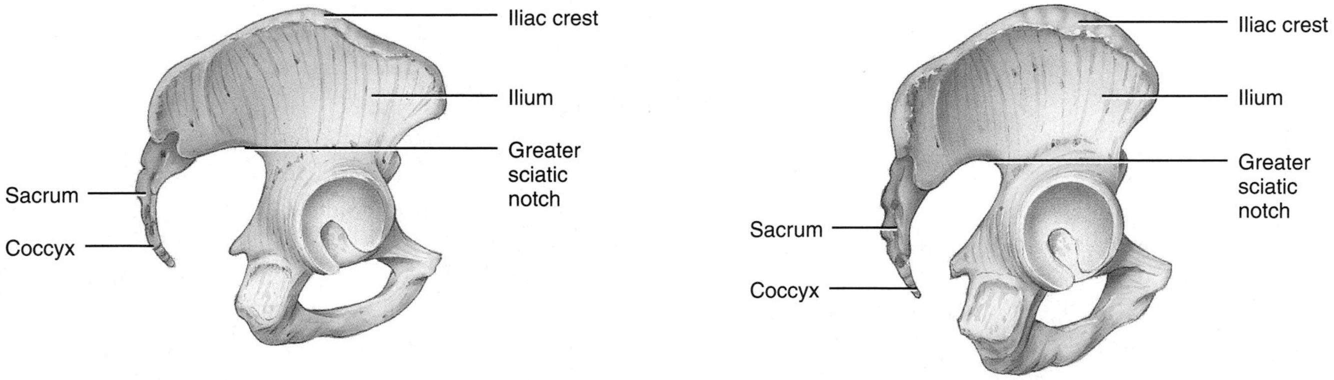

Iliac crest	Less curved.	More curved.
Ilium	Less vertical.	More vertical.
Greater sciatic notch	Wide.	Narrow.
Coccyx	More movable and more curved anteriorly.	Less movable and less curved anteriorly.
Sacrum	Short, wide (see anterior views), and more curved anteriorly.	Long, narrow (see anterior views), and less curved anteriorly.

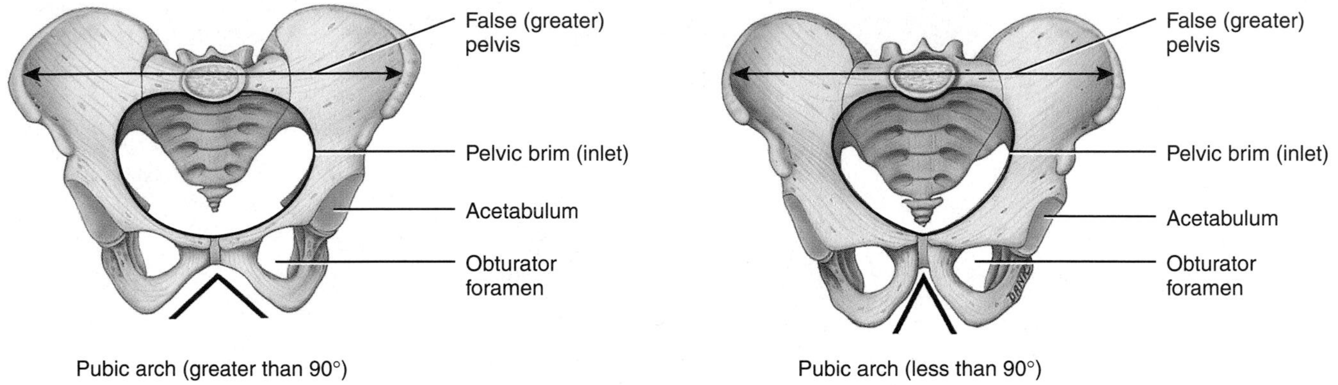

Right lateral views

Table 8.1 (continued)

POINT OF COMPARISON	FEMALE	MALE
Pelvic outlet	Wider.	Narrower.
Ischial tuberosity	Longer, closer together, and more medially projecting.	Shorter, farther apart, and more laterally projecting.

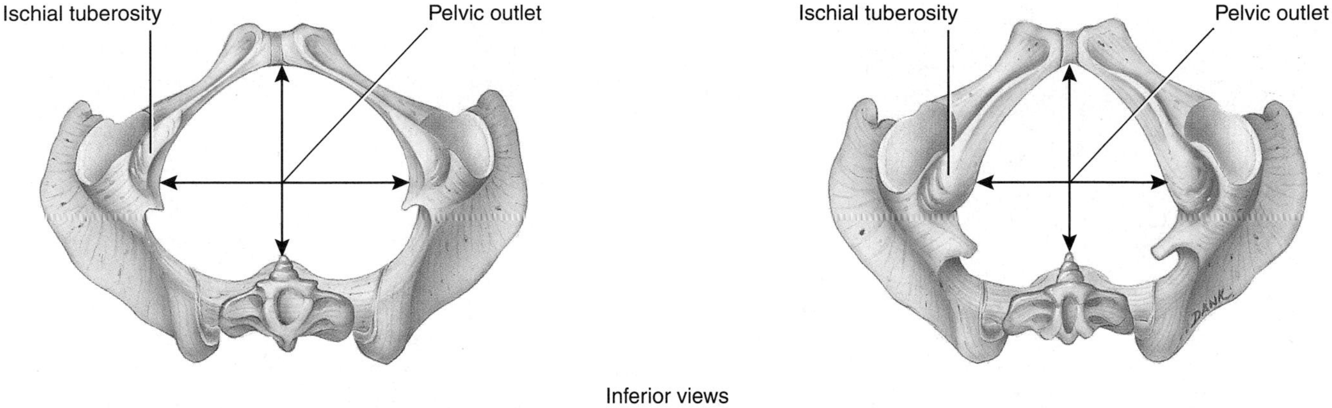

Inferior views

COMPARISON OF FEMALE AND MALE PELVES

OBJECTIVE

• *Compare the principal structural differences between female and male pelves.*

Generally, the bones of a male are larger and heavier than those of a female and possess larger surface markings. Gender-related differences in the features of skeletal bones are readily apparent when comparing the female and male pelves. Most of the structural differences in the pelves are adaptations to the requirements of pregnancy and childbirth. The female's pelvis is wider and shallower than the male's (Table 8.1). Consequently, there is more space in the true pelvis of the female, especially in the pelvic inlet and pelvic outlet, which accommodate the passage of the infant's head at birth. Other significant structural differences between pelves of females and males are listed and illustrated in Table 8.1.

1. Why are structural differences between female and male pelves important?

COMPARISON OF PECTORAL AND PELVIC GIRDLES

OBJECTIVE

• *Describe the differences in the pectoral and pelvic girdles.*

Now that we've studied the structures of the pectoral and pelvic girdles, we can note some of their significant differences. The pectoral girdle does not directly articulate with the vertebral column, whereas the pelvic girdle directly artic-

ulates with the vertebral column via the sacroiliac joint. The sockets (glenoid fossae) for the upper limbs in the pectoral girdle are shallow and maximize movement, whereas the sockets (acetabula) for the lower limbs in the pelvic girdle are deep and allow less movement. Overall, the structure of the pectoral girdle offers more mobility than strength, whereas that of the pelvic girdle offers more strength than mobility.

LOWER LIMB (EXTREMITY)

OBJECTIVE

• *Identify the bones of the lower limb and their principal markings.*

The two **lower limbs (extremities)** are each composed of 30 bones. Each lower limb includes the femur of the thigh; the patella (kneecap); the tibia and fibula of the leg; and the tarsals of the tarsus (ankle), the metatarsals of the metatarsus, and the phalanges (bones of the digits) of the foot (Figure 8.12).

Femur

The **femur,** or thighbone, is the longest, heaviest, and strongest bone in the body (Figure 8.13). Its proximal end articulates with the acetabulum of the hip bone. Its distal end articulates with the tibia and patella. The *body (shaft)* of the femur angles medially and, as a result, the knee joints are brought nearer to the midline. The angle of convergence is greater in females because the female pelvis is broader.

The proximal end of the femur consists of a rounded *head* that articulates with the acetabulum of the hip bone to

Figure 8.12 Right lower limb.

🔑 **The lower limb consists of a femur, patella (kneecap), tibia, fibula, tarsals (ankle bones), metatarsals, and phalanges (bones of the digits).**

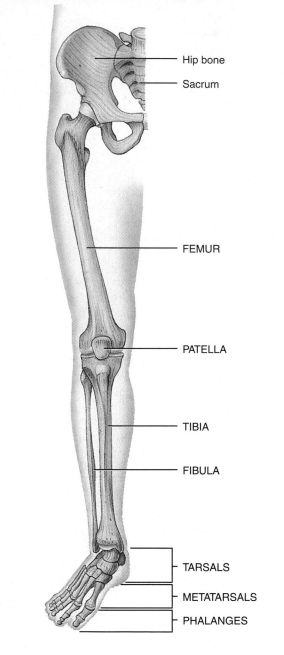

Hip bone
Sacrum
FEMUR
PATELLA
TIBIA
FIBULA
TARSALS
METATARSALS
PHALANGES

Anterior view

Q How many bones make up each lower limb?

form the *hip (coxal) joint.* The ligament of the head of the femur connects the femur to the acetabulum of the hip bone. The *neck* of the femur is a constricted region distal to the head. The *greater trochanter* (trō-KAN-ter) and *lesser trochanter* are projections that serve as points of attachment for the tendons of some of the thigh and buttock muscles. The greater trochanter is the prominence felt and seen anterior to the hollow on the side of the hip. It is a landmark commonly used to locate the site for intramuscular injections into the lateral surface of the thigh. The lesser trochanter is inferior and medial to the greater trochanter. Between the anterior surface of the trochanters is a narrow *intertrochanteric line* (Figure 8.13a); between the posterior surface of the trochanters is an *intertrochanteric crest* (Figure 8.13b).

Inferior to the intertrochanteric crest on the posterior surface of the body of the femur is a vertical ridge called the *gluteal tuberosity.* It blends into another vertical ridge called the *linea aspera* (LIN-ē-a AS-per-a; *asper-* = rough). Both ridges serve as attachment points for the tendons of several thigh muscles.

The distal end of the femur is expanded and includes the *medial condyle* and the *lateral condyle.* These articulate with the medial and lateral condyles of the tibia. Superior to the condyles are the *medial epicondyle* and the *lateral epicondyle.* A depressed area between the condyles on the posterior surface is called the *intercondylar fossa* (in-ter-KON-di-lar). The *patellar surface* is located between the condyles on the anterior surface.

Patella

The **patella** (= little dish), or kneecap, is a small, triangular bone located anterior to the knee joint (Figure 8.14). It is a sesamoid bone that develops in the tendon of the quadriceps femoris muscle. The broad superior end of the patella is called the *base.* The pointed inferior end is the *apex.* The posterior surface contains two *articular facets,* one for the medial condyle and the other for the lateral condyle of the femur. The patellar ligament attaches the patella to the tibial tuberosity. The *patellofemoral joint,* between the posterior surface of the patella and the patellar surface of the femur, is the intermediate component of the *tibiofemoral (knee) joint.* The patella functions to increase the leverage of the tendon of the quadriceps femoris muscle, to maintain the position of the tendon when the knee is bent (flexed), and to protect the knee joint.

CLINICAL APPLICATION
Patellofemoral Stress Syndrome

Patellofemoral stress syndrome ("runner's knee") is one of the most common problems in runners. During normal flexion and extension of the knee, the patella tracks

Figure 8.13 Right femur in relation to the hip bone, patella, tibia, and fibula. (See Tortora, *A Photographic Atlas of the Human Body,* Figure 3.28.)

The acetabulum of the hip bone and head of the femur articulate to form the hip joint.

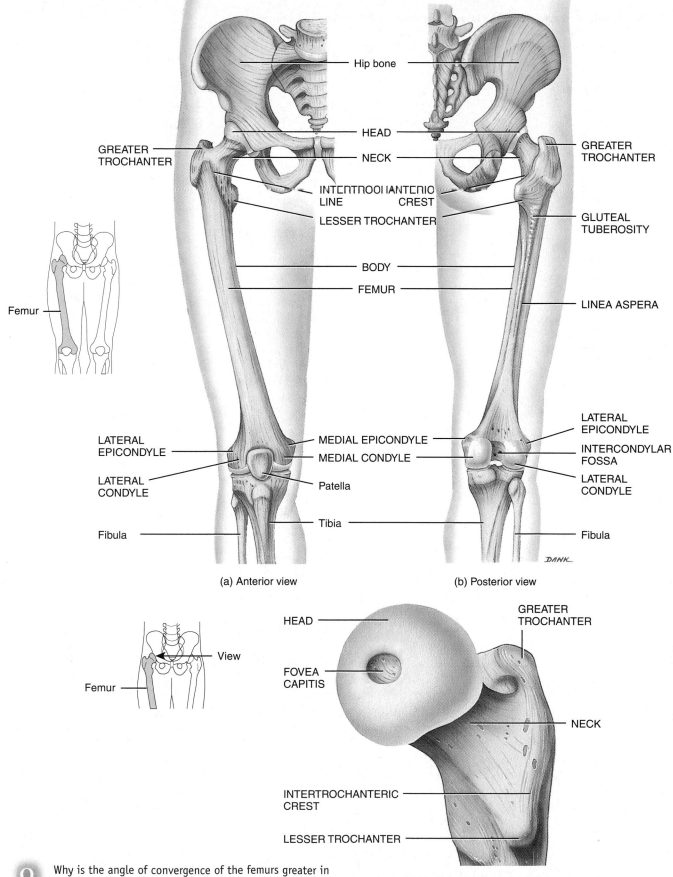

Hip bone

HEAD

NECK

GREATER TROCHANTER

INTERTROCHANTERIC LINE

ANTERIOR CREST

LESSER TROCHANTER

GREATER TROCHANTER

GLUTEAL TUBEROSITY

BODY

FEMUR

LINEA ASPERA

Femur

LATERAL EPICONDYLE

LATERAL CONDYLE

Fibula

MEDIAL EPICONDYLE

MEDIAL CONDYLE

Patella

Tibia

LATERAL EPICONDYLE

INTERCONDYLAR FOSSA

LATERAL CONDYLE

Fibula

DANK

(a) Anterior view

(b) Posterior view

View

Femur

HEAD

FOVEA CAPITIS

GREATER TROCHANTER

NECK

INTERTROCHANTERIC CREST

LESSER TROCHANTER

Q Why is the angle of convergence of the femurs greater in females than males?

(c) Medial view of proximal end of femur

Figure 8.14 Right patella.

🔑 **The patella articulates with the lateral and medial condyles of the femur.**

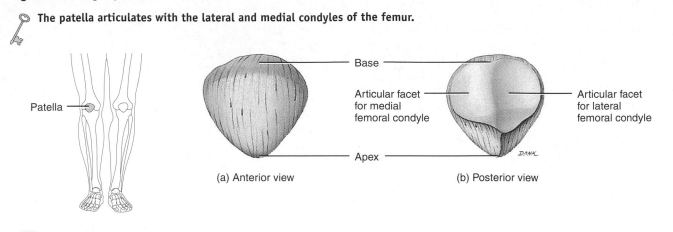

(a) Anterior view

(b) Posterior view

Q Because the patella develops in the tendon of a muscle, it is classified as which type of bone?

(glides) up and down in the groove between the femoral condyles. In patellofemoral stress syndrome, normal tracking does not occur; instead, the patella also tracks laterally, and the increased pressure on the joint causes aching or tenderness around or under the patella. The pain typically occurs after a person has been sitting for a while, especially after exercise. It is worsened by squatting or walking down stairs. One cause of runner's knee is constantly walking, running, or jogging on the same side of the road. Because roads slope down on the sides, the knee that is closer to the center of the road endures greater mechanical stress because it does not fully extend during a stride. Other predisposing factors include having knock-knees, running on hills, and running long distances. ■

Tibia and Fibula

The **tibia,** or shin bone, is the larger, medial, weight-bearing bone of the leg (Figure 8.15). The tibia articulates at its proximal end with the femur and fibula, and at its distal end with the fibula and the talus bone of the ankle. The tibia and fibula, like the ulna and radius, are connected by an interosseous membrane.

The proximal end of the tibia is expanded into a *lateral condyle* and a *medial condyle*. These articulate with the condyles of the femur to form the lateral and medial *tibiofemoral (knee) joints.* The inferior surface of the lateral condyle articulates with the head of the fibula. The slightly concave condyles are separated by an upward projection called the *intercondylar eminence* (Figure 8.15b). The *tibial tuberosity* on the anterior surface is a point of attachment for the patellar ligament. Inferior to and continuous with the tibial tuberosity is a sharp ridge that can be felt below the skin and is known as the *anterior border (crest).*

The medial surface of the distal end of the tibia forms the *medial malleolus* (mal-LĒ-ō-lus; = hammer). This structure articulates with the talus bone of the ankle and forms

the prominence that can be felt on the medial surface of the ankle. The *fibular notch* (Figure 8.15c) articulates with the distal end of the fibula to form the *distal tibiofibular joint.*

The **fibula** is parallel and lateral to the tibia, but it is considerably smaller than the tibia. The *head* of the fibula, the proximal end, articulates with the inferior surface of the lateral condyle of the tibia below the level of the knee joint to form the *proximal tibiofibular joint.* The distal end has a projection called the *lateral malleolus* that articulates with the talus bone of the ankle. This forms the prominence on the lateral surface of the ankle. As noted, the fibula also articulates with the tibia at the fibular notch.

Tarsals, Metatarsals, and Phalanges

The **tarsus** (ankle) is the proximal region of the foot and consists of seven **tarsal bones** (Figure 8.16). They include the **talus** (TĀ-lus; = ankle bone) and **calcaneus** (kal-KĀ-nē-us; = heel), located in the posterior part of the foot. The calcaneus is the largest and strongest tarsal bone. The anterior tarsal bones are the **cuboid** (= cube-shaped), **navicular** (= like a little boat), and three **cuneiform bones** (= wedge-shaped) called the **first (medial), second (intermediate), and third (lateral) cuneiforms.** Joints between tarsal bones are called *intertarsal joints.* The talus, the most superior tarsal bone, is the only bone of the foot that articulates with the fibula and tibia. It articulates on one side with the medial malleolus of the tibia and on the other side with the lateral malleolus of the fibula. These articulations form the *talocrural (ankle) joint.* During walking, the talus transmits about half the weight of the body to the calcaneus. The remainder is transmitted to the other tarsal bones.

The **metatarsus** is the intermediate region of the foot and consists of five **metatarsal bones** numbered I to V (or 1–5) from the medial to lateral position (Figure 8.16). Like the metacarpals of the palm of the hand, each metatarsal

Figure 8.15 Right tibia and fibula in relation to the femur, patella, and talus.
(See Tortora, *A Photographic Atlas of the Human Body,* Figure 3.30)

🔑 **The tibia articulates with the femur and fibula proximally, and with the fibula and talus distally.**

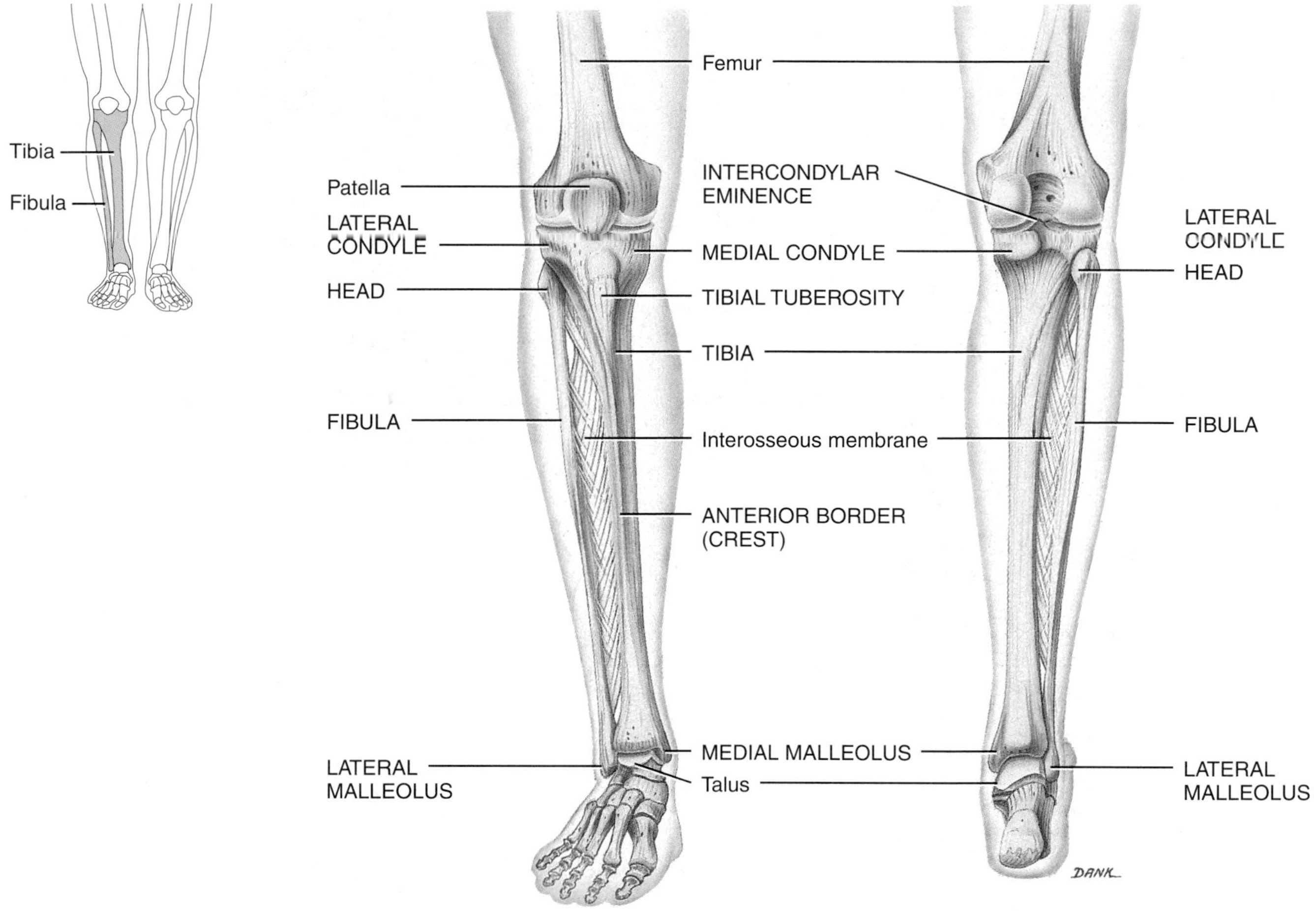

Tibia

Fibula

Femur

Patella

INTERCONDYLAR EMINENCE

LATERAL CONDYLE

MEDIAL CONDYLE

HEAD

TIBIAL TUBEROSITY

TIBIA

FIBULA

Interosseous membrane

ANTERIOR BORDER (CREST)

MEDIAL MALLEOLUS

LATERAL MALLEOLUS

Talus

LATERAL CONDYLE

HEAD

FIBULA

LATERAL MALLEOLUS

(a) Anterior view

(b) Posterior view

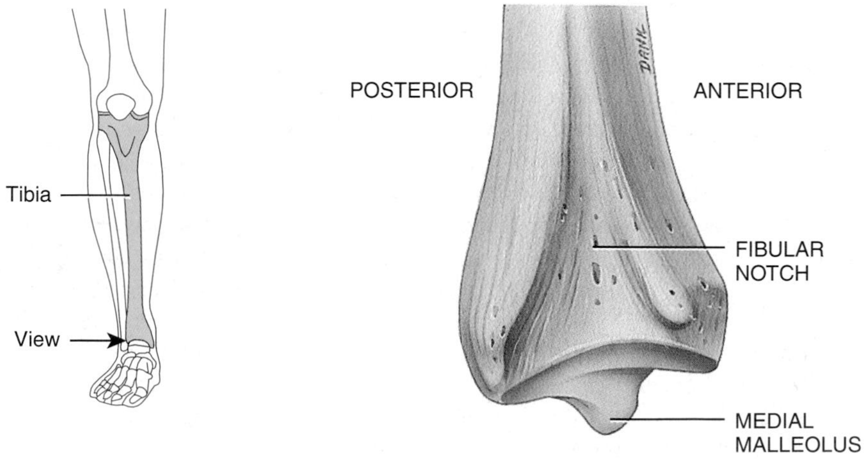

DANK

Tibia

View

POSTERIOR

ANTERIOR

FIBULAR NOTCH

MEDIAL MALLEOLUS

DANK

(c) Lateral view of distal end of tibia

 Which leg bone bears the weight of the body?

Figure 8.16 Right foot. (See Tortora, *A Photographic Atlas of the Human Body,* Figure 3.31.)

 The skeleton of the foot consists of the proximal tarsals, the intermediate metatarsals, and the distal phalanges.

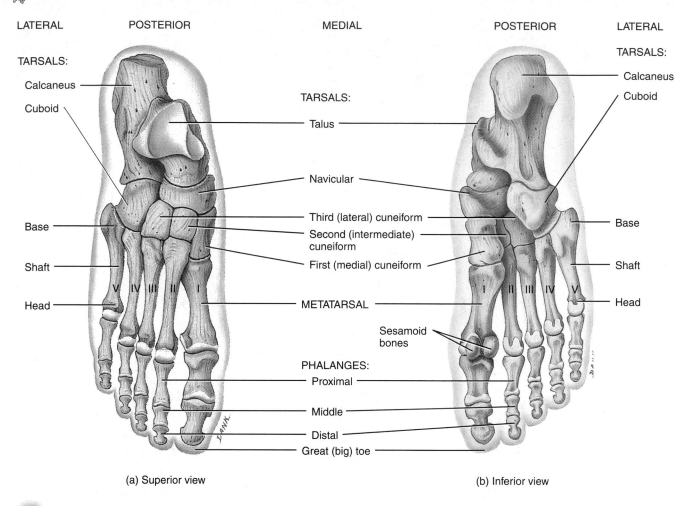

(a) Superior view

(b) Inferior view

Q Which tarsal bone articulates with the tibia and fibula?

consists of a proximal *base,* an intermediate *shaft,* and a distal *head.* The metatarsals articulate proximally with the first, second, and third cuneiform bones and with the cuboid to form the *tarsometatarsal joints.* Distally, they articulate with the proximal row of phalanges to form the *metatarsophalangeal joints.* The first metatarsal is thicker than the others because it bears more weight.

The **phalanges** comprise the distal component of the foot and resemble those of the hand both in number and arrangement. The toes are numbered I to V (or 1–5) beginning with the great toe. Each also consists of a proximal *base,* an intermediate *shaft,* and a distal *head.* The great or big toe (*hallux*) has two large, heavy phalanges called proximal and distal phalanges. The other four toes each have three phalanges—proximal, middle, and distal. Joints between phalanges of the foot, like those of the hand, are called *interphalangeal joints.*

Arches of the Foot

The bones of the foot are arranged in two **arches** (Figure 8.17). The arches enable the foot to support the weight of the body, provide an ideal distribution of body weight over the hard and soft tissues of the foot, and provide leverage when walking. The arches are not rigid; they yield as weight is applied and spring back when the weight is lifted, thus helping to absorb shocks. Usually, the arches are fully developed by the time children reach age 12 or 13.

The *longitudinal arch* has two parts, both of which consist of tarsal and metatarsal bones arranged to form an arch from the anterior to the posterior part of the foot. The *medial part* of the longitudinal arch originates at the calcaneus. It rises to the talus and descends through the navicular, the three cuneiforms, and the heads of the three medial metatarsals. The *lateral part* of the longitudinal arch also begins at the calcaneus. It rises at the cuboid and descends to the heads of the two lateral metatarsals.

Figure 8.17 Arches of the right foot.

🔑 **Arches help the foot support and distribute the weight of the body and provide leverage during walking.**

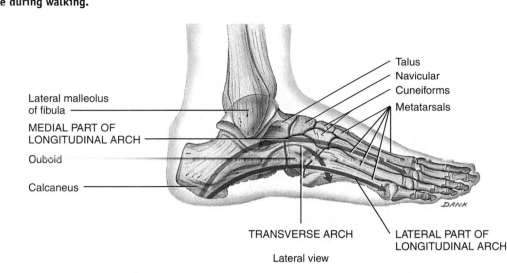

Lateral view

Q   What structural feature of the arches allows them to absorb shocks?

The *transverse arch* is found between the medial and lateral aspects of the foot and is formed by the navicular, three cuneiforms, and the bases of the five metatarsals.

CLINICAL APPLICATION
Flatfoot, Clawfoot, and Clubfoot

The bones composing the arches are held in position by ligaments and tendons. If these ligaments and tendons are weakened, the height of the medial longitudinal arch may decrease or "fall." The result is **flatfoot,** the causes of which include excessive weight, postural abnormalities, weakened supporting tissues, and genetic predisposition. A custom-designed arch support (orthotic) often is prescribed to treat flatfoot.

Clawfoot is a condition in which the medial longitudinal arch is abnormally elevated. It is often caused by muscle deformities, such as may occur in diabetics whose neurological lesions lead to atrophy of muscles of the foot.

Clubfoot or **talipes equinovarus** (*-pes* = foot; *equino-* = horse) is an inherited deformity of the foot that occurs in 1 of every 1000 births. In this malformation, the foot is twisted inferiorly and medially, and the angle of the arch is increased. Treatment consists of manipulating the arch to a normal curvature by casts or adhesive tape, usually soon after birth. Corrective shoes or surgery may also be required. ■

1. Name the bones that form the lower limb, from proximal to distal.
2. Describe the joints the bones of the lower limb form with one another.
3. What are the functions of the arches of the foot?

MEDICAL TERMINOLOGY

Genu valgum (JĒ-noo VAL-gum; *genu-* = knee; *valgum* = bent outward) A decreased space between the knees and increased space between the ankles due to a medial angulation of the legs in relation to the thigh. Also called **knock-knee.**

Genu varum (JĒ-noo VAR-um; *varum* = bent toward the midline) An increased space between the knees due to a lateral angulation of the legs relative to the thigh. Also called **bowleg.**

Hallux valgus (HAL-uks VAL-gus; *hallux* = great toe) Angulation of the great toe away from the midline of the body, typically caused by wearing tightly fitting shoes. Involves lateral deviation of the proximal phalanx of the great toe and medial displacement of metatarsal I. Also called **bunion.**

Pelvimetry (pel-VIM-e-trē) The measurement of the size of the inlet and outlet of the birth canal; may be done by ultrasound or physical examination.

DISORDERS: HOMEOSTATIC IMBALANCE

HIP FRACTURE

Although any region of the hip girdle may fracture, the term **hip fracture** most commonly applies to a break in the bones associated with the hip joint—the head, neck, or trochanteric regions of the femur, or the bones that form the acetabulum. In the United States, 300,000–500,000 people sustain hip fractures each year. The incidence of hip fractures is increasing, due in part to longer life spans. Decreases in bone mass due to osteoporosis (which occurs more often in females) and an increased tendency to fall predispose elderly people to hip fractures.

Hip fractures often require surgical treatment, the goal of which is to repair and stabilize the fracture, increase mobility, and decrease pain. Sometimes the repair is accomplished by using surgical pins, screws, nails, and plates to secure the head of the femur. In severe hip fractures, the femoral head or the acetabulum of the hip bone may be replaced by prostheses (artificial devices). The procedure of replacing either the femoral head or the acetabulum is *hemiarthroplasty* (hem-ē-AR-thrō-plas-tē; *hemi-* = one half; *-arthro-* = joint; *-plasty* = molding). Replacement of both the femoral head and acetabulum is *total hip arthroplasty*. The acetabular prosthesis is made of plastic, whereas the femoral prosthesis is metal; both are designed to withstand a high degree of stress. The prostheses are attached to healthy portions of bone with acrylic cement and screws.

STUDY OUTLINE

PECTORAL (SHOULDER) GIRDLE (p. 218)
1. Each pectoral (shoulder) girdle consists of a clavicle and scapula.
2. Each pectoral girdle attaches an upper limb to the axial skeleton.

UPPER LIMB (p. 221)
1. There are 60 bones in the two upper limbs.
2. The bones of each upper limb include the humerus, the ulna, the radius, the carpals, the metacarpals, and the phalanges.

PELVIC (HIP) GIRDLE (p. 224)
1. The pelvic (hip) girdle consists of two hip bones.
2. Each hip bone consists of three fused components: ilium, pubis, and ischium.
3. The hip bones, sacrum, and pubic symphysis form the bony pelvis. It supports the vertebral column and pelvic viscera and attaches the lower limbs to the axial skeleton.
4. The true pelvis is separated from the false pelvis by the pelvic brim.

COMPARISON OF FEMALE AND MALE PELVES (p. 231)
1. Bones of males are generally larger and heavier than bones of females, with more prominent markings for muscle attachment.

2. The female pelvis is adapted for pregnancy and childbirth. Gender-related differences in pelvic structure are listed and illustrated in Table 8.1.

COMPARISON OF PECTORAL AND PELVIC GIRDLES (p. 231)
1. The pectoral girdle does not directly articulate with the vertebral column; the pelvic girdle does.
2. The glenoid fossae of the scapulae are shallow and maximize movement; the acetabula of the hip bones are deep and allow less movement.

LOWER LIMB (p. 231)
1. There are 60 bones in the two lower limbs.
2. The bones of each lower limb include the femur, the patella, the tibia, the fibula, the tarsals, the metatarsals, and the phalanges.
3. The bones of the foot are arranged in two arches, the longitudinal arch and the transverse arch, to provide support and leverage.

SELF-QUIZ QUESTIONS

Complete the following:
1. Functions of the axial skeleton are related mainly to ___ of internal organs, whereas functions of the appendicular skeleton are related more to ___.
2. A large, triangular, sesamoid bone located anterior to the knee joint is the ___.
3. The bone that initially bears the weight of the entire body during walking is the ___.
4. One of the most frequently broken bones of the body is the ___.

Choose the best answer to the following questions:
5. Which of the following statements are true? (1) The pectoral girdle consists of the scapula, the clavicle, and the sternum. (2) Although the joints of the pectoral girdle are not very stable, they allow free movement in many directions. (3) The anterior component of the pectoral girdle is the scapula. (4) The pectoral girdle articulates directly with the vertebral column. (5) The posterior component of the pectoral girdle is the sternum. (a) 1, 2, and 3, (b) only 2, (c) only 4, (d) 2, 3, and 5, (e) 3, 4, and 5

6. Which of the following bones are carpal bones? (1) phalanges, (2) scaphoid, lunate, and pisiform, (3) trapezium, trapezoid, capitate, (4) hamate and triquetrum, (5) calcaneus and talus. (a) 1, 2, and 3, (b) 2, 3, and 4, (c) 1, 2, 3, and 4, (d) 2, 3, 4, and 5, (e) 1, 2, 3, 4, and 5

7. Which of the following statements is true? (a) The ilium is the smallest component of the hip bone. (b) The greater sciatic notch is found on the ischium. (c) The ischium is the anterior, superior portion of the hip bone. (d) The pubic symphysis is the joint between the two hip bones. (e) The obturator foramen is the smallest foramen in the body.

8. The acetabulum is part of the (a) fibula, (b) sacrum, (c) hip bone, (d) femur, (e) tibia.

9. Which of the following statements are true? (1) Bones of males are generally larger and heavier than those of females. (2) The female pelvis is wider than the male pelvis. (3) The true pelvis of the male contains more space than that of the female. (4) Skeletal bones of males generally possess larger surface features than those of females. (5) Most structural differences between male and female pelves are related to pregnancy and childbirth. (a) 1, 2, and 3, (b) 2, 3, and 4, (c) 3, 4, and 5, (d) 1, 2, 3, and 5, (e) 1, 2, 4, and 5

10. Which of the following is *not* a tarsal bone? (a) patella, (b) navicular, (c) calcaneus, (d) talus, (e) cuneiform

True or false:

11. The arches of the foot enable the foot to support the weight of the body, provide an ideal distribution of body weight over the hard and soft tissues of the foot, and provide leverage for walking.

12. Functionally, the pelvis provides a strong and stable support for the vertebral column and pelvic viscera.

13. Match the following:
    ___ (a) olecranon process   (1) clavicle
    ___ (b) olecranon fossa   (2) scapula
    ___ (c) trochlea   (3) humerus
    ___ (d) greater trochanter   (4) ulna
    ___ (e) medial malleolus   (5) radius
    ___ (f) costal tuberosity   (6) femur
    ___ (g) capitulum   (7) tibia
    ___ (h) acromion   (8) fibula
    ___ (i) radial tuberosity
    ___ (j) lateral malleolus
    ___ (k) glenoid cavity
    ___ (l) coronoid process
    ___ (m) linea aspera

14. Match the following:
    ___ (a) a large, triangular, flat bone found in the posterior part of the thorax   (1) calcaneus
    ___ (b) an S-shaped bone lying horizontally in the superior and anterior part of the thorax   (2) scapula
    ___ (c) articulates proximally with the scapula and distally with the radius and ulna   (3) scaphoid
    ___ (d) located on the medial aspect of the forearm   (4) radius
    ___ (e) located on the lateral aspect of the forearm   (5) femur
    ___ (f) the longest, heaviest, and strongest bone of the body   (6) clavicle
    ___ (g) the larger, medial bone of the leg   (7) ulna
    ___ (h) the smaller, lateral bone of the leg   (8) tibia
    ___ (i) heel bone   (9) humerus
    ___ (j) carpal bone   (10) fibula

15. Match the following:
    ___ (a) caused by narrowing of the carpal tunnel   (1) flatfoot
    ___ (b) condition in which the medial longitudinal arch is abnormally elevated; frequently caused by muscle imbalance   (2) clawfoot
       (3) clubfoot
    ___ (c) also called genu valgum   (4) bunion
    ___ (d) an inherited deformity where the foot is twisted inferiorly and medially and the arch is increased   (5) runner's knee
       (6) knock-knee
       (7) bowleg
    ___ (e) also called genu varum   (8) carpal tunnel syndrome
    ___ (f) patellofemoral stress syndrome
    ___ (g) medial arch height decreases due to weakened ligaments and tendons
    ___ (h) deformity of the great toe

CRITICAL THINKING QUESTIONS

1. Young Tommy watches way too much television, especially cartoons and monster movies. So when he heard that his great uncle had "clawfoot," he couldn't decide if he should be thrilled or terrified! Explain "clawfoot" to Tommy. (HINT: *His great uncle is not changing his species—it's a human condition.*)

2. The local newspaper reported that Farmer White caught his hand in a piece of machinery last Tuesday. He lost the lateral two fingers of his left hand. Science reporter Kent Clark, really a high school junior, reports that Farmer White has three remaining phalanges. Is Kent correct, or does he need a refresher course in anatomy? (HINT: *How many phalanges are in each finger?*)

3. Given that the hips and shoulders are located in the trunk of the body, why aren't they part of the axial skeleton like the rest of the bones that make up the trunk? (HINT: *Have you ever seen a snake with shoulders?*)

ANSWERS TO FIGURE QUESTIONS

8.1 The pectoral girdles attach the upper limbs to the axial skeleton.

8.2 The weakest part of the clavicle is its midregion at the junction of the two curves.

8.3 The highest point of the shoulder is the acromion.

8.4 Each upper limb has 30 bones.

8.5 The radius articulates at the elbow with the capitulum and radial fossa of the humerus. The ulna articulates at the elbow with the trochlea, coronoid fossa, and olecranon fossa of the humerus.

8.6 The "elbow" part of the ulna is the olecranon.

8.7 The radius and ulna form two joints, the proximal and distal radioulnar joints.

8.8 The most frequently fractured wrist bone is the scaphoid.

8.9 The bony pelvis attaches the lower limbs to the axial skeleton and supports the backbone and pelvic viscera.

8.10 The femur articulates with the acetabulum of the hip bone; the sacrum articulates with the auricular surface of the hip bone.

8.11 The pelvic axis is the course taken by a baby's head as it descends through the pelvis during childbirth.

8.12 Each lower limb has 30 bones.

8.13 The angle of convergence of the femurs is greater in females than males because the female pelvis is broader.

8.14 The patella is a sesamoid base—the largest one in the body.

8.15 The tibia is the weight-bearing bone of the leg.

8.16 The talus articulates with the tibia and the fibula.

8.17 The arches are not rigid; they yield when weight is applied and spring back when weight is lifted.

JOINTS

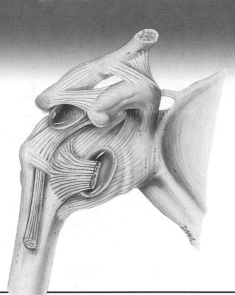

Bones are too rigid to bend without being damaged. Fortunately, flexible connective tissues form joints that hold bones together while still permitting, in most cases, some degree of movement. A **joint,** also called an **articulation** or **arthrosis,** is a point of contact between two bones, between bone and cartilage, or between bone and teeth. When we say one bone *articulates* with another bone, we mean that the bones form a joint. Because most movements of the body occur at joints, you can appreciate their importance if you imagine how a cast over your knee joint makes walking difficult, or how a splint on a finger limits your ability to manipulate small objects. The scientific study of joints is termed **arthrology** (ar-THROL-ō-jē; *arthr-* = joint; *-ology* = study of). The study of motion of the human body is called **kinesiology** (ki-nē-sē′-OL-ō-jē; *kinesi-* = movement).

JOINT CLASSIFICATIONS

OBJECTIVE

• *Describe the structural and functional classifications of joints.*

Joints are classified structurally, based on their anatomical characteristics, and functionally, based on the type of movement they permit.

The structural classification of joints is based on two criteria: (1) the presence or absence of a space between the articulating bones, called a synovial cavity, and (2) the type of connective tissue that binds the bones together. Structurally, joints are classified as one of the following types:

• **Fibrous joints:** The bones are held together by fibrous connective tissue that is rich in collagen fibers; they lack a synovial cavity.
• **Cartilaginous joints:** The bones are held together by cartilage; they lack a synovial cavity.
• **Synovial joints:** The bones forming the joint have a synovial cavity and are united by the dense irregular connective tissue of an articular capsule, and often by accessory ligaments.

The functional classification of joints relates to the degree of movement they permit. Functionally, joints are classified as one of the following types:

• **Synarthrosis** (sin′-ar-THRŌ-sis; *syn-* = together): An immovable joint. The plural is *synarthroses.*
• **Amphiarthrosis** (am′-fē-ar-THRŌ-sis; *amphi-* = on both sides): A slightly movable joint. The plural is *amphiarthroses.*
• **Diarthrosis** (dī-ar-THRŌ-sis; = movable joint): A freely movable joint. The plural is *diarthroses.* All diarthroses are synovial joints. They have a variety of shapes and permit several different types of movements.

The following sections present the joints of the body according to their structural classification. As we examine the structure of each type of joint, we will also explore its functional attributes.

Figure 9.1 Fibrous joints.

🔑 **At a fibrous joint the bones are held together by fibrous connective tissue.**

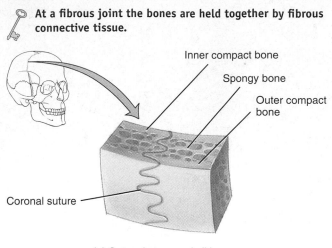

Inner compact bone
Spongy bone
Outer compact bone
Coronal suture

(a) Suture between skull bones

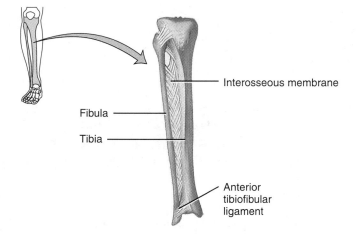

Interosseous membrane
Fibula
Tibia
Anterior tibiofibular ligament

(b) Syndesmoses between tibia and fibula

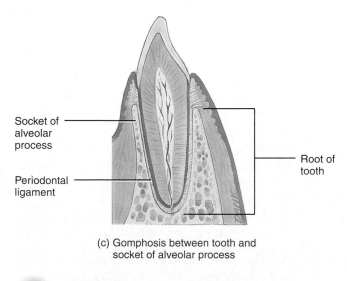

Socket of alveolar process
Periodontal ligament
Root of tooth

(c) Gomphosis between tooth and socket of alveolar process

❓ Functionally, why are sutures classified as synarthroses, and syndesmoses as amphiarthroses?

FIBROUS JOINTS

OBJECTIVE
• *Describe the structure and functions of the three types of fibrous joints.*

Fibrous joints lack a synovial cavity, and the articulating bones are held very closely together by fibrous connective tissue. They permit little or no movement. The three types of fibrous joints are sutures, synedesmoses, and gomphoses.

Sutures

A **suture** (SOO-chur; *sutur-* = seam) is a fibrous joint composed of a thin layer of dense fibrous connective tissue that unites bones of the skull. An example is the coronal suture between the parietal and frontal bones (Figure 9.1a). The irregular, interlocking edges of sutures give them added strength and decrease their chance of fracturing. Because a suture is immovable, it is classified functionally as a synarthrosis.

Some sutures, although present during childhood, are replaced by bone in the adult. Such a suture is called a **synostosis** (sin'-os-TŌ-sis; *os-* = bone), or bony joint—a joint in which there is a complete fusion of bone across the suture line. An example is the frontal suture between the left and right sides of the frontal bone that begins to fuse during infancy.

Syndesmoses

A **syndesmosis** (sin'-dez-MŌ-sis; *syndesmo-* = band or ligament) is a fibrous joint in which, compared to a suture, there is a greater distance between the articulating bones and more fibrous connective tissue. The fibrous connective tissue is arranged either as a bundle (ligament) or as a sheet (interosseous membrane) (Figure 9.1b). One example of a syndesmosis is the distal tibiofibular joint, where the anterior tibiofibular ligament connects the tibia and fibula. Another example is the interosseous membrane between the parallel borders of the tibia and fibula. Because it permits slight movement, a syndesmosis is classified functionally as an amphiarthrosis.

Gomphoses

A **gomphosis** (gom-FŌ-sis; *gompho-* = a bolt or nail) is a type of fibrous joint in which a cone-shaped peg fits into a socket. The only examples of gomphoses are the articulations of the roots of the teeth with the sockets (alveoli) of the alveolar processes of the maxillae and mandible (Figure 9.1c). The dense fibrous connective tissue between a tooth and its socket is the periodontal ligament. A gomphosis is classified functionally as a synarthrosis, an immovable joint.

CARTILAGINOUS JOINTS

OBJECTIVE

• *Describe the structure and functions of the two types of cartilaginous joints.*

Like a fibrous joint, a **cartilaginous joint** lacks a synovial cavity and allows little or no movement. Here the articulating bones are tightly connected by either fibrocartilage or hyaline cartilage. The two types of cartilaginous joints are synchondroses and symphyses.

Synchondroses

A **synchondrosis** (sin′ kon DRŌ sis; *chondro-* = cartilage) is a cartilaginous joint in which the connecting material is hyaline cartilage. An example of a synchondrosis is the epiphyseal plate that connects the epiphysis and diaphysis of a growing bone (Figure 9.2a). Functionally, a synchondrosis is a synarthrosis. When bone elongation ceases, bone replaces the hyaline cartilage, and the synchondrosis becomes a synostosis, a bony joint. Another example of a synchondrosis is the joint between the first rib and the manubrium of the sternum, which also ossifies during adult life and becomes an immovable synostosis.

Symphyses

A **symphysis** (SIM-fi-sis; = growing together) is a cartilaginous joint in which the ends of the articulating bones are covered with hyaline cartilage, but the bones are connected by a broad, flat disc of fibrocartilage. All symphyses occur in the midline of the body. The pubic symphysis between the anterior surfaces of the hip bones is one example of a symphysis (Figure 9.2b). This type of joint is also found at the junction of the manubrium and body of the sternum (see Figure 7.22 on page 211) and at the intervertebral joints between the bodies of vertebrae (see Figure 7.17 on page 204). A portion of the intervertebral disc is fibrocartilage. A symphysis is an amphiarthrosis, a slightly movable joint.

1. List the types of fibrous and cartilaginous joints. For each type, give an example of where in the body it may be found and describe what type of tissue holds the articulating bones together.
2. Which of the fibrous and cartilaginous joints are classified as amphiarthroses?

SYNOVIAL JOINTS

OBJECTIVE

• *Describe the structure of synovial joints.*

Synovial joints (si-NŌ-vē-al) have a space called a **synovial (joint) cavity** between the articulating bones (Figure 9.3). The structure of these joints allows them to be freely moveable. Hence, all synovial joints are classified functionally as diarthroses.

Figure 9.2 Cartilaginous joints.

🔑 **At a cartilaginous joint the bones are held together by cartilage.**

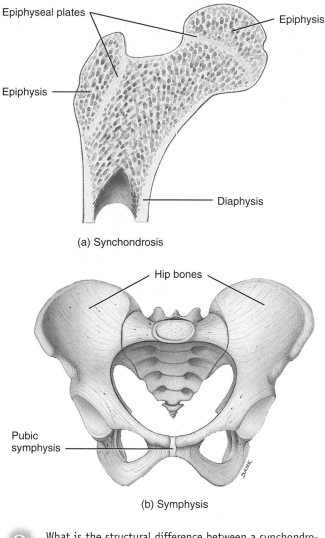

(a) Synchondrosis

(b) Symphysis

Q What is the structural difference between a synchondrosis and a symphysis?

Structure of Synovial Joints

The bones at a synovial joint are covered by **articular cartilage,** which typically is hyaline cartilage and occasionally is fibrocartilage. The cartilage covers the articulating surface of the bones with a smooth, slippery surface but does not bind them together. Articular cartilage reduces friction between bones in the joint during movement and helps to absorb shock.

Articular Capsule

A sleevelike **articular capsule** surrounds a synovial joint, encloses the synovial cavity, and unites the articulating

Figure 9.3 Structure of a typical synovial joint. Note the two layers of the articular capsule—the fibrous capsule and the synovial membrane. Synovial fluid fills the joint cavity between the synovial membrane and the articular cartilage.

🔑 **The distinguishing feature of a synovial joint is the synovial cavity between the articulating bones.**

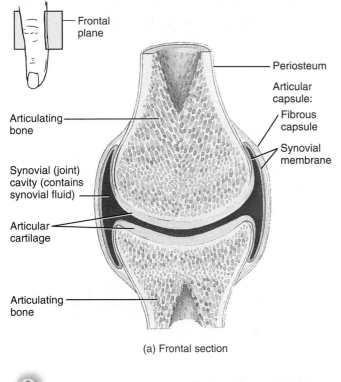

(a) Frontal section

Ⓠ What is the functional classification of synovial joints?

bones. The articular capsule is composed of two layers, an outer fibrous capsule and an inner synovial membrane (see Figure 9.3). The outer layer, the **fibrous capsule,** usually consists of dense, irregular connective tissue that attaches to the periosteum of the articulating bones. The flexibility of the fibrous capsule permits considerable movement at a joint while its great tensile strength (resistance to stretching) helps prevent the bones from dislocating. The fibers of some fibrous capsules are arranged in parallel bundles that are highly adapted for resisting strains. Such fiber bundles are called **ligaments** (*liga-* = bound or tied) and are often designated by individual names. The strength of ligaments is one of the principal mechanical factors that hold bones close together in a synovial joint. The inner layer of the articular capsule, the **synovial membrane,** is composed of areolar connective tissue with elastic fibers. At many synovial joints the synovial membrane includes accumulations of adipose tissue, called **articular fat pads.** An example is the infra-patellar fat pad in the knee (see Figure 9.14c).

Synovial Fluid

The synovial membrane secretes **synovial fluid** (*ov-* = egg), which forms a thin film over the surfaces within the articular capsule. This viscous, clear or pale yellow fluid was named for its similarity in appearance and consistency to uncooked egg white. Synovial fluid consists of hyaluronic acid and interstitial fluid filtered from blood plasma. Its several functions include reducing friction by lubricating the joint, and supplying nutrients to and removing metabolic wastes from the chondrocytes within articular cartilage. (Recall that cartilage is an avascular tissue.) Synovial fluid also contains phagocytic cells that remove microbes and the debris that results from normal wear and tear in the joint. When a synovial joint is immobile for a time, the fluid is quite viscous (gel-like), but as joint movement increases, the fluid becomes less viscous. One of the benefits of a "warm-up" before exercise is that it stimulates the production and secretion of synovial fluid.

Accessory Ligaments and Articular Discs

Many synovial joints also contain **accessory ligaments** called extracapsular ligaments and intracapsular ligaments. *Extracapsular ligaments* lie outside the articular capsule. Examples are the fibular and tibial collateral ligaments of the knee joint (see Figure 9.14d). *Intracapsular ligaments* occur within the articular capsule but are excluded from the synovial cavity by folds of the synovial membrane. Examples are the anterior and posterior cruciate ligaments of the knee joint (see Figure 9.14d).

Inside some synovial joints, such as the knee, are pads of fibrocartilage that lie between the articular surfaces of the bones and are attached to the fibrous capsule. These pads are called **articular discs** or **menisci** (men-IS-ī; the singular is *meniscus*). Figure 9.14d depicts the lateral and medial menisci in the knee joint. The discs usually subdivide the synovial cavity into two separate spaces. By modifying the shape of the joint surfaces of the articulating bones, articular discs allow two bones of different shapes to fit more tightly. Articular discs also help to maintain the stability of the joint and direct the flow of synovial fluid to the areas of greatest friction.

CLINICAL APPLICATION
Torn Cartilage and Arthroscopy

The tearing of articular discs (menisci) in the knee, commonly called **torn cartilage,** occurs often among athletes. Such damaged cartilage will begin to wear and may precipitate arthritis unless surgically removed (meniscectomy). Repair of the torn cartilage may be assisted by **arthroscopy** (ar-THROS-kō-pē; *-scopy* = observation). This procedure involves examination of the interior of a joint, usually the knee, with an arthroscope, a lighted instrument used for visualization. Arthroscopy is used to determine the nature and extent of damage following knee injury; to help remove torn cartilage and repair cruciate ligaments in the knee; to help obtain tissue samples for analysis; to monitor the progression of disease and the effects of therapy; and to help perform surgery on other joints, such as the shoulder, elbow, ankle, and wrist. ■

Nerve and Blood Supply

The nerves that supply a joint are the same as those that supply the skeletal muscles that move the joint. Synovial joints contain many nerve endings that are distributed to the articular capsule and associated ligaments. Some of the nerve endings convey information about pain from the joint to the spinal cord and brain for processing. Other nerve endings are responsive to the degree of movement and stretch at a joint. This information, too, is relayed to the spinal cord and brain, which may respond by sending impulses through different nerves to the muscles to adjust body movements.

Arteries in the vicinity of a synovial joint send out numerous branches that penetrate the ligaments and articular capsule to deliver oxygen and nutrients. Veins remove carbon dioxide and wastes from the joints. The arterial branches from several different arteries typically join together around a joint before penetrating the articular capsule. The articulating portions of a synovial joint receive nourishment from synovial fluid, whereas all other joint tissues are supplied by blood capillaries.

CLINICAL APPLICATION
Sprain and Strain

A **sprain** is the forcible wrenching or twisting of a joint that stretches or tears its ligaments but does not dislocate the bones. It occurs when the ligaments are stressed beyond their normal capacity. Sprains also may damage surrounding blood vessels, muscles, tendons, or nerves. Severe sprains may be so painful that the joint cannot be moved. There is considerable swelling, which results from hemorrhage of ruptured blood vessels. The ankle joint is most often sprained; the lower back is another frequent location for sprains. A less serious injury is a **strain**, which is a stretched or partially torn muscle. It often occurs when a muscle contracts suddenly and powerfully—for example, in sprinters when they accelerate too quickly. ■

1. How does the structure of synovial joints permit them to be diarthroses?
2. What are the functions of articular cartilage, synovial fluid, and articular discs?
3. What types of sensations are perceived at joints, and from what sources do joint components receive their nourishment?

Types of Synovial Joints

OBJECTIVE
• *Describe the six subtypes of synovial joints.*

Although all synovial joints have a generally similar structure, the shapes of the articulating surfaces vary. Ac-

cordingly, synovial joints are divided into six subtypes: planar, hinge, pivot, condyloid, saddle, and ball-and-socket joints.

Planar Joints

The articulating surfaces of bones in a **planar joint** are flat or slightly curved (Figure 9.4a). Some examples of planar joints are the intercarpal joints (between carpal bones at the wrist), intertarsal joints (between tarsal bones at the ankle), sternoclavicular joints (between the sternum and the clavicle), acromioclavicular joints (between the acromion of the scapula and the clavicle), sternocostal joints (between the sternum and ends of the costal cartilages at the tips of the second through seventh pairs of ribs), and vertebrocostal joints (between the heads and tubercles of ribs and transverse processes of thoracic vertebrae). Planar joints primarily permit side-to-side and back-and-forth gliding movements (described shortly). Planar joints are said to be *nonaxial* because the motion they allow does not occur around an axis. However, x-ray films made during wrist and ankle movements reveal some rotation of the small carpal and tarsal bones in addition to their predominant gliding movements.

Hinge Joints

In a **hinge joint,** the convex surface of one bone fits into the concave surface of another bone (Figure 9.4b). Examples of hinge joints are the knee, elbow, ankle, and interphalangeal joints. As the name implies, hinge joints produce an angular, opening-and-closing motion like that of a hinged door. Hinge joints are *monaxial (uniaxial)* because they typically allow motion around a single axis.

Pivot Joints

In a **pivot joint,** the rounded or pointed surface of one bone articulates with a ring formed partly by another bone and partly by a ligament (Figure 9.4c). A pivot joint is monaxial because it allows rotation around its own longitudinal axis only. Examples of pivot joints are the atlanto-axial joint, in which the atlas rotates around the axis and permits the head to turn from side to side as in signifying "no" (see Figure 9.9a on page 251), and the radioulnar joints that enable the palms to turn anteriorly and posteriorly (see Figure 9.10h on page 252).

Condyloid Joints

In a **condyloid joint** (KON-di-loyd; *condyl-* = knuckle) or *ellipsoidal* joint, the convex oval-shaped projection of one bone fits into the oval-shaped depression of another bone (Figure 9.4d). Examples are the wrist and metacarpophalangeal joints for the second through fifth digits. A condyloid joint is *biaxial* because the movement it permits is around two axes. Notice that you can move your index finger both up and down and from side to side.

Figure 9.4 Subtypes of synovial joints. For each subtype, a drawing of the actual joint and a simplified diagram are shown.

🔑 **Synovial joints are classified into subtypes on the basis of the shapes of the articulating bone surfaces.**

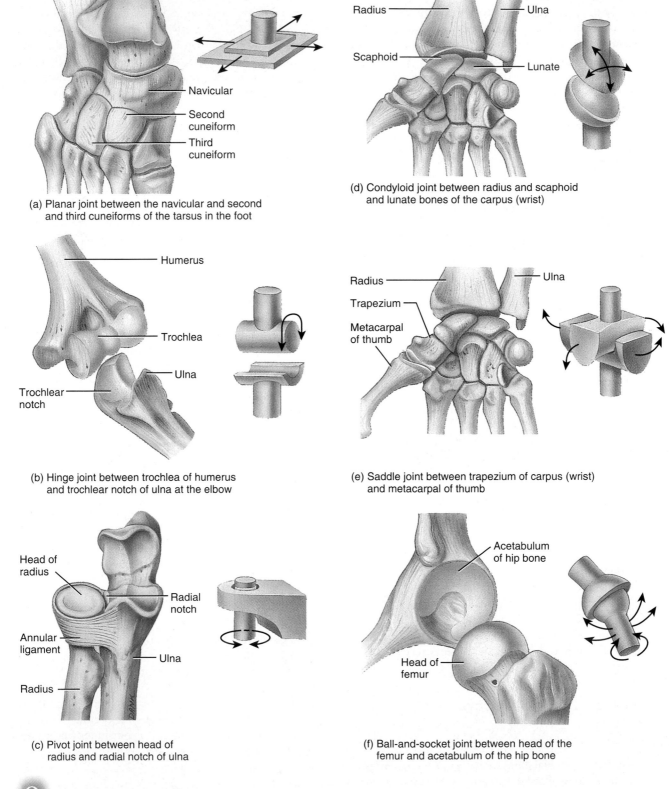

(a) Planar joint between the navicular and second and third cuneiforms of the tarsus in the foot

- Navicular
- Second cuneiform
- Third cuneiform

(b) Hinge joint between trochlea of humerus and trochlear notch of ulna at the elbow

- Humerus
- Trochlea
- Ulna
- Trochlear notch

(c) Pivot joint between head of radius and radial notch of ulna

- Head of radius
- Radial notch
- Annular ligament
- Ulna
- Radius

(d) Condyloid joint between radius and scaphoid and lunate bones of the carpus (wrist)

- Radius
- Ulna
- Scaphoid
- Lunate

(e) Saddle joint between trapezium of carpus (wrist) and metacarpal of thumb

- Radius
- Ulna
- Trapezium
- Metacarpal of thumb

(f) Ball-and-socket joint between head of the femur and acetabulum of the hip bone

- Acetabulum of hip bone
- Head of femur

Ⓠ Which joints shown are biaxial?

Saddle Joints

In a **saddle joint,** the articular surface of one bone is saddle-shaped, and the articular surface of the other bone fits into the "saddle" as a sitting rider would sit (Figure 9.4e). An example of a saddle joint is the carpometacarpal joint between the trapezium of the carpus and metacarpal of the thumb. A saddle joint is a modified condyloid joint in which the movement is somewhat freer. Saddle joints are *biaxial,* producing side to side and up and down movements.

Ball-and-Socket Joints

A ball-and-socket joint consists of the ball-like surface of one bone fitting into a cuplike depression of another bone (Figure 9.4f). Examples of functional ball-and-socket joints are the shoulder and hip joints. Such joints are *multiaxial (polyaxial)* because they permit movement around three axes plus all directions in between.

Bursae and Tendon Sheaths

OBJECTIVE

• *Describe the structure and function of bursae and tendon sheaths.*

The various motions of the body create friction between moving parts. Saclike structures called **bursae** (= purses) are strategically situated to alleviate friction in some joints, such as the shoulder and knee joints (see Figure 9.11 on page 254 and 9.14c on page 260). Bursae (singular is *bursa*) are not strictly part of synovial joints, but do resemble joint capsules because their walls consist of connective tissue lined by a synovial membrane. They are also filled with a fluid similar to synovial fluid. Bursae are located between the skin and bone in places where skin rubs over bone. They are also found between tendons and bones, muscles and bones, and ligaments and bones. The fluid-filled bursal sacs cushion the movement of one body part over another.

In addition to bursae, structures called tendon sheaths also reduce friction at joints. **Tendon sheaths** are tubelike bursae that wrap around tendons where there is considerable friction. This occurs where tendons pass through synovial cavities, such as the tendon of the biceps brachii muscle at the shoulder joint (see Figure 9.11c on page 255). Other areas of considerable friction where tendon sheaths are found are at the wrist and ankle, where many tendons come together in a confined space (see Figure 11.22 on page 369), and in the fingers and toes, where there is a great deal of movement (see Figure 11.18a on page 352).

Table 9.1 summarizes the structural and functional categories of joints.

CLINICAL APPLICATION
Bursitis

An acute or chronic inflammation of a bursa is called **bursitis.** The condition may be caused by trauma, by an acute or chronic infection (including syphilis and tuberculosis), or by rheumatoid arthritis (described on page 264). Re-

Figure 9.5 Gliding movements at synovial joints.

🔑 **Gliding motion consists of side-to-side and back-and-forth movements.**

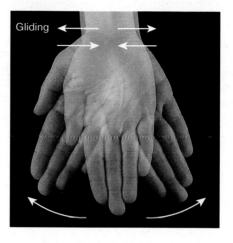

Gliding

Intercarpal joints

Q At which subtype of synovial joints do gliding movements occur?

peated, excessive exertion of a joint often results in bursitis, with local inflammation and the accumulation of fluid. Symptoms include pain, swelling, tenderness, and limited movement. ■

1. Explain which types of joints are nonaxial, monaxial, biaxial, and multiaxial.
2. In what ways are bursae similar to joint capsules? How do they differ?

TYPES OF MOVEMENT AT SYNOVIAL JOINTS

OBJECTIVE

• *Describe the types of movements that can occur at synovial joints.*

Anatomists, physical therapists, and kinesiologists use specific terminology to designate movements that can occur at synovial joints. These precise terms may indicate the form of motion, the direction of movement, or the relationship of one body part to another during movement. Movements at synovial joints are grouped into four main categories: (1) gliding, (2) angular movements, (3) rotation, and (4) special movements. This last category includes movements that occur only at certain joints.

Gliding

Gliding is a simple movement in which relatively flat bone surfaces move back and forth and from side to side with respect to one another (Figure 9.5). There is no significant alteration of the angle between the bones. Gliding

Table 9.1 Summary of Structural Categories and Functional Characteristics of Joints

STRUCTURAL CATEGORY	DESCRIPTION	FUNCTIONAL CHARACTERISTICS	EXAMPLE
Fibrous	Articulating bones held together by fibrous connective tissue; no synovial cavity.		
Suture	Articulating bones united by a thin layer of dense fibrous connective tissue; found between bones of the skull. With age, some sutures are replaced by a synostosis, in which the bones fuse across the former suture.	Synarthrosis (immovable).	Frontal suture.
Syndesmosis	Articulating bones united by dense fibrous connective tissue, either a ligament or an interosseous membrane.	Amphiarthrosis (slightly movable).	Distal tibiofibular joint.
Gomphosis	Articulating bones united by periodontal ligament; cone-shaped peg fits into a socket.	Synarthrosis.	At roots of teeth in alveoli (sockets) of maxillae and mandible.
Cartilaginous	Articulating bones united by cartilage; no synovial cavity.		
Synchondrosis	Connecting material is hyaline cartilage; becomes a synostosis when bone elongation ceases.	Synarthrosis.	Epiphyseal plate at joint between the diaphysis and epiphysis of a long bone.
Symphysis	Connecting material is a broad, flat disc of fibrocartilage.	Amphiarthrosis.	Intervertebral joints, pubic symphysis, and junction of manubrium with body of sternum.
Synovial	Characterized by a synovial cavity, articular cartilage, and an articular capsule; may contain accessory ligaments, articular discs, and bursae.		
Planar	Articulated surfaces are flat or just slightly curved.	Nonaxial diarthrosis (freely movable); gliding motion.	Intercarpal, intertarsal, sternocostal (between sternum and the 2nd–7th pairs of ribs), and vertebrocostal joints.
Hinge	Convex surface fits into a concave surface.	Monaxial diarthrosis; angular motion.	Elbow, ankle, and interphalangeal joints.
Pivot	Rounded or pointed surface fits into a ring formed partly by bone and partly by a ligament.	Monaxial diarthrosis; rotation.	Atlanto-axial and radioulnar joints.
Condyloid	Oval-shaped projection fits into an oval-shaped depression.	Biaxial diarthrosis; angular motion.	Radiocarpal and metacarpophalangeal joints.
Saddle	Articular surface of one bone is saddle-shaped, and the articular surface of the other bone "sits" in the saddle.	Biaxial diarthrosis; angular motion.	Carpometarcarpal joint between trapezium and thumb.
Ball-and-socket	Ball-like surface fits into a cuplike depression.	Multiaxial diarthrosis; angular motion and rotation.	Shoulder and hip joints.

movements are limited in range due to the structure of the articular capsule and associated ligaments and bones. Gliding occurs at planar joints.

Angular Movements

In **angular movements,** there is an increase or a decrease in the angle between articulating bones. The principal angular movements are flexion, extension, lateral extension, hyperextension, abduction, adduction, and circumduction. These movements are discussed with respect to the body in the anatomical position.

Flexion, Extension, Lateral Flexion, and Hyperextension

Flexion and extension are opposite movements. In **flexion** (FLEK-shun; *flex-* = to bend) there is a decrease in the angle between articulating bones; in **extension** (eks-TEN-shun; *exten-* = to stretch out) there is an increase in the angle between articulating bones, often to restore a part of the body to the anatomical position after it has been flexed (Figure 9.6). Both movements usually occur in the sagittal plane. Hinge, pivot, condyloid, saddle, and ball-and-socket joints all permit flexion and extension.

Figure 9.6 Angular movements at synovial joints—flexion, extension, hyperextension, and lateral flexion.

 In angular movements, there is an increase or decrease in the angle between articulating bones.

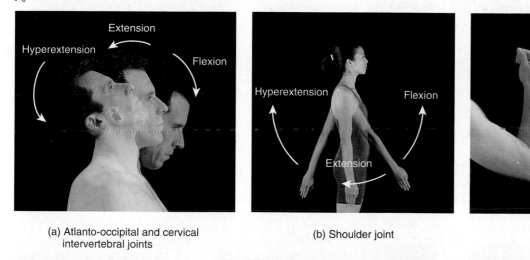

(a) Atlanto-occipital and cervical intervertebral joints

(b) Shoulder joint

(c) Elbow joint

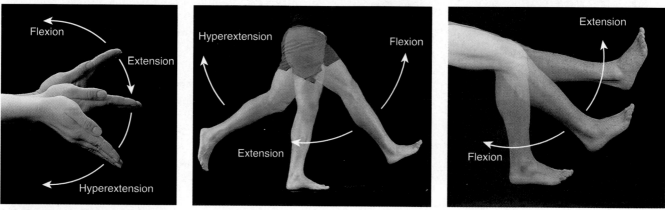

(d) Wrist joint

(e) Hip joint

(f) Knee joint

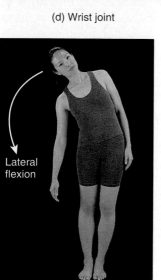

(g) Intervertebral joints

- bending the head toward the chest at the atlanto-occipital joint between the atlas (the first vertebra) and the occipital bone of the skull, and at the cervical intervertebral joints between the cervical vertebrae (Figure 9.6a)
- bending the trunk forward at the intervertebral joints
- moving the humerus forward at the shoulder joint, as in swinging the arms forward while walking (Figure 9.6b)
- moving the forearm toward the arm at the elbow joint between the humerus, ulna, and radius (Figure 9.6c)
- moving the palm toward the forearm at the wrist or radiocarpal joint between the radius and carpals (Figure 9.6d)
- bending the digits of the hand or feet at the interphalangeal joints between phalanges
- moving the femur forward at the hip joint between the femur and hip bone, as in walking (Figure 9.6e)
- moving the leg toward the thigh at the knee or tibiofemoral joint between the tibia, femur, and patella, as occurs when bending the knee (Figure 9.6f)

Q What are two examples of flexion that do not occur in the sagittal plane?

Figure 9.7 Angular movements at synovial joints—abduction and adduction.

🔑 **Abduction and adduction usually occur in the frontal plane.**

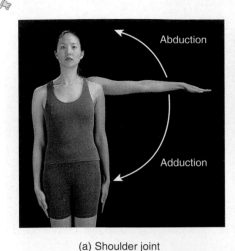

(a) Shoulder joint

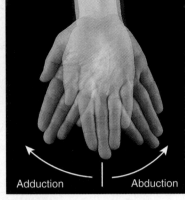

(b) Wrist joint

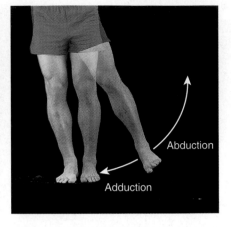

(c) Hip joint

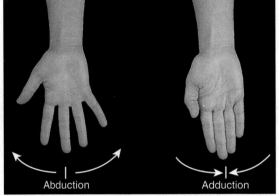

(d) Metacarpophalangeal joints of the fingers (not the thumb)

Q In what way is considering adduction as "adding your limb to your trunk" an effective learning device?

Although flexion and extension usually occur in the sagittal plane, there are a few exceptions. For example, flexion of the thumb involves movement of the thumb medially across the palm at the carpometacarpal joint between the trapezium and metacarpal of the thumb, as in touching the thumb to the opposite side of the palm. Another example is movement of the trunk sideways to the right or left at the waist. This movement, which occurs in the frontal plane and involves the intervertebral joints, is called **lateral flexion** (Figure 9.6g).

Continuation of extension beyond the anatomical position is called **hyperextension** (*hyper-* = beyond or excessive). Examples of hyperextension include:

- bending the head backward at the atlanto-occipital and cervical intervertebral joints (Figure 9.6a)
- bending the trunk backward at the intervertebral joints

- moving the humerus backward at the shoulder joint, as in swinging the arms backward while walking (Figure 9.6b)
- moving the palm backward at the wrist joint (Figure 9.6d)
- moving the femur backward at the hip joint, as in walking (Figure 9.6e)

Hyperextension of other joints, such as the elbow, interphalangeal, and knee joints, is usually prevented by factors such as the arrangement of ligaments and the anatomical alignment of the bones.

Abduction, Adduction, and Circumduction

Abduction (ab-DUK-shun; *ab-* = away; *-duct* = to lead) is the movement of a bone away from the midline, whereas **adduction** (ad-DUK-shun; *ad-* = toward) is the movement of a bone toward the midline. Both movements usually occur in the frontal plane. Condyloid, saddle, and ball-and-socket joints permit abduction and adduction. Examples of abduction include moving the humerus laterally at the shoulder joint, moving the palm laterally at the wrist joint, and moving the femur laterally at the hip joint (Figure 9.7a–c). The movement that returns each of these body parts to the anatomical position is adduction (see Figure 9.7a–c).

With respect to the digits, the midline is not used as a point of reference for abduction and adduction. In abduction of the fingers (but not the thumb), an imaginary line is drawn through the longitudinal axis of the middle (longest) finger, and the fingers move away (spread out) from the middle finger (Figure 9.7d). In abduction of the thumb, the thumb moves away from the palm in the sagittal plane (see Figure 11.18c on page 352). Abduction of the toes is relative to an imaginary line drawn through the second toe. Adduction of the fingers and toes involves returning them to the anatomical position. Adduction of the thumb moves the thumb toward the palm in the sagittal plane (see Figure 11.18c).

Figure 9.8 Angular movements at synovial joints—circumduction.

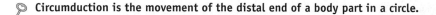

🔑 **Circumduction is the movement of the distal end of a body part in a circle.**

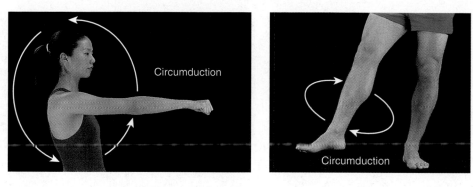

(a) Shoulder joint   (b) Hip joint

Ⓠ Which movements in continuous sequence produce circumduction?

Figure 9.9 Rotation at synovial joints.

🔑 **In rotation, a bone revolves around its own longitudinal axis.**

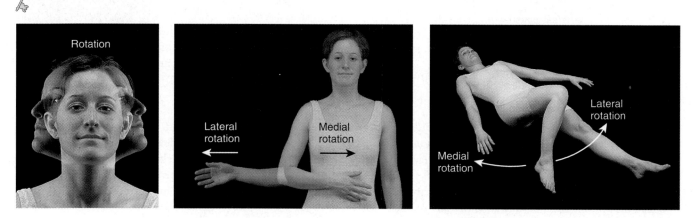

(a) Atlanto-axial joint   (b) Shoulder joint   (c) Hip joint

Ⓠ How do medial and lateral rotation differ?

Circumduction (ser-kum-DUK-shun; *circ-* = circle) is movement of the distal end of a body part in a circle (Figure 9.8). Circumduction occurs as a result of a continuous sequence of flexion, abduction, extension, and adduction. Examples of circumduction are moving the humerus in a circle at the shoulder joint (Figure 9.8a), moving the hand in a circle at the wrist joint, moving the thumb in a circle at the carpometacarpal joint, moving the fingers in a circle at the metacarpophalangeal joints (between the metacarpals and phalanges), and moving the femur in a circle at the hip joint (Figure 9.8b). Both the hip and shoulder joints permit cir-

cumduction, but flexion, abduction, extension, and adduction are more limited in the hip joints due to the tension on certain ligaments and muscles (see Exhibits 9.1 and 9.3).

Rotation

In **rotation** (rō-TĀ-shun; *rota-* = revolve), a bone revolves around its own longitudinal axis. Pivot and ball-and-socket joints permit rotation. One example is turning the head from side to side at the atlanto-axial joint (between the atlas and axis), as in signifying "no" (Figure 9.9a). Another is

Figure 9.10 Special movements at synovial joints.

🔑 **Special movements occur only at certain synovial joints.**

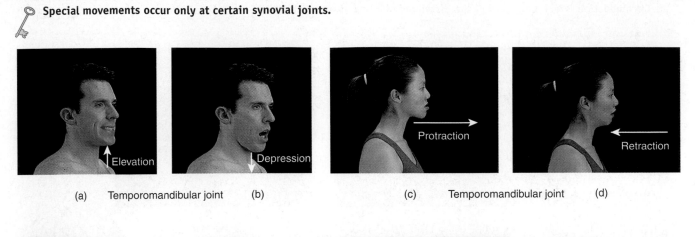

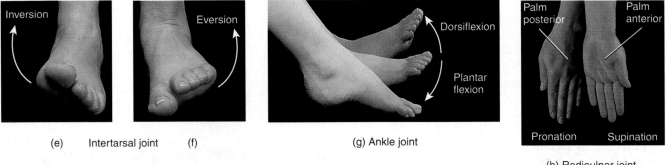

Q What movement of the shoulder girdle is involved in bringing the arms forward until the elbows touch?

turning the trunk from side to side at the intervertebral joints while keeping the hips and lower limbs in the anatomical position. In the limbs, rotation is defined relative to the midline, and specific qualifying terms are used. If the anterior surface of a bone of the limb is turned toward the midline, the movement is called *medial (internal) rotation*. You can medially rotate the humerus at the shoulder joint as follows: Starting in the anatomical position, flex your elbow and then draw your palm across the chest (Figure 9.9b). Medial rotation of the forearm at the radioulnar joints (between the radius and ulna) involves turning the palm medially from the anatomical position (see Figure 9.10h). You can medially rotate the femur at the hip joint as follows: Lie on your back, bend your knee, and then move your leg and foot laterally from the midline. Although you are moving your leg and foot laterally, the femur is rotating medially (Figure 9.9c). Medial rotation of the leg at the knee joint can be produced by sitting on a chair, bending your knee, raising your lower limb off the floor, and turning your toes medially. If the anterior surface of the bone of a limb is turned away from the midline, the movement is called *lateral (external) rotation* (see Figure 9.9b, c).

Special Movements

As noted previously, **special movements** occur only at certain joints. They include elevation, depression, protraction, retraction, inversion, eversion, dorsiflexion, plantar flexion, supination, pronation, and opposition (Figure 9.10):

- **Elevation** (el-e-VĀ-shun; = to lift up) is an upward movement of a part of the body, such as closing the mouth at the temporomandibular joint (between the mandible and temporal bone) to elevate the mandible (Figure 9.10a) or shrugging the shoulders at the acromioclavicular joint to elevate the scapula.

- **Depression** (dē-PRESH-un; = to press down) is a downward movement of a part of the body, such as opening the mouth to depress the mandible (Figure 9.10b) or returning shrugged shoulders to the anatomical position to depress the scapula.

- **Protraction** (prō-TRAK-shun; = to draw forth) is a movement of a part of the body anteriorly in the transverse plane. You can protract your mandible at the temporomandibular joint by thrusting it outward (Figure 9.10c) or protract your clavicles at the acromioclavicular and sternoclavicular joints by crossing your arms.

- **Retraction** (rē-TRAK-shun; = to draw back) is a movement of a protracted part of the body back to the anatomical position (Figure 9.10d).
- **Inversion** (in-VER-zhun; = to turn inward) is movement of the soles medially at the intertarsal joints (between the tarsals) so that the soles face each other (Figure 9.10e).
- **Eversion** (ē-VER-zhun; = to turn outward) is a movement of the soles laterally at the intertarsal joints so that the soles face away from each other (Figure 9.10f).
- **Dorsiflexion** (dor-si-FLEK-shun) refers to bending of the foot at the ankle or talocrural joint (between the tibia, fibula, and talus) in the direction of the dorsum (superior surface) (Figure 9.10g). Dorsiflexion occurs when you stand on your heels.
- **Plantar flexion** involves bending of the foot at the ankle joint in the direction of the plantar or inferior surface (see Figure 9.10g), as when standing on your toes.
- **Supination** (soo-pi-NĀ-shun) is a movement of the forearm at the proximal and distal radioulnar joints in which the palm is turned anteriorly or superiorly (Figure 9.10h). This position of the palms is one of the defining features of the anatomical position.
- **Pronation** (prō-NĀ-shun) is a movement of the forearm at the proximal and distal radioulnar joints in which the distal end of the radius crosses over the distal end of the ulna and the palm is turned posteriorly or inferiorly (see Figure 9.10h).
- **Opposition** (op-ō-ZISH-un) is the movement of the thumb at the carpometacarpal joint (between the trapezium and metacarpal of the thumb) in which the thumb moves across the palm to touch the tips of the fingers on the same hand (see Figure 11.18c on page 352). This is the distinctive digital movement that gives humans and other primates the ability to grasp and manipulate objects very precisely.

A summary of the movements that occur at synovial joints is presented in Table 9.2.

1. On yourself or with a partner, demonstrate each movement listed in Table 9.2.

SELECTED JOINTS OF THE BODY

In this section of the text we examine in detail four selected joints of the body in a series of exhibits. Each exhibit considers a specific synovial joint and contains (1) a definition—a description of the type of joint and the bones that form the joint; (2) the anatomical components—a description of the major connecting ligaments, articular disc, articular capsule, and other distinguishing features of the joint; and (3) the joint's possible movements. Each exhibit also refers you to a figure that illustrates the joint. The joints described are the shoulder (humeroscapular or glenohumeral) joint, elbow joint, hip (coxal) joint, and knee (tibiofemoral) joint.

Table 9.2 Summary of Movements at Synovial Joints

MOVEMENT	DESCRIPTION
Gliding	Movement of relatively flat bone surfaces back and forth and from side to side over one another; little change in the angle between bones occurs.
Angular	Increase or decrease in the angle between bones.
Flexion	Decrease in the angle between articulating bones, usually in the sagittal plane.
Lateral flexion	Movement of the trunk in the frontal plane.
Extension	Increase in the angle between articulating bones, usually in the sagittal plane.
Hyperextension	Extension beyond the anatomical position.
Abduction	Movement of a bone away from the midline, usually in the frontal plane.
Adduction	Movement of a bone toward the midline, usually in the frontal plane.
Circumduction	Flexion, abduction, extension, and adduction in succession, in which the distal end of a body part moves in a circle.
Rotation	Movement of a bone around its longitudinal axis; in the limbs, it may be medial (toward midline) or lateral (away from midline).
Special	Occurs at specific joints.
Elevation	Superior movement of a body part.
Depression	Inferior movement of a body part.
Protraction	Anterior movement of a body part in the transverse plane.
Retraction	Posterior movement of a body part in the transverse plane.
Inversion	Medial movement of the soles so that they face each other.
Eversion	Lateral movement of the soles so that they face away from each other.
Dorsiflexion	Bending the foot in the direction of the dorsum (superior surface).
Plantar flexion	Bending the foot in the direction of the plantar surface (sole).
Supination	Movement of the forearm that turns the palm anteriorly or superiorly.
Pronation	Movement of the forearm that turns the palm posteriorly or inferiorly.
Opposition	Movement of the thumb across the palm to touch fingertips on the same hand.

Text continues on page 261

Exhibit 9.1	*Shoulder (Humeroscapular or Glenohumeral) Joint (Figure 9.11)*

OBJECTIVE

• *Describe the anatomical components of the shoulder joint, and explain which types of movement can occur at this joint.*

DEFINITION

Ball-and-socket joint formed by the head of the humerus and the glenoid cavity of the scapula.

ANATOMICAL COMPONENTS

1. *Articular capsule*. Thin, loose sac that completely envelops the joint. It extends from the glenoid cavity to the anatomical neck of the humerus. The inferior part of the capsule is its weakest area.
2. *Coracohumeral ligament*. Strong, broad ligament that strengthens the superior part of the articular capsule and extends from the coracoid process of the scapula to the greater tubercle of the humerus.
3. *Glenohumeral ligaments*. Three thickenings of the articular capsule over the anterior surface of the joint. They extend from the glenoid cavity to the lesser tubercle and anatomical neck of the humerus. These ligaments are often indistinct or absent and provide only minimal strength.
4. *Transverse humeral ligament*. Narrow sheet extending from the greater tubercle to the lesser tubercle of the humerus.

5. *Glenoid labrum*. Narrow rim of fibrocartilage around the edge of the glenoid cavity. It slightly deepens and enlarges the glenoid cavity.
6. Four *bursae* are associated with the shoulder joint. They are the *subscapular bursa, subdeltoid bursa, subacromial bursa,* and *subcoracoid bursa.*

MOVEMENTS

Produces flexion, extension, abduction, adduction, medial rotation, lateral rotation, and circumduction of the arm (see Figures 9.6–9.9).

The shoulder joint has more freedom of movement than any other joint of the body. This freedom results from the looseness of the articular capsule and shallowness of the glenoid cavity in relation to the large size of the head of the humerus.

Although the ligaments of the shoulder joint strengthen it to some extent, most of the strength results from the muscles that surround the joint, especially the *rotator cuff muscles*. These muscles (supraspinatus, infraspinatus, teres minor, and subscapularis) join the scapula to the humerus (see also Figure 11.14). The tendons of the muscles, together called the **rotator cuff,** encircle the joint (except for the inferior portion) and fuse with the articular capsule. The rotator cuff muscles work as a group to hold the head of the humerus in the glenoid cavity.

Which tendons at the shoulder joint of a baseball pitcher are most likely to be torn due to excessive circumduction?

Figure 9.11 Right shoulder (humeroscapular or glenohumeral) joint. (See Tortora, *A Photographic Atlas of the Human Body,* Figures 4.1 and 4.2)

Most of the stability of the shoulder joints results from the arrangement of the rotator cuff muscles.

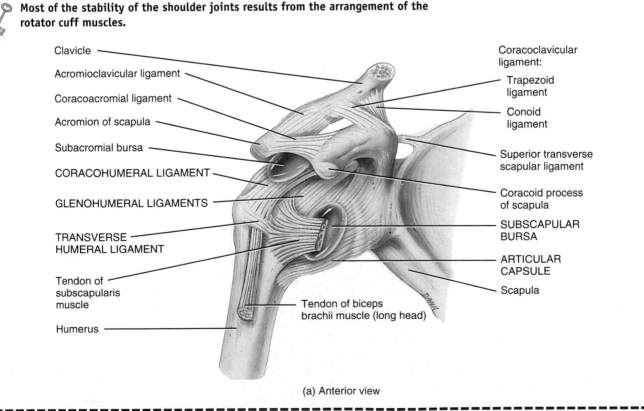

(a) Anterior view

Exhibit 9.1 *(continued)*

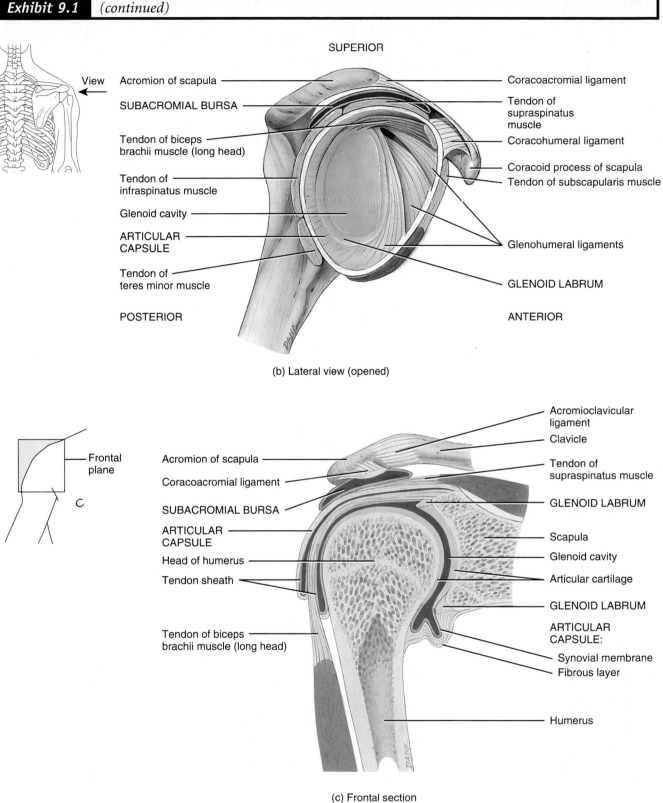

SUPERIOR

View

Acromion of scapula

SUBACROMIAL BURSA

Tendon of biceps brachii muscle (long head)

Tendon of infraspinatus muscle

Glenoid cavity

ARTICULAR CAPSULE

Tendon of teres minor muscle

POSTERIOR

Coracoacromial ligament

Tendon of supraspinatus muscle

Coracohumeral ligament

Coracoid process of scapula
Tendon of subscapularis muscle

Glenohumeral ligaments

GLENOID LABRUM

ANTERIOR

(b) Lateral view (opened)

Frontal plane

Acromion of scapula

Coracoacromial ligament

SUBACROMIAL BURSA

ARTICULAR CAPSULE

Head of humerus

Tendon sheath

Tendon of biceps brachii muscle (long head)

Acromioclavicular ligament

Clavicle

Tendon of supraspinatus muscle

GLENOID LABRUM

Scapula

Glenoid cavity

Articular cartilage

GLENOID LABRUM

ARTICULAR CAPSULE:

Synovial membrane
Fibrous layer

Humerus

(c) Frontal section

Q Why does the shoulder joint have more freedom of movement than any other joint of the body?

Exhibit 9.2	*Elbow Joint (Figure 9.12)*

OBJECTIVE

• *Describe the anatomical components of the elbow joint, and explain which types of movement can occur at this joint.*

DEFINITION

Hinge joint formed by the trochlea of the humerus, the trochlear notch of the ulna, and the head of the radius.

ANATOMICAL COMPONENTS

1. *Articular capsule.* The anterior part covers the anterior part of the joint, from the radial and coronoid fossae of the humerus to the coronoid process of the ulna and the annular ligament of the radius. The posterior part extends from the capitulum, olecranon fossa, and lateral epicondyle of the humerus to the annular ligament of the radius, the olecranon of the ulna, and the ulna posterior to the radial notch.

2. *Ulnar collateral ligament.* Thick, triangular ligament that extends from the medial epicondyle of the humerus to the coronoid process and olecranon of the ulna.

3. *Radial collateral ligament.* Strong, triangular ligament that extends from the lateral epicondyle of the humerus to the annular ligament of the radius and the radial notch of the ulna.

MOVEMENTS

Flexion and extension of the forearm (see Figure 9.6c).

At the elbow joint, which ligaments connect (a) the humerus and the ulna, and (b) the humerus and the radius?

Figure 9.12 Right elbow joint. (See Tortora, *A Photographic Atlas of the Human Body,* Figures 4.3 and 4.4)

The elbow joint is formed by parts of three bones: humerus, ulna, and radius.

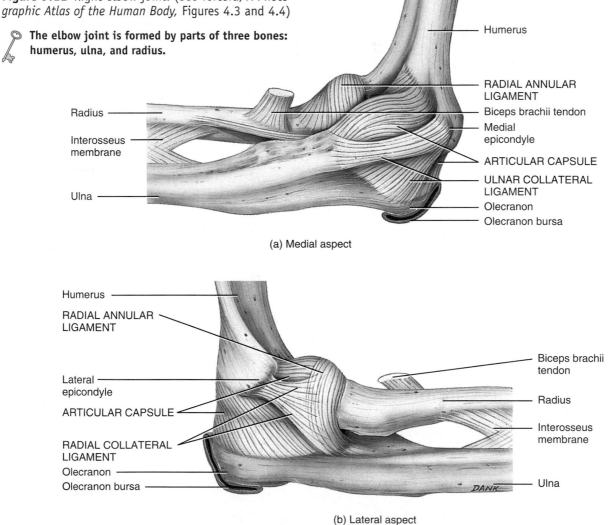

(a) Medial aspect

(b) Lateral aspect

 Which movements are possible at a hinge joint?

Exhibit 9.3   Hip (Coxal) Joint (Figure 9.13)

OBJECTIVE

• *Describe the anatomical components of the hip joint, and explain which types of movement can occur at this joint.*

DEFINITION

Ball-and-socket joint formed by the head of the femur and the acetabulum of the hip bone.

ANATOMICAL COMPONENTS

1. *Articular capsule.* Very dense and strong capsule that extends from the rim of the acetabulum to the neck of the femur. One of the strongest structures of the body, the capsule consists of circular and longitudinal fibers. The circular fibers, called the *zona orbicularis*, form a collar around the neck of the femur. The longitudinal fibers are reinforced by accessory ligaments known as the iliofemoral ligament, pubofemoral ligament, and ischiofemoral ligament.
2. *Iliofemoral ligament.* Thickened portion of the articular capsule that extends from the anterior inferior iliac spine of the hip bone to the intertrochanteric line of the femur.
3. *Pubofemoral ligament.* Thickened portion of the articular capsule that extends from the pubic part of the rim of the acetabulum to the neck of the femur.
4. *Ischiofemoral ligament.* Thickened portion of the articular capsule that extends from the ischial wall of the acetabulum to the neck of the femur.
5. *Ligament of the head of the femur.* Flat, triangular band that extends from the fossa of the acetabulum to the fovea capitis of the head of the femur.

6. *Acetabular labrum.* Fibrocartilage rim attached to the margin of the acetabulum that enhances the depth of the acetabulum. Because the diameter of the acetabular rim is smaller than that of the head of the femur, dislocation of the femur is rare.
7. *Transverse ligament of the acetabulum.* Strong ligament that crosses over the acetabular notch. It supports part of the acetabular labrum and is connected with the ligament of the head of the femur and the articular capsule.

MOVEMENTS

Flexion, extension, abduction, adduction, circumduction, medial rotation, and lateral rotation of thigh (see Figures 9.6–9.9).

The extreme stability of the hip joint is related to the very strong articular capsule at its accessory ligaments, the manner in which the femur fits into the acetabulum, and the muscles surrounding the joint. Although the shoulder and hip joints are both ball-and-socket joints, the movements at the hip joints do not have as wide a range of motion. Flexion is limited by the anterior surface of the thigh coming into contact with the anterior abdominal wall when the knee is flexed and by tension of the hamstring muscles when the knee is extended. Extension is limited by tension of the iliofemoral, pubofemoral, and ischiofemoral ligaments. Abduction is limited by the tension of the pubofemoral ligament, and adduction is limited by contact with the opposite limb and tension in the ligament of the head of the femur. Medial rotation is limited by the tension in the ischiofemoral ligament, and lateral rotation is limited by tension in the iliofemoral and pubofemoral ligaments.

Explain the factors that limit the degree of flexion and abduction at the hip joint.

Figure 9.13 Right hip (coxal) joint. (See Tortora, *A Photographic Atlas of the Human Body,* Figure 4.5)

🔑 **The articular capsule of the hip joint is one of the strongest structures in the body.**

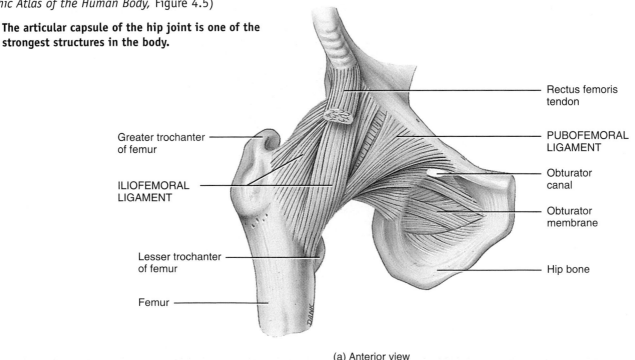

Greater trochanter of femur

ILIOFEMORAL LIGAMENT

Lesser trochanter of femur

Femur

Rectus femoris tendon

PUBOFEMORAL LIGAMENT

Obturator canal

Obturator membrane

Hip bone

(a) Anterior view

Exhibit 9.3 *Hip (Coxal) Joint (continued)*

Figure 9.13 (continued)

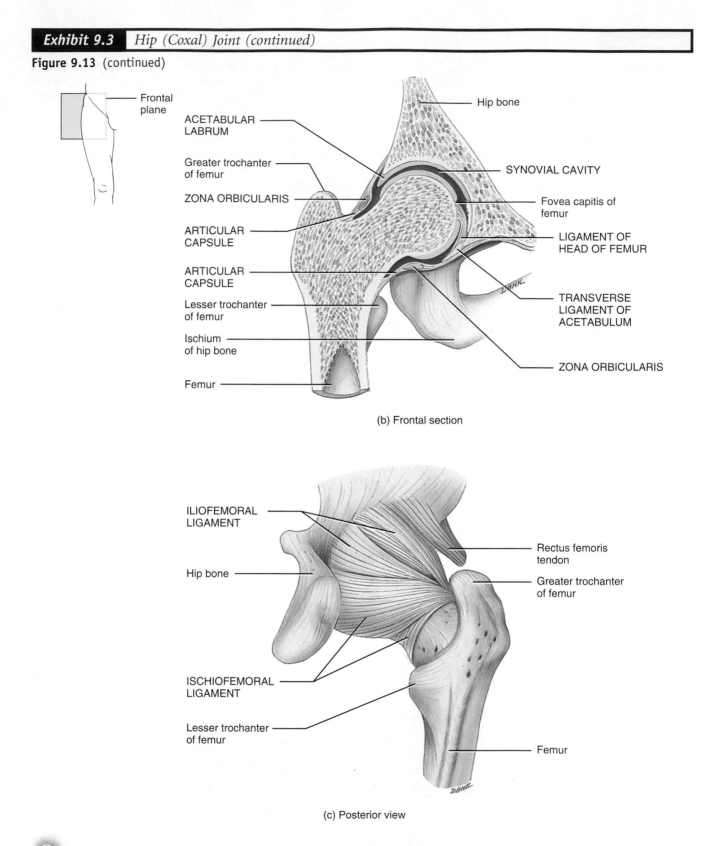

(b) Frontal section

(c) Posterior view

Q   Which ligaments limit the degree of extension that is possible at the hip joint?

Exhibit 9.4 *Knee (Tibiofemoral) Joint (Figure 9.14)*

OBJECTIVE

• *Describe the anatomical components of the knee joint, and explain which types of movement can occur at this joint.*

DEFINITION

The largest and most complex joint of the body, actually consisting of three joints within a single synovial cavity: (1) an intermediate patellofemoral joint between the patella and the patellar surface of the femur, which is a planar joint; (2) a lateral tibiofemoral joint between the lateral condyle of the femur, lateral meniscus, and lateral condyle of the tibia, which is a modified hinge joint; and (3) a medial tibiofemoral joint between the medial condyle of the femur, medial meniscus, and medial condyle of the tibia, which is also a modified hinge joint.

ANATOMICAL COMPONENTS

1. *Articular capsule*. No complete, independent capsule unites the bones. The ligamentous sheath surrounding the joint consists mostly of muscle tendons or expansions of them. There are, however, some capsular fibers connecting the articulating bones.
2. *Medial and lateral patellar retinacula*. Fused tendons of insertion of the quadriceps femoris muscle and the fascia lata (deep fascia of thigh) that strengthen the anterior surface of the joint.
3. *Patellar ligament*. Continuation of the common tendon of insertion of the quadriceps femoris muscle that extends from the patella to the tibial tuberosity. This ligament also strengthens the anterior surface of the joint. The posterior surface of the ligament is separated from the synovial membrane of the joint by an infrapatellar fat pad.
4. *Oblique popliteal ligament*. Broad, flat ligament that extends from the intercondylar fossa of the femur to the head of the tibia. The tendon of the semimembranosus muscle is superficial to the ligament and passes from the medial condyle of the tibia to the lateral condyle of the femur. The ligament and tendon strengthen the posterior surface of the joint.
5. *Arcuate popliteal ligament*. Extends from the lateral condyle of the femur to the styloid process of the head of the fibula. It strengthens the lower lateral part of the posterior surface of the joint.
6. *Tibial collateral ligament*. Broad, flat ligament on the medial surface of the joint that extends from the medial condyle of the femur to the medial condyle of the tibia. The ligament is crossed by tendons of the sartorius, gracilis, and semitendinosus muscles, all of which strengthen the medial aspect of the joint. Because the tibial collateral ligament is firmly attached to the medial meniscus, tearing of the ligament frequently results in tearing of the meniscus and damage to the anterior cruciate ligament, described under 8a.
7. *Fibular collateral ligament*. Strong, rounded ligament on the lateral surface of the joint that extends from the lateral condyle of the femur to the lateral side of the head of the fibula. It strengthens the lateral aspect of the joint. The ligament is covered by the tendon of the biceps femoris muscle. The tendon of the popliteal muscle is deep to the ligament.
8. *Intracapsular ligaments*. Ligaments within the capsule that connect the tibia and femur.
    a. *Anterior cruciate ligament (ACL)*. Extends posteriorly and laterally from the area anterior to the intercondylar eminence of the tibia to the posterior part of the medial surface of the lateral condyle of the femur. This ligament is stretched or torn in about 70% of all serious knee injuries.
    b. *Posterior cruciate ligament (PCL)*. Extends anteriorly and medially from a depression on the posterior intercondylar area of the tibia and lateral meniscus to the anterior part of the medial surface of the medial condyle of the femur.
9. *Articular discs (menisci)*. Two fibrocartilage discs between the tibial and femoral condyles that help compensate for the irregular shapes of the bones and circulate synovial fluid.
    a. *Medial meniscus*. Semicircular piece of fibrocartilage (C-shaped). Its anterior end is attached to the anterior intercondylar fossa of the tibia, anterior to the anterior cruciate ligament. Its posterior end is attached to the posterior intercondylar fossa of the tibia between the attachments of the posterior cruciate ligament and lateral meniscus.
    b. *Lateral meniscus*. Nearly circular piece of fibrocartilage (approaches an incomplete O in shape). Its anterior end is attached anterior to the intercondylar eminence of the tibia, and lateral and posterior to the anterior cruciate ligament. Its posterior end is attached posterior to the intercondylar eminence of the tibia, and anterior to the posterior end of the medial meniscus. The medial and lateral menisci are connected to each other by the *transverse ligament* and to the margins of the head of the tibia by the *coronary ligaments* (not illustrated).
10. The more important *bursae* of the knee include the following:
    a. *Prepatellar bursa* between the patella and skin.
    b. *Intrapatellar bursa* between superior part of tibia and patellar ligament.
    c. *Suprapatellar bursa* between inferior part of femur and deep surface of quadriceps femoris muscle.

MOVEMENTS

Flexion, extension, slight medial rotation, and lateral rotation of leg in flexed position (see Figures 9.6f and 9.9c).

| Explain why the knee joint is said to be "three joints in one."

Exhibit 9.4 *Knee (Tibiofemoral) Joint (continued)*

Figure 9.14 Right knee (tibiofemoral) joint. (See Tortora, *A Photographic Atlas of the Human Body,* Figures 4.6 through 4.8)

🔑 **The knee joint is the largest and most complex joint in the body.**

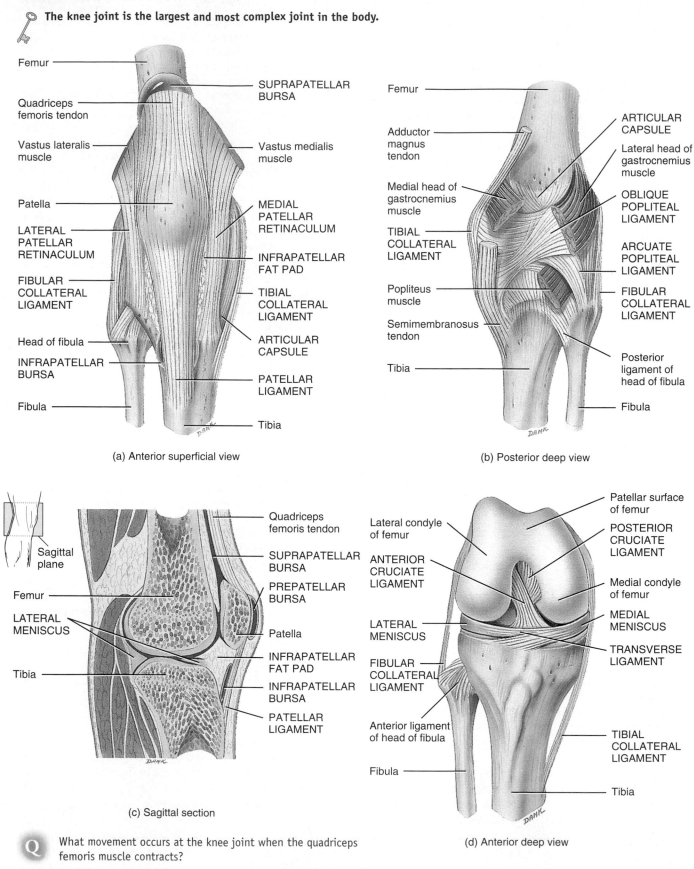

Femur
SUPRAPATELLAR BURSA
Quadriceps femoris tendon
Vastus lateralis muscle
Vastus medialis muscle
Patella
MEDIAL PATELLAR RETINACULUM
LATERAL PATELLAR RETINACULUM
INFRAPATELLAR FAT PAD
FIBULAR COLLATERAL LIGAMENT
TIBIAL COLLATERAL LIGAMENT
Head of fibula
ARTICULAR CAPSULE
INFRAPATELLAR BURSA
PATELLAR LIGAMENT
Fibula
Tibia

(a) Anterior superficial view

Femur
Adductor magnus tendon
ARTICULAR CAPSULE
Medial head of gastrocnemius muscle
Lateral head of gastrocnemius muscle
TIBIAL COLLATERAL LIGAMENT
OBLIQUE POPLITEAL LIGAMENT
Popliteus muscle
ARCUATE POPLITEAL LIGAMENT
Semimembranosus tendon
FIBULAR COLLATERAL LIGAMENT
Tibia
Posterior ligament of head of fibula
Fibula

(b) Posterior deep view

Sagittal plane
Quadriceps femoris tendon
SUPRAPATELLAR BURSA
PREPATELLAR BURSA
Femur
LATERAL MENISCUS
Patella
INFRAPATELLAR FAT PAD
Tibia
INFRAPATELLAR BURSA
PATELLAR LIGAMENT

(c) Sagittal section

Patellar surface of femur
Lateral condyle of femur
POSTERIOR CRUCIATE LIGAMENT
ANTERIOR CRUCIATE LIGAMENT
Medial condyle of femur
LATERAL MENISCUS
MEDIAL MENISCUS
FIBULAR COLLATERAL LIGAMENT
TRANSVERSE LIGAMENT
Anterior ligament of head of fibula
TIBIAL COLLATERAL LIGAMENT
Fibula
Tibia

(d) Anterior deep view

Q What movement occurs at the knee joint when the quadriceps femoris muscle contracts?

Table 9.3 Selected Joints According to Articular Components, Structural and Functional Classification, and Movements

JOINT	ARTICULAR COMPONENTS	CLASSIFICATION	MOVEMENTS
Axial Skeleton			
Suture	Between skull bones.	*Structural:* fibrous. *Functional:* synarthrosis.	None.
Temporomandibular (TMJ)	Between condylar process of mandible and mandibular fossa and articular tubercle of temporal bone.	*Structural:* synovial (combined hinge and planar joint). *Functional:* diarthrosis.	Depression, elevation, protraction, retraction, lateral displacement, and slight rotation of mandible.
Atlanto-occipital	Between superior articular facets of atlas and occipital condyles of occipital bone.	*Structural:* synovial (condyloid). *Functional:* diarthrosis.	Flexion and extension of head and slight lateral flexion of head to either side.
Atlanto-axial	(1) Between dens of axis and anterior arch of atlas and (2) between lateral masses of atlas and axis.	*Structural:* synovial (pivot) between dens and anterior arch, and synovial (planar) between lateral masses. *Functional:* diarthrosis.	Rotation of head.
Intervertebral	(1) Between vertebral bodies and (2) between vertebral arches.	*Structural:* cartilaginous (symphysis) between vertebral bodies, and synovial (planar) between vertebral arches. *Functional:* amphiarthrosis between vertebral bodies, and diarthrosis between vertebral arches.	Flexion, extension, lateral flexion, and rotation of vertebral column.
Vertebrocostal	(1) Between facets of heads of ribs and facets of bodies of adjacent thoracic vertebrae and intervertebral discs between them and (2) between articular part of tubercles of ribs and facets of transverse processes of thoracic vertebrae.	*Structural:* synovial (planar). *Functional:* diarthrosis.	Slight gliding.
Sternocostal	Between sternum and first seven pairs of ribs.	*Structural:* cartilaginous (synchondrosis) between sternum and first pair of ribs, and synovial (planar) between sternum and second through seventh pairs of ribs. *Functional:* synarthrosis between sternum and first pair of ribs, and diarthrosis between sternum and second through seventh pairs of ribs.	None between sternum and first pair of ribs; slight gliding between sternum and second through seventh pairs of ribs.
Lumbosacral	(1) Between body of fifth lumbar vertebra and base of sacrum and (2) between inferior articular facets of fifth lumbar vertebra and superior articular facets of first vertebra of sacrum.	*Structural:* cartilaginous (symphysis) between body and base, and synovial (planar) between articular facets. *Functional:* amphiarthrosis between body and base, and diarthrosis between articular facets.	Flexion, extension, lateral flexion, and rotation of vertebral column.

In previous chapters we discussed the major bones and their markings. In this chapter we have examined how joints are classified according to their structure and function, and we have introduced the movements that occur at joints. Please refer to Table 9.3, which will help you integrate the information you learned about bones with the information about joint classifications and movements. This table lists some of the major joints of the body according to their articular components (the bones that enter into their formation), their structural and functional classification, and the type(s) of movement that occurs at each joint. Because the shoulder, elbow, hip, and knee joints are described in Exhibits 9.1 through 9.4, they are not included in Table 9.3.

Table 9.3 Selected Joints According to Articular Components, Structural and Functional Classification, and Movements (continued)

JOINT	ARTICULAR COMPONENTS	CLASSIFICATION	MOVEMENTS
Girdles and Limbs			
Sternoclavicular	Between sternal end of clavicle, manubrium of sternum, and first costal cartilage.	*Structural:* synovial (planar and pivot). *Functional:* diarthrosis.	Gliding, with limited movements in nearly every direction.
Acromioclavicular	Between acromion of scapula and acromial end of clavicle.	*Structural:* synovial (planar). *Functional:* diarthrosis.	Gliding and rotation of scapula on clavicle.
Radioulnar	Proximal radioulnar joint between head of radius and radial notch of ulna; distal radioulnar joint between ulnar notch of radius and head of ulna.	*Structural:* synovial (pivot). *Functional:* diarthrosis.	Rotation of forearm.
Wrist (radiocarpal)	Between distal end of radius and scaphoid, lunate, and triquetrum of carpus.	*Structural:* synovial (condyloid). *Functional:* diarthrosis.	Flexion, extension, abduction, adduction, and circumduction of wrist.
Intercarpal	Between proximal row of carpal bones, distal row of carpal bones, and between both rows of carpal bones (midcarpal joints).	*Structural:* synovial (planar), except for hamate, scaphoid, and lunate (midcarpal) joint, which is synovial (saddle). *Functional:* diarthrosis.	Gliding plus flexion and abduction at midcarpal joints.
Carpometacarpal	Carpometacarpal joint of thumb between trapezium of carpus and first metacarpal; carpometacarpal joints of remaining digits formed between carpus and second through fifth metacarpals.	*Structural:* synovial (saddle) at thumb and synovial (planar) at remaining digits. *Functional:* diarthrosis.	Flexion, extension, abduction, adduction, and circumduction at thumb, and gliding at remaining digits.
Metacarpophalangeal and metatarsophalangeal	Between heads of metacarpals (or metatarsals) and bases of proximal phalanges.	*Structural:* synovial (condyloid). *Functional:* diarthrosis.	Flexion, extension, abduction, adduction, and circumduction of phalanges.
Interphalangeal	Between heads of phalanges and bases of more distal phalanges.	*Structural:* synovial (hinge). *Functional:* diarthrosis.	Flexion and extension of phalanges.

FACTORS AFFECTING CONTACT AND RANGE OF MOTION AT SYNOVIAL JOINTS

OBJECTIVE

• *Describe six factors that influence the type of movement and range of motion that is possible at a synovial joint.*

The articular surfaces of synovial joints contact one another and determine the type and range of motion that is possible. **Range of motion (ROM)** refers to the range, measured in degrees of a circle, through which the bones of a joint can be moved. The following factors contribute to keeping the articular surfaces in contact and affect range of motion:

1. *Structure or shape of the articulating bones.* The structure or shape of the articulating bones determines how closely they can fit together. The articular surfaces of some bones interlock with one another. This spatial relationship is very obvious at the hip joint, where the head of the femur articulates with the acetabulum of the hip bone. An interlocking fit allows rotational movement.

2. *Strength and tension (tautness) of the joint ligaments.* The different components of a fibrous capsule are tense or taut only when the joint is in certain positions. Tense ligaments not only restrict the range of motion but also direct the movement of the articulating bones with respect to each other. In the knee joint, for example, the anterior cruciate ligament is taut and the posterior cruciate ligament is loose when the knee is straightened, and the reverse occurs when the knee is bent.

3. *Arrangement and tension of the muscles.* Muscle tension reinforces the restraint placed on a joint by its ligaments, and thus restricts movement. A good example of the effect of muscle tension on a joint is seen at the hip joint. When the thigh is raised with the knee straight, the movement is restricted by the tension of the hamstring muscles on the posterior surface of the thigh. But if the knee is bent, the tension on the hamstring muscles is lessened, and the thigh can be raised farther.

Table 9.3 (continued)

JOINT	ARTICULAR COMPONENTS	CLASSIFICATION	MOVEMENTS
Girdles and Limbs			
Sacroiliac	Between auricular surfaces of sacrum and ilia of hip bones.	*Structural:* synovial (planar). *Functional:* diarthrosis.	Slight gliding (even more so during pregnancy).
Pubic symphysis	Between anterior surfaces of hip bones.	*Structural:* cartilaginous (symphysis). *Functional:* amphiarthrosis.	Slight movements (even more so during pregnancy).
Tibiofibular	Proximal tibiofibular joint between lateral condyle of tibia and head of fibula; distal tibiofibular joint between distal end of fibula and fibular notch of tibia	*Structural:* synovial (planar) at proximal joint, and fibrous (syndesmosis) at distal joint. *Functional:* diarthrosis at proximal joint, and amphiarthrosis at distal joint.	Slight gliding at proximal joint, and slight rotation of fibula during dorsiflexion of foot.
Ankle (talocrural)	(1) Between distal end of tibia and its medial malleolus and talus and (2) between lateral malleolus of fibula and talus.	*Structural:* synovial (hinge). *Functional:* diarthrosis.	Dorsiflexion and plantar flexion of foot.
Intertarsal	Subtalar joint between talus and calcaneus of tarsus; talocalcaneonavicular joint between talus and calcaneus and navicular of tarsus; calcaneocuboid joint between calcaneus and cuboid of tarsus.	*Structural:* synovial (planar) at subtalar and calcaneocuboid joints, and synovial at talocalcaneonavicular joint. *Functional:* diarthrosis.	Inversion and eversion of foot.
Tarsometatarsal	Between three cuneiforms of tarsus and bases of five metatarsal bones.	*Structural:* synovial (planar). *Functional:* diarthrosis.	Slight gliding.

4. *Apposition of soft parts.* The point at which one body surface contacts another may limit mobility. For example, if you bend your arm at the elbow, it can move no farther after the anterior surface of the forearm meets with and presses against the biceps brachii muscle of the arm. Joint movement may also be restricted by the presence of adipose tissue.

5. *Hormones.* Joint flexibility may also be affected by hormones. For example, relaxin, a hormone produced by the placenta and ovaries, increases the flexibility of the fibrocartilage of the pubic symphysis and loosens the ligaments between the sacrum, hip bone, and coccyx toward the end of pregnancy. These changes permit expansion of the pelvic outlet, which assists in delivery of the baby.

6. *Disuse.* Movement at a joint may be restricted if a joint has not been used for an extended period. For example, if an elbow joint is immobilized by a cast, only a limited range of motion at the joint may be possible for a time after the cast is removed. Restricted movement due to disuse may result from decreased amounts of synovial fluid, from diminished flexibility of ligaments and tendons, and from muscular atrophy, a reduction in size or wasting away of a muscle.

1. Besides the hip joint, which joints have bones that interlock and allow rotational movement?

AGING AND JOINTS
OBJECTIVE
• *Explain the effects of aging on joints.*

The aging process usually results in decreased production of synovial fluid in joints. In addition, the articular cartilage becomes thinner with age, and ligaments shorten and lose some of their flexibility. The effects of aging on joints, which vary considerably from one person to another, are affected by genetic factors and by wear and tear. Although degenerative changes in joints may begin in individuals as young as 20 years of age, most do not occur until much later. In fact, by age 80, almost everyone develops some type of degeneration in the knees, elbows, hips, and shoulders. Males also commonly develop degenerative changes in the vertebral column, resulting in a hunched-over posture and pressure on nerve roots. As you will see in the following section, one type of arthritis, called osteoarthritis, is at least partially age-related. Nearly everyone over age 70 has evidence of some osteoarthritic changes.

1. Which joints show evidence of degeneration in nearly all individuals as aging progresses?

DISORDERS: HOMEOSTATIC IMBALANCES

RHEUMATISM AND ARTHRITIS

Rheumatism (ROO-ma-tizm) is any of a variety of painful disorders of the supporting structures of the body—bones, ligaments, tendons, or muscles. **Arthritis** is a form of rheumatism in which the joints have become inflamed. Inflammation, pain, and stiffness may also involve adjacent muscles. Arthritis afflicts about 40 million people in the United States.

Three important classes of arthritis are (1) diffuse connective tissue diseases, such as rheumatoid arthritis, (2) degenerative joint disease, such as osteoarthritis, and (3) metabolic and endocrine diseases with associated arthritis, such as gouty arthritis.

Rheumatoid Arthritis

Rheumatoid arthritis (RA) is an autoimmune disease in which the immune system of the body attacks its own tissues—in this case, its own cartilage and joint linings. RA is characterized by inflammation of the joint, which causes swelling, pain, and loss of function. Usually, this form of arthritis occurs bilaterally: If one wrist is affected, the other is also likely to be affected, although usually not to the same degree.

The primary symptom of rheumatoid arthritis is inflammation of the synovial membrane. If untreated, the membrane thickens, and synovial fluid accumulates. The resulting pressure causes pain and tenderness. The membrane then produces an abnormal granulation tissue, called *pannus,* that adheres to the surface of the articular cartilage and sometimes erodes the cartilage completely. When the cartilage is destroyed, fibrous tissue joins the exposed bone ends. The fibrous tissue ossifies and fuses the joint so that it becomes immovable—the ultimate crippling effect of rheumatoid arthritis. The growth of the granulation tissue causes the distortion of the fingers that characterizes hands of RA sufferers.

Osteoarthritis

Osteoarthritis (OA) (os′-tē-ō-ar-THRĪ-tis) is a degenerative joint disease that apparently results from a combination of aging, irritation of the joints, and wear and abrasion. It is commonly known as "wear-and-tear" arthritis and is the leading cause of disability in older individuals.

Osteoarthritis is a progressive disorder of synovial joints, particularly weight-bearing joints. It is characterized by the deterioration of articular cartilage and by the formation of new bone in the subchondral areas and at the margins of the joint. The cartilage slowly degenerates, and as the bone ends become exposed, spurs (small bumps) of new osseous tissue are deposited on them. These spurs decrease the space of the joint cavity and restrict joint movement. Unlike rheumatoid arthritis, osteoarthritis usually affects mainly the articular cartilage, although the synovial membrane often becomes inflamed late in the disease. A major distinction between osteoarthritis and rheumatoid arthritis is that osteoarthritis strikes the larger joints (knees, hips) first, whereas rheumatoid arthritis first strikes smaller joints.

Gouty Arthritis

Uric acid (a substance that gives urine its name) is a waste product produced during the metabolism of nucleic acid (DNA and RNA) subunits. A person who suffers from **gout** (gowt) either produces excessive amounts of uric acid or is not able to excrete as much as normal. The result is a buildup of uric acid in the blood. This excess acid then reacts with sodium to form a salt called sodium urate. Crystals of this salt accumulate in soft tissues such as the kidneys and in the cartilage of the ears and joints.

In **gouty arthritis,** sodium urate crystals are deposited in the soft tissues of the joints. The crystals irritate and erode the cartilage, causing inflammation, swelling, and acute pain. Eventually, the crystals destroy all joint tissues. If the disorder is untreated, the ends of the articulating bones fuse, and the joint becomes immovable.

MEDICAL TERMINOLOGY

Arthralgia (ar-THRAL-jē-a; *arthr-* = joint; *-algia* = pain) Pain in a joint.

Bursectomy (bur-SEK-tō-mē; *-ectomy* = removal of) Removal of a bursa.

Chondritis (kon-DRĪ-tis; *chondr-* = cartilage) Inflammation of cartilage.

Dislocation (dis′-lō-KĀ-shun; *dis-* = apart) The displacement of a bone from a joint, with resultant tearing of ligaments, tendons, and articular capsules; usually caused by a blow or fall. Also termed **luxation** (luks-Ā-shun; *luxatio* = dislocation).

Subluxation (sub-luks-Ā-shun) A partial or incomplete dislocation.

Synovitis (sin′-ō-VĪ-tis) Inflammation of a synovial membrane in a joint.

STUDY OUTLINE

INTRODUCTION (p. 241)

1. A joint (articulation or arthrosis) is a point of contact between two bones, between bone and cartilage, or between bone and teeth.

2. A joint's structure may permit no movement, slight movement, or free movement.

JOINT CLASSIFICATIONS (p. 241)

1. Structural classification is based on the presence or absence of a synovial cavity and the type of connecting tissue. Structurally, joints are classified as fibrous, cartilaginous, or synovial.
2. Functional classification of joints is based on the degree of movement permitted. Joints may be synarthroses (immovable), amphiarthroses (slightly movable), or diarthroses (freely movable).

FIBROUS JOINTS (p. 242)

1. Bones held by fibrous connective tissue are fibrous joints.
2. These joints include immovable sutures (found between skull bones), slightly movable syndesmoses (such as the distal tibiofibular joint), and immovable gomphoses (roots of teeth in alveoli of mandible and maxilla).

CARTILAGINOUS JOINTS (p. 243)

1. Bones held together by cartilage are cartilaginous joints.
2. These joints include immovable synchondroses united by hyaline cartilage (epiphyseal plates between diaphyses and epiphyses) and slightly movable symphyses united by fibrocartilage (pubic symphysis).

SYNOVIAL JOINTS (p. 243)

1. Synovial joints contain a space between bones called the synovial cavity. All synovial joints are diarthroses.
2. Other characteristics of synovial joints are the presence of articular cartilage and an articular capsule, made up of a fibrous capsule and a synovial membrane.
3. The synovial membrane secretes synovial fluid, which forms a thin, viscous film over the surfaces within the articular capsule.
4. Many synovial joints also contain accessory ligaments (extracapsular and intracapsular) and articular discs (menisci).
5. Synovial joints contain an extensive nerve and blood supply. The nerves convey information about pain, joint movements, and the degree of stretch at a joint. Blood vessels penetrate the articular capsule and ligaments.
6. Subtypes of synovial joints are planar, hinge, pivot, condyloid, saddle, and ball-and-socket.
7. Planar joint: articulating surfaces are flat; bone glides back and forth and side to side (nonaxial); found between carpals and tarsals.
8. Hinge joint: convex surface of one bone fits into concave surface of another; motion is angular around one axis (monaxial); examples are the elbow, knee, and ankle joints.
9. Pivot joint: a round or pointed surface of one bone fits into a ring formed by another bone and a ligament; movement is rotation (monaxial); examples are the atlanto-axial and radioulnar joints.
10. Condyloid joint: an oval-shaped projection of one bone fits into an oval cavity of another; motion is angular around two axes (biaxial); examples are the wrist joint and metacarpophalangeal joints for the second through fifth digits.
11. Saddle joint: articular surface of one bone is shaped like a saddle and the other bone fits into the "saddle" like a sitting rider; motion is angular around two axes (biaxial); an example is the carpometacarpal joint between the trapezium and the metacarpal of the thumb.
12. Ball-and-socket joint: ball-shaped surface of one bone fits into cuplike depression of another; motion is angular and rotational around three axes and all directions in between (multiaxial); examples are the shoulder and hip joints.
13. Bursae are saclike structures, similar in structure to joint capsules, that alleviate friction in joints such as the shoulder and knee joints.
14. Tendon sheaths are tubelike bursae that wrap around tendons where there is considerable friction.
15. Table 9.1 on page 248 summarizes the structural and functional categories of joints.

TYPES OF MOVEMENT AT SYNOVIAL JOINTS (p. 247)

1. In a gliding movement, the nearly flat surfaces of bones move back and forth and from side to side. Gliding movements occur at planar joints.
2. In angular movements, a change in the angle between bones occurs. Examples are flexion-extension, lateral flexion, hyperextension, abduction-adduction. Circumduction refers to flexion, abduction, extension, and adduction in succession. Angular movements occur at hinge, condyloid, saddle, and ball-and-socket joints.
3. In rotation, a bone moves around its own longitudinal axis. Rotation can occur at pivot and ball-and-socket joints.
4. Special movements occur at specific synovial joints. Examples are elevation-depression, protraction-retraction, inversion-eversion, dorsiflexion-plantar flexion, supination-pronation, and opposition.
5. Table 9.2 on page 253 summarizes the various types of movements at synovial joints.

SELECTED JOINTS OF THE BODY (p. 253)

1. The shoulder (humeroscapular or glenohumeral) joint is between the head of the humerus and glenoid cavity of the scapula (Exhibit 9.1 on page 254).
2. The elbow joint is between the trochlea of the humerus, the trochlear notch of the ulna, and the head of the radius (Exhibit 9.2 on page 256).
3. The hip (coxal) joint is between the head of the femur and acetabulum of the hip bone (Exhibit 9.3 on page 257).
4. The knee (tibiofemoral) joint is between the patella and patellar surface of femur; lateral condyle of femur, lateral meniscus, and lateral condyle of tibia; and medial condyle of femur, medial meniscus, and medial condyle of tibia (Exhibit 9.4 on page 259).
5. A summary of selected joints of the body, including articular components, structural and functional classifications, and movements, is presented in Table 9.3 on page 261.

FACTORS AFFECTING CONTACT AND RANGE OF MOTION AT SYNOVIAL JOINTS (p. 262)

1. The ways that articular surfaces of synovial joints contact one another determines the type of movement that is possible.
2. Factors that contribute to keeping the surfaces in contact and affect range of motion are structure or shape of the articulating bones, strength and tension of the joint ligaments, arrangement and tension of the muscles, apposition of soft parts, hormones, and disuse.

AGING AND JOINTS (p. 263)

1. With aging, a decrease in synovial fluid, thinning of articular cartilage, and decreased flexibility of ligaments occur.
2. Most individuals experience some degeneration in the knees, elbows, hips, and shoulders due to the aging process.

SELF-QUIZ QUESTIONS

Complete the following:

1. A point of contact between two bones, between bone and cartilage, or between bone and teeth is called a(n) ___.

2. ___ lubricates, reduces friction, and supplies nutrients to and removes wastes from joints.

3. The largest and most complex joint of the body is the ___ joint.

4. ___ or ___ allow two bones of different shapes to fit more tightly, help maintain the stability of the joint, and help direct the flow of synovial fluid to areas of greatest friction.

Choose the best answer to the following questions:

5. A joint with no synovial cavity and bones held together by collagen-rich fibrous connective tissue is a (a) diarthrotic joint, (b) synovial joint, (c) cartilaginous joint, (d) fibrous joint, (e) functional joint.

6. A slightly movable joint is a(n) (a) synarthrosis, (b) amphiarthrosis, (c) diarthrosis, (d) suture, (e) synostosis.

7. Which of the following is a cartilaginous joint? (a) symphysis, (b) gomphosis, (c) suture, (d) syndesmosis, (e) synovial joint.

8. Which of the following statements are true? (1) The bones at a synovial joint are covered by a mucous membrane. (2) The articular capsule surrounds a synovial joint, encloses the synovial cavity, and unites the articulating bones. (3) The fibrous capsule of the articular capsule permits considerable movement at a joint. (4) The tensile strength of the fibrous capsule helps prevent bones from disarticulating. (5) All joints contain a fibrous capsule.
(a) 1, 2, 3, and 4, (b) 2, 3, 4, and 5, (c) 2, 3, and 4, (d) 1, 2, and 3, (e) 2, 4, and 5

9. Which of the following keep the articular surfaces of synovial joints in contact and affect range of motion? (1) structure or shape of the articulating bones, (2) strength and tension of the joint ligaments, (3) arrangement and tension of muscles, (4) vitamins, (5) hormones.
(a) 1, 2, 3, and 5, (b) 2, 3, 4, and 5, (c) 1, 3, 4, and 5, (d) 1, 3, and 5, (e) 1, 2, 3, 4, and 5

10. Which of the following does *not* reduce friction in joints? (a) bursae, (b) articular cartilage, (c) tendon sheaths, (d) synovial fluid, (e) ligaments

11. Match the following:
___ (a) a fibrous joint that unites the bones of the skull; a synarthrosis
___ (b) a fibrous joint between the tibia and fibula; an amphiarthrosis
___ (c) the articulation between bone and teeth
___ (d) the epiphyseal plate
___ (e) joint between the two pubic bones
___ (f) a bony joint

(1) synostosis
(2) synchondrosis
(3) syndesmosis
(4) suture
(5) symphysis
(6) gomphosis

True or false:

12. The fibers of some fibrous capsules are arranged in parallel bundles called ligaments.

13. A sprain is the forcible wrenching or twisting of a joint, with stretching or tearing of its ligaments without dislocation.

14. Match the following:
___ (a) movement in which relatively flat bone surfaces move back and forth and side to side with respect to one another
___ (b) decrease in angle between bones
___ (c) increase in angle between bones
___ (d) movement of bone away from midline
___ (e) movement of bone toward midline
___ (f) movement of distal end of a part of the body in a circle
___ (g) a bone revolves around its own longitudinal axis

(1) flexion
(2) extension
(3) abduction
(4) adduction
(5) gliding
(6) rotation
(7) circumduction

15. Match the following:
___ (a) upward movement of a body part
___ (b) downward movement of a body part
___ (c) movement of a body part anteriorly in the transverse plane
___ (d) movement of an anteriorly projected body part back to the anatomical position
___ (e) movement of the soles medially at the intertarsal joints
___ (f) movement of the soles laterally
___ (g) action that occurs when you stand on your heels
___ (h) action that occurs when you stand on your toes
___ (i) movement of the forearm to turn the palm anteriorly
___ (j) movement of the forearm to turn the palm posteriorly
___ (k) movement of thumb across the palm to touch the tips of the fingers of the same hand

(1) pronation
(2) plantar flexion
(3) eversion
(4) retraction
(5) opposition
(6) elevation
(7) depression
(8) inversion
(9) protraction
(10) dorsiflexion
(11) supination

CRITICAL THINKING QUESTIONS

1. Katie loves pretending that she's a human cannonball. As she jumps off the diving board, she assumes the proper position before she pounds into the water: head and thighs tucked against her chest; back rounded; arms pressed against her sides while her forearms, crossed in front of her shins, hold her legs tightly folded against her chest. Use the proper anatomical terms to describe the position of Katie's back, head, and limbs. (HINT: *She is about as far as she can get from the anatomical position.*)

2. After the cast was removed from his arm, 85-year-old Mr. Singh was referred to physical therapy for a frozen elbow joint. When the therapist said she would be working to free up his "hinge,"

Mr. Singh complained, "You're supposed to be fixing my elbow, not my back door!" Explain the situation to Mr. Singh. (HINT: *Think about how a door moves.*)

3. Kendal was sitting in the cafeteria about 30 minutes before his big A&P exam. "I'm not worried," he told his lab partner while he drank his third cup of coffee. "I've got it all figured out. There are three types of joints: Cartilaginous joints are made out of hyaline cartilage, like the intervertebral discs and the suture; fibrous joints have an articular capsule; and the serous joints have cavities that do not open to the outside of the body." Kendal's lab partner was speechless. Can you explain the three types better? (HINT: *Synovial fluid lubricates diarthroses.*)

ANSWERS TO FIGURE QUESTIONS

9.1 Functionally, sutures are classified as synarthroses because they are immovable; syndesmoses are classified as amphiarthroses because they are slightly movable.

9.2 The difference between a synchondrosis and a symphysis is the type of cartilage that holds the joint together: hyaline cartilage in a synchondrosis and fibrocartilage in a symphysis.

9.3 Functionally, synovial joints are diarthroses, freely movable joints.

9.4 Condyloid and saddle joints are biaxial joints.

9.5 Gliding movements occur at planar joints.

9.6 Two examples of flexion that do not occur in the sagittal plane are flexion of the thumb and lateral flexion of the trunk.

9.7 When you adduct your arm or leg, you bring it closer to the midline of the body, thus "adding" it to the trunk.

9.8 Circumduction involves flexion, abduction, extension, and adduction in continuous sequence.

9.9 The anterior surface of a bone or limb rotates toward the midline in medial rotation, and away from the midline in lateral rotation.

9.10 Bringing the arms forward until the elbows touch is an example of protraction.

9.11 The shoulder joint is so freely movable because of the looseness of its articular capsule and the shallowness of the glenoid cavity in relation to the size of the head of the humerus.

9.12 A hinge joint permits flexion and extension.

9.13 Tension in three ligaments—iliofemoral, pubofemoral, and ischiofemoral—limit the degree of extension at the hip joint.

9.14 Contraction of the quadriceps causes extension at the knee joint.

Although bones provide leverage and form the framework of the body, they cannot move body parts by themselves. Motion results from the alternating contraction and relaxation of muscles, which constitute 40–50% of total body weight. Your muscular strength reflects the prime function of muscle—changing chemical energy into mechanical energy to generate force, perform work, and produce movement. In addition, muscle tissue stabilizes the body's position, regulates organ volume, generates heat, and propels fluids and food matter through various body systems. The scientific study of muscles is known as **myology** (mī-OL-ō-jē; *my-* = muscle; *-ology* = study of).

OVERVIEW OF MUSCLE TISSUE

OBJECTIVE

• *Correlate the three types of muscle tissue with their functions and special properties.*

Types of Muscle Tissue

There are three types of muscle tissue: skeletal, cardiac, and smooth muscle (see the comparison in Table 4.4 on pages 131–132). Whereas the three types of muscle tissue share some properties, they differ from one another in their microscopic anatomy, location, and control by the nervous and endocrine systems.

Skeletal muscle tissue is so-named because the function of most skeletal muscles is to move bones of the skeleton. (A few skeletal muscles attach to tissues other than bone.) Skeletal muscle tissue is termed *striated* because alternating light and dark bands (*striations*) are visible when the tissue is examined under a microscope (see Figure 10.5 on page 275). Skeletal muscle tissue works primarily in a *voluntary* manner because its activity can be consciously controlled by neurons (nerve cells) that are part of the somatic (voluntary) division of the nervous system. (Figure 12.1 on page 379 depicts the divisions of the nervous system.) Most skeletal muscles also are controlled involuntarily to some extent. For instance, we usually are not aware of the alternating contraction and relaxation of the diaphragm, which is the main skeletal muscle that powers breathing; of the continual contraction of postural muscles, which stabilize body positions; or of the stretch reflexes that help adjust muscle tone (see Figure 13.6 on page 421).

Only the heart contains **cardiac muscle tissue,** which forms most of the heart wall. Cardiac muscle is also *striated,* but its action is *involuntary*—its alternating contraction and relaxation cannot be consciously influenced. The heart beats because it has a pacemaker that initiates each contraction; this built-in or intrinsic rhythm is called *autorhythmicity.* Several hormones and neurotransmitters adjust heart rate by speeding or slowing the pacemaker.

Smooth muscle tissue is located in the walls of hollow internal structures, such as blood vessels, airways, and most organs in the abdominopelvic cavity. It is also found in the skin, attached to hair follicles. Under a microscope, this tissue looks *nonstriated* or *smooth.* The action of smooth muscle is usually *involuntary,* and some smooth muscle tissue also has autorhythmicity. Both cardiac muscle and smooth muscle are regulated by neurons that are part of the autonomic (involuntary) division of the nervous system and by hormones released by endocrine glands.

Functions of Muscle Tissue

Through sustained contraction or alternating contraction and relaxation, muscle tissue has five key functions: producing body movements, stabilizing body positions, regulating organ volume, moving substances within the body, and generating heat.

1. *Producing body movements.* Total body movements such as walking and running, and localized movements such as grasping a pencil or nodding the head, rely on the integrated functioning of bones, joints, and skeletal muscles.
2. *Stabilizing body positions.* Skeletal muscle contractions stabilize joints and help maintain body positions, such as standing or sitting. Postural muscles contract continuously when a person is awake; for example, sustained contractions in neck muscles hold the head upright.
3. *Regulating organ volume.* Sustained contractions of ring-like bands of smooth muscles called *sphincters* may prevent outflow of the contents of a hollow organ. Temporary storage of food in the stomach or urine in the urinary bladder is possible because smooth muscle sphincters close off the outlets of these organs.
4. *Moving substances within the body.* Cardiac muscle contractions pump blood through the body's blood vessels. Contraction and relaxation of smooth muscle in the walls of blood vessels help adjust their diameter and thus regulate the rate of blood flow. Smooth muscle contractions also move food and substances such as bile and enzymes through the gastrointestinal tract, push gametes (sperm and oocytes) through the reproductive systems, and propel urine through the urinary system. Skeletal muscle contractions promote flow of lymph and aid the return of blood to the heart.
5. *Producing heat.* As muscle tissue contracts, it also produces heat. Much of the heat released by muscle is used to maintain normal body temperature. Involuntary contractions of skeletal muscle, known as shivering, can increase the rate of heat production several fold.

Properties of Muscle Tissue

Muscle tissue has four special properties that enable it to function and contribute to homeostasis:

1. **Electrical excitability,** a property of both muscle fibers (cells) and neurons, is the ability to respond to certain stimuli by producing electrical signals—for example, *action potentials.* (Chapter 12 describes how action potentials arise; see page 391.) Action potentials propagate (travel) along a cell's plasma membrane due to the presence of specific ion channels. For muscle fibers, the stimuli that trigger action potentials may be autorhythmic electrical signals arising in the muscle tissue itself, such as occurs in the heart's pacemaker, or chemical stimuli, such as neurotransmitters released by neurons, hormones distributed by the blood, or even local changes in pH.
2. **Contractility** is the ability of muscle tissue to contract forcefully when stimulated by an action potential. When muscle contracts, it generates tension (force of contraction) while pulling on its attachment points. In an **isometric contraction** (*iso-* = equal; *-metric* = measure or length), the muscle develops tension but does not shorten. An example is holding a book in an outstretched hand. If the tension generated is great enough to overcome the resistance of the object to being moved, the muscle shortens and movement occurs. In an **isotonic contraction** (*-tonic* = tension), the tension developed by the muscle remains almost constant while the muscle shortens.
3. **Extensibility** is the ability of muscle to stretch without being damaged. Extensibility allows a muscle to contract forcefully even if it is already stretched. Normally, smooth muscle is subject to the greatest amount of stretching. For example, each time the stomach fills with food, the muscle in the wall is stretched. Cardiac muscle also is stretched each time the heart fills with blood. During normal activities, the stretch on skeletal muscle remains fairly constant.
4. **Elasticity** is the ability of muscle tissue to return to its original length and shape after contraction or extension.

This chapter focuses on the structure and function of skeletal muscle tissue. Cardiac muscle and smooth muscle are examined in greater detail later—in discussions of the autonomic nervous system in Chapter 17, of the heart in Chapter 20, and of the various organs that contain smooth muscle in several chapters.

1. What features distinguish the three types of muscle tissue?
2. Summarize the functions of muscle tissue.

SKELETAL MUSCLE TISSUE
OBJECTIVE
• *Explain the relation of connective tissue components, blood vessels, and nerves to skeletal muscles.*

Each skeletal muscle is a separate organ composed of hundreds to thousands of cells called **fibers** because of their elongated shapes. Connective tissues surround muscle fibers and whole muscles, and blood vessels and nerves penetrate into muscle (Figure 10.1). To understand how skeletal muscle contraction can generate tension, one first needs to understand its gross and microscopic anatomy.

Figure 10.1 Organization of skeletal muscle and its connective tissue coverings.

🔑 **A skeletal muscle consists of individual muscle fibers (cells) bundled into fascicles and surrounded by three connective tissue layers that are extensions of the deep fascia.**

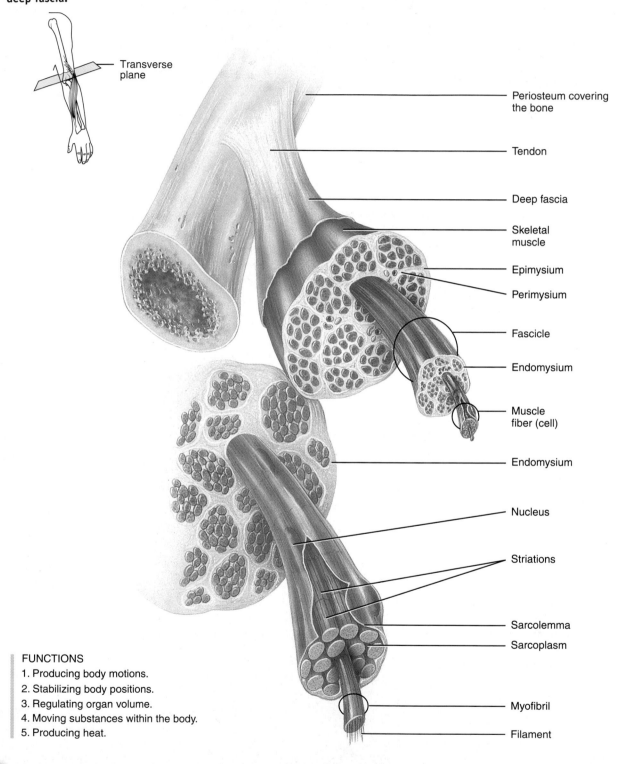

Transverse plane

Periosteum covering the bone

Tendon

Deep fascia

Skeletal muscle

Epimysium

Perimysium

Fascicle

Endomysium

Muscle fiber (cell)

Endomysium

Nucleus

Striations

Sarcolemma

Sarcoplasm

Myofibril

Filament

FUNCTIONS
1. Producing body motions.
2. Stabilizing body positions.
3. Regulating organ volume.
4. Moving substances within the body.
5. Producing heat.

Q Which connective tissue coat surrounds groups of muscle fibers, separating them into fascicles?

Figure 10.2 Blood and nerve supply to skeletal muscle. The site of contact between a somatic motor neuron and a skeletal muscle fiber is the neuromuscular junction.

Branches of one motor neuron typically contact many muscle fibers.

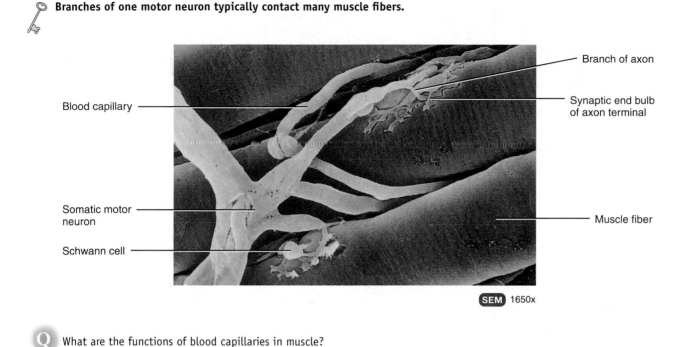

Blood capillary

Somatic motor neuron

Schwann cell

Branch of axon

Synaptic end bulb of axon terminal

Muscle fiber

SEM 1650x

Q What are the functions of blood capillaries in muscle?

Connective Tissue Components

Connective tissue surrounds and protects muscle tissue. **Fascia** (FASH-ē-a; = bandage) is a sheet or broad band of fibrous connective tissue that is deep to the skin and surrounds muscles and other organs of the body. The **superficial fascia** (or **subcutaneous layer**) separates muscle from skin (see Figure 5.1 on page 141). It is composed of areolar connective tissue and adipose tissue and provides a pathway for nerves and blood vessels to enter and exit muscles. The adipose tissue of superficial fascia stores most of the body's triglycerides, serves as an insulating layer that reduces heat loss, and protects muscles from physical trauma. **Deep fascia** is dense, irregular connective tissue that lines the body wall and limbs and holds muscles with similar functions together. Deep fascia allows free movement of muscles, carries nerves and blood and lymphatic vessels, and fills spaces between muscles.

Three layers of connective tissue extend from the deep fascia to further protect and strengthen skeletal muscle (see Figure 10.1). The outermost layer, encircling the whole muscle, is the **epimysium** (ep-i-MĪZ-ē-um; *epi-* = upon). **Perimysium** (per-i-MĪZ-ē-um; *peri-* = around) surrounds groups of 10 to 100 or more individual muscle fibers, separating them into bundles called **fascicles** (FAS-i-kuls; = little bundle). Many fascicles are large enough to be seen with the naked eye. They give a cut of meat its characteristic "grain," and if you tear a piece of meat, it rips apart along the fascicles. Both epimysium and perimysium are dense irregu-

lar connective tissue. Penetrating the interior of each fascicle and separating individual muscle fibers from one another is **endomysium** (en'-dō-MĪZ-ē-um; *endo-* = within), a thin sheath of areolar connective tissue.

Deep fascia, epimysium, perimysium, and endomysium are all continuous with, and contribute collagen fibers to, the connective tissue that attaches skeletal muscle to other structures, such as bone or another muscle. All three connective tissue layers may extend beyond the muscle fibers to form a **tendon**—a cord of dense regular connective tissue that attaches a muscle to the periosteum of a bone. An example is the calcaneal (Achilles) tendon of the gastrocnemius (calf) muscle (see Figure 11.22 on page 370). When the connective tissue elements extend as a broad, flat layer, the tendon is called an **aponeurosis** (*apo-* = from; *neur-* = a sinew). An example of an aponeurosis is the galea aponeurotica on top of the skull (shown in Figure 11.4 on page 314).

Nerve and Blood Supply

Skeletal muscles are well supplied with nerves and blood vessels. Generally, an artery and one or two veins accompany each nerve that penetrates a skeletal muscle (see Figure 10.1). The neurons that stimulate skeletal muscle to contract are the *somatic motor neurons.* Each somatic motor neuron has a threadlike axon that extends from the brain or spinal cord to a group of skeletal muscle fibers (Figure 10.2). The axon is wrapped with *myelin,* which is provided by nearby Schwann cells. Branches of one motor neuron axon typically extend to several different muscle fibers. At the point of contact between the motor neuron and the muscle fiber, called the *neuromuscular junction (NMJ)*, the axon terminals expand into a cluster of *synaptic end bulbs.*

Microscopic blood vessels called capillaries are plentiful in muscle tissue; each muscle fiber is in close contact with one or more capillaries (see Figure 10.2). The blood capillaries bring in oxygen and nutrients and remove heat and the waste products of muscle metabolism. Especially during contraction, a muscle fiber synthesizes and uses considerable ATP (adenosine triphosphate); these reactions require oxygen, glucose, fatty acids, and other substances that are supplied in the blood.

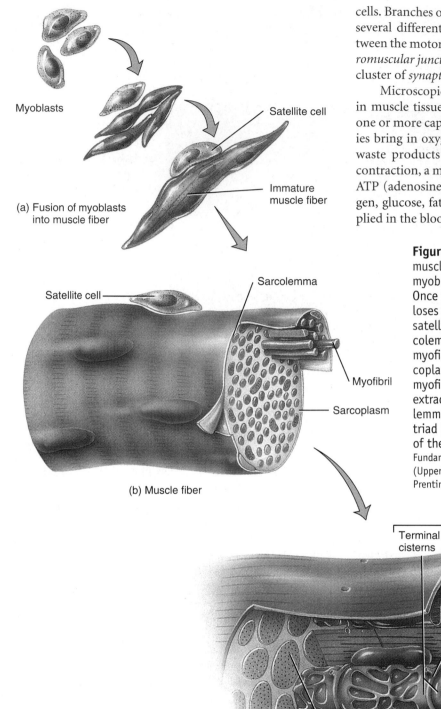

(a) Fusion of myoblasts into muscle fiber

Myoblasts

Satellite cell

Immature muscle fiber

Satellite cell

Sarcolemma

Myofibril

Sarcoplasm

(b) Muscle fiber

Figure 10.3 Microscopic organization of skeletal muscle. (a) During embryonic development, many myoblasts fuse to form one skeletal muscle fiber. Once fusion has occurred, a skeletal muscle fiber loses the ability to undergo cell division, but satellite cells retain this ability. (b) The sarcolemma of the fiber encloses sarcoplasm and myofibrils, which are striated. (c) A sac of sarcoplasmic reticulum (SR) wraps around each myofibril. Thousands of T tubules, filled with extracellular fluid, invaginate from the sarcolemma toward the center of the muscle fiber. A triad is a T tubule and the two terminal cisterns of the SR on either side of it. Adapted from Martini, Fundamentals of Anatomy and Physiology 4e, F10.2, p280, (Upper Saddle River, NJ: Prentice Hall, 1998). ©1998 Prentince Hall, Inc.

Triad

Terminal cisterns

T tubule

Mitochondrion

Sarcolemma

Thick filament

Thin filament

Myofibril

The contractile elements of muscle fibers are the myofibrils, which contain overlapping thick and thin filaments.

Myofibrils

Sarcoplasmic reticulum

(c) Several myofibrils

Q Which structure shown here releases calcium ions to trigger muscle contraction?

Microscopic Anatomy of a Skeletal Muscle Fiber

OBJECTIVE

• *Describe the microscopic anatomy of a skeletal muscle fiber.*

During embryonic development, each skeletal muscle fiber arises from the fusion of a hundred or more small mesodermal cells called *myoblasts* (Figure 10.3a). Hence, each mature skeletal muscle fiber has a hundred or more nuclei. Once fusion has occurred, the muscle fiber loses its ability to undergo cell division. Thus, the number of skeletal muscle fibers is set before birth, and most of these fibers last a lifetime. The dramatic muscle growth that occurs after birth is achieved mainly by enlargement of existing fibers. A few myoblasts do persist in mature skeletal muscle as *satellite cells,* which retain the capacity to fuse with one another or with damaged muscle fibers to regenerate functional muscle fibers. Mature muscle fibers lie parallel to one another and range from 10 to 100 μm* in diameter. Although a typical length is 100 mm, some are up to 30 cm (12 in.) long.

Sarcolemma, T Tubules, and Sarcoplasm

The multiple nuclei of a skeletal muscle fiber are located just beneath the **sarcolemma** (*sarc-* = flesh; *-lemma* = sheath), the fiber's plasma membrane (Figure 10.3b). Thousands of tiny invaginations of the sarcolemma, called **T (transverse) tubules,** tunnel from the surface toward the center of each muscle fiber (Figure 10.3c). T tubules are open to the outside of the fiber and thus are filled with extracellular fluid. Muscle action potentials propagate along the sarcolemma and through the T tubules, quickly spreading throughout the muscle fiber. This arrangement ensures that all parts of the muscle fiber become excited by an action potential virtually simultaneously.

Within the sarcolemma is the **sarcoplasm,** the cytoplasm of a muscle fiber. Sarcoplasm includes a substantial amount of glycogen, which can be split into glucose that is used for ATP synthesis. In addition, the sarcoplasm contains **myoglobin** (mī-ō-GLŌB-in), a red-colored, oxygen-binding protein, found only in muscle fibers, that binds oxygen molecules, which are needed for ATP production within mitochondria. The mitochondria lie in rows throughout the muscle fiber, strategically close to the muscle proteins that use ATP during contraction.

Myofibrils and Sarcoplasmic Reticulum

At high magnification the sarcoplasm appears stuffed with little threads. These small structures are the contractile elements of skeletal muscle, the **myofibrils** (see Figure 10.3b, c). Myofibrils are about 2 μm in diameter and extend the entire length of the muscle fiber. Their prominent striations make the whole muscle fiber look striated.

A fluid-filled system of membranous sacs called the **sarcoplasmic reticulum (SR)** encircles each myofibril (see Figure 10.3c). This elaborate system of sacs is similar to smooth endoplasmic reticulum in nonmuscle cells. Dilated end sacs of the sarcoplasmic reticulum called *terminal cisterns* (= reservoirs) butt against the T tubule from both sides. A transverse tubule and the two terminal cisterns on either side of it form a *triad* (*tri-* = three). In a relaxed muscle fiber, the sarcoplasmic reticulum stores calcium ions (Ca^{2+}). Release of Ca^{2+} from the terminal cisterns of the sarcoplasmic reticulum triggers muscle contraction.

CLINICAL APPLICATION
Muscular Atrophy and Hypertrophy

Muscular atrophy (A-trō-fē; *a-* = without, *-trophy* = nourishment) is a wasting away of muscles. Individual muscle fibers decrease in size as a result of progressive loss of myofibrils. The atrophy that occurs if muscles are not used is termed *disuse atrophy.* Bedridden individuals and people with casts experience disuse atrophy because the flow of nerve impulses to inactive muscle is greatly reduced. If the nerve supply to a muscle is disrupted or cut, the muscle undergoes *denervation atrophy.* In about 6 months to 2 years, the muscle will be one quarter its original size, and the muscle fibers will be replaced by fibrous connective tissue. The transition to connective tissue, when complete, cannot be reversed.

Muscular hypertrophy (hī-PER-trō-fē; *hyper-* = above or excessive) is an increase in the diameter of muscle fibers owing to the production of more myofibrils, mitochondria, sarcoplasmic reticulum, and so forth. It results from very forceful, repetitive muscular activity, such as strength training. Because hypertrophied muscles contain more myofibrils, they are capable of more forceful contractions. ■

Filaments and the Sarcomere

Within myofibrils are two types of even smaller structures called **filaments,** which are only 1–2 μm long (see Figure 10.3c). The diameter of the *thin filaments* is about 8 nm,* whereas that of the *thick filaments* is about 16 nm. The filaments inside a myofibril do not extend the entire length of a muscle fiber; instead, they are arranged in compartments called **sarcomeres** (*-mere* = part), which are the basic functional units of a myofibril (Figure 10.4a). Narrow, plate-shaped regions of dense material called *Z discs* separate one sarcomere from the next.

The thick and thin filaments overlap one another to a greater or lesser extent, depending on whether the muscle is contracted, relaxed, or stretched. The pattern of their overlap, consisting of a variety of zones and bands (Figure 10.4b), creates the striations that can be seen both in single myofibrils and in whole muscle fibers. The darker middle portion of the sarcomere is the **A band,** which extends the entire length of the thick filaments (see Figure 10.4b). Toward each

---

*One micrometer (μm) is 10^{-6} meter (1/25,000 in). One nanometer (nm) is 10^{-9} meter (0.001 μm).

Figure 10.4 The arrangement of filaments within a sarcomere. A sarcomere extends from one Z disc to the next.

🔑 **Myofibrils contain two types of filaments: thick filaments and thin filaments.**

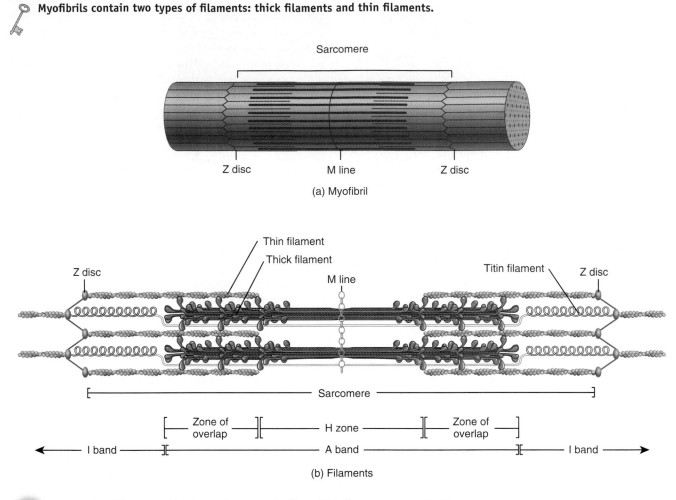

Sarcomere

Z disc   M line   Z disc

(a) Myofibril

Thin filament
Thick filament
M line
Titin filament   Z disc
Z disc

Sarcomere

Zone of overlap   H zone   Zone of overlap

← I band →   A band   I band →

(b) Filaments

Q Among the following, which is smallest: muscle fiber, thick filament, or myofibril? Which is largest?

end of the A band is a *zone of overlap,* where the thick and thin filaments lie side by side. In the zone of overlap, six thin filaments surround each thick filament, and three thick filaments surround each thin filament. Overall, there are two thin filaments for every thick filament. The **I band** is a lighter, less dense area that contains the rest of the thin filaments but no thick filaments (see Figure 10.4b). A Z disc passes through the center of each I band. A narrow *H zone* in the center of each A band contains thick but not thin filaments. Supporting proteins that hold the thick filaments together at the center of the H zone form the *M line,* so-named because it is at the *middle* of the sarcomere. Figure 10.5 shows the relations of the zones, bands, and lines as seen in a transmission electron micrograph.

CLINICAL APPLICATION
Exercise-Induced Muscle Damage

Comparison of electron micrographs of muscle tissue taken from athletes before and after intense exercise reveal considerable exercise-induced muscle damage, including torn sarcolemmas in some muscle fibers, damaged myofibrils, and disrupted Z discs. Microscopic muscle damage after exercise also is indicated by increases in blood levels of proteins, such as myoglobin and the enzyme creatine kinase, that are normally confined within muscle fibers. From 12 to 48 hours after a period of strenuous exercise, skeletal muscles often become sore. Such **delayed onset muscle soreness (DOMS)** is accompanied by stiffness, tenderness, and swelling. Although the causes of DOMS are not completely understood, microscopic muscle damage appears to be a major factor. ■

Muscle Proteins

Myofibrils are built from three kinds of proteins: (1) contractile proteins, which generate force during contraction; (2) regulatory proteins, which help switch the contraction process on and off; and (3) structural proteins, which keep the thick and thin filaments in the proper alignment, give the myofibril elasticity and extensibility, and link the myofibrils to the sarcolemma and extracellular matrix.

The two *contractile proteins* in muscle are myosin and actin, which are the main components of thick and thin filaments, respectively. Myosin functions as a *motor protein* in all three types of muscle tissue. Motor proteins push or pull their cargo to achieve movement by converting the chemical energy in ATP to mechanical energy of motion or force production. About 300 molecules of **myosin** form a single thick filament. Each myosin molecule is shaped like two golf clubs twisted together (Figure 10.6a). The *myosin tail* (golf club handles) points toward the M line in the center of the sarcomere. Tails of neighboring myosin molecules lie parallel to one another, forming the shaft of the thick filament. The two projections of each myosin molecule (golf club heads) are called *myosin heads* or *crossbridges*. The heads project outward from the shaft in a spiraling fashion, each extending toward one of the six thin filaments that surround the thick filament.

Thin filaments extend from anchoring points within the Z discs (see Figure 10.4b). Their main component is the protein **actin.** Individual actin molecules join to form an actin filament that is twisted into a helix (Figure 10.6b). On each actin molecule is a *myosin-binding site*, where a myosin head can attach. Smaller amounts of two *regulatory proteins*— **tropomyosin** and **troponin**—are also part of the thin filament. In relaxed muscle, myosin is blocked from binding to actin because tropomyosin covers the *myosin-binding site* on actin. The tropomyosin strand, in turn, is held in place by troponin.

Besides contractile and regulatory proteins, muscle contains about a dozen *structural proteins,* which contribute to the alignment, stability, elasticity, and extensibility of myofibrils. Several key structural proteins are titin, myomesin, nebulin, and dystrophin. *Titin (titan = gigantic)* is the third most plentiful protein in skeletal muscle (after actin and myosin). This molecule's name reflects its huge size; with a molecular weight of about 3 million daltons, titin is 50 times larger than an average-sized protein. Each titin molecule spans half a sarcomere, from a Z disc to an M line (see Figure 10.4b), a distance of 1–1.2 μm in relaxed muscle. Titin anchors a thick filament to both a Z disc and the M line, thereby helping to stabilize the position of the thick filament. The portion of the titin molecule that extends from the Z disc to the beginning of the thick filament is very elastic. Because it can stretch to at least four times its resting length and then spring back unharmed, titin accounts for much of the elasticity and extensibility of myofibrils. Titin probably helps the sarcomere return to its resting length after a muscle has contracted or been stretched.

Figure 10.5 Transmission electron micrograph showing the characteristic zones and bands of a sarcomere.

🔑 **The striations of skeletal muscle are alternating darker A bands and lighter I bands.**

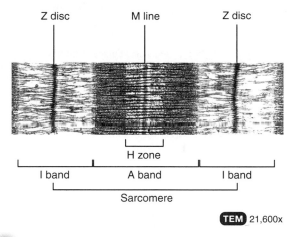

TEM 21,600x

Q What is the ratio between thick and thin filaments in skeletal muscle?

Figure 10.6 Structure of thick and thin filaments. (a) A thick filament (above) contains about 300 myosin molecules, one of which is pictured (below). The myosin tails form the shaft of the thick filament, whereas the myosin heads project outward toward the surrounding thin filaments. (b) Thin filaments contain actin, troponin, and tropomyosin.

🔑 **Contractile proteins (myosin and actin) generate force during contraction, whereas regulatory proteins (troponin and tropomyosin) help switch contraction on and off.**

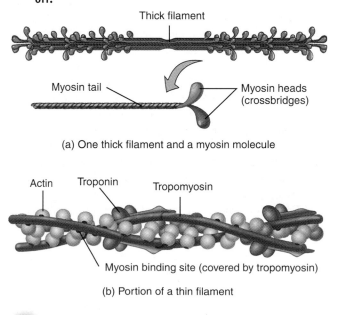

(a) One thick filament and a myosin molecule

(b) Portion of a thin filament

Q Which proteins connect into the Z disc? Which proteins are present in the A band? In the I band?

Table 10.1 Types of Proteins in Skeletal Muscle Fibers

PROTEIN TYPE	NAME*	LOCATION AND FUNCTION
Contractile proteins	Myosin (44%)	Myosin tails form shaft of thick filaments; myosin heads (crossbridges) bind to myosin binding site on actin during muscle contraction.
	Actin (22%)	Forms backbone of thin filament; contains sites where myosin heads bind during muscle contraction.
Regulatory proteins	Tropomyosin (5%)	Part of thin filament; blocks myosin–binding sites when muscle is relaxed.
	Troponin (5%)	Part of thin filament; holds tropomyosin in position.
Structural proteins	Titin (9%)	Extends from Z disc to M line and attaches to myosin; anchors thick filaments to Z discs, stabilizes them during contraction and relaxation, and provides elasticity and extensibility.
	Myomesin	Forms the M line; helps stabilize position of thick filaments.
	Nebulin (3%)	Attaches into Z disc and lies alongside thin filaments; helps maintain alignment of thin filaments.
	Dystrophin	Links thin filaments to integral membrane proteins of sarcolemma; helps transmit muscle tension to tendons.

*Percentage of total protein in myofibrils.

The M line itself is formed by molecules of the protein *myomesin*, which bind to titin and also connect adjacent thick filaments to one another. *Nebulin*, a large but inelastic protein, lies alongside the thin filaments and also attaches into the Z disc. It helps maintain alignment of the thin filaments in the sarcomere. *Dystrophin* is a cytoskeletal protein that links thin filaments of the sarcomere to integral membrane proteins of the sarcolemma. In turn, the membrane proteins attach to proteins in the connective tissue matrix that surrounds muscle fibers. Hence, dystrophin and its associated proteins are thought to reinforce the sarcolemma and help transmit to the tendons the tension generated by the sarcomeres.

Table 10.1 reviews the types of proteins in skeletal muscle fibers.

1. Describe the types of fascia that cover skeletal muscles.
2. Why is a rich blood supply so important to muscle contraction?
3. Describe the components of a sarcomere. How do thin and thick filaments differ structurally?

Figure 10.7 Sliding filament mechanism of muscle contraction, as it occurs in two adjacent sarcomeres.

During muscle contractions, thin filaments move toward the M line of each sarcomere.

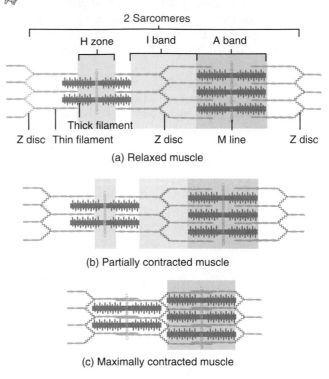

(a) Relaxed muscle

(b) Partially contracted muscle

(c) Maximally contracted muscle

 What happens to the I band and H zone as muscle contracts? Do the lengths of the thick and thin filaments change?

CONTRACTION AND RELAXATION OF SKELETAL MUSCLE FIBERS

When scientists examined the first electron micrographs of skeletal muscle in the mid-1950s, they were surprised to see that the lengths of the thick and thin filaments were the same in both relaxed and contracted muscle. It had been thought that muscle contraction must be a folding process, somewhat like closing an accordion. However, researchers discovered that skeletal muscle shortens during contraction because the thick and thin filaments slide past one another. The model describing the contraction of muscle is known as the **sliding filament mechanism.**

The Sliding Filament Mechanism

OBJECTIVE

• *Outline the steps involved in the sliding filament mechanism of muscle contraction.*

Muscle contraction occurs because myosin heads attach to and "walk" along the thin filaments at both ends of a sarcomere, progressively pulling the thin filaments toward the M line (Figure 10.7). As a result, the thin filaments slide in-

Figure 10.8 The contraction cycle. Sarcomeres shorten through repeated cycles during which the myosin heads (crossbridges) attach to actin, rotate, and detach.

🔑 **During the power stroke of contraction, myosin heads rotate and move the thin filaments past the thick filaments toward the center of the sarcomere.**

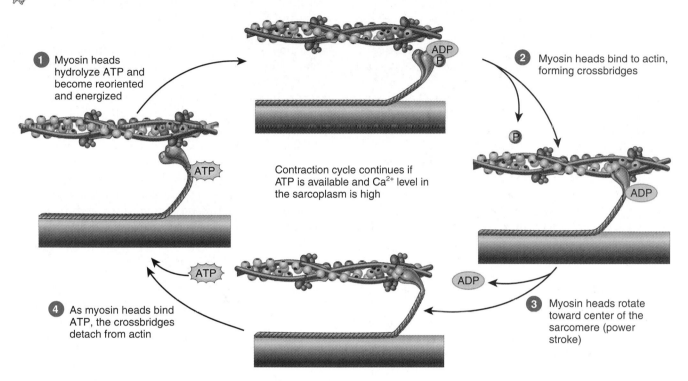

1 Myosin heads hydrolyze ATP and become reoriented and energized

2 Myosin heads bind to actin, forming crossbridges

Contraction cycle continues if ATP is available and Ca²⁺ level in the sarcoplasm is high

4 As myosin heads bind ATP, the crossbridges detach from actin

3 Myosin heads rotate toward center of the sarcomere (power stroke)

Q What would happen if ATP suddenly became unavailable after the sarcomere had started to shorten?

ward and meet at the center of a sarcomere. They may even move so far inward that their ends overlap (Figure 10.7c). As the thin filaments slide inward, the Z discs come closer together, and the sarcomere shortens. However, the lengths of the individual thick and thin filaments do not change. Shortening of the sarcomeres causes shortening of the whole muscle fiber, and ultimately shortening of the entire muscle.

The Contraction Cycle

At the onset of contraction the sarcoplasmic reticulum releases calcium ions (Ca^{2+}), which bind to troponin and cause the troponin–tropomyosin complexes to move away from the myosin-binding sites on actin. Once the binding sites are "free," the **contraction cycle**—the repeating sequence of events that causes the filaments to slide—begins. The contraction cycle consists of four steps (Figure 10.8):

1 *ATP hydrolysis.* The myosin head includes an ATP-binding pocket and an ATPase, an enzyme that hydrolyzes ATP into ADP (adenosine diphosphate) and a phosphate group. This hydrolysis reaction energizes the myosin head. Notice that the products of ATP hydrolysis—ADP and a phosphate group—are still attached to the myosin head.

2 *Attachment of myosin to actin to form crossbridges.* The energized myosin head attaches to the myosin-binding site on actin and then releases the previously hydrolyzed phosphate group.

3 *Power stroke.* The release of the phosphate group triggers the **power stroke** of contraction. During the power stroke, the pocket on the myosin head where ADP is still bound opens, which rotates the myosin head and releases the ADP. The myosin head generates force as it rotates toward the center of the sarcomere, sliding the thin filament past the thick filament toward the M line.

4 *Detachment of myosin from actin.* At the end of the power stroke, the myosin head remains firmly attached to actin until it binds another molecule of ATP. As ATP binds to the ATP-binding pocket on the myosin head, the myosin head detaches from actin.

The contraction cycle repeats as the myosin ATPase again hydrolyzes ATP. The reaction reorients the myosin head and transfers energy from ATP to the myosin head, which is again energized and ready to combine with another myosin-binding site farther along the thin filament. The contraction cycle repeats over and over, as long as ATP is available and the Ca^{2+} level near the thin filament is sufficiently high. The myosin heads keep rotating back and forth with each power stroke, pulling the thin filaments toward the M line. Each of the 600 myosin heads in one thick filament attach and detach about five times per second. At any one instant, some of the myosin heads are attached to actin and generating force, whereas others are detached and available to bind again. Contraction is analogous to running on a nonmotorized treadmill. One foot (myosin head) strikes the belt (thin filament) and pushes it backward (toward the M line). Then the other foot comes down and imparts a second push. The belt (thin filament) moves smoothly while the runner (thick filament) remains stationary. Each myosin head progressively "walks" along a thin filament, coming closer to the Z disc with each "step," while the thin filament moves toward the M line. And like the legs of a runner, to keep going the myosin head needs a constant supply of energy, one ATP for each contraction cycle!

This continual movement of myosin heads applies the force that draws the Z discs toward each other, and the sarcomere shortens. The myofibrils thus contract, and the whole muscle fiber shortens. During a maximal muscle contraction, the distance between Z discs can decrease to half the resting length. But the power stroke does not always result in shortening of the muscle fibers and the whole muscle. In isometric contractions, the myosin heads rotate and generate tension, but the thin filaments do not slide inward because the tension generated is not enough to overcome the resistance.

Excitation–Contraction Coupling

An increase in Ca^{2+} concentration in the cytosol starts muscle contraction, whereas a decrease stops it. When a muscle fiber is relaxed, the concentration of Ca^{2+} in its cytosol is very low, only about 0.1 millimole/liter (10^{-7} M). A huge amount of Ca^{2+}, however, is stored inside the SR (Figure 10.9a). As a muscle action potential propagates along the sarcolemma and into the T tubules, it causes **Ca^{2+} release channels** in the SR membrane to open (Figure 10.9b), permitting Ca^{2+} to diffuse across the SR membrane. As a result, calcium ions flood from the SR into the cytosol around the thick and thin filaments, and the Ca^{2+} concentration in the cytosol rises tenfold or more. The released Ca^{2+} combine with troponin, causing it to change shape. This conformational change moves the troponin–tropomyosin complex away from the myosin-binding sites on actin. Once these binding sites are free, myosin heads bind to them, and the contraction cycle begins. The events just described constitute **excitation–contraction coupling,** the steps that connect excitation (a muscle action potential propagating through the T tubules) to contraction of the muscle fiber.

The sarcoplasmic reticulum membrane also contains **Ca^{2+} active transport pumps** that hydrolyze ATP as they continually move Ca^{2+} from the cytosol into the SR (see Figure 10.9b). As long as muscle action potentials continue to propagate through the T tubules, the Ca^{2+} release channels are open, and calcium ions diffuse into the cytosol more rapidly than they are transported back by the pumps. After the last action potential has propagated throughout the T tubules, the Ca^{2+} release channels close. As the pumps move Ca^{2+} back into the SR, the concentration of calcium ions in the cytosol quickly decreases. Inside the SR, molecules of a calcium-binding protein, appropriately called **calsequestrin,** bind to the Ca^{2+}, enabling even more Ca^{2+} to be sequestered within the SR. As a result, the concentration of Ca^{2+} inside the SR of a relaxed muscle fiber is 10,000 times higher than that in the cytosol. As the Ca^{2+} level drops in the cytosol, the troponin–tropomyosin complexes slide back over and cover the myosin-binding sites, and the muscle fiber relaxes.

CLINICAL APPLICATION
Rigor Mortis

After death, cellular membranes start to become leaky. Calcium ions leak out of the sarcoplasmic reticulum into the cytosol and allow myosin heads to bind to actin. ATP synthesis has ceased, however, so the crossbridges cannot detach from actin. The resulting condition, in which muscles are in a state of rigidity (cannot contract or stretch), is called **rigor mortis** (rigidity of death). Rigor mortis begins 3–4 hours after death and lasts about 24 hours; then it disappears as proteolytic enzymes from lysosomes digest the crossbridges. ■

Length–Tension Relationship

Figure 10.10 plots the **length–tension relationship** for skeletal muscle, which shows how the forcefulness of muscle contraction depends on the length of the sarcomeres within a muscle *before contraction begins.* At a sarcomere length of about 2.0–2.4 μm, the zone of overlap in each sarcomere is optimal, and the muscle fiber can develop maximum tension. Notice in Figure 10.10 that maximum tension (100%) occurs when the zone of overlap between a thick and thin filament extends from the edge of the H zone to one end of a thick filament.

As the sarcomeres of a muscle fiber are stretched to a longer length, the zone of overlap is shorter: Fewer myosin heads can make contact with thin filaments. Consequently, the tension the fiber can produce decreases. When a skeletal muscle fiber is stretched to 170% of its optimal length, there is no overlap between the thick and thin filaments. Because none of the myosin heads can bind to thin filaments, the muscle fiber cannot contract, and tension is zero. As sarcomere lengths become increasingly shorter than the optimum, the tension that can develop again decreases. This is because thick filaments crumple as they are compressed by the Z discs, resulting in fewer myosin heads making contact

Figure 10.9 The role of Ca^{2+} in the regulation of contraction by troponin and tropomyosin. (a) During relaxation, the level of Ca^{2+} in the sarcoplasm is low, only 0.1 mM, because calcium ions are pumped into the sarcoplasmic reticulum (SR) by Ca^{2+} active transport pumps. (b) A muscle action potential propagating along a transverse tubule opens Ca^{2+} release channels in the SR, calcium ions flood into the cytosol, and contraction begins.

 An increase in the Ca^{2+} level in the sarcoplasm starts the sliding of thin filaments; when the level of Ca^{2+} in the sarcoplasm declines, sliding stops.

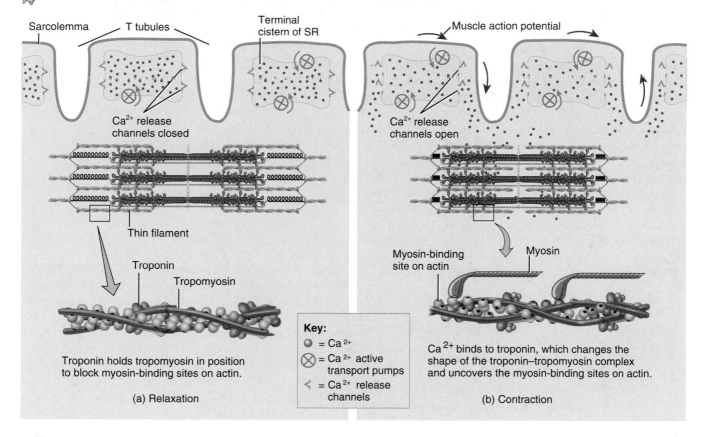

Sarcolemma   T tubules   Terminal cistern of SR   Muscle action potential

Ca^{2+} release channels closed   Ca^{2+} release channels open

Thin filament

Troponin   Myosin-binding site on actin   Myosin

Tropomyosin

Key:
◯ = Ca^{2+}
⊗ = Ca^{2+} active transport pumps
≺ = Ca^{2+} release channels

Troponin holds tropomyosin in position to block myosin-binding sites on actin.

Ca^{2+} binds to troponin, which changes the shape of the troponin–tropomyosin complex and uncovers the myosin-binding sites on actin.

(a) Relaxation

(b) Contraction

Q   What are three functions of ATP in muscle contraction?

Figure 10.10 Length–tension relationship in a skeletal muscle fiber. Maximum tension during contraction occurs when the resting sarcomere length is 2.0–2.4 µm.

 A muscle fiber develops its greatest tension when there is an optimal zone of overlap between thick and thin filaments.

Q   Why is tension maximal at a sarcomere length of 2.2 µm?

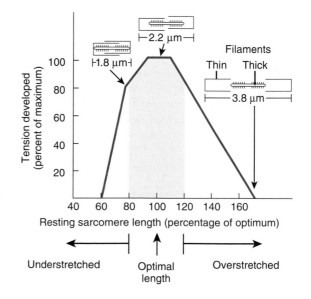

├1.8 µm┤   ├2.2 µm┤   Filaments
Thin   Thick
├──3.8 µm──┤

Tension developed (percent of maximum)

100
80
60
40
20

40 60 80 100 120 140 160
Resting sarcomere length (percentage of optimum)

Understretched   Optimal length   Overstretched

with thin filaments. Normally, resting muscle fiber length is held very close to the optimum by firm attachments of skeletal muscle to bones (via their tendons) and to other inelastic tissues, so that overstretching does not occur.

Active and Passive Tension

As thin filaments start to slide past thick filaments, they pull on the Z discs, which in turn pull on neighboring sarcomeres. Eventually, whole muscle cells pull on their surrounding connective tissue layers. Some of the structural components of muscle are elastic: They stretch slightly before they transfer the tension generated by the sliding filaments. The elastic components include titin molecules, connective tissue around the muscle fibers (endomysium, perimysium, and epimysium), and tendons that attach muscle to bone. Muscle tension that is generated by the contractile components (thin and thick filaments) is called **active tension;** that generated by the elastic components is called **passive tension.** Within limits, the more the elastic components of a muscle are stretched, the greater the passive tension. As a skeletal muscle starts to shorten, it first pulls on its connective tissue coverings and tendons. The coverings and tendons stretch and then become taut, and the tension passed through the tendons pulls on the bones to which they are attached. The result is movement of a part of the body.

The Neuromuscular Junction

OBJECTIVE

• *Describe how muscle action potentials arise at the neuromuscular junction.*

A muscle fiber contracts in response to one or more action potentials propagating along its sarcolemma and through its T tubule system. Muscle action potentials arise at the **neuromuscular junction (NMJ),** the synapse between a somatic motor neuron and a skeletal muscle fiber (Figure 10.11a). A *synapse* is a region where communication occurs between two neurons, or between a neuron and a target cell—for example, between a motor neuron and a muscle fiber. At most synapses a small gap, called the *synaptic cleft,* separates the two cells. Because the cells do not physically touch, the action potential from one cell cannot "jump the gap" to directly excite the next cell. Instead, the first cell communicates with the second indirectly, by releasing a chemical called a **neurotransmitter.**

At a neuromuscular junction, the motor neuron axon terminal divides into a cluster of synaptic end bulbs (Figure 10.11a). Suspended in the cytosol within each bulb are hundreds of membrane-enclosed sacs called **synaptic vesicles.** Inside each synaptic vesicle are thousands of molecules of **acetylcholine** (as′-ē-til-KŌ-lēn), abbreviated **ACh,** the neurotransmitter released at the NMJ. The region of the sarcolemma that is adjacent to the synaptic end bulbs is called the **motor end plate.** It contains 30–40 million *acetylcholine receptors,* which are integral transmembrane proteins that bind specifically to ACh. As you will see, the ACh receptors are gated ion channels. A neuromuscular junction thus includes all the synaptic end bulbs on one side of the synaptic cleft, plus the motor end plate of the muscle fiber on the other side.

A nerve impulse elicits a muscle action potential in the following way (Figure 10.11c).

1. *Release of acetylcholine.* Arrival of the nerve impulse at the synaptic end bulbs triggers exocytosis of many synaptic vesicles. The vesicles fuse with the motor neuron's plasma membrane and liberate ACh, which diffuses across the synaptic cleft between the motor neuron and the motor end plate.

2. *Activation of ACh receptors.* Binding of ACh to its receptor in the sarcolemma opens the gated ion channel portion of the receptor, which allows small cations, most importantly Na^+, to flow across the membrane.

3. *Production of muscle action potential.* The inflow of Na^+ (down its concentration gradient) makes the inside of the muscle fiber more positively charged, which changes the membrane potential and triggers a muscle action potential. The muscle action potential then propagates along the sarcolemma and through the T tubule system. Each nerve impulse normally elicits one muscle action potential.

4. *Termination of ACh activity.* The effect of ACh binding lasts only briefly because the neurotransmitter is rapidly broken down by an enzyme called **acetylcholinesterase (AChE),** which is attached to collagen fibers in the extracellular matrix of the synaptic cleft. AChE breaks down ACh into acetyl and choline, products that cannot activate the ACh receptor. If another nerve impulse releases more acetylcholine, then steps [2] and [3] repeat. When action potentials cease in the motor neuron, ACh release stops and AChE rapidly breaks down the ACh already present in the synaptic cleft. This ends the generation of muscle action potentials, and the Ca^{2+} release channels in the sarcoplasmic reticulum membrane close.

Because skeletal muscle fibers often are very long cells, the NMJ usually is located near the midpoint of the fiber. Muscle action potentials arise at the NMJ and then propagate toward both ends of the fiber. This arrangement permits nearly simultaneous activation (and thus contraction) of all parts of the fiber.

Figure 10.12 summarizes the events that occur during contraction and relaxation of a skeletal muscle fiber.

CLINICAL APPLICATION
Pharmacology of the NMJ

Several plant products and drugs selectively block particular events at the NMJ. *Botulinum toxin,* produced by the bacterium *Clostridium botulinum,* blocks exocytosis of synaptic vesicles at the NMJ. As a result, ACh is not released, and muscle contraction does not occur. The bacteria proliferate in improperly canned foods and their toxin is one of the most lethal chemicals known. A tiny amount can cause death

Figure 10.11 Structure of the neuromuscular junction (NMJ), the synapse between a somatic motor neuron and a skeletal muscle fiber.

🔑 **Synaptic end bulbs at the tips of axon terminals contain synaptic vesicles filled with acetylcholine.**

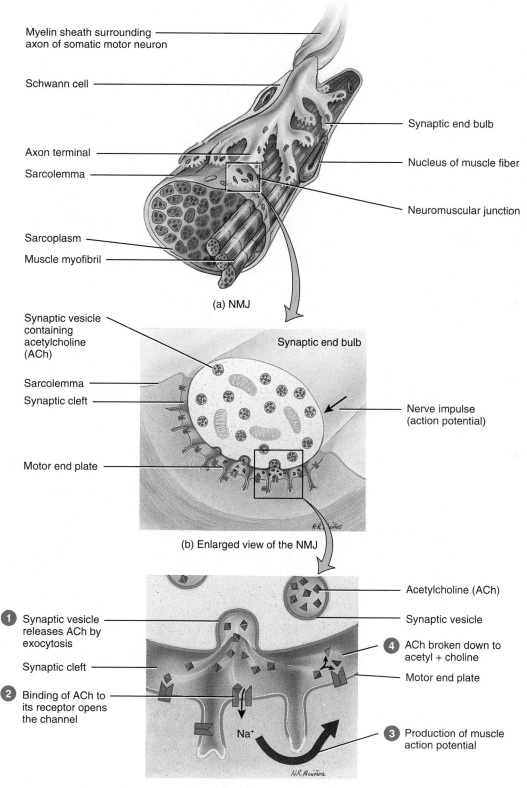

Myelin sheath surrounding axon of somatic motor neuron

Schwann cell

Axon terminal

Sarcolemma

Sarcoplasm

Muscle myofibril

Synaptic end bulb

Nucleus of muscle fiber

Neuromuscular junction

(a) NMJ

Synaptic vesicle containing acetylcholine (ACh)

Sarcolemma

Synaptic cleft

Motor end plate

Synaptic end bulb

Nerve impulse (action potential)

(b) Enlarged view of the NMJ

Acetylcholine (ACh)

❶ Synaptic vesicle releases ACh by exocytosis

Synaptic cleft

❷ Binding of ACh to its receptor opens the channel

Synaptic vesicle

❹ ACh broken down to acetyl + choline

Motor end plate

Na^+

❸ Production of muscle action potential

(c) Binding of acetylcholine to ACh receptors in the motor end plate

 Q What is the term for the portion of the sarcolemma that contains acetylcholine receptors?

Figure 10.12 Summary of the events of contraction and relaxation in a skeletal muscle fiber.

Acetylcholine released at the neuromuscular junction triggers a muscle action potential, which leads to muscle contraction.

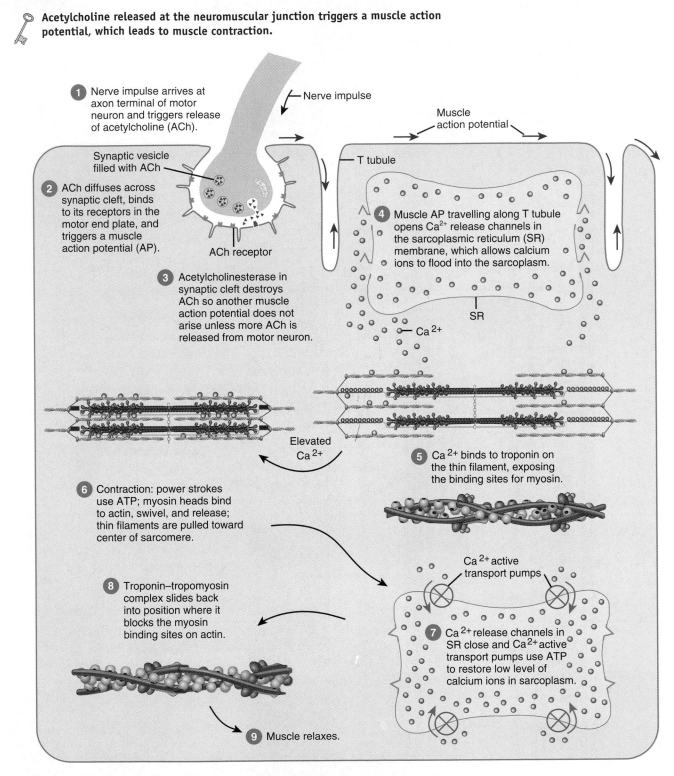

1 Nerve impulse arrives at axon terminal of motor neuron and triggers release of acetylcholine (ACh).

Nerve impulse

Muscle action potential

Synaptic vesicle filled with ACh

T tubule

2 ACh diffuses across synaptic cleft, binds to its receptors in the motor end plate, and triggers a muscle action potential (AP).

4 Muscle AP travelling along T tubule opens Ca^{2+} release channels in the sarcoplasmic reticulum (SR) membrane, which allows calcium ions to flood into the sarcoplasm.

ACh receptor

3 Acetylcholinesterase in synaptic cleft destroys ACh so another muscle action potential does not arise unless more ACh is released from motor neuron.

SR

Ca^{2+}

Elevated Ca^{2+}

5 Ca^{2+} binds to troponin on the thin filament, exposing the binding sites for myosin.

6 Contraction: power strokes use ATP; myosin heads bind to actin, swivel, and release; thin filaments are pulled toward center of sarcomere.

Ca^{2+} active transport pumps

8 Troponin–tropomyosin complex slides back into position where it blocks the myosin binding sites on actin.

7 Ca^{2+} release channels in SR close and Ca^{2+} active transport pumps use ATP to restore low level of calcium ions in sarcoplasm.

9 Muscle relaxes.

Q Which numbered steps in this figure are part of excitation–contraction coupling?

by paralyzing the diaphragm. Yet it is also the first bacterial toxin to be used as a medicine (Botox). Injections of Botox into the affected muscles help patients who have strabismus (crossed eyes) or blepharospasm (uncontrollable blinking).

The plant derivative *curare,* a poison used by South American Indians on arrows and blow-gun darts, causes muscle paralysis by binding to and blocking the ACh receptor such that its gated ion channel does not open. Curare-like drugs are often used during surgery to relax skeletal muscles. A family of chemicals called *anticholinesterase agents* have the property of slowing the enzymatic activity of acetylcholinesterase, thus slowing removal of ACh from the synaptic cleft. At low doses, these agents can strengthen weak muscle contractions. One example is neostigmine, which is used to treat patients with myasthenia gravis (see page 296). Neostigmine is also used as an antidote for curare poisoning and to terminate the effects of curare after surgery. ■

1. What roles do contractile, regulatory, and structural proteins play in muscle contraction and relaxation?
2. How do calcium ions and ATP contribute to muscle contraction and relaxation?
3. Explain how sarcomere length influences the maximum tension that is possible during muscle contraction.
4. Describe the events that occur at the neuromuscular junction and lead to generation of a muscle action potential.

MUSCLE METABOLISM

OBJECTIVE

• *Describe the reactions by which muscle fibers produce ATP.*

Production of ATP in Muscle Fibers

Unlike most cells of the body, skeletal muscle fibers often switch between virtual inactivity, when they are relaxed and using only a modest amount of ATP, and great activity, when they are contracting and using ATP at a rapid pace. Contraction of muscle requires a tremendous amount of ATP for powering the contraction cycle, for pumping Ca^{2+} into the sarcoplasmic reticulum to achieve muscle relaxation, and for other metabolic reactions. However, the ATP present inside muscle fibers is enough to power contraction for only a few seconds. If strenuous exercise is to continue for more than a few seconds, additional ATP must be synthesized. Muscle fibers have three sources for ATP production: (1) creatine phosphate, (2) anaerobic cellular respiration, and (3) aerobic cellular respiration (Figure 10.13). Whereas the first source for ATP production is unique to muscle fibers, all body cells use the second two sources. We consider the events of cellular respiration briefly here and then in greater detail in Chapter 25.

Creatine Phosphate

While at rest, muscle fibers produce more ATP than they need for resting metabolism. Some of the excess ATP is used to synthesize **creatine phosphate,** an energy-rich molecule that is unique to muscle fibers (Figure 10.13a). The enzyme *creatine kinase (CK)* catalyzes the transfer of one of the high-energy phosphate groups of ATP to creatine, forming creatine phosphate and ADP. Creatine phosphate is 3–6 times more plentiful than ATP in the sarcoplasm. When contraction begins and the ADP level starts to rise, CK catalyzes the transfer of a high-energy phosphate group from creatine phosphate back to ADP. This direct phosphorylation reaction quickly forms new ATP molecules. Together, creatine phosphate and ATP provide enough energy for muscles to contract maximally for about 15 seconds. This energy is sufficient for maximal short bursts of activity—for example, to run a 100-meter dash.

CLINICAL APPLICATION
Creatine Supplementation

Creatine is a small, amino-acid–like molecule that is both synthesized in the body and derived from foods. Adults need to synthesize and ingest a total of about 2 grams of creatine daily to make up for the loss in the urine of creatinine, the breakdown product of creatine. Some studies have demonstrated improved performance during intense exercise in subjects who had ingested creatine supplements. For example, college football players who received supplements of 15 grams per day for 28 days gained more muscle mass and had larger gains in lifting power and sprinting performance than the control subjects. Other studies, however, have failed to find a performance-enhancing effect of creatine supplementation. Moreover, ingesting extra creatine tends to shut down the body's own synthesis of creatine, and it is not known whether natural synthesis recovers after long-term creatine supplementation. Further research is needed to determine both the long-term safety and the efficacy of creatine supplementation. ■

Anaerobic Cellular Respiration

Anaerobic cellular respiration is a series of ATP-producing reactions that do not require oxygen. When muscle activity continues and the supply of creatine phosphate is depleted, glucose is catabolized to generate ATP. Glucose easily passes from the blood into contracting muscle fibers via facilitated diffusion, and it is also produced by the breakdown of glycogen within muscle fibers (Figure 10.13b). Then, a series of ten reactions known as *glycolysis* quickly breaks down each glucose molecule into two molecules of pyruvic acid. (Figure 25.4 on page 878 shows the reactions of glycolysis.) These reactions use two ATP but form four ATP for a net gain of two.

Ordinarily, the pyruvic acid formed by glycolysis in the cytosol enters mitochondria, where its oxidation produces a large amount of ATP from ADP through a series of reactions

Figure 10.13 Production of ATP for muscle contraction. (a) Creatine phosphate, formed from ATP while the muscle is relaxed, transfers a high-energy phosphate group to ADP, forming ATP, during muscle contraction. (b) Breakdown of muscle glycogen into glucose and production of pyruvic acid from glucose via glycolysis produce both ATP and lactic acid. Because no oxygen is needed, this is an anaerobic pathway. (c) Within mitochondria, pyruvic acid, fatty acids, and amino acids are used to produce ATP via aerobic cellular respiration, an oxygen-requiring reaction.

🔑 **During a long-term event such as a marathon race, most ATP is produced aerobically.**

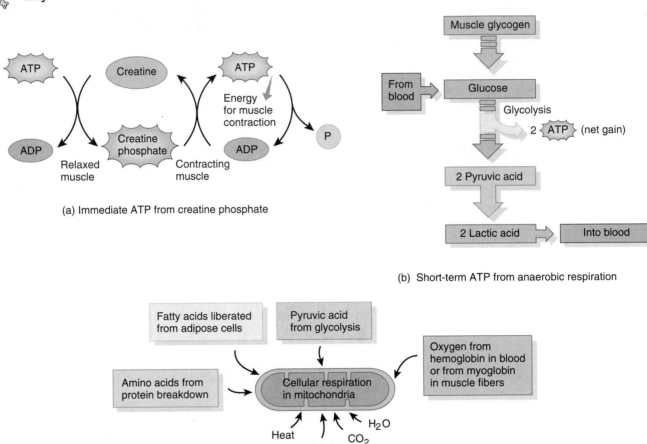

(a) Immediate ATP from creatine phosphate

(b) Short-term ATP from anaerobic respiration

(c) Long-term ATP from aerobic cellular respiration

Q Where inside the muscle fiber are the events shown here occurring?

known as aerobic cellular respiration (described next). During some activities, however, not enough oxygen is available to completely break down pyruvic acid. When this happens, most of the pyruvic acid is converted to lactic acid in the cytosol. About 80% of the lactic acid produced in this way diffuses out of the skeletal muscle fibers into the blood. Liver cells can convert some of the lactic acid back to glucose. This conversion has two benefits: providing new glucose molecules and reducing acidity. In these ways, anaerobic cellular respiration can provide enough energy for about 30–40 seconds of maximal muscle activity—enough, for example, to run a 200-meter race. Eventually, muscle glycogen must also be restored.

Aerobic Cellular Respiration

Muscle activity that lasts longer than half a minute depends increasingly on **aerobic cellular respiration,** a series of oxygen-requiring mitochondrial reactions that produce

ATP. If sufficient oxygen is present, pyruvic acid enters the mitochondria, where it is completely oxidized in reactions that generate ATP, carbon dioxide, water, and heat (Figure 10.13c). Although aerobic cellular respiration is slower than glycolysis, it yields much more ATP—about 36 molecules of ATP from each glucose molecule.

Muscle tissue has two sources of oxygen: (1) oxygen that diffuses into muscle fibers from the blood and (2) oxygen released by myoglobin within muscle fibers. Both myoglobin and hemoglobin (in red blood cells) are oxygen-binding proteins; they bind oxygen when it is plentiful and release oxygen when it is scarce.

Aerobic cellular respiration provides enough ATP for prolonged activity so long as sufficient oxygen and nutrients are available. Besides pyruvic acid obtained from glycolysis of glucose, these nutrients include fatty acids (from the breakdown of triglycerides in adipose cells) and amino acids (from the breakdown of proteins). In activities that last more than 10 minutes, the aerobic system provides more than 90% of the needed ATP. At the end of an endurance event, such as a marathon race, nearly 100% of the ATP is being produced by aerobic cellular respiration.

Muscle Fatigue

The inability of a muscle to contract forcefully after prolonged activity is called **muscle fatigue.** Fatigue results mainly from changes within muscle fibers. Even before actual muscle fatigue occurs, a person may have feelings of tiredness and the desire to cease activity. This response, termed *central fatigue,* may be a protective mechanism in that it might cause a person to stop exercising before muscle becomes too damaged. As you will see, certain types of skeletal muscle fibers fatigue more quickly than others.

Although the precise mechanisms that cause muscle fatigue are still not clear, several factors are thought to contribute. One important factor is inadequate release of calcium ions from the SR, resulting in a decline of Ca^{2+} concentration in the sarcoplasm. Depletion of creatine phosphate also is associated with fatigue. Surprisingly, however, the ATP levels in fatigued muscle often are not much lower than those in resting muscle. Other factors that contribute to muscle fatigue include insufficient oxygen, depletion of glycogen and other nutrients, buildup of lactic acid and ADP, and failure of action potentials in the motor neuron to release enough acetylcholine.

Oxygen Consumption After Exercise

During prolonged periods of muscle contraction, increases in breathing effort and blood flow enhance oxygen delivery to muscle tissue. After muscle contraction has stopped, heavy breathing continues for a period of time, and oxygen consumption remains above the resting level. Depending on the intensity of the exercise, the recovery period may be just a few minutes or several hours. In 1922, A. V. Hill coined the term **oxygen debt** for the added oxygen, over and above the resting oxygen consumption, that is taken into the body after exercise. He proposed that this extra oxygen was used to "pay back" or restore metabolic conditions to the resting level in three ways: (1) to convert lactic acid back into glycogen stores in the liver, (2) to resynthesize creatine phosphate and ATP, and (3) to replace the oxygen removed from myoglobin.

The metabolic changes that occur *during exercise,* however, account for only some of the extra oxygen used *after exercise.* Only a small amount of resynthesis of glycogen occurs from lactic acid. Instead, glycogen stores are replenished much later from dietary carbohydrates. Much of the lactic acid that remains after exercise is converted back to pyruvic acid and used for ATP production via aerobic cellular respiration in the heart, liver, kidneys, and skeletal muscle tissues. Postexercise oxygen use also is boosted by ongoing changes. First, the elevated body temperature after strenuous exercise increases the rate of chemical reactions throughout the body. Faster reactions use ATP more rapidly, and more oxygen is needed to produce ATP. Second, the heart and muscles used in breathing are still working harder than they were at rest, and thus they consume more ATP. Third, tissue repair processes are occurring at an increased pace. For these reasons, **recovery oxygen uptake** is a better term than oxygen debt for the elevated use of oxygen after exercise.

1. Describe which ATP-producing reactions are aerobic and which are anaerobic.
2. Describe the relation between duration of muscle contraction and sources of ATP.
3. What factors contribute to muscle fatigue?
4. Why is the term *oxygen debt* misleading?

CONTROL OF MUSCLE TENSION
OBJECTIVES
• *Describe the structure and function of a motor unit.*

• *Explain the phases of a twitch contraction.*

• *Describe how frequency of stimulation affects muscle tension.*

A single nerve impulse in a motor neuron elicits a single muscle action potential in all the muscle fibers with which it forms synapses. In contrast to action potentials, which always have the same size in a given neuron or muscle fiber, the contraction that results from a single muscle action potential has significantly smaller force than the maximum force the fiber is capable of producing. The total tension that a single fiber can produce depends mainly on the rate at which nerve impulses arrive at the neuromuscular junction. The number of impulses per second is the *frequency of stimulation.* Also, as we saw in Figure 10.10, the amount of stretch before contraction determines the maximum tension

Figure 10.14 Motor units. Shown are two somatic motor neurons, one in purple and one in green, each supplying the muscle fibers of its motor unit.

🔑 **A motor unit consists of a somatic motor neuron plus all the muscle fibers it stimulates.**

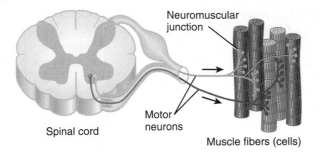

Spinal cord

Neuromuscular junction

Motor neurons

Muscle fibers (cells)

Ⓠ What is the effect of the size of a motor unit on its strength of contraction? (Assume that each muscle fiber generates about the same amount of tension.)

Figure 10.15 Myogram of a twitch contraction. The arrow indicates the time at which the stimulus occurred.

🔑 **A myogram is a record of a muscle contraction.**

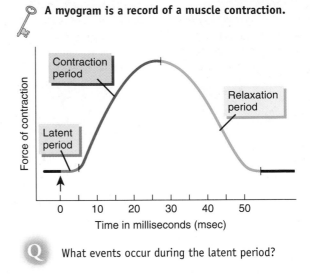

Ⓠ What events occur during the latent period?

that is possible during contraction. Finally, conditions such as nutrient and oxygen availability can influence the tension a fiber can generate. When considering the contraction of a whole muscle, the total tension it can produce depends on the number of fibers that are contracting in unison.

Motor Units

Even though each skeletal muscle fiber has only a single neuromuscular junction, the axon of a motor neuron branches out and forms neuromuscular junctions with many different muscle fibers. A somatic motor neuron plus all the skeletal muscle fibers it stimulates is called a **motor unit** (Figure 10.14). A single motor neuron makes contact with an average of 150 muscle fibers and all muscle fibers in one motor unit contract in unison. Typically, the muscle fibers of a motor unit are dispersed throughout a muscle rather than clustered together.

Muscles that control precise movements consist of many small motor units. For example, muscles of the larynx (voice box) that control voice production have as few as two or three muscle fibers per motor unit, and muscles controlling eye movements may have 10–20 muscle fibers per motor unit. In contrast, some motor units in skeletal muscles responsible for large-scale and powerful movements, such as the biceps brachii muscle in the arm and the gastrocnemius muscle in the leg, may have 2000–3000 muscle fibers each. Remember, all the muscle fibers of a motor unit contract and relax together. Accordingly, the total strength of a contraction depends, in part, on how large the motor units are and how many motor units are activated at the same time.

Twitch Contraction

A **twitch contraction** is a brief contraction of all the muscle fibers in a motor unit in response to a single action

potential in its motor neuron. In the laboratory, a twitch also can be produced by direct electrical stimulation of a motor neuron or its muscle fibers. In the record of a muscle contraction, called a **myogram,** shown in Figure 10.15, we can see that compared to the brief 1–2 msec* duration of an action potential, a twitch is very long lasting—from 20–200 msec.

Note that a brief delay occurs between application of the stimulus (time zero on the graph) and the beginning of contraction; this delay is the **latent period,** which lasts about 2 msec. During this time, calcium ions are being released from the sarcoplasmic reticulum, the filaments start to exert tension, the elastic components stretch, and finally shortening begins. The second phase, the **contraction period,** has a duration of 10–100 msec. The third phase, the **relaxation period,** also lasting 10–100 msec, is caused by the active transport of Ca^{2+} back into the sarcoplasmic reticulum, which results in relaxation. The actual duration of these periods depends on the type of muscle fiber. Some fibers, such as those that move the eyes, are fast-twitch fibers (described shortly); they have a contraction period as brief as 10 msec and an equally brief relaxation period. Others, such as those that move the legs, are slow-twitch fibers, with contraction and relaxation periods of about 100 msec.

If two stimuli are applied, one immediately after the other, the muscle will respond to the first stimulus but not to the second. When a muscle fiber receives enough stimulation to contract, it temporarily loses its excitability and cannot respond for a time. This period of lost excitability, called the **refractory period,** is a characteristic of all muscle and

*One millisecond (msec) is 10^{-3} seconds (0.001 sec).

Figure 10.16 Myograms showing the effects of different frequencies of stimulation. (a) Single twitch. (b) When a second stimulus occurs before the muscle has relaxed, the second contraction is stronger than the first, a phenomenon called wave summation. (The dashed line indicates the force of contraction expected in a single twitch.) (c) In unfused tetanus, the curve looks jagged due to partial relaxation of the muscle between stimuli. (d) In fused tetanus, which occurs when there are 80–100 stimuli per second, the contraction force is steady and sustained.

🔑 **Due to wave summation, the tension produced during a sustained contraction is greater than during a single twitch.**

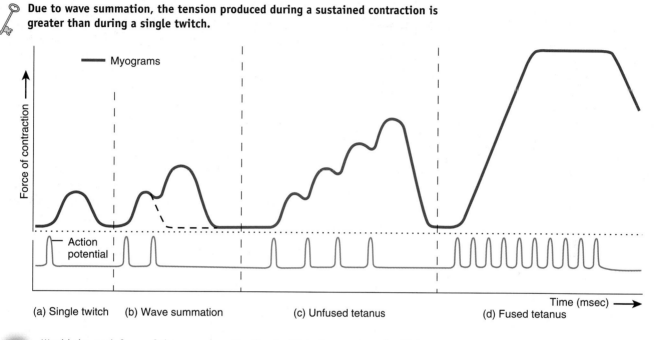

(a) Single twitch   (b) Wave summation   (c) Unfused tetanus   (d) Fused tetanus

Q Would the peak force of the second contraction in (b) be larger or smaller if the second stimulus were applied a few milliseconds later?

nerve cells. The duration of the refractory period varies with the muscle involved. Skeletal muscle has a short refractory period of about 5 msec, whereas cardiac muscle has a long refractory period of about 300 msec.

Frequency of Stimulation

When the second of two stimuli is applied after the refractory period is over, the skeletal muscle will respond to both stimuli. In fact, if the second stimulus occurs after the refractory period but before the muscle fiber has relaxed, the second contraction will actually be stronger than the first (Figure 10.16a, b). This phenomenon, in which stimuli arriving at different times cause larger contractions, is called **wave summation.**

When a skeletal muscle is stimulated at a rate of 20–30 times per second, it can only partially relax between stimuli. The result is a sustained but wavering contraction called **unfused tetanus** (*tetan-* = rigid, tense; Figure 10.16c). Stimulation at a higher rate of 80–100 stimuli per second results in **fused tetanus,** a sustained contraction in which individual twitches cannot be discerned (Figure 10.16d). Wave summation and both kinds of tetanus result from the addition of

Ca^{2+} released from the sarcoplasmic reticulum by the second, and subsequent, stimuli to the Ca^{2+} still in the sarcoplasm from the first stimulus. Because the Ca^{2+} level builds up, the peak tension generated during fused tetanus is 5–10 times larger than the peak tension produced during a single twitch. Nevertheless, smooth, sustained voluntary muscle contractions are achieved mainly by out-of-synchrony unfused tetanus in different motor units.

The stretch of elastic elements is also related to wave summation. During wave summation, elastic elements are not given much time to spring back between contractions, and thus they remain taut. While in this state, the elastic elements do not require very much stretching before the beginning of the next muscular contraction. The combination of the tautness of the elastic elements and the partially contracted state of the filaments enables the force of another contraction to be added more quickly to the one before.

Motor Unit Recruitment

The process in which the number of active motor units is increased is called **motor unit recruitment.** The various motor neurons to a whole muscle fire *asynchronously;* that is, while some motor units are active and contracting, others are inactive and relaxed. This pattern of motor unit activity delays muscle fatigue by allowing alternately contracting motor units to relieve one another, so that the contraction can be sustained for long periods. The weakest motor units are recruited first, with progressively stronger motor units being added if the task requires more force.

Recruitment is one factor responsible for producing smooth movements rather than a series of jerky movements. As mentioned, the number of muscle fibers innervated by one motor neuron varies greatly. Precise movements are brought about by small changes in muscle contraction. Therefore, the muscles that produce precise movements are composed of small motor units. In this way, when a motor unit is recruited or turned off, only slight changes occur in muscle tension. On the other hand, large motor units are active where large tension is needed and precision is less important.

CLINICAL APPLICATION
Endurance Training Versus Strength Training

Regular, repeated activities such as jogging or aerobic dancing increase the supply of oxygen-rich blood available to skeletal muscles for aerobic cellular respiration. By contrast, activities such as weight lifting rely more on anaerobic production of ATP through glycolysis. Such anaerobic activities stimulate synthesis of muscle proteins and result, over a period of time, in increased muscle size (muscle hypertrophy). As a result, aerobic training builds endurance for prolonged activities, whereas anaerobic training builds muscle strength for short-term feats. **Interval training** is a workout regimen that incorporates both types of training—for example, alternating sprints with jogging. ■

Muscle Tone

In a skeletal muscle, a small number of motor units are involuntarily activated to produce a sustained contraction of their muscle fibers, while at the same time the majority of the motor units are not activated and their muscle fibers remain relaxed. This process gives rise to **muscle tone** (*tonos* = tension). To sustain muscle tone, small groups of motor units are alternately active and inactive in a constantly shifting pattern. Muscle tone keeps skeletal muscles firm, but it does not result in a contraction strong enough to produce movement. For example, when the muscles in the back of the neck are in normal tonic contraction, they keep the head upright and prevent it from slumping forward on the chest, but they do not generate enough force to draw the head backward into hyperextension. Muscle tone also is important in smooth muscles, such as those found in the gastrointestinal tract, where the walls of the digestive organs maintain a steady pressure on their contents. The tone of smooth muscles surrounding the walls of blood vessels plays a crucial role in regulating blood pressure.

Isotonic and Isometric Contractions

Isotonic contractions are used for body movements and for moving external objects. There are two types of isotonic contractions: concentric and eccentric. In a **concentric isotonic contraction,** a muscle shortens and pulls on another structure, such as a tendon, to produce movement and to reduce the angle at a joint. Picking a book up off a table involves concentric isotonic contractions of the biceps brachii muscle in the arm (Figure 10.17a). As you lower the book to place it back on the table, the previously shortened biceps gradually lengthens while it continues to contract. When the overall length of a muscle increases during a contraction, it is called an **eccentric isotonic contraction** (Figure 10.17b). For reasons that are not well understood, repeated eccentric isotonic contractions produce more muscle damage and more delayed-onset muscle soreness than do concentric isotonic contractions.

Isometric contractions are important because they stabilize some joints as others are moved. These contractions are important for maintaining posture and for supporting objects in a fixed position. Although isometric contractions do not result in body movement, energy is still expended. In **isometric contractions,** considerable tension is generated without shortening of the muscle. An example would be holding a book steady using an outstretched arm (Figure 10.17c). The book pulls the arm downward, stretching the shoulder and arm muscles. The isometric contraction of the shoulder and arm muscles counteracts the stretch. The two forces—contraction and stretching—applied in opposite directions create the tension. Most activities include both isotonic and isometric contractions.

1. Relate the sizes of motor units to the degree of muscular control they allow.
2. Why is motor unit recruitment important?
3. Define muscle tone and explain why it is important.
4. Define the following terms: concentric isotonic contraction, eccentric isotonic contraction, and isometric contraction.

TYPES OF SKELETAL MUSCLE FIBERS

OBJECTIVE

• *Compare the structure and function of the three types of skeletal muscle fibers.*

Skeletal muscle fibers are not all alike in either composition or function. For example, muscle fibers vary in their content of myoglobin, the red-colored protein that binds oxygen in muscle fibers. Skeletal muscle fibers that have a high myoglobin content are termed *red muscle fibers;* those that have a low content of myoglobin are called *white muscle fibers.* Red muscle fibers also contain more mitochondria for ATP production and are supplied by more blood capillaries than are white muscle fibers.

Figure 10.17 Comparison between isotonic (concentric and eccentric) and isometric contractions. Parts (a) and (b) show isotonic contraction of the biceps brachii muscle in the arm; part (c) shows isometric contraction of shoulder and arm muscles.

🗝 **In an isotonic contraction, tension remains constant as muscle length decreases or increases; in an isometric contraction, tension increases greatly without a change in muscle length.**

(a) Concentric contraction (b) Eccentric contraction (c) Isometric contraction

Q **What kind of contractions are occurring in your neck muscles while you are reading a classroom blackboard?**

Skeletal muscle fibers contract and relax with different velocities. A fiber is categorized as either slow or fast depending on how rapidly the ATPase in its myosin heads hydrolyzes ATP. In addition, we have seen that skeletal muscle fibers vary in the metabolic reactions they use to generate ATP and in how quickly they fatigue. Based on these structural and functional characteristics, skeletal muscle fibers are classified into three main types: (1) slow oxidative fibers, (2) fast oxidative-glycolytic fibers, and (3) fast glycolytic fibers.

Slow Oxidative Fibers

Slow oxidative (SO) fibers are smallest in diameter and thus the least powerful type of muscle fibers. They appear dark red because they contain large amounts of myoglobin and many blood capillaries. Because they have many large mitochondria, SO fibers generate ATP mainly by aerobic cellular respiration, which is why they are called oxidative fibers. These fibers are said to be "slow" because the ATPase in the myosin heads hydrolyzes ATP relatively slowly and the contraction cycle proceeds at a slower pace than in "fast" fibers. As a result, SO fibers have a low contraction velocity. Their twitch contractions last 100–200 msec, and they take longer to reach peak tension. However, slow fibers are very resistant to fatigue and are capable of prolonged, sustained contractions for many hours. These slow twitch, fatigue-resistant fibers are adapted for maintaining posture and for aerobic, endurance-type activities such as running a marathon.

Fast Oxidative-Glycolytic Fibers

Fast oxidative-glycolytic (FOG) fibers are intermediate in diameter between the other two types of fibers. Like slow oxidative fibers, they contain large amounts of myoglobin and many blood capillaries, and thus appear dark red. FOG fibers can generate considerable ATP by aerobic cellular respiration, which gives them a moderately high resistance to fatigue. Because their intracellular glycogen level is high, they also generate ATP by anaerobic glycolysis. These fibers are "fast" because the ATPase in their myosin heads hydrolyzes ATP 3–5 times faster than the myosin ATPase in SO fibers, which makes their contraction velocity faster. Thus, twitches of FOG fibers reach peak tension more quickly than those of SO fibers but are briefer in duration—less than 100 msec. FOG fibers contribute to activities such as walking and sprinting.

Fast Glycolytic Fibers

Fast glycolytic (FG) fibers are largest in diameter, contain the highest number of myofibrils, and thus can generate the most powerful contractions. They have a low myoglobin content, relatively few blood capillaries, few mitochondria, and appear white in color. FG fibers contain large amounts of glycogen and generate ATP mainly by glycolysis. Due to their large size and their ability to hydrolyze ATP rapidly, FG fibers contract strongly and rapidly. These fast-twitch fibers are adapted for intense anaerobic movements of short duration, such as weight lifting or throwing a ball, but they fatigue quickly. Strength training programs that engage a person in activities requiring great strength for short times produce increases in the size, strength, and glycogen content of fast glycolytic fibers. The FG fibers of a weight lifter may be 50% larger than those of a sedentary person or endurance athlete. The increase in size is due to increased synthesis of muscle proteins. The overall result is muscle enlargement due to hypertrophy of the FG fibers.

Distribution and Recruitment of Different Types of Fibers

Most skeletal muscles are a mixture of all three types of skeletal muscle fibers, about half of which are SO fibers. The proportions vary somewhat, depending on the action of the muscle, the person's training regimen, and genetic factors. For example, the continually active postural muscles of the neck, back, and legs have a high proportion of SO fibers. Muscles of the shoulders and arms, in contrast, are not constantly active but are used intermittently and briefly to produce large amounts of tension, such as in lifting and throwing. These muscles have a high proportion of FG fibers. Leg muscles, which not only support the body but are also used for walking and running, have large numbers of both SO and FOG fibers.

Even though most skeletal muscles are a mixture of all three types of skeletal muscle fibers, the skeletal muscle fibers of any given motor unit are all of the same type. However, the different motor units in a muscle are recruited in a specific order, depending on need. For example, if weak contractions suffice to perform a task, only SO motor units are activated. If more force is needed, the motor units of FOG fibers are also recruited. Finally, if maximal force is required, motor units of FG fibers are also called into action. Activation of various motor units is controlled by the brain and spinal cord.

Table 10.2 summarizes the characteristics of the three types of skeletal muscle fibers.

CLINICAL APPLICATION
Anabolic Steroids

The illegal use of **anabolic steroids** by athletes has received widespread attention. These steroid hormones, similar to testosterone, are taken to increase muscle size and thus strength during athletic contests. The large doses needed to produce an effect, however, have damaging, sometimes even devastating side effects, including liver cancer, kidney damage, increased risk of heart disease, stunted growth, wide mood swings, and increased irritability and aggression. Additionally, females may experience atrophy of the breasts and uterus, menstrual irregularities, sterility, facial hair growth,

and deepening of the voice; males may experience diminished testosterone secretion, atrophy of the testes, and baldness. ■

1. What is the basis for classifying skeletal muscle fibers into three types?

CARDIAC MUSCLE TISSUE
OBJECTIVE
• *Describe the main structural and functional characteristics of cardiac muscle tissue.*

The principal tissue in the heart wall is **cardiac muscle tissue** (described in more detail in Chapter 20 and illustrated in Figure 20.9 on page 649). Cardiac muscle fibers have the same arrangement of actin and myosin and the same bands, zones, and Z discs as skeletal muscle fibers. However, the ends of cardiac muscle fibers connect to adjacent fibers by irregular transverse thickenings of the sarcolemma called *intercalated discs* (in-TER-ka-lāt-ed; *intercal-* = to insert between). The discs contain *desmosomes,* which hold the fibers together, and *gap junctions,* which allow muscle action potentials to spread from one cardiac muscle fiber to another (see Figure 4.1 on page 106).

In response to a single action potential, cardiac muscle tissue remains contracted 10–15 times longer than skeletal muscle tissue. This is due to prolonged delivery of Ca^{2+} into the sarcoplasm. In cardiac muscle fibers, calcium ions enter the sarcoplasm both from the sarcoplasmic reticulum (as in skeletal muscle fibers) and from extracellular fluid (ECF). Because the channels that allow inflow of Ca^{2+} from ECF stay open for a relatively long period of time, a cardiac muscle contraction (see Figure 20.11 on page 652) lasts much longer than a skeletal muscle twitch.

We have seen that skeletal muscle tissue contracts only when stimulated by acetylcholine released by a nerve impulse in a motor neuron. In contrast, cardiac muscle tissue contracts when stimulated by its own autorhythmic fibers. Under normal resting conditions, cardiac muscle tissue contracts and relaxes an average of about 75 times a minute. This continuous, rhythmic activity is a major physiological difference between cardiac and skeletal muscle tissue. Accordingly, cardiac muscle tissue requires a constant supply of oxygen, and the mitochondria in cardiac muscle fibers are larger and more numerous than in skeletal muscle fibers. This structural feature correctly suggests that cardiac muscle depends greatly on aerobic cellular respiration to generate ATP. Moreover, cardiac muscle fibers can use lactic acid produced by skeletal muscle fibers to make ATP, a benefit during exercise.

1. Construct a table to compare the properties of skeletal and cardiac muscle.

Table 10.2 Characteristics of the Three Types of Skeletal Muscle Fibers

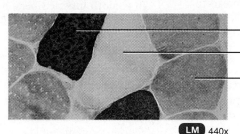

— Slow oxidative fiber
— Fast glycolytic fiber
— Fast oxidative-glycolytic fiber

LM 440x

Transverse section of three types of skeletal muscle fibers

STRUCTURAL CHARACTERISTIC	SLOW OXIDATIVE (SO) FIBERS	FAST OXIDATIVE-GLYCOLYTIC (FOG) FIBERS	FAST GLYCOLYTIC (FG) FIBERS
Fiber diameter	Smallest	Intermediate	Largest
Myoglobin content	Large amount	Large amount	Small amount
Mitochondria	Many	Many	Few
Capillaries	Many	Many	Few
Color	Red	Red-pink	White (pale)
FUNCTIONAL CHARACTERISTIC			
Capacity for generating ATP and method used	High capacity, by aerobic (oxygen-requiring) cellular respiration	Intermediate capacity, by both aerobic (oxygen-requiring) cellular respiration and glycolysis (anaerobic)	Low capacity, by anaerobic cellular respiration (glycolysis)
Rate of ATP hydrolysis by myosin ATPase	Slow	Fast	Fast
Contraction velocity	Slow	Fast	Fast
Fatigue resistance	High	Intermediate	Low
Glycogen stores	Low	Intermediate	High
Order of recruitment	First	Second	Third
Location where fibers are abundant	Postural muscles such as those of the neck	Leg muscles	Arm muscles
Primary functions of fibers	Maintaining posture and endurance-type activities	Walking, sprinting	Rapid, intense movements of short duration

SMOOTH MUSCLE TISSUE

OBJECTIVE

• *Describe the main structural and functional characteristics of smooth muscle tissue.*

Like cardiac muscle tissue, **smooth muscle tissue** is usually activated involuntarily. Of the two types of smooth muscle tissue, the more common type is **visceral (single-unit) smooth muscle tissue** (Figure 10.18a). It is found in wraparound sheets that form part of the walls of small arteries and veins and of hollow viscera such as the stomach, intestines, uterus, and urinary bladder. Like cardiac muscle, visceral smooth muscle is autorhythmic. Because the fibers connect to one another by gap junctions, muscle action potentials spread throughout the network. When a neurotransmitter, hormone, or autorhythmic signal stimulates one fiber, the muscle action potential spreads to neighboring fibers, which then contract in unison, as a single unit.

The second kind of smooth muscle tissue is **multiunit smooth muscle tissue** (Figure 10.18b). It consists of individual fibers, each with its own motor neuron terminals and with few gap junctions between neighboring fibers. Whereas stimulation of one visceral muscle fiber causes contraction of many adjacent fibers, stimulation of one multiunit fiber causes contraction of that fiber only. Multiunit smooth muscle tissue is found in the walls of large arteries, in airways to

Figure 10.18 Two types of smooth muscle tissue. In (a), one autonomic motor neuron synapses with several visceral smooth muscle fibers, and action potentials spread to neighboring fibers through gap junctions. In (b), three autonomic motor neurons synapse with individual multiunit smooth muscle fibers. Stimulation of one multiunit fiber causes contraction of that fiber only.

🔑 **Visceral smooth muscle fibers connect to one another by gap junctions and contract as a single unit; multiunit smooth muscle fibers lack gap junctions and contract independently.**

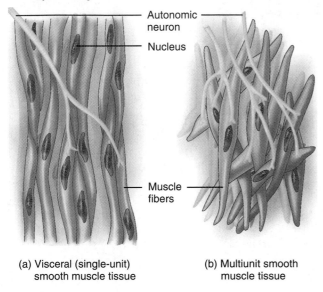

Autonomic neuron

Nucleus

Muscle fibers

(a) Visceral (single-unit) smooth muscle tissue

(b) Multiunit smooth muscle tissue

Q Which type of smooth muscle is more like cardiac muscle than skeletal muscle, with respect to both its structure and function?

the lungs, in the arrector pili muscles that attach to hair follicles, in the muscles of the iris that adjust pupil diameter, and in the ciliary body that adjusts focus of the lens in the eye.

Microscopic Anatomy of Smooth Muscle

Endomysium surrounds smooth muscle fibers, which are considerably smaller than skeletal muscle fibers. A single relaxed smooth muscle fiber is 30–200 μm long, thickest in the middle (3–8 μm), and tapered at each end (Figure 10.19a). Within each fiber is a single, oval, centrally located nucleus (also shown in Table 4.4 on page 132). The sarcoplasm of smooth muscle fibers contains both *thick filaments* and *thin filaments,* in ratios between about 1:10 and 1:15, but they are not arranged in orderly sarcomeres as in striated muscle. Smooth muscle fibers also contain *intermediate filaments.* Because the various filaments have no regular pattern of overlap, smooth muscle fibers do not exhibit striations. This is the reason for the name *smooth.* Smooth mus-

cle fibers also lack transverse tubules and have only scanty sarcoplasmic reticulum for storage of Ca^{2+}.

In smooth muscle fibers, intermediate filaments attach to structures called **dense bodies,** which are functionally similar to Z discs in striated muscle fibers. Some dense bodies are dispersed throughout the sarcoplasm; others are attached to the sarcolemma. Bundles of intermediate filaments stretch from one dense body to another (see Figure 10.19a). During contraction, the sliding filament mechanism involving thick and thin filaments generates tension that is transmitted to intermediate filaments. These, in turn, pull on the dense bodies attached to the sarcolemma, causing a lengthwise shortening of the muscle fiber. Note that shortening of the muscle fiber produces a bubblelike expansion of the sarcolemma (Figure 10.19b). A smooth muscle fiber contracts like a corkscrew turns; the fiber twists in a helix as it contracts and rotates in the opposite direction as it relaxes.

Physiology of Smooth Muscle

Although the principles of contraction are similar in all three types of muscle tissue, smooth muscle tissue exhibits some important physiological differences. Compared with contraction in a skeletal muscle fiber, contraction in a smooth muscle fiber starts more slowly and lasts much longer. Moreover, smooth muscle can both shorten and stretch to a greater extent than other muscle types.

An increase in the concentration of Ca^{2+} in smooth muscle cytosol initiates contraction, just as in striated muscle. Sarcoplasmic reticulum (the reservoir for Ca^{2+} in striated muscle) is scanty in smooth muscle. Calcium ions flow into smooth muscle cytosol from both the extracellular fluid and sarcoplasmic reticulum, but because there are no transverse tubules in smooth muscle fibers, it takes longer for Ca^{2+} to reach the filaments in the center of the fiber and trigger the contractile process. This accounts, in part, for the slow onset and prolonged contraction of smooth muscle.

Several mechanisms regulate contraction and relaxation of smooth muscle cells. In one, a regulatory protein called **calmodulin** (cal-MOD-yoo-lin) binds to Ca^{2+} in the cytosol. (Recall that troponin has this role in striated muscle fibers.) After binding to Ca^{2+}, calmodulin activates an enzyme called *myosin light chain kinase.* This enzyme uses ATP to phosphorylate (add a phosphate group to) a portion of the myosin head. Once the phosphate group is attached, the myosin head can bind to actin, and contraction can occur. Myosin light chain kinase works rather slowly, also contributing to the slowness of smooth muscle contraction.

Not only do calcium ions enter smooth muscle fibers slowly, they also move slowly out of the muscle fiber when excitation declines, which delays relaxation. The prolonged presence of Ca^{2+} in the cytosol provides for **smooth muscle tone,** a state of continued partial contraction. Smooth muscle tissue can thus sustain long-term tone, which is important in the gastrointestinal tract, where the walls maintain a

Figure 10.19 Microscopic anatomy of a smooth muscle fiber.

🔑 **Smooth muscle fibers have thick and thin filaments but no transverse tubules and scanty sarcoplasmic reticulum.**

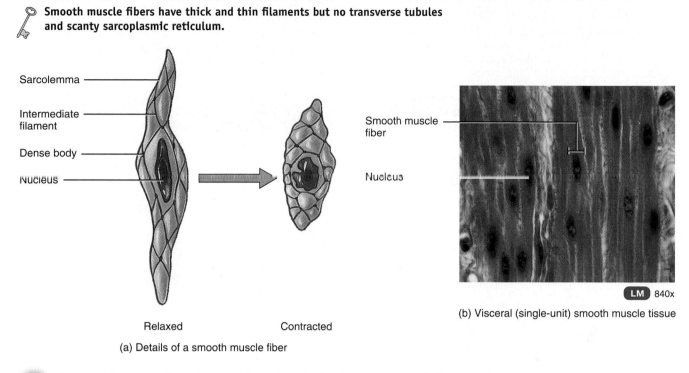

Sarcolemma

Intermediate filament

Dense body

Nucleus

Relaxed   Contracted

(a) Details of a smooth muscle fiber

Smooth muscle fiber

Nucleus

LM 840x

(b) Visceral (single-unit) smooth muscle tissue

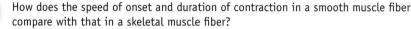

Q How does the speed of onset and duration of contraction in a smooth muscle fiber compare with that in a skeletal muscle fiber?

steady pressure on the contents of the tract, and in the walls of blood vessels called arterioles, which maintain a steady pressure on blood.

Most smooth muscle fibers contract or relax in response to action potentials from the autonomic nervous system. In addition, many smooth muscle fibers contract or relax in response to stretching, hormones, or local factors such as changes in pH, oxygen and carbon dioxide levels, temperature, and ion concentrations. For example, the hormone epinephrine, released by the adrenal medulla, causes relaxation of smooth muscle in the airways and in some blood vessel walls (those that have so-called β_2 receptors; see Exhibit 17.4 on page 554).

Unlike striated muscle fibers, smooth muscle fibers can stretch considerably and still maintain their contractile function. When smooth muscle fibers are stretched, they initially contract, developing increased tension. Within a minute or so, the tension decreases. This phenomenon, termed the **stress–relaxation response,** allows smooth muscle to undergo great changes in length while still retaining the ability to contract effectively. Thus even though smooth muscle in the walls of blood vessels and hollow organs such as the stomach, intestines, and urinary bladder can stretch, the pressure on the contents within them changes very little. Af-

ter the organ empties, however, the smooth muscle in the wall rebounds, and the wall retains its firmness.

1. How do visceral and multiunit smooth muscle differ?
2. Construct a table to compare the properties of skeletal and smooth muscle.

REGENERATION OF MUSCLE TISSUE

OBJECTIVE

• *Explain how muscle fibers regenerate.*

Because mature skeletal muscle fibers have lost the ability to undergo cell division, growth of skeletal muscle after birth is due mainly to hypertrophy, the enlargement of existing cells, rather than to hyperplasia, an increase in the number of fibers. Satellite cells divide slowly and fuse with existing fibers to assist both in muscle growth and in repair of damaged fibers. In response to muscle injury or a disease that causes muscle degeneration, additional cells derived from red bone marrow migrate into the muscle and participate in regeneration of the damaged fibers. The number of new skeletal muscle fibers formed in these ways, however, is not sufficient to compensate for significant skeletal muscle damage or degeneration. In such cases, skeletal muscle tissue

Table 10.3 Summary of the Principal Features of the Three Types of Muscle Tissue

CHARACTERISTIC	SKELETAL MUSCLE	CARDIAC MUSCLE	SMOOTH MUSCLE
Cell appearance and features	Long cylindrical fiber with many peripherally located muclei; striated.	Branched cylindrical fiber with one centrally located nucleus; intercalated discs join neighboring fibers; striated.	Spindle-shaped fiber with one centrally positioned nucleus; not striated.
Location	Attached primarily to bones by tendons.	Heart.	Walls of hollow viscera, airways, blood vessels, iris and ciliary body of eye, arrector pili of hair follicles.
Fiber diameter	Very large (10–100 μm).	Large (14 μm).	Small (3–8 μm).
Connective tissue components	Endomysium, perimysium, and epimysium.	Endomysium.	Endomysium.
Fiber length	100 μm to 30 cm.	50–100 μm.	30–200 μm.
Contractile proteins organized into sarcomeres	Yes.	Yes.	No.
Sarcoplasmic reticulum	Abundant.	Some.	Scanty.
Transverse tubules present?	Yes, aligned with each A-I band junction.	Yes, aligned with each Z disc.	No.
Junctions between fibers	None.	Intercalcated discs contain gap junctions and desmosomes.	Gap junctions in visceral smooth muscle; none in multiunit smooth muscle.
Autorhythmicity	No.	Yes.	Yes, in visceral smooth muscle.
Source of Ca^{2+} for contraction	Sarcoplasmic reticulum.	Sarcoplasmic reticulum and extracellular fluid.	Sarcoplasmic reticulum and extracellular fluid.
Regulator proteins for contraction	Troponin and tropomyosin.	Troponin and tropomyosin.	Calmodulin and myosin light chain kinase.
Speed of contraction	Fast.	Moderate.	Slow.
Nervous control	Voluntary (somatic nervous system).	Involuntary (autonomic nervous system).	Involuntary (autonomic nervous system).
Contraction regulated by	Acetylcholine released by somatic motor neurons.	Acetylcholine, norepinephrine released by autonomic motor neurons; several hormones.	Acetylcholine, norepinephrine released by autonomic motor neurons; several hormones, local chemical changes; stretching.
Capacity for regeneration	Limited, via satellite cells.	None.	Considerable, via pericytes (compared with other muscle tissues, but limited compared with epithelium).

undergoes **fibrosis,** the replacement of muscle fibers by fibrous scar tissue. For this reason, skeletal muscle tissue has only limited powers of regeneration.

Cardiac muscle does not have cells comparable to the satellite cells of skeletal muscle. Damaged cardiac muscle fibers are not repaired or replaced, and healing occurs by fibrosis. These fibers can undergo hypertrophy, however, in response to increased workload. For this reason, many athletes have enlarged hearts.

Smooth muscle tissue, like skeletal and cardiac muscle tissue, can undergo hypertrophy. In addition, certain smooth muscle fibers, such as those in the uterus, retain their capacity for division and thus can grow by hyperplasia. Also, new smooth muscle fibers can arise from cells called *pericytes,* stem cells found in association with blood capillaries and small veins. Smooth muscle fibers can also proliferate in certain pathological conditions, such as occur in the development of atherosclerosis (see page 663). Compared with the other two types of muscle tissue, smooth muscle tissue has considerably greater powers of regeneration. Such powers are still limited when compared with other tissues, such as epithelium.

Table 10.3 summarizes the principal characteristics of the three types of muscle tissue.

DEVELOPMENTAL ANATOMY OF THE MUSCULAR SYSTEM

OBJECTIVE

• *Describe the development of the muscular system.*

In this brief discussion of the development of the human muscular system, we concentrate mostly on skeletal muscles. Except for the muscles of the iris of the eyes and the arrector pili muscles attached to hairs, all muscles of the body are derived from **mesoderm.** As the mesoderm develops, a portion of it becomes arranged in dense columns on either side of the developing nervous system. These columns of mesoderm undergo segmentation into a series of blocks of cells called **somites** (Figure 10.20a). The first pair of somites appears on the 20th day of embryonic development. Eventually, 44 pairs of somites are formed by the 30th day.

With the exception of the skeletal muscles of the head and limbs, *skeletal muscles* develop from the **mesoderm of somites.** Because there are very few somites in the head region of the embryo, most of the skeletal muscles there develop from the **general mesoderm** in the head region. The skeletal muscles of the limbs develop from masses of general mesoderm around developing bones in embryonic limb buds (origins of future limbs; see Figure 6.12a on page 177).

The cells of a somite are differentiated into three regions: (1) **myotome,** which forms some of the skeletal muscles; (2) **dermatome,** which forms the connective tissues, including the dermis; and (3) **sclerotome,** which gives rise to the vertebrae (Figure 10.20b).

Figure 10.20 Location and structure of somites, key structures in the development of the muscular system.

🔑 **Most muscles are derived from mesoderm.**

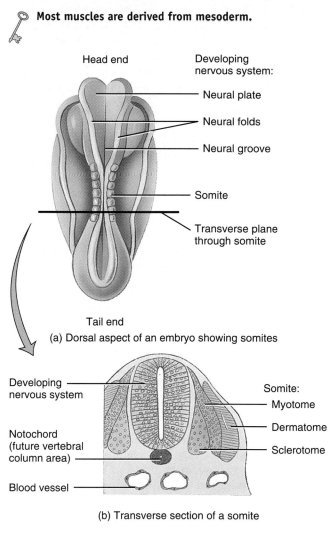

(a) Dorsal aspect of an embryo showing somites

(b) Transverse section of a somite

Q Which portion of a somite differentiates into skeletal muscle?

Cardiac muscle develops from *mesodermal cells* that migrate to and envelop the developing heart while it is still in the form of primitive heart tubes (see in Figure 20.17 on page 662).

Smooth muscle develops from **mesodermal cells** that migrate to and envelop the developing gastrointestinal tract and viscera.

AGING AND MUSCLE TISSUE

OBJECTIVE

• *Explain how aging affects skeletal muscle.*

Beginning at about 30 years of age, humans undergo a progressive loss of skeletal muscle mass that is replaced largely by fibrous connective tissue and adipose tissue. In part, this decline is due to increasing inactivity. Accompanying the loss of muscle mass is a decrease in maximal strength and a slowing of muscle reflexes. In some muscles, a selective loss of muscle fibers of a given type may occur. With aging, the relative number of slow oxidative fibers appears to increase. This could be due either to atrophy of the other fiber types or their conversion into slow oxidative fibers. Whether this is an effect of aging itself or mainly reflects the more limited physical activity of older people is still an unresolved question. Nevertheless, endurance and strength training programs are effective in older people and can slow or even reverse the age-associated decline in muscular performance.

1. Why do skeletal muscle fibers have only limited powers of regeneration?
2. Why does muscle strength decrease with aging?

DISORDERS: HOMEOSTATIC IMBALANCES

Muscle function may be abnormal due to disease or damage of any of the components of a motor unit: somatic motor neurons, neuromuscular junctions, or muscle fibers. The term **neuromuscular disease** encompasses problems at all three sites, whereas the term **myopathy** (mī-OP-a-thē; -*pathy* = disease) signifies a disease or disorder of the skeletal muscle tissue itself.

MYASTHENIA GRAVIS

Myasthenia gravis (mī-as-THĒ-nē-a GRAV-is) is an autoimmune disease that causes chronic, progressive damage of the neuromuscular junction. In people with myasthenia gravis the immune system inappropriately produces antibodies that bind to and block some ACh receptors, thereby decreasing the number of functional ACh receptors at the motor end plates of skeletal muscles (see Figure 10.11). Because 75% of patients with myasthenia gravis have hyperplasia or tumors of the thymus gland, it is possible that thymic abnormalities may cause the disorder. As the disease progresses, more ACh receptors are affected. Muscles become increasingly weaker, fatigue more easily, and may eventually cease to function.

Myasthenia gravis occurs in about 1 in 10,000 people and is more common in women, who typically are ages 20–40 at onset, than in men, who usually are ages 50–60 at onset. The muscles of the face and neck are most often affected. Initial symptoms include a weakness of the eye muscles, which may produce double vision, and difficulty in swallowing. Later, the individual has difficulty chewing and talking. Eventually the muscles of the limbs may become involved. Death may result from paralysis of the respiratory muscles, but often the disorder does not progress to this stage.

Anticholinesterase drugs such as pyridostigmine (Mestinon) or neostigmine are the first line of treatment for myasthenia gravis. They act as inhibitors of acetylcholinesterase, the enzyme that breaks down ACh. Thus, they raise the level of ACh that is available to bind with still-functional receptors. More recently, steroid drugs, such as prednisone, have been used with success to reduce antibody levels. Another treatment is plasmapheresis, a procedure that removes the antibodies from the blood. Often, surgical removal of the thymus gland (thymectomy) is helpful.

MUSCULAR DYSTROPHY

Muscular dystrophy is a group of inherited muscle-destroying diseases that cause progressive degeneration of skeletal muscle fibers. The most common form of muscular dystrophy is *Duchenne muscular dystrophy* (*DMD;* doo-SHĀN). Because the mutated gene is on the X chromosome, and because males have only one copy of it, DMD strikes boys almost exclusively. (Sex-linked inheritance is described in Chapter 29.) Worldwide, about 1 in every 3500 male babies—21,000 in all—are born with DMD each year. The disorder usually becomes apparent between the ages of 2 and 5, when parents notice the child falls often and has difficulty running, jumping, and hopping. By age 12 most boys with DMD are unable to walk. Respiratory or cardiac failure usually causes death between the ages of 20 and 30.

In DMD, the gene that codes for dystrophin is mutated, and little or no dystrophin is present. Without the reinforcing effect of dystrophin, the sarcolemma easily tears during muscle contraction. Because their plasma membrane is damaged, muscle fibers slowly rupture and die. The dystrophin gene was discovered in 1987, and by 1990 the first attempts were made to treat DMD patients with gene therapy. The muscles of three boys with DMD were injected with myoblasts bearing functional dystrophin genes, but only a few muscle fibers gained the ability to produce dystrophin. Similar clinical trials with additional patients have also failed. An alternate approach to the problem is to find a way to induce muscle fibers to produce more utrophin, a structurally similar protein that might be able to substitute for dystrophin. Experiments with dystrophin-deficient mice suggest this approach may work.

ABNORMAL CONTRACTIONS OF SKELETAL MUSCLE

One kind of abnormal muscular contraction is a **spasm,** a sudden involuntary contraction of a single muscle in a large group of muscles. A painful spasmodic contraction is known as a **cramp.** A **tic** is a spasmodic twitching made involuntarily by muscles that are ordinarily under voluntary control. Twitching of the eyelid and facial muscles are examples of tics. A **tremor** is a rhythmic, involuntary, purposeless contraction that produces a quivering or shaking movement. A **fasciculation** is an involuntary, brief twitch of an entire motor unit that is visible under the skin; it occurs irregularly and is not associated with movement of the affected muscle. Fasciculations may be seen in multiple sclerosis (see page 406) or in amyotrophic lateral sclerosis (Lou Gehrig's disease). A **fibrillation** is a spontaneous contraction of a single muscle fiber that is not visible under the skin but can be recorded by electromyography. Fibrillations may signal destruction of motor neurons.

MEDICAL TERMINOLOGY

Charley horse Tearing of a muscle as a result of forceful impact, accompanied by bleeding and severe pain. It often occurs in contact sports and typically affects the quadriceps femoris muscle on the anterior surface of the thigh. The condition is treated by ice immediately after the injury, rest, a supportive wrap, and elevation of the limb.

Hypertonia Increased muscle tone, either rigidity (muscle stiffness) or spasticity (muscle stiffness associated with an increase in tendon reflexes).

Hypotonia (*hypo-* = below or deficient) Decreased or lost muscle tone, usually due to damage to somatic motor neurons.

Myalgia (mī-AL-jē-a; *-algia* = painful condition) Pain in or associated with muscles.

Myoma (mī-Ō-ma; *-oma* = tumor) A tumor consisting of muscle tissue.

Myomalacia (mī'-ō-ma-LĀ-shē-a; *-malacia* = soft) Softening of a muscle.

Myositis (mī'-ō-SĪ-tis; *-itis* = inflammation of) Inflammation of muscle fibers (cells).

Myotonia (mī'-ō-TŌ-nē-a; *-tonia* = tension) Increased muscular excitability and contractility, with decreased power of relaxation; tonic spasm of the muscle.

Volkmann's contracture (FŌLK-manz kon-TRAK-tur; *contra-* = against) Permanent shortening (contracture) of a muscle due to replacement of destroyed muscle fibers by fibrous connective tissue, which lacks extensibility. Destruction of muscle fibers may occur from interference with circulation caused by a tight bandage, a piece of elastic, or a cast.

STUDY OUTLINE

INTRODUCTION (p. 268)

1. Motion results from alternating contraction and relaxation of muscles, which constitute 40–50% of total body weight.
2. The prime function of muscle is changing chemical energy into mechanical energy to perform work.

OVERVIEW OF MUSCLE TISSUE (p. 268)

1. The three types of muscle tissue are skeletal, cardiac, and smooth. Skeletal muscle tissue is primarily attached to bones; it is striated and voluntary. Cardiac muscle tissue forms the wall of the heart; it is striated and involuntary. Smooth muscle tissue is located primarily in internal organs; it is nonstriated (smooth) and involuntary.
2. Through contraction and relaxation, muscle tissue performs five important functions: producing body movements, stabilizing body positions, regulating organ volume, moving substances within the body, and producing heat.
3. Four special properties of muscle tissues are electrical excitability, the property of responding to stimuli by producing action potentials; contractility, the ability to generate tension to do work; extensibility, the ability to be extended (stretched); and elasticity, the ability to return to original shape after contraction or extension.
4. An isotonic contraction occurs when a muscle shortens and moves a constant load; tension remains almost constant. An isometric contraction occurs without shortening of the muscle; tension increases greatly.

SKELETAL MUSCLE TISSUE (p. 269)

1. Connective tissues surrounding muscle are the epimysium, covering the entire muscle; perimysium, covering fascicles; and the endomysium, covering muscle fibers. Superficial fascia separates muscle from skin.
2. Tendons and aponeuroses are extensions of connective tissue beyond muscle fibers that attach the muscle to bone or to other muscle.

3. Skeletal muscles are well supplied with nerves and blood vessels. Generally, an artery and one or two veins accompany each nerve that penetrates a skeletal muscle.
4. Somatic motor neurons provide the nerve impulses that stimulate skeletal muscle to contract.
5. Blood capillaries bring in oxygen and nutrients and remove heat and waste products of muscle metabolism.
6. Each skeletal muscle fiber has 100 or more nuclei because it arises from fusion of many myoblasts. Satellite cells are myoblasts that persist after birth. The sarcolemma is a muscle fiber's plasma membrane; it surrounds the sarcoplasm. T tubules are invaginations of the sarcolemma.
7. Each fiber contains myofibrils, the contractile elements of skeletal muscle. Sarcoplasmic reticulum surrounds each myofibril. Within a myofibril are thin and thick filaments, arranged in compartments called sarcomeres.
8. The overlapping of thick and thin filaments produces striations; darker A bands alternate with lighter I bands.
9. Myofibrils are built from three types of proteins: contractile, regulatory, and structural. The contractile proteins are myosin (thick filament) and actin (thin filament). Regulatory proteins are tropomyosin and troponin (both are part of the thin filament). Structural proteins include titin (links Z disc to M line and stabilizes thick filament), myomesin (M line), nebulin (lies alongside thin filaments), alpha-actinin (Z disc), and dystrophin (links thin filaments to sarcolemma).
10. Projecting myosin heads (crossbridges) contain actin-binding and ATP-binding sites and are the motor proteins that power muscle contraction.
11. Table 10.1 on page 276 reviews the types of proteins in skeletal muscle fibers.

CONTRACTION AND RELAXATION OF SKELETAL MUSCLE FIBERS (p. 276)

1. Muscle contraction occurs because myosin heads attach to and "walk" along the thin filaments at both ends of a sarcomere, progressively pulling the thin filaments toward the center of a sarcomere. As the thin filaments slide inward, the Z discs come closer together, and the sarcomere shortens.

2. The contraction cycle is the repeating sequence of events that causes sliding of the filaments: (1) myosin ATPase hydrolyzes ATP and becomes energized, (2) the myosin head attaches to actin, (3) the myosin head generates force as it rotates toward the center of the sarcomere (power stroke), and (4) binding of ATP to myosin detaches myosin from actin. The myosin head again hydrolyzes the ATP, returns to its original position, and binds to a new site on actin as the cycle continues.

3. An increase in Ca^{2+} concentration in the cytosol starts filament sliding, whereas a decrease turns off the sliding process.

4. The muscle action potential propagating into the T tubule system causes opening of Ca^{2+} release channels in the SR membrane. Calcium ions diffuse from the SR into the cytosol and combine with troponin, causing the troponin–tropomyosin complex to move away from the myosin-binding sites on actin.

5. Ca^{2+} active transport pumps continually remove Ca^{2+} from the sarcoplasm into the SR. When the concentration of calcium ions in the cytosol decreases, the troponin–tropomyosin complexes slide back over and block the myosin-binding sites, and the muscle fiber relaxes.

6. A muscle fiber develops its greatest tension when there is an optimal zone of overlap between thick and thin filaments. This dependence is the length–tension relationship.

7. Tension generated by contractile elements is called active tension; tension generated by elastic elements is called passive tension.

8. The neuromuscular junction (NMJ) is the synapse between a somatic motor neuron and a skeletal muscle fiber. The NMJ includes the axon terminals and synaptic end bulbs of a motor neuron, plus the adjacent motor end plate of the muscle fiber sarcolemma.

9. When a nerve impulse reaches the synaptic end bulbs of a somatic motor neuron, it triggers exocytosis of the synaptic vesicles. ACh diffuses across the synaptic cleft and binds to ACh receptors, initiating a muscle action potential. Acetylcholine esterase then quickly destroys ACh.

MUSCLE METABOLISM (p. 283)

1. Muscle fibers have three sources for ATP production: creatine, anaerobic cellular respiration, and aerobic cellular respiration.

2. Creatine kinase catalyzes the transfer of a high-energy phosphate group from creatine phosphate to ADP to form new ATP molecules. Together, creatine phosphate and ATP provide enough energy for muscles to contract maximally for about 15 seconds.

3. Glucose is converted to pyruvic acid in the reactions of glycolysis, which yield two ATPs without using oxygen. Such anaerobic cellular respiration can provide enough energy for about 30–40 seconds of maximal muscle activity.

4. Muscular activity that lasts longer than half a minute depends on aerobic cellular respiration, mitochondrial reactions that require oxygen to produce ATP. Aerobic cellular respiration yields about 36 molecules of ATP from each glucose molecule.

5. The inability of a muscle to contract forcefully after prolonged activity is muscle fatigue.

6. Elevated oxygen use after exercise is called recovery oxygen uptake.

CONTROL OF MUSCLE TENSION (p. 285)

1. A motor neuron and the muscle fibers it stimulates form a motor unit. A single motor unit may contain as few as two or as many as 3000 muscle fibers.

2. Recruitment is the process of increasing the number of active motor units.

3. A twitch contraction is a brief contraction of all the muscle fibers in a motor unit in response to a single action potential.

4. A record of a contraction is called a myogram. It consists of a latent period, a contraction period, and a relaxation period. The refractory period is the time when a muscle fiber has temporarily lost excitability. Skeletal muscles have a short refractory period, whereas cardiac muscle has a long refractory period.

5. Wave summation is the increased strength of a contraction that occurs when a second stimulus arrives before the muscle has completely relaxed after a previous stimulus.

6. Repeated stimuli can produce unfused tetanus, a sustained muscle contraction with partial relaxation between stimuli; more rapidly repeating stimuli will produce fused tetanus, a sustained contraction without partial relaxation between stimuli.

7. Continual involuntary activation of a small number of motor units produces muscle tone, which is essential for maintaining posture.

8. In a concentric isotonic contraction, the muscle shortens to produce movement and to reduce the angle at a joint. During an eccentric isotonic contraction the muscle lengthens.

9. Isometric contractions, in which tension is generated without muscle shortening, are important because they stabilize some joints as others are moved.

TYPES OF SKELETAL MUSCLE FIBERS (p. 288)

1. On the basis of their structure and function, skeletal muscle fibers are classified as slow oxidative (SO), fast oxidative-glycolytic (FOG), and fast glycolytic (FG) fibers.

2. Most skeletal muscles contain a mixture of all three fiber types; their proportions vary with the typical action of the muscle.

3. The motor units of a muscle are recruited in the following order: first SO fibers, then FOG fibers, and finally FG fibers.

4. Table 10.2 on page 291 summarizes the three types of fibers.

CARDIAC MUSCLE TISSUE (p. 290)

1. Cardiac muscle is found only in the heart. Cardiac muscle fibers have the same arrangement of actin and myosin and the same bands, zones, and Z discs as skeletal muscle fibers. The fibers connect to one another through intercalated discs, which contain both desmosomes and gap junctions.

2. Cardiac muscle tissue remains contracted 10–15 times longer than skeletal muscle tissue due to prolonged delivery of Ca^{2+} into the sarcoplasm.

3. Cardiac muscle tissue contracts when stimulated by its own autorhythmic fibers. Due to its continuous, rhythmic activity cardiac muscle depends greatly on aerobic cellular respiration to generate ATP.

SMOOTH MUSCLE TISSUE (p. 291)

1. Smooth muscle is nonstriated and involuntary.
2. Smooth muscle fibers contain intermediate filaments and dense bodies (whose function is similar to that of the Z discs).
3. Visceral (single-unit) smooth muscle is found in the walls of hollow viscera and of small blood vessels. Many fibers form a network that contracts in unison.
4. Multiunit smooth muscle is found in large blood vessels, large airways to the lungs, arrector pili muscles, and the eye (for adjusting pupil diameter and lens focus). The fibers operate independently rather than in unison.
5. The duration of contraction and relaxation of smooth muscle is longer than in skeletal muscle.
6. Smooth muscle fibers contract in response to nerve impulses, hormones, and local factors.
7. Smooth muscle fibers can stretch considerably and still maintain their contractile function.

REGENERATION OF MUSCLE TISSUE (p. 293)

1. Skeletal muscle fibers cannot divide and have limited powers of regeneration; cardiac muscle fibers can neither divide nor regenerate; smooth muscle fibers have limited capacity for division and regeneration.

2. Table 10.3 on page 294 summarizes the principal characteristics of the three types of muscle tissue.

DEVELOPMENTAL ANATOMY OF THE MUSCULAR SYSTEM (p. 295)

1. With few exceptions, muscles develop from mesoderm.
2. Skeletal muscles of the head and limbs develop from general mesoderm; the remainder of the skeletal muscles develop from the mesoderm of somites.

AGING AND MUSCLE TISSUE (p. 295)

1. Beginning at about 30 years of age, humans undergo a progressive loss of skeletal muscle, which is replaced by fibrous connective tissue and fat.
2. Aging also results in a decrease in muscle strength and in diminished muscle reflexes.

SELF-QUIZ QUESTIONS

True or false:

1. The key functions of muscle are motion of the body, movement of substances within the body, body position stability, organ volume regulation, heat production, and energy formation.

2. The sequence of events resulting in skeletal muscle contraction are (a) generation of a nerve impulse, (b) release of the neurotransmitter acetylcholine, (c) generation of a muscle action potential, (d) release of calcium ions from the sarcoplasmic reticulum, (e) calcium ion binding to the troponin–tropomyosin complex, (f) power stroke with actin and myosin binding and release.

3. Match the following:

___ (a) a sheath of areolar connective tissue that wraps around individual skeletal muscle fibers
___ (b) a sheath of dense irregular connective tissue that surrounds groups of individual muscle fibers
___ (c) bundles of muscle fibers
___ (d) the outermost connective tissue layer of skeletal muscle
___ (e) dense irregular connective tissue that lines the body wall and limbs and hold muscles with similar functions together
___ (f) a cord of dense regular connective tissue that attaches muscle to the periosteum of bone

___ (g) areolar and adipose connective tissue that separates muscle from skin
___ (h) connective tissue elements extended as a broad, flat layer
___ (i) tubes of fibrous connective tissue containing a film of synovial fluid between the two layers of connective tissue; function to reduce friction
___ (j) contractile elements of skeletal muscle

(1) tendon sheaths
(2) myofibrils
(3) aponeurosis
(4) deep fascia
(5) superficial fascia
(6) tendon
(7) endomysium
(8) perimysium
(9) epimysium
(10) fascicles

4. Match the following:

___ (a) point of contact between a motor neuron and a muscle fiber

___ (b) cells that give rise to new smooth muscle fibers

___ (c) myoblasts that persist in mature skeletal muscle

___ (d) plasma membrane of a muscle fiber

___ (e) oxygen-binding protein found only in muscle fibers

___ (f) calcium-ion–storing tubular system similar to smooth endoplasmic reticulum

___ (g) the contracting unit of skeletal muscle

___ (h) area in the contracting unit of a skeletal muscle where the thick filaments are located

___ (i) area in the contracting unit of a skeletal muscle where the thin filaments only are found

___ (j) separates the contracting units from each other

(1) A band
(2) I band
(3) Z disc
(4) sarcomere
(5) neuromuscular junction
(6) myoglobin
(7) satellite cells
(8) pericytes
(9) sarcoplasmic reticulum
(10) sarcolemma

Complete the following:

5. The three types of muscle tissue are ___, ___, and ___.

6. The muscle tissues that show autorythmicity are ___ and ___.

7. The properties of muscle that enable it to carry out its functions and contribute to homeostasis are ___, ___, ___, and ___.

8. The contractile proteins in muscles are ___ and ___.

Choose the best answer to the following questions:

9. In muscle contraction, what starts filament sliding? (a) an increase in Ca^{2+} concentration in the cytosol, (b) a decrease in Na^+ concentration in the cytosol, (c) the bending of the myosin head, (d) the release of energy from the breakdown of ATP, (e) the power stroke

10. What would happen in ATP were suddenly unavailable after the sarcomere had begun to shorten? (a) Nothing. The contraction would proceed normally. (b) The myosin heads would be unable to detach from actin. (c) Troponin would bind with the myosin heads. (d) Actin and myosin myofilaments would separate completely and be unable to recombine. (e) The myosin heads would detach completely from actin and bind to the troponin–tropomyosin complex.

11. Smooth muscle (a) has fibers that are joined by intercalated discs, (b) contains fast oxidative fibers, (c) can both shorten and stretch to a greater extent than either skeletal or cardiac muscle, (d) has the troponin–tropomyosin complex as regulatory proteins, (e) has actin and myosin myofilaments arranged in orderly sarcomeres.

12. The calcium-binding protein in the sarcoplasmic reticulum of skeletal muscle is (a) calmodulin, (b) acetylcholinesterase, (c) troponin, (d) titin, (e) calsequestrin.

13. Which of the following are sources of ATP for muscle contraction? (1) creatine phosphate, (2) glycolysis, (3) anaerobic cellular respiration, (4) aerobic cellular respiration, (5) calmodulin breakdown.
(a) 1, 2, and 3, (b) 2, 3, and 4, (c) 2, 3, and 5, (d) 1, 2, 3, and 4, (e) 2, 3, 4, and 5

14. Which of the following statements are true? (1) Skeletal muscles with a high myoglobin content are termed white muscle fibers. (2) Slow oxidative (SO) fibers generate ATP primarily by aerobic cellular respiration, are resistant to fatigue, and are capable of prolonged, sustained contractions. (3) Fast glycolytic (FG) fibers generate the most powerful contractions, generate ATP mainly by glycolysis, and are adapted for intense anaerobic movements. (4) Fast oxidative-glycolytic (FOG) fibers generate considerable ATP by aerobic cellular respiration but can also generate ATP by anaerobic glycolysis. (5) Most skeletal muscles are a mixture of SO, FOG, and FG fibers in various proportions. (6) Skeletal muscle fibers of any one motor unit vary in muscle fiber types.
(a) 1, 2, 3, and 4, (b) 2, 3, 4, 5, and 6, (c) 2, 3, 4, and 5, (d) 3, 4, 5, and 6, (e) 2, 4, and 6

15. Match the following:

___ (a) the smooth muscle action that allows the fibers to maintain their contractile function even when stretched

___ (b) a brief contraction of all the muscle fibers in a motor unit of a muscle in response to a single action potential in its motor neuron

___ (c) sustained contraction of a muscle

___ (d) larger contractions resulting from stimuli arriving at different times

___ (e) process of increasing the number of activated motor units

___ (f) contraction in which the muscle shortens

___ (g) inability of a muscle to maintain its strength of contraction or tension during prolonged activity

___ (h) continual involuntary activation of a small number of skeletal muscle motor units produces this

___ (i) contraction in which muscle tension is generated without shortening of the muscle

___ (j) contraction in which a muscle lengthens

___ (k) tension generated by actin and myosin resulting in contraction

___ (l) tension generated by the connective tissue around the muscle fibers and not related to contraction

(1) active tension
(2) passive tension
(3) muscle fatigue
(4) twitch contraction
(5) wave summation
(6) tetanus
(7) concentric isotonic contraction
(8) motor unit recruitment
(9) muscle tone
(10) eccentric isotonic contraction
(11) isometric contraction
(12) stress–relaxation response

CRITICAL THINKING QUESTIONS

1. How can smooth muscle contract if it has no striations or sarcomeres? (HINT: *Look for similarities in the structure of smooth and striated muscle.*)
2. Peng is an amateur body builder with a huge collection of weight-lifting equipment in his basement. His little sister Ming tried to lift the 100-pound barbell—it didn't budge. Peng lifted the same barbell over his head, held it there, and then lowered it back to the ground. Ming was very impressed with her big brother. Describe the types of contractions performed by Ming and her brother. (HINT: *Ming's contraction is the same type used by her brother to hold the barbell up.*)
3. As the students from Mr. Klopfer's A&P class filed out of the room, they were massaging their writing hands. "Man, he was on a roll today! Two hours of notes with no break. My hand is killing me!" What's causing the muscle pain? (HINT: *After two hours of A&P, the brain is not the only part of the anatomy that is fatigued.*)

ANSWERS TO FIGURE QUESTIONS

10.1 Perimysium bundles groups of muscle fibers into fascicles.
10.2 Blood capillaries in muscle deliver oxygen and nutrients and remove heat and waste products of muscle metabolism.
10.3 The sarcoplasmic reticulum releases calcium ions to trigger muscle contraction.
10.4 Size, from smallest to largest: thick filament, myofibril, muscle fiber.
10.5 There are two thin filaments for each thick filament in skeletal muscle.
10.6 Actin, titin, and nebulin connect into the Z disc. A bands contain myosin, actin, troponin, tropomyosin, titin, and nebulin; I bands contain actin, troponin, tropomyosin, titin, and nebulin.
10.7 The I bands and H zones disappear. The lengths of the filaments do not change.
10.8 If ATP suddenly became unavailable, the myosin heads would not be able to detach from actin.
10.9 Three functions of ATP in muscle contraction are (1) its hydrolysis by an ATPase activates the myosin head so it can bind to actin and rotate; (2) its binding to myosin causes detachment from actin after the power stroke; and (3) it powers the pumps that transport Ca^{2+} from the cytosol back into the sarcoplasmic reticulum.
10.10 A sarcomere length of 2.2 μm gives a generous zone of overlap between the portions of the thick filaments that have myosin heads and the thin filaments without the overlap being so extensive that sarcomere shortening is limited.
10.11 The portion of the sarcolemma that contains acetylcholine receptors is the motor end plate.
10.12 Steps 4 through 6 are part of excitation–contraction coupling (muscle action potential through binding of myosin heads).
10.13 Glycolysis, exchange of phosphate between creatine phosphate and ADP, and glycogen breakdown occur in the cytosol. Oxidation of pyruvic acid, amino acids, and fatty acids (aerobic cellular respiration) occurs in mitochondria.
10.14 Motor units having many muscle fibers are capable of more forceful contractions than those having only a few fibers.
10.15 Events occurring during the latent period are the events of excitation–contraction coupling: Release of Ca^{2+} from SR and binding to troponin result in attachment of myosin heads to actin and the onset of rotation.
10.16 If the second stimulus were applied a little later, the second contraction would be smaller than the one illustrated in (b).
10.17 Holding your head upright without movement involves mainly isometric contractions.
10.18 Visceral smooth muscle and cardiac muscle are similar in that both include gap junctions, which allow action potentials to spread from one cell to its neighbors.
10.19 Contraction in a smooth muscle fiber starts more slowly and lasts much longer than contraction in a skeletal muscle fiber.
10.20 As its name suggests, the myotome of a somite differentiates into skeletal muscle.

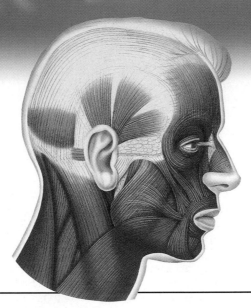

The **muscular system** refers to the voluntary skeletal muscle system: the skeletal muscle tissue and connective tissues that make up individual muscle organs, such as the biceps brachii muscle. To understand how movement occurs at specific joints, you need to be able to identify the body's major skeletal muscle groups. Moreover, it is important to know the origin and insertion of each muscle, and its innervation—the nerve or nerves that stimulate it to contract. Developing a working knowledge of these key aspects of skeletal muscle anatomy will enable you to analyze how normal movements occur. This knowledge is especially important for allied health and physical rehabilitation professionals who work with people whose normal patterns of movement and physical mobility have been disrupted by circumstances such as physical trauma, surgery, or muscular paralysis.

HOW SKELETAL MUSCLES PRODUCE MOVEMENT

Muscle Attachment Sites: Origin and Insertion

OBJECTIVE

• *Describe the relationship between bones and skeletal muscles in producing body movements.*

Skeletal muscles produce movement by exerting force on tendons, which in turn pull on bones or other structures (such as skin). Most muscles cross at least one joint and are usually attached to articulating bones that form the joint (Figure 11.1a).

When a muscle contracts, it draws one of the articulating bones toward the other. The two articulating bones usually do not move equally in response to contraction. One bone remains stationary or near its original position, either because other muscles stabilize that bone by contracting and pulling it in the opposite direction or because its structure makes it less movable. Ordinarily, the attachment of a muscle tendon to the stationary bone is called the **origin;** the attachment of the other muscle tendon to the movable bone is the **insertion.** A good analogy is a spring on a door. In this example, the part of the spring attached to the frame is the origin; the part attached to the door represents the insertion. A good rule of thumb is that the origin is usually proximal and the insertion distal, especially in the limbs.

The fleshy portion of the muscle between the tendons of the origin and insertion is called the **belly** (*gaster*). Keep in mind that muscles that move a body part often do not cover the moving part. Figure 11.1b shows that although one of the functions of the biceps brachii muscle is to move the forearm, the belly of the muscle lies over the humerus, not the forearm. You will also see that muscles that cross two joints, such as the rectus femoris and sartorius, have more complex actions than muscles that cross only one joint.

CLINICAL APPLICATION
Tenosynovitis

Tenosynovitis (ten′-ō-sin-ō-VĪ-tis) is an inflammation of the tendons, tendon sheaths, and synovial membranes surrounding certain joints. The tendons most often

Figure 11.1 Relationship of skeletal muscles to bones. (a) Muscles are attached to bones by tendons known as the origin and insertion. (b) Skeletal muscles produce movements by pulling on bones. Bones serve as levers, and joints act as fulcrums for the levers. Here the lever-fulcrum principle is illustrated by the movement of the forearm. Note where the resistance and effort are applied in this example.

🗝️ **In the limbs, the origin of a muscle is usually proximal and the insertion is usually distal.**

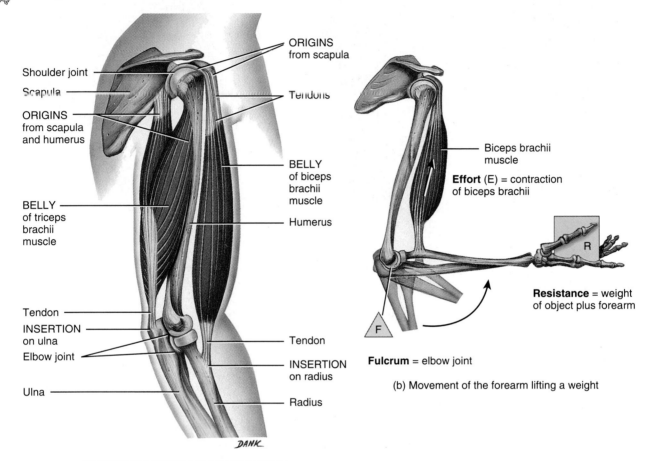

(a) Origin and insertion of a skeletal muscle

(b) Movement of the forearm lifting a weight

Q Where is the belly of the muscle that extends the forearm located?

affected are at the wrists, shoulders, elbows (resulting in a condition called tennis elbow), finger joints (resulting in a condition called trigger finger), ankles, and feet. The affected sheaths sometimes become visibly swollen because of fluid accumulation. Tenderness and pain are frequently associated with movement of the body part. The condition often follows trauma, strain, or excessive exercise. Tenosynovitis of the dorsum of the foot may be caused by tying shoelaces too tightly. Also, gymnasts are prone to developing the condition as a result of chronic, repetitive, and maximum hyperextension at the wrists. ■

1. Using the terms *origin, insertion,* and *belly* in your discussion, describe how skeletal muscles produce body movements by pulling on bones.

Lever Systems and Leverage

OBJECTIVE

• *Define lever and fulcrum, and compare the three types of levers on the basis of location of the fulcrum, effort, and resistance.*

In producing movement, bones act as levers, and joints function as the fulcrums of these levers. A **lever** may be defined as a rigid rod that moves around some fixed point called a **fulcrum,** symbolized by ꜰ . A lever is acted on at two different points by two different forces: the **effort** (E), which causes movement, and the **resistance** ʀ , or *load,* which opposes movement. The effort is the force exerted by muscular contraction, whereas the resistance is typically the weight of

Figure 11.2 Types of levers.

🔑 **Levers are divided into three types based on the placement of the fulcrum, effort, and resistance.**

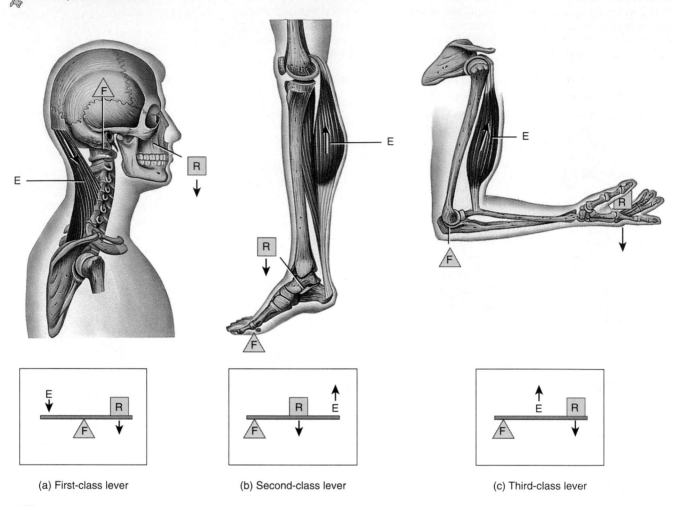

(a) First-class lever

(b) Second-class lever

(c) Third-class lever

Q Which type of lever produces the most force?

the body part that is moved. Motion occurs when the effort applied to the bone at the insertion exceeds the resistance (load). Consider the biceps brachii flexing the forearm at the elbow as an object is lifted (see Figure 11.1b). When the forearm is raised, the elbow is the fulcrum. The weight of the forearm plus the weight of the object in the hand is the resistance. The force of contraction of the biceps brachii pulling the forearm up is the effort.

Levers produce trade-offs between effort and the speed and range of motion. In one situation, a lever operates at a *mechanical advantage*—has **leverage**—when a less forceful effort can move a more forceful resistance. Here the trade-off is that the effort must move farther (must have a longer range of motion) and faster than the resistance. Recall that range of motion refers to the range, measured in degrees of a circle, through which the bones of a joint can be moved. The lever formed by the mandible at the temporomandibular joints (fulcrums) and the effort provided by contraction of the jaw muscles produce a powerful mechanical advantage that crushes food. In a second situation, a lever operates at a *mechanical disadvantage* when a more forceful effort moves a

less forceful resistance. In this case the trade-off is that the effort must move a shorter distance and less quickly than the resistance. The lever formed by the humerus at the shoulder joint (fulcrum) and the effort provided by the back and shoulder muscles produces a mechanical "disadvantage" that enables a big-league pitcher to hurl a baseball at nearly 100 miles per hour!

The positions of the effort, resistance, and fulcrum on the lever determine whether the lever operates at a mechanical advantage or disadvantage. When the resistance is close to the fulcrum and the effort is applied farther away, the lever operates at a mechanical advantage. When you chew food, the resistance (the food) is positioned close to the fulcrums (your temporomandibular joints) while your jaw muscles exert effort farther out from the joints. On the other hand, when the effort is applied close to the fulcrum and the resistance is applied farther away, the lever operates at a mechanical disadvantage. When a pitcher throws a baseball, the back and shoulder muscles apply intense effort very close to the fulcrum (the shoulder joint) while the lighter resistance (the ball) is propelled at the far end of the lever (the arm bone).

Levers are categorized into three types according to the positions of the fulcrum, the effort, and the resistance:

1. The fulcrum is between the effort and the resistance in **first-class levers** (Figure 11.2a). (Think E*F*R.) A seesaw is an example of a first-class lever. A first-class lever can produce either a mechanical advantage or disadvantage depending on whether the effort or the resistance is closer to the fulcrum. As we've seen in the preceding examples, if the effort is farther from the fulcrum than the resistance, a strong resistance can be moved, but not very far or fast. If the effort is closer to the fulcrum than the resistance, only a weaker resistance can be moved, but it moves very far and fast.

   There are few first-class levers in the body. One example is the lever formed by the head resting on the vertebral column (see Figure 11.2a). When the head is raised, the weight of the anterior portion of the skull is the resistance. The joint between the atlas and the occipital bone (atlanto-occipital joint) forms the fulcrum. The contraction of the posterior neck muscles provides the effort.

2. The resistance is between the fulcrum and the effort in **second-class levers** (Figure 11.2b). (Think F*R*E.) They operate like a wheelbarrow. Second-class levers always produce a mechanical advantage because the resistance is always closer to the fulcrum than the effort. In the body, this arrangement sacrifices speed and range of motion for force. Most authorities believe that there are very few examples of second-class levers in the body. The ball of the foot, the tarsal bones, and the calf muscles form a second-class lever. When you rise up on your toes, the ball of the foot is the fulcrum, the body's weight is the resistance, and the contraction of the calf muscles provides the effort, which pulls the heel up off the floor.

3. The effort is between the fulcrum and the resistance in **third-class levers** (Figure 11.2c). (Think F*E*R.) They are the most common levers in the body. Third-class levers always produce a mechanical disadvantage because the effort is always closer to the fulcrum than the resistance. In the body, this arrangement favors speed and range of motion over force. The elbow joint, the bones of the arm and forearm, and the biceps brachii muscle are one example of a third-class lever (see Figure 11.2c). As we have seen, in flexing the forearm at the elbow, the weight of the hand and forearm is the resistance, the elbow joint is the fulcrum, and the contraction of the biceps brachii muscle provides the effort. Another example of a third-class lever is adduction of the thigh, in which the thigh is the resistance, the hip joint is the fulcrum, and the contraction of the adductor muscles is the effort.

Effects of Fascicle Arrangement

OBJECTIVE

• *Identify the various arrangements of muscle fibers in a skeletal muscle, and relate the arrangements to strength of contraction and range of motion.*

Recall from Chapter 10 that skeletal muscle fibers (cells) are arranged within the muscle in bundles called **fascicles.** The muscle fibers are arranged in a parallel fashion within each bundle, but the arrangement of the fascicles with respect to the tendons may take one of five characteristic patterns: parallel, fusiform, circular, triangular, or pennate (Table 11.1).

Fascicular arrangement affects a muscle's power and range of motion. When a muscle fiber contracts, it shortens to a length about 70% of its resting length. Thus, the longer the fibers in a muscle, the greater the range of motion it can produce. By contrast, the strength of a muscle depends on the total number of fibers it contains, because a short fiber can contract as forcefully as a long one. Because a given muscle can contain either a small number of long fibers or a large number of short fibers, fascicular arrangement represents a compromise between power and range of motion. Pennate muscles, for example, have a large number of fascicles distributed over their tendons, giving them greater power but a smaller range of motion. Parallel muscles, in contrast, have comparatively few fascicles that extend the length of the muscle; thus, they have a greater range of motion but less power.

Coordination Within Muscle Groups

OBJECTIVE

• *Explain how the prime mover, antagonist, synergist, and fixator in a muscle group work together to produce movement.*

Most movements are the result of several skeletal muscles acting as a group rather than individually. Most skeletal muscles are arranged in opposing (antagonistic) pairs at joints—that is, flexors-extensors, abductors-adductors, and so on. Within opposing pairs, one muscle, called the **prime mover** or **agonist** (= leader), contracts to cause a desired action while the other muscle, the **antagonist** (*anti-* = against), stretches and yields to the effects of the prime mover. In the process of flexing the forearm at the elbow, for instance, the biceps brachii is the prime mover, and the triceps brachii is the antagonist (see Figure 11.1). The antagonist and prime mover are usually located on opposite sides of the bone or joint, as is the case in this example.

You should not assume, however, that the biceps brachii is always the prime mover and the triceps brachii is always the antagonist. In the process of extending the forearm at the elbow, the triceps brachii serves as the prime mover, and the biceps brachii functions as the antagonist; their roles are reversed. Note that if the prime mover and antagonist contracted simultaneously with equal force, there would be no movement.

There are many examples of a prime mover crossing several joints before it reaches the joint at which its primary action occurs. The biceps brachii, for example, spans both the shoulder and elbow joints, and its primary action is on

Table 11.1 Arrangement of Fascicles

PARALLEL

Fascicles parallel to longitudinal axis of muscle; terminate at either end in flat tendons.

Example: Stylohyoid muscle (see Figure 11.8)

FUSIFORM

Fascicles nearly parallel to longitudinal axis of muscle; terminate in flat tendons; muscle tapers toward tendons, where diameter is less than at belly.

Example: Digastric muscle (see Figure 11.8)

CIRCULAR

Fascicles in concentric circular arrangements form sphincter muscles that enclose an orifice (opening).

Example: Orbicularis oculi muscle (see Figure 11.4)

TRIANGULAR

Fascicles spread over broad area converge at thick central tendon; gives muscle a triangular appearance.

Example: Pectoralis major muscle (see Figure 11.3a)

PENNATE

Short fascicles in relation to total muscle length; tendon extends nearly entire length of muscle.

Unipennate

Fascicles are arranged on only one side of tendon.

Example: Extensor digitorum longus muscle (see Figure 11.22b)

Bipennate

Fascicles are arranged on both sides of centrally positioned tendon.

Example: Rectus femoris muscle (see Figure 11.20a)

Multipennate

Fascicles attach obliquely from many directions to several tendons.

Example: Deltoid muscle (see Figure 11.10a)

the forearm. In order to prevent unwanted movements at intermediate joints or to otherwise aid the movement of the prime mover, muscles called **synergists** (SIN-er-gists; *syn* = together; *ergon* = work) contract and stabilize the intermediate joints. As an example, muscles that flex the fingers (prime movers) cross the intercarpal and radiocarpal joints (intermediate joints). If movement at these intermediate joints was unrestrained, you would not be able to flex your fingers without flexing the wrist at the same time. Synergistic contraction of the wrist extensors stabilizes the wrist joint and prevents it from moving (unwanted movement) while the flexor muscles of the fingers contract to bring about efficient flexion of the fingers (primary action). Synergists are usually located very close to the prime mover.

Some muscles in a group also act as **fixators,** which stabilize the origin of the prime mover so that the prime mover can act more efficiently. Fixators steady the proximal end of a limb while movements occur at the distal end. For example, the scapula in the pectoral (shoulder) girdle is a freely movable bone that serves as the origin for several muscles that move the arm. However, when the arm muscles contract, the scapula must be held steady. This is accomplished by fixator muscles that hold the scapula firmly against the back of the chest. In abduction of the arm, the deltoid muscle serves as the prime mover, whereas fixators (pectoralis minor, trapezius, subclavius, serratus anterior muscles, and others) hold the scapula firmly (see Figure 11.14). These fixators stabilize the scapula, which serves as the attachment site for the origin of the deltoid muscle, whereas the insertion of the muscle pulls on the humerus to abduct the arm. Under different conditions—that is, for different movements—many muscles may act, at various times, as prime movers, antagonists, synergists, or fixators.

1. Describe the three types of levers, and give an example of each in the body.

2. Describe the various arrangements of fascicles.

3. Define the roles of the prime mover (agonist), antagonist, synergist, and fixator in producing movements.

HOW SKELETAL MUSCLES ARE NAMED

OBJECTIVE

• *Explain seven characteristics used in naming skeletal muscles.*

The names of most of the nearly 700 skeletal muscles are based on several characteristics. Learning the terms that refer to these characteristics will help you remember the names of muscles. Although many characteristics may be reflected in a given muscle's name, the most important include the direction in which the muscle fibers run; the size, shape, action, number of origins, and location of the muscle; and the sites of origin and insertion of the muscle (Table 11.2).

1. Select ten muscles presented in Figure 11.3 and identify the bases of their names. (HINT: *Use the prefix, suffix, and root of each muscle's name as a guide.*)

PRINCIPAL SKELETAL MUSCLES

Exhibits 11.1 through 11.20 will assist you in learning the names of the principal skeletal muscles in various regions of the body. The muscles in the exhibits are divided into groups according to the part of the body on which they act. As you study groups of muscles in the exhibits, refer to Figure 11.3 to see how each group is related to the others.

The exhibits contain the following information:

• *Objectives.* These statements describe the principal outcomes that can be expected after studying the exhibit.

• *Overview.* This information provides a general orientation to the muscles under consideration and emphasizes how the muscles are organized within various regions. The discussion also highlights any distinguishing or interesting features about the muscles.

• *Muscle names.* The derivations indicate how the muscles are named. Once you have mastered the naming of the muscles, their actions will have more meaning.

• *Origins, insertions, and actions.* For each muscle, you are also given its origin and insertion (points of attachment to bones or other structures) and actions, the principal movements that occur when the muscle contracts.

• *Innervation.* This section indicates the nerve supply for each muscle. In general, muscles in the head region are served by cranial nerves, which arise from the lower portions of the brain, whereas muscles in the rest of the body are served by spinal nerves, which arise from the spinal cord within the vertebral column. Cranial nerves are designated by both a name and a Roman numeral in parentheses—for example, the facial (VII) cranial nerve. Spinal nerves are numbered in groups according to the portion

of the spinal cord from which they arise: C = cervical (neck region), T = thoracic (chest region), L = lumbar (lower back region), and S = sacral (buttocks region). An example is T1, the first thoracic spinal nerve.

• *"Relating Muscles to Movements."* These exercises will help you organize the muscles in a body region according to the actions they share in common.

• *Figures.* The figures in the exhibits present superficial and deep, anterior and posterior, or medial and lateral views to show each muscle's position as clearly as possible. *The muscle names in all capital letters are specifically referred to in the table portion of the exhibit.* Many of the reference structures (such as blood vessels and nerves) will be discussed more fully in subsequent chapters.

Here are the exhibits and accompanying figures that describe the principal skeletal muscles:

Exhibit 11.1 Muscles of Facial Expression (Figure 11.4), p. 311.

Exhibit 11.2 Muscles that Move the Eyeballs—Extrinsic Muscles (Figure 11.5), p. 315.

Exhibit 11.3 Muscles that Move the Mandible (Lower Jaw) (Figure 11.6), p. 317.

Exhibit 11.4 Muscles that Move the Tongue—Extrinsic Muscles (Figure 11.7), p. 319.

Exhibit 11.5 Muscles of the Floor of the Oral Cavity (Mouth) (Figure 11.8), p. 321.

Exhibit 11.6 Muscles that Move the Head (Figure 11.9), p. 323.

Exhibit 11.7 Muscles that Act on the Abdominal Wall (Figure 11.10), p. 325.

Exhibit 11.8 Muscles Used in Breathing (Figure 11.11), p. 328.

Exhibit 11.9 Muscles of the Pelvic Floor (Figure 11.12), p. 330.

Exhibit 11.10 Muscles of the Perineum (Figures 11.12 and 11.13), p. 332.

Exhibit 11.11 Muscles that Move the Pectoral (Shoulder) Girdle (Figure 11.14), p. 334.

Exhibit 11.12 Muscles that Move the Humerus (Arm) (Figure 11.15), p. 337.

Exhibit 11.13 Muscles that Move the Radius and Ulna (Forearm) (Figure 11.16), p. 341.

Exhibit 11.14 Muscles that Move the Wrist, Hand, and Digits (Figure 11.17), p. 345.

Exhibit 11.15 Intrinsic Muscles of the Hand (Figure 11.18), p. 350.

Exhibit 11.16 Muscles that Move the Vertebral Column (Backbone) (Figure 11.19), p. 353.

Exhibit 11.17 Muscles that Move the Femur (Thigh) (Figure 11.20), p. 358.

Exhibit 11.18 Muscles that Act on the Femur (Thigh) and Tibia and Fibula (Leg) (Figures 11.20 and 11.21), p. 363.

Exhibit 11.19 Muscles that Move the Foot and Toes (Figure 11.22), p. 366.

Exhibit 11.20 Intrinsic Muscles of the Foot (Figure 11.23), p. 371.

Table 11.2 Characteristics Used to Name Muscles

NAME	MEANING	EXAMPLE	FIGURE
DIRECTION: Orientation of muscle fibers relative to the body's midline.			
Rectus	parallel to midline	Rectus abdominis	11.10b
Transverse	perpendicular to midline	Transverse abdominis	11.10b
Oblique	diagonal to midline	External oblique	11.10a
SIZE: Relative size of the muscle.			
Maximus	largest	Gluteus maximus	11.20c
Minimus	smallest	Gluteus minimus	11.20c
Longus	longest	Adductor longus	11.20a
Brevis	shortest	Peroneus brevis	11.22b
Latissimus	widest	Latissimus dorsi	11.15b
Longissimus	longest	Longissimus capitis	11.19a
Magnus	large	Adductor magnus	11.20a
Major	larger	Pectoralis major	11.15a
Minor	smaller	Pectoralis minor	11.14a
Vastus	great	Vastus lateralis	11.20a
SHAPE: Relative shape of the muscle.			
Deltoid	triangular	Deltoid	11.10a
Trapezius	trapezoid	Trapezius	11.3b
Serratus	saw-toothed	Serratus anterior	11.14b
Rhomboideus	diamond-shaped	Rhomboideus major	11.15c
Orbicularis	circular	Orbicularis oculi	11.4a
Pectinate	comblike	Pectineus	11.20a
Piriformis	pear-shaped	Piriformis	11.20c
Platys	flat	Platysma	11.4a
Quadratus	square	Quadratus femoris	11.20c
Gracilis	slender	Gracilis	11.20a
ACTION: Principal action of the muscle.			
Flexor	decreases joint angle	Flexor carpi radialis	11.17a
Extensor	increases joint angle	Extensor carpi ulnaris	11.17c
Abductor	moves bone away from midline	Abductor pollicis longus	11.17c
Adductor	moves bone closer to midline	Adductor longus	11.20a
Levator	produces superior movement	Levator scapulae	11.14a
Depressor	produces inferior movement	Depressor labii inferioris	11.4b
Supinator	turns palm superiorly or anteriorly	Supinator	11.17b
Pronator	turns palm inferiorly or posteriorly	Pronator teres	11.17a
Sphincter	decreases size of opening	External anal sphincter	11.12
Tensor	makes a body part rigid	Tensor fasciae latae	11.20a
Rotator	moves bone around longitudinal axis	Obturator externus	11.20b
NUMBER OF ORIGINS: Number of tendons of origin.			
Biceps	two origins	Biceps brachii	11.16a
Triceps	three origins	Triceps brachii	11.16b
Quadriceps	four origins	Quadriceps femoris	11.20a

LOCATION: Structure near which a muscle is found. *Example:* Frontalis, a muscle near the frontal bone (Figure 11.4a).

ORIGIN AND INSERTION: Sites where muscle originates and inserts. *Example:* Sternocleidomastoid, originating on the sternum and clavicle and inserting on mastoid process of temporal bone (Figure 11.3a).

Figure 11.3 Principal superficial skeletal muscles.

🔑 **Most movements require several skeletal muscles acting in groups rather than individually.**

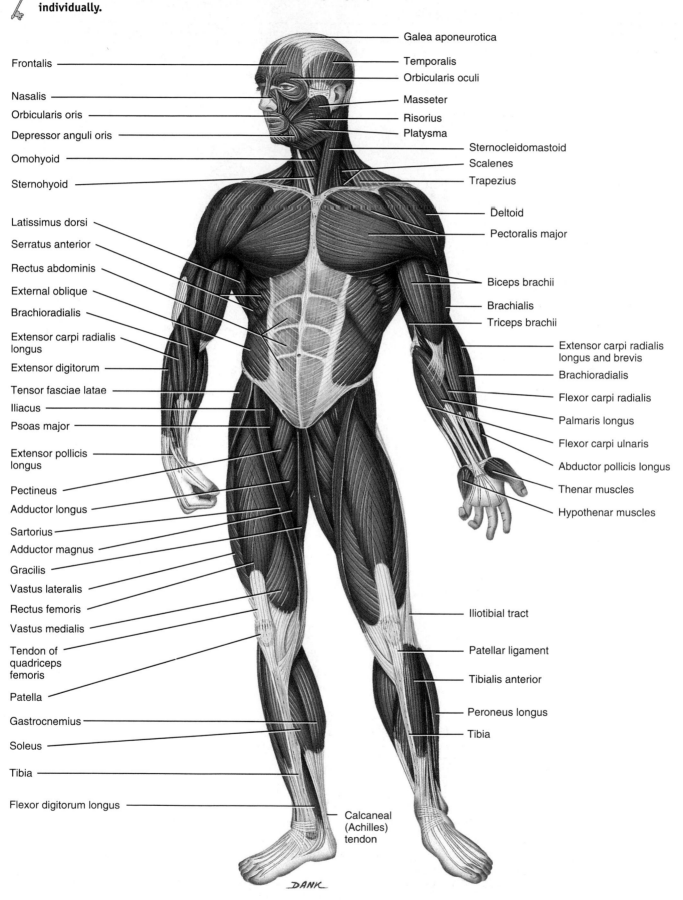

Galea aponeurotica
Frontalis
Temporalis
Orbicularis oculi
Nasalis
Masseter
Orbicularis oris
Risorius
Depressor anguli oris
Platysma
Omohyoid
Sternocleidomastoid
Scalenes
Sternohyoid
Trapezius
Deltoid
Latissimus dorsi
Pectoralis major
Serratus anterior
Rectus abdominis
Biceps brachii
External oblique
Brachialis
Brachioradialis
Triceps brachii
Extensor carpi radialis longus
Extensor carpi radialis longus and brevis
Extensor digitorum
Brachioradialis
Tensor fasciae latae
Flexor carpi radialis
Iliacus
Palmaris longus
Psoas major
Flexor carpi ulnaris
Extensor pollicis longus
Abductor pollicis longus
Pectineus
Thenar muscles
Adductor longus
Hypothenar muscles
Sartorius
Adductor magnus
Gracilis
Vastus lateralis
Rectus femoris
Vastus medialis
Iliotibial tract
Tendon of quadriceps femoris
Patellar ligament
Patella
Tibialis anterior
Gastrocnemius
Peroneus longus
Soleus
Tibia
Tibia
Flexor digitorum longus
Calcaneal (Achilles) tendon

DANK

(a) Anterior view

Figure continues

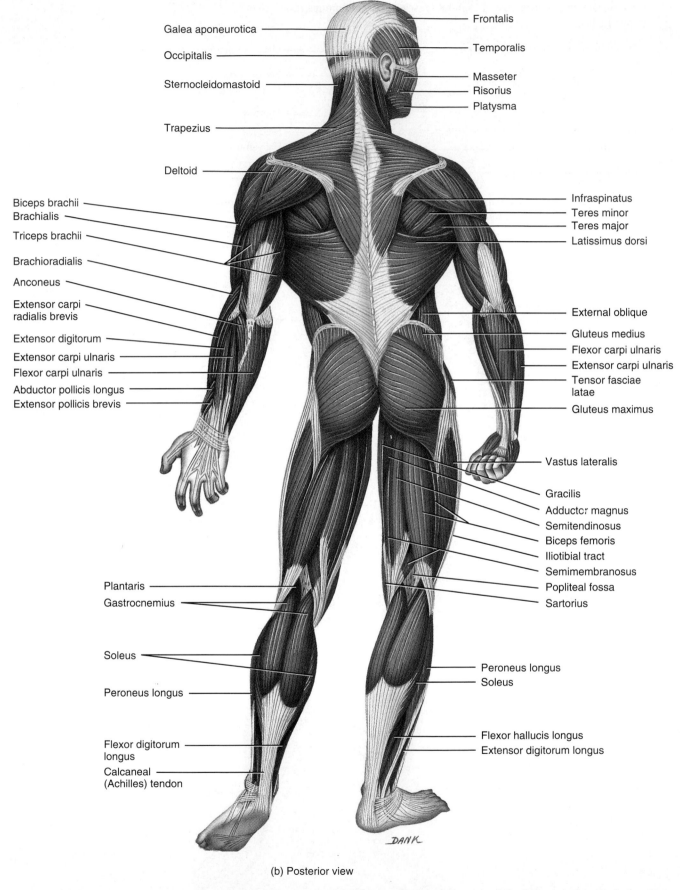

Frontalis

Galea aponeurotica

Temporalis

Occipitalis

Masseter
Risorius

Sternocleidomastoid

Platysma

Trapezius

Deltoid

Infraspinatus
Teres minor
Teres major
Latissimus dorsi

Biceps brachii
Brachialis

Triceps brachii

Brachioradialis

Anconeus

Extensor carpi
radialis brevis

External oblique

Gluteus medius

Extensor digitorum

Flexor carpi ulnaris

Extensor carpi ulnaris

Extensor carpi ulnaris

Flexor carpi ulnaris

Tensor fasciae
latae

Abductor pollicis longus

Extensor pollicis brevis

Gluteus maximus

Vastus lateralis

Gracilis

Adductor magnus

Semitendinosus

Biceps femoris

Iliotibial tract

Semimembranosus

Plantaris

Popliteal fossa

Gastrocnemius

Sartorius

Soleus

Peroneus longus
Soleus

Peroneus longus

Flexor hallucis longus

Extensor digitorum longus

Flexor digitorum
longus

Calcaneal
(Achilles) tendon

(b) Posterior view

 Give an example of a muscle named for each of the following characteristics: direction of fibers, shape, action, size, origin and insertion, location, and number of tendons of origin.

Exhibit 11.1 *Muscles of Facial Expression (Figure 11.4)*

OBJECTIVE

• *Describe the origin, insertion, action, and innervation of the muscles of facial expression.*

The muscles in this group provide humans with the ability to express a wide variety of emotions. The muscles themselves lie within the layers of superficial fascia. They usually originate in the fascia or bones of the skull and insert into the skin. Because of their insertions, the muscles of facial expression move the skin rather than a joint when they contract.

Among the noteworthy muscles in this group are those surrounding the orifices (openings) of the head such as the eyes, nose, and mouth. These muscles function as *sphincters* (SFINGK-ters), which close the orifices, and *dilators,* which open the orifices. For example, the **orbicularis oculi** muscle closes the eye, whereas the **levator palpebrae superioris** muscle opens the eye. The **epicranius** is an unusual muscle in this group because it is made up of two parts: an anterior part called the **frontalis**, which is superficial to the frontal bone, and a posterior part called the **occipitalis**, which is superficial to the occipital bone. The two muscular portions are held together by a strong aponeurosis (sheetlike tendon), the **galea aponeurotica** (GĀ-lē-a ap-ō-noo'-RŌ-ti-ka; *galea* = helmet) or *epicranial aponeurosis,* which covers the superior and lateral surfaces of the skull. The **buccinator** muscle forms the major muscular portion of the cheek. It functions in whistling, blowing, and sucking and also assists in chewing. The duct of the parotid gland (salivary gland) pierces the buccinator muscle to reach the oral cavity. The buccinator muscle is so named because it compresses the cheeks (*bucc* = cheek) during blowing—for example, when a musician plays a wind instrument such as a trumpet.

INNERVATION

All muscles are innervated by the facial (VII) cranial nerve, except for the levator palpebrae superioris muscle, which is innervated by the oculomotor (III) cranial nerve.

CLINICAL APPLICATION
Bell's Palsy

Bell's palsy, also known as **facial paralysis,** is a unilateral paralysis of the muscles of facial expression as a result of damage or disease of the facial (VII) cranial nerve. Although the cause is unknown, inflammation of the facial nerve and a relationship to the herpes simplex virus have been suggested. The paralysis causes the entire side of the face to droop in severe cases, and the person cannot wrinkle the forehead, close the eye, or pucker the lips on the affected side. Difficulty in swallowing and drooling also occur. Eighty percent of patients recover completely within a few weeks to a few months. For others, paralysis is permanent. ■

RELATING MUSCLES TO MOVEMENTS

Arrange the muscles in this exhibit into two groups: (1) those that act on the mouth and (2) those that act on the eyes.

> What muscles would you use to do the following: show surprise, express sadness, bare your upper teeth, pucker your lips, squint, blow up a balloon?

MUSCLE	ORIGIN	INSERTION	ACTION
Epicranius (ep-i-KRĀ-nē-us; *epi* = above; *kranion* = skull)			
Frontalis (fron-TA-lis; *frontalis* = in front)	Galea aponeurotica.	Skin superior to supraorbital margin.	Draws scalp anteriorly, raises eyebrows, and wrinkles skin of forehead horizontally.
Occipitalis (ok-si'-pi-TA-lis; *occipito* = base of skull)	Occipital bone and mastoid process of temporal bone.	Galea aponeurotica.	Draws scalp posteriorly.
Orbicularis oris (or-bi'-kyoo-LAR-is OR-is; *orb* = circular; *or* = mouth)	Muscle fibers surrounding opening of mouth.	Skin at corner of mouth.	Closes and protrudes lips, compresses lips against teeth, and shapes lips during speech.

Exhibit 11.1 *Muscles of Facial Expression (continued)*

MUSCLE	ORIGIN	INSERTION	ACTION
Zygomaticus major (zī-gō-MA-ti-kus; *zygomatic* = cheek bone; *major* = greater)	Zygomatic bone.	Skin at angle of mouth and orbicularis oris.	Draws angle of mouth superiorly and laterally, as in smiling or laughing.
Zygomaticus minor (*minor* = lesser)	Zygomatic bone.	Upper lip.	Elevates upper lip, exposing maxillary teeth.
Levator labii superioris (le-VĀ-tor LĀ-bē-ī soo-per'-ē-OR-is; *levator* = raises or elevates; *labii* = lip; *superioris* = upper)	Superior to infraorbital foramen of maxilla.	Skin at angle of mouth and orbicularis oris.	Elevates upper lip.
Depressor labii inferioris (de-PRE-sor LĀ-bē-ī in-fer'-ē-OR-is; *depressor* = depresses or lowers; *inferioris* = lower)	Mandible.	Skin of lower lip.	Depresses (lowers) lower lip.
Depressor anguli oris (*angul* = angle or corner)	Mandible.	Angle of mouth.	Draws angle of mouth laterally and inferiorly, as in opening mouth.
Buccinator (BUK-si-nā'-tor; *bucca* = cheek)	Alveolar processes of maxilla and mandible and pterygomandibular raphe (fibrous band extending from the pterygoid process to the mandible).	Orbicularis oris.	Presses cheeks against teeth and lips, as in whistling, blowing, and sucking; draws corner of mouth laterally; and assists in mastication (chewing) by keeping food between the teeth (and not between teeth and cheeks).
Mentalis (men-TA-lis; *mentum* = chin)	Mandible.	Skin of chin.	Elevates and protrudes lower lip and pulls skin of chin up, as in pouting.
Platysma (pla-TIZ-ma; *platy* = flat, broad)	Fascia over deltoid and pectoralis major muscles.	Mandible, muscles around angle of mouth, and skin of lower face.	Draws outer part of lower lip inferiorly and posteriorly as in pouting; depresses mandible.
Risorius (ri-ZOR-ē-us; *risor* = laughter)	Fascia over parotid (salivary) gland.	Skin at angle of mouth.	Draws angle of mouth laterally, as in tenseness.
Orbicularis oculi (or-bi'-kyoo-LAR-is OK-yoo-lī; *oculus* = eye)	Medial wall of orbit.	Circular path around orbit.	Closes eye.
Corrugator supercilii (KOR-a-gā'-tor soo-per-SI-lē-ī; *corrugo* = wrinkle; *supercilium* = eyebrow)	Medial end of superciliary arch of frontal bone.	Skin of eyebrow.	Draws eyebrow inferiorly and wrinkles skin of forehead vertically as in frowning.
Levator palpebrae superioris (le-VĀ-tor PAL-pe-brē soo-per'-ē-OR-is; *palpebrae* = eyelids) (see also Figure 11.5a)	Roof of orbit (lesser wing of sphenoid bone).	Skin of upper eyelid.	Elevates upper eyelid (opens eye).

Exhibit 11.1   *(continued)*

Figure 11.4 Muscles of facial expression. (See Tortora, *A Photographic Atlas of the Human Body,* Figures 5.2 through 5.4)

When they contract, muscles of facial expression move the skin rather than a joint.

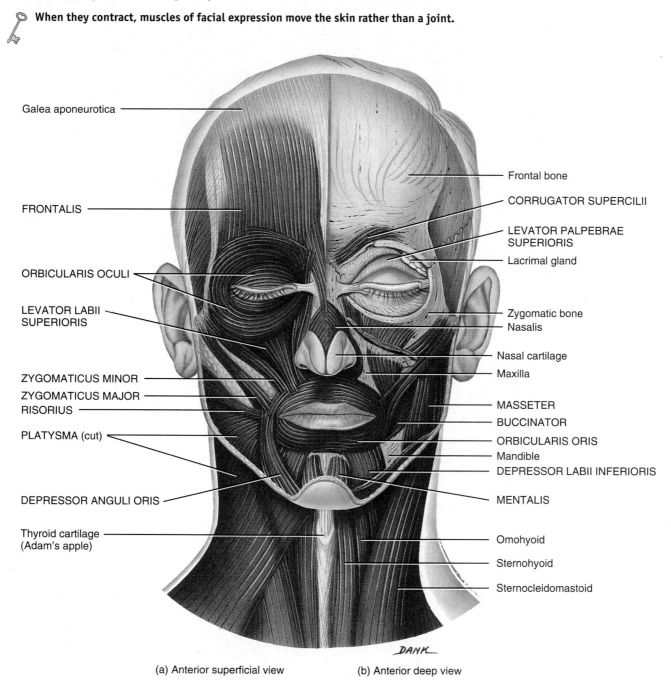

Galea aponeurotica

FRONTALIS

ORBICULARIS OCULI

LEVATOR LABII
SUPERIORIS

ZYGOMATICUS MINOR
ZYGOMATICUS MAJOR
RISORIUS

PLATYSMA (cut)

DEPRESSOR ANGULI ORIS

Thyroid cartilage
(Adam's apple)

Frontal bone

CORRUGATOR SUPERCILII

LEVATOR PALPEBRAE
SUPERIORIS

Lacrimal gland

Zygomatic bone
Nasalis

Nasal cartilage

Maxilla

MASSETER
BUCCINATOR
ORBICULARIS ORIS
Mandible
DEPRESSOR LABII INFERIORIS

MENTALIS

Omohyoid

Sternohyoid

Sternocleidomastoid

DANK

(a) Anterior superficial view   (b) Anterior deep view

Figure continues

Exhibit 11.1 *Muscles of Facial Expression (continued)*

Figure 11.4 (continued)

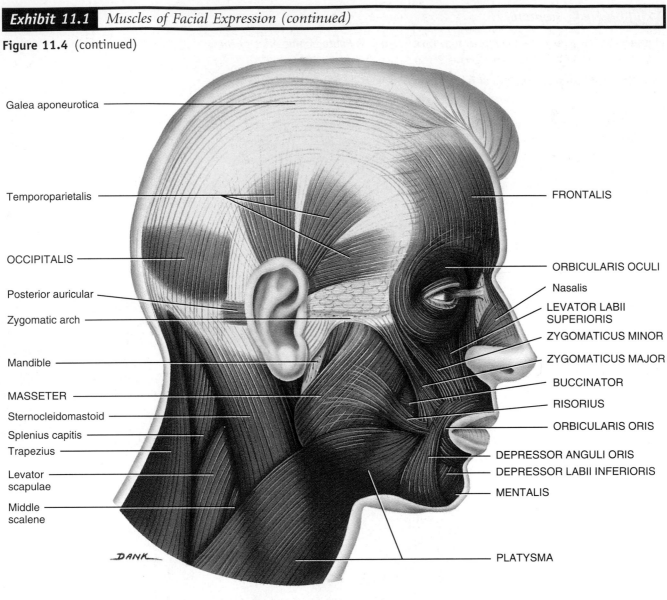

Galea aponeurotica

Temporoparietalis

OCCIPITALIS

Posterior auricular

Zygomatic arch

Mandible

MASSETER

Sternocleidomastoid

Splenius capitis

Trapezius

Levator scapulae

Middle scalene

FRONTALIS

ORBICULARIS OCULI

Nasalis

LEVATOR LABII SUPERIORIS

ZYGOMATICUS MINOR

ZYGOMATICUS MAJOR

BUCCINATOR

RISORIUS

ORBICULARIS ORIS

DEPRESSOR ANGULI ORIS

DEPRESSOR LABII INFERIORIS

MENTALIS

PLATYSMA

DANK

(c) Right lateral superficial view

 Q Which muscles of facial expression cause frowning, smiling, pouting, and squinting?

Exhibit 11.2 | *Muscles that Move the Eyeballs—Extrinsic Muscles (Figure 11.5)*

OBJECTIVE

• *Describe the origin, insertion, action, and innervation of the extrinsic muscles of the eyeballs.*

Muscles that move the eyeballs are called **extrinsic muscles** because they originate outside the eyeballs (in the orbit) and are inserted on the outer surface of the sclera ("white of the eye"). The extrinsic muscles of the eyeballs are among the fastest contracting and most precisely controlled skeletal muscles in the body.

Movements of the eyeballs are controlled by three pairs of extrinsic muscles: (1) superior and inferior recti, (2) lateral and medial recti, and (3) superior and inferior oblique. The four recti muscles (superior, inferior, lateral, and medial) arise from a tendinous ring in the orbit and insert into the sclera of the eye. Whereas the **superior** and **inferior recti** lie in the same vertical plane, the **medial** and **lateral recti** lie in the same horizontal plane. The actions of the recti muscles can be deduced from their insertions on the sclera. The superior and inferior recti move the eyeballs superiorly and inferiorly, respectively; the lateral and medial recti move the eyeballs laterally and medially, respectively. It should be noted that neither the superior nor the inferior rectus muscle pulls directly parallel to the long axis of the eyeballs, and as a result both muscles also move the eyeballs medially.

It is not as easy to deduce the actions of the oblique muscles (superior and inferior) because of their paths through the orbits.

For example, the **superior oblique** muscle originates posteriorly near the tendinous ring and then passes anteriorly and ends in a round tendon, which runs through a pulleylike loop called the *trochlea* (*trochlea* = pulley) in the anterior and medial part of the roof of the orbit. The tendon then turns and inserts on the posterolateral aspect of the eyeballs. Accordingly, the superior oblique muscle moves the eyeballs inferiorly and laterally. The **inferior oblique** muscle originates on the maxilla at the anteromedial aspect of the floor of the orbit. It then passes posteriorly and laterally and inserts on the posterolateral aspect of the eyeballs. Because of this arrangement, the inferior oblique muscle moves the eyeballs superiorly and laterally.

INNERVATION

The superior rectus, inferior rectus, medial rectus, and inferior oblique muscles are innervated by the oculomotor (III) cranial nerve; the lateral rectus is innervated by the abducens (VI) cranial nerve; and the superior oblique is innervated by the trochlear (IV) cranial nerve.

RELATING MUSCLES TO MOVEMENTS

Arrange the muscles in this exhibit according to their actions on the eyeballs: (1) elevation, (2) depression, (3) abduction, (4) adduction, (5) medial rotation, and (6) lateral rotation. The same muscle may be mentioned more than once.

Which muscles contract and relax in each eye as you gaze to your left without moving your head?

MUSCLE	ORIGIN	INSERTION	ACTION
Superior rectus	Common tendinous ring (attached to orbit around optic foramen).	Superior and central part of eyeball.	Moves eyeball superiorly (elevation) and medially (adduction), and rotates it medially.
Inferior rectus	Same as above.	Inferior and central part of eyeball.	Moves eyeball inferiorly (depression) and medially (adduction), and rotates it laterally.
Lateral rectus	Same as above.	Lateral side of eyeball.	Moves eyeball laterally (abduction).
Medial rectus	Same as above.	Medial side of eyeball.	Moves eyeball medially (adduction).
Superior oblique	Sphenoid bone, superior and medial to the tendinous ring in the orbit.	Eyeball between superior and lateral recti. The muscle inserts into the superior and lateral surfaces of the eyeball via a tendon that passes through the trochlea.	Moves eyeball inferiorly (depression) and laterally (abduction), and rotates it medially.
Inferior oblique	Maxilla in floor of orbit.	Eyeball between inferior and lateral recti.	Moves eyeball superiorly (elevation) and laterally (abduction), and rotates it laterally.

- ➤

Exhibit 11.2 *Muscles that Move the Eyeballs—Extrinsic Muscles (continued)*

Figure 11.5 Extrinsic muscles of the eyeball. (See Tortora, *A Photographic Atlas of the Human Body,* Figures 5.4 and 5.5)

 The extrinsic muscles of the eyeball are among the fastest contracting and most precisely controlled skeletal muscles in the body.

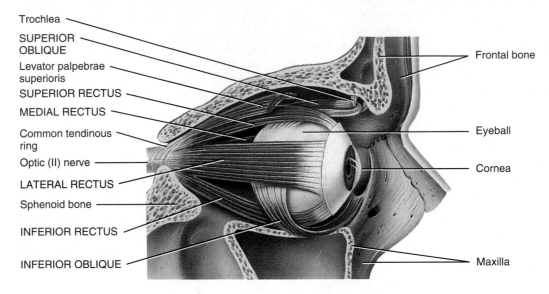

(a) Lateral view of right eyeball

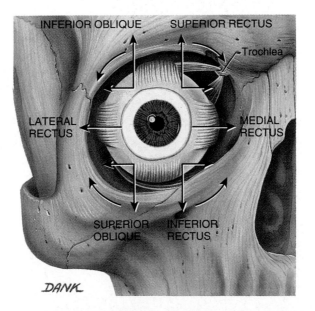

(b) Movements of right eyeball in response to contraction of extrinsic muscles

Q Why does the inferior oblique muscle move the eyeball superiorly and laterally?

| Exhibit 11.3 | *Muscles that Move the Mandible (Lower Jaw) (Figure 11.6)* |

OBJECTIVE

• *Describe the origin, insertion, action, and innervation of the muscles that move the mandible.*

The muscles that move the mandible (lower jaw) at the temporomandibular joint (TMJ) are known as the muscles of mastication because they are involved in chewing (mastication). Of the four pairs of muscles involved in mastication, three are powerful closers of the jaw and account for the strength of the bite: **masseter, temporalis,** and **medial pterygoid.** Of these, the masseter is the strongest muscle of mastication. The medial and **lateral pterygoid** muscles assist in mastication by moving the mandible from side to side to help grind food. Additionally, these muscles protrude the mandible.

In 1996, researchers at the University of Maryland reported that they have identified a new muscle in the skull, tentatively named the *sphenomandibularis muscle.* It extends from the sphenoid bone to the mandible. The muscle is believed to be either a fifth muscle of mastication or a previously unidentified component of an already identified muscle (temporalis or medial pterygoid).

INNERVATION

All muscles are innervated by the mandibular division of the trigeminal (V) cranial nerve, except the sphenomandibularis muscle, which is innervated by the maxillary branch of the trigeminal (V) cranial nerve.

RELATING MUSCLES TO MOVEMENTS

Arrange the muscles in this exhibit according to their actions on the mandible: (1) elevation, (2) depression, (3) retraction, (4) protraction, and (5) side-to-side movement. The same muscle may be mentioned more than once.

What would happen if you lost tone in the masseter and temporalis muscles?

MUSCLE	ORIGIN	INSERTION	ACTION
Masseter (MA-se-ter; *maseter* = chewer) (see Figure 11.4c)	Maxilla and zygomatic arch.	Angle and ramus of mandible.	Elevates mandible, as in closing mouth, and retracts (draws back) mandible.
Temporalis (tem'-por-A-lis; *tempora* = temples)	Temporal and frontal bones.	Coronoid process and ramus of mandible.	Elevates and retracts mandible.
Medial pterygoid (TER-i-goid; *medial* = closer to midline; *pterygoid* = like a wing)	Medial surface of lateral portion of pterygoid process of sphenoid bone; maxilla.	Angle and ramus of mandible.	Elevates and protracts (protrudes) mandible and moves mandible from side to side.
Lateral pterygoid (TER-i-goid; *lateral* = farther from midline)	Greater wing and lateral surface of lateral portion of pterygoid process of sphenoid bone.	Condyle of mandible; temporomandibular joint (TMJ).	Protracts mandible, depresses mandible as in opening mouth, and moves mandible from side to side.

Exhibit 11.3 | *Muscles that Move the Mandible (Lower Jaw) (continued)*

Figure 11.6 Muscles that move the mandible (lower jaw).

 The muscles that move the mandible are also known as muscles of mastication.

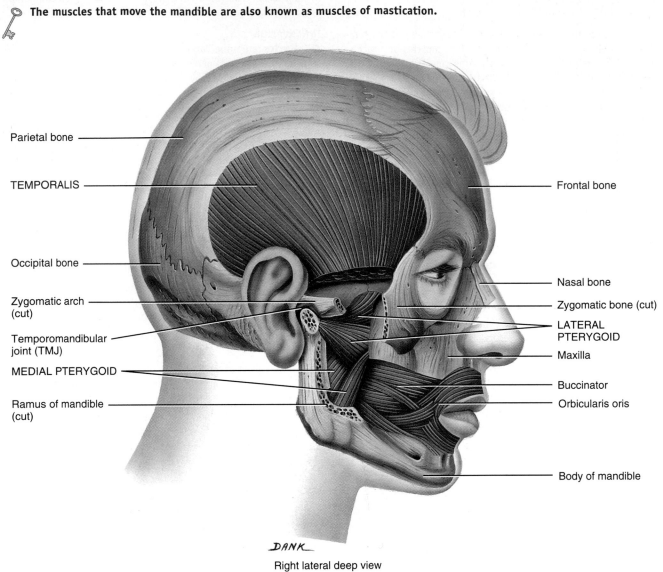

Right lateral deep view

Q Which is the strongest muscle of mastication?

Exhibit 11.4 *Muscles that Move the Tongue—Extrinsic Muscles (Figure 11.7)*

OBJECTIVE

• *Describe the origin, insertion, action, and innervation of the extrinsic muscles of the tongue.*

The tongue is a highly mobile structure that is vital to digestive functions such as mastication, perception of taste, and deglutition (swallowing). It is also important in speech. The tongue's mobility is greatly aided by its suspension from the mandible, styloid process of the temporal bone, and hyoid bone.

The tongue is divided into lateral halves by a median fibrous septum. The septum extends throughout the length of the tongue and is attached inferiorly to the hyoid bone. Muscles of the tongue are of two principal types: extrinsic and intrinsic. **Extrinsic muscles** originate outside the tongue and insert into it. They move the entire tongue in various directions, such as anteriorly, posteriorly, and laterally. **Intrinsic muscles** originate and insert within the tongue. These muscles alter the shape of the tongue rather than moving the entire tongue. The extrinsic and intrinsic muscles of the tongue insert into both lateral halves of the tongue.

When you study the extrinsic muscles of the tongue, you will notice that all of the names end in *glossus,* meaning tongue. You will also notice that the actions of the muscles are obvious, considering the positions of the mandible, styloid process, hyoid bone, and soft palate, which serve as origins for these muscles. For example, the **genioglossus** (originates on the mandible) pulls the tongue downward and forward, the **styloglossus** (originates on the styloid process) pulls the tongue upward and backward, the **hyoglossus** (originates on the hyoid bone) pulls the tongue downward and flattens it, and the **palatoglossus** (originates on the soft palate) raises the back portion of the tongue.

INNERVATION

All muscles are innervated by the hypoglossal (XII) cranial nerve, except the palatoglossus, which is innervated by the pharyngeal plexus, which contains axons from both the vagus (X) and the accessory (XI) cranial nerves.

CLINICAL APPLICATION
Intubation During Anesthesia

When general anesthesia is administered during surgery, a total relaxation of the genioglossus muscle results. This causes the tongue to fall posteriorly, which may obstruct the airway to the lungs. To avoid this, the mandible is either manually thrust forward and held in place, or a tube is inserted from the lips through the laryngopharynx (inferior portion of the throat) into the trachea (endotracheal intubation). ■

RELATING MUSCLES TO MOVEMENTS

Arrange the muscles in this exhibit according to the following actions on the tongue: (1) depression, (2) elevation, (3) protraction, and (4) retraction. The same muscle may be mentioned more than once.

When your physician says, "Open your mouth, stick out your tongue, and say ahh," so she can examine the inside of your mouth for possible signs of infection, which muscles do you contract?

MUSCLE	ORIGIN	INSERTION	ACTION
Genioglossus (jē'-nē-ō-GLOS-us; *geneion* = chin; *glossus* = tongue)	Mandible.	Undersurface of tongue and hyoid bone.	Depresses tongue and thrusts it anteriorly (protraction).
Styloglossus (stī'-lō-GLOS-us; *stylo* = stake or pole; styloid process of temporal bone)	Styloid process of temporal bone.	Side and undersurface of tongue.	Elevates tongue and draws it posteriorly (retraction).
Palatoglossus (pal'-a-tō-GLOS-us; *palato* = palate)	Anterior surface of soft palate.	Side of tongue.	Elevates posterior portion of tongue and draws soft palate down on tongue.
Hyoglossus (hī'-ō-GLOS-us)	Greater horn and body of hyoid bone.	Side of tongue.	Depresses tongue and draws down its sides.

Exhibit 11.4 | *Muscles that Move the Tongue—Extrinsic Muscles (continued)*

Figure 11.7 Muscles that move the tongue.

The extrinsic and intrinsic muscles of the tongue are arranged in both lateral halves of the tongue.

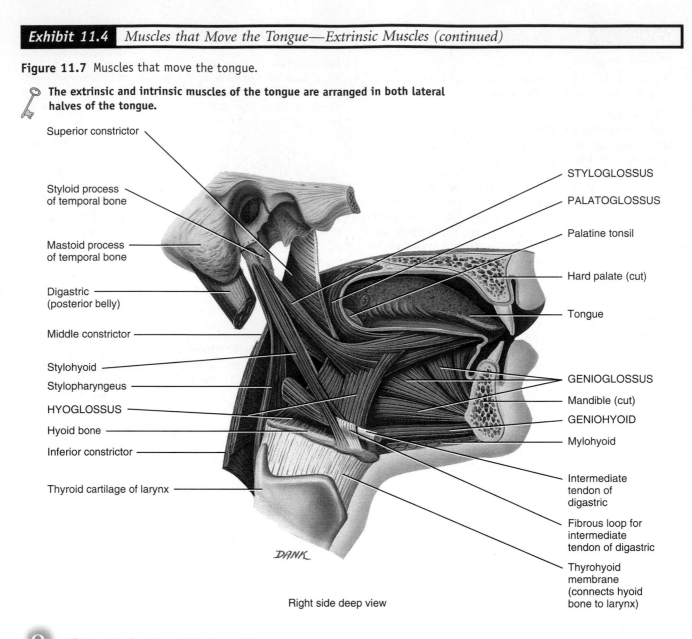

Superior constrictor

Styloid process of temporal bone

Mastoid process of temporal bone

Digastric (posterior belly)

Middle constrictor

Stylohyoid

Stylopharyngeus

HYOGLOSSUS

Hyoid bone

Inferior constrictor

Thyroid cartilage of larynx

DANK

STYLOGLOSSUS

PALATOGLOSSUS

Palatine tonsil

Hard palate (cut)

Tongue

GENIOGLOSSUS

Mandible (cut)

GENIOHYOID

Mylohyoid

Intermediate tendon of digastric

Fibrous loop for intermediate tendon of digastric

Thyrohyoid membrane (connects hyoid bone to larynx)

Right side deep view

Q What are the functions of the tongue?

Exhibit 11.5 — Muscles of the Floor of the Oral Cavity (Mouth) (Figure 11.8)

OBJECTIVE
• Describe the origin, insertion, action, and innervation of the muscles of the floor of the oral cavity.

Two groups of muscles are associated with the anterior aspect of the neck: (1) the **suprahyoid muscles**, so called because they are located superior to the hyoid bone, and (2) the **infrahyoid muscles**, so named because they lie inferior to the hyoid bone. Acting with the infrahyoid muscles, the suprahyoid muscles fix the hyoid bone, thus serving as a firm base upon which the tongue can move. In this exhibit we will consider the suprahyoid muscles, which are associated with the floor of the oral cavity.

As a group, the suprahyoid muscles elevate the hyoid bone, floor of the oral cavity, and tongue during swallowing. As its name suggests, the **digastric** muscle has two bellies, anterior and posterior, united by an intermediate tendon that is held in position by a fibrous loop (see Figure 11.7). This muscle elevates the hyoid bone and larynx (voice box) during swallowing and speech and also depresses the mandible. The **stylohyoid** muscle elevates and draws the hyoid bone posteriorly, thus elongating the floor of the oral cavity during swallowing. The **mylohyoid** muscle elevates the hyoid bone and helps press the tongue against the roof of the oral cavity during swallowing to move food from the oral cavity into the throat. The **geniohyoid** muscle (see Figure 11.7) elevates and draws the hyoid bone anteriorly to shorten the floor of the oral cavity and to widen the throat to receive food that is being swallowed. It also depresses the mandible.

INNERVATION
The anterior belly of the digastric and the mylohyoid are innervated by the mandibular division of the trigeminal (V) cranial nerve; the posterior belly of the digastric and the stylohyoid are innervated by the facial (VII) cranial nerve; and the geniohyoid is innervated by the first cervical spinal nerve.

RELATING MUSCLES TO MOVEMENTS
Arrange the muscles in this exhibit according to the following actions on the hyoid bone: (1) elevating it, (2) drawing it anteriorly, and (3) drawing it posteriorly. The same muscle may be mentioned more than once.

Which tongue, facial, and mandibular muscles would you use when chewing food?

MUSCLE	ORIGIN	INSERTION	ACTION
Digastric (dī′-GAS-trik; di = two; gaster = belly)	Anterior belly from inner side of inferior border of mandible; posterior belly from mastoid process of temporal bone.	Body of hyoid bone via an intermediate tendon.	Elevates hyoid bone and depresses mandible, as in opening the mouth.
Stylohyoid (stī′-lō-HĪ-oid; stylo = stake or pole, styloid process of temporal bone; hyoedes = U-shaped, pertaining to hyoid bone)	Styloid process of temporal bone.	Body of hyoid bone.	Elevates hyoid bone and draws it posteriorly.
Mylohyoid (mī′-lō-HĪ-oid) (myle = mill)	Inner surface of mandible.	Body of hyoid bone.	Elevates hyoid bone and floor of mouth and depresses mandible.
Geniohyoid (jē′-nē-ō-HĪ-oid; geneion = chin) (see also Figure 11.7)	Inner surface of mandible.	Body of hyoid bone.	Elevates hyoid bone, draws hyoid bone and tongue anteriorly, and depresses mandible.

Exhibit 11.5 | *Muscles of the Floor of the Oral Cavity (Mouth) (continued)*

Figure 11.8 Muscles of the floor of the oral cavity and front of the neck.

The suprahyoid muscles elevate the hyoid bone, the floor of the oral cavity, and the tongue during swallowing.

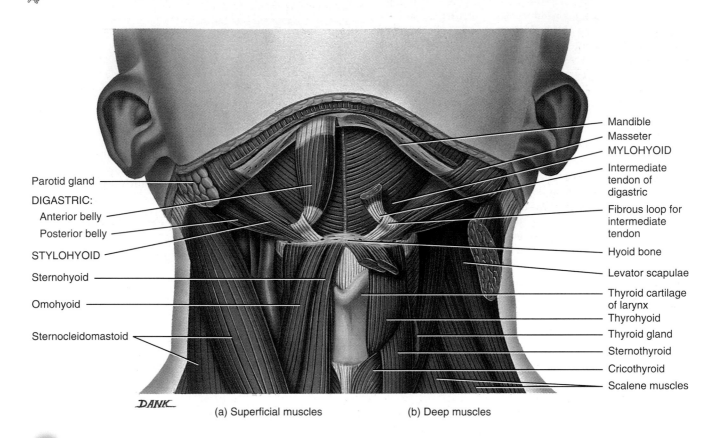

Parotid gland

DIGASTRIC:
  Anterior belly
  Posterior belly

STYLOHYOID

Sternohyoid

Omohyoid

Sternocleidomastoid

Mandible
Masseter
MYLOHYOID
Intermediate tendon of digastric
Fibrous loop for intermediate tendon
Hyoid bone
Levator scapulae
Thyroid cartilage of larynx
Thyrohyoid
Thyroid gland
Sternothyroid
Cricothyroid
Scalene muscles

DANK

(a) Superficial muscles   (b) Deep muscles

Q What is the combined action of the suprahyoid and infrahyoid muscles?

Exhibit 11.6 | Muscles that Move the Head (Figure 11.9)

OBJECTIVE

• *Describe the origin, insertion, action, and innervation of the muscles that move the head.*

The head is attached to the vertebral column at the atlanto-occipital joint formed by the atlas and occipital bone. Balance and movement of the head on the vertebral column involves the action of several neck muscles. For example, contraction of the two **sternocleidomastoid** muscles together (bilaterally) flexes the cervical portion of the vertebral column and extends the head. Acting singly (unilaterally), the sternocleidomastoid muscle laterally flexes and rotates the head. Bilateral contraction of the **semispinalis capitis, splenius capitis,** and **longissimus capitis** muscles extends the head. However, when these same muscles contract unilaterally, their actions are quite different, involving primarily rotation of the head.

The sternocleidomastoid muscle is an important landmark that divides the neck into two major triangles: anterior and posterior. The triangles are important because of the structures that lie within their boundaries.

The **anterior triangle** is bordered superiorly by the mandible, inferiorly by the sternum, medially by the cervical midline, and laterally by the anterior border of the sternocleidomastoid muscle. The anterior triangle is subdivided into an unpaired submental triangle and three paired triangles: submandibular, carotid, and muscular. The anterior triangle contains submental, submandibular,

and deep cervical lymph nodes; the submandibular salivary gland and a portion of the parotid salivary gland; the facial artery and vein; carotid arteries and internal jugular vein; and the glossopharyngeal (IX), vagus (X), accessory (XI), and hypoglossal (XII) cranial nerves.

The **posterior triangle** is bordered inferiorly by the clavicle, anteriorly by the posterior border of the sternocleidomastoid muscle, and posteriorly by the anterior border of the trapezius muscle. The posterior triangle is subdivided into two triangles, occipital and supraclavicular, by the inferior belly of the omohyoid muscle. The posterior triangle contains part of the subclavian artery, external jugular vein, cervical lymph nodes, brachial plexus, and accessory (XI) cranial nerve.

INNERVATION

The sternocleidomastoid is innervated by the accessory (XI) cranial nerve; the capitis muscles are all innervated by cervical spinal nerves.

RELATING MUSCLES TO MOVEMENTS

Arrange the muscles in this exhibit according to the following actions on the head: (1) flexion, (2) lateral flexion, (3) extension, (4) rotation to side opposite contracting muscle, and (5) rotation to same side as contracting muscle. The same muscle may be mentioned more than once.

What muscles would you use to signify "yes" and "no" by moving your head?

MUSCLE	ORIGIN	INSERTION	ACTION
Sternocleidomastoid (ster′-nō-klī′-dō-MAS-toid; *sternum* = breastbone; *cleido* = clavicle; *mastoid* = mastoid process of temporal bone)	Sternum and clavicle.	Mastoid process of temporal bone.	Acting together (bilaterally), flex cervical portion of vertebral column and extend head; acting singly (unilaterally), laterally flex and rotate head to side opposite contracting muscle.
Semispinalis capitis (se′-mē-spi-NA-lis KAP-i-tis; *semi* = half; *spine* = spinous process; *caput* = head) (see Figure 11.19a)	Transverse processes of first six or seven thoracic vertebrae and seventh cervical vertebra, and articular processes of fourth, fifth, and sixth cervical vertebrae.	Occipital bone between superior and inferior nuchal lines.	Acting together, extend head; acting singly, rotate head to side opposite contracting muscle.
Splenius capitis (SPLĒ-nē-us KAP-i-tis; *splenion* = bandage) (see Figure 11.19a)	Ligamentum nuchae and spinous processes of seventh cervical vertebra and first three or four thoracic vertebrae.	Occipital bone and mastoid process of temporal bone.	Acting together, extend head; acting singly, laterally flex and rotate head to same side as contracting muscle.
Longissimus capitis (lon-JIS-i-mus KAP-i-tis; *longissimus* = longest) (see Figure 11.19a)	Transverse processes of upper four thoracic vertebrae and articular processes of last four cervical vertebrae.	Mastoid process of temporal bone.	Acting together, extend head; acting singly, laterally flex and rotate head to same side as contracting muscle.

Exhibit 11.6 | *Muscles that Move the Head (continued)*

Figure 11.9 Triangles of the neck.

The sternocleidomastoid muscle divides the neck into two principal triangles: anterior and posterior.

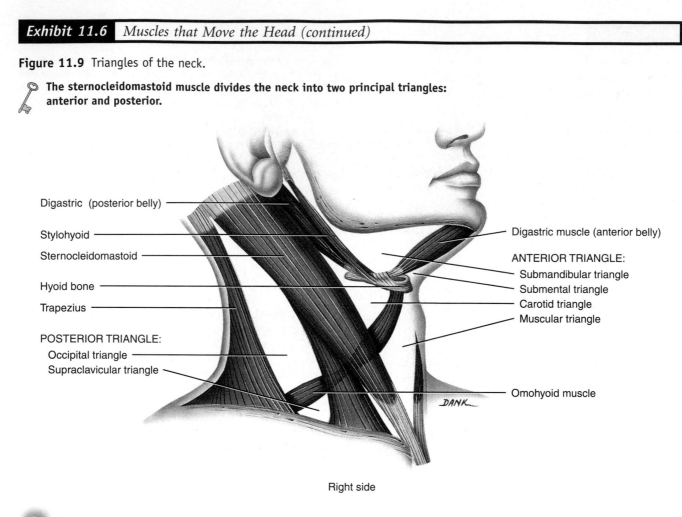

Right side

Q Why are triangles important?

Exhibit 11.7 | Muscles that Act on the Abdominal Wall (Figure 11.10)

OBJECTIVE

• *Describe the origin, insertion, action, and innervation of the muscles that act on the abdominal wall.*

The anterolateral abdominal wall is composed of skin, fascia, and four pairs of muscles: the external oblique, internal oblique, transversus abdominis, and rectus abdominis. The first three muscles are flat muscles; the last is a straplike vertical muscle. The **external oblique** is the external flat muscle with its fibers directed inferiorly and medially. The **internal oblique** is the intermediate flat muscle with its fibers directed at right angles to those of the external oblique. The **transversus abdominis** is the deepest of the flat muscles, with most of its fibers directed horizontally around the abdominal wall. Together, the external oblique, internal oblique, and transversus abdominus form three layers of muscle around the abdomen. The muscle fibers of each layer run cross-directionally to one another, a structural arrangement that affords considerable protection to the abdominal viscera, especially when the muscles have good tone.

The **rectus abdominis** muscle is a long flat muscle that extends the entire length of the anterior abdominal wall, from the pubic crest and pubic symphysis to the cartilages of ribs 5–7 and the xiphoid process of the sternum. The anterior surface of the muscle is interrupted by three transverse fibrous bands of tissue called **tendinous intersections,** believed to be remnants of septa that separated myotomes during embryological development.

As a group, the muscles of the anterolateral abdominal wall help contain and protect the abdominal viscera; flex, laterally flex, and rotate the vertebral column at the intervertebral joints; compress the abdomen during forced expiration; and produce the force required for defecation, urination, and childbirth.

The aponeuroses of the external oblique, internal oblique, and transversus abdominis muscles form the **rectus sheath,** which encloses the rectus abdominis muscles and meet at the midline to form the **linea alba** (= white line), a tough, fibrous band that extends from the xiphoid process of the sternum to the pubic symphysis. In the latter stages of pregnancy, the linea alba stretches to increase the distance between the rectus abdominis muscles. The inferior free border of the external oblique aponeurosis, plus some collagen fibers, forms the **inguinal ligament,** which runs from the anterior superior iliac spine to the pubic tubercle (see Figure 11.20a). Just superior to the medial end of the inguinal lig-

ament is a triangular slit in the aponeurosis referred to as the **superficial inguinal ring,** the outer opening of the **inguinal canal** (see Figure 28.2). The canal contains the spermatic cord and ilioinguinal nerve in males, and the round ligament of the uterus and ilioinguinal nerve in females.

The posterior abdominal wall is formed by the lumbar vertebrae, parts of the ilia of the hip bones, psoas major and iliacus muscles (described in Exhibit 11.17 on page 358), and quadratus lumborum muscle. Whereas the anterolateral abdominal wall can contract and distend, the posterior abdominal wall is bulky and stable by comparison.

INNERVATION

The rectus abdominis is innervated by branches of thoracic nerves T7–T12; the external oblique is innervated by branches of thoracic nerves T7–T12 and the iliohypogastric nerve; the internal oblique and transversus abdominis are innervated by branches of thoracic nerves T8–T12 and the iliohypogastric and ilioinguinal nerves; the quadratus lumborum is innervated by branches of thoracic nerve T12 and lumbar nerves L1–L3 or L1–L4.

CLINICAL APPLICATION
Inguinal Hernia

The inguinal region is a weak area in the abdominal wall. It is often the site of an **inguinal hernia,** a rupture or separation of a portion of the inguinal area of the abdominal wall resulting in the protrusion of a part of the small intestine. Hernia is much more common in males than in females because the inguinal canals in males are larger and represent especially weak points in the abdominal wall. ■

RELATING MUSCLES TO MOVEMENTS

Arrange the muscles in this exhibit according to the following actions on the vertebral column: (1) flexion, (2) lateral flexion, (3) extension, and (4) rotation. The same muscle may be mentioned more than once.

Which muscles do you contract when you "suck in your tummy," thereby compressing the anterior abdominal wall?

Exhibit 11.7 Muscles that Act on the Abdominal Wall (continued)

MUSCLE	ORIGIN	INSERTION	ACTION
Rectus abdominis (REK-tus-ab-DOM-in-is; *rectus* = fibers parallel to midline; *abdomino* = abdomen)	Pubic crest and pubic symphysis.	Cartilage of fifth to seventh ribs and xiphoid process.	Flexes vertebral column, especially lumbar portion, and compresses abdomen to aid in defecation, urination, forced expiration, and childbirth.
External oblique (ō-BLĒK; *external* = closer to surface; *oblique* = fibers diagonal to midline)	Inferior eight ribs.	Iliac crest and linea alba.	Acting together (bilaterally), compress abdomen and flex vertebral column; acting singly (unilaterally), laterally flex vertebral column, especially lumbar portion, and rotate vertebral column.
Internal oblique (ō-BLĒK; *internal* = farther from surface)	Iliac crest, inguinal ligament, and thoracolumbar fascia.	Cartilage of last three or four ribs and linea alba.	Acting together, compress abdomen and flex vertebral column; acting singly, laterally flex vertebral column, especially lumbar portion, and rotate vertebral column.
Transversus abdominis (tranz-VER-sus ab-DOM-in-is; *transverse* = fibers perpendicular to midline)	Iliac crest, inguinal ligament, lumbar fascia, and cartilages of inferior six ribs.	Xiphoid process, linea alba, and pubis.	Compresses abdomen.
Quadratus lumborum (kwod-RĀ-tus lum-BOR-um; *quad* = four; *lumbo* = lumbar region) (see Figure 11.11)	Iliac crest and iliolumbar ligament.	Inferior border of twelfth rib and transverse processes of first four lumbar vertebrae.	Acting together, pull twelfth ribs inferiorly during forced expiration, fix twelfth ribs to prevent their elevation during deep inspiration, and help extend lumbar portion of vertebral column; acting singly, laterally flex vertebral column, especially lumbar portion.

Figure 11.10 (See figure on opposite page.) Muscles of the male anterolateral abdominal wall. (See Tortora, *A Photographic Atlas of the Human Body,* Figure 5.7)

The anterolateral abdominal muscles protect the abdominal viscera, move the vertebral column, and assist in forced expiration, defecation, urination, and childbirth.

Q Which abdominal muscle aids in urination?

Exhibit 11.7 *(continued)*

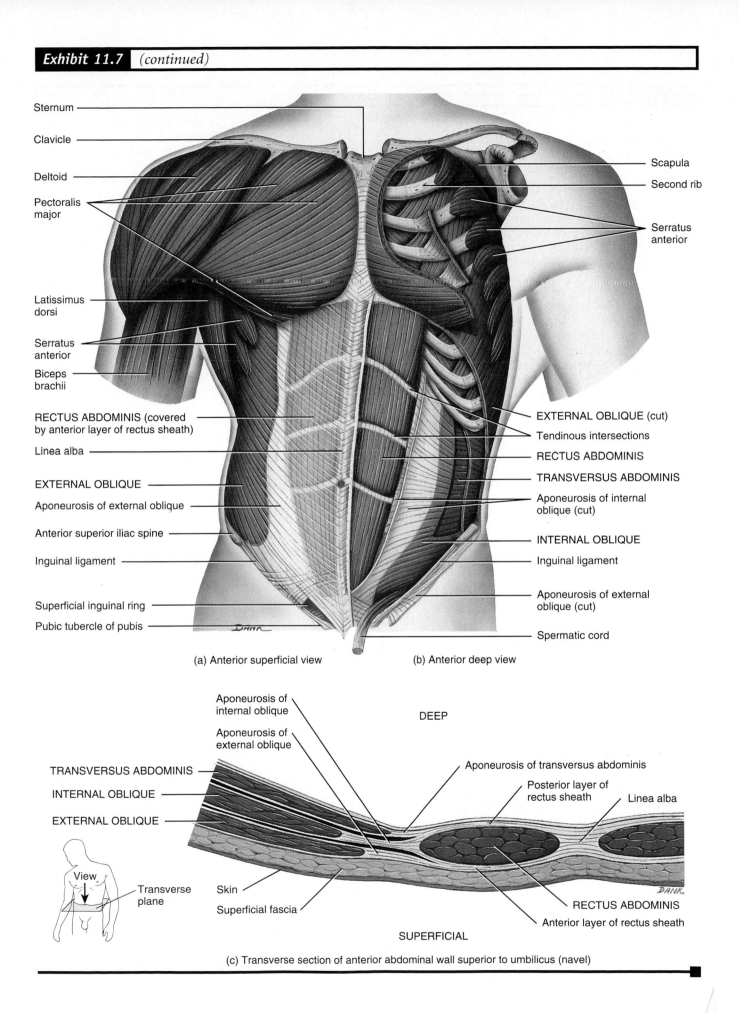

Sternum

Clavicle

Deltoid

Pectoralis major

Latissimus dorsi

Serratus anterior

Biceps brachii

RECTUS ABDOMINIS (covered by anterior layer of rectus sheath)

Linea alba

EXTERNAL OBLIQUE

Aponeurosis of external oblique

Anterior superior iliac spine

Inguinal ligament

Superficial inguinal ring

Pubic tubercle of pubis

Scapula

Second rib

Serratus anterior

EXTERNAL OBLIQUE (cut)

Tendinous intersections

RECTUS ABDOMINIS

TRANSVERSUS ABDOMINIS

Aponeurosis of internal oblique (cut)

INTERNAL OBLIQUE

Inguinal ligament

Aponeurosis of external oblique (cut)

Spermatic cord

(a) Anterior superficial view

(b) Anterior deep view

Aponeurosis of internal oblique

Aponeurosis of external oblique

DEEP

TRANSVERSUS ABDOMINIS

INTERNAL OBLIQUE

EXTERNAL OBLIQUE

Aponeurosis of transversus abdominis

Posterior layer of rectus sheath

Linea alba

View

Transverse plane

Skin

Superficial fascia

RECTUS ABDOMINIS

Anterior layer of rectus sheath

SUPERFICIAL

(c) Transverse section of anterior abdominal wall superior to umbilicus (navel)

Exhibit 11.8 *Muscles Used in Breathing (Figure 11.11)*

OBJECTIVE

• *Describe the origin, insertion, action, and innervation of the muscles used in breathing.*

The muscles described here alter the size of the thoracic cavity so that breathing can occur. Inspiration (breathing in) occurs when the thoracic cavity increases in size, and expiration (breathing out) occurs when the thoracic cavity decreases in size.

The **diaphragm** is the most important muscle in breathing. It is a dome-shaped musculotendinous partition that separates the thoracic and abdominal cavities. The diaphragm is composed of a peripheral muscular portion and a central portion called the central tendon. The **central tendon** is a strong aponeurosis that serves as the tendon of insertion for all the peripheral muscular fibers of the diaphragm. It fuses with the inferior surface of the fibrous pericardium (external covering of the heart) and the parietal pleurae (external coverings of the lungs).

Movements of the diaphragm also help to return to the heart venous blood passing through the abdomen. Together with the anterolateral abdominal muscles, the diaphragm helps to increase intra-abdominal pressure to evacuate the pelvic contents during defecation, urination, and childbirth. This mechanism is further assisted when you take a deep breath and close the rima glottidis (the space between vocal folds). The trapped air in the respiratory system prevents the diaphragm from elevating. The increase in intra-abdominal pressure as just described will also help support the vertebral column and prevent flexion during weight lifting. This greatly assists the back muscles in lifting a heavy weight.

The diaphragm has three major openings through which various structures pass between the thorax and abdomen. These structures include the aorta, along with the thoracic duct and azygos vein, which pass through the **aortic hiatus;** the esophagus with accompanying vagus (X) cranial nerves, which pass through the **esophageal hiatus;** and the inferior vena cava, which passes through the **foramen for the vena cava.** In a condition called a hiatus hernia, the stomach protrudes superiorly through the esophageal hiatus.

The other muscles involved in breathing are called **intercostal** muscles and occupy the intercostal spaces, the spaces between ribs. These muscles are arranged in three layers. The 11 **external intercostal** muscles occupy the superficial layer, and their fibers run obliquely inferiorly and anteriorly from the rib above to the rib below. They elevate the ribs during inspiration to help expand the thoracic cavity. The 11 **internal intercostal** muscles occupy the intermediate layer of the intercostal spaces. The fibers of these muscles run obliquely inferiorly and posteriorly from the inferior border of the rib above to the superior border of the rib below. They draw adjacent ribs together during forced expiration to help decrease the size of the thoracic cavity. The deepest muscle layer is made up of the **transversus thoracis** muscles. These poorly developed muscles (not described in the exhibit) run in the same direction as the internal intercostals, and they may have the same role.

INNERVATION

The diaphragm is innervated by the phrenic nerve, which contains axons from cervical spinal nerves C3–C5; the external and internal intercostals are innervated by thoracic spinal nerves T2–T12.

RELATING MUSCLES TO MOVEMENTS

Arrange the muscles in this exhibit according to the following actions on the size of the thorax: (1) increase in vertical length, (2) increase in lateral and anteroposterior dimensions, and (3) decrease in lateral and anteroposterior dimensions.

What are the names of the three openings in the diaphragm, and which structures pass through each?

MUSCLE	ORIGIN	INSERTION	ACTION
Diaphragm (DĪ-a-fram; *dia* = across; *phragma* = wall)	Xiphoid process of the sternum, costal cartilages of inferior six ribs, and lumbar vertebrae.	Central tendon.	Forms floor of thoracic cavity; pulls central tendon inferiorly during inspiration, and, as dome of diaphragm flattens, increases vertical length of thorax.
External intercostals (in'-ter-KOS-tals; *external* = closer to surface; *inter* = between; *costa* = rib)	Inferior border of rib above.	Superior border of rib below.	Elevate ribs during inspiration and thus increase lateral and anteroposterior dimensions of thorax.
Internal intercostals (in'-ter-KOS-tals; *internal* = farther from surface)	Superior border of rib below.	Inferior border of rib above.	Draw adjacent ribs together during forced expiration and thus decrease lateral and anteroposterior dimensions of thorax.

Exhibit 11.8 *(continued)*

Figure 11.11 Muscles used in breathing, as seen in a male.

Openings in the diaphragm permit the passage of the aorta, esophagus, and inferior vena cava.

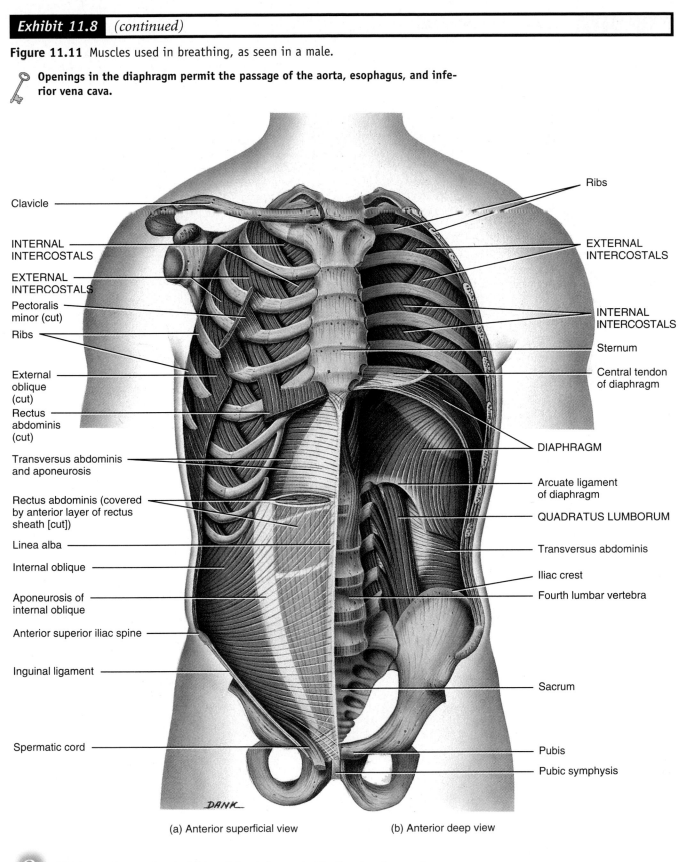

(a) Anterior superficial view

(b) Anterior deep view

Q Which muscle associated with breathing is innervated by the phrenic nerve?

| **Exhibit 11.9** | *Muscles of the Pelvic Floor (Figure 11.12)* |

OBJECTIVE
• *Describe the origin, insertion, action, and innervation of the muscles of the pelvic floor.*

The muscles of the pelvic floor are the levator ani and coccygeus. Together with the fascia covering their internal and external surfaces, these muscles are referred to as the **pelvic diaphragm,** which stretches from the pubis anteriorly to the coccyx posteriorly, and from one lateral wall of the pelvis to the other. This arrangement gives the pelvic diaphragm the appearance of a funnel suspended from its attachments. The pelvic diaphragm is pierced by the anal canal and urethra in both sexes, and also by the vagina in females.

The **levator ani** muscle is divisible into two muscles called the **pubococcygeus** and **iliococcygeus.** These muscles in the female are shown in Figure 11.12 and in the male in Figure 11.13. The levator ani is the largest and most important muscle of the pelvic floor. It supports the pelvic viscera and resists the inferior thrust that accompanies increases in intra-abdominal pressure during functions such as forced expiration, coughing, vomiting, urination, and defecation. The muscle also functions as a sphincter at the anorectal junction, urethra, and vagina. During childbirth, the levator ani muscle supports the head of the fetus, and the muscle may be injured during a difficult childbirth or traumatized during an episiotomy (a cut made with surgical scissors to prevent or direct tearing of the perineum during the birth of a baby). This may cause urinary stress incontinence in which there is a leakage of urine whenever intra-abdominal pressure is increased—for example, during coughing. In addition to assisting the levator ani, the **coccygeus** muscle pulls the coccyx anteriorly after it has been pushed posteriorly following defecation or childbirth. In order to treat urinary stress incontinence, it may be necessary to strengthen and tighten the muscles that support the pelvic viscera. This is accomplished by *Kegel exercises,* the alternate contraction and relaxation of muscles of the pelvic floor.

INNERVATION
The pubococcygeus and iliococcygeus are innervated by sacral spinal nerves S2–S4; the coccygeus is innervated by sacral spinal nerves S4–S5.

RELATING MUSCLES TO MOVEMENTS
Arrange the muscles in this exhibit according to the following actions on the pelvic viscera: supporting and maintaining their position, and resisting an increase in intra-abdominal pressure; and according to the following action on the anus, urethra, and vagina: constriction. The same muscle may be mentioned more than once.

| Which muscles are strengthened by Kegel exercises?

MUSCLE	**ORIGIN**	**INSERTION**	**ACTION**
Levator ani (le-VĀ-tor Ā-nē; *levator* = raises; *ani* = anus)	This muscle is divisible into two parts, the pubococcygeus muscle and the iliococcygeus muscle.		
Pubococcygeus (pu'-bō-kok-SIJ-ē-us; *pubo* = pubis; *coccygeus* = coccyx)	Pubis.	Coccyx, urethra, anal canal, central tendon of perineum, and anococcygeal raphe (narrow fibrous band that extends from anus to coccyx).	Supports and maintains position of pelvic viscera; resists increase in intra-abdominal pressure during forced expiration, coughing, vomiting, urination, and defecation; constricts anus, urethra, and vagina; and supports fetal head during childbirth.
Iliococcygeus (il'-ē-ō-kok-SIJ-ē-us; *ilio* = ilium)	Ischial spine.	Coccyx.	As above.
Coccygeus (kok-SIJ-ē-us)	Ischial spine.	Lower sacrum and upper coccyx.	Supports and maintains position of pelvic viscera; resists increase in intra-abdominal pressure during forced expiration, coughing, vomiting, urination, and defecation; and pulls coccyx anteriorly following defecation or childbirth.

Exhibit 11.9 *(continued)*

Figure 11.12 Muscles of the pelvic floor, as seen in the female perineum.

 The pelvic diaphragm supports the pelvic viscera.

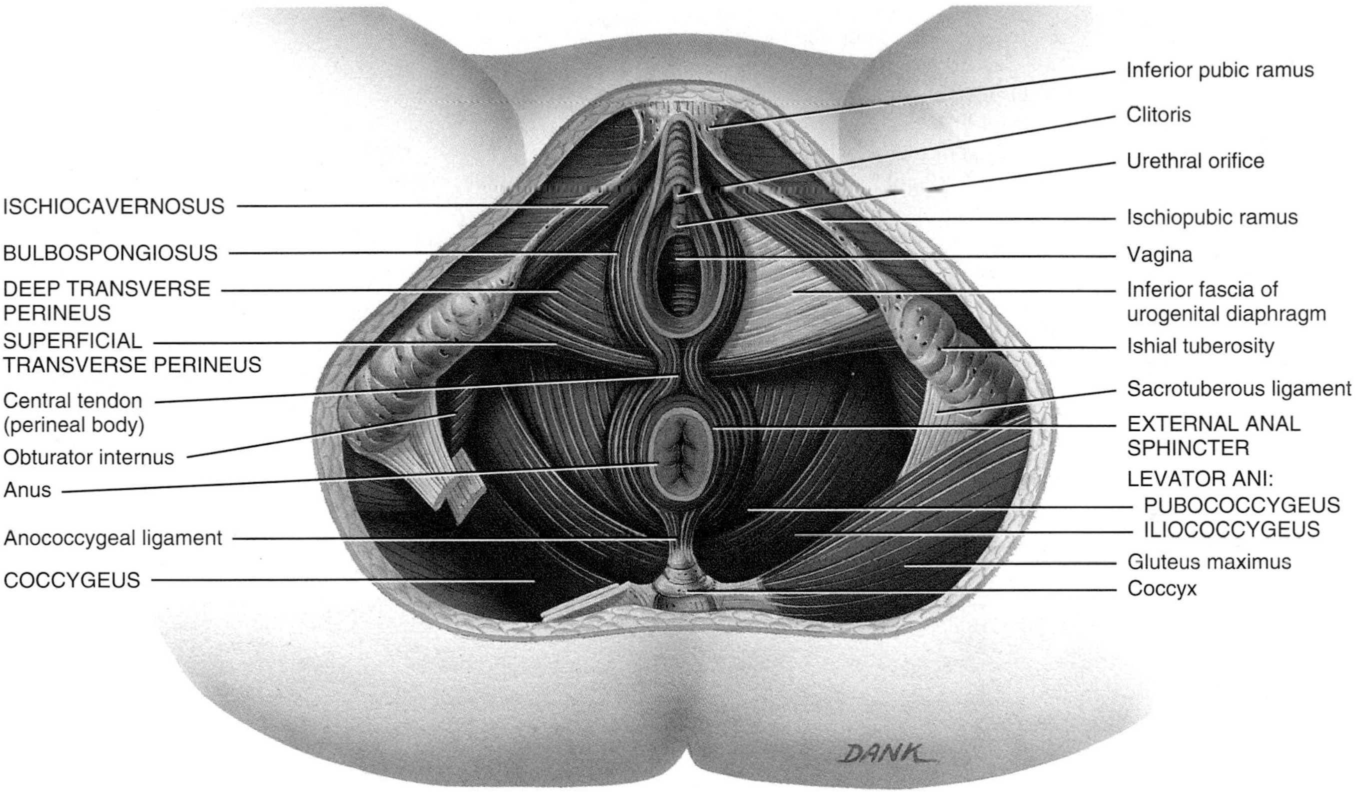

ISCHIOCAVERNOSUS

BULBOSPONGIOSUS

DEEP TRANSVERSE PERINEUS

SUPERFICIAL TRANSVERSE PERINEUS

Central tendon (perineal body)

Obturator internus

Anus

Anococcygeal ligament

COCCYGEUS

Inferior pubic ramus

Clitoris

Urethral orifice

Ischiopubic ramus

Vagina

Inferior fascia of urogenital diaphragm

Ishial tuberosity

Sacrotuberous ligament

EXTERNAL ANAL SPHINCTER

LEVATOR ANI:
PUBOCOCCYGEUS
ILIOCOCCYGEUS

Gluteus maximus

Coccyx

DANK

Inferior superficial view

Q What are the borders of the pelvic diaphragm?

Exhibit 11.10	Muscles of the Perineum (Figures 11.12 and 11.13)

OBJECTIVE

• *Describe the origin, insertion, action, and innervation of the muscles of the perineum.*

The **perineum** is the region of the trunk inferior to the pelvic diaphragm. It is a diamond-shaped area that extends from the pubic symphysis anteriorly, to the coccyx posteriorly, and to the ischial tuberosities laterally. The female and the male perineums may be compared in Figures 11.12 and 11.13, respectively. A transverse line drawn between the ischial tuberosities divides the perineum into an anterior **urogenital triangle** that contains the external genitals and a posterior **anal triangle** that contains the anus (see Figure 28.21). In the center of the perineum is a wedge-shaped mass of fibrous tissue called the **central tendon (perineal body).** It is a strong tendon into which several perineal muscles insert.

The muscles of the perineum are arranged in two layers: **superficial** and **deep.** The muscles of the superficial layer are the **superficial transverse perineus, bulbospongiosus,** and **ischiocavernosus.** The deep muscles of the perineum are the **deep transverse perineus** muscle and **external urethral sphincter** (see Figure 26.24a). The deep transverse perineus, external urethral

sphincter, and their fascia are known as the **urogenital diaphragm.** The muscles of this diaphragm assist in urination and ejaculation in males and urination in females. The **external anal sphincter** closely adheres to the skin around the margin of the anus and keeps the anal canal and anus closed except during defecation.

INNERVATION

Muscles of the perineum are innervated by the pudendal nerve of the sacral plexus (shown in Exhibit 13.4 on page 439). More specifically, all muscles are innervated by the perineal branch of the pudendal nerve, except for the external anal sphincter, which is innervated by sacral spinal nerve S4 and the inferior rectal branch of the pudendal nerve.

RELATING MUSCLES TO MOVEMENTS

Arrange the muscles in this exhibit according to the following actions: (1) expulsion of urine and semen, (2) erection of the clitoris and penis, (3) closing the anal orifice, and (4) constricting the vaginal orifice. The same muscle may be mentioned more than once.

Describe the borders and contents of the urogenital triangle and the anal triangle.

MUSCLE	ORIGIN	INSERTION	ACTION
Superficial Perineal Muscles			
Superficial transverse perineus (per-i-NĒ-us; *superficial* = closer to surface; *transverse* = across; *perineus* = perineum)	Ischial tuberosity.	Central tendon of perineum.	Helps stabilize central tendon of perineum.
Bulbospongiosus (bul'-bō-spon'-jē-Ō-sus; *bulbus* = bulb; *spongia* = sponge)	Central tendon of perineum.	Inferior fascia of urogenital diaphragm, corpus spongiosum of penis, and deep fascia on dorsum of penis in male; pubic arch and root and dorsum of clitoris in female.	Helps expel urine during urination, helps propel semen along urethra, assists in erection of the penis in male; constricts vaginal orifice and assists in erection of clitoris in female.
Ischiocavernosus (is'-kē-ō-ka'-ver-NŌ-sus; *ischion* = hip)	Ischial tuberosity and ischial and pubic rami.	Corpus cavernosum of penis in male and clitoris in female.	Maintains erection of penis in male and clitoris in female.
Deep Perineal Muscles			
Deep transverse perineus (per-i-NĒ-us; *deep* = farther from surface)	Ischial rami.	Central tendon of perineum.	Helps expel last drops of urine and semen in male and urine in female.
External urethral sphincter (yoo-RĒ-thral SFINGK-ter)	Ischial and pubic rami.	Median raphe in male and vaginal wall in female.	Helps expel last drops of urine and semen in male and urine in female.
External anal sphincter (Ā-nal)	Anococcygeal ligament.	Central tendon of perineum.	Keeps anal canal and anus closed.

Exhibit 11.10 (continued)

Figure 11.13 Muscles of the male perineum.

🗝 **The urogenital diaphragm assists in urination in females and males, ejaculation in males, and helps strengthen the pelvic floor.**

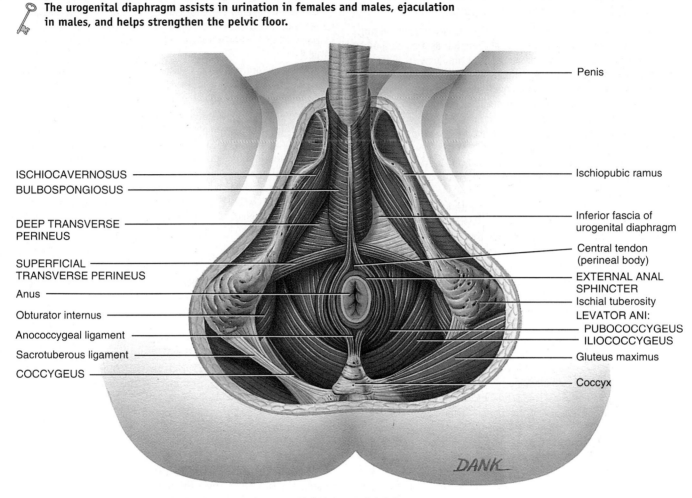

ISCHIOCAVERNOSUS

BULBOSPONGIOSUS

DEEP TRANSVERSE PERINEUS

SUPERFICIAL TRANSVERSE PERINEUS

Anus

Obturator internus

Anococcygeal ligament

Sacrotuberous ligament

COCCYGEUS

Penis

Ischiopubic ramus

Inferior fascia of urogenital diaphragm

Central tendon (perineal body)

EXTERNAL ANAL SPHINCTER

Ischial tuberosity

LEVATOR ANI:

PUBOCOCCYGEUS

ILIOCOCCYGEUS

Gluteus maximus

Coccyx

DANK

Inferior superficial view

Q   What are the borders of the perineum?

Exhibit 11.11 | *Muscles that Move the Pectoral (Shoulder) Girdle (Figure 11.14)*

OBJECTIVE
• *Describe the origin, insertion, action, and innervation of the muscles that move the pectoral girdle.*

The principal action of the muscles that move the pectoral girdle is to stabilize the scapula so it can function as a stable point of origin for most of the muscles that move the humerus. Because scapular movements usually accompany humeral movements in the same direction, the muscles also move the scapula to increase the range of motion of the humerus. For example, it would not be possible to abduct the humerus past the horizontal if the scapula did not move with the humerus. During abduction, the scapula follows the humerus by rotating upward.

Muscles that move the pectoral girdle can be classified into two groups based on their location in the thorax: **anterior** and **posterior thoracic muscles.** The anterior thoracic muscles are the subclavius, pectoralis minor, and serratus anterior. The **subclavius** is a small, cylindrical muscle under the clavicle that extends from the clavicle to the first rib. It steadies the clavicle during movements of the pectoral girdle. The **pectoralis minor** is a thin, flat, triangular muscle that is deep to the pectoralis major. In addition to its role in movements of the scapula, the pectoralis minor muscle assists in forced inspiration. The **serratus anterior** is a large, flat, fan-shaped muscle between the ribs and scapula. It is named because of the saw-toothed appearance of its origins on the ribs.

The posterior thoracic muscles are the trapezius, levator scapulae, rhomboideus major, and rhomboideus minor. The **trapezius** is a large, flat, triangular sheet of muscle extending from the skull and vertebral column medially to the pectoral girdle laterally. It is the most superficial back muscle and covers the posterior neck region and superior portion of the trunk. The two trapezius muscles form a trapezoid (diamond-shaped quadrangle)—hence its name. The **levator scapulae** is a narrow, elongated muscle in the posterior portion of the neck. It is deep to the sternocleidomastoid and trapezious muscles. As its name suggests, one of its actions is to elevate the scapula. The **rhomboideus major** and **rhomboideus minor** lie deep to the trapezius and are not always distinct from each other. They appear as parallel bands that pass inferiolaterally from the vertebrae to the scapula. They are named on the basis of their shape—that is, a rhomboid (an oblique parallelogram). The rhomboideus major is about two times wider than the rhomboideus minor. Both muscles are used when forcibly lowering the raised upper limbs, as in driving a stake with a sledgehammer.

To understand the actions of muscles that move the scapula, it is first helpful to describe the various movements of the scapula:

• **Elevation.** Superior movement of the scapula, such as shrugging the shoulders or lifting a weight over the head.
• **Depression.** Inferior movement of the scapula, as in doing a "pull-up."
• **Abduction (protraction).** Movement of the scapula laterally and anteriorly, as in doing a "push-up" or punching.
• **Adduction (retraction).** Movement of the scapula medially and posteriorly, as in pulling the oars in a rowboat.
• **Upward rotation.** Movement of the inferior angle of the scapula laterally so that the glenoid cavity is moved upward. This movement is required to abduct the humerus past the horizontal.
• **Downward rotation.** Movement of the inferior angle of the scapula medially so that the glenoid cavity is moved downward. This movement is seen when the weight of the body is supported on the hands by a gymnast on parallel bars.

INNERVATION
Muscles that move the shoulder are innervated mainly by nerves that emerge from the cervical and brachial plexuses (shown in Exhibits 13.1 and 13.2 on pages 429–435). More specifically, the subclavius is innervated by the subclavian nerve; the pectoralis minor is innervated by the medial pectoral nerve; the serratus anterior is innervated by the long thoracic nerve; the trapezius is innervated by the accessory (XI) cranial nerve and cervical spinal nerves C3–C5; the levator scapulae is innervated by the dorsal scapular nerve and cervical spinal nerves C3–C5; and the rhomboideus major and rhomboideus minor are innervated by the dorsal scapular nerve.

RELATING MUSCLES TO MOVEMENTS
Arrange the muscles in this exhibit according to the following actions on the scapula: (1) depression, (2) elevation, (3) abduction, (4) adduction, (5) upward rotation, and (6) downward rotation. The same muscle may be mentioned more than once.

What muscles in this exhibit are used to raise your shoulders, lower your shoulders, join your hands behind your back, and join your hands in front of your chest?

Exhibit 11.11 *(continued)*

MUSCLE	ORIGIN	INSERTION	ACTION
Anterior Thoracic Muscles			
Subclavius (sub-KLĀ-vē-us; *sub* = under; *clavius* = clavicle)	First rib.	Clavicle.	Depresses and moves clavicle anteriorly and helps stabilize pectoral girdle.
Pectoralis minor (pek′-tor-A-lis; *pectus* = breast, chest, thorax; *minor* = lesser)	Second through fifth, third through fifth, or second through fourth ribs.	Coracoid process of scapula.	Depresses and abducts scapula and rotates it downward; elevates third through fifth ribs during forced inspiration when scapula is fixed.
Serratus anterior (ser-Ā-tus; *serratus* = saw-toothed; *anterior* = before)	Superior eight or nine ribs.	Vertebral border and inferior angle of scapula.	Abducts scapula and rotates it upward; elevates ribs when scapula is fixed; known as "boxer's muscle."
Posterior Thoracic Muscles			
Trapezius (tra-PĒ-zē-us; *trapezoides* = trapezoid-shaped)	Superior nuchal line of occipital bone, ligamentum nuchae, and spines of seventh cervical and all thoracic vertebrae.	Clavicle and acromion and spine of scapula.	Superior fibers elevate scapula and can help extend head; middle fibers adduct scapula; inferior fibers depress scapula; superior and inferior fibers together rotate scapula upward; stabilizes scapula.
Levator scapulae (le-VĀ-tor SKA-pyoo-lē; *levare* = to raise; *scapulae* = scapula)	Superior four or five cervical vertebrae.	Superior vertebral border of scapula.	Elevates scapula and rotates it downward.
Rhomboideus major (rom-BOID-ē-us; *rhomboides* = rhomboid or diamond-shaped) (see Figure 11.15c)	Spines of second to fifth thoracic vertebrae.	Vertebral border of scapula inferior to spine.	Elevates and adducts scapula and rotates it downward; stabilizes scapula.
Rhomboideus minor (rom-BOID-ē-us) (see Figure 11.15c)	Spines of seventh cervical and first thoracic vertebrae.	Vertebral border of scapula superior to spine.	Elevates and adducts scapula and rotates it downward; stabilizes scapula.

Exhibit 11.11 | *Muscles that Move the Pectoral (Shoulder) Girdle (continued)*

Figure 11.14 Muscles that move the pectoral (shoulder) girdle. (See Tortora, *A Photographic Atlas of the Human Body,* Figure 5.8)

🔑 **Muscles that move the pectoral girdle originate on the axial skeleton and insert on the clavicle or scapula.**

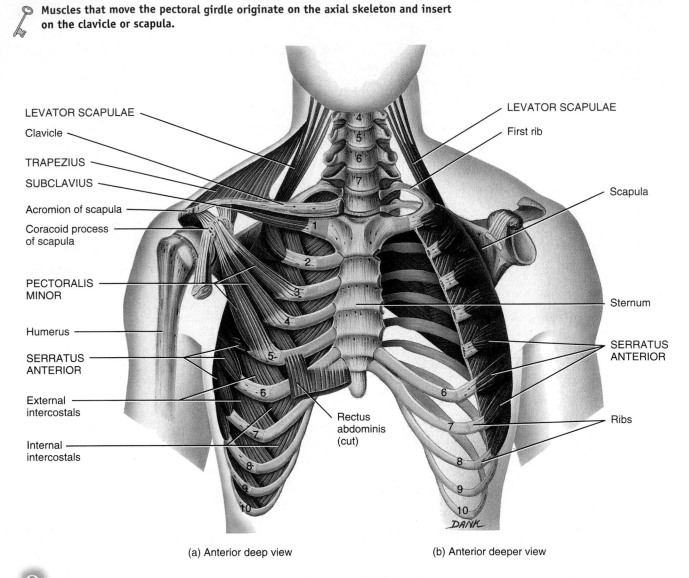

(a) Anterior deep view   (b) Anterior deeper view

❓ What is the main action of the muscles that move the pectoral girdle?

Exhibit 11.12 | Muscles that Move the Humerus (Arm) (Figure 11.15)

OBJECTIVE

• *Describe the origin, insertion, action, and innervation of the muscles that move the humerus.*

Of the nine muscles that cross the shoulder joint, only two of them (pectoralis major and latissimus dorsi) do not originate on the scapula. These two muscles are thus called **axial muscles,** because they originate on the axial skeleton. The remaining seven muscles, the **scapular muscles,** arise from the scapula.

Of the two axial muscles that move the humerus, the **pectoralis major** is a large, thick, fan-shaped muscle that covers the superior part of the thorax. It has two origins: a smaller clavicular head and a larger sternocostal head. The **latissimus dorsi** is a broad, triangular muscle located on the inferior part of the back. It is commonly called the "swimmer's muscle" because its many actions are used while swimming.

Among the scapular muscles, the **deltoid** is a thick, powerful shoulder muscle that covers the shoulder joint and forms the rounded contour of the shoulder. This muscle is a frequent site of intramuscular injections. As you study the deltoid, note that its fibers originate from three different points and that each group of fibers moves the humerus differently. The **subscapularis** is a large triangular muscle that fills the subscapular fossa of the scapula and forms part of the posterior wall of the axilla. The **supraspinatus** is a rounded muscle, named for its location in the supraspinous fossa of the scapula. It lies deep to the trapezius. The **infraspinatus** is a triangular muscle, also named for its location in the infraspinous fossa of the scapula. The **teres major** is a thick, flattened muscle inferior to the teres minor that also helps form part of the posterior wall of the axilla. The **teres minor** is a cylindrical, elongated muscle, often inseparable from the infraspinatus, which lies along its superior border. The **coracobrachialis** is an elongated, narrow muscle in the arm.

The strength and stability of the shoulder joint are not provided by the shape of the articulating bones or its ligaments. Instead, four deep muscles of the shoulder—subscapularis, supraspinatus, infraspinatus, and teres minor—strengthen and stabilize the shoulder joint. These muscles join the scapula to the humerus. Their flat tendons fuse together to form a nearly complete circle around the shoulder joint, like the cuff on a shirt sleeve. This arrangement is referred to as the **rotator (musculotendinous) cuff.** The supraspinatus muscle is especially predisposed to wear and tear because of its location between the head of the humerus and acromion of the scapula, which compress its tendon during shoulder movements.

INNERVATION

Muscles that move the arm are innervated by nerves that emerge from the brachial plexus (shown in Exhibit 13.2 on page 431). More specifically, the pectoralis major is innervated by the medial and lateral pectoral nerves; the latissimus dorsi is innervated by the thoracodorsal nerve; the deltoid and teres minor are innervated by the axillary nerve; the subscapularis is innervated by the upper and lower subscapular nerves; the supraspinatus and infraspinatus are innervated by the suprascapular nerve; the teres major is innervated by the lower subscapular nerve; and the coracobrachialis is innervated by the musculocutaneous nerve.

CLINICAL APPLICATION
Impingement Syndrome

One of the most common causes of shoulder pain and dysfunction in athletes is known as **impingement syndrome.** The repetitive overhead motion that is common in baseball, overhead racquet sports, spiking a volleyball, and swimming puts these athletes at risk for developing this syndrome. Continual pinching of the supraspinatus tendon causes it to become inflamed and results in pain. If the condition persists, the tendon may degenerate near the attachment to the humerus and ultimately may tear away from the bone (rotator cuff injury). ■

RELATING MUSCLES TO MOVEMENTS

Arrange the muscles in this exhibit according to the following actions on the humerus at the shoulder joint: (1) flexion, (2) extension, (3) abduction, (4) adduction, (5) medial rotation, and (6) lateral rotation. The same muscle may be mentioned more than once.

Why are two muscles that cross the shoulder joint called axial muscles, and the seven others called scapular muscles?

Exhibit 11.12 *Muscles that Move the Humerus (Arm) (continued)*

MUSCLE	ORIGIN	INSERTION	ACTION
Axial Muscles that Move the Humerus			
Pectoralis major (pek'-tor-A-lis) (see also Figure 11.10a)	Clavicle (clavicular head), sternum, and costal cartilages of second to sixth ribs (sometimes first to seventh ribs).	Greater tubercle and intertubercular sulcus of humerus.	As a whole, adducts and medially rotates arm at shoulder joint; clavicular head alone flexes arm, and sternocostal head alone extends arm at shoulder joint.
Latissimus dorsi (la-TIS-i-mus DOR-sī; *latissimus* = widest; *dorsum* = back)	Spines of inferior six thoracic vertebrae, lumbar vertebrae, crests of sacrum and ilium, inferior four ribs.	Intertubercular sulcus of humerus.	Extends, adducts, and medially rotates arm at shoulder joint; draws arm inferiorly and posteriorly.
Scapular Muscles that Move the Humerus			
Deltoid (DEL-toyd; *deltoides* = triangular)	Acromial extremity of clavicle (anterior fibers), acromion of scapula (lateral fibers), and spine of scapula (posterior fibers).	Deltoid tuberosity of humerus.	Lateral fibers abduct arm at shoulder joint; anterior fibers flex and medially rotate arm at shoulder joint; posterior fibers extend and laterally rotate arm at shoulder joint.
Subscapularis (sub-scap'-yoo-LA-ris; *sub* = below; *scapularis* = scapula)	Subscapular fossa of scapula.	Lesser tubercle of humerus.	Medially rotates arm at shoulder joint.
Supraspinatus (soo'-pra-spi-NĀ-tus; *supra* = above; *spinatus* = spine of scapula)	Supraspinous fossa of scapula.	Greater tubercle of humerus.	Assists deltoid muscle in abducting arm at shoulder joint.
Infraspinatus (in'-fra-spi-NĀ-tus; *infra* = below)	Infraspinous fossa of scapula.	Greater tubercle of humerus.	Laterally rotates and adducts arm at shoulder joint.
Teres major (TE-rēz; *teres* = long and round)	Inferior angle of scapula.	Intertubercular sulcus of humerus.	Extends arm at shoulder joint and assists in adduction and medial rotation of arm at shoulder joint.
Teres minor (TE-rēz)	Inferior lateral border of scapula.	Greater tubercle of humerus.	Laterally rotates, extends, and adducts arm at shoulder joint.
Coracobrachialis (kor'-a-kō-BRĀ-kē-a'-lis; *coraco* = coracoid process)	Coracoid process of scapula.	Middle of medial surface of shaft of humerus.	Flexes and adducts arm at shoulder joint.

Exhibit 11.12 *(continued)*

Figure 11.15 Muscles that move the humerus (arm). (See Tortora, *A Photographic Atlas of the Human Body,* Figures 5.9 and 5.10)

🗝 The strength and stability of the shoulder joint are provided by the tendons that form the rotator cuff.

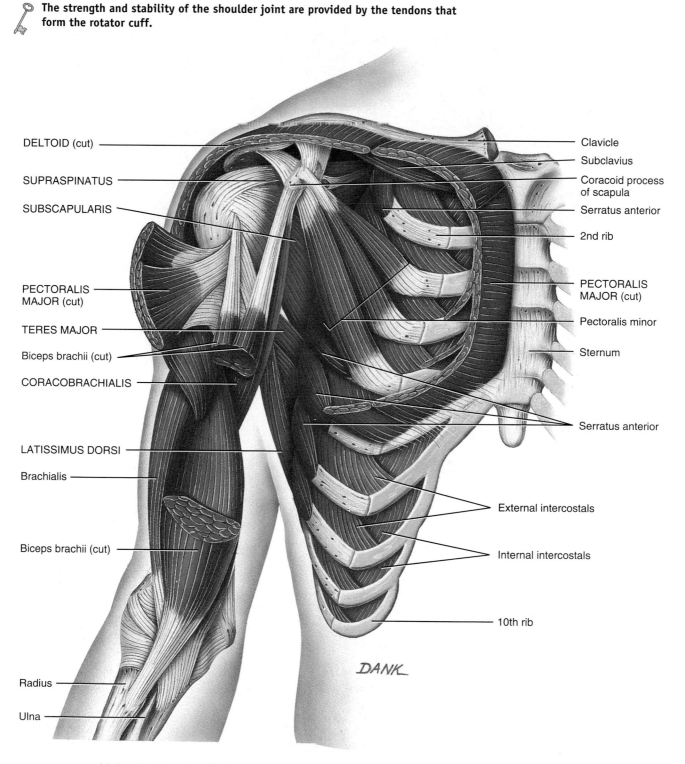

DELTOID (cut)
Clavicle
Subclavius
SUPRASPINATUS
Coracoid process of scapula
SUBSCAPULARIS
Serratus anterior
2nd rib
PECTORALIS MAJOR (cut)
PECTORALIS MAJOR (cut)
TERES MAJOR
Pectoralis minor
Biceps brachii (cut)
Sternum
CORACOBRACHIALIS
Serratus anterior
LATISSIMUS DORSI
Brachialis
External intercostals
Biceps brachii (cut)
Internal intercostals
10th rib
Radius
Ulna

DANK

(a) Anterior deep view (the intact pectoralis major muscle is shown in figure 11.10a)

Figure continues

Exhibit 11.12 *Muscles that Move the Humerus (Arm) (continued)*

Figure 11.15 (continued)

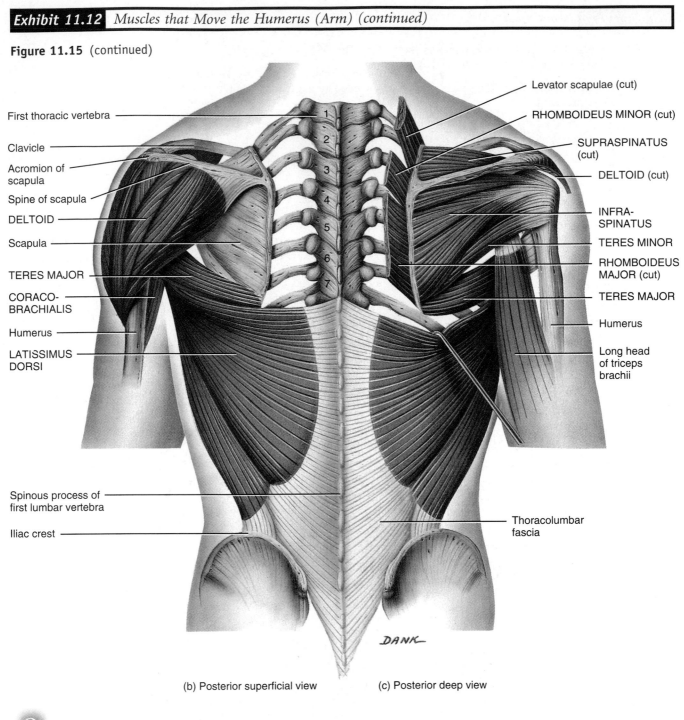

Levator scapulae (cut)

RHOMBOIDEUS MINOR (cut)

SUPRASPINATUS (cut)

DELTOID (cut)

INFRA-SPINATUS

TERES MINOR

RHOMBOIDEUS MAJOR (cut)

TERES MAJOR

Humerus

Long head of triceps brachii

Thoracolumbar fascia

First thoracic vertebra

Clavicle

Acromion of scapula

Spine of scapula

DELTOID

Scapula

TERES MAJOR

CORACO-BRACHIALIS

Humerus

LATISSIMUS DORSI

Spinous process of first lumbar vertebra

Iliac crest

(b) Posterior superficial view

(c) Posterior deep view

Q Which tendons make up the rotator cuff?

Exhibit 11.13 *Muscles that Move the Radius and Ulna (Forearm) (Figure 11.16)*

OBJECTIVE
• *Describe the origin, insertion, action, and innervation of the muscles that move the radius and ulna.*

Most of the muscles that move the radius and ulna (forearm) are involved in flexion and extension at the elbow, which is a hinge joint. The biceps brachii, brachialis, and brachioradialis muscles are the flexor muscles. The extensor muscles are the triceps brachii and the anconeus.

The **biceps brachii** is the large muscle located on the anterior surface of the arm. As indicated by its name, it has two heads of origin (long and short), both from the scapula. The muscle spans both the shoulder and elbow joints. In addition to its role in flexing the forearm at the elbow joint, it also supinates the forearm at the radioulnar joints and flexes the arm at the shoulder joint. The **brachialis** is deep to the biceps brachii muscle. It is the most powerful flexor of the forearm at the elbow joint. For this reason, it is called the "workhorse" of the elbow flexors. The **brachioradialis** flexes the forearm at the elbow joint, especially when a quick movement is required or when a weight is lifted slowly during flexion of the forearm.

The **triceps brachii** is the large muscle located on the posterior surface of the arm. It is the more powerful of the extensors of the forearm at the elbow joint. As its name implies, it has three heads of origin, one from the scapula (long head) and two from the humerus (lateral and medial heads). The long head crosses the shoulder joint; the other heads do not. The **anconeus** is a small muscle located on the lateral part of the posterior aspect of the elbow that assists the triceps brachii in extending the forearm at the elbow joint.

Some muscles that move the radius and ulna are involved in pronation and supination at the radioulnar joints. The pronators,

as suggested by their names, are the **pronator teres** and **pronator quadratus** muscles. The supinator of the forearm is the aptly named **supinator** muscle. You use the powerful action of the supinator when you twist a corkscrew or turn a screw with a screwdriver.

In the limbs, functionally related skeletal muscles and their associated blood vessels and nerves are grouped together by fascia into regions called **compartments.** In the arm, the biceps brachii, brachialis, and coracobrachialis muscles constitute the *flexor compartment;* the triceps brachii muscle forms the *extensor compartment.*

INNERVATION
Muscles that move the radius and ulna are innervated by nerves derived from the brachial plexus (shown in Exhibit 13.2 on page 431). More specifically, the biceps brachii is innervated by the musculocutaneous nerve; the brachialis is innervated by the musculocutaneous and radial nerves; the brachioradialis, triceps brachii, and anconeus are innervated by the radial nerve; the pronator teres and pronator quadratus are innervated by the median nerve; and the supinator is innervated by the deep radial nerve.

RELATING MUSCLES TO MOVEMENTS
Arrange the muscles in this exhibit according to the following actions on the elbow joint: (1) flexion and (2) extension; the following actions on the forearm at the radioulnar joints: (1) supination and (2) pronation; and the following actions on the humerus at the shoulder joint: (1) flexion and (2) extension. The same muscle may be mentioned more than once.

Which muscles are in the anterior and posterior compartments of the arm?

MUSCLE	ORIGIN	INSERTION	ACTION
Forearm Flexors			
Biceps brachii (BĪ-ceps BRĀ-kē-ī; *biceps* = two heads of origin; *brachion* = arm)	*Long head* originates from tubercle above glenoid cavity of scapula (supraglenoid tubercle); *short head* originates from coracoid process of scapula.	Radial tuberosity and bicipital aponeurosis.*	Flexes forearm at elbow joint, supinates forearm at radioulnar joints, and flexes arm at shoulder joint.
Brachialis (brā-kē-A-lis)	Distal, anterior surface of humerus.	Ulnar tuberosity and coronoid process of ulna.	Flexes forearm at elbow joint.
Brachioradialis (bra'-kē-ō-rā-dē-A-lis; *radialis* = radius) (see Figure 11.17a)	Medial and lateral borders of distal end of humerus.	Superior to styloid process of radius.	Flexes forearm at elbow joint and supinates and pronates forearm at radioulnar joints to neutral position.

---

*The bicipital aponeurosis is a broad aponeurosis from the tendon of insertion of the biceps brachii muscle that descends medially across the brachial artery and fuses with deep fascia over the forearm flexor muscles.

Exhibit 11.13 *Muscles that Move the Radius and Ulna (Forearm) (continued)*

MUSCLE	ORIGIN	INSERTION	ACTION
Forearm Extensors			
Triceps brachii (TRĪ-ceps BRĀ-kē-ī; *triceps* = three heads of origin)	*Long head* originates from a projection inferior to glenoid cavity (infraglenoid tubercle) of scapula; *lateral head* originates from lateral and posterior surface of humerus superior to radial groove; *medial head* originates from entire posterior surface of humerus inferior to a groove for the radial nerve.	Olecranon of ulna.	Extends forearm at elbow joint and extends arm at shoulder joint.
Anconeus (an-KŌ-nē-us; *anconeal* = pertaining to elbow) (see Figure 11.17c)	Lateral epicondyle of humerus.	Olecranon and superior portion of shaft of ulna.	Extends forearm at elbow joint.
Forearm Pronators			
Pronator teres (PRŌ-nā-tor TE-rēz; *pronation* = turning palm downward or posteriorly) (see Figure 11.17a)	Medial epicondyle of humerus and coronoid process of ulna.	Midlateral surface of radius.	Pronates forearm at radioulnar joints and weakly flexes forearm at elbow joint.
Pronator quadratus (PRŌ-nā-tor kwod-RĀ-tus; *quadratus* = squared, four-sided) (see Figure 11.17a)	Distal portion of shaft of ulna.	Distal portion of shaft of radius.	Pronates forearm at radioulnar joints.
Forearm Supinator			
Supinator (SOO-pi-nā-tor; *supination* = turning palm upward or anteriorly) (see Figure 11.17b)	Lateral epicondyle of humerus and ridge near radial notch of ulna (supinator crest).	Lateral surface of proximal one-third of radius.	Supinates forearm at radioulnar joints.

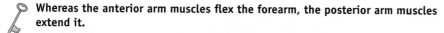

Exhibit 11.13 (continued)

Figure 11.16 Muscles that move the radius and ulna (forearm). (See Tortora, *A Photographic Atlas of the Human Body*, Figure 5.11)

Whereas the anterior arm muscles flex the forearm, the posterior arm muscles extend it.

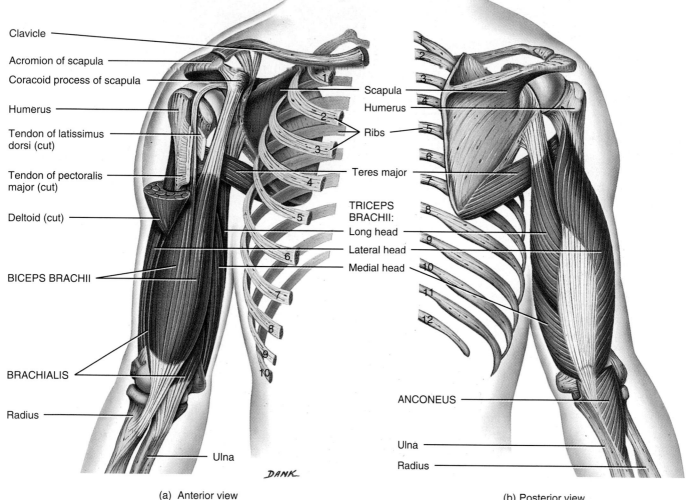

(a) Anterior view

(b) Posterior view

DANK

Figure continues

Exhibit 11.13 *Muscles that Move the Radius and Ulna (Forearm) (continued)*

Figure 11.16 (continued)

View

Transverse plane

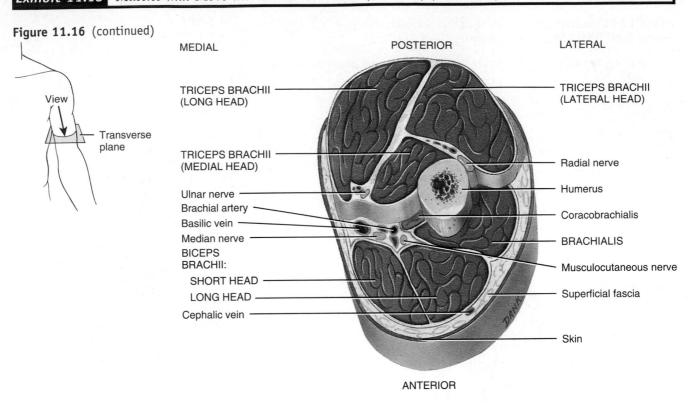

MEDIAL

TRICEPS BRACHII (LONG HEAD)

TRICEPS BRACHII (MEDIAL HEAD)

Ulnar nerve

Brachial artery

Basilic vein

Median nerve

BICEPS BRACHII:

SHORT HEAD

LONG HEAD

Cephalic vein

POSTERIOR

LATERAL

TRICEPS BRACHII (LATERAL HEAD)

Radial nerve

Humerus

Coracobrachialis

BRACHIALIS

Musculocutaneous nerve

Superficial fascia

Skin

ANTERIOR

(c) Superior view of transverse section of arm

Q Which muscles are the most powerful flexor and the most powerful extensor of the forearm?

Exhibit 11.14 | *Muscles that Move the Wrist, Hand, and Digits (Figure 11.17)*

OBJECTIVE

• *Describe the origin, insertion, action, and innervation of the muscles that move the wrist, hand, and digits.*

Muscles that move the wrist, hand, and digits are many and varied. Those in this group that act on the digits are known as **extrinsic muscles** because they originate *outside* the hand and insert within it. As you will see, the names for the muscles that move the wrist, hand, and digits give some indication of their origin, insertion, or action. On the basis of location and function, the muscles are divided into two groups: (1) anterior compartment muscles and (2) posterior compartment muscles. The **anterior compartment** muscles originate on the humerus; typically insert on the carpals, metacarpals, and phalanges; and function as flexors. The bellies of these muscles form the bulk of the forearm. The **posterior compartment** muscles originate on the humerus; insert on the metacarpals and phalanges; and function as extensors. Within each compartment, the muscles are grouped as superficial and deep.

The **superficial anterior compartment** muscles are arranged in the following order from lateral to medial: **flexor carpi radialis**, **palmaris longus** (absent in about 10% of the population), and **flexor carpi ulnaris** (the ulnar nerve and artery are just lateral to the tendon of this muscle at the wrist). The **flexor digitorum superficialis** muscle is actually deep to the other three muscles and is the largest superficial muscle in the forearm.

The **deep anterior compartment** muscles are arranged in the following order from lateral to medial: **flexor pollicis longus** (the only flexor of the distal phalanx of the thumb) and **flexor digitorum profundus** (ends in four tendons that insert into the distal phalanges of the fingers).

The **superficial posterior compartment** muscles are arranged in the following order from lateral to medial: **extensor carpi radialis longus**, **extensor carpi radialis brevis**, **extensor digitorum** (occupies most of the posterior surface of the forearm and divides into four tendons that insert into the middle and distal phalanges of the fingers), **extensor digiti minimi** (a slender muscle generally connected to the extensor digitorum), and the **extensor carpi ulnaris.**

The **deep posterior compartment** muscles are arranged in the following order from lateral to medial: **abductor pollicis longus**, **extensor pollicis brevis**, **extensor pollicis longus**, and **extensor indicis.**

The tendons of the muscles of the forearm that attach to the wrist or continue into the hand, along with blood vessels and nerves, are held close to bones by strong fascial structures. The tendons are also surrounded by tendon sheaths. At the wrist, the deep fascia is thickened into fibrous bands called **retinacula** (*retinere* = retain). The **flexor retinaculum** is located over the palmar surface of the carpal bones. Through it pass the long flexor tendons of the digits and wrist and the median nerve. The **extensor retinaculum** is located over the dorsal surface of the carpal bones and through it pass the extensor tendons of the wrist and digits.

INNERVATION

Muscles that move the wrist, hand, and digits are innervated by nerves derived from the brachial plexus (shown in Exhibit 13.2 on page 431). More specifically, the flexor carpi radialis, palmaris longus, flexor digitorum superficialis, and flexor pollicis longus are innervated by the median nerve; the flexor carpi ulnaris is innervated by the ulnar nerve; the flexor digitorum profundus is innervated by the median and ulnar nerves; the extensor carpi radialis longus, extensor carpi radialis brevis, and extensor digitorum are innervated by the radial nerve; and the extensor digiti minimi, extensor carpi ulnaris, and all of the deep muscles of the posterior compartment are innervated by the deep radial nerve.

RELATING MUSCLES TO MOVEMENTS

Arrange the muscles in this exhibit according to the following actions on the wrist joint: (1) flexion, (2) extension, (3) abduction, and (4) adduction; the following actions on the fingers at the metacarpophalangeal joints: (1) flexion and (2) extension; the following actions on the fingers at the interphalangeal joints: (1) flexion and (2) extension; the following actions on the thumb at the carpometacarpal, metacarpophalangeal, and interphalangeal joints: (1) extension and (2) abduction; and the following action on the thumb at the interphalangeal joint: flexion. The same muscle may be mentioned more than once.

Which muscles and actions of the wrist, hand, and digits are used when writing?

Exhibit 11.14	*Muscles that Move the Wrist, Hand, and Digits (continued)*

MUSCLE	ORIGIN	INSERTION	ACTION
Superficial Anterior Compartment (Flexors)			
Flexor carpi radialis (FLEK-sor KAR-pē rā′-dē-A-lis; *flexor* = decreases angle at joint; *carpus* = wrist; *radialis* = radius)	Medial epicondyle of humerus.	Second and third metacarpals.	Flexes and abducts hand (radial deviation) at wrist joint.
Palmaris longus (pal-MA-ris LON-gus; *palma* = palm; *longus* = long)	Medial epicondyle of humerus.	Flexor retinaculum and palmar aponeurosis (deep fascia in center of palm).	Weakly flexes hand at wrist joint.
Flexor carpi ulnaris (FLEK-sor KAR-pē ul-NAR-is; *ulnaris* = ulna)	Medial epicondyle of humerus and superior posterior border of ulna.	Pisiform, hamate, and base of fifth metacarpal.	Flexes and adducts hand (ulnar deviation) at wrist joint.
Flexor digitorum superficialis (FLEK-sor di′-ji-TOR-um soo′-per-fish′-ē-A-lis; *digit* = finger or toe; *superficialis* = closer to surface)	Medial epicondyle of humerus, coronoid process of ulna, and a ridge along lateral margin of anterior surface (anterior oblique line) of radius.	Middle phalanges of each finger.*	Flexes middle phalanx of each finger at proximal interphalangeal joint, proximal phalanx of each finger at metacarpophalangeal joint, and hand at wrist joint.
Deep Anterior Compartment (Flexors)			
Flexor pollicis longus (FLEK-sor POL-li-kis LON-gus; *pollex* = thumb)	Anterior surface of radius and interosseous membrane (sheet of fibrous tissue that holds shafts of ulna and radius together).	Base of distal phalanx of thumb.	Flexes distal phalanx of thumb at interphalangeal joint.
Flexor digitorum profundus (FLEK-sor di′-ji-TOR-um prō-FUN-dus; *profundus* = deep)	Anterior medial surface of body of ulna.	Bases of distal phalanges of each finger.	Flexes distal and middle phalanges of each finger at interphalangeal joints, proximal phalanx of each finger at metacarpophalangeal joint, and hand at wrist joint.

---

*Reminder: The thumb or pollex is the first digit and has two phalanges: proximal and distal. The remaining digits, the fingers, are numbered 2–5, and each has three phalanges: proximal, middle, and distal.

Exhibit 11.14   *(continued)*

MUSCLE	ORIGIN	INSERTION	ACTION
Superficial Posterior Compartment (Extensors)			
Extensor carpi radialis longus (eks-TEN-sor KAR-pē rā'-dē-A-lis LON-gus; *extensor* = increases angle at joint)	Lateral supracondylar ridge of humerus.	Second metacarpal.	Extends and abducts hand at wrist joint.
Extensor carpi radialis brevis (eks-TEN-sor KAR-pē rā'-dē-A-lis BREV-is; *brevis* = short)	Lateral epicondyle of humerus.	Third metacarpal.	Extends and abducts hand at wrist joint.
Extensor digitorum (eks-TEN-sor di'-ji-TOR um)	Lateral epicondyle of humerus.	Distal and middle phalanges of each finger.	Extends distal and middle phalanges of each finger at interphalangeal joints, proximal phalanx of each finger at metacarpophalangeal joint, and hand at wrist joint.
Extensor digiti minimi (eks-TEN-sor DIJ-i-tē MIN-i-mē; *minimi* = little finger)	Lateral epicondyle of humerus.	Tendon of extensor digitorum on fifth phalanx.	Extends proximal phalanx of little finger at metacarpophalangeal joint and hand at wrist joint.
Extensor carpi ulnaris (eks-TEN-sor KAR-pē ul-NAR-is)	Lateral epicondyle of humerus and posterior border of ulna.	Fifth metacarpal.	Extends and adducts hand at wrist joint.
Deep Posterior Compartment (Extensors)			
Abductor pollicis longus (ab-DUK-tor POL-li-kis LON-gus; *abductor* = moves part away from midline)	Posterior surface of middle of radius and ulna and interosseous membrane.	First metacarpal.	Abducts and extends thumb at carpometacarpal joint and abducts hand at wrist joint.
Extensor pollicis brevis (eks-TEN-sor POL-li-kis BREV-is)	Posterior surface of middle of radius and interosseous membrane.	Base of proximal phalanx of thumb.	Extends proximal phalanx of thumb at metacarpophalangeal joint, first metacarpal of thumb at carpometacarpal joint, and hand at wrist joint.
Extensor pollicis longus (eks-TEN-sor POL-li-kis LON-gus)	Posterior surface of middle of ulna and interosseous membrane.	Base of distal phalanx of thumb.	Extends distal phalanx of thumb at interphalangeal joint, first metacarpal of thumb at carpometacarpal joint, and abducts hand at wrist joint.
Extensor indicis (eks-TEN-sor IN-di-kis; *indicis* = index)	Posterior surface of ulna.	Tendon of extensor digitorum of index finger.	Extends distal and middle phalanges of index finger at interphalangeal joints, proximal phalanx of index finger at metacarpophalangeal joint, and hand at wrist joint.

Exhibit 11.14 *Muscles that Move the Wrist, Hand, and Digits (continued)*

Figure 11.17 Muscles that move the wrist, hand, and digits. (See Tortora, *A Photographic Atlas of the Human Body,* Figures 5.12 and 5.13)

The anterior compartment muscles function as flexors, and the posterior compartment muscles function as extensors.

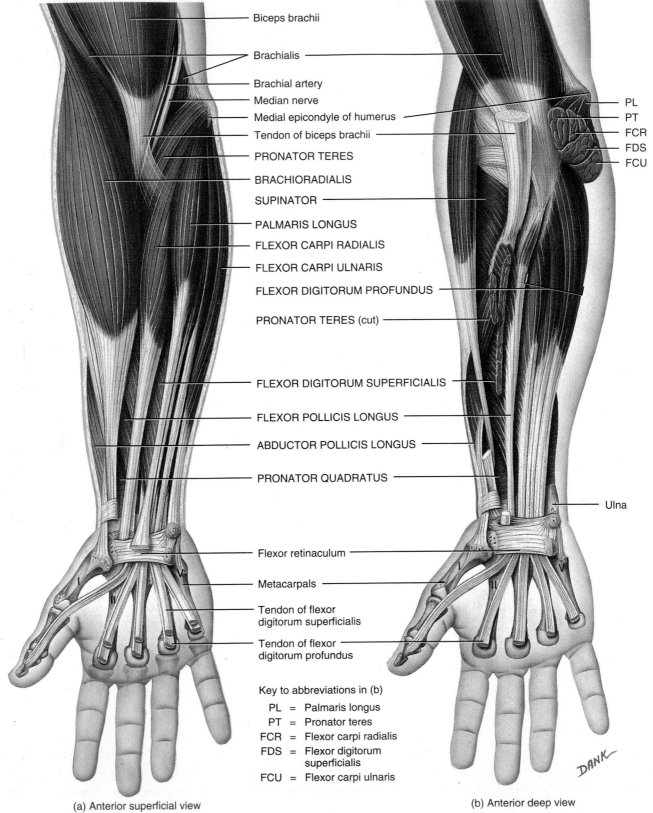

Biceps brachii

Brachialis

Brachial artery

Median nerve

Medial epicondyle of humerus

Tendon of biceps brachii

PRONATOR TERES

BRACHIORADIALIS

SUPINATOR

PALMARIS LONGUS

FLEXOR CARPI RADIALIS

FLEXOR CARPI ULNARIS

FLEXOR DIGITORUM PROFUNDUS

PRONATOR TERES (cut)

FLEXOR DIGITORUM SUPERFICIALIS

FLEXOR POLLICIS LONGUS

ABDUCTOR POLLICIS LONGUS

PRONATOR QUADRATUS

Flexor retinaculum

Metacarpals

Tendon of flexor digitorum superficialis

Tendon of flexor digitorum profundus

PL
PT
FCR
FDS
FCU

Ulna

Key to abbreviations in (b)

PL = Palmaris longus
PT = Pronator teres
FCR = Flexor carpi radialis
FDS = Flexor digitorum superficialis
FCU = Flexor carpi ulnaris

(a) Anterior superficial view

(b) Anterior deep view

348

Exhibit 11.14 (continued)

Triceps brachii

BRACHIORADIALIS

EXTENSOR CARPI RADIALIS LONGUS

Humerus

Medial epicondyle of humerus

Lateral epicondyle of humerus

Olecranon of ulna

ANCONEUS

EXTENSOR CARPI ULNARIS

EXTENSOR DIGITORUM

EXTENSOR CARPI RADIALIS BREVIS

EXTENSOR DIGITI MINIMI

FLEXOR CARPI ULNARIS

FLEXOR DIGITORUM PROFUNDUS

ABDUCTOR POLLICIS LONGUS

EXTENSOR POLLICIS BREVIS

Tendon of extensor carpi ulnaris

Extensor retinaculum

Tendon of extensor digiti minimi

Tendons of extensor digitorum

SUPINATOR

Tendon of pronator teres

EXTENSOR POLLICIS LONGUS

EXTENSOR INDICIS

Carpals

Tendon of extensor indicis

Dorsal interossei

DANK

(c) Posterior superficial view

(d) Posterior deep view

Q What structures pass through the flexor retinaculum?

Exhibit 11.15	*Intrinsic Muscles of the Hand (Figure 11.18)*

OBJECTIVE

- *Describe the origin, insertion, action, and innervation of the intrinsic muscles of the hand.*

Several of the muscles discussed in Exhibit 11.14 move the digits in various ways and are known as extrinsic muscles. They produce the powerful but crude movements of the digits. The **intrinsic muscles** in the palm produce weak but intricate and precise movements of the digits that characterize the human hand. The muscles in this group are so named because their origins and insertions are *within* the hand.

The intrinsic muscles of the hand are divided into three groups: (1) **thenar**, (2) **hypothenar**, and (3) **intermediate**. The four thenar muscles act on the thumb and form the **thenar eminence**, the lateral rounded contour on the palm. The thenar muscles include the **abductor pollicis brevis, opponens pollicis, flexor pollicis brevis**, and **adductor pollicis.**

The three hypothenar muscles act on the little finger and form the **hypothenar eminence**, the rounded contour on the medial aspect of the palm. The hypothenar muscles are the **abductor digiti minimi, flexor digiti minimi brevis**, and **opponens digiti minimi.**

The 12 intermediate (midpalmar) muscles act on all the digits except the thumb. The intermediate muscles include the **lumbricals, palmar interossei**, and **dorsal interossei.** Both sets of interossei muscles are located between the metacarpals and are important in abduction, adduction, flexion, and extension of the fingers, and in movements in skilled activities such as writing, typing, and playing a piano.

The functional importance of the hand is readily apparent when one considers that certain hand injuries can result in permanent disability. Most of the dexterity of the hand depends on movements of the thumb. The general activities of the hand are free motion, power grip (forcible movement of the fingers and thumb against the palm, as in squeezing), precision handling (a change in position of a handled object that requires exact control of finger and thumb positions, as in winding a watch or threading a needle), and pinch (compression between the thumb and index finger or between the thumb and first two fingers).

Movements of the thumb are very important in the precise activities of the hand, and they are defined in different planes from comparable movements of other digits because the thumb is positioned at a right angle to the other digits. The five principal movements of the thumb are illustrated in Figure 11.18c and include *flexion* (movement of the thumb medially across the palm), *extension* (movement of the thumb laterally away from the palm), *abduction* (movement of the thumb in an anteroposterior plane away from the palm), *adduction* (movement of the thumb in an anteroposterior plane toward the palm), and *opposition* (movement of the thumb across the palm so that the tip of the thumb meets the tip of a finger). Opposition is the single most distinctive digital movement that gives humans and other primates the ability to precisely grasp and manipulate objects.

INNERVATION

Intrinsic muscles of the hand are innervated by nerves derived from the brachial plexus (shown in Exhibit 13.2 on page 431). Move specifically, the abductor pollicis brevis and opponens pollicis are innervated by the median nerve. The adductor pollicis, abductor digiti minimi, flexor digiti minimi brevis, opponens digiti minimi, dorsal interossei, and palmar interossei are innervated by the ulnar nerve. The flexor pollicis brevis and lumbricals are innervated by the median and ulnar nerves.

CLINICAL APPLICATION
Carpal Tunnel Syndrome

The **carpal tunnel** is a narrow passageway formed anteriorly by the flexor retinaculum and posteriorly by the carpal bones. Through this tunnel pass the median nerve, the most superficial structure, and the long flexor tendons for the digits. Structures within the carpal tunnel, especially the median nerve, are vulnerable to compression, and the resulting condition is called **carpal tunnel syndrome.** Compression of the median nerve leads to sensory changes over the lateral side of the hand and muscle weakness in the thenar eminence. This results in pain, numbness, and tingling of the fingers. The condition may be caused by inflammation of the digital tendon sheaths, fluid retention, excessive exercise, infection, trauma, and repetitive activities that involve flexion of the wrist, such as keyboarding, cutting hair, and playing a piano. ■

RELATING MUSCLES TO MOVEMENTS

Arrange the muscles in this exhibit according to the following actions on the thumb at the carpometacarpal and metaphalangeal joints: (1) abduction, (2) adduction, (3) flexion, and (4) opposition; and the following actions on the fingers at the metacarpophalangeal and interphalangeal joints: (1) abduction, (2) adduction, (3) flexion, and (4) extension. The same muscle may be mentioned more than once.

Compare the actions of the extrinsic and intrinsic muscles of the hand.

Exhibit 11.15 (continued)

MUSCLE	ORIGIN	INSERTION	ACTION
Thenar			
Abductor pollicis brevis (*abductor* = moves part away from middle; *pollex* = thumb; *brevis* = short)	Flexor retinaculum, scaphoid, and trapezium.	Lateral side of proximal phalanx of thumb.	Abducts thumb at carpometacarpal joint.
Opponens pollicis (*opponens* = opposes)	Flexor retinaculum and trapezium.	Lateral side of first metacarpal (thumb).	Moves thumb across palm to meet little finger (opposition) at the carpometacarpal joint.
Flexor pollicis brevis (*flexor* = decreases angle at joint)	Flexor retinaculum, trapezium, capitate, and trapezoid.	Lateral side of proximal phalanx of thumb.	Flexes thumb at carpometacarpal and metacarpophalangeal joints.
Adductor pollicis (*adductor* = moves part toward midline)	Oblique head; capitate and second and third metacarpals; transverse head: third metacarpal.	Medial side of proximal phalanx of thumb by a tendon containing a sesamoid bone.	Adducts thumb at carpometacarpal and metacarpophalangeal joints.
Hypothenar			
Abductor digiti minimi (*digit* = finger or toe; *minimi* = little)	Pisiform and tendon of flexor carpi ulnaris.	Medial side of proximal phalanx of little finger.	Abducts and flexes little finger at metacarpophalangeal joint.
Flexor digiti minimi brevis	Flexor retinaculum and hamate.	Medial side of proximal phalanx of little finger.	Flexes little finger at carpometacarpal and metacarpophalangeal joints.
Opponens digiti minimi	Flexor retinaculum and hamate.	Medial side of fifth metacarpal (little finger).	Moves little finger across palm to meet thumb (opposition) at the carpometacarpal joint.
Intermediate (Midpalmar)			
Lumbricals (LUM-bri-kals; *lumbricus* = earthworm) (four muscles)	Lateral sides of tendons and flexor digitorum profundus of each finger.	Lateral sides of tendons of extensor digitorum on proximal phalanges of each finger.	Flex each finger at metacarpophalangeal joints and extend each finger at interphalangeal joints.
Palmar interossei (in'-ter-OS-ē-ī (*palmar* = palm; *inter* = between; *ossei* = bones) (four muscles)	Sides of shafts of metacarpals of all digits (except the middle one)	Sides of bases of proximal phalanges of all digits (except the middle one).	Adduct each finger at metacarpophalangeal joints; flex each finger at metacarpophalangeal joints.
Dorsal interossei (in'-ter-OS-ē-ī; *dorsal* = back surface) (four muscles)	Adjacent sides of metacarpals.	Proximal phalanx of each finger.	Abduct fingers 2–4 at metacarpophalangeal joints; flex fingers 2–4 at metacarpophalangeal joints; and extend each finger at interphalangeal joints.

Exhibit 11.15 *Intrinsic Muscles of the Hand (continued)*

Figure 11.18 Intrinsic muscles of the hand. (See Tortora, *A Photographic Atlas of the Human Body,* Figures 5.14 and 5.15)

The intrinsic muscles of the hand produce the intricate and precise movements of the digits that characterize the human hand.

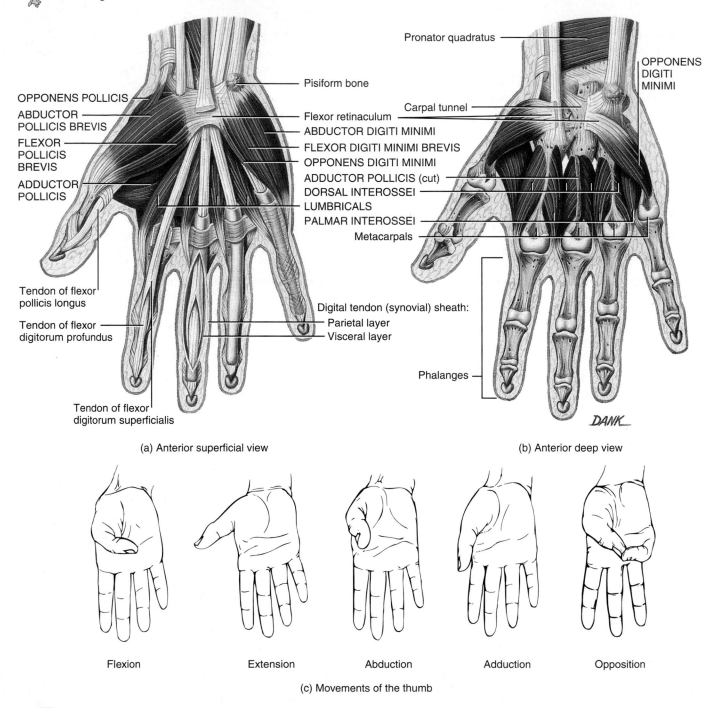

(a) Anterior superficial view

(b) Anterior deep view

(c) Movements of the thumb

Q Muscles of the thenar eminence act on which digit?

| **Exhibit 11.16** | Muscles that Move the Vertebral Column (Backbone) (Figure 11.19) |

OBJECTIVE

• *Describe the origin, insertion, action, and innervation of the muscles that move the vertebral column.*

The muscles that move the vertebral column (backbone) are quite complex because they have multiple origins and insertions and there is considerable overlap among them. One way to group the muscles is on the basis of the general direction of the muscle bundles and their approximate lengths. For example, the splenius muscles arise from the midline and extend laterally and superiorly to their insertions. The erector spinae (sacrospinalis) muscle arises from either the midline or more laterally but usually runs almost longitudinally, with neither a significant lateral nor medial direction as it is traced superiorly. The transversospinalis muscles arise laterally but extend toward the midline as they are traced superiorly. Deep to these three muscle groups are small segmental muscles that extend between spinous processes or transverse processes of vertebrae. Because the scalene muscles also assist in moving the vertebral column, they are included in this exhibit. Note in Exhibit 11.7 that the rectus abdominis, external oblique, internal oblique, and quadratus lumborum muscles also play a role in moving the vertebral column.

The bandage-like **splenius** muscles are attached to the sides and back of the neck. The two muscles in this group are named on the basis of their superior attachments (insertions): **splenius capitis** (head region) and **splenius cervicis** (cervical region). They extend the head and laterally flex and rotate the head.

The **erector spinae** is the largest muscle mass of the back, forming a prominent bulge on either side of the vertebral column. It is the chief extensor of the vertebral column. It is also important in controlling flexion, lateral flexion, and rotation of the vertebral column and in maintaining the lumbar curve, because the main mass of the muscle is in the lumbar region. It consists of three groups: iliocostalis (lateral), longissimus (intermediate), and spinalis (medial). These groups, in turn, consist of a series of overlapping muscles, and the muscles within the groups are named according to the regions of the body with which they are associated. The **iliocostalis group** consists of three muscles: the **iliocostalis cervicis, iliocostalis thoracis,** and **iliocostalis lumborum**. The **longissimus group** resembles a herring bone and consists of three muscles: the **longissimus capitis, longissimus cervicis,** and **longissimus thoracis**. The **spinalis group** also consists of three muscles: the **spinalis capitis, spinalis cervicis,** and **spinalis thoracis**.

The **transversospinalis** muscles are named because their fibers run from the transverse processes to the spinous processes of the vertebrae. The semispinalis muscles in this group are also named according to the region of the body with which they are as-

sociated: **semispinalis capitis, semispinalis cervicis,** and **semispinalis thoracis**. These muscles extend the vertebral column and rotate the head. The **multifidus** muscle in this group, as its name implies, is split into several bundles. It extends and laterally flexes the vertebral column and rotates the head. The **rotatores** muscles of this group are short and run the entire length of the vertebral column. They extend and rotate the vertebral column.

Within the **segmental** muscle group are the **interspinales** and **intertransversarii** muscles, which unite the spinous and transverse processes of consecutive vertebrae. They function primarily in stabilizing the vertebral column during its movements.

Within the **scalene** group, the **anterior scalene** muscle is anterior to the middle scalene muscle, the **middle scalene** muscle is intermediate in placement and is the longest and largest of the scalene muscles, and the **posterior scalene** muscle is posterior to the middle scalene muscle and is the smallest of the scalene muscles. These muscles flex, laterally flex, and rotate the head and assist in deep inspiration.

INNERVATION

The splenius capitis is innervated by the middle cervical nerves; the splenius cervicis is innervated by the inferior cervical nerves; the iliocostalis cervicis and semispinalis capitis are innervated by the cervical and thoracic nerves; the iliocostalis thoracis is innervated by thoracic nerves; the iliocostalis lumborum is innervated by lumbar nerves; the longissimus capitis is innervated by the middle and inferior cervical nerves; the longissimus cervicis, longissimus thoracis, all spinalis muscles, multifidus, rotator, and interspinales are innervated by cervical, thoracic, and lumbar nerves; the semispinalis cervicis and semispinalis thoracis are innervated by cervical and thoracic nerves; the intertransversarii are innervated by spinal nerves; the anterior scalene is innervated by cervical nerves C5–C6; the middle scalene is innervated by cervical nerves C3–C8; and the posterior scalene is innervated by cervical nerves C6–C8.

RELATING MUSCLES TO MOVEMENTS

Arrange the muscles in this exhibit according to the following actions on the head at the atlanto-occipital and intervertebral joints: (1) flexion, (2) extension, (3) lateral flexion, (4) rotation to same side as contracting muscle, and (5) rotation to opposite side as contracting muscle; the following actions on the vertebral column at the intervertebral joints: (1) flexion, (2) extension, (3) lateral flexion, (4) rotation, and (5) stabilization; and the following action on the ribs: elevation during deep inspiration. The same muscle may be mentioned more than once.

What are the four major groups of muscles that move the vertebral column?

Exhibit 11.16 *Muscles that Move the Vertebral Column (Backbone) (continued)*

MUSCLE	ORIGIN	INSERTION	ACTION
Splenius (SPLĒ-nē-us)			
Splenius capitis (KAP-i-tis; *splenium* = bandage; *caput* = head)	Ligamentum nuchae and spinous processes of seventh cervical vertebra and first three or four thoracic vertebrae.	Occipital bone and mastoid process of temporal bone.	Acting together (bilaterally), extend head; acting singly (unilaterally), laterally flex and rotate head to same side as contracting muscle.
Splenius cervicis (SER-vi-kis; *cervix* = neck)	Spinous processes of third through sixth thoracic vertebrae.	Transverse processes of first two or four cervical vertebrae.	Acting together, extend head; acting singly, laterally flex and rotate head to same side as contracting muscle.
Erector Spinae (e-REK-tor SPI-nē) Consists of iliocostalis, longissimus, and spinalis muscles.			
Iliocostalis Group (Lateral)			
Iliocostalis cervicis (il'-ē-ō-kos-TAL-is SER-vi-kis; *ilium* = flank; *costa* = rib)	Superior six ribs.	Transverse processes of fourth to sixth cervical vertebrae.	Acting together, muscles of each region (cervical, thoracic, and lumbar) extend and maintain erect posture of vertebral column of their respective regions; acting singly, laterally flex vertebral column of their respective regions.
Iliocostalis thoracis (il'-ē-ō-kos-TAL-is thō-RA-kis; *thorax* = chest)	Inferior six ribs.	Superior six ribs.	
Iliocostalis lumborum (il'-ē-ō-kos-TAL-is lum-BOR-um)	Iliac crest.	Inferior six ribs.	
Longissimus Group (Intermediate)			
Longissimus capitis (lon-JIS-i-mus KAP-i-tis; *longissimus* = longest)	Transverse processes of superior four thoracic vertebrae and articular processes of inferior four cervical vertebrae.	Mastoid process of temporal bone.	Acting together, both longissimus capitis muscles extend head; acting singly, rotate head to same side as contracting muscle. Acting together, longissimus cervicis and both longissimus thoracis muscles extend vertebral column of their respective regions; acting singly, laterally flex vertebral column of their respective regions.
Longissimus cervicis (lon-JIS-i-mus SER-vi-kis)	Transverse processes of fourth and fifth thoracic vertebrae.	Transverse processes of second to sixth cervical vertebrae.	
Longissimus thoracis (lon-JIS-i-mus thō-RA-kis)	Transverse processes of lumbar vertebrae.	Transverse processes of all thoracic and superior lumbar vertebrae and ninth and tenth ribs.	
Spinalis Group (Medial)			
Spinalis capitis (spi-NA-lis KAP-i-tis; *spinalis* = vertebral column)	Arises with semispinalis capitis.	Occipital bone.	Acting together, muscles of each region (cervical, thoracic, and lumbar) extend vertebral column of their respective regions.
Spinalis cervicis (spi-NA-lis SER-vi-kis)	Ligamentum nuchae and spinous process of seventh cervical vertebra.	Spinous process of axis.	
Spinalis thoracis (spi-NA-lis thō-RA-kis)	Spinous processes of superior lumbar and inferior thoracic vertebrae.	Spinous processes of superior thoracic vertebrae.	

Exhibit 11.16 (continued)

MUSCLE	ORIGIN	INSERTION	ACTION
Transversospinalis (trans-ver'-sō-spi-NA-lis)			
Semispinalis capitis (sem'-ē-spi-NA-lis KAP-i-tis; *semi* = partially or one-half)	Transverse processes of first six or seven thoracic vertebrae and seventh cervical vertebra, and articular processes of fourth, fifth, and sixth cervical vertebrae.	Occipital bone.	Acting together, extend head; acting singly, rotate head to side opposite contracting muscle.
Semispinalis cervicis (sem'-ē-spi-NA-lis SER-vi-kis)	Transverse processes of superior five or six thoracic vertebrae.	Spinous processes of first to fifth cervical vertebrae.	Acting together, both semispinalis cervicis and both semispinalis thoracis muscles extend vertebral column of their respective regions; acting singly, rotate head to side opposite contracting muscle.
Semispinalis thoracis (sem'-ē-spi-NA-lis thō-RA-kis)	Transverse processes of sixth to tenth thoracic vertebrae.	Spinous processes of superior four thoracic and last two cervical vertebrae.	
Multifidus (mul-TIF-i-dus; *multi* = many; *findere* = to split)	Sacrum, ilium, transverse processes of lumbar, thoracic, and inferior four cervical vertebrae.	Spinous process of a more superior vertebra.	Acting together, extend vertebral column; acting singly, laterally flex vertebral column and rotate head to side opposite contracting muscle.
Rotatores (rō'-ta-TŌ-rez; *rotare* = to turn)	Transverse processes of all vertebrae.	Spinous process of vertebra superior to the one of origin.	Acting together, extend vertebral column; acting singly, rotate vertebral column to side opposite contracting muscle.
Segmental (seg-MEN-tal)			
Interspinales (in-ter-SPĪ-nāl-ez; *inter* = between)	Superior surface of all spinous processes.	Inferior surface of spinous process of vertebra superior to the one of origin.	Acting together, extend vertebral column; acting singly, stabilize vertebral column during movement.
Intertransversarii (in'-ter-trans-vers-AR-ē-ī; *inter* = between)	Transverse processes of all vertebrae.	Transverse process of vertebra superior to the one of origin.	Acting together, extend vertebral column; acting singly, laterally flex vertebral column and stabilize it during movements.
Scalene (SKĀ-lēn)			
Anterior scalene (SKĀ-lēn; *anterior* = front; *skalenos* = uneven)	Transverse processes of third through sixth cervical vertebrae.	First rib.	Acting together, both anterior scalene and middle scalene muscles flex head and elevate first ribs during deep inspiration; acting singly, laterally flex head and rotate head to side opposite contracting muscle.
Middle scalene (SKĀ-lēn)	Transverse processes of inferior six cervical vertebrae.	First rib.	
Posterior scalene (SKĀ-lēn)	Transverse processes of fourth through sixth cervical vertebrae.	Second rib.	Acting together, flex head and elevate second ribs during deep inspiration; acting singly, laterally flex head and rotate head to side opposite contracting muscle.

Exhibit 11.16 *Muscles that Move the Vertebral Column (Backbone) (continued)*

Figure 11.19 Muscles that move the vertebral column (backbone).

🔑 The erector spinae group is the largest muscular mass of the body and is the chief extensor of the vertebral column.

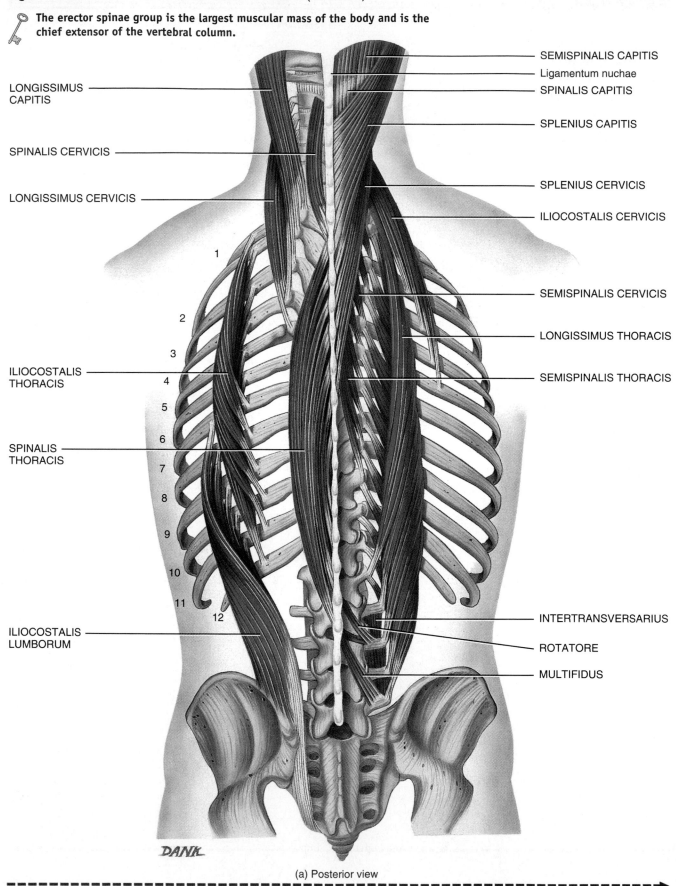

LONGISSIMUS CAPITIS

SPINALIS CERVICIS

LONGISSIMUS CERVICIS

ILIOCOSTALIS THORACIS

SPINALIS THORACIS

ILIOCOSTALIS LUMBORUM

SEMISPINALIS CAPITIS

Ligamentum nuchae

SPINALIS CAPITIS

SPLENIUS CAPITIS

SPLENIUS CERVICIS

ILIOCOSTALIS CERVICIS

SEMISPINALIS CERVICIS

LONGISSIMUS THORACIS

SEMISPINALIS THORACIS

INTERTRANSVERSARIUS

ROTATORE

MULTIFIDUS

DANK

(a) Posterior view

Exhibit 11.16 *(continued)*

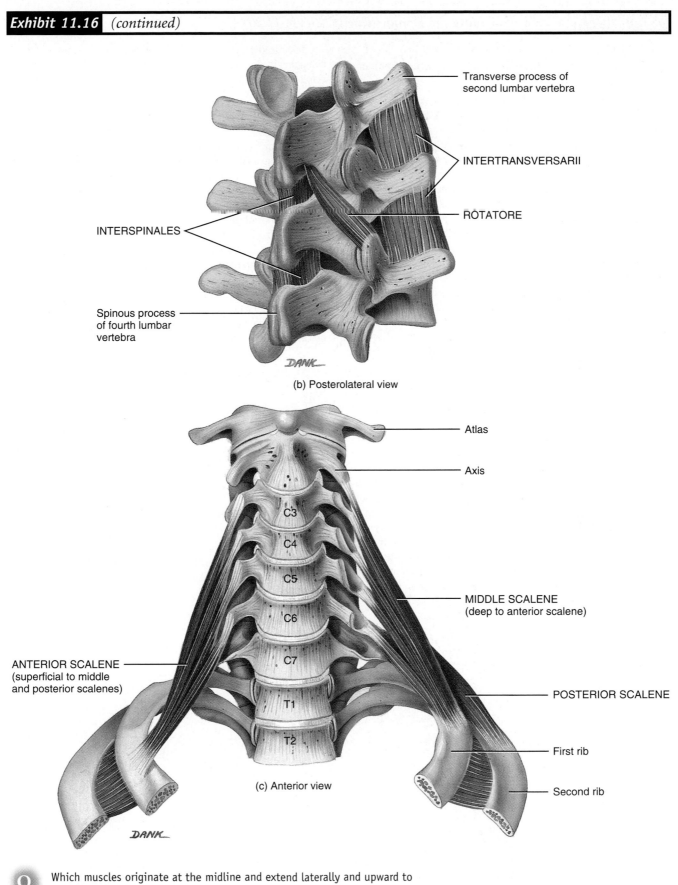

Transverse process of second lumbar vertebra

INTERTRANSVERSARII

ROTATORE

INTERSPINALES

Spinous process of fourth lumbar vertebra

DANK

(b) Posterolateral view

Atlas

Axis

C3

C4

C5

C6

C7

T1

T2

MIDDLE SCALENE (deep to anterior scalene)

ANTERIOR SCALENE (superficial to middle and posterior scalenes)

POSTERIOR SCALENE

First rib

Second rib

(c) Anterior view

DANK

Q Which muscles originate at the midline and extend laterally and upward to their insertion?

Exhibit 11.17 | *Muscles that Move the Femur (Thigh) (Figure 11.20)*

OBJECTIVE

• *Describe the origin, insertion, action, and innervation of the muscles of the femur.*

As you will see, muscles of the lower limbs are larger and more powerful than those of the upper limbs because lower limb muscles function in stability, locomotion, and maintenance of posture. Upper limb muscles are characterized by versatility of movement. In addition, muscles of the lower limbs often cross two joints and act equally on both.

The majority of muscles that move the femur originate on the pelvic girdle and insert on the femur. The **psoas major** and **iliacus** muscles have a common insertion and are together referred to as the **iliopsoas** (il'-ē-ō-SŌ-as) muscle because they share a common insertion (lesser trochanter of femur). There are three gluteal muscles: gluteus maximus, gluteus medius, and gluteus minimus. The **gluteus maximus** is the largest and heaviest of the three muscles and is one of the largest muscles in the body. It is the chief extensor of the femur. The **gluteus medius** is mostly deep to the gluteus maximus and is a powerful abductor of the femur at the hip joint. It is a common site for an intramuscular injection. The **gluteus minimus** is the smallest of the gluteal muscles and lies deep to the gluteus medius.

The **tensor fasciae latae** muscle is located on the lateral surface of the thigh. A layer of deep fascia composed of dense connective tissue encircles the entire thigh and is referred to as the **fascia lata.** It is well-developed laterally where, together with the tendons of the tensor fasciae and gluteus maximus muscles, it forms a structure called the **iliotibial tract**. The tract inserts into the lateral condyle of the tibia.

The **piriformis, obturator internus, obturator externus, superior gemellus, inferior gemellus,** and **quadratus femoris** muscles are all deep to the gluteus maximus muscle and function as lateral rotators of the femur at the hip joint.

Three muscles on the medial aspect of the thigh are the **adductor longus, adductor brevis,** and **adductor magnus.** They originate on the pubic bone and insert on the femur. All three muscles adduct, flex, and medially rotate the femur at the hip joint. The **pectineus** muscle also adducts and flexes the femur at the hip joint.

Technically, the adductor muscles and pectineus muscles are components of the medial compartment of the thigh and could also be included in Exhibit 11.18. However, they are included here because they act on the femur.

INNERVATION

Muscles that move the femur are innervated by nerves derived from the lumar and sacral plexuses (shown in Exhibits 13.3 and 13.4 on pages 436–440). More specifically, the psoas major is innervated by lumbar nerves L2–L3; the iliacus and pectineus are innervated by the femoral nerve; the gluteus maximus is innervated by the inferior gluteal nerve; the gluteus medius and minimus and tensor fasciae latae are innervated by the superior gluteal nerve; the piriformis is innervated by sacral nerves S1 or S2, mainly S1; the obturator internus and superior gemellus are innervated by the nerve to obturator internus; the obturator externus, adductor longus, and adductor brevis are innervated by the obturator nerve; the inferior gemellus and quadratus femoris are innervated by the nerve to quadratus femoris; and the adductor magnus is innervated by the obturator and sciatic nerves.

CLINICAL APPLICATION
Pulled Groin

Certain athletic and other activities may cause a **pulled groin,** a strain, stretching, or tearing of the distal attachments of the medial muscles of the thigh, iliopsoas, and/or adductor muscles. This generally results from activities that involve quick sprints, such as soccer, tennis, football, and running. ■

RELATING MUSCLES TO MOVEMENTS

Arrange the muscles in this exhibit according to the following actions on the thigh at the hip joint: (1) flexion, (2) extension, (3) abduction, (4) adduction, (5) medial rotation, and (6) lateral rotation. The same muscle may be mentioned more than once.

⎮ What forms the iliotibial tract?

Exhibit 11.17 *(continued)*

MUSCLE	ORIGIN	INSERTION	ACTION
Psoas major (SŌ-as; *psoa* = muscle of loin)	Transverse processes and bodies of lumbar vertebrae.	With iliacus into lesser trochanter of femur.	Both psoas major and iliacus muscles acting together flex thigh at hip joint, rotate thigh laterally, and flex trunk on the hip as in sitting up from the supine position.
Iliacus (il'-ē-AK-us; *iliac* = ilium)	Iliac fossa.	With psoas major into lesser trochanter of femur.	
Gluteus maximus (GLOO-tē-us MAK-si-mus; *glutos* = buttock; *maximus* = largest)	Iliac crest, sacrum, coccyx, and aponeurosis of sacrospinalis.	Iliotibial tract of fascia lata and lateral part of linea aspera under greater trochanter (gluteal tuberosity) of femur.	Extends thigh at hip joint and laterally rotates thigh.
Gluteus medius (GLOO-tē-us MĒ-dē-us; *media* = middle)	Ilium.	Greater trochanter of femur.	Abducts thigh at hip joint and medially rotates thigh.
Gluteus minimus (GLOO-tē-us MIN-i-mus; *minimus* = smallest)	Ilium.	Greater trochanter of femur.	Abducts thigh at hip joint and medially rotates thigh.
Tensor fasciae latae (TEN-sor FA-shē-ē LĀ-tē; *tensor* = makes tense; *fascia* = band; *latus* = wide)	Iliac crest.	Tibia by way of the iliotibial tract.	Flexes and abducts thigh at hip joint.
Piriformis (pir-i-FOR-mis; *pirum* = pear; *forma* = shape)	Anterior sacrum.	Superior border of greater trochanter of femur.	Laterally rotates and abducts thigh at hip joint.
Obturator internus (OB-too-rā'-tor in-TER-nus; *obturator* = obturator foramen; *internus* = inside)	Inner surface of obturator foramen, pubis, and ischium.	Greater trochanter of femur.	Laterally rotates and abducts thigh at hip joint.
Obturator externus (OB-too-rā'-tor ex-TER-nus; *externus* = outside)	Outer surface of obturator membrane.	Deep depression inferior to greater trochanter (trochanteric fossa) of femur.	Laterally rotates and abducts thigh at hip joint.
Superior gemellus (jem-EL-lus; *superior* = above; *gemellus* = twins)	Ischial spine.	Greater trochanter of femur.	Laterally rotates and abducts thigh at hip joint.
Inferior gemellus (jem-EL-lus; *inferior* = below)	Ischial tuberosity.	Greater trochanter of femur.	Laterally rotates and abducts thigh at hip joint.
Quadratus femoris (kwod-RĀ-tus FEM-or-is; *quad* = four; *femoris* = femur)	Ischial tuberosity.	Elevation superior to midportion of intertrochanteric crest (quadrate tubercle) on posterior femur.	Laterally rotates and stabilizes hip joint.
Adductor longus (LONG-us; *adductor* = moves part closer to midline; *longus* = long)	Pubic crest and pubic symphysis.	Linea aspera of femur.	Adducts and flexes thigh at hip joint and medially rotates thigh.
Adductor brevis (BREV-is; *brevis* = short)	Inferior ramus of pubis.	Superior half of linea aspera of femur.	Adducts and flexes thigh at hip joint and medially rotates thigh.
Adductor magnus (MAG-nus; *magnus* = large)	Inferior ramus of pubis and ischium to ischial tuberosity.	Linea aspera of femur.	Adducts thigh at hip joint and medially rotates thigh; anterior part flexes thigh at hip joint, and posterior part extends thigh at hip joint.
Pectineus (pek-TIN-ē-us; *pecten* = comb-shaped)	Superior ramus of pubis.	Pectineal line of femur, between lesser trochanter and linea aspera.	Flexes and adducts thigh at hip joint.

Exhibit 11.17 *Muscles that Move the Femur (Thigh) (continued)*

Figure 11.20 Muscles that move the femur (thigh). (See Tortora, *A Photographic Atlas of the Human Body,* Figures 5.16 and 5.17)

🔑 **Most muscles that move the femur originate on the pelvic (hip) girdle and insert on the femur.**

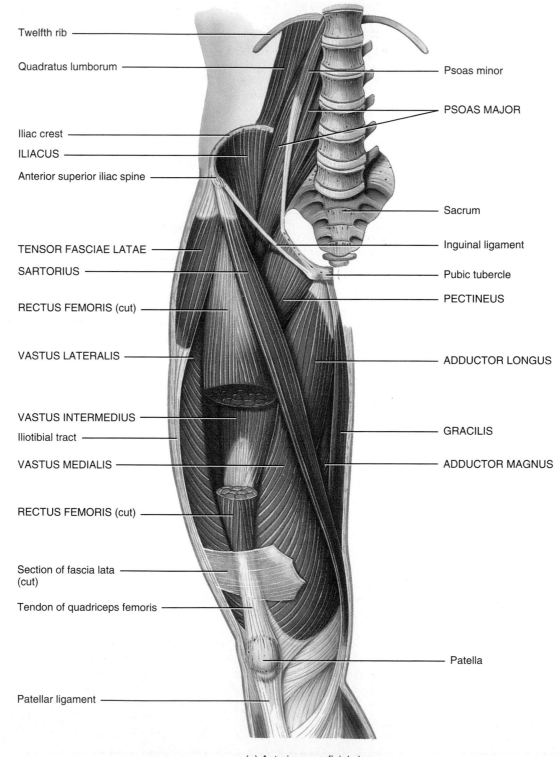

Twelfth rib

Quadratus lumborum

Iliac crest

ILIACUS

Anterior superior iliac spine

TENSOR FASCIAE LATAE

SARTORIUS

RECTUS FEMORIS (cut)

VASTUS LATERALIS

VASTUS INTERMEDIUS

Iliotibial tract

VASTUS MEDIALIS

RECTUS FEMORIS (cut)

Section of fascia lata (cut)

Tendon of quadriceps femoris

Patellar ligament

Psoas minor

PSOAS MAJOR

Sacrum

Inguinal ligament

Pubic tubercle

PECTINEUS

ADDUCTOR LONGUS

GRACILIS

ADDUCTOR MAGNUS

Patella

(a) Anterior superficial view

Exhibit 11.17 (continued)

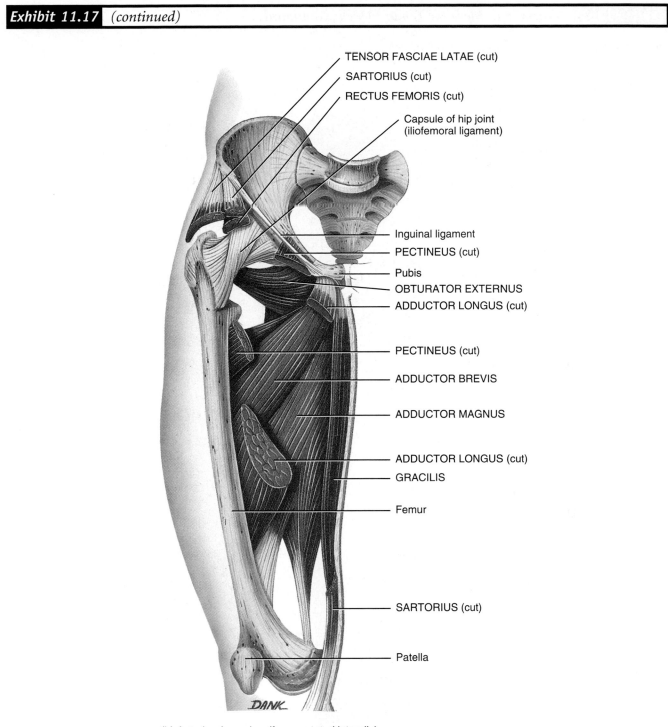

TENSOR FASCIAE LATAE (cut)

SARTORIUS (cut)

RECTUS FEMORIS (cut)

Capsule of hip joint
(iliofemoral ligament)

Inguinal ligament

PECTINEUS (cut)

Pubis

OBTURATOR EXTERNUS

ADDUCTOR LONGUS (cut)

PECTINEUS (cut)

ADDUCTOR BREVIS

ADDUCTOR MAGNUS

ADDUCTOR LONGUS (cut)

GRACILIS

Femur

SARTORIUS (cut)

Patella

DANK

(b) Anterior deep view (femur rotated laterally)

Figure continues

Exhibit 11.17   *Muscles that Move the Femur (Thigh) (continued)*

Figure 11.20 (continued)

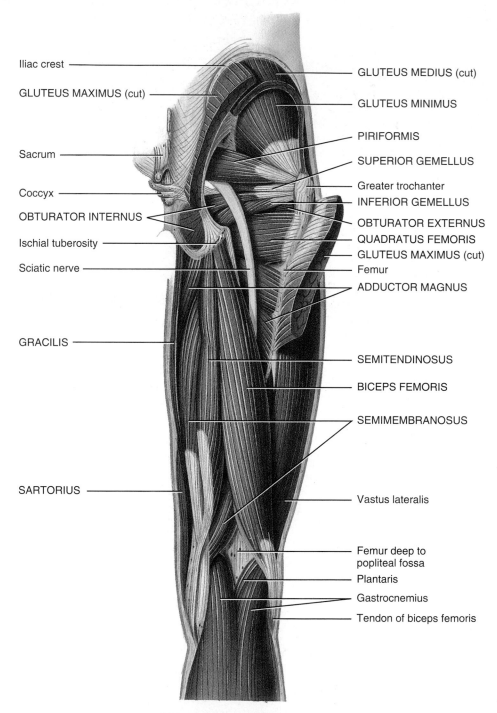

Iliac crest

GLUTEUS MAXIMUS (cut)

Sacrum

Coccyx

OBTURATOR INTERNUS

Ischial tuberosity

Sciatic nerve

GRACILIS

SARTORIUS

GLUTEUS MEDIUS (cut)

GLUTEUS MINIMUS

PIRIFORMIS

SUPERIOR GEMELLUS

Greater trochanter

INFERIOR GEMELLUS

OBTURATOR EXTERNUS

QUADRATUS FEMORIS

GLUTEUS MAXIMUS (cut)

Femur

ADDUCTOR MAGNUS

SEMITENDINOSUS

BICEPS FEMORIS

SEMIMEMBRANOSUS

Vastus lateralis

Femur deep to popliteal fossa

Plantaris

Gastrocnemius

Tendon of biceps femoris

(c) Posterior superficial view

 Q What are the principal differences between the muscles of the upper and lower limbs?

Exhibit 11.18 | *Muscles that Act on the Femur (Thigh) and Tibia and Fibula (Leg) (Figure 11.21)*

OBJECTIVE

• *Describe the origin, insertion, action, and innervation of the muscles that act on the femur and tibia and fibula.*

The muscles that act on the femur (thigh) and tibia and fibula (leg) are separated by deep fascia into medial, anterior, and posterior compartments. The **medial (adductor) compartment** is so named because its muscles adduct the femur at the hip joint. (See the adductor magnus, adductor longus, adductor brevis, and pectineus, which are components of the medial compartment, in Exhibit 11.17.) The **gracilis,** the other muscle in the medial compartment, not only adducts the thigh, but also flexes the leg at the knee joint. For this reason, it is discussed in this exhibit. The gracilis is a long, straplike muscle that lies on the medial aspect of the thigh and knee.

The **anterior (extensor) compartment** is so designated because its muscles extend the leg (and also flex the thigh). This compartment is composed of the quadriceps femoris and sartorius muscles. The **quadriceps femoris** muscle is the biggest muscle in the body, covering most of the anterior surface and sides of the thigh. The muscle is actually a composite muscle, usually described as four separate muscles: (1) **rectus femoris,** on the anterior aspect of the thigh; (2) **vastus lateralis,** on the lateral aspect of the thigh; (3) **vastus medialis,** on the medial aspect of the thigh; and (4) **vastus intermedius,** located deep to the rectus femoris between the vastus lateralis and vastus medialis. The common tendon for the four muscles is known as the **quadriceps tendon,** which inserts into the patella. The tendon continues inferior to the patella as the **patellar ligament,** which attaches to the tibial tuberosity. The quadriceps femoris muscle is the great extensor muscle of the leg. The **sartorius** is a long, narrow muscle that forms a band across the thigh from the ilium of the hip bone to the medial side of the tibia. The various movements it produces help effect the cross-legged sitting position in which the heel of one limb is placed on the knee of the opposite limb. It is known as the tailor's muscle because tailors frequently assume this cross-legged sitting position. (Because the major action of the sartorius muscle is to move the thigh rather than the leg, it could also have been included in Exhibit 11.17.)

The **posterior (flexor) compartment** is so named because its muscles flex the leg (and also extend the thigh). This compartment is composed of three muscles collectively called the **hamstrings:** (1) **biceps femoris,** (2) **semitendinosus,** and (3) **semimembranosus.** The hamstrings are so named because their tendons are long and stringlike in the popliteal area, and from an old practice of butchers in which they hung hams for smoking by these long tendons. Because the hamstrings span two joints (hip and knee), they are both extensors of the thigh and flexors of the leg. The **popliteal fossa** is a diamond-shaped space on the posterior aspect of the knee bordered laterally by the tendons of the biceps femoris muscle and medially by the tendons of the semitendinosus and semimembranosus muscles.

INNERVATION

Muscles that act on the femur and tibia and fibula are innervated by the obturator and femoral nerves derived from the lumbar plexus (shown in Exhibit 13.3 on page 436) and the sciatic nerve derived from the sacral plexus (shown in Exhibit 13.4 on page 439). More specifically, the gracilis is innervated by the obturator nerve; the quadriceps femoris and sartorius are innervated by the femoral nerve; the biceps femoris is innervated by the tibial and common peroneal nerves from the sciatic nerve; and the semimembranosus and semitendinosus are innervated by the tibial nerve from the sciatic nerve.

CLINICAL APPLICATION
Pulled Hamstrings

A strain or partial tear of the proximal hamstring muscles is referred to as **pulled hamstrings** or **hamstring strains.** They are common sports injuries in individuals who run very hard and/or are required to perform quick starts and stops. Sometimes the violent muscular exertion required to perform a feat tears off part of the tendinous origins of the hamstrings, especially the biceps femoris, from the ischial tuberosity. This is usually accompanied by a contusion (bruising), tearing of some of the muscle fibers, and rupture of blood vessels, producing a hematoma (collection of blood) and pain. Adequate training with good balance between the quadriceps femoris and hamstrings and stretching exercises before running or competing are important in preventing this injury. ■

RELATING MUSCLES TO MOVEMENTS

Arrange the muscles in this exhibit according to the following actions on the thigh at the hip joint: (1) abduction, (2) adduction, (3) lateral rotation, (4) flexion, and (5) extension; and according to the following actions on the leg at the knee joint: (1) flexion and (2) extension. The same muscle may be mentioned more than once.

Organize the muscles that act on the femur (thigh) and tibia and fibula (leg) into medial, anterior, and posterior compartments.

Exhibit 11.18 *Muscles that Act on the Femur (Thigh) and Tibia and Fibula (Leg) (continued)*

MUSCLE	ORIGIN	INSERTION	ACTION
Medial (adductor) compartment			
Adductor magnus (MAG-nus)			
Adductor longus (LONG-us)	See Exhibit 11.17.		
Adductor brevis (BREV-is)			
Pectineus (pek-TIN-ē-us)			
Gracilis (gra-SIL-is; *gracilis* = slender)	Pubic symphysis and pubic arch.	Medial surface of body of tibia.	Adducts thigh at hip joint, medially rotates thigh, and flexes leg at knee joint.
Anterior (extensor) compartment			
Quadriceps femoris (KWOD-ri-ceps FEM-or-is; *quadriceps* = four heads of origin; *femoris* = femur)			
Rectus femoris (REK-tus FEM-or-is; *rectus* = fibers parallel to midline)	Anterior inferior iliac spine.	Patella via quadriceps tendon and then tibial tuberosity via patellar ligament.	All four heads extend leg at knee joint; rectus femoris muscle acting alone also flexes thigh at hip joint.
Vastus lateralis (VAS-tus lat′-er-A-lis; *vastus* = large; *lateralis* = lateral)	Greater trochanter and linea aspera of femur.		
Vastus medialis (VAS-tus mē′-dē-A-lis; *medialis* = medial)	Linea aspera of femur.		
Vastus intermedius (VAS-tus in′-ter-MĒ-dē-us; *intermedius* = middle)	Anterior and lateral surfaces of body of femur.		
Sartorius (sar-TOR-ē-us; *sartor* = tailor; longest muscle in body)	Anterior superior iliac spine.	Medial surface of body of tibia.	Flexes leg at knee joint; flexes, abducts, and laterally rotates thigh at hip joint.
Posterior (flexor) compartment			
Hamstrings A collective designation for three separate muscles.			
Biceps femoris (BĪ-ceps FEM-or-is; *biceps* = two heads of origin)	Long head arises from ischial tuberosity; short head arises from linea aspera of femur.	Head of fibula and lateral condyle of tibia.	Flexes leg at knee joint and extends thigh at hip joint.
Semitendinosus (sem′-ē-TEN-di-nō-sus; *semi* = half; *tendo* = tendon)	Ischial tuberosity.	Proximal part of medial surface of shaft of tibia.	Flexes leg at knee joint and extends thigh at hip joint.
Semimembranosus (sem′-ē-MEM-bra-nō-sus; *membran* = membrane)	Ischial tuberosity.	Medial condyle of tibia.	Flexes leg at knee joint and extends thigh at hip joint.

Exhibit 11.18 (continued)

Figure 11.21 Muscles that act on the femur (thigh) and tibia and fibula (leg).

Muscles that act on the leg originate in the hip and thigh and are separated into compartments by deep fascia.

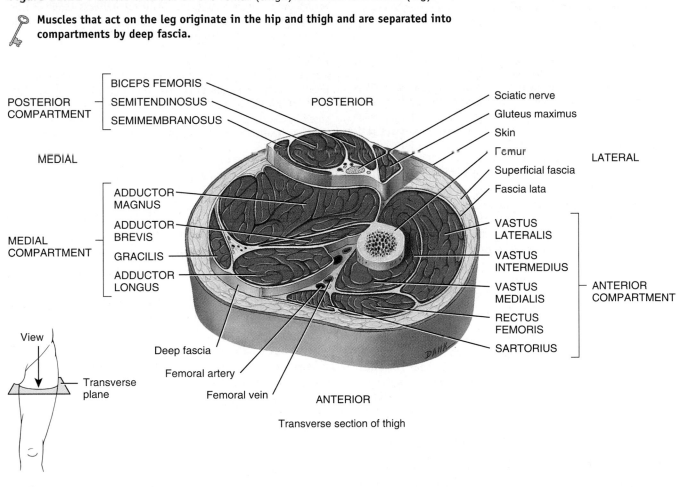

Transverse section of thigh

Q   Which muscles constitute the quadriceps femoris and hamstring muscles?

Exhibit 11.19 Muscles that Move the Foot and Toes (Figure 11.22)

OBJECTIVE

• *Describe the origin, insertion, action, and innervation of the muscles that move the foot and toes.*

Muscles that move the foot and toes are located in the leg. The musculature of the leg, like that of the thigh, is divided by deep fascia into three compartments: anterior, lateral, and posterior. The **anterior compartment** consists of muscles that dorsiflex the foot. In a situation analogous to the wrist, the tendons of the muscles of the anterior compartment are held firmly to the ankle by thickenings of deep fascia called the **superior extensor retinaculum** (*transverse ligament of the ankle*) and **inferior extensor retinaculum** (*cruciate ligament of the ankle*).

Within the anterior compartment, the **tibialis anterior** is a long, thick muscle against the lateral surface of the tibia, where it is easy to palpate. The **extensor hallucis longus** is a thin muscle between and partly deep to the **tibialis anterior** and **extensor digitorum longus** muscles. This latter muscle is featherlike and lies lateral to the tibialis anterior muscle, where it can easily be palpated. The **peroneus tertius** muscle is actually part of the extensor digitorum longus, with which it shares a common origin.

The **lateral (peroneal) compartment** contains two muscles that plantar flex and evert the foot: **peroneus longus** and **peroneus brevis.**

The **posterior compartment** consists of muscles that are divisible into superficial and deep groups. The superficial muscles share a common tendon of insertion, the **calcaneal (Achilles) tendon,** the strongest tendon of the body, that inserts into the calcaneal bone of the ankle. The superficial and most of the deep muscles plantar flex the foot at the ankle joint. The superficial muscles of the posterior compartment are the gastrocnemius, soleus, and plantaris—the so-called calf muscles. The large size of these muscles is directly related to our upright stance, a characteristic of humans. The **gastrocnemius** is the most superficial muscle and forms the prominence of the calf. The **soleus** is broad, flat, and lies deep to the gastrocnemius. It derives its name from its resemblance to a flat fish (sole). The **plantaris** is a small muscle that may be absent; sometimes there are two of them in each leg. It runs obliquely between the gastrocnemius and soleus muscles.

The deep muscles of the posterior compartment are the popliteus, tibialis posterior, flexor digitorum longus, and flexor hallucis longus. The **popliteus** is a triangular muscle that forms the floor of the popliteal fossa. The **tibialis posterior** is the deepest muscle in the posterior compartment. It lies between the flexor digitorum longus and flexor hallucis longus muscles. The **flexor digitorum longus** is smaller than the **flexor hallucis longus,** even though the former flexes four toes, whereas the latter flexes only the great toe at the interphalangeal joint.

INNERVATION

Muscles that move the foot and toes are innervated by nerves derived from the sacral plexes (shown in Exhibit 13.4 on page 439). More specifically, muscles of the anterior compartment are innervated by the deep peroneal nerve; muscles of the lateral compartment are innervated by the superficial peroneal nerve; and muscles of the posterior compartment are innervated by the tibial nerve. The nerves listed here are derived from the sacral plexus, which is described and illustrated in Exhibit 13.4 on page 439.

 CLINICAL APPLICATION
Shinsplint Syndrome

Shinsplint syndrome, or simply **shinsplints,** refers to pain or soreness along the tibia, specifically the medial, distal two-thirds. It may be caused by tendinitis of the anterior compartment muscles, especially the tibialis anterior muscle, inflammation of the periosteum (periostitis) around the tibia, or stress fractures of the tibia. The tendinitis usually occurs when poorly conditioned runners run on hard or banked surfaces with poorly supportive running shoes. The condition may also occur as a result of vigorous activity of the legs following a period of relative inactivity. The muscles in the anterior compartment (mainly the tibialis anterior) can be strengthened to balance the stronger posterior compartment muscles. ■

RELATING MUSCLES TO MOVEMENTS

Arrange the muscles in this exhibit according to the following actions on the foot at the ankle joint: (1) dorsiflexion and (2) plantar flexion; according to the following actions on the foot at the intertarsal joints: (1) inversion and (2) eversion; and according to the following actions on the toes at the metatarsophalangeal and intertarsal joints: (1) flexion and (2) extension. The same muscle may be mentioned more than once.

What is the superior extensor retinaculum? Inferior extensor retinaculum?

Exhibit 11.19 *(continued)*

MUSCLE	ORIGIN	INSERTION	ACTION
Anterior Compartment			
Tibialis anterior (tib′-ē-A-lis; *tibialis* = tibia; *anterior* = front)	Lateral condyle and body of tibia and interosseous membrane (sheet of fibrous tissue that holds shafts of tibia and fibula together).	First metatarsal and first (medial) cuneiform.	Dorsiflexes foot at ankle joint and inverts foot at intertarsal joints.
Extensor hallucis longus (HAL-a-kis LON-gus; *extensor* = increases angle at joint; *hallucis* = hallux or great toe; *longus* = long)	Anterior surface of fibula and interosseous membrane.	Distal phalanx of great toe.	Dorsiflexes foot at ankle joint and extends proximal phalanx of great toe at metatarsophalangeal joint.
Extensor digitorum longus (di′-ji-TOR-um LON-gus)	Lateral condyle of tibia, anterior surface of fibula, and interosseous membrane.	Middle and distal phalanges of toes 2–5.*	Dorsiflexes foot at ankle joint and extends distal and middle phalanges of each toe at interphalangeal joints and proximal phalanx of each toe at metatarsophalangeal joint.
Peroneus tertius (per′-ō-NĒ-us TER-shus; *perone* = fibula; *tertius* = third)	Distal third of fibula and interosseous membrane.	Base of fifth metatarsal.	Dorsiflexes foot at ankle joint and everts foot at intertarsal joints.
Lateral (Peroneal) Compartment			
Peroneus longus (per′-ō-NĒ-us LON-gus)	Head and body of fibula and lateral condyle of tibia.	First metatarsal and first cuneiform.	Plantar flexes foot at ankle joint and everts foot at intertarsal joints.
Peroneus brevis (per′-ō-NĒ-us BREV-is; *brevis* = short)	Body of fibula.	Base of fifth metatarsal.	Plantar flexes foot at ankle joint and everts foot at intertarsal joints.

---

*Reminder: The great toe or hallux is the first toe and has two phalanges: proximal and distal. The remaining toes are numbered 2–5, and each has three phalanges: proximal, middle, and distal.

Exhibit 11.19	*Muscles that Move the Foot and Toes (continued)*

MUSCLE	ORIGIN	INSERTION	ACTION
Superficial Posterior Compartment			
Gastrocnemius (gas'-trok-NĒ-mē-us; *gaster* = belly; *kneme* = leg)	Lateral and medial condyles of femur and capsule of knee.	Calcaneus by way of calcaneal (Achilles) tendon.	Plantar flexes foot at ankle joint and flexes leg at knee joint.
Soleus (SŌ-lē-us; *soleus* = sole of foot)	Head of fibula and medial border of tibia.	Calcaneus by way of calcaneal (Achilles) tendon.	Plantar flexes foot at ankle joint.
Plantaris (plan-TA-ris; *plantar* = sole of foot)	Femur superior to lateral condyle.	Calcaneus by way of calcaneal (Achilles) tendon.	Plantar flexes foot at ankle joint and flexes leg at knee joint.
Deep Posterior Compartment			
Popliteus (pop-LIT-ē-us; *poples* = posterior surface of knee)	Lateral condyle of femur.	Proximal tibia.	Flexes leg at knee joint and medially rotates tibia to unlock the extended knee.
Tibialis (tib'-ē-A-lis) **posterior** (*posterior* = back)	Tibia, fibula, and interosseous membrane.	Second, third, and fourth metatarsals; navicular; all three cuneiforms; and cuboid.	Plantar flexes foot at ankle joint and inverts foot at intertarsal joints.
Flexor digitorum longus (di'-ji-TOR-um LON-gus; *digitorum* = finger or toe)	Posterior surface of tibia.	Distal phalanges of toes 2–5.	Plantar flexes foot at ankle joint; flexes distal and middle phalanges of each toe at interphalangeal joints and proximal phalanx of each toe at metatarsophalangeal joint.
Flexor hallucis longus (HAL-a-kis LON-gus; *flexor* = decreases angle at joint)	Inferior two-thirds of fibula.	Distal phalanx of great toe.	Plantar flexes foot at ankle joint; flexes distal phalanx of great toe at interphalangeal joint and proximal phalanx of great toe at metatarsophalangeal joint.

Exhibit 11.19   *(continued)*

Figure 11.22 Muscles that move the foot and toes. (See Tortora, *A Photographic Atlas of the Human Body,* Figures 5.18 and 5.19)

🔑 **The superficial muscles of the posterior compartment share a common tendon of insertion, the calcaneal (Achilles) tendon, that inserts into the calcaneal bone of the ankle.**

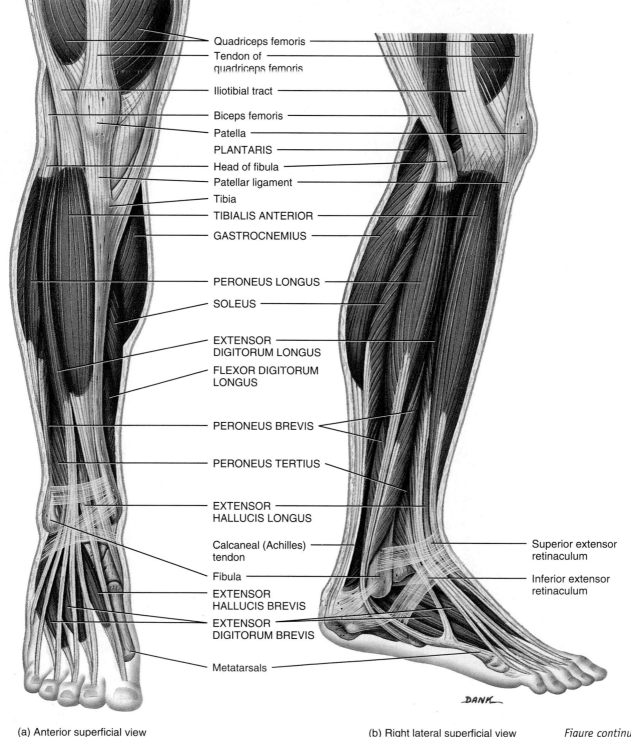

Quadriceps femoris
Tendon of quadriceps femoris
Iliotibial tract
Biceps femoris
Patella
PLANTARIS
Head of fibula
Patellar ligament
Tibia
TIBIALIS ANTERIOR
GASTROCNEMIUS
PERONEUS LONGUS
SOLEUS
EXTENSOR DIGITORUM LONGUS
FLEXOR DIGITORUM LONGUS
PERONEUS BREVIS
PERONEUS TERTIUS
EXTENSOR HALLUCIS LONGUS
Calcaneal (Achilles) tendon
Fibula
EXTENSOR HALLUCIS BREVIS
EXTENSOR DIGITORUM BREVIS
Metatarsals

Superior extensor retinaculum
Inferior extensor retinaculum

DANK

(a) Anterior superficial view

(b) Right lateral superficial view   *Figure continues*

Exhibit 11.19   *Muscles that Move the Foot and Toes (continued)*

Figure 11.22 (continued)

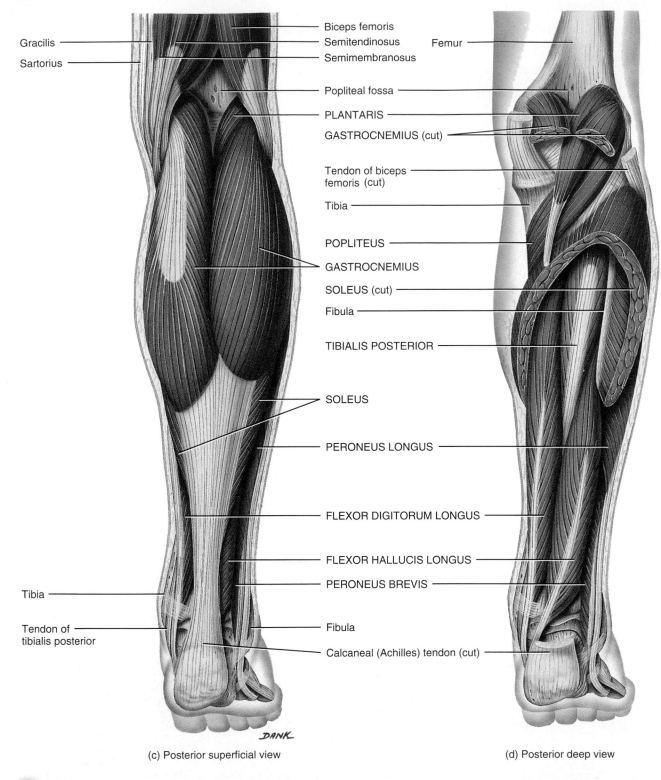

Gracilis
Sartorius

Biceps femoris
Semitendinosus
Semimembranosus

Femur

Popliteal fossa
PLANTARIS
GASTROCNEMIUS (cut)

Tendon of biceps
femoris (cut)

Tibia

POPLITEUS

GASTROCNEMIUS

SOLEUS (cut)

Fibula

TIBIALIS POSTERIOR

SOLEUS

PERONEUS LONGUS

FLEXOR DIGITORUM LONGUS

FLEXOR HALLUCIS LONGUS
PERONEUS BREVIS

Tibia

Tendon of
tibialis posterior

Fibula

Calcaneal (Achilles) tendon (cut)

DANK

(c) Posterior superficial view

(d) Posterior deep view

Q   What structures firmly hold the tendons of the anterior compartment muscles to
the ankle?

Exhibit 11.20 | *Intrinsic Muscles of the Foot (Figure 11.23)*

OBJECTIVE

• *Describe the origin, insertion, action, and innervation of the intrinsic muscles of the foot.*

Several of the muscles discussed in Exhibit 11.19 move the toes in various ways and are known as extrinsic muscles. The muscles in this exhibit are called **intrinsic muscles** because they originate and insert *within* the foot. The intrinsic muscles of the foot are, for the most part, comparable to those in the hand. Whereas the muscles of the hand are specialized for precise and intricate movements, those of the foot are limited to support and locomotion. The deep fascia of the foot forms the **plantar aponeurosis (fascia)** that extends from the calcaneus bone to the phalanges of the toes. The aponeurosis supports the longitudinal arch of the foot and encloses the flexor tendons of the foot.

The intrinsic muscles of the foot are divided into two groups: **dorsal** and **plantar.** There is only one dorsal muscle, the **extensor digitorum brevis,** a four-part muscle deep to the tendons of the extensor digitorum longus muscle, which extends toes 2–5 at the metatarsophalangeal joints.

The plantar muscles are arranged in four layers, the more superficial layer being referred to as the first layer. The muscles in the **first layer** are the **abductor hallucis,** which lies along the medial border of the sole, is comparable to the abductor pollicis brevis in the hand, and abducts the great toe at the metatarsophalangeal joint; the **flexor digitorum brevis,** which lies in the middle of the sole and flexes toes 2–5 at the interphalangeal and metatarsophalangeal joints; and the **abductor digiti minimi,** which lies along the lateral border of the sole, is comparable to the same muscle in the hand, and abducts the small toe.

The **second layer** consists of the **quadratus plantae,** a rectangular muscle that arises by two heads and flexes toes 2–5 at the metatarsophalangeal joints, and the **lumbricals,** four small muscles that are similar to the lumbricals in the hands. They flex the proximal phalanges and extend the distal phalanges of toes 2–5.

The **third layer** is composed of the **flexor hallucis brevis,** which lies adjacent to the plantar surface of the metatarsal of the great toe, is comparable to the same muscle in the hand, and flexes the great toe; the **adductor hallucis,** which has an oblique and transverse head like the adductor pollicis in the hand and adducts the great toe; and the **flexor digiti minimi brevis,** which lies superficial to the metatarsal of the small toe, is comparable to the same muscle in the hand, and flexes the small toe.

The **fourth layer** is the deepest and consists of the **dorsal interossei,** four muscles that adduct toes 2–4, flex the proximal phalanges, and extend the distal phalanges, and the three **plantar interossei,** which adduct toes 3–5, flex the proximal phalanges, and extend the distal phalanges. The interossei of the feet are similar to those of the hand, except that their actions are relative to the midline of the second digit rather than to the third as in the hand.

INNERVATION

Intrinsic muscles of the foot are innervated by nerves derived from the sacral plexus (shown in Exhibit 13.4 on page 439). More specifically, the extensor digitorum brevis is innervated by the deep peroneal nerve; the abductor hallucis and flexor digitorum brevis are innervated by the medial plantar nerve; the abductor digiti minimi, quadratus plantae, adductor hallucis, flexor digiti minimi brevis, and dorsal and plantar interossei are innervated by the lateral plantar nerve; and the lumbricals are innervated by the medial and lateral plantar nerves. The nerves listed here are derived from the sacral plexus, which is described and illustrated in Exhibit 13.4 on page 439.

CLINICAL APPLICATION
Plantar Fasciitis

Plantar fasciitis (fas-ē-Ī-tis) or **painful heel syndrome** is an inflammatory reaction due to chronic irritation of the plantar aponeurosis (fascia) at its origin on the calcaneus (heel bone). The aponeurosis becomes less elastic with age. This condition is also related to weight-bearing activities (walking, jogging, lifting heavy objects), improperly constructed or fitting shoes, excess weight (puts pressure on the feet), and poor biomechanics (flat feet, high arches, and abnormalities in gait may cause uneven distribution of weight on the feet). Plantar fasciitis is the most common cause of heel pain in runners and arises in response to the repeated impact of running. ■

RELATING MUSCLES TO MOVEMENTS

Arrange the muscles in this exhibit according to the following actions on the great toe at the metatarsophalangeal joint: (1) flexion, (2) extension, (3) abduction, and (4) adduction; and according to the following actions on toes 2–5 at the metatarsophalangeal and interphalangeal joints: (1) flexion, (2) extension, (3) abduction, and (4) adduction. The same muscle may be mentioned more than once.

⎪ How do the intrinsic muscles of the hand and foot differ in function?

Exhibit 11.20 *Intrinsic Muscles of the Foot (continued)*

MUSCLE	ORIGIN	INSERTION	ACTION
Dorsal			
Extensor digitorum brevis (*extensor* = increases angle at joint; *digit* = finger or toe; *brevis* = short) (see Figure 11.22a, b)	Calcaneus and inferior extensor retinaculum.	Tendons of extensor digitorum longus on toes 2–4 and proximal phalanx of great toe.*	Extensor hallucis brevis extends great toe at metatarsophalangeal joint and extensor digitorum brevis extends toes 2–4 at interphalangeal joints.
Plantar			
First Layer (most superficial)			
Abductor hallucis (*abductor* = moves part away from midline; *hallucis* = hallux or great toe)	Calcaneus, plantar aponeurosis, and flexor retinaculum.	Medial side of proximal phalanx of great toe with the tendon of the flexor hallucis brevis.	Abducts and flexes great toe at metatarsophalangeal joint.
Flexor digitorum brevis (*flexor* = decreases angle at joint)	Calcaneus and plantar aponeurosis.	Sides of middle phalanx of toes 2–5.	Flexes toes 2–5 at proximal interphalangeal and metatarsophalangeal joints.
Abductor digiti minimi (*minimi* = little)	Calcaneus and plantar aponeurosis.	Lateral side of proximal phalanx of small toe with the tendon of the flexor digiti minimi brevis.	Abducts and flexes small toe at metatarsophalangeal joint.
Second layer			
Quadratus plantae (*quad* = four; *planta* = sole)	Calcaneus.	Tendon of flexor digitorum longus.	Assists flexor digitorum longus to flex toes 2–5 at interphalangeal and metatarsophalangeal joints.
Lumbricals (LUM-bri-kals; *lumbricus* = earthworm)	Tendons of flexor digitorum longus.	Tendons of extensor digitorum longus on proximal phalanges of toes 2–5.	Extend toes 2–5 at interphalangeal joints and flex toes 2–5 at metatarsophalangeal joints.
Third Layer			
Flexor hallucis brevis	Cuboid and third (lateral) cuneiform.	Medial and lateral sides of proximal phalanx of great toe via a tendon containing a sesamoid bone.	Flexes great toe at metatarsophalangeal joint.
Adductor hallucis	Metatarsals 2–4, ligaments of 3–5 metatarsophalangeal joints, and tendon of peroneus longus.	Lateral side of proximal phalanx of great toe.	Adducts and flexes great toe at metatarsophalangeal joint.
Flexor digiti minimi brevis	Metatarsal 5 and tendon of peroneus longus.	Lateral side of proximal phalanx of small toe.	Flexes small toe at metatarsophalangeal joint.
Fourth Layer (deepest)			
Dorsal interossei (not illustrated)	Adjacent side of metatarsals.	Proximal phalanges: both sides of toe 2 and lateral side of toes 3 and 4.	Abduct and flex toes 2–4 at metatarsophalangeal joints and extend toes at interphalangeal joints.
Plantar interossei	Metatarsals 3–5.	Medial side of proximal phalanges of toes 3–5.	Adduct and flex proximal metatarsophalangeal joints and extend toes at interphalangeal joints.

*The tendon that inserts into the proximal phalanx of the great toe, together with its belly, is frequently described as a separate muscle, the extensor hallucis brevis.

Exhibit 11.20 *(continued)*

Figure 11.23 Intrinsic muscles of the foot.

Whereas the muscles of the hand are specialized for precise and intricate movements, those of the foot are limited to support and movement.

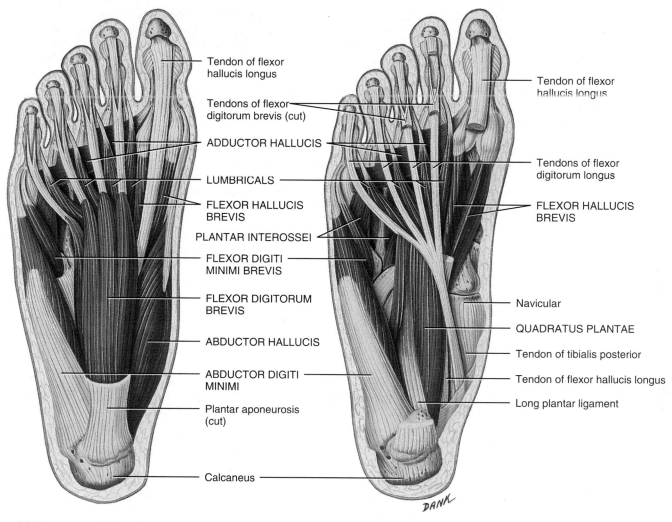

(a) Plantar superficial and deep view

(b) Plantar deep view

Q What structure supports the longitudinal arch and encloses the flexor tendons of the foot?

DISORDERS: HOMEOSTATIC IMBALANCES

RUNNING INJURIES

It is estimated that nearly 70% of those who jog or run will sustain some type of running-related injury. Even though most such injuries are minor, such as sprains and strains, some are quite serious; moreover, untreated or inappropriately treated minor injuries may become chronic. Among runners, common sites of injury include the ankle, knee, calcaneal (Achilles) tendon, hip, groin, foot, and back. Of these, the knee often is the most severely injured area.

Running injuries are frequently related to faulty training techniques. This may involve improper (or lack of) warm-up routines, running too much, or running too soon following an injury. Or it might involve extended running on hard and/or uneven surfaces. Poorly constructed or worn-out running shoes can also contribute to injury. Any biomechanical problem (such as a fallen arch) aggravated by running can also contribute to injuries.

Most sports injuries should be treated initially with RICE therapy, which stands for Rest, Ice, Compression, and Elevation. Immediately apply ice, and rest and elevate the injured part. Then apply an elastic bandage, if possible, to compress the injured tissue. Continue using RICE for 2–3 days, and resist the temptation to apply heat, which may worsen the swelling. Follow-up treatment may include alternating moist heat and ice massage to enhance blood flow in the injured area. Sometimes, nonsteroidal anti-inflammatory drugs (NSAIDs) or local injections of corticosteroids are needed. During the recovery period, it is important to keep active using an alternate fitness program that does not worsen the original injury.

This activity should be determined in consultation with a physician. Finally, careful exercise is needed to rehabilitate the injured area itself.

COMPARTMENT SYNDROME

In several exhibits in this chapter we noted that skeletal muscles in the limbs are organized into units called compartments. This compartmental arrangement causes problems for some individuals under certain conditions. In a disorder called **compartment syndrome,** some external or internal pressure constricts the structures within a compartment, resulting in damaged blood vessels and subsequent reduction of the blood supply (ischemia) to the structures within the compartment. Common causes of compartment syndrome include crushing and penetrating injuries, contusion (damage to subcutaneous tissues without the skin being broken), muscle strain (overstretching of a muscle), or an improperly fitted cast. As a result of hemorrhage, tissue injury, and edema (buildup of interstitial fluid), the pressure increase in the compartment can have serious consequences. Because the fascia that enclose the compartments are very strong, accumulated blood and interstitial fluid cannot escape, and the increased pressure can literally choke off the blood flow and deprive nearby muscles and nerves of oxygen. One treatment option is fasciotomy (fash-ē-OT-ō-me), a surgical procedure in which muscle fascia is cut to relieve the pressure. Without intervention, nerves suffer damage, and muscles develop scar tissue that results in permanent shortening of the muscles, a condition called *contracture.*

STUDY OUTLINE

HOW SKELETAL MUSCLES PRODUCE MOVEMENT (p. 302)

1. Skeletal muscles produce movement by pulling on bones.
2. The attachment to the stationary bone is the origin; the attachment to the movable bone is the insertion.
3. Bones serve as levers, and joints serve as fulcrums. The lever is acted on by two different forces: resistance and effort.
4. Levers are categorized into three types—first-class, second-class, and third-class (most common)—according to the positions of the fulcrum, the effort, and the resistance on the lever.
5. Fascicular arrangements include parallel, fusiform, circular, triangular, and pennate. Fascicular arrangement affects a muscle's power and range of motion.
6. The prime mover produces the desired action; the antagonist produces an opposite action. The synergist assists the prime mover by reducing unnecessary movement. The fixator stabilizes the origin of the prime mover so that it can act more efficiently.

HOW SKELETAL MUSCLES ARE NAMED (p. 307)

1. Skeletal muscles are named on the basis of distinctive criteria: direction of muscle fibers; size, shape, action, number of origins (or heads), and location of the muscle; and sites of origin and insertion of the muscle.

PRINCIPAL SKELETAL MUSCLES (p. 307)

1. Muscles of facial expression move the skin rather than a joint when they contract, and they permit us to express a wide variety of emotions.
2. The extrinsic muscles that move the eyeballs are among the fastest contracting and most precisely controlled skeletal muscles in the body. They permit us to elevate, depress, abduct, adduct, and medially and laterally rotate the eyeballs.
3. Muscles that move the mandible (lower jaw) are also known as the muscles of mastication because they are involved in chewing.
4. The extrinsic muscles that move the tongue are important in chewing, swallowing, and speech.
5. Muscles of the floor of the oral cavity (mouth) are called suprahyoid muscles because they are located above the hyoid bone. They elevate the hyoid bone, oral cavity, and tongue during swallowing.
6. Muscles that move the head alter the position of the head and help balance the head on the vertebral column.
7. Muscles that act on the abdominal wall help contain and protect the abdominal viscera, move the vertebral column, compress the abdomen, and produce the force required for defecation, urination, vomiting, and childbirth.

8. Muscles used in breathing alter the size of the thoracic cavity so that inspiration and expiration can occur.

9. Muscles of the pelvic floor support the pelvic viscera, resist the thrust that accompanies increases in intra-abdominal pressure, and function as sphincters at the anus, urethra, and vagina.

10. Muscles of the perineum assist in urination, erection of the penis and clitoris, ejaculation, and defecation.

11. Muscles that move the pectoral (shoulder) girdle stabilize the scapula so it can function as a stable point of origin for most of the muscles that move the humerus.

12. Muscles that move the humerus (arm) originate for the most part on the scapula (scapular muscles); the remaining muscles originate on the axial skeleton (axial muscles).

13. Muscles that move the radius and ulna (forearm) are involved in flexion and extension at the elbow joint and are organized into flexor and extensor compartments.

14. Muscles that move the wrist, hand, and digits are many and varied; those muscles that act on the digits are called extrinsic muscles.

15. The intrinsic muscles of the hand are important in skilled activities and provide humans with the ability to grasp precisely and manipulate objects.

16. Muscles that move the vertebral column are quite complex because they have multiple origins and insertions and because there is considerable overlap among them.

17. Muscles that move the femur (thigh) originate for the most part on the pelvic girdle and insert on the femur, and these muscles are larger and more powerful than comparable muscles in the upper limb.

18. Muscles that move the femur (thigh) and tibia and fibula (leg) are separated into medial (adductor), anterior (extensor), and posterior (flexor) compartments.

19. Muscles that move the foot and toes are divided into anterior, lateral, and posterior compartments.

20. Intrinsic muscles of the foot, unlike those of the hand, are limited to support and locomotion functions.

SELF-QUIZ QUESTIONS

Complete the following:

1. The less movable end of the muscle is the ___; the more movable end is the ___.

2. The muscle that forms the major portion of the cheek is the ___.

3. The strongest muscle of mastication is the ___.

4. The most important muscle in breathing is the ___.

Choose the best answer to the following questions:

5. Which of the following statements is *false*? (a) Fixators steady the proximal end of a limb whereas movements occur at the distal end. (b) Synergists prevent unwanted movements in intermediate joints. (c) The prime mover, or agonist, causes a desired action. (d) The antagonist supports and aids the agonist's movement. (e) Fascicular arrangement affects a muscle's power and range of motion.

6. Which of the following are important characteristics used in the naming of muscles? (1) muscle shape, (2) muscle size, (3) direction of muscle fibers, (4) muscle action, (5) muscle location, (6) sites of origin and insertion, (7) number of insertions. (a) 1, 2, 3, 4, 5, and 6, (b) 2, 3, 4, 5, 6, and 7, (c) 2, 4, 6, and 7, (d) 1, 3, 5, and 7, (e) 1, 2, 4, and 5

7. The muscle innervated by the oculomotor (III) nerve is the (a) occipitalis, (b) levator palpebrae superioris, (c) levator labii superioris, (d) depressor anguli oris, (e) zygomaticus.

8. The muscle that closes the mouth is the (a) zygomaticus major, (b) levator labii superioris, (c) depressor labii inferioris, (d) mentalis, (e) orbicularis oris.

9. Which of the following are muscle-generated actions on the mandible? (1) elevation, (2) depression, (3) retraction, (4) protraction, (5) abduction, (6) adduction, (7) side-to-side movement. (a) 1, 2, 3, and 5, (b) 2, 3, 4, and 6, (c) 1, 2, 3, 4, and 7, (d) 3, 4, 5, and 7, (e) 1, 3, 5, and 7

10. Which of the following are functions of the anterolateral abdominal muscle group? (1) helps contain and protect the abdominal viscera, (2) helps contain and protect the lower area of the heart and lungs, (3) flexes, laterally flexes, and rotates the vertebral column, (4) expands the abdomen during forced expiration, (5) produces the force required for defecation, urination, and childbirth.
(a) 1, 3, and 5, (b) 2, 4, and 5, (c) 1, 2, and 4, (d) 1 and 5, (e) 1, 2, and 5

11. Match the following:
___(a) most common lever in the body
___(b) lever formed by the head resting on the vertebral column
___(c) always produces a mechanical advantage
___(d) EFR
___(e) FRE
___(f) FER
___(g) formed by the ball of the foot, tarsal bones, and calf muscles when rising on the toes
___(h) adduction of the thigh

(1) first-class lever
(2) second-class lever
(3) third-class lever

True or false:

12. Because of their insertions, the muscles of facial expression move the skin rather than a joint when they contract.

13. Opposition is the single most distinctive digital movement that gives humans and other primates the ability to precisely grasp and manipulate objects.

14. Match the following:

___(a) rectus femoris, vastus later-alis, vastus medialis, vastus intermedius

___(b) biceps femoris, semitendi-nosus, semimembranosus

___(c) erector spinae, iliocostalis, longissimus spinalis

___(d) thenar, hypothenar, inter-mediate

___(e) biceps brachii, brachialis, coracobrachialis

___(f) latissimus dorsi

___(g) subscapularis, supraspinatus, infraspinatus, teres major

___(h) diaphragm, external inter-costals, internal intercostals

(1) breathing muscles
(2) constitute flexor compartment of the arm
(3) hamstrings
(4) intrinsic muscle groups of the hand
(5) muscles that strengthen and

stabilize the shoulder joint; the rotator cuff
(6) quadriceps muscle
(7) largest muscle mass of the back
(8) swimmer's muscle

15. Match the following:

___(a) trapezius
___(b) risorius
___(c) levator ani
___(d) rectus abdominis
___(e) triceps brachii
___(f) gastrocnemius
___(g) temporalis
___(h) external anal sphincter
___(i) external oblique
___(j) splenius capitis
___(k) diagastric
___(l) styloglossus

___(m) medial pterygoid
___(n) diaphragm
___(o) platysma
___(p) latissimus dorsi
___(q) flexor carpi radialis
___(r) biceps brachii
___(s) sternocleidomastoid
___(t) quadriceps femoris
___(u) myohyoid
___(v) tibialis anterior
___(w) external urethral sphincter
___(x) gluteus maximus
___(y) superior rectus

(1) muscle of facial expres-sion
(2) muscle of mastication
(3) muscle that moves the eyeballs
(4) extrinsic muscle that moves the tongue
(5) suprahyoid muscle
(6) muscle of the perineum
(7) muscle that moves the head
(8) abdominal wall muscle
(9) breathing muscle
(10) pelvic floor muscle

(11) pectoral girdle muscle
(12) muscle that moves the humerus
(13) muscle that moves the radius and ulna
(14) muscle that moves the wrist, hand, and digits
(15) muscle that moves the vertebral column
(16) muscle that moves the femur
(17) muscle that acts on the femur, tibia, and fibula
(18) muscle that moves the foot and toes

CRITICAL THINKING QUESTIONS

1. While you proceed through your daily activities, your muscles take turns being the agonist or antagonist—that is, being con-tracted or relaxed. Name some antagonistic pairs of muscles in the arm, thigh, torso, and leg. (HINT: *Antagonistic pairs are often located on opposite sides of a limb or body region.*)

2. Melody was trying to lift a heavy package using the proper back-saving position that she had been taught at work. She squatted down with her knees bent and her back straight. Then she lifted the package using her leg muscles and keeping her back straight. Identify the resistance, fulcrum, and effort involved in lifting the package. What type of lever is this? (HINT: *Determine the rela-tive order of the effort, the fulcrum, and the resistance.*)

3. When Carlos read the first question on his take-home exam, he raised one eyebrow, whistled, then closed his eyes and shook his head. Name the muscles involved in these actions. (HINT: *Try it yourself.*)

ANSWERS TO FIGURE QUESTIONS

11.1 The belly of the muscle that extends the forearm, the triceps brachii, is located posterior to the humerus.

11.2 Second-class levers produce the most force.

11.3 Possible correct responses: direction of fibers: external oblique; shape: deltoid; action: extensor digitorum; size: glu-teus maximus; origin and insertion: sternocleidomastoid; location: tibialis anterior; number of tendons of origin: bi-ceps brachii.

11.4 Frowning: frontalis; smiling: zygomaticus major; pouting: platysma; squinting: orbicularis oculi.

11.5 The inferior oblique muscle moves the eyeball superiorly and laterally because it originates at the anteromedial aspect of the floor of the orbit and inserts on the posterolateral as-pect of the eyeball.

11.6 The masseter is the strongest of the chewing muscles.

11.7 Functions of the tongue include chewing, perception of taste, swallowing, and speech.

11.8 The suprahyoid and infrahyoid muscles fix the hyoid bone to assist in tongue movements.

11.9 Triangles are important because of the structures that lie within their boundaries.

11.10 The rectus abdominis muscle aids in urination.

11.11 The diaphragm is innervated by the phrenic nerve.

11.12 The borders of the pelvic diaphragm are the pubic symphysis anteriorly, the coccyx posteriorly, and the walls of the pelvis laterally.

11.13 The borders of the perineum are the pubic symphysis anteriorly, the coccyx posteriorly, and the ischial tuberosities laterally.

11.14 The main action of the muscles that move the pectoral girdle is to stabilize the scapula to assist in movements of the humerus.

11.15 The rotator cuff consists of the flat tendons of the subscapularis, supraspinatus, infraspinatus, and teres minor muscles that form a nearly complete circle around the shoulder joint.

11.16 Most powerful forearm flexor: brachialis; most powerful forearm extensor: triceps brachii.

11.17 Flexor tendons of the digits and wrist and the median nerve pass through the flexor retinaculum.

11.18 Muscles of the thenar eminence act on the thumb.

11.19 The splenius muscles extend laterally and upward from the ligamentum nuchae at the midline.

11.20 Upper limb muscles exhibit diversity of movement; lower limb muscles function in stability, locomotion, and maintenance of posture. Moreover, lower limb muscles usually cross two joints and act equally on both.

11.21 Quadriceps femoris: rectus femoris, vastus lateralis, vastus medialis, and vastus intermedialius; hamstrings: biceps femoris, semitendinosus, and semimembranosus.

11.22 The superior and inferior extensor retinacula firmly hold the tendons of the anterior compartment muscles to the ankle.

11.23 The plantar aponeurosis supports the longitudinal arch and encloses the flexor tendons of the foot.

NERVOUS TISSUE

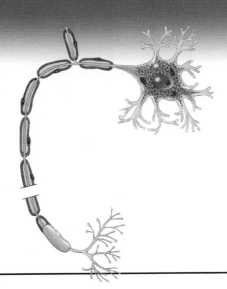

The nervous system and the endocrine system share responsibility for maintaining homeostasis. Their objective is the same—to keep controlled conditions within limits that maintain life—but their means of achieving that objective differ. Whereas the nervous system responds rapidly to stimuli by transmitting nerve impulses (action potentials) to adjust body processes, the endocrine system responds to stimuli more slowly, though no less effectively, by releasing hormones to adjust body processes. The roles of the nervous and endocrine systems in maintaining homeostasis are compared in Chapter 18, on page 566.

Besides helping to maintain homeostasis, the nervous system is responsible for our perceptions, behaviors, and memories, and it initiates all voluntary movements. Because it is quite complex, we will consider different aspects of the nervous system in several related chapters. This chapter focuses on nervous tissue. We will first discuss the organization of the nervous system, and then the basic structure and functions of the nervous system's two principal cell types: neurons (nerve cells) and neuroglia. In chapters that follow we will examine the structure and functions of the spinal cord and spinal nerves, and of the brain and cranial nerves. Then we will discuss the somatic senses—touch, pressure, warmth, cold, pain, and others—and the sensory and motor pathways to understand how nerve impulses pass into the spinal cord and brain or from the spinal cord and brain to muscles and glands. Our exploration continues with a discussion of the special senses: smell, taste, vision, hearing, and equilibrium. Finally, we conclude our study of the nervous system by examining the autonomic nervous system, the part of the nervous system that operates without voluntary control.

The branch of medical science that deals with the normal functioning and disorders of the nervous system is **neurology** (noo-ROL-ō-jē; *neur-* = nerve or nervous system; *-ology* = study of).

OVERVIEW OF THE NERVOUS SYSTEM

OBJECTIVES

• *List the structures and describe the basic functions of the nervous system.*

• *Describe the organization of the nervous system.*

Structure and Functions of the Nervous System

The **nervous system** is a complex, highly organized network of billions of neurons and even more neuroglia. The structures that make up the nervous system include the brain, cranial nerves and their branches, the spinal cord, spinal nerves and their branches, ganglia, enteric plexuses, and sensory receptors (Figure 12.1). The **brain** is housed within the skull and contains about 100 billion neurons. Twelve pairs (right and left) of **cranial nerves,** numbered I through XII, emerge from the base of the brain. A **nerve** is a bundle containing hundreds to thousands of axons plus associated connective tissue and blood vessels. Each nerve follows a defined path and serves a specific region of the body. For example, the right cranial nerve I carries signals for the sense of smell from the right side of the nose to the brain.

Figure 12.1 Major structures of the nervous system.

🔑 **The nervous system includes the brain, cranial nerves, spinal cord, spinal nerves, ganglia, enteric plexuses, and sensory receptors.**

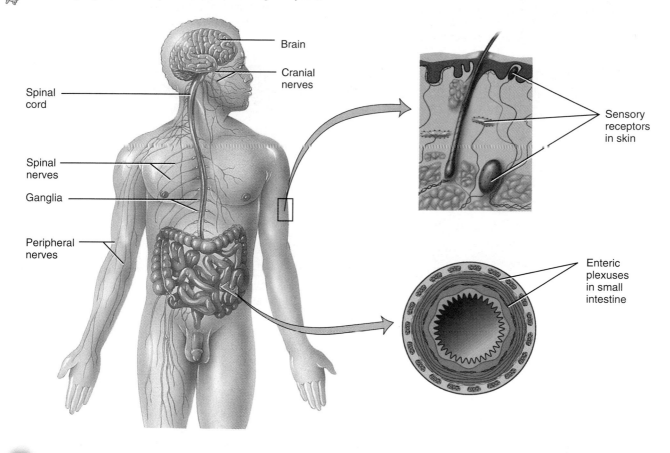

Brain

Cranial nerves

Spinal cord

Spinal nerves

Ganglia

Peripheral nerves

Sensory receptors in skin

Enteric plexuses in small intestine

Q What is the total number of cranial and spinal nerves in the human body?

The **spinal cord** connects to the brain through the foramen magnum of the skull and is encircled by the bones of the vertebral column. It contains about 100 million neurons. Thirty-one pairs of **spinal nerves** emerge from the spinal cord, each serving a specific region on the right or left side of the body. **Ganglia** are small masses of nervous tissue, containing primarily cell bodies of neurons, that are located outside the brain and spinal cord. Ganglia are closely associated with cranial and spinal nerves. In the walls of organs of the gastrointestinal tract are extensive networks of neurons, called **enteric plexuses,** that help regulate the digestive system. **Sensory receptors** are either the dendrites of sensory neurons (described shortly) or separate, specialized cells that monitor changes in the internal or external environment.

The nervous system has three basic functions:

• *Sensory function.* Sensory receptors *detect* internal stimuli, such as an increase in blood acidity, and external stimuli, such as a raindrop landing on your arm. The neurons that carry sensory information into the brain and spinal cord

are **sensory** or **afferent neurons** (AF-er-ent NOO-rons; *af-* = toward; *-ferrent* = carried).

• *Integrative function.* The nervous system *integrates* (processes) sensory information by analyzing and storing some of it and by making decisions regarding appropriate responses. The neurons that serve this function, **interneurons** *(association neurons)*, make up the vast majority of neurons in the body.

• *Motor function.* The nervous system's motor function involves *responding* to integration decisions. The neurons that serve this function are **motor** or **efferent neurons** (EF-er-ent; *ef-* = away from), which carry information out of the brain and spinal cord. The cells and organs innervated by motor neurons are termed **effectors;** examples are muscle fibers and glandular cells.

Organization of the Nervous System

The nervous system is composed of two subsystems: the **central nervous system (CNS)** and the **peripheral** (pe-RIF-er-al) **nervous system (PNS).** The CNS consists of the brain

Figure 12.2 Organization of the nervous system. Subdivisions of the PNS are the somatic nervous system (SNS), the autonomic nervous system (ANS), and the enteric nervous system (ENS).

🔑 **The two main subsystems of the nervous system are (1) the central nervous system (CNS), consisting of the brain and spinal cord, and (2) the peripheral nervous system (PNS), consisting of all nervous tissue outside the CNS.**

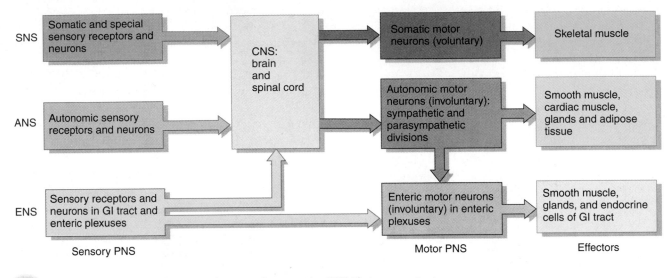

Q What terms are given to neurons that carry input to the CNS? That carry output from the CNS?

and spinal cord, which integrate and correlate many different kinds of incoming sensory information. The CNS is also the source of thoughts, emotions, and memories. Most nerve impulses that stimulate muscles to contract and glands to secrete originate in the CNS. The PNS includes all nervous tissue outside the CNS: cranial nerves and their branches, spinal nerves and their branches, ganglia, and sensory receptors.

The PNS may be subdivided further into a **somatic nervous system (SNS)** (*somat-* = body), an **autonomic nervous system (ANS)** (*auto-* = self; *-nomic* = law), and an **enteric nervous system (ENS)** (*enter-* = intestines) (Figure 12.2). The SNS consists of (1) sensory neurons that convey information from somatic and special sensory receptors primarily in the head, body wall, and limbs to the CNS, and (2) motor neurons from the CNS that conduct impulses to *skeletal muscles* only. Because these motor responses can be consciously controlled, the action of this part of the PNS is *voluntary.*

The ANS consists of (1) sensory neurons that convey information from autonomic sensory receptors, located primarily in the viscera, to the CNS, and (2) motor neurons from the CNS that conduct nerve impulses to *smooth muscle, cardiac muscle, glands,* and *adipose tissue.* As its motor responses are not normally under conscious control, the action of the ANS is *involuntary.* The motor portion of the ANS consists of two branches, the **sympathetic division** and

the **parasympathetic division.** With a few exceptions, effectors are innervated by both, and usually the two divisions have opposing actions. For example, sympathetic neurons increase heart rate, whereas parasympathetic neurons decrease it.

The ENS is the "brain of the gut," and its operation is involuntary. Once considered to be part of the ANS, the ENS consists of approximately 100 million neurons in enteric plexuses that extend the entire length of the gastrointestinal (GI) tract. Many of the neurons of the enteric plexuses function independently of the ANS and CNS to some extent, although they also communicate with the CNS via sympathetic and parasympathetic neurons. Sensory neurons of the ENS monitor chemical changes within the GI tract and the stretching of its walls. Enteric motor neurons govern contraction of GI tract smooth muscle, secretions of the GI tract organs such as acid secretion by the stomach, and activity of GI tract endocrine cells.

1. Discuss the three basic functions of the nervous system.
2. What are the functional differences between typical sensory and motor neurons? Define interneuron.
3. Construct a table to compare the components and operation of the SNS, ANS, and ENS.
4. Relate the terms *voluntary* and *involuntary* to the various subdivisions of the nervous system.

HISTOLOGY OF NERVOUS TISSUE

OBJECTIVES

• *Contrast the histological characteristics and the functions of neurons and neuroglia.*

• *Distinguish between gray matter and white matter.*

Nervous tissue consists of only two principal kinds of cells: neurons and neuroglia. Neurons are responsible for most special functions attributed to the nervous system: sensing, thinking, remembering, controlling muscle activity, and regulating glandular secretions. Neuroglia support, nourish, and protect the neurons and maintain homeostasis in the interstitial fluid that bathes neurons.

Neurons

Like muscle cells, neurons have the property of **electrical excitability,** the capability to produce *action potentials* or *impulses* in response to stimuli. Once they arise, action potentials propagate from one point to the next along the plasma membrane due to the presence of specific types of ion channels.

Parts of a Neuron

Most neurons have three parts: (1) a cell body, (2) dendrites, and (3) an axon (Figure 12.3). The **cell body** contains a nucleus surrounded by cytoplasm that includes typical organelles such as lysosomes, mitochondria, and a Golgi complex. Many neurons also contain *lipofuscin,* a pigment that occurs as clumps of yellowish brown granules in the cytoplasm. Lipofuscin is probably a product of neuronal lysosomes that accumulates as the neuron ages, but it does not seem to harm the neuron. The cell body also contains prominent clusters of rough endoplasmic reticulum, termed *Nissl bodies.* Newly synthesized proteins produced by Nissl bodies are used to replace cellular components, as material for growth of neurons, and to regenerate damaged axons in the PNS. The cytoskeleton includes both *neurofibrils,* composed of bundles of intermediate filaments that provide the cell shape and support, and *microtubules,* which assist in moving materials to and from the cell body and axon.

Two kinds of processes or extensions emerge from the cell body of a neuron: multiple dendrites and a single axon. *Nerve fiber* is a general term for any neuronal process (dendrite or axon). **Dendrites** (= little trees) are the receiving or input portions of a neuron. They usually are short, tapering, and highly branched. In many neurons the dendrites form a tree-shaped array of processes extending off the cell body. Dendrites usually are not myelinated. Their cytoplasm contains Nissl bodies, mitochondria, and other organelles.

The second type of process, the **axon** (= axis), propagates nerve impulses toward another neuron, a muscle fiber, or a gland cell. An axon is a long, thin, cylindrical projection that often joins the cell body at a cone-shaped elevation called the **axon hillock** (= small hill). The first part of the axon is called the **initial segment.** In most neurons, impulses arise at the junction of the axon hillock and the initial segment, which is called the **trigger zone,** and then conduct along the axon. An axon contains mitochondria, microtubules, and neurofibrils. Because rough endoplasmic reticulum is not present, protein synthesis does not occur in the axon. The cytoplasm, called **axoplasm,** is surrounded by a plasma membrane known as the **axolemma** (*lemma* = sheath or husk). Along the length of an axon, side branches called **axon collaterals** may branch off, typically at a right angle to the axon. The axon and its collaterals end by dividing into many fine processes called **axon terminals.**

The site of communication between two neurons or between a neuron and an effector cell is called a **synapse.** The tips of some axon terminals swell into bulb-shaped structures called **synaptic end bulbs,** whereas others exhibit a string of swollen bumps called **varicosities.** Both synaptic end bulbs and varicosities contain many minute membrane-enclosed sacs called **synaptic vesicles** that store a chemical **neurotransmitter.** Neurons were long thought to liberate just one type of neurotransmitter at all synaptic end bulbs. We now know that many neurons contain two or even three neurotransmitters. The neurotransmitter molecules released from synaptic vesicles then influence the activity of other neurons, muscle fibers, or gland cells.

The cell body is where a neuron synthesizes new cell products or recycles old ones. However, because some substances are needed in the axon or at the axon terminals, two types of transport systems carry materials from the cell body to the axon terminals and back. The slower system, which moves materials about 1–5 mm per day, is called **slow axonal transport.** It conveys axoplasm in one direction only—from the cell body toward the axon terminals. It supplies new axoplasm for developing or regenerating axons and replenishes axoplasm in growing and mature axons.

The faster system, which is capable of moving materials a distance of 200–400 mm per day, is called **fast axonal transport.** It uses proteins that function as "motors" to move materials in both directions—away from and toward the cell body—along the surfaces of microtubules. Fast axonal transport moves various organelles and materials that form the membranes of the axolemma, synaptic end bulbs, and synaptic vesicles. Some materials returned to the cell body are degraded or recycled, and others influence its growth.

CLINICAL APPLICATION
Tetanus

Fast axonal transport is the route by which some toxins and disease-causing viruses make their way from axon terminals near skin cuts to cell bodies of neurons, where

Figure 12.3 Structure of a typical neuron. Arrows indicate the direction of information flow: dendrites → cell body → axon → axon terminals. The break in the figure indicates that the axon actually is longer than shown.

🔑 **The basic parts of a neuron are dendrites, a cell body (soma), and an axon.**

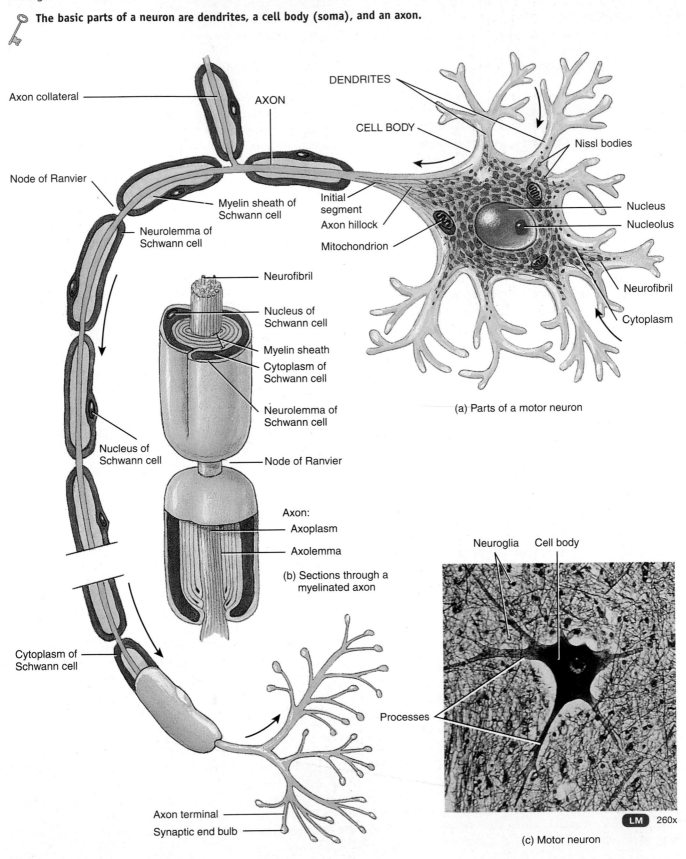

Axon collateral

AXON

Node of Ranvier

Myelin sheath of Schwann cell

Neurolemma of Schwann cell

DENDRITES

CELL BODY

Nissl bodies

Initial segment

Axon hillock

Mitochondrion

Nucleus

Nucleolus

Neurofibril

Cytoplasm

(a) Parts of a motor neuron

Neurofibril

Nucleus of Schwann cell

Myelin sheath

Cytoplasm of Schwann cell

Neurolemma of Schwann cell

Node of Ranvier

Nucleus of Schwann cell

Axon:
Axoplasm
Axolemma

(b) Sections through a myelinated axon

Neuroglia   Cell body

Processes

LM   260x

(c) Motor neuron

Cytoplasm of Schwann cell

Axon terminal

Synaptic end bulb

Q What roles do the dendrites, cell body, and axon play in communication of signals?

Figure 12.4 Structural classification of neurons. Breaks indicate that axons are longer than shown.

🔑 A multipolar neuron has many processes extending from the cell body, whereas a bipolar neuron has two and a unipolar neuron has one.

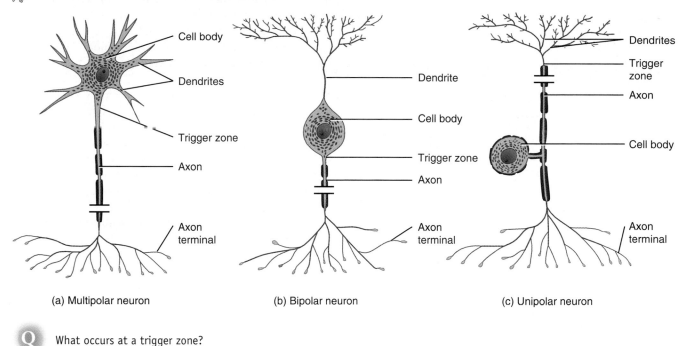

(a) Multipolar neuron (b) Bipolar neuron (c) Unipolar neuron

Q What occurs at a trigger zone?

they can cause damage. For example, the toxin produced by *Clostridium tetani* bacteria is carried by fast axonal transport to the CNS. There it disrupts the actions of motor neurons, causing prolonged painful muscle spasms—a condition called **tetanus.** The delay between the release of the toxin and the first appearance of symptoms is due, in part, to the time required for transport of the toxin to the cell body. For this reason, a laceration or puncture injury to the head or neck is a more serious matter than a similar injury in the leg. The closer the site of injury is to the brain, the shorter the transit time. Thus, treatment must begin quickly to prevent symptoms of tetanus. ■

Structural Diversity in Neurons

Neurons display great diversity in size and shape. For example, their cell bodies range in diameter from 5 micrometers (μm) (smaller than a red blood cell) up to 135 μm (barely large enough to see with the unaided eye). The pattern of dendritic branching is varied and distinctive for neurons in different parts of the nervous system. A few small neurons lack an axon, and many others have very short axons, but the longest neurons have axons that extend for a meter (3.2 ft) or more.

Both structural and functional features are used to classify the various neurons in the body. Structurally, neurons are classified according to the number of processes extending from the cell body (Figure 12.4). **Multipolar neurons**

usually have several dendrites and one axon (see also Figure 12.3). Most neurons in the brain and spinal cord are of this type. **Bipolar neurons** have one main dendrite and one axon; they are found in the retina of the eye, in the inner ear, and in the olfactory area of the brain. **Unipolar neurons** are sensory neurons that originate in the embryo as bipolar neurons. During development, the axon and dendrite fuse into a single process that divides into two branches a short distance from the cell body. Both branches have the characteristic structure and function of an axon: They are long, cylindrical processes that may be myelinated, and they propagate action potentials. However, the axon branch that extends into the periphery has unmyelinated dendrites at its distal tip, whereas the axon branch that extends into the CNS ends in synaptic end bulbs. The dendrites monitor a sensory stimulus such as touch or stretching. The trigger zone for nerve impulses in a unipolar neuron is at the junction of the dendrites and axon (see Figure 12.4c). The impulses then propagate toward the synaptic end bulbs.

Most of the neurons in the body—perhaps 90%—are interneurons of thousands of different types. Interneurons often are named for the histologist who first described them; examples are **Purkinje cells** (pur-KIN-jē) in the cerebellum (Figure 12.5a) and **Renshaw cells** in the spinal cord. Others are named for some aspect of their shape or appearance. For example, **pyramidal cells** (pi-RAM-i-dal), found in the brain, have a cell body shaped like a pyramid (Figure 12.5b).

Figure 12.5 Two examples of interneurons. Arrows indicate the direction of information flow.

🔑 **Interneurons carry nerve impulses from one neuron to another.**

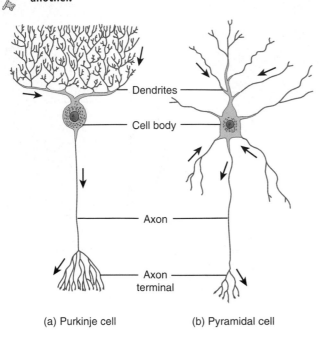

(a) Purkinje cell   (b) Pyramidal cell

Q Do the dendrites of interneurons receive input or provide output?

Neuroglia

Neuroglia (noo-RŌG-lē-a; *-glia* = glue) or **glia** constitute about half the volume of the CNS. Their name derives from early microscopists' idea that they were the "glue" that held nervous tissue together. We now know that neuroglia are not merely passive bystanders but rather active participants in the operation of nervous tissue. Generally, neuroglia are smaller than neurons, and they are 5–50 times more numerous. In contrast to neurons, glia do not generate or propagate action potentials, and they can multiply and divide in the mature nervous system. In cases of injury or disease, neuroglia multiply to fill in the spaces formerly occupied by neurons. Brain tumors derived from glia, called **gliomas,** tend to be highly malignant and grow rapidly. Of the six types of neuroglia, four—astrocytes, oligodendrocytes, microglia, and ependymal cells—are found only in the CNS. The remaining two types—Schwann cells and satellite cells—are present in the PNS. The structure and functions of neuroglia are summarized in Table 12.1 on page 386.

Myelination

The axons of most mammalian neurons are surrounded by a multilayered lipid and protein covering, called the **myelin sheath,** that is produced by neuroglia. The sheath electrically insulates the axon of a neuron and increases the speed of nerve impulse conduction. Axons with such a covering are said to be **myelinated,** whereas those without it are **unmyelinated** (Figure 12.6). Electron micrographs reveal that even unmyelinated axons are surrounded by a thin coat of neuroglial plasma membrane.

Two types of neuroglia produce myelin sheaths: Schwann cells (in the PNS) and oligodendrocytes (in the CNS). In the PNS, **Schwann cells** begin to form myelin sheaths around axons during fetal development. Each Schwann cell wraps about 1 millimeter (1 mm = 0.04 in.) of a single axon's length by spiraling many times around the axon (Figure 12.6b). Eventually, multiple layers of glial plasma membrane surround the axon, with the Schwann cell's cytoplasm and nucleus forming the outermost layer. The inner portion, consisting of up to 100 layers of Schwann cell membrane, is the myelin sheath. The outer nucleated cytoplasmic layer of the Schwann cell, which encloses the myelin sheath, is the **neurolemma (sheath of Schwann).** A neurolemma is found only around axons in the PNS. When an axon is injured, the neurolemma aids regeneration by forming a regeneration tube that guides and stimulates regrowth of the axon. Gaps in the myelin sheath, called **nodes of Ranvier** (RON-vē-ā), appear at intervals along the axon (see Figure 12.3). Each Schwann cell wraps one axon segment between two nodes.

In the CNS, an **oligodendrocyte** myelinates parts of many axons in somewhat the same manner that a Schwann cell myelinates part of a single PNS axon (see Table 12.1). It puts forth an average of 15 broad, flat processes that spiral around CNS axons, forming a myelin sheath. A neurolemma is not present, however, because the oligodendrocyte cell body and nucleus do not envelop the axon. Nodes of Ranvier are present, but they are fewer in number. Axons in the CNS display little regrowth after injury. This is thought to be due, in part, to the absence of a neurolemma, and in part to an inhibitory influence exerted by the oligodendrocytes on axon regrowth.

The amount of myelin increases from birth to maturity, and its presence greatly increases the speed of nerve impulse conduction. An infant's responses to stimuli are neither as rapid nor as coordinated as those of an older child or an adult, in part because myelination is still in progress during infancy. Certain diseases, such as multiple sclerosis (see page 406) and Tay–Sachs disease (see page 831), destroy myelin sheaths.

Gray and White Matter

In a freshly dissected section of the brain or spinal cord, some regions look white and glistening, whereas others appear gray (Figure 12.7). The **white matter** is aggregations of myelinated processes from many neurons. The whitish color of myelin gives white matter its name. The **gray matter** of the nervous system contains neuronal cell bodies, dendrites,

Figure 12.6 Myelinated and unmyelinated axons.

🔑 **Axons of most mammalian neurons are surrounded by a myelin sheath produced by Schwann cells in the PNS and by oligodendrocytes in the CNS.**

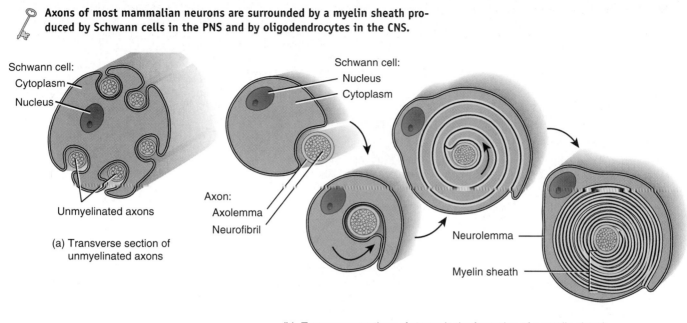

Schwann cell:
Cytoplasm
Nucleus

Unmyelinated axons

(a) Transverse section of unmyelinated axons

Schwann cell:
Nucleus
Cytoplasm

Axon:
Axolemma
Neurofibril

Neurolemma

Myelin sheath

(b) Transverse sections of stages in the formation of a myelin sheath

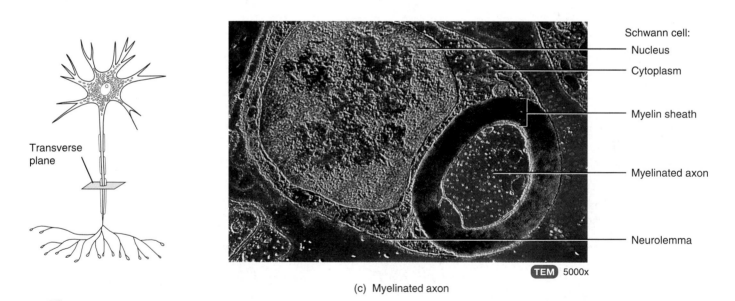

Transverse plane

Schwann cell:
Nucleus
Cytoplasm
Myelin sheath
Myelinated axon
Neurolemma

TEM 5000x

(c) Myelinated axon

Q What is the functional advantage of myelination?

unmyelinated axons, axon terminals, and neuroglia. It looks grayish, rather than white, because there is little or no myelin in these areas. Blood vessels are present in both white and gray matter.

In the spinal cord, the white matter surrounds an inner core of gray matter shaped like a butterfly or the letter H; in the brain, a thin shell of gray matter covers the surface of the largest portions of the brain, the cerebrum and cerebellum (see Figure 12.7). Many nuclei of gray matter also lie deep within the brain. When used to describe nervous tissue, a **nucleus** is a cluster of neuronal cell bodies within the CNS. The arrangements of gray and white matter in the spinal cord and brain are described more extensively in Chapters 13 and 14.

Table 12.1 Neuroglia

TYPE	APPEARANCE	FUNCTIONS
Central Nervous System		
Astrocytes (AS-trō-sīts; *astro-* = star; *-cyte* = cell)	Star-shaped, with many processes. 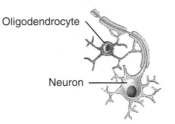	Help maintain appropriate chemical environment for generation of neuron action potentials; provide nutrients to neurons; take up excess neurotransmitters; participate in the metabolism of neurotransmitters; maintain proper balance of Ca^{2+} and K^+; assist with migration of neurons during brain development; help form the blood–brain barrier.
Oligodendrocytes (OL-i-gō-den´-drō-sīts; *oligo-* = few; *dendro-* = tree)	Smaller than astrocytes, with fewer processes; round or oval cell body.	Form supporting network around CNS neurons; produce myelin sheath around several adjacent axons of CNS neurons.
Microglia (mī-KROG-lē-a; *micro-* = small)	Small cells with few processes; derived from mesodermal cells that also give rise to monocytes and macrophages.	Protect CNS cells from disease by engulfing invading microbes; clear away debris of dead cells; migrate to areas of injured nerve tissue.

Figure 12.7 Distribution of gray and white matter in the spinal cord and brain.

🔑 **White matter consists of myelinated processes of many neurons. Gray matter consists of neuron cell bodies, dendrites, axon terminals, bundles of unmyelinated axons, and neuroglia.**

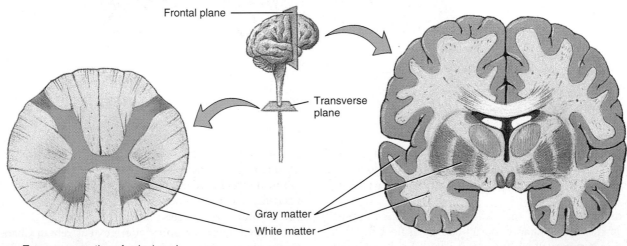

Transverse section of spinal cord

Frontal section of cerebrum of brain

Q What is responsible for the white appearance of white matter?

Table 12.1 (continued)

TYPE	APPEARANCE	FUNCTIONS
Central Nervous System (continued)		
Ependymal cells (ep-EN-di-mal; *epen-* = above; *dym-* = garment)	Epithelial cells arranged in a single layer; range in shape from cuboidal to columnar; many are ciliated.	Line ventricles of the brain (spaces filled with cerebrospinal fluid) and central canal of the spinal cord; form cerebrospinal fluid and assist in its circulation.
Peripheral Nervous System		
Schwann cells (neurolemmocytes)	Flattened cells that encircle PNS axons.	Each cell produces part of the myelin sheath around a single axon of a PNS neuron; participate in regeneration of PNS axons.
Satellite cells (SAT-i-līt)	Flattened cells arranged around the cell bodies of neurons in ganglia.	Support neurons in PNS ganglia.

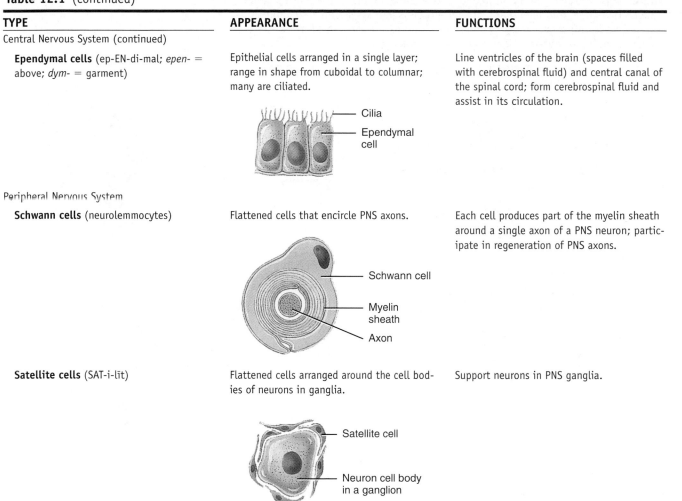

1. Describe the parts of a neuron and the functions of each.
2. Give several examples of the structural diversity of neurons.
3. Define neurolemma. Why is it important?
4. With reference to the nervous system, what is a nucleus?

ELECTRICAL SIGNALS IN NEURONS

OBJECTIVE

• *Describe the cellular properties that permit communication among neurons and effectors.*

Like muscle fibers, neurons are electrically excitable. They communicate with one another using two types of electrical signals: *action potentials,* which allow communication over both short and long distances within the body, and *graded potentials,* which are used for short-distance communication only. Production of both types of signal depends on two basic features of the plasma membrane of excitable cells: the existence of a resting membrane potential and the presence of specific ion channels.

Like most other cells in the body, the plasma membrane of excitable cells exhibits a **membrane potential,** an electrical voltage difference across the membrane. In excitable cells this voltage is termed the **resting membrane potential,** and it is like voltage stored in a battery. If you connect the positive and negative terminals of a battery with a piece of wire, electrons will flow along the wire. This flow of charged particles is called **current.** In living cells, the flow of ions (rather than electrons) constitutes the current.

Graded potentials and action potentials occur because the plasma membranes of neurons contain many different kinds of ion channels that open or close in response to spe-

Figure 12.8 Voltage-gated and ligand-gated ion channels in the plasma membrane. (a) A change in membrane potential opens voltage-gated K⁺ channels during an action potential. (b) A chemical stimulus—here, the neurotransmitter acetylcholine—opens a ligand-gated ion channel.

🔑 **Whereas leakage channels are always open, gated channels open and close in response to a stimulus.**

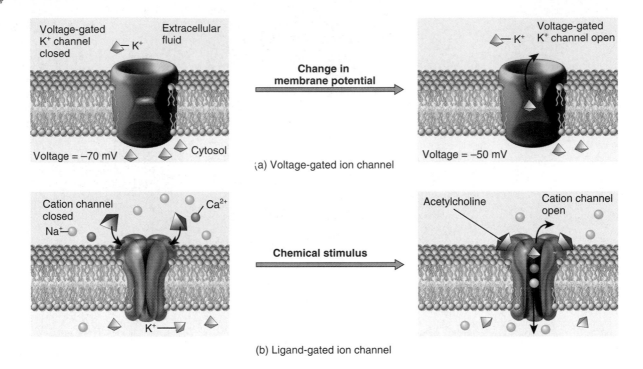

(a) Voltage-gated ion channel

(b) Ligand-gated ion channel

❓ What type of gated ion channel (not shown here) is activated by, for example, a touch on the arm?

cific stimuli. Because the lipid bilayer of the plasma membrane is a good electrical insulator, the main paths for current to flow across the membrane are through the ion channels.

Ion Channels

When they are open, ion channels allow specific ions to diffuse across the plasma membrane, down their electrochemical gradient. Put another way, ions tend to move "downhill" from where they are more concentrated to where they are less concentrated. Similarly, cations tend to move toward a negatively charged area, and anions tend to move toward a positively charged area. In all cases, as ions diffuse across a plasma membrane through ion channels, the result is a flow of current that can change the membrane potential.

There are two basic types of ion channels: leakage channels and gated channels. **Leakage channels** are always open, like a garden hose that has numerous small holes along its length. Because plasma membranes typically have many more potassium ion (K⁺) leakage channels than sodium ion (Na⁺) leakage channels, the membrane's permeability to K⁺ is much higher than its permeability to Na⁺.

Gated channels, in contrast, open and close in response to some sort of stimulus. The presence of gated ion channels in the plasma membranes of neurons and muscle fibers gives these cells their property of electrical excitability. Such excitable cells have three kinds of gated ion channels, based on the stimuli involved: voltage-gated, ligand-gated, and mechanically gated ion channels.

1. A **voltage-gated ion channel** opens in response to a change in membrane potential (voltage) (Figure 12.8a). Voltage-gated channels are used in the generation and conduction of action potentials.

2. A **ligand-gated ion channel** opens and closes in response to a specific chemical stimulus. A wide variety of chemical

Figure 12.9 The distributions of (a) charges and (b) ions that produce the resting membrane potential.

> The resting membrane potential is due to a small buildup of anions, mainly organic phosphates (PO_4^{3-}) and proteins, in the cytosol just inside the membrane and an equal buildup of cations, mainly sodium ions (Na^+), in the extracellular fluid just outside the membrane.

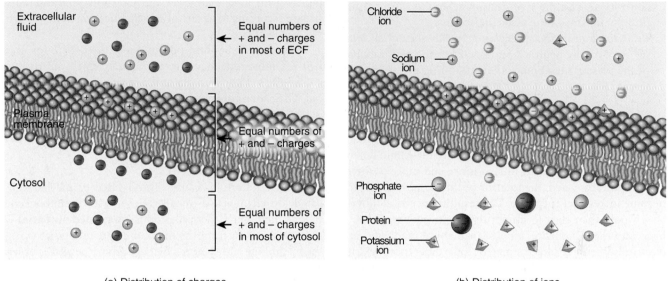

(a) Distribution of charges

(b) Distribution of ions

Q What is a typical value for the resting membrane potential of a neuron?

ligands—including neurotransmitters, hormones, and ions—open or close ligand-gated ion channels. The neurotransmitter acetylcholine, for example, opens cation channels that allow Na^+ and Ca^{2+} to diffuse inward and K^+ to diffuse outward (Figure 12.8b). Ligand-gated ion channels operate in two basic ways. The ligand molecule itself may open or close the channel by binding to a portion of the channel protein, as in the example of acetylcholine. Or the ligand may act *indirectly* via a type of membrane protein called a G protein that activates another molecule, a "second messenger," in the cytosol, which in turn operates the channel's gate. Hormonal ligands often work by such second-messenger systems (see Figure 18.4 on page 571), as do some neurotransmitters.

3. A **mechanically gated ion channel** opens or closes in response to mechanical stimulation in the form of vibration (such as sound waves), pressure (such as touch), or tissue stretching. The force distorts the channel from its resting position, opening the gate. Mechanically gated ion channels are found in auditory receptors in the ears, in receptors that monitor stomach stretching, and in touch receptors in the skin.

Resting Membrane Potential

OBJECTIVE

• *Describe the factors that maintain a resting membrane potential.*

The resting membrane potential exists because of a small buildup of negative ions in the cytosol along the inside of the membrane, and an equal buildup of positive ions in the extracellular fluid along the outside surface of the membrane (Figure 12.9a). Such a separation of positive and negative electrical charges is a form of potential energy, which is measured in volts or millivolts (1 mV = 0.001V). The greater the difference in charge across the membrane, the larger the membrane potential (voltage). Note in Figure 12.9a that the buildup of charge occurs only very close to the membrane. The cytosol or extracellular fluid elsewhere contains equal numbers of positive and negative charges.

In neurons, the resting membrane potential ranges from −40 to −90 mV. A typical value is −70 mV. The minus sign indicates that the inside is negative relative to the outside. A cell that exhibits a membrane potential is said to be **polarized.** Most body cells are polarized; the membrane voltage varies from +5 mV to −100 mV in different types of cells.

The resting membrane potential is maintained primarily by two factors:

1. *Unequal distribution of ions across the plasma membrane.* Recall that the concentrations of major anions and cations are different outside and inside cells (Figure 12.9b). Extracellular fluid is rich in Na^+ and chloride ions (Cl^-). In cytosol, however, the main cation is K^+ (potassium ions), and the two dominant anions are organic phosphates and amino acids in proteins.

2. *Relative permeability of the plasma membrane to Na^+ and K^+.* In a resting neuron or muscle fiber, the permeability of the plasma membrane is 50–100 times greater to K^+ than to Na^+.

To understand how these factors maintain the resting membrane potential, first consider what would happen if the membrane were permeable only to K^+. This cation would tend to leak out of the cell into the extracellular fluid down its concentration gradient. But, as more and more positive potassium ions exited, the interior of the membrane would become increasingly negative. The resulting electrical difference would then start to attract the positively charged K^+ back into the cell. Hence, an electrochemical gradient—an electrical difference plus a concentration difference—exists across the membrane. Eventually, just as many K^+ would be entering because of the electrical difference as would be exiting because of the concentration (chemical) gradient. The membrane potential that just balances the K^+ concentration difference is -90 mV and is called the *potassium equilibrium potential*. The resting membrane potential (-70 mV) is close to, but not exactly equal to, the potassium equilibrium potential, which means that the membrane must be slightly permeable to other ions as well.

Actually, the membrane is moderately permeable to K^+ and Cl^- and very slightly permeable to Na^+. In theory, the electrical effect of K^+ outflow might be canceled if Na^+ could flow inward, in effect exchanging one positive ion for another. But because the membrane permeability to Na^+ is so low, inward leakage of Na^+ is far too slow to keep pace with outward leakage of K^+. A second way to cancel the electrical effect of K^+ outflow might be the simultaneous exit of anions. Most anions in the cell, however, are not free to leave; they are attached either to large proteins or to other organic molecules, such as phosphates in ATP. Finally, inward leakage of Cl^- down its concentration gradient cannot cancel the electrical effect of K^+ outflow. Any chloride ions that enter the cell can only make the inside more negative. As a result of the low permeability of the membrane to Na^+ and to anions inside the cell and the higher permeability to K^+, the fluid along the inner surface of the plasma membrane becomes increasingly more negatively charged as K^+ leaves.

Note that both the electrical and concentration gradients promote Na^+ inflow: The negatively charged interior of the membrane attracts cations, and the concentration of Na^+ is higher outside the cell. Even though membrane permeability to Na^+ is very low, a slow leak would eventually destroy the electrochemical gradient. The small inward Na^+ leak and outward K^+ leak are counteracted by the sodium pumps (Na^+/K^+ ATPase; see Figure 3.11 on page 71); they help maintain the resting membrane potential by pumping out Na^+ as fast as it leaks in. At the same time, the sodium pumps bring in K^+. However, the potassium ions also redistribute according to electrical and chemical gradients, as previously described.

Because the sodium pumps expel three Na^+ for each two K^+ imported, they are *electrogenic,* which means they contribute to the negativity of the resting membrane potential. Their total contribution, however, is very small—about -3 mV of the total -70 mV resting membrane potential in a typical neuron.

Graded Potentials

When a stimulus causes ligand-gated or mechanically gated ion channels to open or close in an excitable cell's plasma membrane, that cell produces a **graded potential**— a small deviation from the membrane potential that makes the membrane either more polarized (more negative) or less polarized (less negative). When the response is a more negative polarization, it is termed a **hyperpolarizing graded potential** (Figure 12.10a). When the response is a less negative polarization, it is termed a **depolarizing graded potential** (Figure 12.10b).

Typically, ligand-gated and mechanically gated ion channels are present in the dendrites of sensory neurons, and ligand-gated ion channels are most numerous in the dendrites and cell bodies of interneurons and motor neurons. Ligand-gated channels are occasionally present in axons. Hence, graded potentials occur most often in the dendrites and cell body of a neuron, and less often in the axon. To say that these electrical signals are *graded* means that they vary in amplitude (size), depending on the strength of the stimulus. They are larger or smaller depending on how many ion channels have opened (or closed) and how long each remains open. The opening or closing of ion channels alters the flow of specific ions across the membrane, producing a flow of current that is *localized,* which means that it spreads along the plasma membrane for only a few micrometers before it dies out. Thus, graded potentials are useful only for short-distance communication.

Graded potentials have different names depending on which type of stimulus causes them and where they occur. For example, when a neurotransmitter binds to its receptors at a ligand-gated ion channel, it produces a graded potential called a *postsynaptic potential* (explained shortly). On the other hand, sensory receptors and sensory neurons produce graded potentials termed *receptor potentials* and *generator potentials* (explained in Chapter 15).

Figure 12.10 Graded potentials. Most graded potentials occur in the dendrites and cell body (as indicated by the blue outline in the inset).

🔑 **In a hyperpolarizing graded potential, the membrane polarization becomes more negative than the resting level, whereas in a depolarizing graded potential the membrane polarization becomes less negative than the resting level.**

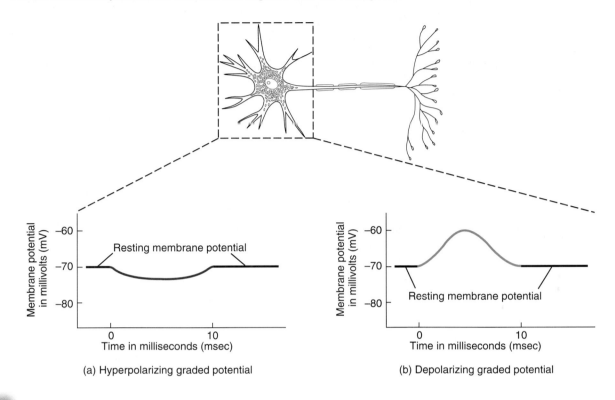

(a) Hyperpolarizing graded potential

(b) Depolarizing graded potential

Q What types of ion channels produce graded potentials when they open or close?

Action Potentials

OBJECTIVE

• *List the sequence of events involved in the generation of a nerve action potential.*

An **action potential (AP)** or **impulse** is a sequence of rapidly occurring events that decrease and eventually reverse the membrane potential, and then restore it to the resting state. During an action potential, two types of voltage-gated ion channels open and then close. These channels are present mainly in the plasma membrane of the axon and axon terminals. The first channels that open allow Na^+ to rush into the cell, which causes depolarization. Then K^+ channels open, allowing K^+ to flow out, which produces repolarization. Together, the depolarization and repolarization phases last about 1 msec (0.001 sec) in a typical neuron.

Action potentials arise according to the **all-or-none principle:** If depolarization reaches a certain level (about −55 mV in many neurons), then the voltage-gated ion channels open, and an action potential that is always the same size (amplitude) occurs. This situation is similar to pushing on the first domino in a long row: When the push on the first domino is strong enough (reaches about −55 mV), that domino falls against the second domino, and the *entire* row topples (an action potential occurs). Stronger pushes on the first domino produce the identical effect— complete toppling. Thus, pushing on the first domino produces an all-or-none event: Either it and all the dominoes fall, or it and all the dominoes remain standing. Because they can travel long distances without dying out, APs function in both short- and long-distance communication.

Depolarizing Phase

If a depolarizing graded potential or some other stimulus causes the membrane to depolarize to a critical level— called **threshold** (typically, about −55 mV)—then voltage-gated Na^+ channels rapidly start to open. Both the electrical and the chemical gradients favor inward diffusion of Na^+ through the opening channels, and the resulting inrush of Na^+ causes the **depolarizing phase** of the action potential (Figure 12.11). The inflow of Na^+ becomes so large that the

Figure 12.11 Action potential (AP) or impulse. When a stimulus depolarizes the membrane to threshold, an AP is generated. The green outline in the inset indicates the presence of voltage-gated Na⁺ and K⁺ channels in the membrane of the axon and axon terminals.

🔑 **An action potential consists of depolarizing and repolarizing phases.**

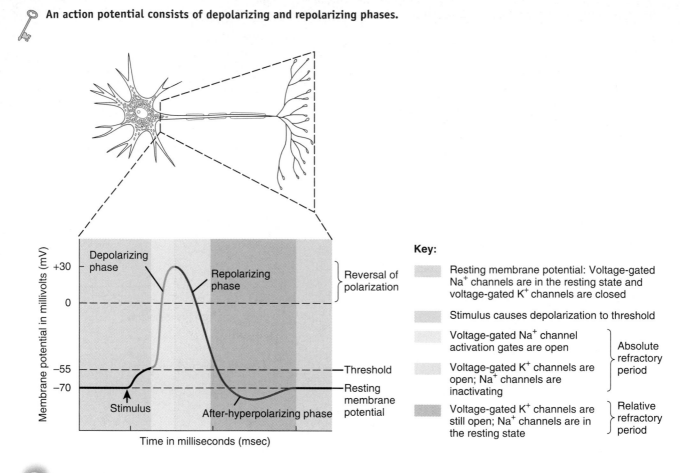

Key:

Resting membrane potential: Voltage-gated Na⁺ channels are in the resting state and voltage-gated K⁺ channels are closed

Stimulus causes depolarization to threshold

Voltage-gated Na⁺ channel activation gates are open ⎫
Voltage-gated K⁺ channels are open; Na⁺ channels are inactivating ⎬ Absolute refractory period
⎭

Voltage-gated K⁺ channels are still open; Na⁺ channels are in the resting state ⎫ Relative refractory period ⎬ ⎭

Q Which channels are open during depolarization? During repolarization?

membrane potential changes from −55 mV, passes 0 mV, and finally reaches +30 mV, at which point the inside of the membrane becomes 30 mV more positive than the outside.

Each voltage-gated Na⁺ channel has two separate gates, an *activation gate* and an *inactivation gate*. In a resting membrane, the inactivation gate is open, but the activation gate is closed (step 1 in Figure 12.12). As a result, Na⁺ cannot diffuse into the cell through these channels. This is the *resting state* of a voltage-gated Na⁺ channel. At threshold, many voltage-gated Na⁺ channels suddenly change from the resting state to the *activated state:* Both the activation and inactivation gates in the channel are open, and Na⁺ moves inward (step 2 in Figure 12.12). As more channels open, more Na⁺ moves inward, the membrane depolarizes further, and so on. Because depolarization causes inflow of Na⁺, which causes further depolarization, which increases Na⁺ inflow, this is a positive feedback system. Different neurons may have different thresholds for generation of an action potential, but the threshold in any one neuron usually is constant.

The same depolarization that opens activation gates also closes inactivation gates for Na⁺ (step 3 in Figure 12.12). This is called the *inactivated state* of the channel. But the *inactivation gate closes* a few ten-thousandths of a second *after the activation gate opens*. Thus, a voltage-gated Na⁺ channel is open for a few ten-thousandths of a second. While a channel is open, about 20,000 Na⁺ flow across the membrane and change the membrane potential considerably. But because this number represents only one of every million Na⁺ in the fluid just outside the membrane at this site, the change in the Na⁺ concentration is very small. Because only a few thousand Na⁺ enter during a single action potential, the sodium pumps easily bail them out and maintain the low concentration of Na⁺ inside the cell.

Repolarizing Phase

A threshold depolarization opens both voltage-gated Na⁺ channels and voltage-gated K⁺ channels (steps 3 and 4 in Figure 12.12). The voltage-gated K⁺ channels open more

Figure 12.12 Changes in ion flow through voltage-gated ion channels during the depolarizing and repolarizing phases of an action potential. (Leakage channels and sodium pumps are not shown.) Adapted from Becker et al., *The World of the Cell* 3e, F22-18, p732 (Menlo Park, CA: Benjamin/ Cummings, 1996) © 1996 The Benjamin/Cummings Publishing Company.

🔑 **Inflow of sodium ions (Na⁺) causes the depolarizing phase, and outflow of potassium ions (K⁺) causes the repolarizing phase.**

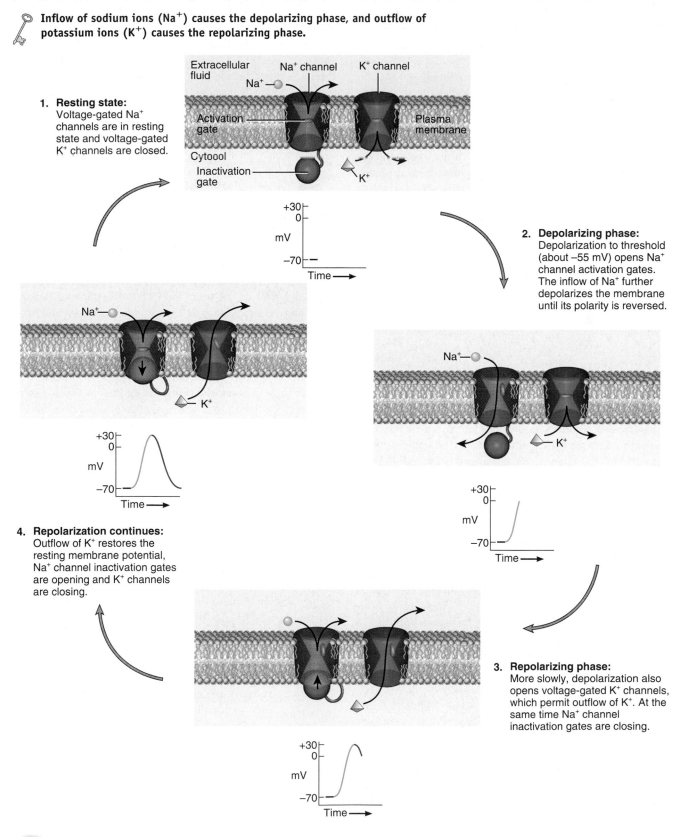

1. **Resting state:**
   Voltage-gated Na⁺ channels are in resting state and voltage-gated K⁺ channels are closed.

2. **Depolarizing phase:**
   Depolarization to threshold (about −55 mV) opens Na⁺ channel activation gates. The inflow of Na⁺ further depolarizes the membrane until its polarity is reversed.

3. **Repolarizing phase:**
   More slowly, depolarization also opens voltage-gated K⁺ channels, which permit outflow of K⁺. At the same time Na⁺ channel inactivation gates are closing.

4. **Repolarization continues:**
   Outflow of K⁺ restores the resting membrane potential, Na⁺ channel inactivation gates are opening and K⁺ channels are closing.

Ⓠ Given the existence of leakage channels for both K⁺ and Na⁺, could the membrane repolarize if the voltage-gated K⁺ channels did not exist?

slowly, however, so their opening occurs at about the same time the voltage-gated Na$^+$ channels are closing. The slower opening of voltage-gated K$^+$ channels, and the closing of previously open Na$^+$ channels, produce the **repolarizing phase** of the action potential, in which the resting membrane potential is restored. As the Na$^+$ channels are inactivated, Na$^+$ inflow slows. At the same time, the K$^+$ channels are opening, and K$^+$ outflow accelerates. Slowing of Na$^+$ inflow and acceleration of K$^+$ outflow causes the membrane potential to change from +30 mV to 0 mV to −70 mV. Repolarization also allows inactivated Na$^+$ channels to revert to their resting state.

While the voltage-gated K$^+$ channels are open, outflow of K$^+$ may be large enough to cause an **after-hyperpolarizing phase** of the action potential (see Figure 12.11). During this phase the membrane is even more permeable to K$^+$ than in the resting state, and the membrane potential drifts toward the potassium equilibrium potential (about −90 mV). As the voltage-gated K$^+$ channels close, however, the membrane potential returns to the resting level of −70 mV. In contrast to voltage-gated Na$^+$ channels, most voltage-gated K$^+$ channels do not exhibit an inactivated state. Instead, they flip back and forth between closed (resting) and open (activated) states.

Refractory Period

The period of time during which an excitable cell cannot generate another action potential is called the **refractory period** (see key in Figure 12.11). During the **absolute refractory period,** a second action potential cannot be initiated, even with a very strong stimulus. This period coincides with the period of Na$^+$ channel activation and inactivation. Inactivated Na$^+$ channels cannot reopen; they first must return to the resting state. In contrast to action potentials, graded potentials do not exhibit a refractory period.

Large-diameter axons have a brief absolute refractory period of about 0.4 msec, and a second nerve impulse can arise very quickly—up to 1000 impulses per second are possible. Small-diameter axons, however, have absolute refractory periods as long as 4 msec, and thus they can transmit a maximum of 250 impulses per second. Under normal body conditions, the maximum frequency of nerve impulses in different axons ranges between 10 and 1000 per second.

The **relative refractory period** is the period of time during which a second action potential can be initiated, but only by a suprathreshold (larger than threshold) stimulus. It coincides with the period when the voltage-gated K$^+$ channels are still open after inactivated Na$^+$ channels have returned to their resting state (see Figure 12.11).

Propagation of Nerve Impulses

To communicate information from one part of the body to another, nerve impulses must travel from where they arise at a trigger zone, often the axon hillock, to axon terminals.

The special mode of impulse travel is called **propagation** or **conduction,** and it depends on positive feedback. As sodium ions flow in, depolarization increases and opens voltage-gated Na$^+$ channels in adjacent portions of membrane. Thus, the nerve impulse self-propagates along the membrane, rather like toppling that long row of dominoes by pushing over the first one in the line. Also, because the membrane is refractory behind the leading edge of a nerve impulse, it normally moves in one direction only—from where it arises at the trigger zone toward the axon terminals.

CLINICAL APPLICATION
Local Anesthetics

Local anesthetics are drugs that block pain and other somatic sensations. Examples include procaine (Novocaine) and lidocaine, which may be used to produce anesthesia in the skin during suturing of a gash, in the mouth during dental work, or in the lower body during childbirth. These drugs act by blocking the opening of voltage-gated Na$^+$ channels. Nerve impulses cannot pass the obstructed region, so pain signals do not reach the CNS. Smaller-diameter axons, such as those carrying pain signals, are more sensitive than larger-diameter fibers to low doses of anesthetic drugs. ■

Continuous and Saltatory Conduction

The type of impulse conduction discussed thus far occurs in muscle fibers and unmyelinated axons. Such a step-by-step depolarization of each adjacent portion of the plasma membrane is called **continuous conduction** (Figure 12.13a). Ionic currents flow across each adjacent portion of the membrane. Note that the impulse has propagated only a relatively short distance after 10 msec.

In myelinated axons, conduction is somewhat different. Wherever the myelin sheath covers the axolemma, there are few voltage-gated ion channels. At the nodes of Ranvier, however, the myelin sheath is interrupted, and the axolemma has a high density of voltage-gated ion channels. Membrane depolarization occurs at the nodes, and current carried by Na$^+$ and K$^+$ flows across the plasma membrane. When a nerve impulse propagates along a myelinated axon, current carried by ions flows through the extracellular fluid surrounding the myelin sheath and through the cytosol from one node to the next (Figure 12.13b). The nerve impulse at the first node generates ionic currents in the cytosol and extracellular fluid that open voltage-gated Na$^+$ channels at the second node. At the second node, the ion flows trigger a nerve impulse. Then the nerve impulse from the second node generates an ionic current that opens voltage-gated Na$^+$ channels at the third node, and so on. Each node repolarizes after it depolarizes. Note that the impulse has propagated much farther along the myelinated axon in 10 msec. Because current flows across the membrane only at the nodes, the impulse appears to leap from node to node as

Figure 12.13 Propagation (conduction) of a nerve impulse after it arises at the trigger zone. Dotted lines indicate ionic current flow. The insets show the path of current flow. (a) In continuous conduction along an unmyelinated axon, ionic currents flow across each adjacent portion of the membrane. (b) In saltatory conduction along a myelinated axon, the nerve impulse at the first node generates ionic currents in the cytosol and extracellular fluid that open voltage-gated Na$^+$ channels at the second node, and so on at each subsequent node.

🔑 **Unmyelinated axons exhibit continuous conduction, whereas myelinated axons exhibit saltatory conduction.**

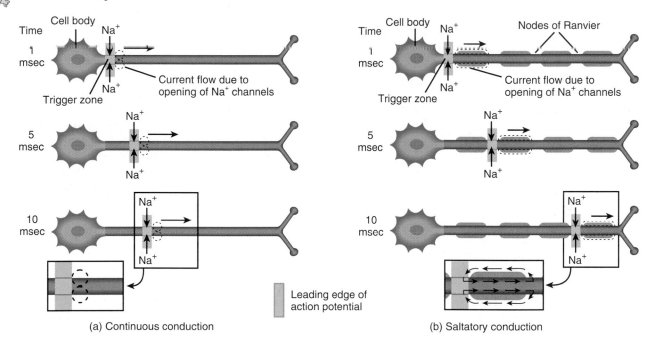

(a) Continuous conduction

(b) Saltatory conduction

Q What factors determine the propagation speed of a nerve impulse?

each nodal area depolarizes to threshold. This type of impulse conduction, which is characteristic of myelinated axons, is called **saltatory conduction** (SAL-ta-tō-rē; *saltat-* = leaping).

Speed of Nerve Impulse Propagation

The propagation speed of a nerve impulse is not related to the strength of the stimulus that triggers it. Rather, the diameter of the axon and the presence or absence of a myelin sheath are the most important factors that determine the speed of nerve impulse propagation. Also, because axons conduct impulses at higher speeds when warmed and at lower speeds when cooled, localized cooling of a nerve can retard impulse conduction. Pain resulting from injured tissue can be reduced by the application of ice because conduction of the pain sensations along axons is partially blocked.

Because an impulse "leaps" long intervals as current flows from one node to the next in saltatory conduction, it travels much faster than it would by continuous conduction in an unmyelinated axon of equal diameter. Saltatory conduction also is more energy efficient. Because only small re-

gions of the membrane depolarize, minimal inflow of Na$^+$ occurs each time a nerve impulse passes by. As a result, less ATP is used by sodium pumps to maintain the low intracellular concentration of Na$^+$.

Larger-diameter axons conduct impulses faster than smaller ones. The largest-diameter axons (about 5–20 μm) are called **A fibers,** and all are myelinated. A fibers have a brief absolute refractory period and conduct impulses at speeds of 12–130 m/sec (27–280 mi/hr). The axons of sensory neurons that conduct impulses associated with touch, pressure, position of joints, and some thermal sensations are A fibers, as are the axons of motor neurons that conduct impulses to skeletal muscles.

B fibers have axon diameters of 2–3 μm and a somewhat longer absolute refractory period than A fibers. They also are myelinated and exhibit saltatory conduction at speeds up to 15 m/sec (32 mi/hr). B fibers conduct sensory nerve impulses from the viscera to the brain and spinal cord. They also constitute all the axons of the autonomic motor neurons that extend from the brain and spinal cord to the ANS relay stations called autonomic ganglia.

Table 12.2 Comparison of Graded Potentials and Action Potentials

CHARACTERISTIC	GRADED POTENTIALS	ACTION POTENTIALS
Origin	Arise mainly in dendrites and cell body (some arise in axons).	Arise at trigger zones and propagate along the axon.
Types of channels	Ligand-gated or mechanically gated ion channels.	Voltage-gated ion channels.
Conduction	Not propagated; localized and thus permit communication over a few micrometers.	Propagate and thus permit communication over longer distances.
Amplitude	Depending on strength of stimulus, varies from less than 1 mV to more than 50 mV.	All-or-none; typically about 100 mV.
Duration	Typically longer, ranging from several msec to several min.	Shorter, ranging from 0.5–2 msec.
Polarity	May be hyperpolarizing (inhibitory to generation of an action potential) or depolarizing (excitatory to generation of an action potential).	Always consist of depolarizing phase followed by repolarizing phase and return to resting membrane potential.
Refractory period present?	No; thus exhibit spatial and temporal summation.	Yes; thus do not exhibit summation.

C fibers have the smallest axon diameters (0.5–1.5 μm) and the longest absolute refractory periods. Nerve impulse conduction along a C fiber ranges from 0.5–2 m/sec (1–4 mi/hr). These unmyelinated axons conduct some sensory impulses for pain, touch, pressure, heat, and cold from the skin, and pain impulses from the viscera. Autonomic motor fibers that extend from autonomic ganglia to stimulate the heart, smooth muscle, and glands are C fibers. Examples of motor functions of B and C fibers are constricting and dilating the pupils, increasing and decreasing the heart rate, and contracting and relaxing the urinary bladder—actions of the autonomic nervous system.

Encoding of Stimulus Intensity

If all nerve impulses are the same size, then how can your sensory systems detect stimuli of differing intensities? Why does a light touch feel different than firmer pressure? The main way that stimulus intensity is conveyed involves the *frequency of impulses*—how often they are generated at the trigger zone. Thus, a light touch generates a low frequency of widely spaced nerve impulses; firmer pressure elicits nerve impulses that pass down the axon at higher frequency. A second factor in the encoding of stimulus intensity is the number of sensory neurons recruited or activated by the stimulus. Firm pressure stimulates more pressure-sensitive neurons than does a light touch.

Comparison of Electrical Signals Produced by Excitable Cells

We have seen that excitable cells—neurons and muscle fibers—produce two types of electrical signals: graded potentials and action potentials. One obvious difference between them is that action potentials permit communication over long distances because they are propagated, whereas graded potentials function only in short-distance communication because they are not propagated. The various differences between graded potentials and action potentials are summarized in Table 12.2.

As we discussed in Chapter 10, propagation of a muscle action potential along the sarcolemma and into the T tubule system initiates the events of muscle contraction. Although action potentials in muscle fibers and in neurons are similar, there are some notable differences. Whereas the typical resting membrane potential of a neuron is −70 mV, it is closer to −90 mV in skeletal and cardiac muscle fibers. The duration of a nerve impulse is 0.5–2 msec, but a muscle action potential is considerably longer—about 1.0–5.0 msec for skeletal muscle fibers and 10–300 msec for cardiac and smooth muscle fibers. Finally, the conduction velocity of action potentials along the largest-diameter, myelinated axons is about 18 times faster than the conduction velocity along the sarcolemma of a skeletal muscle fiber.

1. Compare the basic types of ion channels, and explain how they relate to action potentials and graded potentials.
2. Define the following terms: resting membrane potential, depolarization, repolarization, nerve impulse, and refractory period.
3. How is saltatory conduction different from continuous conduction?
4. What factors determine the speed of propagation of nerve impulses?
5. How is the intensity of a stimulus encoded in the nervous system?

SIGNAL TRANSMISSION AT SYNAPSES

OBJECTIVES

• *Explain the events of signal transmission at a chemical synapse.*

• *Distinguish between spatial and temporal summation.*

• *Give examples of excitatory and inhibitory neurotransmitters, and describe how they act.*

In Chapter 10 we described the events occurring at one type of synapse, the neuromuscular junction. Our focus here is synaptic communication among the billions of neurons in the nervous system. Synapses are essential for homeostasis because they allow information to be filtered and integrated. Learning depends on modifying synapses; your performance on anatomy and physiology exams will be determined by how much some of your synapses have been modified! Certain signals are transmitted while others are blocked. Some diseases and psychiatric disorders result from disruptions of synaptic communication. Synapses also are the sites of action for many therapeutic and addictive chemicals. At a synapse between neurons, the neuron sending the signal is called the **presynaptic neuron,** and the neuron receiving the message is called the **postsynaptic neuron.** Most synapses are either **axodendritic** (from axon to dendrite), **axosomatic** (from axon to soma), or **axoaxonic** (from axon to axon). There are two types of synapses that differ both structurally and functionally: electrical synapses and chemical synapses.

Electrical Synapses

At an **electrical synapse,** ionic current spreads directly between adjacent cells through **gap junctions** (see Figure 4.1 on page 106). Each gap junction contains a hundred or so tubular proteins called *connexons* that form tunnels connecting the cytosol of the two cells. Both ions and molecules are able to flow back and forth through the connexons between adjacent cells. In the case of ions, this provides a path for flow of current. Gap junctions are common in visceral smooth muscle, cardiac muscle, and the developing embryo. They also occur in the CNS.

Electrical synapses have three obvious advantages:

1. *Faster communication.* Electrical synapses allow faster communication than do chemical synapses because action potentials conduct directly across gap junctions. The events that occur at a chemical synapse take time, thus delaying communication for about 0.5 msec.
2. *Synchronization.* Electrical synapses can synchronize the activity of a group of neurons or muscle fibers. The value of synchronized action potentials in the heart or in visceral smooth muscle is that they produce coordinated contraction of these fibers.
3. *Two-way transmission.* Electrical synapses may allow two-way transmission of action potentials; in contrast, chemical synapses function in one-way communication only.

Chemical Synapses

Although the presynaptic and postsynaptic neurons of a **chemical synapse** are in close proximity, their plasma membranes do not touch. They are separated by the **synaptic cleft,** a 20–50-nm space filled with interstitial fluid. Because nerve impulses cannot propagate across the synaptic cleft, an alternate, indirect form of communication occurs across this

space. The presynaptic neuron releases a neurotransmitter that diffuses across the synaptic cleft and acts on receptors in the plasma membrane of the postsynaptic neuron to produce a **postsynaptic potential,** a type of graded potential. In essence, the presynaptic electrical signal (nerve impulse) is converted into a chemical signal (released neurotransmitter). The postsynaptic neuron receives the chemical signal and, in turn, generates an electrical signal (postsynaptic potential). The time required for these processes at a chemical synapse—the **synaptic delay**—is about 0.5 msec. This is why chemical synapses relay signals more slowly than electrical synapses.

A typical chemical synapse transmits a signal as follows (Figure 12.14):

1 An action potential arrives at a synaptic end bulb (or at a varicosity) of a presynaptic axon.

2 The depolarizing phase of the action potential opens **voltage-gated Ca^{2+} channels** in addition to the voltage-gated Na$^+$ channels normally opened. Because it is more concentrated in the extracellular fluid, Ca^{2+} flows inward through the opened channels.

3 An increase in the concentration of Ca^{2+} inside the presynaptic neuron triggers exocytosis of some of the synaptic vesicles. As vesicle membranes merge with the plasma membrane, neurotransmitter molecules within the vesicles are released into the synaptic cleft. Each synaptic vesicle may contain several thousand molecules of neurotransmitter.

4 The neurotransmitter molecules diffuse across the synaptic cleft and bind to **neurotransmitter receptors** in the postsynaptic neuron's plasma membrane. In Figure 12.14 the receptor shown is part of a ligand-gated channel; in other cases the receptor may be a separate protein in the membrane.

5 Binding of neurotransmitter molecules to their receptors on ligand-gated ion channels opens the channels and allows particular ions to flow across the membrane.

6 Depending on which ions the channels admit, the ionic flow causes depolarization or hyperpolarization of the postsynaptic membrane. For example, opening of Na$^+$ channels allows inflow of Na$^+$, which causes depolarization. However, opening of Cl$^-$ channels allows inflow of Cl$^-$, which causes hyperpolarization.

7 If a depolarization reaches threshold, one or more action potentials are generated.

At most chemical synapses, only *one-way information transfer* can occur—from a presynaptic neuron to a postsynaptic neuron or an effector, such as a muscle fiber or a gland cell. The reason is that only synaptic end bulbs of presynaptic neurons can release neurotransmitter, and only the postsynaptic neuron's membrane has the correct receptor proteins to recognize and bind that neurotransmitter. As a result, graded potentials and action potentials move along their pathways in one direction only.

Figure 12.14 Signal transmission at a chemical synapse. Through exocytosis of synaptic vesicles, a presynaptic neuron releases neurotransmitter molecules, which bind to receptors in the plasma membrane of the postsynaptic neuron and produce a postsynaptic potential. From Becker et al., *The World of the Cell* 3e, F22.28, p741 (Menlo Park, CA: Benjamin/ Cummings, 1996) © 1996 The Benjamin/Cummings Publishing Company.

 At a chemical synapse, a presynaptic electrical signal (nerve impulse) is converted into a chemical signal (neurotransmitter release). The chemical signal is then converted back into an electrical signal (postsynaptic potential).

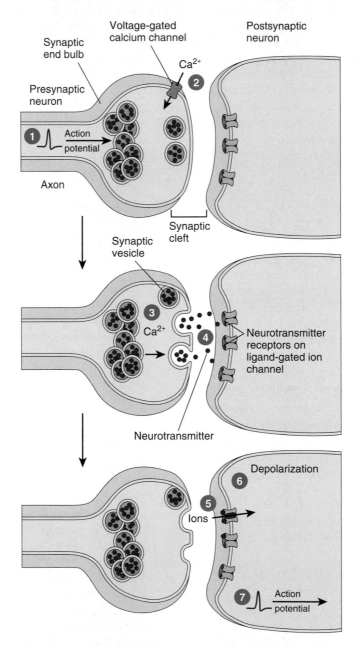

 Why is it that electrical synapses may work in two directions, but chemical synapses can transmit a signal in one direction only?

Excitatory and Inhibitory Postsynaptic Potentials

A neurotransmitter causes either an excitatory or an inhibitory graded potential. If it *depolarizes* the postsynaptic membrane, it is excitatory because it brings the membrane closer to threshold (see Figure 12.10b). A depolarizing postsynaptic potential is called an **excitatory postsynaptic potential (EPSP).** Often, EPSPs result from opening of ligand-gated *cation* channels. These channels allow the three most plentiful cations (Na^+, K^+, and Ca^{2+}) to pass through, but Na^+ inflow is greater than either Ca^{2+} inflow or K^+ outflow. Although a single EPSP normally does not initiate a nerve impulse, the postsynaptic neuron does become more excitable. It is partially depolarized and thus more likely to reach threshold when the next EPSP occurs.

If, in contrast, a neurotransmitter causes *hyperpolarization* of the postsynaptic membrane (see Figure 12.10a), it is inhibitory. It increases the membrane potential by making the inside more negative, and thus generation of a nerve impulse is more difficult than usual because the membrane potential is even farther from threshold than it was in its resting state. A hyperpolarizing postsynaptic potential is inhibitory and is termed an **inhibitory postsynaptic potential (IPSP).** IPSPs often result from the opening of ligand-gated Cl^- or K^+ channels. When Cl^- channels open, chloride ions diffuse inward at a greater pace. When K^+ channels open, the membrane permeability to K^+ increases, more potassium ions diffuse outward, and the inside becomes even more negative (hyperpolarized).

Removal of Neurotransmitter

Removal of the neurotransmitter from the synaptic cleft is essential for normal synaptic function. If a neurotransmitter could linger in the synaptic cleft, it would influence the postsynaptic neuron, muscle fiber, or gland cell indefinitely. Neurotransmitter is removed in three basic ways:

1. *Diffusion.* Some neurotransmitter molecules diffuse out of the synaptic cleft by moving down their concentration gradient.
2. *Enzymatic degradation.* Some neurotransmitters may be inactivated through enzymatic degradation. For example, the enzyme acetylcholinesterase breaks down acetylcholine in the synaptic cleft.
3. *Uptake by cells.* Many neurotransmitters are actively transported back into the neuron that released them (reuptake) or are transported into neighboring neuroglia. Norepinephrine, for example, is rapidly taken up and recycled by the same neurons that release it. The membrane proteins that accomplish such uptake are called *neurotransmitter transporters.* Several therapeutically important drugs selectively block reuptake of specific neurotransmitters by interfering with these transporters. For example, Prozac, which is used to treat some forms of depression, is a *selective serotonin reuptake inhibitor (SSRI).* Thus, it prolongs the synaptic activity of the neurotransmitter serotonin.

Spatial and Temporal Summation of Postsynaptic Potentials

A typical neuron in the CNS receives input from 1000–10,000 synapses. Integration of these inputs, which is known as **summation,** occurs at the trigger zone. The greater the summation of EPSPs, the greater the chance that threshold will be reached and a nerve impulse will be initiated.

When summation results from buildup of neurotransmitter released simultaneously by *several* presynaptic end bulbs, it is called **spatial summation** (Figure 12.15a). When summation results from buildup of neurotransmitter released by a *single* presynaptic end bulb two or more times in rapid succession, it is called **temporal summation** (Figure 12.15b). As a typical EPSP lasts about 15 msec, the second (and subsequent) release of neurotransmitter must occur soon after the first one if temporal summation is to occur.

A single postsynaptic neuron receives input from many presynaptic neurons, some of which release excitatory neurotransmitters and some of which release inhibitory neurotransmitters. The sum of all the excitatory and inhibitory effects at any given time determines the effect on the postsynaptic neuron, which may respond in the following ways:

1. *EPSP.* If the total excitatory effects are greater than the total inhibitory effects but less than the threshold level of stimulation, the result is a subthreshold EPSP. Subsequent stimuli can more easily generate a nerve impulse through summation because the neuron is partially depolarized.
2. *Nerve impulse(s).* If the total excitatory effects are greater than the total inhibitory effects and the threshold level of stimulation is reached or surpassed, the EPSP spreads to the initial segment of the axon and triggers one or more nerve impulses. Impulses continue to be generated as long as the EPSP stays above the threshold level.
3. *IPSP.* If the total inhibitory effects are greater than the excitatory effects, the membrane hyperpolarizes (IPSP). The result is inhibition of the postsynaptic neuron and an inability to generate a nerve impulse.

Table 12.3 summarizes the structural and functional elements of a neuron.

CLINICAL APPLICATION
Strychnine Poisoning

The importance of inhibitory neurons can be appreciated by observing what happens when their activity is blocked. Normally, inhibitory neurons in the spinal cord called *Renshaw cells* release the neurotransmitter glycine (described shortly) at inhibitory synapses with motor neurons. This inhibitory input to motor neurons prevents excessive muscular contraction. Strychnine binds to and blocks glycine receptors; the normal, delicate balance between excitation and inhibition in the CNS is disturbed, and motor neurons generate

Figure 12.15 Spatial and temporal summation. (a) When presynaptic neurons a and b separately cause EPSPs (arrows) in postsynaptic neuron c, the threshold level is not reached in neuron c. Spatial summation occurs only when neurons a and b act simultaneously on neuron c; their EPSPs sum to reach the threshold level and trigger a nerve impulse. (b) Temporal summation occurs when stimuli applied to the same axon in rapid succession (arrows) cause overlapping EPSPs that sum. When depolarization reaches the threshold level, a nerve impulse is triggered.

🔑 **The sum of all excitatory and inhibitory postsynaptic potentials determines whether or not an action potential is generated.**

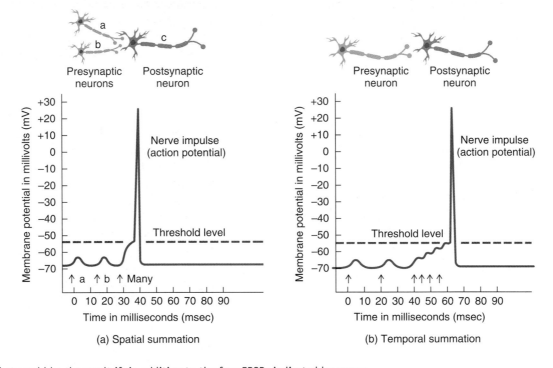

(a) Spatial summation

(b) Temporal summation

Ⓠ What would be the result if, in addition to the four EPSPs indicated by arrows in part (b), an IPSP occurred at time 55 msec?

nerve impulses without restraint, leading to massive tetanic contractions. All skeletal muscles, including the diaphragm, contract fully and remain contracted. Because the diaphragm cannot relax, the victim cannot breathe. ■

1. Construct a table that lists the types and describes the functions of ion channels that participate in signal transmission at a chemical synapse.
2. How is neurotransmitter removed from the synaptic cleft?
3. Distinguish between excitatory and inhibitory postsynaptic potentials.
4. Explain why action potentials are said to be "all-or-none," whereas EPSPs and IPSPs are not.

NEUROTRANSMITTERS

OBJECTIVE

• *Describe the classification and functions of neurotransmitters.*

Even though about 100 substances are either known or suspected neurotransmitters, identifying the function of each is not easy. Dendrites, cell bodies, and axons are tightly packed and closely intermingled in nervous tissue, and the amount of neurotransmitter liberated at a given synapse is tiny. Some neurotransmitters bind to their receptors and act quickly to open or close ion channels in the membrane; others act more slowly via second-messenger systems to influence enzymatic reactions inside cells. The result of either process can be excitation or inhibition of postsynaptic neurons. Many neurotransmitters are also hormones released into the bloodstream by endocrine cells in organs through-

Table 12.3 Summary of Neuronal Structure and Function

	STRUCTURE	FUNCTIONS
	Dendrites	Receive stimuli through activation of ligand- or mechanically gated ion channels; in sensory neurons, produce generator or receptor potentials; in motor neurons and association neurons, produce excitatory and inhibitory postsynaptic potentials (EPSPs and IPSPs).
	Cell body	Receives stimuli and produces EPSPs and IPSPs through activation of ligand- or mechanically gated ion channels.
	Junction of axon hillock and initial segment of axon	Trigger zone in many neurons; integrates EPSPs and IPSPs and, if sum is a depolarization that reaches threshold, initiates action potential (nerve impulse).
	Axon	Propagates (conducts) nerve impulses from initial segment (or from dendrites of sensory neurons) to axon terminals in a self-reinforcing manner; impulse amplitude does not change as it propagates along the axon.
	Axon terminals and synaptic end bulbs (or varicosities)	Inflow of Ca^{2+} caused by depolarizing phase of nerve impulse triggers neurotransmitter release by exocytosis of synaptic vesicles.

▬▬▬ Plasma membrane includes chemically gated channels

▬▬▬ Plasma membrane includes voltage-gated Na^+ and K^+ channels

▬▬▬ Plasma membrane includes voltage-gated Ca^{2+} channels

out the body. Within the brain, certain neurons, called **neurosecretory cells,** also secrete hormones.

The effects of neurotransmitters at chemical synapses can be modified in several ways: (1) *Neurotransmitter synthesis* can be stimulated or inhibited; (2) *neurotransmitter release* can be blocked or enhanced; (3) *neurotransmitter removal* can be stimulated or inhibited; and (4) the *receptor site* can be blocked or activated. An agent that enhances synaptic transmission or mimics the effect of a natural neurotransmitter is an **agonist,** whereas one that blocks the action of a neurotransmitter is an **antagonist.** For example, by being an agonist of the neurotransmitter dopamine, cocaine produces euphoria—intensely pleasurable feelings—by blocking transporters for dopamine uptake. This allows dopamine to linger longer in synaptic clefts, producing excessive stimulation of certain brain regions.

Neurotransmitters can be divided into two classes based on size: small-molecule neurotransmitters and neuropeptides.

Small-Molecule Neurotransmitters

This class of neurotransmitters includes acetylcholine, amino acids, biogenic amines, ATP and other purines, and gases.

Acetylcholine

The best-studied neurotransmitter is **acetylcholine (ACh),** which is released by many PNS neurons and by some CNS neurons. ACh is an excitatory neurotransmitter at some synapses, such as the neuromuscular junction, where it acts directly to open ligand-gated cation channels. It is also known to be an inhibitory neurotransmitter at other synapses, where its effects on ion channels appear to occur indirectly via receptors that link to a G protein. One example is the parasympathetic neurons of the vagus nerve (cranial nerve X) that innervate the heart; ACh slows heart rate via these inhibitory synapses. ACh is inactivated by acetylcholinesterase (AChE), which splits ACh into acetate and choline portions.

Amino Acids

Several amino acids are neurotransmitters in the CNS. **Glutamate** (glutamic acid) and **aspartate** (aspartic acid) have powerful excitatory effects. Nearly all excitatory neurons in the CNS and perhaps half of the synapses in the brain communicate via glutamate. The process for inactivation of glutamate is different from that of ACh. Soon after release from synaptic vesicles, it is actively transported back into the synaptic end bulbs and into neighboring neuroglia by specific glutamate transporters.

Two other amino acids, **gamma aminobutyric** (GAM-ma am-i-nō-byoo-TIR-ik) **acid (GABA)** and **glycine,** are important inhibitory neurotransmitters. Both cause IPSPs by opening Cl⁻ channels. Although GABA is an amino acid, it is not incorporated into proteins; instead, it is found only in the brain, where it is the most common inhibitory neurotransmitter. As many as one-third of all brain synapses use GABA. Antianxiety drugs such as diazepam (Valium) are GABA agonists; they enhance the action of GABA. In the spinal cord, about half of the inhibitory synapses use the amino acid glycine and half use GABA.

CLINICAL APPLICATION
Excitotoxicity

A high level of glutamate in the interstitial fluid of the CNS causes **excitotoxicity**—destruction of neurons through prolonged activation of excitatory synaptic transmission. The most common cause of excitotoxicity is oxygen deprivation of the brain due to ischemia, as happens during a stroke. Lack of oxygen causes the glutamate transporters to fail, and glutamate accumulates in the extracellular spaces between neurons and glia, literally stimulating the neurons to death. Clinical trials are underway to see if antiglutamate drugs can offer some protection from excitotoxicity after a person suffers a "brain attack" or stroke (see page 479). ■

Biogenic Amines

Certain amino acids are modified and decarboxylated (carboxyl group removed) to produce biogenic amines. Those that are prevalent in the nervous system include norepinephrine, epinephrine, dopamine, and serotonin. Depending on the type of receptor (there are three or more different types for each biogenic amine), biogenic amines may cause either excitation or inhibition.

Norepinephrine (NE) has been implicated in maintaining arousal (awakening from deep sleep), dreaming, and regulating mood. A smaller number of neurons in the brain use **epinephrine** as a neurotransmitter. Both epinephrine and norepinephrine also serve as hormones. They are released by the adrenal medulla, the inner portion of the adrenal gland.

Brain neurons containing the neurotransmitter **dopamine (DA)** are involved in emotional responses and in regulating skeletal muscle tone and some aspects of movement due to contraction of skeletal muscles. Degeneration of dopamine-containing axons occurs in Parkinson's disease (see page 507).

Norepinephrine, dopamine, and epinephrine are **catecholamines** (cat-e-KŌL-a-mēns). They all include a catechol ring composed of six carbons and two adjacent hydroxyl (—OH) groups, and all are synthesized from the amino acid tyrosine. Inactivation of catecholamines occurs via reuptake into the synaptic end bulbs. Then they are either recycled back into the synaptic vesicles or destroyed by the enzymes **catechol-*O*-methyltransferase** (kat′-e-kōl-ō-meth-il-TRANS-fer-ās), or **COMT,** and **monoamine oxidase** (mon-ō-AM-ēn OK-si-dās), or **MAO.**

Serotonin, also known as **5-hydroxytryptamine (5-HT),** is concentrated in the neurons in a part of the brain called the raphe nucleus; it is thought to be involved in sensory perception, temperature regulation, control of mood, and the induction of sleep.

ATP and Other Purines

The characteristic ring structure of the adenosine portion of ATP (shown in Figure 2.26 on page 55) is called a purine ring. Adenosine itself, as well as its triphosphate, diphosphate, and monophosphate derivatives (ATP, ADP, and AMP), are all excitatory neurotransmitters in both the CNS and the PNS. Most of the synaptic vesicles that contain ATP also contain another neurotransmitter. In the PNS, ATP and norepinephrine are released together from some sympathetic nerves, whereas ATP and acetylcholine are released together from some parasympathetic nerves.

Gases

A surprising and important newcomer to the ranks of recognized neurotransmitters is the simple gas **nitric oxide (NO),** which has widespread effects throughout the body. It is not to be confused with nitrous oxide (N₂O, laughing gas), which is sometimes used as an anesthetic during dental procedures. Carbon monoxide (CO) may also function as a neurotransmitter.

NO is formed from the amino acid arginine by the enzyme **nitric oxide synthase (NOS).** NO is different from all previously known neurotransmitters because it is not synthesized in advance and packaged into synaptic vesicles. Rather, it is formed on demand and acts immediately. Its action is brief because NO is a highly reactive free radical that lasts less than 10 seconds before it combines with oxygen and water to form inactive nitrates and nitrites. The precise functions of NO released by neurons are still unclear. Because NO is lipid soluble, it diffuses out of cells that produce it and into neighboring cells, where it activates an enzyme for production of a second messenger called cyclic GMP. Some research suggests that NO may play a role in memory and learning.

The first recognition of NO as a regulatory molecule was the discovery in 1987 that EDRF (endothelium-derived relaxing factor) was actually NO. Endothelial cells in blood vessel walls release NO, which diffuses into neighboring smooth muscle cells and causes relaxation. The result is vasodilation, an increase in blood vessel diameter. The effects of such vasodilation range from a lowering of blood pressure to erection of the penis in males. In larger quantities, NO is highly toxic. Phagocytic cells, such as macrophages and certain white blood cells, produce NO to kill microbes and tumor cells. Based on the presence of NOS, it is estimated that more than 2% of the neurons in the brain produce NO. NOS is also highly concentrated in autonomic neurons that cause either relaxation of smooth muscle in the gut or release of epinephrine and norepinephrine from the adrenal medulla.

Neuropeptides

Neurotransmitters consisting of 3–40 amino acids linked by peptide bonds are called **neuropeptides.** They are numerous and widespread in both the CNS and the PNS, and they have both excitatory and inhibitory actions. Neuropeptides are formed in the neuron cell body, packaged into vesicles, and transported to axon terminals. Besides their role as neurotransmitters, many neuropeptides serve as hormones that regulate physiological responses in other parts of the body.

In 1974, scientists discovered that certain brain neurons have plasma membrane receptors for opiate drugs such as morphine and heroin. The quest to find the naturally occurring substances that use these receptors brought to light the first neuropeptides: two molecules, each a chain of five amino acids, named **enkephalins** (en-KEF-a-lins). They have potent analgesic (pain-relieving) effects—200 times stronger than morphine. Besides the enkephalins, other so-called *opioid peptides* include the **endorphins** (en-DOR-fins) and **dynorphins** (dī-NOR-fins). It is thought that opioid peptides are the body's natural painkillers. Acupuncture may produce analgesia (loss of pain sensation) by increasing the release of opioids. They have also been linked to improved memory and learning; feelings of pleasure or euphoria; control of body temperature; regulation of hormones that affect the onset of puberty, sexual drive, and reproduction; and mental illnesses such as depression and schizophrenia.

Another neuropeptide, **substance P,** is released by neurons that transmit pain-related input from peripheral pain receptors into the central nervous system. Enkephalin suppresses the release of substance P, thus decreasing the number of nerve impulses being relayed to the brain for pain sensations. Substance P has also been shown to counter the effects of certain nerve-damaging chemicals, prompting speculation that it might prove useful as a treatment for nerve degeneration.

Table 12.4 provides brief portraits of these neuropeptides, as well as others that are discussed in later chapters.

1. List four ways that synaptic transmission may be modified.
2. List several neurotransmitters, describe their locations and functions in the nervous system, and indicate whether they cause excitation or inhibition.
3. Explain how nitric oxide differs from all previously known neurotransmitters.

Table 12.4 Neuropeptides

SUBSTANCE	COMMENT
Substance P	Found in sensory neurons, spinal cord pathways, and parts of brain associated with pain; enhances perception of pain.
Enkephalins	Inhibit pain impulses by suppressing release of substance P.
Endorphins	Inhibit pain by blocking release of substance P; may have a role in memory and learning, sexual activity, and control of body temperature.
Dynorphins	May be related to controlling pain and registering emotions.
Hypothalamic releasing and inhibiting hormones	Produced by the hypothalamus; regulate the release of hormones by the anterior pituitary gland.
Angiotensin II	Stimulates thirst; may regulate blood pressure in the brain; as a hormone causes vasoconstriction and promotes release of aldosterone, which increases the rate of salt and water reabsorption by the kidneys.
Cholecystokinin (CCK)	Found in the brain and small intestine; may regulate feeding as a "stop eating" signal; as a hormone, regulates pancreatic enzyme secretion during digestion, and contraction of smooth muscle in the gastrointestinal tract.

NEURONAL CIRCUITS IN THE NERVOUS SYSTEM

OBJECTIVE

• *Identify the various types of neuronal circuits in the nervous system.*

The CNS contains billions of neurons organized into complicated networks called **neuronal circuits** over which nerve impulses are conducted. In a **simple series circuit,** a presynaptic neuron stimulates only a single neuron. The second neuron then stimulates another, and so on. Most neuronal circuits, however, are more complex.

A single presynaptic neuron may synapse with several postsynaptic neurons. Such an arrangement, called **divergence,** permits one presynaptic neuron to influence several postsynaptic neurons (or several muscle fibers or gland cells) at the same time. In a **diverging circuit,** the nerve impulse from a single presynaptic neuron causes the stimulation of increasing numbers of cells along the circuit (Figure 12.16a). For example, a small number of neurons in the brain that govern a particular body movement stimulate a much larger number of neurons in the spinal cord. Sensory signals also feed into diverging circuits and are often relayed to several regions of the brain.

Figure 12.16 Examples of neuronal circuits.

🔑 **Neuronal circuits are groups of neurons arranged in patterns over which nerve impulses are conducted.**

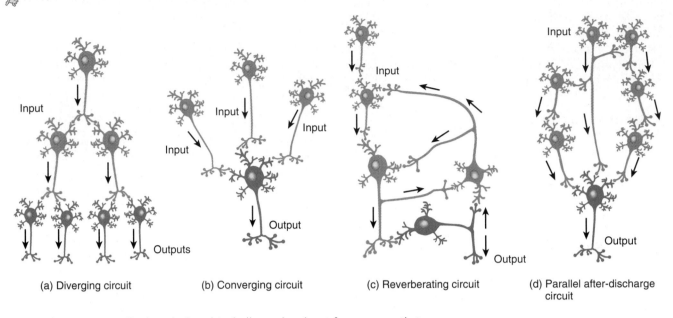

(a) Diverging circuit (b) Converging circuit (c) Reverberating circuit (d) Parallel after-discharge circuit

Ⓠ A motor neuron in the spinal cord typically receives input from neurons that originate in several different regions of the brain. Is this an example of convergence or divergence?

In another arrangement, called **convergence,** several presynaptic neurons synapse with a single postsynaptic neuron. This arrangement permits more effective stimulation or inhibition of the postsynaptic neuron. In one type of **converging circuit** (Figure 12.16b), the postsynaptic neuron receives nerve impulses from several different sources. For example, a single motor neuron that synapses with skeletal muscle fibers at neuromuscular junctions receives input from several pathways that originate in different brain regions.

Some circuits are constructed so that once the presynaptic cell is stimulated, it will cause the postsynaptic cell to transmit a series of nerve impulses. One such circuit is called a **reverberating (oscillatory) circuit** (Figure 12.16c). In this pattern, the incoming impulse stimulates the first neuron, which stimulates the second, which stimulates the third, and so on. Branches from later neurons synapse with earlier ones, however, sending impulses back through the circuit again and again. The output signal may last from a few seconds to many hours, depending on the number of synapses and the arrangement of neurons in the circuit. Inhibitory neurons may turn off a reverberating circuit after a period of time. Among the body responses thought to be the result of output signals from reverberating circuits are breathing, coordinated muscular activities, waking up, sleeping (when reverberation stops), and short-term memory.

A fourth type of circuit is the **parallel after-discharge circuit** (Figure 12.16d). In this circuit, a single presynaptic cell stimulates a group of neurons, each of which synapses with a common postsynaptic cell. A differing number of synapses between the first and last neurons imposes varying synaptic delays, so that the last neuron exhibits multiple EPSPs or IPSPs. If the input is excitatory, the postsynaptic neuron then can send out a stream of impulses in quick succession. It is thought that parallel after-discharge circuits may be involved in precise activities such as mathematical calculations.

1. What is a neuronal circuit?
2. What are the functions of diverging, converging, reverberating, and parallel after-discharge circuits?

REGENERATION AND REPAIR OF NERVOUS TISSUE

OBJECTIVES
• *Define plasticity and neurogenesis.*
• *Describe the events in damage and repair of peripheral nerves.*

Throughout one's life, the nervous system exhibits **plasticity,** the capability to change based on experience. At the level of individual neurons, the changes that can occur include the sprouting of new dendrites, synthesis of new proteins, and changes in synaptic contacts with other neurons. Undoubtedly, both chemical and electrical signals drive the changes that occur. Despite plasticity, however, mammalian neurons have very limited powers of **regeneration,** the capability to replicate or repair themselves. In the PNS, damage to dendrites and myelinated axons may be repaired if the cell body remains intact and if the Schwann cells that produce myelination remain active. In the CNS, little or no repair of damage to neurons occurs; even when the cell body remains intact, a severed axon cannot be repaired or regrown.

Neurogenesis in the CNS

Neurogenesis—the birth of new neurons from undifferentiated stem cells—occurs regularly in some animals. For example, new neurons appear and disappear every year in some songbirds. Until recently, the dogma in humans and other primates was "no new neurons" in the adult brain. Then, in 1992, Canadian researchers published their unexpected finding that **epidermal growth factor (EGF)** stimulated cells taken from the brains of adult mice to proliferate into both neurons and astrocytes. Previously, EGF was known to trigger mitosis in a variety of non-neuronal cells and to promote wound healing and tissue regeneration (see page 144). In 1998 scientists discovered that significant numbers of new neurons do arise in the adult human hippocampus, an area of the brain that is crucial for learning.

The nearly complete lack of neurogenesis in other regions of the brain and spinal cord, however, seems to result from two factors: (1) inhibitory influences from neuroglia, particularly oligodendrocytes, and (2) absence of growth-stimulating cues that were present during fetal development. Axons in the CNS are myelinated by oligodendrocytes that do not form neurolemmas (sheaths of Schwann). In addition, CNS myelin is one of the factors inhibiting regeneration of neurons. Perhaps this is the same mechanism that stops axonal growth once a target region has been reached during development. Also, after axonal damage, nearby astrocytes proliferate rapidly, forming a type of scar tissue and constituting a physical barrier to regeneration. Thus, injury of the brain or spinal cord usually is permanent. Nevertheless, ongoing research seeks ways to stimulate dormant stem cells to replace neurons lost through damage or disease. Also, tissue-cultured neurons might be useful for transplantation purposes.

Damage and Repair in the PNS

Axons and dendrites that are associated with a neurolemma may undergo repair if the cell body is intact, if the Schwann cells are functional, and if scar tissue formation does not occur too rapidly. Most nerves in the PNS consist of processes that are covered with a neurolemma. A person who injures axons of a nerve in an upper limb, for example, has a good chance of regaining nerve function.

Figure 12.17 Damage and repair of a neuron in the PNS.

Myelinated fibers in the peripheral nervous system may be repaired if the cell body remains intact and if Schwann cells remain active.

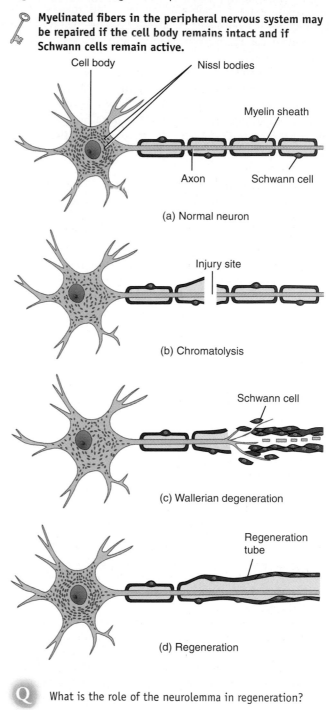

(a) Normal neuron

(b) Chromatolysis

(c) Wallerian degeneration

(d) Regeneration

Q What is the role of the neurolemma in regeneration?

When there is damage to an axon, changes usually occur both in the cell body of the affected neuron and in the portion of the axon distal to the site of injury. Changes also may occur in the portion of the axon proximal to the site of injury.

About 24–48 hours after injury to a process of a normal central or peripheral neuron (Figure 12.17a), the Nissl bodies break up into fine granular masses. This alteration is called **chromatolysis** (krō′-ma-TOL-i-sis; *chromato-* =

color; *-lysis* = dissolution) (Figure 12.17b). It begins between the axon hillock and nucleus and then spreads throughout the cell body. As a result of chromatolysis, the cell body swells, reaching a maximum size between 10 and 20 days after injury.

By the third to fifth day, the part of the process distal to the damaged region becomes slightly swollen and then breaks up into fragments; the myelin sheath also deteriorates (Figure 12.17c). Degeneration of the distal portion of the neuronal process and myelin sheath is called **Wallerian degeneration.** Following degeneration, macrophages phagocytize the debris.

The changes in the proximal portion of the axon, called **retrograde degeneration,** are similar to those that occur during Wallerian degeneration. The main difference in retrograde degeneration is that the changes extend only to the first node of Ranvier.

Following chromatolysis, signs of recovery in the cell body become evident. Synthesis of RNA and protein accelerates, which favors rebuilding or **regeneration** of the axon. Recovery often takes several months. Even though the neuronal process and myelin sheath degenerate, the neu-

rolemma remains. The Schwann cells on either side of the injured site multiply by mitosis, grow toward each other, and may form a **regeneration tube** across the injured area (Figure 12.17d). The tube guides growth of new processes from the proximal area across the injured area into the distal area previously occupied by the original axon. New axons cannot grow if the gap at the site of injury is too large or if the gap becomes filled with collagen fibers.

During the first few days following damage, buds of regenerating axons begin to invade the tube formed by the Schwann cells (see Figure 12.17c). Axons from the proximal area grow at a rate of about 1.5 mm (0.06 in.) per day across the area of damage, find their way into the distal regeneration tubes, and grow toward the distally located receptors and effectors. Thus, some sensory and motor connections are reestablished and some functions restored. In time, the Schwann cells form a new myelin sheath.

1. What factors contribute to a lack of neurogenesis in most parts of the brain?
2. What is the function of the regeneration tube in neuronal repair?

DISORDERS: HOMEOSTATIC IMBALANCES

MULTIPLE SCLEROSIS

Multiple sclerosis (MS) is a progressive destruction of myelin sheaths of neurons in the CNS that afflicts about half a million people in the United States, and 2 million people worldwide. It usually appears in people between the ages of 20 and 40, affecting females twice as often as males. Like rheumatoid arthritis, MS is an autoimmune disease—the body's own immune system spearheads the attack. The condition's name describes the anatomical pathology: Myelin sheaths deteriorate to *scleroses,* which are hardened scars or plaques, in *multiple* regions. Magnetic resonance imaging (MRI) studies reveal numerous plaques in the white matter of the brain and spinal cord. The destruction of myelin sheaths slows and then short-circuits conduction of nerve impulses.

The most common form of the condition is called relapsing-remitting MS. Usually, the first symptoms, including a feeling of heaviness or weakness in the muscles, abnormal sensations, or double vision, occur in early adult life. An attack is followed by a period of remission during which the symptoms temporarily disappear. Sometime later, a new series of plaques develop, and the person suffers a second attack. One attack follows another over the years, usually every year or two. The result is a progressive loss of function interspersed with remission periods, during which symptoms abate. Although the cause of MS is unclear, both genetic susceptibility and exposure to some environmental factor (perhaps a herpesvirus) appear to contribute. Since 1993, many patients with relapsing-remitting MS have been treated with injections of interferon beta, which lengthens the time between relapses, decreases the severity of relapses, and slows formation of new lesions in some cases. Unfortunately, not all MS patients can tolerate interferon beta, and therapy becomes less effective as the disease progresses.

EPILEPSY

The second most common neurological disorder after stroke (rupture or blockage of a brain blood vessel) is **epilepsy,** which afflicts about 1% of the world's population. Epilepsy is characterized by short, recurrent, periodic attacks of motor, sensory, or psychological malfunction. The attacks, called *epileptic seizures,* are initiated by abnormal, synchronous electrical discharges from millions of neurons in the brain, perhaps resulting from abnormal reverberating circuits. The discharges stimulate many of the neurons to send nerve impulses over their conduction pathways; as a result, lights, noise, or smells may be sensed when the eyes, ears, and nose have not been stimulated. Moreover, the skeletal muscles of a person undergoing an attack may contract involuntarily. *Partial seizures* begin in a small focus on one side the brain and produce milder symptoms, whereas *generalized seizures* involve larger areas on both sides of the brain and loss of consciousness.

Epilepsy has many causes, including brain damage at birth (the most common cause); metabolic disturbances (hypoglycemia, hypocalcemia, uremia, hypoxia); infections (encephalitis or meningitis); toxins (alcohol, tranquilizers, hallucinogens); vascular disturbances (hemorrhage, hypotension); head injuries; and tumors and abscesses of the brain. However, most epileptic seizures are idiopathic—they have no demonstrable cause. Epilepsy almost never affects intelligence.

Epileptic seizures often can be eliminated or alleviated by antiepileptic drugs, such as phenytoin, carbamazepine, and valproate sodium. An implantable device that stimulates the vagus nerve (cranial nerve X) also has produced dramatic results in reducing seizures in some patients whose epilepsy was not well-controlled by drugs.

STUDY OUTLINE

OVERVIEW OF THE NERVOUS SYSTEM (p. 378)

1. Structures that make up the nervous system are the brain, 12 pairs of cranial nerves and their branches, the spinal cord, 31 pairs of spinal nerves and their branches, ganglia, enteric plexuses, and sensory receptors.
2. The nervous system helps maintain homeostasis and integrates all body activities by sensing changes (sensory function), interpreting them (integrative function), and reacting to them (motor function).
3. On the basis of their function, neurons are either sensory neurons, interneurons, or motor neurons.
4. The central nervous system (CNS) consists of the brain and spinal cord; the peripheral nervous system (PNS) consists of all nervous tissue outside the CNS—cranial and spinal nerves, ganglia, and sensory receptors.
5. Sensory (afferent) neurons provide input to the CNS; motor (efferent) neurons carry output from the CNS to effectors.
6. The PNS also is subdivided into the somatic nervous system (SNS), autonomic nervous system (ANS), and enteric nervous system (ENS).
7. The SNS consists of (1) neurons that conduct impulses from somatic and special sense receptors to the CNS and (2) motor neurons from the CNS to skeletal muscles.
8. The ANS contains (1) sensory neurons from visceral organs and (2) motor neurons that convey impulses from the CNS to smooth muscle tissue, cardiac muscle tissue, and glands.
9. The ENS consists of neurons in two enteric plexuses that extend the length of the gastrointestinal (GI) tract and function independently of the ANS and CNS to some extent. The ENS monitors sensory changes and controls operation of the GI tract.

HISTOLOGY OF NERVOUS TISSUE (p. 381)

1. Nervous tissue consists of two principal kinds of cells: neurons (nerve cells) and neuroglia. Neurons have the property of electrical excitability and are responsible for most special functions attributed to the nervous system: sensing, thinking, remembering, controlling muscle activity, and regulating glandular secretions. Neuroglia support, nurture, and protect the neurons and maintain homeostasis in the interstitial fluid that bathes neurons.
2. Most neurons consist of many dendrites, which are the main receiving or input region; a cell body that includes typical cellular organelles; and usually a single axon that propagates nerve impulses toward another neuron, a muscle fiber, or a gland cell.
3. Synapses are the site of functional contact between two excitable cells. Axon terminals contain synaptic vesicles filled with neurotransmitter molecules.
4. Slow axonal transport and fast axonal transport are systems for conveying materials to and from the cell body and axon terminals.
5. On the basis of their structure, neurons are either multipolar, bipolar, or unipolar.
6. Neuroglia include astrocytes, oligodendrocytes, microglia, ependymal cells, Schwann cells (neurolemmocytes), and satellite cells (summarized in Table 12.1 on page 386).
7. Two types of neuroglia produce myelin sheaths: oligodendrocytes myelinate axons in the CNS, and Schwann cells myelinate

axons in the PNS.
8. White matter consists of aggregations of myelinated processes, whereas gray matter contains neuron cell bodies, dendrites, and axon terminals or bundles of unmyelinated axons and neuroglia.
9. In the spinal cord, gray matter forms an H-shaped inner core that is surrounded by white matter. In the brain, a thin, superficial shell of gray matter covers the cerebral and cerebellar hemispheres.

ELECTRICAL SIGNALS IN NEURONS (p. 387)

1. The two basic types of ion channels are leakage channels and gated channels.
2. There are three main types of gated ion channels: voltage-gated, ligand-gated, and mechanically gated.
3. The membrane of a resting neuron is positive outside and negative inside owing to the distribution of different ions and the relative greater permeability of the membrane to K^+ than to Na^+.
4. A typical value for the resting membrane potential is -70 mV. A cell that exhibits a membrane potential is said to be polarized.
5. Sodium pumps compensate for slow leakage of Na^+ into the cell by pumping it back out.
6. A graded potential is a small deviation from the resting membrane potential that occurs because ligand-gated or mechanically gated ion channels open or close.
7. Compared to the resting level of polarization, a graded potential makes the membrane potential either more negative (more polarization), termed hyperpolarization, or less negative (less polarization), termed depolarization.
8. The amplitude of a graded potential varies, depending on the strength of the stimulus.
9. During graded potentials, the current caused by the flow of ions is localized; thus, they are useful for short-distance communication only.
10. According to the all-or-none principle, if a stimulus is strong enough to generate an action potential, the impulse generated is of a constant size. A stronger stimulus will not generate a larger impulse.
11. During an action potential, voltage-gated Na^+ and K^+ channels open in sequence. This results first in depolarization—loss and then reversal of membrane polarization (from -70 mV to 0 mV to $+30$ mV). Then repolarization—recovery of the resting membrane potential (from $+30$ mV to -70 mV)—occurs.
12. During the first part of the refractory period (RP), another impulse cannot be generated at all (absolute RP); a little later it can be triggered only by a suprathreshold stimulus (relative RP).
13. An action potential conducts or propagates (travels) from point to point along the membrane, and thus it is useful for long-distance communication.
14. Nerve impulse conduction in which the impulse "leaps" from one node of Ranvier to the next is called saltatory conduction.
15. Fibers with larger diameters conduct impulses faster than those

with smaller diameters; myelinated axons conduct impulses faster than unmyelinated axons.

16. The intensity of a stimulus is encoded in the frequency of action potentials and in the number of sensory neurons that are recruited.

17. Table 12.2 on page 396 compares graded potentials and action potentials.

SIGNAL TRANSMISSION AT SYNAPSES (p. 396)

1. A synapse is the functional junction between one neuron and another, or between a neuron and an effector such as a muscle or a gland.

2. The two types of synapses are electrical and chemical.

3. A chemical synapse produces only one-way information transfer—from a presynaptic neuron to a postsynaptic neuron.

4. An excitatory neurotransmitter is one that can depolarize (make less negative) the postsynaptic neuron's membrane, bringing the membrane potential closer to threshold. An inhibitory neurotransmitter hyperpolarizes the membrane of the postsynaptic neuron.

5. Neurotransmitter is removed from the synaptic cleft in three ways: diffusion, enzymatic degradation, and uptake by cells (neurons and neuroglia).

6. If several presynaptic end bulbs release their neurotransmitter at about the same time, the combined effect may generate a nerve impulse, due to summation. Summation may be spatial or temporal.

7. The postsynaptic neuron is an integrator. It receives excitatory and inhibitory signals, integrates them, and then responds accordingly.

8. Table 12.3 on page 401 summarizes the structural and functional elements of a neuron.

NEUROTRANSMITTERS (p. 400)

1. Both excitatory and inhibitory neurotransmitters are present in the CNS and the PNS. A given neurotransmitter may be excitatory in some locations and inhibitory in others.

2. Neurotransmitters can be divided into two classes based on size: (1) small-molecule neurotransmitters (acetylcholine, amino acids, biogenic amines, ATP and other purines, and gases), and (2) neuropeptides, which are composed of chains of 3–40 amino acids.

3. Chemical synaptic transmission may be modified by affecting neurotransmitter synthesis, neurotransmitter release, transmitter removal, or neurotransmitter receptor sites.

4. Table 12.4 on page 403 describes several important neuropeptides.

NEURONAL CIRCUITS IN THE NERVOUS SYSTEM (p. 403)

1. Neurons in the central nervous system are organized into networks called circuits.

2. Circuits include simple series, diverging, converging, reverberating (oscillatory), and parallel after-discharge circuits.

REGENERATION AND REPAIR OF NERVOUS TISSUE (p. 404)

1. The nervous system exhibits plasticity—the capability to change based on experience—but it has very limited powers of regeneration—the capability to replicate or repair damaged neurons.

2. Neurogenesis—the birth of new neurons from undifferentiated stem cells—is normally very limited. Repair of damaged axons is inhibited in most regions of the CNS.

3. Axons and dendrites that are associated with a neurolemma in the PNS may undergo repair if the cell body is intact, the Schwann cells are functional, and scar tissue formation does not occur too rapidly.

SELF-QUIZ QUESTIONS

Complete the following:

1. The two main divisions of the nervous system are the ___, consisting of the ___ and ___, and the ___, consisting of the ___, ___, ___, and ___.

2. Small masses of nervous tissue, containing primarily cell bodies of neurons, that are located outside the brain and spinal cord are termed ___.

3. The subdivisions of the PNS are the ___, ___, and ___.

4. The nervous system exhibits ___, the capability to change based on experience.

True or false:

5. The nearly complete lack of neurogenesis and regenerative success in the intact CNS seems to result from inhibitory influences from neuroglia and absence of growth cues that were present during development.

6. The two main factors contributing to the resting membrane potential are the equal distribution of ions across the plasma membrane and the relative impermeability of the plasma membrane to sodium and potassium ions.

Choose the best answer to the following questions:

7. Match the following:

___ (a) the part of the neuron that contains the nucleus

___ (b) rough endoplasmic reticulum in neurons

___ (c) store neurotransmitter

___ (d) the process that propagates nerve impulses toward another neuron, muscle fiber, or gland cell

___ (e) the receiving or input portions of a neuron

___ (f) a multilayered lipid and protein covering for axons produced by neuroglia

___ (g) the outer nucleated cytoplasmic layer of the Schwann cell

___ (h) first portion of the axon

___ (i) site of functional contact between two neurons or between a neuron and an effector cell

___ (j) form the cytoskeleton of a neuron

___ (k) gaps in the myelin sheath on an axon

___ (l) general term for any neuronal process

___ (m) area where the axon joins the cell body

___ (n) area where nerve impulses arise

(1) myelin sheath
(2) neurolemma
(3) nodes of Ranvier
(4) cell body
(5) Nissl bodies
(6) neurofibrils
(7) dendrites
(8) axon
(9) axon hillock
(10) initial segment
(11) trigger zone
(12) nerve fiber
(13) synapse
(14) synaptic vesicles

8. Match the following:

___ (a) a sequence of rapidly occurring events that decreases and eventually reverses the membrane potential and then restores it to the resting state; a nerve impulse

___ (b) a small deviation from the resting membrane potential that makes the membrane either more or less negative

___ (c) brought about by rapid opening of voltage-gated sodium ion channels; polarization that is less negative than the resting level

___ (d) the minimum level of depolarization required for a nerve impulse to be generated

___ (e) the recovery of the resting membrane potential

___ (f) a neurotransmitter-caused depolarization of the postsynaptic membrane

___ (g) a neurotransmitter-caused hyperpolarization of the postsynaptic membrane

___ (h) the hyperpolarization that occurs after the repolarizing phase of an action potential

(1) graded potential
(2) action potential
(3) excitatory postsynaptic potential
(4) inhibitory postsynaptic potential
(5) threshold
(6) repolarization
(7) after hyperpolarizing phase
(8) depolarizing graded potential

Choose the best answer to the following questions:

9. Which of the following statements are true? (1) The sensory function of the nervous system involves sensory receptors sensing certain changes in the internal and external environments. (2) Sensory neurons receive electrical signals from sensory receptors. (3) The integrative function of the nervous system involves analyzing sensory information, storing some aspects of it, and making decisions regarding appropriate behaviors. (4) Interneurons carry nerve impulses to effectors. (5) Motor function involves responding to integration decisions. (a) 1, 2, 3, and 4, (b) 2, 4, and 5, (c) 1, 2, 3, and 5, (d) 1, 2, and 4, (e) 2, 3, 4, and 5

10. Which of the following is *not* a neuronal circuit in the body? (a) a diverging circuit, (b) a parallel predischarge circuit, (c) a converging circuit, (d) a reverberating circuit, (e) a parallel afterdischarge circuit

11. Which of the following statements are true? (1) Two or more neurotransmitters may be present in many neurons. (2) Some amino acids, including glutamate and aspartate, are neurotransmitters. (3) The catecholamine neurotransmitter acetylcholine is synthesized from the amino acid tyrosine. (4) Simple gases such as nitric oxide and carbon monoxide can function as neurotransmitters. (5) An excitatory neurotransmitter can never be inhibitory regardless of the neuron that produces it. (a) 1, 2, and 4, (b) 2, 3, and 5, (c) 1, 3, and 5, (d) 1, 3, and 4, (e) 2, 4, and 5

12. Which of the following statements are true? (1) If the excitatory effect is greater than the inhibitory effect but less than the threshold of stimulation, the result is a subthreshold EPSP. (2) If the excitatory effect is greater than the inhibitory effect and reaches or surpasses the threshold level of stimulation, the result is a threshold or suprathreshold EPSP and a nerve impulse. (3) If the inhibitory effect is greater than the excitatory effect, the membrane hyperpolarizes, resulting in inhibition of the postsynaptic neuron and the inability of the neuron to generate a nerve impulse. (4) The greater the summation of hyperpolarizations, the more likely a nerve impulse will be initiated. (5) Inhibitory postsynaptic potentials often result from the opening of ligand-gated Cl^- or K^+ channels. (a) 1, 4, and 5, (b) 2, 4, and 5, (c) 1, 3, and 5, (d) 2, 3, and 4, (e) 1, 2, 3, and 5

13. Which of the following statements are true? (1) The basic types of ion channels are gated, leakage, and electrical. (2) Ion channels allow for the development of graded potentials and action potentials. (3) The major stimuli that operate gated ion channels are voltage changes, ligands (chemicals), and mechanical pressure. (4) Excitable cells that possess ligand-gated or mechanically gated ion channels in the plasma membrane produce action potentials when a stimulus causes those channels to open or close. (5) A graded potential is a small deviation from the resting potential that makes the membrane either more polarized or less polarized.
(a) 1, 2, and 3, (b) 2, 3, and 4, (c) 2, 3, and 5, (d) 2, 3, 4, and 5, (e) 1, 3, and 5

14. Which of the following statements are true? (1) Axons conduct impulses at higher speeds when cooled. (2) Larger-diameter axons conduct nerve impulses faster than smaller ones. (3) Continuous conduction is faster than saltatory conduction. (4) The diameter of an axon and the presence or absence of a myelin sheath are the most important factors that determine the speed of nerve impulse propagation. (5) Action potentials are localized, whereas graded potentials are propagated.
(a) 1, 3, and 5, (b) 3 and 4, (c) 2, 4, and 5, (d) 2 and 4, (e) 1, 2, and 4

15. Match the following:
___ (a) neurons with just one process extending from the cell body; are always sensory neurons
___ (b) small phagocytic neuroglia; serve as the brain's immune cells
___ (c) help maintain an appropriate chemical environment for generation of action potentials by neurons
___ (d) provide myelin sheath for CNS axons
___ (e) contains either neuronal cell bodies, dendrites, and axon terminals or bundles of unmyelinated axons and neuroglia
___ (f) a mass of cell bodies and dendrites of neurons inside the CNS
___ (g) form CSF and assist its circulation
___ (h) neurons showing several dendrites and one axon; most common neuronal type
___ (i) neurons with one main dendrite and one axon; found in the retina of the eye
___ (j) provide myelin sheath for PNS axons
___ (k) aggregation of myelinated processes from many neurons

(1) astrocytes
(2) oligodendroglia
(3) microglia
(4) ependymal cells
(5) Schwann cells
(6) unipolar neurons
(7) bipolar neurons
(8) multipolar neurons
(9) gray matter
(10) white matter
(11) nucleus

CRITICAL THINKING QUESTIONS

1. Elena skipped her A&P class (never again!) and borrowed her friend's notes. Unfortunately, her friend doesn't draw very well, and the notes are out of order. The diagram of the motor neuron looks a lot like the one of the converging circuit, and the bipolar neuron resembles a simple circuit. What do you tell Elena to help her straighten out these notes? (HINT: *Neuronal circuits involve more than one neuron.*)

2. When asked to define "gray matter," Ho answered that it's white matter from a really old person. Explain gray matter to Ho.

(HINT: *At least he's correct that there is a colored substance which collects in the neuron with age.*)

3. The buzzing of the alarm clock woke Carrie. She stretched, yawned, and started to salivate as she smelled the brewing coffee. She could feel her stomach rumble. List the divisions of the nervous system that are involved in each of these actions. (HINT: *Which activities are voluntary?*)

ANSWERS TO FIGURE QUESTIONS

12.1 The total number of cranial and spinal nerves in your body is $(12 \times 2) + (31 \times 2) = 86$.

12.2 Sensory or afferent neurons carry input to the CNS; motor or efferent neurons carry output from the CNS.

12.3 Dendrites receive (motor neurons or interneurons) or generate (sensory neurons) inputs; the cell body also receives input signals; the axon conducts nerve impulses (action potentials) and transmits the message to another neuron or effector cell by releasing a neurotransmitter at its synaptic end bulbs.

12.4 Nerve impulses arise at the trigger zone.

12.5 Dendrites receive input.

12.6 Myelination increases the speed of nerve impulse conduction.

12.7 Myelin makes white matter look shiny and white.

12.8 A touch on the arm activates mechanically gated ion channels.

12.9 A typical value for the resting membrane potential in a neuron is -70 mV.

12.10 Graded potentials result when ligand- or mechanically gated ion channels open or close.

12.11 Voltage-gated Na^+ channels are open during the depolarizing phase, and voltage-gated K^+ channels are open during the repolarizing phase.

12.12 Yes, because the leakage channels would still allow K^+ to exit more rapidly than Na^+ could enter the axon. In fact, some mammalian myelinated axons have only a few voltage-gated K^+ channels.

12.13 The diameter of an axon, presence or absence of a myelin sheath, and temperature determine propagation speed of a nerve impulse.

12.14 In some electrical synapses (gap junctions), ions flow equally well in either direction, so either neuron may be the presynaptic one. At a chemical synapse, one neuron releases neurotransmitter and the other neuron has receptors that bind this chemical. Thus, the signal can proceed in only one direction.

12.15 If an IPSP occurred at 55 msec, threshold depolarization would likely not be reached, and a nerve impulse would not be generated.

12.16 A motor neuron receiving input from several other neurons is an example of convergence.

12.17 The neurolemma provides a regeneration tube that guides regrowth of a severed axon.

THE SPINAL CORD AND SPINAL NERVES

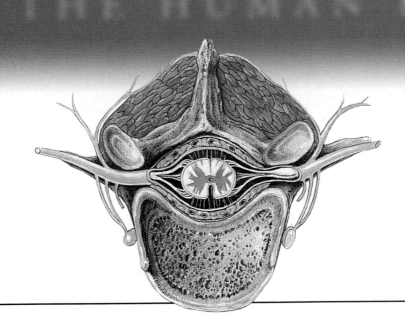

Together, the spinal cord and spinal nerves contain neuronal circuits that mediate some of your quickest reactions to environmental changes. If you pick up something hot, for example, the grasping muscles may relax and you may drop it even before the sensation of extreme heat or pain reaches your conscious perception. This is an example of a spinal cord reflex—a quick, automatic response to certain kinds of stimuli that involves neurons only in the spinal nerves and spinal cord. Besides processing reflexes, the spinal cord also is the site for integration (summing) of excitatory postsynaptic potentials (EPSPs) and inhibitory postsynaptic potentials (IPSPs) that arise locally or are triggered by nerve impulses from the peripheral nervous system (the periphery) and the brain. Moreover, the spinal cord is the highway traveled by sensory nerve impulses headed for the brain, and by motor nerve impulses destined for spinal nerves. Keep in mind that the spinal cord is continuous with the brain and that together they constitute the central nervous system (CNS).

SPINAL CORD ANATOMY

OBJECTIVE
• *Describe the protective structures and gross anatomical features of the spinal cord.*

Protective Structures

Two types of connective tissue coverings—tough meninges and bony vertebrae—plus a cushion of cerebrospinal fluid (produced in the brain) surround and protect the delicate nervous tissue of the spinal cord (and the brain as well).

Meninges

The **meninges** (me-NIN-jēz) are connective tissue coverings that encircle the spinal cord and brain. They are called, respectively, the **spinal meninges** (Figure 13.1a) and the **cranial meninges** (shown in Figure 14.4a on page 451). The most superficial of the three spinal meninges is the **dura mater** (DOO-ra MĀ-ter; = tough mother), which is composed of dense, irregular connective tissue. It forms a sac from the level of the foramen magnum in the occipital bone, where it is continuous with the dura mater of the brain, to the second sacral vertebra, where it is close-ended. The spinal cord is also protected by a cushion of fat and connective tissue located in the **epidural space,** a space between the dura mater and the wall of the vertebral canal (Figure 13.1b).

The middle **meninx** (MĒ-ninks, singular form of meninges) is an avascular covering called the **arachnoid** (a-RAK-noyd; *arachn-* = spider; *-oid* = similar to) because of its spider's web arrangement of delicate collagen fibers and some elastic fibers. It is deep to the dura mater and is also continuous with the arachnoid of the brain. Between the dura mater and the arachnoid is a thin **subdural space,** which contains interstitial fluid.

The innermost meninx is the **pia mater** (PĒ-a MĀ-ter; *pia* = delicate), a thin transparent connective tissue layer that adheres to the surface of the spinal cord and brain. It consists of interlacing bundles of collagen fibers and some

Figure 13.1 Gross anatomy of the spinal cord. The spinal meninges are evident in both views.

🔑 Meninges are connective tissue coverings that surround the spinal cord and brain.

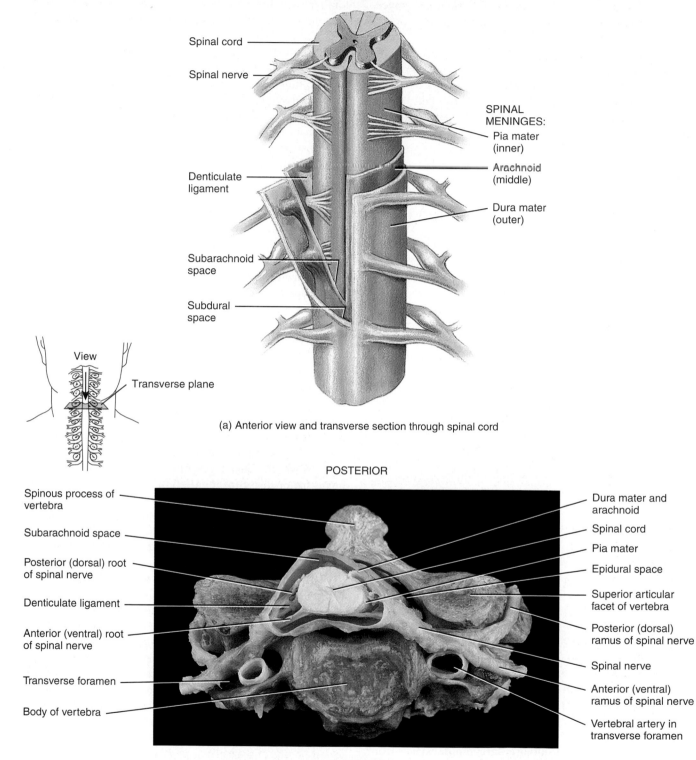

Spinal cord

Spinal nerve

SPINAL MENINGES:
Pia mater (inner)

Arachnoid (middle)

Dura mater (outer)

Denticulate ligament

Subarachnoid space

Subdural space

View

Transverse plane

(a) Anterior view and transverse section through spinal cord

POSTERIOR

Spinous process of vertebra

Subarachnoid space

Posterior (dorsal) root of spinal nerve

Denticulate ligament

Anterior (ventral) root of spinal nerve

Transverse foramen

Body of vertebra

Dura mater and arachnoid

Spinal cord

Pia mater

Epidural space

Superior articular facet of vertebra

Posterior (dorsal) ramus of spinal nerve

Spinal nerve

Anterior (ventral) ramus of spinal nerve

Vertebral artery in transverse foramen

ANTERIOR

(b) Transverse section of the spinal cord within a cervical vertebra

Q What are the superior and inferior boundaries of the spinal dura mater?

fine elastic fibers, and it contains many blood vessels that supply oxygen and nutrients to the spinal cord. Between the arachnoid and the pia mater is the **subarachnoid space,** which contains cerebrospinal fluid. Inflammation of the meninges is known as *meningitis.*

All three spinal meninges cover the spinal nerves up to the point of exit from the spinal column through the intervertebral foramina. Triangular-shaped membranous extensions of the pia mater suspend the spinal cord in the middle of its dural sheath. These extensions, called **denticulate ligaments** (den-TIK-yoo-lāt; = small tooth), are thickenings of the pia mater. They project laterally and fuse with the arachnoid and inner surface of the dura mater between the anterior and posterior nerve roots of spinal nerves on either side. Extending all along the length of the spinal cord, the denticulate ligaments protect the spinal cord against shock and sudden displacement.

Vertebral Column

The spinal cord is located within the vertebral canal of the vertebral column. The vertebral foramina of all the vertebrae, stacked one on top of the other, form the canal. The surrounding vertebrae provide a sturdy shelter for the enclosed spinal cord (see Figure 13.1b). The vertebral ligaments, meninges, and cerebrospinal fluid provide additional protection.

External Anatomy of the Spinal Cord

The **spinal cord,** although roughly cylindrical, is flattened slightly in its anterior-posterior dimension. In adults, it extends from the medulla oblongata, the most inferior part of the brain, to the superior border of the second lumbar vertebra (Figure 13.2). In newborn infants, it extends to the third or fourth lumbar vertebra. During early childhood, both the spinal cord and the vertebral column grow longer as part of overall body growth; around age 4 or 5, however, elongation of the spinal cord stops. Because the vertebral column continues to elongate, the spinal cord does not extend the entire length of the vertebral column in adults. The length of the adult spinal cord ranges from 42 to 45 cm (16–18 in.). Its diameter is about 2 cm (0.75 in.) in the midthoracic region, somewhat larger in the lower cervical and midlumbar regions, and smallest at the inferior tip.

When the spinal cord is viewed externally, two conspicuous enlargements can be seen. The superior enlargement, the **cervical enlargement,** extends from the fourth cervical vertebra to the first thoracic vertebra. Nerves to and from the upper limbs arise from the cervical enlargement. The inferior enlargement, called the **lumbar enlargement,** extends from the ninth to the twelfth thoracic vertebra. Nerves to and from the lower limbs arise from the lumbar enlargement.

Inferior to the lumbar enlargement, the spinal cord tapers to a conical portion known as the **conus medullaris** (KŌ-nus med-yoo-LAR-is; *conus* = cone), which ends at the level of the intervertebral disc between the first and second lumbar vertebrae in adults. Arising from the conus medullaris is the **filum terminale** (FĪ-lum ter-mi-NAL-ē; = terminal filament), an extension of the pia mater that extends inferiorly and anchors the spinal cord to the coccyx.

Nerves that arise from the inferior part of the spinal cord do not leave the vertebral column at the same level as they exit from the cord. The roots (points of attachment to the spinal cord) of these nerves angle inferiorly in the vertebral canal from the end of the spinal cord like wisps of hair. Appropriately, the roots of these nerves are collectively named the **cauda equina** (KAW-da ē-KWĪ-na), meaning "horse's tail."

Spinal cord organization appears to be segmented because the 31 pairs of spinal nerves emerge at regular intervals (see Figure 13.2). Indeed, each pair of spinal nerves is said to arise from a *spinal segment.* Within the spinal cord, however, any obvious segmentation dividing up the white matter or the gray matter is lacking. The naming of spinal nerves and spinal segments is based on their location. There are eight pairs of *cervical nerves* (represented as C1–C8), twelve pairs of *thoracic nerves* (T1–T12), five pairs of *lumbar nerves* (L1–L5), five pairs of *sacral nerves* (S1–S5), and one pair of coccygeal nerves.

Spinal nerves are the paths of communication between the spinal cord and the nerves innervating specific regions of the body. Two bundles of axons, called **roots,** connect each spinal nerve to a segment of the cord (see Figure 13.3a). The **posterior** or **dorsal root** contains only sensory fibers, which conduct nerve impulses from the periphery into the central nervous system. Each posterior root also has a swelling, the **posterior** or **dorsal root ganglion,** which contains the cell bodies of sensory neurons. The **anterior** or **ventral root** contains axons of motor neurons, which conduct impulses from the CNS to effector organs and cells.

CLINICAL APPLICATION
Spinal Tap

In a **spinal tap (lumbar puncture)** a local anesthetic is given, and a long needle is inserted into the subarachnoid space. The procedure is used to withdraw cerebrospinal fluid (CSF) for diagnostic purposes; to introduce antibiotics, contrast media for myelography, or anesthetics; to administer chemotherapy; to measure CSF pressure; and to evaluate the effects of treatment. In adults, a spinal tap is normally performed between the third and fourth or fourth and fifth lumbar vertebrae. Because this region is inferior to the lowest portion of the spinal cord, it provides relatively safe access. (A line drawn across the highest points of the iliac crests, called the *supracristal line,* passes through the spinous process of the fourth lumbar vertebra.) ■

Figure 13.2 External anatomy of the spinal cord and the spinal nerves. (See Tortora, *A Photographic Atlas of the Human Body,* Figures 8.2, 8.3)

The spinal cord extends from the medulla oblongata of the brain to the superior border of the second lumbar vertebra.

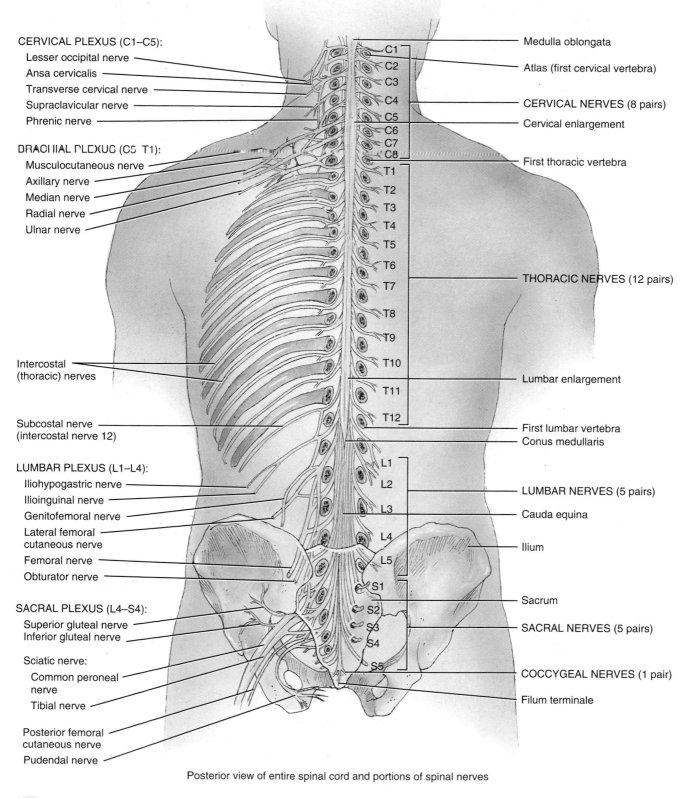

CERVICAL PLEXUS (C1–C5):
Lesser occipital nerve
Ansa cervicalis
Transverse cervical nerve
Supraclavicular nerve
Phrenic nerve

BRACHIAL PLEXUS (C5 T1):
Musculocutaneous nerve
Axillary nerve
Median nerve
Radial nerve
Ulnar nerve

Intercostal
(thoracic) nerves

Subcostal nerve
(intercostal nerve 12)

LUMBAR PLEXUS (L1–L4):
Iliohypogastric nerve
Ilioinguinal nerve
Genitofemoral nerve
Lateral femoral
cutaneous nerve
Femoral nerve
Obturator nerve

SACRAL PLEXUS (L4–S4):
Superior gluteal nerve
Inferior gluteal nerve

Sciatic nerve:
Common peroneal
nerve
Tibial nerve

Posterior femoral
cutaneous nerve
Pudendal nerve

C1
C2
C3
C4
C5
C6
C7
C8
T1
T2
T3
T4
T5
T6
T7
T8
T9
T10
T11
T12
L1
L2
L3
L4
L5
S1
S2
S3
S4
S5

Medulla oblongata

Atlas (first cervical vertebra)

CERVICAL NERVES (8 pairs)

Cervical enlargement

First thoracic vertebra

THORACIC NERVES (12 pairs)

Lumbar enlargement

First lumbar vertebra
Conus medullaris

LUMBAR NERVES (5 pairs)

Cauda equina

Ilium

Sacrum

SACRAL NERVES (5 pairs)

COCCYGEAL NERVES (1 pair)

Filum terminale

Posterior view of entire spinal cord and portions of spinal nerves

Q What portion of the spinal cord connects with sensory and motor nerves of the upper limbs?

Figure 13.3 Internal anatomy of the spinal cord: the organization of gray matter and white matter. For simplicity, dendrites are not shown in this and several other illustrations of transverse sections of the spinal cord. Blue and red arrows in (a) indicate the direction of nerve impulse propagation.

 In the spinal cord, white matter surrounds the gray matter.

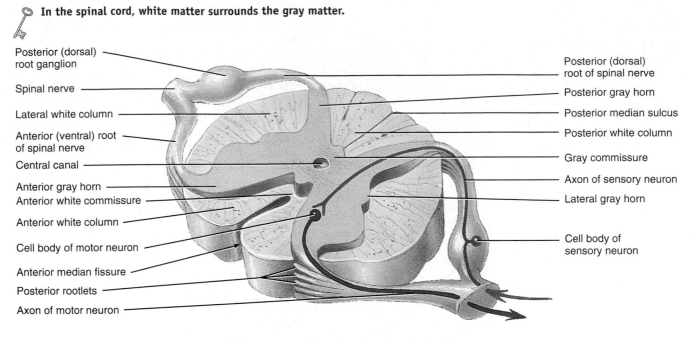

Posterior (dorsal) root ganglion
Spinal nerve
Lateral white column
Anterior (ventral) root of spinal nerve
Central canal
Anterior gray horn
Anterior white commissure
Anterior white column
Cell body of motor neuron
Anterior median fissure
Posterior rootlets
Axon of motor neuron

Posterior (dorsal) root of spinal nerve
Posterior gray horn
Posterior median sulcus
Posterior white column
Gray commissure
Axon of sensory neuron
Lateral gray horn
Cell body of sensory neuron

(a) Transverse section of the thoracic spinal cord

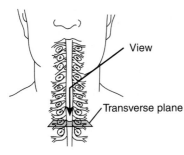

View
Transverse plane

FUNCTIONS

1. White matter tracts propagate sensory impulses from the periphery to the brain and motor impulses from the brain to the periphery.

2. Gray matter receives and integrates incoming and outgoing information.

POSTERIOR

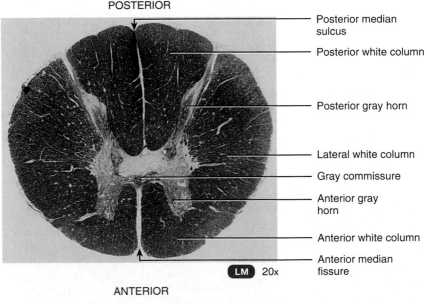

Posterior median sulcus
Posterior white column
Posterior gray horn
Lateral white column
Gray commissure
Anterior gray horn
Anterior white column
Anterior median fissure

LM 20x

ANTERIOR

(b) Transverse section of the thoracic spinal cord

Q What is the difference between a horn and a column?

Internal Anatomy of the Spinal Cord

Two grooves penetrate the white matter of the spinal cord and divide it into right and left sides (Figure 13.3). The **anterior median fissure** is a deep, wide groove on the anterior (ventral) side, whereas the **posterior median sulcus** is a shallower, narrow groove on the posterior (dorsal) surface. The gray matter of the spinal cord is shaped like the letter H or a butterfly and is surrounded by white matter. The gray matter consists primarily of cell bodies of neurons, neuroglia, unmyelinated axons, and dendrites of interneurons and motor neurons. The white matter consists of bundles of myelinated and unmyelinated axons of sensory neurons, interneurons, and motor neurons. The **gray commissure** (KOM-mi-shur) forms the crossbar of the H. In the center of the gray commissure is a small space called the **central canal;** it extends the entire length of the spinal cord. At its superior end, the central canal is continuous with the fourth ventricle (a space that contains cerebrospinal fluid) in the medulla oblongata of the brain. Anterior to the gray commissure is the **anterior (ventral) white commissure,** which connects the white matter of the right and left sides of the spinal cord.

In the gray matter of the spinal cord and brain, clusters of neuronal cell bodies form functional groups called nuclei. *Sensory nuclei* receive input from receptors via sensory neurons, whereas *motor nuclei* provide output to effector tissues via motor neurons. The gray matter on each side of the spinal cord is subdivided into regions called **horns.** The **anterior (ventral) gray horns** contain cell bodies of somatic motor neurons and motor nuclei, which provide nerve impulses for contraction of skeletal muscles. The **posterior (dorsal) gray horns** contain somatic and autonomic sensory nuclei. Between the anterior and posterior gray horns are the **lateral gray horns,** which are present only in the thoracic, upper lumbar, and sacral segments of the cord. Located here are the cell bodies of autonomic motor neurons that regulate activity of smooth muscle, cardiac muscle, and glands.

The white matter, like the gray matter, also is organized into regions. The anterior and posterior gray horns divide the white matter on each side into three broad areas called **columns:** (1) **anterior (ventral) white columns,** (2) **posterior (dorsal) white columns,** and (3) **lateral white columns.** Each column, in turn, contains distinct bundles of nerve axons having a common origin or destination and carrying similar information. These bundles, which may extend long distances up or down the spinal cord, are called **tracts.** **Sensory (ascending) tracts** consist of axons that conduct nerve impulses toward the brain. Tracts consisting of axons that carry nerve impulses down the cord are called **motor (descending) tracts.** Sensory and motor tracts of the spinal cord are continuous with sensory and motor tracts in the brain.

1. Explain the locations and compositions of the spinal meninges. Describe the locations of the epidural, subdural, and subarachnoid spaces.
2. Describe the location of the spinal cord. What are the cervical and lumbar enlargements?
3. Define conus medullaris, filum terminale, and cauda equina. What is a spinal segment? How is the spinal cord partially divided into right and left sides?
4. Based on your knowledge of the structure of the spinal cord in transverse section, define the following: gray commissure, central canal, anterior gray horn, lateral gray horn, posterior gray horn, anterior white column, lateral white column, posterior white column, ascending tract, and descending tract.

SPINAL CORD PHYSIOLOGY

OBJECTIVES

• *Describe the functions of the principal sensory and motor tracts of the spinal cord.*

• *Describe the functional components of a reflex arc and the ways reflexes maintain homeostasis.*

The spinal cord has two principal functions in maintaining homeostasis: nerve impulse propagation and information integration. The *white matter tracts* in the spinal cord are highways for nerve impulse propagation. Along these tracts, sensory impulses flow from the periphery to the brain, and motor impulses flow from the brain to the periphery. The *gray matter* of the spinal cord receives and integrates incoming and outgoing information.

Sensory and Motor Tracts

The first way the spinal cord promotes homeostasis is by conducting nerve impulses along tracts. Often, the name of a tract indicates its position in the white matter, where it begins and ends, and, by extension, the direction of nerve impulse propagation. For example, the anterior spinothalamic tract is located in the *anterior* white column; it begins in the *spinal cord* and ends in the *thalamus* (a region of the brain). Notice that the location of the neuronal cell bodies and dendrites comes first in the name, and the location of the axon terminals comes last. This regularity in naming allows you to determine the direction of information flow along any tract named according to this convention. Thus, because the anterior spinothalamic tract conveys nerve impulses from the cord toward the brain, it is a sensory (ascending) tract. Figure 13.4 highlights the principal sensory and motor tracts in the spinal cord. These tracts are described in detail in Chapter 15 and summarized in Tables 15.3 and 15.4 on pages 498 and 501.

Figure 13.4 The locations of selected sensory and motor tracts, shown in a transverse section of the spinal cord. Sensory tracts are indicated on one half and motor tracts on the other half of the cord, but in fact all tracts are present on both sides.

🔑 The name of a tract often indicates its location in the white matter and where it begins and ends.

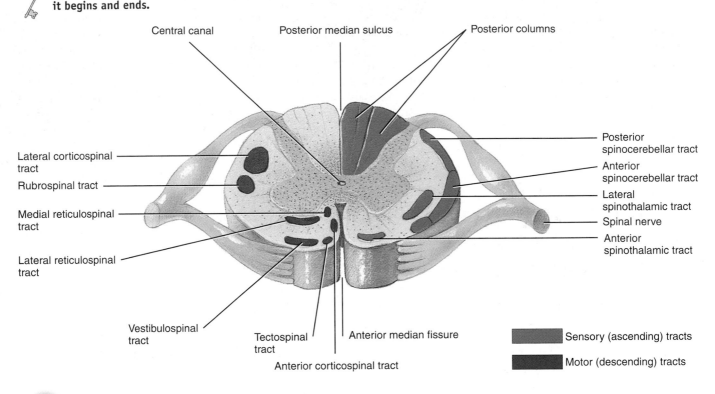

Based on its name, what are the position in the spinal cord, origin, and destination of the anterior corticospinal tract? Is this a sensory or a motor tract?

Sensory information from receptors travels up the spinal cord to the brain along two main routes on each side of the cord: the spinothalamic tracts and the posterior columns. The two **spinothalamic tracts** convey impulses for sensing pain, temperature, deep pressure, and a crude, poorly localized sense of touch. The **posterior columns** carry nerve impulses for sensing (1) proprioception, awareness of the movements of muscles, tendons, and joints; (2) discriminative touch, the ability to feel exactly what part of the body is touched; (3) two-point discrimination, the ability to distinguish the touching of two different points on the skin, even though they are close together; (4) pressure; and (5) vibration.

The sensory systems keep the CNS informed of changes in the external and internal environments. Responses to this information are brought about by motor systems, which enable us to move about and change our physical relationship to the world around us. As sensory information is conveyed to the CNS, it becomes part of a large pool of sensory input. Each piece of incoming information is integrated with all the other information arriving from activated sensory neurons.

Through the activity of interneurons, the integration process occurs not just in one region of the CNS, but in many regions. Integration occurs within the spinal cord and several regions of the brain. As a result, motor responses to make a muscle contract or a gland secrete can be initiated at several levels. Most regulation of involuntary activities of smooth muscle, cardiac muscle, and glands by the autonomic nervous system originates in the brain stem (which is continuous with the spinal cord) and in a nearby brain region called the hypothalamus. More information is given in Chapter 17.

The cerebral cortex (superficial gray matter of the cerebrum) plays a major role in controlling precise, voluntary muscular movements, whereas other brain regions provide important integration for regulation of automatic movements, such as arm swinging during walking. Motor output to skeletal muscles travels down the spinal cord in two types of descending pathways: direct and indirect. **Direct pathways** (the lateral corticospinal, anterior corticospinal, and corticobulbar tracts) convey nerve impulses that originate in the cerebral cortex and are destined to cause precise, *voluntary* movements of skeletal muscles. **Indirect pathways** (rubrospinal, tectospinal, and vestibulospinal tracts) convey nerve impulses that are destined to program automatic

Figure 13.5 General components of a reflex arc. The arrows show the direction of nerve impulse propagation.

🔑 **Reflexes are fast, predictable, automatic responses to changes in the environment.**

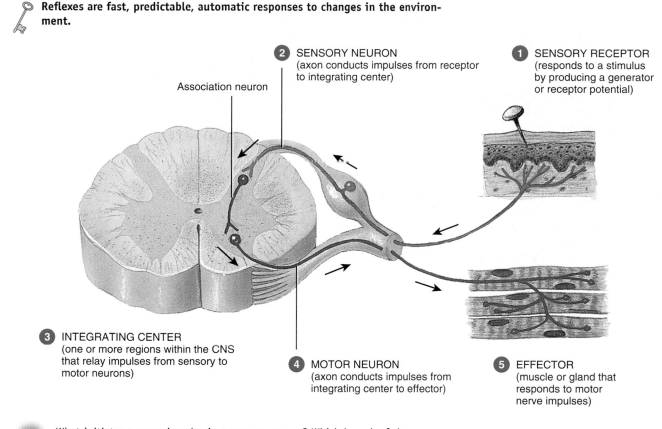

2 SENSORY NEURON
(axon conducts impulses from receptor to integrating center)

1 SENSORY RECEPTOR
(responds to a stimulus by producing a generator or receptor potential)

Association neuron

3 INTEGRATING CENTER
(one or more regions within the CNS that relay impulses from sensory to motor neurons)

4 MOTOR NEURON
(axon conducts impulses from integrating center to effector)

5 EFFECTOR
(muscle or gland that responds to motor nerve impulses)

Q What initiates a nerve impulse in a sensory neuron? Which branch of the nervous system includes all integrating centers for reflexes?

movements, help coordinate body movements with visual stimuli, maintain skeletal muscle tone and posture, and play a major role in equilibrium by regulating muscle tone in response to movements of the head.

Reflexes

The second way the spinal cord promotes homeostasis is by serving as an integrating center for **spinal reflexes.** Integration takes place in the spinal gray matter. **Reflexes** are fast, predictable, automatic responses to changes in the environment. Some reflexes, which occur through the brain stem rather than the spinal cord, involve cranial nerves and are called **cranial reflexes.** You are probably most aware of **somatic reflexes,** which involve contraction of skeletal muscles. Equally important, however, are the **autonomic (visceral) reflexes,** which generally are not consciously perceived. They involve responses of smooth muscle, cardiac muscle, and glands. As we see in Chapter 17, body functions such as heart rate, digestion, urination, and defecation are controlled by the autonomic nervous system through autonomic reflexes.

Reflex Arcs

Nerve impulses propagating into, through, and out of the CNS follow specific pathways, depending on the kind of information, its origin, and its destination. The pathway followed by nerve impulses that produce a reflex is a **reflex arc.** A reflex arc includes the following five functional components (Figure 13.5):

❶ Sensory receptor. The distal end of a sensory neuron (dendrite) or an associated sensory structure serves as a sensory receptor. It responds to a specific **stimulus**—a change in the internal or external environment—by producing a graded potential called a generator (or receptor) potential (described on page 486). If a generator potential reaches the threshold level of depolarization, it will trigger one or more nerve impulses in the sensory neuron.

❷ Sensory neuron. The nerve impulses propagate from the sensory receptor along the axon of the sensory neuron to the axon terminals, which are located in the gray matter of the spinal cord or brain stem.

3 **Integrating center.** One or more regions of gray matter within the CNS act as an integrating center. In the simplest type of reflex, the integrating center is a single synapse between a sensory neuron and a motor neuron. A reflex pathway having only one synapse in the CNS is termed a **monosynaptic reflex arc.** More often the integrating center consists of one or more interneurons, which may relay the impulse to other interneurons as well as to a motor neuron. A **polysynaptic reflex arc** involves more than two types of neurons and more than one CNS synapse.

4 **Motor neuron.** Impulses triggered by the integrating center propagate out of the CNS along a motor neuron to the part of the body that will respond.

5 **Effector.** The part of the body that responds to the motor nerve impulse, such as a muscle or gland, is the effector. Its action is called a reflex. If the effector is skeletal muscle, the reflex is a **somatic reflex.** If the effector is smooth muscle, cardiac muscle, or a gland, the reflex is an **autonomic (visceral) reflex.**

Reflexes enable the body to make exceedingly rapid adjustments to homeostatic imbalances. Because reflexes are normally so predictable, they provide useful information about the health of the nervous system and can greatly aid diagnosis of disease. If a reflex is absent or abnormal, the damage may be anywhere along a particular pathway. For example, tapping the patellar ligament normally causes reflex extension of the knee joint. This is a stretch reflex (described next). Absence of the patellar reflex could indicate damage of the sensory or motor neurons, or a spinal cord injury in the lumbar region. Somatic reflexes generally can be tested simply by tapping or stroking the body surface. Most autonomic reflexes, by contrast, are not practical diagnostic tools because it is difficult to stimulate visceral receptors, which are deep inside the body. An exception is the pupillary light reflex, in which the pupils of both eyes decrease in diameter when either eye is exposed to light. Because the reflex pathway includes synapses in the brain stem and the nearby midbrain, the absence of a normal pupillary light reflex may indicate damage or injury of these brain regions.

Next we examine four important somatic spinal reflexes: the stretch reflex, the tendon reflex, the flexor (withdrawal) reflex, and the crossed extensor reflex.

The Stretch Reflex

The **stretch reflex** is a monosynaptic reflex arc. Only two types of neurons (one sensory and one motor) are involved, and there is only one CNS synapse in the pathway. This reflex results in the contraction of a muscle (effector) when the muscle is stretched. Stretch reflexes can be elicited at the elbow, wrist, knee, and ankle joints.

A stretch reflex operates as follows (Figure 13.6):

1 Slight stretching of a muscle stimulates sensory receptors in the muscle called **muscle spindles** (shown in

more detail in Figure 15.3a on page 492). The spindles monitor changes in the length of the muscle.

2 In response to being stretched, a muscle spindle generates one or more nerve impulses that propagate along a somatic sensory neuron through the posterior root of the spinal nerve and into the spinal cord.

3 In the spinal cord (integrating center), the sensory neuron makes an excitatory synapse with and thereby activates a motor neuron in the anterior gray horn.

4 If the excitation is strong enough, one or more nerve impulses arise in the motor neuron and propagate along its axon, which extends from the spinal cord into the anterior root and through peripheral nerves to the stimulated muscle. The axon terminals of the motor neuron form neuromuscular junctions (NMJs) with skeletal muscle fibers of the stretched muscle.

5 Acetylcholine released by nerve impulses at the NMJs triggers one or more muscle action potentials in the stretched muscle (effector), and the muscle contracts. Thus, muscle stretch is followed by muscle contraction, which relieves the stretching.

In the reflex arc just described, sensory nerve impulses enter the spinal cord on the same side from which motor nerve impulses leave it. This arrangement is called an **ipsilateral reflex arc** (ip′-si-LAT-er-al; = same side). All monosynaptic reflex arcs are ipsilateral.

In addition to the large-diameter motor neurons that innervate typical muscle fibers, smaller-diameter motor neurons innervate smaller, specialized muscle fibers within the muscle spindles themselves. The brain regulates muscle spindle sensitivity through pathways to these smaller motor neurons. This regulation ensures proper muscle spindle signaling over a wide range of muscle lengths during voluntary and reflex contractions. By adjusting how vigorously a muscle spindle responds to stretching, the brain sets an overall level of **muscle tone,** which is the small degree of contraction present while the muscle is at rest. Because the stimulus for the stretch reflex is stretching of muscle, this reflex helps avert injury by preventing overstretching of muscles.

Although the stretch reflex pathway itself is monosynaptic (just two neurons and one synapse), a polysynaptic reflex arc to the antagonistic muscles operates at the same time. This arc involves three neurons and two synapses: An axon collateral (branch) from the muscle spindle sensory neuron also synapses with an inhibitory interneuron in the integrating center; in turn, the interneuron synapses with and inhibits a motor neuron that normally excites the antagonistic muscles (see Figure 13.6). Thus, when the stretched muscle contracts during a stretch reflex, antagonistic muscles that oppose the contraction relax. This type of neural circuit, which simultaneously causes contraction of one muscle and relaxation of its antagonists, is termed **reciprocal innervation.** Reciprocal innervation prevents conflict between opposing muscles and is vital in coordinating body movements.

Figure 13.6 Stretch reflex. This monosynaptic reflex arc has only one synapse in the CNS—between a single sensory neuron from the receptor and a single motor neuron to the effector. A polysynaptic reflex arc to antagonistic muscles that includes two synapses in the CNS and one interneuron is also illustrated. Plus signs (+) indicate excitatory synapses; the minus sign (−) indicates an inhibitory synapse.

🔑 **The stretch reflex causes contraction of a muscle that has been stretched.**

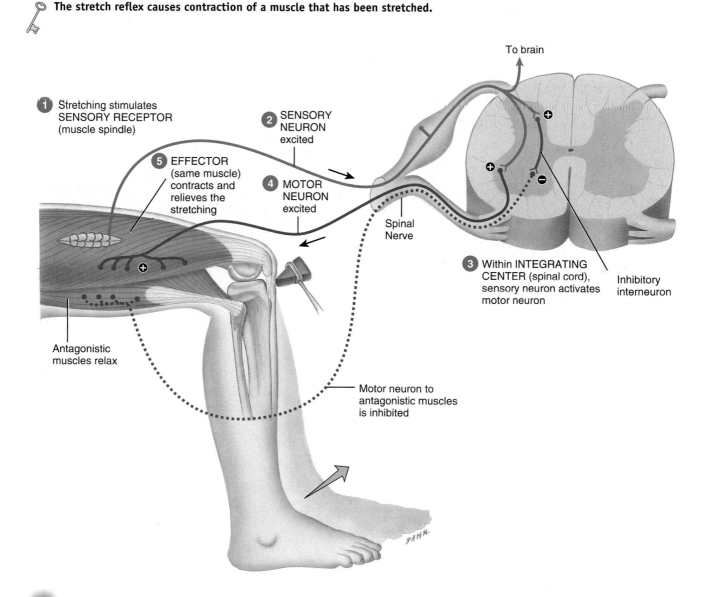

Q   What makes this an ipsilateral reflex?

Axon collaterals of the muscle spindle sensory neuron also relay nerve impulses to the brain over specific ascending pathways. In this way, the brain is given input about the state of stretch or contraction of skeletal muscles, enabling it to coordinate muscular movements and posture. The nerve impulses that pass to the brain also allow conscious perception that the reflex has occurred.

The Tendon Reflex

The stretch reflex operates as a feedback mechanism to control muscle *length* by causing muscle contraction. In contrast, the **tendon reflex** operates as a feedback mechanism to control muscle *tension* by causing muscle relaxation before muscle force becomes so great that tendons might be torn. Like the stretch reflex, the tendon reflex is ipsilateral. The

Figure 13.7 Tendon reflex. This reflex arc is polysynaptic—more than one CNS synapse and more than two different neurons are involved in the pathway. The sensory neuron synapses with two interneurons; an inhibitory interneuron causes relaxation of the effector, and a stimulatory interneuron causes contraction of the antagonistic muscle. Plus signs (+) indicate excitatory synapses; the minus sign (−) indicates an inhibitory synapse.

🔑 **The tendon reflex causes relaxation of the muscle attached to the stimulated tendon organ.**

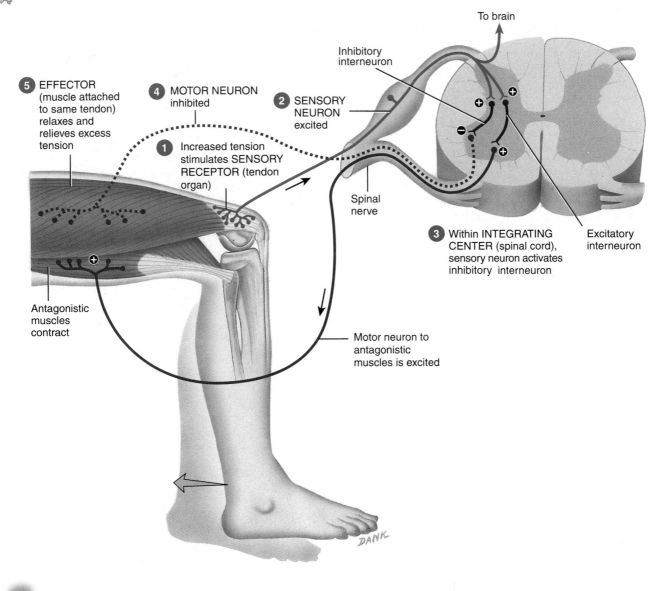

Q What is reciprocal innervation?

sensory receptors for this reflex are called **tendon (Golgi tendon) organs** (shown in more detail in Figure 15.3b on page 492), which lie within a tendon near its junction with a muscle. Whereas muscle spindles are sensitive to changes in muscle length, tendon organs detect and respond to changes in muscle tension that are caused by passive stretch or muscular contraction.

A tendon reflex operates as follows (Figure 13.7):

1 As the tension applied to a tendon increases, the tendon organ (sensory receptor) is stimulated (depolarized to threshold).

2 Nerve impulses are generated and propagate into the spinal cord along a sensory neuron.

③ Within the spinal cord (integrating center), the sensory neuron activates an inhibitory interneuron that synapses with a motor neuron.

④ The inhibitory neurotransmitter inhibits (hyperpolarizes) the motor neuron, which then generates fewer nerve impulses.

⑤ The muscle attached to the same tendon relaxes and relieves excess tension.

Thus, as tension on the tendon organ increases, the frequency of inhibitory impulses increases, and inhibition of the motor neurons to the muscle developing excess tension (effector) causes relaxation of the muscle. In this way, the tendon reflex protects the tendon and muscle from damage due to excessive tension.

Note in Figure 13.7 that the sensory neuron from the tendon organ also synapses with an excitatory interneuron in the spinal cord. The excitatory interneuron, in turn, synapses with motor neurons controlling antagonistic muscles. Thus, while the tendon reflex brings about relaxation of the muscle attached to the tendon organ, it also triggers contraction of antagonists, another example of reciprocal innervation. The sensory neuron also relays nerve impulses to the brain by way of sensory tracts, thus informing the brain about the state of muscle tension throughout the body.

The Flexor and Crossed Extensor Reflexes

Another reflex involving a polysynaptic reflex arc results when, for example, you step on a tack, and in response to this painful stimulus you immediately withdraw your leg. This reflex, called the **flexor** or **withdrawal reflex,** operates as follows (Figure 13.8):

① Stepping on a tack stimulates the dendrites (sensory receptor) of a pain-sensitive neuron.

② This sensory neuron then generates nerve impulses, which propagate into the spinal cord.

③ Within the spinal cord (integrating center), the sensory neuron activates interneurons that extend to several spinal cord segments.

④ The interneurons activate motor neurons in several spinal cord segments. As a result, the motor neurons generate nerve impulses, which propagate toward the axon terminals.

⑤ Acetylcholine released by the motor neurons causes the flexor muscles in the thigh (effectors) to contract, producing withdrawal of the leg. This reflex is protective because contraction of flexor muscles moves a limb away from the source of a painful stimulus.

The flexor reflex, like the stretch reflex, is ipsilateral— the incoming and outgoing impulses propagate into and out of the same side of the spinal cord. The flexor reflex also illustrates another feature of polysynaptic reflex arcs: Because withdrawing your entire lower or upper limb from a painful stimulus involves more than one muscle group, several motor neurons must simultaneously convey impulses to several

lower or upper limb muscles. Because nerve impulses from one sensory neuron ascend and descend in the spinal cord and activate interneurons in several segments of the spinal cord, this type of reflex is called an **intersegmental reflex arc.** Through intersegmental reflex arcs, a single sensory neuron can activate several motor neurons, thereby stimulating more than one effector. The monosynaptic stretch reflex, in contrast, involves muscles receiving nerve impulses from one spinal cord segment only.

Something else may happen when you step on a tack: You may start to lose your balance as your body weight shifts to the other foot. But the pain impulses from stepping on the tack do not merely initiate the flexor reflex that causes you to withdraw the limb; they also initiate a balance-maintaining **crossed extensor reflex,** which operates as follows (Figure 13.9):

① Stepping on a tack stimulates the sensory receptor of a pain-sensitive neuron in the right foot.

② This sensory neuron then generates nerve impulses, which propagate into the spinal cord.

③ Within the spinal cord (integrating center), the sensory neuron activates several interneurons that synapse with motor neurons on the left side of the spinal cord in several spinal cord segments. Thus, incoming pain signals cross to the opposite side through interneurons at that level, and at several levels above and below the point of entry into the spinal cord.

④ The interneurons excite motor neurons that innervate extensor muscles in several spinal cord segments. The motor neurons, in turn, generate more nerve impulses, which propagate toward the axon terminals.

⑤ Acetylcholine released by the motor neurons causes extensor muscles in the thigh (effectors) of the unstimulated left limb to contract, producing extension of the left leg. In this way, weight can be placed on the foot of the limb that must now support the entire body. A comparable reflex occurs with painful stimulation of the left lower limb or either upper limb.

Unlike the flexor reflex, which is an ipsilateral reflex, the crossed extensor reflex involves a **contralateral reflex arc** (kon′-tra-LAT-er-al; = opposite side): Sensory impulses enter one side of the spinal cord and motor impulses exit on the opposite side. Thus, a crossed extensor reflex synchronizes the extension of the contralateral limb with the withdrawal (flexion) of the stimulated limb. Reciprocal innervation also occurs in both the flexor reflex and the crossed extensor reflex. In the flexor reflex, when the flexor muscles of a painfully stimulated lower limb are contracting, the extensor muscles of the same limb are relaxing to some degree. If both sets of muscles contracted at the same time, the two sets of muscles would pull on the bones in opposite directions, which might immobilize the limb. Because of reciprocal innervation, however, one set of muscles contracts while the other relaxes.

Figure 13.8 Flexor (withdrawal) reflex. This reflex arc is polysynaptic and ipsi-lateral. Plus signs (+) indicate excitatory synapses.

🔑 **The flexor reflex causes withdrawal of a part of the body in response to a painful stimulus.**

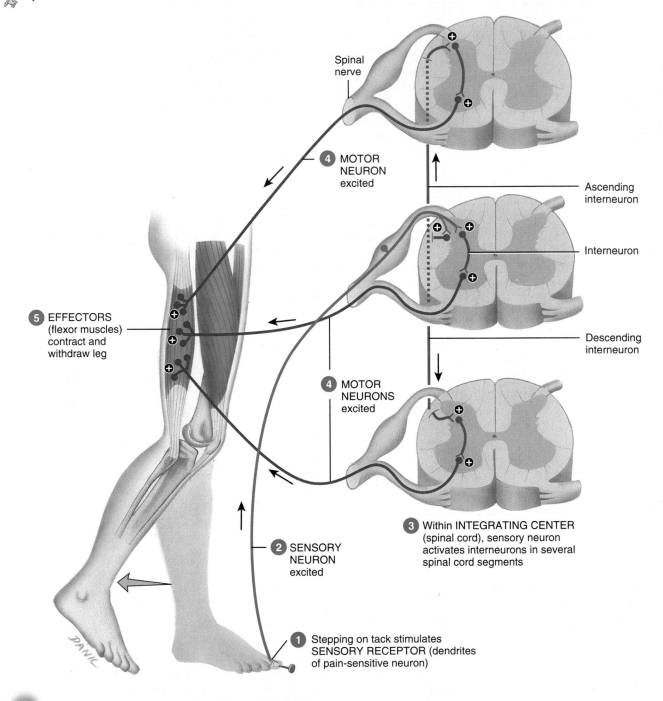

Spinal nerve

4 MOTOR NEURON excited

Ascending interneuron

Interneuron

Descending interneuron

5 EFFECTORS (flexor muscles) contract and withdraw leg

4 MOTOR NEURONS excited

3 Within INTEGRATING CENTER (spinal cord), sensory neuron activates interneurons in several spinal cord segments

2 SENSORY NEURON excited

1 Stepping on tack stimulates SENSORY RECEPTOR (dendrites of pain-sensitive neuron)

Ⓠ Why is the flexor reflex classified as an intersegmental reflex arc?

Figure 13.9 Crossed extensor reflex. The flexor reflex arc is shown (at left) to enable comparison with the crossed extensor reflex arc. Plus signs (+) indicate excitatory synapses.

A crossed extensor reflex causes contraction of muscles that extend joints in the limb opposite a painful stimulus.

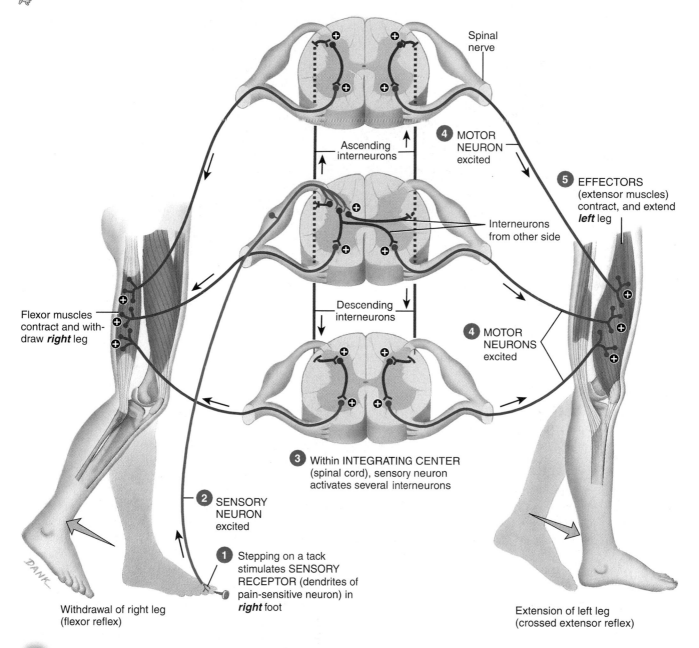

Spinal nerve

Ascending interneurons

4 MOTOR NEURON excited

5 EFFECTORS (extensor muscles) contract, and extend *left* leg

Interneurons from other side

Flexor muscles contract and withdraw *right* leg

Descending interneurons

4 MOTOR NEURONS excited

3 Within INTEGRATING CENTER (spinal cord), sensory neuron activates several interneurons

2 SENSORY NEURON excited

1 Stepping on a tack stimulates SENSORY RECEPTOR (dendrites of pain-sensitive neuron) in *right* foot

Withdrawal of right leg (flexor reflex)

Extension of left leg (crossed extensor reflex)

Q Why is the crossed extensor reflex classified as a contralateral reflex arc?

Figure 13.10 Organization and connective tissue coverings of a spinal nerve. Part b: Richard G. Kessel and Randy H. Kardon from *Tissues and Organs: A Text-Atlas of Scanning Electron Microscopy.* Copyright © 1979 by W. H. Freeman and Company. Reprinted by permission.

Three layers of connective tissue wrappings protect axons: endoneurium surrounds individual axons, perineurium surrounds bundles of axons (fascicles), and epineurium surrounds an entire nerve.

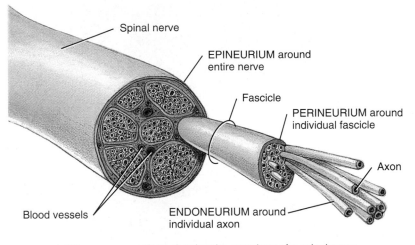

(a) Transverse sections showing the coverings of a spinal nerve

CLINICAL APPLICATION
Plantar Flexion Reflex and Babinski Sign

Several reflexes have clinical significance. Among the most important is the **plantar flexion reflex,** elicited by gently stroking the lateral outer margin of the sole. The normal response is a curling under of all the toes. Damage to descending motor pathways, such as the corticospinal tract, alters this reflex such that stroking the sole causes the great toe to extend, with or without fanning of the other toes, a response called the **Babinski sign.** In children under 18 months of age, the Babinski sign is normal due to incomplete myelination of fibers in the corticospinal tract. The normal response after 18 months of age is the plantar flexion reflex. ■

1. Describe the function of the spinal cord as a conduction pathway.
2. What is a reflex? List and define the components of a reflex arc, and describe how the spinal cord serves as an integrating center for reflexes.
3. Describe the mechanism and function of a stretch reflex, tendon reflex, flexor (withdrawal) reflex, and crossed extensor reflex.
4. Define the following terms related to reflex arcs: monosynaptic, ipsilateral, polysynaptic, intersegmental, contralateral, and reciprocal innervation.

SPINAL NERVES

OBJECTIVES
• *Describe the components, connective tissue coverings, and branching of a spinal nerve.*

• *Define plexus, and identify the distribution of nerves of the cervical, brachial, lumbar, and sacral plexuses.*

• *Describe the clinical significance of dermatomes.*

Spinal nerves and the nerves that branch from them to serve all parts of the body connect the CNS to sensory receptors, muscles, and glands and are part of the peripheral nervous system (PNS). The 31 pairs of spinal nerves are named and numbered according to the region and level of the vertebral column from which they emerge (see Figure 13.2). The first cervical pair emerges between the atlas (first cervical vertebra) and the occipital bone; all other spinal nerves emerge from the vertebral column through the intervertebral foramina between adjoining vertebrae.

Not all spinal cord segments are in line with their corresponding vertebrae. Recall that the spinal cord ends near the level of the superior border of the second lumbar vertebra, and that the roots of the lower lumbar, sacral, and coccygeal nerves descend at an angle to reach their respective foramina before emerging from the vertebral column. This arrangement constitutes the cauda equina (see Figure 13.2).

Figure 13.10 (continued)

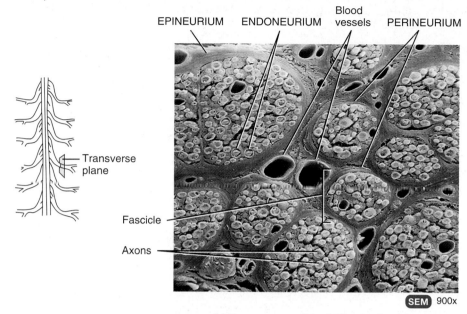

EPINEURIUM   ENDONEURIUM   Blood vessels   PERINEURIUM

Transverse plane

Fascicle

Axons

SEM 900x

(b) Transverse section of 12 nerve fascicles

Q Why are all spinal nerves classified as mixed nerves?

As noted earlier, a typical **spinal nerve** has two connections to the cord: a posterior root and an anterior root (see Figure 13.3a). The posterior and anterior roots unite to form a spinal nerve at the intervertebral foramen. Because the posterior root contains sensory axons and the anterior root contains motor axons, a spinal nerve is a **mixed nerve.** The posterior root contains a ganglion in which cell bodies of sensory neurons are located.

Connective Tissue Coverings of Spinal Nerves

Each cranial nerve and spinal nerve contains layers of protective connective tissue coverings (Figure 13.10). Individual axons, whether myelinated or unmyelinated, are wrapped in **endoneurium** (en'-dō-NOO-rē-um; endo- = within or inner). Groups of axons with their endoneurium are arranged in bundles called **fascicles,** each of which is wrapped in **perineurium** (per'-i-NOO-rē-um; peri- = around). The superficial covering over the entire nerve is the **epineurium** (ep'-i-NOO-rē-um; epi- = over). The dura mater of the spinal meninges fuses with the epineurium as the nerve passes through the intervertebral foramen. Note the presence of many blood vessels, which nourish nerves, within the perineurium and epineurium (Figure 13.10b). You may recall from Chapter 10 that the connective tissue coverings of skeletal muscles—endomysium, perimysium,

and epimysium—are similar in organization to those of nerves.

Distribution of Spinal Nerves

Branches

A short distance after passing through its intervertebral foramen, a spinal nerve divides into several branches (Figure 13.11). These branches are known as **rami** (RĀ-mī = branches). The **posterior (dorsal) ramus** (RĀ-mus; singular form) serves the deep muscles and skin of the dorsal surface of the trunk. The **anterior (ventral) ramus** serves the muscles and structures of the upper and lower limbs and the skin of the lateral and ventral surfaces of the trunk. In addition to posterior and anterior rami, spinal nerves also give off a **meningeal branch,** which reenters the vertebral canal through the intervertebral foramen and supplies the vertebrae, vertebral ligaments, blood vessels of the spinal cord, and meninges. Other branches of a spinal nerve are the **rami communicantes** (kō-myoo-ni-KAN-tēz), components of the autonomic nervous system whose structure and function are discussed in Chapter 17.

Plexuses

The anterior rami of spinal nerves, except for thoracic nerves T2–T12, do not go directly to the body structures

Figure 13.11 Branches of a typical spinal nerve, shown in transverse section through the thoracic portion of the spinal cord. (See Tortora, *A Photographic Atlas of the Human Body,* Figure 8.4)

 The branches of a spinal nerve are the posterior ramus, the anterior ramus, the meningeal branch, and the rami communicantes.

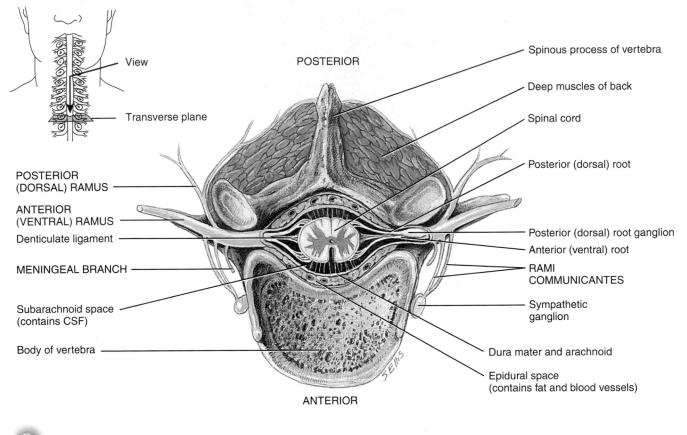

Which spinal nerve branches serve the upper and lower limbs?

they supply. Instead, they form networks on both the left and right sides of the body by joining with various numbers of nerve fibers from anterior rami of adjacent nerves. Such a network is called a **plexus** (= braid). The principal plexuses are the **cervical plexus, brachial plexus, lumbar plexus,** and **sacral plexus.** Refer to Figure 13.2 to see their relationships to one another. Emerging from the plexuses are nerves bearing names that are often descriptive of the general regions they serve or the course they take. Each of the nerves, in turn, may have several branches named for the specific structures they innervate.

Exhibits 13.1–13.4 summarize the principal plexuses. The anterior rami of spinal nerves T2–T12 are called intercostal nerves and will be discussed next.

Intercostal Nerves

The anterior rami of spinal nerves T2–T12 do not enter into the formation of plexuses and are known as **intercostal** or **thoracic nerves.** These nerves directly innervate the structures they supply in the intercostal spaces (see Figure 13.2). After leaving its intervertebral foramen, the anterior ramus of nerve T2 innervates the intercostal muscles of the second intercostal space and supplies the skin of the axilla and posteromedial aspect of the arm. Nerves T3–T6 pass in the costal grooves of the ribs and then to the intercostal muscles and skin of the anterior and lateral chest wall. Nerves T7–T12 supply the intercostal muscles and abdominal muscles, and the overlying skin. The posterior rami of the intercostal nerves supply the deep back muscles and skin of the posterior aspect of the thorax.

Text continues on page 441

Exhibit 13.1 *Cervical Plexus (Figure 13.12)*

OBJECTIVE

• *Describe the origin, distribution, and effects of damage to the cervical plexus.*

The **cervical plexus** (SER-vi-kul) is formed by the anterior (ventral) rami of the first four cervical nerves (C1–C4), with contributions from C5. There is one on each side of the neck alongside the first four cervical vertebrae. The roots of the plexus indicated in Figure 13.12 are the anterior rami.

The cervical plexus supplies the skin and muscles of the head, neck, and superior part of the shoulders and chest. The phrenic nerves arise from the cervical plexuses and supply motor fibers to the diaphragm. Branches of the cervical plexus also run parallel to the accessory (XI) and hypoglossal (XII) cranial nerves.

CLINICAL APPLICATION

Injuries to the Phrenic Nerves

Complete severing of the spinal cord above the origin of the phrenic nerves (C3, C4, and C5) causes respiratory arrest. Breathing stops because the phrenic nerves no longer send impulses to the diaphragm. ■

Which cranial nerves have branches that run alongside branches of the cervical plexus?

NERVE	ORIGIN	DISTRIBUTION
SUPERFICIAL (SENSORY) BRANCHES		
Lesser occipital	C2	Skin of scalp posterior and superior to ear.
Great auricular (aw-RIK-yoo-lar)	C2–C3	Skin anterior, inferior, and over ear, and over parotid glands.
Transverse cervical	C2–C3	Skin over anterior aspect of neck.
Supraclavicular	C3–C4	Skin over superior portion of chest and shoulder.
DEEP (LARGELY MOTOR) BRANCHES		
Ansa cervicalis (AN-sa ser-vi-KAL-is)		This nerve divides into superior and inferior roots.
Superior root	C1	Infrahyoid and geniohyoid muscles of neck.
Inferior root	C2–C3	Infrahyoid muscles of neck.
Phrenic (FREN-ik)	C3–C5	Diaphragm between thorax and abdomen.
Segmental branches	C1–C5	Prevertebral (deep) muscles of neck, levator scapulae, and middle scalene muscles.

Exhibit 13.1 *Cervical Plexus (continued)*

Figure 13.12 Cervical plexus in anterior view. (See Tortora, *A Photographic Atlas of the Human Body*, Figure 8.7)

🔑 The cervical plexus supplies the skin and muscles of the head, neck, superior portion of the shoulders and chest, and diaphragm.

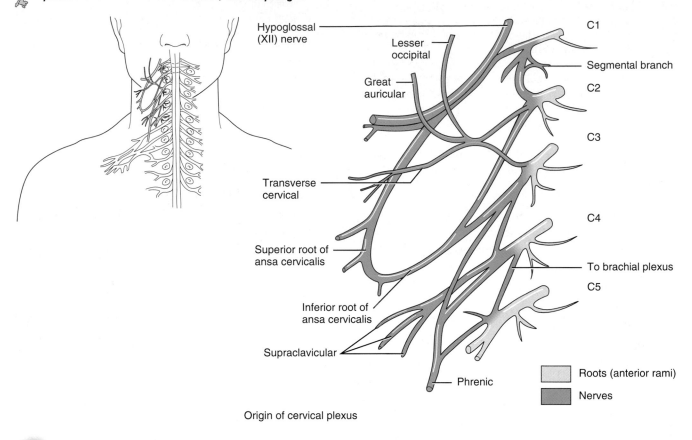

Origin of cervical plexus

Q Why does complete severing of the spinal cord at level C2 cause respiratory arrest?

Exhibit 13.2 | Brachial Plexus (Figures 13.13 and 13.14)

OBJECTIVE

• *Describe the origin, distribution, and effects of damage to the brachial plexus.*

The anterior (ventral) rami of spinal nerves C5–C8 and T1 form the **brachial plexus** (BRĀ-kē-al), which extends inferiorly and laterally on either side of the last four cervical and first thoracic vertebrae (Figure 13.13a). It passes superior to the first rib posterior to the clavicle and then enters the axilla.

The brachial plexus provides the entire nerve supply of the shoulders and upper limbs (Figure 13.13b). Five important nerves arise from the brachial plexus. (1) the **axillary nerve** supplies the deltoid and teres minor muscles. (2) The **musculocutaneous nerve** supplies the flexors of the arm. (3) The **radial nerve** supplies the muscles on the posterior aspect of the arm and forearm. (4) The **median nerve** supplies most of the muscles of the anterior forearm and some of the muscles of the hand. (5) The **ulnar nerve** supplies the anteromedial muscles of the forearm and most of the muscles of the hand.

CLINICAL APPLICATION
Injuries to Nerves Emerging from the Brachial Plexus

Injury to the superior roots of the brachial plexus (C5–C6) may result from forceful pulling away of the head from the shoulder, as might occur from a heavy fall on the shoulder or during childbirth in which the infant's head is excessively stretched. The presentation of this injury is characterized by an upper limb in which the shoulder is adducted, the arm is medially rotated, the elbow is extended, the forearm is pronated, and the wrist is flexed (Figure 13.14a). This condition is called **Erb-Duchene palsy** or **waiter's tip position.** There is loss of sensation along the lateral side of the arm.

Radial (and axillary) **nerve injury** can be caused by improperly administered intramuscular injections into the deltoid muscle. The radial nerve may also be injured when a cast is applied too tightly around the mid-humerus. Radial nerve injury is indicated by **wrist drop,** the inability to extend the wrist and fingers (Figure 13.14b). Sensory loss is minimal due to the overlap of sensory innervation by adjacent nerves.

Median nerve injury is indicated by numbness, tingling, and pain in the palm and fingers. There is also inability to pronate the forearm and flex the proximal interphalangeal joints of all digits and the distal interphalangeal joints of the second and third digits (Figure 13.14c). In addition, wrist flexion is weak and is accompanied by adduction, and thumb movements are weak.

Ulnar nerve injury is indicated by an inability to abduct or adduct the fingers, atrophy of the interosseus muscles of the hand, hyperextension of the metacarpophalangeal joints, and flexion of the interphalangeal joints, a condition called **clawhand** (Figure 13.14d). There is loss of sensation over the little finger.

Long thoracic nerve injury results in paralysis of the serratus anterior muscle. The medial border of the scapula protrudes, giving it the appearance of a wing. When the arm is raised, the vertebral border and inferior angle of the scapula pull away from the thoracic wall, a condition called **winged scapula** (Figure 13.14e). The arm cannot be abducted beyond the horizontal position. ∎

Injury of which nerve could cause paralysis of the serratus anterior muscle?

Exhibit 13.2 | *Brachial Plexus (continued)*

NERVE	ORIGIN	DISTRIBUTION
Dorsal scapular (SKAP-yoo-lar)	C5	Levator scapulae, rhomboideus major, and rhomboideus minor muscles.
Long thoracic (thor-RAS-ik)	C5–C7	Serratus anterior muscle.
Nerve to subclavius (sub-KLĀ-ve-us)	C5–C6	Subclavius muscle.
Suprascapular	C5–C6	Supraspinatus and infraspinatus muscles.
Musculocutaneous (mus'-kyoo-lo-kyoo-TĀN-ē-us)	C5–C7	Coracobrachialis, biceps brachii, and brachialis muscles.
Lateral pectoral (PEK-to-ral)	C5–C7	Pectoralis major muscle.
Upper subscapular	C5–C6	Subscapularis muscle.
Thoracordorsal (tho-RA-ko-dor-sal)	C6–C8	Latissimus dorsi muscle.
Lower subscapular	C5–C6	Subscapularis and teres major muscles.
Axillary (AK-si-lar-ē) or **circumflex** (SER-kum-fleks)	C5–C6	Deltoid and teres minor muscles; skin over deltoid and superior posterior aspect of arm.
Median	C5–T1	Flexors of forearm, except flexor carpi ulnaris and some muscles of the hand (lateral palm); skin of lateral two-thirds of palm of hand and fingers.
Radial	C5–C8, T1	Triceps brachii and other extensor muscles of arm and extensor muscles of forearm; skin of posterior arm and forearm, lateral two-thirds of dorsum of hand, and fingers over proximal and middle phalanges.
Medial pectoral	C8–T1	Pectoralis major and pectoralis minor muscles.
Medial brachial cutaneous (BRĀ-kē-al kyoo'-TĀ-nē-us)	C8–T1	Skin of medial and posterior aspects of distal third of arm.
Medial antebrachial cutaneous (an'-tē-BRĀ-kē-el)	C8–T1	Skin of medial and posterior aspects of forearm.
Ulnar	C8–T1	Flexor carpi ulnaris, flexor digitorum profundus, and most muscles of the hand; skin of medial side of hand, little finger, and medial half of ring finger.

Exhibit 13.2 (continued)

Figure 13.13 Brachial plexus in anterior view. (See Tortora, *A Photographic Atlas of the Human Body,* Figures 8.8, 8.9)

🗝 **The brachial plexus supplies the shoulders and upper limbs.**

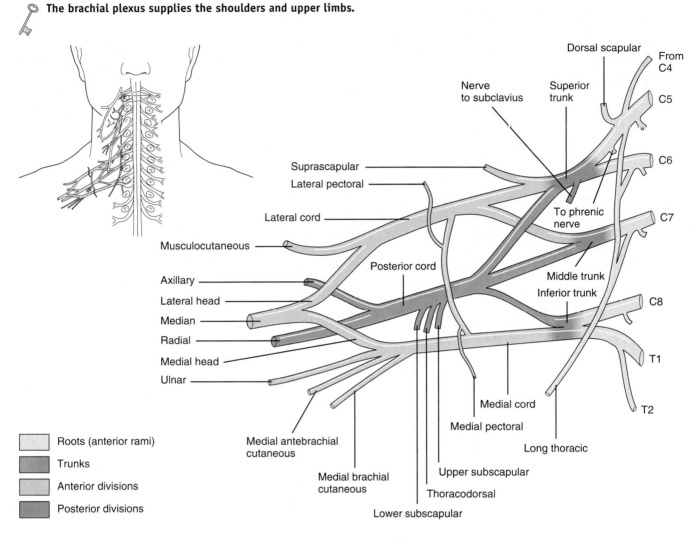

(a) Origin of brachial plexus

Figure continues

Exhibit 13.2 Brachial Plexus *(continued)*

Figure 13.13 *(continued)*

Clavicle
Lateral cord

Posterior cord
Medial cord

Axillary nerve

Musculocutaneous
nerve

Radial nerve

Median nerve

Deep branch
of radial nerve

Superficial
branch of
radial nerve

Ulnar nerve

Median nerve

Radial nerve

Scapula

Ulnar nerve

Humerus

Radius

Ulna

Superficial branch
of ulnar nerve

Digital branch
of median nerve

Digital branch
of ulnar nerve

(b) Distribution of nerves from brachial plexus

Q What five important nerves arise from the brachial plexus?

Exhibit 13.2   *(continued)*

Figure 13.14 Injuries to the brachial plexus.

 Injuries to the brachial plexus affect the sensations and movements of the upper limbs.

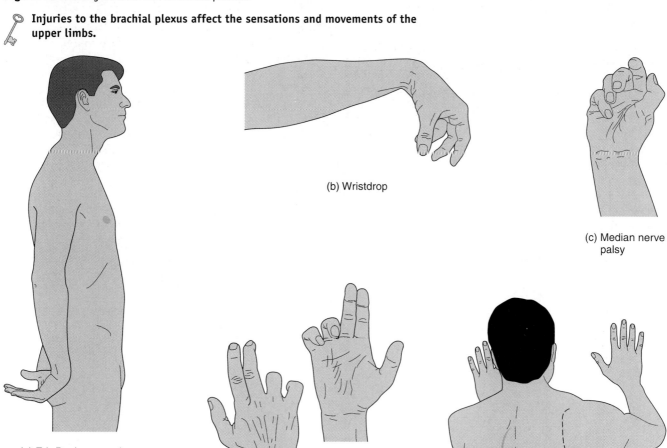

(b) Wristdrop

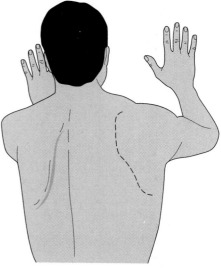

(c) Median nerve palsy

(a) Erb-Duchenne palsy (waiter's tip)

(d) Ulnar nerve palsy

(e) Winging of left scapula

Q Injury to which nerve of the brachial plexus affects sensations on the palm and fingers?

Exhibit 13.3 | Lumbar Plexus (Figure 13.15)

OBJECTIVE

• *Describe the origin, distribution, and effects of damage to the lumbar plexus.*

Anterior (ventral) rami of spinal nerves L1–L4 form the **lumbar** (LUM-bar) **plexus.** It differs from the brachial plexus in that there is no intricate intermingling of fibers. On either side of the first four lumbar vertebrae, the lumbar plexus passes obliquely outward, posterior to the psoas major muscle and anterior to the quadratus lumborum muscle. It then gives rise to its peripheral nerves.

The lumbar plexus supplies the anterolateral abdominal wall, external genitals, and part of the lower limbs.

CLINICAL APPLICATION
Lumbar Plexus Injuries

The largest nerve arising from the lumbar plexus is the femoral nerve. **Femoral nerve injury,** as in stab or gunshot wounds, is indicated by an inability to extend the leg and by loss of sensation in the skin over the anteromedial aspect of the thigh.

Obturator nerve injury, a common complication of childbirth, results in paralysis of the adductor muscles of the leg and loss of sensation over the medial aspect of the thigh. ■

| Injury of which nerve could cause loss of sensation in the buttocks?

NERVE	ORIGIN	DISTRIBUTION
Iliohypogastric (il'-ē-ō-hī-pō-GAS-trik)	L1	Muscles of anterolateral abdominal wall; skin of inferior abdomen and buttock.
Ilioinguinal (il'-ē-ō-IN-gwi-nal)	L1	Muscles of anterolateral abdominal wall; skin of superior medial aspect of thigh, root of penis and scrotum in male, and labia majora and mons pubis in female.
Genitofemoral (jen'-i-tō-FEM-or-al)	L1–L2	Cremaster muscle; skin over middle anterior surface of thigh, scrotum in male, and labia majora in female.
Lateral femoral cutaneous	L2–L3	Skin over lateral, anterior, and posterior aspects of thigh.
Femoral	L2–L4	Flexor muscles of thigh and extensor muscles of leg; skin over anterior and medial aspect of thigh and medial side of leg and foot.
Obturator (OB-too-rā-tor)	L2–L4	Adductor muscles of leg; skin over medial aspect of thigh.

Exhibit 13.3   *(continued)*

Figure 13.15 Lumbar plexus in anterior view. (See Tortora, *A Photographic Atlas of the Human Body,* Figure 8.10)

🔑 **The lumbar plexus supplies the anterolateral abdominal wall, external genitals, and part of the lower limbs.**

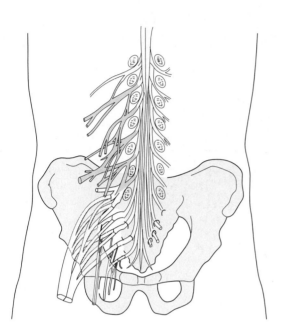

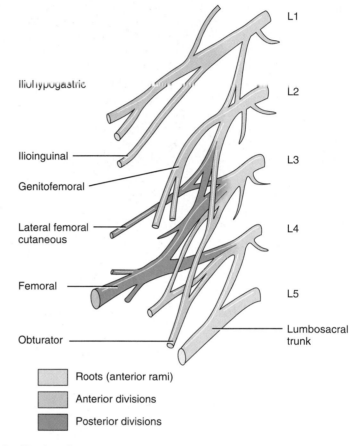

(a) Origin of lumbar plexus

Figure continues

Exhibit 13.3 | *Lumbar Plexus (continued)*

Figure 13.15 *(continued)*

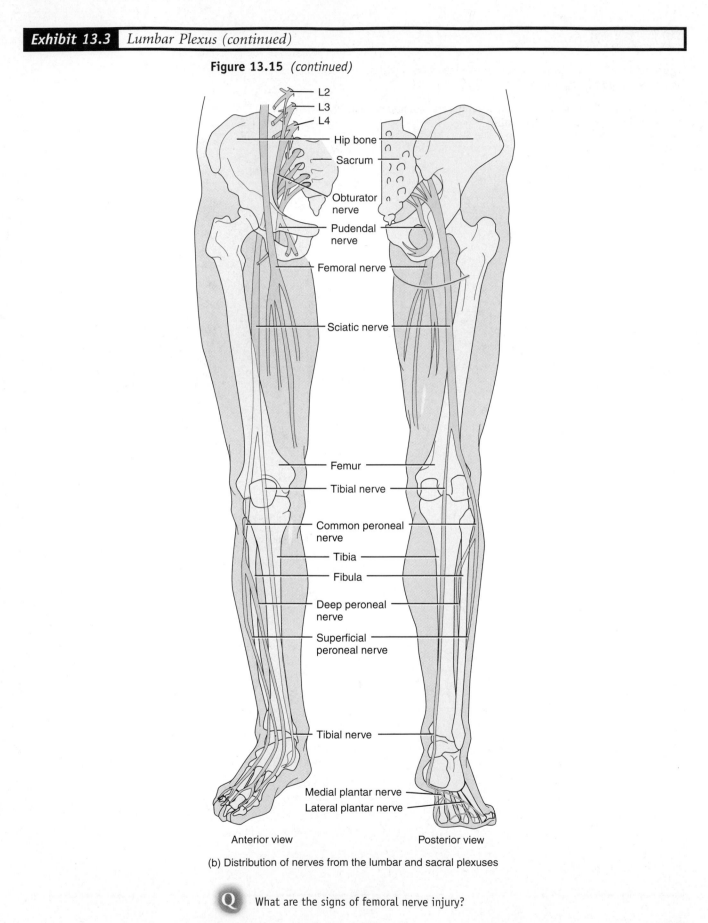

Anterior view   Posterior view

(b) Distribution of nerves from the lumbar and sacral plexuses

Q   What are the signs of femoral nerve injury?

Exhibit 13.4	*Sacral Plexus (Figure 13.16)*

OBJECTIVE
• *Describe the origin, distribution, and effects of damage to the sacral plexus.*

The anterior (ventral) rami of spinal nerves L4–L5 and S1–S4 form the **sacral plexus** (SĀ-kral). It is situated largely anterior to the sacrum. The sacral plexus supplies the buttocks, perineum, and lower limbs. The largest nerve in the body—the sciatic nerve—arises from the sacral plexus.

CLINICAL APPLICATION
Sciatic Nerve Injury

Injury to the sciatic nerve and its branches results in **sciatica**, pain that may extend from the buttock down the posterior and lateral aspect of the leg and the lateral aspect of the foot. The sciatic nerve may be injured because of a herniated (slipped) disc, dislocated hip, osteoarthritis of the lumbosacral spine, pressure from the uterus during pregnancy, or an improperly administered gluteal intramuscular injection.

In the majority of sciatic nerve injuries, the common peroneal portion is the most affected, frequently from fractures of the fibula or by pressure from casts or splints. Damage to the common peroneal nerve causes the foot to be plantar flexed, a condition called **footdrop,** and inverted, a condition called **equinovarus.** There is also loss of function along the anterolateral aspects of the leg and dorsum of the foot and toes. Injury to the tibial portion of the sciatic nerve results in dorsiflexion of the foot plus eversion, a condition called **calcaneovalgus.** Loss of sensation on the sole also occurs. ■

| Injury of which nerve causes footdrop?

NERVE	ORIGIN	DISTRIBUTION
Superior gluteal (GLOO-tē-al)	L4–L5 and S1	Gluteus minimus and gluteus medius muscles and tensor fasciae latae.
Inferior gluteal	L5–S2	Gluteus maximus muscle.
Nerve to piriformis (pir-i-FORM-is)	S1–S2	Piriformis muscle.
Nerve to quadratus femoris (quod-RĀ-tus FEM-or-is) **and inferior gemellus** (jem-EL-us)	L4–L5 and S1	Quadratus femoris and inferior gemellus muscles.
Nerve to obturator internus (OB-too-rā′-tor in-TER-nus) **and superior gemellus**	L5–S2	Obturator internus and superior gemellus muscles.
Perforating cutaneous (kyoo′-TĀ-ne-us)	S2–S3	Skin over inferior medial aspect of buttock.
Posterior femoral cutaneous	S1–S3	Skin over anal region, inferior lateral aspect of buttock, superior posterior aspect of thigh, superior part of calf, scrotum in male, and labia majora in female.
Sciatic (sī-AT-ik)	L4–S3	Actually two nerves—tibial and common peroneal—bound together by common sheath of connective tissue. It splits into its two divisions, usually at the knee. (See below for distributions.) As sciatic nerve descends through the thigh, it sends branches to hamstring muscles and adductor magnus.
Tibial (TIB-ē-al)	L4–S3	Gastrocnemius, plantaris, soleus, popliteus, tibialis posterior, flexor digitorum longus, and flexor hallucis longus muscles. Branches of tibial nerve in foot are medial plantar nerve and lateral plantar nerve.
Medial plantar (PLAN-tar)		Abductor hallucis, flexor digitorum brevis, and flexor hallucis brevis muscles; skin over medial two-thirds of plantar surface of foot.
Lateral plantar		Remaining muscles of foot not supplied by medial plantar nerve; skin over lateral third of plantar surface of foot.
Common peroneal (per′-ō-NĒ-al)	L4–S2	Divides into a superficial peroneal and a deep peroneal branch.
Superficial peroneal		Peroneus longus and peroneus brevis muscles; skin over distal third of anterior aspect of leg and dorsum of foot.
Deep peroneal		Tibialis anterior, extensor hallucis longus, peroneus tertius, and extensor digitorum longus and brevis muscles; skin on adjacent sides of great and second toes.
Pudendal (pyoo-DEN-dal)	S2–S4	Muscles of perineum; skin of penis and scrotum in male and clitoris, labia majora, labia minora, and vagina in female.

Exhibit 13.4 *Sacral Plexus (continued)*

Figure 13.16 Sacral plexus in anterior view. The distribution of the nerves of the sacral plexus is shown in Figure 13.15b. (See Tortora, *A Photographic Atlas of the Human Body*, Figure 8.11)

🔑 **The sacral plexus supplies the buttocks, perineum, and lower limbs.**

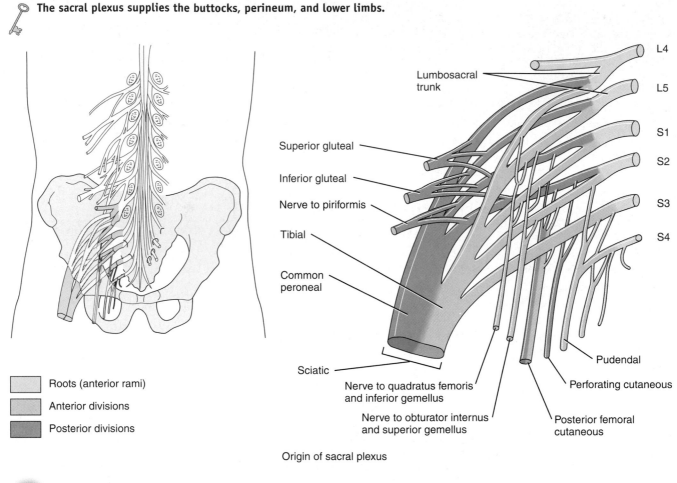

Origin of sacral plexus

Ⓠ What is the origin of the sacral plexus?

Dermatomes

The skin over the entire body is supplied by somatic sensory neurons that carry nerve impulses from the skin into the spinal cord and brain stem. Likewise, the underlying skeletal muscles are innervated by somatic motor neurons that carry impulses out of the spinal cord. Each spinal nerve contains sensory neurons that serve a specific, predictable segment of the body. Most of the skin of the face and scalp is served by cranial nerve V (the trigeminal nerve). The area of the skin that provides sensory input to one pair of spinal nerves or to cranial nerve V (for the face) is called a **dermatome** (*derma-* = skin; *-tome* = thin segment) (Figure 13.17). The nerve supply in adjacent dermatomes overlaps somewhat. In regions in which the overlap is considerable, little loss of sensation may result if only the single nerve supplying one dermatome is damaged.

Knowing which spinal cord segments supply each dermatome makes it possible to locate damaged regions of the spinal cord. If the skin in a particular region is stimulated but the sensation is not perceived, the nerves supplying that dermatome are probably damaged. Information about the innervation patterns of spinal nerves can also be used therapeutically. Cutting roots or infusing local anesthetics can block pain either permanently or transiently. But because dermatomes overlap, deliberate production of a region of complete anesthesia may require that at least three adjacent spinal nerves be cut or blocked by an anesthetic drug.

CLINICAL APPLICATION
Spinal Cord Transection and Muscle Function

An injury that entirely severs the spinal cord is said to cause complete **transection.** After transection, a person will have permanent loss of sensations in dermatomes below the injury because ascending nerve impulses cannot propagate past the transection to reach the brain. At the same time, voluntary muscle contractions will be lost below the transection because nerve impulses descending from the brain also cannot pass. The extent of paralysis of skeletal muscles depends on the level of injury. The following list outlines which muscle functions may be retained at progressively lower levels of spinal cord transection.

- C1–C3: no function maintained from the neck down; ventilator needed for breathing
- C4–C5: diaphragm, which allows breathing
- C6–C7: some arm and chest muscles, which allows feeding, some dressing, and propelling wheelchair
- T1–T3: intact arm function
- T4–T9: control of trunk above the umbilicus
- T10–L1: most thigh muscles, which allows walking with long leg braces
- L1–L2: most leg muscles, which allows walking with short leg braces ■

Figure 13.17 Distribution of dermatomes.

A dermatome is an area of skin that provides sensory input via the posterior roots of one pair of spinal nerves or via cranial nerve V (trigeminal nerve).

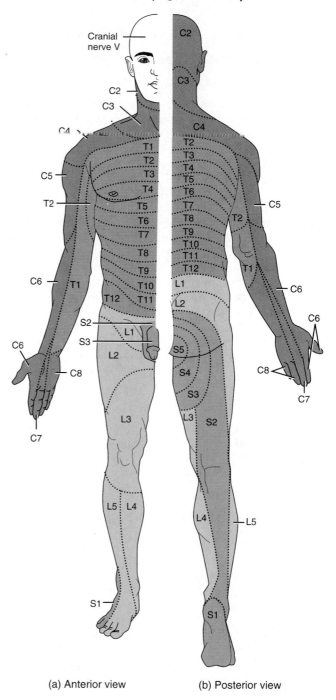

(a) Anterior view   (b) Posterior view

Which is the only spinal nerve that does not have a corresponding dermatome?

1. How are spinal nerves named and numbered? Why are all spinal nerves classified as mixed nerves?
2. Describe how a spinal nerve is connected with the spinal cord.
3. Describe the branches and innervations of a typical spinal nerve.
4. Describe the principal plexuses and the regions they supply.

DISORDERS: HOMEOSTATIC IMBALANCES

NEURITIS

Inflammation of one or several nerves is termed **neuritis** (*neur-* = nerve; *-itis* = inflammation). It may result from irritation to the nerve produced by direct blows, bone fractures, contusions, or penetrating injuries. Additional causes include infections, vitamin deficiency (usually thiamine), and poisons such as carbon monoxide, carbon tetrachloride, heavy metals, and some drugs.

SHINGLES

Shingles is an acute infection of the peripheral nervous system caused by herpes zoster (HER-pēz ZOS-ter), the virus that also causes chickenpox. After a person recovers from chickenpox, the virus retreats to a posterior root ganglion. If the virus is reactivated, the immune system usually prevents it from spreading. From time to time, however, the reactivated virus overcomes a weakened immune system, leaves the ganglion, and travels down sensory neurons of the skin by fast axonal transport (described on page 381). The result is pain, discoloration of the skin, and a characteristic line of skin blisters. The line of blisters marks the distribution (dermatome) of the particular cutaneous sensory nerve belonging to the infected root ganglion.

POLIOMYELITIS

Poliomyelitis, or simply **polio,** is caused by a virus called poliovirus. The onset of the disease is marked by fever, severe headache, a stiff neck and back, deep muscle pain and weakness, and loss of certain somatic reflexes. In its most serious form, the virus produces paralysis by destroying cell bodies of motor neurons, specifically those in the anterior horns of the spinal cord and in the nuclei of the cranial nerves. Polio can cause death from respiratory or heart failure if the virus invades the brain cells of the vital medullary centers that control breathing and heart functions. Even though the availability of polio vaccines has virtually eradicated polio in the United States, outbreaks of polio continue throughout the world. Due to international travel, polio could easily be reintroduced into North America if individuals are not vaccinated appropriately.

Several decades after suffering a severe attack of polio and following their recovery from it, some individuals develop a condition called **post-polio syndrome.** This neurological disorder is characterized by progressive muscle weakness, extreme fatigue, loss of function, and pain, especially in muscles and joints. Post-polio syndrome seems to involve a slow degeneration of motor neurons that innervate muscle fibers. Triggering factors appear to be a fall, a minor accident, surgery, or prolonged bed rest. Possible causes include overuse of surviving motor neurons over time, smaller motor neurons as a result of the initial infection by the virus, reactivation of dormant polio viral particles, immune-mediated responses, hormone deficiencies, and environmental toxins. Treatment consists of muscle-strengthening exercises, administration of pyridostigminine to enhance the action of acetylcholine in stimulating muscle contraction, and administration of nerve growth factors to stimulate both nerve and muscle growth.

STUDY OUTLINE

SPINAL CORD ANATOMY (p. 412)

1. The spinal cord is protected by the meninges, the vertebral column, cerebrospinal fluid, and denticulate ligaments.
2. The meninges are three coverings that run continuously around the spinal cord and brain: dura mater, arachnoid, and pia mater.
3. The spinal cord begins as a continuation of the medulla oblongata and ends at about the second lumbar vertebra in an adult.
4. The spinal cord contains cervical and lumbar enlargements that serve as points of origin for nerves to the limbs.
5. The tapered inferior portion of the spinal cord is the conus medullaris, from which arise the filum terminale and cauda equina.
6. Spinal nerves connect to each segment of the spinal cord by two roots: The posterior or dorsal root contains sensory nerve fibers, and the anterior or ventral root contains motor neuron axons.
7. The spinal cord is partially divided into right and left sides by the anterior median fissure and the posterior median sulcus.
8. The gray matter in the spinal cord is divided into horns, and the white matter into columns. In the center of the spinal cord is the central canal, which runs the length of the spinal cord.
9. Parts of the spinal cord observed in transverse section are the gray commissure; central canal; anterior, posterior, and lateral gray horns; and anterior, posterior, and lateral white columns, which contain ascending and descending tracts. Each part has specific functions.
10. The spinal cord conveys sensory and motor information by way of ascending and descending tracts, respectively.

SPINAL CORD PHYSIOLOGY (p. 417)

1. A major function of the spinal cord is to propagate nerve impulses from the periphery to the brain (sensory tracts) and to conduct motor impulses from the brain to the periphery (motor tracts).

2. Sensory information travels along two main routes in the white matter of the spinal cord: the posterior columns and the spinothalamic tracts.

3. Motor information travels along two main routes in the white matter of the spinal cord: direct pathways and indirect pathways.

4. A second major function of the spinal cord is to serve as an integrating center for spinal reflexes. This integration occurs in the gray matter.

5. A reflex is a fast, predictable, automatic response to changes in the environment that helps maintain homeostasis.

6. Reflexes may be spinal or cranial and somatic or autonomic (visceral).

7. A reflex arc is the simplest type of pathway that connects sensory input to motor output.

8. The components of a reflex arc are sensory receptor, sensory neuron, integrating center, motor neuron, and effector.

9. Somatic spinal reflexes include the stretch reflex, the tendon reflex, the flexor (withdrawal) reflex, and the crossed extensor reflex; all exhibit reciprocal innervation.

10. A two-neuron or monosynaptic reflex arc consists of one sensory neuron and one motor neuron. A stretch reflex, such as the patellar reflex, is an example.

11. The stretch reflex is ipsilateral and is important in maintaining muscle tone.

12. A polysynaptic reflex arc contains sensory neurons, interneurons, and motor neurons. The tendon reflex, flexor (withdrawal) reflex, and crossed extensor reflexes are examples.

13. The tendon reflex is ipsilateral and prevents damage to muscles and tendons when muscle force becomes too extreme; the flexor reflex is ipsilateral and moves a limb away from the source of a painful stimulus; the crossed extensor reflex extends the limb contralateral to a painfully stimulated limb, allowing the weight of the body to shift when a supporting limb is withdrawn.

14. A clinically important somatic reflex is the plantar flexion reflex, which is expressed as the Babinski sign when descending motor pathways are damaged.

SPINAL NERVES (p. 426)

1. The 31 pairs of spinal nerves are named and numbered according to the region and level of the spinal cord from which they emerge.

2. There are 8 pairs of cervical, 12 pairs of thoracic, 5 pairs of lumbar, 5 pairs of sacral, and 1 pair of coccygeal nerves.

3. Spinal nerves typically are connected with the spinal cord by a posterior root and an anterior root. All spinal nerves contain both sensory and motor axons (are mixed nerves).

4. Three connective tissue coverings associated with spinal nerves are endoneurium, perineurium, and epineurium.

5. Branches of a spinal nerve include the posterior ramus, anterior ramus, meningeal branch, and rami communicantes.

6. The anterior rami of spinal nerves, except for T2–T12, form networks of nerves called plexuses.

7. Emerging from the plexuses are nerves bearing names that typically describe the general regions they supply or the route they follow.

8. Nerves of the cervical plexus supply the skin and muscles of the head, neck, and upper part of the shoulders; they connect with some cranial nerves and innervate the diaphragm.

9. Nerves of the brachial plexus supply the upper limbs and several neck and shoulder muscles.

10. Nerves of the lumbar plexus supply the anterolateral abdominal wall, external genitals, and part of the lower limbs.

11. Nerves of the sacral plexus supply the buttocks, perineum, and part of the lower limbs.

12. Anterior rami of nerves T2–T12 do not form plexuses and are called intercostal (thoracic) nerves. They are distributed directly to the structures they supply in intercostal spaces.

13. Sensory neurons within spinal nerves and cranial nerve V (trigeminal) serve specific, constant segments of the skin called dermatomes.

14. Knowledge of dermatomes helps a physician determine which segment of the spinal cord or which spinal nerve is damaged.

SELF-QUIZ QUESTIONS

1. Match the following:
   - ___ (a) the joining together of the anterior rami of adjacent nerves
   - ___ (b) spinal nerve branches that serve the deep muscles and skin of the posterior surface of the trunk
   - ___ (c) spinal nerve branches that serve the muscles and structures of the upper and lower limbs and the lateral and ventral trunk
   - ___ (d) area of the spinal cord from which nerves to and from the upper limbs arise
   - ___ (e) area of the spinal cord from which nerves to and from the lower limbs arise
   - ___ (f) the roots form the nerves that arise from the inferior part of the spinal cord but do not leave the vertebral column at the same level as they exit the cord
   - ___ (g) contains motor neuron axons and conducts impulses from the spinal cord to the periphery
   - ___ (h) contains sensory nerve fibers and conducts impulses from the periphery into the spinal cord
   - ___ (i) an extension of the pia mater that anchors the spinal cord to the coccyx
   - ___ (j) space within the spinal cord filled with cerebrospinal fluid

   (1) cervical enlargement
   (2) lumbar enlargement
   (3) central canal
   (4) filum terminale
   (5) cauda equina
   (6) posterior root
   (7) anterior root
   (8) posterior ramus
   (9) anterior ramus
   (10) plexus

2. Match the following:

___ (a) provides the entire nerve supply of the shoulders and upper limbs

___ (b) provides the nerve supply of the skin and muscles of the head, neck, and superior part of the shoulders and chest

___ (c) provides the nerve supply of the anterolateral abdominal wall, external genitals, and part of the lower limbs

___ (d) innervates the buttocks, perineum, and lower limbs

___ (e) phrenic nerve arises from this plexus

___ (f) median nerve arises from this plexus

___ (g) sciatic nerve arises from this plexus

___ (h) femoral nerve arises from this plexus

___ (i) injury to this plexus can affect breathing

(1) cervical plexus
(2) brachial plexus
(3) lumbar plexus
(4) sacral plexus

True or false:

3. Reflexes permit the body to make exceedingly rapid adjustments to homeostatic imbalances.

4. The two functions of the spinal cord are to serve as an integrating center for thought processes and to serve as a conduction pathway.

Complete the following:

5. ___ are fast, predictable, automatic responses to changes in the environment.

6. ___ reflexes involve responses of smooth muscle, cardiac muscle, and glands.

7. The ___ protect the spinal cord against shock and sudden displacement.

8. The spinal cord ___ serve as conducting pathways, whereas the ___ serves to receive and integrate incoming and outgoing information.

Choose the best answer to the following questions:

9. The connective tissue covering around an entire nerve is the (a) endoneurium, (b) fascicle, (c) perineurium, (d) epineurium, (e) neurolemma.

10. Which of the following is *not* a reflex arc component? (a) sensory receptor, (b) sensory effector, (c) sensory neuron, (d) motor neuron, (e) integrating center

11. Which of the following are functions of the nerves of the posterior columns? (1) proprioception, (2) discriminative touch, (3) pain, (4) temperature variation, (5) pressure, (6) vibration, (7) two-point discrimination.
(a) 1, 2, 4, and 5, (b) 2, 4, 6, and 7, (c) 1, 2, 5, 6, and 7, (d) 3, 4, 5, 6, and 7, (e) 1, 3, 5, 6, and 7

12. Which of the following statements is *false*? (a) The two main spinal cord sensory paths are the spinothalamic and anterior columns. (b) The spinothalamic tracts convey impulses for sensing pain, temperature, touch, and deep pressure. (c) Direct pathways convey nerve impulses destined to cause precise, voluntary movements of skeletal muscles. (d) Indirect pathways convey nerve impulses that program automatic movements, help coordinate body movements with visual stimuli, maintain skeletal muscle tone and posture, and contribute to equilibrium. (e) The direct pathways are motor pathways.

13. The nerves that do *not* enter into the formation of a plexus are (a) C1–C5, (b) T2–T12, (c) C4–T5, (d) L1–L3, (e) T6–L2.

14. Which of the following statements is *false*? (a) The most superficial of the spinal meninges is the dura mater. (b) The epidural space contains fat and connective tissue. (c) The middle spinal meninx is the arachnoid. (d) The innermost spinal meninx is the pia mater. (e) The subarachnoid space contains interstitial fluid and blood cells.

15. Match the following:

___ (a) a reflex resulting in the contraction of a muscle when it is stretched

___ (b) withdrawal reflex

___ (c) a balance-maintaining reflex

___ (d) operates as a feedback mechanism to control muscle tension by causing muscle relaxation when muscle force becomes too extreme

___ (e) maintains proper muscle tone

___ (f) acts as a feedback mechanism to control muscle length by causing muscle contraction

___ (g) protects the tendon and muscle from damage due to excessive tension

(1) stretch reflex
(2) tendon reflex
(3) flexor reflex
(4) crossed extensor reflex

CRITICAL THINKING QUESTIONS

1. Pearl, a lifeguard at the local beach, stepped on a discarded lit cigarette with her bare foot. Trace the reflex arcs set in motion by her accident. Name the reflex types. (HINT: *How will she remain standing when she picks up her foot?*)

2. Why doesn't your spinal cord creep up toward your head every time you bend over? Why doesn't it get all twisted out of position when you exercise? (HINT: *How would you prevent a boat from floating away from a dock?*)

3. Jose's severe headaches and other symptoms were suggestive of meningitis, so his physician ordered a spinal tap. List the structures that the needle will pierce from the most superficial to the deepest. Why would the physician order a test in the spinal region to check a problem in Jose's head? (HINT: *The central nervous system needs protection in both its locations.*)

ANSWERS TO FIGURE QUESTIONS

13.1 The superior boundary of the spinal dura mater is the foramen magnum of the occipital bone; the inferior boundary is the second sacral vertebra.

13.2 The cervical enlargement connects with sensory and motor nerves of the upper limbs.

13.3 A horn is an area of gray matter, and a column is a region of white matter in the spinal cord.

13.4 The anterior corticospinal tract is located on the anterior side of the cord, originates in the cortex of the cerebrum, and ends in the spinal cord. It contains descending fibers and thus is a motor tract.

13.5 A sensory receptor produces a generator potential, which triggers a nerve impulse if the generator potential reaches threshold. Reflex integrating centers are in the CNS.

13.6 In an ipsilateral reflex, the sensory and motor neurons are on the same side of the spinal cord.

13.7 Reciprocal innervation is a type of neural circuit involving simultaneous contraction of one muscle and relaxation of its antagonist.

13.8 The flexor reflex is intersegmental because impulses go out over motor neurons located in several spinal nerves, each arising from a different segment of the spinal cord.

13.9 The crossed extensor reflex is a contralateral reflex arc because the motor impulses leave the spinal cord on the side opposite the entry of sensory impulses.

13.10 All spinal nerves are mixed (have sensory and motor function) because the posterior root containing sensory axons and the anterior root containing motor axons unite to form the spinal nerve.

13.11 The anterior rami serve the upper and lower limbs.

13.12 Severing the spinal cord at level C2 causes respiratory arrest because it prevents descending nerve impulses from reaching the phrenic nerve, which stimulates contraction of the diaphragm.

13.13 The axillary, musculocutaneous, radial, median, and ulnar nerves are five important nerves that arise from the brachial plexus.

13.14 Injury to the median nerve affects sensations on the palm and fingers.

13.15 Signs of femoral nerve injury include inability to extend the leg and loss of sensation in the skin over the anterolateral aspect of the thigh.

13.16 The origin of the sacral plexus is the anterior rami of spinal nerves L4–L5 and S1–S4.

13.17 The only spinal nerve without a corresponding dermatome is C1.

14 THE BRAIN AND CRANIAL NERVES

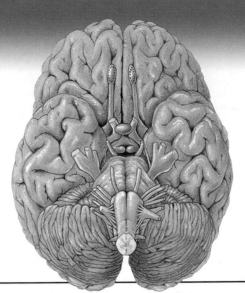

Solving an equation, feeling hungry, breathing—each of these processes is mediated by different regions of the **brain,** that portion of the central nervous system contained within the cranium. The adult brain is made up of about 100 billion neurons and 1000 billion neuroglia; it is one of the largest organs in the body, with a mass of about 1300 g (almost 3 lb.). The brain is the center for registering sensations, correlating them with one another and with stored information, making decisions, and taking actions. It also is the center for the intellect, emotions, behavior, and memory. But the brain encompasses yet a larger domain: It directs our behavior toward others. With ideas that excite, artistry that dazzles, or rhetoric that mesmerizes, one person's thoughts and actions may influence and shape the lives of many others. As you will see, different regions of the brain are specialized for different functions, and many parts of the brain work together to accomplish a particular function. This chapter explores the principal parts of the brain, how the brain is protected and nourished, and how it is related to both the spinal cord and the 12 pairs of cranial nerves.

OVERVIEW OF BRAIN ORGANIZATION AND BLOOD SUPPLY

OBJECTIVES

• *Identify the principal parts of the brain.*

• *Describe how the brain is protected.*

Principal Parts of the Brain

The brain consists of four principal parts: brain stem, cerebellum, diencephalon, and cerebrum (Figure 14.1). The **brain stem** is continuous with the spinal cord and consists of the medulla oblongata, pons, and midbrain. Posterior to the brain stem is the **cerebellum** (ser′-e-BEL-um; = little brain). Superior to the brain stem is the **diencephalon** (dī′-en-SEF-a-lon; *di-* = through; *-encephalon* = brain) consisting primarily of the thalamus and hypothalamus and including the epithalamus and the subthalamus. The **cerebrum** (se-RĒ-brum; = brain) spreads over the diencephalon like a mushroom cap and occupies most of the cranium.

Protective Coverings of the Brain

The brain is protected by the cranial bones (see Figure 7.4 on page 189) and the cranial meninges. The **cranial meninges** surround the brain (Figure 14.2). They are continuous with the spinal meninges, have the same basic structure, and bear the same names: the outer **dura mater,** the middle **arachnoid,** and the inner **pia mater.** Blood vessels that enter brain tissue pass along the surface of the brain, and as they penetrate inward, they are surrounded by a loose-fitting layer of pia mater. Three extensions of the dura mater separate parts of the brain: the **falx cerebri** (FALKS CER-e-brē; *falx* = sickle-shaped) separates the two hemispheres (sides) of the cerebrum; the **falx cerebelli** (cer-e-BEL-ī) separates the two hemispheres of the cerebellum; and the **tentorium cerebelli** (ten-TŌ-rē-um; = tent) separates the cerebrum from the cerebellum.

Figure 14.1 The brain. The infundibulum and pituitary gland are discussed with the endocrine system in Chapter 18. (See Tortora, *A Photographic Atlas of the Human Body,* Figures 8.12, 8.13, 8.15)

🔑 **The four principal parts of the brain are the brain stem, cerebellum, diencephalon, and cerebrum.**

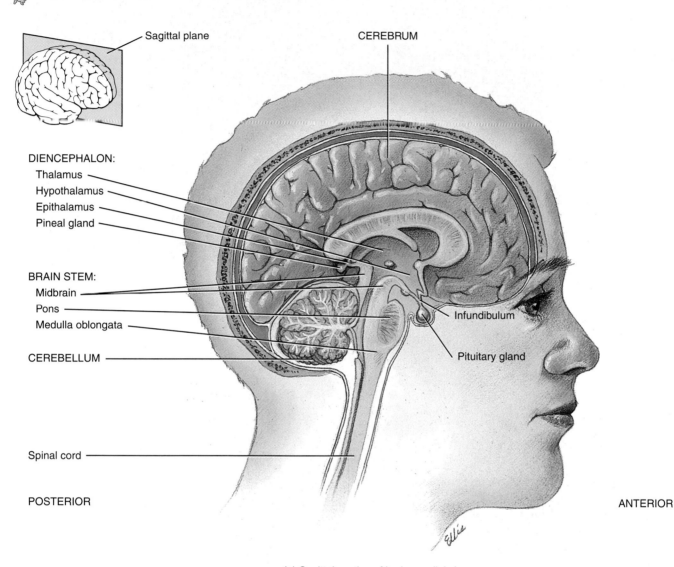

(a) Sagittal section of brain, medial view

Figure continues

Brain Blood Flow and the Blood–Brain Barrier

Blood flows to the brain mainly via blood vessels that branch from the cerebral arterial circle (circle of Willis) at the base of the brain (see Figure 21.20 on page 699); the veins that return blood from the head to the heart are shown in Figure 21.24 (page 711).

In an adult, the brain represents only 2% of total body weight, but it consumes about 20% of the oxygen and glucose used at rest. Neurons synthesize ATP almost exclusively from glucose via reactions that use oxygen (oxidative phosphorylation in mitochondria). When activity of neurons and neuroglia increases in a region of the brain, blood flow to that area also increases. Even a brief slowing of brain blood flow may cause unconsciousness. Typically, an interruption in blood flow for 1 or 2 minutes impairs neuronal function, and total deprivation of oxygen for about 4 minutes causes permanent injury. Because virtually no glucose is stored in the brain, the supply of glucose also must be continuous. If

Figure 14.1 (*continued*)

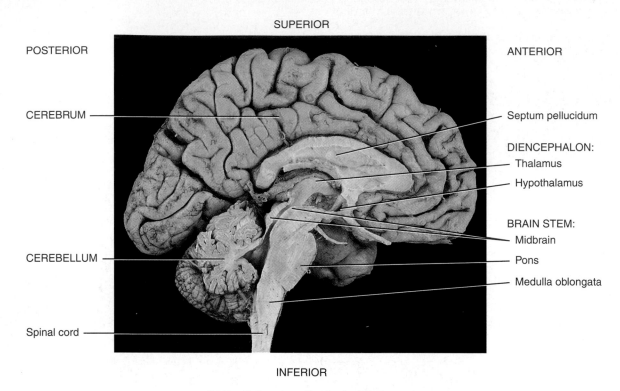

(b) Sagittal section of brain, medial view

 Which part of the brain is the largest?

blood entering the brain has a low level of glucose, mental confusion, dizziness, convulsions, and loss of consciousness may occur.

The existence of a **blood–brain barrier (BBB)** protects brain cells from harmful substances and pathogens by preventing passage of many substances from blood into brain tissue. Tight junctions seal together the endothelial cells of brain capillaries, which also are surrounded by a continuous basement membrane. Also, pressed up against the capillaries are the processes of large numbers of astrocytes (one type of neuroglia), which are thought to selectively pass some substances from the blood but inhibit the passage of others. A few water-soluble substances (for example, glucose) cross the BBB by active transport; other substances, such as creatinine, urea, and most ions, cross the BBB very slowly. Still other substances—proteins and most antibiotic drugs—do not pass at all from the blood into brain tissue. However, the BBB does not prevent passage of lipid-soluble substances, such as oxygen, carbon dioxide, alcohol, and most anesthetic agents, into brain tissue.

CLINICAL APPLICATION
Breaching the Blood–Brain Barrier

We have seen how the BBB prevents the passage into brain tissue of potentially harmful substances. But another consequence of the BBB's efficient protection is that it also prevents the passage of certain drugs that could be therapeutic for brain cancer or other CNS disorders. Researchers are finding ways to move drugs past the BBB. In one method, the drug is injected in a concentrated sugar solution. The high osmotic pressure of the sugar solution causes the endothelial cells of the capillaries to shrink, which opens gaps between their tight junctions. As a result, the drug can enter the brain tissue. ■

1. Compare the sizes and locations of the cerebrum and cerebellum.
2. Describe the locations of the cranial meninges.
3. Explain the blood supply to the brain and the blood–brain barrier.

Figure 14.2 The protective coverings of the brain.

🔑 **Cranial bones and the cranial meninges protect the brain.**

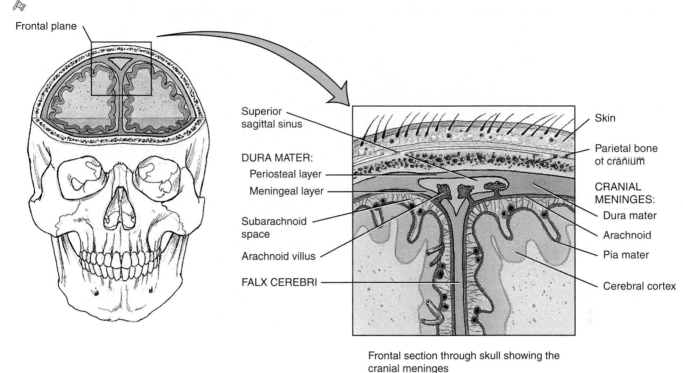

Frontal plane

Superior sagittal sinus

DURA MATER:
Periosteal layer
Meningeal layer

Subarachnoid space

Arachnoid villus

FALX CEREBRI

Skin

Parietal bone of cranium

CRANIAL MENINGES:
Dura mater
Arachnoid
Pia mater

Cerebral cortex

Frontal section through skull showing the cranial meninges

Q What are the three layers of the cranial meninges, from superficial to deep?

CEREBROSPINAL FLUID PRODUCTION AND CIRCULATION IN VENTRICLES

OBJECTIVE
• *Explain the formation and circulation of cerebrospinal fluid.*

Cerebrospinal fluid (CSF) is a clear, colorless liquid that protects the brain and spinal cord against chemical and physical injuries and carries oxygen, glucose, and other needed chemicals from the blood to neurons and neuroglia. This fluid continuously circulates through the subarachnoid space (between the arachnoid and pia mater) around the brain and spinal cord, as well as through cavities within the brain and spinal cord.

Figure 14.3 shows the four CSF-filled cavities within the brain, which are called **ventricles** (VEN-tri-kuls; = little cavities). A **lateral ventricle** is located in each hemisphere of the cerebrum. Anteriorly, the lateral ventricles are separated by a thin membrane, the **septum pellucidum** (SEP-tum pe-LOO-si-dum; *pellucid* = transparent). The **third ventricle** is a narrow cavity along the midline superior to the hypothalamus and between the right and left halves of the thalamus. The **fourth ventricle** lies between the brain stem and the cerebellum.

The total volume of CSF is 80–150 mL (3–5 oz) in an average adult. CSF contains glucose, proteins, lactic acid, urea, cations (Na^+, K^+, Ca^{2+}, Mg^{2+}), and anions (Cl^- and HCO_3^-); it also contains some white blood cells. The CSF contributes to homeostasis in three main ways:

1. *Mechanical protection.* CSF serves as a shock-absorbing medium that protects the delicate tissue of the brain and spinal cord from jolts that would otherwise cause them to hit the bony walls of the cranial and vertebral cavities. The fluid also buoys the brain so that it "floats" in the cranial cavity.

2. *Chemical protection.* CSF provides an optimal chemical environment for accurate neuronal signaling. Even slight changes in the ionic composition of CSF within the brain can seriously disrupt production of action potentials and postsynaptic potentials.

3. *Circulation.* CSF is a medium for exchange of nutrients and waste products between the blood and nervous tissue.

The sites of CSF production are the **choroid plexuses** (KŌ-royd; = membranelike), which are networks of capillaries (microscopic blood vessels) in the walls of the ventricles. The capillaries are covered by ependymal cells that form

Figure 14.3 Locations of ventricles within a "transparent" brain. One interventricular foramen on each side connects a lateral ventricle to the third ventricle, and the cerebral aqueduct connects the third ventricle to the fourth ventricle.

🔑 **Ventricles are cerebrospinal fluid-filled cavities within the brain.**

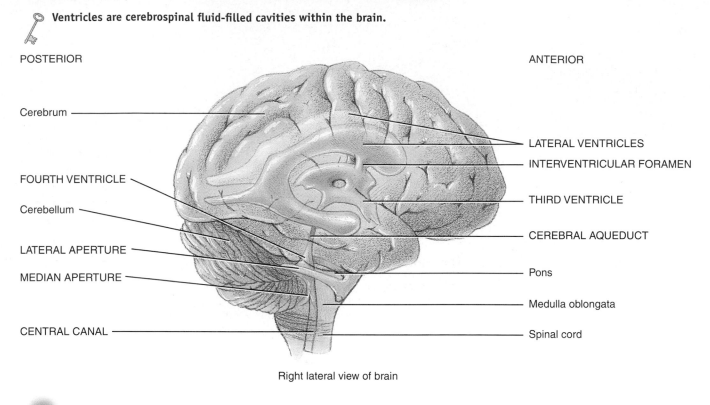

POSTERIOR

ANTERIOR

Cerebrum

LATERAL VENTRICLES

INTERVENTRICULAR FORAMEN

THIRD VENTRICLE

FOURTH VENTRICLE

Cerebellum

CEREBRAL AQUEDUCT

LATERAL APERTURE

MEDIAN APERTURE

Pons

Medulla oblongata

CENTRAL CANAL

Spinal cord

Right lateral view of brain

Q Which brain region is anterior to the fourth ventricle? Which is posterior to it?

cerebrospinal fluid from blood plasma by filtration and secretion. Because the ependymal cells are joined by tight junctions (shown in Figure 4.1 on page 106), materials entering CSF from choroid capillaries cannot leak between these cells; instead, they must pass through the ependymal cells. This **blood–cerebrospinal fluid barrier** permits certain substances to enter the CSF but excludes others, protecting the brain and spinal cord from potentially harmful blood-borne substances.

The CSF formed in the choroid plexuses of each lateral ventricle flows into the third ventricle through a pair of narrow, oval openings, the **interventricular foramina (foramina of Monro)** (Figure 14.4a). More CSF is added by the choroid plexus in the roof of the third ventricle. The fluid then flows through the **cerebral aqueduct (aqueduct of Sylvius),** which passes through the midbrain, into the fourth ventricle. The choroid plexus of the fourth ventricle contributes more fluid. CSF enters the subarachnoid space through three openings in the roof of the fourth ventricle: a **median aperture (of Magendie)** and the paired **lateral apertures (of Luschka),** one on each side. CSF then circulates in the central canal of the spinal cord and in the subarachnoid space around the surface of the brain and spinal

cord. CSF is gradually reabsorbed into the blood through **arachnoid villi,** fingerlike extensions of the arachnoid that project into the dural venous sinuses, especially the **superior sagittal sinus** (Figure 14.4b; also see Figure 14.2). (A sinus is like a vein but has thinner walls.) Normally, CSF is reabsorbed as rapidly as it is formed by the choroid plexuses, at a rate of about 20 mL/hr (480 mL/day). Because the rates of formation and reabsorption are the same, the pressure of CSF normally is constant. Figure 14.4c summarizes the production and flow of CSF.

CLINICAL APPLICATION
Hydrocephalus

Abnormalities in the brain—tumors, inflammation, or developmental malformations—can interfere with the drainage of CSF from the ventricles into the subarachnoid space. When excess CSF accumulates in the ventricles, the CSF pressure rises. Elevated CSF pressure causes a condition called **hydrocephalus** (hī'-drō-SEF-a-lus; *hydro-* = water; *cephal-* = head). In a baby in which the fontanels have not yet closed, the head bulges due to the increased pressure. If the condition persists, the fluid buildup compresses and damages the delicate nervous tissue. Hydrocephalus is relieved by draining the

Figure 14.4 Pathways of circulating cerebrospinal fluid. (See Tortora, *A Photographic Atlas of the Human Body,* Figures 8.15, 8.18)

CSF is formed by ependymal cells that cover the choroid plexuses of the ventricles.

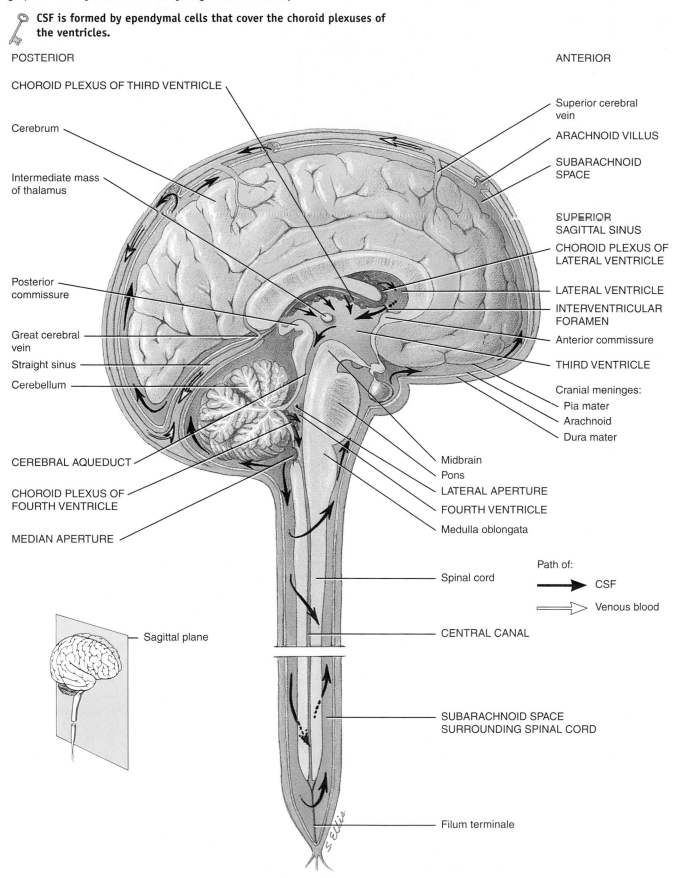

POSTERIOR

ANTERIOR

CHOROID PLEXUS OF THIRD VENTRICLE

Cerebrum

Intermediate mass of thalamus

Posterior commissure

Great cerebral vein

Straight sinus

Cerebellum

CEREBRAL AQUEDUCT

CHOROID PLEXUS OF FOURTH VENTRICLE

MEDIAN APERTURE

Superior cerebral vein

ARACHNOID VILLUS

SUBARACHNOID SPACE

SUPERIOR SAGITTAL SINUS

CHOROID PLEXUS OF LATERAL VENTRICLE

LATERAL VENTRICLE

INTERVENTRICULAR FORAMEN

Anterior commissure

THIRD VENTRICLE

Cranial meninges:
Pia mater
Arachnoid
Dura mater

Midbrain
Pons
LATERAL APERTURE
FOURTH VENTRICLE
Medulla oblongata

Sagittal plane

Spinal cord

CENTRAL CANAL

SUBARACHNOID SPACE SURROUNDING SPINAL CORD

Filum terminale

Path of:

CSF

Venous blood

(a) Sagittal section of brain and spinal cord

Figure continues

Figure 14.4 (*continued*)

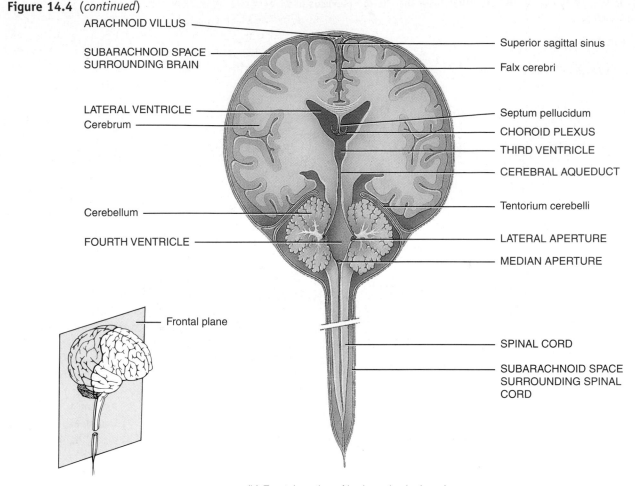

ARACHNOID VILLUS

SUBARACHNOID SPACE SURROUNDING BRAIN

LATERAL VENTRICLE
Cerebrum

Cerebellum

FOURTH VENTRICLE

Frontal plane

Superior sagittal sinus

Falx cerebri

Septum pellucidum

CHOROID PLEXUS

THIRD VENTRICLE

CEREBRAL AQUEDUCT

Tentorium cerebelli

LATERAL APERTURE

MEDIAN APERTURE

SPINAL CORD

SUBARACHNOID SPACE SURROUNDING SPINAL CORD

(b) Frontal section of brain and spinal cord

excess CSF. A neurosurgeon may implant a drain line, called a shunt, into the lateral ventricle to divert CSF into the superior vena cava or abdominal cavity. In adults, hydrocephalus may occur after head injury, meningitis, or subarachnoid hemorrhage. ■

1. What structures are the sites of CSF production, and where are they located?
2. What is the difference between the blood–brain barrier and the blood–cerebrospinal fluid barrier?

THE BRAIN STEM

OBJECTIVE

• *Describe the structures and functions of the brain stem.*

The brain stem is the part of the brain between the spinal cord and the diencephalon; it consists of (1) the medulla oblongata, (2) pons, and (3) midbrain. Extending through the brain stem is the reticular formation, a netlike region of interspersed gray and white matter.

Medulla Oblongata

The **medulla oblongata** (me-DULL-la ob'-long-GA-ta), or more simply the medulla, is a continuation of the superior part of the spinal cord; it forms the inferior part of the brain stem (Figure 14.5; also see Figure 14.1). The medulla begins at the foramen magnum and extends to the inferior border of the pons, a distance of about 3 cm (1.2 in.). Within the medulla are all ascending (sensory) and descending (motor) tracts that connect the spinal cord with the brain, plus many nuclei (gray matter masses of neuronal cell bodies) that regulate various vital body functions. The medulla also contains nuclei that receive sensory input from or provide motor output to five of the twelve cranial nerves (see Table 14.2 on page 475).

The medulla is organized into several major structural and functional regions. On the anterior aspect of the medulla are two conspicuous external bulges called the **pyramids** (Figures 14.5 and 14.6). The pyramids are formed by the largest motor tracts that pass from the cerebrum to the spinal cord. Just superior to the junction of the medulla with

Figure 14.4 (*continued*)

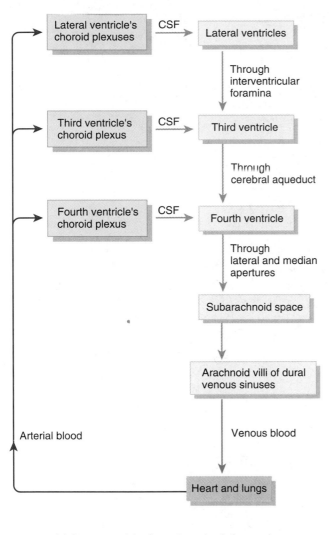

(c) Summary of the formation, circulation, and absorption of cerebrospinal fluid (CSF)

 Where is CSF reabsorbed?

the spinal cord, most of the axons in the left pyramid cross to the right side, and most of the axons in the right pyramid cross over to the left side. This crossing is called the **decussation of pyramids** (dē'-ku-SĀ-shun; *decuss-* = crossing). Thus, neurons in the left cerebral cortex control skeletal muscles on the right side of the body, and neurons in the right cerebral cortex control skeletal muscles on the left side.

Just lateral to each pyramid is an oval-shaped swelling called an **olive** (see Figures 14.5 and 14.6). The swelling is caused mostly by the **inferior olivary nucleus** within the medulla. Neurons here relay impulses from proprioceptors, such as muscle spindles, to the cerebellum. Nuclei associated

with some somatic sensations (touch, vibration, and proprioception) are located on the posterior aspect of the medulla. These nuclei are the right and left **nucleus gracilis** (gras-I-lis; = slender) and **nucleus cuneatus** (kyoo-nē-Ā-tus; = wedge). Many ascending sensory axons form synapses in these nuclei, and postsynaptic neurons then relay the sensory information to the thalamus on the opposite side of the brain (see Figure 15.4 on page 493).

The medulla also contains nuclei that govern several autonomic functions. These nuclei include the **cardiovascular center,** which regulates the rate and force of the heartbeat and the diameter of blood vessels; the **medullary rhythmicity area** of the respiratory center, which adjusts the basic rhythm of breathing; and other centers in the medulla that control reflexes for vomiting, coughing, and sneezing.

Finally, the medulla contains nuclei associated with the following five pairs of cranial nerves (see Figure 14.5).

- *Vestibulocochlear (VIII) nerves.* Several nuclei in the medulla receive sensory input from and provide motor output to the cochlea of the internal ear via the cochlear branches of the vestibulocochlear nerves. These nerves convey impulses related to hearing.
- *Glossopharyngeal (IX) nerves.* Nuclei in the medulla relay sensory and motor nerve impulses related to taste, swallowing, and salivation via the glossopharyngeal nerves.
- *Vagus (X) nerves.* Nuclei in the medulla receive nerve sensory nerve impulses from and provide motor impulses to many thoracic and abdominal viscera via the vagus nerves.
- *Accessory (XI) nerves (cranial portion).* Nuclei in the medulla are the origin for nerve impulses that control swallowing via the cranial portion of the accessory nerves.
- *Hypoglossal (XII) nerves.* Nuclei in the medulla are the origin for nerve impulses that control tongue movements during speech and swallowing via the hypoglossal nerves.

CLINICAL APPLICATION
Injury of the Medulla

Given the many vital activities controlled by the medulla, it is not surprising that a hard blow to the back of the head or upper neck can be fatal. Damage to the respiratory center is particularly serious and can rapidly lead to death. Symptoms of nonfatal medullary injury may include cranial nerve malfunctions on the same side of the body as the injury, paralysis and loss of sensation on the opposite side of the body, and irregularities in breathing or heart rhythm. ■

Pons

The physical relationship of the **pons** (= bridge) to other parts of the brain can be seen in Figures 14.1 and 14.5. The pons lies directly superior to the medulla and anterior to the cerebellum and is about 2.5 cm (1 in.) long. Like the medulla, the pons consists of both nuclei and tracts. As its name implies, the pons is a bridge that connects parts of the

Figure 14.5 Medulla oblongata in relation to the rest of the brain stem. (See Tortora, *A Photographic Atlas of the Human Body,* Figure 8.19)

 The brain stem consists of the medulla oblongata, pons, and midbrain.

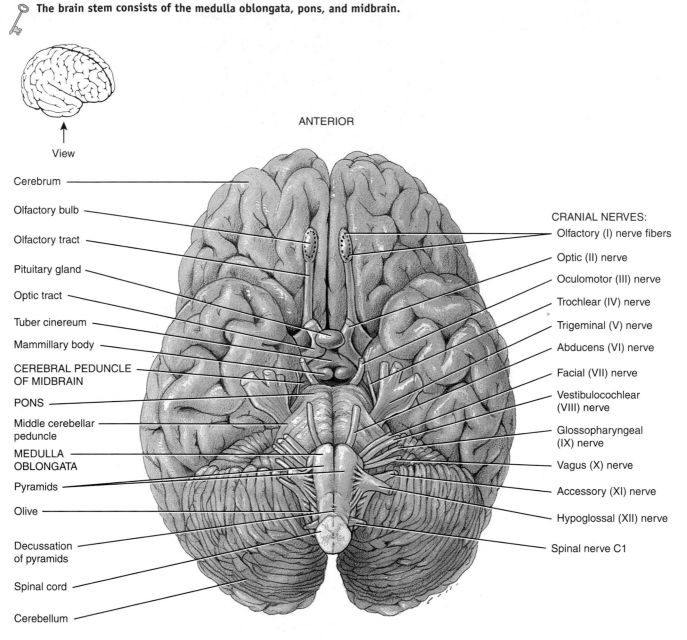

ANTERIOR

View

Cerebrum

Olfactory bulb

Olfactory tract

Pituitary gland

Optic tract

Tuber cinereum

Mammillary body

CEREBRAL PEDUNCLE OF MIDBRAIN

PONS

Middle cerebellar peduncle

MEDULLA OBLONGATA

Pyramids

Olive

Decussation of pyramids

Spinal cord

Cerebellum

CRANIAL NERVES:

Olfactory (I) nerve fibers

Optic (II) nerve

Oculomotor (III) nerve

Trochlear (IV) nerve

Trigeminal (V) nerve

Abducens (VI) nerve

Facial (VII) nerve

Vestibulocochlear (VIII) nerve

Glossopharyngeal (IX) nerve

Vagus (X) nerve

Accessory (XI) nerve

Hypoglossal (XII) nerve

Spinal nerve C1

POSTERIOR

Inferior aspect of brain

Q What part of the brain stem contains the pyramids? The cerebral peduncles? Literally means "bridge"?

Figure 14.6 Internal anatomy of the medulla oblongata.

The pyramids of the medulla contain the largest motor tracts that run from the cerebrum to the spinal cord.

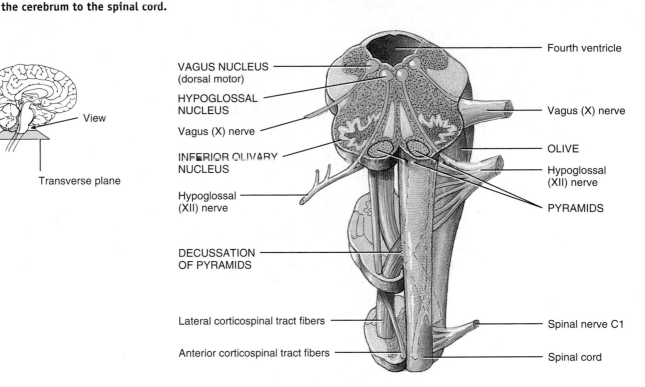

Transverse section and anterior surface of medulla oblongata

Q What does decussation mean? What is the functional consequence of decussation of the pyramids?

brain with one another. These connections are provided by axons that extend in two main directions. Transverse axons of the pons are within paired tracts that connect the right and left sides of the cerebellum; longitudinal axons of the pons are part of ascending sensory tracts and descending motor tracts.

Important nuclei in the pons are the **pneumotaxic area** (noo-mō-TAK-sik) and the **apneustic area** (ap-NOO-stik), shown in Figure 23.25 on page 804. Together with the medullary rhythmicity area, the pneumotaxic and apneustic areas help control breathing. The pons also contains nuclei associated with the following four pairs of cranial nerves (see Figure 14.5):

- *Trigeminal (V) nerves.* Nuclei in the pons receive sensory impulses for somatic sensations from the head and face and provide motor impulses that govern chewing via the trigeminal nerves.
- *Abducens (VI) nerves.* Nuclei in the pons provide motor impulses that control eyeball movement via the abducens nerves.
- *Facial (VII) nerves.* Nuclei in the pons receive sensory impulses for taste and provide motor impulses to regulate

secretion of saliva and tears and contraction of muscles of facial expression via the facial nerves.

- *Vestibulocochlear (VIII) nerves.* Nuclei in the pons receive sensory impulses from and provide motor impulses to the vestibular apparatus via the vestibular branches of the vestibulocochlear nerves. These nerves convey impulses related to balance and equilibrium.

Midbrain

The **midbrain,** or **mesencephalon,** which extends from the pons to the diencephalon (see Figures 14.1 and 14.5), is about 2.5 cm (1 in.) long. The cerebral aqueduct passes through the midbrain, connecting the third ventricle above with the fourth ventricle below. Like the medulla and the pons, the midbrain contains both tracts (white matter) and nuclei (gray matter).

The anterior part of the midbrain contains a pair of tracts called **cerebral peduncles** (pe-DUNG-kulz; = stalk; Figure 14.5 and Figure 14.7b). They contain axons of corticospinal, corticopontine, and corticobulbar motor neurons,

Figure 14.7 Midbrain. (See Tortora, *A Photographic Atlas of the Human Body,* Figure 8.23)

The midbrain connects the pons to the diencephalon.

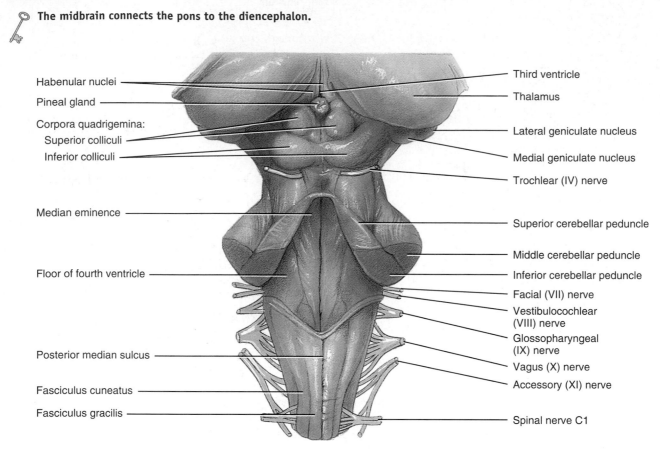

Habenular nuclei

Pineal gland

Corpora quadrigemina:
  Superior colliculi
  Inferior colliculi

Median eminence

Floor of fourth ventricle

Posterior median sulcus

Fasciculus cuneatus

Fasciculus gracilis

Third ventricle

Thalamus

Lateral geniculate nucleus

Medial geniculate nucleus

Trochlear (IV) nerve

Superior cerebellar peduncle

Middle cerebellar peduncle

Inferior cerebellar peduncle

Facial (VII) nerve

Vestibulocochlear (VIII) nerve

Glossopharyngeal (IX) nerve

Vagus (X) nerve

Accessory (XI) nerve

Spinal nerve C1

(a) Posterior view of midbrain in relation to brain stem

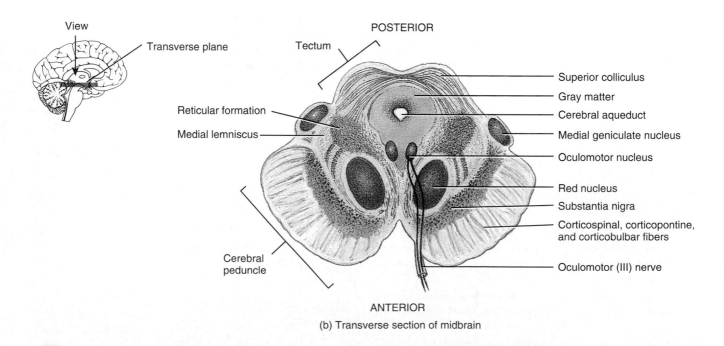

View

Transverse plane

POSTERIOR

Tectum

Reticular formation

Medial lemniscus

Cerebral peduncle

Superior colliculus

Gray matter

Cerebral aqueduct

Medial geniculate nucleus

Oculomotor nucleus

Red nucleus

Substantia nigra

Corticospinal, corticopontine, and corticobulbar fibers

Oculomotor (III) nerve

ANTERIOR

(b) Transverse section of midbrain

Q What is the importance of the cerebral peduncles?

which conduct nerve impulses from the cerebrum to the spinal cord, medulla, and pons. The cerebral peduncles also contain axons of sensory neurons that extend from the medulla to the thalamus.

The posterior part of the midbrain, called the **tectum** (= roof), contains four rounded elevations, the **corpora quadrigemina** (KOR-po-ra kwad-ri-JEM-in-a; *corpora* = bodies; *quadrigeminus* = four twins) (Figure 14.7a). The two superior elevations, known as the **superior colliculi** (ko-LIK-yoo-lī; singular is **colliculus** = small mound), serve as reflex centers that govern movements of the eyes, head, and neck in response to visual and other stimuli. The two inferior elevations, the **inferior colliculi,** are reflex centers for movements of the head and trunk in response to auditory stimuli.

The midbrain contains several nuclei, including the left and right **substantia nigra** (sub-STAN-shē-a; = substance; NĪ-gra; = black), which are large, darkly pigmented nuclei that control subconscious muscle activities (see Figure 14.7b). Also present are the left and right **red nuclei** (see Figure 14.7b), whose name derives from their rich blood supply and an iron-containing pigment in their neuronal cell bodies. Fibers from the cerebellum and cerebral cortex form synapses in the red nuclei, which function with the basal ganglia and cerebellum to coordinate muscular movements.

A band of white matter called the **medial lemniscus** (lem-NIS-kus; = ribbon) extends through the medulla, pons, and midbrain. The medial lemniscus contains axons that convey nerve impulses related to sensations of touch, proprioception (joint and muscle position), pressure, and vibration from the cuneate and gracile nuclei in the medulla to the thalamus.

Finally, nuclei in the midbrain are associated with two cranial nerves (see Figure 14.5):

- *Oculomotor (III) nerves.* Nuclei in the midbrain provide motor impulses that control movements of the eyeball, constriction of the pupil, and changes in shape of the lens via the oculomotor nerves.
- *Trochlear (IV) nerves.* Nuclei in the midbrain provide motor impulses that control movements of the eyeball via the trochlear nerves.

Reticular Formation

The brainstem also contains the **reticular formation** (*ret-* = net), a netlike arrangement of small areas of gray matter interspersed among threads of white matter (see Figure 14.7b). The reticular formation, which also extends into the spinal cord and diencephalon, has both sensory and motor functions. The main sensory function is alerting the cerebral cortex to incoming sensory signals. Part of the reticular formation, called the **reticular activating system (RAS),** consists of fibers that project to the cerebral cortex (see Figure 15.9 on page 503). The RAS is responsible for maintaining consciousness and for awakening from sleep. Incoming impulses from the ears, eyes, and skin are effective stimulators of the RAS. For example, we awaken to the sound of an alarm clock, to a flash of lightning, or to a painful pinch because of RAS activity that arouses the cerebral cortex. The reticular formation's main motor function is to help regulate muscle tone (the slight degree of contraction that characterizes muscles at rest).

The functions of the brain stem are summarized in Table 14.1.

1. Describe the locations of the medulla, pons, and midbrain relative to one another.
2. Define decussation of pyramids. Why is it important?
3. List the body functions that are governed by nuclei in the brain stem.
4. What are two important functions of the reticular formation?

THE CEREBELLUM

OBJECTIVE

- *Describe the structure and functions of the cerebellum.*

The **cerebellum,** the second-largest part of the brain, occupies the inferior and posterior aspects of the cranial cavity. It is posterior to the medulla and pons and inferior to the posterior portion of the cerebrum (see Figure 14.1). The cerebellum is separated from the cerebrum by a deep groove known as the **transverse fissure,** and by the **tentorium cerebelli,** which supports the posterior part of the cerebrum (see Figure 14.4b).

In superior or inferior views, the shape of the cerebellum is somewhat like a butterfly. The central constricted area is the **vermis** (= worm), and the lateral "wings" or lobes are the **cerebellar hemispheres** (Figure 14.8a, b). Each hemisphere consists of lobes separated by deep and distinct fissures. The **anterior lobe** and **posterior lobe** govern subconscious movements of skeletal muscles; the **flocculonodular lobe** (*flocculo-* = wool-like tuft) on the inferior surface is concerned with the sense of equilibrium.

The superficial layer of the cerebellum, called the **cerebellar cortex,** consists of gray matter in a series of slender, parallel ridges called **folia** (= leaves). Deep to the gray matter are tracts called **arbor vitae** (= tree of life) that resemble branches of a tree. Even deeper, within the white matter, are the **cerebellar nuclei,** which give rise to nerve fibers that carry impulses from the cerebellum to other brain centers and to the spinal cord.

The cerebellum is attached to the brain stem by three paired cerebellar peduncles (see Figure 14.7a). The **inferior cerebellar peduncles** connect the medulla oblongata to the cerebellum; they contain axons extending from the inferior olivary nucleus and the spinal cord to the cerebellum. The **middle cerebellar peduncles** connect the pons to the cerebellum; they contain axons extending from the pons to the cerebellum. The **superior cerebellar peduncles** connect the midbrain to the cerebellum; they contain axons that extend mainly from the cerebellum into the midbrain.

Table 14.1 Summary of Functions of Principal Parts of the Brain

PART	FUNCTION	PART	FUNCTION
Brain stem 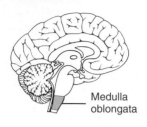	*Medulla oblongata:* Relays motor and sensory impulses between other parts of the brain and the spinal cord. Reticular formation (also in pons, midbrain, and diencephalon) functions in consciousness and arousal. Vital centers regulate heartbeat, breathing (together with pons), and blood vessel diameter. Other centers coordinate swallowing, vomiting, coughing, sneezing, and hiccuping. Contains nuclei of origin for cranial nerves VIII, IX, X, XI, and XII.	**Diencephalon** 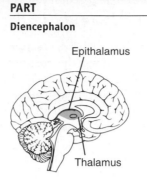 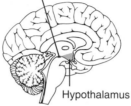	*Epithalamus:* Consists of pineal gland, which secretes melatonin, and the habenular nuclei. *Thalamus:* Relays all sensory input to the cerebral cortex. Provides crude perception of touch, pressure, pain, and temperature. Includes nuclei involved in voluntary motor actions and arousal; anterior nucleus functions in emotions and memory. Also functions in cognition and awareness. *Subthalamus:* Contains the subthalamic nuclei and portions of the red nucleus and the substantia nigra, which are positioned mostly lateral to the midline. These regions communicate with the basal ganglia to help control body movements. *Hypothalamus:* Controls and integrates activities of the autonomic nervous system and pituitary gland. Regulates emotional and behavioral patterns and circadian rhythms. Controls body temperature and regulates eating and drinking behavior. Helps maintain the waking state and establishes patterns of sleep.
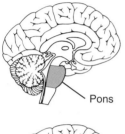	*Pons:* Relays impulses from one side of the cerebellum to the other and between the medulla and midbrain. Contains nuclei of origin for cranial nerves, V, VI, VII, and VIII. Pneumotaxic area and apneustic area, together with the medulla, help control breathing.	**Cerebrum** 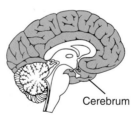	Sensory areas interpret sensory impulses, motor areas control muscular movement, and association areas function in emotional and intellectual processes. Basal ganglia coordinate gross, automatic muscle movements and regulate muscle tone. Limbic system functions in emotional aspects of behavior related to survival.
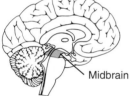	*Midbrain:* Relays motor impulses from the cerebral cortex to the pons and sensory impulses from the spinal cord to the thalamus. Superior colliculi coordinate movements of the eyeballs in response to visual and other stimuli, and the inferior colliculi coordinate movements of the head and trunk in response to auditory stimuli. Most of substantia nigra and red nucleus contribute to control of movement. Contains nuclei of origin for cranial nerves III and IV.		
Cerebellum 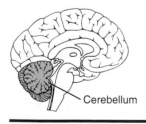	Compares intended movements with what is actually happening to smooth and coordinate complex, skilled movements. Regulates posture and balance.		

A main function of the cerebellum is to evaluate how well movements initiated by motor areas in the cerebrum are actually being carried out. If the movements initiated by the cerebral motor areas are not being carried out, the cerebellum detects the discrepancies and sends feedback signals to the motor areas to correct the errors and modify the movements. This feedback helps to smooth and coordinate complex sequences of skeletal muscle contractions. Besides coordinating skilled movements, the cerebellum is the main brain region that regulates posture and balance. These aspects of cerebellar function make possible all skilled muscular activities, from catching a baseball to dancing.

The functions of the cerebellum are summarized in Table 14.1.

Figure 14.8 Cerebellum. (See Tortora, *A Photographic Atlas of the Human Body*, Figure 8.25)

The cerebellum coordinates skilled movements and regulates posture and balance.

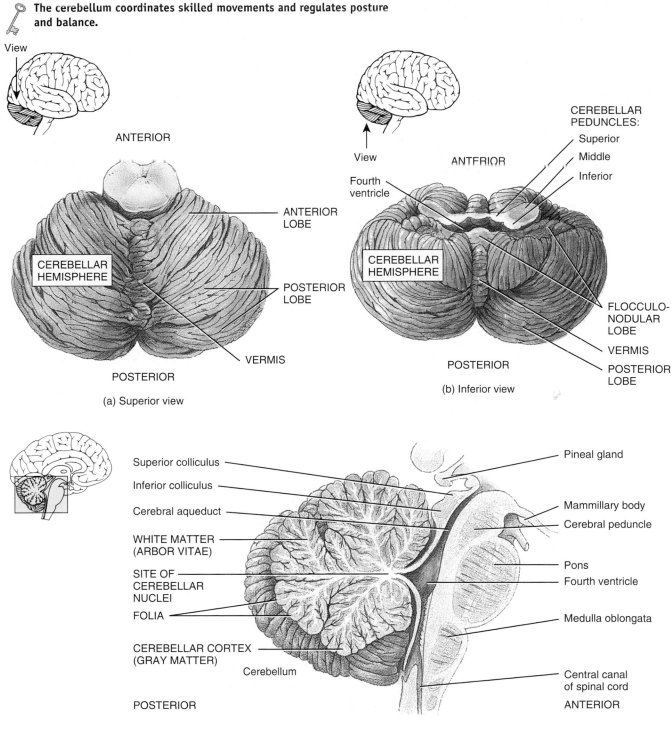

(a) Superior view

ANTERIOR

ANTERIOR LOBE

CEREBELLAR HEMISPHERE

POSTERIOR LOBE

VERMIS

POSTERIOR

(b) Inferior view

CEREBELLAR PEDUNCLES:
Superior
Middle
Inferior

Fourth ventricle

ANTERIOR

CEREBELLAR HEMISPHERE

FLOCCULO-NODULAR LOBE

VERMIS

POSTERIOR LOBE

POSTERIOR

(c) Midsagittal section of cerebellum and brain stem

Superior colliculus
Inferior colliculus
Cerebral aqueduct
WHITE MATTER (ARBOR VITAE)
SITE OF CEREBELLAR NUCLEI
FOLIA
CEREBELLAR CORTEX (GRAY MATTER)
Cerebellum
POSTERIOR

Pineal gland
Mammillary body
Cerebral peduncle
Pons
Fourth ventricle
Medulla oblongata
Central canal of spinal cord
ANTERIOR

Q Which fiber tracts carry information into and out of the cerebellum?

1. Describe the location and principal parts of the cerebellum.
2. What are cerebellar peduncles? List and explain the function of each.

THE DIENCEPHALON

OBJECTIVE

• *Describe the components and functions of the diencephalon.*

The **diencephalon** extends from the brain stem to the cerebrum and surrounds the third ventricle; it includes the thalamus, hypothalamus, epithalamus, and subthalamus.

Thalamus

The **thalamus** (THAL-a-mus; = inner chamber), which measures about 3 cm (1.2 in.) in length and makes up 80% of the diencephalon, consists of paired oval masses of gray matter organized into nuclei with interspersed tracts of white matter (Figure 14.9). Usually, a bridge of gray matter called the **intermediate mass** crosses the third ventricle to join the right and left portions of the thalamus (see Figure 14.10).

The thalamus is the principal relay station for sensory impulses that reach the cerebral cortex from the spinal cord, brain stem, cerebellum, and other parts of the cerebrum. It also allows crude perception of some sensations, such as pain, temperature, and pressure. Precise localization of these sensations depends on the relay of nerve impulses from the thalamus to the cerebral cortex.

The nuclei within each half of the thalamus have various roles. Some nuclei relay impulses to sensory areas of the cerebrum (Figure 14.9a): The **medial geniculate nucleus** (je-NIK-yoo-lāt; = bent like a knee) relays auditory impulses; the **lateral geniculate nucleus** relays visual impulses; and the **ventral posterior nucleus** relays impulses for taste and somatic sensations such as touch, pressure, vibration, heat, cold, and pain. Other nuclei relay impulses to somatic motor areas of the cerebrum: The **ventral lateral nucleus** receives impulses from the cerebellum, and the **ventral anterior nucleus** receives impulses from the basal ganglia (described later in the chapter). The **anterior nucleus** in the floor of the lateral ventricle is concerned with certain emotions and memory.

The thalamus plays an essential role in awareness and in the acquisition of knowledge, which is termed **cognition** (*cogni-* = to get to know). Evidence for this recently appreciated role of the thalamus came from a postmortem examination of the brain of a young woman who had survived in a persistent vegetative state (having waking and sleep cycles but no cognitive functions, awareness, thought, or emotions) for ten years after an automobile accident. An autopsy revealed severe and extensive thalamic damage but only modest abnormalities in the cerebrum and brain stem.

Hypothalamus

The **hypothalamus** (*hypo-* = under) is a small part of the diencephalon located inferior to the thalamus. It is composed of a dozen or so nuclei in four major regions:

• The *mammillary region* (*mammill-* = nipple-shaped), adjacent to the midbrain, is the most posterior part of the hypothalamus. It includes the mammillary bodies and posterior hypothalamic nucleus (Figure 14.10). The **mammillary bodies** are two, small, rounded projections that serve as relay stations for reflexes related to the sense of smell (see also Figure 14.5).

• The *tuberal region,* the widest part of the hypothalamus, includes the dorsomedial, ventromedial, and arcuate nuclei, plus the stalklike **infundibulum,** which connects the pituitary gland to the hypothalamus (see Figure 14.10). The **median eminence** is a slightly raised region that encircles the infundibulum.

• The *supraoptic region* (*supra-* = above; *-optic* = eye) lies superior to the optic chiasm (point of crossing of optic nerves) and contains the paraventricular nucleus, supraoptic nucleus, anterior hypothalamic nucleus, and suprachiasmatic nucleus (see Figure 14.10). Axons from the paraventricular and supraoptic nuclei form the hypothalamohypophyseal tract, which extends through the infundibulum to the posterior pituitary gland.

• The *preoptic region* anterior to the supraoptic region is usually considered part of the hypothalamus because it participates with the hypothalamus in regulating certain autonomic activities. The preoptic region contains the medial and lateral preoptic nuclei (see Figure 14.10).

The hypothalamus controls many body activities and is *one of the major regulators of homeostasis.* Sensory impulses related to both somatic and visceral senses arrive at the hypothalamus via afferent pathways, as do impulses from hearing, taste, and smell receptors. Other receptors within the hypothalamus itself continually monitor osmotic pressure, certain hormone concentrations, and the temperature of blood. The hypothalamus has several very important connections with the pituitary gland and also produces a variety of hormones, which will be discussed in more detail in Chapter 18. Whereas some functions can be attributed to specific nuclei, others are not so precisely localized. The chief functions of the hypothalamus are as follows:

1. *Control of the ANS.* The hypothalamus controls and integrates activities of the autonomic nervous system, which regulates contraction of smooth and cardiac muscle and the secretions of many glands. Axons extend from the hypothalamus to sympathetic and parasympathetic nuclei in the brain stem and spinal cord. Through the ANS, the hypothalamus is a major regulator of visceral activities, including regulation of heart rate, movement of food through the gastrointestinal tract, and contraction of the urinary bladder.

Figure 14.9 Thalamus. The red arrows in (a) indicate some of the connections between the thalamus and the cerebral cortex. (See Tortora, *A Photographic Atlas of the Human Body,* Figure 8.18)

🔑 **The thalamus is the principal relay station for sensory impulses that reach the cerebral cortex from other parts of the brain and the spinal cord.**

POSTERIOR   ANTERIOR

Internal medullary lamina

Ventral posterior nuclei

Anterior nucleus

Ventral anterior nucleus

Medial geniculate nucleus

Ventral lateral nucleus

Lateral geniculate nucleus

(a) Right lateral view showing thalamic nuclei

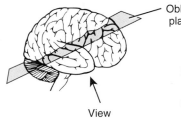

Oblique plane

View

ANTERIOR

Falx cerebri

Cerebrum

Corpus callosum

THALAMUS

Cerebellum

Skin

Caudate nucleus

Lentiform nucleus:
Putamen
Globus pallidus

Third ventricle

Tentorium cerebelli

POSTERIOR

(b) Oblique section of brain

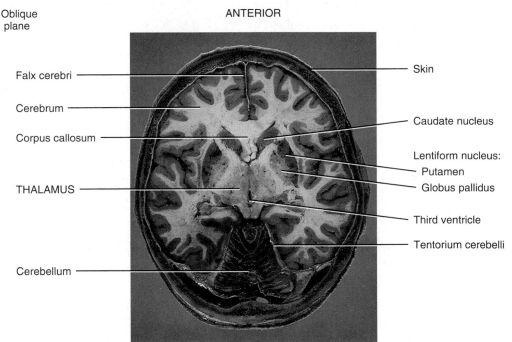

Ⓠ What structure connects the right and left sides of the thalamus?

Figure 14.10 Hypothalamus. Selected portions of the hypothalamus and a three-dimensional representation of hypothalamic nuclei are shown (after Netter).

🔑 **The hypothalamus controls many body activities and is an important regulator of homeostasis.**

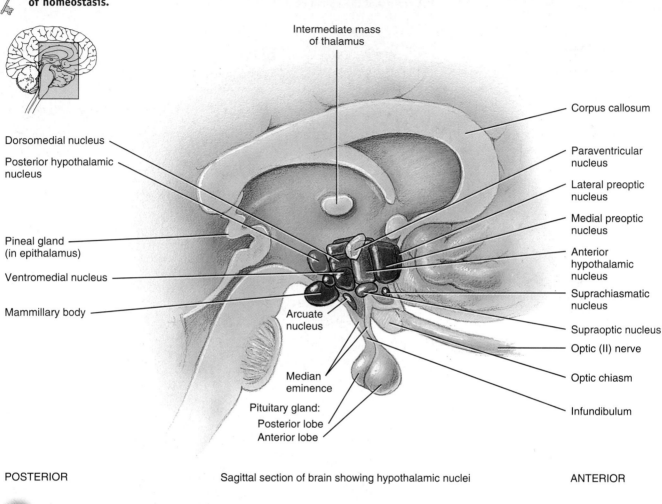

Sagittal section of brain showing hypothalamic nuclei

POSTERIOR   ANTERIOR

Q What are the four major regions of the hypothalamus, from posterior to anterior?

2. *Control of the pituitary gland.* The hypothalamus produces several hormones and has two types of important connections with the pituitary gland, an endocrine gland located inferior to the hypothalamus (see Figure 14.1). First, hypothalamic regulating hormones are released into capillary networks in the median eminence. These hormones then are carried by the bloodstream directly to the anterior pituitary gland, where they stimulate or inhibit secretion of pituitary hormones. Second, axons extend from the paraventricular and supraoptic nuclei through the infundibulum into the posterior pituitary gland. The cell bodies of these neurons make one of two hormones (oxytocin or antidiuretic hormone). Their axons transport the hormones to the posterior pituitary gland, where they are stored and released.

3. *Regulation of emotional and behavioral patterns.* Together with the limbic system (described shortly), the hypothalamus regulates feelings of rage, aggression, pain, and pleasure, and the behavioral patterns related to sexual arousal.

4. *Regulation of eating and drinking.* The hypothalamus regulates food intake through two centers. The **feeding center** is responsible for hunger sensations; when sufficient food has been ingested, the **satiety center** (sa-TĪ-e-tē; *sati-* = full, satisfied) is stimulated and sends out nerve impulses that inhibit the feeding center. The hypothalamus also contains a **thirst center.** When certain cells in the hypothalamus are stimulated by rising osmotic pressure of the extracellular fluid, they cause the sensation of thirst. The intake of water by drinking restores the osmotic pressure to normal, removing the stimulation and relieving the thirst.

5. *Control of body temperature.* If the temperature of blood flowing through the hypothalamus is above normal, the hypothalamus directs the autonomic nervous system to stimulate activities that promote heat loss. If, however, blood temperature is below normal, the hypothalamus generates impulses that promote heat production and retention.

6. *Regulation of circadian rhythms and states of consciousness.* The suprachiasmatic nucleus establishes patterns of sleep that occur on a circadian (daily) schedule.

Epithalamus

The **epithalamus** (*epi-* = above), a small region superior and posterior to the thalamus, consists of the pineal gland and habenular nuclei. The **pineal gland** (PĬN-ē-al; = pinecone-like) is about the size of a small pea and protrudes from the posterior midline of the third ventricle (see Figure 14.1). Although its physiological role is not completely clear, the pineal gland secretes the hormone **melatonin** and is thus an endocrine gland. In part because more melatonin is liberated during darkness than in light, it is thought to promote sleepiness. (Regulation of melatonin release is depicted in Figure 18.20 on page 597.) Melatonin also appears to contribute to the setting of the body's biological clock. The pineal gland is discussed in more detail in Chapter 18.

The **habenular nuclei** (ha-BEN-yoo-lar), shown in Figure 14.7a, are involved in olfaction, especially emotional responses to odors.

Subthalamus

The **subthalamus,** a small area immediately inferior to the thalamus, includes tracts and the paired **subthalamic nuclei,** which connect to motor areas of the cerebrum. Parts of two pairs of midbrain nuclei—the red nucleus and the substantia nigra—also extend into the subthalamus. The subthalamic nuclei, red nuclei, and substantia nigra work together with the basal ganglia, cerebellum, and cerebrum in the control of body movements.

The functions of the four parts of the diencephalon are summarized in Table 14.1 on page 458.

Circumventricular Organs

Parts of the diencephalon, called **circumventricular organs (CVOs)** because they lie in the walls of the third and fourth ventricles, can monitor chemical changes in the blood because they lack the blood–brain barrier. CVOs include part of the hypothalamus, the pineal gland, the pituitary gland, and a few other nearby structures. Functionally, these regions coordinate homeostatic activities of the endocrine and nervous systems, such as the regulation of blood pressure, fluid balance, hunger, and thirst. CVOs are also thought to be the sites of entry into the brain of HIV, the virus that causes AIDS. Once in the brain, HIV may cause dementia (irreversible deterioration of mental state) and other neurological disorders.

1. Explain why the thalamus is considered a "relay station" in the brain.
2. Explain why the hypothalamus is considered to be part of both the nervous system and the endocrine system.

THE CEREBRUM

OBJECTIVE

• *Describe the structures and functions of the cerebrum.*

The **cerebrum** is supported on the diencephalon and brain stem and forms the bulk of the brain (see Figure 14.1). The superficial gray matter layer of the cerebrum is called the **cerebral cortex** (*cortex* = rind or bark) (Figure 14.11a). The cortex, which is only 2–4 mm (0.08–0.16 in.) thick, contains billions of neurons. Deep to the cortex lies the cerebral white matter. The cerebrum is the "seat of intelligence"; it provides us with the ability to read, write, and speak; to make calculations and compose music; and to remember the past, plan for the future, and imagine things that have never existed before.

During embryonic development, when brain size increases rapidly, the gray matter of the cortex enlarges much faster than the deeper white matter. As a result, the cortical region rolls and folds upon itself. The folds are called **gyri** (JĪ-rī; = circles) or **convolutions** (Figure 14.11a, b). The deepest grooves between folds are known as **fissures;** the shallower grooves between folds are termed **sulci** (SUL-sī; = grooves). (The singular terms are gyrus and sulcus.) The most prominent fissure, the **longitudinal fissure,** separates the cerebrum into right and left halves called **cerebral hemispheres.** The hemispheres are connected internally by the **corpus callosum** (kal-LŌ-sum; *corpus* = body; *callosum* = hard), a broad band of white matter containing axons that extend between the hemispheres (see Figure 14.9b).

Lobes of the Cerebrum

Each cerebral hemisphere can be further subdivided into four lobes, named after the bones that cover them: frontal, parietal, temporal, and occipital lobes (see Figure 14.11a, b). The **central sulcus** (SUL-kus) separates the **frontal lobe** from the **parietal lobe.** A major gyrus, the **precentral gyrus**—located immediately anterior to the central sulcus—contains the primary motor area of the cerebral cortex. Another major gyrus, the **postcentral gyrus,** which is located immediately posterior to the central sulcus, contains the primary somatosensory area of the cerebral cortex.

Figure 14.11 Cerebrum. Because the insula cannot be seen externally, it has been projected to the surface in (b). (See Tortora, *A Photographic Atlas of the Human Body,* Figure 8.14)

🔑 **The cerebrum is the "seat of intelligence"; it provides us with the ability to read, write, and speak; to make calculations and compose music; to remember the past and plan for the future; and to create.**

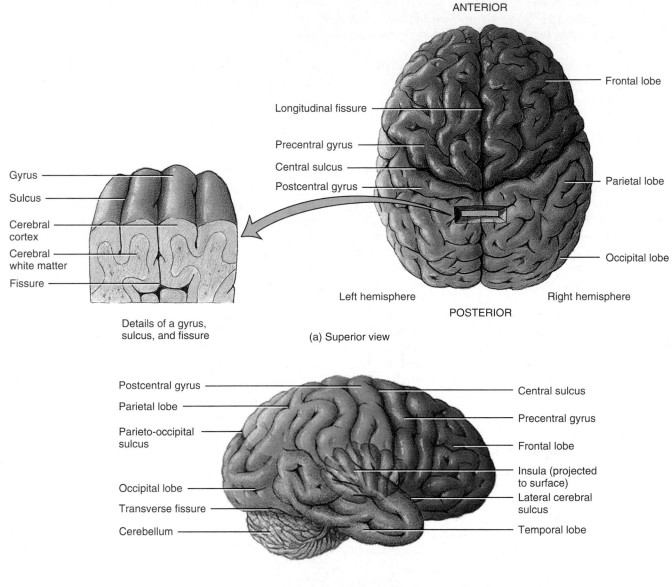

Details of a gyrus, sulcus, and fissure

(a) Superior view

(b) Right lateral view

Ⓠ During development, which part of the brain—gray matter or white matter—enlarges more rapidly? What are the brain folds, shallow grooves, and deep grooves called?

Figure 14.12 Organization of fibers into white matter tracts of the left cerebral hemisphere.

🔑 **Association fibers, commissural fibers, and projection fibers form white matter tracts in the cerebral hemispheres.**

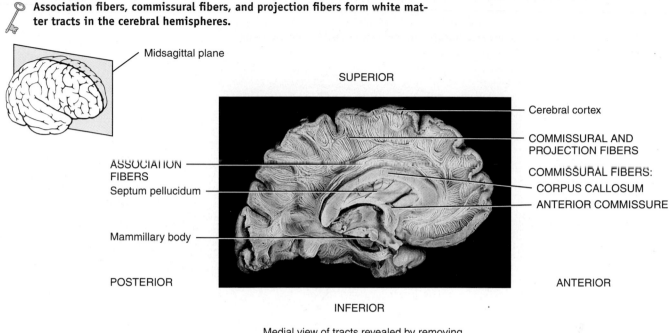

Medial view of tracts revealed by removing gray matter from a midsagittal section

Q Which fibers carry impulses between gyri of the same hemisphere? Between gyri in opposite hemispheres? Between the brain and the spinal cord?

The **lateral cerebral sulcus** separates the **frontal lobe** from the **temporal lobe.** The **parieto-occipital sulcus** separates the **parietal lobe** from the **occipital lobe.** A fifth part of the cerebrum, the **insula,** cannot be seen at the surface of the brain because it lies within the lateral cerebral fissure, deep to the parietal, frontal, and temporal lobes (Figure 14.11b).

Cerebral White Matter

The white matter underlying the cortex consists of myelinated and unmyelinated axons organized into tracts consisting of three principal types of fibers (Figure 14.12):

1. **Association fibers** transmit nerve impulses between gyri in the same hemisphere.
2. **Commissural fibers** transmit impulses from the gyri in one cerebral hemisphere to the corresponding gyri in the other cerebral hemisphere. Three important groups of commissural fibers are the **corpus callosum, anterior commissure,** and **posterior commissure.**
3. **Projection fibers** form descending and ascending tracts that transmit impulses either from the cerebrum and other parts of the brain to the spinal cord, or from the spinal cord to the brain. An example is the **internal capsule,** a thick band of sensory and motor tracts that connect the cerebral cortex with the brain stem and the spinal cord (see Figure 14.13b).

Basal Ganglia

The **basal ganglia** consist of several pairs of nuclei; the two members of each pair are situated in opposite cerebral hemispheres (Figure 14.13). The largest nucleus in the basal ganglia is the **corpus striatum** (strī-Ā-tum; = striped), which consists of the **caudate nucleus** (*caud-* = tail) and the **lentiform nucleus** (*lentiform* = lentil-shaped). Each lentiform nucleus, in turn, is subdivided into a lateral part called the **putamen** (pu-TĀ-men; = shell) and a medial part called the **globus pallidus** (*globus* = ball; *pallidus* = pale). The part of the internal capsule passing between the lentiform nucleus and the caudate nucleus, and between the lentiform nucleus and thalamus, is sometimes considered part of the corpus striatum.

Other structures that are functionally linked to (and sometimes considered part of) the basal ganglia are the substantia nigra and red nuclei of the midbrain (see Figure 14.7b) and the subthalamic nuclei of the diencephalon (see Figure 14.13b). Axons from the substantia nigra terminate in the caudate nucleus and putamen. The subthalamic nuclei connect with the globus pallidus.

The basal ganglia receive input from and provide output to the cerebral cortex, thalamus, and hypothalamus. In addition, many nerve fibers interconnect the nuclei of the basal ganglia. The caudate nucleus and the putamen control automatic movements of skeletal muscles, such as swinging the

Figure 14.13 Basal ganglia. In (a), the basal ganglia have been projected to the surface and are shown in dark green; in (b) they are shown in green and blue. (See Tortora, *A Photographic Atlas of the Human Body*, Figures 8.17, 8.18, 8.22)

🔑 **The basal ganglia control large, automatic movements of skeletal muscles and muscle tone.**

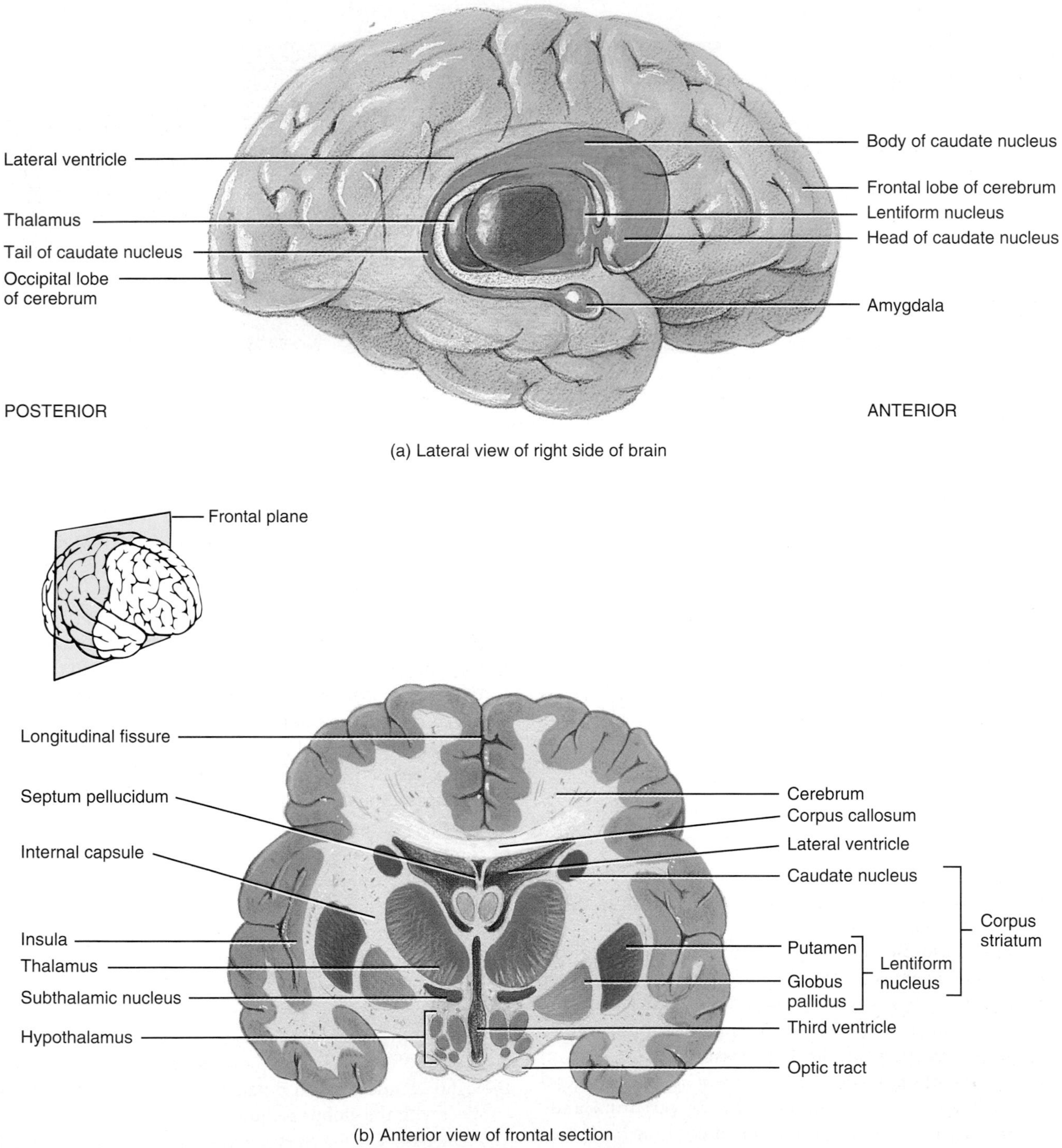

Lateral ventricle

Thalamus

Tail of caudate nucleus

Occipital lobe of cerebrum

Body of caudate nucleus

Frontal lobe of cerebrum

Lentiform nucleus

Head of caudate nucleus

Amygdala

POSTERIOR

ANTERIOR

(a) Lateral view of right side of brain

Frontal plane

Longitudinal fissure

Septum pellucidum

Internal capsule

Insula

Thalamus

Subthalamic nucleus

Hypothalamus

Cerebrum

Corpus callosum

Lateral ventricle

Caudate nucleus

Putamen

Globus pallidus

Third ventricle

Optic tract

Corpus striatum

Lentiform nucleus

(b) Anterior view of frontal section

Q Where are the basal ganglia located relative to the thalamus?

Figure 14.14 Components of the limbic system and surrounding structures.

🔑 **The limbic system governs emotional aspects of behavior.**

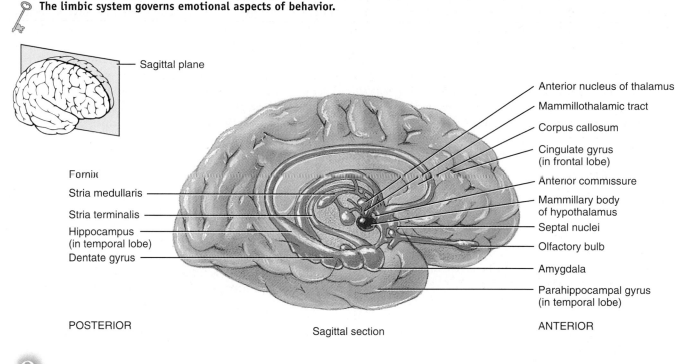

Sagittal plane

Anterior nucleus of thalamus

Mammillothalamic tract

Corpus callosum

Cingulate gyrus (in frontal lobe)

Anterior commissure

Mammillary body of hypothalamus

Septal nuclei

Olfactory bulb

Amygdala

Parahippocampal gyrus (in temporal lobe)

Fornix

Stria medullaris

Stria terminalis

Hippocampus (in temporal lobe)

Dentate gyrus

POSTERIOR

Sagittal section

ANTERIOR

Q Which part of the limbic system functions with the cerebrum in memory?

arms while walking or laughing in response to a joke. The globus pallidus helps regulate the muscle tone required for specific body movements.

The Limbic System

Encircling the upper part of the brain stem and the corpus callosum is a ring of structures on the inner border of the cerebrum and floor of the diencephalon that constitutes the **limbic system** (*limbic* = border). Among the components of the limbic system are the following structures (Figure 14.14):

1. The **parahippocampal** and **cingulate gyri** (*cingul-* = belt), both gyri of the cerebral hemispheres, plus the **hippocampus** (= seahorse), a portion of the parahippocampal gyrus that extends into the floor of the lateral ventricle, make up the *limbic lobe.*
2. The **dentate gyrus** (*dentate* = toothed) lies between the hippocampus and parahippocampal gyrus.
3. The **amygdala** (*amygda-* = almond-shaped) is composed of several groups of neurons located at the tail end of the caudate nucleus.
4. The **septal nuclei** are located within the septal area formed by the regions under the corpus callosum and the paraterminal gyrus (a cerebral gyrus).
5. The **mammillary bodies of the hypothalamus** are two round masses close to the midline near the cerebral peduncles.
6. The **anterior nucleus of the thalamus** is located in the floor of the lateral ventricle.

7. The **olfactory bulbs** are flattened bodies of the olfactory pathway that rest on the cribriform plate.
8. The **fornix, stria terminalis, stria medullaris, medial forebrain bundle,** and **mammillothalamic tract** are linked by bundles of interconnecting myelinated axons.

The limbic system is sometimes called the "emotional brain" because it plays a primary role in a range of emotions, including pain, pleasure, docility, affection, and anger. Experiments have shown that when different areas of animals' limbic system are stimulated, the animals' reactions indicate that they are experiencing intense pain or extreme pleasure. Stimulation of other limbic system areas in animals produces tameness and signs of affection. Stimulation of a cat's amygdala or certain nuclei of the hypothalamus produces a behavioral pattern called rage—the cat extends its claws, raises its tail, opens its eyes wide, and hisses and spits.

The hippocampus of the limbic system, in conjunction with portions of the cerebrum, functions in memory. People with damage to certain limbic system structures forget recent events and cannot commit anything to memory.

The functions of the cerebrum are summarized in Table 14.1 on page 458.

 CLINICAL APPLICATION
Brain Injuries

Brain injuries are commonly associated with head trauma and result in part from displacement and distortion of neuronal tissue at the moment of impact. Secondary effects include decreased blood pressure; sustained, elevated intracranial pressure; infections; and respiratory complications. Additional tissue damage occurs when normal blood flow is restored after a period of ischemia (reduced blood flow) because the sudden increase in oxygen level produces large numbers of oxygen free radicals (charged oxygen molecules with an unpaired electron). Brain cells recovering from the effects of a stroke or cardiac arrest also release free radicals. Free radicals cause damage by disrupting cellular DNA and enzymes and by altering plasma membrane permeability.

Various degrees of brain injury are described by specific terms. A **concussion** is an abrupt, but temporary, loss of consciousness (from seconds to hours) following a blow to the head or the sudden stopping of a moving head. It is the most common brain injury. A concussion produces no obvious bruising of the brain. Signs of a concussion are headache, drowsiness, lack of concentration, confusion, or post-traumatic amnesia (memory loss).

A **contusion** is bruising of the brain due to trauma and includes the leakage of blood from microscopic vessels. It is usually associated with a concussion. In a contusion, the pia mater may be torn, allowing blood to enter the subarachnoid space. The area most commonly affected is the frontal lobe. A contusion usually results in an immediate loss of consciousness (generally lasting no longer than 5 minutes), loss of reflexes, transient cessation of respiration, and decreased blood pressure. Vital signs typically stabilize in a few seconds.

A **laceration** is a tear of the brain, usually from a skull fracture or a gunshot wound. A laceration results in rupture of large blood vessels, with bleeding into the brain and subarachnoid space. Consequences include cerebral hematoma (localized pool of blood, usually clotted, that swells against the brain tissue), edema, and increased intracranial pressure. ∎

FUNCTIONAL ASPECTS OF THE CEREBRAL CORTEX

OBJECTIVE

• *Describe the locations and functions of the sensory, association, and motor areas of the cerebral cortex.*

Specific types of sensory, motor, and integrative signals are processed in certain cerebral regions (Figure 14.15). Generally, **sensory areas** receive and interpret sensory impulses, **motor areas** initiate movements, and **association areas** deal with more complex integrative functions such as memory, emotions, reasoning, will, judgment, personality traits, and intelligence.

Sensory Areas

Sensory impulses arrive mainly in the posterior half of both cerebral hemispheres—that is, in regions posterior to the central sulci. In the cortex, primary sensory areas have the most direct connections with peripheral sensory receptors. Secondary sensory areas and sensory association areas often are adjacent to the primary areas; they usually receive input from the primary areas and from other diverse regions of the brain.

Secondary sensory areas and sensory association areas integrate sensory experiences to generate meaningful patterns of recognition and awareness. For example, whereas a person with damage in the primary visual area would be blind in at least part of his visual field, a person with damage to a visual association area might see normally yet be unable to recognize a friend.

The following are some important sensory areas (see Figure 14.15):

• **Primary Somatosensory Area** (areas 1, 2, and 3) Located directly posterior to the central sulcus of each cerebral hemisphere in the postcentral gyrus of each parietal lobe, the primary somatosensory area extends from the longitudinal fissure on the superior aspect of the cerebrum to the lateral cerebral sulcus.

The primary somatosensory area receives nerve impulses from somatic sensory receptors for touch, proprioception (joint and muscle position), pain, and temperature. Each point within the area receives sensations from a specific part of the body, and the entire body is spatially represented in it. The size of the cortical area receiving impulses from a particular body part depends on the number of receptors present there rather than on the size of the part. For example, a larger portion of the sensory area receives impulses from the lips and fingertips than from the thorax or hip (see Figure 15.5a on page 496). The major function of the primary somatosensory area is to localize exactly the points of the body where sensations originate. Although the thalamus registers sensations in a general way, it cannot distinguish precisely the specific location of stimulation; this capability depends on the primary somatosensory area of the cortex.

• **Primary Visual Area** (area 17) Located on the medial surface of the occipital lobe, it receives impulses that convey visual information. Axons of neurons with cell bodies in the eye form the optic nerves (cranial nerve II), which terminate in the lateral geniculate nucleus of the thalamus. From the thalamus, neurons project to the primary visual area, carrying information concerning shape, color, and movement of visual stimuli.

• **Primary Auditory Area** (areas 41 and 42) Located in the superior part of the temporal lobe near the lateral cerebral sulcus, it interprets the basic characteristics of sound such as pitch and rhythm.

• **Primary Gustatory Area** (area 43) Located at the base of the postcentral gyrus superior to the lateral cerebral sulcus in the parietal cortex, it receives impulses for taste.

• **Primary Olfactory Area** (area 28) Located in the temporal lobe on the medial aspect (and thus not visible in Figure 14.15), it receives impulses for smell.

Figure 14.15 Functional areas of the cerebrum. Broca's speech area is in the left cerebral hemisphere of most people; it is shown here to indicate its relative location. The numbers, still used today, are from K. Brodmann's map of the cerebral cortex, first published in 1909.

🗝 **Particular areas of the cerebral cortex process sensory, motor, and integrative signals.**

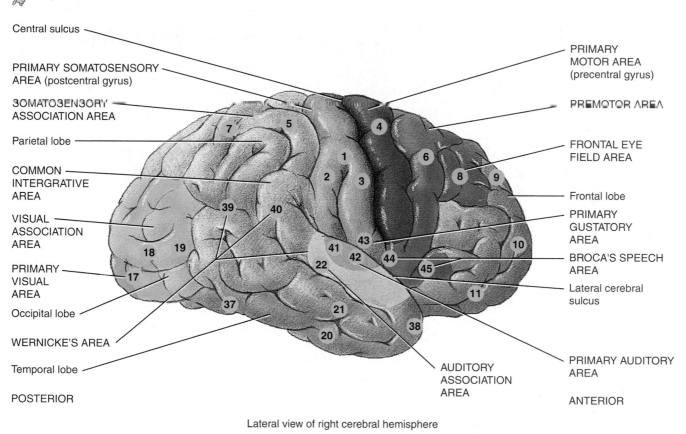

Lateral view of right cerebral hemisphere

Ⓠ What area(s) of the cerebrum integrate(s) interpretation of visual, auditory, and somatic sensations? Translates thoughts into speech? Controls skilled muscular movements? Interprets sensations related to taste? Interprets pitch and rhythm? Interprets shape, color, and movement of objects? Controls voluntary scanning movements of the eyes?

Motor Areas

Motor output from the cerebral cortex flows mainly from the anterior part of each hemisphere. Among the most important motor areas are the following (see Figure 14.15):

- **Primary Motor Area** (area 4) Located in the precentral gyrus of the frontal lobe, each region in the primary motor area controls voluntary contractions of specific muscles or groups of muscles (see Figure 15.5b on page 496). Electrical stimulation of any point in the primary motor area results in contraction of specific skeletal muscle fibers on the opposite side of the body. As is true for the primary somatosensory area, body parts are not repre-

sented in proportion to their size: More cortical area is devoted to those muscles involved in skilled, complex, or delicate movement—as, for example, manipulation of the finger.

- **Broca's Speech Area** (areas 44 and 45) Speaking and understanding language are complex activities that involve several sensory, association, and motor areas of the cortex. In 97% of the population, these language areas are localized in the *left* hemisphere. The production of speech occurs in Broca's (BRŌ-kaz) speech area, located in one frontal lobe—the *left* frontal lobe in most people—just superior to the lateral cerebral sulcus.

Association Areas

The association areas of the cerebrum consist of some motor and sensory areas, plus large areas on the lateral surfaces of the occipital, parietal, and temporal lobes and on the frontal lobes anterior to the motor areas. Association areas are connected with one another by association tracts and include the following (see Figure 14.15):

- **Somatosensory Association Area** (areas 5 and 7) The somatosensory association area is just posterior to and receives input from the primary somatosensory area, as well as from the thalamus and other lower parts of the brain. Its role is to integrate and interpret sensations. This area permits you to determine the exact shape and texture of an object without looking at it, to determine the orientation of one object to another as they are felt, and to sense the relationship of one body part to another. Another role of the somatosensory association area is the storage of memories of past sensory experiences, enabling you to compare current sensations with previous experiences.

- **Visual Association Area** (areas 18 and 19) Located in the occipital lobe, it receives sensory impulses from the primary visual area and the thalamus. It relates present and past visual experiences and is essential for recognizing and evaluating what is seen.

- **Auditory Association Area** (area 22) Located inferior and posterior to the primary auditory area in the temporal cortex, it ascertains whether a sound is speech, music, or noise.

- **Wernicke's (Posterior Language) Area** (area 22, and possibly areas 39 and 40) This area interprets the meaning of speech by recognizing spoken words; it translates words into thoughts. The regions in the *right* hemisphere that correspond to Broca's and Wernicke's areas in the left hemisphere also contribute to verbal communication by adding tonal inflections and emotional content to spoken language. For example, you can tell if a person is angry or joyful by the tone of voice.

- **Common Integrative Area** (areas 5, 7, 39, and 40) This area, bordered by somatosensory, visual, and auditory association areas, receives nerve impulses from these areas, and from the primary gustatory and primary olfactory areas, the thalamus, and parts of the brain stem. It integrates sensory interpretations from the association areas and impulses from other areas, allowing one thought to be formed on the basis of a variety of sensory inputs. It then transmits signals to other parts of the brain to cause the appropriate response to the interpretation of the sensory signals.

- **Premotor Area** (area 6) Immediately anterior to the primary motor area is a motor association area. Neurons in this area communicate with the primary motor cortex, the sensory association areas in the parietal lobe, the basal ganglia, and the thalamus. The premotor area deals with learned motor activities of a complex and sequential nature. It generates nerve impulses that cause specific groups of muscles to contract in specific sequences. For example, this part of the cortex is active when you write a word. The premotor area controls learned skilled movements and serves as a memory bank for such movements.

- **Frontal Eye Field Area** (area 8) This area in the frontal cortex is sometimes included in the premotor area. It controls voluntary scanning movements of the eyes—reading this sentence, for instance.

- **Language Areas** From Broca's speech area, nerve impulses pass to the premotor regions that control the muscles of the larynx, pharynx, and mouth. The impulses from the premotor area result in specific, coordinated muscle contractions that enable you to speak. Simultaneously, impulses are sent from Broca's speech area to the primary motor area. From here, impulses also control the breathing muscles to regulate the proper flow of air past the vocal cords. The coordinated contractions of your speech and breathing muscles enable you to speak your thoughts.

Table 14.1 on page 458 summarizes the various functions of the cerebrum.

CLINICAL APPLICATION
Aphasia

Much of what we know about language areas comes from studies of patients with language or speech disturbances that have resulted from brain damage. Broca's speech area, the auditory association area, and other language areas are located in the left cerebral hemisphere of most people, regardless of whether they are left-handed or right-handed. Injury to the association or motor speech areas results in **aphasia** (a-FĀ-zē-a; *a-* = without; *-phasia* = speech), an inability to use or comprehend words. Damage to Broca's speech area results in nonfluent aphasia, an inability to properly articulate or form words; people with *nonfluent aphasia* know what they wish to say but cannot speak. Damage to the common integrative area or auditory association area (areas 39 and 22) results in *fluent aphasia,* characterized by faulty understanding of spoken or written words. A person experiencing this type of aphasia may fluently produce strings of words that have no meaning. The underlying deficit may be **word deafness,** an inability to understand spoken words; or **word blindness,** an inability to understand written words; or both. ■

Hemispheric Lateralization

Although the brain is fairly symmetrical on its right and left sides, subtle anatomical differences between the two hemispheres exist. For example, in about two-thirds of the population, the planum temporale, a region of the temporal lobe that includes Wernicke's area, is 50% larger on the left side than on the right side. This asymmetry appears in the human fetus at about 30 weeks of gestation. Moreover, although the two hemispheres share performance of many

Figure 14.16 Summary of the principal functional differences between the left and right cerebral hemispheres.

🔑 **Hemispheric lateralization means that each cerebral hemisphere performs unique functions.**

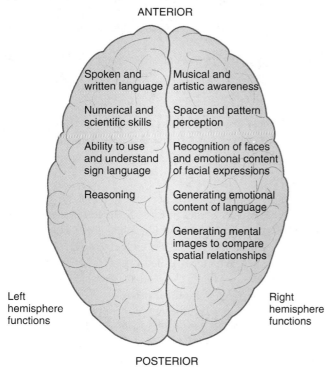

ANTERIOR

Spoken and written language

Musical and artistic awareness

Numerical and scientific skills

Space and pattern perception

Ability to use and understand sign language

Recognition of faces and emotional content of facial expressions

Reasoning

Generating emotional content of language

Generating mental images to compare spatial relationships

Left hemisphere functions

Right hemisphere functions

POSTERIOR

Q Which two areas of the cerebrum are most important for language?

functions, each hemisphere also specializes in performing certain unique functions (Figure 14.16). This functional asymmetry is termed **hemispheric lateralization.**

The left hemisphere receives sensory signals from and controls the right side of the body, whereas the right hemisphere receives sensory signals from and controls the left side of the body. Beyond these differences, however, in most people the left hemisphere is more important for spoken and written language, numerical and scientific skills, ability to use and understand sign language, and reasoning. Patients with damage in the left hemisphere, for example, often exhibit aphasia. Conversely, the right hemisphere is more important for musical and artistic awareness, spatial and pattern perception, recognition of faces, and emotional content of language, and for generating mental images of sight, sound, touch, taste, and smell to compare relationships among them. Patients with damage in the right hemisphere regions that correspond to Broca's and Wernicke's areas in the left hemisphere speak in a monotonous voice, having lost the ability to impart emotional inflection to what they say. In general, lateralization seems less pronounced in females than in males, both for language (left hemisphere) and for visual

Figure 14.17 Types of brain waves recorded in an electroencephalogram (EEG).

🔑 **Brain waves indicate electrical activity of the cerebral cortex.**

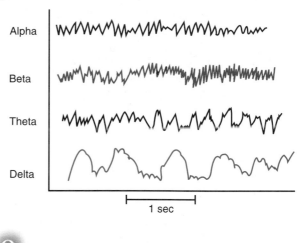

Alpha

Beta

Theta

Delta

1 sec

Q Which type of wave indicates emotional stress?

and spatial skills (right hemisphere). A possibly related observation is that females tend to have a larger anterior commissure and a larger posterior portion of the corpus callosum than males; both of these white matter tracts provide communication between the two hemispheres.

Brain Waves

At any instant, brain cells are generating millions of nerve impulses (nerve action potentials) and graded potentials (excitatory and inhibitory postsynaptic potentials) in individual neurons. Taken together, these electrical signals are called **brain waves.** Brain waves generated by neurons close to the brain surface, mainly neurons in the cerebral cortex, can be detected by sensors called electrodes placed on the forehead and scalp. A record of such waves is called an **electroencephalogram** (e-lek′-trō-en-SEF-a-lō-gram′) or **EEG.** Electroencephalograms are useful both in studying normal brain functions, such as changes that occur during sleep, and in diagnosing a variety of brain disorders, such as epilepsy, tumors, metabolic abnormalities, sites of trauma, and degenerative diseases. An EEG may also be used to establish brain death, the complete absence of brain waves in two EEGs taken 24 hours apart.

Four kinds of waves can be recorded from normal individuals (Figure 14.17):

1. **Alpha waves.** These rhythmic waves occur at a frequency of about 8–13 cycles per second. (The unit commonly used to express frequency is the hertz [Hz]; 1 Hz = 1 cycle per second.) Alpha waves are present in the EEGs of nearly all normal individuals when they are awake and resting with their eyes closed. These waves disappear entirely during sleep.

2. **Beta waves.** The frequency of these waves is between 14 and 30 Hz. Beta waves generally appear when the nervous system is active—that is, during periods of sensory input and mental activity.

3. **Theta waves.** These waves have frequencies of 4–7 Hz. Theta waves normally occur in children and adults experiencing emotional stress. They also occur in many disorders of the brain.

4. **Delta waves.** The frequency of these waves is 1–5 Hz. Whereas delta waves occur during deep sleep in adults, they are normal in awake infants. When produced by an awake adult, they indicate brain damage.

1. Describe the cortex, convolutions, fissures, and sulci of the cerebrum.
2. List and locate the lobes of the cerebrum. How are they separated from one another? What is the insula?
3. Describe the organization of cerebral white matter. Be sure to indicate the function of each major group of fibers.
4. Name the nuclei that form basal ganglia, list the function of each, and describe the effects of basal ganglia damage.
5. Define the limbic system. List several of its functions.
6. Compare the functions of sensory, motor, and association areas of the cerebral cortex.
7. Describe hemispheric lateralization.
8. What is the diagnostic value of an EEG?

CRANIAL NERVES

OBJECTIVE
• *Identify the cranial nerves by name, number, and type, and give the function of each.*

Cranial nerves, like spinal nerves, are part of the peripheral nervous system (PNS). Of the 12 pairs of cranial nerves, ten emerge from the brain stem. The cranial nerves are designated both by roman numerals and names (see Figure 14.5). The roman numerals indicate the order, from anterior to posterior, in which the nerves arise from the brain; the names indicate a nerve's distribution or function.

Cranial nerves emerge from the nose (I), the eyes (II), the brainstem (III–XII), and the spinal cord (a portion of XI). Two cranial nerves (I and II) contain only sensory fibers and thus are called **sensory nerves.** The remainder are **mixed nerves** that contain axons of both sensory and motor neurons. The cell bodies of sensory neurons are located in ganglia outside the brain, whereas the cell bodies of motor neurons lie in nuclei within the brain. Some cranial nerves include both somatic motor fibers and parasympathetic fibers of the autonomic nervous system.

Table 14.2 on page 475 presents a summary of cranial nerves, including clinical applications related to their dysfunction.

1. Define a cranial nerve. How are cranial nerves named and numbered?
2. Distinguish between a mixed and a sensory cranial nerve.
3. For each cranial nerve, devise a test that could reveal if it is damaged.

DEVELOPMENTAL ANATOMY OF THE NERVOUS SYSTEM

OBJECTIVE
• *Describe how the parts of the brain develop.*

The development of the nervous system begins early in the third week with a thickening of the **ectoderm** called the **neural plate** (Figure 14.18). The plate folds inward and forms a longitudinal groove, the **neural groove.** The raised edges of the neural plate are called **neural folds.** As development continues, the neural folds increase in height and meet to form a tube called the **neural tube.**

Three types of cells differentiate from the wall that encloses the neural tube. The outer or **marginal layer** develops into the *white matter* of the nervous system; the middle or **mantle layer** develops into *gray matter;* and the inner or **ependymal layer** eventually forms the *lining of the central canal of the spinal cord and ventricles of the brain.*

The **neural crest** is a mass of tissue between the neural tube and the skin ectoderm (Figure 14.18b). It differentiates and eventually forms the *posterior (dorsal) root ganglia of spinal nerves, spinal nerves, ganglia of cranial nerves, cranial nerves, ganglia of the autonomic nervous system, adrenal medulla,* and *meninges.*

When the neural tube forms from the neural plate, its anterior part develops into three enlarged areas called **primary vesicles** at the end of the fourth week of embryonic development. These are the **prosencephalon** (prōs′-en-SEF-a-lon; *pros-* = before) or forebrain, **mesencephalon** (mes′ en-SEF-a-lon; *mes-* = middle) or midbrain, and **rhombencephalon** (rom′-ben-SEF-a-lon; *rhomb-* = behind) or hindbrain (Figure 14.19a). During the fifth week of development, the prosencephalon develops into two secondary brain vesicles called the **telencephalon** (tel′-en-SEF-a-lon; *tel-* = distant) and the **diencephalon** (dī-en-SEF-a-lon; *di-* = through) (Figure 14.19b). The rhombencephalon also develops into two secondary brain vesicles called the **metencephalon** (met′-en-SEF-a-lon; *met-* = after) and the **myelencephalon** (mī-el-en-SEF-a-lon; *myel-* = marrow). The area of the neural tube inferior to the myelencephalon gives rise to the *spinal cord.*

Figure 14.18 Origin of the nervous system. (a) Dorsal view of an embryo in which the neural folds have partially united, forming the early neural tube. (b) Transverse sections through the embryo showing the formation of the neural tube.

🔑 **The nervous system begins developing in the third week from a thickening of ectoderm called the neural plate.**

Including the mesencephalon, then, there are five secondary brain vesicles, which continue to develop as follows:

- The telencephalon develops into the *cerebral hemispheres* and *basal ganglia* and houses the paired *lateral ventricles*.
- The diencephalon develops into the *epithalamus, thalamus, subthalamus, hypothalamus,* and *pineal gland* and houses the *third ventricle.*
- The mesencephalon develops into the *midbrain* and houses the *cerebral aqueduct.*
- The metencephalon becomes the *pons* and *cerebellum* and houses a portion of the *fourth ventricle.*
- The myelencephalon develops into the *medulla oblongata* and houses a portion of the *fourth ventricle.*

Two neural tube defects—spina bifida (see page 214) and anencephaly (absence of the skull and cerebral hemispheres)—are associated with low levels of folic acid, one of the B vitamins. The incidence of both disorders is greatly reduced when women who may become pregnant take folic acid supplements.

AGING AND THE NERVOUS SYSTEM

OBJECTIVE

• *Describe the effects of aging on the nervous system.*

The brain grows rapidly during the first few years of life. Growth is due mainly to an increase in the size of neurons already present, the proliferation and growth of neuroglia, the development of dendritic branches and synaptic contacts, and myelination of the various fiber tracts. From early adulthood onward, brain weight declines. By the time a person reaches 80, the brain weighs about 7% less than it did in the prime of life. Although the number of neurons present does not decrease very much, the number of synaptic contacts declines. Associated with the decrease in brain mass is a decreased capacity for sending nerve impulses to and from the brain; as a result, processing of information diminishes. Conduction velocity decreases, voluntary motor movements slow down, and reflex times increase.

1. What parts of the brain develop from each primary brain vesicle?
2. How is brain mass related to age?

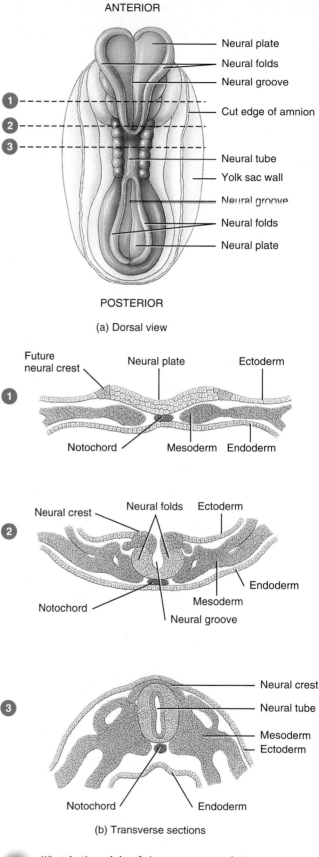

(a) Dorsal view

(b) Transverse sections

Q What is the origin of the gray matter of the nervous system?

Figure 14.19 Development of the brain and spinal cord.

The various parts of the brain develop from the primary brain vesicles.

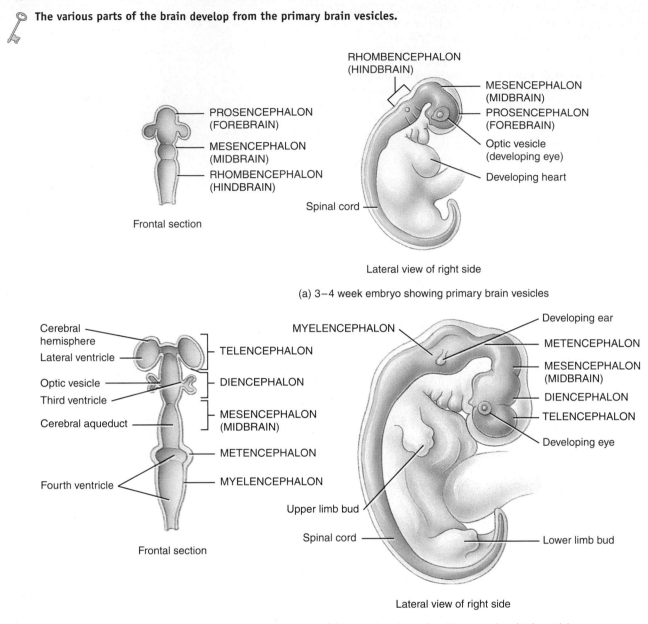

(a) 3–4 week embryo showing primary brain vesicles

(b) 5–week embryo showing secondary brain vesicles

Q   Which primary brain vesicle does not develop into a secondary brain vesicle?

Table 14.2 Summary of Cranial Nerves*

NUMBER AND NAME	TYPE AND LOCATION	FUNCTION AND CLINICAL APPLICATION
Cranial Nerve I: **Olfactory** (ol-FAK-tō-rē; *olfact-* = to smell) 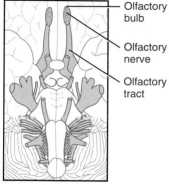	**Sensory** Arises in olfactory mucosa, passes through foramina in the cribriform plate of the ethmoid bone, and ends in the olfactory bulb. The olfactory tract extends via two pathways to olfactory areas of cerebral cortex.	*Function:* Smell. *Clinical application:* Loss of the sense of smell, called *anosmia,* may result from head injuries in which the cribriform plate of the ethmoid bone is fractured and from lesions along the olfactory pathway.
Cranial Nerve II: **Optic** (OP-tik; *opti-* = the eye, vision) 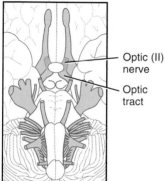	**Sensory** Arises in the retina of the eye, passes through the optic foramen, forms the optic chiasm and then the optic tracts, and terminates in the lateral geniculate nuclei of thalamus. From the thalamus, fibers extend to the primary visual area (area 17) of the cerebral cortex.	*Function:* Vision *Clinical application:* Fractures in the orbit, damage along the visual pathway, and diseases of the nervous system may result in visual field defects and loss of visual acuity. Blindness due to a defect in or loss of one or both eyes is called *anopia.*
Cranial Nerve III: **Oculomotor** (ok′-ū-lō-MŌ-tor; *oculo-* = eye; *-motor* = mover) 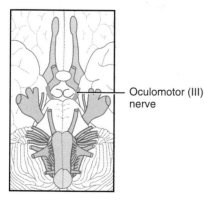	**Mixed (mainly motor)** *Sensory portion:* Consists of fibers from proprioceptors in eyeball muscles that pass through the superior orbital fissure and terminate in the midbrain. *Motor portion:* Originates in the midbrain and passes through the superior orbital fissure. Somatic fibers innervate the levator palpebrae superioris muscle of the upper eyelid and four extrinsic eyeball muscles (superior rectus, medial rectus, inferior rectus, and inferior oblique). Parasympathetic fibers innervate the ciliary muscle of eyeball and the sphincter muscle of the iris.	*Sensory function:* Muscle sense (proprioception). *Motor function:* Movement of eyelid and eyeball, accommodation of lens for near vision, and constriction of pupil. *Clinical application:* Nerve damage causes *strabismus* (a deviation of the eye in which both eyes do not fix on the same object), *ptosis* (drooping) of the upper eyelid, dilation of the pupil, movement of the eyeball downward and outward on the damaged side, loss of accommodation for near vision, and *diplopia* (double vision).

*A mnemonic device that can be used to remember the names of the nerves is: "**O**h, **o**h, **o**h, **t**o **t**ouch and **f**eel **v**ery **g**reen **v**egetables—**A**H!" The boldfaced letters correspond to the initial letter of each pair of cranial nerves.

Table 14.2 Summary of Cranial Nerves (continued)

NUMBER AND NAME	TYPE AND LOCATION	FUNCTION AND CLINICAL APPLICATION
Cranial Nerve IV: **Trochlear** (TRŌK-lē-ar; *trochle-* = a pulley)  — Trochlear (IV) nerve	**Mixed (mainly motor)** *Sensory portion:* Consists of fibers from proprioceptors in the superior oblique muscles, which pass through the superior orbital fissure and terminate in the midbrain. *Motor portion:* Originates in the midbrain and passes through the superior orbital fissure. Innervates the superior oblique muscle, an extrinsic eyeball muscle.	*Sensory function:* Muscle sense (proprioception). *Motor function:* Movement of the eyeball. *Clinical application:* In trochlear nerve paralysis, diplopia and strabismus occur.
Cranial Nerve V: **Trigeminal** (trī-JEM-i-nal; = triple, for its three branches) 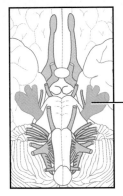 — Trigeminal (V) nerve	**Mixed** *Sensory portion:* Consists of three branches, all of which end in the pons. (1) The **ophthalamic** (*ophthalm-* = the eye) branch contains fibers from the skin over the upper eyelid, eyeball, lacrimal glands, nasal cavity, side of nose, forehead, and anterior half of scalp that pass through superior orbital fissure. (2) The **maxillary** (*maxilla* = upper jaw bone) branch contains fibers from the mucosa of the nose, palate, parts of the pharynx, upper teeth, upper lip, and lower eyelid that pass through the foramen rotundum. (3) The **mandibullar** (*mandibula* = lower jaw bone) branch contains fibers from the anterior two-thirds of the tongue (somatic sensory fibers but not fibers for the special sense of taste), the lower teeth, skin over mandible, cheek and mucosa deep to it, and side of head in front of ear that pass through the foramen ovale. *Motor portion:* Is part of the mandibular branch, which originates in the pons, passes through the foramen ovale, and innervates muscles of mastication (masseter, temporalis, medial pterygoid, lateral pterygoid, anterior belly of digastric, and mylohyoid muscles).	*Sensory function:* Conveys sensations for touch, pain, temperature, and muscle sense (proprioception). *Motor function:* Chewing. *Clinical application:* Injury results in paralysis of the muscles of mastication and a loss of the sensations of touch, temperature, and proprioception. *Neuralgia* (pain) of one or more branches of the trigeminal nerve is called *trigeminal neuralgia (tic douloureux).*

Table 14.2 (continued)

NUMBER AND NAME	TYPE AND LOCATION	FUNCTION AND CLINICAL APPLICATION
Cranial Nerve VI: **Abducens** (ab-DŪ-senz; *ab-* = away; *-ducens* = to lead)  Abducens (VI) nerve	**Mixed (mainly motor)** *Sensory portion:* Consists of fibers from proprioceptors in the lateral rectus muscle that pass through the superior orbital fissure and end in the pons. *Motor portion:* Originates in the pons, passes through the superior orbital fissure, and innervates the lateral rectus muscle, an extrinsic eyeball muscle.	*Sensory function:* Muscle sense (proprioception). *Motor function:* Movement of the eyeball. *Clinical application:* With damage to this nerve, the affected eyeball cannot move laterally beyond the midpoint, and the eye usually is directed medially.
Cranial Nerve VII: **Facial** (FĀ-shal; = face) 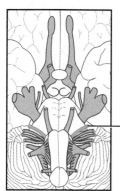 Facial (VII) nerve	**Mixed** *Sensory portion:* Arises from taste buds on the anterior two-thirds of the tongue, passes through the stylomastoid foramen, and ends in the geniculate ganglion, a nucleus in pons. From there, fibers extend to the thalamus and then to the gustatory areas of cerebral cortex. Also contains fibers from proprioceptors in muscles of the face and scalp. *Motor portion:* Originates in the pons and passes through the stylomastoid foramen. Somatic fibers innervate facial, scalp, and neck muscles. Parasympathetic fibers innervate lacrimal, sublingual, submandibular, nasal, and palatine glands.	*Sensory function:* Muscle sense (proprioception) and taste. *Motor function:* Facial expression and secretion of saliva and tears. *Clinical application:* Injury produces *Bell's palsy* (paralysis of the facial muscles), loss of taste, decreased salivation, and loss of ability to close the eyes, even during sleep.
Cranial Nerve VIII: **Vestibulocochlear** (vest-tib-ū-lō-KOK-lē-ar; *vestibulo-* = small cavity; *-cochlear* = a spiral, snail-like)  Vestibulocochlear (VIII) nerve	**Mixed (mainly sensory)** *Vestibular branch, sensory portion:* Arises in the semicircular canals, saccule, and utricle and forms the vestibular ganglion. Fibers end in the pons and cerebellum. *Vestibular branch, motor portion:* Fibers innervate hair cells of the semicircular canals, saccule, and utricle. *Cochlear branch, sensory portion:* Arises in the spiral organ (organ of Corti), forms the spiral ganglion, passes through nuclei in the medulla, and ends in the thalamus. Fibers synapse with neurons that relay impulses to the primary auditory area (areas 41 and 42) of the cerebral cortex. *Cochlear branch, motor portion:* Originates in the pons and terminates on hair cells of the spiral organ.	*Vestibular branch sensory function:* Conveys impulses associated with equilibrium. *Vestibular branch motor function:* Adjusts sensitivity of hair cells. *Cochlear branch sensory function:* Conveys impulses for hearing. *Cochlear branch motor function:* May modify function of hair cells by altering their transmission and mechanical response to sound. *Clinical application:* Injury to the vestibular branch may cause *vertigo* (a subjective feeling of rotation), *ataxia* (muscular incoordination), and *nystagmus* (involuntary rapid movement of the eyeball). Injury to the cochlear branch may cause *tinnitus* (ringing in the ears) or deafness.

Table 14.2 Summary of Cranial Nerves (continued)

NUMBER AND NAME	TYPE AND LOCATION	FUNCTION AND CLINICAL APPLICATION
Cranial Nerve IX: **Glossopharyngeal** (glos'-ō-fah-RIN-jē-al; *glosso-* = tongue; *-pharyngeal* = throat)  Glossopharyngeal (IX) nerve	**Mixed** *Sensory portion:* Consists of fibers from taste buds and somatic sensory receptors on posterior one-third of the tongue, from proprioceptors in swallowing muscles supplied by the motor portion, and from stretch receptors in carotid sinus and chemoreceptors in carotid body near the carotid arteries. Fibers pass through the jugular foramen and end in the medulla. *Motor portion:* Originates in the medulla and passes through the jugular foramen. Somatic fibers innervate the stylopharyngeus muscle. Parasympathetic fibers innervate the parotid (salivary) gland.	*Sensory function:* Taste and somatic sensations (touch, pain, temperature) from posterior third of tongue; muscle sense (proprioception) in swallowing muscles; monitoring of blood pressure; monitoring of O_2 and CO_2 in blood for regulation of breathing rate and depth. *Motor function:* Elevates the pharynx during swallowing and speech; stimulates secretion of saliva. *Clinical application:* Injury causes difficulty in swallowing, reduced secretion of saliva, loss of sensation in the throat, and loss of taste sensation.
Cranial Nerve X: **Vagus** (VĀ-gus; *vagus* = vagrant or wandering) 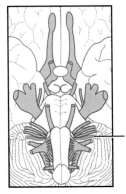 Vagus (X) nerve	**Mixed** *Sensory portion:* Consists of fibers from small number of taste buds in the epiglottis and pharynx, proprioceptors in muscles of the neck and throat, stretch receptors and chemoreceptors in carotid sinus and carotid body near the carotid arteries, chemoreceptors in aortic body near arch of the aorta, and visceral sensory receptors in most organs of the thoracic and abdominal cavities. Fibers pass through the jugular foramen and end in the medulla and pons. *Motor portion:* Originates in medulla and passes through the jugular foramen. Somatic fibers innervate skeletal muscles in the throat and neck. Parasympathetic fibers innervate smooth muscle in the airways, esophagus, stomach, small intestine, most of large intestine, and gallbladder; cardiac muscle in the heart; and glands of the gastrointestinal (GI) tract.	*Sensory function:* Taste and somatic sensations (touch, pain, temperature) from epiglottis and pharynx; monitoring of blood pressure; monitoring of O_2 and CO_2 in blood for regulation of breathing rate and depth; sensations from visceral organs in thorax and abdomen. *Motor function:* Swallowing, coughing, and voice production; smooth muscle contraction and relaxation in organs of the GI tract; slowing of the heart rate; secretion of digestive fluids. *Clinical application:* Injury interrupts sensations from many organs in the thoracic and abdominal cavities, interferes with swallowing, paralyzes vocal cords, and causes heart rate to increase.
Cranial Nerve XI: **Accessory** (ak-SES-ō-rē; = assisting)  Accessory (XI) nerve	**Mixed (mainly motor)** *Sensory portion:* Consists of fibers from proprioceptors in muscles of the pharynx, larynx, and soft palate that pass through the jugular foramen. *Motor portion:* Consists of a cranial portion and a spinal portion. *Cranial portion* originates in the medulla, passes through the jugular foramen, and supplies muscles of the pharynx, larynx, and soft palate. *Spinal portion* originates in the anterior gray horn of the first five cervical segments of the spinal cord, passes through the jugular foramen, and supplies the sternocleidomastoid and trapezius muscles.	*Sensory function:* Muscle sense (proprioception). *Motor function:* Cranial portion mediates swallowing movements; spinal portion mediates movement of head and shoulders. *Clinical application:* If nerves are damaged, the sternocleidomastoid and trapezius muscles become paralyzed, with resulting inability to raise the shoulders and difficulty in turning the head.

Table 14.2 (continued)

NUMBER AND NAME	TYPE AND LOCATION	FUNCTION AND CLINICAL APPLICATION
Cranial Nerve XII: **Hypoglossal** (hī′-pō-GLOS-al; *hypo-* = below; *-glossal* = tongue) 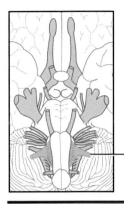 Hypoglossal (XII) nerve	**Mixed (mainly motor)** *Sensory portion:* Consists of fibers from proprioceptors in tongue muscles that pass through the hypoglossal canal and end in the medulla. *Motor portion:* Originates in the medulla, passes through the hypoglossal canal, and supplies muscles of the tongue.	*Sensory function:* Muscle sense (proprioception). *Motor function:* Movement of tongue during speech and swallowing. *Clinical application:* Injury results in difficulty in chewing, speaking, and swallowing. The tongue, when protruded, curls toward the affected side, and the affected side atrophies.

DISORDERS: HOMEOSTATIC IMBALANCES

CEREBROVASCULAR ACCIDENT

The most common brain disorder is a **cerebrovascular accident (CVA),** also called a **stroke** or **brain attack.** CVAs affect 500,000 people a year in the United States and represent the third leading cause of death, behind heart attacks and cancer. A CVA is characterized by abrupt onset of persisting neurological symptoms, such as paralysis or loss of sensation, that arise from destruction of brain tissue. Common causes of CVAs are intracerebral hemorrhage (from a blood vessel in the pia mater or brain), emboli (blood clots), and atherosclerosis (formation of cholesterol-containing plaques that block blood flow) of the cerebral arteries.

Among the risk factors implicated in CVAs are high blood pressure, high blood cholesterol, heart disease, narrowed carotid arteries, transient ischemic attacks (TIAs; discussed next), diabetes, smoking, obesity, and excessive alcohol intake.

A clot-dissolving drug called tissue plasminogen activator (t-PA) is now being used to open up blocked blood vessels in the brain. The drug is most effective when administered within three hours of the onset of the CVA, however, and is helpful only for CVAs due to a blood clot. Use of t-PA can decrease the permanent disability associated with CVAs by 50%.

TRANSIENT ISCHEMIC ATTACK

A **transient ischemic attack (TIA)** is an episode of temporary cerebral dysfunction caused by impaired blood flow to the brain. Symptoms include dizziness, weakness, numbness, or paralysis in a limb or in one side of the body; drooping of one side of the face; headache; slurred speech or difficulty understanding speech; and a partial loss of vision or double vision. Sometimes nausea or vomiting also occurs. The onset of symptoms is sudden and reaches maximum intensity almost immediately. A TIA usually persists for 5–10 minutes and only rarely lasts as long as 24 hours; it leaves no

persistent neurological deficits. The causes of the impaired blood flow that lead to TIAs are blood clots, atherosclerosis, and certain blood disorders.

It is estimated that about one-third of patients who experience a TIA will have a CVA within 5 years. Therapy for TIAs includes drugs such as aspirin that block the aggregation of blood components involved in clotting (platelets) and anticoagulants; cerebral artery bypass grafting; and carotid endarterectomy (removal of the cholesterol-containing plaques and inner lining of an artery).

ALZHEIMER DISEASE

Alzheimer disease (ALTZ-hī-mer) or **AD** is a disabling senile dementia—the loss of reason and ability to care for oneself—that afflicts about 11% of the population over age 65. In the United States, AD afflicts 4 million people and claims over 100,000 lives a year, making it the fourth leading cause of death among the elderly, after heart disease, cancer, and stroke.

Individuals with AD initially have trouble remembering recent events. They then become confused and forgetful, often repeating questions or getting lost while traveling to previously familiar places. Disorientation grows and memories of past events disappear, and episodes of paranoia, hallucination, or violent changes in mood may occur. As their minds continue to deteriorate, they lose their ability to read, write, talk, eat, or walk. The disease ultimately culminates in dementia. A person with AD usually dies of some complication that afflicts bedridden patients, such as pneumonia.

At autopsy, brains of AD victims show three distinct structural abnormalities:
1. *Loss of neurons that liberate acetylcholine.* A major center of neurons that liberate ACh is the nucleus basalis, which is inferior to the globus pallidus. Axons of these neurons project widely throughout the cerebral cortex and limbic system. Their destruction is a hallmark of Alzheimer disease.

2. *Beta-amyloid plaques,* clusters of abnormal proteins deposited outside neurons.
3. *Neurofibrillary tangles,* abnormal bundles of protein filaments inside neurons in affected brain regions.

A variety of risk factors for AD have been identified, although the causes of AD are still not clear. One risk factor for developing AD is a history of head injury. A similar dementia occurs in boxers, probably caused by repeated blows to the head. Hereditary factors also increase one's AD risk. Three forms (alleles) of a gene on chromosome 19 code for a molecule called apolipoprotein E (apoE) that helps transport cholesterol in the blood. People who have one or two copies of the form that codes for **apolipoprotein E4 (apoE4)** have a much higher risk of developing AD (and an earlier age of onset) when compared with people who have genes for the other forms, apoE2 or apoE3. Although the way in which apoE exerts this influence is not known, some scientists suggest that apoE2 and apoE3 may have a protective effect that apoE4 lacks. In a small number of patients, AD is associated with mutations on chromosome 14 or 21. Genetic flaws do not explain all cases, however. Even in identical twins, one may have AD while the other does not. One study of the blood of a small number of Alzheimer patients revealed defects in certain types of plasma membrane K^+ channels in their platelets. A similar abnormality of neuronal ion channels could cause brain dysfunction. Drugs that inhibit acetylcholinesterase (AChE), the enzyme that inactivates ACh, improve alertness and behavior in about 5% of AD patients. Although more studies are needed, some evidence suggests that vitamin E (an antioxidant), estrogen, ibuprofen, and ginko biloba extract may have slight beneficial effects.

MEDICAL TERMINOLOGY

Agnosia (ag-NŌ-zē-a; *a-* =without; *-gnosia* = knowledge) Inability to recognize the significance of sensory stimuli such as sounds, sights, smells, tastes, and touch.

Apraxia (a-PRAK-sē-a; *-praxia* = coordinated) Inability to carry out purposeful movements in the absence of paralysis.

Delirium (de-LIR-ē-um; = off the track) Also called **acute confusional state (ACS).** A transient disorder of abnormal cognition and disordered attention accompanied by disturbances of the sleep-wake cycle and psychomotor behavior (hyperactivity or hypoactivity of movements and speech).

Dementia (de-MEN-shē-a; *de-* = away from; *-mentia* = mind) A mental disorder that results in permanent or progressive general loss of intellectual abilities, including impairment of memory, judgment, and abstract thinking and changes in personality.

Encephalitis (en'-sef-a-LĪ-tis) An acute inflammation of the brain caused by either a direct attack by any of several viruses or an allergic reaction to any of the many viruses that are normally harmless to the central nervous system. If the virus affects the spinal cord as well, the condition is called **encephalomyelitis.**

Lethargy (LETH-ar-jē) A condition of functional sluggishness.

Nerve block Loss of sensation in a region due to injection of a local anesthetic; an example is local dental anesthesia.

Neuralgia (noo-RAL-jē-a; *neur-* = nerve; *-algia* = pain) Attacks of pain along the entire course or a branch of a peripheral sensory nerve.

Stupor (STOO-por) Unresponsiveness from which a patient can be aroused only briefly and only by vigorous and repeated stimulation.

STUDY OUTLINE

OVERVIEW OF BRAIN ORGANIZATION AND BLOOD SUPPLY (p. 446)

1. The principal parts of the brain are the brain stem, cerebellum, diencephalon, and cerebrum.
2. The brain is protected by cranial bones and the cranial meninges.
3. The cranial meninges are continuous with the spinal meninges. From superficial to deep they are the dura mater, arachnoid, and pia mater.
4. Blood flow to the brain is mainly via blood vessels that branch from the cerebral arterial circle (circle of Willis).
5. Any interruption of the oxygen or glucose supply to the brain can result in weakening of, permanent damage to, or death of brain cells.
6. The blood–brain barrier (BBB) causes different substances to move between the blood and the brain tissue at different rates.

CEREBROSPINAL FLUID PRODUCTION AND CIRCULATION IN VENTRICLES (p. 449)

1. Cerebrospinal fluid (CSF) is formed in the choroid plexuses and circulates through the lateral ventricles, third ventricle, fourth ventricle, subarachnoid space, and central canal. Most of the fluid is absorbed into the blood across the arachnoid villi of the superior sagittal blood sinus.
2. Cerebrospinal fluid provides mechanical protection, chemical protection, and circulation of nutrients.

THE BRAIN STEM (p. 452)

1. The medulla oblongata is continuous with the superior part of the spinal cord and contains both motor and sensory tracts. It contains nuclei that are reflex centers for regulation of heart rate, respiratory rate, vasoconstriction, swallowing, coughing, vomiting, and sneezing. It also contains nuclei associated with cranial nerves VIII (cochlear and vestibular branches) through XII.

2. The pons is superior to the medulla. It connects the spinal cord with the brain and links parts of the brain with one another by way of tracts. It relays nerve impulses related to voluntary skeletal movements from the cerebral cortex to the cerebellum. The pons contains the pneumotaxic and apneustic centers, which help control breathing. It contains nuclei associated with cranial nerves V–VII and the vestibular branch of cranial nerve VIII.

3. The midbrain connects the pons and diencephalon and surrounds the cerebral aqueduct. It conveys motor impulses from the cerebrum to the cerebellum and spinal cord, sends sensory impulses from the spinal cord to the thalamus, and regulates auditory and visual reflexes. It also contains nuclei associated with cranial nerves III and IV.

4. A large part of the brain stem consists of small areas of gray matter and white matter called the reticular formation. It helps maintain consciousness, causes awakening from sleep, and contributes to regulating muscle tone.

THE CEREBELLUM (p. 457)

1. The cerebellum occupies the inferior and posterior aspects of the cranial cavity. It consists of two lateral hemispheres and a medial, constricted vermis.

2. It connects to the brain stem by three pairs of cerebellar peduncles.

3. The cerebellum functions to coordinate skeletal muscles and to maintain normal muscle tone, posture, and balance.

THE DIENCEPHALON (p. 460)

1. The diencephalon surrounds the third ventricle and consists of the thalamus, hypothalamus, epithalamus, and subthalamus.

2. The thalamus is superior to the midbrain and contains nuclei that serve as relay stations for all sensory impulses to the cerebral cortex. It also allows crude appreciation of pain, temperature, and pressure.

3. The hypothalamus is inferior to the thalamus. It controls and integrates the autonomic nervous system, connects the nervous and endocrine systems, functions in rage and aggression, controls body temperature, regulates food and fluid intake, and establishes a diurnal sleep pattern.

4. The epithalamus consists of the pineal gland and the habenular nuclei. The pineal gland secretes melatonin, which is thought to promote sleep and to help set the body's biological clock.

5. The subthalamus connects to motor areas of the cerebrum.

6. Circumventricular organs (CVOs) can monitor chemical changes in the blood because they lack the blood–brain barrier.

THE CEREBRUM (p. 463)

1. The cerebrum is the largest part of the brain. Its cortex contains gyri (convolutions), fissures, and sulci.

2. The cerebral lobes are named the frontal, parietal, temporal, and occipital.

3. The white matter is deep to the cortex and consists of myelinated and unmyelinated axons extending to other regions as association, commissural, and projection fibers.

4. The basal ganglia are several groups of nuclei in each cerebral hemisphere. They help control large, automatic movements of skeletal muscles and help regulate muscle tone.

5. The limbic system encircles the upper part of the brain stem and the corpus callosum. It functions in emotional aspects of behavior and memory.

FUNCTIONAL ASPECTS OF THE CEREBRAL CORTEX (p. 468)

1. The sensory areas of the cerebral cortex are concerned with the interpretation of sensory impulses. The motor areas are the regions that govern muscular movement. The association areas are concerned with more complex integrative functions.

2. The primary somatosensory area extends from the longitudinal fissure on the superior aspect of the cerebrum to the lateral cerebral sulcus. It receives nerve impulses from somatic sensory receptors for touch, proprioception, pain, and temperature. Each point within the area receives sensations from a specific part of the body.

3. Areas for the special senses include: the primary visual area (area 17), which receives impulses that convey visual information; the primary auditory area (areas 41 and 42), which interprets the basic characteristics of sound such as pitch and rhythm; the primary gustatory area (area 43), which receives impulses for taste; and the primary olfactory area (area 28), which receives impulses for smell.

4. Motor areas include the primary motor area (area 4), which controls voluntary contractions of specific muscles or groups of muscles, and Broca's speech area (areas 44 and 45), which controls production of speech.

5. Association areas of the cerebrum are connected with one another via tracts.

6. The somatosensory association area (areas 5 and 7) permits you to determine the exact shape and texture of an object without looking at it, to determine the orientation of one object to another as they are felt, and to sense the relationship of one body part to another.

7. The visual association area (areas 18 and 19) relates present to past visual experiences and is essential for recognizing and evaluating what is seen.

8. The auditory association area (area 22) determines if a sound is speech, music, or noise.

9. Wernicke's area (area 22 and possibly 39 and 40) interprets the meaning of speech by translating words into thoughts.

10. The common integrative area (areas 5, 7, 39, and 40) integrates sensory interpretations from the association areas and impulses from other areas, allowing one thought to be formed on the basis of a variety of sensory inputs.

11. The premotor area (area 6) generates nerve impulses that cause specific groups of muscles to contract in specific sequences.

12. The frontal eye field area (area 8) controls voluntary scanning movements of the eyes.

13. Table 14.1 on page 458 summarizes the functions of various parts of the brain.

14. Although the brain is fairly symmetrical on its right and left sides, subtle anatomical differences exist between the two hemispheres, and each has unique functions.

15. The left hemisphere receives sensory signals from and controls the right side of the body; it also is more important for language, numerical and scientific skills, and reasoning.

16. The right hemisphere receives sensory signals from and controls the left side of the body; it also is more important for musical and artistic awareness, spatial and pattern perception, recognition of faces, emotional content of language, and generating mental images of sight, sound, touch, taste, and smell.

17. Brain waves generated by the cerebral cortex are recorded from the surface of the head in an electroencephalogram (EEG).
18. The EEG may be used to diagnose epilepsy, infections, and tumors.

CRANIAL NERVES (p. 472)

1. Twelve pairs of cranial nerves originate from the brain.
2. They are named primarily on the basis of distribution and are numbered I–XII in order of attachment to the brain. Table 14.2 on pages 475–479 summarizes the cranial nerves.

DEVELOPMENTAL ANATOMY OF THE NERVOUS SYSTEM (p. 472)

1. The development of the nervous system begins with a thickening of a region of the ectoderm called the neural plate.

2. During embryological development, primary brain vesicles are formed and serve as forerunners of various parts of the brain.
3. The telencephalon forms the cerebrum, the diencephalon develops into the thalamus and hypothalamus, the mesencephalon develops into the midbrain, the metencephalon develops into the pons and cerebellum, and the myelencephalon forms the medulla.

AGING AND THE NERVOUS SYSTEM (p. 473)

1. The brain grows rapidly during the first few years of life.
2. Age-related effects involve loss of brain mass and decreased capacity for sending nerve impulses.

SELF-QUIZ QUESTIONS

Complete the following:

1. The four principal parts of the brain are the ___, ___, ___, and ___.
2. The ___ functions as a selective anatomical and physiological barrier to protect brain cells from harmful substances and pathogens.
3. The sites of CSF production are the ___.
4. The brain meninx that separates the two hemispheres of the cerebrum is the ___.
5. Another term for the midbrain is the ___.

True or false:

6. The hypothalamus controls many body activities and is one of the major regulators of homeostasis.
7. Delta waves in an EEG produced by an awake adult indicate periods of sensory input and mental activity.

Choose the best answer to the following questions:

8. Which of the following statements is *false*? (a) The blood supply to the brain is provided mainly by blood vessels that form the cerebral circle at the base of the brain. (b) Brain lysosomal membranes are sensitive to decreased oxygen concentration. (c) An interruption of blood flow to the brain for even 30 seconds may impair brain function. (d) Glucose supply to the brain must be continuous. (e) Even a brief interruption of blood flow to the brain may cause unconsciousness.
9. In which of the following ways does cerebrospinal fluid contribute to homeostasis? (1) mechanical protection, (2) chemical protection, (3) electrical protection, (4) transport of chemicals, (5) osmolarity.
   (a) 1, 2, and 3, (b) 2, 3, and 4, (c) 3, 4, and 5, (d) 1, 2, and 4, (e) 2, 4, and 5

10. Which of the following are functions of the hypothalamus? (1) control of the ANS, (2) control of the pituitary gland, (3) regulation of emotional and behavioral patterns, (4) regulation of eating and drinking, (5) control of body temperature, (6) regulation of circadian rhythms and states of consciousness.
    (a) 1, 2, 4, and 6, (b) 2, 3, 5, and 6, (c) 1, 3, 5, and 6, (d) 1, 4, 5, and 6, (e) 1, 2, 3, 4, 5, and 6
11. Which of the following statements is *false*? (a) Association fibers transmit nerve impulses between gyri in the same hemisphere. (b) Commissural fibers transmit impulses from the gyri in one cerebral hemisphere to the corresponding gyri in the other hemisphere. (c) Projection fibers form descending and ascending tracts that transmit impulses from the cerebrum and other parts of the brain to the spinal cord, or from the spinal cord to the brain. (d) The internal capsule is an example of commissural fibers. (e) The corpus callosum is an example of commissural fibers.
12. Which of the following statements is true? (a) The right and left hemispheres of the cerebrum are completely symmetrical. (b) The left hemisphere controls the left side of the body. (c) The right hemisphere is more important for spoken and written language. (d) The left hemisphere is more important for musical and artistic awareness. (e) Hemispheric lateralization is more pronounced in males than in females.
13. Which of the following is not part of the diencephalon? (a) pons, (b) thalamus, (c) mammillary bodies, (d) hypothalamus, (e) pineal gland

14. Match the following:

____ (a) oculomotor
____ (b) trigeminal
____ (c) abducens
____ (d) vestibulocochlear
____ (e) accessory
____ (f) vagus
____ (g) facial
____ (h) glossopharyngeal
____ (i) olfactory
____ (j) trochlear
____ (k) optic
____ (l) hypoglossal
____ (m) functions in sense of smell
____ (n) functions in hearing and equilibrium
____ (o) functions in chewing
____ (p) functions in facial expression and secretion of saliva and tears
____ (q) functions in movement of tongue during speech and swallowing
____ (r) functions in secretion of digestive fluids
____ (s) functions in secretion of saliva, taste, regulation of blood pressure, and muscle sense

(1) cranial nerve I
(2) cranial nerve II
(3) cranial nerve III
(4) cranial nerve IV
(5) cranial nerve V
(6) cranial nerve VI
(7) cranial nerve VII
(8) cranial nerve VIII
(9) cranial nerve IX
(10) cranial nerve X
(11) cranial nerve XI
(12) cranial nerve XII

15. Match the following:

____ (a) emotional brain
____ (b) bridge connecting the spinal cord with the brain and parts of the brain with each other
____ (c) sensory relay area
____ (d) alerts the cerebral cortex to incoming sensory signals and helps regulate muscle tone
____ (e) the motor command center
____ (f) lacks a blood–brain barrier; can monitor chemical changes in the blood
____ (g) site of decussation of pyramids
____ (h) site of pneumotaxic and apneustic areas
____ (i) secretes melatonin
____ (j) contains sensory, motor, and association areas
____ (k) responsible for maintaining consciousness and awakening from sleep
____ (l) controls ANS
____ (m) contains reflex centers for movements of the eyes, head, and neck in response to visual and other stimuli, and reflex center for movements of the head and trunk in response to auditory stimuli
____ (n) plays an essential role in awareness and in the acquisition of knowledge; cognition
____ (o) several groups of nuclei that control large autonomic movements of skeletal muscles and help regulate muscle tone required for specific body movements
____ (p) produces hormones that regulate endocrine gland function
____ (q) contains the vital cardiovascular center and medullary rhythmicity center
____ (r) a thick band of sensory and motor tracts that connect the cerebral cortex with the brain stem and spinal cord

(1) medulla oblongata
(2) pons
(3) midbrain
(4) cerebellum
(5) pineal gland
(6) thalamus
(7) hypothalamus
(8) cerebrum
(9) limbic system
(10) reticular formation
(11) circumventricular organs
(12) reticular activating system
(13) basal ganglia
(14) internal capsule

CRITICAL THINKING QUESTIONS

1. An elderly relative suffered a CVA (stroke) and now has difficulty moving her right arm, and she also has speech problems. What areas of the brain were damaged by the stroke? (HINT: *What results from decussation of the pyramids of the medulla?*)

2. Casey complained to her swim coach that "my bathing cap is so tight, it'll squeeze my brains out my ears!" Her coach is majoring in physical therapy and tells her that that's anatomically impossible. Explain the coach's position. (HINT: *Put your hands on your head and squeeze. What do you feel?*)

3. Ando's first trip to the dentist after a ten-year absence resulted in extensive dental work, so he received numbing injections of anesthetic in several locations during the session. While having lunch right after the appointment, soup dribbles out of his mouth because he has no feeling in his left upper lip, his right lower lip, and the tip of his tongue. What happened to Ando? (HINT: *The dentist wanted to block the sensation of pain from the teeth.*)

ANSWERS TO FIGURE QUESTIONS

14.1 The largest part of the brain is the cerebrum.

14.2 From superficial to deep, the three cranial meninges are the dura mater, arachnoid, and pia mater.

14.3 The brain stem is anterior to the fourth ventricle, and the cerebellum is posterior to it.

14.4 Cerebrospinal fluid is reabsorbed by the arachnoid villi that project into the dural venous sinuses.

14.5 The medulla oblongata contains the pyramids; the midbrain contains the cerebral peduncles; the pons means "bridge."

14.6 Decussation means crossing to the opposite side. Because the pyramids contain motor tracts that extend from the cortex into the spinal cord and convey impulses for contraction of skeletal muscle, the functional consequence of decussation of the pyramids is that one side of the cerebrum controls muscles on the opposite side of the body.

14.7 The cerebral peduncles are the main connections for tracts running between the superior parts of the brain and the inferior parts of the brain and the spinal cord.

14.8 The cerebellar peduncles carry information into and out of the cerebellum.

14.9 The intermediate mass connects the right and left sides of the thalamus.

14.10 From posterior to anterior, the four major regions of the hypothalamus are the mammillary, tuberal, supraoptic, and preoptic regions.

14.11 The gray matter enlarges more rapidly, in the process producing convolutions or gyri (folds), sulci (shallow grooves), and fissures (deep grooves).

14.12 Association fibers connect gyri of the same hemisphere; commissural fibers connect gyri in opposite hemispheres; projection fibers connect the brain and the spinal cord.

14.13 The basal ganglia are lateral, superior, and inferior to the thalamus.

14.14 The hippocampus functions in memory.

14.15 Common integrative area; motor speech area; premotor area; gustatory areas; auditory areas; visual areas; frontal eye field area.

14.16 Broca's and Wernicke's areas are important language centers.

14.17 In an EEG, theta waves indicate emotional stress.

14.18 Gray matter derives from the mantle layer of the neural tube.

14.19 The mesencephalon does not develop into a secondary brain vesicle.

The previous three chapters described the organization of the nervous system; in this chapter we explore the levels and components of sensation. We also examine the pathways that convey somatic sensory nerve impulses from the body to the brain, and the pathways that convey impulses from the brain to skeletal muscles to produce movements. As sensory impulses reach the CNS, they become part of a large pool of sensory input. However, not every bit of input to the CNS elicits a response. Rather, each piece of incoming information is integrated with other arriving and previously stored information. The integration process does not occur just once at a single location, but at many stations along pathways in the CNS, at both conscious and subconscious levels. Integration occurs within the spinal cord, brain stem, cerebellum, basal ganglia, and cerebral cortex, and a motor response to make a muscle contract can be modified at several of these levels. To conclude this chapter we introduce two of the brain's integrative functions: (1) wakefulness and sleep and (2) learning and memory.

SENSATION

OBJECTIVES

• *Define a sensation, and discuss the components of sensation.*

• *Describe the different ways to classify sensory receptors.*

In its broadest definition, **sensation** is the conscious or subconscious awareness of external or internal stimuli. The nature of the sensation and the type of reaction generated vary according to the ultimate CNS destination of nerve impulses that convey sensory information. In the spinal cord, sensory impulses are the afferent portion of spinal reflexes. Sensory impulses that reach the lower brain stem elicit more complex reflexes, such as changes in heart rate or breathing rate. Sensory impulses reaching the thalamus can provide only crude awareness of the location in the body and the *type* of sensation, such as touch, pain, hearing, or taste. When sensory impulses reach the cerebral cortex, we can precisely locate and identify specific sensations. **Perception** is the conscious awareness and the interpretation of meaning of sensations. In addition, memories of previous sensations are stored in the cortex. We have no perception of some sensory impulses because they never reach the thalamus and cerebral cortex. For example, even though one is usually unaware of blood pressure, it is constantly monitored, and the nerve impulses propagate to the cardiovascular center in the medulla oblongata.

Sensory Modalities

Each type of sensation—for example, touch, pain, vision, or hearing—is called a **sensory modality.** Put another way, the distinct quality that makes one sensation different from others is its modality. A given sensory neuron carries information for one modality only. Neurons relaying impulses for touch to the somatosensory area of the cerebral cortex, for example, do not also transmit impulses for pain. Likewise, nerve impulses from the eyes are perceived in the occipital lobe as sight, whereas those from the ears are perceived in the temporal lobes as sounds.

The different sensory modalities can be grouped into two classes: general senses and special senses.

1. The **general senses** include both **somatic senses** (*somat-* = of the body) and **visceral senses.** Somatic modalities include tactile sensations (touch, pressure, and vibration), thermal sensations (warm and cold), pain sensations, and proprioceptive sensations, which allow perception of both the static positions of limbs and body parts (joint and muscle position sense) and movements of the limbs and head. Visceral sensations provide information about conditions within internal organs.
2. The **special senses** include the modalities of smell, taste, vision, hearing, and equilibrium or balance.

In this chapter we discuss the somatic senses and visceral pain. The special senses are the focus of Chapter 16. Visceral senses are discussed further in Chapter 17, and in association with individual organs in later chapters.

The Process of Sensation

The process of sensation begins in a **sensory receptor,** which is either a specialized cell or the dendrites of a sensory neuron that monitors a particular condition in the internal or external environment. Each of the several different types of sensory receptors is sensitive to stimuli for one sensory modality only. Thus, a given sensory receptor responds vigorously to one particular kind of **stimulus,** a change in the environment that can activate certain sensory receptors, but responds only weakly or not at all to other kinds of stimuli. This characteristic of sensory receptors is termed *selectivity.* The stimulus may be in one of three forms of energy: electromagnetic energy, such as light or heat; mechanical energy, such as sound waves or pressure changes; or chemical energy, such as in a molecule of carbon dioxide dissolved in body fluids. Such selectivity is why auditory receptors in the ears, for example, respond to sound waves but not to light.

For a sensation to arise, the following four events typically occur:

1. *Stimulation of the sensory receptor.* The stimulus must occur within the sensory receptor's *receptive field,* the portion of the receptor that is capable of responding to stimulation.
2. *Transduction of the stimulus.* A sensory receptor *transduces* or converts a stimulus into a graded potential. Recall that graded potentials vary in amplitude (size), depending on the strength of the stimulus that evokes them, and are not propagated. (See pages 390–391 to review the differences between action potentials and graded potentials.) Each type of sensory receptor can transduce only one kind of stimulus. For example, odorant molecules in the air stimulate olfactory (smell) receptors in the nose, which transduce the molecules' chemical energy into electrical energy in the form of a graded potential.
3. *Generation of impulses.* When a graded potential in a sensory neuron reaches threshold, it triggers one or more nerve impulses, which then propagate toward the CNS. Sensory neurons that propagate impulses from the PNS into the CNS are called **first-order neurons.**
4. *Integration of sensory input.* A particular region of the CNS receives and integrates the sensory nerve impulses. Conscious sensations or perceptions are integrated in the cerebral cortex. You seem to see with your eyes, hear with your ears, and feel pain in an injured part of your body because sensory impulses from each part of the body arrive in a specific region of the cerebral cortex, which interprets the sensation as coming from the stimulated sensory receptors.

The Nature of Sensory Receptors

Types of Sensory Receptors

Several structural and functional characteristics of sensory receptors can be used to group them into different classes. On a microscopic level, sensory receptors may be (1) free nerve endings, (2) encapsulated nerve endings at the dendrites of first-order sensory neurons, or (3) separate cells that synapse with first-order sensory neurons (Figure 15.1). **Free nerve endings** are bare dendrites that often have no visible structural specializations (Figure 15.1a). Receptors for pain, thermal, tickle, itch, and some touch sensations are free nerve endings. Receptors for other somatic and visceral sensations, such as touch, pressure, and vibration, are **encapsulated nerve endings.** Their dendrites are enclosed in a connective tissue capsule that has a distinctive microscopic structure—for example, lamellated (Pacinian) corpuscles (Figure 15.1b). Sensory receptors for the special senses of vision, hearing, equilibrium, and taste consist of specialized, **separate cells** that synapse with first-order sensory neurons (Figure 15.1c).

Sensory receptors produce different kinds of graded potentials in response to a stimulus. When stimulated, the dendrites of free nerve endings, encapsulated nerve endings, and the receptive part of olfactory receptors produce **generator potentials** (see Figure 15.1a, b). When a generator potential is large enough to reach threshold, it *generates* one or more nerve impulses in its first-order sensory neuron. The resulting nerve impulse propagates along the axon into the CNS. In contrast, the specialized cells that act as receptors for the special senses of vision, hearing, equilibrium, and taste produce **receptor potentials** in response to stimuli (see Figure 15.1c). A receptor potential causes exocytosis of synaptic vesicles. The neurotransmitter molecules liberated from synaptic vesicles diffuse across the synaptic cleft and produce a postsynaptic potential (PSP) in the first-order neuron. In turn, the PSP may trigger one or more nerve impulses, which propagate along the axon into the CNS. The amplitude of both generator potentials and receptor potentials varies with the intensity of the stimulus: An intense stimulus produces a large generator potential or receptor potential, whereas a

Figure 15.1 Types of sensory receptors and their relationship to first-order sensory neurons. (a) Free nerve endings—in this case, a cold-sensitive receptor. These endings are bare dendrites (no apparent structural specialization) of first-order neurons. (b) An encapsulated nerve ending—a pressure-sensitive receptor. Encapsulated nerve endings are dendrites of first-order neurons. (c) A separate receptor cell—here, a gustatory (taste) receptor—and its synapse with a first-order neuron.

Free nerve endings and encapsulated nerve endings produce generator potentials that elicit nerve impulses in first-order neurons. Separate sensory receptors produce a receptor potential that causes release of neurotransmitter, which then stimulates dendrites of a first-order neuron and elicits nerve impulses.

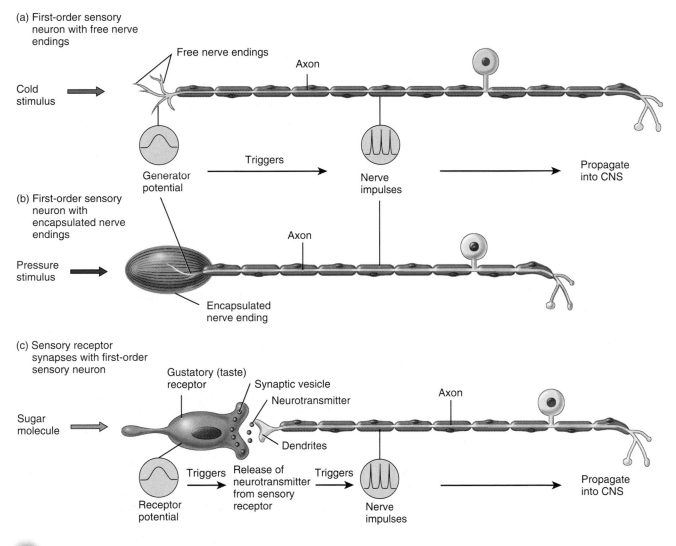

Q Which senses are served by receptors that are separate cells?

weak stimulus elicits a small one. Similarly, large generator potentials or receptor potentials trigger impulses at high frequencies in the first-order neuron, whereas small generator potentials or receptor potentials trigger impulses at lower frequencies.

Another way to group sensory receptors is based on the location of the receptors and the origin of the stimuli that activate them. **Exteroceptors** (EKS-ter-ō-sep′-tors) are located at or near the external surface of the body; they are sensitive to stimuli originating outside the body and provide

Table 15.1 Classification of Sensory Receptors

BASIS OF CLASSIFICATION	DESCRIPTION
Microscopic Features	
Free nerve endings	Bare dendrites associated with pain, thermal, tickle, itch, and some touch sensations.
Encapsulated nerve endings	Dendrites enclosed in a connective tissue capsule, such as a corpuscle of touch.
Separate cells	Receptors synapse with first-order neurons; located in macroscopic structures such as the eye, ear, and tongue.
Receptor Location and Activating Stimuli	
Exteroceptors	Located at or near body surface; sensitive to stimuli originating outside body and provide information about external environment; convey visual, smell, taste, touch, pressure, vibration, thermal, and pain sensations.
Interoceptors	Located in blood vessels, visceral organs, and nervous system; provide information about internal environment; impulses produced usually are not consciously perceived but occasionally may be felt as pain or pressure.
Proprioceptors	Located in muscles, tendons, joints, and inner ear; provide information about body position, muscle length and tension, and position and motion of joints.
Type of Stimulus Detected	
Mechanoreceptors	Detect mechanical pressure; provide sensations of touch, pressure, vibration, proprioception, and hearing and equilibrium; also monitor stretching of blood vessels and internal organs.
Thermoreceptors	Detect changes in temperature.
Nociceptors	Respond to stimuli resulting from physical or chemical damage to tissue.
Photoreceptors	Detect light that strikes the retina of the eye.
Chemoreceptors	Detect chemicals in mouth (taste), nose (smell), and body fluids.

information about the external environment. The sensations of hearing, vision, smell, taste, touch, pressure, vibration, temperature, and pain are conveyed by exteroceptors. **Interoceptors** (IN-ter-ō-sep′-tors) are located in blood vessels, visceral organs, muscles, and the nervous system and monitor conditions in the *internal* environment. The nerve impulses produced by interoceptors usually are not consciously perceived; occasionally, however, activation of interoceptors by strong stimuli may be felt as pain or pressure. **Proprioceptors** (PRŌ-prē-ō-sep′-tors; *proprio-* = one's own) are located in muscles, tendons, joints, and the inner ear; they provide information about body position, muscle tension, and the position and movement of our joints.

A third way to group sensory receptors is according to the type of stimulus they detect.

- **Mechanoreceptors** (MEK-a-nō-rē-sep′-tors) detect mechanical pressure or stretching. Activation of mechanoreceptors gives rise to perceptions of touch, pressure, vibration, proprioception, hearing, and equilibrium.
- **Thermoreceptors** detect changes in temperature.
- **Nociceptors** (NŌ-sē-sep′-tors) respond to stimuli resulting from physical or chemical damage to tissues, giving rise to the sensation of pain.
- **Photoreceptors** detect light that strikes the retina of the eye.
- **Chemoreceptors** detect chemicals in the mouth (taste), nose (smell), and body fluids.

Table 15.1 summarizes the classification of sensory receptors.

Adaptation in Sensory Receptors

A characteristic of most sensory receptors is **adaptation,** in which the generator potential or receptor potential decreases in amplitude during a maintained, constant stimulus. As a result, the frequency of nerve impulses in the first-order neuron also decreases during a prolonged stimulus. Adaptation is caused in part by a decrease in the responsiveness of a receptor. As a result of adaptation, the perception of a sensation may disappear even though the stimulus persists. For example, when you first step into a hot shower, the water may feel very hot, but soon the sensation decreases to one of comfortable warmth even though the stimulus (the high temperature of the water) does not change. Receptors vary in how quickly they adapt. **Rapidly adapting (phasic) receptors** adapt very quickly and are specialized for signaling *changes* in a particular stimulus. Receptors associated with pressure, touch, and smell are rapidly adapting. **Slowly adapting (tonic) receptors,** in contrast, adapt slowly and continue to trigger nerve impulses as long as the stimulus persists. Slowly adapting receptors monitor stimuli associated with pain, body position, and chemical composition of the blood. Even though most adaptation occurs in sensory receptors, further adaptation may also occur as sensory information is processed within the CNS.

1. Distinguish between sensation and perception.
2. What is a sensory modality?
3. What events are needed for a sensation to occur?
4. Distinguish between generator potentials and receptor potentials.
5. What is the difference between rapidly adapting and slowly adapting receptors?

SOMATIC SENSATIONS

OBJECTIVES

• *Describe the location and function of the receptors for tactile, thermal, and pain sensations.*

• *Identify the receptors for proprioception, and describe their functions.*

Somatic sensations arise from stimulation of sensory receptors embedded in the skin or subcutaneous layer; in mucous membranes of the mouth, vagina, and anus; in muscles, tendons, and joints; and in the inner ear. The sensory receptors for somatic sensations are distributed unevenly: Some parts of the body surface are densely populated with receptors, whereas other parts contain only a few. The areas with the highest density of sensory receptors are the tip of the tongue, the lips, and the fingertips. Somatic sensations that result from stimulating the skin surface are called **cutaneous sensations** (kyoo-TĀ-nē-us; *cutane-* = skin).

Somatic sensations are of four modalities: tactile, thermal, pain, and proprioceptive.

Tactile Sensations

The **tactile sensations** (TAK-tĭl; *tact-* = touch) are touch, pressure and vibration, and itch and tickle. Itch and tickle sensations are detected by free nerve endings attached to small-diameter, unmyelinated C fibers. All other tactile sensations are detected by a variety of encapsulated mechanoreceptors attached to large-diameter, myelinated A fibers. Recall that larger-diameter, myelinated fibers propagate nerve impulses more rapidly than do smaller-diameter, unmyelinated fibers (see page 395). Tactile receptors in the skin or subcutaneous layer include corpuscles of touch, hair root plexuses, type I and II cutaneous mechanoreceptors, lamellated corpuscles, and free nerve endings (Figure 15.2).

Touch

Sensations of touch generally result from stimulation of tactile receptors in the skin or subcutaneous layer. **Crude touch** is the ability to perceive that something has contacted the skin, even though its exact location, shape, size, or texture cannot be determined. **Discriminative touch** provides specific information about a touch sensation, such as exactly what point on the body is touched plus the shape, size, and texture of the source of stimulation.

There are two types of rapidly adapting touch receptors. **Corpuscles of touch,** or **Meissner corpuscles** (MĪS-ner), are receptors for discriminative touch that are located in the dermal papillae of hairless skin, especially in the fingertips and palms. Each corpuscle is an egg-shaped mass of dendrites enclosed by a capsule of connective tissue. Because corpuscles of touch are rapidly adapting receptors, they generate nerve impulses mainly at the onset of a touch. They account for 40% of the tactile receptors in the hands and are also abundant in the eyelids, tip of the tongue, lips, nipples, soles, clitoris, and tip of the penis. **Hair root plexuses** are rapidly adapting touch receptors found in hairy skin; they consist of free nerve endings wrapped around hair follicles. Hair root plexuses detect movements on the skin surface that disturb hairs—for example, a flea landing on a hair. Movement of the hair shaft stimulates the free nerve endings.

There are two types of slowly adapting touch receptors. **Type I cutaneous mechanoreceptors,** also known as **tactile** or **Merkel discs,** function in discriminative touch. Merkel discs are saucer-shaped, flattened free nerve endings that contact Merkel cells of the stratum basale (see Figure 5.2 on page 142). These mechanoreceptors are plentiful in the fingertips, lips, and external genitalia; they represent 25% of the tactile receptors in the hands. **Type II cutaneous mechanoreceptors,** or **Ruffini corpuscles,** are elongated, encapsulated receptors located deep in the dermis, and in ligaments and tendons as well. They account for 20% of the sensory receptors in the hands, and they are abundant on the plantar surface of the feet. Although structurally similar to other tactile receptors, the function of Ruffini corpuscles is not well understood. They are most sensitive to stretching that occurs as digits or limbs are moved.

Pressure and Vibration

Pressure is a sustained sensation that is felt over a larger area than is touch. Receptors that contribute to sensations of pressure include corpuscles of touch, type I mechanoreceptors, and lamellated corpuscles. **Lamellated,** or **Pacinian corpuscles** (pa-SIN-ē-an), are large oval structures composed of a connective tissue capsule, layered like an onion, that encloses a dendrite. Like corpuscles of touch, lamellated corpuscles adapt rapidly. They are widely distributed in the body—in the subcutaneous layer (and sometimes the dermis); in submucosal tissues that underlie mucous and serous membranes; around joints, tendons, and muscles; in the periosteum; and in the mammary glands, external genitalia, and certain viscera, such as the pancreas and urinary bladder.

Sensations of **vibration** result from rapidly repetitive sensory signals from tactile receptors. The receptors for vibration sensations are corpuscles of touch and lamellated corpuscles. Whereas corpuscles of touch can detect lower-frequency vibrations, lamellated corpuscles detect higher-frequency vibrations.

Figure 15.2 Structure and location of sensory receptors in the skin and subcutaneous layer.

 The somatic sensations of touch, pressure, vibration, warmth, cold, and pain arise from sensory receptors in the skin, subcutaneous layer, and mucous membranes.

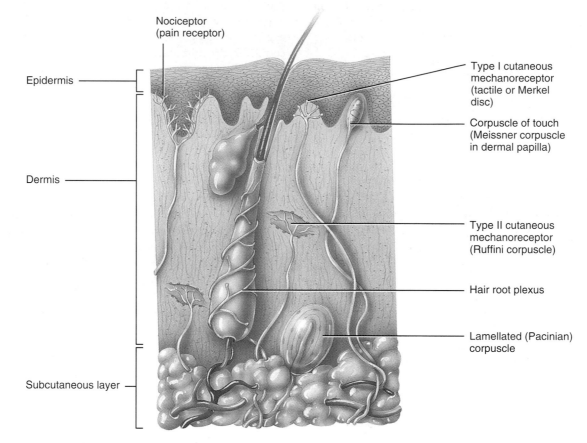

Nociceptor
(pain receptor)

Epidermis

Dermis

Subcutaneous layer

Type I cutaneous
mechanoreceptor
(tactile or Merkel
disc)

Corpuscle of touch
(Meissner corpuscle
in dermal papilla)

Type II cutaneous
mechanoreceptor
(Ruffini corpuscle)

Hair root plexus

Lamellated (Pacinian)
corpuscle

Q   Which sensations can arise when free nerve endings are stimulated?

Itch and Tickle

The **itch** sensation results from stimulation of free nerve endings by certain chemicals, such as bradykinin, often as a result of a local inflammatory response. Free nerve endings and lamellated corpuscles are thought to mediate the **tickle** sensation—the only sensation that you may not be able to elicit on yourself. Why the tickle sensation chiefly arises only when someone else touches you is still a mystery.

Thermal Sensations

Thermoreceptors are free nerve endings that have receptive fields about 1 mm in diameter on the skin surface. Two distinct **thermal sensations** are perceived: cold and warm. **Cold receptors** are located in the stratum basale of the epidermis and are attached to medium-diameter, myelinated A fibers, although a few connect to small-diameter, unmyelinated C fibers. Temperatures between 10° and 40°C

(50–105°F) activate cold receptors. Because skin temperature is about 34°C, the sensation of coolness that occurs when the receptive field of a cold receptor is touched by a small probe having a *higher* temperature is termed *paradoxical cold*. **Warm receptors** are located in the dermis and are attached to small-diameter, unmyelinated C fibers; they are activated by temperatures between 32° and 48°C (90–118°F). Both cold and warm receptors adapt rapidly at the onset of a stimulus but continue to generate impulses at a lower frequency throughout a prolonged stimulus. Temperatures below 10°C and above 48°C stimulate mainly nociceptors, rather than thermoreceptors, producing painful sensations.

Pain Sensations

Pain is indispensable for survival. It serves a protective function by signaling the presence of noxious, tissue-damaging conditions. From a medical standpoint, the subjective description and indication of the location of pain may help pinpoint the underlying cause of disease.

Nociceptors (*noci-* = harmful), the receptors for pain, are free nerve endings found in every tissue of the body except the brain (see Figure 15.2). Intense thermal, mechanical, or chemical stimuli can activate nociceptors. Tissue irritation or injury releases chemicals, such as prostaglandins, kinins, and even potassium ions (K^+), that stimulate nociceptors. Pain may persist even after a pain-producing stimulus is removed because pain-mediating chemicals linger, and because nociceptors adapt only slightly or not at all. Conditions that elicit pain include excessive distension or dilation of a structure, prolonged muscular contractions, muscle spasms, or ischemia (inadequate blood flow to an organ).

Types of Pain

There are two types of pain: fast and slow. The perception of **fast pain** occurs very rapidly, usually within 0.1 second after a stimulus is applied, because the nerve impulses conduct along medium-diameter, myelinated axons called A-delta fibers. This type of pain is also known as acute, sharp, or pricking pain. The pain felt from a needle puncture or knife cut to the skin are examples of fast pain. Fast pain is not felt in deeper tissues of the body. The perception of **slow pain,** by contrast, begins a second or more after a stimulus is applied and then gradually increases in intensity over a period of several seconds or minutes. Impulses for slow pain conduct along small-diameter, unmyelinated C fibers. This type of pain, which may be excruciating, is also referred to as chronic, burning, aching, or throbbing pain. Slow pain can occur both in the skin and in deeper tissues or internal organs. An example is the pain associated with a toothache. You can perceive the difference in onset of these two types of pain best when you injure a body part that is far from the brain because the conduction distance is long. When you stub a toe, for example, there is first the sharp sensation of fast pain and then the slower, aching sensation of slow pain.

Pain that arises from stimulation of receptors in the skin is called **superficial somatic pain,** whereas stimulation of receptors in skeletal muscles, joints, tendons, and fascia causes **deep somatic pain. Visceral pain** results from stimulation of nociceptors in visceral organs. Some stimuli that cause somatic pain do not elicit visceral pain. For example, highly *localized* damage to certain viscera—for example, cutting the intestine in two in a conscious patient—causes little if any pain. But if stimulation is *diffuse* (involves large areas), visceral pain can be severe. Diffuse stimulation of visceral nociceptors might result from organ distension, spasms, or ischemia. For example, a kidney stone or a gallstone might obstruct and distend a ureter or the bile duct and cause severe pain.

Localization of Pain

Fast pain is very precisely localized to the stimulated area. For example, if someone pricks you with a pin, you know exactly which part of your body was stimulated. Slow somatic pain also is well localized but more diffuse; it usually appears to come from a larger area of the skin. In some instances of visceral slow pain, the pain is felt in the stimulated area. If the pleural membranes around the lungs are inflamed, for example, you experience chest pain.

In many instances of visceral pain, however, the pain is felt in or just deep to the skin that overlies the stimulated organ. This phenomenon is called **referred pain.** Referred pain may also be felt in a surface area far from the stimulated organ. In general, the visceral organ involved and the area to which the pain is referred are served by the same segment of the spinal cord. For example, because sensory fibers from the heart, the skin over the heart, and the skin along the medial aspect of the left arm enter spinal cord segments T1–T5 on the left side, the pain of a heart attack is typically felt in the skin over the heart and along the left arm. Figure 15.3 illustrates skin regions to which visceral pain may be referred.

Patients who have had a limb amputated may still experience sensations such as itching, pressure, tingling, or pain as if the limb were still there. This phenomenon is called **phantom limb sensation.** One explanation for phantom sensations is that the cerebral cortex interprets impulses arising in the proximal portions of sensory neurons that previously carried impulses from the limb as coming from the nonexistent (phantom) limb. Another explanation for phantom limb sensations is that the brain itself contains networks of neurons that generate sensations of body awareness. In this view, neurons in the brain that previously received sensory impulses from the missing limb are still active, giving rise to false sensory perceptions.

CLINICAL APPLICATION
Analgesia: Relief from Pain

Some pain sensations are inappropriate; rather than warning of actual or impending damage, they occur out of proportion to minor damage or persist chronically for no obvious reason. In such cases, **analgesia** (*an-* = without; *-algesia* = pain) or pain relief is needed. Analgesic drugs such as aspirin and ibuprofen (for example, Advil) block formation of prostaglandins, which stimulate nociceptors. Local anesthetics, such as Novocaine, provide short-term pain relief by blocking conduction of nerve impulses along first-order pain fibers. Morphine and other opiate drugs alter the quality of pain perception in the brain; pain is still sensed, but it is no longer perceived as being so noxious. ∎

Proprioceptive Sensations

Proprioceptive sensations inform us of the degree to which muscles are contracted, the amount of tension present in tendons, the positions of joints, and the orientation of the head relative to the ground and during movement. Such sensations also report the rate of movement of one body part in relation to others, so that we can walk, type, or dress without

Figure 15.3 Distribution of referred pain. The colored portions of the diagrams indicate skin areas to which visceral pain is referred.

Nociceptors are present in almost every tissue of the body.

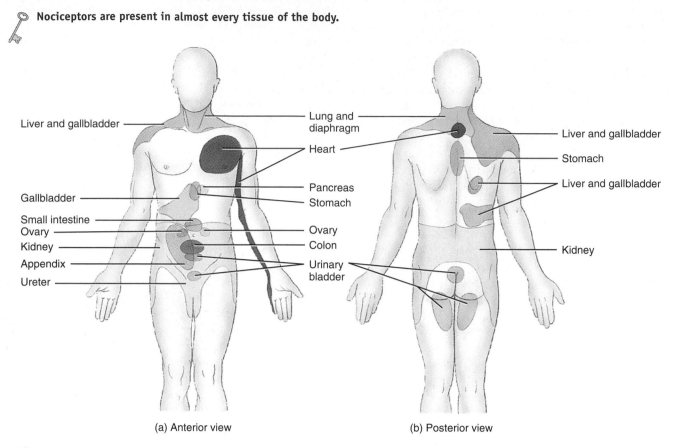

Liver and gallbladder

Gallbladder

Small intestine
Ovary
Kidney
Appendix
Ureter

Lung and diaphragm
Heart
Pancreas
Stomach
Ovary
Colon
Urinary bladder

Liver and gallbladder
Stomach
Liver and gallbladder
Kidney

(a) Anterior view

(b) Posterior view

Q Which visceral organ has the broadest area for referred pain?

using our eyes. **Kinesthesia** (kin′-es-THĒ-zē-a; *kin-* = motion; *-esthesia* = perception) is the perception of body movements. Proprioceptive sensations also allow us to estimate the weight of objects and determine the muscular effort necessary to perform a task. For example, when you pick up a bag you quickly realize whether it contains feathers or books, and you then exert only the amount of effort needed to lift it.

The receptors for proprioception are called **proprioceptors,** and they adapt slowly and only slightly. Thus, the brain continually receives nerve impulses related to the position of different body parts and makes adjustments to ensure coordination. Hair cells of the inner ear are proprioceptors that provide information for maintaining balance and equilibrium (described in Chapter 16). Here we discuss three types of proprioceptors: muscle spindles within skeletal muscles, tendon organs within tendons, and joint kinesthetic receptors within synovial joint capsules.

Muscle Spindles

Muscle spindles are specialized groups of muscle fibers interspersed among and oriented parallel to regular skeletal muscle fibers (Figure 15.4a). The ends of the spindles are anchored to endomysium and perimysium. A muscle spindle consists of 3–10 specialized muscle fibers, called **intrafusal**

muscle fibers, that are enclosed in a spindle-shaped connective tissue capsule. The central region of each intrafusal fiber contains several nuclei but has few or no actin and myosin filaments. Intrafusal fibers contract when stimulated by medium-diameter A fibers called **gamma motor neurons.** Surrounding the muscle spindle are regular skeletal muscle fibers, called **extrafusal muscle fibers,** which are innervated by large-diameter A fibers called **alpha motor neurons.** The cell bodies of both types of motor neurons are located in the anterior gray horn of the spinal cord. The brain regulates muscle spindle sensitivity through pathways to the gamma motor neurons. By adjusting how vigorously a muscle spindle responds to stretching, the brain sets an overall level of muscle tone, the small degree of contraction that is present while the muscle is at rest.

Sensory fibers associated with skeletal muscles are classified according to their axon diameter into four main groups: I (largest diameter), II, III, and IV (smallest diameter); letters are added to designate additional subgroups. The central area of an intrafusal fiber contains two types of slowly adapting sensory fibers. The first is a **type Ia sensory fiber,** which has a large-diameter, rapidly conducting axon. The dendrites of a type Ia sensory fiber spiral around the central area of an intrafusal fiber. The central receptive area of some

Figure 15.4 Two types of proprioceptors: a muscle spindle and a tendon organ. (a) In muscle spindles, which monitor changes in skeletal muscle length, type Ia and type II sensory fibers wrap around the central portion of intrafusal muscle fibers. (b) In tendon organs, which monitor the force of muscle contraction, a type Ib sensory fiber is activated by increasing tension on a tendon.

🔑 **Proprioceptors provide information about movement and the position of the body.**

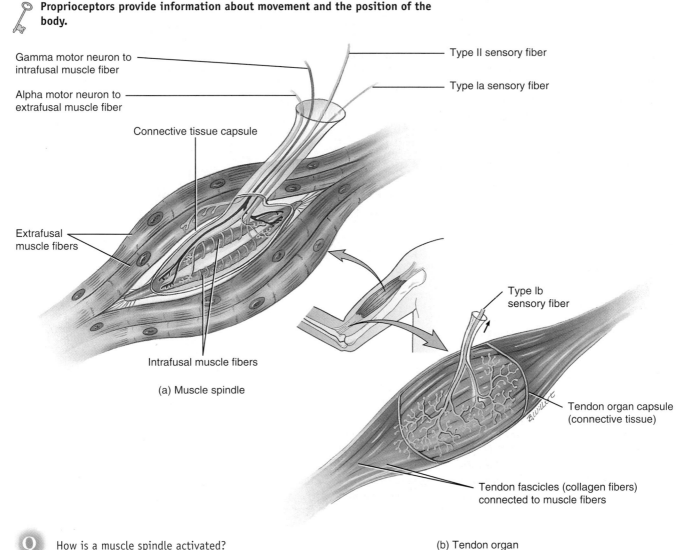

(a) Muscle spindle

(b) Tendon organ

Ⓠ How is a muscle spindle activated?

muscle spindles is also served by a **type II sensory fiber,** whose dendrites are located on either side of the type Ia dendrites. Either sudden or prolonged stretching of the central areas of the intrafusal muscle fibers stimulates the type Ia and type II dendrites. The resulting nerve impulses conduct into the CNS. Muscle spindles thus monitor changes in the length of skeletal muscles. Information from muscle spindles arrives at the cerebral cortex, allowing perception of limb position, and also passes to the cerebellum, where it aids in the coordination of muscle contraction. Impulses from muscle spindles also constitute the afferent part of stretch reflexes (see Figure 13.6 on page 421). Because the stimulus for the reflex is stretching of muscle, the reflex helps prevent injury by preventing overstretching of muscles.

Tendon Organs

Tendon organs are proprioceptors found at the junction of a tendon and a muscle. Each tendon organ consists of a thin capsule of connective tissue that encloses a few collagen fibers (Figure 15.4b). Penetrating the capsule are one or more **type Ib sensory fibers** whose dendrites entwine among and around the collagen fibers. When tension is applied to a tendon, the tendon organs generate nerve impulses that conduct into the CNS, providing information about changes in muscle tension. By initiating tendon reflexes (see Figure 13.7 on page 422), tendon organs protect tendons and their associated muscles from damage due to excessive tension. Tendon reflexes decrease muscle tension by causing muscle relaxation when muscle force becomes too great.

Table 15.2 Summary of Receptors for Somatic Sensations

RECEPTOR TYPE	RECEPTOR STRUCTURE AND LOCATION	SENSATIONS	ADAPTATION RATE
Tactile Receptors			
Corpuscles of touch (Meissner corpuscles)	Capsule surrounds mass of dendrites in dermal papillae of hairless skin.	Touch, pressure, and slow vibrations.	Rapid.
Hair root plexuses	Free nerve endings wrapped around hair follicles in skin.	Touch.	Rapid.
Type I cutaneous mechanoreceptors (tactile or Merkel disc)	Saucer-shaped free nerve endings make contact with Merkel cells in epidermis.	Touch and pressure.	Slow.
Type II cutaneous mechanoreceptors (Ruffini corpuscles)	Elongated capsule surrounds dendrites deep in dermis and in ligaments and tendons.	Stretching of skin.	Slow.
Lamellated (Pacinian) corpuscles	Oval, layered capsule surrounds dendrites; present in subcutaneous layer (and sometimes the dermis), submucosal tissues, joints, periosteum, and some viscera.	Pressure, fast vibrations, and tickling.	Rapid.
Itch and tickle receptors	Free nerve endings and lamellated corpuscles in skin and mucous membranes.	Itching and tickling.	Both slow and rapid.
Thermoreceptors			
Warm receptors and **cold receptors**	Free nerve endings in skin and mucous membranes of mouth, vagina, and anus.	Warmth or cold.	Rapid initially, then slow.
Pain Receptors			
Nociceptors	Free nerve endings in skin and mucous membranes of mouth, vagina, and anus.	Pain.	Slow.
Proprioceptors			
Muscle spindles	Dendrites of types Ia and II fibers wrap around central area of encapsulated intrafusal muscle fibers within most skeletal muscles.	Muscle length.	Slow.
Tendon organs	Capsule encloses collagen fibers and dendrites of type Ib fibers at junction of tendon and muscle.	Muscle tension.	Slow.
Joint kinesthetic receptors	Lamellated corpuscles, Ruffini corpuscles, tendon organs, and free nerve endings.	Joint position and movement.	Rapid.

Joint Kinesthetic Receptors

Several types of **joint kinesthetic receptors** are present within and around the articular capsules of synovial joints. Free nerve endings and type II cutaneous mechanoreceptors (Ruffini corpuscles) in the capsules of joints respond to pressure. Small lamellated (Pacinian) corpuscles in the connective tissue outside articular capsules respond to acceleration and deceleration of joints during movement. Articular ligaments contain receptors similar to tendon organs that adjust reflex inhibition of the adjacent muscles when excessive strain is placed on the joint.

Table 15.2 summarizes the somatic receptors and the sensations they convey.

1. Compare and contrast cutaneous and proprioceptive sensations.

2. Which somatic receptors are encapsulated, and which are not?
3. Contrast somatic, visceral, and phantom pain.
4. What is referred pain, and how is it useful in diagnosing internal disorders?
5. Which somatic receptors are phasic, and which are tonic?

SOMATIC SENSORY PATHWAYS

OBJECTIVE

• *Describe the neuronal components and functions of the posterior column–medial lemniscus, the anterolateral, and the spinocerebellar pathways.*

Somatic sensory pathways relay information from somatic receptors to the primary somatosensory area in the cerebral cortex and to the cerebellum. The pathways to the cerebral cortex are composed of thousands of sets of three neurons: a first-order neuron, a second-order neuron, and a third-order neuron.

1. **First-order neurons** conduct impulses from the somatic receptors into either the brain stem or spinal cord. From the face, mouth, teeth, and eyes, somatic sensory impulses propagate along *cranial nerves* into the brain stem; from the neck, body, and posterior aspect of the head, somatic sensory impulses propagate along *spinal nerves* into the spinal cord.

2. **Second-order neurons** conduct impulses from the spinal cord and brain stem to the thalamus. Axons of second-order neurons decussate (cross over to the opposite side) in the spinal cord or brain stem before ascending to the thalamus. Thus, all somatic sensory information from one side of the body is conveyed to the thalamus on the opposite side.

3. **Third-order neurons** conduct impulses from the thalamus to the primary somatosensory area of the cortex (postcentral gyrus; see Figure 14.15 on page 469), where conscious perception of the sensations results.

Somatic sensory impulses entering the spinal cord ascend to the cerebral cortex by two general pathways: the posterior column–medial lemniscus pathway and the anterolateral (spinothalamic) pathways.

Posterior Column–Medial Lemniscus Pathway to the Cortex

Nerve impulses for conscious proprioception and most tactile sensations ascend to the cortex along a common pathway formed by three-neuron sets (Figure 15.5a). First-order neurons extend from sensory receptors into the spinal cord and up to the medulla oblongata on the same side of the body. The cell bodies of these first-order neurons are in the posterior (dorsal) root ganglia of spinal nerves. In the spinal cord, their axons form the **posterior (dorsal) columns,** which consist of two portions: the **fasciculus gracilis** (fa-SIK-yoo-lus gras-I-lis) and the **fasciculus cuneatus** (kyoo-nē-Ā-tus). The axon terminals synapse with second-order neurons whose cell bodies are located in the nucleus gracilis or nucleus cuneatus of the medulla. Impulses from the neck, upper limbs, and upper chest conducted along axons in the fasciculus cuneatus arrive at the nucleus cuneatus, whereas impulses from the trunk and lower limbs conducted along axons in the fasciculus gracilis arrive at the nucleus gracilis. The axon of the second-order neuron crosses to the opposite side of the medulla and enters the **medial lemniscus,** a projection tract that extends from the medulla to the thalamus. In the thalamus, the axon terminals of second-order neurons synapse with third-order neurons, which project their axons to the primary somatosensory area of the cerebral cortex.

Impulses conducted along the posterior column–medial lemniscus pathway give rise to several highly evolved and refined sensations:

- **Discriminative touch,** the ability to recognize specific information about a touch sensation, such as what point on the body is touched plus the shape, size, and texture of the source of stimulation and to make two-point discriminations.
- **Stereognosis,** the ability to recognize by feel the size, shape, and texture of an object. Examples are reading Braille or identifying a paperclip by feeling it.
- **Proprioception,** the awareness of the precise position of body parts, and **kinesthesia,** the awareness of directions of movement.
- **Weight discrimination,** the ability to assess the weight of an object.
- **Vibratory sensations,** the ability to sense rapidly fluctuating touch.

Anterolateral Pathways to the Cortex

The **anterolateral** or **spinothalamic pathways** (spī-nō-tha-LAM-ik) carry mainly pain and temperature impulses. In addition, they relay the sensations of tickle and itch, as well as some tactile impulses that give rise to a very crude, poorly localized touch or pressure sensation. Like the posterior column–medial lemniscus pathway, the anterolateral pathways are also composed of three-neuron sets (Figure 15.5b). The first-order neuron connects a receptor of the neck, trunk, or limbs with the spinal cord. The cell body of the first-order neuron is in the posterior root ganglion. The axon terminals of the first-order neuron synapse with the second-order neuron, whose cell body is located in the posterior gray horn of the spinal cord. The axon of the second-order neuron crosses to the opposite side of the spinal cord and passes upward to the brain stem in either the **lateral spinothalamic tract** or the **anterior spinothalamic tract.** The axon of the second-order neuron ends in the thalamus, where it synapses with the third-order neuron. The axon of the third-order neuron projects to the primary somatosensory area of the cerebral cortex. The lateral spinothalamic tract conveys sensory impulses for pain and temperature, whereas the anterior spinothalamic tract conveys impulses for tickle, itch, crude touch, and pressure.

Mapping of the Somatosensory Cortex

Researchers have mapped out areas of the primary somatosensory cortex (postcentral gyrus) that receive sensory information from different parts of the body. Figure 15.6a shows the location and areas of representation of the somatosensory cortex of the right cerebral hemisphere. The left cerebral hemisphere has a similar somatosensory cortex.

Figure 15.5 Somatic sensory pathways.

Nerve impulses are conducted along sets of first-order, second-order, and third-order neurons to the primary somatosensory area (postcentral gyrus) of the cerebral cortex.

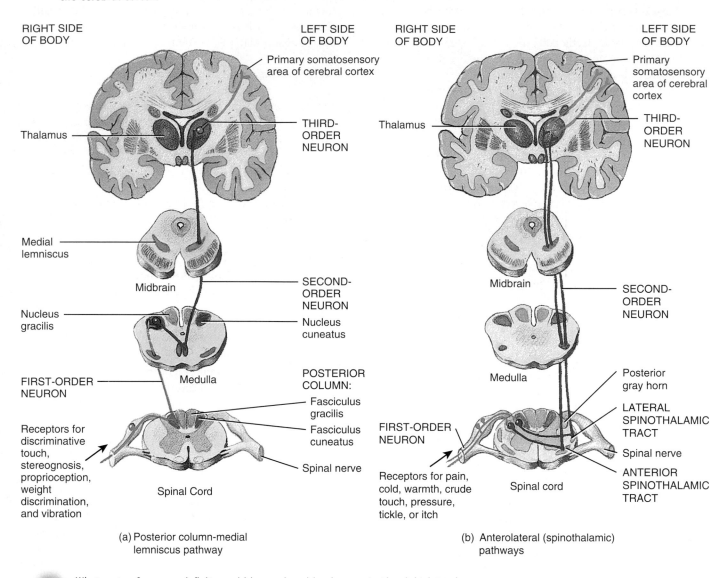

(a) Posterior column-medial lemniscus pathway

(b) Anterolateral (spinothalamic) pathways

Q What sorts of sensory deficits could be produced by damage to the right lateral spinothalamic tract?

Note that some parts of the body—the lips, face, tongue, and thumb, for example—are represented by large areas in the somatosensory cortex. Other parts of the body, such as the trunk and lower limbs, are represented by much smaller areas. The relative sizes of the areas in the somatosensory cortex are proportional to the number of specialized sensory receptors within the corresponding part of the body. For example, there are many sensory receptors in the skin of the lips but few in the skin of the trunk. The size of the cortical area that represents a body part may expand or shrink somewhat, depending on the quantity of sensory impulses received from that body part. For example, people who learn to read Braille ultimately have a larger representation of the fingertips in the somatosensory cortex.

Somatic Sensory Pathways to the Cerebellum

Two tracts in the spinal cord—the **posterior spinocerebellar tract** (spī-nō-ser-e-BEL-ar) and the **anterior spinocerebellar tract**—are the major routes whereby proprioceptive impulses reach the cerebellum. Although not consciously perceived, sensory impulses conveyed to the cerebellum along these two pathways are critical for posture, balance, and coordination of skilled movements. Two other tracts also convey impulses from proprioceptors of the trunk and upper limbs to the cerebellum: the **cuneocerebellar tract** and the **rostral spinocerebellar tract.**

Table 15.3 summarizes the major somatic sensory tracts in the spinal cord and pathways in the brain.

Figure 15.6 (a) Primary somatosensory area (postcentral gyrus) and (b) primary motor area (precentral gyrus) of the right cerebral hemisphere. The left hemisphere has similar representation. (After Penfield and Rasmussen)

Each point on the body surface maps to a specific region in both the primary somatosensory area and the primary motor area.

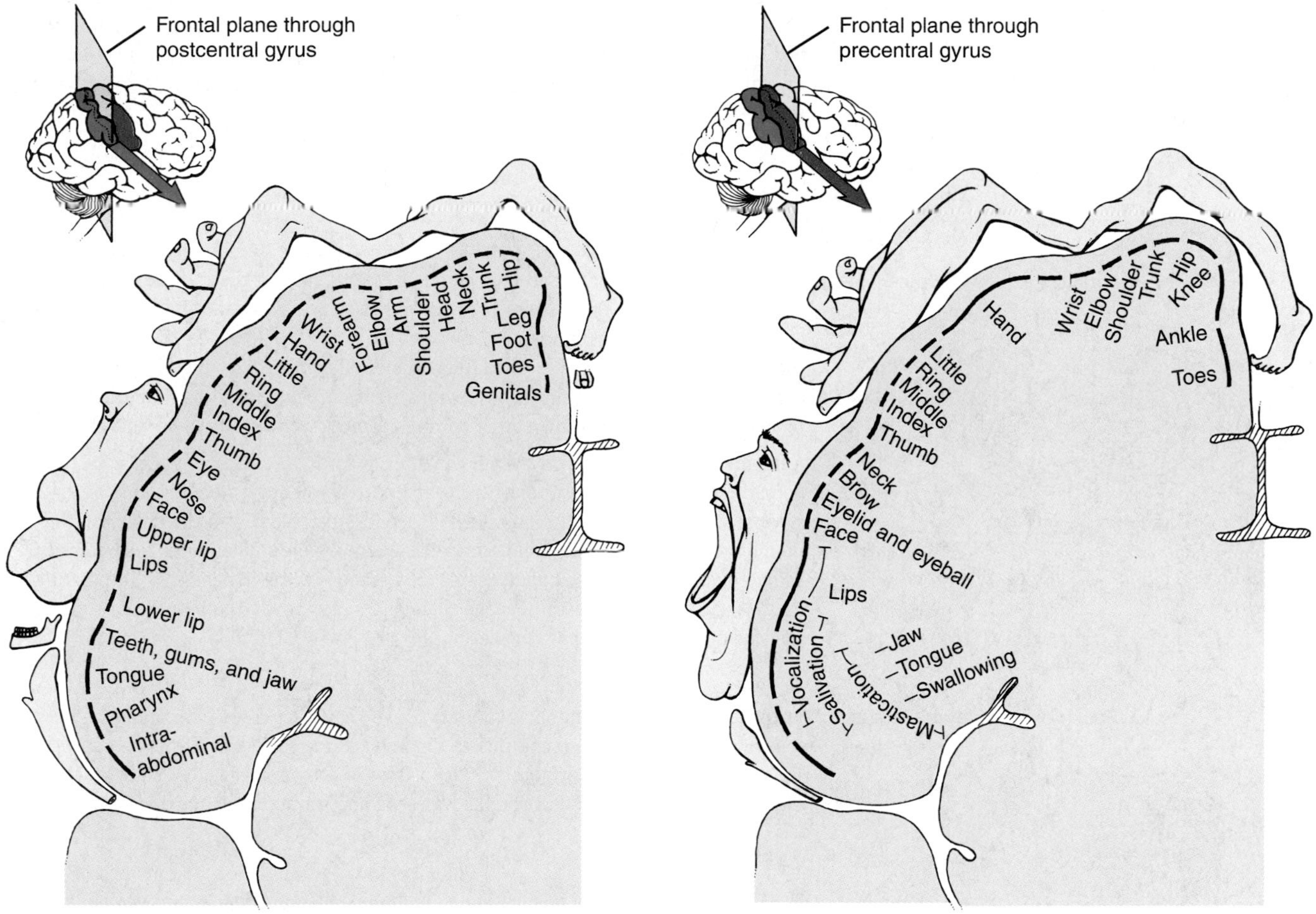

Frontal plane through postcentral gyrus

Frontal plane through precentral gyrus

(a) Frontal section of primary somatosensory area in right cerebral hemisphere

(b) Frontal section of primary motor area in right cerebral hemisphere

Q How do the somatosensory and motor representations compare for the hand, and what does this difference imply?

CLINICAL APPLICATION
Tertiary Syphilis

Syphilis is a common sexually transmitted disease caused by the bacterium *Treponema pallidum*. If the infection is not treated, syphilis progresses through three clinical stages, the third of which—*tertiary syphilis*—typically causes debilitating neurological symptoms. A common outcome is progressive degeneration of the posterior portions of the spinal cord, including the posterior columns, posterior spinocerebellar tracts, and posterior roots. Somatic sensations are lost, and the person's gait becomes uncoordinated because proprioceptive impulses fail to reach the cerebellum. People suffering from tertiary syphilis often watch their feet while walking to maintain their balance. However, vision does not fully compensate for the loss of proprioceptive impulses conducted via the spinocerebellar tracts, and thus body movements are uncoordinated and jerky. ■

1. Describe three differences between the posterior column–medial lemniscus pathway and the anterolateral pathways.
2. How are various parts of the body represented in the somatosensory cortex? Which body parts have the largest representation?
3. What type of sensory information is carried in the spinocerebellar tracts, and what is its usefulness?

Table 15.3 Major Somatic Sensory Tracts in the Spinal Cord and Pathways in the Brain

TRACT AND LOCATION	FUNCTIONS AND PATHWAYS
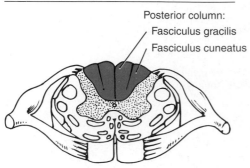 Posterior column: Fasciculus gracilis / Fasciculus cuneatus	**Posterior column:** Conveys nerve impulses for the sensations of discriminative touch, stereognosis, conscious proprioception, kinesthesia, weight discrimination, and vibration. Axons of first-order neurons from one side of the body form the posterior column on the same side and end in the medulla, where they synapse with dendrites and cell bodies of second-order neurons. Axons of second-order neurons decussate, enter the medial lemniscus on the opposite side, and extend to the thalamus. Third-order neurons transmit nerve impulses from the thalamus to the primary somatosensory cortex on the side opposite the site of stimulation.
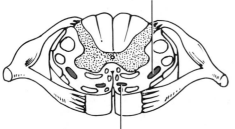 Lateral spinothalamic tract / Anterior spinothalamic tract	**Lateral spinothalamic:** Conveys nerve impulses for pain and thermal sensations. Axons of first-order neurons from one side of the body synapse with dendrites and cell bodies of second-order neurons in the posterior gray horn on the same side. Axons of second-order neurons decussate, enter the lateral spinothalamic tract on the opposite side, and extend to the thalamus. Third-order neurons transmit nerve impulses from the thalamus to the primary somatosensory cortex on the side opposite the site of stimulation. **Anterior spinothalamic:** Conveys nerve impulses for itch, tickle, pressure, and crude, poorly localized touch sensations. Axons of first-order neurons from one side of the body synapse with dendrites and cell bodies of second-order neurons in the posterior gray horn on the same side. Axons of second-order neurons decussate, enter the anterior spinothalamic tract on the opposite side, and extend to the thalamus. Third-order neurons transmit nerve impulses from the thalamus to the primary somatosensory cortex on the side opposite the site of stimulation.
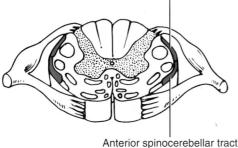 Posterior spinocerebellar tract / Anterior spinocerebellar tract	**Anterior and posterior spinocerebellar:** Convey nerve impulses from proprioceptors in the trunk and lower limb of one side of the body to the same side of the cerebellum. The proprioceptive input informs the cerebellum of actual movements, allowing it to coordinate, smooth, and refine skilled movements and maintain posture and balance.

SOMATIC MOTOR PATHWAYS

OBJECTIVES

• *Compare the locations and functions of the direct and indirect motor pathways.*

• *Explain how the basal ganglia and cerebellum contribute to motor responses.*

Control of body movements involves several regions of the brain. Motor portions of the cerebral cortex play the major role for initiating and controlling precise, discrete muscular movements. The basal ganglia help establish a normal level of muscle tone and integrate semivoluntary, automatic movements, whereas the cerebellum assists the motor cortex and basal ganglia by making body movements smooth and coordinated and by helping to maintain normal posture and balance. Two main types of somatic motor pathways extend from the brain to the skeletal muscles: Direct somatic motor pathways extend from the cerebral cortex into the spinal cord and out to skeletal muscles; indirect somatic motor pathways include synapses in the basal ganglia, thalamus, reticular formation, and cerebellum.

Before we examine these pathways we consider the role of the motor cortex in voluntary movement.

Mapping of the Motor Cortex

The **primary motor area (precentral gyrus)** of the cerebral cortex is the major control region for initiation of voluntary movements. The adjacent **premotor area** and even the **primary somatosensory area** in the postcentral gyrus also contribute fibers to the descending motor pathways. As is true for somatic sensory representation in the somatosensory area, different muscles are represented unequally in the primary motor area (see Figure 15.6b). The degree of representation is proportional to the number of motor units in a particular muscle of the body. For example, the muscles in the thumb, fingers, lips, tongue, and vocal cords have large representations, whereas the trunk has a much smaller representation. By comparing Figures 15.6a and 15.6b, you can see that somatosensory and motor representations are similar but not identical for any given part of the body.

Direct Motor Pathways

Nerve impulses for voluntary movements propagate from the motor cortex to somatic motor neurons that innervate skeletal muscles via the **direct** or **pyramidal pathways** (Figure 15.7). The simplest of these pathways consists of sets of two neurons: upper motor neurons and lower motor neurons. About 1 million cell bodies of direct pathway **upper motor neurons (UMNs)** are located in the cortex, 60% in the precentral gyrus and 40% in the postcentral gyrus. Their axons descend through the internal capsule of the cerebrum. In the medulla oblongata, the axon bundles form the ventral bulges known as the pyramids. About 90% of the axons of upper motor neurons decussate (cross over) to the contralateral (opposite) side in the medulla oblongata. The 10% that remain on the ipsilateral (same) side eventually decussate at lower levels. Thus, the motor cortex of the right side of the brain controls muscles on the left side of the body, and the motor cortex of the left side of the brain controls muscles on the right side of the body. The upper motor neuron axons terminate in nuclei of cranial nerves in the brainstem or in the anterior gray horn of the spinal cord.

Lower motor neurons (LMNs) extend from the motor nuclei of cranial nerves to skeletal muscles of the face and head, and from the anterior horn of each spinal cord segment to skeletal muscle fibers of the trunk and limbs. Close to their termination point, most upper motor neurons synapse with an interneuron, which, in turn, synapses with a lower motor neuron. Thus, most impulses from the brain are conveyed to interneurons before being received by lower motor neurons. (A few upper motor neurons synapse directly with lower motor neurons.) Lower motor neurons are the only neurons that carry information from the CNS to skeletal muscle fibers, and in that role each of them receives and integrates an enormous amount of excitatory and inhibitory input from many presynaptic neurons, both upper motor neurons and interneurons. For this reason, lower motor neurons are also called the **final common pathway.**

Figure 15.7 Direct motor pathways whereby signals initiated by the primary motor area in the right hemisphere control skeletal muscles on the left side of the body. Spinal cord tracts carrying impulses of direct motor pathways are the lateral corticospinal tract and anterior corticospinal tract.

🔑 **Direct pathways convey impulses that result in precise, voluntary movements.**

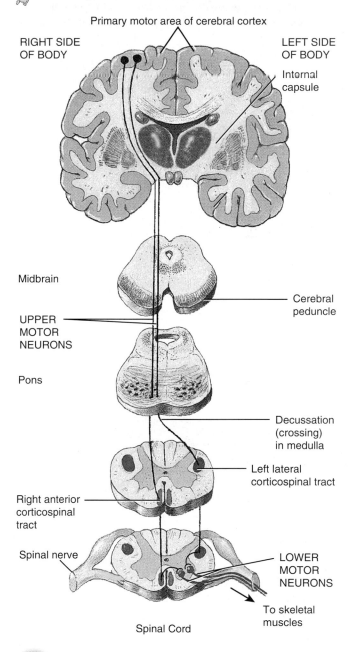

Q What other tracts (not shown in the figure) convey impulses that result in precise, voluntary movements?

Figure 15.8 Indirect motor pathways for coordination and control of movement. Red arrows show connections between brain regions by neurons of the indirect motor pathways; green arrows indicate indirect motor pathway tracts in the spinal cord. For comparison, connections made by upper motor neurons of the direct pathway are shown at the right of the figure.

🔑 **Because lower motor neurons receive input from all direct and indirect pathways and from spinal cord interneurons, they are called the final common pathway.**

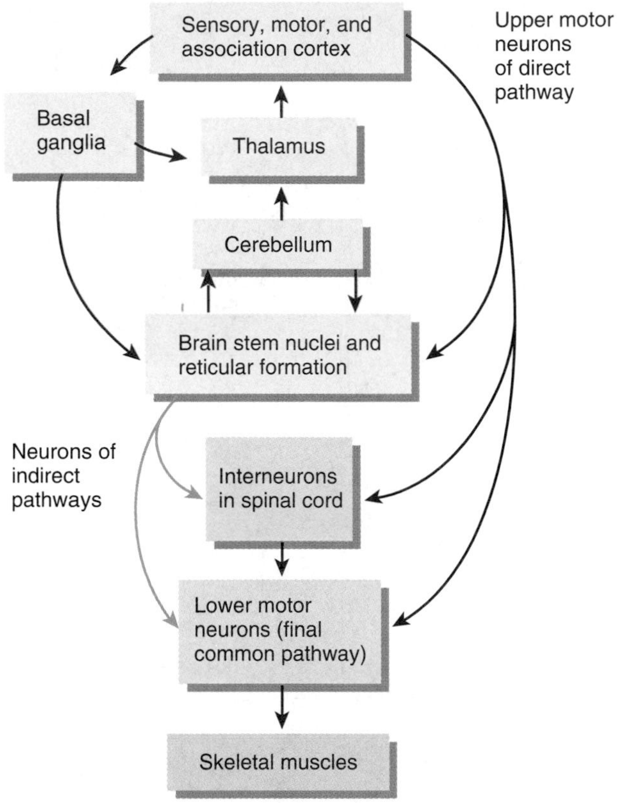

[Diagram: Sensory, motor, and association cortex; Basal ganglia; Thalamus; Cerebellum; Brain stem nuclei and reticular formation; Interneurons in spinal cord; Lower motor neurons (final common pathway); Skeletal muscles. Labels: Upper motor neurons of direct pathway; Neurons of indirect pathways]

Q For signals that originate in the cortex, what brain regions participate in the two circular feedback loops that pass through the thalamus?

The direct pathways convey impulses from the cortex that result in precise, voluntary movements. Three tracts contain axons of upper motor neurons:

1. *Lateral corticospinal tracts.* Axons of upper motor neurons that decussate in the medulla form the **lateral corticospinal tracts** (kor'-ti-kō-SPI-nal) in the right and left lateral white columns of the spinal cord (see Figure 15.7). Axons of lower motor neurons emerge from all levels of the spinal cord in the anterior roots of spinal nerves and terminate in skeletal muscles. These motor neurons control skilled movements of the limbs, hands, and feet.
2. *Anterior corticospinal tracts.* Axons of upper motor neurons that do not decussate in the medulla form the **anterior corticospinal tracts** in the right and left anterior

white columns (see Figure 15.7). At each spinal cord level at which they terminate, some of the axons of these upper motor neurons decussate. After crossing over to the opposite side, they synapse with interneurons or lower motor neurons in the anterior gray horn of the spinal cord. Axons of these lower motor neurons exit the cervical and upper thoracic segments of the cord in the anterior roots of spinal nerves. They terminate in skeletal muscles that control movements of the neck and part of the trunk, thus coordinating movements of the axial skeleton.
3. *Corticobulbar tracts.* The axons of upper motor neurons that conduct impulses for the control of skeletal muscles in the head extend through the internal capsule to the midbrain, where they form the **corticobulbar tracts** (kor'-ti-kō-BUL-bar) in the right and left cerebral peduncles (see Table 15.4). Some of the axons in the corticobulbar tracts have decussated, whereas others have not. The axons terminate in the nuclei of nine pairs of cranial nerves in the pons and medulla oblongata: the oculomotor (III), trochlear (IV), trigeminal (V), abducens (VI), facial (VII), glossopharyngeal (IX), vagus (X), accessory (XI), and hypoglossal (XII). The lower motor neurons of cranial nerves convey impulses that control precise, voluntary movements of the eyes, tongue, and neck, plus chewing, facial expression, and speech.

Table 15.4 summarizes the functions and pathways of the tracts in the direct motor pathways.

CLINICAL APPLICATION
Paralysis

During a neurological exam, assessment of muscle tone, reflexes, and the ability to perform voluntary movements helps pinpoint certain types of motor system dysfunction. Damage or disease of *lower* motor neurons, either of their cell bodies in the anterior horn or of their axons in the anterior root or spinal nerve, produces **flaccid paralysis** of muscles on the same side of the body. There is neither voluntary nor reflex action of the innervated muscle fibers, muscle tone is decreased or lost, and the muscle remains limp or flaccid. Injury or disease of *upper* motor neurons causes **spastic paralysis** of muscles on the opposite side of the body. This condition is characterized by varying degrees of spasticity (increased muscle tone), exaggerated reflexes, and pathological reflexes such as the Babinski sign. ■

Indirect Motor Pathways

The **indirect** or **extrapyramidal pathways** include all somatic motor tracts other than the corticospinal and corticobulbar tracts. Nerve impulses conducted along the indirect pathways follow complex, polysynaptic circuits that involve the motor cortex, basal ganglia, limbic system, thalamus, cerebellum, reticular formation, and nuclei in the brain stem (Figure 15.8). Axons of neurons that carry nerve

Table 15.4 Major Somatic Motor Pathways in the Brain and Tracts in the Midbrain and Spinal Cord

TRACT AND LOCATION	FUNCTIONS AND PATHWAYS
Direct (pyramidal) tracts 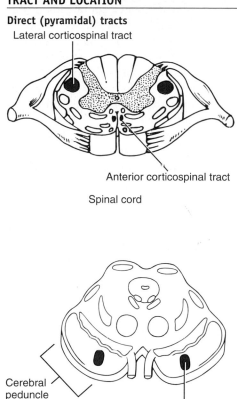 Lateral corticospinal tract / Anterior corticospinal tract / Spinal cord	**Lateral corticospinal:** Conveys nerve impulses from the motor cortex to skeletal muscles on opposite side of body for precise, voluntary movements of the limbs, hands, and feet. Axons of upper motor neurons (UMNs) descend from the precentral and postcentral gyri of the cortex into the medulla. Here 90% decussate (cross over to the opposite side) and then enter the contralateral side of the spinal cord to form this tract. At their level of termination, these UMNs end in the anterior gray horn on the same side. They provide input to lower motor neurons, which innervate skeletal muscles. **Anterior corticospinal:** Conveys nerve impulses from the motor cortex to skeletal muscles on opposite side of body for movements of the axial skeleton. Axons of UMNs descend from the cortex into the medulla. Here the 10% that do not decussate enter the spinal cord and form this tract. At their level of termination, these UMNs decussate and end in the anterior gray horn on the opposite side. They provide input to lower motor neurons, which innervate skeletal muscles.
Cerebral peduncle / Corticobulbar tract / Midbrain	**Corticobulbar:** Conveys nerve impulses from the motor cortex to skeletal muscles of the head and neck to coordinate precise, voluntary movements. Axons of UMNs descend from the cortex into the brain stem, where some decussate and others do not. They provide input to lower motor neurons in the nuclei of cranial nerves III, IV, V, VI, VII, IX, X, XI, and XII, which control voluntary movements of the eyes, tongue and neck; chewing; facial expression; and speech.
Indirect (extrapyramidal) tracts 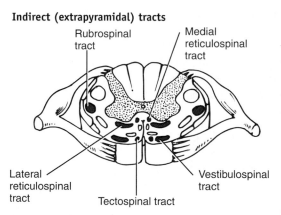 Rubrospinal tract / Medial reticulospinal tract / Lateral reticulospinal tract / Tectospinal tract / Vestibulospinal tract	**Rubrospinal:** Conveys nerve impulses from the red nucleus (which receives input from the cortex and cerebellum) to contralateral skeletal muscles that govern precise movements of the limbs, hands, and feet. **Tectospinal:** Conveys nerve impulses from the superior colliculus to contralateral skeletal muscles that move the head and eyes in response to visual stimuli. **Vestibulospinal:** Conveys nerve impulses from the vestibular nucleus (which receives input about head movements from the vestibular apparatus in the inner ear) to regulate ipsilateral muscle tone for maintaining balance in response to head movements. **Lateral reticulospinal:** Conveys from the reticular formation nerve impulses that facilitate flexor reflexes, inhibit extensor reflexes, and decrease muscle tone in muscles of the axial skeleton and proximal portions of the limbs. **Medial reticulospinal:** Conveys from the reticular formation nerve impulses that facilitate extensor reflexes, inhibit flexor reflexes, and increase muscle tone in muscles of the axial skeleton and proximal portions of the limbs.

impulses from the indirect pathways descend from various nuclei of the brain stem into five major tracts of the spinal cord and terminate on interneurons or lower motor neurons. These tracts are the **rubrospinal** (ROO-brō-spī-nal), **tectospinal** (TEK-tō-spī-nal), **vestibulospinal** (ves-TIB-yoo-lō-spī-nal), **lateral reticulospinal** (re-TIK-yoo-lō-spī-nal), and **medial reticulospinal tracts.**

Table 15.4 summarizes the functions and pathways of the tracts in the indirect motor pathways.

Roles of the Basal Ganglia

The basal ganglia have many connections with other parts of the brain. Through these connections they help to program habitual or automatic movement sequences such as arm swinging during walking or laughing in response to a joke and to set an appropriate level of muscle tone. The basal ganglia also selectively inhibit other motor neuron circuits that are intrinsically active or excitatory. This effect is revealed in certain types of basal ganglia damage, which often are characterized by abnormal, uncontrollable movements or tremors caused by loss of inhibitory signals from the basal ganglia.

The caudate nucleus and the putamen receive input from sensory, motor, and association areas of the cortex and the substantia nigra. Output from the basal ganglia comes mainly from the globus pallidus. (Figure 14.13b on page 466 shows these parts of the basal ganglia.) The globus pallidus, in turn, sends feedback signals to the motor cortex by way of the thalamus. This feedback circuit—from cortex to basal ganglia to thalamus to cortex—appears to function in planning and programming movements. The globus pallidus also sends into the reticular formation impulses that reduce muscle tone. Damage or destruction of certain basal ganglia connections causes a generalized increase in muscle tone that produces abnormal muscle rigidity, as occurs in Parkinson disease (see page 507).

Roles of the Cerebellum

The cerebellum is active in both learning and performing rapid, coordinated, highly skilled movements such as hitting a golf ball, speaking, and swimming. It also functions to maintain proper posture and equilibrium (balance). Cerebellar function involves four activities (Figure 15.9):

❶ The cerebellum *monitors intentions for movement* by receiving impulses from the motor cortex and basal ganglia via the pontine nuclei in the pons regarding what movements are planned (red lines).

❷ The cerebellum *monitors actual movement* by receiving input from proprioceptors in joints and muscles that reveals what actually is happening (blue lines). These nerve impulses travel in the anterior and posterior spinocerebellar tracts. Nerve impulses from the vestibular (equilibrium-sensing) apparatus in the inner ear and from the eyes also enter the cerebellum.

❸ The cerebellum *compares the command signals* (intentions for movement) *with sensory information* (actual performance).

❹ The cerebellum *sends out corrective signals*, both to nuclei in the brain stem and to the motor cortex via the thalamus (green lines).

In summary, the cerebellum receives information from higher brain centers about what the muscles should be doing, and from the peripheral nervous system about what the muscles are doing. If there is a discrepancy between the two, corrective feedback signals are sent from the cerebellum via the thalamus to the cerebrum, where new commands are initiated to decrease the discrepancy and smooth the motion.

Skilled activities such as tennis or volleyball provide good examples of the contribution of the cerebellum to movement. To make a good stop volley or to block a spike, you must bring your racket or arms forward just far enough to make solid contact. How do you stop at exactly the right point? Before you even hit the ball, the cerebellum has sent nerve impulses to the cerebral cortex and basal ganglia informing them where your swing must stop. In response to impulses from the cerebellum, the cortex and basal ganglia transmit motor impulses to opposing body muscles to stop the swing.

1. Which parts of the body have the largest and smallest representation in the motor cortex?
2. Explain why the two main somatic motor pathways are named "direct" and "indirect."
3. Given the normal functions of the basal ganglia and the cerebellum, what differences might you see with disease of the basal ganglia versus damage to the cerebellum?

INTEGRATIVE FUNCTIONS OF THE CEREBRUM

OBJECTIVE

• *Compare the integrative cerebral functions of wakefulness and sleep, and learning and memory.*

We turn now to a fascinating, though incompletely understood, function of the cerebrum: integration. The **integrative functions** include cerebral activities such as sleep and wakefulness, learning and memory, and emotional responses. (The role of the limbic system in emotional behavior was discussed in Chapter 14.)

Wakefulness and Sleep

Humans sleep and awaken in a fairly constant 24-hour cycle called a **circadian rhythm** (ser-KĀ-dē-an; *circa-* = about; *-dia* = a day) that is established by the suprachiasmatic nucleus of the hypothalamus (see Figure 14.10 on page 462). A person who is awake is in a state of readiness and is able to react consciously to various stimuli. Moreover, EEG recordings show that the cerebral cortex is very active during wakefulness, whereas fewer impulses arise during most stages of sleep.

The Role of the Reticular Activating System in Awakening

Stimulation of some parts of the reticular formation results in increased cortical activity. Thus, a portion of the reticular formation is known as the **reticular activating sys-**

Figure 15.9 Input to and output from the cerebellum.

The cerebellum coordinates and smoothes contractions of skeletal muscles during skilled movements and helps maintain posture and balance.

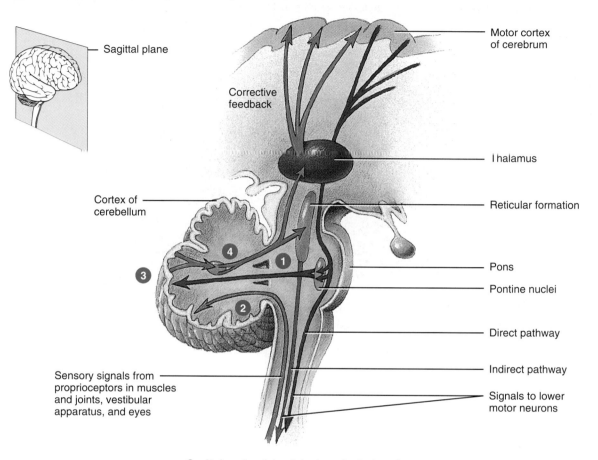

Sagittal section through brain and spinal cord

Q Which tracts carry information from proprioceptors in joints and muscles to the cerebellum?

tem or **RAS** (Figure 15.10). When this area is active, many nerve impulses pass upward to widespread areas of the cerebral cortex, both directly and via the thalamus. The effect is a generalized increase in cortical activity.

Arousal, or awakening from sleep, also involves increased activity in the RAS. For arousal to occur, the RAS must be stimulated. Many sensory stimuli can activate the RAS: painful stimuli detected by nociceptors, touch and pressure on the skin, movement of the limbs, bright light, or the buzz of an alarm clock. Once the RAS is activated, the cerebral cortex is also activated, and arousal occurs. The result is a state of wakefulness called **consciousness.** Notice in Figure 15.10 that even though the RAS receives input from somatic sensory systems and several of the special senses, there is little or no input from the olfactory system, so even strong odors may fail to cause arousal. People who die in

house fires usually succumb to smoke inhalation without awakening. For this reason, all sleeping areas should have a nearby smoke detector that emits a loud alarm (or a flashing light for hearing-impaired individuals).

 CLINICAL APPLICATION
Coma

Damage to the RAS or other parts of the brain can produce **coma,** a state of deep unconsciousness from which a person cannot be aroused. A comatose patient lies in a sleeplike state with eyes closed. In the lightest stages of coma, brain stem and spinal cord reflexes persist, but in the deepest stages, even reflexes are lost. If respiratory and cardiovascular controls are lost, the patient dies. ■

Figure 15.10 The reticular activating system (RAS), which consists of fibers that project from the reticular formation through the thalamus to the cerebral cortex.

🔑 **Awakening from sleep (arousal) involves increased activity of the RAS.**

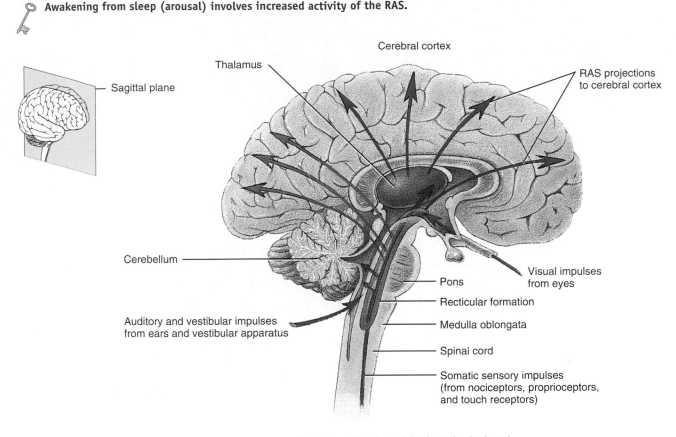

Sagittal section through brain and spinal cord

Ⓠ Why should every sleeping area have a smoke detector?

Sleep

Sleep is a state of altered consciousness or partial unconsciousness from which an individual can be aroused by many different stimuli. Prolonged wakefulness or sleep-deprivation causes fatigue, increasing both the urge to sleep and the deepness of sleep once it occurs. Although the exact triggers for sleep are still unclear, several lines of evidence suggest the existence of sleep-inducing chemicals in the brain. One apparent sleep-inducer is adenosine, which accumulates during periods of high ATP (adenosine triphosphate) use by the nervous system. Adenosine binds to specific receptors, called A1 receptors, and inhibits certain cholinergic (acetylcholine-releasing) neurons of the RAS that participate in arousal. Thus, activity in the RAS during sleep is low due to the inhibitory effect of adenosine. Caffeine (in coffee) and theophylline (in tea)—substances known for their ability to maintain wakefulness—bind to and block the A1 receptors, preventing adenosine from binding and inducing sleep.

Just as there are different levels of awareness when a person is awake, there are different levels of sleep. Normal sleep consists of two components: non-rapid eye movement (NREM) sleep and rapid eye movement (REM) sleep. **NREM sleep** consists of four gradually merging stages, each of which is characterized by a different kind of EEG activity (Figure 15.11a):

1. *Stage 1* is a transition stage between wakefulness and sleep that normally lasts 1–7 minutes. The person is relaxed with eyes closed and has fleeting thoughts. People awakened during this stage often say they have not been sleeping. Alpha waves, which are present in people awake with eyes closed, diminish and are replaced by lower frequency, slightly larger amplitude EEG waves.

2. *Stage 2* or *light sleep* is the first stage of true sleep. In it, a person is a little more difficult to awaken. Fragments of dreams may be experienced, and the eyes may slowly roll from side to side. The EEG shows *sleep spindles*—bursts of sharply pointed waves that occur at 12–14 Hz (cycles per second) and last 1–2 seconds.

3. *Stage 3* is a period of moderately deep sleep. Body temperature and blood pressure decrease. It is difficult to awaken the person, and the EEG shows a mixture of sleep spindles and larger, lower frequency waves. This stage occurs about 20 minutes after falling asleep.

4. *Stage 4* or *slow-wave sleep* is the deepest level of sleep. The person responds slowly if awakened, and the EEG is dominated by slow, large amplitude delta waves. Although brain metabolism decreases significantly during slow-wave sleep and body temperature drops slightly, most reflexes are intact, and muscle tone is decreased only slightly. When sleepwalking occurs, it does so during this stage.

Typically, a person goes from stage 1 to stage 4 of NREM sleep in less than an hour (Figure 15.11b). During a typical 7- or 8-hour sleep period, there are three to five episodes of **REM sleep,** during which the eyes move rapidly back and forth under closed eyelids. The person may rapidly ascend through stages 3 and 2 before entering REM sleep. The first episode of REM sleep lasts 10–20 minutes and is followed by another interval of NREM sleep.

Most dreaming occurs during REM sleep, and the EEG readings are similar to those of an awake person. With the exception of motor neurons that govern breathing and eye movements, most somatic motor neurons are inhibited during REM sleep, which decreases muscle tone and even paralyzes the skeletal muscles. Many people experience a momentary feeling of paralysis if they are awakened during REM sleep. The autonomic nervous system becomes more active during REM sleep; heart rate and blood pressure increase. In men, autonomic excitation causes erection of the penis during most REM intervals, even when dream content is not sexual. Presence of penile erections during REM sleep in a man with erectile dysfunction (inability to attain an erection while awake) indicates a psychological, rather than a physical, cause of his problem.

REM and NREM sleep alternate throughout the night. REM periods, which occur approximately every 90 minutes, gradually lengthen, until the final one lasts about 50 minutes. In adults, REM sleep totals 90–120 minutes during a typical sleep period. As a person ages, the average total time spent sleeping decreases; moreover, the percentage of REM sleep declines. As much as 50% of an infant's sleep is REM sleep, as opposed to 35% for 2 year olds and 25% for adults. Although we do not yet understand the function of REM sleep, the high percentage of REM sleep in infants and children is thought to be important for the maturation of the brain. Neuronal activity is high during REM sleep—brain oxygen use is higher during REM sleep than during intense mental or physical activity while awake.

Learning and Memory

Without memory, we would repeat mistakes and be unable to learn. Similarly, we would not be able to repeat our successes or accomplishments, except by chance. Although

Figure 15.11 The stages of sleep. (a) EEG recordings during sleep stages. In an awake person whose eyes are closed, alpha waves predominate; slow, high-amplitude delta waves characterize stage 4 sleep. (b) Alternating intervals of NREM and REM sleep during a typical sleep period.

🔑 **As non-rapid eye movement (NREM) sleep progresses from stage 1 to stage 4, the EEG waves become slower (lower frequency) and larger (higher amplitude).**

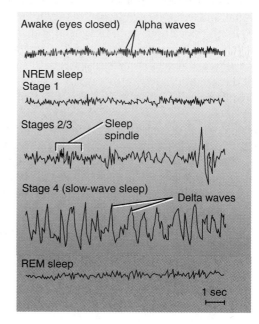

(a) EEG waves during sleep stages

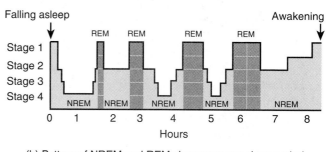

(b) Pattern of NREM and REM sleep over one sleep period

Q In which phase of sleep do dreaming and paralysis of skeletal muscles occur?

both learning and memory have been extensively studied, we still have no completely satisfactory explanation for how we recall information or how we remember events. However, we do know something about how information is acquired and stored, and it is clear that there are different categories of memory.

Learning is the ability to acquire new knowledge or skills through instruction or experience. **Memory** is the process by which that knowledge is retained over time. For an experience to become part of memory, it must produce persistent functional changes that represent the experience in the brain. This capability for change associated with learning is termed **plasticity.** Nervous system plasticity underlies our ability to change our behavior in response to stimuli from the external and internal environments. It involves changes in individual neurons—for example, synthesis of different proteins or sprouting of new dendrites—as well as changes in the strengths of synaptic connections among neurons. The portions of the brain known to be involved with memory include the association cortex of the frontal, parietal, occipital, and temporal lobes; parts of the limbic system, especially the hippocampus and amygdala; and the diencephalon. The primary somatosensory and primary motor areas in the brain also exhibit plasticity. If a particular body part is used more intensively or in a newly learned activity, such as reading Braille, the cortical areas devoted to that body part gradually expand.

Memory occurs in stages over a period of time. **Short-term memory** is the temporary ability to recall a few pieces of information. One example is when you look up an unfamiliar telephone number, cross the room to the phone, and then dial the new number. If the number has no special significance, it is usually forgotten within a few seconds. Information in short-term memory may later be transformed into a more permanent type of memory, called **long-term memory,** that lasts from days to years. If you use that new telephone number often enough, it becomes part of long-term memory. Information in long-term memory usually can be retrieved for use whenever needed. The reinforcement that results from the frequent retrieval of a piece of information is called **memory consolidation.**

Although the brain receives many stimuli, we pay attention to only a few of them at any one time. It has been estimated that of all the information that comes to our consciousness, only about 1% is stored as long-term memory. Moreover, much of what goes into long-term memory is eventually forgotten. Memory does not record every detail as if it were magnetic tape. Even when details are lost, we can often explain the idea or concept using our own words and ways of viewing things.

Some evidence supports the notion that short-term memory depends more on electrical and chemical events in the brain than on structural changes, such as the formation of new synapses. Several conditions that inhibit the electrical activity of the brain, such as anesthesia, coma, electroconvulsive shock, and ischemia of the brain, disrupt retention of recently acquired information without altering previously established long-term memories. It is a common experience that people who suffer retrograde amnesia (loss of past memories) cannot remember anything that occurred during the 30 minutes or so before the amnesia developed. As a person recovers from amnesia, the most recent memories return last.

One theory of short-term memory holds that memories may be caused by reverberating neuronal circuits—an incoming nerve impulse stimulates the first neuron, which stimulates the second, which stimulates the third, and so on. Branches from the second and third neurons synapse with the first, sending the impulse back through the circuit again and again. Once fired, the output signal may last from a few seconds to many hours, depending on the arrangement of neurons in the circuit. If this pattern is applied to short-term memory, an incoming thought (say, a phone number) continues in the brain even after the initial stimulus (looking at the number in the phone book) is gone. Thus, you can recall the thought only for as long as the reverberation continues.

Most research on long-term memory focuses on anatomical or biochemical changes that might enhance facilitation at synapses. Any of the events in synaptic transmission could be responsible for enhanced communication between neurons. For example, there could be an increase in the number of receptor molecules in the postsynaptic plasma membrane or a decrease in the rate of removal of a neurotransmitter. Alternatively, disease may adversely affect memory by interfering with synthesis of specific neurotransmitters. In Alzheimer disease, which eventually wipes out the ability to remember, a key enzyme needed to synthesize ACh becomes depleted in specific brain regions.

Anatomical changes occur in neurons when they are stimulated. For example, electron micrographs of neurons subjected to prolonged, intense activity reveal an increase in the number of presynaptic terminals and enlargement of synaptic end bulbs in presynaptic neurons, as well as an increase in the number of dendritic branches in postsynaptic neurons. Moreover, neurons grow new synaptic end bulbs with increasing age, presumably as a result of increased use. Opposite changes occur when neurons are inactive. For example, in animals that have lost their eyesight, the cerebral cortex in the visual area becomes thinner.

Believed to underlie some aspects of memory is a phenomenon called **long-term potentiation (LTP),** in which transmission at some synapses within the hippocampus is enhanced (potentiated) for hours or weeks after a brief period of high-frequency stimulation. The neurotransmitter released is glutamate, which acts on NMDA* glutamate receptors on the postsynaptic neurons. In some cases, induction of LTP depends on the release of nitric oxide (NO) from the postsynaptic neurons after they have been activated by glutamate. The NO, in turn, diffuses into the presynaptic neurons and causes LTP.

1. Describe how sleep and wakefulness are related to the reticular activating system (RAS).
2. What are the four stages of non-rapid eye movement (NREM) sleep? How is NREM sleep distinguished from rapid eye movement (REM) sleep?
3. Define memory. What are the two kinds of memory? What is memory consolidation?

---

*Named after the chemical *N*-methyl D-aspartate, which is used to detect this type of glutamate receptor.

DISORDERS: HOMEOSTATIC IMBALANCES

SPINAL CORD INJURY

The spinal cord may be damaged by a tumor either within or adjacent to the spinal cord, herniated intervertebral discs, blood clots, penetrating wounds caused by projectile fragments, or other trauma. In many cases of traumatic injury of the spinal cord, the patient has an improved outcome if an anti-inflammatory corticosteroid drug called methylprednisolone is given within 8 hours of the injury.

Depending on the location and extent of spinal cord damage, paralysis may occur. **Monoplegia** (*mono-* = one; *-plegia* = blow or strike) is paralysis of one limb only. **Diplegia** (*di-* = two) is paralysis of both upper limbs or both lower limbs. **Paraplegia** (*para-* = beyond) is paralysis of both lower limbs. **Hemiplegia** (*hemi-* = half) is paralysis of the upper limb, trunk, and lower limb on one side of the body, and **quadriplegia** (*quad-* = four) is paralysis of all four limbs.

Complete transection of the spinal cord means that the cord is severed from one side to the other, thus cutting all sensory and motor tracts. It results in a loss of all sensations and voluntary movement *below* the level of the transection. **Hemisection** is a partial transection of the cord on either the right or left side. Below the level of the hemisection there is a loss of proprioception and discriminative touch sensations on the same side as the injury if the posterior column is cut, paralysis on the same side if the lateral corticospinal tract is cut, and loss of pain, temperature, and crude tactile sensations on the opposite side if the spinothalamic tracts are cut.

Following complete transection, and to varying degrees after hemisection, spinal shock occurs. **Spinal shock** is an immediate response to spinal cord injury characterized by temporary loss of reflex function, called **areflexia** (a'-rē-FLEK-sē-a), below the level of the injury. Signs of acute spinal shock include slow heart rate, low blood pressure, flaccid paralysis of skeletal muscles, loss of somatic sensations, and urinary bladder dysfunction. Spinal shock may begin within 1 hour after injury and may last from several minutes to several months, after which reflex activity gradually returns.

CEREBRAL PALSY

Cerebral palsy (CP) is a group of motor disorders that cause loss of muscle control and coordination. It is due to damage of the motor areas of the brain during fetal life, birth, or infancy and occurs in about 2 of every 1000 children. One cause is infection of the mother by the German measles (rubella) virus during her first 3 months of pregnancy. Radiation during fetal life, temporary lack of oxygen during birth, and hydrocephalus during infancy may also cause cerebral palsy. Cerebral palsy is not a progressive disease; it does not worsen as time elapses. Once the damage is done, however, it is irreversible.

PARKINSON DISEASE

Parkinson disease (PD) is a progressive disorder of the CNS that typically affects its victims around age 60. The cause is un-

known, but toxic environmental factors are suspected, in part because only 5% of PD patients have a family history of the disease. Neurons that extend from the substantia nigra to the basal ganglia, where they release the neurotransmitter dopamine (DA), degenerate in PD. Also in the basal ganglia—in the caudate nucleus—are neurons that liberate the neurotransmitter acetylcholine (ACh). Although the level of ACh does not change as the level of DA declines, the imbalance of neurontransmitter activity—too little DA and too much ACh—is thought to bring about most of the symptoms.

In PD patients, involuntary skeletal muscle contractions often interfere with voluntary movement. For instance, the muscles of the upper limb may alternately contract and relax, causing the hand to shake. This shaking, called **tremor,** is the most common symptom of PD. Also, muscle tone may increase greatly, causing rigidity of the involved body part. Rigidity of the facial muscles gives the face a masklike appearance. The expression is characterized by a wide-eyed, unblinking stare and a slightly open mouth with uncontrolled drooling.

Motor performance is also impaired by **bradykinesia** (*brady-* = slow), in which activities such as shaving, cutting food, and buttoning a blouse take longer and become increasingly more difficult as the disease progresses. Muscular movements are performed not only slowly but with decreasing range of motion, or **hypokinesia** (*hypo-* = under). For example, handwritten letters get smaller, become poorly formed, and eventually become illegible. Often, walking is impaired; steps become shorter and shuffling, and arm swing diminishes. Even speech may be affected.

Treatment of PD is directed toward increasing levels of DA and decreasing levels of ACh. Although people with PD do not manufacture enough dopamine, taking it orally is useless because DA cannot cross the blood–brain barrier. Even though symptoms are partially relieved by a drug developed in the 1960s called levodopa (L-dopa), a precursor of DA, the drug does not slow the progression of the disease. As more and more affected brain cells die, the drug becomes useless. In an alternative approach, a drug called selegiline (deprenyl) inhibits monoamine oxidase, which is one of the enzymes that degrades catecholamine neurotransmitters, including dopamine. This drug slows progression of PD and may be used together with levodopa.

For more than a decade, surgeons have sought to reverse the effects of Parkinson disease by transplanting dopamine-rich fetal nerve tissue into the basal ganglia (usually the putamen) of patients with severe PD. Only a few postsurgical patients have shown any degree of improvement, such as less rigidity and improved quickness of motion. A more recent surgical technique that has produced improvement for some patients is *pallidotomy,* in which a part of the globus pallidus that generates tremors and produces muscle rigidity is destroyed.

MEDICAL TERMINOLOGY

Ataxia (a-TAK-sē-a; *a-* = without; *-taxia* = coordination) Failure of muscular coordination; a common sign of cerebellar trauma or disease. Blindfolded people with ataxia cannot touch the tip of their nose with a finger because they cannot coordinate movements with their sense of where a body part is located. Often, speech is slurred due to a lack of coordination of speech muscles.

Huntington's chorea (KŌ-rē-a; = a dance) A hereditary disorder of the basal ganglia in which a person has purposeless and involuntary jerky contractions of skeletal muscles and progressive mental deterioration. Symptoms often do not appear until age 30 or 40.

Insomnia (in-SOM-nē-a; *in-* = not; *-somnia* = sleep) Difficulty in falling asleep and staying asleep.

Narcolepsy (NAR-kō-lep-sē; *narco-* = numbness; *-lepsy* = a seizure) A condition in which REM sleep cannot be inhibited during waking periods. As a result, involuntary periods of sleep that last about 15 minutes occur throughout the day.

Sleep apnea (AP-nē-a; *a-* = without; *-pnea* = breath) A disorder in which a person repeatedly stops breathing for 10 or more seconds while sleeping. Most often it occurs because loss of muscle tone in pharyngeal muscles allows the airway to collapse.

STUDY OUTLINE

SENSATION (p. 485)

1. Sensation is a conscious or subconscious awareness of external or internal stimuli.
2. The nature of a sensation and the type of reaction generated vary according to the destination of sensory impulses in the CNS.
3. Each different type of sensation is a sensory modality; usually, a given sensory neuron serves only one modality.
4. General senses include somatic senses (touch, pressure, vibration, warmth, cold, pain, itch, tickle, and proprioception) and visceral senses; special senses include the modalities of smell, taste, vision, hearing, and equilibrium.
5. For a sensation to arise, four events typically occur: stimulation, transduction, generation of impulses, and integration.
6. Simple receptors, consisting of free nerve endings and encapsulated nerve endings, are associated with the general senses; complex receptors are associated with the special senses.
7. Sensory receptors respond to stimuli by producing receptor or generator potentials.
8. Table 15.1 on page 488 summarizes the classification of sensory receptors.
9. Adaptation is a decrease in sensitivity during a long-lasting stimulus. Receptors are either rapidly adapting (phasic) or slowly adapting (tonic).

SOMATIC SENSATIONS (p. 489)

1. Somatic sensations include tactile sensations (touch, pressure, vibration, itch, and tickle), thermal sensations (warmth and cold), pain, and proprioception.
2. Receptors for tactile, thermal, and pain sensations are located in the skin, subcutaneous layer, and mucous membranes of the mouth, vagina, and anus.
3. Receptors for proprioceptive sensations (position and movement of body parts) are located in muscles, tendons, joints, and the inner ear.
4. Receptors for touch are (a) hair root plexuses and corpuscles of touch (Meissner corpuscles), which are rapidly adapting, and (b) slowly adapting type I cutaneous mechanoreceptors (tactile or Merkel discs). Type II cutaneous mechanoreceptors (Ruffini

corpuscles), which are slowly adapting, are sensitive to stretching. Receptors for pressure include corpuscles of touch, type I mechanoreceptors, and lamellated (Pacinian) corpuscles. Receptors for vibration are corpuscles of touch and lamellated corpuscles. Itch receptors are free nerve endings, whereas both free nerve endings and lamellated corpuscles mediate the tickle sensation.
5. Thermoreceptors are free nerve endings. Cold receptors are located in the stratum basale of the epidermis; warm receptors are located in the dermis.
6. Pain receptors (nociceptors) are free nerve endings that are located in nearly every body tissue.
7. Nerve impulses for fast pain propagate along medium-diameter, myelinated A-delta fibers, whereas those for slow pain conduct along small-diameter unmyelinated C fibers.
8. Proprioceptors include muscle spindles, tendon organs, joint kinesthetic receptors, and hair cells of the inner ear.
9. Table 15.2 on page 494 summarizes the somatic sensory receptors and the sensations they convey.

SOMATIC SENSORY PATHWAYS (p. 494)

1. Somatic sensory pathways from receptors to the cerebral cortex involve three-neuron sets: first-order, second-order, and third-order neurons.
2. Axon collaterals (branches) of somatic sensory neurons simultaneously carry signals into the cerebellum and the reticular formation of the brain stem.
3. Impulses propagating along the posterior column–medial lemniscus pathway relay discriminative touch, stereognosis, proprioception, weight discrimination, and vibratory sensations.
4. The neural pathway for pain and temperature sensations is the lateral spinothalamic tract.
5. The neural pathway for tickle, itch, crude touch, and pressure sensations is the anterior spinothalamic pathway.
6. Specific regions of the primary somatosensory area (postcentral gyrus) of the cerebral cortex receive information from different parts of the body.

7. The pathways to the cerebellum are the anterior and posterior spinocerebellar tracts, which transmit impulses for subconscious muscle and joint position sense from the trunk and lower limbs.

8. Table 15.3 on page 498 summarizes the major somatic sensory pathways.

SOMATIC MOTOR PATHWAYS (p. 498)

1. The primary motor area (precentral gyrus) of the cortex is the major control region for initiation of voluntary movement.

2. Impulses governing voluntary movements propagate from the motor cortex to somatic motor neurons that innervate skeletal muscles via the direct pathways. The simplest pathways consist of upper and lower motor neurons.

3. The direct (pyramidal) pathways include the lateral and anterior corticospinal tracts and corticobulbar tracts.

4. Indirect (extrapyramidal) pathways involve the motor cortex, basal ganglia, limbic system, thalamus, cerebellum, reticular formation, and nuclei in the brain stem.

5. The basal ganglia program automatic movements, regulate muscle tone, and inhibit other motor neuron circuits.

6. The cerebellum is active in learning and performing rapid, coordinated, highly skilled movements. It also contributes to maintaining balance and posture.

7. Table 15.4 on page 501 summarizes the major somatic motor pathways.

INTEGRATIVE FUNCTIONS OF THE CEREBRUM (p. 502)

1. Sleep and wakefulness are integrative functions that are controlled by the suprachiasmatic nucleus and the reticular activating system (RAS).

2. Non-rapid eye movement (NREM) sleep consists of four stages identified by characteristic waves in EEG recordings.

3. Most dreaming occurs during rapid eye movement (REM) sleep.

4. Memory, the ability to store and recall thoughts, involves persistent changes in the brain, a capability called plasticity.

5. Memory is generally classified into two kinds: short-term memory and long-term memory.

6. Short-term memory is thought to be related to electrical and chemical events; long-term memory is more likely to depend on anatomical and biochemical changes at synapses.

SELF-QUIZ QUESTIONS

Complete the following:

1. ___ is the conscious or subconscious awareness of external or internal stimuli.

2. ___ is a decline in sensitivity during a maintained, constant stimulus.

3. Touch, pressure, and vibration are all examples of ___ sensations.

Choose the best answer to the following questions:

4. Which of the following is a modality of the general senses? (a) smell, (b) taste, (c) touch, (d) hearing, (e) vision

5. Which of the following events typically occur if a sensation is to arise? (1) stimulation of the sensory receptor, (2) generation of a nerve impulse, (3) transduction of the stimulus, (4) integration of sensory input, (5) effector response to the stimulus.
(a) 1, 2, 3, and 4, (b) 2, 3, 4, and 5, (c) 1, 3, 4, and 5, (d) 1, 4, and 5, (e) 2, 4, and 5

6. Which of the following statement is *false*? (a) First-order sensory neurons carry signals from the somatic receptors into either the brain stem or spinal cord. (b) Second-order neurons carry signals from the spinal cord and brain stem to the thalamus. (c) Third-order neurons project to the primary somatosensory area of the cortex, where conscious perception of the sensation results. (d) The somatic sensory pathways to the cerebellum are the posterior column–medial lemniscus pathway and the anterolateral pathway. (e) Axons of second-order neurons decussate in the spinal cord or brain stem before ascending to the thalamus.

7. Which of the following is *not* an aspect of cerebellar function? (a) monitoring movement intention, (b) monitoring actual movement, (c) comparing intent with actual performance, (d) sending out corrective signals, (e) directing sensory input to effectors

8. The autonomic nervous system is most active during (a) stage 1 NREM sleep, (b) REM sleep, (c) stage 4 NREM sleep, (d) stage 2 NREM sleep, (e) stage 3 NREM sleep

9. Which of the following statements is *false*? (a) Learning is the ability to acquire new knowledge or skills through instruction or experience. (b) Memory is the process of knowledge retention over time. (c) For an experience to become a memory, it must produce persistent functional changes in the brain that represent the experience. (d) Human memory records everything as if it were a magnetic tape. (e) The capacity for change associated with learning is termed plasticity.

10. Which of the following statements is *incorrect*? (a) The graded potentials produced by receptors that serve the senses of touch, pressure, stretching, vibration, pain, proprioception, and smell are generator potentials. (b) Graded potentials produced by receptors that serve the special senses of vision, hearing, equilibrium, and taste are receptor potentials. (c) When a generator potential is large enough to reach threshold, it generates one or more nerve impulses in its first-order sensory neuron. (d) A receptor potential generates a nerve impulse in the second-order neuron. (e) The amplitude of both generator and receptor potentials varies with the intensity of the stimulus.

True or false:

11. Sensory input is filtered to ensure that only a small portion reaches the conscious perception at any given time.

12. Sleep walking can occur during stage 4 NREM sleep.

13. Match the following:

____ (a) receptors located in muscles, tendons, joints, and the inner ear

____ (b) receptors located in blood vessels, visceral organs, muscles, and the nervous system

____ (c) receptors that detect temperature changes

____ (d) receptors that detect light that strikes the retina of the eye

____ (e) receptors located at or near the external surface of the body

____ (f) receptors that provide information about body position, muscle tension, and position and activity of joints

____ (g) receptors that detect chemicals in the mouth, nose, and body fluids

____ (h) receptors that detect mechanical pressure or stretching

____ (i) receptors that respond to stimuli resulting from physical or chemical damage to tissues

(1) exteroceptors
(2) interoceptors
(3) proprioceptors
(4) mechanoreceptors
(5) thermoreceptors
(6) nociceptors
(7) photoreceptors
(8) chemoreceptors

14. Match the following:

____ (a) the major control region of the cerebral cortex for initiation of voluntary movements

____ (b) direct pathways conveying impulses from the cortex to the spinal cord that result in precise, voluntary movements

____ (c) contain motor neurons that control skilled movements of the hands and feet

____ (d) tracts include rubrospinal, tectospinal, vestibulospinal, lateral reticulospinal, and medial reticulospinal

____ (e) contain neurons that help program habitual or automatic movement sequences, such as arm swinging during walking

____ (f) mainly carry pain and temperature impulses

____ (g) the major routes whereby proprioceptive input reaches the cerebellum

____ (h) include the fasciculus gracilis and fasciculus cuneatus

____ (i) contain motor neurons that coordinate movements of the axial skeleton

____ (j) contain axons that convey impulses for precise, voluntary movements of the eyes, tongue, and neck, plus chewing, facial expression, and speech

____ (k) convey sensations of discriminative touch, stereognosis, proprioception, and weight discrimination to the cerebral cortex.

(1) posterior column
(2) anterolateral pathways
(3) spinocerebellar pathways
(4) lateral corticospinal tracts
(5) anterior corticospinal tracts
(6) corticobulbar pathways
(7) extrapyramidal pathways
(8) pyramidal pathways
(9) precentral gyrus
(10) basal ganglia
(11) posterior column–medial lemniscus pathways

15. Match the following:

____ (a) specialized groupings of muscle fibers interspersed among regular skeletal muscle fibers and oriented parallel to them; monitor changes in the length of a skeletal muscle

____ (b) inform the CNS about changes in muscle tension

____ (c) free nerve ending receptors for pain

____ (d) encapsulated receptors for touch located in the dermal papillae; found in hairless skin, eyelids, tip of the tongue, and lips

____ (e) lamellated corpuscles that detect pressure

____ (f) type II cutaneous mechanoreceptors; most sensitive to stretching that occurs as digits or limbs are moved

____ (g) located in the stratum basale and activated by low temperatures

____ (h) located in the dermis and activated by high temperatures

____ (i) found within and around the articular capsules of synovial joints; respond to pressure and acceleration and deceleration of joints

____ (j) type I cutaneous mechanoreceptors that function in discriminative touch

(1) Meissner corpuscles
(2) Merkel discs
(3) Ruffini corpuscles
(4) Pacinian corpuscles
(5) cold receptors
(6) warm receptors
(7) nociceptors
(8) Golgi tendon organs
(9) joint kinesthetic receptors
(10) muscle spindles

CRITICAL THINKING QUESTIONS

1. When Joni first stepped onto the sailboat, she smelled the tangy sea air and felt the motion of water beneath her feet. After a few minutes, she no longer noticed the smell, but unfortunately she was aware of the rolling motion for hours. What types of receptors are involved in smell and detection of motion? Why did her sensation of smell fade but the rolling sensation remain? (HINT: *One way to classify receptors is by the type of stimulus that they detect.*)

2. Damara was laughing uncontrollably; her brother had her by the leg and was tickling her foot. How does her brain know that her foot is being tickled? (HINT: *If you blocked the pathway for tickle, you'd still be able to feel pain.*)

3. Yoshio used to predict the weather by how much the bunion (abnormally swollen joint on big toe) on his left foot was bothering him. Last year, Yoshio's left foot was amputated due to complications from diabetes, but sometimes he still thinks he feels that bunion. Explain his experience. (HINT: *How did Yoshio's toe communicate with his brain before the amputation?*)

ANSWERS TO FIGURE QUESTIONS

15.1 The special senses of vision, hearing, taste, and equilibrium are served by separate sensory cells.

15.2 Pain, thermal sensations, and tickle and itch involve free nerve endings.

15.3 The kidneys have the broadest area for referred pain.

15.4 Muscle spindles are activated when the central areas of their intrafusal fibers are stretched.

15.5 Damage to the right lateral spinothalamic tract could result in loss of pain and thermal sensations on the left side of the body.

15.6 The hand has a larger representation in the motor area than in the somatosensory area, which implies greater precision in the hand's movement control than discriminative ability in its sensation.

15.7 The corticobulbar and rubrospinal tracts (see Table 15.4 on page 501) convey impulses that result in precise, voluntary movements.

15.8 Feedback loops: cortex to basal ganglia to thalamus to cortex; cortex to brain stem nuclei to cerebellum to thalamus to cortex.

15.9 The anterior and posterior spinocerebellar tracts carry information from proprioceptors in joints and muscles to the cerebellum.

15.10 A smoke detector responds to smoke by sounding a loud bell or buzzer, which awakens sleepers by providing auditory input that stimulates the RAS.

15.11 Dreaming and paralysis of skeletal muscles occur during REM sleep.

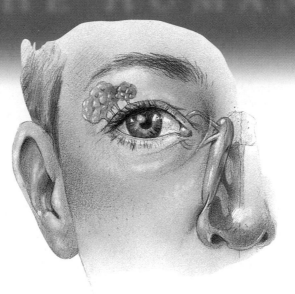

Receptors for the special senses—smell, taste, vision, hearing, and equilibrium—are housed in complex sensory organs such as the eyes and ears. Moreover, receptors for special senses are separate receptor cells that are usually embedded in epithelial tissue within the special sense organs. In this chapter we examine the structure and function of the special sense organs, and the pathways involved in conveying information from them to the central nervous system. **Ophthalmology** (of-thal-MOL-ō-jē; *ophthalmo-* = eye; *-ology* = study of) is the science that deals with the eye and its disorders. The other special senses are, in large part, the concern of **otolaryngology** (ō-tō-lar-in-GOL-ō-jē; *oto-* = ear; *laryngo-* = larynx).

OLFACTION: SENSE OF SMELL

OBJECTIVE

• *Describe the olfactory receptors and the neural pathway for olfaction.*

Both smell and taste are chemical senses; the sensations arise from the interaction of molecules with smell or taste receptors. Because impulses for smell and taste propagate to the limbic system (and to higher cortical areas as well), certain odors and tastes can evoke strong emotional responses or a flood of memories.

Anatomy of Olfactory Receptors

The nose contains 10–100 million receptors for the sense of smell or **olfaction** (ol-FAK-shun; *olfact-* = smell).

The total area of olfactory epithelium is 5 cm² (a little less than 1 in.²). It occupies the superior portion of the nasal cavity, covering the inferior surface of the cribriform plate and extending along the superior nasal concha and upper part of the middle nasal concha (Figure 16.1a). The olfactory epithelium consists of three kinds of cells: olfactory receptors, supporting cells, and basal stem cells (Figure 16.1b).

Olfactory receptors, which are the first-order neurons of the olfactory pathway, are bipolar neurons whose exposed tip is a knob-shaped dendrite. The sites of olfactory transduction are the **olfactory hairs,** which are cilia that project from the dendrite. Olfactory receptors respond to the chemical stimulation of an odorant molecule by producing a generator potential, thus initiating the olfactory response. From the base of each olfactory receptor, a single axon projects through the cribriform plate and into the olfactory bulb.

Supporting cells are columnar epithelial cells of the mucous membrane lining the nose. They provide physical support, nourishment, and electrical insulation for the olfactory receptors, and they help detoxify chemicals that come in contact with the olfactory epithelium. **Basal stem cells** lie between the bases of the supporting cells and continually undergo cell division to produce new olfactory receptors, which live for only a month or so before being replaced. This process is remarkable because olfactory receptors are neurons, and in general, mature neurons are not replaced.

Within the connective tissue that supports the olfactory epithelium are **olfactory (Bowman's) glands,** which produce mucus that is carried to the surface of the epithelium by ducts. The secretion moistens the surface of the olfactory ep-

Figure 16.1 Olfactory epithelium and olfactory receptors. (a) Location of olfactory epithelium in nasal cavity. (b) Anatomy of olfactory receptors, consisting of first-order neurons whose axons extend through the cribriform plate and terminate in the olfactory bulb. (See Tortora, *A Photographic Atlas of the Human Body,* Figures 9.1b and 9.1c)

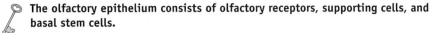

The olfactory epithelium consists of olfactory receptors, supporting cells, and basal stem cells.

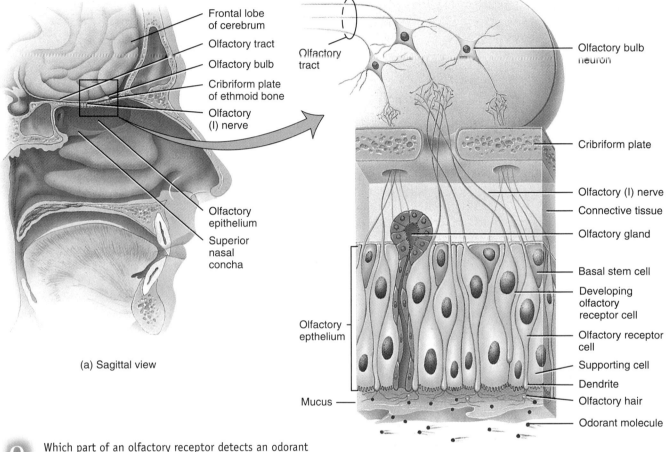

(a) Sagittal view

(b) Enlarged aspect of olfactory receptors

Q Which part of an olfactory receptor detects an odorant molecule?

ithelium and dissolves odorants. Both supporting cells of the nasal epithelium and olfactory glands are innervated by branches of the facial (VII) nerve, which can be stimulated by certain chemicals. Impulses in these nerves, in turn, stimulate the lacrimal glands in the eyes and nasal mucous glands. The result is tears and a runny nose after inhaling substances such as pepper or the vapors of household ammonia.

Physiology of Olfaction

Many attempts have been made to distinguish among and classify "primary" sensations of smell. Genetic evidence now suggests the existence of hundreds of primary scents. Our ability to recognize about 10,000 different scents probably depends on patterns of activity in the brain that arise from activation of many different combinations of olfactory receptors.

Olfactory receptors react to odorant molecules in the same way that most sensory receptors react to their specific stimuli: A generator potential (depolarization) develops and triggers one or more nerve impulses. How the generator potential arises is known in some cases. Some odorants bind to receptors that are linked to G proteins in the plasma membrane and activate the enzyme adenylate cyclase (see page 571). The result is the following chain of events: opening of sodium ion (Na^+) channels → inflow of Na^+ → depolarizing generator potential → nerve impulses.

Odor Thresholds and Adaptation

Olfaction, like all the special senses, has a low threshold: Only a few molecules of certain substances need be present in air to be perceived as an odor. A good example is the chemical methyl mercaptan, which smells like rotten cabbage and can be detected in concentrations as low as 1/25 billionth of a milligram per milliliter of air. Because the natural gas used for cooking and heating is odorless but lethal and potentially explosive if it accumulates, a small amount of methyl mercaptan is added to provide olfactory warning of gas leaks.

Adaptation (decreasing sensitivity) to odors occurs rapidly. Olfactory receptors adapt by about 50% in the first second or so after stimulation and adapt very slowly thereafter. Still, complete insensitivity to certain strong odors occurs in about a minute after exposure; apparently, reduced sensitivity also involves some sort of adaptation process in the central nervous system as well.

The Olfactory Pathway

On each side of the nose, bundles of the slender, unmyelinated axons of olfactory receptors extend through about 20 olfactory foramina in the cribriform plate of the ethmoid bone (see Figure 16.1b). Collectively, these 40 or so bundles of axons, termed the **olfactory (I) nerves,** terminate in the brain in paired masses of gray matter called the **olfactory bulbs,** which are located inferior to the frontal lobes of the cerebrum and lateral to the crista galli of the ethmoid bone. Within the olfactory bulbs the axon terminals of olfactory receptors—the first-order neurons—form synapses with the dendrites and cell bodies of second-order neurons in the olfactory pathway.

Axons of olfactory bulb neurons extend posteriorly and form the **olfactory tract** (see Figure 16.1b). These axons project to the lateral olfactory area, which is located at the inferior and medial surface of the temporal lobe. This cortical region is part of the limbic system and includes some of the amygdala (see Figure 14.14 on page 467). Because many olfactory tract axons terminate in the lateral olfactory area, it is considered the primary olfactory area, where conscious awareness of smells begins. Connections to other limbic system regions and to the hypothalamus probably account for our emotional and memory-evoked responses to odors. Examples include sexual excitement upon smelling a certain perfume, nausea upon smelling a food that once made you violently ill, or an odor-evoked memory of a childhood experience. From the lateral olfactory area, pathways also extend to the frontal lobe, both directly and indirectly via the thalamus. An important region for odor identification and discrimination is the orbitofrontal area, corresponding to Brodmann's area 11 (see Figure 14.15 on page 469). People who suffer damage in this area have difficulty identifying different odors. Positron emission tomography (PET) studies suggest some degree of hemispheric lateralization: The orbitofrontal area of the *right* hemisphere exhibits greater activity during olfactory processing.

1. How do basal stem cells contribute to olfaction?
2. Describe the sequence of events from the binding of an odorant molecule to an olfactory hair to the arrival of a nerve impulse in an olfactory bulb.

GUSTATION: SENSE OF TASTE
OBJECTIVE
• *Describe the gustatory receptors and the neural pathway for gustation.*

Like olfaction, taste is a chemical sense; to be detected by either sense, stimulating molecules must be dissolved. Taste or **gustation** (GUS-tā-shun; *gust-* = taste) is much simpler than olfaction in that only four major classes of stimuli can be distinguished: sour, sweet, bitter, and salty. All other "tastes," such as chocolate, pepper, and coffee, are combinations of these four, plus the accompanying olfactory sensations. Odors from food can pass upward from the mouth into the nasal cavity, where they stimulate olfactory receptors. Because olfaction is much more sensitive than taste, a given concentration of a food substance may stimulate the olfactory system thousands of times more strongly than it stimulates the gustatory system. When persons with colds or allergies complain that they cannot taste their food, they are reporting blockage of olfaction, not of taste.

Anatomy of Gustatory Receptors

The receptors for sensations of taste or are located in the taste buds (Figure 16.2). The nearly 10,000 taste buds of a young adult are mainly on the tongue, but they are also found on the soft palate (posterior portion of roof of mouth), pharynx (throat), and larynx (voice box). The number of taste buds declines with age. Each **taste bud** is an oval body consisting of three kinds of epithelial cells: supporting cells, gustatory receptor cells, and basal cells (see Figure 16.2c). The **supporting cells** surround about 50 **gustatory receptor cells.** A single, long microvillus, called a **gustatory hair,** projects from each gustatory receptor cell to the external surface through the **taste pore,** an opening in the taste bud. **Basal cells,** found at the periphery of the taste bud near the connective tissue layer, produce supporting cells, which then develop into gustatory receptor cells with a life span of about 10 days. At their base, the receptor cells synapse with dendrites of first-order neurons that form the first part of the gustatory pathway. The dendrites of each first-order neuron branch profusely and contact many receptors in several taste buds.

Taste buds are found in elevations on the tongue called **papillae** (pa-PIL-ē), which give the upper surface of the tongue its rough appearance (see Figure 16.2a, b). **Circumvallate papillae** (ser-kum-VAL-āt), the largest type, are circular and form an inverted V-shaped row at the posterior portion of the tongue. **Fungiform papillae** (FUN-ji-form) are mushroom-shaped elevations scattered over the entire surface of the tongue. All circumvallate and most fungiform papillae contain taste buds. In addition, the entire surface of the tongue has **filiform papillae** (FIL-i-form), pointed, threadlike structures that rarely contain taste buds.

Figure 16.2 The relationship of gustatory receptors in taste buds to tongue papillae. (See Tortora, *A Photographic Atlas of the Human Body,* Figure 9.2)

 Gustatory (taste) receptors are located in taste buds.

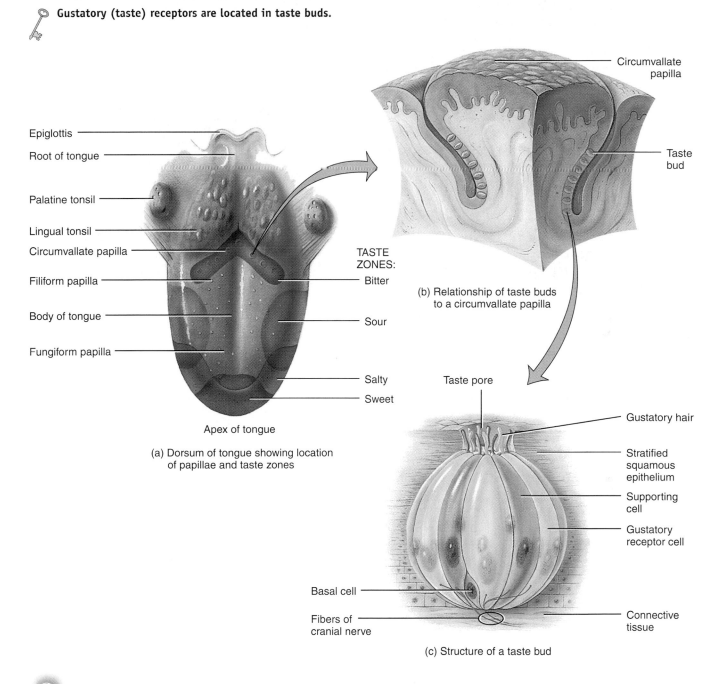

(a) Dorsum of tongue showing location of papillae and taste zones

(b) Relationship of taste buds to a circumvallate papilla

(c) Structure of a taste bud

Q   Beginning at the receptors, what structures form the gustatory pathway?

Physiology of Gustation

Once a chemical is dissolved in saliva, it can make contact with the plasma membrane of the gustatory hairs, which are the sites of taste transduction. The result is a receptor potential that stimulates exocytosis of neurotransmitter-containing synaptic vesicles from the gustatory receptor cell. Nerve impulses may then arise in the first-order sensory neurons that synapse with gustatory receptor cells.

Individual gustatory receptor cells may respond to more than one of the four primary tastes, but receptors in certain regions of the tongue are more sensitive than others to the primary taste sensations (see Figure 16.2a). Receptors in the tip of the tongue are highly sensitive to sweet and salty substances, receptors in the posterior portion of the tongue are highly sensitive to bitter substances, and those on the lateral aspects of the tongue are most sensitive to sour substances.

Taste Thresholds and Adaptation

The threshold for taste varies for each of the primary tastes; the threshold for bitter substances, as measured by using quinine, is lowest. Because poisonous substances often are bitter, the low threshold (or high sensitivity) may have a protective function. The threshold for sour substances, as measured by using hydrochloric acid, is somewhat higher. The thresholds for salty substances, represented by sodium chloride, and for sweet substances, as measured by using sucrose, are about the same and are higher than those for bitter or sour substances.

Complete adaptation to a specific taste can occur in 1–5 minutes of continuous stimulation. Taste adaptation is due to changes that occur in the taste receptors, in olfactory receptors, and in neurons of the gustatory pathway in the CNS.

The Gustatory Pathway

Three cranial nerves include first-order gustatory fibers from taste buds. The facial (VII) nerve serves the anterior two-thirds of the tongue, the glossopharyngeal (IX) nerve serves the posterior one-third of the tongue, and the vagus (X) nerve serves the throat and epiglottis (cartilage lid over the voice box). From taste buds, impulses propagate along these cranial nerves to the medulla oblongata. From the medulla, some taste fibers project to the limbic system and the hypothalamus, whereas others project to the thalamus. Taste fibers extend from the thalamus to the primary gustatory area in the parietal lobe of the cerebral cortex (see area 43 in Figure 14.15 on page 469), giving rise to the conscious perception of taste.

1. How do olfactory receptors and taste receptors differ in structure and function?
2. Compare the olfactory and gustatory pathways.

VISION

OBJECTIVES

• *List and describe the accessory structures of the eye and the structural components of the eyeball.*

• *Discuss image formation by describing refraction, accommodation, and constriction of the pupil.*

More than half the sensory receptors in the human body are located in the eyes, and a large part of the cerebral cortex is devoted to processing visual information. In this section of the chapter we examine the accessory structures of the eye, the eyeball, how visual images are formed, the physiology of vision, and the visual pathway.

Accessory Structures of the Eye

The **accessory structures** of the eye are the eyelids, eyelashes, eyebrows, the lacrimal (tearing) apparatus, and extrinsic eye muscles.

Figure 16.3 Surface anatomy of the right eye.

🔑 **The palpebral fissure is the space between the upper and lower eyelids that exposes the eyeball.**

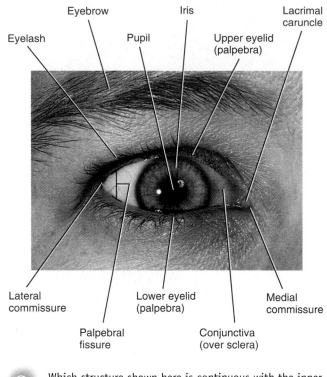

Eyebrow — Iris — Lacrimal caruncle
Eyelash — Pupil — Upper eyelid (palpebra)
Lateral commissure — Lower eyelid (palpebra) — Medial commissure
Palpebral fissure — Conjunctiva (over sclera)

Q Which structure shown here is continuous with the inner lining of the eyelids?

Eyelids

The upper and lower **eyelids,** or **palpebrae** (PAL-pe-brē), shade the eyes during sleep, protect the eyes from excessive light and foreign objects, and spread lubricating secretions over the eyeballs (Figure 16.3). The upper eyelid is more movable than the lower and contains in its superior region the **levator palpebrae superioris muscle.** The space between the upper and lower eyelids that exposes the eyeball is the **palpebral fissure.** Its angles are known as the **lateral commissure** (KOM-i-shur), which is narrower and closer to the temporal bone, and the **medial commissure,** which is broader and nearer the nasal bone. In the medial commissure is a small, reddish elevation, the **lacrimal caruncle** (KAR-ung-kul), that contains sebaceous (oil) glands and sudoriferous (sweat) glands. The whitish material that sometimes collects in the medial commissure comes from these glands.

From superficial to deep, each eyelid consists of epidermis, dermis, subcutaneous tissue, fibers of the orbicularis oculi muscle, a tarsal plate, tarsal glands, and conjunctiva (Figure 16.4a). The **tarsal plate** is a thick fold of connective tissue that gives form and support to the eyelids. Embedded in each tarsal plate is a row of elongated modified sebaceous

Figure 16.4 Accessory structures of the eye.

🔑 **Accessory structures of the eye are the eyelids, eyelashes, eyebrows, the lacrimal apparatus, and extrinsic eye muscles.**

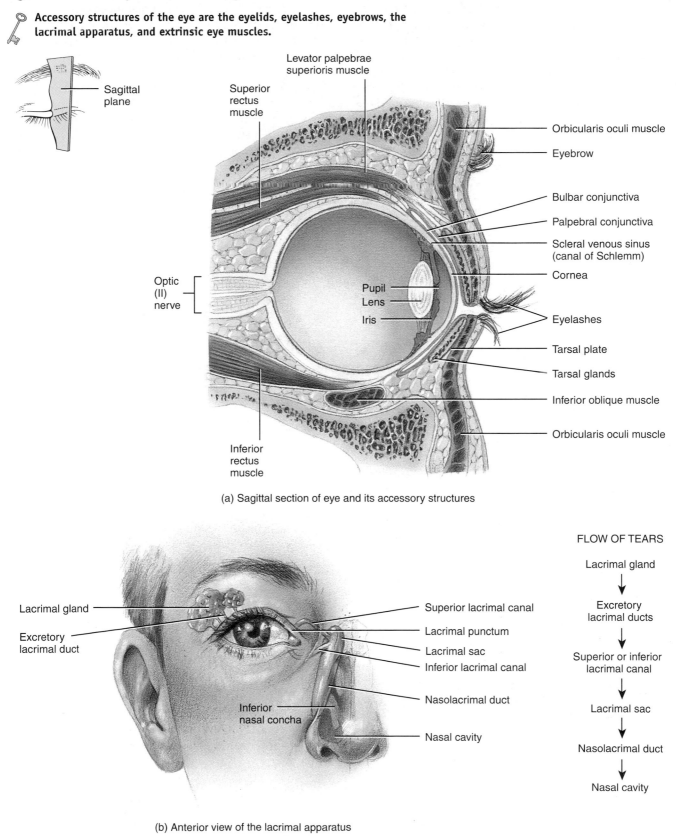

Sagittal plane

Levator palpebrae superioris muscle

Superior rectus muscle

Orbicularis oculi muscle

Eyebrow

Bulbar conjunctiva

Palpebral conjunctiva

Scleral venous sinus (canal of Schlemm)

Cornea

Pupil
Lens
Iris

Eyelashes

Tarsal plate

Tarsal glands

Inferior oblique muscle

Orbicularis oculi muscle

Optic (II) nerve

Inferior rectus muscle

(a) Sagittal section of eye and its accessory structures

Lacrimal gland

Excretory lacrimal duct

Inferior nasal concha

Superior lacrimal canal

Lacrimal punctum

Lacrimal sac

Inferior lacrimal canal

Nasolacrimal duct

Nasal cavity

(b) Anterior view of the lacrimal apparatus

FLOW OF TEARS

Lacrimal gland
↓
Excretory lacrimal ducts
↓
Superior or inferior lacrimal canal
↓
Lacrimal sac
↓
Nasolacrimal duct
↓
Nasal cavity

Q What is lacrimal fluid, and what are its functions?

glands, known as **tarsal** or **Meibomian glands** (mī-BŌ-mē-an), that secrete a fluid that helps keep the eyelids from adhering to each other. Infection of the tarsal glands produces a tumor or cyst on the eyelid called a **chalazion** (ka-LĀ-zē-on). The **conjunctiva** (kon′-junk-TĪ-va) is a thin, protective mucous membrane composed of stratified columnar epithelium with numerous goblet cells that is supported by areolar connective tissue. The **palpebral conjunctiva** lines the inner aspect of the eyelids, and the **bulbar conjunctiva** passes from the eyelids onto the anterior surface of the eyeball lateral to the cornea. Dilation and congestion of the blood vessels of the bulbar conjunctiva due to local irritation or infection are the cause of bloodshot eyes.

Eyelashes and Eyebrows

The **eyelashes,** which project from the border of each eyelid, and the **eyebrows,** which arch transversely above the upper eyelids, help protect the eyeballs from foreign objects, perspiration, and the direct rays of the sun. Sebaceous glands at the base of the hair follicles of the eyelashes, called **sebaceous ciliary glands,** release a lubricating fluid into the follicles. Infection of these glands is called a **sty.**

The Lacrimal Apparatus

The **lacrimal apparatus** (*lacrim-* = tears) is a group of structures that produces and drains **lacrimal fluid** or **tears.** The **lacrimal glands,** each about the size and shape of an almond, secrete lacrimal fluid, which drains into 6–12 **excretory lacrimal ducts** that empty tears onto the surface of the conjunctiva of the upper lid (Figure 16.4b). From here the tears pass medially over the anterior surface of the eyeball to enter two small openings called **lacrimal puncta.** Tears then pass into two ducts, the **lacrimal canals,** which lead into the **lacrimal sac** and then into the **nasolacrimal duct.** This duct carries the lacrimal fluid into the nasal cavity just inferior to the inferior nasal concha.

Lacrimal fluid is a watery solution containing salts, some mucus, and **lysozyme,** a protective bactericidal enzyme. The fluid protects, cleans, lubricates, and moistens the eyeball. After being secreted, lacrimal fluid is spread medially over the surface of the eyeball by the blinking of the eyelids. Each gland produces about 1 ml of lacrimal fluid per day.

Normally, tears are cleared away as fast as they are produced, either by evaporation or by passing into the lacrimal canals and then into the nasal cavity. If an irritating substance contacts the conjunctiva, however, the lacrimal glands are stimulated to oversecrete, and tears accumulate (watery eyes). Lacrimation is a protective mechanism, as the tears dilute and wash away the irritating substance. Watery eyes also occur when an inflammation of the nasal mucosa, such as occurs with a cold, obstructs the nasolacrimal ducts and blocks drainage of tears. Humans are unique in expressing emotions, both happiness and sadness, by **crying.** In response to parasympathetic stimulation, the lacrimal glands produce excessive lacrimal fluid that may spill over the edges of the eyelids and even fill the nasal cavity with fluid.

Extrinsic Eye Muscles

The six extrinsic eye muscles that move each eye receive their innervation from cranial nerves III, IV, or VI. These muscles are the **superior rectus, inferior rectus, lateral rectus, medial rectus, superior oblique,** and **inferior oblique** (see Figures 16.4 and 16.5). In general, the motor units in these muscles are small. Some motor neurons serve only two or three muscle fibers—fewer than in any other part of the body except the larynx (voice box)—which permits smooth, precise, and rapid movement of the eyes. As indicated in Exhibit 11.2 on page 316, the extrinsic eye muscles move the eyeball in various directions (laterally, medially, superiorly, or inferiorly). Looking to the right, for example, requires simultaneous contraction of the right lateral rectus and left medial rectus muscles and relaxation of the left lateral rectus and right medial rectus. The oblique muscles preserve rotational stability of the eyeball. Circuits in the brain stem and cerebellum coordinate and synchronize the movements of the eyes.

Anatomy of the Eyeball

The adult **eyeball** measures about 2.5 cm (1 in.) in diameter. Of its total surface area, only the anterior one-sixth is exposed; the remainder is recessed and protected by the orbit, into which it fits. Anatomically, the wall of the eyeball consists of three layers: fibrous tunic, vascular tunic, and retina.

Fibrous Tunic

The **fibrous tunic,** the superficial coat of the eyeball, is avascular and consists of the anterior cornea and posterior sclera (Figure 16.5). The **cornea** (KOR-nē-a) is a transparent coat that covers the colored iris. Because it is curved, the cornea helps focus light onto the retina. Its outer surface consists of nonkeratinized stratified squamous epithelium. The middle coat of the cornea consists of collagen fibers and fibroblasts, and the inner surface is simple squamous epithelium. The **sclera** (SKLE-ra; *scler-* = hard), the "white" of the eye, is a layer of dense connective tissue made up mostly of collagen fibers and fibroblasts. The sclera covers all of the eyeball except the cornea; it gives shape to the eyeball, makes it more rigid, and protects its inner parts. At the junction of the sclera and cornea is an opening known as the **scleral venous sinus (canal of Schlemm)** (see Figure 16.4a).

CLINICAL APPLICATION
Corneal Transplants

Corneal transplants are the most common organ transplant operation—and the most successful type of transplant as well. Because the cornea is avascular, blood-borne antibodies that might cause rejection do not enter the transplanted tissue, and rejection rarely occurs. The defective cornea is removed and a donor cornea of similar diameter is sewn in. The shortage of donor corneas has been partially overcome by the development of artificial corneas made of plastic. ■

Figure 16.5 Gross structure of the eyeball. (See Tortora, *A Photographic Atlas of the Human Body,* Figure 9.3b)

519

🔑 The wall of the eyeball consists of three layers: the fibrous tunic, the vascular tunic, and the retina.

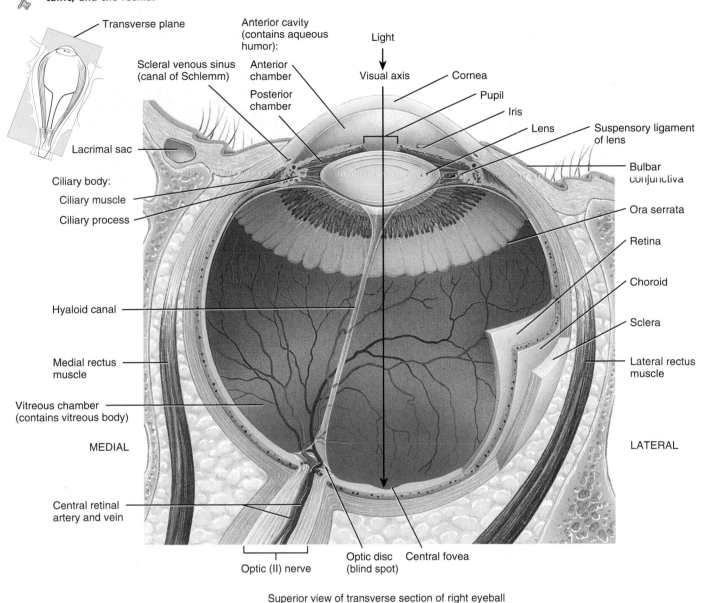

Superior view of transverse section of right eyeball

Q What are the components of the fibrous tunic and vascular tunic?

Vascular Tunic

The **vascular tunic** or **uvea** (YOO-vē-a) is the middle layer of the eyeball and has three parts: choroid, ciliary body, and iris (see Figure 16.5). The highly vascularized **choroid** (KŌ-royd), which is the posterior portion of the vascular tunic, lines most of the internal surface of the sclera. It provides nutrients to the posterior surface of the retina.

In the anterior portion of the vascular tunic, the choroid becomes the **ciliary body** (SIL-ē-ar′-ē). It extends from the **ora serrata** (Ō-ra ser-RĀ-ta), the jagged anterior margin of the retina, to a point just posterior to the junction of the sclera and cornea. The ciliary body consists of the ciliary

processes and the ciliary muscle. The **ciliary processes** are protrusions or folds on the internal surface of the ciliary body; they contain blood capillaries that secrete aqueous humor, and they also attach to suspensory ligaments, which connect to the lens. The **ciliary muscle** is a circular band of smooth muscle that alters the shape of the lens, adapting it for near or far vision. When the muscle contracts, it pulls on the ciliary body, which then pulls on the suspensory ligament and alters the shape of the lens.

The **iris,** the colored portion of the eyeball, is shaped like a flattened donut. It is suspended between the cornea

Figure 16.6 Responses of the pupil to light of varying brightness.

Contraction of the circular muscles cause constriction of the pupil; contraction of the radial muscles cause dilation of the pupil.

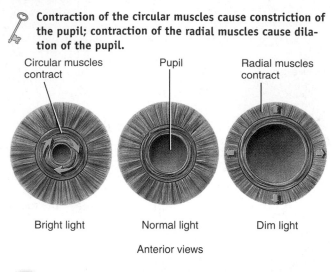

Circular muscles contract   Pupil   Radial muscles contract

Bright light   Normal light   Dim light

Anterior views

Q Which division of the autonomic nervous system causes pupillary constriction? Which causes pupillary dilation?

Figure 16.7 A normal retina, as seen through an ophthalmoscope.

Blood vessels in the retina can be viewed directly and examined for pathological changes.

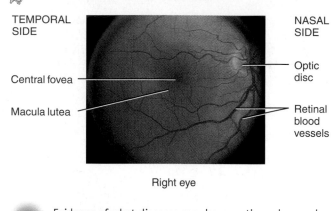

TEMPORAL SIDE   NASAL SIDE

Central fovea   Optic disc

Macula lutea   Retinal blood vessels

Right eye

Q Evidence of what diseases may be seen through an ophthalmoscope?

and the lens and is attached at its outer margin to the ciliary processes. It consists of circular and radial smooth muscle fibers. A principal function of the iris is to regulate the amount of light entering the vitreous chamber of the eyeball through the **pupil,** the hole in the center of the iris. Autonomic reflexes regulate pupil diameter in response to light levels (Figure 16.6). When bright light stimulates the eye, parasympathetic neurons stimulate the **circular muscles (constrictor pupillae)** of the iris to contract, causing a decrease in the size of the pupil (constriction). In dim light, sympathetic neurons stimulate the **radial muscles (dilator pupillae)** of the iris to contract, causing an increase in the pupil's size (dilation).

Retina

The third and inner coat of the eyeball, the **retina,** lines the posterior three-quarters of the eyeball and is the beginning of the visual pathway (see Figure 16.5). An ophthalmoscope allows an observer to peer through the pupil, providing a magnified image of the retina and the blood vessels that course across the retina's anterior surface (Figure 16.7). The surface of the retina is the only place in the body where blood vessels can be viewed directly and examined for pathological changes, such as those that occur with hypertension or diabetes mellitus. Several landmarks are visible. The **optic disc** is the site where the optic nerve exits the eyeball. Bundled together with the optic nerve are the **central retinal artery,** a branch of the ophthalmic artery, and the **central retinal vein.** Branches of the central retinal artery fan out to nourish the anterior surface of the retina; the central retinal vein drains blood from the retina through the optic disc.

The retina consists of a pigment epithelium (nonvisual portion) and a neural portion (visual portion). The **pigment epithelium** is a sheet of melanin-containing epithelial cells that lies between the choroid and the neural portion of the retina. (Some histologists classify it as part of the choroid, not part of the retina.) Melanin in the choroid and in the pigment epithelium absorbs stray light rays, which prevents reflection and scattering of light within the eyeball. As a result, the image cast on the retina by the cornea and lens remains sharp and clear. Albinos lack melanin in all parts of the body, including the eye. They often need to wear sunglasses, even indoors, because even moderately bright light is perceived as bright glare due to light scattering.

The **neural portion** of the retina is a multilayered outgrowth of the brain that extensively processes visual data before transmitting nerve impulses to the thalamus. Three distinct layers of retinal neurons—the **photoreceptor layer,** the **bipolar cell layer,** and the **ganglion cell layer**—are separated by two zones, the outer and inner synaptic layers, where synaptic contacts are made (Figure 16.8). Note that light passes through the ganglion and bipolar cell layers and both synaptic layers before reaching the photoreceptor layer. Two other types of cells present in the retina are called **horizontal cells** and **amacrine cells.** These cells form laterally directed pathways that modify the signals being transmitted along the pathway from photoreceptors to bipolar cells to ganglion cells.

Two types of photoreceptors are specialized to transduce light rays into receptor potentials: rods and cones. Each retina has about 6 million cones and 120 million rods. **Rods** have a low light threshold, allowing us to see in dim light, such as moonlight. Because they do not provide color vision, in dim light we see only shades of gray. Brighter lights stimulate the **cones,** which have a higher threshold and produce

Figure 16.8 Microscopic structure of the retina. The downward blue arrow at left indicates the direction of the signals passing through the neural portion of the retina. Eventually, nerve impulses arise in ganglion cells and propagate along their axons, which make up the optic (II) nerve. (See Tortora, *A Photographic Atlas of the Human Body,* Figure 9.3c)

🔑 In the retina, visual signals pass from photoreceptors to bipolar cells to ganglion cells.

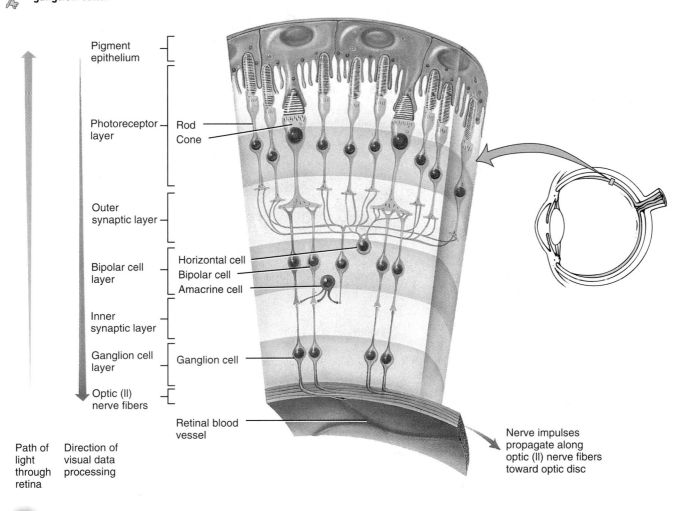

Q What are the two types of photoreceptors, and how do their functions differ?

color vision. Most of our visual experiences are mediated by the cone system, the loss of which produces legal blindness. In contrast, a person who loses rod vision mainly has difficulty seeing in dim light and thus should not, for example, drive at night.

The **macula lutea** (MAK-yoo-la LOO-tē-a; *macula* = a small, flat spot; *lute-* = yellowish) is in the exact center of the posterior portion of the retina, at the visual axis of the eye. The **central fovea** (see Figure 16.5), a small depression in the center of the macula lutea, contains only cones. In addition, the layers of bipolar and ganglion cells, which scatter light to some extent, do not cover the cones here; these layers are displaced to the periphery of the fovea. As a result, the central fovea is the area of highest **visual acuity** or **resolution** (sharpness of vision). A main reason that you move your

head and eyes while looking at something is to place images of interest on your fovea—as you do to read each of the words in this sentence! Rods are absent from the fovea and are more plentiful toward the periphery of the retina. Because rod vision is more sensitive than cone vision, you can see a faint object (such as a dim star) better if you gaze slightly to one side rather than looking directly at it.

From photoreceptors, information flows to bipolar cells through the outer synaptic layer, and then from bipolar cells through the inner synaptic layer to ganglion cells. The axons of ganglion cells extend posteriorly to the optic disc and exit the eyeball as the optic nerve. The optic disc is also called the **blind spot.** Because it contains no rods or cones, we cannot see an image that strikes the blind spot. Normally, you are not aware of having a blind spot, but you can easily demon-

Figure 16.9 The anterior and posterior chambers of the eye, seen in a section through the anterior portion of the eyeball at the junction of the cornea and sclera. Arrows indicate the flow of aqueous humor.

🔑 **The lens separates the posterior chamber of the anterior cavity from the vitreous chamber.**

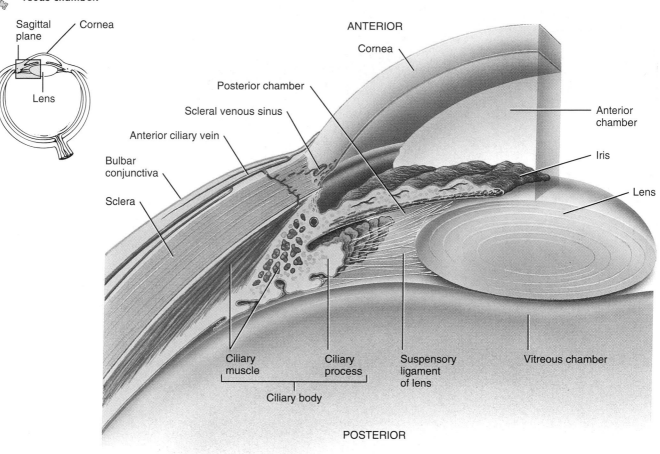

Sagittal plane
Cornea
Lens

ANTERIOR
Cornea
Posterior chamber
Scleral venous sinus
Anterior ciliary vein
Bulbar conjunctiva
Sclera

Anterior chamber
Iris
Lens

Ciliary muscle
Ciliary process
Ciliary body
Suspensory ligament of lens
Vitreous chamber

POSTERIOR

Q Where is aqueous humor produced, what is its circulation path, and where does it drain from the eyeball?

strate its presence. Cover your left eye and gaze directly at the cross below. Then increase or decrease the distance between the book and your eye. At some point the square will disappear as its image falls on the blind spot.

+   ■

Lens

Just posterior to the pupil and iris, within the cavity of the eyeball, is the avascular **lens** (see Figure 16.5). Proteins called **crystallins,** arranged like the layers of an onion, make up the lens, which normally is perfectly transparent. It is enclosed by a clear connective tissue capsule and held in position by encircling **suspensory ligaments,** which are attached to the ciliary process. The lens fine-tunes focusing of light rays onto the retina to facilitate clear vision.

Interior of the Eyeball

The interior of the eyeball is divided by the lens into two cavities: the anterior cavity and vitreous chamber. The **anterior cavity**—the space anterior to the lens—consists of the **anterior chamber,** which lies behind the cornea and in front of the iris, and the **posterior chamber,** which lies behind the iris and in front of the suspensory ligaments and lens (Figure 16.9). The anterior cavity is filled with **aqueous humor** (*aqua* = water), a watery fluid that is continually filtered from blood capillaries in the ciliary processes and that nourishes the lens and cornea. Aqueous humor flows into the posterior chamber, then forward between the iris and the lens, through the pupil, and into the anterior chamber. From the anterior chamber, aqueous humor drains into the scleral venous sinus (canal of Schlemm) and then into the blood. Normally, aqueous humor is completely replaced about every 90 minutes.

The pressure in the eye, called **intraocular pressure,** is produced mainly by the aqueous humor and partly by the

Table 16.1 Summary of Structures Associated with the Eyeball

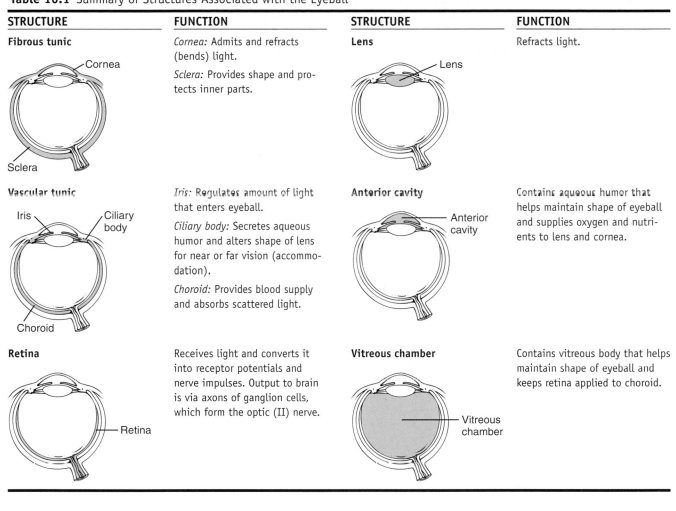

STRUCTURE	FUNCTION	STRUCTURE	FUNCTION
Fibrous tunic	*Cornea:* Admits and refracts (bends) light. *Sclera:* Provides shape and protects inner parts.	**Lens**	Refracts light.
Vascular tunic	*Iris:* Regulates amount of light that enters eyeball. *Ciliary body:* Secretes aqueous humor and alters shape of lens for near or far vision (accommodation). *Choroid:* Provides blood supply and absorbs scattered light.	**Anterior cavity**	Contains aqueous humor that helps maintain shape of eyeball and supplies oxygen and nutrients to lens and cornea.
Retina	Receives light and converts it into receptor potentials and nerve impulses. Output to brain is via axons of ganglion cells, which form the optic (II) nerve.	**Vitreous chamber**	Contains vitreous body that helps maintain shape of eyeball and keeps retina applied to choroid.

vitreous body (described shortly); normally it is about 16 mm of Hg. The intraocular pressure maintains the shape of the eyeball and prevents the eyeball from collapsing.

The second, and larger, cavity of the eyeball is the **vitreous chamber (posterior cavity),** which lies between the lens and the retina. Within the vitreous chamber is the **vitreous body,** a jellylike substance that contributes to intraocular pressure and holds the retina flush against the choroid, so that the retina provides an even surface for the reception of clear images. Unlike the aqueous humor, the vitreous body does not undergo constant replacement. It is formed during embryonic life and is not replaced thereafter. The vitreous body also contains phagocytic cells that remove debris, keeping this part of the eye clear for unobstructed vision. Occasionally, collections of debris may cast a shadow on the retina and create the appearance of specks that dart in and out of the field of vision. These "vitreal floaters," which are more common in older individuals, are usually harmless and do not require treatment. The **hyaloid canal** is a narrow channel that runs through the vitreous body from the optic disc to the posterior aspect of the lens. In the fetus, it was occupied by the hyaloid artery.

Table 16.1 summarizes the structures associated with the eyeball.

Image Formation

In some ways the eye is like a camera: Its optical elements focus an image of some object on a light-sensitive "film"—the retina—while ensuring the correct amount of light makes the proper "exposure." To understand how the eye forms clear images of objects on the retina, we must examine three processes: (1) the refraction or bending of light by the lens and cornea; (2) accommodation, the change in shape of the lens; and (3) constriction or narrowing of the pupil.

Refraction of Light Rays

When light rays traveling through a transparent substance (such as air) pass into a second transparent substance with a different density (such as water), they bend at the junction between the two. This bending is called **refraction** (Figure 16.10a). As light rays enter the eye, they are refracted at the anterior and posterior surfaces of the cornea. Both surfaces of the lens of the eye further refract the light rays so that they come into exact focus on the retina.

Images focused on the retina are inverted; they are upside down (Figure 16.10b, c). They also undergo right to left

Figure 16.10 Refraction of light rays. In accommodation (c), the lens changes shape to increase the refraction of light.

🔑 **Refraction is the bending of light rays at the junction of two transparent substances with different densities.**

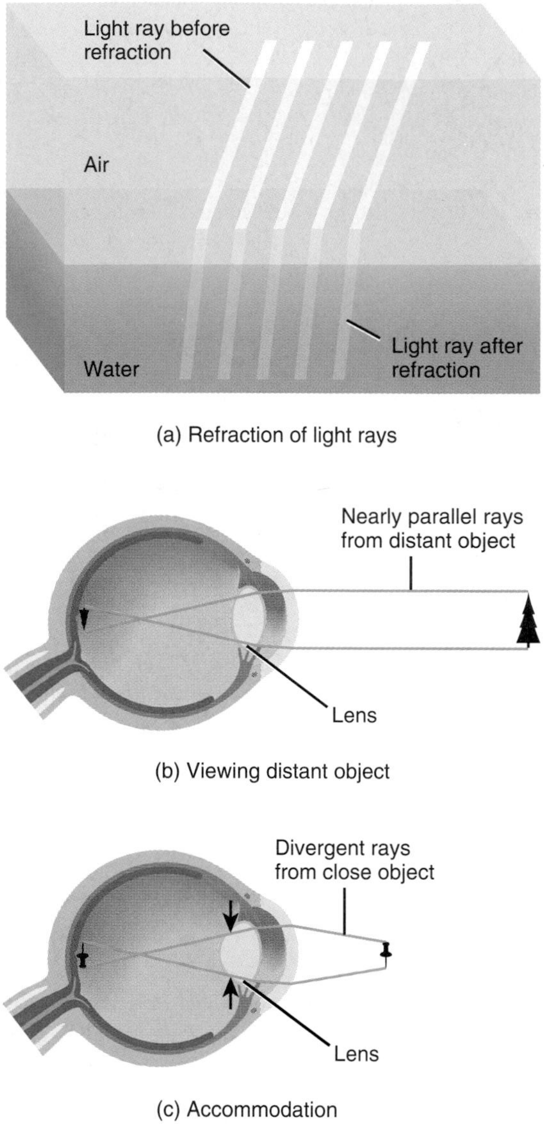

(a) Refraction of light rays

(b) Viewing distant object

(c) Accommodation

Q What sequence of events occur during accommodation?

reversal; that is, light from the right side of an object strikes the left side of the retina, and vice versa. The reason the world does not look inverted and reversed is that the brain "learns" early in life to coordinate visual images with the orientations of objects. The brain stores the inverted and reversed images we acquired when we first reached for and touched objects and interprets those visual images as being correctly oriented in space.

About 75% of the total refraction of light occurs at the cornea. The lens then fine-tunes image focus and changes the focus for near or distant objects. When an object is 6 m (20 ft) or more away from the viewer, the light rays reflected from the object are nearly parallel to one another (see Figure 16.10b). These parallel rays must be bent sufficiently to fall exactly focused on the central fovea, where vision is sharpest. However, because light rays that are reflected from objects closer than 6 m (20 ft) are divergent rather than parallel (see Figure 16.10c), the rays must be refracted more if they are to be focused on the retina. This additional refraction is accomplished in the process called accommodation.

Accommodation and the Near Point of Vision

A surface that curves outward, like the surface of a ball, is said to be *convex*. When the surface of a lens is convex, that lens will refract incoming light rays toward each other, such that they eventually intersect. If the surface of a lens curves inward, like the inside of a hollow ball, the lens is said to be *concave* and causes light rays to refract away from each other. The lens of the eye is convex on both its anterior and posterior surfaces, and its focusing power increases as its curvature becomes greater. When the eye is focusing on a close object, the lens becomes more curved and refracts the light rays more. This increase in the curvature of the lens for near vision is called **accommodation** (see Figure 16.10c).

How does accommodation occur? When you are viewing distant objects, the ciliary muscle is relaxed and the lens is fairly flat because it is stretched in all directions by taut suspensory ligaments. When you view a close object, the ciliary muscle contracts, which pulls the ciliary process and choroid forward toward the lens. This action releases tension on the lens and suspensory ligaments. Because it is elastic, the lens becomes rounder (more convex), which increases its focusing power and causes greater convergence of the light rays.

The **near point of vision** is the minimum distance from the eye that an object can be clearly focused with maximum effort. This is about 10 cm (4 in.) in a young adult. With aging, the lens loses elasticity and thus its ability to accommodate. As a consequence, older people cannot read print at the same close range as can youngsters. This condition is called **presbyopia** (prez-bē-Ō-pē-a; *presby-* = old; *-opia* = pertaining to the eye or vision). By age 40 the near point of vision may have increased to 20 cm (8 in.), and at age 60 to 80 cm (31 in.). Presbyopia usually begins in the mid-forties. At about that age, people who already wear glasses typically start to need bifocals, and those who have not previously worn glasses begin to require them for reading.

Refraction Abnormalities

The normal eye, known as an **emmetropic eye** (em'-e-TROP-ik), can sufficiently refract light rays from an object 6 m (20 ft) away so that a clear image is focused on the retina. Many people, however, lack this ability because of refraction abnormalities. Among these abnormalities are **myopia** (mī-Ō-pē-a), or nearsightedness, and **hypermetropia** (hī'-per-me-TRŌ-pē-a), or farsightedness (also known as **hyperopia**). Figure 16.11 illustrates these conditions and explains how they are corrected. Another refraction abnormality is **astigmatism** (a-STIG-ma-tizm), in which either the cornea or the lens has an irregular curvature. As a result, parts of the image are out of focus, and thus vision is blurred or distorted. Most errors of vision can be corrected by eyeglasses or contact lenses. A contact lens floats on a film of tears over the cornea; the anterior outer surface of the contact lens corrects the visual defect, whereas its posterior surface matches the curvature of the cornea.

Constriction of the Pupil

The circular muscle fibers of the iris also have a role in the formation of clear retinal images. Part of the accommodation mechanism consists of the contraction of the circular muscles of the iris to constrict the pupil. **Constriction of the pupil** is a narrowing of the diameter of the hole through which light enters the eye. This autonomic reflex occurs simultaneously with accommodation and prevents light rays from entering the eye through the periphery of the lens. Light rays entering at the periphery would not be brought to focus on the retina and would result in blurred vision. The pupil, as noted earlier, also constricts in bright light.

Convergence

Because of the position of their eyes in their heads, many animals see one set of objects off to the left through one eye, and an entirely different set of objects off to the right through the other. In humans, both eyes focus on only one set of objects—a characteristic called **binocular vision.** This feature of our visual system allows the perception of depth and an appreciation of the three-dimensional nature of objects.

Binocular vision occurs when light rays from an object strike corresponding points on the two retinas. When we stare straight ahead at a distant object, the incoming light rays are aimed directly at both pupils and are refracted to comparable spots on the retinas of both eyes. But as we move closer to the object, our eyes must rotate medially if the light rays from the object are to strike the same points on both retinas. The term **convergence** refers to this medial movement of the two eyeballs so that both are directed toward the object being viewed. The nearer the object, the greater the degree of convergence needed to maintain binocular vision. The coordinated action of the extrinsic eye muscles brings about convergence.

Physiology of Vision

OBJECTIVE

• *Describe how photoreceptors and photopigments function in vision.*

Photoreceptors and Photopigments

Rods and cones were named for the different appearance of each's *outer segment*—the distal end of the photoreceptor next to the pigment epithelium. The outer segments

Figure 16.11 Refraction abnormalities in the eyeball and their correction. (a) Normal (emmetropic) eye. (b) In the nearsighted or myopic eye, the image is focused in front of the retina. The condition may result from an elongated eyeball or thickened lens. (c) Correction of myopia is by use of a concave lens that diverges entering light rays so that they come into focus directly on the retina. (d) In the farsighted or hypermetropic eye, the image is focused behind the retina. The condition results from a shortened eyeball or a thin lens. (e) Correction of hypermetropia is by a convex lens that converges entering light rays so that they focus directly on the retina.

🔑 **In myopia (nearsightedness), distant objects can't be seen clearly; in hypermetropia (farsightedness), nearby objects can't be seen clearly.**

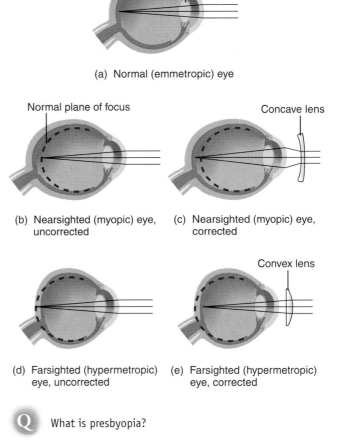

(a) Normal (emmetropic) eye

(b) Nearsighted (myopic) eye, uncorrected

(c) Nearsighted (myopic) eye, corrected

(d) Farsighted (hypermetropic) eye, uncorrected

(e) Farsighted (hypermetropic) eye, corrected

Q What is presbyopia?

of rods are cylindrical or rod-shaped, whereas those of cones are tapered or cone-shaped (Figure 16.12). Transduction of light energy into a receptor potential occurs in the outer segment. The photopigments are integral proteins in the plasma membrane of the outer segment; in cones the plasma membrane folds back and forth in a pleated fashion, whereas in rods the pleats pinch off from the plasma membrane to form discs. The outer segment of each rod contains a stack of about 1000 discs, piled up like coins inside a wrapper.

Figure 16.12 Structure of rod and cone photoreceptors. The inner segments contain the metabolic machinery for synthesis of photopigments and production of ATP. The photopigments are embedded in the membrane discs or folds of the outer segments. New discs in rods and new folds in cones form at the junction of the outer and inner segments. Pigment epithelial cells phagocytize old discs and the folds that slough off the distal tip of the outer segments.

🔑 **Transduction of light into a receptor potential occurs in the outer segments of rods and cones.**

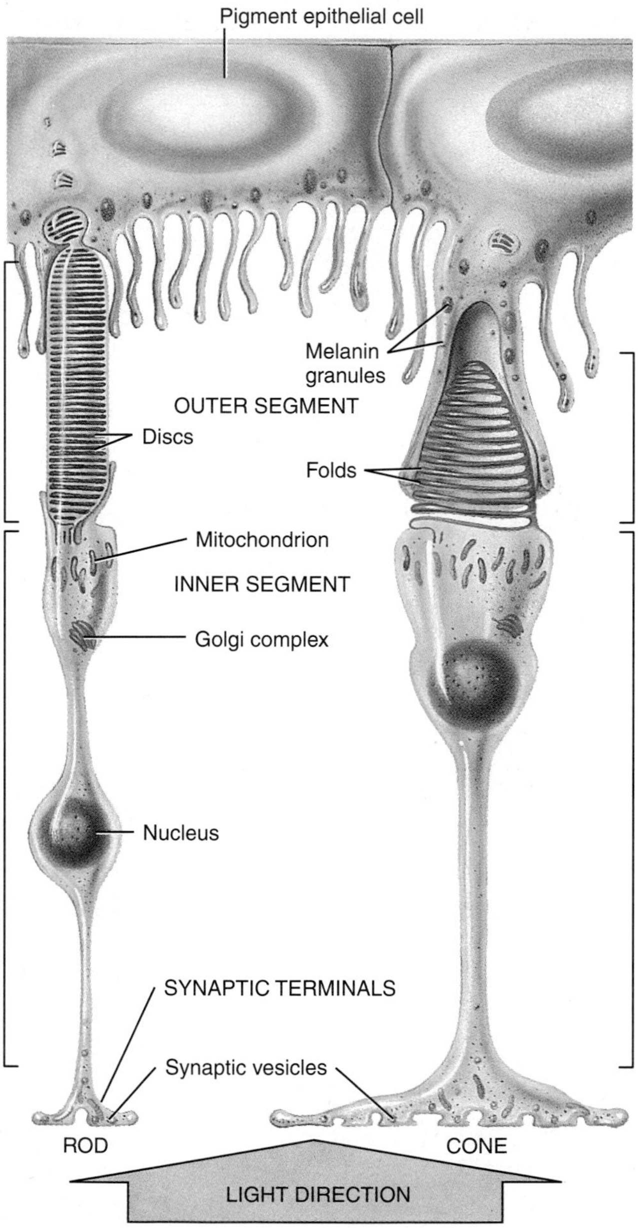

Pigment epithelial cell

Melanin granules

OUTER SEGMENT

Discs

Folds

Mitochondrion

INNER SEGMENT

Golgi complex

Nucleus

SYNAPTIC TERMINALS

Synaptic vesicles

ROD   CONE

LIGHT DIRECTION

❓ What are the functional similarities between rods and cones?

Photoreceptor outer segments are renewed at an astonishingly rapid pace. In rods, one to three new discs are added to the base of the outer segment every hour while old discs slough off at the tip and are phagocytized by pigment epithelial cells. The *inner segment* contains the cell nucleus, Golgi complex, and many mitochondria. At its proximal end, the photoreceptor expands into bulblike synaptic terminals filled with synaptic vesicles.

The first step in visual transduction is absorption of light by a **photopigment,** a colored protein that undergoes structural changes when it absorbs light, initiating events that lead to the production of a receptor potential. The single type of photopigment in rods is **rhodopsin** (*rhodo-* = rose; *opsin-* = related to vision). Three different **cone photopigments** are present in the retina, one in each of three types of cones. Color vision results from different colors of light selectively activating the different cone photopigments. Most forms of **color blindness,** an inability to distinguish between certain colors, result from the absence or a deficiency of one of the three cone photopigments. The most common type is red-green color blindness, in which a photopigment sensitive to orange-red light or green light is missing. As a result, the person cannot distinguish between red and green.

All photopigments associated with vision contain two parts: a glycoprotein known as **opsin** and a derivative of vitamin A called **retinal.** Vitamin A derivatives are formed from carotenoids, the plant pigments that give carrots their orange color. Good vision depends on adequate dietary intake of carotenoid-rich vegetables such as carrots, spinach, broccoli, and yellow squash, or foods that contain vitamin A, such as liver. Prolonged vitamin A deficiency and the consequent inability to synthesize a normal amount of rhodopsin may cause **night blindness** or **nyctalopia** (nik′-ta-LŌ-pē-a), an inability to see well at low light levels.

Retinal is the light-absorbing part of all visual photopigments. In the human retina, there are four different opsins, one for each cone photopigment and another for rhodopsin. Small variations in the amino acid sequences of the different opsins permit the rods and cones to absorb different colors (wavelengths) of incoming light. Rhodopsin absorbs blue to green light most effectively, whereas the three different cone photopigments most effectively absorb blue, green, or yellow-orange light.

Photopigments respond to light in the following cyclical process (Figure 16.13):

❶ In darkness, retinal has a bent shape, called *cis*-retinal, which fits snugly into the opsin portion of the photopigment. When *cis*-retinal absorbs a photon of light, it straightens out to a shape called *trans*-retinal. This *cis* to *trans* conversion is called **isomerization** and is the first step in visual transduction. After retinal isomerizes, several unstable chemical intermediates form and disappear. These chemical changes lead to production of a receptor potential (described shortly).

② In about a minute, *trans*-retinal completely separates from opsin. The final products look colorless, so this portion of the cycle is called **bleaching** of photopigment.

③ An enzyme called **retinal isomerase** converts *trans*-retinal back to *cis*-retinal.

④ The *cis*-retinal then can bind to opsin, reforming a functional photopigment. This portion of the cycle—resynthesis of a photopigment—is called **regeneration.**

The pigment epithelium adjacent to the photoreceptors stores a large quantity of vitamin A and contributes to the regeneration process in rods. The extent of rhodopsin regeneration decreases drastically if the retina detaches from the pigment epithelium. Cone photopigments regenerate much more quickly than does rhodopsin and are less dependent on the pigment epithelium. After complete bleaching, it takes 5 minutes to regenerate half of the rhodopsin, but only 90 seconds to regenerate half of the cone photopigments. Full regeneration of bleached rhodopsin takes 30–40 minutes.

Light and Dark Adaptation

When you emerge from dark surroundings (say, a tunnel) into the sunshine, **light adaptation** occurs—your visual system adjusts in seconds to the brighter environment by decreasing its sensitivity. On the other hand, when you enter a darkened room such as a theater, your visual system undergoes **dark adaptation**—its sensitivity increases slowly over many minutes. Bleaching and regeneration of the photopigments account for much (but not all) of the sensitivity changes during light and dark adaptation.

As the light level increases, more and more photopigment is bleached. While light is bleaching some photopigment molecules, however, others are being regenerated. In daylight, regeneration of rhodopsin cannot keep up with the bleaching process, so rods contribute little to daylight vision. In contrast, cone pigments regenerate rapidly enough that some of the *cis* form is always present, even in very bright light.

If the light level decreases abruptly, sensitivity increases rapidly at first and then more slowly. In complete darkness, full regeneration of cone photopigments occurs during the first 8 minutes of dark adaptation. During this time, a threshold (barely perceptible) light flash is seen as having color. Rhodopsin regenerates more slowly, and our visual sensitivity increases until even a single photon (the smallest unit of light) can be detected. In that situation, although much dimmer light can be detected, threshold flashes appear gray-white, regardless of their color. At very low light levels, such as starlight, objects appear as shades of gray because only the rods are functioning.

Release of Neurotransmitter by Photoreceptors

The absorption of light and isomerization of retinal initiates chemical changes in the photoreceptor outer segment that lead to production of a receptor potential. To under-

Figure 16.13 The cyclical bleaching and regeneration of photopigment. Blue arrows indicate bleaching steps; black arrows indicate regeneration steps.

🔑 **Retinal, a derivative of vitamin A, is the light-absorbing portion of all visual photopigments.**

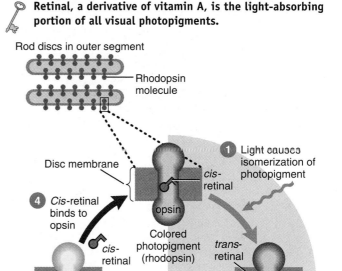

❓ What is the conversion of *cis*-retinal to *trans*-retinal called?

stand how the receptor potential arises, however, we first need to examine the operation of photoreceptors in the absence of light. In darkness, sodium ions (Na^+) flow into photoreceptor outer segments through ligand-gated Na^+ channels (Figure 16.14a). The ligand that holds these channels open is **cyclic GMP (guanosine monophosphate)** or **cGMP.** The inflow of Na^+, called the "dark current," partially depolarizes the photoreceptor. As a result, the membrane potential of a photoreceptor is about −30 mV, much closer to zero than a typical neuron's −70 mV membrane potential. This partial depolarization during darkness triggers continual release of neurotransmitter at the synaptic terminals. The neurotransmitter in rods, and perhaps in cones also, is the amino acid glutamate (glutamic acid). At synapses between rods and some bipolar cells, glutamate is an inhibitory neurotransmitter: It triggers inhibitory postsynaptic potentials (IPSPs) that hyperpolarize the bipolar cells and prevent them from sending signals on to the ganglion cells.

When light strikes the retina and *cis*-retinal undergoes isomerization, enzymes are activated that break down

Figure 16.14 Operation of rod photoreceptors.

 Light causes a hyperpolarizing receptor potential in photoreceptors, which decreases release of an inhibitory neurotransmitter (glutamate).

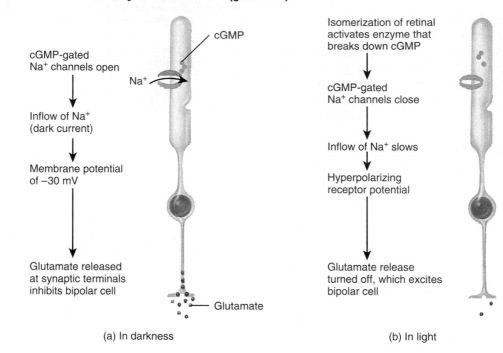

(a) In darkness   (b) In light

cGMP-gated Na⁺ channels open → Inflow of Na⁺ (dark current) → Membrane potential of −30 mV → Glutamate released at synaptic terminals inhibits bipolar cell → Glutamate

cGMP

Na⁺

Isomerization of retinal activates enzyme that breaks down cGMP → cGMP-gated Na⁺ channels close → Inflow of Na⁺ slows → Hyperpolarizing receptor potential → Glutamate release turned off, which excites bipolar cell

Q What is the function of cGMP in photoreceptors?

cGMP. As a result, some cGMP-gated Na$^+$ channels close, Na$^+$ inflow decreases, the membrane potential becomes more negative—approaching -70 mV (Figure 16.14b). This sequence of events produces a hyperpolarizing receptor potential that decreases the release of glutamate. Dim lights cause small and brief receptor potentials that partially turn off glutamate release; brighter lights elicit larger and longer receptor potentials that more completely shut down neurotransmitter release. The surprising result is this: Light excites the bipolar cells that synapse with rods by turning off the release of an inhibitory neurotransmitter!

The Visual Pathway

OBJECTIVE

• *Describe the processing of visual signals in the retina and the neural pathway for vision.*

After considerable processing of visual signals in the retina—at synapses among the various types of neurons (see Figure 16.8)—the axons of retinal ganglion cells provide output from the retina to the brain. They exit the eyeball via the **optic (II) nerve.**

Processing of Visual Input in the Retina

Within the retina, certain features of visual input are enhanced while other features may be discarded. Input from several cells may either converge upon a smaller number of postsynaptic neurons or diverge to a large number. On the whole, however, convergence predominates because there are only 1 million ganglion cells receiving input from about 126 million photoreceptors.

Once receptor potentials arise in rods and cones, they spread through the inner segments to the synaptic terminals. Neurotransmitter molecules released by rods and cones induce local graded potentials in both bipolar cells and horizontal cells. Between 6 and 600 rods synapse with a single bipolar cell in the outer synaptic layer, whereas a cone more often synapses with just one bipolar cell. The convergence of many rods onto a single bipolar cell increases the light sensitivity of rod vision but slightly blurs the image that is perceived. Cone vision, although less sensitive, has higher acuity because of the one-to-one synapses between cones and their bipolar cells. Stimulation of rods by light excites their bipolar cells, whereas cone bipolar cells may be either excited or inhibited when a light is turned on.

Horizontal cells transmit inhibitory signals to bipolar cells in the areas lateral to excited rods and cones. This lateral inhibition enhances contrasts in the visual scene between areas of the retina that are strongly stimulated and adjacent areas that are more weakly stimulated. Horizontal cells also assist in the differentiation of various colors. Amacrine cells, which are excited by bipolar cells, synapse with ganglion cells and transmit information to them that signals a change in the level of illumination of the retina. When bipolar or amacrine cells transmit excitatory signals to ganglion cells, the ganglion cells become depolarized and initiate nerve impulses.

Brain Pathway and Visual Fields

The axons of the optic (II) nerve pass through the **optic chiasm** (kī-AZ-m; = a crossover, as in the letter X), a crossing point of the optic nerves (Figure 16.15a, b). Some fibers cross to the opposite side, whereas others remain uncrossed. After passing through the optic chiasm, the fibers, now part of the **optic tract,** enter the brain and terminate in the lateral geniculate nucleus of the thalamus. Here the fibers synapse with neurons whose axons form the **optic radiations,** which project to the primary visual areas in the occipital lobes of the cerebral cortex (area 17 in Figure 14.15 on page 469).

Everything that can be seen with an eye is that eye's **visual field.** Because our eyes are located anteriorly in the head, there is considerable overlap of the visual fields of the two eyes (see Figure 16.15b). We have binocular vision due to the large region where the visual fields of the two eyes overlap—the **binocular visual field.** The visual field of each eye is divided into two regions: the **nasal** or **central half** and the **temporal** or **peripheral half** (Figure 16.15c, d). For each eye, light rays from an object in the nasal half of the visual field fall on the temporal half of the retina, and light rays from an object in the temporal half of the visual field fall on the nasal half of the retina. Moreover, visual information from the *right* half of each visual field is conveyed to the *left* side of the brain, whereas visual information from the *left* half of each visual field is conveyed to the *right* side of the brain, as follows (see Figure 16.15c, d):

❶ The axons of all retinal ganglion cells in one eye exit the eyeball at the optic disc and form the optic nerve on that side.

❷ At the optic chiasm, nerve fibers from the temporal half of each retina do not cross but continue directly to the lateral geniculate nucleus of the thalamus on the same side.

❸ In contrast, nerve fibers from the nasal half of each retina cross and continue to the opposite thalamus.

❹ Each optic tract consists of crossed and uncrossed axons that project from the optic chiasm to the thalamus on one side.

❺ Axon collaterals (branches) of the retinal ganglion cells also project to the midbrain, where they participate in circuits that govern constriction of the pupils in response to light and coordination of head and eye movements, and to the suprachiasmatic nucleus of the hypothalamus, which establishes patterns of sleep and other activities that occur on a circadian or daily schedule in response to intervals of light and darkness.

❻ The axons of thalamic neurons form the optic radiations as they project to the primary visual area of the cortex on the same side.

Although we have just described the visual pathway as a more or less single pathway, visual signals are thought to be processed by at least three separate systems in the cerebral cortex, each with its own function. One system processes information related to the shape of objects, another system processes information regarding color of objects, and a third system processes information about movement, location, and spatial organization.

1. Describe the structure and importance of the eyelids, eyelashes, and eyebrows.
2. What is the function of the lacrimal apparatus?
3. Describe the histology of the neural portion of the retina.
4. Explain how each of the following events is related to the physiology of vision: (a) refraction of light, (b) accommodation of the lens, and (c) constriction of the pupil.
5. Describe the structure of rods and cones, and explain how photopigments respond to light and recover in darkness.
6. How do receptor potentials develop in photoreceptors?
7. Describe the pathway for seeing an object in the nasal half of the visual field of the left eye.

HEARING AND EQUILIBRIUM
OBJECTIVE
• *Describe the anatomy of the structures in the three principal regions of the ear.*
• *List the principal events in the physiology of hearing.*
• *Identify the receptor organs for equilibrium, and describe how they function.*

The ear is an engineering marvel because its sensory receptors can transduce sound vibrations with amplitudes as small as the diameter of an atom of gold (0.3 nm) into electrical signals 1000 times faster than photoreceptors can respond to light. Besides receptors for sound waves, the ear also contains receptors for equilibrium.

Anatomy of the Ear
The ear is divided into three principal regions: the external ear, which collects sound waves and channels them inward; the middle ear, which conveys sound vibrations to the oval window; and the internal ear, which houses the receptors for hearing and equilibrium.

External (Outer) Ear
The **external (outer) ear** consists of the auricle, external auditory canal, and eardrum (Figure 16.16). The **auricle (pinna)** is a flap of elastic cartilage shaped like the flared end of a trumpet and covered by skin. The rim of the auricle is the **helix;** the inferior portion is the **lobule.** The auricle is attached to the head by ligaments and muscles. The **external**

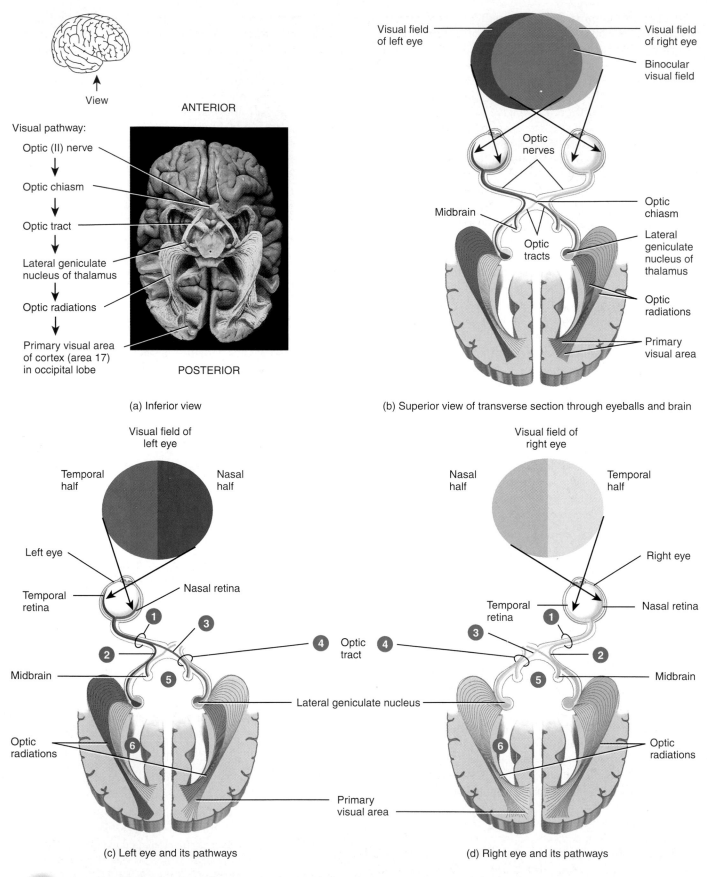

View

Visual pathway:

Optic (II) nerve
↓
Optic chiasm
↓
Optic tract
↓
Lateral geniculate nucleus of thalamus
↓
Optic radiations
↓
Primary visual area of cortex (area 17) in occipital lobe

ANTERIOR

POSTERIOR

(a) Inferior view

Visual field of left eye

Visual field of right eye

Binocular visual field

Optic nerves

Midbrain

Optic chiasm

Optic tracts

Lateral geniculate nucleus of thalamus

Optic radiations

Primary visual area

(b) Superior view of transverse section through eyeballs and brain

Visual field of left eye

Temporal half

Nasal half

Left eye

Temporal retina

Nasal retina

Midbrain

Optic radiations

Optic tract

Lateral geniculate nucleus

Primary visual area

(c) Left eye and its pathways

Visual field of right eye

Nasal half

Temporal half

Right eye

Temporal retina

Nasal retina

Midbrain

Optic radiations

(d) Right eye and its pathways

 On which half of the retina do light rays from an object in the temporal half of the visual field fall?

Figure 16.15 (See figure opposite.) The visual pathway. (a) Partial dissection of the brain reveals the optic radiations (axons extending from the thalamus to the occipital lobe). (b) An object in the binocular visual field can be seen with both eyes. In (c) and (d), note that information from the right side of the visual field of each eye projects to the left side of the brain, and information from the left side of the visual field of each eye projects to the right side of the brain. Adapted from Seeley et al., *Anatomy and Physiology* 4e, F15.22, p480; (New York: WCB McGraw-Hill, 1998) © 1998 The McGraw-Hill Companies.

🔑 **The axons of ganglion cells in the temporal half of the retina extend to the thalamus on the same side; the axons of ganglion cells in the nasal half of the retina extend to the thalamus on the opposite side.**

auditory canal (*audit-* = hearing) is a curved tube about 2.5 cm (1 in.) long that lies in the temporal bone and leads from the auricle to the eardrum. The **eardrum** or **tympanic membrane** (tim-PAN-ik; *tympan-* = a drum) is a thin, semitransparent partition between the external auditory canal and middle ear. The eardrum is covered by epidermis and lined by simple cuboidal epithelium. Between the epithelial layers is connective tissue composed of collagen, elastic fibers, and fibroblasts.

Near the exterior opening, the external auditory canal contains a few hairs and specialized sebaceous (oil) glands called **ceruminous glands** (se-ROO-mi-nus) that secrete

Figure 16.16 Structure of the ear, illustrated in a frontal section through the right ear and skull. (See Tortora, *A Photographic Atlas of the Human Body,* Figure 9.4a)

🔑 **The ear has three principal regions: the external (outer) ear, the middle ear, and the internal (inner) ear.**

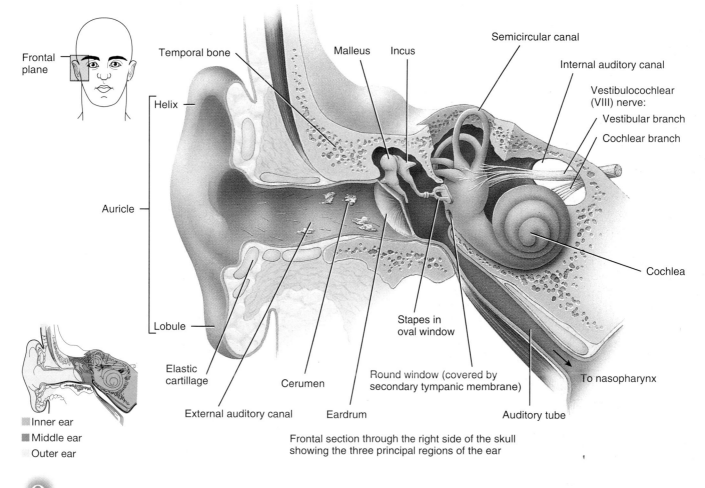

Frontal section through the right side of the skull showing the three principal regions of the ear

■ Inner ear
■ Middle ear
▫ Outer ear

Q To which structure of the external ear does the malleus of the middle ear attach?

Figure 16.17 The right middle ear containing the auditory ossicles. (See Tortora, *A Photographic Atlas of the Human Body,* Figure 3.14)

Common names for the malleus, incus, and stapes are the hammer, anvil, and stirrup, respectively.

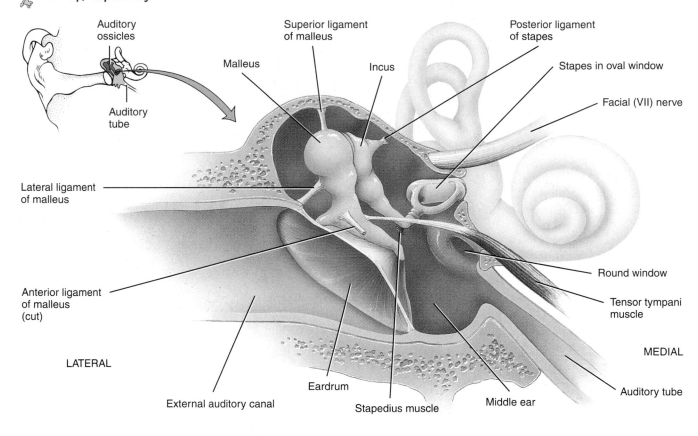

Frontal section showing location of auditory ossicles

Q What structures separate the middle ear from the external ear and from the internal ear?

earwax or **cerumen** (se-ROO-min). The combination of hairs and cerumen helps prevent dust and foreign objects from entering the ear. Cerumen usually dries up and falls out of the ear canal. Some people, however, produce a large amount of cerumen, which can become impacted and can muffle incoming sounds.

Middle Ear

The **middle ear** is a small, air-filled cavity in the temporal bone that is lined by epithelium (Figure 16.17). It is separated from the external ear by the eardrum and from the internal ear by a thin bony partition that contains two small membrane-covered openings: the oval window and the round window. Extending across the middle ear and attached to it by ligaments are the three smallest bones in the body, the **auditory ossicles** (OS-si-kuls), which are connected by synovial joints. The bones, named for their shapes, are the malleus, incus, and stapes—commonly called the hammer, anvil, and stirrup, respectively. The "handle" of the **malleus** (MAL-ē-us) is attached to the internal surface of the eardrum. The head of the malleus articulates with the body of the incus. The **incus** (ING-kus), the middle bone in the

series, articulates with the head of the stapes. The base or footplate of the **stapes** (STĀ-pēz) fits into the **oval window.** Directly below the oval window is another opening, the **round window,** that is enclosed by a membrane called the **secondary tympanic membrane.**

Besides the ligaments, two tiny skeletal muscles also attach to the ossicles (see Figure 16.17). The **tensor tympani muscle,** which is innervated by the mandibular branch of the trigeminal (V) nerve, limits movement and increases tension on the eardrum to prevent damage to the inner ear from loud noises. The **stapedius muscle,** which is innervated by the facial (VII) nerve, is the smallest of all skeletal muscles. By dampening large vibrations of the stapes due to loud noises, it protects the oval window, but it also decreases the sensitivity of hearing. For this reason, paralysis of the stapedius muscle is associated with **hyperacusia** (abnormally sensitive hearing). Because it takes a fraction of a second for the tensor tympani and stapedius muscles to contract, they can protect the inner ear from prolonged loud noises, but not from brief ones such as a gunshot.

The anterior wall of the middle ear contains an opening that leads directly into the **auditory (Eustachian) tube.** The

Figure 16.18 The right internal ear. The outer, cream-colored area is part of the bony labyrinth; the inner, pink-colored area is the membranous labyrinth.

🔑 **The bony labyrinth contains perilymph, and the membranous labyrinth contains endolymph.**

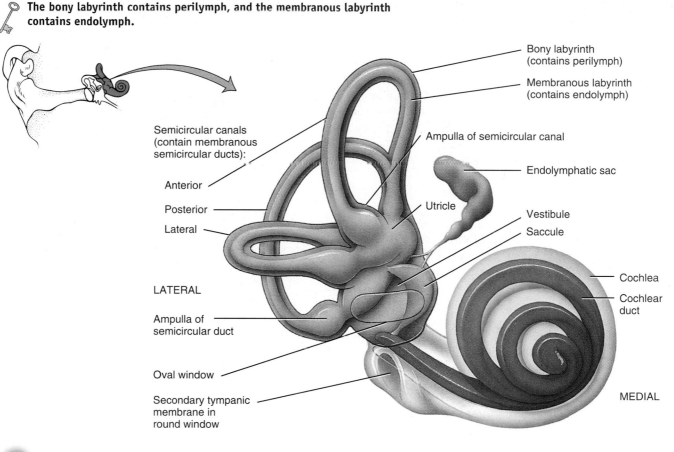

Semicircular canals (contain membranous semicircular ducts):

Anterior

Posterior

Lateral

LATERAL

Ampulla of semicircular duct

Oval window

Secondary tympanic membrane in round window

Bony labyrinth (contains perilymph)

Membranous labyrinth (contains endolymph)

Ampulla of semicircular canal

Endolymphatic sac

Utricle

Vestibule

Saccule

Cochlea

Cochlear duct

MEDIAL

Q What are the names of the two sacs that lie in the vestibule?

auditory tube, which consists of both bone and hyaline cartilage, connects the middle ear with the nasopharynx (upper portion of the throat). It is normally closed at its medial (pharyngeal) end; during swallowing and yawning, it opens, allowing air to enter or leave the middle ear until the pressure in the middle ear equals the atmospheric pressure. When the pressures are balanced, the eardrum vibrates freely as sound waves strike it. If the pressure is not equalized, intense pain, hearing impairment, ringing in the ears, and vertigo could develop. The auditory tube also is a route whereby pathogens may travel from the nose and throat to the middle ear.

Internal (Inner) Ear

The **internal (inner) ear** is also called the **labyrinth** (LAB-i-rinth) because of its complicated series of canals (Figure 16.18). Structurally, it consists of two main divisions: an outer bony labyrinth that encloses an inner membranous labyrinth. The **bony labyrinth** is a series of cavities in the temporal bone divided into three areas: (1) the semicircular canals and (2) the vestibule, both of which contain

receptors for equilibrium; and (3) the cochlea, which contains receptors for hearing. The bony labyrinth is lined with periosteum and contains **perilymph.** This fluid, which is chemically similar to cerebrospinal fluid, surrounds the **membranous labyrinth,** a series of sacs and tubes inside the bony labyrinth and having the same general form. The membranous labyrinth is lined by epithelium and contains **endolymph.** The level of K^+ in endolymph is unusually high for an extracellular fluid, and potassium ions play a role in the generation of auditory signals (described shortly).

The **vestibule** (VES-ti-būl) is the oval central portion of the bony labyrinth. The membranous labyrinth in the vestibule consists of two sacs called the **utricle** (YOO-tri-kul; = little bag) and the **saccule** (SAK-yool; = little sac), which are connected by a small duct. Projecting superiorly and posteriorly from the vestibule are the three bony **semicircular canals,** each of which lies at approximately right angles to the other two. Based on their positions, they are called the anterior, posterior, and lateral semicircular canals. The anterior and posterior semicircular canals are oriented vertically; the lateral one is oriented horizontally. At one end of each canal is a swollen enlargement called the **ampulla** (am-POOL-la; = saclike duct). The portions of the membranous labyrinth that lie inside the bony semicircular canals are called the **membranous semicircular ducts.** These structures communicate with the utricle of the vestibule.

The vestibular branch of the vestibulocochlear (VIII) nerve consists of *ampullary, utricular,* and *saccular nerves.* These nerves contain both first-order sensory neurons and motor neurons that synapse with receptors for equilibrium. The first-order sensory neurons carry sensory information from the receptors, and the motor neurons carry feedback signals to the receptors, apparently to modify their sensitivity. Cell bodies of the sensory neurons are located in the **vestibular ganglia** (see Figure 16.19b).

Anterior to the vestibule is the **cochlea** (KŌK-lē-a; = snail-shaped), a bony spiral canal (Figure 16.19a) that resembles a snail's shell and makes almost three turns around a central bony core called the **modiolus** (mŌ-DĪ-ō-lus; Figure 16.19b). Sections through the cochlea (see Figure 16.19a–c) show that it is divided into three channels. Together, the partitions that separate the channels are shaped like the letter Y. The stem of the Y is a bony shelf that protrudes into the canal; the wings of the Y are composed mainly of membranous labyrinth. The channel above the bony partition is the **scala vestibuli,** which ends at the oval window; the channel below is the **scala tympani,** which ends at the round window.

The scala vestibuli and scala tympani both contain perilymph and are completely separated, except for an opening at the apex of the cochlea, the **helicotrema** (hel-i-kō-TRĒ-ma). See Figure 16.20. The cochlea adjoins the wall of the vestibule, into which the scala vestibuli opens. The perilymph in the vestibule is continuous with that of the scala vestibuli. The third channel (between the wings of the Y) is the **cochlear duct** or **scala media.** The **vestibular membrane** separates the cochlear duct from the scala vestibuli, and the **basilar membrane** separates the cochlear duct from the scala tympani.

Resting on the basilar membrane is the **spiral organ** or **organ of Corti** (see Figure 16.19c, d). The spiral organ is a coiled sheet of epithelial cells, including supporting cells and about 16,000 **hair cells,** which are the receptors for hearing. There are two groups of hair cells: The *inner hair cells* are arranged in a single row and extend the entire length of the cochlea; the *outer hair cells* are arranged in three rows. At the apical tip of each hair cell is a **hair bundle,** consisting of 30–100 *stereocilia* that extend into the endolymph of the cochlear duct. Despite their name, stereocilia are actually long, hairlike microvilli arranged in several rows of graded height.

At their basal ends, inner and outer hair cells synapse both with first-order sensory neurons and with motor neurons from the cochlear branch of the vestibulocochlear (VIII) nerve. Cell bodies of the sensory neurons are located in the **spiral ganglion** (see Figure 16.19b, c). Although outer hair cells outnumber them by 3 to 1, the inner hair cells

synapse with 90–95% of the first-order sensory neurons in the cochlear nerve that relay auditory information to the brain. By contrast, 90% of the motor neurons in the cochlear nerve synapse with outer hair cells. Projecting over and in contact with the hair cells of the spiral organ is the **tectorial membrane** (*tector-* = covering), a flexible gelatinous membrane.

The Nature of Sound Waves

Sound waves are a series of alternating high- and low-pressure regions traveling in the same direction through some medium (such as air). They originate from a vibrating object in much the same way that ripples arise and travel over the surface of a pond when you toss a stone into it. The sounds heard most acutely by human ears are those from sources that vibrate at frequencies between 500 and 5000 hertz (Hz; 1 Hz = 1 cycle per second). The entire audible range extends from 20 to 20,000 Hz. Sounds of speech contain frequencies mainly between 100 and 3000 Hz, and the "high C" sung by a soprano has a dominant frequency at 1048 Hz. The sounds from a jet plane several miles away range from 20 to 100 Hz.

The *frequency* of a sound vibration is its *pitch.* The greater the frequency of vibration, the higher the pitch. Also, the greater the *intensity* (size or amplitude) of the vibration, the *louder* the sound. Sound intensity is measured in units called **decibels (dB).** Each ten decibels represents a tenfold increase in sound intensity. The hearing threshold—the point at which an average young adult can just distinguish sound from silence—is defined as 0 dB at 1000 Hz. Rustling leaves have a decibel level of 15; whispered speech, 30; normal conversation, 60; a vacuum cleaner, 75; shouting, 80; and a nearby motorcycle or jackhammer, 90. Sound becomes uncomfortable to a normal ear at about 120 dB, and painful above 140 dB. Because hearing loss results from prolonged noise exposure, employers in the United States must require workers to use hearing protectors when occupational noise levels exceed 90 dB. Rock concerts and even inexpensive headphones can easily produce sounds over 110 dB.

CLINICAL APPLICATION
Loud Sounds and Hair Cell Damage

Exposure to loud music, the engine roar of jet planes, revved-up motorcycles, lawn mowers, and vacuum cleaners damages hair cells of the cochlea. Continued exposure to high-intensity sounds causes **deafness,** a significant or total hearing loss. The louder the sounds, the quicker the loss. Deafness usually begins with loss of sensitivity for high-pitched sounds. If you are listening to music through headphones and bystanders can hear it, the dB level is in the damaging range. Most people fail to notice their progressive hearing loss until destruction is extensive and they begin having difficulty understanding speech. Wearing earplugs with a noise-reduction rating of 30 dB while engaging in noisy activities can protect the sensitivity of your ears. ∎

Figure 16.19 Semicircular canals, vestibule, and cochlea of the right ear. Note that the cochlea makes nearly three complete turns. (See Tortora, *A Photographic Atlas of the Human Body,* Figure 9.4b)

🔑 **The three channels in the cochlea are the scala vestibuli, the scala tympani, and the cochlear duct.**

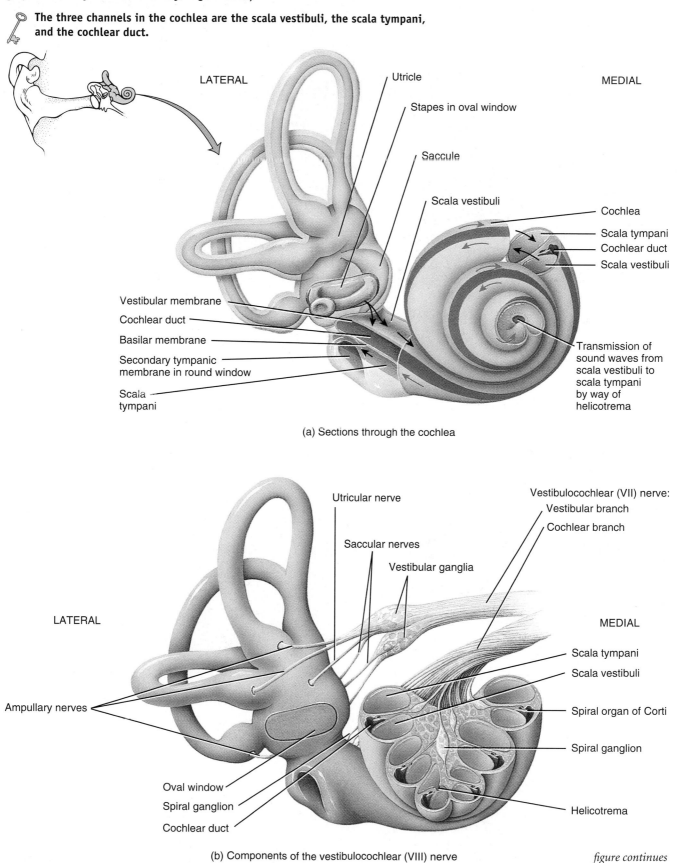

(a) Sections through the cochlea

(b) Components of the vestibulocochlear (VIII) nerve

figure continues

Figure 16.19 *(continued)*

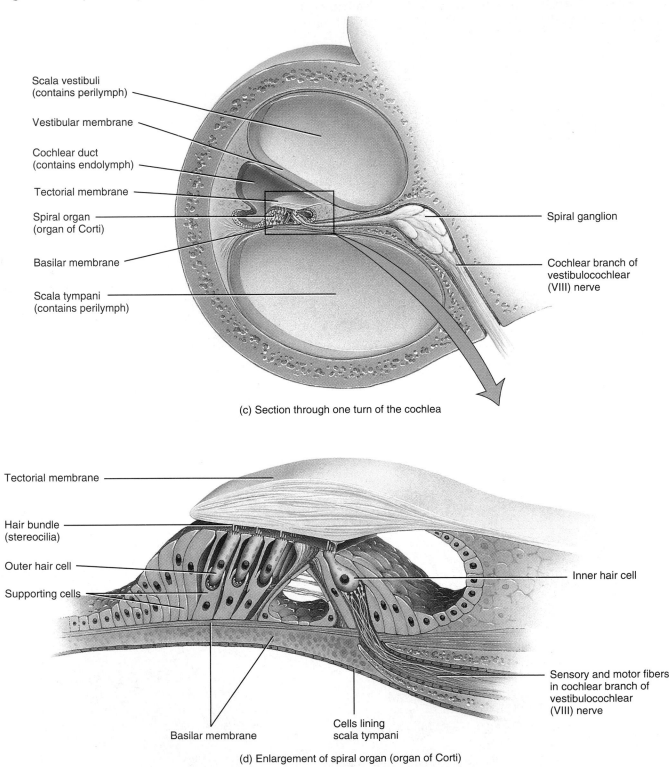

Scala vestibuli
(contains perilymph)

Vestibular membrane

Cochlear duct
(contains endolymph)

Tectorial membrane

Spiral organ
(organ of Corti)

Basilar membrane

Scala tympani
(contains perilymph)

Spiral ganglion

Cochlear branch of
vestibulocochlear
(VIII) nerve

(c) Section through one turn of the cochlea

Tectorial membrane

Hair bundle
(stereocilia)

Outer hair cell

Supporting cells

Inner hair cell

Sensory and motor fibers
in cochlear branch of
vestibulocochlear
(VIII) nerve

Basilar membrane

Cells lining
scala tympani

(d) Enlargement of spiral organ (organ of Corti)

Q What are the three subdivisions of the bony labyrinth?

Figure 16.20 Events in the stimulation of auditory receptors in the right ear. The numbers correspond to the events listed in the text. The cochlea has been uncoiled to more easily visualize the transmission of sound waves and their distortion of the vestibular and basilar membranes of the cochlear duct.

🔑 **The function of hair cells of the spiral organ (organ of Corti) is to convert a mechanical vibration (stimulus) into an electrical signal (receptor potential).**

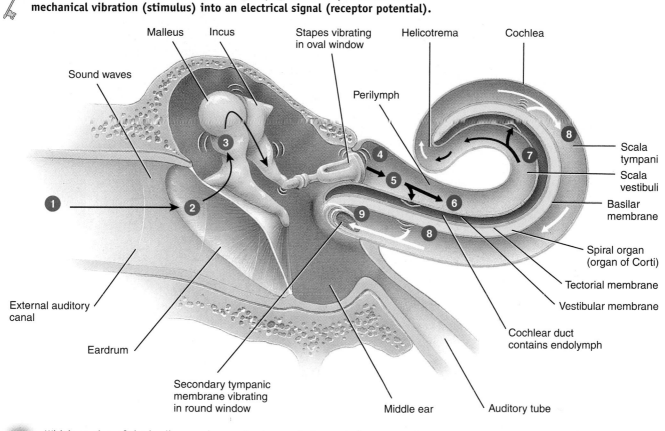

Q Which portion of the basilar membrane vibrates most vigorously in response to high-frequency (high-pitched) sounds?

Physiology of Hearing

The following events are involved in hearing (Figure 16.20):

1 The auricle directs sound waves into the external auditory canal.

2 When sound waves strike the eardrum, the alternating high- and low-pressure of the air cause the eardrum to vibrate back and forth. The distance it moves, which is very small, depends on the intensity and frequency of the sound waves. The eardrum vibrates slowly in response to low-frequency (low-pitched) sounds and rapidly in response to high-frequency (high-pitched) sounds.

3 The central area of the eardrum connects to the malleus, which also starts to vibrate. The vibration is transmitted from the malleus to the incus and then to the stapes.

4 As the stapes moves back and forth, it pushes the membrane of the oval window in and out. The oval window vibrates about 20 times more vigorously than the eardrum because the ossicles efficiently transmit small vibrations spread over a large surface area (eardrum) into larger vibrations of a smaller surface (oval window).

5 The movement of the oval window sets up fluid pressure waves in the perilymph of the cochlea. As the oval window bulges inward, it pushes on the perilymph of the scala vestibuli.

6 Pressure waves are transmitted from the scala vestibuli to the scala tympani and eventually to the round window, causing it to bulge outward into the middle ear. (See **9**.)

7 As the pressure waves deform the walls of the scala vestibuli and scala tympani, they also push the vestibular membrane back and forth, creating pressure waves in the endolymph inside the cochlear duct.

8 The pressure waves in the endolymph cause the basilar membrane to vibrate, which moves the hair cells of the spiral organ against the tectorial membrane. Bending of the stereocilia produces receptor potentials that ultimately lead to the generation of nerve impulses in cochlear nerve fibers.

Sound waves of various frequencies cause specific regions of the basilar membrane to vibrate more intensely than others. In other words, each segment of the basilar membrane is "tuned" for a particular pitch. The membrane is narrower and stiffer at the base of the cochlea (portion closer to the oval window); high-frequency (high-pitched) sounds near 20,000 Hz induce maximal vibrations in this region. Toward the apex of the cochlea near the helicotrema, the basilar membrane is wider and more flexible; low-frequency (low-pitched) sounds near 20 Hz cause maximal vibration of the basilar membrane there. Loudness is determined by the intensity of sound waves. High-intensity sound waves cause greater vibration of the basilar membrane, which leads to a higher frequency of nerve impulses reaching the brain. Louder sounds also may stimulate a larger number of hair cells.

The hair cells transduce mechanical vibrations into electrical signals. As the basilar membrane vibrates, the hair bundles at the apex of the hair cell bend back and forth and slide against one another. A *tip link* protein connects the tip of each stereocilium to a mechanically gated ion channel called the **transduction channel** in its taller stereocilium neighbor. As the stereocilia bend in the direction of the taller stereocilia, the tip links tug on the transduction channels and open them. These channels allow cations in the endolymph, primarily K^+, to enter the hair cell cytosol. As cations enter, they produce a depolarizing receptor potential. Depolarization quickly spreads along the plasma membrane and opens voltage-gated Ca^{2+} channels in the base of the hair cell. The resulting inflow of Ca^{2+} triggers exocytosis of synaptic vesicles containing a neurotransmitter, which is probably glutamate. As more neurotransmitter is released, the frequency of nerve impulses in the first-order sensory nerve fibers that synapse with the base of the hair cell increases. Bending of the stereocilia in the opposite direction closes the transduction channels, allows repolarization or even hyperpolarization to occur, and reduces neurotransmitter release from the hair cells. This decreases the frequency of nerve impulses in sensory nerve fibers.

Besides its role in detecting sounds, the cochlea has the surprising ability to produce sounds. These usually inaudible sounds, called **otoacoustic emissions,** can be picked up by placing a sensitive microphone next to the eardrum. They are caused by vibrations of the outer hair cells that occur in response to sound waves and to signals from motor neurons. As they depolarize and repolarize, the outer hair cells rapidly shorten and lengthen. This vibratory behavior appears to change the stiffness of the tectorial membrane and is thought to enhance the displacement of the basilar membrane, which amplifies the responses of the inner hair cells. At the same time, the outer hair cell vibrations set up a traveling wave that goes back toward the stapes and leaves the ear as a sound. Detection of otoacoustic emissions is a fast, inexpensive, and noninvasive way to screen newborns for hearing defects.

The Auditory Pathway

First-order sensory neurons in the cochlear branch of each vestibulocochlear (VIII) nerve terminate in the cochlear nuclei of the medulla oblongata on the same side. From there, fibers carrying auditory signals project to the superior olivary nuclei, also in the medulla, on both sides. Slight differences in the timing of impulses arriving from the two ears at the olivary nuclei allow us to locate the source of a sound. From both the cochlear nuclei and the olivary nuclei, fibers ascend to the inferior colliculus in the midbrain, and then to the medial geniculate body of the thalamus. From the thalamus, auditory signals project to the primary auditory area in the superior temporal gyrus of the cerebral cortex (areas 41 and 42 in Figure 14.15 on page 469). Because many auditory axons decussate (cross over) in the medulla while others remain on the same side, the right and left primary auditory areas receive nerve impulses from both ears.

CLINICAL APPLICATION
Cochlear Implants

Cochlear implants translate sounds into electronic signals that can be interpreted by the brain. They are useful for people with deafness caused by a disorder or an injury that has destroyed hair cells in the cochlea. Sound waves are picked up by a tiny microphone and converted into electrical signals by a microprocessor. The signals then travel to electrodes implanted in the cochlea, where they trigger nerve impulses in fibers of the cochlear branch of the vestibulocochlear nerve. These artificially induced nerve impulses propagate over their normal pathways to the brain. The sounds perceived are crude compared to normal hearing, but they provide a sense of rhythm and loudness; information about certain noises, such as those made by telephones and automobiles; and the pitch and cadence of speech. A few patients hear well enough with a cochlear implant to use the telephone. ■

Physiology of Equilibrium

There are two kinds of **equilibrium** (balance). One kind, called **static equilibrium,** refers to the maintenance of the position of the body (mainly the head) relative to the force of gravity. The second kind, **dynamic equilibrium,** is the maintenance of body position (mainly the head) in response to sudden movements such as rotation, acceleration, and deceleration. Collectively, the receptor organs for equilibrium are called the **vestibular apparatus** (ves-TIB-yoo-lar), which includes the saccule, utricle, and membranous semicircular ducts.

Otolithic Organs: Saccule and Utricle

The walls of both the utricle and the saccule contain a small, thickened region called a **macula** (MAK-yoo-la; Figure 16.21). The two maculae (plural), which are perpendicular to one another, are the receptors for static equilibrium, and they also contribute to some aspects of dynamic equilibrium. For static equilibrium, they provide sensory information on the position of the head in space and are essential for maintaining appropriate posture and balance. For dynamic

Figure 16.21 Location and structure of receptors in the maculae of the right ear. Both first-order sensory neurons (blue) and motor neurons (red) synapse with the hair cells.

🔑 **The movement of stereocilia initiates depolarizing receptor potentials.**

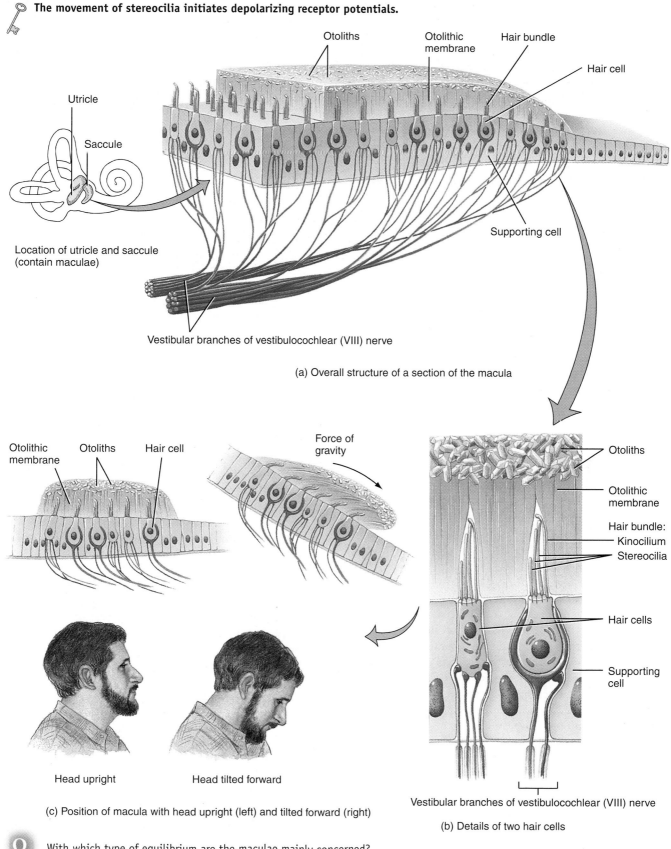

Otoliths

Otolithic membrane

Hair bundle

Hair cell

Utricle

Saccule

Supporting cell

Location of utricle and saccule (contain maculae)

Vestibular branches of vestibulocochlear (VIII) nerve

(a) Overall structure of a section of the macula

Otolithic membrane

Otoliths

Hair cell

Force of gravity

Otoliths

Otolithic membrane

Hair bundle:
Kinocilium
Stereocilia

Hair cells

Supporting cell

Head upright

Head tilted forward

(c) Position of macula with head upright (left) and tilted forward (right)

Vestibular branches of vestibulocochlear (VIII) nerve

(b) Details of two hair cells

Q With which type of equilibrium are the maculae mainly concerned?

equilibrium, they detect linear acceleration and deceleration—for example, the sensations you feel while in an elevator or a car that is speeding up or slowing down.

The two maculae consist of two kinds of cells: **hair cells,** which are the sensory receptors, and **supporting cells.** Hair cells feature **hair bundles** that consist of 70 or more *stereocilia,* which are actually microvilli, plus one *kinocilium,* a conventional cilium anchored firmly to its basal body and extending beyond the longest stereocilia. As in the cochlea, the stereocilia are connected by tip links. Scattered among the hair cells are columnar supporting cells that probably secrete the thick, gelatinous, glycoprotein layer, called the **otolithic membrane,** that rests on the hair cells. A layer of dense calcium carbonate crystals, called **otoliths** (*oto-* = ear; *-liths* = stones) extends over the entire surface of the otolithic membrane.

Because the otolithic membrane sits on top of the macula, if you tilt your head forward, the otolithic membrane (and the otoliths as well) is pulled by gravity and slides downhill over the hair cells in the direction of the tilt, bending the hair bundles. However, if you are sitting upright in a car that suddenly jerks forward, the otolithic membrane, due to its inertia, lags behind the head movement. As the otolithic membrane lags behind, it pulls on the hair bundles and makes them bend in the other direction. Bending of the hair bundles in one direction stretches the tip links, which pull open transduction channels thereby producing depolarizing receptor potentials; bending in the opposite direction closes the transduction channels and produces repolarization.

As the hair cells depolarize and repolarize, they release neurotransmitter at a faster or slower rate. The hair cells synapse with first-order sensory neurons in the vestibular branch of the vestibulocochlear (VIII) nerve (see Figure 16.21a). These neurons fire impulses at a slow or rapid pace depending on how much neurotransmitter is present. Motor fibers also synapse with the hair cells and sensory neurons. Evidently, they regulate the sensitivity of the hair cells and sensory neurons.

Membranous Semicircular Ducts

The three membranous semicircular ducts, together with the saccule and the utricle, function in dynamic equilibrium. The ducts lie at right angles to one another in three planes (Figure 16.22): The two vertical ones are the anterior and posterior membranous semicircular ducts, and the horizontal one is the lateral membranous semicircular duct (see also Figure 16.18). This positioning permits detection of rotational acceleration or deceleration. In the ampulla, the dilated portion of each duct, is a small elevation called the **crista.** Each crista contains a group of **hair cells** and **supporting cells** covered by a mass of gelatinous material called the **cupula.** When the head moves, the attached membranous semicircular ducts and hair cells move with it. The endolymph, however, is not attached and lags behind due to its inertia. As the moving hair cells drag along the stationary fluid, the hair bundles bend. Bending of the hair bundles produces receptor potentials, which lead to nerve impulses that pass along the vestibular branch of the vestibulocochlear (VIII) nerve.

Equilibrium Pathways

Most of the vestibular branch fibers of the vestibulocochlear (VIII) nerve enter the brain stem and terminate in several vestibular nuclei in the medulla and pons. The remaining fibers enter the cerebellum through the inferior cerebellar peduncle (see Figure 14.5b). Bidirectional pathways connect the vestibular nuclei and cerebellum. Fibers from all the vestibular nuclei extend to the nuclei of cranial nerves that control eye movements—oculomotor (III), trochlear (IV), and abducens (VI)—and to the accessory (XI) nerve nucleus that helps control head and neck movements. In addition, fibers from the lateral vestibular nucleus form the vestibulospinal tract, which conveys impulses to skeletal muscles that regulate muscle tone in response to head movements. Various pathways between the vestibular nuclei, cerebellum, and cerebrum enable the cerebellum to play a key role in maintaining static and dynamic equilibrium. The cerebellum continuously receives updated sensory information from the utricle and saccule. The cerebellum monitors this information and makes corrective adjustments in the motor activities that originate in the cerebral cortex. Essentially, in response to input from the utricle, saccule, and membranous semicircular ducts, the cerebellum sends continuous nerve impulses to the motor areas of the cerebrum. This feedback allows correction of signals from the motor cortex to specific skeletal muscles to maintain equilibrium.

Table 16.2 summarizes the structures of the ear related to hearing and equilibrium.

1. Explain the events involved in the transmission of sound from the auricle to the spiral organ of Corti.
2. How do hair cells in the cochlea and vestibular apparatus operate?
3. What is the pathway for auditory impulses from the cochlea to the cerebral cortex?
4. Compare the function of the maculae in maintaining static equilibrium with the role of the cristae in maintaining dynamic equilibrium.
5. Describe the role of vestibular input to the cerebellum.

Figure 16.22 Location and structure of the membranous semicircular ducts of the right ear. Both first-order sensory neurons (blue) and motor neurons (red) synapse with the hair cells. The ampullary nerves are branches of the vestibular division of the vestibulocochlear (VIII) nerve.

🔑 **The positions of the membranous semicircular ducts permit detection of rotational movements.**

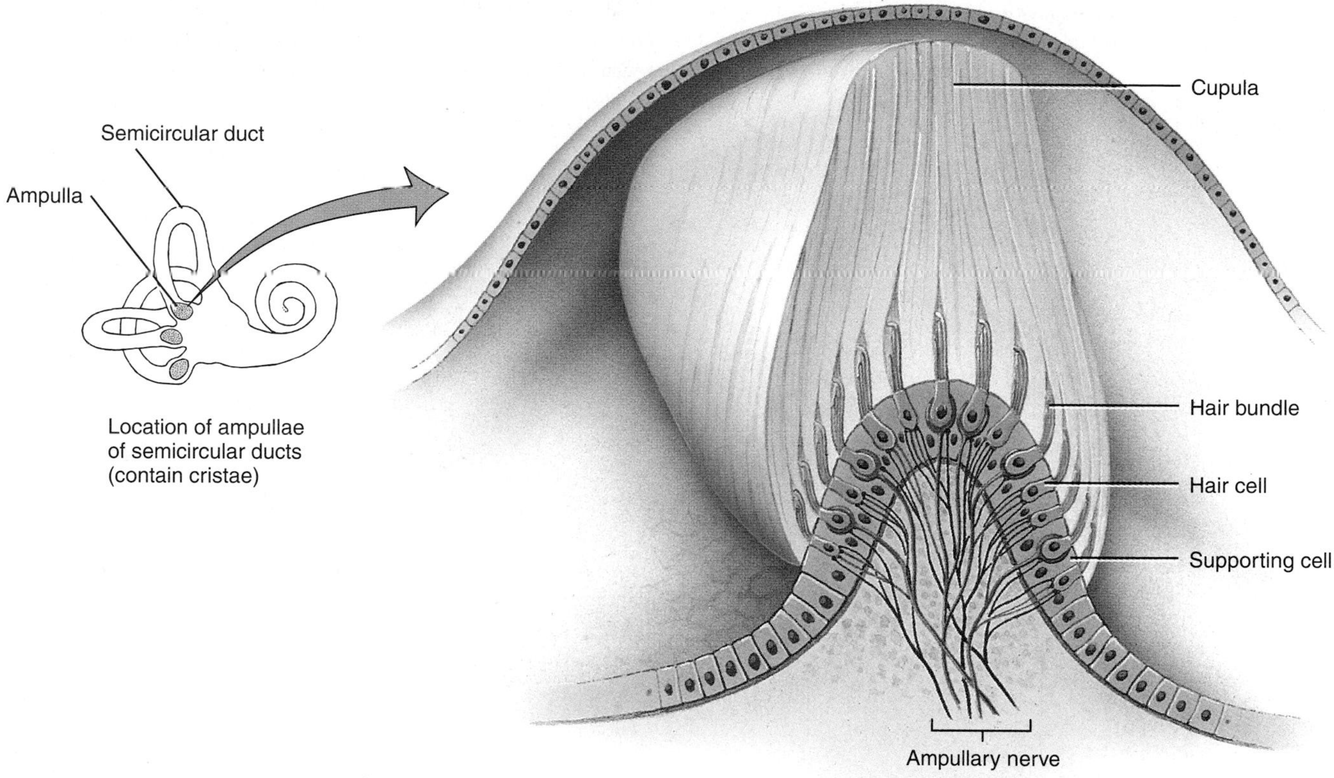

Semicircular duct

Ampulla

Cupula

Hair bundle

Hair cell

Supporting cell

Ampullary nerve

Location of ampullae of semicircular ducts (contain cristae)

(a) Details of a crista

Section of ampulla of membranous labyrinth in semicircular duct

Crista

Ampulla

Ampullary nerve

Cupula sensing movement and direction of flow of endolymph

Head in still position

Head rotating

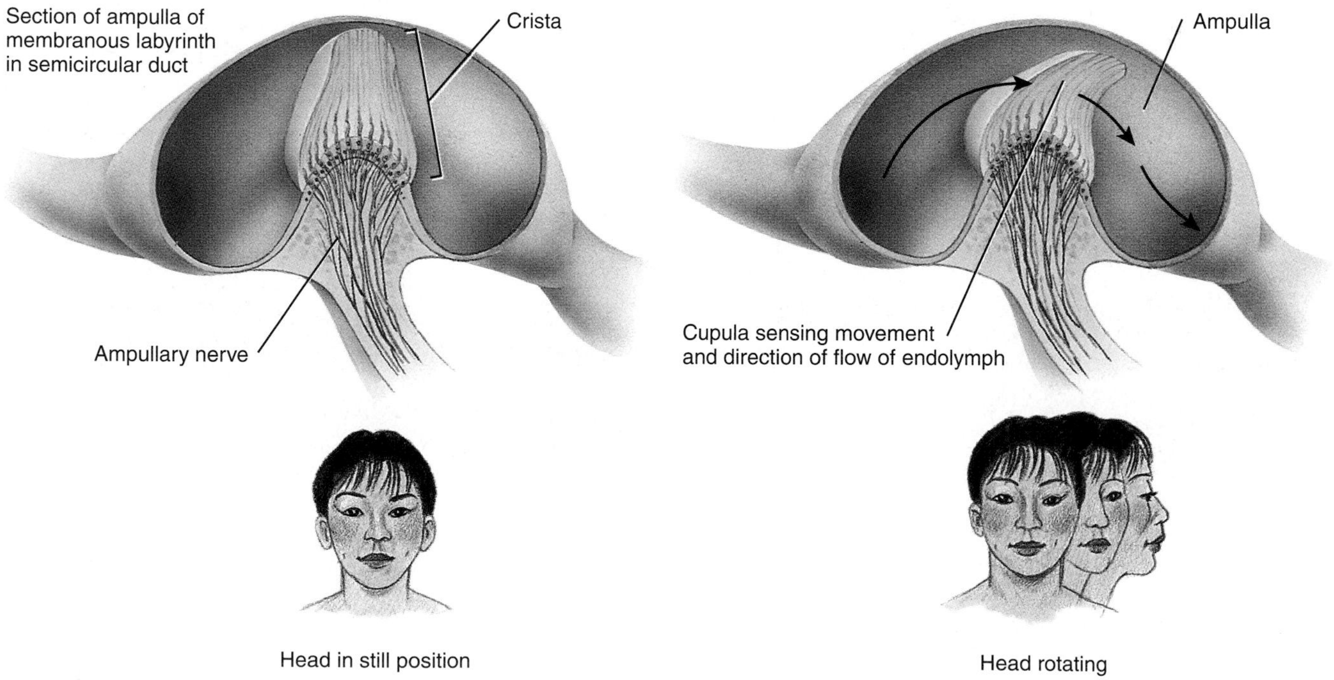

(b) Position of a crista with the head in the still position (left) and when the head rotates (right)

Q With which type of equilibrium are the membranous semicircular ducts, the utricle, and the saccule associated?

Table 16.2 Summary of Structures of the Ear Related to Hearing and Equilibrium

REGIONS OF THE EAR AND KEY STRUCTURES	FUNCTION
External (outer) ear 	*Auricle (pinna):* Collects sound waves. *External auditory canal (meatus):* Directs sound waves to eardrum. *Eardrum (tympanic membrane):* Sound waves cause it to vibrate, which, in turn, causes the malleus to vibrate.
Middle ear 	*Auditory ossicles:* Transmit and amplify vibrations from tympanic membrane to oval window. *Auditory (Eustachian) tube:* Equalizes air pressure on both sides of the tympanic membrane.
Internal (inner) ear 	*Cochlea:* Contains a series of fluids, channels, and membranes that transmit vibrations to the spiral organ (organ of Corti), the organ of hearing; hair cells in the spiral organ produce receptor potentials, which elicit nerve impulses in the cochlear branch of the vestibulo-cochlear (VIII) nerve. *Semicircular ducts:* Contain cristae, site of hair cells for dynamic equilibrium. *Utricle:* Contains macula, site of hair cells for static and dynamic equilibrium. *Saccule:* Contains macula, site of hair cells for static and dynamic equilibrium.

DISORDERS: HOMEOSTATIC IMBALANCES

CATARACTS

A common cause of blindness is a loss of transparency of the lens known as a **cataract.** The lens becomes cloudy (less transparent) due to changes in the structure of the lens proteins. Cataracts often occur with aging but may also be caused by injury, excessive exposure to ultraviolet rays, certain medications (such as long-term use of steroids), or complications of other diseases (for example, diabetes). People who smoke also have increased risk of developing cataracts. Fortunately, sight can usually be restored by surgical removal of the old lens and implantation of an artificial one.

GLAUCOMA

Glaucoma is the most common cause of blindness in the United States, afflicting about 2% of the population over age 40. Glaucoma is an abnormally high intraocular pressure due to a buildup of aqueous humor within the anterior chamber. The fluid compresses the lens into the vitreous body and puts pressure on the neurons of the retina. Persistent pressure results in a progression from mild visual impairment to irreversible destruction of neurons of the retina, damage to the optic (II) nerve, and blindness. Glaucoma is painless, and the other eye compensates to a large extent, so a person may experience considerable retinal damage and loss of vision before the condition is diagnosed. Because glaucoma occurs more often with advancing age, regular measurement of intraocular pressure is an increasingly important part of an eye exam as people grow older. Risk factors include race (blacks are more susceptible), increasing age, family history, and past eye injuries and disorders.

MACULAR DEGENERATION

Macular degeneration, the leading cause of blindness in individuals over age 75, is a deterioration of the retina in the region of the macula, which is ordinarily the area of most acute vision. Initially, a person may experience blurring and distortion at the center of the visual field. In "dry" macular degeneration, central vision gradually diminishes because the pigment epithelium atrophies and degenerates. There is no effective treatment. "Wet" macular degeneration is due to the growth of new blood vessels that form under the retina and leak plasma or blood. Smokers have a threefold greater risk of developing macular degeneration than nonsmokers. Vision loss can be slowed by using laser surgery to destroy leaking blood vessels.

DEAFNESS

Deafness is significant or total hearing loss. **Sensorineural deafness** is caused by either impairment of hair cells in the cochlea or damage of the cochlear branch of the vestibulocochlear (VIII) nerve. This type of deafness may be caused by atherosclerosis, which reduces blood supply to the ears; by repeated exposure to loud noise, which destroys hair cells of the spiral organ; and by certain drugs such as aspirin and streptomycin. **Conduction deafness** is caused by impairment of the external and middle ear mechanisms for transmitting sounds to the cochlea. Conduction deafness may be caused by otosclerosis, the deposition of new bone around the oval window; by impacted cerumen; by injury to the eardrum; and by aging, which often results in thickening of the eardrum and stiffening of the joints of the auditory ossicles.

MÉNIÈRE'S DISEASE

Ménière's disease (men'-ē-ĀRZ) results from an increased amount of endolymph that enlarges the membranous labyrinth. Among the symptoms are fluctuating hearing loss (caused by distortion of the basilar membrane of the cochlea) and roaring tinnitus (ringing). Spinning or whirling vertigo is characteristic of Ménière's disease. Almost total destruction of hearing may occur over a period of years.

OTITIS MEDIA

Otitis media is an acute infection of the middle ear caused primarily by bacteria and associated with infections of the nose and throat. Symptoms include pain, malaise, fever, and a reddening and outward bulging of the eardrum, which may rupture unless prompt treatment is received (this may involve draining pus from the middle ear). Bacteria from the nasopharynx passing into the auditory tube are the primary cause of all middle ear infections. Children are more susceptible than adults to middle ear infections because their auditory tubes are shorter, wider, and almost horizontal, which decreases drainage.

MEDICAL TERMINOLOGY

Amblyopia (am'-blē-Ō-pē-a; *ambly-* = dull or dim) Term used to describe the loss of vision in an otherwise normal eye that, because of muscle imbalance, cannot focus in synchrony with the other eye.

Conjunctivitis (pinkeye) An inflammation of the conjunctiva; when caused by bacteria such as pneumococci, staphylococci, or *Hemophilus influenzae*, it is very contagious and more common in children. Conjunctivitis may also be caused by irritants, such as dust, smoke, or pollutants in the air, in which case it is not contagious.

Keratitis (ker'-a-TĪ-tis; *kerat-* = cornea) An inflammation or infection of the cornea.

Mydriasis (mi-DRĪ-a-sis) Dilated pupil.

Nystagmus (nis-TAG-mus; *nystagm-* = nodding or drowsy) A rapid involuntary movement of the eyeballs, possibly caused by a disease of the central nervous system. It is associated with conditions that cause vertigo.

Otalgia (ō-TAL-jē-a; *ot-* = ear; *-algia* = pain) Earache.

Scotoma (skō-TŌ-ma; = darkness) An area of reduced or lost vision in the visual field.

Strabismus (stra-BIZ-mus) An imbalance in the extrinsic eye muscles that produces a squint.

Tinnitus (ti-NĪ-tus) A ringing, roaring, or clicking in the ears.

Trachoma (tra-KŌ-ma) A serious form of conjunctivitis and the greatest single cause of blindness in the world. It is caused by the bacterium *Chlamydia trachomatis*. The disease produces an excessive growth of subconjunctival tissue and invasion of blood vessels into the cornea, which progresses until the entire cornea is opaque, causing blindness.

Vertigo (VER-ti-gō; = dizziness) A sensation of spinning or movement in which the world seems to revolve or the person seems to revolve in space.

OLFACTION: SENSE OF SMELL (p. 512)

1. The receptors for olfaction, which are bipolar neurons, are in the nasal epithelium.
2. In olfactory reception, a generator potential develops and triggers one or more nerve impulses.
3. The threshold of smell is low, and adaptation to odors occurs quickly.
4. Axons of olfactory receptors form the olfactory (I) nerves, which convey nerve impulses to the olfactory bulbs, olfactory tracts, limbic system, and cerebral cortex (temporal and frontal lobes).

GUSTATION: SENSATION OF TASTE (p. 514)

1. The receptors for gustation, the gustatory receptor cells, are located in taste buds.
2. Before a substance can be tasted, it must be dissolved.
3. Receptor potentials developed in gustatory receptor cells cause the release of neurotransmitter, which can generate nerve impulses in first-order sensory neurons.
4. The threshold varies with the taste involved, and adaptation to taste occurs quickly.
5. Gustatory receptor cells trigger nerve impulses in cranial nerves VII, IX, and X. Taste signals then pass to the medulla oblongata, thalamus, and cerebral cortex (parietal lobe).

VISION (p. 516)

1. Accessory structures of the eyes include the eyebrows, eyelids, eyelashes, lacrimal apparatus, and extrinsic eye muscles.
2. The lacrimal apparatus consists of structures that produce and drain tears.
3. The eye is constructed of three layers: (a) fibrous tunic (sclera and cornea), (b) vascular tunic (choroid, ciliary body, and iris), and (c) retina.
4. The retina consists of pigment epithelium and a neural portion (photoreceptor layer, bipolar cell layer, ganglion cell layer, horizontal cells, and amacrine cells).
5. The anterior cavity contains aqueous humor; the vitreous chamber contains the vitreous body.
6. Image formation on the retina involves refraction of light rays by the cornea and lens, which focus an inverted image on the central fovea of the retina.
7. For viewing close objects, the lens increases its curvature (accommodation) and the pupil constricts to prevent light rays from entering the eye through the periphery of the lens.
8. The near point of vision is the minimum distance from the eye at which an object can be clearly focused with maximum effort.
9. In convergence, the eyeballs move medially so they are both directed toward an object being viewed.

10. The first step in vision is the absorption of light by photopigments in rods and cones (photoreceptors) and isomerization of *cis*-retinal.
11. Receptor potentials in rods and cones decrease the release of inhibitory neurotransmitter, which induces graded potentials in bipolar cells and horizontal cells.
12. Horizontal cells transmit inhibitory signals to bipolar cells; bipolar or amacrine cells transmit excitatory signals to ganglion cells, which depolarize and initiate nerve impulses.
13. Impulses from ganglion cells are conveyed into the optic (II) nerve, through the optic chiasm and optic tract, to the thalamus. From the thalamus, impulses for vision propagate to the cerebral cortex (occipital lobe). Axon collaterals of retinal ganglion cells extend to the midbrain and hypothalamus.

HEARING AND EQUILIBRIUM (p. 529)

1. The external (outer) ear consists of the auricle, external auditory canal, and eardrum (tympanic membrane).
2. The middle ear consists of the auditory or Eustachian tube, ossicles, oval window, and round window.
3. The internal (inner) ear consists of the bony labyrinth and membranous labyrinth. The internal ear contains the spiral organ (organ of Corti), the organ of hearing.
4. Sound waves enter the external auditory canal, strike the eardrum, pass through the ossicles, strike the oval window, set up waves in the perilymph, strike the vestibular membrane and scala tympani, increase pressure in the endolymph, vibrate the basilar membrane, and stimulate hair bundles on the spiral organ (organ of Corti).
5. Hair cells convert mechanical vibrations into a receptor potential, which releases neurotransmitter that can initiate nerve impulses in first-order sensory neurons.
6. Sensory fibers in the cochlear branch of the vestibulocochlear (VIII) nerve terminate in the medulla oblongata. Auditory signals then pass to the inferior colliculus, thalamus, and temporal lobes of the cerebral cortex.
7. Static equilibrium is the orientation of the body relative to the pull of gravity. The maculae of the utricle and saccule are the sense organs of static equilibrium.
8. Dynamic equilibrium is the maintenance of body position in response to movement. The cristae in the membranous semicircular ducts are the principal sense organs of dynamic equilibrium.
9. Most vestibular branch fibers of the vestibulocochlear (VIII) nerve enter the brain stem and terminate in the medulla and pons; other fibers enter the cerebellum.

SELF-QUIZ QUESTIONS

True or false:

1. Of all of the special senses, only smell and taste sensations project both to higher cortical areas and to the limbic system.
2. Rods are most important for color vision, whereas cones are most important for seeing shades of gray in dim light.

Complete the following:

3. The four primary taste sensations are ___, ___, ___, and ___.
4. A small depression in the center of the macula lutea that contains only cone photoreceptors and is the area of highest visual acuity is the ___.
5. The pressure equalization tube that connects the middle ear to the nasopharynx is the ___ tube.
6. ___ equilibrium refers to the maintenance of the position of the body relative to the force gravity; ___ equilibrium refers to the maintenance of body position in response to sudden movements such as rotation, acceleration, and deceleration.

Choose the best answer to the following questions:

7. Which of the following statements is *incorrect*? (a) Olfactory receptors respond to the chemical stimulation of an odorant molecule by producing a receptor potential. (b) Basal stem cells continually produce new olfactory receptors. (c) Adaptation to odors is rapid and occurs in both olfactory receptors and the CNS. (d) Olfaction has a low threshold. (e) The lateral olfactory area is considered to be the primary olfaction area, where conscious awareness of smells begins.
8. Which of the following statements is *incorrect*? (a) Taste is a chemical sense. (b) The receptors for taste sensations are found in taste buds located on the tongue, the soft palate, the pharynx, and the larynx. (c) Taste receptors generate a receptor potential. (d) The threshold for bitter substances is the highest. (e) Adaptation to taste is rapid.
9. Which of the following are functions of the visual processing systems in the cerebral cortex? (1) information processing regarding shapes of objects, (2) information processing regarding weights of objects, (3) information processing regarding colors of objects, (4) information processing regarding ages of objects, (5) information processing regarding movement, location, and spatial organization, (a) 1, 2, and 3, (b) 2, 3, and 4, (c) 1, 3, and 5, (d) 1, 3, and 4, (e) 2, 4, and 5
10. Which of the following statements is *incorrect*? (a) Retinal is the light-absorbing portion of all visual photopigments. (b) Whereas the photopigment in rods is rhodopsin, three different cone opsins are present in the retina. (c) Retinal is a derivative of vitamin C. (d) Rhodopsin absorbs blue to green light most effectively; the cone opsins absorb blue, green, or yellow-orange light. (e) Bleaching and regeneration of the photopigments account for much but not all of the sensitivity changes during light and dark adaptation.
11. Which of the following is the correct sequence for the visual pathway? (a) cornea, pupil, iris, photoreceptors, ganglion cells, bipolar cells, thalamus, visual cortex, (b) cornea, pupil, photoreceptors, bipolar cells, ganglion cells, thalamus, visual cortex, (c) sclera, iris, photoreceptors, bipolar cells, ganglion cells, thalamus, visual cortex, (d) cornea, vitreous body, aqueous humor, ciliary body, photoreceptors, thalamus, visual cortex, (e) cornea, aqueous humor, pupil, vitreous body, photoreceptors, bipolar cells, ganglion cells, thalamus, visual cortex
12. Which of the following is the correct sequence for sound waves? (a) external auditory canal, tympanic membrane, auditory ossicles, oval window, cochlea and organ of Corti, (b) tympanic membrane, external auditory canal, ossicles, cochlea and organ of Corti, round window, (c) ossicles, tympanic membrane, cochlea and organ of Corti, round window, oval window, external auditory canal, (d) auricle, tympanic membrane, round window, cochlea and organ of Corti, oval window, (e) external auditory canal, tympanic membrane, ossicles, internal auditory canal, organ or Corti, oval window
13. Match the following:

    ___ (a) upper and lower eyelids
    ___ (b) produces and drains tears
    ___ (c) arch transversely above the eyeballs and help protect the eyeballs from foreign objects, perspiration, and the direct rays of the sun
    ___ (d) move the eyeball medially, laterally, superiorly, or inferiorly
    ___ (e) shade the eyes during sleep, spread lubricating secretions over the eyeballs
    ___ (f) a thick fold of connective tissue that gives form and support to the eyelids
    ___ (g) modified sebaceous glands; secretion helps keep eyelids from adhering to each other
    ___ (h) project from the border of each eyelid; help protect the eyeballs from foreign objects, perspiration, and direct rays of the sun
    ___ (i) a thin, protective mucous membrane that lines the inner aspect of the eyelids and passes from the eyelids onto the anterior surface of the eyeball

    (1) palpebrae
    (2) tarsal or Meibomian glands
    (3) conjunctiva
    (4) eyelashes
    (5) lacrimal apparatus
    (6) extrinsic eye muscles
    (7) eyebrows
    (8) tarsal plate

14. Match the following:

____ (a) posterior portion of the vascular tunic; provides nutrients to the posterior surface of the retina

____ (b) colored portion of the eyeball; regulates the amount of light entering the posterior part of the eyeball

____ (c) beginning of the visual pathway; contains rods and cones

____ (d) fine tunes focusing of light rays for clear vision

____ (e) transparent part of fibrous tunic; helps focus light

____ (f) circular band of smooth muscle that alters the shape of the lens for near or far vision

____ (g) site where the optic nerve exits the eyeball; the blind spot

____ (h) fluid in the anterior cavity that helps nourish the lens and cornea; helps maintain shape of the eyeball

____ (i) the hole in the center of the iris

____ (j) jelly-like substance in the vitreous chamber that helps prevent the eyeball from collapsing and holds the retina flush against the internal portions of the eyeball

____ (k) white of the eye; gives shape to the eyeball, makes it more rigid, protects its inner parts; a part of the fibrous tunic

____ (l) contain blood capillaries that secrete aqueous humor; attach to suspensory ligaments of lens

(1) cornea
(2) sclera
(3) choroid
(4) ciliary processes
(5) ciliary muscle
(6) iris
(7) pupil
(8) retina
(9) optic disc
(10) lens
(11) aqueous humor
(12) vitreous body

15. Match the following:

____ (a) eardrum

____ (b) oval central portion of the bony labyrinth

____ (c) receptor for static equilibrium; also contributes to some aspects of dynamic equilibrium; consists of hair cells and supporting cells

____ (d) organ of Corti; organ for hearing

____ (e) ear bones: anvil, malleus, stapes

____ (f) opening into the middle ear; is enclosed by a membrane called the secondary tympanic membrane

____ (g) contains the organ of Corti

____ (h) the flap of elastic cartilage covered by skin that captures sound waves; the pinna

____ (i) receptor organs for equilibrium; the saccule, utricle, and semicircular canals

____ (j) opening between the middle and inner ear; receives base of stapes

(1) auricle
(2) tympanic membrane
(3) auditory ossicles
(4) vestibular apparatus
(5) vestibule
(6) cochlea
(7) spiral organ
(8) oval window
9) round window
(10) macula

CRITICAL THINKING QUESTIONS

1. Mother held a dish of steaming food under Brenna's nose. "How can you say it tastes terrible? You haven't even tried it!" Brenna's lips remained clamped shut. How can Brenna be so sure she won't like the food? Trace the pathway of the food's smell to the brain. (HINT: *How would Brenna know if she had tried this food before?*)

2. Fred will be 81 years old on his next birthday. He started working on the fishing boat when he was 15, but lately he's had to let his grandson do the piloting because his vision has gotten so cloudy. Although he's a smoker, his health is generally good, and he's had his blood pressure and intraocular pressure checked yearly. What could be the problem with his eyes? (HINT: *Fishermen are constantly exposed to bright sunlight and glare off the water.*)

3. Mari's mother always says to her, "Anything I say goes in one ear and out the other!" (At least that's what Mari thinks her mom says.) Follow the path of sound from the external ear to the organ of hearing. (HINT: *The inner ear on one side of the head does not connect to the ear on the other side.*)

ANSWERS TO FIGURE QUESTIONS

16.1 The olfactory hairs detect odorant molecules.

16.2 Gustatory receptors → cranial nerves VII, IX, or X → medulla oblongata → either limbic system and hypothalamus, or thalamus → primary gustatory area in the parietal lobe of the cerebral cortex.

16.3 The conjunctiva is continuous with the inner lining of the eyelids.

16.4 Lacrimal fluid, or tears, is a watery solution containing salts, some mucus, and lysozyme that protects, cleans, lubricates, and moistens the eyeball.

16.5 The fibrous tunic consists of the cornea and sclera; the vascular tunic consists of the choroid, ciliary body, and lens.

16.6 The parasympathetic division of the ANS causes pupillary constriction, whereas the sympathetic division causes pupillary dilation.

16.7 An ophthalmoscopic examination can reveal evidence of hypertension, diabetes mellitus, cataract, and macular degeneration.

16.8 The two types of photoreceptors are rods and cones. Rods provide black-and-white vision in dim light, whereas cones provide high visual acuity and color vision in bright light.

16.9 After its secretion by the ciliary process, aqueous humor flows into the posterior chamber, around the iris, into the anterior chamber, and out of the eyeball through the scleral venous sinus.

16.10 During accommodation the ciliary muscle contracts, causing the suspensory ligaments to slacken. The lens then becomes more convex, increasing its focusing power.

16.11 Presbyopia is the loss of lens elasticity that occurs with aging.

16.12 Both rods and cones transduce light into receptor potentials, use a photopigment embedded in outer segment membrane discs or folds, and release neurotransmitter at synapses with bipolar cells and horizontal cells.

16.13 The conversion of *cis*-retinal to *trans*-retinal is called isomerization.

16.14 Cyclic GMP is the ligand that opens Na^+ channels in photoreceptors.

16.15 Light rays from an object in the temporal half of the visual field fall on the nasal half of the retina.

16.16 The malleus is attached to the eardrum.

16.17 The eardrum separates the middle ear from the external ear; the oval and round windows separate the middle ear from the internal ear.

16.18 The two sacs in the vestibule are the utricle and saccule.

16.19 The bony labyrinth consists of semicircular canals, vestibule, and cochlea.

16.20 The region of the basilar membrane close to the oval and round windows vibrates most vigorously in response to high-frequency sounds.

16.21 Maculae mainly operate in static equilibrium.

16.22 The membranous semicircular ducts, utricle, and saccule are mainly associated with dynamic equilibrium.

The **autonomic nervous system (ANS)** regulates the activity of smooth muscle, cardiac muscle, and certain glands. The ANS has often been described as a specific *motor output* portion of the peripheral nervous system. To operate, however, the ANS depends on a continual flow of *sensory input* from visceral organs and blood vessels into *integrating centers* in the central nervous system (CNS). Structurally, then, the ANS includes autonomic sensory neurons, integrating centers in the CNS, and autonomic motor neurons.

Functionally, the ANS usually operates without conscious control. The system was originally named *autonomic* because it was thought to function autonomously or in a self-governing manner, without control by the CNS. However, the ANS is regulated by centers in the brain, mainly in the hypothalamus and brain stem, that receive input from the limbic system and other regions of the cerebrum.

In this chapter we compare structural and functional features of the somatic and autonomic nervous systems, then we discuss the anatomy of the motor portion of the ANS and compare the organization and actions of its two major branches, the sympathetic and parasympathetic divisions.

COMPARISON OF SOMATIC AND AUTONOMIC NERVOUS SYSTEMS

OBJECTIVE

• *Compare the structural and functional differences of the somatic and autonomic portions of the nervous system.*

The somatic nervous system includes both sensory and motor neurons. The sensory neurons convey input from re-

ceptors for the special senses (vision, hearing, taste, smell, and equilibrium) and from receptors for somatic senses (pain, temperature, tactile, and proprioceptive sensations). All these sensations normally are consciously perceived. In turn, somatic motor neurons innervate skeletal muscle—the effector tissue of the somatic nervous system—and produce conscious, voluntary movements. In the somatic nervous system, the effect of a motor neuron always is excitation. When a somatic motor neuron stimulates a skeletal muscle, the muscle contracts. If somatic motor neurons cease to stimulate a muscle, the result is a paralyzed, limp muscle that has no muscle tone. In addition, even though we are generally not conscious of breathing, the muscles that generate respiratory movements are skeletal muscles controlled by somatic motor neurons. If the respiratory motor neurons become inactive, breathing stops. A few skeletal muscles, such as those in the middle ear, cannot be voluntarily controlled.

The main input to the ANS comes from **autonomic sensory neurons.** Mostly, these neurons are associated with interoceptors, such as chemoreceptors that monitor blood CO_2 level and mechanoreceptors that detect degree of stretch in the walls of organs or blood vessels. These sensory signals are not consciously perceived most of the time, although intense activation of interoceptors may give rise to conscious sensations. Two examples of visceral sensations that are perceived are sensations of pain or nausea from damaged viscera and angina pectoris (chest pain) from inadequate blood flow to the heart. Input that influences the ANS also includes some sensations monitored by somatic sensory and special sensory neurons. For example, pain can produce dramatic changes in some autonomic activities.

Figure 17.1 Motor neuron pathways in the ANS (a) and the somatic nervous system (b). Note that autonomic motor neurons release either acetylcholine (ACh) or norepinephrine (NE); somatic motor neurons release ACh.

🔑 **Stimulation by the autonomic nervous system either excites or inhibits visceral effectors; somatic nervous system stimulation always excites its effectors.**

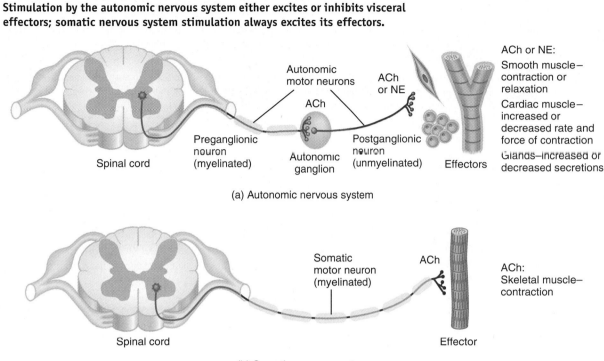

(a) Autonomic nervous system

(b) Somatic nervous system

ⓠ What does dual innervation mean?

Autonomic motor neurons regulate visceral activities by either increasing (exciting) or decreasing (inhibiting) ongoing activities in their effector tissues, which are cardiac muscle, smooth muscle, and glands. Unlike skeletal muscle, these tissues generally function even if their nerve supply is interrupted. The heart continues to beat, for instance, when it is removed for transplantation into another person. Examples of autonomic responses are changes in the diameter of the pupil, dilation and constriction of blood vessels, and adjustment of the rate and force of the heartbeat.

Most autonomic responses cannot be consciously altered or suppressed to any great degree—you probably cannot voluntarily slow your heartbeat to half its normal rate. For this reason, some autonomic responses are the basis for polygraph ("lie detector") tests. Nevertheless, practitioners of yoga or other techniques of meditation may through long practice learn how to modulate at least some of their autonomic activities. Input from the general somatic and special senses, acting via the limbic system, also influences responses of autonomic motor neurons. Seeing a bike about to hit you, hearing squealing brakes of a nearby car, or being grabbed by an attacker, for example, would increase the rate and force of your heartbeat.

All autonomic motor pathways consist of two motor neurons in series, one following the other (Figure 17.1a). The first neuron has its cell body in the CNS; its myelinated axon extends from the CNS to an **autonomic ganglion.** (Recall that a ganglion is a collection of neuronal cell bodies outside the CNS.) The cell body of the second neuron is also in that autonomic ganglion; its unmyelinated axon extends directly from the ganglion to the effector (smooth muscle, cardiac muscle, or a gland). In contrast, to cause voluntary contraction of a skeletal muscle fiber, a single myelinated somatic motor neuron extends from the CNS to the effector (Figure 17.1b). Also, whereas all somatic motor neurons release only acetylcholine (ACh) as their neurotransmitter, autonomic motor neurons release either ACh or norepinephrine (NE).

The output (motor) part of the ANS has two principal branches: the **sympathetic division** and the **parasympathetic division.** Most organs have **dual innervation**—they receive impulses from both sympathetic and parasympathetic neurons. In general, nerve impulses from one division stimulate the organ to increase its activity (excitation), whereas impulses from the other division decrease the organ's activity (inhibition). For example, an increased rate of nerve impulses from the sympathetic division increases heart

Table 17.1 Summary of Autonomic and Somatic Nervous Systems

	AUTONOMIC NERVOUS SYSTEM	SOMATIC NERVOUS SYSTEM
Sensory input	Mainly from interoceptors; some from special senses and somatic senses.	Special senses and somatic senses.
Control of motor output	Involuntary control from limbic system, hypothalamus, brain stem, and spinal cord; limited control from cerebral cortex.	Voluntary control from cerebral cortex, with contributions from basal ganglia, cerebellum, brain stem, and spinal cord.
Motor neuron pathway	Two-neuron pathway: preganglionic neurons extending from CNS synapse with postganglionic neurons in an autonomic ganglion, and postganglionic neurons extending from ganglion synapse with a visceral effector. Also, preganglionic neurons extending from CNS synapse with cells of adrenal medulla.	One-neuron pathway: somatic motor neurons extending from CNS synapse directly with effector.
Neurotransmitters and hormones	Preganglionic axons release acetylcholine (ACh); postganglionic axons release ACh (parasympathetic division and sympathetic fibers to sweat glands) or norepinephrine (NE; remainder of sympathetic division); adrenal medulla releases epinephrine and norepinephrine.	All somatic motor neurons release ACh.
Effectors	Smooth muscle, cardiac muscle, glands.	Skeletal muscle.
Responses	Contraction or relaxation of smooth muscle; increased or decreased rate and force of contraction of cardiac muscle; increased or decreased secretions of glands.	Contraction of skeletal muscle.

rate, whereas an increased rate of nerve impulses from the parasympathetic division decreases heart rate. Table 17.1 summarizes the similarities and differences between the somatic and autonomic nervous systems.

1. Explain why the autonomic nervous system is so named.
2. What are the principal input and output components of the autonomic nervous system?

ANATOMY OF AUTONOMIC MOTOR PATHWAYS

OBJECTIVES

• *Describe preganglionic and postganglionic neurons of the autonomic nervous system.*

• *Compare the anatomical structure of the sympathetic and parasympathetic divisions of the autonomic nervous system.*

Anatomical Components

The first of the two motor neurons in any autonomic motor pathway is called a **preganglionic neuron** (see Figure 17.1a). Its cell body is in the brain or spinal cord, and its axon exits the CNS as part of a cranial or spinal nerve. The axon of a preganglionic neuron is a small-diameter, myelinated type B fiber that extends to an autonomic ganglion, where it synapses with the **postganglionic neuron,** the second neuron in the autonomic motor pathway. Notice that the postganglionic neuron lies entirely outside the CNS; its

cell body and dendrites are located in an autonomic ganglion, where it synapses with one or more preganglionic fibers. The axon of a postganglionic neuron is a small-diameter, unmyelinated type C fiber that terminates in a visceral effector. Thus, preganglionic neurons convey motor impulses from the CNS to autonomic ganglia, and postganglionic neurons relay the impulses from autonomic ganglia to visceral effectors.

Preganglionic Neurons

In the sympathetic division, the preganglionic neurons have their cell bodies in the lateral horns of the gray matter in the 12 thoracic segments and the first two lumbar segments of the spinal cord (Figure 17.2). For this reason, the sympathetic division is also called the **thoracolumbar division** (thō′-ra-kō-LUM-bar), and the axons of the sympathetic preganglionic neurons are known as the **thoracolumbar outflow.**

The cell bodies of the preganglionic neurons of the parasympathetic division are located in the nuclei of the four

Figure 17.2 (See figure opposite.) Structure of the sympathetic and parasympathetic divisions of the autonomic nervous system (ANS). The parasympathetic division is shown only on one side, and the sympathetic division only on the other side, for diagrammatic purposes; each division innervates tissues and organs on both sides of the body.

Sympathetic and parasympathetic stimulation have opposing effects on organs that receive dual innervation.

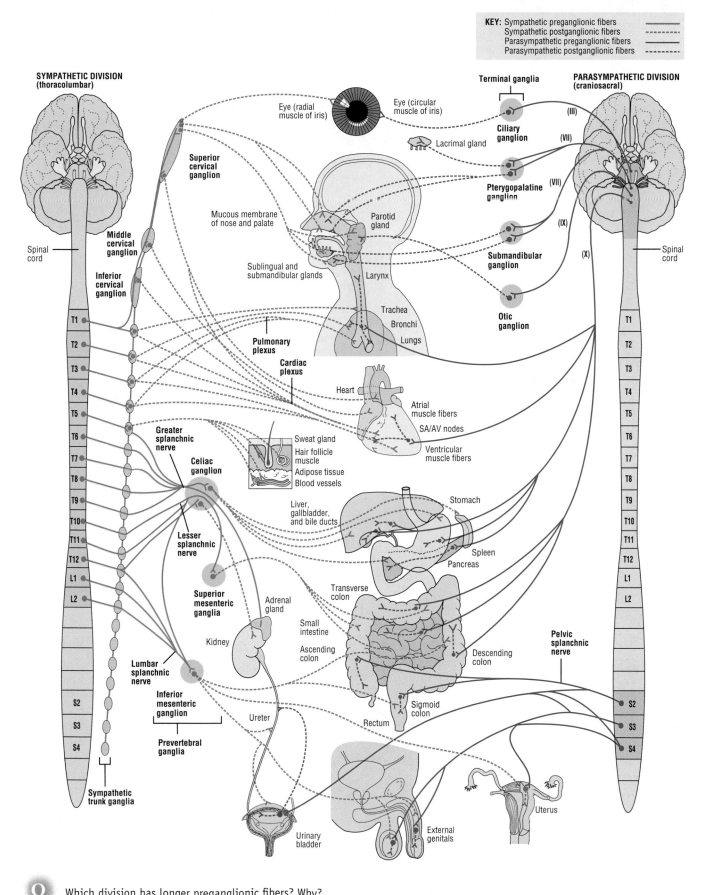

KEY: Sympathetic preganglionic fibers
Sympathetic postganglionic fibers
Parasympathetic preganglionic fibers
Parasympathetic postganglionic fibers

SYMPATHETIC DIVISION (thoracolumbar)

Eye (radial muscle of iris)

Eye (circular muscle of iris)

Terminal ganglia

PARASYMPATHETIC DIVISION (craniosacral)

(III)

Ciliary ganglion

(VII)

Lacrimal gland

Superior cervical ganglion

Mucous membrane of nose and palate

Parotid gland

(VII)

Pterygopalatine ganglion

Spinal cord

Middle cervical ganglion

Inferior cervical ganglion

Sublingual and submandibular glands

Larynx

(IX)

Submandibular ganglion

(X)

Spinal cord

T1

T2

T3

T4

T5

T6

T7

T8

T9

T10

T11

T12

L1

L2

Trachea

Bronchi

Lungs

Otic ganglion

Pulmonary plexus

Cardiac plexus

Heart

Atrial muscle fibers

SA/AV nodes

Ventricular muscle fibers

Greater splanchnic nerve

Celiac ganglion

Sweat gland
Hair follicle muscle
Adipose tissue
Blood vessels

Liver, gallbladder, and bile ducts

Stomach

Lesser splanchnic nerve

Spleen

Pancreas

Superior mesenteric ganglia

Adrenal gland

Transverse colon

T1

T2

T3

T4

T5

T6

T7

T8

T9

T10

T11

T12

L1

L2

Kidney

Small intestine

Ascending colon

Descending colon

Pelvic splanchnic nerve

Lumbar splanchnic nerve

Inferior mesenteric ganglion

Ureter

Sigmoid colon

Rectum

Prevertebral ganglia

S2

S3

S4

S2

S3

S4

Sympathetic trunk ganglia

Urinary bladder

External genitals

Uterus

Q Which division has longer preganglionic fibers? Why?

Figure 17.3 Autonomic plexuses in the thorax, abdomen, and pelvis.

 An autonomic plexus is a network of sympathetic and parasympathetic axons that sometimes also includes autonomic sensory axons and sympathetic ganglia.

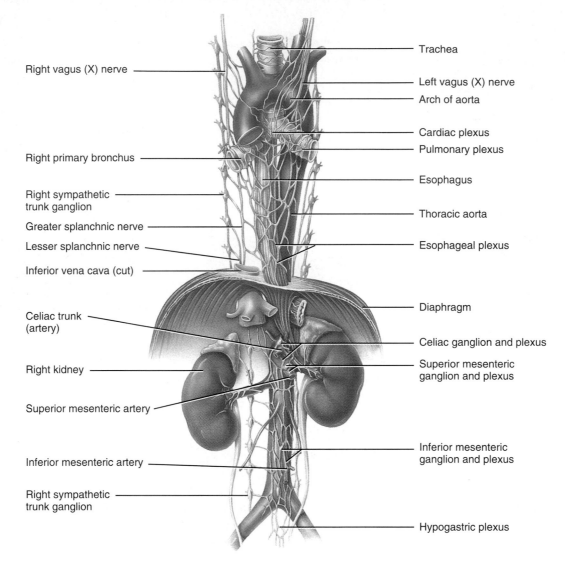

Right vagus (X) nerve

Right primary bronchus

Right sympathetic trunk ganglion

Greater splanchnic nerve

Lesser splanchnic nerve

Inferior vena cava (cut)

Celiac trunk (artery)

Right kidney

Superior mesenteric artery

Inferior mesenteric artery

Right sympathetic trunk ganglion

Trachea

Left vagus (X) nerve

Arch of aorta

Cardiac plexus

Pulmonary plexus

Esophagus

Thoracic aorta

Esophageal plexus

Diaphragm

Celiac ganglion and plexus

Superior mesenteric ganglion and plexus

Inferior mesenteric ganglion and plexus

Hypogastric plexus

Q Which is the largest autonomic plexus?

cranial nerves in the brain stem and in the lateral gray horns of the second through fourth sacral segments of the spinal cord. Hence, the parasympathetic division is also known as the **craniosacral division,** and the axons of the parasympathetic preganglionic neurons are referred to as the **craniosacral outflow.**

Autonomic Ganglia

The autonomic ganglia may be divided into three general groups: Two of the groups are components of the sympathetic division, and one group is a component of the parasympathetic division.

Sympathetic Ganglia The sympathetic ganglia are the sites of synapses between sympathetic preganglionic and postganglionic neurons. The two groups of sympathetic ganglia are sympathetic trunk ganglia and prevertebral ganglia. **Sympathetic trunk (chain) ganglia** (also called *vertebral chain ganglia* or *paravertebral ganglia*) lie in a vertical row on either side of the vertebral column, extending from the base of the skull to the coccyx (see Figure 17.2). Because the sympathetic trunk ganglia are near the spinal cord, most sympathetic preganglionic axons are short. Postganglionic axons from sympathetic trunk ganglia, in general, innervate organs above the diaphragm. Examples of sympathetic trunk ganglia are the **superior, middle,** and **inferior cervical ganglia.**

The second group of sympathetic ganglia, the **prevertebral** *(collateral)* **ganglia,** lie anterior to the spinal column and close to the large abdominal arteries. In general, postganglionic fibers from prevertebral ganglia innervate organs below the diaphragm. Examples of prevertebral ganglia are the **celiac ganglion** (SĒ-lē-ak), on either side of the celiac artery just inferior to the diaphragm; the **superior mesenteric ganglion,** near the beginning of the superior mesenteric artery in the upper abdomen; and the **inferior mesenteric ganglion,** near the beginning of the inferior mesenteric artery in the middle of the abdomen (see Figures 17.2 and 17.3).

Parasympathetic Ganglia Preganglionic fibers of the parasympathetic division synapse with postganglionic neurons in **terminal** *(intramural)* **ganglia.** These ganglia are located close to or actually within the wall of a visceral organ. Because the axons of parasympathetic preganglionic neurons extend from the CNS to a terminal ganglion in an innervated organ, they are longer than most of the axons of sympathetic preganglionic neurons. Examples of terminal ganglia include the ciliary ganglion, pterygopalatine ganglion, submandibular ganglion, and otic ganglion (see Figure 17.2).

Autonomic Plexuses

In the thorax, abdomen, and pelvis, axons of both sympathetic and parasympathetic neurons form tangled networks called **autonomic plexuses,** many of which lie along major arteries. The autonomic plexuses also may contain sympathetic ganglia and axons of autonomic sensory neurons. Often plexuses are named after the artery along which they are distributed. The major plexuses in the thorax are the **cardiac plexus,** which is at the base of the heart, surrounding the large blood vessels emerging from the heart, and the **pulmonary plexus,** located mostly posterior to each lung (Figure 17.3).

The abdomen and pelvis also contain major autonomic plexuses. The **celiac (solar) plexus** is found at the level of the last thoracic and first lumbar vertebrae. It is the largest autonomic plexus and surrounds the celiac and superior mesenteric arteries. It contains two large celiac ganglia and a dense network of autonomic fibers. Secondary plexuses from or connected to the celiac plexus are distributed to the diaphragm, liver, gallbladder, stomach, pancreas, spleen, kidneys, medulla (inner region) of the adrenal gland, testes, and ovaries. The **superior mesenteric plexus,** a secondary plexus of the celiac plexus, contains the superior mesenteric ganglion and supplies the small and large intestine. The **inferior mesenteric plexus,** also a secondary plexus of the celiac plexus, contains the inferior mesenteric ganglion and innervates the large intestine. The **hypogastric plexus** is anterior to the fifth lumbar vertebra and supplies pelvic viscera.

Postganglionic Neurons

Once axons of preganglionic neurons of the sympathetic division pass to sympathetic trunk ganglia, they may connect with postganglionic neurons in one of the following ways (Figure 17.4):

1 An axon may synapse with postganglionic neurons in the ganglion it first reaches.

2 An axon may ascend or descend to a higher or lower ganglion before synapsing with postganglionic neurons.

3 An axon may continue, without synapsing, through the sympathetic trunk ganglion to end at a prevertebral ganglion and synapse with postganglionic neurons there.

A single sympathetic preganglionic fiber has many axon collaterals (branches) and may synapse with 20 or more postganglionic fibers. This pattern of projection is an example of divergence and helps explain why many sympathetic responses affect almost the entire body simultaneously. After exiting their ganglia, the postganglionic fibers typically innervate several visceral effectors (see Figure 17.2).

Axons of preganglionic neurons of the parasympathetic division pass to terminal ganglia near or within a visceral effector (see Figure 17.2). In the ganglion, the presynaptic neuron usually synapses with only four or five postsynaptic neurons, all of which supply a single visceral effector. Thus, parasympathetic responses can be localized to a single effector. With this background in mind, we can now examine some specific structural features of the sympathetic and parasympathetic divisions of the ANS.

Structure of the Sympathetic Division

Cell bodies of sympathetic preganglionic neurons are part of the lateral horns of all thoracic segments and of the first two lumbar segments of the spinal cord (see Figure 17.2). The preganglionic axons leave the spinal cord through the anterior root of a spinal nerve along with the somatic motor fibers at the same segmental level. After exiting through the intervertebral foramina, the myelinated preganglionic sympathetic axons enter a short pathway called a **white ramus** before passing to the nearest sympathetic trunk ganglion on the same side (see Figure 17.4).

Collectively, the white rami are called the **white rami communicantes** (kō-myoo-ni-KAN-tēz; the singular is **ramus communicans**). The "white" in their name indicates that they contain myelinated fibers. Only the thoracic and first two or three lumbar nerves have white rami communicantes. The white rami communicantes connect the anterior ramus of the spinal nerve with the ganglia of the sympathetic trunk. The paired sympathetic trunk ganglia are arranged anterior and lateral to the vertebral column, one on either side. Typically, there are 3 cervical, 11 or 12 thoracic, 4 or 5 lumbar, and 4 or 5 sacral sympathetic trunk ganglia. Although the sympathetic trunk ganglia extend inferiorly from the neck, chest, and abdomen to the coccyx, they receive preganglionic fibers only from the thoracic and lumbar segments of the spinal cord (see Figure 17.2).

The cervical portion of each sympathetic trunk is located in the neck and is subdivided into superior, middle,

Figure 17.4 Types of connections between ganglia and postganglionic neurons in the sympathetic division of the ANS. Numbers correspond to descriptions in the text. Also illustrated are the gray and white rami communicantes.

🔑 **Sympathetic ganglia lie in two chains on either side of the vertebral column (sympathetic trunk ganglia) and near large abdominal arteries anterior to the vertebral column (prevertebral ganglia).**

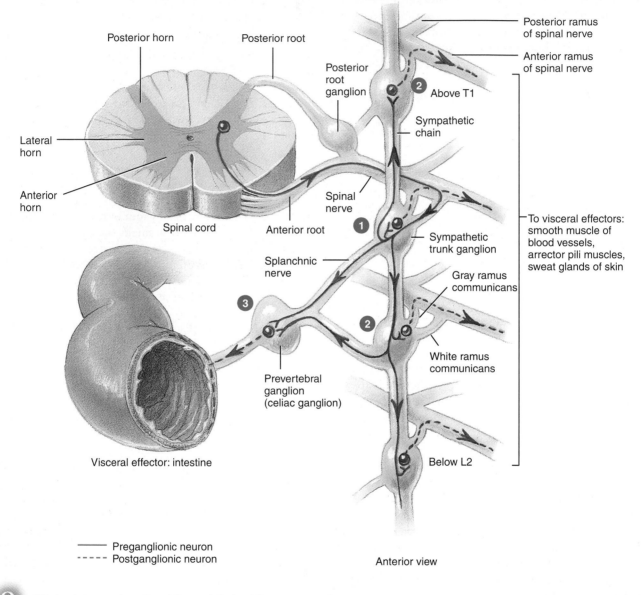

Posterior horn

Posterior root

Posterior root ganglion

② Above T1

Posterior ramus of spinal nerve

Anterior ramus of spinal nerve

Sympathetic chain

Lateral horn

Anterior horn

Spinal cord

Spinal nerve

Anterior root

①

Splanchnic nerve

③

Prevertebral ganglion (celiac ganglion)

Visceral effector: intestine

Sympathetic trunk ganglion

Gray ramus communicans

②

White ramus communicans

Below L2

To visceral effectors: smooth muscle of blood vessels, arrector pili muscles, sweat glands of skin

—— Preganglionic neuron
- - - - Postganglionic neuron

Anterior view

Q What substance gives the white rami their white appearance?

and inferior ganglia (see Figure 17.2). The **superior cervical ganglion** is anterior to a transverse process of the second cervical vertebra. Postganglionic fibers leaving the superior cervical ganglion serve the head. They are distributed to sweat glands, smooth muscle of the eye, blood vessels of the face, nasal mucosa, and the submandibular, sublingual, and parotid salivary glands. Gray rami communicantes (described shortly) from the ganglion also pass to the upper two

to four cervical spinal nerves. The **middle cervical ganglion** lies near the sixth cervical vertebra, and the **inferior cervical ganglion** is located near the first rib, anterior to the transverse processes of the seventh cervical vertebra. Postganglionic fibers from the middle and inferior cervical ganglia innervate the heart.

The thoracic portion of each sympathetic trunk lies anterior to the necks of the corresponding ribs. This portion of the sympathetic trunk receives most of the sympathetic preganglionic fibers. Postganglionic fibers from the thoracic sympathetic trunk innervate the heart, lungs, bronchi, and other thoracic viscera. In the skin, these fibers also innervate sweat glands, blood vessels, and arrector pili muscles of hair follicles.

The lumbar portion of each sympathetic trunk lies lateral to the corresponding lumbar vertebrae. The sacral portion of the sympathetic trunk lies in the pelvic cavity on the medial side of the sacral foramina. Unmyelinated postganglionic fibers from the lumbar and sacral sympathetic trunk ganglia enter a short pathway called a **gray ramus** and then merge with a spinal nerve or join the hypogastric plexus via direct visceral branches. The **gray rami communicantes** are structures containing the postganglionic fibers that connect the ganglia of the sympathetic trunk to spinal nerves (see Figure 17.4). The fibers are unmyelinated. Gray rami communicantes outnumber the white rami because there is a gray ramus leading to each of the 31 pairs of spinal nerves.

As preganglionic fibers extend from a white ramus communicans into the sympathetic trunk, they give off several axon collaterals (branches). These collaterals terminate and synapse in several ways. Some synapse in the first ganglion at the level of entry. Others pass up or down the sympathetic trunk for a variable distance to form the **sympathetic chains,** the fibers on which the ganglia are strung (see Figure 17.4). Many postganglionic fibers rejoin the spinal nerves through gray rami and supply peripheral visceral effectors such as sweat glands, smooth muscle in blood vessels, and arrector pili muscles of hair follicles.

Some preganglionic fibers pass through the sympathetic trunk without terminating in the trunk. Beyond the trunk, they form nerves known as **splanchnic nerves** (SPLANK-nik; see Figure 17.4), which extend to and terminate in the outlying prevertebral ganglia. Splanchnic nerves from the thoracic area terminate in the **celiac ganglion,** where the preganglionic fibers synapse with postganglionic cell bodies. The *greater splanchnic nerve* enters the celiac ganglion of the celiac plexus. From here, postganglionic fibers extend to the stomach, spleen, liver, kidney, and small intestine (see Figure 17.2). The *lesser splanchnic nerve* passes through the celiac plexus to enter the superior mesenteric ganglion of the superior mesenteric plexus. Postganglionic fibers from this ganglion innervate the small intestine and colon. The *lowest splanchnic nerve,* not always present, enters the renal plexus near the kidney. Postganglionic fibers supply kidney arterioles and the ureter.

Preganglionic fibers that form the *lumbar splanchnic nerve* enter the inferior mesenteric plexus and terminate in the inferior mesenteric ganglion, where they synapse with postganglionic neurons. Axons of postganglionic neurons extend through the hypogastric plexus and supply the distal colon and rectum, urinary bladder, and genital organs. Postganglionic fibers leaving the prevertebral ganglia follow the course of various arteries to abdominal and pelvic visceral effectors.

Sympathetic *preganglionic fibers* also extend to the adrenal medullae. Developmentally, the adrenal medullae are modified sympathetic ganglia, and their cells are similar to sympathetic postganglionic neurons. Rather than extending to another organ, however, these cells release hormones into the blood. Upon stimulation by sympathetic preganglionic neurons, the adrenal medullae release a mixture of catecholamine hormones—about 80% **epinephrine,** 20% **norepinephrine,** and a trace amount of **dopamine.**

CLINICAL APPLICATION
Horner's Syndrome

In **Horner's syndrome** the sympathetic innervation to one side of the face is lost due to an inherited mutation, an injury, or a disease that affects sympathetic outflow through the superior cervical ganglion. Symptoms occur on the affected side and include ptosis (drooping of the upper eyelid), miosis (constricted pupil), and anhidrosis (lack of sweating). ■

Structure of the Parasympathetic Division

Cell bodies of parasympathetic preganglionic neurons are found in nuclei in the brain stem and in the lateral horns of the second through fourth sacral segments of the spinal cord (see Figure 17.2). Their axons emerge as part of a cranial nerve or as part of the anterior root of a spinal nerve. The **cranial parasympathetic outflow** consists of preganglionic axons that extend from the brain stem in four cranial nerves. The **sacral parasympathetic outflow** consists of preganglionic axons in anterior roots of the second through fourth sacral nerves. The preganglionic axons of both the cranial and sacral outflows end in terminal ganglia, where they synapse with postganglionic neurons.

The cranial outflow has five components: four pairs of ganglia and the plexuses associated with the vagus (X) nerve. The four pairs of cranial parasympathetic ganglia innervate structures in the head and are located close to the organs they innervate (see Figure 17.2). The **ciliary ganglia** lie lateral to each optic (II) nerve near the posterior aspect of the orbit. Preganglionic axons pass with the oculomotor (III) nerves to the ciliary ganglia. Postganglionic axons from the ganglia innervate smooth muscle fibers in the eyeball. The **pterygopalatine ganglia** (ter'-i-gō-PAL-a-tīn) are lateral to the sphenopalatine foramen, between the sphenoid and palatine bones; they receive preganglionic axons from the facial (VII) nerve and send postganglionic axons to the nasal mucosa, palate, pharynx, and lacrimal glands. The **submandibular ganglia** are found near the ducts of the submandibular salivary glands; they receive preganglionic axons from the facial (VII) nerves and send postganglionic axons to the submandibular and sublingual salivary glands. The **otic ganglia** are situated just inferior to each foramen ovale; they receive preganglionic axons from the glossopharyngeal (IX) nerves and send postganglionic axons to the parotid salivary glands.

Table 17.2 Comparative Anatomy of the Sympathetic and Parasympathetic Divisions

	SYMPATHETIC	PARASYMPATHETIC
Distribution	Bodywide: skin, sweat glands, arrector pili muscles of hair follicles, adipose tissue, smooth muscle of blood vessels.	Limited mainly to head and to viscera of thorax, abdomen, and pelvis; some blood vessels.
Outflow	Thoracolumbar (T1–L2).	Craniosacral (cranial nerves III, VII, IX, and X; S2–S4).
Associated ganglia	Two types: sympathetic trunk and prevertebral ganglia.	One type: terminal ganglia.
Ganglia locations	Close to CNS and distant from visceral effectors.	Typically near or within wall of visceral effectors.
Fiber length and divergence	Short preganglionic fibers synapse with many long postganglionic neurons that pass to many visceral effectors.	Long preganglionic fibers usually synapse with four to five short postganglionic neurons that pass to a single visceral effector.
Rami communicantes	Both present; white rami communicantes contain myelinated preganglionic fibers, and gray rami communicantes contain unmyelinated postganglionic fibers.	Neither present.

Preganglionic axons that leave the brain as part of the vagus (X) nerves carry nearly 80% of the total craniosacral outflow. Vagal fibers extend to many terminal ganglia in the thorax and abdomen. Because the terminal ganglia are close to or in the walls of their visceral effectors, postganglionic parasympathetic axons are very short. As the vagus (X) nerve passes through the thorax, it sends axons to the heart and the airways of the lungs. In the abdomen, it supplies the liver, gallbladder, stomach, pancreas, small intestine, and part of the large intestine.

The sacral parasympathetic outflow consists of preganglionic axons from the anterior roots of the second through fourth sacral nerves. Collectively, they form the **pelvic splanchnic nerves** (see Figure 17.2). These nerves synapse with parasympathetic postganglionic neurons located in terminal ganglia in the walls of the innervated viscera. From the ganglia, parasympathetic postganglionic axons innervate smooth muscle and glands in the walls of the colon, ureters, urinary bladder, and reproductive organs.

The anatomical features of the sympathetic and parasympathetic divisions are compared in Table 17.2.

1. Explain why the sympathetic division is called the thoracolumbar division even though its ganglia extend from the cervical to the sacral region.
2. Prepare a list of the organs served by each sympathetic and parasympathetic ganglion.
3. Describe the locations of sympathetic trunk ganglia, prevertebral ganglia, and terminal ganglia. Which types of autonomic fibers synapse in each type of ganglion?
4. Explain in anatomical terms why the sympathetic division may produce simultaneous effects throughout the body, whereas parasympathetic effects typically are localized to specific organs.

ANS NEUROTRANSMITTERS AND RECEPTORS

OBJECTIVE

• *Describe the neurotransmitters and receptors involved in autonomic responses.*

Based on the neurotransmitter they produce and liberate, autonomic neurons are classified as either cholinergic or adrenergic. The receptors for the neurotransmitters are integral membrane proteins located in the plasma membrane of the postsynaptic neuron or effector cell.

Cholinergic Neurons and Receptors

Cholinergic neurons (kō′-lin-ER-jik) release the neurotransmitter **acetylcholine (ACh).** In the ANS, the cholinergic neurons include (1) all sympathetic and parasympathetic preganglionic neurons, (2) sympathetic postganglionic neurons that innervate most sweat glands, and (3) all parasympathetic postganglionic neurons (Figure 17.5).

ACh is stored in synaptic vesicles and released by exocytosis. It then diffuses across the synaptic cleft and binds with specific **cholinergic receptors,** which are integral membrane proteins in the *postsynaptic* plasma membrane. The two types of cholinergic receptors, both of which bind ACh, are nicotinic receptors and muscarinic receptors. **Nicotinic receptors** are present in the plasma membrane of dendrites and cell bodies of both sympathetic and parasympathetic postganglionic neurons (see Figure 17.5) and in the motor end plate at the neuromuscular junction. They are so named because nicotine mimics the action of ACh by binding to these receptors. (Nicotine, a natural substance in tobacco leaves, is not normally present in nonsmoking humans.) **Muscarinic receptors** are present in the plasma membranes of all effectors (smooth muscle, cardiac muscle, and glands) innervated by parasympathetic postganglionic axons. In addition, most sweat glands receive their innervation from *cholinergic* sympathetic postganglionic fibers and possess muscarinic receptors (see Figure 17.5b). These receptors are so named because a mushroom poison called muscarine mimics the actions of ACh by binding to them. Nicotine does not activate muscarinic receptors, and muscarine does not activate nicotinic receptors, but ACh does activate both types of cholinergic receptor.

Activation of nicotinic receptors by ACh causes depolarization and thus excitation of the postsynaptic cell—either a postganglionic neuron, an autonomic effector, or a skeletal

Figure 17.5 Cholinergic neurons (aqua) and adrenergic neurons (orange) in the sympathetic and parasympathetic divisions. Cholinergic neurons release acetylcholine, whereas adrenergic neurons release norepinephrine. Cholinergic and adrenergic receptors all are integral membrane proteins located in the plasma membrane of a postsynaptic neuron or an effector cell.

🔑 **Most sympathetic postganglionic neurons are adrenergic; other autonomic neurons are cholinergic.**

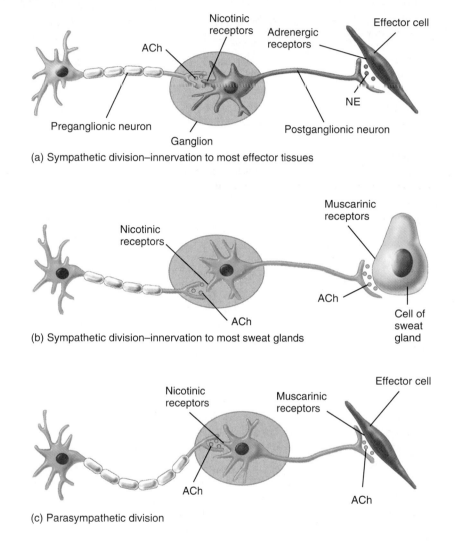

(a) Sympathetic division—innervation to most effector tissues

(b) Sympathetic division—innervation to most sweat glands

(c) Parasympathetic division

Q Which neurons are cholinergic and possess nicotinic ACh receptors? What type of receptors for ACh do the effector tissues innervated by these neurons possess?

muscle fiber. Activation of muscarinic receptors by ACh sometimes causes depolarization (excitation) and sometimes causes hyperpolarization (inhibition), depending on which particular cell bears the muscarinic receptors. The binding of ACh to muscarinic receptors inhibits (relaxes) smooth muscle sphincters in the gastrointestinal tract, for example, but excites smooth muscle fibers in the circular muscles of the iris of the eye, causing them to contract. Because acetylcholine is quickly inactivated by the enzyme **acetylcholinesterase (AChE),** effects triggered by cholinergic fibers are brief.

Adrenergic Neurons and Receptors

In the ANS, **adrenergic neurons** (ad′-ren-ER-jik) release **norepinephrine (NE),** also known as **noradrenalin** (see Figure 17.5a). Most sympathetic postganglionic neurons are adrenergic. Like ACh, NE is synthesized and stored in synaptic vesicles and released by exocytosis. The molecules diffuse across the synaptic cleft and bind to specific adrenergic receptors on the postsynaptic membrane, causing either excitation or inhibition of the effector cell.

Table 17.3 Locations and Responses of Adrenergic and Cholinergic Receptors

TYPE OF RECEPTOR	MAJOR LOCATIONS	EFFECTS OF RECEPTOR ACTIVATION
Cholinergic	Integral proteins in postsynaptic plasma membranes; activated by the neurotransmitter acetylcholine.	
Nicotinic	Plasma membrane of postganglionic sympathetic and parasympathetic neurons.	Excitation → impulses in postganglionic neurons.
	Cells of adrenal medullae.	Epinephrine and norepinephrine secretion.
	Sarcolemma of skeletal muscle fibers (motor end plate).	Excitation → contraction.
Muscarinic	Effectors innervated by parasympathetic postganglionic neurons.	In some receptors, excitation; in others, inhibition.
	Sweat glands innervated by cholinergic sympathetic postganglionic neurons.	Increased sweating.
Adrenergic	Integral proteins in postsynaptic plasma membranes; activated by the neurotransmitter norepinephrine, and by hormonal norepinephrine and epinephrine.	
α_1	Smooth muscle fibers in blood vessels that serve salivary glands, skin, mucosal membranes, kidneys, and abdominal viscera; radial muscle in iris of eye; sphincter muscles of stomach and urinary bladder.	Excitation → contraction, which causes vasoconstriction, dilation of pupil, and closing of sphincters.
	Salivary gland cells.	Secretion of K^+ and water.
	Sweat glands on palms and soles.	Increased sweating.
α_2	Smooth muscle fibers in some blood vessels.	Inhibition → relaxation → vasodilation.
	Cells of pancreatic islets that secrete the hormone insulin (beta cells).	Decreased insulin secretion.
	Platelets in blood.	Aggregation to form platelet plug.
β_1	Cardiac muscle fibers.	Excitation → increased force and rate of contraction.
	Juxtaglomerular cells of kidneys.	Renin secretion.
	Posterior pituitary gland.	Secretion of antidiuretic hormone.
	Adipose cells.	Breakdown of triglycerides → release of fatty acids into blood.
β_2	Smooth muscle in walls of airways; in blood vessels that serve the heart, skeletal muscle, adipose tissue, and liver; and in walls of visceral organs.	Inhibition → relaxation, which causes dilation of airways, vasodilation, and relaxation of organ walls.
	Ciliary muscle in eye.	Inhibition → relaxation.
	Hepatocytes in liver.	Glycogenolysis (breakdown of glycogen into glucose).
β_3	Brown adipose tissue.	Thermogenesis (heat production).

Adrenergic receptors bind both norepinephrine and epinephrine. Moreover, the receptors are activated by norepinephrine released as a neurotransmitter by sympathetic postganglionic neurons and by epinephrine and norepinephrine released as hormones into the blood by the adrenal medulla. The two main types of adrenergic receptors are **alpha (α) receptors** and **beta (β) receptors,** which are found on visceral effectors innervated by most sympathetic postganglionic axons. These receptors are further classified into subtypes—α_1, α_2, β_1, β_2, and β_3—based on the specific responses they elicit and by their selective binding of drugs that activate or block them. Although there are some exceptions, activation of α_1 and β_1 receptors generally produces excitation, whereas activation of α_2 and β_2 receptors causes inhibi-

tion. β_3 receptors are present only on cells of brown adipose tissue, where their activation causes thermogenesis (heat production). Cells of most effectors contain either alpha or beta receptors; some visceral effector cells contain both. Norepinephrine stimulates alpha receptors more strongly than beta receptors, whereas epinephrine is a potent stimulator of both alpha and beta receptors.

The activity of norepinephrine at a synapse is terminated either when the NE is taken up by the axon that released it or when the NE is enzymatically inactivated by either **catechol-O-methyltransferase (COMT)** or **monoamine oxidase (MAO).** Compared to ACh, norepinephrine lingers in the synaptic cleft for a longer period of time. Thus, effects triggered by adrenergic neurons typically are longer lasting than those triggered by cholinergic neurons.

Table 17.3 describes the locations of cholinergic and adrenergic receptors and summarizes the responses that occur when each type of receptor is activated.

Receptor Agonists and Antagonists

A large variety of drugs and natural products can selectively activate or block specific cholinergic or adrenergic receptors. An **agonist** is a substance that binds to and activates a receptor, in the process mimicking the effect of a natural neurotransmitter or hormone. Phenylephrine, which activates α_1 receptors, is a common ingredient in cold and sinus medications. Because it constricts blood vessels in the nasal mucosa, phenylephrine reduces production of mucus, thus relieving nasal congestion. An **antagonist** is a substance that binds to and blocks a receptor, thereby preventing a natural neurotransmitter or hormone from exerting its effect. For example, atropine, which blocks muscarinic ACh receptors, dilates the pupils, reduces glandular secretions, and relaxes smooth muscle in the gastrointestinal tract. As a result, it is used to dilate the pupils during eye examinations, in the treatment of smooth muscle disorders such as iritis and intestinal hypermotility, and as an antidote for chemical warfare agents that inactivate acetylcholinesterase.

Propranolol (Inderal) often is prescribed for patients with hypertension (high blood pressure). It is a nonselective beta blocker, meaning it binds to all types of beta receptors and prevents their activation by epinephrine and norepinephrine. The desired effects of propranolol are due to its blockade of β_1 receptors—namely, decreased heart rate and force of contraction and a consequent decrease in blood pressure. Undesired effects due to blockade of β_2 receptors may include hypoglycemia (low blood glucose), resulting from decreased glycogen breakdown and decreased gluconeogenesis (the conversion of a noncarbohydrate into glucose in the liver); and mild bronchoconstriction (narrowing of the airways). If these side effects pose a threat to the patient, a selective β_1 blocker such as metoprolol (Lopressor) can be prescribed instead of propranolol.

PHYSIOLOGICAL EFFECTS OF THE ANS

As noted earlier, most body organs are innervated by both divisions of the ANS, which typically work in opposition to one another. The balance between sympathetic and parasympathetic activity or "tone" is regulated by the hypothalamus. Typically, the hypothalamus turns up sympathetic tone at the same time it turns down parasympathetic tone, and vice versa. The two divisions can affect body organs differently because their postganglionic neurons release different neurotransmitters and because the effector organs possess different adrenergic and cholinergic receptors. A few structures receive only sympathetic innervation—sweat

glands, arrector pili muscles attached to hair follicles in the skin, the kidneys, most blood vessels, and the adrenal medullae (see Figure 17.2). In these cases there is no opposition from the parasympathetic division. Still, an increase in sympathetic tone has a given effect, whereas a decrease in sympathetic tone produces an opposite effect.

Sympathetic Responses

During physical or emotional stress, the sympathetic division dominates the parasympathetic division. High sympathetic tone favors body functions that can support vigorous physical activity and rapid production of ATP. At the same time, the sympathetic division reduces body functions that favor the storage of energy. Besides physical exertion, a variety of emotions—such as fear, embarrassment, or rage—stimulate the sympathetic division. Visualizing body changes that occur during "E situations" (exercise, emergency, excitement, embarrassment) will help you remember most of the sympathetic responses. Activation of the sympathetic division and release of hormones by the adrenal medullae set in motion a series of physiological responses collectively called the **fight-or-flight response,** which produces the following effects:

1. The pupils of the eyes dilate.
2. Heart rate, force of heart contraction, and blood pressure increase.
3. The airways dilate, allowing faster movement of air into and out of the lungs.
4. The blood vessels that supply nonessential organs such as the kidneys and gastrointestinal tract constrict.
5. Blood vessels that supply organs involved in exercise or fighting off danger—skeletal muscles, cardiac muscle, liver, and adipose tissue—dilate, allowing greater blood flow through these tissues.
6. Hepatocytes in the liver perform glycogenolysis (breakdown of glycogen to glucose), and adipose tissue cells perform lipolysis (breakdown of triglycerides to fatty acids and glycerol).
7. Release of glucose by the liver increases blood glucose level.
8. Processes that are not essential for meeting the stressful situation are inhibited. For example, muscular movements of the gastrointestinal tract and digestive secretions slow down or even stop.

The effects of sympathetic stimulation are longer lasting and more widespread than the effects of parasympathetic stimulation for three reasons:

- Sympathetic postganglionic fibers diverge more extensively; as a result, many tissues are activated simultaneously.
- Acetylcholinesterase quickly inactivates acetylcholine, whereas norepinephrine lingers in the synaptic cleft for a longer period of time.

Table 17.4 Activities of Sympathetic and Parasympathetic Divisions of the ANS

VISCERAL EFFECTOR	EFFECT OF SYMPATHETIC STIMULATION (α OR β ADRENERGIC RECEPTORS, EXCEPT AS NOTED)[*]	EFFECT OF PARASYMPATHETIC STIMULATION (MUSCARINIC ACh RECEPTORS)
Glands		
Adrenal medullae	Secretion of epinephrine and norepinephrine (nicotinic ACh receptors).	No known effect.
Lacrimal (tear)	Slight secretion of tears (α).	Secretion of tears.
Pancreas	Inhibits secretion of digestive enzymes and the hormone insulin (α_2); promotes secretion of the hormone glucagon (β_2).	Secretion of digestive enzymes and the hormone insulin.
Posterior pituitary	Secretion of antidiuretic hormone (ADH) (β_1).	No known effect.
Sweat	Increases sweating in most body regions (muscarinic ACh receptors); sweating on palms of hands and soles of feet (α_1).	No known effect.
Adipose tissue[†]	Lipolysis (breakdown of triglycerides into fatty acids and glycerol) (β_1); release of fatty acids into blood (β_1 and β_3).	No known effect.
Liver[†]	Glycogenolysis (conversion of glycogen into glucose); gluconeogenesis (conversion of noncarbohydrates into glucose); decreased bile secretion (α and β_2).	Glycogen synthesis; increased bile secretion.
Kidney, juxtaglomerular cells[†]	Secretion of renin (β_1).	No known effect.
Cardiac (heart) muscle	Increased heart rate and force of atrial and ventricular contractions (β_1).	Decreased heart rate; decreased force of atrial contraction.
Smooth muscle		
Iris, radial muscle	Contraction $\rightarrow$ dilation of pupil (α_1).	No known effect.
Iris, circular muscle	No known effect.	Contraction $\rightarrow$ constriction of pupil.
Ciliary muscle of eye	Relaxation for far vision (β_2).	Contraction for near vision.
Lungs, bronchial muscle	Relaxation $\rightarrow$ airway dilation (β_2).	Contraction $\rightarrow$ airway constriction.
Gallbladder and ducts	Relaxation (β_2).	Contraction $\rightarrow$ increased release of bile into small intestine.

[*]Subcategories of α and β receptors are listed if known.
[†]Grouped with glands because they release substances into the blood.

• Epinephrine and norepinephrine secreted into the blood from the adrenal medulla intensify and prolong the responses caused by NE liberated from sympathetic postganglionic axons. These blood-borne hormones circulate throughout the body, affecting all tissues that have alpha and beta receptors. In time, blood-borne NE and epinephrine are inactivated by enzymatic destruction in the liver.

CLINICAL APPLICATION
Raynaud's Disease

Raynaud's disease is due to excessive sympathetic stimulation of arterioles within the fingers and toes. Its cause is unknown. Because arterioles in the digits vasoconstrict in response to sympathetic stimulation, blood flow is greatly diminished. The digits may even become deprived of blood for minutes to hours and, in extreme cases, may become necrotic from lack of oxygen and nutrients. The disease is most common in young women and is worsened by cold climates. ■

Parasympathetic Responses

In contrast to the "fight-or-flight" activities of the sympathetic division, the parasympathetic division enhances "rest-and-digest" activities. Parasympathetic responses support body functions that conserve and restore body energy during times of rest and recovery. In the quiet intervals between periods of exercise, parasympathetic impulses to the digestive glands and the smooth muscle of the gastrointesti-

Table 17.4 (continued)

VISCERAL EFFECTOR	EFFECT OF SYMPATHETIC STIMULATION (α OR β ADRENERGIC RECEPTORS, EXCEPT AS NOTED)[a]	EFFECT OF PARASYMPATHETIC STIMULATION (MUSCARINIC ACH RECEPTORS)
Smooth muscle (continued)		
Stomach and intestines	Decreased motility and tone (α_1, α_2, β_2); contraction of sphincters (α_1).	Increased motility and tone; relaxation of sphincters.
Spleen	Contraction and discharge of stored blood into general circulation (α_1).	No known effect.
Ureter	Increases motility (α_1)	Increases motility (?).
Urinary bladder	Relaxation of muscular wall (β_2); contraction of sphincter (α_1).	Contraction of muscular wall; relaxation of sphincter.
Uterus	Inhibits contraction in nonpregnant woman (β_2); promotes contraction in pregnant woman (α_1).	Minimal effect.
Sex organs	In males: contraction of smooth muscle of ductus (vas) deferens, seminal vesicle, prostate → ejaculation of semen (α_1).	Vasodilation; erection of clitoris (females) and penis (males).
Hair follicles, arrector pili muscle	Contraction → erection of hairs (α_1).	No known effect.
Vascular smooth muscle		
Salivary gland arterioles	Vasoconstriction, which decreases secretion (β_2).	Vasodilation, which increases K^+ and water secretion.
Gastric gland arterioles	Vasoconstriction, which inhibits secretion (α_1).	Secretion of gastric juice.
Intestinal gland arterioles	Vasoconstriction, which inhibits secretion (α_1).	Secretion of intestinal juice.
Heart arterioles	Relaxation → vasodilation (β_2).	Contraction → vasoconstriction.
Skin and mucosal arterioles	Contraction → vasoconstriction (α_1).	Vasodilation, which may not be physiologically significant.
Skeletal muscle arterioles	Contraction → vasoconstriction (α_1); relaxation → vasodilation (β_2).	No known effect.
Abdominal viscera arterioles	Contraction → vasoconstriction (α_1, β_2).	No known effect.
Brain arterioles	Slight contraction → vasoconstriction (α_1).	No known effect.
Kidney arterioles	Constriction of blood vessels → decreases urine volume (α_1).	No known effect.
Systemic veins	Contraction → constriction (α_1); relaxation → dilation (β_2).	No known effect.

nal tract predominate over sympathetic impulses, allowing energy-supplying food to be digested and absorbed. At the same time, parasympathetic responses reduce body functions that support physical activity.

The acronym "SLUDD" can be helpful in remembering five parasympathetic responses. It stands for salivation (S), lacrimation (L), urination (U), digestion (D), and defecation (D). All of these activities are stimulated mainly by the parasympathetic division. Besides the increasing "SLUDD" responses, other important parasympathetic responses are three "decreases": decreased heart rate, decreased diameter of airways (bronchoconstriction), and decreased diameter (constriction) of the pupils.

Fear typically elicits sympathetic responses, but in the case of "paradoxical fear" there is massive activation of the parasympathetic division. This type of fear occurs when one is backed into a corner, with no escape route or no way to win. It may happen to soldiers in a losing battle, to unprepared students taking an exam, or to athletes before competition. The high level of parasympathetic tone in such situations can cause loss of control over urination or defecation.

Table 17.4 summarizes the responses of glands, cardiac muscle, and smooth muscle to stimulation by the sympathetic and parasympathetic divisions of the ANS.

1. Define cholinergic and adrenergic neurons and receptors.
2. Give examples of the antagonistic effects of the sympathetic and parasympathetic divisions of the autonomic nervous system.

3. Describe what happens during the fight-or-flight response.
4. Why is the parasympathetic division of the ANS called an energy conservation/restoration system?
5. Describe the sympathetic response in a frightening situation for each of the following body parts: hair follicles, iris of eye, lungs, spleen, adrenal medullae, urinary bladder, stomach, intestines, gallbladder, liver, heart, arterioles of the abdominal viscera, and arterioles of skeletal muscles.

INTEGRATION AND CONTROL OF AUTONOMIC FUNCTIONS

OBJECTIVES
• *Describe the components of an autonomic reflex.*
• *Explain the relationship of the hypothalamus to the ANS.*

Autonomic Reflexes

Autonomic reflexes are responses that occur when nerve impulses pass over an autonomic reflex arc. These reflexes play a key role in regulating controlled conditions in the body, such as *blood pressure,* by adjusting heart rate, force of ventricular contraction, and blood vessel diameter; *respiration,* by regulating the diameter of bronchial tubes; *digestion,* by adjusting the motility (movement) and muscle tone of the gastrointestinal tract; and *defecation* and *urination,* by regulating the opening and closing of sphincters.

The components of an autonomic reflex arc are as follows:

1. *Receptor.* Like the receptor in a somatic reflex arc (see Figure 13.5 on page 419), the receptor in an autonomic reflex arc is the distal end of a sensory neuron, which responds to a stimulus and produces a change that will ultimately trigger nerve impulses. Autonomic sensory receptors are mostly associated with interoceptors.
2. *Sensory neuron.* Conducts nerve impulses from receptors to the CNS.
3. *Integrating center.* Interneurons within the CNS relay signals from sensory neurons to motor neurons. The main integrating centers for most autonomic reflexes are located in the hypothalamus and brain stem. Some autonomic reflexes, such as those for urination and defecation, have integrating centers in the spinal cord.
4. *Motor neurons.* Nerve impulses triggered by the integrating center propagate out of the CNS along motor neurons to an effector. In an autonomic reflex arc, two motor neurons connect the CNS to an effector: The preganglionic neuron conducts motor impulses from the CNS to an autonomic ganglion, and the postganglionic neuron conducts motor impulses from an autonomic ganglion to an effector (see Figure 17.1).
5. *Effector.* In an autonomic reflex arc, the effectors are smooth muscle, cardiac muscle, and glands, and the reflex is called an autonomic reflex.

Autonomic Control by Higher Centers

Normally, we are not aware of muscular contractions of the digestive organs, heartbeat, changes in the diameter of blood vessels, and pupil dilation and constriction because the integrating centers for these autonomic responses are in the spinal cord or the lower regions of the brain. Somatic or autonomic sensory neurons deliver input to these centers, and autonomic motor neurons provide output that adjusts activity in the visceral effector, usually without our conscious perception.

The hypothalamus is the major control and integration center of the ANS. The hypothalamus receives sensory input related to visceral functions, olfaction (smell), and gustation (taste), as well as input related to changes in temperature, osmolarity, and levels of various substances in blood. In addition, it receives input relating to emotions from the limbic system. Output from the hypothalamus influences autonomic centers both in the brain stem (such as the cardiovascular, salivation, swallowing, and vomiting centers) and in the spinal cord (such as the defecation and urination reflex centers in the sacral spinal cord).

Anatomically, the hypothalamus is connected to both the sympathetic and parasympathetic divisions of the ANS by axons of neurons whose dendrites and cell bodies are in various hypothalamic nuclei. The axons form tracts from the hypothalamus to sympathetic and parasympathetic nuclei in the brain stem and spinal cord through relays in the reticular formation. The posterior and lateral portions of the hypothalamus control the sympathetic division. Stimulation of these areas produces an increase in heart rate and force of contraction, a rise in blood pressure due to constriction of blood vessels, an increase in body temperature, dilation of the pupils, and inhibition of the gastrointestinal tract. In contrast, the anterior and medial portions of the hypothalamus control the parasympathetic division. Stimulation of these areas results in a decrease in heart rate, lowering of blood pressure, constriction of the pupils, and increased secretion and motility of the gastrointestinal tract.

 CLINICAL APPLICATION
Autonomic Dysreflexia

Autonomic dysreflexia is an exaggerated response of the sympathetic division of the ANS that occurs in about 85% of individuals with spinal cord injury at or above the level of T6. The condition is seen after recovery from spinal shock (see page 507) and occurs due to interruption of the control of the ANS by higher centers. When certain sensory impulses, such as those resulting from stretching of a full urinary bladder, are unable to ascend the spinal cord, mass stimulation of the sympathetic nerves inferior to the level of injury occurs. Other triggers include stimulation of pain receptors and visceral contractions. Among the effects of increased sympathetic activity is severe vasoconstriction, which elevates blood pressure. In response, the cardiovascular center in the medulla oblongata sends out parasympathetic signals via the vagus (X) nerve that decrease heart rate and dilate blood vessels superior to the level of the injury.

Autonomic dysreflexia is characterized by pounding headache; hypertension; flushed, warm skin with profuse sweating above the injury level; pale, cold, and dry skin below the injury level; and anxiety. It is an emergency condition that requires immediate intervention. If untreated, autonomic dys-reflexia can cause seizures, stroke, or heart attack.

1. Give three examples of controlled conditions in the body that are kept in homeostatic balance by autonomic reflexes.
2. How does an autonomic reflex arc differ from a somatic reflex arc?

STUDY OUTLINE

COMPARISON OF SOMATIC AND AUTONOMIC NERVOUS SYSTEMS (p. 548)

1. The somatic nervous system operates under conscious control; the ANS usually operates without conscious control.
2. Sensory input for the somatic nervous system is mainly from the special senses and somatic senses; sensory input for the ANS is from interoceptors, in addition to special senses and somatic senses.
3. The axons of somatic motor neurons extend from the CNS and synapse directly with an effector. Autonomic motor pathways consist of two motor neurons in series. The axon of the first motor neuron extends from the CNS and synapses in a ganglion with the second motor neuron; the second neuron synapses with an effector.
4. The output (motor) portion of the ANS has two divisions: sympathetic and parasympathetic. Most body organs receive dual innervation; usually one ANS division causes excitation and the other causes inhibition.
5. Somatic nervous system effectors are skeletal muscles; ANS effectors include cardiac muscle, smooth muscle, and glands.
6. Table 17.1 on page 550 compares the somatic and autonomic nervous systems.

ANATOMY OF AUTONOMIC MOTOR PATHWAYS (p. 550)

1. Preganglionic neurons are myelinated; postganglionic neurons are unmyelinated.
2. The cell bodies of sympathetic preganglionic neurons are in the lateral gray horns of the 12 thoracic and the first two or three lumbar segments of the spinal cord; the cell bodies of parasympathetic preganglionic neurons are in four cranial nerve nuclei (III, VII, IX, and X) in the brain stem and lateral gray horns of the second through fourth sacral segments of the spinal cord.
3. Autonomic ganglia are classified as sympathetic trunk ganglia (on both sides of vertebral column), prevertebral ganglia (anterior to vertebral column), and terminal ganglia (near or inside visceral effectors).
4. Sympathetic preganglionic neurons synapse with postganglionic neurons in ganglia of the sympathetic trunk or in prevertebral ganglia; parasympathetic preganglionic neurons synapse with postganglionic neurons in terminal ganglia.
5. Table 17.2 on page 556 compares anatomical features of the sympathetic and parasympathetic divisions.

ANS NEUROTRANSMITTERS AND RECEPTORS (p. 556)

1. Cholinergic neurons release acetylcholine, which binds to nicotinic or muscarinic cholinergic receptors.
2. In the ANS, the cholinergic neurons include all sympathetic and parasympathetic preganglionic neurons, all parasympathetic postganglionic neurons, and sympathetic postganglionic neurons that innervate most sweat glands.
3. In the ANS, adrenergic neurons release norepinephrine. Both epinephrine and norepinephrine bind to alpha and beta adrenergic receptors.
4. Most sympathetic postganglionic neurons are adrenergic.
5. Table 17.3 on page 558 summarizes the types of cholinergic and adrenergic receptors.
6. An agonist is a substance that binds to and activates a receptor, mimicking the effect of a natural neurotransmitter or hormone. An antagonist is a substance that binds to and blocks a receptor, thereby preventing a natural neurotransmitter or hormone from exerting its effect.

PHYSIOLOGICAL EFFECTS OF THE ANS (p. 559)

1. The sympathetic division favors body functions that can support vigorous physical activity and rapid production of ATP (fight-or-flight response); the parasympathetic division regulates activities that conserve and restore body energy.
2. The effects of sympathetic stimulation are longer lasting and more widespread than the effects of parasympathetic stimulation.
3. Table 17.4 on pages 560–561 summarizes sympathetic and parasympathetic responses.

INTEGRATION AND CONTROL OF AUTONOMIC FUNCTIONS (p. 562)

1. An autonomic reflex adjusts the activities of smooth muscle, cardiac muscle, and glands.
2. An autonomic reflex arc consists of a receptor, a sensory neuron, an integrating center, two autonomic motor neurons, and a visceral effector.
3. The hypothalamus is the major control and integration center of the ANS. It is connected to both the sympathetic and the parasympathetic divisions.

SELF-QUIZ QUESTIONS

1. Match the following:
   ___ (a) plexus located mostly posterior to each lung
   ___ (b) plexus containing the inferior mesenteric ganglion and supplying the large intestine
   ___ (c) plexus located anterior to the fifth lumbar vertebra and the superior portion of the sternum; supplies pelvic viscera
   ___ (d) plexus at the base of the heart surrounding the large blood vessels emerging from the heart
   ___ (e) the largest autonomic plexus; found at the level of the last thoracic and first lumbar vertebrae
   ___ (f) plexus containing the superior mesenteric ganglion and supplying the small and large intestine

   (1) cardiac plexus
   (2) pulmonary plexus
   (3) celiac (solar) plexus
   (4) superior mesenteric plexus
   (5) inferior mesenteric plexus
   (6) hypogastric plexus

Complete the following:

2. The part of the nervous system that regulates the activity of smooth muscle, cardiac muscle, and certain glands is the ___ nervous system.

3. The sympathetic division of the ANS is also called the ___ division, whereas the parasympathetic division is also called the ___ division.

4. Adrenergic neurons of the ANS release ___ as their neurotransmitter.

5. Match the following:
   ___ (a) also known as intramural ganglia
   ___ (b) includes the celiac, superior mesenteric, and inferior mesenteric ganglia
   ___ (c) also called vertebral chain or paravertebral ganglia
   ___ (d) lie in a vertical row on either side of the vertebral column
   ___ (e) postganglionic fibers, in general, innervate organs below the diaphragm
   ___ (f) ganglia located at the end of an autonomic motor pathway close to or actually within the wall of a visceral organ
   ___ (g) includes ciliary, pterygopalatine, submandibular, and otic ganglia

   (1) sympathetic trunk ganglia
   (2) prevertebral ganglia
   (3) terminal ganglia

True or false:

6. Structurally, the ANS includes autonomic sensory neurons and autonomic motor neurons.

7. Functionally, the ANS operates primarily without conscious control but can operate under conscious control when an individual is awake, alert, and mentally active.

8. Match the following:
   ___ (a) stimulates urination and defecation
   ___ (b) prepares the body for emergency situations
   ___ (c) fight-or-flight response
   ___ (d) promotes digestion and absorption of food
   ___ (e) concerned primarily with processes involving the expenditure of energy
   ___ (f) controlled by the posterior and lateral portions of the hypothalamus
   ___ (g) controlled by the anterior and medial portions of the hypothalamus
   ___ (h) causes a decrease in heart rate

   (1) increased activity of the sympathetic division of the ANS
   (2) increased activity of the parasympathetic division of the ANS

Choose the best answer to the following questions:

9. Which of the following statements are true? (1) The somatic nervous system and the ANS both include sensory and motor neurons. (2) Somatic sensory receptors are interoceptors. (3) The effect of an autonomic motor neuron is either excitation or inhibition, whereas that of a somatic motor neuron is always excitation. (4) Autonomic sensory neurons are mostly associated with interoceptors. (5) Autonomic motor pathways always consist of two motor neurons in series. (6) Somatic motor pathways may or may not consist of two motor neurons in series.
   (a) 1, 2, 3, 4, and 5, (b) 1, 3, 4, and 5, (c) 2, 3, 5, and 6, (d) 1, 3, 5, and 6, (e) 2, 4, 5, and 6

10. Which of the following statements is *false*? (a) The first neuron in an autonomic pathway is the preganglionic neuron. (b) The preganglionic neuron's cell body is within the CNS. (c) The postganglionic neuron's cell body is within the CNS. (d) Postganglionic neurons relay impulses from autonomic ganglia to visceral effectors. (e) A major difference between autonomic ganglia and posterior root ganglia is that only autonomic ganglia contain synapses.

11. Which of the following statements is *false*? (a) A single sympathetic preganglionic fiber may synapse with 20 or more postganglionic fibers, which partly explains why sympathetic responses tend to be widespread throughout the body. (b) Parasympathetic effects tend to be localized because parasympathetic neurons usually synapse in the terminal ganglia with only four or five postsynaptic neurons (all of which supply a single effector). (c) Only the thoracic and first lumbar nerves have white rami communicantes. (d) The white rami communicantes connect the posterior ramus of the spinal nerve with the ganglia of the sympathetic trunk. (e) White rami communicantes contain myelinated fibers.

12. Which of the following are cholinergic neurons? (1) all sympathetic preganglionic neurons, (2) all parasympathetic preganglionic neurons, (3) all parasympathetic postganglionic neurons, (4) all sympathetic postganglionic neurons, (5) some sympathetic postganglionic neurons.
   (a) 1, 2, 3, and 5, (b) 1, 2, 3, and 4, (c) 2, 3, and 5, (d) 2 and 5, (e) 1, 3, and 5

13. Which of the following statements are true? (1) Most sympathetic postganglionic axons are adrenergic. (2) Cholinergic receptors include nicotinic and muscarinic receptors. (3) Adrenergic receptors include alpha and beta receptors. (4) Muscarinic receptors are present on all effectors innervated by parasympathetic postganglionic axons. (5) In general, norepinephrine stimulates alpha receptors more vigorously than beta receptors, whereas epinephrine is a potent stimulator of both alpha and beta receptors.
(a) 1, 2, 3, 4, and 5, (b) 2, 3, 4, and 5, (c) 1, 3, 4, and 5, (d) 3, 4, and 5, (e) 1, 2, 3, and 4

14. Which of the following are reasons why the effects of sympathetic stimulation are longer lasting and more widespread than those of parasympathetic stimulation? (1) There is greater divergence of sympathetic postganglionic fibers. (2) There is less divergence of sympathetic postganglionic fibers. (3) AChE quickly inactivates ACh, whereas norepinephrine lingers in the synaptic cleft for a longer time. (4) Norepinephrine and epinephrine secreted into the blood by the adrenal medullae intensify the actions of the sympathetic division. (5) ACh remains in the synaptic cleft until norepinephrine is produced.
(a) 1, 2, and 3, (b) 1, 3, and 5, (c) 1, 3, and 4, (d) 2, 3, and 4, (e) 2, 3, and 5

15. Which of the following are components of an autonomic reflex arc? (1) receptor, (2) sensory neuron, (3) integrating center, (4) two motor neurons, (5) effector.
(a) 3, 4, and 5, (b) 1, 3, and 4, (c) 2, 4, and 5, (d) 1, 4, and 5, (e) 1, 2, 3, 4, and 5

CRITICAL THINKING QUESTIONS

1. It's Thanksgiving and you've just eaten a huge turkey dinner with all the trimmings. Now you're going to watch the football game on the television. Which division of the nervous system will be handling your body's after-dinner activities? List several organs involved and the effects of the nervous system on their functions. (HINT: *The most strenuous activity of the afternoon will be reaching for the TV's remote control.*)

2. After his lunch, baby Bobby was looking uncomfortable, squirming and fretting. Suddenly, a pungent odor wafted from his diaper and he looked very content. Trace the reflex arc responsible for Bobby's present condition. (HINT: *Baby Bobby's body responds to the stimuli automatically—say that three times fast.*)

3. The advertisements made cigarette smoking sound cool and relaxing, so Jamal decided to try it. After smoking half a pack, Jamal felt nervous and edgy, his muscles felt "twitchy," and his hands were shaking. What's happened to Jamal? (HINT: *How does your nervous system cause a muscle twitch?*)

ANSWERS TO FIGURE QUESTIONS

17.1 Dual innervation means that a body organ innervated by the ANS receives both sympathetic and parasympathetic fibers.

17.2 Most parasympathetic preganglionic axons are longer than most sympathetic preganglionic axons because most parasympathetic ganglia are in the walls of visceral organs, whereas most sympathetic ganglia are close to the spinal cord in the sympathetic trunk.

17.3 The largest autonomic plexus is the celiac plexus.

17.4 White rami look white due to the presence of myelin.

17.5 Cholinergic neurons that have nicotinic ACh receptors include sympathetic postganglionic neurons innervating sweat glands and all parasympathetic postganglionic neurons. The effectors innervated by these cholinergic neurons possess muscarinic receptors.

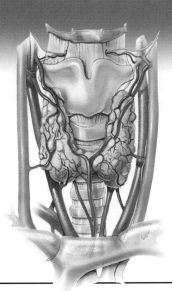

Together, the nervous and endocrine systems coordinate functions of all body systems. The nervous system controls body activities through nerve impulses conducted along axons of neurons. At synapses, nerve impulses trigger the release of mediator molecules called neurotransmitters. In contrast, the glands of the endocrine system release mediator molecules called **hormones** (*hormon* = to excite or get moving) into the bloodstream. The circulating blood then delivers hormones to virtually all cells throughout the body. The science of the structure and function of the endocrine glands and the diagnosis and treatment of disorders of the endocrine system is **endocrinology** (en′-dō-kri-NOL-ō-jē; *endo-* = within; *-crin* = to secrete; *-ology* = study of).

The nervous and endocrine systems are coordinated as an interlocking supersystem termed the **neuroendocrine system.** Certain parts of the nervous system stimulate or inhibit the release of hormones, which in turn may promote or inhibit the generation of nerve impulses. The nervous system causes muscles to contract and glands to secrete either more or less of their product. The endocrine system not only helps regulate the activity of smooth muscle, cardiac muscle, and some glands; it affects virtually all other tissues as well. Hormones alter metabolism, regulate growth and development, and influence reproductive processes.

The nervous and endocrine systems respond to a stimulus at different rates. Nerve impulses most often produce an effect within a few milliseconds; some hormones can act within seconds, whereas others can take several hours or more to cause a response. Moreover, the effects of activating the nervous system are generally briefer than effects produced by the endocrine system. Table 18.1 compares the characteristics of the nervous and endocrine systems.

The focus of this chapter is the endocrine system; in it we examine the principal endocrine glands and hormone-producing tissues, and their roles in coordinating body activities.

Table 18.1 Comparison of the Nervous and Endocrine Systems

CHARACTERISTIC	NERVOUS SYSTEM	ENDOCRINE SYSTEM
Mediator molecules	Neurotransmitters released in response to nerve impulses.	Hormones delivered to tissues throughout the body by the blood.
Cells affected	Muscle cells, gland cells, other neurons.	Virtually all body cells.
Time to onset of action	Typically within milliseconds.	Seconds to hours or days.
Duration of action	Generally briefer.	Generally longer.

Figure 18.1 Location of many endocrine glands. Also shown are other organs that contain endocrine tissue, and associated structures.

🔑 Endocrine glands secrete hormones, which circulating blood delivers to target tissues.

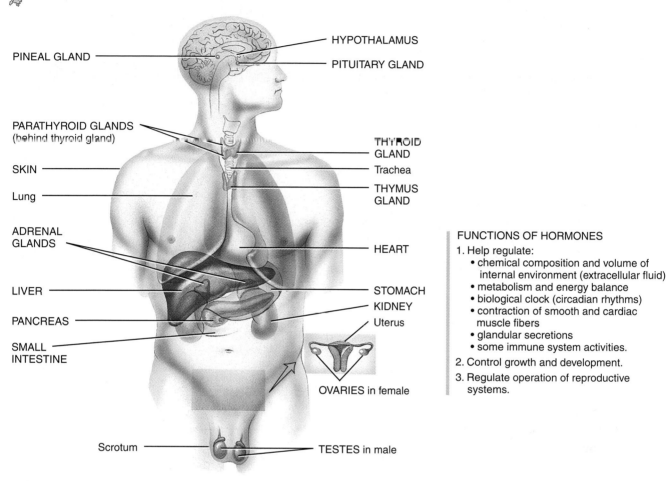

FUNCTIONS OF HORMONES
1. Help regulate:
   • chemical composition and volume of internal environment (extracellular fluid)
   • metabolism and energy balance
   • biological clock (circadian rhythms)
   • contraction of smooth and cardiac muscle fibers
   • glandular secretions
   • some immune system activities.
2. Control growth and development.
3. Regulate operation of reproductive systems.

Q What is the basic difference between endocrine glands and exocrine glands?

ENDOCRINE GLANDS DEFINED

OBJECTIVE

• *Distinguish between exocrine glands and endocrine glands.*

The body contains two kinds of glands: exocrine glands and endocrine glands. **Exocrine glands** (*exo-* = outside) secrete their products into ducts that carry the secretions into body cavities, into the lumen of an organ, or to the outer surface of the body. Exocrine glands include sudoriferous (sweat), sebaceous (oil), mucous, and digestive glands. **Endocrine glands,** by contrast, secrete their products (hormones) into the interstitial fluid surrounding the secretory cells, rather than into ducts. The secretion then diffuses into capillaries and is carried away by the blood. The endocrine glands of the body, which constitute the **endocrine system,** include the pituitary, thyroid, parathyroid, adrenal, and pineal glands (Figure 18.1). In addition, several organs and

tissues of the body contain cells that secrete hormones but are not endocrine glands exclusively. These include the hypothalamus, thymus gland, pancreas, ovaries, testes, kidneys, stomach, liver, small intestine, skin, heart, adipose tissue, and placenta.

HORMONE ACTIVITY

OBJECTIVES

• *Describe how hormones interact with target-cell receptors.*

• *Compare the two chemical classes of hormones based on their solubility.*

Hormones have powerful effects when present even in very low concentrations. As a rule, most of the 50 or so hormones affect only a few types of cells. The reason that some cells respond to a particular hormone and others do not involves hormone receptors.

The Role of Hormone Receptors

Although a given hormone travels throughout the body in the blood, it affects only specific **target cells.** Hormones, like neurotransmitters, influence their target cells by chemically binding to specific protein or glycoprotein **receptors.** Only the target cells for a given hormone have receptors that bind and recognize that hormone. For example, thyroid-stimulating hormone (TSH) binds to receptors on cells of the thyroid gland, but it does not bind to cells of the ovaries because ovarian cells do not have TSH receptors.

Receptors, like other cellular proteins, are constantly being synthesized and broken down. Generally, a target cell has 2000–100,000 receptors for a particular hormone. When a hormone is present in excess, the number of target-cell receptors may decrease—an effect called **down-regulation.** For example, when certain cells of the testes are exposed to a high concentration of luteinizing hormone (LH), the number of LH receptors decreases. In down-regulation, the receptors undergo endocytosis in clathrin-coated vesicles, and then are degraded by lysosomes. Down-regulation decreases the responsiveness of target cells to the hormone. In contrast, when a hormone (or neurotransmitter) is deficient, the number of receptors may increase. This phenomenon, known as **up-regulation,** makes a target tissue more sensitive to a hormone.

CLINICAL APPLICATION
Blocking Hormone Receptors

Synthetic hormones that block the receptors for certain naturally occurring hormones are available as drugs. For example, RU486 (mifepristone), which is used to induce abortion, binds to the receptors for progesterone (a female sex hormone) and prevents progesterone from exerting its normal effect. When RU486 is given to a pregnant woman, the uterine conditions needed for nurturing an embryo are not maintained, embryonic development stops, and the embryo is sloughed off along with the uterine lining. This example illustrates an important endocrine principle: If a hormone is prevented from interacting with its receptors, the hormone cannot perform its normal functions. ■

Circulating and Local Hormones

Hormones can be classified into two groups according to how far from their site of production they act. Hormones that pass into the blood and act on distant target cells are called **circulating hormones** or **endocrines;** those that act locally without first entering the bloodstream are called **local hormones.** Among local hormones are those that act on neighboring cells, called **paracrines** (*para-* = beside or near), and those that act on the same cell that secreted them, termed **autocrines** (*auto-* = self). Figure 18.2 compares the sites of action of circulating and local hormones. Local hormones usually are inactivated quickly; circulating hormones may linger in the blood and exert their effects for a few minutes or occasionally for a few hours. In time, circulating hormones are inactivated by the liver and excreted by the kidneys. In cases of kidney or liver failure, excessive levels of hormones may build up in the blood. An example of a local

Figure 18.2 Comparison of circulating hormones (endocrines) and local hormones (autocrines and paracrines).

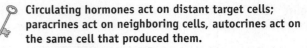

Circulating hormones act on distant target cells; paracrines act on neighboring cells, autocrines act on the same cell that produced them.

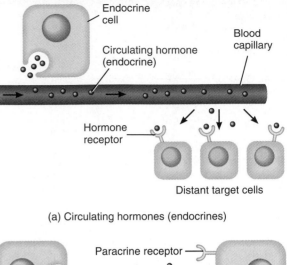

(a) Circulating hormones (endocrines)

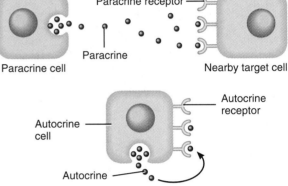

(b) Local hormones (paracrines and autocrines)

In the stomach, one stimulus for secretion of hydrochloric acid by parietal cells is the release of histamine by neighboring mast cells. Is histamine an endocrine, autocrine, or a paracrine in this situation?

hormone is the gas nitric oxide (NO), which is released by endothelial cells lining blood vessels. NO causes relaxation of nearby vascular smooth muscle fibers, which in turn causes vasodilation and increases the flow of blood in that region.

Chemical Classes of Hormones

Chemically, hormones can be divided into two broad classes: those that are soluble in lipids, and those that are soluble in water.

Lipid-soluble Hormones

1. **Steroid hormones** are derived from cholesterol and are synthesized on smooth endoplasmic reticulum. Each steroid hormone is unique due to the presence of different functional groups attached at various sites on the four rings at the core of its structure. These small differences allow for a large diversity of functions. Endocrine tissues that secrete steroid hormones all are derived from mesoderm.

Table 18.2 Chemical Classes of Hormones

EXAMPLES	WHERE PRODUCED	HORMONES
Lipid-soluble **Steroids** Aldosterone	Adrenal cortex.	Aldosterone, cortisol, and androgens.
	Kidneys.	Calcitriol.
	Testes.	Testosterone.
	Ovaries.	Estrogens and progesterone.
Thyroid hormones Triiodothyronine (T_3)	Thyroid gland (follicular cells).	T_3 (triiodothyronine) and T_4 (thyroxine).
Gas (NO) $N = O$	Endothelial cells lining blood vessels.	Nitric oxide (NO).
Water-soluble **Amines** Norepinephrine	Adrenal medulla.	Epinephrine and norepinephrine (catecholamines).
	Pineal gland.	Melatonin.
	Mast cells in connective tissues.	Histamine.
	Platelets in blood.	Serotonin.
Peptides and proteins Oxytocin	Hypothalamic neurosecretory cells.	All hypothalamic releasing and inhibiting hormones, oxytocin, antidiuretic hormone.
	Anterior pituitary gland.	Human growth hormone, thyroid stimulating hormone, adrenocorticotropic hormone, follicle stimulating hormone, luteinizing hormone, prolactin, melanocyte stimulating hormone.
	Pancreas.	Insulin, glucagon, somatostatin, pancreatic polypeptide.
	Parathyroid glands.	Parathyroid hormone.
	Thyroid gland (parafollicular cells).	Calcitonin.
	Stomach and small intestine (enteroendocrine cells).	Gastrin, secretin, cholecystokinin, glucose-dependent insulinotropic peptide.
	Kidneys.	Erythropoietin.
	Adipose tissue.	Leptin.
Eicosanoids A leukotriene (LTB_4)	All cells except red blood cells.	Prostaglandins, leukotrienes.

2. Two **thyroid hormones** (T_3 and T_4) are synthesized by iodination and coupling of two molecules of tyrosine, an amino acid. The benzene ring of tyrosine plus the attached iodines make T_3 and T_4 very lipid-soluble.

3. The gas **nitric oxide (NO)** is both a hormone and a neurotransmitter. Its synthesis is catalyzed by the enzyme nitric oxide synthase.

Water-soluble Hormones

1. **Amine hormones** are synthesized by decarboxylating and modifying certain amino acids. They are called amines because they retain an amino group ($-NH_3^+$). The catecholamines—epinephrine, norepinephrine, and dopamine—are synthesized by modifying the amino acid tyrosine. Histamine is synthesized from histidine by mast cells and platelets. Serotonin and melatonin are derived from tryptophan.

2. **Peptide hormones** and **protein hormones** are synthesized on rough endoplasmic reticulum and consist of chains of 3–200 amino acids. Several of the protein hormones—for example, thyroid-stimulating hormone (TSH)—have attached carbohydrate groups and thus are **glycoprotein hormones.**

3. A more recently discovered group of chemical mediators are the **eicosanoid hormones** (ī-KŌ-sa-noid; *eicos-* = twenty forms; *-oid* = resembling), which are derived from arachidonic acid, a 20-carbon fatty acid. The two major types of eicosanoids are **prostaglandins** and **leukotrienes.** The eicosanoids are important local hormones, and they may act as circulating hormones as well.

Figure 18.3 Mechanism of action of the lipid-soluble steroid hormones and thyroid hormones.

🔑 **Lipid-soluble hormones bind to receptors inside target cells.**

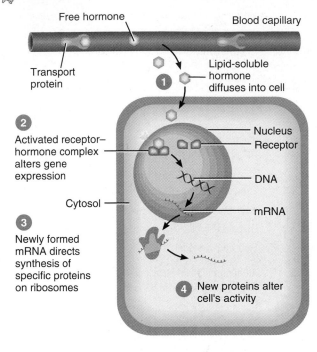

Q Approximately how long does it take a steroid hormone to begin to exert its effects?

Some examples of lipid-soluble and water-soluble hormones, and their sites of production, are listed in Table 18.2.

Hormone Transport in the Blood

Most water-soluble hormone molecules circulate in the watery blood plasma in a "free" form (not attached to plasma proteins), whereas most lipid-soluble hormone molecules bind to **transport proteins.** The transport proteins, which are synthesized by cells in the liver, have three functions:

- They improve the transportability of the lipid-soluble hormones by making them temporarily water-soluble.
- They retard passage of the small hormone molecules through the filtering mechanism in the kidneys, thus slowing the rate of hormone loss in the urine.
- They provide a ready reserve of hormone, already present in the bloodstream.

In general, 0.1–10% of the molecules of a lipid-soluble hormone are not bound to a transport protein. It is this **free fraction** that diffuses out of capillaries, binds to receptors, and triggers responses. As free hormone molecules leave the blood and bind to their receptors, transport proteins release new ones to replenish the free fraction.

1. Contrast the control of homeostasis by the nervous and endocrine systems.

2. Distinguish down-regulation from up-regulation.
3. Identify the chemical classes of hormones, and give an example of each.
4. How are hormones transported in the blood?

MECHANISMS OF HORMONE ACTION

OBJECTIVE

• *Describe the two general mechanisms of hormonal action.*

The response to a hormone depends on both the hormone and the target cell. Various target cells respond differently to the same hormone. Insulin, for example, stimulates synthesis of glycogen in liver cells but synthesis of triglycerides in adipose cells.

The response to a hormone is not always the synthesis of new molecules, as is the case for insulin. Other hormonal effects include changing the permeability of the plasma membrane, stimulating transport of a substance into or out of the target cells, altering the rate of specific metabolic reactions, or causing contraction of smooth muscle or cardiac muscle. In part, these varied effects of hormones are possible because a single hormone can set in motion several different cellular responses. First, however, a hormone must "announce its arrival" to a target cell by binding to receptors. Depending on whether a hormone is lipid-soluble or water-soluble, its target-cell receptors are located either inside the cell or in the plasma membrane.

Action of Lipid-soluble Hormones

Lipid-soluble hormones, including steroid hormones and thyroid hormones, bind to receptors within target cells. Their mechanism of action is as follows (Figure 18.3):

1️⃣ A lipid-soluble hormone diffuses from the blood, through interstitial fluid, and through the lipid bilayer of the plasma membrane into a cell.

2️⃣ If the cell is a target cell, the hormone will bind to and activate receptors located within the cytosol or nucleus. An activated receptor then alters gene expression: It turns specific genes of the nuclear DNA on or off.

3️⃣ As the DNA is transcribed, new messenger RNA (mRNA) forms, leaves the nucleus, and enters the cytosol, where it directs synthesis of new proteins, usually enzymes, on the ribosomes.

4️⃣ The new proteins alter the cell's activity and cause the typical physiological responses of that hormone.

Action of Water-soluble Hormones

Catecholamine, peptide, protein, and eicosanoid hormones are not lipid-soluble, and thus they cannot diffuse through the lipid bilayer of the plasma membrane to attach to receptors inside target cells. Instead, the receptors for the water-soluble hormones are integral proteins in the plasma membrane that protrude into the interstitial fluid. Because a water-soluble hormone binds to a receptor only at the extracellular surface of the plasma membrane, it acts as the **first messenger.** A **second messenger** is then released inside the cell, where hormone-stimulated responses can take place. Neurotransmitters, neuropeptides, and several sensory transduction mechanisms (for example, vision; see Figure 16.14 on page 528) also act by means of second-messenger systems.

One common second messenger is **cyclic AMP (cAMP),** which is synthesized from ATP. The enzyme that catalyzes formation of cAMP from ATP is **adenylate cyclase,** which is attached to the inner surface of the plasma membrane. Cyclic AMP and other second messengers alter a cell's function in specific ways. For example, an increase in cAMP causes adipose cells to break down triglycerides and release fatty acids more rapidly, but cAMP stimulates thyroid cells to secrete more thyroid hormone.

The action of a typical water-soluble hormone occurs as follows (Figure 18.4):

1 A water-soluble hormone diffuses from the blood through interstitial fluid and then binds to its receptor in a target cell's plasma membrane. This binding activates another membrane protein, called a G protein (described shortly), which activates adenylate cyclase.

2 Adenylate cyclase then converts ATP into cyclic AMP (cAMP) in the cytosol of the cell.

3 Cyclic AMP (the second messenger) activates one or several **protein kinases,** which may be free in the cytosol or bound to the plasma membrane. A protein kinase is an enzyme that phosphorylates (adds a phosphate group to) cellular proteins. The donor of the phosphate group is ATP, which is converted to ADP.

4 Activated protein kinases phosphorylate one or several other enzymes, here called Enzyme 1 and Enzyme 2. Phosphorylation activates some enzymes and inactivates others, rather like an on-off switch. The result of phosphorylating a particular enzyme could be regulation of other enzymes, secretion, protein synthesis, or a change in plasma membrane permeability.

5 Enzymes activated by phosphorylation, in turn, catalyze reactions that produce physiological responses. Different protein kinases exist within different target cells and within different organelles of the same target cell. Thus, one protein kinase may trigger glycogen synthesis, a second triglyceride breakdown, a third protein synthesis, and so forth. Moreover, phosphorylation by a protein kinase can inhibit certain enzymes. For example, some of the kinases unleashed when epinephrine binds to liver cells inactivate an enzyme needed for glycogen synthesis.

After a brief period of time, an enzyme called **phosphodiesterase** inactivates cAMP. Thus, the cell's response is turned off unless new hormone molecules continue to bind to their receptors in the plasma membrane.

Many hormones exert at least some of their physiological effects through the *increased* synthesis of cAMP. Examples

Figure 18.4 Mechanism of action of the water-soluble hormones (catecholamines, peptides, and proteins).

🔑 **Water-soluble hormones bind to receptors embedded in the plasma membrane of target cells.**

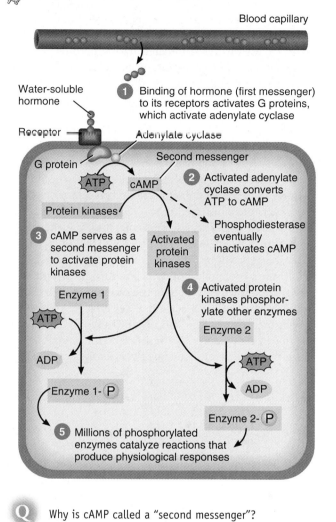

Q Why is cAMP called a "second messenger"?

include antidiuretic hormone (ADH), thyroid-stimulating hormone (TSH), adrenocorticotropic hormone (ACTH), glucagon, epinephrine, and hypothalamic releasing hormones. In other cases, such as growth hormone inhibiting hormone (GHIH), the level of cyclic AMP *decreases* in response to the binding of a hormone to its receptor. Besides cAMP, several other substances are known second messengers, including calcium ions (Ca^{2+}), cGMP (cyclic guanosine monophosphate, a cyclic nucleotide similar to cAMP), inositol trisphosphate (IP_3), and diacylglycerol (DAG). Although it is a lipid-soluble hormone, nitric oxide exerts its effect inside smooth muscle fibers by activating guanylyl cyclase. This enzyme, in turn, converts guanosine triphosphate (GTP) to cGMP, which causes calcium ions to enter storage areas of the smooth muscle fiber. The lowered Ca^{2+} level in the cytosol then causes muscle relaxation. A given hormone (or neurotransmitter) may use different second messengers in different target cells.

In the cyclic AMP second-messenger system, hormone receptors do not connect *directly* to adenylate cyclase. Instead, molecules called **G proteins** link receptors on the outer surface of the plasma membrane to the adenylate cyclase molecules on the inner surface (see Figure 18.4). The binding of a hormone to its receptor activates many G-protein molecules, which in turn activate molecules of adenylate cyclase. Unless further stimulated by the binding of more hormone molecules to receptors, G proteins slowly become inactivated, thus helping to stop the hormone response. G proteins are a common feature of most second-messenger systems.

Hormones that bind to plasma membrane receptors can induce their effects at very low concentrations because they initiate a cascade or chain reaction, each step of which multiplies or amplifies the initial effect. For example, the binding of a single molecule of epinephrine to its receptor on a liver cell may activate a hundred or so G proteins, each of which activates an adenylate cyclase molecule. If each adenylate cyclase produces even 1000 cAMP, then 100,000 of these second messengers will be liberated inside the cell. Each cAMP may activate a protein kinase, which in turn can act on hundreds or thousands of substrate molecules. Some of the kinases phosphorylate and activate a key enzyme needed for glycogen breakdown. The end result of the binding of epinephrine to its receptor is the breakdown of millions of glycogen molecules into glucose (glycogenolysis).

CLINICAL APPLICATION
Cholera Toxin and G Proteins

The toxin produced by cholera bacteria is deadly. It produces massive watery diarrhea, and an infected person can rapidly die from the resulting dehydration. The cholera toxin modifies the G proteins in intestinal epithelial cells such that they become locked in an activated state. As a result, the intracellular cAMP concentration skyrockets. One of the effects of cAMP in these cells is to stimulate an active transport pump that ejects chloride ions (Cl^-) from the cells into the lumen of the intestines; water follows the Cl^- by osmosis, and positively charged sodium ions follow the negatively charged Cl^-. Thus, cholera toxin causes a huge outflow of Na^+, Cl^-, and water into the fecal material. Treatment consists of replacement of lost fluids, either intravenously or by mouth (oral rehydration therapy), plus antibiotic (tetracycline) therapy. ■

Hormone Interactions

The responsiveness of a target cell to a hormone depends on (1) the hormone's concentration, (2) the abundance of the target cell's hormone receptors, and (3) influences exerted by other hormones. A target cell responds more vigorously when the level of a hormone rises or when it has more receptors (up-regulation). Also, the actions of some hormones on target cells require a simultaneous or recent exposure to a second hormone. In such cases, the second hormone is said to have a **permissive effect.** For example, epinephrine alone only weakly stimulates lipolysis (the breakdown of triglycerides), but when small amounts of thyroid hormones (T_3 and T_4) are also present, the same amount of epinephrine stimulates lipolysis much more powerfully. Sometimes the permissive hormone up-regulates receptors for the other hormone, and sometimes it promotes the synthesis of an enzyme required for the expression of the other hormone's effects.

When the effect of two hormones acting together is greater or more extensive than the sum of each hormone acting alone, the two hormones are said to have a **synergistic effect.** For example, secretion of neither estrogens from the ovaries nor follicle stimulating hormone (FSH) from the anterior pituitary gland alone is sufficient for normal production of oocytes by the ovaries. When they act together, however, the ovaries normally produce oocytes.

When one hormone opposes the actions of another hormone, the two hormones are said to have **antagonistic effects.** An example of an antagonistic pair of hormones is insulin, which promotes synthesis of glycogen by liver cells, and glucagon, which stimulates breakdown of glycogen in the liver.

CONTROL OF HORMONE SECRETION

OBJECTIVE

• *Describe the three types of signals that can control hormone secretion.*

Most hormones are released in short bursts, with little or no secretion between bursts. When increasingly stimulated, an endocrine gland will release its hormone in more frequent bursts, and thus the hormone's overall blood concentration increases; in the absence of stimulation, bursts are minimal or inhibited, and the blood level of the hormone decreases. Regulation of secretion normally maintains homeostasis and prevents overproduction or underproduction of any given hormone.

Hormone secretion is regulated by (1) signals from the nervous system, (2) chemical changes in the blood, and (3) other hormones. Thus, nerve impulses to the adrenal medullae regulate the release of epinephrine; blood Ca^{2+} level regulates the secretion of parathyroid hormone; and a hormone from the anterior pituitary gland (adrenocorticotropic hormone) stimulates the release of cortisol by the adrenal cortex. As we will see when we examine more examples of each of these three mechanisms of hormonal regulation later in this chapter, most hormonal regulatory systems work via negative feedback.

Disorders of the endocrine system often involve either **hyposecretion** (*hypo-* = too little or under), inadequate release of a hormone, or **hypersecretion** (*hyper-* = too much or above), excessive release of a hormone. In most cases the problem is faulty control of secretion, but in other cases the problem is faulty hormone receptors or an inadequate number of receptors. Several endocrine disorders are described in more detail at the end of the chapter (see page 603).

Occasionally a positive feedback system contributes to the regulation of hormone secretion. One example occurs during childbirth. The hormone oxytocin stimulates contractions of the uterus. Uterine contractions, in turn, stimulate more oxytocin release (see Figure 1.4 on page 9). Another positive feedback system is the luteinizing hormone (LH) surge that results in ovulation (described on page 577). In both cases, the response intensifies the initiating stimulus. A part of our discussion of the effects of various hormones in this chapter describes how the secretions of each hormone are controlled.

1. Distinguish among permissive effects, synergistic effects, and antagonistic effects.
2. How do negative and positive feedback systems differ from one another?

Now that we have an overview of the roles of hormones in the endocrine system, we turn to discussions of the various endocrine glands and the hormones they secrete.

HYPOTHALAMUS AND PITUITARY GLAND

OBJECTIVES

• *Explain why the hypothalamus is an endocrine gland.*
• *Describe the location, histology, hormones, and functions of the anterior and posterior pituitary glands.*

For many years the **pituitary gland** or **hypophysis** (hī-POF-i-sis) was called the "master" endocrine gland because it secretes several hormones that control other endocrine glands. We now know that the pituitary gland itself has a master—the **hypothalamus.** This small region of the brain, inferior to the thalamus, is the major integrating link between the nervous and endocrine systems. It receives input from several other regions of the brain, including the limbic system, cerebral cortex, thalamus, and reticular activating system. It also receives sensory signals from internal organs and from the retina.

Painful, stressful, and emotional experiences all cause changes in hypothalamic activity. In turn, the hypothalamus controls the autonomic nervous system and regulates body temperature, thirst, hunger, sexual behavior, and defensive reactions such as fear and rage. Not only is the hypothalamus an important regulatory center in the nervous system; it is also a crucial endocrine gland. Cells in the hypothalamus synthesize at least nine different hormones, and the pituitary gland secretes seven more. Together, the hormones play important roles in the regulation of virtually all aspects of growth, development, metabolism, and homeostasis.

The pituitary gland is a pea-shaped structure measuring about 1–1.5 cm (0.5 in.) in diameter that lies in the sella tur-

cica of the sphenoid bone and attaches to the hypothalamus by a stalk, the **infundibulum** (= a funnel; Figure 18.5). The pituitary gland has two anatomically and functionally separate portions. The **anterior pituitary gland (anterior lobe)** accounts for about 75% of the total weight of the gland. It develops from an outgrowth of ectoderm called the hypophyseal pouch in the roof of the mouth (see Figure 18.22b). The **posterior pituitary gland (posterior lobe)** also develops from an ectodermal outgrowth, called the neurohypophyseal bud (see Figure 18.22b). It contains axons and axon terminals of more than 10,000 neurons whose cell bodies are located in the supraoptic and paraventricular nuclei of the hypothalamus (see Figure 18.8). The axon terminals in the posterior pituitary gland are associated with specialized neuroglia called **pituicytes** (pi-TOO-i-sītz).

A third region called the **pars intermedia** atrophies during fetal development and ceases to exist as a separate lobe in adults (see Figure 18.22b). However, some of its cells migrate into adjacent parts of the anterior pituitary gland, where they persist.

Anterior Pituitary Gland

The **anterior pituitary gland** or **adenohypophysis** (ad′-e-nō-hī-POF-i-sis; *adeno-* = gland; *-physis* = organ suspended from the hypothalamus) secretes hormones that regulate a wide range of bodily activities, from growth to reproduction. Release of anterior pituitary gland hormones is stimulated by **releasing hormones** and suppressed by **inhibiting hormones** from the hypothalamus. These hypothalamic hormones are an important link between the nervous and endocrine systems.

Hypothalamic hormones reach the anterior pituitary gland through a portal system. A portal system carries blood between two capillary networks without passing through the heart. Most often, blood passes from the heart through an artery to a capillary to a vein and back to the heart. In the **hypophyseal portal system** (hī′-pō-FIZ-ē-al) blood flows from the median eminence of the hypothalamus into the infundibulum and anterior pituitary gland principally from several **superior hypophyseal arteries** (see Figure 18.5). These arteries are branches of the internal carotid and posterior communicating arteries. The superior hypophyseal arteries form the **primary plexus of the hypophyseal portal system,** a capillary network at the base of the hypothalamus. Near the median eminence and above the optic chiasm are two groups of specialized neurons, called **neurosecretory cells,** that secrete hypothalamic releasing and inhibiting hormones into the primary plexus. These hormones are synthesized in the neuronal cell bodies and packaged inside vesicles, which reach the axon terminals by axonal transport. When nerve impulses reach the axon terminals, they stimulate the vesicles to undergo exocytosis. The hormones then diffuse into the primary plexus of the hypophyseal portal system.

Figure 18.5 Hypothalamus and pituitary gland, and their blood supply. The small figure at right indicates that releasing and inhibiting hormones synthesized by hypothalamic neurons are transported within axons and released from the axon terminals. The hormones diffuse into capillaries of the primary plexus of the hypophyseal portal system and are carried by the hypophyseal portal veins to the secondary plexus of the hypophyseal portal system for distribution to target cells in the anterior pituitary gland. (See Tortora, *A Photographic Atlas of the Human Body,* Figure 10.1b)

🔑 **Hypothalamic hormones are an important link between the nervous and endocrine systems.**

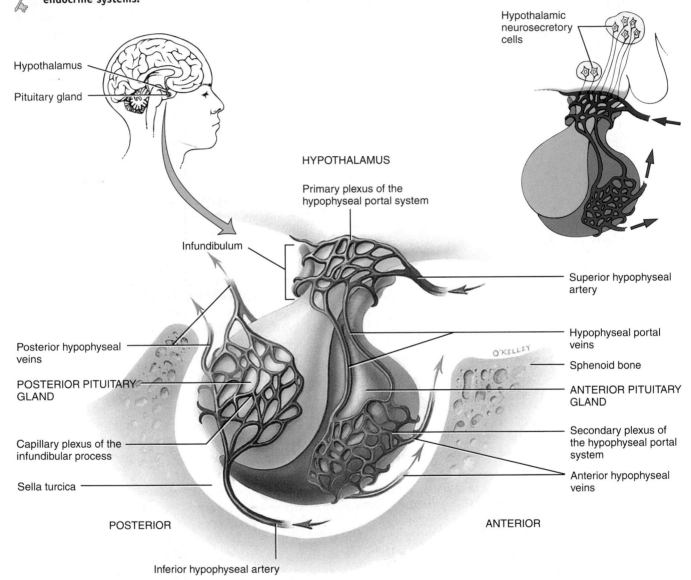

Hypothalamic neurosecretory cells

Hypothalamus

Pituitary gland

HYPOTHALAMUS

Primary plexus of the hypophyseal portal system

Infundibulum

Superior hypophyseal artery

Posterior hypophyseal veins

Hypophyseal portal veins

POSTERIOR PITUITARY GLAND

Sphenoid bone

O'KELLEY

ANTERIOR PITUITARY GLAND

Capillary plexus of the infundibular process

Secondary plexus of the hypophyseal portal system

Sella turcica

Anterior hypophyseal veins

POSTERIOR

ANTERIOR

Inferior hypophyseal artery

Q What is the functional importance of the hypophyseal portal veins?

From the primary plexus, blood drains into the **hypophyseal portal veins** that pass down the outside of the infundibulum. In the anterior pituitary gland, the hypophyseal portal veins redivide into another capillary network called the **secondary plexus of the hypophyseal portal system.**

This direct route permits hypothalamic hormones to act quickly on anterior pituitary gland cells, before the hormones are diluted or destroyed in the systemic circulation. Hormones secreted by anterior pituitary gland cells pass into the secondary plexus of the hypophyseal portal system and

Table 18.3 Hormones of the Anterior Pituitary Gland

HORMONE	SECRETED BY	RELEASING HORMONES (STIMULATE SECRETION)	INHIBITING HORMONES (SUPPRESS SECRETION)
Human growth hormone (hGH) or **somatotropin**	Somatotrophs.	Growth hormone releasing hormone (GHRH), also known as somatocrinin.	Growth hormone inhibiting hormone (GHIH), also known as somatostatin.
Thyroid-stimulating hormone (TSH) or **thyrotropin**	Thyrotrophs.	Thyrotropin releasing hormone (TRH).	Growth hormone inhibiting hormone (GHIH).
Follicle-stimulating hormone (FSH)	Gonadotrophs.	Gonadotropic releasing hormone (GnRH).	—
Luteinizing hormone (LH)	Gonadotrophs.	Gonadotropic releasing hormone (GnRH).	—
Prolactin (PRL)	Lactotrophs.	Prolactin releasing hormone (PRH); TRH.	Prolactin inhibiting hormone (PIH), which is dopamine.
Adrenocorticotropic hormone (ACTH) or **corticotropin**	Corticotrophs.	Corticotropin releasing hormone (CRH).	—
Melanocyte stimulating hormone	Corticotrophs.	Corticotropin releasing hormone (CRH).	Dopamine.

then into the **anterior hypophyseal veins** for distribution to target tissues throughout the body.

The following list describes the seven major hormones secreted by five types of anterior pituitary gland cells:

- **Human growth hormone (hGH)** or **somatotropin** (sō′-ma-tō-TRŌ-pin; *somato-* = body; *-tropin* = change) is secreted by cells called **somatotrophs.** Human growth hormone in turn stimulates several tissues to secrete **insulinlike growth factors,** hormones that stimulate general body growth and regulate aspects of metabolism.
- **Thyroid-stimulating hormone (TSH)** or **thyrotropin** (thī-rō-TRŌ-pin; *thyro-* = shield), which controls the secretions and other activities of the thyroid gland, is secreted by **thyrotrophs.**
- **Follicle-stimulating hormone (FSH)** and **luteinizing hormone (LH)** (LOO-tē-in′-īz-ing) are secreted by **gonadotrophs** (*gonado-* = seed). FSH and LH both act on the gonads: They stimulate secretion of estrogens and progesterone and the maturation of oocytes in the ovaries, and they stimulate secretion of testosterone and sperm production in the testes.
- **Prolactin (PRL),** which initiates milk production in the mammary glands, is released by **lactotrophs** (*lacto-* = milk).
- **Adrenocorticotropic hormone (ACTH)** or **corticotropin** (*cortico-* = rind or bark), which stimulates the adrenal cortex to secrete glucocorticoids, is synthesized by **corticotrophs.** Some corticotrophs, remnants of the pars intermedia, also secrete **melanocyte-stimulating hormone (MSH).**

Hormones that influence another endocrine gland are called **tropic hormones** or **tropins.** Several of the anterior pituitary gland hormones are tropins. The two **gonadotropins,** FSH and LH, regulate the functions of the gonads (ovaries and testes). Thyrotropin stimulates the thyroid gland, whereas corticotropin acts on the cortex of the adre-

nal gland. The hormones of the anterior pituitary gland are summarized in Table 18.3.

Secretion of anterior pituitary gland hormones is regulated in two ways. First, neurosecretory cells in the hypothalamus secrete five releasing hormones, which stimulate secretion of anterior pituitary gland hormones, and two inhibiting hormones, which suppress secretion of anterior pituitary gland hormones (see Table 18.3). Second, negative feedback in the form of hormones released by target glands adjusts secretions of anterior pituitary gland cells (Figure 18.6). In such negative feedback loops, the secretory activity of thyrotrophs, gonadotrophs, and corticotrophs decreases when blood levels of their target gland hormones rise. For example, corticotropin (ACTH) stimulates the cortex of the adrenal gland to secrete glucocorticoids, mainly cortisol. In turn, an elevated blood level of cortisol decreases secretion of both corticotropin and corticotropin releasing hormone (CRH) due to negative feedback suppression of the anterior pituitary corticotrophs and hypothalamic neurosecretory cells.

Human Growth Hormone and Insulinlike Growth Factors

Somatotrophs are the most abundant cells in the anterior pituitary gland, and human growth hormone (hGH) is the most plentiful anterior pituitary hormone. For the most part, hGH acts indirectly on tissues by promoting the synthesis and secretion of small protein hormones called **insulinlike growth factors (IGFs).** In response to human growth hormone, cells in the liver, skeletal muscle, cartilage, bone, and other tissues secrete IGFs, which may either enter the bloodstream from the liver or act locally in other tissues as autocrine or paracrine substances. IGFs cause cells to grow and multiply by increasing the rate at which amino acids enter cells and are used to synthesize proteins. IGFs also decrease the breakdown of proteins and the use of amino acids for ATP production. Due to these effects of the IGFs, human growth hormone increases the growth rate of

Figure 18.6 Negative feedback regulation of hypothalamic neurosecretory cells and anterior pituitary corticotrophs.

🔑 **Cortisol secreted by the adrenal cortex suppresses secretion of CRH and ACTH.**

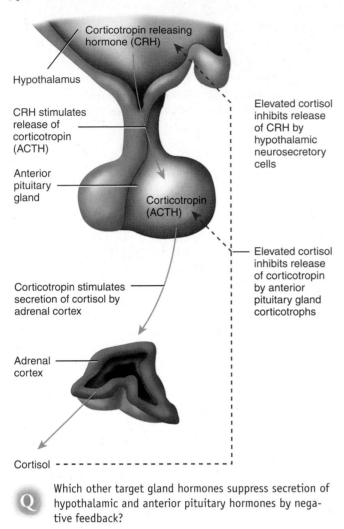

Q Which other target gland hormones suppress secretion of hypothalamic and anterior pituitary hormones by negative feedback?

Figure 18.7 Effects of human growth hormone (hGH) and insulinlike growth factors (IGFs).

🔑 **Secretion of hGH is stimulated by growth hormone releasing hormone (GHRH) and inhibited by growth hormone inhibiting hormone (GHIH).**

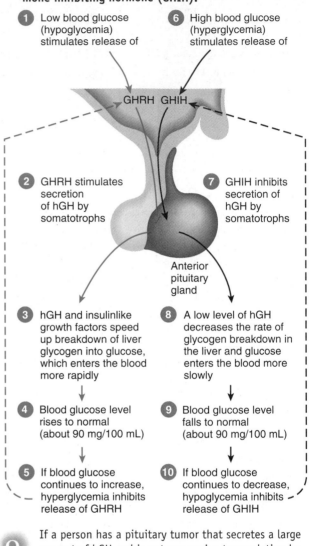

Q If a person has a pituitary tumor that secretes a large amount of hGH and is not responsive to regulation by GHRH and GHIH, will hyperglycemia or hypoglycemia be more likely?

the skeleton and skeletal muscles during childhood and the teenage years. In adults, human growth hormone and IGFs help maintain muscle and bone mass and promote healing of injuries and tissue repair.

IGFs also enhance lipolysis in adipose tissue, which results in increased use of the released fatty acids for ATP production by body cells. Besides affecting protein and lipid metabolism, human growth hormone and IGFs influence carbohydrate metabolism by decreasing glucose uptake, thereby decreasing the use of glucose for ATP production by most body cells. This action spares glucose so that neurons may continue to use it for ATP production in times of glucose scarcity. IGFs and human growth hormone may also stimulate liver cells to release glucose into the blood.

Somatotrophs in the anterior pituitary gland release bursts of human growth hormone every few hours, especially during sleep. Their secretory activity is controlled

mainly by growth hormone releasing hormone (GHRH) and growth hormone inhibiting hormone (GHIH). GHRH promotes secretion of human growth hormone, whereas GHIH suppresses it. Blood glucose level is a major regulator of GHRH and GHIH secretion, and thus of the effects of human growth hormone (Figure 18.7):

1 Low blood glucose level (hypoglycemia) stimulates the hypothalamus to secrete GHRH, which flows toward the anterior pituitary gland in the portal veins.

2 Upon reaching the anterior pituitary gland, GHRH stimulates somatotrophs to release human growth hormone.

3 Together, human growth hormone and insulinlike growth factors speed up breakdown of liver glycogen into glucose, which enters the blood.

4 As a result, blood glucose rises to the normal level (about 90 mg/100 ml of blood plasma).

5 An increase in blood glucose above the normal level inhibits release of GHRH.

6 Abnormally high blood glucose (hyperglycemia) stimulates the hypothalamus to secrete GHIH (while reducing the secretion of GHRH).

7 Upon reaching the anterior pituitary gland in portal blood, GHIH inhibits secretion of human growth hormone by somatotrophs.

8 A low level of human growth hormone slows liver glycogen breakdown, and glucose is released into the blood more slowly.

9 Blood glucose falls to the normal level.

10 A decrease in blood glucose below the normal level (hypoglycemia) inhibits release of GHIH.

Other stimuli that promote secretion of human growth hormone include decreased fatty acids and increased amino acids in the blood; deep sleep (stages 3 and 4 of non-rapid eye movement sleep); increased activity of the sympathetic division of the autonomic nervous system, such as might occur with stress or vigorous physical exercise; and other hormones, including glucagon, estrogens, cortisol, and insulin. Factors that inhibit human growth hormone secretion are increased fatty acids and decreased amino acids in the blood; rapid eye movement sleep; emotional deprivation; obesity; low levels of thyroid hormones; and human growth hormone itself (through negative feedback).

CLINICAL APPLICATION
Diabetogenic Effect of Human Growth Hormone

One symptom of excess human growth hormone is **hyperglycemia** (hī'-per-glī-SĒ-mē-a), high blood glucose concentration. Persistent hyperglycemia, in turn, stimulates the pancreas to secrete insulin continually. Such excessive stimulation, if it lasts for weeks or months, may cause "beta-cell burnout," a greatly decreased capacity of pancreatic beta cells to synthesize and secrete insulin. Eventually, excess secretion of human growth hormone may have a **diabetogenic effect;** that is, it causes diabetes mellitus (lack of insulin activity). ■

Thyroid-stimulating Hormone

Thyroid-stimulating hormone (TSH) stimulates the synthesis and secretion of the two thyroid hormones, triiodothyronine (T_3) and thyroxine (T_4), both produced by the thyroid gland. TSH secretion is controlled by thyrotropin releasing hormone (TRH) from the hypothalamus. Release of TRH, in turn, depends on blood levels of TSH and T_3, blood glucose level, and the body's metabolic rate, among other factors. The release of TRH operates according to a negative feedback system that is explained later in the chapter and depicted in Figure 18.12.

Follicle-stimulating Hormone

In females, follicle-stimulating hormone (FSH) is transported by the blood from the anterior pituitary gland to the ovaries, where each month it initiates the development of follicles—saclike arrangements of secretory cells that surround a developing oocyte. FSH also stimulates follicular cells to secrete estrogens (female sex hormones). In males, FSH stimulates sperm production in the testes. Gonadotropin releasing hormone (GnRH) from the hypothalamus stimulates FSH release. GnRH and FSH release is suppressed by estrogens in the female, and by testosterone (the principal male sex hormone in males), through negative feedback systems.

Luteinizing Hormone

In females, luteinizing hormone (LH), together with follicle-stimulating hormone, stimulates secretion of estrogen by ovarian cells, resulting in the release of a secondary oocyte (future ovum) by the ovary, a process called ovulation. LH also stimulates formation of the corpus luteum (structure formed after ovulation) in the ovary and the secretion of progesterone (another female sex hormone) by the corpus luteum. Estrogens and progesterone prepare the uterus for implantation of a fertilized ovum and help prepare the mammary glands for milk secretion. In males, LH stimulates the interstitial cells in the testes to secrete testosterone. Secretion of LH, like that of FSH, is controlled by gonadotropin releasing hormone.

Prolactin

Prolactin (PRL), together with other hormones, initiates and maintains milk secretion by the mammary glands. By itself, prolactin has only a weak effect. Only after the mammary glands have been primed by estrogens, progesterone, glucocorticoids, human growth hormone, thyroxine, and insulin, which exert permissive effects, does PRL bring about milk secretion. Ejection of milk from the mammary glands depends on the hormone oxytocin, which is released from the posterior pituitary gland. Together, milk secretion and ejection constitute *lactation.*

The hypothalamus secretes both inhibitory and excitatory hormones that regulate prolactin secretion. Prolactin inhibiting hormone (PIH), which is dopamine, inhibits the release of prolactin from the anterior pituitary gland. As the levels of estrogens and progesterone fall just before menstruation begins, the secretion of PIH diminishes and the blood level of prolactin rises. Breast tenderness just before menstruation may be caused by elevated prolactin, but because the high prolactin level does not last long enough and glandular cells are not fully developed, milk production does not start. As the menstrual cycle begins anew and the level of estrogens rises, PIH is again secreted and the prolactin level drops. Prolactin level rises during pregnancy, stimulated by

Table 18.4 Summary of the Principal Actions of Anterior Pituitary Gland Hormones

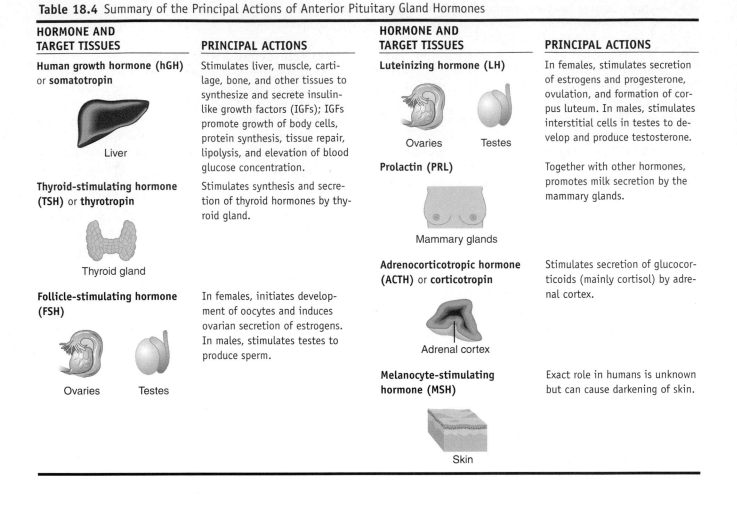

HORMONE AND TARGET TISSUES	PRINCIPAL ACTIONS	HORMONE AND TARGET TISSUES	PRINCIPAL ACTIONS
Human growth hormone (hGH) or **somatotropin** Liver	Stimulates liver, muscle, cartilage, bone, and other tissues to synthesize and secrete insulin-like growth factors (IGFs); IGFs promote growth of body cells, protein synthesis, tissue repair, lipolysis, and elevation of blood glucose concentration.	**Luteinizing hormone (LH)** Ovaries Testes	In females, stimulates secretion of estrogens and progesterone, ovulation, and formation of corpus luteum. In males, stimulates interstitial cells in testes to develop and produce testosterone.
Thyroid-stimulating hormone (TSH) or **thyrotropin** Thyroid gland	Stimulates synthesis and secretion of thyroid hormones by thyroid gland.	**Prolactin (PRL)** Mammary glands	Together with other hormones, promotes milk secretion by the mammary glands.
Follicle-stimulating hormone (FSH) Ovaries Testes	In females, initiates development of oocytes and induces ovarian secretion of estrogens. In males, stimulates testes to produce sperm.	**Adrenocorticotropic hormone (ACTH)** or **corticotropin** Adrenal cortex	Stimulates secretion of glucocorticoids (mainly cortisol) by adrenal cortex.
		Melanocyte-stimulating hormone (MSH) Skin	Exact role in humans is unknown but can cause darkening of skin.

prolactin releasing hormone (PRH) from the hypothalamus. The sucking action of a nursing infant causes a reduction in hypothalamic secretion of PIH.

The function of prolactin is not known in males, but its hypersecretion causes erectile dysfunction or impotence (inability to have an erection of the penis). In females, hypersecretion of prolactin causes galactorrhea (inappropriate lactation) and amenorrhea (absence of menstrual cycles).

Adrenocorticotropic Hormone

Corticotrophs secrete mainly adrenocorticotropic hormone (ACTH) or adrenocorticotropin (ad-rē′-nō-kor′-ti-kō-TRŌ-pin). ACTH controls the production and secretion of hormones called glucocorticoids by the cortex (outer portion) of the adrenal glands. Corticotropin releasing hormone (CRH) from the hypothalamus stimulates secretion of ACTH by corticotrophs. Stress-related stimuli, such as low blood glucose or physical trauma, and interleukin-1 (IL-1), a substance produced by macrophages, also stimulate release of ACTH. Glucocorticoids cause negative feedback inhibition of both CRH and ACTH release.

Melanocyte-stimulating Hormone

Melanocyte-stimulating hormone (MSH) increases skin pigmentation in amphibians by stimulating the dispersion of melanin granules in melanocytes. Its exact role in humans is unknown. However, continued administration of MSH for several days does produce a darkening of the skin, and without MSH the skin may be pallid. Corticotropin releasing hormone (CRH) stimulates MSH release, whereas dopamine inhibits MSH release.

Table 18.4 summarizes the principal actions of the anterior pituitary hormones.

Posterior Pituitary Gland

Although the **posterior pituitary gland** or **neurohypophysis** does not *synthesize* hormones, it does *store* and *release* two hormones. As noted earlier, it consists of pituicytes and axon terminals of hypothalamic neurosecretory cells. The cell bodies of the neurosecretory cells are in the paraventricular and supraoptic nuclei of the hypothalamus; their axons form the **hypothalamohypophyseal tract** (hī′-pō-

Figure 18.8 Axons of hypothalamic neurosecretory cells form the hypothalamohypophyseal tract, which extends from the paraventricular and supraoptic nuclei to the posterior pituitary gland. Hormone molecules synthesized in the cell body of a neurosecretory cell are packaged into secretory vesicles that move down to the axon terminals. Nerve impulses trigger exocytosis of the vesicles and release of the hormone.

🔑 **Oxytocin and antidiuretic hormone are synthesized in the hypothalamus and released into capillaries of the posterior pituitary gland.**

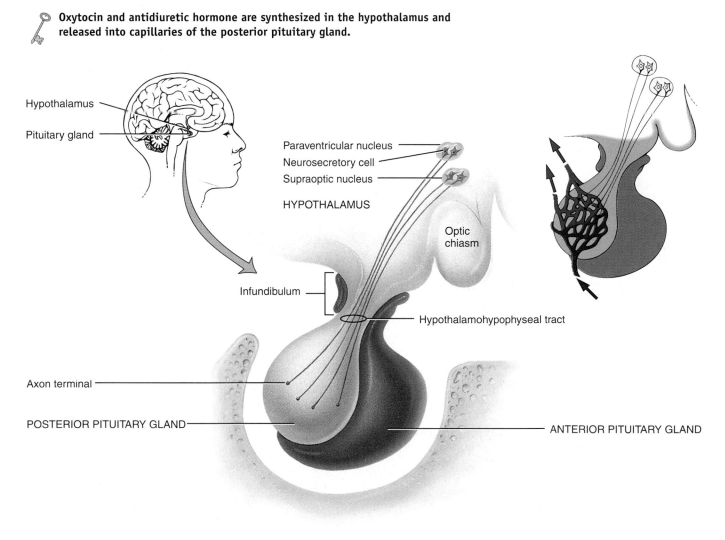

Hypothalamus

Pituitary gland

Paraventricular nucleus

Neurosecretory cell

Supraoptic nucleus

HYPOTHALAMUS

Optic chiasm

Infundibulum

Hypothalamohypophyseal tract

Axon terminal

POSTERIOR PITUITARY GLAND

ANTERIOR PITUITARY GLAND

Q Functionally, how are the hypothalamohypophyseal tract and the hypophyseal portal veins similar? Structurally, how are they different?

thal′-a-mō-hī-pō-FIZ-ē-al), which begins in the hypothalamus and ends near blood capillaries in the posterior pituitary gland (Figure 18.8). Different neurosecretory cells produce two hormones: **oxytocin** (**OT;** ok′-sē-TŌ-sin; *oxytoc-* = quick birth) and **antidiuretic hormone (ADH),** also called **vasopressin.**

After their production in the cell bodies of neurosecretory cells, oxytocin and antidiuretic hormone are packed into vesicles, which move by fast axonal transport (described on page 381) to the axon terminals in the posterior pituitary gland. Nerve impulses that propagate along the axon and

reach the axon terminals trigger exocytosis of these secretory vesicles.

Blood is supplied to the posterior pituitary gland by the **inferior hypophyseal arteries** (see Figure 18.5), which branch from the internal carotid arteries. In the posterior pituitary gland, the inferior hypophyseal arteries drain into the **capillary plexus of the infundibular process,** a capillary network that receives secreted oxytocin and antidiuretic hormone (see Figure 18.5). From this plexus, hormones pass into the **posterior hypophyseal veins** for distribution to target cells in other tissues.

Figure 18.9 Regulation of secretion and actions of antidiuretic hormone (ADH).

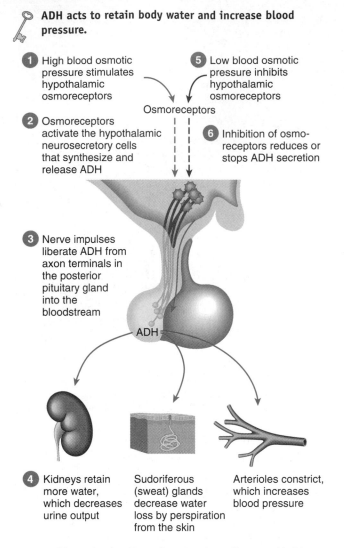

🔑 **ADH acts to retain body water and increase blood pressure.**

1 High blood osmotic pressure stimulates hypothalamic osmoreceptors

5 Low blood osmotic pressure inhibits hypothalamic osmoreceptors

Osmoreceptors

2 Osmoreceptors activate the hypothalamic neurosecretory cells that synthesize and release ADH

6 Inhibition of osmoreceptors reduces or stops ADH secretion

3 Nerve impulses liberate ADH from axon terminals in the posterior pituitary gland into the bloodstream

ADH

4 Kidneys retain more water, which decreases urine output

Sudoriferous (sweat) glands decrease water loss by perspiration from the skin

Arterioles constrict, which increases blood pressure

Q If you drank a liter of water, what effect would this have on the osmotic pressure of your blood, and how would the level of ADH change in your blood?

Oxytocin

During and after delivery of a baby, oxytocin has two target tissues: the mother's uterus and breasts. During delivery, oxytocin enhances contraction of smooth muscle cells in the wall of the uterus; after delivery, it stimulates milk ejection ("letdown") from the mammary glands in response to the mechanical stimulus provided by a suckling infant. The function of oxytocin in males and in nonpregnant females is not clear. Experiments with animals have suggested that it has actions within the brain that foster parental caretaking behavior toward young offspring. It may also be responsible, in part, for the feelings of sexual pleasure during and after intercourse.

Antidiuretic Hormone

An **antidiuretic** is a substance that decreases urine production. ADH causes the kidneys to return more water to the blood, thus decreasing urine volume. In the absence of ADH, urine output increases more than tenfold, from the normal 1–2 liters to about 20 liters a day. ADH also decreases the water lost through sweating and causes constriction of arterioles, which increases blood pressure. This hormone's other name, vasopressin, reflects its effect on blood pressure. Hyposecretion of ADH or nonfunctional ADH receptors causes diabetes insipidus (see page 603).

The amount of ADH secreted varies with blood osmotic pressure and blood volume. Regulation of the secretion and actions of ADH occurs as follows (Figure 18.9):

1 High blood osmotic pressure—due to dehydration or decline in blood volume because of hemorrhage, diarrhea, or excessive sweating—stimulates **osmoreceptors,** neurons in the hypothalamus that monitor blood osmotic pressure. Elevated blood osmotic pressure activates the osmoreceptors directly; they also receive excitatory input from other brain areas when blood volume decreases.

2 Osmoreceptors activate the hypothalamic neurosecretory cells that synthesize and release ADH.

3 When neurosecretory cells receive excitatory input from the osmoreceptors, they generate nerve impulses that cause exocytosis of ADH-containing vesicles from their axon terminals in the posterior pituitary gland. This liberates ADH, which diffuses into blood capillaries of the posterior pituitary gland.

4 The blood carries ADH to three target tissues: the kidneys, sudoriferous (sweat) glands, and smooth muscle in blood vessel walls. The kidneys respond by retaining more water, which decreases urine output. Secretory activity of sweat glands decreases, which lowers the rate of water loss by perspiration from the skin. Smooth muscle in the walls of arterioles (small arteries) contracts in response to high levels of ADH, which constricts (narrows) the lumen of these blood vessels and increases blood pressure.

5 Low osmotic pressure of blood or increased blood volume inhibits the osmoreceptors.

6 Inhibition of osmoreceptors reduces or stops ADH secretion. The kidneys then retain less water by forming a larger volume of urine, secretory activity of sweat glands increases, and arterioles dilate. The blood volume and osmotic pressure of body fluids return to normal.

Secretion of ADH can also be altered in other ways. Pain, stress, trauma, anxiety, acetylcholine, nicotine, and drugs such as morphine, tranquilizers, and some anesthetics stimulate ADH secretion. Alcohol inhibits ADH secretion, thereby increasing urine output. The resulting dehydration may cause both the thirst and the headache typical of a hangover.

Table 18.5 Summary of Posterior Pituitary Gland Hormones

HORMONE AND TARGET TISSUES	PRINCIPAL ACTIONS	CONTROL OF SECRETION
Oxytocin (OT)  Uterus  Mammary glands	Stimulates contraction of smooth muscle cells of uterus during childbirth; stimulates contraction of myoepithelial cells in mammary glands to cause milk ejection.	Neurosecretory cells of hypothalamus secrete OT in response to uterine distention and stimulation of nipples.
Antidiuretic hormone (ADH) or **vasopressin** 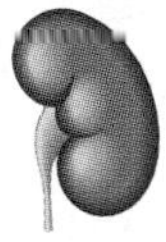 Kidneys 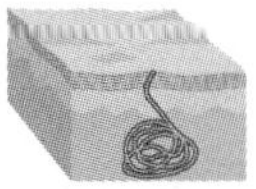 Sudoriferous (sweat) glands 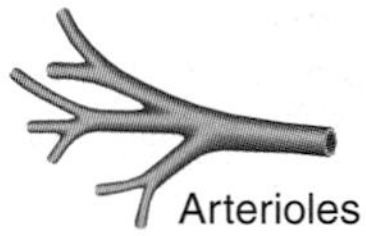 Arterioles	Conserves body water by decreasing urine volume; decreases water loss through perspiration; raises blood pressure by constricting aterioles.	Neurosecretory cells of hypothalamus secrete ADH in response to elevated blood osmotic pressure, dehydration, loss of blood volume, pain, or stress; low blood osmotic pressure, high blood volume, and alcohol inhibit ADH secretion.

Table 18.5 lists the posterior pituitary gland hormones and summarizes their principal actions and the control of their secretion.

1. In what respect is the pituitary gland actually two glands?
2. How do hypothalamic releasing and inhibiting hormones influence secretions of the anterior pituitary gland?
3. Describe the structure and importance of the hypothalamohypophyseal tract.
4. Explain how blood levels of T_3/T_4, TSH, and TRH would change in a lab animal that has undergone a thyroidectomy (complete removal of its thyroid gland).

THYROID GLAND

OBJECTIVE

• *Describe the location, histology, hormones, and functions of the thyroid gland.*

The butterfly-shaped **thyroid gland** is located just inferior to the larynx (voice box); the right and left **lateral lobes** lie one on either side of the trachea (Figure 18.10a). Connecting the lobes is a mass of tissue called an **isthmus** (IS-mus) that lies anterior to the trachea. A small, pyramidal-shaped lobe sometimes extends upward from the isthmus. The gland usually weighs about 30 g (1 oz) and has a rich blood supply, receiving 80–120 mL of blood per minute.

Microscopic spherical sacs called **thyroid follicles** (Figure 18.10b; also see Figure 18.13c) make up most of the thyroid gland. The wall of each follicle consists primarily of cells called **follicular cells,** most of which extend to the lumen (internal space) of the follicle. When the follicular cells are inactive, their shape is low cuboidal to squamous, but under the influence of TSH they become cuboidal or low columnar and actively secretory. The follicular cells produce two hormones: **thyroxine** (thī-ROK-sin), which is also called **tetraiodothyronine** (tet-ra-ī-ō-dō-THĪ-rō-nēn) or T_4 because it contains four atoms of iodine, and **triiodothyronine** (trī-ī'-ō-dō-THĪ-rō-nēn) or T_3, which contains three atoms of iodine. T_3 and T_4 are also known as **thyroid hormones.** A few cells called **parafollicular cells** or **C cells** may be embedded within a follicle or lie between follicles. They produce the hormone **calcitonin** (kal-si-TŌ-nin), which helps regulate calcium homeostasis.

Formation, Storage, and Release of Thyroid Hormones

The thyroid gland is the only endocrine gland that stores its secretory product in large quantity—normally about a 100-day supply. In essence, under the stimulation of TSH, T_3 and T_4 are synthesized by attaching iodine atoms to the amino acid tyrosine, then stored for some period of time, and finally secreted into the blood, as follows (Figure 18.11):

1 *Iodide trapping.* Thyroid follicular cells trap iodide ions (I^-) by actively transporting them from the blood into the cytosol. The I^- concentration inside follicular cells is 20–40 times that of blood plasma. As a result, the thyroid gland normally contains most of the iodide in the body.

Figure 18.10 Location, blood supply, and histology of the thyroid gland. (See Tortora, *A Photographic Atlas of the Human Body,* Figure 10.3a)

🔑 **Thyroid hormones regulate (1) oxygen use and basal metabolic rate, (2) cellular metabolism, and (3) growth and development.**

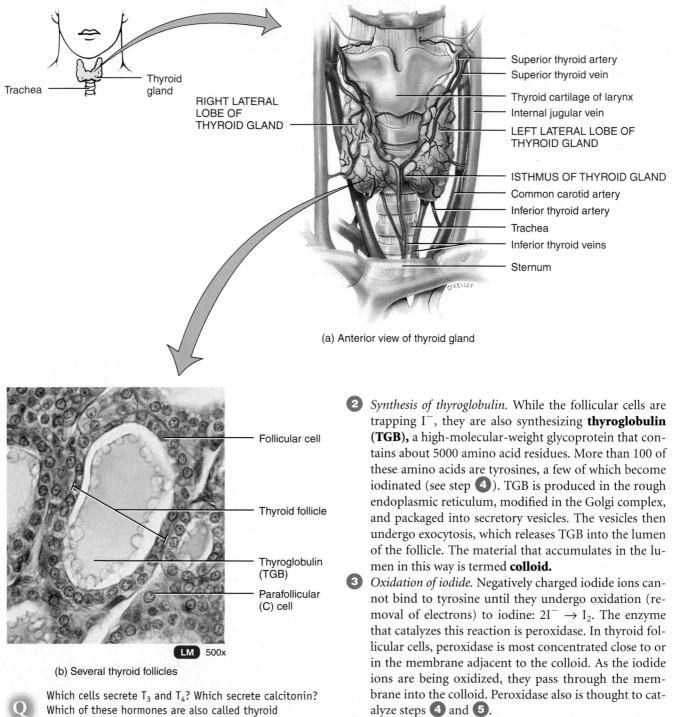

Trachea — Thyroid gland

RIGHT LATERAL LOBE OF THYROID GLAND

Superior thyroid artery
Superior thyroid vein
Thyroid cartilage of larynx
Internal jugular vein
LEFT LATERAL LOBE OF THYROID GLAND
ISTHMUS OF THYROID GLAND
Common carotid artery
Inferior thyroid artery
Trachea
Inferior thyroid veins
Sternum

O'KELLEY

(a) Anterior view of thyroid gland

Follicular cell

Thyroid follicle

Thyroglobulin (TGB)

Parafollicular (C) cell

LM 500x

(b) Several thyroid follicles

Q Which cells secrete T_3 and T_4? Which secrete calcitonin? Which of these hormones are also called thyroid hormones?

❷ *Synthesis of thyroglobulin.* While the follicular cells are trapping I^-, they are also synthesizing **thyroglobulin (TGB),** a high-molecular-weight glycoprotein that contains about 5000 amino acid residues. More than 100 of these amino acids are tyrosines, a few of which become iodinated (see step ❹). TGB is produced in the rough endoplasmic reticulum, modified in the Golgi complex, and packaged into secretory vesicles. The vesicles then undergo exocytosis, which releases TGB into the lumen of the follicle. The material that accumulates in the lumen in this way is termed **colloid.**

❸ *Oxidation of iodide.* Negatively charged iodide ions cannot bind to tyrosine until they undergo oxidation (removal of electrons) to iodine: $2I^- \rightarrow I_2$. The enzyme that catalyzes this reaction is peroxidase. In thyroid follicular cells, peroxidase is most concentrated close to or in the membrane adjacent to the colloid. As the iodide ions are being oxidized, they pass through the membrane into the colloid. Peroxidase also is thought to catalyze steps ❹ and ❺.

❹ *Iodination of tyrosine.* As iodine molecules (I_2) form, they react with tyrosine amino acids that are part of thyroglobulin molecules in the colloid. Binding of one iodine atom yields monoiodotyrosine (T_1), and a second iodination produces diiodotyrosine (T_2).

5 *Coupling of T_1 and T_2.* During the last step in the synthesis of thyroid hormone, two T_2 molecules join to form T_4, or one T_1 and one T_2 join to form T_3. Each TGB ultimately contains about six T_1, five T_2, and one to five T_4. There is a single T_3 in one of every four thyroglobulins. Thus, thyroid hormones are stored as part of thyroglobulin molecules within the colloid.

6 *Pinocytosis and digestion of colloid.* Droplets of colloid reenter follicular cells by pinocytosis and merge with lysosomes. Digestive enzymes break down TGB, cleaving off molecules of T_3 and T_4. T_1 and T_2 are also released, but they undergo deiodination (removal of iodine). The iodine is reused to synthesize more T_3 and T_4.

7 *Secretion of thyroid hormones.* Because T_3 and T_4 are lipid-soluble, they diffuse through the plasma membrane to enter the blood.

8 *Transport in the blood.* More than 99% of both the T_3 and the T_4 combine with transport proteins in the blood, mainly **thyroxine-binding globulin (TBG).**

T_4 normally is secreted in greater quantity than T_3, but T_3 is several times more potent. Moreover, as it circulates in the blood and enters cells throughout the body, most T_4 is converted to T_3 by removal of one iodine.

Actions of Thyroid Hormones

The thyroid hormones regulate (1) oxygen use and basal metabolic rate, (2) cellular metabolism, and (3) growth and development.

Thyroid hormones increase basal metabolic rate or BMR (rate of oxygen consumption at rest after an overnight fast) by stimulating the use of cellular oxygen to produce ATP. The active transport pumps that continually eject sodium ions (Na^+) from the cytosol into the extracellular fluid use a large portion of the ATP produced by most cells. A major effect of the thyroid hormones is to stimulate synthesis of the enzyme that runs the pump, Na^+/K^+ ATPase. As cells use more oxygen to produce ATP, more heat is given off, and body temperature rises. This phenomenon is called the **calorigenic effect** of the thyroid hormones. In this way, they play an important role in the maintenance of normal body temperature. Normal mammals can survive in freezing temperatures, but those whose thyroid glands have been removed cannot.

In the regulation of metabolism, the thyroid hormones stimulate protein synthesis and increase the use of glucose for ATP production. They also increase lipolysis and enhance cholesterol excretion in bile (a substance made in the liver that aids lipid digestion), thus reducing blood cholesterol level.

The thyroid hormones enhance some actions of the catecholamines (norepinephrine and epinephrine) because they up-regulate beta (β) receptors. For this reason, symptoms of hyperthyroidism include increased heart rate, more forceful heartbeats, and increased blood pressure.

Figure 18.11 Steps in the synthesis and secretion of thyroid hormones.

🔑 **Thyroid hormones are synthesized by attaching iodine atoms to the amino acid tyrosine.**

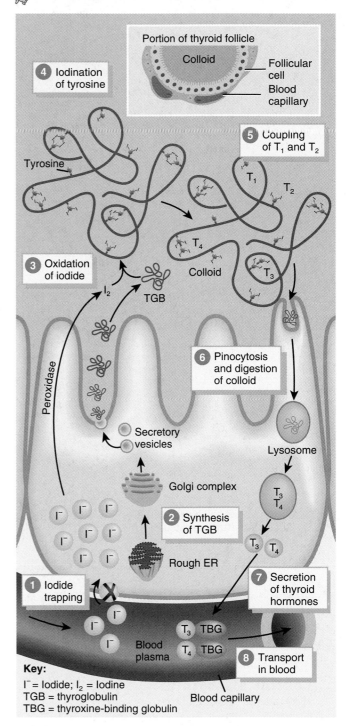

Key:
I^- = Iodide; I_2 = Iodine
TGB = thyroglobulin
TBG = thyroxine-binding globulin

Q What is the storage form of thyroid hormones?

Figure 18.12 Negative feedback regulation of thyroid hormone secretion.

🔑 **Thyroid-stimulating hormone (TSH) promotes release of thyroid hormones (T₃ and T₄) by the thyroid gland.**

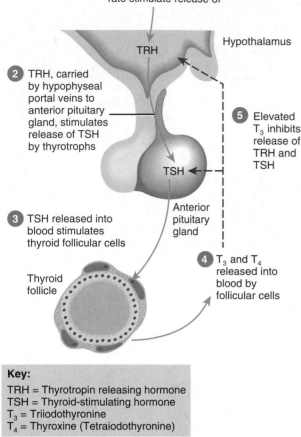

1 Low blood levels of T_3 and T_4 or low metabolic rate stimulate release of

TRH

Hypothalamus

2 TRH, carried by hypophyseal portal veins to anterior pituitary gland, stimulates release of TSH by thyrotrophs

TSH

5 Elevated T_3 inhibits release of TRH and TSH

3 TSH released into blood stimulates thyroid follicular cells

Anterior pituitary gland

Thyroid follicle

4 T_3 and T_4 released into blood by follicular cells

Key:
TRH = Thyrotropin releasing hormone
TSH = Thyroid-stimulating hormone
T_3 = Triiodothyronine
T_4 = Thyroxine (Tetraiodothyronine)

Q How could an iodine-deficient diet lead to goiter, an enlargement of the thyroid gland?

Together with human growth hormone and insulin, thyroid hormones accelerate body growth, particularly the growth of nervous tissue. Deficiency of thyroid hormones during fetal development, infancy, or childhood results in *cretinism,* which is characterized by small stature and mental retardation.

Control of Thyroid Hormone Secretion

The size and the secretory activity (and thus the size) of the thyroid gland are controlled in two main ways. First, although iodine is needed for synthesis of thyroid hormones, an abnormally high blood iodine concentration suppresses release of thyroid hormones. Second, negative feedback systems involving thyroptin releasing hormone (TRH) from the hypothalamus and TSH from the anterior pituitary gland

stimulate synthesis and release of thyroid hormones, as follows (Figure 18.12):

1 Low blood levels of T_3 and T_4 or low metabolic rate stimulate the hypothalamus to secrete thyrotropin releasing hormone.

2 TRH enters the hypophyseal portal veins and is carried to the anterior pituitary gland, where it stimulates thyrotrophs to secrete thyroid-stimulating hormone (TSH).

3 TSH stimulates virtually all aspects of thyroid follicular cell activity, including iodide trapping, hormone synthesis and secretion, and growth of the follicular cells.

4 The thyroid follicular cells release T_3 and T_4 into the blood until the metabolic rate returns to normal.

5 An elevated level of T_3 inhibits release of TRH and TSH.

Conditions that increase ATP demand—a cold environment, hypoglycemia, high altitude, and pregnancy—also affect this negative feedback system and increase the secretion of the thyroid hormones.

Calcitonin

The hormone produced by the parafollicular cells of the thyroid gland is **calcitonin** (**CT;** kal-si-TŌ-nin). Although it acts rapidly in experimental situations, calcitonin's importance in normal physiology is not clear because it can be present in excess or completely absent without causing clinical symptoms. When given as a drug, calcitonin lowers the amount of blood calcium and phosphates by inhibiting bone resorption (breakdown of bone matrix) and by accelerating uptake of calcium and phosphates into bone matrix. Calcitonin exerts its effects by inhibiting the action of osteoclasts (bone-destroying cells). Miacalcin, a calcitonin extract derived from salmon, is used as a drug to treat osteoporosis.

Table 18.6 on page 586 summarizes the hormones produced by the thyroid gland, their principal actions, and control of secretion.

1. How are the thyroid hormones made, stored, and secreted?

2. Discuss the physiological effects of the thyroid hormones.

3. How is the secretion of T_3 and T_4 regulated?

PARATHYROID GLANDS

OBJECTIVE
• *Describe the location, histology, hormone, and functions of the parathyroid glands.*

Attached to the posterior surface of the lateral lobes of the thyroid gland are small, round masses of tissue called the **parathyroid glands.** Usually, one superior and one inferior parathyroid gland is attached to each lateral thyroid lobe (Figure 18.13a).

Figure 18.13 Location, blood supply, and histology of the parathyroid glands.
(See Tortora, *A Photographic Atlas of the Human Body,* Figure 10.4a)

The parathyroid glands, normally four in number, are embedded in the posterior surface of the thyroid gland.

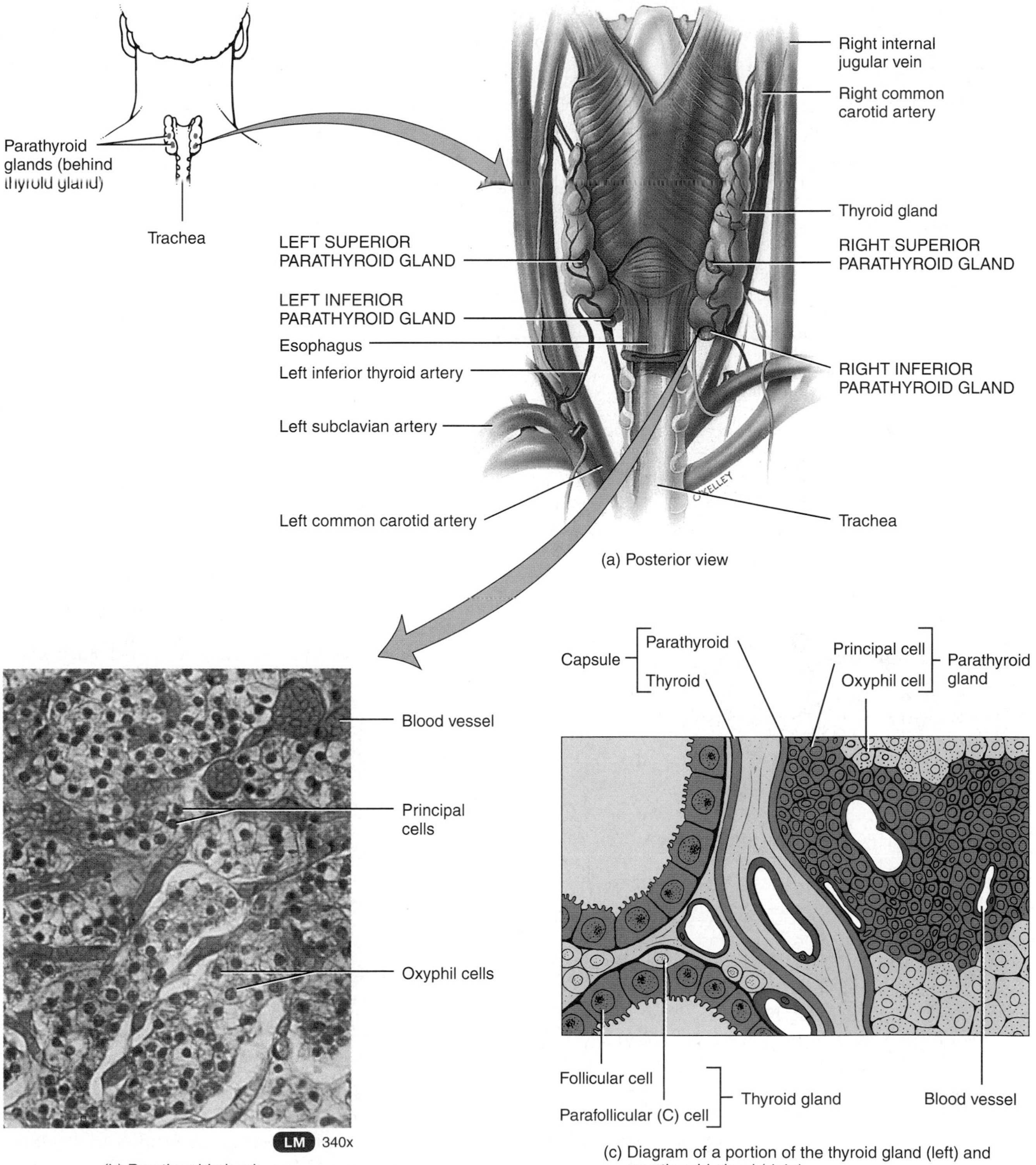

(a) Posterior view

(b) Parathyroid gland

LM 340x

(c) Diagram of a portion of the thyroid gland (left) and parathyroid gland (right)

Q What are the secretory products of (1) parafollicular cells of the thyroid gland and (2) principal cells of the parathyroid glands?

Table 18.6 Summary of Thyroid Gland Hormones

HORMONE	PRINCIPAL ACTIONS	CONTROL OF SECRETION
T_3 (triiodothyronine) and T_4 (thyroxine) or **thyroid hormones** from follicular cells	Increase basal metabolic rate, stimulate synthesis of proteins, increase use of glucose for ATP production, increase lipolysis, enhance cholesterol excretion in bile, accelerate body growth, and contribute to development of the nervous system.	Secretion increased by thyrotropin releasing hormone (TRH), which stimulates release of thyroid-stimulating hormone (TSH) in response to low thyroid hormone levels, low metabolic rate, cold, pregnancy, and high altitudes; TRH and TSH secretion is inhibited in response to high thyroid hormone levels; high iodine level suppresses T_3/T_4 secretion.
Calcitonin (CT) from parafollicular cells	Lowers blood levels of ionic calcium and phosphates by inhibiting bone resorption and accelerating uptake of calcium and phosphates into bone matrix.	High blood Ca^{2+} levels stimulate secretion; low blood Ca^{2+} levels inhibit secretion.

Microscopically, the parathyroids contain two kinds of epithelial cells (Figure 18.13b, c). The more numerous cells, called **principal cells,** probably are the major source of **parathyroid hormone (PTH),** or **parathormone.** The function of the other kind of cell, called an *oxyphil cell,* is not known.

Parathyroid Hormone

PTH increases the number and activity of osteoclasts. The result is elevated bone resorption, which releases ionic calcium (Ca^{2+}) and phosphates (HPO_4^{2-}) into the blood. PTH also produces two changes in the kidneys: (1) It increases the rate at which the kidneys remove Ca^{2+} and magnesium (Mg^{2+}) from urine that is being formed and returns them to the blood, and (2) it inhibits the reabsorption of HPO_4^{2-} filtered by the kidneys, so that more of it is excreted in urine. More HPO_4^{2-} is lost in the urine than is gained from the bones. Overall, then, PTH decreases blood HPO_4^{2-} level and increases blood Ca^{2+} and Mg^{2+} levels. With respect to blood Ca^{2+} level, PTH and calcitonin are antagonists; that is, they have opposite actions (see Figure 18.14).

A third effect of PTH on the kidneys is to promote formation of the hormone **calcitriol,** which is the active form of vitamin D. Calcitriol, also known as *1,25-dihydroxy vitamin D_3,* increases the rate of Ca^{2+}, HPO_4^{2-}, and Mg^{2+} absorption from the gastrointestinal tract into the blood.

The blood calcium level directly controls the secretion of calcitonin and parathyroid hormone via negative feedback loops that do not involve the pituitary gland (Figure 18.14):

1 A higher than normal level of calcium ions (Ca^{2+}) in the blood stimulates parafollicular cells of the thyroid gland.

2 Parafollicular cells release more calcitonin as blood Ca^{2+} level rises.

3 Calcitonin promotes deposition of blood Ca^{2+} into the matrix of bone tissue, thereby decreasing blood Ca^{2+} level.

4 A lower than normal level of Ca^{2+} in the blood stimulates principal cells of the parathyroid gland.

5 Principal cells release more parathyroid hormone (PTH) as blood Ca^{2+} level falls.

6 PTH promotes release of Ca^{2+} from bone matrix into the blood and retards loss of Ca^{2+} in the urine, raising the blood level of Ca^{2+}.

7 PTH also stimulates the kidneys to release calcitriol.

8 Calcitriol stimulates increased absorption of Ca^{2+} from foods in the gastrointestinal tract, which helps increase the blood level of Ca^{2+}.

Table 18.7 summarizes the principal actions and control of secretion of parathyroid hormone.

1. How is secretion of parathyroid hormone regulated?

ADRENAL GLANDS

OBJECTIVE

• *Describe the location, histology, hormones, and functions of the adrenal glands.*

The paired **adrenal (suprarenal) glands,** one of which lies superior to each kidney (Figure 18.15a on page 588), have a flattened pyramidal shape. In an adult, each adrenal gland is 3–5 cm in height, 2–3 cm in width, and a little less than 1 cm thick; it weighs 3.5–5 g, only half its weight at birth. During embryonic development, the adrenal glands

Figure 18.14 The roles of calcitonin (green arrows), parathyroid hormone (blue arrows), and calcitriol (red arrows) in calcium homeostasis.

🗝 **With respect to regulation of blood Ca²⁺ level, calcitonin and PTH are antagonists.**

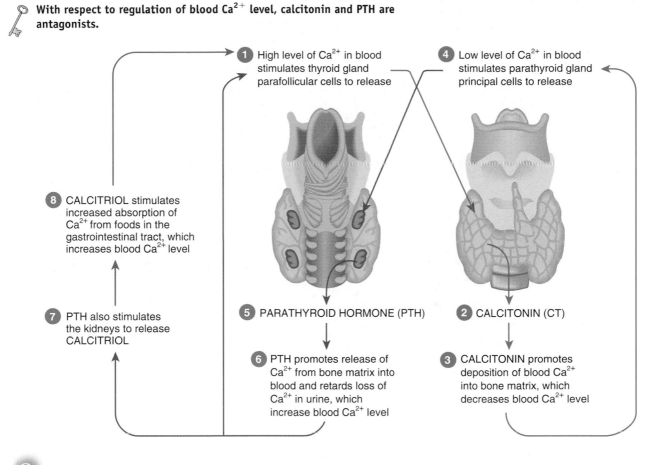

1 High level of Ca²⁺ in blood stimulates thyroid gland parafollicular cells to release

4 Low level of Ca²⁺ in blood stimulates parathyroid gland principal cells to release

8 CALCITRIOL stimulates increased absorption of Ca²⁺ from foods in the gastrointestinal tract, which increases blood Ca²⁺ level

7 PTH also stimulates the kidneys to release CALCITRIOL

5 PARATHYROID HORMONE (PTH)

2 CALCITONIN (CT)

6 PTH promotes release of Ca²⁺ from bone matrix into blood and retards loss of Ca²⁺ in urine, which increase blood Ca²⁺ level

3 CALCITONIN promotes deposition of blood Ca²⁺ into bone matrix, which decreases blood Ca²⁺ level

Ⓠ What are the primary target tissues for PTH, CT, and calcitriol?

Table 18.7 Summary of Parathyroid Gland Hormone

HORMONE	PRINCIPAL ACTIONS	CONTROL OF SECRETION
Parathyroid hormone (PTH) from principal cells Principal cell	Increases blood Ca²⁺ and Mg²⁺ levels and decreases blood phosphate level; increases rate of dietary Ca²⁺ and Mg²⁺ absorption; increases bone resorption by osteoclasts; increases Ca²⁺ reabsorption and phosphate excretion by kidneys; and promotes formation of calcitriol (active form of vitamin D).	Low blood Ca²⁺ levels stimulate secretion. High blood Ca²⁺ levels inhibit secretion.

Figure 18.15 Location, blood supply, and histology of the adrenal (suprarenal) glands. (See Tortora, *A Photographic Atlas of the Human Body,* Figure 10.5a)

🔑 **The adrenal cortex secretes steroid hormones that are essential for life; the adrenal medulla secretes norepinephrine and epinephrine.**

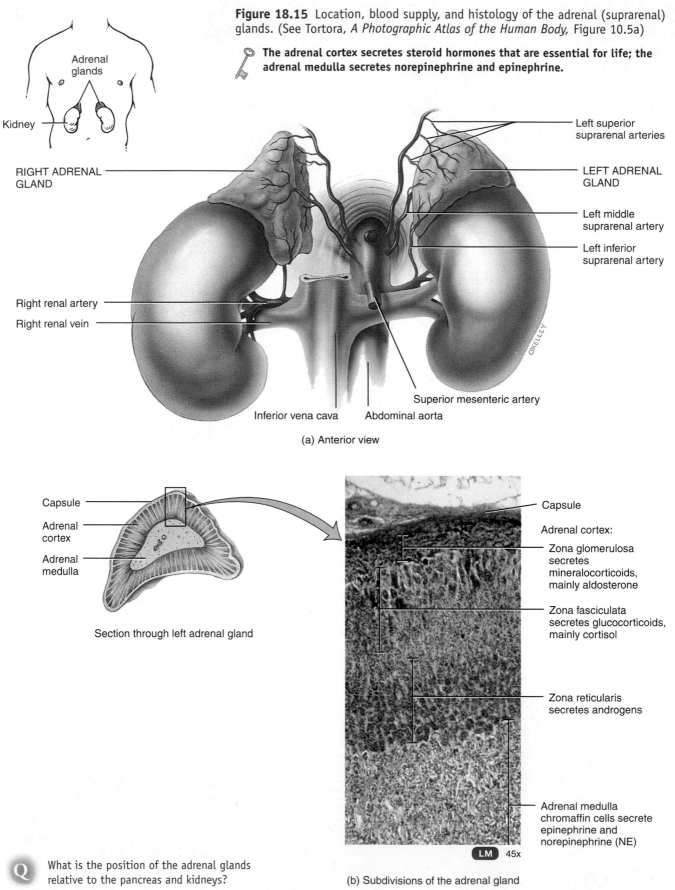

Adrenal glands

Kidney

RIGHT ADRENAL GLAND

Right renal artery

Right renal vein

Left superior suprarenal arteries

LEFT ADRENAL GLAND

Left middle suprarenal artery

Left inferior suprarenal artery

Superior mesenteric artery

Inferior vena cava

Abdominal aorta

OKELLEY

(a) Anterior view

Capsule

Adrenal cortex

Adrenal medulla

Section through left adrenal gland

Capsule

Adrenal cortex:

Zona glomerulosa secretes mineralocorticoids, mainly aldosterone

Zona fasciculata secretes glucocorticoids, mainly cortisol

Zona reticularis secretes androgens

Adrenal medulla chromaffin cells secrete epinephrine and norepinephrine (NE)

LM 45x

(b) Subdivisions of the adrenal gland

Q What is the position of the adrenal glands relative to the pancreas and kidneys?

differentiate into two structurally and functionally distinct regions: A large, peripherally located **adrenal cortex,** representing 80–90% of the gland by weight, develops from mesoderm; a small, centrally located **adrenal medulla** develops from ectoderm (Figure 18.15b). The adrenal cortex produces steroid hormones that are essential for life. Complete loss of adrenocortical hormones leads to death due to dehydration and electrolyte imbalances in a few days to a week, unless hormone replacement therapy begins promptly. The adrenal medulla produces two catecholamine hormones: norepinephrine and epinephrine. Covering the gland is a connective tissue capsule. The adrenal glands, like the thyroid gland, are highly vascularized.

Adrenal Cortex

The adrenal cortex is subdivided into three zones, each of which secretes different hormones (Figure 18.15b). The outer zone, just deep to the connective tissue capsule, is called the **zona glomerulosa** (*zona* = belt; *glomerul-* = little ball). Its cells, which are closely packed and arranged in spherical clusters and arched columns, secrete hormones called **mineralocorticoids** (min′-er-al-ō-KOR-ti-koyds) because they affect homeostasis of the minerals sodium and potassium. The middle zone, or **zona fasciculata** (*fascicul-* = little bundle), is the widest of the three zones and consists of cells arranged in long, straight cords. The cells of the zona fasciculata secrete mainly **glucocorticoids** (gloo′-kō-KOR-ti-koyds), so named because they affect glucose homeostasis. The cells of the inner zone, the **zona reticularis** (*reticul-* = network), are arranged in branching cords. They synthesize small amounts of weak **androgens** (*andro-* = a man), steroid hormones that have masculinizing effects.

Mineralocorticoids

Mineralocorticoids help control water and electrolyte homeostasis, particularly the concentrations of sodium ions (Na^+) and potassium ions (K^+). Although the adrenal cortex secretes at least three different hormones classified as mineralocorticoids, about 95% of mineralocorticoid activity is due to **aldosterone** (al-DA-ster-ōn). Aldosterone acts on certain tubule cells in the kidneys to increase their reabsorption of Na^+. By stimulating return of Na^+ to the blood, aldosterone prevents depletion of Na^+ from the body. The Na^+ reabsorption also leads to reabsorption of Cl^- (chloride ions), HCO_3^- (bicarbonate ions), and water molecules. At the same time, aldosterone promotes secretion of K^+, thereby increasing K^+ excretion in the urine. Aldosterone also promotes secretion of H^+ into the urine; this removal of acids from the body can help prevent acidosis (blood pH below 7.35).

The most important mechanism for control of aldosterone secretion is the **renin–angiotensin pathway** (RĒ-nin an′je-ō-TEN-sin; Figure 18.16):

1 Stimuli that initiate the renin–angiotensin pathway include dehydration, Na^+ deficiency, or hemorrhage.

2 These conditions cause a decrease in blood volume.

3 Decreased blood volume causes a decrease in blood pressure.

4 Lowered blood pressure stimulates certain cells of the kidneys, called juxtaglomerular cells, to secrete the enzyme **renin.**

5 The level of renin in the blood increases.

6 Renin converts **angiotensinogen,** a plasma protein produced by the liver, into **angiotensin I.**

7 Blood containing increased levels of angiotensin I circulates to the lungs.

8 As blood flows through capillaries, particularly those of the lungs, an enzyme called **angiotensin converting enzyme (ACE)** converts angiotensin I into the hormone **angiotensin II.**

9 Blood level of angiotensin II increases.

10 Angiotensin II has two main target tissues, one of which is the adrenal cortex, which it stimulates to secrete aldosterone.

11 Blood containing increased levels of aldosterone circulates to the kidneys.

12 In the kidneys, aldosterone increases Na^+ reabsorption, and water follows by osmosis. Aldosterone also stimulates the kidneys to increase secretion of K^+ into the urine.

13 As a result of increased water reabsorption by the kidneys, blood volume increases.

14 As blood volume increases, blood pressure increases to normal.

15 The second target tissue of angiotensin II is smooth muscle in the walls of arterioles, which responds by contracting to produce vasoconstriction. Vasoconstriction of arterioles also increases blood pressure and thus helps raise blood pressure to normal.

16 A second mechanism for the control of aldosterone secretion involves the blood K^+ level. An increase in the K^+ concentration of blood (and thus of interstitial fluid) directly stimulates aldosterone secretion by the adrenal cortex, causing the kidneys to eliminate excess K^+. A decrease in the blood K^+ level has the opposite effect.

Glucocorticoids

The glucocorticoids, which regulate metabolism and resistance to stress, include **cortisol (hydrocortisone), corticosterone,** and **cortisone.** Of the three, cortisol is the most abundant and is responsible for about 95% of glucocorticoid activity. Glucocorticoids have the following effects:

1. *Breakdown of proteins.* Glucocorticoids increase the rate of protein breakdown, mainly in muscle fibers, and thus increase the liberation of amino acids into the bloodstream. The amino acids may be used by liver cells for synthesis of new plasma proteins (including the enzymes needed for metabolic reactions), or they may be used by other cells for ATP production.

Figure 18.16 Regulation of aldosterone secretion by the renin–angiotensin pathway.

🔑 **Aldosterone helps regulate blood volume, blood pressure, and levels of Na⁺, K⁺, and H⁺ in the blood.**

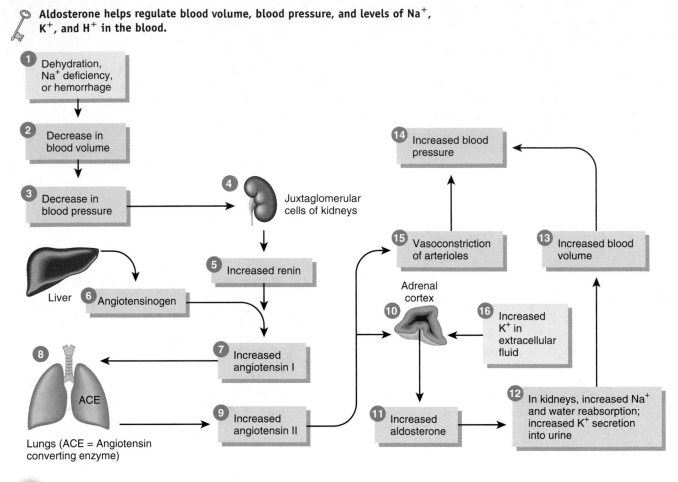

Q In what two ways can angiotensin II increase blood pressure, and what are its target tissues in each case?

2. *Formation of glucose.* Liver cells also may convert certain amino acids or lactate (lactic acid) to glucose. This conversion of a substance other than glycogen or another monosaccharide into glucose is called **gluconeogenesis** (gloo′-ko-nē′-ō-JEN-e-sis).

3. *Lipolysis.* Glucocorticoids stimulate **lipolysis,** the breakdown of triglycerides and release of fatty acids from adipose tissue.

4. *Resistance to stress.* Glucocorticoids work in many ways to provide resistance to stress. Additional glucose provides tissues with a ready source of ATP to combat a range of stresses, including exercise, fasting, fright, temperature extremes, high altitude, bleeding, infection, surgery, trauma, and disease. Because glucocorticoids also make blood vessels more sensitive to other mediators that cause vasoconstriction, they raise blood pressure. This effect is an advantage if the stress happens to be significant blood loss, which makes blood pressure fall.

5. *Anti-inflammatory effects.* Glucocorticoids are anti-inflammatory compounds that inhibit cells participating in inflammatory responses. Glucocorticoids (a) reduce the number of mast cells, thereby reducing release of histamine; (b) stabilize lysosomal membranes, thereby slowing release of destructive enzymes; (c) decrease blood capillary permeability; and (d) depress phagocytosis. Unfortunately, they also retard connective tissue repair, and as a result they slow wound healing. Although high doses can cause severe mental disturbances, glucocorticoids are very useful in the treatment of chronic inflammatory disorders such as rheumatoid arthritis.

6. *Depression of immune responses.* High doses of glucocorticoids depress immune responses. For this reason, glucocorticoids are taken by organ transplant recipients to retard tissue rejection by the immune system.

The control of glucocorticoid secretion is via a typical negative feedback system (Figure 18.17). Low blood levels of glucocorticoids, mainly cortisol, stimulate neurosecretory cells in the hypothalamus to secrete **corticotropin releasing hormone (CRH).** CRH coupled with a low level of cortisol promotes the release of ACTH from the anterior pituitary.

ACTH is carried by the blood to the adrenal cortex, where it stimulates glucocorticoid secretion. The discussion of stress at the end of the chapter describes how the hypothalamus also increases CRH release in response to a variety of physical and emotional stresses.

Androgens

In both males and females, the adrenal cortex secretes small amounts of androgens, which are also released in much greater quantity by the testes in males. The major androgen secreted by the adrenal gland is **dehydroepiandrosterone** (**DHEA;** dē-hī-drō-ep′-ē-an-DROS-ter-ōn). In adult males the amount of androgens secreted by the adrenal gland is usually so low that their effects are insignificant. In females, however, androgens from the adrenal glands play important roles: They contribute to libido (sex drive) and also are converted into estrogens (feminizing sex steroids) by other body tissues. After menopause, when ovarian secretion of estrogens ceases, the small amount of estrogens present comes from conversion of adrenal androgens. Adrenal androgens also stimulate growth of axillary and pubic hair in boys and girls and contribute to the prepubertal growth spurt. Although control of adrenal androgen secretion is not fully understood, the main hormone that stimulates its secretion is ACTH.

Adrenal Medulla

The adrenal medulla consists of hormone-producing cells, called **chromaffin cells** (krō-MAF-in; *chrom-* = color; *-affin* = affinity for; see Figure 18.15b), which surround large blood vessels. Chromaffin cells receive direct innervation from preganglionic neurons of the sympathetic division of the autonomic nervous system (ANS) and develop from the same embryonic tissue as all other sympathetic postganglionic cells. Thus, they are sympathetic postganglionic cells that are specialized to secrete hormones instead of a neurotransmitter. Because the ANS controls the chromaffin cells directly, hormone release can occur very quickly.

The two principal hormones synthesized by the adrenal medulla are **epinephrine** and **norepinephrine (NE),** also called adrenaline and noradrenaline, respectively. Epinephrine constitutes about 80% of the total secretion of the gland. Both hormones are **sympathomimetic** (sim′-pa-thō-mi-MET-ik)—their effects mimic those brought about by the sympathetic division of the ANS. To a large extent, they are responsible for the fight-or-flight response. Like the glucocorticoids of the adrenal cortex, these hormones help resist stress. Unlike the hormones of the adrenal cortex, however, the medullary hormones are not essential for life. By increasing heart rate and force of contraction, epinephrine and norepinephrine increase the pumping output of the heart, which increases blood pressure. They also increase blood flow to the heart, liver, skeletal muscles, and adipose tissue; dilate airways to the lungs; and increase blood levels of glucose and fatty acids.

Figure 18.17 Negative feedback regulation of glucocorticoid secretion.

🔑 **A high level of CRH and a low level of glucocorticoids promote the release of ACTH, which stimulates glucocorticoid secretion by the adrenal cortex.**

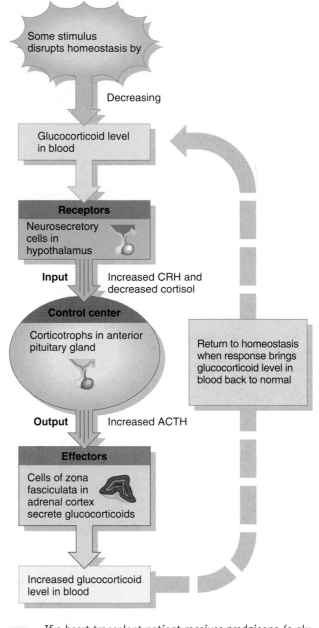

Q If a heart transplant patient receives prednisone (a glucocorticoid) to help prevent rejection of the transplanted tissue, will blood levels of ACTH and CRH be high or low? Explain.

In stressful situations and during exercise, impulses received by the hypothalamus are conveyed to sympathetic preganglionic neurons, which release the neurotransmitter acetylcholine. Acetylcholine causes the chromaffin cells to increase their output of epinephrine and norepinephrine.

Table 18.8 Summary of Adrenal Gland Hormones

HORMONE	PRINCIPAL ACTIONS	CONTROL OF SECRETION
Adrenal cortical hormones		
Mineralocorticoids (mainly **aldosterone**) from zona glomerulosa cells	Increase blood levels of Na$^+$ and water and decrease blood level of K$^+$.	Increased blood K$^+$ level and angiotensin II stimulate secretion.
Glucocorticoids (mainly **cortisol**) from zona fasciculata cells	Increase protein breakdown (except in liver), stimulate gluconeogenesis and lipolysis, provide resistance to stress, dampen inflammation, and depress immune responses.	ACTH stimulates release; corticotropin releasing hormone (CRH) promotes ACTH secretion in response to stress and low blood levels of glucocorticoids.
Androgens (mainly **dehydroepiandrosterone** or **DHEA**) from zona reticularis cells	Assist in early growth of axillary and pubic hair in both sexes; in females, contribute to libido and are source of estrogens after menopause.	ACTH stimulates secretion.
—Adrenal cortex		
Adrenal medullary hormones		
Epinephrine and **norepinephrine** from chromaffin cells.	Produce effects that enhance those of the sympathetic division of the autonomic nervous system (ANS) during stress.	Sympathetic preganglionic neurons release acetylcholine, which stimulates secretion.
— Adrenal medulla		

Hypoglycemia also stimulates medullary secretion of epinephrine and norepinephrine.

Table 18.8 summarizes the hormones produced by the adrenal glands, their principal actions, and control of secretion.

1. Compare the adrenal cortex and the adrenal medulla with regard to location and histology.
2. How is secretion of adrenal cortex hormones regulated?
3. Describe the relationship of the adrenal medulla to the autonomic nervous system.

PANCREAS

OBJECTIVE

• *Describe the location, histology, hormones, and functions of the pancreas.*

The **pancreas** (*pan-* = all; *-creas* = flesh) is both an endocrine gland and an exocrine gland. We discuss its endocrine functions here and its exocrine functions in Chapter 24 on the digestive system. The pancreas is a flattened organ that measures about 12.5–15 cm (4.5–6 in.) in length. It is located posterior and slightly inferior to the stomach and consists of a head, a body, and a tail (Figure 18.18a). Roughly 99% of the pancreatic cells are arranged in clusters called **acini** (singular is acinus); these cells produce digestive enzymes, which flow into the gastrointestinal tract through a network of ducts. Scattered among the exocrine acini are 1–2 million tiny clusters of endocrine tissue called **pancreatic islets** or **islets of Langerhans** (LAHNG-er-hanz; Figure 18.19b, c). Abundant capillaries serve both the exocrine and endocrine portions of the pancreas.

Cell Types in the Pancreatic Islets

Each pancreatic islet includes four types of hormone-secreting cells: (1) **alpha** or **A cells** constitute about 20% of pancreatic islet cells and secrete **glucagon** (GLOO-ka-gon); (2) **beta** or **B cells** constitute about 70% of pancreatic islet cells and secrete **insulin** (IN-soo-lin); (3) **delta** or **D cells** constitute about 5% of pancreatic islet cells and secrete **somatostatin** (identical to growth hormone inhibiting hormone secreted by the hypothalamus); and (4) **F cells** constitute the remainder of pancreatic islet cells and secrete **pancreatic polypeptide.**

The interactions of the four pancreatic hormones are complex and not completely understood. Whereas glucagon raises blood glucose level, insulin lowers it. Somatostatin acts in a paracrine manner to inhibit both insulin and glucagon release from neighboring beta and alpha cells, and it is also thought to slow absorption of nutrients from the gastrointestinal tract. Pancreatic polypeptide inhibits somatostatin secretion, gallbladder contraction, and secretion of digestive enzymes by the pancreas.

Regulation of Glucagon and Insulin Secretion

The principal action of glucagon is to increase blood glucose level when it falls below normal. Insulin, on the other hand, helps lower blood glucose level when it is too

Figure 18.18 Location, blood supply, and histology of the pancreas. (See Tortora, *A Photographic Atlas of the Human Body,* Figure 10.6a)

🔑 **Pancreatic hormones regulate blood glucose level.**

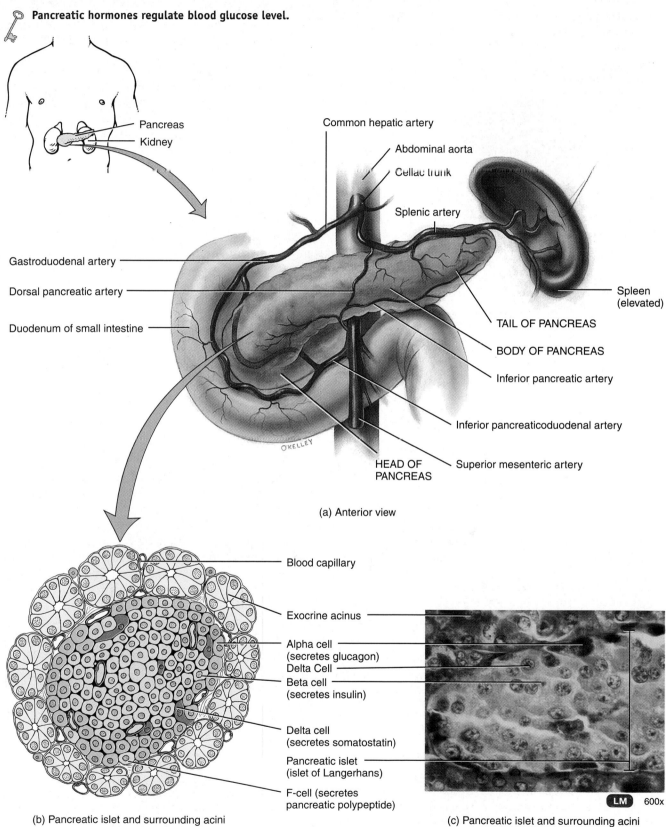

Pancreas
Kidney

Common hepatic artery
Abdominal aorta
Celiac trunk
Splenic artery

Gastroduodenal artery
Dorsal pancreatic artery
Duodenum of small intestine

Spleen (elevated)
TAIL OF PANCREAS
BODY OF PANCREAS
Inferior pancreatic artery

Inferior pancreaticoduodenal artery

O'KELLEY

HEAD OF PANCREAS
Superior mesenteric artery

(a) Anterior view

Blood capillary

Exocrine acinus

Alpha cell (secretes glucagon)
Delta Cell
Beta cell (secretes insulin)

Delta cell (secretes somatostatin)

Pancreatic islet (islet of Langerhans)

F-cell (secretes pancreatic polypeptide)

LM 600x

(b) Pancreatic islet and surrounding acini

(c) Pancreatic islet and surrounding acini

Q Is the pancreas an exocrine gland or an endocrine gland?

Figure 18.19 Negative feedback regulation of the secretion of glucagon (blue arrows) and insulin (orange arrows).

Low blood glucose stimulates release of glucagon, whereas high blood glucose stimulates secretion of insulin.

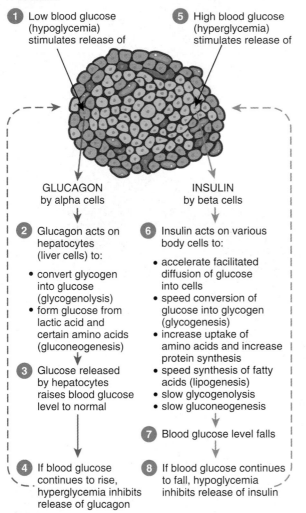

Why is glucagon sometimes called an "anti-insulin" hormone?

---

④ If blood glucose continues to rise, high blood glucose level (hyperglycemia) inhibits release of glucagon (negative feedback).

⑤ At the same time, however, high blood glucose (hyperglycemia) stimulates release of insulin from beta cells of the pancreatic islets.

⑥ Insulin acts on various cells in the body to accelerate facilitated diffusion of glucose into cells, especially skeletal muscle fibers; to speed conversion of glucose into glycogen (glycogenesis); to increase uptake of amino acids by cells and to increase protein synthesis; to speed synthesis of fatty acids (lipogenesis); to slow glycogenolysis; and to slow gluconeogenesis.

⑦ As a result, blood glucose level falls.

⑧ If blood glucose level continues to fall, low blood glucose inhibits release of insulin (negative feedback).

Although rising blood glucose is the most important stimulator of insulin release, several hormones and neurotransmitters also regulate the release of insulin and glucagon. Insulin secretion is stimulated by (1) acetylcholine, the neurotransmitter liberated from axon terminals of parasympathetic vagus nerve fibers that innervate the pancreatic islets; (2) the amino acids arginine and leucine; (3) glucagon; and (4) glucose-dependent insulinotropic peptide (GIP)*, a hormone released by enteroendocrine cells of the small intestine in response to the presence of glucose in the gastrointestinal tract. Thus, digestion and absorption of food containing both carbohydrates and proteins provide strong stimulation for insulin release.

Increased activity of the sympathetic division of the ANS, as occurs during exercise, enhances glucagon release. Also, a rise in blood amino acids stimulates glucagon secretion if blood glucose level is low, which could occur after a meal that contained mainly protein. Whereas glucagon stimulates insulin release, insulin suppresses glucagon secretion. Thus, as blood glucose level declines and less insulin is secreted, the alpha cells are released from the inhibitory effect of insulin and secrete more glucagon. Indirectly, human growth hormone (hGH) and adrenocorticotropic hormone (ACTH) stimulate secretion of insulin because they act to elevate blood glucose.

Table 18.9 summarizes the hormones produced by the pancreas, their principal actions, and control of secretion.

---

1. How are blood levels of glucagon and insulin controlled?

2. Compare the effects on secretion of insulin and glucagon of exercise versus eating a carbohydrate- and protein-rich meal.

---

*GIP—previously called gastric inhibitory peptide—was renamed because at physiological concentration its inhibitory effect on stomach function is negligible.

---

high. The level of blood glucose controls secretion of glucagon and insulin via negative feedback systems (Figure 18.19):

① Low blood glucose level (hypoglycemia) stimulates release of glucagon from alpha cells of the pancreatic islets.

② Glucagon acts on hepatocytes (liver cells) to accelerate the conversion of glycogen into glucose (glycogenolysis) and to promote formation of glucose from lactic acid (lactate) and certain amino acids (gluconeogenesis).

③ As a result, hepatocytes release glucose into the blood more rapidly, and blood glucose level rises.

Table 18.9 Summary of Hormones Produced by the Pancreas

HORMONE	PRINCIPAL ACTIONS	CONTROL OF SECRETION
Glucagon from alpha or A cells of pancreatic islets 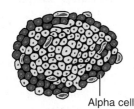 Alpha cell	Raises blood glucose level by accelerating breakdown of glycogen into glucose in liver (glycogenolysis), converting other nutrients into glucose in liver (gluconeogenesis), and releasing glucose into the blood.	Decreased blood level of glucose, exercise, and mainly protein meals stimulate secretion; somatostatin and insulin inhibit secretion.
Insulin from beta or B cells of pancreatic islets 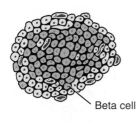 Beta cell	Lowers blood glucose level by accelerating transport of glucose into cells, converting glucose into glycogen (glycogenesis), and decreasing glycogenolysis and gluconeogenesis; also increases lipogenesis and stimulates protein synthesis.	Increased blood level of glucose, acetylcholine (released by parasympathetic vagus nerve fibers), arginine and leucine (two amino acids), glucagon, GIP, hGH, and ACTH stimulate secretion; somatostatin inhibits secretion.
Somatostatin from delta or D cells of pancreatic islets 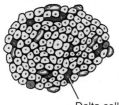 Delta cell	Inhibits secretion of insulin and glucagon and slows absorption of nutrients from the gastrointestinal tract.	Pancreatic polypeptide inhibits secretion.
Pancreatic polypeptide from F cells of pancreatic islets 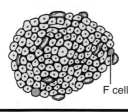 F cell	Inhibits somatostatin secretion, gallbladder contraction, and secretion of pancreatic digestive enzymes.	Meals containing protein, fasting, exercise, and acute hypoglycemia stimulate secretion; somatostatin and elevated blood glucose level inhibit secretion.

OVARIES AND TESTES

OBJECTIVE

• *Describe the location, hormones, and functions of the male and female gonads.*

The female gonads, called the **ovaries,** are paired oval bodies located in the pelvic cavity. The ovaries produce female sex hormones called **estrogens** and **progesterone.** Along with the gonadotropic hormones of the pituitary gland, the sex hormones regulate the female reproductive cycle, maintain pregnancy, and prepare the mammary glands for lactation. These hormones are also responsible for the development and maintenance of feminine secondary sex characteristics. The ovaries also produce **inhibin,** a protein hormone that inhibits secretion of follicle-stimulating hormone (FSH). During pregnancy, the ovaries and placenta produce a peptide hormone called **relaxin,** which increases the flexibility of the pubic symphysis during pregnancy and helps dilate the uterine cervix during labor and delivery. These actions help ease the baby's passage by enlarging the birth canal.

The male has two oval gonads, called **testes,** that produce **testosterone,** the primary androgen. Testosterone regulates production of sperm and stimulates the development

Table 18.10 Summary of Hormones of the Ovaries and Testes

HORMONE	PRINCIPAL ACTIONS
Ovarian hormones	
Estrogens and **progesterone**	Together with gonadotropic hormones of the anterior pituitary gland, regulate the female reproductive cycle, maintain pregnancy, prepare the mammary glands for lactation, regulate oogenesis, and promote development and maintenance of feminine secondary sex characteristics.
Relaxin	Increases flexibility of pubic symphysis during pregnancy and helps dilate uterine cervix during labor and delivery.
Inhibin	Inhibits secretion of FSH from anterior pituitary gland.

Ovaries

Testicular hormones	
Testosterone	Stimulates descent of testes before birth, regulates spermatogenesis, and promotes development and maintenance of masculine secondary sex characteristics.
Inhibin	Inhibits secretion of FSH from anterior pituitary gland.

Testes

and maintenance of masculine secondary sex characteristics such as beard growth. The testes also produce inhibin, which inhibits secretion of FSH. The specific roles of gonadotropic hormones and sex hormones are discussed in Chapter 28.

Table 18.10 summarizes the hormones produced by the ovaries and testes and their principal actions.

| **1.** Explain why the ovaries and testes are endocrine glands.

PINEAL GLAND

OBJECTIVE

• *Describe the location, histology, hormone, and functions of the pineal gland.*

The **pineal gland** (PĪN-ē-al; = pinecone shape) is a small endocrine gland attached to the roof of the third ven-

tricle of the brain at the midline (see Figure 18.1). It is part of the epithalamus, positioned between the two superior colliculi, and it weighs 0.1–0.2 g. The gland, which is covered by a capsule formed by the pia mater, consists of masses of neuroglia and secretory cells called **pinealocytes** (pin-ē-AL-ō-sīts). Sympathetic postganglionic fibers from the superior cervical ganglion terminate in the pineal gland.

Although many anatomical features of the pineal gland have been known for years, its physiological role is still unclear. One hormone secreted by the pineal gland is **melatonin,** an amine hormone derived from serotonin. More melatonin is released in darkness, and less melatonin is liberated in strong sunlight, according to the following sequence (Figure 18.20):

1 Light enters the eyes, strikes the retina, and stimulates photoreceptors.

2 Retinal neurons activated by photoreceptors transmit impulses to the suprachiasmatic nucleus of the hypothalamus.

3 From the suprachiasmatic nucleus, nerve impulses are transmitted to the superior cervical ganglion.

4 Sympathetic postganglionic fibers from the superior cervical ganglion extend to the pineal gland and form synaptic contacts with cells of the pineal gland.

5 In darkness, retinal neurons transmit fewer impulses to the pineal gland by way of the suprachiasmatic nucleus and superior cervical ganglion.

6 Lack of norepinephrine stimulates secretion of melatonin by cells of the pineal gland, and the result is sleepiness.

7 In bright light, norepinephrine released by the sympathetic fibers inhibits secretion of melatonin by cells of the pineal gland.

8 Inhibition of melatonin secretion results in lack of sleepiness. Thus, the release of melatonin is governed by the circadian (daily) dark-light cycle.

Melatonin contributes to setting the body's biological clock, which is controlled from the suprachiasmatic nucleus. During sleep, plasma levels of melatonin increase tenfold and then decline to a low level again before awakening. Small doses of melatonin given orally can induce sleep and reset daily rhythms, which might benefit workers whose shifts alternate between daylight and nighttime hours. Melatonin also is a potent antioxidant that may provide some protection against damaging oxygen free radicals. In animals that breed during specific seasons, melatonin inhibits reproductive functions. Whether melatonin influences human reproductive function, however, is still unclear. Melatonin levels are higher in children and decline with age into adulthood, but there is no evidence that changes in melatonin secretion correlate with the onset of puberty and sexual maturation. Nevertheless, because melatonin causes atrophy of the gonads in several animal species, the possibility of adverse effects on human reproduction must be studied before its use to reset daily rhythms can be recommended.

Figure 18.20 Pathway whereby light slows release of melatonin from the pineal gland.

🔑 **In darkness, secretion of melatonin by the pineal gland promotes sleepiness.**

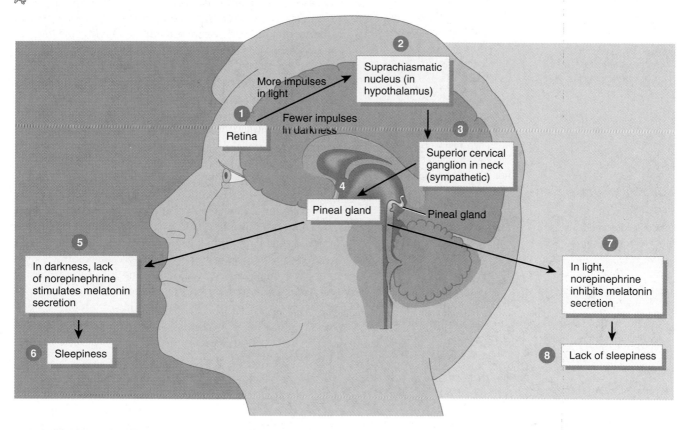

Q To reduce the effects of jet lag by resetting her biological clock, should a New Yorker arriving in London at 9 A.M. (3 A.M. NY time) expose herself to 3 hours of bright light before leaving or after arriving?

CLINICAL APPLICATION
Seasonal Affective Disorder and Jet Lag

Seasonal affective disorder (SAD) is a type of depression that afflicts some people during the winter months, when day length is short. It is thought to be due, in part, to overproduction of melatonin. Bright light therapy—repeated doses of several hours of exposure to artificial light as bright as sunlight—provides relief for some people. Three to six hours of exposure to bright light also appears to speed recovery from jet lag, the fatigue suffered by travelers who cross several time zones. ■

1. What are the relationships among light exposure, melatonin secretion, and sleep?

THYMUS GLAND

Because of its role in immunity, the details of the structure and functions of the **thymus gland** are discussed in Chapter 22, which examines the lymphatic system and im-

munity. The hormones produced by the thymus gland, called **thymosin, thymic humoral factor (THF), thymic factor (TF),** and **thymopoietin,** promote the proliferation and maturation of T cells (a type of white blood cell), which destroy microbes and foreign substances, and may retard the aging process.

MISCELLANEOUS HORMONES

Hormones from Other Endocrine Cells

OBJECTIVE

• *List the hormones secreted by cells in tissues and organs other than endocrine glands, and describe their functions.*

Some tissues and organs other than those normally classified as endocrine glands also contain endocrine cells and thus secrete hormones. These hormones and their actions are summarized in Table 18.11.

Table 18.11 Summary of Hormones Produced by Organs and Tissues that Contain Endocrine Cells

HORMONES	PRINCIPAL ACTIONS
Gastrointestinal tract	
Gastrin	Promotes secretion of gastric juice and increases motility of the stomach.
Glucose-dependent insulinotropic peptide (GIP)	Stimulates release of insulin by pancreatic beta cells.
Secretin	Stimulates secretion of pancreatic juice and bile.
Cholecystokinin (CCK)	Stimulates secretion of pancreatic juice, regulates release of bile from the gallbladder, and brings about a feeling of fullness after eating.
Placenta	
Human chorionic gonadotropin (hCG)	Stimulates the corpus luteum in the ovary to continue the production of estrogens and progesterone to maintain pregnancy.
Estrogens and progesterone	Maintain pregnancy and prepare mammary glands to secrete milk.
Human chorionic somatomammotropin (hCS)	Stimulates the development of the mammary glands for lactation.
Kidneys	
Erythropoietin (EPO)	Increases rate of red blood cell production.
Calcitriol[a] (active form of vitamin D)	Aids in the absorption of dietary calcium and phosphorus.
Heart	
Atrial nutriuretic peptide (ANP)	Decreases blood pressure.
Adipose tissue	
Leptin	Suppresses appetite and may have a permissive effect on activity of GnRH and gonadotropins.

[a]Synthesis begins in the skin, continues in the liver, and ends in the kidneys.

Eicosanoids

OBJECTIVE

• *Explain the actions of eicosanoids.*

Two families of eicosanoid molecules—the **prostaglandins** (pros'-ta-GLAN-dins), or **PGs,** and the **leukotrienes** (loo-kō-TRĪ-ēns), or **LTs**—act as local hormones in most tissues of the body. They are synthesized by clipping a 20-carbon fatty acid called **arachidonic acid** from membrane phospholipid molecules. From arachidonic acid, different enzymatic reactions produce PGs or LTs. **Thromboxane (TX)** is a modified PG that constricts blood vessels and promotes platelet activation. In response to chemical and mechanical stimuli, virtually all body cells except red blood cells release these local hormones, which act mainly as potent paracrines and autocrines and appear in the blood only in minute quantities. They are present only briefly due to rapid inactivation.

To exert their effects, eicosanoids bind to receptors on target-cell plasma membranes and stimulate or inhibit the synthesis of second messengers such as cyclic AMP. Leukotrienes stimulate chemotaxis of white blood cells and mediate inflammation. The importance of prostaglandins in both normal physiology and pathology is apparent in their broad range of biological activities: They alter smooth muscle contraction, glandular secretions, blood flow, reproductive processes, platelet function, respiration, nerve impulse transmission, fat metabolism, and immune responses. They also have roles in inflammation, in promoting fever, and in intensifying pain. What has intrigued investigators even more than the physiological role of prostaglandins is their potential therapeutic effects, including lowering or raising blood pressure, reducing gastric secretion, dilating or constricting airways in the lungs, stimulating or inhibiting blood platelet aggregation, contracting or relaxing intestinal and uterine smooth muscle, inducing labor in pregnancy, and promoting diuresis.

CLINICAL APPLICATION
Nonsteroidal Anti-inflammatory Drugs

In 1971, scientists solved the long-standing puzzle of how aspirin works. Aspirin and related **nonsteroidal anti-inflammatory drugs (NSAIDs),** such as ibuprofen (Motrin), inhibit a key enzyme in prostaglandin synthesis without affecting synthesis of leukotrienes. They are used to treat a wide variety of inflammatory disorders, from rheumatoid arthritis to tennis elbow. The success of NSAIDs in reducing fever, pain, and inflammation points to prostaglandins as factors in these woes. ■

Growth Factors

OBJECTIVE

• *List six important growth factors.*

Several of the hormones we've described—insulinlike growth factor, thymosin, insulin, thyroid hormones, human growth hormone, prolactin, and erythropoietin—stimulate cell growth and division. In addition, several more recently discovered hormones called **growth factors** play important roles in tissue development, growth, and repair. Growth factors are *mitogenic* substances—they cause growth by stimulating cell division. Many growth factors act locally, as autocrines or paracrines. A summary of sources and actions of six important growth factors is presented in Table 18.12.

1. List the hormones secreted by the gastrointestinal tract, placenta, kidneys, skin, adipose tissue, and heart.

2. What are some functions of prostaglandins and leukotrienes?

3. What are growth factors? List several and explain their actions.

STRESS AND THE GENERAL ADAPTATION SYNDROME

OBJECTIVE

• *Describe the general adaptation syndrome.*

Homeostatic mechanisms *attempt* to counteract the everyday stresses of living. When they are successful, the internal environment remains within normal physiological limits of chemistry, temperature, and pressure. If stress is extreme, unusual, or long-lasting, however, the normal mechanisms may not produce sufficient results. In 1936, Hans Selye, a pioneer in stress research, showed that a variety of stressful conditions or noxious agents elicit a similar sequence of bodily changes. This wide-ranging set of changes is now called the **stress response** or **general adaptation syndrome (GAS).** Unlike homeostatic mechanisms, the GAS does not maintain the normal internal environment. It resets the levels of various controlled conditions to prepare the body to meet an emergency. For instance, blood pressure and blood glucose are raised above normal levels.

It is impossible to remove all stress from our everyday lives. Some stress, called **eustress,** prepares us to meet certain challenges and thus is productive. Other stress, called **distress,** is harmful. Distress may lower resistance to infection by temporarily inhibiting certain components of the immune system.

Any stimulus that produces a stress response is called a **stressor.** A stressor may be almost any disturbance—heat or cold, environmental poisons, toxins given off by bacteria during a raging infection, heavy bleeding from a wound or surgery, or a strong emotional reaction. Stressors may be pleasant or unpleasant, and they vary among people and even within the same person at different times.

Stages of the General Adaptation Syndrome

If exposure to a stressor is prolonged, the response may occur in three stages: (1) an initial *alarm reaction,* (2) a slower *resistance reaction,* and eventually (3) *exhaustion.*

The Alarm Reaction

The **alarm reaction** or **fight-or-flight response** is a complex of reactions initiated by hypothalamic stimulation of the sympathetic division of the ANS and the adrenal medulla (Figure 18.21a). If, for example, a person is being attacked by a dog, the response is rapid, mobilizing the body's resources for immediate physical activity. In essence, the alarm reaction brings huge amounts of glucose and oxygen

Table 18.12 Summary of Selected Growth Factors

GROWTH FACTOR	COMMENT
Epidermal growth factor (EGF)	Produced in submaxillary (salivary) glands; stimulates proliferation of epithelial cells, fibroblasts, neurons, and astrocytes; suppresses some cancer cells and secretion of gastric juice by the stomach.
Platelet-derived growth factor (PDGF)	Produced in blood platelets; stimulates proliferation of neuroglia, smooth muscle fibers, and fibroblasts, appears to have a role in wound healing; may contribute to the development of atherosclerosis.
Fibroblast growth factor (FGF)	Found in pituitary gland and brain; stimulates proliferation of many cells derived from embryonic mesoderm (fibroblasts, adrenocortical cells, smooth muscle fibers, chondrocytes, and endothelial cells); also stimulates formation of new blood vessels (angiogenesis).
Nerve growth factor (NGF)	Produced in submaxillary (salivary) glands and hippocampus of brain; stimulates the growth of ganglia in embryonic life, maintains sympathetic nervous system; stimulates hypertrophy and differentiation of neurons.
Tumor angiogenesis factors (TAFs)	Produced by normal and tumor cells; stimulate growth of new capillaries, organ regeneration, and wound healing.
Transforming growth factors (TGFs)	Produced by various cells as separate molecules called TGF-alpha and TGF-beta. TGF-alpha has activities similar to epidermal growth factor, and TGF-beta inhibits proliferation of many cell types.

to the organs that are most active in warding off danger: the brain, which must become highly alert; the skeletal muscles, which may have to fight off an attacker or flee; and the heart, which must work vigorously to pump enough blood to the brain and muscles. Overall, the stress responses of the alarm stage increase circulation and promote ATP synthesis. During the alarm reaction, nonessential body functions such as digestive, urinary, and reproductive activities are inhibited. If the stress is great enough, the body's mechanisms may not be able to cope, and death can result.

Figure 18.21 Responses to stressors during the general adaptation syndrome (GAS). Red arrows (hormonal responses) and green arrows (neural responses) in (a) indicate immediate alarm ("fight-or-flight") reactions; black arrows in (b) indicate long-term resistance reactions.

🔑 **Stressors stimulate the hypothalamus to initiate the GAS through the alarm reaction and the resistance reaction.**

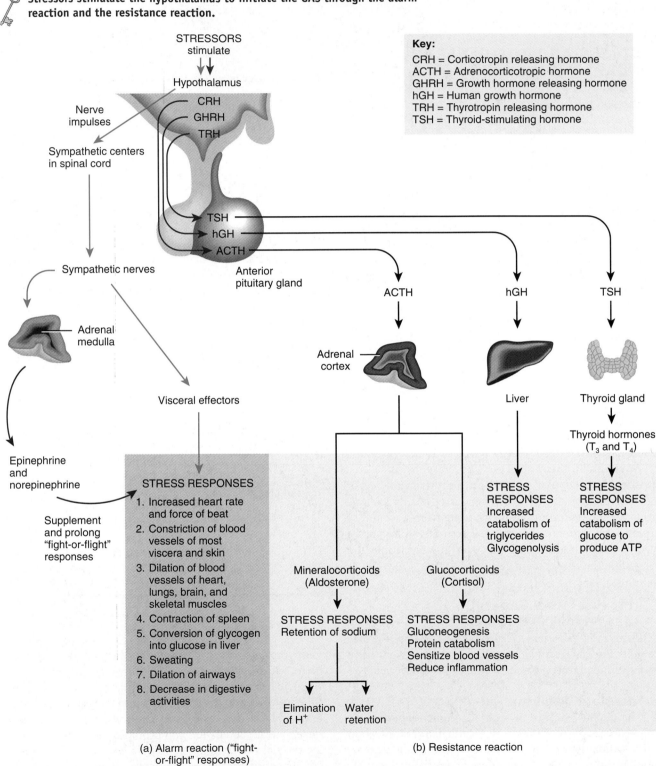

(a) Alarm reaction ("fight-or-flight" responses)

(b) Resistance reaction

Q What is the basic difference between the GAS and homeostasis?

The Resistance Reaction

The second stage in the stress response is the **resistance reaction** (Figure 18.21b). Unlike the short-lived alarm reaction, which is initiated by nerve impulses from the hypothalamus, the resistance reaction is initiated in large part by hypothalamic releasing hormones and is a long-term reaction. The hormones involved are corticotropin releasing hormone (CRH), growth hormone releasing hormone (GHRH), and thyrotropin releasing hormone (TRH).

CRH stimulates the anterior pituitary gland to increase secretion of ACTH, which stimulates the adrenal cortex to secrete more mineralocortoids (aldosterone). Aldosterone acts to conserve Na^+ and eliminate H^+. Thus, a lowering of body pH is prevented during stress. Na^+ retention also leads to water retention by the kidneys, thus maintaining the high blood pressure of the alarm reaction. Water retention also would help preserve body fluid volume if fluid has been lost through severe bleeding. ACTH also stimulates the adrenal cortex to secrete more glucocorticoids (cortisol). Cortisol in turn stimulates the conversion of noncarbohydrates into glucose (gluconeogenesis) and enhances protein catabolism (breakdown). It also makes blood vessels more sensitive to stimuli that bring about their constriction. This response counteracts a drop in blood pressure caused by bleeding. In addition, cortisol reduces inflammation and prevents it from becoming disruptive rather than protective. Because cortisol also discourages formation of new connective tissue, wound healing is slow during a prolonged resistance reaction.

GHRH causes the anterior pituitary gland to secrete human growth hormone (hGH). hGH stimulates the catabolism of triglycerides and the conversion of glycogen to glucose (glycogenolysis).

TRH causes the anterior pituitary gland to secrete thyroid-stimulating hormone (TSH). TSH stimulates the thyroid gland to secrete the thyroid hormones T_3 and T_4, which stimulate the increased catabolism of glucose to produce ATP. The combined actions of hGH and TSH thereby supply additional ATP for metabolically active cells.

The resistance stage of the stress response allows the body to continue fighting a stressor long after the alarm reaction dissipates. It also provides the ATP, enzymes, and circulatory changes needed to meet emotional crises, perform strenuous tasks, or resist the threat of bleeding to death. During the resistance reaction, blood chemistry returns to nearly normal. The cells use glucose at the same rate it enters the bloodstream, and thus blood glucose level returns to normal.

Generally, the resistance stage is successful in seeing us through a stressful episode, and our bodies then return to normal. Occasionally, however, the resistance stage fails to combat the stressor, and the GAS moves into the state of exhaustion.

Exhaustion

The resources of the body may eventually become depleted such that the resistance stage cannot be sustained, and a state termed **exhaustion** ensues. Prolonged exposure to high levels of cortisol and other resistance reaction hormones causes wasting of muscle, suppression of the immune system, ulceration of the gastrointestinal tract, and failure of the pancreatic beta cells. In addition, pathological changes may occur because resistance reactions persist after the stressor has been removed.

Stress and Disease

Although the exact role of stress in human diseases is not known, it is clear that stress can lead to particular diseases by temporarily inhibiting certain components of the immune system. Stress-related disorders include gastritis, ulcerative colitis, irritable bowel syndrome, hypertension, asthma, rheumatoid arthritis (RA), migraine headaches, anxiety, and depression. It has also been shown that people under stress are at a greater risk of developing chronic disease or dying prematurely.

Interleukin-1, a cytokine secreted by macrophages of the immune system (see page 757), is an important link between stress and immunity. In response to infection, inflammation, and other stressors, interleukin-1 stimulates production of immune substances by the liver and other tissues, increases the number of circulating neutrophils (phagocytic white blood cells), activates other cells that participate in immunity, and induces fever. Together, these responses result in a powerful immune response. However, interleukin-1 also stimulates the secretion of corticotropin (ACTH). ACTH, in turn, stimulates the production of cortisol, which not only provides resistance to stress and inflammation but also suppresses further production of interleukin-1. Thus, the immune system turns on the stress response. This negative feedback system keeps the immune response in check once it has accomplished its goal. Because glucocorticoids, such as cortisol, suppress certain aspects of the immune response, they are used as immunosuppressive drugs after organ transplantation.

---

1. How do homeostatic responses differ from stress responses?
2. What is the central role of the hypothalamus during stress?
3. Outline the body's reactions during the alarm reaction, the resistance reaction, and exhaustion.
4. Explain how stress and immunity are related.

DEVELOPMENTAL ANATOMY OF THE ENDOCRINE SYSTEM

OBJECTIVE

• *Describe the development of endocrine glands.*

The development of the endocrine system is not as localized as the development of other systems because endocrine organs develop in widely separated parts of the embryo.

Figure 18.22 Development of the endocrine system.

🔑 **Glands of the endocrine system develop from all three primary germ layers.**

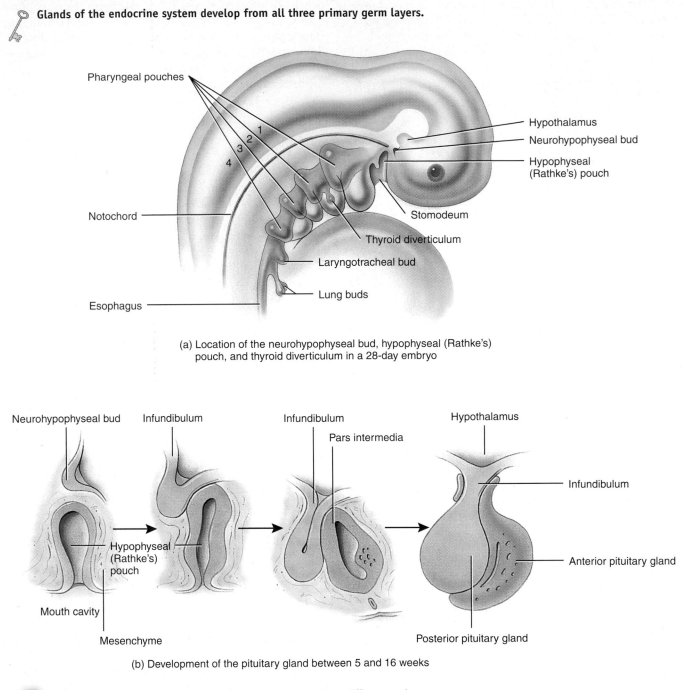

(a) Location of the neurohypophyseal bud, hypophyseal (Rathke's) pouch, and thyroid diverticulum in a 28-day embryo

(b) Development of the pituitary gland between 5 and 16 weeks

Q Which two endocrine glands develop from tissues with two different embryological origins?

The *pituitary gland (hypophysis)* originates from two different regions of the ectoderm. The *posterior pituitary gland (neurohypophysis)* derives from an outgrowth of ectoderm called the **neurohypophyseal bud,** located on the floor of the hypothalamus (Figure 18.22a). The *infundibulum,* also an outgrowth of the neurohypophyseal bud, connects the posterior pituitary gland to the hypothalamus. The *anterior pituitary gland (adenohypophysis)* is derived from an

outgrowth of ectoderm from the roof of the mouth called the **hypophyseal (Rathke's) pouch.** The pouch grows toward the neurohypophyseal bud and eventually loses its connection with the roof of the mouth.

The *thyroid gland* develops as a midventral outgrowth of endoderm, called the **thyroid diverticulum,** from the floor of the pharynx at the level of the second pair of pharyngeal pouches. The outgrowth projects inferiorly and differentiates into the right and left lateral lobes and the isthmus of the gland.

The *parathyroid glands* develop from endoderm as outgrowths from the third and fourth **pharyngeal pouches.**

The adrenal cortex and adrenal medulla have completely different embryological origins. The *adrenal cortex* is derived from intermediate mesoderm from the same region that produces the gonads. The *adrenal medulla* is ectodermal in origin and derives from the **neural crest,** which also gives rise to sympathetic ganglia and other structures of the nervous system (see Figure 14.18b on page 473).

The *pancreas* develops from two outgrowths of endoderm from the part of the **foregut** that later becomes the duodenum (see Figure 24.29 on page 463). The two outgrowths eventually fuse to form the pancreas. The origin of the ovaries and testes is discussed in the section on the reproductive system.

The *pineal gland* arises as an outgrowth between the thalamus and colliculi from ectoderm associated with the **diencephalon** (see Figure 14.19b on page 473).

The *thymus gland* arises from endoderm of the third **pharyngeal pouches.**

AGING AND THE ENDOCRINE SYSTEM

OBJECTIVE

• *Describe the effects of aging on the endocrine system.*

Although some endocrine glands shrink as we get older, their performance may or may not be compromised. With respect to the pituitary gland, production of human growth hormone decreases, which is one cause of muscle atrophy as aging proceeds. But production of gonadotropins and thyroid-stimulating hormone actually increases with age. Output of adrenocorticotropic hormone by the pituitary gland is apparently unaffected by age. The thyroid gland decreases its output of thyroxine with age, causing a decrease in metabolic rate, an increase in body fat, and hypothyroidism, which is seen more often in older people.

The thymus gland is largest in infancy. After puberty its size begins to decrease, and thymic tissue is replaced by adipose and areolar connective tissue. In older adults the thymus has atrophied significantly.

The adrenal glands contain increasingly more fibrous tissue and produce less cortisol and aldosterone. However, production of epinephrine and norepinephrine remains normal. The pancreas releases insulin more slowly with age, and receptor sensitivity to glucose declines. As a result, blood glucose levels in older people increase faster and return to normal more slowly than in younger individuals.

The ovaries experience a drastic decrease in size with age, and they no longer respond to gonadotropins. The resultant decreased output of estrogen by the ovaries leads to conditions such as osteoporosis and atherosclerosis. Although testosterone production by the testes decreases with age, the effects are not usually apparent until very old age, and many elderly males can still produce active sperm in normal numbers.

DISORDERS: HOMEOSTATIC IMBALANCES

Most disorders of the endocrine system are due to secretion of too much or too little of a given hormone. However, because hormones cannot function until they bind to their receptors, some endocrine disorders are due to receptor aberrations or defects in second-messenger systems. Because hormones are distributed in the blood to target tissues throughout the body, problems associated with endocrine dysfunction may also be widespread.

PITUITARY GLAND DISORDERS

Pituitary Dwarfism, Giantism, and Acromegaly

Hyposecretion of human growth hormone (hGH) during the growth years slows bone growth, and the epiphyseal plates close before normal height is reached. This condition is called **pituitary dwarfism.** Other organs of the body also fail to grow, and a pituitary dwarf is childlike in many physical respects. Treatment requires administration of hGH during childhood, before the epiphyseal plates close.

Hypersecretion of hGH during childhood results in **giantism,** an abnormal increase in the length of long bones. A person with this condition is unusually tall, but has approximately normal body proportions. Hypersecretion of hGH during adulthood causes **acromegaly** (ak′-rō-MEG-a-lē). Further lengthening of the long bones cannot occur because the epiphyseal plates are closed. Instead, the bones of the hands, feet, cheeks, and jaw thicken. Also, the eyelids, lips, tongue, and nose enlarge, and the skin thickens and develops furrows, especially on the forehead and soles.

Diabetes Insipidus

The most common abnormality associated with dysfunction of the posterior pituitary gland is **diabetes insipidus** (dī-a-BĒ-tēs in-SIP-i-dus; *diabetes* = overflow; *insipidus* = tasteless) or **DI.** An inability either to secrete or to respond to antidiuretic hormone (ADH) can cause diabetes insipidus. Neurogenic diabetes insipidus results from hyposecretion of ADH, usually caused by a brain tumor, head trauma, or brain surgery that damages the posterior pituitary gland and the hypothalamic paraventricular and supraoptic nuclei. In nephrogenic diabetes insipidus, the kidneys do not respond to ADH. The ADH receptors may be nonfunctional, or the kidneys may be damaged. A common symptom of both forms of the disorder is excretion of large volumes of urine, with resulting dehydration and thirst. Bed-wetting is common in afflicted children. Because so much water is lost in the urine, a person with severe diabetes insipidus may die of dehydration if deprived of water for only a day or so. Treatment of neurogenic diabetes insipidus involves hormone replacement, usually for life. Either subcutaneous injection or nasal spray application of ADH analogs is effective. Treatment of nephrogenic diabetes insipidus is more complex and depends on the nature of the kidney dysfunction. Restriction of salt in the diet and, paradoxically, the use of certain diuretic drugs, are helpful.

THYROID GLAND DISORDERS

Thyroid gland disorders affect all major body systems and are among the most common endocrine disorders. Hyposecretion of thyroid hormones during fetal life or infancy results in **cretinism** (KRĒ-tin-izm). Cretins exhibit dwarfism because the skeleton fails to grow and mature, and they are severely mentally retarded because the brain fails to develop fully. Because lipid-soluble maternal thyroid hormones cross the placenta and allow normal development, newborn babies usually appear normal, even if they can't produce their own thyroid hormones. Most states require testing of all newborns to ensure adequate thyroid function. If hypothyroidism is detected early, cretinism can be prevented by giving oral thyroid hormone.

Hypothyroidism during the adult years produces **myxedema** (mix-e-DĒ-ma). A hallmark of this disorder is edema (accumulation of interstitial fluid) that causes the facial tissues to swell and look puffy. A person with myxedema has a slow heart rate, low body temperature, sensitivity to cold, dry hair and skin, muscular weakness, general lethargy, and a tendency to gain weight easily. Because the brain has already reached maturity, mental retardation does not occur, but the person's mental functions may be dulled, such that the person is less alert. Myxedema occurs about five times more often in females than in males. Oral thyroid hormones reduce the symptoms.

The most common form of hyperthyroidism is **Graves' disease,** which is an autoimmune disorder. It occurs seven to ten times more often in females than in males, usually before age 40. Because the person produces antibodies that mimic the action of thyroid-stimulating hormone (TSH), the thyroid gland is continually stimulated to grow and produce thyroid hormones. A primary sign is an enlarged thyroid, which may be two to three times its normal size. Graves' patients often have a peculiar edema behind the eyes, called **exophthalmos** (ek′-sof-THAL-mos), which causes the eyes to protrude. Treatment may include surgically removing part or all of the thyroid gland (thyroidectomy), using radioactive iodine (^{131}I) to selectively destroy thyroid tissue, and using antithyroid drugs to block synthesis of thyroid hormones.

A **goiter** (GOY-ter; *guttur* = throat) is simply an enlarged thyroid gland, and it may be associated with hyperthyroidism, hypothyroidism, or euthyroidism (*eu-* = good), which means normal secretion of thyroid hormone. In some places in the world, dietary iodine intake is inadequate; the resultant low level of thyroid hormone in the blood stimulates secretion of TSH, which causes thyroid gland enlargement.

PARATHYROID GLAND DISORDERS

Hypoparathyroidism—too little parathyroid hormone—leads to a deficiency of Ca^{2+}, which causes neurons and muscle fibers to depolarize and produce action potentials spontaneously. This leads to twitches, spasms, and **tetany** of skeletal muscle. The leading cause of hypoparathyroidism is accidental damage to the parathyroid glands or to their blood supply during thyroidectomy surgery.

Hyperparathyroidism, usually due to a tumor of the parathyroid gland, produces **osteitis fibrosa cystica,** which causes demineralization of bone. The bones become brittle and easily fractured due to excessive loss of Ca^{2+}.

ADRENAL GLAND DISORDERS

Disorders of the adrenal cortex involve hyperfunction or hypofunction and result in complex physical, psychological, and metabolic changes that may be life threatening. Disorders of the adrenal medulla involve hyperfunction and are not life threatening.

Cushing's Syndrome

Hypersecretion of cortisol by the adrenal cortex produces **Cushing's syndrome.** Causes include a tumor of the adrenal gland that secretes cortisol, or a tumor elsewhere that secretes adrenocorticotropic hormone (ACTH), which in turn stimulates excessive secretion of cortisol. The condition is characterized by breakdown of muscle proteins and redistribution of body fat, resulting in spindly arms and legs accompanied by a rounded "moon face," "buffalo hump" on the back, and pendulous (hanging) abdomen. Facial skin is flushed, and the skin covering the abdomen develops stretch marks. The person also bruises easily, and wound healing is poor. The elevated level of cortisol causes hyperglycemia, osteoporosis, weakness, hypertension, increased susceptibility to infection, decreased resistance to stress, and mood swings. People who need long-term, glucocorticoid therapy—for instance, to prevent rejection of a transplanted organ—may develop a cushinoid appearance.

Addison's Disease

Progressive destruction of the adrenal cortex leads to hyposecretion of glucocorticoids and aldosterone and causes **Addison's disease (primary adrenocortical insufficiency).** The majority of cases are thought to be autoimmune disorders in which antibodies either cause adrenal cortex destruction or block binding of ACTH to its receptors. Pathogens, such as the bacterium that causes tuberculosis, also may trigger adrenal cortex destruction. Symptoms, which typically do not appear until 90% of the adrenal cortex has been destroyed, include mental lethargy, anorexia, nausea and vomiting, weight loss, hypoglycemia, and muscular weakness. Loss of aldosterone leads to elevated potassium and decreased sodium in the blood, low blood pressure, dehydration, decreased cardiac output, arrhythmias, and potential cardiac arrest. Excessive skin pigmentation, especially in sun-exposed areas and in mucous membranes, also occurs. Treatment consists of replacing glucocorticoids and mineralocorticoids and increasing sodium in the diet.

Pheochromocytomas

Usually benign tumors of the chromaffin cells of the adrenal medulla, called **pheochromocytomas** (fē-ō-krō′-mō-si-TŌ-mas; *pheo-* = dusky; *chromo-* = color; *cyto-* = cell), cause hypersecretion of the medullary hormones. Hypersecretion of epinephrine and norepinephrine causes a prolonged version of the fight-or-flight response: rapid heart rate, headache, high blood pressure, high levels of glucose in blood and urine, an elevated basal metabolic rate (BMR), flushed face, nervousness, sweating, and decreased gastrointestinal motility. Treatment is surgical removal of the tumor.

PANCREATIC DISORDERS

Among the pancreatic disorders is the most common endocrine disorder, **diabetes mellitus** (MEL-i-tus; *melli-* = honey sweetened), a chronic disease that affects about 12 million Americans and is the fourth leading cause of death by disease in the United States, primarily because of its widespread cardiovascular effects. Diabetes mellitus is actually a group of disorders caused by an inability to produce or use insulin. The result is an elevation of blood glucose (hyperglycemia) and loss of glucose in the urine (glucosuria). Hallmarks of diabetes mellitus are the three "polys": *polyuria,* excessive urine production due to an inability of the kidneys to reabsorb water; *polydipsia,* excessive thirst; and *polyphagia,* excessive eating.

There are two major types of diabetes mellitus. **Type I diabetes** is caused by an absolute deficiency of insulin. Consequently, Type I diabetes is also called **insulin-dependent diabetes mellitus (IDDM)** because regular injections of insulin are required to prevent death. Most commonly, IDDM develops in people younger than age 20, although the condition persists throughout life. IDDM is an autoimmune disorder in which a person's immune system destroys the pancreatic beta cells. The drug cyclosporine, which suppresses the immune system, shows promise in being able to interrupt the destruction of beta cells.

The cellular metabolism of an untreated type I diabetic is similar to that of a starving person. Because insulin is not present to aid the entry of glucose into body cells, most cells use fatty acids to produce ATP. The accumulation of by-products of fatty acid breakdown—organic acids called ketones (ketone bodies)—cause a form of acidosis called **ketoacidosis,** which lowers the pH of the blood and can result in death. The breakdown of stored triglycerides and proteins also causes weight loss. As lipids are transported by the blood from storage depots to cells, lipid particles are deposited on the walls of blood vessels, leading to atherosclerosis and a multitude of cardiovascular problems, including cerebrovascular insufficiency, ischemic heart disease, peripheral vascular disease, and gangrene. One of the major complications of diabetes is loss of vision due either to cataracts (excessive glucose attaches to lens proteins, causing cloudiness) or to damage to blood vessels of the retina. Severe kidney problems also may result from damage to renal blood vessels.

Type II diabetes, also called **non-insulin-dependent diabetes mellitus (NIDDM),** is much more common than type I, representing more than 90% of all cases. Type II diabetes most often occurs in people who are over 35 and overweight. Clinical symptoms are mild, and the high glucose levels in the blood often can be controlled by diet, exercise, and weight loss. Sometimes, an antidiabetic drug such as *glyburide* (DiaBeta) is used to stimulate secretion of insulin by beta cells of the pancreas. Although some type II diabetics need insulin, many have a sufficient amount (or even a surplus) of insulin in the blood. For these people, diabetes arises not from a shortage of insulin but because target cells become less sensitive to it due to down-regulation of insulin receptors.

Hyperinsulinism most often results when a diabetic injects too much insulin. The principal symptom is **hypoglycemia,** decreased blood glucose level, which occurs because the excess insulin stimulates excessive uptake of glucose by many cells of the body. The resulting hypoglycemia stimulates the secretion of epinephrine, glucagon, and human growth hormone. As a consequence, anxiety, sweating, tremor, increased heart rate, hunger, and weakness occur. When blood glucose falls, brain cells are deprived of the steady supply of glucose they need to function effectively. This condition leads to mental disorientation, convulsions, unconsciousness, and shock and is termed **insulin shock.** Death can occur quickly unless blood glucose is raised.

STUDY OUTLINE

INTRODUCTION (p. 566)

1. The nervous system controls homeostasis through nerve impulses; the endocrine system uses hormones.
2. The nervous system causes muscles to contract and glands to secrete; the endocrine system affects virtually all body tissues.

ENDOCRINE GLANDS DEFINED (p. 567)

1. Exocrine (sudoriferous, sebaceous, and digestive) glands secrete their products through ducts into body cavities or onto body surfaces.
2. Endocrine glands secrete hormones into the blood.
3. The endocrine system consists of endocrine glands and several organs that contain endocrine tissue (see Figure 18.1).
4. Hormones regulate the internal environment, metabolism, and energy balance.
5. They also help regulate muscular contraction, glandular secretion, and certain immune responses.
6. Hormones affect growth, development, and reproduction.

HORMONE ACTIVITY (p. 567)

1. Hormones affect only specific target cells that have receptors to recognize (bind) a given hormone.
2. The number of hormone receptors may decrease (down-regulation) or increase (up-regulation).
3. Chemically, hormones are either lipid-soluble (steroids, thyroid hormones, and nitric oxide) or water-soluble (amines; peptides, proteins, and glycoproteins; and eicosanoids).

4. Water-soluble hormones circulate in "free" form in the blood; lipid-soluble steroid and thyroid hormones attach to specific transport proteins synthesized by the liver.

MECHANISMS OF HORMONE ACTION (p. 570)

1. Lipid-soluble steroid hormones and thyroid hormones affect cell function by altering gene expression.
2. Water-soluble hormones alter cell function by activating plasma membrane receptors, which initiate a cascade of events inside the cell.
3. Nitric oxide activates guanylyl cyclase in smooth muscle fibers.
4. Three hormonal interactions are a permissive effect, a synergistic effect, and an antagonistic effect.

CONTROL OF HORMONE SECRETION (p. 572)

1. Regulation of secretion is designed to prevent oversecretion or undersecretion.
2. Hormone secretion is controlled by signals from the nervous system, chemical changes in blood, and other hormones.
3. Most often, negative feedback systems regulate hormonal secretions.

HYPOTHALAMUS AND PITUITARY GLAND (p. 573)

1. The hypothalamus is the major integrating link between the nervous and endocrine systems.

2. The hypothalamus and pituitary gland regulate virtually all aspects of growth, development, metabolism, and homeostasis.

3. The pituitary gland is located in the sella turcica and is divided into the anterior pituitary and posterior pituitary gland.

4. Secretion of anterior pituitary gland hormones is stimulated by releasing hormones and suppressed by inhibiting hormones from the hypothalamus.

5. The blood supply to the anterior pituitary gland is from the superior hypophyseal arteries. Hypothalamic releasing and inhibiting hormones enter the primary plexus and are carried to the anterior pituitary gland by the hypophyseal portal veins.

6. The anterior pituitary gland consists of somatotrophs that produce human growth hormone (hGH); lactotrophs that produce prolactin (PRL); corticotrophs that secrete adrenocorticotropic hormone (ACTH) and melanocyte-stimulating hormone (MSH); thyrotrophs that secrete thyroid-stimulating hormone (TSH); and gonadotrophs that synthesize follicle-stimulating hormone (FSH) and luteinizing hormone (LH).

7. Human growth hormone (hGH) stimulates body growth through insulinlike growth factors (IGFs). Secretion of hGH is inhibited by GHIH (growth hormone inhibiting hormone, or somatostatin) and promoted by GHRH (growth hormone releasing hormone).

8. TSH regulates thyroid gland activities. Its secretion is stimulated by TRH (thyrotropin releasing hormone) and suppressed by GHIH.

9. FSH and LH regulate the activities of the gonads—ovaries and testes—and are controlled by GnRH (gonadotropin releasing hormone).

10. Prolactin (PRL) helps initiate milk secretion. Prolactin inhibiting hormone suppresses secretion of PRL, whereas prolactin releasing hormone and TRH stimulate PRL secretion.

11. ACTH regulates the activities of the adrenal cortex and is controlled by CRH (corticotropin releasing hormone).

12. Secretion of MSH is stimulated by corticotropin releasing hormone (CRH) and inhibited by dopamine.

13. The neural connection between the hypothalamus and posterior pituitary gland is via the hypothalamohypophyseal tract.

14. Hormones made by the hypothalamus and stored in the posterior pituitary gland are oxytocin (OT), which stimulates contraction of the uterus and ejection of milk from the breasts, and antidiuretic hormone (ADH), which stimulates water reabsorption by the kidneys and constriction of arterioles. ADH is also known as vasopressin.

15. OT secretion is stimulated by uterine distension and suckling during nursing; ADH secretion is controlled by osmotic pressure of the blood and blood volume.

THYROID GLAND (p. 581)

1. The thyroid gland is located inferior to the larynx.

2. It consists of thyroid follicles composed of follicular cells, which secrete the thyroid hormones thyroxine (T_4) and triiodothyronine (T_3), and parafollicular cells, which secrete calcitonin (CT).

3. Thyroid hormones are synthesized from iodine and tyrosine within thyroglobulin (TGB).

4. They are transported in the blood bound to plasma proteins, mostly thyroxine-binding globulin (TBG).

5. Thyroid hormones regulate the rate of metabolism, growth, and development. Their secretion is controlled by the level of iodine in the blood and TSH.

6. Calcitonin (CT) can lower the blood level of calcium ions (Ca^{2+}) and promote deposition of Ca^{2+} into bone matrix. Secretion of CT is controlled by Ca^{2+} level in the blood.

PARATHYROID GLANDS (p. 584)

1. The parathyroid glands are embedded on the posterior surfaces of the lateral lobes of the thyroid.

2. The parathyroids consist of principal cells and oxyphil cells.

3. Parathyroid hormone (PTH) regulates the homeostasis of calcium and phosphate ions by increasing blood calcium level and decreasing blood phosphate level. Secretion is controlled by Ca^{2+} level in the blood.

ADRENAL GLANDS (p. 586)

1. The adrenal glands are located superior to the kidneys. They consist of an outer adrenal cortex and inner adrenal medulla.

2. The adrenal cortex is divided into a zona glomerulosa, a zona fasciculata, and a zona reticularis; the adrenal medulla consists of chromaffin cells and large blood vessels.

3. Cortical secretions are mineralocorticoids, glucocorticoids, and androgens.

4. Mineralocorticoids (mainly aldosterone) increase sodium and water reabsorption and decrease potassium reabsorption. Secretion is controlled by the renin–angiotensin pathway and K^+ level in the blood.

5. Glucocorticoids (mainly cortisol) promote protein breakdown, gluconeogenesis, and lipolysis; help resist stress; and serve as anti-inflammatory substances. Their secretion is controlled by ACTH.

6. Androgens secreted by the adrenal cortex stimulate growth of axillary and pubic hair, aid the prepubertal growth spurt, and contribute to libido.

7. Medullary secretions are epinephrine and norepinephrine (NE), which produce effects similar to sympathetic responses. They are released during stress.

PANCREAS (p. 592)

1. The pancreas is posterior and slightly inferior to the stomach.

2. It consists of pancreatic islets or islets of Langerhans (endocrine cells) and clusters of enzyme-producing cells (acini). Four types of cells in the endocrine portion are alpha cells, beta cells, delta cells, and F cells.

3. Alpha cells secrete glucagon, beta cells secrete insulin, delta cells secrete somatostatin, and F cells secrete pancreatic polypeptide.

4. Glucagon increases blood glucose, and its secretion is stimulated by a low blood glucose level.

5. Insulin decreases blood glucose, and its secretion is stimulated by a high blood glucose level.

OVARIES AND TESTES (p. 595)

1. The ovaries are located in the pelvic cavity and produce estrogens, progesterone, and inhibin. These sex hormones govern the development and maintenance of feminine secondary sex characteristics, reproductive cycles, pregnancy, lactation, and normal reproductive functions.

2. The testes lie inside the scrotum and produce testosterone and inhibin. These sex hormones govern the development and maintenance of masculine secondary sex characteristics and normal reproductive functions.

PINEAL GLAND (p. 596)

1. The pineal gland is attached to the roof of the third ventricle.
2. It consists of secretory cells called pinealocytes, neuroglia, and scattered postganglionic sympathetic fibers.
3. The pineal gland secretes melatonin, which contributes to setting the body's biological clock (controlled in the suprachiasmatic nucleus). During sleep, plasma levels of melatonin increase tenfold and then decline to a low level again before awakening.

THYMUS GLAND (p. 597)

1. The thymus gland secretes several hormones related to immunity.
2. Thymosin, thymic humoral factor (THF), thymic factor (TF), and thymopoietin promote the maturation of T cells.

MISCELLANEOUS HORMONES (p. 597)

1. The gastrointestinal (GI) tract synthesizes several hormones, including gastrin, glucose-dependent insulinotropic peptide (GIP), secretin, and cholecystokinin (CCK).
2. The placenta produces human chorionic gonadotropin (hCG), estrogens, progesterone, relaxin, and human chorionic somatomammotropin.
3. The kidneys release erythropoietin and calcitriol, the active form of vitamin D.
4. The atria of the heart produce atrial natriuretic peptide (ANP).
5. Adipose tissue produces leptin.
6. Prostaglandins and leukotrienes are eicosanoids that act as paracrines and autocrines in most body tissues by altering the production of second messengers, such as cyclic AMP.
7. Prostaglandins have a wide range of biological activity in normal physiology and pathology.
8. Growth factors are local hormones that stimulate cell growth and division. They include epidermal growth factor (EGF), platelet-derived growth factor (PDGF), fibroblast growth factor (FGF), nerve growth factor (NGF), tumor angiogenesis factors (TAFs), and transforming growth factors (TGFs).

STRESS AND THE GENERAL ADAPTATION SYNDROME (p. 599)

1. If a stress is extreme or unusual, it triggers a wide-ranging set of bodily changes called the general adaptation syndrome (GAS).

The stimuli that produce the general adaptation syndrome are called stressors.
2. Stressors include surgical operations, poisons, infections, fever, and strong emotional responses.
3. Productive stress is termed eustress, and harmful, nonproductive stress is termed distress.
4. The alarm reaction is initiated by nerve impulses from the hypothalamus to the sympathetic division of the autonomic nervous system and adrenal medulla.
5. Alarm responses are the immediate and brief fight-or-flight responses that increase circulation, promote ATP production, and decrease nonessential activities.
6. The resistance reaction is initiated by hormones secreted by the hypothalamus, most importantly CRH, TRH, and GHRH.
7. Resistance reactions are long-term and accelerate breakdown reactions to provide ATP for counteracting stress.
8. Exhaustion results from depletion of body resources during the resistance stage.
9. Stress appears to trigger certain diseases by inhibiting the immune system.
10. An important link between stress and immunity is interleukin-1 produced by macrophages; it stimulates secretion of ACTH.

DEVELOPMENTAL ANATOMY OF THE ENDOCRINE SYSTEM (p. 601)

1. The development of the endocrine system is not as localized as other systems because endocrine organs develop in widely separated parts of the embryo.
2. The pituitary gland, adrenal medulla, and pineal gland develop from ectoderm; the adrenal cortex develops from mesoderm; and the thyroid gland, parathyroid glands, pancreas, and thymus gland develop from endoderm.

AGING AND THE ENDOCRINE SYSTEM (p. 603)

1. Although endocrine glands shrink with age, their performance may not necessarily be compromised.
2. With advancing age the pituitary gland produces less human growth hormone but more gonadotropins and thyroid-stimulating hormone.

SELF-QUIZ QUESTIONS

Choose the best answer to the following questions:
1. Which of the following comparisons are true? (1) Nerve impulses produce their effects quickly, whereas hormones can act either quickly or slowly. (2) Nervous system effects are brief, whereas endocrine system effects are longer lasting. (3) The nervous system controls homeostasis through nerve impulses; the endocrine system works through hormones. (4) The nervous system causes muscles to contract; the endocrine system affects virtually all body tissues. (5) The nervous system includes organs scattered throughout the body; the endocrine system consists of only the pituitary gland, thyroid gland, adrenal gland, and pancreas.
(a) 1, 2, 3, 4, and 5, (b) 1, 2, 3, and 4, (c) 2, 3, 4, and 5, (d) 2, 4, and 5, (e) 1, 4, and 5
2. Which of the following is *not* a class of hormone? (a) steroids, (b) amines, (c) alcohols, (d) peptides and proteins, (e) eicosanoids

3. Which of the following factors affect the responsiveness of a target cell to a hormone? (1) hormone concentration, (2) target cell type, (3) abundance of hormone receptors, (4) influences of other hormones, (5) age of the target cell.
(a) 1, 2, 3, and 4, (b) 2, 3, 4, and 5, (c) 1, 2, and 3, (d) 1, 3, and 4, (e) 2, 4, and 5

4. The class of adrenal gland hormones that provide resistance to stress, produce anti-inflammatory effects, and promote normal metabolism to ensure adequate quantities of ATP is (a) glucocorticoids, (b) mineralocorticoids, (c) androgens, (d) catecholamines, (e) sympathomimetics.

5. Which of the following regulate secretion of hormones? (1) signals from the environment, (2) signals from the nervous system, (3) chemical changes in the blood, (4) other hormones, (5) chemical changes in target cells.
(a) 1, 2, and 3, (b) 2, 3, and 4, (c) 2, 4, and 5, (d) 1, 3, and 5, (e) 2, 3, and 5

6. The body system that turns on the stress response is the (a) circulatory system, (b) nervous system, (c) immune system, (d) lymphatic system, (e) respiratory system.

Complete the following:

7. A water-soluble hormone binds to a receptor on the target cell membrane, which activates another membrane protein called a ___, which in turn acts to turn on an enzyme called ___. This enzyme then converts ATP into cyclic AMP, the second messenger. Cyclic AMP activates ___, which ___ cellular proteins. These cellular proteins catalyze reactions that produce the physiological response attributed to the hormone.

8. The three stages of the stress response or general adaptation syndrome, in order of occurrence, are ___, ___, and ___.

9. The major integrating link between the nervous and endocrine systems and an endocrine gland itself is the ___.

10. ___ glands secrete their products into ducts that carry the product to the destination site. ___ glands secrete their products into the interstitial fluid, where they diffuse into capillaries.

True or false:

11. The posterior pituitary gland synthesizes oxytocin and ADH.

12. In the direct gene activation method of hormone action, the hormone enters the target cell and binds to an intracellular receptor. The activated receptor then alters gene expression to produce the protein that causes the physiological responses that are characteristic of the hormone.

13. Match the following:
___ (a) increases blood Ca^{2+} level
___ (b) only increases blood glucose level
___ (c) decreases blood Ca^{2+} level
___ (d) decreases blood glucose level
___ (e) initiates and maintains milk secretion by the mammary glands
___ (f) stimulates gamete formation
___ (g) stimulates sex hormone production
___ (h) stimulates milk ejection
___ (i) regulates metabolism and resistance to stress
___ (j) helps control water and electrolyte homeostasis
___ (k) suppresses release of FSH
___ (l) inhibits water loss through the kidneys
___ (m) determines anterior pituitary gland function
___ (n) regulates oxygen use and basal metabolic rate, cellular metabolism, and growth and development
___ (o) stimulates protein synthesis, inhibits protein breakdown, stimulates lipolysis, and retards use of glucose for ATP production
___ (p) increases skin pigmentation

(1) insulin
(2) glucagon
(3) inhibin
(4) FSH
(5) LH
(6) thyroxine and triiodothyronine
(7) calcitonin
(8) parathormone
(9) melanocyte-stimulating hormone
(10) oxytocin
(11) ADH
(12) prolactin
(13) hGH
(14) hypothalamic regulating hormones
(15) aldosterone
(16) cortisol

14. Match the following hormone secreting cells to the hormone(s) they secrete:
___ (a) ACTH
___ (b) TSH
___ (c) glucagon
___ (d) calcitonin
___ (e) insulin
___ (f) progesterone
___ (g) FSH and LH
___ (h) hGH
___ (i) testosterone
___ (j) thyroxine and triiodothyronine

(1) beta cells of pancreatic islet
(2) alpha cells of pancreatic islet
(3) follicular cells of thyroid gland
(4) parafollicular cells of thyroid gland
(5) interstitial cells of testis
(6) corpus luteum of ovary
(7) somatotrophs
(8) thyrotrophs
(9) gonadotrophs
(10) corticotrophs

15. Match the endocrine disorder to the problem that produced the disorder:
___ (a) hyposecretion of insulin or insufficient insulin receptors
___ (b) hypersecretion of hGH before epiphyseal plate closure
___ (c) hyposecretion of thyroid hormone
___ (d) hypersecretion of glucocorticoids
___ (e) hyposecretion of hGH before epiphyseal plate closure
___ (f) hyperparathyroidism
___ (g) hypersecretion of hGH after epiphyseal plate closure
___ (h) hyposecretion of glucocorticoids and aldosterone
___ (i) hyposecretion of ADH
___ (j) hyperthyroidism, an autoimmune disease

(1) giantism
(2) acromegaly
(3) pituitary dwarfism
(4) diabetes insipidus
(5) cretinism
(6) Graves' disease
(7) Cushing's syndrome
(8) osteitis fibrosa cystica
(9) Addison's disease
(10) diabetes mellitus

CRITICAL THINKING QUESTIONS

1. A possible treatment being investigated for type I diabetes is the transplantation of only one type of cell into the patient, instead of the entire source organ. What cell type could this be, and from what organ? How could just one cell type provide a cure for diabetes? (HINT: *The organ has both exocrine and endocrine functions.*)

2. Amanda hates her new student ID photo. Her hair looks dry, the extra weight that she's gained shows, and her neck looks fat. In fact, there's an odd butterfly shaped swelling across the front of her neck, under her chin. Amanda's also been feeling very tired and mentally "dull" lately, but she figures all new A&P students feel that way. Should she visit the clinic or just wear turtlenecks? (HINT: *She also feels constantly cold.*)

3. Jamal is a 15-year-old computer whiz with an interest in medicine. He likes to practice virtual surgery using the computer program he wrote himself. Today he's transplanting an adrenal gland. Describe the location and structure of the adrenal glands. (HINT: *It's really two glands in one.*)

ANSWERS TO FIGURE QUESTIONS

18.1 Secretions of endocrine glands diffuse into the blood; exocrine secretions flow into ducts that lead into body cavities or to the body surface.

18.2 In the stomach, histamine is a paracrine because it acts on nearby parietal cells without entering the blood.

18.3 Synthesis of new proteins takes several minutes to half an hour, due to the time needed for activation of DNA transcription and the translation of the mRNA code of nucleotides into a chain of amino acids. The effects of steroid and thyroid hormones thus appear slowly.

18.4 Cyclic AMP is termed a second messenger because it translates the presence of the first messenger, the water-soluble hormone, into a response inside the cell.

18.5 The hypophyseal portal veins carry blood from the median eminence of the hypothalamus, where hypothalamic releasing and inhibiting hormones are secreted, to the anterior pituitary gland, where these hormones act.

18.6 Thyroid hormones suppress secretion of TSH by thyrotrophs and of TRH by hypothalamic neurosecretory cells; gonadal hormones suppress secretion of FSH and LH by gonadotrophs and of GnRH by hypothalamic neurosecretory cells.

18.7 Excess hGH causes hyperglycemia.

18.8 Functionally, both the hypothalamohypophyseal tract and the hypophyseal portal veins carry hypothalamic hormones to the pituitary gland. Structurally, the tract is composed of axons of neurons that extend from the hypothalamus to the posterior pituitary gland, whereas the portal veins are blood vessels that extend to the anterior pituitary gland.

18.9 Absorption of a liter of water in the intestines would decrease the osmotic pressure of your blood plasma, turning off secretion of ADH and decreasing the ADH level in your blood.

18.10 Follicular cells secrete T_3 and T_4, also known as thyroid hormones. Parafollicular cells secrete calcitonin.

18.11 The storage form of thyroid hormones is thyroglobulin.

18.12 Lack of iodine in the diet → diminished production of T_3 and T_4 → increased release of TSH → growth (enlargement) of the thyroid gland → goiter.

18.13 Parafollicular cells of the thyroid gland secrete calcitonin; principal cells of the parathyroid gland secrete PTH.

18.14 Target tissues for PTH are bone and kidneys; target tissue for CT is bone; target tissue for calcitriol is the GI tract.

18.15 The adrenal glands are superior to the kidneys in the retroperitoneal space.

18.16 Angiotensin II acts to constrict blood vessels by causing contraction of smooth muscle, and it stimulates secretion of aldosterone (by zona glomerulosa cells of the adrenal cortex), which in turn causes the kidneys to conserve water and increase blood volume.

18.17 A transplant recipient who takes prednisone will have low blood levels of ACTH and CRH due to negative feedback suppression of the anterior pituitary and hypothalamus by the prednisone.

18.18 The pancreas is both an endocrine and an exocrine gland.

18.19 Glucagon is considered an anti-insulin hormone because it has several effects that are opposite to those of insulin.

18.20 Exposure to bright light after arriving may help reset the biological clock to earlier times for falling asleep and awakening.

18.21 Homeostasis maintains controlled conditions typical of a normal internal environment; the GAS resets controlled conditions at a different level to cope with various stressors.

18.22 The pituitary gland and the adrenal glands both include tissues having two different embryological origins.

THE CARDIOVASCULAR SYSTEM: THE BLOOD

The **cardiovascular system** (*cardio-* = heart; *vascular* = blood or blood vessels) consists of three interrelated components: blood, the heart, and blood vessels. The focus of this chapter is blood; the next two chapters will examine the heart and blood vessels, respectively.

Most cells of a multicellular organism cannot move around to obtain oxygen and nutrients and escape carbon dioxide and other wastes. Instead, to accomplish these needs the cells must be serviced by two fluids: blood and interstitial fluid. **Blood** is a connective tissue composed of a liquid portion called plasma and a cellular portion consisting of various cells and cell fragments. **Interstitial fluid** is the fluid that bathes body cells. Oxygen breathed into the lungs and nutrients absorbed in the gastrointestinal tract are transported by the blood to body tissues. The oxygen and nutrients diffuse from the blood into the interstitial fluid and then into body cells. Carbon dioxide and other wastes move in the reverse direction, from cells into the interstitial fluid and then into the blood. Blood then transports the wastes to various organs—the lungs, kidneys, skin, and digestive system—for elimination from the body. As we will see, blood not only transports various substances; it also helps regulate several life processes and affords protection against disease. For all of its similarities in origin, composition, and functions, blood is as unique from one person to another as are skin, bone, and hair. Health-care professionals routinely examine and analyze its differences through various blood tests when attempting to determine the cause of different diseases. The branch of science concerned with the study of blood, blood-forming tissues, and the disorders associated with them is **hematology** (hēm-a-TOL-ō-jē; *hem-* or *hemat-* = blood; *-ology* = study of).

FUNCTIONS OF BLOOD

OBJECTIVE

• *List and describe the functions of blood.*

Blood, the only liquid connective tissue, has three general functions:

1. *Transportation.* Blood transports oxygen from the lungs to the cells of the body and carbon dioxide from the cells to the lungs. It also carries nutrients from the gastrointestinal tract to body cells, heat and waste products away from cells, and hormones from endocrine glands to other body cells.
2. *Regulation.* Blood helps regulate pH through buffers. It also participates in adjusting body temperature through the heat-absorbing and coolant properties of the water in plasma and its variable rate of flow through the skin, where excess heat can be lost from the blood to the environment. Blood osmotic pressure also influences the water content of cells, mainly through interactions involving dissolved ions and proteins.
3. *Protection.* Blood can clot, which protects against its excessive loss from the cardiovascular system after an injury. In addition, white blood cells protect against disease by carrying on phagocytosis and producing proteins

called antibodies. Blood also contains additional proteins, called interferons and complement, that help protect against disease.

PHYSICAL CHARACTERISTICS OF BLOOD

OBJECTIVE

• *List the principal physical characteristics of blood.*

Blood is denser and more viscous than water, which is part of the reason it flows more slowly than water. The temperature of blood is about 38°C (100.4°F), which is slightly higher than normal body temperature, and it has a slightly alkaline pH ranging from 7.35 to 7.45. Blood constitutes about 8% of the total body weight. The blood volume is 5–6 liters (1.5 gal) in an average-sized adult male and 4–5 liters (1.2 gal) in an average-sized adult female. Several hormonal negative feedback systems ensure that blood volume and osmotic pressure remain relatively constant. Especially important systems are those involving aldosterone, antidiuretic hormone, and atrial natriuretic peptide, which regulate how much water is excreted in the urine (see pages 936–938).

CLINICAL APPLICATION
Withdrawing Blood

Blood samples for laboratory testing may be obtained in several ways. The most frequently used procedure is **venipuncture,** withdrawal of blood from a vein using a hypodermic needle and syringe. A tourniquet is wrapped around the arm above the venipuncture site to cause the blood to accumulate in the vein. This increased blood volume causes the vein to stand out. Opening and closing the fist further causes it to stand out, making the venipuncture more successful. Another method of drawing blood is through a **finger** or **heel stick.** This is frequently used by diabetic patients who monitor their daily blood sugar as well as for drawing blood from infants and children. A common site for venipuncture is the median cubital vein anterior to the elbow (see Figure 21.26b on page 717). ■

COMPONENTS OF BLOOD

OBJECTIVE

• *Describe the principal components of blood.*

Whole blood is composed of two components: (1) blood plasma, a watery liquid that contains dissolved substances, and (2) formed elements, which are cells and cell fragments. If a sample of blood is centrifuged (spun) in a small glass tube, the cells sink to the bottom of the tube while the lighter-weight plasma forms a layer on top (Figure 19.1a). Blood is about 45% formed elements and 55% plasma. Normally, more than 99% of the formed elements are red-colored red blood cells (RBCs). Pale, colorless white blood cells (WBCs) and platelets occupy less than 1% of total blood volume. They form a very thin layer, called the

buffy coat, between the packed RBCs and plasma in centrifuged blood. Figure 19.1b shows the composition of blood plasma and the numbers of the various types of formed elements in blood.

Blood Plasma

When the formed elements are removed from blood, a straw-colored liquid called **blood plasma** (or simply **plasma**) is left. Plasma is about 91.5% water and 8.5% solutes, most of which (7% by weight) are proteins. Some of the proteins in plasma are also found elsewhere in the body, but those confined to blood are called **plasma proteins.** Among other functions, these proteins play a role in maintaining proper blood osmotic pressure, which is an important factor in the exchange of fluids across capillary walls (discussed in Chapter 21).

Hepatocytes (liver cells) synthesize most of the plasma proteins, which include the **albumins** (54% of plasma proteins), **globulins** (38%), and **fibrinogen** (7%). Certain blood cells develop into cells that produce gamma globulins, an important type of globulin. These plasma proteins are also called **antibodies** or **immunoglobulins** because they are produced during certain immune responses. Foreign invaders such as bacteria and viruses stimulate production of millions of different antibodies. An antibody binds specifically to the foreign substance, called an **antigen,** that provoked its production. When the two come together they form an **antigen–antibody complex,** which disables the invading antigen (see Figure 22.18).

Besides proteins, other solutes in plasma include electrolytes, nutrients, regulatory substances such as enzymes and hormones, gases, and waste products such as urea, uric acid, creatinine, ammonia, and bilirubin.

Table 19.1 describes the chemical composition of blood plasma.

Formed Elements

The **formed elements** of the blood include three principal components: **red blood cells (RBCs), white blood cells (WBCs),** and **platelets** (Figure 19.2). Although RBCs and WBCs are living cells, platelets are cell fragments. Unlike RBCs and platelets, each of which performs limited roles, WBCs have a number of specialized roles. For this reason, WBCs exist in several distinct cell types—neutrophils, lymphocytes, monocytes, eosinophils, and basophils—each having a unique microscopic appearance. The roles of each type of WBC are discussed later in this chapter.

The percentage of total blood volume occupied by RBCs is called the **hematocrit** (he-MAT-ō-krit). For example, a hematocrit of 40 means that 40% of the volume of blood is composed of RBCs. The normal range of hematocrit for adult females is 38–46% (average = 42); for adult males it is 40–54% (average = 47). The hormone testosterone, which is present in much higher concentration in males, stimulates synthesis of erythropoietin by the kidneys (the hormone that

Figure 19.1 Components of blood in a normal adult.

🔑 **Blood is a connective tissue that consists of plasma (liquid) plus formed elements (red blood cells, white blood cells, and platelets).**

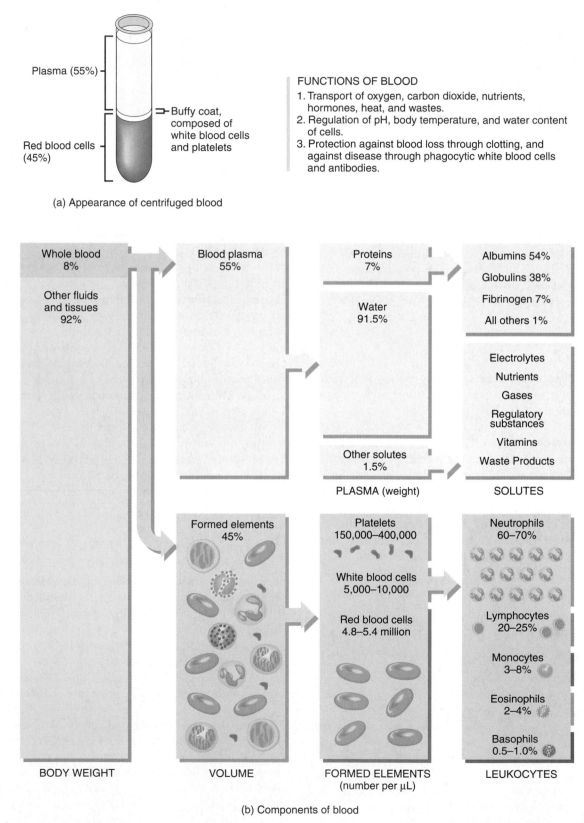

FUNCTIONS OF BLOOD
1. Transport of oxygen, carbon dioxide, nutrients, hormones, heat, and wastes.
2. Regulation of pH, body temperature, and water content of cells.
3. Protection against blood loss through clotting, and against disease through phagocytic white blood cells and antibodies.

(a) Appearance of centrifuged blood

(b) Components of blood

Q What is the approximate volume of blood in your body?

Table 19.1 Substances in Plasma

CONSTITUENT	DESCRIPTION
Water	Liquid portion of blood; constitutes about 91.5% of plasma. Acts as solvent and suspending medium for components of blood; absorbs, transports, and releases heat.
Proteins	Constitute about 7.0% (by weight) of plasma. Exert colloid osmotic pressure, which helps maintain water balance between blood and tissues, and regulate blood volume.
Albumins	Smallest and most numerous plasma proteins; produced by liver. Function as transport proteins for several steroid hormones and for fatty acids.
Globulins	Protein group to which antibodies (immunoglobulins) belong. Produced by liver and by plasma cells, which develop from B lymphocytes. Antibodies help attack viruses and bacteria. Alpha and beta globulins transport iron, lipids, and fat-soluble vitamins.
Fibrinogen	Produced by liver. Plays essential role in blood clotting.
Other Solutes	Constitute about 1.5% (by weight) of plasma.
Electrolytes	Inorganic salts. Positively charged ions (cations) include Na^+, K^+, Ca^{2+}, Mg^{2+}; negatively charged ions (anions) include Cl^-, HPO_4^{2-}, SO_4^{2-}, and HCO_3^-. Help maintain osmotic pressure and play essential roles in the function of cells.
Nutrients	Products of digestion pass into blood for distribution to all body cells. Include amino acids (from proteins), glucose (from carbohydrates), fatty acids and glycerol (from triglycerides), vitamins, and minerals.
Gases	Oxygen (O_2), carbon dioxide (CO_2), and nitrogen (N_2). Whereas more O_2 is associated with hemoglobin inside red blood cells, more CO_2 is dissolved in plasma. N_2 has no known function in the body.
Regulatory substances	Enzymes, produced by body cells, catalyze chemical reactions. Hormones, produced by endocrine glands, regulate growth and development in body.
Wastes	Most are breakdown products of protein metabolism and are carried by blood to organs of excretion. Include urea, uric acid, creatine, creatinine, bilirubin, and ammonia.

stimulates production of RBCs), which contributes to higher hematocrits in males. Lower values in women during their reproductive years may be due to excessive loss of blood during menstruation. A significant drop in hematocrit indicates anemia, a lower than normal number of RBCs. In *polycythemia* the percentage of RBCs is abnormally high, and the hematocrit may be 65% or higher. Polycythemia may be caused by conditions such as an unregulated increase in RBC production, tissue hypoxia, dehydration, and blood doping by athletes.

CLINICAL APPLICATION
Blood Doping

In recent years, some athletes have been tempted to try **blood doping** in efforts to enhance their performance. In this procedure, blood cells are removed from the body, stored for a month or so, and then reinjected a few days before an athletic event. Because delivery of oxygen to muscle is a limiting factor in muscular feats, and given that red blood cells carry oxygen, increasing the oxygen-carrying capacity of the blood was expected to increase muscular performance. Indeed, blood doping can improve athletic performance in endurance events. The practice is dangerous, however, because it increases the workload of the heart. The increased numbers of RBCs raises the viscosity of the blood, making the blood more difficult for the heart to pump. Moreover, the procedure is banned by the International Olympics Committee. ■

Figure 19.2 Scanning electron micrograph of the formed elements of blood.

🔑 **The formed elements of blood are red blood cells (RBCs), white blood cells (WBCs), and platelets.**

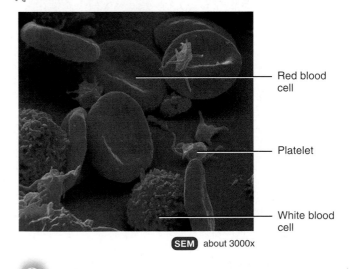

Red blood cell

Platelet

White blood cell

SEM about 3000x

Q Which formed elements of the blood are cell fragments?

1. Distinguish between plasma and formed elements.
2. What are the major constituents of plasma? What does each do?

Figure 19.3 Origin, development, and structure of blood cells. Some of the generations of some cell lines have been omitted.

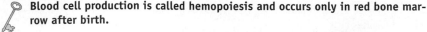 Blood cell production is called hemopoiesis and occurs only in red bone marrow after birth.

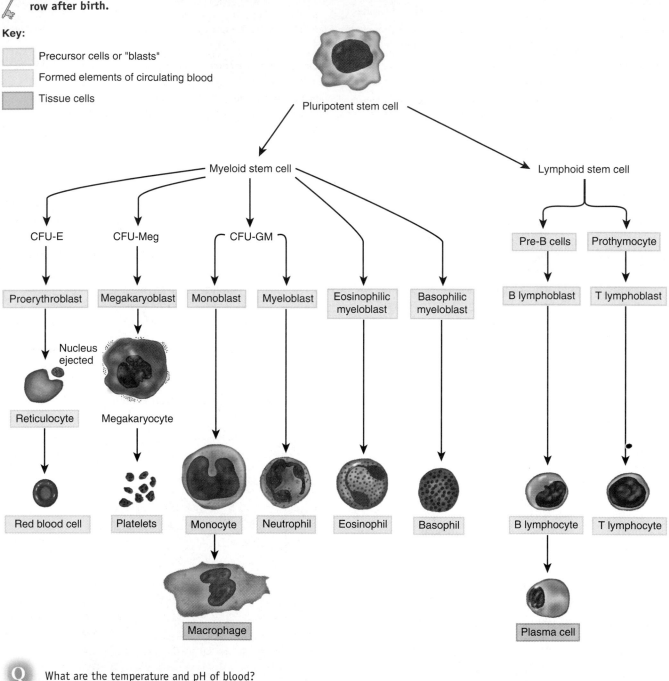

Key:

☐ Precursor cells or "blasts"

☐ Formed elements of circulating blood

☐ Tissue cells

Pluripotent stem cell

Myeloid stem cell — Lymphoid stem cell

CFU-E — CFU-Meg — CFU-GM — Pre-B cells — Prothymocyte

Proerythroblast — Megakaryoblast — Monoblast — Myeloblast — Eosinophilic myeloblast — Basophilic myeloblast — B lymphoblast — T lymphoblast

Nucleus ejected

Reticulocyte — Megakaryocyte

Red blood cell — Platelets — Monocyte — Neutrophil — Eosinophil — Basophil — B lymphocyte — T lymphocyte

Macrophage — Plasma cell

Q What are the temperature and pH of blood?

FORMATION OF BLOOD CELLS

OBJECTIVE

• *Explain the origin of blood cells.*

Although some lymphocytes have a lifetime measured in years, most formed elements of the blood are continually dying and being replaced within hours, days, or weeks. Neg-ative feedback systems regulate the total number of RBCs and platelets in circulation, and the numbers normally remain steady. The abundance of different types of WBCs, however, varies in response to challenges by invading pathogens and foreign antigens.

The process by which the formed elements of blood develop is called **hemopoiesis** (hē-mō-poy-Ē-sis; *-poiesis* =

making) or *hematopoiesis*. About 0.05–0.1% of red bone marrow cells are **pluripotent stem cells** (Figure 19.3), which are derived from mesenchyme. These cells replenish themselves, proliferate, and differentiate into cells that give rise to *all* the formed elements of the blood.

Originating from the pluripotent stem cells are two other types of stem cells, called *myeloid stem cells* and *lymphoid stem cells*. They also have the ability to replenish themselves, and additionally they can differentiate into more specific formed elements of blood. Myeloid stem cells begin their development in red bone marrow and give rise to red blood cells, platelets, monocytes, neutrophils, eosinophils, and basophils. Lymphoid stem cells begin their development in red bone marrow but complete it in lymphatic tissues; they give rise to lymphocytes. Although the various stem cells have distinctive cell identity markers in their plasma membranes, they cannot be distinguished histologically and resemble lymphocytes.

During hemopoiesis, the myeloid stem cells differentiate into **progenitor cells**—cells that no longer are capable of replenishing themselves and are committed to giving rise to more specific elements of blood. Some progenitor cells are referred to as *colony-forming units (CFUs)*. After the CFU designation is an abbreviation that designates the mature elements in blood they will produce: CFU-E ultimately produces red blood cells (erythrocytes), CFU-Meg ultimately produces platelets, and CFU-GM ultimately produces neutrophils and monocytes. Progenitor cells, like stem cells, resemble lymphocytes and cannot be distinguished by their microscopic appearance alone. Other myeloid stem cells develop directly into cells called precursor cells (described next). Lymphoid stem cells differentiate into pre-B cells and prothymocytes, which ultimately develop into B lymphocytes and T lymphocytes, respectively.

In the next generation, the cells are called **precursor cells,** also known as **blasts.** They develop over several cell divisions into the actual formed elements of blood. For example, monoblasts develop into monocytes, eosinophilic myeloblasts develop into eosinophils, and so on. Precursor cells exhibit recognizable histological characteristics.

During embryonic and fetal life, the yolk sac, liver, spleen, thymus gland, lymph nodes, and red bone marrow all participate at various times in producing the formed elements. After birth, however, hemopoiesis takes place *in red bone marrow only,* although small numbers of stem cells circulate in the bloodstream. Red bone marrow is found in the epiphyses (ends) of long bones such as the humerus and femur; in flat bones such as the sternum, ribs, and cranial bones; in vertebrae; and in the pelvis. With the exception of lymphocytes, formed elements do not divide once they leave the bone marrow.

Several **hemopoietic growth factors** regulate the differentiation and proliferation of particular progenitor cells. **Erythropoietin** (e-rith′-rō-POY-e-tin) or **EPO,** a hormone produced mainly by the kidneys, increases the numbers of red blood cell precursors. **Thrombopoietin** (throm′-bō-poy-Ē-tin) or **TPO** is a hormone produced by the liver that stimulates formation of platelets (thrombocytes). Several different cytokines regulate development of different blood cell types. **Cytokines** are small glycoproteins, produced by red bone marrow cells, leukocytes, macrophages, fibroblasts, and endothelial cells, that typically act as local hormones (autocrines or paracrines) to maintain normal cell functions and to stimulate proliferation. Two important families of cytokines that stimulate white blood cell formation are **colony-stimulating factors (CSFs)** and **interleukins.**

CLINICAL APPLICATION
Medical Uses of Hemopoietic Growth Factors

Hemopoietic growth factors made available through recombinant DNA technology hold tremendous potential for medical uses when a person's natural ability to form blood cells is diminished or defective. Recombinant erythropoietin (EPO) is very effective in treating the diminished red blood cell production that accompanies end-stage kidney disease. Granulocyte–macrophage colony-stimulating factor (GM-CSF) and granulocyte CSF (G-CSF) are given to stimulate white blood cell formation in cancer patients who are receiving chemotherapy, which tends to kill their red bone marrow cells as well as the cancer cells. Thrombopoietin shows great promise for preventing platelet depletion during chemotherapy. CSFs and thrombopoietin also improve the outcome of patients who receive bone marrow transplants. ■

1. Describe the hemopoietic growth factors that regulate differentiation and proliferation of particular progenitor cells.

RED BLOOD CELLS
OBJECTIVE
• *Describe the structure, functions, life cycle, and production of red blood cells.*

Red blood cells (RBCs) or **erythrocytes** (e-RITH-rō-sīts; *erythro-* = red; *-cyte* = cell) contain the oxygen-carrying protein **hemoglobin,** which is a pigment that gives whole blood its red color. A healthy adult male has about 5.4 million red blood cells per microliter (μL) of blood[*], and a healthy adult female has about 4.8 million. (One drop of blood is about 50 μL.) To maintain normal quantities of RBCs, new mature cells must enter the circulation at the astonishing rate of at least 2 million per second, a pace that balances the equally high rate of RBC destruction.

---

[*]$1 \mu L = 1 mm^3 = 10^{-6}$ liter

Figure 19.4 The shapes of a red blood cell (RBC) and a hemoglobin molecule, and the structure of a heme group. In (b), note that each of the four polypeptide chains of a hemoglobin molecule has one heme group, which contains an iron ion (Fe^{2+}).

🔑 **The iron portion of a heme group binds oxygen for transport by hemoglobin.**

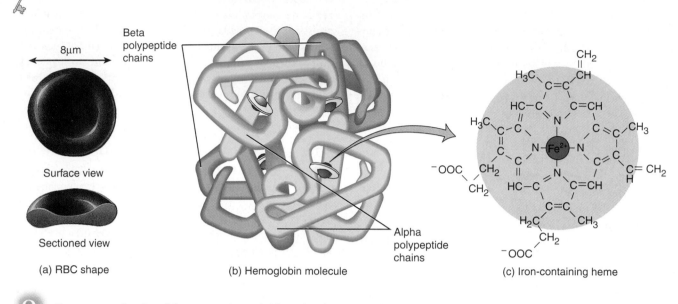

(a) RBC shape   (b) Hemoglobin molecule   (c) Iron-containing heme

Q   How many molecules of O_2 can one hemoglobin molecule transport?

RBC Anatomy

RBCs are biconcave discs with a diameter of 7–8 μm (Figure 19.4a). Mature red blood cells have a simple structure. Their plasma membrane is both strong and flexible, which allows RBCs to deform without rupturing as they squeeze through narrow capillaries. As you will see later, certain glycolipids in the plasma membrane of RBCs are antigens that account for the various blood groups such as the ABO and Rh groups. RBCs lack a nucleus and other organelles and can neither reproduce nor carry on extensive metabolic activities. Dissolved in the cytosol are the hemoglobin molecules, which were synthesized before loss of the nucleus during RBC production and which constitute about 33% of the cell's weight.

RBC Physiology

Red blood cells are highly specialized for their oxygen transport function. Each one contains about 280 million hemoglobin molecules. Because RBCs have no nucleus, all their internal space is available for oxygen transport. Moreover, RBCs lack mitochondria and generate ATP anaerobically (without oxygen); as a result, they do not consume any of the oxygen they transport. Even the shape of an RBC facilitates its function. A biconcave disc has a much greater surface area for its volume than, say, a sphere or a cube. This shape provides a large surface area for the diffusion of gas molecules into and out of the RBC.

A hemoglobin molecule consists of a protein called **globin,** composed of four polypeptide chains (two alpha and two beta chains), plus four nonprotein pigments called **hemes** (Figure 19.4b). Each heme is associated with one polypeptide chain and contains an iron ion (Fe^{2+}) that can combine reversibly with one oxygen molecule (Figure 19.4c). The oxygen picked up in the lungs is transported in this state to other tissues of the body. In the tissues, the iron–oxygen reaction reverses. Hemoglobin releases oxygen, which diffuses first into the interstitial fluid and then into cells.

Hemoglobin also transports about 23% of the total carbon dioxide, a waste product of metabolism. Blood flowing through tissue capillaries picks up carbon dioxide, some of which combines with amino acids in the globin portion of hemoglobin. As blood flows through the lungs, the carbon dioxide is released from hemoglobin and then exhaled.

In addition to its key role in transporting oxygen and carbon dioxide, hemoglobin also assumes a function in blood pressure regulation, a recent and surprising discovery. The iron ions in the heme portion of hemoglobin have a strong affinity for *nitric oxide (NO)*, a gas produced by the endothelial cells that line blood vessels. Hemoglobin that has already circulated through the body and given up its oxygen to tissues passes into the lungs, where it releases both carbon dioxide to be exhaled and NO. Then the hemoglobin picks up a fresh supply of oxygen and a different form of NO called *super nitric oxide (SNO),* probably produced by lung cells. In

Figure 19.5 Formation and destruction of red blood cells, and the recycling of hemoglobin components.

 The rate of RBC formation by red bone marrow roughly equals the rate of RBC destruction by macrophages.

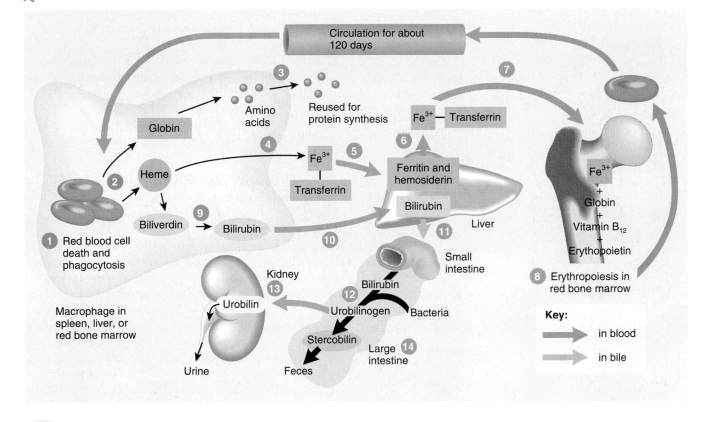

Q What is the function of transferrin?

body tissues, hemoglobin releases oxygen and SNO and picks up carbon dioxide. The released SNO causes *vasodilation,* an increase in blood vessel diameter that occurs when the smooth muscle in its wall relaxes. At the tissue level, hemoglobin also picks up excess NO, which tends to cause *vasoconstriction,* a decrease in blood vessel diameter that occurs when the smooth muscle in its wall contracts. Vasodilation decreases blood pressure, whereas vasoconstriction increases it. By ferrying NO and SNO throughout the body, hemoglobin helps regulate blood pressure by adjusting the amount of NO or SNO to which blood vessels are exposed. This newly appreciated role of hemoglobin may influence development of drugs to treat hypertension (high blood pressure).

RBC Life Cycle

Red blood cells live only about 120 days because of the wear and tear their plasma membranes undergo as they squeeze through blood capillaries. Without a nucleus and other organelles, RBCs cannot synthesize new components to replace damaged ones. The plasma membranes become more fragile with age, and the cells are more likely to burst, especially as they squeeze through narrow channels in the spleen. Worn-out red blood cells are removed from circula-

tion and destroyed by fixed phagocytic macrophages in the spleen and liver, and the breakdown products are recycled, as follows (Figure 19.5):

1 Macrophages in the spleen, liver, or red bone marrow phagocytize worn-out red blood cells.

2 The globin and heme portions of hemoglobin are split apart.

3 Globin is broken down into amino acids, which can be reused to synthesize other proteins.

4 Iron removed from the heme portion is in the form of Fe^{3+}, which associates with the plasma protein **transferrin** (trans-FER-in; *trans-* = across; *ferr-* = iron), a transporter for Fe^{3+} in the bloodstream.

5 In muscle fibers, liver cells, and macrophages of the spleen and liver, Fe^{3+} detaches from transferrin and attaches to iron-storage proteins called **ferritin** and **hemosiderin** (hē-mō-SID-er-in).

6 Upon release from a storage site or absorption from the gastrointestinal tract, Fe^{3+} reattaches to transferrin.

7 The Fe^{3+}-transferrin complex is then carried to bone marrow, where RBC precursor cells take it up through receptor-mediated endocytosis (see Figure 3.13 on page 73) for use in hemoglobin synthesis. Iron is needed for

Figure 19.6 Negative feedback regulation of erythropoiesis (red blood cell formation).

🔑 The main stimulus for erythropoiesis is a decrease in the oxygen-carrying capacity of the blood.

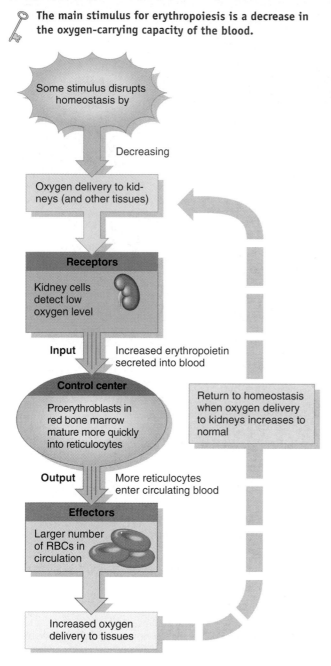

Some stimulus disrupts homeostasis by

Decreasing

Oxygen delivery to kidneys (and other tissues)

Receptors

Kidney cells detect low oxygen level

Input — Increased erythropoietin secreted into blood

Control center

Proerythroblasts in red bone marrow mature more quickly into reticulocytes

Return to homeostasis when oxygen delivery to kidneys increases to normal

Output — More reticulocytes enter circulating blood

Effectors

Larger number of RBCs in circulation

Increased oxygen delivery to tissues

Q How might your hematocrit change if you moved from a town at sea level to a high-mountain village?

the heme portion of the hemoglobin molecule, whereas amino acids are needed for the globin portion. Vitamin B_{12} is also needed for synthesis of hemoglobin.

8 Erythropoiesis in red bone marrow results in the production of red blood cells, which enter the circulation.

9 When iron is removed from heme, the non-iron portion of heme is converted to **biliverdin** (bil′-i-VER-din), a green pigment, and then into **bilirubin** (bil′-i-ROO-bin), a yellow-orange pigment.

10 Bilirubin enters the blood and is transported to the liver.

11 Within the liver, bilirubin is secreted by liver cells into bile, which passes into the small intestine and then into the large intestine.

12 In the large intestine, bacteria convert bilirubin into **urobilinogen** (yoo′-rō-bi-LIN-ō-jen).

13 Some urobilinogen is absorbed back into the blood, converted to a yellow pigment called **urobilin** (yoo′-rō-BĪ-lin), and excreted in urine.

14 Most urobilinogen is eliminated in feces in the form of a brown pigment called **stercobilin** (ster′-kō-BĪ-lin), which gives feces its characteristic color.

Erythropoiesis: Production of RBCs

Erythropoiesis (e-rith′-rō-poy-Ē-sis), the production of RBCs, starts in the red bone marrow with a precursor cell called a proerythroblast (see Figure 19.3). The proerythroblast gives rise to several cells that begin to synthesize hemoglobin. Ultimately, a cell near the end of the developmental sequence ejects its nucleus and becomes a **reticulocyte** (re-TIK-yoo-lō-sīt). Loss of the nucleus causes the center of the cell to indent, producing a distinctive biconcave shape. Reticulocytes, which are about 34% hemoglobin and retain some mitochondria, ribosomes, and endoplasmic reticulum, pass from red bone marrow into the bloodstream by squeezing between the endothelial cells of blood capillaries. Reticulocytes usually develop into **erythrocytes,** or mature red blood cells, within 1–2 days after their release from bone marrow.

Normally, erythropoiesis and red blood cell destruction proceed at roughly the same pace. If the oxygen-carrying capacity of the blood falls because erythropoiesis is not keeping up with RBC destruction, a negative feedback system steps up RBC production (Figure 19.6). The controlled condition is the amount of oxygen delivered to body tissues. Cellular oxygen deficiency, called **hypoxia** (hī-POKS-ē-a), may occur if not enough oxygen enters the blood. For example, the reduced oxygen content of air at high altitudes results in a reduced level of oxygen in the blood. Oxygen delivery may also fall due to anemia, which has many causes; lack of iron, lack of certain amino acids, and lack of vitamin B_{12} are but a few (see page 630). Circulatory problems that reduce blood flow to tissues may also reduce oxygen delivery. Whatever the cause, hypoxia stimulates the kidneys to step up the release of erythropoietin. This hormone circulates through the blood to the red bone marrow, where it speeds the development of proerythroblasts into reticulocytes. Premature newborns often exhibit anemia, due in part to inadequate production of erythropoietin. During the first weeks after birth, the liver, not the kidneys, produces most EPO. Because the liver is less sensitive than the kidneys to hypoxia, newborns have a smaller EPO response to anemia than do adults.

CLINICAL APPLICATION
Reticulocyte Count

The rate of erythropoiesis is measured by a **reticulocyte count.** Normally, a little less than 1% of the oldest RBCs are replaced by newcomer reticulocytes on any given day. It then takes 1–2 days for the reticulocytes to lose the last vestiges of endoplasmic reticulum and become mature RBCs. Thus, reticulocytes account for about 0.5–1.5% of all RBCs in a normal blood sample. A low "retic" count in a person who is anemic might indicate an inability of the red bone marrow to respond to erythropoietin, perhaps because of a nutritional deficiency or leukemia. A high "retic" count might indicate a good red bone marrow response to previous blood loss or to iron therapy in someone who had been iron deficient. ■

1. Describe the size and microscopic appearance of RBCs.
2. Explain how hemoglobin is recycled.
3. Define erythropoiesis. Relate erythropoiesis to hematocrit. What factors accelerate and slow erythropoiesis?

WHITE BLOOD CELLS

OBJECTIVE

• *Describe the structure, functions, and production of white blood cells.*

WBC Anatomy and Types

Unlike red blood cells, white blood cells or **leukocytes** (LOO-kō-sīts; *leuko-* = white) have a nucleus and do not contain hemoglobin (Figure 19.7). WBCs are classified as either granular or agranular, depending on whether they contain conspicuous chemical-filled cytoplasmic vesicles (originally called granules) that are made visible by staining. *Granular leukocytes* include neutrophils, eosinophils, and basophils; *agranular leukocytes* include lymphocytes and monocytes. As shown in Figure 19.3, monocytes and granular leukocytes develop from a myeloid stem cell. In contrast, lymphocytes develop from a lymphoid stem cell.

Granular Leukocytes

After staining, each of the three types of granular leukocytes display conspicuous granules with distinctive coloration that can be recognized under a light microscope. The large, uniform-sized granules of an **eosinophil** (ē-ō-SIN-ō-fil) are *eosinophilic* (= eosin-loving)—they stain red-orange with acidic dyes (Figure 19.7a). The granules usually do not cover or obscure the nucleus, which most often has two or three lobes connected by either a thin strand or a thick strand. The round, variable-sized granules of a **basophil** (BĀ-sō-fil) are *basophilic* (= basic loving)—they stain blue-purple with basic dyes (Figure 19.7b). The granules commonly obscure the nucleus, which has two lobes. The granules of a **neutrophil** (NOO-trō-fil) are smaller, evenly distributed, and pale lilac in color (Figure 19.7c); the nucleus

Figure 19.7 Structure of white blood cells.

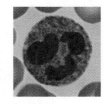

White blood cells are distinguished from one another by the shape of their nuclei and the staining properties of their cytoplasmic granules.

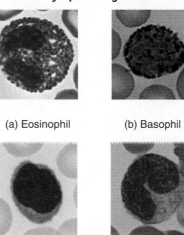

(a) Eosinophil   (b) Basophil   (c) Neutrophil

 LM   all at 1500x

(d) Small lymphocyte   (e) Monocyte

 Which WBCs are called granular leukocytes? Why?

has two to five lobes, connected by very thin strands of chromatin. As the cells age, the number of nuclear lobes increases. Because older neutrophils thus have several differently shaped nuclear lobes, they are often called *polymorphonuclear leukocytes (PMNs)*, polymorphs, or "polys." Younger neutrophils are often called *bands* because their nucleus is more rod-shaped.

Agranular Leukocytes

Even though so-called agranular leukocytes possess cytoplasmic granules, the granules are not visible under a light microscope because of their small size and poor staining qualities.

The nucleus of a **lymphocyte** (LIM-fō-sīt) stains dark and is round or slightly indented. The cytoplasm stains sky blue and forms a rim around the nucleus. The larger the cell, the more cytoplasm is visible. Lymphocytes are classified as large or small based on cell diameter: 6–9 μm in small lymphocytes and 10–14 μm in large lymphocytes (Figure 19.7d). (Although the functional significance of the size difference between small and large lymphocytes is unclear, the distinction is still useful clinically because an increase in the number of large lymphocytes has diagnostic significance in acute viral infections and in some immunodeficiency diseases.)

Monocytes (MON-ō-sīts) are 12–20 μm in diameter (Figure 19.7e). The nucleus of a monocyte is usually kidney-shaped or horseshoe-shaped, and the cytoplasm is blue-gray

Figure 19.8 Emigration of white blood cells.

🔑 **Adhesion molecules (selectins and integrins) assist the emigration of WBCs from the bloodstream into interstitial fluid.**

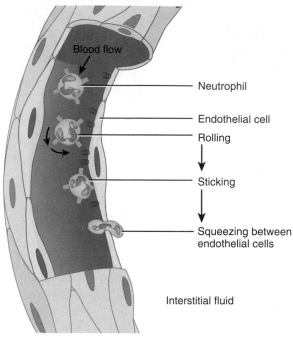

Blood flow

Neutrophil

Endothelial cell

Rolling

Sticking

Squeezing between endothelial cells

Interstitial fluid

Key:

⑧ Selectins on endothelial cells

▇ Integrins on neutrophil

Q In what way is the "traffic pattern" of lymphocytes in the body different from those of other WBCs?

and has a foamy appearance. The blood is merely a conduit for monocytes, which migrate out into the tissues, enlarge, and differentiate into **macrophages** (= large eaters). Some are **fixed macrophages,** which means they reside in a particular tissue; examples are alveolar macrophages in the lungs, macrophages in the spleen, or stellate reticuloendothelial (Kupffer) cells in the liver. Others are **wandering macrophages,** which roam the tissues and gather at sites of infection or inflammation.

White blood cells and other nucleated body cells have proteins, called *major histocompatibility (MHC) antigens,* protruding from their plasma membrane into the extracellular fluid. These "cell identity markers" are unique for each person (except identical twins). Although RBCs possess blood group antigens, they lack the MHC antigens.

WBC Physiology

In a healthy body, some WBCs, especially lymphocytes, can live for several months or years, but most live only a few days. During a period of infection, phagocytic WBCs may

live only a few hours. WBCs are far less numerous than red blood cells, about 5000–10,000 cells per μL of blood. RBCs therefore outnumber white blood cells by about 700:1. **Leukocytosis** (loo′-kō-sī-TŌ-sis), an increase in the number of WBCs, is a normal, protective response to stresses such as invading microbes, strenuous exercise, anesthesia, and surgery. An abnormally low level of white blood cells (below 5000/μL) is termed **leukopenia** (loo-kō-PĒ-nē-a). It is never beneficial and may be caused by radiation, shock, and certain chemotherapeutic agents.

The skin and mucous membranes of the body are continuously exposed to microbes and their toxins. Some of these microbes can invade deeper tissues to cause disease. Once pathogens enter the body, the general function of white blood cells is to combat them by phagocytosis or immune responses. To accomplish these tasks, many WBCs leave the bloodstream and collect at points of pathogen invasion or inflammation. Once granulocytes and monocytes leave the bloodstream to fight injury or infection, they never return to it. Lymphocytes, on the other hand, continually recirculate—from blood to interstitial spaces of tissues to lymphatic fluid and back to blood. Only 2% of the total lymphocyte population is circulating in the blood at any given time; the rest are in lymphatic fluid and organs such as skin, lungs, lymph nodes, and spleen.

WBCs leave the bloodstream by a process termed **emigration** (em′-i-GRĀ-shun; *e-* = out; *migra-* = wander) in which they roll along the endothelium, stick to it, and then squeeze between endothelial cells (Figure 19.8). The precise signals that stimulate emigration through a particular blood vessel vary for the different types of WBCs. Molecules known as **adhesion molecules** help WBCs stick to the endothelium. For example, endothelial cells display adhesion molecules called *selectins* in response to nearby injury and inflammation. Selectins stick to carbohydrates on the surface of neutrophils, causing them to slow down and roll along the endothelial surface. On the neutrophil surface are other adhesion molecules called *integrins,* which tether neutrophils to the endothelium and assist their movement through the blood vessel wall and into the interstitial fluid of the injured tissue.

Neutrophils and macrophages are active in **phagocytosis;** they can ingest bacteria and dispose of dead matter (see Figure 3.14 on page 74). Several different chemicals released by microbes and inflamed tissues attract phagocytes, a phenomenon called **chemotaxis.** Among the substances that provide stimuli for chemotaxis are toxins produced by microbes; kinins, which are specialized products of damaged tissues; and some of the colony-stimulating factors. The CSFs also enhance the phagocytic activity of neutrophils and macrophages.

Among WBCs, neutrophils respond most quickly to tissue destruction by bacteria. After engulfing a pathogen during phagocytosis, a neutrophil unleashes several destructive chemicals. These chemicals include the enzyme **lysozyme,**

Table 19.2 Significance of Elevated and Depressed White Blood Cell Counts

WBC TYPE	HIGH COUNT MAY INDICATE	LOW COUNT MAY INDICATE
Neutrophils	Bacterial infection, burns, stress, inflammation.	Radiation exposure, drug toxicity, vitamin B_{12} deficiency, systemic lupus erythematosus (SLE).
Lymphocytes	Viral infections, some leukemias.	Prolonged illness, immunosuppression, treatment with cortisol.
Monocytes	Viral or fungal infections, tuberculosis, some leukemias, other chronic diseases.	Bone marrow depression, treatment with cortisol.
Eosinophils	Allergic reactions, parasitic infections, autoimmune diseases.	Drug toxicity, stress.
Basophils	Allergic reactions, leukemias, cancers, hypothyroidism.	Pregnancy, ovulation, stress, hyperthyroidism.

which destroys certain bacteria, and **strong oxidants,** such as the superoxide anion (O_2^-), hydrogen peroxide (H_2O_2), and the hypochlorite anion (OCl^-), which is similar to household bleach. Neutrophils also contain **defensins,** proteins that exhibit a broad range of antibiotic activity against bacteria and fungi. Defensins form peptide "spears" that poke holes in microbe membranes; the resulting loss of cellular contents kills the invader.

Monocytes take longer to reach a site of infection than do neutrophils, but they arrive in larger numbers and destroy more microbes. Upon arrival they enlarge and differentiate into wandering macrophages, which clean up cellular debris and microbes following an infection.

Eosinophils leave the capillaries and enter tissue fluid. They are believed to release enzymes, such as histaminase, that combat the effects of histamine and other mediators of inflammation in allergic reactions. Eosinophils also phagocytize antigen–antibody complexes and are effective against certain parasitic worms. A high eosinophil count often indicates an allergic condition or a parasitic infection.

Basophils are also involved in inflammatory and allergic reactions. They leave capillaries, enter tissues, and develop into mast cells, which can liberate heparin, histamine, and serotonin. These substances intensify the inflammatory reaction and are involved in hypersensitivity (allergic) reactions (see Chapter 22).

The major types of lymphocytes are B cells, T cells, and natural killer cells, which are the major combatants in immune responses (described in detail in Chapter 22). B cells are particularly effective in destroying bacteria and inactivating their toxins. T cells attack viruses, fungi, transplanted cells, cancer cells, and some bacteria. Immune responses carried out by B cells and T cells help combat infection and provide protection against some diseases. T cells also are responsible for transfusion reactions, allergies, and rejection of transplanted organs. Natural killer cells attack a wide variety of infectious microbes and certain spontaneously arising tumor cells.

An increase in the number of circulating WBCs usually indicates inflammation or infection. A physician may order a **differential white blood cell count** to detect infection or inflammation, determine the effects of possible poisoning by chemicals or drugs, monitor blood disorders (for example, leukemia) and effects of chemotherapy, or detect allergic reactions and parasitic infections. Because each type of white blood cell plays a different role, determining the *percentage* of each type in the blood assists in diagnosing the condition. Table 19.2 lists the significance of both elevated and depressed WBC counts.

CLINICAL APPLICATION
Bone Marrow Transplant

A **bone marrow transplant** is the intravenous transfer of red bone marrow from a healthy donor to a recipient, with the goal of establishing normal hemopoiesis and consequently normal blood cell counts in the recipient. In patients with cancer or certain genetic diseases, the defective red bone marrow first must be destroyed by high doses of chemotherapy and whole body radiation. The donor marrow must be very closely matched to that of the recipient, or rejection occurs. When a transplant is successful, stem cells from the donor marrow reseed and grow in the recipient's red bone marrow cavities. Bone marrow transplants have been used to treat aplastic anemia, certain types of leukemia, severe combined immunodeficiency disease (SCID), Hodgkin's disease, non-Hodgkin's lymphoma, multiple myeloma, thalassemia, sickle-cell disease, breast cancer, ovarian cancer, testicular cancer, and hemolytic anemia. ■

1. Explain the importance of emigration, chemotaxis, and phagocytosis in fighting bacterial invaders.
2. Distinguish between leukocytosis and leukopenia.
3. What is a differential white blood cell count?
4. What functions are performed by B cells and T cells?

PLATELETS

OBJECTIVE

• *Describe the structure, function, and origin of platelets.*

Besides the immature cell types that develop into erythrocytes and leukocytes, hemopoietic stem cells also differentiate into cells that produce platelets. Under the influence of the hormone **thrombopoietin,** myeloid stem cells develop into megakaryocyte-colony-forming cells that, in turn, develop into precursor cells called megakaryoblasts (see Figure 19.3). Megakaryoblasts transform into megakaryocytes, huge cells that splinter into 2000–3000 fragments. Each fragment, enclosed by a piece of the cell membrane, is a **platelet (thrombocyte).** Platelets break off from the megakaryocytes in red bone marrow and then enter the blood circulation. Between 150,000 and 400,000 platelets are present in each μL of blood. They are disc-shaped, 2–4 μm in diameter, and exhibit many granules but no nucleus. Platelets help stop blood loss from damaged blood vessels by forming a platelet plug. Their granules also contain chemicals that, once released, promote blood clotting. Platelets have a short life span, normally just 5–9 days. Aged and dead platelets are removed by fixed macrophages in the spleen and liver.

Table 19.3 summarizes the formed elements in blood.

CLINICAL APPLICATION
Complete Blood Count

A **complete blood count (CBC)** is a valuable test that screens for anemia and various infections. Usually included are counts of RBCs, WBCs, and platelets per μL of whole blood; hematocrit; and differential white blood cell count. The amount of hemoglobin in grams per mL of blood also is determined. Normal hemoglobin ranges are: infants, 14–20 g/100 mL of blood; adult females, 12–16 g/100 mL of blood; and adult males, 13.5–18 g/100 mL of blood.

1. Compare RBCs, WBCs, and platelets with respect to size, number per μL, and life span.

HEMOSTASIS

OBJECTIVES

• *Describe the mechanisms that contribute to hemostasis.*

• *Identify the stages involved in blood clotting, and explain the various factors that promote and inhibit blood clotting.*

Hemostasis (hē-mō-STĀ-sis) is a sequence of responses that stops bleeding. When blood vessels are damaged or ruptured, the hemostatic response must be quick, localized to the region of damage, and carefully controlled. Three mechanisms reduce blood loss: (1) vascular spasm, (2) platelet plug formation, and (3) blood clotting (coagulation). When successful, hemostasis prevents **hemorrhage** (HEM-or-ij; *-rhage* = burst forth), the loss of a large amount of blood from the vessels. Hemostatic mechanisms can prevent hemorrhage from smaller blood vessels, but extensive hemorrhage from larger vessels usually requires medical intervention.

Vascular Spasm

When arteries or arterioles are damaged, the circularly arranged smooth muscle in their walls contracts immediately. Such a **vascular spasm** reduces blood loss for several minutes to several hours, during which time the other hemostatic mechanisms go into operation. The spasm is probably caused by damage to the smooth muscle and by reflexes initiated by pain receptors.

Platelet Plug Formation

In their unstimulated state, platelets are disc-shaped. Considering their small size, platelets store an impressive array of chemicals. Two types of granules are present in the cytoplasm: (1) **Alpha granules** contain clotting factors and **platelet-derived growth factor (PDGF),** which can cause proliferation of vascular endothelial cells, vascular smooth muscle fibers, and fibroblasts to help repair damaged blood vessel walls; (2) **dense granules** contain ADP, ATP, Ca^{2+}, and serotonin. Also present are enzymes that produce thromboxane A2, a prostaglandin; *fibrin-stabilizing factor,* which helps to strengthen a blood clot; lysosomes; some mitochondria; membrane systems that take up and store calcium and provide channels for release of the contents of granules; and glycogen.

Platelet plug formation occurs as follows (Figure 19.9):

1 Initially, platelets contact and stick to parts of a damaged blood vessel, such as collagen fibers of the connective tissue underlying the damaged endothelial cells. This process is called **platelet adhesion.**

2 As a result of adhesion, the platelets become activated, and their characteristics change dramatically. They extend many projections that enable them to contact and interact with one another, and they begin to liberate the contents of their granules. This phase is called the **platelet release reaction.** Liberated ADP and thromboxane A2 play a major role by activating nearby platelets. Serotonin and thromboxane A2 function as vasoconstrictors, causing and sustaining contraction of vascular smooth muscle, which decreases blood flow through the injured vessel.

3 The release of ADP makes other platelets in the area sticky, and the stickiness of the newly recruited and activated platelets causes them to adhere to the originally activated platelets. This gathering of platelets is called **platelet aggregation.** Eventually, the accumulation and attachment of large numbers of platelets form a mass called a **platelet plug.**

A platelet plug is very effective in preventing blood loss in a small vessel. Although initially the platelet plug is loose, it becomes quite tight when reinforced by fibrin threads formed during clotting (see Figure 19.10). A platelet plug can stop blood loss completely if the hole in a blood vessel is small enough.

Figure 19.9 Platelet plug formation.

🔑 **A platelet plug can stop blood loss completely if the hole in a blood vessel is small enough.**

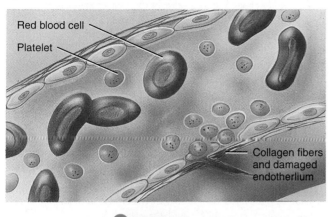

Red blood cell
Platelet
Collagen fibers and damaged endotherlium

1 Platelet adhesion

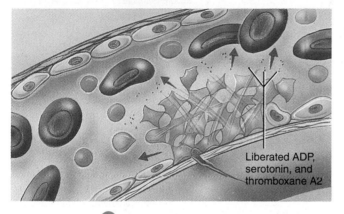

Liberated ADP, serotonin, and thromboxane A2

2 Platelet release reaction

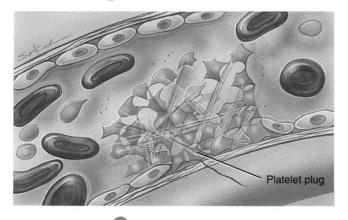

Platelet plug

3 Platelet aggregation

Q Besides platelet plug formation, which two mechanisms also contribute to hemostasis?

Figure 19.10 Scanning electron micrograph of a portion of a blood clot, showing a platelet and red blood cells trapped by fibrin threads.

🔑 **A blood clot is a gel that contains formed elements of the blood entangled in fibrin threads.**

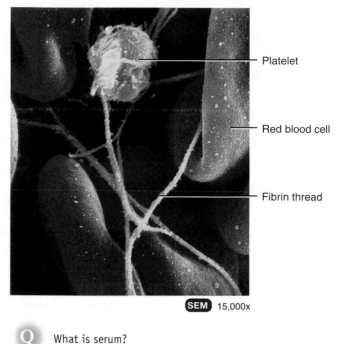

Platelet

Red blood cell

Fibrin thread

SEM 15,000x

Q What is serum?

Blood Clotting

Normally, blood remains in its liquid form as long as it stays within its vessels. If it is drawn from the body, however, it thickens and forms a gel. Eventually, the gel separates from the liquid. The straw-colored liquid, called **serum,** is simply plasma minus the clotting proteins. The gel is called a **clot** and consists of a network of insoluble protein fibers called fibrin in which the formed elements of blood are trapped (Figure 19.10).

The process of gel formation, called **clotting** or **coagulation,** is a series of chemical reactions that culminates in formation of fibrin threads. If blood clots too easily, the result can be **thrombosis**—clotting in an unbroken blood vessel. If the blood takes too long to clot, hemorrhage can result.

Clotting involves several substances known as **clotting (coagulation) factors.** These factors include calcium ions (Ca^{2+}), several inactive enzymes that are synthesized by hepatocytes (liver cells) and released into the bloodstream, and various molecules associated with platelets or released by damaged tissues. Many clotting factors are identified by Roman numerals that indicate the order of their discovery, and not necessarily the order of their participation in the clotting process.

Clotting is a complex cascade of reactions in which each clotting factor activates the next one in a fixed sequence.

Table 19.3 Summary of the Formed Elements in Blood

NAME AND APPEARANCE	NUMBER	CHARACTERISTICS[*]	FUNCTIONS
Red blood cells (RBCs) *or erythrocytes*	4.8 million/μL in females; 5.4 million/μL in males.	7–8 μm diameter, biconcave discs, without a nucleus; live for about 120 days.	Hemoglobin within RBCs transports most of the oxygen and part of the carbon dioxide in the blood. It also transports NO and SNO, which help adjust blood pressure.
White blood cells (WBCs) *or leukocytes*	5000–10,000/μL	Most live for a few hours to a few days.[†]	Combat pathogens and other foreign substances that enter the body.
Granular leukocytes			
Neutrophils	60–70% of all WBCs.	10–12 μm diameter; nucleus has 2–5 lobes connected by thin strands of chromatin; cytoplasm has very fine, pale lilac granules.	Phagocytosis. Destruction of bacteria with lysozyme, defensins, and strong oxidants, such as superoxide anion, hydrogen peroxide, and hypochlorite anion.
Eosinophils	2–4% of all WBCs.	10–12 μm diameter; nucleus has 2 or 3 lobes; large, red-orange granules fill the cytoplasm.	Combat the effects of histamine in allergic reactions, phagocytize antigen–antibody complexes, and destroy certain parasitic worms.
Basophils	0.5–1% of all WBCs.	8–10 μm diameter; nucleus has 2 lobes; large cytoplasmic granules appear deep blue-purple.	Liberate heparin, histamine, and serotonin in allergic reactions that intensify the overall inflammatory response.
Agranular leukocytes			
Lymphocytes (T cells, B cells, and natural killer cells)	20–25% of all WBCs.	Small lymphocytes are 6–9 μm in diameter; large lymphocytes are 10–14 mm in diameter; nucleus is round or slightly indented; cytoplasm forms a rim around the nucleus that looks sky blue; the larger the cell, the more cytoplasm is visible.	Mediate immune responses, including antigen–antibody reactions. B cells develop into plasma cells, which secrete antibodies. T cells attack invading viruses, cancer cells, and transplanted tissue cells. Natural killer cells attack a wide variety of infectious microbes and certain spontaneously arising tumor cells.
Monocytes	3–8% of all WBCs.	12–20 μm diameter; nucleus is kidney-shaped or horseshoe-shaped; cytoplasm is blue-gray and has foamy appearance.	Phagocytosis (after transforming into fixed or wandering macrophages).
Platelets (thrombocytes)	150,000–400,000/μL.	2–4 μm diameter cell fragments that live for 5–9 days; contain many granules but no nucleus.	Form platelet plug in hemostasis; release chemicals that promote vascular spasm and blood clotting.

[*]Colors are those seen when using Wright's stain.

[†]Some lymphocytes, called T and B memory cells, can live for many years once they are established.

Each clotting factor activates many molecules of the next one, and so on, so that at the end of the process a large quantity of product (fibrin) is formed. Clotting can be divided into three basic stages (Figure 19.11):

Stage ❶ The formation of prothrombinase (prothrombin activator), is initiated by either (or both) of two pathways called the extrinsic pathway and the intrinsic pathway.

Once prothrombinase is formed, the steps involved in the next two stages of clotting are the same for both the extrinsic and intrinsic pathways, and together these two stages are referred to as the common pathway.

Stage ❷ Prothrombinase and Ca^{2+} convert prothrombin (a plasma protein formed by the liver) into the enzyme thrombin.

Stage ❸ Soluble fibrinogen (another plasma protein formed by the liver) is converted into insoluble fibrin by thrombin. Fibrin forms the threads of the clot.

Stage ❶: Formation of Prothrombinase

As we just noted, the formation of prothrombinase is initiated by either or both of two mechanisms: the extrinsic and intrinsic pathways of blood clotting.

The Extrinsic Pathway The **extrinsic pathway** of blood clotting has fewer steps than the intrinsic pathway and occurs rapidly—within a matter of seconds if trauma is severe. It is so named because a tissue protein called **tissue factor (TF),** also known as **thromboplastin,** leaks into the blood from cells *outside (extrinsic to)* blood vessels and initiates the formation of prothrombinase. TF is a complex mixture of lipoproteins and phospholipids released from the surfaces of damaged cells. In the presence of Ca^{2+}, TF begins a sequence of reactions that ultimately activates clotting factor X (see Figure 19.11a). Once factor X is activated, it combines with factor V in the presence of Ca^{2+} to form the active enzyme prothrombinase, completing the extrinsic pathway.

The Intrinsic Pathway The **intrinsic pathway** of blood clotting is more complex than the extrinsic pathway, and it occurs more slowly, usually requiring several minutes. The intrinsic pathway is so named because its activators are either in direct contact with blood or contained *within (intrinsic to)* the blood; outside tissue damage is not needed. If endothelial cells become roughened or damaged, blood can come in contact with collagen fibers in the connective tissue beneath the endothelium of the blood vessel. In addition, trauma to endothelial cells causes damage to platelets, resulting in the release of phospholipids by the platelets. Contact with collagen fibers (or with the slippery glass sides of a blood collection tube) activates clotting factor XII (see Figure 19.11b), which begins a sequence of reactions that ultimately activates clotting factor X. Platelet phospholipids and Ca^{2+} can also participate in the activation of factor X. Once factor X is activated, it combines with factor V to form the active enzyme prothrombinase (just as occurs in the extrinsic pathway), completing the intrinsic pathway.

Stages ❷ and ❸: The Common Pathway

Once prothrombinase is formed, the common pathway follows. In the second stage of blood clotting (see Figure 19.11c), prothrombinase and Ca^{2+} catalyze the conversion of prothrombin to thrombin. In the third stage, thrombin, in the presence of Ca^{2+}, converts fibrinogen, which is soluble, to loose fibrin threads, which are insoluble. Thrombin also activates factor XIII (fibrin stabilizing factor), which strengthens and stabilizes the fibrin threads into a sturdy clot. Plasma contains some factor XIII, which is also released by platelets trapped in the clot.

Thrombin has two positive feedback effects. In the first positive feedback loop, which involves factor V, it accelerates

Figure 19.11 The blood clotting cascade.

 In clotting, coagulation factors activate subsequent factors in sequence, resulting in a cascade of reactions that includes positive feedback cycles.

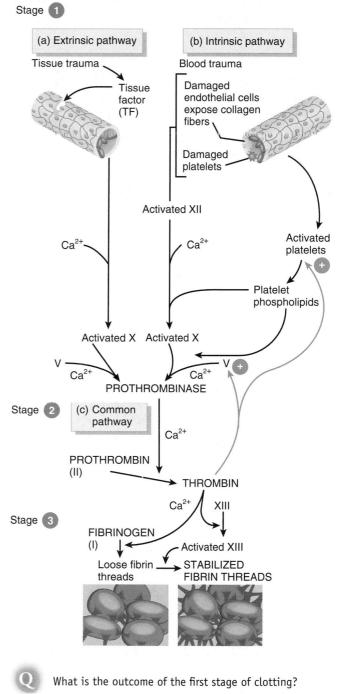

Q What is the outcome of the first stage of clotting?

the formation of prothrombinase. Prothrombinase, in turn, accelerates the production of more thrombin, and so on. In the second positive feedback loop, thrombin activates platelets, which reinforces their aggregation and the release of platelet phospholipids.

Table 19.4 Clotting (Coagulation) Factors

NUMBER*	NAME(S)	SOURCE	PATHWAY(S) OF ACTIVATION
I	Fibrinogen.	Liver.	Common.
II	Prothrombin.	Liver.	Common.
III	Tissue factor (thromboplastin).	Damaged tissues and activated platelets.	Extrinsic.
IV	Calcium ions (Ca^{2+}).	Diet, bones, and platelets.	All.
V	Proaccelerin, labile factor, or accelerator globulin (AcG).	Liver and platelets.	Extrinsic and intrinsic.
VII	Serum prothrombin conversion accelerator (SPCA), stable factor, or proconvertin.	Liver.	Extrinsic.
VIII	Antihemophilic factor (AHF), antihemophilic factor A, or antihemophilic globulin (AHG).	Platelets and endothelial cells.	Intrinsic.
IX	Christmas factor, plasma thromboplastin component (PTC), or antihemophilic factor B.	Liver.	Intrinsic.
X	Stuart factor, Prower factor, or thrombokinase.	Liver.	Extrinsic and intrinsic.
XI	Plasma thromboplastin antecedent (PTA) or antihemophilic factor C.	Liver.	Intrinsic.
XII	Hageman factor, glass factor, contact factor, or antihemophiliac factor D.	Liver.	Intrinsic.
XIII	Fibrin-stabilizing factor (FSF).	Liver and platelets.	Common.

*There is no factor VI. Prothrombinase (prothrombin activator) is a combination of activated factors V and X.

Clot Retraction and Blood Vessel Repair

Once a clot is formed, it plugs the ruptured area of the blood vessel and thus stops blood loss. **Clot retraction** is the consolidation or tightening of the fibrin clot. The fibrin threads attached to the damaged surfaces of the blood vessel gradually contract as platelets pull on them. As the clot retracts, it pulls the edges of the damaged vessel closer together, decreasing the risk of further damage. During retraction, some serum can escape between the fibrin threads, but the formed elements in blood cannot. Normal retraction depends on an adequate number of platelets in the clot, which release factor XIII and other factors, thereby strengthening and stabilizing the clot. Permanent repair of the blood vessel can then take place. In time, fibroblasts form connective tissue in the ruptured area, and new endothelial cells repair the vessel lining.

Role of Vitamin K in Clotting

Normal clotting depends on adequate vitamin K in the body. Although vitamin K is not involved in actual clot formation, it is required for the synthesis of four clotting factors by hepatocytes: factors II (prothrombin), VII, IX, and X. Vitamin K, which is normally produced by bacteria that inhabit the large intestine, is a fat-soluble vitamin, and it can be absorbed through the lining of the intestine and into the blood only if absorption of lipids is normal. People suffering from disorders that slow absorption of lipids (for example, inadequate release of bile into the small intestine) often experience uncontrolled bleeding as a consequence of vitamin K deficiency.

The various clotting factors, their sources, and the pathways of activation are summarized in Table 19.4.

Hemostatic Control Mechanisms

Many times a day little clots start to form, often at a site of minor roughness or at a developing atherosclerotic plaque inside a blood vessel. Because blood clotting involves amplification and positive feedback cycles, a clot has a tendency to enlarge, creating the potential for impairment of blood flow through undamaged vessels. The **fibrinolytic system** (fī-bri-nō-LIT-ik) dissolves small, inappropriate clots; it also dissolves clots at a site of damage once the damage is repaired. Dissolution of a clot is called **fibrinolysis** (fī-bri-NOL-i-sis). When a clot is formed, an inactive plasma enzyme called **plasminogen** is incorporated into the clot. Both body tissues and blood contain substances that can activate plasminogen to **plasmin (fibrinolysin),** an active plasma enzyme. Among these substances are thrombin, activated factor XII, and tissue plasminogen activator (t-PA), which is synthesized in endothelial cells of most tissues and liberated into the blood. Once plasmin is formed, it can dissolve the clot by digesting fibrin threads and inactivating substances such as fibrinogen, prothrombin, and factors V, VIII, and XII.

Even though thrombin has a positive feedback effect on blood clotting, clot formation normally remains localized at the site of damage. A clot does not extend beyond a wound site into the general circulation, in part because fibrin absorbs thrombin into the clot. Another reason for localized clot formation is that as a result of the dispersal of some of the clotting factors by the blood, their concentrations are not high enough to bring about widespread clotting.

Several other mechanisms also operate to control blood clotting. For example, endothelial cells and white blood cells produce a prostaglandin called **prostacyclin** that opposes the actions of thromboxane A2. Prostacyclin is a powerful inhibitor of platelet adhesion and release.

Moreover, substances that inhibit clotting, called **anticoagulants,** are present in blood. These include **antithrombin III (AT-III),** which blocks the action of factors XII, XI, IX, X, and II (thrombin); **protein C,** which inactivates factors V and VIII, the two major clotting factors not blocked by AT-III, and enhances activity of plasminogen activators; **alpha-2-macroglobulin,** which inactivates thrombin and plasmin; and **alpha-1-antitrypsin,** which inhibits factor XI. **Heparin,** another anticoagulant that is produced by mast cells and basophils, combines with AT-III and increases its effectiveness in blocking thrombin. Heparin is also a pharmacologic anticoagulant extracted from animal lung tissue and intestinal mucosa.

Intravascular Clotting

Despite the anticoagulating and fibrinolytic mechanisms, blood clots sometimes form within the cardiovascular system. Such clots may be initiated by roughened endothelial surfaces of a blood vessel resulting from atherosclerosis, trauma, or infection. These conditions induce adhesion of platelets. Intravascular clots may also form when blood flows too slowly (stasis), allowing clotting factors to accumulate locally in high enough concentrations to initiate coagulation. Clotting in an unbroken blood vessel (usually a vein) is called **thrombosis** (*thromb-* = clot; *-osis* = a condition of). The clot itself is a **thrombus,** and it may dissolve spontaneously. If it remains intact, however, the thrombus may become dislodged and be swept away in the blood. A blood clot, bubble of air, fat from broken bones, or a piece of debris transported by the bloodstream is called an **embolus** (*em-* = in; *bolus* = a mass). An embolus that breaks away from an arterial wall may lodge in a smaller-diameter artery downstream and block blood flow to a vital organ. When an embolus lodges in the lungs, the condition is called **pulmonary embolism.**

At low doses, aspirin inhibits vasoconstriction and platelet aggregation by blocking synthesis of thromboxane A2, and it also reduces the chance of inappropriate clot formation. In patients with heart and blood vessel disease, the events of hemostasis may occur even without injury to a blood vessel. Aspirin reduces the risk of transient ischemic attacks (TIA) and strokes (see page 479), myocardial infarction (see page 664), and blockage of peripheral arteries.

CLINICAL APPLICATION
Anticoagulants and Thrombolytic Agents

Patients who are at increased risk of forming blood clots may receive an **anticoagulant drug,** a substance that delays, suppresses, or prevents blood clotting. Examples are heparin or warfarin. **Heparin** is often administered during hemodialysis and open heart surgery. **Warfarin (Coumadin)** acts as an antagonist to vitamin K and thus blocks synthesis of four clotting factors (II, VII, IX, and X). Warfarin is slower acting than heparin. To prevent clotting in donated blood, blood banks and laboratories often add a substance that removes Ca^{2+}; one example is CPD (citrate phosphate dextrose).

Thrombolytic agents are chemical substances that are injected into the body to dissolve blood clots and restore circulation. They either directly or indirectly activate plasminogen. The first thrombolytic agent, approved in 1982 for dissolving clots in the coronary arteries of the heart, was **streptokinase,** which is produced by streptococcal bacteria. A genetically engineered version of human **tissue plasminogen activator (t-PA)** is also used to treat both heart attacks and brain attacks (strokes) that are caused by blood clots. ■

1. Define hemostasis. Explain the mechanism involved in vascular spasm and platelet plug formation.
2. What is fibrinolysis? Why does blood rarely remain clotted inside blood vessels?
3. How do the extrinsic and intrinsic pathways of blood clotting differ?
4. Define the following: thrombus, embolus, anticoagulant, and thrombolytic agent.

BLOOD GROUPS AND BLOOD TYPES

OBJECTIVE
• *Explain the ABO and Rh blood groups.*

The surfaces of erythrocytes contain a genetically determined assortment of glycoproteins and glycolipids that can act as antigens. Called **isoantigens** or **agglutinogens** (agloo-TIN-ō-jens), these antigens occur in characteristic combinations. Based on the presence or absence of various isoantigens, blood is categorized into different **blood groups.** Within a given blood group there may be two or more different **blood types.** There are at least 24 blood groups and more than 100 isoantigens that can be detected on the surface of red blood cells. Here we discuss two major blood groups—ABO and Rh; among other blood groups are the Lewis, Kell, Kidd, and Duffy systems.

Figure 19.12 Antigens and antibodies of the ABO blood types.

🔑 **The antibodies in your plasma do not react with the antigens on your red blood cells.**

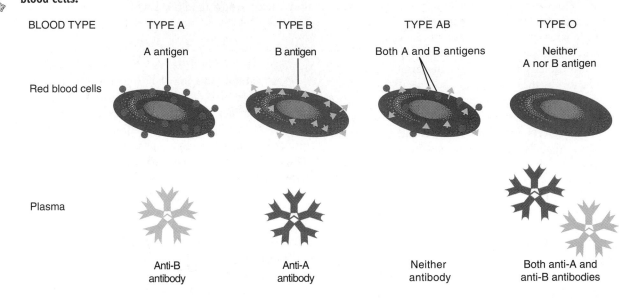

Q Which antibodies are usually present in type O blood?

Table 19.5 Incidence of Blood Types in the United States

POPULATION GROUP	BLOOD TYPE (PERCENTAGE)				
	O	A	B	AB	Rh⁺
White	45	40	11	4	85
Black	49	27	20	4	95
Korean	32	28	30	10	100
Japanese	31	38	21	10	100
Chinese	42	27	25	6	100
Native American	79	16	4	1	100

ABO Blood Group

The **ABO blood group** is based on two glycolipid isoantigens called A and B (Figure 19.12). People whose RBCs display *only antigen A* are said to have **type A** blood. Those who have *only antigen B* are **type B.** Individuals who have *both A and B antigens* are **type AB,** whereas those who have neither *antigen A nor B* are **type O.** The incidence of ABO blood types varies among different population groups, as indicated in Table 19.5.

In addition to isoantigens on RBCs, blood plasma usually contains **isoantibodies** or **agglutinins** (a-GLOO-ti-nins) that react with the A or B antigens if the two are mixed. These are **anti-A antibody,** which reacts with antigen A, and **anti-B antibody,** which reacts with antigen B. The antibodies present in each of the four blood types are shown in Figure 19.12. You do not have antibodies that react with the antigens of your own RBCs, but most likely you do have an-

tibodies for any antigens that your RBCs lack. Although isoantibodies start to appear in the blood within a few months after birth, the reason for their presence is not clear. Perhaps they are formed in response to bacteria that normally inhabit the gastrointestinal tract. Because the isoantibodies are large IgM type antibodies (see Table 22.3 on page 763) that do not cross the placenta, ABO incompatibility between a mother and her fetus rarely causes problems.

Rh Blood Group

The **Rh blood group** is so named because the antigen was discovered in the blood of the *Rhesus* monkey. The alleles of three genes may code for the Rh antigen. People whose RBCs have Rh antigens are designated Rh⁺ (Rh positive); those who lack Rh antigens are designated Rh⁻ (Rh negative). The incidence of Rh⁺ and Rh⁻ individuals in various populations is shown in Table 19.5. Normally, plasma does not contain anti-Rh antibodies. If an Rh⁻ person receives an Rh⁺ blood transfusion, however, the immune system starts to make anti-Rh antibodies that will remain in the blood. If a second transfusion of Rh⁺ blood is given later, the previously formed anti-Rh antibodies will cause **hemolysis** (rupture) of the RBCs in the donated blood, and a severe reaction may occur.

Hemolytic Disease of the Newborn

The most common problem with Rh incompatibility, **hemolytic disease of the newborn (HDN),** may arise during pregnancy (Figure 19.13). Normally, no direct contact occurs between maternal and fetal blood while a woman is pregnant. However, if a small amount of Rh$^+$ blood leaks from the fetus through the placenta into the bloodstream of an Rh$^-$ mother, the mother will start to make anti-Rh antibodies. Because the greatest possibility of fetal blood transfer occurs at delivery, the firstborn baby will be unaffected. If the mother becomes pregnant again, however, her anti-Rh antibodies can cross the placenta and enter the bloodstream of the fetus. If the fetus is Rh$^-$, there is no problem, because Rh$^-$ blood does not have the Rh antigen. If, however, the fetus is Rh$^+$, hemolysis brought on by fetal–maternal incompatibility may occur in the fetal blood. HDN is prevented by giving all Rh$^-$ women an injection of anti-Rh antibodies called anti-Rh gamma globulin (RhoGAM) soon after every delivery, miscarriage, or abortion. These antibodies bind to and inactivate the fetal Rh antigens, if they are present, and the mother's immune system does produce antibodies.

Transfusions

Despite the differences in RBC antigens reflected in the blood group systems, blood is the most easily shared of human tissues, saving many thousands of lives every year through transfusions. A **transfusion** (trans-FYOO-shun) is the transfer of whole blood or blood components (red blood cells only or plasma only) into the bloodstream. A transfusion is most often given to alleviate anemia or when blood volume is low—for example, after a severe hemorrhage. However, the normal components of one person's RBC plasma membrane can trigger damaging antigen–antibody responses in a transfusion recipient. In an incompatible blood transfusion, isoantibodies in the recipient's plasma bind to the isoantigens on the donated RBCs. When these antigen–antibody complexes form, they activate plasma proteins of the complement family (described on page 762). In essence, complement molecules make the plasma membrane of the donated RBCs leaky, causing them to hemolyze (burst) and release hemoglobin into the plasma. The liberated hemoglobin may cause kidney damage.

As an example of an incompatible blood transfusion, consider what happens if a person with type A blood receives a transfusion of type B blood. The recipient's blood (type A) contains A antigens on the red blood cells and anti-B antibodies in the plasma. The donor's blood (type B) contains B antigens and anti-A antibodies. In this situation, two things can happen. First, the anti-B antibodies in the recipient's plasma can bind to the B antigens on the donor's erythrocytes, causing hemolysis of the red blood cells. Second, the anti-A antibodies in the donor's plasma can bind to the A antigens on the recipient's red blood cells. However, the second reaction is usually not serious because the donor's anti-A antibodies become so diluted in the recipient's plasma that they do not cause any significant hemolysis of the recipient's

Figure 19.13 Development of hemolytic disease of the newborn (HDN). (a) At birth, a small quantity of fetal blood usually leaks across the placenta into the maternal bloodstream. (b) A problem can arise when the mother is Rh$^-$ and the baby is Rh$^+$, having inherited an allele for one of the Rh antigens from the father. Upon exposure to Rh antigen, the mother's immune system responds by making anti-Rh antibodies. (c) During a subsequent pregnancy, the maternal antibodies cross the placenta into the fetal blood. If the second fetus is Rh$^+$, the ensuing antigen–antibody reaction causes hemolysis of fetal RBCs. The result is HDN.

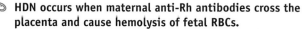

HDN occurs when maternal anti-Rh antibodies cross the placenta and cause hemolysis of fetal RBCs.

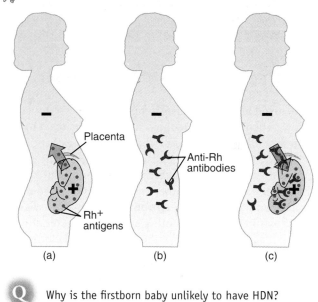

(a)   (b)   (c)

Q Why is the firstborn baby unlikely to have HDN?

RBCs. The interactions of the four blood types of the ABO system are summarized in Table 19.6.

People with type AB blood do not have any anti-A or anti-B antibodies in their plasma. They are sometimes called "universal recipients" because theoretically they can receive blood from donors of all four blood types. They have no antibodies to attack antigens on donated RBCs (see Table 19.6). People with type O blood have neither A nor B antigens on their RBCs and are sometimes called "universal donors" because theoretically they can donate blood to all four ABO blood types. Type O persons requiring blood may receive only type O blood (see Table 19.6). In practice, use of the terms *universal recipient* and *universal donor* is misleading and dangerous. Blood contains antigens and antibodies other than those associated with the ABO system, and they can cause transfusion problems. Thus, blood should be carefully cross-matched or screened before transfusion. In about 80% of the population, soluble antigens of the ABO type appear in saliva and other body fluids, in which case blood type can be identified from a sample of saliva.

Table 19.6 Summary of ABO Blood Group Interactions

BLOOD TYPE	A	B	AB	O
Antigen on RBCs	A	B	Both A and B	Neither A nor B
Antibody in plasma	anti-B	anti-A	Neither anti-A nor anti-B	Both anti-A and anti-B
Compatible donor blood types (no hemolysis)	A, O	B, O	A, B, AB, O	O
Incompatible donor blood types (hemolysis)	B, AB	A, AB	—	A, B, AB

Typing and Cross-Matching Blood for Transfusion

To avoid blood-type mismatches, laboratory technicians type the patient's blood and then either cross-match it to potential donor blood or screen it for the presence of antibodies. Outside the body, at room temperature, mixing of incompatible blood causes **agglutination** (clumping) that is visible to the naked eye, rather than hemolysis. Agglutination is an antigen–antibody response, whereby the cells become cross-linked to one another and form a visible clump. (Note that agglutination is not the same as blood clotting.)

In the procedure for ABO blood typing, single drops of blood are mixed with different antisera, solutions that contain antibodies. One drop of blood is mixed with anti-A serum, which contains anti-A antibodies that will agglutinate red blood cells that possess A antigens. Another drop is mixed with anti-B serum, which contains anti-B antibodies that will agglutinate red blood cells that possess B antigens. If the red blood cells agglutinate only when mixed with anti-A serum, the blood is type A. If the red blood cells agglutinate only when mixed with anti-B serum, the blood is type B. The blood is type AB if both drops agglutinate; if neither drop agglutinates, the blood is type O.

In the procedure for determining Rh factor, a drop of blood is mixed with antiserum containing antibodies that will agglutinate RBCs displaying Rh antigens. If the blood agglutinates, it is Rh^+; no agglutination indicates Rh^-.

Once the patient's blood type is known, donor blood of the same ABO and Rh type is selected. In a **cross-match,** the possible donor RBCs are mixed with the recipient's serum. If agglutination does not occur, the recipient does not have antibodies that will attack the donor RBCs. Alternatively, the recipient's serum can be **screened** against a test panel of RBCs having antigens known to cause blood transfusion reactions to detect any antibodies that may be present.

1. What is the basis for ABO blood grouping?
2. What is the basis for the Rh system?
3. What precautions must be taken before a person receives a blood transfusion?

DISORDERS: HOMEOSTATIC IMBALANCES

ANEMIA

Anemia is a condition in which the oxygen-carrying capacity of blood is reduced. Many kinds of anemia exist; all are characterized by reduced numbers of RBCs or a decreased amount of hemoglobin in the blood. The person feels fatigued and is intolerant of cold, both of which are related to lack of oxygen needed for ATP and heat production. Also, the skin appears pale, due to the low content of red-colored hemoglobin circulating in skin blood vessels. Among the most important types of anemia are the following:

- *Iron-deficiency anemia,* the most prevalent kind of anemia, is caused by inadequate absorption of iron, excessive loss of iron, increased iron requirement, or insufficient intake of iron. Women are at greater risk for iron-deficiency anemia due to menstrual blood losses and increased iron demands of the growing fetus during pregnancy. Gastrointestinal losses such as occurs with malignancy or ulceration also contribute to this type of anemia.
- *Pernicious anemia* is caused by insufficient hemopoiesis resulting from an inability of the stomach to produce intrinsic factor, which is needed for absorption of vitamin B_{12} in the small intestine.
- *Hemorrhagic anemia* is due to an excessive loss of RBCs through bleeding resulting from large wounds, stomach ulcers, or especially heavy menstruation.

- In *hemolytic anemia,* RBC plasma membranes rupture prematurely, and their hemoglobin pours into the plasma. The condition may result from inherited defects such as abnormal red blood cell enzymes, or from outside agents such as parasites, toxins, or antibodies from incompatible transfused blood.
- *Thalassemia* (thal′-a-SĒ-mē-a) is a group of hereditary hemolytic anemias associated with deficient synthesis of hemoglobin. The RBCs are small (microcytic), pale (hypochromic), and short-lived. Thalassemia occurs primarily in populations from countries bordering the Mediterranean Sea.
- *Aplastic anemia* results from destruction of the red bone marrow. Toxins, gamma radiation, and certain medications that inhibit enzymes needed for hemopoiesis are causes.

SICKLE-CELL DISEASE

The RBCs of a person with **sickle cell disease (SCD)** contain Hb-S, an abnormal kind of hemoglobin. When Hb-S gives up oxygen to the interstitial fluid, it forms long, stiff, rodlike structures that bend the erythrocyte into a sickle shape (Figure 19.14). The sickled cells rupture easily. Even though erythropoiesis is stimulated by the loss of the cells, it cannot keep pace with hemolysis; hemolytic anemia is the result. Prolonged oxygen reduction may eventually cause extensive tissue damage.

Sickle-cell disease is inherited. People with two sickle-cell genes have severe anemia, whereas those with only one defective gene may have minor problems. Sickle-cell genes are found primarily among populations, or descendants of populations, that live in the malaria belt around the world, including parts of Mediterranean Europe, sub-Saharan Africa, and tropical Asia. Some 1–2% of African Americans have SCD. The gene responsible for the tendency of the RBCs to sickle also alters the permeability of the plasma membranes of sickled cells, causing potassium ions to leak out. Low levels of potassium kill the malaria parasites that infect sickled cells. Because of this effect, a person with one normal gene and one sickle-cell gene has a high resistance to malaria. The possession of a single sickle-cell gene thus confers a survival advantage.

Treatment consists of administration of analgesics to relieve pain, fluid therapy to maintain hydration, oxygen to reduce the stimulus for the crisis, antibiotics to counter infections, and blood transfusions. People who suffer sickle-cell disease have normal fetal hemoglobin (Hb-F), a slightly different form of hemoglobin that predominates at birth and normally is present in small amounts thereafter. In some patients with sickle-cell disease, a drug called hydroxyurea promotes transcription of the normal Hb-F gene, elevates the level of Hb-F, and reduces the chance that the RBCs will sickle. Unfortunately, this drug also has toxic effects on the bone marrow; thus, its safety for long-term use is questionable.

HEMOPHILIA

Hemophilia (-*philia* = loving) is an inherited deficiency of clotting in which bleeding may occur spontaneously or after only minor trauma. Different types of hemophilia are due to deficiencies of different blood clotting factors and exhibit varying degrees of severity, ranging from mild to severe bleeding tendencies. The most common type (classic hemophilia) is hemophilia A, in which factor VIII is absent. People with hemophilia B lack factor IX. Hemophilia A and B occur primarily among males because these are sex-linked recessive disorders, whereas hemophilia C affects both males and females. Hemophilia C is caused by a lack of factor XI (which in turn activates factor IX) and is much less severe than hemophilia A or B because an alternate activator of factor IX is present—namely, factor VII. Hemophilia is characterized by spontaneous or traumatic subcutaneous and intramuscular hemorrhaging, nosebleeds, blood in the urine, and hemorrhages in joints that produce pain and tissue damage. Treatment involves transfusions of fresh plasma or concentrates of the deficient clotting factor to relieve the tendency to bleed.

DISSEMINATED INTRAVASCULAR CLOTTING

Disseminated intravascular clotting (DIC) is a disorder of hemostasis characterized by simultaneous and unregulated blood

Figure 19.14 Red blood cells from a patient with sickle-cell disease.

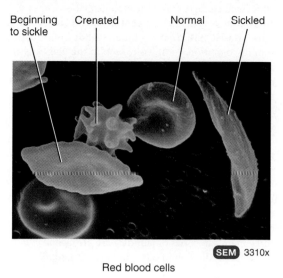

Beginning to sickle   Crenated   Normal   Sickled

SEM 3310x

Red blood cells

clotting and hemorrhage throughout the body. DIC is associated primarily with endothelial damage, tissue damage, and direct activation of clotting factor X, all of which activate the blood clotting sequence. Common causes of DIC are infections, hypoxia, low blood flow rates, trauma, tumors, hypotension, and hemolysis. DIC may be severe and life threatening. The clots decrease blood flow and eventually cause ischemia, infarction, and necrosis, resulting in multisystem organ dysfunction. A peculiar feature of DIC is that the patient often begins to bleed despite forming clots. This occurs because so many clotting factors are removed by the widespread clotting that too few remain to allow normal clotting of the remaining blood. Thus, DIC presents the paradoxical presence of blood clotting and bleeding at the same time.

LEUKEMIA

Acute leukemia is a malignant disease of blood-forming tissues characterized by uncontrolled production and accumulation of immature leukocytes. In **chronic leukemia,** mature leukocytes accumulate in the bloodstream because they do not die at the end of their normal life span. The *human T cell leukemia-lymphoma virus-1 (HTLV-1)* is strongly associated with some types of leukemia. The abnormal accumulation of immature leukocytes may be reduced by treatment with x-rays and antileukemic drugs. In some cases, a bone marrow transplant can cure the leukemia.

MEDICAL TERMINOLOGY

Acute normovolemic hemodilution (nor-mō-vō-LĒ-mik hē-mō-di-LOO-shun) Removal of blood immediately before surgery and its replacement with a cell-free solution to maintain sufficient blood volume for adequate circulation. At the end of surgery, once bleeding has been controlled, the collected blood is returned to the body.

Autologous preoperative transfusion (aw-TOL-o-gus trans-FYOO-zhun; *auto-* = self) Donating one's own blood; can be done up to 6 weeks before elective surgery. Also called *predonation.*

Blood bank A facility that collects and stores a supply of blood for future use by the donor or others. Because blood banks have now assumed additional and diverse functions (immunohematology reference work, continuing medical education, bone and tissue storage, and clinical consultation), they are more appropriately referred to as **centers of transfusion medicine.**

Cyanosis (sī-a-NŌ-sis; *cyano-* = blue) Slightly bluish/dark-purple skin discoloration, most easily seen in the nail beds and mucous membranes, due to an increased quantity of reduced hemoglobin (hemoglobin not combined with oxygen) in systemic blood.

Hemochromatosis (hē-mō-krō-ma-TŌ-sis; *chroma* = color) Disorder of iron metabolism characterized by excess deposits of iron in tissues (especially the liver, heart, pituitary gland, gonads, and pancreas) that result in bronze discoloration of the skin, cirrhosis, diabetes mellitus, and bone and joint abnormalities.

Jaundice (*jaund-* = yellow) An abnormal yellowish discoloration of the sclerae of the eyes, skin, and mucous membranes due to excess bilirubin (yellow-orange pigment) in the blood. The three main categories of jaundice are *prehepatic jaundice,* due to excess production of bilirubin; *hepatic jaundice,* due to ab-

normal bilirubin processing by the liver caused by congenital liver disease, cirrhosis (scar tissue formation) of the liver, or hepatitis (liver inflammation); and *extrahepatic jaundice,* due to blockage of bile drainage by gallstones or cancer of the bowel or pancreas.

Phlebotomist (fle-BOT-ō-mist; *phlebo-* = vein; *-tom* = cut) A technician who specializes in withdrawing blood.

Septicemia (sep'-ti-SĒ-mē-a; *septic-* = decay; *-emia* = condition of blood) Toxins or disease-causing bacteria in the blood. Also called "blood poisoning."

Thrombocytopenia (throm'-bō-sī'-tō-PĒ-nē-a; *-penia* = poverty) Very low platelet count that results in a tendency to bleed from capillaries.

Venesection (vē'-ne-SEK-shun; *ven-* = vein) Opening of a vein for withdrawal of blood. Although **phlebotomy** (fle-BOT-ō-me) is a synonym for venesection, in clinical practice phlebotomy refers to therapeutic bloodletting, such as the removal of some blood to lower its viscosity in a patient with polycythemia.

Whole blood Blood containing all formed elements, plasma, and plasma solutes in natural concentrations.

STUDY OUTLINE

INTRODUCTION (p. 610)
1. The cardiovascular system consists of the blood, heart, and blood vessels.
2. Blood is a connective tissue composed of plasma (liquid portion) and cells and cell fragments (cellular portion).

FUNCTIONS OF BLOOD (p. 610)
1. Blood transports oxygen, carbon dioxide, nutrients, wastes, and hormones.
2. It helps regulate pH, body temperature, and water content of cells.
3. It provides protection through clotting and by combating toxins and microbes through certain phagocytic white blood cells or specialized plasma proteins.

PHYSICAL CHARACTERISTICS OF BLOOD (p. 611)
1. Physical characteristics of blood include a viscosity greater than that of water; a temperature of 38°C (100.4°F); and a pH of 7.35–7.45.
2. Blood constitutes about 8% of body weight, and its volume is 4–6 liters in adults.

COMPONENTS OF BLOOD (p. 611)
1. Blood consists of 55% plasma and 45% formed elements.
2. The hematocrit is the percentage of total blood volume occupied by red blood cells.
3. Plasma consists of 91.5% water and 8.5% solutes.
4. Principal solutes include proteins (albumins, globulins, fibrinogen), nutrients, vitamins, hormones, respiratory gases, electrolytes, and waste products.
5. The formed elements in blood include red blood cells (erythrocytes), white blood cells (leukocytes), and platelets.

FORMATION OF BLOOD CELLS (p. 614)
1. Hemopoiesis is the formation of blood cells from hemopoietic stem cells in red bone marrow.
2. Myeloid stem cells form RBCs, platelets, granulocytes, and monocytes. Lymphoid stem cells give rise to lymphocytes.
3. Several hemopoietic growth factors stimulate differentiation and proliferation of the various blood cells.

RED BLOOD CELLS (p. 615)
1. Mature RBCs are biconcave discs that lack nuclei and contain hemoglobin.
2. The function of the hemoglobin in red blood cells is to transport oxygen and some carbon dioxide.
3. RBCs live about 120 days. A healthy male has about 5.4 million RBCs/μL of blood; a healthy female, about 4.8 million/ μL.
4. After phagocytosis of aged RBCs by macrophages, hemoglobin is recycled.
5. RBC formation, called erythropoiesis, occurs in adult red bone marrow of certain bones. It is stimulated by hypoxia, which stimulates release of erythropoietin by the kidneys.
6. A reticulocyte count is a diagnostic test that indicates the rate of erythropoiesis.

WHITE BLOOD CELLS (p. 619)
1. WBCs are nucleated cells. The two principal types are granulocytes (neutrophils, eosinophils, and basophils) and agranulocytes (lymphocytes and monocytes).
2. The general function of WBCs is to combat inflammation and infection. Neutrophils and macrophages (which develop from monocytes) do so through phagocytosis.
3. Eosinophils combat the effects of histamine in allergic reactions, phagocytize antigen–antibody complexes, and combat parasitic worms; basophils develop into mast cells that liberate heparin, histamine, and serotonin in allergic reactions that intensify the inflammatory response.

4. B lymphocytes, in response to the presence of foreign substances called antigens, differentiate into plasma cells that produce antibodies. Antibodies attach to the antigens and render them harmless. This antigen–antibody response combats infection and provides immunity. T lymphocytes destroy foreign invaders directly.

5. Except for lymphocytes, which may live for years, WBCs usually live for only a few hours or a few days. Normal blood contains 5000–10,000 WBCs /μL.

PLATELETS (p. 621)

1. Platelets (thrombocytes) are disc-shaped structures without nuclei.

2. They are fragments derived from megakaryocytes and are involved in clotting.

3. Normal blood contains 150,000–400,000 platelets/μL.

HEMOSTASIS (p. 622)

1. Hemostasis refers to the stoppage of bleeding.

2. It involves vascular spasm, platelet plug formation, and blood clotting (coagulation).

3. In vascular spasm, the smooth muscle of a blood vessel wall contracts, which slows blood loss.

4. Platelet plug formation involves the aggregation of platelets to stop bleeding.

5. A clot is a network of insoluble protein fibers (fibrin) in which formed elements of blood are trapped.

6. The chemicals involved in clotting are known as clotting (coagulation) factors.

7. Blood clotting involves a cascade of reactions that may be divided into three stages: formation of prothrombinase, conversion of prothrombin into thrombin, and conversion of soluble fibrinogen into insoluble fibrin.

8. Clotting is initiated by the interplay of the extrinsic and intrinsic pathways of blood clotting.

9. Normal coagulation requires vitamin K and is followed by clot retraction (tightening of the clot) and ultimately fibrinolysis (dissolution of the clot).

10. Clotting in an unbroken blood vessel is called thrombosis. A thrombus that moves from its site of origin is called an embolus.

11. Anticoagulants (for example, heparin) prevent clotting.

BLOOD GROUPS AND BLOOD TYPES (p. 627)

1. ABO and Rh blood groups are genetically determined and based on antigen–antibody responses.

2. In the ABO blood group, the presence or absence of A and B antigens on the surface of RBCs determines blood type.

3. In the Rh system, individuals whose RBCs have Rh antigens are classified as Rh^+; those who lack the antigen are Rh^-.

4. Hemolytic disease of the newborn (HDN) can occur when an Rh^- mother is pregnant with an Rh^+ fetus.

5. Before blood is transfused, a recipient's blood is typed and then either cross-matched to potential donor blood or screened for the presence of antibodies.

SELF-QUIZ QUESTIONS

Choose the best answer to the following questions:

1. Which of the following are functions of blood? (1) transportation, (2) excretion, (3) regulation, (4) protection, (5) secretion, (a) 1, 2, 3, and 4, (b) 1, 2, 4, and 5, (c) 1, 2, and 4, (d) 2, 4, and 5, (e) 1, 3, and 4

2. The plasma proteins called antibodies are (a) globulins, (b) albumins, (c) fibrinogen, (d) thrombin, (e) fibrin.

3. Which of the following are events of hemostasis? (1) vascular spasm, (2) enlargement of the rupture site, (3) hemorrhage, (4) platelet plug formation, (5) blood clotting.
(a) 1, 2, and 3, (b) 2, 3, and 4, (c) 2, 4, and 5, (d) 1, 4, and 5, (e) 1, 3, and 4

4. Which of the following statements explain why red blood cells are highly specialized for oxygen transport? (1) Red blood cells contain hemoglobin. (2) Red blood cells lack a nucleus. (3) Red blood cells have many mitochondria and thus generate ATP aerobically. (4) The biconcave shape of RBCs provides a large surface area for the inward and outward diffusion of gas molecules. (5) Red blood cells can carry up to four oxygen molecules for each hemoglobin molecule.
(a) 1, 2, 3, and 5, (b) 1, 2, 4, and 5, (c) 2, 3, 4, and 5, (d) 1, 3, and 5, (e) 2, 4, and 5

5. Which of the following are true statements? (1) White blood cells leave the bloodstream by emigration. (2) Adhesion molecules help white blood cells stick to the endothelium, which

aids emigration. (3) Neutrophils and macrophages are active in phagocytosis. (4) The attraction of phagocytes to microbes and inflamed tissue is termed chemotaxis. (5) White blood cells are the most numerous formed elements during disease processes.
(a) 1, 2, 4, and 5, (b) 2, 3, 4, and 5, (c) 1, 2, 3, and 4, (d) 1, 3, and 5, (e) 1, 2, and 4

6. A person with type A Rh negative blood can receive a transfusion with blood of which of the following types? (1) A positive, (2) B negative, (3) AB negative, (4) O negative, (5) A negative.
(a) 1 only, (b) 3 only, (c) 4 only, (d) 4 and 5, (e) 1 and 5

Complete the following:

7. The components of blood are ___ and ___, which includes ___, ___, and ___.

8. Plasma minus its clotting proteins is termed ___.

9. The stages of blood clotting are as follows: (1) the formation of ___ by initiation of either the ___ or ___ pathway, or both; (2) the conversion of ___ into ___ by the action of the chemical formed in stage 1; and (3) the conversion of ___ into ___ by the action of the chemical formed in stage 2.

10. Cell fragments enclosed by a piece of the cell membrane of megakaryocytes and containing clotting factors are called ___ or ___.

True or false:

11. Hemoglobin functions in both oxygen and carbon dioxide transport and blood pressure regulation.

12. Blood groups include the ABO, Rh, Lewis, Kell, Kidd, and Duffy systems, among others.

13. Match the following:

____ (a) a tissue protein that leaks into the blood from cells outside blood vessels and initiates the formation of prothrombin activator

____ (b) an anticoagulant

____ (c) required for normal blood clotting

____ (d) its formation is initiated by either the extrinsic or intrinsic pathway or both; catalyzes the conversion of prothrombin to thrombin

____ (e) glycoproteins and glycolipids on the surfaces of red blood cells that can act as antigens

____ (f) forms the threads of a clot; produced from fibrinogen

____ (g) can dissolve a clot by digesting fibrin threads

____ (h) serves as the catalyst to form fibrin; formed from prothrombin

(1) prothrombinase
(2) thrombin
(3) fibrin
(4) thromboplastin
(5) plasmin
(6) heparin
(7) agglutinogens
(8) vitamin K

14. Match the following:

____ (a) contain hemoglobin and function in gas transport

____ (b) young neutrophils showing a rod-shaped nucleus

____ (c) white blood cell showing a kidney-shaped nucleus

____ (d) monocytes that roam the tissues and gather at sites of infection or inflammation

____ (e) occur as B cells, T cells, and natural killer cells

____ (f) combat the effects of histamine and other mediators of inflammation in allergic reaction; also phagocytize antigen–antibody complexes

____ (g) respond to tissue destruction by bacteria; release lysozyme, strong oxidants, and defensins

____ (h) older neutrophils with several different shaped nuclear lobes

____ (i) monocytes that leave the blood and reside in a particular tissue such as Kupffer cells in the liver

____ (j) involved in inflammatory and allergic reactions; are involved in hypersensitivity reactions

(1) neutrophils
(2) lymphocytes
(3) monocytes
(4) eosinophils
(5) basophils
(6) red blood cells
(7) polymorphs
(8) bands
(9) fixed macrophages
(10) wandering macrophages

15. Match the following:

____ (a) individual forms of progenitor cells; named on the basis of the mature elements in blood they will ultimately produce

____ (b) cells that give rise to all the formed elements of blood; derived from mesenchyme

____ (c) give rise to red blood cells

____ (d) give rise to lymphocytes

____ (e) hormone that increases the numbers of red blood cell precursors

____ (f) cells no longer capable of replenishing themselves; can only give rise to more specific formed elements of blood

____ (g) stimulate white blood cell formation

____ (h) hormone that stimulates formation of platelets

(1) pluripotent (hemopoietic) stem cells
(2) myeloid cells
(3) lymphoid stem cells
(4) progenitor cells
(5) colony forming units
(6) erythropoietin
(7) thrombopoietin
(8) cytokines

CRITICAL THINKING QUESTIONS

1. During his anatomy class, Josef was holding a heart in the palm of his hand while the laboratory instructor started to explain the dissection procedure. "Put the heart in the pan before you make the first cut." Too late—Josef cut through the heart and into his hand. Describe the composition of the red fluid now flowing onto his palm. (HINT: *It's a liquid connective tissue.*)

2. "We may as well check your blood type," said Josef's instructor, "since you're bleeding anyway." Josef's blood type is B negative. Explain how they determined Josef's blood type. (HINT: *The "B" and the "negative" refer to two different antigens.*)

3. When Josef removed the bandage from his hand, the bleeding had stopped. How did this happen? (HINT: *A cell fragment plays a major role in this process.*)

ANSWERS TO FIGURE QUESTIONS

19.1 Blood volume is about 6 liters in males and 4–5 liters in females, representing about 8% of body weight.

19.2 Platelets are cell fragments.

19.3 Blood temperature is about 38°C (100.4°F); its pH is 7.35–7.45.

19.4 One hemoglobin molecule can transport four O_2 molecules —one bound to each heme group.

19.5 Transferrin is an iron-carrying plasma protein.

19.6 Your hematocrit once you moved to high altitude would increase due to increased secretion of erythropoietin.

19.7 Neutrophils, eosinophils, and basophils are called granulocytes because all have cytoplasmic granules that stains make visible through a light microscope.

19.8 Lymphocytes recirculate between blood and tissues, whereas after leaving the blood other WBCs remain in the tissues until they die.

19.9 Besides platelet plug formation, vascular spasm and blood clotting contribute to hemostasis.

19.10 Serum is blood plasma minus the clotting proteins.

19.11 The outcome of the first stage of clotting is the formation of prothrombinase.

19.12 Type O blood usually contains both anti-A and anti-B antibodies.

19.13 Because the mother is most likely to start making anti-Rh antibodies after the baby is already born, that baby suffers no damage.

THE CARDIOVASCULAR SYSTEM: THE HEART

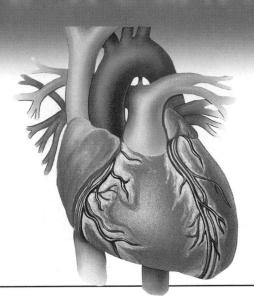

The cardiovascular system consists of the blood, heart, and blood vessels. In the previous chapter we examined the composition and functions of blood. For blood to reach body cells and exchange materials with them, it must be constantly propelled through blood vessels. The heart is the pump that circulates the blood through an estimated 100,000 km (60,000 mi) of blood vessels. Even while you are sleeping, your heart pumps 30 times its own weight each minute, about 5 liters (5.3 qt) to the lungs and the same volume to the rest of the body. At this rate, the heart pumps more than 14,000 liters (3,600 gal) of blood in a day, or 10 million liters (2.6 million gal) in a year. You don't spend all your time sleeping, however, and your heart pumps more vigorously when you are active. Thus, the actual blood volume the heart pumps in a single day is much larger. The study of the normal heart and the diseases associated with it is **cardiology** (kar-dē-OL-ō-jē; *cardio-* = heart; *-ology* = study of). This chapter explores the design of the heart and the unique properties that permit it to pump for a lifetime without rest.

LOCATION AND SURFACE PROJECTION OF THE HEART

OBJECTIVE

• *Describe the location of the heart, and trace its outline on the surface of the chest.*

For all its might, the cone-shaped heart is relatively small, roughly the same size as a closed fist—about 12 cm (5

in.) long, 9 cm (3.5 in.) wide at its broadest point, and 6 cm (2.5 in.) thick. Its mass averages 250 g (8 oz) in adult females and 300 g (10 oz) in adult males. The heart rests on the diaphragm, near the midline of the thoracic cavity in the **mediastinum** (mē-dē-a-STĪ-num), a mass of tissue that extends from the sternum to the vertebral column and between the coverings (pleurae) of the lungs (Figure 20.1a). About two-thirds of the mass of the heart lies to the left of the body's midline. The position of the heart in the mediastinum is more readily appreciated by examining its ends, surfaces, and borders (Figure 20.1b). Visualize the heart as a cone lying on its side. The pointed end of the heart is the **apex,** which is directed anteriorly, inferiorly, and to the left. The broad portion of the heart opposite the apex is the **base,** which is directed posteriorly, superiorly, and to the right. In addition to the apex and base, the heart has several surfaces and borders (margins) that are useful in determining its surface projection (described shortly). The **anterior surface** is deep to the sternum and ribs. The **inferior surface** is the portion of the heart that rests mostly on the diaphragm and is found between the apex and right border (see Figure 20.1b). The **right border** faces the right lung and extends from the inferior surface to the base; the **left border,** also called the pulmonary border, faces the left lung and extends from the base to the apex.

Determining an organ's **surface projection** involves outlining its dimensions with respect to landmarks on the surface of the body. This practice is useful when conducting diagnostic procedures (for example, a lumbar puncture), auscultation (for example, listening to heart and lung

636

Figure 20.1 Position of the heart and associated structures in the medi-astinum (dashed outline), and the points of the heart that correspond to its surface projection. (See Tortora, *A Photographic Atlas of the Human Body,* Figures 6.5 and 6.6)

🔑 **The heart is located in the mediastinum; two-thirds of its mass is to the left of the midline.**

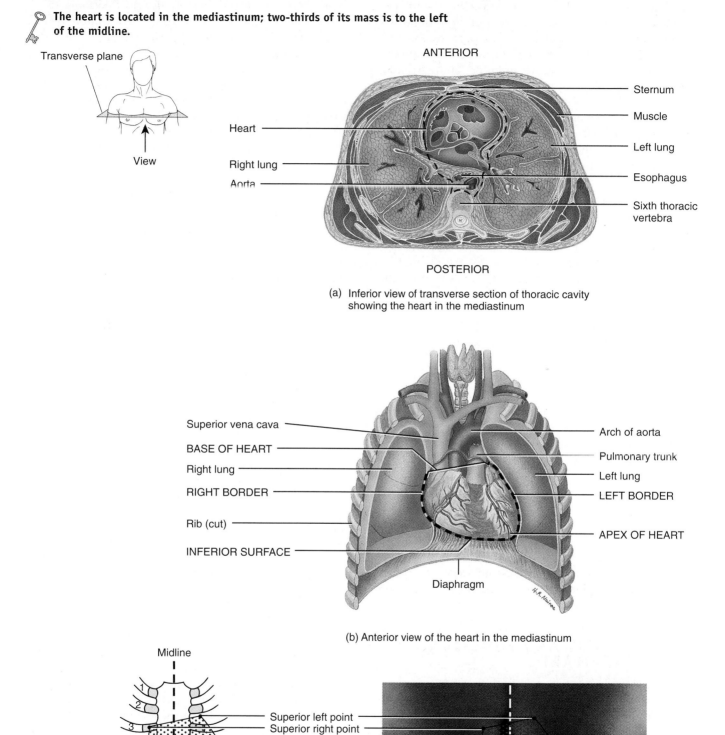

(a) Inferior view of transverse section of thoracic cavity showing the heart in the mediastinum

(b) Anterior view of the heart in the mediastinum

(c) Surface projection of the heart

Ⓠ What is the mediastinum?

sounds), and anatomical studies. We can project the heart on the anterior surface of the chest by locating the following landmarks (Figure 20.1c): The **superior right point** is located at the superior border of the third right costal cartilage, about 3 cm (1 in.) to the right of the midline. The **superior left point** is located at the inferior border of the second left costal cartilage, about 3 cm to the left of the midline. A line connecting these two points corresponds to the base of the heart. The **inferior left point** is located at the apex of the heart in the fifth left intercostal space, about 9 cm (3.5 in.) to the left of the midline. A line connecting the superior and inferior left points corresponds to the left border of the heart. The **inferior right point** is located at the superior border of the sixth right costal cartilage, about 3 cm to the right of the midline. A line connecting the inferior left and right points corresponds to the inferior surface of the heart, and a line connecting the inferior and superior right points corresponds to the right border of the heart. When all four points are connected, they form a shape that roughly reveals the size and shape of the heart.

CLINICAL APPLICATION
Cardiopulmonary Resuscitation

Because the heart lies between two rigid structures —the vertebral column and the sternum (see Figure 20.1a)—external pressure (compression) on the chest can be used to force blood out of the heart and into the circulation. In cases in which the heart suddenly stops beating, **cardiopulmonary resuscitation (CPR)**—properly applied cardiac compressions, performed in conjunction with artificial ventilation of the lungs—saves lives by keeping oxygenated blood circulating until the heart can be restarted. ■

1. Describe the position of the heart in the mediastinum by defining its apex, base, anterior and posterior surfaces, and right and left borders.
2. Explain the location of the superior right point, superior left point, inferior left point, and inferior right point. Why are these points significant?

STRUCTURE AND FUNCTION OF THE HEART

OBJECTIVES
- *Describe the structure of the pericardium and the heart wall.*
- *Discuss the external and internal anatomy of the chambers of the heart.*
- *Describe the structure and function of the valves of the heart.*

Pericardium

The membrane that surrounds and protects the heart is the **pericardium** (*peri-* = around). It confines the heart to its position in the mediastinum, while allowing sufficient freedom of movement for vigorous and rapid contraction. The pericardium consists of two principal portions: the fibrous pericardium and the serous pericardium (Figure 20.2a). The superficial **fibrous pericardium** is a tough, inelastic, dense irregular connective tissue. It resembles a bag that rests on and attaches to the diaphragm; its open end is fused to the connective tissues of the blood vessels entering and leaving the heart. The fibrous pericardium prevents overstretching of the heart, provides protection, and anchors the heart in the mediastinum.

The deeper **serous pericardium** is a thinner, more delicate membrane that forms a double layer around the heart (see Figure 20.2a). The outer **parietal layer** of the serous pericardium is fused to the fibrous pericardium. The inner **visceral layer** of the serous pericardium, also called the **epicardium** (*epi-* = on top of), adheres tightly to the surface of the heart. Between the parietal and visceral layers of the serous pericardium is a thin film of serous fluid. This fluid, known as **pericardial fluid,** is a slippery secretion of the pericardial cells that reduces friction between the membranes as the heart moves. The space that contains the few milliliters of pericardial fluid is called the **pericardial cavity.**

CLINICAL APPLICATION
Pericarditis and Cardiac Tamponade

Inflammation of the pericardium is known as **pericarditis.** If production of pericardial fluid diminishes, painful rubbing together of the parietal and visceral serous pericardial layers may result. A buildup of pericardial fluid (which may also occur in pericarditis) or extensive bleeding into the pericardium are life-threatening conditions. Because the pericardium cannot stretch, the buildup of fluid or blood compresses the heart. This compression, known as **cardiac tamponade** (tam'-pon-ĀD), can stop the beating of the heart. ■

Layers of the Heart Wall

The wall of the heart consists of three layers (see Figure 20.2a): the epicardium (external layer), the myocardium (middle layer), and the endocardium (inner layer). The outermost **epicardium,** also called the *visceral layer of the serous pericardium,* is the thin, transparent outer layer of the wall. It is composed of mesothelium and delicate connective tissue that imparts a smooth, slippery texture to the outermost surface of the heart. The middle **myocardium** (*myo-* = muscle), which is cardiac muscle tissue, makes up the bulk of the heart and is responsible for its pumping action. Although it is striated like skeletal muscle, cardiac muscle is involuntary like smooth muscle. The cardiac muscle fibers swirl diagonally around the heart in interlacing bundles (Figure 20.2b). The innermost **endocardium** (*endo-* = within) is a thin layer of endothelium overlying a thin layer of connective tissue. It provides a smooth lining for the chambers of the heart and covers the valves of the heart. The endocardium is con-

Figure 20.2 Pericardium and heart wall.

🔑 **The pericardium is a triple-layered sac that surrounds and protects the heart.**

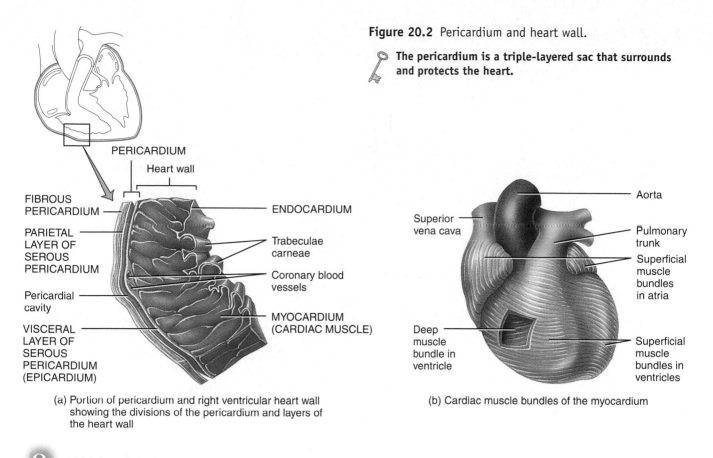

PERICARDIUM

Heart wall

FIBROUS PERICARDIUM

PARIETAL LAYER OF SEROUS PERICARDIUM

Pericardial cavity

VISCERAL LAYER OF SEROUS PERICARDIUM (EPICARDIUM)

ENDOCARDIUM

Trabeculae carneae

Coronary blood vessels

MYOCARDIUM (CARDIAC MUSCLE)

(a) Portion of pericardium and right ventricular heart wall showing the divisions of the pericardium and layers of the heart wall

Aorta

Superior vena cava

Pulmonary trunk

Superficial muscle bundles in atria

Deep muscle bundle in ventricle

Superficial muscle bundles in ventricles

(b) Cardiac muscle bundles of the myocardium

Q Which layer is both a part of the pericardium and a part of the heart wall?

tinuous with the endothelial lining of the large blood vessels attached to the heart.

Chambers of the Heart

The heart contains four chambers. The two upper chambers are the **atria** (= entry halls or chambers), and the two lower chambers are the **ventricles** (= little bellies). On the anterior surface of each atrium is a wrinkled pouchlike structure called an **auricle** (OR-i-kul; *auri-* = ear), so named because of its resemblance to a dog's ear (Figure 20.3). Each auricle slightly increases the capacity of an atrium so that it can hold a greater volume of blood. Also on the surface of the heart are a series of grooves, called **sulci** (SUL-sē), that contain coronary blood vessels and a variable amount of fat. Each sulcus (SUL-kus) marks the external boundary between two chambers of the heart. The deep **coronary sulcus** (*coron-* = resembling a crown) encircles most of the heart and marks the boundary between the superior atria and inferior ventricles. The **anterior interventricular sulcus** is a shallow groove on the anterior surface of the heart that marks the boundary between the right and left ventricles. This sulcus continues around to the posterior surface of the heart as the **posterior interventricular sulcus,** which marks

the boundary between the ventricles on the posterior aspect of the heart (see Figure 20.3c).

Right Atrium

The **right atrium** forms the right border of the heart (see Figure 20.1b). It receives blood from three veins: *superior vena cava, inferior vena cava,* and *coronary sinus* (Figure 20.4a). The anterior and posterior walls within the right atrium differ considerably. Whereas the posterior wall is smooth, the anterior wall is rough due to the presence of muscular ridges called **pectinate muscles** (*pectin* = comb), which also extend into the auricle (Figure 20.4b). Between the right atrium and left atrium is a thin partition called the **interatrial septum** (*inter-* = between; *septum* = a dividing wall or partition). A prominent feature of this septum is an oval depression called the **fossa ovalis,** which is the remnant of the foramen ovale, an opening in the interatrial septum of the fetal heart that normally closes soon after birth (see Figure 21.31 on page 730). Blood passes from the right atrium into the right ventricle through a valve called the **tricuspid valve** (tri-KUS-pid; *tri-* = three; *cuspid* = point) because it consists of three leaflets or cusps (see Figure 20.4a). The valves of the heart are composed of dense connective tissue covered by endocardium.

Figure 20.3 Structure of the heart: surface features.

🗝 **Sulci are grooves that contain blood vessels and fat and mark the boundaries between the various chambers.**

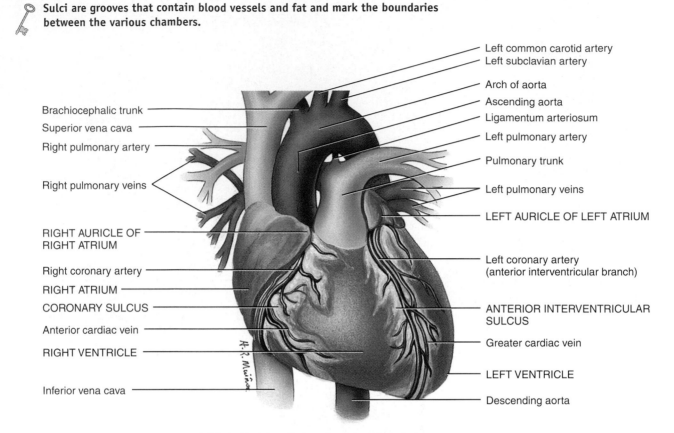

Left common carotid artery
Left subclavian artery
Arch of aorta
Ascending aorta
Ligamentum arteriosum
Left pulmonary artery
Pulmonary trunk
Left pulmonary veins
LEFT AURICLE OF LEFT ATRIUM
Left coronary artery (anterior interventricular branch)
ANTERIOR INTERVENTRICULAR SULCUS
Greater cardiac vein
LEFT VENTRICLE
Descending aorta

Brachiocephalic trunk
Superior vena cava
Right pulmonary artery
Right pulmonary veins
RIGHT AURICLE OF RIGHT ATRIUM
Right coronary artery
RIGHT ATRIUM
CORONARY SULCUS
Anterior cardiac vein
RIGHT VENTRICLE
Inferior vena cava

(a) Anterior external view showing surface features

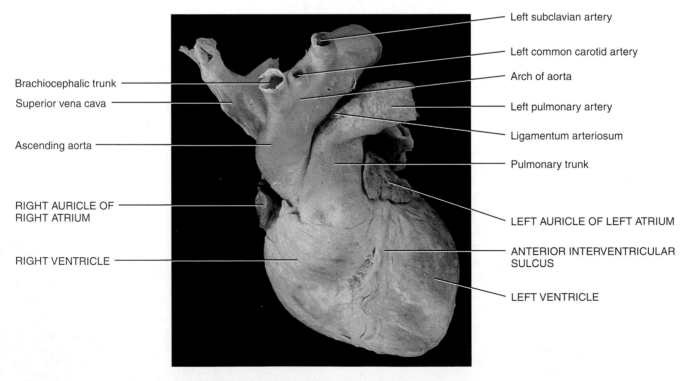

Left subclavian artery
Left common carotid artery
Arch of aorta
Left pulmonary artery
Ligamentum arteriosum
Pulmonary trunk
LEFT AURICLE OF LEFT ATRIUM
ANTERIOR INTERVENTRICULAR SULCUS
LEFT VENTRICLE

Brachiocephalic trunk
Superior vena cava
Ascending aorta
RIGHT AURICLE OF RIGHT ATRIUM
RIGHT VENTRICLE

(b) Anterior external view showing surface features

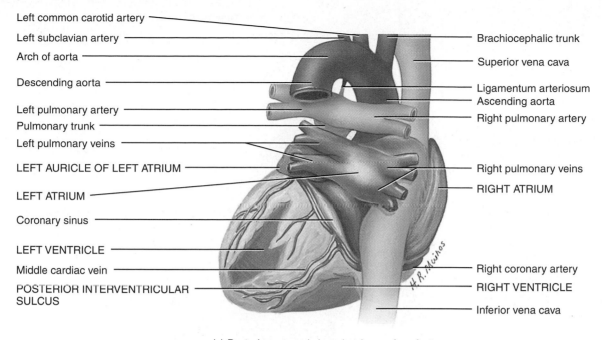

(c) Posterior external view showing surface features

Q   The coronary sulcus forms a boundary between which chambers of the heart?

Right Ventricle

The **right ventricle** forms most of the anterior surface of the heart. The inside of the right ventricle contains a series of ridges formed by raised bundles of cardiac muscle fibers called **trabeculae carneae** (tra-BEK-yoo-lē KAR-nē-ē; *trabeculae* = little beams; *carneae* = fleshy). Some of the trabeculae carneae convey part of the conduction system of the heart (described on page 648). The cusps of the tricuspid valve are connected to tendonlike cords, the **chordae tendineae** (KOR-dē ten-DIN-ē-ē; *chord-* = cord; *tend-* = tendon), which, in turn, are connected to cone-shaped trabeculae carneae called **papillary muscles** (*papill-* = nipple). The right ventricle is separated from the left ventricle by a partition called the **interventricular septum.** Blood passes from the right ventricle through the **pulmonary semilunar (SL) valve** (*semi-* = half; *lunar* = moon-shaped) into a large artery called the *pulmonary trunk,* which divides into right and left *pulmonary arteries.*

Left Atrium

The **left atrium** forms most of the base of the heart (see Figure 20.1b). It receives blood from the lungs through four *pulmonary veins.* Like the right atrium, the inside of the left atrium has a smooth posterior wall. Because pectinate muscles are confined to the auricle of the left atrium, the anterior wall of the left atrium also is smooth. Blood passes from the left atrium into the left ventricle through the **bicuspid (mitral) valve** (*bi-* = two), which has two cusps.

Left Ventricle

The **left ventricle** forms the apex of the heart (see Figure 20.1b). Like the right ventricle, the left ventricle also contains trabeculae carneae and has chordae tendinae that anchor the cusps of the bicuspid valve to papillary muscles. Blood passes from the left ventricle through the **aortic semilunar valve** into the largest artery of the body, the *ascending aorta* (*aorte* = to suspend, because the aorta once was believed to lift up the heart). From here, some of the blood flows into the *coronary arteries,* which branch from the ascending aorta and carry blood to the heart wall; the remainder of the blood passes into the *arch of the aorta* and *descending aorta* (*thoracic aorta* and *abdominal aorta*). Branches of the arch of the aorta and descending aorta carry blood throughout the body.

During fetal life, a temporary blood vessel, called the *ductus arteriosus,* shunts blood from the pulmonary trunk into the aorta, so that only a small amount of blood enters the nonfunctioning fetal lungs (see Figure 21.31 on page 730). The ductus arteriosus normally closes shortly after birth, leaving a remnant known as the **ligamentum arteriosum,** which connects the arch of the aorta and pulmonary trunk (see Figure 20.4a).

Myocardial Thickness and Function

The thickness of the myocardium of the four chambers varies according to each chamber's function. The atria are thin-walled because they deliver blood into the adjacent ventricles; because the ventricles pump blood greater distances,

Figure 20.4 Structure of the heart: internal anatomy.

🔑 The thickness of the four chambers varies according to their functions.

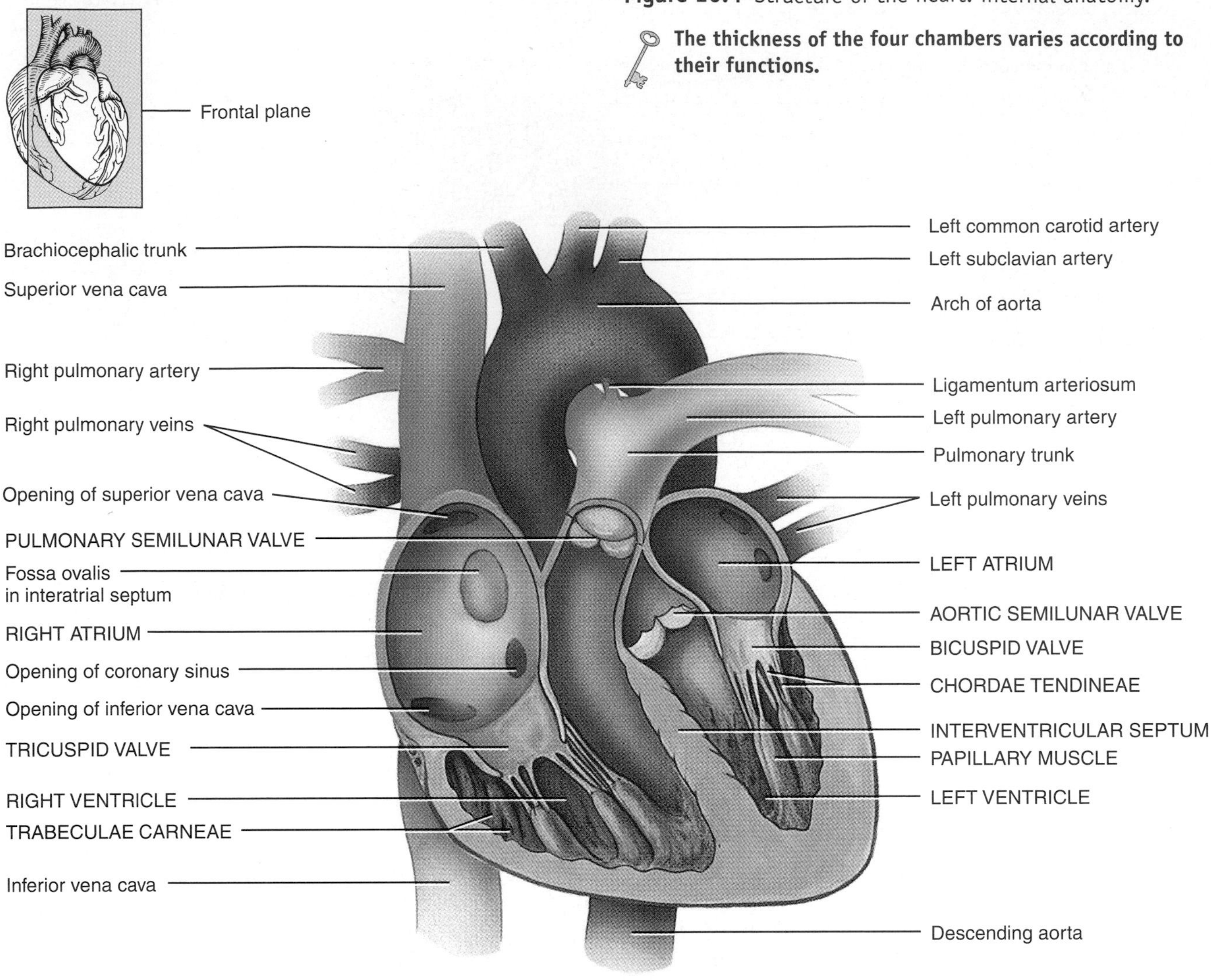

Frontal plane

Brachiocephalic trunk

Superior vena cava

Right pulmonary artery

Right pulmonary veins

Opening of superior vena cava

PULMONARY SEMILUNAR VALVE

Fossa ovalis in interatrial septum

RIGHT ATRIUM

Opening of coronary sinus

Opening of inferior vena cava

TRICUSPID VALVE

RIGHT VENTRICLE

TRABECULAE CARNEAE

Inferior vena cava

Left common carotid artery

Left subclavian artery

Arch of aorta

Ligamentum arteriosum

Left pulmonary artery

Pulmonary trunk

Left pulmonary veins

LEFT ATRIUM

AORTIC SEMILUNAR VALVE

BICUSPID VALVE

CHORDAE TENDINEAE

INTERVENTRICULAR SEPTUM

PAPILLARY MUSCLE

LEFT VENTRICLE

Descending aorta

(a) Anterior view of frontal section showing internal anatomy

their walls are thicker (see Figure 20.4a). Even though the right and left ventricles act as two separate pumps that simultaneously eject equal volumes of blood, the right side has a much smaller workload. This is because the right ventricle pumps blood into the lungs, which are nearby and offer only small resistance to blood flow, whereas the left ventricle pumps blood to all other parts of the body, where the resistance to blood flow is higher. Thus, the left ventricle works harder than the right ventricle to maintain the same rate of blood flow. The anatomy of the two ventricles confirms this functional difference: The muscular wall of the left ventricle is considerably thicker than that of the right ventricle (see Figure 20.4c). Note also that the perimeter of the lumen (space) of the left ventricle is circular, whereas that of the right ventricle is crescent-shaped.

Fibrous Skeleton of the Heart

In addition to cardiac muscle tissue, the heart wall also contains dense connective tissue that forms the **fibrous skeleton of the heart** (Figure 20.5). Essentially, the fibrous skeleton consists of dense connective tissue rings that sur-round the valves of the heart, fuse with one another, and merge with the interventricular septum. Four fibrous rings support the four valves of the heart and are fused to each other. The skeleton forms the foundation to which the heart valves attach, serves as a point of insertion for cardiac muscle bundles, prevents overstretching of the valves as blood passes through them, and acts as an electrical insulator that prevents the direct spread of action potentials from the atria to the ventricles.

Operation of Heart Valves

As each chamber of the heart contracts, it pushes a volume of blood into a ventricle or out of the heart into an artery. Valves open and close in response to pressure changes as the heart contracts and relaxes. Each of the four valves helps to ensure one-way flow of blood by opening to let blood through and closing to prevent its backflow.

Atrioventricular Valves

Because they are located between an atrium and a ventricle, the tricuspid and bicuspid valves are termed **atrioventricular (AV) valves.** When an AV valve is open, the pointed

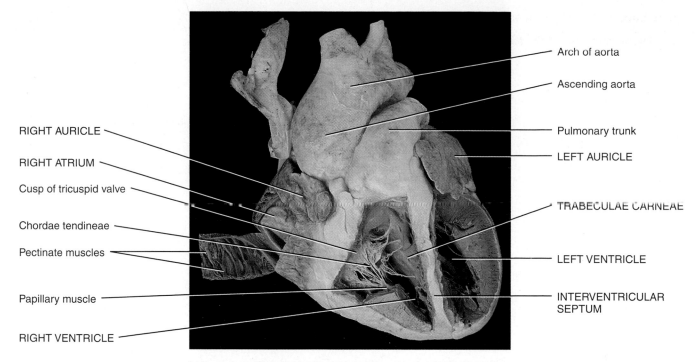

RIGHT AURICLE

RIGHT ATRIUM

Cusp of tricuspid valve

Chordae tendineae

Pectinate muscles

Papillary muscle

RIGHT VENTRICLE

Arch of aorta

Ascending aorta

Pulmonary trunk

LEFT AURICLE

TRABECULAE CARNEAE

LEFT VENTRICLE

INTERVENTRICULAR SEPTUM

(b) Partially sectioned heart in anterior view showing internal anatomy

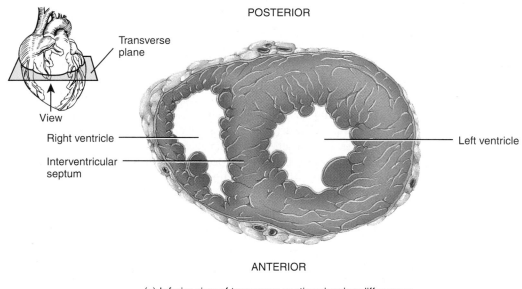

POSTERIOR

Transverse plane

View

Right ventricle

Interventricular septum

Left ventricle

ANTERIOR

(c) Inferior view of transverse section showing differences in thickness of ventricular walls

Q Which chamber has the thickest wall?

ends of the cusps project into the ventricle. Blood moves from the atria into the ventricles through open AV valves when ventricular pressure is lower than atrial pressure (Figure 20.6a, c). At this time, the papillary muscles are relaxed, and the chordae tendineae are slack. When the ventricles contract, the pressure of the blood drives the cusps upward until their edges meet and close the opening (Figure 20.6b, d). At the same time, the papillary muscles are also contracting, which pulls on and tightens the chordae tendineae, preventing the valve cusps from everting (being forced to open in the opposite direction into the atria due to the high ventricular pressure). If the AV valves or chordae tendineae are damaged, blood may regurgitate (flow back) into the atria when the ventricles contract.

Figure 20.5 Fibrous skeleton of the heart (depicted in blue).

🔑 **Fibrous rings support the four valves of the heart and are fused to each other.**

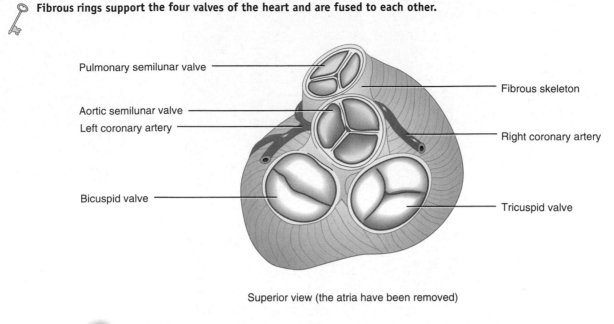

Superior view (the atria have been removed)

Q In what two ways does the fibrous skeleton contribute to the functioning of heart valves?

Semilunar Valves

The two **semilunar (SL) valves** allow ejection of blood from the heart and into arteries but prevent backflow of blood into the ventricles. Both SL valves consist of three crescent-shaped cusps (see Figure 20.6c), each of which is attached to the artery wall by its convex outer margin. The free borders of the cusps curve outward and project into the lumen of the artery. When the ventricles contract, pressure builds up within them. The semilunar valves open when pressure in the ventricles exceeds the pressure in the arteries, permitting ejection of blood from the ventricles into the pulmonary trunk and aorta (see Figure 20.6d). As the ventricles relax, blood starts to flow back toward the heart. This backflowing blood fills the valve cusps, which tightly closes the semilunar valves (see Figure 20.6c).

 CLINICAL APPLICATION
Rheumatic Fever

Certain infectious diseases can damage or destroy the heart valves. One example is **rheumatic fever,** an acute systemic inflammatory disease that usually occurs after a streptococcal infection of the throat and can affect many of the body's connective tissues. The bacteria trigger an immune response in which antibodies that are produced to destroy the bacteria attack and inflame the connective tissues in joints, heart valves, and other organs. Even though the entire heart wall may be weakened, rheumatic fever most often damages the bicuspid (mitral) and aortic semilunar valves, which then may fail to open and close properly or may constantly leak. The damage to valves produced by rheumatic fever is permanent. ■

1. Define each of the following external features of the heart: auricle, coronary sulcus, anterior interventricular sulcus, and posterior interventricular sulcus.
2. Describe the characteristic internal features of each chamber of the heart.
3. For each chamber of the heart, list the blood vessels that deliver blood to it or receive ejected blood, and name the valve that blood passes through on its way to the next heart chamber or blood vessel.
4. Describe the relationship between wall thickness and function for each heart chamber.
5. How does the fibrous skeleton of the heart assist the operation of heart valves?
6. What causes the heart valves to open and to close?

CIRCULATION OF BLOOD

OBJECTIVES

• *Describe the flow of blood through the chambers of the heart and through the systemic and pulmonary circulations.*

• *Discuss the coronary circulation.*

Systemic and Pulmonary Circulations

With each beat, the heart pumps blood into two closed circuits—the **systemic circulation** and the **pulmonary circulation** (*pulmon-* = lung). The left side of the heart is the pump for the systemic circulation; it receives freshly oxygenated blood from the lungs. The left ventricle ejects blood into the *aorta* (Figure 20.7a). From the aorta, the blood di-

Figure 20.6 Responses of the valves to the pumping of the heart. (See Tortora, *A Photographic Atlas of the Human Body,* Figure 6.7)

🔑 **Heart valves prevent backflow of blood.**

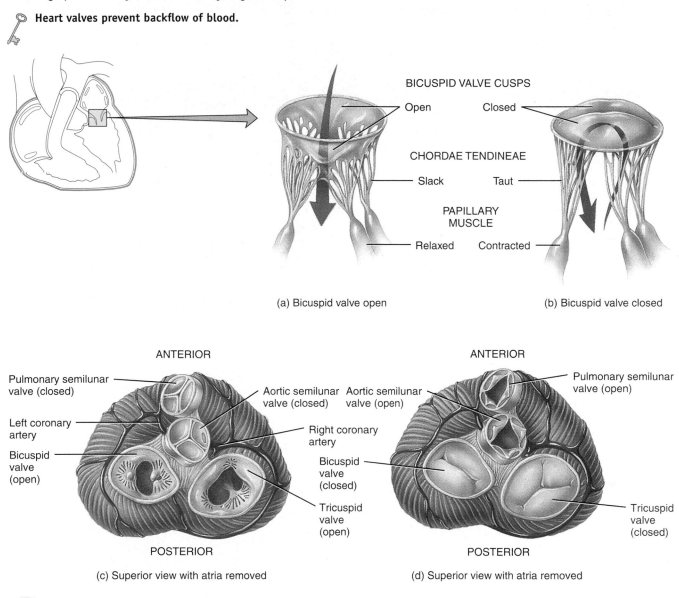

(a) Bicuspid valve open

(b) Bicuspid valve closed

(c) Superior view with atria removed

(d) Superior view with atria removed

Q How do papillary muscles prevent AV valve cusps from everting or swinging upward into the atria?

vides into separate streams, entering progressively smaller and smaller *systemic arteries* that carry it to all organs throughout the body—except for the air sacs (alveoli) of the lungs, which are supplied by the pulmonary circulation. In systemic tissues, arteries give rise to smaller-diameter *arterioles*, which finally lead into extensive beds of *systemic capillaries*. Exchange of nutrients and gases occurs across the thin capillary walls: Blood unloads O_2 (oxygen) and picks up CO_2 (carbon dioxide). In most cases, blood flows through only one capillary and then enters a *systemic venule*. Venules carry deoxygenated (O_2-poor) blood away from tissues and merge

to form larger *systemic veins,* and ultimately the blood flows back to the right atrium.

The right side of the heart is the pump for the pulmonary circulation; it receives all the deoxygenated blood returning from the systemic circulation. Blood ejected from the right ventricle flows into the *pulmonary trunk,* which branches into *pulmonary arteries* that carry blood to the right and left lungs. In pulmonary capillaries, then, blood unloads CO_2, which is exhaled, and picks up O_2. The freshly oxygenated blood then flows into pulmonary veins and returns to the left atrium. Figure 20.7b reviews the route of blood

Figure 20.7 Systemic and pulmonary circulations. Throughout this book, blood vessels that carry oxygenated blood are colored red, whereas those that carry deoxygenated blood are colored blue.

🔑 **The left side of the heart pumps freshly oxygenated blood into the systemic circulation to all tissues of the body except the air sacs (alveoli) of the lungs; the right side of the heart pumps deoxygenated (oxygen-poor) blood into the pulmonary circulation to the air sacs (alveoli) of the lungs.**

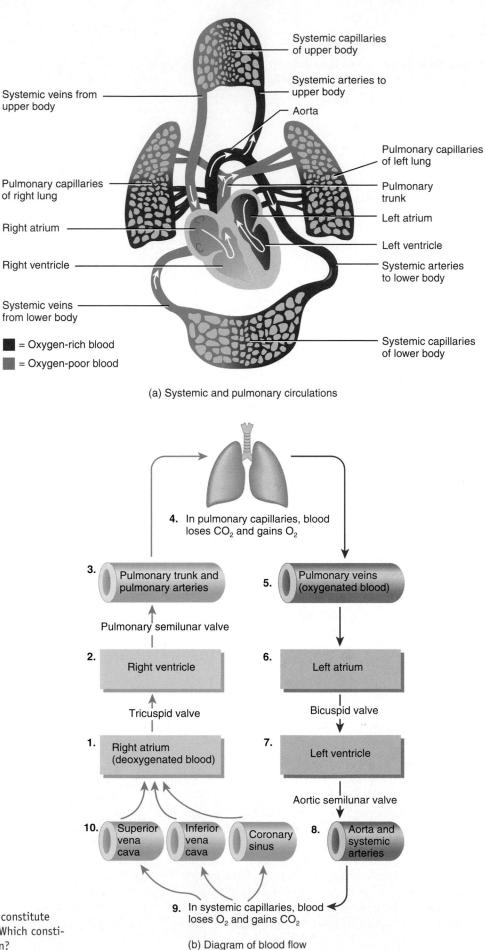

(a) Systemic and pulmonary circulations

= Oxygen-rich blood
= Oxygen-poor blood

4. In pulmonary capillaries, blood loses CO_2 and gains O_2

3. Pulmonary trunk and pulmonary arteries

5. Pulmonary veins (oxygenated blood)

Pulmonary semilunar valve

2. Right ventricle

6. Left atrium

Tricuspid valve

Bicuspid valve

1. Right atrium (deoxygenated blood)

7. Left ventricle

Aortic semilunar valve

10. Superior vena cava Inferior vena cava Coronary sinus

8. Aorta and systemic arteries

9. In systemic capillaries, blood loses O_2 and gains CO_2

Ⓠ In part (b), which numbers constitute the pulmonary circulation? Which constitute the systemic circulation?

(b) Diagram of blood flow

Figure 20.8 Coronary (cardiac) circulation. This view of the heart from the anterior aspect is drawn as if the heart were transparent to reveal blood vessels on the posterior aspect. (See Tortora, *A Photographic Atlas of the Human Body*, Figures 6.8 and 6.9)

🔑 **The right and left coronary arteries deliver blood to the heart; the coronary veins drain blood from the heart into the coronary sinus.**

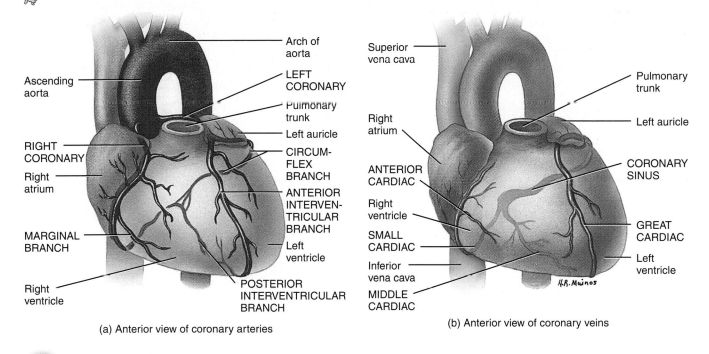

(a) Anterior view of coronary arteries

(b) Anterior view of coronary veins

Q   Which blood vessel delivers oxygenated blood to the left atrium and left ventricle?

flow through the chambers and valves of the heart and the pulmonary and systemic circulations.

Coronary Circulation

Nutrients could not possibly diffuse from the chambers of the heart through all the layers of cells that make up the heart tissue. For this reason, the wall of the heart has its own blood vessels. The flow of blood through the many vessels that pierce the myocardium is called the **coronary (cardiac) circulation.** The arteries of the heart encircle it like a crown encircles the head. While it is contracting, the heart receives little oxygenated blood by way of the **coronary arteries,** which branch from the ascending aorta (Figure 20.8a). When the heart relaxes, however, the high pressure of blood in the aorta propels blood through the coronary arteries, into capillaries, and then into **coronary veins** (Figure 20.8b).

Coronary Arteries

Two coronary arteries, the right and left coronary arteries, branch from the ascending aorta and supply oxygenated blood to the myocardium (see Figure 20.8a). The **left coronary artery** passes inferior to the left auricle and divides into the anterior interventricular and circumflex branches. The **anterior interventricular branch** or **left anterior descending (LAD) artery** is in the anterior interventricular sulcus and supplies oxygenated blood to the walls of both ventricles. The **circumflex branch** lies in the coronary sulcus and distributes oxygenated blood to the walls of the left ventricle and left atrium.

The **right coronary artery** supplies small branches (atrial branches) to the right atrium. It continues inferior to the right auricle and divides into the posterior interventricular and marginal branches. The **posterior interventricular branch** follows the posterior interventricular sulcus and supplies the walls of the two ventricles with oxygenated blood. The **marginal branch** in the coronary sulcus transports oxygenated blood to the myocardium of the right ventricle.

Most parts of the body receive blood from branches of more than one artery, and where two or more arteries supply the same region, they usually connect. These connections, called **anastomoses** (a-nas′-tō-MŌ-sēs), provide alternate routes for blood to reach a particular organ or tissue. The myocardium contains many anastomoses that connect branches of a given coronary artery or extend between branches of different coronary arteries. They provide detours for arterial blood if a main route becomes obstructed. Thus, heart muscle may receive sufficient oxygen even if one of its coronary arteries is partially blocked.

Coronary Veins

After blood passes through the arteries of the coronary circulation, where it delivers oxygen and nutrients to the heart muscle, it passes into veins, where it collects carbon dioxide and wastes. The deoxygenated blood then drains into a large vascular sinus on the posterior surface of the heart, called the **coronary sinus** (see Figure 20.8b), which empties into the right atrium. A vascular sinus is a thin-walled vein that has no smooth muscle to alter its diameter. The principal tributaries carrying blood into the coronary sinus are the **great cardiac vein,** which drains the anterior aspect of the heart, and the **middle cardiac vein,** which drains the posterior aspect of the heart.

1. In correct sequence, list the heart chambers, heart valves, and blood vessels encountered by a drop of blood as it flows out of the right atrium until it reaches the aorta.
2. Which arteries deliver oxygenated blood to the myocardium of the left and right ventricles?

CARDIAC MUSCLE AND THE CARDIAC CONDUCTION SYSTEM

OBJECTIVES
• *Describe the structural and functional characteristics of cardiac muscle tissue.*

• *Explain the structural and functional features of the conduction system of the heart.*

• *Describe how an action potential occurs in cardiac contractile fibers.*

Histology of Cardiac Muscle

Compared to skeletal muscle fibers, cardiac muscle fibers are shorter in length, larger in diameter, and not as circular in transverse section (Figure 20.9). They also exhibit branching, which gives an individual fiber a Y-shaped appearance (see Table 4.4 on page 131). A typical cardiac muscle fiber is 50–100 μm long and has a diameter of about 14 μm. Usually there is only one centrally located nucleus, although an occasional cell may have two nuclei. The sarcolemma of cardiac muscle fibers is similar to that of skeletal muscle, but the sarcoplasm is more abundant and the mitochondria are larger and more numerous. Cardiac muscle fibers have the same arrangement of actin and myosin, and the same bands, zones, and Z discs, as skeletal muscle fibers. The transverse tubules of cardiac muscle are wider but less abundant than those of skeletal muscle; there is only one transverse tubule per sarcomere, located at the Z disc. Also, the sarcoplasmic reticulum of cardiac muscle fibers is scanty compared with the SR of skeletal muscle fibers. As a result, cardiac muscle has a limited intracellular reserve of Ca^{2+}. During contraction, a substantial amount of Ca^{2+} enters cardiac muscle fibers from extracellular fluid.

Although cardiac muscle fibers branch and interconnect with each other, they form two separate functional networks. The muscular walls and partition of the atria compose one network, whereas the muscular walls and partition of the ventricles compose the other network. The ends of each fiber in a network connect to its neighbors by irregular transverse thickenings of the sarcolemma called **intercalated discs** (in-TER-ka-lāt-ed; *intercalat-* = to insert between). The discs contain **desmosomes,** which hold the fibers together, and **gap junctions,** which allow muscle action potentials to spread from one muscle fiber to another. As a consequence, when a single fiber of either network is stimulated, all the other fibers in the network become stimulated as well. Thus, each network contracts as a functional unit. When the fibers of the atria contract as a unit, blood moves into the ventricles; when the ventricular fibers contract as a unit, they eject blood out of the heart into arteries.

Autorhythmic Cells: The Conduction System

An inherent and rhythmical electrical activity is the reason for the heart's continuous beating. The source of stimulation is a network of specialized cardiac muscle fibers that are called **autorhythmic cells** because they are self-excitable. Autorhythmic cells repeatedly generate spontaneous action potentials that trigger heart contractions. The autorhythmic cells explain why a heart that has been removed from the body—for example, to be transplanted into another person—continues to beat even though all of its nerves have been cut. Nerve impulses from the autonomic nervous system and blood-borne hormones (such as epinephrine) modify the heartbeat, but they *do not establish the fundamental rhythm*.

During embryonic development, about 1% of the cardiac muscle fibers become autorhythmic cells that repeatedly and rhythmically generate action potentials. Autorhythmic cells have two important functions: They act as a **pacemaker,** setting the rhythm for the entire heart, and they form the **conduction system,** the route for propagating action potentials throughout the heart muscle. The conduction system assures that cardiac chambers become stimulated to contract in a coordinated manner, which makes the heart an effective pump. Cardiac action potentials propagate through the following components of the conduction system (Figure 20.10):

1 Normally, cardiac excitation begins in the **sinoatrial (SA) node,** located in the right atrial wall just inferior to the opening of the superior vena cava. Each action potential from the SA node propagates throughout both atria via gap junctions in the intercalated discs of atrial fibers. In the wake of the action potential, the atria contract.

Figure 20.9 Histology of cardiac muscle.

🔑 **Muscle fibers of the atria form one functional network, whereas muscle fibers of the ventricles form a second functional network.**

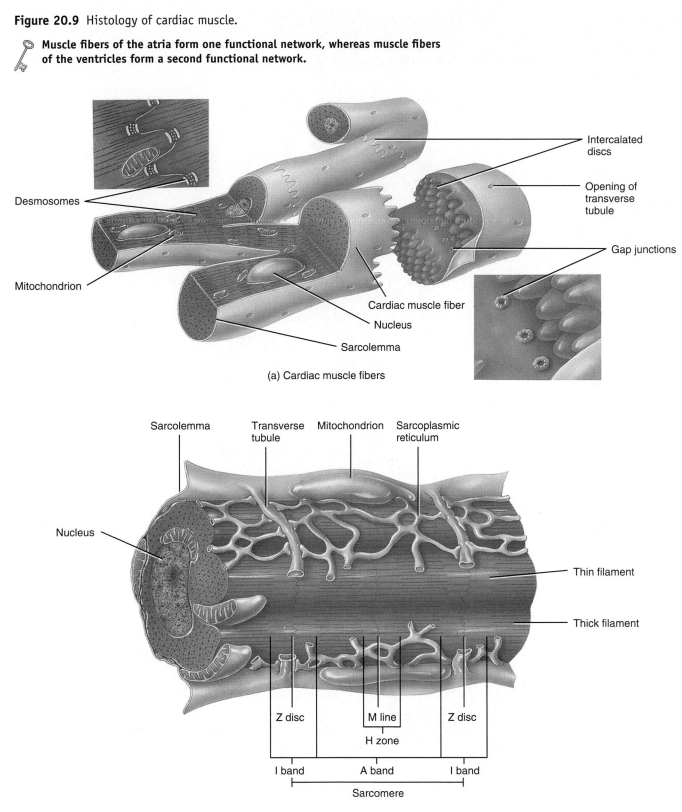

Desmosomes

Mitochondrion

Intercalated discs

Opening of transverse tubule

Gap junctions

Cardiac muscle fiber

Nucleus

Sarcolemma

(a) Cardiac muscle fibers

Sarcolemma Transverse tubule Mitochondrion Sarcoplasmic reticulum

Nucleus

Thin filament

Thick filament

Z disc M line Z disc

H zone

I band A band I band

Sarcomere

(b) Cardiac myofibrils based on an electron micrograph

Q What are the functions of intercalated discs in cardiac muscle fibers?

Figure 20.10 The conduction system of the heart. Autorhythmic fibers in the SA node, located in the right atrial wall, act as the heart's pacemaker, initiating cardiac action potentials that cause contraction of the heart's chambers. The route of action potentials through the numbered components of the conduction system is described in the text.

🔑 **The conduction system ensures that cardiac chambers contract in a coordinated manner.**

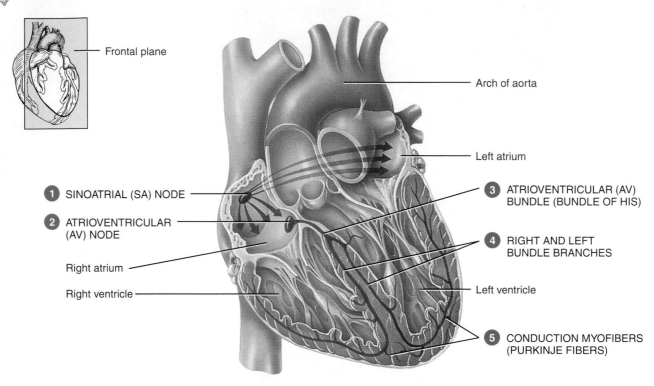

Frontal plane

Arch of aorta

Left atrium

① SINOATRIAL (SA) NODE

③ ATRIOVENTRICULAR (AV) BUNDLE (BUNDLE OF HIS)

② ATRIOVENTRICULAR (AV) NODE

④ RIGHT AND LEFT BUNDLE BRANCHES

Right atrium

Right ventricle

Left ventricle

⑤ CONDUCTION MYOFIBERS (PURKINJE FIBERS)

Anterior view of frontal section

Q Which component of the conduction system provides the only electrical connection between the atria and the ventricles?

② Propagating along atrial muscle fibers, the action potential reaches the **atrioventricular (AV) node,** located in the septum between the two atria, just anterior to the opening of the coronary sinus.

③ From the AV node, the action potential enters the **atrioventricular (AV) bundle** (also known as the **bundle of His**), the only electrical connection between the atria and the ventricles. (Elsewhere, we find that the fibrous skeleton of the heart electrically insulates the atria from the ventricles.)

④ After conducting along the AV bundle, the action potential then enters both the **right** and **left bundle branches** that course through the interventricular septum toward the apex of the heart.

⑤ Finally, the large-diameter **conduction myofibers (Purkinje fibers)** rapidly conduct the action potential, first to the apex of the ventricular myocardium and then upward to the remainder of the ventricular myocardium. About 0.20 sec (200 milliseconds or msec) after the atria contract, the ventricles contract.

On their own, autorhythmic fibers in the SA node initiate action potentials 90–100 times per minute, faster than any other region. As a result, action potentials from the SA node spread to other areas of the conduction system and stimulate them before they are able to generate an action potential at their own, intrinsically slower rate. Thus, the normal *pacemaker* of the heart is the SA node. Various hormones and neurotransmitters can speed or slow pacing of the heart by SA node fibers. In a person at rest, for example, acetylcholine released by the parasympathetic division of the ANS slows SA node pacing to about 75 action potentials per minute.

Sometimes, a site other than the SA node becomes the pacemaker because it develops abnormal self-excitability. Such a site is called an **ectopic pacemaker** (ek-TOP-ik; *ectop-* = displaced). An ectopic pacemaker may operate only occasionally, producing extra beats, or it may pace the heart for some period of time. Triggers of ectopic activity include caffeine and nicotine, electrolyte imbalances, hypoxia, and toxic reactions to drugs such as digitalis.

Timing of Atrial and Ventricular Excitation

From the SA node, a cardiac action potential propagates throughout the atrial muscle and down to the AV node in about 50 msec. The action potential slows considerably at the AV node because the fibers there have much smaller diameters. (Recall what happens to automobile traffic when a four-lane highway narrows to just two lanes.) The resulting 100-msec delay has a beneficial function: It gives the atria time to contract fully, thus adding to the volume of blood in the ventricles before ventricular contraction begins. The action potential propagates rapidly again after entering the AV bundle. About 200 msec after the action potential arises in the SA node, it has propagated throughout the entire myocardium.

If the SA node becomes diseased or damaged, the slower AV node fibers can take over the pacemaking chores. With pacing by the AV node, however, heart rate is 40–50 beats/min. If the activity of both nodes is suppressed, the heartbeat may still be maintained by autorhythmic fibers in the ventricles—the AV bundle, a bundle branch, or conduction myofibers. These fibers generate action potentials very slowly, only about 20–40 times per minute. At such a low heart rate, blood flow to the brain is inadequate. When this condition occurs, normal heart rhythm can be restored and maintained by surgically implanting an **artificial pacemaker,** a device that sends out small electrical currents to stimulate the heart to maintain adequate cardiac output. Many of the newer pacemakers, called activity-adjusted pacemakers, automatically speed up the heartbeat during exercise.

Physiology of Cardiac Muscle Contraction

The action potential initiated by the SA node travels along the conduction system and spreads out to excite the "working" atrial and ventricular muscle fibers, which are called **contractile fibers.** An action potential occurs in a contractile fiber as follows (Figure 20.11a):

1 *Depolarization.* Contractile fibers have a resting membrane potential close to −90 mV. When they are brought to threshold by excitation in neighboring fibers, certain sodium ion (Na^+) channels open very rapidly; these are called **voltage-gated fast Na^+ channels.** When these channels open, the permeability of the sarcolemma (plasma membrane) to sodium ions (P_{Na^+}) increases (Figure 20.11b). Because the cytosol is electrically more negative than extracellular fluid and Na^+ concentration is higher in extracellular fluid, the result is an inflow of Na^+ along the electrochemical gradient that produces a **rapid depolarization.** Within a few milliseconds, the fast Na^+ channels automatically inactivate and P_{Na^+} decreases.

2 *Plateau.* The next phase, called the **plateau,** involves opening of **voltage-gated slow Ca^{2+} channels** in the sarcolemma and sarcoplasmic reticulum membrane. This increases permeability to calcium ions (P_{Ca2+}), allowing the level of Ca^{2+} to increase in the cytosol (see Figure 20.11b). Some calcium ions pass through the sarcolemma from the extracellular fluid (which has a higher Ca^{2+} concentration) while others pour out of the sarcoplasmic reticulum within the fiber. At the same time, the membrane permeability to potassium ions (P_{K+}) decreases due to closing of K^+ channels. For about 250 msec the membrane potential stays close to 0 mV as a small outflow of K^+ just balances the inflow of Ca^{2+}. (By comparison, depolarization in a neuron or skeletal muscle fiber lasts about 1 msec.)

3 *Repolarization.* The recovery of the resting membrane potential during the **repolarization** phase of a cardiac action potential resembles that of other excitable tissues. After a delay (which is particularly prolonged in cardiac muscle), **voltage-gated K^+ channels** open, thereby increasing the membrane permeability to potassium ions, which then diffuse out more rapidly due to the concentration difference. At the same time, the calcium channels are closing. As more K^+ leave the fiber and fewer Ca^{2+} enter, the negative resting membrane potential (−90 mV) is restored.

The mechanism of contraction is similar in cardiac and skeletal muscle: The electrical activity (action potential) leads to the mechanical response (contraction) after a short lag time. As Ca^{2+} concentration rises inside a contractile fiber, Ca^{2+} binds to the regulator protein troponin, which allows the actin and myosin filaments to begin sliding past one another, and tension starts to develop. Substances that alter the movement of Ca^{2+} through slow Ca^{2+} channels influence the strength of heart contractions. Epinephrine, for example, increases contraction force by enhancing Ca^{2+} inflow.

In muscle, the **refractory period** is the time interval during which a second contraction cannot be triggered. The refractory period of a cardiac fiber lasts longer than the contraction itself (see Figure 20.11). As a result, another contraction cannot begin until relaxation is well underway. For this reason, tetanus (maintained contraction) cannot occur in cardiac muscle tissue. The advantage is apparent if you consider how the ventricles work. Their pumping function depends on alternating contraction (when they eject blood) and relaxation (when they refill). If heart muscle could undergo tetanus, blood flow would cease.

CLINICAL APPLICATION
Help for Failing Hearts

Physicians are using or investigating a variety of surgical techniques and medical devices that can aid a failing heart. For some patients, even a 10% increase in the volume of blood ejected from the ventricles can mean the difference between being bedridden and having limited mobility. **Heart transplants** are fairly common today and produce good

Figure 20.11 Action potential in a ventricular contractile fiber. The resting membrane potential is about −90 mV.

🔑 **The "working" atrial and ventricular muscle fibers are called contractile fibers.**

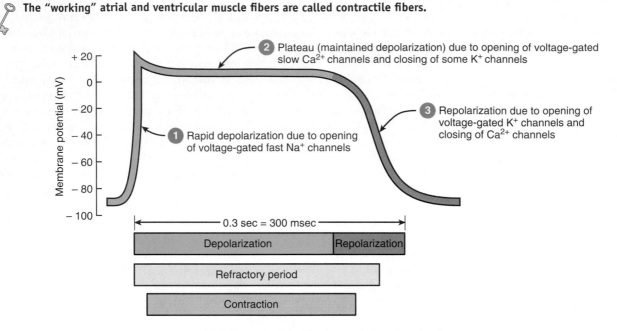

② Plateau (maintained depolarization) due to opening of voltage-gated slow Ca^{2+} channels and closing of some K^+ channels

③ Repolarization due to opening of voltage-gated K^+ channels and closing of Ca^{2+} channels

① Rapid depolarization due to opening of voltage-gated fast Na^+ channels

(a) Action potential, refractory period, and contraction

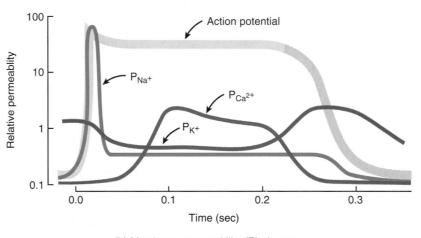

(b) Membrane permeability (P) changes

Ⓠ How does the duration of depolarization and repolarization in a ventricular contractile fiber compare with that in a skeletal muscle fiber?

results, but the availability of donor hearts is very limited. In the United States there are about 50 potential candidates for each available donor heart. Another approach is to use **cardiac assist devices** and surgical procedures that augment heart function without removing the heart. Table 20.1 describes several of these. ■

1. How do cardiac muscle fibers differ structurally and functionally from skeletal muscle fibers?
2. In what ways are autorhythmic fibers similar to and different from contractile fibers?

3. Describe the phases of an action potential in ventricular contractile fibers.

ELECTROCARDIOGRAM

OBJECTIVE

• *Explain the meaning of an electrocardiogram (ECG) and its diagnostic importance.*

Action potential propagation through the heart generates electrical currents that can be detected at the surface of

Table 20.1 Cardiac Assist Devices and Procedures

DEVICE	DESCRIPTION
Intra-aortic balloon pump (IABP)	A 40-ml polyurethane balloon mounted on a catheter is inserted into an artery in the groin and threaded into the thoracic aorta. An external pump inflates the balloon with gas at the beginning of ventricular diastole. As the balloon inflates, it pushes blood both backward toward the heart, which improves coronary blood flow, and forward toward peripheral tissues. The balloon then is rapidly deflated just before the next ventricular systole, making it easier for the left ventricle to eject blood. Because the balloon is inflated between heartbeats, this technique is called intra-aortic balloon counterpulsation.
Hemopump	This propeller-like pump is threaded through an artery in the groin and then into the left ventricle. There, the blades of the pump whirl at about 25,000 revolutions per minute, pulling blood out of the left ventricle and pushing it into the aorta.
Left ventricular assist device (LVAD)	The LVAD is a completely portable assist device. It is implanted within the abdomen and powered by a battery pack worn in a shoulder holster. The LVAD is connected to the patient's weakened left ventricle and pumps blood into the aorta. The pumping rate increases automatically during exercise.
Cardiomyoplasty	A large piece of the patient's own skeletal muscle (left latissimus dorsi) is partially freed from its connective tissue attachments and wrapped around the heart, leaving the blood and nerve supply intact. An implanted pacemaker stimulates the skeletal muscle's motor neurons to cause contraction 10–20 times per minute, in synchrony with some of the heartbeats.
Skeletal muscle assist device	A piece of the patient's own skeletal muscle is used to fashion a pouch that is inserted between the heart and the aorta, functioning as a booster heart. A pacemaker stimulates the muscle's motor neurons to elicit contraction.

the body. A recording of these electrical changes is called an **electrocardiogram** (e-lek′-trō-KAR-dē-ō-gram), abbreviated either **ECG** or **EKG** (from the German word *Elektrokardiogram*). The ECG is a composite recording of action potentials produced by all the heart muscle fibers during each heartbeat. The instrument used to record the changes is an **electrocardiograph.** In clinical practice, the ECG is recorded by placing electrodes on the arms and legs (limb leads) and at six positions on the chest (chest leads). The electrocardiograph amplifies the heart's electrical activity and produces 12 different tracings from different combinations of limb and chest leads. Each limb and chest electrode records slightly different electrical activity because it is in a different position relative to the heart. By comparing these records with one another and with normal records, it is possible to determine (1) if the conduction pathway is abnormal, (2) if the heart is enlarged, and (3) if certain regions are damaged.

In a typical Lead II record (right arm to left leg), three clearly recognizable waves accompany each heartbeat. The first, called the **P wave,** is a small upward deflection on the ECG (Figure 20.12); it represents **atrial depolarization,** which spreads from the SA node throughout both atria. About 100 msec after the P wave begins, the atria contract. The second wave, called the **QRS complex,** begins as a downward deflection, continues as a large, upright, triangular wave, and ends as a downward wave. The QRS complex represents the onset of **ventricular depolarization,** as the wave of electrical excitation spreads through the ventricles. Shortly after the QRS complex begins, the ventricles start to contract. The third wave is a dome-shaped upward deflection called the **T wave;** it indicates **ventricular repolarization** and occurs just before the ventricles start to relax. The T

Figure 20.12 Electrocardiogram or ECG (Lead II). P wave = atrial depolarization; QRS complex = onset of ventricular depolarization; T wave = ventricular repolarization.

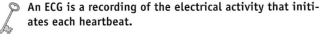
An ECG is a recording of the electrical activity that initiates each heartbeat.

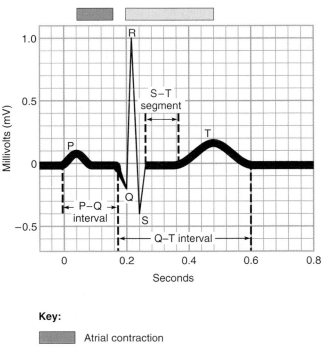

Key:

▨	Atrial contraction
▨	Ventricular contraction

Q What is the significance of an enlarged Q wave?

wave is smaller and wider than the QRS complex because repolarization occurs more slowly than depolarization. Repolarization of the atria is not usually evident in an ECG because it is masked by the larger QRS complex.

In reading an ECG, the size of the waves can provide clues to abnormalities. Larger P waves, for example, indicate enlargement of an atrium; an enlarged Q wave may indicate a myocardial infarction (heart attack), and an enlarged R wave generally indicates enlarged ventricles. The T wave is flatter than normal when the heart muscle is receiving insufficient oxygen—as, for example, in coronary artery disease. The T wave may be elevated in hyperkalemia (increased blood K^+ level).

Analysis of an ECG also involves examination of the time spans between waves, which are called **intervals** or **segments.** For example, the **P–Q interval** is measured from the beginning of the P wave to the beginning of the QRS complex; it represents the conduction time from the beginning of atrial excitation to the beginning of ventricular excitation. Put another way, the P–Q interval is the time required for an impulse to travel through the atria, atrioventricular node, and the remaining fibers of the conduction system. In coronary artery disease and rheumatic fever, scar tissue may form in the heart. As the impulse detours around scar tissue, the P–Q interval lengthens.

The **S–T segment** begins at the end of the S wave and ends at the beginning of the T wave; it represents the time when the ventricular contractile fibers are fully depolarized during the plateau phase of the action potential. The S–T segment is elevated (above the baseline) in acute myocardial infarction and depressed (below the baseline) when the heart muscle receives insufficient oxygen. The **Q–T interval** extends from the start of the QRS complex to the end of the T wave; it is the time from the beginning of ventricular depolarization to the end of ventricular repolarization. The Q–T interval may be lengthened by myocardial damage, coronary ischemia, or conduction abnormalities.

Sometimes it is necessary to evaluate the heart's response to the stress of physical exercise. Such a test is called a **stress electrocardiogram,** or **stress test.** Although narrowed coronary arteries may carry adequate oxygenated blood while a person is at rest, during strenuous exercise they will be unable to meet the heart's increased need for oxygen, creating changes that can be noted on an electrocardiogram.

CLINICAL APPLICATION
Arrhythmias

Arrhythmia (a-RITH-mē-a) or *dysrhythmia* is a general term referring to an irregularity in heart rhythm resulting from a defect in the conduction system of the heart. Arrhythmias are caused by such factors as caffeine, nicotine, alcohol, and certain other drugs; anxiety; hyperthyroidism; potassium deficiency; and certain heart diseases. One serious arrhythmia is *heart block,* in which propagation of action potentials along the conduction system is either slowed or blocked. The most common site of blockage is the atrioventricular node, a condition called *atrioventricular (AV) block.* In *atrial flutter,* the atria contract abnormally rapidly, about 300 times per minute. *Atrial fibrillation* is asynchronous contraction of atrial fibers (about 400–600 times per minute), whereas *ventricular fibrillation* is asynchronous contractions of ventricular contractile fibers. In both cases, the fibrillating chambers fail to pump blood because some muscle fibers are contracting while others are relaxing. In a strong heart, atrial fibrillation reduces the pumping effectiveness of the heart by only 20–30%, which is compatible with life. In ventricular fibrillation, however, death occurs quickly because blood is not being ejected from the ventricles. ■

1. Draw, label, and define the waves of a normal electrocardiogram (ECG).
2. Describe the significance of the P–Q interval and the S–T segment.

THE CARDIAC CYCLE
OBJECTIVE
• *Describe the phases, timing, and heart sounds associated with a cardiac cycle.*

A single **cardiac cycle** includes all the events associated with one heartbeat. In each cardiac cycle, the atria and ventricles alternately contract and relax, forcing blood from areas of higher pressure to areas of lower pressure. As a chamber of the heart contracts, blood pressure within it increases.

Figure 20.13 shows the relation between the heart's electrical signals (ECG) and mechanical events (contraction and relaxation), including changes in atrial pressure, ventricular pressure, aortic pressure, and ventricular volume during the cardiac cycle. The pressures given in Figure 20.13 apply to the left side of the heart; pressures on the right side are considerably lower. Each ventricle, however, expels the same volume of blood per beat, and the same pattern exists for both pumping chambers.

In a normal cardiac cycle, the two atria contract while the two ventricles relax; then, while the two ventricles contract, the two atria relax. The term **systole** (SIS-tō-lē; = contraction) refers to the phase of contraction; the phase of relaxation is **diastole** (dī-AS-tō-lē; *diastole* = dilation or expansion). A cardiac cycle consists of systole and diastole of both atria plus systole and diastole of both ventricles.

Phases of the Cardiac Cycle

When resting heart rate is 75 beats/min, each cardiac cycle lasts about 0.8 seconds (800 msec). For the purposes of our discussion, we can divide the cardiac cycle of a resting adult into three phases (see Figure 20.13).

❶ *Isovolumetric relaxation.* Repolarization of the ventricular muscle fibers (T wave in the ECG) initiates relaxation. As the ventricles relax, pressure within the chambers drops, and blood starts to flow from the pulmonary

Figure 20.13 Cardiac cycle. (a) ECG. (b) Changes in left atrial pressure (green solid line), left ventricular pressure (green dashed line), and aortic pressure (red line) as they relate to the opening and closing of valves. (c) Changes in left ventricular volume. (d) Heart sounds. (e) Phases of the cardiac cycle; steps 1–3 are discussed in the text.

A cardiac cycle is composed of all the events associated with one heartbeat.

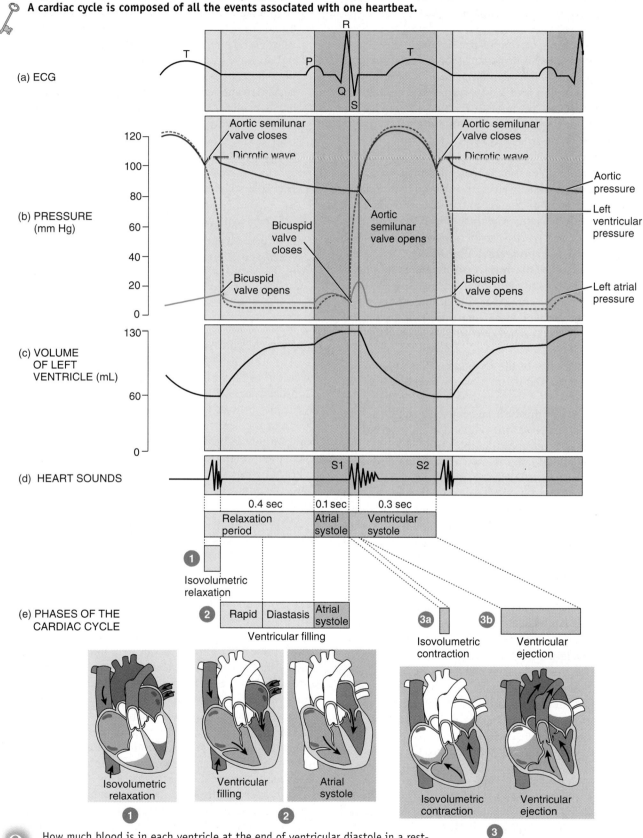

Q How much blood is in each ventricle at the end of ventricular diastole in a resting person? What is this volume called?

trunk and aorta back toward the ventricles. The back-flowing blood becomes trapped in the semilunar cusps, however, and closes the valves. Rebound of blood off the closed cusps produces the *dicrotic wave* on the aortic pressure curve.

Following the closing of the semilunar valves is a brief interval when ventricular blood volume does not change because both semilunar and AV valves are closed. This period is called **isovolumetric relaxation.** As the ventricles continue to relax, the space within them expands, and the pressure falls quickly. When ventricular pressure drops below atrial pressure, the AV valves open and ventricular filling begins.

② *Ventricular filling.* The major part of ventricular filling occurs just after the AV valves open. Blood that has been flowing into and building up in the atria while the ventricles were contracting now rushes into the ventricles. The first third of ventricular filling time thus is known as the period of **rapid ventricular filling.** During the middle third, called **diastasis,** a much smaller volume of blood flows into the ventricles. During isovolumetric relaxation, rapid ventricular filling, and diastasis, all four heart chambers are in diastole. This interval is known as the **relaxation period** and lasts about 400 msec.

Depolarization of the SA node results in atrial depolarization, marked by the P wave in the ECG. **Atrial systole** follows the P wave and lasts about 100 msec. It occurs in the last third of the ventricular filling period and forces a final 20–25 mL of blood into the ventricles. At the end of ventricular diastole, each ventricle contains about 130 mL; this volume of blood is called **end-diastolic volume (EDV).** Because atrial systole contributes only 20–30% of the total blood volume in the ventricles, atrial contraction is not absolutely necessary for adequate blood circulation. Throughout the period of ventricular filling, the AV valves are open and the semilunar valves are closed.

③ *Ventricular systole.* For the next 0.3 seconds (300 msec), the atria are relaxing and the ventricles are contracting. Near the end of atrial systole, the action potential from the SA node has passed through the AV node and into the ventricles, causing them to depolarize. Onset of ventricular depolarization is indicated by the QRS complex in the ECG. Then, **ventricular systole** begins, and blood is pushed up against the AV valves, forcing them shut. For about 0.05 seconds (50 msec), all four valves are closed again; this period is called **isovolumetric contraction ③ₐ**. During this interval, cardiac muscle fibers are contracting and exerting force but are not yet shortening. Thus, the muscle contraction is isometric (same length). Moreover, because all four valves are closed, ventricular volume remains the same (isovolumic).

As ventricular contraction continues, pressure inside the chambers rises sharply. When left ventricular pressure surpasses aortic pressure at about 80 millimeters of mercury (mm Hg) and right ventricular pressure rises above the pressure in the pulmonary trunk (about 20 mm Hg), both semilunar valves open, and ejection of blood from the heart begins. The pressure in the left ventricle continues to rise to about 120 mm Hg, whereas the pressure in the right ventricle climbs to about 30 mm Hg. The period when the semilunar valves are open is called **ventricular ejection ③ᵇ** and lasts for about 0.25 seconds (250 msec). When the ventricles begin to relax, ventricular pressure drops, the semilunar valves close, and another relaxation period begins. As the heart beats faster, the relaxation period becomes shorter and shorter, whereas the durations of atrial systole and ventricular systole shorten only slightly.

The volume of blood left in a ventricle at the end of systole is called **end-systolic volume (ESV).** At rest, ESV is about 60 mL. **Stroke volume,** the volume ejected per beat from each ventricle, equals end-diastolic volume minus end-systolic volume: SV = EDV − ESV. At rest, the stroke volume is about 130 mL − 60 mL = 70 mL (a little more than 2 oz).

Heart Sounds

The act of listening to sounds within the body is called **auscultation** (aws-kul-TĀ-shun; *ausculta-* = listening), and it is usually done with a stethoscope. The sound of the heartbeat comes primarily from blood turbulence caused by the closing of the heart valves. During each cardiac cycle, four **heart sounds** are generated, but in a normal heart only the first and second heart sounds (S1 and S2) are loud enough to be heard by listening through a stethoscope. Figure 20.13d shows the timing of S1 and S2 relative to other events in the cardiac cycle.

The first sound (S1), which can be described as a **lubb** sound, is louder and a bit longer than the second sound. The lubb is the sound created by blood turbulence associated with closure of the AV valves soon after ventricular systole begins. The second sound (S2), which is shorter and not as loud as the first, can be described as a **dupp** sound. S2 is caused by blood turbulence associated with closure of the semilunar valves at the beginning of ventricular diastole. Although these heart sounds are due to blood turbulence associated with the closure of valves, they are best heard at the surface of the chest in locations that differ slightly from the actual locations of the valves (Figure 20.14). The sound associated with closure of the aortic semilunar valve (first part of S2) is best heard near the superior right point, the sound associated with closure of the pulmonary semilunar valve (second part of S2) near the superior left point, the sound associated with closure of the bicuspid valve (first part of S1) near the inferior left point, and the sound associated with closure of the tricuspid valve (second part of S1) near the inferior right point. Although normally not loud enough to be heard, S3 is due to blood turbulence during rapid ventricular

Figure 20.14 Location of valves (green) and auscultation sites (red) for heart sounds.

 Listening to sounds within the body is called auscultation; it is usually done with a stethoscope.

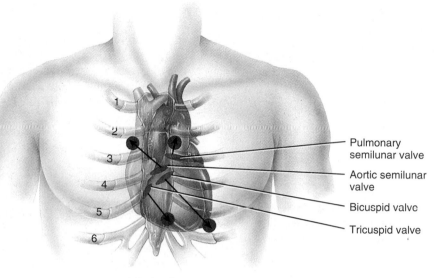

Pulmonary semilunar valve

Aortic semilunar valve

Bicuspid valve

Tricuspid valve

Anterior view

 Which heart sound is related to blood turbulence associated with closure of the AV valves?

filling, and S4 is due to blood turbulence during atrial contraction.

CLINICAL APPLICATION
Heart Murmurs

Heart sounds provide valuable information about the mechanical operation of the heart. A **heart murmur** is an abnormal sound consisting of a rushing or gurgling noise that is heard before, between, or after the normal heart sounds, or that may mask the normal heart sounds. Although some heart murmurs are "innocent," meaning they are not associated with a significant heart problem, most often a murmur indicates a valve disorder.

Among the valvular abnormalities that may contribute to murmurs are **mitral stenosis** (narrowing of the mitral valve by scar formation or a congenital defect), **mitral insufficiency** (backflow or regurgitation of blood from the left ventricle into the left atrium due to a damaged mitral valve or ruptured chordae tendineae), **aortic stenosis** (narrowing of the aortic semilunar valve), and **aortic insufficiency** (backflow of blood from the aorta into the left ventricle). Another cause of a heart murmur is **mitral valve prolapse (MVP)**, an inherited disorder in which one or both cusps of the mitral valve protrude into the left atrium during ventricular contraction. Although a small volume of blood may flow back into the left atrium during ventricular contraction and result in a murmur, mitral valve pro-

lapse does not always pose a serious threat. MVP occurs in 10–15% of the population, more often in females (65%). ■

1. List in sequence and describe the events of the cardiac cycle, including atrial systole, ventricular systole, isovolumetric relaxation, isovolumetric contraction, relaxation period, diastasis, ventricular ejection, and rapid ventricular filling.
2. Draw and label a diagram that relates the events of the cardiac cycle to time.
3. Describe the source of the four normal heart sounds. Which ones can usually be heard through a stethoscope?

CARDIAC OUTPUT

OBJECTIVE

• *Define cardiac output, and describe the factors that affect it.*

Although the heart has autorhythmic fibers that enable it to beat independently, its operation is governed by events occurring in the rest of the body. All body cells must receive a certain amount of oxygenated blood each minute to maintain health and life. When cells are metabolically active, as during exercise, they must take up more oxygen from the blood. During rest periods, cellular metabolic need is reduced, and the workload of the heart decreases.

Cardiac output (CO) is the volume of blood ejected from the left ventricle (or the right ventricle) into the aorta (or pulmonary trunk) each minute. Cardiac output equals the **stroke volume (SV),** the volume of blood ejected by the ventricle with each contraction, multiplied by the **heart rate (HR),** the number of heartbeats per minute:

$$\underset{\text{(mL/min)}}{\text{CO}} = \underset{\text{(mL/beat)}}{\text{SV}} \times \underset{\text{(beats/min)}}{\text{HR}}$$

In a typical resting adult male, stroke volume averages 70 mL/beat, and heart rate is about 75 beats/min. Thus, average cardiac output is

$$
\begin{aligned}
\text{CO} &= 70 \text{ mL/beat} \times 75 \text{ beats/min} \\
&= 5250 \text{ mL/min} \\
&= 5.25 \text{ L/min}
\end{aligned}
$$

This volume is close to the total blood volume, which is about 5 liters in a typical adult male. Approximately the entire blood volume thus flows through the pulmonary and systemic circulations each minute. When body tissues use more or less oxygen, cardiac output changes to meet the need; factors that increase stroke volume or heart rate normally increase CO. During mild exercise, for example, stroke volume may increase to 100 mL/beat, and heart rate to 100 beats/min. Cardiac output then would be 10 L/min. During intense (but still not maximal) exercise, the heart rate may accelerate to 150 beats/min, and stroke volume may rise to 130 mL/beat, providing a cardiac output of 19.5 L/min.

Cardiac reserve is the ratio between a person's maximum cardiac output and cardiac output at rest. A cardiac reserve that is four or five times the resting value is average; top endurance athletes may have a cardiac reserve seven or eight times their resting CO. People with severe heart disease may have little or no cardiac reserve, which limits their ability to carry out even the simple tasks of daily living.

Regulation of Stroke Volume

A healthy heart pumps out all the blood that has entered its chambers during the previous diastole. The more blood that returns to the heart during diastole, the more blood that is ejected during the next systole. At rest, the stroke volume is 50–60% of the end-diastolic volume because 40–50% remains in the ventricles after each contraction (end-systolic volume). Three important factors regulate stroke volume and ensure that the left and right ventricles pump equal volumes of blood: (1) **preload,** the degree of stretch in the heart before it contracts; (2) **contractility,** the forcefulness of contraction of individual ventricular muscle fibers; and (3) **afterload,** the pressure that must be exceeded if ejection of blood from the ventricles is to occur.

Preload: Effect of Stretching

A greater preload (stretch) on cardiac muscle fibers prior to contraction increases their force of contraction.

Within limits, the more the heart is filled during diastole, the greater the force of contraction during systole—a relationship known as the **Frank–Starling law of the heart.** In the body, the preload is the volume of blood that fills the ventricles at the end of diastole, the end-diastolic volume (EDV). Normally, the greater the EDV (preload), the more forceful the ensuing contraction.

The duration of ventricular diastole and venous pressure are the two key factors that determine EDV. When heart rate increases, the duration of diastole is shorter. Less filling time means a smaller EDV, and the ventricles may contract before they are adequately filled. Conversely, when venous pressure increases, a greater volume of blood is forced into the ventricles, and the EDV is increased.

When heart rate exceeds about 160 beats/min, stroke volume usually declines due to the short filling time. At such rapid heart rates, EDV is less, and the preload is lower. People who have slow resting heart rates, in contrast, usually have large resting stroke volumes because filling time is prolonged and preload is larger.

The Frank–Starling law of the heart equalizes the output of the right and left ventricles and keeps the same volume of blood flowing to both the systemic and pulmonary circulations. If the left side of the heart pumps a little more blood than the right side, for example, the volume of blood returning to the right ventricle (venous return) increases. With increased EDV, then, the right ventricle contracts more forcefully on the next beat, and the two sides are again in balance.

Contractility

The second factor that influences stroke volume is myocardial **contractility,** which is the strength of contraction at any given preload. Substances that increase contractility are called **positive inotropic agents,** whereas those that decrease contractility are called **negative inotropic agents.** Thus, for a constant preload, the stroke volume increases when a positive inotropic substance is present. Positive inotropic agents often promote Ca^{2+} inflow during cardiac action potentials, which strengthens the force of the subsequent muscle fiber contraction. Stimulation of the sympathetic division of the autonomic nervous system (ANS), hormones such as epinephrine and norepinephrine, increased Ca^{2+} level in the extracellular fluid, and the drug digitalis all have positive inotropic effects. In contrast, inhibition of the sympathetic division of the ANS, anoxia, acidosis, some anesthetics (for example, halothane), and increased K^+ level in the extracellular fluid have negative inotropic effects. A class of drugs called calcium channel blockers exert a negative inotropic effect by reducing Ca^{2+} inflow, thereby decreasing the strength of the heartbeat.

Afterload

Ejection of blood from the heart begins when pressure in the right ventricle exceeds the pressure in the pulmonary trunk (about 20 mm Hg), and when the pressure in the left ventricle exceeds the pressure in the aorta (about 80 mm Hg). At that point, the higher pressure in the ventricles causes blood to push the semilunar valves open. The pressure that must be overcome before a semilunar valve can open is termed the **afterload.** At any given preload, an increase in afterload causes stroke volume to decrease, and more blood remains in the ventricles at the end of systole. Conditions that can increase afterload include hypertension (elevated blood pressure) and narrowing of arteries by atherosclerosis (see page 663).

CLINICAL APPLICATION
Congestive Heart Failure

In **congestive heart failure (CHF),** the heart is a failing pump. Causes of CHF include coronary artery disease (see page 663), congenital defects, long-term high blood pressure (which increases the afterload), myocardial infarctions (regions of dead heart tissue due to a previous heart attack), and valve disorders. As the pump becomes less effective, more blood remains in the ventricles at the end of each cycle, and gradually the end-diastolic volume (preload) increases. Initially, increased preload may promote increased force of contraction (the Frank–Starling law of the heart), but as the preload increases further, the heart is overstretched and contracts less forcefully. The result is a positive feedback loop: Less-effective pumping leads to even lower pumping capability.

Often, one side of the heart starts to fail before the other. If the left ventricle fails first, it can't pump out all the blood it receives. As a result, blood backs up in the lungs. The result is *pulmonary edema,* fluid accumulation in the lungs that can suffocate an untreated person. If the right ventricle fails first, blood backs up in the systemic vessels. In this case, the resulting *peripheral edema* is usually most noticeable in the feet and ankles. ■

Regulation of Heart Rate

As we have seen, cardiac output depends on heart rate as well as stroke volume. Adjustments in heart rate are important in the short-term control of cardiac output and blood pressure. The sinoatrial (SA) node initiates contraction and, if left to itself, would set a constant heart rate of 90–100 beats/min. However, tissues require a different volume of blood flow under different conditions. During exercise, for example, cardiac output rises to supply working tissues with increased amounts of oxygen and nutrients. Stroke volume may fall if the ventricular myocardium is damaged or if blood volume is reduced by bleeding. In these cases, homeostatic mechanisms act to maintain adequate cardiac output by increasing the heart rate and contractility. Among the several factors that contribute to regulation of heart rate, the most important are the autonomic nervous system and hormones released by the adrenal medullae (epinephrine and norepinephrine).

Autonomic Regulation of Heart Rate

Nervous system regulation of the heart originates in the **cardiovascular center** in the medulla oblongata. This region of the brain stem receives input from a variety of sensory receptors and from higher brain centers, such as the limbic system and cerebral cortex. The cardiovascular center then directs appropriate output by increasing or decreasing the frequency of nerve impulses via both the sympathetic and parasympathetic branches of the ANS (Figure 20.15).

Even before physical activity begins, especially in competitive situations, heart rate may climb. This anticipatory increase occurs because the limbic system in the brain sends nerve impulses to the cardiovascular center in the medulla. Then, as physical activity begins, **proprioceptors** that are monitoring the position of limbs and muscles send an increased frequency of nerve impulses to the cardiovascular center. This proprioceptor input is a major stimulus for the quick rise in heart rate that occurs at the onset of physical activity. Other sensory receptors that provide input to the cardiovascular center include **chemoreceptors,** which monitor chemical changes in the blood, and **baroreceptors,** which monitor blood pressure in major arteries and veins. Important baroreceptors located in the arch of the aorta and in the carotid arteries (see Figure 21.13 on page 684) detect changes in blood pressure and relay the information to the cardiovascular center. Baroreceptor reflexes, which are also important for the regulation of blood pressure, are discussed in detail in Chapter 21. Here we focus specifically on the innervation of the heart by the sympathetic and parasympathetic branches of the ANS.

Sympathetic fibers extend from the medulla oblongata into the spinal cord. From the thoracic region of the spinal cord, **cardiac accelerator nerves** extend out to the SA node, AV node, and most portions of the myocardium. Impulses in the cardiac accelerator nerves trigger the release of norepinephrine, which binds to beta-1 receptors on cardiac muscle fibers. This interaction has two separate effects: (1) In SA (and AV) node fibers, norepinephrine speeds the rate of spontaneous depolarization so that these pacemakers fire impulses more rapidly and heart rate increases. (2) In contractile fibers throughout the atria and ventricles, norepinephrine enhances Ca^{2+} entry through the voltage-gated slow Ca^{2+} channels, thereby increasing contractility and resulting in greater ejection of blood during systole. With a moderate increase in heart rate, stroke volume does not decline because the increased contractility offsets the decreased preload. With maximal sympathetic stimulation, however, heart rate may reach 200 beats/min in a 20-year-old person. At such a high heart rate, stroke volume is lower than at rest due to the very short filling time. The maximal heart rate declines with age; as a rule, subtracting one's age from 220 provides a good estimate of maximal heart rate in beats per minute.

Parasympathetic nerve impulses reach the heart via the right and left **vagus (X) nerves.** These fibers innervate the SA

Figure 20.15 Nervous system control of the heart.

🔑 The cardiovascular center in the medulla oblongata controls both sympathetic and parasympathetic nerves that innervate the heart.

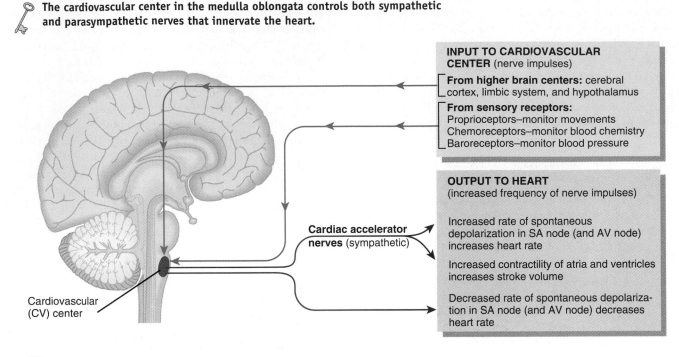

INPUT TO CARDIOVASCULAR CENTER (nerve impulses)

From higher brain centers: cerebral cortex, limbic system, and hypothalamus

From sensory receptors:
Proprioceptors–monitor movements
Chemoreceptors–monitor blood chemistry
Baroreceptors–monitor blood pressure

Cardiac accelerator nerves (sympathetic)

Cardiovascular (CV) center

OUTPUT TO HEART
(increased frequency of nerve impulses)

Increased rate of spontaneous depolarization in SA node (and AV node) increases heart rate

Increased contractility of atria and ventricles increases stroke volume

Decreased rate of spontaneous depolarization in SA node (and AV node) decreases heart rate

Q What region of the heart is innervated by the sympathetic division but not by the parasympathetic division?

node, AV node, and atrial myocardium. They release acetylcholine, which decreases heart rate by causing hyperpolarization and slowing the rate of spontaneous depolarization in autorhythmic fibers. As only a few vagal fibers innervate ventricular muscle, changes in parasympathetic activity have little or no effect on contractility of the ventricles.

A continually shifting balance exists between sympathetic and parasympathetic stimulation of the heart. At rest, the parasympathetic stimulation predominates. The resting heart rate—about 75 beats/min—is usually lower than the autorhythmic rate of the SA node (90–100 beats/min). With maximal stimulation by the parasympathetic division, the heart can slow to 20 or 30 beats/min, or can even stop momentarily.

Chemical Regulation of Heart Rate

Certain chemicals influence both the basic physiology of cardiac muscle and heart rate. For example, hypoxia (lowered oxygen level), acidosis (low pH), and alkalosis (high pH) all depress cardiac activity. Two broad types of chemicals have major effects on the heart:

1. *Hormones.* Epinephrine and norepinephrine (from the adrenal medullae) enhance the heart's pumping effectiveness. These hormones affect cardiac muscle fibers in much the same way as does norepinephrine released by cardiac accelerator nerves—they increase both heart rate and contractility. Exercise, stress, and excitement cause

the adrenal medulla to release more hormones. Thyroid hormones also enhance cardiac contractility and increase heart rate. One sign of hyperthyroidism (excessive thyroid hormone) is tachycardia (elevated heart rate at rest).

2. *Ions.* Given that differences between intracellular and extracellular concentrations of several ions (for example, Na^+ and K^+) are crucial for the production of action potentials in all nerve and muscle fibers, it is not surprising that ionic imbalances can quickly compromise the pumping effectiveness of the heart. In particular, the relative concentrations of three cations—K^+, Ca^{2+}, and Na^+—have a large effect on cardiac function. Elevated blood levels of K^+ or Na^+ decrease heart rate and contractility. Excess Na^+ blocks Ca^{2+} inflow during cardiac action potentials, thereby decreasing the force of contraction, whereas excess K^+ blocks generation of action potentials. A moderate increase in extracellular (and thus intracellular) Ca^{2+} speeds heart rate and strengthens the heartbeat.

Other Factors in Heart Rate Regulation

Age, gender, physical fitness, and body temperature also influence resting heart rate. A newborn baby is likely to have a resting heart rate over 120 beats/min; the rate then declines through childhood. Senior citizens may develop a more rapid heartbeat, however. Adult females generally have slightly higher resting heart rates than adult males, although regular

Figure 20.16 Factors that increase cardiac output.

661

Cardiac output equals stroke volume multiplied by heart rate.

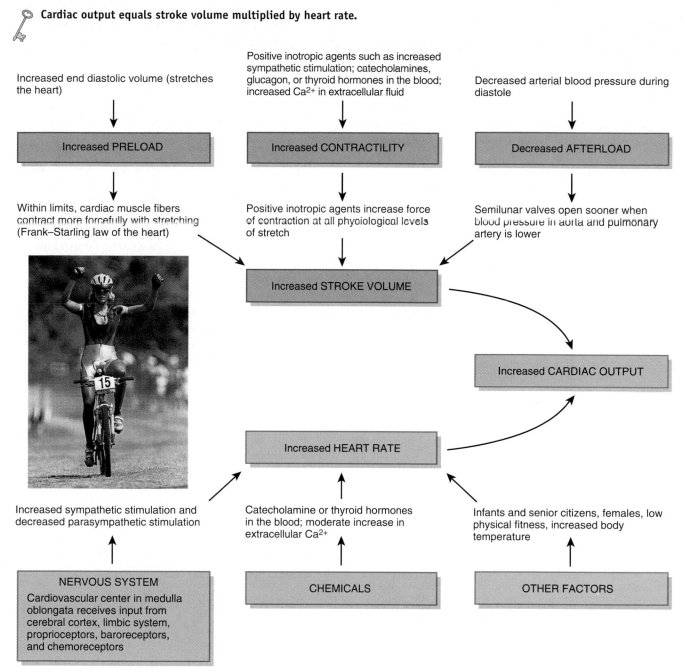

Increased end diastolic volume (stretches the heart)

Positive inotropic agents such as increased sympathetic stimulation; catecholamines, glucagon, or thyroid hormones in the blood; increased Ca^{2+} in extracellular fluid

Decreased arterial blood pressure during diastole

Increased PRELOAD

Increased CONTRACTILITY

Decreased AFTERLOAD

Within limits, cardiac muscle fibers contract more forcefully with stretching (Frank–Starling law of the heart)

Positive inotropic agents increase force of contraction at all physiological levels of stretch

Semilunar valves open sooner when blood pressure in aorta and pulmonary artery is lower

Increased STROKE VOLUME

Increased CARDIAC OUTPUT

Increased HEART RATE

Increased sympathetic stimulation and decreased parasympathetic stimulation

Catecholamine or thyroid hormones in the blood; moderate increase in extracellular Ca^{2+}

Infants and senior citizens, females, low physical fitness, increased body temperature

NERVOUS SYSTEM
Cardiovascular center in medulla oblongata receives input from cerebral cortex, limbic system, proprioceptors, baroreceptors, and chemoreceptors

CHEMICALS

OTHER FACTORS

Q When you are exercising, contraction of skeletal muscles helps return blood to the heart more rapidly. Would this effect tend to increase or decrease stroke volume?

exercise tends to bring resting heart rate down in both sexes. A physically fit person may even exhibit bradycardia, a resting heart rate under 60 beats/min. This is a beneficial effect of endurance-type training because a slowly beating heart is more energy efficient than one that beats more rapidly.

Increased body temperature, such as occurs during fever or strenuous exercise, causes the SA node to discharge impulses faster, thereby increasing heart rate. Decreased body temperature decreases heart rate and strength of contraction. During surgical repair of certain heart abnormalities, it

is helpful to slow a patient's heart rate by **hypothermia** (hī-pō-THER-mē-a), in which the person's body is deliberately cooled to a low core temperature. Hypothermia also slows metabolism, which reduces the oxygen needs of the tissues, allowing the heart and brain to withstand short periods of interrupted or reduced blood flow during the procedure.

Figure 20.16 summarizes the factors that can increase stroke volume and heart rate to achieve an increase in cardiac output.

1. How is cardiac output calculated?
2. Define stroke volume (SV), and explain the factors that regulate it.
3. What is the Frank–Starling law of the heart? What is its significance?
4. Define cardiac reserve. How does it vary with training?
5. Explain how the sympathetic and parasympathetic divisions of the autonomic nervous system adjust heart rate.

EXERCISE AND THE HEART

OBJECTIVE

- *Explain the relationship between exercise and the heart.*

No matter what a person's level of fitness, it can be improved at any age with regular exercise. Of the various types of exercise, some are more effective than others for improving the health of the cardiovascular system. **Aerobics,** any activity that works large body muscles for at least 20 minutes, elevates cardiac output and accelerates metabolic rate. Three to five such sessions a week are usually recommended for improving the health of the cardiovascular system. Brisk walking, running, bicycling, cross-country skiing, and swimming are examples of aerobic activities.

Sustained exercise increases the oxygen demand of the muscles. Whether the demand is met depends primarily on the adequacy of cardiac output and proper functioning of the respiratory system. After several weeks of training, a healthy person increases maximal cardiac output, thereby increasing the maximal rate of oxygen delivery to the tissues. Oxygen delivery also rises because hemoglobin level increases and skeletal muscles develop more capillary networks in response to long-term training.

During strenuous activity, a well-trained athlete can achieve a cardiac output double that of a sedentary person, in part because training causes hypertrophy (enlargement) of the heart. Even though the heart of a well-trained athlete is larger, *resting* cardiac output is about the same as in a healthy untrained person, because stroke volume is increased while heart rate is decreased. The resting heart rate of a trained athlete often is only 40–60 beats per minute (resting bradycardia). Regular exercise also helps to reduce blood pressure, anxiety, and depression; control weight; and increase the body's ability to dissolve blood clots by increasing fibrinolytic activity.

1. What are some of the cardiovascular benefits of regular exercise?

DEVELOPMENTAL ANATOMY OF THE HEART

OBJECTIVE

- *Describe the development of the heart.*

The *heart,* a derivative of **mesoderm,** begins to develop before the end of the third week of gestation. It begins its development in the ventral region of the embryo inferior to the foregut (see Figure 24.28b). The first step in its development is the formation of a pair of tubes, the **endothelial (endocardial) tubes,** which develop from mesodermal cells (Figure 20.17). These tubes then unite to form a common tube, referred to as the **primitive heart tube.** Next, the primitive heart tube develops into five distinct regions: (1) **truncus arteriosus,** (2) **bulbus cordis,** (3) **ventricle,** (4) **atrium,** and (5) **sinus venosus.** Because the bulbus cordis and ventricle grow most rapidly, and because the heart enlarges more rapidly than its superior and inferior attachments, the heart first assumes a U-shape and then an S-shape. The flexures of the heart reorient the regions so that the atrium and sinus venosus eventually come to lie superior to the bulbus cordis, ventricle, and truncus arteriosus. Contractions of the primitive heart begin by day 22; they originate in the sinus venosus and force blood through the tubular heart.

At about the seventh week of development, a partition called the **interatrial septum** forms in the atrial region. This septum divides the atrial region into a *right atrium* and *left atrium.* The opening in the partition is the **foramen ovale,** which normally closes at birth and later forms a depression called the *fossa ovalis.* An **interventricular septum** also develops; it partitions the ventricular region into a *right ventricle* and *left ventricle.* The bulbus cordis and truncus arteriosus divide into two vessels, the *aorta* (arising from the left ventricle) and the *pulmonary trunk* (arising from the right ventricle). The great veins of the heart, the *superior vena cava* and the *inferior vena cava,* develop from the venous end of the primitive heart tube.

1. What structures develop from the bulbus cordis and truncus arteriosus?

Figure 20.17 Development of the heart. Arrows within the structures indicate the direction of blood flow.

 The heart begins its development from mesoderm before the end of the third week of gestation.

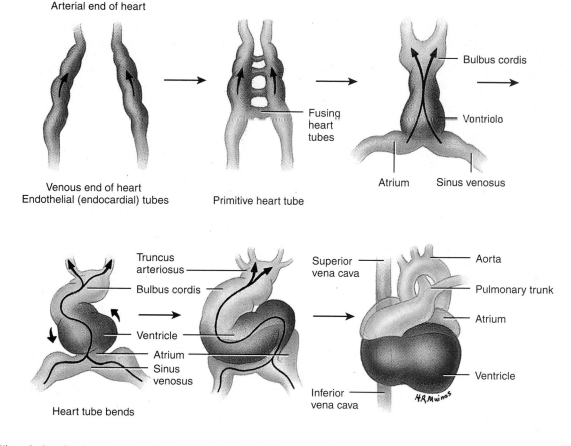

Heart tube bends

Q When during development does the primitive heart begin to contract?

DISORDERS: HOMEOSTATIC IMBALANCES

CORONARY ARTERY DISEASE

Coronary artery disease (CAD) is a serious medical problem that annually affects about 7 million people and causes nearly half a million deaths in the United States. CAD is defined as the effects of the accumulation of atherosclerotic plaques (described shortly) in coronary arteries that lead to a reduction in blood flow to the myocardium. Some individuals have no signs or symptoms; others experience angina pectoris (chest pain), and still others suffer a heart attack.

Risk Factors for CAD

People who possess combinations of certain risk factors are more likely to develop CAD. *Risk factors* are characteristics, symptoms, or signs present in a disease-free person that are statistically associated with a greater chance of developing a disease. Some risk factors in CAD are modifiable; that is, they can be altered by changing diet and other habits or can be controlled by taking medications. Among these risk factors for CAD are high blood cholesterol level, high blood pressure, cigarette smoking, obesity, diabetes,

"type A" personality, and sedentary lifestyle. It has been found, for example, that moderate exercise and reducing fat in the diet can not only stop the progression of CAD, but actually reverse the process. Other risk factors are unmodifiable—that is, beyond our control—including genetic predisposition (family history of CAD at an early age), age, and gender. (For example, even though adult males are more likely than adult females to develop CAD, after age 70 their risks are roughly equal.)

Development of Atherosclerotic Plaques

Although the following discussion of atherosclerosis applies to coronary arteries, the process can also occur in arteries outside the heart. Thickening of the walls of arteries and loss of elasticity are the main characteristics of a group of diseases called **arteriosclerosis** (ar-tē-rē-ō-skle-RŌ-sis; *sclero-* = hardening). One form of arteriosclerosis is **atherosclerosis** (ath-er-ō-skle-RŌ-sis), a progressive disease characterized by the formation in the walls of large and medium-sized arteries of lesions called **atherosclerotic plaques**

Figure 20.18 Photomicrograph of an artery partially obstructed by an atherosclerotic plaque.

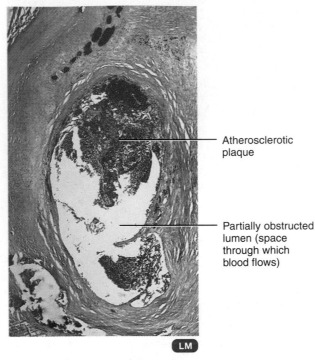

Atherosclerotic plaque

Partially obstructed lumen (space through which blood flows)

LM

Transverse section

(Figure 20.18). Atherosclerosis is initiated by one or more unknown factors that cause damage to the endothelial lining of the arterial wall. Factors that may initiate the process include high circulating LDL levels, cytomegalovirus (a common herpes virus), prolonged high blood pressure, carbon monoxide in cigarette smoke, and diabetes mellitus. Atherosclerosis is thought to begin when one of these factors injures the endothelium of an artery, promoting the aggregation of platelets and also attracting phagocytes.

At the injury site, cholesterol and triglycerides collect in the inner layer of the arterial wall. As part of the inflammatory process, macrophages also arrive at the site. Contact with platelets, lipids, and other components of blood stimulates smooth muscle cells and collagen fibers in the arterial wall to proliferate abnormally. In response to the proliferation of smooth muscle cells and collagen fibers and to the buildup of lipids, an atherosclerotic plaque develops, progressively obstructing blood flow as it enlarges. An additional danger is that a plaque may provide a roughened surface that attracts platelets, initiating clot formation, further obstructing blood flow. Moreover, a thrombus or piece of a thrombus may dislodge to become an embolus and obstruct blood flow in other vessels.

Diagnosis of CAD

Cardiac catheterization (kath'-e-ter-i-ZĀ-shun) is an invasive procedure used to visualize the heart's coronary arteries, chambers, valves, and great vessels. It may also be used to measure pressure in the heart and blood vessels; to assess function, cardiac output, and diastolic properties of the left ventricle; to measure the flow of blood through the heart and blood vessels, the oxygen content of blood, and the status of heart valves and conduction system; and to identify the exact location of septal and valvular defects. The basic procedure involves inserting a long, flexible, radiopaque

catheter (plastic tube) into a peripheral vein (for right heart catheterization) or a peripheral artery (for left heart catheterization) and guiding it under fluoroscopy (x-ray observation).

Cardiac angiography (an'-jē-OG-ra-fē) is another invasive procedure in which a cardiac catheter is used to inject a radiopaque contrast medium into blood vessels or heart chambers. The procedure may be used to visualize coronary arteries, the aorta, pulmonary blood vessels, and the ventricles to assess structural abnormalities in blood vessels (such as atherosclerotic plaques and emboli), ventricular volume, wall thickness, and wall motion. Angiography can also be used to inject clot-dissolving drugs, such as streptokinase or tissue plasminogen activator (t-PA), into a coronary artery to dissolve an obstructing thrombus.

Treatment of CAD

Treatment options for CAD include drugs (nitroglycerine, beta blockers, and cholesterol-lowering and clot-dissolving agents) and various surgical and nonsurgical procedures designed to increase the blood supply to the heart.

Coronary artery bypass grafting (CABG) is a surgical procedure in which a blood vessel from another part of the body is attached ("grafted") to a coronary artery so as to bypass an area of blockage. A piece of the grafted blood vessel is sutured between the aorta and the unblocked portion of the coronary artery (Figure 20.19a).

A nonsurgical procedure used to treat CAD is termed **percutaneous transluminal coronary angioplasty (PTCA)** (*percutaneous* = through the skin; *trans-* = across; *lumen* = an opening or channel in a tube; *angio-* = blood vessel; *-plasty* = to mold or to shape). In this procedure, a balloon catheter (plastic tube) is inserted into an artery of an arm or leg and gently guided into a coronary artery (Figure 20.19b). Then, while dye is released, angiograms (x-rays of blood vessels) are taken to locate the plaques. Next, the catheter is advanced to the point of obstruction, and a balloonlike device is inflated with air to squash the plaque against the blood vessel wall. Because about 30–50% of PTCA-opened arteries fail due to restenosis (renarrowing) within six months after the procedure is done, a special device called a stent may be inserted via a catheter. A *stent* is a stainless steel device, resembling a spring coil, that is permanently placed in an artery to keep the artery patent (open), permitting blood to circulate (Figure 20.19c). Restenosis may be due to damage from the procedure itself, for PTCA may damage the coronary artery wall, leading to platelet activation, proliferation of smooth muscle fibers, and plaque formation.

MYOCARDIAL ISCHEMIA AND INFARCTION

Partial obstruction of blood flow in the coronary arteries may cause **myocardial ischemia** (is-KĒ-mē-a; *ische-* = to obstruct; *-emia* = in the blood). Usually, ischemia causes **hypoxia** (reduced oxygen supply), which may weaken cells without killing them. **Angina pectoris** (an-JĪ-na, or AN-ji-na, PEK-to-ris), literally meaning "strangled chest," is a severe pain that usually accompanies myocardial ischemia. Typically, sufferers describe it as a tightness or squeezing sensation, as though the chest were in a vise. Angina pectoris often occurs during exertion, when the heart demands more oxygen, and then disappears with rest. The pain associated with angina pectoris is often referred to the neck, chin, or down the left arm to the elbow. In some people, especially people with diabetes who suffer from neuropathy (disease of the peripheral or autonomic nervous system), ischemic episodes occur without producing pain. This is known as **silent myocardial ischemia** and is particularly dangerous because the person has no forewarning of an impending heart attack.

A complete obstruction to blood flow may result in a **myocardial infarction** (in-FARK-shun), or **MI**, commonly called a heart attack. *Infarction* means the death of an area of tissue because of interrupted blood supply. Because the heart tissue distal to the obstruction dies and is replaced by noncontractile scar tissue, the heart muscle loses at least some of its strength. The aftereffects depend partly on the size and location of the infarcted (dead) area. Besides killing normal heart tissue, the infarction may disrupt the conduction system of the heart and cause sudden death by triggering ventricular fibrillation. Treatment for a myocardial infarction may involve injection of a thrombolytic (clot-dissolving) agent such as streptokinase or t-PA, plus heparin (an anticoagulant), or performing coronary angioplasty or coronary artery bypass grafting. Fortunately, heart muscle can remain alive in a resting person if it receives as little as 10–15% of its normal blood supply, but such a person may have little ability to engage in physical activity. Furthermore, the extensive anastomoses of coronary blood vessels are a significant factor in helping some people to survive myocardial infarctions.

CONGENITAL HEART DEFECTS

A defect that is present at birth, and usually before, is called a **congenital defect.** Many such defects are not serious and may go unnoticed for a lifetime, and others heal themselves, but some are life-threatening and must be repaired by surgical techniques ranging from simple suturing to replacement of malfunctioning parts with synthetic materials. Among the several congenital defects that affect the heart are the following:

- **Coarctation** (kō′-ark-TĀ-shun) **of the aorta.** In this condition, a segment of the aorta is too narrow, and thus the flow of oxygenated blood to the body is reduced, the left ventricle is forced to pump harder, and high blood pressure develops.
- **Patent ductus arteriosus.** In some babies, the ductus arteriosus, a temporary blood vessel between the aorta and the pulmonary trunk, remains open rather than closing shortly after birth. As a result, aortic blood flows into the lower-pressure pulmonary trunk, thus increasing the pulmonary trunk blood pressure and overworking both ventricles.
- **Septal defect.** This defect is an opening in the septum that separates the interior of the heart into left and right sides. In an **interatrial septal defect** the fetal foramen ovale between the two atria fails to close after birth. An **interventricular septal defect** is caused by incomplete closure of the interventricular septum, which permits oxygenated blood to flow directly from the left ventricle into the right ventricle, where it mixes with deoxygenated blood.
- **Tetralogy of Fallot** (tet-RAL-ō-jē of fal-Ō). This condition is actually a combination of four defects: an interventricular septal defect, an aorta that emerges from both ventricles instead of from the left ventricle only, a stenosed pulmonary semilunar valve, and an enlarged right ventricle. Deoxygenated blood from the right ventricle enters the left ventricle through the interventricular septum, mixes with oxygenated blood, and is pumped into the systemic circulation. Moreover, because the aorta emerges from the right ventricle and the pulmonary trunk is stenosed, very little blood reaches the pulmonary circulation. This causes cyanosis, the bluish discoloration most easily seen in nail beds and mucous membranes when the level of deoxygenated hemoglobin is high. For this reason, tetralogy of Fallot is one of the conditions that cause a "blue baby."

Figure 20.19 Three procedures for reestablishing blood flow in occluded coronary arteries.

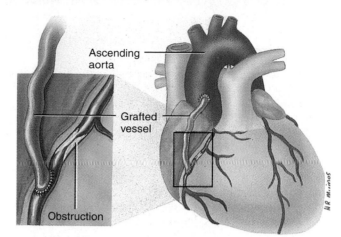

(a) Coronary artery bypass grafting (CABG)

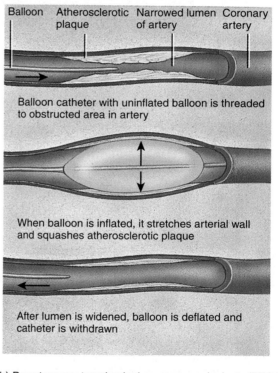

(b) Percutaneous transluminal coronary angioplasty (PTCA)

(c) Stent in an artery

MEDICAL TERMINOLOGY

Cardiac arrest (KAR-dē-ak a-REST) A clinical term meaning cessation of an effective heartbeat. The heart may be completely stopped or in ventricular fibrillation.

Cardiomegaly (kar'-dē-ō-MEG-a-lē; *mega* = large) Heart enlargement.

Corpulmonale (**CP**) (kor pul-mōn-ALE; *cor-* = heart; *pulmon-* = lung) A term referring to right ventricular hypertrophy from disorders that bring about hypertension (high blood pressure) in the pulmonary circulation.

Cardiopulmonary resuscitation (kar'-dē-ō-PUL-mō-ner-ē re-sus'-i-TĀ-shun) (**CPR**) The artificial establishment of normal or near-normal respiration and circulation. The **A, B, C's** of cardiopulmonary resuscitation are **Airway, Breathing,** and **Circulation,** meaning the rescuer must establish an airway,

provide artificial ventilation if breathing has stopped, and reestablish circulation if there is inadequate cardiac action. The procedure should be performed in that order (A, B, C).

Incompetent valve (in-KOM-pe-tent) Any valve that does not close properly and thus permits a backflow of blood; also called *valvular insufficiency.*

Palpitation (pal'-pi-TĀ-shun) A fluttering of the heart or an abnormal rate or rhythm of the heart.

Paroxysmal tachycardia (par'-ok-SIZ-mal tak'-e-KAR-dē-a) A period of rapid heartbeats that begins and ends suddenly.

Sudden cardiac death The unexpected cessation of circulation and breathing due to an underlying heart disease such as ischemia, myocardial infarction, or a disturbance in cardiac rhythm.

STUDY OUTLINE

LOCATION AND SURFACE PROJECTION OF THE HEART (p. 636)

1. The heart is located in the mediastinum; about two-thirds of its mass is to the left of the midline.
2. The heart is basically a cone lying on its side; it consists of an apex, a base, anterior and inferior surfaces, and right and left borders.
3. Four points are used to project the heart's location to the surface of the chest.

STRUCTURE AND FUNCTION OF THE HEART (p. 638)

1. The pericardium is the membrane that surrounds and protects the heart; it consists of an outer fibrous layer and an inner serous pericardium, which is composed of a parietal and a visceral layer.
2. Between the parietal and visceral layers of the serous pericardium is the pericardial cavity, a potential space filled with a few milliliters of pericardial fluid that reduces friction between the two membranes.
3. The wall of the heart has three layers: epicardium (visceral layer of the serous pericardium), myocardium, and endocardium.
4. The epicardium consists of mesothelium and connective tissue, the myocardium is composed of cardiac muscle tissue, and the endocardium consists of endothelium and connective tissue.
5. The heart chambers include two superior chambers, the right and left atria, and two inferior chambers, the right and left ventricles.
6. External features of the heart include the auricles (flaps on each atrium that increase their volume), the coronary sulcus between the atria and ventricles, and the anterior and posterior sulci between the ventricles on the anterior and posterior surfaces of the heart, respectively.
7. The right atrium receives blood from the superior vena cava, inferior vena cava, and coronary sinus. It is separated from the left atrium by the interatrial septum, which contains the fossa ovalis. Blood exits the right atrium through the tricuspid valve.

8. The right ventricle receives blood from the right atrium. It is separated from the left ventricle by the interventricular septum and pumps blood to the lungs through the pulmonary semilunar valve and pulmonary trunk.
9. Oxygenated blood enters the left atrium from the pulmonary veins and exits through the bicuspid (mitral) valve.
10. The left ventricle pumps oxygenated blood into the systemic circulation through the aortic semilunar valve and aorta.
11. The thickness of the myocardium of the four chambers varies according to the chamber's function. The left ventricle has the thickest wall because of its high workload.
12. The fibrous skeleton of the heart is dense connective tissue that surrounds and supports the valves of the heart.
13. Heart valves prevent backflow of blood within the heart.
14. The atrioventricular (AV) valves, which lie between atria and ventricles, are the tricuspid valve on the right side of the heart and the bicuspid (mitral) valve on the left. The chordae tendineae and papillary muscles stabilize the flaps of the AV valves and stop blood from backing into the atria.
15. Each of the two arteries that leaves the heart has a semilunar valve (aortic and pulmonary).

CIRCULATION OF BLOOD (p. 644)

1. The left side of the heart is the pump for the systemic circulation, the circulation of blood throughout the body except for the air sacs of the lungs. The left ventricle ejects blood into the aorta, and blood then flows into systemic arteries, arterioles, capillaries, venules, and veins, which carry it back to the right atrium.
2. The right side of the heart is the pump for pulmonary circulation, the circulation of blood through the lungs. The right ventricle ejects blood into the pulmonary trunk, and blood then flows into pulmonary arteries, pulmonary capillaries, and pulmonary veins, which carry it back to the left atrium.
3. The flow of blood through the heart is called the coronary (cardiac) circulation.
4. The principal arteries are left and right coronary arteries; the principal veins are the cardiac vein and the coronary sinus.

CARDIAC MUSCLE AND THE CARDIAC CONDUCTION SYSTEM (p. 648)

1. Cardiac muscle fibers usually contain a single centrally located nucleus. Compared to skeletal muscle fibers, cardiac muscle fibers have more sarcoplasm, more mitochondria, less well-developed sarcoplasmic reticulum, and wider transverse tubules, which are located at Z discs rather than at A–I band junctions.
2. Cardiac muscle fibers branch and are connected via end-to-end intercalated discs, which provide strength and aid in conduction of muscle action potentials by way of gap junctions located in the discs.
3. Autorhythmic cells form the conduction system, cardiac muscle fibers that spontaneously generate action potentials.
4. Components of the conduction system are the sinoatrial (SA) node (pacemaker), atrioventricular (AV) node, atrioventricular (AV) bundle (bundle of His), bundle branches, and conduction myofibers (Purkinje fibers).
5. An action potential in a ventricular contractile fiber is characterized by rapid depolarization, a long plateau, and repolarization.
6. Cardiac muscle tissue has a long refractory period, which prevents tetanus.

ELECTROCARDIOGRAM (p. 652)

1. The record of electrical changes during each cardiac cycle is called an electrocardiogram (ECG).
2. A normal ECG consists of a P wave (atrial depolarization), a QRS complex (onset of ventricular depolarization), and a T wave (ventricular repolarization).
3. The P–Q interval represents the conduction time from the beginning of atrial excitation to the beginning of ventricular excitation. The S–T segment represents the time when ventricular contractile fibers are fully depolarized.

THE CARDIAC CYCLE (p. 654)

1. A cardiac cycle consists of the systole (contraction) and diastole (relaxation) of both atria, plus the systole and diastole of both ventricles.
2. The phases of the cardiac cycle are (a) isovolumetric relaxation, (b) ventricular filling, and (c) ventricular systole.
3. With an average heartbeat of 75 beats/min, a complete cardiac cycle requires 0.8 seconds (800 msec).

4. S1, the first heart sound (lubb), is caused by blood turbulence associated with the closing of the atrioventricular valves. S2, the second sound (dupp), is caused by blood turbulence associated with the closing of semilunar valves.

CARDIAC OUTPUT (p. 657)

1. Cardiac output (CO) is the amount of blood ejected by the left ventricle (or right ventricle) into the aorta (or pulmonary trunk) per minute. It is calculated as follows: CO (mL/min) = stroke volume (SV) in mL/beat × heart rate (HR) in beats per minute.
2. Stroke volume (SV) is the amount of blood ejected by a ventricle during each systole.
3. Cardiac reserve is the ratio between a person's maximum cardiac output and his or her cardiac output at rest.
4. Stroke volume is related to preload (stretch on the heart before it contracts), contractility (forcefulness of contraction), and afterload (pressure that must be exceeded before ventricular ejection can begin).
5. According to the Frank–Starling law of the heart, a greater preload (stretch) on cardiac muscle fibers just before they contract increases their force of contraction until the stretching becomes excessive.
6. Nervous control of the cardiovascular system stems from the cardiovascular center in the medulla oblongata.
7. Sympathetic impulses increase heart rate and force of contraction; parasympathetic impulses decrease heart rate.
8. Heart rate is affected by hormones (epinephrine, norepinephrine, thyroid hormones), ions (Na^+, K^+, Ca^{2+}), age, gender, physical fitness, and body temperature.

EXERCISE AND THE HEART (p. 662)

1. Sustained exercise increases oxygen demand on muscles.
2. Among the benefits of aerobic exercise are increased cardiac output, decreased blood pressure, weight control, and increased fibrinolytic activity.

DEVELOPMENTAL ANATOMY OF THE HEART (p. 663)

1. The heart develops from mesoderm.
2. The endothelial tubes develop into the four-chambered heart and great vessels of the heart.

SELF-QUIZ QUESTIONS

1. Match the following:

___ (a) collects oxygenated blood from the pulmonary circulation
___ (b) pumps deoxygenated blood to the lungs for oxygenation
___ (c) their contraction pulls on and tightens the chordae tendinae, preventing the valve cusps from everting
___ (d) prevents backflow of blood from the right ventricle into the right atrium

___ (e) tendonlike cords connected to the AV valve cusps which, along with the papillary muscles, prevent valve eversion
___ (f) pumps oxygenated blood to all body cells, except the air sacs of the lungs
___ (g) collects deoxygenated blood from the systemic circulation
___ (h) left atrioventricular valve

(1) right atrium
(2) right ventricle
(3) left atrium
(4) left ventricle
(5) tricuspid valve
(6) bicuspid valve
(7) chordae tendineae
(8) papillary muscles

Choose the best answer to the following questions:

2. Which of the following is the correct route of blood through the heart from the systemic circulation to the pulmonary circulation and back to the systemic circulation? (a) right atrium, tricuspid valve, right ventricle, pulmonary semilunar valve, left atrium, mitral valve, left ventricle, aortic semilunar valve, (b) left atrium, tricuspid valve, left ventricle, pulmonary semilunar valve, right atrium, mitral valve, right ventricle, aortic semilunar valve, (c) left atrium, pulmonary semilunar valve, right atrium, tricuspid valve, left ventricle, aortic semilunar valve, right ventricle, mitral valve, (d) left ventricle, mitral valve, left atrium, pulmonary semilunar valve, right ventricle, tricuspid valve, right atrium, aortic semilunar valve, (e) right atrium, mitral valve, right ventricle, pulmonary semilunar valve, left atrium, tricuspid valve, left ventricle, aortic semilunar valve

3. Which of the following represents the correct pathway of impulse conduction through the heart? (a) AV node, SA node, conduction myofibers, bundle branches, (b) AV node, bundle branches, SA node, conduction myofibers, (c) SA node, AV node, bundle branches, conduction myofibers, (d) SA node, bundle branches, AV node, conduction myofibers, (e) SA node, AV node, conduction myofibers, bundle branches

4. Which of the following factors regulate stroke volume? (1) preload, (2) contractility, (3) extensibility, (4) afterload, (5) relaxation.
   (a) 1, 2, and 3, (b) 2, 4, and 5, (c) 1, 3, and 5, (d) 1, 2, and 4, (e) 2, 3, and 5

5. Which of the following statements is *incorrect*? (a) Cardiac output is the amount of blood ejected from the left ventricle into the aorta each minute. (b) Cardiac output equals stroke volume divided by heart rate. (c) A greater stretch on cardiac muscle fibers just before they contract increases their force of contraction. (d) The pressure that must be overcome before the semilunar valves can open is termed afterload. (e) Myocardial contractility is the strength of contraction at any given load.

6. Which of the following does *not* contribute to regulation of heart rate? (a) the autonomic nervous system, (b) hormones, (c) ions, (d) age, (e) blood volume

7. Which of the following are benefits of regular exercise sessions on heart activity? (1) an increase in maximal cardiac output, (2) an increase in hemoglobin, (3) an increase in fibrinolytic activity, (4) a reduction in blood pressure, (5) a decrease in endorphins.
   (a) 1, 2, and 3, (b) 2, 3, and 4, (c) 1, 3, 4, and 5, (d) 1, 2, 3, and 4, (e) 1, 2, 4, and 5

8. Match the following:
   ___ (a) cardiac muscle tissue
   ___ (b) blood vessels that pierce the heart muscle and supply blood to the heart muscle cells
   ___ (c) separate the upper and lower heart chambers, preventing backflow of blood from the lower chambers back into the upper chambers
   ___ (d) endothelial cells lining the interior of the heart; are continuous with the endothelium of the blood vessels
   ___ (e) inner visceral layer of the serous pericardium; adheres tightly to the surface of the heart
   ___ (f) the remnant of the foramen ovale, an opening in the interatrial septum of the fetal heart
   ___ (g) outer layer of the serous pericardium; is fused to the fibrous pericardium
   ___ (h) the gap junction and desmosome connections between individual heart muscle cells
   ___ (i) the superficial dense irregular connective tissue covering of the heart
   ___ (j) prevent backflow of blood from the arteries into the lower chambers

   (1) fibrous pericardium
   (2) parietal pericardium
   (3) epicardium
   (4) myocardium
   (5) endocardium
   (6) atrioventricular valves
   (7) semilunar valves
   (8) intercalated discs
   (9) fossa ovalis
   (10) coronary vessels

Complete the following:

9. The period of the cardiac cycle during which ventricular blood volume does not change because both semilunar and atrioventricular valves are closed is called the ___ period.

10. The phase of heart contraction is called ___; the phase of relaxation is called ___.

11. The ___ heart sound occurs when the atrioventricular valves ___. The ___ heart sound occurs when the semilunar valves ___.

12. Cardiac output equals ___ multiplied by heart rate.

True or false:

13. Positive inotropic agents decrease myocardial contractility.

14. The Frank–Starling law of the heart equalizes the output of the right and left ventricles and keeps the same volume of blood flowing to both the systemic and pulmonary circulations.

15. Match the following:
    ___ (a) indicates ventricular repolarization
    ___ (b) represents the time from the beginning of ventricular depolarization to the end of ventricular repolarization
    ___ (c) represents atrial depolarization
    ___ (d) represents the time when the ventricular contractile fibers are fully depolarized; occurs during the plateau phase of the action potential
    ___ (e) represents the onset of ventricular depolarization
    ___ (f) represents the conduction time from the beginning of atrial excitation to the beginning of ventricular excitation

    (1) P wave
    (2) QRS complex
    (3) T wave
    (4) P–Q interval
    (5) S–T segment
    (6) Q–T interval

CRITICAL THINKING QUESTIONS

1. Arian, a member of the college track team, volunteered for a study that the exercise physiology class was conducting on the cardiovascular system. His resting heart rate was measured at 55 beats per minute and he had a normal cardiac output (CO). After strenuous exercise, his cardiac reserve was calculated to be six times the resting value. Arian is in excellent physical condition. Predict Arian's resting CO and stroke volume. What would his CO be during strenuous exercise? (HINT: *At rest, each ventricle pumps a volume of blood per minute that is about equal to the body's total blood volume.*)

2. At family gatherings, Aunt Mercy likes to tell the story of when she had rheumatic fever as a child and how her mother treated her with home remedies. Now Aunt Mercy complains of a "weak heart." Could the childhood illness be related to her heart problem? (HINT: *The immune system may mistakenly attack healthy body tissue along with the bacteria.*)

3. Mr. Perkins is a large, 62-year-old man with a weakness for sweets and fried foods. His idea of exercise is to get up to change the channel while watching sports on television. Lately, he's been troubled by chest pains when he walks up stairs. His doctor told him to quit smoking and scheduled a cardiac angiography for next week. What is involved in performing this procedure? Why did the doctor order this test? (HINT: *Mr Perkins' stress electrocardiogram revealed changes that suggest myocardial ischemia.*)

ANSWERS TO FIGURE QUESTIONS

20.1 The mediastinum is the mass of tissue that extends from the sternum to the vertebral column between the pleurae of the lungs.

20.2 The visceral layer of the serous pericardium (epicardium) is both a part of the pericardium and a part of the heart wall.

20.3 The coronary sulcus forms a boundary between the atria and ventricles.

20.4 The left ventricle has the thickest wall.

20.5 The skeleton attaches to the heart valves and prevents overstretching of the valves as blood passes through them.

20.6 The papillary muscles contract, which pulls on the chordae tendineae and prevents valve cusps from everting.

20.7 Numbers 2 (right ventricle) through 6 depict the pulmonary circulation, whereas numbers 7 (left ventricle) through 10 and 1 (right atrium) depict the systemic circulation.

20.8 The circumflex artery delivers oxygenated blood to the left atrium and left ventricle.

20.9 The intercalated discs hold the cardiac muscle fibers together and enable action potentials to propagate from one muscle fiber to another.

20.10 The only electrical connection between the atria and the ventricles is the atrioventricular bundle.

20.11 The duration of an action potential is much longer in a ventricular contractile fiber (300 msec) than in a skeletal muscle fiber (1–2 msec).

20.12 An enlarged Q wave may indicate a myocardial infarction (heart attack).

20.13 The amount of blood in each ventricle at the end of ventricular diastole—the end-diastolic volume—is about 130 mL in a resting person.

20.14 The first sound (S1), or lubb, is associated with the closure of AV valves.

20.15 The ventricular myocardium is innervated by the sympathetic division only.

20.16 The skeletal muscle "pump" increases stroke volume by increasing preload (end-diastolic volume).

20.17 The heart begins to contract by the 22nd day of gestation.

THE CARDIOVASCULAR SYSTEM: BLOOD VESSELS AND HEMODYNAMICS

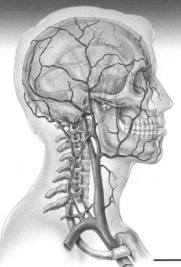

This chapter focuses on the structure and functions of the various types of blood vessels; on **hemodynamics** (hē-mō-dī-NAM-ics; *hemo-* = blood; *dynamics* = power), the forces involved in circulating blood throughout the body; and on the blood vessels that constitute the major circulatory routes.

ANATOMY OF BLOOD VESSELS

OBJECTIVE
• *Contrast the structure and function of arteries, arterioles, capillaries, venules, and veins.*

Blood vessels form a closed system of tubes that carries blood away from the heart, transports it to the tissues of the body, and then returns it to the heart. **Arteries** are vessels that carry blood from the heart to the tissues. Large, elastic arteries leave the heart and divide into medium-sized, muscular arteries that branch out into the various regions of the body. Medium-sized arteries then divide into small arteries, which, in turn, divide into still smaller arteries called **arterioles** (ar-TER-ē-ōls). As the arterioles enter a tissue, they branch into countless microscopic vessels called **capillaries** (KAP-i-lar′-ēs). Substances are exchanged between the blood and body tissues through the thin walls of capillaries. Before leaving the tissue, groups of capillaries unite to form small veins called **venules** (VEN-yools), which merge to form progressively larger blood vessels called veins. **Veins** then convey blood from the tissues back to the heart. Because blood vessels require oxygen (O_2) and nutrients just like other tissues of the body, larger blood vessels are served by their own blood vessels, called **vasa vasorum** (literally, vasculature of vessels), located within their walls.

Arteries

In ancient times, **arteries** (*ar-* = air; *ter-* = to carry) were found empty at death and thus were thought to contain only air. The wall of an artery has three coats or tunics: (1) tunica interna, (2) tunica media, and (3) tunica externa (Figure 21.1). The innermost coat, the **tunica interna (intima),** is composed of a lining of simple squamous epithelium called *endothelium,* a *basement membrane,* and a layer of elastic tissue called the *internal elastic lamina.* The endothelium is a continuous layer of cells that line the inner surface of the entire cardiovascular system (the heart and all blood vessels). Normally, the only tissue that blood comes in contact with is endothelium. The tunica interna is closest to the **lumen,** the hollow center through which blood flows. The middle coat, or **tunica media,** is usually the thickest layer; it consists of elastic fibers and smooth muscle fibers (cells), which are arranged circularly (in rings) around the lumen. Due to their plentiful elastic fibers, arteries normally have high *compliance,* which means that their walls easily stretch or expand without tearing in response to a small increase in pressure. The outer coat, the **tunica externa,** is composed principally of elastic and collagen fibers. In muscular arteries (described shortly), an *external elastic lamina* composed of elastic tissue separates the tunica externa from the tunica media.

Sympathetic fibers of the autonomic nervous system innervate vascular smooth muscle. An increase in sympathetic

Figure 21.1 Comparative structure of blood vessels. The relative size of the capillary in (c) is enlarged.

Arteries carry blood from the heart to tissues; veins carry blood from tissues to the heart.

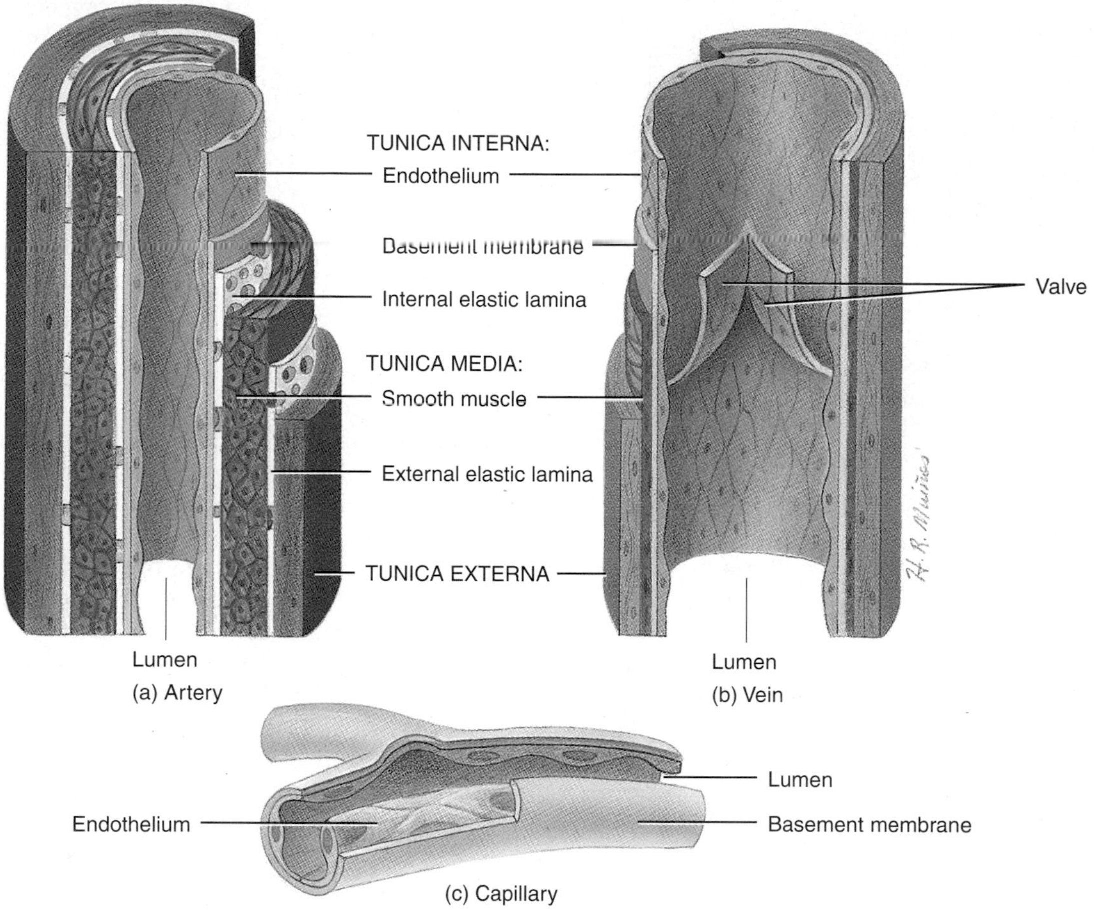

TUNICA INTERNA:
- Endothelium
- Basement membrane
- Internal elastic lamina

TUNICA MEDIA:
- Smooth muscle
- External elastic lamina

TUNICA EXTERNA

Valve

Lumen
(a) Artery

Lumen
(b) Vein

Endothelium

Lumen

Basement membrane

(c) Capillary

Q Which vessel—the femoral artery or the femoral vein—has a thicker wall? Which has a wider lumen?

stimulation typically stimulates the smooth muscle to contract, squeezing the vessel wall and narrowing the lumen. Such a decrease in the diameter of the lumen of a blood vessel is called **vasoconstriction.** In contrast, when sympathetic stimulation decreases, or in the presence of certain chemicals (such as nitric oxide, K^+, H^+, and lactic acid), smooth muscle fibers relax. The resulting increase in lumen diameter is called **vasodilation.** Additionally, when an artery or arteriole is damaged, its smooth muscle contracts, producing vascular spasm of the vessel. Such a vasospasm limits blood flow through the damaged vessel and helps reduce blood loss if the vessel is small.

Elastic Arteries

The largest-diameter arteries, termed **elastic arteries** because the tunica media contains a high proportion of elastic fibers, have walls that are relatively thin in proportion to their overall diameter. Elastic arteries perform an important

function: They help propel blood onward while the ventricles are relaxing. As blood is ejected from the heart into elastic arteries, their highly elastic walls stretch, accommodating the surge of blood. By stretching, the elastic fibers momentarily store mechanical energy, functioning as a **pressure reservoir** (Figure 21.2a). Then, the elastic fibers recoil and convert stored (potential) energy in the vessel into kinetic energy of the blood. Thus, blood continues to move through the arteries even while the ventricles are relaxed (Figure 21.2b). Because they conduct blood from the heart to medium-sized, more-muscular arteries, elastic arteries also are called *conducting arteries.* The aorta and the brachiocephalic, common carotid, subclavian, vertebral, pulmonary, and common iliac arteries are elastic arteries (see Figure 21.18).

Muscular Arteries

Medium-sized arteries are called **muscular arteries** because their tunica media contains more smooth muscle and

Figure 21.2 Pressure reservoir function of elastic arteries.

🔑 **Recoil of elastic arteries keeps blood moving during ventricular relaxation (diastole).**

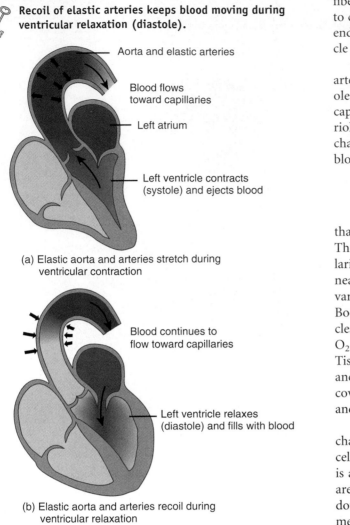

Aorta and elastic arteries

Blood flows toward capillaries

Left atrium

Left ventricle contracts (systole) and ejects blood

(a) Elastic aorta and arteries stretch during ventricular contraction

Blood continues to flow toward capillaries

Left ventricle relaxes (diastole) and fills with blood

(b) Elastic aorta and arteries recoil during ventricular relaxation

Q In atherosclerosis, the walls of elastic arteries become less compliant (stiffer). What effect does reduced compliance have on the pressure reservoir function of arteries?

fewer elastic fibers than elastic arteries. Thus, muscular arteries are capable of greater vasoconstriction and vasodilation to adjust the rate of blood flow. The large amount of smooth muscle makes the walls of muscular arteries relatively thick. Muscular arteries also are called *distributing arteries* because they distribute blood to various parts of the body. Examples include the brachial artery in the arm and radial artery in the forearm.

Arterioles

An **arteriole** (= small artery) is a very small (almost microscopic) artery that delivers blood to capillaries (Figure 21.3). Arterioles near the arteries from which they branch have a tunica interna like that of arteries, a tunica media composed of smooth muscle and very few elastic fibers, and

a tunica externa composed mostly of elastic and collagen fibers. In the smallest-diameter arterioles, which are closest to capillaries, the tunics consist of little more than a ring of endothelial cells surrounded by a few scattered smooth muscle fibers.

Arterioles play a key role in regulating blood flow from arteries into capillaries. When the smooth muscle of arterioles contracts, causing vasoconstriction, blood flow into capillaries decreases; when the smooth muscle relaxes, arterioles vasodilate and blood flow into capillaries increases. A change in diameter of arterioles can also significantly affect blood pressure.

Capillaries

Capillaries (*capillar-* = hairlike) are microscopic vessels that usually connect arterioles and venules (see Figure 21.3). The flow of blood from arterioles to venules through capillaries is called the **microcirculation.** Capillaries are found near almost every cell in the body, but their distribution varies with the metabolic activity of the tissue they serve. Body tissues with high metabolic requirements, such as muscles, the liver, the kidneys, and the nervous system, use more O_2 and nutrients and thus have extensive capillary networks. Tissues with lower metabolic requirements, such as tendons and ligaments, contain fewer capillaries. A few tissues—all covering and lining epithelia, the cornea and lens of the eye, and cartilage—lack capillaries.

The primary function of capillaries is to permit the exchange of nutrients and wastes between the blood and tissue cells through the interstitial fluid. The structure of capillaries is admirably suited to this purpose: Because capillary walls are composed of only a single layer of epithelial cells (endothelium) and a basement membrane and have no tunica media or tunica externa (see Figure 21.1c), a substance in the blood must pass through just one cell layer to reach the interstitial fluid and tissue cells. Exchange of materials occurs only through the walls of capillaries and the beginning of venules; the walls of arteries, arterioles, most venules, and veins present too thick a barrier. Capillaries form extensive branching networks that increase the surface area available for rapid exchange of materials. In most tissues, blood flows through only a small portion of the capillary network when metabolic needs are low. However, when a tissue is active, such as contracting muscle, the entire capillary network fills with blood.

A **metarteriole** (*met-* = beyond) is a vessel that emerges from an arteriole and supplies a group of 10–100 capillaries that constitute a **capillary bed** (see Figure 21.3). The proximal portion of a metarteriole is surrounded by scattered smooth muscle fibers whose contraction and relaxation help regulate blood flow through the capillary bed. The distal portion of a metarteriole, which empties into a venule, has no smooth muscle fibers and is called a **thoroughfare channel.** Blood flowing through a thoroughfare channel bypasses the capillary bed.

Figure 21.3 Arteriole, capillaries, and venule.

🔑 Arterioles regulate blood flow into capillaries, where nutrients, gases, and wastes are exchanged between the blood and interstitial fluid.

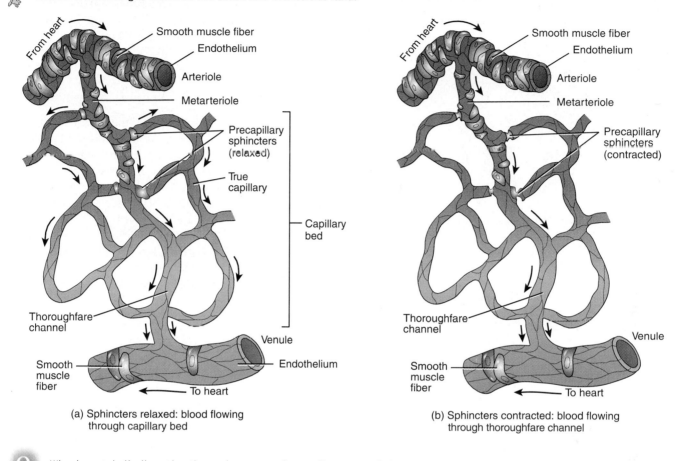

(a) Sphincters relaxed: blood flowing through capillary bed

(b) Sphincters contracted: blood flowing through thoroughfare channel

Q Why do metabolically active tissues have extensive capillary networks?

True capillaries emerge from arterioles or metarterioles. At their sites of origin, a ring of smooth muscle fibers called a **precapillary sphincter** controls the flow of blood into a true capillary. When the precapillary sphincters are relaxed (open), blood flows into the capillary bed (see Figure 21.3a); when precapillary sphincters contract (close or partially close), blood flow through the capillary bed ceases or decreases (see Figure 21.3b). Typically, blood flows intermittently through a capillary bed due to alternating contraction and relaxation of the smooth muscle of metarterioles and the precapillary sphincters. This intermittent contraction and relaxation, which may occur 5–10 times per minute, is called **vasomotion.** In part, vasomotion is due to chemicals released by the endothelial cells; nitric oxide is one example. At any given time, blood flows through only about 25% of a capillary bed.

The body contains different types of capillaries (Figure 21.4). Many capillaries are **continuous capillaries,** in which the plasma membranes of endothelial cells form a continuous tube that is interrupted only by **intercellular clefts,** which are gaps between neighboring endothelial cells (Fig-ure 21.4a). Continuous capillaries are found in skeletal and smooth muscle, connective tissues, and the lungs. Other capillaries of the body are **fenestrated capillaries** (*fenestr- = window*), which differ from continuous capillaries in that the plasma membranes of their endothelial cells have many **fenestrations,** small holes ranging from 70–100 nm in diameter (Figure 21.4b). Fenestrated capillaries are found in the kidneys, villi of the small intestine, choroid plexuses of the ventricles in the brain, ciliary processes of the eyes, and endocrine glands.

Sinusoids are wider and more winding than other capillaries. Their endothelial cells may have unusually large fenestrations. In addition to an incomplete or absent basement membrane (Figure 21.4c), sinusoids have very large intercellular clefts that allow proteins and in some cases even blood cells to pass from a tissue into the bloodstream. For example, newly formed blood cells enter the bloodstream through the sinusoids of red bone marrow. In addition, sinusoids contain specialized lining cells that are adapted to the function of the tissue. Sinusoids in the liver, for example, contain phagocytic cells that remove bacteria and other debris from

Figure 21.4 Types of capillaries (shown in transverse sections).

🔑 **Capillaries are microscopic blood vessels that connect arterioles and venules.**

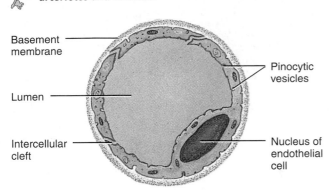

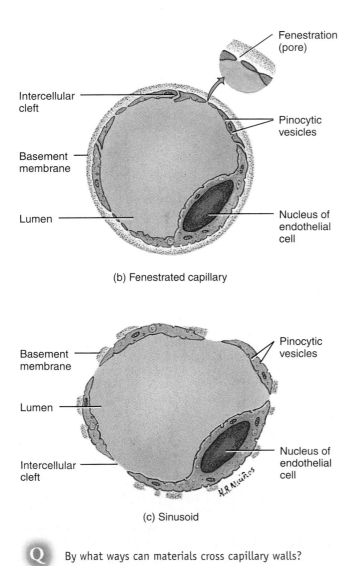

(a) Continuous capillary formed by endothelial cells

(b) Fenestrated capillary

(c) Sinusoid

Q By what ways can materials cross capillary walls?

the blood. Besides the liver and red bone marrow, sinusoids are also present in the spleen, anterior pituitary gland, and parathyroid glands.

Materials that are exchanged between blood and interstitial fluid can move across capillary walls in four basic ways: through intercellular clefts, through fenestrations, within pinocytic vesicles that undergo transcytosis, and through the plasma membranes of the endothelial cells.

Venules

When several capillaries unite, they form small veins called **venules** (= little vein). Venules collect blood from capillaries and drain into veins. The smallest venules, those closest to the capillaries, consist of a tunica interna of endothelium and a tunica media that has only a few scattered smooth muscle fibers and fibroblasts (see Figure 21.3). Like capillaries, the walls of the smallest venules are very porous and are the site where many phagocytic white blood cells emigrate from the bloodstream into an inflamed or infected tissue. As venules become larger and converge to form veins, they contain the tunica externa characteristic of veins.

Veins

Although **veins** (VĀNZ) are composed of essentially the same three coats as arteries, the relative thicknesses of the layers are different. The tunica interna of veins is thinner than that of arteries; the tunica media of veins is much thinner than in arteries, with relatively little smooth muscle and elastic fibers. The tunica externa of veins is the thickest layer and consists of collagen and elastic fibers; the tunica externa of the inferior vena cava also contains longitudinal fibers of smooth muscle. Veins lack the external or internal elastic laminae found in arteries (see Figure 21.1b). Despite these differences, veins are still distensible enough to adapt to variations in the volume and pressure of blood passing through them, although they are not designed to withstand high pressure. Furthermore, the lumen of a vein is larger than that of a comparable artery, and veins frequently appear collapsed (flattened) when sectioned.

Many veins, especially those in the limbs, also feature abundant **valves,** which are thin folds of tunica interna that form flaplike cusps and project into the lumen and toward the heart (Figure 21.5). Because the low blood pressure in veins allows returning blood to slow and even back up, the valves aid in venous return by preventing the backflow of blood.

A **vascular (venous) sinus** is a vein with a thin endothelial wall that has no smooth muscle to alter its diameter. In a vascular sinus the surrounding dense connective tissue replaces the tunica media and tunica externa in providing support. For example, dural venous sinuses, which are supported by the dura mater, convey deoxygenated blood from the brain to the heart. Another example of a vascular sinus is the coronary sinus of the heart.

Figure 21.5 Venous valves.

 Valves in veins allow blood to flow in one direction only—toward the heart.

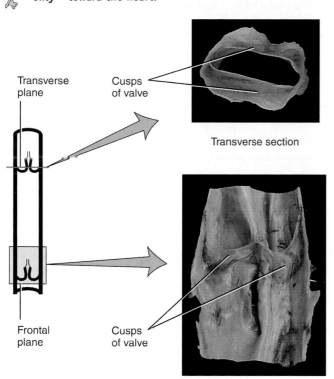

Transverse plane

Cusps of valve

Transverse section

Frontal plane

Cusps of valve

Longitudinally cut

 Why is it more important for valves to be present in arm veins and leg veins than in neck veins?

 CLINICAL APPLICATION
Varicose Veins

Leaky venous valves can cause veins to become dilated and twisted in appearance, a condition called **varicose veins** (*varic-* = a swollen vein). The condition may occur in the veins of almost any body part, but it is most common in the esophagus and in superficial veins of the lower limbs. The valvular defect may be congenital or may result from mechanical stress (prolonged standing or pregnancy) or aging. The leaking venous valves allow the backflow of blood, which causes pooling of blood. This, in turn, creates pressure that distends the vein and allows fluid to leak into surrounding tissue. As a result, the affected vein and the tissue around it may become inflamed and painfully tender. Veins close to the surface of the legs, especially the saphenous vein, are highly susceptible to varicosities, whereas deeper veins are not as vulnerable because surrounding skeletal muscles prevent their walls from stretching excessively. ■

Figure 21.6 Blood distribution in the cardiovascular system at rest.

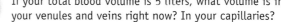 **Because systemic veins and venules contain more than half the total blood volume, they are called blood reservoirs.**

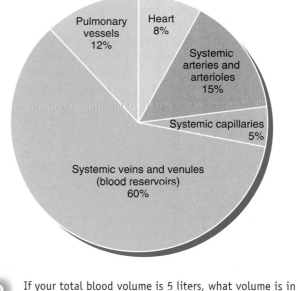

Pulmonary vessels 12%

Heart 8%

Systemic arteries and arterioles 15%

Systemic capillaries 5%

Systemic veins and venules (blood reservoirs) 60%

Q If your total blood volume is 5 liters, what volume is in your venules and veins right now? In your capillaries?

Anastomoses

Most tissues of the body receive blood from more than one artery. The union of the branches of two or more arteries supplying the same body region is called an **anastomosis** (a-nas-tō-MŌ-sis; = connecting; plural is **anastomoses**). Anastomoses between arteries provide alternate routes for blood to reach a tissue or organ. If blood flow stops momentarily when normal movements compress a vessel, or if a vessel is blocked by disease, injury, or surgery, then circulation to a part of the body is not necessarily stopped. The alternate route of blood flow to a body part through an anastomosis is known as **collateral circulation.** Anastomoses may also occur between veins and between arterioles and venules. Arteries that do not anastomose are known as **end arteries.** Obstruction of an end artery interrupts the blood supply to a whole segment of an organ, producing necrosis (death) of that segment. Alternate blood routes may also be provided by nonanastomosing vessels that supply the same region of the body.

Blood Distribution

The largest portion of your blood volume at rest—about 60%—is in systemic veins and venules (Figure 21.6).

Systemic capillaries hold only about 5% of the blood volume, and arteries and arterioles about 15%. Because systemic veins and venules contain a large percentage of the blood volume, they function as **blood reservoirs** from which blood can be diverted quickly if the need arises. For example, when there is increased muscular activity, the cardiovascular center in the brain stem sends more sympathetic impulses to veins. The result is *venoconstriction,* constriction of veins, which reduces the volume of blood in reservoirs and allows a greater blood volume to flow to skeletal muscles, where it is needed most. A similar mechanism operates in cases of hemorrhage, when blood volume and pressure decrease; in this case, venoconstriction helps counteract the drop in blood pressure. Among the principal blood reservoirs are the veins of the abdominal organs (especially the liver and spleen) and the veins of the skin.

1. Discuss the importance of elastic fibers and smooth muscle in the tunica media of arteries.
2. Distinguish between elastic and muscular arteries in terms of location, histology, and function.
3. Describe the structural features of capillaries that allow exchange of materials between blood and body cells.
4. Distinguish between pressure reservoirs and blood reservoirs. Why is each important?
5. Describe the relationship between anastomoses and collateral circulation.

CAPILLARY EXCHANGE

OBJECTIVE

• *Discuss the various pressures involved in the movement of fluids between capillaries and interstitial spaces.*

The mission of the entire cardiovascular system is to keep blood flowing through capillaries so that **capillary exchange**—the movement of substances into and out of capillaries—can occur. Although only 5% of the blood is in systemic capillaries, it is mainly this blood that exchanges materials with interstitial fluid. Substances enter and leave capillaries by three basic mechanisms: diffusion, transcytosis, and bulk flow.

Diffusion

The most important method of capillary exchange is simple diffusion. Substances such as O_2, carbon dioxide (CO_2), glucose, amino acids, and hormones diffuse through capillary walls down their concentration gradients. All plasma solutes except larger proteins pass freely across most capillary walls. Lipid-soluble materials, such as O_2, CO_2, and steroid hormones, may pass directly through the lipid bilayer of endothelial cell plasma membranes. Water-soluble substances, such as glucose and amino acids, pass through either fenestrations or intercellular clefts (see Figure 21.4). The intercellular clefts between endothelial cells that line liver sinu-

soids are so large that even proteins such as fibrinogen and albumin may enter the bloodstream. The prime exception to diffusion of water-soluble materials across capillary walls is in the brain, where the endothelial cells in most regions are nonfenestrated and sealed together by tight junctions. (The blood–brain barrier is described on page 448.)

Transcytosis

A small quantity of material crosses capillary membranes by **transcytosis** (*trans-* = across). In this process, substances in blood plasma become enclosed within tiny vesicles that first enter endothelial cells by endocytosis and then exit on the other side by exocytosis. This method of transport is important mainly for large, lipid-insoluble molecules that cannot cross capillary walls in any other way. The hormone insulin enters the bloodstream by transcytosis, and certain antibody proteins pass from the maternal circulation into the fetal circulation by transcytosis.

Bulk Flow: Filtration and Reabsorption

Bulk flow is a passive process by which *large* numbers of ions, molecules, or particles that are dissolved or suspended in a fluid move together in the same direction. The substances move in unison in response to pressure, and they move at rates far greater than can be accounted for by diffusion or osmosis alone. Bulk flow is always from an area of higher pressure to an area of lower pressure, and it continues as long as a pressure difference exists.

Substances move back and forth across capillary walls between blood and interstitial fluid by bulk flow. Whereas diffusion is more important for *solute exchange* between plasma and interstitial fluid, bulk flow is more important for regulation of the *relative volumes of blood and interstitial fluid.* Pressure-driven movement of fluid and solutes from blood capillaries *into* interstitial fluid is termed **filtration,** whereas pressure-driven movement *from* interstitial fluid into blood capillaries is called **reabsorption.** Two pressures promote filtration: **blood hydrostatic pressure (BHP),** the pressure generated by the pumping action of the heart, and **interstitial fluid osmotic pressure.** The main pressure promoting reabsorption of fluid is **blood colloid osmotic pressure.** The balance of these pressures, called **net filtration pressure (NFP),** determines whether blood volume remains steady or changes. Overall, the volume of fluid and solutes reabsorbed normally is almost as large as the volume filtered; this near equilibrium is known as **Starling's law of the capillaries.** Let's see how these hydrostatic and osmotic pressures balance.

Within vessels, the hydrostatic pressure is due to the pressure that water in plasma exerts against blood vessel walls. The blood hydrostatic pressure (BHP) is about 35 millimeters of mercury (mm Hg) at the arterial end of a capillary, and about 16 mm Hg at the venous end (Figure 21.7).

Figure 21.7 Dynamics of capillary exchange (Starling's law of the capillaries).

🔑 **Blood hydrostatic pressure pushes fluid out of capillaries (filtration), whereas blood colloid osmotic pressure pulls fluid into capillaries (reabsorption).**

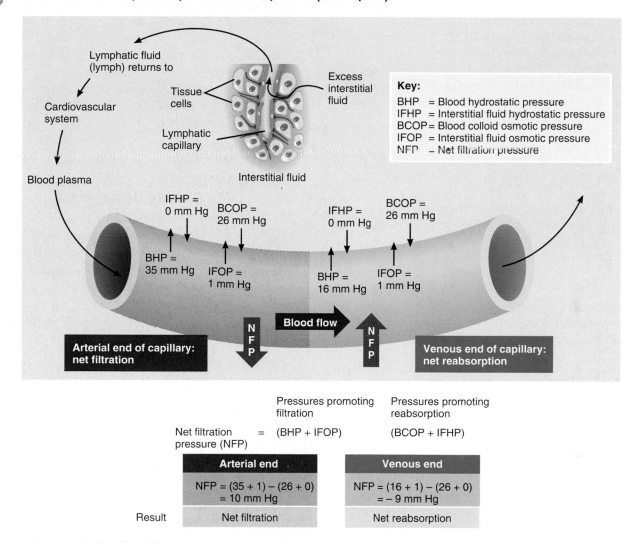

A person who has liver failure cannot synthesize the normal amount of plasma proteins. How does this affect blood colloid osmotic pressure, and what is the effect on capillary filtration and reabsorption?

The pressure of the interstitial fluid, called **interstitial fluid hydrostatic pressure (IFHP),** is close to zero. (IFHP is difficult to measure, and its reported values vary from small positive values to small negative values. For our discussion we assume that IFHP equals 0 mm Hg all along the capillaries). The difference between BHP and IFHP tends to force fluid out of capillaries and into interstitial fluid.

The difference in osmotic pressure across a capillary wall is due almost entirely to the presence in blood of plasma proteins, which are too large to pass through either fenestrations or gaps between endothelial cells. **Blood colloid osmotic pressure (BCOP)** is a force caused by the colloidal suspension of these large plasma proteins. The effect of

BCOP is to pull fluid from interstitial spaces into capillaries; it averages 26 mm Hg in most capillaries. Opposing BCOP is **interstitial fluid osmotic pressure (IFOP),** which tends to move fluid out of capillaries into interstitial fluid. Normally, IFOP is very small—0.1–5 mm Hg—because only tiny amounts of protein are present in interstitial fluid. The small amount of protein that leaks from plasma into interstitial fluid does not accumulate there because it enters lymphatic fluid and is returned to the blood. For discussion, we can use a value of 1 mm Hg for IFOP.

Whether fluids leave or enter capillaries depends on the net balance of pressures. If the pressures that push fluid out of capillaries exceed the pressures that pull fluid into capillaries, fluid will move from capillaries into interstitial spaces (filtration). If, however, the pressures that move fluid out of interstitial spaces into capillaries exceed the pressures that move fluid out of capillaries, then fluid will move from interstitial spaces into capillaries (reabsorption).

The **net filtration pressure (NFP),** which indicates the direction of fluid movement, is calculated as follows:

$$\underset{\substack{\text{Pressures that promote}\\\text{filtration}}}{\text{NFP} = (\text{BHP} + \text{IFOP})} - \underset{\substack{\text{Pressures that promote}\\\text{reabsorption}}}{(\text{BCOP} + \text{IFHP})}$$

At the arterial end of a capillary,

$$\text{NFP} = (35 + 1) - (26 + 0) = 36 - 26 = 10 \text{ mm Hg}$$

Thus, at the arterial end of a capillary, there is a *net outward pressure* of 10 mm Hg, and fluid moves out of the capillary into interstitial spaces (filtration).

At the venous end of a capillary,

$$\text{NFP} = (16 + 1) - (26 + 0) = 17 - 26 = -9 \text{ mm Hg}$$

At the venous end of a capillary, the negative value (-9 mm Hg) represents a *net inward pressure,* and fluid moves into the capillary from tissue spaces (reabsorption).

On average, about 85% of the fluid filtered out of capillaries is reabsorbed. Some of the filtered fluid and any proteins that escape from blood into interstitial fluid flow into lymphatic capillaries and return to the blood via the lymphatic system. Every day about 20 liters of fluid filters out of capillaries, 17 liters are reabsorbed, and 3 liters enter lymphatic capillaries. (Excluded from these numbers are the 180 liters of fluid that are filtered from capillaries in the kidneys and the 178 liters that are reabsorbed during the formation of urine each day.)

CLINICAL APPLICATION
Edema

If filtration greatly exceeds reabsorption, the result is an abnormal increase in interstitial fluid volume, which is termed **edema** (= swelling). Edema is not usually detectable in tissues until interstitial fluid volume has risen to 30% above normal. Edema can result from either excess filtration or inadequate reabsorption.

Two situations may cause excess filtration:

• *Increased blood pressure (hypertension)* causes more fluid to be filtered from capillaries.
• *Increased permeability of capillaries* raises interstitial fluid osmotic pressure by allowing some plasma proteins to escape. Such leakiness may be caused by the destructive effects of chemical, bacterial, thermal, or mechanical agents on capillary walls.

One situation commonly causes inadequate reabsorption:

• *Decreased concentration of plasma proteins* lowers the blood colloid osmotic pressure. Inadequate synthesis or loss of plasma proteins is associated with liver disease, burns, malnutrition, and kidney disease. ■

1. Describe how substances enter and leave the blood in capillaries.
2. Describe how hydrostatic and osmotic pressures govern fluid movement across the walls of capillaries.
3. Write an equation to describe Starling's law of the capillaries.

HEMODYNAMICS: FACTORS AFFECTING CIRCULATION

OBJECTIVES
• *Explain the factors that regulate the velocity and volume of blood flow.*
• *Explain how blood pressure changes throughout the cardiovascular system, and describe the factors that determine mean arterial blood pressure.*
• *Describe the factors that determine systemic vascular resistance.*

In Chapter 20 we saw that total cardiac output depends on heart rate and stroke volume. The distribution of the cardiac output to various tissues depends, however, on the interplay of (1) the *pressure difference* that drives the blood flow and (2) the *resistance* to blood flow, which is the opposition to flow of blood through specific blood vessels.

Velocity of Blood Flow

The *volume* of blood that flows through any tissue in a given period of time (in mL/min) is called **blood flow.** The *velocity* (speed) of blood flow (in cm/sec) is inversely related to cross-sectional area; that is, velocity is slowest where the total cross-sectional area is greatest (Figure 21.8). Each time an artery branches, the total cross-sectional area of all its branches is greater than that of the original vessel, and thus blood flow is slower in the branches. Conversely, when branches combine—for example, as venules merge to form veins—the total cross-sectional area becomes smaller and flow becomes faster. In an adult, the cross-sectional area of the aorta is only 3–5 cm^2, and the average velocity of the blood there is 40 cm/sec; in capillaries, the total cross-sectional area is 4500–6000 cm^2, and the velocity of blood flow is less than 0.1 cm/sec. In the two venae cavae combined, the cross-sectional area is about 14 cm^2, and the velocity is 5–20 cm/sec. Thus, the velocity of blood flow decreases as blood flows from the aorta to arteries to arterioles to capillaries, and increases as it leaves capillaries and returns to the heart. The relatively slow rate of flow through capillaries facilitates the exchange of materials between blood and adjacent interstitial fluid.

Circulation time is the time required for a drop of blood to pass from the right atrium, through the pulmonary circulation, back to the left atrium, through the systemic circulation down to the foot, and back again to the right atrium. In a resting person, circulation time normally is about 1 minute.

Volume of Blood Flow

We saw in Chapter 20 that cardiac output (CO)—the volume of blood that circulates through systemic (or pulmonary) blood vessels each minute—equals stroke volume (SV) multiplied by heart rate (HR):

$$CO = SV \times HR$$

Two other factors influence cardiac output: (1) blood pressure and (2) resistance, which is due mainly to friction between blood and blood vessel walls. Blood flows from regions of higher to lower pressure; the greater the pressure difference, the greater the blood flow. The higher the resistance, on the other hand, the lower the blood flow.

Blood Pressure

Blood pressure (BP) is the hydrostatic pressure exerted by blood on the walls of a blood vessel. Blood pressure is generated by contraction of the ventricles and is highest in the aorta and large systemic arteries, where in a resting, young adult, BP rises to about 120 mm Hg during systole (contraction) and drops to about 80 mm Hg during diastole (relaxation). **Mean arterial blood pressure (MABP),** the average pressure, is calculated as follows:

$$MABP = \text{diastolic BP} + \tfrac{1}{3}(\text{systolic BP} - \text{diastolic BP})$$

Thus, in a person whose BP is 120/80 mm Hg, MABP is about 93 mm Hg.

Cardiac output equals mean arterial blood pressure (MABP) divided by resistance (R):

$$CO = MABP \div R$$

If cardiac output rises due to an increase in stroke volume or heart rate, then the mean arterial blood pressure rises so long as resistance remains steady. Likewise, a decrease in cardiac output causes a decrease in blood pressure if resistance does not change.

As blood leaves the aorta and flows through the systemic circulation, its pressure falls progressively as the distance from the left ventricle increases (Figure 21.9). Mean blood pressure decreases from 93 mm Hg to about 35 mm Hg as blood passes from arteries through arterioles and into capillaries, where the pressure fluctuations disappear. At the venous end of capillaries, blood pressure has dropped to about 16 mm Hg. Blood pressure continues to drop as blood enters venules and then veins because these vessels are farthest from the left ventricle. Finally, blood pressure reaches

Figure 21.8 Relationship between velocity of blood flow and total cross-sectional area in different types of blood vessels.

🔑 **Velocity of blood flow is slowest in the capillaries because they have the largest total cross-sectional area.**

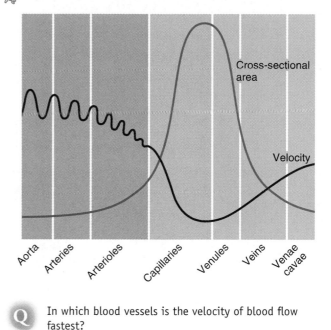

Ⓠ In which blood vessels is the velocity of blood flow fastest?

Figure 21.9 Blood pressures in various parts of the cardiovascular system. The dashed line is the mean (average) blood pressure.

🔑 **Blood pressure rises and falls with each heartbeat in blood vessels leading to capillaries.**

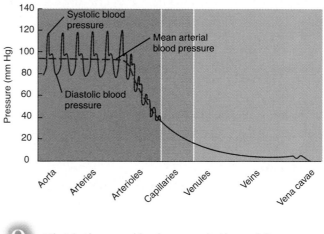

Ⓠ What is the mean blood pressure in the aorta?

Figure 21.10 Action of the skeletal muscle pump in returning blood to the heart. Steps **1** through **3** are described in the text on page 681.

🔑 *Milking* refers to skeletal muscle contractions that drive venous blood toward the heart.

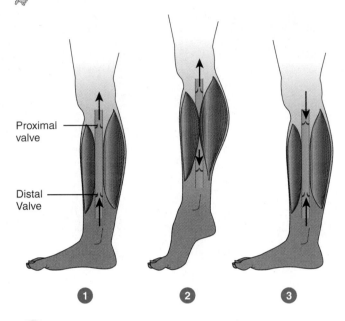

Proximal valve

Distal Valve

1　　**2**　　**3**

Q What mechanisms, besides cardiac contractions, act as pumps to boost venous return?

0 mm Hg as blood flows into the right ventricle. Blood always flows through vessels down a pressure gradient (difference); if there is no pressure difference, there is no flow.

Blood pressure also depends on the total volume of blood in the cardiovascular system. The normal volume of blood in an adult is about 5 liters (5.3 qt). Any decrease in this volume, as from hemorrhage, decreases the amount of blood that is circulated through the arteries each minute. A modest decrease can be compensated for by homeostatic mechanisms that help maintain blood pressure (described on page 687), but if the decrease in blood volume is greater than 10% of total blood volume, blood pressure drops. Conversely, anything that increases blood volume, such as water retention in the body, tends to increase blood pressure.

Resistance

As noted earlier, **resistance** refers to the opposition to blood flow principally as a result of friction between blood and the walls of blood vessels. The friction, and thus the resistance, depends on (1) average blood vessel radius, (2) blood viscosity, and (3) total blood vessel length.

1. *Average blood vessel radius.* Resistance is inversely proportional to the fourth power of the radius of the blood vessel ($R \propto 1/r^4$). The smaller the radius of the blood vessel, the greater the resistance it offers to blood flow. If, for ex-

ample, the radius of a blood vessel decreases by one-half, its resistance to blood flow increases 16 times:

$$R = 1/(\tfrac{1}{2})^4 = 2^4 = 2 \times 2 \times 2 \times 2 = 16$$

Normally, moment-to-moment fluctuations in blood pressure are due to changes in blood vessel radius.

2. *Blood viscosity.* The viscosity ("thickness") of blood depends mostly on the ratio of red blood cells to plasma (fluid) volume, and to a smaller extent on the concentration of proteins in plasma. Resistance to blood flow is directly proportional to the viscosity of blood; any condition that increases the viscosity of blood, such as dehydration or polycythemia (an unusually high number of red blood cells), increases resistance and thus blood pressure. A depletion of plasma proteins or red blood cells, as a result of anemia or hemorrhage, decreases resistance and thus blood pressure.

3. *Total blood vessel length.* Resistance to blood flow through a vessel is directly proportional to the length of the blood vessel. The longer a blood vessel, the greater the resistance to blood that flows through it. An obese person may have hypertension (elevated blood pressure) due to increase in total blood vessel length caused by the additional blood vessels in adipose tissue. An estimated 300 km (about 200 miles) of additional blood vessels develop for each extra pound of fat.

Systemic vascular resistance (SVR), also known as *total peripheral resistance (TPR)*, refers to all the vascular resistances offered by systemic blood vessels. Most resistance is contributed by the smallest vessels—arterioles, capillaries, and venules. The diameters of arteries and veins are large, and thus their resistance is very small because most of the blood does not come into physical contact with the walls of the blood vessel. A major function of arterioles is to control SVR—and therefore blood pressure and blood flow to particular tissues—by changing their diameters. Arterioles need vasodilate or vasoconstrict only slightly to have a large effect on SVR. The principal center for regulation of SVR is the vasomotor center in the brain stem (described shortly).

Venous Return

Venous return, the volume of blood flowing back to the heart from the systemic veins, depends on the pressure difference from venules (averaging about 16 mm Hg) to the right ventricle (0 mm Hg). Although this pressure difference is small, venous return to the right atrium keeps pace with output from the left ventricle because resistance of veins is so low. If pressure increases in the right atrium, however, venous return will decrease. One cause of increased pressure in the right atrium is an incompetent (leaky) tricuspid valve, which lets blood regurgitate (flow backward) as the ventricles contract. The result is buildup of blood on the venous side of the systemic circulation.

Figure 21.11 Summary of factors that increase blood pressure. Changes noted within green boxes increase cardiac output, whereas changes noted within blue boxes increase systemic vascular resistance.

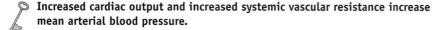

Increased cardiac output and increased systemic vascular resistance increase mean arterial blood pressure.

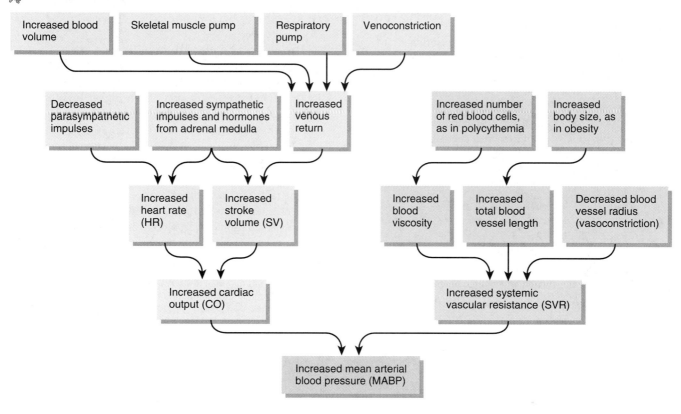

Q   Which type of blood vessel exerts the major control of systemic vascular resistance on a moment-to-moment basis, and how does it achieve this?

Besides the heart, two other mechanisms act as pumps to boost venous return: the contraction of skeletal muscles in the lower limbs and the pressure changes in the thorax and abdomen during respiration (breathing). The presence of valves in veins allows both of these pumps to contribute to venous return.

The **skeletal muscle pump** operates as follows (Figure 21.10):

1 While standing at rest, both venous valves in this segment of the leg are open, and blood flows upward toward the heart.

2 Contraction of leg muscles compresses the vein, which pushes blood through the proximal valve, an action called *milking,* and closes the distal valve in the uncompressed segment of the vein just below. People who are immobilized through injury or disease lack these contractions. As a result, the return of venous blood to the heart is slower.

3 Just after muscle relaxation, pressure falls in the previously compressed section of vein, which causes the proximal valve to close. The distal valve now opens because blood pressure in the foot is higher than in the leg, and the vein fills with blood from the foot.

The **respiratory pump** is also based on alternating compression and decompression of veins. During inhalation the diaphragm moves inferiorly, which causes a decrease in pressure in the thoracic cavity and an increase in pressure in the abdominal cavity. As a result, abdominal veins are compressed, and a greater volume of blood moves from the compressed abdominal veins into the decompressed thoracic veins and then into the right atrium. When the pressures reverse during exhalation, the valves in the veins prevent backflow of blood.

Figure 21.11 summarizes the factors that increase blood pressure through increasing cardiac output or systemic vascular resistance.

CLINICAL APPLICATION
Syncope

Syncope (SIN-kō-pē), or fainting, is a sudden, temporary loss of consciousness that is not due to head trauma, followed by spontaneous recovery. It is most commonly due to cerebral ischemia, lack of sufficient blood flow to the brain. Syncope may occur for several reasons:

• *Vasodepressor syncope* is due to sudden emotional stress or real, threatened, or fantasized injury.
• *Situational syncope* is caused by pressure stress associated with urination, defecation, or severe coughing.
• *Drug-induced syncope* may be caused by drugs such as antihypertensives, diuretics, vasodilators, and tranquilizers.
• *Orthostatic hypotension,* an excessive decrease in blood pressure that occurs upon standing up, may cause fainting. ■

1. Why is the velocity of blood flow faster in arteries and veins than in capillaries?
2. Explain how blood pressure and resistance determine volume of blood flow.
3. Define resistance, and explain the factors that contribute to it.
4. Explain how the return of venous blood to the heart is accomplished.

CONTROL OF BLOOD PRESSURE AND BLOOD FLOW

OBJECTIVE
• *Describe how blood pressure is regulated.*

From moment to moment and day to day, several interconnected negative feedback systems control blood pressure by adjusting heart rate, stroke volume, systemic vascular resistance, and blood volume. Some systems allow rapid adjustment of blood pressure to cope with sudden changes, such as the drop in blood pressure in the brain that occurs when you get out of bed; others act more slowly to provide long-term regulation of blood pressure. Even if blood pressure is steady, the distribution of blood flow may require adjustment, which is accomplished mainly by altering the diameter of arterioles. During exercise, for example, a greater percentage of blood flow is diverted to skeletal muscles.

Role of the Cardiovascular Center

In Chapter 20 we noted how the **cardiovascular center** in the medulla oblongata helps regulate heart rate and stroke volume. Now we extend that account by describing the neural, hormonal, and local negative feedback systems that regulate blood pressure and blood flow to specific tissues. Groups of neurons scattered within the cardiovascular (CV) center regulate heart rate, contractility (force of contraction) of the ventricles, and blood vessel diameter. Some of the CV center's neurons stimulate the heart (cardiostimulatory cen-

ter), whereas others inhibit the heart (cardioinhibitory center); still others control blood vessel diameter (vasomotor center), either by causing constriction (vasoconstrictor center) or dilation (vasodilator center). Because these clusters of neurons communicate with one another, function together, and are not clearly separated anatomically, we discuss them here as a group.

Input to the Cardiovascular Center

The cardiovascular center receives input both from higher brain regions and from sensory receptors (Figure 21.12). Nerve impulses descend from higher brain regions, including the cerebral cortex, limbic system, and hypothalamus, to affect the cardiovascular center. For example, even before you start to run a race, your heart rate may increase due to nerve impulses conveyed from the limbic system to the cardiovascular center. If your body temperature rises during a race, the thermoregulatory center of the hypothalamus sends nerve impulses to the cardiovascular center; the result is vasodilation of skin blood vessels, which allows heat to dissipate more rapidly from the surface of the skin. The three main types of sensory receptors that provide input to the cardiovascular center are proprioceptors, baroreceptors, and chemoreceptors. *Proprioceptors,* which monitor movements of joints and muscles, provide input to the cardiovascular center during physical activity and account for the rapid increase in heart rate at the beginning of exercise. *Baroreceptors* monitor changes in pressure and stretch in the walls of blood vessels, and *chemoreceptors* monitor the concentration of various chemicals in the blood.

Output from the Cardiovascular Center

Output from the cardiovascular center flows along sympathetic and parasympathetic fibers of the ANS (see Figure 21.12). Sympathetic impulses reach the heart via the **cardiac accelerator nerves.** An increase in sympathetic stimulation increases heart rate and contractility, whereas a decrease in sympathetic stimulation decreases heart rate and contractility. Parasympathetic stimulation, conveyed along the **vagus (X) nerves,** decreases heart rate. Thus, autonomic control of the heart is the result of opposing sympathetic (stimulatory) and parasympathetic (inhibitory) influences.

The cardiovascular center also continually sends impulses to smooth muscle in blood vessel walls via sympathetic fibers called **vasomotor nerves.** Autonomic control of blood vessel diameter is mostly via the sympathetic division. Sympathetic vasomotor nerve fibers exit the spinal cord through all thoracic and the first one or two lumbar spinal nerves and then pass into the sympathetic trunk ganglia (see Figure 17.2 on page 551). From there, impulses propagate along sympathetic nerves that innervate blood vessels in viscera and peripheral areas. The vasomotor region of the cardiovascular center continually sends impulses over these routes to arterioles throughout the body, but especially to those in the skin and abdominal viscera. The result is a mod-

Figure 21.12 Location and function of the cardiovascular (CV) center in the medulla oblongata. The CV center receives input from higher brain centers, proprioceptors, baroreceptors, and chemoreceptors, and it provides output to both the sympathetic and parasympathetic divisions of the autonomic nervous system.

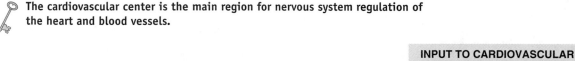

The cardiovascular center is the main region for nervous system regulation of the heart and blood vessels.

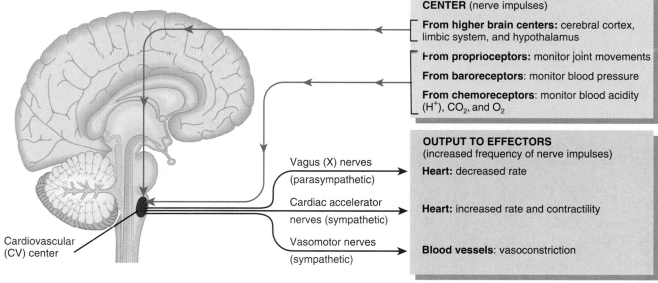

INPUT TO CARDIOVASCULAR CENTER (nerve impulses)

From higher brain centers: cerebral cortex, limbic system, and hypothalamus

From proprioceptors: monitor joint movements

From baroreceptors: monitor blood pressure

From chemoreceptors: monitor blood acidity (H^+), CO_2, and O_2

OUTPUT TO EFFECTORS (increased frequency of nerve impulses)

Heart: decreased rate

Heart: increased rate and contractility

Blood vessels: vasoconstriction

Vagus (X) nerves (parasympathetic)

Cardiac accelerator nerves (sympathetic)

Vasomotor nerves (sympathetic)

Cardiovascular (CV) center

Q What types of effector tissues are regulated by the cardiovascular center?

erate state of tonic contraction or vasoconstriction, called **vasomotor tone,** that sets the resting level of systemic vascular resistance. Sympathetic stimulation of most veins results in constriction that moves blood out of venous blood reservoirs and increases blood pressure.

Neural Regulation of Blood Pressure

The nervous system regulates blood pressure via negative feedback loops in two types of reflexes: baroreceptor reflexes and chemoreceptor reflexes.

Baroreceptor Reflexes

Baroreceptors in the walls of some arteries and veins monitor blood pressure. The two most important negative feedback systems involving baroreceptors are the carotid sinus reflex and the aortic reflex.

The **carotid sinus reflex,** which helps maintain normal blood pressure in the brain, is initiated by baroreceptors in the wall of the carotid sinuses. The **carotid sinuses** are small widenings of the right and left internal carotid arteries just above the point where they branch from the common carotid artery (Figure 21.13). Any increase in blood pressure stretches the wall of the carotid sinus, which stimulates the

baroreceptors. Nerve impulses propagate from the carotid sinus baroreceptors over sensory fibers in the glossopharyngeal (IX) nerves to the cardiovascular center in the medulla oblongata. The **aortic reflex** governs general systemic blood pressure; it is initiated by baroreceptors in the wall of the ascending aorta and arch of the aorta. Nerve impulses from aortic baroreceptors reach the cardiovascular center via sensory fibers of the vagus (X) nerves.

If blood pressure falls, however, the baroreceptors are stretched less, and they send nerve impulses at a slower rate to the cardiovascular center (Figure 21.14). In response, the cardiovascular center decreases parasympathetic stimulation of the heart via motor fibers of the vagus (X) nerves and increases sympathetic stimulation of the heart via cardiac accelerator nerves. Another effect of increased sympathetic stimulation is increased epinephrine and norepinephrine secretion by the adrenal medulla. The effects on the heart and blood vessels are to accelerate heart rate, increase force of contraction, and promote vasoconstriction. As the heart beats faster and more forcefully, and as systemic vascular resistance increases, blood pressure increases, promoting a return to homeostasis as blood pressure returns to the normal level.

Figure 21.13 ANS innervation of the heart and the baroreceptor reflexes that help regulate blood pressure.

🔑 **Baroreceptors are pressure-sensitive neurons that monitor stretching.**

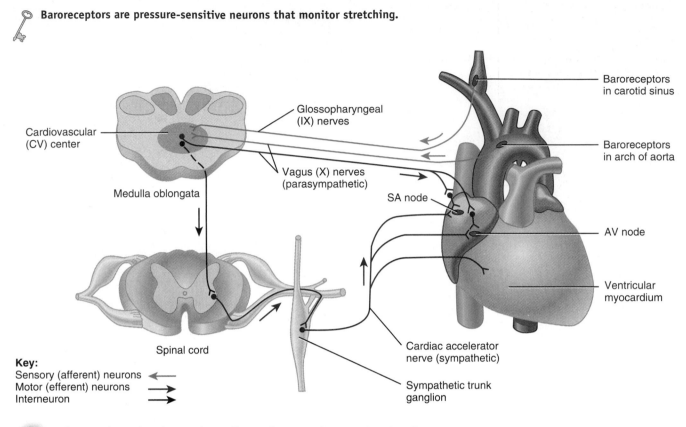

Key:
Sensory (afferent) neurons ⟵
Motor (efferent) neurons ⟶
Interneuron ⟶

Q What are the paths taken to the cardiovascular center by nerve impulses from baroreceptors in the carotid sinuses and the arch of the aorta?

Conversely, when an increase in aortic and carotid artery pressures is detected by the respective baroreceptors, the cardiovascular center responds by increasing parasympathetic stimulation and decreasing sympathetic stimulation. The resulting decreases in heart rate and force of contraction lower cardiac output. Moreover, the cardiovascular center slows the sympathetic impulses it sends along vasomotor fibers that normally cause vasoconstriction. The result is vasodilation, which lowers systemic vascular resistance. Both decreased cardiac output and decreased systemic vascular resistance lower systemic arterial blood pressure.

The ability of the aortic and carotid sinus reflexes to correct a drop in blood pressure is very important when a person sits up or stands from a prone (lying down) position. Moving from a prone to an erect position decreases blood pressure and blood flow in the head and upper part of the body. The drop in pressure, however, is quickly counteracted by the aortic and carotid sinus reflexes. Sometimes these reflexes operate more slowly than normal, especially in older people, in which case a person can faint when standing up too quickly because of an inadequate blood supply to the brain.

CLINICAL APPLICATION
*Carotid Sinus Massage
and Carotid Sinus Syncope*

Because the carotid sinus is close to the anterior surface of the neck, it is possible to stimulate the baroreceptors there by putting pressure on the neck. Physicians sometimes use **carotid sinus massage**, which involves carefully massaging the neck over the carotid sinus, to slow heart rate in a person who has paroxysmal supraventricular tachycardia, a type of tachycardia that originates in the atria. Anything that stretches or puts pressure on the carotid sinus, such as hyperextension of the head, tight collars, or carrying heavy shoulder loads, may also slow heart rate and can cause **carotid sinus syncope,** fainting due to inappropriate stimulation of the carotid sinus baroreceptors. ▪

Chemoreceptor Reflexes

Chemoreceptors that monitor the chemical composition of blood are located close to the baroreceptors of the carotid sinus and arch of the aorta in small structures called **carotid bodies** and **aortic bodies,** respectively. These chemoreceptors detect changes in blood level of O_2, CO_2, and H^+. *Hypoxia* (lowered O_2 availability), *acidosis* (an increase in H^+ concentration), or *hypercapnia* (excess CO_2)

stimulate the chemoreceptors to send impulses to the cardiovascular center. In response, the CV center increases sympathetic stimulation to arterioles and veins, producing vasoconstriction and an increase in blood pressure. As we will see in Chapter 23, these chemoreceptors also provide input to the respiratory center in the brain stem for adjustments in the rate of breathing.

Hormonal Regulation of Blood Pressure

The following hormones and hormone systems help regulate blood pressure and blood flow by altering cardiac output, changing systemic vascular resistance, or adjusting the total blood volume:

1. *Renin–angiotensin–aldosterone (RAA) system.* When blood volume falls or blood flow to the kidneys decreases, juxtaglomerular cells in the kidneys release increased amounts of the enzyme *renin* into the bloodstream. In sequence, renin and angiotensin converting enzyme (ACE) act on their substrates to produce the active hormone **angiotensin II,** which raises blood pressure in two ways: First, angiotensin II is a potent vasoconstrictor that raises systemic vascular resistance. Second, it stimulates secretion of *aldosterone,* which increases reabsorption of sodium ions (Na^+) and water by the kidneys. This action increases total blood volume, which increases blood pressure.
2. *Epinephrine and norepinephrine.* These adrenal medulla hormones increase cardiac output by increasing the rate and force of heart contractions; they also bring about vasoconstriction of arterioles and veins in the skin and abdominal organs. In addition, epinephrine causes vasodilation of arterioles in cardiac and skeletal muscle.
3. *Antidiuretic hormone (ADH).* One of the actions of ADH, produced by the hypothalamus and released from the posterior pituitary gland, is to cause vasoconstriction. For this reason ADH is also called **vasopressin.**
4. *Atrial natriuretic peptide (ANP).* Released by cells in the atria of the heart, ANP lowers blood pressure by causing vasodilation and by promoting the loss of salt and water in the urine, which reduces blood volume.

Table 21.1 summarizes the regulation of blood pressure by hormones.

Local Regulation of Blood Pressure

In each capillary bed, localized changes can regulate vasomotion. By causing vasodilation or vasoconstriction, local factors influence systemic vascular resistance and thus blood pressure. Vasodilators produce local dilation of arterioles and relaxation of precapillary sphincters; the result is an increased flow of blood into capillary beds, which restores O_2 level to normal. Vasoconstrictors have opposite effects. The ability of a tissue to automatically adjust the blood flow

Figure 21.14 Negative feedback regulation of blood pressure via baroreceptor reflexes.

🗝 **According to Marey's law of the heart, as heart rate increases blood pressure increases.**

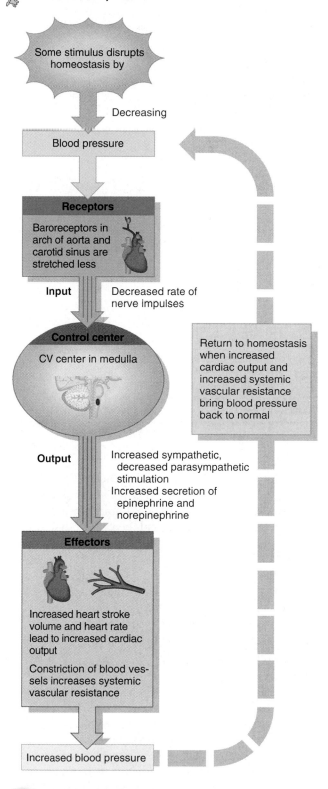

Q Does this negative feedback cycle represent the changes that occur when you lie down or when you stand up?

Table 21.1 Blood Pressure Regulation by Hormones

FACTOR INFLUENCING BLOOD PRESSURE	HORMONE	EFFECT ON BLOOD PRESSURE
Cardiac Output		
Increased heart rate and contractility	Norepinephrine Epinephrine	Increase
Systemic Vascular Resistance		
Vasoconstriction	Angiotensin II	Increase
	Antidiuretic hormone (vasopressin)	
	Norepinephrine*	
	Epinephrine*	
Vasodilation	Atrial natriuretic peptide	Decrease
	Epinephrine†	
	Nitric oxide	
Blood Volume		
Blood volume increase	Aldosterone	Increase
	Antidiuretic hormone	
Blood volume decrease	Atrial natriuretic peptide	Decrease

*Acts at α_1 receptors in arterioles of abdomen and skin.

†Acts at β_2 receptors in arterioles of cardiac and skeletal muscle; norepinephrine has a much smaller vasodilating effect.

through it to match its metabolic demand for O_2 and nutrients and the removal of wastes is called **autoregulation.** In tissues such as the heart and skeletal muscle, where the demand can increase as much as tenfold during physical activity, autoregulation is an important contributor to increased blood flow through the tissue; it also is the major regulator of regional blood flow in the brain. Even though total blood flow to the brain remains almost constant regardless of the degree of physical or mental activity, blood distribution to various parts of the brain changes dramatically for different activities. During a conversation, for example, blood flow increases to the motor speech areas when one is talking and increases to the auditory areas when one is listening.

Two general types of stimuli cause autoregulatory changes in blood flow:

1. *Physical changes.* Warming promotes vasodilation, whereas cooling causes vasoconstriction. Furthermore, smooth muscle in arteriole walls exhibits a **myogenic response**—it contracts more forcefully when it is stretched and relaxes when stretching lessens. In an arteriole, the degree

to which smooth muscle is stretched depends on its blood flow: If blood flow decreases, then stretch decreases; the smooth muscle relaxes and produces vasodilation, which increases blood flow.

2. *Chemical mediators.* Several types of cells—including white blood cells, platelets, smooth muscle fibers, macrophages, and endothelial cells—release a wide variety of **vasoactive factors,** chemicals that alter blood-vessel diameter. Vasodilating chemicals released by metabolically active tissue cells include K^+, H^+, lactic acid (lactate), and adenosine (from ATP). Another important vasodilator released by endothelial cells is nitric oxide, named endothelium-derived relaxation factor (EDRF) before its chemical identity was known. Tissue trauma or inflammation causes release of vasodilating kinins and histamine. Vasoconstrictors include certain eicosanoids such as thromboxane A2 and prostaglandin $F_{2\alpha}$, superoxide radicals, serotonin (from platelets), and endothelins (from endothelial cells).

An important difference between the pulmonary and systemic circulations is their autoregulatory response to changes in O_2 level. In the systemic circulation, blood vessels *dilate* in response to low O_2 concentration, but in the pulmonary circulation, blood vessels *constrict* in response to low levels of O_2. This mechanism is very important in distributing blood to those areas of the lungs where it can pick up the most O_2. If some alveoli (air sacs) are not well ventilated by fresh air, the blood vessels in the affected area constrict so that blood will largely bypass the poorly functioning areas. As a result, most of the blood flows to other, better ventilated areas of the lung.

1. What are the principal inputs to and outputs from the cardiovascular center?
2. Describe the operation of the carotid sinus reflex and the aortic reflex.
3. Explain the role of chemoreceptors in the regulation of blood pressure.
4. Describe the hormonal regulation of blood pressure.
5. What is autoregulation? Describe the physical and chemical changes that cause vasodilation and vasoconstriction.

SHOCK AND HOMEOSTASIS

OBJECTIVE

• *Define shock, and describe the four types of shock.*

Shock is a failure of the cardiovascular system to deliver enough O_2 and nutrients to meet cellular metabolic needs. The causes of shock are many and varied, but all are characterized by inadequate perfusion of (blood flow to) body tissues. With inadequate oxygen delivery, cells switch from aerobic to anaerobic production of ATP, and lactic acid accumulates in body fluids. If shock persists, cells and organs become damaged, and cells may die unless proper treatment begins quickly.

Types of Shock

Shock can be of four different types: (1) **hypovolemic shock** (hī-pō-vō-LĒ-mik; *hypo-* = low; *-volemic* = volume) due to decreased blood volume, (2) **cardiogenic shock** due to poor heart function, (3) **vascular shock** due to inappropriate vasodilation, and (4) **obstructive shock** due to obstruction of blood flow.

A common cause of hypovolemic shock is acute (sudden) hemorrhage. The blood loss may be external as occurs in trauma, or internal, as in rupture of an aortic aneurysm. Loss of body fluids through excessive sweating, diarrhea, or vomiting can also cause hypovolemic shock. Other conditions—for instance, diabetes mellitus—may cause excessive loss of fluid in the urine. Sometimes, hypovolemic shock is due to inadequate intake of fluid. Whatever the cause, when the volume of body fluids falls, venous return to the heart declines such that filling of the heart is diminished, stroke volume decreases, and cardiac output decreases.

In cardiogenic shock, the pumping function of the heart is compromised, most often because of a myocardial infarction (heart attack). Other causes of cardiogenic shock include poor perfusion of the heart (ischemia), heart valve problems, excessive preload or afterload, impaired contractility of heart muscle fibers, and arrhythmias.

Even with normal blood volume and cardiac output, shock may occur if blood pressure drops due to a decrease in systemic vascular resistance. A variety of conditions can cause inappropriate dilation of arterioles or venules. In *anaphylactic shock,* a severe allergic reaction—for example, to a bee sting—releases histamine and other mediators that cause vasodilation. In *neurogenic shock,* vasodilation may occur following trauma to the head that causes malfunction of the cardiovascular center in the medulla. Shock stemming from certain bacterial toxins that produce vasodilation is termed *septic shock.* In the United States, septic shock causes more than 100,000 deaths per year and is the most common cause of death in hospital critical care units.

Obstructive shock occurs when blood flow through a portion of the circulation is blocked. The most common cause is *pulmonary embolism,* a blood clot lodged in a blood vessel of the lungs.

Homeostatic Responses to Shock

The major mechanisms of compensation in shock are *negative feedback systems* that work to return cardiac output and arterial blood pressure to normal. When shock is mild, compensation by homeostatic mechanisms prevents serious damage. In an otherwise healthy person, for example, compensatory mechanisms can maintain adequate blood flow and blood pressure despite an acute blood loss of as much as 10% of total volume, as follows:

1. *Activation of the renin–angiotensin–aldosterone system.* Decreased blood flow to the kidneys causes the kidneys to secrete renin and initiates the renin–angiotensin–aldosterone system (see Figure 18.16 on page 590). Recall that angiotensin II is a potent vasoconstrictor that also stimulates the adrenal cortex to secrete aldosterone—a hormone that increases reabsorption of Na^+ and water by the kidneys. The increases in systemic vascular resistance and blood volume help raise blood pressure.

2. *Secretion of antidiuretic hormone.* In response to decreased blood pressure, the posterior pituitary gland releases more antidiuretic hormone (ADH). ADH causes vasoconstriction, which increases systemic vascular resistance, and enhances water reabsorption by the kidneys, which conserves remaining blood volume.

3. *Activation of the sympathetic division of the ANS.* As blood pressure decreases, the aortic and carotid baroreceptors initiate powerful sympathetic responses throughout most of the body. One result is marked vasoconstriction of arterioles and veins of the skin, kidneys, and other abdominal viscera. (Vasoconstriction does not occur in the brain or heart.) The vasoconstriction increases systemic vascular resistance, which helps maintain adequate venous return. Sympathetic stimulation also increases heart rate and force of contraction and increases secretion of epinephrine and norepinephrine by the adrenal medulla. These hormones intensify vasoconstriction and increase heart rate and contractility, all of which help raise blood pressure.

4. *Release of local vasodilators.* In response to *hypoxia,* cells liberate vasodilators—including K^+, H^+, lactic acid, adenosine, and nitric oxide—that dilate arterioles and relax precapillary sphincters. Such vasodilation increases local blood flow and may restore O_2 level to normal in an underperfused part of the body. However, vasodilation also has the potentially harmful effect of decreasing systemic vascular resistance and thus lowering the blood pressure.

The negative feedback mechanisms that respond to hypovolemic shock are diagrammed in Figure 21.15.

If blood volume drops more than 10–20%, or if the heart cannot bring blood pressure up sufficiently, compensatory mechanisms may fail to maintain adequate perfusion of tissues. At this point, shock becomes life threatening as damaged cells start to die.

Signs and Symptoms of Shock

Even though the signs and symptoms of shock vary with the severity of the condition, most can be predicted in light of the responses generated by the negative feedback systems

Figure 21.15 Negative feedback systems that can restore normal blood pressure during hypovolemic shock.

🔑 **Homeostatic mechanisms can compensate for an acute blood loss of as much as 10% of total blood volume.**

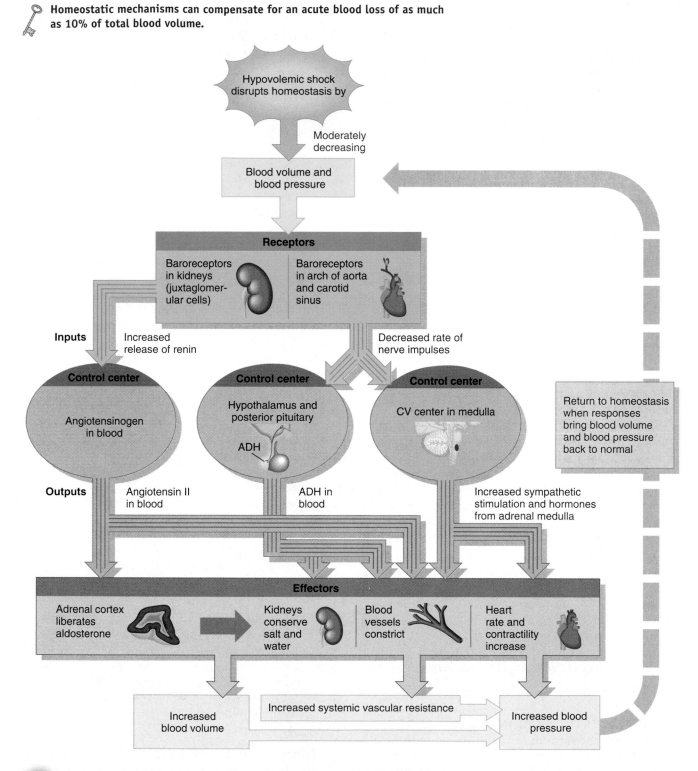

Q Does almost-normal blood pressure in a person who has lost blood indicate that the patient's tissues are receiving adequate perfusion (blood flow)?

Table 21.2 Pulse Points

STRUCTURE		LOCATION	STRUCTURE		LOCATION
Superficial temporal artery		Lateral to orbit of eye.	Femoral artery		Inferior to inguinal ligament.
Facial artery		Mandible (lower jaw-bone) on a line with the corners of the mouth.	Popliteal artery		Posterior to the knee.
Common carotid artery		Lateral to larynx (voice box).	Radial artery		Distal half of wrist.
Brachial artery		Medial side of biceps brachii muscle.	Dorsalis pedis artery		Superior to instep of foot.

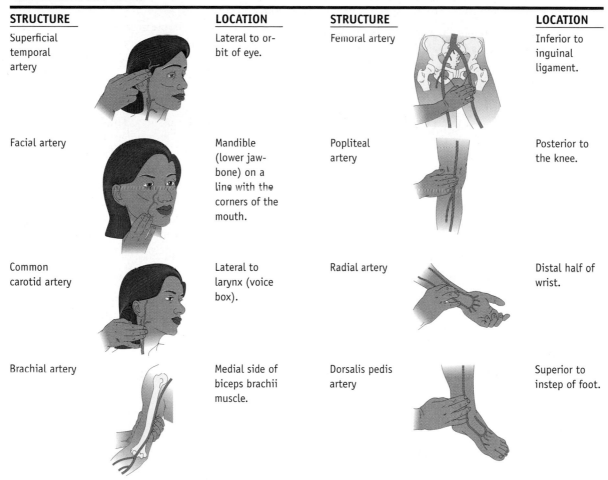

that attempt to correct the problem. Among these signs and symptoms are the following:

- *Rapid resting heart rate* due to sympathetic stimulation and increased blood levels of epinephrine and norepinephrine produced by the adrenal medulla
- *Weak, rapid pulse* due to reduced cardiac output and fast heart rate
- *Clammy, cool, pale skin* due to sympathetic vasoconstriction of skin blood vessels
- *Sweating* due to sympathetic stimulation
- *Altered mental state* due to cerebral ischemia
- *Reduced urine formation* due to sympathetic vasoconstriction of renal blood vessels and increased levels of aldosterone and antidiuretic hormone (ADH)
- *Thirst* due to loss of extracellular fluid
- *Acidosis* due to buildup of lactic acid
- *Nausea* due to impaired circulation to the digestive system due to sympathetic vasoconstriction

1. Explain which symptoms of hypovolemic shock relate to actual body fluid loss, and which relate to the negative feedback systems that attempt to maintain blood pressure and blood flow.

EVALUATING CIRCULATION

OBJECTIVE

- *Define pulse, and define systolic, diastolic, and pulse pressures.*

Pulse

The alternate expansion and recoil of elastic arteries after each systole of the left ventricle create a traveling pressure wave that is called the **pulse.** Pulse is strongest in the arteries closest to the heart, becomes weaker in the arterioles, and disappears altogether in the capillaries. The pulse may be felt in any artery that lies near the surface of the body and runs over a bone or other firm structure. Table 21.2 depicts some common pulse points.

The pulse rate normally is the same as the heart rate. Normal resting pulse rate is between 70 and 80 beats per minute. **Tachycardia** (tak'-i-KAR-dē-a; *tachy-* = fast) means a rapid resting heart or pulse rate—over 100 beats/min. **Bradycardia** (brād'-i-KAR-dē-a; *brady-* = slow) indicates a slow resting heart or pulse rate—under 60 beats/min. Endurance-trained athletes normally exhibit bradycardia.

Figure 21.16 Relationship of blood pressure changes to cuff pressure.

🔑 As the cuff is deflated, sounds first occur at the systolic blood pressure; the sounds suddenly become faint at the diastolic blood pressure.

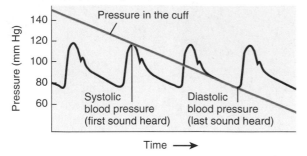

Q If a blood pressure is reported as "142 over 95," what are the diastolic, systolic, and pulse pressures? Does this person have hypertension as defined on page 732?

Measuring Blood Pressure

Blood pressure is usually measured in the brachial artery using a **sphygmomanometer** (sfig'-mō-ma-NOM-e-ter; *sphygmo-* = pulse; *manometer* = instrument used to measure pressure). A common sphygmomanometer consists of a rubber cuff attached by a rubber tube to a compressible hand pump or bulb used to inflate the cuff. Another tube attaches to the cuff to a column of mercury or a pressure dial marked off in millimeters of mercury to measure the pressure. The cuff is wrapped around the arm over the brachial artery and inflated until the pressure in the cuff exceeds the pressure in the artery. At this point, the walls of the brachial artery are compressed tightly against each other, and no blood can flow through. Two signs can confirm that the artery is occluded: (1) No sounds can be heard if a stethoscope is placed over the artery below the cuff, and (2) no pulse can be felt by placing the fingers over the radial artery at the wrist.

Next, the cuff is deflated gradually until the pressure in the cuff is slightly less than the maximal pressure in the brachial artery. At this point, the artery opens, a spurt of blood passes through, and a sound resulting from the turbulence of blood flow may be heard through the stethoscope. When this first sound is heard, the reading on the mercury column or dial corresponds to **systolic blood pressure (SBP)**—the highest pressure attained by blood as a result of ventricular contraction (Figure 21.16). As cuff pressure is further reduced, the sounds suddenly become faint as blood turbulence declines significantly. The pressure reading when the sounds suddenly become faint is **diastolic blood pressure (DBP);** it corresponds to the lowest blood pressure that occurs in arteries during ventricular relaxation. Whereas systolic pressure reflects the force of left ventricular contraction, diastolic pressure provides information about systemic vascular resistance. At pressures below diastolic blood pressure, sounds disappear altogether. The various sounds that are heard while taking blood pressure are called **Korotkoff sounds** (kō-ROT-kof).

Although some people may have a lower or higher blood pressure, the normal blood pressure of a young adult male is about 120 mm Hg systolic and 80 mm Hg diastolic, reported as "120 over 80" and written as 120/80. In young adult females, the pressures are 8–10 mm Hg lower. People who exercise regularly and are in good physical condition also tend to have lower blood pressures. Thus, blood pressure slightly lower than 120/80 may be a sign of good health and fitness.

The difference between systolic and diastolic pressure is called **pulse pressure.** This pressure, which averages 40 mm Hg, provides information about the condition of the cardiovascular system. For example, conditions such as atherosclerosis and patent (open) ductus arteriosus greatly increase pulse pressure. The normal ratio of systolic pressure to diastolic pressure to pulse pressure is about 3:2:1.

1. Where may the pulse be felt?
2. Define tachycardia and bradycardia.
3. Explain how systolic and diastolic blood pressures are measured with a sphygmomanometer.

CIRCULATORY ROUTES

OBJECTIVE
• *Describe the two main circulatory routes in adults.*

Arteries, arterioles, capillaries, venules, and veins are organized into routes that deliver blood throughout the body. Figure 21.17 shows these **circulatory routes** of blood flow. The routes are parallel—in most cases a portion of the cardiac output flows separately to each tissue of the body. Each organ receives its own supply of freshly oxygenated blood. The two basic routes for blood flow are the systemic and pulmonary circulations. The **systemic circulation** includes all the arteries and arterioles that carry oxygenated blood from the left ventricle to systemic capillaries, plus the veins and venules that return deoxygenated blood to the right atrium. The bronchial arteries, which carry nutrients to the lungs, also are part of the systemic circulation. Blood leaving the aorta and flowing through the systemic arteries is bright red. As it moves through capillaries, it loses some of its oxygen and picks up carbon dioxide, becoming darker red.

When blood returns to the heart from the systemic circulation, it is pumped out of the right ventricle through the **pulmonary circulation** to the lungs (see Figure 21.30). In pulmonary capillaries surrounding air sacs (alveoli) of the lungs, blood loses some of its carbon dioxide and takes up oxygen. Bright red again, it returns to the left atrium of the heart and reenters the systemic circulation as it is pumped out by the left ventricle.

Another major route—the **fetal circulation**—exists only in the fetus and contains special structures that allow the developing fetus to exchange materials with its mother (see Figure 21.31).

Figure 21.17 Circulatory routes. Heavy black arrows indicate the systemic circulation (detailed in Exhibits 21.3–21.12), thin black arrows the pulmonary circulation (detailed in Figure 21.30), and red arrows the hepatic portal circulation (detailed in Figure 21.29). Refer to Figure 20.8 on page 647 for details of the coronary circulation, and to Figure 21.31 for details of the fetal circulation.

🔑 **Blood vessels are organized into various routes that deliver blood to tissues of the body.**

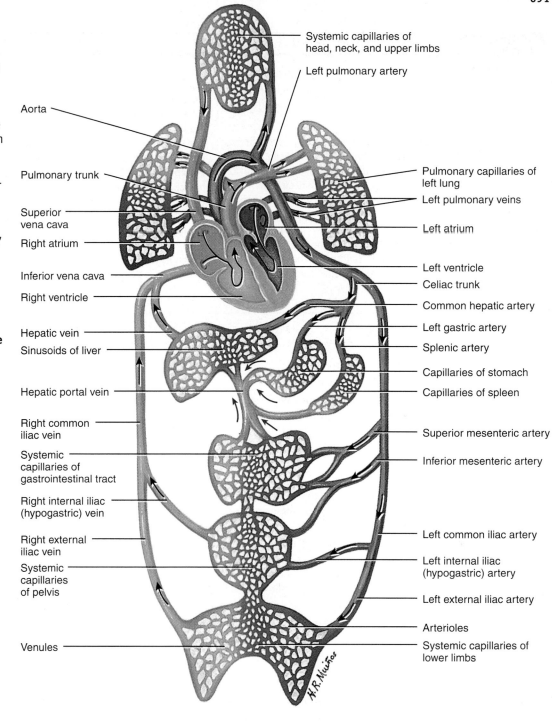

Q What are the two main circulatory routes?

The Systemic Circulation

All systemic arteries branch from the **aorta.** Completing the circuit, all the veins of the systemic circulation drain into the **superior vena cava,** the **inferior vena cava,** or the **coronary sinus,** which in turn empty into the right atrium.

The principal arteries and veins of the systemic circulation are described and illustrated in Exhibits 21.1–21.12 and Figures 21.18–21.28. The blood vessels are organized in the exhibits according to body regions. Figure 21.18a presents an overview of the major arteries, and Figure 21.24 provides an overview of the major veins. As you study the various blood vessels in the exhibits, refer to these two figures to see the relationships of the blood vessels under consideration to other regions of the body.

Exhibit 21.1   The Aorta and Its Branches (Figure 21.18)

OBJECTIVE

• *Identify the four principal divisions of the aorta and locate the major arterial branches arising from each division.*

The **aorta** (*aortae* = to lift up) is the largest artery of the body, with a diameter of 2–3 cm (about 1 in.). Its four principal divisions are the ascending aorta, arch of the aorta, thoracic aorta, and abdominal aorta. The portion of the aorta that emerges from the left ventricle posterior to the pulmonary trunk is the **ascending aorta.** The beginning of the aorta contains the aortic semilunar valve (see Figure 20.2b on page 639). The ascending aorta gives off two coronary artery branches that supply the myocardium of the heart. Then it turns to the left, forming the **arch of the aorta**, which descends and ends at the level of the intervertebral

disc between the fourth and fifth thoracic vertebrae. As the aorta continues to descend, it lies close to the vertebral bodies, passes through the diaphragm, and divides at the level of the fourth lumbar vertebra into two **common iliac arteries,** which carry blood to the lower limbs. The section of the aorta between the arch of the aorta and the diaphragm is called the **thoracic aorta;** the section between the diaphragm and the common iliac arteries is the **abdominal aorta.** Each division of the aorta gives off arteries that branch into distributing arteries that lead to organs. Within the organs, the arteries divide into arterioles and then into capillaries that service the systemic tissues (all tissues except the alveoli of the lungs).

What general regions are supplied by each of the four principal divisions of the aorta?

DIVISION AND BRANCHES	REGION SUPPLIED
Ascending Aorta	
Right and left coronary arteries	Heart
Arch of the Aorta	
Bracheocephalic trunk (brā′-kē-ō-se-FAL-ik)	
Right common carotid artery (ka-ROT-id)	Right side of head and neck.
Right subclavian artery (sub-KLĀ-vē-an)	Right upper limb.
Left common carotid artery	Left side of head and neck.
Left subclavian artery	Left upper limb.
Thoracic Aorta (*thorac-* = chest)	
Intercostal arteries (in′-ter-KOS-tal)	Intercostal and chest muscles and plurae.
Superior phrenic arteries (FREN-ik)	Posterior and superior surfaces of diaphragm.
Bronchial arteries (BRONG-kē-al)	Bronchi of lungs.
Esophageal arteries (e-sof′-a-JĒ-al)	Esophagus.
Abdominal Aorta	
Inferior phrenic arteries (FREN-ik)	Inferior surface of diaphragm.
Celiac trunk (SĒ-lē-ak)	
Common hepatic artery (he-PAT-ik)	Liver.
Left gastric artery (GAS-trik)	Stomach and esophagus.
Splenic artery (SPĒN-ik)	Spleen, pancreas, and stomach.
Superior mesenteric artery (MES-en-ter′-ik)	Small intestine, cecum, ascending and transverse colons, and pancreas.
Suprarenal arteries (soo′-pra-RĒ-nal)	Adrenal (suprarenal) glands.
Renal arteries (RĒ-nal)	Kidneys.
Gonadal arteries (gō-NAD-al)	
Testicular arteries (tes-TIK-yoo-lar)	Testes (male).
Ovarian arteries (ō-VAR-ē-an)	Ovaries (female).
Inferior mesenteric artery	Transverse, descending, and sigmoid colons; rectum.
Common iliac arteries (IL-ē-ak)	
External iliac arteries	Lower limbs.
Interal iliac (hypogastric) arteries	Uterus (female), prostate gland (male), muscles of buttocks, and urinary bladder.

Exhibit 21.1 *(continued)*

Figure 21.18 Aorta and its principal branches.

🔑 **All systemic arteries branch from the aorta.**

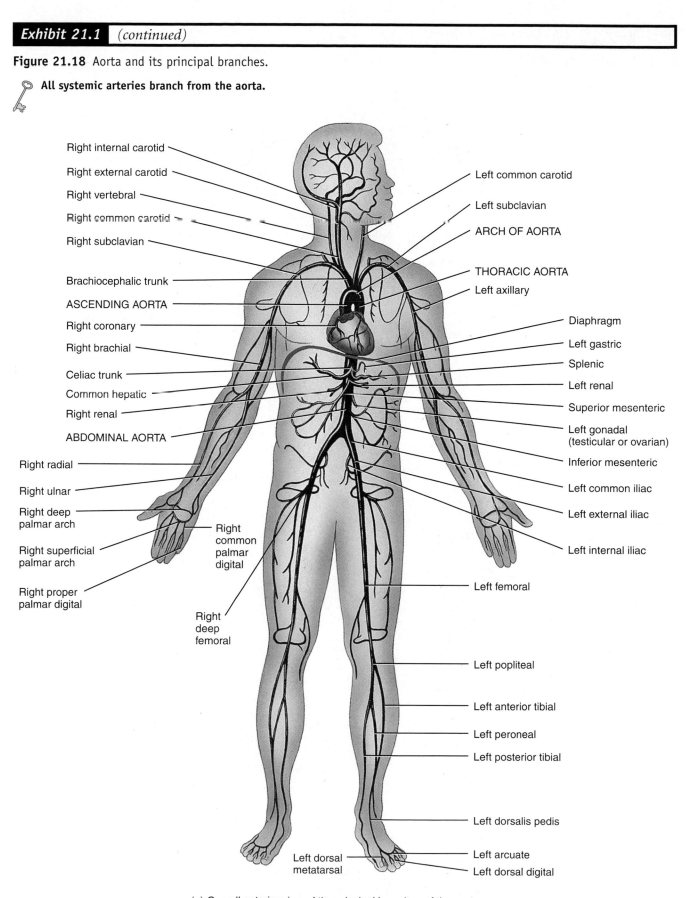

(a) Overall anterior view of the principal branches of the aorta

Exhibit 21.1 | *The Aorta and Its Branches (continued)*

Figure 21.18 (continued)

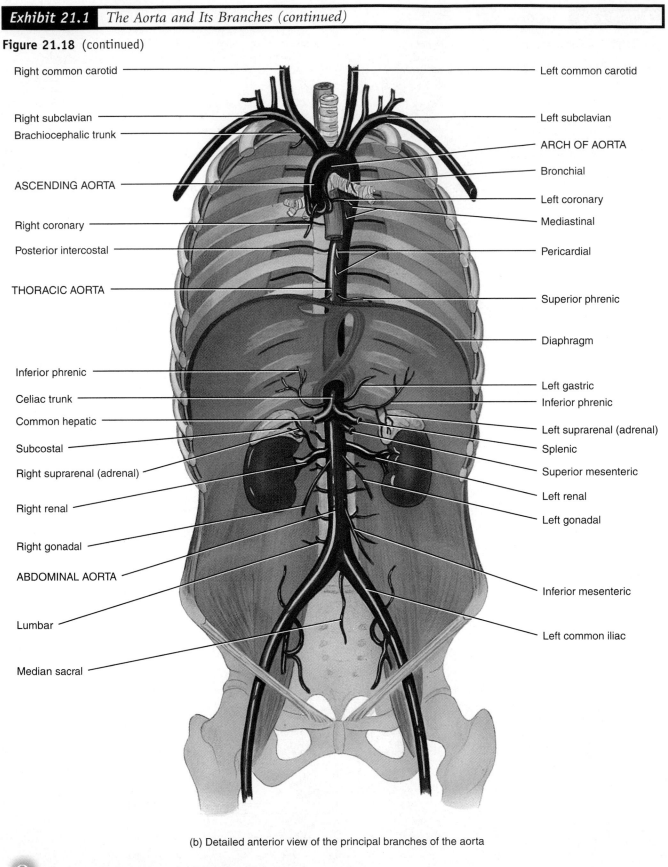

(b) Detailed anterior view of the principal branches of the aorta

Q What are the four principal subdivisions of the aorta?

Exhibit 21.2 | *Ascending Aorta (Figure 21.19)*

OBJECTIVE

- *Identify the two primary arterial branches of the ascending aorta.*

The **ascending aorta** is about 5 cm (2 in.) in length and begins at the aortic semilunar valve. It is directed superiorly, slightly anteriorly, and to the right; it ends at the level of the sternal angle, where it becomes the arch of the aorta. The beginning of the ascending aorta is posterior to the pulmonary trunk and right auricle; the right pulmonary artery is posterior to it. At its origin, the ascending aorta contains three dilations called aortic sinuses. Two of these, the right and left sinuses, give rise to the right and left coronary arteries, respectively.

The right and left **coronary arteries** (*coron-* = crown) arise from the ascending aorta just superior to the aortic semilunar valve. They form a crownlike ring around the heart, giving off

branches to the atrial and ventricular myocardium. The **posterior interventricular branch** (in-ter-ven-TRIK-yoo-lar; *inter-* = between) of the right coronary artery supplies both ventricles, and the **marginal branch** supplies the right ventricle. The **anterior interventricular branch** of the left coronary artery supplies both ventricles, and the **circumflex branch** (SER-kum-flex; *circum-* = around; *-flex* = to bend) supplies the left atrium and left ventricle.

Which branches of the coronary arteries supply the left ventricle? Why does the left ventricle have such an extensive arterial blood supply?

Figure 21.19 Ascending aorta and its branches.

🔑 **The ascending aorta is the first division of the aorta.**

SCHEME OF DISTRIBUTION

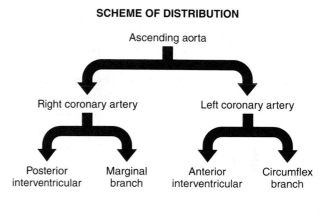

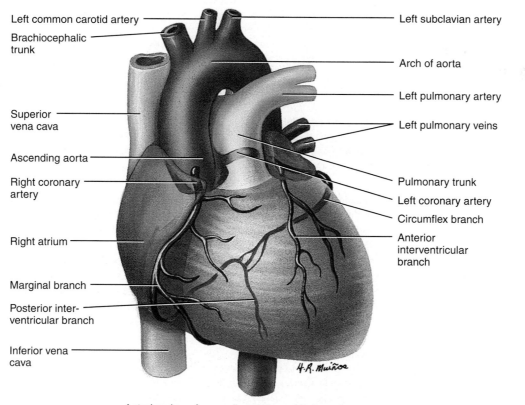

Anterior view of ascending aorta and its branches

Q Which arteries arise from the ascending aorta?

Exhibit 21.3   *The Arch of the Aorta (Figure 21.20)*

OBJECTIVE

• *Identify the three principal arteries that branch from the arch of the aorta.*

The arch of the aorta is 4–5 cm (almost 2 in.) in length and is the continuation of the ascending aorta. It emerges from the pericardium posterior to the sternum at the level of the sternal angle. The arch is directed superiorly and posteriorly to the left and then inferiorly; it ends at the intervertebral disc between the fourth and fifth thoracic vertebrae, where it becomes the thoracic aorta. Three major arteries branch from the superior aspect of the arch of the aorta: the brachiocephalic trunk, the left common carotid, and the left subclavian. The first and largest branch from the arch of the aorta is the **brachiocephalic trunk** (brā-kē-ō-se-FAL-ik; *brachio-* = arm; *-cephalic* = head). It extends superiorly, bending slightly to the right, and divides at the right sternoclavicular joint to form the right subclavian artery and right common carotid artery. The second branch from the arch of the aorta is the **left common carotid artery** (ka-ROT-id), which divides into basically the same branches with the same names as the right common carotid artery. The third branch from the arch of the aorta is the **left subclavian artery** (sub-KLĀ-vē-an), which distributes blood to the left vertebral artery and vessels of the left upper limb. Arteries branching from the left subclavian artery are similar in distribution and name to those branching from the right subclavian artery. The accompanying table focuses on the principal arteries originating from the brachiocephalic trunk.

> What general regions are supplied by the arteries that arise from the arch of the aorta?

BRANCH	DESCRIPTION AND REGION SUPPLIED
Brachiocephalic trunk	The **brachiocephalic trunk** divides to form the right subclavian artery and right common carotid artery (Figure 21.20a).
Right subclavian artery (sub-KLĀ-vē-an)	The **right subclavian artery** extends from the brachiocephalic trunk to the first rib and then passes into the armpit (axilla). The general distribution of the artery is to the brain and spinal cord, neck, shoulder, thoracic viscera and wall, and scapular muscles.
Axillary artery (AK-si-ler-ē; = armpit)	Continuation of the right subclavian artery into the axilla is called the **axillary artery.** (Note that the right subclavian artery, which passes deep to the clavicle, is a good example of the practice of giving the same vessel different names as it passes through different regions.) Its general distribution is the shoulder, thoracic and scapular muscles, and humerus.
Brachial artery (BRĀ-kē-al; = arm)	The **brachial artery** is the continuation of the axillary artery into the arm. The brachial artery provides the main blood supply to the arm and is superficial and palpable along its course. It begins at the tendon of the teres major muscle and ends just distal to the bend of the elbow. At first, the brachial artery is medial to the humerus, but as it descends it gradually curves laterally and passes through the cubital fossa, a triangular depression anterior to the elbow where you can easily detect the pulse of the brachial artery and listen to the various sounds when taking a person's blood pressure. Just distal to the bend in the elbow, the brachial artery divides into the radial artery and ulnar artery.
Radial artery (RĀ-dē-al; = radius)	The **radial artery** is the smaller branch and is a direct continuation of the brachial artery. It passes along the lateral (radial) aspect of the forearm and then through the wrist and hand, supplying these structures with blood. At the wrist, the radial artery comes into contact with the distal end of the radius, where it is covered only by fascia and skin. Because of its superficial location at this point, it is a common site for measuring radial pulse.
Ulnar artery (UL-nar; = ulna)	The **ulnar artery,** the larger branch of the brachial artery, passes along the medial (ulnar) aspect of the forearm and then into the wrist and hand, supplying these structures with blood. In the palm, branches of the radial and ulnar arteries anastomose to form the superficial palmar arch and the deep palmar arch.
Superficial palmar arch (*palma* = palm)	The **superficial palmar arch** is formed mainly by the ulnar artery, with a contribution from a branch of the radial artery. The arch is superficial to the long flexor tendons of the fingers and extends across the palm at the bases of the metacarpals. It gives rise to **common palmar digital arteries,** which supply the palm. Each divides into a pair of **proper palmar digital arteries,** which supply the fingers.
Deep palmar arch	The **deep palmar arch** is formed mainly by the radial artery, with a contribution from a branch of the ulnar artery. The arch is deep to the long flexor tendons of the fingers and extends across the palm, just distal to the bases of the metacarpals. Arising from the deep palmar arch are **palmar metacarpal arteries,** which supply the palm and anastomose with the common palmar digital arteries of the superficial palmar arch.

Exhibit 21.3 *(continued)*

BRANCH	DESCRIPTION AND REGION SUPPLIED
Vertebral artery (VER-te-bral)	Before passing into the axilla, the right subclavian artery gives off a major branch to the brain called the **right vertebral artery** (Figure 21.20b). The right vertebral artery passes through the foramina of the transverse processes of the sixth through first cervical vertebrae and enters the skull through the foramen magnum to reach the inferior surface of the brain. Here it unites with the left vertebral artery to form the **basilar** (BAS-i-lar) **artery.** The vertebral artery supplies the posterior portion of the brain with blood. The basilar artery passes along the midline of the anterior aspect of the brain stem and supplies the cerebellum and pons of the brain and the inner ear.
Right common carotid artery	The **right common carotid artery** begins at the bifurcation of the brachiocephalic trunk, posterior to the right sternoclavicular joint, and passes superiorly in the neck to supply structures in the head (Figure 21.20b). At the superior border of the larynx (voice box), it divides into the right external and right internal carotid arteries.
External carotid artery	The **external carotid artery** begins at the superior border of the larynx and terminates near the temporomandibular joint in the substance of the parotid gland, where it divides into two branches: the superficial temporal and maxillary arteries. The carotid pulse can be detected in the external carotid artery just anterior to the sternocleidomastoid muscle at the superior border of the larynx. The general distribution of the external carotid artery is to structures *external* to the skull.
Internal carotid artery	The **internal carotid artery** has no branches in the neck and supplies structures *internal* to the skull. It enters the cranial cavity through the carotid foramen in the temporal bone. The internal carotid artery supplies blood to the eyeball and other orbital structures, ear, most of the cerebrum of the brain, pituitary gland, and external nose. The terminal branches of the internal carotid artery are the **anterior cerebral artery,** which supplies most of the medial surface of the cerebrum, and the **middle cerebral artery,** which supplies most of the lateral surface of the cerebrum (Figure 21.20c). Inside the cranium, anastomoses of the left and right internal carotid arteries along with the basilar artery form an arrangement of blood vessels at the base of the brain near the sella turcica called the **cerebral arterial circle (circle of Willis).** From this circle (Figure 21.20c) arise arteries supplying most of the brain. Essentially, the cerebral arterial circle is formed by the union of the **anterior cerebral arteries** (branches of internal carotids) and **posterior cerebral arteries** (branches of basilar artery). The posterior cerebral arteries are connected with the internal carotid arteries by the **posterior communicating arteries** (ko-MYOO-ni-kā'-ting). The anterior cerebral arteries are connected by the **anterior communicating arteries.** The **internal carotid arteries** are also considered part of the cerebral arterial circle. The functions of the cerebral arterial circle are to equalize blood pressure to the brain and provide alternate routes for blood flow to the brain, should the arteries become damaged.
Left common carotid artery	See description above, on page 696.
Left subclavian artery	See description above, on page 696.

Exhibit 21.3 *The Arch of the Aorta (continued)*

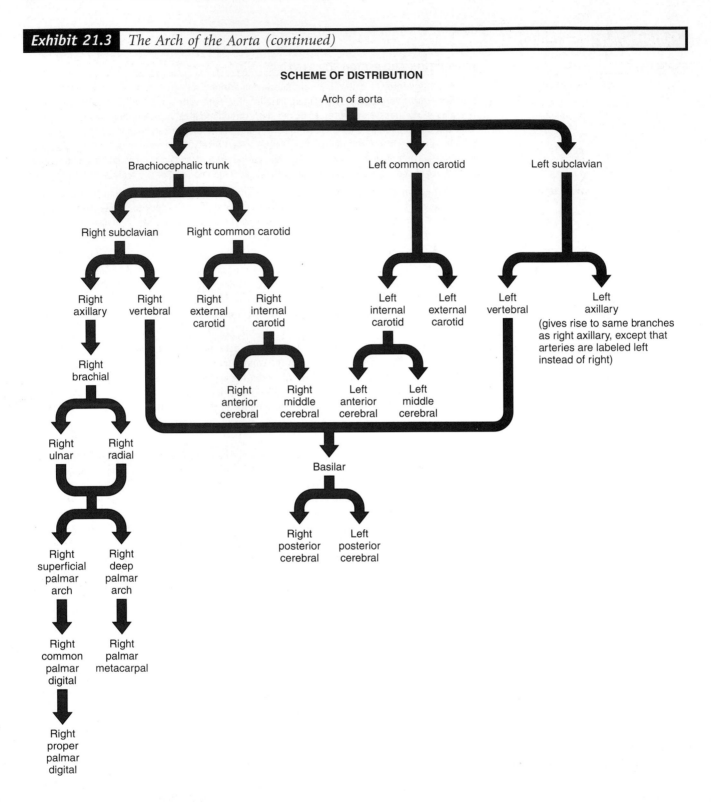

Exhibit 21.3 (continued)

Figure 21.20 Arch of the aorta and its branches. Note in (c) the arteries that constitute the cerebral arterial circle (circle of Willis).

🔑 **The arch of the aorta ends at the level of the intervertebral disc between the fourth and fifth thoracic vertebrae.**

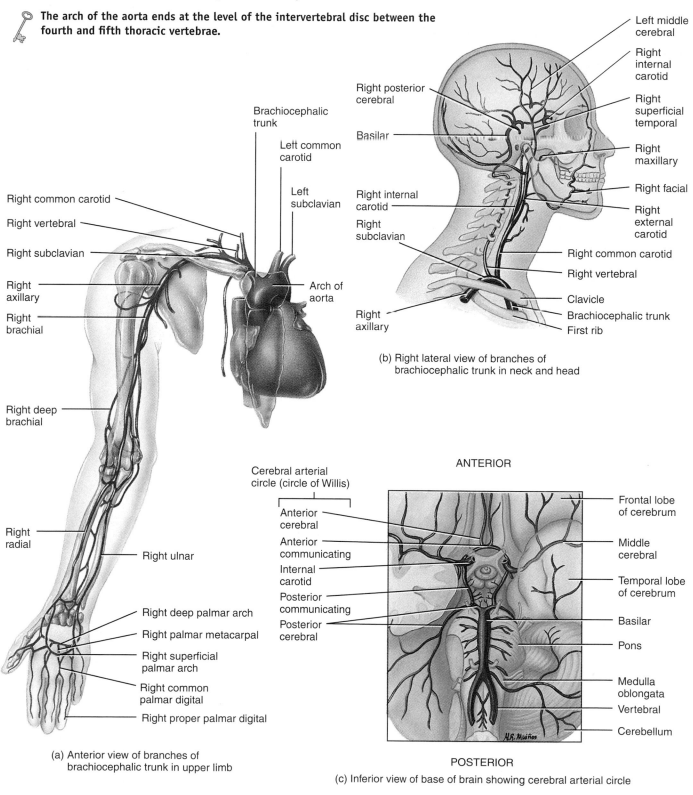

(a) Anterior view of branches of brachiocephalic trunk in upper limb

(b) Right lateral view of branches of brachiocephalic trunk in neck and head

(c) Inferior view of base of brain showing cerebral arterial circle

Ⓠ What are the three major branches of the arch of the aorta, in order of their origination?

Exhibit 21.4 | Thoracic Aorta (Figure 21.21)

OBJECTIVE

• *Identify the visceral and parietal branches of the thoracic aorta.*

The **thoracic aorta** is about 20 cm (8 in.) long and is a continuation of the arch of the aorta. It begins at the level of the intervertebral disc between the fourth and fifth thoracic vertebrae, where it lies to the left of the vertebral column. As it descends, it moves closer to the midline and ends at an opening in the diaphragm (aortic hiatus) anterior to the vertebral column at the level of the intervertebral disc between the twelfth thoracic and first lumbar vertebrae.

Along its course the thoracic aorta sends off numerous small arteries, **visceral branches** to viscera and **parietal branches** to body wall structures.

What general regions are supplied by the visceral and parietal branches of the thoracic aorta?

BRANCH	DESCRIPTION AND REGION SUPPLIED
Visceral	
Pericardial arteries (per'-i-KAR-dē-al); *peri-* = around; *cardia-* = heart)	Two or three minute **pericardial arteries** supply blood to the pericardium.
Bronchial arteries (BRONG-kē-al; = windpipe)	One right and two left **bronchial arteries** supply the bronchial tubes, pleurae, bronchial lymph nodes, and esophagus. (Whereas the right bronchial artery arises from the third posterior intercostal artery, the two left bronchial arteries arise from the thoracic aorta).
Esophageal arteries (e-sof'-a-JĒ-al; *oisein* = to carry; *phage-* = food)	Four or five **esophageal arteries** supply the esophagus.
Mediastinal arteries (mē'-dē-as-TĪ-nal)	Numerous small **mediastinal arteries** supply blood to structures in the mediastinum.
Parietal	
Posterior intercostal arteries (in'-ter-KOS-tal; *inter-* = between; *costa* = rib)	Nine pairs of **posterior intercostal arteries** supply the intercostal, pectoralis major and minor, and serratus anterior muscles; overlying subcutaneous tissue and skin; mammary glands; and vertebrae, meninges, and spinal cord.
Subcostal arteries (sub-KOS-tal; *sub-* = under)	The left and right **subcostal arteries** have a distribution similar to that of the posterior intercostals.
Superior phrenic arteries (FREN-ik; = diaphragm)	Small **superior phrenic arteries** supply the superior and posterior surfaces of the diaphragm.

SCHEME OF DISTRIBUTION

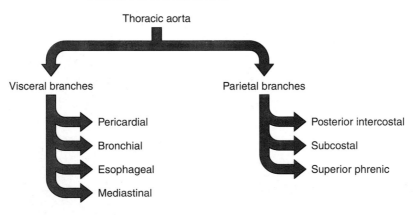

Exhibit 21.4 *(continued)*

Figure 21.21 Thoracic and abdominal aorta and their principal branches.

The thoracic aorta is the continuation of the ascending aorta.

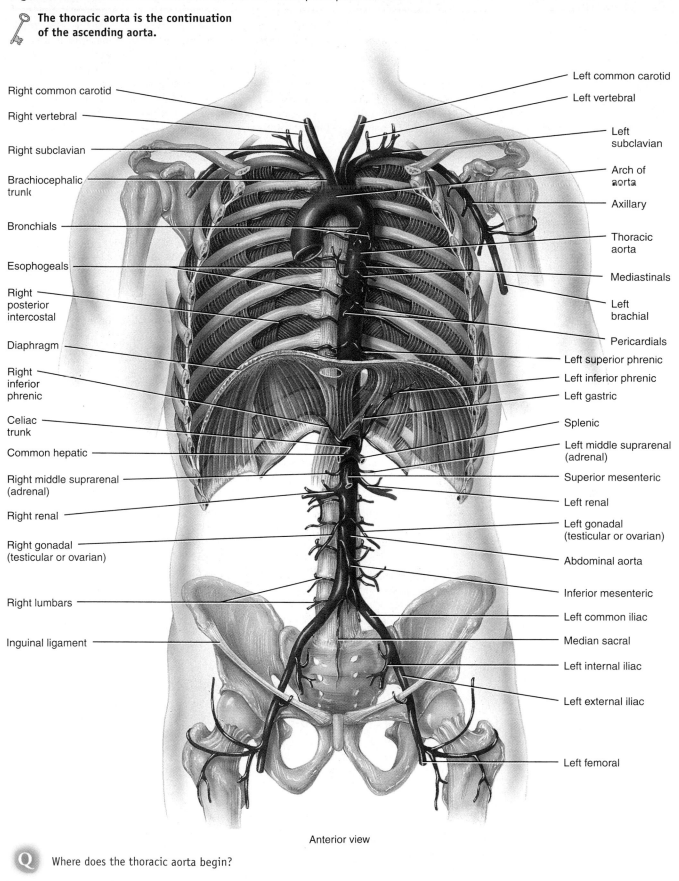

Right common carotid

Right vertebral

Right subclavian

Brachiocephalic trunk

Bronchials

Esophogeals

Right posterior intercostal

Diaphragm

Right inferior phrenic

Celiac trunk

Common hepatic

Right middle suprarenal (adrenal)

Right renal

Right gonadal (testicular or ovarian)

Right lumbars

Inguinal ligament

Left common carotid

Left vertebral

Left subclavian

Arch of aorta

Axillary

Thoracic aorta

Mediastinals

Left brachial

Pericardials

Left superior phrenic

Left inferior phrenic

Left gastric

Splenic

Left middle suprarenal (adrenal)

Superior mesenteric

Left renal

Left gonadal (testicular or ovarian)

Abdominal aorta

Inferior mesenteric

Left common iliac

Median sacral

Left internal iliac

Left external iliac

Left femoral

Anterior view

Q Where does the thoracic aorta begin?

Exhibit 21.5   *Abdominal Aorta (Figure 21.22)*

OBJECTIVE

• *Identify the visceral and parietal branches of the abdominal aorta.*

The **abdominal aorta** (ab-DOM-i-nal) is the continuation of the thoracic aorta. It begins at the aortic hiatus in the diaphragm and ends at about the level of the fourth lumbar vertebra, where it divides into the right and left common iliac arteries. The abdominal aorta lies anterior to the vertebral column.

As with the thoracic aorta, the abdominal aorta gives off visceral and parietal branches. The unpaired visceral branches arise from the anterior surface of the aorta and include the **celiac trunk** and the **superior mesenteric** and **inferior mesenteric arteries** (see Figure 21.21). The paired visceral branches arise from the lateral surfaces of the aorta and include the **suprarenal, renal,** and **gonadal arteries.** The unpaired parietal branch is the **median sacral artery.** The paired parietal branches arise from the posterolateral surfaces of the aorta and include the **inferior phrenic** and **lumbar arteries.**

Name the paired visceral and parietal branches and the unpaired visceral and parietal branches of the abdominal aorta, and indicate the general regions they supply.

BRANCH	DESCRIPTION AND REGION SUPPLIED
Unpaired Visceral Branches	
Celiac trunk (SĒ-lē-ak)	The **celiac trunk (artery)** is the first visceral branch from the aorta inferior to the diaphragm, at about the level of the twelfth thoracic vertebra (Figure 21.22a). Almost immediately, the celiac trunk divides into three branches—the left gastric, splenic, and common hepatic arteries (Figure 21.22a).

1. The **left gastric artery** (GAS-trik; = stomach) is the smallest of the three branches. It passes superiorly to the left toward the esophagus and then turns to follow the lesser curvature of the stomach. It supplies the stomach and esophagus.
2. The **splenic artery** (SPLĒN-ik) is the largest branch of the celiac trunk. It arises from the left side of the celiac trunk distal to the left gastric artery, passes horizontally to the left along the pancreas to reach the spleen, and gives rise to three arteries:
   • **Pancreatic artery** (pan-krē-AT-ik), which supplies the pancreas.
   • **Left gastroepiploic artery** (gas'-trō-ep'-i-PLŌ-ik; *epiplo-* = omentum), which supplies the stomach and greater omentum.
   • **Short gastric artery,** which supplies the stomach.
3. The **common hepatic artery** (he-PAT-ik; = liver) is intermediate in size between the left gastric and splenic arteries. Unlike the other two branches of the celiac trunk, the common hepatic artery arises from the right side; it gives rise to three arteries:
   • **Proper hepatic artery,** which supplies the liver, gallbladder, and stomach.
   • **Right gastric artery,** which supplies the stomach.
   • **Gastroduodenal artery** (gas'-trō-doo'-ō-DĒ-nal), which supplies the stomach, duodenum of the small intestine, pancreas, and greater omentum.

Superior mesenteric (MES-en-ter'-ik; *meso-* = middle; *-enteric* = intestines)

The **superior mesenteric artery** (Figure 21.22b) arises from the anterior surface of the abdominal aorta about 1 cm inferior to the celiac trunk at the level of the first lumbar vertebra. It extends inferiorly and anteriorly and between the layers of mesentery, which is a portion of the peritoneum that attaches the small intestine to the posterior abdominal wall. It anastomoses extensively and has five branches:

1. The **inferior pancreaticoduodenal artery** (pan'-krē-at'-i-kō-doo'-ō-DĒ-nal) supplies the pancreas and duodenum.
2. The **jejunal** (je-JOO-nal) and **ileal arteries** (IL-ē-al) supply the jejunum and ileum of the small intestine, respectively.
3. The **ileocolic artery** (il'ē-ō-KŌL-ik) supplies the ileum and ascending colon of the large intestine.
4. The **right colic artery** (KŌL-ik) supplies the ascending colon.
5. The **middle colic artery** supplies the transverse colon of the large intestine.

| **Exhibit 21.5** | *(continued)* |

BRANCH	DESCRIPTION AND REGION SUPPLIED
Inferior mesenteric	The **inferior mesenteric artery** (Figure 21.22c) arises from the anterior aspect of the abdominal aorta at the level of the third lumbar vertebra and then passes inferiorly to the left of the aorta. It anastomoses extensively and has three branches:

1. The **left colic artery** supplies the transverse colon and descending colon of the large intestine.

2. The **sigmoid arteries** (SIG-moyd) supply the descending colon and sigmoid colon of the large intestine.

3. The **superior rectal artery** (REK-tal) supplies the rectum of the large intestine.

Paired Visceral Branches

Suprarenal arteries (soo'-pra-RĒ-nal; *supra-* = above; *ren-* = kidney)	Although there are three pairs of **suprarenal (adrenal) arteries** that supply the adrenal (suprarenal) glands (superior, middle, and inferior), only the middle pair originates directly from the abdominal aorta (see Figure 21.21). The middle suprarenal arteries arise at the level of the first lumbar vertebra at or superior to the renal arteries. The superior suprarenal arteries arise from the inferior phrenic artery, and the inferior suprarenal arteries originate from the renal arteries.
Renal arteries (RĒ-nal; = kidney)	The right and left **renal arteries** usually arise from the lateral aspects of the abdominal aorta at the superior border of the second lumbar vertebra, about 1 cm inferior to the superior mesenteric artery (see Figure 21.21). The right renal vein, which is longer than the left, arises slightly lower than the left and passes posterior to the right renal vein and inferior vena cava. The left renal artery is posterior to the left renal vein and is crossed by the inferior mesenteric vein. The renal arteries carry blood to the kidneys, adrenal (suprarenal) glands, and ureters. Their distribution within the kidneys is discussed in Chapter 26.
Gonadal (gō-NAD-al; *gon-* = seed) **[testicular** (test-TIK-yoo-lar) **or ovarian arteries** (ō-VA-rē-an)**]**	The **gonadal arteries** arise from the abdominal aorta at the level of the second lumbar vertebra just inferior to the renal arteries (see Figure 21.21). In males, the gonadal arteries are specifically referred to as the **testicular arteries.** They pass through the inguinal canal and supply the testes, epididymis, and ureters. In females, the gonadal arteries are called the **ovarian arteries.** They are much shorter than the testicular arteries and supply the ovaries, uterine (Fallopian) tubes, and ureters.

Unpaired Parietal Branch

| **Median sacral artery** (SĀ-kral; = sacrum) | The **median sacral artery** arises from the posterior surface of the abdominal aorta about 1 cm superior to the bifurcation of the aorta into the right and left common iliac arteries (see Figure 21.21). The median sacral artery supplies the sacrum and coccyx. |

Paired Parietal Branches

| **Inferior phrenic arteries** (FREN-ik; = diaphragm) | The **inferior phrenic arteries** are the first paired branches of the abdominal aorta, immediately superior to the origin of the celiac trunk (see Figure 21.21). (They may also arise from the renal arteries.) The inferior phrenic arteries are distributed to the inferior surface of the diaphragm and adrenal (suprarenal) glands. |
| **Lumbar arteries** (LUM-bar; = loin) | The four pairs of **lumbar arteries** arise from the posterolateral surface of the abdominal aorta (see Figure 21.21). They supply the lumbar vertebrae, spinal cord and its meninges, and the muscles and skin of the lumbar region of the back. |

Exhibit 21.5   *Abdominal Aorta (continued)*

SCHEME OF DISTRIBUTION

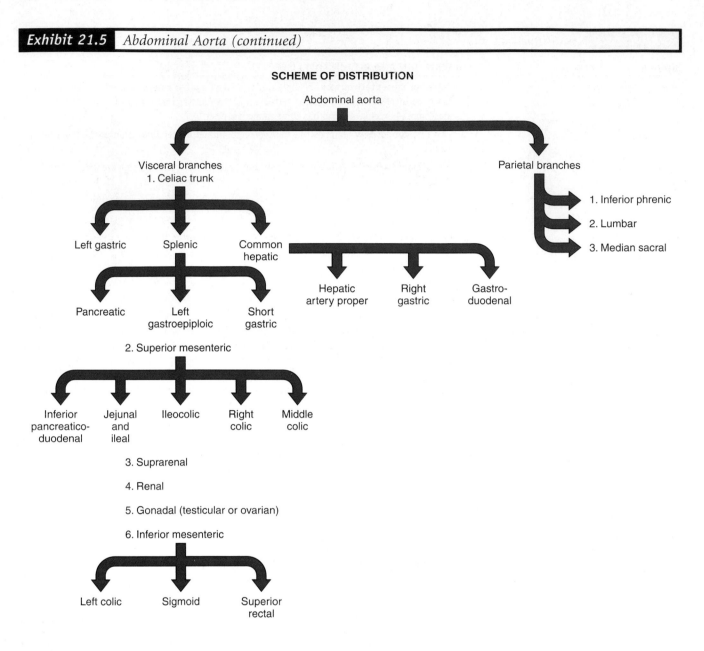

Exhibit 21.5   *(continued)*

Figure 21.22 Abdominal aorta and its principal branches.

🔑 **The abdominal aorta is the continuation of the thoracic aorta.**

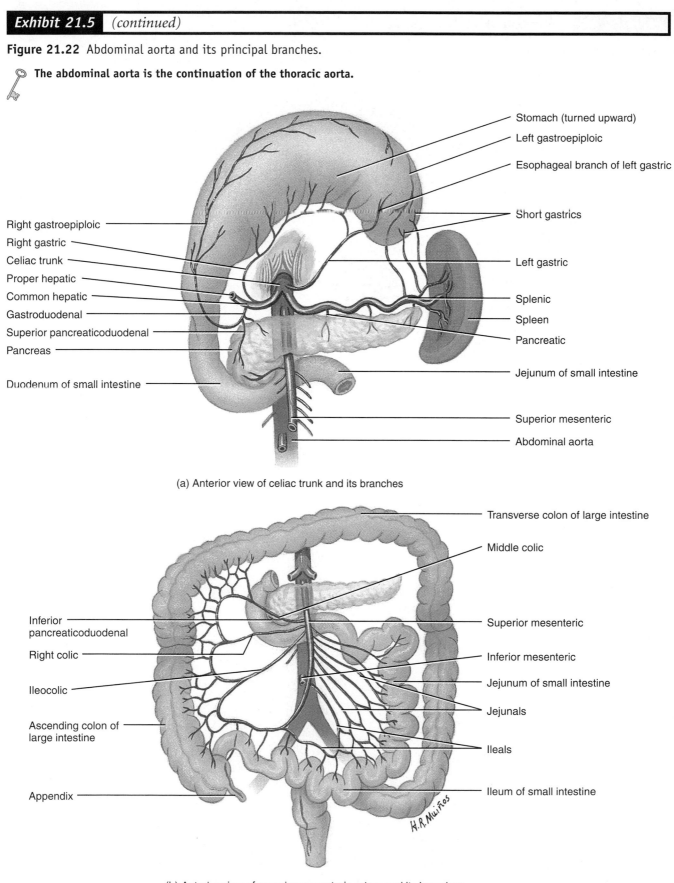

(a) Anterior view of celiac trunk and its branches

(b) Anterior view of superior mesenteric artery and its branches

Exhibit 21.5 *Abdominal Aorta (continued)*

Figure 21.22 (continued)

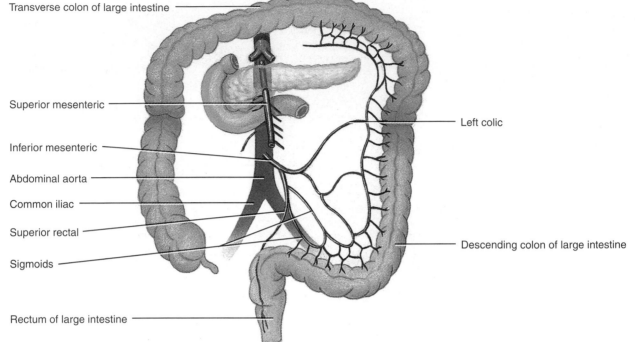

Transverse colon of large intestine

Superior mesenteric

Inferior mesenteric

Abdominal aorta

Common iliac

Superior rectal

Sigmoids

Rectum of large intestine

Left colic

Descending colon of large intestine

(c) Anterior view of inferior mesenteric artery and its branches

 Q Where does the abdominal aorta begin?

Exhibit 21.6 | Arteries of the Pelvis and Lower Limbs (Figure 21.23)

OBJECTIVE

• *Identify the two major branches of the common iliac arteries.*

The abdominal aorta ends by dividing into the right and left **common iliac arteries.** These, in turn, divide into the **internal** and **external iliac arteries.** In sequence, the external iliacs become the **femoral arteries** in the thighs, the **popliteal arteries** posterior to the knee, and the **anterior** and **posterior tibial arteries** in the legs.

What general regions are supplied by the internal and external iliac arteries?

BRANCH	DESCRIPTION AND REGION SUPPLIED
Common iliac arteries (IL-ē-ak; = ilium)	At about the level of the fourth lumbar vertebra, the abdominal aorta divides into the right and left **common iliac arteries,** the terminal branches of the abdominal aorta. Each passes inferiorly about 5 cm (2 in.) and gives rise to two branches: internal iliac and external iliac arteries. The general distribution of the common iliac arteries is to the pelvis, external genitals, and lower limbs.
Internal iliac arteries	The **internal iliac (hypogastric) arteries** are the primary arteries of the pelvis. They begin at the bifurcation of the common iliac arteries anterior to the sacroiliac joint at the level of the lumbosacral intervertebral disc. They pass posteromedially as they descend in the pelvis and divide into anterior and posterior divisions. The general distribution of the internal iliac arteries is to the pelvis, buttocks, external genitals, and thigh.
External iliac arteries	The **external iliac arteries** are larger than the internal iliac arteries. Like the internal iliac arteries, they begin at the bifurcation of the common iliac arteries. They descend along the medial border of the psoas major muscles following the pelvic brim, pass posterior to the midportion of the inguinal ligaments, and become the femoral arteries. The general distribution of the external iliac arteries is to the lower limbs. Specifically, branches of the external iliac arteries supply the muscles of the anterior abdominal wall, the cremasteric muscle in males and the round ligament of the uterus in females, and the lower limbs.
Femoral arteries (FEM-o-ral; = thigh)	The **femoral arteries** descend along the anteriomedial aspects of the thighs to the junction of the middle and lower third of the thighs. Here they pass through an opening in the tendon of the adductor magnus muscle, where they emerge posterior to the femurs as the popliteal arteries. A pulse may be felt in the femoral artery just inferior to the inguinal ligament. The general distribution of the femoral arteries is to the lower abdominal wall, groin, external genitals, and muscles of the thigh.
Popliteal arteries (pop'-li-TĒ-al; = posterior surface of the knee)	The **popliteal arteries** are the continuation of the femoral arteries through the popliteal fossa (space behind the knee). They descend to the inferior border of the popliteus muscles, where they divide into the anterior and posterior tibial arteries. A pulse may be detected in the popliteal arteries. In addition to supplying the adductor magnus and hamstring muscles and the skin on the posterior aspect of the legs, branches of the popliteal arteries also supply the gastrocnemius, soleus, and plantaris muscles of the calf, knee joint, femur, patella, and fibula.
Anterior tibial arteries (TIB-ē-al; = shin bone)	The **anterior tibial arteries** descend from the bifurcation of the popliteal arteries. They are smaller than the posterior tibial arteries. The anterior tibial arteries descend through the anterior muscular compartment of the leg. They pass through the interosseous membrane that connects the tibia and fibula, lateral to the tibia. The anterior tibial arteries supply the knee joints, anterior compartment muscles of the legs, skin over the anterior aspects of the legs, and ankle joints. At the ankles, the anterior tibial arteries become the **dorsalis pedis arteries** (PED-is; *ped* = foot), also arteries from which a pulse may be detected. The dorsalis pedis arteries supply the muscles, skin, and joints on the dorsal aspects of the feet. On the dorsum of the feet, the dorsalis pedis arteries give off a transverse branch at the first (medial) cuneiform bone called the **arcuate arteries** (*arcuat-* = bowed) that run laterally over the bases of the metatarsals. From the arcuate arteries branch **dorsal metatarsal arteries,** which supply the feet. The dorsal metatarsal arteries terminate by dividing into the **dorsal digital arteries,** which supply the toes.
Posterior tibial arteries	The **posterior tibial arteries,** the direct continuations of the popliteal arteries, descend from the bifurcation of the popliteal arteries. They pass down the posterior muscular compartment of the legs posterior to the medial malleolus of the tibia. They terminate by dividing into the medial and lateral plantar arteries. Their general distribution is to the muscles, bones, and joints of the leg and foot. Major branches of the posterior tibial arteries are the **peroneal arteries** (per'-o-NĒ-al; *perone* = fibula). They supply the peroneal, soleus, tibialis posterior, and flexor hallucis muscles, fibula, tarsus, and lateral aspect of the heel. The bifurcation of the posterior tibial arteries into the medial and lateral plantar arteries occurs deep to the flexor retinaculum on the medial side of the feet. The **medial plantar arteries** (PLAN-tar = sole of foot) supply the abductor hallucis and flexor digitorum brevis muscles and the toes. The **lateral plantar arteries** unite with a branch of the dorsalis pedis arteries to form the **plantar arch.** The arch begins at the base of the fifth metatarsal and extends medially across the metacarpals. As the arch crosses the foot, it gives off **plantar metatarsal arteries,** which supply the feet. These terminate by dividing into **plantar digital arteries,** which supply the toes.

Exhibit 21.6 *Arteries of the Pelvis and Lower Limbs (continued)*

SCHEME OF DISTRIBUTION

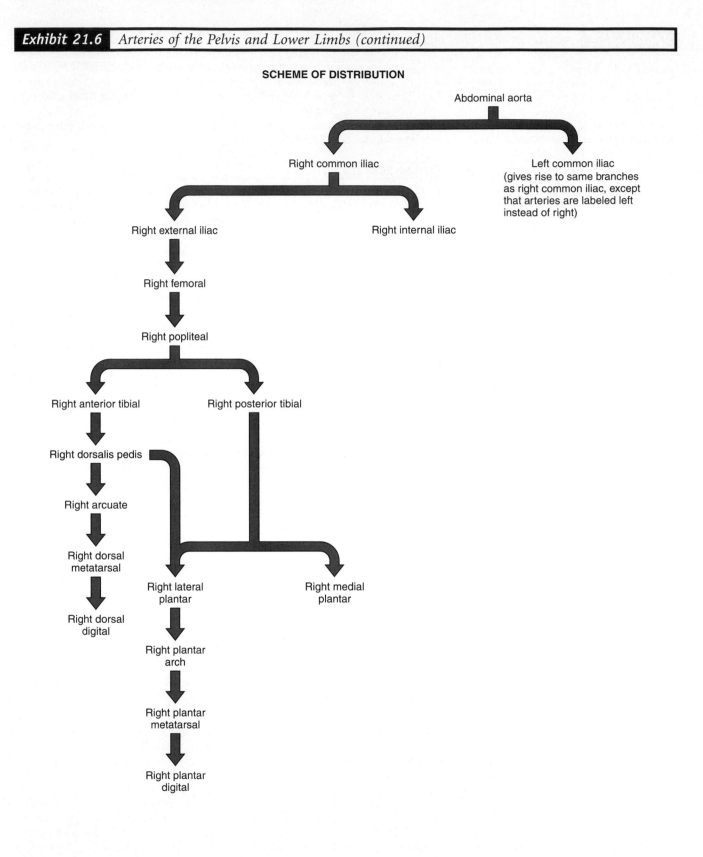

Exhibit 21.6 *(continued)*

Figure 21.23 Arteries of the pelvis and right lower limb.

🔑 **The internal iliac arteries carry most of the blood supply to the pelvic viscera and wall.**

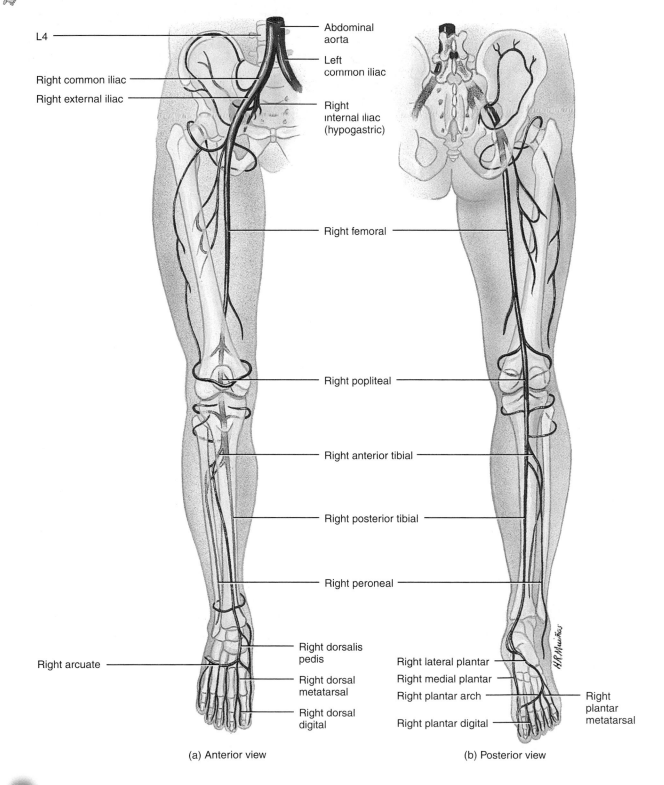

(a) Anterior view

(b) Posterior view

Q At what point does the abdominal aorta divide into the common iliac arteries?

| **Exhibit 21.7** | *Veins of the Systemic Circulation (Figure 21.24)* |

OBJECTIVE

• *Identify the three systemic veins that return deoxygenated blood to the heart.*

Whereas arteries distribute blood to various parts of the body, veins drain blood away from them. For the most part, arteries are deep, whereas veins may be superficial or deep. Superficial veins are located just beneath the skin and can easily be seen. Because there are no large superficial veins, the names of superficial veins do not correspond to those of arteries. Superficial veins are clinically important as sites for withdrawing blood or giving injections. Deep veins generally travel alongside arteries and usually bear the same name. Arteries usually follow definite pathways; veins are more difficult to follow because they connect in irregular networks in which many tributaries merge to form a large vein. Although only one systemic artery, the aorta, takes oxygenated blood away from the heart (left ventricle), three systemic veins, the **coronary sinus, superior vena cava,** and **inferior vena cava,** deliver deoxygenated blood to the heart (right atrium). The coronary sinus receives blood from the cardiac veins; the superior vena cava receives blood from other veins superior to the diaphragm, except the air sacs (alveoli) of the lungs; the inferior vena cava receives blood from veins inferior to the diaphragm.

| What are the three tributaries of the coronary sinus?

VEIN	DESCRIPTION AND REGION SUPPLIED
Coronary sinus (KOR-ō-nar-ē; *corona* = crown)	The **coronary sinus** is the main vein of the heart; it receives almost all venous blood from the myocardium. It is located in the coronary sulcus (see Figure 20.3c) and opens into the right atrium between the orifice of the inferior vena cava and the tricuspid valve. It is a wide venous channel into which three veins drain. It receives the **great cardiac vein** (in the anterior interventricular sulcus) into its left end, and the **middle cardiac vein** (in the posterior interventricular sulcus) and the **small cardiac vein** into its right end. Several **anterior cardiac veins** drain directly into the right atrium.
Superior vena cava (SVC) (VĒ-na CĀ-va; *vena* = vein; *cava* = cavelike)	The **SVC** is about 7.5 cm. (3 in.) long and 2 cm (1 in.) in diameter and empties its blood into the superior part of the right atrium. It begins posterior to the right first costal cartilage by the union of the right and left brachiocephalic veins and ends at the level of the right third costal cartilage, where it enters the right atrium. The SVC drains the head, neck, chest, and upper limbs.
Inferior vena cava (IVC)	The **IVC** is the largest vein in the body, about 3.5 cm (1½ in.) in diameter. It begins anterior to the fifth lumbar vertebra by the union of the common iliac veins, ascends behind the peritoneum to the right of the midline, pierces the costal tendon of the diaphragm at the level of the eighth thoracic vertebra, and enters the inferior part of the right atrium. The IVC drains the abdomen, pelvis, and lower limbs. The inferior vena cava is commonly compressed during the later stages of pregnancy by the enlarging uterus, producing edema of the ankles and feet and temporary varicose veins.

Exhibit 21.7 (continued)

Figure 21.24 Principal veins.

Deoxygenated blood returns to the heart via the superior and inferior venae cavae and the coronary sinus.

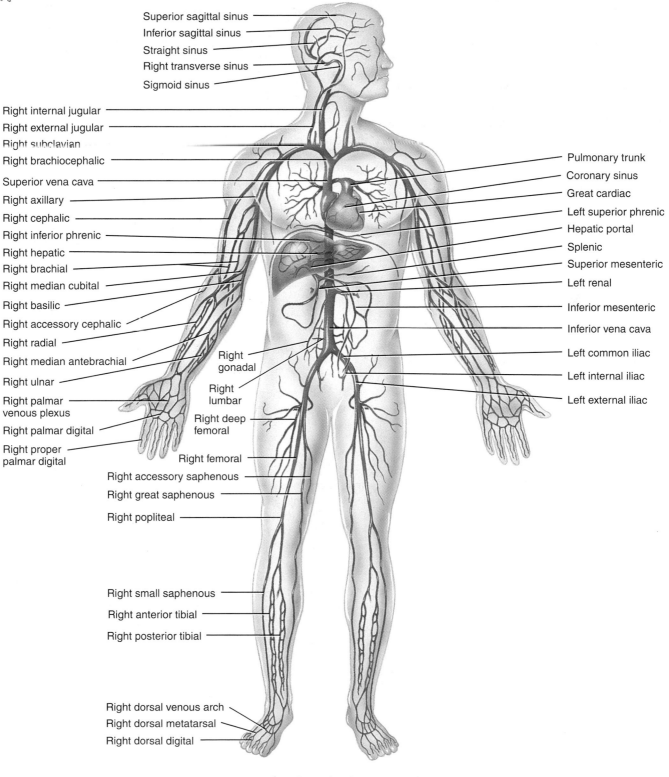

Superior sagittal sinus
Inferior sagittal sinus
Straight sinus
Right transverse sinus
Sigmoid sinus

Right internal jugular
Right external jugular
Right subclavian
Right brachiocephalic
Superior vena cava
Right axillary
Right cephalic
Right inferior phrenic
Right hepatic
Right brachial
Right median cubital
Right basilic
Right accessory cephalic
Right radial
Right median antebrachial
Right ulnar
Right palmar venous plexus
Right palmar digital
Right proper palmar digital

Right gonadal
Right lumbar
Right deep femoral
Right femoral
Right accessory saphenous
Right great saphenous
Right popliteal

Right small saphenous
Right anterior tibial
Right posterior tibial

Right dorsal venous arch
Right dorsal metatarsal
Right dorsal digital

Pulmonary trunk
Coronary sinus
Great cardiac
Left superior phrenic
Hepatic portal
Splenic
Superior mesenteric
Left renal
Inferior mesenteric
Inferior vena cava
Left common iliac
Left internal iliac
Left external iliac

Overall anterior view

Q Which general regions of the body are drained by the superior vena cava and the inferior vena cava?

Exhibit 21.8 | *Veins of the Head and Neck (Figure 21.25)*

OBJECTIVE
• *Identify the three major veins that drain blood from the head.*

The majority of blood draining from the head passes into three pairs of veins: the **internal jugular**, **external jugular**, and **verte-** bral veins. Within the brain, all veins drain into dural venous sinuses and then into the internal jugular veins. **Dural venous sinuses** are endothelial-lined venous channels between layers of the cranial dura mater.

Which general areas are drained by the internal jugular, external jugular, and vertebral veins?

VEIN	DESCRIPTION AND REGION DRAINED
Internal jugular veins (JUG-yoo-lar; *jugular* = throat)	The flow of blood from the dural venous sinuses into the internal jugular veins is as follows (Figure 21.25): The **superior sagittal sinus** (SAJ-i-tal; *sagitallis* = straight) begins at the frontal bone, where it receives a vein from the nasal cavity, and passes posteriorly to the occipital bone. Along its course, it receives blood from the superior, medial, and lateral aspects of the cerebral hemispheres, meninges, and cranial bones. The superior sagittal sinus usually turns to the right and drains into the right transverse sinus.
	The **inferior sagittal sinus** is much smaller than the superior sagittal sinus; it begins posterior to the attachment of the falx cerebri and receives the great cerebral vein to become the straight sinus. The great cerebral vein drains the deeper parts of the brain. Along its course the inferior sagittal sinus also receives tributaries from the superior and medial aspects of the cerebral hemispheres.
	The **straight sinus** runs in the tentorium cerebelli and is formed by the union of the inferior sagittal sinus and the great cerebral vein. The straight sinus also receives blood from the cerebellum and usually drains into the left transverse sinus.
	The **transverse (lateral) sinuses** begin near the occipital bone, pass laterally and anteriorly, and become the sigmoid sinuses near the temporal bone. The transverse sinuses receive blood from the cerebrum, cerebellum, and cranial bones.
	The **sigmoid sinuses** (SIG-moid = S-shaped) are located along the temporal bone. They pass through the jugular foramina, where they terminate in the internal jugular veins. The sigmoid sinuses drain the transverse sinuses.
	The cavernous sinuses (KAV-er-nus = cavelike) are located on either side of the sphenoid bone. They receive blood from the ophthalmic veins from the orbits, and from the cerebral veins from the cerebral hemispheres. They ultimately empty into the transverse sinuses and internal jugular veins. The cavernous sinuses are unique because they have nerves and a major blood vessel passing through them on their way to the orbit and face. Through the sinuses pass the oculomotor (III), trochlear (IV), and ophthalmic and maxillary divisions of the trigeminal (V) nerve, as well as the internal carotid arteries.
	The right and left **internal jugular veins** pass inferiorly on either side of the neck lateral to the internal carotid and common carotid arteries. They then unite with the subclavian veins posterior to the clavicles at the sternoclavicular joints to form the right and left **brachiocephalic veins** (brā-kē-ō-se-FAL-ik; *bracium* = arm; *cephalic* = head). From here blood flows into the superior vena cava. The general structures drained by the internal jugular veins are the brain (through the dural venous sinuses), face, and neck.
External jugular veins	The right and left **external jugular veins** begin in the parotid glands near the angle of the mandible. They are superficial veins that descend through the neck across the sternocleidomastoid muscles. They terminate at a point opposite the middle of the clavicle, where they empty into the subclavian veins. The general structures drained by the external jugular veins are external to the cranium, such as the scalp and superficial and deep regions of the face.
Vertebral veins (VER-te-bral; *vertebra* = vertebrae)	The right and left **vertebral veins** originate inferior to the occipital condyles. They descend through successive transverse foramina of the first six cervical vertebrae and emerge from the foramina of the sixth cervical vertebra to enter the brachiocephalic veins in the root of the neck. The vertebral veins drain deep structures in the neck such as the cervical vertebrae, cervical spinal cord, and some neck muscles.

- ➝

Exhibit 21.8 *(continued)*

SCHEME OF DRAINAGE

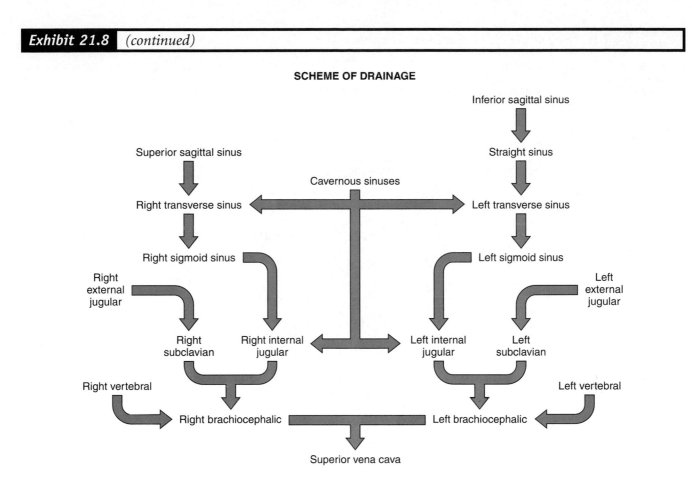

Exhibit 21.8 | *Veins of the Head and Neck (continued)*

Figure 21.25 Principal veins of the head and neck.

🔑 **Blood draining from the head passes into the internal jugular, external jugular, and vertebral veins.**

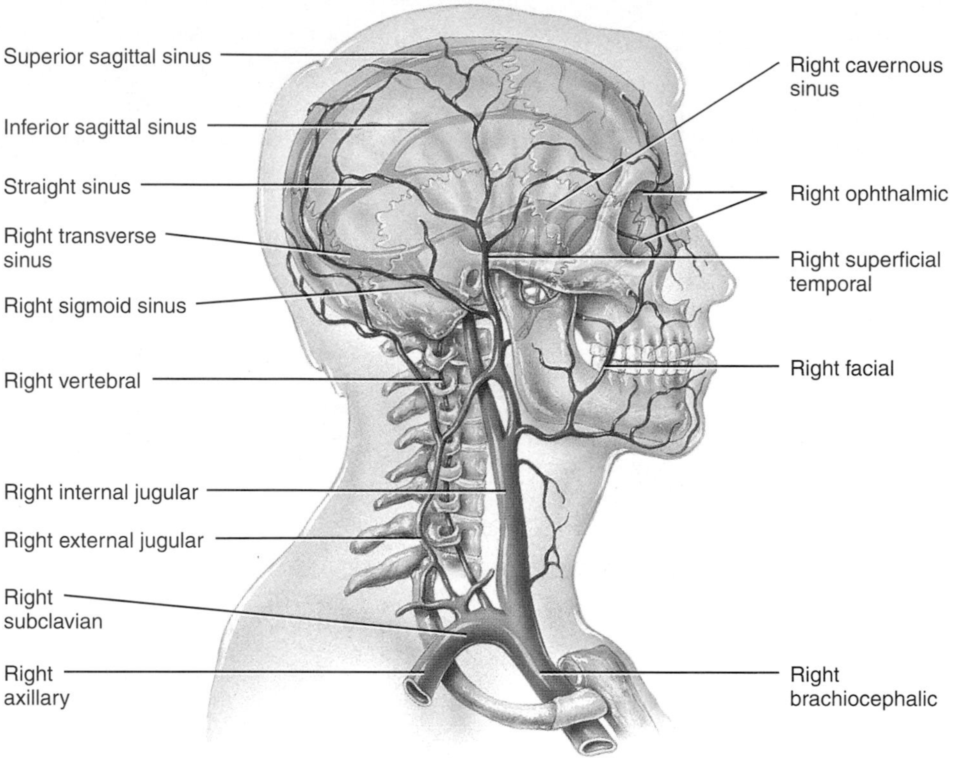

Right lateral view

Q Into which veins in the neck does all venous blood in the brain drain?

Exhibit 21.9 | Veins of the Upper Limbs (Figure 21.26)

OBJECTIVE

- *Identify the principal superficial veins that drain the upper limbs.*

Blood from the upper limbs is returned to the heart by both superficial and deep veins. Both sets of veins have valves, which are more numerous in the deep veins. **Superficial veins** are located just deep to the skin and are often visible. They anastomose extensively with one another and with deep veins, and they do not accompany arteries. Superficial veins are larger than deep veins and return most of the blood from the upper limbs. **Deep veins** are located deep in the body. They usually accompany arteries and have the same names as the corresponding arteries.

| Where do the cephalic, basilic, median antebrachial, radial, and ulnar veins originate?

VEIN	DESCRIPTION AND REGION DRAINED
Superficial	
Cephalic veins (se-FAL-ik; *cephalic* = head)	The principal superficial veins that drain the upper limbs are the cephalic and basilic veins. They originate in the hand and convey blood from the smaller superficial veins into the axillary veins. The **cephalic veins** begin on the lateral aspect of the **dorsal venous arches** (VĒ-nus), networks of veins on the dorsum of the hands formed by the **dorsal metacarpal veins** (Figure 21.26a). These veins, in turn, drain the **dorsal digital veins,** which pass along the sides of the fingers. Following their formation from the dorsal venous arches, the cephalic veins arch around the radial side of the forearms to the anterior surface and ascend through the entire limbs along the anterolateral surface. The cephalic veins end where they join the axillary veins, just inferior to the clavicles. **Accessory cephalic veins** originate either from a venous plexus on the dorsum of the forearms or from the medial aspects of the dorsal venous arches and unite with the cephalic veins just inferior to the elbow. The cephalic veins drain blood from the lateral aspect of the upper limbs.
Basilic veins (ba-SIL-ik; *basilikos* = royal)	The **basilic veins** begin on the medial aspects of the dorsal venous arches and ascend along the posteromedial surface of the forearm and anteromedial surface of the arm (Figure 21.26b). They drain blood from the medial aspects of the upper limbs. Anterior to the elbow, the basilic veins are connected to the cephalic veins by the **median cubital veins** (*cubitus* = elbow), which drain the forearm. If veins must be punctured for an injection, transfusion, or removal of a blood sample, the medial cubital veins are preferred. After receiving the median cubital veins, the basilic veins continue ascending until they reach the middle of the arm. There they penetrate the tissues deeply and run alongside the brachial arteries until they join the brachial veins. As the basilic and brachial veins merge in the axillary area, they form the axillary veins.
Median antebrachial veins (an'-tē-BRĀ-kē-al; *ante-* = before, in front of; *brachium* = arm)	The **median antebrachial veins (median veins of the forearm)** begin in the **palmar venous plexuses,** networks of veins on the palms. The plexuses drain the **palmar digital veins** in the fingers. The median antebrachial veins ascend anteriorly in the forearms to join the basilic or median cubital veins, sometimes both. They drain the palms and forearms.
Deep	
Radial veins (RĀ-dē-al = radius)	The paired radial veins begin at the **deep venous palmar arches** (Figure 21.26c). These arches drain the **palmar metacarpal veins** in the palms. The radial veins drain the lateral aspects of the forearms and pass alongside the radial arteries. Just inferior to the elbow joint, the radial veins unite with the ulnar veins to form the brachial veins.
Ulnar veins (UL-nar = ulna)	The paired **ulnar veins,** which are larger than the radial veins, begin at the **superficial palmar venous arches.** These arches drain the **common palmar digital veins** and the **proper palmar digital veins** in the fingers. The ulnar veins drain the medial aspect of the forearms, pass alongside the ulnar arteries, and join with the radial veins to form the brachial veins.
Brachial veins (BRĀ-kē-al; *brachium* = arm)	The paired **brachial veins** accompany the brachial arteries. They drain the forearms, elbow joints, arms, and humerus. They pass superiorly and join with the basilic veins to form the axillary veins.
Axillary veins (AK-si-ler'-ē; *axilla* = armpit)	The **axillary veins** ascend to the outer borders of the first ribs, where they become the subclavian veins. The axillary veins receive tributaries that correspond to the branches of the axillary arteries. The axillary veins drain the arms, axillas, and superolateral chest wall.
Subclavian veins (sub-KLĀ-vē-an; *sub-* = under; *clavicula* = clavicle)	The **subclavian veins** are continuations of the axillary veins that terminate at the sternal end of the clavicles, where they unite with the internal jugular veins to form the brachiocephalic veins. The subclavian veins drain the arms, neck, and thoracic wall. The thoracic duct of the lymphatic system delivers lymph into the left subclavian vein at the junction with the internal jugular. The right lymphatic duct delivers lymph into the right subclavian vein at the corresponding junction (see Figure 22.3a).

Exhibit 21.9 *Veins of the Upper Limbs (continued)*

SCHEME OF DRAINAGE

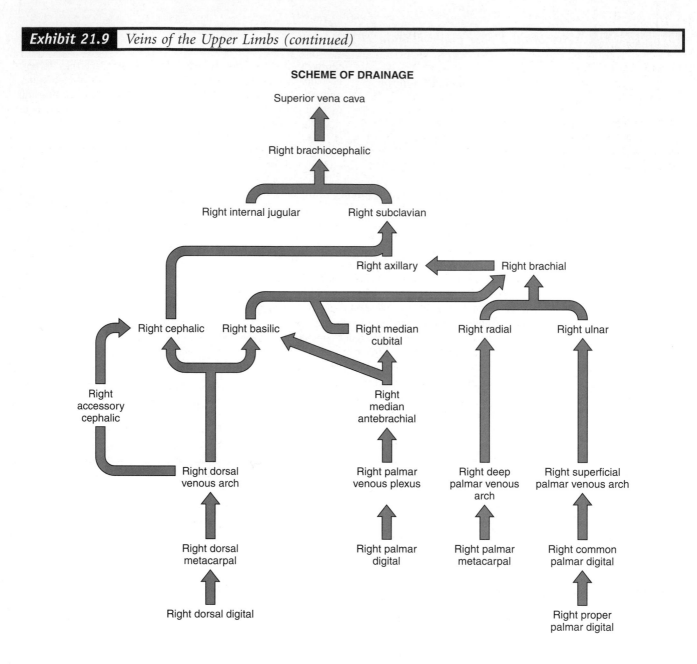

Exhibit 21.9   *(continued)*

Figure 21.26 Principal veins of the right upper limb.

🗝 **Deep veins usually accompany arteries that have similar names.**

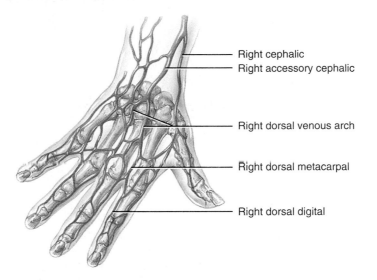

Right cephalic
Right accessory cephalic
Right dorsal venous arch
Right dorsal metacarpal
Right dorsal digital

(a) Posterior view of superficial veins of the hand

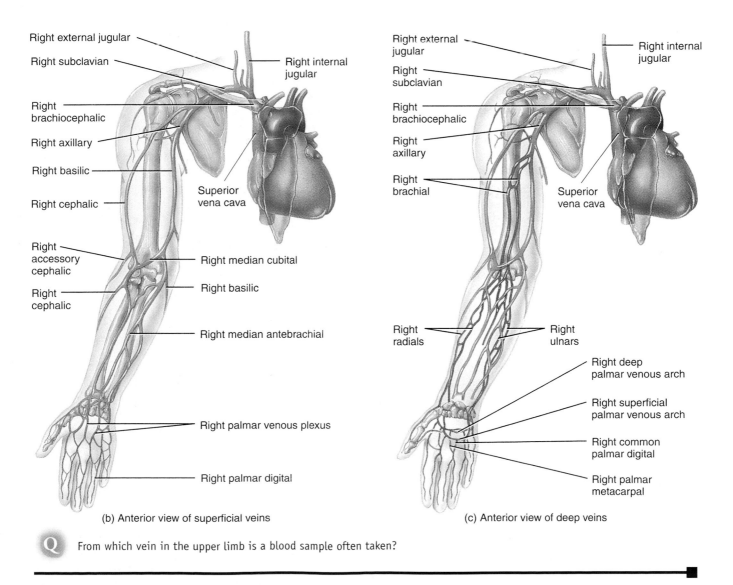

Right external jugular
Right subclavian
Right brachiocephalic
Right axillary
Right basilic
Right cephalic
Right internal jugular
Superior vena cava
Right accessory cephalic
Right cephalic
Right median cubital
Right basilic
Right median antebrachial
Right palmar venous plexus
Right palmar digital

(b) Anterior view of superficial veins

Right external jugular
Right subclavian
Right brachiocephalic
Right axillary
Right brachial
Right internal jugular
Superior vena cava
Right radials
Right ulnars
Right deep palmar venous arch
Right superficial palmar venous arch
Right common palmar digital
Right palmar metacarpal

(c) Anterior view of deep veins

Q From which vein in the upper limb is a blood sample often taken?

Exhibit 21.10 Veins of the Thorax (Figure 21.27)

OBJECTIVE
• *Identify the components of the azygos system of veins.*

Although the brachiocephalic veins drain some portions of the thorax, most thoracic structures are drained by a network of veins, called the **azygos system,** that runs on either side of the vertebral column. The system consists of three veins—the **azygous, hemiazygos,** and **accessory hemiazygos veins**—that show considerable variation in origin, course, tributaries, anastomoses, and termination. Ultimately they empty into the superior vena cava.

What is the importance of the azygos system relative to the inferior vena cava?

VEIN	DESCRIPTION AND REGION DRAINED
Brachiocephalic vein (brā′-kē-ō-se-FAL-ik; *brachium* = arm; *cephalic* = head)	The right and left **brachiocephalic veins,** formed by the union of the subclavian veins and internal jugular veins, drain blood from the head, neck, upper limbs, mammary glands, and superior thorax. The brachiocephalic veins unite to form the superior vena cava. Because the superior vena cava is to the right, the left brachiocephalic vein is longer than the right. The right brachiocephalic vein is anterior and to the right of the brachiocephalic trunk. The left brachiocephalic vein is anterior to the brachiocephalic trunk, the left common carotid and left subclavian arteries, the trachea, and the left vagus (X) and phrenic nerves.
Azygos system (az-Ī-gos = unpaired)	The **azygos system,** besides collecting blood from the thorax and abdominal wall, may serve as a bypass for the inferior vena cava that drains blood from the lower body. Several small veins directly link the azygos system with the inferior vena cava. Large veins that drain the lower limbs and abdomen conduct blood into the azygos system. If the inferior vena cava or hepatic portal vein becomes obstructed, the azygos system can return blood from the lower body to the superior vena cava.
Azygos vein	The **azygos vein** lies anterior to the vertebral column, slightly to the right of the midline. It usually begins at the junction of the right ascending lumbar and right subcostal veins near the diaphragm. At the level of the fourth thoracic vertebra, it arches over the root of the right lung to end in the superior vena cava. Generally, the azygos vein drains the right side of the thoracic wall, thoracic viscera, and abdominal wall. Specifically, the azygos vein receives blood from most of the **right posterior intercostal, hemiazygos, accessory hemiazygos, esophageal, mediastinal, pericardial,** and **bronchial veins.**
Hemiazygos vein (HEM-ē-az-ī-gos; *hemi-* = half)	The **hemiazygos vein** is anterior to the vertebral column and slightly to the left of the midline. It frequently begins at the junction of the left ascending lumbar and left subcostal veins. It terminates by joining the azygos vein at about the level of the ninth thoracic vertebra. Generally, the hemiazygos vein drains the left side of the thoracic wall, thoracic viscera, and abdominal wall. Specifically, the hemiazygos vein receives blood from the ninth through eleventh **left posterior intercostal, esophageal, mediastinal,** and sometimes the **accessory hemiazygos veins.**
Accessory hemiazygos vein	The **accessory hemiazygos vein** is also anterior to the vertebral column and to the left of the midline. It begins at the fourth or fifth intercostal space and descends from the fifth to the eighth thoracic vertebra or ends in the hemiazygos vein. It terminates by joining the azygos vein at about the level of the eighth thoracic vertebra. The accessory hemiazygos vein drains the left side of the thoracic wall. It receives blood from the fourth through eighth **left posterior intercostal veins** (the first through third left posterior intercostal veins open into the left brachiocephalic vein), **left bronchial,** and **mediastinal veins.**

Exhibit 21.10 *(continued)*

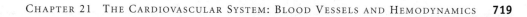

SCHEME OF DRAINAGE

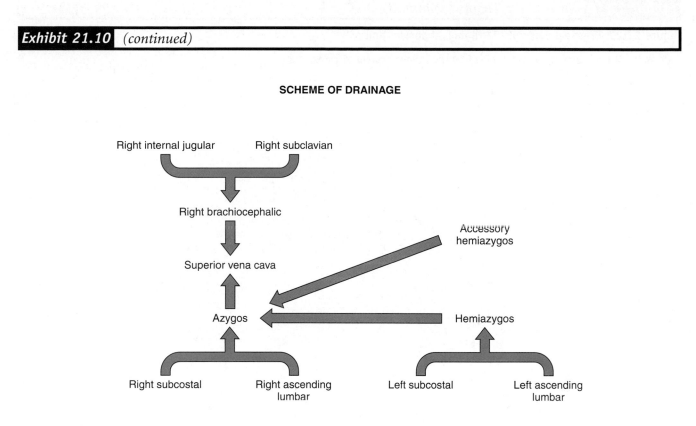

Exhibit 21.10 *Veins of the Thorax (continued)*

Figure 21.27 Principal veins of the thorax, abdomen, and pelvis.

🔑 **Most thoracic structures are drained by the azygos system of veins.**

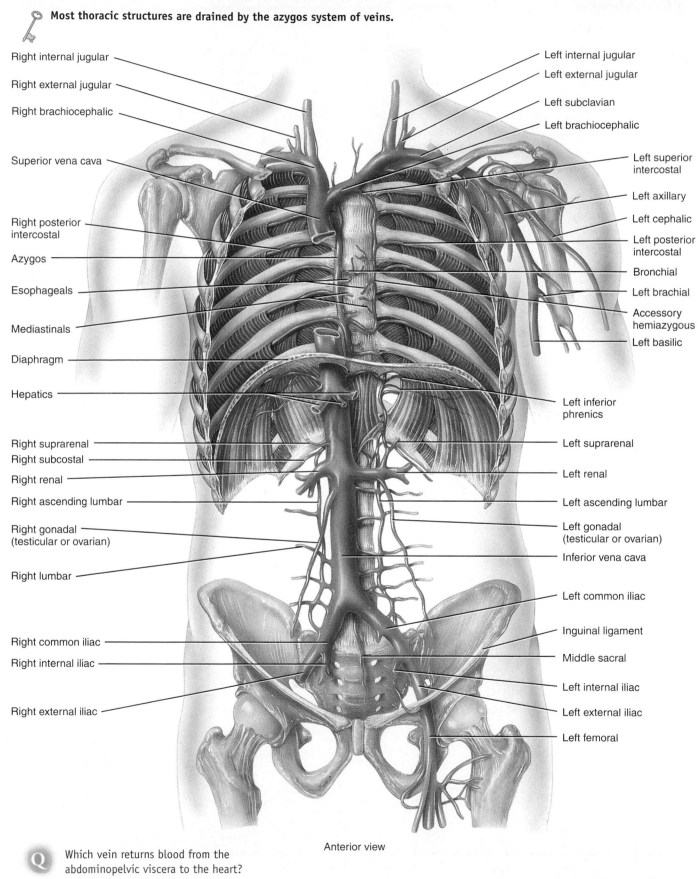

Right internal jugular

Right external jugular

Right brachiocephalic

Superior vena cava

Right posterior intercostal

Azygos

Esophageals

Mediastinals

Diaphragm

Hepatics

Right suprarenal

Right subcostal

Right renal

Right ascending lumbar

Right gonadal (testicular or ovarian)

Right lumbar

Right common iliac

Right internal iliac

Right external iliac

Left internal jugular

Left external jugular

Left subclavian

Left brachiocephalic

Left superior intercostal

Left axillary

Left cephalic

Left posterior intercostal

Bronchial

Left brachial

Accessory hemiazygous

Left basilic

Left inferior phrenics

Left suprarenal

Left renal

Left ascending lumbar

Left gonadal (testicular or ovarian)

Inferior vena cava

Left common iliac

Inguinal ligament

Middle sacral

Left internal iliac

Left external iliac

Left femoral

Anterior view

Q Which vein returns blood from the abdominopelvic viscera to the heart?

Exhibit 21.11 | *Veins of the Abdomen and Pelvis (Figure 21.27)*

OBJECTIVE

• *Identify the principal veins that drain the abdomen and pelvis.*

Blood from the abdominopelvic viscera and abdominal wall returns to the heart via the **inferior vena cava.** Many small veins enter the inferior vena cava. Most carry return flow from parietal branches of the abdominal aorta, and their names correspond to the names of the arteries.

The inferior vena cava does not receive veins directly from the gastrointestinal tract, spleen, pancreas, and gallbladder. These or-

gans pass their blood into a common vein, the **hepatic portal vein,** which delivers the blood to the liver. The hepatic portal vein is formed by the union of the superior mesenteric and splenic veins (see Figure 21.29). This special flow of venous blood is called **hepatic portal circulation,** which is described shortly. After passing through the liver for processing, blood drains into the hepatic veins, which empty into the inferior vena cava.

What structures are drained by the lumbar, gonadal, renal, suprarenal, inferior phrenic, and hepatic veins?

VEIN	DESCRIPTION AND REGION DRAINED
Inferior vena cava (VĒ-na CĀ-va; *vena* = vein; *cava* = cavelike)	The **inferior vena cava** is formed by the union of two common iliac veins that drain the lower limbs, pelvis, and abdomen (see Figure 21.27). The inferior vena cava extends superiorly thorough the abdomen and thorax to the right atrium.
Common iliac veins (IL-ē-ak; *iliac* = pertaining to the ilium)	The **common iliac veins** are formed by the union of the internal and external iliac veins anterior to the sacroiliac joint and represent the distal continuation of the inferior vena cava at their bifurcation. The right common iliac vein is much shorter than the left and is also more vertical. Generally, the common iliac veins drain the pelvis, external genitals, and lower limbs.
Internal iliac veins	The **internal (hypogastric) iliac veins** begin near the superior portion of the greater sciatic notch and run medial to their corresponding arteries (see Figure 21.27). Generally, the veins drain the thigh, buttocks, external genitals, and pelvis.
External iliac veins	The **external iliac veins** are companions of the internal iliac arteries and begin at the inguinal ligaments as continuations of the femoral veins. They end anterior to the sacroiliac joint where they join with the internal iliac veins to form the common iliac veins. The external iliac veins drain the lower limbs, cremasteric muscle in males, and the abdominal wall.
Lumbar veins (LUM-bar; *lumbar* = loin)	A series of parallel **lumbar veins,** usually four on each side, drain blood from both sides of the posterior abdominal wall, vertebral canal, spinal cord, and meninges (see Figure 21.27). The lumbar veins run horizontally with the lumbar arteries. The lumbar veins connect at right angles with the right and left **ascending lumbar veins,** which form the origin of the corresponding azygos or hemiazygos vein. The lumbar veins drain blood into the ascending lumbars and then run to the inferior vena cava, where they release the remainder of the flow.
Gonadal veins (gō-NAD-al; *gono* = seed) **[testicular** (test-TIK-yoo-lor) or **ovarian** (ō-VA-rē-an)]	The **gonadal veins** ascend with the gonadal arteries along the posterior abdominal wall (see Figure 21.27). In the male, the gonadal veins are called the testicular veins. The **testicular veins** drain the testes (the left testicular vein empties into the left renal vein, and the right testicular vein drains into the inferior vena cava). In the female, the gonadal veins are called the ovarian veins. The **ovarian veins** drain the ovaries (the left ovarian vein empties into the left renal vein, and the right ovarian vein drains into the inferior vena cava).
Renal veins (RĒ-nal; *renal* = kidney)	The **renal veins** are large and pass anterior to the renal arteries (see Figure 21.27). The left renal vein is longer than the right renal vein and passes anterior to the abdominal aorta. It receives the left testicular (or ovarian), left inferior phrenic, and usually left suprarenal veins. The right renal vein empties into the inferior vena cava posterior to the duodenum. The renal veins drain the kidneys.
Suprarenal veins (soo′-pra-RĒ-nal; *supra* = above)	The **suprarenal veins** drain the adrenal (suprarenal) glands (the left suprarenal vein empties into the left renal vein, and the right suprarenal vein empties into the inferior vena cava). See Figure 21.27.
Inferior phrenic veins (FREN-ik; *phrenic* = diaphragm)	The **inferior phrenic veins** drain the diaphragm (the left inferior phrenic vein usually sends one tributary to the left suprarenal vein, which empties into the left renal vein, and another tributary that empties into the inferior vena cava; the right inferior phrenic vein empties into the inferior vena cava). See Figure 21.27.
Hepatic veins (he-PAT-ik; *hepatic* = liver)	The **hepatic veins** drain the liver. (See Figure 21.27).

Exhibit 21.11 *Veins of the Abdomen and Pelvis (continued)*

SCHEME OF DRAINAGE

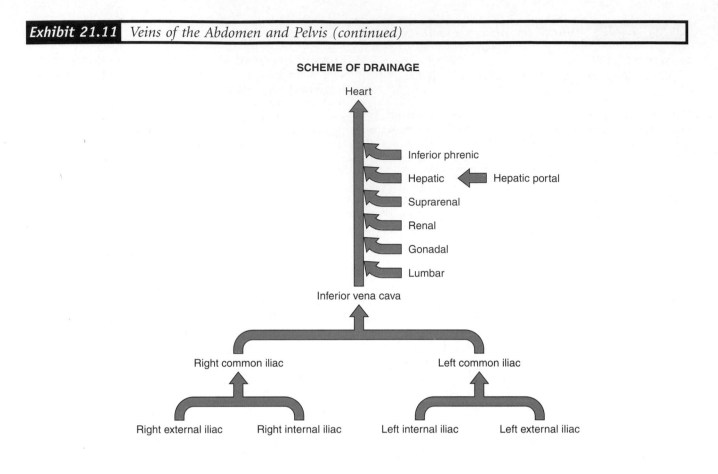

Exhibit 21.12 | *Veins of the Lower Limbs (Figure 21.28)*

OBJECTIVE

• *Identify the principal superficial and deep veins that drain the lower limbs.*

As with the upper limbs, blood from the lower limbs is drained by both **superficial** and **deep veins.** The superficial veins often anastomose with each other and with deep veins along their length. Deep veins, for the most part, have the same names as corresponding arteries. All veins of the lower limbs have valves, which are more numerous than in veins of the upper limbs.

| Why are the great saphenous veins clinically important?

VEIN	DESCRIPTION AND REGION DRAINED
Superficial Veins	
Great saphenous veins (sa-FĒ-nus; *saphenes* = clearly visible)	The **great (long) saphenous veins,** the longest veins in the body, ascend from the foot to the groin in the subcutaneous layer. They begin at the medial end of the dorsal venous arches of the foot. The **dorsal venous arches** (VĒ-nus) are networks of veins on the dorsum of the foot formed by the **dorsal digital veins,** which collect blood from the toes, and then unite in pairs to form the **dorsal metatarsal veins,** which parallel the metatarsals. As the dorsal metatarsal veins approach the foot, they combine to form the dorsal venous arches. The great saphenous veins pass anterior to the medial malleolus of the tibia and then superiorly along the medial aspect of the leg and thigh just deep to the skin. They receive tributaries from superficial tissues and connect with the deep veins as well. They empty into the femoral veins at the groin. Generally, the great saphenous veins drain mainly the medial side of the leg and thigh, the groin, external genitals, and abdominal wall. Along their length, the great saphenous veins have from 10 to 20 valves, with more located in the leg than the thigh. These veins are more likely to be subject to varicosities than other veins in the lower limbs because they must support a long column of blood and are not well supported by skeletal muscles. The great saphenous veins are often used for prolonged administration of intravenous fluids. This is particularly important in very young children and in patients of any age who are in shock and whose veins are collapsed. The great saphenous veins are also often used as a source of vascular grafts, especially for coronary bypass surgery. In the procedure, the vein is removed and then reversed so that the valves do not obstruct the flow of blood.
Small saphenous veins	The **small (short) saphenous veins** begin at the lateral aspect of the dorsal venous arches of the foot. They pass posterior to the lateral malleolus of the fibula and ascend deep to the skin along the posterior aspect of the leg. They empty into the popliteal veins in the popliteal fossa, posterior to the knee. Along their length, the small saphenous veins have from 9 to 12 valves. The small saphenous veins drain the foot and posterior aspect of the leg. They may communicate with the great saphenous veins in the proximal thigh.
Deep Veins	
Posterior tibial veins (TIB-ē-al)	The **plantar digital veins** on the plantar surfaces of the toes unite to form the **plantar metatarsal veins,** which parallel the metatarsals. They unite to form the **deep plantar venous arches.** From each arch emerges the **medial** and **lateral plantar veins.** The paired **posterior tibial veins,** which sometimes unite to form a single vessel, are formed by the medial and lateral plantar veins, posterior to the medial malleolus of the tibia. They accompany the posterior tibial artery through the leg. They ascend deep to the muscles in the posterior aspect of the leg and drain the foot and posterior compartment muscles. About two-thirds the way up the leg, the posterior tibial veins drain blood from the **peroneal veins** (per'-ō-NĒ-al; *perone* = fibula), which drain the lateral and posterior leg muscles. The posterior tibial veins unite with the anterior tibial veins just inferior to the popliteal fossa to form the popliteal veins.
Anterior tibial veins	The paired **anterior tibial veins** arise in the dorsal venous arch and accompany the anterior tibial artery. They ascend in the interosseous membrane between the tibia and fibula and unite with the posterior tibial veins to form the popliteal vein. The anterior tibial veins drain the ankle joint, knee joint, tibiofibular joint, and anterior portion of leg.

Exhibit 21.12 *Veins of the Lower Limbs (continued)*

VEIN	DESCRIPTION AND REGION DRAINED
Deep Veins (continued)	
Popliteal veins (pop′-li-TĒ-al; *popliteus* = hollow behind knee)	The **popliteal veins** are formed by the union of the anterior and posterior tibial veins. The popliteal veins also receive blood from the small saphenous veins and tributaries that correspond to branches of the popliteal artery. The popliteal veins drain the knee joint and the skin, muscles, and bones of portions of the calf and thigh around the knee joint.
Femoral veins (FEM-o-ral)	The **femoral veins** accompany the femoral arteries and are the continuations of the popliteal veins just superior to the knee. The femoral veins extend up the posterior surface of the thighs and drain the muscles of the thighs, femurs, external genitals, and superficial lymph nodes. The largest tributaries of the femoral veins are the **deep femoral veins.** Just before penetrating the abdominal wall, the femoral veins receive the deep femoral veins and the great saphenous veins. The veins formed from their union penetrate the body wall and enter the pelvic cavity. Here they are known as the **external iliac veins.**

SCHEME OF DRAINAGE

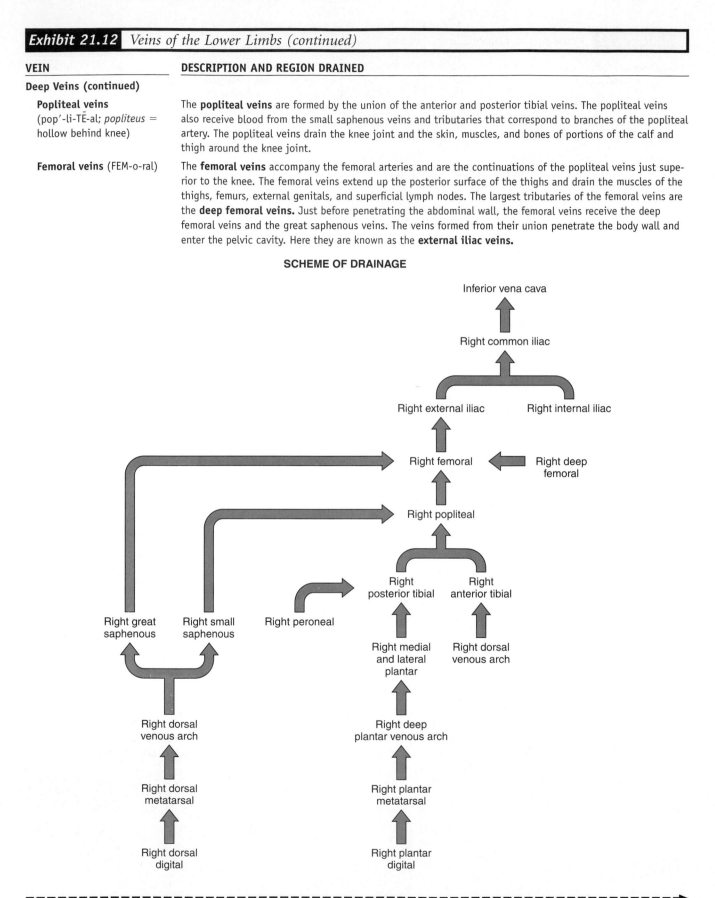

Exhibit 21.12 *(continued)*

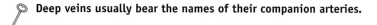

Figure 21.28 Principal veins of the pelvis and lower limbs.

🔑 **Deep veins usually bear the names of their companion arteries.**

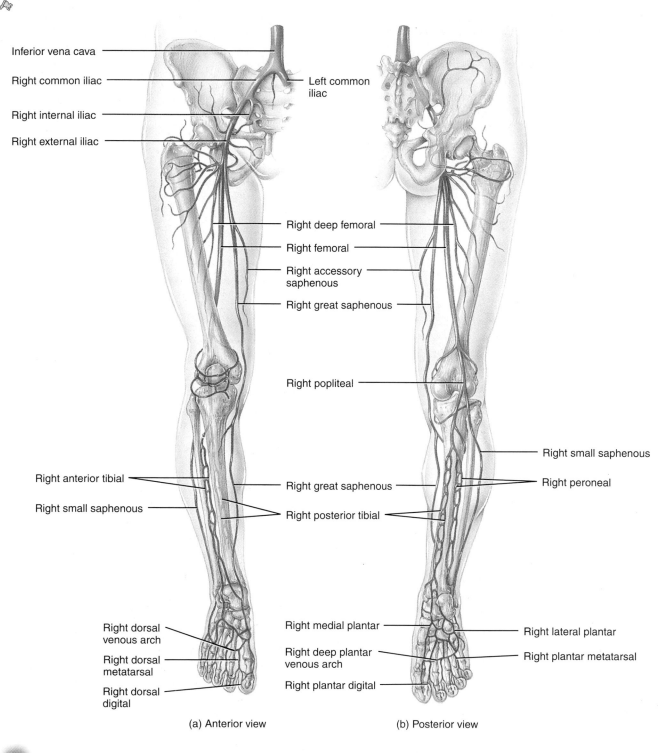

Inferior vena cava

Right common iliac

Right internal iliac

Right external iliac

Left common iliac

Right deep femoral

Right femoral

Right accessory saphenous

Right great saphenous

Right popliteal

Right small saphenous

Right peroneal

Right anterior tibial

Right great saphenous

Right small saphenous

Right posterior tibial

Right dorsal venous arch

Right dorsal metatarsal

Right dorsal digital

Right medial plantar

Right deep plantar venous arch

Right plantar digital

Right lateral plantar

Right plantar metatarsal

(a) Anterior view

(b) Posterior view

 Which veins of the lower limb are superficial?

The Hepatic Portal Circulation

OBJECTIVE

• *Identify the blood vessels and route of the hepatic portal circulation.*

We noted earlier that the two principal circulatory routes are the systemic and pulmonary circulations. Before looking at the pulmonary circulation, we first examine the hepatic portal circulation, a subdivision of the systemic circulation. The **hepatic portal circulation** (*hepat-* = liver) detours venous blood from the gastrointestinal organs and spleen through the liver before it returns to the heart (Figure 21.29). A *portal system* carries blood between two capillary networks, from one location in the body to another, without passing through the heart. In this case, blood flows from capillaries of the gastrointestinal tract to sinusoids of the liver. After a meal, hepatic portal blood is rich in substances absorbed from the gastrointestinal tract. The liver stores some of them and modifies others before they pass into the general circulation. For example, the liver converts glucose into glycogen for storage, thereby helping to maintain homeostasis of blood glucose; the liver also detoxifies harmful substances that have been absorbed from the gastrointestinal tract and destroys bacteria by phagocytosis.

The **hepatic portal vein** is formed by the union of the superior mesenteric and splenic veins. The **superior mesenteric vein** drains blood from the small intestine and portions of the large intestine, stomach, and pancreas through the *jejunal, ileal, ileocolic, right colic, middle colic, pancreaticoduodenal,* and *right gastroepiploic veins*. The **splenic vein** drains blood from the stomach, pancreas, and portions of the large intestine through the *short gastric, left gastroepiploic, pancreatic,* and *inferior mesenteric veins*. The inferior mesenteric vein, which passes into the splenic vein, drains portions of the large intestine through the superior *rectal, sigmoidal,* and *left colic veins*. The *right* and *left gastric veins*, which open directly into the hepatic portal vein, drain the stomach. The *cystic vein*, which also opens into the hepatic portal vein, drains the gallbladder.

At the same time that the liver receives deoxygenated blood via the hepatic portal system, it also receives oxygenated blood via the proper hepatic artery, a branch of the celiac artery. Ultimately, all blood leaves the liver through the **hepatic veins,** which drain into the inferior vena cava.

The other portal veins in the body—the hypophyseal portal veins that carry blood from the hypothalamus to the anterior pituitary gland—may be reviewed in Figure 18.5 on page 574.

The Pulmonary Circulation

OBJECTIVE

• *Identify the blood vessels and route of the pulmonary circulation.*

The **pulmonary circulation** (*pulmo-* = lung) carries deoxygenated blood from the right ventricle to the air sacs (alveoli) within the lungs and returns oxygenated blood from the air sacs to the left atrium (Figure 21.30). The **pulmonary trunk** emerges from the right ventricle and passes superiorly, posteriorly, and to the left. It then divides into two branches: the **right pulmonary artery** to the right lung and the **left pulmonary artery** to the left lung. The pulmonary arteries are the only postnatal (after birth) arteries that carry deoxygenated blood. On entering the lungs, the branches divide and subdivide until finally they form capillaries around the air sacs within the lungs. CO_2 passes from the blood into the air sacs and is exhaled. Inhaled O_2 passes from the air within the lungs into the blood. The pulmonary capillaries unite to form venules and eventually **pulmonary veins,** which exit the lungs and transport the oxygenated blood to the left atrium. Two left and two right pulmonary veins enter the left atrium. After birth, the pulmonary veins are the only veins that carry oxygenated blood. Contractions of the left ventricle then eject the oxygenated blood into the systemic circulation.

The pulmonary and systemic circulations are different in two important ways. First, blood in the pulmonary circulation need not be pumped as far as blood in the systemic circulation. Second, compared to systemic arteries, pulmonary arteries have larger diameters, thinner walls, and less elastic tissue; as a result, the resistance to pulmonary blood flow is very low, which means that less pressure is needed to move blood through the lungs. The peak systolic pressure in the right ventricle is about 20% of that in the left ventricle.

The Fetal Circulation

OBJECTIVE

• *Identify the blood vessels and route of the fetal circulation.*

The circulatory system of a fetus, called the **fetal circulation,** differs from the postnatal circulation because the lungs, kidneys, and gastrointestinal organs do not begin to function until birth. The fetus obtains its O_2 and nutrients by diffusion from the maternal blood and eliminates its CO_2 and wastes by diffusion into the maternal blood.

The exchange of materials between fetal and maternal circulations occurs through the **placenta** (pla-SEN-ta), which forms inside the mother's uterus and attaches to the umbilicus (navel) of the fetus by the **umbilical cord** (um-BIL-i-kal). The placenta communicates with the mother's cardiovascular system through many small blood vessels that emerge from the uterine wall. The umbilical cord contains blood vessels that branch into capillaries in the placenta. Wastes from the fetal blood diffuse out of the capillaries, into spaces containing maternal blood (intervillous spaces) in the placenta, and finally into the mother's uterine veins (see Figure 29.8a on page 1032). Nutrients travel the opposite route—from the maternal blood vessels to the intervillous spaces to the fetal capillaries. Normally, there is no direct mixing of maternal and fetal blood because all exchanges occur by diffusion through capillary walls.

Figure 21.29 Hepatic portal circulation. A schematic diagram of blood flow through the liver, including arterial circulation, is shown in (b); deoxygenated blood is indicated in blue, oxygenated blood in red.

The hepatic portal circulation delivers venous blood from the organs of the gastrointestinal tract and spleen to the liver.

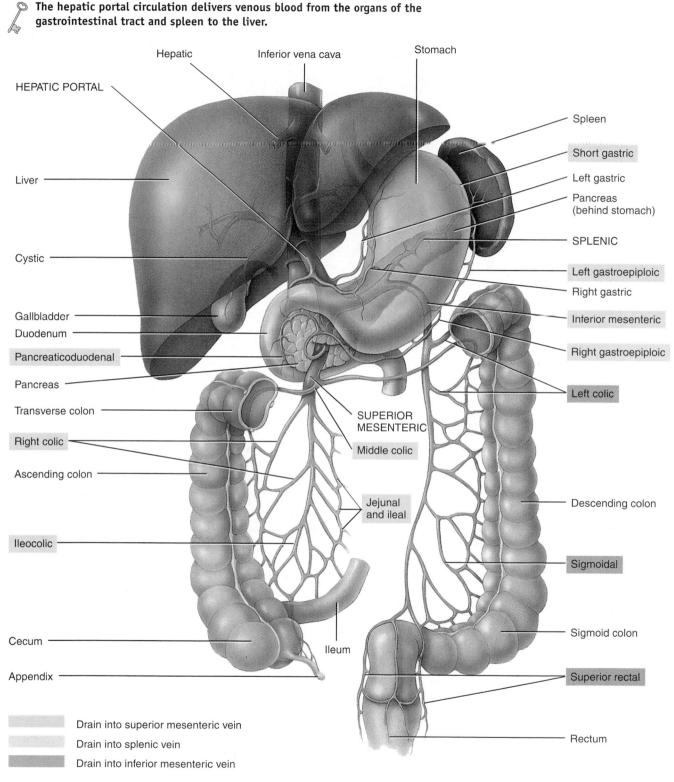

Drain into superior mesenteric vein

Drain into splenic vein

Drain into inferior mesenteric vein

(a) Anterior view of veins draining into the hepatic portal vein

figure continues

Figure 21.29 (continued)

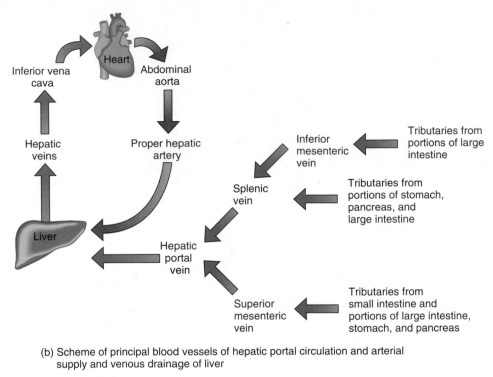

(b) Scheme of principal blood vessels of hepatic portal circulation and arterial supply and venous drainage of liver

 Which veins carry blood away from the liver?

Blood passes from the fetus to the placenta via two **umbilical arteries** (Figure 21.31a, c). These branches of the internal iliac (hypogastric) arteries are within the umbilical cord. At the placenta, fetal blood picks up O_2 and nutrients and eliminates CO_2 and wastes. The oxygenated blood returns from the placenta via a single **umbilical vein.** This vein ascends to the liver of the fetus, where it divides into two branches. Whereas some blood flows through the branch that joins the hepatic portal vein and enters the liver, most of the blood flows into the second branch, the **ductus venosus** (DUK-tus ve-NŌ-sus), which drains into the inferior vena cava.

Deoxygenated blood returning from the inferior regions mingles with oxygenated blood from the ductus venosus in the inferior vena cava. This mixed blood then enters the right atrium. Deoxygenated blood returning from the superior regions of the fetus enters the superior vena cava and passes into the right atrium.

Most of the fetal blood does not pass from the right ventricle to the lungs, as it does in postnatal circulation, because an opening called the **foramen ovale** (fō-RĀ-men ō-VAL-ē) exists in the septum between the right and left atria. About one third of the blood passes through the foramen ovale directly into the systemic circulation. The blood that does pass into the right ventricle is pumped into the pulmonary trunk, but little of this blood reaches the nonfunctioning fetal lungs. Instead, most is sent through the **ductus arteriosus** (ar-tē-rē-Ō-sus), a vessel that connects the pulmonary trunk with the aorta, so that most blood bypasses the fetal lungs. The blood in the aorta is carried to all fetal tissues through the systemic circulation. When the common iliac arteries branch into the external and internal iliacs, part of the blood flows into the internal iliacs, into the umbilical arteries, and back to the placenta for another exchange of materials. The only fetal vessel that carries fully oxygenated blood is the umbilical vein.

After birth, when pulmonary (lung), renal, and digestive functions begin, the following vascular changes occur (Figure 21.31b).

1. When the umbilical cord is tied off, blood no longer flows through the umbilical arteries, they fill with connective tissue, and the distal portions of the umbilical arteries become fibrous cords called the **medial umbilical ligaments.** Although the arteries are closed functionally only a few minutes after birth, complete obliteration of the lumens may take 2–3 months.

2. The umbilical vein collapses but remains as the **ligamentum teres (round ligament),** a structure that attaches the umbilicus to the liver.

3. The ductus venosus collapses but remains as the **ligamentum venosum,** a fibrous cord on the inferior surface of the liver.

Figure 21.30 Pulmonary circulation.

The pulmonary circulation brings deoxygenated blood from the right ventricle to the lungs and returns oxygenated blood from the lungs to the left atrium.

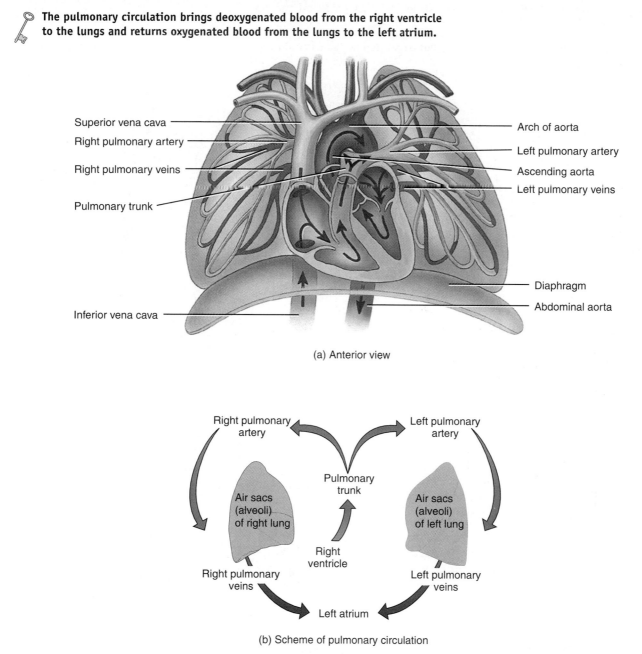

Superior vena cava

Right pulmonary artery

Right pulmonary veins

Pulmonary trunk

Inferior vena cava

Arch of aorta

Left pulmonary artery

Ascending aorta

Left pulmonary veins

Diaphragm

Abdominal aorta

(a) Anterior view

Right pulmonary artery

Left pulmonary artery

Pulmonary trunk

Air sacs (alveoli) of right lung

Air sacs (alveoli) of left lung

Right ventricle

Right pulmonary veins

Left pulmonary veins

Left atrium

(b) Scheme of pulmonary circulation

Q After birth, which are the only arteries that carry deoxygenated blood?

4. The placenta is expelled as the **"afterbirth."**

5. The foramen ovale normally closes shortly after birth to become the **fossa ovalis,** a depression in the interatrial septum. When an infant takes its first breath, the lungs expand and blood flow to the lungs increases. Blood returning from the lungs to the heart increases pressure in the left atrium. This closes the foramen ovale by pushing the valve that guards it against the interatrial septum. Permanent closure occurs in about a year.

6. The ductus arteriosus closes by vasoconstriction almost immediately after birth and becomes the **ligamentum arteriosum.** Complete anatomical obliteration of the lumen takes 1–3 months.

1. Prepare a diagram to show the hepatic portal circulation. Why is this route important?

2. Prepare a diagram to show the route of the pulmonary circulation.

Figure 21.31 Fetal circulation and changes at birth. The boxes between parts (a) and (b) describe the fate of certain fetal structures once postnatal circulation is established.

🔑 **The lungs and gastrointestinal organs do not begin to function until birth.**

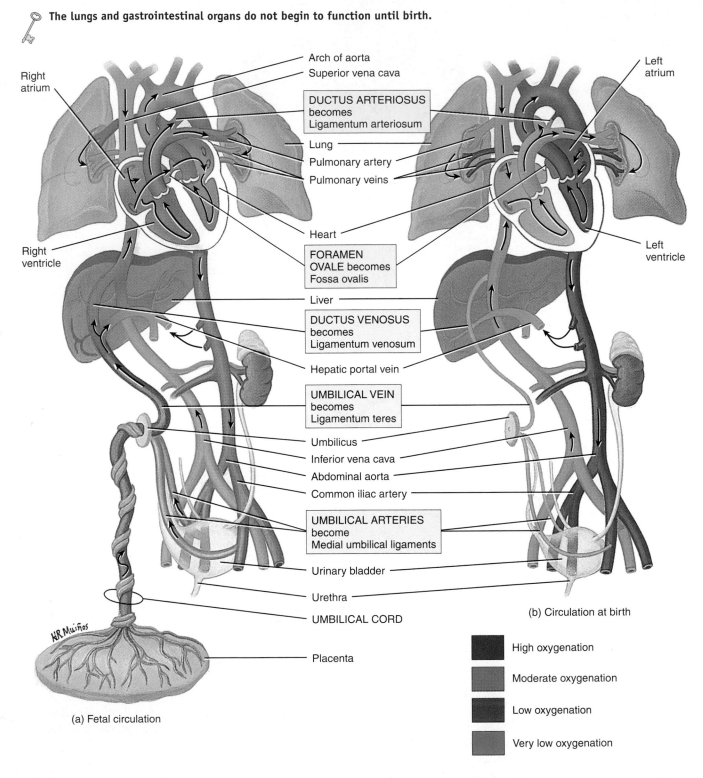

Right atrium

Right ventricle

Left atrium

Left ventricle

Arch of aorta

Superior vena cava

DUCTUS ARTERIOSUS becomes Ligamentum arteriosum

Lung

Pulmonary artery

Pulmonary veins

Heart

FORAMEN OVALE becomes Fossa ovalis

Liver

DUCTUS VENOSUS becomes Ligamentum venosum

Hepatic portal vein

UMBILICAL VEIN becomes Ligamentum teres

Umbilicus

Inferior vena cava

Abdominal aorta

Common iliac artery

UMBILICAL ARTERIES become Medial umbilical ligaments

Urinary bladder

Urethra

UMBILICAL CORD

Placenta

(a) Fetal circulation

(b) Circulation at birth

HR Muiños

High oxygenation

Moderate oxygenation

Low oxygenation

Very low oxygenation

Figure 21.31 (continued)

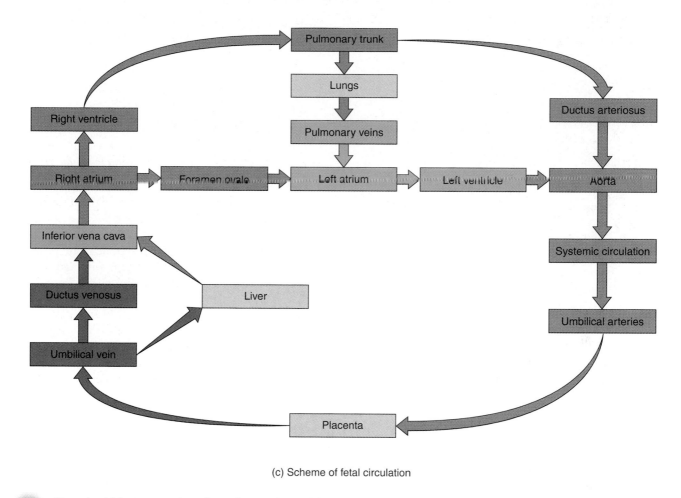

(c) Scheme of fetal circulation

Q Through which structure does the exchange of materials between mother and fetus occur?

---

3. Discuss the anatomy and physiology of the fetal circulation. Indicate the function of the umbilical arteries, umbilical vein, ductus venosus, foramen ovale, and ductus arteriosus.

 ### DEVELOPMENTAL ANATOMY OF BLOOD VESSELS AND BLOOD

OBJECTIVE

• *Describe the development of blood vessels and blood.*

The human yolk sac has little yolk to nourish the developing embryo. Blood and blood vessel formation starts as early as 15–16 days in the **mesoderm** of the yolk sac, chorion, and body stalk.

Blood vessels develop from isolated masses and cords of mesenchyme in the mesoderm called **blood islands** (Figure

21.32). Spaces soon appear in the islands and become the lumens of the blood vessels. Some of the mesenchymal cells immediately around the spaces give rise to the *endothelial lining of the blood vessels*. Mesenchyme around the endothelium forms the *tunics* (intima, media, externa) of the larger blood vessels. Growth and fusion of blood islands form an extensive network of blood vessels throughout the embryo.

Blood plasma and *blood cells* are produced by the endothelial cells and appear in the blood vessels of the yolk sac and allantois quite early. Blood formation in the embryo itself begins at about the second month in the liver and spleen, a little later in red bone marrow, and much later in lymph nodes.

1. Compare the origin of blood vessels and blood.

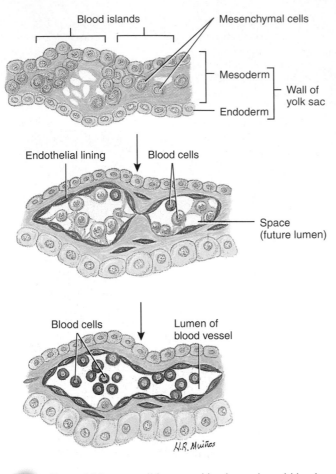

Figure 21.32 Development of blood vessels and blood cells from blood islands.

🔑 **Blood vessel development begins in the embryo on about the 15th or 16th day.**

Q From which germ cell layer are blood vessels and blood derived?

AGING AND THE CARDIOVASCULAR SYSTEM

OBJECTIVE

• *Explain the effects of aging on the cardiovascular system.*

General changes in the cardiovascular system associated with aging include decreased compliance of the aorta, reduction in cardiac muscle fiber size, progressive loss of cardiac muscular strength, reduced cardiac output, a decline in maximum heart rate, and an increase in systolic blood pressure. Total blood cholesterol tends to increase with age, as does low-density lipoprotein (LDL); high-density lipoprotein (HDL) tends to decrease. There is an increase in the incidence of coronary artery disease (CAD), the major cause of heart disease and death in older Americans. Congestive heart failure, a set of symptoms associated with impaired pumping of the heart, is also prevalent in older individuals. Changes in blood vessels that serve brain tissue—for example, atherosclerosis—reduce nourishment to the brain and result in the malfunction or death of brain cells. By age 80, cerebral blood flow is 20% less and renal blood flow is 50% less than in the same person at age 30.

DISORDERS: HOMEOSTATIC IMBALANCES

HYPERTENSION

Hypertension, or persistently high blood pressure, is defined as systolic blood pressure of 140 mm Hg or greater and diastolic blood pressure of 90 mm Hg or greater. Recall that a blood pressure of 120/80 is normal and desirable in a healthy adult. In industrialized societies, hypertension is the most common disorder affecting the heart and blood vessels; it is a major cause of heart failure, kidney disease, and stroke. The classification system adopted in 1997 ranks blood pressure values for adults as follows:

Optimal	Systolic less than 120
	diastolic less than 80
Normal	Systolic less than 130
	diastolic less than 85
High-normal	Systolic 130–139; diastolic 85–89
Hypertension	Systolic 140 or greater; diastolic 90 or greater
Stage 1	Systolic 140–159; diastolic 90–99
Stage 2	Systolic 160–179; diastolic 100–109
Stage 3	Systolic 180 or greater; diastolic 110 or greater

Types and Causes of Hypertension

Amazingly, 90–95% of all cases of hypertension are classified as **primary hypertension,** which is a persistently elevated blood pressure that cannot be attributed to any identifiable cause. The remaining 5–10% of cases are **secondary hypertension,** which has an identifiable underlying cause. Several disorders cause secondary hypertension:

• *Obstruction of renal blood flow* or disorders that damage renal tissue may cause the kidneys to release excessive amounts of renin into the blood. The resulting high level of angiotensin II causes vasoconstriction, thus increasing systemic vascular resistance.

• *Hypersecretion of aldosterone*—resulting, for instance, from a tumor of the adrenal cortex—stimulates excess reabsorption of salt and water by the kidneys, which increases the volume of body fluids.

• *Hypersecretion of epinephrine and norepinephrine* by a **pheochromocytoma** (fē-ō-krō′-mō-sī-TŌ-ma), a tumor of the adrenal medulla. Epinephrine and norepinephrine increase heart rate and contractility and increase systemic vascular resistance.

Damaging Effects of Untreated Hypertension

High blood pressure is known as the "silent killer" because it can cause considerable damage to the blood vessels, heart, brain, and kidneys before it causes pain or other noticeable symptoms. In blood vessels, hypertension causes thickening of the tunica media, accelerates development of atherosclerosis and coronary artery disease (CAD), and increases systemic vascular resistance. In the heart, hypertension increases the afterload, which forces the ventricles to work harder to eject blood. The response to an increased workload is hypertrophy of the myocardium, especially in the wall of the left ventricle. If the coronary blood flow cannot meet the increased demand for oxygen, angina pectoris or even myocardial infarction may result. When myocardial hypertrophy can no longer compensate for the increased afterload, the left ventricle becomes dilated and weakened. Because arteries in the brain are usually less protected by surrounding tissues than are the major arteries in other parts of the body, prolonged hypertension can eventually cause them to rupture, and a stroke occurs due to a brain hemorrhage. Hypertension also damages kidney arterioles, causing them to thicken, which narrows the lumen; because the blood supply to the kidneys is thereby reduced, the kidneys secrete more renin, which elevates the blood pressure even more.

Lifestyle Changes to Reduce Hypertension

Although several categories of drugs (described next) can reduce elevated blood pressure, the following lifestyle changes are also effective in managing hypertension.

- *Lose weight.* This is the best treatment for high blood pressure short of using drugs. Loss of even a few pounds helps reduce blood pressure in overweight hypertensive individuals.
- *Limit alcohol intake.* Drink less than 2 ounces of 100-proof liquor a day, or avoid alcohol altogether.
- *Exercise.* Becoming more physically fit by engaging in moderate activity (such as brisk walking) several times a week for 30–45 minutes can lower systolic blood pressure by about 10 mm Hg.
- *Reduce intake of sodium (salt).* Roughly half the people with hypertension are "salt sensitive." For them, a high-salt diet appears to promote hypertension, and a low-salt diet can lower their blood pressure.
- *Maintain recommended dietary intake of potassium, calcium, and magnesium.* Higher levels of potassium, calcium, and magnesium in the diet are associated with a lower risk of hypertension.
- *Don't smoke.* Smoking has devastating effects on the heart and can augment the damaging effects of high blood pressure by promoting vasoconstriction.
- *Manage stress.* Various meditation and biofeedback techniques help some people reduce high blood pressure. These methods may work by decreasing the daily release of epinephrine and norepinephrine by the adrenal medulla.

Drug Treatment of Hypertension

Drugs having several different mechanisms of action are effective in lowering blood pressure. Many people are successfully treated with *diuretics*, agents that decrease blood pressure by decreasing blood volume because they increase elimination of water and salt in the urine. *ACE (angiotensin converting enzyme) inhibitors* block formation of angiotensin II and thereby promote vasodilation and decrease the liberation of aldosterone. *Beta blockers* reduce blood pressure by inhibiting the secretion of renin and by decreasing heart rate and contractility. *Vasodilators* relax the smooth muscle in arterial walls, causing vasodilation and lowering blood pressure by lowering systemic vascular resistance. An important category of vasodilators are the *calcium channel blockers,* which slow the inflow of Ca^{2+} into vascular smooth muscle cells. They reduce the heart's work load by slowing Ca^{2+} entry into myocardial fibers, thereby decreasing the force of myocardial contraction.

MEDICAL TERMINOLOGY

Aneurysm (AN-yoo-rizm) A thin, weakened section of the wall of an artery or a vein that bulges outward, forming a balloonlike sac. Common causes are atherosclerosis, syphilis, congenital blood vessel defects, and trauma. If untreated, the aneurysm enlarges and the blood vessel wall becomes so thin that it bursts. The result is massive hemorrhage with shock, severe pain, stroke, or death.

Angiogenesis (an′-jē-ō-JEN-e-sis) Formation of new blood vessels.

Aortography (ā′-or-TOG-ra-fē) X-ray examination of the aorta and its main branches after injection of a radiopaque dye.

Arteritis (ar′-te-RĪ-tis; *-itis* = inflammation of) Inflammation of an artery, probably due to an autoimmune response.

Carotid endarterectomy (ka-ROT-id end′-ar-ter-EK-tō-mē) The removal of atherosclerotic plaque from the carotid artery to restore greater blood flow to the brain.

Claudication (klaw′-di-KĀ-shun) Pain and lameness or limping caused by defective circulation of the blood in the vessels of the limbs.

Deep venous thrombosis The presence of a thrombus (blood clot) in a deep vein of the lower limbs. It may lead to (1) pulmonary embolism, if the thrombus dislodges and then lodges within the pulmonary arterial blood flow, and (2) postphlebitic syndrome, which consists of edema, pain, and skin changes due to destruction of venous valves.

Hypotension (hī-pō-TEN-shun) Low blood pressure; most commonly used to describe an acute drop in blood pressure, as occurs during excessive blood loss.

Normotensive (nor′-mō-TEN-siv) Characterized by normal blood pressure.

Occlusion (ō-KLOO-shun) The closure or obstruction of the lumen of a structure such as a blood vessel. An example is an atherosclerotic plaque in an artery.

Orthostatic hypotension (or′-thō-STAT-ik; *ortho-* = straight; *-static* = causing to stand) An excessive lowering of systemic blood pressure when a person assumes an erect or semierect posture; it is usually a sign of a disease. May be caused by excessive fluid loss, certain drugs, and cardiovascular or neurogenic factors. Also called **postural hypotension.**

Phlebitis (fle-BĪ-tis; *phleb-* = vein) Inflammation of a vein, often in a leg.

Raynaud's disease (rā-NŌZ) A vascular disorder, primarily of females, characterized by bilateral attacks of ischemia, usually of the fingers and toes, in which the skin becomes pale and exhibits burning and pain. It is brought on by cold temperatures or emotional stimuli.

Thrombectomy (throm-BEK-tō-mē; *thrombo-* = clot) An operation to remove a blood clot from a blood vessel.

Thrombophlebitis (throm′-bō-fle-BĪ-tis) Inflammation of a vein involving clot formation. Superficial thrombophlebitis occurs in veins under the skin, especially in the calf.

White coat (office) hypertension A stress-induced syndrome found in patients who have elevated blood pressure when being examined by health-care personnel, but otherwise have normal blood pressure.

STUDY OUTLINE

ANATOMY OF BLOOD VESSELS (p. 670)

1. Arteries carry blood away from the heart. The wall of an artery consists of a tunica interna, a tunica media (which maintains elasticity and contractility), and a tunica externa.
2. Large arteries are termed elastic (conducting) arteries, and medium-sized arteries are called muscular (distributing) arteries.
3. Many arteries anastomose: The distal ends of two or more vessels unite. An alternate blood route from an anastomosis is called collateral circulation. Arteries that do not anastomose are called end arteries.
4. Arterioles are small arteries that deliver blood to capillaries.
5. Through constriction and dilation, arterioles assume a key role in regulating blood flow from arteries into capillaries and in altering arterial blood pressure.
6. Capillaries are microscopic blood vessels through which materials are exchanged between blood and tissue cells; some capillaries are continuous, whereas others are fenestrated.
7. Capillaries branch to form an extensive network throughout a tissue. This network increases surface area, allowing a rapid exchange of large quantities of materials.
8. Precapillary sphincters regulate blood flow through capillaries.
9. Microscopic blood vessels in the liver are called sinusoids.
10. Venules are small vessels that continue from capillaries and merge to form veins.
11. Veins consist of the same three tunics as arteries but have a thinner tunica interna and media. The lumen of a vein is also larger than that of a comparable artery.
12. Veins contain valves to prevent backflow of blood.
13. Weak valves can lead to varicose veins.
14. Vascular (venous) sinuses are veins with very thin walls.
15. Systemic veins are collectively called blood reservoirs because they hold a large volume of blood. If the need arises, this blood can be shifted into other blood vessels through vasoconstriction of veins.
16. The principal blood reservoirs are the veins of the abdominal organs (liver and spleen) and skin.

CAPILLARY EXCHANGE (p. 676)

1. Substances enter and leave capillaries by diffusion, vesicular transport, or bulk flow.
2. The movement of water and solutes (except proteins) through capillary walls depends on hydrostatic and osmotic pressures.
3. The near equilibrium between filtration and reabsorption in capillaries is called Starling's law of the capillaries.
4. Edema is an abnormal increase in interstitial fluid.

HEMODYNAMICS: FACTORS AFFECTING CIRCULATION (p. 678)

1. The velocity of blood flow is inversely related to the cross-sectional area of blood vessels; blood flows slowest where cross-sectional area is greatest.
2. The velocity of blood flow decreases from the aorta to arteries to capillaries and increases as blood returns to the heart.
3. Blood flow is determined by blood pressure and resistance.
4. Blood flows from regions of higher to lower pressure; the higher the resistance the lower the blood flow.
5. Cardiac output equals the mean arterial blood pressure divided by total resistance ($CO = MABP \div R$).
6. Blood pressure is the pressure exerted on the walls of a blood vessel.
7. Factors that affect blood pressure are cardiac output, blood volume, viscosity, resistance, and the elasticity of arteries.
8. As blood leaves the aorta and flows through the systemic circulation, its pressure progressively falls to 0 mm Hg by the time it reaches the right ventricle.
9. Resistance depends on blood viscosity, blood vessel length, and blood vessel radius.
10. Venous return depends on pressure differences between the venules and the right ventricle.
11. Blood return to the heart is maintained by several factors, including skeletal muscular contractions, valves in veins (especially in the limbs), and pressure changes associated with breathing.

CONTROL OF BLOOD PRESSURE AND BLOOD FLOW (p. 682)

1. The cardiovascular (CV) center is a group of neurons in the medulla oblongata that regulates heart rate, contractility, and blood vessel diameter.
2. The cardiovascular center receives input from higher brain regions and sensory receptors (baroreceptors and chemoreceptors).
3. Output from the cardiovascular center flows along sympathetic and parasympathetic fibers. Sympathetic impulses propagated along cardioaccelerator nerves increase heart rate and contractility, whereas parasympathetic impulses propagated along vagus nerves decrease heart rate.
4. Baroreceptors monitor blood pressure, and chemoreceptors monitor blood levels of O_2, CO_2, and hydrogen ions.
5. The carotid sinus reflex helps regulate blood pressure in the brain.

6. The aortic reflex is concerned with general systemic blood pressure.

7. Hormones that help regulate blood pressure are epinephrine, norepinephrine, ADH (vasopressin), angiotensin II, and ANP.

8. Autoregulation refers to local, automatic adjustments of blood flow in a given region to meet a particular tissue's need.

9. O_2 level is the principal stimulus for autoregulation.

SHOCK AND HOMEOSTASIS (p. 686)

1. Shock is a failure of the cardiovascular system to deliver enough O_2 and nutrients to meet the metabolic needs of cells.

2. Types of shock include hypovolemic, cardiogenic, vascular, and obstructive.

3. Signs and symptoms include rapid resting heart rate; weak, rapid pulse; clammy, cool, pale skin; sweating; hypotension; altered mental state; decreased urinary output; thirst; and acidosis.

EVALUATING CIRCULATION (p. 689)

1. Pulse is the alternate expansion and elastic recoil of an artery wall with each heartbeat. It may be felt in any artery that lies near the surface or over a hard tissue.

2. A normal resting pulse (heart) rate is 70–80 beats/min.

3. Blood pressure is the pressure exerted by blood on the wall of an artery when the left ventricle undergoes systole and then diastole. It is measured by the use of a sphygmomanometer.

4. Systolic blood pressure (SBP) is the force of blood recorded during ventricular contraction. Diastolic blood pressure (DBP) is the force of blood recorded during ventricular relaxation. Normal blood pressure is 120/80 mm Hg.

5. Pulse pressure is the difference between systolic and diastolic pressure. It normally is about 40 mm Hg.

CIRCULATORY ROUTES (p. 690)

1. The two basic postnatal circulatory routes are the systemic and pulmonary circulations.

2. Among the subdivisions of the systemic circulation are the coronary (cardiac) circulation and the hepatic portal circulation.

3. Fetal circulation exists only in the fetus.

4. The systemic circulation carries oxygenated blood from the left ventricle through the aorta to all parts of the body (including some lung tissue, but *not* the air sacs (alveoli) of the lungs) and returns the deoxygenated blood to the right atrium.

5. The aorta is divided into the ascending aorta, the arch of the aorta, and the descending aorta. Each section gives off arteries that branch to supply the whole body.

6. Blood returns to the heart through the systemic veins. All veins of the systemic circulation drain into the superior or inferior venae cavae or the coronary sinus, which, in turn, empty into the right atrium.

7. The principal blood vessels of the systemic circulation may be reviewed in Exhibits 21.1–21.12.

8. The hepatic portal circulation detours venous blood from the gastrointestinal organs and spleen and directs it into the hepatic portal vein of the liver before it is returned to the heart. It enables the liver to utilize nutrients and detoxify harmful substances in the blood.

9. The pulmonary circulation takes deoxygenated blood from the right ventricle to the alveoli within the lungs and returns oxygenated blood from the alveoli to the left atrium. It allows blood to be oxygenated for systemic circulation.

10. The fetal circulation involves the exchange of materials between fetus and mother.

11. The fetus derives O_2 and nutrients and eliminates CO_2 and wastes through the maternal blood supply via the placenta.

12. At birth, when pulmonary (lung), digestive, and liver functions begin, the special structures of fetal circulation are no longer needed.

DEVELOPMENTAL ANATOMY OF BLOOD VESSELS AND BLOOD (p. 731)

1. Blood vessels develop from isolated masses of mesenchyme in mesoderm called blood islands.

2. Blood is produced by the endothelium of blood vessels.

AGING AND THE CARDIOVASCULAR SYSTEM (p. 732)

1. General changes associated with aging include reduced elasticity of blood vessels, reduction in cardiac muscle size, reduced cardiac output, and increased systolic blood pressure.

2. The incidence of coronary artery disease (CAD), congestive heart failure (CHF), and atherosclerosis increases with age.

SELF-QUIZ QUESTIONS

Choose the best answer to the following questions:

1. The blood vessels that function as pressure reservoirs are (a) muscular arteries, (b) elastic arteries, (c) anastomoses, (d) veins, (e) arterioles.

2. The main vessels that permit the exchange of nutrients and wastes between the blood and tissue cells through interstitial spaces are (a) arteries, (b) arterioles, (c) capillaries, (d) venules, (e) veins.

3. Which of the following are causes of edema? (1) increased blood colloid osmotic pressure, (2) increased blood hydrostatic pressure in capillaries, (3) decreased plasma protein concentration, (4) increased capillary permeability, (5) blockage of lymph vessels.

(a) 1, 2, and 3, (b) 2, 3, and 4, (c) 3, 4, and 5, (d) 1, 3, 4, and 5, (e) 2, 3, 4, and 5

4. Systemic vascular resistance depends on which of the following factors? (1) blood viscosity, (2) total blood vessel length, (3) average blood vessel radius, (4) type of blood vessel, (5) oxygen concentration of the blood.

(a) 1, 2, and 3, (b) 2, 3, and 4, (c) 3, 4, and 5, (d) 1, 3, and 5, (e) 2, 4, and 5

5. Which of the following help regulate blood pressure? (1) baroreceptor and chemoreceptor reflexes, (2) hormones, (3) autoregulation, (4) H^+ concentration of blood, (5) oxygen concentration of the blood.

(a) 1, 2, and 4, (b) 2, 4, and 5, (c) 1, 4, and 5, (d) 1, 2, 3, 4, and 5, (e) 3, 4, and 5

6. Which of the following statements are correct? (1) Hypertension most commonly affects the heart. (2) Hypertension compromises kidney function. (3) Primary hypertension is persistently elevated blood pressure that cannot be attributed to any identifiable cause. (4) Continual hypertension can cause a stroke. (5) Lifestyle changes such as weight loss, exercise, and reduced salt intake can help manage hypertension. (a) 1, 3, 4, and 5, (b) 2, 3, 4, and 5, (c) 1, 2, 3, 4, and 5, (d) 1, 4, and 5, (e) 2, 3, and 5

7. Match the following:

____ (a) a traveling pressure wave created by the alternate expansion and recoil of elastic arteries after each systole of the left ventricle

____ (b) the lowest force of blood in arteries during ventricular relaxation

____ (c) a slow resting heart rate or pulse rate

____ (d) an inadequate cardiac output that results in a failure of the cardiovascular system to deliver enough oxygen and nutrients to meet the metabolic needs of body cells

____ (e) a rapid resting heart rate or pulse rate

____ (f) the highest force with which blood pushes against arterial walls as a result of ventricular contraction

(1) shock
(2) pulse
(3) tachycardia
(4) bradycardia
(5) systolic blood pressure
(6) diastolic blood pressure

8. Match the following:

____ (a) supplies blood to the kidney

____ (b) drains blood from the small intestine, portions of the large intestine, stomach, and pancreas

____ (c) supply and drain blood from the heart muscle

____ (d) supply blood to the lower limbs

____ (e) drains oxygenated blood from the lungs and sends it to the left atrium

____ (f) supplies blood to the stomach, liver, and pancreas

____ (g) supply blood to the brain

____ (h) supplies blood to the large intestine

____ (i) drain blood from the head

____ (j) detours venous blood from the gastrointestinal organs and spleen through the liver before it returns to the heart

____ (k) supplies blood to the small intestine

____ (l) a part of the venous circulation of the leg; a vessel used in heart bypass surgery

(1) superior mesenteric vein
(2) inferior mesenteric artery
(3) pulmonary vein
(4) coronary vessels
(5) hepatic portal circulation
(6) carotid arteries
(7) jugular veins
(8) celiac trunk
(9) common iliac arteries
(10) superior mesenteric artery
(11) renal artery
(12) saphenous vein

True or false:

9. The most important method of capillary exchange is simple diffusion.

10. The overall function of the cardiovascular system is to ensure adequate circulation of blood to all body tissues to enable capillary exchange between blood plasma, interstitial fluid, and tissue cells.

Complete the following:

11. Substances enter and leave capillaries by ____, ____, and ____.

12. For bulk flow to occur, two pressures, ____ and ____, promote filtration; ____ is the main pressure promoting reabsorption.

13. The distribution of the cardiac output to various tissues depends on the interplay of the ____ that drives the blood flow, the ____ to blood flow, ____ and ____.

14. The ____ reflex helps maintain normal blood pressure in the brain; the ____ reflex governs general systemic blood pressure.

15. Match the following:

____ (a) returns oxygenated blood from the placenta

____ (b) an opening in the septum between the right and left atria

____ (c) becomes the ligamentum venosum after birth

____ (d) passes blood from the fetus to the placenta

____ (e) bypasses the nonfunctioning lungs; becomes the ligamentum arteriosum at birth

____ (f) become the medial umbilical ligament at birth

____ (g) becomes the ligamentum teres at birth

(1) ductus venosus
(2) ductus arteriosus
(3) foramen ovale
(4) umbilical arteries
(5) umbilical vein

CRITICAL THINKING QUESTIONS

1. Which structures that are present in the fetal circulation are absent in the adult circulatory system? Why do these changes occur? (HINT: *A mother may want to do everything for her baby after he's born, but she can't breathe for him.*)

2. Josef is thinking about the route that his blood follows on its way from his heart to his hand. He realizes that the route is a bit different for blood flowing to the left hand as compared to the right hand. Trace the pathway of blood from the heart to the right and left hands. (HINT: *The lack of symmetry is closer to the heart than the hand.*)

3. Lance spent the week before the exam lying on a beach studying the back of his eyelids. "What was that question about sinuses doing on the circulatory test?" he complained. "We did that back with bones!" How would you enlighten Lance? (HINT: *In bone, sinuses are filled with air; these aren't.*)

ANSWERS TO FIGURE QUESTIONS

21.1 The femoral artery has the thicker wall; the femoral vein has the wider lumen.

21.2 As a result of atherosclerosis, less energy is stored in the less-compliant elastic arteries during systole; thus, the heart must pump harder to maintain the same rate of blood flow.

21.3 Metabolically active tissues use O_2 and produce wastes more rapidly than inactive tissues.

21.4 Materials cross capillary walls through intercellular clefts and fenestrations, via transcytosis in pinocytic vesicles, and through the plasma membranes of endothelial cells.

21.5 When you are standing, gravity tends to cause pooling of blood in the veins of the limbs. The valves prevent backflow as the blood proceeds toward the right atrium after each heartbeat. When you are erect, gravity aids the flow of blood in neck veins back toward the heart.

21.6 Volume in veins is about 60% of 5 liters, or 3 liters; volume in capillaries is about 5% of 5 liters, or 250 mL.

21.7 Blood colloid osmotic pressure is lower than normal in a person whose plasma protein level is depressed, and therefore capillary reabsorption is low. The result is edema (see page 678).

21.8 Velocity of blood flow is fastest in the aorta and arteries.

21.9 Mean blood pressure in the aorta is about 93 mm Hg.

21.10 Venous return is aided by the skeletal muscle and respiratory pumps.

21.11 Systemic vascular resistance is regulated mainly by vasodilation and vasoconstriction of arterioles.

21.12 The cardiovascular center regulates cardiac muscle in the heart and smooth muscle in blood vessel walls.

21.13 Impulses to the cardiovascular center pass from baroreceptors in the carotid sinuses via the glossopharyngeal (IX) nerves and from baroreceptors in the arch of the aorta via the vagus (X) nerves.

21.14 It represents a change that occurs when you stand up because gravity causes increased pooling of blood in leg veins once you are upright, decreasing the blood pressure in your upper body.

21.15 Not necessarily; if systemic vascular resistance has increased greatly, perfusion may be inadequate.

21.16 Diastolic blood pressure = 95 mm Hg; systolic blood pressure = 142 mm Hg; pulse pressure = 47 mm Hg. This person has stage I hypertension because the systolic blood pressure is greater than 140 mm Hg and the diastolic blood pressure is greater than 90 mm Hg.

21.17 The two main circulatory routes are the systemic and pulmonary circulations.

21.18 The subdivisions of the aorta are the ascending aorta, arch of the aorta, thoracic aorta, and abdominal aorta.

21.19 The coronary arteries arise from the ascending aorta.

21.20 Branches of the arch of aorta are the brachiocephalic trunk, left common carotid artery, and left subclavian artery.

21.21 The thoracic aorta begins at the level of the intervertebral disc between T4 and T5.

21.22 The abdominal aorta begins at the aortic hiatus in the diaphragm.

21.23 The abdominal aorta divides into the common iliac arteries at about the level of L4.

21.24 The superior vena cava drains regions above the diaphragm, and the inferior vena cava drains regions below the diaphragm.

21.25 All venous blood in the brain drains into the internal jugular veins.

21.26 The median cubital vein is often used for withdrawing blood.

21.27 The inferior vena cava returns blood from abdominopelvic viscera to the heart.

21.28 Superficial veins of the lower limbs are the dorsal venous arch and the great saphenous and small saphenous veins.

21.29 The hepatic veins carry blood away from the liver.

21.30 The pulmonary arteries carry deoxygenated blood.

21.31 Exchange of materials between mother and fetus occurs across the placenta.

21.32 Blood vessels and blood derive from mesoderm.

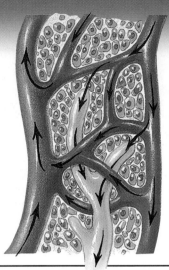

The ability of the body to ward off disease is called **resistance;** conversely, vulnerability or lack of resistance is termed **susceptibility.** Resistance to disease is of two types: nonspecific and specific. **Nonspecific resistance** to disease includes defense mechanisms that provide immediate but general protection against invasion by a wide range of **pathogens,** which are disease-producing microbes such as bacteria, viruses, and parasites. The first line of defense in nonspecific resistance is provided by mechanical and chemical barriers of the skin and mucous membranes; the acidity of the stomach contents, for example, kills many bacteria ingested in food. **Specific resistance** or **immunity** develops more slowly and involves activation of specific lymphocytes that combat a particular pathogen or other foreign substance. The body system responsible for immunity is the lymphatic system. This chapter describes the mechanisms that provide defenses against intruders and promote the repair of damaged body tissues.

THE LYMPHATIC SYSTEM

OBJECTIVES
• *Describe the general components of the lymphatic system, and list its functions.*
• *Describe the organization of lymphatic vessels.*
• *Describe the formation and flow of lymph.*
• *List and describe the primary and secondary lymphatic organs and tissues.*

The **lymphatic system** (lim-FAT-ik) consists of a fluid called lymph flowing within lymphatic vessels, several structures and organs that contain lymphatic tissue, and red bone marrow, which houses stem cells that develop into lymphocytes (Figure 22.1). The composition of interstitial fluid and lymph are basically the same; the major difference between the two is location. After fluid passes from interstitial spaces into lymphatic vessels, it is called **lymph** (LIMF; = clear fluid). Lymphatic tissue is a specialized type of connective tissue that contains large numbers of lymphocytes.

Functions of the Lymphatic System

The lymphatic system has three primary functions:

1. *Draining interstitial fluid.* Lymphatic vessels drain excess interstitial fluid from tissue spaces.
2. *Transporting dietary lipids.* Lymphatic vessels transport the lipids and lipid-soluble vitamins (A, D, E, and K) absorbed by the gastrointestinal tract to the blood.
3. *Facilitating immune responses.* Lymphatic tissue initiates highly specific responses directed against particular microbes or abnormal cells. Lymphocytes, aided by macrophages, recognize foreign cells, microbes, toxins, and cancer cells and respond to them in two basic ways: Lymphocytes called T cells destroy the intruders by causing them to rupture or by releasing cytotoxic (cell-killing) substances; lymphocytes called B cells differentiate into plasma cells, which secrete antibodies—proteins that combine with and cause destruction of specific foreign substances.

Figure 22.1 Components of the lymphatic system.

739

 The lymphatic system consists of lymph, lymphatic vessels, lymphatic tissues, and red bone marrow.

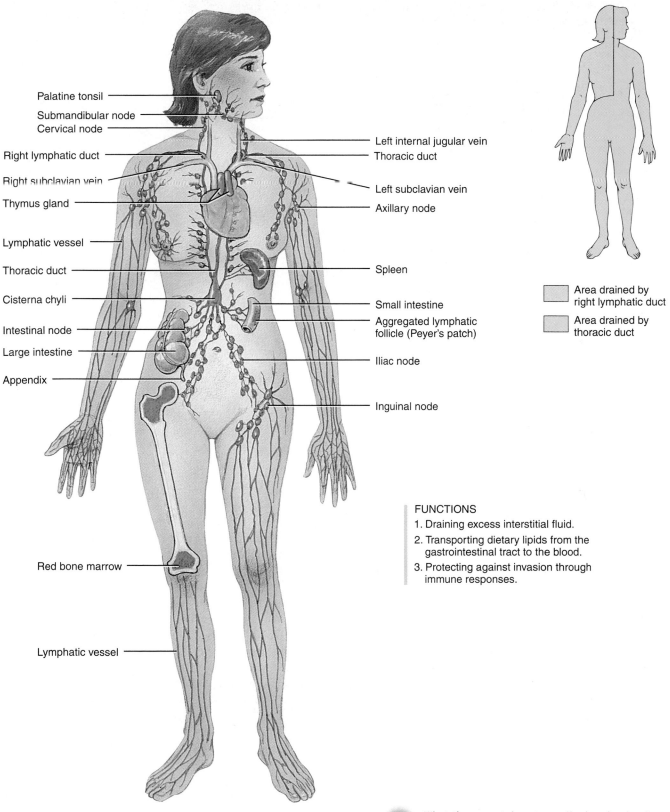

Palatine tonsil

Submandibular node

Cervical node

Right lymphatic duct

Right subclavian vein

Thymus gland

Lymphatic vessel

Thoracic duct

Cisterna chyli

Intestinal node

Large intestine

Appendix

Red bone marrow

Lymphatic vessel

Left internal jugular vein

Thoracic duct

Left subclavian vein

Axillary node

Spleen

Small intestine

Aggregated lymphatic follicle (Peyer's patch)

Iliac node

Inguinal node

Area drained by right lymphatic duct

Area drained by thoracic duct

FUNCTIONS

1. Draining excess interstitial fluid.
2. Transporting dietary lipids from the gastrointestinal tract to the blood.
3. Protecting against invasion through immune responses.

Anterior view of principal components of lymphatic system

 What tissue contains stem cells that develop into lymphocytes?

Figure 22.2 Lymphatic capillaries.

Lymphatic capillaries are found throughout the body except in avascular tissues, the central nervous system, portions of the spleen, and red bone marrow.

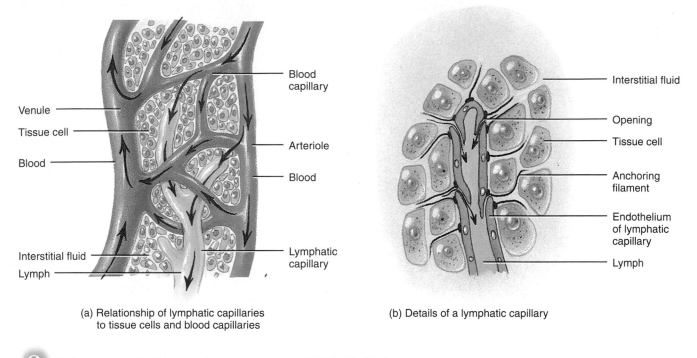

(a) Relationship of lymphatic capillaries to tissue cells and blood capillaries

(b) Details of a lymphatic capillary

Q Is lymph more similar to blood plasma or to interstitial fluid? Why?

Lymphatic Vessels and Lymph Circulation

Lymphatic vessels begin as **lymphatic capillaries,** closed-ended tubes located in the spaces between cells (Figure 22.2). Just as blood capillaries converge to form venules and veins, lymphatic capillaries unite to form larger **lymphatic vessels** (see Figure 22.1), which resemble veins in structure but have thinner walls and more valves. At various intervals along the lymphatic vessels, lymph flows through lymphatic tissue structures called **lymph nodes.** In the skin, lymphatic vessels lie in subcutaneous tissue and generally follow veins; lymphatic vessels of the viscera generally follow arteries, forming plexuses (networks) around them.

Lymphatic Capillaries

Lymphatic capillaries are slightly larger in diameter than blood capillaries and are found throughout the body, except in avascular tissues (such as cartilage, the epidermis, and the cornea of the eye), the central nervous system, portions of the spleen, and red bone marrow. The unique structure of lymphatic capillaries permits interstitial fluid to flow into but not out of them. The margins of endothelial cells that make up the wall of a lymphatic capillary overlap; when pressure is greater in the interstitial fluid than in lymph, the cells separate slightly, like a one-way valve opening, and fluid

enters the lymphatic capillary (see Figure 22.2b). When pressure is greater inside the lymphatic capillary, the cells adhere more closely, so lymph cannot flow back into interstitial fluid. *Anchoring filaments,* which lie at right angles to the lymphatic capillary, attach lymphatic endothelial cells to surrounding tissues. When excess interstitial fluid accumulates and causes tissue swelling, the anchoring filaments are pulled, making the openings between cells even larger so that more fluid can flow into the lymphatic capillary.

In the small intestine, specialized lymphatic capillaries called **lacteals** (LAK-tē-als; *lact-* = milky) carry dietary lipids into lymphatic vessels and ultimately into the blood. The presence of these lipids causes the lymph draining the small intestine to appear creamy white; such lymph is referred to as **chyle** (KĪL; = juice). Elsewhere, lymph is a clear, pale yellow fluid.

Lymph Trunks and Ducts

Lymph passes from lymphatic capillaries into lymphatic vessels and then through lymph nodes. Lymphatic vessels exiting lymph nodes pass lymph either toward another node within the same group or on to another group of nodes. From the most proximal group of each chain of nodes, the exiting vessels unite to form **lymph trunks.** The principal

Figure 22.3 Routes for drainage of lymph from lymph trunks into thoracic and right lymphatic ducts.

🔑 **All lymph returns to the bloodstream through the thoracic (left) lymphatic duct and right lymphatic duct.**

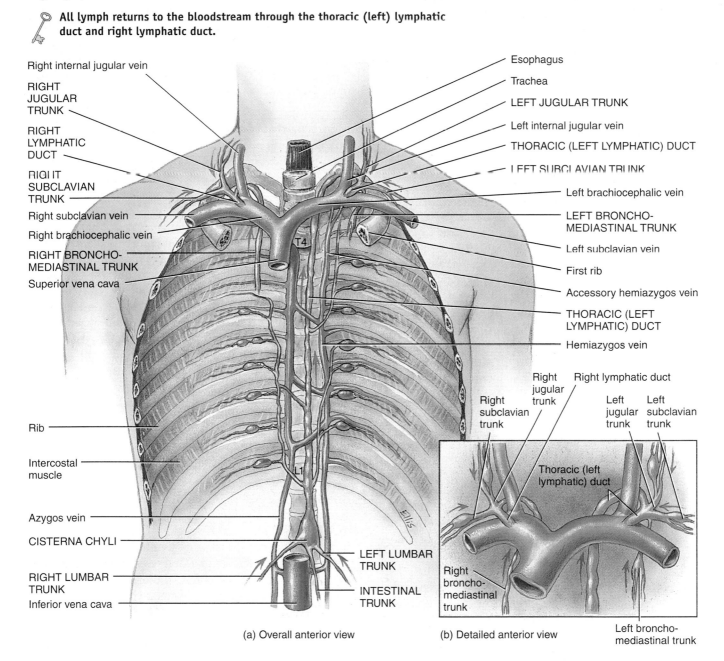

(a) Overall anterior view

(b) Detailed anterior view

Q Which lymphatic vessels empty into the cisterna chyli, and which duct receives lymph from the cisterna chyli?

trunks are the **lumbar, intestinal, bronchomediastinal, subclavian,** and **jugular trunks** (Figure 22.3). The principal trunks pass their lymph into two main channels, the thoracic duct and the right lymphatic duct. Lymph passes from these ducts into venous blood.

The **thoracic (left lymphatic) duct** is about 38–45 cm (15–18 in.) long and begins as a dilation called the **cisterna chyli** (sis-TER-na KĪ-lē; *cisterna* = cavity or reservoir) ante-

rior to the second lumbar vertebra. The thoracic duct, the main collecting duct of the lymphatic system, receives lymph from the left side of the head, neck, and chest, the left upper limb, and the entire body inferior to the ribs. The thoracic duct drains lymph into venous blood via the **left subclavian vein.**

The cisterna chyli receives lymph from the right and left lumbar trunks and from the intestinal trunk. The lumbar trunks drain lymph from the lower limbs, the wall and viscera of the pelvis, the kidneys, the adrenal glands, and the deep lymphatic vessels that drain lymph from most of the abdominal wall. The intestinal trunk drains lymph from the stomach, intestines, pancreas, spleen, and part of the liver.

In the neck, the thoracic duct also receives lymph from the left jugular, left subclavian, and left bronchomediastinal trunks. The left jugular trunk drains lymph from the left side of the head and neck, and the left subclavian trunk drains lymph from the left upper limb. The left bronchomediastinal trunk drains lymph from the left side of the deeper parts of the anterior thoracic wall, the superior part of the anterior abdominal wall, the anterior part of the diaphragm, the left lung, and the left side of the heart.

The **right lymphatic duct** (see Figure 22.3) is about 1.25 cm (0.5 in.) long and drains lymph from the upper right side of the body into venous blood via the **right subclavian vein.** Three lymphatic trunks drain into the right lymphatic duct: the right jugular trunk, which drains the right side of the head and neck; the right subclavian trunk, which drains the right upper limb; and the right bronchomediastinal trunk, which drains the right side of the thorax, the right lung, the right side of the heart, and part of the liver.

Formation and Flow of Lymph

Most components of blood plasma freely filter through the capillary walls to form interstitial fluid. More fluid filters out of blood capillaries, however, than returns to them by reabsorption (see Figure 21.7 on page 677). The excess filtered fluid—about 3 liters per day—drains into lymphatic vessels and becomes lymph. Because most plasma proteins are too large to leave blood vessels, interstitial fluid contains only small amounts of proteins. Proteins that do leave plasma, however, cannot return to the blood directly by diffusion because the concentration gradient (high level of proteins inside blood capillaries, low level outside) prevents such movement. Thus, an important function of lymphatic vessels is to return lost plasma proteins to the bloodstream.

Ultimately, lymph drains into venous blood through the right lymphatic duct and the thoracic duct at the junction of the internal jugular and subclavian veins (see Figure 22.3). Thus, the sequence of fluid flow is blood capillaries (blood) → interstitial spaces (interstitial fluid) → lymphatic capillaries (lymph) → lymphatic vessels (lymph) → lymphatic ducts (lymph) → subclavian veins (blood). This sequence, as well as the relationship of the lymphatic and cardiovascular systems, is illustrated in Figure 22.4.

Recall that the skeletal muscle and respiratory pumps promote return of venous blood to the heart (see page 681). These two pumps similarly promote the flow of lymph from tissue spaces to the large lymphatic ducts to the subclavian veins. Muscle contractions compress lymphatic vessels and force lymph toward the subclavian veins in a kind of milking action. Lymphatic vessels contain one-way valves, similar to those found in veins, that prevent backflow of lymph. During respirations, the pressure present in the lymphatic sys-

tem changes. With each inhalation, lymph flows from the abdominal region, where the pressure is higher, toward the thoracic region, where it is lower. Likewise, the fall in abdominal pressure during exhalation promotes the flow of lymph from more distal vessels into the abdominal vessels. In addition, when a lymphatic vessel distends, the smooth muscle in its wall contracts, which helps move lymph from one segment of the vessel to the next.

Lymphatic Organs and Tissues

The organs and tissues of the lymphatic system, which are widely distributed throughout the body, are classified into two groups based on their functions. **Primary lymphatic organs** provide the appropriate environment for stem cells to divide and mature into B cells and T cells, the lymphocytes that carry out immune responses. The primary lymphatic organs are the **red bone marrow** (in flat bones and the epiphyses of long bones of adults) and the **thymus gland.** Pluripotent stem cells in red bone marrow give rise to mature B cells and to pre-T cells, the latter of which migrate to and mature in the thymus gland. The **secondary lymphatic organs** and **tissues,** which are the sites where most immune responses occur, include **lymph nodes,** the **spleen,** and **lymphatic nodules.** The thymus gland, lymph nodes, and spleen are considered organs because each is surrounded by a connective tissue capsule; lymphatic nodules, in contrast, are not organs because they lack a capsule.

Thymus Gland

The **thymus gland** usually has two lobes and is located in the mediastinum, posterior to the sternum (Figure 22.5a). An enveloping layer of connective tissue holds the two **thymic lobes** closely together, but a connective tissue **capsule** encloses each lobe separately. Extensions of the capsule, called **trabeculae** (tra-BEK-yoo-lē; = little beams) penetrate inward and divide the lobes into **lobules** (Figure 22.5b).

Each lobule consists of a deeply staining outer **cortex** and a lighter staining central **medulla.** The cortex is composed of tightly packed lymphocytes, epithelial cells called **reticular epithelial cells** that surround clusters of lymphocytes, and macrophages. The medulla consists mostly of reticular epithelial cells and more widely scattered lymphocytes. Although only some of their functions are known, the reticular epithelial cells produce thymic hormones, which are thought to aid in the maturation of T cells. In addition, the medulla contains characteristic **thymic (Hassall's) corpuscles,** concentric layers of flattened reticular epithelial cells filled with keratohyalin granules and keratin (Figure 22.5c).

The thymus gland is large in infants, having a mass of about 70 g (a little more than 2 oz). After puberty, adipose and areolar connective tissue begin to replace the thymic tissue. By the time a person reaches maturity, the gland has atrophied considerably, and in old age it may weigh only 3 g.

Figure 22.4 Schematic diagram showing the relationship of the lymphatic system to the cardiovascular system.

 The sequence of fluid flow is blood capillaries (blood) → interstitial spaces (interstitial fluid) → lymphatic capillaries (lymph) → lymphatic vessels (lymph) → lymphatic ducts (lymph) → subclavian veins (blood).

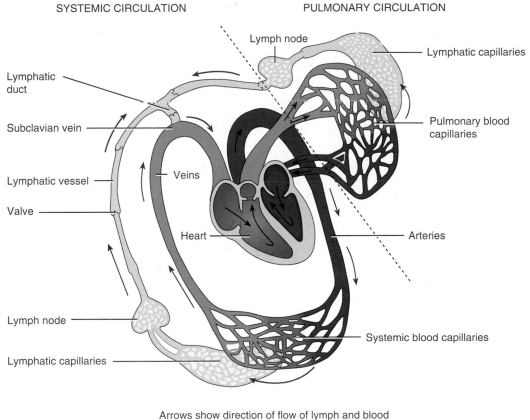

Arrows show direction of flow of lymph and blood

Q   Does inhalation promote or hinder the flow of lymph?

Lymph Nodes

The approximately 600 bean-shaped organs located along lymphatic vessels are called **lymph nodes.** They are scattered throughout the body, both superficially and deep, and usually in groups (see Figure 22.1). Lymph nodes are heavily concentrated near the mammary glands and in the axillae and groin.

Lymph nodes are 1–25 mm (0.04–1 in.) long and are covered by a **capsule** of dense connective tissue that extends into the node (Figure 22.6). The capsular extensions, called the **trabeculae,** divide the node into compartments, provide support, and provide a route for blood vessels into the interior of a node. Internal to the capsule is a supporting network of reticular fibers and fibroblasts. The capsule, trabeculae, reticular fibers, and fibroblasts constitute the *stroma,* or framework, of a lymph node. The *parenchyma* of a lymph node is specialized into two regions: a superficial cortex and a deep medulla. The cortex of a lymph node is divided into outer and inner regions. The **outer cortex** contains lymphatic nodules of B cells (B lymphocytes). Lighter-staining

areas in the nodules, called **germinal centers,** are where B cells proliferate into antibody-secreting plasma cells. The reticular cells of the germinal centers, called **dendritic cells,** serve as antigen-presenting cells that help initiate immune responses. The germinal centers also contain macrophages. Deep to the outer cortex is the **inner cortex,** which contains T cells (T lymphocytes). The **medulla** of a lymph node contains B cells and plasma cells in tightly packed strands called **medullary cords.**

Lymph flows through a node in one direction only. It enters through **afferent lymphatic vessels** (*afferent* = to carry toward), which penetrate the convex surface of the node at several points. The afferent vessels contain valves that open toward the center of the node, such that the lymph is directed *inward.* Within the node, lymph enters **sinuses,** which are a series of irregular channels that contain branching reticular fibers, lymphocytes, and macrophages. From the afferent lymphatic vessels, lymph flows into the **subcapsular sinus,** immediately beneath the capsule. Then lymph flows into **trabecular sinuses,** which extend through the cortex and run parallel to the trabeculae. From there, lymph

Figure 22.5 Thymus gland. (See Tortora, *A Photographic Atlas of the Human Body*, Figure 7.2)

 The bilobed thymus gland is largest at puberty and then atrophies as individuals age.

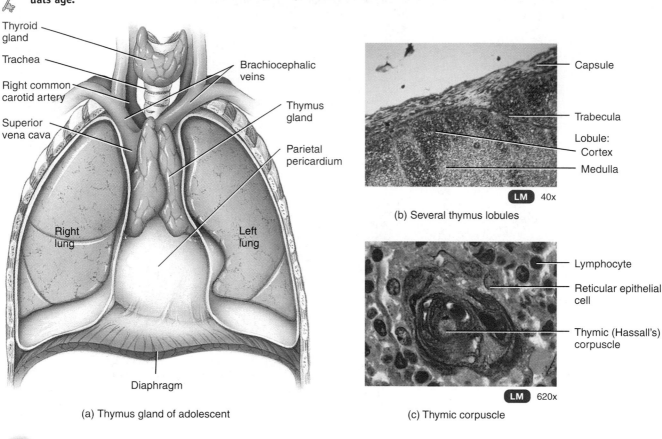

(a) Thymus gland of adolescent

(b) Several thymus lobules

(c) Thymic corpuscle

Q Which lymphocytes mature in the thymus gland?

flows into **medullary sinuses,** which extend through the medulla. The medullary sinuses drain into one or two **efferent lymphatic vessels** (*efferent* = to carry away), which are wider than afferent vessels and fewer in number. They contain valves that open away from the center of the node to convey lymph *out* of the node. Efferent lymphatic vessels emerge from one side of the lymph node at a slight depression called a **hilus** (HĪ-lus). Blood vessels also enter and leave the node at the hilus.

Only lymph nodes filter lymph. As lymph enters one end of a node, foreign substances are trapped by the reticular fibers within the sinuses of the node. Then macrophages destroy some foreign substances by phagocytosis while lymphocytes destroy others by a variety of immune responses. Filtered lymph then leaves the other end of the node. Plasma cells and T cells that have proliferated within a lymph node can also leave in lymph and circulate to other parts of the body.

 CLINICAL APPLICATION
Metastasis Through the Lymphatic System

Metastasis (me-TAS-ta-sis; *meta-* = beyond; *stasis* = to stand) is the spread of disease from one organ

to another not directly connected to it. It is a characteristic of all malignant tumors. Cancer cells may travel via the blood or lymphatic system and establish new tumors where they lodge. When metastasis occurs via the lymphatic system, secondary tumor sites can be predicted according to the direction of lymph flow from the primary tumor site. Cancerous lymph nodes feel enlarged, firm, nontender, and fixed to underlying structures. By contrast, most lymph nodes that are enlarged due to an infection are not firm, are moveable, and are very tender. ■

Spleen

The oval **spleen** is the largest single mass of lymphatic tissue in the body, measuring about 12 cm (5 in.) in length (Figure 22.7a). It is located in the left hypochondriac region between the stomach and diaphragm. The superior surface of the spleen is smooth and convex and conforms to the concave surface of the diaphragm. Neighboring organs make indentations in the visceral surface of the spleen—the gastric impression (stomach), the renal impression (left kidney), and the colic impression (left flexure of colon). Like lymph nodes, the spleen has a hilus through which pass the splenic artery, splenic vein, and efferent lymphatic vessels.

Figure 22.6 Structure of a lymph node. Arrows indicate the direction of lymph flow through a lymph node. (See Tortora, *A Photographic Atlas of the Human Body,* Figure 7.4)

🔑 **Lymph nodes are present throughout the body, usually clustered in groups.**

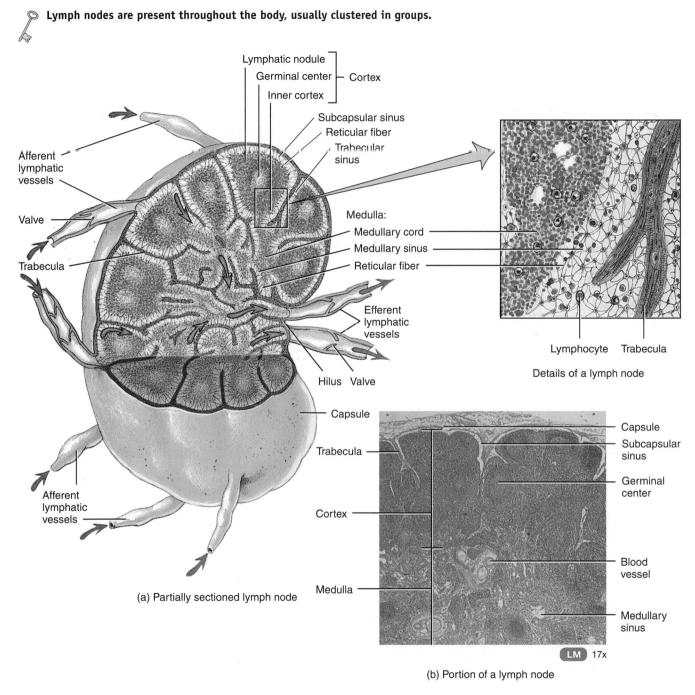

(a) Partially sectioned lymph node

Details of a lymph node

(b) Portion of a lymph node

Q   What happens to foreign substances that enter a lymph node in lymph?

A capsule of dense connective tissue surrounds the spleen. Trabeculae extend inward from the capsule, which, in turn, is covered by a serous membrane, the visceral peritoneum. The capsule plus trabeculae, reticular fibers, and fibroblasts constitute the stroma of the spleen; the parenchyma of the spleen consists of two different kinds of tissue called white pulp and red pulp (Figure 22.7b). **White pulp** is lymphatic tissue, mostly lymphocytes and macrophages, arranged around branches of the splenic artery called central arteries. The **red pulp** consists of **venous sinuses** filled with blood and cords of splenic tissue called **splenic (Billroth's) cords.** Splenic cords consist of red blood cells, macrophages, lymphocytes, plasma cells, and granulocytes. Veins are closely associated with the red pulp.

Figure 22.7 Structure of the spleen. (See Tortora, *A Photographic Atlas of the Human Body,* Figure 7.3)

🔑 **The spleen is the largest single mass of lymphatic tissue in the body.**

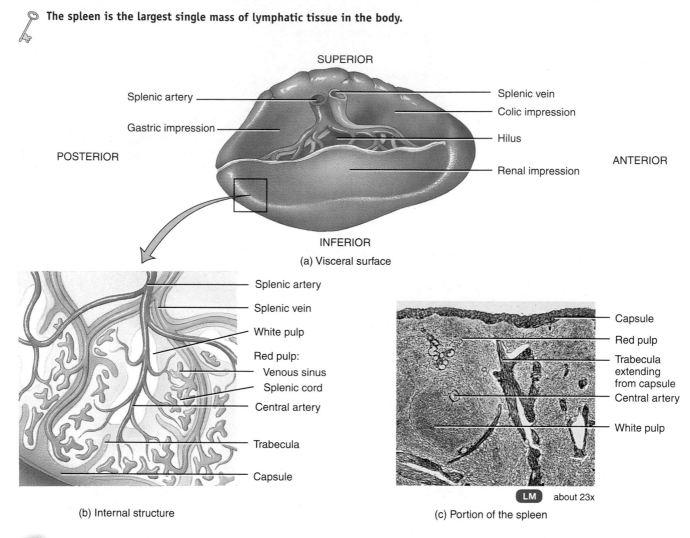

(a) Visceral surface

(b) Internal structure

(c) Portion of the spleen

Q After birth, what are the main functions of the spleen?

Blood flowing into the spleen through the splenic artery enters the central arteries of the white pulp. Within the white pulp, B cells and T cells carry out immune functions while spleen macrophages destroy blood-borne pathogens by phagocytosis. Within the red pulp, the spleen performs three functions related to blood cells: (1) removal by macrophages of worn out or defective blood cells and platelets; (2) storage of platelets, perhaps up to one-third of the body's supply; and (3) production of blood cells (hemopoiesis) during fetal life.

The spleen is the organ most often damaged in cases of abdominal trauma. A ruptured spleen causes severe intraperitoneal hemorrhage and shock. Prompt removal of the spleen *(splenectomy)* is needed to prevent the patient from bleeding to death. Other structures, particularly red bone marrow and the liver, can take over functions normally carried out by the spleen.

Lymphatic Nodules

Lymphatic nodules are oval-shaped concentrations of lymphatic tissue that are not surrounded by a capsule. Because they are scattered throughout the lamina propria (connective tissue) of mucous membranes lining the gastrointestinal, urinary, and reproductive tracts and the respiratory airways, lymphatic nodules are also referred to as **mucosa-associated lymphoid tissue (MALT).**

Although many lymphatic nodules are small and solitary, some occur in multiple large aggregations in specific parts of the body. Among these are the tonsils in the pharyngeal region and the aggregated lymphatic follicles (Peyer's patches) in the ileum of the small intestine. Aggregations of lymphatic nodules also occur in the appendix. Usually there are five **tonsils** that form a ring at the junction of the oral cavity and oropharynx and at the junction of the nasal cavity and nasopharynx (see Figure 23.2b on page 778). Thus, the tonsils are strategically positioned to participate in immune responses against foreign substances that are inhaled or ingested. The single **pharyngeal tonsil** (fa-RIN-jē-al) or **adenoid** is embedded in the posterior wall of the nasopharynx. The two **palatine tonsils** (PAL-a-tīn) lie at the posterior region of the oral cavity, one on either side; these are the tonsils commonly removed in a tonsillectomy. The paired **lingual tonsils** (LIN-gwal), located at the base of the tongue, may also require removal during a tonsillectomy.

1. How are interstitial fluid and lymph similar, and how do they differ?
2. How do lymphatic vessels differ in structure from veins?
3. Construct a diagram that shows the route of lymph circulation.
4. What is the role of the thymus gland in immunity?
5. What functions do lymph nodes serve?
6. Describe the functions of the spleen.

DEVELOPMENTAL ANATOMY OF THE LYMPHATIC SYSTEM

OBJECTIVE

• *Describe the development of the lymphatic system.*

The lymphatic system begins to develop by the end of the fifth week of embryonic life. *Lymphatic vessels* develop from **lymph sacs** that arise from developing veins, which are derived from **mesoderm.**

The first lymph sacs to appear are the paired **jugular lymph sacs** at the junction of the internal jugular and subclavian veins (Figure 22.8). From the jugular lymph sacs, capillary plexuses spread to the thorax, upper limbs, neck, and head. Some of the plexuses enlarge and form lymphatic vessels in their respective regions. Each jugular lymph sac retains at least one connection with its jugular vein, the left one developing into the superior portion of the thoracic duct (left lymphatic duct).

The next lymph sac to appear is the unpaired **retroperitoneal lymph sac** at the root of the mesentery of the intestine. It develops from the primitive vena cava and mesonephric (primitive kidney) veins. Capillary plexuses and lymphatic vessels spread from the retroperitoneal lymph sac to the abdominal viscera and diaphragm. The sac establishes connections with the cisterna chyli but loses its connections with neighboring veins.

At about the time the retroperitoneal lymph sac is developing, another lymph sac, the **cisterna chyli,** develops inferior to the diaphragm on the posterior abdominal wall. It gives rise to the inferior portion of the *thoracic duct* and the *cisterna chyli* of the thoracic duct. Like the retroperitoneal lymph sac, the cisterna chyli also loses its connections with surrounding veins.

The last of the lymph sacs, the paired **posterior lymph sacs,** develop from the iliac veins. The posterior lymph sacs

Figure 22.8 Development of the lymphatic system.

 The lymphatic system is derived from mesoderm.

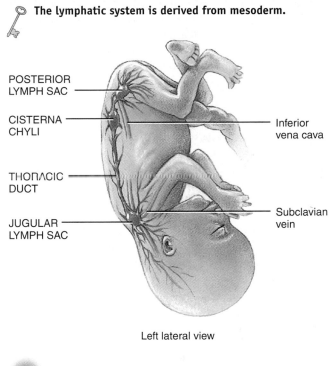

POSTERIOR LYMPH SAC

CISTERNA CHYLI

THORACIC DUCT

JUGULAR LYMPH SAC

Inferior vena cava

Subclavian vein

Left lateral view

Q When does the lymphatic system begin to develop?

produce capillary plexuses and lymphatic vessels of the abdominal wall, pelvic region, and lower limbs. The posterior lymph sacs join the cisterna chyli and lose their connections with adjacent veins.

With the exception of the anterior part of the sac from which the cisterna chyli develops, all lymph sacs become invaded by **mesenchymal cells** and are converted into groups of *lymph nodes.*

The *spleen* develops from **mesenchymal cells** between layers of the dorsal mesentery of the stomach. The *thymus gland* arises as an outgrowth of the **third pharyngeal pouch** (see Figure 18.22a on page 602).

1. Name the four lymph sacs from which lymphatic vessels develop.

NONSPECIFIC RESISTANCE TO DISEASE

OBJECTIVE

• *Describe the mechanisms of nonspecific resistance to disease.*

Although several mechanisms contribute to nonspecific resistance to disease, all of them have one thing in common: They offer immediate protection against a wide variety of pathogens and foreign substances. Nonspecific resistance, as its name suggests, lacks specific responses to specific invaders; instead, its protective mechanisms function the same way, regardless of the type of invader. Mechanisms of nonspecific resistance include the external mechanical and chemical barriers provided by the skin and mucous membranes, and various internal nonspecific defenses, including antimicrobial proteins, natural killer cells and phagocytes, inflammation, and fever.

First Line of Defense: Skin and Mucous Membranes

The skin and mucous membranes of the body are the first line of defense against pathogens. Both mechanical and chemical barriers discourage pathogens and foreign substances from penetrating the body and causing disease.

With its many layers of closely packed, keratinized cells, the outer epithelial layer of the skin—the **epidermis**—provides a formidable physical barrier to the entrance of microbes (see Figure 5.1 on page 141). In addition, periodic shedding of epidermal cells helps remove microbes at the skin surface. Bacteria rarely penetrate the intact surface of healthy epidermis. But if the epithelial surface is broken by cuts, burns, or punctures, pathogens can penetrate the epidermis and then invade adjacent tissues or circulate in the blood to other parts of the body.

The epithelial layer of **mucous membranes,** which line body cavities, secretes a fluid called **mucus** that lubricates and moistens the cavity surface. Because mucus is slightly viscous, it traps many microbes and foreign substances. The mucous membrane of the nose has mucus-coated **hairs** that trap and filter microbes, dust, and pollutants from inhaled air. The mucous membrane of the upper respiratory tract contains **cilia,** microscopic hairlike projections on the surface of the epithelial cells. The waving action of cilia propels inhaled dust and microbes that have become trapped in mucus toward the throat. Coughing and sneezing accelerate movement of mucus.

Other fluids produced by various organs also help protect epithelial surfaces of the skin and mucous membranes. For example, the **lacrimal apparatus** (LAK-ri-mal) of the eyes (see Figure 16.4 on page 518) manufactures and drains away tears in response to irritants. Blinking spreads tears over the surface of the eyeball, and the continual washing action of tears helps to dilute microbes and keep them from settling on the surface of the eye.

Saliva, produced by the salivary glands, washes microbes from the surfaces of the teeth and from the mucous membrane of the mouth, much as tears wash the eyes. The flow of saliva reduces colonization of the mouth by microbes.

The cleansing of the urethra by the **flow of urine** retards microbial colonization of the urinary system. Vaginal secretions likewise move microbes out of the body in females. **Defecation** and **vomiting** also are processes that can expel microbes. For example, in response to microbial toxins that

irritate the lining of the lower gastrointestinal tract, the smooth muscle of the tract contracts vigorously; the resulting diarrhea rapidly expels many of the microbes.

Certain chemicals also contribute to the high degree of resistance of the skin and mucous membranes to microbial invasion. Sebaceous (oil) glands of the skin secrete an oily substance called **sebum** that forms a protective film over the surface of the skin; the unsaturated fatty acids in sebum inhibit the growth of certain pathogenic bacteria and fungi. The acidity of the skin (pH 3–5) is caused in part by the secretion of fatty acids and lactic acid. **Perspiration** helps flush microbes from the surface of the skin, and it also contains **lysozyme,** an enzyme capable of breaking down the cell walls of certain bacteria. Lysozyme is also found in tears, saliva, nasal secretions, and tissue fluids, where it also exhibits antimicrobial activity. **Gastric juice,** produced by the glands of the stomach, is a mixture of hydrochloric acid, enzymes, and mucus. The strong acidity of gastric juice (pH 1.2–3.0) destroys many bacteria and most bacterial toxins. **Vaginal secretions** also are slightly acidic, which discourages bacterial growth.

Second Line of Defense: Internal Defenses

When pathogens penetrate the mechanical and chemical barriers of the skin and mucous membranes, they encounter a second line of defense: internal antimicrobial proteins, phagocytic and natural killer cells, inflammation, and fever.

Antimicrobial Proteins

Blood and interstitial fluids contain three main types of **antimicrobial proteins** that discourage microbial growth:

1. *Interferons.* Lymphocytes, macrophages, and fibroblasts infected with viruses produce proteins called **interferons** (in′-ter-FĒR-ons), or **IFNs.** Once produced by and released from virus-infected cells, IFNs diffuse to uninfected neighboring cells and bind to surface receptors, where they induce synthesis of antiviral proteins that interfere with viral replication. Although IFNs do not prevent viruses from attaching to and penetrating host cells, they do stop replication, and viruses can cause disease only if they can replicate within body cells. IFNs are an important defense against infection by many different viruses. The three types of interferon are alpha-, beta-, and gamma-IFN.

2. *Complement.* A group of normally inactive proteins in blood plasma and on plasma membranes make up the **complement system.** When activated, these proteins "complement" or enhance certain immune, allergic, and inflammatory reactions.

3. *Transferrins.* Iron-binding proteins called **transferrins** inhibit the growth of certain bacteria by reducing the amount of available iron.

Figure 22.9 Phagocytosis of a microbe. The first phase of phagocytosis, chemotaxis, is not depicted.

🔑 **The major types of phagocytes are neutrophils and macrophages.**

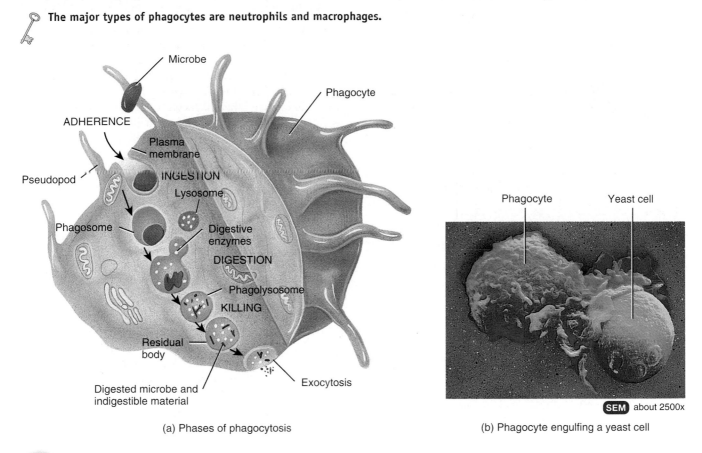

(a) Phases of phagocytosis

(b) Phagocyte engulfing a yeast cell

SEM about 2500x

Q What chemicals are responsible for killing ingested microbes?

Natural Killer Cells and Phagocytes

If microbes penetrate the skin and mucous membranes and survive the action of antimicrobial proteins in blood, they may be attacked by natural killer cells or phagocytes. **Natural killer (NK) cells** are lymphocytes that lack the membrane molecules that identify B cells and T cells yet have the ability to kill a wide variety of infectious microbes plus certain spontaneously arising tumor cells. About 5–10% of lymphocytes in the blood are NK cells; they are also present in the spleen, lymph nodes, and red bone marrow. NK cells attack cells that display abnormal plasma membrane proteins called major histocompatibility complex (MHC) antigens, which are described on page 754. NK cells achieve cell destruction in at least two ways: either by releasing **perforins**, chemicals that when inserted into the plasma membrane of a microbe make the membrane so leaky that cytolysis occurs, or by binding to a target cell and inflicting damage by direct contact.

Phagocytes (*phago-* = eat; *-cytes* = cells) are cells specialized to perform **phagocytosis** (*-osis* = process), the ingestion of microbes or other particulate matter (see Figure 3.14 on page 74). The two major types of phagocytes are

neutrophils and **macrophages** (MAK-rō-fā-jez); the latter are scavenger cells that develop from monocytes. Mobile macrophages, called **wandering macrophages,** are present in most tissues. Other macrophages, called **fixed macrophages,** stand guard in specific tissues. Among the fixed macrophages are histiocytes in the skin and subcutaneous layer, stellate reticuloendothelial cells (Kupffer cells) in the liver, alveolar macrophages in the lungs, microglia in the nervous system, and tissue macrophages in the spleen, lymph nodes, and red bone marrow. In addition to being a nonspecific defense mechanism, phagocytosis plays a vital role in immunity, as discussed later in the chapter.

Phagocytosis has several phases: chemotaxis, adherence, ingestion, digestion, and killing (Figure 22.9):

1. *Chemotaxis.* The chemical attraction of phagocytes to a particular location is termed **chemotaxis** (kē'-mō-TAK-sis; *chemo-* = chemical; *-taxis* = arrangement). Among the chemotactic chemicals that attract phagocytes are certain microbial products, components of white blood cells and damaged tissue cells, and activated complement proteins.

Figure 22.10 Inflammation.

🔑 **The three stages of inflammation are (1) vasodilation and increased permeability of blood vessels, (2) phagocyte emigration, and (3) tissue repair.**

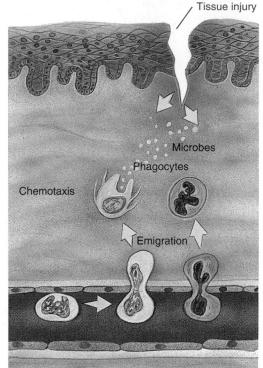

Phagocytes migrate from blood to site of tissue injury

Q What causes each of the following signs and symptoms of inflammation: redness, pain, heat, and swelling?

2. *Adherence.* The attachment of the plasma membrane of a phagocyte to the surface of a microorganism or other foreign material is called **adherence.**
3. *Ingestion.* Following adherence, **ingestion** occurs. The cell membrane of the phagocyte extends projections, called pseudopods, that engulf the microorganism. Once the microorganism is surrounded, the pseudopods meet and fuse, enclosing the microorganism within a vesicle called a **phagosome.**
4. *Digestion.* The phagosome, which forms when a portion of the membrane completely pinches off, enters the cytoplasm and merges with lysosomes to form a single, larger structure called a **phagolysosome.** The lysosome contributes lysozyme, which breaks down microbial cell walls, and digestive enzymes, which degrade carbohydrates, proteins, lipids, and nucleic acids. The phagocyte also forms lethal oxidants, such as superoxide anion (O_2^-), hypochlorite anion (OCl^-), and hydrogen peroxide (H_2O_2), in a process called an **oxidative burst.**
5. *Killing.* Within a phagolysosome, the chemical onslaught provided by lysozyme, digestive enzymes, and oxidants

quickly kills many types of microbes. Any materials that cannot be degraded further remain in structures called **residual bodies,** which eventually undergo exocytosis.

Some microbes, such as the toxin-producing staphylococci that cause one kind of food poisoning, may be ingested but not killed; instead, the toxins may kill the phagocytes. Other microbes—such as the tubercle bacillus, which causes tuberculosis—may multiply within the phagolysosome and eventually destroy the phagocyte. Still other microbes, such as those that cause tularemia and brucellosis, may remain dormant in phagocytes for months or years at a time.

Inflammation

Cells damaged by microbes, physical agents, or chemical agents initiate a defensive response called **inflammation.** The four characteristic signs and symptoms of inflammation are **redness, pain, heat,** and **swelling.** Inflammation can also cause the **loss of function** in the injured area, depending on the site and extent of the injury. Inflammation traps microbes, toxins, and foreign material at the site of injury and prepares the site for tissue repair. Thus, it helps restore tissue homeostasis.

Because inflammation is one of the body's nonspecific defenses, the response of a tissue to, say, a cut is similar to the response to damage caused by burns, radiation, or bacterial or viral invasion. In each case, inflammation has three basic stages: (1) vasodilation and increased permeability of blood vessels, (2) phagocyte emigration, and, ultimately, (3) tissue repair.

Immediately after tissue damage, blood vessels in the area of the injury dilate and become more permeable. **Vasodilation,** an increase in blood vessel diameter, allows more blood to flow through the damaged area; **increased permeability** means that substances normally retained in the bloodstream, including antibodies, clotting proteins, and phagocytes, are allowed to pass more easily out of blood vessels (Figure 22.10). The increased blood flow also helps remove toxic products released by the invading microorganisms and dead cells.

Among the substances that contribute to vasodilation, increased permeability, and other aspects of the inflammatory response are the following:

- *Histamine.* In response to injury, mast cells in connective tissue and basophils and platelets in blood release **histamine.** Neutrophils and macrophages attracted to the site of injury also stimulate the release of histamine, which causes vasodilation and increased permeability of blood vessels.
- *Kinins.* These polypeptides, formed in blood from inactive precursors called kininogens, induce vasodilation and increased permeability and serve as chemotactic agents for phagocytes.
- *Prostaglandins (PGs).* These lipids, especially those of the E series, are released by damaged cells and intensify the

effects of histamine and kinins. PGs also may stimulate the emigration of phagocytes through capillary walls.

- *Leukotrienes (LTs).* Produced by basophils and mast cells by breakdown of membrane phospholipids, LTs cause increased permeability; they also function in adherence of phagocytes to pathogens and as chemotactic agents that attract phagocytes.

- *Complement.* Different components of the complement system stimulate histamine release, attract neutrophils by chemotaxis, and promote phagocytosis; some components can also destroy bacteria.

Within minutes after an injury, dilation of arterioles and increased permeability of capillaries produce heat, redness, and edema (swelling) in the affected area. The large amount of warm blood flowing through the area produces both heat and redness (erythema). As the local temperature rises slightly, metabolic reactions proceed more rapidly and release additional heat. Edema results from increased permeability of blood vessels, which permits more fluid to move from blood into tissue spaces. Pain, whether immediate or delayed, is a cardinal symptom of inflammation; it can result from injury of nerve fibers or from irritation by toxic chemicals from microorganisms. Kinins affect some nerve endings, causing much of the pain associated with inflammation. Prostaglandins intensify and prolong the pain associated with inflammation. Pain may also be due to increased pressure from edema.

The increased permeability of capillaries allows leakage of blood-clotting factors into tissues. The clotting cascade is set into motion, and fibrinogen is ultimately converted to an insoluble, thick network of fibrin threads that localizes and traps invading microbes and blocks their spread.

Within an hour after the inflammatory process starts, phagocytes appear on the scene, first neutrophils and then monocytes (see Figure 22.10). They leave the bloodstream at an inflamed site by a process termed **emigration** (formerly termed *diapedesis*). Neutrophil emigration depends on chemotaxis; these phagocytes are attracted by microbes, kinins, complement, and other neutrophils, and they attempt to destroy the invaders by phagocytosis. A steady stream of neutrophils is ensured by the production and release of additional cells from bone marrow. Such an increase in white blood cells in the blood is termed **leukocytosis.**

Neutrophils predominate in the early stages of inflammation but die off rapidly. As the inflammatory response continues hours later, monocytes follow the neutrophils into the infected area. Once in the inflamed tissue, monocytes transform into wandering macrophages that augment the phagocytic activity of fixed macrophages. They are more powerful phagocytes than neutrophils and engulf damaged tissue, worn-out neutrophils, and invading microbes.

Eventually, these phagocytes also die. Within a few days, a pocket of dead phagocytes and damaged tissue forms; this collection of dead cells and fluid is called **pus.** Pus formation occurs in most inflammatory responses and usually contin-

ues until the infection subsides. At times, pus reaches the surface of the body or drains into an internal cavity and is dispersed; on other occasions the pus remains even after the infection is terminated. In this case, the pus is gradually destroyed over a period of days and is absorbed.

CLINICAL APPLICATION
Abscesses and Ulcers

If pus cannot drain out of an inflamed region, the result is an **abscess**—an excessive accumulation of pus in a confined space. Common examples are pimples and boils. When superficial inflamed tissue sloughs off the surface of an organ or tissue, the resulting open sore is called an **ulcer.** People with poor circulation—for instance, diabetics with advanced atherosclerosis—are susceptible to ulcers in the tissues of their legs. These ulcers, which are called stasis ulcers, develop because of poor oxygen and nutrient supply to tissues that then become very susceptible to even a very mild injury or an infectious process. ■

Fever

Fever is an abnormally high body temperature that occurs because the hypothalamic thermostat is reset. Although its significance is still not fully understood, fever commonly occurs during infection and inflammation. Many bacterial toxins elevate body temperature, sometimes by triggering release of fever-causing cytokines such as interleukin-1. Elevated body temperature intensifies the effects of interferons, inhibits the growth of some microbes, and speeds up body reactions that aid repair. (Fever is discussed in more detail on page 908.)

Table 22.1 summarizes the components of nonspecific resistance.

1. List the mechanical and chemical factors that provide protection from disease in the skin and mucous membranes.
2. What internal defenses provide protection against microbes that penetrate the skin and mucous membranes?
3. Compare the activities of natural killer cells and phagocytes.
4. Define inflammation, and describe its principal signs and symptoms and its stages.

SPECIFIC RESISTANCE: IMMUNITY

OBJECTIVES

- *Define immunity, and describe how T cells and B cells arise.*

- *Explain the relationship between an antigen and an antibody.*

- *Describe the role of antigen-presenting cells in processing exogenous and endogenous antigens.*

Table 22.1 Summary of Nonspecific Resistance

COMPONENT	FUNCTIONS
First Line of Defense: Skin and Mucous Membranes	
Mechanical Factors	
Epidermis of skin	Forms a physical barrier to the entrance of microbes.
Mucous membranes	Inhibit the entrance of many microbes, but not as effective as intact skin.
Mucus	Traps microbes in respiratory and gastrointestinal tracts.
Hairs	Filter out microbes and dust in nose.
Cilia	Together with mucus, trap and remove microbes and dust from upper respiratory tract.
Lacrimal apparatus	Tears dilute and wash away irritating substances and microbes.
Saliva	Washes microbes from surfaces of teeth and mucous membranes of mouth.
Urine	Washes microbes from urethra.
Defecation and vomiting	Expel microbes from body.
Chemical Factors	
Acid pH of skin	Discourages growth of many microbes.
Unsaturated fatty acids	Antibacterial substance in sebum.
Lysozyme	Antimicrobial substance in perspiration, tears, saliva, nasal secretions, and tissue fluids.
Gastric juice	Destroys bacteria and most toxins in stomach.
Vaginal secretions	Slight acidity discourages bacterial growth.
Second Line of Defense: Internal Defenses	
Antimicrobial Proteins	
Interferons (IFNs)	Protect uninfected host cells from viral infection.
Complement system	Causes cytolysis of microbes, promotes phagocytosis, and contributes to inflammation.
Natural Killer (NK) Cells	Kill a wide variety of microbes and certain tumor cells.
Phagocytes	Ingest foreign particulate matter.
Inflammation	Confines and destroys microbes and initiates tissue repair.
Fever	Intensifies the effects of interferons, inhibits growth of some microbes, and speeds up body reactions that aid repair.

The ability of the body to defend itself against specific invading agents such as bacteria, toxins, viruses, and foreign tissues is called **specific resistance** or **immunity.** Substances that are recognized as foreign and provoke immune responses are called **antigens (Ags).** Two properties distinguish immunity from nonspecific defenses: (1) *specificity* for particular foreign molecules (antigens), which also involves distinguishing self from nonself molecules, and (2) *memory* for most previously encountered antigens such that a second encounter prompts an even more rapid and vigorous response. The branch of science that deals with the responses of the body when challenged by antigens is called **immunology** (im´-yoo-NOL-ō-jē; *immun-* = free from service; *-ology* = study of). The **immune system** includes the cells and tissues that carry out immune responses.

Maturation of T Cells and B Cells

The cells that develop **immunocompetence,** the ability to carry out immune responses if properly stimulated, are lymphocytes called B cells and T cells. Both develop from pluripotent stem cells that originate in red bone marrow (see Figure 19.3 on page 614). B cells complete their development into mature, immunocompetent cells in bone marrow, a process that continues throughout one's lifetime; T cells develop from pre-T cells that migrate from bone marrow into the thymus (Figure 22.11). Although most T cells arise before puberty, some continue to mature throughout life.

Before T cells leave the thymus or B cells leave red bone marrow, they acquire several distinctive surface proteins. Some function as **antigen receptors**—molecules capable of recognizing specific antigens (see Figure 22.11). In addition, T cells exit the thymus as either CD4+ or CD8+ cells, which means they display either a protein called CD4 or one called CD8 on their plasma membrane. As we will see later in this chapter, these two types of T cells, called T4 and T8 cells, have very different functions.

Types of Immune Responses

Immunity consists of two kinds of closely allied responses, both triggered by antigens. In the first kind, called **cell-mediated immune responses,** CD8+ T cells proliferate into "killer" T cells and directly attack the invading antigen. In the second kind, called **antibody-mediated (humoral) immune responses,** B cells transform into plasma cells, which synthesize and secrete specific proteins called **antibodies (Abs)** or **immunoglobulins** (im´-yoo-nō-GLOB-yoo-lins). Antibodies bind to and inactivate a specific antigen. Most CD4+ T cells become helper T cells that aid both cell-mediated and antibody-mediated immune responses.

To some extent, each type of immune response specializes in dealing with certain types of invaders. Cell-mediated immunity is particularly effective against (1) intracellular pathogens that reside within host cells (primarily fungi, parasites, and viruses); (2) some cancer cells; and (3) foreign tis-

Figure 22.11 Maturation of lymphocytes and their roles in the two types of immune responses.

 B cells and pre-T cells develop from pleuripotent stem cells in red bone marrow.

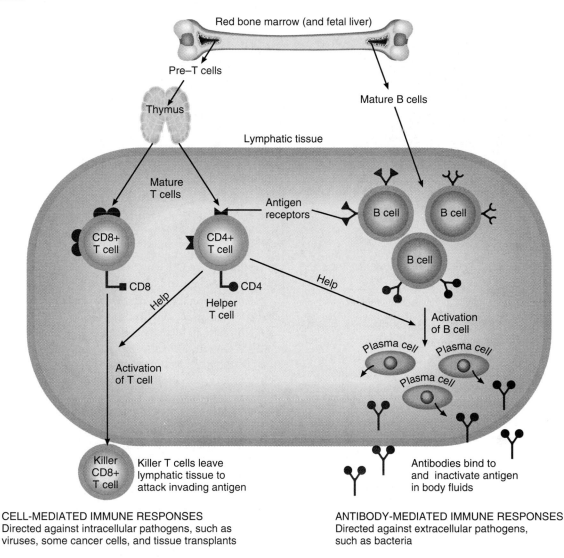

CELL-MEDIATED IMMUNE RESPONSES
Directed against intracellular pathogens, such as viruses, some cancer cells, and tissue transplants

ANTIBODY-MEDIATED IMMUNE RESPONSES
Directed against extracellular pathogens, such as bacteria

Q Which type of T cell participates in both cell-mediated and antibody-mediated immune responses?

sue transplants. Thus, cell-mediated immunity always involves cells attacking cells. Antibody-mediated immunity works mainly against (1) antigens present in body fluids and (2) extracellular pathogens that multiply in body fluids but rarely enter body cells (primarily bacteria). Often, however, a given pathogen provokes both types of immune responses.

Antigens

Antigens have two important characteristics: immunogenicity and reactivity. **Immunogenicity** (im′-yoo-nō-jen-

IS-it-ē; -*genic* = producing) is the ability to provoke an immune response by stimulating the production of specific antibodies, the proliferation of specific T cells, or both. The term *antigen* derives from its function as an *anti*body *gener*-ator. **Reactivity** is the ability of the antigen to react specifically with the antibodies or cells it provoked. Strictly speaking, immunologists define antigens as substances that have reactivity; substances with both immunogenicity and reactivity are considered **complete antigens.** Commonly, however, the term *antigen* implies both immunogenicity and reactivity, and we use the word in this way.

Figure 22.12 Epitopes (antigenic determinants).

🔑 **Most antigens have several epitopes that induce the production of different antibodies or activate different T cells.**

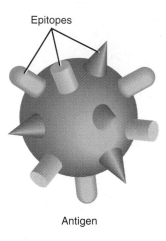

Epitopes

Antigen

Q What is the difference between an epitope and a hapten?

Entire microbes or parts of microbes may act as antigens. Bacterial structures such as flagella, capsules, and cell walls are antigenic, as are bacterial toxins. Nonmicrobial examples of antigens include pollen, egg white, incompatible blood cells, and transplanted tissues and organs. The huge variety of antigens in the environment provides myriad opportunities for provoking immune responses.

Antigens that get past the nonspecific defenses generally follow one of three routes into lymphatic tissue: (1) Most antigens that enter the bloodstream (for example, through an injured blood vessel) are deposited in the spleen; (2) antigens that penetrate the skin enter lymphatic vessels and reach lymph nodes; and (3) antigens that penetrate mucous membranes lodge in the mucosa-associated lymphoid tissue (MALT).

Chemical Nature of Antigens

Antigens are large, complex molecules, most frequently proteins. However, nucleic acids, lipoproteins, glycoproteins, and certain large polysaccharides may also act as antigens. T cells respond only to antigens that include protein; B cells respond to antigens made of proteins, certain lipids, carbohydrates, and nucleic acids. Complete antigens usually have large molecular weights of 10,000 daltons or more, but large molecules that have simple, repeating subunits—for example, cellulose and most plastics—are not usually antigenic. This is why plastic materials can be used in artificial heart valves or joints.

A smaller substance that has reactivity but lacks immunogenicity is called a **hapten** (HAP-ten = to grasp). A hapten can stimulate an immune response only if it is at-

tached to a larger carrier molecule. An example is the small lipid toxin in poison ivy, which triggers an immune response after combining with a body protein. Likewise, some drugs, such as penicillin, may combine with proteins in the body to form immunogenic complexes. Such hapten-stimulated immune responses are responsible for some allergic reactions to drugs and other substances in the environment (see page 769).

Epitopes

The small, specific portions of antigen molecules that trigger immune responses are called **epitopes** (EP-i-tōps), or *antigenic determinants* (Figure 22.12). Most antigens have many epitopes, each of which induces production of a specific antibody or activates a specific T cell. As a rule, antigens are foreign substances; they are not usually part of body tissues. However, sometimes the immune system fails to distinguish "friend" (self) from "foe" (nonself), and the result is an autoimmune disorder (see page 770) in which self molecules or cells are attacked as though they were foreign.

Diversity of Antigen Receptors

An amazing feature of the human immune system is its ability to recognize and bind to at least a billion (10^9) different epitopes. Before a particular antigen ever enters the body, T cells and B cells that can recognize and respond to that intruder are ready and waiting. Cells of the immune system can even recognize artificially made molecules that do not exist in nature. The basis for the ability to recognize so many epitopes is an equally large diversity of antigen receptors. Given that human cells contain only about 100,000 genes, how could a billion or more different antigen receptors possibly be generated?

The answer to this puzzle turned out to be simple in concept. The diversity of antigen receptors in both B cells and T cells and the diversity of antibodies produced result from the shuffling and rearranging of a few hundred versions of several small gene segments within. This process is called **genetic recombination.** The gene segments are put together in different combinations as the lymphocytes are developing from stem cells in red bone marrow and the thymus gland. The situation is similar to shuffling a deck of 52 cards and then dealing out three cards. If you did this over and over, you could generate many more than 52 different sets of three cards. As a result of genetic recombination, each B cell or T cell has a unique set of gene segments that codes for its unique antigen receptor. After transcription and translation, the receptor molecules are inserted into the plasma membrane.

Major Histocompatibility Complex Antigens

At the plasma membrane surface of most body cells are "self-antigens," the **major histocompatibility complex**

(MHC) antigens. These integral membrane glycoproteins are also called *human leukocyte associated (HLA) antigens* because they were first identified on white blood cells. Unless you have an identical twin, your MHC antigens are unique. Thousands to several hundred thousand MHC molecules mark the surface of all your body cells (except red blood cells). Although MHC antigens are the reason that tissues are rejected when they are transplanted from one person to another, their normal function is to help T cells to recognize that an antigen is foreign, not self, an important first step in any immune response.

The two types of major histocompatibility complex antigens are class I and class II. Class I MHC (MHC-I) molecules are built into the plasma membranes of all body cells except red blood cells. Class II MHC (MHC-II) molecules appear only on the surface of antigen-presenting cells (described in the next section), thymus cells, and T cells that have been activated by exposure to an antigen.

CLINICAL APPLICATION
Histocompatibility Testing

The success of an organ or tissue transplant depends on **histocompatibility** (his'-tō-kom-pat-i-BIL-i-tē)— that is, the tissue compatibility between the donor and the recipient. The more similar the MHC antigens, the greater the histocompatibility, and thus the greater the probability that the transplant will not be rejected. **Tissue typing (histocompatibility testing)** is done before any organ transplant. A nationwide computerized registry helps physicians select the most histocompatible and needy organ transplant recipients whenever donor organs become available. Also, in cases of disputed parentage, tissue typing can be used to identify the biological parents. ■

Pathways of Antigen Processing

For an immune response to occur, B cells and T cells must recognize that a foreign antigen is present. Whereas B cells can recognize and bind to antigens in extracellular fluid, T cells can only recognize fragments of antigenic proteins that first have been processed and presented in association with major histocompatibility complex self-antigens. As proteins inside body cells are being broken down, some peptide fragments associate with a peptide-binding groove of newly synthesized MHC molecules. This association stabilizes the MHC molecule and aids its proper folding, so that the MHC molecule can be inserted into the plasma membrane. When a peptide fragment from a *self-protein* is associated with an MHC antigen on the surface of a cell, T cells ignore it, but when the fragment is from a *foreign protein*, a few T cells recognize it as an intruder, and an immune response ensues. Preparation of a foreign antigen for display at the cell surface is termed *processing and presenting* of antigen; it occurs in two ways, depending on whether the antigen is an exogenous or endogenous antigen.

Processing of Exogenous Antigens
Foreign antigens that are present in fluids outside body cells are termed *exogenous antigens.* They include intruders such as bacteria and bacterial toxins, parasites, inhaled pollen and dust, and viruses that have not yet infected a body cell. A special class of cells called **antigen-presenting cells (APCs)** process and present exogenous antigens. APCs include macrophages, B cells, and dendritic cells (so-named for their long, branchlike projections). APCs are strategically located in places where antigens are likely to penetrate nonspecific defenses and enter the body, such as the epidermis and dermis of the skin (Langerhans cells are a type of dendritic cell); the mucous membranes that line the respiratory, gastrointestinal, urinary, and reproductive tracts; and lymph nodes. After processing an antigen, APCs migrate from tissues via lymphatic vessels to lymph nodes.

The steps in the processing and presenting of an exogenous antigen by an antigen-presenting cell occur as follows (Figure 22.13):

1 *Ingestion of the antigen.* Antigen-presenting cells ingest antigens by phagocytosis or endocytosis. Ingestion could occur almost anywhere in the body that invaders, such as microbes, have penetrated the nonspecific defenses.

2 *Digestion of antigen into peptide fragments.* Within the phagosome or endosome, digestive enzymes split large antigens into short peptide fragments. At the same time, the antigen-presenting cell is synthesizing MHC-II molecules and packaging them into vesicles. The MCH-II molecules are anchored to the inner membrane surface of the vesicles.

3 *Fusion of vesicles.* The vesicles containing antigen peptide fragments and MHC-II molecules merge and fuse.

4 *Binding of peptide fragments to MHC-II molecules.* After fusion of the two types of vesicles, antigen peptide fragments bind to MHC-II molecules.

5 *Insertion of antigen–MHC-II complex into the plasma membrane.* The combined vesicle containing antigen–MHC-II complexes undergoes exocytosis. As a result, the antigen–MHC-II complexes are inserted into the plasma membrane.

After processing an antigen, the antigen-presenting cell migrates to lymphatic tissue to present the antigen to T cells. Within lymphatic tissue, a small number of T cells that have compatibly shaped receptors recognize and bind to the antigen fragment–MHC II complex, triggering either a cell-mediated or an antibody-mediated immune response. The presentation of exogenous antigen together with MHC-II molecules by antigen-presenting cells informs T cells that intruders are present in the body and that combative action should begin.

Processing of Endogenous Antigens
Foreign antigens that are synthesized within a body cell are termed *endogenous antigens.* Such antigens may be viral

Figure 22.13 Processing and presenting of exogenous antigen by an antigen-presenting cell (APC).

Except for identical twins, major histocompatibility complex (MHC) molecules are unique in each person. They help T cells recognize foreign invaders.

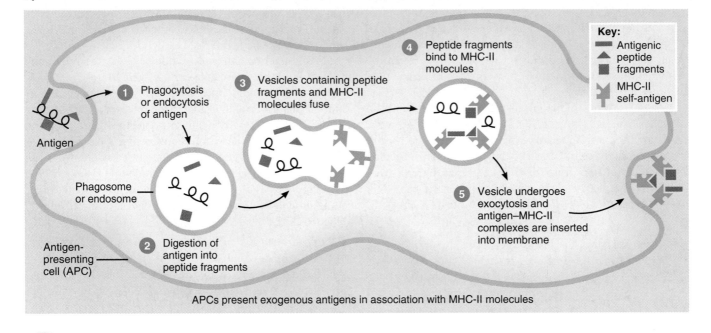

APCs present exogenous antigens in association with MHC-II molecules

Q   What cells are APCs, and where in the body are they found?

proteins produced after a virus infects the cell and takes over the cell's metabolic machinery, or they may be abnormal proteins synthesized by a cancerous cell. Fragments of endogenous antigens associate with major histocompatibility complex-I molecules inside infected cells. The resulting endogenous antigen fragment–MHC-I complex then moves to the plasma membrane, where it is displayed at the surface of the cell. Most cells of the body can process and present endogenous antigens. Presentation of endogenous antigen bound to an MHC-I molecule signals that a cell has been infected and needs help.

Cytokines

Cytokines are small protein hormones that stimulate or inhibit many normal cell functions, such as cell growth and differentiation. Lymphocytes and antigen-presenting cells secrete cytokines, as do fibroblasts, endothelial cells, monocytes, hepatocytes, and kidney cells. Some cytokines stimulate proliferation of progenitor blood cells in bone marrow. Others regulate activities of cells involved in nonspecific defenses or immune responses, as described in Table 22.2.

CLINICAL APPLICATION
Cytokine Therapy

Cytokine therapy is the use of cytokines to treat medical conditions. Interferons were the first cyto-

kines shown to be effective against a human cancer. Alpha-interferon (Intron A) is approved in the United States for treating Kaposi's (KAP-ō-sēs) sarcoma, a cancer that often occurs in patients infected with HIV (the virus that causes AIDS). Other approved uses for alpha-interferon include treating genital herpes caused by herpes virus; treating hepatitis B and C, caused by the hepatitis B and C viruses; and treating hairy cell leukemia. A form of beta-interferon (Betaseron) slows the progression of multiple sclerosis (MS) and lessens the frequency and severity of MS attacks. Of the interleukins, the one most widely used to fight cancer is interleukin-2. Although this treatment is effective in causing tumor regression in some patients, it also can be very toxic. Among the adverse effects are high fever, severe weakness, difficulty breathing due to pulmonary edema, and hypotension leading to shock. ■

1. What is immunocompetence, and which body cells display it?
2. Compare the functions of the major histocompatibility complex class I and class II "self-antigens."
3. How do antigens arrive at lymphatic tissues?
4. What are cytokines, what is their origin, and how do they function?

CELL-MEDIATED IMMUNITY
OBJECTIVE
• Describe the steps of a cell-mediated immune response.

Table 22.2 Summary of Cytokines Participating in Immune Responses

CYTOKINE	ORIGINS AND FUNCTIONS
Interleukin-1 (IL-1)	Produced by monocytes and macrophages; costimulator of T cell and B cell proliferation; acts on hypothalamus to cause fever.
Interleukin-2 (IL-2) (T cell growth factor)	Secreted by helper T cells to costimulate the proliferation of helper T cells, cytotoxic T cells, and B cells; activates NK cells.
Interleukin-4 (IL-4) (B cell-stimulating factor)	Produced by activated helper T cells; costimulator for B cells; causes plasma cells to secrete IgE antibodies (see Table 22.3); promotes growth of T cells.
Interleukin-5 (IL-5)	Produced by certain activated CD4+ T cells and activated mast cells; costimulator for B cells; causes plasma cells to secrete IgA antibodies.
Tumor necrosis factor (TNF)	Produced mainly by macrophages; stimulates accumulation of neutrophils and macrophages at sites of inflammation and stimulates their killing of microbes; stimulates macrophages to produce IL-1; induces synthesis of colony-stimulating factors by endothelial cells and fibroblasts; exerts an interferon-like protective effect against viruses; and functions as an endogenous pyrogen to induce fever (see page 908).
Transforming growth factor beta (TGF-β)	Secreted by T cells and macrophages; has some positive effects but is thought to be important for turning off immune responses; inhibits proliferation of T cells and activation of macrophages.
Gamma-interferon (γ-IFN)	Secreted by helper and cytotoxic T cells and NK cells; strongly stimulates phagocytosis by neutrophils and macrophages; activates NK cells; enhances both cell-mediated and antibody-mediated immune responses.
Alpha- and beta-interferons (α-IFN and β-IFN)	Produced by virus-infected cells to inhibit viral replication in uninfected cells; produced by antigen-stimulated macrophages to stimulate T cell growth; activate NK cells, inhibit cell growth and suppress formation of some tumors.
Lymphotoxin (LT)	Secreted by cytotoxic T cells; kills cells by causing fragmentation of DNA.
Perforin	Secreted by cytotoxic T cells and perhaps by NK cells; perforates cell membranes of target cells, which causes cytolysis.
Macrophage migration inhibiting factor	Produced by T cells; prevents macrophages from leaving site of infection.

A cell-mediated immune response begins with *activation* of a small number of T cells by a specific antigen. Once a T cell has been activated, it undergoes *proliferation* and *differentiation* into a clone of **effector cells,** a population of identical cells that can recognize the same antigen and carry out some aspect of the immune attack. Finally, the immune response results in *elimination* of the intruder.

Activation, Proliferation, and Differentiation of T Cells

Antigen receptors on the surface of T cells, called **T cell receptors (TCRs),** recognize and bind to specific foreign antigen fragments that are presented in antigen–MHC complexes. There are millions of different T cells; each has its own unique TCRs that can recognize a specific antigen–MHC complex. At any given time, most T cells are inactive. When an antigen enters the body, only a few T cells have TCRs that can recognize and bind to the antigen. Antigen recognition by a TCR is the *first signal* in activation of a T cell.

Successful activation of a T cell also requires a *second signal,* called a **costimulator.** Of the more than 20 known costimulators, some are cytokines, such as **interleukin-1** and **interleukin-2.** Other costimulators include pairs of plasma membrane molecules, one on the surface of the T cell and a

second on the surface of an antigen-presenting cell, that enable the two cells to adhere to one another for a period of time. The need for two signals is a little like starting and driving a car: When you insert the correct key (antigen) in the ignition (TCR) and turn it, the car starts (recognition of specific antigen), but it cannot move forward until you move the gear shift into drive (costimulation). The need for costimulation may prevent an immune response from occurring accidentally. It is thought that different costimulators affect the activated T cell in different ways, just as shifting a car into reverse has a different effect than shifting it into drive. Moreover, recognition (antigen binding to a receptor) without costimulation is thought to lead to a prolonged *state of inactivity* called **anergy** in both T cells and B cells—rather like a car in neutral with its engine running until it's out of gas!

Once a T cell has received two signals (antigen recognition and costimulation), it is said to be **activated.** It then enlarges and begins to **proliferate** (divide several times) and to **differentiate** (form more highly specialized cells). The result is a **clone,** or identical population of cells that can recognize the same specific antigen. Before the first exposure to a given antigen, only a handful of T cells might be able to recognize it, but once an immune response has begun, there are thousands. Activation, differentiation, and proliferation of T cells occur in the secondary lymphatic organs and tissues. If you

Figure 22.14 Activation, proliferation, and differentiation of T cells.

The binding of CD4 to MHC-II and CD8 to MHC-I helps anchor the TCR–antigen interaction so that antigen recognition can occur.

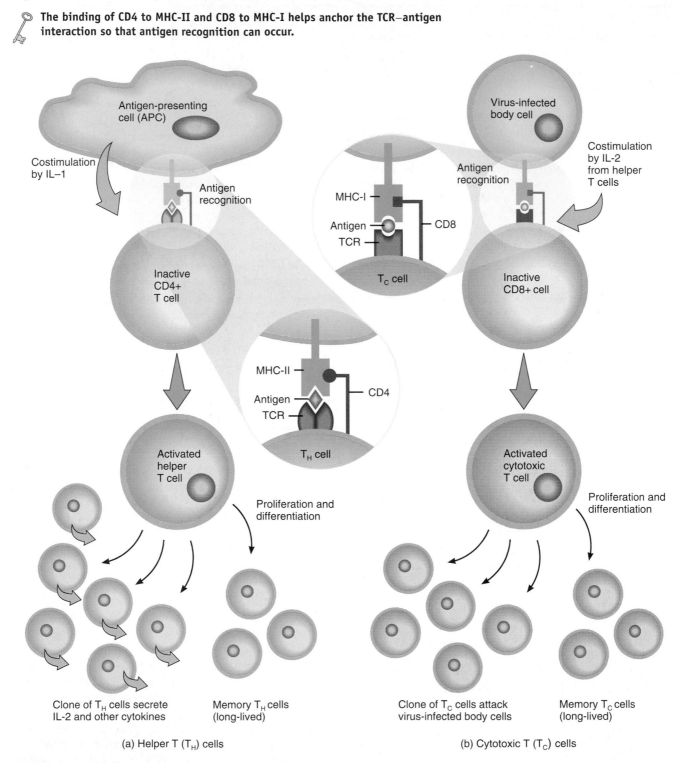

Clone of T$_H$ cells secrete IL-2 and other cytokines

Memory T$_H$ cells (long-lived)

(a) Helper T (T$_H$) cells

Clone of T$_C$ cells attack virus-infected body cells

Memory T$_C$ cells (long-lived)

(b) Cytotoxic T (T$_C$) cells

Q What are the first and second signals in activation of a T cell?

have ever had swollen tonsils or lymph nodes in your neck, the proliferation of lymphocytes participating in an immune response was likely the cause.

Types of T Cells

Three main types of differentiated T cells exist: helper T cells, cytotoxic (killer) T cells, and memory T cells.

Helper T Cells

Most T cells that display CD4 develop into **helper T (T$_H$) cells** or **T4 cells.** Resting (inactive) T$_H$ cells recognize antigen fragments associated with major histocompatibility complex class II (MHC-II) molecules and are costimulated by interleukin-1, which is secreted by macrophages (Figure 22.14a). For this reason, helper T cells are activated mainly by antigen-presenting cells.

Within hours after costimulation, helper T cells start secreting a variety of cytokines (see Table 22.2). Different subsets of T$_H$ cells specialize in the production of particular cytokines. Moreover, particular subtypes appear in certain diseases, such as asthma, multiple sclerosis, and Lyme arthritis. One very important cytokine produced by T$_H$ cells is interleukin-2 (IL-2), which is needed for virtually all immune responses and is the prime trigger of T cell proliferation. It can act as a costimulator for resting T$_H$ or cytotoxic T cells, and it enhances activation and proliferation of T cells, B cells, and natural killer cells.

Some actions of interleukin-2 provide a good example of a beneficial positive feedback system. As noted earlier, activation of a helper T cell stimulates it to start secreting IL-2, which then acts in an autocrine manner by binding to IL-2 receptors on the plasma membrane of the cell that secreted it. One effect is stimulation of cell division. As the T$_H$ cells proliferate, a positive feedback effect occurs because they secrete more IL-2, which causes further cell division. IL-2 may also act in a paracrine manner by binding to IL-2 receptors on neighboring T$_H$ cells, cytotoxic T cells, or B cells. If any of these cells have already bound an antigen, IL-2 serves as a costimulator to activate them.

Cytotoxic T Cells

T cells that display CD8 develop into **cytotoxic T (T$_C$) cells** or **T8 cells,** also known as **killer T cells.** T$_C$ cells recognize foreign antigens combined with major histocompatibility complex class I (MHC-I) molecules on the surfaces of (1) body cells infected by viruses, (2) some tumor cells, and (3) cells of a tissue transplant (Figure 22.14b). However, to become cytolytic (able to lyse cells) they need costimulation by interleukin-2 or other cytokines produced by helper T cells. (Recall that T$_H$ cells are activated by antigen associated with MHC-II molecules.) Thus, maximal activation of T$_C$ cells requires presentation of antigen associated with both MHC-I and MHC-II molecules.

Memory T Cells

T cells that remain from a proliferated clone after a cell-mediated immune response are termed **memory T cells.** Should a pathogen bearing the same foreign antigen invade the body at a later date, thousands of memory cells are available to initiate a far swifter reaction than occurred during the first invasion. The second response usually is so fast and so vigorous that the pathogens are destroyed before any signs or symptoms of disease can occur.

Figure 22.15 Activity of cytotoxic T cells. After delivering a "lethal hit," a cytotoxic T cell can detach and attack another target cell displaying the same antigen.

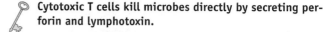

🔑 **Cytotoxic T cells kill microbes directly by secreting perforin and lymphotoxin.**

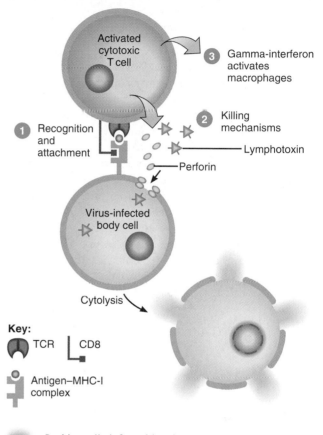

Key:
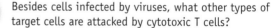

Q Besides cells infected by viruses, what other types of target cells are attacked by cytotoxic T cells?

Elimination of Invaders

Cytotoxic T cells are the soldiers that march forth to do battle with foreign invaders in cell-mediated immune responses. They leave secondary lymphatic organs and tissues and migrate to the site of invasion, infection, or tumor formation. They recognize and attach to target cells bearing the antigen that stimulated activation and proliferation of their progenitor cells. Then the cytotoxic T cells deliver a "lethal hit" that kills the target cells without damaging themselves (Figure 22.15). After detaching from a target cell, a cytotoxic T cell can seek out and destroy another invader that displays the same antigen.

Cytotoxic T cells use two killing mechanisms. In the first, granules containing the protein perforin undergo exocytosis from the cytotoxic T cell. **Perforin** forms holes in the plasma membrane of the target cell, which allow extracellular fluid to flow in, and the cell bursts—a process known as **cytolysis.** In the second mechanism, the cytotoxic T cell secretes a toxic molecule known as **lymphotoxin** that activates enzymes within the target cell. These enzymes cause the target cell's DNA to fragment, and the cell dies. In these ways, cytotoxic T cells destroy infected cells. Also, cytotoxic T cells secrete gamma-interferon, which activates phagocytic cells at the scene of the battle. Cytotoxic T cells are especially effective against slowly developing bacterial diseases (such as tuberculosis and brucellosis), some viruses, fungi, cancer cells associated with viral infection, and transplanted cells.

Immunological Surveillance

When a normal cell transforms into a cancerous cell, it often displays novel cell surface components called **tumor antigens,** molecules that are rarely, if ever, displayed on the surface of normal cells. If the immune system recognizes a tumor antigen as nonself, it can destroy the cancer cells carrying them. Such an immune response, called **immunological surveillance,** is carried out by cytotoxic T cells, macrophages, and natural killer cells. Immunological surveillance is most effective in eliminating tumor cells due to cancer-causing viruses. As a result, transplant recipients who are taking immunosuppressive drugs to prevent transplant rejection, for example, do not have a higher than normal incidence of most cancers, but they do have a greatly increased incidence of virus-associated cancers.

CLINICAL APPLICATION
Graft Rejection

Organ transplantation involves the replacement of an injured or diseased organ, such as the heart, liver, kidney, lungs, or pancreas, with an organ donated by some other individual. Usually, the immune system recognizes the proteins in the transplanted organ as foreign and mounts both cell-mediated and antibody-mediated immune responses against them. This phenomenon is known as **graft rejection.** The more closely matched are the major histocompatibility complex antigens of the donor and recipient, the weaker the graft rejection response. To reduce the risk of graft rejection, organ transplant recipients receive immunosuppressive drugs. One such drug is *cyclosporine,* derived from a fungus, which inhibits secretion of interleukin-2 by helper T cells but has only a minimal effect on B cells. Thus, the risk of rejection is diminished while maintaining resistance to some diseases. ■

1. Describe the functions of helper, cytotoxic, and memory T cells.
2. How do cytotoxic T cells kill their targets?
3. Explain the usefulness of immunological surveillance.

ANTIBODY-MEDIATED IMMUNITY
OBJECTIVES
• *Describe the steps in an antibody-mediated immune response.*
• *Describe the chemical characteristics and actions of antibodies.*

The body contains not only millions of different T cells but also millions of different B cells, each capable of responding to a specific antigen. Whereas cytotoxic T cells leave lymphatic tissues to seek out and destroy a foreign antigen, B cells stay put. In the presence of a foreign antigen, specific B cells in lymph nodes, the spleen, or lymphatic tissue in the gastrointestinal tract become activated. They then differentiate into plasma cells that secrete specific antibodies, which in turn circulate in the lymph and blood to reach the site of invasion.

Activation, Proliferation, and Differentiation of B Cells

During activation of a B cell, antigen receptors on the cell surface bind to an antigen (Figure 22.16). B cell antigen receptors are chemically similar to the antibodies that will eventually be secreted by the cell's progeny. Although B cells can respond to an unprocessed antigen present in lymph or interstitial fluid, their response is much more intense when nearby dendritic cells also process and present antigen to them. Some antigen is then taken into the B cell, broken down into peptide fragments and combined with MHC-II self-antigen, and moved to the B cell surface. Helper T cells recognize the antigen–MHC-II complex and deliver the costimulation needed for B cell proliferation and differentiation. The helper T cell produces interleukin-2 and other cytokines that function as costimulators to activate B cells. Interleukin-1, secreted by macrophages, also enhances B cell proliferation and differentiation into plasma cells.

Some of the activated B cells enlarge, divide, and differentiate into a clone of antibody-secreting **plasma cells.** The phenomenal rate of antibody secretion by plasma cells is about 2000 molecules per second for each cell, and secretion occurs for about 4 or 5 days, until the plasma cell dies. The activated B cells that do not differentiate into plasma cells remain as **memory B cells** that are ready to respond more rapidly and forcefully should the same antigen reappear at a future time.

Different antigens stimulate different B cells to develop into plasma cells and their accompanying memory B cells. The B cells of a particular clone are capable of secreting only one kind of antibody, which is identical in specificity to the antigen receptor displayed by the B cell that first responded to the antigen. Each specific antigen activates only those B cells that are predestined (by the combination of gene segments they carry) to secrete antibody specific to that antigen. Antibodies produced by a clone of plasma cells enter the circulation and form antigen–antibody complexes with the antigen that initiated their production.

Antibodies

An **antibody (Ab)** can combine specifically with the epitope on the antigen that triggered its production. Its structure matches its antigen much as a lock fits a specific key. In theory, B cells could produce as many different antibodies as there are antigen receptors on the cells; the same recombined gene segments code for both the antigen receptors on B cells and the antibodies eventually secreted by plasma cells.

Antibody Structure

Antibodies belong to a group of glycoproteins called globulins, and for this reason they are also known as **immunoglobulins (Igs).** Most antibodies contain four polypeptide chains (Figure 22.17). Two of the chains are identical to each other and are called **heavy (H) chains;** each consists of about 450 amino acids. Short carbohydrate chains are attached to each heavy polypeptide chain. The other two polypeptide chains, also identical to each other, are called **light (L) chains,** and each consists of about 220 amino acids. A disulfide bond (S–S) holds each light chain to a heavy chain. Two disulfide bonds also link the midregion of the two heavy chains; this part of the antibody displays considerable flexibility and is called the **hinge region.** Because the antibody "arms" can move somewhat as the hinge region bends, an antibody can assume either a T shape (Figure 22.17a) or a Y shape (Figure 22.17b).

Within each H and L chain are two distinct regions. The tips of the H and L chains, called the **variable (V) regions,** constitute the **antigen binding site.** The variable region, which is different for each kind of antibody, is the part of the antibody that recognizes and attaches specifically to a particular antigen. Because most antibodies have two antigen binding sites, they are said to be bivalent. Flexibility at the hinge allows the antibody to simultaneously bind to two epitopes that are some distance apart—for example, on the surface of a microbe.

The remainder of each H and L chain, called the **constant (C) region,** is nearly the same in all antibodies of the same class and is responsible for the type of antigen–antibody reaction that occurs. However, the constant region of the H chain differs from one class of antibody to another, and its structure serves as a basis for distinguishing five different classes, designated IgG, IgA, IgM, IgD, and IgE. Each class has a distinct chemical structure and a specific biological role. Because they appear first and are relatively short-lived, the presence of IgM antibodies indicates a recent invasion. In a sick patient, the responsible pathogen may be suggested by the presence of high levels of IgM specific to a particular organism. Resistance of the fetus and newborn baby to infection stems mainly from maternal IgG antibodies that cross the placenta before birth and IgA antibodies that are absorbed from breast milk after birth. Table 22.3 summarizes the structures and functions of the five classes of antibodies.

Antibody Actions

Even though the five classes of immunoglobulins have actions that differ somewhat, all act to disable antigens. Actions of antibodies include:

- *Neutralization of antigen.* The reaction of antibody with antigen blocks or neutralizes the damaging effect of some

Figure 22.16 Activation, proliferation, and differentiation of B cells into plasma cells and memory cells. Plasma cells are actually much larger than B cells.

 **Plasma cells secrete antibodies.**

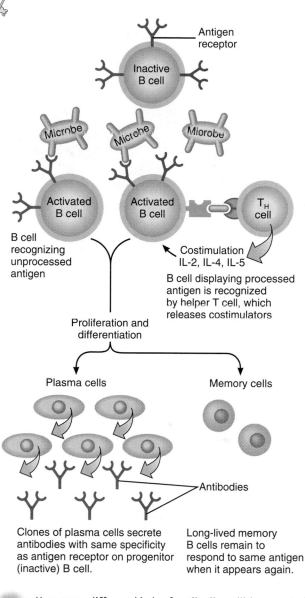

Q How many different kinds of antibodies will be secreted by the plasma cells in the clone shown here?

bacterial toxins and prevents attachment of some viruses to body cells.

- *Immobilization of bacteria.* If antibodies form against antigens on the cilia or flagella of motile bacteria, the antigen–antibody reaction may cause the bacteria to lose their motility, which limits their spread into nearby tissues.

- *Agglutination and precipitation of antigen.* Because antibodies have two or more sites for binding to antigen, the antigen–antibody reaction may cross-link pathogens to one another, causing agglutination (clumping together).

Figure 22.17 Chemical structure of the immunoglobulin G (IgG) class of antibody. Each molecule is composed of four polypeptide chains (two heavy and two light) plus a short carbohydrate chain attached to each heavy chain. In (a), each circle represents one amino acid. In (b), V_L = variable region of light chain, C_L = constant region of light chain, V_H = variable region of heavy chain, and C_H = constant region of heavy chain.

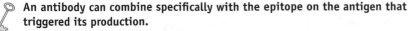

An antibody can combine specifically with the epitope on the antigen that triggered its production.

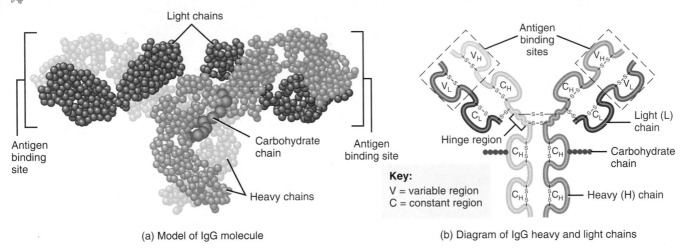

(a) Model of IgG molecule

(b) Diagram of IgG heavy and light chains

Q What is the function of the variable regions?

Likewise, soluble antigens may come out of solution and form a more-easily phagocytized precipitate when cross-linked by antibodies.

- *Activation of complement.* Antigen–antibody complexes initiate the classical pathway of the complement system (to be discussed shortly).

- *Enhancement of phagocytosis.* Antibodies enhance the activity of phagocytes by causing agglutination and precipitation, by activating complement, and by coating microbes so that they are more susceptible to phagocytosis, a process known as **opsonization** (op-sō-ni-ZĀ-shun).

CLINICAL APPLICATION
Monoclonal Antibodies

The antibodies produced against a given antigen by plasma cells can be harvested from an individual's blood. However, because an antigen typically has many epitopes, a variety of antibodies against the antigen are produced by many different clones of plasma cells. If a single plasma cell could be isolated and induced to proliferate into a clone of identical plasma cells, then a large quantity of identical antibodies could be produced. Unfortunately, lymphocytes and plasma cells are difficult to grow in culture, so scientists side-stepped this difficulty by fusing B cells with tumor cells that grow easily and proliferate endlessly. The resulting hybrid cell is called a **hybridoma** (hī-bri-DŌ-ma). Hybridomas are long-term sources of large quantities of pure, identical antibodies,

called **monoclonal antibodies (MAbs)** because they come from a single clone of identical cells. Such antibodies combine with just one epitope. One clinical use of monoclonal antibodies is for measuring levels of a drug in a patient's blood; other uses include the diagnosis of strep throat, pregnancy, allergies, and diseases such as hepatitis, rabies, and some sexually transmitted diseases. MAbs have also been used to detect cancer at an early stage and to ascertain the extent of metastasis. ■

Role of the Complement System in Immunity

The **complement system** is a defensive system consisting of plasma proteins that attack and destroy microbes. The system can be activated in one of two pathways (classical and alternative), both of which are an ordered sequence or cascade of reactions. Both pathways lead to the same events: inflammation, enhancement of phagocytosis, and bursting of microbes.

The complement system consists of more than 20 different plasma proteins. Included are proteins called C1 through C9 (the C stands for complement) and proteins called factors B, D, and P (properdin). The **classical pathway** (Figure 22.18) begins by the binding of antibody to antigen. The antigen could be a bacterium or other foreign cell. The antigen-antibody attachment activates complement protein C1 and the cascade begins. The **alternative pathway** does not involve antibodies. It begins by the interaction between

Figure 22.18 The classical and alternative pathways of the complement system.

When activated, complement proteins enhance certain immune, allergic, and inflammatory reactions.

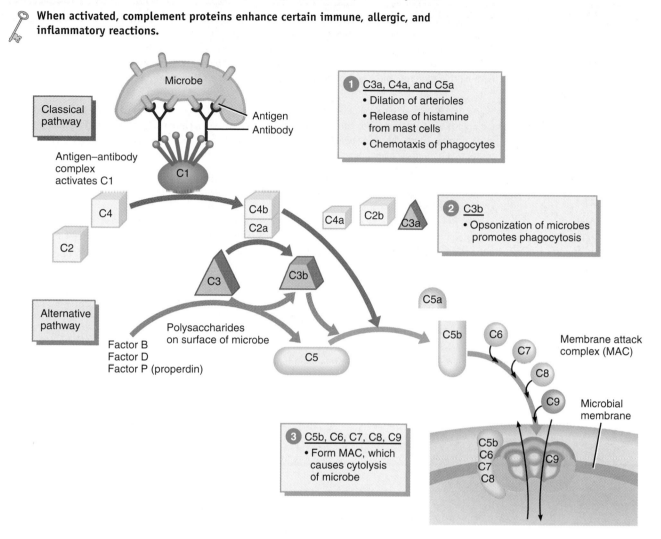

Q   Which pathway for activation of complement is linked to the immune system? Explain why.

polysaccharides on the surface of a microbe and factors B, D, and P (Figure 22.18). This interaction activates complement protein C3 and the cascade begins.

The consequences of activation of the classical and alternative pathways are as follows:

1 *Activation of inflammation.* Some complement proteins (C3a, C4a, and C5a) contribute to the development of inflammation: They dilate arterioles, which increases blood flow to the area, and cause the release of histamine from mast cells, basophils, and platelets. Because histamine increases the permeability of blood capillaries, white blood cells can more easily move into tissues to combat infection or allergy. Other complement proteins serve as chemotactic agents that attract phagocytes to the site of microbial invasion.

2 *Opsonization.* Complement fragment C3b binds to the surface of a microbe and then interacts with receptors

on phagocytes to promote phagocytosis, an example of opsonization.

3 *Cytolysis.* Several complement proteins (C5b, C6, C7, C8, and C9) come together to form a **membrane attack complex (MAC)** that inserts into the plasma membrane of the microbe, forming large holes. The ensuing leakiness allows fluid to flow into the cell's interior, and the microbe swells and bursts (cytolysis).

Immunological Memory

A hallmark of immune responses is memory for specific antigens that have triggered immune responses in the past. Immunological memory is due to the presence of long-lasting antibodies and very long-lived lymphocytes that arise during proliferation and differentiation of antigen-stimulated B cells and T cells.

Table 22.3 Classes of Immunoglobulins (Igs)

NAME AND STRUCTURE	CHARACTERISTICS AND FUNCTIONS
IgG	Most abundant, about 80% of all antibodies in the blood; found in blood, lymph, and the intestines; monomer (one-unit)) structure. Protects against bacteria and viruses by enhancing phagocytosis, neutralizing toxins, and triggering the complement system. It is the only class of antibody to cross the placenta from mother to fetus, thereby conferring considerable immune protection in newborns.
IgA	Makes up 10–15% of all antibodies in the blood; occurs as monomers and dimers (two units). Found mainly in sweat, tears, saliva, mucus, milk and gastrointestinal secretions. Smaller quantities are present in blood and lymph. Levels decrease during stress, lowering resistance to infection. Provides localized protection on mucous membranes against bacteria and viruses.
IgM	About 5–10% of all antibodies in the blood; occurs as pentamers (five units); first antibody class to be secreted by plasma cells after an initial exposure to any antigen; found in blood and lymph. Activates complement and causes agglutination and lysis of microbes. Also present as monomers on the surfaces of B cells, where they serve as antigen receptors. In blood plasma, the anti-A and anti-B antibodies of the ABO blood group, which bind to A and B antigens during incompatible blood transfusions, are also IgM antibodies (see Figure 19.11 on page 625).
IgD	About 0.2% of all antibodies in the blood; occurs as monomers; found in blood, in lymph, and on the surfaces of B cells as antigen receptors. Involved in activation of B cells.
IgE	Less than 0.1% of all antibodies in the blood; occurs as monomers; located on mast cells and basophils. Involved in allergic and hypersensitivity reactions; provides protection against parasitic worms.

Immune responses, whether cell-mediated or antibody-mediated, are much quicker and more intense after a second or subsequent exposure to an antigen than after the first exposure. Initially, only a few cells have the correct specificity to respond, and the immune response may take several days to build to maximum intensity. Because thousands of memory cells exist after an initial encounter with an antigen, the next time the same antigen appears they can proliferate and differentiate into plasma cells or cytotoxic T cells within hours.

One measure of immunological memory is the amount of antibody in serum, called the *antibody titer* (TĪ-ter). Following an initial contact with an antigen, no antibodies are present for a period of several days; then a slow rise in the antibody titer occurs, first IgM and then IgG, followed by a gradual decline in antibody titer (Figure 22.19). This is called the **primary response.**

Memory cells may remain for decades. Every new encounter with the same antigen results in a rapid proliferation of memory cells. The antibody titer after subsequent encounters is far greater than during a primary response and consists mainly of IgG antibodies. This accelerated, more intense response is called the **secondary response.** Antibodies produced during a secondary response have an even higher affinity for the antigen than those produced during a primary response, and thus they are more successful in disposing of it.

Primary and secondary responses occur during microbial infection. When you recover from an infection without taking antimicrobial drugs, it is usually because of the primary response. If at a later time you are infected by the same microbe, the secondary response could be so swift that the microbes are destroyed before you exhibit any signs or symptoms of infection.

Immunological memory provides the basis for immunization by vaccination against certain diseases (for example, polio). When you receive the vaccine, which may contain weakened or killed whole microbes or portions of microbes, your B cells and T cells are activated. Should you subsequently encounter the living pathogen as an infecting microbe, your body initiates a secondary response.

Table 22.4 summarizes the various types of naturally and artificially acquired immunity.

Figure 22.19 Production of antibodies in the primary (after first exposure) and secondary (after second exposure) responses to a given antigen.

🔑 **Immunological memory is the basis for successful immunization by vaccination.**

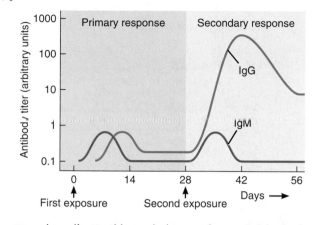

🇶 According to this graph, how much more IgG is circulating in the blood in the secondary response than in the primary response?

Table 22.4 Types of Immunity

TYPE OF IMMUNITY	HOW ACQUIRED
Naturally acquired active immunity	Antigen recognition by B cells and T cells and costimulation lead to antibody-secreting plasma cells, cytotoxic T cells, and B and T memory cells.
Naturally acquired passive immunity	Transfer of IgG antibodies from mother to fetus across placenta, or of IgA antibodies from mother to baby in milk during breast-feeding.
Artificially acquired active immunity	Antigens introduced during a vaccination stimulate cell-mediated and antibody-mediated immune responses, leading to production of memory cells. The antigens are pretreated to be immunogenic but not pathogenic; that is, they will trigger an immune response but not cause significant illness.
Artificially acquired passive immunity	Intravenous injection of immunoglobulins (antibodies).

1. How do the five classes of antibodies differ in structure and function?
2. Compare and contrast the cell-mediated and antibody-mediated immune responses.
3. In what ways does the complement system augment antibody-mediated immune responses?
4. Discuss the importance of the secondary response to an antigen.

SELF-RECOGNITION AND IMMUNOLOGICAL TOLERANCE

OBJECTIVE

• *Describe how self-recognition and immunological tolerance develop.*

To function properly, each of your T cells must have two traits: (1) They must be able to recognize your own major histocompatibility complex (MHC) molecules, a process known as **self-recognition,** and (2) they must lack reactivity to peptide fragments from your own proteins, a condition known as **immunological tolerance** (Figure 22.20). B cells also display immunological tolerance. Loss of immunological tolerance leads to the development of autoimmune diseases (see page 770).

While residing in the thymus gland, immature T cells that become capable of recognizing self-MHC molecules survive, whereas those lacking self-recognition undergo apoptosis (programmed cell death). This aspect of development of immunocompetence is termed **positive selection** (see Figure 22.20a). The T cells selected to survive *can recognize* the MHC part of an antigen–MHC complex.

The development of immunological tolerance, in contrast, occurs by a weeding-out process called **negative selection** in which T cells with receptors that recognize peptide fragments from self-proteins are eliminated or inactivated (see Figure 22.20a). The T cells selected to survive *do not respond* to fragments of molecules that are normally present in the body. Negative selection occurs in two ways: deletion and anergy. In **deletion,** self-reactive T cells undergo apoptosis and die, whereas in **anergy** they remain alive but are unresponsive to antigenic stimulation. It is estimated that only 1 in 100 immature T cells in the thymus receives the proper signals to survive apoptosis during both positive and negative selection and emerges as a mature, immunocompetent T cell.

Once T cells have emerged from the thymus, they may still come in contact with an unfamiliar self-protein; in such cases they may also become anergic if there is no costimulator (see Figure 22.20b). Some evidence suggests that deletion of self-reactive T cells may also occur after they leave the thymus. B cells also develop tolerance through deletion and anergy (see Figure 22.20c). While B cells are developing in bone marrow, those cells exhibiting antigen receptors that recognize common self-antigens (such as MHC antigens or blood group antigens) are deleted. Once B cells are released into the blood, however, anergy appears to be the main mechanism for preventing responses to self-proteins. When B cells encounter antigen not associated with an antigen-presenting cell, the necessary costimulation signal often is missing. In this case, the B cell is likely to become anergic (inactivated) rather than activated.

Table 22.5 summarizes the activities of cells involved in immune responses.

Figure 22.20 Development of self-recognition and immunological tolerance.

Positive selection allows recognition of self-MHC-I and self-MHC-II molecules; negative selection provides immunological tolerance of self-peptides.

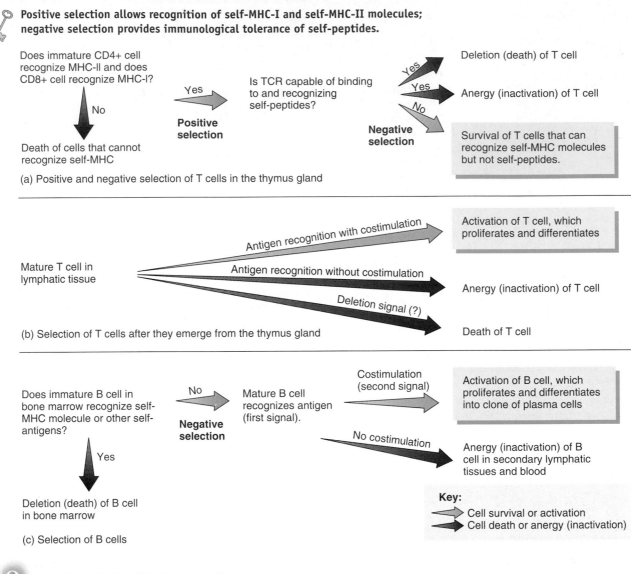

(a) Positive and negative selection of T cells in the thymus gland

(b) Selection of T cells after they emerge from the thymus gland

(c) Selection of B cells

Key:
Cell survival or activation
Cell death or anergy (inactivation)

Q How does deletion differ from anergy?

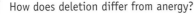

CLINICAL APPLICATION
Tumor Immunotherapy

Researchers have been trying for years to induce the immune system to mount an attack against cancer. This approach, called **tumor immunotherapy,** is effective against some types of cancer. In *adoptive cellular immunotherapy,* cells that have antitumor activity are injected into the bloodstream of a cancer patient in the hope that these "adopted" cells will seek out and destroy tumor cells. One method of adoptive cellular immunotherapy uses a patient's own inactive cytotoxic T cells and natural killer cells that have been removed in a blood sample and cultured with interleukin-2, which activates them. Such cells, called *lymphokine-activated killer (LAK) cells,* are then transfused back into the patient's blood. Although LAK cells cause some tumor regression, severe complications affect most patients. ■

1. Define the following terms: positive selection, negative selection, and anergy.

AGING AND THE IMMUNE SYSTEM
OBJECTIVE

• *Describe the effects of aging on the immune system.*

With advancing age, elderly individuals become more susceptible to all types of infections and malignancies. Their response to vaccines is decreased, and they tend to produce more autoantibodies (antibodies against their body's own molecules). In addition, the immune system exhibits lowered levels of function. For example, T cells become less responsive to antigens, and fewer T cells respond to infections. This may result from age-related atrophy of the thymus

Table 22.5 Summary of Functions of Cells Participating in Immune Responses

CELL	FUNCTIONS
Antigen-presenting Cells (APCs)	
Macrophage	Phagocytosis; processing and presentation of foreign antigens to T cells; secretion of interleukin-1, which stimulates secretion of interleukin-2 by helper T cells and induces proliferation of B cells; secretion of interferons that stimulate T cell growth.
Dendritic cell	Processes and presents antigen to T cells and B cells; found in mucous membranes, skin, and lymph nodes.
B cell	Processes and presents antigen to helper T cells.
Lymphocytes	
Cytotoxic T cell (T$_C$ or T8 cell)	Causes lysis and death of foreign cells by releasing perforin and lymphotoxin; releases other cytokines that attract macrophages and increase their phagocytic activity (gamma-IFN) and prevent macrophage migration from site of action (macrophage migration inhibition factor).
Helper T cell (T$_H$ or T4 cell)	Cooperates with B cells to amplify antibody production by plasma cells and secretes interleukin-2, which stimulates proliferation of T cells and B cells. May secrete gamma-IFN and tumor necrosis factor (TNF), which stimulate inflammatory response.
Memory T cell	Remains in lymphatic tissue and recognizes original invading antigens, even years after the first encounter.
B cell	Differentiates into antibody-producing plasma cell.
Plasma cell	Descendant of B cell that produces and secretes antibodies.
Memory B cell	Ready to respond more rapidly and forcefully than initially should the same antigen enter the body in the future.

gland or decreased production of thymic hormones. Because the T cell population decreases with age, B cells are also less responsive. Consequently, antibody levels do not increase in response to a challenge by an antigen, resulting in increased susceptibility to various infections. It is for this key reason that elderly individuals are encouraged to get influenza (flu) vaccinations each year.

1. What are the general responses of the lymphatic system as aging occurs?

DISORDERS: HOMEOSTATIC IMBALANCES

AIDS: ACQUIRED IMMUNODEFICIENCY SYNDROME

Acquired immunodeficiency syndrome (AIDS) is a condition in which a person experiences a telltale assortment of infections as a result of the progressive destruction of immune system cells by the **human immunodeficiency virus (HIV).** As we will see, AIDS represents the end stage of infection by HIV. A person who is infected with HIV may be symptom free for many years, even while the virus is actively attacking the immune system. HIV infection is serious and usually fatal because it takes control of and destroys the very cells that the body deploys to attack the virus.

Epidemiology

Because HIV is present in the blood and some body fluids, it is most effectively transmitted by actions or practices that involve the exchange of blood or body fluids between people. Within most populations, HIV is transmitted in semen or vaginal fluid during unprotected sexual intercourse or oral sex. HIV also is transmitted by direct blood-to-blood contact, such as occurs among intravenous drug users who share hypodermic needles. People at high risk are the sexual partners of HIV-infected individuals; those at lesser risk include health-care professionals who may be accidentally stuck by HIV-contaminated hypodermic needles. In addition,

HIV may be transmitted from a mother to her fetus or suckling infant. In the United States and Europe prior to 1985, HIV was unknowingly spread by the transfusion of blood and blood products containing the virus. Effective HIV screening of blood instituted after 1985 has largely eliminated this mode of HIV transmission in the United States and other developed nations. By contrast, in sub-Saharan Africa, where two-thirds of those infected with HIV live, 25% of transfused blood is not screened for HIV.

Most people in economically advantaged, industrial nations who are diagnosed as having AIDS are either homosexual men who have engaged in unprotected anal intercourse or intravenous drug users. The rate of new HIV infections in the United States paints a different and disturbing picture: The greatest increases in new HIV infections are among people of color, women, and teenagers. In developing nations, HIV is largely transmitted during unprotected heterosexual intercourse, but the problem is compounded by a contaminated blood supply. Of the 40 million people infected with HIV worldwide in the year 2000, about half are women and one-quarter are children.

HIV is a very fragile virus; it cannot survive for long outside the human body. The virus is not transmitted by insect bites. It is important to understand that one cannot become infected by casual physical contact with an HIV-infected person, such as by hug-

Figure 22.21 Human immunodeficiency virus (HIV), the causative agent of AIDS. The core contains RNA and reverse transcriptase, plus several other enzymes. The protein coat (capsid) around the core consists of a protein called P24. The envelope consists of a lipid bilayer studded with glycoproteins (GP120 and GP41) that play a vital role when HIV binds to and enters certain target cells.

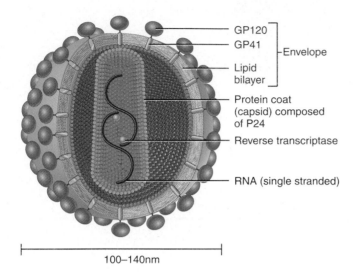

GP120 ⎤
GP41 ⎥ Envelope
Lipid bilayer ⎦

Protein coat (capsid) composed of P24

Reverse transcriptase

RNA (single stranded)

100–140nm

ging or sharing household items. The virus can be eliminated from personal care items and medical equipment by exposing them to heat (135°F for 10 minutes) or by cleaning them with common disinfectants such as hydrogen peroxide, rubbing alcohol, Lysol, household bleach, or germicidal cleansers such as Betadine or Hibiclens. Standard dishwashing and clothes washing also kills HIV.

The epidemiology of HIV also suggests ways the disease can be prevented. The chance of transmitting or of being infected by HIV during vaginal or anal intercourse can be greatly minimized—although not entirely eliminated—by the use of latex condoms. Public health programs aimed at encouraging intravenous drug users not to share needles have proved effective at checking the increase in new HIV infections among this population. Also, the prophylactic administration of certain drugs such as AZT (discussed shortly) to pregnant HIV-infected women has proved remarkably effective in minimizing the transmission of the virus to their newborn babies.

Pathogenesis of HIV Infection

HIV consists of a single strand of genetic material surrounded by a protective protein coat. To replicate, a virus must enter a host cell, where it uses the cell's enzymes, ribosomes, and nutrients to reproduce copies of its genetic information. Unlike most viruses, however, HIV is a form of **retrovirus,** a virus whose genetic information is carried in RNA instead of DNA. Once inside the host cell, a viral enzyme called **reverse transcriptase** reads the viral RNA strand and makes a DNA copy.

The coat that surrounds HIV's RNA and reverse transcriptase is composed of many molecules of a protein termed P24. In addition, the coat is wrapped by an envelope composed of a lipid bilayer penetrated by a distinctive assortment of glycoproteins (Figure 22.21). One of these glycoproteins, GP120, functions as the docking

protein that binds to CD4 molecules on T cells, macrophages, and dendritic cells. In addition, HIV must simultaneously attach to a coreceptor in the host cell's plasma membrane—a molecule designated CCR5 in antigen-presenting cells and CXCR4 in T cells—before it can gain entry into the cell. Another glycoprotein, GP41, helps the viral lipid bilayer fuse with the host cell's lipid bilayer.

The binding of HIV's docking proteins with the receptors and coreceptors in the host cell's plasma membrane causes receptor-mediated endocytosis, which brings the virus into the cell's cytoplasm. Once inside the cell, HIV sheds its protein coat. The reverse transcriptase enzyme makes a DNA copy of the viral RNA, and the viral DNA copy becomes integrated into the cell's DNA. Thus, the viral DNA is duplicated along with the host cell's DNA during normal cell division. In addition, under circumstances that are still not well understood, the viral DNA can cause the infected cell to begin producing millions of copies of viral RNA and to assemble new protein coats for each copy. The new HIV copies bud off from the cell's plasma membrane and circulate in the blood to infect other cells.

HIV mainly damages T4 (helper T) cells, and it does so in various ways. Over 100 billion viral copies may be synthesized each day. The viruses bud so rapidly from an infected cell's plasma membrane that cell lysis eventually occurs. In addition, the body's defenses attack the infected cells, killing them as well as the viruses they harbor.

As we saw in this chapter, T4 cells are instrumental in orchestrating the actions of the immune system. In most HIV-infected individuals, the body is able to replace HIV-infected T4 cells at about the same rate that they are destroyed. This process continues for many years while the body's ability to replace T4 cells is slowly exhausted; the number of T4 cells in circulation progressively declines, at an estimated rate of about 20 million cells per day.

Signs, Symptoms, and Diagnosis of HIV Infection

Immediately following infection with HIV, most people experience a brief flu-like illness. Common signs and symptoms are fever, fatigue, rash, headache, joint pain, sore throat, and swollen lymph nodes. In addition, about 50% of infected people have night sweats. After three to four weeks, plasma cells begin secreting antibodies to components of the HIV protein coat. These antibodies are detectable in blood plasma and form the basis for some of the screening tests for HIV. When people test "HIV-positive," this usually means they have antibodies to HIV in their bloodstream. If an acute HIV infection is suspected but the antibody test is negative, laboratory tests based on detection of HIV's RNA or P24 coat protein in blood plasma can confirm the presence of HIV.

Progression to AIDS

After a period of 2–10 years, the virus destroys enough T4 cells that most infected people begin to experience symptoms of immunodeficiency. HIV-infected people commonly have enlarged lymph nodes and experience persistent fatigue, involuntary weight loss, night sweats, skin rashes, diarrhea, and various lesions of the mouth and gums. In addition, the virus may begin to infect neurons in the brain, affecting the person's memory and producing visual disturbances.

As the immune system slowly collapses, an HIV-infected person becomes susceptible to a host of *opportunistic infections,* diseases caused by microorganisms that are normally held in check but now proliferate because of the defective immune system. AIDS is diagnosed when the T4 cell count drops below 200 cells per nanoliter (= cubic millimeter) of blood or when opportunistic infections arise, whichever occurs first. Typically, it is these opportunistic infections that eventually cause the death of the person.

About 5% of individuals infected with HIV have not developed AIDS; these people are called long-term nonprogressors. Their T4 count is stable, and they are symptom free. Their survival may be due to infection by a weak strain of HIV or the presence of potent natural killer cells and strong antibodies against HIV. Some of these individuals have mutations in both of the genes that code for the coreceptor CCR5, which likely prevents certain strains of HIV from entering their antigen-presenting cells.

Treatment of HIV Infection

At present, infection with HIV cannot be cured, and despite intensive research, no effective vaccine is yet available to provide immunity against HIV. However, two categories of drugs have proved successful in extending the life of many HIV-infected individuals. The first category, *reverse transcriptase inhibitors,* interferes with the action of the reverse transcriptase enzyme that the virus uses to convert its RNA into a DNA copy. Among the drugs in this category are azidovudine (AZT), didanosine (ddI), dideoxycytidine (ddC), and stavudine (d4T). The second and more recently discovered category is the *protease inhibitors.* These drugs interfere with the action of protease, a viral enzyme that cuts proteins into pieces that are assembled into the coat of newly produced HIV particles. Drugs in this category include nelfinavir, saquinavir, ritonavir, and indinavir.

Beginning in late 1995, researchers discovered that most HIV-infected individuals who receive *triple therapy*—a combination of two differently acting reverse transcriptase inhibitors and one protease inhibitor—experience a drastic reduction in viral load (the number of copies of HIV RNA in a milliliter of plasma) and an increase in the number of T4 cells in whole blood. Not only does triple therapy delay the progression of HIV infection to AIDS, but many individuals with AIDS have seen the remission or disappearance of opportunistic infections and an apparent return to health. Unfortunately, triple therapy is very costly (exceeding $10,000 per year), the dosing schedule is grueling, and not all people can tolerate the harsh side effects of these drugs. It is currently thought that people who find the drugs helpful must keep taking them for a long time, perhaps for life.

ALLERGIC REACTIONS

A person who is overly reactive to a substance that is tolerated by most other people is said to be **hypersensitive (allergic).** Whenever an allergic reaction occurs, some tissue injury results. The antigens that induce an allergic reaction are called **allergens.** Common allergens include certain foods (milk, peanuts, shellfish, eggs), antibiotics (penicillin, tetracycline), vaccines (pertussis, typhoid), venoms (honeybee, wasp, snake), cosmetics, chemicals in plants such as poison ivy, pollens, dust, molds, iodine-containing dyes used in certain x-ray procedures, and even microbes.

There are four basic types of hypersensitivity reactions: type I (anaphylactic), type II (cytotoxic), type III (immune-complex), and type IV (cell-mediated). The first three are antibody-mediated immune responses; the last is a cell-mediated immune response.

Type I (anaphylactic) reactions are the most common and occur within a few minutes after a person sensitized to an allergen is reexposed to it. **Anaphylaxis** (an′-a-fi-LAK-sis) results from the interaction of allergens with IgE antibodies on the surface of mast cells and basophils. In response to certain allergens, some people produce IgE antibodies that bind to the surface of mast cells and basophils. The next time the same allergen enters the body, it attaches to the IgE antibodies already present. In response, the mast cells and basophils release histamine, prostaglandins, leukotrienes, and kinin. Collectively, these mediators cause vasodilation, increased blood capillary permeability, increased smooth muscle contraction in the airways of the lungs, and increased mucus secretion. As a result, a person may experience inflammatory responses, difficulty in breathing through the constricted airways, and a runny nose from excess mucus secretion. In **anaphylactic shock,** which may occur in a susceptible individual who has just received a triggering drug or been stung by a wasp, wheezing and shortness of breath as airways constrict are usually accompanied by shock due to vasodilation and fluid loss from blood. This life-threatening emergency is usually treated by injecting epinephrine to dilate the airways and strengthen the heartbeat.

Type II (cytotoxic) reactions are caused by antibodies (IgG or IgM) directed against antigens on a person's blood cells (red blood cells, lymphocytes, or platelets) or tissue cells. The reaction of antibodies and antigens usually leads to activation of complement. Type II reactions, which may occur in incompatible blood transfusion reactions, damage cells by causing lysis.

Type III (immune-complex) reactions involve antigens, antibodies (IgA or IgM), and complement. When certain ratios of antigen to antibody occur, the immune complexes are small enough to escape phagocytosis, but they become trapped in the basement membrane under the endothelium of blood vessels, activate complement, and cause inflammation. Conditions that so arise include glomerulonephritis and rheumatoid arthritis (RA).

Type IV (cell-mediated) reactions or **delayed hypersensitivity reactions** usually appear 12–72 hours after exposure to an allergen. Type IV reactions occur when allergens are taken up by antigen-presenting cells (such as Langerhans cells in the skin) that migrate to lymph nodes and present the allergen to T cells, which then proliferate. Some of the new T cells return to the site of allergen entry into the body, where they produce gamma-interferon, which activates macrophages, and tumor necrosis factor, which stimulates an inflammatory response. Intracellular bacteria such as *Mycobacterium tuberculosis* trigger this type of cell-mediated immune response, as do certain haptens, such as poison ivy toxin. The skin test for tuberculosis also is a delayed hypersensitivity reaction.

INFECTIOUS MONONUCLEOSIS

Infectious mononucleosis (IM) is a contagious disease caused by the *Epstein–Barr virus (EBV).* It occurs mainly in children and young adults, and more often in females than in males by a 3:1 ratio. The virus most commonly enters the body through intimate oral contact such as kissing; it then multiplies in lymphatic tissues and spreads into the blood, where it infects and multiplies in B lymphocytes, the primary host cells. As a result of this infection, the B cells become enlarged and abnormal in appearance such that they resemble monocytes, the primary reason for the term *mononucleosis.* Signs and symptoms include an elevated white blood cell count with an abnormally high percentage of lymphocytes, fatigue, headache, dizziness, sore throat, enlarged and tender lymph nodes, and fever. There is no cure for infectious mononucleosis, but the disease usually runs its course in a few weeks.

MEDICAL TERMINOLOGY

Adenitis (ad′-e-NĪ-tis; *aden-* = gland; *-itis* = inflammation of) Enlarged, tender, and inflamed lymph nodes resulting from an infection.

Allograft (AL-ō-graft; *allo-* = other) A transplant between genetically distinct individuals of the same species. Skin transplants from other people and blood transfusions are allografts.

Autograft (AW-tō-graft; *auto-* = self) A transplant in which one's own tissue is grafted to another part of the body (such as skin grafts for burn treatment or plastic surgery).

Autoimmune disease (aw-tō-i-MYOON) A disease in which the immune system fails to recognize self-antigens and attacks the person's own cells. Examples are rheumatoid arthritis (RA), systemic lupus erythematosus (SLE), rheumatic fever, hemolytic and pernicious anemias, Addison's disease, Graves' disease, insulin-dependent diabetes mellitus, myasthenia gravis, multiple sclerosis (MS), and ulcerative colitis. Also called **autoimmunity.**

Chronic fatigue syndrome (CFS) A disorder, usually occurring in young adults and primarily in females, characterized by (1) extreme fatigue that impairs normal activities for at least 6 months and (2) the absence of other known diseases (cancer, infections, drug abuse, toxicity, or psychiatric disorders) that might produce similar symptoms.

Gamma globulin (GLOB-yoo-lin) Suspension of immunoglobulins from blood consisting of antibodies that react with a specific pathogen. It is prepared by injecting the pathogen into animals, removing blood from the animals after antibodies have been produced, isolating the antibodies, and injecting them into a human to provide short-term immunity.

Lymphadenopathy (lim-fad′-e-NOP-a-thē; *lymph-* = clear fluid; *-pathy* = disease) Enlarged, sometimes tender lymph glands.

Lymphedema (lim′-fe-DĒ-ma; *edema* = swelling) Accumulation of lymph producing subcutaneous tissue swelling.

Lymphomas (lim-FŌ-mas; *-omas* = tumors) Cancers of the lymphatic system, especially the lymph nodes. The two principal types of lymphomas are Hodgkin's disease and non-Hodgkin's lymphoma.

Splenomegaly (splē′-nō-MEG-a-lē; *mega-* = large) Enlarged spleen.

Systemic lupus erythematosus (er-e′-thēm-a-TŌ-sus), **SLE,** or **lupus** (*lupus* = wolf) An autoimmune, noncontagious, inflammatory disease of connective tissue, occurring mostly in young women. In SLE, damage to blood vessel walls results in the release of chemicals that mediate inflammation. Symptoms of SLE include joint pain, slight fever, fatigue, oral ulcers, weight loss, enlarged lymph nodes and spleen, photosensitivity, rapid loss of large amounts of scalp hair, and sometimes an eruption across the bridge of the nose and cheeks called a "butterfly rash."

Tonsillectomy (ton′-si-LEK-tō-mē; *-ectomy* = excision) Removal of a tonsil.

Xenograft (ZEN-ō-graft; *xeno-* = strange or foreign) A transplant between animals of different species. Xenografts from porcine (pig) or bovine (cow) tissue may be used in a human as a physiological dressing for severe burns.

STUDY OUTLINE

INTRODUCTION (p. 738)

1. The ability to ward off disease is called resistance. Lack of resistance is called susceptibility.
2. Nonspecific resistance refers to a wide variety of body responses against a wide range of pathogens; specific resistance or immunity involves activation of specific lymphocytes to combat a particular foreign substance.

THE LYMPHATIC SYSTEM (p. 738)

1. The lymphatic system carries out immune responses and consists of lymph, lymphatic vessels, and structures and organs that contain lymphatic tissue (specialized reticular tissue containing many lymphocytes).
2. The lymphatic system drains interstitial fluid, transports dietary lipids, and protects against invasion through immune responses.
3. Lymphatic vessels begin as closed-ended lymph capillaries in tissue spaces between cells.
4. Interstitial fluid drains into lymphatic capillaries, thus forming lymph.
5. Lymph capillaries merge to form larger vessels, called lymphatic vessels, which convey lymph into and out of structures called lymph nodes.

6. The route of lymph flow is from lymph capillaries to lymphatic vessels to lymph trunks to the thoracic duct (or right lymphatic duct) to the subclavian veins.
7. Lymph flows as a result of skeletal muscle contractions and respiratory movements. It is also aided by valves in lymphatic vessels.
8. The primary lymphatic organs are red bone marrow and the thymus gland. Secondary lymphatic organs are lymph nodes, spleen, and lymphatic nodules.
9. The thymus gland lies between the sternum and the large blood vessels above the heart. It is the site of T cell maturation.
10. Lymph nodes are encapsulated, oval structures located along lymphatic vessels.
11. Lymph enters nodes through afferent lymphatic vessels, is filtered, and exits through efferent lymphatic vessels.
12. Lymph nodes are the site of proliferation of plasma cells and T cells.
13. The spleen is the largest single mass of lymphatic tissue in the body. It is a site of B cell proliferation into plasma cells and phagocytosis of bacteria and worn-out red blood cells.
14. Lymphatic nodules are scattered throughout the mucosa of the gastrointestinal, respiratory, urinary, and reproductive tracts. This lymphatic tissue is termed mucosa-associated lymphoid tissue (MALT).

DEVELOPMENTAL ANATOMY OF THE LYMPHATIC SYSTEM (p. 747)

1. Lymphatic vessels develop from lymph sacs, which arise from developing veins. Thus, they are derived from mesoderm.
2. Lymph nodes develop from lymph sacs that become invaded by mesenchymal cells.

NONSPECIFIC RESISTANCE TO DISEASE (p. 747)

1. Mechanisms of nonspecific resistance include mechanical factors, chemical factors, antimicrobial proteins, natural killer cells, phagocytes, inflammation, and fever.
2. The skin and mucous membranes are the first line of defense against entry of pathogens.
3. Antimicrobial proteins include interferons, the complement system, and transferrins.
4. Natural killer cells and phagocytes attack and kill pathogens and defective cells in the body.
5. Inflammation aids disposal of microbes, toxins, or foreign material at the site of an injury, and prepares the site for tissue repair.
6. Fever intensifies the antiviral effects of interferons, inhibits growth of some microbes, and speeds up body reactions that aid repair.
7. Table 22.1 on page 752 summarizes the components of nonspecific resistance.

SPECIFIC RESISTANCE: IMMUNITY (p. 751)

1. Specific resistance to disease involves the production of a specific lymphocyte or antibody against a specific antigen and is called immunity.
2. B cells and T cells derive from stem cells in red bone marrow.
3. T cells complete their maturation and develop immunocompetence in the thymus gland.
4. In cell-mediated immune responses, "killer" T cells directly attack the invading antigen, whereas in antibody-mediated (humoral) immune responses, plasma cells secrete antibodies.
5. Antigens (Ags) are chemical substances that are recognized as foreign by the immune system.
6. Antigen receptors exhibit great diversity due to genetic recombination.
7. "Self-antigens" termed major histocompatibility complex (MHC) antigens are unique to each person's body cells. All cells except red blood cells display MHC-I molecules; some cells also display MHC-II molecules.
8. Cells called antigen-presenting cells (APCs), which include macrophages, B cells, and dendritic cells, process antigens.
9. Exogenous antigens (formed outside the body) are presented together with MHC-II molecules to T cells, whereas endogenous (formed inside a body cell) antigens are presented together with MHC-I molecules.
10. Cytokines are small protein hormones needed for many normal cell functions. Some of them regulate immune responses (see Table 22.2 on page 757).

CELL-MEDIATED IMMUNITY (p. 756)

1. In a cell-mediated immune response, an antigen is recognized, specific T cells proliferate and differentiate into effector cells, and the antigen is eliminated.
2. T cell receptors (TCRs) recognize antigen fragments associated with MHC molecules on the surface of a body cell.

3. Proliferation of T cells requires costimulation, either by cytokines such as interleukin-1 and interleukin-2 or by pairs of plasma membrane molecules.
4. T cells consist of several subpopulations. Helper T cells display CD4 protein, recognize antigen fragments associated with MHC-II molecules, and secrete several cytokines, most importantly interleukin-2, which acts as a costimulator for other helper T cells, cytotoxic T cells, and B cells. Cytotoxic T cells display CD8 protein and recognize antigen fragments associated with MHC-I molecules. Memory T cells remain after a cell-mediated immune response and initiate a faster response should a pathogen bearing the same foreign antigen invade the body at a later date.
5. Cytotoxic T cells eliminate invaders by secreting perforin, which causes cytolysis, and lymphotoxin, which causes fragmentation of the DNA of a target cell.
6. Cytotoxic T cells, macrophages, and natural killer cells carry out immunological surveillance, recognizing and destroying cancerous cells that display tumor antigens.

ANTIBODY-MEDIATED IMMUNITY (p. 760)

1. B cells can respond to unprocessed antigens, but their response is more intense when dendritic cells present antigen to them. Interleukin-2 and other cytokines secreted by helper T cells provide costimulation for the proliferation of B cells.
2. An activated B cell develops into a clone of antibody-producing plasma cells.
3. An antibody (Ab) is a protein that combines specifically with the antigen that triggered its production.
4. Antibodies consist of heavy and light chains and variable and constant regions.
5. Based on chemistry and structure, antibodies are grouped into five principal classes, (IgG, IgA, IgM, IgD, and IgE), each with specific biological roles.
6. Actions of antibodies include neutralization of antigen, immobilization of bacteria, agglutination and precipitation of antigen, activation of complement, and enhancement of phagocytosis.
7. Complement is a group of proteins whose functions complement immune responses and help clear antigens from the body.
8. Immunization against certain microbes is possible because memory B cells and memory T cells remain after a primary response to an antigen. The secondary response provides protection should the same microbe enter the body again.

SELF-RECOGNITION AND IMMUNOLOGICAL TOLERANCE (p. 765)

1. T cells undergo positive selection to ensure that they can recognize self-MHC antigens (self-recognition), and negative selection to ensure that they do not react to other self-proteins (tolerance). Negative selection involves both deletion and anergy.
2. B cells develop tolerance through deletion and anergy.

AGING AND THE IMMUNE SYSTEM (p. 766)

1. With advancing age, individuals become more susceptible to infections and malignancies, respond less well to vaccines, and produce more autoantibodies.
2. Immune responses also diminish with age.

Choose the best answer to the following questions:

1. Which of the following are functions of the lymphatic system? (1) draining interstitial fluid, (2) draining intracellular fluid, (3) transporting dietary lipids, (4) transporting nucleic acids, (5) protecting against invasion.
   (a) 1, 2, and 3, (b) 2, 3, and 4, (c) 3, 4, and 5, (d) 1, 2, and 4, (e) 1, 3, and 5

2. Which of the following statements are correct? (1) Lymphatic vessels are found throughout the body, except in avascular tissues, the CNS, portions of the spleen, and red bone marrow. (2) Lymphatic capillaries allow interstitial fluid to flow into them but not out of them. (3) Anchoring filaments attach lymphatic endothelial cells to surrounding tissues. (4) Lymphatic vessels freely receive all the components of blood, including the formed elements. (5) Lymph vessels directly connect to blood vessels by way of the subclavian veins.
   (a) 1, 3, 4, and 5, (b) 2, 3, 4, and 5, (c) 1, 2, 3, and 4, (d) 1, 2, 4, and 5, (e) 1, 2, 3, and 5

3. Which of the following are mechanical factors that help fight pathogens and disease? (1) tight junctions of epidermal cells, (2) mucus of mucous membranes, (3) saliva, (4) interferons, (5) complement.
   (a) 1, 3, and 4, (b) 2, 4, and 5, (c) 1, 4, and 5, (d) 1, 2, and 3, (e) 1, 2, and 4

4. Which of the following are phases of inflammation? (1) vasodilation and increased permeability of blood vessels, (2) phagocyte emigration, (3) tissue repair, (4) opsonization, (5) adherence.
   (a) 1, 2, and 3, (b) 2, 3, and 4, (c) 3, 4, and 5, (d) 1, 3, and 5, (e) 2, 4, and 5

5. Antibody-mediated immunity works mainly against (a) foreign tissue transplants, (b) intracellular pathogens, (c) extracellular pathogens, (d) cancer cells, (e) viruses

6. Which of the following are functions of antibodies? (1) neutralization of antigens, (2) immobilization of bacteria, (3) agglutination and precipitation of antigen, (4) activation of complement, (5) enhancement of phagocytosis.
   (a) 1, 3, and 4, (b) 2, 4, and 5, (c) 1, 2, 3, and 4, (d) 1, 2, 3, and 5, (e) 1, 2, 3, 4, and 5

Complete the following:

7. The factors that maintain lymph flow are ___, ___, and ___.

8. The first three phases of phagocytosis are ___, ___, and ___.

9. The first line of nonspecific defense against pathogens is the ___ and ___; the second line of nonspecific defense is the ___ and ___.

10. The two distinguishing features of immunity are ___ and ___.

True or false:

11. The sequence of fluid flow from blood vessel to blood vessel by way of the lymphatic system is arteries (blood) to blood capillaries (blood) to interstitial spaces (interstitial fluid) to lymphatic capillaries (lymph) to lymphatic vessels (lymph) to lymphatic ducts (lymph) to subclavian veins (blood).

12. A person's T cells must be able to recognize the person's own MHC molecules, a process known as self-recognition, and lack reactivity to peptide fragments from the person's own proteins, a condition known as immunological tolerance.

13. Match the following:
    ___ (a) oval or bean-shaped structures located along the length of lymphatic vessels; contain T cells, macrophages, and follicular dendritic cells; filter lymph
    ___ (b) produces pre-T cells and B cells; found in flat bones and epiphyses of long bones
    ___ (c) the single largest mass of lymphatic tissue in the body
    ___ (d) responsible for the maturation of T cells
    ___ (e) clusters of lymphocytes that stand guard in all mucous membranes

    (1) red bone marrow
    (2) thymus gland
    (3) lymph nodes
    (4) spleen
    (5) lymphatic nodules

14. Match the following:
    ___ (a) recognize foreign antigens combined with MHC-I molecules on the surface of body cells infected by viruses, some tumor cells, and cells of a tissue transplant
    ___ (b) are programmed to recognize the reappearance of the original invading antigen
    ___ (c) differentiate into plasma cells that secrete specific antibodies
    ___ (d) process and present exogenous antigens; include macrophages, B cells, and dendritic cells
    ___ (e) secrete cytokines as costimulators
    ___ (f) ingest microbes or any foreign particulate matter; include neutrophils and macrophages
    ___ (g) may down-regulate or dampen parts of the immune response
    ___ (h) lymphocytes that have the ability to kill a wide variety of infectious microbes plus certain spontaneously arising tumor cells; lack antigen receptors

    (1) helper T cells
    (2) cytotoxic (killer) T cells
    (3) suppressor T cells
    (4) memory T cells
    (5) B cells
    (6) NK cells
    (7) phagocytes
    (8) antigen-presenting cells

15. Match the following:

___ (a) participate in inflammation, opsinization, and cytolysis

___ (b) stimulate histamine release, attract neutrophils by chemotaxis, promote phagocytosis, and destroy bacteria

___ (c) glycoproteins that mark the surface of all body cells except for RBCs; distinguish self from nonself

___ (d) foreign antigens present in fluids outside body cells

___ (e) foreign antigens synthesized within a body cell

___ (f) small protein hormones that stimulate or inhibit many normal cell functions; serve as costimulators for B cell and T cell activity

___ (g) a substance that has reactivity but lacks immunogenicity

___ (h) causes vasodilation and increased permeability of blood vessels; is found in mast cells in connective tissue and in basophils and platelets in blood

___ (i) polypeptides formed in blood; induce vasodilation and increased permeability of blood vessels; serve as chemotaxic agents for phagocytes

___ (j) glycoproteins that contain four polypeptide chains and bind the antigen

(1) exogenous antigens
(2) endogenous antigens
(3) complement proteins
(4) hapten
(5) cytokines
(6) antibodies
(7) histamine
(8) major histocompatibility complex (MHC) antigens
(9) kinins
(10) complement proteins

CRITICAL THINKING QUESTIONS

1. Three-year-old Tariq was running barefoot through the grass when he felt a sharp pain and ran crying to his mother about a stepping on a "buzzy bug." His mother pulled an insect stinger out of his foot. Thirty minutes later, the sole of Tariq's foot was swollen around the stinger hole and itchy. What type of immune response is Tariq exhibiting? (HINT: *Ice and topical cortisone ointment relieved the symptoms.*)

2. On the second day of Cara's wilderness backpacking trip, she got a splinter stuck in her right thumb. It didn't bother her too much (her feet hurt more than her thumb), so she put a Band-

Aid on it and ignored it. At the end of the trip, Cara had red streaks running along her right arm and tender swollen bumps in her right armpit. What happened to her arm? (HINT: *She should have paid more attention to that splinter.*)

3. Esperanza watched as her mother got her "flu shot." "Why do you need a shot if you're not sick?" she asked. "So I won't get sick," answered her mom. Explain how the influenza vaccination prevents illness. (HINT: *A flu shot won't protect against measles, chicken pox, or pneumonia; it only works against the flu.*)

ANSWERS TO FIGURE QUESTIONS

22.1 Red bone marrow contains stem cells that develop into lymphocytes.

22.2 Lymph is more similar to interstitial fluid than to plasma because the protein content of lymph is low.

22.3 The left and right lumbar trunks and the intestinal trunk empty into the cisterna chyli, which then drains into the thoracic duct.

22.4 Inhalation promotes the movement of lymph from abdominal lymphatic vessels toward the thoracic region.

22.5 T cells mature in the thymus gland.

22.6 Foreign substances that enter a lymph node in lymph may be phagocytized by macrophages or attacked by lymphocytes that mount immune responses.

22.7 White pulp of the spleen functions in immunity; red pulp of the spleen performs functions related to blood cells.

22.8 The lymphatic system begins to develop by the end of the fifth week of gestation.

22.9 Microbes are killed by lysozyme, digestive enzymes, and oxidants.

22.10 *Redness* results from increased blood flow due to vasodilation; *pain*, from injury of nerve fibers, irritation by microbial toxins, kinins, and prostaglandins, and pressure due to edema; *heat*, from increased blood flow and heat released by locally increased metabolic reactions; *swelling*, from leakage of fluid from capillaries due to increased permeability.

22.11 Helper T cells participate in both cell-mediated and antibody-mediated immune responses.

22.12 Epitopes are small immunogenic parts of a larger antigen; haptens are small molecules that become immunogenic only when attached to a body protein.

22.13 APCs include macrophages in tissues throughout the body, B cells in blood and lymphatic tissue, and dendritic cells in mucous membranes and the skin.

22.14 The first signal in T cell activation is antigen binding to TCR; the second signal is a costimulator, such as a cytokine or another pair of plasma membrane molecules.

22.15 Cytotoxic T cells attack some tumor cells and transplanted tissue cells, as well as cells infected by viruses.

22.16 A clone of plasma cells secretes just one kind of antibody.

22.17 The variable regions recognize and bind to a specific antigen.

22.18 The classical pathway for the activation of complement is linked to antibody–mediated immunity because Ag–Ab complexes activate C1.

22.19 At peak secretion, approximately 1000 times more IgG is produced in the secondary response than in the primary response.

22.20 In deletion, self-reactive T cells or B cells die; in anergy, T cells or B cells are alive but are unresponsive to antigenic stimulation.

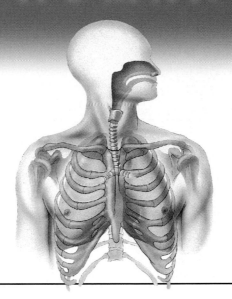

Cells continually use oxygen (O_2) for the metabolic reactions that release energy from nutrient molecules and produce ATP. At the same time, these reactions release carbon dioxide (CO_2). Because an excessive amount of CO_2 produces acidity that can be toxic to cells, excess CO_2 must be eliminated quickly and efficiently. The two systems that cooperate to supply O_2 and eliminate CO_2 are the cardiovascular and respiratory systems. The respiratory system provides for gas exchange—intake of O_2 and elimination of CO_2—whereas the cardiovascular system transports blood containing the gases between the lungs and body cells. Failure of either system disrupts homeostasis by causing rapid death of cells from oxygen starvation and the buildup of waste products. In addition to functioning in gas exchange, the respiratory system also participates in regulating blood pH, contains receptors for the sense of smell, filters inspired air, produces sounds, and rids the body of some water and heat in exhaled air.

The process of gas exchange in the body, called **respiration,** is composed of three basic steps:

1. **Pulmonary ventilation** (*pulmon-* = lung), or breathing, is the mechanical flow of air into (inspiration) and out of (expiration) the lungs.
2. **External respiration** is the exchange of gases between the air spaces of the lungs and the blood in pulmonary capillaries. In this process, pulmonary capillary blood gains O_2 and loses CO_2.
3. **Internal respiration** is the exchange of gases between blood in systemic capillaries and tissue cells. The blood

loses O_2 and gains CO_2. Within cells, the metabolic reactions that consume O_2 and give off CO_2 during the production of ATP are termed *cellular respiration* (discussed in Chapter 25).

RESPIRATORY SYSTEM ANATOMY

OBJECTIVE

• *Describe the anatomy and histology of the nose, pharynx, larynx, trachea, bronchi, and lungs.*

• *Identify the functions of each respiratory system structure.*

The **respiratory system** consists of the nose, pharynx (throat), larynx (voice box), trachea (windpipe), bronchi, and lungs (Figure 23.1). Structurally, the respiratory system consists of two portions: (1) the **upper respiratory system,** which includes the nose, pharynx, and associated structures, and (2) the **lower respiratory system,** which includes the larynx, trachea, bronchi, and lungs. Functionally, the respiratory system also consists of two portions: (1) the **conducting portion,** which consists of a series of interconnecting cavities and tubes both outside and within the lungs—the nose, pharynx, larynx, trachea, bronchi, bronchioles, and terminal bronchioles—that filter, warm, and moisten air and conduct it into the lungs, and (2) the **respiratory portion,** which consists of tissues within the lungs where gas exchange occurs—the respiratory bronchioles, alveolar ducts, alveolar sacs, and alveoli, the main sites of gas exchange between air and blood. The volume of the conducting portion

Figure 23.1 Structures of the respiratory system.

🔑 **The upper respiratory system includes the nose, pharynx, and associated structures; the lower respiratory system includes the larynx, trachea, bronchi, and lungs.**

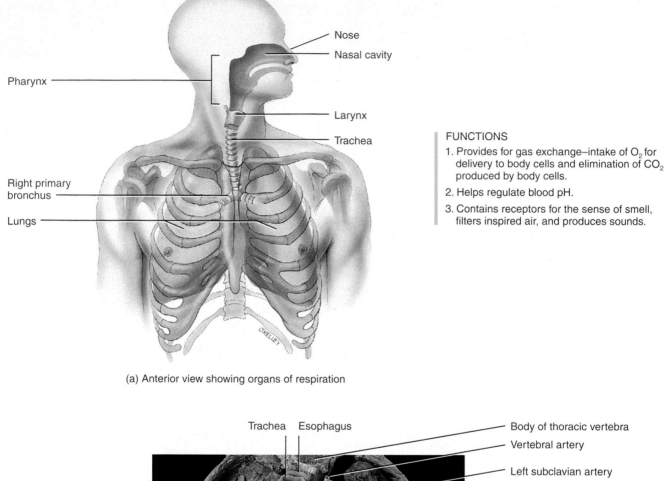

FUNCTIONS

1. Provides for gas exchange–intake of O_2 for delivery to body cells and elimination of CO_2 produced by body cells.
2. Helps regulate blood pH.
3. Contains receptors for the sense of smell, filters inspired air, and produces sounds.

(a) Anterior view showing organs of respiration

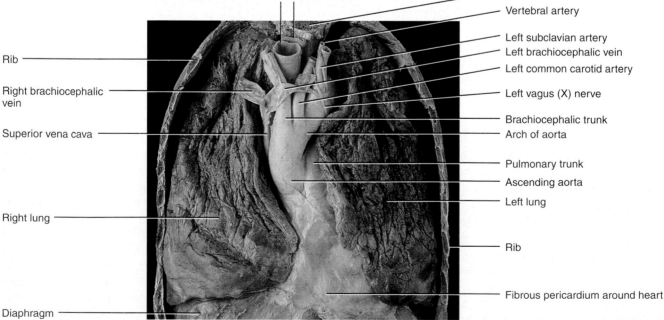

(b) Anterior view of the lungs after removal of the anterolateral thoracic wall and parietal pleura

 Which structures are part of the conducting portion of the respiratory system?

in an adult is about 150 mL; that of the respiratory portion is 5–6 liters.

The branch of medicine that deals with the diagnosis and treatment of diseases of the ears, nose, and throat is called **otorhinolaryngology** (ō′-tō-rī′-nō-lar′-in-GOL-ō-jē; *oto-* = ear; *rhino-* = nose; *laryngo-* = voice box; *-ology* = study of). A **pulmonologist** is a specialist in the diagnosis and treatment of diseases of the lungs.

Nose

The **nose** can be divided into external and internal portions. The external portion consists of a supporting framework of bone and hyaline cartilage covered with muscle and skin and lined by a mucous membrane. The bony framework of the nose is formed by the frontal bone, nasal bones, and maxillae (Figure 23.2a). The cartilaginous framework consists of the **septal cartilage,** which forms the anterior portion of the nasal septum; the **lateral nasal cartilages** inferior to the nasal bones; and the **alar cartilages,** which form a portion of the walls of the nostrils. Because it has a framework of pliable hyaline cartilage, the rest of the external nose is somewhat flexible. On the undersurface of the external nose are two openings called the **external nares** (NA-rēz; singular is **naris**) or **nostrils.** The surface anatomy of the nose is shown in Figure 23.3. The interior structures of the external portion of the nose have three functions: (1) warming, moistening, and filtering incoming air; (2) detecting olfactory stimuli; and (3) modifying speech vibrations as they pass through the large, hollow resonating chambers.

The internal portion of the nose is a large cavity in the anterior aspect of the skull that lies inferior to the nasal bone and superior to the mouth; it also includes muscle and mucous membrane. Anteriorly, the internal nose merges with the external nose, and posteriorly it communicates with the pharynx through two openings called the **internal nares** or **choanae** (kō-Ā-nē) (see Figure 23.2b). Ducts from the paranasal sinuses (frontal, sphenoidal, maxillary, and ethmoidal sinuses) and the nasolacrimal ducts also open into the internal nose. The lateral walls of the internal nose are formed by the ethmoid, maxillae, lacrimal, palatine, and inferior nasal conchae bones (see Figure 7.9 on page 195); the ethmoid also forms the roof. The floor of the internal nose is formed mostly by the palatine bones and palatine processes of the maxillae, which together constitute the hard palate.

The space inside the internal nose is called the **nasal cavity;** it is divided into right and left sides by a vertical partition called the **nasal septum.** The anterior portion of the septum consists primarily of hyaline cartilage; the remainder is formed by the vomer, perpendicular plate of the ethmoid, maxillae, and palatine bones (see Figure 7.14 on page 200). The anterior portion of the nasal cavity, just inside the nostrils, is called the **vestibule** and is surrounded by cartilage; the superior part of the nasal cavity is surrounded by bone.

When air enters the nostrils, it passes first through the vestibule, which is lined by skin containing coarse hairs that filter out large dust particles. Three shelves formed by projections of the superior, middle, and inferior nasal conchae extend out of each lateral wall of the cavity. The conchae, almost reaching the septum, subdivide each side of the nasal cavity into a series of groovelike passageways—the **superior, middle,** and **inferior meatuses** (mē-Ā-tes-ez; = openings or passages). Mucous membrane lines the cavity and its shelves. The arrangement of conchae and meatuses increases surface area in the internal nose and prevents dehydration by acting as a baffle that traps water droplets during exhalation.

The olfactory receptors lie in the membrane lining the superior nasal conchae and adjacent septum. This region is called the **olfactory epithelium.** Inferior to the olfactory epithelium, the mucous membrane contains capillaries and pseudostratified ciliated columnar epithelium with many goblet cells. As inspired air whirls around the conchae and meatuses, it is warmed by blood in the capillaries. Mucus secreted by the goblet cells moistens the air and traps dust particles. Drainage from the nasolacrimal ducts and perhaps secretions from the paranasal sinuses also help moisten the air. The cilia move the mucus and trapped dust particles toward the pharynx, at which point they can be swallowed or spit out, thus removing particles from the respiratory tract.

 CLINICAL APPLICATION
Rhinoplasty

Rhinoplasty (RĪ-nō-plas′-tē; *-plasty* = to mold or to shape), commonly called a "nose job," is a surgical procedure in which the structure of the external nose is altered. Although it is often done for cosmetic reasons, it is sometimes performed to repair a fractured nose or a deviated nasal septum. In the procedure, both a local and general anesthetic are given, and with instruments inserted through the nostrils, the nasal cartilage is reshaped, and the nasal bones are fractured and repositioned, to achieve the desired shape. An internal packing and splint are inserted to keep the nose in the desired position while it heals. ■

Pharynx

The **pharynx** (FAIR-inks), or throat, is a funnel-shaped tube about 13 cm (5 in.) long that starts at the internal nares and extends to the level of the cricoid cartilage, the most inferior cartilage of the larynx (voice box) (Figure 23.4). The pharynx lies just posterior to the nasal and oral cavities, superior to the larynx, and just anterior to the cervical vertebrae. Its wall is composed of skeletal muscles and is lined with a mucous membrane. The pharynx functions as a passageway for air and food, provides a resonating chamber for speech sounds, and houses the tonsils, which participate in immunological reactions against foreign invaders.

The pharynx can be divided into three anatomical regions: (1) nasopharynx, (2) oropharynx, (3) laryngopharynx. (See the lower orientation diagram in Figure 23.4.) The

Figure 23.2 Respiratory structures in the head and neck. (See Tortora, *A Photographic Atlas of the Human Body,* Figures 11.2 and 11.3)

 As air passes through the nose, it is warmed, filtered, and moistened.

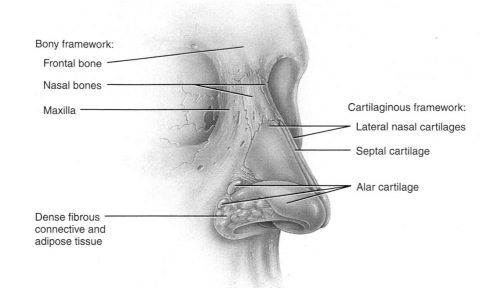

Bony framework:
Frontal bone
Nasal bones
Maxilla

Cartilaginous framework:
Lateral nasal cartilages
Septal cartilage
Alar cartilage

Dense fibrous connective and adipose tissue

(a) Anterolateral view of external portion of nose showing cartilaginous and bony framework

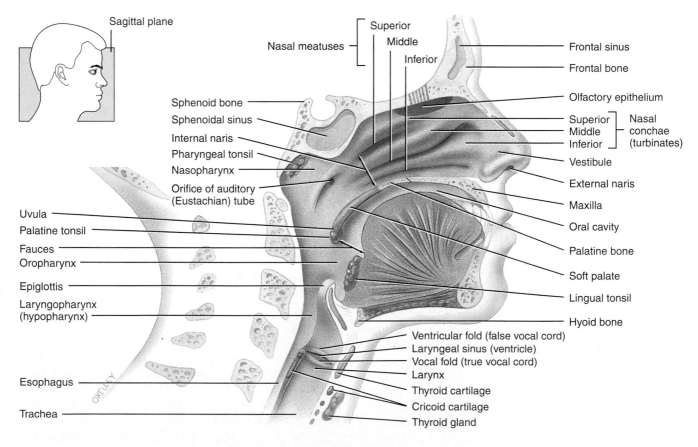

Sagittal plane

Nasal meatuses —
Superior
Middle
Inferior

Frontal sinus
Frontal bone

Olfactory epithelium

Sphenoid bone
Sphenoidal sinus
Internal naris
Pharyngeal tonsil
Nasopharynx
Orifice of auditory (Eustachian) tube

Superior
Middle   Nasal conchae (turbinates)
Inferior

Vestibule
External naris
Maxilla

Uvula
Palatine tonsil
Fauces
Oropharynx
Epiglottis
Laryngopharynx (hypopharynx)

Oral cavity
Palatine bone
Soft palate
Lingual tonsil

Esophagus

Hyoid bone
Ventricular fold (false vocal cord)
Laryngeal sinus (ventricle)
Vocal fold (true vocal cord)
Larynx
Thyroid cartilage

Trachea

Cricoid cartilage
Thyroid gland

(b) Sagittal section of the left side of the head and neck showing the location of respiratory structures

Q What is the path taken by air molecules into and through the nose?

muscles of the entire pharynx are arranged in two layers, an outer circular layer and an inner longitudinal layer.

The superior portion of the pharynx, called the **nasopharynx,** lies posterior to the nasal cavity and extends to the plane of the soft palate. There are five openings in its wall: two internal nares, two openings that lead into the auditory (Eustachian) tubes, and the opening into the oropharynx. The posterior wall also contains the **pharyngeal tonsil.** Through the internal nares, the nasopharynx receives air from the nasal cavity and receives packages of dust-laden mucus. It is lined with pseudostratified ciliated columnar epithelium, and the cilia move the mucus down toward the most inferior part of the pharynx. The nasopharynx also exchanges small amounts of air with the auditory (Eustachian) tubes to equalize air pressure between the pharynx and the middle ear.

The intermediate portion of the pharynx, the **oropharynx,** lies posterior to the oral cavity and extends from the soft palate inferiorly to the level of the hyoid bone. It has only one opening, the **fauces** (FAW-sēz; = throat), the opening from the mouth. This portion of the pharynx has both respiratory and digestive functions because it is a common passageway for air, food, and drink. Because the oropharynx

Figure 23.3 Surface anatomy of the nose.

🔑 **The external nose has a cartilaginous and bony framework.**

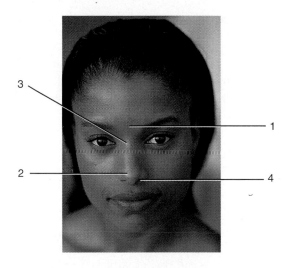

Anterior view

1. **Root**: Superior attachment of the nose to the frontal bone
2. **Apex**: Tip of nose
3. **Bridge**: Bony framework of nose formed by nasal bones
4. **External naris**: Nostril; external opening into nasal cavity

Q   At what point is the nose attached to the frontal bone?

Figure 23.4 Pharynx. (See Tortora, *A Photographic Atlas of the Human Body,* Figure 11.4)

🔑 **The three subdivisions of the pharynx are the (1) nasopharynx, (2) oropharynx, and (3) laryngopharynx.**

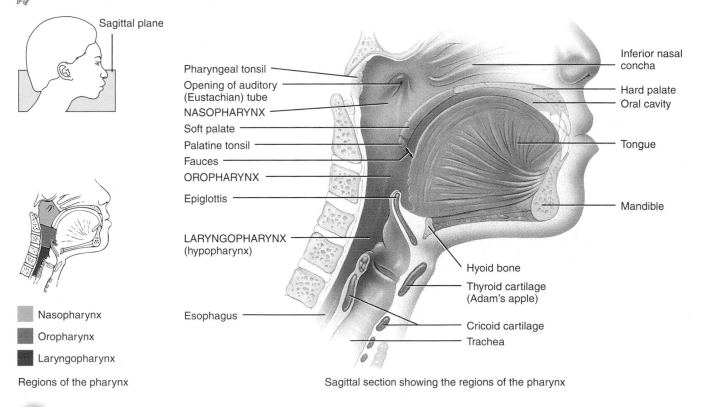

Regions of the pharynx

Sagittal section showing the regions of the pharynx

Q   What are the superior and inferior borders of the pharynx?

is subject to abrasion by food particles, it is lined with nonkeratinized stratified squamous epithelium. Two pairs of tonsils, the **palatine** and **lingual tonsils,** are found in the oropharynx.

The inferior portion of the pharynx, the **laryngopharynx** (la-rin′-gō-FAIR-inks), or **hypopharynx,** begins at the level of the hyoid bone and connects the esophagus (food tube) with the larynx (voice box). Like the oropharynx, the laryngopharynx is both a respiratory as well as a digestive pathway and is lined by nonkeratinized stratified squamous epithelium.

Larynx

The **larynx** (LAIR-inks), or voice box, is a short passageway that connects the laryngopharynx with the trachea. It lies in the midline of the neck anterior to the fourth through sixth cervical vertebrae (C4–C6).

The wall of the larynx is composed of nine pieces of cartilage (Figure 23.5). Three occur singly (thyroid cartilage, epiglottic cartilage, and cricoid cartilage), and three occur in pairs (arytenoid, cuneiform, and corniculate cartilages). Of the paired cartilages, the arytenoid cartilages are the most important because they influence the positions and tensions of the vocal folds (true vocal cords). Whereas the extrinsic muscles of the larynx connect the cartilages to other structures in the throat, the intrinsic muscles connect the cartilages to each other.

The **thyroid cartilage (Adam's apple)** consists of two fused plates of hyaline cartilage that form the anterior wall of the larynx and give it a triangular shape. It is usually larger in males than in females due to the influence of male sex hormones on its growth during puberty. The ligament that connects the thyroid cartilage to the hyoid bone is called the **thyrohyoid membrane.**

The **epiglottis** (*epi-* = over; *glottis* = tongue) is a large, leaf-shaped piece of elastic cartilage that is covered with epithelium (see also Figure 23.4). The "stem" of the epiglottis is attached to the anterior rim of the thyroid cartilage, but the "leaf" portion is unattached and free to move up and down like a trap door. During swallowing, the pharynx and larynx rise. Elevation of the pharynx widens it to receive food or drink; elevation of the larynx causes the free edge of the epiglottis to move down and form a lid over the glottis, closing it off. The **glottis** consists of a pair of folds of mucous membrane, the vocal folds (true vocal cords) in the larynx, and the space between them called the **rima glottidis** (RĪ-ma GLOT-ti-dis). The closing of the larynx in this way during swallowing routes liquids and foods into the esophagus and keeps them out of the larynx and airways inferior to it. When small particles of dust, smoke, food, or liquids pass into the larynx, a cough reflex occurs, usually expelling the material.

The **cricoid cartilage** (KRĪ-koyd; = ringlike) is a ring of hyaline cartilage that forms the inferior wall of the larynx.

It is attached to the first ring of cartilage of the trachea by the **cricotracheal ligament.** The thyroid cartilage is connected to the cricoid cartilage by the **cricothyroid ligament.** The cricoid cartilage is the landmark for making an emergency airway (a tracheostomy; see page 783).

The paired **arytenoid cartilages** (ar′-i-TĒ-noyd; = ladlelike) are triangular pieces of mostly hyaline cartilage located at the posterior, superior border of the cricoid cartilage. They attach to the vocal folds and intrinsic pharyngeal muscles. Supported by the arytenoid cartilages, the intrinsic pharyngeal muscles contract and thus move the vocal folds.

The paired **corniculate cartilages** (kor-NIK-yoo-lāt; = shaped like a small horn), which are horn-shaped pieces of elastic cartilage, are located at the apex of each arytenoid cartilage. The paired **cuneiform cartilages** (kyoo-NĒ-i-form; = wedge-shaped), which are club-shaped elastic cartilages anterior to the corniculate cartilages, support the vocal folds and lateral aspects of the epiglottis.

The lining of the larynx superior to the vocal folds is nonkeratinized stratified squamous epithelium. The lining of the larynx inferior to the vocal folds is pseudostratified ciliated columnar epithelium consisting of ciliated columnar cells, goblet cells, and basal cells. Its mucus helps trap dust not removed in the upper passages. Whereas the cilia in the upper respiratory tract move mucus and trapped particles *down* toward the pharynx, the cilia in the lower respiratory tract move them *up* toward the pharynx.

The Structures of Voice Production

The mucous membrane of the larynx forms two pairs of folds (see Figure 23.5c): a superior pair called the **ventricular folds (false vocal cords)** and an inferior pair called simply the **vocal folds (true vocal cords).** The space between the ventricular folds is known as the **rima vestibuli.** The **laryngeal sinus (ventricle)** is a lateral expansion of the middle portion of the laryngeal cavity between the ventricular folds above and the vocal folds below (see Figure 23.2b).

When the ventricular folds are brought together, they function in holding the breath against pressure in the thoracic cavity, such as might occur when a person strains to lift a heavy object. Deep to the mucous membrane of the vocal folds, which is lined by nonkeratinized stratified squamous epithelium, are bands of elastic ligaments stretched between pieces of rigid cartilage like the strings on a guitar. Skeletal muscles of the larynx, called intrinsic muscles, attach to both the rigid cartilage and the vocal folds. When the muscles contract, they pull the elastic ligaments tight and stretch the vocal folds out into the airways so that the rima glottidis is narrowed. If air is directed against the vocal folds, they vibrate and set up sound waves in the column of air in the pharynx, nose, and mouth. The greater the pressure of air, the louder the sound.

When the intrinsic muscles of the larynx contract, they pull on the arytenoid cartilages, which causes them to pivot.

Figure 23.5 Larynx. (See Tortora, *A Photographic Atlas of the Human Body,* Figures 11.5 and 11.6)

🔑 **The larynx is composed of nine pieces of cartilage.**

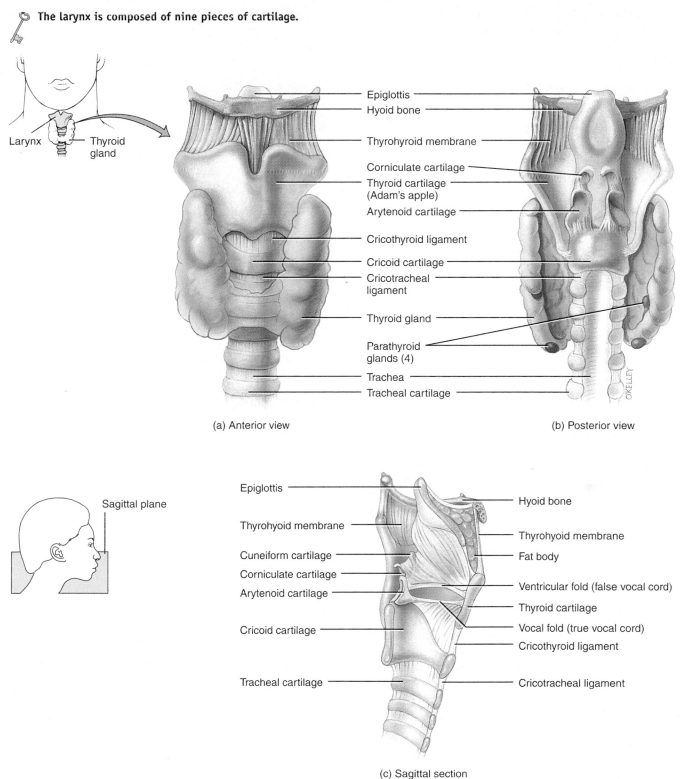

(a) Anterior view

(b) Posterior view

(c) Sagittal section

Q How does the epiglottis prevent aspiration of foods and liquids?

Figure 23.6 Movement of the vocal folds. (See Tortora, *A Photographic Atlas of the Human Body,* Figure 11.7)

🔑 **The glottis consists of a pair of folds of mucous membrane, the vocal folds in the larynx, and the space between them (the rima glottidis).**

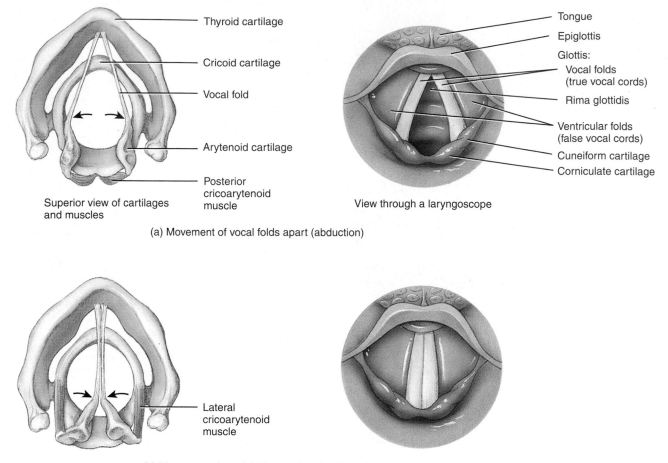

Thyroid cartilage

Cricoid cartilage

Vocal fold

Arytenoid cartilage

Posterior cricoarytenoid muscle

Superior view of cartilages and muscles

Tongue

Epiglottis

Glottis:
Vocal folds (true vocal cords)
Rima glottidis
Ventricular folds (false vocal cords)
Cuneiform cartilage
Corniculate cartilage

View through a laryngoscope

(a) Movement of vocal folds apart (abduction)

Lateral cricoarytenoid muscle

(b) Movement of vocal folds together (adduction)

Ⓠ What is the main function of the vocal folds?

Contraction of the posterior cricoarytenoid muscles, for example, moves the vocal folds apart (abduction), thereby opening the rima glottidis (Figure 23.6a). By contrast, contraction of the lateral cricoarytenoid muscles moves the vocal folds together (adduction), thereby closing the rima glottidis (Figure 23.6b). Other intrinsic muscles can elongate (and place tension on) or shorten (and relax) the vocal folds.

Pitch is controlled by the tension on the vocal folds. If they are pulled taut by the muscles, they vibrate more rapidly, and a higher pitch results. Lower sounds are produced by decreasing the muscular tension on the vocal folds. Due to the influence of androgens (male sex hormones), vocal folds are usually thicker and longer in males than in females, and therefore they vibrate more slowly. Thus, men's voices generally have a lower range of pitch than women's.

Sound originates from the vibration of the vocal folds, but other structures are necessary for converting the sound into recognizable speech. The pharynx, mouth, nasal cavity, and paranasal sinuses all act as resonating chambers that give

the voice its human and individual quality. We produce the vowel sounds by constricting and relaxing the muscles in the wall of the pharynx. Muscles of the face, tongue, and lips help us enunciate words.

Whispering is accomplished by closing all but the posterior portion of the rima glottidis. Because the vocal folds do not vibrate during whispering, there is no pitch to this form of speech. However, we can still produce intelligible speech while whispering by changing the shape of the oral cavity as we enunciate. As the size of the oral cavity changes, its resonance qualities change, which imparts a vowel-like pitch to the air as it rushes toward the lips.

Trachea

The **trachea** (TRĀ-kē-a; = sturdy), or windpipe, is a tubular passageway for air that is about 12 cm (5 in.) long and 2.5 cm (1 in.) in diameter. It is located anterior to the esophagus (Figure 23.7) and extends from the larynx to the supe-

Figure 23.7 Location of the trachea in relation to the esophagus.

The trachea is anterior to the esophagus and extends from the larynx to the superior border of the fifth thoracic vertebra.

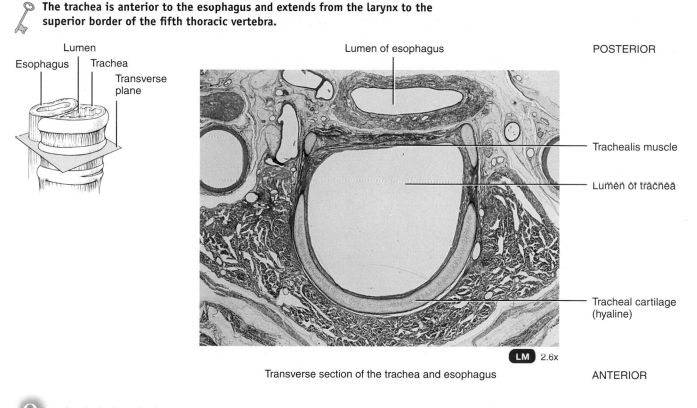

Lumen
Esophagus | Trachea
Transverse plane

Lumen of esophagus

POSTERIOR

Trachealis muscle

Lumen of trachea

Tracheal cartilage (hyaline)

LM 2.6x

Transverse section of the trachea and esophagus

ANTERIOR

Q What is the benefit of not having cartilage between the trachea and the esophagus?

rior border of the fifth thoracic vertebra (T5), where it divides into right and left primary bronchi (see Figure 23.8).

The layers of the tracheal wall, from deep to superficial, are (1) mucosa, (2) submucosa, (3) hyaline cartilage, and (4) adventitia, which is composed of areolar connective tissue. The mucosa of the trachea consists of an epithelial layer of pseudostratified ciliated columnar epithelium and an underlying layer of lamina propria that contains elastic and reticular fibers (see Table 4.1 on page 113). The epithelium consists of ciliated columnar cells and goblet cells that reach the luminal surface, plus basal cells that do not. The epithelium provides the same protection against dust as the membrane lining the nasal cavity and larynx. The submucosa consists of areolar connective tissue that contains seromucous glands and their ducts. The 16–20 incomplete, horizontal rings of hyaline cartilage resemble the letter C and are stacked one on top of another. They may be felt through the skin inferior to the larynx. The open part of each C-shaped cartilage ring faces the esophagus (see Figure 23.7), an arrangement that accommodates slight expansion of the esophagus into the trachea during swallowing. Transverse smooth muscle fibers, called the **trachealis muscle,** and elastic connective tissue stabilize the open ends of the cartilage rings. The solid C-shaped cartilage rings provide a semirigid support so that the tracheal wall does not collapse inward (especially during in-

spiration) and obstruct the air passageway. The adventitia of the trachea consists of areolar connective tissue that joins the trachea to surrounding tissues.

CLINICAL APPLICATION
Tracheostomy and Intubation

Several conditions may block airflow by obstructing the trachea. For example, the rings of cartilage that support the trachea may collapse due to a crushing injury to the chest, inflammation of the mucous membrane may cause it to swell so much that the airway closes, or vomit or a foreign object may be aspirated into it. Two methods are used to reestablish airflow past a tracheal obstruction. If the obstruction is superior to the level of the larynx, a **tracheostomy** (trā-kē-OS-tō-mē) may be performed. In this procedure, a skin incision is followed by a short longitudinal incision into the trachea inferior to the cricoid cartilage. The patient can then breathe through a metal or plastic tracheal tube inserted through the incision. The second method is **intubation,** in which a tube is inserted into the mouth or nose and passed inferiorly through the larynx and trachea. The firm wall of the tube pushes aside any flexible obstruction, and the lumen of the tube provides a passageway for air; any mucus clogging the trachea can be suctioned out through the tube. ■

Figure 23.8 Branching of airways from the trachea: the bronchial tree. (See Tortora, *A Photographic Atlas of the Human Body,* Figure 11.8)

🔑 **The bronchial tree begins at the trachea and ends at the terminal bronchioles.**

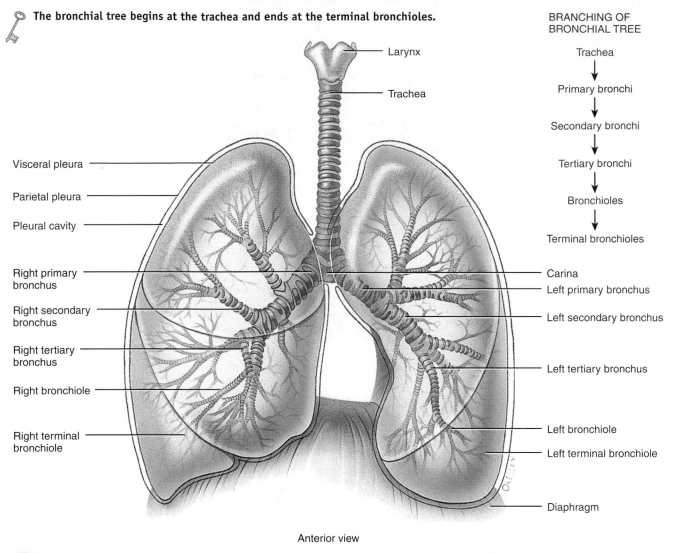

BRANCHING OF
BRONCHIAL TREE

Trachea
↓
Primary bronchi
↓
Secondary bronchi
↓
Tertiary bronchi
↓
Bronchioles
↓
Terminal bronchioles

Larynx

Trachea

Visceral pleura

Parietal pleura

Pleural cavity

Right primary bronchus

Right secondary bronchus

Right tertiary bronchus

Right bronchiole

Right terminal bronchiole

Carina

Left primary bronchus

Left secondary bronchus

Left tertiary bronchus

Left bronchiole

Left terminal bronchiole

Diaphragm

Anterior view

Q How many lobes and secondary bronchi are present in each lung?

Bronchi

At the superior border of the fifth thoracic vertebra, the trachea divides into a **right primary bronchus** (BRON-kus; = windpipe), which goes into the right lung, and a **left primary bronchus,** which goes into the left lung (Figure 23.8). The right primary bronchus is more vertical, shorter, and wider than the left. As a result, an aspirated object is more likely to enter and lodge in the right primary bronchus than the left. Like the trachea, the primary bronchi (BRON-kē) contain incomplete rings of cartilage and are lined by pseudostratified ciliated columnar epithelium.

At the point where the trachea divides into right and left primary bronchi is an internal ridge called the **carina** (ka-RĪ-na; = keel of a boat). It is formed by a posterior and somewhat inferior projection of the last tracheal cartilage. The mucous membrane of the carina is one of the most sensitive areas of the entire larynx and trachea for triggering a

cough reflex. Widening and distortion of the carina is a serious sign because it usually indicates a carcinoma of the lymph nodes around the region where the trachea divides.

On entering the lungs, the primary bronchi divide to form smaller bronchi—the **secondary (lobar) bronchi,** one for each lobe of the lung. (The right lung has three lobes; the left lung has two.) The secondary bronchi continue to branch, forming still smaller bronchi, called **tertiary (segmental) bronchi,** that divide into **bronchioles.** Bronchioles, in turn, branch repeatedly, and the smallest ones branch into even smaller tubes called **terminal bronchioles.** This extensive branching from the trachea resembles an inverted tree and is commonly referred to as the **bronchial tree.**

As the branching becomes more extensive in the bronchial tree, several structural changes may be noted. First, the epithelium gradually changes from pseudostratified ciliated columnar epithelium in the bronchi to nonciliated simple cuboidal epithelium in the terminal bronchioles. (In regions where nonciliated cuboidal epithelium is present, inhaled particles are removed by macrophages.) Second, the incomplete rings of cartilage in primary bronchi are gradually replaced by plates of cartilage that finally disappear. Third, as the amount of cartilage decreases, the amount of smooth muscle increases. Smooth muscle encircles the lumen in spiral bands. Because there is no supporting cartilage, however, muscle spasms can close off the airways; this is what happens during an asthma attack, and it can be a life-threatening situation. During exercise, activity in the sympathetic division of the ANS increases and the adrenal medulla releases the hormones epinephrine and norepinephrine, both of which cause relaxation of smooth muscle in the bronchioles, which dilates the airways. The result is improved lung ventilation because air reaches the alveoli more quickly. The parasympathetic division of the ANS and mediators of allergic reactions such as histamine cause contraction of bronchiolar smooth muscle and result in constriction of distal bronchioles.

1. What functions do the respiratory and cardiovascular systems have in common?
2. Distinguish between the upper and lower respiratory systems.
3. Compare the structure and functions of the external nose and the internal nose.
4. What are the three anatomical regions of the pharynx? List the roles of each in respiration.
5. Explain how the larynx functions in respiration and voice production.
6. Describe the location, structure, and function of the trachea.
7. What is the bronchial tree? Describe its structure.

Lungs

OBJECTIVE

• *Describe the anatomy and histology of the lungs.*

The **lungs** (= lightweights, because they float) are paired cone-shaped organs lying in the thoracic cavity. They are separated from each other by the heart and other structures in the mediastinum, which separates the thoracic cavity into two anatomically distinct chambers. As a result, should trauma cause one lung to collapse, the other may remain expanded. Two layers of serous membrane, collectively called the **pleural membrane** (*pleur-* = side), enclose and protect each lung. The superficial layer lines the wall of the thoracic cavity and is called the **parietal pleura;** the deep layer, the **visceral pleura,** covers the lungs themselves (Figure 23.9). Between the visceral and parietal pleurae is a small space, the **pleural cavity,** which contains a small amount of lubricating fluid secreted by the membranes. This fluid reduces friction between the membranes, allowing them to slide easily over one another during breathing. Pleural fluid

also causes the two membranes to adhere to one another just as a film of water causes two glass slides to stick together. Separate pleural cavities surround the left and right lungs. Inflammation of the pleural membrane, called **pleurisy** or **pleuritis,** may in its early stages cause pain due to friction between the parietal and visceral layers of the pleura. If the inflammation persists, excess fluid accumulates in the pleural space, a condition known as **pleural effusion.**

The lungs extend from the diaphragm to just slightly superior to the clavicles and lie against the ribs anteriorly and posteriorly (Figure 23.10). The broad inferior portion of the lung, the **base,** is concave and fits over the convex area of the diaphragm. The narrow superior portion of the lung is the **apex.** The surface of the lung lying against the ribs, the **costal surface,** matches the rounded curvature of the ribs. The **mediastinal (medial) surface** of each lung contains a region, the **hilus,** through which bronchi, pulmonary blood vessels, lymphatic vessels, and nerves enter and exit. These structures are held together by the pleura and connective tissue and constitute the **root** of the lung. Medially, the left lung also contains a concavity, the **cardiac notch,** in which the heart lies. Due to the space occupied by the heart, the left lung is about 10% smaller than the right lung. Although the right lung is thicker and broader, it is also somewhat shorter than the left lung because the diaphragm is higher on the right side, accommodating the liver that lies inferior to it.

The lungs almost fill the thorax (see Figure 23.10a). The apex of the lungs lies superior to the medial third of the clavicles and is the only area that can be palpated. The anterior, lateral, and posterior surfaces of the lungs lie against the ribs. The base of the lungs extends from the sixth costal cartilage anteriorly to the spinous process of the tenth thoracic vertebra posteriorly. The pleura extends about 5 cm below the base from the sixth costal cartilage anteriorly to the twelfth rib posteriorly. Thus, the lungs do not completely fill the pleural cavity in this area, and removal of excessive fluid in the pleural cavity can be accomplished without injuring lung tissue by inserting the needle posteriorly through the seventh intercostal space, a procedure termed **thoracentesis** (thor′-a-sen-TĒ-sis; -*centesis* = puncture).

Lobes, Fissures, and Lobules

Each lung is divided into lobes by one or more fissures (see Figure 23.10b–e). Both lungs have an **oblique fissure,** which extends inferiorly and anteriorly; the right lung also has a **horizontal fissure.** The oblique fissure in the left lung separates the **superior lobe** from the **inferior lobe.** In the right lung, the superior part of the oblique fissure separates the superior lobe from the inferior lobe, whereas the inferior part of the oblique fissure separates the inferior lobe from the **middle lobe.** The horizontal fissure of the right lung subdivides the superior lobe, thus forming a middle lobe.

Each lobe receives its own secondary (lobar) bronchus. Thus, the right primary bronchus gives rise to three secondary (lobar) bronchi called the **superior, middle,** and **inferior secondary (lobar) bronchi,** whereas the left primary

Figure 23.9 Relationship of the pleural membranes to the lungs. The arrow in the inset indicates the direction from which the lungs are viewed (superior).

 The parietal pleura lines the thoracic cavity, whereas the visceral pleura covers the lungs.

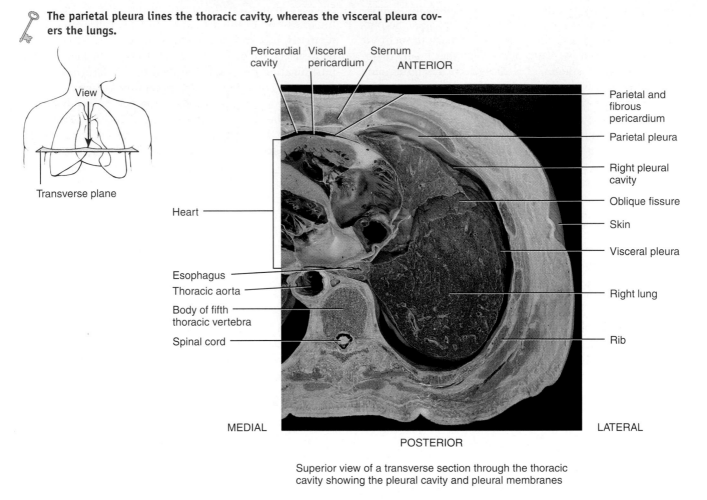

Superior view of a transverse section through the thoracic cavity showing the pleural cavity and pleural membranes

Q What type of membrane is the pleural membrane?

bronchus gives rise to **superior** and **inferior secondary (lobar) bronchi.** Within the substance of the lung, the secondary bronchi give rise to the **tertiary (segmental) bronchi,** which are constant in both origin and distribution—there are ten tertiary bronchi in each lung. The segment of lung tissue that each supplies is called a **bronchopulmonary segment.** Bronchial and pulmonary disorders (such as tumors or abscesses) that are localized in a bronchopulmonary segment may be surgically removed without seriously disrupting the surrounding lung tissue.

Each bronchopulmonary segment of the lungs has many small compartments called **lobules,** each of which is wrapped in elastic connective tissue and contains a lymphatic vessel, an arteriole, a venule, and a branch from a terminal bronchiole (Figure 23.11a). Terminal bronchioles subdivide into microscopic branches called **respiratory bronchioles** (Figure 23.11b). As the respiratory bronchioles penetrate more deeply into the lungs, the epithelial lining changes from simple cuboidal to simple squamous. Respiratory bronchioles, in turn, subdivide into several (2–11) **alve-**

olar ducts. The respiratory passages from the trachea to the alveolar ducts contain about 25 orders of branching; that is, branching—from the trachea into primary bronchi (first-order branching) into secondary bronchi (second-order branching) and so on down to the alveolar ducts—occurs about 25 times.

Alveoli

Around the circumference of the alveolar ducts are numerous alveoli and alveolar sacs. An **alveolus** (al-VĒ-ō-lus) is a cup-shaped outpouching lined by simple squamous epithelium and supported by a thin elastic basement membrane; **alveolar sacs** are two or more alveoli that share a common opening (see Figure 23.11a, b). The walls of alveoli consist of two types of alveolar epithelial cells (Figure 23.12). **Type I alveolar cells** are simple squamous epithelial cells that form a mostly continuous lining of the alveolar wall that is interrupted by occasional **type II alveolar cells,** also called **septal cells.** The thin type I alveolar cells are the main site of gas exchange. Type II alveolar cells, which are rounded or

Figure 23.10 Surface anatomy of the lungs. (See Tortora, *A Photographic Atlas of the Human Body,* Figures 11.12 and 11.14)

The subdivisions of the lungs are lobes, bronchopulmonary segments, and lobules.

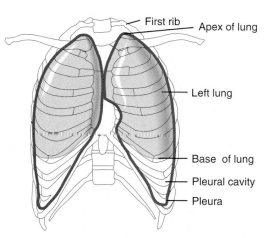

First rib
Apex of lung
Left lung
Base of lung
Pleural cavity
Pleura

(a) Anterior view of lungs and pleurae in thorax

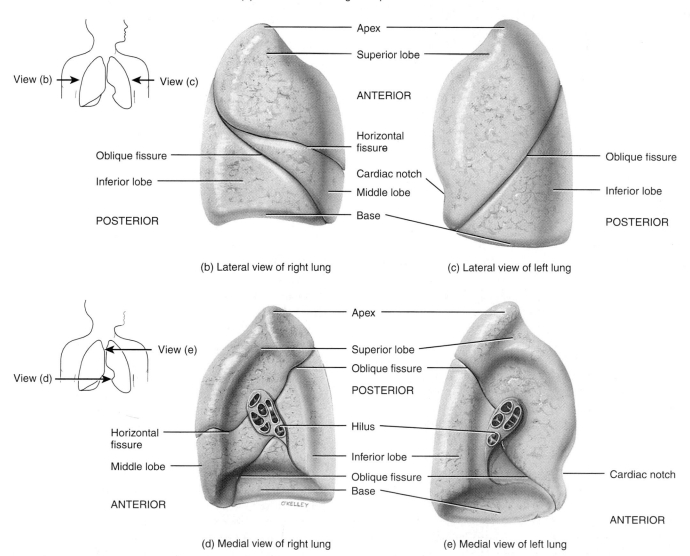

View (b)   View (c)

Apex
Superior lobe
ANTERIOR
Oblique fissure
Horizontal fissure
Inferior lobe
Cardiac notch
Oblique fissure
Middle lobe
Inferior lobe
POSTERIOR
Base
POSTERIOR

(b) Lateral view of right lung

(c) Lateral view of left lung

View (e)
View (d)

Apex
Superior lobe
Oblique fissure
POSTERIOR
Horizontal fissure
Hilus
Middle lobe
Inferior lobe
Oblique fissure
Cardiac notch
Base
ANTERIOR
ANTERIOR

O'KELLEY

(d) Medial view of right lung

(e) Medial view of left lung

Q Why are the right and left lungs slightly different in size and shape?

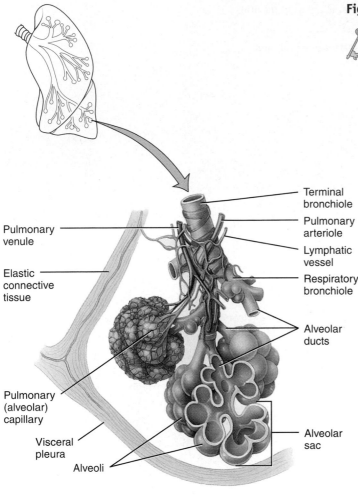

(a) Diagram of a portion of a lobule of the lung

Figure 23.11 Microscopic anatomy of a lobule of the lungs.

🔑 **Alveolar sacs are two or more alveoli that share a common opening.**

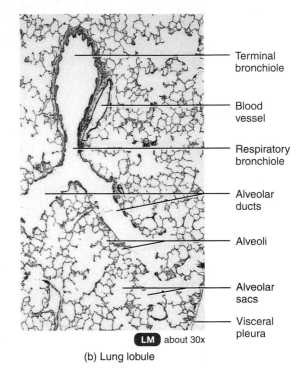

LM about 30x

(b) Lung lobule

Q What type of epithelium makes up the wall of an alveolus?

---

cuboidal epithelial cells whose free surfaces contain microvilli, secrete alveolar fluid, which keeps the surface between the cells and the air moist. Included in the alveolar fluid is **surfactant** (sur-FAK-tant), a complex detergentlike mixture of phospholipids and lipoproteins. Surfactant lowers the surface tension of alveolar fluid, which reduces the tendency of alveoli to collapse. Associated with the alveolar wall are **alveolar macrophages (dust cells),** wandering phagocytes that remove fine dust particles and other debris in the alveolar spaces. Also present are fibroblasts that produce reticular and elastic fibers. Underlying the layer of type I alveolar cells is an elastic basement membrane. Around the alveoli, the lobule's arteriole and venule disperse into a network of blood capillaries that consist of a single layer of endothelial cells and basement membrane.

The exchange of O_2 and CO_2 between the air spaces in the lungs and the blood takes place by diffusion across alveolar and capillary walls. The gases diffuse through the **respiratory membrane,** which consists of four layers (see Figure 23.12b):

1. A layer of type I and type II alveolar cells and associated alveolar macrophages that constitutes the **alveolar wall.**

2. An **epithelial basement membrane** underlying the alveolar wall.
3. A **capillary basement membrane** that is often fused to the epithelial basement membrane.
4. The **endothelial cells** of the capillary.

Despite having several layers, the respiratory membrane is very thin—only 0.5 μm thick, about one-sixteenth the diameter of a red blood cell. This thinness allows rapid diffusion of gases. Moreover, it has been estimated that the lungs contain 300 million alveoli, providing an immense surface area of 70 m² (750 ft²)—about the size of a handball court—for the exchange of gases.

Blood Supply to the Lungs

The lungs receive blood via two sets of arteries: pulmonary arteries and bronchial arteries. Deoxygenated blood passes through the pulmonary trunk, which divides into a left pulmonary artery that enters the left lung and a right pulmonary artery that enters the right lung. Return of the oxygenated blood to the heart occurs by way of the pulmonary veins, which drain into the left atrium (see Figure 21.30 on

Figure 23.12 Structural components and function of an alveolus.

🔑 **The exchange of respiratory gases occurs by diffusion across the respiratory membrane.**

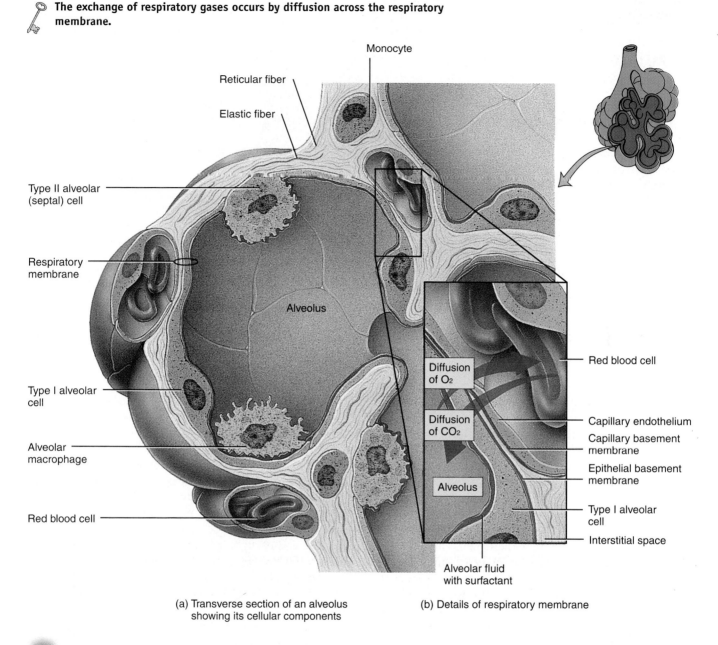

(a) Transverse section of an alveolus
showing its cellular components

(b) Details of respiratory membrane

Q What is the thickness of the respiratory membrane?

page 729). A unique feature of pulmonary blood vessels is their constriction in response to localized hypoxia (low O_2 level). In all other body tissues, hypoxia causes dilation of blood vessels, which serves to increase blood flow to a tissue that is not receiving adequate O_2. In the lungs, however, vasoconstriction in response to hypoxia diverts pulmonary blood from poorly ventilated areas to well-ventilated regions of the lungs.

Bronchial arteries, which branch from the aorta, deliver oxygenated blood to the lungs. This blood mainly perfuses the walls of the bronchi and bronchioles. Connections exist

between branches of the bronchial arteries and branches of the pulmonary arteries, however, and most blood returns to the heart via pulmonary veins. Some blood, however, drains into bronchial veins, branches of the azygos system, and returns to the heart via the superior vena cava.

1. Where are the lungs located? Distinguish the parietal pleura from the visceral pleura.
2. Define each of the following parts of a lung: base, apex, costal surface, medial surface, hilus, root, cardiac notch, lobe, and lobule.

3. What is a bronchopulmonary segment?

4. Describe the histology and function of the respiratory membrane.

PULMONARY VENTILATION

OBJECTIVE

• *Describe the events that cause inspiration and expiration.*

Pulmonary ventilation (breathing) is the process by which gases are exchanged between the atmosphere and lung alveoli. Air flows between the atmosphere and the lungs because of alternating pressure differences created by contraction and relaxation of respiratory muscles. The rate of airflow and the amount of effort needed for breathing is also influenced by alveolar surface tension, compliance of the lungs, and airway resistance.

Pressure Changes During Pulmonary Ventilation

Air moves into the lungs when the air pressure inside the lungs is less than the air pressure in the atmosphere and out of the lungs when the pressure inside the lungs is greater than the pressure in the atmosphere.

Inspiration

Breathing in is called **inspiration (inhalation).** Just before each inspiration, the air pressure inside the lungs is equal to the pressure of the atmosphere, which at sea level is about 760 millimeters of mercury (mm Hg), or 1 atmosphere (atm). For air to flow into the lungs, the pressure inside the alveoli must become lower than the atmospheric pressure. This condition is achieved by increasing the volume of the lungs.

The pressure of a gas in a closed container is inversely proportional to the volume of the container. This means that if the size of a closed container is increased, the pressure of the gas inside the container decreases, and that if the size of the container is decreased, then the pressure inside it increases. This inverse relationship between volume and pressure, called **Boyle's law,** may be demonstrated as follows (Figure 23.13): Suppose we place a gas in a cylinder that has a movable piston and a pressure gauge, and that the initial pressure created by the gas molecules striking the wall of the container is 1 atm. If the piston is pushed down, the gas is compressed into a smaller volume, so that the same number of gas molecules strike less wall area. The gauge shows that the pressure doubles as the gas is compressed to half its original volume. In other words, the same number of molecules in half the volume produces twice the pressure. Conversely, if the piston is raised to increase the volume, the pressure decreases. Thus, the pressure of a gas varies inversely with volume. Boyle's law also applies to everyday activities such as the operation of a bicycle pump or the blowing up of a balloon.

Differences in pressure caused by changes in lung vol-

Figure 23.13 Boyle's law.

🔑 **The volume of a gas varies inversely with the pressure.**

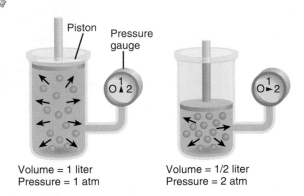

Volume = 1 liter
Pressure = 1 atm

Volume = 1/2 liter
Pressure = 2 atm

Q If the volume is decreased from 1 liter to ¼ liter, how does the pressure change?

ume force air into our lungs when we inhale and out when we exhale. For inspiration to occur, the lungs must expand, which increases lung volume and thus decreases the pressure in the lungs to below atmospheric pressure. The first step in expanding the lungs involves contraction of the main inspiratory muscle—the diaphragm (Figure 23.14).

The diaphragm is a dome-shaped skeletal muscle that forms the floor of the thoracic cavity. It is innervated by fibers of the phrenic nerves, which emerge from the spinal cord at cervical levels 3, 4, and 5. Contraction of the diaphragm causes it to flatten and increases the vertical dimension of the thoracic cavity. During normal quiet breathing, the diaphragm descends about 1 cm (0.4 in.), producing a pressure difference of 1–3 mm Hg and the inhalation of about 500 mL of air. In strenuous breathing, the diaphragm may descend 10 cm (4 in.), which produces a pressure difference of 100 mm Hg and the inhalation of 2–3 liters of air.

During normal breathing, the pressure between the two pleural layers, called **intrapleural pressure,** is always subatmospheric (lower than atmospheric pressure). Just before inspiration, it is about 4 mm Hg less than the atmospheric pressure, or about 756 mm Hg if the atmospheric pressure is 760 mm Hg (Figure 23.15). As the diaphragm contracts and the overall size of the thoracic cavity increases, the volume of the pleural cavity also increases, which causes intrapleural pressure to decrease to about 754 mm Hg. During expansion of the thorax the parietal and visceral pleurae normally adhere strongly to each other because of the subatmospheric pressure between them and because of the surface tension created by their moist adjoining surfaces. As the thoracic cavity expands, the parietal pleura lining the cavity is pulled outward in all directions, and the visceral pleura and lungs are pulled along with it.

As the volume of the lungs increases in this way, the pressure inside the lungs, called the **alveolar (intrapul-**

Figure 23.14 Muscles of inspiration and expiration and their roles in pulmonary ventilation. The pectoralis minor muscle, an accessory inspiratory muscle, is illustrated in Figure 11.15a.

🔑 **During deep, labored inspiration, accessory muscles of inspiration (sternocleidomastoids, scalenes, and pectoralis minors) participate.**

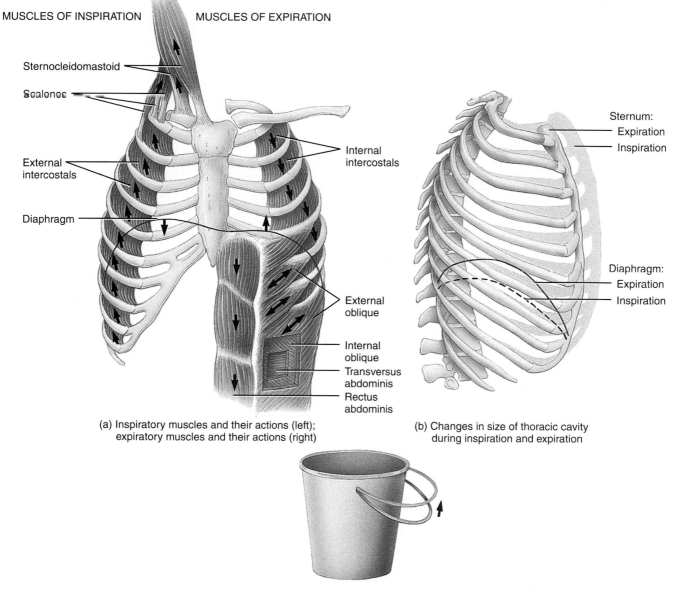

MUSCLES OF INSPIRATION   MUSCLES OF EXPIRATION

Sternocleidomastoid

Scalenes

External intercostals

Diaphragm

Internal intercostals

Sternum:
Expiration
Inspiration

Diaphragm:
Expiration
Inspiration

External oblique

Internal oblique

Transversus abdominis

Rectus abdominis

(a) Inspiratory muscles and their actions (left); expiratory muscles and their actions (right)

(b) Changes in size of thoracic cavity during inspiration and expiration

(c) During inspiration, the ribs move upward and outward like the handle on a bucket

Q Right now, what muscle is mainly responsible for your breathing?

monic) pressure, drops from 760 to 758 mm Hg. A pressure difference is thus established between the atmosphere and the alveoli. Because air always flows from a region of higher pressure to a region of lower pressure, inspiration takes place. Air continues to flow into the lungs as long as a pressure difference exists. During deep, forceful inspirations, accessory muscles of inspiration also participate in increasing the size of the thoracic cavity (see Figure 23.14a) in the following ways: The external intercostals elevate the ribs, the sternocleidomastoid elevates the sternum, the scalenes elevate the superior two ribs, and the pectoralis minor elevates the third through fifth ribs.

Figure 23.16a summarizes the events of inspiration.

Figure 23.15 Pressure changes in pulmonary ventilation. During inspiration, the diaphragm contracts, the chest expands, the lungs are pulled outward, and alveolar pressure decreases. During expiration, the diaphragm relaxes, the lungs recoil inward and alveolar pressure increases, forcing air out of the lungs.

🔑 **Air moves into the lungs when alveolar pressure is less than atmospheric pressure, and out of the lungs when alveolar pressure is greater than atmospheric pressure.**

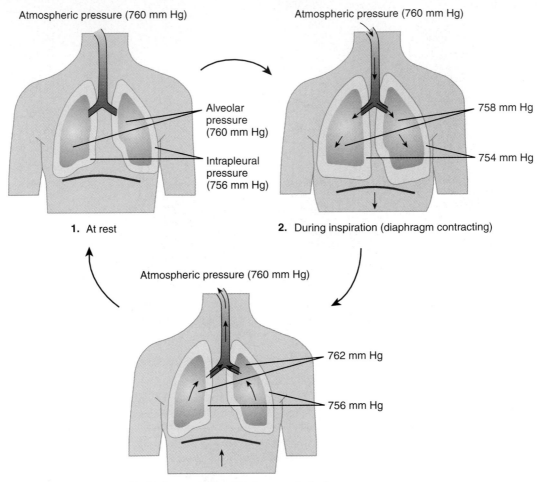

Atmospheric pressure (760 mm Hg)

Alveolar pressure (760 mm Hg)

Intrapleural pressure (756 mm Hg)

1. At rest

Atmospheric pressure (760 mm Hg)

758 mm Hg

754 mm Hg

2. During inspiration (diaphragm contracting)

Atmospheric pressure (760 mm Hg)

762 mm Hg

756 mm Hg

3. During expiration (diaphragm relaxing)

Q How does the intrapleural pressure change during a normal, quiet breath?

Expiration

Breathing out, called **expiration (exhalation),** is also due to a pressure gradient, but in this case the gradient is reversed: The pressure in the lungs is greater than the pressure of the atmosphere. Normal expiration during quiet breathing, unlike inspiration, is a *passive process* because no muscular contractions are involved. Instead, expiration results from **elastic recoil** of the chest wall and lungs, both of which have a natural tendency to spring back after they have been stretched. Two inwardly directed forces contribute to elastic recoil: (1) the recoil of elastic fibers that were stretched during inspiration and (2) the inward pull of surface tension due to the film of alveolar fluid.

Expiration starts when the inspiratory muscles relax. As the diaphragm relaxes, its dome moves superiorly owing to its elasticity. These movements decrease the vertical and anterior-posterior dimensions of the thoracic cavity, which decreases lung volume. In turn, the alveolar pressure increases, to about 762 mm Hg. Air then flows from the area of higher pressure in the alveoli to the area of lower pressure in the atmosphere (see Figure 23.15).

Expiration becomes active only during more forceful breathing, such as in playing a wind instrument or during exercise. During these times, muscles of expiration—the abdominals and internal intercostals (see Figure 23.14a)—contract, which increases pressure in the abdominal region

Figure 23.16 Summary of events of inspiration and expiration.

🔑 **Inspiration and expiration are caused by changes in alveolar pressure.**

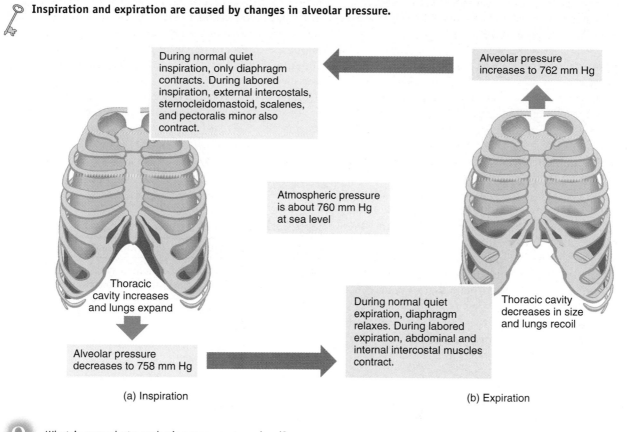

During normal quiet inspiration, only diaphragm contracts. During labored inspiration, external intercostals, sternocleidomastoid, scalenes, and pectoralis minor also contract.

Alveolar pressure increases to 762 mm Hg

Atmospheric pressure is about 760 mm Hg at sea level

Thoracic cavity increases and lungs expand

During normal quiet expiration, diaphragm relaxes. During labored expiration, abdominal and internal intercostal muscles contract.

Thoracic cavity decreases in size and lungs recoil

Alveolar pressure decreases to 758 mm Hg

(a) Inspiration

(b) Expiration

Q What is normal atmospheric pressure at sea level?

and thorax. Contraction of the abdominal muscles moves the inferior ribs downward and compresses the abdominal viscera, thereby forcing the diaphragm superiorly. Contraction of the internal intercostals, which extend inferiorly and posteriorly between adjacent ribs, pulls the ribs inferiorly. Although intrapleural pressure is always less than alveolar pressure, it may briefly exceed atmospheric pressure during a forceful expiration, such as during a cough.

Figure 23.16b summarizes the events of expiration.

Other Factors Affecting Pulmonary Ventilation

Whereas air pressure differences drive airflow during inspiration and expiration, three other factors affect the rate of airflow and the ease of pulmonary ventilation: surface tension of the alveolar fluid, compliance of the lungs, and airway resistance.

Surface Tension of the Alveolar Fluid

As noted earlier, a thin layer of alveolar fluid coats the luminal surface of alveoli and exerts a force known as **surface tension.** Surface tension arises at all air–water interfaces because the polar water molecules are more strongly attracted to each other than they are to gas molecules in the air.

When liquid surrounds a sphere of air, as in an alveolus or a soap bubble, surface tension produces an inwardly directed force. Soap bubbles "burst" because they collapse inward due to surface tension. In the lungs, surface tension causes the alveoli to assume the smallest possible diameter. During breathing, surface tension must be overcome to expand the lungs during each inspiration. Surface tension also accounts for two-thirds of lung elastic recoil, which decreases the size of alveoli during expiration.

The surfactant present in alveolar fluid reduces its surface tension below the surface tension of pure water. A deficiency of surfactant in premature infants causes *respiratory distress syndrome,* in which the surface tension of alveolar fluid is greatly increased, so that many alveoli collapse at the end of each exhalation. Great effort is then needed at the next inhalation to reopen the collapsed alveoli.

 CLINICAL APPLICATION
Pneumothorax

The pleural cavities are sealed off from the outside environment, which prevents them from equalizing their pressure with that of the atmosphere. Injuries of the chest wall that allow air to enter the intrapleural space either from the outside or from the alveoli cause **pneumothorax** (*pneumo-* = air), filling of the pleural cavity with air. Because the intrapleural pressure is then equal to atmospheric pressure rather than subatmospheric, surface tension and recoil of elastic fibers cause the lung to collapse. ■

Compliance of the Lungs

Compliance refers to how much effort is required to stretch the lungs and thoracic wall. High compliance means that the lungs and thoracic wall expand easily; low compliance means that they resist expansion. By analogy, a thin balloon that is easy to inflate has high compliance, whereas a heavy and stiff balloon that takes a lot of effort to inflate has low compliance. In the lungs, compliance is related to two principal factors: elasticity and surface tension. The lungs normally have high compliance and expand easily because elastic fibers in lung tissue are easily stretched and surfactant in alveolar fluid reduces surface tension. Decreased compliance is a common feature in pulmonary conditions that (1) scar lung tissue (for example, tuberculosis), (2) cause lung tissue to become filled with fluid (pulmonary edema), (3) produce a deficiency in surfactant, or (4) impede lung expansion in any way (for example, paralysis of the intercostal muscles). An increase in lung compliance occurs in emphysema due to destruction of elastic fibers in alveolar walls.

Airway Resistance

As is true for blood flow through blood vessels, the rate of airflow through the airways depends on both the pressure difference and the resistance: Airflow equals the pressure difference between the alveoli and the atmosphere divided by the resistance. The walls of the airways, especially the bronchioles, offer some resistance to the normal flow of air into and out of the lungs. As the lungs expand during inhalation, the bronchioles enlarge because their walls are pulled outward in all directions. Larger diameter airways have decreased resistance. Airway resistance then increases during exhalation as the diameter of bronchioles decreases. Also, the degree of contraction or relaxation of smooth muscle in the walls of the airways regulates airway diameter and thus resistance. Increased signals from the sympathetic division of the autonomic nervous system cause relaxation of this smooth muscle, which results in bronchodilation and decreased resistance.

Any condition that narrows or obstructs the airways increases resistance, and more pressure is required to maintain the same airflow. The hallmark of asthma or chronic obstructive pulmonary disease (COPD)—emphysema or chronic bronchitis—is increased airway resistance due to obstruction or collapse of airways.

Breathing Patterns and Modified Respiratory Movements

The term for the normal pattern of quiet breathing is **eupnea** (yoop-NĒ-a; *eu-* = good, easy, or normal; *-pnea* =

breath). Eupnea can consist of shallow, deep, or combined shallow and deep breathing. A pattern of shallow (chest) breathing, called **costal breathing,** consists of an upward and outward movement of the chest as a result of contraction of the external intercostal muscles. A pattern of deep (abdominal) breathing, called **diaphragmatic breathing,** consists of the outward movement of the abdomen as a result of the contraction and descent of the diaphragm.

Respirations also provide humans with methods for expressing emotions such as laughing, sighing, and sobbing. Moreover, respiratory air can be used to expel foreign matter from the lower air passages through actions such as sneezing

Table 23.1 Modified Respiratory Movements

MOVEMENT	DESCRIPTION
Coughing	A long-drawn and deep inspiration followed by a complete closure of the rima glottidis, which results in a strong expiration that suddenly pushes the rima glottidis open and sends a blast of air through the upper respiratory passages. Stimulus for this reflex act may be a foreign body lodged in the larynx, trachea, or epiglottis.
Sneezing	Spasmodic contraction of muscles of expiration that forcefully expels air through the nose and mouth. Stimulus may be an irritation of the nasal mucosa.
Sighing	A long-drawn and deep inspiration immediately followed by a shorter but forceful expiration.
Yawning	A deep inspiration through the widely opened mouth producing an exaggerated depression of the mandible. It may be stimulated by drowsiness, fatigue, or someone elses's yawning, but precise cause is unknown.
Sobbing	A series of convulsive inspirations followed by a single prolonged expiration. The rima glottidis closes earlier than normal after each inspiration, so only a little air enters the lungs with each inspiration.
Crying	An inspiration followed by many short convulsive expirations, during which the rima glottidis remains open and the vocal folds vibrate; accompanied by characteristic facial expressions and tears.
Laughing	The same basic movements as crying, but the rhythm of the movements and the facial expressions usually differ from those of crying. Laughing and crying are sometimes indistinguishable.
Hiccuping	Spasmodic contraction of the diaphragm followed by a spasmodic closure of the rima glottidis to produce a sharp inspiratory sound. Stimulus is usually irritation of the sensory nerve endings of the gastrointestinal tract.
Valsalva maneuver	Forced expiration against a closed rima glottidis as may occur during periods of straining while defecating.

Figure 23.17 Spirogram of lung volumes and capacities (average values for a healthy adult).

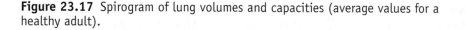
Lung capacities are combinations of various lung volumes.

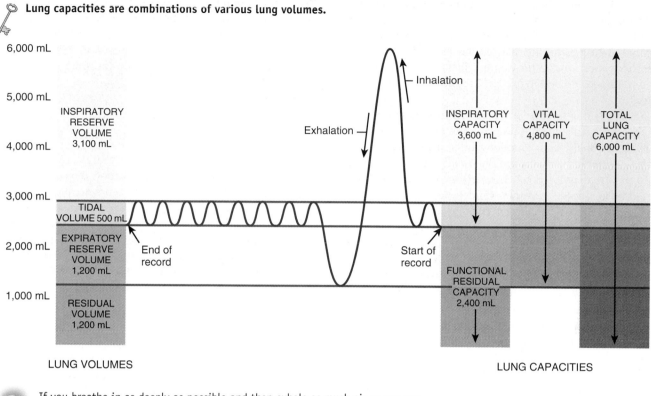

LUNG VOLUMES   LUNG CAPACITIES

Q If you breathe in as deeply as possible and then exhale as much air as you can, which lung capacity have you demonstrated?

and coughing. Respiratory movements are also modified and controlled during talking and singing. Some of the modified respiratory movements that express emotion or clear the airways are listed in Table 23.1. All these movements are reflexes, but some of them also can be initiated voluntarily.

1. What are the basic differences among pulmonary ventilation, external respiration, and internal respiration?
2. Compare what happens during quiet versus forceful ventilation.
3. Describe how alveolar surface tension, compliance, and airway resistance affect pulmonary ventilation.
4. Define the various kinds of modified respiratory movements.

LUNG VOLUMES AND CAPACITIES

OBJECTIVE

• *Define the various lung volumes and capacities.*

While at rest, a healthy adult averages 12 breaths a minute, with each inspiration and expiration moving about 500 mL of air into and out of the lungs. The volume of one breath is called the **tidal volume (V_T).** Thus, the **minute ventilation (MV)**—the total volume of air inhaled and ex-

haled each minute—is respiratory rate multiplied by tidal volume:

$$MV = 12 \text{ breaths/min} \times 500 \text{ mL}$$
$$= 6 \text{ liters/min}$$

A lower-than-normal minute ventilation usually is a sign of pulmonary malfunction. The apparatus commonly used to measure the volume of air exchanged during breathing and the respiratory rate is a **spirometer** (*spiro-* = breathe; *meter* = measuring device) or **respirometer.** The record is called a **spirogram.** Inspiration is recorded as an upward deflection, and expiration is recorded as a downward deflection, and the recording pen usually moves from right to left (Figure 23.17).

Tidal volume varies considerably from one person to another and in the same person at different times. In an average adult, about 70% of the tidal volume (350 mL) actually reaches the respiratory portion of the respiratory system—the respiratory bronchioles, alveolar ducts, alveolar sacs, and alveoli—and participates in external respiration; the other 30% (150 mL) remains in the conducting airways of the nose, pharynx, larynx, trachea, bronchi, bronchioles, and terminal bronchioles. Collectively, these conducting airways are known as the **anatomic dead space.** (An easy rule of thumb for determining the volume of your anatomic dead space in milliliters is that it is about the same as your ideal weight in pounds.) Not all of the minute ventilation can be used in gas exchange because some of it remains in the anatomic dead space. The **alveolar ventilation rate** is the volume of air per minute that reaches the alveoli and other respiratory portions. In the example just given, alveolar ventilation rate would be 350 mL × 12 respirations/min = 4200 mL/min.

Several other lung volumes are defined relative to forceful breathing. In general, these volumes are larger in males, taller individuals, and younger adults, and smaller in females, shorter individuals, and the elderly. Various disorders also may be diagnosed by comparison of actual and predicted normal values for one's gender, height, and age. The values given here are averages for young adults.

By taking a very deep breath, you can inspire a good deal more than 500 mL. This additional inhaled air, called the **inspiratory reserve volume,** is about 3100 mL (see Figure 23.17). Even more air can be inhaled if inspiration follows forced expiration. If you inhale normally and then exhale as forcibly as possible, you should be able to push out 1200 mL of air in addition to the 500 mL of tidal volume. The extra 1200 mL is called the **expiratory reserve volume.** The **FEV$_{1.0}$** is the **forced expiratory volume in 1 second,** which is the volume of air that can be expelled from the lungs in 1 second with maximal effort following a maximal inhalation. Typically, chronic obstructive pulmonary disease (COPD) greatly reduces FEV$_{1.0}$. That is because COPD increases airway resistance.

Even after the expiratory reserve volume is expelled, considerable air remains in the lungs because the subatmospheric intrapleural pressure keeps the alveoli slightly inflated, and some air also remains in the noncollapsible airways. This volume, which cannot be measured by spirometry, is called the **residual volume** and amounts to about 1200 mL.

If the thoracic cavity is opened, the intrapleural pressure rises to equal the atmospheric pressure and forces out some of the residual volume. The air remaining is called the **minimal volume.** Minimal volume provides a medical and legal tool for determining whether a baby was born dead or died after birth. The presence of minimal volume can be demonstrated by placing a piece of lung in water and observing if it floats. Fetal lungs contain no air, and so the lung of a stillborn baby will not float in water.

Lung capacities are combinations of specific lung volumes (see Figure 23.17). **Inspiratory capacity** is the sum of tidal volume and inspiratory reserve volume (500 mL + 3100 mL = 3600 mL). **Functional residual capacity** is the sum of residual volume and expiratory reserve volume (1200 mL + 1200 mL = 2400 mL). **Vital capacity** is the sum of inspiratory reserve volume, tidal volume, and expiratory reserve volume (4800 mL). Finally, **total lung capacity** is the sum of all volumes (6000 mL).

1. What is a spirometer?
2. Distinguish between lung volumes and lung capacities.
3. How is minute ventilation calculated?
4. Define alveolar ventilation rate and FEV$_{1.0}$.

EXCHANGE OF OXYGEN AND CARBON DIOXIDE

OBJECTIVE

• *Explain Dalton's law and Henry's law.*

• *Describe the exchange of oxygen and carbon dioxide in external and internal respiration.*

The exchange of oxygen and carbon dioxide between alveolar air and pulmonary blood occurs via passive diffusion, which is governed by the behavior of gases as described by Dalton's law and Henry's law. Whereas Dalton's law is important for understanding how gases move down their pressure differences by diffusion, Henry's law helps explain how the solubility of a gas relates to its diffusion.

Gas Laws: Dalton's Law and Henry's Law

According to **Dalton's law,** each gas in a mixture of gases exerts its own pressure as if all the other gases were not present. The pressure of a specific gas in a mixture is called its *partial pressure* and is denoted as P_x, where the subscript is the formula of the gas. The total pressure of the mixture is calculated simply by adding all the partial pressures. Atmospheric air is a mixture of gases—nitrogen (N_2), oxygen, water vapor (H_2O), and carbon dioxide, plus other gases present in small quantities. Atmospheric pressure is the sum of the pressures of all these gases:

$$\text{Atmospheric pressure (760 mm Hg)} = P_{N_2} + P_{O_2} + P_{H_2O} + P_{CO_2} + P_{\text{other gases}}$$

We can determine the partial pressure exerted by each component in the mixture by multiplying the percentage of the gas in the mixture by the total pressure of the mixture. Atmospheric air is 78.6% nitrogen, 20.9% oxygen, 0.04% carbon dioxide, and 0.06% other gases; a variable amount of water vapor is also present, about 0.4% on a cool, dry day. Thus, the partial pressures of the gases in inhaled air are as follows:

$$P_{N_2} = 0.786 \times 760 \text{ mm Hg} = 597.4 \text{ mm Hg}$$
$$P_{O_2} = 0.209 \times 760 \text{ mm Hg} = 158.8 \text{ mm Hg}$$
$$P_{H_2O} = 0.004 \times 760 \text{ mm Hg} = 3.0 \text{ mm Hg}$$
$$P_{CO_2} = 0.0004 \times 760 \text{ mm Hg} = 0.3 \text{ mm Hg}$$
$$P_{\text{other gases}} = 0.0006 \times 760 \text{ mm Hg} = \underline{0.5 \text{ mm Hg}}$$
$$\text{Total} = 760 \text{ mm Hg}$$

These partial pressures are important in determining the movement of O_2 and CO_2 between the atmosphere and lungs, between the lungs and blood, and between the blood and body cells. When a mixture of gases diffuses across a permeable membrane, each gas diffuses from the area where its partial pressure is greater to the area where its partial pressure is less. The greater the difference in partial pressure, the faster the rate of diffusion. Each gas behaves as if the other gases in the mixture were not present and diffuses at a rate determined by its own partial pressure.

Henry's law states that the quantity of a gas that will dissolve in a liquid is proportional to the partial pressure of the gas and its solubility coefficient. In body fluids, the ability of a gas to stay in solution is greater when its partial pressure is higher and when it has a high solubility coefficient in water. The higher the partial pressure of a gas over a liquid and the higher the solubility coefficient, the more gas will stay in solution. In comparison to oxygen, much more CO_2 is dissolved in plasma because the solubility coefficient of CO_2 is 24 times greater than that of O_2. Even though the air we breathe contains almost 79% N_2, this gas has no known effect on bodily functions, and at sea level pressure very little of it dissolves in blood plasma because its solubility coefficient is very low.

An everyday experience gives a demonstration of Henry's law. You have probably noticed that a soft drink makes a hissing sound when the top of the container is removed, and bubbles rise to the surface for some time afterward. The gas dissolved in carbonated beverages is CO_2. Because the soft drink is bottled or canned under high pressure and capped, the CO_2 remains dissolved so long as the container is unopened. Once you remove the cap, the pressure decreases and the gas begins to bubble out of solution.

CLINICAL APPLICATION
Hyperbaric Oxygenation

A major clinical application of Henry's law is **hyperbaric oxygenation** (*hyper* = over; *baros* = pressure). Using pressure to cause more O_2 to dissolve in the blood is an effective technique in treating patients infected by anaerobic bacteria, such as those that cause tetanus and gangrene. (Anaerobic bacteria cannot live in the presence of free O_2.) A person undergoing hyperbaric oxygenation is placed in a hyperbaric chamber, which contains O_2 at a pressure of 3–4 atmospheres (2280–3040 mm Hg). As body tissues pick up the O_2, the bacteria are killed. Hyperbaric chambers may also be used for treating certain heart disorders, carbon monoxide poisoning, gas embolisms, crush injuries, cerebral edema, certain hard-to-treat bone infections caused by anaerobic bacteria, smoke inhalation, near-drowning, asphyxia, vascular insufficiencies, and burns. ■

External and Internal Respiration

External (pulmonary) respiration is the exchange of O_2 and CO_2 between air in the alveoli of the lungs and blood in pulmonary capillaries (Figure 23.18a). External respiration in the lungs results in the conversion of **deoxygenated blood** (depleted of some O_2) coming from the right side of

Figure 23.18 Changes in partial pressures of oxygen and carbon dioxide (in mm Hg) during external and internal respiration.

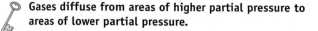
Gases diffuse from areas of higher partial pressure to areas of lower partial pressure.

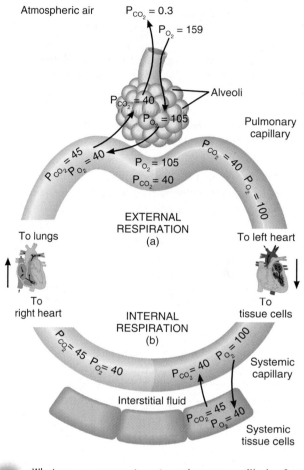

What causes oxygen to enter pulmonary capillaries from alveoli and to enter tissue cells from systemic capillaries?

the heart to **oxygenated blood** (saturated with O_2) returning to the left side of the heart (see Figure 20.7 on page 646). During inspiration, atmospheric air containing O_2 enters the alveoli, and during expiration CO_2 is exhaled into the atmosphere. Deoxygenated blood is pumped by the right ventricle through the pulmonary arteries into the pulmonary capillaries surrounding the alveoli. The partial pressures of gases in pulmonary capillaries come into equilibrium with the partial pressures of gases in alveolar air.

The P_{O_2} of alveolar air is 105 mm Hg. If you are at rest, the P_{O_2} of deoxygenated blood entering your pulmonary capillaries is about 40 mm Hg; if you have been exercising, it will be even lower because contracting muscle fibers are using more O_2. Due to the difference in P_{O_2}, there is a net diffusion of O_2 from the alveoli into the deoxygenated blood until equilibrium is reached. The P_{O_2} of the now oxygenated blood increases to 105 mm Hg. Because blood leaving capillaries near alveoli mixes with a small volume of blood that has flowed through conducting portions of the respiratory system, where gas exchange does not occur, the P_{O_2} of blood in the pulmonary veins is slightly less than the P_{O_2} in pulmonary capillaries—about 100 mm Hg.

While O_2 is diffusing from the alveoli into deoxygenated blood, CO_2 is diffusing in the opposite direction. The P_{CO_2} of deoxygenated blood is 45 mm Hg in a resting person, whereas P_{CO_2} of alveolar air is 40 mm Hg. Because of this difference in P_{CO_2}, carbon dioxide diffuses from deoxygenated blood into the alveoli until the P_{CO_2} of the blood decreases to 40 mm Hg.

The *rate* of gas exchange during external respiration depends on several factors:

- *Partial pressure difference of the gases.* As long as alveolar P_{O_2} is higher than P_{O_2} in pulmonary capillaries, O_2 diffuses from the alveoli into the blood. The rate of diffusion is faster when the difference between P_{O_2} in alveolar air and pulmonary capillary blood is larger; diffusion is slower when the difference is smaller, as occurs at high altitude. With increasing altitude, the total atmospheric pressure decreases, as does the partial pressure of O_2— from 159 mm Hg at sea level, to 110 mm Hg at 10,000 ft, to 73 mm Hg at 20,000 ft. Although O_2 still is 20.9% of the total, the P_{O_2} of inhaled air decreases with increasing altitude. Alveolar P_{O_2} decreases correspondingly, and O_2 diffuses into the blood more slowly. The common symptoms of **high altitude sickness**—shortness of breath, headache, fatigue, insomnia, nausea, dizziness, and disorientation—are due to the low O_2 content of the blood. The differences between P_{O_2} and P_{CO_2} in alveolar air versus pulmonary blood increase during exercise. As more-active tissues use more O_2 and give off more CO_2, the P_{O_2} of blood returning to the right side of the heart decreases, whereas its P_{CO_2} increases. The larger partial pressure differences accelerate the rates of gas diffusion. The partial pressures of O_2 and CO_2 in the alveoli also depend on the rate of airflow into and out of the lungs. Certain drugs (such as morphine) slow ventilation, thereby decreasing the amount of O_2 and CO_2 that can be exchanged between the alveoli and the blood.

- *Surface area for gas exchange.* Any pulmonary disorder that decreases the functional surface area formed by the respiratory membranes (about 70 m^2 or 750 ft^2) decreases the rate of external respiration. In emphysema, for example, alveolar walls disintegrate, and surface area is smaller than normal.

- *Diffusion distance.* If the total thickness of the respiratory membranes (only 0.5 μm) were greater, the rate of diffusion would slow. Also, the capillaries are so narrow that the red blood cells must pass through them in single file, which minimizes the diffusion distance from an alveolar air space to hemoglobin inside red blood cells. Buildup of intersitial fluid between alveoli, as occurs in pulmonary edema, slows the rate of gas exchange because it increases diffusion distance.

- *Solubility and molecular weight of the gases.* Because O_2 has a lower molecular weight than CO_2, it could be expected to diffuse across the respiratory membrane about 1.2 times faster. However, the solubility of CO_2 in the fluid portions of the respiratory membrane is about 24 times greater than that of O_2. Taking both of these factors into account, net outward CO_2 diffusion occurs 20 times more rapidly than net inward O_2 diffusion. Consequently, when diffusion is slower than normal, for example, in emphysema or pulmonary edema, O_2 insufficiency (hypoxia) typically occurs before there is significant retention of CO_2 (hypercapnia).

The number of capillaries near alveoli in the lungs is very large, and blood flows slowly enough through these capillaries that it becomes completely saturated with O_2. During vigorous exercise, when cardiac output is increased and blood flows more rapidly through both the systemic and pulmonary circulations, blood's transit time through the pulmonary capillaries is shorter. Still, the P_{O_2} of blood in the pulmonary veins normally is 100 mm Hg. In diseases that decrease the rate of gas diffusion, however, the blood may not come into full equilibrium with alveolar air, especially when transit time is reduced during exercise; as a result, P_{O_2} declines and P_{CO_2} rises.

The left ventricle pumps oxygenated blood into the aorta and through the systemic arteries to systemic capillaries, and finally to tissue cells. The exchange of O_2 and CO_2 between systemic capillaries and tissue cells is called **internal (tissue) respiration** (see Figure 23.18b), and it results in the conversion of oxygenated blood into deoxygenated blood. Oxygenated blood entering tissue capillaries has a P_{O_2} of 100 mm Hg, whereas tissue cells have an average P_{O_2} of 40 mm Hg. Because of this difference in P_{O_2}, oxygen diffuses from the oxygenated blood through interstitial fluid and into tissue cells until the P_{O_2} in the blood declines to 40 mm Hg, the average P_{O_2} of deoxygenated blood entering tissue venules at rest.

Entry into the tissues of only about 25% of the available O_2 in oxygenated blood is sufficient to support the needs of resting cells. Thus, deoxygenated blood in a resting individual still retains 75% of its O_2 content. During exercise, more O_2 diffuses from the blood into active cells, which are generating CO_2 as they use O_2 for ATP production, and the O_2 content of deoxygenated blood drops below 75%.

While O_2 diffuses from the tissue capillaries into tissue cells, CO_2 diffuses in the opposite direction. The average P_{CO_2} of tissue cells is 45 mm Hg, whereas the P_{CO_2} of tissue capillary oxygenated blood is 40 mm Hg. As a result, CO_2 diffuses from tissue cells through interstitial fluid into the oxygenated blood until the P_{CO_2} in the blood increases to 45 mm Hg, the P_{CO_2} of tissue capillary deoxygenated blood. The deoxygenated blood returns to the heart and is pumped to the lungs for another cycle of external respiration.

In summary, the partial pressures of O_2 and CO_2 in mm Hg in alveolar air, oxygenated blood, tissue cells, and deoxygenated blood are as follows:

Alveoli: $P_{O_2} = 105$ mm Hg; $P_{CO_2} = 40$ mm Hg
Oxygenated blood: $P_{O_2} = 100$ mm Hg; $P_{CO_2} = 40$ mm Hg
Tissue cells (average): $P_{O_2} = 40$ mm Hg; $P_{CO_2} = 45$ mm Hg
Deoxygenated blood: $P_{O_2} = 40$ mm Hg; $P_{CO_2} = 45$ mm Hg

The relative amounts of O_2 and CO_2 differ in inspired (atmospheric) air, alveolar air, and expired air, as follows:

Inspired air: 20.9% O_2, 0.04% CO_2
Alveolar air: 13.6% O_2, 5.2% CO_2
Expired air: 16% O_2, 4.5% CO_2

Compared with inspired air, alveolar air has less O_2 (20.9% versus 13.6%) and more CO_2 (0.04% versus 5.2%) because gas exchange is occurring in the alveoli. In contrast, expired air contains more O_2 than alveolar air (16% versus 14%) and less CO_2 (4.5% versus 5.2%) because some of the expired air was in the anatomic dead space and did not participate in gas exchange. Expired air is a mixture of alveolar air and inspired air that was in the anatomic dead space. Also, expired air and alveolar air have higher water vapor content than inspired air because the moist mucosal linings humidify air once it enters the respiratory tract.

1. Define the partial pressure of a gas. How is the partial pressure of oxygen related to altitude?
2. Construct a diagram to show how and why the respiratory gases diffuse during external and internal respiration.

TRANSPORT OF OXYGEN AND CARBON DIOXIDE IN THE BLOOD

OBJECTIVE

• *Explain how oxygen and carbon dioxide are transported in the blood.*

The transport of gases between the lungs and body tissues is accomplished by the blood. When O_2 and CO_2 enter the blood, certain physical and chemical changes occur that aid in gas transport and exchange. First we examine the dynamics of oxygen transport.

Oxygen Transport

Oxygen does not dissolve easily in water, and therefore very little O_2—only about 1.5%—is carried in the dissolved state in blood plasma. The remainder of the O_2—about 98.5%—is transported in chemical combination with hemoglobin inside red blood cells (Figure 23.19). Each 100 mL of oxygenated blood contains the equivalent of 20 mL of gaseous O_2—0.3 mL is dissolved in the plasma and 19.7 mL is bound to hemoglobin.

Figure 23.19 Transport of oxygen (O_2) and carbon dioxide (CO_2) in the blood.

🔑 **Most O_2 is transported by hemoglobin as oxyhemoglobin within red blood cells; most CO_2 is transported in blood plasma as bicarbonate ions.**

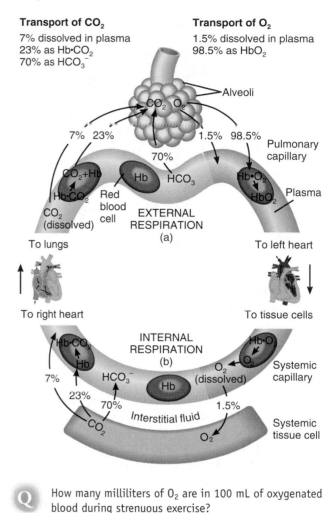

Transport of CO_2
7% dissolved in plasma
23% as $Hb \cdot CO_2$
70% as HCO_3^-

Transport of O_2
1.5% dissolved in plasma
98.5% as HbO_2

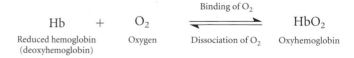

Q How many milliliters of O_2 are in 100 mL of oxygenated blood during strenuous exercise?

Hemoglobin consists of a protein portion called globin and an iron-containing pigment portion called heme (see Figure 19.4b, c on page 616). A hemoglobin molecule has four heme groups, each of which can combine with one molecule of O_2. Oxygen and hemoglobin combine in an easily reversible reaction to form **oxyhemoglobin** as follows:

$$Hb + O_2 \underset{\text{Dissociation of } O_2}{\overset{\text{Binding of } O_2}{\rightleftarrows}} HbO_2$$

Reduced hemoglobin (deoxyhemoglobin)   Oxygen   Oxyhemoglobin

Because 98.5% of the O_2 is bound to hemoglobin and therefore trapped inside RBCs, only the dissolved O_2 (1.5%) can diffuse out of tissue capillaries into tissue cells. Thus, it is

Figure 23.20 Oxygen–hemoglobin dissociation curve showing the relationship between hemoglobin saturation and P_{O_2} at normal body temperature.

🔑 **As P_{O_2} increases, more O_2 combines with hemoglobin.**

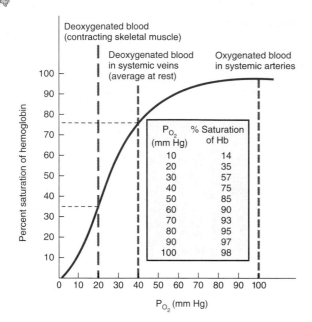

Q What point on the curve represents blood in your pulmonary veins right now? In your pulmonary veins if you were jogging?

important to understand the factors that promote O_2 binding to and dissociation (separation) from hemoglobin.

The Relation Between Hemoglobin and Oxygen Partial Pressure

The most important factor that determines how much O_2 combines with hemoglobin is the P_{O_2}; the higher the P_{O_2}, the more O_2 combines with Hb. When reduced hemoglobin (deoxyhemoglobin) is completely converted to HbO_2, the hemoglobin is said to be **fully saturated;** when hemoglobin consists of a mixture of Hb and HbO_2, it is **partially saturated.** The **percent saturation of hemoglobin** expresses the average saturation of hemoglobin with oxygen. For instance, if each hemoglobin molecule has bound two O_2 molecules, then the hemoglobin is 50% saturated because each Hb can bind a maximum of four O_2. The relation between the percent saturation of hemoglobin and P_{O_2} is illustrated in the oxygen–hemoglobin dissociation curve in Figure 23.20. Note that when the P_{O_2} is high, hemoglobin binds with large amounts of O_2 and is almost 100% saturated. When P_{O_2} is low, hemoglobin is only partially saturated. In other words, the greater the P_{O_2}, the more O_2 will combine with hemoglobin, until all the available hemoglobin molecules are saturated. Therefore, in pulmonary capillaries, where P_{O_2} is high, a lot of O_2 binds with hemoglobin. In tissue capillaries,

where the P_{O_2} is lower, hemoglobin does not hold as much O_2, and the O_2 is unloaded via diffusion into tissue cells. Note that hemoglobin is still 75% saturated with O_2 at a P_{O_2} of 40 mm Hg, the average P_{O_2} of tissue cells at rest. This is the basis for the earlier statement that only 25% of the available O_2 unloads from hemoglobin and is used by tissue cells under resting conditions.

When the P_{O_2} is between 60 and 100 mm Hg, hemoglobin is 90% or more saturated with O_2 (see Figure 23.20). Thus, blood picks up a nearly full load of O_2 from the lungs even when the P_{O_2} of alveolar air is as low as 60 mm Hg. The $Hb-P_{O_2}$ curve explains why people can still perform well at high altitudes or when they have certain cardiac and pulmonary diseases, even though P_{O_2} may drop as low as 60 mm Hg. Note also in the curve that at a considerably lower P_{O_2} of 40 mm Hg, hemoglobin is still 75% saturated with O_2. However, oxygen saturation of Hb drops to 35% at 20 mm Hg. Between 40 and 20 mm Hg, large amounts of O_2 are released from hemoglobin in response to only small decreases in P_{O_2}. In active tissues such as contracting muscles, P_{O_2} may drop well below 40 mm Hg. Then, a large percentage of the O_2 is released from hemoglobin, providing more O_2 to metabolically active tissues.

Other Factors Affecting Hemoglobin's Affinity for Oxygen

Although P_{O_2} is the most important factor that determines the percent O_2 saturation of hemoglobin, several other factors influence the tightness or **affinity** with which hemoglobin binds O_2. In effect, these factors shift the entire curve either to the left (higher affinity) or to the right (lower affinity). The changing affinity of hemoglobin for O_2 is another example of how homeostatic mechanisms adjust body activities to cellular needs. Each one makes sense if you keep in mind that metabolically active tissue cells need O_2 and produce acids, CO_2, and heat as wastes. The following four factors affect hemoglobin's affinity for O_2:

1. *Acidity (pH).* As acidity increases (pH decreases), the affinity of hemoglobin for O_2 decreases, and O_2 dissociates more readily from hemoglobin (Figure 23.21a). In other words, increasing acidity enhances the unloading of oxygen from hemoglobin. The main acids produced by metabolically active tissues are lactic acid and carbonic acid. When pH decreases, the entire oxygen–hemoglobin dissociation curve shifts to the right—at any given P_{O_2}, Hb is less saturated with O_2—a change termed the **Bohr effect.** The explanation for the Bohr effect is that hemoglobin can act as a buffer of hydrogen ions (H^+). But when H^+ bind to amino acids in hemoglobin, they slightly alter its structure and thereby decrease its oxygen-carrying capacity. Thus, lowered pH drives O_2 off hemoglobin, making more O_2 available for tissue cells. By contrast, elevated pH increases the affinity of hemoglobin for O_2 and shifts the oxygen–hemoglobin dissociation curve to the left.

Figure 23.21 Oxygen–hemoglobin dissociation curves showing the relationship of (a) pH and (b) P_{CO_2} to hemoglobin saturation at normal body temperature. As pH increases or P_{CO_2} decreases, O_2 combines more tightly with hemoglobin, so that less is available to tissues. These relationships are emphasized by the broken lines.

🔑 **As pH decreases or P_{CO_2} increases, the affinity of hemoglobin for O_2 declines, so less O_2 combines with hemoglobin and more is available to tissues.**

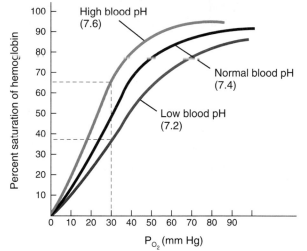

(a) Effect of pH on affinity of hemoglobin for oxygen

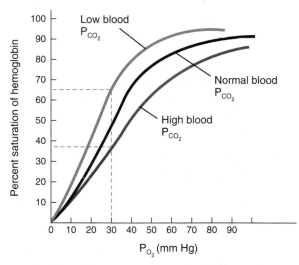

(b) Effect of P_{CO_2} on affinity of hemoglobin for oxygen

Ⓠ In comparison to the value when you are sitting, is the affinity of your hemoglobin for O_2 higher or lower when you are exercising? What is the benefit of this change to you?

2. *Partial pressure of carbon dioxide.* CO_2 also can bind to hemoglobin, and the effect is similar to that of H^+ (shifting the curve to the right). As P_{CO_2} rises, hemoglobin releases O_2 more readily (Figure 23.21b). P_{CO_2} and pH are related factors because low blood pH (acidity) results from high P_{CO_2}. As CO_2 enters the blood, much of it is temporarily converted to carbonic acid (H_2CO_3), a reaction catalyzed by an enzyme in red blood cells called *carbonic anhydrase* (CA):

Figure 23.22 Oxygen–hemoglobin dissociation curves showing the relationship between temperature and hemoglobin saturation with O_2.

🔑 **As temperature increases, the affinity of hemoglobin for O_2 decreases.**

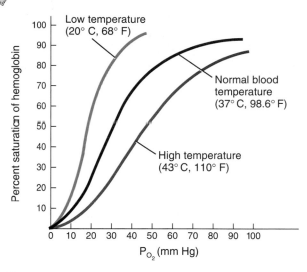

Ⓠ Is O_2 more available or less available to tissue cells when you have a fever? Why?

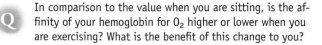

$$CO_2 + H_2O \underset{}{\overset{CA}{\rightleftharpoons}} H_2CO_3 \rightleftharpoons H^+ + HCO_3^-$$

Carbon dioxide · Water · Carbonic acid · Hydrogen ion · Bicarbonate ion

The carbonic acid thus formed in red blood cells dissociates into hydrogen ions and bicarbonate ions. As the H^+ concentration increases, pH decreases. Thus, an increased P_{CO_2} produces a more acidic environment, which helps release O_2 from hemoglobin. During exercise, lactic acid—a by-product of anaerobic metabolism within muscles—also decreases blood pH. Decreased P_{CO_2} (and elevated pH) shifts the saturation curve to the left.

3. *Temperature.* Within limits, as temperature increases, so does the amount of O_2 released from hemoglobin (Figure 23.22). Heat is a by-product of the metabolic reactions of all cells, and the heat released by contracting muscle fibers tends to raise body temperature. Metabolically active cells require more O_2 and liberate more acids and heat. The acids and heat, in turn, promote release of O_2 from oxyhemoglobin. Fever produces a similar result. In contrast, during hypothermia (lowered body temperature) cellular metabolism slows, the need for O_2 is reduced, and more O_2 remains bound to hemoglobin (a shift to the left in the saturation curve).

4. *BPG.* A substance found in red blood cells called **2,3-bisphosphoglycerate (BPG),** previously called diphosphoglycerate (DPG), decreases the affinity of hemoglobin for O_2 and thus helps unload O_2 from hemoglobin. BPG is

Figure 23.23 Oxygen–hemoglobin dissociation curves comparing fetal and maternal hemoglobin.

 Fetal hemoglobin has a higher affinity for O_2 than does adult hemoglobin.

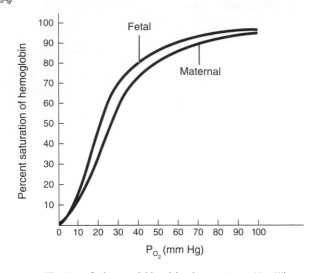

 The P_{O_2} of placental blood is about 40 mm Hg. What are the O_2 saturations of maternal and fetal hemoglobin at this P_{O_2}?

formed in red blood cells when they break down glucose to produce ATP in a process called glycolysis (described on page 874). When BPG combines with hemoglobin, the hemoglobin binds O_2 less tightly. The greater the level of BPG, the more O_2 is unloaded from hemoglobin. Certain hormones, such as thyroxine, human growth hormone, epinephrine, norepinephrine, and testosterone, increase the formation of BPG. The level of BPG also is higher in people living at higher altitudes.

Oxygen Affinity of Fetal and Adult Hemoglobin

Fetal hemoglobin (Hb-F) differs from **adult hemoglobin (Hb-A)** in structure and in its affinity for O_2. Hb-F has a higher affinity for O_2 because it binds BPG less strongly. Thus, when P_{O_2} is low, Hb-F can carry up to 30% more O_2 than maternal Hb-A (Figure 23.23). As the maternal blood enters the placenta, O_2 is readily transferred to fetal blood. This is very important because the O_2 saturation in maternal blood in the placenta is quite low, and the fetus might suffer hypoxia were it not for the greater affinity of fetal hemoglobin for O_2.

CLINICAL APPLICATION
Carbon Monoxide Poisoning

Carbon monoxide (CO) is a colorless and odorless gas found in exhaust fumes from automobiles, gas furnaces, and space heaters and in tobacco smoke. It is a by-product of the combustion of carbon-containing materials such as coal, gas, and wood. CO combines with the heme group of hemoglobin, just as O_2 does, except that the binding of carbon monoxide to hemoglobin is over 200 times as strong as the binding of O_2 to hemoglobin. Thus, at a concentration as small as 0.1% (P_{CO} = 0.5 mm Hg), CO will combine with half the available hemoglobin molecules and reduce the oxygen-carrying capacity of the blood by 50%. Elevated blood levels of CO cause **carbon monoxide poisoning,** one sign of which is a bright, cherry-red color of the lips and oral mucosa. The condition may be treated by administering pure O_2, which hastens the separation of carbon monoxide from hemoglobin. ■

Carbon Dioxide Transport

Under normal resting conditions, each 100 mL of deoxygenated blood contains the equivalent of 53 mL of gaseous CO_2, which is transported in the blood in three main forms (see Figure 23.19):

1. *Dissolved CO_2.* The smallest percentage—about 7%—is dissolved in plasma. Upon reaching the lungs, it diffuses into the alveoli.

2. *Carbamino compounds.* A somewhat higher percentage, about 23%, combines with the amino groups of amino acids and proteins in blood to form **carbamino compounds.** Because the most prevalent protein in blood is hemoglobin, most of the CO_2 transported in this manner is bound to amino acids in the globin portion of hemoglobin. Hemoglobin that has bound CO_2 is termed **carbaminohemoglobin (Hb·CO_2):**

$$\underset{\text{Hemoglobin}}{\text{Hb}} + \underset{\text{Carbon dioxide}}{CO_2} \rightleftharpoons \underset{\text{Carbaminohemoglobin}}{\text{Hb·}CO_2}$$

The formation of carbaminohemoglobin is greatly influenced by P_{CO_2}. For example, in tissue capillaries P_{CO_2} is relatively high, which promotes formation of carbaminohemoglobin. But in pulmonary capillaries, P_{CO_2} is relatively low, and the CO_2 readily splits apart from globin and enters the alveoli by diffusion.

3. *Bicarbonate ions.* The greatest percentage of CO_2—about 70%—is transported in plasma as **bicarbonate ions (HCO_3^-)** as a result of a reaction we noted earlier:

$$\underset{\substack{\text{Carbon}\\\text{dioxide}}}{CO_2} + \underset{\text{Water}}{H_2O} \overset{\text{CA}}{\rightleftharpoons} \underset{\substack{\text{Carbonic}\\\text{acid}}}{H_2CO_3} \rightleftharpoons \underset{\substack{\text{Hydrogen}\\\text{ion}}}{H^+} + \underset{\substack{\text{Bicarbonate}\\\text{ion}}}{HCO_3^-}$$

As CO_2 diffuses into tissue capillaries and enters red blood cells, it reacts with water in the presence of the enzyme carbonic anhydrase (CA) to form carbonic acid, which dissociates into H^+ and HCO_3^-. As HCO_3^- accumulates inside RBCs, some diffuses out into the plasma, down its concentration gradient. In exchange, chloride ions (Cl^-) diffuse from plasma into RBCs. This exchange of negative ions, which maintains the electrical balance between plasma and RBCs, is known as the **chloride**

Figure 23.24 Summary of gas exchange and transport.

 As CO_2 leaves tissue cells and enters blood cells, it causes more O_2 to dissociate from hemoglobin (Bohr effect), and thus more CO_2 combines with hemoglobin and more bicarbonate ions (HCO_3^-) are produced. As O_2 passes from alveoli into blood cells, hemoglobin becomes saturated with O_2 and becomes more acidic. The more acidic hemoglobin releases more hydrogen ions (H^+), which bind to HCO_3^- to form carbonic acid (H_2CO_3). The H_2CO_3 dissociates into H_2O and CO_2, and the CO_2 diffuses from blood into alveoli.

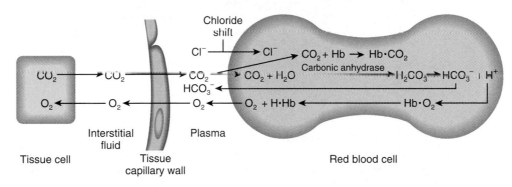

(a) Exchange of O_2 and CO_2 in the tissues (internal respiration)

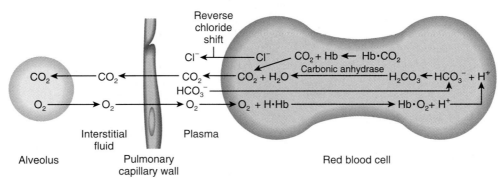

(b) Exchange of O_2 and CO_2 in the lungs (external respiration)

Q Would you expect the concentration of HCO_3^- to be greater in plasma taken from an artery of the arm or a vein of the arm?

---

shift. The net effect of these reactions is that CO_2 is removed from tissue cells and transported in plasma as HCO_3^-. As blood passes through pulmonary capillaries in the lungs, all these reactions reverse as CO_2 is exhaled.

The amount of CO_2 that can be transported in the blood is influenced by the percent saturation of hemoglobin with oxygen. The lower the oxyhemoglobin (HbO_2), the higher the CO_2-carrying capacity of the blood, a relationship known as the **Haldane effect.** Two characteristics of deoxyhemoglobin give rise to the Haldane effect. (1) Deoxyhemoglobin binds to and thus transports more CO_2 than does HbO_2. (2) Deoxyhemoglobin also buffers more H^+ than does HbO_2, thereby removing H^+ from solution and promoting conversion of CO_2 to HCO_3^- via the reaction catalyzed by carbonic anhydrase.

Summary of Gas Exchange and Transport in Lungs and Tissues

Deoxygenated blood returning to the lungs (Figure 23.24a) contains CO_2 dissolved in plasma, CO_2 combined with globin as carbaminohemoglobin, and CO_2 incorporated in HCO_3^- within RBCs. The RBCs have also picked up H^+, some of which is buffered by hemoglobin ($H \cdot Hb$). In the pulmonary capillaries, the chemical reactions reverse (Figure 23.24b). CO_2 dissolved in plasma and CO_2 that dissociates from the globin portion of hemoglobin diffuse into alveoli and are exhaled. At the same time, inhaled O_2 is diffusing from alveoli into RBCs and is attaching to hemoglobin. The Bohr effect is reversible: Just as an increase in H^+ in blood causes O_2 to unload from hemoglobin, the binding of O_2 to hemoglobin causes unloading of H^+ from hemoglo-

Figure 23.25 Locations of areas of the respiratory center.

🔑 The respiratory center is composed of neurons in the medullary rhythmicity area in the medulla oblongata plus the pneumotaxic and apneustic areas in the pons.

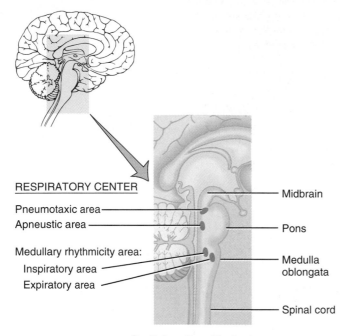

RESPIRATORY CENTER

Pneumotaxic area

Apneustic area

Medullary rhythmicity area:

Inspiratory area

Expiratory area

Midbrain

Pons

Medulla oblongata

Spinal cord

Sagittal section of brain stem

Q Which area contains autorhythmic neurons that are active and then inactive in a repeating cycle?

bin. Carbon dioxide in HCO_3^- is released when H^+ combines with HCO_3^- inside RBCs to form H_2CO_3, which then splits into CO_2 and H_2O. The direction of the carbonic acid reaction depends mostly on P_{CO_2}. In tissue capillaries, where P_{CO_2} is high, H^+ and HCO_3^- are formed; in pulmonary capillaries, where P_{CO_2} is low, CO_2 and H_2O are formed. As the concentration of HCO_3^- declines inside RBCs in pulmonary capillaries, HCO_3^- diffuses in from the plasma, again in exchange for Cl^-. Carbon dioxide diffuses out of RBCs into alveoli and is exhaled. Thus, oxygenated blood leaving the lungs has increased O_2 content and decreased CO_2 and H^+.

1. In a resting person, how many O_2 molecules are attached to each hemoglobin molecule, on average, in blood in the pulmonary arteries? In blood in the pulmonary veins?
2. Describe the relationship between hemoglobin and P_{O_2}. Explain how temperature, H^+, P_{CO_2}, and BPG influence the affinity of Hb for O_2.
3. Explain why hemoglobin can unload more oxygen as blood flows through metabolically active tissues, such as skeletal muscle during exercise, than is unloaded at rest.

REGULATION OF RESPIRATION

OBJECTIVE

• *Describe the various factors that regulate the rate and depth of respiration.*

At rest, about 200 mL of O_2 are used each minute by body cells. During strenuous exercise, however, O_2 use typically increases 15–20-fold in normal healthy adults, and as much as 30-fold in elite endurance-trained athletes. Thus, certain mechanisms exist to match respiratory effort to metabolic demand. The basic rhythm of respiration is controlled by groups of neurons in the medulla oblongata and pons. We first examine the principal mechanisms involved in nervous system control of the rhythm of respiration.

Role of the Respiratory Center

The size of the thorax is altered by the action of the respiratory muscles, which contract and relax as a result of nerve impulses transmitted to them from centers in the brain. The area from which nerve impulses are sent to respiratory muscles consists of clusters of neurons located bilaterally in the medulla oblongata and pons of the brain stem. This area, called the **respiratory center,** consists of a widely dispersed group of neurons functionally divided into three areas: (1) the medullary rhythmicity (rith-MIS-i-tē) area in the medulla oblongata; (2) the pneumotaxic (noo-mō-TAK-sik) area in the pons; and (3) the apneustic (ap-NOO-stik) area, also in the pons (Figure 23.25).

Medullary Rhythmicity Area

The function of the **medullary rhythmicity area** is to control the basic rhythm of respiration. In the basic rhythm of respiration in the normal resting state, inspiration usually lasts for about 2 seconds and expiration for about 3 seconds. Within the medullary rhythmicity area are both inspiratory and expiratory neurons that constitute inspiratory and expiratory areas, respectively. We first consider the role of the inspiratory neurons in respiration.

The basic rhythm of respiration is determined by nerve impulses generated in the inspiratory area (Figure 23.26a). At the beginning of expiration, the inspiratory area is inactive, but after 3 seconds it automatically becomes active due to impulses generated by autorhythmic neurons. Even when all incoming nerve connections to the inspiratory area are cut or blocked, neurons in this area still rhythmically discharge impulses that result in inspiration. Nerve impulses from the active inspiratory area last for about 2 seconds and reach the diaphragm via the phrenic nerves. When the nerve impulses reach the diaphragm, it contracts and inspiration occurs. At the end of 2 seconds, the inspiratory muscles relax for about 3 seconds, and then the cycle repeats.

The neurons of the expiratory area remain inactive during most normal, quiet respirations. During quiet breathing, inspiration is accomplished by active contraction of the diaphragm, and expiration results from passive elastic recoil of

Figure 23.26 Proposed roles of the medullary rhythmicity area in controlling (a) the basic rhythm of respiration and (b) labored breathing.

🔑 **During normal, quiet respirations, the expiratory area is inactive; during forceful respirations, the expiratory area is activated by the inspiratory area.**

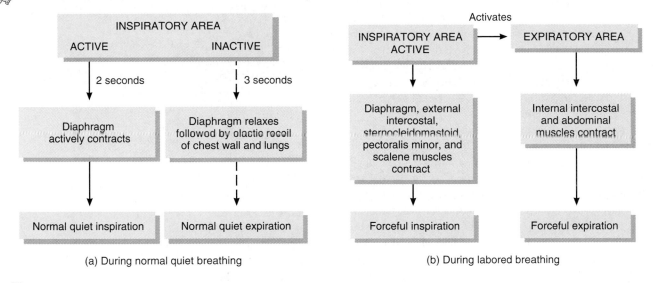

(a) During normal quiet breathing

(b) During labored breathing

ⓆWhich nerves convey impulses from the respiratory center to the diaphragm?

the lungs and thoracic wall as the diaphragm relaxes. However, during forceful ventilation, nerve impulses from the inspiratory area activate the expiratory area (Figure 23.26b). Impulses from the expiratory area cause contraction of the internal intercostals and abdominal muscles, which decreases the size of the thoracic cavity and causes forceful expiration.

Pneumotaxic Area

Although the medullary rhythmicity area controls the basic rhythm of respiration, other sites in the brain stem help coordinate the transition between inspiration and expiration. One of these sites is the **pneumotaxic area** (*pneumo-* = breath; *-taxic* = arrangement) in the superior portion of the pons (see Figure 23.25), which transmits inhibitory impulses to the inspiratory area. The major effect of these nerve impulses is to help turn off the inspiratory area before the lungs become too full of air. In other words, the impulses limit the duration of inspiration and thus facilitate the onset of expiration. When the pneumotaxic area is more active, breathing rate is more rapid.

Apneustic Area

Another part of the brain stem that coordinates the transition between inspiration and expiration is the **apneustic area** in the inferior portions of the pons (see Figure 23.25). This area sends stimulatory impulses to the inspiratory area that activate it and prolong inspiration, thus inhibiting expiration. Such stimulation occurs when the pneu-

motaxic area is inactive; when the pneumotaxic area is active, it overrides the apneustic area.

Regulation of the Respiratory Center

Although the basic rhythm of respiration is set and coordinated by the inspiratory area, the rhythm can be modified in response to inputs from other brain regions and receptors in the peripheral nervous system. Next we discuss several factors that influence the regulation of respiration.

Cortical Influences on Respiration

Because the cerebral cortex has connections with the respiratory center, we can voluntarily alter our pattern of breathing. We can even refuse to breathe at all for a short time. Voluntary control is protective because it enables us to prevent water or irritating gases from entering the lungs. The ability to not breathe, however, is limited by the buildup of CO_2 and H^+ in the blood. When P_{CO_2} and the concentration of H^+ increase to a certain level, the inspiratory area is strongly stimulated, nerve impulses are sent along the phrenic and intercostal nerves to inspiratory muscles, and breathing resumes, whether the person wants it or not. It is impossible for people to kill themselves by holding their breath. Even if a person faints, breathing resumes when consciousness is lost. Nerve impulses from the hypothalamus and limbic system also stimulate the respiratory center, allowing emotional stimuli to alter respirations (as, for example, in crying).

Chemical Regulation of Respiration

Certain chemical stimuli modulate how fast and deeply we breathe. The respiratory system functions to maintain proper levels of CO_2 and O_2, and the system is, not surprisingly, highly responsive to changes in the blood levels of either. Chemoreceptors in two locations monitor levels of CO_2 and O_2 and provide input to the respiratory center: **Central chemoreceptors** are located in the medulla oblongata (*central* nervous system), whereas **peripheral chemoreceptors** are located in the walls of systemic arteries and relay impulses to the respiratory center over two cranial nerves of the *peripheral* nervous system.

Central chemoreceptors respond to changes in H^+ concentration or P_{CO_2}, or both, in cerebrospinal fluid. Peripheral chemoreceptors are especially sensitive to changes in P_{O_2}, as well as H^+ and P_{CO_2}, in the blood; they are in the **aortic body,** a cluster of chemoreceptors located in the wall of the arch of the aorta, and in the **carotid bodies,** which are oval nodules in the wall of the left and right common carotid arteries where they divide into the internal and external carotid arteries. Sensory fibers from the aortic body join the vagus (X) nerve, whereas those from the carotid bodies join the right and left glossopharyngeal (IX) nerves.

Because CO_2 is lipid-soluble, it easily diffuses across plasma membranes, including those that form the blood–brain barrier. Carbonic anhydrase is present within cells, so CO_2 can combine with water (H_2O) to form carbonic acid (H_2CO_3), which quickly breaks down into H^+ and HCO_3^-. Any increase in CO_2 will thus cause an increase in H^+, and any decrease in CO_2 will cause a decrease in H^+.

Under normal circumstances, the P_{CO_2} in arterial blood is 40 mm Hg. If even a slight increase in P_{CO_2} occurs—a condition called **hypercapnia**—the central chemoreceptors are stimulated and respond vigorously to the increase in cerebrospinal fluid H^+ concentration that accompanies hypercapnia. H^+ and CO_2 concentrations fluctuate more readily in cerebrospinal fluid than in blood plasma because cerebrospinal fluid contains fewer buffers than blood. The peripheral chemoreceptors in the carotid and aortic bodies also are stimulated by both the high P_{CO_2} and the rise in H^+ concentration. In addition, the peripheral chemoreceptors respond to deficiency of O_2. If arterial P_{O_2} falls from a normal of 100 mm Hg to about 50 mm Hg, the peripheral chemoreceptors are strongly stimulated.

As a result of increased P_{CO_2}, increased H^+ concentration, and decreased P_{O_2}, input from the central and peripheral chemoreceptors causes the inspiratory area to become highly active, and the rate and depth of breathing increase (Figure 23.27). Rapid and deep breathing, called **hyperventilation,** allows the exhalation of more CO_2 until P_{CO_2} and H^+ are lowered to normal. Slow and shallow breathing is called **hypoventilation.**

If arterial P_{CO_2} is lower than 40 mm Hg—a condition called **hypocapnia**—the central and peripheral chemore-

Figure 23.27 Regulation of breathing in response to changes in blood P_{CO_2}, P_{O_2}, and pH (H^+ concentration) via negative feedback control.

An increase in arterial blood P_{CO_2} stimulates the inspiratory center.

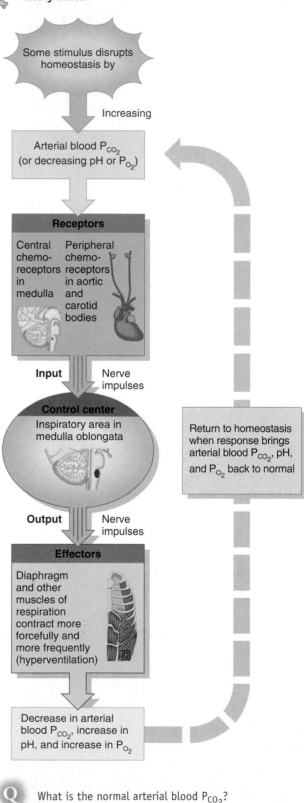

Some stimulus disrupts homeostasis by

Increasing

Arterial blood P_{CO_2} (or decreasing pH or P_{O_2})

Receptors

Central chemoreceptors in medulla

Peripheral chemoreceptors in aortic and carotid bodies

Input   Nerve impulses

Control center

Inspiratory area in medulla oblongata

Return to homeostasis when response brings arterial blood P_{CO_2}, pH, and P_{O_2} back to normal

Output   Nerve impulses

Effectors

Diaphragm and other muscles of respiration contract more forcefully and more frequently (hyperventilation)

Decrease in arterial blood P_{CO_2}, increase in pH, and increase in P_{O_2}

Q   What is the normal arterial blood P_{CO_2}?

ceptors are not stimulated, and stimulatory impulses are not sent to the inspiratory area. Consequently, the area sets its own moderate pace until CO_2 accumulates and the P_{CO_2} rises to 40 mm Hg. People who hyperventilate voluntarily and cause hypocapnia can hold their breath for an unusually long period of time. Swimmers were once encouraged to hyperventilate just before diving in to compete, but this practice is risky because the O_2 level may fall dangerously low and cause fainting before the P_{CO_2} rises high enough to stimulate inspiration. A person who faints on land may suffer bumps and bruises, but one who faints in the water may drown.

Severe deficiency of O_2 depresses activity of the central chemoreceptors and inspiratory area which then do not respond well to any inputs and send fewer impulses to the muscles of respiration (Figure 23.28). As the respiration rate decreases or breathing ceases altogether, P_{O_2} falls lower and lower, thereby establishing a positive feedback cycle.

CLINICAL APPLICATION
Hypoxia

Hypoxia (hī-POK-sē-a; *hypo-* = under) is a deficiency of O_2 at the tissue level. Based on the cause, we can classify hypoxia into four types, as follows:

1. **Hypoxic hypoxia** is caused by a low P_{O_2} in arterial blood as a result of high altitude, airway obstruction, or fluid in the lungs.
2. In **anemic hypoxia,** too little functioning hemoglobin is present in the blood, which reduces O_2 transport to tissue cells. Among the causes are hemorrhage, anemia, and failure of hemoglobin to carry its normal complement of O_2, as in carbon monoxide poisoning.
3. In **ischemic hypoxia,** blood flow to a tissue is so reduced that too little O_2 is delivered to it, even though P_{O_2} and oxyhemoglobin level are normal.
4. In **histotoxic hypoxia,** the blood delivers adequate O_2 to tissues, but the tissues are unable to use it properly because of the action of some toxic agent. One cause is cyanide poisoning, in which cyanide blocks an enzyme needed for O_2 utilization during ATP synthesis. ∎

Proprioceptor Input and Respiration

As soon as you start exercising, your rate and depth of breathing increase, even before changes in P_{O_2}, P_{CO_2}, or H^+ concentration occur. The main stimulus for these quick changes in respiratory effort is thought to be input from proprioceptors, which monitor movement of joints and muscles. Nerve impulses from the proprioceptors stimulate the inspiratory area of the medulla oblongata. At the same time, axon collaterals (branches) of upper motor neurons that originate in the primary motor cortex (precentral gyrus) also feed excitatory impulses into the inspiratory area.

Figure 23.28 Reduction of P_{O_2} in blood via positive feedback.

🔑 **A severe decrease in arterial blood P_{O_2} depresses activity of the central chemoreceptors and inspiratory area.**

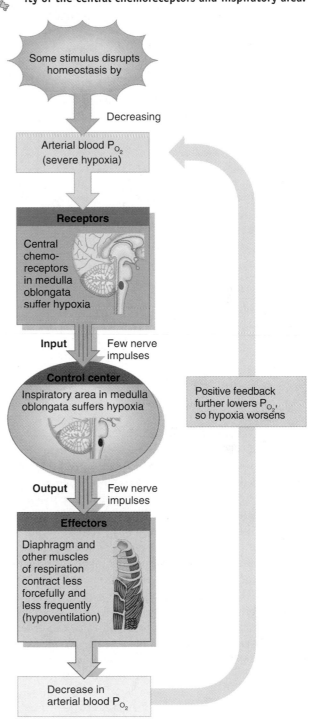

Q How might one intervene to break this positive feedback loop?

Table 23.2 Summary of Regulation of Ventilation Rate and Depth

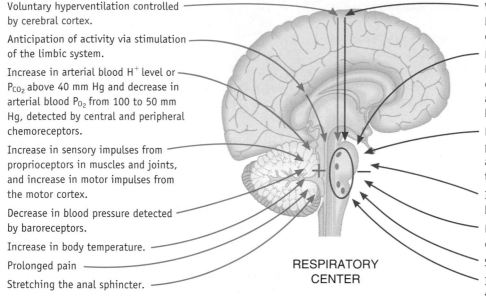

VENTILATION RATE AND DEPTH INCREASE WITH:	VENTILATION RATE AND DEPTH DECREASE WITH:
Voluntary hyperventilation controlled by cerebral cortex.	Voluntary hypoventilation controlled by cerebral cortex (limited by buildup of CO_2 and H^+).
Anticipation of activity via stimulation of the limbic system.	Decrease in arterial blood H^+ level or P_{CO_2} below 40 mm Hg detected by central and peripheral chemoreceptors, and decrease in arterial blood P_{O_2} below 50 mm Hg.
Increase in arterial blood H^+ level or P_{CO_2} above 40 mm Hg and decrease in arterial blood P_{O_2} from 100 to 50 mm Hg, detected by central and peripheral chemoreceptors.	Decrease in sensory impulses from proprioceptors in muscles and joints, and decrease in motor impulses from the motor cortex.
Increase in sensory impulses from proprioceptors in muscles and joints, and increase in motor impulses from the motor cortex.	Increase in blood pressure detected by baroreceptors.
Decrease in blood pressure detected by baroreceptors.	Decrease in body temperature (sudden cold stimulus causes apnea).
Increase in body temperature.	Severe pain causes apnea.
Prolonged pain	Irritation of pharynx or larynx by touch or chemicals causes apnea followed by coughing or sneezing.
Stretching the anal sphincter.	

RESPIRATORY CENTER

The Inflation Reflex

Located in the walls of bronchi and bronchioles are stretch-sensitive receptors called **baroreceptors** or **stretch receptors.** When these receptors become stretched during overinflation of the lungs, nerve impulses are sent along the vagus (X) nerves to the inspiratory and apneustic areas. In response, the inspiratory area is inhibited, and the apneustic area is inhibited from activating the inspiratory area. As a result, expiration begins. As air leaves the lungs during expiration, the lungs deflate and the stretch receptors are no longer stimulated. Thus, the inspiratory and apneustic areas are no longer inhibited, and a new inspiration begins. Some evidence suggests that this reflex, referred to as the **inflation (Hering–Breuer) reflex,** is mainly a protective mechanism for preventing excessive inflation of the lungs rather than a key component in the normal regulation of respiration.

Other Influences on Respiration

Among the other factors that contribute to regulation of respiration are the following:

- *Limbic system stimulation.* Anticipation of activity or emotional anxiety may stimulate the limbic system, which then sends excitatory input to the inspiratory area to increase the rate and depth of ventilation.

- *Temperature.* An increase in body temperature, as during a fever or vigorous muscular exercise, increases the rate of respiration; a decrease in body temperature decreases respiratory rate. A sudden cold stimulus (such as plunging into cold water) causes **apnea** (AP-nē-a; *a-* = without; *-pnea* = breath), a temporary cessation of breathing.

- *Pain.* A sudden, severe pain brings about brief apnea, but a prolonged somatic pain increases respiratory rate. Visceral pain may slow respiratory rate.

- *Stretching the anal sphincter muscle.* This action increases the respiratory rate and is sometimes used to stimulate respiration in a newborn baby or a person who has stopped breathing.

- *Irritation of airways.* Mechanical or chemical irritation of the pharynx or larynx brings about an immediate cessation of breathing followed by coughing or sneezing.

- *Blood pressure.* The carotid and aortic sinuses, which are near the carotid and aortic bodies, contain baroreceptors (stretch receptors) that detect changes in blood pressure. Although these baroreceptors are active mainly in the control of blood pressure, they have a small effect on respiration. A sudden rise in blood pressure decreases the rate of respiration, and a drop in blood pressure increases the respiratory rate.

Table 23.2 summarizes the changes that increase or decrease ventilation rate and depth.

1. How does the medullary rhythmicity area function in regulating respiration? How are the apneustic and pneumotaxic areas related to the control of respiration?
2. Explain how each of the following modifies respiration: cerebral cortex, inflation reflex, CO_2, O_2, proprioceptors, temperature, pain, and irritations of the respiratory mucosa.

EXERCISE AND THE RESPIRATORY SYSTEM

OBJECTIVE
• Describe the effects of exercise on the respiratory system.

During exercise, the respiratory and cardiovascular systems make adjustments in response to both the intensity and duration of the exercise. The effects of exercise on the heart are discussed in Chapter 20; here we focus on how exercise affects the respiratory system.

Recall that the heart pumps the same amount of blood to the lungs as to all the rest of the body. Thus, as cardiac output rises, the blood flow to the lungs, termed **pulmonary perfusion,** increases as well. In addition, the **O_2 diffusing capacity,** a measure of the rate at which O_2 can diffuse from alveolar air into the blood, may increase threefold during maximal exercise because more pulmonary capillaries become maximally perfused. As a result, there is a greater surface area available for diffusion of O_2 into pulmonary blood capillaries.

When muscles contract during exercise, they consume large amounts of O_2 and produce large amounts of CO_2. During vigorous exercise, O_2 consumption and pulmonary ventilation both increase dramatically. At the onset of exercise, an abrupt increase in pulmonary ventilation is followed by a more gradual increase. With moderate exercise, the increase is due mostly to an increase in the depth of ventilation rather than to increased breathing rate. When exercise is more strenuous, the frequency of breathing also increases.

The abrupt increase in ventilation at the start of exercise is due to *neural* changes that send excitatory impulses to the inspiratory area in the medulla oblongata. These changes include (1) anticipation of the activity, which stimulates the limbic system; (2) sensory impulses from proprioceptors in muscles, tendons, and joints; and (3) motor impulses from the primary motor cortex (precentral gyrus). The more gradual increase in ventilation during moderate exercise is due to *chemical* and *physical* changes in the bloodstream, including (1) slightly decreased P_{O_2}, due to increased O_2 consumption; (2) slightly increased P_{CO_2}, due to increased CO_2 production by contracting muscle fibers; and (3) increased temperature due to liberation of more heat as more O_2 is utilized. Moreover, during strenuous exercise, HCO_3^- buffers H^+ released by lactic acid in a reaction that liberates CO_2, which further increases P_{CO_2}.

At the end of an exercise session, an abrupt decrease in pulmonary ventilation is followed by a more gradual decline to the resting level. The initial decrease is due mainly to changes in neural factors when movement stops or slows, whereas the more gradual phase reflects the slower return of blood chemistry levels and temperature to the resting state.

CLINICAL APPLICATION
Why Smokers Have Lowered Respiratory Efficiency

Smoking may cause a person to become easily "winded" with even moderate exercise because several factors decrease respiratory efficiency in smokers: (1) Nicotine constricts terminal bronchioles, which decreases airflow into and out of the lungs. (2) Carbon monoxide in smoke binds to hemoglobin and reduces its oxygen-carrying capability. (3) Irritants in smoke cause increased mucus secretion by the mucosa of the bronchial tree and swelling of the mucosal lining, both of which impede airflow into and out of the lungs. (4) Irritants in smoke also inhibit the movement of cilia and destroy cilia in the lining of the respiratory system. Thus, excess mucus and foreign debris are not easily removed, which further adds to the difficulty in breathing. (5) With time, smoking leads to destruction of elastic fibers in the lungs and is the prime cause of emphysema (described on page 811). These changes cause collapse of small bronchioles and the trapping of air in alveoli at the end of exhalation; as a result, gas exchange is less efficient. ■

DEVELOPMENTAL ANATOMY OF THE RESPIRATORY SYSTEM

OBJECTIVE
• Describe the development of the respiratory system.

The development of the mouth and pharynx is discussed in Chapter 24; here we consider the remainder of the respiratory system. At about 4 weeks of fetal development, the respiratory system begins as an outgrowth of the **endoderm** of the foregut (precursor of some digestive organs) just posterior to the pharynx. This outgrowth is called the **laryngotracheal bud** (see Figure 18.22 on page 602). As the bud grows, it elongates and differentiates into the future epithelial lining of the *larynx* and other structures. Its proximal end maintains a slitlike opening into the pharynx called the *rima glottis*. The middle portion of the bud gives rise to the epithelial lining of the *trachea*. The distal portion divides into two **lung buds,** which grow into the epithelial lining of the *bronchi* and *lungs* (Figure 23.29).

As the lung buds develop, they branch and rebranch and give rise to all the *bronchial tubes*. After the sixth month, the closed terminal portions of the tubes dilate and become the *alveoli* of the lungs. The smooth muscle, cartilage, and connective tissues of the bronchial tubes and the pleural sacs of the lungs are contributed by **mesenchymal (mesodermal) cells.**

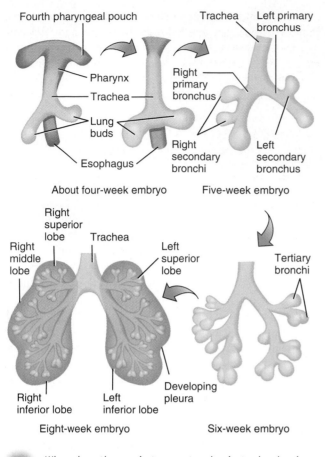

Figure 23.29 Development of the bronchial tubes and lungs.

🔑 **The respiratory system develops from endoderm and mesoderm.**

AGING AND THE RESPIRATORY SYSTEM

OBJECTIVE
• *Describe the effects of aging on the respiratory system.*

With advancing age, the airways and tissues of the respiratory tract, including the alveoli, become less elastic and more rigid; the chest wall becomes more rigid as well. The result is a decrease in lung capacity. In fact, vital capacity (the maximum amount of air that can be expired after maximal inspiration) can decrease as much as 35% by age 70. Moreover, a decrease in blood levels of O_2, decreased activity of alveolar macrophages, and diminished ciliary action of the epithelium lining the respiratory tract occur. Owing to all these age-related factors, elderly people are more susceptible to pneumonia, bronchitis, emphysema, and other pulmonary disorders.

1. How does exercise affect the inspiratory area?
2. What structures develop from the laryngotracheal bud?
3. What accounts for the decrease in lung capacity with aging?

Q When does the respiratory system begin to develop in an embryo?

DISORDERS: HOMEOSTATIC IMBALANCES

ASTHMA

Asthma (AZ-ma; = panting) is a disorder characterized by chronic airway inflammation, airway hypersensitivity to a variety of stimuli, and airway obstruction that is at least partially reversible, either spontaneously or with treatment. It affects 3–5% of the U.S. population and is more common in children than in adults. Airway obstruction may be due to smooth muscle spasms in the walls of smaller bronchi and bronchioles, edema of the mucosa of the airways, increased mucus secretion, and damage to the epithelium of the airway.

Individuals with asthma typically react to low concentrations of agents that do not normally cause symptoms in people without asthma. Sometimes the trigger is an allergen such as pollen, house dust mites, molds, or a particular food. Other common triggers of asthma attacks are emotional upset, aspirin, sulfiting agents (used in wine and beer and to keep greens fresh in salad bars), exercise, and breathing cold air or cigarette smoke. In the early phase (acute) response, smooth muscle spasm is accompanied by excessive secretion of mucus that may clog the bronchi and bronchioles and worsen the attack. The late phase (chronic) response is characterized by inflammation, fibrosis, edema, and necrosis (death) of bronchial epithelial cells. A host of mediator chemicals, including

leukotrienes, prostaglandins, thromboxane, platelet-activating factor, and histamine, take part.

Symptoms include difficult breathing, coughing, wheezing, chest tightness, tachycardia, fatigue, moist skin, and anxiety. An acute attack is treated by giving an inhaled beta$_2$-adrenergic agonist (albuterol) to help relax smooth muscle in the bronchioles and open up the airways. However, long-term therapy of asthma strives to suppress the underlying inflammation. The anti-inflammatory drugs that are used most often are inhaled corticosteroids (glucocorticoids), cromolyn sodium (Intal), and leukotriene blockers (Accolate).

CHRONIC OBSTRUCTIVE PULMONARY DISEASE

Chronic obstructive pulmonary disease (COPD) is a type of respiratory disorder characterized by chronic and recurrent obstruction of airflow, which increases airway resistance. COPD affects about 30 million Americans and is the fourth leading cause of death behind heart disease, cancer, and cerebrovascular disease. The principal types of COPD are emphysema and chronic bronchitis. In most cases, COPD is preventable because its most common cause is cigarette smoking or breathing secondhand smoke. Other causes include air pollution, pulmonary infection, occupational exposure to dusts and gases, and genetic factors. Because men, on average, have more years of exposure to cigarette smoke than women, men are twice as likely as women to suffer from COPD; still, the incidence of COPD in women has risen six-fold in the past 50 years, a reflection of increased smoking among women.

Emphysema

Emphysema (em′-fi-SĒ-ma′; = blown up or full of air) is a disorder characterized by destruction of the walls of the alveoli, producing abnormally large air spaces that remain filled with air during expiration. With less surface area for gas exchange, O_2 diffusion across the damaged respiratory membrane is reduced. Blood O_2 level is somewhat lowered, and any mild exercise that raises the O_2 requirements of the cells leaves the patient breathless. As increasing numbers of alveolar walls are damaged, lung elastic recoil decreases due to loss of elastic fibers, and an increasing amount of air becomes trapped in the lungs at the end of exhalation. Over several years, added inspiratory exertion increases the size of the chest cage, resulting in a "barrel chest."

Emphysema is generally caused by a long-term irritation; cigarette smoke, air pollution, and occupational exposure to industrial dust are the most common irritants. Some destruction of alveolar sacs may be caused by an imbalance between enzymes called proteases (for example, elastase) and *alpha-1-antitrypsin,* a molecule that inhibits proteases. When decreased production by the liver of the plasma protein alpha-1-antitrypsin occurs, elastase is not inhibited and is free to attack the connective tissue in the walls of alveolar sacs. Cigarette smoke not only deactivates a protein that is apparently crucial in preventing emphysema; it also prevents the repair of affected lung tissue.

Treatment consists of cessation of smoking and removing other environmental irritants, exercise training under careful medial supervision, breathing exercises, use of bronchodilators, and oxygen therapy.

Chronic Bronchitis

Chronic bronchitis is a disorder characterized by excessive secretion of bronchial mucus and accompanied by a productive cough (sputum is raised) that lasts for at least three months of the year for two successive years. Cigarette smoking is the leading cause of chronic bronchitis. Inhaled irritants lead to chronic inflammation with an increase in the size and number of mucous glands and goblet cells in the airway epithelium. The thickened and excessive mucus produced narrows the airway and impairs ciliary function. Thus, inhaled pathogens become embedded in airway secretions and multiply rapidly. Besides a productive cough, symptoms of chronic bronchitis are shortness of breath, wheezing, cyanosis, and pulmonary hypertension. Treatment for chronic bronchitis is similar to that for emphysema.

Lung Cancer

In the United States **lung cancer** is the leading cause of cancer death in males and females, accounting for 160,000 deaths annually. At the time of diagnosis, lung cancer is usually well advanced, with distant metastases present in about 55% of patients, and regional lymph node involvement in an additional 25%. Most people with lung cancer die within a year of the initial diagnosis; the overall survival rate is only 10–15%. Cigarette smoke is the most important cause of lung cancer. Roughly 85% of lung cancer cases are related to smoking, and the disease is 10–30 times more common in smokers than nonsmokers. Exposure to secondhand smoke is also associated with lung cancer and heart disease. In the United

States, secondhand smoke causes an estimated 4000 deaths a year from lung cancer, and nearly 40,000 deaths a year from heart disease. Other causes of lung cancer are ionizing radiation and inhaled irritants, such as asbestos and radon gas. Emphysema is a common precursor to the development of lung cancer.

The most common type of lung cancer, **bronchogenic carcinoma,** starts in the epithelium of the bronchial tubes. Bronchogenic tumors are named on the basis of where they arise. For example, *adenocarcinomas* develop in peripheral areas of the lungs from bronchial glands and alveolar cells, *squamous cell carcinomas* develop from the epithelium of larger bronchial tubes, and *small (oat) cell carcinomas* develop from epithelial cells in primary bronchi near the hilus of the lungs and tend to involve the mediastinum early on. Depending on the type of bronchogenic tumors, they may be aggressive, locally invasive, and undergo widespread metastasis. The tumors begin as epithelial lesions that grow to form masses that obstruct the bronchial tubes or invade adjacent lung tissue. Bronchogenic carcinomas metastasize to lymph nodes, the brain, bones, liver, and other organs.

Symptoms of lung cancer are related to the location of the tumor. These may include a chronic cough, spitting blood from the respiratory tract, wheezing, shortness of breath, chest pain, hoarseness, difficulty swallowing, weight loss, anorexia, fatigue, bone pain, confusion, problems with balance, headache, anemia, thrombocytopenia, and jaundice.

Treatment consists of partial or complete surgical removal of a diseased lung (pulmonectomy), radiation therapy, and chemotherapy.

PNEUMONIA

Pneumonia or **pneumonitis** is an acute infection or inflammation of the alveoli. It is the most common infectious cause of death in the United States, where an estimated 4 million cases occur annually. When certain microbes enter the lungs of susceptible individuals, they release damaging toxins, stimulating inflammation and immune responses that have damaging side effects. The toxins and immune response damage alveoli and bronchial mucous membranes; inflammation and edema cause the alveoli to fill with debris and exudate, interfering with ventilation and gas exchange.

The most common cause of pneumonia is the pneumococcal bacterium *Streptococcus pneumoniae,* but other microbes may also cause pneumonia. Those who are most susceptible to pneumonia are the elderly, infants, immunocompromised individuals (those having AIDS or a malignancy, or those taking immunosuppressive drugs), cigarette smokers, and individuals with an obstructive lung disease. Most cases of pneumonia are preceded by an upper respiratory infection that is frequently viral. Individuals then develop fever, chills, productive or dry cough, malaise, chest pain, and sometimes dyspnea and hemoptysis.

Treatment may involve antibiotics, bronchodilators, oxygen therapy, increased fluid intake, and chest physiotherapy (percussion, vibration, and postural drainage).

TUBERCULOSIS

The bacterium *Mycobacterium tuberculosis* produces an infectious, communicable disease called **tuberculosis (TB)** that most often affects the lungs and the pleurae but may involve other parts of the body. Once the bacteria are inside the lungs, they multiply and cause inflammation, which stimulates neutrophils and macrophages to migrate to the area and engulf the bacteria to prevent their spread. If the immune system is not impaired, the bacteria remain dormant for life, but impaired immunity may enable the bacteria to escape into blood and lymph to infect other organs. In many people, symptoms—fatigue, weight loss, lethargy, anorexia, a low grade fever, night sweats, cough, dyspnea, chest pain, and hemoptysis—do not develop until the disease is advanced.

During the past several years, the incidence of TB in the United States has risen dramatically. Perhaps the single most important factor related to this increase is the presence of the human immunodeficiency virus (HIV). People infected with HIV are much more likely to develop tuberculosis because their immune systems are impaired. Among the other factors that have contributed to the increased number of cases are homelessness, increased drug abuse, increased immigration from countries with a high prevalence of tuberculosis, increased crowding in housing among the poor, and airborne transmission of tuberculosis in prisons and shelters. In addition, recent outbreaks of tuberculosis involving multi-drug-resistant strains of *Mycobacterium tuberculosis* have occurred because patients fail to complete their antibiotic and other treatment regimens.

CORYZA AND INFLUENZA

Hundreds of viruses are responsible for **coryza** (ko-RĪ-za) or the **common cold.** A group of viruses called rhinoviruses is responsible for about 40% of all colds in adults. Typical symptoms include sneezing, excessive nasal secretion, dry cough, and congestion. The uncomplicated common cold is not usually accompanied by a fever. Complications include sinusitis, asthma, bronchitis, ear infections, and laryngitis. Recent investigations suggest an association between emotional stress and the common cold: The higher the stress level, the greater the frequency and duration of colds.

Influenza (flu) is also caused by a virus. Its symptoms include chills, fever (usually higher than 101°F = 39°C), headache, and muscular aches. Coldlike symptoms appear as the fever subsides.

PULMONARY EDEMA

Pulmonary edema is an abnormal accumulation of interstitial fluid in the interstitial spaces and alveoli of the lungs. The edema may arise from increased pulmonary capillary permeability (pulmonary origin) or increased pulmonary capillary pressure (cardiac origin); the latter cause may coincide with congestive heart failure. The most common symptom is dyspnea. Others include wheezing, tachypnea (rapid breathing rate), restlessness, a feeling of suffocation, cyanosis, pallor (paleness), and diaphoresis (excessive perspiration). Treatment consists of administering oxygen, drugs that dilate the bronchioles and lower blood pressure, diuretics to rid the body of excess fluid, and drugs that correct acid–base imbalance; suctioning of airways; and mechanical ventilation.

CYSTIC FIBROSIS

Cystic fibrosis (CF) is an inherited disease of secretory epithelia that affects the airways, liver, pancreas, small intestine, and sweat glands. It is the most common lethal genetic disease in whites: 5% of the population are thought to be genetic carriers. The cause of cystic fibrosis is a genetic mutation affecting a transporter protein that carries chloride ions across the plasma membranes of many epithelial cells. Because dysfunction of sweat glands causes perspiration to contain excessive sodium chloride (salt), measurement of the excess chloride is one index for diagnosing CF. The mutation also disrupts the normal functioning of several organs by causing ducts within them to become obstructed by thick mucus secretions that do not drain easily from the passageways. Buildup of these secretions leads to inflammation and replacement of injured cells with connective tissue that further blocks the ducts. Clogging and infection of the airways leads to difficulty in breathing and eventual destruction of lung tissue. Lung disease accounts for most deaths from CF. Obstruction of small bile ducts in the liver interferes with digestion and disrupts liver function; clogging of pancreatic ducts prevents digestive enzymes from reaching the small intestine. Because pancreatic juice contains the main fat-digesting enzyme, the person fails to absorb fats or fat-soluble vitamins and thus suffers from vitamin A, D, and K deficiency diseases. With respect to the reproductive systems, blockage of the vas deferens leads to infertility in males; the formation of dense mucus plugs in the vagina restricts the entry of sperm into the uterus and can lead to infertility in females.

A child suffering from cystic fibrosis is given pancreatic extract and large doses of vitamins A, D, and K. The recommended diet is high in calories, fats, and proteins, with vitamin supplementation and liberal use of salt.

MEDICAL TERMINOLOGY

Acute respiratory distress syndrome (ARDS) A form of respiratory failure characterized by excessive leakiness of the respiratory membranes and severe hypoxia. Situations that can cause ARDS include near-drowning, aspiration of acidic gastric juice, drug reactions, inhalation of an irritating gas such as ammonia, allergic reactions, various lung infections such as pneumonia or TB, and pulmonary hypertension. ARDS strikes about 250,000 people a year in the United States, and about 50% of them die despite intensive medical care.

Asphyxia (as-FIK-sē-a; *sphyxia* = pulse) Oxygen starvation due to low atmospheric oxygen or interference with ventilation, external respiration, or internal respiration.

Aspiration (as′-pi-RĀ-shun) Inhalation of a foreign substance such as water, food, or a foreign body into the bronchial tree; also, the drawing of a substance in or out by suction.

Atelectasis (at′-ē-LEK-ta-sis; *atel-* = incomplete; *ectasis* = expansion) Incomplete expansion of a lung or a portion of a lung caused by airway obstruction, lung compression, or inadequate pulmonary surfactant.

Bronchiectasis (bron′-kē-EK-ta-sis) A chronic dilation of the bronchi or bronchioles.

Bronchography (bron-KOG-ra-fē) An imaging technique used to visualize the bronchial tree using x-rays. After an opaque contrast medium is inhaled through an intratracheal catheter, radiographs of the chest in various positions are taken, and the developed film, a **bronchogram** (BRON-kō-gram), provides a picture of the bronchial tree.

Bronchoscopy (bron-KOS-kō-pē) Visual examination of the bronchi through a **bronchoscope,** an illuminated, flexible tubular instrument that is passed through the mouth (or nose), larynx, and trachea into the bronchi. The examiner can view the interior of the trachea and bronchi to biopsy a tumor, clear an obstructing object or secretions from an airway, take cultures or smears for microscopic examination, stop bleeding, or deliver drugs.

Cheyne–Stokes respiration (CHĀN STŌKS res′-pi-RĀ-shun) A repeated cycle of irregular breathing that begins with shallow breaths that increase in depth and rapidity and then decrease and cease altogether for 15–20 seconds. Cheyne–Stokes is normal in infants; it is also often seen just before death from pulmonary, cerebral, cardiac, and kidney disease.

Dyspnea (DISP-nē-a; *dys-* = painful, difficult) Painful or labored breathing.

Epistaxis (ep′-i-STAK-sis) Loss of blood from the nose due to trauma, infection, allergy, malignant growths, or bleeding disorders. It can be arrested by cautery with silver nitrate, electrocautery, or firm packing. Also called **nosebleed.**

Heimlich (abdominal thrust) maneuver (HĪM-lik ma-NOO-ver) First-aid procedure designed to clear the airways of obstructing objects. It is performed by applying a quick upward thrust between the navel and costal margin that causes sudden elevation of the diaphragm and forceful, rapid expulsion of air in the lungs; this action forces air out the trachea to eject the ob-

structing object. The Heimlich maneuver is also used to expel water from the lungs of near-drowning victims before resuscitation is begun.

Hemoptysis (hē-MOP-ti-sis; *hemo-* = blood; *-ptysis* = spit) Spitting of blood from the respiratory tract.

Rales (RĀLS) Sounds sometimes heard in the lungs that resemble bubbling or rattling. Rales are to the lungs what murmurs are to the heart. Different types are due to the presence of an abnormal type or amount of fluid or mucus within the bronchi or alveoli, or to bronchoconstriction that causes turbulent airflow.

Respirator (RES-pi-rā′-tor) An apparatus fitted to a mask over the nose and mouth, or hooked directly to an endotracheal or tracheotomy tube, that is used to assist or support ventilation or to provide nebulized medication to the air passages.

Respiratory failure A condition in which the respiratory system either cannot supply sufficient O_2 to maintain metabolism or cannot eliminate enough CO_2 to prevent respiratory acidosis (a lower-than-normal pH in extracellular fluid).

Rhinitis (rī-NĪ-tis; *rhin-* = nose) Chronic or acute inflammation of the mucous membrane of the nose.

Sudden infant death syndrome (SIDS) Death of infants between the ages of 1 week and 12 months thought to be due to hypoxia while sleeping in a prone position (on the stomach) and the rebreathing of exhaled air trapped in a depression of the mattress. It is now recommended that normal newborns be placed on their backs for sleeping.

Tachypnea (tak′-ip-NĒ-a; *tachy-* = rapid) Rapid breathing rate.

STUDY OUTLINE

RESPIRATORY SYSTEM ANATOMY (p. 775)

1. The respiratory system consists of the nose, pharynx, larynx, trachea, bronchi, and lungs. They act with the cardiovascular system to supply oxygen (O_2) and remove carbon dioxide (CO_2) from the blood.
2. The external portion of the nose is made of cartilage and skin and is lined with a mucous membrane. Openings to the exterior are the external nares.
3. The internal portion of the nose communicates with the paranasal sinuses and nasopharynx through the internal nares.
4. The nasal cavity is divided by a septum. The anterior portion of the cavity is called the vestibule. The nose warms, moistens, and filters air and functions in olfaction and speech.
5. The pharynx (throat) is a muscular tube lined by a mucous membrane. The anatomic regions are the nasopharynx, oropharynx, and laryngopharynx.
6. The nasopharynx functions in respiration. The oropharynx and laryngopharynx function both in digestion and in respiration.
7. The larynx (voice box) is a passageway that connects the pharynx with the trachea. It contains the thyroid cartilage (Adam's apple); the epiglottis, which prevents food from entering the larynx; the cricoid cartilage, which connects the larynx and tra-

chea; and the paired arytenoid, corniculate, and cuneiform cartilages.
8. The larynx contains vocal folds, which produce sound as they vibrate. Taut folds produce high pitches, and relaxed ones produce low pitches.
9. The trachea (windpipe) extends from the larynx to the primary bronchi. It is composed of C-shaped rings of cartilage and smooth muscle and is lined with pseudostratified ciliated columnar epithelium.
10. The bronchial tree consists of the trachea, primary bronchi, secondary bronchi, tertiary bronchi, bronchioles, and terminal bronchioles. Walls of bronchi contain rings of cartilage; walls of bronchioles contain increasingly smaller plates of cartilage and increasing amounts of smooth muscle.
11. Lungs are paired organs in the thoracic cavity enclosed by the pleural membrane. The parietal pleura is the superficial layer that lines the thoracic cavity; the visceral pleura is the deep layer that covers the lungs.
12. The right lung has three lobes separated by two fissures; the left lung has two lobes separated by one fissure and a depression, the cardiac notch.
13. Secondary bronchi give rise to branches called segmental bronchi, which supply segments of lung tissue called bronchopulmonary segments.

14. Each bronchopulmonary segment consists of lobules, which contain lymphatics, arterioles, venules, terminal bronchioles, respiratory bronchioles, alveolar ducts, alveolar sacs, and alveoli.

15. Alveolar walls consist of type I alveolar cells, type II alveolar cells, and associated alveolar macrophages.

16. Gas exchange occurs across the respiratory membranes.

PULMONARY VENTILATION (p. 790)

1. Pulmonary ventilation, or breathing, consists of inspiration and expiration.

2. The movement of air into and out of the lungs depends on pressure changes governed in part by Boyle's law, which states that the volume of a gas varies inversely with pressure, assuming that temperature remains constant.

3. Inspiration occurs when alveolar pressure falls below atmospheric pressure. Contraction of the diaphragm increases the size of the thorax, thereby decreasing the intrapleural pressure so that the lungs expand. Expansion of the lungs decreases alveolar pressure so that air moves down a pressure gradient from the atmosphere into the lungs.

4. During forceful inspiration, accessory muscles of inspiration (external intercostals, sternocleidomastoids, scalenes, and pectoralis minors) are also used.

5. Expiration occurs when alveolar pressure is higher than atmospheric pressure. Relaxation of the diaphragm results in elastic recoil of the chest wall and lungs, which increases intrapleural pressure; lung volume decreases and alveolar pressure increases, so air moves from the lungs to the atmosphere.

6. Forceful expiration involves contraction of the internal intercostal and abdominal muscles.

7. The surface tension exerted by alveolar fluid is decreased by the presence of surfactant.

8. Compliance is the ease with which the lungs and thoracic wall can expand.

9. The walls of the airways offer some resistance to breathing.

10. Normal quiet breathing is termed eupnea; other patterns are costal breathing and diaphragmatic breathing. Modified respiratory movements, such as coughing, sneezing, sighing, yawning, sobbing, crying, laughing, and hiccuping, are used to express emotions and to clear the airways.

LUNG VOLUMES AND CAPACITIES (p. 795)

1. Lung volumes exchanged during breathing and the rate of respiration are measured with a spirometer.

2. Lung volumes measured by spirometry include tidal volume, minute ventilation, alveolar ventilation rate, inspiratory reserve volume, expiratory reserve volume, and $FEV_{1.0}$. Other lung volumes are anatomic dead space, residual volume, and minimal volume.

3. Lung capacities, the sum of two or more volumes, include inspiratory, functional residual, vital, and total lung capacities.

EXCHANGE OF OXYGEN AND CARBON DIOXIDE (p. 796)

1. The partial pressure of a gas is the pressure exerted by that gas in a mixture of gases. It is symbolized by P_x, where the subscript is the formula of the gas.

2. According to Dalton's law, each gas in a mixture of gases exerts its own pressure as if all the other gases were not present.

3. Henry's law states that the quantity of a gas that will dissolve in a liquid is proportional to the partial pressure of the gas and its solubility coefficient (given that the temperature remains constant).

4. In internal and external respiration, O_2 and CO_2 diffuse from areas of higher partial pressures to areas of lower partial pressures.

5. External respiration is the exchange of gases between alveoli and pulmonary blood capillaries. It depends on partial pressure differences, a large surface area for gas exchange, a small diffusion distance across the respiratory membrane, and the rate of airflow into and out of the lungs.

6. Internal respiration is the exchange of gases between tissue blood capillaries and tissue cells.

TRANSPORT OF OXYGEN AND CARBON DIOXIDE IN THE BLOOD (p. 799)

1. In each 100 mL of oxygenated blood, 1.5% of the O_2 is dissolved in plasma and 98.5% is bound to hemoglobin as oxyhemoglobin (HbO_2).

2. The association of O_2 and hemoglobin is affected by P_{O_2}, acidity (pH), P_{CO_2}, temperature, and BPG.

3. Fetal hemoglobin differs from adult hemoglobin in structure and has a higher affinity for O_2.

4. In each 100 mL of deoxygenated blood, 7% of CO_2 is dissolved in plasma, 23% combines with hemoglobin as carbaminohemoglobin ($Hb \cdot CO_2$), and 70% is converted to bicarbonate ions (HCO_3^-).

5. In an acidic environment, hemoglobin's affinity for O_2 is lower, and O_2 dissociates more readily from it (Bohr effect).

6. In the presence of O_2, less CO_2 binds to hemoglobin (Haldane effect).

REGULATION OF RESPIRATION (p. 804)

1. The respiratory center consists of a medullary rhythmicity area, a pneumotaxic area, and an apneustic area.

2. The inspiratory area has an intrinsic excitability (autorhythmicity) that sets the basic rhythm of respiration.

3. The pneumotaxic and apneustic areas coordinate the transition between inspiration and expiration.

4. Respirations may be modified by a number of factors, including cortical influences; the inflation reflex; chemical stimuli, such as O_2 and CO_2 and H^+ levels; proprioceptor input; blood pressure changes; the limbic system stimulation; temperature; pain; and irritation to the airways.

EXERCISE AND THE RESPIRATORY SYSTEM (p. 809)

1. The rate and depth of ventilation change in response to both the intensity and duration of exercise.

2. An increase in pulmonary perfusion and O_2 diffusing capacity occurs during exercise.

3. The abrupt increase in ventilation at the start of exercise is due to neural changes that send excitatory impulses to the inspiratory area in the medulla oblongata. The more gradual increase in ventilation during moderate exercise is due to chemical and physical changes in the bloodstream.

DEVELOPMENTAL ANATOMY OF THE RESPIRATORY SYSTEM (p. 809)

1. The respiratory system begins as an outgrowth of endoderm called the laryngotracheal bud.
2. Smooth muscle, cartilage, and connective tissue of the bronchial tubes and pleural sacs develop from mesoderm.

AGING AND THE RESPIRATORY SYSTEM (p. 810)

1. Aging results in decreased vital capacity, decreased blood level of O_2, and diminished alveolar macrophage activity.
2. Elderly people are more susceptible to pneumonia, emphysema, bronchitis, and other pulmonary disorders.

SELF-QUIZ QUESTIONS

1. Match the following:
   ___ (a) functions as a passageway for air and food, provides a resonating chamber for speech sounds, and houses the tonsils
   ___ (b) site of external respiration
   ___ (c) connects the laryngopharynx with the trachea; houses the vocal cords
   ___ (d) serous membrane that surrounds the lungs
   ___ (e) functions in warming, moistening, and filtering air; receives olfactory stimuli; is a resonating chamber for sound
   ___ (f) form a continuous lining of the alveolar wall
   ___ (g) a tubular passageway for air connecting the larynx to the bronchi
   ___ (h) secretes alveolar fluid, which keeps the alveolar cell moist; secretes surfactant
   ___ (i) prevents food or fluid from entering the airways
   ___ (j) air passageways entering the lungs

   (1) nose
   (2) pharynx
   (3) larynx
   (4) epiglottis
   (5) trachea
   (6) bronchi
   (7) pleura
   (8) alveoli
   (9) type I alveolar cells
   (10) type II alveolar cells

Complete the following:

2. The three basic steps of respiration are ___, ___, and ___.
3. The process by which gases are exchanged between the atmosphere and lung alveoli is termed ___. Breathing in is called ___, and breathing out is called ___.
4. Oxygen in blood is carried primarily in the form of ___; carbon dioxide is carried primarily in the form of ___.
5. The chemical reaction that applies to the transport of carbon dioxide in blood is ___.

True or false:

6. Pulmonary circulation differs from systemic circulation in that the pulmonary blood vessels provide more resistance to blood flow, so more pressure is required to move blood through the pulmonary circulation.
7. The most important muscle of respiration is the diaphragm.

8. Match the following:
   ___ (a) a deficiency of oxygen at the tissue level
   ___ (b) a slight increase in the partial pressure of carbon dioxide
   ___ (c) normal quiet breathing
   ___ (d) deep, abdominal breathing
   ___ (e) the ease with which the lungs and thoracic wall can be expanded
   ___ (f) a temporary cessation of breathing
   ___ (g) rapid breathing
   ___ (h) painful or labored breathing
   ___ (i) shallow, chest breathing
   ___ (j) collapse or incomplete expansion of lung tissue

   (1) eupnea
   (2) apnea
   (3) dyspnea
   (4) hyperventilation
   (5) costal breathing
   (6) diaphragmatic breathing
   (7) atelectasis
   (8) compliance
   (9) hypoxia
   (10) hypercapnia

Choose the best answer to the following questions:

9. Which of the following statements are correct? (1) Normal expiration during quiet breathing is an active process involving intensive muscle contraction. (2) Passive expiration results from elastic recoil of the chest wall and lungs. (3) Ventilation is due to a pressure gradient between the lungs and the atmospheric air. (4) During normal breathing, the pressure between the two pleural layers (intrapleural pressure) is always subatmospheric. (5) Boyle's law—the pressure of a gas in a closed container is directly proportional to the volume of the container—explains ventilation.
   (a) 1, 2, and 3, (b) 2, 3, and 4, (c) 3, 4, and 5, (d) 1, 3, and 5, (e) 2, 3, and 5
10. Which of the following factors affect the rate of external respiration? (1) partial pressure differences of the gases, (2) surface area for gas exchange, (3) diffusion distance, (4) solubility and molecular weight of the gases, (5) presence of surfactant.
   (a) 1, 2, and 3, (b) 2, 4, and 5, (c) 1, 2, 4, and 5, (d) 1, 2, 3, and 4, (e) 2, 3, 4, and 5
11. The most important factor in determining the percent oxygen saturation of hemoglobin is (a) the partial pressure of oxygen, (b) acidity, (c) the partial pressure of carbon dioxide, (d) temperature, (e) BPG
12. The lung volume that provides a medical or legal tool for determining whether a baby was born dead or died after birth is the (a) tidal volume, (b) residual volume, (c) minimal volume, (d) expiratory reserve volume, (e) inspiratory reserve volume
13. Which gas law that states that each gas in a mixture exerts its own pressure as if all the other gases were not present? (a) Boyle's law, (b) Henry's law, (c) Dalton's law, (d) the Haldane effect, (e) the Bohr effect

14. Which of the following statements are true? (1) It is impossible for people to kill themselves by holding their breath. (2) The cerebral cortex provides voluntary control over breathing. (3) Emotional stimuli can alter respiration. (4) Certain chemical stimuli determine the rate and depth of breathing. (5) At the onset of exercise, an abrupt increase in pulmonary ventilation is followed by a more gradual increase in ventilation.
(a) 1, 2, 4, and 5, (b) 2, 3, and 5, (c) 1, 3, and 5, (d) 2, 3, 4, and 5, (e) 1, 2, 3, 4, and 5

15. Match the following:
___ (a) prevents excessive inflation of the lungs
___ (b) controls the basic rhythm of respiration
___ (c) sends stimulatory impulses to the inspiratory area that activate it and prolong inspiration, thus inhibiting expiration
___ (d) as acidity increases, the affinity of hemoglobin for oxygen decreases and oxygen dissociates more readily from hemoglobin
___ (e) transmits inhibitory impulses to the inspiratory area of the medulla oblongata to turn off the inspiratory area before the lungs become too full of air
___ (f) relates to the partial pressure of a gas in a mixture of gases

(1) Bohr effect
(2) Dalton's law
(3) medullary rhythmicity area
(4) pneumotaxic area
(5) apneustic area
(6) Hering–Breuer reflex

CRITICAL THINKING QUESTIONS

1. Leung, a member of the college swim team, was measuring his lung volumes during his A & P lab. Although his tidal volume was the average value for a male of his age and size, his inspiratory reserve volume was the highest in the class. List the normal male values for lung volumes that can be measured with spirometer. Would you expect the average female volumes to be the same? Why or why not? (HINT: *Total lung capacity includes a volume that cannot be measured by spirometry.*)

2. Samir stepped up to the plate and lifted his bat. The pitch was wild, and the ball hit him right in the nose! The x-ray revealed a break in both medial bones of the external nose. Describe the structure of the external nose, and identify the location of the break. (HINT: *Samir may need a rhinoplasty to repair his nose.*)

3. Her mother calls Kirsten a "challenging personality," but you have some other adjectives in mind. She's threatening to hold her breath until she, as she puts it, "turns blue, falls down, and dies—and won't you be sorry!" Do you need to worry about the "dies" part? (HINT: *Only some aspects of respiration may be partly under voluntary control.*)

ANSWERS TO FIGURE QUESTIONS

23.1 The conducting portion of the respiratory system includes the nose, pharynx, larynx, trachea, bronchi, and bronchioles (except the respiratory bronchioles).

23.2 Air's path is external nares → vestibule → nasal cavity → internal nares.

23.3 The root of the nose attaches it to the frontal bone.

23.4 The superior border of the pharynx is the internal nares; the inferior border of the pharynx is the cricoid cartilage.

23.5 During swallowing, the epiglottis closes over the rima glottidis, the entrance to the trachea.

23.6 The main function of the vocal folds is voice production.

23.7 Because the tissues between the esophagus and trachea are soft, the esophagus can bulge and press against the trachea during swallowing.

23.8 The left lung has two lobes and secondary bronchi; the right lung has three of each.

23.9 The pleural membrane is a serous membrane.

23.10 Because two-thirds of the heart lies to the left of the midline, the left lung contains a cardiac notch to accommodate the presence of the heart.

23.11 The wall of an alveolus is made up of simple squamous epithelium.

23.12 The respiratory membrane averages 0.5 μm in thickness.

23.13 The pressure would increase to 4 atm.

23.14 If you are at rest while reading, your diaphragm is the most active muscle of respiration.

23.15 At the start of inspiration, intrapleural pressure is about 756 mm Hg. With contraction of the diaphragm, it decreases to about 754 mm Hg as the volume of the space between the two pleural layers expands. With relaxation of the diaphragm, it increases back to 756 mm Hg.

23.16 Normal atmospheric pressure at sea level is 760 mm Hg.

23.17 Breathing in and then exhaling as much air as possible demonstrates vital capacity.

23.18 In both places, a difference in P_{O_2} promotes oxygen diffusion.

23.19 During strenuous exercise, oxygenated blood contains 20 mL of O_2 per 100 mL of blood (the same as at rest).

23.20 In both cases, hemoglobin in your pulmonary veins would be fully saturated with O_2, a point which is at the upper right of the curve.

23.21 Because lactic acid (lactate) and CO_2 are produced by active skeletal muscles, blood pH decreases slightly and P_{CO_2} increases when you are actively exercising. The result is lowered affinity of hemoglobin for O_2, so more O_2 is available to the working muscles.

23.22 O_2 is more available when you have a fever because the affinity of hemoglobin for O_2 decreases with increasing temperature.

23.23 Fetal Hb is 80% saturated with O_2, whereas maternal Hb is about 75% saturated at a P_{O_2} of 40 mm Hg.

23.24 Blood in a vein would have a higher concentration of HCO_3^-.

23.25 The medullary inspiratory area contains autorhythmic neurons.

23.26 The phrenic nerves innervate the diaphragm.

23.27 Normal arterial P_{CO_2} is 40 mm Hg.

23.28 One could administer air enriched in O_2, with artificial ventilation if the person has stopped breathing. Even mouth-to-mouth resuscitation might save the person.

23.29 The respiratory system begins to develop at about 4 weeks of gestation.

THE DIGESTIVE SYSTEM

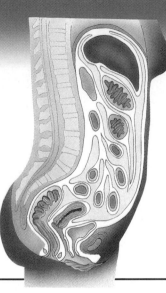

Food contains a variety of nutrients—molecules needed for building new body tissues, repairing damaged tissues, and sustaining needed chemical reactions. Food is also vital for life because it is the source of energy that drives the chemical reactions occurring in every cell. As consumed, however, most food cannot be used as a source of cellular energy. First, it must be broken down into molecules small enough to cross the plasma membranes of cells, a process known as **digestion.** The passage of these smaller molecules through cells into the blood and lymph is termed **absorption.** The organs that collectively perform these functions—the **digestive system**—are the focus of this chapter.

The medical specialty that deals with the structure, function, diagnosis, and treatment of diseases of the stomach and intestines is called **gastroenterology** (gas′-trō-en′-ter-OL-ō-jē; *gastro-* = stomach; *enter-* = intestines; *-ology* = study of). The medical specialty that deals with the diagnosis and treatment of disorders of the rectum and anus is called **proctology** (prok-TOL-ō-jē; *proct-* = rectum).

OVERVIEW OF THE DIGESTIVE SYSTEM

OBJECTIVES

• *Identify the organs of the digestive system.*

• *Describe the basic processes performed by the digestive system.*

The digestive system (Figure 24.1) is composed of two groups of organs: the gastrointestinal (GI) tract and the ac-

cessory digestive organs. The **gastrointestinal (GI) tract,** or **alimentary canal** (*alimentary* = nourishment), is a continuous tube that extends from the mouth to the anus through the ventral body cavity. Organs of the gastrointestinal tract include the mouth, most of the pharynx, esophagus, stomach, small intestine, and large intestine. The length of the GI tract taken from a cadaver is about 9 m (30 ft). In a living person it is shorter because the muscles along the walls of GI tract organs are in a state of tone (sustained contraction). The **accessory digestive organs** are the teeth, tongue, salivary glands, liver, gallbladder, and pancreas. Teeth aid in the physical breakdown of food, and the tongue assists in chewing and swallowing. The other accessory digestive organs, however, never come into direct contact with food. They produce or store secretions that flow into the GI tract through ducts and aid in the chemical breakdown of food.

The GI tract contains food from the time it is eaten until it is digested and absorbed or eliminated. Muscular contractions in the wall of the GI tract physically break down the food by churning it. The contractions also help to dissolve foods by mixing them with fluids secreted into the tract. Enzymes secreted by accessory structures and cells that line the tract break down the food chemically. Wavelike contractions of the smooth muscle in the wall of the GI tract propel the food along the tract, from the esophagus to the anus.

Overall, the digestive system has six basic functions:

1. *Ingestion.* This process involves taking foods and liquids into the mouth (eating).

Figure 24.1 Organs of the digestive system.

🔑 **Organs of the gastrointestinal (GI) tract are the mouth, pharynx, esophagus, stomach, small intestine, and large intestine. Accessory digestive organs are the teeth, tongue, salivary glands, liver, gallbladder, and pancreas.**

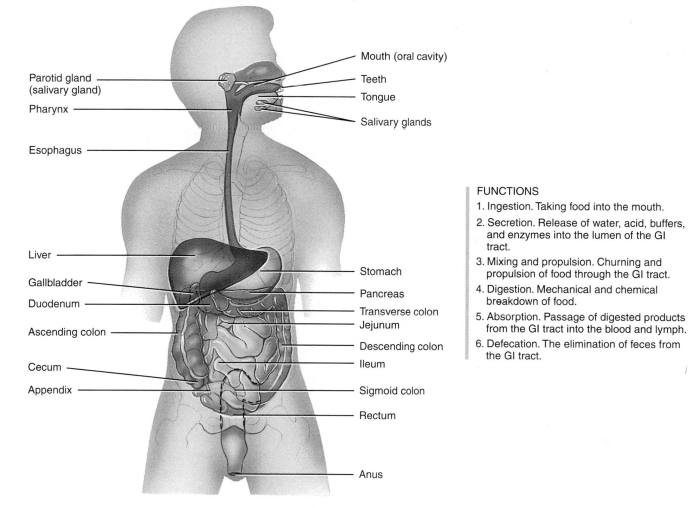

Right lateral view of head and neck and anterior view of trunk

FUNCTIONS
1. Ingestion. Taking food into the mouth.
2. Secretion. Release of water, acid, buffers, and enzymes into the lumen of the GI tract.
3. Mixing and propulsion. Churning and propulsion of food through the GI tract.
4. Digestion. Mechanical and chemical breakdown of food.
5. Absorption. Passage of digested products from the GI tract into the blood and lymph.
6. Defecation. The elimination of feces from the GI tract.

Q Which structures of the digestive system secrete digestive enzymes?

2. *Secretion.* Each day, cells within the walls of the GI tract and accessory organs secrete a total of about 7 liters of water, acid, buffers, and enzymes into the lumen of the tract.
3. *Mixing and propulsion.* Alternating contraction and relaxation of smooth muscle in the walls of the GI tract mix food and secretions and propel them toward the anus. This capability of the GI tract to mix and move material along its length is termed **motility.**
4. *Digestion.* Mechanical and chemical processes break down ingested food into small molecules. In **mechanical digestion** the teeth cut and grind food before it is swallowed, and then smooth muscles of the stomach and small intestine churn the food. As a result, food molecules

become dissolved and thoroughly mixed with digestive enzymes. In **chemical digestion** the large carbohydrate, lipid, protein, and nucleic acid molecules in food are split into smaller molecules by hydrolysis (see Figure 2.16 on page 44). Digestive enzymes produced by the salivary glands, tongue, stomach, pancreas, and small intestine catalyze these catabolic reactions. A few substances in food can be absorbed without chemical digestion, including amino acids, cholesterol, glucose, vitamins, minerals, and water.
5. *Absorption.* During absorption the secreted fluids and the small molecules and ions that are products of digestion enter the epithelial cells lining the lumen of the GI tract

Figure 24.2 Three-dimensional depiction of the various layers of the gastrointestinal tract.

🗝 **The four layers of the GI tract, from deep to superficial, are the mucosa, submucosa, muscularis, and serosa.**

Nerve

Artery

Mesentery

Myenteric plexus (plexus of Auerbach)

Submucosal plexus (plexus of Meissner)

Gland in mucosa

Duct of gland outside tract (such as pancreas)

Mucosa associated lymphoid tissue (MALT)

Lumen

MUCOSA:
Epithelium
Lamina propia
Muscularis mucosae

SUBMUCOSA

Gland in submucosa

MUSCULARIS:
Longitudinal muscle
Circular muscle

SEROSA:
Connective tissue
Epithelium

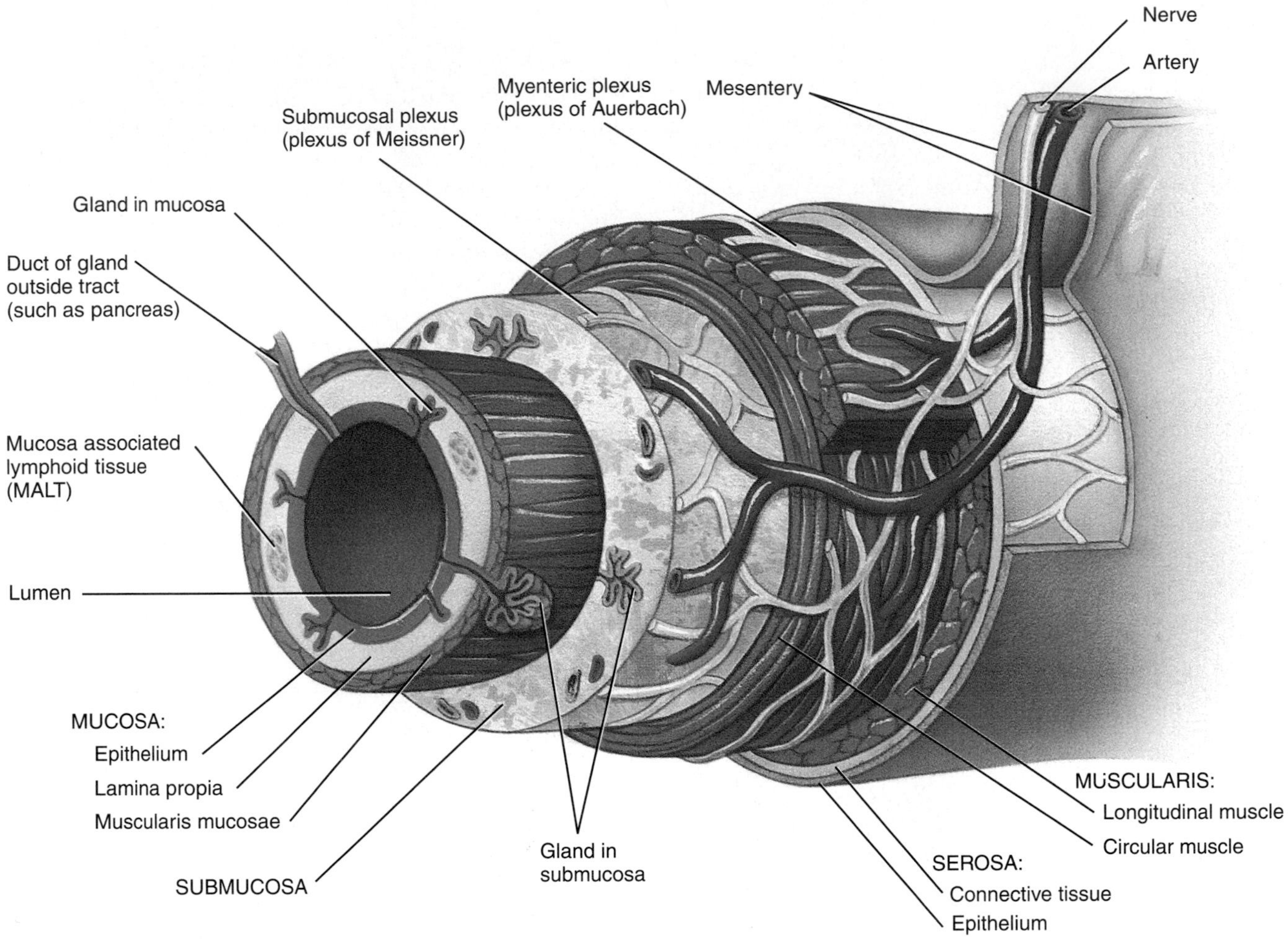

Q What is the function of the enteric plexuses in the wall of the gastrointestinal tract?

either via active transport or via passive diffusion. The absorbed substances pass into blood or lymph and circulate to cells throughout the body.

6. *Defecation.* Wastes, indigestible substances, bacteria, cells sloughed from the lining of the GI tract, and digested materials that were not absorbed leave the body through the anus in a process called **defecation.** The eliminated material is termed **feces.**

1. List the GI tract organs and the accessory digestive organs.
2. Which organs of the digestive system come in contact with food, and what are some of their digestive functions?

3. Which kinds of food molecules undergo chemical digestion, and which do not?

LAYERS OF THE GI TRACT

OBJECTIVE

• *Describe the layers that form the wall of the gastrointestinal tract.*

The wall of the GI tract, from the esophagus to the anal canal, has the same basic, four-layered arrangement of tissues. The four layers of the tract, from deep to superficial, are the mucosa, submucosa, muscularis, and serosa (Figure 24.2).

Mucosa

The lumen of the GI tract is lined by a mucous membrane, the **mucosa.** GI tract mucosa contains three layers: (1) a lining of epithelium in direct contact with the contents of the tract, (2) an underlying layer of areolar connective tissue, and (3) a thin layer of smooth muscle.

1. The **epithelium** in the mouth, pharynx, esophagus, and anal canal is mainly nonkeratinized stratified squamous epithelium that serves a protective function. Simple columnar epithelium, which functions in secretion and absorption, lines the stomach and intestines. Neighboring simple columnar epithelial cells are firmly sealed to each other by tight junctions that restrict leakage between the cells. The rate of renewal of GI tract epithelial cells is rapid: Every 5–7 days they slough off and are replaced by new cells. Located among the absorptive epithelial cells are exocrine cells that secrete mucus and fluid into the lumen of the tract, and several types of endocrine cells, collectively called **enteroendocrine cells,** that secrete hormones into the bloodstream.

2. The **lamina propria** (*lamina* = thin, flat plate) is areolar connective tissue containing many blood and lymphatic vessels, which are the routes by which nutrients absorbed into the GI tract reach the other tissues of the body. This layer supports the epithelium and binds it to the muscularis mucosae (discussed next). The lamina propria also contains most of the cells of the **mucosa-associated lymphoid tissue (MALT).** These prominent lymphatic nodules contain immune system cells that protect against disease. MALT is present all along the GI tract, especially in the tonsils, small intestine, appendix, and large intestine, and it contains about as many immune cells as are present in all the rest of the body. The lymphocytes and macrophages in MALT mount immune responses against microbes, such as bacteria, that may penetrate the epithelium.

3. A thin layer of smooth muscle fibers called the **muscularis mucosae** throws the mucous membrane of the stomach and small intestine into many small folds, which increase the surface area for digestion and absorption. Movements of the muscularis mucosae ensure that all absorptive cells are fully exposed to the contents of the GI tract.

Submucosa

The **submucosa** consists of areolar connective tissue that binds the mucosa to the third layer, the muscularis. It is highly vascular and contains the **submucosal plexus** or *plexus of Meissner,* a portion of the **enteric nervous system (ENS).** The ENS is the "brain of the gut" and consists of approximately 100 million neurons in two enteric plexuses that extend the entire length of the GI tract. The submucosal plexus contains sensory and motor enteric neurons, plus parasympathetic and sympathetic postganglionic fibers that innervate the mucosa and submucosa. The plexus regulates movements of the mucosa and vasoconstriction of blood vessels. Because it also innervates secretory cells of mucosal glands, it is important in controlling secretions by the GI tract. The submucosa may also contain glands and lymphatic tissue.

Muscularis

The **muscularis** of the mouth, pharynx, and superior and middle parts of the esophagus contains *skeletal muscle* that produces voluntary swallowing. Skeletal muscle also forms the external anal sphincter, which permits voluntary control of defecation. Throughout the rest of the tract, the muscularis consists of *smooth muscle* that is generally found in two sheets: an inner sheet of circular fibers and an outer sheet of longitudinal fibers. Involuntary contractions of the smooth muscles help break down food physically, mix it with digestive secretions, and propel it along the tract. The muscularis also contains the second plexus of the enteric nervous system—the **myenteric plexus** (*my-* = muscle) or *plexus of Auerbach,* which contains enteric neurons, parasympathetic ganglia and postganglionic fibers, and sympathetic postganglionic fibers that innervate the muscularis. This plexus mostly controls GI tract motility, in particular the frequency and strength of contraction of the muscularis.

Serosa

The **serosa** is the superficial layer of those portions of the GI tract that are suspended in the abdominopelvic cavity. It is a serous membrane composed of connective tissue and simple squamous epithelium. As we will see shortly, the esophagus, which passes through the mediastinum, has a superficial layer called the *adventitia* composed of areolar connective tissue. Inferior to the diaphragm, the serosa is also called the **visceral peritoneum;** it forms a portion of the peritoneum, which we examine in detail next.

1. Where along the GI tract is the muscularis composed of skeletal muscle? Is control of this skeletal muscle voluntary or involuntary?
2. What two plexuses form the enteric nervous system, and where are they located?

PERITONEUM

OBJECTIVE

• *Describe the peritoneum and its folds.*

The **peritoneum** (per′-i-tō-NĒ-um; *peri-* = around) is the largest serous membrane of the body; it consists of a layer of simple squamous mesothelium with an underlying supporting layer of connective tissue. Whereas the **parietal peritoneum** lines the wall of the abdominopelvic cavity, the **visceral peritoneum** covers some of the organs in the cavity and is their serosa (Figure 24.3a). The slim space between the parietal and visceral portions of the peritoneum is called the

Figure 24.3 Relationship of the peritoneal folds to each other and to organs of the digestive system. The size of the peritoneal cavity is exaggerated for emphasis.

🔑 **The peritoneum is the largest serous membrane in the body.**

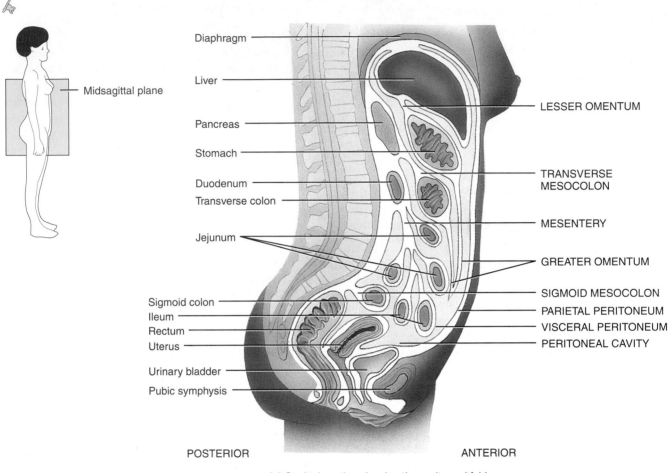

(a) Sagittal section showing the peritoneal folds

peritoneal cavity, which contains serous fluid. In certain diseases, the peritoneal cavity may become distended by the accumulation of several liters of fluid, a condition called **ascites** (a-SĪ-tēz).

As we will see, some organs lie on the posterior abdominal wall and are covered by peritoneum on their anterior surfaces only. Such organs, including the kidneys and pancreas, are said to be **retroperitoneal** (*retro-* = behind).

Unlike the pericardium and pleurae, which smoothly cover the heart and lungs, the peritoneum contains large folds that weave between the viscera. The folds bind the organs to each other and to the walls of the abdominal cavity and contain blood and lymphatic vessels and nerves that supply the abdominal organs. One fold of the peritoneum, called the **mesentery** (MEZ-en-ter′-ē; *mes-* = middle), is an outward fold of the serous coat of the small intestine (see Figure 24.3a, d); the tip of the fold binds the small intestine to the posterior abdominal wall. A fold of peritoneum called the **mesocolon** (mez′-ō-KŌ-lon) binds the large intestine to

the posterior abdominal wall; it also carries blood and lymphatic vessels to the intestines. The mesentery and mesocolon hold the intestines loosely in place, allowing for a great amount of movement as muscular contractions mix and move the luminal contents along the GI tract.

Other important peritoneal folds are the falciform ligament, lesser omentum, and greater omentum. The **falciform ligament** (FAL-si-form; *falc-* = sickle-shaped) attaches the liver to the anterior abdominal wall and diaphragm (Figure 24.3b). (The liver is the only digestive organ that is attached to the anterior abdominal wall.) The **lesser omentum** (ō-MENT-um; = fat skin) arises as two folds in the serosa of the stomach and duodenum, and it suspends the stomach and duodenum from the liver (Figure 24.3c). It contains some lymph nodes. The **greater omentum,** the largest peritoneal fold, hangs loosely like a "fatty apron" over the transverse colon and coils of the small intestine (see Figure 24.3b, d). It is a double sheet that folds back upon itself, and thus it is a four-layered structure. From at-

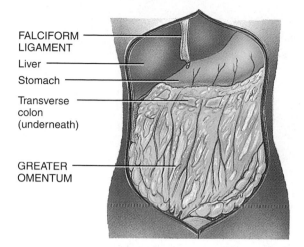

FALCIFORM LIGAMENT
Liver
Stomach
Transverse colon (underneath)
GREATER OMENTUM

(b) Greater omentum, anterior view

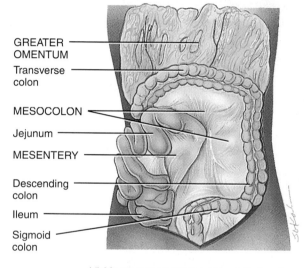

GREATER OMENTUM
Transverse colon
MESOCOLON
Jejunum
MESENTERY
Descending colon
Ileum
Sigmoid colon

(d) Mesentery and mesocolon, anterior view
(greater omentum lifted and small intestine reflected to right side)

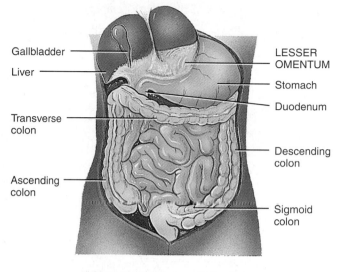

Gallbladder
Liver
Transverse colon
Ascending colon
LESSER OMENTUM
Stomach
Duodenum
Descending colon
Sigmoid colon

(c) Lesser omentum, anterior view
(liver and gallbladder lifted)

Q Which peritoneal fold binds the small intestine to the posterior abdominal wall?

example, bacteria gain access to the peritoneal cavity through an intestinal perforation or rupture of the appendix, they can produce an acute, life-threatening form of peritonitis. A less serious (but still painful) form of peritonitis can result from the rubbing together of inflamed peritoneal surfaces. ■

1. Describe the locations of the visceral peritoneum and parietal peritoneum.
2. Describe the attachment sites and functions of the mesentery, mesocolon, falciform ligament, lesser omentum, and greater omentum.

MOUTH

OBJECTIVES

- *Identify the locations of the salivary glands, and describe the functions of their secretions.*
- *Describe the structure and functions of the tongue.*
- *Identify the parts of a typical tooth, and compare deciduous and permanent dentitions.*

The **mouth,** also referred to as the **oral** or **buccal cavity** (BUK-al; *bucca* = cheeks), is formed by the cheeks, hard and soft palates, and tongue (Figure 24.4). Forming the lateral walls of the oral cavity are the **cheeks**—muscular structures covered externally by skin and internally by nonkeratinized stratified squamous epithelium. The anterior portions of the cheeks end at the lips.

The **lips** or **labia** (= fleshy borders) are fleshy folds sur-

tachments along the stomach and duodenum, the greater omentum extends downward anterior to the small intestine, then turns and extends upward and attaches to the transverse colon. The greater omentum contains considerable adipose tissue and many lymph nodes. It contributes macrophages and antibody-producing plasma cells that help combat an infection of the GI tract and prevent the infection from spreading.

CLINICAL APPLICATION
Peritonitis

Peritonitis is an acute inflammation of the peritoneum. A common cause of the condition is contamination of the peritoneum by infectious microbes, which can result from accidental or surgical wounds in the abdominal wall, or from perforation or rupture of abdominal organs. If, for

Figure 24.4 Structures of the mouth (oral cavity).

🔑 **The mouth is formed by the cheeks, hard and soft palates, and tongue.**

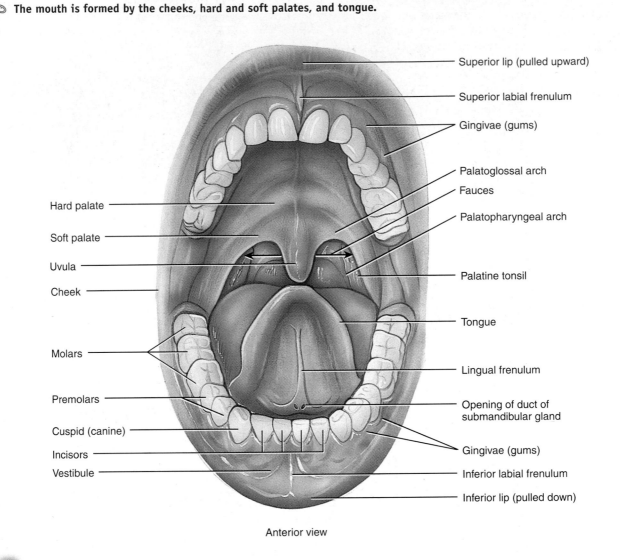

Superior lip (pulled upward)

Superior labial frenulum

Gingivae (gums)

Palatoglossal arch

Fauces

Palatopharyngeal arch

Palatine tonsil

Tongue

Lingual frenulum

Opening of duct of submandibular gland

Gingivae (gums)

Inferior labial frenulum

Inferior lip (pulled down)

Hard palate

Soft palate

Uvula

Cheek

Molars

Premolars

Cuspid (canine)

Incisors

Vestibule

Anterior view

Q What is the function of the uvula?

rounding the opening of the mouth. They are covered externally by skin and internally by a mucous membrane. There is a transition zone where the two kinds of covering tissue meet. This portion of the lips is nonkeratinized, and the color of the blood in the underlying blood vessels is visible through the transparent surface layer. The inner surface of each lip is attached to its corresponding gum by a midline fold of mucous membrane called the **labial frenulum** (LĀ-bē-al FREN-yoo-lum; *frenulum* = small bridle).

The orbicularis oris muscle and connective tissue lie between the skin and the mucous membrane of the oral cavity. During chewing, contraction of the buccinator muscles in the cheeks and orbicularis oris muscle in the lips help keep food between the upper and lower teeth. These muscles also assist in speech.

The **vestibule** (= entrance to a canal) of the oral cavity

is a space bounded externally by the cheeks and lips and internally by the gums and teeth. The **oral cavity proper** is a space that extends from the gums and teeth to the **fauces** (FAW-sēs; = passages), the opening between the oral cavity and the pharynx or throat.

The **hard palate**—the anterior portion of the roof of the mouth—is formed by the maxillae and palatine bones, is covered by mucous membrane, and forms a bony partition between the oral and nasal cavities. The **soft palate,** which forms the posterior portion of the roof of the mouth, is an arch-shaped muscular partition between the oropharynx and nasopharynx that is lined by mucous membrane.

Hanging from the free border of the soft palate is a conical muscular process called the **uvula** (YOU-vyoo-la; = little grape). During swallowing, the soft palate and uvula are drawn superiorly, closing off the nasopharynx and preventing swallowed foods and liquids from entering the nasal cavity. Lateral to the base of the uvula are two muscular folds that run down the lateral sides of the soft palate: Anteriorly, the **palatoglossal arch** extends to the side of the base of the tongue; posteriorly, the **palatopharyngeal arch** (PAL-a-tō-fa-rin'-jē-al) extends to the side of the pharynx. The palatine tonsils are situated between the arches, and the lingual tonsils are situated at the base of the tongue. At the posterior border of the soft palate, the mouth opens into the oropharynx through the fauces (see Figure 24.4).

Structure and Function of the Salivary Glands

A **salivary gland** is any cell or organ that releases a secretion called saliva into the oral cavity. Ordinarily, just enough saliva is secreted to keep the mucous membranes of the mouth and pharynx moist and to cleanse the mouth and teeth. When food enters the mouth, however, secretion of saliva increases, and it lubricates, dissolves, and begins the chemical breakdown of the food.

The mucous membrane of the mouth and tongue contains many small salivary glands that open directly, or indirectly via short ducts, to the oral cavity. These glands include *labial, buccal,* and *palatal glands* in the lips, cheeks, and palate, respectively, and *lingual glands* in the tongue, all of which make a small contribution to saliva. However, most saliva is secreted by the **major salivary glands,** which lie beyond the oral mucosa. Their secretions empty into ducts that lead to the oral cavity.

There are three pairs of major salivary glands: the parotid, submandibular, and sublingual glands (Figure 24.5a). The **parotid glands** (*par-* = near; *ot-* = ear) are located inferior and anterior to the ears, between the skin and the masseter muscle. Each secretes saliva into the oral cavity via a **parotid duct** that pierces the buccinator muscle to open into the vestibule opposite the second maxillary (upper) molar tooth. The **submandibular glands** are found beneath the base of the tongue in the posterior part of the floor of the mouth. Their ducts, the **submandibular ducts,** run under the mucosa on either side of the midline of the floor of the mouth and enter the oral cavity proper lateral to the lingual frenulum. The **sublingual glands** are superior to the submandibular glands. Their ducts, the **lesser sublingual ducts,** open into the floor of the mouth in the oral cavity proper.

Composition and Functions of Saliva

Chemically, **saliva** is 99.5% water and 0.5% solutes. Among the solutes are ions, including sodium, potassium, chloride, bicarbonate, and phosphate. Also present are some dissolved gases and various organic substances, including urea and uric acid, mucus, immunoglobulin A, the bacteriolytic enzyme lysozyme, and two digestive enzymes—salivary amylase, which acts on starch, and lingual lipase, which acts on triglycerides.

Each major salivary gland supplies different proportions of ingredients to saliva. The parotid glands contain cells that secrete a watery (serous) liquid containing salivary amylase. Because the submandibular glands contain cells similar to those found in the parotid glands, plus some mucous cells, they secrete a fluid that contains amylase but is thickened with mucus. The sublingual glands contain mostly mucous cells, so they secrete a much thicker fluid that contributes only a small amount of amylase to the saliva.

The water in saliva provides a medium for dissolving foods so that they can be tasted and digestive reactions can begin. Chloride ions in the saliva activate salivary amylase. Bicarbonate and phosphate ions buffer acidic foods that enter the mouth; as a result, saliva is only slightly acidic (pH 6.35–6.85). Urea and uric acid are found in saliva because salivary glands (like the sweat glands of the skin) help remove waste molecules from the body. Mucus lubricates the food so it can easily be moved about in the mouth, formed into a ball, and swallowed. Immunoglobulin A is a type of antibody that inhibits bacterial growth, and the enzyme lysozyme kills bacteria. Even though these substances help protect the mucous membrane from infection and the teeth from decay, they are not present in large enough quantities to eliminate all oral bacteria.

Salivation

Secretion of saliva, or **salivation** (sal-i-VĀ-shun), is controlled by the nervous system. Amounts of saliva secreted daily vary considerably but average 1000–1500 mL (1–1.6 qt). Normally, parasympathetic stimulation promotes continuous secretion of a moderate amount of saliva, which keeps the mucous membranes moist and lubricates the movements of the tongue and lips during speech. The saliva is then swallowed and helps moisten the esophagus. Eventually, most components of saliva are reabsorbed, which prevents fluid loss. Sympathetic stimulation dominates during stress, resulting in dryness of the mouth. During dehydration, the salivary glands stop secreting saliva to conserve water; the resulting dryness in the mouth contributes to the sensation of thirst. Drinking will then not only restore the homeostasis of body water but also moisten the mouth.

The touch and taste of food also are potent stimulators of salivary gland secretions. Chemicals in the food stimulate receptors in taste buds on the tongue, and impulses are conveyed from the taste buds to two salivary nuclei in the brain stem. Returning parasympathetic impulses in fibers of the facial (VII) and glossopharyngeal (IX) nerves stimulate the secretion of saliva. Saliva continues to be heavily secreted for some time after food is swallowed; this flow of saliva washes out the mouth and dilutes and buffers the remnants of irritating chemicals.

The smell, sight, sound, or thought of food may also stimulate secretion of saliva. These stimuli constitute psychological activation and involve learned behavior. When memories that associate the stimuli with food occur, nerve impulses propagate from the cortex to the nuclei in the brain

Figure 24.5 Major salivary glands. The submandibular gland shown in the photomicrograph in (b) consists mostly of serous acini (serous-fluid–secreting portions of gland) and a few mucous acini (mucus-secreting portions of gland); the parotid glands consist of serous acini only, and the sublingual glands consist of mostly mucous acini and a few serous acini. (See Tortora, *A Photographic Atlas of the Human Body,* Figure 12.6)

🔑 **Saliva lubricates and dissolves foods and begins the chemical breakdown of carbohydrates and lipids.**

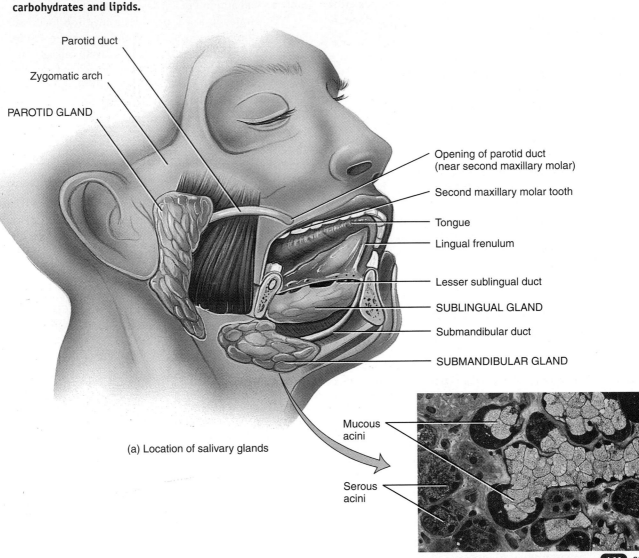

(a) Location of salivary glands

(b) Submandibular gland

LM 350x

Q What is the function of the chloride ions in saliva?

stem, and the salivary glands are activated. Psychological activation of the glands has some benefit to the body because it allows chemical digestion to start in the mouth as soon as the food is ingested. Salivation also occurs in response to swallowing irritating foods or during nausea due to reflexes originating in the stomach and upper small intestine. This mechanism presumably helps dilute or neutralize the irritating substance.

CLINICAL APPLICATION
Mumps

Although any of the salivary glands may be the target of a nasopharyngeal infection, the mumps virus (myxovirus) typically attacks the parotid glands. **Mumps** is an inflammation and enlargement of the parotid glands accompanied by moderate fever, malaise (general discomfort), and extreme pain in the throat, especially when swallowing sour foods or acidic juices. Swelling occurs on one or both sides of the face, just anterior to the ramus of the mandible. In about 30% of males past puberty, the testes may also become inflamed; sterility rarely occurs because testicular involvement is usually unilateral (one testis only). Since a vaccine became available for mumps in 1967, the incidence of the disease has declined. ■

Structure and Function of the Tongue

The **tongue** is an accessory digestive organ composed of skeletal muscle covered with mucous membrane. Together with its associated muscles, it forms the floor of the oral cavity. The tongue is divided into symmetrical lateral halves by a median septum that extends its entire length, and it is attached inferiorly to the hyoid bone, styloid process of the temporal bone, and mandible. Each half of the tongue consists of an identical complement of extrinsic and intrinsic muscles.

The **extrinsic muscles** of the tongue, which originate outside the tongue (attach to bones in the area) and insert into connective tissues in the tongue, include the hyoglossus, genioglossus, and styloglossus muscles (see Figure 11.7 on page 320). The extrinsic muscles move the tongue from side to side and in and out to maneuver food for chewing, shape the food into a rounded mass, and force the food to the back of the mouth for swallowing. They also form the floor of the mouth and hold the tongue in position. The **intrinsic muscles** originate in and insert into connective tissue within the tongue and alter the shape and size of the tongue for speech and swallowing. The intrinsic muscles include the longitudinalis superior, longitudinalis inferior, transversus linguae, and verticalis linguae muscles. The **lingual frenulum** (*lingua* = the tongue), a fold of mucous membrane in the midline of the undersurface of the tongue, is attached to the floor of the mouth and aids in limiting the movement of the tongue posteriorly (see Figures 24.4 and 24.5). If a person's lingual frenulum is abnormally short or rigid—a condition called **ankyloglossia** (ang'-kē-lō-GLOSS-ē-a)—then eating and speaking are impaired such that the person is said to be "tongue-tied."

The dorsum (upper surface) and lateral surfaces of the tongue are covered with **papillae** (pa-PIL-ē; = nipple-shaped projections), projections of the lamina propria covered with keratinized epithelium (see Figure 16.2a on page 515). Many papillae contain taste buds, the receptors for gustation (taste). **Fungiform papillae** (FUN-ji-form; = shaped like a mushroom) are mushroomlike elevations distributed among the filiform papillae that are more numerous near the tip of the tongue. They appear as red dots on the surface of the tongue, and most of them contain taste buds. **Circumvallate papillae** (*circum-* = around; *vall-* = wall) are arranged in an inverted V shape on the posterior surface of the tongue; all of them contain taste buds. **Filiform papillae** (FIL-i-form; = threadlike) are whitish, conical projections distributed in parallel rows over the anterior two-thirds of the tongue. Although filiform papillae lack taste buds, they increase friction between the tongue and food,

making it easier for the tongue to move food particles within the oral cavity. **Lingual glands** in the lamina propria secrete both mucus and a watery serous fluid that contains lingual lipase into saliva.

Structure and Function of the Teeth

The **teeth,** or **dentes** (Figure 24.6), are accessory digestive organs located in sockets of the alveolar processes of the mandible and maxillae. The alveolar processes are covered by the **gingivae** (jin-JI-vē), or gums, which extend slightly into each socket to form the gingival sulcus. The sockets are lined by the **periodontal ligament** (*odont-* = tooth), which consists of dense fibrous connective tissue and is attached to the socket walls and the cemental surface of the roots. Thus, it anchors the teeth in position and acts as a shock absorber during chewing.

A typical tooth consists of three principal regions. The **crown** is the visible portion above the level of the gums. Embedded in the socket are one to three **roots.** The **neck** is the constricted junction of the crown and root near the gum line.

Teeth are composed primarily of **dentin,** a calcified connective tissue that gives the tooth its basic shape and rigidity. It is harder than bone because of its higher content of calcium salts (70% of dry weight). The dentin encloses a cavity. The enlarged part of the cavity, the **pulp cavity,** lies within the crown and is filled with **pulp,** a connective tissue containing blood vessels, nerves, and lymphatic vessels. Narrow extensions of the pulp cavity, called **root canals,** run through the root of the tooth. Each root canal has an opening at its base, the **apical foramen,** through which blood vessels, lymphatic vessels, and nerves extend.

The dentin of the crown is covered by **enamel** that consists primarily of calcium phosphate and calcium carbonate. Enamel, the hardest substance in the body and the richest in calcium salts (about 95% of dry weight), protects the tooth from the wear of chewing. It is also a barrier against acids that easily dissolve the dentin. The dentin of the root is covered by **cementum,** another bonelike substance, which attaches the root to the periodontal ligament.

The branch of dentistry that is concerned with the prevention, diagnosis, and treatment of diseases that affect the pulp, root, periodontal ligament, and alveolar bone is known as **endodontics** (en'-dō-DON-tiks; *endo-* = within). **Orthodontics** (or'-thō-DON-tiks; *ortho-* = straight) is a branch of dentistry that is concerned with the prevention and correction of abnormally aligned teeth, whereas **periodontics** (per'-ē-ō-DON-tiks) is a branch of dentistry concerned with the treatment of abnormal conditions of the tissues immediately surrounding the teeth.

Humans have two **dentitions,** or sets of teeth: deciduous and permanent. The first of these—the **deciduous teeth** (*decidu-* = falling out), also called **primary teeth, milk teeth,** or **baby teeth**—begin to erupt at about 6 months of age, and one pair of teeth appears at about each month

Figure 24.6 A typical tooth and surrounding structures.

 Teeth are anchored in sockets of the alveolar processes of the mandible and maxillae.

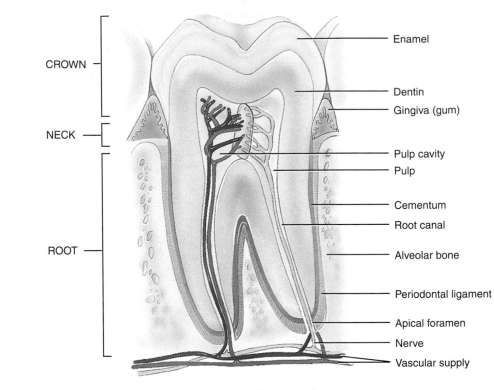

Sagittal section of a mandibular (lower) molar

Q What type of tissue is the main component of teeth?

thereafter, until all 20 are present (Figure 24.7a). The incisors, which are closest to the midline, are chisel-shaped and adapted for cutting into food. They are referred to as either **central** or **lateral incisors** on the basis of their position. Next to the incisors, moving posteriorly, are the **cuspids (canines),** which have a pointed surface called a cusp. Cuspids are used to tear and shred food. Incisors and cuspids have only one root apiece. Posterior to them lie the **first** and **second molars,** which have four cusps. Maxillary (upper) molars have three roots; mandibular (lower) molars have two roots. The molars crush and grind food.

All the deciduous teeth are lost—generally between ages 6 and 12 years—and are replaced by the **permanent (secondary) teeth** (Figure 24.7b). The permanent dentition contains 32 teeth that erupt between age 6 and adulthood. The pattern resembles the deciduous dentition, with the following exceptions. The deciduous molars are replaced by the **first** and **second premolars (bicuspids),** which have two cusps and one root (upper first bicuspids have two roots) and are used for crushing and grinding. The permanent molars, which erupt into the mouth posterior to the bicuspids, do not replace any deciduous teeth and erupt as the jaw grows to accommodate them—the **first molars** at age 6, the

second molars at age 12, the **third molars (wisdom teeth)** after age 17.

Often the human jaw does not have enough room posterior to the second molars to accommodate the eruption of the third molars. In this case, the third molars remain embedded in the alveolar bone and are said to be "impacted." They often cause pressure and pain and must be removed surgically. In some people, third molars may be dwarfed in size or may not develop at all.

Mechanical and Chemical Digestion in the Mouth

Mechanical digestion in the mouth results from chewing, or **mastication** (mas′-ti-KĀ-shun; = to chew), in which food is manipulated by the tongue, ground by the teeth, and mixed with saliva. As a result, the food is reduced to a soft, flexible, easily swallowed mass called a **bolus** (= lump). Food molecules begin to dissolve in the water in saliva, an important activity because enzymes can react with food molecules in a liquid medium only.

Two enzymes contribute to chemical digestion in the mouth: salivary amylase and lingual lipase. **Salivary amylase**

Figure 24.7 Dentitions and times of eruptions (indicated in parentheses). (See Tortora, *A Photographic Atlas of the Human Body,* Figure 12.7)

🔑 **Deciduous teeth begin to erupt at 6 months of age, and one pair of teeth appears about each month thereafter, until all 20 are present.**

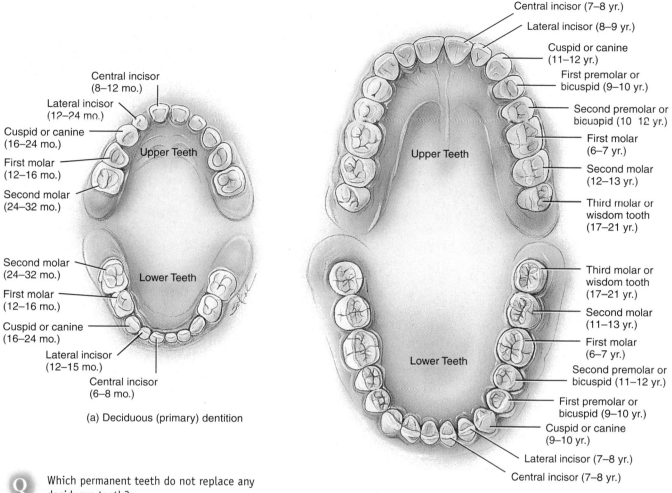

(a) Deciduous (primary) dentition

(b) Permanent (secondary) dentition

Q Which permanent teeth do not replace any deciduous teeth?

---

initiates the breakdown of starch. Dietary carbohydrates are either monosaccharide and disaccharide sugars or complex polysaccharides such as starches (see page 44). Most of the carbohydrates we eat are starches, but only monosaccharides can be absorbed into the bloodstream. Thus, ingested disaccharides and starches must be broken down into monosaccharides. The function of salivary amylase is to break certain chemical bonds between glucose units in the starches, which reduces the long-chain polysaccharides to the disaccharide maltose, the trisaccharide maltotriose, and short-chain glucose polymers called α-dextrins. Even though food is usually swallowed too quickly for all the starches to be reduced to disaccharides in the mouth, salivary amylase in the swallowed food continues to act on the starches for about another hour, at which time stomach acids inactivate it. Saliva also contains **lingual lipase,** which starts the digestion of di-

etary triglycerides into fatty acids and monoglycerides. This enzyme becomes activated in the acidic environment of the stomach and thus starts to work after food is swallowed.

Table 24.1 summarizes the digestive activities in the mouth.

1. What structures form the mouth (oral cavity)?
2. How are the major salivary glands distinguished histologically?
3. Describe the composition of saliva and the role of each of its components.
4. How is saliva secretion regulated?
5. Contrast the functions of incisors, cuspids, premolars, and molars.
6. What is a bolus? How is it formed?

Table 24.1 Summary of Digestive Activities in the Mouth

STRUCTURE	ACTIVITY	RESULT
Cheeks and Lips	Keep food between teeth during mastication.	Foods uniformly chewed.
Salivary Glands	Secrete saliva.	Lining of mouth and pharynx moistened and lubricated. Saliva softens, moistens, and dissolves food and cleanses mouth and teeth. Salivary amylase splits starch into smaller fragments.
Tongue		
Extrinsic muscles	Move tongue from side to side and in and out.	Food maneuvered for mastication, shaped into bolus, and maneuvered for swallowing.
Intrinsic muscles	Alter shape of tongue.	Swallowing and speech.
Taste buds	Serve as receptors for gustation (taste) and presence of food in mouth.	Secretion of saliva stimulated by nerve impulses from taste buds to salivary nuclei in brain stem to salivary glands.
Lingual Glands	Secrete lingual lipase.	Triglycerides broken down.
Teeth	Cut, tear, and pulverize food.	Solid foods reduced to smaller particles for swallowing.

PHARYNX

OBJECTIVE
• *Describe the location and function of the pharynx.*

When food is first swallowed, it passes from the mouth into the **pharynx** (= throat), a funnel-shaped tube that extends from the internal nares to the esophagus posteriorly and to the larynx anteriorly (see Figure 23.4 on page 779). The pharynx is composed of skeletal muscle and lined by mucous membrane. Whereas the nasopharynx functions only in respiration, both the oropharynx and laryngopharynx have digestive as well as respiratory functions. Swallowed food passes from the mouth into the oropharynx and laryngopharynx, the muscular contractions of which help propel food into the esophagus.

The movement of food from the mouth into the stomach is achieved by the act of swallowing, or **deglutition** (dē-gloo-TISH-un) (Figure 24.8). Deglutition is facilitated by saliva and mucus and involves the mouth, pharynx, and

esophagus. Swallowing occurs in three stages: (1) the voluntary stage, in which the bolus is passed into the oropharynx; (2) the pharyngeal stage, which is the involuntary passage of the bolus through the pharynx into the esophagus; and (3) the esophageal stage (described in the section on the esophagus), which is the involuntary passage of the bolus through the esophagus into the stomach.

Swallowing starts when the bolus is forced to the back of the oral cavity and into the oropharynx by the movement of the tongue upward and backward against the palate; these actions constitute the **voluntary stage** of swallowing. With the passage of the bolus into the oropharynx, the involuntary **pharyngeal stage** of swallowing begins (Figure 24.8b). The respiratory passageways close, and breathing is temporarily interrupted. The bolus stimulates receptors in the oropharynx, which send impulses to the **deglutition center** in the medulla oblongata and lower pons of the brain stem. The returning impulses cause the soft palate and uvula to move upward to close off the nasopharynx, and the larynx is pulled forward and upward under the tongue. As the larynx rises, the epiglottis moves backward and downward and seals off the rima glottidis. The movement of the larynx also pulls the vocal cords together, further sealing off the respiratory tract, and widens the opening between the laryngopharynx and esophagus. The bolus passes through the laryngopharynx and enters the esophagus in 1–2 seconds. The respiratory passageways then reopen, and breathing resumes.

1. Define deglutition. List the sequence of events involved in passing a bolus from the mouth into the stomach.
2. Discuss the voluntary and pharyngeal stages of swallowing.

ESOPHAGUS

OBJECTIVE
• *Describe the location, anatomy, histology, and function of the esophagus.*

The **esophagus** (e-SOF-a-gus; = eating gullet) is a collapsible muscular tube that lies posterior to the trachea. It is about 25 cm (10 in.) long. It begins at the inferior end of the laryngopharynx, passes through the mediastinum anterior to the vertebral column, pierces the diaphragm through an opening called the **esophageal hiatus,** and ends in the superior portion of the stomach (see Figure 24.1).

Histology of the Esophagus

The **mucosa** of the esophagus consists of nonkeratinized stratified squamous epithelium, lamina propria (areolar connective tissue), and a muscularis mucosae (smooth muscle) (Figure 24.9). Near the stomach, the mucosa of the esophagus also contains mucous glands. The stratified squamous epithelium associated with the lips, mouth, tongue, oropharynx, laryngopharynx, and esophagus affords considerable protection against abrasion and wear-and-tear from

Figure 24.8 Deglutition (swallowing). During the pharyngeal stage of deglutition (b), the tongue rises against the palate, the nasopharynx is closed off, the larynx rises, the epiglottis seals off the larynx, and the bolus is passed into the esophagus.

🗝 **Deglutition is a mechanism that moves food from the mouth into the stomach.**

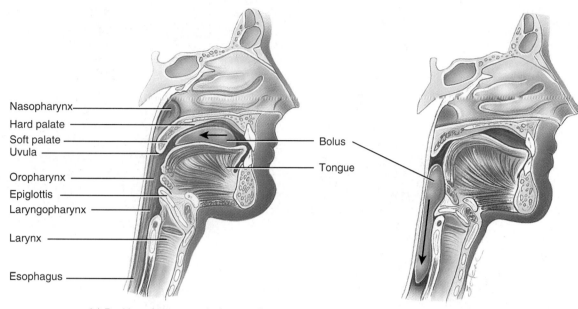

Nasopharynx
Hard palate
Soft palate
Uvula
Oropharynx
Epiglottis
Laryngopharynx
Larynx
Esophagus

Bolus
Tongue

(a) Position of structures before swallowing

(b) During the pharyngeal stage of swallowing

Q Is swallowing a voluntary action or an involuntary action?

food particles that are chewed, mixed with secretions, and swallowed. The **submucosa** contains areolar connective tissue, blood vessels, and mucous glands. The **muscularis** of the superior third of the esophagus is skeletal muscle, the intermediate third is skeletal and smooth muscle, and the inferior third is smooth muscle. The superficial layer is known as the **adventitia** (ad-ven-TISH-ya), rather than the serosa, because the areolar connective tissue of this layer is not covered by mesothelium and because the connective tissue merges with the connective tissue of surrounding structures of the mediastinum, through which it passes. The adventitia attaches the esophagus to surrounding structures.

Physiology of the Esophagus

The esophagus secretes mucus and transports food into the stomach. It does not produce digestive enzymes, and it does not carry on absorption. The passage of food from the laryngopharynx into the esophagus is regulated at the entrance to the esophagus by a sphincter (a circular band or ring of muscle that is normally contracted) called the **upper esophageal sphincter** (e-sof'-a-JĒ-al). It consists of the cricopharyngeus muscle attached to the cricoid cartilage. The elevation of the larynx during the pharyngeal stage of swallowing causes the sphincter to relax, and the bolus enters the esophagus. This sphincter also relaxes during exhalation.

During the **esophageal stage** of swallowing, a progression of coordinated contractions and relaxations of the circular and longitudinal layers of the muscularis—called **peristalsis** (per'-is-STAL-sis; *stalsis* = constriction)—push the food bolus onward (Figure 24.10). (Peristalsis occurs in other tubular structures, including other portions of the GI tract and the ureters, bile ducts, and uterine tubes; in the esophagus it is controlled by the medulla oblongata.) In the section of the esophagus lying just superior to the bolus, the circular muscle fibers contract, constricting the esophageal wall and squeezing the bolus toward the stomach. Meanwhile, longitudinal fibers inferior to the bolus also contract, which shortens this inferior section and pushes its walls outward so it can receive the bolus. The contractions are repeated in a wave that pushes the food toward the stomach. Mucus secreted by esophageal glands lubricates the bolus and reduces friction. The passage of solid or semisolid food from the mouth to the stomach takes 4–8 seconds; very soft foods and liquids pass through in about 1 second.

Just superior to the diaphragm, the esophagus narrows slightly due to a maintained contraction of the muscularis at the lowest part of the esophagus. This physiological sphincter, known as the **lower esophageal (cardiac) sphincter,** relaxes during swallowing and thus allows the bolus to pass from the esophagus into the stomach.

Figure 24.9 Histology of the esophagus. An enlarged view of nonkeratinized, stratified squamous epithelium is shown in Table 4.1 on page 108. (See Tortora, *A Photographic Atlas of the Human Body,* Figure 12.8)

🔑 **The esophagus secretes mucus and transports food to the stomach.**

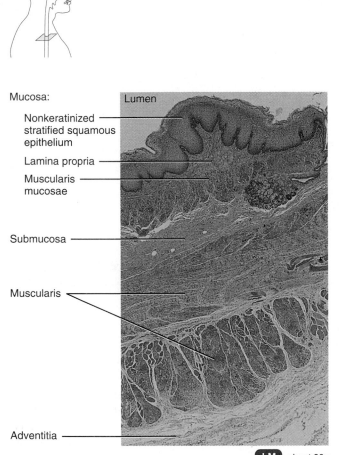

Wall of the esophagus

LM about 90x

Q In which layers of the esophagus are the glands that secrete lubricating mucus located?

Table 24.2 summarizes the digestive activities of the pharynx and esophagus.

 CLINICAL APPLICATION
Gastroesophageal Reflux Disease

If the lower esophageal sphincter fails to close adequately after food has entered the stomach, the stomach contents can reflux, or back up, into the inferior portion of the esophagus. This condition is known as **gastroesophageal reflux disease (GERD).** Hydrochloric acid (HCl) from the stomach contents can irritate the esophageal wall, re-

Figure 24.10 Peristalsis during the esophageal stage of deglutition (swallowing).

🔑 **Peristalsis consists of progressive, wavelike contractions of the muscularis.**

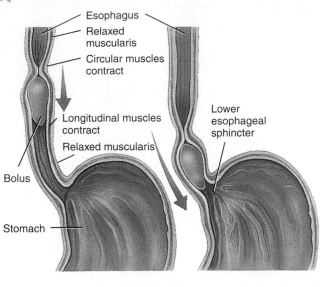

Anterior view of peristalsis in esophagus

Q Does peristalsis "push" or "pull" food along the gastrointestinal tract?

sulting in a burning sensation called **heartburn** because it is experienced in a region very near the heart, even though it is unrelated to any cardiac problem. Drinking alcohol and smoking can cause the sphincter to relax, worsening the problem. The symptoms of GERD often can be controlled by avoiding foods that strongly stimulate stomach acid secretion (coffee, chocolate, tomatoes, fatty foods, peppermint, spearmint, and onions). Other acid-reducing strategies include taking over-the-counter histamine-2 (H_2) blockers such as Tagamet HB or Pepcid AC 30–60 minutes before eating, and neutralizing already secreted acid with antacids such as Tums or Maalox. Symptoms are less likely to occur if food is eaten in smaller amounts and if the person does not lie down immediately after a meal. GERD may be associated with cancer of the esophagus. ■

1. Describe the location and histology of the esophagus. What is its role in digestion?
2. Explain the operation of the upper and lower esophageal sphincters.

STOMACH
OBJECTIVE
• *Describe the location, anatomy, histology, and function of the stomach.*

The **stomach** is a typically J-shaped enlargement of the GI tract directly inferior to the diaphragm in the epigastric, umbilical, and left hypochondriac regions of the abdomen (see Figure 1.12a on page 19). The stomach connects the esophagus to the duodenum, the first part of the small intestine (Figure 24.11). Because a meal can be eaten much more quickly than the intestines can digest and absorb it, one of the functions of the stomach is to serve as a mixing vat and holding reservoir. At appropriate intervals after food is ingested, the stomach forces a small quantity of material into the first portion of the small intestine. The position and size of the stomach vary continually; the diaphragm pushes it inferiorly with each inspiration and pulls it superiorly with each expiration. Empty, it is about the size of a large sausage, but it is the most distensible portion of the GI tract and can accommodate a large quantity of food. In the stomach, digestion of starch continues, digestion of proteins and triglycerides begins, the semisolid bolus is converted to a liquid, and certain substances are absorbed.

Anatomy of the Stomach

The stomach has four main regions: the cardia, fundus, body, and pylorus (see Figure 24.11). The **cardia** (CAR-dē-a) surrounds the superior opening of the stomach. The rounded portion superior to and to the left of the cardia is the **fundus** (FUN-dus). Inferior to the fundus is the large central portion of the stomach, called the **body.** The region of the stomach that connects to the duodenum is the **pylorus** (pī-LOR-us; *pyl-* = gate; *-orus* = guard); it has two parts, the **pyloric antrum** (AN-trum; = cave), which connects to the body of the stomach, and the **pyloric canal,** which leads into the duodenum. When the stomach is empty, the mucosa lies in large folds, called **rugae** (ROO-gē; = wrinkles), that can be seen with the unaided eye. The pylorus communicates with the duodenum of the small intestine via a sphincter called the **pyloric sphincter.** The concave medial border of the stomach is called the **lesser curvature,** and the convex lateral border is called the **greater curvature.**

CLINICAL APPLICATION
Pylorospasm and Pyloric Stenosis

Two abnormalities of the pyloric sphincter can occur in infants. In **pylorospasm** the muscle fibers of the sphincter fail to relax normally, so food does not pass easily from the stomach to the small intestine, the stomach becomes overly full, and the infant vomits often to relieve the pressure. Pylorospasm is treated by drugs that relax the muscle fibers of the sphincter. **Pyloric stenosis** is a narrowing of the pyloric sphincter, and it must be corrected surgically. The hallmark symptom is projectile vomiting—the spraying of liquid vomitus some distance from the infant. ■

Histology of the Stomach

The stomach wall is composed of the same four basic layers as the rest of the GI tract, with certain modifications (Figure 24.12a). The surface of the **mucosa** is a layer of sim-

Table 24.2 Summary of Digestive Activities in the Pharynx and Esophagus

STRUCTURE	ACTIVITY	RESULT
Pharynx	Pharyngeal stage of deglutition.	Moves bolus from oropharynx to laryngopharynx and into esophagus; closes air passageways.
Esophagus	Relaxation of upper esophageal sphincter.	Permits entry of bolus from laryngopharynx into esophagus.
	Esophageal stage of deglutition (peristalsis).	Pushes bolus down esophagus.
	Relaxation of lower esophageal sphincter.	Permits entry of bolus into stomach.
	Secretion of mucus.	Lubricates esophagus for smooth passage of bolus.

ple columnar epithelial cells called **mucous surface cells.** The mucosa contains a **lamina propria** (areolar connective tissue) and a **muscularis mucosae** (smooth muscle). Epithelial cells extend down into the lamina propria, where they form columns of secretory cells called **gastric glands** that line many narrow channels called **gastric pits.** Secretions from several gastric glands flow into each gastric pit and then into the lumen of the stomach.

The gastric glands contain three types of *exocrine gland cells* that secrete their products into the stomach lumen: mucous neck cells, chief cells, and parietal cells. Both mucous surface cells and **mucous neck cells** secrete mucus (Figure 24.12b). The **chief (zymogenic) cells** secrete pepsinogen and gastric lipase. **Parietal cells** produce hydrochloric acid and intrinsic factor (needed for absorption of vitamin B$_{12}$). The secretions of the mucous, chief, and parietal cells form **gastric juice,** which totals 2000–3000 mL (roughly 2–3 qt.) per day. In addition, gastric glands include a type of enteroendocrine cell, the **G cell,** which is located mainly in the pyloric antrum and secretes the hormone gastrin into the bloodstream. As we will see shortly, this hormone stimulates several aspects of gastric activity.

Three additional layers lie deep to the mucosa. The **submucosa** of the stomach is composed of areolar connective tissue. The **muscularis** has three (rather than two) layers of smooth muscle: an outer longitudinal layer, a middle circular layer, and an inner oblique layer. The oblique layer is limited primarily to the body of the stomach. The **serosa** (simple squamous mesothelium and areolar connective tissue) covering the stomach is part of the visceral peritoneum. At

Figure 24.11 External and internal anatomy of the stomach. (See Tortora, *A Photographic Atlas of the Human Body*, Figure 12.9)

🗝 **The four regions of the stomach are the cardia, fundus, body, and pylorus.**

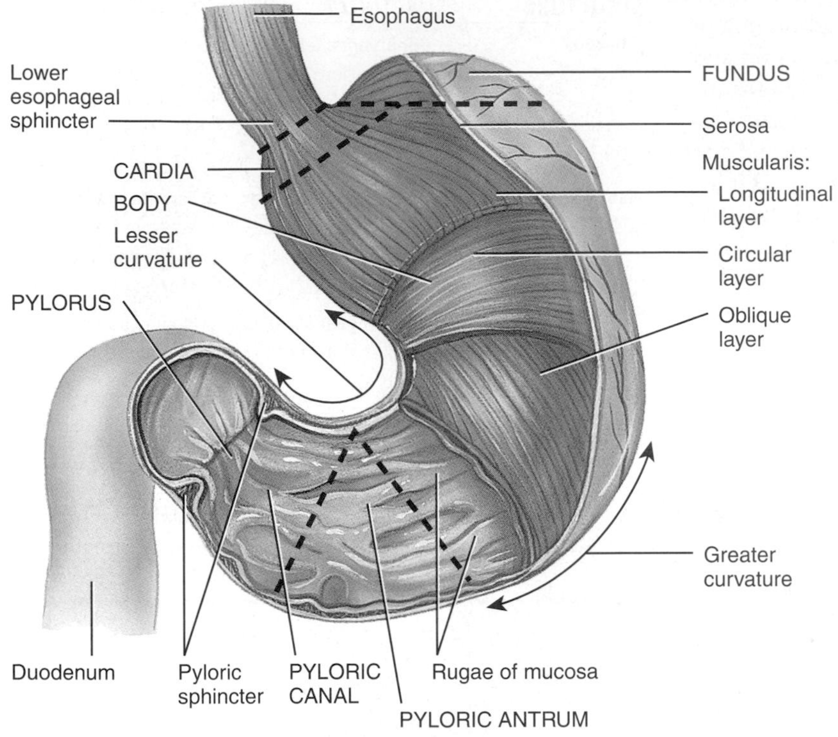

(a) Anterior view of regions of the stomach

FUNCTIONS OF THE STOMACH

1. Mixes saliva, food, and gastric juice to form chyme.
2. Reservoir for holding food before release into small intestine.
3. Secretes gastric juice, which contains HCl, pepsin, intrinsic factor, and gastric lipase.
4. HCl kills bacteria and denatures proteins. Pepsin begins the digestion of proteins. Intrinsic factor aids absorption of vitamin B_{12}. Gastric lipase aids digestion of triglycerides.
5. Secretes gastrin into blood.

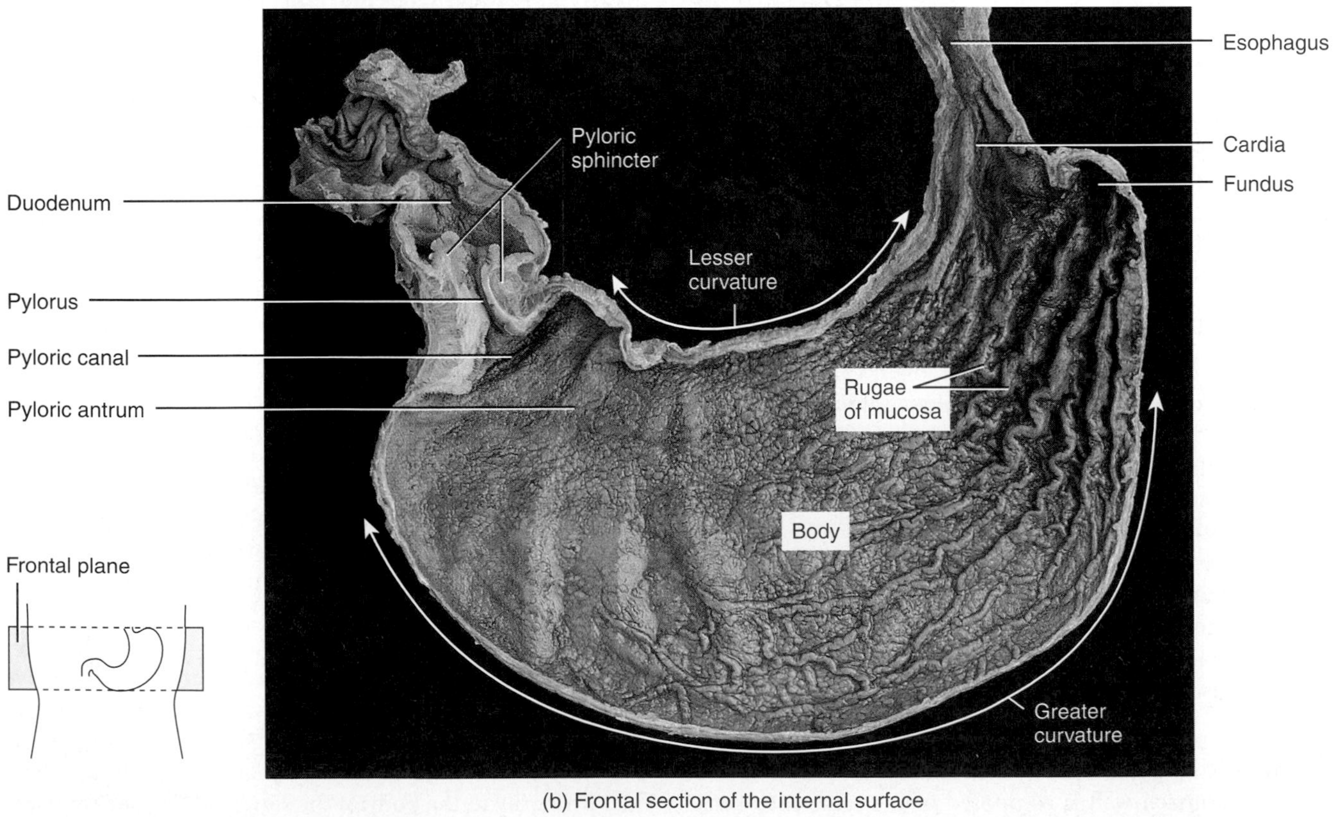

(b) Frontal section of the internal surface

Q After a very large meal, does your stomach have rugae?

Figure 24.12 Histology of the stomach.

The muscularis of the stomach has three layers of smooth muscle tissue.

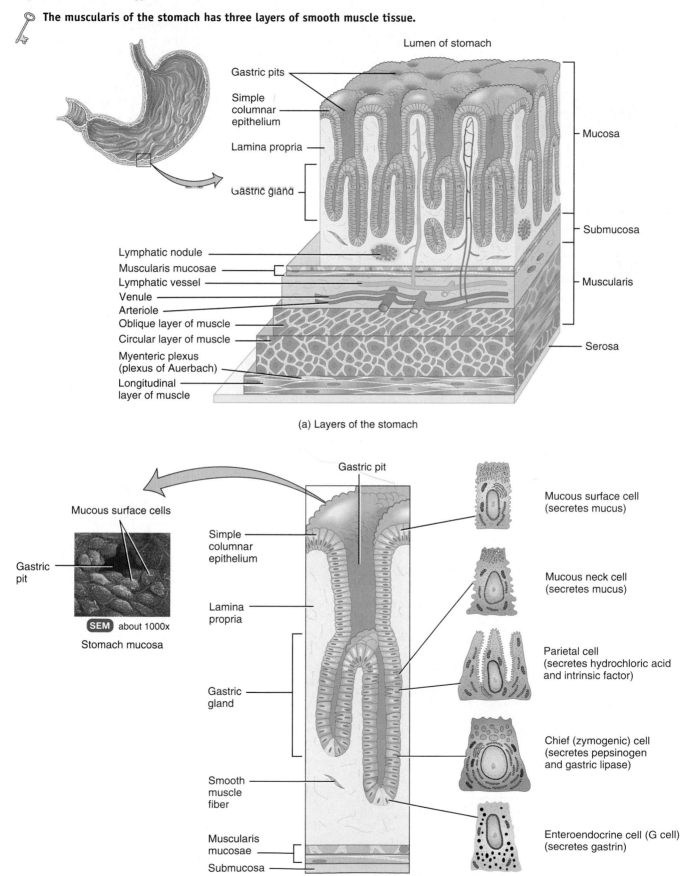

(a) Layers of the stomach

(b) Stomach mucosa showing gastric glands and cell types

figure continues

Figure 24.12 *(continued)*

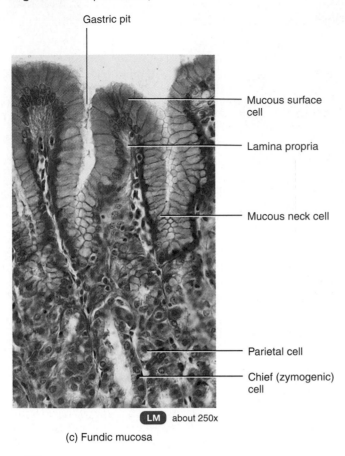

Gastric pit

Mucous surface cell

Lamina propria

Mucous neck cell

Parietal cell

Chief (zymogenic) cell

LM about 250x

(c) Fundic mucosa

 What types of cells are found in gastric glands, and what does each secrete?

the lesser curvature, the visceral peritoneum extends superiorly to the liver as the lesser omentum. At the greater curvature, the visceral peritoneum continues inferiorly as the greater omentum and drapes over the intestines.

Mechanical and Chemical Digestion in the Stomach

Several minutes after food enters the stomach, gentle, rippling, peristaltic movements called **mixing waves** pass over the stomach every 15–25 seconds. These waves macerate food, mix it with secretions of the gastric glands, and reduce it to a soupy liquid called **chyme** (kīm; = juice). Few mixing waves are observed in the fundus, which primarily has a storage function. As digestion proceeds in the stomach, more vigorous mixing waves begin at the body of the stomach and intensify as they reach the pylorus. The pyloric sphincter normally remains almost, but not completely, closed; as food reaches the pylorus, each mixing wave forces several milliliters of chyme into the duodenum through the pyloric sphincter. Most of the chyme is forced back into the body of the stomach, where mixing continues. The next wave pushes the chyme forward again and forces a little more into the duodenum. These forward and backward

movements of the gastric contents are responsible for most mixing in the stomach.

Foods may remain in the fundus for about an hour without becoming mixed with gastric juice. During this time, digestion by salivary amylase continues. Soon, however, the churning action mixes chyme with acidic gastric juice, inactivating salivary amylase and activating lingual lipase, which starts to digest triglycerides into fatty acids and diglycerides.

Even though parietal cells secrete hydrogen ions (H^+) and chloride ions (Cl^-) separately into the stomach lumen, the net effect is secretion of hydrochloric acid (HCl). **Proton pumps** powered by H^+/K^+ ATPases actively transport H^+ into the lumen while bringing potassium ions (K^+) into the cell. At the same time Cl^- and K^+ diffuse out through Cl^- and K^+ leakage channels in the apical membrane (next to the lumen). In the basolateral membrane, which faces the lamina propria, Cl^-/HCO_3^- antiporters bring Cl^- into the parietal cell in exchange for HCO_3^-, which diffuses into blood capillaries. The result is an "alkaline tide" of bicarbonate ions entering the bloodstream after a meal. Additionally, the strongly acidic fluid of the stomach kills many microbes in food, and HCl partially denatures (unfolds) proteins in food and stimulates the secretion of hormones that promote the flow of bile and pancreatic juice.

The enzymatic digestion of proteins begins in the stomach. The only proteolytic (protein-digesting) enzyme in the stomach is **pepsin,** which is secreted by chief cells. Because pepsin breaks certain peptide bonds between the amino acids making up proteins, a protein chain of many amino acids is broken down into smaller peptide fragments. Pepsin is most effective in the very acidic environment of the stomach (pH 2); it becomes inactive at higher pH.

What keeps pepsin from digesting the protein in stomach cells along with the food? First, pepsin is secreted in an inactive form called *pepsinogen;* in this form it cannot digest the proteins in the chief cells that produce it. Pepsinogen is not converted into active pepsin until it comes in contact with active pepsin molecules or hydrochloric acid secreted by parietal cells. Second, the stomach epithelial cells are protected from gastric juices by a 1–3 mm thick layer of alkaline mucus secreted by mucous surface cells and mucous neck cells.

Another enzyme of the stomach is **gastric lipase,** which splits the short-chain triglycerides in fat molecules found in milk into fatty acids and monoglycerides. This enzyme, which has a limited role in the adult stomach, operates best at a pH of 5–6. More important than either lingual lipase or gastric lipase is pancreatic lipase, an enzyme secreted by the pancreas into the small intestine.

Only a little absorption occurs in the stomach because its epithelial cells are impermeable to most materials. However, mucous cells of the stomach absorb some water, ions, and short-chain fatty acids, as well as certain drugs (especially aspirin) and alcohol.

Table 24.3 summarizes the digestive activities of the stomach.

Table 24.3 Summary of Digestive Activities in the Stomach

STRUCTURE	ACTIVITY	RESULT(S)
Mucosa		
Chief cells	Secrete pepsinogen.	Pepsin, the activated form, breaks certain peptide bonds in proteins.
	Secrete gastric lipase.	Splits short-chain triglycerides into fatty acids and monoglycerides.
Parietal cells	Secrete hydrochloric acid.	Kills microbes in food; denatures proteins; converts pepsinogen into pepsin.
	Secrete intrinsic factor.	Needed for absorption of vitamin B_{12}, which is required for normal red blood cell formation (erythropoiesis).
Mucous surface cells and **mucous neck cells**	Secrete mucus.	Forms a protective barrier that prevents digestion of stomach wall.
G cells	Secrete gastrin.	Stimulates parietal cells to secrete HCl and chief cells to secrete pepsinogen; contracts lower esophageal sphincter, increases motility of the stomach, and relaxes pyloric sphincter.
Muscularis	Mixing waves.	Macerate food and mix it with gastric juice, forming chyme.
	Peristalsis.	Forces chyme through pyloric sphincter.
Pyloric Sphincter	Opens to permit passage of chyme into duodenum.	Regulates passage of chyme from stomach to duodenum; prevents backflow of chyme from duodenum to stomach.

Regulation of Gastric Secretion and Motility

Both neural and hormonal mechanisms control the secretion of gastric juice and the contraction of smooth muscle in the stomach wall. Events in gastric digestion occur in three overlapping phases: the cephalic, gastric, and intestinal phases (Figure 24.13).

Cephalic Phase

The **cephalic phase** of gastric digestion consists of reflexes initiated by sensory receptors in the head. Even before food enters the stomach, the sight, smell, taste, or thought of food initiates this reflex. The cerebral cortex and feeding center in the hypothalamus send nerve impulses to the medulla oblongata, which then transmits impulses to parasympathetic preganglionic fibers in the vagus (X) nerves, which stimulate parasympathetic postganglionic fibers in the submucosal plexus. In turn, parasympathetic fibers transmit impulses to parietal cells, chief cells, and mucous cells and increase secretions from all the gastric glands. These impulses stimulate the gastric glands to secrete pepsinogen, hydrochloric acid, and mucus into stomach chyme, and gastrin into the blood. Impulses from parasympathetic fibers also increase stomach motility. Emotions such as anger, fear, and anxiety may slow digestion in the stomach because they stimulate the sympathetic nervous system, which inhibits gastric activity.

Gastric Phase

Once food reaches the stomach, sensory receptors in the stomach initiate both neural and hormonal mechanisms to ensure that gastric secretion and motility continue. This is the **gastric phase** of gastric digestion (see Figure 24.13). Food of any kind distends (stretches) the stomach and stimulates stretch receptors in its walls, and chemoreceptors monitor the pH of the stomach chyme. When the stomach walls are distended or pH increases because proteins have entered the stomach and buffered some of the stomach acid, the stretch receptors and chemoreceptors are activated, and a neural negative feedback loop is set in motion (Figure 24.14). From the receptors, nerve impulses propagate to the submucosal plexus, where they activate parasympathetic and enteric fibers. The resulting nerve impulses cause waves of peristalsis and continue to stimulate the flow of gastric juice from parietal cells, chief cells, and mucous cells.

The peristaltic waves mix the food with gastric juice, and when the waves become strong enough, a small quantity of chyme—about 10–15 mL (2–3 teaspoons)—squirts past the pyloric sphincter into the duodenum. As the pH of the stomach chyme becomes more acidic again and the stomach walls are distended less because chyme has passed into the small intestine, this negative feedback cycle suppresses secretion of gastric juice.

Hormonal negative feedback also regulates gastric secretion during the gastric phase (see Figure 24.13). Partially digested proteins buffer H^+, thus increasing pH, and ingested food distends the stomach. Chemoreceptors and stretch receptors monitoring these changes stimulate parasympathetic fibers to release acetylcholine. In turn, acetylcholine stimulates secretion of the hormone **gastrin** by G cells, enteroendocrine cells in the mucosa of the pyloric antrum. (A small amount of gastrin is also secreted by enteroendocrine cells in the small intestine; moreover, some chemicals in food—for example, caffeine—directly stimulate gastrin release.) Gastrin enters the bloodstream and finally reaches its target cells, the gastric glands.

Gastrin stimulates growth of the gastric glands and secretion of large amounts of gastric juice. It also strengthens contraction of the lower esophageal sphincter, increases motility of the stomach, and relaxes the pyloric and ileocecal

Figure 24.13 The cephalic, gastric, and intestinal phases of gastric digestion.

🔑 **Reflexes initiated by sensory receptors in the head, stomach, and small intestine constitute the cephalic, gastric, and intestinal phases of gastric digestion, respectively.**

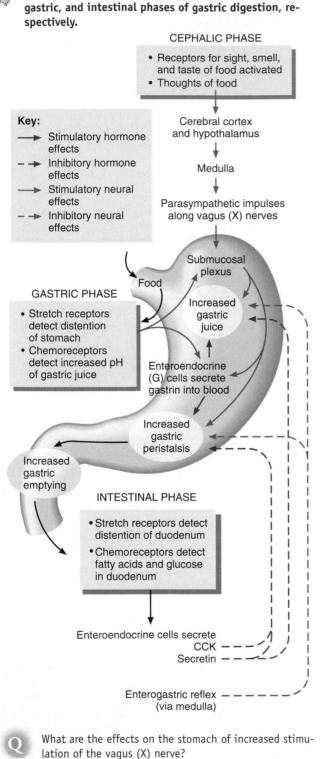

What are the effects on the stomach of increased stimulation of the vagus (X) nerve?

Figure 24.14 Neural negative feedback regulation of pH of gastric juice and gastric motility during the gastric phase of digestion.

🔑 **What are the effects on the stomach of increased stimulation of the vagus (X) nerves?**

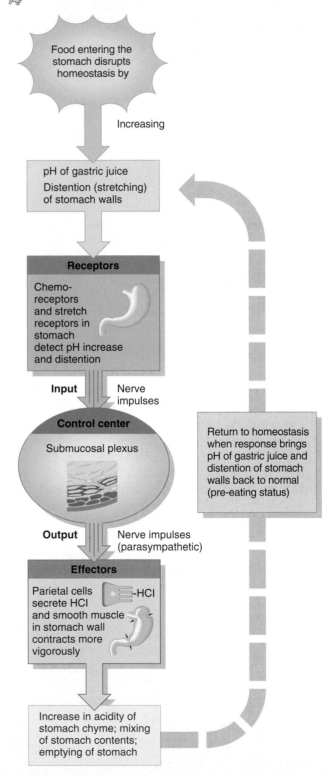

Why is this negative feedback loop called a "local reflex"?

sphincters (described later). Gastrin secretion is inhibited when the pH of gastric juice drops below 2.0 and is stimulated when the pH rises. This negative feedback mechanism helps provide an optimal low pH for the functioning of pepsin, the killing of microbes, and the denaturing of proteins in the stomach.

Acetylcholine (ACh) released by parasympathetic fibers and gastrin secreted by G cells stimulate parietal cells to secrete more HCl in the presence of histamine. In other words, histamine, a paracrine substance that is released by mast cells in the lamina propria and acts on neighboring parietal cells, acts synergistically with acetylcholine and gastrin to enhance their effects. Receptors for all three substances are present in the plasma membrane of parietal cells. The histamine receptors on parietal cells are called H_2 receptors; they mediate different responses than do the H_1 receptors involved in allergic responses.

Intestinal Phase

The **intestinal phase** of gastric digestion is due to activation of receptors in the small intestine. Whereas reflexes initiated during the cephalic and gastric phases stimulate stomach secretory activity and motility, those occurring during the intestinal phase have inhibitory effects (see Figure 24.13) that slow the exit of chyme from the stomach and prevent overloading of the duodenum with more chyme than it can handle. In addition, responses occurring during the intestinal phase promote the continued digestion of foods that have reached the small intestine. When chyme containing fatty acids and glucose leaves the stomach and enters the small intestine, it triggers enteroendocrine cells in the small intestinal mucosa to release into the blood two hormones that affect the stomach—**secretin** (se-KRĒ-tin) and **cholecystokinin** (kō-′-lē-sis′-tō-KĪN-in) or **CCK.** With respect to the stomach, secretin mainly decreases gastric secretions, whereas CCK mainly inhibits stomach emptying. Both hormones have other important effects on the pancreas, liver, and gallbladder (explained shortly) that contribute to the regulation of digestive processes.

Regulation of Gastric Emptying

Gastric emptying, the periodic release of chyme from the stomach into the duodenum, is regulated by both neural and hormonal reflexes, as follows (Figure 24.15a):

1 Stimuli such as distention of the stomach and the presence of partially digested proteins, alcohol, and caffeine initiate gastric emptying.

2 These stimuli increase the secretion of gastrin and generate parasympathetic impulses in the vagus (X) nerves.

3 Gastrin and nerve impulses stimulate contraction of the lower esophageal sphincter, increase motility of the stomach, and relax the pyloric sphincter.

4 The net effect of these actions is gastric emptying.

Figure 24.15 Neural and hormonal regulation of gastric emptying.

🔑 **Foods rich in carbohydrates are the first to leave the stomach.**

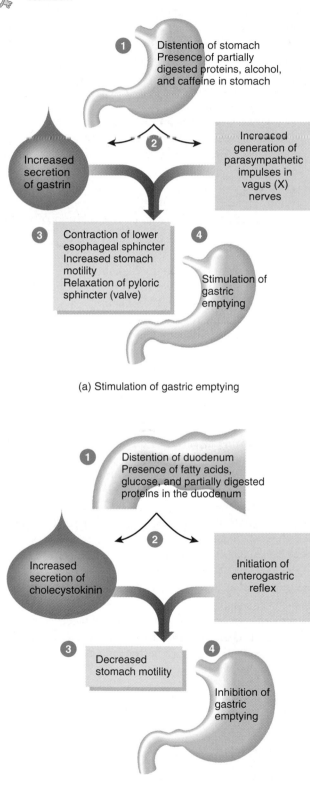

(a) Stimulation of gastric emptying

(b) Inhibition of gastric emptying

Q What effect would vagotomy (cutting of vagal nerve fibers) have on gastric emptying?

Neural and hormonal reflexes also help ensure that the stomach does not release to the small intestine more chyme than the latter can process. The neural reflex called the **enterogastric reflex** and cholecystokinin inhibit gastric emptying, as follows (Figure 24.15b):

1 Stimuli such as distention of the duodenum and the presence of fatty acids, glucose, and partially digested proteins in the duodenal chyme inhibit gastric emptying.

2 These stimuli initiate the enterogastric reflex: Nerve impulses propagate from the duodenum to the medulla oblongata, where they inhibit parasympathetic stimulation and stimulate sympathetic activity in the stomach. The same stimuli also increase secretion of cholecystokinin.

3 Increased sympathetic impulses and cholecystokinin both decrease gastric motility.

4 The net effect of these actions is inhibition of gastric emptying.

Within 2–4 hours after eating a meal, the stomach has emptied its contents into the duodenum. Foods rich in carbohydrate spend the least time in the stomach; high-protein foods remain somewhat longer, and emptying is slowest after a fat-laden meal containing large amounts of triglycerides. The reason for slower emptying after eating triglycerides is that fatty acids in chyme stimulate release of cholecystokinin, which slows stomach emptying.

CLINICAL APPLICATION
Vomiting

Vomiting or *emesis* is the forcible expulsion of the contents of the upper GI tract (stomach and sometimes duodenum) through the mouth. The strongest stimuli for vomiting are irritation and distension of the stomach; other stimuli include unpleasant sights, general anesthesia, dizziness, and certain drugs such as morphine and derivatives of digitalis. Nerve impulses are transmitted to the vomiting center in the medulla oblongata, and returning impulses propagate to the upper GI tract organs, diaphragm, and abdominal muscles. Vomiting basically involves squeezing the stomach between the diaphragm and abdominal muscles and expelling the contents through open esophageal sphincters. Prolonged vomiting, especially in infants and elderly people, can be serious because the loss of acidic gastric juice can lead to alkalosis (higher than normal blood pH). ■

1. Compare the epithelium of the esophagus with that of the stomach. How is each adapted to the function of the organ?

2. What is the importance of rugae, mucous surface cells, mucous neck cells, chief cells, parietal cells, and G cells in the stomach?

3. What is the role of pepsin? Why is it secreted in an inactive form?

4. What are the functions of gastric lipase and lingual lipase in the stomach?

5. Describe the factors that stimulate and inhibit gastric secretion and motility. Why are the three phases of gastric digestion named the cephalic, gastric, and intestinal phases?

PANCREAS

OBJECTIVE
• *Describe the location, anatomy, histology, and function of the pancreas.*

From the stomach, chyme passes into the small intestine. Because chemical digestion in the small intestine depends on activities of the pancreas, liver, and gallbladder, we first consider the activities of these accessory digestive organs and their contributions to digestion in the small intestine.

Anatomy of the Pancreas

The **pancreas** (*pan-* = all; *-creas* = flesh), a retroperitoneal gland that is about 12–15 cm (5–6 in.) long and 2.5 cm (1 in.) thick, lies posterior to the greater curvature of the stomach. The pancreas consists of a head, a body, and a tail and is connected to the duodenum usually by two ducts (Figure 24.16). The **head** is the expanded portion of the organ near the curve of the duodenum; superior to and to the left of the head are the central **body** and the tapering **tail.**

Pancreatic secretions pass from the secreting cells in the pancreas into small ducts that ultimately unite to form two larger ducts that convey the secretions into the small intestine. The larger of the two ducts is called the **pancreatic duct (duct of Wirsung).** In most people, the pancreatic duct joins the common bile duct from the liver and gallbladder and enters the duodenum as a common duct called the **hepatopancreatic ampulla (ampulla of Vater).** The ampulla opens on an elevation of the duodenal mucosa known as the **major duodenal papilla,** that lies about 10 cm (4 in.) inferior to the pyloric sphincter of the stomach. The smaller of the two ducts, the **accessory duct (duct of Santorini),** leads from the pancreas and empties into the duodenum about 2.5 cm (1 in.) superior to the hepatopancreatic ampulla.

Histology of the Pancreas

The pancreas is made up of small clusters of glandular epithelial cells, about 99% of which are arranged in clusters called **acini** (AS-i-nē) and constitute the *exocrine* portion of the organ (see Figure 18.18b, c on page 593). The cells within acini secrete a mixture of fluid and digestive enzymes called **pancreatic juice.** The remaining 1% of the cells are organized into clusters called **pancreatic islets (islets of Langerhans),** the *endocrine* portion of the pancreas. These cells secrete the hormones glucagon, insulin, somatostatin, and

Figure 24.16 Relation of the pancreas to the liver, gallbladder, and duodenum. The inset shows details of the common bile duct and pancreatic duct forming the hepatopancreatic ampulla (ampulla of Vater) and emptying into the duodenum. (See Tortora, *A Photographic Atlas of the Human Body,* Figures 12.10, 12.11)

🔑 **Pancreatic enzymes digest starches (polysaccharides), proteins, triglycerides, and nucleic acids.**

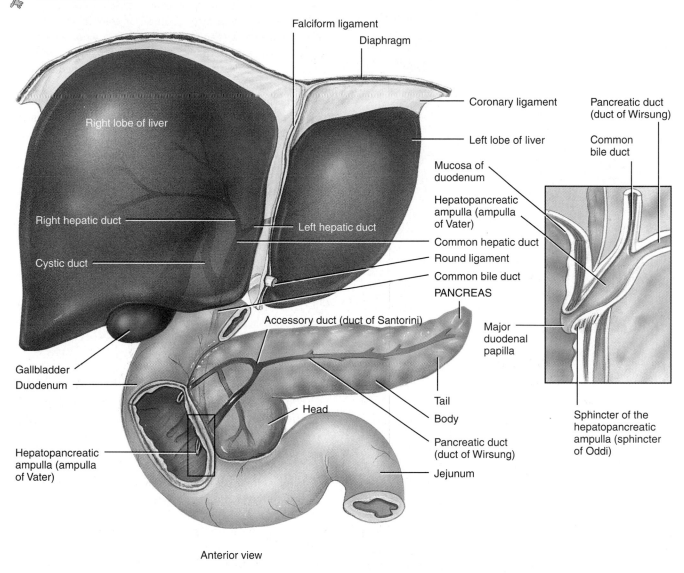

Anterior view

Q What type of fluid is found in the pancreatic duct? The common bile duct? The hepatopancreatic ampulla?

pancreatic polypeptide. The functions of these hormones are discussed in Chapter 18.

Composition and Functions of Pancreatic Juice

Each day the pancreas produces 1200–1500 mL (about 1.2–1.5 qt) of pancreatic juice, which is a clear, colorless liq-

uid consisting mostly of water, some salts, sodium bicarbonate, and several enzymes. The sodium bicarbonate gives pancreatic juice a slightly alkaline pH (7.1–8.2) that buffers acidic gastric juice in chyme, stops the action of pepsin from the stomach, and creates the proper pH for action of digestive enzymes in the small intestine. The enzymes in pancreatic juice include a carbohydrate-digesting enzyme called **pancreatic amylase;** several protein-digesting enzymes

Figure 24.17 Neural and hormonal enhancement of the secretion of pancreatic juice.

🔑 **Parasympathetic stimulation (via the vagus nerves) and acidic chyme in the small intestine stimulate secretin release into the blood. Vagal stimulation and fatty acids and amino acids in chyme within the small intestine stimulate cholecystokinin (CCK) release into the blood.**

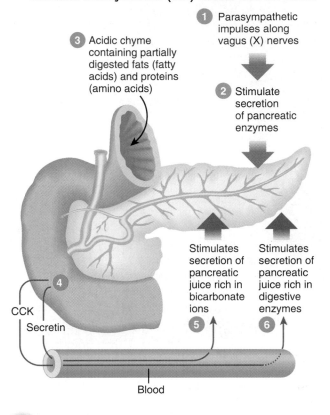

1 Parasympathetic impulses along vagus (X) nerves

3 Acidic chyme containing partially digested fats (fatty acids) and proteins (amino acids)

2 Stimulate secretion of pancreatic enzymes

Stimulates secretion of pancreatic juice rich in bicarbonate ions **5**

Stimulates secretion of pancreatic juice rich in digestive enzymes **6**

CCK
Secretin

4

Blood

❓ Why is secretion of pancreatic juice rich in bicarbonate ions helpful at this point in digestion?

called **trypsin** (TRIP-sin), **chymotrypsin** (kī′-mō-TRIP-sin), **carboxypeptidase** (kar-bok′-sē-PEP-ti-dās), and **elastase** (ē-LAS-tās); the principal triglyceride-digesting enzyme in adults, called **pancreatic lipase;** and nucleic acid-digesting enzymes called **ribonuclease** and **deoxyribonuclease.**

Just as pepsin is produced in the stomach in an inactive form (pepsinogen), so too are the protein-digesting enzymes of the pancreas. Because they are inactive, the enzymes do not digest cells of the pancreas itself. Trypsin is secreted in an inactive form called **trypsinogen** (trip-SIN-ō-jen). Pancreatic acinar cells also secrete a protein called **trypsin inhibitor,** that combines with any trypsin formed accidentally in the pancreas or in pancreatic juice and blocks its enzymatic activity. When trypsinogen reaches the lumen of the small intestine, it comes in contact with an activating brush-border enzyme called **enterokinase** (en′-ter-ō-KĪ-nās), which splits off part of the trypsinogen molecule to form trypsin. In turn, trypsin acts on the inactive precursors

(called **chymotrypsinogen, procarboxypeptidase,** and **proelastase**) to produce chymotrypsin, carboxypeptidase, and elastase, respectively.

🩺 **CLINICAL APPLICATION**
Pancreatitis

Inflammation of the pancreas, as may occur in association with alcohol abuse or chronic gallstones, is called **pancreatitis** (pan′-krē-a-TĪ-tis). In a more severe condition known as **acute pancreatitis,** which is associated with heavy alcohol intake or biliary tract obstruction, the pancreatic cells may release either trypsin instead of trypsinogen or insufficient amounts of trypsin inhibitor, and the trypsin begins to digest the pancreatic cells. Patients with acute pancreatitis usually respond to treatment, but recurrent attacks are the rule. ■

Regulation of Pancreatic Secretions

Pancreatic secretion, like gastric secretion, is regulated by both neural and hormonal mechanisms as follows (Figure 24.17).

1 During the cephalic and gastric phases of gastric digestion, parasympathetic impulses are transmitted along the vagus (X) nerves to the pancreas.

2 These parasympathetic nerve impulses stimulate increased secretion of pancreatic enzymes.

3 Acidic chyme containing partially digested fats and proteins enters the small intestine.

4 In response to fatty acids and amino acids, some enteroendocrine cells in the small intestine secrete cholecystokinin (CCK) into the blood. In response to acidic chyme, other enteroendocrine cells in the small intestinal mucosa liberate secretin into the blood.

5 Secretin stimulates the flow of pancreatic juice that is rich in bicarbonate ions.

6 Cholecystokinin (CCK) stimulates a pancreatic secretion that is rich in digestive enzymes.

1. Describe the duct system connecting the pancreas to the duodenum.
2. What are pancreatic acini? Contrast their functions with those of the pancreatic islets (islets of Langerhans).
3. Describe the composition of pancreatic juice and the digestive functions of each component.
4. How is pancreatic juice secretion regulated?

LIVER AND GALLBLADDER

OBJECTIVE

• *Describe the location, anatomy, histology, and functions of the liver and gallbladder.*

The **liver** is the heaviest gland of the body, weighing about 1.4 kg (about 3 lb) in an average adult, and after the skin it is the second largest organ of the body. It is inferior to the diaphragm and occupies most of the right hypochondriac and part of the epigastric regions of the abdominopelvic cavity (see Figure 1.12a on page 19).

The **gallbladder** (*gall-* = bile) is a pear-shaped sac that is located in a depression of the posterior surface of the liver. It is 7–10 cm (3–4 in.) long and typically hangs from the anterior inferior margin of the liver (see Figure 24.16).

Anatomy of the Liver and Gallbladder

The liver is almost completely covered by visceral peritoneum and is completely covered by a dense irregular connective tissue layer that lies deep to the peritoneum. The liver is divided into two principal lobes—a large **right lobe** and a smaller **left lobe**—by the **falciform ligament** (see Figure 24.16). Even though the right lobe is considered by many anatomists to include an inferior **quadrate lobe** and a posterior **caudate lobe,** on the basis of internal morphology (primarily the distribution of blood vessels), the quadrate and caudate lobes more appropriately belong to the left lobe. The falciform ligament, a fold of the parietal peritoneum, extends from the undersurface of the diaphragm between the two principal lobes of the liver to the superior surface of the liver, helping to suspend the liver. In the free border of the falciform ligament is the **ligamentum teres (round ligament),** a fibrous cord that is a remnant of the umbilical vein of the fetus (see Figure 21.31a, b on page 730); it extends from the liver to the umbilicus. The right and left **coronary ligaments** are narrow reflections of the parietal peritoneum that suspend the liver from the diaphragm.

The parts of the gallbladder are the broad *fundus,* which projects downward beyond the inferior border of the liver; the *body,* the central portion; and the *neck,* the tapered portion. The body and neck project upward.

Histology of the Liver and Gallbladder

The lobes of the liver are made up of many functional units called **lobules** (Figure 24.18). A lobule consists of specialized epithelial cells, called **hepatocytes** (*hepat-* = liver; *-cytes* = cells), arranged in irregular, branching, interconnected plates around a **central vein.** Instead of capillaries, the liver has larger, endothelium-lined spaces called **sinusoids,** through which blood passes. Also present in the sinusoids are fixed phagocytes called **stellate reticuloendothelial (Kupffer's) cells,** which destroy worn-out leukocytes and red blood cells, bacteria, and other foreign matter in the venous blood draining from the gastrointestinal tract.

Bile, which is secreted by hepatocytes, enters **bile canaliculi** (kan'-a-LIK-yoo-lī; = small canals), which are narrow intercellular canals that empty into small *bile ductules* (see Figure 24.18a). The ductules pass bile into *bile ducts* at the periphery of the lobules. The bile ducts merge and eventually form the larger **right** and **left hepatic ducts,** which unite and exit the liver as the **common hepatic duct** (see Figure 24.16). Farther on, the common hepatic duct joins the **cystic duct** (*cystic* = bladder) from the gallbladder to form the **common bile duct.** Bile enters the cystic duct and is temporarily stored in the gallbladder.

The mucosa of the gallbladder consists of simple columnar epithelium arranged in rugae resembling those of the stomach. The gallbladder lacks a submucosa. The middle, muscular coat of the wall consists of smooth muscle fibers, the contraction of which ejects the contents of the gallbladder into the **cystic duct.** The gallbladder's outer coat is the visceral peritoneum. The functions of the gallbladder are to store and concentrate bile (up to tenfold) until it is needed in the small intestine. In the concentration process, water and ions are absorbed by the gallbladder mucosa.

Blood Supply of the Liver

The liver receives blood from two sources (Figure 24.19). From the hepatic artery it obtains oxygenated blood, and from the hepatic portal vein it receives deoxygenated blood containing newly absorbed nutrients, drugs, and possibly microbes and toxins from the gastrointestinal tract (see Figure 21.29 on page 727). Branches of both the hepatic artery and the hepatic portal vein carry blood into liver sinusoids, where oxygen, most of the nutrients, and certain toxic substances are taken up by the hepatocytes. Products manufactured by the hepatocytes and nutrients needed by other cells are secreted back into the blood, which then drains into the central vein and eventually passes into a hepatic vein. Because blood from the gastrointestinal tract passes through the liver as part of the hepatic portal circulation, the liver is often a site for metastasis of cancer that originates in the GI tract. Branches of the hepatic portal vein, hepatic artery, and bile duct typically accompany each other in their distribution through the liver. Collectively, these three structures are called a **portal triad** (see Figure 24.18).

Role and Composition of Bile

Each day, hepatocytes secrete 800–1000 mL (about 1 qt) of **bile,** a yellow, brownish, or olive-green liquid. It has a pH of 7.6–8.6 and consists mostly of water and bile acids, bile salts, cholesterol, a phospholipid called lecithin, bile pigments, and several ions.

Bile is partially an excretory product and partially a digestive secretion. Bile salts, which are sodium salts and potassium salts of bile acids (mostly cholic acid and chenodeoxycholic acid), play a role in **emulsification,** the breakdown of large lipid globules into a suspension of droplets about 1 μm in diameter, and in absorption of lipids following their digestion. The tiny lipid droplets present a very large surface area which allows pancreatic lipase to more rapidly accomplish digestion of triglycerides. Cholesterol is made soluble in bile by bile salts and lecithin.

Figure 24.18 Histology of a lobule, the functional unit of the liver.

🔑 **A lobule consists of hepatocytes arranged around a central vein.**

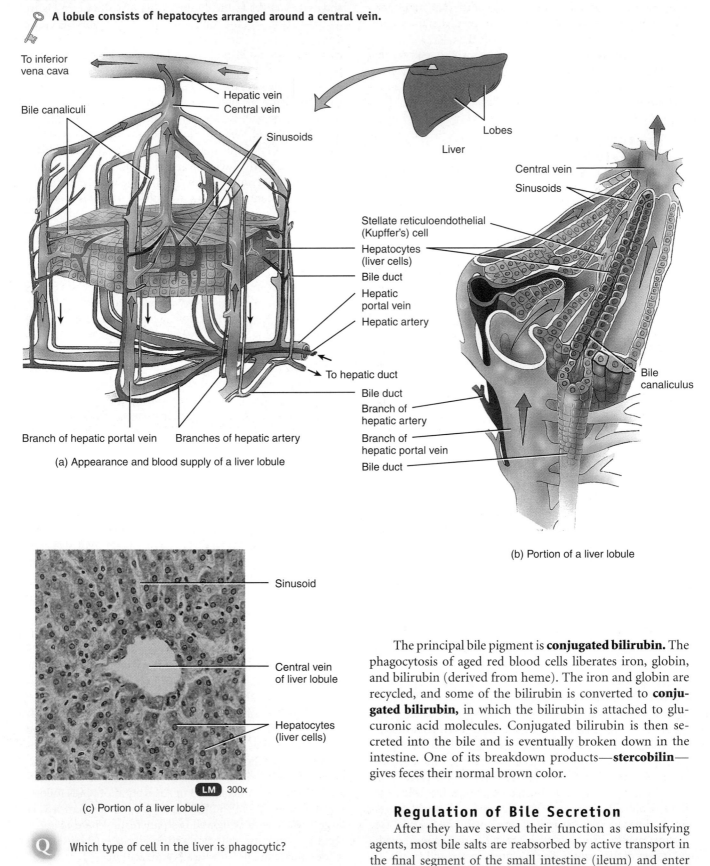

(a) Appearance and blood supply of a liver lobule

(b) Portion of a liver lobule

(c) Portion of a liver lobule

Ⓠ Which type of cell in the liver is phagocytic?

The principal bile pigment is **conjugated bilirubin.** The phagocytosis of aged red blood cells liberates iron, globin, and bilirubin (derived from heme). The iron and globin are recycled, and some of the bilirubin is converted to **conjugated bilirubin,** in which the bilirubin is attached to glucuronic acid molecules. Conjugated bilirubin is then secreted into the bile and is eventually broken down in the intestine. One of its breakdown products—**stercobilin**—gives feces their normal brown color.

Regulation of Bile Secretion

After they have served their function as emulsifying agents, most bile salts are reabsorbed by active transport in the final segment of the small intestine (ileum) and enter portal blood flowing toward the liver. Although hepatocytes continually release bile, they increase production and secretion when the portal blood contains more bile acids; thus, as digestion and absorption continue in the small intestine, bile release increases. Between meals, after most absorption has occurred, bile flows into the gallbladder for storage because the **sphincter of the hepatopancreatic ampulla** (*sphincter of Oddi;* see Figure 24.16) closes off the entrance to the duodenum. After a meal, several neural and hormonal stimuli promote production and release of bile (Figure 24.20):

1 Parasympathetic impulses along the vagus (X) nerve fibers can stimulate the liver to increase bile production to more than twice the baseline rate.

2 Fatty acids and amino acids in chyme entering the duodenum stimulate some duodenal enteroendocrine cells to secrete the hormone cholecystokinin (CCK) into the blood. Acidic chyme entering the duodenum stimulates other enteroendocrine cells to secrete the hormone secretin into the blood.

3 CCK causes contraction of the wall of the gallbladder, which squeezes stored bile out of the gallbladder into the cystic duct and through the common bile duct. CCK also causes relaxation of the sphincter of the hepatopancreatic ampulla, which allows bile to flow into the duodenum.

4 Secretin, which stimulates secretion of pancreatic juice that is rich in HCO_3^-, also stimulates the secretion of HCO_3^- by hepatocytes into bile.

Functions of the Liver

Besides secreting bile, which is needed for absorption of dietary fats, the liver performs many other vital functions:

- *Carbohydrate metabolism.* The liver is especially important in maintaining a normal blood glucose level. When blood glucose is low, the liver can break down glycogen to glucose and release glucose into the bloodstream. The liver can also convert certain amino acids and lactic acid to glucose, and it can convert other sugars, such as fructose and galactose, into glucose. When blood glucose is high, as occurs just after eating a meal, the liver converts glucose to glycogen and triglycerides for storage.

- *Lipid metabolism.* Hepatocytes store some triglycerides; break down fatty acids to generate ATP; synthesize lipoproteins, which transport fatty acids, triglycerides, and cholesterol to and from body cells; synthesize cholesterol; and use cholesterol to make bile salts.

- *Protein metabolism.* Hepatocytes deaminate (remove the amino group, NH_2, from) amino acids so that the amino acids can be used for ATP production or converted to carbohydrates or fats. The resulting toxic ammonia (NH_3) is then converted into the much less toxic urea, which is excreted in urine. Hepatocytes also synthesize most plasma proteins, such as alpha and beta globulins, albumin, prothrombin, and fibrinogen.

Figure 24.19 Hepatic blood flow: sources, path through the liver, and return to the heart.

🔑 **The liver receives oxygen-rich blood via the hepatic artery and nutrient-rich deoxygenated blood via the hepatic portal vein.**

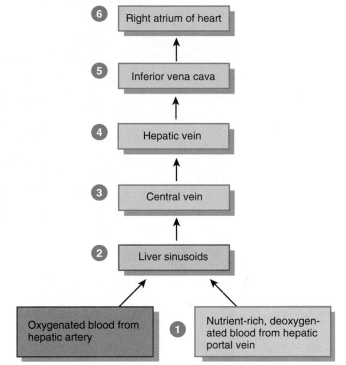

Q During the first few hours after a meal, how does the chemical composition of blood change as it flows through the liver sinusoids?

- *Processing of drugs and hormones.* The liver can detoxify substances such as alcohol or excrete drugs such as penicillin, erythromycin, and sulfonamides into bile. It can also chemically alter or excrete thyroid hormones and steroid hormones such as estrogens and aldosterone.

- *Excretion of bilirubin.* As previously noted, bilirubin, derived from the heme of aged red blood cells, is absorbed by the liver from the blood and secreted into bile. Most of the bilirubin in bile is metabolized in the small intestine by bacteria and eliminated in feces.

- *Synthesis of bile salts.* Bile salts are used in the small intestine for the emulsification and absorption of lipids, cholesterol, phospholipids, and lipoproteins.

- *Storage.* In addition to glycogen, the liver is a prime storage site for certain vitamins (A, B_{12}, D, E, and K) and minerals (iron and copper), which are released from the liver when needed elsewhere in the body.

- *Phagocytosis.* The stellate reticuloendothelial (Kupffer's) cells of the liver phagocytize aged red blood cells and white blood cells and some bacteria.

Figure 24.20 Neural and hormonal stimuli that promote production and release of bile.

🔑 **Bile salts in bile function in emulsification and lipid absorption.**

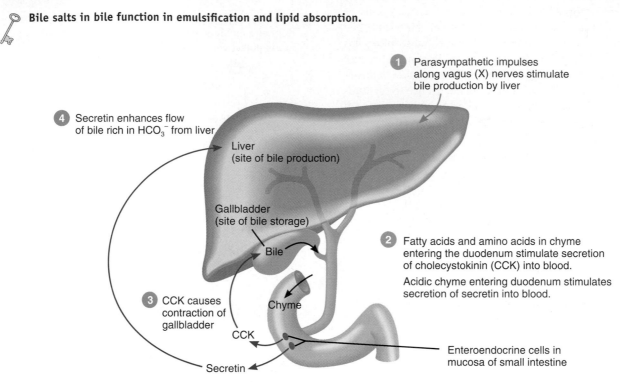

1 Parasympathetic impulses along vagus (X) nerves stimulate bile production by liver

4 Secretin enhances flow of bile rich in HCO_3^- from liver

Liver (site of bile production)

Gallbladder (site of bile storage)

Bile

3 CCK causes contraction of gallbladder

Chyme

CCK

Secretin

2 Fatty acids and amino acids in chyme entering the duodenum stimulate secretion of cholecystokinin (CCK) into blood.

Acidic chyme entering duodenum stimulates secretion of secretin into blood.

Enteroendocrine cells in mucosa of small intestine

Q In what respect are the effects of secretin on the liver and on the pancreas similar?

• *Activation of vitamin D.* The skin, liver, and kidneys participate in synthesizing the active form of vitamin D (see page 151).

The liver functions related to metabolism are discussed more fully in Chapter 25.

CLINICAL APPLICATION
Gallstones

If bile contains either insufficient bile salts or lecithin or excessive cholesterol, the cholesterol may crystallize to form **gallstones.** As they grow in size and number, gallstones may cause minimal, intermittent, or complete obstruction to the flow of bile from the gallbladder into the duodenum. Treatment consists of using gallstone-dissolving drugs, lithotripsy (shock-wave therapy), or surgery. For people with recurrent gallstones or for whom drugs or lithotripsy is not indicated, *cholecystectomy*—the removal of the gallbladder and its contents—is necessary. More than half a million cholecystectomies are performed each year in the United States. ■

1. Draw and label a diagram of a liver lobule.
2. Describe the pathways of blood flow into, through, and out of the liver.

3. How are the liver and gallbladder connected to the duodenum?
4. Once bile has been formed by the liver, how is it collected and transported to the gallbladder for storage?
5. What is the function of bile?
6. How is bile secretion regulated?

SUMMARY: DIGESTIVE HORMONES
OBJECTIVE
• *Describe the site of secretion and actions of gastrin, secretin, and cholecystokinin.*

The effects of the three major digestive hormones—gastrin, secretin, and cholecystokinin—and the stimuli that promote release of each are summarized in Table 24.4. All three are secreted into the blood by enteroendocrine cells located in the GI tract mucosa. Gastrin exerts its major effects on the stomach, whereas secretin and cholecystokinin affect the pancreas, liver, and gallbladder most strongly.

Stretching of the stomach as it receives food and the buffering of gastric acids by proteins in food trigger the release of gastrin. Gastrin, in turn, promotes secretion of gastric juice and increases gastric motility so that ingested food becomes well mixed into a thick, soupy chyme. Reflux of the

Table 24.4 Major Hormones that Control Digestion

HORMONE	STIMULUS AND SITE OF SECRETION	ACTIONS
Gastrin	Distension of stomach, partially digested proteins and caffeine in stomach, and high pH of stomach chyme stimulate gastrin secretion by enteroendocrine G cells, located mainly in the mucosa of pyloric antrum of stomach.	*Major effects:* Promotes secretion of gastric juice, increases gastric motility, and promotes growth of gastric mucosa. *Minor effects:* Constricts lower esophageal sphincter; relaxes pyloric sphincter and ileocecal sphincter.
Secretin	Acidic (high H^+ level) chyme that enters the small intestine stimulates secretion of secretin by enteroendocrine S cells in the mucosa of the small intestine.	*Major effects:* Stimulates secretion of pancreatic juice and bile that are rich in HCO_3^- (bicarbonate ions). *Minor effects:* Inhibits secretion of gastric juice, promotes normal growth and maintenance of the pancreas, and enhances effects of CCK.
Cholecystokinin (CCK)	Partially digested proteins (amino acids), triglycerides, and fatty acids that enter the small intestine stimulate secretion of CCK by enteroendocrine CCK cells in the mucosa of the small intestine; CCK is also released in the brain.	*Major effects:* Stimulates secretion of pancreatic juice rich in digestive enzymes, causes ejection of bile from the gallbladder and opening of the sphincter of the hepatopancreatic ampulla (sphincter of Oddi), and induces satiety (feeling full to satisfaction). *Minor effects:* Inhibits gastric emptying, promotes normal growth and maintenance of the pancreas, and enhances effects of secretin.

acidic chyme into the esophagus is prevented by tonic contraction of the lower esophageal sphincter, which is enhanced by gastrin.

The major stimulus for secretion of secretin is acidic chyme (high concentration of H^+) entering the small intestine. In turn, secretin promotes secretion of bicarbonate ions (HCO_3^-) into pancreatic juice and bile (see Figures 24.17 and 24.20). HCO_3^- acts to buffer ("soak up") excess H^+. Besides these major effects, secretin inhibits secretion of gastric juice, promotes normal growth and maintenance of the pancreas, and enhances the effects of CCK. Overall, secretin causes buffering of acid in chyme that reaches the duodenum and slows production of acid in the stomach.

Amino acids from partially digested proteins and fatty acids from partially digested triglycerides stimulate secretion of cholecystokinin by enteroendocrine cells in the mucosa of the small intestine. CCK stimulates secretion of pancreatic juice that is rich in digestive enzymes (see Figure 24.17) and ejection of bile into the duodenum (see Figure 24.20), slows gastric emptying by promoting contraction of the pyloric sphincter, and produces satiety (feeling full to satisfaction) by acting on the hypothalamus in the brain. Like secretin, CCK promotes normal growth and maintenance of the pancreas; it also enhances the effects of secretin.

Besides these three hormones, at least ten other so-called "gut hormones" are secreted by and have effects on the GI tract, including *motilin, substance P,* and *bombesin,* which stimulate motility of the intestines; *vasoactive intestinal polypeptide (VIP),* which stimulates secretion of ions and water by the intestines and inhibits gastric acid secretion; *gastrin-releasing peptide,* which stimulates release of gastrin;

and *somatostatin,* which inhibits gastrin release. Some of these hormones are thought to act as local hormones (paracrines), whereas others are secreted into the blood or even into the lumen of the GI tract. The physiological roles of these and other gut hormones are still under investigation.

1. What are the stimuli that trigger secretion of gastrin, secretin, and cholecystokinin?

SMALL INTESTINE

OBJECTIVE

• *Describe the location, anatomy, histology, and function of the small intestine.*

Now that we have examined the sources of the main digestive enzymes and hormones, we can proceed to the next part of the GI tract and pick up the story of digestion and absorption. The major events of digestion and absorption occur in a long tube called the **small intestine.** Because almost all digestion and absorption of nutrients occur in the small intestine, its structure is specially adapted for this function. Its length alone provides a large surface area for digestion and absorption, and that area is further increased by circular folds, villi, and microvilli. The small intestine begins at the pyloric sphincter of the stomach, coils through the central and inferior part of the abdominal cavity, and eventually opens into the large intestine. It averages 2.5 cm (1 in.) in diameter; its length is about 3 m (10 ft) in a living person and about 6.5 m (21 ft) in a cadaver due to the loss of smooth muscle tone after death.

Figure 24.21 Regions of the small intestine. (See Tortora, *A Photographic Atlas of the Human Body*, Figure 12.12)

Most digestion and absorption occur in the small intestine.

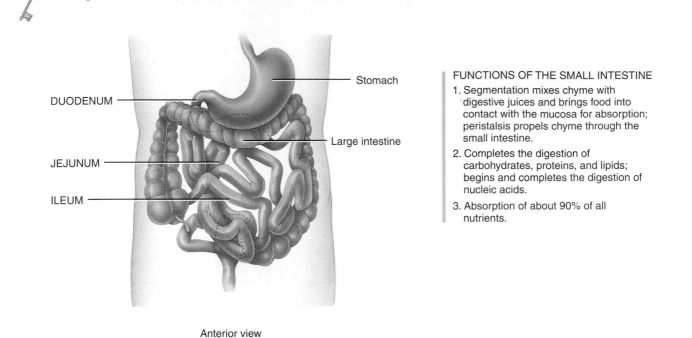

FUNCTIONS OF THE SMALL INTESTINE

1. Segmentation mixes chyme with digestive juices and brings food into contact with the mucosa for absorption; peristalsis propels chyme through the small intestine.
2. Completes the digestion of carbohydrates, proteins, and lipids; begins and completes the digestion of nucleic acids.
3. Absorption of about 90% of all nutrients.

Anterior view

Q Which portion of the small intestine is the longest?

Anatomy of the Small Intestine

The small intestine is divided into three regions (Figure 24.21). The **duodenum** (doo'-ō-DĒ-num), the shortest region, is retroperitoneal. It starts at the pyloric sphincter of the stomach and extends about 25 cm (10 in.) until it merges with the jejunum. *Duodenum* means "12"; it is so named because it is about as long as the width of 12 fingers. The **jejunum** (jē-JOO-num) is about 1 m (3 ft) long and extends to the ileum. *Jejunum* means "empty," which is how it is found at death. The final and longest region of the small intestine, the **ileum** (IL-ē-um; = twisted), measures about 2 m (6 ft) and joins the large intestine at the **ileocecal** (il'-ē-ō-SĒ-kal) **sphincter.**

Projections called **circular folds** are 10 mm (0.4 in.) high permanent ridges in the mucosa (see Figure 24.22c). The circular folds begin near the proximal portion of the duodenum and end at about the midportion of the ileum; some extend all the way around the circumference of the intestine, and others extend only part of the way around. They enhance absorption by increasing surface area and causing the chyme to spiral, rather than move in a straight line, as it passes through the small intestine.

Histology of the Small Intestine

Even though the wall of the small intestine is composed of the same four coats that make up most of the GI tract, special features of both the mucosa and the submucosa facilitate the processes of digestion and absorption. The mucosa forms a series of fingerlike **villi** (= tufts of hair), projections that are 0.5–1 mm long (Figure 24.22a). The large number of villi (20–40 per square millimeter) vastly increases the surface area of the epithelium available for absorption and digestion and gives the intestinal mucosa a velvety appearance. Each villus (singular form) has a core of lamina propria (areolar connective tissue); embedded in this connective tissue are an arteriole, a venule, a blood capillary network, and a **lacteal** (LAK-tē-al; = milky), which is a lymphatic capillary. Nutrients absorbed by the epithelial cells covering the villus pass through the wall of a capillary or a lacteal to enter blood or lymph, respectively.

The epithelium of the mucosa consists of simple columnar epithelium that contains absorptive cells, goblet cells, enteroendocrine cells, and Paneth cells (Figure 24.22b). The apical (free) membrane of absorptive cells features **microvilli** (mī'-krō-VIL-ī; *micro-* = small); each microvillus is a 1 μm-long cylindrical, membrane-covered projection that contains a bundle of 20–30 actin filaments. In a photomicrograph taken through a light microscope, the microvilli are too small to be seen individually; instead they form a fuzzy line, called the **brush border,** extending into the lumen of the small intestine (see Figure 24.23b). There are an estimated 200 million microvilli per square millimeter of small intestine. Because the microvilli greatly increase the surface area of the plasma membrane, larger amounts of

Figure 24.22 Anatomy of the small intestine.

Circular folds, villi, and microvilli increase the surface area of the small intestine for digestion and absorption.

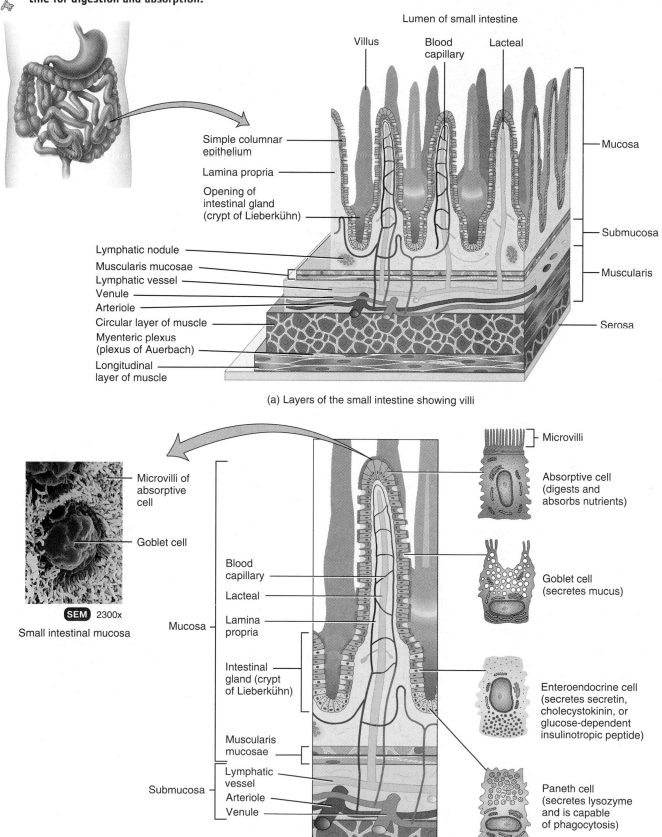

(a) Layers of the small intestine showing villi

(b) Enlarged villus showing lacteal, capillaries, and intestinal gland and cell types   *figure continues*

Figure 24.22 *(continued)*

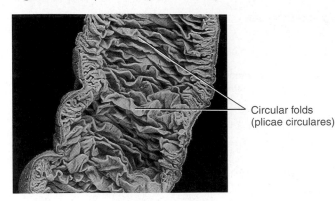

Circular folds
(plicae circulares)

(c) Jejunum cut open to expose the circular folds

 What is the functional significance of the blood capillary network and lacteal in the center of each villus?

digested nutrients can diffuse into absorptive cells in a given period of time. The brush border also contains several, brush-border enzymes that have digestive functions (discussed shortly).

The mucosa contains many deep crevices lined with glandular epithelium. Cells lining the crevices form the **intestinal glands (crypts of Lieberkühn)** and secrete intestinal juice. Many of the epithelial cells in the mucosa are goblet cells, which secrete mucus. **Paneth cells,** found in the deepest parts of the intestinal glands, secrete lysozyme, a bactericidal enzyme, and are also capable of phagocytosis. They may have a role in regulating the microbial population in the intestines. Three types of enteroendocrine cells, also in the deepest part of the intestinal glands, secrete hormones: secretin (by S cells), cholecystokinin (by CCK cells), and glucose-dependent insulinotropic peptide (by K cells). The lamina propria of the small intestine has an abundance of mucosa-associated lymphoid tissue (MALT). **Solitary lymphatic nodules** are most numerous in the distal part of the ileum; groups of lymphatic nodules referred to as **aggregated lymphatic follicles (Peyer's patches)** are also numerous in the ileum. The muscularis mucosae consists of smooth muscle. The submucosa of the duodenum contains **duodenal (Brunner's) glands** (Figure 24.23a), which secrete an alkaline mucus that helps neutralize gastric acid in the chyme.

The **muscularis** of the small intestine consists of two layers of smooth muscle. The outer, thinner layer contains longitudinal fibers; the inner, thicker layer contains circular fibers. Except for a major portion of the duodenum, the serosa (or visceral peritoneum) completely surrounds the small intestine.

Roles of Intestinal Juice and Brush-Border Enzymes

Intestinal juice is a clear yellow fluid secreted in amounts of 1–2 liters (about 1–2 qt) a day. It contains water and mucus and is slightly alkaline (pH 7.6). Together, pancreatic and intestinal juices provide a liquid medium that aids the absorption of substances from chyme as they come in contact with the microvilli. The absorptive epithelial cells synthesize several digestive enzymes, called **brush-border enzymes,** and insert them in the plasma membrane of the microvilli. Thus, some enzymatic digestion occurs at the surface of the epithelial cells that line the villi, rather than in the lumen exclusively, as occurs in other parts of the GI tract. Among the brush-border enzymes are four carbohydrate-digesting enzymes called **α-dextrinase, maltase, sucrase,** and **lactase;** protein-digesting enzymes called **peptidases** (**aminopeptidase** and **dipeptidase**); and two types of nucleotide-digesting enzymes, **nucleosidases** and **phosphatases.** Also, as cells slough off into the lumen of the small intestine, they break apart and release enzymes that help digest nutrients in the chyme.

Mechanical Digestion in the Small Intestine

The two types of movements of the small intestine— segmentations and a type of peristalsis called migrating motility complexes—are governed mainly by the myenteric plexus. **Segmentations** are a localized, mixing type of contraction that occurs in portions of intestine distended by a large volume of chyme. Segmentations mix chyme with the digestive juices and bring the particles of food into contact with the mucosa for absorption; they do not push the intestinal contents along the tract. A segmentation starts with the contractions of circular muscle fibers in a portion of the small intestine, an action that constricts the intestine into segments. Next, muscle fibers that encircle the middle of each segment also contract, dividing each segment again. Finally, the fibers that first contracted relax, and each small segment unites with an adjoining small segment so that large segments are formed again. As this sequence of events repeats, the chyme sloshes back and forth. Segmentations occur most rapidly in the duodenum, about 12 times per minute, and progressively slow to about 8 times per minute in the ileum. This movement is similar to alternately squeezing the middle and then the ends of a capped tube of toothpaste.

After most of a meal has been absorbed, which lessens distention of the wall of the small intestine, segmentation stops and peristalsis begins. The type of peristalsis that occurs in the small intestine, termed a **migrating motility complex (MMC),** begins in the lower portion of the stomach and pushes chyme forward along a short stretch of small intestine before dying out. The MMC slowly migrates down the small intestine, reaching the end of the ileum in 90–120 minutes. Then another MMC begins in the stomach. Altogether, chyme remains in the small intestine for 3–5 hours.

Figure 24.23 Histology of the duodenum.

🔑 **Microvilli greatly increase the surface area of the small intestine for digestion and absorption.**

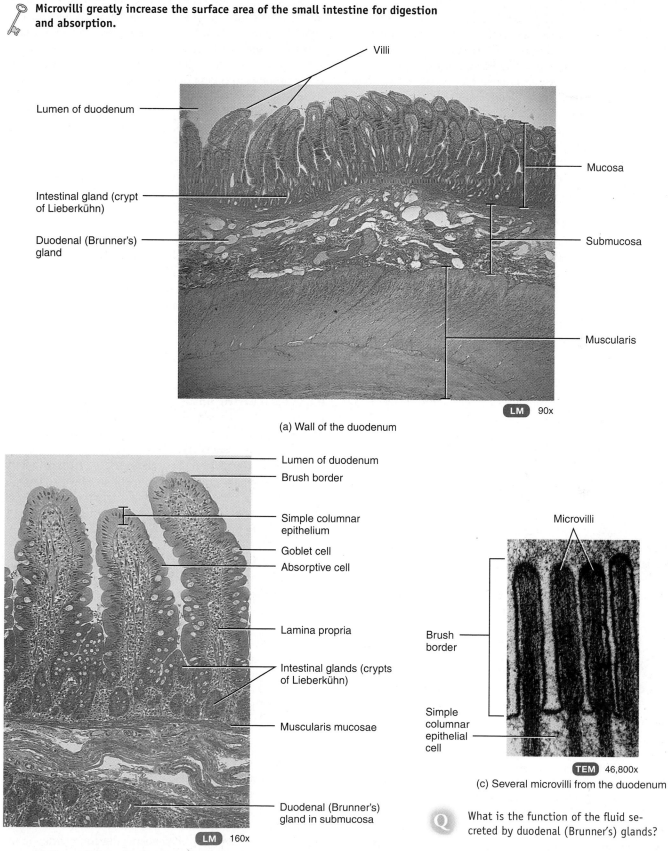

(a) Wall of the duodenum

LM 90x

(b) Three villi from the duodenum of the small intestine

LM 160x

(c) Several microvilli from the duodenum

TEM 46,800x

Q What is the function of the fluid secreted by duodenal (Brunner's) glands?

Chemical Digestion in the Small Intestine

In the mouth, salivary amylase converts starch (a polysaccharide) to maltose (a disaccharide), maltotriose (a trisaccharide), and α-dextrins (short-chain, branched fragments of starch with five to ten glucose units). In the stomach, pepsin converts proteins to peptides (small fragments of proteins), and lingual and gastric lipases convert some triglycerides into fatty acids, diglycerides, and monoglycerides. Thus, chyme entering the small intestine contains partially digested carbohydrates, proteins, and lipids. The completion of the digestion of carbohydrates, proteins, and lipids is a collective effort of pancreatic juice, bile, and intestinal juice in the small intestine.

Digestion of Carbohydrates

Even though the action of **salivary amylase** may continue in the stomach for awhile, the acidic pH of the stomach destroys salivary amylase and ends its activity. Thus, only a few starches are reduced to maltose by the time chyme leaves the stomach. Those starches not already broken down into maltose, maltotriose, and α-dextrins are cleaved by **pancreatic amylase,** an enzyme in pancreatic juice that acts in the small intestine. Although amylase acts on both glycogen and starches, it does not act on another polysaccharide called cellulose, which is an indigestible plant fiber. After amylase (either salivary or pancreatic) has split starch into smaller fragments, a brush-border enzyme called **α-dextrinase** acts on the resulting α-dextrins, clipping off one glucose unit at a time.

Ingested molecules of sucrose, lactose, and maltose—three disaccharides—are not acted on until they reach the small intestine. Three brush-border enzymes digest the disaccharides into monosaccharides: **Sucrase** breaks sucrose into a molecule of glucose and a molecule of fructose; **lactase** digests lactose into a molecule of glucose and a molecule of galactose; and **maltase** splits maltose and maltotriose into two or three molecules of glucose, respectively. Digestion of carbohydrates ends with the production of monosaccharides, as mechanisms exist for their absorption.

CLINICAL APPLICATION
Lactose Intolerance

In some people the mucosal cells of the small intestine fail to produce enough lactase, which is essential for the digestion of lactose. This results in a condition called **lactose intolerance,** in which undigested lactose in chyme retains fluid in the feces, and bacterial fermentation of lactose results in the production of gases. Symptoms of lactose intolerance include diarrhea, gas, bloating, and abdominal cramps after consumption of milk and other dairy products. The severity of symptoms varies from relatively minor to sufficiently serious to require medical attention. Persons with lactose intolerance can take dietary supplements to aid in the digestion of lactose. ■

Digestion of Proteins

Protein digestion starts in the stomach, where proteins are fragmented into peptides by the action of **pepsin.** Enzymes in pancreatic juice—**trypsin, chymotrypsin, carboxypeptidase,** and **elastase**—continue to break down proteins into peptides. Although all these enzymes convert whole proteins into peptides, their actions differ somewhat because each splits peptide bonds between different amino acids. Trypsin, chymotrypsin, and elastase all cleave the peptide bond between a specific amino acid and its neighbor; carboxypeptidase breaks the peptide bond that attaches the terminal amino acid to the carboxyl (acid) end of the peptide. Protein digestion is completed by two **peptidases** in the brush border: aminopeptidase and dipeptidase. **Aminopeptidase** acts on peptides by breaking the peptide bond that attaches the terminal amino acid to the amino end of the peptide. **Dipeptidase** splits dipeptides (two amino acids joined by a peptide bond) into single amino acids.

Digestion of Lipids

The most abundant lipids in the diet are triglycerides, which consist of a molecule of glycerol bonded to three molecules of fatty acid (see Figure 2.17 on page 46). Enzymes that split triglycerides and phospholipids are called **lipases.** In adults, most lipid digestion occurs in the small intestine, although some occurs in the stomach through the action of **lingual** and **gastric lipases.** When chyme enters the small intestine, bile salts emulsify the globules of triglycerides into droplets about 1 μm in diameter, which increases the surface area exposed to **pancreatic lipase,** another enzyme in pancreatic juice. This enzyme hydrolyzes triglycerides into fatty acids and monoglycerides, the main end products of triglyceride digestion. Pancreatic and gastric lipases remove two of the three fatty acids from glycerol; the third remains attached to the glycerol, forming a monoglyceride.

Digestion of Nucleic Acids

Pancreatic juice contains two nucleases: **ribonuclease,** which digests RNA, and **deoxyribonuclease,** which digests DNA. The nucleotides that result from the action of the two nucleases are further digested by brush-border enzymes called **nucleosidases** and **phosphatases** into pentoses, phosphates, and nitrogenous bases. These products are absorbed via active transport.

Table 24.5 summarizes the sources, substrates, and products of the digestive enzymes.

Regulation of Intestinal Secretion and Motility

The most important mechanisms that regulate small intestinal secretion and motility are enteric reflexes that respond to the presence of chyme; vasoactive intestinal polypeptide (VIP) also stimulates the production of intestinal juice. Segmentation movements depend mainly on in-

Table 24.5 Summary of Digestive Enzymes

ENZYME	SOURCE	SUBSTRATES	PRODUCTS
Saliva			
Salivary amylase	Salivary glands.	Starches (polysaccharides).	Maltose (disaccharide), maltotriose (trisaccharide), and α-dextrins.
Lingual lipase	Glands in the tongue.	Triglycerides (fats and oils) and other lipids.	Fatty acids and diglycerides.
Gastric Juice			
Pepsin (activated from pepsinogen by pepsin and hydrochloric acid)	Stomach chief (zymogenic) cells.	Proteins.	Peptides.
Gastric lipase	Stomach chief (zymogenic) cells.	Short-chain triglycerides (fats and oils) in fat molecules in milk.	Fatty acids and monoglycerides.
Pancreatic Juice			
Pancreatic amylase	Pancreatic acinar cells.	Starches (polysaccharides).	Maltose (disaccharide), maltotriose (trisaccharide), and α-dextrins.
Trypsin (activated from trypsinogen by enterokinase)	Pancreatic acinar cells.	Proteins.	Peptides.
Chymotrypsin (activated from chymotrypsinogen by trypsin)	Pancreatic acinar cells.	Proteins.	Peptides.
Elastase (activated from proelastase by trypsin)	Pancreatic acinar cells.	Proteins.	Peptides.
Carboxypeptidase (activated from procarboxypeptidase by trypsin)	Pancreatic acinar cells.	Terminal amino acid at carboxyl (acid) end of peptides.	Peptides and amino acids.
Pancreatic lipase	Pancreatic acinar cells.	Triglycerides (fats and oils) that have been emulsified by bile salts.	Fatty acids and monoglycerides.
Nucleases			
Ribonuclease	Pancreatic acinar cells.	Ribonucleic acid.	Nucleotides.
Deoxyribonuclease	Pancreatic acinar cells.	Deoxyribonucleic acid.	Nucleotides.
Brush Border			
α-Dextrinase	Small intestine.	α-Dextrins.	Glucose.
Maltase	Small intestine.	Maltose.	Glucose.
Sucrase	Small intestine.	Sucrose.	Glucose and fructose.
Lactase	Small intestine.	Lactose.	Glucose and galactose.
Enterokinase	Small intestine.	Trypsinogen.	Trypsin.
Peptidases			
Aminopeptidase	Small intestine.	Terminal amino acid at amino end of peptides.	Peptides and amino acids.
Dipeptidase	Small intestine.	Dipeptides.	Amino acids.
Nucleosidases and phosphatases	Small intestine.	Nucleotides.	Nitrogenous bases, pentoses, and phosphates.

testinal distention, which initiates nerve impulses to the enteric plexuses and the central nervous system. Enteric reflexes and returning parasympathetic impulses from the CNS increase motility; sympathetic impulses decrease intestinal motility. Migrating motility complexes strengthen when most nutrients and water have been absorbed—that is, when the walls of the small intestine are less distended. With more vigorous peristalsis, the chyme moves along toward the

Figure 24.24 Absorption of digested nutrients in the small intestine. For simplicity, all digested foods are shown in the lumen of the small intestine, even though some nutrients are digested by brush-border enzymes.

🔑 **Long-chain fatty acids and monoglycerides are absorbed into lacteals; other products of digestion enter blood capillaries.**

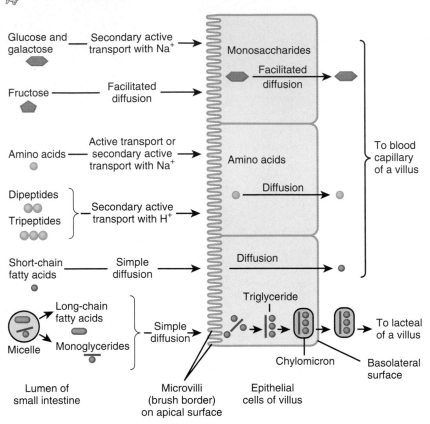

(a) Mechanisms for movement of nutrients through epithelial cells of the villi

large intestine as fast as 10 cm/sec. The first remnants of a meal reach the beginning of the large intestine in about 4 hours.

Absorption in the Small Intestine

All the chemical and mechanical phases of digestion from the mouth through the small intestine are directed toward changing food into forms that can pass through the epithelial cells lining the mucosa and into the underlying blood and lymphatic vessels. These forms are monosaccharides (glucose, fructose, and galactose) from carbohydrates; single amino acids, dipeptides, and tripeptides from proteins; and fatty acids, glycerol, and monoglycerides from triglycerides. Passage of these digested nutrients from the gastrointestinal tract into the blood or lymph is called **absorption.**

Absorption of materials occurs via diffusion, facilitated diffusion, osmosis, and active transport. About 90% of all absorption of nutrients occurs in the small intestine; the other 10% occurs in the stomach and large intestine. Any

undigested or unabsorbed material left in the small intestine passes on to the large intestine.

Absorption of Monosaccharides

All carbohydrates are absorbed as monosaccharides. The capacity of the small intestine to absorb monosaccharides is huge—an estimated 120 grams per hour. As a result, all dietary carbohydrates that are digested normally are absorbed, leaving only indigestible cellulose and fibers in the feces. Monosaccharides pass from the lumen through the apical membrane via *facilitated diffusion* or *active transport*. Fructose, a monosaccharide found in fruits, is transported via *facilitated diffusion;* glucose and galactose are transported into epithelial cells of the villi via *secondary active transport* that is coupled to the active transport of Na^+ (Figure 24.24a). The transporter has binding sites for one glucose molecule and two sodium ions; unless all three sites are filled, neither substance is transported. Galactose competes with glucose to ride the same transporter. (Because both Na^+ and glucose or galactose move in the same direction,

Figure 24.24 *(continued)*

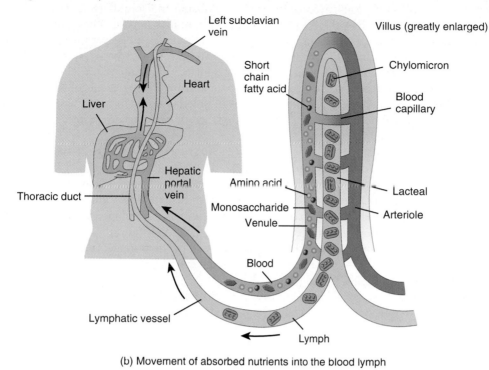

(b) Movement of absorbed nutrients into the blood lymph

Q A monoglyceride may be larger than an amino acid. Why can monoglycerides be absorbed by simple diffusion, whereas amino acids cannot?

this is a *symporter.* The same type of Na$^+$–glucose symporter reabsorbs filtered blood glucose in the tubules of the kidneys; see Figure 26.12 on page 932.) Monosaccharides then move out of the epithelial cells through their basolateral surfaces via *facilitated diffusion* and enter the capillaries of the villi (see Figure 24.24a, b).

Absorption of Amino Acids, Dipeptides, and Tripeptides

Most proteins are absorbed as amino acids via *active transport* processes that occur mainly in the duodenum and jejunum. About half of the absorbed amino acids are present in food, whereas the other half comes from proteins in digestive juices and dead cells that slough off the mucosal surface! Normally, 95–98% of the protein present in the small intestine is digested and absorbed. Several transporters carry different types of amino acids. Some amino acids enter epithelial cells of the villi via Na$^+$-dependent secondary active transport processes that are similar to the glucose transporter; other amino acids are actively transported by themselves. At least one symporter brings in dipeptides and tripeptides together with H$^+$; the peptides then are hydrolyzed to single amino acids inside the epithelial cells.

Amino acids move out of the epithelial cells via diffusion and enter capillaries of the villus (see Figure 24.24a, b). Both monosaccharides and amino acids are transported in the blood to the liver by way of the hepatic portal system. If not removed by hepatocytes, they enter the general circulation.

Absorption of Lipids

All dietary lipids are absorbed via *simple diffusion.* Adults absorb about 95% of the lipids present in the small intestine; due to their lower production of bile, newborn infants absorb only about 85% of lipids. As a result of their emulsification and digestion, triglycerides are broken down into monoglycerides and fatty acids. Recall that lingual and pancreatic lipases remove two of the three fatty acids from glycerol during digestion of a triglyceride; the other fatty acid remains attached to glycerol, thus forming a monoglyceride. The small amount of short-chain fatty acids (having fewer than 10–12 carbon atoms) in the diet passes into the epithelial cells via simple diffusion and follows the same route taken by monosaccharides and amino acids into a blood capillary of a villus (see Figure 24.24a ,b).

Most dietary fatty acids, however, are long-chain fatty acids. They and monoglycerides reach the bloodstream by a different route and require bile for adequate absorption. Bile salts are amphipathic; they have both polar (hydrophilic) and nonpolar (hydrophobic) portions. Thus, they can form tiny spheres called **micelles** (mī-SELZ; = small morsels), which are 2–10 nm in diameter and include 20–50 bile salt molecules. Because they are small and have the polar portions of bile salt molecules at their surface, micelles can dissolve in the water of intestinal fluid. In contrast, partially digested dietary lipids can dissolve in the nonpolar central core of micelles. It is in this form that fatty acids and monoglycerides reach the epithelial cells of the villi.

At the apical surface of the epithelial cells, fatty acids and monoglycerides diffuse into the cells, leaving the micelles behind in chyme. The micelles continually repeat this ferrying function. When chyme reaches the ileum, 90–95% of the bile salts are reabsorbed and returned by the blood to the liver through the hepatic portal system for recycling. This cycle of bile salt secretion by hepatocytes into bile, reabsorption by the ileum, and resecretion into bile is called the **enterohepatic circulation.** Insufficient bile salts, due either to obstruction of the bile ducts or removal of the gallbladder, can result in the loss of up to 40% of dietary lipids in feces due to diminished lipid absorption. Moreover, when lipids are not absorbed properly, the fat-soluble vitamins—A, D, E, and K—are not adequately absorbed.

Within the epithelial cells, many monoglycerides are further digested by lipase to glycerol and fatty acids. The fatty acids and glycerol are then recombined to form triglycerides, which aggregate into globules along with phospholipids and cholesterol and become coated with proteins. These large (about 80 nm in diameter) spherical masses are called **chylomicrons.** The hydrophilic protein coat keeps the chylomicrons suspended and prevents them from sticking to each other. Chylomicrons leave the epithelial cell via exocytosis. Because they are so large and bulky, chylomicrons cannot enter blood capillaries in the small intestine; instead, they enter the much leakier lacteals. From there they are transported by way of lymphatic vessels to the thoracic duct and enter the blood at the left subclavian vein (see Figure 24.24b).

Within 10 minutes after their absorption, about half of the chylomicrons have already been removed from the blood as they pass through blood capillaries in the liver and adipose tissue. This removal is accomplished by an enzyme in capillary endothelial cells, called **lipoprotein lipase,** that breaks down triglycerides in chylomicrons and other lipoproteins into fatty acids and glycerol. The fatty acids diffuse into hepatocytes and adipose cells and combine with glycerol during resynthesis of triglycerides. Two or three hours after a meal, few chylomicrons remain in the blood.

Absorption of Electrolytes

Many of the electrolytes absorbed by the small intestine come from gastrointestinal secretions, and some are part of ingested foods and liquids. Sodium ions are actively transported out of intestinal epithelial cells by sodium pumps

(Na^+/K^+ ATPase) after they have moved into epithelial cells via diffusion and secondary active transport. Thus, most of the sodium ions in gastrointestinal secretions are reclaimed and not lost in the feces. Negatively charged chloride, iodide, and nitrate ions can passively follow Na^+ or be actively transported. Calcium ions also are absorbed actively in a process stimulated by calcitriol. Other electrolytes such as iron, potassium, magnesium, and phosphate ions also are absorbed via active transport mechanisms.

Absorption of Vitamins

The fat-soluble vitamins A, D, E, and K are included with ingested dietary lipids in micelles and are absorbed via simple diffusion. Most water-soluble vitamins, such as most B vitamins and vitamin C, also are absorbed via simple diffusion. Vitamin B_{12}, however, combines with intrinsic factor produced by the stomach, and the combination is absorbed in the ileum via an active transport mechanism.

Absorption of Water

The total volume of fluid that enters the small intestine each day—about 9.3 liters (9.8 qt)—comes from ingestion of liquids (about 2.3 liters) and from various gastrointestinal secretions (about 7.0 liters). Figure 24.25 depicts the amounts of fluid ingested, secreted, absorbed, and excreted by the GI tract. The small intestine absorbs about 8.3 liters of the fluid; the remainder passes into the large intestine, where most of the rest of it—about 0.9 liter—is also absorbed. Only 0.1 liter (100 mL) of water is excreted in the feces each day.

All water absorption in the GI tract occurs via *osmosis* from the lumen of the intestines through epithelial cells and into blood capillaries. Because water can move across the intestinal mucosa in both directions, the absorption of water from the small intestine depends on the absorption of electrolytes and nutrients to maintain an osmotic balance with the blood. The absorbed electrolytes, monosaccharides, and amino acids establish a concentration gradient for water that promotes water absorption via osmosis.

Table 24.6 summarizes the digestive activities of the pancreas, liver, gallbladder, and small intestine.

1. What are the regions of the small intestine?
2. In what ways are the mucosa and submucosa of the small intestine adapted for digestion and absorption?
3. Describe the types of movement in the small intestine.
4. Explain the function of each digestive enzyme.
5. How is small intestinal secretion regulated?
6. Define absorption. How are the end products of carbohydrate and protein digestion absorbed? How are the end products of lipid digestion absorbed?
7. By what routes do absorbed nutrients reach the liver?
8. Describe the absorption of electrolytes, vitamins, and water by the small intestine.

Figure 24.25 Daily volumes of fluid ingested, secreted, absorbed, and excreted from the GI tract.

🔑 **All water absorption in the GI tract occurs via osmosis.**

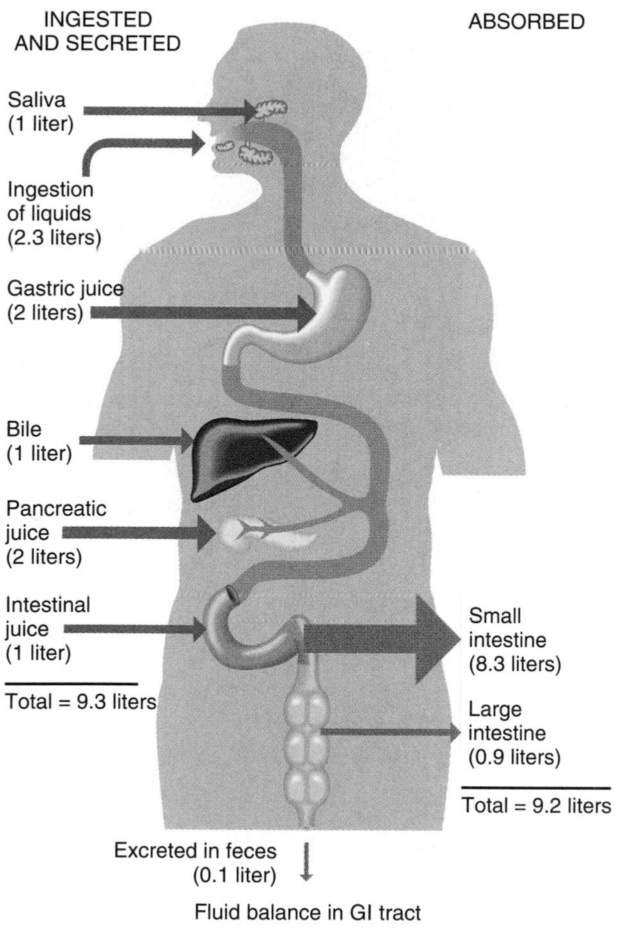

INGESTED
AND SECRETED

ABSORBED

Saliva
(1 liter)

Ingestion
of liquids
(2.3 liters)

Gastric juice
(2 liters)

Bile
(1 liter)

Pancreatic
juice
(2 liters)

Intestinal
juice
(1 liter)

Total = 9.3 liters

Small
intestine
(8.3 liters)

Large
intestine
(0.9 liters)

Total = 9.2 liters

Excreted in feces
(0.1 liter)

Fluid balance in GI tract

Q Which two organs of the digestive system secrete the most fluid?

LARGE INTESTINE

OBJECTIVE

• *Describe the anatomy, histology, and functions of the large intestine.*

The large intestine is the terminal portion of the GI tract and is divided into four principal regions. The overall functions of the large intestine are the completion of absorption, the production of certain vitamins, the formation of feces, and the expulsion of feces from the body.

Anatomy of the Large Intestine

The **large intestine,** which is about 1.5 m (5 ft) long and 6.5 cm (2.5 in.) in diameter, extends from the ileum to the anus and is attached to the posterior abdominal wall by its **mesocolon,** which is a double layer of peritoneum. Structurally, the four principal regions of the large intestine are the cecum, colon, rectum, and anal canal (Figure 24.26a).

Table 24.6 Summary of Digestive Activities in the Pancreas, Liver, Gallbladder, and Small Intestine

STRUCTURE	ACTIVITIES
Pancreas	Delivers pancreatic juice into the duodenum via the pancreatic duct (see Table 24.5 for pancreatic enzymes and their functions).
Liver	Produces bile (bile salts) necessary for emulsification and absorption of lipids.
Gallbladder	Stores, concentrates, and delivers bile into the duodenum via the common bile duct.
Small Intestine	Major site of digestion and absorption of nutrients and water in the gastrointestinal tract.
Mucosa/ submucosa	
Intestinal glands	Secrete intestinal juice.
Duodenal (Brunner's) glands	Secrete alkaline fluid to buffer stomach acids, and mucus for protection and lubrication.
Microvilli	Microscopic, membrane-covered projections of epithelial cells that contain brush-border enzymes (listed in Table 24.5) and increase surface area for absorption and digestion.
Villi	Fingerlike projections of mucosa that are the sites of absorption of digested food and that increase the surface area for digestion and absorption.
Circular folds	Folds of mucosa and submucosa that increase surface area for absorption and digestion.
Muscularis	
Segmentation	Consists of alternating contractions of circular smooth muscle fibers that produce segmentation and resegmentation of portions of the small intestine; mixes chyme with digestive juices and brings food into contact with the mucosa for absorption.
Migrating motility complex (MMC)	A type of peristalsis consisting of waves of contraction and relaxation of circular and longitudinal smooth muscle fibers passing the length of the small intestine; moves chyme toward ileocecal sphincter.

The opening from the ileum into the large intestine is guarded by a fold of mucous membrane called the **ileocecal sphincter,** which allows materials from the small intestine to pass into the large intestine. Hanging inferior to the ileocecal valve is the **cecum,** a blind pouch about 6 cm (2.4 in.) long. Attached to the cecum is a twisted, coiled tube, measuring about 8 cm (3 in.) in length, called the **appendix** or **vermiform appendix** (*vermiform* = worm-shaped; *appendix* = appendage). The mesentery of the appendix, called the **mesoappendix,** attaches the appendix to the inferior part of the mesentery of the ileum.

Figure 24.26 Anatomy of large intestine. (See Tortora, *A Photographic Atlas of the Human Body,* Figure 12.13)

🔑 **The regions of the large intestine are the cecum, colon, rectum, and anal canal.**

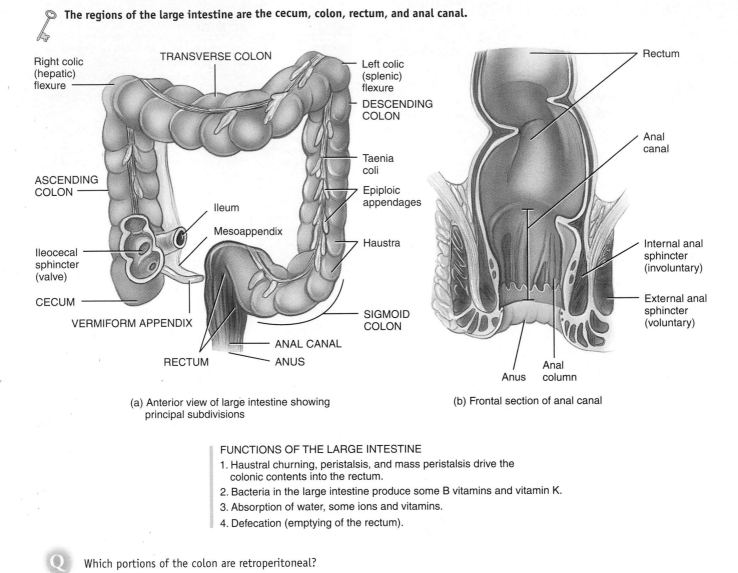

(a) Anterior view of large intestine showing principal subdivisions

(b) Frontal section of anal canal

FUNCTIONS OF THE LARGE INTESTINE
1. Haustral churning, peristalsis, and mass peristalsis drive the colonic contents into the rectum.
2. Bacteria in the large intestine produce some B vitamins and vitamin K.
3. Absorption of water, some ions and vitamins.
4. Defecation (emptying of the rectum).

Q Which portions of the colon are retroperitoneal?

The open end of the cecum merges with a long tube called the **colon** (= food passage), which is divided into ascending, transverse, descending, and sigmoid portions. Both the ascending and descending colon are retroperitoneal, whereas the transverse and sigmoid colon are not. The **ascending colon** ascends on the right side of the abdomen, reaches the inferior surface of the liver, and turns abruptly to the left to form the **right colic (hepatic) flexure.** The colon continues across the abdomen to the left side as the **transverse colon.** It curves beneath the inferior end of the spleen on the left side as the **left colic (splenic) flexure** and passes inferiorly to the level of the iliac crest as the **descending colon.** The **sigmoid colon** (*sigm-* = S-shaped) begins near the left iliac crest, projects medially to the midline, and terminates as the rectum at about the level of the third sacral vertebra.

The **rectum,** the last 20 cm (8 in.) of the GI tract, lies anterior to the sacrum and coccyx. The terminal 2–3 cm (1 in.) of the rectum is called the **anal canal** (Figure 24.26b). The mucous membrane of the anal canal is arranged in longitudinal folds called **anal columns** that contain a network of arteries and veins. The opening of the anal canal to the exterior, called the **anus,** is guarded by an internal sphincter of smooth muscle (involuntary) and an external sphincter of skeletal muscle (voluntary). Normally the anus is closed except during the elimination of feces.

CLINICAL APPLICATION
Appendicitis

Inflammation of the appendix, termed **appendicitis,** is preceded by obstruction of the lumen of the appendix by chyme, inflammation, a foreign body, a carcinoma of the cecum, stenosis, or kinking of the organ. It is characterized by high fever, elevated white cell count, and a neutrophil count higher than 75%. The infection that follows may result in edema and ischemia and may progress to gangrene and perforation within 24 to 36 hours. Typically, appendicitis begins with referred pain in the umbilical region of the abdomen, followed by anorexia (loss of appetite for food), nausea, and vomiting. After several hours the pain localizes in the right lower quadrant (RLQ) and is continuous, dull or severe, and intensified by coughing, sneezing, or body movements. Early appendectomy (removal of the appendix) is recommended because it is safer to operate than to risk rupture, peritonitis, and gangrene. ■

Histology of the Large Intestine

The wall of the large intestine differs from that of the small intestine in several respects. No villi or permanent circular folds are found in the **mucosa,** which consists of simple columnar epithelium, lamina propria (areolar connective tissue), and muscularis mucosae (smooth muscle) (Figure 24.27a). The epithelium contains mostly absorptive and goblet cells (Figure 24.27b and c). The absorptive cells function primarily in water absorption, whereas the goblet cells secrete mucus that lubricates the passage of the colonic contents. Both absorptive and goblet cells are located in long, straight, tubular intestinal glands that extend the full thickness of the mucosa. Solitary lymphatic nodules are also found in the mucosa. The **submucosa** of the large intestine is similar to that found in the rest of the GI tract. The **muscularis** consists of an external layer of longitudinal smooth muscle and an internal layer of circular smooth muscle. Unlike other parts of the GI tract, portions of the longitudinal muscles are thickened, forming three conspicuous longitudinal bands called **taeniae coli** (TĒ-nē-ē KŌ-lī; *taenia* = flat band), that run most of the length of the large intestine (see Figure 24.26a). The taeniae coli are separated by portions of the wall with less or no longitudinal muscle. Tonic contractions of the bands gather the colon into a series of pouches called **haustra** (HAWS-tra; = shaped like pouches), which give the colon a puckered appearance. A single layer of circular smooth muscle lies between taeniae coli. The **serosa** of the large intestine is part of the visceral peritoneum. Small pouches of visceral peritoneum filled with fat are attached to taeniae coli and are called **epiploic appendages.**

Mechanical Digestion in the Large Intestine

The passage of chyme from the ileum into the cecum is regulated by the action of the ileocecal sphincter. Normally, the valve remains partially closed so that the passage of chyme into the cecum usually occurs slowly. Immediately after a meal, a **gastroileal reflex** intensifies ileal peristalsis and forces any chyme in the ileum into the cecum. The hormone gastrin also relaxes the sphincter. Whenever the cecum is distended, the degree of contraction of the ileocecal sphincter intensifies.

Movements of the colon begin when substances pass the ileocecal sphincter. Because chyme moves through the small intestine at a fairly constant rate, the time required for a meal to pass into the colon is determined by gastric emptying time. As food passes through the ileocecal sphincter, it fills the cecum and accumulates in the ascending colon.

One movement characteristic of the large intestine is **haustral churning.** In this process, the haustra remain relaxed and become distended while they fill up. When the distension reaches a certain point, the walls contract and squeeze the contents into the next haustrum. **Peristalsis** also occurs, although at a slower rate (3–12 contractions per minute) than in more proximal portions of the tract. A final type of movement is **mass peristalsis,** a strong peristaltic wave that begins at about the middle of the transverse colon and quickly drives the colonic contents into the rectum. Because food in the stomach initiates this **gastrocolic reflex** in the colon, mass peristalsis usually takes place three or four times a day, during or immediately after a meal.

Chemical Digestion in the Large Intestine

The final stage of digestion occurs in the colon through the activity of bacteria that inhabit the lumen. Mucus is secreted by the glands of the large intestine, but no enzymes are secreted. Chyme is prepared for elimination by the action of bacteria, which ferment any remaining carbohydrates and release hydrogen, carbon dioxide, and methane gases. These gases contribute to flatus (gas) in the colon, termed *flatulence* when it is excessive. Bacteria also convert remaining proteins to amino acids and break down the amino acids into simpler substances: indole, skatole, hydrogen sulfide, and fatty acids. Some of the indole and skatole is eliminated in the feces and contributes to their odor; the rest is absorbed and transported to the liver, where these compounds are converted to less toxic compounds and excreted in the urine. Bacteria also decompose bilirubin to simpler pigments, including stercobilin, which give feces their brown color. Several vitamins needed for normal metabolism, including some B vitamins and vitamin K, are bacterial products that are absorbed in the colon.

Absorption and Feces Formation in the Large Intestine

By the time chyme has remained in the large intestine 3–10 hours, it has become solid or semisolid as a result of water absorption and is now called **feces.** Chemically, feces consist of water, inorganic salts, sloughed-off epithelial cells from the mucosa of the gastrointestinal tract, bacteria, products of bacterial decomposition, unabsorbed digested materials, and indigestible parts of food.

Although most water absorption occurs in the small intestine, the large intestine absorbs enough to make it an important organ in maintaining the body's water balance. Of

Figure 24.27 Histology of the large intestine.

🔑 **Intestinal glands formed by simple columnar epithelial cells and goblet cells extend the full thickness of the mucosa.**

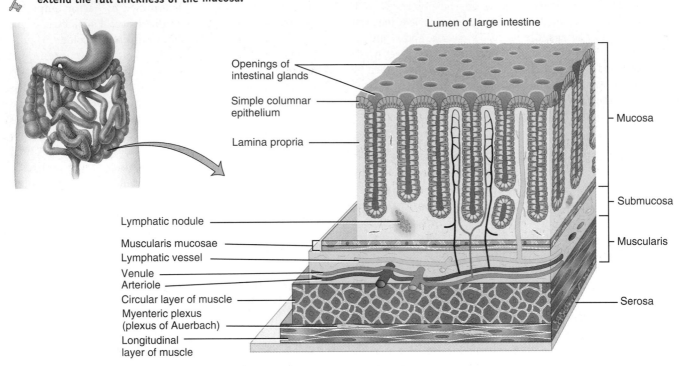

(a) Layers of the large intestine

the 0.5–1.0 liter of water that enters the large intestine, all but about 100–200 mL is absorbed via osmosis. The large intestine also absorbs electrolytes, including sodium and chloride, and some vitamins.

The Defecation Reflex

Mass peristaltic movements push fecal material from the sigmoid colon into the rectum. The resulting distention of the rectal wall stimulates stretch receptors, which initiates a **defecation reflex** that empties the rectum. The defecation reflex occurs as follows: In response to distention of the rectal wall, the receptors send sensory nerve impulses to the sacral spinal cord. Motor impulses from the cord travel along parasympathetic nerves back to the descending colon, sigmoid colon, rectum, and anus. The resulting contraction of the longitudinal rectal muscles shortens the rectum, thereby increasing the pressure within it. This pressure, along with voluntary contractions of the diaphragm and abdominal muscles, plus parasympathetic stimulation, open the internal sphincter.

The external sphincter is voluntarily controlled. If it is voluntarily relaxed, defecation occurs and the feces are expelled through the anus; if it is voluntarily constricted, defecation can be postponed. Voluntary contractions of the diaphragm and abdominal muscles aid defecation by increasing the pressure within the abdomen, which pushes the walls of the sigmoid colon and rectum inward. If defecation does not

occur, the feces back up into the sigmoid colon until the next wave of mass peristalsis again stimulates the stretch receptors, again creating the urge to defecate. In infants, the defecation reflex causes automatic emptying of the rectum because voluntary control of the external anal sphincter has not yet developed.

Diarrhea (dī-a-RĒ-a; *dia-* = through; *rrhea* = flow) is an increase in the frequency, volume, and fluid content of the feces caused by increased motility of and decreased absorption by the intestines. When chyme passes too quickly through the small intestine and feces pass too quickly through the large intestine, there is not enough time for absorption. Frequent diarrhea can result in dehydration and electrolyte imbalances. Excessive motility may be caused by lactose intolerance, stress, and microbes that irritate the gastrointestinal mucosa.

Constipation (kon-sti-PĀ-shun; *con-* = together; *stip-* = to press) refers to infrequent or difficult defecation caused by decreased motility of the intestines. Because the feces remain in the colon for prolonged periods of time, excessive water absorption occurs, and the feces become dry and hard. Constipation may be caused by poor habits (delaying defecation), spasms of the colon, insufficient fiber in the diet, inadequate fluid intake, lack of exercise, emotional stress, and certain drugs. A common treatment is a mild laxative, such as milk of magnesia, which induces defecation. However,

Figure 24.27 *(continued)*

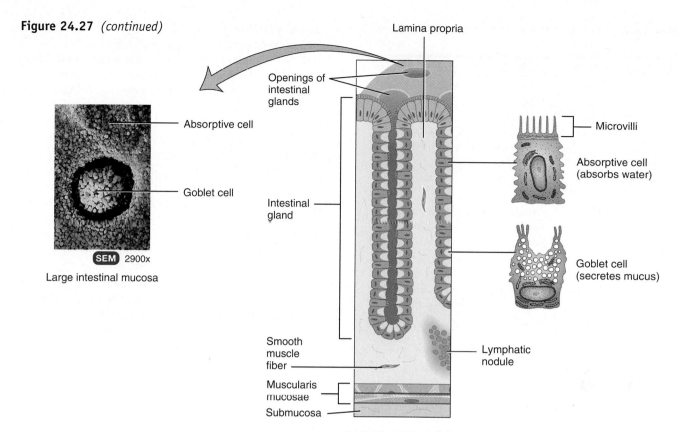

 What is the function of goblet cells?

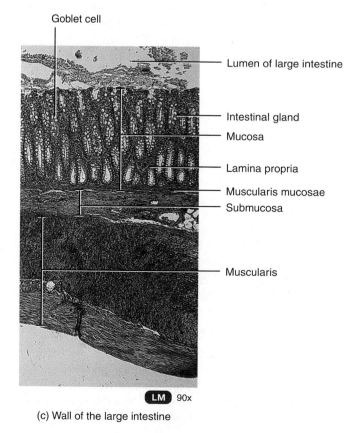

(b) Sectional view of the large intestinal mucosa showing intestinal glands and cell types

(c) Wall of the large intestine

many physicians maintain that laxatives are habit-forming, and that adding fiber to the diet, increasing the amount of exercise, and increasing fluid intake are safer ways of controlling this common problem.

Table 24.7 summarizes the digestive activities in the large intestine.

CLINICAL APPLICATION
Dietary Fiber

Dietary fiber consists of indigestible plant carbohydrates—such as cellulose, lignin, and pectin—found in fruits, vegetables, grains, and beans. **Insoluble fiber,** which does not dissolve in water, includes the woody or structural parts of plants such as the skins of fruits and vegetables and the bran coating around wheat and corn kernels. Insoluble fiber passes through the GI tract largely unchanged and speeds up the passage of material through the tract. **Soluble fiber** dissolves in water and forms a gel, which slows the passage of material through the tract; it is found in abundance in beans, oats, barley, broccoli, prunes, apples, and citrus fruits.

People who choose a fiber-rich diet may reduce their risk of developing obesity, diabetes, atherosclerosis, gallstones,

Table 24.7 Summary of Digestive Activities in the Large Intestine

STRUCTURE	ACTIVITY	FUNCTION(S)
Lumen	Bacterial activity.	Breaks down undigested carbohydrates, proteins, and amino acids into products that can be expelled in feces or absorbed and detoxified by liver; synthesizes certain B vitamins and vitamin K.
Mucosa	Secretes mucus.	Lubricates colon and protects mucosa.
	Absorbs water and other soluble compounds.	Maintains water balance; solidifies feces; absorbs vitamins and some ions.
Muscularis	Haustral churning.	Moves contents from haustrum to haustrum by muscular contractions.
	Peristalsis.	Moves contents along length of colon by contractions of circular and longitudinal muscles.
	Mass peristalsis.	Forces contents into sigmoid colon and rectum.
	Defecation reflex.	Eliminates feces by contractions in sigmoid colon and rectum.

hemorrhoids, diverticulitis, appendicitis, and colorectal cancer. Soluble fiber also may help lower blood cholesterol because the fiber binds bile salts and prevents their reabsorption; as a result, more cholesterol is used to replace the bile salts lost in the feces. ■

1. What are the principal regions of the large intestine?
2. How does the muscularis of the large intestine differ from that of the rest of the gastrointestinal tract? What are haustra?
3. Describe the mechanical movements that occur in the large intestine.
4. Define defecation. How does it occur?
5. Explain the activities of the large intestine that change its contents into feces.

DEVELOPMENTAL ANATOMY OF THE DIGESTIVE SYSTEM

OBJECTIVE

• *Describe the development of the digestive system.*

About the 14th day after fertilization, the cells of the endoderm form a cavity referred to as the **primitive gut** (Figure 24.28a). Soon after the mesoderm forms and splits into two layers (somatic and splanchnic), the splanchnic mesoderm associates with the endoderm of the primitive gut; as a result, the primitive gut has a double-layered wall. The **endodermal layer** gives rise to the *epithelial lining* and *glands* of most of the gastrointestinal tract; the **mesodermal layer** produces the *smooth muscle* and *connective tissue* of the tract.

The primitive gut elongates, and during the third week it differentiates into an anterior **foregut,** an intermediate **midgut,** and a posterior **hindgut** (Figure 24.28b). Until the fifth week of development, the midgut opens into the yolk sac; after that time, the yolk sac constricts and detaches from the midgut, and the midgut seals. In the region of the foregut, a depression consisting of ectoderm, the **stomodeum,** appears (Figure 24.28c). This develops into the *oral cavity.* The **oral membrane** that separates the foregut from the stomodeum ruptures during the fourth week of development, so that the foregut is continuous with the outside of the embryo through the oral cavity. Another depression consisting of ectoderm, the **proctodeum,** forms in the hindgut and goes on to develop into the *anus.* The **cloacal membrane,** which separates the hindgut from the proctodeum, ruptures, so that the hindgut is continuous with the outside of the embryo through the anus. Thus, the GI tract forms a continuous tube from mouth to anus.

The foregut develops into the *pharynx, esophagus, stomach,* and a *portion of the duodenum.* The midgut is transformed into the *remainder of the duodenum,* the *jejunum,* the *ileum,* and *portions of the large intestine* (cecum, appendix, ascending colon, and most of the transverse colon). The hindgut develops into the *remainder of the large intestine,* except for a portion of the anal canal that is derived from the proctodeum.

As development progresses, the endoderm at various places along the foregut develops into hollow buds that grow into the mesoderm. These buds will develop into the *salivary glands, liver, gallbladder,* and *pancreas* (Figure 24.28d). Each of the glands retains a connection with the gastrointestinal tract through ducts.

1. What structures develop from the foregut, midgut, and hindgut?

AGING AND THE DIGESTIVE SYSTEM

OBJECTIVE

• *Describe the effects of aging on the digestive system.*

Overall changes of the digestive system associated with aging include decreased secretory mechanisms, decreased motility of the digestive organs, loss of strength and tone of the muscular tissue and its supporting structures, changes in neurosensory feedback regarding enzyme and hormone release, and diminished response to pain and internal sensations. In the upper portion of the GI tract, common changes include reduced sensitivity to mouth irritations and sores, loss of taste, periodontal disease, difficulty in swallowing, hi-

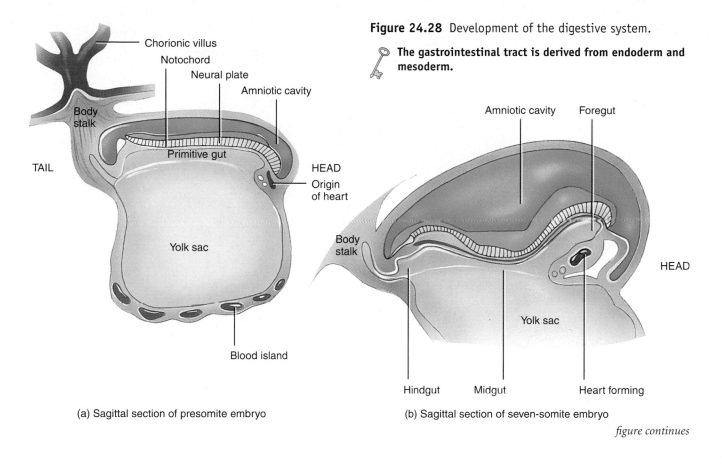

Figure 24.28 Development of the digestive system.

The gastrointestinal tract is derived from endoderm and mesoderm.

(a) Sagittal section of presomite embryo

(b) Sagittal section of seven-somite embryo

figure continues

atal hernia, gastritis, and peptic ulcer disease. Changes that may appear in the small intestine include duodenal ulcers, appendicitis, malabsorption, and maldigestion. Other pathologies that increase in incidence with age are gallblad-

der problems, jaundice, cirrhosis, and acute pancreatitis. Large intestinal changes such as constipation, hemorrhoids, and diverticular disease may also occur. Cancer of the colon or rectum is quite common.

DISORDERS: HOMEOSTATIC IMBALANCES

DENTAL CARIES

Dental caries, or tooth decay, involve a gradual demineralization (softening) of the enamel and dentin. If untreated, microorganisms may invade the pulp, causing inflammation and infection, with subsequent death of the pulp and abscess of the alveolar bone surrounding the root's apex. Such teeth are treated by root canal therapy.

Dental caries begin when bacteria, acting on sugars, produce acids that demineralize the enamel. **Dextran,** a sticky polysaccharide produced from sucrose, causes the bacteria to stick to the teeth. Masses of bacterial cells, dextran, and other debris adhering to teeth constitute **dental plaque.** Saliva cannot reach the tooth surface to buffer the acid because the plaque covers the teeth. Brushing the teeth immediately after eating removes the plaque from flat surfaces before the bacteria can produce acids. Dentists also recommend that the plaque between the teeth be removed every 24 hours with dental floss.

PERIODONTAL DISEASE

Periodontal disease is a collective term for a variety of conditions characterized by inflammation and degeneration of the gingivae, alveolar bone, periodontal ligament, and cementum. One such condition is called **pyorrhea,** the initial symptoms of which are enlargement and inflammation of the soft tissue and bleeding of the gums. Without treatment, the soft tissue may deteriorate and the alveolar bone may be resorbed, causing loosening of the teeth and recession of the gums. Periodontal diseases are often caused by poor oral hygiene; by local irritants, such as bacteria, impacted food, and cigarette smoke; or by a poor "bite."

PEPTIC ULCER DISEASE

In the United States, 5–10% of the population develops **peptic ulcer disease (PUD).** An **ulcer** is a craterlike lesion in a membrane; ulcers that develop in areas of the GI tract exposed to acidic gastric juice are called **peptic ulcers.** The most common complication of peptic ulcers is bleeding, which can lead to anemia if enough

Figure 24.28 (continued)

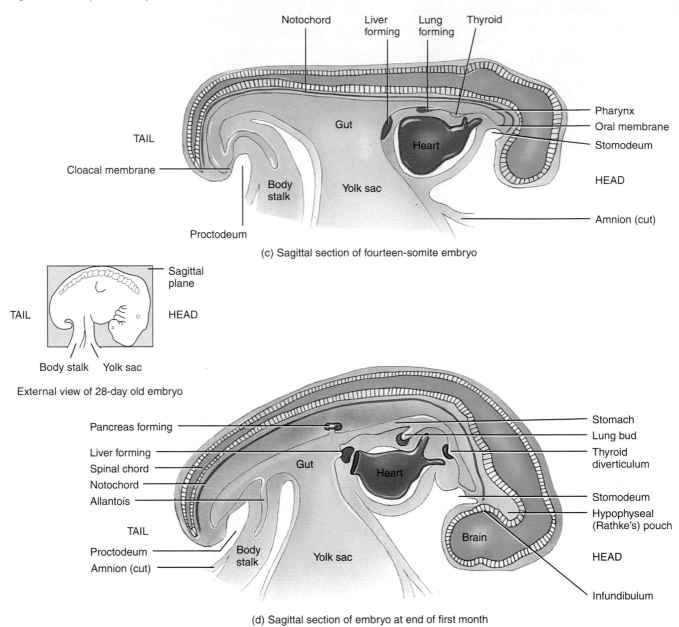

(c) Sagittal section of fourteen-somite embryo

External view of 28-day old embryo

(d) Sagittal section of embryo at end of first month

Q When does the gastrointestinal tract begin to develop?

blood is lost. In acute cases, peptic ulcers can lead to shock and death. Three distinct causes of PUD are recognized: (1) the bacterium *Helicobacter pylori;* (2) nonsteroidal anti-inflammatory drugs (NSAIDs) such as aspirin; and (3) hypersecretion of HCl, as occurs in Zollinger–Ellison syndrome, a gastrin-producing tumor usually of the pancreas.

Helicobacter pylori (previously named *Campylobacter pylori*) is the most frequent cause of PUD. The bacterium produces an enzyme called urease, which splits urea into ammonia and carbon dioxide. While shielding the bacterium from the acidity of the stomach, the ammonia also damages the protective mucous layer of

the stomach and the underlying gastric cells. *H. pylori* also produces catalase, an enzyme that may protect the microbe from phagocytosis by neutrophils, plus several adhesion proteins that allow the bacterium to attach itself to gastric cells.

Several therapeutic approaches are helpful in the treatment of PUD. Because cigarette smoke, alcohol, caffeine, and NSAIDs can impair mucosal defensive mechanisms, in the process increasing mucosal susceptibility to the damaging effects of HCl, these substances should be avoided. In cases associated with *H. pylori*, treatment with an antibiotic drug often resolves the problem. Oral antacids such as Tums or Maalox can help temporarily by buffering gastric acid. When hypersecretion of HCl is the cause of PUD, H_2 blockers (such as Tagamet) or proton pump inhibitors such as omeprazole (Prilosec), which block secretion of H^+ from parietal cells, may be used.

DIVERTICULITIS

Diverticulosis is the development of diverticula, saclike outpouchings of the wall of the colon in places where the muscularis has become weak. Many people who develop diverticulosis have no symptoms and experience no complications, but about 15% of people with diverticulosis will eventually develop an inflammation known as **diverticulitis.** This condition may be characterized by pain, either constipation or increased frequency of defecation, nausea, vomiting, and low-grade fever. Because diets low in fiber contribute to development of diverticulitis, patients who change to high-fiber diets show marked relief of symptoms. In severe cases, affected portions of the colon may require surgical removal. If diverticula rupture, the release of bacteria into the abdominal cavity can cause peritonitis.

COLORECTAL CANCER

Colorectal cancer is among the deadliest of malignancies, ranking second to lung cancer in males and third after lung cancer and breast cancer in females. Genetics plays a very important role in that an inherited predisposition contributes to more than half of all cases of colorectal cancer. Intake of alcohol and diets high in animal fat and protein are associated with increased risk of colorectal cancer, whereas dietary fiber, retinoids, calcium, and selenium may be protective. Signs and symptoms of colorectal cancer include diarrhea, constipation, cramping, abdominal pain, and rectal bleeding, either visible or occult. Screening for colorectal cancer includes testing for blood in the feces, digital rectal examination, sigmoidoscopy, colonoscopy, and barium enema. Tumors may be removed endoscopically or surgically.

HEPATITIS

Hepatitis is an inflammation of the liver that can be caused by viruses, drugs, and chemicals, including alcohol. Clinically, several types of viral hepatitis are recognized. **Hepatitis A (infectious hepatitis)** is caused by hepatitis A virus and is spread via fecal contamination of objects such as food, clothing, toys, and eating utensils (fecal–oral route). It is generally a mild disease of children and young adults characterized by loss of appetite, malaise, nausea, diarrhea, fever, and chills. Eventually, jaundice appears. This type of hepatitis does not cause lasting liver damage. Most people recover in 4–6 weeks.

Hepatitis B is caused by hepatitis B virus and is spread primarily by sexual contact and contaminated syringes and transfusion equipment. It can also be spread via saliva and tears. Hepatitis B virus can be present for years or even a lifetime, and it can produce cirrhosis and possibly cancer of the liver. Individuals who harbor the active hepatitis B virus are at risk for cirrhosis and also become carriers. Vaccines produced through recombinant DNA technology (for example, Recombivax HB) are available to prevent hepatitis B infection.

Hepatitis C, caused by hepatitis C virus, is clinically similar to hepatitis B and is often spread by blood transfusions. Hepatitis C can cause cirrhosis and possibly liver cancer.

Hepatitis D is caused by hepatitis D virus. It is transmitted like hepatitis B and, in fact, a person must be coinfected with hepatitis B before contracting hepatitis D. Hepatitis D results in severe liver damage and has a higher fatality rate than infection with hepatitis B virus alone.

Hepatitis E is caused by hepatitis E virus and is spread like hepatitis A. Although it does not cause chronic liver disease, hepatitis E virus is responsible for a very high mortality rate in pregnant women.

ANOREXIA NERVOSA

Anorexia nervosa is a chronic disorder characterized by self-induced weight loss, negative perception of body image, and physiological changes that result from nutritional depletion. Patients with anorexia nervosa have a fixation on weight control and often insist on having a bowel movement every day despite inadequate food intake. They abuse laxatives, which worsens the fluid and electrolyte imbalances and nutrient deficiencies. The disorder is found predominantly in young, single females, and it may be inherited. Abnormal patterns of menstruation, amenorrhea (absence of menstruation), and a lowered basal metabolic rate reflect the depressant effects of starvation. Individuals may become emaciated and may ultimately die of starvation or one of its complications. Also associated with the disorder are osteoporosis, depression, and brain abnormalities coupled with impaired mental performance. Treatment consists of psychotherapy and dietary regulation.

MEDICAL TERMINOLOGY

Achalasia (ak′-a-LĀ-zē-a; *a-* = without; *chalasis* = relaxation) A condition, caused by malfunction of the myenteric plexus, in which the lower esophageal sphincter fails to relax normally as food approaches. A whole meal may become lodged in the esophagus and enter the stomach very slowly. Distension of the esophagus results in chest pain that is often confused with pain originating from the heart.

Borborygmus (bor′-bō-RIG-mus) A rumbling noise caused by the propulsion of gas through the intestines.

Bulimia (*bu-* = ox; *limia* = hunger) or **binge–purge syndrome** A disorder that typically affects young, single, middle-class, white females, characterized by overeating at least twice a week followed by purging by self-induced vomiting, strict dieting or fasting, vigorous exercise, or use of laxatives or diuretics; it occurs in response to fears of being overweight, stress, depression, and physiological disorders such as hypothalamic tumors.

Canker sore (KANG-ker) Painful ulcer on the mucous membrane of the mouth that affects females more often than males, usually between ages 10 and 40; may be an autoimmune reaction or a food allergy.

Cholecystitis (kō′-lē-sis-TĪ-tis; *chole-* = bile; *cyst-* = bladder; *-itis* = inflammation of) In some cases, an autoimmune inflammation of the gallbladder; other cases are caused by obstruction of the cystic duct by bile stones.

Cirrhosis Distorted or scarred liver as a result of chronic inflammation due to hepatitis, chemicals that destroy hepatocytes, parasites that infect the liver, and alcoholism; the hepatocytes are replaced by fibrous or adipose connective tissue. Symptoms include jaundice, edema in the legs, uncontrolled bleeding, and increased sensitivity to drugs.

Colitis (ko-LĪ-tis) Inflammation of the mucosa of the colon and rectum in which absorption of water and salts is reduced, producing watery, bloody feces and, in severe cases, dehydration and salt depletion. Spasms of the irritated muscularis produce cramps. It is thought to be an autoimmune condition.

Colostomy (kō-LOS-tō-mē; -*stomy* = provide an opening) The diversion of the fecal stream through an opening in the colon, creating a surgical "stoma" (artificial opening) that is affixed to the exterior of the abdominal wall. This opening serves as a substitute anus through which feces are eliminated into a bag worn on the abdomen.

Dysphagia (dis-FĀ-jē-a; *dys-* = abnormal; *phagia* = to eat) Difficulty in swallowing that may be caused by inflammation, paralysis, obstruction, or trauma.

Enteritis (en′-ter-Ī-tis; *enter-* = intestine) An inflammation of the intestine, particularly the small intestine.

Flatus (FLĀ-tus) Air (gas) in the stomach or intestine, usually expelled through the anus. If the gas is expelled through the mouth, it is called **eructation** or **belching** (burping). Flatus may result from gas released during the breakdown of foods in the stomach or from swallowing air or gas-containing substances such as carbonated drinks.

Gastrectomy (gas-TREK-tō-mē; *gastr-* = stomach; *-ectomy* = to cut out) Removal of all or a portion of the stomach.

Gastroscopy (gas-TROS-kō-pē; *-scopy* = to view with a lighted instrument) Endoscopic examination of the stomach in which the examiner can view the interior of the stomach directly to evaluate an ulcer, tumor, inflammation, or source of bleeding.

Hernia (HER-nē-a) Protrusion of all or part of an organ through a membrane or cavity wall, usually the abdominal cavity. *Diaphragmatic (hiatal) hernia* is the protrusion of the lower esophagus, stomach, or intestine into the thoracic cavity through the esophageal hiatus. *Inguinal hernia* is the protrusion of the hernial sac into the inguinal opening; it may contain a portion of the bowel in an advanced stage and may extend into the scrotal compartment in males, causing strangulation of the herniated part.

Inflammatory bowel disease (in-FLAM-a-tō′-rē BOW-el) Disorder that exists in two forms: (1) Crohn's disease, an inflammation of the gastrointestinal tract, especially the distal ileum and proximal colon, in which the inflammation may extend from the mucosa through the serosa, and (2) ulcerative colitis, an inflammation of the mucosa of the gastrointestinal tract, usually limited to the large intestine and usually accompanied by rectal bleeding.

Irritable bowel syndrome (IBS) Disease of the entire gastrointestinal tract in which a person reacts to stress by developing symptoms (such as cramping and abdominal pain) associated with alternating patterns of diarrhea and constipation. Excessive amounts of mucus may appear in feces, and other symptoms include flatulence, nausea, and loss of appetite. The condition is also known as **irritable colon** or **spastic colitis.**

Malocclusion (mal′-ō-KLOO-zhun; *mal-* = bad; *occlusion* = to fit together) Condition in which the surfaces of the maxillary (upper) and mandibular (lower) teeth fit together poorly.

Traveler's diarrhea Infectious disease of the gastrointestinal tract that results in loose, urgent bowel movements, cramping, abdominal pain, malaise, nausea, and occasionally fever and dehydration. It is acquired through ingestion of food or water contaminated with fecal material typically containing bacteria (especially *Escherichia coli*); viruses or protozoan parasites are a less common cause.

STUDY OUTLINE

INTRODUCTION (p. 818)

1. The breaking down of larger food molecules into smaller molecules is called digestion; the passage of these smaller molecules into blood and lymph is termed absorption.
2. The organs that collectively perform digestion and absorption constitute the digestive system and are usually composed of two main groups: the gastrointestinal (GI) tract and accessory digestive organs.
3. The GI tract is a continuous tube extending from the mouth to the anus.
4. The accessory structures include the teeth, tongue, salivary glands, liver, gallbladder, and pancreas.

OVERVIEW OF THE DIGESTIVE SYSTEM (p. 818)

1. Digestion includes six basic processes: ingestion, secretion, mixing and propulsion, mechanical and chemical digestion, absorption, and defecation.
2. Mechanical digestion consists of mastication and movements of the gastrointestinal tract that aid chemical digestion.

3. Chemical digestion is a series of hydrolysis reactions that break down large carbohydrates, lipids, proteins, and nucleic acids in foods into smaller molecules that are usable by body cells.

LAYERS OF THE GI TRACT (p. 820)

1. The basic arrangement of layers in most of the gastrointestinal tract, from deep to superficial, is the mucosa, submucosa, muscularis, and serosa.
2. Associated with the lamina propria of the mucosa are extensive patches of lymphatic tissue called mucosa-associated lymphoid tissue (MALT).

PERITONEUM (p. 821)

1. The peritoneum is the largest serous membrane of the body; it lines the wall of the abdominal cavity and covers some abdominal organs.
2. Folds of the peritoneum include the mesentery, mesocolon, falciform ligament, lesser omentum, and greater omentum.

MOUTH (p. 823)

1. The mouth is formed by the cheeks, hard and soft palates, lips, and tongue.
2. The vestibule is the space bounded externally by the cheeks and lips and internally by the teeth and gums.
3. The oral cavity proper extends from the vestibule to the fauces.
4. The tongue, together with its associated muscles, forms the floor of the oral cavity. It is composed of skeletal muscle covered with mucous membrane.
5. The upper surface and sides of the tongue are covered with papillae, some of which contain taste buds.
6. The major portion of saliva is secreted by the salivary glands, which lie outside the mouth and pour their contents into ducts that empty into the oral cavity.
7. There are three pairs of salivary glands: parotid, submandibular (submaxillary), and sublingual glands.
8. Saliva lubricates food and starts the chemical digestion of carbohydrates.
9. Salivation is controlled by the nervous system.
10. The teeth (dentes) project into the mouth and are adapted for mechanical digestion.
11. A typical tooth consists of three principal regions: crown, root, and neck.
12. Teeth are composed primarily of dentin and are covered by enamel, the hardest substance in the body.
13. There are two dentitions: deciduous and permanent.
14. Through mastication, food is mixed with saliva and shaped into a soft, flexible mass called a bolus.
15. Salivary amylase begins the digestion of starches, and lingual lipase acts on triglycerides.

PHARYNX (p. 830)

1. Deglution, or swallowing, moves a bolus from the mouth to the stomach.
2. Swallowing consists of a voluntary stage, a pharyngeal stage (involuntary), and an esophageal stage (involuntary).

ESOPHAGUS (p. 830)

1. The esophagus is a collapsible, muscular tube that connects the pharynx to the stomach.
2. It passes a bolus into the stomach by peristalsis.
3. It contains an upper and a lower esophageal sphincter.

STOMACH (p. 832)

1. The stomach connects the esophagus to the duodenum.
2. The principal anatomic regions of the stomach are the cardia, fundus, body, and pylorus.
3. Adaptations of the stomach for digestion include rugae; glands that produce mucus, hydrochloric acid, pepsin, gastric lipase, and intrinsic factor; and a three-layered muscularis.
4. Mechanical digestion consists of mixing waves.
5. Chemical digestion consists mostly of the conversion of proteins into peptides by pepsin.
6. The stomach wall is impermeable to most substances.
7. Among the substances the stomach can absorb are water, certain ions, drugs, and alcohol.
8. Gastric secretion is regulated by neural and hormonal mechanisms.
9. Stimulation of gastric secretion occurs in three overlapping phases: the cephalic, gastric, and intestinal phases.

10. During the cephalic and gastric phases, peristalsis is stimulated; during the intestinal phase, motility is inhibited.
11. Gastric emptying is stimulated in response to distension, and gastrin is released in response to the presence of certain types of foods.
12. Gastric emptying is inhibited by the enterogastric reflex and by hormones (CCK and GIP).

PANCREAS (p. 840)

1. The pancreas consists of a head, a body, and a tail and is connected to the duodenum via the pancreatic duct and accessory duct.
2. Endocrine pancreatic islets (islets of Langerhans) secrete hormones, and exocrine acini secrete pancreatic juice.
3. Pancreatic juice contains enzymes that digest starch (pancreatic amylase), proteins (trypsin, chymotrypsin, carboxypeptidase, and elastase), triglycerides (pancreatic lipase), and nucleic acids (ribonuclease and deoxyribonuclease).
4. Pancreatic secretion is regulated by neural and hormonal mechanisms.

LIVER AND GALLBLADDER (p.842)

1. The liver has left and right lobes; the right lobe includes a quadrate and caudate lobe. The gallbladder is a sac located in a depression on the posterior surface of the liver that stores and concentrates bile.
2. The lobes of the liver are made up of lobules that contain hepatocytes (liver cells), sinusoids, stellate reticuloendothelial (Kupffer's) cells, and a central vein.
3. Hepatocytes produce bile that is carried by a duct system to the gallbladder for concentration and temporary storage. Cholecystokinin (CCK) stimulates ejection of bile into the common bile duct.
4. Bile's contribution to digestion is the emulsification of dietary lipids.
5. The liver also functions in carbohydrate, lipid, and protein metabolism; processing of drugs and hormones; excretion of bilirubin; synthesis of bile salts; storage of vitamins and minerals; phagocytosis; and activation of vitamin D.
6. Bile secretion is regulated by neural and hormonal mechanisms.

SUMMARY: DIGESTIVE HORMONES (p. 846)

1. Enteroendocrine cells in the mucosa secrete several hormones that regulate digestive processes. The major hormones are gastrin (from the stomach), secretin (from the small intestine), and cholecystokinin (from the small intestine).
2. Gastrin exerts its major effects on the stomach, whereas secretin and cholecystokinin affect the pancreas, liver, and gallbladder most strongly.

SMALL INTESTINE (p. 847)

1. The small intestine extends from the pyloric sphincter to the ileocecal sphincter.
2. It is divided into duodenum, jejunum, and ileum.
3. Its glands secrete fluid and mucus, and the circular folds, villi, and microvilli of its wall provide a large surface area for digestion and absorption.
4. Brush-border enzymes digest α-dextrins, maltose, sucrose, lactose, peptides, and nucleotides at the surface of mucosal epithelial cells.

5. Pancreatic and intestinal brush-border enzymes break down starches into maltose, maltotriose, and α-dextrins (pancreatic amylase), α-dextrins into glucose (α-dextrinase), maltose to glucose (maltase), sucrose to glucose and fructose (sucrase), lactose to glucose and galactose (lactase), and proteins into peptides (trypsin, chymotrypsin, and elastase). Also, enzymes break peptide bonds that attach terminal amino acids to carboxyl ends of peptides (carboxypeptidases) and peptide bonds that attach terminal amino acids to amino ends of peptides (aminopeptidases). Finally, enzymes split dipeptides to amino acids (dipeptidase), triglycerides to fatty acids and monoglycerides (lipases), and nucleotides to pentoses and nitrogenous bases (nucleosidases and phosphatases).

6. Mechanical digestion in the small intestine involves segmentation and migrating motility complexes.

7. The most important regulators of intestinal secretion and motility are enteric reflexes and digestive hormones.

8. Parasympathetic impulses increase motility; sympathetic impulses decrease motility.

9. Absorption occurs via diffusion, facilitated diffusion, osmosis, and active transport; most absorption occurs in the small intestine.

10. Monosaccharides, amino acids, and short-chain fatty acids pass into the blood capillaries.

11. Long-chain fatty acids and monoglycerides are absorbed from micelles, resynthesized to triglycerides, and formed into chylomicrons.

12. Chylomicrons move into lymph in the lacteal of a villus.

13. The small intestine also absorbs electrolytes, vitamins, and water.

LARGE INTESTINE (p. 857)

1. The large intestine extends from the ileocecal sphincter to the anus.

2. Its regions include the cecum, colon, rectum, and anal canal.

3. The mucosa contains many goblet cells, and the muscularis consists of taeniae coli and haustra.

4. Mechanical movements of the large intestine include haustral churning, peristalsis, and mass peristalsis.

5. The last stages of chemical digestion occur in the large intestine through bacterial action. Substances are further broken down, and some vitamins are synthesized.

6. The large intestine absorbs water, electrolytes, and vitamins.

7. Feces consist of water, inorganic salts, epithelial cells, bacteria, and undigested foods.

8. The elimination of feces from the rectum is called defecation.

9. Defecation is a reflex action aided by voluntary contractions of the diaphragm and abdominal muscles and relaxation of the external anal sphincter.

DEVELOPMENTAL ANATOMY OF THE DIGESTIVE SYSTEM (p. 862)

1. The endoderm of the primitive gut forms the epithelium and glands of most of the gastrointestinal tract.

2. The mesoderm of the primitive gut forms the smooth muscle and connective tissue of the gastrointestinal tract.

AGING AND THE DIGESTIVE SYSTEM (p. 862)

1. General changes include decreased secretory mechanisms, decreased motility, and loss of tone.

2. Specific changes may include loss of taste, pyorrhea, hernias, peptic ulcer disease, constipation, hemorrhoids, and diverticular diseases.

SELF-QUIZ QUESTIONS

1. Match the following:
   ___ (a) collapsed, muscular tube involved in deglutition and peristalsis
   ___ (b) vestigial organ containing a large amount of lymphatic tissue
   ___ (c) contains Brunner's glands in the submucosa
   ___ (d) produces and secretes bile
   ___ (e) contains Peyer's patches in the submucosa
   ___ (f) responsible for ingestion, mastication, and deglutition

   ___ (g) responsible for churning, peristalsis, storage, and chemical digestion with the enzyme pepsin
   ___ (h) storage area for bile
   ___ (i) responsible for forming a semi-solid waste material through haustral churning and peristalsis
   ___ (j) passageway for food, fluid, and air; involved in deglutition

   (1) mouth
   (2) pharynx
   (3) esophagus
   (4) stomach
   (5) duodenum
   (6) ileum
   (7) colon
   (8) liver
   (9) gallbladder
   (10) appendix

2. Match the following:
___ (a) an activating brush-border enzyme that splits off part of the trypsinogen molecule to form trypsin, a protease
___ (b) an enzyme that initiates carbohydrate digestion in the mouth
___ (c) the principal triglyceride-digesting enzyme in adults
___ (d) stimulates growth of gastric glands and secretion of gastric juices
___ (e) secreted by chief cells in the stomach; a protease
___ (f) stimulates the flow of pancreatic juice rich in bicarbonates
___ (g) a nonenzymatic fat-emulsifying agent
___ (h) causes contraction of the gallbladder and stimulates the production of pancreatic juice rich in digestive enzymes
___ (i) inhibits gastric juice release
___ (j) stimulates secretion of ions and water by the intestines and inhibits gastric acid secretion

(1) gastrin
(2) cholecystokinin
(3) secretin
(4) enterokinase
(5) pepsin
(6) salivary amylase
(7) pancreatic lipase
(8) bile
(9) vasoactive intestinal polypeptide
(10) somatostatin

Choose the best answer to the following questions:

3. Which of the following are basic digestive processes? (1) ingestion, (2) secretion, (3) mixing and propulsion, (4) ventilation, (5) absorption, (6) defecation, (7) excretion, (8) digestion. (a) 1, 2, 3, 5, 6, and 8, (b) 2, 3, 4, 6, 7, and 8, (c) 1, 3, 5, and 8, (d) 2, 3, 6, and 8, (e) 1, 3, 6, and 8

4. Which of the following are functions of the liver? (1) carbohydrate, lipid, and protein metabolism, (2) nucleic acid metabolism, (3) excretion of bilirubin, (4) synthesis of bile salts, (5) activation of vitamin D. (a) 1, 2, 3, and 5, (b) 1, 2, 3, and 4, (c) 1, 3, 4, and 5, (d) 2, 3, 4, and 5, (e) 1, 2, 4, and 5

5. The part of the wall of the GI tract that contains the mucosa-associated lymphoid tissue is the (a) muscularis mucosa, (b) submucosa, (c) muscularis, (d) serosa, (e) lamina propria.

6. The digestive organ that contains three layers in the muscularis is the (a) esophagus, (b) colon, (c) small intestine, (d) stomach, (e) secum.

7. Which of the following statements regarding the regulation of gastric secretion and motility are true? (1) The sight, smell, taste, or thought of food can initiate the cephalic phase of gastric activity. (2) The gastric phase begins when food enters the small intestine. (3) Once activated, stretch receptors in the stomach trigger the flow of gastric juice and peristalsis. (4) The intestinal phase reflexes inhibit gastric activity. (5) The enterogastric reflex is initiated during the gastric phase. (a) 1, 3, and 4, (b) 2, 4, and 5, (c) 1, 4, and 5, (d) 1, 2, and 5, (e) 1, 2, 3, and 4

8. The major movement of the small intestine is (a) peristalsis, (b) churning, (c) segmentation, (d) deglutition, (e) stretching.

Complete the following:

9. The four layers of the GI tract, from deep to superficial, are the ___, ___, ___, and ___.

10. The largest serous membrane of the body is the ___. Folds of the membrane are the ___ and ___.

11. The ___ regulates movements of the mucosa and vasoconstriction of blood vessels. The ___ controls GI tract motility, the frequency and strength of contraction of the muscularis.

12. The end-products of chemical digestion of carbohydrates are ___, of proteins are ___, of lipids are ___ and ___, and of nucleic acids are ___, ___, and ___.

True or false:

13. All digestive end-produces are absorbed from the lumen of the GI tract directly into the blood capillaries of the cardiovascular system and are then taken to the liver for processing.

14. The last stage of digestion occurs in the colon through the activity of the haustra.

15. Match the following:
___ (a) microvilli of the small intestine that increase surface area for absorption; also contain some digestive enzymes
___ (b) finger-like projections of the mucosa of the small intestine that increase surface area
___ (c) produce hydrochloric acid in the stomach
___ (d) secrete lysozyme; help regulate microbial population in the intestines
___ (e) forces the food to the back of the mouth for swallowing; places food in contact with the teeth
___ (f) aids in limiting the movement of the tongue posteriorly
___ (g) produce a fluid in the mouth that helps cleanse the mouth and teeth; lubricates, dissolves, and begins the chemical breakdown of food
___ (h) the primary structures of mastication
___ (i) secrete pepsinogen and gastric lipase in the stomach
___ (j) permanent ridges in the mucosa of the small intestine; enhance absorption by increasing surface area and causing chyme to spiral rather than move in a straight line
___ (k) stellate reticuloendothelial cells of the liver; destroy worn-out leukocytes and red blood cells, bacteria, and other foreign matter in the blood draining the GI tract

(1) salivary glands
(2) parietal cells
(3) chief cells
(4) brush border
(5) Kupffer's cells
(6) teeth
(7) tongue
(8) villi
(9) circular folds
(10) Paneth cells
(11) lingual frenulum

CRITICAL THINKING QUESTIONS

1. Katie's first two permanent teeth came in just in time to fill the upper gap in her smile before the start of school. Unfortunately, the day before her class picture, she fell off her bike and re-created the gap in her smile. Use the proper anatomical terms to describe the teeth Katie lost and the ones she has remaining. (HINT: *A complete set of deciduous teeth includes a different number than a complete set of permanent teeth.*)

2. Obesity is a health concern for many people. Suppose that a CCK (cholecystokinin) nasal spray is developed as a weight-loss treatment. How would the CCK affect the digestive system and the appetite? (HINT: *CCK is produced by enteroendocrine cells.*)

3. The surgeon explained to the woman that he was going to remove the smaller lobe of her liver and transplant it into her child to replace the child's diseased liver. Use proper anatomical terminology to describe the location of this section of the woman's liver. (HINT: *If the hands are positioned incorrectly when performing CPR, the xiphoid process may pierce the liver.*)

ANSWERS TO FIGURE QUESTIONS

24.1 Digestive enzymes are produced by the salivary glands, tongue, stomach, pancreas, and small intestine.

24.2 The enteric plexuses help regulate secretions and motility of the GI tract.

24.3 Mesentery binds the small intestine to the posterior abdominal wall.

24.4 The uvula helps prevent foods and liquids from entering the nasal cavity during swallowing.

24.5 Chloride ions in saliva activate salivary amylase.

24.6 The main component of teeth is connective tissue, specifically dentin.

24.7 The first, second, and third molars do not replace any deciduous teeth.

24.8 Both. Initiation of swallowing is voluntary and the action is carried out by skeletal muscles. Completion of swallowing—moving a bolus along the esophagus and into the stomach—is involuntary and involves peristalsis by smooth muscle.

24.9 The esophageal mucosa and submucosa contain mucus-secreting glands.

24.10 Food is pushed along by contraction of smooth muscle behind the bolus and relaxation of smooth muscle in front of it.

24.11 After a large meal, the rugae stretch and disappear as the stomach fills.

24.12 Mucous surface cells and mucous neck cells secrete mucus; chief cells secrete pepsinogen and gastric lipase; parietal cells secrete HCl and intrinsic factor; G cells secrete gastrin.

24.13 Stimulation of the vagus (X) nerve causes secretion of gastric juice and increased gastric motility.

24.14 This negative feedback loop is a "local reflex" because all parts of the loop occur within the stomach; impulses from the brain or spinal cord are not needed.

24.15 A vagotomy would slow the rate of gastric emptying.

24.16 The pancreatic duct contains pancreatic juice (fluid and digestive enzymes); the common bile duct contains bile; the hepatopancreatic ampulla contains pancreatic juice and bile.

24.17 The HCO_3^- in pancreatic juice helps buffer gastric acid and raises the pH of chyme so that pancreatic enzymes can work effectively.

24.18 The phagocytic cell in the liver is the stellate reticuloendothelial (Kupffer's) cell.

24.19 While a meal is being absorbed, hepatocytes remove nutrients from blood flowing through liver sinusoids. They also remove O_2 and extract certain toxic substances.

24.20 Secretin promotes liberation of fluid rich in HCO_3^- from both the pancreas and the liver.

24.21 The ileum is the longest part of the small intestine.

24.22 Nutrients being absorbed enter the blood via capillaries or the lymph via lacteals.

24.23 The fluid secreted by Brunner's glands—alkaline mucus—neutralizes gastric acid and protects the mucosal lining of the duodenum.

24.24 Because monoglycerides are nonpolar (hydrophobic) molecules, they can dissolve in and diffuse through the lipid bilayer of the plasma membrane.

24.25 The stomach and pancreas are the two digestive system organs that secrete the largest volumes of fluid.

24.26 The ascending and descending portions of the colon are retroperitoneal.

24.27 Goblet cells secrete mucus to lubricate colonic contents.

24.28 The digestive system begins to develop about 14 days after fertilization.

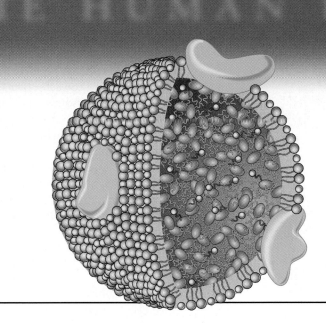

Whereas plants use the green pigment chlorophyll to trap energy in sunlight, we don't have a similarly functioning pigment in our skin, so the food we eat is our only source of energy for performing biological work. Many molecules needed to maintain cells and tissues can be made from simpler precursors by the body's metabolic reactions; others—the essential amino acids, essential fatty acids, vitamins, and minerals—must be obtained in food because we cannot synthesize them. Carbohydrates, lipids, and proteins in food are digested by enzymes in the gastrointestinal tract. The end products of digestion that reach body cells are monosaccharides, fatty acids, glycerol, monoglycerides, and amino acids. Some minerals and many vitamins are part of enzyme systems that catalyze the breakdown and synthesis of carbohydrates, lipids, and proteins. Food molecules absorbed by the gastrointestinal (GI) tract have three main fates:

1. Most food molecules are used to *supply energy* for sustaining life processes, such as active transport, DNA replication, protein synthesis, muscle contraction, and mitosis.
2. Some food molecules *serve as building blocks* for the synthesis of more complex structural or functional molecules, such as muscle proteins, hormones, and enzymes.
3. Other food molecules are *stored for future use*. For example, glycogen is stored in liver cells, and triglycerides are stored in adipose cells.

In this chapter we discuss how metabolic reactions harvest the chemical energy in foods, how each group of food molecules contributes to the body's growth, repair, and energy needs, and how heat and energy balance is maintained in the body. Finally, we explore some aspects of nutrition.

METABOLIC REACTIONS

OBJECTIVE

• *Explain the role of ATP in anabolism and catabolism.*

Metabolism (me-TAB-ō-lizm; *metabol-* = change) refers to all the chemical reactions of the body. Metabolism is an energy-balancing act between catabolic reactions, which are decomposition reactions, and anabolic reactions, which are synthesis reactions. Overall, catabolic reactions are *exergonic;* they produce more energy than they consume. By contrast, anabolic reactions are *endergonic;* they consume more energy than they produce. The molecule that participates most often in energy exchanges in living cells is **ATP (adenosine triphosphate),** which couples energy-releasing catabolic reactions to energy-requiring anabolic reactions.

The metabolic reactions that occur depend on which enzymes are active in a particular cell at a particular time. Often, catabolic reactions occur in one compartment of a cell, for example, the mitochondria, whereas synthetic reactions take place in another location such as the endoplasmic reticulum.

A molecule synthesized in an anabolic reaction has a limited lifetime. With few exceptions, it will eventually be broken down and its component atoms recycled into other molecules or excreted from the body. Recycling of biological molecules occurs continuously in living tissues, rapidly in

Figure 25.1 Role of ATP in linking anabolic and catabolic reactions. When complex molecules and polymers are split apart (catabolism, at left), some of the energy is transferred to ATP and the rest is given off as heat. When simple molecules and monomers are combined to form complex molecules (anabolism, at right), ATP provides the energy for synthesis, and again some energy is given off as heat.

🔑 **The coupling of energy-releasing and energy-requiring reactions is achieved through ATP.**

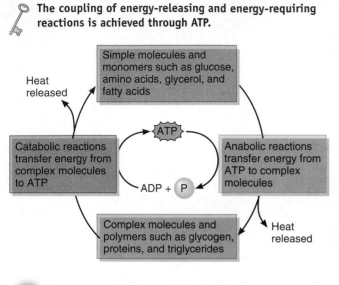

Q In a pancreatic cell producing digestive enzymes, does anabolism or catabolism predominate?

some tissues and very slowly in others. Individual cells may be refurbished molecule by molecule, or a whole tissue may be rebuilt cell by cell.

Catabolism and Anabolism Defined

The chemical reactions that break down complex organic molecules into simpler ones are collectively known as **catabolism** (ka-TAB-ō-lizm; *cata-* = downward). Catabolic reactions release the chemical energy stored in organic molecules. Important sets of catabolic reactions occur in glycolysis, the Krebs cycle, and the electron transport chain, each of which will be discussed later in the chapter.

Chemical reactions that combine simple molecules and monomers to form the body's complex structural and functional components are collectively known as **anabolism** (a-NAB-ō-lizm; *ana-* = upward). Examples of anabolic reactions are the formation of peptide bonds between amino acids during protein synthesis, the building of fatty acids into phospholipids that form the plasma membrane bilayer, and the linkage of glucose monomers to form glycogen.

Coupling of Catabolism and Anabolism by ATP

The chemical reactions of living systems depend on efficiently transferring manageable amounts of energy from one

molecule to another. The molecule that most often performs this task is ATP, the "energy currency" of a living cell. Like money, it is readily available to "buy" cellular activities; it is spent and remade over and over. A typical cell has about a billion molecules of ATP, each of which typically lasts for less than a minute before being used. Thus, ATP is not a long-term storage form of currency, like gold in a vault, but rather convenient cash for moment-to-moment transactions.

Recall from Chapter 2 that a molecule of ATP consists of an adenine molecule, a ribose molecule, and three phosphate groups bonded to each other (see Figure 2.26 on page 55). Figure 25.1 shows how ATP links anabolic and catabolic reactions. When the terminal phosphate group is split off of ATP in anabolic reactions, adenosine diphosphate (ADP) and a phosphate group (symbolized (P)) are formed. Some of the energy released is used to drive anabolic reactions such as the formation of glycogen from glucose. Subsequently, energy from complex molecules is used in catabolic reactions to combine ADP and a phosphate group to resynthesize ATP:

$$\text{ADP} + (\text{P}) + \text{energy} \longrightarrow \text{ATP}$$

About 40% of the energy released in catabolism is used for cellular functions; the rest is converted to heat, some of which helps maintain normal body temperature. Excess heat is lost to the environment. Compared with machines, which typically convert only 10–20% of energy into work, the 40% efficiency of the body's metabolism is impressive. Still, the body has a continuous need to take in and process external sources of energy so that cells can synthesize enough ATP to sustain life.

1. What is metabolism? Distinguish between anabolism and catabolism, and give examples of each.
2. How does ATP couple anabolism and catabolism?

ENERGY TRANSFER

OBJECTIVES

• *Describe oxidation–reduction reactions.*

• *Explain the role of ATP in metabolism.*

Various catabolic reactions transfer energy into the "high-energy" phosphate bonds of ATP. Although the amount of energy in these bonds is not exceptionally large, it can be released quickly and easily. Before discussing metabolic pathways, we first consider two important aspects of energy transfer: oxidation–reduction reactions and mechanisms of ATP generation.

Oxidation–Reduction Reactions

Oxidation is the *removal of electrons* from an atom or molecule; the result is a *decrease* in the potential energy of the atom or molecule. Because most biological oxidation reactions involve the loss of hydrogen atoms, they are called *dehydrogenation reactions*. An example of an oxidation reaction is the conversion of lactic acid into pyruvic acid:

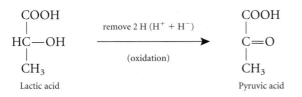

Reduction is the opposite of oxidation; it is the *addition of electrons* to a molecule. Reduction results in an *increase* in the potential energy of the molecule. An example of a reduction reaction is the conversion of pyruvic acid into lactic acid:

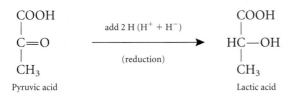

When a substance is oxidized, the liberated hydrogen atoms do not remain free in the cell but are transferred immediately by coenzymes to another compound. Two coenzymes are commonly used by animal cells to carry hydrogen atoms: **nicotinamide adenine dinucleotide (NAD$^+$),** a derivative of the B vitamin niacin, and **flavin adenine dinucleotide (FAD),** a derivative of vitamin B$_2$ (riboflavin). The oxidation and reduction states of NAD$^+$ and FAD can be represented as follows:

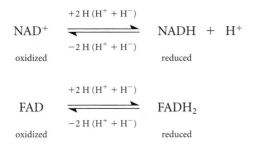

In the preceding equations, 2 H (H$^+$ + H$^-$) means that two neutral hydrogen atoms (2 H) are equivalent to one hydrogen ion (H$^+$) plus one hydride ion (H$^-$). When NAD$^+$ is reduced to NADH + H$^+$, the NAD$^+$ gains a hydride ion (H$^-$), and the H$^+$ is released into the surrounding solution. The addition of a hydride ion to NAD$^+$ neutralizes the charge on NAD$^+$ and adds a hydrogen atom so that the reduced form is NADH. When NADH is oxidized to NAD$^+$, a hydride ion is lost from NADH, which results in one less hydrogen atom and an additional positive charge. Thus, the oxidized form is NAD$^+$. FAD is reduced to FADH$_2$ when it gains a hydrogen ion and a hydride ion, and FADH$_2$ is oxidized to FAD when it loses the same two ions.

Oxidation and reduction reactions are always coupled; each time one substance is oxidized, another is simultaneously reduced. Such paired reactions are called **oxidation–reduction** or **redox reactions.** For example, when lactic acid is *oxidized* to form pyruvic acid, the two hydrogen atoms removed in the reaction are used to *reduce* NAD$^+$. This coupled redox reaction may be written as follows:

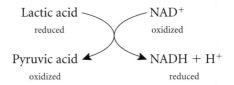

An important point to remember about oxidation–reduction reactions is that oxidation is usually an exergonic (energy-releasing) reaction. Cells use multistep biochemical reactions to release energy from energy-rich, highly reduced compounds (with many hydrogen atoms) to lower-energy, highly oxidized compounds (with many oxygen atoms or multiple bonds). For example, when a cell oxidizes a molecule of glucose (C$_6$H$_{12}$O$_6$), the energy in the glucose molecule is removed in a stepwise manner. Ultimately, some of the energy is captured by transferring it to ATP, which then serves as an energy source for energy-requiring reactions within the cell. Compounds that have many hydrogen atoms (for example, glucose) are highly reduced compounds that contain more chemical potential energy than oxidized compounds. For this reason, glucose is a valuable nutrient.

Mechanisms of ATP Generation

Some of the energy released during oxidation reactions is captured within a cell when ATP is formed. Briefly, a phosphate group (P) is added to ADP, with an input of energy, to form ATP. The two high-energy phosphate bonds that can be used to transfer energy are indicated by "squiggles" (~):

$$Adenosine—\textcircled{P}\sim\textcircled{P}+\textcircled{P}+energy \longrightarrow$$
ADP

$$Adenosine—\textcircled{P}\sim\textcircled{P}\sim\textcircled{P}$$
ATP

The high-energy phosphate bond that attaches the third phosphate group contains the energy stored in this reaction. The addition of a phosphate group to a molecule, called **phosphorylation** (fos′-for-i-LĀ-shun), increases its potential energy. Organisms use three mechanisms of phosphorylation to generate ATP:

1. **Substrate-level phosphorylation** generates ATP by transferring a high-energy phosphate group from an intermediate phosphorylated metabolic compound—a substrate—directly to ADP. In human cells this process occurs in the cytosol.

2. **Oxidative phosphorylation** removes electrons from organic compounds and passes them through a series of electron acceptors, called the **electron transport chain,** to molecules of oxygen (O_2). This process occurs in the inner mitochondrial membrane of cells.

3. **Photophosphorylation** occurs only in chlorophyll-containing plant cells or in certain bacteria that contain other light-absorbing pigments.

1. How is a hydride ion different from a hydrogen ion? Describe the involvement of both ions in redox reactions.
2. Describe the three ways that ATP can be generated.

CARBOHYDRATE METABOLISM

OBJECTIVE

• *Describe the fate, metabolism, and functions of carbohydrates.*

During digestion, polysaccharides and disaccharides are hydrolyzed into the monosaccharides glucose (about 80%), fructose, and galactose. (Some fructose is converted into glucose as it is absorbed through the intestinal epithelial cells.) The three monosaccharides are absorbed into the capillaries of the villi of the small intestine and then carried through the hepatic portal vein to the liver. Hepatocytes (liver cells) convert most of the remaining fructose and practically all the galactose to glucose. Thus, the story of carbohydrate metabolism is really the story of glucose metabolism. Because negative feedback systems maintain blood glucose at about 90 mg/100 mL of plasma (5 mmol/liter), a total of 2–3 g of glucose normally circulates in the blood.

The Fate of Glucose

Because glucose is the body's preferred source for synthesizing ATP, the fate of glucose absorbed from the diet—the use to which it is put—depends on the needs of body cells, which use glucose for:

• *ATP production.* In body cells that require immediate energy, glucose is oxidized to produce ATP. Glucose not needed for immediate ATP production can enter one of several other metabolic pathways.
• *Amino acid synthesis.* Cells throughout the body can use glucose to form several amino acids, which then can be incorporated into proteins.
• *Glycogen synthesis.* Hepatocytes and muscle fibers can perform **glycogenesis** (glī'-kō-JEN-e-sis; *glyco-* = sugar or sweet; *-genesis* = to generate), in which hundreds of glucose monomers are combined to form the polysaccharide glycogen. Total storage capacity of glycogen is about 125 g in the liver and 375 g in skeletal muscles.
• *Triglyceride synthesis.* When the glycogen storage areas are filled up, hepatocytes can transform the glucose to glycerol and fatty acids that can be used for **lipogenesis** (lip'-ō-JEN-ē-sis), the synthesis of triglycerides. Triglycerides then are deposited in adipose tissue, which has virtually unlimited storage capacity.

Glucose Movement into Cells

Before glucose can be used by body cells, it must first pass through the plasma membrane and enter the cytosol. Whereas glucose absorption in the GI tract (and kidney tubules) is accomplished via secondary active transport (Na^+–glucose symporters), glucose movement from blood into most other body cells occurs via GluT molecules, a family of facilitated diffusion glucose transporters (see page 70). Insulin increases the insertion of GluT4 into the plasma membranes of most body cells, thereby increasing the rate of facilitated diffusion of glucose into cells. In neurons and hepatocytes, however, numerous GluT molecules are always in the plasma membrane, so glucose entry is always "turned on." Upon entering a cell, glucose becomes phosphorylated. Because GluT cannot transport phosphorylated glucose, this reaction traps glucose within the cell.

Glucose Catabolism

The oxidation of glucose to produce ATP is also known as **cellular respiration,** and it involves four sets of reactions: glycolysis, the formation of acetyl coenzyme A, the Krebs cycle, and the electron transport chain (Figure 25.2).

1 *Glycolysis* is a set of reactions in which one glucose molecule is oxidized and two molecules of pyruvic acid are produced. The reactions also produce two molecules of ATP and two energy-containing NADH + H^+. Because glycolysis does not require oxygen, it is a way to produce ATP anaerobically (without oxygen) and is known as **anaerobic cellular respiration.**

2 *Formation of acetyl coenzyme A* is a transition step that prepares pyruvic acid for entrance into the Kreb's cycle. This step also produces energy-containing NADH + H^+ plus carbon dioxide (CO_2).

3 *Krebs cycle* reactions oxidize acetyl coenzyme A and produce CO_2, ATP, energy-containing NADH + H^+, and $FADH_2$.

4 *Electron transport chain* reactions oxidize NADH + H^+ and $FADH_2$ and transfer their electrons to a series of electron carriers. The Krebs cycle and the electron transport chain together require oxygen to produce ATP and are known as **aerobic cellular respiration.**

Glycolysis

Glycolysis (glī-KOL-i-sis; *-lysis* = breakdown) consists of ten chemical reactions that split a six-carbon molecule of glucose into two three-carbon molecules of pyruvic acid. (Caution: Do not confuse *glycolysis* with *glycogenolysis,* which is the breakdown of glycogen to glucose.)

Figure 25.2 Overview of cellular respiration (oxidation of glucose). A modified version of this figure appears in several places in this chapter to indicate the relationships of particular reactions to the overall process of cellular respiration.

🔑 **The oxidation of glucose involves glycolysis, the formation of acetyl coenzyme A, the Krebs cycle, and the electron transport chain.**

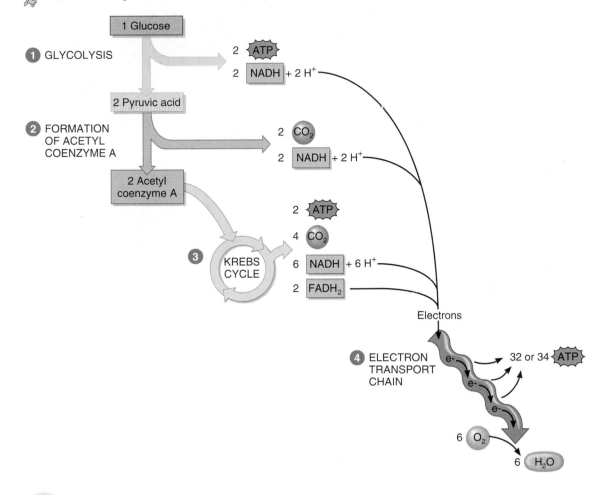

Q Which of the four processes shown here is also called anaerobic cellular respiration?

Figure 25.3a on page 876 shows the steps of glycolysis; each of the reactions is catalyzed by a specific enzyme. The ten steps of glycolysis proceed as follows:

1 The first reaction phosphorylates glucose, using a phosphate group from an ATP molecule.

2 The second reaction converts glucose 6-phosphate to fructose 6-phosphate.

3 The third reaction uses a second ATP to add a second phosphate group. *Phosphofructokinase*, the enzyme that catalyzes this step, is the key regulator of the rate of glycolysis. Its activity is high when ADP concentration is high, in which case ATP is produced rapidly, and its activity is low when ATP concentration is high. When the activity of phosphofructokinase is low, most glucose does not enter the reactions of glycolysis but instead undergoes conversion to glycogen for storage.

4 and **5** The doubly phosphorylated molecule of fructose splits into two three-carbon compounds, glyceraldehyde 3-phosphate (G 3-P) and dihydroxyacetone phosphate. These two compounds are interconvertible, but it is G 3-P that undergoes further oxidation to pyruvic acid.

6 Oxidation occurs as two molecules of NAD^+ accept two pairs of electrons and hydrogen ions from two molecules of G 3-P, forming two molecules each of NADH and 1,3-bisphosphoglyceric acid (BPG). Many body cells use the two NADH produced in this step to generate four ATPs in the electron transport chain. A few types of cells, such as hepatocytes, kidney cells, and cardiac muscle fibers, can generate six ATPs from the two NADH.

Figure 25.3 Glycolysis. Part (a) shows each of the ten steps of glycolysis; part (b) provides a simplified summary.

🔑 **Glycolysis results in a net gain of two ATP, two NADH, and two H⁺.**

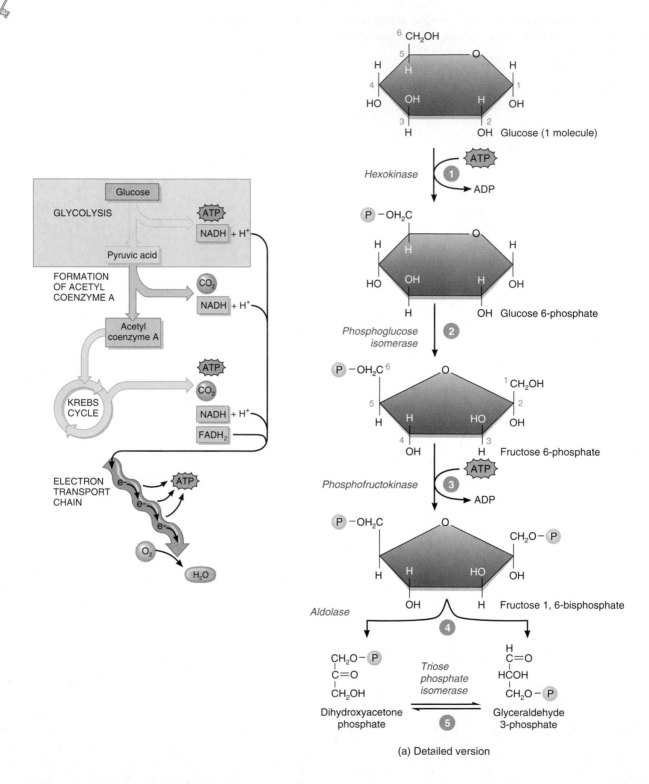

(a) Detailed version

figure continues

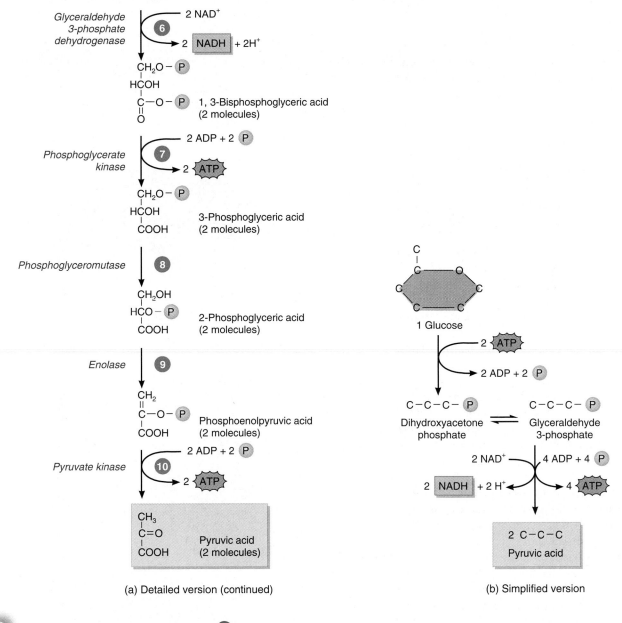

(a) Detailed version (continued)

(b) Simplified version

Q Why is the enzyme that catalyzes step ❶ named hexokinase?

7–**10** These reactions generate four molecules of ATP and produce two molecules of pyruvic acid (pyruvate*).

To summarize, even though glycolysis uses two ATP molecules, it produces four ATP molecules, for a net gain of two ATP molecules for each glucose molecule that is oxidized (Figure 25.3b).

The Fate of Pyruvic Acid

The fate of pyruvic acid produced during glycolysis depends on the availability of oxygen (Figure 25.4). If oxygen is scarce (anaerobic conditions)—for example, in skeletal muscle fibers during strenuous exercise—then pyruvic acid is reduced via an aerobic pathway by the addition of two hydrogen atoms to form lactic acid (lactate):

$$2 \text{ Pyruvic acid } + \text{ 2 NADH } + \text{ 2 H}^+ \longrightarrow$$
(oxidized)

$$2 \text{ Lactic acid } + \text{ 2 NAD}^+$$
(reduced)

This reaction regenerates the NAD^+ that was used in the oxidation of glyceraldehyde 3-phosphate (step **6** in glycolysis) and thus allows glycolysis to continue. As lactic acid is produced, it rapidly diffuses out of the cell and enters the blood. Hepatocytes remove lactic acid from the blood and convert it back to pyruvic acid.

When oxygen is plentiful (aerobic conditions), most cells convert pyruvic acid to acetyl coenzyme A. This molecule links glycolysis, which occurs in the cytosol, with the Krebs cycle, which occurs in the matrix of mitochondria. Pyruvic acid enters the mitochondrial matrix with the help of a special transporter protein. Because they lack mitochondria, red blood cells can produce ATP through glycolysis only.

Formation of Acetyl Coenzyme A

Each step in the oxidation of glucose requires a different enzyme, and often a coenzyme as well. The coenzyme used at this point in cellular respiration is **coenzyme A (CoA)**, which is derived from pantothenic acid, a B vitamin. During the transitional step between glycolysis and the Krebs cycle, pyruvic acid is prepared for entrance into the cycle. The enzyme *pyruvate dehydrogenase,* which is located exclusively in the mitochondrial matrix, converts pyruvate to a two-carbon fragment by removing a molecule of carbon dioxide (see

---

*The carboxyl groups (—COOH) of intermediates in glycolysis and in the citric acid cycle are mostly ionized at the pH of body fluids to —COO⁻. The suffix "-ic acid" indicates the nonionized form, whereas the ending "-ate" indicates the ionized form. Although the "-ate" names are more correct, we will use the "acid" names because these terms are more familiar.

Figure 25.4 Fate of pyruvic acid.

🔑 **When oxygen is plentiful, pyruvic acid enters mitochondria, is converted to acetyl coenzyme A, and enters the Krebs cycle (aerobic pathway). When oxygen is scarce, most pyruvic acid is converted to lactic acid via an anaerobic pathway.**

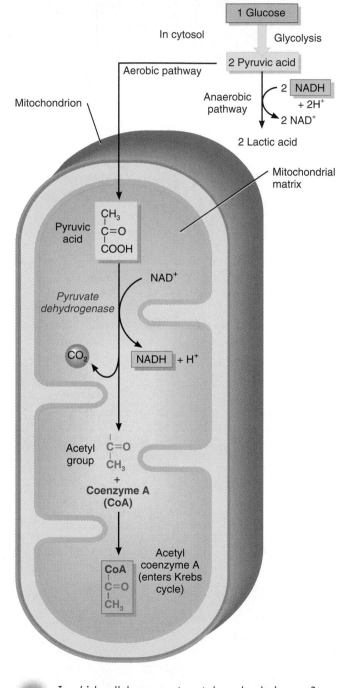

Q In which cellular compartment does glycolysis occur? The Krebs cycle?

Figure 25.4). The loss of a molecule of CO_2 by a substance is called **decarboxylation** (dē-kar-bok′-si-LĀ-shun). This is the first reaction in cellular respiration that releases CO_2. During this reaction, pyruvic acid is also oxidized. Each pyruvic acid loses two hydrogen atoms in the form of one hydride ion (H^-) plus one hydrogen ion (H^+). The coenzyme NAD^+ is reduced as it picks up the H^- from pyruvic acid; the H^+ is released into the mitochondrial matrix. The reduction of NAD^+ to $NADH + H^+$ is indicated in Figure 25.4 by the curved arrow entering and then leaving the reaction. Recall that the oxidation of one glucose molecule produces two molecules of pyruvic acid, so for each molecule of glucose, two molecules of carbon dioxide are lost and two $NADH + H^+$ are produced. The two-carbon fragment, called an **acetyl group,** attaches to coenzyme A, and the whole complex is called **acetyl coenzyme A (acetyl CoA).** Once the pyruvic acid has undergone decarboxylation and the remaining acetyl group has attached to CoA, the resulting compound (acetyl CoA) is ready to enter the Krebs cycle.

The Krebs Cycle

The **Krebs cycle**—named for the biochemist Hans Krebs, who described these reactions in the 1930s—is also known as the **citric acid cycle,** for the first molecule formed when an acetyl group joins the cycle. For the most part, the Krebs cycle is a series of oxidation–reduction reactions and decarboxylation reactions, each catalyzed by a specific enzyme in the matrix of mitochondria.

In the Krebs cycle, the large amount of chemical potential energy stored in intermediate substances derived from pyruvic acid is released in stepwise fashion. The oxidation–reduction reactions transfer chemical energy, in the form of electrons, to two coenzymes—NAD^+ and FAD. The pyruvic acid derivatives are oxidized, whereas the coenzymes are reduced. In addition, one step generates ATP. As each reaction is described in the text, refer to the numbered reactions in Figure 25.5a. The enzyme that catalyzes each reaction is indicated in parentheses.

1 *Entry of the acetyl group.* The chemical bond that attaches the acetyl group to coenzyme A (CoA) breaks, and the two-carbon acetyl group attaches to a four-carbon molecule of oxaloacetic acid to form a six-carbon molecule called citric acid. CoA is free to combine with another acetyl group from pyruvic acid and repeat the process. *(citric synthase)*

2 *Isomerization.* Citric acid undergoes isomerization to isocitric acid, which has the same molecular formula as citrate. Notice, however, that the hydroxyl group (—OH) is attached to a different carbon. *(aconitase)*

3 *Oxidative decarboxylation.* Isocitric acid is oxidized and loses a molecule of CO_2, forming a five-carbon molecule called alpha-ketoglutaric acid. The H^- from the oxidation is passed on to NAD^+, which is reduced to $NADH + H^+$. This is the second reaction in cellular respiration that releases CO_2. *(isocitric dehydrogenase)*

4 *Oxidative decarboxylation.* Alpha-ketoglutaric acid is oxidized, loses a molecule of CO_2, and picks up CoA to form a four-carbon molecule called succinyl CoA. This is the third reaction in cellular respiration that releases CO_2. *(alpha-ketoglutarate dehydrogenase)*

5 *Substrate-level phosphorylation.* At this point in the cycle, a substrate-level phosphorylation occurs. CoA is displaced by a phosphate group, which is then transferred to guanosine diphosphate (GDP) to form guanosine triphosphate (GTP). GTP is similar to ATP and can donate a phosphate group to ADP to form ATP. These reactions also convert succinyl CoA to succinic acid. *(succinyl-CoA synthetase)*

6 *Dehydrogenation.* Succinic acid is oxidized to fumaric acid as two of its hydrogen atoms are transferred to the coenzyme flavin adenine nucleotide (FAD), which is reduced to $FADH_2$. *(succinate dehydrogenase)*

7 *Hydration.* Fumaric acid is converted to malic acid by the addition of a molecule of water. *(fumarase)*

8 *Dehydrogenation.* In the final step in the cycle, malic acid is oxidized to re-form oxaloacetic acid. In the process, two hydrogen atoms are removed and one is transferred to NAD^+, which is reduced to $NADH + H^+$. The regenerated oxaloacetic acid can combine with another molecule of acetyl CoA, beginning a new cycle. *(malic dehydrogenase)*

The reduced coenzymes ($NADH$ and $FADH_2$) are the most important outcome of the Krebs cycle because they contain the energy originally stored in glucose and then in pyruvic acid. Overall, for every acetyl CoA that enters the Krebs cycle, three NADH, three H^+, and one $FADH_2$ are produced by oxidation–reduction reactions, and one molecule of ATP is generated by substrate-level phosphorylation (Figure 25.5b). In the electron transport chain, the three $NADH +$ three H^+ will later yield nine ATP molecules, and the $FADH_2$ will later yield two ATP molecules. Thus, each "turn" of the Krebs cycle eventually generates 12 molecules of ATP. Because each glucose provides two acetyl CoA, glucose catabolism via the Krebs cycle and the electron transport chain yields 24 molecules of ATP per glucose molecule.

Liberation of CO_2 occurs as pyruvic acid is converted to acetyl CoA and during the two decarboxylation reactions of the Krebs cycle (see Figure 25.5b). But because each molecule of glucose generates two molecules of pyruvic acid, a total of six molecules of CO_2 are liberated from each original glucose molecule catabolized along this pathway. The molecules of CO_2 leave the mitochondria, diffuse through the cytosol to the plasma membrane, and then diffuse into the blood. Eventually, the CO_2 is transported by the blood to the lungs and is exhaled.

The Electron Transport Chain

The **electron transport chain** is a series of **electron carriers** that are integral membrane proteins in the inner mitochondrial membrane. This membrane is folded into

Figure 25.5 The Krebs cycle. Part (a) shows each of the nine steps of the cycle; part (b) provides a simplified summary.

The net results of the Krebs cycle are (1) the production of reduced coenzymes (NADH + H$^+$ and FADH$_2$), which contain stored energy; (2) the generation of GTP, a high-energy compound that is used to produce ATP; and (3) the formation of CO$_2$, which is transported to the lungs and exhaled.

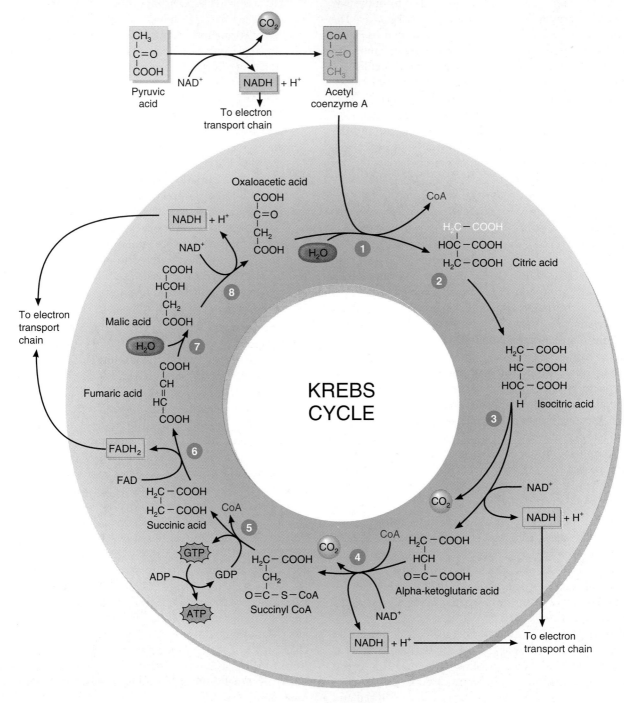

(a) Detailed version

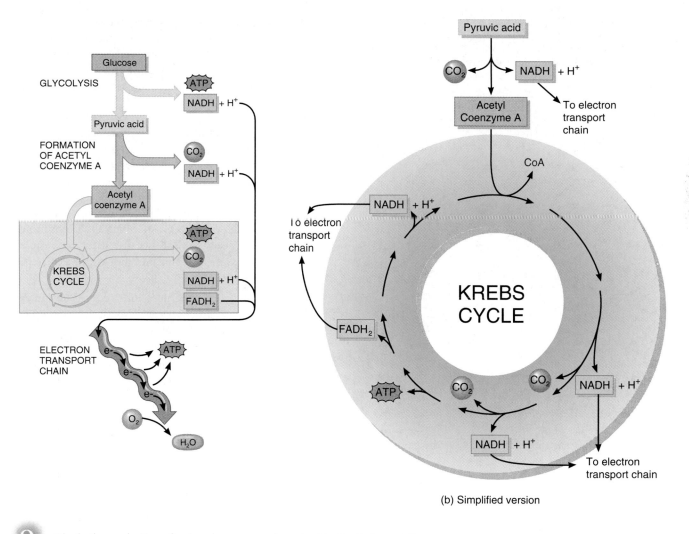

(b) Simplified version

Q Why is the production of reduced coenzymes important in the Krebs cycle?

cristae that increase its surface area, thus accommodating thousands of copies of the transport chain in each mitochondrion. Each carrier in the chain is reduced as it picks up electrons and is oxidized as it gives up electrons. As electrons pass through the chain, exergonic reactions release small amounts of energy in a series of steps, and this energy is used to form ATP. In aerobic cellular respiration, the final electron acceptor of the chain is oxygen. Because this mechanism of ATP generation links chemical reactions (the passage of electrons along the transport chain) with the pumping of hydrogen ions, it is called **chemiosmosis** (kem'-ē-oz-MŌ-sis; *chemi-* = chemical; *-osmosis* = pushing). Briefly, chemiosmosis works as follows (Figure 25.6):

1 Energy from NADH + H$^+$ passes along the electron transport chain and is used to pump H$^+$ (protons) from the matrix of the mitochondrion into the space between the inner and outer mitochondrial membranes.

2 A high concentration of H$^+$ accumulates between the inner and outer mitochondrial membranes.

3 ATP synthesis then occurs as hydrogen ions diffuse back into the mitochondrial matrix through a special type of H$^+$ channel in the inner membrane.

First, we examine the electron carriers, and then we explore chemiosmosis in greater detail.

Electron Carriers Several types of molecules and atoms serve as electron carriers:

• **Flavin mononucleotide (FMN),** like FAD (flavin adenine dinucleotide), is a flavoprotein derived from riboflavin (vitamin B$_2$).

• **Cytochromes** (SĪ-tō-krōmz) are proteins with an iron-containing group (heme) capable of existing alternately in a reduced form (Fe^{2+}) and an oxidized form (Fe^{3+}). The several cytochromes involved in the electron transport chain are cytochrome *b* (cyt *b*), cytochrome *c*$_1$ (cyt

Figure 25.6 Chemiosmosis.

🔑 **In chemiosmosis, ATP is produced when hydrogen ions diffuse back into the mitochondrial matrix.**

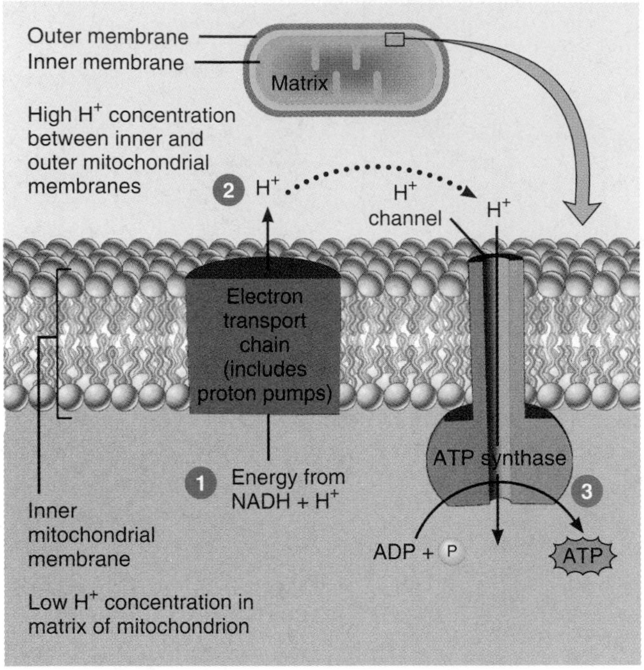

Outer membrane
Inner membrane
Matrix
High H^+ concentration between inner and outer mitochondrial membranes
② H^+
H^+ channel
H^+
Electron transport chain (includes proton pumps)
ATP synthase
① Energy from $NADH + H^+$
③
Inner mitochondrial membrane
$ADP + P$
ATP
Low H^+ concentration in matrix of mitochondrion

❓ What is the energy source that powers the proton pumps?

c_1), cytochrome c (cyt c), cytochrome a (cyt a), and cytochrome a_3 (cyt a_3).

- **Iron–sulfur (Fe–S) centers** contain either two or four iron atoms bound to sulfur atoms that form an electron transfer center within a protein.
- **Copper (Cu) atoms** bound to two proteins in the chain also participate in electron transfer.
- **Coenzyme Q,** symbolized **Q,** is a nonprotein, low-molecular-weight carrier that is mobile in the lipid bilayer of the inner membrane.

Steps in Electron Transport and the Chemiosmotic Generation of ATP

Within the inner mitochondrial membrane, the carriers of the electron transport chain are clustered into three complexes, each of which acts as a **proton pump** that expels H^+ from the mitochondrial matrix and helps create an electrochemical gradient of H^+. Each of the three proton pumps is a complex that transports electrons and pumps H^+, as follows (Figure 25.7):

① The first proton pump is the **NADH dehydrogenase complex,** which contains flavin mononucleotide (FMN) and five or more Fe–S centers. $NADH + H^+$ is oxidized to NAD^+, and FMN is reduced to $FMNH_2$, which in turn is oxidized as it passes electrons to the iron–sulfur centers. Q, which is mobile in the membrane, shuttles electrons to the second pump complex.

② The second proton pump is the **cytochrome b–c_1 complex,** which contains cytochromes and an iron–sulfur center. Electrons are passed successively from Q to cyt b, to Fe–S, to cyt c_1. The mobile shuttle that passes electrons from the second pump complex to the third is cytochrome c (cyt c).

③ The third proton pump is the **cytochrome oxidase complex,** which contains cytochromes a and a_3 and two copper atoms. Electrons pass from cyt c, to Cu, to cyt a, and finally to cyt a_3. Cyt a_3 passes its electrons to one-half of a molecule of oxygen (O_2), which becomes negatively charged and then picks up two H^+ from the surrounding medium to form H_2O. This is the only part of aerobic cellular respiration in which O_2 is consumed. Cyanide is a deadly poison because it binds to the cytochrome oxidase complex and blocks this last step in electron transport.

The pumping of H^+ produces both a concentration gradient of protons and an electrical gradient. The buildup of H^+ on one side of the inner mitochondrial membrane makes that side positively charged compared with the other side. The resulting electrochemical gradient has potential energy, called the *proton motive force.* Included in the inner mitochondrial membrane are proton channels that allow H^+ to diffuse back across the membrane, driven by the proton motive force. As H^+ diffuse back, they generate ATP because the H^+ channels also include an enzyme called **ATP synthase.** The enzyme uses the proton motive force to synthesize ATP from ADP and P. This process of chemiosmosis is responsible for most of the ATP produced during cellular respiration.

Summary of Cellular Respiration

The various electron transfers in the electron transport chain generate either 32 or 34 ATP molecules from each molecule of glucose that is oxidized: either 28 or 30* from the ten molecules of $NADH + H^+$ and two from each of the two molecules of $FADH_2$ (four total). Thus, during cellular respiration, 36 or 38 ATPs can be generated from one molecule of glucose. Note that two of those ATPs come from substrate-level phosphorylation in glycolysis, and two come from substrate-level phosphorylation in the Krebs cycle. The overall reaction is:

$$C_6H_{12}O_6 + 6\ O_2 + 36 \text{ or } 38 \text{ ADPs} + 36 \text{ or } 38\ P$$

Glucose   Oxygen

$$\longrightarrow 6\ CO_2 + 6\ H_2O + 36 \text{ or } 38 \text{ ATPs}$$

Carbon   Water
dioxide

---

*The two NADH produced in the cytosol during glycolysis cannot enter mitochondria. Instead, they donate their electrons to one of two transfer molecules, known as the malate shuttle and the glycerol phosphate shuttle. In cells of the liver, kidneys, and heart, use of the malate shuttle results in three ATP synthesized for each NADH. In other body cells, such as skeletal muscle fibers and neurons, use of the glycerol phosphate shuttle results in two ATP synthesized for each NADH.

Figure 25.7 The actions of the three proton pumps and ATP synthase in the inner membrane of mitochondria. Each pump is a complex of three or more electron carriers. As H^+ diffuse back into the mitochondrial matrix, ATP is synthesized.

🔑 As the three proton pumps pass electrons from one carrier to the next, they also move H^+ from the matrix into the space between the inner and outer mitochondrial membranes.

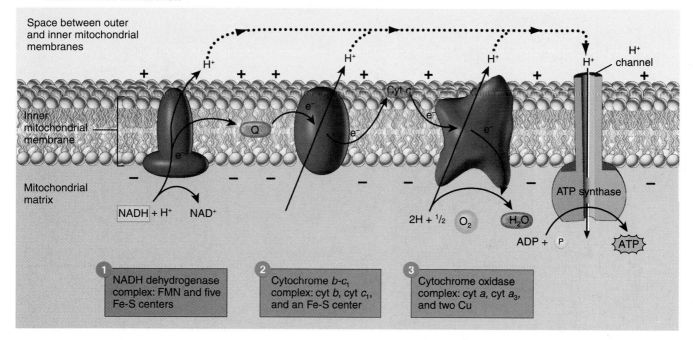

Q Where is the concentration of H^+ highest?

Table 25.1 summarizes the ATP yield during cellular respiration. A schematic depiction of the principal reactions of cellular respiration is presented in Figure 25.8. The actual ATP yield may be lower than 36 or 38 ATPs per glucose. One uncertainty is the exact number of H^+ that must be pumped out to generate one ATP during chemiosmosis. Moreover, the ATP generated in mitochondria must be transported out of these organelles into the cytosol for use elsewhere in a cell. Exporting ATP in exchange for the inward movement of ADP that is formed from metabolic reactions in the cytosol uses up a portion of the proton motive force.

Glycolysis, the Krebs cycle, and especially the electron transport chain provide all the ATP for cellular activities, and because the Krebs cycle and electron transport chain are aerobic processes, cells cannot carry on their activities for long without sufficient oxygen.

CLINICAL APPLICATION
Carbohydrate Loading

The amount of glycogen stored in the liver and skeletal muscles varies and can be completely exhausted during long-term athletic endeavors. Thus, many marathon

Table 25.1 Summary of ATP Produced in Cellular Respiration

SOURCE	ATP YIELD (PROCESS)
Glycolysis	
Oxidation of one glucose molecule to two pyruvic acid molecules	2 ATPs (substrate-level phosphorylation)
Production of 2 NADH + H^+	4 or 6 ATPs (oxidative phosphorylation in electron transport chain)
Formation of Acetyl Coenzyme A	
2 NADH + 2 H^+	6 ATPs (oxidative phosphorylation in electron transport chain)
Krebs Cycle and Electron Transport Chain	
Oxidation of succinyl CoA to succinic acid	2 GTPs that are converted to 2 ATPs (substrate-level phosphorylation)
Production of 6 NADH + 6 H^+	18 ATPs (oxidative phosphorylation in electron transport chain)
Production of 2 FADH$_2$	4 ATPs (oxidative phosphorylation in electron transport chain)
Total:	36 or 38 ATPs per glucose molecule (theoretical maximum)

Figure 25.8 Summary of the principal reactions of cellular respiration. ETC = electron transport chain and chemiosmosis.

🔑 **Except for glycolysis, which occurs in the cytosol, all other reactions of cellular respiration occur within mitochondria.**

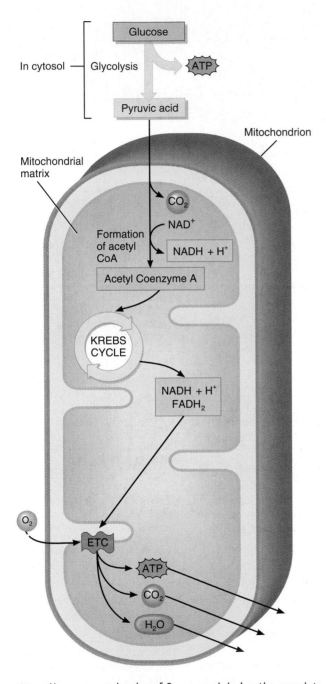

🔍 How many molecules of O_2 are used during the complete oxidation of one glucose molecule? How many molecules of CO_2 are produced?

Figure 25.9 Glycogenesis and glycogenolysis.

🔑 **The glycogenesis pathway converts glucose into glycogen, whereas the glycogenolysis pathway breaks down glycogen into glucose.**

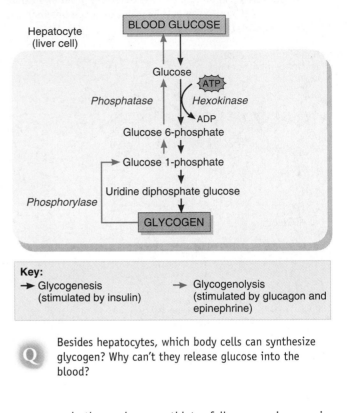

Key:
→ Glycogenesis (stimulated by insulin)

→ Glycogenolysis (stimulated by glucagon and epinephrine)

🔍 Besides hepatocytes, which body cells can synthesize glycogen? Why can't they release glucose into the blood?

runners and other endurance athletes follow a precise exercise and dietary regimen that includes eating large amounts of complex carbohydrates, such as pasta and potatoes, in the three days before an event. This practice, called **carbohydrate loading,** is intended to maximize muscle levels of glycogen available for ATP production. For athletic events lasting more than an hour, carbohydrate loading has been shown to increase an athlete's endurance. ∎

Glucose Anabolism

Even though most of the glucose in the body is catabolized to generate ATP, glucose may take part in several anabolic reactions. One is the synthesis of glycogen; another is the synthesis of new glucose molecules from some of the products of protein and lipid breakdown.

Glucose Storage: Glycogenesis

If glucose is not needed immediately for ATP production, it is combined with many other molecules of glucose to form the polysaccharide glycogen. In the process of **glycogenesis** (Figure 25.9), glucose that enters hepatocytes or skeletal muscle cells is first phosphorylated to glucose 6-phosphate by hexokinase. Glucose 6-phosphate is converted to glucose 1-phosphate, then to uridine diphosphate glucose, and finally to glycogen. Glycogenesis is stimulated by insulin from pancreatic beta cells.

Figure 25.10 Gluconeogenesis, the conversion of noncarbohydrate molecules (amino acids, lactic acid, and glycerol) into glucose.

🔑 **About 60% of the amino acids in the body can undergo gluconeogenesis.**

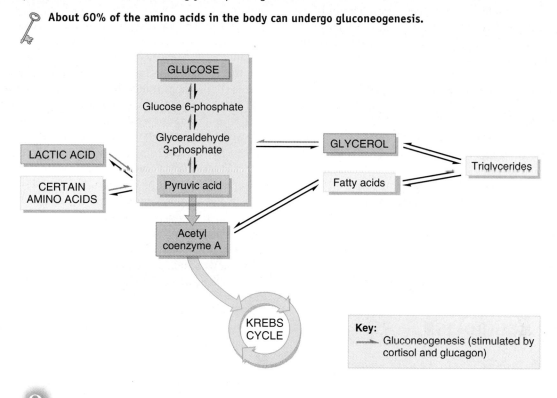

Q What cells can carry out gluconeogenesis and glycogenesis?

Glucose Release: Glycogenolysis

When body activities require ATP, glycogen stored in hepatocytes is broken down into glucose and released into the blood to be transported to cells, where it will be catabolized. The process of converting glycogen back to glucose is called **glycogenolysis** (glī′-kō-je-NOL-e-sis).

Glycogenolysis is not a simple reversal of the steps of glycogenesis (see Figure 25.9). It begins by splitting glucose molecules off of the branched glycogen molecule via phosphorylation to form glucose 1-phosphate. Phosphorylase, the enzyme that catalyzes this reaction, is activated by the hormones glucagon from the pancreatic alpha cells and epinephrine from the adrenal medulla. Glucose 1-phosphate is then converted to glucose 6-phosphate and finally to glucose. Because phosphatase, the enzyme that converts glucose 6-phosphate into glucose, is present in hepatocytes but absent in skeletal muscle cells, hepatocytes can release glucose derived from glycogen to the bloodstream, whereas skeletal muscle cells cannot. In skeletal muscle cells, glycogen is broken down into glucose 1-phosphate, which is then catabolized for ATP production via glycolysis and the Krebs cycle. However, the lactic acid produced by glycolysis in muscle cells can be converted to glucose in the liver. In this way, muscle glycogen can be an indirect source of blood glucose.

Formation of Glucose from Proteins and Fats: Gluconeogenesis

When the supply of liver glycogen runs low, body cells start catabolizing more triglycerides and proteins. Although some catabolism of triglycerides and proteins occurs normally, large-scale triglyceride and protein breakdown does not occur unless an individual is fasting, starving, eating meals that contain very few carbohydrates, or suffering from an endocrine disorder.

Certain molecules may be broken down and converted to glucose in the liver. The process by which new glucose is formed from noncarbohydrate sources is called **gluconeogenesis** (gloo′-kō-nē′-ō-JEN-e-sis; *neo-* = new). Lactic acid, certain amino acids, and the glycerol portion of triglyceride molecules can be used to form new glucose molecules through gluconeogenesis (Figure 25.10). About 60% of the amino acids in the body can undergo this conversion. Amino acids such as alanine, cysteine, glycine, serine, and threonine, and lactic acid are converted to pyruvic acid, which then may be synthesized into glucose or enter the Krebs cycle. Glycerol may be converted into glyceraldehyde 3-phosphate, which may form pyruvic acid or be used to synthesize glucose.

Gluconeogenesis is stimulated by cortisol, the main glucocorticoid hormone of the adrenal cortex, and by glucagon from the pancreas. In addition, cortisol stimulates the breakdown of proteins into amino acids, thus expanding the pool of amino acids available for gluconeogenesis. Thyroid hormones (thyroxine and triiodothyronine) also mobilize proteins and may mobilize triglycerides from adipose tissue, thereby making glycerol available for gluconeogenesis.

1. Explain how glucose moves into body cells.
2. Define glycolysis, and describe its principal events and outcome. What is the fate of pyruvic acid?
3. Describe how acetyl coenzyme A is formed.
4. Outline the principal events and outcomes of the Krebs cycle.
5. Explain what happens in the electron transport chain and why this process is called chemiosmosis.
6. Summarize the sources of ATP during the complete oxidation of a molecule of glucose.
7. Define glycogenesis and glycogenolysis. Under what circumstances does each occur?
8. What is gluconeogenesis, and why is it important?

LIPID METABOLISM

OBJECTIVES

• *Describe the lipoproteins that transport lipids in the blood.*

• *Describe the fate, metabolism, and functions of lipids.*

Transport of Lipids by Lipoproteins

Most lipids, such as cholesterol and triglycerides, are nonpolar—and therefore very hydrophobic—molecules. To be transported in watery blood, such molecules first must be made water-soluble by combining them with proteins produced by the liver and intestine. The combinations thus formed are **lipoproteins,** spherical particles that contain hundreds of molecules (Figure 25.11). In a lipoprotein, an outer shell of polar proteins plus amphipathic phospholipid and cholesterol molecules surrounds an inner core of hydrophobic triglyceride and cholesterol ester molecules. The proteins in the outer shell are called **apoproteins (apo)** and are designated by the letters A, B, C, D, and E plus a number. Besides helping to solubilize the lipoprotein in body fluids, each apoprotein also has specific functions.

There are several types of lipoproteins, each having different functions, but all essentially are transport vehicles: They provide a kind of delivery and pick-up service so that the various types of lipids can be available to cells that need them or removed from circulation if not needed. Lipoproteins are categorized and named mainly according to their density, which varies with the ratio of lipids (which have a low density) to proteins (which have a high density). From largest and lightest to smallest and heaviest, the four major classes of lipoproteins are chylomicrons, very low-density lipoproteins (VLDLs), low-density lipoproteins (LDLs), and high-density lipoproteins (HDLs).

Chylomicrons form in mucosal epithelial cells of the small intestine and contain *exogenous* (dietary) lipids. They contain about 1–2% proteins, 85% triglycerides, 7% phospholipids, and 6–7% cholesterol, plus a small amount of fat-soluble vitamins. Chylomicrons enter lacteals of intestinal villi and are carried by lymph into venous blood and then into the systemic circulation. Their presence gives blood plasma a milky appearance, but they remain in the blood for only a few minutes. As chylomicrons circulate through the capillaries of adipose tissue, one of their apoproteins, **apo C-2,** activates *endothelial lipoprotein lipase,* an enzyme that removes fatty acids from chylomicron triglycerides. The free fatty acids are then taken up by adipocytes for storage and by muscle cells for ATP production. Hepatocytes remove chylomicron remnants from the blood via receptor-mediated endocytosis, in which another chylomicron apoprotein, **apo E,** is the docking protein.

Very low-density lipoproteins (VLDLs) form in hepatocytes and contain *endogenous* triglycerides. VLDLs contain about 10% proteins, 50% triglycerides, 20% phospholipids, and 20% cholesterol. VLDLs transport triglycerides synthesized in hepatocytes to adipocytes for storage. Like chylomicrons, they lose triglycerides as their **apo C-2** activates endothelial lipoprotein lipase, and the resulting fatty acids are taken up by adipocytes for storage and by muscle cells for ATP production. After depositing some of their triglycerides in adipose cells, VLDLs are converted to LDLs.

Low-density lipoproteins (LDLs) contain 25% proteins, 5% triglycerides, 20% phospholipids, and 50% cholesterol. They carry about 75% of the total cholesterol in blood and deliver it to cells throughout the body for use in repair of cell membranes and synthesis of steroid hormones and bile salts. The only apoprotein LDLs contain is **apo B100,** which is the docking protein that binds to LDL receptors for receptor-mediated endocytosis of the LDL into a body cell. Within the cell, the LDL is broken down, and the cholesterol is released to serve the cell's needs. Once a cell has sufficient cholesterol for its activities, a negative feedback system inhibits the cell's synthesis of new LDL receptors.

When present in excessive numbers, LDLs also deposit cholesterol in and around smooth muscle fibers in arteries, forming fatty atherosclerotic plaques that increase the risk of coronary artery disease (see page 663). For this reason, the cholesterol in LDLs, called LDL-cholesterol, is known as "bad" cholesterol. Because some people have too few LDL receptors, their body cells cannot remove LDL from the blood as effectively; as a result, their plasma LDL level is abnormally high, and they are more likely to develop fatty plaques. Furthermore, eating a high-fat diet increases the production of VLDLs, which elevates the LDL level and increases the formation of fatty plaques.

High-density lipoproteins (HDLs), which contain 40–45% proteins, 5–10% triglycerides, 30% phospholipids, and 20% cholesterol, remove excess cholesterol from body cells and transport it to the liver for elimination. Because HDLs prevent accumulation of cholesterol in the blood, a

Figure 25.11 A lipoprotein. Shown here is a VLDL.

🔑 **A single layer of amphipathic phospholipids, cholesterol, and proteins surrounds a core of nonpolar lipids.**

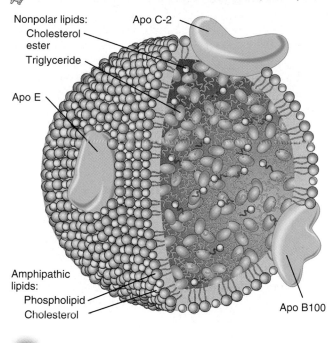

Nonpolar lipids:
Cholesterol ester
Triglyceride

Apo C-2

Apo E

Amphipathic lipids:
Phospholipid
Cholesterol

Apo B100

Q Which type of lipoprotein delivers cholesterol to body cells?

high HDL level is associated with decreased risk of coronary artery disease. For this reason, HDL-cholesterol is known as "good" cholesterol.

Sources and Significance of Blood Cholesterol

There are two sources of cholesterol in the body. Some is present in foods (eggs, dairy products, organ meats, beef, pork, and processed luncheon meats), but most is synthesized by the liver. Fatty foods that don't contain any cholesterol at all can still dramatically increase blood cholesterol level in two ways: First, a high intake of dietary fats stimulates reabsorption of cholesterol-containing bile back into the blood, so less cholesterol is lost in the feces. Second, when saturated fats are broken down in the body, hepatocytes use some of the breakdown products to produce cholesterol.

A lipid profile test usually measures total cholesterol (TC), HDL-cholesterol, and triglycerides (VLDLs). LDL-cholesterol then is calculated by using the following formula: LDL = TC − HDL − (triglycerides/5). In the United States, blood cholesterol is usually measured in milligrams per deciliter (mg/dL); a deciliter is 0.1 liter or 100 mL. For adults, desirable levels of blood cholesterol are total cholesterol under 200 mg/dL, LDL under 130 mg/dL, and HDL over 40 mg/dL. Normally, triglycerides are in the range of 10–190 mg/dL.

As total cholesterol level increases, the risk of coronary artery disease begins to rise. When total cholesterol is above 200 mg/dL (5.2 mmol/liter), the risk of a heart attack doubles with every 50 mg/dL (1.3 mmol/liter) increase in total cholesterol. Total cholesterol of 200–239 mg/dL and LDL of 130–159 mg/dL are borderline-high, whereas total cholesterol above 239 mg/dL and LDL above 159 are classified as high blood cholesterol. The ratio of total cholesterol to HDL-cholesterol predicts the risk of developing coronary artery disease. For example, a person with a total cholesterol of 180 and HDL of 60 has a risk ratio of 3. Ratios above 4 are considered undesirable; the higher the ratio, the greater the risk of developing coronary artery disease.

Among the therapies used to reduce blood cholesterol level are exercise, diet, and drugs. Regular physical activity at aerobic and nearly aerobic levels raises HDL level. Dietary changes are aimed at reducing the intake of total fat, saturated fats, and cholesterol. Drugs used to treat high blood cholesterol levels include cholestyramine (Questran) and colestipol (Colestid), which promote excretion of bile in the feces; nicotinic acid (Lipo-nicin); and lovastatin (Mevacor) and simvastatin (Zocor), which block a key enzyme (HMG-CoA reductase) needed for cholesterol synthesis.

The Fate of Lipids

Lipids, like carbohydrates, may be oxidized to produce ATP. If the body has no immediate need to use lipids in this way, they are stored in adipose tissue (fat depots) throughout the body and in the liver. A few lipids are used as structural molecules or to synthesize other essential substances. Some examples include phospholipids, which are constituents of plasma membranes; lipoproteins, which are used to transport cholesterol throughout the body; thromboplastin, which is needed for blood clotting; and myelin sheaths, which speed up nerve impulse conduction. The various functions of lipids in the body may be reviewed in Table 2.7 on page 45.

Triglyceride Storage

A major function of adipose tissue is to remove triglycerides from chylomicrons and VLDLs and store them until they are needed for ATP production in other parts of the body. Adipocytes in the subcutaneous layer contain about 50% of the stored triglycerides. Other adipose tissues account for the other half—about 12% around the kidneys, 10–15% in the omenta, 15% in genital areas, 5–8% between muscles, and 5% behind the eyes, in the sulci of the heart, and attached to the outside of the large intestine. Triglycerides in adipose tissue are continually broken down and resynthesized. Thus, the triglycerides stored in adipose tissue today are not the same molecules that were present last month because they are continually released from storage, transported in the blood, and redeposited in other adipose tissue cells.

Figure 25.12 Pathways of lipid metabolism. Glycerol may be converted to glyceraldehyde 3-phosphate, which can then be converted to glucose or enter the Krebs cycle for oxidation. Fatty acids undergo beta oxidation and enter the Krebs cycle via acetyl coenzyme A. The synthesis of lipids from glucose or amino acids is called lipogenesis.

🔑 **Glycerol and fatty acids are catabolized in separate pathways.**

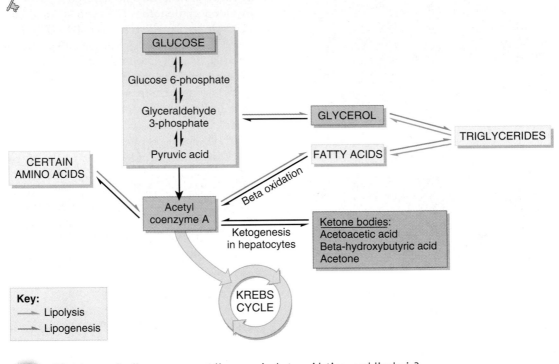

Q What types of cells can carry out lipogenesis, beta oxidation, and lipolysis? What type of cell can carry out ketogenesis?

Lipid Catabolism: Lipolysis

Triglycerides stored in adipose tissue constitute 98% of all body energy reserves. They are stored more readily than is glycogen, in part because triglycerides are hydrophobic and do not exert osmotic pressure on cell membranes. Several tissues (muscle, liver, and adipose tissue) routinely oxidize fatty acids derived from triglycerides to produce ATP. Before triglycerides can be catabolized to produce ATP, they must be split into glycerol and fatty acids—a process called **lipolysis** (li-POL-i-sis) that is catalyzed by enzymes called **lipases.** Two hormones that enhance triglyceride breakdown into fatty acids and glycerol are epinephrine and norepinephrine, which are released when sympathetic tone increases, as occurs, for example, during exercise. Other lipolytic hormones are cortisol, thyroid hormones, and insulinlike growth factors. By contrast, insulin itself inhibits lipolysis.

The glycerol and fatty acids that result from lipolysis are catabolized via different pathways (Figure 25.12). Glycerol is converted by many cells of the body to glyceraldehyde 3-phosphate, one of the compounds also formed during the catabolism of glucose. If ATP supply in a cell is high, glyceraldehyde 3-phosphate is converted into glucose, an example

of gluconeogenesis. When a cell needs to produce more ATP, however, glyceraldehyde 3-phosphate enters the catabolic pathway to pyruvic acid.

Fatty acids are catabolized differently than is glycerol and yield more ATP. The first stage in fatty acid catabolism is a series of reactions, called **beta oxidation,** that occurs in the matrix of mitochondria. Enzymes remove two carbon atoms at a time from the long chain of carbon atoms composing a fatty acid and attach the resulting two-carbon fragment to coenzyme A, forming acetyl CoA. In the second stage of fatty acid catabolism, the acetyl CoA formed as a result of beta oxidation enters the Krebs cycle (see Figure 25.12). A 16-carbon fatty acid such as palmitic acid can yield a net gain of 129 ATPs upon its complete oxidation via beta oxidation, the Krebs cycle, and the electron transport chain.

As part of normal fatty acid catabolism, hepatocytes can take two acetyl CoA molecules at a time and condense them to form **acetoacetic acid.** This reaction liberates the bulky CoA portion, which cannot diffuse out of cells. Some acetoacetic acid is converted into **beta-hydroxybutyric acid** and **acetone.** The formation of these three substances, collectively known as **ketone bodies,** is called **ketogenesis** (see

Figure 25.12). Because ketone bodies freely diffuse through plasma membranes, they leave hepatocytes and enter the bloodstream.

Other cells take up acetoacetic acid and attach its four carbons to two coenzyme A molecules to form two acetyl CoA molecules, which can then enter the Krebs cycle for oxidation. Heart muscle and the cortex (outer part) of the kidneys use acetoacetic acid in preference to glucose for generating ATP. Hepatocytes, which make acetoacetic acid, cannot use it for ATP production because they lack the enzyme that transfers acetoacetic acid back to coenzyme A.

Lipid Anabolism: Lipogenesis

Liver cells and adipose cells can synthesize lipids from glucose or amino acids through **lipogenesis** (see Figure 25.12), which is stimulated by insulin. Lipogenesis occurs when individuals consume more calories than are needed to satisfy their ATP needs. Excess dietary carbohydrates, proteins, and fats all have the same fate—they are converted into triglycerides. Many amino acids can be converted into acetyl CoA, which can then be converted into triglycerides. The steps in the conversion of glucose to lipids involve the formation of glyceraldehyde 3-phosphate, which can be converted to glycerol, and acetyl CoA, which can be converted to fatty acids. The resulting glycerol and fatty acids can undergo anabolic reactions to become triglycerides that can be stored, or they can go through a series of anabolic reactions to produce other lipids such as lipoproteins, phospholipids, and cholesterol.

CLINICAL APPLICATION
Ketosis

The level of ketone bodies in the blood normally is very low because other tissues use them for ATP production as fast as they are generated from the breakdown of fatty acids by hepatocytes. During periods of excessive beta oxidation, however, the production of ketone bodies exceeds their uptake and use by body cells. Excessive beta oxidation might occur after a meal rich in triglycerides, or during fasting or starvation, because few carbohydrates are available for catabolism. Excessive beta oxidation may also occur in poorly controlled or untreated diabetes mellitus for two reasons: (1) Because adequate glucose cannot get into cells, triglycerides are used for ATP production, and (2) because insulin normally inhibits lipolysis, a lack of insulin accelerates the pace of lipolysis. When the concentration of ketone bodies in the blood rises above normal—a condition called **ketosis**— the ketone bodies, most of which are acids, must be buffered. If too many accumulate, they decrease the concentration of buffers such as bicarbonate ions, and blood pH falls. Thus, extreme or prolonged ketosis can lead to **acidosis (ketoacidosis),** an abnormally low blood pH. When a diabetic becomes seriously insulin deficient, one of the telltale signs is the sweet smell on the breath of the ketone body acetone. ■

1. What are the functions of apoproteins in lipoproteins?
2. Explain which lipoprotein particles contain "good" and "bad" cholesterol, and explain why these terms are used.
3. Where are triglycerides stored in the body?
4. Explain the principal events of the catabolism of glycerol and fatty acids.
5. What are ketone bodies? What is ketosis?
6. Define lipogenesis and explain its importance.

PROTEIN METABOLISM
OBJECTIVE
• *Describe the fate, metabolism, and functions of proteins.*

During digestion, proteins are broken down into their constituent amino acids, which are then absorbed into blood capillaries in intestinal villi and transported to the liver via the hepatic portal vein. Unlike carbohydrates and triglycerides, which are stored, proteins are not warehoused for future use; instead, amino acids are either oxidized to produce ATP or used to synthesize new proteins for body growth and repair. Excess dietary amino acids are not excreted in the urine or feces but instead are converted into glucose (gluconeogenesis) or triglycerides (lipogenesis).

The Fate of Proteins

The active transport of amino acids into body cells is stimulated by insulinlike growth factors (IGFs) and insulin. Almost immediately after entry into the body, amino acids are incorporated into proteins. Many proteins function as enzymes; other proteins are involved in transportation (hemoglobin) or serve as antibodies, clotting chemicals (fibrinogen), hormones (insulin), or contractile elements in muscle fibers (actin and myosin). Several proteins serve as structural components of the body (collagen, elastin, and keratin). The various functions of proteins in the body may be reviewed in Table 2.8 on page 48.

Protein Catabolism

A certain amount of protein catabolism occurs in the body each day. Proteins are extracted from worn-out cells (such as red blood cells) and broken down into free amino acids. Some amino acids are converted into other amino acids, peptide bonds are re-formed, and new proteins are synthesized as part of the constant state of turnover in all cells. A significant fraction of the amino acids absorbed by the GI tract are from proteins in worn-out cells that have sloughed off the intestinal mucosa into the lumen.

Proteins being recycled are first broken down into amino acids. Hepatocytes, then, either convert amino acids to fatty acids, ketone bodies, or glucose, or oxidize them to carbon dioxide and water. However, before amino acids can be catabolized, they must first be converted to various substances that can enter the Krebs cycle. One such conversion

Figure 25.13 Various points at which amino acids (shown in yellow boxes) enter the Krebs cycle for oxidation.

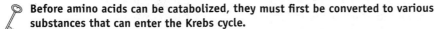

Before amino acids can be catabolized, they must first be converted to various substances that can enter the Krebs cycle.

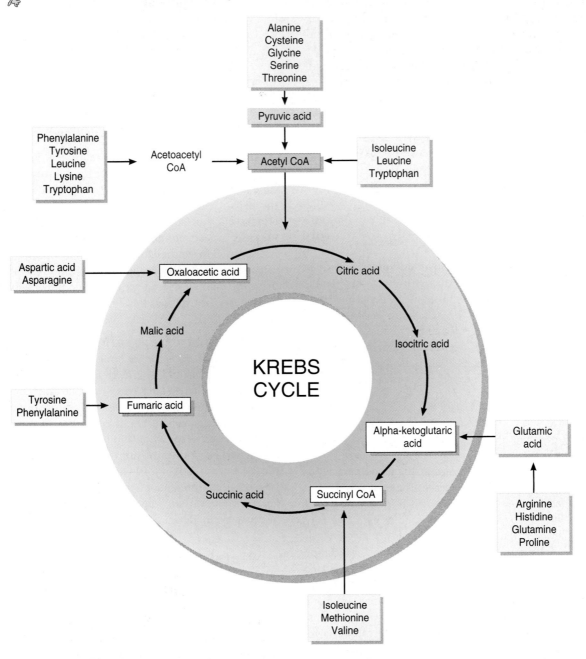

Q What group is removed from an amino acid before it can enter the Krebs cycle, and what is this process called?

consists of removing the amino group (NH₂) from the amino acid—a process called **deamination** (dē-am′-i-NĀ-shun)—and converting it to ammonia (NH₃). Hepatocytes then convert ammonia to urea, which is excreted in the urine. Other conversions are decarboxylation and dehydro-genation. Figure 25.13 shows that particular amino acids enter the Krebs cycle at different points. The conversion of amino acids into glucose (gluconeogenesis) may be reviewed in Figure 25.10; the conversion of amino acids into fatty acids (lipogenesis) or ketone bodies (ketogenesis) is shown in Figure 25.12.

Protein Anabolism

Protein anabolism involves the formation of peptide bonds between amino acids to produce new proteins. Protein anabolism, or protein synthesis, occurs on the ribosomes in almost every cell in the body, as directed by the cell's DNA and RNA (see Figure 3.30 on page 90). Insulinlike growth factors, thyroid hormones (T3 and T4), insulin, estrogen, and testosterone stimulate protein synthesis. Because proteins are a main component of most cell structures, adequate dietary protein is especially essential during the growth years, during pregnancy, and when tissue has been damaged by disease or injury. Once dietary intake of protein is adequate, however, eating more protein will not increase bone or muscle mass; only a regular program of forceful, weight-bearing muscular activity accomplishes that goal.

Of the 20 amino acids in the human body, ten are **essential amino acids**—they must be present in the diet because they cannot be synthesized in the body (at least in adequate amounts). Humans are unable to synthesize eight amino acids (isoleucene, leucine, methionine, phenylalanine, threonine, tryptophan, and valine) and synthesize two others (arginine and histidine) in inadequate amounts (especially in childhood). Because plants synthesize them, humans can obtain the essential amino acids by eating plants or animals (which eat plants). **Nonessential amino acids** can be synthesized by a process called **transamination,** the transfer of an amino group from an amino acid to pyruvic acid or to an acid in the Krebs cycle. Once the appropriate essential and nonessential amino acids are present in cells, protein synthesis occurs rapidly.

CLINICAL APPLICATION
Phenylketonuria

Phenylketonuria (fen′-il-kē′-tō-NOO-rē-a) or **PKU** is a genetic error of protein metabolism characterized by elevated blood levels of the amino acid phenylalanine. Most children with phenylketonuria have a mutation in the gene that codes for the enzyme phenylalanine hydroxylase; this enzyme is needed to convert phenylalanine into the amino acid tyrosine, which can enter the Krebs cycle. Because the enzyme is deficient, phenylalanine cannot be metabolized, and what is not used in protein synthesis builds up in the blood. If untreated, the disorder causes vomiting, rashes, seizures, growth deficiency, and severe mental retardation. Newborns are screened for PKU, and mental retardation can be prevented by restricting the child to a diet that supplies only the amount of phenylalanine needed for growth, although learning disabilities may still ensue. Because the artificial sweetener aspartame (NutraSweet) contains phenylalanine, its consumption must be restricted in children with PKU. ■

1. Relate deamination to amino acid catabolism.
2. Summarize the major steps of protein synthesis.
3. Define essential and nonessential amino acids.

KEY MOLECULES AT METABOLIC CROSSROADS

OBJECTIVE

• *Identify the key molecules in metabolism, and describe the reactions and the products they may form.*

Although there are thousands of different chemicals in cells, three molecules—glucose 6-phosphate, pyruvic acid, and acetyl coenzyme A—play pivotal roles in metabolism (Figure 25.14). These molecules stand at "metabolic crossroads"; the reactions that occur depend on the nutritional or activity status of the individual. Reactions **1** through **7** of Figure 25.14 occur in the cytosol, whereas reactions **8** through **10** occur inside mitochondria.

The Role of Glucose 6-Phosphate

Shortly after glucose enters a body cell, a kinase converts it to **glucose 6-phosphate.** Four possible fates await glucose 6-phosphate (see Figure 25.14):

1 *Synthesis of glycogen.* When glucose is abundant in the bloodstream, such as occurs just after a meal, considerable glucose 6-phosphate is used to synthesize glycogen, the storage form of carbohydrate in animals. Subsequent breakdown of glycogen into glucose 6-phosphate occurs through a slightly different series of reactions. Synthesis and breakdown of glycogen occur mainly in skeletal muscle fibers and hepatocytes.

2 *Release of glucose into the bloodstream.* If the enzyme glucose 6-phosphatase is present and active, glucose 6-phosphate can be dephosphorylated. Once glucose is released from the phosphate group, it can leave the cell and enter the bloodstream. Hepatocytes are the main cells that can provide glucose to the bloodstream in this way.

3 *Synthesis of nucleic acids.* Glucose 6-phosphate is the precursor used by cells throughout the body to make ribose 5-phosphate, a five-carbon sugar that is needed for synthesis of RNA (ribonucleic acid) and DNA (deoxyribonucleic acid). The same sequence of reactions that produces ribose 5-phosphate also produces NADPH, which is needed for certain reduction reactions.

4 *Glycolysis.* Some ATP is produced anaerobically via glycolysis, in which glucose 6-phosphate is converted to pyruvic acid, another key molecule in metabolism. Most body cells carry out glycolysis.

The Role of Pyruvic Acid

Each 6-carbon molecule of glucose that undergoes glycolysis yields two 3-carbon molecules of **pyruvic acid.** This molecule stands at a metabolic crossroads: Given enough oxygen, the aerobic (oxygen-consuming) reactions of cellular respiration can proceed; if oxygen is in short supply, anaerobic reactions can occur (see Figure 25.14):

Figure 25.14 Summary of the roles of the key molecules in metabolic pathways. Double-headed arrows indicate that reactions between two molecules may proceed in either direction, if the appropriate enzymes are present and the conditions are favorable; single-headed arrows signify the presence of an irreversible step.

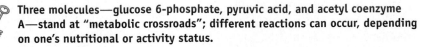

Three molecules—glucose 6-phosphate, pyruvic acid, and acetyl coenzyme A—stand at "metabolic crossroads"; different reactions can occur, depending on one's nutritional or activity status.

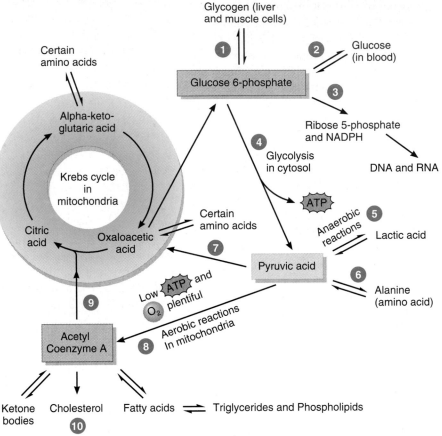

Q Which substance is the gateway into the Krebs cycle for fuel molecules that are being oxidized to generate ATP?

5 *Production of lactic acid.* When oxygen is in short supply in a tissue, as in actively contracting skeletal or cardiac muscle, some pyruvic acid is changed to lactic acid. The lactic acid then diffuses into the bloodstream and is taken up by hepatocytes, which eventually convert it back to pyruvic acid.

6 *Production of alanine.* One link between carbohydrate and protein metabolism occurs at pyruvic acid. Through transamination, an amino group ($-NH_3$) can either be added to pyruvic acid (a carbohydrate) to produce the amino acid alanine, or be removed from alanine to generate pyruvic acid.

7 *Gluconeogenesis.* Pyruvic acid and certain amino acids also can be converted to oxaloacetic acid, one of the Krebs cycle intermediates, which, in turn, can be used to

form glucose 6-phosphate. This sequence of gluconeogenesis reactions bypasses certain one-way reactions of glycolysis.

The Role of Acetyl Coenzyme A

8 When the ATP level in a cell is low but oxygen is plentiful, most pyruvic acid streams toward ATP-producing reactions—the Krebs cycle and electron transport chain—via conversion to **acetyl coenzyme A.**

9 *Entry into the Krebs cycle.* Acetyl CoA is the vehicle for two-carbon acetyl groups to enter the Krebs cycle. Oxidative Krebs cycle reactions convert acetyl CoA to CO_2 and produce reduced coenzymes (NADH and $FADH_2$) that transfer electrons into the electron transport chain. Oxidative reactions in the electron transport chain, in

Table 25.2 Summary of Metabolism

PROCESS	COMMENTS
Carbohydrates	
Glucose catabolism	Complete oxidation of glucose (cellular respiration) is the chief source of ATP in cells and consists of glycolysis, Krebs cycle, and electron transport chain. Complete oxidation of 1 molecule of glucose yields a maximum of 36 or 38 molecules of ATP.
Glycolysis	Conversion of glucose into pyruvic acid results in the production of some ATP. Reactions do not require oxygen (anaerobic cellular respiration).
Krebs cycle	Cycle includes series of oxidation–reduction reactions in which coenzymes (NAD^+ and FAD) pick up hydrogen ions and hydride ions from oxidized organic acids, and some ATP is produced. CO_2 and H_2O are by-products. Reactions are aerobic.
Electron transport chain	Third set of reactions in glucose catabolism is another series of oxidation–reduction reactions, in which electrons are passed from one carrier to the next, and most of the ATP is produced. Reactions require oxygen (aerobic cellular respiration).
Glucose anabolism	Some glucose is converted into glycogen (glycogenesis) for storage if not needed immediately for ATP production. Glycogen can be reconverted to glucose (glycogenolysis). The conversion of amino acids, glycerol, and lactic acid into glucose is called gluconeogenesis.
Lipids	
Triglyceride catabolism	Triglycerides are broken down into glycerol and fatty acids. Glycerol may be converted into glucose (gluconeogenesis) or catabolized via glycolysis. Fatty acids are catabolized via beta oxidation into acetyl coenzyme A that can enter the Krebs cycle for ATP production or be converted into ketone bodies (ketogenesis).
Triglyceride anabolism	The synthesis of triglycerides from glucose and fatty acids is called lipogenesis. Triglycerides are stored in adipose tissue.
Proteins	
Protein catabolism	Amino acids are oxidized via the Krebs cycle after deamination. Ammonia resulting from deamination is converted into urea in the liver, passed into blood, and excreted in urine. Amino acids may be converted into glucose (gluconeogenesis), fatty acids, or ketone bodies.
Protein anabolism	Protein synthesis is directed by DNA and utilizes cells' RNA and ribosomes.

turn, generate ATP. Most fuel molecules that will be oxidized to generate ATP—glucose, fatty acids, and ketone bodies—are first converted to acetyl CoA.

10 *Synthesis of lipids.* Acetyl CoA also is used for synthesis of certain lipids, including fatty acids, ketone bodies, and cholesterol. Because pyruvic acid can be converted to acetyl CoA, carbohydrates can be turned into triglycerides; this is the metabolic pathway for storage of most excess calories as fat. Mammals, including humans, cannot reconvert acetyl CoA to pyruvic acid, however, so fatty acids cannot be used to generate glucose or other carbohydrate molecules.

Table 25.2 summarizes carbohydrate, lipid, and protein metabolism.

1. Explain the importance of glucose 6-phosphate, pyruvic acid, and acetyl coenzyme A in metabolism.
2. Where within cells does each reaction in Figure 25.14 occur?

METABOLIC ADAPTATIONS

OBJECTIVE

• *Compare metabolism during the absorptive and postabsorptive states.*

Regulation of metabolic reactions depends both on the chemical environment within body cells, such as the levels of ATP and oxygen, and on signals from the nervous and endocrine systems. Some aspects of metabolism depend on how much time has passed since the last meal. During the **absorptive state,** ingested nutrients are entering the bloodstream, and glucose is readily available for ATP production. During the **postabsorptive state,** absorption of nutrients from the GI tract is complete, and energy needs must be met by fuels already in the body. A typical meal requires about 4 hours for complete absorption, and given three meals a day, the absorptive state exists for about 12 hours each day. Assuming no between-meal snacks, the other 12 hours—typically late morning, late afternoon, and most of the night—are spent in the postabsorptive state.

Because the nervous system and red blood cells continue to depend on glucose for ATP production during the postabsorptive state, maintaining a steady blood glucose

Figure 25.15 Principal metabolic pathways during the absorptive state.

🔑 **During the absorptive state, most body cells produce ATP by oxidizing glucose to CO₂ and H₂O.**

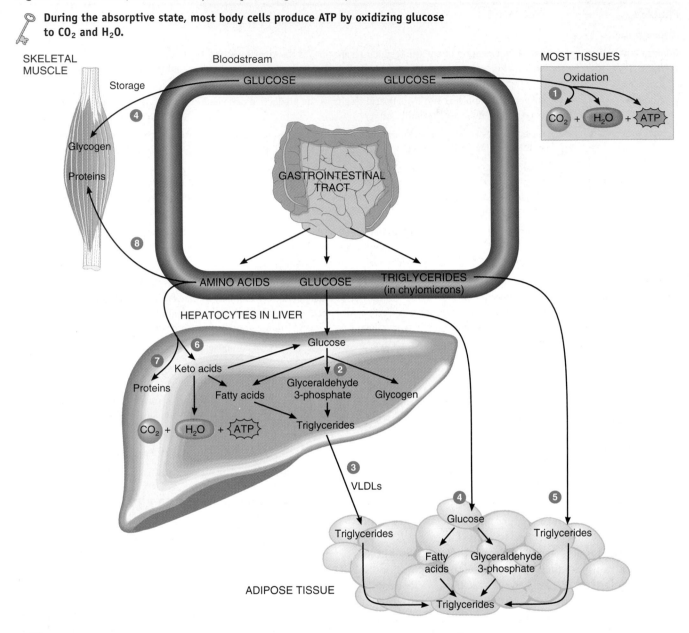

❓ Are the reactions shown in this figure mainly anabolic or catabolic?

level is critical during this period. Hormones are the major regulators of metabolism in each state. The effects of insulin dominate in the absorptive state, whereas several other hormones regulate metabolism in the postabsorptive state. During fasting and starvation, many body cells increasingly use ketone bodies for ATP production.

Metabolism During the Absorptive State

Soon after a meal, nutrients start to enter the blood. Recall that ingested food reaches the bloodstream mainly as glucose, amino acids, and triglycerides (in chylomicrons).

Two metabolic hallmarks of the absorptive state are the oxidation of glucose for ATP production, which occurs in most body cells, and the storage of excess fuel molecules for future between-meal use, which occurs mainly in hepatocytes, adipocytes, and skeletal muscle fibers.

Absorptive State Reactions

The following reactions dominate during the absorptive state (Figure 25.15):

❶ About 50% of the glucose absorbed from a typical meal is oxidized by cells throughout the body to produce ATP

Table 25.3 Hormonal Regulation of Metabolism in the Absorptive State

PROCESS	LOCATION(S)	MAIN STIMULATING HORMONE(S)
Facilitated diffusion of glucose into cells	Most cells.	Insulin.*
Active transport of amino acids into cells	Most cells.	Insulin.
Glycogenesis (glycogen synthesis)	Hepatocytes and muscle fibers.	Insulin.
Protein synthesis	All body cells.	Insulin, thyroid hormones, and insulinlike growth factors.
Lipogenesis (triglyceride synthesis)	Adipose cells and hepatocytes.	Insulin.

*Facilitated diffusion of glucose into hepatocytes (liver cells) and neurons is always "turned on" and does not require insulin.

via glycolysis, the Krebs cycle, and the electron transport chain.

2 Most glucose that enters hepatocytes is converted to triglycerides or glycogen.

3 Some fatty acids and triglycerides synthesized in the liver remain there, but hepatocytes package most into VLDLs, which carry lipids to adipose tissue for storage.

4 Adipocytes also take up glucose not picked up by the liver and convert it into triglycerides for storage. Overall, about 40% of the glucose absorbed from a meal is converted to triglycerides, and about 10% is stored as glycogen in skeletal muscles and hepatocytes.

5 Most dietary lipids (mainly triglycerides and fatty acids) are stored in adipose tissue; only a small portion is used for synthesis reactions. Adipocytes obtain the lipids from chylomicrons, from VLDLs, and from their own synthesis reactions.

6 Many absorbed amino acids that enter hepatocytes are deaminated to keto acids, which can either enter the Krebs cycle for ATP production or be used to synthesize glucose or fatty acids.

7 Some amino acids that enter hepatocytes are used to synthesize proteins (for example, plasma proteins).

8 Amino acids not taken up by hepatocytes are used in other body cells (such as muscle cells) for synthesis of proteins or regulatory chemicals such as hormones or enzymes.

Regulation of Metabolism During the Absorptive State

Soon after a meal, glucose-dependent insulinotropic peptide (GIP), plus the rising blood levels of glucose and certain amino acids, stimulate pancreatic beta cells to release insulin. In general, insulin increases the activity of enzymes needed for anabolism and the synthesis of storage molecules; at the same time it decreases the activity of enzymes needed for catabolic or breakdown reactions. Insulin promotes the entry of glucose and amino acids into cells of many tissues, and it stimulates the phosphorylation of glucose in hepatocytes and the conversion of glucose 6-phosphate to glycogen in both liver and muscle cells. In liver and adipose tissue, insulin enhances the synthesis of triglycerides, and in cells

throughout the body insulin stimulates protein synthesis. (See page 595 to review the effects of insulin.) Insulinlike growth factors and the thyroid hormones (T3 and T4) also stimulate protein synthesis. Table 25.3 summarizes the hormonal regulation of metabolism in the absorptive state.

Metabolism During the Postabsorptive State

About 4 hours after the last meal, absorption of nutrients from the small intestine is nearly complete, and blood glucose level starts to fall because glucose continues to leave the bloodstream and enter body cells while none is being absorbed from the GI tract. Thus, the main metabolic challenge during the postabsorptive state is to maintain the normal blood glucose level of 70–110 mg/100 mL (3.9–6.1 mmol/liter). Homeostasis of blood glucose concentration is especially important for the nervous system and for red blood cells for the following reasons:

- The dominant fuel molecule for ATP production in the nervous system is glucose, because fatty acids are unable to pass the blood–brain barrier.
- Red blood cells derive all their ATP from glycolysis of glucose because they have no mitochondria, and thus the Krebs cycle and the electron transport chain are not available to them.

Postabsorptive State Reactions

During the postabsorptive state, both *glucose production* and *glucose conservation* help maintain blood glucose level: Hepatocytes produce glucose molecules and export them into the blood, and other body cells switch from glucose to alternative fuels for ATP production to conserve scarce glucose. The major reactions of the postabsorptive state that produce glucose are the following (Figure 25.16):

1 *Breakdown of liver glycogen.* During fasting, a major source of blood glucose is liver glycogen, which can provide about a 4-hour supply of glucose. Liver glycogen is continually being formed and broken down as needed.

2 *Lipolysis.* Glycerol, produced by breakdown of triglycerides in adipose tissue, is also used to form glucose.

3 *Gluconeogenesis using lactic acid.* During exercise, skeletal muscle tissue breaks down stored glycogen (see step

Figure 25.16 Principal metabolic pathways during the postabsorptive state.

🔑 **The principal function of postabsorptive state reactions is to maintain a normal blood glucose level.**

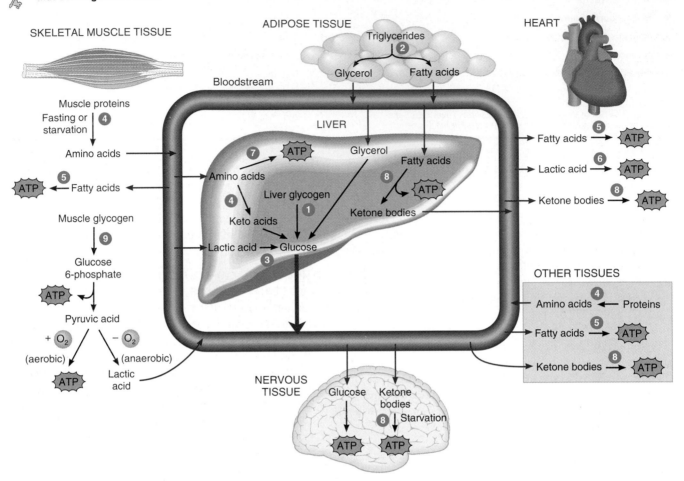

Q What processes directly elevate blood glucose during this state, and where does each occur?

❾) and produces some ATP anaerobically via glycolysis. Some of the pyruvic acid that results is converted to acetyl CoA, and some is converted to lactic acid, which diffuses into the blood. In the liver, lactic acid can be used for gluconeogenesis, and the resulting glucose is released into the blood.

❹ *Gluconeogenesis using amino acids.* Modest breakdown of proteins in skeletal muscle and other tissues releases large amounts of amino acids, which then can be converted to glucose by gluconeogenesis in the liver.

Despite these ways to produce glucose, blood glucose level cannot be maintained for very long without further metabolic changes. Thus, a major adjustment must be made during the postabsorptive state to produce ATP while conserving glucose. The following reactions produce ATP without using glucose:

❺ *Oxidation of fatty acids.* The fatty acids released by lipolysis of triglycerides cannot be used for glucose production because acetyl CoA cannot readily be converted to pyruvic acid. But most cells can oxidize the fatty acids directly, feed them into the Krebs cycle as acetyl CoA, and produce ATP through the electron transport chain.

❻ *Oxidation of lactic acid.* Cardiac muscle can produce ATP aerobically from lactic acid.

❼ *Oxidation of amino acids.* In hepatocytes, amino acids may be oxidized directly to produce ATP.

❽ *Oxidation of ketone bodies.* Hepatocytes also convert fatty acids to ketone bodies, which can be used by the heart, kidneys, and other tissues for ATP production.

❾ *Breakdown of muscle glycogen.* Skeletal muscle cells break down glycogen to glucose 6-phosphate, which undergoes glycolysis and provides ATP for muscle contraction.

Regulation of Metabolism During the Postabsorptive State

Both hormones and the sympathetic division of the autonomic nervous system (ANS) regulate metabolism during the postabsorptive state. The hormones that regulate postabsorptive state metabolism sometimes are called anti-insulin hormones because they counter the effects of insulin that dominate during the absorptive state. As blood glucose level declines, the secretion of insulin falls and the release of anti-insulin hormones rises.

When blood glucose concentration starts to drop, the pancreatic alpha cells release glucagon at a faster rate, and the beta cells secrete insulin more slowly. The primary target tissue of glucagon is the liver; the major effect is increased release of glucose into the bloodstream due to gluconeogenesis and glycogenolysis.

Low blood glucose level also activates the sympathetic branch of the ANS. Glucose-sensitive neurons in the hypothalamus detect low blood glucose and increase sympathetic output. As a result, sympathetic nerve endings release the neurotransmitter norepinephrine, and the adrenal medulla releases two catecholamine hormones—epinephrine and norepinephrine—into the bloodstream. Like glucagon, epinephrine stimulates glycogen breakdown. Epinephrine and norepinephrine are both potent stimulators of lipolysis. These actions of the catecholamines help to increase glucose and free fatty acid levels in the blood. As a result, muscle uses more fatty acids for ATP production, and more glucose is available to the nervous system. Table 25.4 summarizes the hormonal regulation of metabolism in the postabsorptive state.

Metabolism During Fasting and Starvation

The term **fasting** means going without food for many hours or a few days, whereas **starvation** implies weeks or months of food deprivation or inadequate food intake. People can survive without food for two months or more if they drink enough water to prevent dehydration. Although glycogen stores are depleted within a few hours of beginning a fast, catabolism of stored triglycerides and structural proteins can provide energy for several weeks. The amount of adipose tissue the body contains determines the lifespan possible without food.

During fasting and starvation, nervous tissue and RBCs continue to use glucose for ATP production. There is a ready supply of amino acids for gluconeogenesis because lowered insulin and increased cortisol levels slow the pace of protein synthesis and promote protein catabolism. Most cells in the body, especially skeletal muscle cells because they contain so much protein to begin with, can spare a fair amount of protein before their performance is adversely affected. During the first few days of fasting, protein catabolism outpaces protein synthesis by about 75 grams daily as some of the "old"

Table 25.4 Hormonal Regulation of Metabolism in the Postabsorptive State

PROCESS	LOCATION(S)	MAIN STIMULATING HORMONE(S)
Glycogenolysis (glycogen breakdown)	Hepatocytes and skeletal muscle fibers.	Glucagon and epinephrine.
Lipolysis (triglyceride breakdown)	Adipocytes.	Epinephrine, norepinephrine, cortisol, insulinlike growth factors, thyroid hormones, and others.
Protein breakdown	Most body cells, but especially skeletal muscle fibers.	Cortisol.
Gluconeogenesis (synthesis of glucose from noncarbohydrates)	Hepatocytes and kidney cortex cells.	Glucagon and cortisol.

amino acids are being deaminated and used for gluconeogenesis and "new" (that is, dietary) amino acids are lacking.

By the second day of a fast, blood glucose level has stabilized at about 65 mg/100 mL (3.6 mmol/liter), whereas the level of fatty acids in plasma has risen fourfold. Lipolysis of triglycerides in adipose tissue releases glycerol, which is used for gluconeogenesis, and fatty acids. The fatty acids diffuse into muscle fibers and other body cells, where they are used to produce acetyl-CoA, which enters the Krebs cycle. ATP then is synthesized as oxidation proceeds via the Krebs cycle and the electron transport chain.

The most dramatic metabolic change that occurs with fasting and starvation is the increase in the formation of ketone bodies by hepatocytes. During fasting, only small amounts of glucose undergo glycolysis to pyruvic acid, which in turn can be converted to oxaloacetic acid. Acetyl-CoA enters the Krebs cycle by combining with oxaloacetic acid (see Figure 25.14); when oxaloacetic acid is scarce due to fasting, only some of the available acetyl-CoA can enter the Krebs cycle. Surplus acetyl-CoA is used for ketogenesis, mainly in hepatocytes. Ketone body production thus increases as catabolism of fatty acids rises. Lipid-soluble ketone bodies can diffuse through plasma membranes and across the blood–brain barrier and be used as an alternate fuel for ATP production, especially by cardiac and skeletal muscle fibers and neurons. Normally, only a trace of ketone bodies (0.01 mmol/liter) are present in the blood, and thus they are a negligible fuel source. After two days of fasting, however, the level of ketones is 100–300 times higher and supplies roughly a third of the brain's fuel for ATP production. By 40 days of starvation, ketones provide up to two-thirds of the brain's energy needs. The presence of ketones actually reduces the use of glucose for ATP production, which in turn decreases the demand for gluconeogenesis and slows the catabolism of muscle proteins later in starvation to about 20 grams daily.

CLINICAL APPLICATION
Absorption of Alcohol

The intoxicating and incapacitating effects of alcohol depend on the blood alcohol level. Because it is lipid soluble, alcohol begins to be absorbed in the stomach. However, the surface area available for absorption is much greater in the small intestine than in the stomach, so when alcohol passes into the duodenum, it is absorbed more rapidly. Thus, the longer the alcohol remains in the stomach, the more slowly blood alcohol level rises. Because fatty acids in chyme slow gastric emptying, blood alcohol level will rise more slowly when fat-rich foods, such as pizza, hamburgers, or nachos, are consumed with alcoholic beverages. Also, the enzyme alcohol dehydrogenase, which is present in gastric mucosa cells, breaks down some of the alcohol to acetaldehyde, which is not intoxicating. When the rate of gastric emptying is slower, proportionally more alcohol will be absorbed and converted to acetaldehyde in the stomach, and thus less alcohol will reach the bloodstream. Given identical consumption of alcohol, females often develop higher blood alcohol levels (and therefore experience greater intoxication) than males of comparable size for two reasons: (1) Females typically have a smaller blood volume, and (2) the activity of gastric alcohol dehydrogenase is up to 60% lower in females than in males. Asian males may also have lower levels of this gastric enzyme. ■

1. Describe the roles of the following hormones in the regulation of metabolism: insulin, glucagon, epinephrine, insulinlike growth factors, thyroxine, cortisol, estrogen, and testosterone.
2. Why is ketogenesis more significant during fasting than during alternative absorptive and postabsorptive states?

HEAT AND ENERGY BALANCE

OBJECTIVES

• *Define basal metabolic rate (BMR), and explain several factors that affect it.*

• *Describe the factors that influence body heat production.*

• *Explain how normal body temperature is maintained by negative feedback loops involving the hypothalamic thermostat.*

The body produces more or less heat depending on the rates of metabolic reactions. Homeostasis of body temperature can be maintained only if the rate of heat loss from the body equals the rate of heat production by metabolism. Thus, it is important to understand the ways in which heat can be lost, gained, or conserved.

Metabolic Rate

The overall rate at which metabolic reactions use energy is termed the **metabolic rate.** Some of the energy is used to produce ATP, and some is released as heat. **Heat** is a form of kinetic energy that can be measured as **temperature** and expressed in units called calories. A **calorie,** spelled with a lowercase c, is the amount of heat energy required to raise the temperature of 1 gram of water from 14°C to 15°C. Because the calorie is a relatively small unit, the **kilocalorie (kcal)** or **Calorie (Cal),** spelled with an uppercase C, is often used to measure the body's metabolic rate and to express the energy content of foods. A kilocalorie equals 1000 calories.

Because many factors affect metabolic rate, it is measured under standard conditions, with the body in a quiet, resting, and fasting condition called the **basal state.** The measurement obtained is the **basal metabolic rate (BMR).** The most common way to determine BMR is by measuring the amount of oxygen used per kilocalorie of food metabolized. When the body uses 1 liter of oxygen to oxidize a typical dietary mixture of triglycerides, carbohydrates, and proteins, about 4.8 Cal of energy is released. BMR is 1200–1800 Cal/day in adults, or about 24 Cal/kg of body mass in adult males and 22 Cal/kg in adult females. The added calories needed to support daily activities, such as digestion and walking, range from 500 Cal for a small, relatively sedentary person to over 3000 Cal for a person in training for Olympic-level competitions or mountain climbing.

Body Temperature Homeostasis

Despite wide fluctuations in environmental temperature, homeostatic mechanisms can maintain a normal range for internal body temperature. If the rate of body heat production equals the rate of heat loss, the body maintains a constant core temperature near 37°C (98.6°F). **Core temperature** is the temperature in body structures deep to the skin and subcutaneous layer. **Shell temperature** is the temperature near the body surface—in the skin and subcutaneous layer. Depending on the environmental temperature, shell temperature is 1–6°C lower than core temperature. Too high a core temperature kills by denaturing body proteins, whereas too low a core temperature causes cardiac arrhythmias that result in death.

Heat Production

The production of body heat is proportional to metabolic rate. Several factors affect the metabolic rate and thus the rate of heat production:

• *Exercise.* During strenuous exercise, the metabolic rate may increase to as much as 15 times the basal rate. In well-trained athletes, the rate may increase up to 20 times.
• *Hormones.* Thyroid hormones (thyroxine and triiodothyronine) are the main regulators of BMR, which increases as the blood levels of thyroid hormones rise. The response

to changing levels of thyroid hormones is slow, however, taking several days to appear. Thyroid hormones increase BMR in part by stimulating aerobic cellular respiration. As cells use more oxygen to produce ATP, more heat is given off, and body temperature rises. Other hormones have minor effects on BMR. Testosterone, insulin, and human growth hormone can increase the metabolic rate by 5–15%.

- *Nervous system.* During exercise or in a stressful situation, the sympathetic division of the autonomic nervous system is stimulated. Its postganglionic neurons release norepinephrine (NE), and it also stimulates release of the hormones epinephrine and norepinephrine by the adrenal medulla. Both epinephrine and norepinephrine increase the metabolic rate of body cells.

- *Body temperature.* The higher the body temperature, the higher the metabolic rate. Each 1°C rise in core temperature increases the rate of biochemical reactions by about 10%. As a result, metabolic rate may be substantially increased during a fever.

- *Ingestion of food.* The ingestion of food can raise metabolic rate by 10–20%. This effect, *food-induced thermogenesis,* is greatest after eating a high-protein meal and is less after eating carbohydrates and lipids.

- *Age.* The metabolic rate of a child, in relation to its size, is about double that of an elderly person due to the high rates of reactions related to growth.

- *Miscellaneous factors.* Other factors that affect metabolic rate are gender (lower in females, except during pregnancy and lactation), climate (lower in tropical regions), sleep (lower), and malnutrition (lower).

Mechanisms of Heat Transfer

Maintaining normal body temperature depends on the ability to lose heat to the environment at the same rate as it is produced by metabolic reactions. Heat can be transferred from the body to its surroundings in four ways: via conduction, convection, radiation, and evaporation.

1. **Conduction** is the heat exchange that occurs between molecules of two materials that are in direct contact with each other. At rest, about 3% of body heat is lost via conduction to solid materials in contact with the body, such as a chair, clothing, and jewelry. Heat can also be gained via conduction—for example, while soaking in a hot tub. Because water conducts heat 20 times more effectively than air, heat loss or heat gain via conduction is much greater when the body is submerged in cold or hot water.

2. **Convection** is the transfer of heat by the movement of a fluid (a gas or a liquid) between areas of different temperature. The contact of air or water with your body results in heat transfer by both conduction and convection. When cool air makes contact with the body, it becomes warmed and therefore less dense and is carried away by convection currents created as the less dense air rises. The faster the air moves—for example, by a breeze or a fan—the faster the rate of convection. At rest, about 15% of body heat is lost to the air via conduction and convection.

3. **Radiation** is the transfer of heat in the form of infrared rays between a warmer object and a cooler one without physical contact. Your body loses heat by radiating more infrared waves than it absorbs from cooler objects. If surrounding objects are warmer than you are, you absorb more heat than you lose by radiation. In a room at 21°C (70°F), about 60% of heat loss occurs via radiation in a resting person.

4. **Evaporation** is the conversion of a liquid to a vapor. Every milliliter of evaporating water takes with it a great deal of heat—about 0.58 Cal/mL. Under typical resting conditions, about 22% of heat loss occurs through evaporation of about 700 mL of water per day—300 mL in exhaled air and 400 mL from the skin surface. Because we are not normally aware of this water loss through the skin and mucous membranes of the mouth and respiratory system, it is termed **insensible water loss.** The rate of evaporation is inversely related to relative humidity, the ratio of the actual amount of moisture in the air to the maximum amount it can hold at a given temperature. The higher the relative humidity, the lower the rate of evaporation. At 100% humidity, heat is gained via condensation of water on the skin surface as fast as heat is lost via evaporation. Evaporation provides the main defense against overheating during exercise. Under extreme conditions, a maximum of about 3 liters of sweat can be produced each hour, removing more than 1700 Calories of heat if all of it evaporates. (Note: Sweat that drips off the body rather than evaporating removes very little heat.)

Hypothalamic Thermostat

The control center that functions as the body's thermostat is a group of neurons in the anterior part of the hypothalamus, the **preoptic area.** This area receives impulses from thermoreceptors in the skin and mucous membranes and in the hypothalamus. Neurons of the preoptic area generate nerve impulses at a higher frequency when blood temperature increases, and at a lower frequency when blood temperature decreases.

Nerve impulses from the preoptic area propagate to two other parts of the hypothalamus known as the **heat-losing center** and the **heat-promoting center,** which, when stimulated by the preoptic area, set into operation a series of responses that lower body temperature and raise body temperature, respectively.

Thermoregulation

If core temperature declines, mechanisms that help conserve heat and increase heat production act via several negative feedback loops to raise the body temperature to normal (Figure 25.17). Thermoreceptors in the skin and hypothalamus send nerve impulses to the preoptic area and the heat-promoting center in the hypothalamus, and to hypothalamic

Figure 25.17 Negative feedback mechanisms that conserve heat and increase heat production.

🔑 **Core temperature is the temperature in body structures deep to the skin and subcutaneous layer; shell temperature is the temperature near the body surface.**

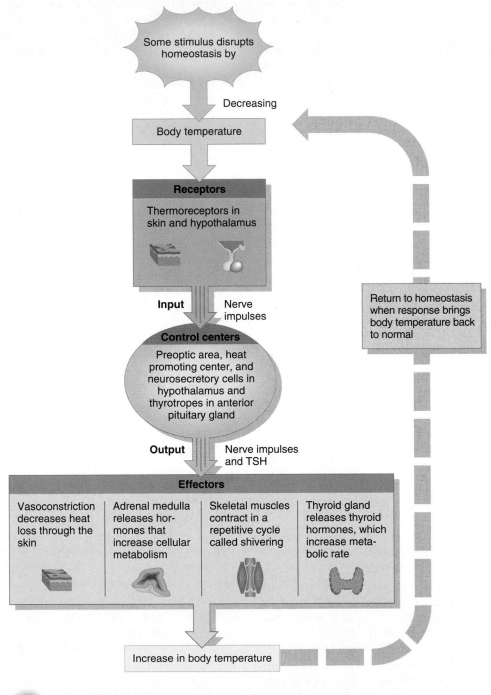

Q What factors can increase metabolic rate and thus increase the rate of heat production?

neurosecretory cells that produce thyrotropin-releasing hormone or TRH. In response, the hypothalamus discharges nerve impulses and secretes TRH, which in turn stimulates thyrotrophs in the anterior pituitary gland to release thyroid stimulating hormone (TSH). Nerve impulses from the hypothalamus and TSH then activate several effectors.

Each effector responds in a way that helps increase core temperature to the normal value:

- Nerve impulses from the heat-promoting center stimulate sympathetic nerves that cause blood vessels of the skin to constrict. Vasoconstriction decreases the flow of warm blood, and thus the transfer of heat, from the internal organs to the skin. Slowing the rate of heat loss allows the internal body temperature to increase as metabolic reactions continue to produce heat.

- Nerve impulses in sympathetic nerves leading to the adrenal medulla stimulate the release of epinephrine and norepinephrine into the blood. The hormones, in turn, bring about an increase in cellular metabolism, which increases heat production.

- The heat-promoting center stimulates parts of the brain that increase muscle tone and hence heat production. As muscle tone increases in one muscle (the agonist), the small contractions stretch muscle spindles in its antagonist, initiating a stretch reflex. The resulting contraction in the antagonist stretches muscle spindles in the agonist, and it too develops a stretch reflex. This repetitive cycle—called **shivering**—greatly increases the rate of heat production. During maximal shivering, body heat production can rise to about four times the basal rate in just a few minutes.

- The thyroid gland responds to TSH by releasing more thyroid hormones into the blood. As increased levels of thyroid hormones slowly increase the metabolic rate, body temperature rises.

If core body temperature rises above normal, a negative feedback loop opposite to the one depicted in Figure 25.17 goes into action. The higher temperature of the blood stimulates thermoreceptors that send nerve impulses to the preoptic area, which, in turn, stimulate the heat-losing center and inhibit the heat-promoting center. Nerve impulses from the heat-losing center cause dilation of blood vessels in the skin. The skin becomes warm, and the excess heat is lost to the environment via radiation and conduction as an increased volume of blood flows from the warmer core of the body into the cooler skin. At the same time, metabolic rate decreases, and shivering does not occur. The high temperature of the blood stimulates sweat glands of the skin via hypothalamic activation of sympathetic nerves. As the water in perspiration evaporates from the surface of the skin, the skin is cooled. All these responses counteract heat-promoting effects and help return body temperature to normal.

CLINICAL APPLICATION
Hypothermia

Hypothermia is a lowering of core body temperature to 35°C (95°F) or below. Causes of hypothermia include an overwhelming cold stress (immersion in icy water), metabolic diseases (hypoglycemia, adrenal insufficiency, or hypothyroidism), drugs (alcohol, antidepressants, sedatives, or tranquilizers), burns, and malnutrition. Hypothermia is characterized by the following as core body temperature falls: sensation of cold, shivering, confusion, vasoconstriction, muscle rigidity, bradycardia, acidosis, hypoventilation, hypotension, loss of spontaneous movement, coma, and death (usually caused by cardiac arrhythmias). Because the elderly have reduced metabolic protection against a cold environment coupled with a reduced perception of cold, they are at greater risk for developing hypothermia. ■

Regulation of Food Intake

When the energy content of food balances the energy used by all the cells of the body, constant body weight is maintained. This is true unless there is a gain or loss of water. In the more affluent nations, however, a significant fraction of the population is overweight, a condition that increases one's risk of dying from a variety of cardiovascular and metabolic disorders, for example, diabetes mellitus.

Most mature animals and many men and women maintain a stable body weight for long periods of time. Weight stability may persist despite large day-to-day variations in activity and food intake. But there are no sensory receptors that monitor one's weight or size. How, then, is food intake regulated? The answer to this question remains elusive. It seems to depend on many factors, including levels of certain nutrients in the blood, certain hormones, psychological elements such as stress or depression, signals from the GI tract and the special senses, and neural connections between the hypothalamus and other parts of the brain.

Within the hypothalamus are nuclei that play key roles in regulating food intake (see Figure 14.10 on page 462). When one cluster of neurons—the **feeding center,** located in the lateral hypothalamic nuclei—is stimulated in animals, they begin to eat heartily, even if they are already full. A second center, a cluster of neurons called the **satiety center** (sa-TĪ-i-tē), is located in the ventromedial nuclei of the hypothalamus. Satiety is the opposite of appetite; it is a sensation of fullness accompanied by a lack of desire to eat. Stimulation of this center causes animals to stop eating, even if they have been starved for days.

How do neurons in the hypothalamus sense whether a person is well-fed or in need of food? One possibility is that changes in the chemical composition of the blood after eating and during fasting might alert the hypothalamic neurons.

Signaling molecules in the blood that act on the hypothalamus to decrease appetite and increase energy expenditure include several hormones (glucagon, cholecystokinin, and epinephrine acting via beta receptors), glucose, and leptin, which is released by adipocytes as they synthesize triglycerides. Signaling molecules that increase appetite and decrease energy expenditure include the following: opioids, growth hormone-releasing hormone (GHRH), glucocorticoids, epinephrine acting via alpha receptors, insulin, progesterone, and somatostatin.

Food intake is also regulated by distention of the GI tract, particularly the stomach and duodenum. The stretching of these organs initiates a reflex that activates the satiety center and depresses the feeding center. Psychological factors may override the usual intake mechanisms, as occurs, for example, in anorexia nervosa, bulimia, and obesity (see pages 865 and 908).

1. Define a kilocalorie (kcal). How is the unit used? How does it relate to a calorie?
2. Distinguish between core temperature and shell temperature.
3. In what ways can a person lose heat to or gain heat from the surroundings? How is it possible for a person to lose heat on a sunny beach when the temperature is 40°C (104°F) and the humidity is 85%?
4. Discuss how food intake is regulated.

NUTRITION

OBJECTIVE

• *Describe how to select foods to maintain a healthy diet.*

• *Compare the sources, functions, and importance of minerals and vitamins in metabolism.*

Foods contain the substances, called **nutrients,** that the body uses for growth, maintenance, and repair of its tissues. Nutrients include water, carbohydrates, lipids, proteins, vitamins, and minerals. Water is the nutrient needed in the largest amount—about 2,000–3,000 grams (2–3 liters) per day. As the most abundant compound in the body, water provides the medium in which most metabolic reactions occur, and it also participates in some reactions (for example, hydrolysis reactions). The important roles of water in the body can be reviewed on page 39. Three organic nutrients—carbohydrates, lipids, and proteins—provide the energy needed for metabolic reactions and serve as building blocks for body structures. *Essential nutrients* are specific nutrient molecules that the body cannot make in sufficient quantity to meet its needs and thus must obtain from the diet. Some amino acids, some fatty acids, vitamins, and minerals are essential nutrients.

Next we discuss some guidelines for healthy eating and the roles of minerals and vitamins in metabolism.

Guidelines for Healthy Eating

On a daily basis, many women and older people need about 1600 Calories; children, teenage girls, active women, and most men need about 2200 Calories; and teenage boys and active men need about 2800 Calories. Each gram of protein or carbohydrate in food provides about 4 Calories, whereas a gram of fat (lipids) provides about 9 Calories.

We still do not know with certainty what levels and types of carbohydrate, fat, and protein are optimal in the diet, for different populations the world over eat radically different diets that are adapted to their particular lifestyles. Experts recommend the following distribution of calories: 50–60% from carbohydrates, with less than 15% from simple sugars; less than 30% from fats (triglycerides are the main type of dietary fat), with no more than 10% as saturated fats; and about 12–15% from proteins.

The guidelines for healthy eating are:

• Eat a variety of foods.
• Maintain healthy weight.
• Choose foods low in fat, saturated fat, and cholesterol.
• Eat plenty of vegetables, fruits, and grain products.
• Use sugars in moderation only.
• Use salt and sodium in moderation only (less than 6 grams daily).
• If you drink alcoholic beverages, do so in moderation (less than 1 ounce of the equivalent of pure alcohol per day).

To help people achieve a good balance of vitamins, minerals, carbohydrates, fats, and proteins in their food, the U.S. Department of Agriculture developed the food guide pyramid (Figure 25.18). The sections of the pyramid indicate how many servings of each of the five major food groups to eat each day. The smallest number of servings corresponds to a 1600 Cal/day diet, whereas the largest number of servings corresponds to a 2800 Cal/day diet. Because they should be consumed in largest quantity, foods rich in complex carbohydrates—the bread, cereal, rice, and pasta group—form the base of the pyramid. Vegetables and fruits form the next level. The health benefits of eating generous amounts of these foods are well documented. To be eaten in smaller quantities are foods on the next level up—the milk, yogurt, and cheese group and the meat, poultry, fish, dry beans, eggs, and nuts group. These two food groups have higher fat and protein content than the food groups below them. To lower your daily intake of fats, choose low-fat foods from these groups—nonfat milk and yogurt, low-fat cheese, fish, and poultry (remove the skin).

The apex of the pyramid is not a food group but rather a caution to use fats, oils, and sweets sparingly. Even though the food guide pyramid does not distinguish among the different types of fatty acids—saturated, polyunsaturated, and monounsaturated—in dietary fats, note that atherosclerosis and coronary artery disease are prevalent in populations that consume large amounts of saturated fats and cholesterol.

Figure 25.18 The food guide pyramid. The smallest number of servings corresponds to 1600 Calories per day, whereas the largest number of servings corresponds to 2800 Calories per day. Each example given equals one serving.

The sections of the pyramid show how many servings of five major food groups to eat each day.

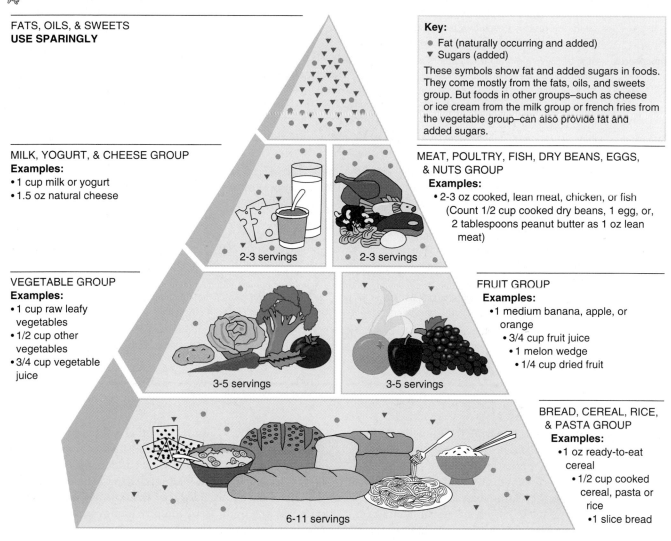

FATS, OILS, & SWEETS
USE SPARINGLY

Key:
- Fat (naturally occurring and added)
- ▼ Sugars (added)

These symbols show fat and added sugars in foods. They come mostly from the fats, oils, and sweets group. But foods in other groups—such as cheese or ice cream from the milk group or french fries from the vegetable group—can also provide fat and added sugars.

MILK, YOGURT, & CHEESE GROUP
Examples:
- 1 cup milk or yogurt
- 1.5 oz natural cheese

2-3 servings

MEAT, POULTRY, FISH, DRY BEANS, EGGS, & NUTS GROUP
Examples:
- 2-3 oz cooked, lean meat, chicken, or fish (Count 1/2 cup cooked dry beans, 1 egg, or, 2 tablespoons peanut butter as 1 oz lean meat)

2-3 servings

VEGETABLE GROUP
Examples:
- 1 cup raw leafy vegetables
- 1/2 cup other vegetables
- 3/4 cup vegetable juice

3-5 servings

FRUIT GROUP
Examples:
- 1 medium banana, apple, or orange
- 3/4 cup fruit juice
- 1 melon wedge
- 1/4 cup dried fruit

3-5 servings

BREAD, CEREAL, RICE, & PASTA GROUP
Examples:
- 1 oz ready-to-eat cereal
- 1/2 cup cooked cereal, pasta or rice
- 1 slice bread

6-11 servings

Q Which foods shown contain cholesterol and most of the saturated fatty acids in the diet?

Populations living around the Mediterranean Sea, in contrast, have low rates of coronary artery disease despite eating a diet that contains up to 40% of the calories as fats. Most of their fat comes from olive oil, however, which is rich in monounsaturated fatty acids and has no cholesterol. Canola oil, avocados, nuts, and peanut oil are also rich in monounsaturated fatty acids.

Minerals

Some minerals and many vitamins are components of the enzyme systems that catalyze metabolic reactions. **Minerals** are inorganic elements that occur naturally in the Earth's crust. In the body they appear in combination with each other, in combination with organic compounds, or as ions in solution. Minerals constitute about 4% of total body weight, and they are concentrated most heavily in the skeleton. The minerals present in largest quantity in the body are calcium, phosphorus, potassium, sulfur, sodium, chloride, magnesium, iron, and iodide (see Table 2.1 on page 27). Other minerals known to have functions essential to life include manganese, copper, cobalt, zinc, fluoride, selenium, and chromium. The minerals aluminum, silicon, arsenic,

Table 25.5 Minerals Vital to the Body

MINERAL	COMMENTS	IMPORTANCE
Calcium	Most abundant mineral in body. Appears in combination with phosphates. About 99% is stored in bone and teeth. Blood Ca^{2+} level is controlled by calcitonin (CT) and parathyroid hormone (PTH). Absorption occurs only in the presence of vitamin D. Excess is excreted in feces and urine. Sources are milk, egg yolk, shellfish, green leafy vegetables.	Formation of bones and teeth, blood clotting, normal muscle and nerve activity, endocytosis and exocytosis, cellular motility, chromosome movement before cell division, glycogen metabolism, and synthesis and release of neurotransmitters.
Phosphorus	About 80% found in bones and teeth as phosphate salts. Blood phosphate level is controlled by calcitonin (CT) and parathyroid hormone (PTH). Excess is excreted in urine; small amount is eliminated in feces. Sources are dairy products, meat, fish, poultry, nuts.	Formation of bones and teeth. Phosphates ($H_2PO_4^-$, HPO_4^{2-}, and PO_4^{3-}) constitute a major buffer system of blood. Plays important role in muscle contraction and nerve activity. Component of many enzymes. Involved in energy transfer (ATP). Component of DNA and RNA.
Potassium	Principal cation (K^+) in intracellular fluid. Excess excreted in urine. Normal food intake supplies required amounts.	Functions in nerve and muscle action potential conduction.
Sulfur	Constituent of many proteins (such as insulin), electron carriers in oxidative phosphorylation, and some vitamins (thiamine and biotin). Excreted in urine. Sources include beef, liver, lamb, fish, poultry, eggs, cheese, beans.	As components of hormones and vitamins, regulates various body activities. Needed for ATP production by aerobic cellular respiration.
Sodium	Most abundant cation (Na^+) in extracellular fluids; some found in bones. Excreted in urine and perspiration. Normal intake of NaCl (table salt) supplies more than the required amounts.	Strongly affects distribution of water through osmosis. Part of bicarbonate buffer system. Functions in nerve and muscle action potential conduction.
Chloride	Principal anion (Cl^-) in extracellular fluid. Excess excreted in urine. Normal intake of NaCl supplies required amounts.	Plays role in acid–base balance of blood, water balance, and formation of HCl in stomach.
Magnesium	Important cation (Mg^{2+}) in intracellular fluid. Excreted in urine and feces. Widespread in various foods, such as green leafy vegetables, seafood, and whole-grain cereals.	Required for normal functioning of muscle and nervous tissue. Participates in bone formation. Constituent of many coenzymes.

and nickel are present in the body, but their functions, if any, are not known.

Calcium and phosphorus form part of the matrix of bone. Because minerals do not form long-chain compounds, they are otherwise poor building materials. A major role of minerals is to help regulate enzymatic reactions. Calcium, iron, magnesium, and manganese are constituents of some coenzymes. Magnesium also serves as a catalyst for the conversion of ADP to ATP. Minerals such as sodium and phosphorus work in buffer systems. Sodium also helps regulate the osmosis of water and, along with other ions, is involved in the generation of nerve impulses. Table 25.5 describes the vital functions of selected minerals. Note that the body generally uses the ions of the minerals rather than the nonionized form. Some minerals, such as chlorine, are toxic or even fatal if ingested in the nonionized form.

Vitamins

Organic nutrients required in small amounts to maintain growth and normal metabolism are called **vitamins.** Unlike carbohydrates, lipids, or proteins, vitamins do not provide energy or serve as the body's building materials. Of the vitamins whose functions are known, most serve as coenzymes.

Most vitamins cannot be synthesized by the body; they must be ingested. Other vitamins, such as vitamin K, are produced by bacteria in the GI tract and then absorbed. The body can assemble some vitamins if the raw materials, called **provitamins,** are provided. For example, vitamin A is produced by the body from the provitamin beta-carotene, a chemical present in yellow vegetables such as carrots and in dark green vegetables such as spinach. No single food contains all the required vitamins—one of the best reasons to eat a varied diet.

Vitamins are divided into two main groups: fat-soluble vitamins and water-soluble vitamins. **Fat-soluble vitamins** are absorbed along with other dietary lipids in the small intestine and packaged into chylomicrons. They cannot be absorbed in adequate quantity unless they are ingested with other lipids. Fat-soluble vitamins may be stored in cells, particularly hepatocytes. Examples of the fat-soluble vitamins include vitamins A, D, E, and K. **Water-soluble vitamins,** by contrast, are absorbed along with water in the GI tract and dissolve in body fluids. Excess quantities of these vitamins are not stored but instead are excreted in the urine. Examples of water-soluble vitamins are the B vitamins and vitamin C.

Table 25.5 (continued)

MINERAL	COMMENTS	IMPORTANCE
Iron	About 66% found in hemoglobin of blood. Normal losses of iron occur by shedding of hair, epithelial cells, and mucosal cells, and in sweat, urine, feces, bile, and blood lost during menstruation. Sources are meat, liver, shellfish, egg yolk, beans, legumes, dried fruits, nuts, cereals.	As component of hemoglobin, reversibly binds O_2. Component of cytochromes involved in electron transport chain.
Iodide	Essential component of thyroid hormones. Excreted in urine. Sources are seafood, iodized salt, and vegetables grown in iodine-rich soils.	Required by thyroid gland to synthesize thyroid hormones, which regulate metabolic rate.
Manganese	Some stored in liver and spleen. Most excreted in feces.	Activates several enzymes. Needed for hemoglobin synthesis, urea formation, growth, reproduction, lactation, bone formation, and possibly production and release of insulin, and inhibition of cell damage.
Copper	Some stored in liver and spleen. Most excreted in feces. Sources include eggs, whole-wheat flour, beans, beets, liver, fish, spinach, asparagus.	Required with iron for synthesis of hemoglobin. Component of coenzymes in electron transport chain and enzyme necessary for melanin formation.
Cobalt	Constituent of vitamin B_{12}.	As part of vitamin B_{12}, required for erythropoiesis.
Zinc	Important component of certain enzymes. Widespread in many foods, especially meats.	As a component of carbonic anhydrase, important in carbon dioxide metabolism. Necessary for normal growth and wound healing, normal taste sensations and appetite, and normal sperm counts in males. As a component of peptidases, it is involved in protein digestion.
Fluoride	Components of bones, teeth, other tissues.	Appears to improve tooth structure and inhibit tooth decay.
Selenium	Found in seafood, meat, chicken, grain cereals, egg yolk, milk, mushrooms, and garlic.	An antioxidant. Prevents chromosome breakage and may play a role in preventing certain birth defects.
Chromium	Found in high concentrations in brewer's yeast. Also found in wine and some brands of beer.	Needed for normal activity of insulin in carbohydrate and lipid metabolism.

Besides their other functions, three vitamins—C, E, and beta-carotene (a provitamin)—are termed **antioxidant vitamins** because they inactivate oxygen free radicals, which are highly reactive particles that carry an unpaired electron (see Figure 2.3 on page 30). Free radicals damage cell membranes, DNA, and other cellular structures and contribute to the formation of artery-narrowing atherosclerotic plaques. Some free radicals arise naturally in the body, and others derive from environmental hazards such as tobacco smoke and radiation. Antioxidant vitamins are thought to play a role in protecting against some kinds of cancer, reducing the buildup of atherosclerotic plaque, delaying some effects of aging, and decreasing the chance of cataract formation in the lens of the eyes. Table 25.6 lists the principal vitamins, their sources, their functions, and related deficiency disorders.

CLINICAL APPLICATION
Vitamin and Mineral Supplements

Most nutritionists recommend eating a balanced diet that includes a variety of foods rather than taking vitamin or mineral supplements, except in special circumstances. Common examples of warranted supplementations include: iron for women who have excessive menstrual bleeding; iron and calcium for women who are pregnant or breast-feeding; folic acid (folate) for all women who may become pregnant, to reduce the risk of fetal neural tube defects; calcium for most adults, because they do not receive the recommended amount in their diets; and vitamin B_{12} for strict vegetarians, who eat no meat. Because most North Americans do not ingest in their food the high levels of antioxidant vitamins thought to have beneficial effects, some experts recommend supplementing vitamins C and E. More is not always better, however; megadoses of vitamins or minerals can be very harmful. ■

1. Define a nutrient, and list the different types of nutrients.
2. Describe the food guide pyramid and give examples of foods from each food group.
3. What is a mineral? Briefly describe the functions of the following minerals: calcium, phosphorus, potassium, sulfur, sodium, chloride, magnesium, iron, iodine, copper, zinc, fluoride, manganese, cobalt, chromium, and selenium.
4. Define a vitamin. Explain how we obtain vitamins. Distinguish between a fat-soluble vitamin and a water-soluble vitamin.
5. For each of the following vitamins, indicate its principal function and the effect(s) of deficiency: A, D, E, K, B_1, B_2, niacin, B_6, B_{12}, pantothenic acid, folic acid, biotin, and C.

Table 25.6 The Principal Vitamins

VITAMIN	COMMENT AND SOURCE	FUNCTIONS	DEFICIENCY SYMPTOMS AND DISORDERS
Fat-soluble	All require bile salts and some dietary lipids for adequate absorption.		
A	Formed from provitamin beta-carotene (and other provitamins) in GI tract. Stored in liver. Sources of carotene and other provitamins include yellow and green vegetables; sources of vitamin A include liver and milk.	Maintains general health and vigor of epithelial cells. Beta-carotene acts as an antioxidant to inactivate free radicals.	Deficiency results in atrophy and keratinization of epithelium, leading to dry skin and hair; increased incidence of ear, sinus, respiratory, urinary, and digestive system infections; inability to gain weight; drying of cornea; and skin sores.
		Essential for formation of photopigments, light-sensitive chemicals in photoreceptors of retina.	**Night blindness** or decreased ability for dark adaptation.
		Aids in growth of bones and teeth, apparently by helping to regulate activity of osteoblasts and osteoclasts.	Slow and faulty development of bones and teeth.
D	Sunlight converts 7-dehydrocholesterol in the skin to cholecalciferol (vitamin D_3). A liver enzyme then converts cholecalciferol to 25-hydroxycholecalciferol. A second enzyme in the kidneys converts 25-hydroxycholecalciferol to calcitriol (1,25-dihydroxycalciferol), which is the active form of vitamin D. Most excreted via bile. Dietary sources include fish-liver oils, egg yolk, fortified milk.	Essential for absorption and utilization of calcium and phosphorus from GI tract. Works with parathyroid hormone (PTH) to maintain Ca^{2+} homeostasis.	Defective utilization of calcium by bones leads to **rickets** in children and **osteomalacia** in adults. Possible loss of muscle tone.
E (tocopherols)	Stored in liver, adipose tissue, and muscles. Sources include fresh nuts and wheat germ, seed oils, green leafy vegetables.	Inhibits catabolism of certain fatty acids that help form cell structures, especially membranes. Involved in formation of DNA, RNA, and red blood cells. May promote wound healing, contribute to the normal structure and functioning of the nervous system, and prevent scarring. May help protect liver from toxic chemicals such as carbon tetrachloride. Acts as an antioxidant to inactivate free radicals.	May cause the oxidation of monounsaturated fats, resulting in abnormal structure and function of mitochondria, lysosomes, and plasma membranes. A possible consequence is hemolytic anemia. Deficiency also causes muscular dystrophy in monkeys and sterility in rats.
K	Produced by intestinal bacteria. Stored in liver and spleen. Dietary sources include spinach, cauliflower, cabbage, liver.	Coenzyme essential for synthesis of several clotting factors by liver, including prothrombin.	Delayed clotting time results in excessive bleeding.
Water-soluble	Absorbed along with water in GI tract and dissolved in body fluids.		
B_1 (thiamine)	Rapidly destroyed by heat. Not stored in body. Excess intake eliminated in urine. Sources include whole-grain products, eggs, pork, nuts, liver, yeast.	Acts as coenzyme for many different enzymes that break carbon-to-carbon bonds and are involved in carbohydrate metabolism of pyruvic acid to CO_2 and H_2O. Essential for synthesis of acetylcholine.	Improper carbohydrate metabolism leads to buildup of pyruvic and lactic acids and insufficient production of ATP for muscle and nerve cells. Deficiency leads to: (1) **beriberi**—partial paralysis of smooth muscle of GI tract, causing digestive disturbances; skeletal muscle paralysis; and atrophy of limbs; (2) **polyneuritis**—due to degeneration of myelin sheaths; impaired reflexes, impaired sense of touch, stunted growth in children, and poor appetite.

Table 25.6 (continued)

VITAMIN	COMMENT AND SOURCE	FUNCTIONS	DEFICIENCY SYMPTOMS AND DISORDERS
B_2 (riboflavin)	Not stored in large amounts in tissues. Most is excreted in urine. Small amounts supplied by bacteria of GI tract. Dietary sources include yeast, liver, beef, veal, lamb, eggs, whole-grain products, asparagus, peas, beets, peanuts.	Component of certain coenzymes (for example, FAD and FMN) in carbohydrate and protein metabolism, especially in cells of eye, integument, mucosa of intestine, blood.	Deficiency may lead to improper utilization of oxygen resulting in blurred vision, cataracts, and corneal ulcerations. Also dermatitis and cracking of skin, lesions of intestinal mucosa, and development of one type of anemia.
Niacin (nicotinamide)	Derived from amino acid tryptophan. Sources include yeast, meats, liver, fish, whole-grain products, peas, beans, nuts.	Essential component of NAD and NADP, coenzymes in oxidation–reduction reactions. In lipid metabolism, inhibits production of cholesterol and assists in triglyceride breakdown.	Principal deficiency is **pellagra,** characterized by dermatitis, diarrhea, and psychological disturbances.
B_6 (pyridoxine)	Synthesized by bacteria of GI tract. Stored in liver, muscle, brain. Other sources include salmon, yeast, tomatoes, yellow corn, spinach, whole-grain products, liver, yogurt.	Essential coenzyme for normal amino acid metabolism. Assists production of circulating antibodies. May function as coenzyme in triglyceride metabolism.	Most common deficiency symptom is dermatitis of eyes, nose, and mouth. Other symptoms are retarded growth and nausea.
B_{12} (cyanocobalamin)	Only B vitamin not found in vegetables; only vitamin containing cobalt. Absorption from GI tract depends on intrinsic factor secreted by gastric mucosa. Sources include liver, kidney, milk, eggs, cheese, meat.	Coenzyme necessary for red blood cell formation, formation of the amino acid methionine, entrance of some amino acids into Krebs cycle, and manufacture of choline (used to synthesize acetylcholine).	Pernicious anemia, neuropsychiatric abnormalities (ataxia, memory loss, weakness, personality and mood changes, and abnormal sensations), and impaired osteoblast activity.
Pantothenic acid	Stored primarily in liver and kidneys. Some produced by bacteria of GI tract. Other sources include kidney, liver, yeast, green vegetables, cereal.	Constituent of coenzyme A essential for transfer of acetyl group from pyruvic acid into Krebs cycle, conversion of lipids and amino acids into glucose, and synthesis of cholesterol and steroid hormones.	Experimental deficiency tests indicate fatigue, muscle spasms, neuromuscular degeneration, insufficient production of adrenal steroid hormones.
Folic acid (folate, folacin)	Synthesized by bacteria of GI tract. Dietary sources include green leafy vegetables, broccoli, asparagus, breads, dried beans, and citrus fruits.	Component of enzyme systems synthesizing purines and pyrimidines built into DNA and RNA. Essential for normal production of red and white blood cells.	Production of abnormally large red blood cells (macrocytic anemia). Higher risk of neural tube defects in babies born to folate-deficient mothers.
Biotin	Synthesized by bacteria of GI tract. Dietary sources include yeast, liver, egg yolk, kidneys.	Essential coenzyme for conversion of pyruvic acid to oxaloacetic acid and synthesis of fatty acids and purines.	Mental depression, muscular pain, dermatitis, fatigue, nausea.
C (ascorbic acid)	Rapidly destroyed by heat. Some stored in glandular tissue and plasma. Sources include citrus fruits, tomatoes, green vegetables.	Promotes many metabolic reactions, particularly protein metabolism, including laying down of collagen in formation of connective tissue. As coenzyme, may combine with poisons, rendering them harmless until excreted. Works with antibodies, promotes wound healing, and functions as an antioxidant.	Scurvy; anemia; many symptoms related to poor connective tissue growth and repair, including tender swollen gums, loosening of teeth (alveolar processes also deteriorate), poor wound healing, bleeding (vessel walls fragile because of connective tissue degeneration), and retardation of growth.

DISORDERS: HOMEOSTATIC IMBALANCES

BODY TEMPERATURE ABNORMALITIES

Fever

A **fever** is an elevation of core temperature that results from a resetting of the hypothalamic thermostat. The most common cause of fever is a viral or bacterial infection (or bacterial toxins); other causes are ovulation, excessive secretion of thyroid hormones, tumors, and reactions to vaccines. A fever-producing substance is called a **pyrogen** (PĪ-rō-gen; *pyro-* = fire; *-gen* = produce) and can induce a fever in the following way: When phagocytes ingest certain bacteria, they are stimulated to secrete interleukin-1, which acts as a pyrogen. It circulates to the hypothalamus and induces neurons of the preoptic area to secrete prostaglandins, particularly of the E series. Prostaglandins reset the hypothalamic thermostat at a higher temperature, and temperature-regulating reflex mechanisms then act to bring the core body temperature up to this new setting. *Antipyretics* are agents that relieve or reduce fever; examples are aspirin, acetaminophen (Tylenol), and ibuprofen (Advil), which reduce fever by inhibiting synthesis of prostaglandins.

Suppose that as a result of pyrogens the thermostat is reset at 39°C (103°F). Now the heat-promoting mechanisms (vasoconstriction, increased metabolism, shivering) are operating at full force. Thus, even though core temperature is climbing higher than normal—say, 38°C (101°F)—the skin remains cold, and shivering occurs. This condition, called a **chill**, is a definite sign that core temperature is rising. After several hours, core temperature reaches the setting of the thermostat, and the chills disappear. But now the body will continue to regulate temperature at 39°C (103°F). When the pyrogens disappear, the thermostat is reset at normal—37.0°C (98.6°F). Because core temperature is high in the beginning, the heat-losing mechanisms (vasodilation and sweating) go into operation to decrease core temperature. The skin becomes warm, and the person begins to sweat. This phase of the fever is called the **crisis,** and it indicates that core temperature is falling.

Up to a point, fever is beneficial. A higher temperature intensifies the effect of interferon and the phagocytic activities of macrophages while hindering replication of some pathogens. Because fever increases heart rate, infection-fighting white blood cells are delivered to sites of infection more rapidly. In addition, antibody production and T cell proliferation increase. Moreover, heat speeds up the rate of chemical reactions, which may help body cells repair themselves more quickly during a disease. Among the complications of fever are dehydration, acidosis, and permanent brain damage. As a rule, death results if core temperature rises above 44–46°C (112–114°F).

Heat Cramps, Heat Exhaustion, and Heatstroke

Heat cramps occur as a result of profuse sweating that removes water and salt (NaCl) from the body. The salt loss causes painful contractions of muscles; such cramps tend to occur in muscles used while working but do not appear until the person relaxes once the work is done. Salted liquids usually lead to rapid improvement.

In **heat exhaustion (heat prostration),** the core temperature is generally normal, or a little below, and the skin is cool and moist due to profuse perspiration. Heat exhaustion is normally characterized by fluid and electrolyte loss, especially salt. The salt loss results

in muscle cramps, dizziness, vomiting, and fainting; fluid loss may cause low blood pressure. Complete rest and salt tablets are recommended.

When **heatstroke (sunstroke)** occurs, the temperature and relative humidity are high, making it difficult for the body to lose heat by radiation, conduction, or evaporation, and neurological impairment has resulted. Blood flow to the skin is decreased, perspiration is greatly reduced, and core temperature rises sharply. The skin is thus dry and hot—its temperature may reach 43°C (110°F). Because brain cells are affected, the hypothalamic thermostat fails to operate. Treatment, which must be undertaken immediately, consists of cooling the body by immersing the victim in cool water and by administering fluids and electrolytes.

OBESITY

Obesity—body weight more than 20% above some desirable standard due to an excessive accumulation of adipose tissue—affects one-third of the adult population in the United States. (An athlete may be *overweight* due to higher-than-normal amounts of muscle tissue without being obese.) Even moderate obesity is hazardous to health; it is implicated as a risk factor in cardiovascular disease, hypertension, pulmonary disease, non-insulin-dependent diabetes mellitus, arthritis, certain cancers (breast, uterus, and colon), varicose veins, and gallbladder disease. Also, loss of body fat in obese individuals has been shown to elevate HDL cholesterol, the type associated with prevention of cardiovascular disease.

In a few cases, obesity may result from trauma of or tumors in the food-regulating centers in the hypothalamus. In most cases of obesity, no specific cause can be identified. Contributing factors include genetic factors, eating habits taught early in life, overeating to relieve tension, and social customs. Studies indicate that some obese people burn fewer calories during digestion and absorption of a meal. Additionally, obese people who lose weight require about 15% fewer calories to maintain normal body weight than do people who have never been obese. Although leptin suppresses appetite and produces satiety in experimental animals, it is not deficient in most obese people.

MALNUTRITION

Malnutrition (*mal-* = bad) is an imbalance of nutrient intake—either inadequate or excessive total caloric intake or deficiency of specific nutrients. One cause of malnutrition—undernutrition, or inadequate food intake—can result from conditions such as fasting, anorexia nervosa, food deprivation, cancer, gastrointestinal obstructions, inability to swallow, renal disease, and poor dentition. Other causes of malnutrition are an imbalance of nutrients, malabsorption of nutrients, improper distribution of nutrients, inability to use nutrients (for example, in diabetes mellitus), increased nutrient requirements (due to fever, infections, burns, fractures, stress, and exposure to heat or cold), increased nutrient losses (diarrhea, bleeding, or glucosuria), and overnutrition (excess vitamins, minerals, and calories).

One of the major types of undernutrition results from inadequate intake of protein and/or calories needed to meet a person's nutritional requirements. Such protein–calorie undernutrition may be classified into two types based on what is lacking in the diet. In one type, called **kwashiorkor** (kwash-ē-OR-kor), protein intake is deficient despite normal or nearly normal caloric intake. Because the main protein in corn, zein, lacks the essential amino acids tryptophan and lysine, which are needed for growth and tissue repair, many African children whose diet consists largely of cornmeal develop kwashiorkor. The main signs of kwashiorkor are edema of the abdomen, enlarged liver, decreased blood pressure, bradycardia, hypothermia, anorexia, lethargy, dry and hyperpigmented skin, easily pluckable hair, and sometimes mental retardation.

Another type of protein–calorie undernutrition, called **marasmus** (mar-AZ-mus), results from inadequate intake of both protein and calories. Its characteristics include retarded growth, low weight, muscle wasting, emaciation, dry skin, and thin, dry, dull hair.

STUDY OUTLINE

INTRODUCTION (p. 871)

1. Our only source of energy for performing biological work is the food we eat. Food also provides essential substances that we cannot synthesize.
2. Most food molecules absorbed by the gastrointestinal tract are used to supply energy for life processes, serve as building blocks during synthesis of complex molecules, or are stored for future use.

METABOLIC REACTIONS (p. 871)

1. Metabolism refers to all chemical reactions of the body and is of two types: catabolism and anabolism.
2. Catabolism is the term for reactions that break down complex organic compounds into simple ones. Overall, catabolic reactions are exergonic; they produce more energy than they consume.
3. Chemical reactions that combine simple molecules into more complex ones that form the body's structural and functional components are collectively known as anabolism. Overall, anabolic reactions are endergonic; they consume more energy than they produce.
4. The coupling of anabolism and catabolism occurs via ATP.

ENERGY TRANSFER (p. 872)

1. Oxidation is the removal of electrons from a substance; reduction is the addition of electrons to a substance.
2. Two coenzymes used to carry hydrogen atoms during coupled oxidation–reduction reactions are nicotinamide adenine dinucleotide (NAD^+) and flavin adenine dinucleotide (FAD).
3. ATP can be generated via substrate-level phosphorylation, oxidative phosphorylation, and photophosphorylation.

CARBOHYDRATE METABOLISM (p. 874)

1. During digestion, polysaccharides and disaccharides are hydrolyzed into the monosaccharides glucose (about 80%), fructose, and galactose; the latter two are then converted to glucose.
2. Some glucose is oxidized by cells to provide ATP. Glucose also can be used to synthesize amino acids, glycogen, and triglycerides.
3. Glucose moves into most cells via facilitated diffusion and becomes phosphorylated to glucose 6-phosphate; this process is stimulated by insulin. Glucose entry into neurons and hepatocytes is always "turned on."

4. Cellular respiration, the complete oxidation of glucose to CO_2 and H_2O, involves glycolysis, the Krebs cycle, and the electron transport chain.
5. Glycolysis is the breakdown of glucose into two molecules of pyruvic acid; there is a net production of two molecules of ATP.
6. When oxygen is in short supply, pyruvic acid is reduced to lactic acid; under aerobic conditions, pyruvic acid enters the Krebs cycle.
7. Pyruvic acid is prepared for entrance into the Krebs cycle by conversion to a two-carbon acetyl group followed by the addition of coenzyme A to form acetyl coenzyme A.
8. The Krebs cycle involves decarboxylations, oxidations, and reductions of various organic acids.
9. Each molecule of pyruvic acid that is converted to acetyl coenzyme A and then enters the Krebs cycle produces three molecules of CO_2, four molecules of NADH and four H^+, one molecule of $FADH_2$, and one molecule of ATP.
10. The energy originally stored in glucose and then in pyruvic acid is transferred primarily to the reduced coenzymes NADH and $FADH_2$.
11. The electron transport chain involves a series of oxidation–reduction reactions in which the energy in NADH and $FADH_2$ is liberated and transferred to ATP.
12. The electron carriers include FMN, cytochromes, iron–sulfur centers, copper atoms, and coenzyme Q.
13. The electron transport chain yields a maximum of 32 or 34 molecules of ATP and six molecules of H_2O.
14. Table 25.1 on page 883 summarizes the ATP yield during cellular respiration. The complete oxidation of glucose can be represented as follows:

$$C_6H_{12}O_6 + 6\ O_2 + 36 \text{ or } 38 \text{ ADPs} + 36 \text{ or } 38 \enclose{circle}{P}$$
$$\longrightarrow 6\ CO_2 + 6\ H_2O + 36 \text{ or } 38 \text{ ATPs}$$

15. The conversion of glucose to glycogen for storage in the liver and skeletal muscle is called glycogenesis. It is stimulated by insulin.
16. The conversion of glycogen to glucose is called glycogenolysis. It occurs between meals and is stimulated by glucagon and epinephrine.
17. Gluconeogenesis is the conversion of noncarbohydrate molecules into glucose. It is stimulated by cortisol and glucagon.

LIPID METABOLISM (p. 886)

1. Lipoproteins transport lipids in the bloodstream. Types of lipoproteins include chylomicrons, which carry dietary lipids to adipose tissue; very low-density lipoproteins (VLDLs), which carry triglycerides from the liver to adipose tissue; low-density lipoproteins (LDLs), which deliver cholesterol to body cells; and high-density lipoproteins (HDLs), which remove excess cholesterol from body cells and transport it to the liver for elimination.
2. Cholesterol in the blood comes from two sources: from food and from synthesis by the liver.
3. Lipids may be oxidized to produce ATP or stored in adipose tissue as triglycerides.
4. A few lipids are used as structural molecules or to synthesize essential molecules.
5. Triglycerides are stored in adipose tissue, mostly in the subcutaneous layer.
6. Adipose tissue contains lipases that catalyze the deposition of triglycerides from chylomicrons and hydrolyze triglycerides into fatty acids and glycerol.
7. In lipolysis, triglycerides are split into fatty acids and glycerol and released from adipose tissue under the influence of epinephrine, norepinephrine, cortisol, thyroid hormones, and insulinlike growth factors.
8. Glycerol can be converted into glucose by conversion into glyceraldehyde 3-phosphate.
9. In beta oxidation of fatty acids, carbon atoms are removed in pairs from fatty acid chains; the resulting molecules of acetyl coenzyme A enter the Krebs cycle.
10. The conversion of glucose or amino acids into lipids is called lipogenesis; it is stimulated by insulin.

PROTEIN METABOLISM (p. 889)

1. During digestion, proteins are hydrolyzed into amino acids, which enter the liver via the hepatic portal vein.
2. Amino acids, under the influence of insulinlike growth factors and insulin, enter body cells via active transport.
3. Inside cells, amino acids are synthesized into proteins that function as enzymes, hormones, structural elements, and so forth; stored as fat or glycogen; or used for energy.
4. Before amino acids can be catabolized, they must be converted to substances that can enter the Krebs cycle; these conversions involve deamination, decarboxylation, and hydrogenation.
5. Amino acids may also be converted into glucose, fatty acids, and ketone bodies.
6. Protein synthesis is stimulated by insulinlike growth factors, thyroid hormones, insulin, estrogen, and testosterone.
7. Table 25.2 on page 893 summarizes carbohydrate, lipid, and protein metabolism.

KEY MOLECULES AT METABOLIC CROSSROADS (p. 891)

1. Three molecules play a key role in metabolism: glucose 6-phosphate, pyruvic acid, and acetyl coenzyme A.
2. Glucose 6-phosphate may be converted to glucose, glycogen, ribose 5-phosphate, and pyruvic acid.
3. When ATP is low and oxygen is plentiful, pyruvic acid is converted to acetyl coenzyme A; when oxygen supply is low, pyruvic acid is converted to lactic acid. One link between carbohydrate and protein metabolism occurs via pyruvic acid.

4. Acetyl coenzyme A is the molecule that enters the Krebs cycle; it is also used to synthesize fatty acids, ketone bodies, and cholesterol.

METABOLIC ADAPTATIONS (p. 893)

1. During the absorptive state, ingested nutrients enter the blood and lymph from the GI tract.
2. During the absorptive state, blood glucose is oxidized to form ATP, and glucose transported to the liver is converted to glycogen or triglycerides. Most triglycerides are stored in adipose tissue. Amino acids in hepatocytes are converted to carbohydrates, fats, and proteins. Table 25.3 on page 895 summarizes the hormonal regulation of metabolism during the absorptive state.
3. During the postabsorptive state, absorption is complete and the ATP needs of the body are satisfied by nutrients already present in the body.
4. The major task during the postabsorptive state is to maintain normal blood glucose level. This involves conversion of glycogen in the liver and skeletal muscle into glucose, conversion of glycerol into glucose, conversion of amino acids into glucose, and oxidation of fatty acids, ketone bodies, and amino acids to supply ATP. Table 25.4 on page 897 summarizes the hormonal regulation of metabolism during the postabsorptive state.
5. Fasting is going without food for a few days; starvation implies weeks or months of inadequate food intake.
6. During fasting and starvation, fatty acids and ketone bodies are increasingly utilized for ATP production.

HEAT AND ENERGY BALANCE (p. 898)

1. Measurement of the metabolic rate under basal conditions is called the basal metabolic rate (BMR).
2. A kilocalorie (kcal) or Calorie is the amount of energy required to raise the temperature of 1000 g of water from 14°C to 15°C.
3. Normal core temperature is maintained by a delicate balance between heat-producing and heat-losing mechanisms.
4. Metabolic rate is affected by exercise, hormones, the nervous system, body temperature, ingestion of food, age, gender, climate, sleep, and malnutrition.
5. Mechanisms of heat transfer are conduction, convection, radiation, and evaporation.
6. Conduction is the transfer of heat between two substances or objects in contact with each other.
7. Convection is the transfer of heat by a liquid or gas between areas of different temperatures.
8. Radiation is the transfer of heat from a warmer object to a cooler object without physical contact.
9. Evaporation is the conversion of a liquid to a vapor; in the process, heat is lost.
10. The hypothalamic thermostat is in the preoptic area.
11. Responses that produce, conserve, or retain heat when core temperature falls are vasoconstriction; release of epinephrine, norepinephrine, and thyroid hormones; and shivering.
12. Responses that increase heat loss when core temperature increases include vasodilation, decreased metabolic rate, and evaporation of perspiration.
13. Two centers in the hypothalamus related to regulation of food intake are the feeding center and satiety center.

NUTRITION (p. 902)

1. Nutrients include water, carbohydrates, lipids, proteins, minerals, and vitamins.
2. Most teens and adults need between 1600 and 2800 Calories per day.
3. Nutrition experts suggest dietary calories be 50–60% from carbohydrates, 30% or less from fats, and 12–15% from proteins, although the optimal levels of these nutrients may vary.
4. The food guide pyramid indicates how many servings of five food groups are recommended each day to attain the number of calories and variety of nutrients needed for wellness.
5. Minerals known to perform essential functions are calcium, phosphorus, potassium, sulfur, sodium, chloride, magnesium, iron, iodide, manganese, cobalt, copper, zinc, fluoride, selenium, and chromium. Their functions are summarized in Table 25.5 on page 904.
6. Vitamins are organic nutrients that maintain growth and normal metabolism. Many function in enzyme systems.
7. Fat-soluble vitamins are absorbed with fats and include vitamins A, D, E, and K; water-soluble vitamins are absorbed with water and include the B vitamins and vitamin C.
8. The functions and deficiency disorders of the principal vitamins are summarized in Table 25.6 on page 906.

SELF-QUIZ QUESTIONS

Complete the following:

1. The three fates of food molecules absorbed by the GI tract are to ___, ___, and ___.
2. The thermostat and food intake regulating center of the body is in the ___ of the brain.
3. The three key molecules of metabolism are ___, ___, and ___.
4. ___ are the primary regulators of metabolism.

True or false:

5. The energy-transferring molecule of the body is glucose.
6. Most of the glucose in the body is used to produce ATP.

Choose the best answer to the following questions:

7. Which of the following is *not* considered a nutrient? (a) water, (b) carbohydrates, (c) nucleic acids, (d) proteins, (e) lipids
8. Which of the following factors affect metabolic rate and thus the production of body heat? (1) exercise, (2) hormones, (3) minerals, (4) food intake, (5) waste product removal, (6) age. (a) 1, 2, 3, and 4, (b) 2, 3, 4, and 5, (c) 3, 4, 5, and 6, (d) 1, 2, 4, and 6, (e) 1, 3, 5, and 6
9. Which of the following are fates of glucose in the body? (1) ATP production, (2) amino acid synthesis, (3) glycogen synthesis, (4) triglyceride synthesis, (5) vitamin synthesis. (a) 1, 3, 4, and 5, (b) 2, 3, 4, and 5, (c) 1, 3, and 5, (d) 2, 4, and 5, (e) 1, 2, 4, and 5
10. Which of the following is the correct sequence for the oxidation of glucose to produce ATP? (a) electron transport chain, Krebs cycle, glycolysis, formation of acetyl CoA, (b) Krebs cycle, formation of acetyl CoA, electron transport chain, glycolysis, (c) glycolysis, electron transport chain, Krebs cycle, formation of acetyl CoA, (d) glycolysis, formation of acetyl CoA, Krebs cycle, electron transport chain, (e) formation of acetyl CoA, Krebs cycle, glycolysis, electron transport chain
11. Which of the following are absorptive state reactions? (1) aerobic cellular respiration, (2) glycogenesis, (3) glycogenolysis, (4) gluconeogenesis using lactic acid, (5) lipolysis. (a) 1 and 2, (b) 2 and 3, (c) 3 and 4, (d) 4 and 5, (e) 1 and 5
12. Which of the following are heat-loss mechanisms? (1) shivering of skeletal muscles, (2) release of thyroid hormones, (3) conduction, (4) evaporation, (5) vasoconstriction, (6) convection, (7) radiation. (a) 1, 3, 5, and 7, (b) 3, 4, 6, and 7, (c) 2, 4, 5, and 7, (d) 1, 2, 4, and 6, (e) 1, 3, 5, and 6

13. Match the following:
    ___ (a) the mechanism of ATP generation that links chemical reactions with pumping of hydrogen ions
    ___ (b) the removal of electrons from an atom or molecule resulting in a decrease in energy
    ___ (c) refers to all the chemical reactions in the body
    ___ (d) the oxidation of glucose to produce ATP
    ___ (e) the addition of electrons to a molecule resulting in an increase in energy content of the molecule
    ___ (f) chemical reactions that break down complex organic molecules and polymers into simpler ones
    ___ (g) overall rate at which heat is produced in the body
    ___ (h) the formation of ketone bodies
    ___ (i) chemical reactions that combine simple molecules and monomers to make more complex ones
    ___ (j) the addition of a phosphate group to a chemical compound

    (1) metabolism
    (2) catabolism
    (3) anabolism
    (4) metabolic rate
    (5) oxidation
    (6) reduction
    (7) phosphorylation
    (8) chemiosmosis
    (9) ketogenesis
    (10) cellular respiration

14. Match the following:
___ (a) the conversion of glycogen back to glucose
___ (b) the removal of the amino group from an amino acid
___ (c) the splitting of a triglyceride into glycerol and fatty acids
___ (d) the cleavage of one pair of carbon atoms at a time from a fatty acid
___ (e) the formation of ketone bodies
___ (f) the synthesis of lipids from glucose or amino acids
___ (g) the conversion of glucose into glycogen
___ (h) the transfer of an amino group from an amino acid to a substance such as pyruvic acid, resulting in the formation of a nonessential amino acid
___ (i) the formation of glucose from noncarbohydrate sources
___ (j) the breakdown of glucose into two molecules of pyruvic acid

(1) glycolysis
(2) glycogenolysis
(3) glycogenesis
(4) gluconeogenesis
(5) lipolysis
(6) ketogenesis
(7) lipogenesis
(8) deamination
(9) transamination
(10) beta oxidation

15. Match the following:
___ (a) deliver cholesterol to body cells for use in repair of cell membranes and synthesis of steroid hormones and bile salts; in excess, deposit cholesterol in blood vessels
___ (b) remove excess cholesterol from body cells and transport it to the liver for elimination
___ (c) organic nutrients required in minute amounts to maintain growth and normal metabolism
___ (d) the energy molecule of the body
___ (e) nutrient molecules that can be oxidized to produce ATP, stored in adipose tissue, used as structural molecules in cell membranes, and used to transport cholesterol in the blood
___ (f) a steroid molecule that is commonly used as an indicator of risk for coronary artery disease
___ (g) the body's preferred source for synthesizing ATP
___ (h) composed of amino acids and are the primary regulatory molecules in the body
___ (i) acetoacetic acid, beta-hydroxybutyric acid, and acetone
___ (j) inorganic substances that perform many vital functions in the body

(1) vitamins
(2) minerals
(3) glucose
(4) lipids
(5) proteins
(6) ATP
(7) cholesterol
(8) ketone bodies
(9) low-density lipoproteins
(10) high-density lipoproteins

CRITICAL THINKING QUESTIONS

1. Deb was training for a marathon. With only one day to go before the big event, she had shortened her morning training run and was loading up on bread and pasta for her last meal. Why is Deb eating mounds of bread and spaghetti? (HINT: *Bread and pasta are high in carbohydrates.*)

2. Mr. Hernandez believed that the perfect lawn needed to be cut according to a strict schedule. As he was pushing the lawn mower around the yard on a scorching hot Saturday morning, he began to feel dizzy and nauseous. When he sat down, he was bothered by cramps in his legs, and he nearly fainted. What's wrong with Mr. Hernandez? (HINT: *His shirt was soaked with sweat, and his forehead actually felt cool.*)

3. Three-year-old Sara's mother was concerned about her child's eating habits. She complained that "one day Sara will only eat a few bites, and then the next day she eats like a horse!" The pediatrician told Sara's mom that it sounded like Sara had perfectly normal control over her food intake. How is food intake controlled? (HINT: *Sara is too young to count calories.*)

ANSWERS TO FIGURE QUESTIONS

25.1 In pancreatic acinar cells, anabolism predominates because they are synthesizing complex molecules (digestive enzymes).

25.2 Glycolysis is also called anaerobic cellular respiration.

25.3 The substrate is glucose, a hexose; kinases are enzymes that phosphorylate (add phosphate to) their substrate. Thus, the enzyme is named hexokinase.

25.4 Glycolysis occurs in the cytosol; the Krebs cycle occurs within mitochondria.

25.5 The production of reduced coenzymes is important in the Krebs cycle because they will subsequently yield ATP in the electron transport chain.

25.6 The energy source that powers the proton pumps is electrons provided by NADH + H$^+$.

25.7 The concentration of H$^+$ is highest in the space between the inner and outer mitochondrial membranes.

25.8 During the complete oxidation of one glucose molecule, six molecules of O$_2$ are used and six molecules of CO$_2$ are produced.

25.9 Skeletal muscle fibers can synthesize glycogen, but they cannot release glucose into the blood because they lack the enzyme phosphatase.

25.10 Hepatocytes can carry out gluconeogenesis and glycogenesis.

25.11 LDLs deliver cholesterol to body cells.

25.12 Hepatocytes and adipose cells carry out lipogenesis, beta oxidation, and lipolysis; hepatocytes carry out ketogenesis.

25.13 An amino group is removed from an amino acid via deamination.

25.14 Acetyl coenzyme A is the gateway into the Krebs cycle.

25.15 Reactions of the absorptive state are mainly anabolic.

25.16 Processes that directly elevate blood glucose during the postabsorptive state include lipolysis (in adipocytes and hepatocytes), gluconeogenesis (in hepatocytes), and glycogenolysis (in hepatocytes).

25.17 Exercise, the sympathetic nervous system, hormones (epinephrine, norepinephrine, thyroxine, testosterone, human growth hormone), elevated body temperature, and ingestion of food increase metabolic rate.

25.18 Foods that contain cholesterol and most of the saturated fatty acids in the diet are milk, yogurt, cheeses, and meats.

THE URINARY SYSTEM

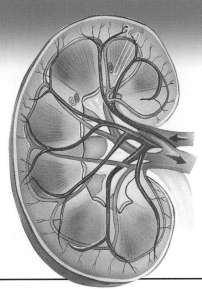

The **urinary system** consists of two kidneys, two ureters, one urinary bladder, and one urethra (Figure 26.1). After the kidneys filter blood and return most of the water and many solutes to the bloodstream, the remaining water and solutes constitute **urine.** Urine is excreted from each kidney through its ureter and is stored in the urinary bladder until it is expelled from the body through the urethra. In males, the urethra is also the route through which semen exits the body. **Nephrology** (nef-ROL-ō-jē; *nephr-* = kidney; *-ology* = study of) is the scientific study of the anatomy, physiology, and pathology of the kidney. The branch of medicine that deals with the male and female urinary systems and the male reproductive system is **urology** (yoo-ROL-ō-jē; *uro-* = urine).

OVERVIEW OF KIDNEY FUNCTIONS

OBJECTIVE

- *List the functions of the kidneys.*

The kidneys do the major work of the urinary system, as the other parts of the system are primarily passageways and storage areas. In filtering blood and forming urine, the kidneys contribute to homeostasis of body fluids in several ways. Functions of the kidneys include:

- *Regulation of blood ionic composition.* The kidneys help regulate the blood levels of several ions, most importantly sodium ions (Na^+), potassium ions (K^+), calcium ions (Ca^{2+}), chloride ions (Cl^-), and phosphate ions (HPO_4^{2-}).

- *Maintenance of blood osmolarity.* By separately regulating loss of water and loss of solutes in the urine, the kidneys maintain a relatively constant blood osmolarity close to 290 milliosmoles per liter (mOsm/liter).*

- *Regulation of blood volume.* By conserving or eliminating water, the kidneys adjust blood volume and thus in effect regulate the volume of interstitial fluid. Also, an increase in blood volume increases blood pressure, whereas a decrease in blood volume decreases blood pressure.

- *Regulation of blood pressure.* Besides adjusting blood volume, the kidneys help regulate blood pressure in two other ways: by secreting the enzyme renin, which activates the renin–angiotensin pathway (see Figure 18.16 on page 590), and by adjusting renal resistance, the resistance encountered by blood flowing through the kidneys, which in turn affects systemic vascular resistance (see Figure 21.11 on page 681). The result of an increase in renin or an increase in renal resistance is an increase in blood pressure.

---

*The **osmolarity** of a solution is a measure of the total number of dissolved particles per liter of solution. The particles may be molecules, ions, or a mixture of both. To calculate osmolarity, multiply molarity (see page 39) by the number of particles per molecule, once the molecule dissolves. A similar term, *osmolality*, is the number of particles of solute per *kilogram* of water. Because it is easier to measure volumes of solutions than to determine the mass of water they contain, osmolarity is used more commonly than osmolality. Most body fluids and solutions used clinically are dilute, in which case there is less than a 1% difference between the two measures.

Figure 26.1 Organs of the urinary system, shown in relation to the surrounding structures in a female. (See Tortora, *A Photographic Atlas of the Human Body,* Figure 13.2)

 Urine formed by the kidneys passes first into the ureters, then to the urinary bladder for storage, and finally through the urethra for elimination from the body.

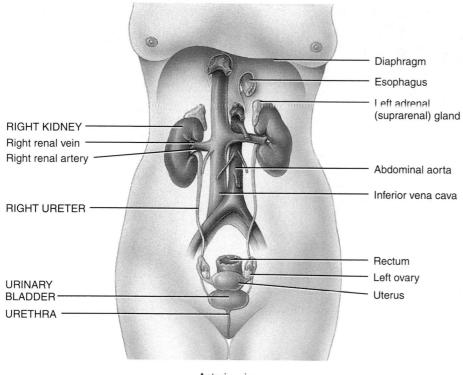

Anterior view

FUNCTIONS OF THE URINARY SYSTEM

1. The kidneys regulate blood volume and composition, help regulate blood pressure, synthesize glucose, release erythropoietin, and participate in vitamin D synthesis.
2. The ureters transport urine from the kidneys to the urinary bladder.
3. The urinary bladder stores urine.
3. The urethra discharges urine from the body.

Q   Which organs constitute the urinary system?

• *Regulation of blood pH.* The kidneys excrete a variable amount of H$^+$ into the urine and conserve bicarbonate ions (HCO$_3^-$), an important buffer of H$^+$. Both of these activities contribute to the regulation of blood pH.

• *Release of hormones.* The kidneys release two hormones: *calcitriol,* the active form of vitamin D, which helps regulate calcium homeostasis (see Figure 18.14 on page 587), and *erythropoietin,* which stimulates production of red blood cells (see Figure 19.5 on page 617).

• *Regulation of blood glucose level.* The kidneys can deaminate the amino acid glutamine, use it for *gluconeogenesis* (the synthesis of new glucose molecules), and release glucose into the blood.

• *Excretion of wastes and foreign substances.* By forming urine, the kidneys help excrete wastes—substances that have no useful function in the body. Some wastes excreted in urine result from metabolic reactions in the body, such as ammonia and urea from the deamination of amino acids; bilirubin from the catabolism of hemoglobin; creatinine from the breakdown of creatine phosphate in muscle fibers; and uric acid from the catabolism

of nucleic acids. Other wastes excreted in urine are foreign substances, such as drugs and environmental toxins.

1. Describe three ways that the kidneys help regulate blood pressure.
2. What is the osmolarity of a 0.2 molar solution of CaCl$_2$? Of a 0.2 molar solution of glucose? Of a 0.2 molar solution of NaCl? (HINT: *Count the number of particles that result when each solute dissolves in water.*)
3. What are wastes, and how do the kidneys contribute to their removal from the body?

ANATOMY AND HISTOLOGY OF THE KIDNEYS

OBJECTIVES

• *Describe the external and internal gross anatomical features of the kidneys.*

• *Trace the path of blood flow through the kidneys.*

• *Describe the structure of renal corpuscles and renal tubules.*

Figure 26.2 Position and coverings of the kidneys. (See Tortora, *A Photographic Atlas of the Human Body,* Figure 13.3)

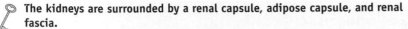

The kidneys are surrounded by a renal capsule, adipose capsule, and renal fascia.

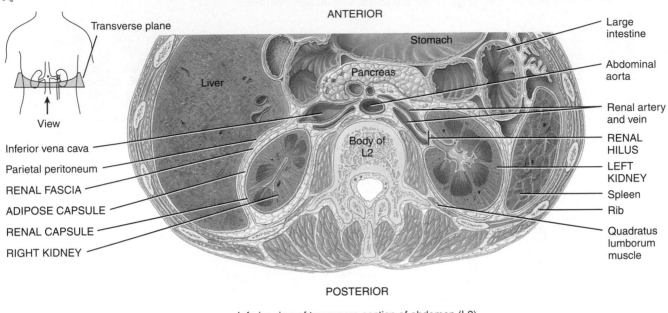

Inferior view of transverse section of abdomen (L2)

Q Why are the kidneys said to be retroperitoneal?

The paired **kidneys** are reddish, kidney-bean-shaped organs located just above the waist between the peritoneum and the posterior wall of the abdomen. Because their position is posterior to the peritoneum of the abdominal cavity, they are said to be **retroperitoneal** (re′-trō-per-i-tō-NĒ-al; *retro-* = behind) organs (Figure 26.2). (Other retroperitoneal structures include the ureters and adrenal glands.) The kidneys are located between the levels of the last thoracic and third lumbar vertebrae, a position where they are partially protected by the eleventh and twelfth pairs of ribs. The right kidney is slightly lower than the left (see Figure 26.1) because the liver occupies considerable space on the right side superior to the kidney.

External Anatomy of the Kidneys

A typical kidney in an adult is 10–12 cm (4–5 in.) long, 5–7 cm (2–3 in.) wide, and 3 cm (1 in.) thick—about the size of a bar of bath soap—and has a mass of 135–150 g. The concave medial border of each kidney faces the vertebral column (see Figure 26.1). Near the center of the concave border is a deep vertical fissure called the **renal hilus** (see Figure 26.3) through which the ureter leaves the kidney, as do blood vessels, lymphatic vessels, and nerves.

Three layers of tissue surround each kidney (see Figure 26.2). The deep layer, the **renal capsule** (*ren-* = kidney), is a smooth, transparent, fibrous membrane that is continuous with the outer coat of the ureter. It serves as a barrier against

trauma and helps maintain the shape of the kidney. The intermediate layer, the **adipose capsule,** is a mass of fatty tissue surrounding the renal capsule. It also protects the kidney from trauma and holds it firmly in place within the abdominal cavity. The superficial layer, the **renal fascia,** is a thin layer of dense, irregular connective tissue that anchors the kidney to the surrounding structures and to the abdominal wall.

Internal Anatomy of the Kidneys

A frontal section through a kidney reveals two distinct regions: a superficial, smooth-textured reddish area called the **renal cortex** (*cortex* = rind or bark) and a deep, reddish-brown region called the **renal medulla** (*medulla* = inner portion) (Figure 26.3). The medulla consists of 8–18 cone-shaped **renal pyramids.** The base (wider end) of each pyramid faces the renal cortex, and its apex (narrower end), called a **renal papilla,** points toward the center of the kidney. The portions of the renal cortex that extend between renal pyramids are called **renal columns.**

Together, the renal cortex and renal pyramids of the renal medulla constitute the functional portion or **parenchyma** of the kidney. Within the parenchyma are the functional units of the kidney—about 1 million microscopic structures called **nephrons** (NEF-rons). Urine formed by the nephrons drains into large **papillary ducts,** which extend

Figure 26.3 Internal anatomy of the kidneys. (See Tortora, *A Photographic Atlas of the Human Body,* Figures 13.4, 13.5)

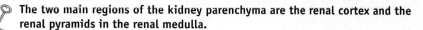

The two main regions of the kidney parenchyma are the renal cortex and the renal pyramids in the renal medulla.

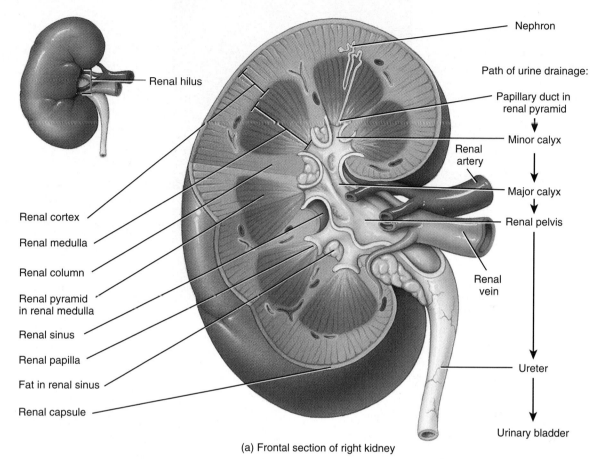

(a) Frontal section of right kidney

figure continues

through the renal papillae of the pyramids. The papillary ducts drain into cuplike structures called **minor** and **major calyces** (KĀ-li-sēz; = cups; singular is *calyx*). Each kidney has 8–18 minor calyces and 2–3 major calyces. A minor calyx receives urine from the papillary ducts of one renal papilla and delivers it to a major calyx. From the major calyces, urine drains into a single large cavity called the **renal pelvis** (*pelv-* = basin) and then out through the ureter to the urinary bladder.

The hilus expands into a cavity within the kidney called the **renal sinus,** which contains part of the renal pelvis, the calyces, and branches of the renal blood vessels and nerves. Adipose tissue helps stabilize the position of these structures in the renal sinus.

Blood and Nerve Supply of the Kidneys

Because the kidneys remove wastes from the blood and regulate its volume and ionic composition, it is not surprising that they are abundantly supplied with blood vessels. Although the kidneys constitute less than 0.5% of total body mass, they receive 20–25% of the resting cardiac output via the right and left **renal arteries** (Figure 26.4). In adults, renal blood flow is about 1200 mL per minute.

Within the kidney, the renal artery divides into several **segmental arteries,** each of which gives off several branches that enter the parenchyma and pass through the renal columns between the renal pyramids as the **interlobar arteries.** At the bases of the renal pyramids, the interlobar arteries arch between the renal medulla and cortex; here they are known as the **arcuate arteries** (*arcuat-* = shaped like a bow). Divisions of the arcuate arteries produce a series of **interlobular arteries,** which enter the renal cortex and give off branches called **afferent arterioles** (*af-* = toward; *-ferrent* = to carry).

Each nephron receives one afferent arteriole, which divides into a tangled, ball-shaped capillary network called the **glomerulus** (glō-MER-yoo-lus; = little ball; plural is *glomeruli*). The glomerular capillaries then reunite to form an **efferent arteriole** (*ef-* = out) that drains blood out of the glomerulus. Glomerular capillaries are unique because they are positioned between two arterioles, rather than between

Figure 26.3 (continued)

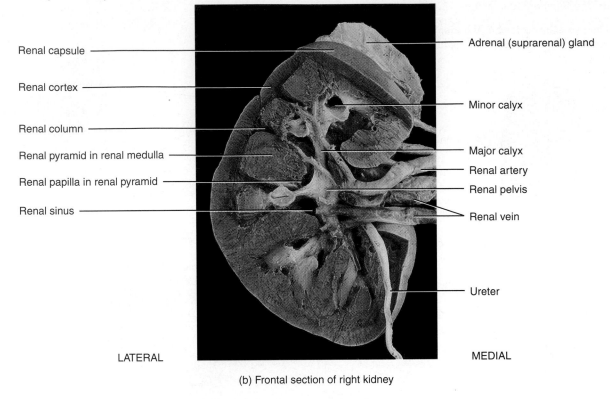

Renal capsule

Renal cortex

Renal column

Renal pyramid in renal medulla

Renal papilla in renal pyramid

Renal sinus

Adrenal (suprarenal) gland

Minor calyx

Major calyx

Renal artery

Renal pelvis

Renal vein

Ureter

LATERAL

MEDIAL

(b) Frontal section of right kidney

Q What structures pass through the renal hilus?

an arteriole and a venule. Coordinated vasodilation and vasoconstriction of the afferent and efferent arterioles can produce large changes in both renal blood flow and renal vascular resistance, which in turn affects systemic vascular resistance. Because they are capillary networks, the glomeruli are part of both the cardiovascular and the urinary systems.

The efferent arterioles divide to form a network of capillaries, called the **peritubular capillaries** (*peri-* = around), that surrounds tubular portions of the nephron in the renal cortex. Extending from some efferent arterioles are long loop-shaped capillaries called **vasa recta** (VĀ-sa REK-ta; *vasa* = vessels; *recta* = straight) that supply tubular portions of the nephron in the renal medulla (see Figure 26.5a).

The peritubular capillaries eventually reunite to form **peritubular venules** and then **interlobular veins.** (The interlobular veins also receive blood from the vasa recta.) Then the blood drains through the **arcuate veins** to the **interlobar veins** running between the renal pyramids, and on to the **segmental veins.** Blood leaves the kidney through a single **renal vein** that exits at the renal hilus.

Most renal nerves originate in the *celiac ganglion* and pass through the *renal plexus* into the kidneys along with the renal arteries. All these nerves are part of the sympathetic di-

vision of the autonomic nervous system. Most are vasomotor nerves that innervate blood vessels; that is, they regulate the flow of blood through the kidney and renal resistance by altering the diameters of arterioles.

The Nephron

Nephrons are the functional units of the kidneys that engage in three basic processes: filtering blood, returning useful substances to blood so that they are not lost from the body, and removing substances from the blood that are not needed by the body. As a result of these processes, nephrons maintain the homeostasis of blood and urine is produced.

Parts of a Nephron

Each nephron (Figure 26.5) consists of two portions: a **renal corpuscle** (KOR-pus-sul; = tiny body), where plasma is filtered, and a **renal tubule** into which the filtered fluid passes. Each renal corpuscle has two components: the **glomerulus** and the **glomerular (Bowman's) capsule,** a double-walled epithelial cup that surrounds the glomerulus. From the glomerular capsule, fluid filtered from the plasma passes into the renal tubule, which has three main sections. In the order that fluid passes through them, the renal tubule consists of a (1) **proximal convoluted tubule,** (2) **loop of**

Figure 26.4 Blood supply of the kidneys. (See Tortora, *A Photographic Atlas of the Human Body,* Figure 13.6)

The renal arteries deliver 20–25% of the resting cardiac output to the kidneys.

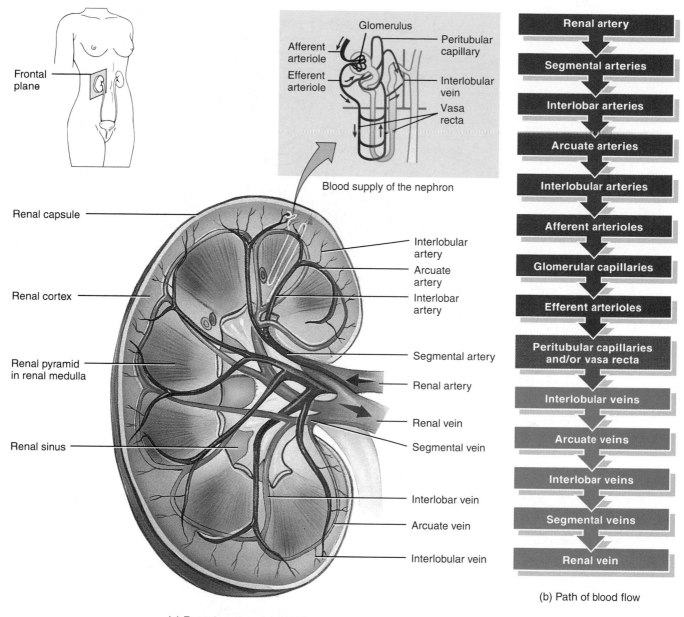

Blood supply of the nephron

(a) Frontal section of right kidney

(b) Path of blood flow

Q What volume of blood enters the renal arteries per minute?

Henle **(nephron loop),** and (3) **distal convoluted tubule.** *Proximal* denotes the portion of the tubule attached to the glomerular capsule, and *distal* denotes the portion that is farther away. *Convoluted* means the tubule is tightly coiled rather than straight. The renal corpuscle and both convo-

luted tubules lie within the renal cortex, whereas the loop of Henle extends into the renal medulla, makes a hairpin turn, and then returns to the renal cortex.

The distal convoluted tubules of several nephrons empty into a single **collecting duct.** Collecting ducts then

Figure 26.5 The structure of nephrons (colored gold) and associated blood vessels. (a) A cortical nephron. (b) A juxtamedullary nephron.

🔑 **Nephrons are the functional units of the kidneys.**

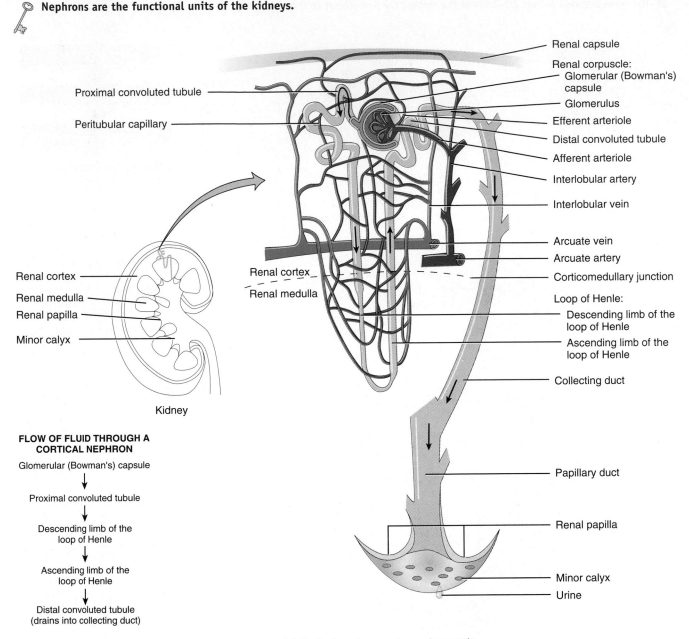

Renal capsule
Renal corpuscle:
  Glomerular (Bowman's) capsule
  Glomerulus
Efferent arteriole
Distal convoluted tubule
Afferent arteriole
Interlobular artery
Interlobular vein
Arcuate vein
Arcuate artery
Corticomedullary junction
Loop of Henle:
  Descending limb of the loop of Henle
  Ascending limb of the loop of Henle
Collecting duct
Papillary duct
Renal papilla
Minor calyx
Urine

Proximal convoluted tubule
Peritubular capillary

Renal cortex
Renal medulla

Renal cortex
Renal medulla
Renal papilla
Minor calyx

Kidney

FLOW OF FLUID THROUGH A CORTICAL NEPHRON

Glomerular (Bowman's) capsule
↓
Proximal convoluted tubule
↓
Descending limb of the loop of Henle
↓
Ascending limb of the loop of Henle
↓
Distal convoluted tubule (drains into collecting duct)

(a) Cortical nephron and vascular supply

unite and converge until eventually there are only several hundred large **papillary ducts,** which drain into the minor calyces. The collecting ducts and papillary ducts extend from the renal cortex through the renal medulla to the renal pelvis. Although a kidney has about 1 million nephrons, it has a much smaller number of collecting ducts and even fewer papillary ducts.

In a nephron, the loop of Henle connects the proximal and distal convoluted tubules. The first portion of the loop of Henle dips into the renal medulla, where it is called the **descending limb of the loop of Henle** (see Figure 26.5). It

then makes that hairpin curve and returns to the renal cortex as the **ascending limb of the loop of Henle.** About 80–85% of the nephrons are termed **cortical nephrons;** their renal corpuscles lie in the outer portion of the renal cortex, and they have *short* loops of Henle that lie mainly in the cortex and penetrate only into the superficial region of the renal medulla (see Figure 26.5a). The short loops of Henle receive their blood supply from peritubular capillaries that arise from efferent arterioles. The other 15–20% of the nephrons are called **juxtamedullary nephrons** (*juxta-* = near to); their renal corpuscles are deep in the cortex, close to the

Figure 26.5 (continued)

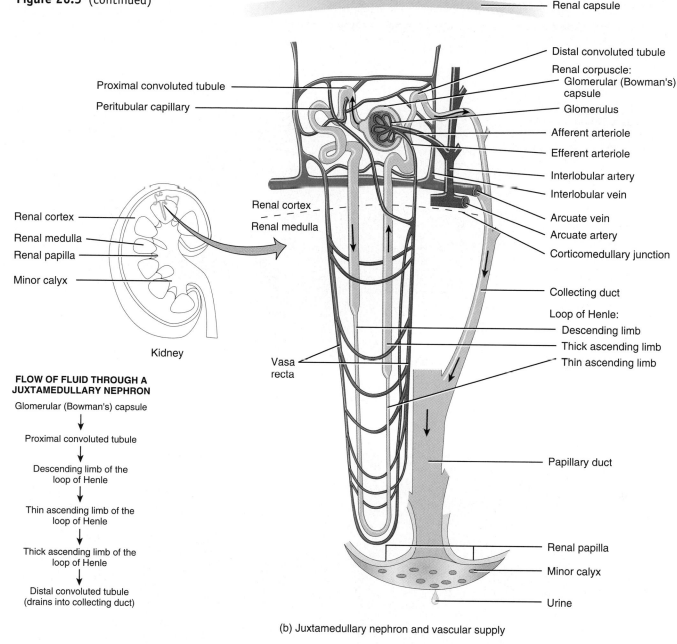

FLOW OF FLUID THROUGH A
JUXTAMEDULLARY NEPHRON

Glomerular (Bowman's) capsule
↓
Proximal convoluted tubule
↓
Descending limb of the
loop of Henle
↓
Thin ascending limb of the
loop of Henle
↓
Thick ascending limb of the
loop of Henle
↓
Distal convoluted tubule
(drains into collecting duct)

(b) Juxtamedullary nephron and vascular supply

Q What are the basic differences between cortical and juxtamedullary nephrons?

medulla, and they have a *long* loop of Henle that extends into the deepest region of the medulla (see Figure 26.5b). Long loops of Henle receive their blood supply from peritubular capillaries and vasa recta that arise from efferent arterioles. In addition, the ascending limb of the loop of Henle of juxtamedullary nephrons consists of two portions: a **thin ascending limb** followed by a **thick ascending limb** (see Figure 26.5b). Nephrons with long loops of Henle enable the kidneys to excrete very dilute or very concentrated urine (described on page 938).

Histology of the Nephron and Collecting Duct

Beginning at the glomerular capsule, a single layer of epithelial cells forms the entire wall of the glomerular capsule, renal tubule, and ducts. Each part, however, has distinctive histological features that reflect its particular functions. In the order that fluid flows through them, the parts are the glomerular capsule, the renal tubule, and the collecting duct.

Glomerular Capsule The glomerular (Bowman's) capsule consists of visceral and parietal layers (Figure 26.6a). The

Figure 26.6 Histology of a renal corpuscle.

🔑 **A renal corpuscle consists of a glomerular (Bowman's) capsule and a glomerulus.**

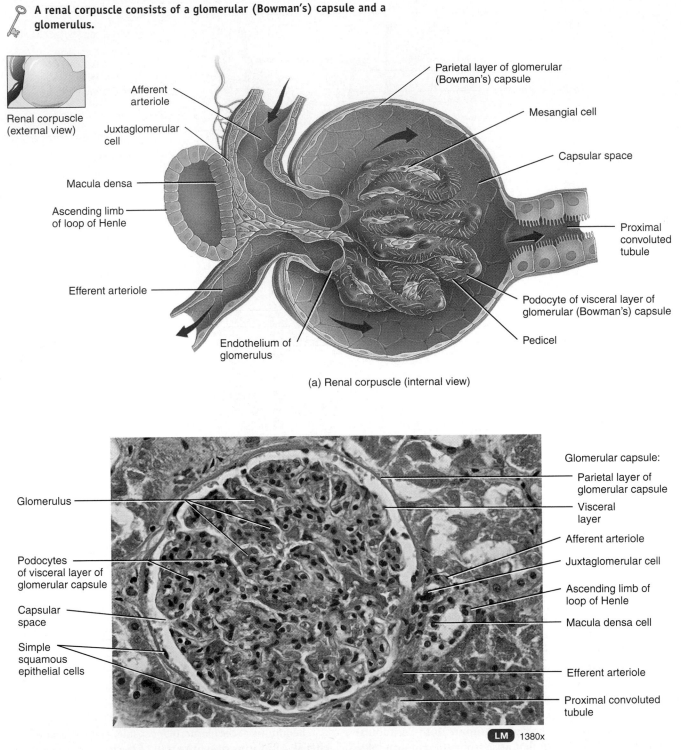

Renal corpuscle (external view)

Afferent arteriole

Juxtaglomerular cell

Macula densa

Ascending limb of loop of Henle

Efferent arteriole

Endothelium of glomerulus

Parietal layer of glomerular (Bowman's) capsule

Mesangial cell

Capsular space

Proximal convoluted tubule

Podocyte of visceral layer of glomerular (Bowman's) capsule

Pedicel

(a) Renal corpuscle (internal view)

Glomerulus

Podocytes of visceral layer of glomerular capsule

Capsular space

Simple squamous epithelial cells

Glomerular capsule:

Parietal layer of glomerular capsule

Visceral layer

Afferent arteriole

Juxtaglomerular cell

Ascending limb of loop of Henle

Macula densa cell

Efferent arteriole

Proximal convoluted tubule

LM 1380x

(b) Renal corpuscle

 Is the photomicrograph in (b) from a section through the renal cortex or renal medulla? How can you tell?

Table 26.1 Histological Features of the Renal Tubule and Collecting Duct

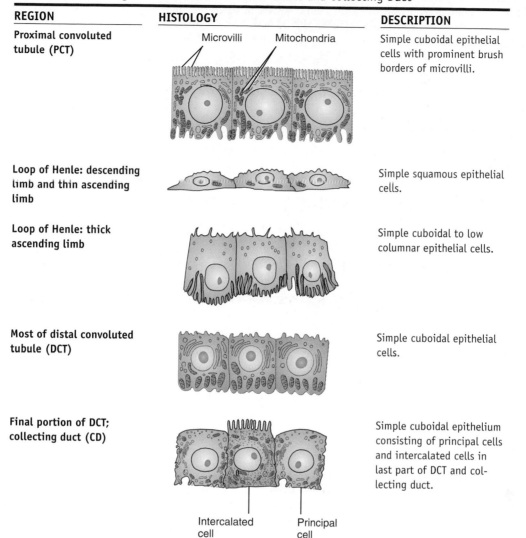

REGION	HISTOLOGY	DESCRIPTION
Proximal convoluted tubule (PCT)	Microvilli Mitochondria	Simple cuboidal epithelial cells with prominent brush borders of microvilli.
Loop of Henle: descending limb and thin ascending limb		Simple squamous epithelial cells.
Loop of Henle: thick ascending limb		Simple cuboidal to low columnar epithelial cells.
Most of distal convoluted tubule (DCT)		Simple cuboidal epithelial cells.
Final portion of DCT; collecting duct (CD)	Intercalated cell Principal cell	Simple cuboidal epithelium consisting of principal cells and intercalated cells in last part of DCT and collecting duct.

visceral layer consists of modified simple squamous epithelial cells called **podocytes** (PŌ-dō-cīts; *podo-* = foot; *-cytes* = cells). The many footlike projections of these cells (pedicels) wrap around the single layer of endothelial cells of the glomerular capillaries and form the inner wall of the capsule. The parietal layer of the glomerular capsule consists of simple squamous epithelium and forms the outer wall of the capsule. Fluid filtered from the glomerular capillaries enters the **capsular (Bowman's) space,** the space between the two layers of the glomerular capsule.

Renal Tubule and Collecting Duct Table 26.1 illustrates the histology of the cells that form the renal tubule and collecting duct. In the proximal convoluted tubule, the cells are simple cuboidal epithelial cells and have a prominent brush border of microvilli on their apical surface (surface facing the lumen). These microvilli, like those of the small intestine, increase the surface area for reabsorption and secretion. The

descending limb of the loop of Henle and the first part of the ascending limb of the loop of Henle (the thin ascending limb) are simple squamous epithelium. (Recall that cortical or short-loop nephrons lack the thin ascending limb.) The thick ascending limb of the loop of Henle is composed of simple cuboidal to low columnar epithelium.

In each nephron, the final portion of the ascending limb of the loop of Henle makes contact with the afferent arteriole serving that renal corpuscle (see Figure 26.6a). Because the tubule cells in this region are columnar and crowded together, they are known as the **macula densa** (*macula* = spot; *densa* = dense). Alongside the macula densa, the wall of the afferent arteriole (and sometimes the efferent arteriole) contains modified smooth muscle fibers called **juxtaglomerular (JG) cells.** Together with the macula densa, they constitute the **juxtaglomerular apparatus,** or **JGA.** The distal convoluted tubule (DCT) begins a short distance past the macula densa. Even though the cells of the distal convoluted tubule

Figure 26.7 Relation of a nephron's structure to its three basic functions: glomerular filtration, tubular reabsorption, and tubular secretion. Excreted substances remain in the urine and subsequently leave the body. For any substance S, excretion rate of S = filtration rate of S − reabsorption rate of S + secretion rate of S.

🔑 **Glomerular filtration occurs in the renal corpuscle, whereas tubular reabsorption and tubular secretion occur all along the renal tubule and collecting duct.**

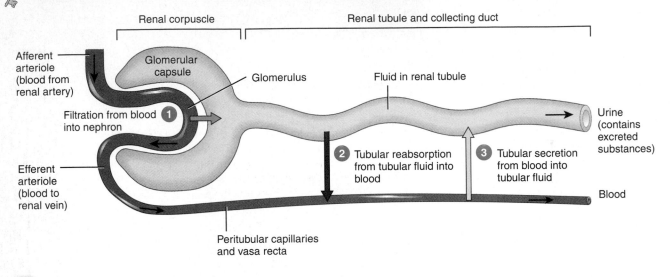

Q When cells of the renal tubules secrete the drug penicillin, is the drug being added to or removed from the bloodstream?

and collecting ducts all are simple cuboidal epithelial cells that have only a few microvilli, beginning in the last portion of the DCT and continuing into the collecting ducts two different types of cells are present. Most are **principal cells,** which have infoldings of the basal membrane; a smaller number are **intercalated cells,** which have microvilli at the apical surface and a large number of mitochondria. Principal cells have receptors for both antidiuretic hormone (ADH) and aldosterone, two hormones that regulate their functions. Intercalated cells play a role in the homeostasis of blood pH. The collecting ducts drain into large papillary ducts, which are lined by simple columnar epithelium.

CLINICAL APPLICATION
Number of Nephrons

The number of nephrons remains constant from birth; any increase in kidney size is due solely to the growth of individual nephrons. If nephrons are injured or become diseased, new ones do not form. Signs of kidney dysfunction do not usually become apparent until function declines to less than 25% of normal because the remaining functional nephrons gradually adapt to handle a larger-than-normal load. Surgical removal of one kidney, for example, stimulates hypertrophy (enlargement) of the remaining kidney, which eventually is able to filter blood at 80% of the rate of two normal kidneys. ■

1. Describe the location of the kidneys. Why are they said to be retroperitoneal?
2. Which branch of the autonomic nervous system innervates renal blood vessels?
3. How do cortical nephrons and juxtamedullary nephrons differ structurally?
4. Describe the histology of the various portions of a nephron and collecting duct.
5. Describe the structure of the juxtaglomerular apparatus (JGA).

OVERVIEW OF RENAL PHYSIOLOGY
OBJECTIVE
• *Identify the three basic tasks performed by nephrons and collecting ducts, and indicate where each task occurs.*

In the process of producing urine, nephrons and collecting ducts perform three basic processes—glomerular filtration, tubular secretion, and tubular reabsorption (Figure 26.7):

❶ *Glomerular filtration.* In the first step of urine production, water and most solutes in plasma pass from blood across the wall of glomerular capillaries into the glomerular capsule, which empties into the renal tubule.

❷ *Tubular reabsorption.* As filtered fluid flows along the renal tubule and through the collecting duct, most filtered water and many useful solutes are reabsorbed by the tubule cells and returned to the blood as it flows through the peritubular capillaries and vasa recta. Note that *reabsorption* refers to the return of substances to the bloodstream, as distinguished from absorption, which means entry of new substances into the body.

❸ *Tubular secretion.* As fluid flows along the tubule and through the collecting duct, the tubule and duct cells secrete additional materials, such as wastes, drugs, and excess ions, into the fluid. Tubular secretion *removes* a substance from the blood; in other instances of secretion—for instance, secretion of hormones—a substance often is released into the blood.

Solutes in the fluid that drains into the renal pelvis remain in the urine and are excreted. The rate of urinary excretion of any solute is equal to its rate of glomerular filtration, plus its rate of secretion, minus its rate of reabsorption.

By filtering, reabsorbing, and secreting, nephrons maintain homeostasis of the blood. The situation is somewhat analogous to a recycling center: Garbage trucks dump refuse into an input hopper, where the smaller refuse passes onto a conveyor belt (glomerular filtration of plasma). As the conveyor belt carries the garbage along, workers remove useful items, such as aluminum cans, plastics, and glass containers (reabsorption). Other workers place additional garbage left at the center and larger items onto the conveyor belt (secretion). At the end of the belt, all remaining garbage falls into a truck for transport to the landfill (excretion of wastes in urine).

GLOMERULAR FILTRATION

OBJECTIVES
• *Describe the filtration membrane.*

• *Discuss the pressures that promote and oppose glomerular filtration.*

The fluid that enters the capsular space is **glomerular filtrate.** The fraction of plasma in the afferent arterioles of the kidneys that becomes glomerular filtrate is the *filtration fraction.* Although a filtration fraction of 0.16–0.20 (16–20%) is typical, the value varies considerably in both health and disease. On average, the daily volume of glomerular filtrate in adults is 150 liters in females and 180 liters in males, a volume that represents about 65 times the entire blood plasma volume. More than 99% of the glomerular filtrate returns to the bloodstream via tubular reabsorption, however, so only 1–2 liters (about 1–2 qt) are excreted as urine.

The Filtration Membrane

Together, endothelial cells of glomerular capillaries and podocytes, which completely encircle the capillaries, form a

leaky barrier known as the **filtration membrane** or **endothelial–capsular membrane.** This sandwichlike assembly permits filtration of water and small solutes but prevents filtration of most plasma proteins, blood cells, and platelets. Filtered substances move from the bloodstream through three barriers—a glomerular endothelial cell, the basal lamina, and a filtration slit formed by a podocyte (Figure 26.8):

❶ Glomerular endothelial cells are quite leaky because they have large **fenestrations** (pores) that are 70–100 nm (0.07–0.1μm) in diameter. This size permits all solutes in blood plasma to exit glomerular capillaries but prevents filtration of blood cells and platelets. Located among the glomerular capillaries and in the cleft between afferent and efferent arterioles are **mesangial cells** (*mes-* = in the middle; *-angi* = blood vessel), contractile cells that help regulate glomerular filtration (see Figure 26.6a).

❷ The **basal lamina,** a layer of acellular material between the endothelium and the podocytes, consists of fibrils in a glycoprotein matrix; it prevents filtration of larger plasma proteins.

❸ Extending from each podocyte are thousands of footlike processes termed **pedicels** (PED-i-sels; = little feet) that wrap around glomerular capillaries. The spaces between pedicels are the **filtration slits.** A thin membrane, the **slit membrane,** extends across each filtration slit; it permits the passage of molecules having a diameter smaller than 6–7 nm (0.006–0.007 μm), including water, glucose, vitamins, amino acids, very small plasma proteins, ammonia, urea, and ions. Because the most plentiful plasma protein—albumin—has a diameter of 7.1 nm, less than 1% of it passes the slit membrane.

The principle of filtration—the use of pressure to force fluids and solutes through a membrane—is the same in glomerular capillaries as in capillaries elsewhere in the body (see Starling's law of the capillaries, page 676). However, the volume of fluid filtered by the renal corpuscle is much larger than in other capillaries of the body for three reasons:

1. Because they are long and extensive, *the glomerular capillaries present a large surface area for filtration.* How much of this surface area is available for filtration is regulated by the mesangial cells. When they are relaxed, surface area is maximal, and glomerular filtration is very high; when they contract, the available surface area is reduced, and glomerular filtration decreases.

2. *The filtration membrane is thin and porous.* Despite having several layers, the thickness of the filtration membrane is only 0.1 μm. Glomerular capillaries also are about 50 times leakier than capillaries in most other tissues, mainly because of their large fenestrations.

3. *Glomerular capillary blood pressure is high.* Because the efferent arteriole is smaller in diameter than the afferent arteriole, resistance to the outflow of blood from the glomerulus is high. As a result, blood pressure in

Figure 26.8 The filtration (endothelial–capsular) membrane. The size of the endothelial fenestrations and filtration slits in (a) have been exaggerated for emphasis. Richard K. Kessel and Randy H. Kardon, *Tissues and Organs: A Text-Atlas of Scanning Electron Microscopy.* © 1979 by W. H. Freeman and Company. Reprinted by permission.

🔑 **During glomerular filtration, water and solutes pass from blood plasma into the capsular space.**

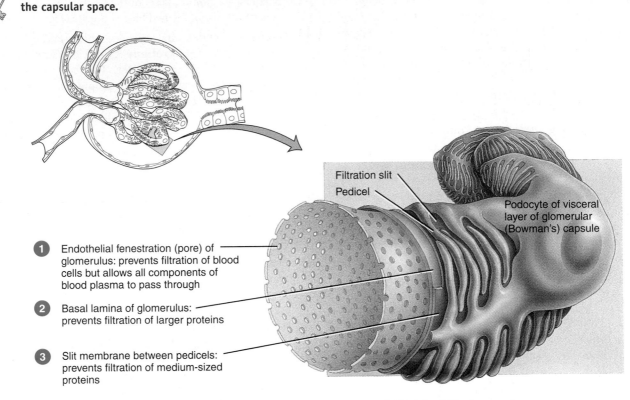

① Endothelial fenestration (pore) of glomerulus: prevents filtration of blood cells but allows all components of blood plasma to pass through

② Basal lamina of glomerulus: prevents filtration of larger proteins

③ Slit membrane between pedicels: prevents filtration of medium-sized proteins

Filtration slit
Pedicel
Podocyte of visceral layer of glomerular (Bowman's) capsule

(a) Details of filtration membrane

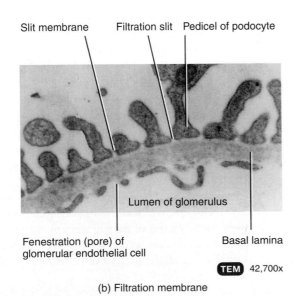

Slit membrane   Filtration slit   Pedicel of podocyte

Lumen of glomerulus

Fenestration (pore) of glomerular endothelial cell

Basal lamina

TEM 42,700x

(b) Filtration membrane

Q Which part of the filtration membrane prevents red blood cells from entering the capsular space?

glomerular capillaries is considerably higher than in capillaries elsewhere in the body; a higher pressure produces more filtrate.

Net Filtration Pressure

Glomerular filtration depends on three main pressures—one that *promotes* filtration and two that *oppose* filtration (Figure 26.9):

① **Glomerular blood hydrostatic pressure (GBHP)** promotes filtration—it forces water and solutes in blood plasma through the filtration membrane. Glomerular blood hydrostatic pressure is the blood pressure in glomerular capillaries, which is about 55 mm Hg.

② **Capsular hydrostatic pressure (CHP)** opposes filtration. It is the hydrostatic pressure exerted against the filtration membrane by fluid already in the capsular space and renal tubule. CHP is about 15 mm Hg.

③ **Blood colloid osmotic pressure (BCOP),** which is due to proteins such as albumin, globulins, and fibrinogen in blood plasma, also opposes filtration. The average BCOP in glomerular capillaries is about 30 mm Hg.

Figure 26.9 The pressures that drive glomerular filtration. Taken together, these pressures determine net filtration pressure (NFP).

🔑 **Glomerular blood hydrostatic pressure promotes filtration, whereas capsular hydrostatic pressure and blood colloid osmotic pressure oppose filtration.**

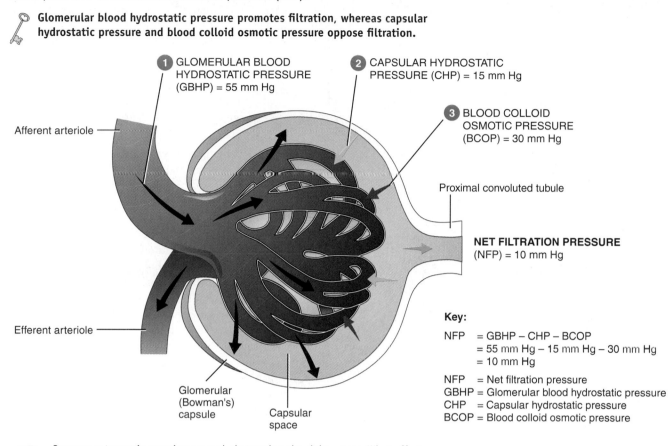

① GLOMERULAR BLOOD HYDROSTATIC PRESSURE (GBHP) = 55 mm Hg

② CAPSULAR HYDROSTATIC PRESSURE (CHP) = 15 mm Hg

③ BLOOD COLLOID OSMOTIC PRESSURE (BCOP) = 30 mm Hg

Afferent arteriole

Proximal convoluted tubule

NET FILTRATION PRESSURE (NFP) = 10 mm Hg

Efferent arteriole

Key:

NFP = GBHP – CHP – BCOP
      = 55 mm Hg – 15 mm Hg – 30 mm Hg
      = 10 mm Hg

NFP = Net filtration pressure
GBHP = Glomerular blood hydrostatic pressure
CHP = Capsular hydrostatic pressure
BCOP = Blood colloid osmotic pressure

Glomerular (Bowman's) capsule

Capsular space

Q Suppose a tumor is pressing on and obstructing the right ureter. What effect might this have on CHP and thus on NFP in the right kidney? Would the left kidney also be affected?

Net filtration pressure (NFP), the total pressure that promotes filtration, is determined as follows:

Net Filtration Pressure (NFP) = GBHP − CHP − BCOP

By substituting the values just given, normal NFP may be calculated:

$$NFP = 55 \text{ mm Hg} - 15 \text{ mm Hg} - 30 \text{ mm Hg}$$
$$= 10 \text{ mm Hg}$$

Thus, a pressure of only 10 mm Hg causes a normal amount of plasma (minus plasma proteins) to filter from the glomerulus into the capsular space.

Glomerular Filtration Rate

The amount of filtrate formed in all the renal corpuscles of both kidneys each minute is called the **glomerular filtration rate (GFR).** In adults, the GFR averages 125 mL/min in males and 105 mL/min in females. Homeostasis of body fluids requires that the kidneys maintain a relatively constant GFR. If the GFR is too high, needed substances may pass so quickly through the renal tubules that some are not reabsorbed and are lost in the urine. If the GFR is too low, nearly all the filtrate may be reabsorbed and certain waste products may not be adequately excreted.

GFR is directly related to the pressures that determine net filtration pressure; any change in net filtration pressure will affect GFR. Severe blood loss, for example, reduces systemic blood pressure, which also decreases the glomerular blood hydrostatic pressure. Filtration ceases if glomerular blood hydrostatic pressure drops to 45 mm Hg because the opposing pressures add up to 45 mm Hg. Amazingly, when systemic blood pressure rises above normal, net filtration pressure and GFR increase very little. GFR is nearly constant when the mean arterial blood pressure (MABP) is anywhere between 80 and 180 mm Hg.

Regulation of GFR

The mechanisms that regulate GFR operate in two main ways: (1) by adjusting blood flow into and out of the glomerulus and (2) by altering the glomerular capillary surface area

Figure 26.10 Tubuloglomerular feedback.

🔑 Macula densa cells of the juxtaglomerular apparatus provide negative feedback regulation of glomerular filtration rate.

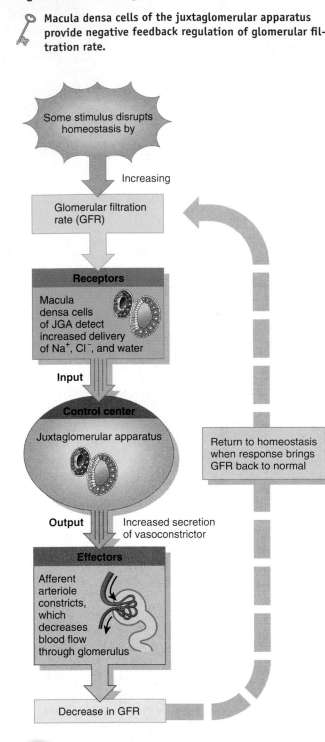

Q Why is this process termed *auto*regulation?

available for filtration. GFR increases when blood flow into the glomerular capillaries increases. Coordinated control of the diameter of both afferent and efferent arterioles regulates glomerular blood flow. Constriction of the afferent arteriole, for example, decreases blood flow into the glomerulus, whereas dilation of the afferent arteriole increases blood flow into the glomerulus. Three mechanisms control GFR: renal autoregulation, neural regulation, and hormonal regulation.

Renal Autoregulation of GFR The kidneys themselves help maintain a constant renal blood flow and GFR despite normal, everyday changes in systemic arterial pressure (as occur, for instance, during exercise). This capability is called **renal autoregulation** and consists of two mechanisms—the myogenic mechanism and tubuloglomerular feedback. Working together, they can maintain nearly constant GFR over a wide range of systemic blood pressures.

The **myogenic mechanism** (*myo-* = muscle; *-genic* = producing) occurs when stretching triggers contraction of smooth muscle cells in the wall of afferent arterioles. When blood pressure rises, GFR also rises because renal blood flow increases. However, the elevated blood pressure also stretches the walls of the afferent arterioles. In response, smooth muscle fibers in the wall of the afferent arteriole contract, the arteriole's lumen narrows, and renal blood flow decreases, which reduces GFR to its previous level. Conversely, when arterial blood pressure drops, the smooth muscle cells are stretched less and thus relax. The afferent arterioles dilate, renal blood flow increases, and GFR increases. The myogenic mechanism normalizes renal blood flow and GFR within seconds after blood pressure changes occur.

The second contributor to renal autoregulation, **tubuloglomerular feedback,** is so named because part of the renal tubule—the macula densa—provides feedback to the glomerulus (Figure 26.10). When GFR is above normal due to elevated systemic blood pressure, filtered fluid flows more rapidly along the renal tubules. As a result, the proximal convoluted tubule and loop of Henle have less time to reabsorb Na^+, Cl^-, and water. Macula densa cells are thought to detect the increased delivery of Na^+, Cl^-, and water and to stimulate the release of an as yet unidentified vasoconstrictor from cells in the juxtaglomerular apparatus (JGA). The vasoconstrictor causes afferent arterioles to constrict, less blood flows into the glomerular capillaries, and GFR decreases. If blood pressure falls and GFR is lower than normal, the opposite sequence of events occurs, although to a lesser degree. Tubuloglomerular feedback operates more slowly than the myogenic mechanism.

Neural Regulation of GFR Like most blood vessels of the body, those of the kidneys are supplied by sympathetic ANS fibers that release norepinephrine. Norepinephrine causes vasoconstriction through the activation of α_1 receptors, which are particularly plentiful in the smooth muscle fibers of afferent arterioles. At rest, sympathetic stimulation is

Table 26.2 Regulation of Glomerular Filtration Rate (GFR)

TYPE OF REGULATION	MAJOR STIMULUS	MECHANISM AND SITE OF ACTION	EFFECT ON GFR
Renal Autogregulation			
Myogenic mechanism	Increased stretching of smooth muscle fibers in afferent arteriole walls due to increased blood pressure.	Stretched smooth muscle fibers contract, thereby narrowing the lumen of the afferent arterioles.	Decrease.
Tubuloglomerular feedback	Rapid delivery of Na^+ and Cl^- to the macula densa due to high systemic blood pressure.	Increased release of a vasoconstrictor by the juxtaglomerular apparatus causes constriction of afferent arterioles.	Decrease.
Neural Regulation	Increase in level of activity of renal sympathetic nerves releases norepinephrine.	Constriction of afferent arterioles through activation of α_1 receptors and increased release of renin.	Decrease.
Hormonal Regulation			
Angiotensin II	Decreased blood volume or blood pressure stimulates production of angiotensin II.	Constriction of both afferent and efferent arterioles.	Decrease.
Atrial natriuretic peptide (ANP)	Stretching of the heart stimulates secretion of ANP.	Relaxation of mesangial cells in glomerulus increases capillary surface area available for filtration.	Increase.

moderately low, the afferent and efferent arterioles are dilated, and renal autoregulation of GFR prevails. With moderate sympathetic stimulation, both afferent and efferent arterioles constrict to the same degree. Blood flow into and out of the glomerulus is restricted to the same extent, which decreases GFR only slightly. With greater sympathetic stimulation, however, as occurs during exercise or hemorrhage, vasoconstriction of the afferent arterioles predominates. As a result, blood flow into glomerular capillaries is greatly decreased, and GFR drops. This lowering of renal blood flow has two consequences: It reduces urine output, which helps conserve blood volume, and it permits greater blood flow to other body tissues. Sympathetic stimulation also stimulates release of the enzyme renin from juxtaglomerular cells, which in turn accelerates production of the hormone angiotensin II (see Figure 18.6 on page 576).

Hormonal Regulation of GFR Angiotensin II reduces GFR, whereas atrial natriuretic peptide (ANP) increases GFR. **Angiotensin II** is a very potent vasoconstrictor that narrows both afferent and efferent arterioles and reduces renal blood flow, thereby decreasing GFR. As its name suggests, **atrial natriuretic peptide (ANP)** is secreted by cells in the atria of the heart. Stretching of the atria, as occurs when blood volume increases, stimulates secretion of ANP. By causing relaxation of the glomerular mesangial cells, ANP increases the capillary surface area available for filtration, which increases glomerular filtration rate.

Table 26.2 summarizes the ways in which glomerular filtration rate is regulated.

1. If the urinary excretion rate of a drug such as penicillin is greater than the rate at which it is filtered at the glomerulus, how else is it getting into the urine?

2. What is the major chemical difference between plasma and glomerular filtrate?
3. Describe the factors that allow much greater filtration through glomerular capillaries than through capillaries elsewhere in the body.
4. Write the equation for the calculation of net filtration pressure (NFP).
5. Discuss the ways in which glomerular filtration rate is regulated.

TUBULAR REABSORPTION AND TUBULAR SECRETION

OBJECTIVES
• *Describe the routes and mechanisms of tubular reabsorption and secretion.*
• *Describe how specific segments of the renal tubule and collecting duct reabsorb water and solutes.*
• *Describe how specific segments of the renal tubule and collecting duct secrete solutes into the urine.*

Principles of Tubular Reabsorption and Secretion

The normal rate of glomerular filtration is so high that the volume of fluid entering the proximal convoluted tubules in half an hour is greater than the total plasma volume. Reabsorption—returning most of the filtered water and many of the filtered solutes to the bloodstream—is the second function of the nephron and collecting duct. Epithelial cells all along the renal tubule and duct carry out reabsorption, but proximal convoluted tubule cells make the largest contribution. Solutes that are reabsorbed by both active and passive processes include glucose, amino acids, urea, and ions such as Na^+, K^+, Ca^{2+}, Cl^-, HCO_3^- (bicarbonate),

Table 26.3 Substances in Plasma and Amounts Filtered, Reabsorbed, and Excreted in Urine

SUBSTANCE	TOTAL AMOUNT IN PLASMA*	FILTERED[†] (ENTERS GLOMERULAR CAPSULE PER DAY)	REABSORBED (RETURNED TO BLOOD PER DAY)	URINE (EXCRETED PER DAY)
Water	3 liters	180 liters	178–179 liters	1–2 liters
Proteins	200 g	2.0 g	1.9 g	0.1 g
Sodium ions (Na^+)	9.7 g (420 mmol)	579 g (25,174 mmol)	575g (25,000 mmol)	4 g (174 mmol)
Chloride ions (Cl^-)	10.7 g (300 mmol)	640 g (18,000 mmol)	633.7 g (17,850 mmol)	6.3 g (150 mmol)
Bicarbonate ions (HCO_3^-)	4.6 g (75 mmol)	275 g (4500 mmol)	275 g (4500 mmol)	0.03 g (2 mmol)
Glucose	2.7 g (15 mmol)	162 g (900 mmol)	162 g (900 mmol)	0
Urea	0.9 g (15 mmol)	54 g (900 mmol)	27 g (450 mmol)	27 g (450 mmol)[‡]
Potassium ions (K^+)	0.5 g (12.6 mmol)	29.6 g (756 mmol)	29.6 g (756 mmol)	2.0 g (50 mmol)[§]
Uric acid	0.15 g	8.5 g	7.7 g	0.8 g
Creatinine	0.03 g	1.6 g	0	1.6 g

*Assuming total plasma volume is 3 liters.
[†]Assuming GFR is 180 liters per day.
[‡]In addition to being filtered and reabsorbed, urea is secreted.
[§]After virtually all filtered K^+ is reabsorbed in the convoluted tubules and loop of Henle, a variable amount of K^+ is secreted by principal cells in the collecting duct.

and HPO_4^{2-} (phosphate). Cells located more distally fine-tune the reabsorption processes to maintain homeostatic balances of water and selected ions. Most small proteins and peptides that pass through the filter also are reabsorbed, usually via pinocytosis. To appreciate the huge extent of tubular reabsorption, look at Table 26.3 and compare the amounts of substances that are filtered, reabsorbed, and excreted in urine with the amounts present in blood plasma.

The third function of nephrons and collecting ducts is tubular secretion, the transfer of materials from the blood and tubule cells into tubular fluid. Secreted substances include H^+, K^+, ammonium ions (NH_4^+), creatinine, and certain drugs such as penicillin. Tubular secretion has two important outcomes: The secretion of H^+ helps control blood pH, and the secretion of other substances helps eliminate them from the body.

Reabsorption Routes

A substance being reabsorbed from the fluid in the tubule lumen can take one of two routes before entering a peritubular capillary: It can move *between* adjacent tubule cells or *through* an individual tubule cell (Figure 26.11). Throughout the renal tubule, tight junctions surround and join neighboring cells to one another, much like the plastic rings that hold a six-pack of soda cans together. The **apical membrane** (like the tops of the soda cans) contacts the tubular fluid, and the **basolateral membrane** (like the bottoms and sides of the soda cans) contacts interstitial fluid at the sides and base of the cell. Although the tight junctions prevent mixing of membrane proteins in the apical and basolat-

eral membrane compartments, they do not completely seal off the interstitial fluid from the fluid in the tubule lumen.

Some fluid can leak *between* the cells in a passive process known as **paracellular reabsorption** (*para-* = beside). In some parts of the renal tubule, the paracellular route is thought to account for up to 50% of the reabsorption of certain ions and the water that osmotically accompanies them. In **transcellular reabsorption** (*trans-* = across), a substance passes from the fluid in the tubular lumen through the apical membrane of a tubule cell, across the cytosol, and out into interstitial fluid through the basolateral membrane.

Transport Mechanisms

When renal cells transport solutes out of or into tubular fluid, they move specific substances in one direction only. Not surprisingly, the apical and basolateral membranes have different types of transport proteins. Moreover, the tight junctions form a barrier that prevents diffusion of proteins between these two membrane compartments.

Reabsorption of Na^+ by the renal tubules is especially important because more sodium ions pass the glomerular filters than any other substance except water. Furthermore, solute reabsorption drives water reabsorption because water is reabsorbed only passively via osmosis. Several different transport systems accomplish Na^+ reabsorption in each portion of the renal tubule and collecting duct. These mechanisms recover not only filtered Na^+ but also other electrolytes, nutrients, and water. Some of the transporters also work to secrete excess hydrogen and potassium ions.

Cells lining the renal tubules, like other cells throughout the body, have a low concentration of Na^+ in their cytosol

Figure 26.11 Reabsorption routes: paracellular (between cells) reabsorption and transcelluar (through cells) reabsorption.

🔑 **In paracellular reabsorption, water and solutes in tubular fluid return to the bloodstream by moving between tubule cells; in transcellular reabsorption, solutes and water in tubular fluid return to the bloodstream by passing through a tubule cell.**

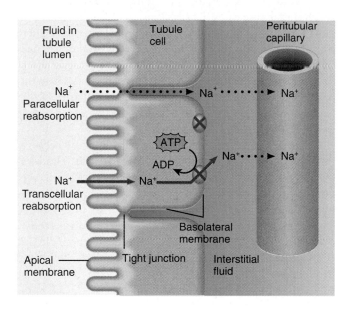

Key:

•••••▶ Diffusion

──────▶ Active transport

 Sodium pump (Na⁺/K⁺ ATPase)

Q What is the main function of the tight junctions between tubule cells?

due to the activity of sodium pumps (Na^+/K^+ ATPases), which eject Na^+ from the cells through the basolateral membrane (see Figure 26.11). The absence of sodium pumps in the apical membrane ensures that reabsorption of Na^+ is a one-way process: Most sodium ions that cross the apical membrane will be pumped into interstitial fluid at the base and sides of the cell. The amount of ATP used by sodium pumps in the renal tubules is substantial—an estimated 6% of the total ATP consumption at rest. For comparison, this is about the same energy used by the diaphragm as it contracts during quiet breathing.

As we noted in Chapter 3, transport of materials across membranes may be either active or passive. Recall that in **primary active transport** the energy derived from hydrolysis of ATP is used to "pump" a substance across a membrane. Because the sodium pump uses ATP in this manner, it is a primary active transport pump. In **secondary active transport** the energy stored in an ion's electrochemical gradient,

rather than hydrolysis of ATP, drives another substance across a membrane. Secondary active transport couples the "downhill" movement of an ion along its electrochemical gradient to the "uphill" movement of a second substance against its electrochemical gradient. The membrane proteins that perform secondary active transport are called *symporters* when they move two or more substances in the same direction across a membrane, and *antiporters* when they move two or more substances in opposite directions across a membrane. Each type of transporter has an upper limit on how fast it can work, just as an escalator has a limit on how many people it can carry from one level to another in a given period. This limit is called the **transport maximum (T_m)** and is measured in mg/min.

The mechanism for reabsorption of water by the renal tubules and collecting duct is osmosis. About 90% of the reabsorption of water filtered by the kidneys occurs together with the reabsorption of solutes such as Na^+, Cl^-, and glucose. Water reabsorbed together with solutes in tubular fluid is termed **obligatory water reabsorption** (ob-LIG-a-tor'-ē) because the water is "obliged" to follow the solutes when they are reabsorbed. This type of water reabsorption occurs in the proximal convoluted tubule and the descending limb of the loop of Henle because these segments of the nephron are always permeable to water. Reabsorption of the final 10% of the water, a total of 10–20 liters per day, is termed **facultative water reabsorption** (FAK-ul-tā'-tiv). The word *facultative* means "capable of adapting to a need." Facultative water reabsorption occurs mainly in the collecting ducts and is regulated by antidiuretic hormone.

CLINICAL APPLICATION
Glucosuria

When the blood concentration of glucose is above 200 mg/mL, the renal symporters cannot work fast enough to reabsorb all the glucose that enters the glomerular filtrate. As a result, some glucose remains in the urine, a condition called **glucosuria** (gloo'-kō-SOO-rē-a). The most common cause of glucosuria is diabetes mellitus, in which the blood glucose level may rise far above normal because insulin activity is deficient. Rare genetic mutations in the renal Na^+–glucose symporter that greatly reduces its T_m also cause glucosuria. In these cases, glucose appears in the urine even though the blood glucose level is normal. ■

Now that we have discussed the principles of renal transport, we can follow the filtered fluid from the glomerular capsule into the proximal convoluted tubule, loop of Henle, distal convoluted tubule, and collecting ducts to see where and how specific substances are reabsorbed and secreted. Due to reabsorption and secretion, the fluid's composition changes as it flows along the nephron tubule and through the collecting duct. The filtered fluid becomes *tubular fluid* once it enters the proximal convoluted tubule, and the fluid finally excreted by the kidneys is *urine*, which drains from papillary ducts into the renal pelvis.

Figure 26.12 Reabsorption of glucose by Na$^+$–glucose symporters in cells of the proximal convoluted tubule (PCT).

🔑 **Normally, all filtered glucose is reabsorbed in the PCT.**

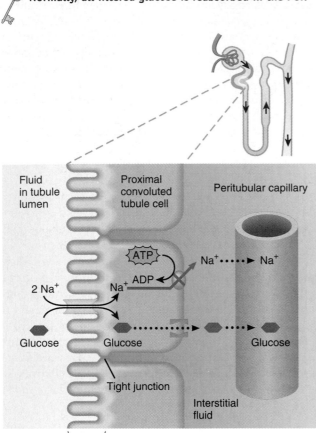

Fluid in tubule lumen

Proximal convoluted tubule cell

Peritubular capillary

ATP

ADP

Na$^+$ ⋯⋯▶ Na$^+$

2 Na$^+$

Na$^+$

Glucose

Glucose

Glucose

Tight junction

Interstitial fluid

Brush border (microvilli)

Key:

▬ Na$^+$– glucose symporter

⟶ Secondary active transport

⟩ Glucose facilitated diffusion transporter

•••▶ Diffusion

✕ Sodium pump

Q How does filtered glucose enter and leave a PCT cell?

Reabsorption in the Proximal Convoluted Tubule

The majority of solute and water reabsorption from filtered fluid occurs in the proximal convoluted tubules, and most absorptive processes involve Na$^+$. Two different types of Na$^+$ transporters are located in the proximal convoluted tubule: (1) several different Na$^+$ symporters effect reabsorption of Na$^+$ together with a variety of other solutes, and (2) Na$^+$/H$^+$ antiporters effect reabsorption of Na$^+$ in exchange for secretion of H$^+$. These proximal convoluted tubule Na$^+$ transporters promote reabsorption from the filtrate of 100% of most organic solutes, such as glucose and amino acids;

80–90% of the HCO$_3^-$; 65% of the water, Na$^+$, and K$^+$; 50% of the Cl$^-$; and a variable amount of Ca^{2+}, Mg^{2+}, and HPO$_4^{2-}$.

Normally, filtered glucose, amino acids, lactic acid, water-soluble vitamins, and other nutrients are not lost in the urine because they are completely reabsorbed in the first half of the proximal convoluted tubule (PCT) by **Na$^+$ symporters** located in the apical membrane. Figure 26.12 depicts the operation of the main Na$^+$–glucose symporter in the apical membrane of PCT cells. Two Na$^+$ and a molecule of glucose attach to the symporter protein, which carries them from the tubular fluid into the tubule cell. Glucose brought into proximal convoluted tubule cells by Na$^+$ symporters exits via facilitated diffusion through the basolateral membrane and then diffuses into peritubular capillaries. Other Na$^+$ symporters in the PCT reclaim phosphate and sulfate ions, all amino acids, and lactic acid.

Another secondary active transport process achieves Na$^+$ reabsorption and also returns filtered HCO$_3^-$ and water to the peritubular capillaries. **Na$^+$/H$^+$ antiporters** carry filtered Na$^+$ down its concentration gradient into PCT cells in exchange for H$^+$ (Figure 26.13a). Thus, Na$^+$ is reabsorbed and H$^+$ is secreted. PCT cells continually produce the H$^+$ needed to keep the antiporters running. Carbon dioxide (CO$_2$) diffuses from peritubular blood or tubular fluid or is produced by metabolic reactions within the cells. In the presence of the enzyme *carbonic anhydrase* (CA), CO$_2$ combines with water (H$_2$O) to form carbonic acid (H$_2$CO$_3$), which then dissociates into H$^+$ and HCO$_3^-$:

$$CO_2 + H_2O \xrightarrow{\text{Carbonic anhydrase}} H_2CO_3 \longrightarrow H^+ + HCO_3^-$$

This mechanism also is responsible for the reabsorption of 80–90% of the filtered bicarbonate ions, an important body buffer. HCO$_3^-$ reabsorption is depicted in Figure 26.13b. After H$^+$ is secreted into the fluid within the lumen of the proximal convoluted tubule, it combines with filtered HCO$_3^-$. This reaction, catalyzed by carbonic anhydrase present in the brush border, forms H$_2$CO$_3$, which in turn dissociates into CO$_2$ and H$_2$O. Carbon dioxide then diffuses into the tubule cells and joins with H$_2$O to form H$_2$CO$_3$, which dissociates into H$^+$ and HCO$_3^-$. As the level of HCO$_3^-$ rises in the cytosol, HCO$_3^-$ exits the cell via a facilitated diffusion transporter in the basolateral membrane and diffuses into the blood together with Na$^+$. Thus, for every H$^+$ secreted into the tubular fluid, one filtered HCO$_3^-$ eventually returns to the blood.

Besides achieving reabsorption of sodium ions, the Na$^+$ symporters and Na$^+$/H$^+$ antiporters promote passive reabsorption of other solutes. As Na$^+$, HCO$_3^-$, organic solutes, and water leave the tubular fluid due to the operation of Na$^+$ symporters and Na$^+$/H$^+$ antiporters, the concentrations of the remaining filtered solutes increase. In the second half of the PCT, electrochemical gradients for Cl$^-$, K$^+$, Ca^{2+}, Mg^{2+}, and urea promote their passive diffusion into peritubular capillaries via both paracellular and transcellular routes (Figure 26.14). Among these ions, Cl^- is present in the highest concentration. Diffusion of negatively charged Cl^- into interstitial fluid via the paracellular route makes the interstitial fluid electrically more negative than the tubular fluid. This electrical potential difference promotes passive paracellular reabsorption of filtered cations, namely Na^+, K^+, Ca^{2+}, and Mg^{2+}.

Reabsorption of Na^+ and other solutes also promotes reabsorption of water via osmosis (see Figure 26.14). Each reabsorbed solute increases the osmolarity, first inside the tubule cell, then in interstitial fluid, and finally in the blood. Water thus moves rapidly from the tubular fluid, via both the paracellular and transcellular routes, into the peritubular capillaries and restores osmotic balance. In other words, reabsorption of the solutes creates an osmotic gradient that promotes the reabsorption of water via osmosis. Cells lining the proximal convoluted tubule and the descending limb of the loop of Henle are especially permeable to water because they have numerous molecules of *aquaporin-1*, a membrane protein that functions as a water channel.

Secretion of NH_3 and NH_4^+ in the Proximal Convoluted Tubule

Ammonia (NH_3) is a poisonous waste product derived from the deamination (removal of an amino group) of various amino acids, which occurs mainly in hepatocytes. In addition, hepatocytes convert most ammonia to urea, which is a less-toxic compound. Although tiny amounts of urea and ammonia are present in sweat, most excretion of these nitrogen-containing waste products occurs via the urine. Urea and ammonia in blood are both filtered at the glomerulus and secreted by proximal convoluted tubule cells into the tubular fluid.

Proximal convoluted tubule cells produce additional ammonia by deaminating the amino acid glutamine in a reaction that generates both NH_3 and new HCO_3^-. Most NH_3 quickly binds H^+ to become an ammonium ion (NH_4^+), which can substitute for H^+ aboard Na^+/H^+ antiporters in the apical membrane and be secreted into the tubular fluid. The HCO_3^- moves through the basolateral membrane and diffuses into the bloodstream. Ammonia production and secretion in the proximal convoluted tubule increases during acidosis (blood pH below 7.35), and the resulting increase in blood HCO_3^- helps raise blood pH. During alkalosis (blood pH above 7.45), by contrast, ammonia formation decreases, and the resulting decrease in blood HCO_3^- helps lower blood pH. The renal contribution to acid–base balance is discussed more fully in Chapter 27.

Reabsorption in the Loop of Henle

Because the proximal convoluted tubules reabsorb about 65% of the filtered water (about 80 mL/min), fluid enters the loop of Henle (LOH) at a rate of 40–45 mL/min. The chemical composition of the tubular fluid now is quite different from that of blood plasma (and that of glomerular fil-

Figure 26.13 Actions of Na^+/H^+ antiporters in proximal convoluted tubule cells. (a) Reabsorption of sodium ions (Na^+) and secretion of hydrogen ions (H^+) via secondary active transport through the apical membrane; (b) reabsorption of bicarbonate ions (HCO_3^-) via facilitated diffusion through the basolateral membrane. CO_2 = carbon dioxide; H_2CO_3 = carbonic acid; CA = carbonic anhydrase.

🔑 Na^+/H^+ antiporters promote transcellular reabsorption of Na^+, HCO_3^-, and water in the proximal convoluted tubule.

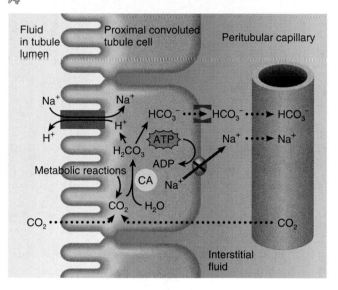

(a) Na^+ reabsorption and H^+ secretion

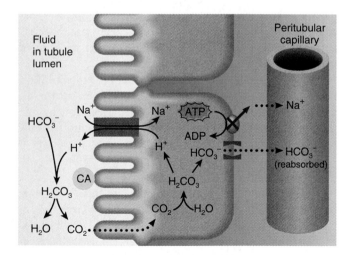

(b) HCO_3^- reabsorption

Key:

▇	Na^+/H^+ antiporter
→	Secondary active transport
⊃⊂	HCO_3^- facilitated diffusion transporter
·····▶	Diffusion
⊗	Sodium pump

Q Which step in Na^+ movement in part (a) is promoted by the electrochemical gradient?

Figure 26.14 Passive reabsorption of Cl⁻, K⁺, Ca²⁺, Mg²⁺, urea, and water in the second half of the proximal convoluted tubule.

🔑 **Electrochemical gradients promote passive reabsorption of solutes via both paracellular and transcellular routes.**

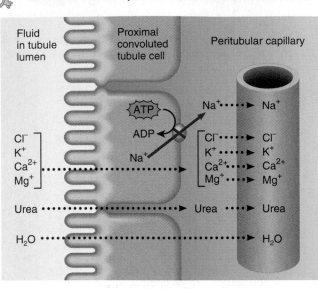

Q By what mechanism is water reabsorbed from tubular fluid?

trate as well) because glucose, amino acids, and other nutrients are no longer present. The osmolarity of the tubular fluid is still close to the osmolarity of blood, however, because reabsorption of water by osmosis keeps pace with reabsorption of solutes all along the proximal convoluted tubule.

The LOH reabsorbs about 20–30% of the filtered Na⁺, K⁺, Ca²⁺; 10–20% of the filtered HCO₃⁻; 35% of the filtered Cl⁻; and 15% of the filtered water. Here, for the first time, reabsorption of water via osmosis is *not* automatically coupled to reabsorption of filtered solutes. The loop of Henle thus sets the stage for *independent* regulation of both the *volume* and *osmolarity* of body fluids.

The apical membranes of cells in the thick ascending limb of the LOH have **Na⁺–K⁺–2Cl⁻ symporters** that simultaneously reclaim one Na⁺, one K⁺, and two Cl⁻ from the fluid in the tubular lumen (Figure 26.15). Na⁺ that is actively transported into interstitial fluid at the base and sides of the cell passively diffuses into the vasa recta. Cl⁻ diffuses through leakage channels in the basolateral membrane. Because numerous K⁺ leakage channels are present in the apical membrane, most K⁺ brought in by the symporters diffuses down its concentration gradient back into the tubular fluid. Thus, the main effect of the Na⁺–K⁺–2Cl⁻ symporters is reabsorption of Na⁺ and Cl⁻.

Positively charged K⁺ returning to the tubular fluid through the apical membrane channels gives the interstitial fluid and blood a net negative charge relative to fluid in the

ascending limb of the LOH. This relative negativity draws cations—Na⁺, K⁺, Ca²⁺, and Mg²⁺—from tubular fluid into the vasa recta via the paracellular route.

Although about 15% of the filtered water is reabsorbed in the *descending* limb of the LOH, little or no water is reabsorbed in the *ascending* limb because the apical membranes of ascending limb cells are virtually impermeable to water. Because ions but not water molecules are reabsorbed, the osmolarity of the tubular fluid progressively decreases as fluid flows along the ascending limb.

Reabsorption in the Distal Convoluted Tubule

Fluid enters the distal convoluted tubules (DCT) at a rate of about 25 mL/min because 80% of the filtered water (100 mL/min) has now been reabsorbed. As fluid flows along the DCT, reabsorption of Na⁺ and Cl⁻ continues by means of **Na⁺–Cl⁻ symporters** in the apical membranes. Sodium pumps and Cl⁻ leakage channels in the basolateral membranes then permit reabsorption of Na⁺ and Cl⁻ into the peritubular capillaries. The extent of Na⁺ and Cl⁻ reabsorption here depends on how fast the ions are delivered to the DCT; higher delivery rates lead to enhanced reabsorption in the DCT. The DCT also is the major site where parathyroid hormone stimulates reabsorption of Ca²⁺. Like the thick ascending limb cells, DCT cells are not very permeable to water, so solutes are reabsorbed with little accompanying water.

Reabsorption and Secretion in the Collecting Duct

By the time fluid reaches the end of the distal convoluted tubule, 90–95% of the filtered solutes and water have been returned to the bloodstream. Recall that two different types of cells—principal cells and intercalated cells—are present at the end of the distal convoluted tubule and in the collecting duct. Whereas principal cells reabsorb Na⁺ and secrete K⁺, intercalated cells reabsorb K⁺ and HCO₃⁻ and secrete H⁺.

Reabsorption of Na⁺ and Secretion of K⁺ by Principal Cells

In contrast to earlier segments of the nephron, Na⁺ passes through the apical membrane of principal cells via Na⁺ leakage channels rather than via symporters or antiporters (Figure 26.16). The concentration of Na⁺ in the cytosol remains low, as usual, because the sodium pumps actively transport Na⁺ across the basolateral membranes. Then Na⁺ passively diffuses into the peritubular capillaries from the interstitial spaces around the tubule cells.

Normally, transcellular and paracellular reabsorption in the proximal convoluted tubule and loop of Henle return most filtered K⁺ to the bloodstream. To adjust for varying dietary intake of potassium and to maintain a stable level of K⁺ in body fluids, principal cells secrete a variable amount of

Figure 26.15 Na^+–K^+ – $2Cl^-$ symporter in the thick ascending limb of the loop of Henle.

🗝 **Cells in the thick ascending limb have symporters that simultaneously reabsorb one Na^+, one K^+, and two Cl^-.**

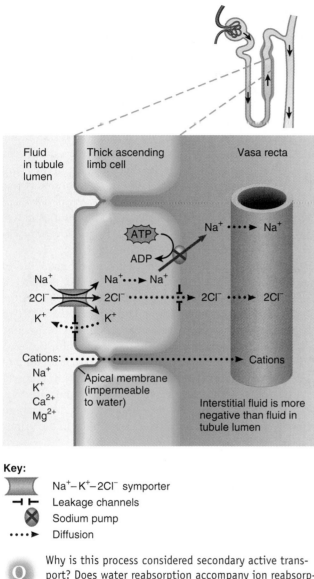

Key:

▨	Na^+–K^+–$2Cl^-$ symporter
⊣ ⊢	Leakage channels
⊗	Sodium pump
••••►	Diffusion

Q Why is this process considered secondary active transport? Does water reabsorption accompany ion reabsorption in this region of the nephron?

Figure 26.16 Reabsorption of Na^+ and secretion of K^+ by principal cells in the last part of the distal convoluted tubule and in the collecting duct.

🗝 **In the apical membrane of principal cells, Na^+ leakage channels allow entry of Na^+ while K^+ leakage channels allow exit of K^+ into the tubular fluid.**

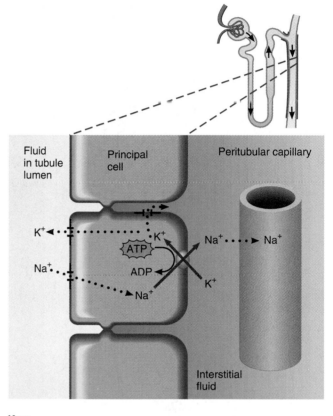

Key:

••••►	Diffusion
⊣ ⊢	Leakage channels
⊗	Sodium pump

Q What hormone stimulates reabsorption and secretion by principal cells, and through which mechanism does this hormone act?

K^+ (see Figure 26.16). Because the basolateral sodium pumps continually bring K^+ into principal cells, its intracellular concentration remains high. K^+ leakage channels are present in both the apical and basolateral membranes. Thus, some K^+ diffuses down its concentration gradient into the tubular fluid, where the K^+ concentration is very low. This secretion mechanism is the main source of K^+ excreted in the urine.

The hormone aldosterone increases Na^+ and water reabsorption and K^+ secretion by principal cells because it both increases the activity of existing sodium pumps and leakage channels and stimulates the synthesis of new pumps

and channels. If the level of aldosterone is low, the principal cells reabsorb few Na^+ and secrete few K^+, which can cause the blood level of K^+ to increase dangerously. As the plasma level of K^+ rises, disturbances in cardiac rhythm may develop, and at higher concentrations cardiac arrest may occur.

The amount of K^+ secreted by principal cells is increased by:

1. *High K^+ level in plasma.* Elevated levels of K^+ in plasma stimulate the adrenal cortex to release aldosterone.
2. *Increased aldosterone.* Aldosterone stimulates the principal cells to secrete more K^+ into tubular fluid.
3. *Increased delivery of Na^+.* A high level of Na^+ in the fluid delivered to the collecting ducts increases the rate of Na^+ absorption and K^+ secretion.

Figure 26.17 Secretion of H$^+$ by intercalated cells in the collecting duct. HCO$_3^-$ = bicarbonate ion; CO$_2$ = carbon dioxide; H$_2$O = water; H$_2$CO$_3$ = carbonic acid; Cl$^-$ = chloride ion; NH$_3$ = ammonia; NH$_4^+$ = ammonium ion; HPO$_4^{2-}$ = monohydrogen phosphate ion; H$_2$PO$_4^-$ = dihydrogen phosphate ion.

🔑 Urine can be up to 1000 times more acidic than blood due to the operation of these proton pumps.

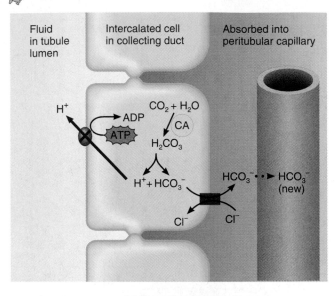

(a) Secretion of H$^+$

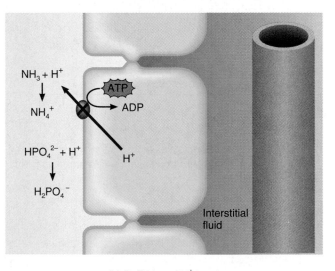

(b) Buffering of H$^+$ in urine

Key:

 Proton pump (H$^+$ ATPase)

 HCO$_3^-$/Cl$^-$ antiporter

•• ► Diffusion

Q What are the effects of a drug that blocks the activity of carbonic anhydrase?

Secretion of H$^+$ and Absorption of HCO$_3^-$ by Intercalated Cells

The apical membranes of some intercalated cells include **proton pumps (H$^+$ ATPases)** that secrete H$^+$ into the tubular fluid (Figure 26.17). Intercalated cells can secrete H$^+$ against a concentration gradient so effectively that urine can be up to 1000 times (3 pH units) more acidic than blood. HCO$_3^-$ produced by dissociation of H$_2$CO$_3$ inside intercalated cells crosses the basolateral membrane using **Cl$^-$/HCO$_3^-$ antiporters** and then diffuses into peritubular capillaries (see Figure 26.17a). The HCO$_3^-$ that enters the blood in this way is *new* (not filtered); for this reason, blood leaving the kidney in the renal vein may have a higher HCO$_3^-$ concentration than blood entering the kidney in the renal artery. Interestingly, a second type of intercalated cells has proton pumps in their basolateral membranes and Cl$^-$/HCO$_3^-$ antiporters in their apical membranes; these cells secrete HCO$_3^-$ and reabsorb H$^+$. Thus, the two types of intercalated cells help maintain the pH of body fluids in two ways—by excreting excess H$^+$ when pH is too low or by excreting excess HCO$_3^-$ when pH is too high.

Some hydrogen ions secreted into the tubular fluid of the collecting duct are buffered. By now, most of the filtered bicarbonate ions have been reabsorbed; few are left in the tubule lumen to combine with and buffer secreted H$^+$. Two other buffers, however, are present and available to combine with H$^+$ (see Figure 26.17b). The most plentiful buffer is HPO$_4^{2-}$ (monohydrogen phosphate ion); a small amount of NH$_3$ (ammonia) also is present. H$^+$ combines with HPO$_4^{2-}$ to form H$_2$PO$_4^-$ (dihydrogen phosphate ion) and with NH$_3$ to form NH$_4^+$ (ammonium ion). Because these ions cannot diffuse back into tubule cells, they are excreted in the urine.

Hormonal Regulation of Tubular Reabsorption and Tubular Secretion

Four hormones affect the extent of Na$^+$, Cl$^-$, and water reabsorption and K$^+$ secretion by the renal tubules. The most important hormonal regulators of electrolyte reabsorption and secretion are angiotensin II and aldosterone. The major hormone that regulates water reabsorption is antidiuretic hormone. Atrial natriuretic peptide plays a minor role in inhibiting both electrolyte and water reabsorption.

Renin–Angiotensin–Aldosterone System

When blood volume and blood pressure decrease, the walls of the afferent arterioles are stretched less, and the juxtaglomerular cells secrete the enzyme **renin** into the blood. Sympathetic stimulation also directly stimulates release of renin from juxtaglomerular cells. Renin clips off a 10–amino-acid peptide called angiotensin I from angiotensinogen, which is synthesized by hepatocytes. By clipping off two more amino acids, *angiotensin converting enzyme (ACE)* converts angiotensin I to **angiotensin II,** which is the active hormone.

Angiotensin II affects renal physiology in four ways:

1. It decreases the glomerular filtration rate by causing vasoconstriction of the afferent arterioles.
2. It enhances reabsorption of Na^+, Cl^-, and water in the proximal convoluted tubule by stimulating the activity of Na^+/H^+ antiporters.
3. It stimulates the adrenal cortex to release **aldosterone,** a hormone that in turn stimulates the principal cells in the collecting ducts to reabsorb more Na^+ and Cl^-. The osmotic consequence of excreting fewer Na^+ and Cl^- is excreting less water, which increases blood volume.
4. It stimulates release of antidiuretic hormone, which increases reabsorption of water in the collecting duct.

Antidiuretic Hormone

Antidiuretic hormone (ADH or **vasopressin)** regulates facultative water reabsorption by increasing the water permeability of principal cells. In the absence of ADH, the apical membranes of principal cells have a very low permeability to water. Within principal cells are tiny vesicles containing many copies of a water channel protein known as **aquaporin-2.*** ADH stimulates insertion of the aquaporin-2–containing vesicles into the apical membranes via exocytosis. As a result, the water permeability of the principal cell's apical membrane increases, and water molecules move more rapidly from the tubular fluid into the cells. Because the basolateral membranes are always permeable to water, water molecules then move rapidly into the blood. The kidneys produce as little as 400–500 mL of very concentrated urine each day when ADH concentration is maximal. When ADH level declines, the aquaporin-2 channels are removed from the apical membrane via endocytosis, and then a larger volume of more dilute urine is excreted.

A negative feedback system involving ADH regulates facultative water reabsorption (Figure 26.18). When the osmolarity or osmotic pressure of plasma and interstitial fluid increases—that is, when water concentration decreases—by as little as 1%, osmoreceptors in the hypothalamus detect the change. Their nerve impulses stimulate secretion of more ADH into the blood, and the principal cells become more permeable to water. As facultative water reabsorption increases, plasma osmolarity decreases to normal. In the absence of ADH activity—a condition known as diabetes insipidus—a person may excrete up to 20 liters of very dilute urine daily.

Atrial Natriuretic Peptide

A large increase in blood volume promotes release of atrial natriuretic peptide (ANP). Although the importance of

---

*The previously mentioned water channel—aquaporin-1—is not governed by ADH.

Figure 26.18 Negative feedback regulation of facultative water reabsorption by ADH.

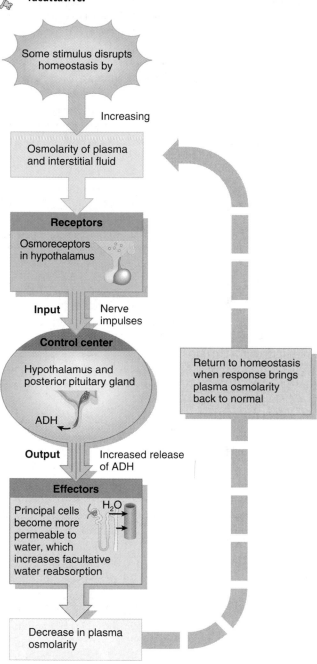

🔑 **Most water reabsorption (90%) is obligatory; 10% is facultative.**

Q Besides ADH, which other hormones contribute to the regulation of water reabsorption?

Table 26.4 Hormonal Regulation of Tubular Reabsorption and Tubular Secretion

HORMONE	MAJOR STIMULI THAT TRIGGER RELEASE	MECHANISM AND SITE OF ACTION	EFFECTS
Angiotensin II	Low blood volume or low blood pressure stimulates renin-induced production of angiotensin II.	Stimulates activity of Na^+/H^+ antiporters in proximal tubule cells.	Increases reabsorption of Na^+, other solutes, and water, which increases blood volume.
Aldosterone	Increased angiotensin II level and increased level of plasma K^+ promote release of aldosterone by adrenal cortex.	Enhances activity and synthesis of sodium pumps in basolateral membrane and Na^+ channels in apical membrane of principal cells in collecting duct.	Increases secretion of K^+ and reabsorption of Na^+, Cl^-, and water, which increases blood volume.
Antidiuretic hormone (ADH) or vasopressin	Increased osmolarity of extracellular fluid or increased angiotensin II level promote release of ADH from the posterior pituitary gland.	Stimulates insertion of water-channel proteins, called aquaporin-2, into the apical membranes of principal cells.	Increases facultative reabsorption of water, which decreases osmolarity of body fluids.
Atrial natriuretic peptide (ANP)	Stretching of atria of heart stimulates secretion of ANP.	Suppresses reabsorption of Na^+ and water in proximal tubule and collecting duct; also inhibits secretion of aldosterone and ADH.	Increases excretion of Na^+ in urine (natriuresis); increases urine output (diuresis) and thus decreases blood volume.

ANP in normal regulation of tubular function is unclear it can inhibit reabsorption of Na^+ and water in the proximal convoluted tubule and collecting duct, and it suppresses the secretion of aldosterone and ADH. These effects increase the excretion of Na^+ in urine (natriuresis) and increase urine output (diuresis), which decreases blood volume.

Table 26.4 summarizes hormonal regulation of tubular reabsorption and tubular secretion.

1. Draw a diagram that depicts how substances are reabsorbed via the transcellular and paracellular routes. Label the apical membrane and the basolateral membrane Where are the sodium pumps located?
2. For reabsorption of Na^+, describe two mechanisms in the PCT, one in the LOH, one in the DCT, and one in the collecting duct. What other solutes are reabsorbed or secreted together with Na^+ in each case?
3. How are hydrogen ions secreted by intercalated cells?
4. Prepare a table that indicates the proportions of filtered water and filtered Na^+ that are reabsorbed in the PCT, LOH, DCT, and collecting duct. Indicate which hormones, if any, regulate reabsorption in each segment.

Production of Dilute and Concentrated Urine

OBJECTIVE

• *Describe how the renal tubule produces dilute and concentrated urine.*

Even though a person's fluid intake can be highly variable, the total volume of fluid in the body remains fairly stable. Homeostasis of body fluid volume depends in large part on the ability of the kidneys to regulate the rate of water loss in urine. Normally functioning kidneys produce a large volume of dilute urine when fluid intake is excessive, and a small volume of concentrated urine when fluid intake is meager or fluid loss is large. ADH controls whether dilute urine or concentrated urine is formed. In the absence of ADH, urine contains a high ratio of water to solutes (dilute urine); however, in the presence of ADH, much water is reabsorbed back into blood, leaving a lower ratio of water to solutes in the urine (concentrated urine).

Formation of Dilute Urine

Glomerular filtrate has the same ratio of water and solute particles as blood; its osmolarity is about 300 mOsm/liter. As previously noted, fluid leaving the proximal convoluted tubule is still isotonic to plasma. When *dilute* urine is being formed (Figure 26.19), the osmolarity of the fluid in the tubular lumen *increases* as it flows down the descending limb of the loop of Henle, *decreases* as it flows up the ascending limb, and *decreases* still more as it flows through the rest of the nephron and collecting duct. These changes in osmolarity result from the following conditions along the path of tubular fluid:

1. Because the osmolarity of the interstitial fluid of the renal medulla becomes progressively greater, more and more water is reabsorbed by osmosis as tubular fluid flows along the descending limb toward the tip of the loop. (The source of this medullary osmotic gradient is explained shortly.) As a result, the fluid remaining in the lumen becomes progressively more concentrated.
2. Cells lining the thick ascending limb of the loop have symporters that actively reabsorb Na^+, K^+, and Cl^- from the tubular fluid (see Figure 26.15). The ions pass from the tubular fluid into thick ascending limb cells, then into

interstitial fluid, and finally some diffuse into the blood in the vasa recta.

3. Although solutes are being reabsorbed in the thick ascending limb, the water permeability of this portion of the nephron is always quite low, so water cannot follow by osmosis. As solutes—but not water molecules—are leaving the tubular fluid, its osmolarity drops to about 150 mOsm/liter. The fluid entering the distal convoluted tubule is thus more dilute than plasma.

4. While the fluid continues flowing along the distal convoluted tubule, additional solutes but essentially no water molecules are reabsorbed because the distal convoluted tubule also is impermeable to water, independent of the presence or absence of ADH.

5. Finally, because the principal cells of the collecting ducts are impermeable to water when the ADH level is very low, tubular fluid becomes progressively more dilute as it flows onward. By the time the tubular fluid drains into the renal pelvis, its concentration can be as low as 65–70 mOsm/liter, four times more dilute than blood plasma or glomerular filtrate.

Formation of Concentrated Urine

When water intake is low or water loss is high (such as during heavy sweating) the kidneys must conserve water while still eliminating wastes and excess ions. Under the influence of ADH the kidneys produce a small volume of highly concentrated urine. Urine can be four times more concentrated (up to 1200 mOsm/liter) than blood plasma or glomerular filtrate (300 mOsm/liter).

The ability of ADH to cause excretion of concentrated urine depends on the presence of an **osmotic gradient** of solutes in the interstitial fluid of the renal medulla. Notice in Figure 26.20 that the solute concentration of the interstitial fluid in the kidney increases from about 300 mOsm/liter in the renal cortex to about 1200 mOsm/liter deep in the renal medulla. The major solutes that contribute to this high osmolarity are Na^+, Cl^-, and urea. Two main factors contribute to building and maintaining this osmotic gradient: (1) differences in solute and water permeability and reabsorption in different sections of the long loops of Henle and the collecting duct, and (2) the countercurrent flow (flow in opposite directions) of fluid in neighboring descending and ascending limbs of the loop of Henle.

Production of concentrated urine occurs as follows (Figure 26.20):

1 *In long-loop nephrons, symporters in thick ascending limb cells of the loop of Henle establish the osmotic gradient in the renal medulla.* In the *thick* ascending limb of the LOH, the $Na^+-K^+-2Cl^-$ symporters reabsorb Na^+ and Cl^- (but not water) from the tubular fluid (Figure 26.20a). As a result, these ions become increasingly concentrated in the interstitial fluid of the outer medulla. Although the mechanism is not clear, cells in the *thin* ascending limb of the LOH also appear to contribute to

Figure 26.19 Formation of dilute urine. Numbers indicate osmolarity in milliosmoles per liter (mOsm/liter). Heavy brown lines in the ascending limb of the loop of Henle and in the distal convoluted tubule indicate impermeability to water; heavy blue lines indicate the last part of the distal convoluted tubule and the collecting duct, which are impermeable to water in the absence of ADH; light blue areas around the nephron represent interstitial fluid. When ADH is absent, the osmolarity of urine can be as low as 65 mOsm/liter.

🔑 When ADH level is low, urine is dilute and has an osmolarity less than the osmolarity of blood.

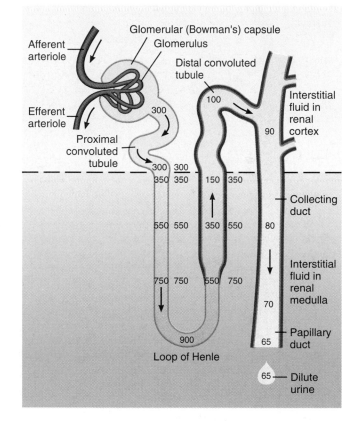

Q Which portions of the renal tubule and collecting duct reabsorb more solutes than water to produce dilute urine?

building the osmotic gradient in the inner medulla. Those ions that diffuse into the vasa recta are carried deep into the inner medulla by the blood flow (Figure 26.20b). Because blood flow is sluggish in the vasa recta, however, there is time for diffusion of solutes to occur among tubular fluid, interstitial fluid, and blood at each level in the medulla. Thus, fluid in the descending limb, interstitial fluid, and plasma attain the same osmolarity.

2 *Cells in the collecting ducts reabsorb more water and urea.* When ADH increases the water permeability of the principal cells, water quickly moves via osmosis out of

Figure 26.20 Mechanism of urine concentration in long-loop juxtamedullary nephrons. The green line indicates the presence of $Na^+-K^+-2Cl^-$ symporters that simultaneously reabsorb these ions into the interstitial fluid of the renal medulla; this portion of the nephron is also relatively impermeable to water and urea. All concentrations are in milliosmoles per liter (mOsm/liter).

🔑 **The formation of concentrated urine depends on high concentrations of solutes in interstitial fluid in the renal medulla.**

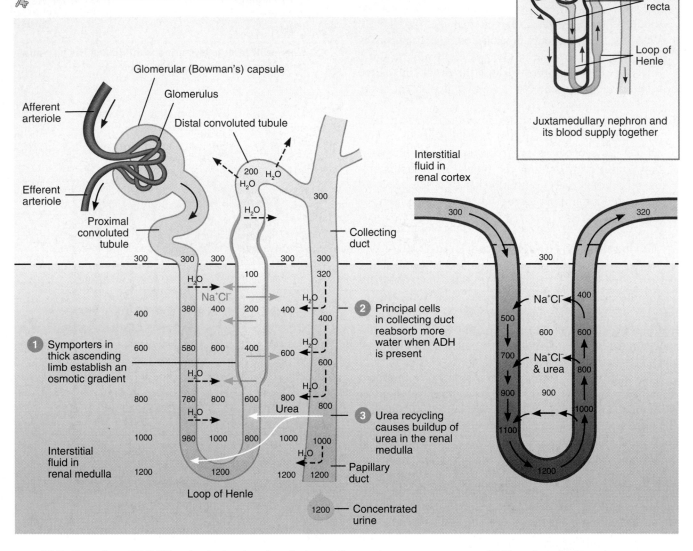

(a) Reabsorption of Na⁺, Cl⁻ and water in a long-loop juxtamedullary nephron

(b) Recycling of salts and urea in the vasa recta

Q Which solutes are the main contributors to the high osmolarity of interstitial fluid in the renal medulla?

the collecting duct tubular fluid, into the interstitial fluid of the inner medulla, and then into the vasa recta. With loss of water, the urea left behind in the tubular fluid of the collecting duct becomes increasingly concentrated. Because duct cells deep in the medulla are permeable to urea, it diffuses from the fluid in the duct into the interstitial fluid of the medulla.

❸ *Urea recycling causes a buildup of urea in the renal medulla.* As urea accumulates in the interstitial fluid,

some of it diffuses into the tubular fluid in the descending and thin ascending limbs of the long loops of Henle, which also are permeable to urea (see Figure 26.20a). However, while the fluid flows through the thick ascending limb, DCT, connecting tubule, and cortical portion of the collecting duct, urea remains in the lumen because cells in these segments are quite impermeable to it. As fluid flows along the collecting ducts, water reabsorption continues via osmosis because ADH is present. This water reabsorption *further increases* the concentration of urea in the tubular fluid, *more* urea diffuses into the interstitial fluid of the inner renal medulla, and the cycle repeats itself. Repeated transfer of urea between segments of the renal tubule and the interstitial fluid of the medulla is termed *urea recycling.* In this way, reabsorption of water from the tubular fluid of the ducts promotes the buildup of urea in the interstitial fluid of the renal medulla, which in turn promotes water reabsorption. The solutes left behind in the lumen thus become very concentrated, and a small volume of concentrated urine is excreted.

The second contributor to the osmotic gradient in the renal medulla is the **countercurrent mechanism,** which has its basis in the hairpin shape of the long loops of Henle of juxtamedullary nephrons. Note in Figure 26.20a that the descending limb of the loop of Henle carries tubular fluid from the renal cortex deep into the medulla, and the ascending limb carries it in the opposite direction. Thus, fluid flowing in one tube runs counter (opposite) to fluid flowing in a nearby parallel tube, an arrangement called *countercurrent flow.*

The descending limb of the loop of Henle is very permeable to water but quite impermeable to solutes except urea. Because the osmolarity of the interstitial fluid outside the descending limb is higher than the tubular fluid within it, water moves out of the descending limb via osmosis, which causes the osmolarity of the tubular fluid to increase. As the fluid continues along the descending limb, its osmolarity increases even more: At the hairpin turn of the loop, the osmolarity can reach 1200 mOsm/liter.

As previously noted, the ascending limb of the loop is impermeable to water, but its symporters reabsorb Na^+ and Cl^- from the tubular fluid into the interstitial fluid of the renal medulla, so the osmolarity of the tubular fluid progressively decreases as it flows through the ascending limb. At the junction of the medulla and cortex, the osmolarity of the tubular fluid has fallen to about 100 mOsm/liter. Overall, tubular fluid becomes progressively more concentrated as it flows along the descending limb and progressively more dilute as it moves along the ascending limb.

Note in Figure 26.20b that the vasa recta also consists of descending and ascending portions that are parallel to each other and to the loop of Henle. Just as tubular fluid flows in opposite directions in the loop of Henle, blood flows in opposite directions in the ascending and descending portions of the vasa recta. Blood entering the vasa recta has an osmolarity of about 300 mOsm/liter. As it flows along the descending portion into the renal medulla, where the interstitial fluid becomes increasingly concentrated, Na^+, Cl^-, and urea diffuse from interstitial fluid into the blood. But after its osmolarity increases, the blood flows into the ascending portion of the vasa recta; here blood flows through a region where the interstitial fluid becomes increasingly less concentrated. As a result, ions and urea diffuse from the blood into

interstitial fluid, and reabsorbed water diffuses from interstitial fluid into the vasa recta. The osmolarity of blood leaving the vasa recta is only slightly higher than the osmolarity of blood entering the vasa recta. Thus, the vasa recta provides oxygen and nutrients to the renal medulla without diminishing the osmotic gradient.

Figure 26.21 summarizes the processes of filtration, reabsorption, and secretion in each segment of the nephron and collecting duct.

CLINICAL APPLICATION
Diuretics

Diuretics are substances that slow renal reabsorption of water and thereby cause *diuresis,* an elevated urine flow rate. Naturally occurring diuretics include *caffeine* in coffee, tea, and cola sodas, which inhibits Na^+ reabsorption, and *alcohol* in beer, wine, and mixed drinks, which inhibits secretion of ADH. Most diuretics act by interfering with a mechanism for reabsorption of filtered Na^+. For example, loop diuretics, such as furosamide (Lasix), selectively inhibit the $Na^+-K^+-2Cl^-$ symporters in the thick ascending limb of the loop of Henle (see Figure 26.15). The thiazide diuretics, such as chlorthiazide (Diuril), act in the distal convoluted tubule, where they promote loss of NaCl in the urine by inhibiting Na^+-Cl^- symporters.

Diuretic drugs that decrease Na^+ reabsorption in the proximal convoluted tubule or loop of Henle cause a greater than normal amount of fluid and solutes to reach more distal parts of the nephron. In response to increased delivery of Na^+, Cl^-, and water, the distal convoluted tubule increases its reabsorption of Na^+ and Cl^- and its secretion of K^+. Although the elevated reabsorption fails to completely make up for the previous diminished reabsorption in the tubule, the enhanced secretion of K^+ often causes excessive loss of K^+ in the urine. So-called potassium-sparing diuretics can counteract this effect; they act on principal cells in the collecting duct either to inhibit the action of aldosterone—for example, spironolactone (Aldactone)—or to block the apical membrane Na^+ leakage channels (for example, amiloride). By slowing reabsorption of Na^+, other solutes, and water while also inhibiting secretion of K^+ in principal cells, these drugs promote mild diuresis without excessive urinary loss of K^+. ■

1. How do symporters in the ascending limb of the LOH and principal cells in the collecting duct contribute to the formation of concentrated urine?
2. Explain how ADH regulates facultative water reabsorption.
3. What is the countercurrent mechanism? Why is it important?
4. What is diuresis? Describe the mechanisms of action of furosamide, thiazide, and potassium-sparing diuretics.

Figure 26.21 Summary of filtration, reabsorption, and secretion in the nephron and collecting duct.

Filtration occurs in the renal corpuscle; reabsorption occurs all along the renal tubule and collecting ducts.

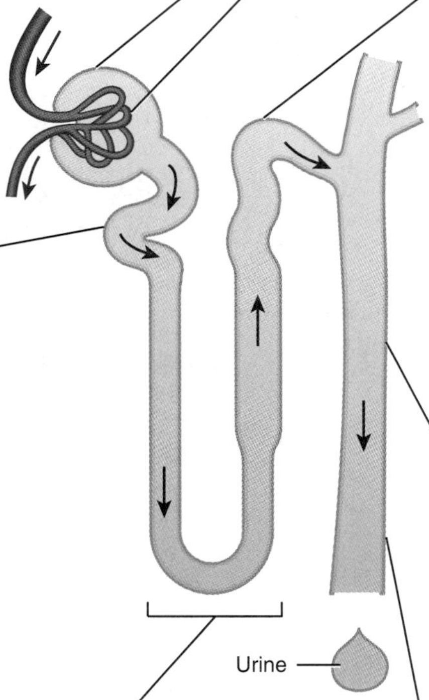

PROXIMAL CONVOLUTED TUBULE

Reabsorption (into blood) of filtered:

Water	65% (osmosis)
Na^+	65% (sodium pumps, symporters, antiporters)
K^+	65% (diffusion)
Glucose	100% (symporters and facilitated diffusion)
Amino acids	100% (symporters and facilitated diffusion)
Cl^-	50% (diffusion)
HCO_3^-	80–90% (facilitated diffusion)
Urea	50% (diffusion)
Ca^{2+}, Mg^{2+}	variable (diffusion)

Secretion (into urine) of:

H^+	variable (antiporters)
NH_4^+	variable, increases in acidosis (antiporters)
Urea	variable (diffusion)
Creatinine	small amount

At end of PCT, tubular fluid is still isotonic to blood (300 mOsm/liter).

LOOP OF HENLE

Reabsorption (into blood) of:

Water	15% (osmosis in descending limb)
Na^+	20–30% (symporters in ascending limb)
K^+	20–30% (symporters in ascending limb)
Cl^-	35% (symporters in ascending limb)
HCO_3^-	10–20% (facilitated diffusion)
Ca^{2+}, Mg^{2+}	variable (diffusion)

Secretion (into urine) of:

Urea	variable (recycling from collecting duct)

At end of loop of Henle, tubular fluid is hypotonic (100–150 mOsm/liter).

RENAL CORPUSCLE

Glomerular filtration rate: 105–125 ml/min of fluid that is isotonic to blood

Filtered substances: water and all solutes present in blood (except proteins) including ions, glucose, amino acids, creatinine, uric acid

DISTAL CONVOLUTED TUBULE

Reabsorption (into blood) of:

Water	10–15% (osmosis)
Na^+	5% (symporters)
Cl^-	5% (symporters)
Ca^{2+}	variable (stimulated by parathyroid hormone)

PRINCIPAL CELLS IN LATE DISTAL TUBULE AND COLLECTING DUCT

Reabsorption (into blood) of:

Water	5–9% (insertion of water channels stimulated by ADH)
Na^+	1–4% (sodium pumps)
Urea	variable (recycling to loop of Henle)

Secretion (into urine) of:

K^+	variable amount to adjust for dietary intake (leakage channels)

Tubular fluid leaving the collecting duct is dilute when ADH level is low and concentrated when ADH level is high.

INTERCALATED CELLS IN LATE DISTAL TUBULE AND COLLECTING DUCT

Reabsorption (into blood) of:

HCO_3^- (new)	varible amount, depends on H^+ secretion (antiporters)
Urea	variable (recycling to loop of Henle)

Secretion (into urine) of:

H^+	variable amounts to maintain acid-base homeostasis (H^+ pumps)

Urine

Q In which parts of the nephron and collecting duct does secretion occur?

EVALUATION OF KIDNEY FUNCTION

OBJECTIVES

- *Define urinalysis and describe its importance.*
- *Define renal plasma clearance and describe its importance.*

Routine assessment of kidney function involves evaluating both the quantity and quality of urine and the levels of wastes in the blood.

Urinalysis

An analysis of the volume and physical, chemical, and microscopic properties of urine, called a **urinalysis,** reveals much about the state of the body. The principal characteristics of normal urine are summarized in Table 26.5. The volume of urine eliminated per day in a normal adult is 1–2 liters (about 1–2 qt). Urine volume is influenced by fluid intake, blood pressure, blood osmolarity, diet, body temperature, diuretics, mental state, and general health. Low blood pressure triggers the renin–angiotensin–aldosterone pathway, which increases reabsorption of water and salts in the renal tubules and decreases urine volume. By contrast, when blood osmolarity decreases—for example, after drinking a large volume of water—secretion of ADH is inhibited and a larger volume of urine is excreted.

Water accounts for about 95% of the total volume of urine; the remaining 5% consists of electrolytes, solutes derived from cellular metabolism, and exogenous substances such as drugs. Normal urine is virtually protein free. Typical solutes normally present in urine include filtered and secreted electrolytes that are not reabsorbed, urea (from breakdown of proteins), creatinine (from breakdown of creatine phosphate in muscle fibers), uric acid (from breakdown of nucleic acids), urobilinogen (from breakdown of hemoglobin), and small quantities of other substances, such as fatty acids, pigments, enzymes, and hormones.

If disease alters body metabolism or kidney function, traces of substances not normally present may appear in the urine, or normal constituents may appear in abnormal amounts. Table 26.6 lists several abnormal constituents in urine that may be detected as part of a urinalysis.

Blood Tests

Two blood screening tests can provide information about kidney function. One is the **blood urea nitrogen (BUN)** test, which measures the blood nitrogen that is part of the urea resulting from catabolism and deamination of amino acids. When glomerular filtration rate decreases severely, as may occur with renal disease or obstruction of the urinary tract, BUN rises steeply. One strategy in treating such patients is to minimize their protein intake, thereby reducing the rate of urea production.

Table 26.5 Characteristics of Normal Urine

CHARACTERISTIC	DESCRIPTION
Volume	One to two liters in 24 hours but varies considerably.
Color	Yellow or amber but varies with urine concentration and diet. Color is due to urochrome (pigment produced from breakdown of bile). Concentrated urine is darker in color. Diet (reddish colored urine from beets), medications, and certain diseases affect color. Kidney stones may produce blood in urine.
Turbidity	Transparent when freshly voided but becomes turbid (cloudy) upon standing.
Odor	Mildly aromatic but becomes ammonia-like upon standing. Some people inherit the ability to form methylmercaptan from digested asparagus that gives urine a characteristic odor. Urine of diabetics has a fruity odor due to presence of ketone bodies.
pH	Ranges between 4.6 and 8.0; average 6.0; varies considerably with diet. High-protein diets increase acidity; vegetarian diets increase alkalinity.
Specific gravity	Specific gravity (density) is the ratio of the weight of a volume of a substance to the weight of an equal volume of distilled water. In urine, it ranges from 1.001 to 1.035. The higher the concentration of solutes, the higher the specific gravity.

Another test often used to evaluate kidney function is measurement of **plasma creatinine,** which results from catabolism of creatine phosphate in skeletal muscle. Normally, the blood creatinine level remains steady because the rate of creatinine excretion in the urine equals its discharge from muscle. A creatinine level above 1.5 mg/dL (about 135 mmol/liter) usually is an indication of poor renal function. Normal values for selected blood tests are listed in Appendix C.

Renal Plasma Clearance

Even more useful than BUN and blood creatinine values in the diagnosis of kidney problems is a measure of how effectively the kidneys are removing a given substance from blood plasma. **Renal plasma clearance** is the volume of blood that is "cleaned" or cleared of a substance per unit of time, usually expressed in units of *milliliters per minute.* High renal plasma clearance indicates efficient excretion of a substance in the urine; low clearance indicates inefficient excretion. For example, the clearance of glucose is normally zero because it is not excreted at all; instead, 100% of the filtered glucose is returned to the blood via tubular reabsorption. In drug therapy, knowing a drug's clearance is necessary for

Table 26.6 Summary of Abnormal Constituents in Urine

ABNORMAL CONSTITUENT	COMMENTS
Albumin	A normal constituent of plasma, but it usually appears in only very small amounts in urine because it is too large to pass through capillary fenestrations. The presence of excessive albumin in the urine—**albuminuria** (al'-byoo-mi-NOO-rē-a)—indicates an increase in the permeability of filtration membranes due to injury or disease, increased blood pressure, or irritation of kidney cells by substances such as bacterial toxins, ether, or heavy metals.
Glucose	The presence of glucose in the urine is called **glucosuria** (gloo-kō-SOO-rē-a) and usually indicates diabetes mellitus. Occasionally it may be caused by stress, which can cause excessive amounts of epinephrine to be secreted. Epinephrine stimulates the breakdown of glycogen and liberation of glucose from the liver.
Red blood cells (erythrocytes)	The presence of red blood cells in the urine is called **hematuria** (hēm-a-TOO-rē-a) and generally indicates a pathological condition. One cause is acute inflammation of the urinary organs as a result of disease or irritation from kidney stones. Other causes include tumors, trauma, and kidney disease. One should make sure the urine sample was not contaminated with menstrual blood from the vagina.
White blood cells (leukocytes)	The presence of white blood cells and other components of pus in the urine, referred to as **pyuria** (pī-YOO-rē-a), indicates infection in the kidney or other urinary organs.
Ketone bodies	High levels of ketone bodies in the urine, called **ketonuria** (kē-tō-NOO-rē-a), may indicate diabetes mellitus, anorexia, starvation, or simply too little carbohydrate in the diet.
Bilirubin	When red blood cells are destroyed by macrophages, the globin portion of hemoglobin is split off and the heme is converted to biliverdin. Most of the biliverdin is converted to bilirubin, which gives bile its major pigmentation. An above-normal level of bilirubin in urine is called **bilirubinuria** (bil'-ē-roo-bi-NOO-rē-a).
Urobilinogen	The presence of urobilinogen (breakdown product of hemoglobin) in urine is called **urobilinogenuria** (yoo'-rō-bi-lin'-ō-je-NOO-rē-a). Traces are normal, but elevated urobilinogen may be due to hemolytic or pernicious anemia, infectious hepatitis, biliary obstruction, jaundice, cirrhosis, congestive heart failure, or infectious mononucleosis.
Casts	**Casts** are tiny masses of material that have hardened and assumed the shape of the lumen of the tubule in which they formed. They are then flushed out of the tubule when filtrate builds up behind them. Casts are named after the cells or substances that compose them or on the basis of their appearance. For example, there are white blood cell casts, red blood cell casts, and epithelial cell casts that contain cells from the walls of the tubules.
Microbes	The number and type of bacteria vary with specific infections in the urinary tract. One of the most common is *E. coli*. The most common fungus to appear in urine is *Candida albicans*, a cause of vaginitis. The most frequent protozoan seen is *Trichomonas vaginalis*, a cause of vaginitis in females and urethritis in males.

determining the correct dosage. If clearance is high (one example is penicillin), then the dosage must also be high, and the drug must be given several times a day to maintain an adequate therapeutic level in the blood.

The following equation is used to calculate clearance:

$$\text{Renal plasma clearance of substance } S = \frac{U \times V}{P}$$

where U and P are the concentrations of the substance in urine and plasma, respectively (both expressed in the same units, such as mg/mL); V is the urine flow rate in mL/min.

The clearance of a solute depends on the three basic processes of a nephron: glomerular filtration, tubular reabsorption, and tubular secretion. If a substance is filtered but neither reabsorbed nor secreted, then its clearance equals the glomerular filtration rate, because all the molecules that pass the filtration membrane appear in the urine. This is very nearly the situation for creatinine; it easily passes the filter, it is not reabsorbed, and it is secreted only to a very small extent. Measuring the creatinine clearance, which normally is

120–140 mL/min, is the easiest way to assess glomerular filtration rate. The waste product urea is filtered, reabsorbed, and secreted to varying extents. Its clearance typically is less than the GFR, about 70 mL/min.

CLINICAL APPLICATION
Dialysis

If a person's kidneys are so impaired by disease or injury that they are unable to function adequately, then blood must be cleansed artificially by **dialysis,** the separation of large solutes from smaller ones through use of a selectively permeable membrane. One method of dialysis is the artificial kidney machine, which performs **hemodialysis** (*hemo-* = blood) because it directly filters the patient's blood. As blood flows through tubing made of selectively permeable dialysis membrane, waste products diffuse from the blood into the dialysis solution surrounding the membrane. The dialysis solution is continuously replaced to maintain favorable concentration gradients for diffusion of solutes into and out of the blood. After passing through the dialysis tubing, the cleansed blood flows back into the body.

Continuous ambulatory peritoneal dialysis (CAPD) uses the peritoneal lining of the peritoneal cavity as the dialysis membrane. The tip of a catheter is surgically placed in the patient's peritoneal cavity and connected to a sterile dialysis solution. The dialysate flows into the peritoneal cavity from its plastic container by gravity. The solution remains in the cavity until metabolic waste products, excess electrolytes, and extracellular fluid diffuse into the dialysis solution. The solution is then drained from the cavity by gravity into a sterile bag that is discarded. ■

1. Describe the following characteristics of normal urine: color, turbidity, odor, pH, and specific gravity.
2. Describe the chemical composition of normal urine.
3. How may kidney function be evaluated?
4. Explain why the renal plasma clearances of glucose, urea, and creatinine are different. How does each clearance compare to glomerular filtration rate?

URINE TRANSPORTATION, STORAGE, AND ELIMINATION

OBJECTIVE

• *Describe the anatomy, histology, and physiology of the ureters, urinary bladder, and urethra.*

Urine drains through papillary ducts into the minor calyces, which join to become major calyces that unite to form the renal pelvis (see Figure 26.3). From the renal pelvis, urine first drains into the ureters and then into the urinary bladder; urine is then discharged from the body through the single urethra (see Figure 26.1).

Ureters

Each of the two **ureters** (YOO-re-ters) transports urine from the renal pelvis of one kidney to the urinary bladder. Peristaltic contractions of the muscular walls of the ureters push urine toward the urinary bladder, but hydrostatic pressure and gravity also contribute. Peristaltic waves that pass from the renal pelvis to the urinary bladder vary in frequency from one to five per minute, depending on how fast urine is being formed.

The ureters are 25–30 cm (10–12 in.) long and are thick-walled, narrow tubes that vary in diameter from 1 mm to 10 mm along their course between the renal pelvis and the urinary bladder. Like the kidneys, the ureters are retroperitoneal. At the base of the urinary bladder the ureters curve medially and pass obliquely through the wall of the posterior aspect of the urinary bladder (Figure 26.22).

Even though there is no anatomical valve at the opening of each ureter into the urinary bladder, there is a physiological one that is quite effective. As the urinary bladder fills with urine, pressure within it compresses the oblique openings into the ureters and prevents the backflow of urine. When this physiological valve is not operating properly, it is possible for microbes to travel up the ureters from the urinary bladder to infect one or both kidneys.

Three coats of tissue form the wall of the ureters. The deepest coat, or **mucosa**, is a mucous membrane with **transitional epithelium** (see Table 4.1E on page 112) and an underlying **lamina propria** of areolar connective tissue with considerable collagen, elastic fibers, and lymphatic tissue. Transitional epithelium is able to stretch—a marked advantage for any organ that must accommodate a variable volume of fluid. Mucus secreted by the mucosa prevents the cells from coming in contact with urine, the solute concentration and pH of which may differ drastically from the cytosol of cells that form the wall of the ureters. Throughout most of the length of the ureters, the intermediate coat, the **muscularis**, is composed of inner longitudinal and outer circular layers of smooth muscle fibers, an arrangement opposite that of the gastrointestinal tract which contains inner circular and outer longitudinal layers; the muscularis of the distal third of the ureters also contains an outer layer of longitudinal muscle fibers. Thus, the muscularis in the distal third of the ureter is inner longitudinal, middle circular, and outer longitudinal. Peristalsis is the major function of the muscularis. The superficial coat of the ureters is the **adventitia**, a layer of areolar connective tissue containing blood vessels, lymphatic vessels, and nerves that serve the muscularis and mucosa. The adventitia blends in with surrounding connective tissue and anchors the ureters in place.

Urinary Bladder

The **urinary bladder** is a hollow, distensible muscular organ situated in the pelvic cavity posterior to the pubic symphysis. In males, it is directly anterior to the rectum; in females it is anterior to the vagina and inferior to the uterus (see Figure 28.13b on p. 991). It is held in position by folds of the peritoneum. The shape of the urinary bladder depends on how much urine it contains. Empty, it is collapsed; when slightly distended it becomes spherical; as urine volume increases it becomes pear-shaped and rises into the abdominal cavity. Urinary bladder capacity averages 700–800 mL; it is smaller in females because the uterus occupies the space just superior to the bladder.

Anatomy and Histology of the Urinary Bladder

In the floor of the urinary bladder is a small triangular area called the **trigone** (TRĪ-gōn; = triangle). The two posterior corners of the trigone contain the two ureteral openings, whereas the opening into the urethra, the **internal urethral orifice**, lies in the anterior corner (see Figure 26.22). Because its mucosa is firmly bound to the muscularis, the trigone has a smooth appearance.

Three coats make up the wall of the urinary bladder. The deepest is the **mucosa**, a mucous membrane composed of **transitional epithelium** and an underlying **lamina propria** similar to that of the ureters. Rugae (the folds in the

Figure 26.22 Ureters, urinary bladder, and urethra (shown in a female). (See Tortora, *A Photographic Atlas of the Human Body,* Figures 13.8, 13.10)

🔑 **Urine is stored in the urinary bladder before being expelled by micturition.**

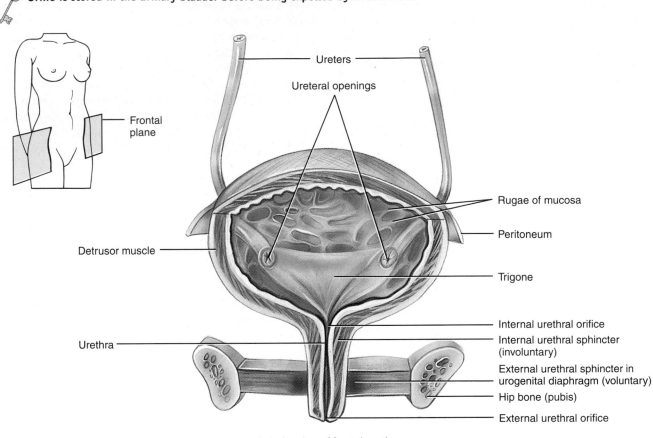

Frontal plane

Ureters

Ureteral openings

Rugae of mucosa

Peritoneum

Detrusor muscle

Trigone

Urethra

Internal urethral orifice

Internal urethral sphincter (involuntary)

External urethral sphincter in urogenital diaphragm (voluntary)

Hip bone (pubis)

External urethral orifice

Anterior view of frontal section

Q What is a lack of voluntary control over micturition called?

mucosa) are also present. Surrounding the mucosa is the intermediate **muscularis,** also called the **detrusor muscle** (de-TROO-ser; = to push down), which consists of three layers of smooth muscle fibers: the inner longitudinal, middle circular, and outer longitudinal layers. Around the opening to the urethra the circular fibers form an **internal urethral sphincter;** inferior to it is the **external urethral sphincter,** which is composed of skeletal muscle and is a modification of the urogenital diaphragm muscle (see Figure 11.12 on page 331). The most superficial coat of the urinary bladder on the posterior and inferior surfaces is the **adventitia,** a layer of areolar connective tissue that is continuous with that of the ureters. Over the superior surface of the urinary bladder is the **serosa,** a layer of visceral peritoneum.

The Micturition Reflex

Discharge of urine from the urinary bladder, called **micturition** (mik'-too-RISH-un; *mictur-* = urinate), is also known as *urination* or *voiding.* Micturition occurs via a combination of involuntary and voluntary muscle contractions.

When the volume of urine in the bladder exceeds 200–400 mL, pressure within the bladder increases considerably, and stretch receptors in its wall transmit nerve impulses into the spinal cord. These impulses propagate to the **micturition center** in sacral spinal cord segments S2 and S3 and trigger a spinal reflex called the **micturition reflex.** In this reflex arc, parasympathetic impulses from the micturition center propagate to the urinary bladder wall and internal urethral sphincter. The nerve impulses cause *contraction* of the detrusor muscle and *relaxation* of the internal urethral sphincter muscle. Simultaneously, the micturition center inhibits somatic motor neurons that innervate skeletal muscle in the external urethral sphincter. Upon contraction of the bladder wall and relaxation of the sphincters, urination takes place. Bladder filling causes a sensation of fullness that initiates a conscious desire to urinate before the micturition reflex actually occurs. Although emptying of the urinary bladder is a reflex, in early childhood we learn to initiate it and stop it voluntarily. Through learned control of the external urethral sphincter muscle and certain muscles of the pelvic floor, the cerebral cortex can initiate micturition or delay its occurrence for a limited period of time.

Urethra

The **urethra** is a small tube leading from the internal urethral orifice in the floor of the urinary bladder to the exterior of the body (see Figure 26.1). In both males and females, the urethra is the terminal portion of the urinary system and the passageway for discharging urine from the body; in males it discharges semen as well.

In females, the urethra lies directly posterior to the pubic symphysis, is directed obliquely inferiorly and anteriorly, and has a length of 4 cm (1.5 in.) (see Figure 28.13b on page 991). The opening of the urethra to the exterior, the **external urethral orifice,** is located between the clitoris and the vaginal opening (see Figure 28.22 on page 999). The wall of the female urethra consists of a deep **mucosa** and a superficial **muscularis.** The mucosa is a mucous membrane composed of **epithelium** and **lamina propria** (areolar connective tissue with elastic fibers and a plexus of veins). The muscularis consists of circularly arranged smooth muscle fibers and is continuous with that of the urinary bladder. Near the urinary bladder, the mucosa contains transitional epithelium that is continuous with that of the urinary bladder; near the external urethral orifice the epithelium is nonkeratinized stratified squamous epithelium. Between these areas, the mucosa contains stratified columnar or pseudostratified columnar epithelium.

In males, the urethra also extends from the internal urethral orifice to the exterior, but its length and passage through the body are considerably different than in females (see Figure 28.3 on page 977 and Figure 28.12 on page 988). The male urethra first passes through the prostate gland, then through the urogenital diaphragm, and finally through the penis, a distance of 15–20 cm (6–8 in.).

The male urethra, which also consists of a deep **mucosa** and a superficial **muscularis,** is subdivided into three anatomical regions: (1) The **prostatic urethra** passes through the prostate gland; (2) the **membranous urethra,** the shortest portion, passes through the urogenital diaphragm; and (3) the **spongy urethra,** the longest portion, passes through the penis. The epithelium of the prostatic urethra is continuous with that of the urinary bladder and consists of transitional epithelium that becomes stratified columnar or pseudostratified columnar epithelium more distally. The mucosa of the membranous urethra contains stratified columnar or pseudostratified columnar epithelium. The epithelium of the spongy urethra is stratified columnar or pseudostratified columnar epithelium, except near the external urethral orifice, which is nonkeratinized stratified squamous epithelium. The **lamina propria** of the male urethra is areolar connective tissue with elastic fibers and a plexus of veins.

The muscularis of the prostatic urethra is composed of wisps of mostly circular smooth muscle fibers superficial to the lamina propria; these circular fibers help form the internal urethral sphincter of the urinary bladder. The muscularis of the membranous urethra consists of circularly arranged skeletal muscle fibers of the urogenital diaphragm that help form the external urethral sphincter of the urinary bladder.

CLINICAL APPLICATION
Urinary Incontinence

A lack of voluntary control over micturition is called **urinary incontinence.** In infants and children under 2–3 years old, incontinence is normal because neurons to the external urethral sphincter muscle are not completely developed, and voiding occurs whenever the urinary bladder is sufficiently distended to stimulate the micturition reflex. Urinary incontinence also occurs in adults. Of the more than 10 million U.S. adults who suffer from urinary incontinence, 1–2 million have **stress incontinence.** In this condition, physical stresses that increase abdominal pressure, such as coughing, sneezing, laughing, exercising, pregnancy, or simply walking cause leakage of urine from the bladder. Other causes of incontinence in adults are injury to the nerves controlling the urinary bladder, loss of bladder flexibility with age, disease or irritation of the bladder or urethra, damage to the external urethral sphincter, or certain drugs. And, compared with nonsmokers, those who smoke have twice the risk of developing incontinence. ■

1. What forces help propel urine from the renal pelvis to the bladder?
2. What is micturition? Describe the micturition reflex.
3. Compare the location, length, and histology of the urethra in males and females.

WASTE MANAGEMENT IN OTHER BODY SYSTEMS

OBJECTIVE

• *Describe the ways that body wastes are handled.*

As we have seen, just one of the many functions of the urinary system is to help rid the body of some kinds of waste materials. Besides the kidneys, several other tissues, organs, and processes contribute to the temporary confinement of wastes, the transport of waste materials for disposal, the recycling of materials, and the excretion of excess or toxic substances in the body. These waste management systems include the following:

• *Body buffers.* Buffers bind excess hydrogen ions (H^+), thereby preventing an increase in the acidity of body fluids. Buffers are like wastebaskets in that they have a limited capacity; eventually the H^+, like the paper in a wastebasket, must be eliminated from the body by excretion.
• *Blood.* The bloodstream provides pick-up and delivery services for the transport of wastes, in much the same way that garbage trucks and sewer lines serve a community.
• *Liver.* The liver is the primary site for metabolic recycling,

Figure 26.23 Development of the urinary system.

🔑 **Three pairs of kidneys form within intermediate mesoderm in successive time periods: pronephros, mesonephros, and metanephros.**

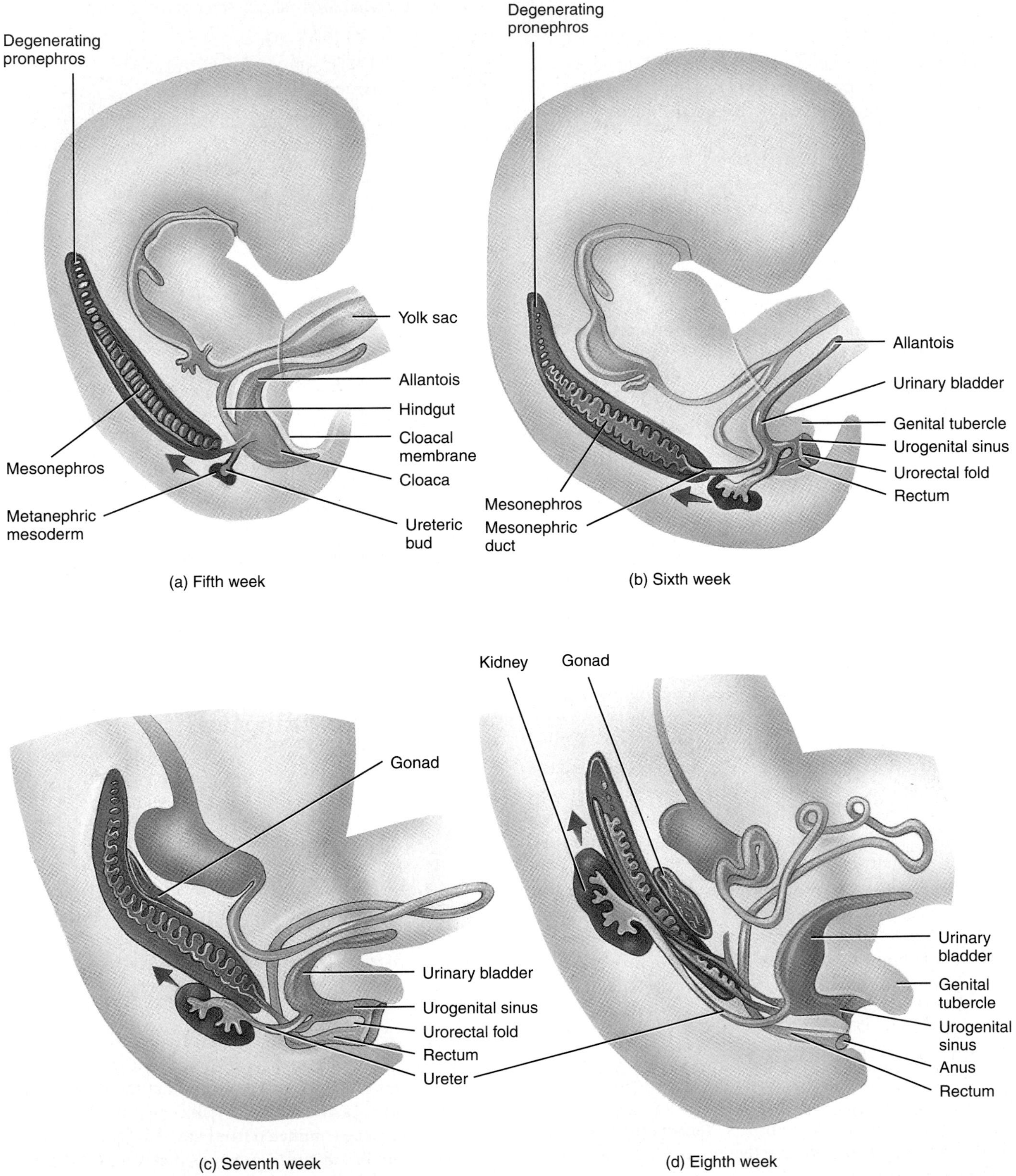

Degenerating pronephros

Yolk sac
Allantois
Hindgut
Cloacal membrane
Cloaca

Mesonephros

Metanephric mesoderm

Ureteric bud

(a) Fifth week

Degenerating pronephros

Allantois
Urinary bladder
Genital tubercle
Urogenital sinus
Urorectal fold
Rectum

Mesonephros
Mesonephric duct

(b) Sixth week

Gonad

Urinary bladder
Urogenital sinus
Urorectal fold
Rectum
Ureter

(c) Seventh week

Kidney   Gonad

Urinary bladder
Genital tubercle
Urogenital sinus
Anus
Rectum

(d) Eighth week

Q When do the kidneys begin to develop?

as occurs, for example, in the conversion of amino acids into glucose or of glucose into fatty acids. The liver also converts toxic substances into less toxic ones, such as ammonia into urea. These functions of the liver are described in Chapters 24 and 25.

- *Lungs.* With each exhalation, the lungs excrete CO_2, and expel heat and a little water vapor.
- *Sweat (sudoriferous) glands.* Especially during exercise, sweat glands in the skin help eliminate excess heat, water, and CO_2, plus small quantities of salts and urea as well.
- *Gastrointestinal tract.* Through defecation, the gastrointestinal tract eliminates solid, undigested foods; wastes; some CO_2; water; salts; and heat.

DEVELOPMENTAL ANATOMY OF THE URINARY SYSTEM

OBJECTIVE
• *Describe the development of the urinary system.*

Starting in the third week of fetal development, a portion of the mesoderm along the posterior aspect of the embryo, the **intermediate mesoderm,** differentiates into the kidneys. Three pairs of kidneys form within the intermediate mesoderm in successive time periods: pronephros, mesonephros, and metanephros (Figure 26.23). Only the last pair remains as the functional kidneys of the newborn.

The first kidney to form, the **pronephros,** is the most superior of the three and has an associated **pronephric duct.** This duct empties into the **cloaca,** which functions as a common outlet of the urinary, digestive, and reproductive ducts. The pronephros begins to degenerate during the fourth week and is completely gone by the sixth week; the pronephric ducts, however, remain.

The second kidney, the **mesonephros,** replaces the pronephros. The retained portion of the pronephric duct, which connects to the mesonephros, develops into the **mesonephric duct.** The mesonephros begins to degenerate by the sixth week and is almost gone by the eighth week.

At about the fifth week, a mesodermal outgrowth, called a **ureteric bud,** develops from the distal end of the mesonephric duct near the cloaca. The **metanephros,** or ultimate kidney, develops from the ureteric bud and metanephric mesoderm. The ureteric bud forms the *collecting ducts, calyces, renal pelvis,* and *ureter.* The **metanephric mesoderm** forms the nephrons of the kidneys. By the third month, the fetal kidneys begin excreting urine into the surrounding amniotic fluid; indeed, fetal urine makes up most of the amniotic fluid.

During development, the cloaca divides into a **urogenital sinus,** into which urinary and genital ducts empty, and a *rectum* that discharges into the anal canal. The *urinary bladder* develops from the urogenital sinus. In females, the *urethra* develops as a result of lengthening of the short duct that extends from the urinary bladder to the urogenital sinus. The *vestibule,* into which the urinary and genital ducts empty, is also derived from the urogenital sinus. In males, the urethra is considerably longer and more complicated, but it is also derived from the urogenital sinus.

AGING AND THE URINARY SYSTEM

OBJECTIVE
• *Describe the effects of aging on the urinary system.*

With aging, the kidneys shrink in size, have lowered blood flow, and filter less blood. The mass of the two kidneys decreases from an average of 260 g in 20 year olds to less than 200 g by age 80. Likewise, renal blood flow and filtration rate decline by 50% between ages 40 and 70. Kidney diseases that become more common with age include acute and chronic kidney inflammations and renal calculi. Because the sensation of thirst diminishes with age, older individuals are susceptible to dehydration. Urinary tract infections are more common among the elderly, as are polyuria (excessive urine production), nocturia (excessive urination at night), increased frequency of urination, dysuria (painful urination), urinary retention or incontinence, and hematuria (blood in the urine). Cancer of the prostate is the most common malignancy in elderly males (see page 1014).

DISORDERS: HOMEOSTATIC IMBALANCES

RENAL CALCULI

The crystals of salts present in urine occasionally precipitate and solidify into insoluble stones called **renal calculi** (*calculi* = pebbles) or **kidney stones.** They commonly contain crystals of calcium oxalate, uric acid, or calcium phosphate. Conditions leading to calculus formation include the ingestion of excessive calcium, scanty water intake, abnormally alkaline or acidic urine, and overactivity of the parathyroid glands. When a stone lodges in a narrow passage, such as a ureter, the pain can be excruciating. **Shock wave lithotripsy** (LITH-ō-trip′-sē; *litho-* = stone) offers an alternative to surgical removal of kidney stones. A device, called a lithotripter,

delivers brief, high-intensity sound waves through a water-filled cushion. Over a period of 30–60 minutes, 1000 or more hydraulic shock waves pulverize the stone until the fragments are small enough to wash out in the urine.

URINARY TRACT INFECTIONS

The term **urinary tract infection (UTI)** is used to describe either an infection of a part of the urinary system or the presence of large numbers of microbes in urine. UTIs are more common in females due to their shorter urethra. Symptoms include painful or burning urination, urgent and frequent urination, low back pain, and bed-wetting. UTIs include *urethritis* (inflammation of the urethra), *cystitis* (inflammation of the urinary bladder), and *pyelonephritis* (inflammation of the kidneys). If pyelonephritis becomes chronic, scar tissue forms in the kidneys and severely impairs their function.

GLOMERULAR DISEASES

A variety of conditions may damage the kidney glomeruli, either directly or indirectly as a consequence of disease elsewhere in the body. Typically, the filtration membrane sustains damage, and its permeability increases.

Glomerulonephritis is an inflammation of the kidney that involves the glomeruli. One of the most common causes is an allergic reaction to the toxins produced by streptococcal bacteria that have recently infected another part of the body, especially the throat. The glomeruli become so inflamed, swollen, and engorged with blood that the filtration membranes allow blood cells and plasma proteins to enter the filtrate. As a result, the urine contains many erythrocytes (hematuria) and much protein. The glomeruli may be permanently damaged, leading to chronic renal failure.

Nephrotic syndrome is a condition characterized by *proteinuria* (protein in the urine) and *hyperlipidemia* (high blood levels of cholesterol, phospholipids, and triglycerides). The proteinuria is due to an increased permeability of the filtration membrane, which permits proteins, especially albumin, to escape from blood into urine. Loss of albumin results in *hypoalbuminemia* (low blood albumin level) once liver production of albumin fails to meet increased urinary losses. Edema, usually seen around the eyes, ankles, feet, and abdomen, occurs in nephrotic syndrome because loss of albumin from the blood decreases blood colloid osmotic pressure. Nephrotic syndrome is associated with several glomerular diseases of unknown cause, as well as with systemic disorders such as diabetes mellitus, systemic lupus erythematosus (SLE), a variety of cancers, and AIDS.

RENAL FAILURE

Renal failure is a decrease or cessation of glomerular filtration. In **acute renal failure (ARF)** the kidneys abruptly stop working entirely (or almost entirely). The main feature of ARF is the suppression of urine flow, usually characterized either by *oliguria* (*olig-* = scanty; *-uria* = urine production), which is daily urine output less than 250 mL, or by *anuria*, daily urine output less than 50 mL. Causes include low blood volume (for example, due to hemorrhage), decreased cardiac output, damaged renal tubules, kidney stones, the dyes used to visualize blood vessels in angiograms, nonsteroidal anti-inflammatory drugs, and some antibiotic drugs. Renal failure causes edema due to salt and water retention; acidosis due to inability of the kidneys to excrete acidic substances; increased levels of urea due to impaired renal excretion of metabolic waste products; elevated potassium levels that can lead to cardiac arrest; anemia, because the kidneys no longer produce enough erythropoietin for adequate red blood cell production; and osteomalacia, because the kidneys are no longer able to convert vitamin D to calcitriol, which is needed for adequate calcium absorption from the small intestine.

Chronic renal failure (CRF) refers to a progressive and usually irreversible decline in glomerular filtration rate (GFR). CRF may result from chronic glomerulonephritis, pyelonephritis, polycystic kidney disease, or traumatic loss of kidney tissue. CRF develops in three stages. In the first stage, *diminished renal reserve*, nephrons are destroyed until about 75% of the functioning nephrons are lost. At this stage, a person may have no signs or symptoms because the remaining nephrons enlarge and take over the function of those that have been lost. Once 75% of the nephrons are lost, the person enters the second stage, called *renal insufficiency*, characterized by a decrease in GFR and increased blood levels of nitrogen-containing wastes and creatinine. Also, the kidneys cannot effectively concentrate or dilute the urine. The final stage, called *end-stage renal failure*, occurs when about 90% of the nephrons have been lost. At this stage, GFR diminishes to 10–15% of normal, oliguria is present, and blood levels of nitrogen-containing wastes and creatinine increase further. People with end-stage renal failure need dialysis therapy and are possible candidates for a kidney transplant operation.

POLYCYSTIC KIDNEY DISEASE

Polycystic kidney disease (PKD) is one of the most common inherited disorders. In infants it results in death at birth or shortly thereafter; in adults it accounts for 6–12% of kidney transplants. In PKD, the kidney tubules become riddled with hundreds or thousands of cysts (fluid-filled cavities). In addition, inappropriate apoptosis (programmed cell death) of cells in noncystic tubules leads to progressive impairment of renal function and eventually to end-stage renal failure.

People with PKD also may have cysts and apoptosis in the liver, pancreas, spleen, and gonads; increased risk of cerebral aneurysms; heart valve defects; and diverticuli in the colon. Typically, symptoms are not noticed until adulthood, when patients may have back pain, urinary tract infections, blood in the urine, hypertension, and large abdominal masses. Progression to renal failure may be slowed by using drugs to restore normal blood pressure, restricting protein and salt in the diet, and controlling urinary tract infections.

MEDICAL TERMINOLOGY

Azotemia (az-ō-TĒ-mē-a; *azot-* = nitrogen; *-emia* = condition of blood) Presence of urea or other nitrogen-containing substances in the blood.

Cystocele (SIS-tō-sēl; *cysto-* = bladder; *-cele* = hernia or rupture) Hernia of the urinary bladder.

Enuresis (en′-yoo-RĒ-sis; = to void urine) Involuntary voiding of urine after the age at which voluntary control has typically been attained.

Intravenous pyelogram (in′-tra-VĒ-nus PĪ-e-lō-gram′; *intra-* = within; *veno-* = vein; *pyelo-* = pelvis of kidney; *-gram* = record), or **IVP** Radiograph (x-ray) of the kidneys after venous injection of a dye.

Nocturnal enuresis (nok-TUR-nal en′-yoo-RĒ-sis) Discharge of urine during sleep, resulting in bed-wetting; occurs in about 15% of 5-year-old children and generally resolves spontaneously, afflicting only about 1% of adults. It may have a genetic basis, as bed-wetting occurs more often in identical twins than in fraternal twins and more often in children whose parents or siblings were bed-wetters. Possible causes include smaller-than-normal bladder capacity, failure to awaken in response to a full bladder, and above-normal production of urine at night. Also referred to as **nocturia.**

Polyuria (pol′-ē-YOO-rē-a; *poly-* = too much) Excessive urine formation.

Stricture (STRIK-chur) Narrowing of the lumen of a canal or hollow organ, as may occur in the ureter, urethra, or any other tubular structure in the body.

Urinary retention A failure to completely or normally void urine; may be due to an obstruction in the urethra or neck of the urinary bladder, to nervous contraction of the urethra, or to lack of urge to urinate. In men, an enlarged prostate may constrict the urethra and cause urinary retention.

STUDY OUTLINE

INTRODUCTION (p. 914)

1. The organs of the urinary system are the kidneys, ureters, urinary bladder, and urethra.
2. After the kidneys filter blood and return most water and many solutes to the bloodstream, the remaining water and solutes constitute urine.

OVERVIEW OF KIDNEY FUNCTIONS (p. 914)

1. The kidneys regulate blood ionic composition, blood osmolarity, blood volume, blood pressure, and blood pH.
2. The kidneys also perform gluconeogenesis, release calcitriol and erythropoietin, and excrete wastes and foreign substances.

ANATOMY AND HISTOLOGY OF THE KIDNEYS (p. 915)

1. The kidneys are retroperitoneal organs attached to the posterior abdominal wall.
2. Three layers of tissue surround the kidneys: renal capsule, adipose capsule, and renal fascia.
3. Internally, the kidneys consist of a renal cortex, a renal medulla, renal pyramids, renal papillae, renal columns, calyces, and a renal pelvis.
4. Blood flows into the kidney through the renal artery and successively into segmental, interlobar, arcuate, and interlobular arteries; afferent arterioles; glomerular capillaries; efferent arterioles; peritubular capillaries and vasa recta; and interlobular, arcuate, interlobar, and segmental veins before flowing out of the kidney through the renal vein.
5. Vasomotor nerves from the sympathetic division of the autonomic nervous system supply kidney blood vessels; they help regulate flow of blood through the kidney.
6. The nephron is the functional unit of the kidneys. A nephron consists of a renal corpuscle (glomerulus and glomerular or Bowman's capsule) and a renal tubule.
7. A renal tubule consists of a proximal convoluted tubule, a loop of Henle, and a distal convoluted tubule, which drains into a collecting duct (shared by several nephrons). The loop of Henle consists of a descending limb and an ascending limb.
8. A cortical nephron has a short loop that dips only into the superficial region of the renal medulla; a juxtamedullary nephron has a long loop of Henle that stretches through the renal medulla almost to the renal papilla.
9. The wall of the entire glomerular capsule, renal tubule, and ducts consists of a single layer of epithelial cells. The epithelium has distinctive histological features in different parts of

the tubule. Table 26.1 on page 923 summarizes the histological features of the renal tubule and collecting duct.
10. The juxtaglomerular apparatus (JGA) consists of the juxtaglomerular cells of an afferent arteriole and the macula densa of the final portion of the ascending limb of the loop of Henle.

OVERVIEW OF RENAL PHYSIOLOGY (p. 924)

1. Nephrons perform three basic tasks: glomerular filtration, tubular secretion, and tubular reabsorption.

GLOMERULAR FILTRATION (p. 925)

1. Fluid that enters the capsular space is glomerular filtrate.
2. The filtration (endothelial–capsular) membrane consists of the glomerular endothelium, basal lamina, and filtration slits between pedicels of podocytes.
3. Most substances in plasma easily pass through the glomerular filter. However, blood cells and most proteins normally are not filtered.
4. Glomerular filtrate amounts to up to 180 liters of fluid per day. This large amount of fluid is filtered because the filter is porous and thin, the glomerular capillaries are long, and the capillary blood pressure is high.
5. Glomerular blood hydrostatic pressure (GBHP) promotes filtration, whereas capsular hydrostatic pressure (CHP) and blood colloid osmotic pressure (BCOP) oppose filtration. Net filtration pressure (NFP) = GBHP − CHP − BCOP. NFP is about 10 mm Hg.
6. Glomerular filtration rate (GFR) is the amount of filtrate formed in both kidneys per minute; it is normally 105–125 mL/min.
7. Glomerular filtration rate depends on renal autoregulation, neural regulation, and hormonal regulation. Table 26.2 on page 929 summarizes regulation of GFR.

TUBULAR REABSORPTION AND TUBULAR SECRETION (p. 929)

1. Tubular reabsorption is a selective process that reclaims materials from tubular fluid and returns them to the bloodstream. Reabsorbed substances include water, glucose, amino acids, urea, and ions, such as sodium, chloride, potassium, bicarbonate, and phosphate (see Table 26.3 on page 930).
2. Some substances not needed by the body are removed from the blood and discharged into the urine via tubular secretion. Included are ions (K^+, H^+, and NH_4^+), urea, creatinine, and certain drugs.

3. Reabsorption routes include both paracellular (between tubule cells) and transcellular (across tubule cells) routes.

4. The maximum amount of a substance that can be reabsorbed per unit time is called the transport maximum (T_m).

5. About 90% of water reabsorption is obligatory; it occurs via osmosis, together with reabsorption of solutes. The remaining 10% is facultative water reabsorption, which varies according to body needs and is regulated by ADH.

6. Na^+ are reabsorbed throughout the basolateral membrane via primary active transport.

7. In the proximal convoluted tubule, sodium ions are reabsorbed through the apical membranes via Na^+–glucose symporters and Na^+/H^+ antiporters; water is reabsorbed via osmosis; Cl^-, K^+, Ca^{2+}, Mg^{2+}, and urea are reabsorbed via passive diffusion; and NH_3 and NH_4^+ are secreted.

8. The loop of Henle reabsorbs 20–30% of the filtered Na^+, K^+, Ca^{2+}, and HCO_3^-; 35% of the filtered Cl^-, and 15% of the filtered water.

9. The distal convoluted tubule reabsorbs sodium and chloride ions via Na^+–Cl^- symporters.

10. In the collecting duct, principal cells reabsorb Na^+ and secrete K^+. Some intercalated cells reabsorb K^+ and HCO_3^- and secrete H^+, whereas other intercalated cells secrete HCO_3^-.

11. Angiotensin II, aldosterone, antidiuretic hormone, and atrial natriuretic peptide regulate solute and water reabsorption, as summarized in Table 26.4 on page 938.

12. In the absence of ADH, the kidneys produce dilute urine; renal tubules absorb more solutes than water.

13. In the presence of ADH, the kidneys produce concentrated urine; large amounts of water are reabsorbed from the tubular fluid into interstitial fluid, increasing solute concentration of the urine.

14. The countercurrent mechanism establishes an osmotic gradient in the interstitial fluid of the renal medulla that enables production of concentrated urine when ADH is present.

EVALUATION OF KIDNEY FUNCTION (p. 943)

1. A urinalysis is an analysis of the volume and physical, chemical, and microscopic properties of a urine sample. Table 26.5 on page 943 summarizes the principal physical characteristics of urine.

2. Chemically, normal urine contains about 95% water and 5% solutes. The solutes normally include urea, creatinine, uric acid, urobilinogen, and various ions.

3. Table 26.6 on page 944 lists several abnormal components that can be detected in a urinalysis, including albumin, glucose, red and white blood cells, ketone bodies, bilirubin, excessive urobilinogen, casts, and microbes.

4. Renal clearance refers to the ability of the kidneys to clear (remove) a specific substance from blood.

URINE TRANSPORTATION, STORAGE, AND ELIMINATION (p. 945)

1. The ureters are retroperitoneal and consist of a mucosa, muscularis, and adventitia. They transport urine from the renal pelvis to the urinary bladder, primarily via peristalsis.

2. The urinary bladder is located in the pelvic cavity posterior to the pubic symphysis; its function is to store urine prior to micturition.

3. The urinary bladder consists of a mucosa with rugae, a muscularis (detrusor muscle), and an adventitia (serosa over the superior surface).

4. The micturition reflex discharges urine from the urinary bladder via parasympathetic impulses that cause contraction of the detrusor muscle and relaxation of the internal urethral sphincter muscle and via inhibition of impulses in somatic motor neurons to the external urethral sphincter.

5. The urethra is a tube leading from the floor of the urinary bladder to the exterior. Its anatomy and histology differ in females and males. In both sexes the urethra functions to discharge urine from the body; in males it discharges semen as well.

WASTE MANAGEMENT IN OTHER BODY SYSTEMS (p. 947)

1. Besides the kidneys, several other tissues, organs, and processes temporarily confine wastes, transport waste materials for disposal, recycle materials, and excrete excess or toxic substances.

2. Buffers bind excess H^+, the blood transports wastes, the liver converts toxic substances into less toxic ones, the lungs exhale CO_2, sweat glands help eliminate excess heat, and the gastrointestinal tract eliminates solid wastes.

DEVELOPMENTAL ANATOMY OF THE URINARY SYSTEM (p. 949)

1. The kidneys develop from intermediate mesoderm.

2. They develop in the following sequence: pronephros, mesonephros, metanephros. Only the metanephros remains and develops into a functional kidney.

AGING AND THE URINARY SYSTEM (p. 949)

1. With aging, the kidneys shrink in size, have lowered blood flow, and filter less blood.

2. Common problems related to aging include urinary tract infections, increased frequency of urination, urinary retention or incontinence, and renal calculi.

SELF-QUIZ QUESTIONS

1. Match the following:
   ___ (a) cells in the last portion of the distal convoluted tubule and in the collecting ducts; regulated by ADH and aldosterone
   ___ (b) a capillary network lying in Bowman's capsule and functioning in filtration
   ___ (c) the functional unit of the kidney
   ___ (d) drains into a collecting duct
   ___ (e) combined glomerulus and glomerular capsule; where plasma is filtered
   ___ (f) the visceral layer of the glomerular capsule consisting of modified simple squamous epithelial cells
   ___ (g) cells of the final portion of the ascending limb of the loop of Henle that make contact with the afferent arteriole
   ___ (h) site of obligatory water reabsorption
   ___ (i) can secrete H^+ against a concentration gradient
   ___ (j) modified smooth muscle cells in the wall of the afferent arteriole

   (1) podocytes
   (2) glomerulus
   (3) renal corpuscle
   (4) proximal convoluted tubule
   (5) distal convoluted tubule
   (6) juxtaglomerular cells
   (7) macula densa
   (8) principal cells
   (9) intercalated cells
   (10) nephron

2. Match the following:
   ___ (a) the primary site for metabolic recycling
   ___ (b) excrete excess water and solutes, H^+ and other ions, some heat, and CO_2
   ___ (c) excretes solid, undigested foods; wastes; some CO_2; water; salts; and heat
   ___ (d) picks up wastes from body cells
   ___ (e) excrete CO_2, heat, and water vapor
   ___ (f) bind excess hydrogen ions
   ___ (g) eliminate excess heat, water, CO_2, and small quantities of salts and urea

   (1) buffers
   (2) blood
   (3) liver
   (4) GI tract
   (5) lungs
   (6) sweat glands
   (7) kidneys

Choose the best answer to the following questions:

3. Which of the following are features of the renal corpuscle that enhance its filtering capacity? (1) large glomerular capillary surface area, (2) thick, selectively permeable filtration membrane, (3) low glomerular capillary pressure, (4) high glomerular capillary pressure, (5) thin and porous filtration. (a) 1, 2, and 3, (b) 2, 4, and 5, (c) 1, 4, and 5, (d) 2, 3, and 4, (e) 2, 3, and 5

4. The pressure that promotes glomerular filtration is (a) glomerular blood hydrostatic pressure, (b) capsular hydrostatic pressure, (c) blood colloid pressure, (d) capsular osmotic pressure, (e) combined capsular hydrostatic and osmotic pressures

5. Which of the following statements are correct? (1) Glomerular filtration rate (GFR) is directly related to the pressures that determine net filtration pressure. (2) Any change in net filtration rate affects GFR. (3) Mechanisms that regulate GFR work by adjusting blood flow into and out of the glomerulus and by altering the glomerular capillary surface area available for filtration. (4) GFR increases when blood flow into glomerular capillaries decreases. (5) Normally, GFR increases very little when systemic blood pressure rises.
   (a) 1, 2, and 3, (b) 2, 3, and 4, (c) 3, 4, and 5, (d) 1, 2, 3, and 5, (e) 2, 3, 4, and 5

6. Which of the following are mechanisms that control GFR? (1) renal autoregulation, (2) neural regulation, (3) hormonal regulation, (4) chemical regulation of ions, (5) presence or absence of a transporter.
   (a) 1, 2, and 3, (b) 2, 3, and 4, (c) 3, 4, and 5, (d) 1, 3, and 5, (e) 1, 3, and 4

7. Which of the following hormones affect Na^+, Cl^-, and water reabsorption and K^+ secretion by the renal tubules? (1) angiotensin II, (2) aldosterone, (3) ADH, (4) atrial natriuretic peptide, (5) thyroid hormone.
   (a) 1, 3, and 5, (b) 2, 3, and 4, (c) 2, 4, and 5, (d) 1, 2, 4, and 5, (e) 1, 2, 3, and 4

8. Which of the following is the correct sequence of tubular fluid flow? (a) proximal convoluted tubule, descending limb of loop of Henle, ascending limb of loop of Henle, collecting duct, proximal convoluted tubule, papillary duct, (b) proximal convoluted tubule, descending limb of loop of Henle, ascending limb of loop of Henle, distal convoluted tubule, collecting duct, papillary duct, (c) papillary duct, distal convoluted tubule, proximal convoluted tubule, collecting duct, ascending limb of loop of Henle, descending limb of loop of Henle, (d) collecting duct, proximal convoluted tubule, ascending limb of loop of Henle, descending limb of loop of Henle, papillary duct, distal convoluted tubule, (e) distal convoluted tubule, ascending limb of loop of Henle, descending limb of loop of Henle, proximal convoluted tubule, papillary duct, collecting duct.

True or false:

9. When dilute urine is being formed, the osmolarity of the fluid in the tubular lumen increases as it flows down the descending limb of the loop of Henle, decreases as it flows up the ascending limb, and continues to decrease as it flows through the rest of the nephron and collecting duct.

10. The route of blood flow from the renal artery to the renal vein is renal artery, segmental arteries, interlobar arteries, arcuate arteries, interlobular arteries, afferent arterioles, glomerulus, efferent arterioles, peritubular capillaries, vasa recta, peritubular venules, interlobular veins, arcuate veins, interlobar veins, segmental veins, and renal vein.

Complete the following:

11. The ___ cells of the ureters and urinary bladder allow distension for urine flow and storage.

12. Micturition is both a ___ and an ___ reflex.

13. The three nephron functions are, ___, ___, and ___.

14. Eight functions of the kidney are ___, ___, ___, ___, ___, ___, ___, and ___.

15. Match the following:
   ___ (a) membrane proteins that function as water channels; found in cells of the proximal convoluted tubule and descending limb of the loop of Henle
   ___ (b) a secondary active transport process that achieves Na^+ reabsorption, returns filtered HCO_3^- and water to the peritubular capillaries, and secretes H^+
   ___ (c) H^+ ATPases that secrete H^+ into the tubular fluid

   ___ (d) stimulates principal cells to secrete more K^+ into tubular fluid
   ___ (e) reduces glomerular filtration rate; increases blood volume and pressure
   ___ (f) increases glomerular filtration rate
   ___ (g) regulates facultative water reabsorption by increasing the water permeability of principal cells
   ___ (h) reabsorb Na^+ together with a variety of other solutes

   (1) angiotensin II
   (2) atrial natriuretic peptide
   (3) Na^+ symporters
   (4) Na^+/H^+ antiporters
   (5) aquaporins
   (6) aldosterone
   (7) ADH
   (8) proton pumps

CRITICAL THINKING QUESTIONS

1. Imagine the discovery of a new toxin that blocks renal tubule reabsorption but does not affect filtration. Predict the short-term effects of this toxin. (HINT: *Normally about 99% of glomerular filtrate is reabsorbed.*)

2. While traveling down the interstate, little Caitlin told her father to stop the car "NOW! I'm so full I'm gonna burst!" Her dad thinks she can make it to the next rest stop. What structures of the urinary bladder will help her "hold it"? (HINT: *Children often visit the emergency room with broken bones, but not with exploding bladders.*)

3. Although urinary catheters come in one length only, the number of centimeters of the catheter that must be inserted to reach the bladder and release urine differs significantly in males and females. Why? (HINT: *Why do men and women usually assume different positions when they urinate?*)

ANSWERS TO FIGURE QUESTIONS

26.1 The kidneys, ureters, urinary bladder, and urethra are the components of the urinary system.

26.2 The kidneys are retroperitoneal because they are posterior to the peritoneum.

26.3 Blood vessels, lymphatic vessels, nerves, and a ureter pass through the renal hilus.

26.4 About 1200 mL of blood enters the renal arteries each minute.

26.5 Cortical nephrons have glomeruli in the superficial renal cortex, and their short loops of Henle penetrate only into the superficial renal medulla; juxtamedullary nephrons have glomeruli deep in the renal cortex, and their long loops of Henle extend through the renal medulla nearly to the renal papilla.

26.6 This section must pass through the renal cortex because there are no renal corpuscles in the renal medulla.

26.7 Secreted penicillin is being removed from the bloodstream.

26.8 Endothelial fenestrations (pores) in glomerular capillaries are too small for red blood cells to pass through.

26.9 Obstruction of the right ureter would increase CHP and thus decrease NFP in the right kidney; the obstruction would have no effect on the left kidney.

26.10 *Auto* means self; tubuloglomerular feedback is an example of autoregulation because it takes place entirely within the kidneys.

26.11 The tight junctions form a barrier that prevents diffusion of transporter, channel, and pump proteins between the apical and basolateral membranes.

26.12 Glucose enters a PCT cell via a Na^+–glucose symporter in the apical membrane and leaves via facilitated diffusion through the basolateral membrane.

26.13 The electrochemical gradient promotes movement of Na^+ into the tubule cell through the apical membrane antiporters.

26.14 Reabsorption of the solutes creates an osmotic gradient that promotes the reabsorption of water via osmosis.

26.15 This is secondary active transport because the symporter uses the energy stored in the concentration gradient of Na^+

between extracellular fluid and the cytosol. No water is reabsorbed here because the thick ascending limb of the loop of Henle is virtually impermeable to water.

26.16 In principal cells, aldosterone stimulates secretion of K^+ and reabsorption of Na^+ by increasing the activity of existing sodium pumps and by stimulating the synthesis of new sodium pumps and leakage channels for Na^+ and K^+.

26.17 A carbonic anhydrase inhibitor reduces secretion of H^+ into the urine and reduces reabsorption of Na^+ and HCO_3^- into the blood. It has a diuretic effect, and it also tends to cause acidosis (lowered pH of the blood) due to loss of HCO_3^- in the urine.

26.18 Aldosterone and atrial natriuretic peptide also influence renal water reabsorption.

26.19 Dilute urine is produced when the thick ascending limb of the loop of Henle, the distal convoluted tubule, and the collecting duct reabsorb more solutes than water.

26.20 The high osmolarity of interstitial fluid in the renal medulla is due mainly to Na^+, Cl^-, and urea.

26.21 Secretion occurs in the proximal convoluted tubule, the loop of Henle, and the collecting duct.

26.22 Lack of voluntary control over micturition is termed urinary incontinence.

26.23 The kidneys start to develop during the 3rd week of gestation.

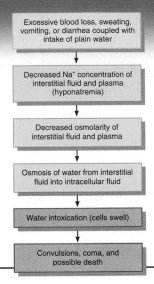

Excessive blood loss, sweating, vomiting, or diarrhea coupled with intake of plain water

Decreased Na⁺ concentration of interstitial fluid and plasma (hyponatremia)

Decreased osmolarity of interstitial fluid and plasma

Osmosis of water from interstitial fluid into intracellular fluid

Water intoxication (cells swell)

Convulsions, coma, and possible death

The water and dissolved solutes in each of the body's fluid compartments constitute **body fluids.** Water is the main component of all body fluids. Regulatory mechanisms function to maintain homeostasis of body fluids in health; their malfunction may seriously endanger the functioning of organs throughout the body, including the nervous system. This chapter explores the mechanisms that regulate the total volume of body water and its distribution among the body's fluid compartments. In addition, it examines the factors that determine the concentrations of solutes and the pH of body fluids.

FLUID COMPARTMENTS AND FLUID BALANCE

OBJECTIVES

- *Compare the locations of intracellular fluid (ICF) and extracellular fluid (ECF), and describe the various fluid compartments of the body.*

- *Describe the sources of water and solute gain and loss, and explain how each is regulated.*

In lean adults, body fluid constitutes about 55% and 60% total body mass in females and males, respectively, and is partitioned into two main compartments—the intracellular fluid compartment and the extracellular fluid compartment (Figure 27.1). About two-thirds of body fluid is within cells and is termed **intracellular fluid** (ICF; *intra-* = within). The other third, called **extracellular fluid** (ECF; *extra-* = outside), is outside cells and includes all other body fluids. About 80% of the ECF is **interstitial fluid** (*inter-* = between), which occupies the microscopic spaces between tissue cells, and 20% of the ECF is **plasma,** the liquid portion of the blood. In addition to interstitial fluid and plasma, the ECF also includes lymph in lymphatic vessels; cerebrospinal fluid in the nervous system; fluids in the gastrointestinal tract; synovial fluid in joints; aqueous humor and vitreous body in the eyes; endolymph and perilymph in the ears; pleural, pericardial, and peritoneal fluids between serous membranes; and glomerular filtrate in the kidneys.

Plasma membranes of individual cells separate their intracellular fluid from the interstitial fluid, and blood vessel walls divide interstitial fluid from plasma. Only in the smallest blood vessels, the capillaries, are the walls thin enough and leaky enough to permit the exchange of water and solutes between plasma and interstitial fluid.

To say that the body is in **fluid balance** means that the required amounts of water and solutes are present and are correctly proportioned among the various compartments. Despite continual exchange of water and solutes between fluid compartments, by filtration, reabsorption, diffusion, and osmosis, the volume of fluid in each compartment remains fairly stable. The pressures that promote filtration and reabsorption can be reviewed in Figure 21.7 on page 677. Because osmosis is the primary means of water movement between intracellular fluid and interstitial fluid, the concentration of solutes in these fluids determines the *direction* of water movement. Most solutes in body fluids are **electrolytes,** inorganic compounds that dissociate into ions. Fluid balance depends on electrolyte balance; the two are in-

Figure 27.1 Body fluid compartments.

🔑 **The term *body fluid* refers to body water and its dissolved substances.**

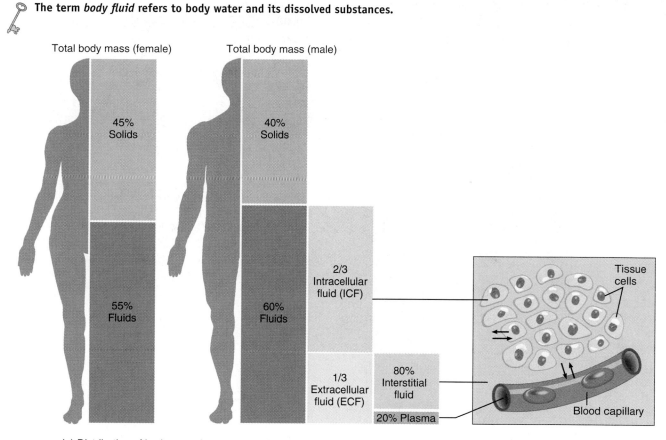

(a) Distribution of body water in an average lean, adult female and male

(b) Exchange of water among body fluid compartments

Q What is the approximate volume of plasma in a lean 60-kg male? In a lean 60-kg female? (NOTE: *One liter of body fluid has a mass of 1 kilogram.*)

terrelated. Because intake of water and electrolytes rarely occurs in exactly the same proportions as their presence in body fluids, the ability of the kidneys to excrete excess water by producing dilute urine, or to excrete excess electrolytes by producing concentrated urine, is of utmost importance.

Avenues of Body Water Gain and Loss

Water is by far the largest single component of the body, constituting 45–75% of total body mass, depending on age and gender. Infants have the highest percentage of water, up to 75% of body mass; the percentage decreases until about 2 years of age. Until puberty, body mass of both boys and girls is about 60% water. Because adipose tissue contains almost no water, fatter people have a smaller proportion of water than do leaner people. In lean adult males, water accounts for about 60% of body mass. Even lean females, on average, have more subcutaneous fat than do males, so their total body water is lower, accounting for about 55% of body mass.

The body can gain water in two ways: via ingestion and metabolic synthesis (Figure 27.2). The main sources of body water are ingested liquids (1600 mL) and moist foods (700 mL) absorbed from the gastrointestinal tract, which amount to about 2300 mL/day. The other source of water is **metabolic water**—about 200 mL/day—which is produced in the body mainly when electrons are accepted by oxygen during aerobic cellular respiration (see Figure 25.7 on page 883) and to a smaller extent during dehydration synthesis reactions (see Figure 2.16 on page 44). Daily water gain totals about 2500 mL.

Normally, water loss equals water gain, so body fluid volume remains constant. Water loss occurs via four avenues (see Figure 27.2). Each day the kidneys excrete about 1500 mL in urine, the skin evaporates about 600 mL (400 mL through insensible perspiration and 200 mL as sweat), the lungs exhale about 300 mL as water vapor, and the gastrointestinal tract excretes about 100 mL in feces. In women of reproductive age, menstrual flow is an additional route for loss

Figure 27.2 Sources of daily water gain and loss under normal conditions. Numbers are average volumes for adults.

🔑 **Normally, daily water loss equals daily water gain.**

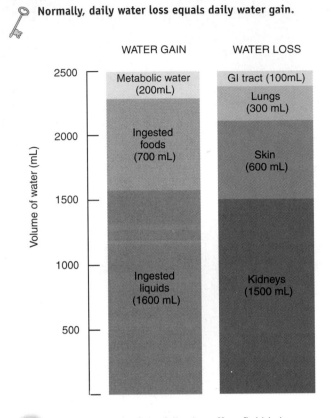

Q How does each of the following affect fluid balance: Hyperventilation? Vomiting? Fever? Diuretics?

Figure 27.3 Pathways through which dehydration stimulates thirst.

🔑 **Dehydration occurs when water loss is greater than water gain.**

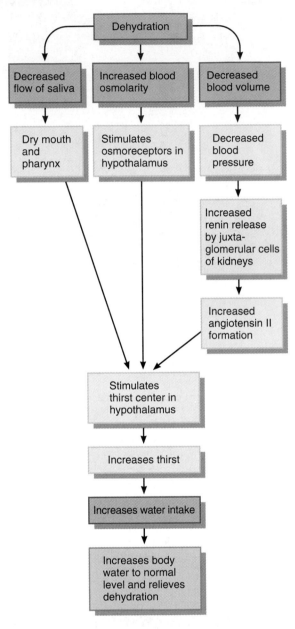

Q Does regulation of these pathways occur via negative or positive feedback? Why?

of body water. On average, daily water loss totals about 2500 mL. The amount of water lost by a given route can vary considerably over time. For example, water may literally pour from the skin in the form of sweat during strenuous exertion; in other instances water may be lost in diarrhea during a GI tract infection.

Regulation of Water Gain

The rate of formation of metabolic water is not regulated to maintain homeostasis of body water. Rather, metabolic water volume is primarily a function of the level of aerobic cellular respiration, which reflects the level of demand for ATP in body cells. Body water gain is regulated principally by adjusting the volume of water intake, mainly by drinking more or less fluid. An area in the hypothalamus known as the **thirst center** governs the urge to drink.

When water loss is greater than water gain, **dehydration**—a decrease in volume and an increase in osmolarity of body fluids—stimulates thirst (Figure 27.3). Mild dehydration occurs when body mass decreases by 2% due to fluid loss. A decrease in blood volume causes blood pressure to

fall. This change stimulates the kidneys to release renin, which promotes the formation of angiotensin II. Increased nerve impulses from osmoreceptors in the hypothalamus, triggered by increased blood osmolarity, and increased angiotensin II in the blood both stimulate the thirst center in the hypothalamus. Other signals that stimulate thirst are inputs from neurons in the mouth that detect dryness due to a decreased flow of saliva and from baroreceptors that detect lowered pressure in the heart and blood vessels. As a result, the sensation of thirst increases, which usually leads to increased fluid intake (if fluids are available) and restoration of normal fluid volume. The net effect of the cycle is that fluid gain balances fluid loss. Sometimes, however, either the sensation of thirst does not occur quickly enough or access to fluids is restricted, and significant dehydration ensues. This happens most often in elderly people, in infants, and in those who are in a confused mental state. In situations in which heavy sweating or fluid loss from diarrhea or vomiting occurs, it is wise to start replacing body fluids by drinking even before thirst becomes apparent.

Regulation of Water and Solute Loss

Even though the loss of water and solutes through sweating and exhalation increases during exercise, elimination of *excess* body water or solutes occurs mainly via the urine. As we will see, the extent of *urinary NaCl loss* is the main determinant of body fluid *volume* because "water follows salt." In contrast, the extent of *urinary water loss* is the main determinant of body fluid *osmolarity*.

Because our daily diet contains a highly variable amount of NaCl (salt), urinary excretion of Na^+ and Cl^- must also vary if homeostasis is to be maintained. Hormonal changes regulate the urinary loss of these ions, which in turn affects blood volume. The sequence of changes that occur after a salty meal are depicted in Figure 27.4. The increased intake of NaCl produces an increase in plasma levels of Na^+ and Cl^- (the major contributors to osmolarity of extracellular fluid), and the resultant increased osmolarity of interstitial fluid causes osmosis of water from intracellular fluid into interstitial fluid and then into plasma, which increases blood volume.

The three most important hormones that regulate the extent of renal Na^+ and Cl^- reabsorption (and thus how much is lost in the urine) are **angiotensin II, aldosterone, and atrial natriuretic peptide (ANP).** Angiotensin II and aldosterone promote reabsorption (reduce urinary loss) of Na^+ and Cl^- and thereby increase the volume of body fluids. ANP promotes **natriuresis,** elevated urinary excretion of Na^+ (and Cl^-), which decreases blood volume. An increase in blood volume stretches the atria of the heart and promotes release of atrial natriuretic peptide, which promotes natriuresis. The increase in blood volume also slows release of renin from juxtaglomerular cells of the kidneys. When renin level declines, less angiotensin II is formed. Decline in angiotensin II from a moderate level to a low level increases glomerular filtration rate and reduces Na^+, Cl^-, and water reabsorption in the kidney tubules. Reduced angiotensin II also decreases the secretion of aldosterone, and as aldosterone level declines, Na^+ and Cl^- reabsorption slows in the renal collecting ducts; more of the filtered Na^+ and Cl^- thus remains in the tubular fluid, and these ions are excreted in the urine. The osmotic consequence of excreting more Na^+ and Cl^- is loss of more water in urine which decreases blood volume.

Figure 27.4 Hormonal changes following increased NaCl intake. The resulting reduced blood volume is achieved by increasing urinary loss of Na^+ and Cl^-.

🔑 **The three main hormones that regulate renal Na^+ and Cl^- reabsorption (and thus the amount lost in the urine) are angiotensin II, aldosterone, and atrial natriuretic peptide.**

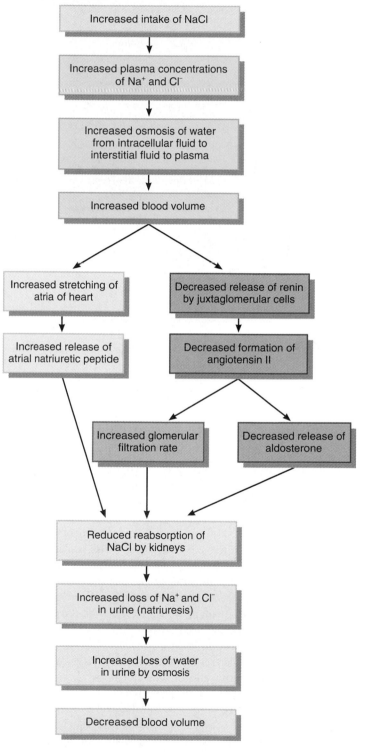

Q How does hyperaldosteronism (excessive aldosterone secretion) cause edema?

Figure 27.5 Series of events in water intoxication.

🔑 **Water intoxication is a state in which excessive body water causes cells to become hypotonic and swell.**

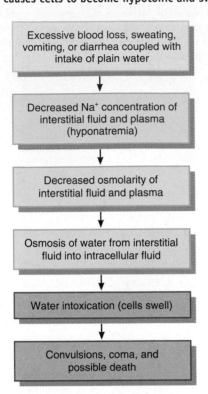

Excessive blood loss, sweating, vomiting, or diarrhea coupled with intake of plain water

⬇

Decreased Na⁺ concentration of interstitial fluid and plasma (hyponatremia)

⬇

Decreased osmolarity of interstitial fluid and plasma

⬇

Osmosis of water from interstitial fluid into intracellular fluid

⬇

Water intoxication (cells swell)

⬇

Convulsions, coma, and possible death

Q Why do solutions used for oral rehydration therapy contain a small amount of table salt (NaCl)?

The major hormone that regulates water loss is **antidiuretic hormone (ADH),** which is produced by neurosecretory cells that extend from the hypothalamus to the posterior pituitary gland. Besides stimulating the thirst mechanism, an increase in the osmolarity of plasma and interstitial fluid stimulates release of ADH (see Figure 26.18 on page 937), which promotes the insertion of water-channel proteins (aquaporin-2) into the apical membranes of principal cells in the collecting ducts of the kidneys. As a result, the water permeability of these cells increases. Water molecules move by osmosis from the renal tubular fluid into the cells and then from the cells into the bloodstream; the result is production of a small volume of very concentrated urine. By contrast, intake of plain water decreases the osmolarity of blood and interstitial fluid. Within minutes, ADH secretion shuts down, and soon its blood level is close to zero. When the principal cells are not stimulated by ADH, aquaporin-2 molecules are removed from the apical membrane by endocytosis. The principal cells' water permeability is thereby decreased, and more water is lost in the urine.

Under some conditions, factors other than blood osmolarity influence urinary water loss. A large decrease in blood volume, which is detected by baroreceptors (stretch receptors) in the left atrium and in blood vessel walls, also stimu-

lates ADH release. In severe dehydration, glomerular filtration rate decreases because blood pressure falls, and less water is lost in the urine. Conversely, the intake of too much water increases blood pressure, the rate of glomerular filtration rises, and more water is lost in the urine. Hyperventilation (abnormally fast and deep breathing) increases fluid loss through the exhalation of more water vapor. Vomiting and diarrhea result in fluid loss from the gastrointestinal tract. Finally, fever, heavy sweating, and destruction of extensive areas of the skin from burns result in excessive water loss through the skin.

Movement of Water Between Fluid Compartments

Intracellular and interstitial fluids normally have the same osmolarity, so cells neither shrink nor swell, but fluid imbalance between these two compartments can be caused by a change in their osmolarity. An increase in the osmolarity of interstitial fluid draws water out of cells, and they shrink slightly; a decrease in the osmolarity of interstitial fluid, by contrast, causes cells to swell. Changes in osmolarity most often result from changes in the concentration of Na⁺.

A decrease in the osmolarity of interstitial fluid normally inhibits secretion of ADH. Normally functioning kidneys then excrete excess water in the urine, which raises the osmotic pressure of body fluids to normal. As a result, body cells swell only slightly, and only for a brief period of time. But when a person steadily consumes water faster than the kidneys can excrete it (the maximum urine flow rate is about 15 mL/min) or when renal function is poor, the result may be **water intoxication,** a state in which excessive body water causes cells to become hypotonic and to swell dangerously (Figure 27.5). If the body water and Na⁺ lost during blood loss or excessive sweating, vomiting, or diarrhea is replaced by drinking plain water, body fluids become more dilute. This dilution can cause the Na⁺ concentration of plasma and then of interstitial fluid to fall below the normal range (hyponatremia). When the Na⁺ concentration of interstitial fluid decreases, its osmolarity also falls. The net result is osmosis of water from interstitial fluid into intracellular fluid. Water entering the cells causes them to become hypotonic and swell, producing convulsions, coma, and possibly death. To prevent this dire sequence of events in cases of severe electrolyte and water loss, solutions given for intravenous or oral rehydration therapy (ORT) include a small amount of table salt (NaCl).

 CLINICAL APPLICATION
Enemas and Fluid Balance

An **enema** (EN-e-ma) is the introduction of a solution into the rectum and colon to stimulate bowel activity and evacuate feces. Repeated enemas, especially in young children, increase the risk of fluid and electrolyte imbalances. Because tap water (which is hypotonic) and commercially prepared hypertonic solutions may cause rapid movement of fluid between compartments, an isotonic solution is preferred. ■

1. List the body fluid compartments and calculate the approximate volume of each.
2. Which routes of water gain and loss from the body are most susceptible to regulation?
3. Discuss the role of thirst in regulating water intake.
4. Describe how angiotensin II, aldosterone, atrial natriuretic peptide, and antidiuretic hormone regulate the volume and osmolarity of body fluids.
5. Describe the factors involved in the movement of water between interstitial fluid and intracellular fluid.

ELECTROLYTES IN BODY FLUIDS

OBJECTIVES

• *Compare the electrolyte composition of the three major fluid compartments: plasma, interstitial fluid, and intracellular fluid.*

• *Discuss the functions of sodium, chloride, potassium, bicarbonate, calcium, phosphate, and magnesium ions, and explain the regulation of their concentrations.*

The ions formed when electrolytes dissolve serve four general functions in the body: (1) Because they are largely confined to particular fluid compartments and are more numerous than nonelectrolytes, certain ions *control the osmosis of water between fluid compartments.* (2) Ions *help maintain the acid–base balance* required for normal cellular activities. (3) Ions *carry electrical current*, which allows production of action potentials and graded potentials and controls secretion of some hormones and neurotransmitters. (4) Several ions *serve as cofactors* needed for optimal activity of enzymes.

Concentrations of Electrolytes in Body Fluids

To compare the charge carried by ions in different solutions, the concentration of ions is typically expressed in **milliequivalents/liter (mEq/liter),** which gives the concentration of cations or anions in a given volume of solution. One equivalent is the positive or negative charge equal to the amount of charge in 1 mole of H^+; a milliequivalent is one-thousandth of an equivalent. Recall that a mole of a substance is its molecular weight expressed in grams. For ions such as sodium (Na^+), potassium (K^+), and bicarbonate (HCO_3^-), which have a single positive or negative charge, the number of mEq/liter is equal to the number of mmol/liter. For ions such as calcium (Ca^{2+}) or phosphate (HPO_4^{2-}), which have two positive or negative charges, the number of mEq/liter is twice the number of mmol/liter.

Figure 27.6 compares the concentrations of the main electrolytes and protein anions in plasma, interstitial fluid, and intracellular fluid. The chief difference between the two extracellular fluids—plasma and interstitial fluid—is that plasma contains many protein anions, whereas interstitial fluid has very few. Because normal capillary membranes are virtually impermeable to protein, only a few plasma proteins leak out of blood vessels into the interstitial fluid. This difference in protein concentration is largely responsible for the blood colloid osmotic pressure exerted by plasma. In other respects the two fluids are similar.

The electrolyte content of intracellular fluid, however, differs considerably from that of extracellular fluid. In extracellular fluid, the most abundant cation is Na^+, and the most abundant anion is Cl^-. In intracellular fluid, the most abundant cation is K^+, and the most abundant anions are proteins and phosphates (HPO_4^{2-}). By actively transporting Na^+ out of cells and K^+ into cells, sodium pumps (Na^+/K^+ ATPase) play a major role in maintaining the high intracellular concentration of K^+ and high extracellular concentration of Na^+.

Sodium

Sodium ions (Na^+) are the most abundant extracellular ions, representing about 90% of extracellular cations. Normal plasma Na^+ concentration is 136–148 mEq/liter. As we have already seen, Na^+ plays a pivotal role in fluid and electrolyte balance because it accounts for almost half of the osmolarity (142 of about 290 mOsm/liter) of extracellular fluid (ECF). The flow of Na^+ through voltage-gated channels in the plasma membrane also is necessary for the generation and conduction of action potentials in neurons and muscle fibers. The typical daily intake of Na^+ in North America often far exceeds the body's normal daily requirements due largely to excess dietary salt. The kidneys excrete excess Na^+, but they also can conserve it during periods of shortage.

The Na^+ level in the blood is controlled by aldosterone, antidiuretic hormone (ADH), and atrial natriuretic peptide (ANP). Aldosterone increases renal reabsorption of Na^+. When the plasma concentration of Na^+ drops below about 135 mEq/liter, ADH release ceases. The lack of ADH, in turn, permits greater excretion of water in urine and restoration of the normal Na^+ level in ECF. Atrial natriuretic peptide (ANP) increases Na^+ and water excretion by the kidneys when Na^+ level is too high.

CLINICAL APPLICATION
Edema and Hypovolemia Indicate Na^+ Imbalance

If excess sodium ions remain in the body because the kidneys fail to excrete enough of them, water is also osmotically retained. The result is **edema,** an abnormal accumulation of interstitial fluid. Renal failure and hyperaldosteronism (excessive aldosterone secretion) are two causes of Na^+ retention. Excessive urinary loss of Na^+, by contrast, has the osmotic effect of causing excessive water loss, which results in

Figure 27.6 Electrolyte and protein anion concentrations in plasma, interstitial fluid, and intracellular fluid. The height of each column represents the milliequivalents/liter (mEq/liter).

The electrolytes present in extracellular fluids are different from those present in intracellular fluid.

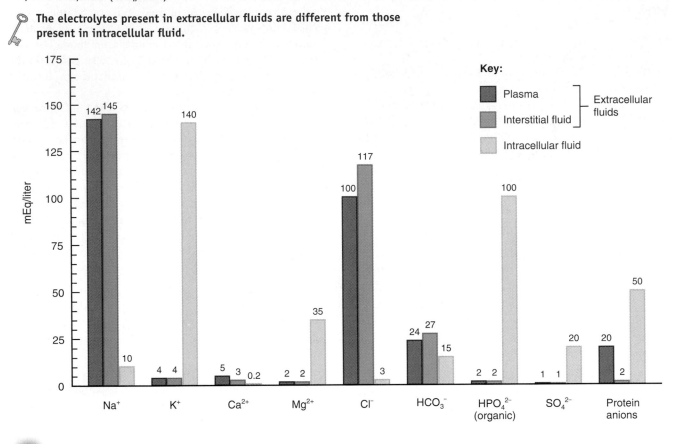

Q What are the major cation and the two major anions in ECF? In ICF?

hypovolemia, an abnormally low blood volume. Hypovolemia related to Na^+ loss is most often due to inadequate secretion of aldosterone associated with either adrenal insufficiency or overly vigorous therapy with diuretic drugs. ■

Chloride

Chloride ions (Cl^-) are the most prevalent extracellular anions. Normal plasma Cl^- concentration is 95–105 mEq/liter. Cl^- diffuses relatively easily between the extracellular and intracellular compartments because most plasma membranes contain numerous Cl^- leakage channels. For this reason, Cl^- can help balance the level of anions in different fluid compartments. One example is the chloride shift that occurs between red blood cells and plasma as the blood level of carbon dioxide either increases or decreases (see Figure 23.24 on page 803). In this case, the exchange of Cl^- for HCO_3^- maintains the correct balance of anions in ECF and ICF. Chloride ions also are part of the hydrochloric acid secreted into gastric juice. ADH helps regulate Cl^- balance in body fluids because it governs the extent of water loss in

urine. Processes that increase or decrease renal reabsorption of sodium ions also affect reabsorption of chloride ions, which follow sodium ions passively by electrical attraction —the negatively charged Cl^- follows the positively charged Na^+.

Potassium

Potassium ions (K^+) are the most abundant cations in intracellular fluid (140 mEq/liter). K^+ plays a key role in establishing the resting membrane potential and in the repolarization phase of action potentials in neurons and muscle fibers; K^+ also helps maintain normal intracellular fluid volume. When K^+ moves in or out of cells, it often is exchanged for H^+. This shift of H^+ helps regulate pH.

Normal plasma K^+ concentration is 3.5–5.0 mEq/liter. The plasma level of K^+ is controlled mainly by aldosterone. When plasma K^+ concentration is high, more aldosterone is secreted into the blood and subsequently stimulates principal cells of the renal collecting ducts to secrete more K^+, with the result that more K^+ is lost in the urine. Conversely, when plasma K^+ concentration is low, aldosterone secretion decreases and less K^+ is excreted in urine. Abnormal plasma K^+ levels adversely affect neuromuscular and cardiac function.

Bicarbonate

Bicarbonate ions (HCO_3^-) are the second most prevalent extracellular anions. Normal plasma HCO_3^- concentration is 22–26 mEq/liter in systemic arterial blood and 23–27 mEq/liter in systemic venous blood. HCO_3^- concentration increases as blood flows through systemic capillaries because carbon dioxide being released by metabolically active cells combines with water to form carbonic acid, which dissociates into H^+ and HCO_3^-. As blood flows through pulmonary capillaries, however, the concentration of HCO_3^- decreases again as carbon dioxide is exhaled. (Figure 23.24 on page 803 shows these reactions.) Intracellular fluid also contains a small amount of HCO_3^-. As previously noted, the exchange of Cl^- for HCO_3^- helps maintain the correct balance of anions in extracellular fluid and intracellular fluid.

The kidneys are the main regulators of blood HCO_3^- concentration. The intercalated cells of the renal tubule can either form HCO_3^- and release it into the blood when the blood level is low (see Figure 26.17 on page 936) or excrete excess HCO_3^- in the urine when the level in blood is too high. Changes in the blood level of HCO_3^- are considered later in this chapter in the section on acid–base balance.

Calcium

Because such a large amount of calcium is stored in bone, it is the most abundant mineral in the body. About 98% of the calcium in adults is in the skeleton and teeth, where it is combined with phosphates to form a crystal lattice of mineral salts. In body fluids, calcium is mainly an extracellular cation (Ca^{2+}). The normal concentration of free or unattached Ca^{2+} in plasma is 4.5–5.5 mEq/liter; about the same amount of Ca^{2+} is attached to various plasma proteins. Besides contributing to the hardness of bones and teeth, Ca^{2+} plays important roles in blood clotting, neurotransmitter release, maintenance of muscle tone, and excitability of nervous and muscle tissue.

The two main regulators of Ca^{2+} concentration in plasma are parathyroid hormone (PTH) and calcitriol (1,25-dihydroxy vitamin D_3), the form of vitamin D that acts as a hormone (see Figure 18.14 on page 587). A low plasma Ca^{2+} level promotes release of more parathyroid hormone, which stimulates osteoclasts in bone tissue to release calcium (and phosphate) from mineral salts of bone matrix. Thus, parathyroid hormone increases bone *resorption*. Parathyroid hormone also increases production of calcitriol (which in turn increases Ca^{2+} *absorption* from the gastrointestinal tract) and enhances *reabsorption* of Ca^{2+} from glomerular filtrate through renal tubule cells and back into blood.

Phosphate

About 85% of the phosphate in adults is present as calcium phosphate salts, which are structural components of bone and teeth; the remaining 15% is ionized. Three phosphate ions ($H_2PO_4^-$, HPO_4^{2-}, and PO_4^{3-}) are important intracellular anions, but at the normal pH of body fluids, HPO_4^{2-} is the most prevalent form. Phosphates contribute about 100 mEq/liter of anions to intracellular fluid. HPO_4^{2-} is an important buffer of H^+, both in body fluids and in the urine. Although some are "free," most phosphate ions are covalently bound to organic molecules such as lipids (phospholipids), proteins, carbohydrates, nucleic acids (DNA and RNA), and adenosine triphosphate (ATP).

The normal plasma concentration of ionized phosphate is only 1.7–2.6 mEq/liter. The same two hormones that govern calcium homeostasis—parathyroid hormone (PTH) and calcitriol—also regulate the level of HPO_4^{2-} in blood plasma. PTH stimulates resorption of bone matrix by osteoclasts, which releases both phosphate and calcium ions into the bloodstream. In the kidneys, however, PTH inhibits reabsorption of phosphate ions while stimulating reabsorption of calcium ions by renal tubular cells. Thus, PTH increases urinary excretion of phosphate and lowers blood phosphate level. Calcitriol promotes absorption of both phosphates and calcium from the digestive tract.

Magnesium

In adults, about 54% of the total body magnesium is deposited in bone matrix as magnesium salts; the remaining 46% occurs as magnesium ions (Mg^{2+}) in intracellular fluid (45%) and extracellular fluid (1%). Mg^{2+} is the second most common intracellular cation (35 mEq/liter). Functionally, Mg^{2+} is a cofactor for enzymes involved in the metabolism of carbohydrates and proteins and for Na^+/K^+ ATPase (the sodium pump enzyme). Mg^{2+} is also important in neuromuscular activity, nerve impulse transmission, and myocardial functioning, and is needed for the secretion of parathyroid hormone.

Normal plasma Mg^{2+} concentration is low, only 1.3–2.1 mEq/liter. Several factors regulate the plasma level of Mg^{2+} by varying the rate at which it is excreted in the urine. The kidneys increase urinary excretion of Mg^{2+} in hypercalcemia, hypermagnesemia, increases in extracellular fluid volume, decreases in parathyroid hormone, and acidosis. Opposite conditions decrease renal excretion of Mg^{2+}.

Table 27.1 on page 964 describes the imbalances that result from the deficiency or excess of several electrolytes.

CLINICAL APPLICATION
Individuals at Risk for Fluid and Electrolyte Imbalances

Due to a number of situations, a variety of individuals are at risk for fluid and electrolyte imbalances. Those

Table 27.1 Blood Electrolyte Imbalances

| ELECTROLYTE* | DEFICIENCY | | EXCESS | |
	NAME & CAUSES	SYMPTOMS	NAME & CAUSES	SYMPTOMS
Sodium (Na^+) 136–148 mEq/liter	**Hyponatremia** (hī-pō-na-TRĒ-mē-a) may be due to decreased sodium intake; increased sodium loss through vomiting, diarrhea, aldosterone deficiency, or taking certain diuretics; and excessive water intake.	Muscular weakness; dizziness, headache, and hypotension; tachycardia and shock; mental confusion, stupor, and coma.	**Hypernatremia** may occur with dehydration, water deprivation, or excessive sodium in diet or intravenous fluids; causes hypertonicity of ECF, which pulls water out of body cells into ECF, causing cellular dehydration.	Intense thirst, hypertension, edema, agitation, and convulsions.
Chloride (Cl^-) 95–105 mEq/liter	**Hypochloremia** (hī-pō-klō-RĒ-mē-a) may be due to excessive vomiting, overhydration, aldosterone deficiency, congestive heart failure, and therapy with certain diuretics such as furosemide (Lasix).	Muscle spasms, metabolic alkalosis, shallow respirations, hypotension, and tetany.	**Hyperchloremia** may result from dehydration due to water loss or water deprivation; excessive chloride intake; or severe renal failure, hyperaldosteronism, certain types of acidosis, and some drugs.	Lethargy, weakness, metabolic acidosis, and rapid, deep breathing.
Potassium (K^+) 3.5–5.0 mEq/liter	**Hypokalemia** (hī-pō-ka-LĒ-mē-a) may result from excessive loss due to vomiting or diarrhea, decreased potassium intake, hyperaldosteronism, kidney disease, and therapy with some diuretics.	Muscle fatigue, flaccid paralysis, mental confusion, increased urine output, shallow respirations, and changes in the electrocardiogram, including flattening of the T wave.	**Hyperkalemia** may be due to excessive intake, renal failure, aldosterone deficiency, crushing injuries to body tissues, or transfusion of hemolyzed blood.	Irritability, nausea, vomiting, diarrhea, muscular weakness; can cause death by inducing ventricular fibrillation.
Calcium (Ca^{2+}) total = 9–10.5 mg/dL; ionized = 4.5–5.5 mEq/liter	**Hypocalcemia** (hī-pō-kal-SĒ-mē-a) may be due to increased calcium loss, reduced calcium intake, elevated levels of phosphate, or hypoparathyroidism.	Numbness and tingling of the fingers; hyperactive reflexes, muscle cramps, tetany, and convulsions; bone fractures; spasms of laryngeal muscles that can cause death by asphyxiation.	**Hypercalcemia** may result from hyperparathyroidism, some cancers, excessive intake of vitamin D, and Paget's disease of bone.	Lethargy, weakness, anorexia, nausea, vomiting, polyuria, itching, bone pain, depression, confusion, paresthesia, stupor, and coma.
Phosphate (HPO_4^{2-}) 1.7–2.6 mEq/liter	**Hypophosphatemia** (hī-pō-fos′-fa-TĒ-mē-a) may occur through increased urinary losses, decreased intestinal absorption, or increased utilization.	Confusion, seizures, coma, chest and muscle pain, numbness and tingling of the fingers, uncoordination, memory loss, and lethargy.	**Hyperphosphatemia** occurs when the kidneys fail to excrete excess phosphate, as happens in renal failure; can also result from increased intake of phosphates or destruction of body cells, which releases phosphate into the blood.	Anorexia, nausea, vomiting, muscular weakness, hyperactive reflexes, tetany, and tachycardia.
Magnesium (Mg^{2+}) 1.3–2.1 mEq/liter	**Hypomagnesemia** (hī′-pō-mag′-ne-SĒ-mē-a) may be due to inadequate intake or excessive loss in urine or feces; also occurs in alcoholism, malnutrition, diabetes mellitus, and diuretic therapy.	Weakness, irritability, tetany, delirium, convulsions, confusion, anorexia, nausea, vomiting, paresthesia, and cardiac arrhythmias.	**Hypermagnesemia** occurs in renal failure or due to increased intake of magnesium, such as magnesium-containing antacids; also occurs in aldosterone deficiency and hypothyroidism.	Hypotension, muscular weakness or paralysis, nausea, vomiting, and altered mental functioning.

*Values are normal ranges of plasma levels in adults.

at risk include people who depend on others for fluid and food, such as infants, the elderly, and the hospitalized; and individuals undergoing medical treatment that involves intravenous infusions, drainages or suctions, and urinary catheters. People who receive diuretics, experience excessive fluid losses and require increased fluid intake, or experience fluid retention and have fluid restrictions are at risk. Finally at risk are postoperative individuals, severe burn or trauma cases, individuals with chronic diseases (congestive heart failure, diabetes, chronic obstructive lung disease, and cancer), people in confinement, and individuals with altered levels of consciousness who may be unable to communicate needs or respond to thirst. ■

1. Describe the functions of electrolytes in the body.
2. Name three important extracellular electrolytes and three important intracellular electrolytes.

ACID–BASE BALANCE

OBJECTIVES

• *Compare the roles of buffers, exhalation of carbon dioxide, and kidney excretion of H^+ in maintaining pH of body fluids.*

• *Define acid–base imbalances, describe their effects on the body, and explain how they are treated.*

From our discussion thus far, it is clear that various ions play different roles in helping to maintain homeostasis. A major homeostatic challenge is keeping the H^+ concentration (pH) of body fluids at an appropriate level. This task—the maintenance of acid–base balance—is of critical importance because the three-dimensional shape of all body proteins, which enables them to perform specific functions, is very sensitive to pH changes. When the diet contains a large amount of protein, as is typical in North America, cellular metabolism produces more acids than bases and thus tends to acidify the blood. (You may wish to review the discussion of acids, bases, and pH in Chapter 2.)

In a healthy person, the pH of systemic arterial blood remains between 7.35 and 7.45. (A pH of 7.4 corresponds to a H^+ concentration of 0.00004 mEq/liter = 40 nEq/liter.) Because metabolic reactions often produce a huge excess of H^+, the lack of any mechanism for the disposal of H^+ would cause H^+ level in body fluids to rise and quickly lead to death. Homeostasis of H^+ concentration within a narrow pH range is thus essential to survival. The removal of H^+ from body fluids and its subsequent elimination from the body depend on the following three major mechanisms:

1. *Buffer systems.* Buffers act quickly to temporarily bind H^+, removing the highly reactive, excess H^+ from solution, but not from the body.
2. *Exhalation of carbon dioxide.* By increasing the rate and depth of breathing, more carbon dioxide can be exhaled. Within minutes this reduces the level of carbonic acid in blood, which raises blood pH (reduces blood H^+ level).

3. *Kidney excretion of H^+.* The slowest mechanism, but the only way to eliminate acids other than carbonic acid, is through their excretion in urine.

We next examine each of these mechanisms in turn.

The Actions of Buffer Systems

Most buffer systems in the body consist of a weak acid and the salt of that acid, which functions as a weak base. Buffers prevent rapid, drastic changes in the pH of a body fluid by converting strong acids and bases into weak acids and bases. Buffers work within fractions of a second. Strong acids lower pH more than weak acids because strong acids release H^+ more readily and thus contribute more free hydrogen ions. Similarly, strong bases raise pH more than weak ones. The principal buffer systems of the body fluids are the protein buffer system, the carbonic acid–bicarbonate buffer system, and the phosphate buffer system.

Protein Buffer System

The **protein buffer system** is the most abundant buffer in intracellular fluid and plasma. The protein hemoglobin is an especially good buffer within red blood cells, and albumin is the main protein buffer in plasma. Proteins are composed of amino acids, organic molecules that contain at least one carboxyl group (—COOH) and at least one amino group (—NH$_2$); these groups are the functional components of the protein buffer system. The free carboxyl group at one end of a protein acts like an acid by releasing H^+ when pH rises; it can dissociate as follows:

$$\underset{\underset{H}{|}}{\overset{\overset{R}{|}}{NH_2-C-COOH}} \longrightarrow \underset{\underset{H}{|}}{\overset{\overset{R}{|}}{NH_2-C-COO^-}} + H^+$$

The H^+ is then able to react with any excess OH^- in the solution to form water. The free amino group at the other end of a protein can act as a base by combining with H^+ when pH falls, as follows:

$$\underset{\underset{H}{|}}{\overset{\overset{R}{|}}{NH_2-C-COOH}} + H^+ \longrightarrow \underset{\underset{H}{|}}{\overset{\overset{R}{|}}{{}^+NH_3-C-COOH}}$$

Thus, proteins can buffer both acids and bases. Besides the terminal carboxyl and amino groups, side chains that can buffer H^+ are present on seven of the 20 amino acids.

Hemoglobin effectively buffers H^+ in red blood cells in the following way: As blood flows through the systemic capillaries, carbon dioxide (CO_2) passes from tissue cells into red blood cells, where it combines with water (H_2O) to form carbonic acid. Once formed, H_2CO_3 dissociates into H^+ and

HCO_3^-. At the same time that CO_2 is entering red blood cells, oxyhemoglobin ($Hb \cdot O_2$) is giving up its oxygen to tissue cells. Reduced hemoglobin (deoxyhemoglobin) is an excellent buffer of H^+, so it picks up most of the H^+. For this reason, reduced hemoglobin usually is written as $Hb \cdot H$. The following reactions summarize these relations:

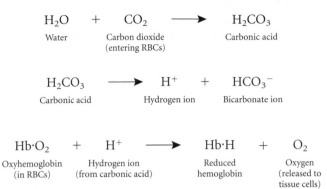

$$H_2O \quad + \quad CO_2 \quad \longrightarrow \quad H_2CO_3$$
Water   Carbon dioxide   Carbonic acid
(entering RBCs)

$$H_2CO_3 \quad \longrightarrow \quad H^+ \quad + \quad HCO_3^-$$
Carbonic acid   Hydrogen ion   Bicarbonate ion

$$Hb \cdot O_2 \quad + \quad H^+ \quad \longrightarrow \quad Hb \cdot H \quad + \quad O_2$$
Oxyhemoglobin   Hydrogen ion   Reduced   Oxygen
(in RBCs)   (from carbonic acid)   hemoglobin   (released to tissue cells)

Carbonic Acid–Bicarbonate Buffer System

The **carbonic acid–bicarbonate buffer system** is based on the *bicarbonate ion* (HCO_3^-), which can act as a weak base, and *carbonic acid* (H_2CO_3), which can act as a weak acid. HCO_3^- is a significant anion in both intracellular and extracellular fluids (see Figure 27.6), and its reaction with the CO_2 constantly released during cellular respiration produces H_2CO_3. If there is an excess of H^+, HCO_3^- can function as a weak base and remove the excess H^+, as follows:

$$H^+ \quad + \quad HCO_3^- \quad \longrightarrow \quad H_2CO_3$$
Hydrogen ion   Bicarbonate ion   Carbonic acid
(weak base)

Subsequently, H_2CO_3 dissociates into water and carbon dioxide in the lungs, and the CO_2 is exhaled.

Conversely, if there is a shortage of H^+, H_2CO_3 can function as a weak acid and provide H^+, as follows:

$$H_2CO_3 \quad \longrightarrow \quad H^+ \quad + \quad HCO_3^-$$
Carbonic acid   Hydrogen ion   Bicarbonate ion
(weak acid)

At a pH of 7.4, HCO_3^- concentration is about 24 mEq/liter, whereas H_2CO_3 concentration is about 1.2 mmol/liter; thus, bicarbonate ions outnumber carbonic acid molecules by a factor of 20. Because CO_2 and H_2O combine to form H_2CO_3, this buffer system cannot protect against pH changes due to respiratory problems in which there is an excess or shortage of CO_2.

Phosphate Buffer System

The **phosphate buffer system** acts via essentially the same mechanism as the carbonic acid–bicarbonate buffer system. The components of the phosphate buffer system are the *dihydrogen phosphate ion* ($H_2PO_4^-$) and the *monohydrogen phosphate ion* (HPO_4^{2-}). Recall that phosphates are major anions in intracellular fluid and minor ones in extracellu-

lar fluids (see Figure 27.6). The dihydrogen phosphate ion acts as a weak acid and is capable of buffering strong bases such as OH^-, as follows:

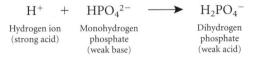

$$OH^- \quad + \quad H_2PO_4^- \quad \longrightarrow \quad H_2O \quad + \quad HPO_4^{2-}$$
Hydroxide ion   Dihydrogen   Water   Monohydrogen
(strong base)   phosphate   phosphate
  (weak acid)   (weak base)

The monohydrogen phosphate ion, in contrast, acts as a weak base and is capable of buffering the H^+ released by a strong acid such as hydrochloric acid (HCl):

$$H^+ \quad + \quad HPO_4^{2-} \quad \longrightarrow \quad H_2PO_4^-$$
Hydrogen ion   Monohydrogen   Dihydrogen
(strong acid)   phosphate   phosphate
  (weak base)   (weak acid)

Because the concentration of phosphates is highest in intracellular fluid, the phosphate buffer system is an important regulator of pH in the cytosol. It also acts to a smaller degree in extracellular fluids, and it acts to buffer acids in urine. $H_2PO_4^-$ is formed when excess H^+ in the kidney tubule fluid combine with HPO_4^{2-} (see Figure 26.17 on page 936). The H^+ that becomes part of the $H_2PO_4^-$ passes into the urine. This reaction is one means by which the kidneys help maintain blood pH by excreting H^+ in the urine.

Exhalation of Carbon Dioxide

Breathing also plays a role in maintaining the pH of body fluids. An increase in the carbon dioxide (CO_2) concentration in body fluids increases H^+ concentration and thus lowers the pH (makes it more acidic). Because H_2CO_3 can be eliminated by exhaling CO_2, it is called a **volatile acid.** Conversely, a decrease in the CO_2 concentration of body fluids raises the pH (makes it more alkaline). This chemical interaction is illustrated by the following reversible reactions:

$$CO_2 \quad + \quad H_2O \quad \rightleftharpoons \quad H_2CO_3 \quad \rightleftharpoons \quad H^+ \quad + \quad HCO_3^-$$
Carbon   Water   Carbonic   Hydrogen   Bicarbonate
dioxide   acid   ion   ion

The pH of body fluids may be adjusted, usually in 1–3 minutes, by a change in the rate and depth of breathing. With increased ventilation, more CO_2 is exhaled, the preceding reaction is driven to the left, H^+ concentration falls, and blood pH rises. Doubling the ventilation increases pH by about 0.23 units, from 7.4 to 7.63. If ventilation slows, less carbon dioxide is exhaled, and the blood pH falls. Reducing ventilation to one-quarter of normal lowers the pH by 0.4 units, from 7.4 to 7.0. These examples show the powerful effect of alterations in breathing on the pH of body fluids.

The pH of body fluids and the rate and depth of breathing interact via a negative feedback loop (Figure 27.7). If, for example, the blood becomes more acidic, the decrease in pH (increase in concentration of H^+) is detected by central chemoreceptors in the medulla oblongata and peripheral chemoreceptors in the aortic and carotid bodies, both of which stimulate the inspiratory area in the medulla oblongata. As a result, the diaphragm and other muscles that drive ventilation contract more forcefully and frequently, so more CO_2 is exhaled. As less H_2CO_3 forms and fewer H^+ are present, blood pH increases. When the response brings blood pH (H^+ concentration) back to normal, there is a return to homeostasis. The same negative feedback loop operates if the blood level of CO_2 increases. Ventilation increases, which removes more CO_2 from the blood and reduces the H^+ concentration, and thus blood pH increases.

By contrast, if the pH of the blood increases, the respiratory center is inhibited and ventilation decreases. A decrease in the CO_2 concentration of the blood has the same effect. When ventilation decreases, CO_2 accumulates in the blood and the H^+ concentration increases. This respiratory mechanism is a powerful eliminator of acid, but it can only get rid of the sole volatile acid—carbonic acid.

Kidney Excretion of H^+

Metabolic reactions produce **nonvolatile acids** such as sulfuric acid at a rate of about 1 mEq of H^+ per day for every kilogram of body mass. The only way to eliminate this huge acid load is to excrete H^+ in the urine. Because the kidneys also synthesize new HCO_3^- and reabsorb filtered HCO_3^-, this important buffer is not lost in the urine. Given the magnitude of these contributions to acid–base balance, it's not surprising that renal failure can quickly cause death. The role of the kidneys in maintaining pH is discussed in Chapter 26; carefully review Figure 26.17 on page 936.

Table 27.2 summarizes the mechanisms that maintain pH of body fluids.

Acid–Base Imbalances

The normal pH range of systemic arterial blood is between 7.35 (= 45 nEq of H^+/liter) and 7.45 (= 35 nEq of H^+/liter). **Acidosis** (or **acidemia**) is a condition in which blood pH is below 7.35; **alkalosis** (or **alkalemia**) is a condition in which blood pH is higher than 7.45.

The principal physiological effect of acidosis is depression of the central nervous system through depression of synaptic transmission. If the systemic arterial blood pH falls below 7, depression of the nervous system is so severe that the individual becomes disoriented, then comatose, and may die. Patients with severe acidosis usually die in a state of coma. A major physiological effect of alkalosis, by contrast, is overexcitability in both the central nervous system and peripheral nerves. Neurons conduct impulses repetitively, even when not stimulated by normal stimuli; the results are nervousness, muscle spasms, and even convulsions and death.

A change in blood pH that leads to acidosis or alkalosis may be countered by **compensation,** the physiological response to an acid–base imbalance that acts to normalize arterial blood pH. Compensation may be either *complete,* if pH

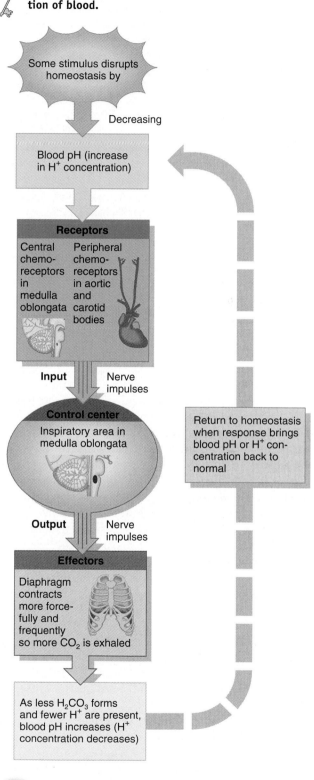

Figure 27.7 Negative feedback regulation of blood pH by the respiratory system.

🔑 **Exhalation of carbon dioxide lowers the H^+ concentration of blood.**

Some stimulus disrupts homeostasis by

Decreasing

Blood pH (increase in H^+ concentration)

Receptors

Central chemo-receptors in medulla oblongata

Peripheral chemo-receptors in aortic and carotid bodies

Input Nerve impulses

Control center
Inspiratory area in medulla oblongata

Return to homeostasis when response brings blood pH or H^+ concentration back to normal

Output Nerve impulses

Effectors
Diaphragm contracts more forcefully and frequently so more CO_2 is exhaled

As less H_2CO_3 forms and fewer H^+ are present, blood pH increases (H^+ concentration decreases)

Q If you hold your breath for 30 seconds, what is likely to happen to your blood pH?

Table 27.2 Mechanisms that Maintain pH of Body Fluids

MECHANISM	COMMENTS
Buffer Systems	Most consist of a weak acid and the salt of that acid, which functions as a weak base. They prevent drastic changes in body fluid pH.
Proteins	The most abundant buffers in body cells and blood. Histidine and cysteine are the two amino acids that contribute most of the buffering capacity of proteins. Hemoglobin inside red blood cells is a good buffer.
Carbonic acid–bicarbonate	Important regulator of blood pH. The most abundant buffers in extracellular fluid (ECF).
Phosphates	Important buffers in intracellular fluid and in urine.
Exhalation of CO_2	With increased exhalation of CO_2, pH rises (fewer H^+). With decreased exhalation of CO_2, pH falls (more H^+).
Kidneys	Renal tubules secrete H^+ into the urine and reabsorb HCO_3^- so it is not lost in the urine.

indeed is brought within the normal range, or *partial,* if systemic arterial blood pH is still lower than 7.35 or higher than 7.45. If a person has altered blood pH due to metabolic causes, hyperventilation or hypoventilation can help bring blood pH back toward the normal range; this form of compensation, termed **respiratory compensation,** occurs within minutes and reaches its maximum within hours. If, however, a person has altered blood pH due to respiratory causes, then **renal compensation**—changes in secretion of H^+ and reabsorption of HCO_3^- by the kidney tubules—can help reverse the change. Renal compensation may begin in minutes, but it takes days to reach maximum effectiveness.

In the discussion that follows, note that both respiratory acidosis and respiratory alkalosis are disorders resulting from changes in the partial pressure of CO_2 (P_{CO_2}) in systemic arterial blood (normal range is 35–45 mm Hg). By contrast, both metabolic acidosis and metabolic alkalosis are disorders resulting from changes in HCO_3^- concentration (normal range is 22–26 mEq/liter in systemic arterial blood).

Respiratory Acidosis

The hallmark of **respiratory acidosis** is a P_{CO_2} of systemic arterial blood above 45 mm Hg. Inadequate exhalation of CO_2 causes the blood pH to drop; any condition that decreases the movement of CO_2 from the blood to the alveoli of the lungs to the atmosphere causes a buildup of CO_2, H_2CO_3, and H^+. Such conditions include emphysema, pulmonary edema, injury to the respiratory center of the

medulla oblongata, airway obstruction, or disorders of the muscles involved in breathing. If the respiratory problem is not too severe, the kidneys can help raise the blood pH into the normal range by increasing excretion of H^+ and reabsorption of HCO_3^- (renal compensation). The goal in treatment of respiratory acidosis is to increase the exhalation of CO_2, as, for instance, by providing ventilation therapy. In addition, intravenous administration of HCO_3^- may be helpful.

Respiratory Alkalosis

In **respiratory alkalosis,** arterial blood P_{CO_2} falls below 35 mm Hg. The cause of the drop in P_{CO_2} and the resulting increase in pH is hyperventilation, which occurs in conditions that stimulate the inspiratory area in the brain stem. Such conditions include oxygen deficiency due to high altitude or pulmonary disease, cerebrovascular accident (stroke), or severe anxiety. Again, renal compensation may bring blood pH into the normal range if the kidneys decrease excretion of H^+ and reabsorption of HCO_3^-. Treatment of respiratory alkalosis is aimed at increasing the level of CO_2 in the body. One simple treatment is to have the person inhale and exhale into a paper bag for a short period of time; as a result, the person inhales air containing a higher than normal concentration of CO_2.

Metabolic Acidosis

In **metabolic acidosis,** the systemic arterial plasma HCO_3^- level drops below 22 mEq/liter. Such a decline in this important buffer causes the blood pH to decrease. Three situations may lower the plasma level of HCO_3^-: (1) actual loss of HCO_3^-, such as may occur with severe diarrhea or renal dysfunction; (2) accumulation of an acid other than carbonic acid, as may occur in ketosis (described on page 889); or (3) failure of the kidneys to excrete H^+ from metabolism of dietary proteins. If the problem is not too severe, hyperventilation can help bring blood pH into the normal range (respiratory compensation). Treatment of metabolic acidosis consists of administering intravenous solutions of sodium bicarbonate and correcting the cause of the acidosis.

Metabolic Alkalosis

In **metabolic alkalosis,** the blood HCO_3^- concentration is above 26 mEq/liter. A nonrespiratory loss of acid by the body or excessive intake of alkaline drugs causes the pH to increase above 7.45. Excessive vomiting of gastric contents, which results in a substantial loss of hydrochloric acid, is probably the most frequent cause of metabolic alkalosis. Other causes include gastric suctioning, use of certain diuretics, endocrine disorders, excessive intake of alkaline drugs, and severe dehydration. Respiratory compensation through hypoventilation may bring blood pH into the normal range. Treatment of metabolic alkalosis consists of fluid therapy to correct chloride, potassium, and other electrolyte deficiencies and correcting the cause of alkalosis.

Table 27.3 Summary of Acidosis and Alkalosis

CONDITION	DEFINITION	COMMON CAUSES	COMPENSATORY MECHANISM
Respiratory acidosis	Increased P_{CO_2} (above 45 mm Hg) and decreased pH (below 7.35) if there is no compensation.	Hypoventilation due to emphysema, pulmonary edema, trauma to respiratory center, airway obstructions, or dysfunction of muscles of respiration.	*Renal:* increased excretion of H^+; increased reabsorption of HCO_3^-. If compensation is complete, pH will be within the normal range but P_{CO_2} will be high.
Respiratory alkalosis	Decreased P_{CO_2} (below 35 mm Hg) and increased pH (above 7.45) if there is no compensation.	Hyperventilation due to oxygen deficiency, pulmonary disease, cerebrovascular accident (CVA), anxiety, or aspirin overdose.	*Renal:* decreased excretion of H^+; decreased reabsorption of HCO_3^-. If compensation is complete, pH will be within the normal range but P_{CO_2} will be low.
Metabolic acidosis	Decreased HCO_3^- (below 22 mEq/liter) and decreased pH (below 7.35) if there is no compensation.	Loss of bicarbonate ions due to diarrhea, accumulation of acid (ketosis), renal dysfunction.	*Respiratory:* hyperventilation, which increases loss of CO_2. If compensation is complete, pH will be within the normal range but HCO_3^- will be low.
Metabolic alkalosis	Increased HCO_3^- (above 26 mEq/liter) and increased pH (above 7.45) if there is no compensation.	Loss of acid due to vomiting, gastric suctioning, or use of certain diuretics; excessive intake of alkaline drugs.	*Respiratory:* hypoventilation, which slows loss of CO_2. If compensation is complete, pH will be within the normal range but HCO_3^- will be high.

Table 27.3 summarizes respiratory and metabolic acidosis and alkalosis.

CLINICAL APPLICATION
Diagnosis of Acid–Base Imbalances

One can often pinpoint the cause of an acid–base imbalance by careful evaluation of three factors in a sample of systemic arterial blood: pH, concentration of HCO_3^-, and P_{CO_2}. These three blood chemistry values are examined in the following four-step sequence:

1. Note whether the pH is high (alkalosis) or low (acidosis).
2. Then decide which value—P_{CO_2} or HCO_3^-—is out of the normal range and could be the *cause* of the pH change. For example, *elevated pH* could be caused by *low P_{CO_2}* or *high HCO_3^-*.
3. If the cause is a change in P_{CO_2}, the problem is *respiratory;* if the cause is a *change in HCO_3^-*, the problem is *metabolic.*
4. Now look at the value that doesn't correspond with the observed pH change. If it is within its normal range, there is no compensation. If it is outside the normal range, compensation is occurring and is partially correcting the pH imbalance.

1. Explain how each of the following buffer systems helps to maintain the pH of body fluids: proteins, carbonic acid–bicarbonate buffers, and phosphates.
2. Define acidosis and alkalosis. Distinguish among respiratory and metabolic acidosis and alkalosis.
3. What are the principal physiological effects of acidosis and alkalosis?
4. Given that you know pH, HCO_3^-, and P_{CO_2} values, explain how this information can help you determine the cause of an acid–base imbalance.

AGING AND FLUID, ELECTROLYTE, AND ACID–BASE BALANCE

OBJECTIVE

• *Describe the changes in fluid, electrolyte, and acid–base balance that may occur with aging.*

There are significant differences between adults and infants, especially premature infants, with respect to fluid distribution, regulation of fluid and electrolyte balance, and acid–base homeostasis. Accordingly, infants experience more problems than adults in these areas. The differences are related to the following conditions:

• *Proportion and distribution of water.* Whereas a newborn's total body mass is about 75% water (and can be as high as 90% in a premature infant), an adult's is about 55–60% water. (The "adult" percentage is achieved at about 2 years of age.) Moreover, whereas adults have twice as much water in ICF as ECF, the opposite is true in premature infants. Because ECF is subject to more changes than ICF, rapid losses or gains of body water are much more critical in infants. Given that the rate of fluid intake and output is about seven times higher in infants than in adults, the slightest changes in fluid balance can result in severe abnormalities.

• *Metabolic rate.* The metabolic rate of infants is about double that of adults. This results in the production of more metabolic wastes and acids, which can lead to the development of acidosis in infants.

• *Functional development of the kidneys.* In infants, the kidneys are only about half as efficient in concentrating urine as those of adults. (Functional development is not complete until about the end of the first month after birth.) As a result, the kidneys of newborns can neither concentrate urine nor rid the body of excess acids produced as a result of high metabolic rate as effectively as those of adults.

- *Body surface area.* The ratio of body surface area to body volume of infants is about three times greater than that of adults. This significantly increases water loss through the skin in infants.
- *Breathing rate.* The higher breathing rate of infants (about 30–80 times a minute) causes greater water loss from the lungs. Moreover, respiratory alkalosis may occur because greater ventilation eliminates more CO_2 and lowers the P_{CO_2}.
- *Ion concentrations.* Newborns have higher K^+ and Cl^- concentrations than adults. This creates a tendency toward metabolic acidosis.

By comparison with children and younger adults, older adults often have an impaired ability to maintain fluid, electrolyte, and acid–base balance. With increasing age, many people have a decreased volume of intracellular fluid and decreased total body potassium due to declining skeletal muscle mass and increasing mass of adipose tissue (which contains very little water). Age-related decreases in respiratory and renal functioning may compromise acid–base balance by slowing the exhalation of CO_2 and the excretion of excess acids in urine. Other kidney changes, such as decreased blood flow, decreased glomerular filtration rate, and reduced sensitivity to antidiuretic hormone, have an adverse effect on the ability to maintain fluid and electrolyte balance. Due to a decrease in the number and efficiency of sweat glands, both sensible and insensible water loss from the skin declines with age. Because of these age-related changes, older adults are susceptible to several fluid and electrolyte disorders:

- *Dehydration* and *hypernatremia* often occur due to inadequate fluid intake or loss of more water than sodium in vomit, feces, or urine.
- *Hyponatremia* may occur due to inadequate intake of sodium; elevated loss of sodium in urine, vomit, or diarrhea; or impaired ability of the kidneys to produce dilute urine.
- *Hypokalemia* often occurs in older adults who chronically use laxatives to relieve constipation or who take potassium-depleting diuretic drugs for treatment of hypertension or heart disease.
- *Acidosis* may occur due to impaired ability of the lungs and kidneys to compensate for acid–base imbalances. One cause of acidosis is that renal tubule cells produce less ammonia, which then is not available to combine with H^+ and be excreted in urine as NH_4^+; another cause is reduced exhalation of CO_2.

STUDY OUTLINE

FLUID COMPARTMENTS AND FLUID BALANCE (p. 956)

1. Body fluid includes water and dissolved solutes.
2. About two-thirds of the body's fluid is located within cells and is called intracellular fluid (ICF). The other one-third, called extracellular fluid (ECF), includes interstitial fluid; plasma and lymph; cerebrospinal fluid; gastrointestinal tract fluids; synovial fluid; fluids of the eyes and ears; pleural, pericardial, and peritoneal fluids; and glomerular filtrate.
3. Fluid balance means that the various body compartments contain the normal amount of water.
4. An inorganic substance that dissociates into ions in solution is called an electrolyte. Fluid balance and electrolyte balance are interrelated.
5. Water is the largest single constituent in the body—45–75% of total body mass depending on age and the amount of adipose tissue present.
6. Daily water gain and loss are each about 2500 ml. Sources of water gain are ingested liquids and foods, and water produced by cellular respiration and dehydration synthesis reactions (metabolic water). Water is lost from the body via urination, evaporation from the skin surface, exhalation of water vapor, and defecation. In women, menstrual flow is an additional route for loss of body water.

7. The main way to regulate body water gain is by adjusting the volume of water intake, mainly by drinking more or less fluid. The thirst center in the hypothalamus governs the urge to drink.
8. Although increased amounts of water and solutes are lost through sweating and exhalation during exercise, loss of excess body water or excess solutes depends mainly on regulating excretion in the urine. The extent of urinary NaCl loss is the main determinant of body fluid volume, whereas the extent of urinary water loss is the main determinant of body fluid osmolarity.
9. Angiotensin II and aldosterone reduce urinary loss of Na^+ and Cl^- and thereby increase the volume of body fluids. ANP promotes natriuresis, elevated excretion of Na^+ (and Cl^-), which decreases blood volume.
10. The major hormone that regulates water loss and thus body fluid osmolarity is antidiuretic hormone (ADH).
11. An increase in the osmolarity of interstitial fluid draws water out of cells, and they shrink slightly. A decrease in the osmolarity of interstitial fluid causes cells to swell. Most often a change in osmolarity is due to a change in the concentration of Na^+, the dominant solute in interstitial fluid.
12. When a person consumes water faster than the kidneys can excrete it or when renal function is poor, the result may be water intoxication, in which cells swell dangerously.

ELECTROLYTES IN BODY FLUIDS (p. 961)

1. Ions formed when electrolytes dissolve in body fluids control the osmosis of water between fluid compartments, help maintain acid base balance, and carry electrical current.
2. The concentrations of cations and anions is expressed in units of milliequivalents/liter (mEq/liter).
3. Plasma, interstitial fluid, and intracellular fluid contain varying kinds and amounts of ions.
4. Sodium ions (Na^+) are the most abundant extracellular ions. They are involved in impulse transmission, muscle contraction, and fluid and electrolyte balance. Na^+ level is controlled by aldosterone, antidiuretic hormone, and atrial natriuretic peptide.
5. Chloride ions (Cl^-) are the major extracellular anions. They play a role in regulating osmotic pressure and forming HCl in gastric juice. Cl^- level is controlled indirectly by antidiuretic hormone and by processes that increase or decrease renal reabsorption of Na^+.
6. Potassium ions (K^+) are the most abundant cations in intracellular fluid. They play a key role in the resting membrane potential and action potential of neurons and muscle fibers; help maintain intracellular fluid volume; and contribute to regulation of pH. K^+ level is controlled by aldosterone.
7. Bicarbonate ions (HCO_3^-) are the second most abundant anions in extracellular fluid. They are the most important buffer in plasma.
8. Calcium is the most abundant mineral in the body. Calcium salts are structural components of bones and teeth. Ca^{2+}, which are principally extracellular cations, function in blood clotting, neurotransmitter release, and contraction of muscle. Ca^{2+} level is controlled mainly by parathyroid hormone and calcitriol.
9. Phosphate ions ($H_2PO_4^-$, HPO_4^{2-}, and PO_4^{3-}) are principally intracellular anions, and their salts are structural components of bones and teeth. They are also required for the synthesis of nucleic acids and ATP and participate in buffer reactions. Their level is controlled by parathyroid hormone and calcitriol.
10. Magnesium ions (Mg^{2+}) are primarily intracellular cations. They act as cofactors in several enzyme systems.
11. Table 27.1 on page 964 describes the imbalances that result from deficiency or excess of important body electrolytes.

ACID–BASE BALANCE (p. 965)

1. The overall acid–base balance of the body is maintained by controlling the H^+ concentration of body fluids, especially extracellular fluid.
2. The normal pH of systemic arterial blood is 7.35–7.45.
3. Homeostasis of pH is maintained by buffer systems, via exhalation of carbon dioxide, and via kidney excretion of H^+ and reabsorption of HCO_3^-.
4. The important buffer systems include proteins, carbonic acid–bicarbonate buffers, and phosphates.
5. An increase in exhalation of carbon dioxide increases blood pH; a decrease of CO_2 in exhalation decreases blood pH.
6. The kidneys excrete H^+ and reabsorb HCO_3^-.
7. Table 27.2 on page 968 summarizes the mechanisms that maintain pH of body fluids.
8. Acidosis is a systemic arterial blood pH below 7.35; its principal effect is depression of the central nervous system (CNS). Alkalosis is a systemic arterial blood pH above 7.45; its principal effect is overexcitability of the CNS.
9. Respiratory acidosis and alkalosis are disorders due to changes in blood P_{CO_2}, whereas metabolic acidosis and alkalosis are disorders associated with changes in blood HCO_3^- concentration.
10. Metabolic acidosis or alkalosis can be compensated by respiratory mechanisms; respiratory acidosis or alkalosis can be compensated by renal mechanisms.
11. Table 27.3 on page 969 summarizes respiratory and metabolic acidosis and alkalosis.
12. By examining pH, HCO_3^-, and P_{CO_2} values, it is possible to pinpoint the cause of an acid–base imbalance.

SELF-QUIZ QUESTIONS

Complete the following:

1. The largest single component of body fluids is ___.
2. The principal physiological effect of acidosis is ___ of the ___ through depression of ___. A major physiological effect of alkalosis is ___ in both the ___ and ___.
3. Most buffers consist of a ___ and a ___.
4. The principal buffer systems of the body are the ___ systems, the ___ systems, and the ___ systems.

True or false:

5. The most abundant buffer in body cells and plasma is the protein buffer system.
6. Normally, water loss equals water gain, so body fluid volume is constant.

Choose the best answer to the following questions:

7. The primary means of regulating body water gain is adjusting (a) the volume of water intake, (b) the rate of cellular respiration, (c) the formation of metabolic water, (d) the volume of metabolic water, (e) the metabolic use of water.
8. Which of the following are ways that dehydration stimulates thirst? (1) It decreases the production of saliva. (2) It increases the production of saliva. (3) It increases osmolarity of body fluids. (4) It decreases osmolarity of body fluids. (5) It decreases blood volume. (6) It increases blood volume.
   (a) 1, 2, 4, and 6, (b) 1, 3, 5, and 6, (c) 1, 3, and 5, (d) 2, 4, and 6, (e) 1, 4, 5, and 6
9. Which of the following hormones regulate fluid loss? (1) antidiuretic hormone, (2) aldosterone, (3) atrial natriuretic peptide, (4) thyroxine, (5) cortisol.
   (a) 1, 3, and 5, (b) 1, 2, and 3, (c) 2, 4, and 5, (d) 2, 3, and 4, (e) 1, 3, and 4
10. Which of the following are true concerning ions in the body? (1) They control osmosis of water between fluid compartments. (2) They help maintain acid–base balance. (3) They

carry electrical current. (4) They serve as cofactors for enzyme activity. (5) They serve as local hormones under special circumstances.

(a) 1, 3, and 5, (b) 2, 4, and 5, (c) 1, 4, and 5, (d) 1, 2, and 4, (e) 1, 2, 3, and 4

11. Which of the following statements are true? (1) Buffers prevent rapid, drastic changes in pH of a body fluid. (2) Buffers work slowly. (3) Strong acids lower pH more than weak acids because strong acids contribute fewer H^+. (4) Proteins can buffer both acids and bases. (5) Hemoglobin is an important buffer.

(a) 1, 2, 3, and 5, (b) 1, 3, 4, and 5, (c) 1, 3, and 5, (d) 1, 4, and 5, (e) 2, 3, and 5

12. Which of the following statements are true? (1) An increase in the carbon dioxide concentration in body fluids increases H^+ concentration and thus lowers pH. (2) Breath holding results in a decline in blood pH. (3) The respiratory buffer mechanism can eliminate a single volatile acid: carbonic acid. (4) The only way to eliminate fixed acids is to excrete H^+ in the urine. (5) When the diet contains a large amount of protein, normal metabolism produces more acids than bases.

(a) 1, 2, 3, 4, and 5, (b) 1, 3, 4, and 5, (c) 1, 2, 3, and 4, (d) 1, 2, 4, and 5, (e) 1, 3, and 4

13. Match the following:

___ (a) negatively charged ions
___ (b) positively charged ions
___ (c) compounds with covalent bonds that do not form ions when dissolved in water
___ (d) inorganic substances that dissociate into ions when in solution
___ (e) substances that act to prevent rapid, drastic changes in the pH of a body fluid

(1) electrolytes
(2) anions
(3) cations
(4) nonelectrolytes
(5) buffers

14. Match the following:

___ (a) the most abundant cation in extracellular fluid; plays a key role in establishing the resting membrane potential
___ (b) the most abundant mineral in the body; plays important roles in blood clotting, neurotransmitter release, maintenance of muscle tone, and excitability of nervous and muscle tissue
___ (c) second most common intracellular cation; is a cofactor for enzymes involved in carbohydrate, protein, and Na^+/K^+ ATPase metabolism
___ (d) the most abundant extracellular ion; essential in fluid and electrolyte balance
___ (e) ions that are mostly combined with lipids, proteins, carbohydrates, nucleic acids, and ATP inside cells
___ (f) most prevalent extracellular anion; can help balance the level of anions in different fluid compartments
___ (g) second most prevalent extracellular anion; mainly regulated by the kidneys; important for acid–base balance

(1) sodium
(2) chloride
(3) potassium
(4) bicarbonate
(5) calcium
(6) phosphate
(7) magnesium

15. Match the following:

___ (a) an abnormal increase in the volume of interstitial fluid
___ (b) the swelling of cells due to water moving from plasma into interstitial fluid and then into cells
___ (c) occurs when water loss is greater than water gain
___ (d) condition that can occur as water moves out of plasma into interstitial fluid and blood volume decreases
___ (e) can be caused by emphysema, pulmonary edema, injury to the respiratory center of the medulla oblongata, airway destruction, or disorders of the muscles involved in breathing
___ (f) can be caused by actual loss of bicarbonate ions, ketosis, or failure of kidneys to excrete H^+
___ (g) can be caused by excessive vomiting of gastric contents, gastric suctioning, use of certain diuretics, severe dehydration, or excessive intake of alkaline drugs
___ (h) can be caused by oxygen deficiency at high altitude, stroke, severe anxiety, or aspirin overdose

(1) respiratory acidosis
(2) respiratory alkalosis
(3) metabolic acidosis
(4) metabolic alkalosis
(5) dehydration
(6) hypovolemic shock
(7) water intoxication
(8) edema

CRITICAL THINKING QUESTIONS

1. After a late-night party, Gary ended up in the emergency room. According to his friends, Gary had said he was so thirsty that he could drink a tub full of water, so someone dared him to try it. His friends called for an ambulance when Gary suffered a convulsion. What has happened to Gary? (HINT: *Gary was intoxicated, but not with alcohol.*)

2. "I'm so mad at Jacques," Aimee told her girlfriend. "We're about the same height and the same weight, but when we had our body fat measured in A&P lab, he had less fat than I did!" How do you explain this to Aimee? (HINT: *Fat and water don't mix.*)

3. When Carlos ate some of his roommate's leftovers, he thought the food tasted a little odd, but he didn't feel like cooking so he ate it anyway. Awhile later, Carlos was vomiting the entire contents of his stomach. Carlos then tried to settle his queasy stomach by eating an entire bottle of antacids. What effect will this episode have on Carlos's fluid and electrolyte balance? (HINT: *What is the pH of gastric juice?*)

4. Henry is in the intensive care unit because he suffered a severe myocardial infarction three days ago. The lab reports the following values from an arterial blood sample: pH = 7.30, $HCO_3^- =$ 20 mEq/liter, P_{CO_2} = 32 mm Hg. Diagnose Henry's acid-base status and decide whether or not compensation is occurring. (HINT: Use the four-step method described on page 969.)

ANSWERS TO FIGURE QUESTIONS

27.1 Plasma volume equals body mass × percent of body mass that is body fluid, × proportion of body fluid that is ECF, × proportion of ECF that is plasma, × a conversion factor (1 liter/kg). For males:

$$\text{plasma volume} = 60\,\text{kg} \times 0.60 \times \tfrac{1}{3} \times 0.20 \times 1\,\text{liter/kg} = 2.4\,\text{liters}$$

For females, plasma volume is 2.2 liters.

27.2 Hyperventilation, vomiting, fever, and diuretics all increase fluid loss.

27.3 Negative feedback is in operation because the result (an increase in fluid intake) is opposite to the initiating stimulus (dehydration).

27.4 Elevated aldosterone promotes abnormally high renal reabsorption of NaCl and water, which expands blood volume and increases blood pressure. Increased blood pressure causes more fluid to filter out of capillaries and accumulate in the interstitial fluid; this is edema.

27.5 When both the salt and water are absorbed in the digestive tract, blood volume increases without a decrease in osmolarity, so water intoxication does not occur.

27.6 In ECF, the major cation is Na^+, and the major anions are Cl^- and HCO_3^-. In ICF, the major cation is K^+, and the major anions are proteins and organic phosphates (for example, ATP).

27.7 Breath holding causes blood pH to decrease slightly as CO_2 and H^+ accumulate.

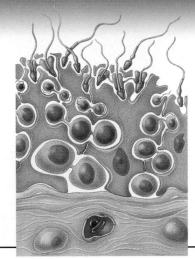

Sexual reproduction is a process in which organisms produce offspring by means of germ cells called **gametes** (GAM-ēts; = spouses). After the male gamete (sperm cell) unites with the female gamete (secondary oocyte)—an event called **fertilization**—the resulting cell contains one set of chromosomes from each parent. Males and females have anatomically distinct reproductive organs that are adapted for producing gametes, facilitating fertilization, and in females sustaining the growth of the embryo and fetus.

The male and female reproductive organs can be grouped by function. The **gonads**—testes in males and ovaries in females—produce gametes and secrete sex hormones. Various **ducts** then store and transport the gametes, and **accessory sex glands** produce substances that protect the gametes and facilitate their movement. Finally, *supporting structures,* such as the penis and the uterus, assist the delivery and joining of gametes and, in females, the growth of the fetus during pregnancy.

Gynecology (gī-ne-KOL-ō-jē; *gynec-* = woman; *-ology* = study of) is the specialized branch of medicine concerned with the diagnosis and treatment of diseases of the female reproductive system. As noted in Chapter 26, **urology** (u-ROL-ō-jē) is the study of the urinary system. Urologists also diagnose and treat diseases and disorders of the male reproductive system.

THE CELL CYCLE IN THE GONADS

OBJECTIVE

• *Describe the process of meiosis.*

Before we can discuss the cell cycle in the gonads, we must first understand the distribution of genetic material in cells.

Number of Chromosomes in Somatic Cells and Gametes

In humans, somatic cells, such as brain cells, stomach cells, kidney cells, and so forth, contain 23 pairs of chromosomes, or a total of 46 chromosomes; one member of each pair is inherited from each parent. The chromosomes that make up each pair are called **homologous chromosomes** (hō-MOL-ō-gus; *homo-* = same) or **homologs;** they contain similar genes arranged in the same (or almost the same) order. When examined under a light microscope (see Figure 29.18 on page 1047), homologous chromosomes generally look very similar; the exception to this rule is a pair of chromosomes called the **sex chromosomes,** designated X and Y. In females the homologous pair of sex chromosomes consists of two X chromosomes; in males the pair consists of an X and a Y chromosome. The other 22 pairs of chromosomes are called **autosomes.** Because somatic cells contain two sets of chromosomes, they are called **diploid cells** (DIP-loid; *dipl-* = double; *-loid* = form). Geneticists use the symbol n to denote the number of different chromosomes in an organism; in humans, $n = 23$. Diploid cells are $2n$.

In sexual reproduction, each new organism is the result of the union and fusion of two different gametes, one produced by each parent. But if each gamete had the same number of chromosomes as somatic cells, then the number of chromosomes would double each time fertilization oc-

Figure 28.1 Meiosis, reproductive cell division. Details of events are discussed in the text.

🔑 **In reproductive cell division, a single diploid parent cell undergoes meiosis I and meiosis II to produce four haploid gametes that are genetically different from the parent cell.**

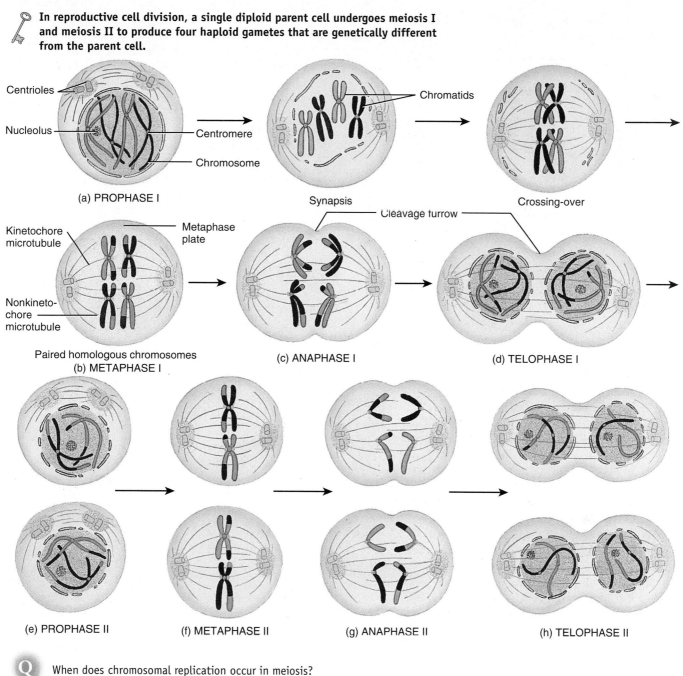

(a) PROPHASE I — Centrioles, Nucleolus, Centromere, Chromosome — Synapsis — Chromatids — Crossing-over

(b) METAPHASE I — Kinetochore microtubule, Metaphase plate, Nonkineto-chore microtubule, Paired homologous chromosomes

(c) ANAPHASE I — Cleavage furrow

(d) TELOPHASE I

(e) PROPHASE II

(f) METAPHASE II

(g) ANAPHASE II

(h) TELOPHASE II

Q When does chromosomal replication occur in meiosis?

curred. Chromosome number does not double at fertilization, however, because the cell cycle in the gonads produces gametes in which the number of chromosomes is reduced by half; gametes contain a single set of chromosomes and thus are **haploid cells** (HAP-loyd; *hapl-* = single). Put another way, gametes receive a single set of chromosomes (1*n*) via a special type of cell division called **meiosis** (mī-Ō-sis; *mei-* = lessening; *-osis* = condition of).

Meiosis

During development of gametes, meiosis results in the production of haploid cells that contain only 23 chromo-

somes. Meiosis occurs in two successive stages: **meiosis I** and **meiosis II.** During the interphase that precedes meiosis I, the chromosomes replicate in a manner similar to that in the interphase before mitosis in somatic cell division.

Meiosis I

Meiosis I, which begins once chromosomal replication is complete, consists of four phases: prophase I, metaphase I, anaphase I, and telophase I (Figure 28.1a–d). Prophase I is an extended phase in which the chromosomes shorten and thicken, the nuclear envelope and nucleoli disappear, and the mitotic spindle appears. In contrast to prophase of mitosis, the chromosomes become arranged in homologous

Figure 28.2 Crossing-over within a tetrad during prophase I of meiosis.

🔑 **Crossing-over permits an exchange of genes between homologous chromosomes.**

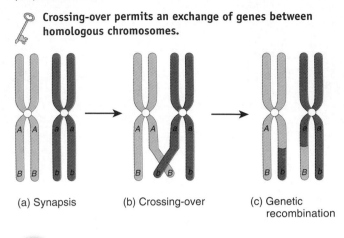

(a) Synapsis   (b) Crossing-over   (c) Genetic recombination

Q How does crossing-over affect the genetic content of daughter cells?

pairs. In metaphase I, the homologous pairs of chromosomes line up along the metaphase plate of the cell, with homologous chromosomes side by side. (No such pairing of homologous chromosomes occurs during metaphase of mitosis.) The pericentriolar area of a centrosome forms kinetochore microtubules that attach the centromeres to opposite poles of the cell. During anaphase I, the members of each homologous pair separate, with one member of each pair moving to an opposite pole of the cell. The centromeres do not split, and the paired chromatids, held by a centromere, remain together. (During anaphase of mitosis, the centromeres split and the sister chromatids separate.) Telophase I and cytokinesis are similar to telophase and cytokinesis of mitosis. The net effect of meiosis I is that each resulting daughter cell contains the haploid number of chromosomes; each cell contains only one member of each pair of the homologous chromosomes present in the parent cell.

Two events that are not seen in prophase I of mitosis (or in prophase II of meiosis) occur during prophase I of meiosis. First, the two chromatids of each pair of homologous chromosomes pair off, an event called **synapsis.** The resulting four chromatids form a **tetrad.** Second, portions of one chromatid may be exchanged with portions of another; such an exchange is termed **crossing-over** (Figure 28.2). This process, among others, permits an exchange of genes between homologous chromatids so that the resulting daughter cells are genetically unlike each other and genetically unlike the parent cell that produced them. Crossing-over results in **genetic recombination**—the formation of new combinations of genes—and accounts for part of the great genetic variation among humans and other organisms that form gametes via meiosis.

Meiosis II

The second stage of meiosis, meiosis II, also consists of four phases: prophase II, metaphase II, anaphase II, and

telophase II (see Figure 28.1e–h). These phases are similar to those that occur during mitosis; the centromeres split, and the sister chromatids separate and move toward opposite poles of the cell.

In summary, meiosis I begins with a diploid parent cell and ends with two daughter cells, each with the haploid number of chromosomes. During meiosis II, each haploid cell formed during meiosis I divides, and the net result is four genetically different haploid cells.

1. Distinguish between haploid (*n*) and diploid (*2n*) cells.
2. What are homologous chromosomes?
3. Prepare a table to compare meiosis with mitosis. (HINT: *To review mitosis, refer to Figure 3.33 on page 93.*)

THE MALE REPRODUCTIVE SYSTEM

OBJECTIVE

• *Describe the structure and functions of the organs of the male reproductive system.*

The organs of the male reproductive system are the testes, a system of ducts (including the ductus deferens, ejaculatory ducts, and urethra), accessory sex glands (seminal vesicles, prostate gland, and bulbourethral gland), and several supporting structures, including the scrotum and the penis (Figure 28.3). The testes (male gonads) produce sperm and secrete hormones. A system of ducts transports and stores sperm, assists in their maturation, and conveys them to the exterior. Semen contains sperm plus the secretions provided by the accessory sex glands.

Scrotum

The **scrotum** (SKRŌ-tum; = bag), the supporting structure for the testes, is a sac consisting of loose skin and superficial fascia that hangs from the root (attached portion) of the penis (see Figure 28.3a). Externally, the scrotum looks like a single pouch of skin separated into lateral portions by a median ridge called the **raphe** (RĀ-fē; = seam); internally, the scrotal septum divides the scrotum into two sacs, each containing a single testis (Figure 28.4). The septum consists of superficial fascia and muscle tissue called the **dartos muscle** (DAR-tōs; = skinned), which consists of bundles of smooth muscle fibers; the dartos muscle is also found in the subcutaneous tissue of the scrotum and is directly continuous with the subcutaneous tissue of the abdominal wall. When it contracts, the dartos muscle causes wrinkling of the skin of the scrotum.

The location of the scrotum and the contraction of its muscle fibers regulate the temperature of the testes. A temperature about 2–3°C below core temperature, which is required for normal sperm production, is maintained within the scrotum because it is outside the pelvic cavity. The **cremaster muscle** (krē-MAS-ter; = suspender), a small band of skeletal muscle in the spermatic cord that is a continuation of the internal oblique muscle, elevates the testes upon expo-

Figure 28.3 Male organs of reproduction and surrounding structures. (See Tortora, *A Photographic Atlas of the Human Body,* Figure 14.2)

🔑 **Reproductive organs are adapted for producing new individuals and passing on genetic material from one generation to the next.**

FUNCTIONS OF THE MALE REPRODUCTIVE SYSTEM
1. Testes: produce sperm and the male sex hormone testosterone.
2. Ducts: transport, store, and assist in maturation of sperm.
3. Accessory sex glands: secrete most of the liquid portion of semen.
4. Penis: contains the urethra, a passageway for ejaculation of semen and excretion of urine.

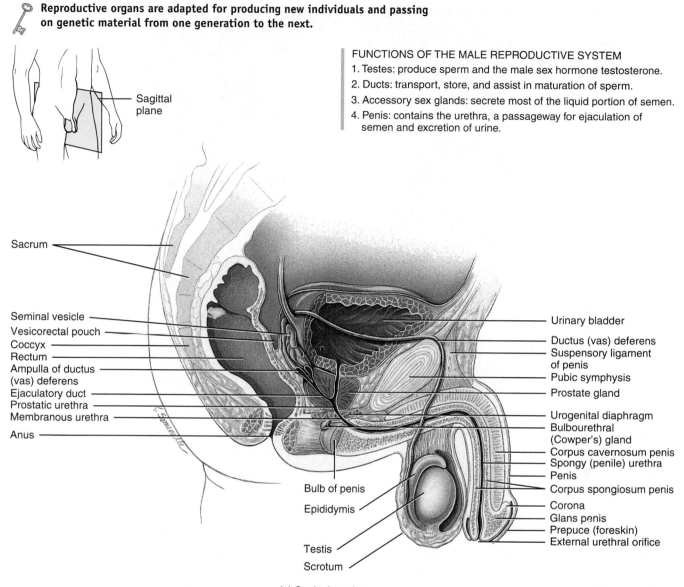

(a) Sagittal section

figure continues

sure to cold (and during sexual arousal). This action moves the testes closer to the pelvic cavity, where they can absorb body heat. Exposure to warmth reverses the process. The dartos muscle also contracts in response to cold and relaxes in response to warmth.

Testes

The **testes** (TES-tēz), or **testicles,** are paired oval glands measuring about 5 cm (2 in.) long and 2.5 cm (1 in.) in diameter (Figure 28.5 on page 980). Each **testis** (singular) weighs 10–15 grams. The testes develop near the kidneys, in the posterior portion of the abdomen, and they usually begin their descent into the scrotum through the inguinal canals

(passageways in the anterior abdominal wall; see Figure 28.4) during the latter half of the seventh month of fetal development. A serous membrane called the **tunica vaginalis** (*tunica* = sheath), which is derived from the peritoneum and forms during the descent of the testes, partially covers the testes. Internal to the tunica vaginalis is a dense white fibrous capsule, the **tunica albuginea** (al′-byoo-JIN-ē-a; *albu-* = white); it extends inward, forming septa that divide each testis into a series of internal compartments called **lobules.** Each of the 200–300 lobules contains one to three tightly coiled tubules, the **seminiferous tubules** (*semin-* = seed; *fer-* = to carry), where sperm are produced (Figure 28.6 on page 981).

Figure 28.3 (continued)

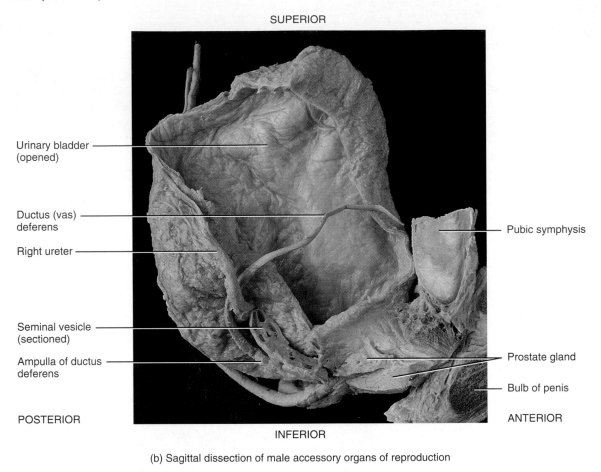

SUPERIOR

Urinary bladder (opened)

Ductus (vas) deferens

Right ureter

Pubic symphysis

Seminal vesicle (sectioned)

Ampulla of ductus deferens

Prostate gland

Bulb of penis

POSTERIOR

ANTERIOR

INFERIOR

(b) Sagittal dissection of male accessory organs of reproduction

Q What are the groups of reproductive organs in males, and what are the functions of each group?

Spermatogenic cells are cells at any of the sperm-forming stages. Sperm production begins in stem cells called **spermatogonia** (sper′-ma-tō-GŌ-nē-a; *-gonia* = offspring) that line the periphery of the seminiferous tubules (see Figure 28.6). These cells develop from **primordial germ cells** (*primordi-* = primitive or early form) that arise from yolk sac endoderm and enter the testes early in development. In the embryonic testes, the primordial germ cells differentiate into spermatogonia, which remain dormant during childhood. At puberty, they begin to undergo mitosis, then meiosis, and finally differentiation to eventually produce sperm. Toward the lumen of the tubule are layers of progressively more mature cells. In order of advancing maturity, these are primary spermatocytes, secondary spermatocytes, spermatids, and sperm. By the time a **sperm cell, or spermatozoon** (sper′-ma-tō-ZŌ-on; *-zoon* = life) has nearly reached maturity, it is released into the lumen of the seminiferous tubule. (The plural terms are **sperm** and **spermatozoa.**)

Embedded among the spermatogenic cells in the tubules are large **Sertoli cells** or *sustentacular cells* (sus′-ten-TAK-yoo-lar), which extend from the basement membrane

to the lumen of the tubule. Just internal to the basement membrane, neighboring Sertoli cells are joined to one another by tight junctions that form the **blood–testis barrier.** To reach the developing gametes, substances must first pass through the Sertoli cells. By isolating the spermatogenic cells from the blood, the barrier prevents an immune response against the spermatogenic cell's surface antigens, which are recognized as foreign by the immune system.

Sertoli cells support and protect developing spermatogenic cells; nourish spermatocytes, spermatids, and sperm; phagocytize excess spermatid cytoplasm as development proceeds; and control movements of spermatogenic cells and the release of sperm into the lumen of the seminiferous tubule. They also produce fluid for sperm transport, secrete androgen-binding protein and the hormone inhibin, and mediate the effects of other hormones.

In the spaces between adjacent seminiferous tubules are clusters of cells called **Leydig cells** or *interstitial endocrinocytes* (see Figure 28.6b). These cells secrete testosterone, the most important androgen (male sex hormone).

Figure 28.4 The scrotum, the supporting structure for the testes.

The scrotum consists of loose skin and superficial fascia and supports the testes.

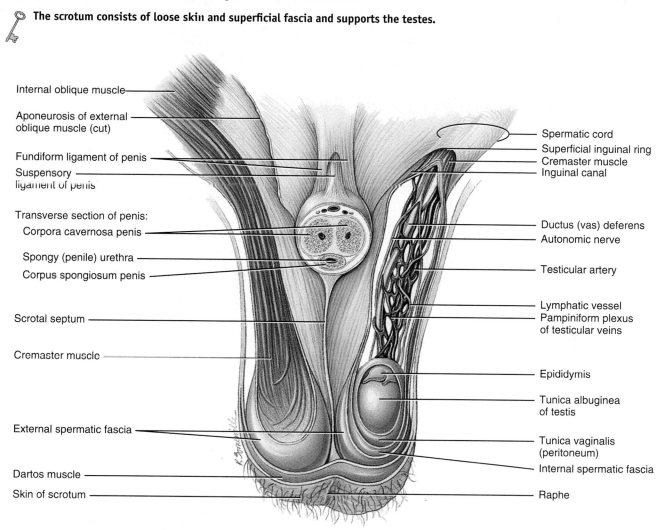

Internal oblique muscle

Aponeurosis of external oblique muscle (cut)

Fundiform ligament of penis

Suspensory ligament of penis

Transverse section of penis:

　Corpora cavernosa penis

　Spongy (penile) urethra

　Corpus spongiosum penis

Scrotal septum

Cremaster muscle

External spermatic fascia

Dartos muscle

Skin of scrotum

Spermatic cord

Superficial inguinal ring

Cremaster muscle

Inguinal canal

Ductus (vas) deferens

Autonomic nerve

Testicular artery

Lymphatic vessel

Pampiniform plexus of testicular veins

Epididymis

Tunica albuginea of testis

Tunica vaginalis (peritoneum)

Internal spermatic fascia

Raphe

Anterior view of scrotum and testes and transverse section of penis

Q Which muscles help regulate the temperature of the testes?

CLINICAL APPLICATION
Cryptorchidism

The condition in which the testes do not descend into the scrotum is called **cryptorchidism** (krip-TOR-ki-dizm; *crypt-* = hidden; *orchid* = testis); it occurs in about 3% of full-term infants and about 30% of premature infants. Untreated bilateral cryptorchidism results in sterility because the cells involved in the initial stages of spermatogenesis are destroyed by the higher temperature of the pelvic cavity. The chance of testicular cancer is 30–50 times greater in cryptorchid testes. The testes of about 80% of boys with cryptorchidism will descend spontaneously during the first year of life. When the testes remain undescended, the condition can be corrected surgically, ideally before 18 months of age. ■

The Process of Spermatogenesis

The process by which the seminiferous tubules of the testes produce haploid sperm is called **spermatogenesis** (sper′-ma-tō-JEN-e-sis; *genesis* = beginning process or production). In humans, this process takes about 65–75 days. Spermatogenesis begins in the spermatogonia, which contain the diploid (2*n*) chromosome number (Figure 28.7 on page 982). Spermatogonia are *stem cells* because when they undergo mitosis, some of the daughter cells remain near the basement membrane of the seminiferous tubule in an undifferentiated state to serve as a reservoir of cells for future mitosis and subsequent sperm production. The rest of the daughter cells lose contact with the basement membrane, undergo developmental changes, and differentiate into **primary spermatocytes** (SPER-ma-tō-sītz′). Primary spermatocytes, like spermatogonia, are diploid (2*n*); that is, they have 46 chromosomes.

Figure 28.5 Internal and external anatomy of a testis.

The testes are the male gonads, which produce haploid sperm.

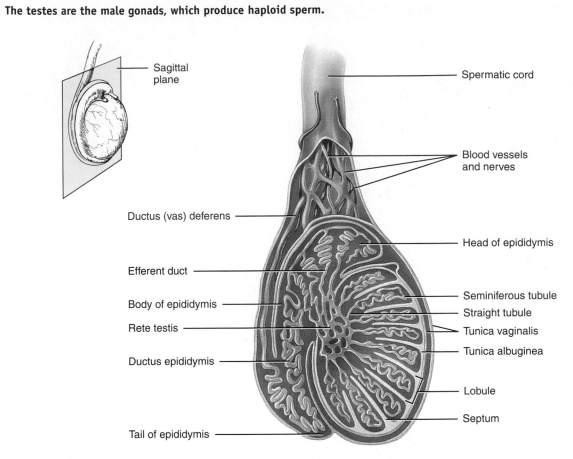

(a) Sagittal section of a testis showing seminiferous tubules

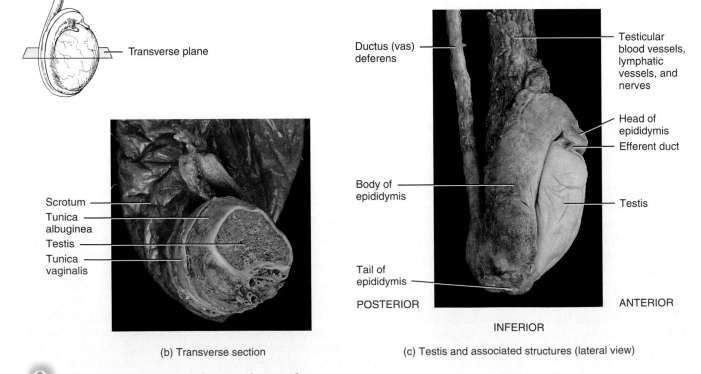

(b) Transverse section

(c) Testis and associated structures (lateral view)

Q What tissue layers cover and protect the testes?

Figure 28.6 Microscopic anatomy of the seminiferous tubules and stages of sperm production (spermatogenesis). Arrows in (b) indicate the progression of spermatogenic cells from least mature to most mature. The *(n)* and (2*n*) refer to haploid and diploid chromosome number, respectively.

🔑 **Spermatogenesis occurs in the seminiferous tubules of the testes.**

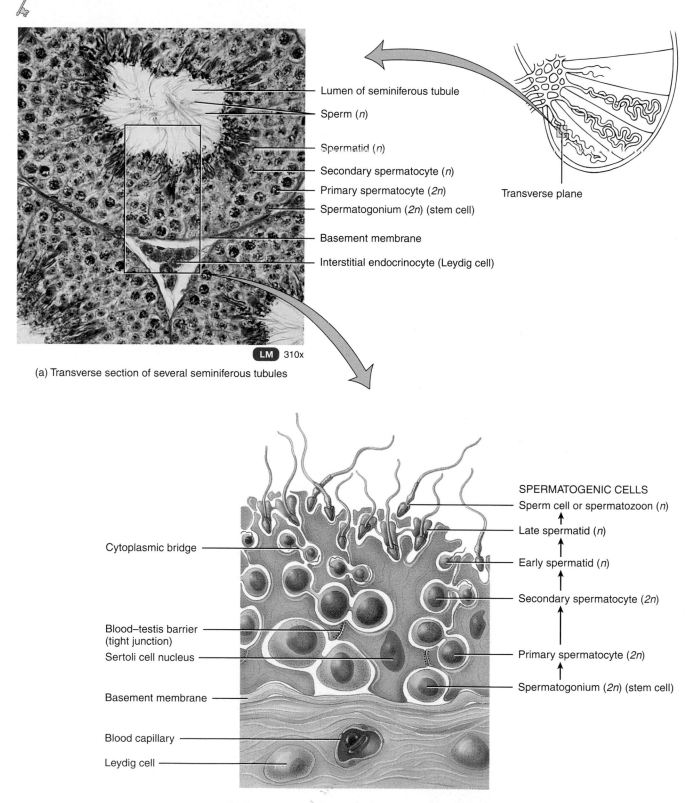

- Lumen of seminiferous tubule
- Sperm (*n*)
- Spermatid (*n*)
- Secondary spermatocyte (*n*)
- Primary spermatocyte (2*n*)
- Spermatogonium (2*n*) (stem cell)
- Basement membrane
- Interstitial endocrinocyte (Leydig cell)

Transverse plane

LM 310x

(a) Transverse section of several seminiferous tubules

SPERMATOGENIC CELLS
- Sperm cell or spermatozoon (*n*)
   ↑
- Late spermatid (*n*)
   ↑
- Early spermatid (*n*)
   ↑
- Secondary spermatocyte (2*n*)
   ↑
- Primary spermatocyte (2*n*)
   ↑
- Spermatogonium (2*n*) (stem cell)

Cytoplasmic bridge

Blood–testis barrier (tight junction)

Sertoli cell nucleus

Basement membrane

Blood capillary

Leydig cell

(b) Transverse section of a portion of a seminiferous tubule

Q Which cells produce testosterone?

Figure 28.7 Events in spermatogenesis. Diploid cells (*2n*) have 46 chromosomes; haploid cells (*n*) have 23 chromosomes.

🔑 **Spermiogenesis involves the maturation of spermatids into sperm.**

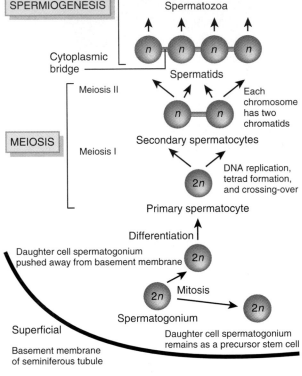

Q What is "reduced" during meiosis I?

Figure 28.8 A sperm cell (spermatozoon).

🔑 **About 300 million sperm mature each day.**

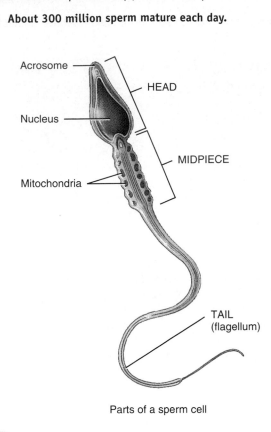

Parts of a sperm cell

Q What are the functions of each part of a sperm cell?

Each primary spermatocyte enlarges before dividing. Then two nuclear divisions occur as part of meiosis (see Figure 28.7). In meiosis I, DNA replicates, homologous pairs of chromosomes line up at the metaphase plate, and crossing-over occurs. Then, the meiotic spindle forms and pulls one (duplicated) chromosome of each pair to an opposite pole of the dividing cell. This random assortment of maternally derived chromosomes and paternally derived chromosomes toward opposite poles is another reason for genetic variation among gametes. The cells formed by meiosis I are called **secondary spermatocytes,** and each cell has 23 chromosomes —the haploid number. Each chromosome within a secondary spermatocyte, however, is made up of two chromatids (two copies of the DNA) still attached by a centromere.

In meiosis II, no replication of DNA occurs. The chromosomes line up in single file along the metaphase plate, and the chromatids of each chromosome separate from each other. The cells resulting from meiosis II are called **sper-**

matids; each spermatid is haploid. A primary spermatocyte therefore produces four spermatids through two rounds of cell division (meiosis I and meiosis II).

A unique and very interesting process occurs during spermatogenesis. As the sperm cells proliferate, they fail to complete cytoplasmic separation (cytokinesis). The four daughter cells remain in contact via cytoplasmic bridges through their entire development (see Figures 28.6b and 28.7). This pattern of development most likely accounts for the synchronized production of sperm in any given area of seminiferous tubule. It may have survival value in that half of the sperm contain an X chromosome and half contain a Y chromosome. The larger X chromosome may carry genes needed for spermatogenesis that are lacking on the smaller Y chromosome.

The final stage of spermatogenesis, called **spermiogenesis** (sper´-mē-ō-JEN-e-sis), is the maturation of spermatids into sperm. Because no cell division occurs in spermiogenesis, each spermatid develops into a single **sperm cell.** The release of a sperm cell from its connection to a Sertoli cell is known as **spermiation.** Sperm then enter the lumen of the seminiferous tubule and flow toward ducts of the testes.

Sperm mature at the rate of about 300 million per day and, once ejaculated, most probably do not survive more than 48 hours within the female reproductive tract. A sperm cell consists of structures highly adapted for reaching and penetrating a secondary oocyte: a head, a midpiece, and a tail (Figure 28.8). The **head** contains the nuclear material (DNA) and an **acrosome** (*acro-* = atop), a vesicle that contains hyaluronidase and proteinases, enzymes that aid penetration of the sperm cell into a secondary oocyte. Numerous mitochondria in the **midpiece** carry on the metabolism that provides ATP for locomotion. The **tail,** a typical flagellum, propels the sperm cell along its way.

Hormonal Control of Spermatogenesis

At the onset of puberty, the anterior pituitary gland increases its secretion of the gonadotropic hormones: **luteinizing hormone (LH)** and **follicle-stimulating hormone (FSH).** Their release is controlled by **gonadotropin releasing hormone (GnRH)** from the hypothalamus. Figure 28.9 summarizes the hormonal control of testicular functions.

LH stimulates the Leydig cells to secrete the hormone **testosterone** (tes-TOS-te-rōn), which is synthesized from cholesterol in the testes and is the principal androgen. It is lipid soluble and diffuses out of Leydig cells into interstitial fluid, and then into the bloodstream. In some target cells, such as those in the prostate gland and seminal vesicles, an enzyme called 5 alpha-reductase converts testosterone to an even more potent androgen called **dihydrotestosterone (DHT).**

FSH acts indirectly to stimulate spermatogenesis (see Figure 28.9). FSH and testosterone act synergistically on the Sertoli cells to stimulate secretion of **androgen-binding protein (ABP)** into the lumen of the seminiferous tubules and into the interstitial fluid around the spermatogenic cells. ABP binds to testosterone, thereby keeping the concentration of testosterone high near the seminiferous tubules. Testosterone stimulates the final steps of spermatogenesis.

Both androgens (testosterone and dihydrotestosterone) bind to the same androgen receptors, which are found within the nuclei of target cells. The hormone–receptor complex acts to regulate gene transcription, turning on some genes and turning others off. As a result of these changes, the androgens produce several effects:

- *Prenatal development.* Before birth, testosterone stimulates the male pattern of development of reproductive system ducts and the descent of the testes. Dihydrotestosterone, by contrast, stimulates development of the external genitals (described on page 1012). Testosterone also is converted in the brain to estrogens (female sex hormones), which may play a role in the development of certain regions of the brain in males.

- *Development of male sexual characteristics.* At puberty, testosterone and dihydrotestosterone bring about development and enlargement of the male sex organs and the development of masculine secondary sexual characteristics. These include muscular and skeletal growth that results in wide shoulders and narrow hips; pubic, axillary,

Figure 28.9 Hormonal control of testicular functions. In response to stimulation by FSH and testosterone, Sertoli cells secrete androgen-binding protein (ABP).

🔑 **Release of FSH is stimulated by GnRH and inhibited by inhibin; release of LH is stimulated by GnRH and inhibited by testosterone.**

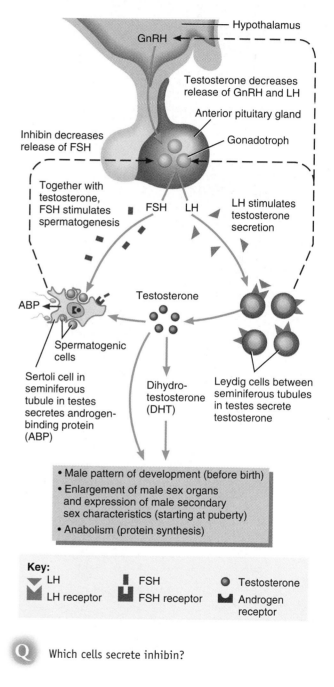

Q Which cells secrete inhibin?

facial, and chest hair (within hereditary limits); thickening of the skin; increased sebaceous (oil) gland secretion; and enlargement of the larynx and consequent deepening of the voice.

Figure 28.10 Negative feedback control of blood level of testosterone.

Gonadotrophs of the anterior pituitary gland produce leutinizing hormone (LH).

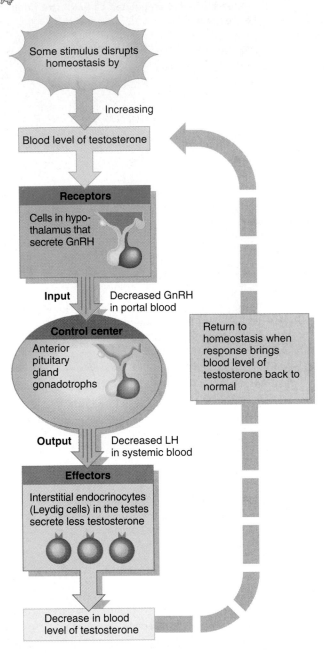

Which hormones inhibit secretion of FSH and LH by the anterior pituitary gland?

fect is obvious in the heavier muscle and bone mass of most men as compared to women. Androgens also stimulate closure of the epiphyseal plates.

A negative feedback system regulates testosterone production (Figure 28.10). When testosterone concentration in the blood increases to a certain level, it inhibits the release of GnRH by cells in the hypothalamus. As a result, there is less GnRH in the portal blood that flows from the hypothalamus to the anterior pituitary gland. Gonadotrophs in the anterior pituitary gland then release less LH, so the concentration of LH in systemic blood falls. With less stimulation by LH, the Leydig cells in the testes secrete less testosterone, and there is a return to homeostasis. If the testosterone concentration in the blood falls too low, however, GnRH is again released by the hypothalamus and stimulates secretion of LH by the anterior pituitary gland, which stimulates testosterone production by the testes.

Once the degree of spermatogenesis required for male reproductive functions has been achieved, Sertoli cells release **inhibin,** a protein hormone named for its inhibition of FSH secretion by the anterior pituitary gland (see Figure 28.9). Inhibin thus ultimately decreases spermatogenesis. If spermatogenesis is proceeding too slowly, less inhibin is released, which permits more FSH secretion and an increased rate of spermatogenesis.

1. Describe the function of the scrotum in protecting the testes from temperature fluctuations.
2. Describe the internal structure of a testis. Where are sperm cells produced? What are the functions of Sertoli cells and Leydig cells?
3. Describe the principal events of spermatogenesis.
4. Identify the principal parts of a sperm cell, and list the functions of each.
5. Explain the effects of FSH and LH on the male reproductive system. How are these hormones controlled by GnRH?
6. Describe the physiological effects of testosterone and inhibin on the male reproductive system. How is the blood level of testosterone controlled?

Reproductive System Ducts in Males

Ducts of the Testis

After their release into the lumen of the very convoluted seminiferous tubules, sperm and fluid are propelled toward the straight tubules by pressure generated by the continual release of sperm and fluid secreted by Sertoli cells. The straight tubules lead to a network of ducts in the testis called the **rete testis** (RĒ-tē; = network) (see Figure 28.5a). From the rete testis, sperm move into a series of coiled **efferent ducts** in the epididymis that empty into a single tube called the **ductus epididymis.**

- *Development of sexual function.* Androgens contribute to male sexual behavior and spermatogenesis and to sex drive (libido) in both males and females. Recall that the adrenal cortex is the main source of androgens in females.
- *Stimulation of anabolism.* Androgens are anabolic hormones; that is, they stimulate protein synthesis. This ef-

Epididymis

The **epididymis** (ep′-i-DID-i-mis; *epi-* = over; *-didymis* = testis) is a comma-shaped organ about 4 cm (1.5 in.) long that lies along the posterior border of each testis (see Figure 28.5a). The plural is **epididymides** (ep′-i-did-ĪM-i-dēs). Each epididymis consists mostly of the tightly coiled **ductus epididymis.** The larger, superior portion of the epididymis, the **head,** is where the efferent ducts from the testis join the ductus epididymis. The **body** is the narrow midportion of the epididymis, and the **tail** is the smaller, inferior portion. At its distal end, the tail of the epididymis continues as the ductus (vas) deferens (discussed shortly).

The ductus epididymis is a tightly coiled structure that would measure about 6 m (20 ft) in length if it were straightened out. It is lined with pseudostratified columnar epithelium and encircled by layers of smooth muscle. The free surfaces of the columnar cells contain long, branching microvilli called **stereocilia** that increase surface area for the reabsorption of degenerated sperm.

Functionally, the ductus epididymis is the site where sperm motility increases over a 10–14 day period. The ductus epididymis also stores sperm and helps propel them by peristaltic contraction of its smooth muscle into the ductus (vas) deferens. Sperm may remain in storage in the ductus epididymis for a month or more.

Ductus Deferens

Within the tail of the epididymis, the ductus epididymis becomes less convoluted, and its diameter increases; beyond this point, the duct is referred to as the **ductus deferens** or **vas deferens** (see Figure 28.5a). The ductus deferens, which is about 45 cm (18 in.) long, ascends along the posterior border of the epididymis, passes through the inguinal canal (see Figure 28.4), and enters the pelvic cavity; there it loops over the ureter and passes over the side and down the posterior surface of the urinary bladder (see Figure 28.3a). The dilated terminal portion of the ductus deferens is known as the **ampulla** (am-POOL-la; = little jar) (see Figure 28.11). The ductus deferens is lined with pseudostratified columnar epithelium and contains a heavy coat of three layers of muscle; the inner and outer layers are longitudinal, and the middle layer is circular.

Functionally, the ductus deferens stores sperm; they can remain viable here for up to several months. The ductus deferens also conveys sperm from the epididymis toward the urethra by peristaltic contractions of the muscular coat. Sperm that are not ejaculated are ultimately reabsorbed.

The **spermatic cord** is a supporting structure of the male reproductive system that ascends out of the scrotum (see Figure 28.4). It consists of the ductus (vas) deferens as it ascends through the scrotum, the testicular artery, autonomic nerves, veins that drain the testes and carry testosterone into circulation (pampiniform plexus), lymphatic vessels, and the cremaster muscle. The spermatic cord and ilioinguinal nerve pass through the **inguinal canal** (IN-gwin-al; = groin), an oblique passageway in the anterior abdominal wall just superior and parallel to the medial half of the inguinal ligament. The canal, which is about 4–5 cm (about 2 in.) long, originates at the **deep (abdominal) inguinal ring,** a slitlike opening in the aponeurosis of the transversus abdominis muscle; the canal ends at the **superficial (subcutaneous) inguinal ring** (see Figure 28.4), a somewhat triangular opening in the aponeurosis of the external oblique muscle. In females, the round ligament of the uterus and ilioinguinal nerve pass through the inguinal canal.

CLINICAL APPLICATION
Inguinal Hernias

Because the inguinal region is a weak area in the abdominal wall, it is often the site of an **inguinal hernia**—a rupture or separation of a portion of the inguinal area of the abdominal wall resulting in the protrusion of a part of the small intestine. In an *indirect inguinal hernia,* a part of the small intestine protrudes through the deep inguinal ring and enters the scrotum. In a *direct inguinal hernia,* a portion of the small intestine pushes into the posterior wall of the inguinal canal, usually causing a localized bulging in the wall of the canal. Inguinal hernias are much more common in males than in females because the larger inguinal canals in males represent larger weak points in the abdominal wall. ■

Ejaculatory Ducts

Each **ejaculatory duct** (e-JAK-yoo-la-tō′-rē; *ejacul-* = to expel) is about 2 cm (1 in.) long and is formed by the union of the duct from the seminal vesicle and the ampulla of the ductus (vas) deferens (Figure 28.11). The ejaculatory ducts form just above the base (superior portion) of the prostate gland and pass inferiorly and anteriorly through the prostate gland. They terminate in the prostatic urethra, where they eject sperm and seminal vesicle secretions just before **ejaculation,** the powerful propulsion of semen from the urethra to the exterior. They also transport and eject secretions of the seminal vesicles (described shortly).

Urethra

In males, the **urethra** is the shared terminal duct of the reproductive and urinary systems; it serves as a passageway for both semen and urine. The urethra, which is about 20 cm (8 in.) long, passes through the prostate gland, the urogenital diaphragm, and the penis, and is subdivided into three parts (see Figures 28.3a and 28.11). The **prostatic urethra** is 2–3 cm (1 in.) long and passes through the prostate gland. As this duct continues inferiorly, it passes through the urogenital diaphragm (a muscular partition between the two ischial and pubic rami; see Figure 11.13 on page 333), where it is known as the membranous urethra. The **membranous urethra** is about 1 cm (0.5 in.) in length. As this duct passes through the corpus spongiosum of the penis, it is known as the **spongy (penile) urethra,** which is about 15–20 cm

Figure 28.11 Locations of several accessory reproductive organs in males. The prostate gland, urethra, and penis have been sectioned to show internal details.

🗝 **The male urethra has three subdivisions: the prostatic, membranous, and spongy (penile) urethra.**

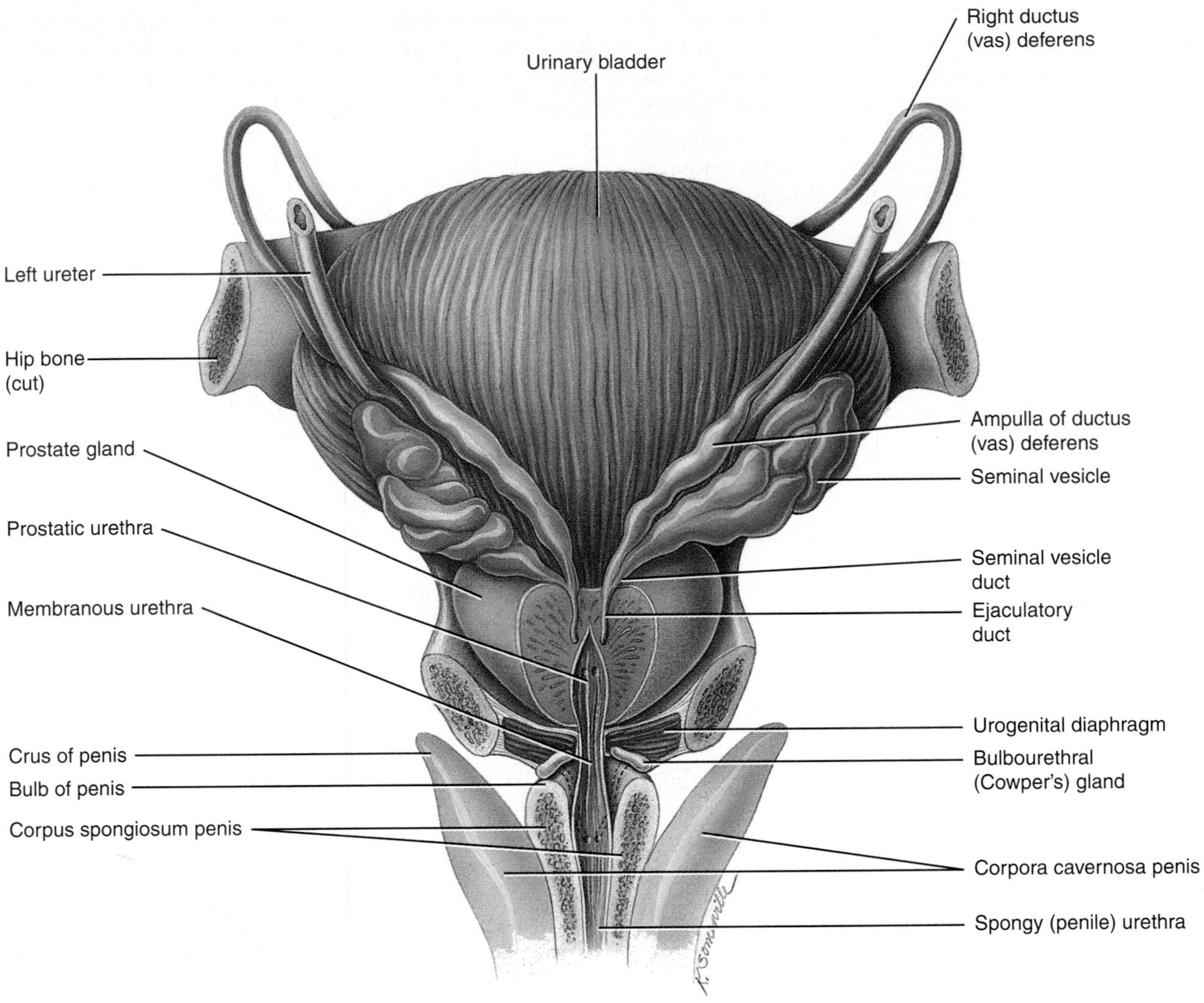

Urinary bladder

Right ductus (vas) deferens

Left ureter

Hip bone (cut)

Prostate gland

Prostatic urethra

Membranous urethra

Crus of penis

Bulb of penis

Corpus spongiosum penis

Ampulla of ductus (vas) deferens

Seminal vesicle

Seminal vesicle duct

Ejaculatory duct

Urogenital diaphragm

Bulbourethral (Cowper's) gland

Corpora cavernosa penis

Spongy (penile) urethra

Posterior view of male accessory organs of reproduction

FUNCTIONS OF ACCESSORY SEX GLAND SECRETIONS

1. Seminal vesicles: secrete alkaline, viscous fluid that helps neutralize acid in the female reproductive tract, provides fructose for ATP production by sperm, contributes to sperm motility and viability, and helps semen coagulate after ejaculation.
2. Prostate gland: secretes a milky, slightly acidic fluid that helps semen coagulate after ejaculation and subsequently breaks down the clot.
3. Bulbourethral (Cowper's) glands: secrete alkaline fluid that neutralizes the acidic environment of the uretha and mucus that lubricates lining of the urethra and tip of the penis during sexual intercourse.

Q   What accessory sex gland contributes the majority of the seminal fluid?

(6–8 in.) long. The spongy urethra ends at the **external urethral orifice.** The histology of the male urethra may be reviewed on page 947.

1. Which ducts transport sperm within the testes?
2. Describe the location, structure, and functions of the ductus epididymis, ductus (vas) deferens, and ejaculatory duct.
3. List the structures within the spermatic cord.
4. Give the locations of the three subdivisions of the male urethra.
5. Trace the course of sperm through the system of ducts from the seminiferous tubules through the urethra.

Accessory Sex Glands

Whereas the ducts of the male reproductive system store and transport sperm cells, the **accessory sex glands** secrete most of the liquid portion of semen. Among the accessory sex glands are the seminal vesicles, the prostate gland, and the bulbourethral glands.

Seminal Vesicles

The paired **seminal vesicles** (VES-i-kuls) are convoluted pouchlike structures, about 5 cm (2 in.) in length, lying posterior to and at the base of the urinary bladder anterior to the rectum (see Figure 28.11). They secrete an alkaline, viscous fluid that contains fructose (a monosaccharide sugar), prostaglandins, and clotting proteins unlike those found in blood. The alkaline nature of the fluid helps to neutralize the acidic environment of the male urethra and female reproductive tract that otherwise would inactivate and kill sperm. The fructose is used for ATP production by sperm. Prostaglandins contribute to sperm motility and viability and may also stimulate muscular contraction within the female reproductive tract. Fluid secreted by the seminal vesicles normally constitutes about 60% of the volume of semen.

Prostate Gland

The **prostate gland** (PROS-tāt) is a single, doughnut-shaped gland about the size of a chestnut that lies inferior to the urinary bladder and surrounds the prostatic urethra (see Figure 28.11). The prostate secretes a milky, slightly acidic fluid (pH about 6.5) that contains (1) *citric acid,* which can be used by sperm for ATP production via the Krebs cycle (see page 879); (2) acid phosphatase (the function of which is unknown); and (3) several proteolytic enzymes, such as *prostate-specific antigen (PSA),* pepsinogen, lysozyme, amylase, and hyaluronidase.

Secretions of the prostate gland enter the prostatic urethra through many prostatic ducts. Prostatic secretions make up about 25% of the volume of semen and contribute to sperm motility and viability. The prostate gland slowly increases in size from birth to puberty, and then expands rapidly. The size attained by age 30 remains stable until about age 45, when further enlargement may occur.

Bulbourethral Glands

The paired **bulbourethral glands** (bul'-bō-yoo-RĒ-thral), or **Cowper's glands,** each about the size of a pea, lie inferior to the prostate gland on either side of the membranous urethra within the urogenital diaphragm; their ducts open into the spongy urethra (see Figure 28.11). During sexual arousal, the bulbourethral glands secrete an alkaline substance that protects the passing sperm by neutralizing acids from urine in the urethra. At the same time, they also secrete mucus that lubricates the end of the penis and the lining of the urethra, thereby decreasing the number of sperm damaged during ejaculation.

Semen

Semen (= seed) is a mixture of sperm and **seminal fluid,** a liquid that consists of the secretions of the seminiferous tubules, seminal vesicles, prostate gland, and bulbourethral glands. The volume of semen in a typical ejaculation is 2.5–5 mL, with a sperm count (concentration) of 50–150 million sperm per milliliter. A male whose sperm count falls below 20 million/mL is likely to be infertile. The very large number is required because only a tiny fraction ever reach the secondary oocyte.

Despite the slight acidity of prostatic fluid, semen has a slightly alkaline pH of 7.2–7.7 due to the higher pH and larger volume of fluid from the seminal vesicles. The prostatic secretion gives semen a milky appearance, whereas fluids from the seminal vesicles and bulbourethral glands give it a sticky consistency. Seminal fluid provides sperm with a transportation medium and nutrients, and it neutralizes the hostile acidic environment of the male's urethra and of the vagina. Semen also contains an antibiotic, *seminalplasmin,* that can destroy certain bacteria. Because both semen and the lower female reproductive tract contain bacteria, the antibiotic activity of seminalplasmin may help control the abundance of these bacteria.

Once ejaculated, liquid semen coagulates within 5 minutes due to the presence of clotting proteins from the seminal vesicles. The functional role of semen coagulation is not known, but the proteins involved are different from those that cause blood coagulation. After about 10–20 minutes, semen reliquefies because prostate-specific antigen and other proteolytic enzymes produced by the prostate gland break down the clot. Abnormal or delayed liquefaction of clotted semen may cause complete or partial immobilization of sperm, thereby inhibiting their movement through the cervix of the uterus.

Penis

The **penis** contains the urethra and is a passageway for the ejaculation of semen and the excretion of urine. It is cylindrical in shape and consists of a body, root, and glans penis. The **body** of the penis is composed of three cylindrical masses of tissue, each surrounded by fibrous tissue called the

Figure 28.12 Internal structure of the penis. The inset in (b) shows details of the skin and fascia. (See Tortora, *A Photographic Atlas of the Human Body,* Figure 14.6)

🔑 **The penis contains the urethra, a pathway for the ejaculation of semen and the excretion of urine.**

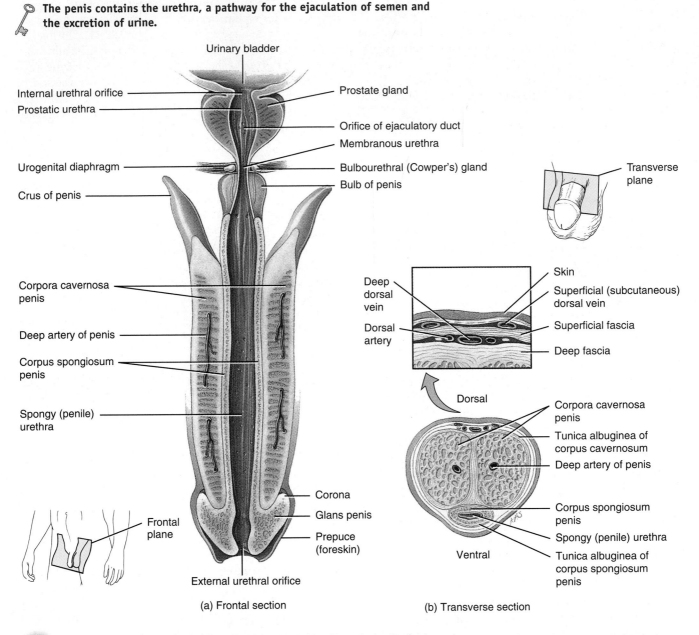

(a) Frontal section

(b) Transverse section

Q Which tissue masses form the erectile tissue in the penis, and why do they become rigid during sexual arousal?

tunica albuginea (Figure 28.12b). The paired dorsolateral masses are called the **corpora cavernosa penis** (*corpora* = main bodies; *cavernosa* = hollow); the smaller midventral mass, the **corpus spongiosum penis,** contains the spongy urethra and functions in keeping the spongy urethra open during ejaculation. All three masses are enclosed by fascia and skin and consist of erectile tissue permeated by blood sinuses.

Upon sexual stimulation, which may be visual, tactile, auditory, olfactory, or from the imagination, the arteries supplying the penis dilate, and large quantities of blood enter the blood sinuses. Expansion of these spaces compresses the veins draining the penis, so blood outflow is slowed.

These vascular changes, due to a parasympathetic reflex, result in an **erection.** The penis returns to its flaccid state when the arteries constrict and pressure on the veins is relieved.

Ejaculation is a sympathetic reflex. As part of the reflex, the smooth muscle sphincter at the base of the urinary bladder closes. As a result, urine is not expelled during ejaculation, and semen does not enter the urinary bladder. Even before ejaculation occurs, peristaltic contractions in the ampulla of the ductus deferens, seminal vesicles, ejaculatory ducts, and prostate gland propel semen into the penile portion of the urethra (spongy urethra). Typically, this leads to **emission** (ē-MISH-un), the discharge of a small volume of semen before ejaculation. Emission may also occur during sleep (nocturnal emission).

The root of the penis is the attached portion (proximal portion) and consists of the **bulb of the penis,** the expanded portion of the base of the corpus spongiosum penis, and the **crura of the penis** (singular is **crus** = resembling a leg), the two separated and tapered portions of the corpora cavernosa penis (Figure 28.12a). The bulb of the penis is attached to the inferior surface of the urogenital diaphragm and enclosed by the bulbospongiosus muscle. Each crus of the penis is attached to the ischial and inferior pubic rami and is surrounded by the ischiocavernosus muscle (see Figure 11.13 on page 333). Contraction of these skeletal muscles aids ejaculation.

The distal end of the corpus spongiosum penis is a slightly enlarged, acorn-shaped region called the **glans penis;** its margin is the **corona.** The distal urethra enlarges within the glans penis and forms a terminal slitlike opening, the **external urethral orifice.** Covering the glans in an uncircumcised penis is the loosely fitting **prepuce** (PRĒ-pyoos), or **foreskin.** The weight of the penis is supported by two ligaments that are continuous with the fascia of the penis: the **fundiform ligament,** which arises from the inferior part of the linea alba, and the **suspensory ligament of the penis,** which arises from the pubic symphysis.

CLINICAL APPLICATION
Circumcision

Circumcision (= to cut around) is a surgical procedure in which part or all of the prepuce is removed. It is usually performed just after delivery, 3–4 days after birth, or on the eighth day as part of a Jewish religious rite. Although some health-care professionals can find no medical justification for circumcision, others feel that it has benefits, such as a lower risk of urinary tract infections, protection against penile cancer, and possibly a lower risk for sexually transmitted diseases. ■

1. Briefly explain the locations and functions of the seminal vesicles, the prostrate gland, and the bulbourethral (Cowper's) glands.
2. What is semen? What is its function?
3. How does an erection occur?

THE FEMALE REPRODUCTIVE SYSTEM
OBJECTIVE

• *Describe the location, structure, and functions of the organs of the female reproductive system.*

The organs of reproduction in females (Figure 28.13) include the ovaries, which produce secondary oocytes and hormones, such as progesterone and estrogens (the female sex hormones), inhibin, and relaxin; the uterine (Fallopian) tubes, or oviducts, which transport secondary oocytes and fertilized ova to the uterus; the uterus, in which embryonic and fetal development occur; the vagina; and external organs that constitute the vulva, or pudendum. The mammary glands also are considered part of the female reproductive system.

Ovaries

The **ovaries** (= egg receptacles) are paired glands that resemble unshelled almonds in size and shape; they are homologous to the testes. (*Homologous* means that two organs have the same embryonic origin.) The ovaries, one on either side of the uterus, descend to the brim of the superior portion of the pelvic cavity during the third month of development. A series of ligaments holds them in position (Figure 28.14 on page 992). The **broad ligament** of the uterus (see also Figure 28.13b), which is itself part of the parietal peritoneum, attaches to the ovaries by a double-layered fold of peritoneum called the **mesovarium.** The **ovarian ligament** anchors the ovaries to the uterus, and the **suspensory ligament** attaches them to the pelvic wall. Each ovary contains a **hilus,** the point of entrance and exit for blood vessels and nerves and along which the mesovarium is attached.

Histology of the Ovary

Each ovary consists of the following parts (Figure 28.15 on page 993):

• The **germinal epithelium** is a layer of simple epithelium (low cuboidal or squamous) that covers the surface of the ovary and is continuous with the mesothelium that covers the mesovarium. The term *germinal epithelium* is a misnomer because it does not give rise to ova, although at one time people believed that it did. Now we know that the progenitors of ova arise from the endoderm of the yolk sac and migrate to the ovaries during embryonic development.
• The **tunica albuginea** is a whitish capsule of dense, irregular connective tissue immediately deep to the germinal epithelium.
• The **ovarian cortex** is a region just deep to the tunica albuginea that consists of dense connective tissue and contains ovarian follicles (described shortly).
• The **ovarian medulla** is a region deep to the ovarian cortex that consists of loose connective tissue and contains blood vessels, lymphatics, and nerves.

Figure 28.13 Organs of reproduction and surrounding structures in females.

🔑 **The organs of reproduction in females include the ovaries, uterine (Fallopian) tubes, uterus, vagina, vulva, and mammary glands.**

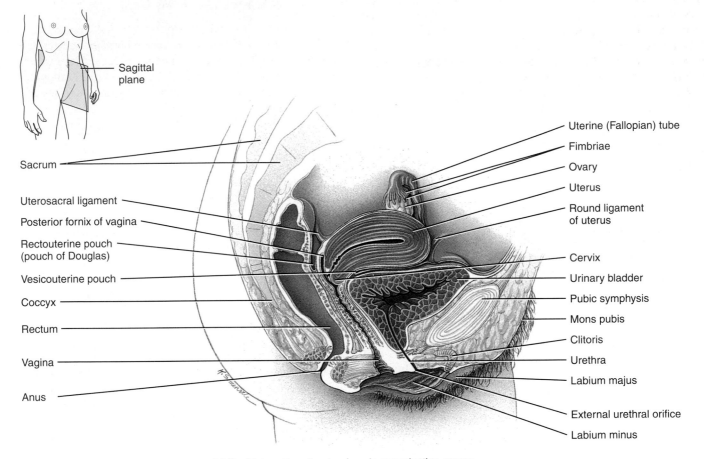

(a) Sagittal section showing female reproductive organs

FUNCTIONS OF THE FEMALE REPRODUCTIVE SYSTEM

1. Ovaries: produce secondary oocytes and hormones, including progesterone and estrogens (female sex hormones), inhibin, and relaxin.

2. Uterine tubes: transport a secondary oocyte to the uterus and normally are the sites where fertilization occurs.

3. Uterus: site of implantation of a fertilized ovum, development of the fetus during pregnancy, and labor.

4. Vagina: receives the penis during sexual intercourse and is a passageway for childbirth.

5. Mammary glands: synthesize, secrete, and eject milk for nourishment of the newborn.

- **Ovarian follicles** (*folliculus* = little bag) lie in the cortex and consist of **oocytes** in various stages of development, plus the surrounding cells. When the surrounding cells form a single layer, they are called **follicular cells;** later in development, when they form several layers, they are referred to as **granulosa cells.** The surrounding cells nourish the developing oocyte and begin to secrete estrogens as the follicle grows larger.

- A **mature (Graafian) follicle** is a large, fluid-filled follicle that soon will rupture and expel a secondary oocyte, a process called **ovulation.**

- A **corpus luteum** (= yellow body) contains the remnants of an ovulated mature follicle. The corpus luteum produces progesterone, estrogens, relaxin, and inhibin until it degenerates and turns into fibrous tissue called a **corpus albicans** (= white body).

Figure 28.13 (continued)

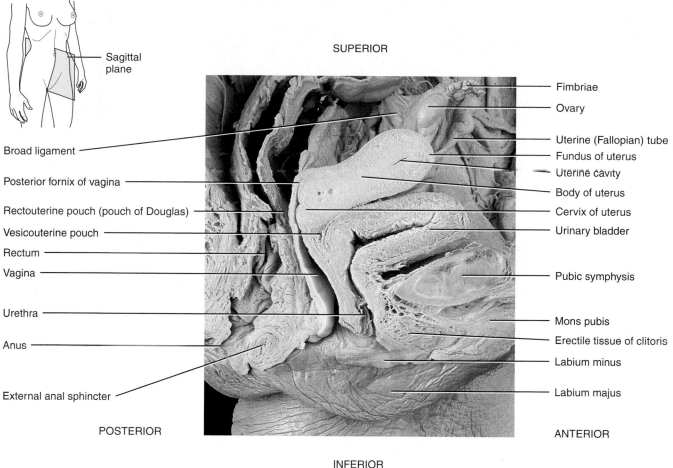

(b) Sagittal section showing female reproductive organs

Q Which structures in males are homologous to the ovaries, the clitoris, the paraurethral glands, and the greater vestibular glands?

Oogenesis

Formation of gametes in the ovaries is termed **oogenesis** (ō-ō-JEN-e-sis; *oo-* = egg), and like spermatogenesis it involves meiosis. During early fetal development, primordial (primitive) germ cells migrate from the endoderm of the yolk sac to the ovaries. There, germ cells differentiate within the ovaries into **oogonia** (ō'-o-GO-nē-a; singular is **oogonium;** ō'-o-GŌ-nē-um), which are diploid (2*n*) cells that divide mitotically to produce millions of germ cells. Even before birth, most of these germ cells degenerate, a process known as **atresia.** A few, however, develop into larger cells called **primary oocytes** (Ō-ō-sītz) that enter prophase of meiosis I during fetal development but do not complete that phase until after puberty. At birth, 200,000–2,000,000 oogonia and primary oocytes remain in each ovary. Of these, about 40,000 remain at puberty, and only 400 will mature and ovulate during a woman's reproductive lifetime; the remainder undergo atresia.

Each primary oocyte is surrounded by a single layer of follicular cells, and the entire structure is called a **primordial follicle** (Figure 28.16a on page 994). Although the stimulating mechanism is unclear, a few primordial follicles periodically start to grow, even during childhood. They become **primary follicles,** which are surrounded first by one layer of cuboidal-shaped follicular cells and then by six to seven layers of cuboidal and low-columnar cells called **granulosa cells.** As a follicle grows, it forms a clear glycoprotein layer, called the **zona pellucida** (pe-LOO-si-da), between the primary oocyte and the granulosa cells. The innermost layer of granulosa cells becomes firmly attached to the zona pellucida and is called the **corona radiata** (*corona* = crown; *radiata* = radiation) (Figure 28.16b). The outermost granulosa cells rest on a basement membrane that separates them from the surrounding ovarian stroma; this outer region is called the **theca folliculi.** As a primary follicle continues to grow, the theca differentiates into two layers: (1) the **theca interna,** a

Figure 28.14 Relative positions of the ovaries, the uterus, and the ligaments that support them. (See Tortora, *A Photographic Atlas of the Human Body*, Figure 14.9)

🔑 **Ligaments holding the ovaries in position are the mesovarium, the ovarian ligament, and the suspensory ligament.**

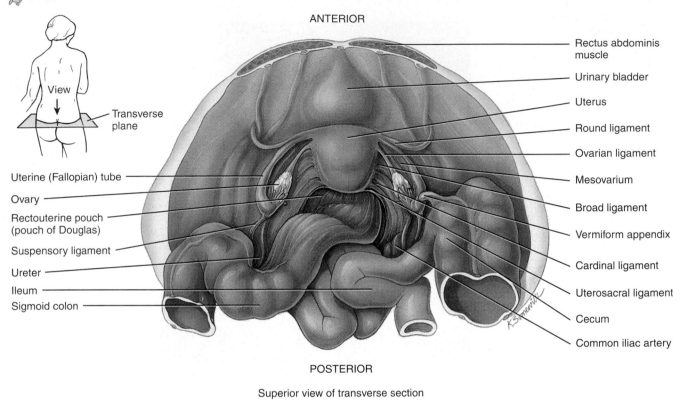

ANTERIOR

Rectus abdominis muscle
Urinary bladder
Uterus
Round ligament
Ovarian ligament
Mesovarium
Broad ligament
Vermiform appendix
Cardinal ligament
Uterosacral ligament
Cecum
Common iliac artery

View
Transverse plane

Uterine (Fallopian) tube
Ovary
Rectouterine pouch (pouch of Douglas)
Suspensory ligament
Ureter
Ileum
Sigmoid colon

POSTERIOR

Superior view of transverse section

Q To which structures do the mesovarium, ovarian ligament, and suspensory ligament anchor the ovary?

vascularized internal layer of secretory cells, and (2) the **theca externa,** an outer layer of connective tissue cells. The granulosa cells begin to secrete follicular fluid, which builds up in a cavity called the antrum in the center of the follicle, which is now termed a **secondary follicle.** During childhood, primordial and developing follicles continue to undergo atresia.

Each month after puberty, the gonadotropic hormones secreted by the anterior pituitary gland stimulate the resumption of oogenesis (Figure 28.17 on page 994). Meiosis I resumes in several secondary follicles, although only one will eventually reach the maturity needed for ovulation. The diploid primary oocyte completes meiosis I, and two haploid cells of unequal size—both with 23 chromosomes (*n*) of two chromatids each—are produced. The smaller cell produced by meiosis I, called the **first polar body,** is essentially a packet of discarded nuclear material; the larger cell, known as the **secondary oocyte,** receives most of the cytoplasm. Once a secondary oocyte is formed, it proceeds to the metaphase of meiosis II and then stops. The follicle in which these events are taking place—the **mature (Graafian) follicle**—soon ruptures and releases its secondary oocyte, a process known as **ovulation.**

At ovulation, usually one secondary oocyte (with the first polar body and corona radiata) is expelled into the pelvic cavity. Normally these cells are swept into the uterine tube. If fertilization does not occur, the secondary oocyte degenerates. If sperm are present in the uterine tube and one penetrates the secondary oocyte, however, then meiosis II resumes. The secondary oocyte splits into two haploid (*n*) cells, again of unequal size. The larger cell is the **ovum,** or mature egg; the smaller one is the **second polar body.** The nuclei of the sperm cell and the ovum then unite, forming a diploid (2*n*) **zygote.** If the first polar body undergoes another division to produce two polar bodies, then the primary oocyte ultimately gives rise to a single haploid (*n*) ovum and three haploid (*n*) polar bodies, which all degenerate. Thus, one oogonium gives rise to a single gamete (an ovum), whereas one spermatogonium produces four gametes (sperm).

1. How are the ovaries held in position in the pelvic cavity?
2. Describe the microscopic structure and function of an ovary.
3. Describe the principal events of oogenesis.

Figure 28.15 Histology of the ovary. The arrows indicate the sequence of developmental stages that occur as part of the maturation of an ovum during the ovarian cycle.

🔑 **The ovaries are the female gonads; they produce haploid oocytes.**

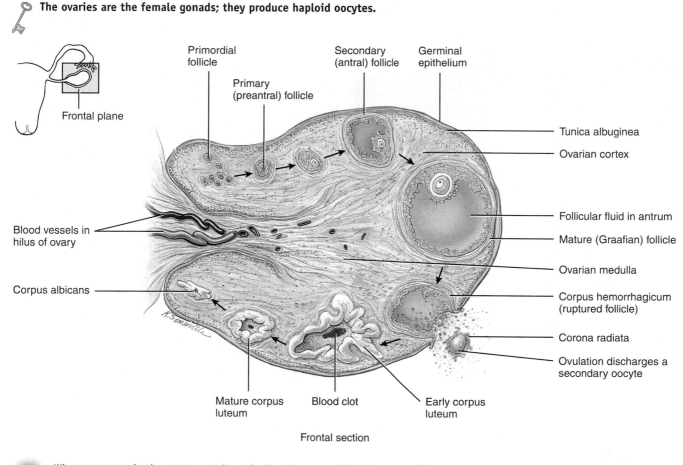

Frontal section

Q What structures in the ovary contain endocrine tissue, and what hormones do they secrete?

Uterine Tubes

Females have two **uterine (Fallopian) tubes,** also called **oviducts,** that extend laterally from the uterus (Figure 28.18). The tubes, which measure about 10 cm (4 in.) long and lie between the folds of the broad ligaments of the uterus, transport secondary oocytes and fertilized ova from the ovaries to the uterus. The funnel-shaped portion of each tube, called the **infundibulum,** is close to the ovary but is open to the pelvic cavity. It ends in a fringe of fingerlike projections called **fimbriae** (FIM-brē-ē; = fringe), one of which is attached to the lateral end of the ovary. From the infundibulum, the uterine tube extends medially and eventually inferiorly and attaches to the superior lateral angle of the uterus. The **ampulla** of the uterine tube is the widest, longest portion, making up about the lateral two-thirds of its length. The **isthmus** of the uterine tube is the more medial, short, narrow, thick-walled portion that joins the uterus.

Histologically, the uterine tubes are composed of three layers. The internal mucosa contains ciliated columnar epithelial cells, which help move the fertilized ovum (or secondary oocyte) along the tube, and secretory cells, which have microvilli and may provide nutrition for the ovum (Figure 28.19 on page 996). The middle layer, the muscularis, is composed of an inner, thick, circular ring of smooth muscle and an outer, thin, region of longitudinal smooth muscle. Peristaltic contractions of the muscularis and the ciliary action of the mucosa help move the oocyte or fertilized ovum toward the uterus. The outer layer of the uterine tubes is a serous membrane, the serosa.

After ovulation, local currents produced by movements of the fimbriae, which surround the surface of the mature follicle just before ovulation occurs, sweep the secondary oocyte into the uterine tube. A sperm cell usually encounters and fertilizes a secondary oocyte in the ampulla of the uterine tube, although fertilization in the abdominopelvic cavity is not uncommon. Fertilization may occur within the tube at any time up to about 24 hours after ovulation. Some hours after fertilization, the nuclear materials of the haploid ovum and sperm unite; the diploid fertilized ovum is now called a zygote. After several cell divisions, it typically arrives at the uterus about 7 days after ovulation.

Figure 28.16 Ovarian follicles. (a) Primordial and primary follicles in the ovarian cortex. (b) A secondary follicle.

🔑 **As an ovarian follicle enlarges, follicular fluid accumulates in a cavity called the antrum.**

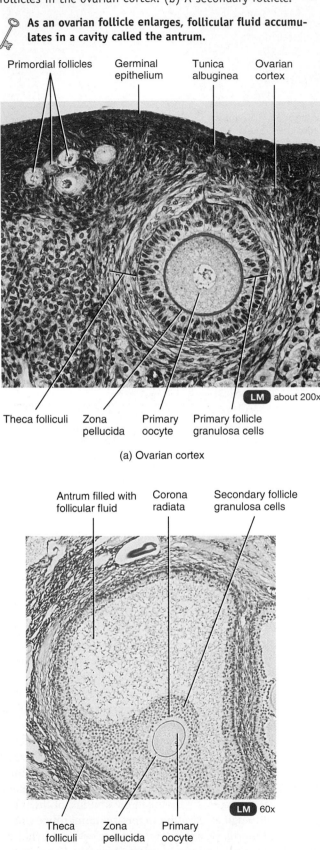

Primordial follicles   Germinal epithelium   Tunica albuginea   Ovarian cortex

LM about 200x

Theca folliculi   Zona pellucida   Primary oocyte   Primary follicle granulosa cells

(a) Ovarian cortex

Antrum filled with follicular fluid   Corona radiata   Secondary follicle granulosa cells

LM 60x

Theca folliculi   Zona pellucida   Primary oocyte

(b) Secondary follicle

Q What happens to most ovarian follicles?

Figure 28.17 Oogenesis. Diploid cells (2n) have 46 chromosomes; haploid cells (n) have 23 chromosomes.

🔑 **In an oocyte, meiosis II is completed only if fertilization occurs.**

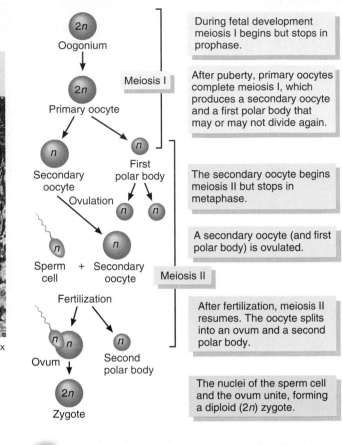

2n
Oogonium

2n
Primary oocyte

Meiosis I

During fetal development meiosis I begins but stops in prophase.

After puberty, primary oocytes complete meiosis I, which produces a secondary oocyte and a first polar body that may or may not divide again.

n
Secondary oocyte

n
First polar body

Ovulation

n   n

The secondary oocyte begins meiosis II but stops in metaphase.

n
Sperm cell

+

n
Secondary oocyte

A secondary oocyte (and first polar body) is ovulated.

Meiosis II

Fertilization

n n

n

Ovum

Second polar body

After fertilization, meiosis II resumes. The oocyte splits into an ovum and a second polar body.

2n
Zygote

The nuclei of the sperm cell and the ovum unite, forming a diploid (2n) zygote.

Q How does the age of a primary oocyte in a female compare with the age of a primary spermatocyte in a male?

Uterus

The **uterus** (womb) serves as part of the pathway for sperm to reach the uterine tubes (see Figure 28.18). It is also the site of menstruation, implantation of a fertilized ovum, development of the fetus during pregnancy, and labor. Situated between the urinary bladder and the rectum, the uterus is the size and shape of an inverted pear. In females who have never been pregnant, it is about 7.5 cm (3 in.) long, 5 cm (2 in.) wide, and 2.5 cm (1 in.) thick; it is larger in females who have recently been pregnant, and smaller (atrophied) when sex hormone levels are low, as occurs after menopause.

Anatomical subdivisions of the uterus include: (1) a dome-shaped portion superior to the uterine tubes called the **fundus,** (2) a tapering central portion called the **body,** and (3) an inferior narrow portion called the **cervix** that opens into the vagina. Between the body of the uterus and the cervix is the **isthmus** (IS-mus), a constricted region about 1 cm (0.5 in.) long. The interior of the body of the uterus is called the uterine cavity, and the interior of the narrow cervix is called the cervical canal. The cervical canal opens into the uterine cavity at the **internal os** (*os* = mouthlike opening) and into the vagina at the **external os.**

Figure 28.18 Relationship of the uterine (Fallopian) tubes to the ovaries, uterus, and associated structures. In the left side of the drawing the uterine tube and uterus have been sectioned to show internal structures. (See Tortora, *A Photographic Atlas of the Human Body,* Figure 14.9)

After ovulation, a secondary oocyte and its corona radiata move from the pelvic cavity into the infundibulum of the uterine tube. The uterus is the site of menstruation, implantation of a fertilized ovum, development of the fetus, and labor.

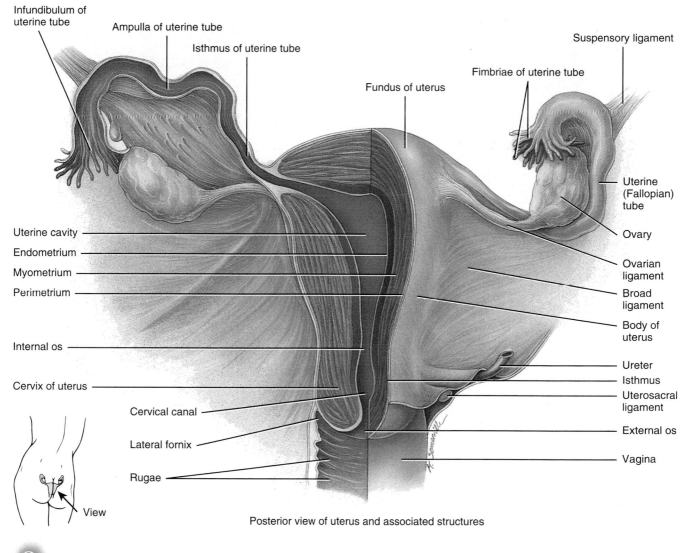

Posterior view of uterus and associated structures

Q Where does fertilization usually occur?

Normally, the body of the uterus projects anteriorly and superiorly over the urinary bladder in a position called **anteflexion.** The cervix projects inferiorly and posteriorly and enters the anterior wall of the vagina at nearly a right angle (see Figure 28.13). Several ligaments that are either extensions of the parietal peritoneum or fibromuscular cords maintain the position of the uterus (see Figure 28.14). The paired **broad ligaments** are double folds of peritoneum attaching the uterus to either side of the pelvic cavity. The paired **uterosacral ligaments,** also peritoneal extensions, lie on either side of the rectum and connect the uterus to the sacrum. The **cardinal (lateral cervical) ligaments** extend inferior to the bases of the broad ligaments between the

pelvic wall and the cervix and vagina. The **round ligaments** are bands of fibrous connective tissue between the layers of the broad ligament; they extend from a point on the uterus just inferior to the uterine tubes to a portion of the labia majora of the external genitalia. Although the ligaments normally maintain the anteflexed position of the uterus, they also afford the uterine body enough movement such that the uterus may become malpositioned. A posterior tilting of the uterus is called **retroflexion** (*retro-* = backward or behind).

Histologically, the uterus consists of three layers of tissue: the perimetrium, myometrium, and endometrium (Figure 28.20). The outer layer—the **perimetrium** (*peri-* = around; *-metrium* = uterus) or serosa—is part of the vis-

Figure 28.19 Mucosal cells of the uterine (Fallopian) tube.

Peristaltic contractions of the uterine tube muscularis and ciliary action of the mucosa help move the oocyte or fertilized ovum toward the uterus.

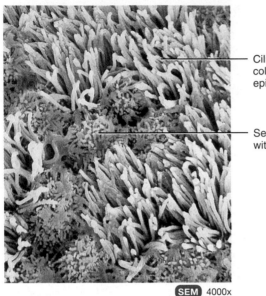

Cilia of ciliated columnar epithelial cell

Secretory cell with microvilli

SEM 4000x

Cilia and secretory cells lining the uterine (Fallopian) tube

 What types of cells line the uterine tubes?

Figure 28.20 Histology of the uterus; the superficial perimetrium (serosa) is not shown.

The three layers of the uterus from superficial to deep are the perimetrium (serosa), the myometrium, and the endometrium.

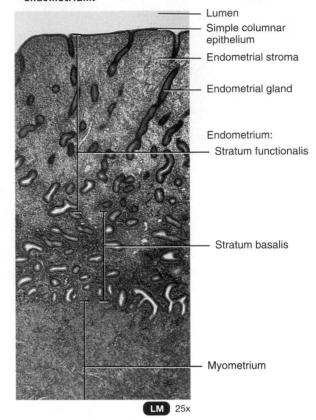

Lumen

Simple columnar epithelium

Endometrial stroma

Endometrial gland

Endometrium:
Stratum functionalis

Stratum basalis

Myometrium

LM 25x

Portion of endometrium and myometrium

 What structural features of the endometrium and myometrium contribute to their functions?

ceral peritoneum; it is composed of simple squamous epithelium and areolar connective tissue. Laterally, it becomes the broad ligament; anteriorly, it covers the urinary bladder and forms a shallow pouch, the **vesicouterine pouch** (ves'-i-kō-YOO-ter-in; *vesico-* = bladder; see Figure 28.13); posteriorly, it covers the rectum and forms a deep pouch, the **rectouterine pouch** (rek-tō-YOO-ter-in; *recto-* = rectum) or *pouch of Douglas*—the most inferior point in the pelvic cavity.

The middle layer of the uterus, the **myometrium** (*myo-* = muscle), consists of three layers of smooth muscle fibers and is thickest in the fundus and thinnest in the cervix. The thicker middle layer is circular, whereas the inner and outer layers are longitudinal or oblique. During labor and childbirth, coordinated contractions of the myometrium in response to oxytocin from the posterior pituitary gland help expel the fetus from the uterus.

The inner layer of the uterus, the **endometrium** (*endo-* = within), is highly vascular and is composed of an innermost layer of simple columnar epithelium (ciliated and secretory cells) that lines the lumen; an underlying endometrial stroma that is a very thick region of lamina propria (areolar connective tissue); and endometrial (uterine) glands that develop as invaginations of the luminal epithe-

lium and extend almost to the myometrium. The endometrium is divided into two layers: The **stratum functionalis** *(functional layer)* lines the uterine cavity and is shed during menstruation; a deeper layer, the **stratum basalis** *(basal layer)*, is permanent and gives rise to a new stratum functionalis after each menstruation.

The secretory cells of the mucosa of the cervix produce a secretion called **cervical mucus,** a mixture of water, glycoprotein, serum-type proteins, lipids, enzymes, and inorganic salts. During their reproductive years, females secrete 20–60 mL of cervical mucus per day. Cervical mucus is more hospitable to sperm at or near the time of ovulation because it is then less viscous and more alkaline (pH 8.5); at other times, viscous mucus forms a cervical plug that physically impedes sperm penetration. Cervical mucus supplements the energy needs of sperm, and both the cervix and cervical mucus serve as a sperm reservoir, protect sperm from the hostile environment of the vagina, and protect sperm from phagocytes.

Figure 28.21 Blood supply of the uterus. The inset shows histological details of the blood vessels of the endometrium.

🔑 Straight arterioles supply the materials needed for regeneration of the stratum functionalis.

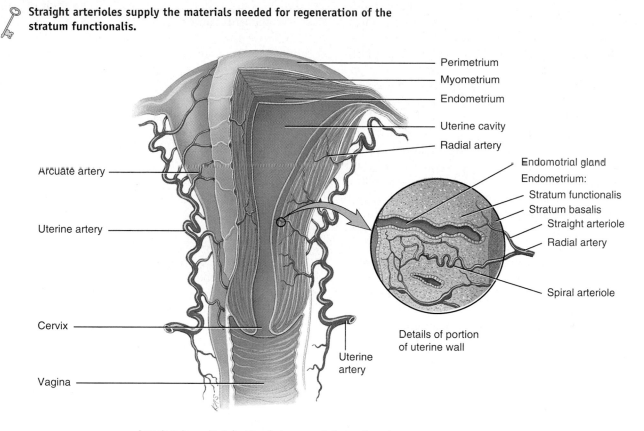

Anterior view with left side of uterus partially sectioned

Q What is the functional significance of the stratum basalis of the endometrium?

They may also play a role in *capacitation*—a functional change that sperm undergo in the female reproductive tract before they are able to fertilize a secondary oocyte.

Blood is supplied to the uterus by branches of the internal iliac artery called **uterine arteries** (Figure 28.21). Branches called **arcuate arteries** (= shaped like a bow) are arranged in a circular fashion in the myometrium and give off **radial arteries** that penetrate deeply into the myometrium. Just before the branches enter the endometrium, they divide into two kinds of arterioles: **straight arterioles** supply the stratum basalis with the materials needed to regenerate the stratum functionalis; **spiral arterioles** supply the stratum functionalis and change markedly during the menstrual cycle. Blood leaving the uterus is drained by the **uterine veins** into the internal iliac veins. The extensive blood supply of the uterus is essential to support regrowth of a new stratum functionalis after menstruation, implantation of a fertilized ovum, and development of the placenta.

🩺 **CLINICAL APPLICATION**
Hysterectomy

Hysterectomy (hiss-te-RECT-tō-mē; *hyster-* = uterus), the surgical removal of the uterus, is the most common gynecological operation. It may be indicated in conditions such as endometriosis, pelvic inflammatory disease, recurrent ovarian cysts, excessive uterine bleeding, and cancer of the cervix, uterus, or ovaries. In a partial or subtotal hysterectomy, the body of the uterus is removed but the cervix is left in place; a complete hysterectomy is the removal of the body and cervix of the uterus. A radical hysterectomy includes removal of the body and cervix of the uterus, uterine tubes, possibly the ovaries, superior portion of the vagina, pelvic lymph nodes, and supporting structures, such as ligaments. ∎

1. Where are the uterine tubes located? What is their function?
2. Diagram and label the principal parts of the uterus.
3. Describe the arrangement of ligaments that hold the uterus in its normal position.
4. Describe the histology of the uterus.
5. Discuss the blood supply to the uterus. Why is an abundant blood supply important?

Vagina

The **vagina** (= sheath) serves as a passageway for menstrual flow, for childbirth, and for semen from the penis during sexual intercourse. It is a tubular, 10-cm (4-in.) long fibromuscular organ lined with mucous membrane (see Figures 28.13 and 28.18). Situated between the urinary bladder and the rectum, the vagina is directed superiorly and posteriorly, where it attaches to the uterus. A recess called the **fornix** (= arch or vault) surrounds the vaginal attachment to the cervix.

The mucosa of the vagina is continuous with that of the uterus. Histologically, it consists of nonkeratinized stratified squamous epithelium and areolar connective tissue that lies in a series of transverse folds called **rugae.** Dendritic cells in the mucosa are antigen-presenting cells (described on page 755); they participate in the transmission of viruses—for example, HIV (the virus that causes AIDS)—to a female during intercourse with an infected male.

The **mucosa** of the vagina contains large stores of glycogen, the decomposition of which produces organic acids. The resulting acidic environment retards microbial growth, but it also is harmful to sperm. Alkaline components of semen, mainly from the seminal vesicles, neutralize the acidity of the vagina and increase viability of the sperm.

The **muscularis** is composed of an outer circular layer and an inner longitudinal layer of smooth muscle that can stretch considerably to accommodate the penis during sexual intercourse and a child during birth.

The **adventitia,** the superficial layer of the vagina, consists of areolar connective tissue; it anchors the vagina to adjacent organs such as the urethra and urinary bladder anteriorly and the rectum and anal canal posteriorly.

At the inferior end of the vaginal opening to the exterior, the **vaginal orifice,** there may be a thin fold of vascularized mucous membrane, called the **hymen** (= membrane), that forms a border around the orifice, partially closing it (see Figure 28.22). Sometimes the hymen completely covers the orifice, a condition called **imperforate hymen** (im-PER-fō-rāt); surgery may be required to open the orifice and permit the discharge of menstrual flow.

Vulva

The term **vulva** (VUL-va; = to wrap around), or **pudendum** (pyoo-DEN-dum), refers to the external genitals of the female (Figure 28.22). The vulva comprises the following components:

- Anterior to the vaginal and urethral openings is the **mons pubis** (MONZ PŪ-bis; *mons* = mountain), an elevation of adipose tissue covered by skin and coarse pubic hair that cushions the pubic symphysis.
- From the mons pubis, two longitudinal folds of skin, the **labia majora** (LĀ-bē-a ma-JŌ-ra; *labia* = lips; *majora* = larger), extend inferiorly and posteriorly. The singular term is *labium majus*. The labia majora are covered by pubic hair and contain an abundance of adipose tissue, se-

baceous (oil) glands, and apocrine sudoriferous (sweat) glands. They are homologous to the scrotum.
- Medial to the labia majora are two smaller folds of skin called the **labia minora** (mī-NŌ-ra; *minora* = smaller). The singular term is *labium minus.* Unlike the labia majora, the labia minora are devoid of pubic hair and fat and have few sudoriferous glands, but they do contain many sebaceous glands. The labia minora are homologous to the spongy (penile) urethra.
- The **clitoris** (KLI-to-ris) is a small cylindrical mass of erectile tissue and nerves located at the anterior junction of the labia minora. A layer of skin called the **prepuce** (foreskin) is formed at the point where the labia minora unite and covers the body of the clitoris. The exposed portion of the clitoris is the glans. The clitoris is homologous to the glans penis in males; like the male structure, it is capable of enlargement upon tactile stimulation and has a role in sexual excitement in the female.
- The region between the labia minora is the **vestibule.** Within the vestibule are the hymen (if still present), the vaginal orifice, the external urethral orifice, and the openings of the ducts of several glands. The vestibule is homologous to the membranous urethra of males. The **vaginal orifice,** the opening of the vagina to the exterior, occupies the greater portion of the vestibule and is bordered by the hymen. Anterior to the vaginal orifice and posterior to the clitoris is the **external urethral orifice,** the opening of the urethra to the exterior. On either side of the external urethral orifice are the openings of the ducts of the **paraurethral** *(Skene's)* **glands,** which are embedded in the wall of the urethra and secrete mucus. The paraurethral glands are homologous to the prostate gland. On either side of the vaginal orifice itself are the **greater vestibular** *(Bartholin's)* **glands** (see Figure 28.23), which open by ducts into a groove between the hymen and labia minora and produce a small quantity of mucus during sexual arousal and intercourse that adds to cervical mucus and provides lubrication. The greater vestibular glands are homologous to the bulbourethral glands in males. Several **lesser vestibular glands** also open into the vestibule.
- The **bulb of the vestibule** (see Figure 28.23) consists of two elongated masses of erectile tissue just deep to the labia on either side of the vaginal orifice. The bulb of the vestibule becomes engorged with blood during sexual arousal, narrowing the vaginal orifice and placing pressure on the penis during intercourse. The bulb of the vestibule is homologous to the corpus spongiosum penis and bulb of the penis in males.

Table 28.1 summarizes the homologous structures of the female and male reproductive systems.

Perineum

The **perineum** (per'-i-NĒ-um) is the diamond-shaped area medial to the thighs and buttocks of both males and fe-

Figure 28.22 Components of the vulva (pudendum). (See Tortora, *A Photographic Atlas of the Human Body,* Figure 14.7)

🔑 **The vulva refers to the external genitals of the female.**

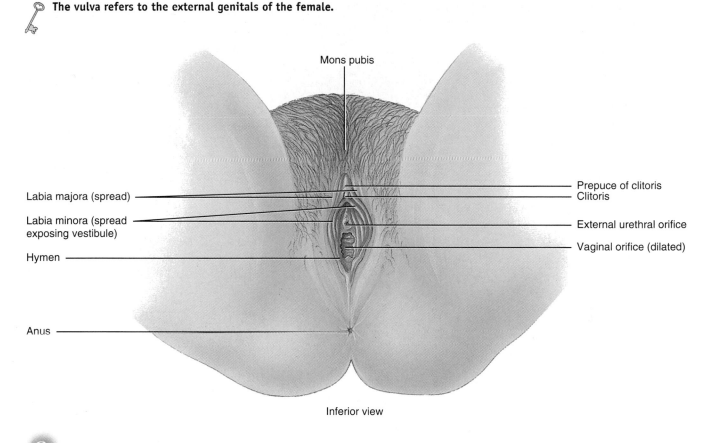

Mons pubis

Labia majora (spread)

Labia minora (spread exposing vestibule)

Hymen

Anus

Prepuce of clitoris
Clitoris

External urethral orifice

Vaginal orifice (dilated)

Inferior view

Q What surface structures are anterior to the vaginal opening? Lateral to it?

males that contains the external genitals and anus (Figure 28.23). The perineum is bounded anteriorly by the pubic symphysis, laterally by the ischial tuberosities, and posteriorly by the coccyx. A transverse line drawn between the ischial tuberosities divides the perineum into an anterior **urogenital triangle** (yoo′-rō-JEN-i-tal) that contains the external genitalia and a posterior anal triangle that contains the anus.

Mammary Glands

The two **mammary glands** (*mamma* = breast) are modified sudoriferous (sweat) glands that produce milk. They lie over the pectoralis major and serratus anterior muscles and are attached to them by a layer of deep fascia that is dense irregular connective tissue (Figure 28.24).

Each breast has one pigmented projection, the **nipple,** that has a series of closely spaced openings of ducts called **lactiferous ducts,** where milk emerges. The circular pigmented area of skin surrounding the nipple is called the **areola** (a-RĒ-ō-la; = small space); it appears rough because it contains modified sebaceous (oil) glands. Strands of connective tissue called the **suspensory ligaments of the breast**

Table 28.1 Summary of Homologous Structures of the Female and Male Reproductive Systems

FEMALE STRUCTURES	MALE STRUCTURES
Ovaries	Testes
Ovum	Sperm cell
Labia majora	Scrotum
Labia minora	Spongy (penile) urethra
Vestibule	Membranous urethra
Bulb of vestibule	Corpus spongiosum penis and bulb of penis
Clitoris	Glans penis
Paraurethral glands	Prostate gland
Greater vestibular glands	Bulbourethral (Cowper's) glands

(Cooper's ligaments) run between the skin and deep fascia and support the breast. These ligaments become looser with age or with excessive strain, as occurs in long-term jogging or high-impact aerobics. Wearing a supportive bra slows the appearance of "Cooper's droop."

Figure 28.23 Perineum of a female. (Figure 11.13 shows the perineum of a male.)

🗝 **The perineum is a diamond-shaped area medial to the thighs and buttocks that contains the external genitals and anus.**

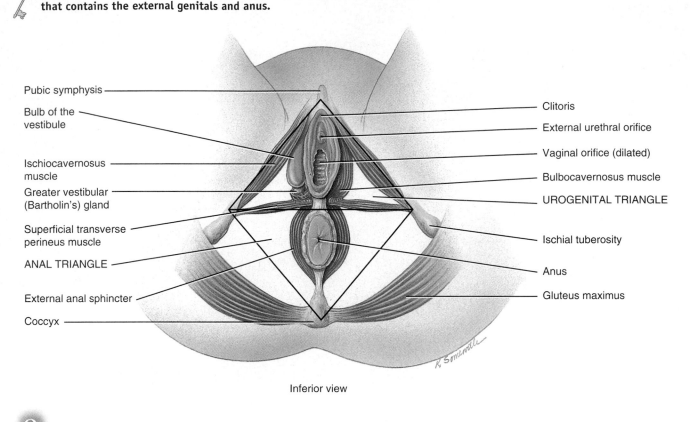

Pubic symphysis

Bulb of the vestibule

Ischiocavernosus muscle

Greater vestibular (Bartholin's) gland

Superficial transverse perineus muscle

ANAL TRIANGLE

External anal sphincter

Coccyx

Clitoris

External urethral orifice

Vaginal orifice (dilated)

Bulbocavernosus muscle

UROGENITAL TRIANGLE

Ischial tuberosity

Anus

Gluteus maximus

Inferior view

Q   Why is the anterior portion of the perineum called the urogenital triangle?

Internally, the mammary gland consists of 15–20 lobes, or compartments, separated by adipose tissue. The amount of adipose tissue, not the amount of milk produced, determines the size of the breasts. In each lobe are several smaller compartments called **lobules,** composed of grapelike clusters of milk-secreting glands termed **alveoli** (= small cavities) embedded in connective tissue. Surrounding the alveoli are spindle-shaped cells called **myoepithelial cells,** the contraction of which helps propel milk toward the nipples. When milk is being produced, it passes from the alveoli into a series of **secondary tubules** and then into the **mammary ducts.** Near the nipple, the mammary ducts expand to form sinuses called **lactiferous sinuses** (*lact-* = milk), where some milk may be stored before draining into a lactiferous duct. Each lactiferous duct typically carries milk from one of the lobes to the exterior.

The essential functions of the mammary glands are the synthesis, secretion, and ejection of milk; these functions, called **lactation,** are associated with pregnancy and childbirth. Milk production is stimulated largely by the hormone prolactin, with contributions from progesterone and estrogens. The ejection of milk is stimulated by oxytocin, which is released from the posterior pituitary gland in response to the sucking of an infant on the mother's nipple (suckling).

 CLINICAL APPLICATION
Fibrocystic Disease of the Breasts

The breasts of females are highly susceptible to cysts and tumors. In **fibrocystic disease,** the most common cause of breast lumps in females, one or more cysts (fluid-filled sacs) and thickening of alveoli (clusters of milk-secreting cells) develop. The condition, which occurs mainly in females between the ages of 30 and 50, is probably due to a hormonal imbalance—a relative excess of estrogens or a deficiency of progesterone—in the postovulatory (luteal) phase of the reproductive cycle (discussed shortly). Fibrocystic disease usually causes one or both breasts to become lumpy, swollen, and tender a week or so before menstruation begins. ■

1. What is the function of the vagina? Describe its histology.
2. List the parts of the vulva, and explain the functions of each.

Figure 28.24 Mammary glands. (See Tortora, *A Photographic Atlas of the Human Body,* Figure 14.14)

🔑 **The mammary glands function in the synthesis, secretion, and ejection of milk (lactation).**

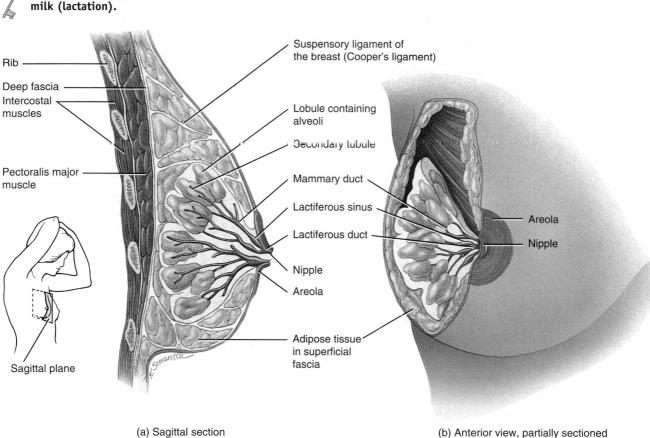

Rib

Deep fascia

Intercostal muscles

Pectoralis major muscle

Sagittal plane

Suspensory ligament of the breast (Cooper's ligament)

Lobule containing alveoli

Secondary tubule

Mammary duct

Lactiferous sinus

Lactiferous duct

Nipple

Areola

Adipose tissue in superficial fascia

Areola

Nipple

(a) Sagittal section

(b) Anterior view, partially sectioned

Q What hormones regulate the synthesis and ejection of milk?

3. Describe the structure of the mammary glands. How are they supported?
4. Describe the passage of milk from the alveoli of the mammary gland to the nipple.

THE FEMALE REPRODUCTIVE CYCLE

OBJECTIVE

• *Compare the principal events of the ovarian and uterine cycles.*

During their reproductive years, nonpregnant females normally experience a cyclical sequence of changes in the ovaries and uterus. Each cycle takes about a month and involves both oogenesis and preparation of the uterus to receive a fertilized ovum. Hormones secreted by the hypothalamus, anterior pituitary gland, and ovaries control the principal events. The **ovarian cycle** is a series of events in the ovaries that occur during and after the maturation of an

oocyte. The **uterine (menstrual) cycle** is a concurrent series of changes in the endometrium of the uterus to prepare it for the arrival of a fertilized ovum that will develop in the uterus until birth. If fertilization does not occur, the stratum functionalis of the endometrium is shed. The general term **female reproductive cycle** encompasses the ovarian and uterine cycles, the hormonal changes that regulate them, and the related cyclical changes in the breasts and cervix.

Hormonal Regulation of the Female Reproductive Cycle

The ovarian and uterine cycles are controlled by **gonadotropin releasing hormone (GnRH)** secreted by the hypothalamus (Figure 28.25). GnRH stimulates the release of **follicle-stimulating hormone (FSH)** and **luteinizing hormone (LH)** from the anterior pituitary gland. FSH, in turn, initiates follicular growth and the secretion of estrogens by the growing follicles. LH stimulates the further development of ovarian follicles and their full secretion of estrogens,

Figure 28.25 Secretion and physiological effects of estrogens, progesterone, relaxin, and inhibin in the female reproductive cycle.

 The uterine and ovarian cycles are controlled by gonadotropin releasing hormone (GnRH) and ovarian hormones (estrogens and progesterone).

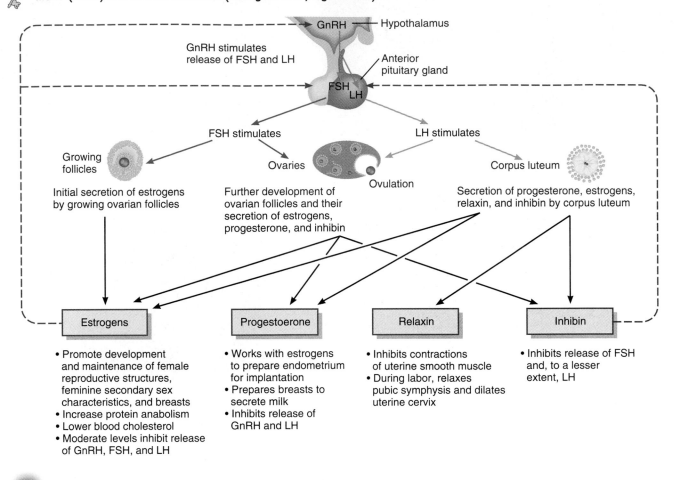

Q Of the several estrogens, which one exerts the major effect?

brings about ovulation, promotes formation of the corpus luteum, and stimulates the production of estrogens, progesterone, relaxin, and inhibin by the corpus luteum.

Even though at least six different estrogens have been isolated from the plasma of human females, only three are present in significant quantities: *beta (β)-estradiol, estrone,* and *estriol.* In a nonpregnant woman, the principal estrogen is β-estradiol, which is synthesized from cholesterol in the ovaries.

Estrogens secreted by follicular cells have several important functions: (1) Estrogens promote the development and maintenance of female reproductive structures, secondary sex characteristics, and the breasts. The secondary sex characteristics include distribution of adipose tissue in the breasts, abdomen, mons pubis, and hips; voice pitch; a broad pelvis; and pattern of hair growth on the head and body. (2) Estrogens increase protein anabolism; in this regard, estro-

gens are synergistic with human growth hormone (hGH). (3) Estrogens lower blood cholesterol level, which is probably the reason that women under age 50 have a much lower risk of coronary artery disease than do men of comparable age. Moderate levels of estrogens in the blood inhibit both the release of GnRH by the hypothalamus and secretion of LH and FSH by the anterior pituitary gland.

Progesterone, secreted mainly by cells of the corpus luteum, acts synergistically with estrogens to prepare the endometrium for the implantation of a fertilized ovum and the mammary glands for milk secretion. High levels of progesterone also inhibit secretion of GnRH and LH.

The small quantity of relaxin, produced by the corpus luteum during each monthly cycle, relaxes the uterus by inhibiting contractions; presumably, implantation of a fertilized ovum occurs more readily in a relaxed uterus. During pregnancy, the placenta produces much more relaxin, and it

Figure 28.26 The female reproductive cycle. Events in the ovarian and uterine cycles and the release of anterior pituitary gland hormones are correlated with the sequence of the cycle's four phases. In the cycle shown, fertilization and implantation have not occurred.

🔑 **The length of the female reproductive cycle typically is 24–36 days; the pre-ovulatory phase is more variable in length than the other phases.**

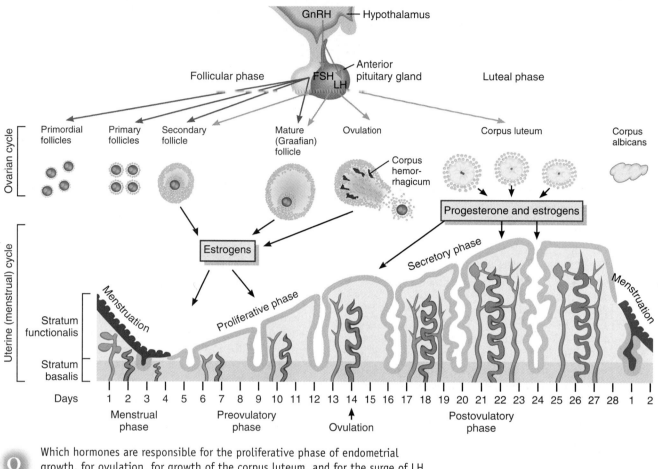

Q Which hormones are responsible for the proliferative phase of endometrial growth, for ovulation, for growth of the corpus luteum, and for the surge of LH at midcycle?

continues to relax uterine smooth muscle. At the end of pregnancy, relaxin also increases the flexibility of the pubic symphysis and may help dilate the uterine cervix, both of which ease delivery of the baby.

Inhibin is secreted by granulosa cells of growing follicles and by the corpus luteum of the ovary. It inhibits secretion of FSH and, to a lesser extent, LH.

Phases of the Female Reproductive Cycle

The duration of the female reproductive cycle typically is 24–35 days. For this discussion we assume a duration of 28 days, divided into four phases: the menstrual phase, the pre-

ovulatory phase, ovulation, and the postovulatory phase (Figure 28.26).

Menstrual Phase

The **menstrual phase** (MEN-stroo-al), also called **menstruation** (men′-stroo-Ā-shun) or **menses** (= month), lasts for roughly the first 5 days of the cycle. (By convention, the first day of menstruation marks the first day of a new cycle.)

Events in the Ovaries During the menstrual phase, 20 or so small secondary follicles, some in each ovary, begin to enlarge. Follicular fluid, secreted by the granulosa cells and oozing from blood capillaries, accumulates in the enlarging antrum while the oocyte remains near the edge of the follicle (see Figure 28.16b).

Figure 28.27 Relative concentrations of anterior pituitary gland hormones (FSH and LH) and ovarian hormones (estrogens and progesterone) during the phases of a normal female reproductive cycle.

🔑 Estrogens are the primary ovarian hormones before ovulation; after ovulation, both progesterone and estrogens are secreted by the corpus luteum.

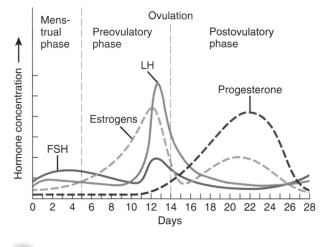

Q Which hormones initiate rebuilding of the stratum functionalis?

Events in the Uterus Menstrual flow from the uterus consists of 50–150 mL of blood, tissue fluid, mucus, and epithelial cells derived from the endometrium. This discharge occurs because the declining level of ovarian hormones, especially progesterone, stimulates release of prostaglandins that cause the uterine spiral arterioles to constrict. As a result, the cells they supply become oxygen-deprived and start to die. Eventually, the entire stratum functionalis sloughs off. At this time the endometrium is very thin, about 2–5mm, because only the stratum basalis remains. The menstrual flow passes from the uterine cavity to the cervix and through the vagina to the exterior.

Preovulatory Phase

The **preovulatory phase,** the second phase of the female reproductive cycle, is the time between menstruation and ovulation. The preovulatory phase of the cycle is more variable in length than the other phases and accounts for most of the difference when cycles are shorter or longer than 28 days. It lasts from days 6 to 13 in a 28-day cycle.

Events in the Ovaries Under the influence of FSH, the group of about 20 secondary follicles continues to grow and begins to secrete estrogens and inhibin. By about day 6, one follicle in one ovary has outgrown all the others to become the dominant follicle. Estrogens and inhibin secreted by the dominant follicle decrease the secretion of FSH, which causes the other less well-developed follicles to stop growing and undergo atresia.

The one dominant follicle becomes the **mature (Graafian) follicle,** which continues to enlarge until it is more than 20 mm in diameter and ready for ovulation (see Figure 28.15). This follicle forms a blisterlike bulge on the surface of the ovary. Fraternal (nonidentical) twins may result if two secondary follicles achieve codominance, and both ovulate and are fertilized. During the final maturation process, the dominant follicle continues to increase its estrogen production under the influence of an increasing level of LH (Figure 28.27). Although estrogens are the primary ovarian hormones before ovulation, small amounts of progesterone are produced by the mature follicle a day or two before ovulation.

With reference to the ovarian cycle, the menstrual and preovulatory phases together are termed the **follicular phase** (fō-LIK-yoo-lar) because ovarian follicles are growing and developing.

Events in the Uterus Estrogens liberated into the blood by growing ovarian follicles stimulate the repair of the endometrium; cells of the stratum basalis undergo mitosis and produce a new stratum functionalis. As the endometrium thickens, the short, straight endometrial glands develop, and the arterioles coil and lengthen as they penetrate the stratum functionalis. The thickness of the endometrium approximately doubles, to about 4–10 mm. With reference to the uterine cycle, the preovulatory phase is also termed the **proliferative phase** because the endometrium is proliferating.

Ovulation

Ovulation, the rupture of the mature follicle and the release of the secondary oocyte into the pelvic cavity, usually occurs on day 14 in a 28–day cycle. During ovulation, the secondary oocyte remains surrounded by its zona pellucida and corona radiata. Development of a secondary follicle into a fully mature follicle generally takes a total of about 20 days (spanning the last 6 days of the previous cycle and the first 14 days of the current cycle). During this time the primary oocyte completes meiosis I to become a secondary oocyte, which begins meiosis II but then halts in metaphase.

The *high* levels of estrogens during the last part of the preovulatory phase exert a positive feedback effect on both LH and gonadotropic releasing hormone and cause ovulation, as follows (Figure 28.28):

1 When estrogens are present in high enough concentrations, they stimulate the hypothalamus to release more gonadotropic releasing hormone, and the anterior pituitary gland to produce more LH.

2 Gonadotropic releasing hormone promotes the release of FSH and more LH by the anterior pituitary gland.

3 The LH surge brings about rupture of the dominant follicle and expulsion of a secondary oocyte. The ovulated oocyte and its corona radiata cells are usually swept into the uterine tube, but some oocytes are lost into the pelvic cavity and disintegrate.

An over-the-counter home test that detects the LH surge associated with ovulation can be used to predict ovulation a day in advance. FSH also increases at this time, but not as dramatically as LH because FSH is stimulated only by the increase in gonadotropic releasing hormone. The positive feedback effect of estrogens on the hypothalamus and anterior pituitary gland does not occur if progesterone is present at the same time.

After ovulation, the mature follicle collapses; once the blood resulting from minor bleeding during rupture of the follicle clots, the follicle becomes the **corpus hemorrhagicum** (*hemo-* = blood; *rrhagic-* = bursting forth) (see Figure 28.15). The clot is absorbed by the remaining follicular cells, which enlarge and form the corpus luteum under the influence of LH. Stimulated by LH, the corpus luteum secretes progesterone, estrogen, relaxin, and inhibin.

Postovulatory Phase

The postovulatory phase of the female reproductive cycle is the most constant in duration: It lasts for 14 days, from days 15 to 28 in a 28–day cycle (see Figure 28.26). It represents the time between ovulation and the onset of the next menses.

Events in One Ovary After ovulation, LH stimulates the remnants of the mature follicle to develop into the corpus luteum, which secretes increasing quantities of progesterone and some estrogens. With reference to the ovarian cycle, this phase is also called the **luteal phase.** Subsequent events in the ovary that ovulated an oocyte depend on whether or not the oocyte becomes fertilized. If the oocyte is not fertilized, the corpus luteum has a lifespan of only 2 weeks; its secretory activity then declines, and it degenerates into a corpus albicans (see Figure 28.15). As the levels of progesterone, estrogens, and inhibin decrease, release of GnRH, FSH, and LH rise due to loss of negative feedback suppression by the ovarian hormones. Then, follicular growth resumes and a new ovarian cycle begins.

If the secondary oocyte is fertilized and begins to divide, the corpus luteum persists past its normal 2-week life span. It is "rescued" from degeneration by **human chorionic gonadotropin** (kō-rē-ON-ik) **(hCG,)** a hormone produced by the chorion of the embryo as early as 8–12 days after fertilization; hCG acts like LH in stimulating the secretory activity of the corpus luteum. The presence of hCG in maternal blood or urine is an indicator of pregnancy.

Events in the Uterus Progesterone and estrogens produced by the corpus luteum promote growth and coiling of the endometrial glands (which begin to secrete glycogen), vascularization of the superficial endometrium, thickening

Figure 28.28 Role of high levels of estrogens in increasing secretion of GnRH and LH and initiating ovulation.

 At midcycle, a surge of LH triggers ovulation.

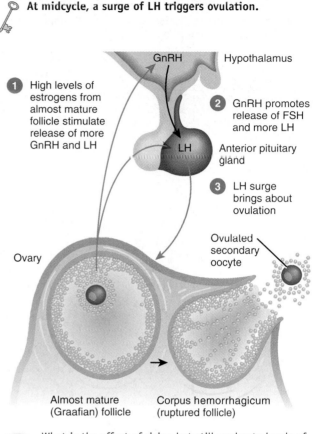

1 High levels of estrogens from almost mature follicle stimulate release of more GnRH and LH

GnRH

Hypothalamus

2 GnRH promotes release of FSH and more LH

LH

Anterior pituitary gland

3 LH surge brings about ovulation

Ovary

Ovulated secondary oocyte

Almost mature (Graafian) follicle

Corpus hemorrhagicum (ruptured follicle)

 What is the effect of rising but still moderate levels of estrogens on the secretion of gonadotropin releasing hormone, LH, and FSH?

of the endometrium to 12–18 mm, and an increase in the amount of tissue fluid. These preparatory changes peak about one week after ovulation, corresponding to the time that a fertilized ovum might arrive. With reference to the uterine cycle, this phase is called the **secretory phase** because of the secretory activity of the endometrial glands. If fertilization does not occur, the level of progesterone declines due to degeneration of the corpus luteum, and menstruation ensues. Figure 28.29 summarizes the hormonal interactions during the ovarian and uterine cycles.

CLINICAL APPLICATION
Menstrual Abnormalities

Amenorrhea (ā-men′-ō-RĒ-a; *a-* = without; *men-* = month; *-rrhea* = a flow) is the absence of menstruation; it may be caused by a hormone imbalance, obesity, extreme weight loss, or very low body fat as may occur during rigorous athletic training. **Dysmenorrhea** (dis′-men-ō-RĒ-a; *dys* = difficult or painful) refers to pain associated with menstru-

Figure 28.29 Summary of hormonal interactions in the ovarian and uterine cycles.

Hormones from the anterior pituitary gland regulate ovarian function, and hormones from the ovaries regulate the changes in the endometrial lining of the uterus.

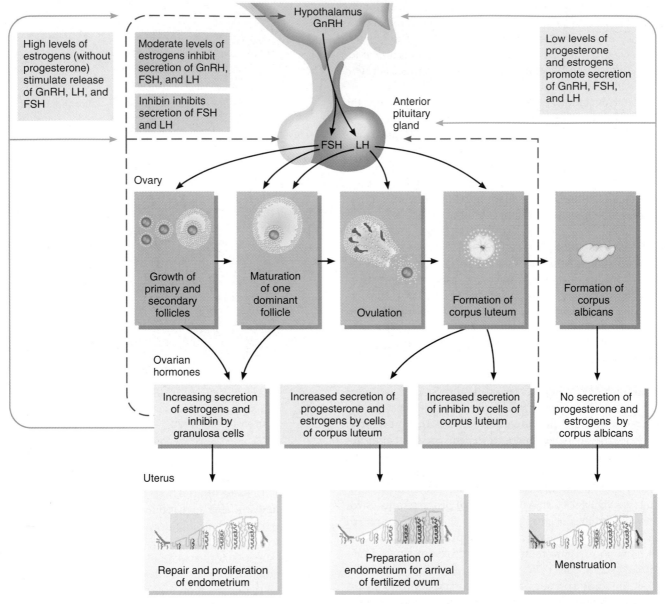

Q When declining levels of estrogens and progesterone stimulate secretion of GnRH, is this a positive or a negative feedback effect? Why?

ation; the term is usually reserved to describe menstrual symptoms that are severe enough to prevent a woman from functioning normally for one or more days each month. Some cases are caused by uterine tumors, ovarian cysts, pelvic inflammatory disease, or intrauterine devices.

Abnormal uterine bleeding includes menstruation of excessive duration or excessive amount, diminished menstrual flow, too frequent menstruation, intermenstrual bleeding, and postmenopausal bleeding. These abnormalities may be caused by disordered hormonal regulation, emotional factors, fibroid tumors of the uterus, and systemic diseases. ■

1. Describe the function of each of the following hormones in the uterine and ovarian cycles: GnRH, FSH, LH, estrogens, progesterone, and inhibin.

2. Briefly outline the major events of each phase of the uterine cycle, and correlate them with the events of the ovarian cycle.

3. Prepare a labeled diagram of the principal hormonal interactions involved in the uterine and ovarian cycles.

THE HUMAN SEXUAL RESPONSE

OBJECTIVE

• *Describe the similarities and differences in the sexual responses of males and females.*

During sexual intercourse—also called **coitus** (KŌ-i-tus; = a coming together) when it is between a male and a female—sperm are ejaculated from the male urethra into the vagina. The similar sequence of physiological and emotional changes experienced by both males and females before, during, and after intercourse is termed the *human sexual response.*

Stages of the Human Sexual Response

William Masters and Virginia Johnson, who began their pioneering research on human sexuality in the late 1950s, divided the human sexual response into four stages: excitement, plateau, orgasm, and resolution. During sexual **excitement,** also known as **arousal,** various physical and psychological stimuli trigger parasympathetic reflexes in which nerve impulses propagate from the gray matter of the second, third, and fourth sacral segments of the spinal cord along the pelvic splanchnic nerves. Some of these parasympathetic postganglionic fibers produce relaxation of vascular smooth muscle, which allows **vasocongestion**—engorgement with blood—of genital tissues. Parasympathetic impulses also stimulate the secretion of lubricating fluids, most of which are contributed by the female. Without satisfactory lubrication, sexual intercourse is difficult and painful for both partners and inhibits orgasm.

Other changes include increases in heart rate and blood pressure, increased tone in skeletal muscles throughout the body, and hyperventilation. Excitement can be induced by direct physical contact (as in kissing or rubbing), especially of the glans penis, clitoris, nipples of the breasts, and earlobes. However, anticipation or fear, memories, visual, olfactory, and auditory sensations, and fantasies can enhance or diminish the likelihood that excitement will occur. Due to the tactile stimulation of the breast while nursing an infant, feelings of sexual arousal during breastfeeding are fairly common and should not be cause for concern.

The changes that begin during excitement are sustained at an intense level in the **plateau** stage, which may last for only a few seconds or for many minutes. At this stage, many females and some males display a sex flush—a rashlike redness of the face and chest. Generally, the briefest stage is **orgasm** *(climax),* during which the male ejaculates and mem-bers of both sexes experience several rhythmic muscular contractions about 0.8 sec apart, accompanied by intense, pleasurable sensations. The sex flush also is most prominent at this time. During orgasm, bursts of *sympathetic* nerve impulses leave the spinal cord at the first and second lumbar levels and cause rhythmic contractions of smooth muscle in genital organs. At the same time, *somatic motor neurons* from lumbar and sacral segments of the spinal cord trigger powerful, rhythmic contractions of perineal skeletal muscles, particularly the bulbocavernosus muscles, ischiocavernosus muscles, and anal sphincter (see Figure 11.13 on page 333 for males and Figure 28.23 for females).

In males, orgasm usually accompanies ejaculation. Reception of the ejaculate, however, provides little stimulus for a female, especially if she is not already at the plateau stage; this is why a female does not automatically experience orgasm simultaneously with her partner. On the other hand, females may experience two or more orgasms in rapid succession, whereas males enter a **refractory period,** a recovery time during which a second ejaculation and orgasm is physiologically impossible. In some males the refractory period lasts only a few minutes; in others it lasts for several hours.

In the final stage—**resolution,** which begins with a sense of profound relaxation—genital tissues, heart rate, blood pressure, breathing, and muscle tone return to the unaroused state. If sexual excitement has been intense but orgasm has not occurred, resolution takes place more slowly. The four phases of the human sexual response are not always clearly separated from one another and may vary considerably among different people, and even in the same person at different times.

Changes in Males

Most of the time, the penis is flaccid (limp) because sympathetic impulses cause vasoconstriction of its arteries, which limits blood inflow. The first sign of sexual excitement is **erection,** the enlargement and stiffening of the penis. Parasympathetic impulses cause release of neurotransmitters and local hormones, including the gas nitric oxide, which relaxes vascular smooth muscle in the penis. The arteries of the penis dilate, and blood fills the blood sinuses of the three corpora. Expansion of these erectile tissues compresses the superficial veins that normally drain the penis, resulting in engorgement and rigidity (erection). Parasympathetic impulses also cause the bulbourethral (Cowper's) glands to secrete mucus, which flows through the urethra and provides a small amount of lubrication for intercourse.

During the plateau stage, the head of the penis increases in diameter, and vasocongestion causes the testes to swell. As orgasm begins, rhythmic sympathetic impulses cause peristaltic contractions of smooth muscle in the ducts of each testis, epididymis, and ductus (vas) deferens, as well as in the walls of the seminal vesicles and prostate gland. These contractions propel sperm and fluid into the urethra (emission). During this time, contraction of urethral sphincter muscles prevents flow of semen into the urinary bladder, or of urine into the urethra. Ejaculation quickly follows emission. During ejaculation, peristaltic contractions in the ducts and urethra combine with rhythmic contractions of skeletal muscles in the perineum and at the base of the penis to propel the semen from the urethra to the exterior.

Changes in Females

Just as engorgement with blood causes erection of the penis, the first signs of sexual excitement in females are also due to vasocongestion. Within a few seconds to a minute, parasympathetic impulses stimulate release of fluids that lubricate the walls of the vagina. Although the vaginal mucosa lacks glands, engorgement of its connective tissue with blood during sexual excitement causes lubricating fluid to ooze from capillaries and seep through the epithelial layer via a process called **transudation** (trans'-yoo-DĀ-shun; *trans-* = across or through; *sud-* = sweat). Glands within the cervical mucosa and the greater vestibular (Bartholin's) glands contribute a small quantity of lubricating mucus. During excitement, parasympathetic impulses also trigger erection of the clitoris, engorgement of the labia, and relaxation of vaginal smooth muscle. The breasts also may swell due to vasocongestion, and erection may occur in the nipples of the breasts.

Late in the plateau stage, pronounced vasocongestion of the distal third of the vagina swells the tissue and narrows the opening. Because of this response, the vagina grips the penis more firmly. If effective sexual stimulation continues, orgasm may occur, associated with 3–15 rhythmic contractions of the vagina, uterus, and perineal muscles. In both males and females, orgasm is a total body response that may produce milder sensations on some occasions and more intense, explosive sensations at other times.

CLINICAL APPLICATION
Erectile Dysfunction

Erectile dysfunction (ED), previously termed *impotence,* is the consistent inability of an adult male to ejaculate or to attain or hold an erection long enough for sexual intercourse. Many cases of impotence are caused by insufficient release of nitric oxide, which relaxes the smooth muscle of the penile arteries. The drug Viagra (sildenafil) enhances effects stimulated by nitric oxide in the penis. Other causes of erectile dysfunction include diabetes mellitus, physical abnormalities of the penis, systemic disorders such as syphilis, vascular disturbances (arterial or venous obstructions), neurological disorders, surgery, testosterone deficiency, and drugs (alcohol, antidepressants, antihistamines, antihypertensives, narcotics, nicotine, and tranquilizers). Psychological factors such as anxiety or depression, fear of causing pregnancy, fear of sexually transmitted diseases, religious inhibitions, and emotional immaturity may also cause impotence. ■

1. Describe the stages of the human sexual response.
2. How are the physiological changes during the human sexual response similar and different in males and females?

BIRTH CONTROL METHODS
OBJECTIVE
• *Compare the various kinds of birth control methods and their effectiveness.*

Although no single, ideal method of **birth control** exists, several methods are available, each with advantages and disadvantages. The only method of preventing pregnancy that is 100% reliable is total **abstinence,** the avoidance of sexual intercourse. Other methods discussed here are surgical sterilization, hormonal methods, intrauterine devices, spermatocides, barrier methods, periodic abstinence, coitus interruptus, and induced abortion.

Table 28.2 provides the failure rates for each method.

Surgical Sterilization

Sterilization is a procedure that renders an individual incapable of reproduction. The most common means of sterilization of males is **vasectomy** (vas-EK-tō-mē; *-ectomy* = cut out), in which a portion of each ductus deferens is removed. An incision is made in the posterior side of the scrotum, the ducts are located, each is tied in two places, and the portion between the ties is removed. Although sperm production continues in the testes, sperm can no longer reach the exterior; instead, they degenerate and are destroyed by phagocytosis. Because the blood vessels are not cut, testosterone levels in the blood remain normal, so a vasectomy has no effect on sexual desire or performance. Sterilization in females most often is achieved by performing a **tubal ligation** (lī-GĀ-shun), in which the uterine tubes are tied closed and then cut. As a result, the secondary oocyte cannot pass into the uterus, and sperm cannot reach the oocyte. Tubal ligation reduces the risk of pelvic inflammatory disease in women who are exposed to sexually transmitted infections; it may also reduce the risk of ovarian cancer.

Hormonal Methods

Except for total abstinence or surgical sterilization, hormonal methods are the most effective means of birth control. By using **oral contraceptives** *("the pill")* to adjust hormone levels, it is possible to interfere with the production of gametes or implantation of a fertilized ovum in the uterus. The pills used most often contain a higher concentration of a progestin (a substance similar to progesterone) and a lower concentration of estrogens (combination pills). These two hormones act via negative feedback on the anterior pituitary gland to decrease the secretion of FSH and LH, and on the hypothalamus to inhibit secretion of gonadotropic releasing hormone. The low levels of FSH and LH usually prevent both follicular development and ovulation; as a result, pregnancy cannot occur because there is no secondary oocyte to fertilize. Even if ovulation does occur, as it does in some cases, oral contraceptives also alter cervical mucus such that it is more hostile to sperm.

Among the noncontraceptive benefits of oral contraceptives are regulation of the length of menstrual cycles and decreased menstrual flow (and therefore decreased risk of anemia). The pill also provides protection against endometrial and ovarian cancers and reduces the risk of endometriosis. However, oral contraceptives may not be advised for women with a history of blood clotting disorders, cerebral blood vessel damage, migraine headaches, hypertension, liver malfunction, or heart disease. Women who take the pill and smoke face far higher odds of having a heart attack or stroke than do nonsmoking pill users. Smokers should quit smoking or use an alternative method of birth control. Oral contraceptives also may be used for **emergency contraception (EC),** the so-called "morning-after pill." When two pills are taken within 72 hours after unprotected intercourse and another two pills are taken 12 hours later, the chance of pregnancy is reduced by 75%.

Other hormonal methods of contraception are Norplant, Depo-provera, and the vaginal ring. **Norplant** consists of six slender hormone-containing capsules that are surgically implanted under the skin of the arm using local anesthesia. They slowly and continually release a progestin, which inhibits ovulation and thickens the cervical mucus. The effects last for 5 years, and Norplant is about as reliable as sterilization. Removing the Norplant capsules restores fertility. **Depo-provera,** which is given as an intramuscular injection once every 3 months, contains a hormone similar to progesterone that prevents maturation of the ovum and causes changes in the uterine lining that make pregnancy less likely. Expected on the market at the beginning of the new millennium, the **vaginal ring** is a doughnut-shaped ring that fits in the vagina and releases either a progestin alone or a progestin and an estrogen. It is worn for 3 weeks and removed for 1 week to allow menstruation to occur.

The quest for an efficient oral contraceptive for males has been disappointing. The challenge is to find substances that block production of functional sperm without causing erectile dysfunction.

Intrauterine Devices

An **intrauterine device (IUD)** is a small object made of plastic, copper, or stainless steel that is inserted into the cavity of the uterus. IUDs cause changes in the uterine lining that prevent implantation of a fertilized ovum. The IUD most commonly used in the United States today is the Copper T 380A, which is approved for up to 10 years of use and has long-term effectiveness comparable to that of tubal ligation. Some women cannot use IUDs because of expulsion, bleeding, or discomfort.

Table 28.2 Failure Rates of Several Birth Control Methods

| METHOD | FAILURE RATES* | |
	Perfect Use[†]	Typical Use
None	85%	85%
Complete abstinence	0%	0%
Surgical sterilization		
Vasectomy	0.10%	0.15%
Tubal ligation	0.5%	0.5%
Hormonal methods		
Oral contraceptives	0.1%	3%[‡]
Norplant	0.3%	0.3%
Depo-provera	0.05%	0.05%
Intrauterine device		
Copper T 380A	0.6%	0.8%
Spermatocides	6%	26%[‡]
Barrier methods		
Male condom	3%	14%[‡]
Vaginal pouch	5%	21%[‡]
Diaphragm	6%	20%[‡]
Periodic abstinence		
Rhythm	9%	25%[‡]
Sympto-thermal	2%	20%[‡]
Coitus interruptus	4%	18%[‡]

*Defined as percent of women having an unintended pregnancy during the first year of use.
[†]Failure rate when the method is used correctly and consistently.
[‡]Includes couples who forgot to use the method.

Spermatocides

Various foams, creams, jellies, suppositories, and douches that contain sperm-killing agents, or **spermatocides,** make the vagina and cervix unfavorable for sperm survival and are available without prescription. The most widely used spermatocide is nonoxynol-9, which kills sperm by disrupting the plasma membrane. It also inactivates the AIDS virus and decreases the incidence of gonorrhea (described on page 1014). A spermatocide is more effective when used together with a diaphragm or a condom.

Barrier Methods

Barrier methods are designed to prevent sperm from gaining access to the uterine cavity and uterine tubes. In addition to preventing pregnancy, barrier methods may also provide some protection against sexually transmitted diseases (STDs) such as HIV infection, in contrast to oral contraceptives and IUDs, which confer no such protection.

Among the barrier methods are use of a condom, a vaginal pouch, or a diaphragm.

A **condom** is a nonporous, latex covering placed over the penis that prevents deposition of sperm in the female reproductive tract. A **vaginal pouch,** sometimes called a female condom, is made of two flexible rings connected by a polyurethane sheath. One ring lies inside the sheath and is inserted to fit over the cervix; the other ring remains outside the vagina and covers the female external genitals.

A **diaphragm** is a rubber dome-shaped structure that fits over the cervix and is used in conjunction with a spermatocide. It can be inserted up to 6 hours before intercourse. The diaphragm stops most sperm from passing into the cervix; the spermatocide kills most sperm that do get by. Although a diaphragm use does decrease the risk of some STDs, it does not fully protect against HIV infection.

Periodic Abstinence

A couple can use their knowledge of the physiological changes that occur during the female reproductive cycle to decide either to abstain from intercourse on those days when pregnancy is a likely result, or to plan intercourse on those days if they wish to conceive a child. In females with normal and regular menstrual cycles, these physiological events help to predict the day on which ovulation is likely to occur.

The first physiologically based method, developed in the 1930s, is known as the **rhythm method.** It takes advantage of the fact that a secondary oocyte is fertilizable for only 24 hours and is available for only 3–5 days in each reproductive cycle. During this time (3 days before ovulation, the day of ovulation, and 3 days after ovulation) the couple abstains from intercourse. The effectiveness of the rhythm method for birth control is poor in many women due to their irregular cycles.

Another system is the **sympto-thermal method,** in which couples are instructed to know and understand certain signs of fertility and infertility. The signs of ovulation include increased basal body temperature; the production of clear, stretchy cervical mucus; abundant cervical mucus; and pain associated with ovulation. If a couple abstains from sexual intercourse when the signs of ovulation are present and for 3 days afterward, the chance of pregnancy is decreased. A problem with this method is that fertilization is quite likely if intercourse occurred up to 2 days *before* ovulation.

Coitus Interruptus

Coitus interruptus is withdrawal of the penis from the vagina just before ejaculation. Failures with this method are due either to failure to withdraw before ejaculation or to preejaculatory emission of sperm-containing fluid from the urethra. In addition, this method offers no protection against transmission of STDs.

Induced Abortion

Abortion refers to the premature expulsion from the uterus of the products of conception, usually before the 20th week of pregnancy. An abortion may be spontaneous (naturally occurring; also called a miscarriage) or induced (intentionally performed). When birth control methods are not used or fail to prevent an unwanted pregnancy, an **induced abortion** may be performed. Induced abortions may involve vacuum aspiration (suction), infusion of a saline solution, or surgical evacuation (scraping).

Certain drugs, most notably RU 486, can induce a so-called nonsurgical abortion. **RU 486 (mifepristone)** is an antiprogestin; it blocks the action of progesterone. Progesterone prepares the uterine endometrium for implantation and then maintains the uterine lining after implantation. If progesterone levels fall during pregnancy or if the action of the hormone is blocked, menstruation occurs, and the embryo is sloughed off along with the uterine lining. RU 486 occupies the endometrial receptor sites for progesterone. Within 12 hours, the endometrium starts to degenerate, and within 72 hours it begins to slough off. A form of prostaglandin E (misoprostol), which stimulates uterine contractions, is given after RU 486 to aid in expulsion of the endometrium. RU 486 can be taken up to 5 weeks after conception. One side effect of the drug is uterine bleeding. RU 486 has been used for several years in France, Sweden, the United Kingdom, and China.

1. Explain how oral contraceptives reduce the likelihood of pregnancy.
2. Why do some methods of birth control protect against sexually transmitted diseases, whereas others do not?

DEVELOPMENTAL ANATOMY OF THE REPRODUCTIVE SYSTEMS

OBJECTIVE
•*Describe the development of the male and female reproductive systems.*

The *gonads* develop from the **intermediate mesoderm.** By the sixth week of gestation, they appear as bulges that protrude into the ventral body cavity (Figure 28.30). Adjacent to the gonads are the **mesonephric (Wolffian) ducts,** which eventually develop into structures of the reproductive system in males. A second pair of ducts, the **paramesonephric (Müllerian) ducts,** develop lateral to the mesonephric ducts and eventually form structures of the reproductive system in females. Both sets of ducts empty into the urogenital sinus. An early embryo has the potential to follow either the male or the female developmental pattern because it contains both sets of ducts and primitive gonads that can differentiate into either testes or ovaries.

Cells of male embryo have one X chromosome and one Y chromosome. The male pattern of differentiation is initiated by a Y chromosome "master switch" gene named *SRY,*

Figure 28.30 Development of the internal reproductive systems.

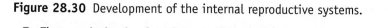

The gonads develop from intermediate mesoderm.

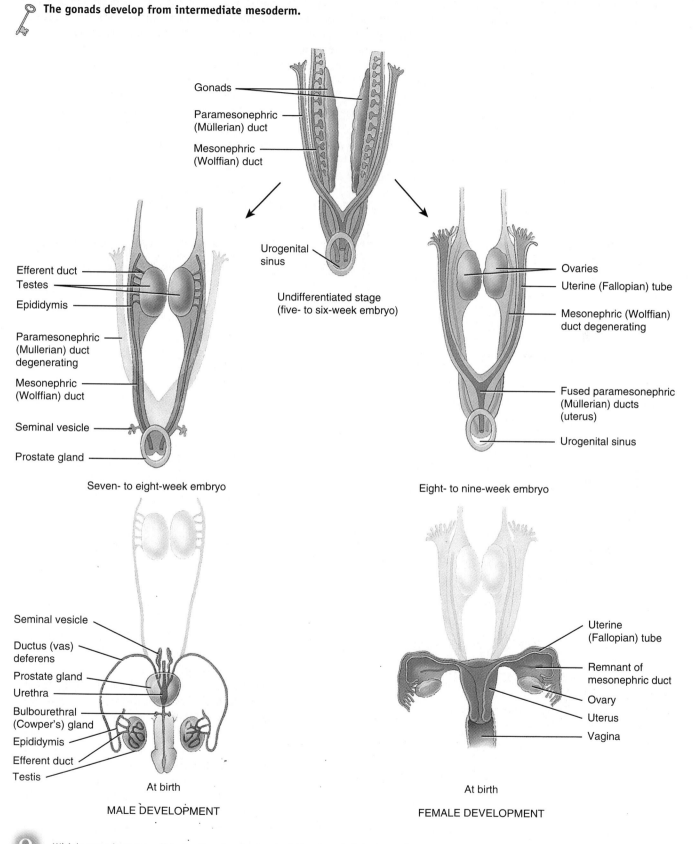

Gonads

Paramesonephric (Müllerian) duct

Mesonephric (Wolffian) duct

Urogenital sinus

Undifferentiated stage (five- to six-week embryo)

Efferent duct

Testes

Epididymis

Paramesonephric (Mullerian) duct degenerating

Mesonephric (Wolffian) duct

Seminal vesicle

Prostate gland

Seven- to eight-week embryo

Ovaries

Uterine (Fallopian) tube

Mesonephric (Wolffian) duct degenerating

Fused paramesonephric (Müllerian) ducts (uterus)

Urogenital sinus

Eight- to nine-week embryo

Seminal vesicle

Ductus (vas) deferens

Prostate gland

Urethra

Bulbourethral (Cowper's) gland

Epididymis

Efferent duct

Testis

At birth

MALE DEVELOPMENT

Uterine (Fallopian) tube

Remnant of mesonephric duct

Ovary

Uterus

Vagina

At birth

FEMALE DEVELOPMENT

Q Which gene is responsible for the development of the gonads into testes?

which stands for **S**ex-determining **R**egion of the **Y** chromosome. When *SRY* is activated, primitive Sertoli cells begin to differentiate in the gonadal tissues of male embryos during the seventh week. The developing Sertoli cells secrete a hormone called **Müllerian-inhibiting substance (MIS),** which causes the death of cells within the paramesonephric (Müllerian) ducts. As a result, those cells do not contribute any functional structures to the male reproductive system. Stimulated by human chorionic gonadotropin (hCG), primitive Leydig cells in the gonadal tissue begin to secrete the androgen **testosterone** during the eighth week. Testosterone then stimulates development of the mesonephric duct on each side into the *epididymis, ductus (vas) deferens, ejaculatory duct,* and *seminal vesicle.* The *testes* connect to the mesonephric duct through a series of tubules that eventually become the *seminiferous tubules.* The *prostate* and *bulbourethral glands* are **endodermal** outgrowths of the urethra.

Cells of a female embryo have two X chromosomes and no Y chromosome. Because *SRY* is absent, the gonads develop into *ovaries,* and because MIS is not produced the paramesonephric ducts flourish. The distal ends of the paramesonephric ducts fuse to form the *uterus* and *vagina* whereas the unfused proximal portions become the *uterine (Fallopian) tubes.* The mesonephric ducts degenerate without contributing any functional structures to the female reproductive system because testosterone is absent. The *greater* and *lesser vestibular glands* develop from **endodermal** outgrowths of the vestibule.

The *external genitals* of both male and female embryos also remain undifferentiated until about the eighth week. Before differentiation, all embryos have an elevated midline swelling called the **genital tubercle** (Figure 28.31). The tubercle consists of the **urethral groove** (opening into the urogenital sinus), paired **urethral folds,** and paired **labioscrotal swellings.**

In male embryos, some testosterone is converted to a second androgen called **dihydrotestosterone (DHT).** DHT stimulates development of the urethra, prostate gland, and external genitals (scrotum and penis). A portion of the genital tubercle elongates and develops into a penis. Fusion of the urethral folds forms the *spongy (penile) urethra* and leaves an opening to the exterior only at the distal end of the penis, the *external urethral orifice.* The labioscrotal swellings develop into the *scrotum.* In the absence of DHT, the genital tubercle gives rise to the *clitoris* in female embryos. The urethral folds remain open as the *labia minora,* and the labioscrotal swellings become the *labia majora.* The urethral groove becomes the *vestibule.* After birth, androgen levels decline because hCG is no longer present to stimulate secretion of testosterone.

CLINICAL APPLICATION
Deficiency of 5 Alpha-Reductase

A rare genetic defect leads to a deficiency in 5 alpha-reductase, the enzyme that converts testosterone to dihydrotestosterone. At birth such a baby looks like a female

externally because of the absence of dihydrotestosterone during development. At puberty, however, testosterone level rises, masculine characteristics start to appear; and the breasts fail to develop. Internal examination reveals testes and structures that normally develop from the mesonephric duct (epididymis, ductus deferens, seminal vesicle, and ejaculatory duct) rather than ovaries and a uterus. ■

1. Describe the role of hormones in differentiation of the gonads, the mesonephric ducts, the paramesonephric ducts, and the external genitals.

AGING AND THE REPRODUCTIVE SYSTEMS
OBJECTIVE
• *Describe the effects of aging on the reproductive systems.*

During the first decade of life, the reproductive system is in a juvenile state. At about age 10, hormone-directed changes start to occur in both sexes. **Puberty** (PŪ-ber-tē; = a ripe age) is the period when secondary sexual characteristics begin to develop and the potential for sexual reproduction is reached.

In females, the reproductive cycle normally occurs once each month from **menarche** (me-NAR-kē), the first menses, to **menopause,** the permanent cessation of menses. Between the ages of 40 and 50 the pool of remaining ovarian follicles becomes exhausted; as a result, the ovaries become less responsive to hormonal stimulation and the production of estrogens declines, despite copious secretion of FSH and LH by the anterior pituitary gland. In addition, changes in the pattern of gonadotropin-releasing hormone (GnRH) release may also contribute to menopausal changes. Some women experience hot flashes and copious sweating, which appear to coincide with increased release of GnRH. Other symptoms of menopause are headache, hair loss, muscular pains, vaginal dryness, insomnia, depression, weight gain, and mood swings. Some atrophy of the ovaries, uterine tubes, uterus, vagina, external genitalia, and breasts occurs in postmenopausal women. Osteoporosis is also a possible occurrence due to the diminished level of estrogens. Sexual desire (libido) does not show a parallel decline; it may be maintained by adrenal sex steroids.

The female reproductive system has a time-limited span of fertility between menarche and menopause. Fertility declines with age, due to the small number of follicles left in the ovaries, less frequent ovulation, and possibly the declining ability of the uterine tubes and uterus to support the young embryo. Uterine cancer peaks at about 65 years of age, but cervical cancer is more common in younger women.

In males, declining reproductive function is much more subtle than in females. Healthy men often retain reproductive capacity into their eighties or nineties. At about age 55 a decline in testosterone synthesis leads to reduced muscle

Figure 28.31 Development of the external genitals.

🔑 **The external genitals of male and female embryos remain undifferentiated until about the eighth week.**

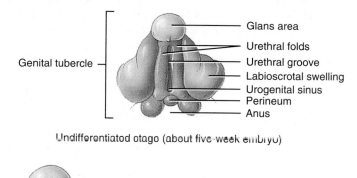

Undifferentiated stage (about five-week embryo)

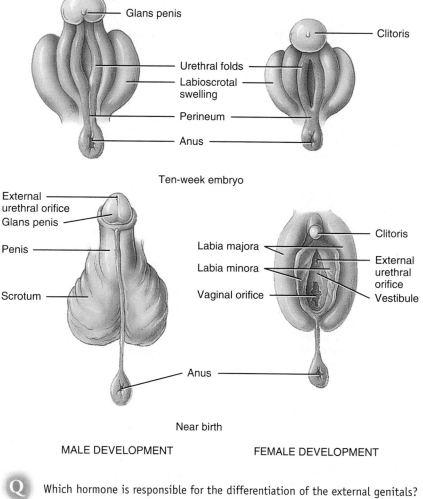

Ten-week embryo

Near birth

MALE DEVELOPMENT   FEMALE DEVELOPMENT

Q Which hormone is responsible for the differentiation of the external genitals?

strength, fewer viable sperm, and decreased sexual desire. However, abundant sperm may be present even in old age.

Enlargement of the prostate gland to two to four times its normal size occurs in approximately one-third of all males over age 60. This condition, called **benign prostatic hyperplasia (BPH),** is characterized by frequent urination, nocturia (bed-wetting), hesitancy in urination, decreased force of urinary stream, postvoiding dribbling, and a sensation of incomplete emptying.

1. List some of the changes that occur in males and females at puberty.
2. Distinguish between menarche and menopause.

SEXUALLY TRANSMITTED DISEASES

A **sexually transmitted disease (STD)** is one that is spread by sexual contact. In most developed countries of the world, such as those of the European Community, Japan, Australia, and New Zealand, the incidence of STDs has declined markedly during the past 25 years. In the United States, by contrast, STDs have been rising to near epidemic proportions, especially among urban populations. AIDS and hepatitis B, which are sexually transmitted diseases that also may be contracted in other ways, are discussed in Chapters 22 and 24, respectively.

Chlamydia

Chlamydia (kla-MID-ē-a) is a sexually transmitted disease caused by the bacterium *Chlamydia trachomatis* (*chlamy-* = cloak). This unusual bacterium cannot reproduce outside body cells; it "cloaks" itself inside cells, where it divides. At present, chlamydia is the most prevalent sexually transmitted disease, affecting 3–5 million individuals in the United States and causing sterility in more than 20,000 young men and women annually.

In most cases the initial infection is asymptomatic and thus difficult to recognize clinically. In males, urethritis is the principal result, causing a clear discharge, burning on urination, frequent urination, and painful urination. Without treatment, the epididymides may also become inflamed, leading to sterility. In 70% of females with chlamydia, symptoms are absent, but chlamydia is the leading cause of pelvic inflammatory disease. Moreover, the uterine tubes may also become inflamed, which increases the risk of ectopic pregnancy (implantation of a fertilized ovum outside the uterus) and infertility due to the formation of scar tissue in the tubes.

Gonorrhea

Gonorrhea (gon-ō-RĒ-a) or **"the clap"** is caused by the bacterium *Neisseria gonorrhoeae*. In the United States, 1–2 million new cases of gonorrhea appear each year, most among individuals aged 15–19 years. Discharges from infected mucus membranes are the source of transmission of the bacteria either during sexual contact or during the passage of a newborn through the birth canal. The infection site can be in the mouth and throat after oral-genital contact, in the vagina and penis after genital intercourse, or in the rectum after recto-genital contact.

Males usually experience urethritis with profuse pus drainage and painful urination. The prostate gland and epididymis may also become infected. In females, infection typically occurs in the vagina, often with a discharge of pus. Both infected males and females may harbor the disease without any symptoms, however, until it has progressed to a more advanced stage; about 5–10% of males and 50% of females are asymptomatic. In females, the infection and consequent inflammation can proceed from the vagina into the uterus, uterine tubes, and pelvic cavity. An estimated 50,000–80,000 women in the United States are made infertile by gonorrhea every year as a result of scar tissue formation that closes the uterine tubes. If bacteria in the birth canal are transmitted to the eyes of a newborn, blindness can result. Administration of a 1% silver nitrate solution in the infant's eyes prevents infection.

Syphilis

Syphilis, caused by the bacterium *Treponema pallidum*, is transmitted through sexual contact or exchange of blood, or through the placenta to a fetus. The disease progresses through several stages. During the *primary stage,* the chief sign is a painless open sore, called a **chancre** (SHANG-ker), at the point of contact. The chancre heals within 1 to 5 weeks. From 6 to 24 weeks later, signs and symptoms such as a skin rash, fever, and aches in the joints and muscles usher in the *secondary state,* which is systemic— the infection spreads to all major body systems. When signs of organ degeneration appear, the disease is said to be in the *tertiary stage.* If the nervous system is involved, the tertiary stage is called **neurosyphilis.** As motor areas become extensively damaged, victims may be unable to control urine and bowel movements; eventually they may become bedridden, unable even to feed themselves. Damage to the cerebral cortex produces memory loss and personality changes that range from irritability to hallucinations.

Genital Herpes

Genital herpes is an incurable STD. Type II herpes simplex virus (HSV-2) causes genital infections, producing painful blisters on the prepuce, glans penis, and penile shaft in males and on the vulva or sometimes high up in the vagina in females. The blisters disappear and reappear in most patients, but the virus itself remains in the body. A related virus, type I herpes simplex virus (HSV-1), causes cold sores on the mouth and lips. Infected individuals typically experience recurrences of symptoms several times a year.

REPRODUCTIVE SYSTEM DISORDERS IN MALES

Testicular Cancer

Testicular cancer is the most common cancer in males between the ages of 20 and 35; it is also one of the most curable cancers in males. More than 95% of testicular cancers arise from spermatogenic cells within the seminiferous tubules. An early sign of testicular cancer is a mass in the testis, often associated with a sensation of testicular heaviness or a dull ache in the lower abdomen; pain usually does not occur. All males should perform regular testicular self-examinations.

Prostate Disorders

Because the prostate surrounds a portion of the urethra, any infection, enlargement, or tumor in it can obstruct the flow of urine. Acute and chronic infections of the prostate gland are common in postpubescent males, often in association with inflammation of the urethra. In **acute prostatitis,** the prostate gland becomes swollen and tender. **Chronic prostatitis** is one of the most common chronic infections in men of the middle and later years; on examination, the prostate gland feels enlarged, soft, and very tender, and its surface outline is irregular.

Prostate cancer is the leading cause of death from cancer in men in the United States, having surpassed lung cancer in 1991. Each year it strikes almost 200,000 U.S. men, causing 40,000 deaths. A blood test can measure the level of prostate-specific antigen (PSA) in the blood. The amount of PSA, which is produced only by prostate epithelial cells, increases with enlargement of the prostate gland and may indicate infection, benign enlargement, or prostate cancer. For males over 40, the American Cancer Society recommends annual examination of the prostate gland by a **digital rectal exam,** in which a physician palpates the gland through the rectum with the fingers (digits). Many physicians also recommend an annual PSA test for males over 50. Treatment for prostate cancer may involve surgery, radiation, hormonal therapy, and chemotherapy. Because many prostate cancers grow very slowly, some urologists recommend "watchful waiting" before treating small tumors in men over age 70.

REPRODUCTIVE SYSTEM DISORDERS IN FEMALES

Premenstrual Syndrome

Premenstrual syndrome (PMS) refers to severe physical and emotional distress that occurs during the postovulatory (luteal) phase of the female reproductive cycle. Signs and symptoms usually increase in severity until the onset of menstruation and then dramatically disappear. The signs and symptoms are highly variable from one woman to another and include edema, weight gain, breast swelling and tenderness, abdominal distension, backache, joint pain, constipation, skin eruptions, fatigue and lethargy, greater need for sleep, depression or anxiety, irritability, mood swings, headache, poor coordination and clumsiness, and cravings for sweet or salty foods. The pathophysiology of PMS is unknown. For some women, getting regular exercise, avoiding caffeine, salt, and alcohol, and eating a diet that is high in complex carbohydrates and lean proteins can bring considerable relief. Because PMS occurs only after ovulation has occurred, oral contraceptives are an effective treatment for women whose symptoms are incapacitating.

Endometriosis

Endometriosis (en-dō-mē-trē-Ō-sis; endo- = within; metri- = uterus; osis = condition) is characterized by the growth of endometrial tissue outside the uterus. The tissue enters the pelvic cavity via the open uterine tubes and may be found in any of several sites—on the ovaries, the rectouterine pouch, the outer surface of the uterus, the sigmoid colon, pelvic and abdominal lymph nodes, the cervix, the abdominal wall, the kidneys, and the urinary bladder. Endometrial tissue responds to hormonal fluctuations—whether it is inside or outside the uterus—by first proliferating and then breaking down and bleeding, which causes inflammation, pain, scarring, and infertility. Symptoms include premenstrual pain or unusually severe menstrual pain.

Breast Cancer

One in eight U.S. women faces the prospect of **breast cancer.** After lung cancer, it is the second-leading cause of death from cancer in U.S. women; it seldom occurs in men. In females, breast cancer is rarely seen before age 30, and its incidence rises rapidly after menopause. An estimated 5% of the 180,000 cases diagnosed each year in the United States, particularly those that arise in younger women, stem from inherited genetic mutations (changes in the DNA). Researchers have now identified two genes that increase susceptibility to breast cancer: BRCA1 (breast cancer 1) and BRCA2. Mutation of BRCA1 also confers a high risk for ovarian cancer. In addition, mutations of the p53 gene increase the risk of breast cancer in both males and females, and mutations of the androgen receptor gene are associated with the occurrence of breast cancer in some males. Despite the fact that breast cancer generally is not painful until it becomes quite advanced, any lump, no matter how small, should be reported to a physician at once. Early detection—by breast self-examination and mammograms—is the best way to increase the chance of survival.

The most effective technique for detecting tumors less than 1 cm (about 0.5 in.) in diameter is **mammography** (mam-OG-ra-fē; -graphy = to record), a type of radiography using very sensitive x-ray film. The image of the breast, called a **mammogram,** is best obtained by compressing the breasts, one at a time, using flat plates. A supplementary procedure for evaluating breast abnormalities is **ultrasound.** Although ultrasound cannot detect tumors smaller than 1 cm in diameter, it can be used to determine whether a lump is a benign, fluid-filled cyst or a solid (and therefore possibly malignant) tumor.

Among the factors that increase the risk of developing breast cancer are (1) a family history of breast cancer, especially in a mother or sister; (2) nulliparity (never having borne a child) or having a first child after age 35; (3) previous cancer in one breast; (4) exposure to ionizing radiation, such as x-rays; (5) excessive alcohol intake; and (6) cigarette smoking.

The American Cancer Society recommends the following steps to help diagnose breast cancer as early as possible:

- All women over 20 should develop the habit of monthly breast self-examination.
- A physician should examine the breasts every 3 years when a woman is between the ages of 20 and 40, and every year after age 40.
- A mammogram should be taken in women between the ages of 35 and 39, to be used later for comparison (baseline mammogram).
- Women with no symptoms should have a mammogram every year or two between ages 40 and 49, and every year after age 50.
- Women of any age with a history of breast cancer, a strong family history of the disease, or other risk factors should consult a physician to determine a schedule for mammography.

Treatment for breast cancer may involve hormone therapy, chemotherapy, radiation therapy, **lumpectomy** (removal of the tumor and the immediate surrounding tissue), a modified or radical mastectomy, or a combination of these approaches. A **radical mastectomy** (mast- = breast) involves removal of the affected breast along with the underlying pectoral muscles and the axillary lymph nodes. (Lymph nodes are removed because metastasis of cancerous cells usually occurs through lymphatic or blood vessels.) Radiation treatment and chemotherapy may follow the surgery to ensure the destruction of any stray cancer cells. In some cases of metastatic breast cancer, Herceptin—a monoclonal antibody drug that targets an antigen on the surface of breast-cancer cells—can cause regression of the tumors and retard progression of the disease. Finally, two promising drugs for breast cancer prevention now on the market—Nolvadex (tamoxifen) and Evista (raloxifene)—reduce the incidence of breast cancer.

Ovarian Cancer

Even though ovarian cancer is the sixth most common form of cancer in females, it is the leading cause of death from all gynecological malignancies (excluding breast cancer) because it is difficult to detect before it metastasizes (spreads) beyond the ovaries. Risk factors associated with ovarian cancer include age (usually over age 50); race (whites are at highest risk); family history of ovarian cancer; more than 40 years of active ovulation; nulliparity or first pregnancy after age 30; a high-fat, low-fiber, vitamin A-deficient diet; and prolonged exposure to asbestos and talc. Early ovarian cancer has no symptoms or only mild ones associated with other common problems, such as abdominal discomfort, heartburn, nausea, loss of appetite, bloating, and flatulence. Later-stage signs and symptoms include an enlarged abdomen, abdominal and/or pelvic pain, persistent gastrointestinal disturbances, urinary complications, menstrual irregularities, and heavy menstrual bleeding.

Cervical Cancer

Cervical cancer, carcinoma of the cervix of the uterus, starts with **cervical dysplasia** (dis-PLĀ-sē-a), a change in the shape, growth, and number of cervical cells. If the condition is minimal, the cells may return to normal; if it is severe, it may progress to cancer. In most cases cervical cancer may be detected in its earliest stages by a Pap smear. Some evidence links cervical cancer to the virus that causes genital warts (human papilloma virus). Increased risk is associated with having a large number of sexual partners, having first intercourse at a young age, and smoking cigarettes.

Vulvovaginal Candidiasis

Candida albicans is a yeastlike fungus that commonly grows on mucous membranes of the gastrointestinal and genitourinary tracts. The organism is responsible for **vulvovaginal candidiasis** (vul-vō-VAJ-i-nal can-di-DĪ-a-sis), the most common form of **vaginitis** (vaj-i-NĪ-tis), inflammation of the vagina. Candidiasis is characterized by severe itching; a thick, yellow, cheesy discharge; a yeasty odor; and pain. The disorder, experienced at least once by about 75% of females, is usually a result of proliferation of the fungus following antibiotic therapy for another condition. Predisposing conditions include the use of oral contraceptives or cortisone-like medications, pregnancy, and diabetes.

MEDICAL TERMINOLOGY

Castration (kas-TRĀ-shun; = to prune) Removal, inactivation, or destruction of the gonads; commonly used in reference to removal of the testes only.

Culdoscopy (kul-DOS-kō-pē; -*scopy* = to examine) A procedure in which a culdoscope (endoscope) is inserted through the vagina to view the pelvic cavity.

Endocervical curettage (ku-re-TAZH; *curette* = scraper) A procedure in which the cervix is dilated and the endometrium of the uterus is scraped with a spoon-shaped instrument called a curette; commonly called a D and C (dilation and curettage).

Episiotomy (e-piz-ē-OT-ō-mē; *episi-* = vulva or pubic region; -*otomy* = incision) A perineal cut made with surgical scissors during childbirth to prevent excessive stretching and even tearing of this region. The cut enlarges the vaginal opening to make more room for the fetus to pass. In effect, a straight, more easily sutured cut is substituted for a jagged tear.

Hermaphroditism (her-MAF-rō-dī-tizm′) Presence of both male and female sex organs in one individual.

Hypospadias (hī′-pō-SPĀ-dē-as; *hypo-* = below) A displaced urethral opening. In males, the displaced opening may be on the underside of the penis, at the penoscrotal junction, between the scrotal folds, or in the perineum; in females, the urethra opens into the vagina.

Leukorrhea (loo′-kō-RĒ-a; *leuko-* = white) A whitish (nonbloody) vaginal discharge that may occur at any age and affects most women at some time.

Oophorectomy (ō′-of-ō-REK-tō-mē; *oophor-* = bearing eggs) Removal of the ovaries.

Ovarian cyst The most common form of ovarian tumor, in which a fluid-filled follicle or corpus luteum persists and continues growing.

Pelvic inflammatory disease (PID) A collective term for any extensive bacterial infection of the pelvic organs, especially the uterus, uterine tubes, or ovaries, which is characterized by pelvic soreness, lower back pain, abdominal pain, and urethritis. Often the early symptoms of PID occur just after menstruation. As infection spreads and cases advance, fever may develop, along with painful abscesses of the reproductive organs.

Salpingectomy (sal′-pin-JEK-tō-mē; *salpingo* = tube) Removal of a uterine tube (oviduct).

Smegma (SMEG-ma) The secretion, consisting principally of desquamated epithelial cells, found chiefly around the external genitalia and especially under the foreskin of the male.

STUDY OUTLINE

THE CELL CYCLE IN THE GONADS (p. 974)

1. Reproduction is the process by which new individuals of a species are produced and the genetic material is passed from generation to generation.
2. The organs of reproduction are grouped as gonads (produce gametes), ducts (transport and store gametes), accessory sex glands (produce materials that support gametes), and supporting structures (have various roles in reproduction).
3. Gametes contain the haploid (*n*) chromosome number, and most somatic cells contain the diploid (2*n*) chromosome number.
4. Meiosis is the process that produces haploid gametes; it consists of two successive nuclear divisions called meiosis I and meiosis II.
5. During meiosis I, homologous chromosomes undergo synapsis (pairing) and crossing-over; the net result is two haploid daughter cells that are genetically unlike each other and unlike the parent cell that produced them.
6. During meiosis II, the two haploid daughter cells divide to form four haploid cells.

THE MALE REPRODUCTIVE SYSTEM (p. 976)

1. The male structures of reproduction include the testes, ductus epididymis, ductus (vas) deferens, ejaculatory duct, urethra, seminal vesicles, prostate gland, bulbourethral (Cowper's) glands, and penis.
2. The scrotum is a sac that hangs from the root of the penis and consists of loose skin and superficial fascia; it supports the testes.
3. The temperature of the testes is regulated by contraction of the cremaster muscle and dartos muscle, which either elevates them and brings them closer to the pelvic cavity or relaxes and moves them farther from the pelvic cavity.
4. The testes are paired oval glands (gonads) in the scrotum containing seminiferous tubules, in which sperm cells are made; Sertoli cells (sustentacular cells), which nourish sperm cells and secrete inhibin; and Leydig cells (interstitial endocrinocytes), which produce the male sex hormone testosterone.
5. The testes descend into the scrotum through the inguinal canals during the seventh month of fetal development. Failure of the testes to descend is called cryptorchidism.
6. Secondary oocytes and sperm, both of which are called gametes, are produced in the gonads.
7. Spermatogenesis, which occurs in the testes, is the process whereby immature spermatogonia develop into mature sperm. The spermatogenesis sequence, which includes meiosis I, meiosis II, and spermiogenesis, results in the formation of four haploid sperm (spermatozoa) from each primary spermatocyte.
8. Mature sperm consist of a head, a midpiece, and a tail. Their function is to fertilize a secondary oocyte.
9. At puberty, gonadotropin releasing hormone (GnRH) stimulates anterior pituitary gland secretion of FSH and LH. LH stimulates production of testosterone; FSH and testosterone stimulate spermatogenesis. Sertoli cells secrete androgen-binding protein (ABP), which binds to testosterone and keeps its concentration high in the seminiferous tubule.
10. Testosterone controls the growth, development, and maintenance of sex organs; stimulates bone growth, protein anabolism, and sperm maturation; and stimulates development of masculine secondary sex characteristics.
11. Inhibin is produced by Sertoli cells; its inhibition of FSH helps regulate the rate of spermatogenesis.
12. The duct system of the testes includes the seminiferous tubules, straight tubules, and rete testis. Sperm flow out of the testes through the efferent ducts.
13. The ductus epididymis is the site of sperm maturation and storage.
14. The ductus (vas) deferens stores sperm and propels them toward the urethra during ejaculation.
15. Each ejaculatory duct, formed by the union of the duct from the seminal vesicle and ductus (vas) deferens, is the passageway for ejection of sperm and secretions of the seminal vesicles into the first portion of the urethra, the prostatic urethra.
16. The urethra in males is subdivided into three portions: the prostatic, membranous, and spongy (penile) urethra.
17. The seminal vesicles secrete an alkaline, viscous fluid that constitutes about 60% of the volume of semen and contributes to sperm viability.
18. The prostate gland secretes a slightly acidic fluid that constitutes about 25% of the volume of semen and contributes to sperm motility.
19. The bulbourethral (Cowper's) glands secrete mucus for lubrication and an alkaline substance that neutralizes acid.
20. Semen is a mixture of sperm and seminal fluid; it provides the fluid in which sperm are transported, supplies nutrients, and neutralizes the acidity of the male urethra and the vagina.
21. The penis consists of a root, a body, and a glans penis.
22. Engorgement of the penile blood sinuses under the influence of sexual excitation is called erection.

THE FEMALE REPRODUCTIVE SYSTEM (p. 989)

1. The female organs of reproduction include the ovaries (gonads), uterine (Fallopian) tubes or oviducts, uterus, vagina, and vulva.
2. The mammary glands are considered part of the reproductive system in females.
3. The ovaries, the female gonads, are located in the superior portion of the pelvic cavity, lateral to the uterus.
4. Ovaries produce secondary oocytes, discharge secondary oocytes (the process of ovulation), and secrete estrogens, progesterone, relaxin, and inhibin.
5. Oogenesis (production of haploid secondary oocytes) begins in the ovaries. The oogenesis sequence includes meiosis I and meiosis II, which goes to completion only after an ovulated secondary oocyte is fertilized by a sperm cell.
6. The uterine (Fallopian) tubes transport secondary oocytes from the ovaries to the uterus and are the normal sites of fertilization. Ciliated cells and peristaltic contractions help move a secondary oocyte or fertilized ovum toward the uterus.
7. The uterus is an organ the size and shape of an inverted pear that functions in menstruation, implantation of a fertilized ovum, development of a fetus during pregnancy, and labor. It also is part of the pathway for sperm to reach the uterine tubes to fertilize a secondary oocyte. Normally, the uterus is held in position by a series of ligaments.
8. Histologically, the layers of the uterus are an outer perimetrium (serosa), a middle myometrium, and an inner endometrium.
9. The vagina is a passageway for sperm and the menstrual flow, the receptacle of the penis during sexual intercourse, and the inferior portion of the birth canal. It is capable of considerable distension.
10. The vulva, a collective term for the external genitals of the female, consists of the mons pubis, labia majora, labia minora, clitoris, vestibule, vaginal and urethral orifices, hymen, bulb of the vestibule, and the paraurethral (Skene's), greater vestibular (Bartholin's), and lesser vestibular glands.
11. The perineum is a diamond-shaped area at the inferior end of the trunk medial to the thighs and buttocks.
12. The mammary glands are modified sweat glands lying superficial to the pectoralis major muscles. Their function is to synthesize, secrete, and eject milk (lactation).
13. Mammary gland development depends on estrogens and progesterone.
14. Milk production is stimulated by prolactin, estrogen, and progesterone; milk ejection is stimulated by oxytocin.

THE FEMALE REPRODUCTIVE CYCLE (p. 1001)

1. The function of the ovarian cycle is to develop a secondary oocyte, whereas that of the uterine (menstrual) cycle is to prepare the endometrium each month to receive a fertilized egg. The female reproductive cycle includes both the ovarian and uterine cycles.
2. The uterine and ovarian cycles are controlled by GnRH from the hypothalamus, which stimulates the release of FSH and LH by the anterior pituitary gland.
3. FSH stimulates development of secondary follicles and initiates secretion of estrogens by the follicles. LH stimulates further development of the follicles, secretion of estrogens by follicular cells, ovulation, formation of the corpus luteum, and the secretion of progesterone and estrogens by the corpus luteum.
4. Estrogens stimulate the growth, development, and maintenance of female reproductive structures; stimulate the development of secondary sex characteristics; and stimulate protein synthesis.
5. Progesterone works with estrogens to prepare the endometrium for implantation and the mammary glands for milk synthesis.
6. Relaxin relaxes the pubic symphysis and helps dilate the uterine cervix to facilitate delivery.
7. During the menstrual phase, the stratum functionalis of the endometrium is shed, discharging blood, tissue fluid, mucus, and epithelial cells.
8. During the preovulatory phase, a group of follicles in the ovaries begin to undergo final maturation. One follicle outgrows the others and becomes dominant while the others degenerate. At the same time, endometrial repair occurs in the uterus. Estrogens are the dominant ovarian hormones during the preovulatory phase.
9. Ovulation is the rupture of the dominant mature (Graafian) follicle and the release of a secondary oocyte into the pelvic cavity. It is brought about by a surge of LH. Signs and symptoms of ovulation include increased basal body temperature; clear, stretchy cervical mucus; changes in the uterine cervix; and ovarian pain.
10. During the postovulatory phase, both progesterone and estrogens are secreted in large quantity by the corpus luteum of the ovary, and the uterine endometrium thickens in readiness for implantation.
11. If fertilization and implantation do not occur, the corpus luteum degenerates, and the resulting low level of progesterone allows discharge of the endometrium followed by the initiation of another reproductive cycle.
12. If fertilization and implantation occur, the corpus luteum is maintained by placental hCG, and the corpus luteum and later the placenta secrete progesterone and estrogens to support pregnancy and breast development for lactation.

THE HUMAN SEXUAL RESPONSE (p. 1007)

1. The similar sequence of changes experienced by both males and females before, during, and after intercourse is termed the human sexual response; it occurs in four stages: excitement (arousal), plateau, orgasm, and resolution.
2. During excitement and plateau, parasympathetic nerve impulses produce genital vasocongestion, engorgement of tissues with blood, and secretion of lubricating fluids. Heart rate, blood pressure, breathing rate, and muscle tone increase.
3. During orgasm, sympathetic and somatic motor nerve impulses cause rhythmical contractions of smooth and skeletal muscles.
4. During resolution, the body relaxes and returns to the unaroused state.

BIRTH CONTROL METHODS (p. 1008)

1. Methods include surgical sterilization (vasectomy, tubal ligation), hormonal methods, intrauterine devices, spermatocides, barrier methods (condom, vaginal pouch, diaphragm), periodic abstinence (rhythm and sympto-thermal methods), coitus interruptus, and induced abortion. See Table 28.2 on page 1009 for failure rates for these methods.
2. Contraceptive pills of the combination type contain estrogens and progestins in concentrations that decrease the secretion of FSH and LH and thereby inhibit development of ovarian follicles and ovulation.
3. An abortion is the premature expulsion from the uterus of the products of conception; it may be spontaneous or induced. RU 486 can induce abortion by blocking the action of progesterone.

DEVELOPMENTAL ANATOMY OF THE REPRODUCTIVE SYSTEMS (p. 1010)

1. The gonads develop from intermediate mesoderm. In the presence of the *SRY* gene, the gonads begin to differentiate into testes during the seventh week. The gonads differentiate into ovaries when the *SRY* gene is absent.
2. In males, testosterone stimulates development of each mesonephric duct into an epididymis, ductus (vas) deferens, ejaculatory duct, and seminal vesicle, and Müllerian-inhibiting substance (MIS) causes the paramesonephric duct cells to die. In females, testosterone and MIS are absent; the paramesonephric ducts develop into the uterine tubes, uterus, and vagina and the mesonephric ducts degenerate.
3. The external genitals develop from the genital tubercle and are stimulated to develop into typical male structures by the hormone dihydrotestosterone (DHT). The external genitals develop into female structures when DHT is not produced, the normal situation in female embryos.

AGING AND THE REPRODUCTIVE SYSTEMS (p. 1012)

1. Puberty refers to the period of time when secondary sex characteristics begin to develop and the potential for sexual reproduction is reached.
2. The onset of puberty in males is signaled by increased levels of LH, FSH, and testosterone.
3. The onset of puberty in females is signaled by increased levels of LH, FSH, and estrogens.
4. In older males, decreased levels of testosterone are associated with decreased muscle strength, waning sexual desire, and fewer viable sperm; prostate disorders are common.
5. In older females, levels of progesterone and estrogens decrease, resulting in changes in menstruation and then menopause; uterine and breast cancer increase in incidence with age.

SELF-QUIZ QUESTIONS

1. Match the following:

____ (a) a small, cylindrical mass of erectile tissue and nerves in the female; homologue of the male glans penis

____ (b) produce mucus in the female during sexual arousal and intercourse; homologous to the male bulbourethral glands

____ (c) the group of cells that nourish the developing oocyte and begin to secrete estrogens

____ (d) a pathway for sperm to reach the uterine tubes; the site of menstruation; the site of implantation of a fertilized ovum; the womb

____ (e) produces progesterone, estrogens, relaxin, and inhibin

____ (f) draw the ovum into the uterine tube

____ (g) the opening between the uterus and vagina

____ (h) mucous-secreting glands in the female that are homologous to the prostate gland

____ (i) the female copulatory organ; the birth canal

____ (j) passageway for the ovum to the uterus; usual site of fertilization

____ (k) refers to the external genitals of the female

____ (l) the layer of the uterine lining that is partially shed during each monthly cycle

(1) follicle
(2) corpus luteum
(3) uterine tube
(4) fimbriae
(5) uterus
(6) cervix
(7) endometrium
(8) vagina
(9) vulva
(10) clitoris
(11) paraurethral glands
(12) greater vestibular glands

2. Match the following:

____ (a) site of sperm maturation

____ (b) the male copulatory organ; a passageway for ejaculation of sperm and excretion of urine

____ (c) sperm-forming cells

____ (d) during sexual arousal, produce an alkaline substance that protects sperm by neutralizing acids in the urethra

____ (e) ejects sperm into the urethra just before ejaculation

____ (f) the supporting structure for the testis

____ (g) carries the sperm from the scrotum into the abdominopelvic cavity for release by ejaculation; is cut and tied as a means of sterilization

____ (h) the shared terminal duct of the reproductive and urinary systems in the male

____ (i) surrounds the urethra at the base of the bladder; produces secretions that contribute to sperm motility and viability

____ (j) produce testosterone

____ (k) support and protect developing spermatogenic cells; secrete inhibin; form the blood–testis barrier

____ (l) secrete an alkaline fluid to help neutralize acids in the female reproductive tract; secrete fructose for use in ATP production by sperm

(1) spermatogenic cells
(2) Sertoli cells
(3) Leydig cells
(4) penis
(5) scrotum
(6) epididymis
(7) ductus deferens
(8) ejaculatory duct
(9) urethra
(10) seminal vesicles
(11) prostrate gland
(12) bulbourethral glands

Choose the best answer to the following questions:

3. Which of the following are functions of Sertoli cells? (1) protection of developing spermatogenic cells, (2) nourishment of spermatocytes, spermatids, and sperm, (3) phagocytosis of excess sperm cytoplasm as development proceeds, (4) mediation of the effects of testosterone and FSH, (5) control of movements of spermatogenic cells and release of sperm into the lumen of seminiferous tubules
(a) 1, 2, 4, and 5, (b) 1, 2, 3, and 5, (c) 2, 3, 4, and 5, (d) 1, 2, 3, and 4, (e) 1, 2, 3, 4, and 5

4. Which of the following are true concerning androgens? (1) They stimulate the male pattern of development. (2) They stimulate the male and female patterns of development. (3) They contribute to sex drive in males and females. (4) They contribute to male secondary sex characteristics. (5) They stimulate protein synthesis.
(a) 1, 2, and 3, (b) 1, 3, 4, and 5, (c) 1, 2, 4, and 5, (d) 1, 2, 3, and 4, (e) 1, 3, and 5

5. Which of the following are true concerning estrogens? (1) They promote development and maintenance of female reproductive structures and secondary sex characteristics. (2) They help control fluid and electrolyte balance. (3) They increase protein catabolism. (4) They lower blood cholesterol. (5) In moderate levels, they inhibit the release of GnRH and the secretion of LH and FSH.
(a) 1, 2, 4, and 5, (b) 1, 3, 4, and 5, (c) 1, 2, 3, and 5, (d) 1, 2, 3, and 4, (e) 1, 2, 3, 4, and 5

6. Which of the following statements are correct? (1) A sperm head contains DNA and an acrosome. (2) An acrosome is a specialized lysosome that contains enzymes that enable sperm to produce the ATP needed to propel themselves out of the male reproductive tract. (3) Mitochondria in the midpiece of a sperm produce ATP for sperm motility. (4) A sperm's tail, a flagellum, propels it along its way. (5) Once ejaculated, sperm are viable and normally are able to fertilize a secondary oocyte for 5 days.
(a) 1, 2, 3, and 4, (b) 2, 3, 4, and 5, (c) 1, 3, and 4, (d) 2, 4, and 5, (e) 2, 3, and 4

7. Which of the following statements are correct? (1) Spermatogonia are stem cells because when they undergo mitosis, some of the daughter cells remain to serve as a reservoir of cells for future mitosis. (2) Meiosis I is a division of pairs of chromosomes resulting in daughter cells with only one member of each chromosome pair. (3) Meiosis II separates the chromatids of each chromosome. (4) Spermiogenesis involves the maturation of spermatids into sperm. (5) The process by which the seminiferous tubules produce haploid sperm is called spermatogenesis.
(a) 1, 2, 3, and 5, (b) 1, 2, 3, 4, and 5, (c) 1, 3, 4, and 5, (d) 1, 2, 3, and 4, (e) 1, 3, and 5

8. Which of the following statements are correct? (1) Cells from the endoderm of the yolk sac give rise to oogonia. (2) Ova arise from the germinal epithelium of the ovary. (3) Primary oocytes enter prophase of meiosis I during fetal development but do not complete it until after puberty. (4) Once a secondary oocyte is formed, it proceeds to metaphase of meiosis II and stops at this stage. (5) The secondary oocyte resumes meiosis II and forms the ovum and a polar body only if fertilization occurs. (6) A primary oocyte gives rise to an ovum and four polar bodies.
(a) 1, 3, 4, and 5, (b) 1, 3, 4, and 6, (c) 1, 2, 4, and 6, (d) 1, 2, 4, and 5, (e) 1, 2, 5, and 6

9. Which of the following statements are correct? (1) The female reproductive cycle consists of a menstrual phase, a preovulatory phase, ovulation, and a postovulatory phase. (2) During the menstrual phase, small secondary follicles in the ovary begin to enlarge while the uterus is shedding its lining. (3) During the preovulatory phase, a dominant follicle continues to grow and begins to secrete estrogens and inhibin while the uterine lining begins to rebuild. (4) Ovulation results in the release of an ovum and the shedding of the uterine lining to nourish and support the released ovum. (5) After ovulation, a corpus luteum forms from the ruptured follicle and begins to secrete progesterone and estrogens. which it will continue to do throughout pregnancy if the egg is fertilized. (6) If pregnancy does not occur, then the corpus luteum degenerates into a scar called the corpus albicans, and the uterine lining is prepared to be shed again.
(a) 1, 2, 4, and 5, (b) 2, 4, 5, and 6, (c) 1, 4, 5, and 6, (d) 1, 3, 4, and 6, (e) 1, 2, 3, and 6

10. Match the following:
____ (a) relaxes the uterus by inhibiting contractions during monthly cycles; increases flexibility of the pubic symphysis during childbirth
____ (b) stimulates Leydig cells to secrete testosterone
____ (c) inhibits production of FSH by the anterior pituitary gland
____ (d) stimulates male pattern of development; stimulates protein synthesis; contributes to sex drive
____ (e) maintains the corpus luteum during the first trimester of pregnancy
____ (f) promotes development of female reproductive structures; helps control fluid and electrolyte balance; lowers blood cholesterol
____ (g) stimulates the initial secretion of estrogen by growing follicles; promotes follicle growth
____ (h) is secreted by the corpus luteum to maintain the uterine lining during the first trimester of pregnancy

(1) inhibin
(2) LH
(3) FSH
(4) testosterone
(5) estrogen
(6) progesterone
(7) relaxin
(8) human chorionic gonadotropin

11. Match the following:
____ (a) the process during meiosis when portions of a chromatid from the maternal homologue may be exchanged with the corresponding portion of the paternal homologue
____ (b) refers to cells containing one-half the chromosome number
____ (c) the cell produced by the union of an egg and a sperm
____ (d) all chromosomes except the sex chromosomes
____ (e) two chromosomes that belong to a pair (one maternal and one paternal)
____ (f) the degeneration of oogonia before and after birth
____ (g) a packet of discarded nuclear material from the first or second meiotic division of the egg
____ (h) refers to cells containing the full chromosome number

(1) autosomes
(2) zygote
(3) homologous chromosomes
(4) haploid
(5) diploid
(6) crossing-over
(7) polar body
(8) atresia

Complete the following:
12. The female gonads are the ____; the male gonads are the ____. The female gamete is the ____; the male gamete is the ____.
13. The ____ produced by the tight junctions connecting the ____ prevents an immune response against the surface antigens of sperm by isolating the spermatogenic cells from the blood.

14. The period of time when secondary sexual characteristics begin to develop and the potential for sexual reproduction is reached is called ____. The first menses is called ____, and the permanent cessation of menses is called ____.

True or false:
15. Spermatogenesis does not occur at normal core body temperature.

CRITICAL THINKING QUESTIONS

1. Melissa skipped A&P class when the reproductive system was covered because, as she said, "I already know all that stuff." On the exam she answered "false" to the statement "LH is a hormone of the male reproductive system." Melissa should have attended class. Explain her error. (HINT: *The corpus luteum secretes sex hormones.*)

2. Why is the germinal epithelium of the ovaries superficial to the tunica albuginea, whereas that of the testes is deep? Where is the

tunica vaginalis located? (HINT: *Male and female structures develop from common embryonic origins.*)

3. If two babies were produced following the fertilization of an ovum (actually, the secondary oocyte) and a polar body, would the babies be identical twins? (HINT: *Identical twins have an identical chromosomal makeup.*)

ANSWERS TO FIGURE QUESTIONS

28.1 In meiosis, chromosomal replication occurs during the interphase that precedes meiosis I.

28.2 The result of crossing-over is that daughter cells are genetically unlike each other and genetically unlike the parent cell that produced them.

28.3 The gonads (testes) produce gametes (sperm) and hormones; the ducts transport, store, and receive gametes; and the accessory sex glands secrete materials that support gametes.

28.4 The cremaster and dartos muscles help regulate the temperature of the testes.

28.5 The tunica vaginalis and tunica albuginea are tissue layers that cover and protect the testes.

28.6 Leydig cells secrete testosterone.

28.7 During meiosis I, the number of chromosomes in each cell is reduced—by half.

28.8 The sperm head contains DNA and enzymes for penetration of a secondary oocyte; the midpiece contains mitochondria for ATP production; the tail consists of a flagellum that provides propulsion.

28.9 Sertoli cells secrete inhibin.

28.10 Testosterone inhibits secretion of LH, and inhibin inhibits secretion of FSH.

28.11 Seminal vesicles contribute the largest volume to seminal fluid.

28.12 Two corpora cavernosa penis and one corpus spongiosum penis contain blood sinuses that fill with blood that cannot flow out of the penis as quickly as it flows in. The trapped blood engorges and stiffens the tissue. The corpus spongiosum penis keeps the spongy urethra open so that ejaculation can occur.

28.13 The testes are homologous to the ovaries; the glans penis is homologous to the clitoris; the prostate gland is homologous to the paraurethral glands; and the bulbourethral gland is homologous to the greater vestibular glands.

28.14 The mesovarium anchors the ovary to the broad ligament of the uterus and the uterine tube; the ovarian ligament anchors it to the uterus; the suspensory ligament anchors it to the pelvic wall.

28.15 Ovarian follicles secrete estrogens; the corpus luteum secretes progesterone, estrogens, relaxin, and inhibin.

28.16 Most ovarian follicles undergo atresia (degeneration).

28.17 Primary oocytes are present in the ovary at birth, so they are

as old as the woman is. In males, primary spermatocytes are continually being formed from stem cells (spermatogonia) and thus are only a few days old.

28.18 Fertilization most often occurs in the ampulla of the uterine tube.

28.19 Ciliated columnar epithelial cells and secretory cells with microvilli line the uterine tubes.

28.20 The endometrium is a highly vascular, secretory epithelium that provides the oxygen and nutrients needed to sustain a fertilized egg; the myometrium is a thick smooth muscle layer that supports the uterine wall during pregnancy and contracts to expel the fetus at birth.

28.21 The stratum basalis of the endometrium provides cells to replace those shed (the stratum functionalis) during each menstruation.

28.22 Anterior to the vaginal opening are the mons pubis, clitoris, and prepuce. Lateral to the vaginal opening are the labia minora and labia majora.

28.23 The anterior portion of the perineum is called the urogenital triangle because its borders form a triangle that encloses the urethral (uro-) and vaginal (-genital) orifices.

28.24 Prolactin, estrogens, and progesterone regulate the synthesis of milk. Oxytocin regulates the ejection of milk.

28.25 The principal estrogen is β-estradiol.

28.26 The hormones responsible for the proliferative phase of endometrial growth are estrogens; for ovulation, LH; for growth of the corpus luteum, LH; and for the midcycle surge of LH, estrogens.

28.27 Estrogens initiate rebuilding of the stratum functionalis.

28.28 The effect of rising but moderate levels of estrogens is negative feedback inhibition of the secretion of gonadotropin-releasing hormone, LH, and FSH.

28.29 This is negative feedback, because the response is opposite to the stimulus. A reduced amount of negative feedback due to declining levels of estrogens and progesterone stimulates release of GnRH, which in turn increases the production and release of FSH and LH, which ultimately stimulate the secretion of estrogens.

28.30 The *SRY* gene on the Y chromosome is responsible for the development of the gonads into testes.

28.31 The presence of dihydrotestosterone (DHT) stimulates differentiation of the external genitals in males; its absence allows differentiation of the external genitals in females.

DEVELOPMENT AND INHERITANCE

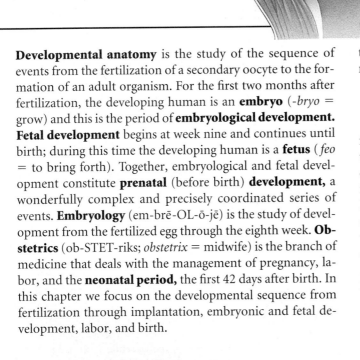

Developmental anatomy is the study of the sequence of events from the fertilization of a secondary oocyte to the formation of an adult organism. For the first two months after fertilization, the developing human is an **embryo** (-*bryo* = grow) and this is the period of **embryological development.** **Fetal development** begins at week nine and continues until birth; during this time the developing human is a **fetus** (*feo* = to bring forth). Together, embryological and fetal development constitute **prenatal** (before birth) **development,** a wonderfully complex and precisely coordinated series of events. **Embryology** (em-brē-OL-ō-jē) is the study of development from the fertilized egg through the eighth week. **Obstetrics** (ob-STET-riks; *obstetrix* = midwife) is the branch of medicine that deals with the management of pregnancy, labor, and the **neonatal period,** the first 42 days after birth. In this chapter we focus on the developmental sequence from fertilization through implantation, embryonic and fetal development, labor, and birth.

FROM FERTILIZATION TO IMPLANTATION

OBJECTIVE

• *Explain the processes associated with fertilization, morula formation, blastocyst development, and implantation.*

Once sperm and a secondary oocyte have developed through meiosis and maturation, and the sperm have been deposited in the vagina, pregnancy can occur. **Pregnancy** is a sequence of events that begins with fertilization, proceeds to implantation, embryonic development, and fetal development, and normally ends with birth about 38 weeks later.

Fertilization

During **fertilization** (fer-til-i-ZĀ-shun; *fertil-* = fruitful) the genetic material from a haploid sperm cell (spermatozoon) and a haploid secondary oocyte merges into a single diploid nucleus. Of the 300–500 million sperm introduced into the vagina, fewer than 1% reach the secondary oocyte. Fertilization normally occurs in the uterine (Fallopian) tube about 12–24 hours after ovulation. Because sperm remain viable in the vagina for about 48 hours and a secondary oocyte is viable for about 24 hours after ovulation, there typically is a 3-day window during which pregnancy is most likely to occur—from 2 days before ovulation to 1 day after ovulation.

The process leading to fertilization begins when peristaltic contractions and the action of cilia transport the oocyte through the uterine tube. Sperm swim up the uterus and into the uterine tubes, propelled by the whiplike movements of their tails (flagella) and possibly guided by chemical attractants released by the oocyte. Additionally, muscular contractions of the uterus, stimulated by prostaglandins in semen, probably aid the movement of sperm toward the uterine tube. Although 100 or so sperm may reach the vicinity of the oocyte within minutes after ejaculation, they *do not become capable* of fertilizing it until several hours later. During this time in the female reproductive tract, sperm undergo **capacitation,** a series of functional changes that cause

Figure 29.1 Selected structures and events in fertilization. (a) A sperm cell penetrating the corona radiata and zona pellucida around a secondary oocyte. (b) A sperm cell in contact with a secondary oocyte. (c) Male and female pronuclei.

🗝 **During fertilization, genetic material from a sperm cell and an oocyte merge to form a single diploid nucleus.**

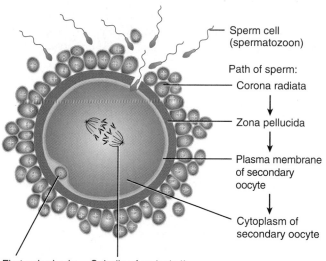

Sperm cell
(spermatozoon)

Path of sperm:

Corona radiata

↓

Zona pellucida

↓

Plasma membrane
of secondary
oocyte

↓

Cytoplasm of
secondary oocyte

First polar body   Spindle of meiosis II

(a) Sperm cell penetrating a secondary oocyte

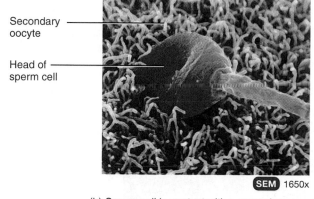

Secondary
oocyte

Head of
sperm cell

SEM 1650x

(b) Sperm cell in contact with a secondary oocyte

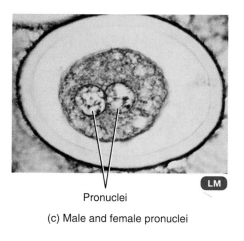

Pronuclei

LM

(c) Male and female pronuclei

Q   What is capacitation?

the sperm's tail to beat even more vigorously and enable its plasma membrane to fuse with the oocyte's plasma membrane. For fertilization to occur, a sperm cell first must penetrate the **corona radiata,** the cloud of granulosa cells that surround the oocyte, and then the **zona pellucida,** the clear glycoprotein layer between the corona radiata and the oocyte's plasma membrane (Figure 29.1a). A glycoprotein in the zona pellucida called ZP3 acts as a sperm receptor; it binds to specific membrane proteins in the sperm head and triggers the **acrosomal reaction,** the release of the contents

of the acrosome. The proteolytic acrosomal enzymes digest a path through the zona pellucida as the lashing sperm tail pushes the sperm cell onward. Many sperm bind to ZP3 molecules and undergo acrosomal reactions, but only the first sperm cell to penetrate the entire zona pellucida and reach the oocyte's plasma membrane fuses with the oocyte.

Fusion of a sperm with a secondary oocyte, called **syngamy** (*syn-* = coming together; *-gamy* = marriage), sets in motion events that block **polyspermy,** fertilization by more than one sperm cell. Within 1–3 seconds the cell membrane of the oocyte depolarizes, which acts as a *fast block to polyspermy*—a depolarized oocyte cannot fuse with another sperm. Depolarization also triggers the intracellular release of calcium ions, which stimulate exocytosis of secretory vesicles from the oocyte. The molecules released by exocytosis inactivate ZP3 and harden the entire zona pellucida, events that constitute the *slow block to polyspermy.*

Once a sperm cell enters a secondary oocyte, the oocyte completes meiosis II. It divides into a larger ovum (mature egg) and a smaller second polar body that fragments and disintegrates (see Figure 28.17 on page 994). The nucleus in the head of the sperm develops into the **male pronucleus,** and the nucleus of the fertilized ovum develops into the **female pronucleus** (Figure 29.1c). After the pronuclei form, they fuse, producing a single diploid nucleus that contains 23 chromosomes from each pronucleus. Thus, the fusion of the haploid (*n*) pronuclei restores the diploid number (2*n*) of 46 chromosomes. The fertilized ovum now is called a **zygote** (ZĪ-gōt; *zygosis* = a joining).

Dizygotic (fraternal) twins are produced from the independent release of two secondary oocytes and the subsequent fertilization of each by different sperm. They are the

Figure 29.2 Cleavage and the formation of the morula and blastocyst.

🔑 **Cleavage refers to the early, rapid mitotic divisions of a zygote.**

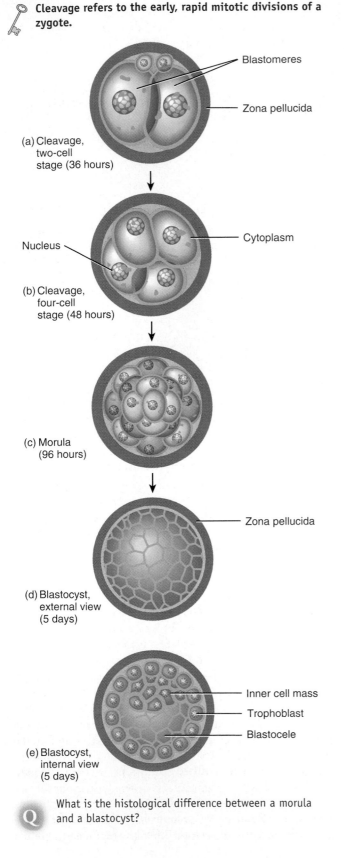

(a) Cleavage, two-cell stage (36 hours)

Blastomeres

Zona pellucida

Nucleus

Cytoplasm

(b) Cleavage, four-cell stage (48 hours)

(c) Morula (96 hours)

Zona pellucida

(d) Blastocyst, external view (5 days)

Inner cell mass

Trophoblast

Blastocele

(e) Blastocyst, internal view (5 days)

Q What is the histological difference between a morula and a blastocyst?

same age and in the uterus at the same time, but they are genetically as dissimilar as any other siblings. Dizygotic twins may or may not be the same sex. Because **monozygotic (identical) twins** develop from a single fertilized ovum, they contain exactly the same genetic material and are always the same sex. Monozygotic twins arise from separation of the developing cells into two embryos, which occurs before 8 days after fertilization 99% of the time. Separations that occur later are likely to produce **conjoined twins,** in which the twins are joined together and share some body structures.

Formation of the Morula

After fertilization, rapid mitotic cell divisions of the zygote called **cleavage** take place (Figure 29.2). The first division of the zygote begins about 24 hours after fertilization and is completed about 30 hours after fertilization. Each succeeding division takes slightly less time. By the second day after fertilization, the second cleavage is completed and there are four cells (see Figure 29.2b); by the end of the third day, there are 16 cells. The progressively smaller cells produced by cleavage are called **blastomeres** (BLAS-tō-mērz; *blasto-* = germ or sprout; *-meres* = parts). Successive cleavages eventually produce a solid sphere of cells, still surrounded by the zona pellucida, called the **morula** (MOR-yoo-la; *morula* = mulberry), which is about the same size as the original zygote (see Figure 29.2c).

Development of the Blastocyst

By the end of the fourth day, the number of cells in the morula increases as it continues to move through the uterine tube toward the uterine cavity. At 4.5–5 days, the dense cluster of cells has developed into a hollow ball of cells that enters the uterine cavity; it is now called a **blastocyst** (*-cyst* = bag) (see Figure 29.2d, e).

The blastocyst has an outer covering of cells called the **trophoblast** (TRŌF-ō-blast; *tropho-* = develop or nourish), an **inner cell mass,** and an internal fluid-filled cavity called the **blastocele** (BLAS-tō-sēl; *-cele* = hollow space). The trophoblast and part of the inner cell mass ultimately form the membranes composing the fetal portion of the placenta; the rest of the inner cell mass develops into the embryo.

Implantation

The blastocyst remains free within the cavity of the uterus for about 2 days before it attaches to the uterine wall. The endometrium is in its secretory phase, and the blastocyst receives nourishment from the glycogen-rich secretions of the endometrial glands, sometimes called uterine milk. During this time, the zona pellucida disintegrates and the blastocyst enlarges. About 6 days after fertilization the blastocyst attaches to the endometrium, a process called **implantation** (Figure 29.3).

As the blastocyst implants, usually in either the posterior portion of the fundus or the body of the uterus, it is ori-

Figure 29.3 Relation of a blastocyst to the endometrium of the uterus at the time of implantation.

Implantation, the attachment of a blastocyst to the endometrium, occurs about 6 days after fertilization.

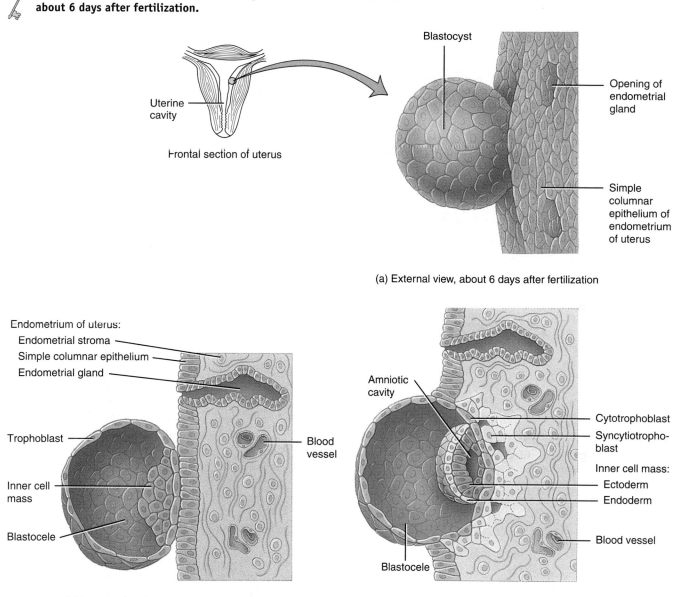

(a) External view, about 6 days after fertilization

(b) Internal view, about 6 days after fertilization

(c) Internal view, about 7 days after fertilization

Q How does the blastocyst merge with and burrow into the endometrium?

ented such that the inner cell mass is toward the endometrium (Figure 29.3b). In the region of contact between the blastocyst and endometrium, the trophoblast develops two layers: a **syncytiotrophoblast** (sin-sīt′-ē-ō-TRŌF-ō-blast) that contains no cell boundaries, and a **cytotrophoblast** (sī-tō-TRŌF-ō-blast) between the inner cell mass and syncytiotrophoblast that is composed of distinct cells (Figure 29.3c). These two layers of trophoblast become part

of the chorion (one of the fetal membranes) as they undergo further growth (see Figure 29.5c). During implantation, the syncytiotrophoblast secretes enzymes that enable the blastocyst to penetrate the uterine lining by digesting and liquefying the endometrial cells. The endometrial secretions further nourish the burrowing blastocyst for about a week after implantation. Eventually, the blastocyst becomes buried in the endometrium. Another secretion of the trophoblast is human

Figure 29.4 Summary of events associated with fertilization and implantation.

 Fertilization usually occurs in the uterine tube.

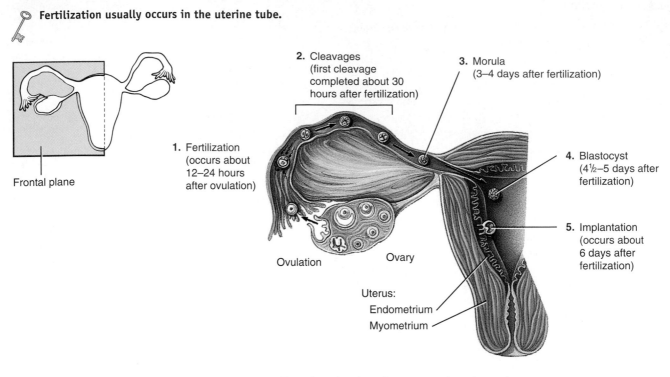

Frontal section through uterus, uterine tube, and ovary

Q In which phase of the uterine cycle does implantation occur?

chorionic gonadotropin (hCG), which has actions similar to LH. Human chorionic gonadotropin rescues the corpus luteum from degeneration and sustains its secretion of progesterone and estrogens. These hormones maintain the uterine lining in a secretory state and thereby prevent menstruation.

The principal events associated with fertilization and implantation are summarized in Figure 29.4.

CLINICAL APPLICATION
Ectopic Pregnancy

Ectopic pregnancy (*ec-* = out of; *-topic* = place) is the development of an embryo or fetus outside the uterine cavity. An ectopic pregnancy usually occurs when passage of the fertilized ovum through the uterine tube is impaired, typically either by decreased motility of the uterine tube smooth muscle or abnormal anatomy. Although the most common sites of ectopic pregnancies are the ampullar and infundibular portions of the uterine tube, ectopic pregnancies may also occur in the abdominal cavity or uterine cervix. Compared to nonsmokers, women who smoke and become pregnant are twice as likely to have an ectopic pregnancy because nicotine in cigarette smoke paralyzes the cilia in the lining of the uterine tube (as it does those in the airways). Scars from pelvic inflammatory disease, previous uterine tube surgery, and previous ectopic pregnancy may hinder movement of the fertilized ovum.

The signs and symptoms of ectopic pregnancy include one or two missed menstrual cycles followed by bleeding and acute abdominal and pelvic pain. Unless removed, the developing embryo can rupture the tube, often resulting in death of the mother. ■

1. Define developmental anatomy.
2. Where does fertilization normally occur?
3. How is polyspermy prevented?
4. What is a morula, and how is it formed?
5. Describe the components of a blastocyst.

EMBRYONIC AND FETAL DEVELOPMENT

OBJECTIVES

• *Discuss the formation of the primary germ layers and embryonic membranes as the principal events of the embryonic period.*

• *List representative body structures produced by the primary germ layers.*

• *Describe the formation of the placenta and umbilical cord.*

The time span from fertilization to birth is the **gestation period** (jes-TĀ-shun; *gest-* = to bear). The human gestation period is about 38 weeks, counted from the estimated day of fertilization (or 2 weeks after the first day of the last menstruation). By the end of the **embryonic period,** the first two months of development, the rudiments of all the principal adult organs are present, and the embryonic membranes are developed. During the **fetal period,** after the second month, organs established by the primary germ layers grow rapidly, and the fetus takes on a human appearance. By the end of the third month, the placenta, which is the site of exchange of nutrients and wastes between the mother and the fetus, is functioning.

The Beginnings of Organ Systems

The first major event of the embryonic period is **gastrulation** (gas'-troo-LĀ-shun), in which the inner cell mass of the blastocyst differentiates into three **primary germ layers:** ectoderm, endoderm, and mesoderm. These germ layers are the major embryonic tissues from which all tissues and organs of the body develop.

Within 8 days after fertilization, the cells of the inner cytotrophoblast proliferate and form the amnion (a fetal membrane) and a space, the **amniotic cavity** (am-nē-OT-ik; *amnio-* = lamb), adjacent to the inner cell mass (Figure 29.5a).

The layer of cells of the inner cell mass that is closer to the amniotic cavity develops into the **ectoderm** (*ecto-* = outside; *derm-* = skin); the layer of cells of the inner cell mass that borders the blastocele develops into the **endoderm** (*endo-* = inside). As the amniotic cavity forms, the inner cell mass at this stage is called the **embryonic disc** and contains ectodermal and endodermal cells; mesodermal cells are scattered external to the disc.

About the 12th day after fertilization, formation of the primary germ layers and associated structures produces striking changes. The cells of the endodermal layer have been dividing rapidly, such that groups of them now extend around in a hollow sphere, forming the yolk sac, another fetal membrane (described shortly). Cells of the cytotrophoblast give rise to a loose connective tissue, the **extraembryonic mesoderm** (*meso-* = middle), which completely fills the space between the cytotrophoblast and yolk sac. Soon large spaces develop in the extraembryonic mesoderm and come together to form a single, larger cavity called the **extraembryonic coelom** (SĒ-lōm; = cavity), the future ventral body cavity (Figure 29.5b).

About the 14th day, differentiation of the cells of the embryonic disc produces three distinct layers: the ectoderm, mesoderm, and endoderm (Figure 29.5b). As the embryo develops, the endoderm becomes the epithelial lining of the

Figure 29.5 Formation of the primary germ layers and associated structures.

The primary germ layers (ectoderm, mesoderm, and endoderm) are the embryonic tissues from which all tissues and organs develop.

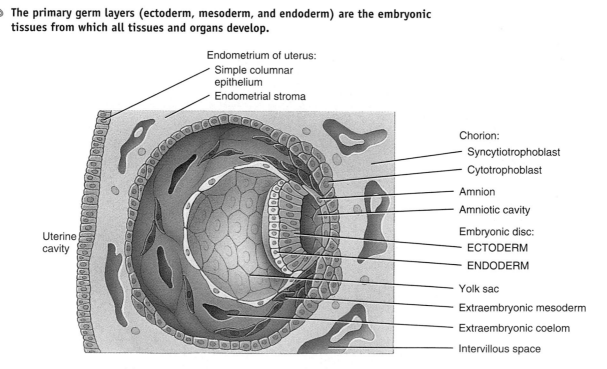

(a) Internal view, about 12 days after fertilization

figure continues

Figure 29.5 *(continued)*

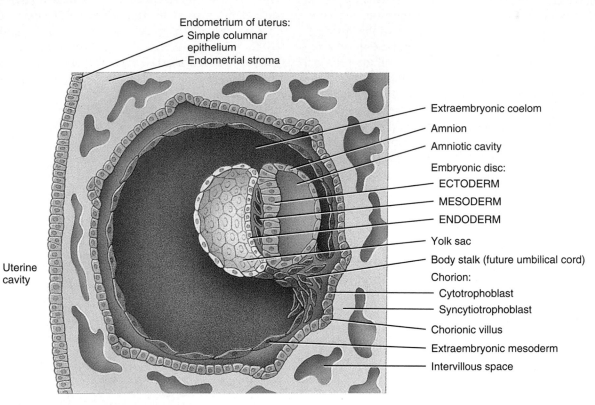

Endometrium of uterus:
Simple columnar epithelium
Endometrial stroma

Extraembryonic coelom

Amnion

Amniotic cavity

Embryonic disc:
ECTODERM
MESODERM
ENDODERM

Yolk sac

Body stalk (future umbilical cord)

Chorion:
Cytotrophoblast
Syncytiotrophoblast

Chorionic villus

Extraembryonic mesoderm

Intervillous space

Uterine cavity

(b) Internal view, about 14 days after fertilization

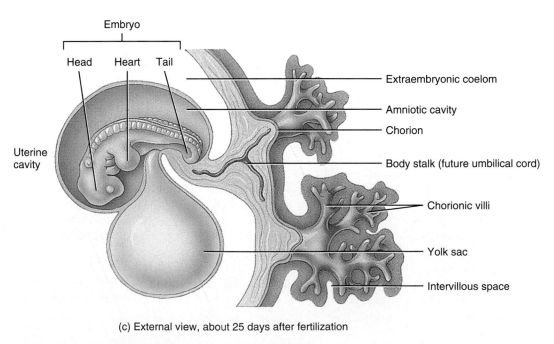

Embryo

Head   Heart   Tail

Extraembryonic coelom

Amniotic cavity

Chorion

Body stalk (future umbilical cord)

Chorionic villi

Yolk sac

Intervillous space

Uterine cavity

(c) External view, about 25 days after fertilization

Q Which cells of the blastocyst give rise to the embryonic disc?

Table 29.1 Structures Produced by the Three Primary Germ Layers

ENDODERM	MESODERM	ECTODERM
Epithelial lining of gastrointestinal tract (except the oral cavity and anal canal) and the epithelium of its glands.	All skeletal, most smooth, and all cardiac muscle.	All nervous tissue.
Epithelial lining of urinary bladder, gallbladder, and liver.	Cartilage, bone, and other connective tissues.	Epidermis of skin.
Epithelial lining of pharynx, auditory (Eustachian) tubes, tonsils, larynx, trachea, bronchi, and lungs.	Blood, red bone marrow, and lymphatic tissue.	Hair follicles, arrector pili muscles, nails, and epithelium of skin glands (sebaceous and sudoriferous).
	Endothelium of blood vessels and lymphatic vessels.	Lens, cornea, and internal eye muscles.
Epithelium of thyroid gland, parathyroid glands, pancreas, and thymus gland.	Dermis of skin.	Internal and external ear.
	Fibrous tunic and vascular tunic of eye.	Neuroepithelium of sense organs.
Epithelial lining of prostate and bulbourethral (Cowper's) glands, vagina, vestibule, urethra, and associated glands such as the greater (Bartholin's) vestibular and lesser vestibular glands.	Middle ear.	Epithelium of oral cavity, nasal cavity, paranasal sinuses, salivary glands, and anal canal.
	Mesothelium of ventral body cavity.	
	Epithelium of kidneys and ureters.	Epithelium of pineal gland, pituitary gland (hypophysis), and adrenal medullae.
	Epithelium of adrenal cortex.	
	Epithelium of gonads and genital ducts.	

gastrointestinal and respiratory tracts, and of several other organs. The mesoderm forms muscle, bone and other connective tissues, and the peritoneum. The ectoderm develops into the epidermis of the skin and the nervous system. Table 29.1 provides more details about the fates of these primary germ layers.

Formation of Embryonic Membranes

A second major event that occurs during the embryonic period is the formation of the **embryonic membranes.** These membranes lie outside the embryo and protect and nourish the embryo and, later, the fetus. (Recall that the developing embryo becomes a fetus after the second month.) The embryonic membranes are the yolk sac, the amnion, the chorion, and the allantois (Figure 29.6).

In species whose young develop inside a shelled egg (such as birds), the **yolk sac** is the primary source of blood vessels that transport nutrients to the embryo (see Figure 29.5c). However, human embryos receive their nutrients from the endometrium; the yolk sac remains small and functions as an early site of blood formation. The yolk sac also contains cells that migrate into the gonads and differentiate into the primitive germ cells (spermatogonia and oogonia).

The amnion is a thin, protective membrane that forms by the eighth day after fertilization and initially overlies the embryonic disc (see Figure 29.5a, b). As the embryo grows, the amnion entirely surrounds the embryo, creating a cavity that becomes filled with amniotic fluid (see Figure 29.6a). Most amniotic fluid is initially derived from a filtrate of maternal blood; later, the fetus makes daily contributions to the fluid by excreting urine into the amniotic cavity. Amniotic fluid serves as a shock absorber for the fetus, helps regulate fetal body temperature, and prevents adhesions between the

skin of the fetus and surrounding tissues. Embryonic cells are sloughed off into amniotic fluid; they can be examined in the procedure called amniocentesis (am´-nē-ō-sen-TĒ-sis), which is described on page 1033. The amnion usually ruptures just before birth; it and its fluid constitutes the "bag of waters."

The **chorion** (KŌR-ē-on) is derived from the trophoblast of the blastocyst and the mesoderm that lines the trophoblast. It surrounds the embryo and, later, the fetus. Eventually the chorion becomes the principal embryonic part of the placenta, the structure for exchange of materials between mother and fetus. It also produces human chorionic gonadotropin (hCG). The inner layer of the chorion eventually fuses with the amnion.

The **allantois** (a-LAN-tō-is; *allant-* = sausage) is a small, vascularized structure that serves as another early site of blood formation. Later its blood vessels form part of the link between mother and fetus.

Placenta and Umbilical Cord

Development of the **placenta** (pla-SEN-ta; = flat cake), which is the site of exchange of nutrients and wastes between the mother and the fetus, occurs during the third month of pregnancy; it is formed by the chorion of the embryo and a portion of the endometrium of the mother. When fully developed, the placenta is shaped like a pancake (see Figure 29.8b). Functionally, the placenta allows oxygen and nutrients to diffuse from maternal blood into fetal blood while carbon dioxide and wastes diffuse from fetal blood into maternal blood.

The placenta also is a protective barrier because most microorganisms cannot pass through it. However, certain viruses, such as those that cause AIDS, German measles, chickenpox, measles, encephalitis, and poliomyelitis, may

Figure 29.6 Embryonic membranes.

🔑 **Embryonic membranes are outside the embryo; they protect and nourish the embryo and, later, the fetus.**

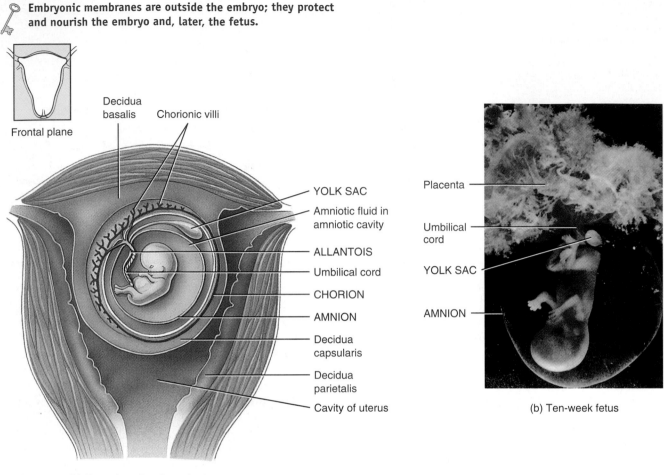

Frontal plane

Decidua basalis

Chorionic villi

YOLK SAC

Amniotic fluid in amniotic cavity

ALLANTOIS

Umbilical cord

CHORION

AMNION

Decidua capsularis

Decidua parietalis

Cavity of uterus

(a) Frontal section through uterus

Placenta

Umbilical cord

YOLK SAC

AMNION

(b) Ten-week fetus

Q How do the amnion and the chorion differ in function?

cross the placenta. The placenta also stores nutrients such as carbohydrates, proteins, calcium, and iron, which are released into fetal circulation as required, and it produces several hormones that are necessary to maintain pregnancy. Almost all drugs, including alcohol and many other substances that can cause birth defects, pass freely through the placenta.

If implantation occurs, a portion of the endometrium becomes modified and is known as the **decidua** (dē-SID-yoo-a; = falling off). The decidua includes all but the stratum basalis layer of the endometrium; it separates from the endometrium after the fetus is delivered just as it does in normal menstruation. Different regions of the decidua, all of which are areas of the stratum functionalis, are named based on their positions relative to the site of the implanted blastocyst (Figure 29.7). The **decidua basalis** is the portion of the endometrium between the chorion and the stratum basalis of the uterus; it becomes the maternal part of the placenta. The **decidua capsularis** is the portion of the endometrium that is located between the embryo and the uterine cavity. The **decidua parietalis** (par-rī-e-TAL-is) is the remaining

modified endometrium that lines the noninvolved areas of the rest of the uterus. As the embryo and later the fetus enlarges, the decidua capsularis bulges into the uterine cavity and initially fuses with the decidua parietalis, thereby obliterating the uterine cavity. By about 27 weeks, the decidua capsularis degenerates and disappears.

Connections between mother and developing child are established via the developing placenta and umbilical cord (Figure 29.8a). During embryonic life, fingerlike projections of the chorion, called **chorionic villi** (kō′-rē-ON-ik VIL-ī), grow into the decidua basalis of the endometrium. These projections, which will contain fetal blood vessels of the allantois, continue growing until they are bathed in maternal blood sinuses called **intervillous spaces** (in-ter-VIL-us). The result is that maternal and fetal blood vessels are brought into close proximity. Note, however, that maternal and fetal blood vessels do not join, and the blood they carry does not normally mix. Instead, oxygen and nutrients in the blood of the mother's intervillous spaces diffuse across the cell membranes into the capillaries of the villi while waste products diffuse in the opposite direction. From the capillaries of the villi, nutrients and oxygen enter the fetus through the umbilical vein. Wastes leave the fetus through the umbilical arteries, pass into the capillaries of the villi, and diffuse into the maternal blood. A few materials, such as IgG antibodies, pass from the blood of the mother into the capillaries of the villi via transcytosis (described on page 676) in which endocytosis is receptor-mediated.

The **umbilical cord** (um-BIL-i-kul), the vascular connection between mother and fetus, consists of two umbilical arteries that carry deoxygenated fetal blood to the placenta, one umbilical vein that carries oxygenated blood into the fetus, and supporting mucous connective tissue called Wharton's jelly derived from the allantois. The entire umbilical cord is surrounded by a layer of amnion (see Figure 29.8a).

After the birth of the baby, the placenta detaches from the uterus and is termed the **afterbirth.** At this time, the umbilical cord is tied off and then severed, leaving the baby on its own. The small portion (about an inch) of the cord that remains still attached to the infant begins to wither and eventually falls off, usually within 12–15 days after birth. The area where the cord was attached becomes covered by a thin layer of skin, and scar tissue forms. The scar is the **umbilicus (navel).**

Pharmaceutical companies use human placentas as a source of hormones, drugs, and blood; portions of placentas are also used for burn coverage. The placental and umbilical cord veins can also be used in blood vessel grafts, and cord blood can be frozen to provide a future source of pluripotent stem cells.

Throughout the text we have discussed the developmental anatomy of the various body systems in their respective chapters. The following list of these sections is presented here for your review.

A summary of changes associated with embryonic and fetal growth is presented in Table 29.2 on page 1034.

CLINICAL APPLICATION
Placenta Previa

In some cases, part or all of the placenta becomes implanted in the inferior portion of the uterus, near or covering the internal os of the cervix. This condition is called **placenta previa** (PRĒ-vē-a; = before or in front of). Al-

Figure 29.7 Regions of the decidua.

🔑 **The decidua is a modified portion of the endometrium that develops after implantation.**

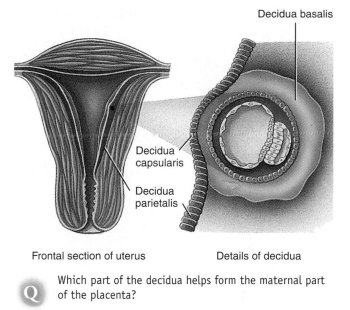

Decidua basalis

Decidua capsularis

Decidua parietalis

Frontal section of uterus   Details of decidua

Q Which part of the decidua helps form the maternal part of the placenta?

though placenta previa may lead to spontaneous abortion, it also occurs in approximately 1 in 250 live births. It is dangerous to the fetus because it may cause premature birth and intrauterine hypoxia due to maternal bleeding. Maternal mortality is increased due to hemorrhage and infection. The most important symptom is sudden, painless, bright red vaginal bleeding in the third trimester. Cesarean section is the preferred method of delivery in placenta previa. ■

Prenatal Diagnostic Tests

Several tests are available to detect genetic disorders and assess fetal well-being. Here we describe fetal ultrasonography, amniocentesis, and chorionic villi sampling (CVS).

Fetal Ultrasonography

If there is a question about the normal progress of a pregnancy, **fetal ultrasonography** (ul′-tra-son-OG-ra-fē) may be performed. By far the most common use of diagnostic ultrasound is to determine a more accurate fetal age when the date of conception is unclear; it is also used to evaluate fetal viability and growth, determine fetal position, identify multiple pregnancies, identify fetal–maternal abnormalities, and serve as an adjunct to special procedures such as amniocentesis. Ultrasound is not used routinely to determine the gender of a fetus; it is performed only for a specific medical indication.

An instrument (transducer) that emits high-frequency sound waves is passed back and forth over the abdomen. The reflected sound waves from the developing fetus are picked up by the transducer and converted to an on-screen image called a **sonogram** (see Table 1.4 on page 20). Because the

Figure 29.8 Placenta and umbilical cord.

The placenta is formed by the chorion of the embryo and part of the endometrium of the mother.

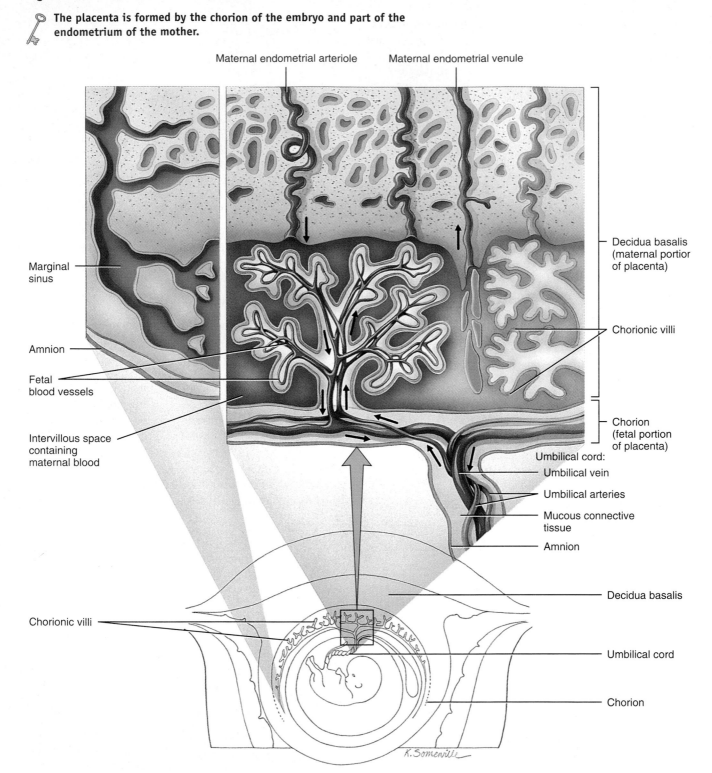

(a) Overall structure

figure continues

Figure 29.8 *(continued)*

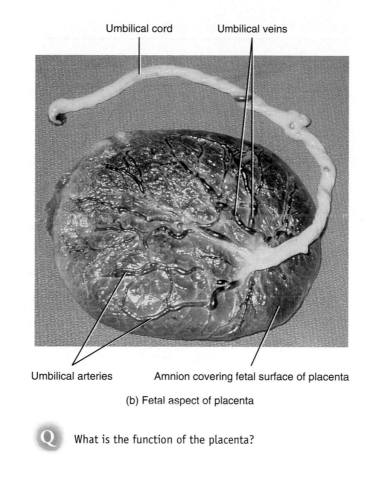

Umbilical cord   Umbilical veins

Umbilical arteries   Amnion covering fetal surface of placenta

(b) Fetal aspect of placenta

Q What is the function of the placenta?

urinary bladder serves as a landmark during the procedure, the patient needs to drink liquids and not void urine so as to maintain a full bladder.

Amniocentesis

Amniocentesis (am'-nē-ō-sen-TĒ-sis; *amnio-* = amnion; *-centesis* = puncture to remove fluid) involves withdrawing some of the amniotic fluid that bathes the developing fetus and analyzing the fetal cells and dissolved substances. It is used either to test for the presence of certain genetic disorders, such as Down syndrome (DS), spina bifida, hemophilia, Tay–Sachs disease, sickle-cell disease, and certain muscular dystrophies, or to determine fetal maturity and well-being near the time of delivery. To detect suspected genetic abnormalities, the test is usually done at 14–16 weeks of gestation; to assess fetal maturity, it is usually performed after the 35th week of gestation. About 300 chromosomal disorders and over 50 biochemical defects can be detected through amniocentesis. It can also reveal gender, which is important information for the diagnosis of sex-linked disorders, in which an abnormal gene carried by the mother affects her male offspring only (described later in the chapter). If the fetus is female, it will not be afflicted unless the father also carries the defective gene.

During amniocentesis, the position of the fetus and placenta is first identified using ultrasound and palpation. After the skin is prepared with an antiseptic and a local anesthetic is given, a hypodermic needle is inserted through the mother's abdominal wall and uterus into the amniotic cavity, and about 10 mL of fluid are aspirated (Figure 29.9a). The fluid and suspended cells are subjected to microscopic examination and biochemical testing. Elevated levels of alphafetoprotein (AFP) and acetylcholinesterase may indicate failure of the nervous system to develop properly, as occurs in spina bifida or anencephaly (absence of the cerebrum). Chromosome studies, which require growing the cells for 2–4 weeks in a culture medium, may reveal rearranged, missing, or extra chromosomes. Amniocentesis is performed only when a risk for genetic defects is suspected because there is about a 0.5% chance of spontaneous abortion after the procedure.

Chorionic Villi Sampling

In **chorionic villi sampling** (ko-rē-ON-ik VIL-ī) or **CVS,** a catheter is guided through the vagina and cervix of the uterus and then advanced to the chorionic villi under ultrasound guidance (Figure 29.9b). About 30 milligrams of tissue are suctioned out and prepared for chromosomal

Table 29.2 Changes Associated with Embryonic and Fetal Growth

END OF MONTH	APPROXIMATE SIZE AND WEIGHT	REPRESENTATIVE CHANGES
1	0.6 cm (³/₁₆ in.)	Eyes, nose, and ears are not yet visible. Vertebral column and vertebral canal form. Small buds that will develop into limbs form. Heart forms and starts beating. Body systems begin to form. The central nervous system appears at the start of the third week.
2	3 cm (1¼ in.) 1 g (¹/₃₀ oz)	Eyes are far apart, eyelids fused. Nose is flat. Ossification begins. Limbs become distinct, and digits are well formed. Major blood vessels form. Many internal organs continue to develop.
3	7½ cm (3 in.) 30 g (1 oz)	Eyes are almost fully developed but eyelids are still fused, nose develops a bridge, and external ears are present. Ossification continues. Limbs are fully formed and nails develop. Heartbeat can be detected. Urine starts to form. Fetus begins to move, but it cannot be felt by mother. Body systems continue to develop.
4	18 cm (6½–7 in.) 100 g (4 oz)	Head is large in proportion to rest of body. Face takes on human features, and hair appears on head. Many bones are ossified, and joints begin to form. Rapid development of body systems occurs.
5	25–30 cm (10–12 in.) 200–450 g (½–1 lb)	Head is less disproportionate to rest of body. Fine hair (lanugo) covers body. Brown fat forms and is the site of heat production. Fetal movements are commonly felt by mother (quickening). Rapid development of body systems occurs.
6	27–35 cm (11–14 in.) 550–800 g (1¼–1½ lb)	Head becomes even less disproportionate to rest of body. Eyelids separate and eyelashes form. Substantial weight gain occurs. Skin in wrinkled. Type II alveolar cells begin to produce surfactant.
7	32–42 cm (13–17 in.) 110–1350 g (2½–3 lb)	Head and body are more proportionate. Skin is wrinkled. Seven-month fetus (premature baby) is capable of survival. Fetus assumes an upside-down position. Testes start to descend into scrotum.
8	41–45 cm (16½–18 in.) 2000–2300 g (4½–5 lb)	Subcutaneous fat is deposited. Skin is less wrinkled.
9	50 cm (20 in.) 3200–3400 g (7–7½ lb)	Additional subcutaneous fat accumulates. Lanugo is shed. Nails extend to tips of fingers and maybe even beyond.

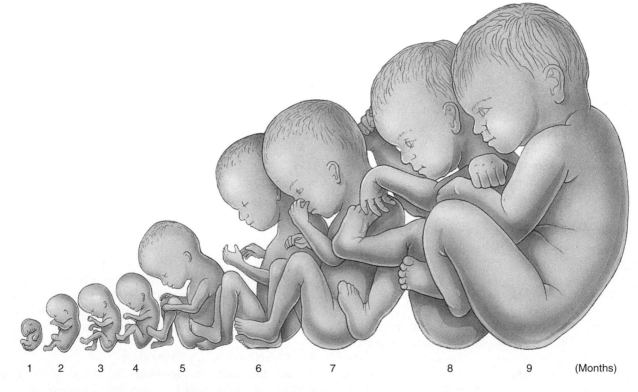

1   2   3   4   5   6   7   8   9   (Months)

Figure 29.9 Amniocentesis and chorionic villi sampling.

To detect genetic abnormalities, amniocentesis is performed at 14–16 weeks of gestation, whereas chorionic villi sampling may be performed as early as 8 weeks of gestation.

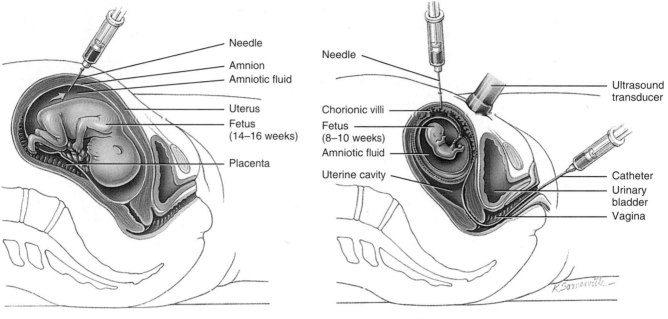

(a) Amniocentesis

(b) Chorionic villi sampling (CVS)

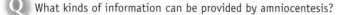

Q What kinds of information can be provided by amniocentesis?

analysis. Alternatively, the chorionic villi can be sampled by inserting a needle through the abdominal cavity, as performed in amniocentesis.

CVS can identify the same defects as amniocentesis because chorion cells and fetal cells contain the same genome. Moreover, CVS offers several advantages over amniocentesis: It can be performed as early as 8 weeks of gestation, and test results are available in only a few days, permitting an earlier decision on whether or not to continue the pregnancy. In addition, the procedure does not require penetration of the abdomen, uterus, or amniotic cavity by a needle. However, CVS is slightly riskier than amniocentesis; it carries a 1–2% chance of spontaneous abortion after the procedure.

1. Define the embryonic period and the fetal period.
2. Explain the importance of the placenta and umbilical cord to fetal growth.
3. Discuss the principal body changes associated with fetal growth.
4. Describe the following prenatal diagnostic tests: fetal ultrasonography, amniocentesis, and chorionic villi sampling.

MATERNAL CHANGES DURING PREGNANCY

OBJECTIVES

• Describe the sources and functions of the hormones secreted during pregnancy.
• Describe the hormonal, anatomical, and physiological changes in the mother during pregnancy.

Hormones of Pregnancy

During the first 3–4 months of pregnancy, the corpus luteum continues to secrete **progesterone** and **estrogens,** which maintain the lining of the uterus during pregnancy and prepare the mammary glands to secrete milk. The amounts secreted by the corpus luteum, however, are only slightly more than those produced after ovulation in a normal menstrual cycle. From the third month through the remainder of the pregnancy, the placenta itself provides the high levels of progesterone and estrogens required. As noted previously, the chorion of the placenta secretes **human chorionic gonadotropin (hCG)** into the blood. In turn,

Figure 29.10 Hormones during pregnancy.

🔑 **Whereas the corpus luteum produces progesterone and estrogens during the first 3−4 months of pregnancy, the placenta assumes this function from the third month on.**

Placenta

| Human chorionic gonadotropin (hCG) | Relaxin | Human chorionic somatomammotropin (hCS) | Corticotropin-releasing hormone |

Rescues corpus luteum from degeneration until the 3rd or 4th month of pregnancy

Corpus luteum (in ovary)

Progesterone Estrogens

1. Maintain endometrium of uterus during pregnancy

2. Help prepare mammary glands for lactation

3. Prepare mother's body for birth of baby

1. Increases flexibility of pubic symphysis

2. Helps dilate uterine cervix during labor

1. Helps prepare mammary glands for lactation

2. Enhances growth by increasing protein synthesis

3. Decreases glucose use and increases fatty acid use for ATP production

1. Establishes the timing of birth

2. Increases secretion of cortisol

(a) Sources and functions of hormones

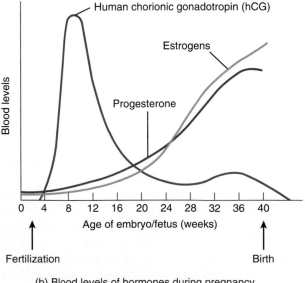

(b) Blood levels of hormones during pregnancy

Q Which hormone is detected by early pregnancy tests?

hCG stimulates the corpus luteum to continue production of progesterone and estrogens—an activity required to prevent menstruation and for the continued attachment of the embryo and fetus to the lining of the uterus (Figure 29.10a). By the eighth day after fertilization, hCG can be detected in the blood of a pregnant woman. Peak secretion of hCG occurs at about the ninth week of pregnancy (Figure 29.10b); the hCG level decreases sharply during the fourth and fifth months and then levels off until childbirth.

The chorion of the placenta begins to secrete estrogens after the first 3 or 4 weeks of pregnancy and progesterone by the sixth week. These hormones are secreted in increasing quantities until the time of birth (see Figure 29.10b). By the fourth month, when the placenta is fully established, the secretion of hCG has been greatly reduced because the secretions of the corpus luteum are no longer essential. Thus, from the third month to the ninth month, the placenta supplies the levels of estrogens and progesterone needed to maintain the pregnancy. A high level of progesterone ensures that the uterine myometrium is relaxed and that the cervix is tightly closed. After delivery, estrogens and progesterone in the blood decrease to normal levels.

Relaxin, a hormone produced first by the corpus luteum of the ovary and later by the placenta, increases the flexibility of the pubic symphysis and ligaments of the sacroiliac and sacrococcygeal joints and helps dilate the uterine cervix during labor. Both of these actions ease delivery of the baby.

A third hormone produced by the chorion of the placenta is **human chorionic somatomammotropin (hCS),** also known as **human placental lactogen (hPL).** The rate of secretion of hCS increases in proportion to placental mass, reaching maximum levels after 32 weeks and remaining relatively constant after that. It is thought to help prepare the mammary glands for lactation, enhance maternal growth by increasing protein synthesis, and regulate certain aspects of metabolism in mother and fetus alike. For example, hCS causes decreased use of glucose by the mother, thus making more available for the fetus. Additionally, hCS promotes the release of fatty acids from adipose tissue, providing an alternative to glucose for the mother's ATP production.

The hormone most recently found to be produced by the placenta is **corticotropin-releasing hormone (CRH),** which in nonpregnant people is secreted only by neurosecretory cells in the hypothalamus. CRH is now thought to be the "clock" that establishes the timing of birth. Secretion of CRH by the placenta begins at about 12 weeks and increases enormously toward the end of pregnancy. Women who have higher levels of CRH earlier in pregnancy are more likely to deliver prematurely, whereas those who have low levels are more likely to deliver after their due date. CRH from the placenta has a second important effect: It increases secretion of cortisol, which is needed for maturation of the fetal lungs and the production of surfactant.

CLINICAL APPLICATION
Early Pregnancy Tests

Early pregnancy tests detect the tiny amounts of human chorionic gonadotropin (hCG) in the urine that begin to be excreted about 8 days after fertilization. The test kits can detect pregnancy as early as the first day of a missed menstrual period—that is, at about 14 days after fertilization. Chemicals in the kits produce a color change if a reaction occurs between hCG in the urine and hCG antibodies included in the kit.

Several of the test kits available at pharmacies are as sensitive and accurate as test methods used in many hospitals. Still, false-negative and false-positive results can occur. A false-negative result (the test is negative, even though the woman is pregnant) may be due to testing too soon or to an ectopic pregnancy. A false-positive result (the test is positive, but the woman is not pregnant) may be due to excess protein or blood in the urine or to hCG production due to a rare type of uterine cancer. Thiazide diuretics, hormones, steroids, and thyroid drugs may also affect the outcome of an early pregnancy test. ■

Anatomical and Physiological Changes During Pregnancy

By about the end of the third month of pregnancy, the uterus occupies most of the pelvic cavity; as the fetus continues to grow, the uterus extends higher and higher into the abdominal cavity. Toward the end of a full-term pregnancy, the uterus fills nearly all of the abdominal cavity, reaching above the costal margin nearly to the xiphoid process of the sternum (Figure 29.11). It pushes the maternal intestines, liver, and stomach superiorly, elevates the diaphragm, and widens the thoracic cavity. Pressure on the stomach may force the stomach contents superiorly into the esophagus, resulting in heartburn. In the pelvic cavity, compression of the ureters and urinary bladder occurs.

Besides the anatomical changes associated with pregnancy, pregnancy-induced physiological changes also occur, including weight gain due to the fetus, amniotic fluid, the placenta, uterine enlargement, and increased total body water; increased storage of proteins, triglycerides, and minerals; marked breast enlargement in preparation for lactation; and lower back pain due to lordosis (swayback).

Changes in the maternal cardiovascular system include an increase in stroke volume by about 30%; a rise in cardiac output of 20–30% due to increased maternal blood flow to the placenta and increased metabolism; an increase in heart rate of about 10–15%; and an increase in blood volume of 30–50%, mostly during the second half of pregnancy. These increases are necessary to meet the additional demands of the fetus for nutrients and oxygen. When a pregnant woman is lying on her back, the enlarged uterus may compress the aorta, resulting in diminished blood flow to the uterus. Compression of the inferior vena cava also decreases venous return, which leads to edema in the lower limbs and may produce varicose veins. Compression of the renal artery can lead to renal hypertension.

Pulmonary function is also altered during pregnancy to meet the added oxygen demands of the fetus. Tidal volume can increase by 30–40%, expiratory reserve volume can be reduced by up to 40%, functional residual capacity can decline by up to 25%, minute ventilation (the total volume of air inhaled and exhaled each minute) can increase by up to 40%, airway resistance in the bronchial tree can decline by 30–40%, and total body oxygen consumption can increase by about 10–20%. Dyspnea (difficult breathing) also occurs.

With regard to the gastrointestinal tract, pregnant women experience an increase in appetite, but a general decrease in GI tract motility can result in constipation and a delay in gastric emptying time. Nausea, vomiting, and heartburn can also occur.

Figure 29.11 Normal fetal location and position at the end of a full-term pregnancy.

⚷ **The gestation period is the time interval (about 38 weeks) from fertilization to birth.**

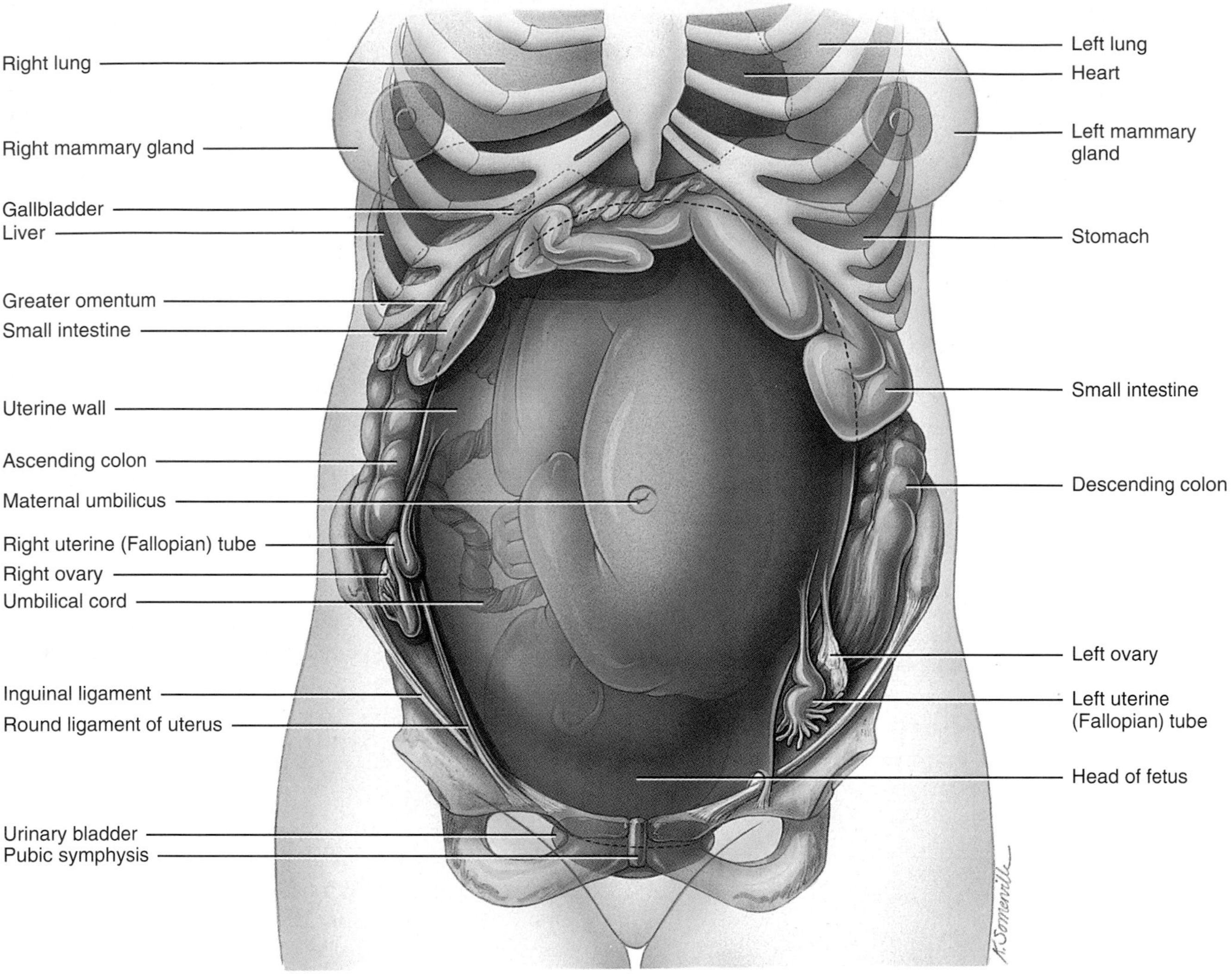

Right lung

Right mammary gland

Gallbladder
Liver

Greater omentum
Small intestine

Uterine wall

Ascending colon

Maternal umbilicus

Right uterine (Fallopian) tube
Right ovary
Umbilical cord

Inguinal ligament
Round ligament of uterus

Urinary bladder
Pubic symphysis

Left lung
Heart

Left mammary gland

Stomach

Small intestine

Descending colon

Left ovary
Left uterine (Fallopian) tube

Head of fetus

Anterior view of position of organs at end of full-term pregnancy

Q What hormone increases the flexibility of the pubic symphysis and helps dilate the cervix of the uterus to ease delivery of the baby?

Pressure on the urinary bladder by the enlarging uterus can produce urinary symptoms, such as increased frequency and urgency of urination, and stress incontinence. An increase in renal plasma flow up to 35% and an increase in glomerular filtration rate up to 40% increase the renal filtering capacity, which allows faster elimination of the extra wastes produced by the fetus.

Changes in the skin during pregnancy are more apparent in some women than others. Included are increased pigmentation around the eyes and cheekbones in a masklike pattern (chloasma), in the areolae of the breasts, and in the linea alba of the lower abdomen (linea nigra). Striae (stretch marks) over the abdomen can occur as the uterus enlarges, and hair loss also increases.

Changes in the reproductive system include edema and increased vascularity of the vulva and increased pliability and vascularity of the vagina. The uterus increases from its nonpregnant mass of 60–80 g to 900–1200 g at term as a result of hyperplasia of muscle fibers in the myometrium in early pregnancy and hypertrophy of muscle fibers during the second and third trimesters.

CLINICAL APPLICATION
Pregnancy-Induced Hypertension

About 10–15% of all pregnant women in the United States experience **pregnancy-induced hypertension (PIH),** elevated blood pressure associated with pregnancy. The major cause is **preeclampsia** (prē-ē-KLAMP-sē-a), an abnormal condition of pregnancy characterized by sudden hypertension, large amounts of protein in the urine, and generalized edema that typically appears after the 20th week of pregnancy. Other signs and symptoms are generalized edema, blurred vision, and headaches. Preeclampsia might be related to an autoimmune or allergic reaction resulting from the presence of a fetus. When the condition is also associated with convulsions and coma, it is termed **eclampsia.** Other forms of PIH are not associated with protein in the urine. ■

1. List the hormones involved in pregnancy, and describe the functions of each.
2. Describe several structural and functional changes that occur in the mother during pregnancy.

EXERCISE AND PREGNANCY

OBJECTIVE

• *Explain the effects of pregnancy on exercise, and of exercise on pregnancy.*

Only a few changes in early pregnancy affect exercise. A pregnant woman may tire more easily than usual, or morning sickness may interfere with regular exercise. As the pregnancy progresses, weight is gained and posture changes, so more energy is needed to perform activities, and certain maneuvers (sudden stopping, changes in direction, rapid movements) are more difficult to execute. In addition, certain joints, especially the pubic symphysis, become less stable in response to the increased level of the hormone relaxin. As compensation, many mothers-to-be walk with widely spread legs and a shuffling motion.

Although blood shifts from viscera (including the uterus) to the muscles and skin during exercise, there is no evidence of inadequate blood flow to the placenta. The heat generated during exercise may cause dehydration and further increase body temperature. During early pregnancy especially, excessive exercise and heat buildup should be avoided because elevated body temperature has been implicated in neural tube defects. Exercise has no known effect on lactation, provided a woman remains hydrated and wears a bra that provides good support. Moderate physical activity does not endanger the fetuses of healthy women who have a normal pregnancy.

Among the benefits of exercise during pregnancy are improvement in oxygen capacity, a greater sense of well-being, and fewer minor complaints.

LABOR

OBJECTIVE

• *Explain the events associated with the three stages of labor.*

Labor is the process by which the fetus is expelled from the uterus through the vagina to the outside. **Parturition** (par′-too-RISH-un; *parturit-* = childbirth) also means giving birth.

The onset of labor is determined by complex interactions of several placental and fetal hormones. Because progesterone inhibits uterine contractions, labor cannot take place until its effects are diminished. Toward the end of gestation, the levels of estrogens in the mother's blood rise sharply, producing changes that overcome the inhibiting effects of progesterone. The rise in estrogens results from increasing secretion by the placenta of corticotropin-releasing hormone, which stimulates the fetal anterior pituitary gland to secrete ACTH (adrenocorticotropic hormone). In turn, ACTH stimulates the fetal adrenal gland to secrete both cortisol and dehydroepiandrosterone (DHEA), the major adrenal androgen. The placenta then converts DHEA into an estrogen. High levels of estrogens cause uterine muscle fibers to display receptors for oxytocin and to form gap junctions with one another. Oxytocin stimulates uterine contractions, and relaxin assists by increasing the flexibility of the pubic symphysis and helping dilate the uterine cervix. Estrogen also stimulates the placenta to release prostaglandins, which induce production of enzymes that digest collagen fibers in the cervix, causing it to soften.

Control of labor contractions during parturition occurs via a positive feedback loop (see Figure 1.4 on page 9). Contractions of the uterine myometrium force the baby's head or body into the cervix, thereby distending (stretching) the cervix. Stretch receptors in the cervix send nerve impulses to neurosecretory cells in the hypothalamus, causing them to release oxytocin into blood capillaries of the posterior pituitary gland. Oxytocin then is carried by the blood to the uterus, where it stimulates the myometrium to contract more forcefully. As the contractions intensify, the baby's body stretches the cervix still more, and the resulting nerve impulses stimulate the secretion of yet more oxytocin. With birth of the infant, the positive feedback cycle is broken because cervical distention suddenly lessens.

Uterine contractions occur in waves (quite similar to peristaltic waves) that start at the top of the uterus and move downward, eventually expelling the fetus. **True labor** begins when uterine contractions occur at regular intervals, usually producing pain. As the interval between contractions shortens, the contractions intensify. Another symptom of true labor in some women is localization of pain in the back that is intensified by walking. The reliable indicator of true labor is dilation of the cervix and the "show," a discharge of a blood-containing mucus that appears in the cervical canal during

Figure 29.12 Stages of true labor.

🔑 **The term *parturition* refers to birth.**

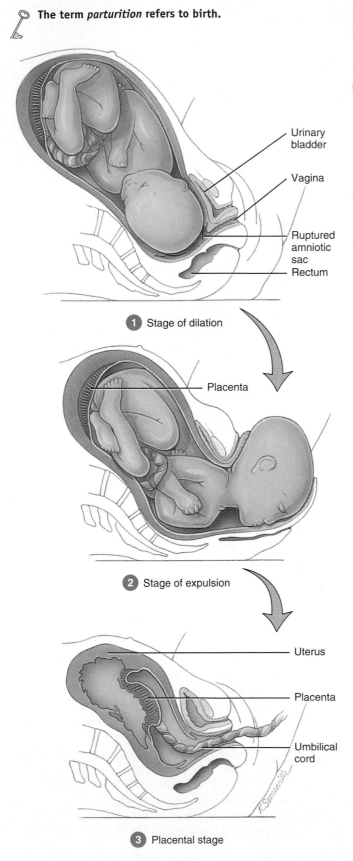

Urinary
bladder

Vagina

Ruptured
amniotic
sac

Rectum

1 Stage of dilation

Placenta

2 Stage of expulsion

Uterus

Placenta

Umbilical
cord

3 Placental stage

Ⓠ What event marks the beginning of the stage of expulsion?

labor. In **false labor,** pain is felt in the abdomen at irregular intervals, and it does not intensify and it is not altered significantly by walking. There is no "show" and no cervical dilation.

True labor can be divided into three stages (Figure 29.12):

1 *Stage of dilation.* The time from the onset of labor to the complete dilation of the cervix is the **stage of dilation.** This stage, which typically lasts 6–12 hours, features regular contractions of the uterus, usually a rupturing of the amniotic sac, and complete dilation (to 10 cm) of the cervix. If the amniotic sac does not rupture spontaneously, it is ruptured intentionally.

2 *Stage of expulsion.* The time (10 minutes to several hours) from complete cervical dilation to delivery of the baby is the **stage of expulsion.**

3 *Placental stage.* The time (5–30 minutes or more) after delivery until the placenta or "afterbirth" is expelled by powerful uterine contractions is the **placental stage.** These contractions also constrict blood vessels that were torn during delivery, thereby reducing the likelihood of hemorrhage.

As a rule, labor lasts longer with first babies (typically about 14 hours); for women who have previously given birth, the average duration of labor is about 8 hours—although the time varies enormously among births. Because the fetus may be squeezed through the birth canal (cervix and vagina) for up to several hours, the fetus is stressed during childbirth: The fetal head is compressed, and the fetus undergoes some degree of intermittent hypoxia due to compression of the umbilical cord and the placenta during uterine contractions. In response to this stress, the fetal adrenal medullae secrete very high levels of epinephrine and norepinephrine, the "fight-or-flight" hormones. Much of the protection against the stresses of parturition, as well as preparation of the infant for surviving extrauterine life, are provided by the adrenal medullary hormones. Among other functions, the hormones clear the lungs and alter their physiology in readiness for breathing air, mobilize readily usable nutrients for cellular metabolism, and promote an increased blood flow to the brain and heart.

About 7% of pregnant women have not delivered by 2 weeks after their due date. Such cases carry an increased risk of brain damage to the fetus, and even fetal death, due to inadequate supplies of oxygen and nutrients from an aging placenta. Post-term deliveries may be facilitated by inducing labor (initiated by administration of oxytocin) or by surgical delivery.

Following the delivery of the baby and placenta is a 6-week period during which the maternal reproductive organs and physiology return to the prepregnancy state. This period is called the **puerperium** (pyoo′-er-PE-rē-um). Through a process of tissue catabolism, the uterus undergoes a remarkable reduction in size (especially in lactating women), called **involution** (in′-vō-LOO-shun). The cervix loses its elasticity and regains its prepregnancy firmness. For 2–4 weeks after delivery, women have a uterine discharge called **lochia** (LŌ-kē-a), which consists initially of blood and later of serous fluid derived from the former site of the placenta.

CLINICAL APPLICATION
Dystocia and Cesarean Section

Dystocia (dis-TŌ-sē-a; *dys-* = painful or difficult; *toc-* = birth), or difficult labor, may result either from an abnormal position (presentation) of the fetus or a birth canal of inadequate size to permit vaginal birth. In a **breech presentation,** for example, the fetal buttocks or lower limbs, rather than the head, enter the birth canal first; this occurs most often in premature births. If fetal or maternal distress prevents a vaginal birth, the baby may be delivered surgically through an abdominal incision. A low, horizontal cut is made through the abdominal wall and lower portion of the uterus, through which the baby and placenta are removed. Even though it is popularly associated with the birth of Julius Caesar, the procedure for performing this operation is termed a **cesarean section (C-section)** because it was described in Roman Law, *lex cesarea,* 600 years before Julius Caesar was born. Even a history of multiple C-sections need not exclude a pregnant woman from attempting a vaginal delivery. ■

1. Describe the hormonal changes that induce labor.
2. Distinguish between false labor and true labor.
3. Describe what happens during the stage of dilation, the stage of expulsion, and the placental stage of true labor.

ADJUSTMENTS OF THE INFANT AT BIRTH

OBJECTIVE
• *Explain the respiratory and cardiovascular adjustments that occur in an infant at birth.*

During pregnancy, the embryo (and later the fetus) is totally dependent on the mother for its existence. The mother supplies the fetus with oxygen and nutrients, eliminates its carbon dioxide and other wastes, protects it against shocks and temperature changes, and provides antibodies that confer protection against certain harmful microbes. At birth, a physiologically mature baby becomes much more self-supporting, and the newborn's body systems must make various adjustments. Next we examine some of the changes that occur in the respiratory and cardiovascular systems.

Respiratory Adjustments

The fetus depends entirely on the mother for obtaining oxygen and eliminating carbon dioxide. The fetal lungs are either collapsed or partially filled with amniotic fluid, which is absorbed at birth. The production of surfactant begins by the end of the sixth month of development, and because the respiratory system is fairly well developed at least 2 months before birth, premature babies delivered at 7 months are able to breathe and cry. After delivery, the baby's supply of oxygen from the mother ceases. Circulation in the baby continues, and as the blood level of carbon dioxide increases, the respiratory center in the medulla oblongata is stimulated, causing the respiratory muscles to contract, and the baby draws its first breath. Because the first inspiration is unusually deep as the lungs contain no air, the baby exhales vigorously and naturally cries. A full-term baby may breathe 45 times a minute for the first 2 weeks after birth. Breathing rate gradually declines until it approaches a normal rate of 12 breaths per minute.

Cardiovascular Adjustments

After the baby's first inspiration, the cardiovascular system must make several adjustments (see Figure 21.31 on page 730). Closure of the foramen ovale between the atria of the fetal heart, which occurs at the moment of birth, diverts deoxygenated blood to the lungs for the first time. The foramen ovale is closed by two flaps of septal heart tissue that fold together and permanently fuse. The remnant of the foramen ovale is the fossa ovalis.

Once the lungs begin to function, the ductus arteriosus is shut off by contractions of the muscles in its wall and becomes the ligamentum arteriosum. The muscle contraction is probably mediated by the polypeptide bradykinin, released from the lungs during their initial inflation. The ductus arteriosus generally does not completely and irreversibly close for about 3 months after birth. Incomplete closure results in a condition called **patent ductus arteriosus.**

After the umbilical cord is tied off and severed and blood no longer flows through the umbilical arteries, they fill with connective tissue, and their distal portions become the medial umbilical ligaments. After the umbilical cord is severed, the umbilical vein becomes the ligamentum teres (round ligament) of the liver.

In the fetus, the ductus venosus connects the umbilical vein directly with the inferior vena cava, allowing blood from the placenta to bypass the fetal liver. When the umbilical cord is severed, the ductus venosus collapses, and venous blood from the viscera of the fetus flows into the hepatic portal vein to the liver and then via the hepatic vein to the inferior vena cava. The remnant of the ductus venosus becomes the ligamentum venosum.

At birth, an infant's pulse may range from 120 to 160 beats per minute and may go as high as 180 upon excitation. After birth, oxygen use increases, which stimulates an increase in the rate of red blood cell and hemoglobin production. Moreover, the white blood cell count at birth is very high—sometimes as much as 45,000 cells per cubic microliter—but the count decreases rapidly by the seventh day.

Figure 29.13 The milk ejection reflex, a positive feedback cycle.

 Oxytocin stimulates contraction of myoepithelial cells in the breasts, which squeezes the glandular and duct cells and causes milk ejection.

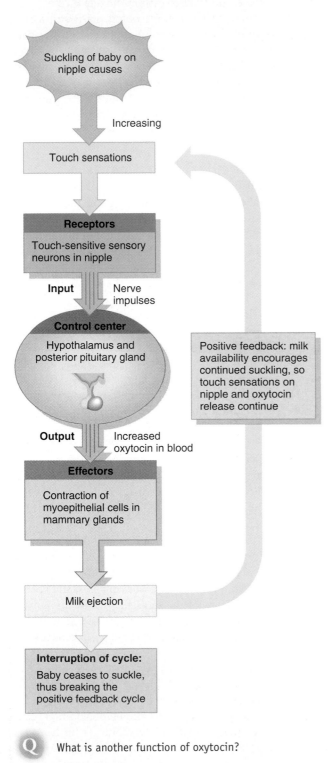

Q What is another function of oxytocin?

 CLINICAL APPLICATION
Premature Infants

Delivery of a physiologically immature baby carries certain risks. A **premature infant** or "preemie" is generally considered to be a baby who weighs less than 2500 g (5.5 lb) at birth. Poor prenatal care, drug abuse, history of a previous premature delivery, and mother's age below 16 or above 35 increase the chance of premature delivery. The body of a premature infant is not yet ready to sustain some critical functions, and thus its survival is uncertain without medical intervention. The major problem after delivery of an infant under 36 weeks of gestation is respiratory distress syndrome (RDS) of the newborn due to insufficient surfactant. RDS can be eased by use of artificial surfactant and a ventilator that delivers oxygen until the lungs can operate on their own. ∎

1. Explain the importance of surfactant at birth and afterward.
2. What cardiovascular adjustments are made by an infant at birth?

THE PHYSIOLOGY OF LACTATION

OBJECTIVE

• *Discuss the physiology and hormonal control of lactation.*

Lactation (lak′-TĀ-shun) is the secretion and ejection of milk from the mammary glands. A principal hormone in promoting milk secretion is **prolactin (PRL),** which is secreted from the anterior pituitary gland. Even though prolactin levels increase as the pregnancy progresses, no milk secretion occurs because progesterone inhibits the effects of prolactin. After delivery, the levels of estrogens and progesterone in the mother's blood decrease, and the inhibition is removed. The principal stimulus in maintaining prolactin secretion during lactation is the sucking action of the infant. Suckling initiates nerve impulses from stretch receptors in the nipples to the hypothalamus; the impulses decrease hypothalamic release of prolactin inhibiting hormone (PIH) and increase release of prolactin releasing hormone (PRH), so more prolactin is released by the anterior pituitary gland.

Oxytocin causes release of milk into the mammary ducts via the **milk ejection reflex** (Figure 29.13). Milk formed by the glandular cells of the breasts is stored until the baby begins active suckling. Stimulation of touch receptors in the nipple initiates sensory nerve impulses that are relayed to the hypothalamus. In response, secretion of oxytocin from the posterior pituitary gland increases. Carried by the bloodstream to the mammary glands, oxytocin stimulates contraction of myoepithelial (smooth-muscle-like) cells surrounding the glandular cells and ducts. The resulting compression moves the milk from the alveoli of the mammary glands into the mammary ducts, where it can be suckled. This process is termed **milk ejection (let-down).** Even though the actual ejection of milk does not occur until 30–60 seconds after nursing begins (the latent period), some milk stored in lactiferous sinuses near the nipple is available during the latent period. Stimuli other than suckling, such as hearing a baby's cry or touching the genitals, also can trigger oxytocin release and milk ejection. The suckling stimulation that produces the release of oxytocin also inhibits the release of PIH; this results in increased secretion of prolactin, which maintains lactation.

During late pregnancy and the first few days after birth, the mammary glands secrete a cloudy fluid called **colostrum.** Although it is not as nutritious as milk—it contains less lactose and virtually no fat—colostrum serves adequately until the appearance of true milk on about the fourth day. Colostrum and maternal milk contain antibodies that protect the infant during the first few months of life.

Following birth of the infant, the prolactin level starts to return to the nonpregnant level. However, each time the mother nurses the infant, nerve impulses from the nipples to the hypothalamus increase the release of PRH (and decrease the release of PIH), resulting in a tenfold increase in prolactin secretion by the anterior pituitary gland that lasts about an hour. Prolactin acts on the mammary glands to provide milk for the next nursing period. If this surge of prolactin is blocked by injury or disease, or if nursing is discontinued, the mammary glands lose their ability to secrete milk in only a few days. Even though milk secretion normally decreases considerably within 7–9 months after birth, it can continue for several years if nursing continues. A woman who maintains lactation by nursing other women's infants is called a wet-nurse.

Lactation often blocks ovarian cycles for the first few months following delivery, if the frequency of sucking is about 8–10 times a day. This effect is inconsistent, however, and ovulation normally precedes the first menstrual period after delivery of a baby. As a result, the mother can never be certain she is not fertile. Breast-feeding is therefore not a very reliable birth control measure. The suppression of ovulation during lactation is believed to occur as follows: During breast-feeding, neural input from the nipple reaches the hypothalamus and causes it to produce neurotransmitters that suppress the release of gonadotropin releasing hormone (GnRH). As a result, production of LH and FSH decreases, and ovulation is inhibited.

A primary benefit of **breast-feeding** is nutritional: Human milk is a sterile solution that contains amounts of fatty acids, lactose, amino acids, minerals, vitamins, and water that are ideal for the baby's digestion, brain development, and growth. Breast-feeding also benefits infants by providing the following:

- *Beneficial cells.* Several types of white blood cells are present in breast milk. Neutrophils serve as phagocytes, ingesting bacteria in the baby's gastrointestinal tract. Macrophages also kill microbes in the baby's gastrointestinal tract; additionally they produce lysozyme and other immune system components. Plasma cells, which develop from B lymphocytes, produce antibodies against specific microbes, and T lymphocytes kill microbes directly or help mobilize other defenses.

- *Beneficial molecules.* Breast milk also contains an abundance of helpful molecules. Maternal IgA antibodies in breast milk bind to microbes in the baby's gastrointestinal tract and prevent their migration into other body tissues. Because a mother produces antibodies to whatever disease-causing microbes are present in her environment, her breast milk affords protection against the specific infectious agents to which her baby is also exposed. Additionally, two milk proteins bind to nutrients that many bacteria need to grow and survive, thereby making the nutrients unavailable: B_{12} binding protein ties up vitamin B_{12}, and lactoferrin ties up iron. Some fatty acids can kill certain viruses by disrupting their membranes, and lysozyme kills bacteria by disrupting their cell walls. Finally, interferons enhance the antimicrobial activity of immune cells.

- *Decreased incidence of diseases later in life.* Breast-feeding provides children with a slight reduction in risk of lymphoma, heart disease later in life, allergies, respiratory and gastrointestinal infections, ear infections, diarrhea, diabetes mellitus, and meningitis. Breast-feeding also protects the mother against osteoporosis and breast cancer.

- *Miscellaneous benefits.* Breast-feeding supports optimal infant growth, enhances intellectual and neurological development, and fosters mother-infant relations by establishing early and prolonged contact between them. Compared to cow's milk, the fats and iron in breast milk are more easily absorbed, the proteins in breast milk are more readily metabolized, and the lower sodium content of breast milk is more suited to an infant's needs. Premature infants benefit from breast-feeding because the milk produced by mothers of premature infants seems to be specially adapted to the infant's needs by having a higher protein content than the milk of mothers of full-term infants. Finally, a baby is less likely to have an allergic reaction to its mother's milk than to milk from another source.

CLINICAL APPLICATION
Nursing and Childbirth

Years before oxytocin was discovered, it was common practice in midwifery to let a first-born twin nurse at the mother's breast to speed the birth of the second child. Now we know why this practice is helpful—it stimulates the release of oxytocin. Even after a single birth, nursing promotes expulsion of the placenta (afterbirth) and helps the uterus return to its normal size. Synthetic oxytocin (Pitocin) is often given to induce labor or to increase uterine tone and control hemorrhage just after parturition. ∎

Figure 29.14 Inheritance of phenylketonuria (PKU).

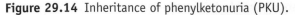

Whereas genotype refers to genetic makeup, phenotype refers to the physical or outward expression of a gene.

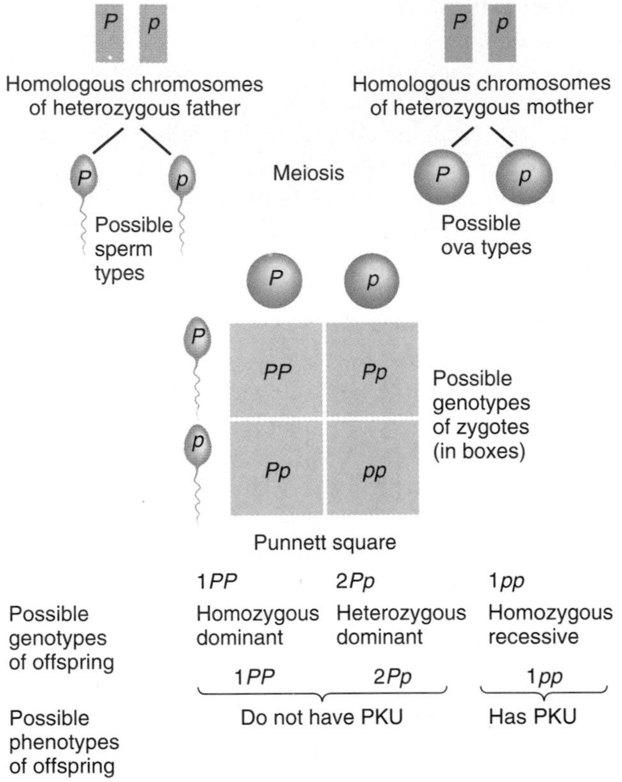

Homologous chromosomes of heterozygous father

Homologous chromosomes of heterozygous mother

Meiosis

Possible sperm types

Possible ova types

Possible genotypes of zygotes (in boxes)

	P	p
P	PP	Pp
p	Pp	pp

Punnett square

Possible genotypes of offspring

1*PP* — Homozygous dominant

2*Pp* — Heterozygous dominant

1*pp* — Homozygous recessive

Possible phenotypes of offspring

1*PP* 2*Pp* — Do not have PKU

1*pp* — Has PKU

Q If parents have the genotypes shown here, what is the chance (expressed as percentage) that their first child will have PKU? What is the chance of PKU occurring in their second child?

1. Name the hormones that contribute to lactation, and describe the function of each.
2. What are the benefits of breast-feeding over bottle-feeding?

INHERITANCE

OBJECTIVE

• *Define inheritance, and explain the inheritance of dominant, recessive, polygenic, and sex-linked traits.*

As previously indicated, the genetic material of a father and a mother unite when a sperm cell fuses with a secondary oocyte to form a zygote. Children resemble their parents because they inherit traits passed down from both parents. We now examine some of the principles involved in that process, called inheritance.

Inheritance is the passage of hereditary traits from one generation to the next. It is the process by which you acquired your characteristics from your parents and may

transmit some of your traits to your children. The branch of biology that deals with inheritance is called **genetics** (je-NET-iks). The area of health care that offers advice on genetic problems (or potential problems) is called **genetic counseling.**

Genotype and Phenotype

The nuclei of all human cells except gametes contain 23 pairs of chromosomes—the diploid number. One chromosome in each pair came from the mother, and the other came from the father. Each homologue—one of the two chromosomes that make up a pair—contains genes that control the same traits. If a chromosome contains a gene for body hair, for example, its homologue will also contain a gene for body hair in the same position on the chromosome. Such alternative forms of a gene that code for the same trait and are at the same location on homologous chromosomes are called **alleles** (ah-LĒLZ). For example, one allele of a body hair gene might code for coarse hair, whereas another allele codes for fine hair. A **mutation** (myoo-TĀ-shun; *muta-* = change) is a permanent heritable change in an allele that produces a different varient of the same trait.

The relationship of genes to heredity is illustrated by examining the alleles involved in a disorder called **phenylketonuria** or **PKU**. People with PKU (see page 891) are unable to manufacture the enzyme phenylalanine hydroxylase. The allele that codes for phenylalanine hydroxylase is symbolized as *P* whereas the mutated allele that fails to produce a functional enzyme is symbolized as *p*. The chart in Figure 29.14, which shows the possible combinations of gametes from two parents who each have one *P* and one *p* allele, is called a **Punnett square.** In constructing a Punnett square, the possible paternal alleles in sperm are written at the left side and the possible maternal alleles in ova (or secondary oocytes) are written at the top. The four spaces on the chart show how the alleles can combine in zygotes formed by the union of these sperm and ova to produce the three different genetic makeups, or **genotypes** (JĒ - nō - tīps): *PP, Pp,* or *pp*. Notice from the Punnett square that 25 percent of the offspring will have the *PP* genotype, 50 percent will have the *Pp* genotype, and 25 percent will have the *pp* genotype. People who inherit *PP* or *Pp* genotypes do not have PKU; those with a *pp* genotype suffer from the disorder. Although people with a *Pp* genotype have one PKU allele (*p*), the allele that codes for the normal trait (*P*) is more dominant. An allele that dominates or masks the presence of another allele and is fully expressed (*P* in this example) is said to be a **dominant allele,** and the trait expressed is called a dominant trait. The allele whose presence is completely masked (*p* in this example) is said to be a **recessive allele,** and the trait it controls is called a recessive trait.

By tradition, the symbols for genes are written in italics, with dominant alleles written in capital letters and recessive alleles in lowercase letters. A person with the same alleles on homologous chromosomes (for example, *PP* or *pp*) is said to be **homozygous** for the trait. *PP* is homozygous dominant, and *pp* is homozygous recessive. An individual with different alleles on homologous chromosomes (for example, *Pp*) is said to be **heterozygous** for the trait.

Phenotype (FĒ-nō-tīp; *pheno-* = showing) refers to how the genetic makeup is expressed in the body; it is the physical or outward expression of a gene. A person with *Pp* (a heterozygote) has a different genotype from a person with *PP* (a homozygote), but both have the same phenotype—normal production of phenylalanine hydroxylase. Heterozygous individuals who carry a recessive gene but do not express it *(Pp)* can pass the gene on to their offspring. Such individuals are called **carriers** of the recessive gene.

Most genes give rise to the same phenotype whether they are inherited from the mother or the father. In a few cases, however, the phenotype is dramatically different, depending on the parental origin. This surprising phenomenon, first appreciated in the 1980s, is called **genomic imprinting.** In humans, the abnormalities most clearly associated with mutation of an imprinted gene are *Angelman syndrome,* which results when the gene for a particular abnormal trait is inherited from the mother, and *Prader–Willi syndrome,* which results when it is inherited from the father.

Alleles that code for normal traits do not always dominate over those that code for abnormal ones, but dominant alleles for severe disorders usually are lethal; they cause death of the embryo or fetus. One exception is Huntington's disease (HD), which is caused by a dominant allele whose effects do not become manifest until adulthood. Both homozygous dominant and heterozygous people exhibit the disease, whereas homozygous recessive people are normal. HD causes progressive degeneration of the nervous system and eventual death, but because symptoms typically do not appear until after age 30 or 40, many afflicted individuals have already passed the allele for the condition on to their children.

Occasionally an error in meiosis called **nondisjunction** results in an abnormal number of chromosomes. In this situation, homologous chromosomes fail to separate properly during anaphase of meiosis I. A cell that has one or more chromosomes of a set added or deleted is called an **aneuploid** (an′-yoo-PLOID). A monosomic cell $(2n-1)$ is missing a chromosome; a trisomic cell $(2n+1)$ has an extra chromosome. Down syndrome (see page 1050) is one example of an aneuploid disorder in which there is trisomy of chromosome 21. Another error in meiosis is a **translocation,** in which the location of a chromosome segment is changed; it is moved either to another chromosome or to another location within the same chromosome. A translocation may result from crossing-over between two chromosomes that are not homologues. One form of Down syndrome results from a translocation of a portion of chromosome 21 to another chromosome. Affected individuals then have three, rather than two, copies of that portion of chromosome 21.

Table 29.3 lists some dominant and recessive inherited structural and functional traits in humans.

Table 29.3 Selected Hereditary Traits in Humans

DOMINANT	RECESSIVE
Coarse body hair	Fine body hair
Male pattern baldness	Baldness
Normal skin pigmentation	Albinism
Freckles	Absence of freckles
Astigmatism	Normal vision
Near- or farsightedness	Normal vision
Normal hearing	Deafness
Broad lips	Thin lips
Tongue roller	Inability to roll tongue into a U shape
PTC taster*	PTC nontaster
Large eyes	Small eyes
Polydactylism (extra digits)	Normal digits
Brachydactylism (short digits)	Normal digits
Syndactylism (webbed digits)	Normal digits
Feet with normal arches	Flat feet
Hypertension	Normal blood pressure
Diabetes insipidus	Normal excretion
Huntington's disease	Normal nervous system
Normal mentality	Schizophrenia
Migraine headaches	Normal
Widow's peak	Straight hairline
Curved (hyperextended) thumb	Straight thumb
Normal Cl⁻ transport	Cystic fibrosis
Hypercholesterolemia (familial)	Normal cholesterol level

*Ability to taste a chemical compound called phenylthiocarbamide (PTC).

Variations on Dominant–Recessive Inheritance

Most patterns of inheritance are not simple **dominant–recessive inheritance** in which only dominant and recessive alleles interact. The phenotypic expression of a particular gene may be influenced not only by which alleles are present, but also by other genes and the environment. Moreover, most inherited traits are influenced by more than one gene, and most genes can influence more than a single trait. Next we discuss three variations on dominant–recessive inheritance.

Incomplete Dominance

In **incomplete dominance,** neither member of an allelic pair is dominant over the other, and the heterozygote has a phenotype intermediate between the homozygous dominant

Figure 29.15 Inheritance of sickle-cell disease.

🔑 **Sickle-cell disease is an example of incomplete dominance.**

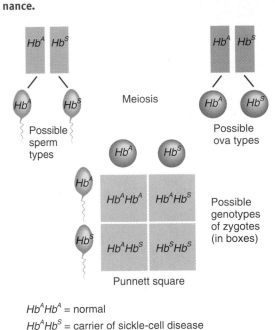

Meiosis

Possible sperm types

Possible ova types

Possible genotypes of zygotes (in boxes)

Punnett square

$Hb^A Hb^A$ = normal
$Hb^A Hb^S$ = carrier of sickle-cell disease
$Hb^S Hb^S$ = has sickle-cell disease

❓ What are the distinguishing features of incomplete dominance?

Figure 29.16 The ten possible combinations of parental ABO blood types and the blood types their offspring could (and could not) inherit. For each possible set of parents, the blue letters represent the blood types their offspring could inherit, whereas the red letters represent blood types their offspring could not inherit.

🔑 **Inheritance of ABO blood types is an example of multiple-allele inheritance.**

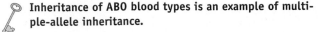
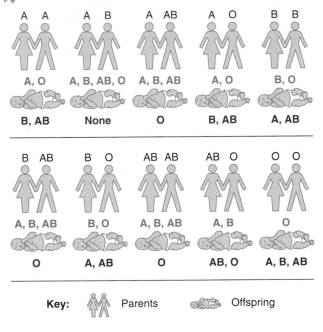

Key: Parents   Offspring

❓ Is it possible for a baby to have type O blood if neither parent is type O?

and the homozygous recessive phenotypes. An example of incomplete dominance in humans is the inheritance of **sickle-cell disease (SCD)** (Figure 29.15). People with the homozygous dominant genotype $Hb^A Hb^A$ form normal hemoglobin; those with the homozygous recessive genotype $Hb^S Hb^S$ have sickle-cell disease and severe anemia. Although they are usually healthy, those with the heterozygous genotype $Hb^A Hb^S$ have minor problems with anemia because half their hemoglobin is normal and half is not. Heterozygotes are carriers, and they are said to have *sickle-cell trait.*

Another example of incomplete dominance in humans is the inheritance of hair texture. In this example, *HH* represents the homozygous genotype for curly hair; $H^1 H^1$ represents the homozygous genotype for straight hair; and HH^1 represents the heterozygous genotype for wavy hair.

Multiple-Allele Inheritance

Although a single individual inherits only two alleles for each gene, some genes may have more than two alternate forms, and this is the basis for **multiple-allele inheritance.**

One example of multiple-allele inheritance in humans is the inheritance of the ABO blood group. The four blood types (phenotypes) of the ABO group—A, B, AB, and O—result from the inheritance of six combinations of three different alleles of a single gene called the *I* gene: (1) allele I^A produces the A antigen, (2) allele I^B produces the B antigen, and (3) allele *i* produces neither A nor B antigen. Each person inherits two *I*-gene alleles, one from each parent, that give rise to the various phenotypes. The six possible genotypes produce four blood types, as follows:

Genotype	Blood type
$I^A I^A$ or $I^A i$	A
$I^B I^B$ or $I^B i$	B
$I^A I^B$	AB
ii	O

Notice that both I^A and I^B are inherited as dominant traits, whereas *i* is inherited as a recessive trait. Because an individual with type AB blood has characteristics of both type A and type B red blood cells expressed in the phenotype, alleles I^A and I^B are said to be **codominant.** In other words, both genes are expressed equally in the heterozygote. Depending on the parental blood types, different offspring may have blood types different from each other. Figure 29.16

Figure 29.17 Polygenic inheritance of skin color.

🔑 **In polygenic inheritance, a trait is controlled by the combined effects of several genes.**

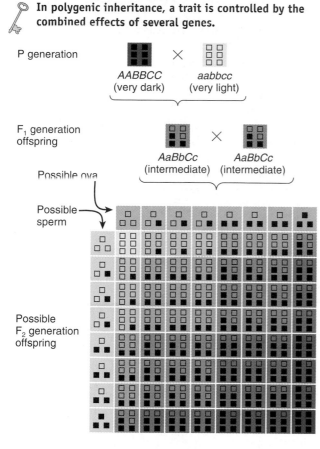

P generation

AABBCC (very dark) × aabbcc (very light)

F₁ generation offspring

AaBbCc (intermediate) × AaBbCc (intermediate)

Possible ova

Possible sperm

Possible F₂ generation offspring

Q What other traits are transmitted by polygenic inheritance?

Figure 29.18 Autosomes and sex chromosomes.

🔑 **Human somatic cells contain 23 different pairs of chromosomes.**

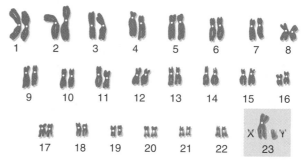

1 2 3 4 5 6 7 8
9 10 11 12 13 14 15 16
17 18 19 20 21 22 X 23 Y

Q What are the two sex chromosomes in females and males?

individual with the genotype *aabbcc* is very light skinned, and a person with the genotype *AaBbCc* has an intermediate skin color. Parents having an intermediate skin color may have children with very light, very dark, or intermediate skin color. Figure 29.17 illustrates the inheritance of skin color and the various gradations of skin tone. Note that the **P generation** (parental generation) is the starting generation, the **F₁ generation** (first filial generation) is produced from the P generation, and the **F₂ generation** (second filial generation) is produced from the F₁ generation.

Autosomes, Sex Chromosomes, and Sex Determination

When viewed under a microscope, the 46 human chromosomes in a normal somatic cell can be identified by their size, shape, and staining pattern to be a member of 23 different pairs of chromosomes. In 22 of the pairs, the homologous chromosomes look alike and have the same appearance in both males and females; these 22 pairs are called **autosomes.** The two members of the 23rd pair are termed the **sex chromosomes;** they look different in males and females (Figure 29.18). In females, the pair consists of two chromosomes called X chromosomes. One X chromosome is also present in males, but its mate is a much smaller chromosome called a Y chromosome.

When a spermatocyte undergoes meiosis to reduce its chromosome number, one daughter cell (spermatid) will contain the X chromosome, and the other will contain the Y chromosome. Oocytes have no Y chromosomes and produce only X-containing gametes. If the secondary oocyte is fertilized by an X-bearing sperm, the offspring normally is female (XX). Fertilization by a Y-bearing sperm produces a male (XY). Thus, an individual's sex is determined by the father's chromosomes (Figure 29.19).

Both female and male embryos develop identically until about 7 weeks after fertilization. At that point, one or more

shows which blood types offspring could and could not inherit given the blood types of their parents.

Polygenic Inheritance

Most inherited traits are not controlled by one gene, but instead by the combined effects of many genes, a situation referred to as **polygenic inheritance** (*poly-* = many). A polygenic trait shows a continuous gradation of small differences between extremes among individuals. Examples of polygenic traits include skin color, hair color, eye color, height, and body build. Whereas it is relatively easy to predict the risk of passing on an undesirable trait that is due to a single dominant or recessive gene, it is very difficult to make this prediction when the trait is polygenic. Such traits are difficult to follow in a family because the range of variation is so large, and because the number of different genes involved usually is not known.

As an example of polygenic inheritance we consider skin color. Suppose that skin color is controlled by three separate genes, each having two alleles: *A, a; B, b;* and *C, c.* Whereas a person with the genotype *AABBCC* is very dark skinned, an

Figure 29.19 Sex determination.

🔑 **Sex is determined at the time of fertilization by the presence or absence of a Y chromosome in the sperm.**

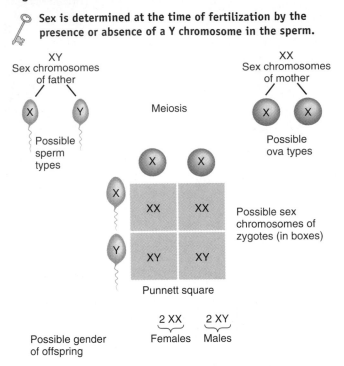

Q What are all the chromosomes except the sex chromosomes called?

genes set into motion a cascade of events that leads to the development of a male; in the absence of the gene or genes, the female pattern of development occurs. It has been known since 1959 that the Y chromosome is needed to initiate male development. Experiments published in 1991 established that the prime male-determining gene is one called **SRY (sex-determining region of the Y chromosome).** When a small DNA fragment containing this gene was inserted into 11 female mouse embryos, three of them developed as males. (The researchers suspected that the gene failed to be integrated into the genetic material in the other eight.) *SRY* acts as a molecular switch to turn on the male pattern of development. Only if the *SRY* gene is present and functional in a fertilized ovum will the fetus develop testes and differentiate into a male; in the absence of *SRY,* the fetus will develop ovaries and differentiate into a female.

An experiment of nature provided evidence confirming the key role of *SRY* in directing the male pattern of development in humans. In two cases, phenotypic females with an XY genotype were found to have mutated *SRY* genes. In other words, these individuals failed to develop normally as males because their *SRY* gene was defective.

Sex-Linked Inheritance

The sex chromosomes also are responsible for the transmission of several nonsexual traits. Genes for these traits appear on X chromosomes but are absent from Y chromo-

somes; this feature produces a pattern of heredity that is different from the patterns described previously.

Red–Green Color Blindness

Consider the most common type of color blindness, called **red–green color blindness.** This condition is characterized by a deficiency in either red- or green-sensitive cones, so red and green are seen as the same color (either red or green, depending on which cone is present). The gene for red–green color blindness is a recessive one designated *c.* Normal color vision, designated *C,* dominates. The *C/c* genes are located only on the X chromosome, and thus the ability to see colors depends entirely on the X chromosomes. The possible combinations are as follows:

Genotype	Phenotype
$X^C X^C$	Normal female
$X^C X^c$	Normal female (but a carrier of the recessive gene)
$X^c X^c$	Red–green color-blind female
$X^C Y$	Normal male
$X^c Y$	Red–green color-blind male

Only females who have two X^c genes are red–green color blind. This rare situation can result only from the mating of a color-blind male and a color-blind or carrier female. (In $X^C X^c$ females the trait is masked by the normal, dominant gene.) Because males, in contrast, do not have a second X chromosome that could mask the trait, all males with an X^c gene will be red–green color blind. Figure 29.20 illustrates the inheritance of red–green color blindness in the offspring of a normal male and a carrier female.

Traits inherited in the manner just described are called **sex-linked traits.** The most common type of **hemophilia**—a condition in which the blood fails to clot or clots very slowly after an injury—is a sex-linked trait. Like the trait for red–green color blindness, hemophilia is caused by a recessive gene. Other sex-linked traits in humans are fragile X syndrome (described on page 1051), nonfunctional sweat glands, certain forms of diabetes, some types of deafness, uncontrollable rolling of the eyeballs, absence of central incisors, night blindness, one form of cataract, juvenile glaucoma, and juvenile muscular dystrophy.

X-Chromosome Inactivation

Whereas females have two X chromosomes in every cell (except developing oocytes), males have only one. Females thus have a double set of all genes on the X chromosome. A mechanism termed **X-chromosome inactivation** or **lyonization** in effect reduces the X-chromosome genes to a single set in females. In each cell of a female's body, one X chromosome is randomly and permanently inactivated early in development, and the genes of the inactivated X chromosome are not expressed (transcribed and translated). The nuclei of cells in female mammals contain a dark-staining body, called a **Barr body,** that is not present in the nuclei of cells in males. Geneticist Mary Lyon correctly predicted in 1961 that the Barr body is the inactivated X chromosome. During inactivation, chemical groups that prevent transcription into RNA are added to the X chromosome's DNA. As a result, an inactivated X chromosome reacts differently to histological stains and has a different appearance than the rest of the DNA. In nondividing (interphase) cells, it remains tightly coiled and can be seen as a dark-staining body within the nucleus. In a blood smear, the Barr body of neutrophils is termed a "drumstick" because it looks like a tiny drumstick-shaped projection of the nucleus.

Environmental Influences on Genotype and Phenotype

A given phenotype often is the result of both genotype and environment. Factors in the environment seem to be particularly influential in the case of polygenic traits. For example, even though a person inherits several genes for tallness, full height potential may not be reached due to environmental factors, such as disease or malnutrition during the growth years. Similarly, the risk of having a child with a neural tube defect is greater in pregnant women who lack adequate folic acid in their diet (an environmental factor); the fact that neural tube defects are more prevalent in some families than in others suggests the presence of a genetic component.

Exposure of a developing embryo or fetus to certain environmental factors can damage the developing organism or even cause death. A **teratogen** (*terato-* = monster) is any agent or influence that causes developmental defects in the embryo. In the following sections we briefly discuss several examples.

Chemicals and Drugs

Because the placenta is not an absolute barrier between the maternal and fetal circulations, any drug or chemical that is dangerous to an infant should be considered potentially dangerous to the fetus when given to the mother. Alcohol is by far the number one fetal teratogen. Intrauterine exposure to even a small amount of alcohol may result in **fetal alcohol syndrome (FAS),** one of the most common causes of mental retardation and the most common preventable cause of birth defects in the United States. The symptoms of FAS may include slow growth before and after birth, characteristic facial features (short palpebral fissures, a thin upper lip, and sunken nasal bridge), defective heart and other organs, malformed limbs, genital abnormalities, and central nervous system damage. Behavioral problems, such as hyperactivity, extreme nervousness, reduced ability to concentrate, and an inability to appreciate cause-and-effect relationships, are common. Acetaldehyde, one of the toxic products of alcohol metabolism, may cross the placenta and cause fetal damage.

Other teratogens include pesticides; defoliants (chemicals that cause plants to shed their leaves prematurely); in-

Figure 29.20 An example of the inheritance of red–green color blindness.

🔑 **Red–green color blindness and hemophilia are examples of sex-linked traits.**

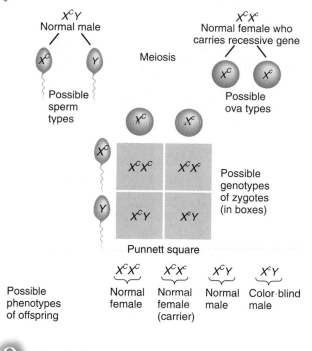

Q What is the genotype of a red–green color-blind female?

dustrial chemicals; some hormones; antibiotics; oral anticoagulants, anticonvulsants, antitumor agents, thyroid drugs, thalidomide, diethylstilbestrol (DES), and numerous other prescription drugs; LSD; marijuana; and cocaine. A pregnant woman who uses cocaine, for example, subjects the fetus to higher risk of retarded growth, attention and orientation problems, hyperirritability, a tendency to stop breathing, malformed or missing organs, strokes, and seizures. The risks of spontaneous abortion, premature birth, and stillbirth also increase with fetal exposure to cocaine.

Cigarette Smoking

Strong evidence implicates cigarette smoking during pregnancy as a cause of low infant birth weight; a strong association between smoking and a higher fetal and infant mortality rate also exists. Women who smoke have a much higher risk of an ectopic pregnancy. Cigarette smoke may be teratogenic and may cause cardiac abnormalities and anencephaly (a developmental defect characterized by the absence of a cerebrum). Maternal smoking also appears to be a significant factor in the development of cleft lip and palate and has tentatively been linked with sudden infant death syndrome (SIDS). Infants nursing from smoking mothers have also been found to have an increased incidence of gastrointestinal disturbances. Even a mother's exposure to secondhand cigarette smoke (breathing air containing tobacco smoke) predisposes her baby to increased incidence of respiratory problems, including bronchitis and pneumonia, during the first year of life.

Irradiation

Ionizing radiation of various kinds is a potent teratogen. Exposure of pregnant mothers to x-rays or radioactive isotopes during the embryo's susceptible period of development may cause microcephaly (small head size relative to the rest of the body), mental retardation, and skeletal malformations. Caution is advised, especially during the first trimester of pregnancy.

1. Define the following terms: genotype, phenotype, dominant, recessive, homozygous, and heterozygous.
2. Describe genomic imprinting and nondisjunction.
3. Define incomplete dominance. Give an example.
4. What is multiple-allele inheritance? Give an example.
5. Define polygenic inheritance and give an example.
6. Set up a Punnett square to show the inheritance of the following: sex, red–green color blindness, and hemophilia.
7. Why does X-chromosome inactivation occur?
8. How can environmental factors influence genotype and phenotype?

DISORDERS: HOMEOSTATIC IMBALANCES

INFERTILITY

Female infertility, or the inability to conceive, occurs in about 10 percent of all women of reproductive age in the United States. Female infertility may be caused by ovarian disease, obstruction of the uterine tubes, or conditions in which the uterus is not adequately prepared to receive a fertilized ovum. **Male infertility (sterility)** is an inability to fertilize a secondary oocyte; it does not imply impotence. Male fertility requires production of adequate quantities of viable, normal sperm by the testes, unobstructed transport of sperm though the ducts, and satisfactory deposition in the vagina. The seminiferous tubules of the testes are sensitive to many factors—x-rays, infections, toxins, malnutrition, and higher-than-normal scrotal temperatures—that may cause degenerative changes and produce male sterility.

One cause of infertility in females is inadequate body fat. To begin and maintain a normal reproductive cycle, a female must have a minimum amount of body fat. A moderate deficiency of fat—10–15% below normal weight for height—may delay the onset of menstruation (menarche), inhibit ovulation during the reproductive cycle, or cause amenorrhea, cessation of menstruation. Both dieting and intensive exercise may reduce body fat below the minimum amount and lead to infertility that is reversible, if weight gain or reduction of intensive exercise or both occurs. Studies of very obese women indicate that they, like very lean ones, experience problems with amenorrhea and infertility. Males also experience reproductive problems in response to undernutrition and weight loss. For example, they produce less prostatic fluid and reduced numbers of sperm having decreased motility.

Many fertility-expanding techniques now exist for assisting infertile couples to have a baby. The birth of Louise Joy Brown on July 12, 1978, near Manchester, England, was the first recorded case of **in vitro fertilization (IVF)**—fertilization in a laboratory dish. In the IVF procedure, the mother-to-be is given follicle-stimulating hormone (FSH) soon after menstruation, so that several secondary oocytes, rather than the typical single oocyte, will be produced (superovulation). When several follicles have reached the appropriate size, a small incision is made near the umbilicus, and the secondary oocytes are aspirated from the stimulated follicles and transferred to a solution containing sperm, where the oocytes undergo fertil-

ization. Alternatively, an oocyte may be fertilized in vitro by suctioning a sperm or even a spermatid obtained from the testis into a tiny pipette and then injecting it into the oocyte's cytoplasm. This procedure, termed **intracytoplasmic sperm injection (ICSI),** has been used when infertility is due to impairments in sperm motility or to the failure of spermatids to develop into spermatozoa. When the zygote achieved by IVF reaches the 8-cell or 16-cell stage, it is introduced into the uterus for implantation and subsequent growth.

In **embryo transfer,** a man's semen is used to artificially inseminate a fertile secondary oocyte donor. After fertilization in the donor's uterine tube, the morula or blastocyst is transferred from the donor to the infertile woman, who then carries it (and subsequently the fetus) to term. Embryo transfer is indicated for women who are infertile or who do not want to pass on their own genes because they are carriers of a serious genetic disorder.

In **gamete intrafallopian transfer (GIFT)** the goal is to mimic the normal process of conception by uniting sperm and secondary oocyte in the prospective mother's uterine tubes. It is an attempt to bypass conditions in the female reproductive tract that might prevent fertilization, such as high acidity or inappropriate mucus. In this procedure, a woman is given FSH and LH to stimulate the production of several secondary oocytes, which are aspirated from the mature follicles with a laparoscope fitted with a suction device, mixed outside the body, with a solution containing sperm, and then immediately inserted into the uterine tubes.

DOWN SYNDROME

Down syndrome (DS) is a disorder that most often results from nondisjunction of chromosome 21 during meiosis, which causes an extra chromosome to pass to one of the gametes. For this reason, DS is also called trisomy 21. Most of the time the extra chromosome comes from the mother, a not too surprising finding given that all her oocytes began meiosis when she herself was a fetus; they may have been exposed to chromosome-damaging chemicals and radiation for years. Undoubtedly, the kinetochore microtubules responsible for pulling sister chromatids to opposite poles of the cell (see Figure 28.1 on page 975) sustain increasing damage with the passage of time. (Sperm, by contrast, usually are less than 10 weeks old at the time they fertilize a secondary oocyte.) Overall, 1 infant in 800 is born with Down syndrome. However, older women are more likely to have a DS baby. The chance of conceiving a baby with this syndrome, which is less than 1 in 3000 for women under age 30, increases to 1 in 300 in the 35–39 age group and to 1 in 9 at age 48.

Down syndrome is characterized by mental retardation, retarded physical development (short stature and stubby fingers), distinctive facial structures (large tongue, flat profile, broad skull, slanting eyes, and round head), and malformations of the heart, ears, hands, and feet. Sexual maturity is rarely attained.

FRAGILE X SYNDROME

Fragile X syndrome, a recently recognized disorder caused by a defective gene on the X chromosome, is so named because a small portion of the tip of the X chromosome seems susceptible to breakage. It is the leading cause of mental retardation among newborns.

Fragile X syndrome affects males more than females and results in learning difficulties, mental retardation, and physical abnormalities such as oversized ears, elongated forehead, enlarged testes, and double jointedness. The syndrome may also be involved in autism in which the individual exhibits extreme withdrawal and refusal to communicate.

Fragile X syndrome can be diagnosed via amniocentesis. Although the syndrome is largely sex-linked, 20–50% of males who inherit the trait are unaffected but can pass it on to their daughters. Although the daughters also are unaffected, their children, male and female alike, may suffer mental retardation. Researchers have suggested that this pattern of inheritance could be explained by maternal imprinting of the gene. According to this theory, the father's fragile X gene becomes chemically changed as it passes through the daughter so that the gene is expressed in the daughter's offspring.

MEDICAL TERMINOLOGY

Breech presentation A malpresentation in which the fetal buttocks or lower limbs present into the maternal pelvis; the most common cause is prematurity.

Emesis gravidarum (EM-e-sis gra-VID-ar-um; *emeo* = to vomit; *gravida* = a pregnant woman) Episodes of nausea and possibly vomiting that are most likely to occur in the morning during the early weeks of pregnancy; also called **morning sickness.** Its cause is unknown, but the high levels of human chorionic gonadotropin (hCG) secreted by the placenta, and of progesterone secreted by the ovaries, have been implicated. In some women the severity of these symptoms requires hospitalization for intravenous feeding, and the condition is then known as **hyperemesis gravidarum.**

Karyotype (KAR-ē-ō-tīp; *karyo-* = nucleus) The chromosomal characteristics of an individual presented as a systematic arrangement of pairs of metaphase chromosomes arrayed in descending order of size and according to the position of the centromere; useful in judging whether or not chromosomes are normal in number and structure.

Klinefelter's syndrome A sex chromosome aneuploidy, usually due to trisomy XXY, that occurs once in every 500 births. Such individuals are somewhat mentally disadvantaged, sterile males with undeveloped testes, scant body hair, and enlarged breasts.

Lethal gene (LĒ-thal jēn; *lethum* = death) A gene that, when expressed, results in death either in the embryonic state or shortly after birth.

Metafemale syndrome A sex chromosome aneuploidy characterized by at least three X chromosomes (XXX) that occurs about once in every 700 births. These females have underdeveloped genital organs and limited fertility and generally are mentally retarded.

Puerperal fever (pyoo-ER-per-al; *puer* = child, *parere* = to bring forth) An infectious disease of childbirth, also called puerperal sepsis and childbed fever. The disease, which results from an infection originating in the birth canal, affects the endometrium; it may spread to other pelvic structures and lead to septicemia.

Turner's syndrome A sex chromosome aneuploidy caused by the presence of a single X chromosome (designated XO); occurs about once in every 5000 births and produces a sterile female with virtually no ovaries and limited development of secondary sex characteristics. Other features include short stature, webbed neck, underdeveloped breasts, and widely spaced nipples. Intelligence usually is normal.

STUDY OUTLINE

FROM FERTILIZATION TO IMPLANTATION (p. 1022)

1. Pregnancy is a sequence of events that starts with fertilization and normally ends in birth.
2. Its various events are hormonally controlled.
3. Fertilization refers to the penetration of a secondary oocyte by a sperm cell and the subsequent union of their pronuclei to form a zygote.
4. Penetration of the zona pellucida is facilitated by enzymes in the sperm's acrosome.
5. Normally, only one sperm cell fertilizes a secondary oocyte because both fast and slow blocks to polyspermy exist.
6. Early rapid cell division of a zygote is called cleavage, and the cells produced by cleavage are called blastomeres.
7. The solid sphere of cells produced by cleavage is a morula.
8. The morula develops into a blastocyst, a hollow ball of cells differentiated into a trophoblast and an inner cell mass.
9. The attachment of a blastocyst to the endometrium is called implantation; it occurs via enzymatic degradation of the endometrium.

EMBRYONIC AND FETAL DEVELOPMENT (p. 1026)

1. During embryonic growth, the primary germ layers and embryonic membranes are formed.
2. The primary germ layers—ectoderm, mesoderm, and endoderm—form all the tissues of the developing organism. Table 29.1 on page 1029 summarizes the structures produced by the three germ layers.
3. The embryonic membranes are the yolk sac, the amnion, the chorion, and the allantois.
4. Fetal and maternal materials are exchanged through the placenta.
5. During the fetal period, organs established by the primary germ layers grow rapidly.
6. The principal changes associated with embryonic and fetal growth are summarized in Table 29.2 on page 1034.
7. Several prenatal diagnostic tests are used to detect genetic disorders and to assess fetal well-being, including fetal ultrasonography, in which an image of a fetus is displayed on a screen; amniocentesis, the withdrawal and analysis of amniotic fluid and the fetal cells within it; chorionic villi sampling (CVS), which involves withdrawal of chorionic villi tissue for chromosomal analysis. CVS can be done earlier than amniocentesis, and the results are available more quickly, but it is also slightly riskier than amniocentesis.

MATERNAL CHANGES DURING PREGNANCY (p. 1035)

1. Pregnancy is maintained by human chorionic gonadotropin (hCG), estrogens, and progesterone.
2. Human chorionic somatomammotropin (hCS) assumes a role in breast development, protein anabolism, and glucose and fatty acid catabolism.
3. Relaxin relaxes the pubic symphysis and helps dilate the uterine cervix near the end of pregnancy.
4. Corticotropin-releasing hormone, produced by the placenta, is thought to establish the timing of birth and also stimulates the secretion of cortisol by the fetal adrenal gland.
5. During pregnancy, several anatomical and physiological changes occur in the mother.

EXERCISE AND PREGNANCY (p. 1039)

1. During pregnancy, some joints become less stable, and certain maneuvers are more difficult to execute.
2. Moderate physical activity does not endanger the fetus in a normal pregnancy.

LABOR (p. 1039)

1. Labor is the process by which the fetus is expelled from the uterus through the vagina to the outside. True labor involves dilation of the cervix, expulsion of the fetus, and delivery of the placenta.
2. Oxytocin stimulates uterine contractions via a positive feedback cycle.

ADJUSTMENTS OF THE INFANT AT BIRTH (p. 1041)

1. The fetus depends on the mother for oxygen and nutrients, the removal of wastes, and protection.
2. Following birth, an infant's respiratory and cardiovascular systems undergo changes in adjusting to becoming self-supporting during postnatal life.

THE PHYSIOLOGY OF LACTATION (p. 1042)

1. Lactation refers to the production and ejection of milk by the mammary glands.
2. Milk production is influenced by prolactin (PRL), estrogens, and progesterone.
3. Milk ejection is stimulated by oxytocin.
4. A few of the many benefits of breast-feeding include ideal nutrition for the infant, protection from disease, and decreased likelihood of developing allergies.

INHERITANCE (p. 1044)

1. Inheritance is the passage of hereditary traits from one generation to the next.
2. The genetic makeup of an organism is called its genotype; the traits expressed are called its phenotype.
3. Dominant genes control a particular trait; expression of recessive genes is masked by dominant genes.
4. Many patterns of inheritance do not conform to the simple dominant–recessive patterns.
5. In incomplete dominance, neither member of an allelic pair dominates; phenotypically, the heterozygote is intermediate between the homozygous dominant and the homozygous recessive. An example is sickle-cell disease.
6. In multiple-allele inheritance, genes have more than two alternate forms. An example is the inheritance of ABO blood groups.
7. In polygenic inheritance, an inherited trait is controlled by the combined effects of many genes. An example is skin color.
8. Each somatic cell has 46 chromosomes—22 pairs of autosomes and 1 pair of sex chromosomes.
9. In females, the sex chromosomes are two X chromosomes; in males, they are one X chromosome and a much smaller Y chromosome, which normally includes the prime male-determining gene, called *SRY*.
10. If the *SRY* gene is present and functional in a fertilized ovum, the fetus will develop testes and differentiate into a male. In the absence of *SRY*, the fetus will develop ovaries and differentiate into a female.
11. Red–green color blindness and hemophilia result from recessive genes located on the X chromosome; they occur primarily in males because of the absence of any counterbalancing dominant genes on the Y chromosome.
12. A mechanism termed X-chromosome inactivation or lyonization balances the difference in number of X chromosomes between males (one X) and females (two Xs). In each cell of a female's body, one X chromosome is randomly and permanently inactivated early in development and becomes a Barr body.
13. A given phenotype is the result of the interactions of genotype and the environment.
14. Teratogens, which are agents that cause physical defects in developing embryos, include chemicals and drugs, alcohol, nicotine, and ionizing radiation.

SELF-QUIZ QUESTIONS

1. Match the following:

___ (a) the penetration of a secondary oocyte by a single sperm cell

___ (b) fertilization of a secondary oocyte by more than one sperm

___ (c) the attachment of a blastocyst to the endometrium

___ (d) the merging of the genetic material from a haploid sperm and a haploid secondary oocyte into a single diploid nucleus

___ (e) the inactivation of one X chromosome in the female

___ (f) the induction by the female reproductive tract of functional changes in sperm that allow them to fertilize a secondary oocyte

___ (g) the examination of embryonic cells sloughed off into the amniotic fluid

___ (h) an abnormal condition of pregnancy characterized by sudden hypertension, large amounts of protein in urine, and generalized edema

___ (i) the process of giving birth

___ (j) the period of time (about 6 weeks) during which the maternal reproductive organs and physiology return to the prepregnancy state

(1) fertilization
(2) capacitation
(3) syngamy
(4) polyspermy
(5) implantation
(6) amniocentesis
(7) lyonization
(8) preeclampsia
(9) parturition
(10) puerperium

2. Match the following:

___ (a) a hollow sphere of cells that enters the uterine cavity

___ (b) cells produced by cleavage

___ (c) the developing individual from the third month of pregnancy until birth

___ (d) the outer covering of cells of the blastocyst

___ (e) membrane derived from trophoblast

___ (f) early divisions of the zygote

___ (g) a solid sphere of cells still surrounded by the zona pellucida

___ (h) event in which differentiation into the three primary germ layers occurs

(1) cleavage
(2) blastomeres
(3) morula
(4) trophoblast
(5) blastocyst
(6) gastrulation
(7) chorion
(8) fetus

3. Match the following:

___ (a) the control of inherited traits by the combined effects of many genes; an example is skin color

___ (b) the two alternative forms of a gene that code for the same trait and are at the same location on homologous chromosomes

___ (c) inheritance based on genes that have more than two alternate forms; an example is the inheritance of blood type

___ (d) a cell in which one or more chromosomes of a set is added or deleted

___ (e) refers to an individual with different alleles on homologous chromosomes

___ (f) traits linked to the X chromosome

___ (g) neither member of the allelic pair is dominant over the other, and the heterozygote has a phenotype intermediate between the homozygous dominant and homozygous recessive

___ (h) refers to how the genetic makeup is expressed in the body

___ (i) a homozygous dominant, homozygous recessive, or heterozygous genetic makeup; the actual gene arrangement

___ (j) refers to a person with the same alleles on homologous chromosomes

___ (k) individuals who possess a recessive gene (but do not express it) and can pass the gene on to their offspring

___ (l) an allele that masks the presence of another allele and is fully expressed

(1) genotype
(2) phenotype
(3) alleles
(4) aneuploid
(5) incomplete dominance
(6) multiple-allele inheritance
(7) polygenic inheritance
(8) sex-linked inheritance
(9) homozygous
(10) heterozygous
(11) carriers
(12) dominant trait

Choose the best answer to the following questions:

4. The group of cells that secrete enzymes that enable the blastocyst to penetrate the uterine lining is the (a) synctiotrophoblast, (b) cytotrophoblast, (c) corona radiata, (d) zona pellucida, (e) trophoblast.

5. The only tissue of the developing embryo that contacts the mother's uterus is the (a) ectoderm, (b) corona radiata, (c) trophoblast, (d) zona pellucida, (e) acrosome.

6. Which of the following statements are correct? (1) The embryonic period is the first four months of development. (2) The inner cell mass differentiates into the three primary germ layers. (3) The three primary germ layers are the endoderm, mesoderm, and ectoderm. (4) The embryonic membranes are the yolk sac, amnion, chorion, and allantois. (5) Placenta development is accomplished during the fourth month of pregnancy. (a) 1, 2, and 3, (b) 2, 3, and 4, (c) 3, 4, and 5, (d) 1, 4, and 5, (e) 2, 3, and 5

7. The hormone of pregnancy that mimics LH and acts to rescue the corpus luteum from degeneration and to stimulate its continued production of progesterone and estrogens is (a) relaxin, (b) human chorionic somatomammotropin, (c) human placental lactogen, (d) inhibin, (e) human chorionic gonadotropin.

8. Which of the following are maternal changes that occur during pregnancy? (1) altered pulmonary function, (2) increased stroke volume, cardiac output, and heart rate, and decreased blood volume, (3) weight gain, (4) increased gastric motility, a delay in gastric emptying time, (5) edema and possible varicose veins.
(a) 1, 2, 3, and 4, (b) 2, 3, 4, and 5, (c) 1, 3, 4, and 5, (d) 1, 3, and 5, (e) 2, 4, and 5

9. Which of the following statements is correct? (a) Normal traits always dominate over abnormal traits. (b) Occasionally an error in meiosis called nondisjunction results in an abnormal number of chromosomes. (c) The mother always determines the sex of the child because she has either an X or a Y gene in her oocytes. (d) Most patterns of inheritance are simple dominant-recessive inheritances. (e) Genes are expressed normally regardless of any outside influence such as chemicals or radiation.

Complete the following:

10. The three stages of true labor, in order of occurrence, are ___, ___, and ___.

11. The first major event of the embryonic period is ___. The second major event of the embryonic period is the formation of ___. The third major event of the embryonic period is the development of the ___.

12. Hormones produced by the ___ are responsible for maintaining the pregnancy during the first 3–4 months.

True or false:

13. The principal stimulus in maintaining prolactin secretion during lactation is the sucking action of the infant.

14. Male and female embryos develop identically until the *SRY* gene sets into motion a cascade of events that lead to the development of a female.

15. Match the following:

___ (a) the embryonic membrane that entirely surrounds the embryo	(1) decidua
___ (b) functions as an early site of blood formation; contains cells that migrate into the gonads and differentiate into the primitive germ cells	(2) placenta (3) amnion (4) chorion
___ (c) becomes the principal part of the placenta; produces human chorionic gonadotropin	(5) allantois (6) yolk sac (7) chorionic villi
___ (d) includes all but the stratum basalis layer of the endometrium and separates from the endometrium after the fetus is delivered	(8) umbilical cord
___ (e) contains the vascular connections between mother and fetus	
___ (f) allows oxygen and nutrients to diffuse from maternal blood into fetal blood	
___ (g) serves as an early site of blood formation; its blood vessels serve as the umbilical connection in the placenta between mother and fetus	
___ (h) finger-like projections of the chorion that are bathed in maternal blood sinuses, thereby bringing maternal and fetal blood vessels into close proximity	

CRITICAL THINKING QUESTIONS

1. Gladys, the neighborhood busybody, watched from her window as 2-year-old Angelica's parents and grandmother took her for a walk. "I still don't think it's biologically possible for two brown-eyed, brown-haired parents to give birth to a little blue-eyed blond like Angelica," she announced. "There's your proof," answered her husband. Who's right? (HINT: *Everyone says that Angelica is the "spitting image" of her grandmother.*)

2. When pregnant Phoebe learned that Fragile X syndrome was present in her husband's family, she said, "It's one of those boy problems, like color blindness. We won't have to worry about it." Explain the inheritance of Fragile X syndrome more clearly. (HINT: *Most cases of Fragile X syndrome are sex-linked.*)

3. Usually, when you think about arteries, you think about oxygenated blood, but this textbook states that the umbilical arteries carry deoxygenated blood in the fetus. Could the authors, Tortora and Grabowski, be wrong about this? Please explain. (HINT: *Think about the pulmonary arteries in an adult.*)

ANSWERS TO FIGURE QUESTIONS

29.1 Capacitation entails the functional changes in sperm after they have been deposited in the female reproductive tract that enable them to fertilize a secondary oocyte.

29.2 A morula is a solid ball of cells, whereas a blastocyst consists of a rim of cells (trophoblast) surrounding a cavity (blastocele) and an inner cell mass.

29.3 The blastocyst secretes digestive enzymes that eat away the endometrial lining at the site of implantation.

29.4 Implantation occurs during the secretory phase of the uterine cycle.

29.5 The inner cell mass of the blastocyst gives rise to the embryonic disc.

29.6 The amnion functions as a shock absorber, whereas the chorion forms the placenta.

29.7 The decidua basalis helps form the maternal part of the placenta.

29.8 The placenta functions in exchange of materials between fetus and mother.

29.9 Amniocentesis is used mainly to detect genetic disorders, but it also provides information concerning the maturity (and survivability) of the fetus.

29.10 Early pregnancy tests detect human chorionic gonadotropin (hCG).

29.11 Relaxin increases the flexibility of the pubic symphysis and helps dilate the cervix of the uterus to ease delivery.

29.12 Complete dilation of the cervix marks the onset of the stage of expulsion.

29.13 Oxytocin also stimulates contraction of the uterus during delivery of a baby.

29.14 The odds that a child will have PKU are the same for each child—25%.

29.15 In incomplete dominance, neither member of an allelic pair is dominant; the heterozygote has a phenotype intermediate between the homozygous dominant and the homozygous recessive phenotypes.

29.16 Yes, a baby can have blood type O if each parent has one *i* allele and passes it on to the offspring.

29.17 Hair color, height, and body build, among others, are polygenic traits.

29.18 The female sex chromosomes are XX, and the male sex chromosomes are XY.

29.19 The chromosomes that are not sex chromosomes are called autosomes.

29.20 A red–green color-blind female has an $X^c X^c$ genotype.

A | MEASUREMENTS

U.S. CUSTOMARY SYSTEM

	Unit	Relation to Other U.S. Units	SI (Metric) Equivalent
Length	inch	1/12 foot	2.54 centimeters
	foot	12 inches	0.305 meter
	yard	36 inches	9.144 meters
	mile	5,280 feet	1.609 kilometers
Mass	grain	1/1000 pound	64.799 milligrams
	dram	1/16 ounce	1.772 grams
	ounce	16 drams	28.350 grams
	pound	16 ounces	453.6 grams
	ton	2,000 pounds	907.18 kilograms
Volume (Liquid)	ounce	1/16 pint	29.574 milliliters
	pint	16 ounces	0.473 liter
	quart	2 pints	0.946 liter
	gallon	4 quarts	3.785 liters
Volume (Dry)	pint	1/2 quart	0.551 liter
	quart	2 pints	1.101 liters
	peck	8 quarts	8.810 liters
	bushel	4 pecks	35.239 liters

INTERNATIONAL SYSTEM (SI)

BASE UNITS

Unit	Quantity	Symbol
meter	length	m
kilogram	mass	kg
second	time	s
liter	volume	L
mole	amount of matter	mol

PREFIXES

Prefix	Multiplier	Symbol
tera-	$10^{12} = 1,000,000,000,000$	T
giga-	$10^{9} = 1,000,000,000$	G
mega-	$10^{6} = 1,000,000$	M
kilo-	$10^{3} = 1,000$	K
hecto-	$10^{2} = 100$	h
deca-	$10^{1} = 10$	da
deci-	$10^{-1} = 0.1$	d
centi-	$10^{-2} = 0.01$	c
milli-	$10^{-3} = 0.001$	m
micro-	$10^{-6} = 0.000,001$	μ
nano-	$10^{-9} = 0.000,000,001$	n
pico-	$10^{-12} = 0.000,000,000,001$	p

APOTHECARY MEASURES

	Unit	US Equivalent	SI Equivalent
Mass	grain	1 grain	64.8 milligrams
	dram	2.194 drams	3.888 grams
	ounce	1.097 ounces	31.104 grams
	pound	0.823 pound	373.242 grams
Volume	minims	1/60 ounce	0.06 milliliter
	fluid dram	1/8 ounce	3.697 milliliters
	fluid ounce	1 ounce	29.573 milliliters
	pint	1 pint	0.473 liter

U.S. TO SI (METRIC) CONVERSION

When you know	Multiply by	To find
inches	2.54	centimeters
feet	30.48	centimeters
yards	0.91	meters
miles	1.61	kilometers
ounces	28.35	grams
pounds	0.45	kilograms
tons	0.91	metric tons
fluid ounces	29.57	milliliters
pints	0.47	liters
quarts	0.95	liters
gallons	3.79	liters

TEMPERATURE CONVERSION

FAHRENHEIT (F) TO CELSIUS (C)

$$°C = (°F - 32) \div 1.8$$

CELSIUS (C) TO FAHRENHEIT (F)

$$°F = (°C \times 1.8) + 32$$

SI (METRIC) TO U.S. CONVERSION

When you know	Multiply by	To find
millimeters	0.04	inches
centimeters	0.39	inches
meters	3.28	feet
kilometers	0.62	miles
liters	1.06	quarts
cubic meters	35.32	cubic feet
grams	0.035	ounces
kilograms	2.21	pounds

The periodic table lists the known **chemical elements,** the basic units of matter. The elements in the table are arranged left-to-right in rows in order of their **atomic number,** the number of protons in the nucleus. Each horizontal row, numbered from 1 to 7, is a **period.** All elements in a given period have the same number of electron shells as their period number. For example, an atom of hydrogen or helium each has one electron shell, while an atom of potassium or calcium each has four electron shells. The elements in each column, or **group,** share chemical properties. For example, the elements in column IA are very chemically reactive, whereas the elements in column VIIIA have full electron shells and thus are chemically inert.

Scientists now recognize 112 different elements; 92 occur naturally on Earth, and the rest are produced from the natural elements using devices such as particle accelerators or nuclear reactors. Elements are designated by **chemical symbols,** which are the first one or two letters of the element's name in English, Latin, or another language.

Twenty-six of the 92 naturally occurring elements normally are present in your body. Of these, just four elements—oxygen, carbon, hydrogen, and nitrogen (coded blue)—constitute about 96% of the body's mass. Nine others—calcium, phosphorus, potassium, sulfur, sodium, chlorine, magnesium, iodine, and iron (coded red)— contribute an additional 3.9% of the body's mass. An additional 13 elements, called **trace elements** because they are present in tiny concentrations, account for the remaining 0.1% of the body's mass. The trace elements are aluminum, boron, chromium, cobalt, copper, fluorine, manganese, molybdenum, selenium, silicon, tin, vanadium, and zinc (coded yellow).

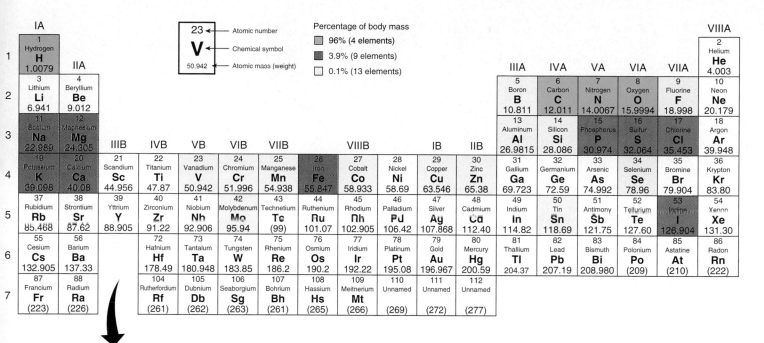

57–71, Lanthanides

57	58	59	60	61	62	63	64	65	66	67	68	69	70	71
Lanthanum	Cerium	Praseodymium	Neodymium	Promethium	Samarium	Europium	Gadolinium	Terbium	Dysprosium	Holmium	Erbium	Thulium	Ytterbium	Lutetium
La	**Ce**	**Pr**	**Nd**	**Pm**	**Sm**	**Eu**	**Gd**	**Tb**	**Dy**	**Ho**	**Er**	**Tm**	**Yb**	**Lu**
138.91	140.12	140.907	144.24	144.913	150.35	151.96	157.25	158.925	162.50	164.930	167.26	168.934	173.04	174.97

89–103, Actinides

89	90	91	92	93	94	95	96	97	98	99	100	101	102	103
Actinium	Thorium	Protactinium	Uranium	Neptunium	Plutonium	Americium	Curium	Berkelium	Californium	Einsteinium	Fermium	Mendelevium	Nobelium	Lawrencium
Ac	**Th**	**Pa**	**U**	**Np**	**Pu**	**Am**	**Cm**	**Bk**	**Cf**	**Es**	**Fm**	**Md**	**No**	**Lr**
(227)	232.038	(231)	238.03	(237)	244.064	(243)	(247)	(247)	242.058	(254)	257.095	258.10	259.10	260.105

The system of international (SI) units (Système Internationale d'Unités) is used in most countries and in many medical and scientific journals. Clinical laboratories in the United States, by contrast, usually report values for blood and urine tests in conventional units. The laboratory values in this Appendix give conventional units first, followed by SI equivalents in parentheses. Values listed for various blood tests should be viewed as reference values rather than absolute "normal" values for all well people. Values may vary due to age, gender, diet, and environment of the subject or the equipment, methods, and standards of the lab performing the measurement.

KEY TO SYMBOLS

g = gram
mg = milligram = 10^{-3} gram
μg = microgram = 10^{-6} gram
U = units
L = liter
dL = deciliter
mL = milliliter
μL = microliter
mEq/L = milliequivalents per liter
mmol/L = millimoles per liter
μmol/L = micromoles per liter
$>$ = greater than; $<$ = less than

TEST (SPECIMEN)	REFERENCE VALUES (SI UNITS)	VALUES INCREASE IN	VALUES DECREASE IN
Aminotransferases (serum)			
Alanine amino-transferase (ALT)	0–35 U/L (same)	Liver disease or liver damage due to toxic drugs.	
Aspartate amino-transferase (AST)	0–35 U/L (same)	Myocardial infarction, liver disease, trauma to skeletal muscles, severe burns.	Beriberi, uncontrolled diabetes mellitus with acidosis, pregnancy.
Ammonia (plasma)	20–120 μg/dL (12–55 μmol/L)	Liver disease, heart failure, emphysema, pneumonia, hemolytic disease of newborn.	Hypertension.
Bilirubin (serum)	Conjugated: <0.5 mg/dL (<5.0 μmol/L) Unconjugated: 0.2–1.0 mg/dL (18–20 μmol/L) Newborn: 1.0–12.0 mg/dL (<200 μmol/L)	Conjugated bilirubin: liver dysfunction or gallstones. Unconjugated bilirubin: excessive hemolysis of red blood cells.	
Blood urea nitrogen (BUN) (serum)	8–26 mg/dL (2.9–9.3 mmol/L)	Kidney disease, urinary tract obstruction, shock, diabetes, burns, dehydration, myocardial infarction.	Liver failure, malnutrition, overhydration, pregnancy.
Carbon dioxide content (bicarbonate + dissolved CO_2) (whole blood)	Arterial: 19–24 mEq/L (19–24 mmol/L) Venous: 22–26 mEq/L (22–26 mmol/L)	Severe diarrhea, severe vomiting, starvation, emphysema, aldosteronism.	Renal failure, diabetic ketoacidosis, shock.
Cholesterol, total (plasma)	<200 mg/dL (<5.2 mmol/L) is desirable	Hypercholesterolemia, uncontrolled diabetes mellitus, hypothyroidism, hypertension, atherosclerosis, nephrosis.	Liver disease, hyperthyroidism, fat malabsorption, pernicious or hemolytic anemia, severe infections.
HDL cholesterol (plasma)	>40 mg/dL (>1.0 mmol/L) is desirable		
LDL cholesterol (plasma)	<130 mg/dL (<3.2 mmol/L) is desirable		

TEST (SPECIMEN)	REFERENCE VALUES (SI UNITS)	VALUES INCREASE IN	VALUES DECREASE IN
Creatine (serum)	Males: 0.15–0.5 mg/dL (10–40 μmol/L) Females: 0.35–0.9 mg/dL (30–70 μmol/L)	Muscular dystrophy, damage to muscle tissue, electric shock, chronic alcoholism.	
Creatine kinase (CK), also known as **Creatine phosphokinase (CPK)** (serum)	0–130 U/L (same)	Myocardial infarction, progressive muscular dystrophy, hypothyroidism, pulmonary edema.	
Creatinine (serum)	0.5–1.2 mg/dL (45–105 μmol/L)	Impaired renal function, urinary tract obstruction, giantism, acromegaly.	Decreased muscle mass, as occurs in muscular dystrophy or myasthenia gravis.
Electrolytes (plasma)	See Table 27.1 on page 964.		
Gamma-glutamyl transferase (GGT) (serum)	0–30 U/L (same)	Bile duct obstruction, cirrhosis, alcoholism, metastatic liver cancer, congestive heart failure.	
Glucose (plasma)	70–110 mg/dL (3.9–6.1 mmol/L)	Diabetes mellitus, acute stress, hyperthyroidism, chronic liver disease, Cushing's disease.	Addison's disease, hypothyroidism, hyperinsulinism.
Hemoglobin (whole blood)	Males: 14–18 g/100 mL (140–180 g/L) Females: 12–16 g/100 mL (120–160 g/L) Newborns: 14–20 g/100 mL (140–200 g/L)	Polycythemia, congestive heart failure, chronic obstructive pulmonary disease, living at high altitude.	Anemia, severe hemorrhage, cancer, hemolysis, Hodgkin's disease, nutritional deficiency of vitamin B_{12}, systemic lupus erythematosus, kidney disease.
Iron, total (serum)	Males: 80–180 μg/dL (14–32 μmol/L) Females: 60–160 μg/dL (11–29 μmol/L)	Liver disease, hemolytic anemia, iron poisoning.	Iron-deficiency anemia, chronic blood loss, pregnancy (late), chronic heavy menstruation.
Lactic dehydrogenase (LDH) (serum)	71–207 U/L (same)	Myocardial infarction, liver disease, skeletal muscle necrosis, extensive cancer.	
Lipids (serum) Total	400–850 mg/dL (4.0–8.5 g/L)	Hyperlipidemia, diabetes mellitus, hypothyroidism.	Fat malabsorption.
Triglycerides	10–190 mg/dL (0.1–1.9 g/L)		
Platelet (thrombocyte) count (whole blood)	150,000–400,000/μL	Cancer, trauma, leukemia, cirrhosis.	Anemias, allergic conditions, hemorrhage.
Protein (serum) Total Albumin Globulin	6–8 g/dL (60–80 g/L) 4–6 g/dL (40–60 g/L) 2.3–3.5 g/dL (23–35 g/L)	Dehydration, shock, chronic infections.	Liver disease, poor protein intake, hemorrhage, diarrhea, malabsorption, chronic renal failure, severe burns.
Red blood cell (erythrocyte) count (whole blood)	Males: 4.5–6.5 million/μL Females: 3.9–5.6 million/μL	Polycythemia, dehydration, living at high altitude.	Hemorrhage, hemolysis, anemias, cancer, overhydration.
Uric acid (urate) (serum)	2.0–7.0 mg/dL (120–420 μmol/L)	Impaired renal function, gout, metastatic cancer, shock, starvation.	
White blood cell (leukocyte) count, total (whole blood)	5,000–10,000/μL (See Table 19.3 on page 623 for relative percentages of different types of WBCs.)	Acute infections, trauma, malignant diseases, cardiovascular diseases. (See also Table 19.2 on page 621.)	Diabetes mellitus, anemia. (See also Table 19.2 on page 621.)

ANSWERS TO SELF-QUIZ QUESTIONS

Chapter 1
1. cell 2. epithelium, connective tissue, muscle, nervous tissue 3. catabolism 4. homeostasis 5. (1) f, (2) a, (3) b, (4) e, (5) c, (6) h, (7) g, (8) d 6. false 7. true 8. d 9. b 10. a 11. b 12. e 13. c 14. (1) d, (2) f, (3) h, (4) a, (5) e, (6) b, (7) g, (8) c 15. (1) c, (2) g, (3) a, (4) e, (5) f, (6) d, (7) h, (8) b

Chapter 2
1. protons, electrons, neutrons 2. 8 3. hydrogen 4. concentration, temperature 5. (a) 2, (b) 4, (c) 6, (d) 7, (e) 5, (f) 8, (g) 3, (h) 1 6. true 7. false 8. b 9. d 10. a 11. c 12. e 13. a 14. (a) 3, (b) 4, (c) 6, (d) 8, (e) 2, (f) 5, (g) 1, (h) 7 15. (a) 1, (b) 2 (c) 1, (d) 4, (e) 5

Chapter 3
1. plasma membrane, cytoplasm, nucleus 2. apoptosis, necrosis 3. cytosol 4. nucleolus 5. (a) 2, (b) 3, (c) 5, (d) 7, (e) 6, (f) 8, (g) 1, (h) 4 6. false 7. true 8. a 9. d 10. b 11. c 12. c 13. (a) 4, (b) 10, (c) 1, (d) 2, (e) 5, (f) 9, (g) 8, (h) 7, (i) 6, (j) 3 14. (a) 2, (b) 9, (c) 3, (d) 5, (e) 6, (f) 8, (g) 1 (h) 4 (i) 7 15. (a) 3, (b) 9, (c) 1, (d) 5, (e) 6, (f) 4, (g) 8, (h) 7, (i) 2, (j) 10

Chapter 4
1. (a) 2, (b) 5, (c) 3, (d) 4, (e) 1 2. true 3. false 4. a 5. c 6. b 7. e 8. b 9. d 10. (a) 3, (b) 5, (c) 8, (d) 12, (e) 9, (f) 7, (g) 11, (h) 6, (i) 2, (j) 4, (k) 10, (l) 1 11. (a) 4, (b) 2, (c) 5, (d) 6, (e) 8, (f) 3, (g) 1, (h) 7 12. tissue 13. endocrine 14. cells, ground substance, fibers 15. regeneration, fibrosis

Chapter 5
1. epidermis, dermis 2. strength, extensibility, elasticity 3. papillary, reticular 4. melanin, carotene, hemoglobin 5. true 6. false 7. (a) 3, (b) 1, (c) 2, (d) 5, (e) 4 8. a 9. b 10. d 11. c 12. a 13. e 14. (a) 3, (b) 5, (c) 4, (d) 1, (e) 2, (f) 6, (g) 7, (h) 8 15. (a) 5, (b) 13, (c) 3, (d) 7, (e) 4, (f) 2, (g) 6, (h) 8, (i) 10, (j) 12, (k) 11, (l) 1, (m) 14, (n) 9

Chapter 6
1. hardness, tensile strength 2. osteons, trabeculae 3. hyaline cartilage, mesenchyme 4. parathyroid hormone 5. (a) 3, (b) 9, (c) 8, (d) 1, (e) 5, (f) 4, (g) 6, (h) 7, (i) 2, (j) 10 6. true 7. false 8. (a) 1, (b) 2, (c) 7, (d) 10, (e) 8, (f) 3, (g) 6, (h) 4, (i) 9 (j) 5 9. d 10. b 11. a 12. c 13. a 14. b 15. (a) 10, (b) 6, (c) 5, (d) 1, (e) 3, (f) 2, (g) 4, (h) 9, (i) 7, (j) 8

Chapter 7
1. false 2. false 3. pituitary gland 4. fontanels 5. hyoid 6. intervertebral discs 7. (a) 8, (b) 6, (c) 1, (d) 7, (e) 2, (f) 5, (g) 9, (h) 3, (i) 10, (j) 4, (k) 11, (l) 12, (m) 15, (n) 13, (o) 14 8. (a) 2, (b) 3, (c) 5, (d) 6, (e) 4, (f) 1, (g) 5, (h) 4, (i) 2, (j) 4, (k) 3 9. a 10. c 11. b 12. b 13. d 14. e 15. (a) 1, (b) 3, (c) 6, (d) 9, (e) 13, (f) 12, (g) 2, (h) 4, (i) 5, (j) 7, (k) 10, (l) 8, (m) 11, (n) 14

Chapter 8
1. protection, movement 2. patella 3. talus 4. clavicle 5. b 6. b 7. d 8. c 9. e 10. a 11. true 12. true 13. (a) 4, (b) 3, (c) 3, (d) 6, (e) 7, (f) 1, (g) 3, (h) 2, (i) 5, (j) 8, (k) 2, (l) 4, (m) 6 14. (a) 2, (B) 6, (c) 9, (d) 7, (e) 4, (f) 5, (g) 8, (h) 10, (i) 1, (j) 3 15. (a) 8, (b) 2, (c) 6, (d) 3, (e) 7, (f) 5, (g) 1, (h) 4

Chapter 9
1. joint or articulation or arthrosis 2. synovial fluid 3. knee 4. articular discs, menisci 5. d 6. b 7. a 8. c 9. a 10. e 11. (a) 4, (b) 3, (c) 6, (d) 2, (e) 5, (f) 1 12. true 13. true 14. (a) 5, (b) 1, (c) 2, (d) 3, (e) 4, (f) 7, (g) 6 15. (a) 6, (b) 7, (c) 9, (d) 4, (e) 8, (f) 3, (g) 10, (h) 2, (i) 11, (j) 1, (k) 5

Chapter 10
1. false 2. true 3. (a) 7, (b) 8, (c) 10, (d) 9, (e) 4, (f) 6, (g) 5, (h) 3, (i) 1, (j) 2 4. (a) 5, (b) 8, (c) 7, (d) 10, (e) 6, (f) 9, (g) 4, (h) 1, (i) 2, (j) 3 5. skeletal, cardiac, smooth 6. cardiac and visceral smooth muscle 7. electrical excitability, contractility, extensibility, elasticity 8. actin, myosin 9. a 10. b 11. c 12. e 13. d 14. c 15. (a) 12, (b) 4, (c) 6, (d) 5, (e) 8, (f) 7, (g) 3, (h) 9, (i) 11, (j) 10, (k) 1, (l) 2

Chapter 11
1. origin, insertion 2. buccinator 3. masseter 4. diaphragm 5. d 6. a 7. b 8. e 9. c 10. a 11. (a) 3, (b) 1, (c) 2, (d) 1, (e) 2, (f) 3, (g) 2, (h) 3 12. true 13. true 14. (a) 6, (b) 3, (c) 7, (d) 4, (e) 2, (f) 8, (g) 5, (h) 1 15. (a) 11, (b) 1, (c) 10, (d) 8, (e) 13, (f) 18, (g) 2, (h) 6, (i) 8, (j) 15, (k) 5, (l) 4, (m) 2, (n) 9, (o) 1, (p) 12, (q) 14, (r) 13, (s) 7, (t) 17, (u) 5, (v) 18, (w) 6, (x) 16, (y) 3

Chapter 12
1. CNS, brain, spinal cord; PNS, cranial nerves, spinal nerves, ganglia, sensory receptors 2. ganglia 3. somatic nervous system, autonomic nervous system, enteric nervous system 4. plasticity 5. true 6. false 7. (a) 4, (b) 5, (c) 14, (d) 8, (e) 7, (f) 1, (g) 2, (h) 10, (i) 13, (j) 6, (k) 3, (l) 12, (m) 9, (n) 11 8. (a) 2, (b) 1, (c) 8, (d) 5, (e) 6, (f) 3, (g) 4, (h) 7 9. c 10. b 11. a 12. e 13. c 14. d 15. (a) 6, (b) 3, (c) 1, (d) 2, (e) 9, (f) 11, (g) 4, (h) 8, (i) 7, (j) 5, (k) 10

Chapter 13
1. (a) 10, (b) 8, (c) 9, (d) 1, (e) 2, (f) 5, (g) 7, (h) 6, (i) 4, (j) 3 2. (a) 2, (b) 1, (c) 3, (d) 4, (e) 1, (f) 2, (g) 4, (h) 3, (i) 1 3. true 4. false 5. reflexes 6. autonomic 7. denticulate ligaments 8. white matter tracts, gray matter 9. d 10. b 11. c 12. a 13. b 14. e 15. (a) 1, (b) 3, (c) 4, (d) 2, (e) 1, (f) 1, (g) 2

Chapter 14
1. brain stem, cerebellum, diencephalon, cerebrum 2. blood–brain barrier 3. choroid plexuses 4. falx cerebri 5. mesencephalon 6. true 7. false 8. c 9. d 10. e 11. d 12. e 13. a 14. (a) 3, (b) 5, (c) 6, (d) 8, (e) 11, (f) 10, (g) 7, (h) 9, (i) 1, (j) 4, (k) 2, (l) 12, (m) 1, (n) 8, (o) 5, (p) 7, (q) 12, (r) 10, (s) 9 15. (a) 9, (b) 2, (c) 6, (d) 10, (e) 4, (f) 11, (g) 1, (h) 2, (i) 5, (j) 8, (k) 12, (l) 7, (m) 3, (n) 6, (o) 13, (p) 7, (q) 1, (r) 14

Chapter 15
1. sensation 2. adaptation 3. tactile 4. c 5. a 6. d 7. e 8. b 9. d 10. d 11. true 12. true 13. (a) 3, (b) 2, (c) 5, (d) 7, (e) 1, (f) 3, (g) 8, (h) 4, (i) 6 14. (a) 9, (b) 8, (c) 4, (d) 7, (e) 10, (f) 2, (g) 3, (h) 1, (i) 5, (j) 6, (k) 11 15. (a) 10, (b) 8, (c) 7, (d) 1, (e) 4, (f) 3, (g) 5, (h) 6, (i) 9, (j) 2

Chapter 16
1. true 2. false 3. sweet, sour, salty, bitter 4. central fovea 5. auditory or Eustachian 6. static, dynamic 7. a 8. d 9. c 10. c 11. e 12. a 13. (a) 1, (b) 5, (c) 7, (d) 6, (e) 1, (f) 8, (g) 2, (h) 4, (i) 3 14. (a) 3, (b) 6, (c) 8, (d) 10, (e) 1, (f) 5, (g) 9, (h) 11, (i) 7, (j) 12, (k) 2, (l) 4 15. (a) 2, (b) 5, (c) 10, (d) 7, (e) 3, (f) 9, (g) 6, (h) 1, (i) 4, (j) 8

Chapter 17
1. (a) 2, (b) 5, (c) 6, (d) 1, (e) 3, (f) 4 2. autonomic 3. thoracolumbar, craniosacral 4. norepinephrine 5. (a) 3, (b) 2, (c) 1, (d) 1, (e) 2, (f) 3, (g) 3 6. true 7. false 8. (a) 2, (b) 1, (c) 1, (d) 2, (e) 1, (f) 1, (g) 2, (h) 2 9. b 10. c 11. d 12. a 13. a 14. c 15. e

Chapter 18
1. b 2. c 3. d 4. a 5. b 6. c 7. G protein, adenylate cyclase, protein kinase, phosphorylates 8. alarm reaction, resistance reaction, exhaustion 9. hypothalamus 10. Exocrine, Endocrine 11. false 12. true 13. (a) 8, (b) 2, (c) 7, (d) 1, (e) 12, (f) 4, (g) 5, (h) 10, (i) 16, (j) 15, (k) 3, (l) 11, (m) 14, (n) 6, (o) 13, (p) 9 14. (a) 10, (b) 8, (c) 2, (d) 4, (e) 1, (f) 6, (g) 9, (h) 7, (i) 5, (j) 3 15. (a) 10, (b) 1, (c) 5, (d) 7, (e) 3, (f) 8, (g) 2, (h) 9, (i) 4, (j) 6

Chapter 19
1. e 2. a 3. d 4. b 5. c 6. d 7. plasma, formed elements, red blood cells, white blood cells, platelets 8. serum 9. prothrombinase, intrinsic, extrinsic, prothrombin, thrombin, fibrinogen, fibrin 10. platelets, thrombocytes 11. true 12. false 13. (a) 4, (b) 6, (c) 8, (d) 1, (e) 7, (f) 3, (g) 5, (h) 2 14. (a) 6, (b) 8, (c) 3, (d) 10, (e) 2, (f) 4, (g) 1, (h) 7, (i) 9, (j) 5 15. (a) 5, (b) 1, (c) 2, (d) 3, (e) 6, (f) 4, (g) 8, (h) 7

Chapter 20
1. (a) 3, (b) 2, (c) 8, (d) 5, (e) 7, (f) 4, (g) 1, (h) 6 2. a 3. c 4. d 5. b 6. e 7. d 8. (a) 4, (b) 10, (c) 6, (d) 5, (e) 3, (f) 9, (g) 2, (h) 8, (i) 1, (j) 7 9. isovolumetric 10. systole, diastole 11. lubb, close; dupp, close 12. stroke volume 13. false 14. true 15. (a) 3, (b) 6, (c) 1, (d) 5, (e) 2, (f) 4

Chapter 21
1. b 2. c 3. e 4. a 5. d 6. c 7. (a) 2, (b) 6, (c) 4, (d) 1, (e) 3, (f) 5 8. (a) 11, (b) 1, (c) 4, (d) 9, (e) 3, (f) 8, (g) 6, (h) 2, (i) 7, (j) 5, (k) 10, (l) 12 9. true 10. true 11. diffusion, vesicular transport, bulk flow 12. blood hydrostatic pressure, interstitial fluid hydrostatic pressure, blood colloidal osmotic pressure 13. pressure difference, resistance, heart rate, stroke volume 14. carotid sinus, aortic 15. (a) 5, (b) 3, (c) 1, (d) 4, (e) 2, (f) 4, (g) 5

Chapter 22
1. e 2. e 3. d 4. a 5. c 6. e 7. skeletal muscle contraction, one-way valves, breathing movements 8. chemotaxis, adherence, ingestion 9. skin and mucous membranes; antimicrobial proteins, and NK cells and phagocytes 10. specificity, memory 11. true 12. true 13. (a) 3, (b) 1, (c) 4, (d) 2, (e) 5 14. (a) 2, (b) 4, (c) 5, (d) 8, (e) 1, (f) 7, (g) 3, (h) 6 15. (a) 3 or 10, (b) 10 or 3, (c) 8, (d) 1, (e) 2, (f) 5, (g) 4, (h) 7, (i) 9, (j) 6

Chapter 23
1. (a) 2, (b) 8, (c) 3, (d) 7, (e) 1, (f) 9, (g) 5, (h) 10, (i) 4, (j) 6 2. pulmonary ventilation, external respiration, internal respiration 3. pulmonary ventilation, inspiration (inhalation), expiration (exhalation) 4. oxyhemoglobin, bicarbonate ion 5. $CO_2 + H_2O \rightarrow H_2CO_3 \rightarrow H^+ + HCO_3^-$ 6. false 7. true 8. (a) 9, (b) 10, (c) 1, (d) 6, (e) 8, (f) 2, (g) 4, (h) 3, (i) 5, (j) 7 9. b 10. d 11. a 12. c 13. c 14. e 15. (a) 6, (b) 3, (c) 5, (d) 1, (e) 4, (f) 2

Chapter 24
1. (a) 3, (b) 10, (c) 5, (d) 8, (e) 6, (f) 1, (g) 4, (h) 9, (i) 7, (j) 2 2. (a) 4, (b) 6, (c) 7, (d) 1, (e) 5, (f) 3, (g) 8, (h) 2, (i) 10, (j) 9 3. a 4. c 5. e 6. d 7. a 8. c 9. mucosa, submucosa, muscularis, serosa 10. peritoneum; omenta, mesentery 11. submucosal plexus or plexus of Meissner; myenteric plexus or plexus of Auerbach 12. simple sugars; amino acids; monoglycerides and fatty acids; pentoses, phosphates, and nitrogenous bases 13. false 14. false 15. (a) 4, (b) 8, (c) 2, (d) 10, (e) 7, (f) 11, (g) 1, (h) 6, (i) 3, (j) 9, (k) 5

Chapter 25
1. supply energy, serve as building blocks, be stored for future use 2. hypothalamus 3. glucose 6-phosphate, pyruvic acid, acetyl coenzyme A 4. hormones 5. false 6. true 7. c 8. d 9. a 10. d 11. a 12. b 13. (a) 8, (b) 5, (c) 1, (d) 10, (e) 6, (f) 2, (g) 4, (h) 9, (i) 3, (j) 7 14. (a) 2, (b) 8, (c) 5, (d) 10, (e) 6, (f) 7, (g) 3, (h) 9, (i) 4, (j) 1 15. (a) 9, (b) 10, (c) 1, (d) 6, (e) 4, (f) 7, (g) 3, (h) 5, (i) 8, (j) 2

Chapter 26
1. (a) 8, (b) 2, (c) 10, (d) 5, (e) 3, (f) 1, (g) 7, (h) 4, (i) 9, (j) 6 2. (a) 3, (b) 7, (c) 4, (d) 2, (e) 5, (f) 1, (g) 6 3. c 4. a 5. d 6. a 7. e 8. b 9. true 10. true 11. transitional epithelial 12. voluntary, involuntary 13. glomerular filtration, tubular reabsorption, tubular secretion 14. regulation of blood ionic composition, maintenance of blood osmolarity, regulation of blood volume, regulation of blood pressure, regulation of blood pH, endocrine secretion, regulation of blood glucose levels, excretion of wastes and foreign substances 15. (a) 5, (b) 4, (c) 8, (d) 6, (e) 1, (f) 2, (g) 7, (h) 3

Chapter 27
1. water 2. depression, CNS, synaptic transmission; overexcitability, CNS, peripheral nerves 3. weak acid, weak base 4. protein buffer, carbonic acid–bicarbonate, phosphate 5. true 6. true 7. a 8. c 9. b 10. e 11. d 12. a 13. (a) 2, (b) 3, (c) 4, (d) 1, (e) 5 14. (a) 3, (b) 5, (c) 7, (d) 1, (e) 6, (f) 2, (g) 4 15. (a) 8, (b) 7, (c) 5, (d) 6, (e) 1, (f) 3, (g) 4, (h) 2

Chapter 28
1. (a) 10, (b) 12, (c) 1, (d) 5, (e) 2, (f) 4, (g) 6, (h) 11, (i) 8, (j) 3, (k) 9, (l) 7 2. (a) 6, (b) 4, (c) 1, (d) 12, (e) 8, (f) 5, (g) 7, (h) 9, (i) 11, (j) 3, (k) 2, (l) 10 3. e 4. b 5. e 6. c 7. b 8. d 9. e 10. (a) 7, (b) 2, (c) 1, (d) 4, (e) 8, (f) 5, (g) 3, (h) 6 11. (a) 6, (b) 4, (c) 2, (d) 1, (e) 3, (f) 8, (g) 7, (h) 5 12. ovaries, testes, secondary oocyte, sperm cell 13. blood–testis barrier, Sertoli cells 14. puberty, menarche, menopause 15. true

Chapter 29
1. (a) 3, (b) 4, (c) 5, (d) 1, (e) 7, (f) 2, (g) 6, (h) 8, (i) 9, (j) 10 2. (a) 5, (b) 2, (c) 8, (d) 4, (e) 7, (f) 1, (g) 3, (h) 6 3. (a) 7, (b) 3, (c) 6, (d) 4, (e) 10, (f) 8, (g) 5, (h) 2, (i) 1, (j) 9, (k) 11, (l) 12 4. a 5. c 6. b 7. e 8. d 9. b 10. dilation, expulsion, placental 11. gastrulation; embryonic membranes; placenta 12. corpus luteum 13. true 14. false 15. (a) 3, (b) 6, (c) 4, (d) 1, (e) 8, (f) 2, (g) 5, (h) 7

ANSWERS TO CRITICAL THINKING QUESTIONS

Chapter 1
1. Homeostasis is the relative constancy of the body's internal environment. Body temperature should vary within a narrow range around normal body temperature (38 degrees C or 98.6 degrees F).

2. Bilateral means two sides or both arms in this case. The carpal region is the wrist area.
3. A x-ray provides a good image of dense tissue such as bones. An MRI is used for imaging soft tissues, not bones. An MRI cannot be used when metal is present due to the magnetic field used.

Chapter 2

1. Fatty acids are in all lipids including plant oils and the phospholipids which compose cell membranes. Sugars (monosaccharides) are needed for energy production, are a component of nucleotides, and are the basic units of disaccharides and polysaccharides including starch, glycogen, and cellulose.
2. DNA = deoxyribonucleic acid. DNA is composed of 4 different nucleotides. The order of the nucleotides is unique in every person. The 20 different amino acids form proteins.
3. One pH unit equals a tenfold change in H^+ concentration. Pure water has a pH of 7, which equals 1×10^{-7} moles of H^+ per liter. Blood has a pH of 7.4, which equals 0.4×10^{-7} moles H^+ per liter. Thus, blood has less than half as many H^+ as water.

Chapter 3

1. The tissues are destroyed due to autolysis of the cells caused by the release of acids and digestive enzymes from the lysosomes.
2. Synthesis of mucin by the ribosomes on rough endoplasmic reticulum, to transport vesicle, to entry face of Golgi, to transfer vesicle, to medial cisterns, where protein is modified, to transfer vesicle, to exit face, to secretory vesicle, to plasma membrane, where it undergoes exocytosis.
3. The sense strand is composed of DNA codons transcribed into mRNA. Only the exons will be found in the mRNA. The introns are clipped out.

Chapter 4

1. Many possible adaptations including: more adipose tissue for insulation, thicker bone for support, more red blood cells for oxygen transport, and so on.
2. The surface layer of the skin is keratinized, stratified squamous epithelium. It is avascular. If the pin were stuck straight in, the connective tissue would be pierced and the finger would bleed.
3. The gelatin portion of the salad represents the ground substance of the connective tissue matrix. The grapes are cells such as fibroblasts. The shredded carrots and coconut are fibers embedded in the ground substance.

Chapter 5

1. The dust particles are mostly keratinocytes shed from the stratum corneum of the skin.
2. It would be neither wise nor feasible to remove the exocrine glands. Essential exocrine glands include the sudoriferous glands (sweat helps control body temperature), sebaceous glands (sebum lubricates the skin), and ceruminous glands (provide protective lubricant for ear canal).
3. The superficial layer of the epidermis of the skin is the stratum corneum. Its cells are full of intermediate filaments of keratin, keratohyalin, and lipids from lamellar granules, making this layer a water-repellent barrier. The epidermis also contains an abundance of desmosomes.

Chapter 6

1. In greenstick fracture, which occurs only in children due to the greater flexibility of their bones, the bone breaks on one side but bends on the other side, resembling what happens when one tries to break a green (not dry) stick. The child probably broke her fall with her extended arm, breaking the radius or ulna.
2. At Aunt Edith's age, the production of several hormones (such as estrogens and human growth hormone) necessary for bone remodeling is likely to be decreased. Her decreased height results, in part, from compression of her vertebrae due to flattening of the discs between them. She may also have osteoporosis and an increased susceptibility to fractures resulting in damage to the vertebrae and loss of height.
3. Exercise causes mechanical stress on bones, but because there is effectively zero gravity in space, the pull of gravity on bones is missing. The lack of stress from gravity results in bone demineralization and weakness.

Chapter 7

1. Fontanels, the soft spots between cranial bones, are fibrous connective tissue membranes that will be replaced by bone as the baby matures. They allow the infant's head to be molded during its passage through the birth canal and allow for brain and skull growth in the infant.
2. Barbara probably broke her coccyx or tailbone. The coccygeal vertebrae usually fuse by age 30. The coccyx points inferiorly in females.
3. Infants are born with a single concave curve in the vertebral column. Adults have four curves in their spinal cord—at the cervical (convex), thoracic (concave), lumbar (convex), and sacral regions. Buying a mattress from this company and sleeping on it would result in potentially severe back problems.

Chapter 8

1. Clawfoot is an abnormal elevation of the medial longitudinal arch of the foot.
2. There are 14 phalanges in each hand: two bones in the thumb and three in each of the other fingers. Farmer White has lost five phalanges on his left hand so he has nine remaining on his left and 14 remaining on his right for a total of 23.
3. Snakes don't have appendages, so they don't need shoulders and hips. Pelvic and pectoral girdles connect the upper and lower limbs (appendages) to the axial skeleton, and thus are considered part of the appendicular skeleton.

Chapter 9

1. Katie's vertebral column, head, thigh, lower leg, and lower arm are all flexed. Her lower arm and shoulder are medially rotated.
2. The elbow is a hinge-type synovial joint. The trochlea of the humerus articulates with the trochlear notch of the ulna. The elbow has monaxial (opening and closing) movement similar to a door.
3. Cartilaginous joints can be composed of hyaline cartilage, such as at the epiphyseal plate, or fibrocartilage, such as the intervertebral disc. Sutures are an example of fibrous joints, which are composed of dense fibrous tissue. Synovial (diarthrotic) joints are held together by an articular capsule, which has an inner synovial membrane layer.

Chapter 10

1. Smooth muscle contains both thick and thin filaments as well as intermediate filaments attached to dense bodies. The smooth muscle fiber contracts like a corkscrew turns; the fiber twists like a helix as it contracts and shortens.
2. Ming's muscles performed isometric contractions. Peing used primarily isotonic concentric contraction to lift, isometric to hold, and then isotonic eccentric contractions to lower the barbell.
3. Mr. Klopfer's students have muscle fatigue in their hands. Fatigue has many causes including an increase in lactic acid and ADP, and a decrease in Ca^{2+}, oxygen, creatine phosphate and glycogen.

Chapter 11

1. Some of the antagonistic pairs for the regions are: upper arm: biceps brachii vs. triceps brachii; upper leg: quadriceps femoris vs. hamstrings; torso: rectus abdominus vs. erector spinae; lower leg: gastrocnemius vs. tibialis anterior.
2. The fulcrum is the knee joint, the resistance is the weight of the upper body and the package, and the effort is the thigh muscles. This is the third–class lever.
3. Lifting eyebrow uses frontalis; whistling uses buccinator and orbicularis oris; closing eyes uses orbicularis oculi; shaking head can use several muscles, including sternocleidomastoid, semispinalis capitis, splenius capitis, and longissimus capitis.

Chapter 12

1. The motor neuron is multipolar in structure with an axon and several dendrites projecting from the cell body. The converging circuit has several neurons which converge to form synapses with one common neuron. The bipolar neuron has one dendrite and an axon projecting from the cell body. In a simple circuit, each presynaptic neuron makes a single synapse with one postsynaptic neuron.

2. Gray matter appears gray in color due to the absence of myelin. It is composed of cell bodies, and unmyelinated axons and dendrites. White matter conatains many myelinated axons. Lipofuscin, a yellowing pigment, collects in the neuron with age.

3. Smelling coffee and hearing alarm are somatic sensory, stretching and yawning are somatic motor, salivating is autonomic (parasympathetic) motor, stomach rumble is enteric motor.

Chapter 13

1. A withdrawal/flexor reflex is ipsilateral and polysynaptic. The route is pain receptor of sensory neuron to spinal cord (integrating center) interneurons to motor neurons to flexor muscles in the leg. Also present is a crossed extensor reflex, which is contralateral and polysynaptic. The route diverges at the spinal cord: Interneurons cross to the other side to motor neurons to extensor muscles in the opposite leg.

2. The spinal cord is anchored in place by the filum terminale and the denticulate ligaments.

3. The needle will pierce the epidermis, the dermis, the hypodermis, then go between the vertebrae through the epidural space, the dura matter, the subdural space, the arachnoid, and into the CSF in the subarachnoid space. CSF is produced in the brain, and the spinal meninges are continuous with cranial meninges.

Chapter 14

1. Movement of the right arm is controlled by the left hemisphere's primary motor area, located in the precentral gyrus. Speech is controlled by Broca's area in the left hemisphere's frontal lobe just superior to the lateral cerebral sulcus.

2. The brain is enclosed by the cranial bones and meninges. The temporal bone houses the middle and inner ear, separating these from the brain.

3. The dentist has injected anesthetic into the inferior alveolar nerve, a branch of the mandibular nerve which numbs the lower teeth and the lower lip. The tongue is numbed by blocking the lingual nerve. The upper teeth and lip are numbed by injecting the superior alveolar nerve, a branch of the maxillary nerve.

Chapter 15

1. Chemoreceptors, a type of exteroceptor, detect odor. Proprioceptors detect body position and are involved in equilibrium. The receptors for smell are rapidly adapting (phasic), whereas proprioceptors are slowly adapting (tonic).

2. The tickle receptors are free-nerve endings in the foot. The impulses travel along the first-order sensory neuron to the posterior gray horn, then along the second-order neuron, crossing to the other side of the cord and up the anterior spinothalamic tract to the thalamus. The third-order neuron extends from the thalamus to the "foot" region of the somatosensory area of the cerebral cortex.

3. Yoshio's perception of feeling in his amputated foot is called phantom limb sensation. Impulses from the remaining proximal section of the sensory neuron are perceived by the brain as still coming from the amputated foot.

Chapter 16

1. Because smell and taste have ties to the cortex and limbic areas, Brenna may be recalling a memory of this or similar food. Pathway: Olfactory receptors to cranial nerve I to olfactory bulbs to olfactory tracts to lateral olfactory area in temporal lobe of cerebral cortex.

2. Fred may have cataracts, a loss of transparency in the lens of the eye. Cataracts are often associated with age, smoking, and exposure to UV light.

3. Auricle to external auditory meatus to tympanic membrane to maleus to incus to stapes to oval window to perilymph of scala vestibuli and tympani to vestibular membrane to endolymph of cochlear duct to basilar membrane to spiral organ.

Chapter 17

1. Digestion and relaxation are controlled by the parasympathetic division of the ANS. The salivary glands, pancreas, and liver will show increased secretion; the stomach and intestines will have increased activity; the gallbladder will have increased contractions; heart contractions will have decreased force and rate.

2. The interoceptor in the colon (stimulated) autonomic sensory neuron, sacral region of spinal cord, parasympathetic preganglionic neuron, terminal ganglion, parasympathetic postganglionic neuron, smooth muscle in colon, rectum, and sphincter.

3. Nicotine from the cigarette smoke binds nicotinic receptors on skeletal muscles, mimicking the effect of acetylcholine and causing increased contractions ("twitches"). Nicotine also binds to nicotinic receptors on cells in the adrenal medulla, stimulating release of epinephrine and NE, which mimics "fight-or-flight" response.

Chapter 18

1. The beta cells are one of the cell types found in the pancreatic islets of the pancreas. In Type 1 diabetes, only the beta cells are destroyed; the rest of the pancreas is not affected. A successful transplantation of the beta cells would enable the recipient to produce the hormone insulin and would cure the diabetes.

2. Amanda has an enlarged thyroid gland or goiter. The goiter is probably due to hypothyroidism, which is causing the weight gain, fatigue, mental dullness, and other symptoms.

3. The two adrenal glands are located superior to the two kidneys. The adrenals are about 4 cm high, 2 cm wide, and 1 cm thick. The outer adrenal cortex is composed of three layers: zona glomerulosa, zona fasiculata, and zona reticularis. The inner adrenal medulla is composed of chromaffin cells.

Chapter 19

1. Blood is composed of liquid plasma and the formed elements: erythrocytes, leukocytes, and platelets. Plasma contains water, proteins, electrolytes, gases, and nutrients. There are about 5 million RBCs, 5000 WBCs, and 250,000 platelets per μL of blood. Leukocytes include neutrophils, eosinophils, basophils, monocytes, and lymphocytes.

2. To determine the blood type, a drop of each of three different antibody solutions (antisera) are added to three separate blood drops. The solutions are anti-A, anti-B, and anti-Rh. In Josef's test, anti-B caused agglutination (clump formation) when added to his blood, indicating the presence of antigen B on the RBCs. Anti-A and anti-Rh did not cause agglutination, indicating the absence of these antigens.

3. Hemostasis occurred. The stages involved are vascular spasm (if an arteriole was cut), platelet plug formation, and coagulation (clotting).

Chapter 20

1. With a normal resting CO of about 5.25 liters/min, and a heart rate of 55 beats per minute, Arian's stroke volume would be 95 mL/beat. His CO during strenuous exercise is 6 times his resting CO, about 31,500 mL/min.

2. Rheumatic fever is caused by inflammation of the bicuspid and the aortic semilunar valves following a streptococcal infection. Antibodies produced by the immune system to destroy the bacteria may also attack and damage the heart valves. Heart problems later in life may be related to this damage.

3. Mr. Perkins is suffering from angina pectoris and has several risk factors for coronary artery disease such as smoking, obesity, and male gender. The cardiac angiography involves the use of a cardiac catheter to inject a radiopaque medium into the heart and its vessels. It may reveal blockages such as atherosclerotic plaques in his coronary arteries.

Chapter 21

1. The foramen ovale and ductus arteriosus close to establish the pulmonary circulation. The umbilical artery and veins close because the placenta is no longer functioning. The ductus venosus closes so that the liver is no longer bypassed.

2. Right hand: Left ventricle to ascending aorta to arch of aorta to brachiocephalic trunk to right subclavian to right axillary to right brachial to right radial and ulnar arteries to right superficial palmar arch. Left hand: Left ventricle to ascending aorta to arch of aorta to left subclavian to left axillary to left brachial to left radial and ulnar arteries to left superficial palmar arch.

3. A vascular (venous) sinus is a vein that lacks smooth muscle in its tunica medica. Dense connective tissue replaces the tunica media and externa.

Chapter 22

1. Tariq had a hypersensitivity reaction to an insect sting; he was probably allergic to the venom. He had a localized anaphylactic or type I reaction.

2. The site of the embedded splinter had probably become infected with bacteria. The red streaks are the lymphatic vessels that drain the infected area; the swollen tender bumps are the axillary lymph nodes, which are swollen due to the immune response to the infection.

3. Influenza vaccination introduces a weakened or killed virus (which will not cause the disease) to the body. The immune system recognizes the antigen and mounts a primary immune response. Upon exposure to the same flu virus that was in the vaccine, the body will produce a secondary response, which will usually prevent a case of the flu. This is artificially acquired active immunity.

Chapter 23

1. Normal average male volumes are 500 mL for tidal volume during quiet breathing, 3100 mL for inspiratory reserve volume, and 1200 mL for expiratory reserve volume. Average female volumes would be expected to be less than average male volumes, because the average female is generally smaller than the average male.

2. The bones of the external nose are the frontal bone, the two nasal bones (the location of the fracture), and the maxillae. The external nose is also composed of the septal, lateral nasal, and alar cartilages, as well as skin, muscle, and mucous membrane.

3. The cerebral cortex can temporarily allow voluntary breath holding. P_{CO_2} and H^+ levels in the blood and CSF increase with breath holding, strongly stimulating the inspiratory area, which stimulates inspiration to begin again. Breathing will begin again even if the person loses consciousness.

Chapter 24

1. Katie lost the two upper central permanent incisors. The remaining deciduous teeth include the lower central incisors, an upper and lower pair of lateral incisors, an upper and lower pair of cuspids, an upper and lower pair of first molars, and upper and lower pair of second molars.

2. CCK promotes ejection of bile, secretion of pancreatic juice, and contraction of the pyloric sphincter. It also promotes pancreatic growth and enhances the effects of secretin. CCK acts on the hypothalamus to induce satiety (fullness) and should therefore decrease the appetite.

3. The smaller left lobe of the liver is separated from the larger right lobe by the falciform ligament. The left lobe is inferior to the diaphragm in the epigastric region of the abdominopelvic cavity.

Chapter 25

1. Deb is eating a diet high in carbohydrates in order to store maximum amounts of glycogen in her skeletal muscles and liver. This practice is called carbohydrate loading. Gycogenolysis of stored glycogen supplies the muscles with the glucose needed for cellular respiration.

2. Mr. Hernandez was suffering from heat exhaustion caused by loss of fluids and electrolytes. Lack of NaCl causes muscles cramps, nausea and vomiting, dizziness, and fainting. Low blood pressure may also result.

3. A stable body weight may be maintained despite day-to-day fluctuations in food intake. Food intake is controlled by many factors, including the feeding and satiety centers in the hypothalamus, blood glucose level, amount of adipose tissue, CCK, distention of the GI tract, and body temperature.

Chapter 26

1. Without reabsorption, initially 105–125 mL of filtrate would be lost per minute, assuming normal GFR. Fluid loss from the blood would cause a decrease in blood pressure, and therefore a decrease in GBHP. When GBHP dropped below 45mm Hg, filtration would stop (assuming normal CHP and BCOP) because NFP would be zero.

2. The urinary bladder can stretch considerably due to the presence of transitional epithelium, rugae, and three layers of smooth muscle in the detrusor muscle.

3. In females the urethra is about 4 cm long; in males the urethra is about 15–20 cm long, including its passage through the penis, urogenital diaphragm, and prostate gland.

Chapter 27

1. Gary has water intoxication. The Na^+ concentration of his plasma and interstitial fluid is below normal. Water moved by osmosis into the cells, resulting in hypotonic intracellular fluid and water intoxication. Decreased plasma volume, due to water movement into the interstitial fluid, caused hypovolemic shock.

2. The average female body contains about 55% water, versus 60% water in the average male. Due to the influence of male and female hormones, the average female has relatively more subcutaneous fat (which is very low in water) and relatively less muscle and other tissues (which are high in water) than the average male.

3. Excessive vomiting causes a loss of hydrochloric acid in gastric juice, and intake of antacids increases the amount of alkali in the body fluids, resulting in metabolic alkalosis. Vomiting also causes a loss of fluid and may cause dehydration.

4. (Step 1) pH = 7.30 indicates slight acidosis, which could be caused by elevated P_{CO_2} or lowered HCO_3^-. (Step 2) The HCO_3^- is lower than normal (20 mEq/liter), so (step 3) the cause is metabolic. (Step 4) the P_{CO_2} is lower than normal (32 mm Hg), so hyperventilation is providing some compensation. Diagnosis: Henry has partially compensated metabolic acidosis. A possible cause is kidney damage that resulted from interruption of blood flow during the heart attack.

Chapter 28

1. LH (luteinizing hormone) is a hormone of the male and female reproductive systems. In males, LH stimulates the Leydig cells of the testes to secrete testosterone, the principal sex hormone in males.

2. The germinal epithelium, located on the surface of the ovaries, is a misnomer. The oocytes are located in the ovarian cortex, deep to the tunica albuginea. The tunica vaginalis covers the testes, superficial to the tunica albuginea. It forms from the peritoneum during the descent of the testes from the abdomen into the scrotum during fetal development.

3. No. The first polar body, formed by meoisis I, would contain half the homologous chromosome pairs, the other half being in the secondary oocyte. The second polar body, formed by meoisis II, would contain chromosomes identical to the ovum; however, fertilization would be by two different sperm, so the babies would still not be identical.

Chapter 29

1. Blond hair and blue eyes are both autosomal recessive traits. Angelica's genotype is homozygous recessive for both traits. Her parents are heterozygous for both traits. They exhibit the dominant phenotype for brown hair and brown eyes and are carriers of the recessive traits.

2. Fragile X syndrome is due to a defective gene on the tip of the X chromosome. Up to 50% of male carriers of the trait are unaffected. If a carrier father passes the defective X to a daughter, she will be unaffected but the X will undergo maternal imprinting. Both the male and female offspring of the daughter may be affected.

3. All arteries carry blood away from the heart. The umbilical arteries, located in the fetus, carry deoxygenated blood away from the fetal heart towards the placenta where the blood will be oxygenated as it passes through the placenta.

GLOSSARY

Pronunciation Key

1. The most strongly accented syllable appears in capital letters, for example, bilateral (bī-LAT-er-al) and diagnosis (dī-ag-NŌ-sis).

2. If there is a secondary accent, it is noted by a prime ('), for example, constitution (kon'-sti-TOO-shun) and physiology (fiz'-ē-OL-ō-jē). Any additional secondary accents are also noted by a prime, for example, decarboxylation (dē'-kar-bok'-si-LĀ-shun).

3. Vowels marked by a line above the letter are pronounced with the long sound, as in the following common words:

    ā as in *māke*
    ē as in *bē*
    ī as in *īvy*
    ō as in *pōle*

4. Vowels not so marked are pronounced with the short sound, as in the following words:

    a as in *above*
    e as in *bet*
    i as in *sip*
    o as in *not*
    u as in *bud*

5. Other phonetic symbols are used to indicate the following sounds:

    oo as in *sue*
    yoo as in *cute*
    oy as in *oil*

Abdomen (ab-DŌ-men or AB-dō-men) The area between the diaphragm and pelvis.

Abdominal (ab-DŌM-i-nal) **cavity** Superior portion of the abdominopelvic cavity that contains the stomach, spleen, liver, gallbladder, pancreas, small intestine, and most of the large intestine.

Abdominopelvic (ab-dom'-i-nō-PEL-vic) **cavity** Inferior component of the ventral body cavity that is subdivided into a superior abdominal cavity and an inferior pelvic cavity.

Abduction (ab-DUK-shun) Movement away from the axis or midline of the body.

Abortion (a-BOR-shun) The premature loss (spontaneous) or removal (induced) of the embryo or nonviable fetus; any failure in the normal process of developing or maturing.

Abscess (AB-ses) A localized collection of pus and liquefied tissue in a cavity.

Absorption (ab-SORP-shun) Intake of fluids or other substances by cells of the skin or mucous membranes; the passage of digested foods from the gastrointestinal tract into blood or lymph.

Absorptive (fed) state Metabolic state during which ingested nutrients are being absorbed into the blood or lymph from the gastrointestinal tract.

Accessory duct A duct of the pancreas that empties into the duodenum about 2.5 cm (1 in.) superior to the ampulla of Vater (hepatopancreatic ampulla). Also called the **duct of Santorini** (san'-tō-RĒ-nē).

Accommodation (a-kom-ō-DĀ-shun) A change in the curvature of the eye lens to adjust for vision at various distances.

Acetabulum (as'-e-TAB-yoo-lum) The rounded cavity on the external surface of the hip bone that receives the head of the femur.

Acetylcholine (as′-ē-til-KŌ-lēn) **(ACh)** A neurotransmitter liberated by many peripheral nervous system neurons and some central nervous system neurons. It is excitatory at neuromuscular junctions but inhibitory at some other synapses (e.g. it slows heart rate).

Acid (AS-id) A proton donor, or a substance that dissociates into hydrogen ions (H^+) and anions; characterized by an excess of hydrogen ions and a pH less than 7.

Acidosis (as-i-DŌ-sis) A condition in which blood pH is below 7.35. Also known as **acidemia.**

Acini (AS-i-nē) Masses of cells in the pancreas that secrete digestive enzymes.

Acoustic (a-KOOS-tik) Pertaining to sound or the sense of hearing.

Acrosome (AK-rō-sōm) A dense lysosomelike body in the head of a sperm cell that contains enzymes that facilitate the penetration of a sperm cell into a secondary oocyte.

Actin (AK-tin) The contractile protein that is part of thin filaments in muscle fibers.

Action potential An electrical signal that propagates along the membrane of a neuron or muscle fiber (cell); a rapid change in membrane potential that involves a depolarization followed by a repolarization. Also called a **nerve action potential** or **nerve impulse** as it relates to a neuron, and a **muscle action potential** as it relates to a muscle fiber (cell).

Activation (ak′-ti-VĀ-shun) **energy** The minimum amount of energy required for a chemical reaction to occur.

Active transport The movement of substances across cell membranes against a concentration gradient, requiring the expenditure of cellular energy (ATP).

Acute (a-KYOOT) Having rapid onset, severe symptoms, and a short course; not chronic.

Adaptation (ad′-ap-TĀ-shun) The adjustment of the pupil of the eye to light variations. The property by which a neuron relays a decreased frequency of action potentials from a receptor, even though the strength of the stimulus remains constant; the decrease in perception of a sensation over time while the stimulus is still present.

Adduction (ad-DUK-shun) Movement toward the axis or midline of the body.

Adenoids (AD-e-noyds) The pharyngeal tonsils.

Adenosine triphosphate (a-DEN-ō-sēn trī-FOS-fāt) **(ATP)** The energy-carrying molecule manufactured in all living cells as a means of capturing and storing energy. ATP consists of the purine base *adenine* and the five-carbon sugar *ribose,* to which are added, in linear array, three *phosphate* groups.

Adenylate cyclase (a-DEN-i-lāt SĪ-klās) An enzyme in the postsynaptic membrane that is activated when certain neurotransmitters (or hormones) bind to their receptors; the enzyme that converts ATP into cyclic AMP.

Adhesion (ad-HĒ-zhun) Abnormal joining of parts to each other.

Adipocyte (AD-i-pō-sīt) Fat cell, derived from a fibroblast.

Adipose (AD-i-pōz) **tissue** Tissue composed of adipocytes specialized for triglyceride storage and present in the form of soft pads between various organs for support, protection, and insulation.

Adrenal cortex (a-DRĒ-nal KOR-teks) The outer portion of an adrenal gland, divided into three zones: the zona glomerulosa, the zona fasciculata, and the zona reticularis.

Adrenal glands Two glands located superior to each kidney. Also called the **suprarenal** (soo′-pra-RĒ-nal) **glands.**

Adrenal medulla (me-DUL-a) The inner portion of an adrenal gland, consisting of cells that secrete epinephrine and norepi-

nephrine (NE) in response to the stimulation of preganglionic sympathetic neurons.

Adrenergic (ad′-ren-ER-jik) **fiber** A nerve fiber that when stimulated releases epinephrine (adrenaline) or norepinephrine (noradrenaline) at a synapse.

Adrenocorticotropic (ad-rē′-nō-kor-ti-kō-TRŌP-ik) **hormone (ACTH)** A hormone produced by the anterior pituitary gland that influences the production and secretion of certain hormones of the adrenal cortex.

Adventitia (ad-ven-TISH-ya) The outermost covering of a structure or organ.

Aerobic (air-Ō-bik) Requiring molecular oxygen.

Afferent arteriole (AF-er-ent ar-TĒ-rē-ōl) A blood vessel of a kidney that divides into the capillary network called a glomerulus; there is one afferent arteriole for each glomerulus.

Agglutination (a-gloo′-ti-NĀ-shun) Clumping of microorganisms or blood cells, typically due to an antigen–antibody reaction.

Aggregated lymphatic follicles Clusters of lymph nodules that are most numerous in the ileum. Also called **Peyer's** (PĪ-erz) **patches.**

Albumin (al-BYOO-min) The most abundant (60%) and smallest of the plasma proteins, which is the main contributor to blood colloid osmotic pressure (BCOP).

Aldosterone (al-do-STĒR-ōn) A mineralocorticoid produced by the adrenal cortex that brings about sodium and water reabsorption and potassium excretion.

Alkaline (AL-ka-līn) Containing more hydroxide ions (OH^-) than hydrogen ions (H^+) to produce a pH higher than 7.

Alkalosis (al-ka-LŌ-sis) A condition in which blood pH is higher than 7.45. Also known as **alkalemia.**

Allantois (a-LAN-tō-is) A small, vascularized membrane between the chorion and amnion of the fetus that serves as an early site for blood formation.

Alleles (a-LĒLZ) Alternate forms of a single gene that control the same inherited trait (such as height or eye color) and are located at the same position on homologous chromosomes.

Allergen (AL-er-jen) An antigen that evokes a hypersensitivity reaction.

Alpha (AL-fa) **cell** A cell in the pancreatic islets (islets of Langerhans) in the pancreas that secretes glucagon. Also termed an **A cell.**

Alpha receptor Receptor found on visceral effectors innervated by most sympathetic postganglionic axons.

Alveolar duct Branch of a respiratory bronchiole around which alveoli and alveolar sacs are arranged.

Alveolar macrophage (MAK-rō-fāj) Highly phagocytic cell found in the alveolar walls of the lungs. Also called a **dust cell.**

Alveolar (al-VĒ-ō-lar) **pressure** Air pressure within the lungs. Also called **intrapulmonic pressure.**

Alveolar sac A collection or cluster of alveoli that share a common opening.

Alveolus (al-VĒ-ō-lus) A small hollow or cavity; an air sac in the lungs; milk-secreting portion of a mammary gland. *Plural is* **alveoli** (al-VĒ-ol-ī).

Amenorrhea (ā-men-ō-RĒ-a) Absence of menstruation.

Amino (ah-MĒ-nō) **acid** An organic acid, containing an acidic carboxyl group (COOH) and a basic amino group (NH_2), that is the building unit from which proteins are formed.

Amnion (AM-nē-on) The deepest fetal membrane; a thin transparent sac that holds the fetus suspended in amniotic fluid. Also called the **"bag of waters."**

Amniotic (am'-nē-OT-ik) **fluid** Fluid in the amniotic cavity, the space between the developing embryo (or fetus) and amnion; the fluid is initially produced as a filtrate from maternal blood and later includes fetal urine.

Amphiarthrosis (am'-fē-ar-THRŌ-sis) A slightly movable articulation, in which the articulating bony surfaces are separated by fibrous connective tissue or fibrocartilage to which both are attached; types are syndesmosis and symphysis.

Ampulla (am-POOL-la) A saclike dilation of a canal.

Anabolism (a-NAB-ō-lizm) Synthetic energy-requiring reactions whereby small molecules are built up into larger ones.

Anaerobic (an-air-Ō-bik) Not requiring molecular oxygen.

Anal (Ā-nal) **canal** The terminal 2 or 3 cm (1 in.) of the rectum; opens to the exterior through the anus.

Anal column A longitudinal fold in the mucous membrane of the anal canal that contains a network of arteries and veins.

Anal triangle The subdivision of the female or male perineum that contains the anus.

Analgesia (an-al-JĒ-zē-a) Pain relief.

Anaphase (AN-a-fāz) The third stage of mitosis in which the chromatids that have separated at the centromeres move to opposite poles of the cell.

Anastomosis (a-nas-tō-MŌ-sis) An end-to-end union or joining together of blood vessels, lymphatic vessels, or nerves.

Anatomical (an'-a-TOM-i-kal) **position** A position of the body universally used in anatomical descriptions in which the body is erect, facing the observer, the upper limbs are at the sides, the palms are facing forward, and the feet are flat on the floor.

Anatomic dead space Spaces of the nose, pharynx, larynx, trachea, bronchi, and bronchioles that contain 150 mL of tidal volume; the air does not reach the alveoli to participate in gas exchange.

Anatomy (a-NAT-ō-mē) The structure or study of structure of the body and the relation of its parts to each other.

Androgen (AN-drō-jen) Substance producing or stimulating masculine characteristics, such as the male hormone testosterone.

Anemia (a-NĒ-mē-a) Condition of the blood in which the number of functional red blood cells or their hemoglobin content is below normal.

Aneuploid (an'-yoo-PLOID) A cell that has one or more chromosomes of a set added or deleted.

Anesthesia (an'-es-THĒ-zē-a) A total or partial loss of feeling or sensation, usually defined with respect to loss of pain sensation; may be general or local.

Angiotensin (an-jē-ō-TEN-sin) Either of two forms of a protein associated with regulation of blood pressure. Angiotensin I is produced by the action of renin on angiotensinogen and is converted by the action of ACE (angiotensin converting enzyme) into angiotensin II, which stimulates aldosterone secretion by the adrenal cortex, stimulates the sensation of thirst, and causes vasoconstriction with resulting increase in systemic vascular resistance.

Anion (AN-ī-on) A negatively charged ion. An example is the chloride ion (Cl^-).

Ankylosis (ang'-ki-LŌ-sus) Severe or complete loss of movement at a joint.

Anomaly (a-NOM-a-lē) An abnormality that may be a developmental (congenital) defect; a variant from the usual standard.

Anoxia (an-OK-sē-a) Deficiency of oxygen.

Antagonist (an-TAG-ō-nist) A muscle that has an action opposite that of the prime mover (agonist) and yields to the movement of the prime mover.

Antagonistic (an-tag-ō-NIST-ik) **effect** A hormonal interaction in which the effect of one hormone on a target cell is opposed by another hormone. For example, calcitonin (CT) lowers blood calcium level, whereas parathormone (PTH) raises it.

Anterior (an-TĒR-ē-or) Nearer to or at the front of the body. Equivalent to **ventral** in bipeds.

Anterior pituitary gland Anterior lobe of the pituitary gland. Also called the **adenohypophysis** (ad'-e-nō-hī-POF-i-sis).

Anterior root The structure composed of axons of motor (efferent) fibers that emerges from the anterior aspect of the spinal cord and extends laterally to join a posterior root, forming a spinal nerve. Also called a **ventral root.**

Antibody (AN-ti-bod'-ē) A protein produced by certain cells in response to a specific antigen; the antibody combines with that antigen to neutralize, inhibit, or destroy it. Also called an **immunoglobulin** (im-yoo-nō-GLOB-yoo-lin) or **Ig.**

Antibody-mediated immunity That component of immunity in which lymphocytes (B cells) develop into plasma cells that produce antibodies that destroy antigens. Also called **humoral** (YOO-mor-al) **immunity.**

Anticoagulant (an-tīcō-AG-yoo-lant) A substance that is able to delay, suppress, or prevent the clotting of blood.

Antidiuretic hormone (ADH) Hormone produced by neurosecretory cells in the paraventricular and supraoptic nuclei of the hypothalamus that stimulates water reabsorption from kidney cells into the blood and vasoconstriction of arterioles. Also called **vasopressin** (vāz-ō-PRES-in).

Antigen (AN-ti-jen) A substance that has immunogenicity—the ability to provoke an immune response—and reactivity—the ability to react with the antibodies or cells that result from the immune response. Also termed a **complete antigen.**

Antigen-presenting cell (APC) Special class of migratory cells that process and present antigens to T cells during an immune response; APCs include macrophages, B cells, and dendritic cells, which are present in the skin and in mucous membranes.

Antiport Process by which two substances, often Na^+ and another substance, move in opposite directions across a plasma membrane. Also called **countertransport.**

Anulus fibrosus (AN-yoo-lus fī-BRŌ-sus) A ring of fibrous tissue and fibrocartilage that encircles the pulpy substance (nucleus pulposus) of an intervertebral disc.

Anuria (a-NOO-rē-a) A daily urine output of less that 50 mL.

Anus (Ā-nus) The distal end and outlet of the rectum.

Aorta (ā-OR-ta) The main systemic trunk of the arterial system of the body that emerges from the left ventricle.

Aortic (ā-OR-tik) **body** Cluster of receptors on or near the arch of the aorta that respond to alterations in blood levels of oxygen, carbon dioxide, and hydrogen ions (H^+).

Aortic reflex A reflex concerned with maintaining normal systemic blood pressure.

Aperture (AP-er-chur) An opening or orifice.

Apex (Ā-peks) The pointed end of a conical structure, such as the apex of the heart.

Apnea (AP-nē-a) Temporary cessation of breathing.

Apneustic (ap-NOO-stik) **area** Portion of the respiratory center in the pons that sends stimulatory nerve impulses to the inspiratory area that activate and prolong inspiration and inhibit expiration.

Apocrine (AP-ō-krin) **gland** A type of gland in which the secretory products gather at the free end of the secreting cell and are pinched off, along with some of the cytoplasm, to become the secretion, as in mammary glands.

Aponeurosis (ap′-ō-noo-RŌ-sis) A sheetlike tendon joining one muscle with another or with bone.

Apoptosis (ap-ō-TŌ-sis) A normal type of cell death that removes unneeded cells during embryological development, regulates the number of cells in tissues, and eliminates many potentially dangerous cells such as cancer cells. During apoptosis, the DNA fragments, the nucleus condenses, mitochondria cease to function, and the cytoplasm shrinks, but the plasma membrane remains intact. Phagocytes engulf and digest the apoptotic cells, and an inflammatory response does not occur.

Appendage (a-PEN-dij) A structure attached to the body.

Appositional (ap′-ō-ZISH-o-nal) **growth** Growth due to surface deposition of material, as in the growth in diameter of cartilage and bone. Also called **exogenous** (eks-OJ-e-nus) **growth.**

Aqueous humor (AK-wē-us HYOO-mor) The watery fluid, similar in composition to cerebrospinal fluid, that fills the anterior cavity of the eye.

Arachnoid (a-RAK-noyd) The middle of the three coverings (meninges) of the brain and spinal cord.

Arachnoid villus (VIL-us) Berrylike tuft of arachnoid that protrudes into the superior sagittal sinus and through which cerebrospinal fluid is reabsorbed into the bloodstream.

Arbor vitae (AR-bor VĪ-tē) The white matter tracts of the cerebellum, which have a treelike appearance when seen in midsagittal section.

Arch of the aorta The most superior portion of the aorta, lying between the ascending and descending segments of the aorta.

Areola (a-RĒ-ō-la) Any tiny space in a tissue. The pigmented ring around the nipple of the breast.

Arm The portion of the upper limb from the shoulder to the elbow.

Arousal (a-ROW-zal) Awakening from sleep, a response due to stimulation of the reticular activating system (RAS).

Arrector pili (a-REK-tor PI-lē) Smooth muscles attached to hairs; contraction pulls the hairs into a more vertical position, resulting in "goose bumps."

Arrhythmia (a-RITH-mē-a) Irregular heart rhythm. Also called a **dysrhythmia**.

Arteriole (ar-TĒ-rē-ōl) A small, almost microscopic, artery that delivers blood to a capillary.

Artery (AR-ter-ē) A blood vessel that carries blood away from the heart.

Arthrology (ar-THROL-ō-jē) The study or description of joints.

Arthrosis (ar-THRŌ-sis) A joint or articulation.

Articular (ar-TIK-yoo-lar) **capsule** Sleevelike structure around a synovial joint composed of a fibrous capsule and a synovial membrane.

Articular cartilage (KAR-ti-lij) Hyaline cartilage attached to articular bone surfaces.

Articular disc Fibrocartilage pad between articular surfaces of bones of some synovial joints. Also called a **meniscus** (men-IS-cus).

Articulation (ar-tik′-yoo-LĀ-shun) A joint; a point of contact between bones, cartilage and bones, or teeth and bones.

Arytenoid (ar′-i-TĒ-noyd) **cartilages** A pair of small, pyramidal cartilages of the larynx that attach to the vocal folds and intrinsic pharyngeal muscles and can move the vocal folds.

Ascending colon (KŌ-lon) The portion of the large intestine that passes superiorly from the cecum to the inferior edge of the liver, where it bends at the right colic (hepatic) flexure to become the transverse colon.

Ascites (as-SĪ-tē z) Abnormal accumulation of serous fluid in the peroneal cavity.

Association area A portion of the cerebral cortex connected by many motor and sensory fibers to other parts of the cortex. The association areas are concerned with motor patterns, memory, concepts of word-hearing and word-seeing, reasoning, will, judgment, and personality traits.

Astigmatism (a-STIG-ma-tizm) An irregularity of the lens or cornea of the eye causing the image to be out of focus and producing faulty vision.

Astrocyte (AS-trō-sīt) A neuroglial cell having a star shape that participates in brain development and the metabolism of neurotransmitters, helps form the blood–brain barrier and maintain the proper balance of K^+ for generation of nerve impulses, and provides a link between neurons and blood vessels.

Atherosclerotic (ath′-er-ō-skle-RO-tic) **plaque** (PLAK) A lesion that results from accumulated cholesterol and smooth muscle fibers (cells) of the tunica media of an artery; may become obstructive.

Atom Unit of matter that makes up a chemical element; consists of a nucleus and electrons.

Atomic mass (weight) Average mass of all stable atoms of an element, reflecting the relative proportion of atoms with different mass numbers.

Atomic number Number of protons in an atom.

Atresia (a-TRĒ-zē-a) Degeneration and reabsorption of an ovarian follicle before it fully matures and ruptures; abnormal closure of a passage, or absence of a normal body opening.

Atrial fibrillation (Ā-trē-al fib-ri-LĀ-shun) Asynchronous contraction of cardiac muscle fibers in the atria that results in the cessation of atrial pumping.

Atrial natriuretic (na′-trē-yoo-RET-ik) **peptide (ANP)** Peptide hormone, produced by the atria of the heart in response to stretching, that inhibits aldosterone production and thus lowers blood pressure.

Atrioventricular (AV) (ā′-trē-ō-ven-TRIK-yoo-lar) **bundle** The portion of the conduction system of the heart that begins at the atrioventricular (AV) node, passes through the cardiac skeleton separating the atria and the ventricles, then extends a short distance down the interventricular septum before splitting into right and left bundle branches. Also called the **bundle of His** (HISS).

Atrioventricular (AV) node The portion of the conduction system of the heart made up of a compact mass of conducting cells located in the septum between the two atria.

Atrioventricular (AV) valve A heart valve made up of membranous flaps or cusps that allows blood to flow in one direction only, from an atrium into a ventricle.

Atrophy (AT-rō-f ē) Wasting away or decrease in size of a part, due to a failure, abnormality of nutrition, or lack of use.

Atrium (Ā-trē-um) A superior chamber of the heart.

Auditory ossicle (AW-di-tō-rē OS-si-kul) One of the three small bones of the middle ear called the **malleus, incus,** and **stapes.**

Auditory tube The tube that connects the middle ear with the nose and nasopharynx region of the throat. Also called the **Eustachian** (yoo-STĀ-kē-an) **tube.**

Auscultation (aws-kul-TĀ-shun) Examination by listening to sounds in the body.

Autocrine (AW-tō-krin) Local hormone, such as interleukin-2, that acts on the same cell that secreted it.

Autoimmunity An immunologic response against a person's own tissues.

Autolysis (aw-TOL-i-sis) Self-destruction of cells by their own lysosomal digestive enzymes after death or in a pathological process.

Autonomic ganglion (aw′-tō-NOM-ik GANG-lē-on) A cluster of sympathetic or parasympathetic cell bodies located outside the central nervous system.

Autonomic nervous system (ANS) Visceral sensory (afferent) and motor (efferent) neurons, both sympathetic and parasympathetic. Motor neurons transmit nerve impulses from the central nervous system to smooth muscle, cardiac muscle, and glands; so named because this portion of the nervous system was thought to be self-governing or spontaneous.

Autonomic plexus (PLEK-sus) An extensive network of sympathetic and parasympathetic fibers, the cardiac, celiac, and pelvic plexuses are located in the thorax, abdomen, and pelvis, respectively.

Autophagy (aw-TOF-a-jē) Process by which worn-out organelles are digested within lysosomes.

Autoregulation (aw-tō-reg-yoo-LĀ-shun) A local, automatic adjustment of blood flow in a given region of the body in response to tissue needs.

Autorhythmic cells Cardiac or smooth muscle fibers that are self-excitable (generate impulses without an external stimulus); act as the heart's pacemaker and conduct the pacing impulse through the conduction system of the heart; self-excitable neurons in the central nervous system, as in the inspiratory area of the brain stem.

Autosome (AW-tō-sōm) Any chromosome other than the pair of sex chromosomes.

Axilla (ak-SIL-a) The small hollow beneath the arm where it joins the body at the shoulders. Also called the **armpit.**

Axon (AK-son) The usually single, long process of a nerve cell that propagates a nerve impulse toward the axon terminals.

Axon terminal Terminal branch of an axon where synaptic vesicles undergo exocytosis to release neurotransmitter molecules.

Azygos (AZ-ī-gos) An anatomical structure that is not paired; occurring singly.

Back The posterior part of the body; the dorsum.

Ball-and-socket joint A synovial joint in which the rounded surface of one bone moves within a cup-shaped depression or fossa of another bone, as in the shoulder or hip joint. Also called a **spheroid** (SFĒ-roid) **joint.**

Baroreceptor (bar′-ō-re-SEP-tor) Nerve cell capable of responding to changes in blood, air, or fluid pressure. Also called a **pressoreceptor.**

Basal ganglia (GANG-glē-a) Paired clusters of cell bodies that make up the central gray matter in each cerebral hemisphere, including the caudate nucleus, lentiform nucleus, claustrum, and amygdala. Also called **cerebral nuclei** (SER-e-bral NOO-klē-ī).

Basal metabolic (BĀ-sal met′-a-BOL-ik) **rate (BMR)** The rate of metabolism measured under standard or basal conditions (awake, at rest, fasting).

Base A nonacid or a proton acceptor, characterized by excess of hydroxide ions (OH^-) and a pH greater than 7. A ring-shaped, nitrogen-containing organic molecule that is one of the components of a nucleotide, namely, adenine, guanine, cytosine, thymine, and uracil; also known as nitrogenous base.

Basement membrane Thin, extracellular layer between epithelium and connective tissue consisting of a basal lamina and a reticular lamina.

Basilar (BAS-i-lar) **membrane** A membrane in the cochlea of the inner ear that separates the cochlear duct from the scala tympani and on which the spiral organ (organ of Corti) rests.

Basophil (BĀ-sō-fil) A type of white blood cell characterized by a pale nucleus and large granules that stain blue-purple with basic dyes.

B cell A lymphocyte that can develop into an antibody-producing plasma cell or a memory cell.

Belly The abdomen. The gaster or prominent, fleshy part of a skeletal muscle.

Beta (BĀ-ta) **cell** A cell in the pancreatic islets (islets of Langerhans) in the pancreas that secretes insulin. Also termed a **B cell.**

Beta receptor Receptor for epinephrine and norepinephrine found on most visceral effectors innervated by sympathetic postganglionic axons.

Bicuspid (bī-KUS-pid) **valve** Atrioventricular (AV) valve on the left side of the heart. Also called the **mitral valve.**

Bilateral (bī-LAT-er-al) Pertaining to two sides of the body.

Bile (bīl) A secretion of the liver consisting of water, bile salts, bile pigments, cholesterol, lecithin, and several ions; it emulsifies lipids prior to their digestion.

Bilirubin (bil-ē-ROO-bin) An orange pigment that is one of the end products of hemoglobin breakdown in the hepatocytes and is excreted as a waste material in the bile.

Blastocele (BLAS-tō-sēl) The fluid-filled cavity within the blastocyst.

Blastocyst (BLAS-tō-sist) In the development of an embryo, a hollow ball of cells that consists of a blastocele (the internal cavity), trophoblast (outer cells), and inner cell mass.

Blastomere (BLAS-tō-mēr) One of the cells resulting from the cleavage of a fertilized ovum.

Blind spot Area in the retina at the end of the optic (II) nerve in which there are no photoreceptors.

Blood The fluid that circulates through the heart, arteries, capillaries, and veins and that constitutes the chief means of transport within the body.

Blood–brain barrier (BBB) A barrier consisting of specialized brain capillaries and astrocytes that prevents the passage of materials from the blood to the cerebrospinal fluid and brain.

Blood island Isolated mass and cord of mesenchyme in the mesoderm from which blood vessels develop.

Blood pressure (BP) Force exerted by blood against the walls of blood vessels due to contraction of the heart and influenced by the elasticity of the vessel walls; clinically, a measure of the pressure in arteries during ventricular systole and ventricular diastole. *See also* **mean arterial blood pressure.**

Blood reservoir (REZ-er-vwar) Systemic veins that contain large amounts of blood that can be moved quickly to parts of the body requiring the blood.

Blood–testis barrier (BTB) A barrier formed by Sertoli cells that prevents an immune response against antigens produced by spermatogenic cells by isolating the cells from the blood.

Body cavity A space within the body that contains various internal organs.

Body fluid Body water and its dissolved substances; constitutes about 60% of total body weight.

Bohr (BŌR) **effect** In an acidic environment, oxygen unloads more readily from hemoglobin because when hydrogen ions (H^+) bind to hemoglobin, they alter the structure of hemoglobin, thereby reducing its oxygen-carrying capacity.

Bolus (BŌ-lus) A soft, rounded mass, usually food, that is swallowed.

Bony labyrinth (LAB-i-rinth) A series of cavities within the petrous portion of the temporal bone forming the vestibule, cochlea, and semicircular canals of the inner ear.

Brachial plexus (BRĀ-kē-al PLEK-sus) A network of nerve fibers of the ventral rami of spinal nerves C5, C6, C7, C8, and T1. The nerves that emerge from the brachial plexus supply the upper limb.

Bradycardia (brād′-i-KAR-de-a) A slow resting heart or pulse rate (under 60/min).

Brain A mass of nervous tissue located in the cranial cavity.

Brain stem The portion of the brain immediately superior to the spinal cord, made up of the medulla oblongata, pons, and midbrain.

Brain waves Electrical activity produced as a result of action potentials of brain neurons.

Broad ligament A double fold of parietal peritoneum attaching the uterus to the side of the pelvic cavity.

Broca's (BRŌ-kaz) **area** Motor area of the brain in the frontal lobe that translates thoughts into speech. Also called the **motor speech area.**

Bronchi (BRONG-kē) Branches of the respiratory passageway including primary bronchi (the two divisions of the trachea), secondary or lobar bronchi (divisions of the primary that are distributed to the lobes of the lung), and tertiary or segmental bronchi (divisions of the secondary that are distributed to bronchopulmonary segments of the lung). *Singular is* **bronchus.**

Bronchial tree The trachea, bronchi, and their branching structures up to and including the terminal bronchioles.

Bronchiole (BRONG-kē-ōl) Branch of a tertiary bronchus further dividing into terminal bronchioles (distributed to lobules of the lung), which divide into respiratory bronchioles (distributed to alveolar sacs).

Bronchopulmonary (brong′-kō-PUL-mō-ner-ē) **segment** One of the smaller divisions of a lobe of a lung supplied by its own branches of a bronchus.

Buccal (BUK-al) Pertaining to the cheek or mouth.

Buffer (BUF-er) **system** A pair of chemicals—one a weak acid and the other the salt of the weak acid, which functions as a weak base—that resists changes in pH.

Bulb of penis Expanded portion of the base of the corpus spongiosum penis.

Bulbourethral (bul′-bō-yoo-RĒ-thral) **gland** One of a pair of glands located inferior to the prostate gland on either side of the urethra that secretes an alkaline fluid into the cavernous urethra. Also called a **Cowper's** (KOW-perz) **gland.**

Bulk flow The movement of large numbers of ions, molecules, or particles in the same direction as a result of pressure differences (osmotic, hydrostatic, or air pressure).

Bundle branch One of the two branches of the atrioventricular (AV) bundle made up of specialized muscle fibers (cells) that transmit electrical impulses to the ventricles.

Bursa (BUR-sa) A sac or pouch of synovial fluid located at friction points, especially about joints.

Buttocks (BUT-oks) The two fleshy masses on the posterior aspect of the inferior trunk, formed by the gluteal muscles.

Calcaneal (kal-KĀ-nē-al) **tendon** The tendon of the soleus, gastrocnemius, and plantaris muscles at the back of the heel. Also called the **Achilles'** (a-KIL-ēz) **tendon.**

Calcification (kal-si-fi-KĀ-shun) Deposition of mineral salts, primarily hydroxyapatite, in a framework formed by collagen fibers in which the tissue hardens. Also called **mineralization** (min′-e-ral-i-ZĀ-shun).

Calcitonin (kal-si-TŌ-nin) **(CT)** A hormone produced by the thyroid gland that lowers the calcium and phosphate levels of the blood by inhibiting bone breakdown and accelerating calcium absorption by bones.

Calculus (KAL-kyoo-lus) A stone, or insoluble mass of crystallized salts or other material, formed within the body, as in the gallbladder, kidney, or urinary bladder.

Callus (KAL-lus) A growth of new bone tissue in and around a fractured area, ultimately replaced by mature bone. An acquired, localized thickening.

Calmodulin (kal-MOD-yoo-lin) An intracellular protein that binds with calcium ions and activates or inhibits enzymes, many of which are protein kinases, to elicit physiological responses of hormones.

Calorie (KAL-ō-rē) A unit of heat. A calorie (cal) is the standard unit and is the amount of heat necessary to raise 1 g of water from 14°C to 15°C. The kilocalorie (kcal), used in metabolic and nutrition studies, is equal to 1000 cal.

Calyx (KĀL-iks) Any cuplike division of the kidney pelvis. *Plural is* **calyces** (KĀ-li-sēz).

Canal (ka-NAL) A narrow tube, channel, or passageway.

Canaliculus (kan′-a-LIK-yoo-lus) A small channel or canal, as in bones, where they connect lacunae. *Plural is* **canaliculi** (kan′-a-LIK-yoo-lī).

Cancellous (KAN-sel-us) Having a reticular or latticework structure, as in spongy tissue of bone.

Capacitation (ka′-pas-i-TĀ-shun) The functional changes that sperm undergo in the female reproductive tract that allow them to fertilize a secondary oocyte.

Capillary (KAP-i-lar′-ē) A microscopic blood vessel located between an arteriole and venule through which materials are exchanged between blood and body cells.

Carbohydrate (kar′-bō-HĪ-drāt) An organic compound containing carbon, hydrogen, and oxygen in a particular amount and arrangement and composed of monosaccharide subunits; usually has the formula $(CH_2O)n$.

Carcinogen (kar-SIN-ō-jen) Any substance that causes cancer.

Cardiac (KAR-dē-ak) **arrest** Cessation of an effective heartbeat in which the heart is completely stopped or in ventricular fibrillation.

Cardiac cycle A complete heartbeat consisting of systole (contraction) and diastole (relaxation) of both atria plus systole and diastole of both ventricles.

Cardiac muscle Striated muscle fibers (cells) that form the wall of the heart; stimulated by an intrinsic conduction system and autonomic motor neurons.

Cardiac notch An angular notch in the anterior border of the left lung into which a portion of the heart fits.

Cardiac output (CO) The volume of blood pumped from one ventricle of the heart (usually measured from the left ventricle) in 1 min; about 5.2 liters/min under normal resting conditions.

Cardiac reserve The maximum percentage that cardiac output can increase above normal.

Cardinal ligament A ligament of the uterus, extending laterally from the cervix and vagina as a continuation of the broad ligament.

Cardiology (kar-dē-OL-ō-jē) The study of the heart and diseases associated with it.

Cardiovascular (kar-dē-ō-VAS-kyoo-lar) **center** Groups of neurons scattered within the medulla oblongata that regulate heart rate, force of contraction, and blood vessel diameter.

Carotene (KAR-o-tēn) Antioxidant vitamin; yellow-orange pigment present in the stratum corneum of the epidermis. Accounts for the yellowish coloration of skin. Also termed **beta carotene.**

Carotid (ka-ROT-id) **body** Cluster of receptors on or near the carotid sinus that respond to alterations in blood levels of oxygen, carbon dioxide, and hydrogen ions.

Carotid sinus A dilated region of the internal carotid artery immediately superior to the branching of the common carotid artery containing receptors that monitor blood pressure.

Carotid sinus reflex A reflex concerned with maintaining normal blood pressure in the brain.

Carpus (KAR-pus) A collective term for the eight bones of the wrist.

Cartilage (KAR-ti-lij) A type of connective tissue consisting of chondrocytes in lacunae embedded in a dense network of collagen and elastic fibers and a matrix of chondroitin sulfate.

Cartilaginous (kar′-ti-LAJ-i-nus) **joint** A joint without a synovial (joint) cavity where the articulating bones are held tightly together by cartilage, allowing little or no movement.

Cast A small mass of hardened material formed within a cavity in the body and then discharged from the body; can originate in different areas and can be composed of various materials.

Catabolism (ka-TAB-o-lizm) Chemical reactions that break down complex organic compounds into simple ones, with the net release of energy.

Catalyst (KAT-a-list) A substance that speeds up a chemical reaction without itself being altered; enzyme.

Cataract (KAT-a-rakt) Loss of transparency of the lens of the eye or its capsule or both.

Cation (KAT-ī-on) A positively charged ion. An example is a sodium ion (Na^+).

Cauda equina (KAW-da ē-KWĪ-na) A tail-like array of roots of spinal nerves at the inferior end of the spinal cord.

Caudal (KAW-dal) Pertaining to any tail-like structure; inferior in position.

Cecum (SĒ-kum) A blind pouch at the proximal end of the large intestine to which the ileum is attached.

Celiac plexus (PLEK-sus) A large mass of ganglia and nerve fibers located at the level of the superior part of the first lumbar vertebra. Also called the **solar plexus.**

Cell The basic structural and functional unit of all organisms; the smallest structure capable of performing all the activities vital to life.

Cell cycle Growth and division of a single cell into daughter cells; consists of interphase and cell division.

Cell division Process by which a cell reproduces itself that consists of a nuclear division (mitosis) and a cytoplasmic division (cytokinesis); types include somatic and reproductive cell division.

Cell inclusion Principally organic substance produced by a cell that is not enclosed by a membrane and may appear or disappear at various times in the life of a cell; an example is glycogen granules.

Cell-mediated immunity That component of immunity in which specially sensitized lymphocytes (T cells) attach to antigens to destroy them. Also called **cellular immunity.**

Cementum (se-MEN-tum) Calcified tissue covering the root of a tooth.

Center of ossification (os′-i-fi-KĀ-shun) An area in the cartilage model of a future bone where the cartilage cells hypertrophy and then secrete enzymes that result in the calcification of their matrix, resulting in the death of the cartilage cells, followed by the invasion of the area by osteoblasts that then lay down bone.

Central canal A microscopic tube running the length of the spinal cord in the gray commissure. A circular channel running longitudinally in the center of an osteon (Haversian system) of mature compact bone, containing blood and lymphatic vessels and nerves. Also called a **Haversian** (ha-VĒR-shun) **canal.**

Central nervous system (CNS) That portion of the nervous system that consists of the brain and spinal cord.

Centrioles (SEN-trē-ōlz) Paired, cylindrical structures within a centrosome, each consisting of a ring of microtubules and arranged at right angles to each other.

Centromere (SEN-trō-mēr) The clear, constricted portion of a chromosome where the two chromatids are joined; serves as the point of attachment for the chromosomal microtubules.

Centrosome (SEN-trō-sōm) A rather dense area of cytoplasm, near the nucleus of a cell, containing a pair of centrioles. During prophase, it forms the mitotic spindle.

Cephalic (se-FAL-ik) Pertaining to the head; superior in position.

Cerebellar peduncle (ser-e-BEL-ar pe-DUNG-kul) A bundle of nerve fibers connecting the cerebellum with the brain stem.

Cerebellum (ser-e-BEL-um) The portion of the brain lying posterior to the medulla oblongata and pons; governs coordination of skilled movements and balance.

Cerebral aqueduct (SER-ē-bral AK-we-dukt) A channel through the midbrain connecting the third and fourth ventricles and containing cerebrospinal fluid. Also termed the **aqueduct of Sylvius.**

Cerebral arterial circle A ring of arteries forming an anastomosis at the base of the brain between the internal carotid and basilar arteries and arteries supplying the brain. Also called the **circle of Willis.**

Cerebral cortex The surface of the cerebral hemispheres, 2–4 mm thick, consisting of six layers of neuronal cell bodies (gray matter) in most areas.

Cerebral peduncle One of a pair of nerve fiber bundles located on the anterior surface of the midbrain, conducting nerve impulses between the pons and the cerebral hemispheres.

Cerebrospinal (se-rē′-brō-SPĪ-nal) **fluid (CSF)** A fluid produced by ependymal cells that cover choroid plexuses in the ventricles of the brain; the fluid circulates in the ventricles, the central canal, and the subarachnoid space around the brain and spinal cord.

Cerebrovascular (se rē′-brō-VAS-kyoo-lar) **accident (CVA)** Destruction of brain tissue (infarction) resulting from disorders of blood vessels that supply the brain. Also called a **stroke.**

Cerebrum (SER-ē-brum *or* ser-Ē-brum) The two hemispheres of the forebrain, making up the largest part of the brain.

Cerumen (se-ROO-men) Waxlike secretion produced by ceruminous glands in the external auditory meatus (ear canal).

Ceruminous (se-ROO-mi-nus) **gland** A modified sudoriferous (sweat) gland in the external auditory meatus that secretes cerumen (ear wax).

Cervical ganglion (SER-vi-kul GANG-glē-on) A cluster of cell bodies of postganglionic sympathetic neurons located in the neck, near the vertebral column.

Cervical plexus (PLEK-sus) A network of nerve fibers formed by the ventral rami of the first four cervical nerves.

Cervix (SER-viks) Neck; any constricted portion of an organ, such as the inferior cylindrical part of the uterus.

Chemical bond Force of attraction in a molecule or compound that holds its atoms together. Examples include ionic and covalent bonds.

Chemical element Unit of matter that cannot be decomposed into a simpler substance by ordinary chemical reactions. Examples include hydrogen (H), carbon (C), and oxygen (O).

Chemically gated ion channel An ion channel that opens and closes in response to a specific chemical stimulus, such as a neurotransmitter, a hormone, or certain ions, for example, H^+ or Ca^{2+}.

Chemical reaction The combination or breaking apart of atoms in which chemical bonds are formed or broken and new products with different properties are produced.

Chemiosmosis (kem′-ē-oz-MŌ-sis) Mechanism for ATP generation that links chemical reactions (electrons passing along the electron transport chain) with pumping of H^+ out of the mitochondrial matrix. ATP synthesis occurs as H^+ diffuse back into the mitochondrial matrix through special H^+ channels in the membrane.

Chemoreceptor (kē′-mō-rē-SEP-tor) Receptor that detects the presence of chemicals.

Chemotaxis (kē-mō-TAK-sis) Attraction of phagocytes to microbes by a chemical stimulus.

Chiasm (KĪ-azm) A crossing; especially the crossing of the optic (II) nerve fibers.

Chief cell The secreting cell of a gastric gland that produces pepsinogen, the precursor of the enzyme pepsin, and the enzyme gastric lipase. Also called a **zymogenic** (zī′-mō-JEN-ik) **cell.**

Chiropractic (kī′-rō-PRAK-tik) A system of treating disease by using one's hands to manipulate body parts, mostly the vertebral column.

Chloride shift Exchange of bicarbonate ions ($HCO3^-$) for chloride ions (Cl^-) between red blood cells and plasma; maintains electrical balance inside red blood cells as bicarbonate ions are produced or eliminated during respiration.

Cholesterol (kō-LES-te-rol) Classified as a lipid, the most abundant steroid in animal tissues; located in cell membranes and used for the synthesis of steroid hormones and bile salts.

Cholinergic (kō′-lin-ER-jik) **fiber** A neuron that liberates acetylcholine at its synapses.

Chondrocyte (KON-drō-sīt) Cell of mature cartilage.

Chondroitin (kon-DROY-tin) **sulfate** An amorphous matrix material found outside connective tissue cells.

Chordae tendineae (KOR-dē TEN-di-nē-ē) Tendonlike, fibrous cords that connect heart AV valves with papillary muscles.

Chorion (KŌ-rē-on) The most superficial fetal membrane that becomes the principal embryonic portion of the placenta; serves a protective and nutritive function.

Chorionic villi (kō′-rē-ON-ik VIL-lī) Fingerlike projections of the chorion that grow into the decidua basalis of the endometrium and contain fetal blood vessels.

Choroid (KŌ-royd) One of the vascular coats of the eyeball.

Choroid plexus (PLEK-sus) A network of capillaries located in the roof of each of the four ventricles of the brain; ependymal cells around choroid plexuses produce cerebrospinal fluid.

Chromaffin (krō-MAF-in) **cell** Cell that has an affinity for chrome salts, due in part to the presence of the precursors of the neurotransmitter epinephrine; found, among other places, in the adrenal medulla.

Chromatid (KRŌ-ma-tid) One of a pair of identical connected nucleoprotein strands that are joined at the centromere and separate during cell division, each becoming a chromosome of one of the two daughter cells.

Chromatin (KRŌ-ma-tin) The threadlike mass of genetic material, consisting principally of DNA, which is present in the nucleus of a nondividing or interphase cell.

Chromatolysis (krō′-ma-TOL-i-sis) The breakdown of chromatophilic substance (Nissl bodies) into finely granular masses in the cell body of a central or peripheral neuron whose process (axon or dendrite) has been damaged.

Chromatophilic substance Rough endoplasmic reticulum in the cell bodies of neurons that functions in protein synthesis. Also called **Nissl bodies.**

Chromosome (KRŌ-mō-sōm) One of the small, threadlike structures in the nucleus of a cell, normally 46 in a human diploid cell, that bears the genetic material; composed of DNA and proteins (histones) that form a delicate chromatin thread during interphase; becomes packaged into compact rodlike structures that are visible under the light microscope during cell division.

Chronic (KRON-ik) Long term or frequently recurring; applied to a disease that is not acute.

Chronic obstructive pulmonary disease (COPD) A disease, such as bronchitis or emphysema, in which there is some degree of obstruction of air passageways and consequent increase in airway resistance.

Chyle (KĪL) The milky-appearing fluid found in the lacteals of the small intestine after digestion.

Chylomicron (kī-lō-MĪK-ron) Protein-coated spherical structure that contains triglycerides, phospholipids, and cholesterol and is absorbed into the lacteal of a villus in the small intestine.

Chyme (KĪM) The semifluid mixture of partly digested food and digestive secretions found in the stomach and small intestine during digestion of a meal.

Ciliary (SIL-ē-ar′-ē) **body** One of the three portions of the vascular tunic of the eyeball, the others being the choroid and the iris; includes the ciliary muscle and the ciliary processes.

Ciliary ganglion (GANG-glē-on) A very small parasympathetic ganglion whose preganglionic fibers come from the oculomotor (III) nerve and whose postganglionic fibers carry nerve impulses to the ciliary muscle and the sphincter muscle of the iris.

Cilium (SIL-ē-um) A hair or hairlike process projecting from a cell that may be used to move the entire cell or to move substances along the surface of the cell. *Plural is* **cilia.**

Circadian (ser-KĀ-dē-an) **rhythm** A cycle of active and nonactive periods in organisms determined by internal mechanisms and repeating about every 24 hours.

Circular folds Permanent, deep, transverse folds in the mucosa and submucosa of the small intestine that increase the surface area for absorption. Also called **plicae circulares** (PLĪ-kē SER-kyoo-lar-ēs).

Circulation time Time required for blood to pass from the right atrium, through pulmonary circulation, back to the left ventricle, through systemic circulation to the foot, and back again to the right atrium; normally about 1 min.

Circumduction (ser′-kum-DUK-shun) A movement at a synovial joint in which the distal end of a bone moves in a circle while the proximal end remains relatively stable.

Circumvallate papilla (ser′-kum-VAL-āt pa-PIL-a) One of the circular projections that is arranged in an inverted V-shaped row at the back of the tongue; the largest of the elevations on the upper surface of the tongue containing taste buds.

Cisterna chyli (sis-TER-na KĪ-lē) The origin of the thoracic duct.

Cleavage The rapid mitotic divisions following the fertilization of a secondary oocyte, resulting in an increased number of progressively smaller cells, called blastomeres.

Climacteric (klī-mak-TER-ik) Cessation of the reproductive function in the female or diminution of testicular activity in the male.

Climax The peak period or moments of greatest intensity during sexual excitement.

Clitoris (KLI-to-ris) An erectile organ of the female, located at the anterior junction of the labia minora, that is homologous to the male penis.

Clone (KLŌN) A population of identical cells.

Clot The end result of a series of biochemical reactions that changes liquid plasma into a gelatinous mass; specifically, the conversion of fibrinogen into a tangle of polymerized fibrin molecules.

Clot retraction (rē-TRAK-shun) The consolidation of a fibrin clot to pull damaged tissue together.

Coagulation (cō-ag-yoo-LĀ-shun) Process by which a blood clot is formed. Also known as **clotting.**

Coccyx (KOK-six) The fused bones at the inferior end of the vertebral column.

Cochlea (KŌK-lē-a) A winding, cone-shaped tube forming a portion of the inner ear and containing the spiral organ (organ of Corti).

Cochlear duct The membranous cochlea consisting of a spirally arranged tube enclosed in the bony cochlea and lying along its outer wall. Also called the **scala media** (SCA-la MĒ-dē-a).

Coenzyme A type of cofactor; a nonprotein organic molecule that is associated with and activates an enzyme; many are derived from vitamins. An example is nicotinamide adenine dinucleotide (NAD), derived from the B vitamin niacin.

Coitus (KŌ-i-tus) Sexual intercourse. Also called **copulation** (cop-yoo-LĀ-shun).

Collagen (KOL-a-jen) A protein that is the main organic constituent of connective tissue.

Collateral circulation The alternate route taken by blood through an anastomosis.

Colliculus (ko-LIK-yoo-lus) A small elevation.

Colon The division of the large intestine consisting of ascending, transverse, descending, and sigmoid portions.

Colony-stimulating factor (CSF) One of a group of molecules that stimulates development of white blood cells. Examples are macrophage CSF and granulocyte CSF.

Colostrum (kō-LOS-trum) A thin, cloudy fluid secreted by the mammary glands a few days prior to or after delivery before true milk is produced.

Column (KOL-um) Group of white matter tracts in the spinal cord.

Common bile duct A tube formed by the union of the common hepatic duct and the cystic duct that empties bile into the duodenum at the hepatopancreatic ampulla (ampulla of Vater).

Compact (dense) bone tissue Bone tissue that contains few spaces between osteons (Haversian systems); forms the external portion of all bones and the bulk of the diaphysis (shaft) of long bones; is found immediately deep to the periosteum and external to spongy bone.

Complement (KOM-ple-ment) A group of at least 20 normally inactive proteins found in plasma that forms a component of nonspecific resistance and immunity by bringing about cytolysis, inflammation, and opsonization.

Compliance The ease with which the lungs and thoracic wall or blood vessels can be expanded.

Compound A substance that can be broken down into two or more other substances by chemical means.

Concha (KONG-ka) A scroll-like bone found in the skull. *Plural is* **conchae** (KONG-kē).

Concussion (kon-KUSH-un) Traumatic injury to the brain that produces no visible bruising but may result in abrupt, temporary loss of consciousness.

Conduction myofiber Muscle fiber (cell) in the ventricular tissue of the heart specialized for conducting an action potential to the myocardium; part of the conduction system of the heart. Also called a **Purkinje** (pur-KIN-jē) **fiber.**

Conduction system A series of autorhythmic cardiac muscle fibers that generates and distributes electrical impulses to stimulate coordinated contraction of the heart chambers; includes the sinoatrial (SA) node, the atrioventricular (AV) node, the atrioventricular (AV) bundle, the right and left bundle branches, and the conduction myofibers (Purkinje fibers).

Conductivity (kon′-duk-TIV-i-tē) The ability of a cell to propagate (conduct) action potentials along its plasma membrane; characteristic of neurons and muscle fibers (cells).

Condyloid (KON-di-loid) **joint** A synovial joint structured so that an oval-shaped condyle of one bone fits into an elliptical cavity of another bone, permitting side-to-side and back-and-forth movements, such as the joint at the wrist between the radius and carpals. Also called an **ellipsoidal** (e-lip-soy-dal) **joint.**

Cone The type of photoreceptor in the retina that is specialized for highly acute, color vision in bright light.

Congenital (kon-JEN-i-tal) Present at the time of birth.

Conjunctiva (kon′-junk-TĪ-va) The delicate membrane covering the eyeball and lining the eyes.

Connective tissue The most abundant of the four basic tissue types in the body, performing the functions of binding and supporting; consists of relatively few cells in a generous matrix (the ground substance and fibers between the cells).

Consciousness (KON-shus-nes) A state of wakefulness in which an individual is fully alert, aware, and oriented, partly as a result of feedback between the cerebral cortex and reticular activating system.

Contact inhibition Phenomenon by which migration of a growing cell is stopped when it makes contact with another cell of its own kind.

Continuous conduction (kon-DUK-shun) Propagation of an action potential (nerve impulse) in a step-by-step depolarization of each adjacent area of an axon membrane.

Contraception (kon′-tra-SEP-shun) The prevention of fertilization or impregnation without destroying fertility.

Contractility (kon′-trak-TIL-i-tē) The ability of cells or parts of cells to actively generate force to undergo shortening and change form for purposeful movements. Muscle fibers (cells) exhibit a high degree of contractility.

Contralateral (kon′-tra-LAT-er-al) On the opposite side; affecting the opposite side of the body.

Control center The component of a feedback system, such as the brain, that determines the point at which a controlled condition, such as body temperature, is maintained.

Conus medullaris (KŌ-nus med-yoo-LAR-is) The tapered portion of the spinal cord inferior to the lumbar enlargement.

Convergence (con-VER-jens) An anatomical arrangement in which the synaptic end bulbs of several presynaptic neurons terminate on one postsynaptic neuron. The medial movement of the two eyeballs so that both are directed toward a near object being viewed in order to produce a single image.

Convulsion (con-VUL-shun) Violent, involuntary, tetanic contractions of an entire group of muscles.

Cornea (KOR-nē-a) The nonvascular, transparent fibrous coat through which the iris can be seen.

Coronal (kō-RŌ-nal) **plane** A plane that divides the body into anterior and posterior portions. Also called **frontal plane.**

Corona radiata The innermost layer of granulosa cells that is firmly attached to the zona pellucida around a secondary oocyte.

Coronary artery disease (CAD) A condition such as atherosclerosis that causes narrowing of coronary arteries so that blood flow to the heart is reduced. The result is **coronary heart disease (CHD)** in which the heart muscle receives inadequate blood flow due to an interruption of its blood supply.

Coronary circulation The pathway followed by the blood from the ascending aorta through the blood vessels supplying the heart and returning to the right atrium. Also called **cardiac circulation.**

Coronary sinus (SĪ-nus) A wide venous channel on the posterior surface of the heart that collects the blood from the coronary circulation and returns it to the right atrium.

Corpora quadrigemina (KOR-por-a kwad-ri-JEM-in-a) Four small elevations (superior and inferior colliculi) in the posterior portion of the midbrain concerned with visual and auditory reflexes.

Corpus (KOR-pus) The principal part of any organ; any mass or body.

Corpus albicans (KOR-pus AL-bi-kanz) A white fibrous patch in the ovary that forms after the corpus luteum regresses.

Corpus callosum (kal-LŌ-sum) The great commissure of the brain between the cerebral hemispheres.

Corpuscle of touch The sensory receptor for the sensation of touch; found in the dermal papillae, especially in palms and soles. Also called a **Meissner** (MĪS-ner) **corpuscle.**

Corpus luteum (LOO-tē-um) A yellow endocrine gland in the ovary formed when a follicle has discharged its secondary oocyte; secretes estrogens, progesterone, relaxin, and inhibin.

Corpus striatum (strī-Ā-tum) An area in the interior of each cerebral hemisphere composed of the caudate and lentiform nuclei of the basal ganglia and white matter of the internal capsule, arranged in a striated manner.

Cortex (KOR-teks) An outer layer of an organ. The convoluted layer of gray matter covering each cerebral hemisphere.

Costal (KOS-tal) Pertaining to a rib.

Costal cartilage (KAR-ti-lij) Hyaline cartilage that attaches a rib to the sternum.

Countercurrent mechanism One mechanism involved in the ability of the kidneys to produce hypertonic urine.

Cramp A spasmodic, usually painful contraction of a muscle.

Cranial (KRĀ-nē-al) **cavity** A subdivision of the dorsal body cavity formed by the cranial bones and containing the brain.

Cranial nerve One of 12 pairs of nerves that leave the brain, pass through foramina in the skull, and supply the head, neck, and part of the trunk; each is designated by a Roman numeral and a name.

Craniosacral (krā-nē-ō-SĀ-kral) **outflow** The fibers of parasympathetic preganglionic neurons, which have their cell bodies located in nuclei in the brain stem and in the lateral gray matter of the sacral portion of the spinal cord.

Cranium (KRĀ-nē-um) The skeleton of the skull that protects the brain and the organs of sight, hearing, and balance; includes the frontal, parietal, temporal, occipital, sphenoid, and ethmoid bones.

Creatine phosphate (KRĒ-a-tin FOS-fāt) Molecule in skeletal muscle fibers that contains high-energy phosphate bonds; used to generate ATP rapidly from ADP by transfer of phosphate group. Also called **phosphocreatine** (fos′-fō-KRĒ-a-tin).

Crenation (krē-NĀ-shun) The shrinkage of red blood cells into knobbed, starry forms when they are placed in a hypertonic solution.

Crista (KRIS-ta) A crest or ridged structure. A small elevation in the ampulla of each semicircular duct that contains receptors for dynamic equilibrium.

Crossed extensor reflex A reflex in which extension of the joints in one limb occurs in conjunction with contraction of the flexor muscles of the opposite limb.

Crossing-over The exchange of a portion of one chromatid with another in a tetrad during meiosis. It permits an exchange of genes among chromatids and is one factor that results in genetic variation of progeny.

Crus (KRUS) **of penis** Separated, tapered portion of the corpora cavernosa penis. *Plural is* **crura** (KROO-ra).

Cryptorchidism (krip-TOR-ki-dizm) The condition of undescended testes.

Cupula (KUP-yoo-la) A mass of gelatinous material covering the hair cells of a crista; a receptor in the ampulla of a semicircular canal stimulated when the head moves.

Cutaneous (kyoo-TĀ-nē-us) Pertaining to the skin.

Cyanosis (sī-a-NŌ-sis) Reduced hemoglobin (deoxygenated) concentration of blood of more than 5 g/dL that results in a blue or dark purple discoloration that is most easily seen in nail beds and mucous membranes.

Cyclic AMP (cyclic adenosine-3′, 5′-monophosphate) Molecule formed from ATP by the action of the enzyme adenylate cyclase; serves as an intracellular messenger (second messenger) for some hormones.

Cyst (SIST) A sac with a distinct connective tissue wall, containing a fluid or other material.

Cystic (SIS-tik) **duct** The duct that transports bile from the gallbladder to the common bile duct.

Cytolysis (sī-TOL-i-sis) The rupture of living cells in which the contents leak out.

Cytochrome (SĪ-tō-krōm) A protein with an iron-containing group (heme) capable of alternating between a reduced form (Fe^{2+}) and an oxidized form (Fe^{3+}).

Cytokines (SĪ-to-kīns) Small protein hormones produced by lymphocytes, fibroblasts, endothelial cells, and antigen-presenting cells that act as autocrine or paracrine substances to stimulate or inhibit cell growth and differentiation, regulate immune responses, or aid nonspecific defenses.

Cytokinesis (sī′-tō-ki-NĒ-sis) Distribution of the cytoplasm into two separate cells during cell division; coordinated with nuclear division (mitosis).

Cytology (sī-TOL-ō-jē) The study of cells.

Cytoplasm (SĪ-tō-plazm) Cytosol, all organelles (except the nucleus), and inclusions.

Cytoskeleton Complex internal structure of cytoplasm consisting of microfilaments, microtubules, and intermediate filaments.

Cytosol (SĪ-tō-sol) Semifluid portion of cytoplasm in which organelles and inclusions are suspended and solutes are dissolved. Also called **intracellular fluid.**

Dartos (DAR-tōs) The contractile tissue deep to the skin of the scrotum.

Decibel (DES-i-bel) **(dB)** A unit that measures relative sound intensity (loudness).

Decidua (de-SID-yoo-a) That portion of the endometrium of the uterus (all but the deepest layer) that is modified during pregnancy and shed after childbirth.

Deciduous (dē-SID-yoo-us) Falling off or being shed seasonally or at a particular stage of development. In the body, referring to the first set of teeth.

Decussation (dē′-ku-SĀ-shun) A crossing-over; usually refers to the crossing of 90% of the fibers in the large motor tracts to opposite sides in the medullary pyramids.

Deep Away from the surface of the body or an organ.

Deep fascia (FASH-ē-a) A sheet of connective tissue wrapped around a muscle to hold it in place.

Deep inguinal (IN-gwi-nal) **ring** A slitlike opening in the aponeurosis of the transversus abdominis muscle that represents the origin of the inguinal canal.

Defecation (def-e-KĀ-shun) The discharge of feces from the rectum.

Deglutition (dē-gloo-TISH-un) The act of swallowing.

Dehydration (dē-hī-DRĀ-shun) Excessive loss of water from the body or its parts.

Delta cell A cell in the pancreatic islets (islets of Langerhans) in the pancreas that secretes somatostatin. Also termed a **D cell.**

Demineralization (de-min′-er-al-i-ZĀ-shun) Loss of calcium and phosphorus from bones.

Denaturation (de-nā-chur-Ā-shun) Disruption of the tertiary structure of a protein by agents such as heat, changes in pH, or other physical or chemical methods, in which the protein loses its physical properties and biological activity.

Dendrite (DEN-drīt) A neuronal process that carries a nerve impulse toward the cell body.

Dendritic (den-DRIT-ik) **cell** One type of antigen-presenting cell with long branchlike projections that commonly is present in epithelial linings (such as the vagina) and in the skin (for example, Langerhans cells in the epidermis).

Dental caries (KA-rēz) Gradual demineralization of the enamel and dentin of a tooth that may invade the pulp and alveolar bone. Also called **tooth decay.**

Denticulate (den-TIK-yoo-lāt) Finely toothed or serrated; characterized by a series of small, pointed projections.

Dentin (DEN-tin) The bony tissues of a tooth enclosing the pulp cavity.

Dentition (den-TI-shun) The eruption of teeth. The number, shape, and arrangement of teeth.

Deoxyribonucleic (dē-ok′-sē-ri′-bō-nyoo-KLĒ-ik) **acid (DNA)** A nucleic acid constructed of nucleotides consisting of one of four nitrogenous bases (adenine, cytosine, guanine, or thymine), deoxyribose, and a phosphate group; encoded in the nucleotides is genetic information.

Depolarization (dē-pō-lar-i-ZĀ-shun) Used in neurophysiology to describe the reduction of voltage across a plasma membrane; expressed as a change toward less negative (more positive) voltages on the interior surface of the plasma membrane.

Depression (de-PRESS-shun) Movement in which a part of the body moves inferiorly.

Dermal papilla (pa-PILL-a) Fingerlike projection of the papillary region of the dermis that may contain blood capillaries or corpuscles of touch (Meissner corpuscles).

Dermatology (der-ma-TOL-ō-jē) The medical specialty dealing with diseases of the skin.

Dermatome (DER-ma-tōm) The cutaneous area developed from one embryonic spinal cord segment and receiving most of its sensory innervation from one spinal nerve. An instrument for incising the skin or cutting thin transplants of skin.

Dermis (DER-mis) A layer of dense irregular connective tissue lying deep to the epidermis.

Descending colon (KŌ-lon) The part of the large intestine descending from the left colic (splenic) flexure to the level of the left iliac crest.

Detritus (de-TRĪ-tus) Particulate matter produced by or remaining after the wearing away or disintegration of a substance or tissue; scales, crusts, or loosened skin.

Detrusor (de-TROO-ser) **muscle** Muscle in the wall of the urinary bladder.

Developmental anatomy The study of development from the fertilized egg to the adult form. The branch of anatomy called embryology is generally restricted to the study of development from the fertilized egg through the eighth week in utero.

Diagnosis (dī-ag-NŌ-sis) Distinguishing one disease from another or determining the nature of a disease from signs and symptoms by inspection, palpation, laboratory tests, and other means.

Diaphragm (DĪ-a-fram) Any partition that separates one area from another, especially the dome-shaped skeletal muscle between the thoracic and abdominal cavities. Also a dome-shaped device that is placed over the cervix, usually with a spermicide, to prevent conception.

Diaphysis (dī-AF-i-sis) The shaft of a long bone.

Diarthrosis (dī-ar-THRŌ-sis) A freely movable joint; types are gliding, hinge, pivot, condyloid, saddle, and ball-and-socket.

Diastole (dī-AS-tō-lē) In the cardiac cycle, the phase of relaxation or dilation of the heart muscle, especially of the ventricles.

Diastolic (dī-as-TOL-ik) **blood pressure** The force exerted by blood on arterial walls during ventricular relaxation; the lowest blood pressure measured in the large arteries, about 80 mm Hg under normal conditions for a young adult.

Diencephalon (dī′-en-SEF-a-lon) A part of the brain consisting of the thalamus, hypothalamus, epithalamus, and subthalamus.

Digestion (dī-JES-chun) The mechanical and chemical breakdown of food to simple molecules that can be absorbed and used by body cells.

Dilate (DĪ-lāt) To expand or swell.

Diploid (DIP-loyd) Having the number of chromosomes characteristically found in the somatic cells of an organism. Symbolized 2*n*.

Direct motor pathways Collections of upper motor neurons with cell bodies in the motor cortex that project axons into the spinal cord, where they synapse with lower motor neurons or association neurons in the anterior horns. Also called the **pyramidal pathways.**

Disease Any change from a state of health.

Dislocation (dis-lō-KĀ-shun) Displacement of a bone from a joint with tearing of ligaments, tendons, and articular capsules. Also called **luxation** (luks-Ā-shun).

Dissect (di-SEKT) To separate tissues and parts of a cadaver (corpse) or an organ for anatomical study.

Distal (DIS-tal) Farther from the attachment of a limb to the trunk; farther from the point of origin or attachment.

Diuretic (dī-yoo-RET-ik) A chemical that inhibits sodium reabsorption, reduces antidiuretic hormone (ADH) concentration, and increases urine volume by inhibiting facultative reabsorption of water.

Divergence (dī-VER-jens) An anatomical arrangement in which the synaptic end bulbs of one presynaptic neuron terminate on several postsynaptic neurons.

Diverticulum (dī-ver-TIK-yoo-lum) A sac or pouch in the wall of a canal or organ, especially in the colon.

Dominant gene A gene that is able to override the influence of the complementary gene on the homologous chromosome; the gene that is expressed.

Dorsal body cavity Cavity near the dorsal (posterior) surface of the body that consists of a cranial cavity and vertebral canal.

Dorsal ramus (RĀ-mus) A branch of a spinal nerve containing motor and sensory fibers supplying the muscles, skin, and bones of the posterior part of the head, neck, and trunk.

Dorsiflexion (dor'-si-FLEK-shun) Bending the foot in the direction of the dorsum (upper surface).

Down-regulation Phenomenon in which there is a decrease in the number of receptors in response to an excess of a hormone or neurotransmitter.

Ductus arteriosus (DUK-tus ar-tē-rē-Ō-sus) A small vessel connecting the pulmonary trunk with the aorta; found only in the fetus.

Ductus (vas) deferens (DEF-er-ens) The duct that carries sperm from the epididymis to the ejaculatory duct. Also called the **seminal duct.**

Ductus epididymis (ep'-i-DID-i-mis) A tightly coiled tube inside the epididymis, distinguished into a head, body, and tail, in which sperm undergo maturation.

Ductus venosus (ve-NŌ-sus) A small vessel in the fetus that helps the circulation bypass the liver.

Duodenal (doo-ō-DĒ-nal) **gland** Gland in the submucosa of the duodenum that secretes an alkaline mucus to protect the lining of the small intestine from the action of enzymes and to help neutralize the acid in chyme. Also called **Brunner's** (BRUN-erz) **gland.**

Duodenum (doo'-ō-DĒ-num) The first 25 cm (10 in.) of the small intestine, which connects the stomach and the ileum.

Dura mater (DYOO-ra MĀ-ter) The outer membrane (meninx) covering the brain and spinal cord.

Dynamic equilibrium (ē-kwi-LIB-rē-um) The maintenance of body position, mainly the head, in response to sudden movements such as rotation.

Dysfunction (dis-FUNK-shun) Absence of completely normal function.

Dysmenorrhea (dis'-men-ō-RĒ-a) Painful menstruation.

Dysplasia (dis-PLĀ-zē-a) Change in the size, shape, and organization of cells due to chronic irritation or inflammation; may either revert to normal if stress is removed or progress to neoplasia.

Dyspnea (DISP-nē-a) Shortness of breath.

Eardrum A thin, semitransparent partition of fibrous connective tissue between the external auditory meatus and the middle ear. Also called the **tympanic membrane.**

Ectoderm The primary germ layer that gives rise to the nervous system and the epidermis of skin and its derivatives.

Ectopic (ek-TOP-ik) Out of the normal location, as in ectopic pregnancy.

Edema (e-DĒ-ma) An abnormal accumulation of interstitial fluid.

Effector (e-FEK-tor) An organ of the body, either a muscle or a gland, that responds to a motor neuron impulse.

Efferent arteriole (EF-er-ent ar-TĒ-rē-ōl) A vessel of the renal vascular system that transports blood from a glomerulus to a peritubular capillary.

Efferent (EF-er-ent) **ducts** A series of coiled tubes that transport sperm cells from the rete testis to the epididymis.

Effusion (e-FYOO-zhun) The escape of fluid from the lymphatic vessels or blood vessels into a cavity or into tissues.

Eicosanoids (ī-KŌ-sa-noyds) Local hormones derived from a 20-carbon fatty acid (arachidonic acid); two important types are prostaglandins and leukotrienes.

Ejaculation (e-jak-yoo-LĀ-shun) The reflex ejection or expulsion of semen from the penis.

Ejaculatory (e-JAK-yoo-la-tō-rē) **duct** A tube that transports sperm cells from the ductus (vas) deferens to the prostatic urethra.

Elasticity (e-las-TIS-i-tē) The ability of tissue to return to its original shape after contraction or extension.

Electrocardiogram (e-lek'-trō-KAR-dē-ō-gram) **(ECG or EKG)** A recording of the electrical changes that accompany the cardiac cycle that can be detected at the surface of the body; may be resting, stress, or ambulatory.

Electroencephalogram (e-lek'-trō-en-SEF-a-lō-gram) **(EEG)** A recording of the electrical impulses of the brain from the scalp surface; used to diagnose certain diseases (such as epilepsy), furnish information regarding sleep and wakefulness, and confirm brain death.

Electrolyte (ē-LEK-trō-līt) Any compound that separates into ions when dissolved in water and is able to conduct electricity.

Electromyography (e-lek'-trō-mī-OG-ra-fē) Evaluation of the electrical activity of resting and contracting muscle to ascertain causes of muscular weakness, paralysis, involuntary twitching, and abnormal levels of muscle enzymes; also used as part of biofeedback studies.

Electron transport chain A sequence of electron carrier molecules on the inner mitochondrial membrane that undergo oxidation and reduction as they pump hydrogen ions (H^+) through the membrane. ATP synthesis then occurs as H^+ diffuse back into the mitochondrial matrix through special H^+ channels. *See also* **Chemiosmosis**.

Elevation (el-e-VĀ-shun) Movement in which a part of the body moves superiorly.

Embolism (EM-bō-lizm) Obstruction or closure of a vessel by an embolus.

Embolus (EM-bō-lus) A blood clot, bubble of air or fat from broken bones, mass of bacteria, or other debris or foreign material transported by the blood.

Embryo (EM-brē-ō) The young of any organism in an early stage of development; in humans, the developing organism from fertilization to the end of the eighth week in utero.

Embryology (em'-brē-OL-ō-jē) The study of development from the fertilized egg to the end of the eighth week in utero.

Emesis (EM-e-sis) Vomiting.

Emigration (em′-e-GRĀ-shun) Process whereby white blood cells (WBCs) leave the bloodstream by slowing down, rolling along the endothelium, and squeezing between the endothelial cells. Adhesion molecules help WBCs stick to the endothelium. Also known as **migration** or **extravasation.**

Emission (ē-MISH-un) Propulsion of sperm into the urethra due to peristaltic contractions of the ducts of the testes, epididymides, and ductus (vas) deferens as a result of sympathetic stimulation.

Emmetropia (em′-e-TRŌ-pē-a) Normal vision in which light rays are focused exactly on the retina.

Emphysema (em′-fi′SE̅-ma) A lung disorder in which alveolar walls disintegrate, producing abnormally large air spaces and loss of elasticity in the lungs; typically caused by exposure to cigarette smoke.

Emulsification (ē-mul′-si-fi-KĀ-shun) The dispersion of large lipid globules to smaller, uniformly distributed particles in the presence of bile.

Enamel (e-NAM-el) The hard, white substance covering the crown of a tooth.

End-diastolic (dī-as-TO-lik) **volume (EDV)** The volume of blood, about 130 ml, remaining in a ventricle at the end of its diastole (relaxation).

Endergonic (end′-er-GON-ik) **reaction** Type of chemical reaction in which the energy released as new bonds form is less than the energy needed to break apart old bonds; an energy-requiring reaction.

Endocardium (en-dō-KAR-dē-um) The layer of the heart wall, composed of endothelium and smooth muscle, that lines the inside of the heart and covers the valves and tendons that hold the valves open.

Endochondral ossification (en′-dō-KON-dral os′-i-fi-KĀ-shun) The replacement of cartilage by bone. Also called **intracartilaginous** (in′-tra-kar′-ti-LAJ-i-nus) **ossification.**

Endocrine (EN-dō-krin) **gland** A gland that secretes hormones into the blood; a ductless gland.

Endocrinology (en′-dō-kri-NOL-ō-jē) The science concerned with the structure and functions of endocrine glands and the diagnosis and treatment of disorders of the endocrine system.

Endocytosis (en′-dō-sī-TŌ-sis) The uptake into a cell of large molecules and particles in which a segment of plasma membrane surrounds the substance, encloses it, and brings it in; includes phagocytosis, pinocytosis, and receptor-mediated endocytosis.

Endoderm (EN-dō-derm) The primary germ layer of the developing embryo that gives rise to the gastrointestinal tract, urinary bladder and urethra, and respiratory tract.

Endodontics (en′-dō-DON-tiks) The branch of dentistry concerned with the prevention, diagnosis, and treatment of diseases that affect the pulp, root, periodontal ligament, and alveolar bone.

Endogenous (en-DOJ-e-nus) Growing from or beginning within the organism.

Endolymph (EN-dō-lymf′) The fluid within the membranous labyrinth of the inner ear.

Endometrium (en′-dō-ME̅-trē-um) The mucous membrane lining the uterus.

Endomysium (en′-dō-MIZ-ē-um) Invagination of the perimysium separating each individual muscle fiber (cell).

Endoneurium (en′-dō-NYOO-rē-um) Connective tissue wrapping around individual nerve fibers (cells).

Endoplasmic reticulum (en′-do-PLAZ-mik re-TIK-yoo-lum) **(ER)** A network of channels running through the cytoplasm of a cell that serves in intracellular transportation, support, storage, synthesis, and packaging of molecules. Portions of ER where ribosomes are attached to the outer surface are called **rough ER;** portions that have no ribosomes are called **smooth ER.**

Endorphin (en-DOR-fin) A neuropeptide in the central nervous system that acts as a painkiller.

Endosteum (en-DOS-tē-um) The membrane that lines the medullary (marrow) cavity of bones, consisting of osteoprogenitor cells and scattered osteoclasts.

Endothelial–capsular membrane A filtration membrane in a nephron of a kidney consisting of the endothelium and basement membrane of the glomerulus and the epithelium of the visceral layer of the glomerular (Bowman's) capsule.

Endothelium (en′-dō-THE̅-lē-um) The layer of simple squamous epithelium that lines the cavities of the heart, blood vessels, and lymphatic vessels.

End-systolic (sis-TO-lik) **volume (ESV)** The volume of blood, about 60 mL, remaining in a ventricle after its systole (contraction).

Energy The capacity to do work.

Enkephalin (en-KEF-a-lin) A peptide found in the central nervous system that acts as a painkiller.

Enteroendocrine (en-ter-ō-EN-dō-krin) **cell** A cell of the mucosa of the gastrointestinal tract that secretes the hormones gastrin, cholecystokinin, gastric inhibitory peptide, or secretin.

Enterogastric (en-te-rō-GAS-trik) **reflex** A reflex that inhibits gastric secretion; initiated by food in the small intestine.

Enzyme (EN-zīm) A substance that affects the speed of chemical changes; an organic catalyst, usually a protein.

Eosinophil (ē′-ō-SIN-ō-fil) A type of white blood cell characterized by granules that stain red or pink with acid dyes.

Ependymal (e-PEN-de-mal) **cells** Neuroglial cells that cover choroid plexuses and produce cerebrospinal fluid (CSF); they also line the ventricles of the brain and probably assist in the circulation of CSF.

Epicardium (ep′-i-KAR-dē-um) The thin outer layer of the heart wall, composed of serous tissue and mesothelium. Also called the **visceral pericardium.**

Epidemiology (ep′-i-dē-mē-OL-ō-jē) Medical science concerned with the occurrence and distribution of diseases and disorders in human populations.

Epidermis (ep-i-DERM-is) The superficial, thinner layer of skin, composed of keratinized stratified squamous epithelium.

Epididymis (ep′-i-DID-i-mis) A comma-shaped organ that lies along the posterior border of the testis and contains the ductus epididymis, in which sperm undergo maturation. *Plural is* **epididymides** (ep′-i-DID-i-mi-dēz).

Epidural (ep′-i-DOO-ral) **space** A space between the spinal dura mater and the vertebral canal, containing areolar connective tissue and a plexus of veins.

Epiglottis (ep′-i-GLOT-is) A large, leaf-shaped piece of cartilage lying on top of the larynx, with its "stem" attached to the thyroid cartilage and its "leaf" portion unattached and free to move up and down to cover the glottis (vocal folds and rima glottidis).

Epimysium (ep´-i-MĪZ-ē-um) Fibrous connective tissue around muscles.

Epinephrine (ep-ē-NEF-rin) Hormone secreted by the adrenal medulla that produces actions similar to those that result from sympathetic stimulation. Also called **adrenaline** (a-DREN-a-lin).

Epineurium (ep´-i-NYOO-rē-um) The superficial covering around the entire nerve.

Epiphyseal (ep´-i-FIZ-ē-al) **line** The remnant of the epiphyseal plate in a long bone.

Epiphyseal (ep´-i-FIZ-ē-al) **plate** The hyaline cartilage plate between the epiphysis and diaphysis that is responsible for the lengthwise growth of long bones.

Epiphysis (ē-PIF-i-sis) The end of a long bone, usually larger in diameter than the shaft (diaphysis).

Epithalamus Uppermost portion of the diencephalon of the brain.

Epithelial (ep´-i-THĒ-lē-al) **tissue** The tissue that forms glands or the superficial part of the skin and lines blood vessels, hollow organs, and passages that lead externally from the body.

Eponychium (ep´-ō-NIK-ē-um) Narrow band of stratum corneum at the proximal border of a nail that extends from the margin of the nail wall. Also called the **cuticle.**

Erectile dysfunction Failure to maintain an erection long enough for sexual intercourse. Also known as **impotence** (IM-pō-tens).

Erection (ē-REK-shun) The enlarged and stiff state of the penis or clitoris resulting from the engorgement of the spongy erectile tissue with blood.

Eructation (e-ruk´-TĀ-shun) The forceful expulsion of gas from the stomach. Also called **belching.**

Erythema (er´-e-THĒ-ma) Skin redness usually caused by engorgement of the capillaries in the deeper layers of the skin.

Erythrocyte (e-RITH-rō-sīt) Red blood cell.

Erythropoiesis (e-rith´-rō-poy-Ē-sis) The process by which erythrocytes (red blood cells) are formed.

Erythropoietin (e-rith´-rō-POY-ē-tin) A hormone released by the kidneys that stimulates erythrocyte (red blood cell) production.

Esophagus (e-SOF-a-gus) A hollow muscular tube connecting the pharynx and the stomach.

Essential amino acids Those ten amino acids that cannot be synthesized by the human body at an adequate rate to meet its needs and therefore must be obtained from the diet.

Estrogens (ES-tro-jens) Female sex hormones produced by the ovaries concerned with the development and maintenance of female reproductive structures and secondary sex characteristics, fluid and electrolyte balance, and protein anabolism. Examples are b-estradiol, estrone, and estriol.

Etiology (ē´-tē-OL-ō-jē) The study of the causes of disease, including theories of the origin and organisms (if any) involved.

Eupnea (YOOP-nē-a) Normal quiet breathing.

Eversion (ē-VER-zhun) The movement of the sole laterally at the ankle joint or of an atrioventricular valve into an atrium during ventricular contraction.

Exacerbation (eg-zas´-er-BĀ-shun) An increase in the severity of symptoms or of disease.

Excitability (ek-sīt´-a-BIL-i-tē) The ability of muscle tissue to receive and respond to stimuli; the ability of nerve cells to respond to stimuli and convert them into nerve impulses.

Excitatory postsynaptic potential (EPSP) A small depolarization of the postsynaptic membrane when it is stimulated by an excitatory neurotransmitter. The EPSP is a localized event that decreases in strength from the point of excitation.

Excrement (EKS-kre-ment) Material eliminated from the body as waste, especially fecal matter.

Excretion (eks-KRĒ-shun) The process of eliminating waste products from the body; also the products excreted.

Exergonic (eks´-er-GON-ik) **reaction** Type of chemical reaction in which the energy released as new bonds form is greater than the energy needed to break apart old bonds; an energy-releasing reaction.

Exocrine (EK-sō-krin) **gland** A gland that secretes substances into ducts that empty at covering or lining epithelium or directly onto a free surface.

Exocytosis (ex´-ō-sī-TŌ-sis) A process of discharging large particles through the plasma membrane. The particles are enclosed in vesicles, which fuse with the plasma membrane, expelling the particles from the cell.

Exogenous (ex-SOJ-e-nus) Originating outside an organ or part.

Exon (EX-on) A region of DNA that codes for synthesis of a protein.

Expiration (ek-spi-RĀ-shun) Breathing out; expelling air from the lungs into the atmosphere. Also called **exhalation.**

Expiratory (eks-PĪ-ra-tō-rē) **reserve volume** The volume of air in excess of tidal volume that can be exhaled forcibly; about 1200 mL.

Extensibility (ek-sten´-si-BIL-i-tē) The ability of muscle tissue to be stretched when pulled.

Extension (ek-STEN-shun) An increase in the angle between two bones; restoring a body part to its anatomical position after flexion.

External Located on or near the surface.

External auditory (AW-di-tōr-ē) **canal** or **meatus** (mē-Ā-tus) A curved tube in the temporal bone that leads to the middle ear.

External ear The outer ear, consisting of the pinna, external auditory canal, and tympanic membrane (eardrum).

External nares (NA-rēz) The external nostrils, or the openings into the nasal cavity on the exterior of the body.

External respiration The exchange of respiratory gases between the lungs and blood. Also called **pulmonary respiration.**

Exteroceptor (eks´-ter-ō-SEP-tor) A receptor adapted for the reception of stimuli from outside the body.

Extracellular fluid (ECF) Fluid outside body cells, such as interstitial fluid and plasma.

Extravasation (eks-trav-a-SĀ-shun) The escape of fluid, especially blood, lymph, or serum, from a vessel into the tissues.

Extrinsic (ek-STRIN-sik) Of external origin.

Extrinsic pathway (of blood clotting) Sequence of reactions leading to blood clotting that is initiated by the release of tissue factor (TF), also known as thromboplastin, that leaks into the blood from damaged cells *outside* the blood vessels.

Exudate (EKS-yoo-dāt) Escaping fluid or semifluid material that oozes from a space and that may contain serum, pus, and cellular debris.

Eyebrow The hairy ridge superior to the eye.

Face The anterior aspect of the head.

Facilitated diffusion (fa-SIL-i-tā-ted dif-YOO-zhun) Diffusion in which a substance not soluble by itself in lipids diffuses across a selectively permeable membrane with the help of a transporter (carrier) protein.

Falciform ligament (FAL-si-form LIG-a-ment) A sheet of parietal peritoneum between the two principal lobes of the liver. The ligamentum teres, or remnant of the umbilical vein, lies within its fold.

Falx cerebelli (FALKS ser′-e-BEL-lē) A small triangular process of the dura mater attached to the occipital bone in the posterior cranial fossa and projecting inward between the two cerebellar hemispheres.

Falx cerebri (FALKS SER-e-brē) A fold of the dura mater extending deep into the longitudinal fissure between the two cerebral hemispheres.

Fascia (FASH-ē-a) A fibrous membrane covering, supporting, and separating muscles.

Fascicle (FAS-i-kul) A small bundle or cluster, especially of nerve or muscle fibers (cells). Also called a **fasciculus** (fa-SIK-yoo-lus). *Plural is* **fasciculi** (fa-SIK-yoo-lī).

Fasciculation (fa-sik′-yoo-LĀ-shun) Abnormal, spontaneous twitch of all skeletal muscle fibers in one motor unit that is visible at the skin surface; not associated with movement of the affected muscle; present in progressive diseases of motor neurons, for example, poliomyelitis and amyotrophic lateral sclerosis (ALS).

Fauces (FAW-sēz) The opening from the mouth into the pharynx.

F cell A cell in the pancreatic islets (islets of Langerhans) that secretes pancreatic polypeptide.

Feces (FĒ-sēz) Material discharged from the rectum and made up of bacteria, excretions, and food residue. Also called **stool.**

Feedback system A sequence of events in which information about the status of a situation is continually reported (fed back) to a control center.

Feeding (hunger) center A cluster of neurons in lateral nuclei of the hypothalamus that, when stimulated, brings about feeding.

Female reproductive cycle General term for the ovarian and uterine cycles, the hormonal changes that accompany them, and cyclic changes in the breasts and cervix; includes changes in the endometrium of a nonpregnant female that prepares the lining of the uterus to receive a fertilized ovum. Less correctly termed **menstrual cycle.**

Fertilization (fer′-ti-li-ZĀ-shun) Penetration of a secondary oocyte by a sperm cell and subsequent union of the nuclei of the gametes.

Fetal circulation The cardiovascular system of the fetus, including the placenta and special blood vessels involved in the exchange of materials between fetus and mother.

Fetus (FĒ-tus) The latter stages of the developing young of an animal; in humans, the developing organism *in utero* from the beginning of the third month to birth.

Fever An elevation in body temperature above the normal temperature of 37°C (98.6°F) due to a resetting of the hypothalamic thermostat.

Fibrillation (fi-bri-LĀ-shun) Abnormal, spontaneous twitch of a single skeletal muscle fiber (cell) that can be detected with electromyography but is not visible at the skin surface; not associated with movement of the affected muscle; present in certain disorders of motor neurons, for example, amyotrophic lateral sclerosis (ALS). With reference to cardiac muscle, *see* **Atrial fibrillation** and **Ventricular fibrillation.**

Fibrin (FĪ-brin) An insoluble protein that is essential to blood clotting; formed from fibrinogen by the action of thrombin.

Fibrinogen (fi-BRIN-ō-jen) A clotting factor in blood plasma that by the action of thrombin is converted to fibrin.

Fibrinolysis (fī-bri-NOL-i-sis) Dissolution of a blood clot by the action of a proteolytic enzyme, such as plasmin (fibrinolysin), that dissolves fibrin threads and inactivates fibrinogen and other blood clotting factors.

Fibroblast (FĪ-brō-blast) A large, flat cell that secretes most of the matrix (extracellular) material of areolar and dense connective tissues.

Fibrous (FĪ-brus) **joint** A joint that allows little or no movement, such as a suture or a syndesmosis.

Fibrous tunic (TOO-nik) The superficial coat of the eyeball, made up of the posterior sclera and the anterior cornea.

Fight-or-flight response The effects produced upon stimulation of the sympathetic division of the autonomic nervous system.

Filiform papilla (FIL-i-form pa-PIL-a) One of the conical projections that are distributed in parallel rows over the anterior two-thirds of the tongue and contain no taste buds.

Filtration (fil-TRĀ-shun) The flow of a liquid through a filter (or membrane that acts like a filter) due to hydrostatic pressure, as occurs in capillaries.

Filtration fraction The percentage of plasma entering the kidneys that becomes glomerular filtrate.

Filum terminale (FĪ-lum ter-mi-NAL-ē) Nonnervous fibrous tissue of the spinal cord that extends inferiorly from the conus medullaris to the coccyx.

Fimbriae (FIM-brē-ē) Fingerlike structures, especially the lateral ends of the uterine (Fallopian) tubes.

Fissure (FISH-ur) A groove, fold, or slit that may be normal or abnormal.

Fistula (FIS-choo-la) An abnormal passage between two organs or between an organ cavity and the outside.

Fixator A muscle that stabilizes the origin of the prime mover so that the prime mover can act more efficiently.

Fixed macrophage (MAK-rō-fāj) Stationary phagocytic cell found in the liver, lungs, brain, spleen, lymph nodes, subcutaneous tissue, and red bone marrow. Also called a **histiocyte** (HIS-tē-ō-sīt).

Flagellum (fla-JEL-um) A hairlike, motile process on the extremity of a bacterium, protozoan, or sperm cell. *Plural is* **flagella** (fla-JEL-a).

Flatus (FLĀ-tus) Gas in the stomach or intestines, commonly used to denote expulsion of gas through the rectum.

Flexion (FLEK-shun) Movement in which there is a decrease in the angle between two bones.

Flexor reflex A protective reflex in which flexor muscles are stimulated while extensor muscles are inhibited.

Fluid mosaic (mō-ZĀ-ik) **model** Model of plasma membrane structure that describes the molecular arrangement of the plasma membrane and other membranes in living organisms.

Follicle (FOL-i-kul) A small secretory sac or cavity; the group of cells that contains a developing oocyte in the ovaries.

Follicle-stimulating hormone (FSH) Hormone secreted by the anterior pituitary gland that initiates development of ova and stimulates the ovaries to secrete estrogens in females, and initiates sperm production in males.

Fontanel (fon′-ta-NEL) A membrane-covered spot where bone formation is not yet complete, especially between the cranial bones of an infant's skull.

Foot The terminal part of the lower limb.

Foramen (fō-RĀ-men) A passage or opening; a communication between two cavities of an organ, or a hole in a bone for passage of vessels or nerves. *Plural is* **foramina** (fō-RAM-i-na).

Foramen ovale (fō-RĀ-men ō-VAL-ē) An opening in the fetal heart in the septum between the right and left atria. A hole in the greater wing of the sphenoid bone that transmits the mandibular branch of the trigeminal (V) nerve.

Forearm (FOR-arm) The part of the upper limb between the elbow and the wrist.

Fornix (FOR-niks) An arch or fold; a tract in the brain made up of association fibers, connecting the hippocampus with the mammillary bodies; a recess around the cervix of the uterus where it protrudes into the vagina.

Fossa (FOS-a) A furrow or shallow depression.

Fourth ventricle (VEN-tri-kul) A cavity filled with cerebrospinal fluid within the brain lying between the cerebellum and the medulla oblongata and pons.

Fovea (FŌ-vē-ah) A cuplike depression in the center of the macula lutea of the retina, containing cones only; the area of clearest vision.

Fracture (FRAK-chur) Any break in a bone.

Frenulum (FREN-yoo-lum) A small fold of mucous membrane that connects two parts and limits movement.

Frontal plane A plane at a right angle to a midsagittal plane that divides the body or organs into anterior and posterior portions. Also called a **coronal** (kō-RŌ-nal) **plane.**

Functional residual (re-ZID-yoo-al) **capacity** The sum of residual volume plus expiratory reserve volume; about 2400 mL.

Fundus (FUN-dus) The part of a hollow organ farthest from the opening.

Fungiform papilla (FUN-ji-form pa-PIL-a) A mushroomlike elevation on the upper surface of the tongue appearing as a red dot; most contain taste buds.

Gallbladder A small pouch, located inferior to the liver, that stores bile and empties via the cystic duct.

Gallstone A solid mass, usually containing cholesterol, in the gallbladder or a bile-containing duct; formed anywhere between bile canaliculi in the liver and the hepatopancreatic ampulla (ampulla of Vater), where bile enters the duodenum. Also called a **biliary calculus.**

Gamete (GAM-ēt) A male or female reproductive cell; a sperm cell or secondary oocyte.

Ganglion (GANG-glē-on) Usually, a group of nerve cell bodies lying outside the central nervous system (CNS); also used for one group of nerve cell bodies within the CNS—the basal ganglia. *Plural is* **ganglia** (GANG-glē-a).

Gastric (GAS-trik) **glands** Glands in the mucosa of the stomach composed of cells that empty their secretions into narrow channels called gastric pits. Types of cells are chief cells (secrete pepsinogen), parietal cells (secrete hydrochloric acid and intrinsic factor), mucous surface and mucous neck cells (secrete mucus), and G cells (secrete gastrin).

Gastroenterology (gas'-trō-en'-ter-OL-ō-jē) The medical specialty that deals with the structure, function, diagnosis, and treatment of diseases of the stomach and intestines.

Gastrointestinal (gas-trō-in-TES-ti-nal) **(GI) tract** A continuous tube running through the ventral body cavity extending from the mouth to the anus. Also called the **alimentary** (al'-i-MEN-tar-ē) **canal.**

Gastrulation (gas'-troo-LĀ-shun) The migration of groups of cells that transform a blastula into a gastrula and establish the primary germ layers.

Gene (jēn) Biological unit of heredity; a segment of DNA located in a definite position on a particular chromosome; a sequence of DNA that codes for a particular mRNA, rRNA, or tRNA.

General adaptation syndrome (GAS) Wide-ranging set of bodily changes, triggered by a stressor, that gears the body to meet an emergency.

Generator potential The graded depolarization that results in a change in the resting membrane potential in a receptor (specialized neuronal ending); may trigger a nerve action potential (nerve impulse) if depolarization reaches threshold.

Genetic engineering The manufacture and manipulation of genetic material.

Genetics The study of genes and heredity.

Genitalia (jen'-i-TĀL-ya) Reproductive organs.

Genome (JĒ-nōm) The complete set of genes of an organism.

Genotype (JĒ-nō-tīp) The total hereditary information carried by an individual; the genetic makeup of an organism.

Geriatrics (jer'-ē-AT-riks) The branch of medicine devoted to the medical problems and care of elderly persons.

Gestation (jes-TĀ-shun) The period of development from fertilization to birth.

Gingivae (jin-JI-vē) Gums. They cover the alveolar processes of the mandible and maxilla and extend slightly into each socket.

Gland Single or group of specialized epithelial cells that secrete substances.

Glans penis (glanz PĒ-nis) The slightly enlarged region at the distal end of the penis.

Glaucoma (glaw-KŌ-ma) An eye disorder in which there is increased intraocular pressure due to an excess of aqueous humor.

Gliding joint A synovial joint having articulating surfaces that are usually flat, permitting only side-to-side and back-and-forth movements, as between carpal bones, tarsal bones, and the scapula and clavicle. Also called an **arthrodial** (ar-THRŌ-dē-al) **joint.**

Glomerular (glō-MER-yoo-lar) **capsule** A double-walled globe at the proximal end of a nephron that encloses the glomerular capillaries. Also called **Bowman's** (BŌ-manz) **capsule.**

Glomerular filtrate (glō-MER-yoo-lar FIL-trāt) The fluid produced when blood is filtered by the endothelial–capsular membrane in the glomeruli of the kidneys.

Glomerular filtration The first step in urine formation in which substances in blood are filtered at the endothelial–capsular membrane and the filtrate enters the proximal convoluted tubule of a nephron.

Glomerular filtration rate (GFR) The total volume of fluid that enters all the glomerular (Bowman's) capsules of the kidneys in 1 min; about 125 mL/min.

Glomerulus (glō-MER-yoo-lus) A rounded mass of nerves or blood vessels, especially the microscopic tuft of capillaries that is surrounded by the glomerular (Bowman's) capsule of each kidney tubule.

Glottis (GLOT-is) The vocal folds (true vocal cords) in the larynx plus the space between them (rima glottidis).

Glucagon (GLOO-ka-gon) A hormone, produced by the alpha cells of the pancreatic islets (islets of Langerhans), that increases blood glucose level.

Glucocorticoids (gloo-kō-KOR-ti-koyds) Hormones secreted by the cortex of the adrenal gland, especially cortisol, that influence glucose metabolism.

Gluconeogenesis (gloo'-kō-nē-ō-JEN-e-sis) The synthesis of glucose from certain amino acids or lactic acid.

Glucose (GLOO-kōs) A six-carbon sugar, $C_6H_{12}O_6$; the major energy source for the production of ATP by body cells.

Glucosuria (gloo'-kō-SOO-rē-a) The presence of glucose in the urine; may be temporary or pathological. Also called **glycosuria.**

Glycogen (GLĪ-kō-jen) A highly branched polymer of glucose containing thousands of subunits; functions as a compact store of glucose molecules in liver and muscle fibers (cells).

Glycogenesis (glī′-kō-JEN-e-sis) The process by which many molecules of glucose combine to form a molecule called glycogen.

Glycogenolysis (glī-kō-je-NOL-i-sis) The breakdown of glycogen into glucose.

Glycolysis (glī-KOL-i-sis) Series of chemical reactions in the cytosol of a cell in which a molecule of glucose is split into two molecules of pyruvic acid with production of two ATPs.

Goblet cell A goblet-shaped unicellular gland that secretes mucus; present in epithelium of airways and intestines.

Goiter (GOY-ter) An enlargement of the thyroid gland.

Golgi (GOL-jē) **complex** An organelle in the cytoplasm of cells consisting of four to six flattened sacs (cisterns), stacked on one another, with expanded areas at their ends; functions in processing, sorting, packaging, and delivering proteins and lipids to the plasma membrane, lysosomes, and secretory vesicles.

Gomphosis (gom-FŌ-sis) A fibrous joint in which a cone-shaped peg fits into a socket.

Gonad (GŌ-nad) A gland that produces gametes and hormones; the ovary in the female and the testis in the male.

Gonadotropic hormone Anterior pituitary hormone affecting the gonads.

Gray commissure (KOM-i-shur) A narrow strip of gray matter connecting the two lateral gray masses within the spinal cord.

Gray matter Area in the central nervous system and ganglia consisting of nonmyelinated nerve tissue.

Gray ramus communicans (RĀ-mus kō-MYOO-ni-kans) A short nerve containing postganglionic sympathetic fibers; the cell bodies of the fibers are in a sympathetic chain ganglion, and the nonmyelinated axons extend by way of the gray ramus to a spinal nerve and then to the periphery to supply smooth muscle in blood vessels, arrector pili muscles, and sweat glands. *Plural is* **rami communicantes** (RĀ-mē kō-myoo-ni-KAN-tēz).

Greater omentum (ō-MEN-tum) A large fold in the serosa of the stomach that hangs down like an apron anterior to the intestines.

Greater vestibular (ves-TIB-yoo-lar) **glands** A pair of glands on either side of the vaginal orifice that open by a duct into the space between the hymen and the labia minora. Also called **Bartholin's** (BAR-to-linz) **glands.**

Groin (GROYN) The depression between the thigh and the trunk; the inguinal region.

Gross anatomy The branch of anatomy that deals with structures that can be studied without using a microscope. Also called **macroscopic anatomy.**

Growth An increase in size due to an increase in (1) the number of cells, (2) the size of existing cells as internal components increase in size, or (3) the size of intercellular substances.

Gustatory (GUS-ta-tō′-rē) Pertaining to taste.

Gynecology (gī′-ne-KOL-ō-jē) The branch of medicine dealing with the study and treatment of disorders of the female reproductive system.

Gyrus (JĪ-rus) One of the folds of the cerebral cortex of the brain. *Plural is* **gyri** (JĪ-rī). Also called a **convolution.**

Hair A threadlike structure produced by hair follicles that develops in the dermis. Also called **pilus** (PĪ-lus).

Hair follicle (FOL-li-kul) Structure, composed of epithelium and surrounding the root of a hair, from which hair develops.

Hair root plexus (PLEK-sus) A network of dendrites arranged around the root of a hair as free or naked nerve endings that are stimulated when a hair shaft is moved.

Haldane (HAWL-dān) **effect** In the presence of oxygen, less carbon dioxide binds in the blood because when oxygen combines with hemoglobin, the hemoglobin becomes a stronger acid, which combines with less carbon dioxide.

Hand The terminal portion of an upper limb, including the carpus, metacarpus, and phalanges.

Haploid (HAP-loyd) Having half the number of chromosomes characteristically found in the somatic cells of an organism; characteristic of mature gametes. Symbolized *n.*

Hard palate (PAL-at) The anterior portion of the roof of the mouth, formed by the maxillae and palatine bones and lined by mucous membrane.

Haustra (HAWS-tra) The sacculated elevations of the colon. *Singular is* **haustrum.**

Head The superior part of a human, cephalic to the neck. The superior or proximal part of a structure.

Heart A hollow muscular organ lying slightly to the left of the midline of the chest that pumps the blood through the cardiovascular system.

Heart block An arrhythmia (dysrhythmia) of the heart in which the atria and ventricles contract independently because of a blocking of electrical impulses through the heart at some point in the conduction system.

Heart murmur (MER-mer) An abnormal sound that consists of a flow noise that is heard before, between, or after the normal lubb–dupp or that may mask normal heart sounds.

Hematocrit (hē-MAT-ō-krit) **(Hct)** The percentage of blood made up of red blood cells. Usually calculated by centrifuging a blood sample in a graduated tube and then reading the volume of red blood cells and total blood.

Hematology (hē′-ma-TOL-ō-jē) The study of blood.

Hemiplegia (hem-i-PLĒ-jē-a) Paralysis of the upper limb, trunk, and lower limb on one side of the body.

Hemodynamics (hē-mō-dī-NA-miks) The study of factors and forces that govern the flow of blood through blood vessels.

Hemoglobin (hē′-mō-GLŌ-bin) **(Hb)** A substance in red blood cells consisting of the protein globin and the iron-containing red pigment heme and constituting about 33% of the cell volume; involved in the transport of oxygen and carbon dioxide.

Hemolysis (hē-MOL-i-sis) The escape of hemoglobin from the interior of an erythrocyte (red blood cell) into the surrounding medium; results from disruption of the cell membrane by toxins or drugs, freezing or thawing, or hypotonic solutions.

Hemophilia (hē′-mō-FĒL-ē-a) A hereditary blood disorder where there is a deficient production of certain factors involved in blood clotting, resulting in excessive bleeding into joints, deep tissues, and elsewhere.

Hemopoiesis (hē-mō-poy-Ē-sis) Blood cell production occurring in red bone marrow. Also called **hematopoiesis** (hem′-a-tō-poy-Ē-sis).

Hemorrhage (HEM-or-rij) Bleeding; the escape of blood from blood vessels, especially when the loss is profuse.

Hemorrhoids (HEM-ō-royds) Dilated or varicosed blood vessels (usually veins) in the anal region. Also called **piles.**

Hemostasis (hē-MOS-tā-sis) The stoppage of bleeding.

Heparin (HEP-a-rin) An anticoagulant given to slow the conversion of prothrombin to thrombin, thus reducing the risk of blood clot formation; also found naturally in basophils and most cells.

Hepatic (he-PAT-ik) Refers to the liver.

Hepatic duct A duct that receives bile from the bile capillaries. Small hepatic ducts merge to form the larger right and left hepatic ducts that unite to leave the liver as the common hepatic duct.

Hepatic portal circulation The flow of blood from the gastrointestinal organs to the liver before returning to the heart.

Hepatocyte (he-PAT-ō-cyte) A liver cell.

Hepatopancreatic (hep′-a-tō-pan′-krē-A-tik) **ampulla** A small, raised area in the duodenum where the combined common bile duct and main pancreatic duct empty into the duodenum. Also called the **ampulla of Vater** (VA-ter).

Hernia (HER-nē-a) The protrusion or projection of an organ or part of an organ through a membrane or cavity wall, usually the abdominal cavity.

Herniated (HER-nē-Ā′-ted) **disc** A rupture of an intervertebral disc so that the nucleus pulposus protrudes into the vertebral cavity. Also called a **slipped disc.**

Heterozygous (he-ter-ō-ZĪ-gus) Possessing a pair of different genes on homologous chromosomes for a particular hereditary characteristic.

Hiatus (hī-Ā-tus) An opening; a foramen.

Hilus (HĪ-lus) An area, depression, or pit where blood vessels and nerves enter or leave an organ. Also called a **hilum.**

Hinge joint A synovial joint in which a convex surface of one bone fits into a concave surface of another bone, such as the elbow, knee, ankle, and interphalangeal joints. Also called a **gingly-mus** (JIN-gli-mus) **joint.**

Histamine (HISS-ta-mēn) Substance found in many cells, especially mast cells, basophils, and platelets, released when the cells are injured; results in vasodilation, increased permeability of blood vessels, and constriction of bronchioles.

Histology (hiss-TOL-ō-jē) Microscopic study of the structure of tissues.

Holocrine (HŌL-ō-krin) **gland** A type of gland in which entire secretory cells, along with their accumulated secretions, make up the secretory product of the gland, as in the sebaceous (oil) glands.

Homeostasis (hō′-mē-ō-STĀ-sis) The condition in which the body's internal environment remains relatively constant, within physiological limits.

Homologous chromosomes Two chromosomes that belong to a pair. Also called **homologues.**

Homozygous (hō-mō-ZĪ-gus) Possessing a pair of similar genes on homologous chromosomes for a particular hereditary characteristic.

Hormone (HOR-mōn) A secretion of endocrine cells that alters the physiological activity of target cells of the body.

Horn Principal area of gray matter in the spinal cord.

Human chorionic gonadotropin (hCG) (kō-rē-ON-ik gō-nad-ō-TRŌ-pin) A hormone produced by the developing placenta that maintains the corpus luteum.

Human chorionic somatomammotropin (sō-mat-ō-mam-ō-TRŌ-pin) **(hCS)** A hormone produced by the chorion of the placenta that may stimulate breast tissue for lactation, enhance body growth, and regulate metabolism. Also called **human placental lactogen (hPL).**

Human growth hormone (hGH) Hormone secreted by the anterior pituitary gland that stimulates growth of body tissues, especially skeletal and muscular. Also known as **somatotropin** and **somatotropic hormone (STH).**

Hyaluronic (hī′-a-loo-RON-ik) **acid** A viscous, amorphous extracellular material that binds cells together, lubricates joints, and maintains the shape of the eyeballs.

Hyaluronidase (hī′-a-loo-RON-i-dās) An enzyme that breaks down hyaluronic acid, increasing the permeability of connective tissues by dissolving the substances that hold body cells together.

Hydride ion (HI-drīd Ī-on) A hydrogen nucleus with two orbiting electrons (H⁻), in contrast to the more common **hydrogen ion,** which has no orbiting electrons (H⁺).

Hymen (HĪ-men) A thin fold of vascularized mucous membrane at the vaginal orifice.

Hypercalcemia (hī′-per-kal-SĒ-mē-a) An excess of calcium in the blood.

Hypercapnia (hī′-per-KAP-nē-a) An abnormal increase in the amount of carbon dioxide in the blood.

Hyperextension (hī′-per-ek-STEN-shun) Continuation of extension beyond the anatomical position, as in bending the head backward.

Hyperglycemia (hī′-per-glī-SĒ-mē-a) An elevated blood glucose level.

Hyperkalemia (hī′-per-kā-LĒ-mē-a) An excess of potassium ions in the blood.

Hypermagnesemia (hī′-per-mag′-ne-SĒ-mē-a) An excess of magnesium ions in the blood.

Hypermetropia (hī′-per-mē-TRŌ-pē-a) A condition in which visual images are focused behind the retina, with resulting defective vision of near objects; farsightedness.

Hyperphosphatemia (hī-per-fos′-fa-TĒ-mē-a) An abnormally high level of phosphates in the blood.

Hyperplasia (hī′-per-PLĀ-zē-a) An abnormal increase in the number of normal cells in a tissue or organ, increasing its size.

Hyperpolarization (hī′-per-PŌL-a-ri-zā′-shun) Increase in the internal negativity across a cell membrane, thus increasing the voltage and moving it farther away from the threshold value.

Hypersecretion (hī′-per-se-KRĒ-shun) Overactivity of glands resulting in excessive secretion.

Hypersensitivity (hī′-per-sen-si-TI-vi-tē) Overreaction to an allergen that results in pathological changes in tissues. Also called **allergy.**

Hypertension (hī′-per-TEN-shun) High blood pressure.

Hyperthermia (hī′-per-THERM-ē-a) An elevated body temperature.

Hypertonia (hī-per-TŌ-nē-a) Increased muscle tone that is expressed as spasticity or rigidity.

Hypertonic (hī′-per-TON-ik) Solution that causes cells to shrink due to loss of water by osmosis.

Hypertrophy (hī-PER-trō-fē) An excessive enlargement or overgrowth of tissue without cell division.

Hyperventilation (hī′-per-ven-ti-LĀ-shun) A rate of respiration higher than that required to maintain a normal level of plasma PCO₂.

Hypocalcemia (hī′-pō-kal-SĒ-mē-a) A below normal level of calcium in the blood.

Hypochloremia (hī′-pō-klō-RĒ-mē-a) Deficiency of chloride ions in the blood.

Hypoglycemia (hī′-pō-glī-SĒ-mē-a) An abnormally low concentration of glucose in the blood; can result from excess insulin (injected or secreted).

Hypokalemia (hī′-pō-ka-LĒ-mē-a) Deficiency of potassium ions in the blood.

Hypomagnesemia (hī′-pō-mag′-ne-SĒ-mē-a) Deficiency of magnesium ions in the blood.

Hyponatremia (hī′-pō-na-TRĒ-mē-a) Deficiency of sodium ions in the blood.

Hyponychium (hī-pō-NIK-ē-um) Free edge of the fingernail.

Hypophosphatemia (hī-pō-fos′-fa-TĒ-mē-a) An abnormally low level of phosphates in the blood.

Hypophyseal (hī′-pō-FIZ-ē-al) **pouch** An outgrowth of ectoderm from the roof of the mouth from which the anterior pituitary gland develops.

Hypophysis (hī-POF-i-sis) Pituitary gland.

Hyposecretion (hī′-pō-se-KRĒ-shun) Underactivity of glands resulting in diminished secretion.

Hypothalamohypophyseal (hī-pō-tha-lam′-ō-hī-pō-FIZ-ē-al) **tract** A bundle of axons that have their cell bodies in the hypothalamus but release their neurosecretions in the posterior pituitary gland.

Hypothalamus (hī-pō-THAL-a-mus) A portion of the diencephalon, lying beneath the thalamus and forming the floor and part of the wall of the third ventricle.

Hypothermia (hī-pō-THER-mē-a) Lowering of body temperature below 35°C (95°F); in surgical procedures, it refers to deliberate cooling of the body to slow down metabolism and reduce oxygen needs of tissues.

Hypotonia (hī′-pō-TŌ-nē-a) Decreased or lost muscle tone in which muscles appear flaccid.

Hypotonic (hī′-pō-TON-ik) Solution that causes cells to swell and perhaps rupture due to gain of water by osmosis.

Hypoventilation (hī-pō-ven-ti-LĀ-shun) A rate of respiration lower than that required to maintain a normal level of plasma $p\text{CO}_2$.

Hypovolemic (hī-pō-vō-LĒ-mik) **shock** A type of shock characterized by decreased intravascular volume resulting from blood loss; may be caused by acute hemorrhage or excessive fluid loss.

Hypoxia (hī-POKS-ē-a) Lack of adequate oxygen at the tissue level.

Hysterectomy (hiss-te-REK-tō-mē) The surgical removal of the uterus.

Ileocecal (il-ē-ō-SĒ-kal) **sphincter** A fold of mucous membrane that guards the opening from the ileum into the large intestine. Also called the **ileocecal valve.**

Ileum (IL-ē-um) The terminal portion of the small intestine.

Immunity (im-YOO-ni-tē) The state of being resistant to injury, particularly by poisons, foreign proteins, and invading pathogens.

Immunogenicity (im-yoo-nō-jen-IS-it-ē) Ability of an antigen to provoke an immune response.

Immunoglobulin (im-yoo-nō-GLOB-yoo-lin) **(Ig)** An antibody synthesized by plasma cells derived from B lymphocytes in response to the introduction of antigen. Immunoglobulins are divided into five kinds (IgG, IgM, IgA, IgD, IgE) based primarily on the larger protein component present in the immunoglobulin.

Immunology (im′-yoo-NOL-ō-jē) The branch of science that deals with the responses of the body when challenged by antigens.

Implantation (im-plan-TĀ-shun) The insertion of a tissue or a part into the body. The attachment of the blastocyst to the lining of the uterus 7–8 days after fertilization.

Incontinence (in-KON-ti-nens) Inability to retain urine, semen, or feces through loss of sphincter control.

Indirect motor pathways Motor tracts that convey information from the brain down the spinal cord for automatic movements, coordination of body movements with visual stimuli, skeletal muscle tone and posture, and balance. Also known as **extrapyramidal pathways.**

Infarction (in-FARK-shun) A localized area of necrotic tissue, produced by inadequate oxygenation of the tissue.

Infection (in-FEK-shun) Invasion and multiplication of microorganisms in body tissues, which may be inapparent or characterized by cellular injury.

Inferior (in-FĒR-ē-or) Away from the head or toward the lower part of a structure. Also called **caudad** (KAW-dad).

Inferior vena cava (VĒ-na CĀ-va) **(IVC)** Large vein that collects blood from parts of the body inferior to the heart and returns it to the right atrium.

Infertility Inability to conceive or to cause conception. Also called **sterility.**

Inflammation (in′-fla-MĀ-shun) Localized, protective response to tissue injury designed to destroy, dilute, or wall off the infecting agent or injured tissue; characterized by redness, pain, heat, swelling, and sometimes loss of function.

Inflation reflex Reflex that prevents overinflation of the lungs. Also called **Hering–Breuer reflex**.

Infundibulum (in′-fun-DIB-yoo-lum) The stalklike structure that attaches the pituitary gland to the hypothalamus of the brain. The funnel-shaped, open, distal end of the uterine (Fallopian) tube.

Ingestion (in-JES-chun) The taking in of food, liquids, or drugs, by mouth.

Inguinal (IN-gwi-nal) Pertaining to the groin.

Inguinal canal An oblique passageway in the anterior abdominal wall just superior and parallel to the medial half of the inguinal ligament that transmits the spermatic cord and ilioinguinal nerve in the male and round ligament of the uterus and ilioinguinal nerve in the female.

Inheritance The acquisition of body characteristics and qualities by transmission of genetic information from parents to offspring.

Inhibin A hormone secreted by the gonads that inhibits release of follicle-stimulating hormone (FSH) by the anterior pituitary gland.

Inhibiting hormone Chemical secretion of the hypothalamus that can suppress secretion of hormones by the anterior pituitary gland.

Inhibitory postsynaptic potential (IPSP) A small hyperpolarization caused by neurotransmitter at a synapse in which the membrane potential becomes more negative.

Inner cell mass A region of cells of a blastocyst that differentiates into the three primary germ layers—ectoderm, mesoderm, and endoderm—from which all tissues and organs develop; also called an **embryoblast.**

Inorganic (in′-or-GAN-ik) **compound** Compound that usually lacks carbon, usually is small, and often contains ionic bonds. Examples include water and many acids, bases, and salts.

Insertion (in-SER-shun) The attachment of a muscle tendon to a movable bone or the end opposite the origin.

Inspiration (in-spi-RĀ-shun) The act of drawing air into the lungs. Also termed **inhalation.**

Inspiratory (in-SPĪ-ra-tor-ē) **capacity** Total inspiratory capacity of the lungs; the total of tidal volume plus inspiratory reserve volume; averages 3600 mL.

Inspiratory (in-SPĪ-ra-tor-ē) **reserve volume** Additional inspired air over and above tidal volume; averages 3100 mL.

Insula (IN-su-la) A triangular area of cerebral cortex that lies deep within the lateral cerebral fissue, under the parietal, frontal, and temporal lobes.

Insulin (IN-su-lin) A hormone produced by the beta cells of a pancreatic islet (islet of Langerhans) that decreases the blood glucose level.

Insulinlike growth factor (IGF) Small protein, produced by the liver and other tissues in response to stimulation by human growth hormone (hGH), that mediates most of the effects of human growth hormone. Previously called **somatomedin** (sō′-ma-tō-MĒ-din).

Integrin (IN-te-grin) Receptor on a plasma membrane that interacts with an adhesion protein found in intercellular material and blood.

Integumentary (in-teg′-yoo-MEN-tar-ē) Relating to the skin.

Intercalated (in-TER-ka-lāt-ed) **disc** An irregular transverse thickening of sarcolemma that contains desmosomes that hold cardiac muscle fibers (cells) together and gap junctions that aid in conduction of muscle action potentials.

Intercostal (in′-ter-KOS-tal) **nerve** A nerve supplying a muscle located between the ribs.

Interferons (in′-ter-FĒR-ons) **(IFNs)** Three principal types of protein (alpha, beta, gamma) naturally produced by virus-infected host cells that induce uninfected cells to synthesize antiviral proteins that inhibit intracellular viral replication in uninfected host cells.

Intermediate Between two structures, one of which is medial and one of which is lateral.

Intermediate filament Protein filament, ranging from 8 to 12 nm in diameter, that may provide structural reinforcement, hold organelles in place, and give shape to a cell.

Internal Away from the surface of the body.

Internal capsule A tract of projection fibers connecting various parts of the cerebral cortex and lying between the thalamus and the caudate and lentiform nuclei of the basal ganglia.

Internal ear The inner ear or labyrinth, lying inside the temporal bone, containing the organs of hearing and balance.

Internal nares (NA-rēz) The two openings posterior to the nasal cavities opening into the nasopharynx. Also called the **choanae** (kō-A-nē).

Internal respiration The exchange of respiratory gases between blood and body cells. Also called **tissue respiration.**

Interneuron (in′-ter-NOO-ron) A nerve cell lying completely within the central nervous system. Also called an **association neuron.**

Interoceptor (in′-ter-ō-SEP-tor) Receptor located in blood vessels and viscera that provides information about the body's internal environment. Also termed **visceroceptor** (vis′-er-ō-SEP-tor).

Interphase (IN-ter-fāz) The period of the cell cycle between cell divisions, consisting of the G₁- (gap or growth) phase, when the cell is engaged in growth, metabolism, and production of substances required for division; S- (synthesis) phase, during which chromosomes are replicated; and G₂-phase.

Interstitial (in′-ter-STISH-al) **fluid** The portion of extracellular fluid that fills the microscopic spaces between the cells of tissues; the internal environment of the body. Also called **intercellular** or **tissue fluid.**

Interstitial growth Growth from within, as in the growth of cartilage. Also called **endogenous** (en-DOJ-e-nus) **growth.**

Interventricular (in′-ter-ven-TRIK-yoo-lar) **foramen** A narrow, oval opening through which the lateral ventricles of the brain communicate with the third ventricle. Also called the **foramen of Monro.**

Intervertebral (in′-ter-VER-te-bral) **disc** A pad of fibrocartilage located between the bodies of two vertebrae.

Intestinal gland A gland that opens onto the surface of the intestinal mucosa and secretes digestive enzymes. Also called a **crypt of Lieberkühn** (LĒ-ber-kyoon).

Intracellular (in′-tra-SEL-yoo-lar) **fluid (ICF)** Fluid located within cells. Also called **cytosol** (SĪ-tō-sol).

Intrafusal (in′-tra-FYOO-zal) **fibers** Three to ten specialized muscle fibers (cells), partially enclosed in a spindle-shaped connective tissue capsule; these fibers make up a muscle spindle.

Intramembranous ossification (in′-tra-MEM-bra-nus os′-i′-fi-KĀ-shun) The method of bone formation in which the bone is formed directly in membranous tissue.

Intraocular (in-tra-OK-yoo-lar) **pressure (IOP)** Pressure in the eyeball, produced mainly by aqueous humor.

Intrapleural pressure Air pressure between the two pleural layers of the lungs, usually subatmospheric. Also called **intrathoracic pressure.**

Intrinsic (in-TRIN-sik) Of internal origin.

Intrinsic pathway (of blood clotting) Sequence of reactions leading to blood clotting that is initiated by damage to blood vessel endothelium or platelets; activators of this pathway are contained *within* blood itself or are in direct contact with blood.

Intrinsic factor (IF) A glycoprotein, synthesized and secreted by the parietal cells of the gastric mucosa, that facilitates vitamin B₁₂ absorption in the small intestine.

Intron (IN-tron) A region of DNA that does not code for the synthesis of a protein.

In utero (YOO-ter-ō) Within the uterus.

Invagination (in-vaj′-i-NĀ-shun) The pushing of the wall of a cavity into the cavity itself.

Inversion (in-VER-zhun) The movement of the sole medially at the ankle joint.

In vitro (VĒ-trō) Literally, in glass; outside the living body and in an artificial environment such as a laboratory test tube.

In vivo (VĒ-vō) In the living body.

Ion (Ī-on) Any charged particle or group of particles; usually formed when a substance, such as a salt, dissolves and dissociates.

Ionization (ī′-on-i-ZĀ-shun) Separation of inorganic acids, bases, and salts into ions when dissolved in water. Also called **dissociation.**

Ipsilateral (ip′-si-LAT-er-al) On the same side, affecting the same side of the body.

Iris The colored portion of the eyeball seen through the cornea that consists of circular and radial smooth muscle; the hole in the center of the iris is the pupil.

Ischemia (is-KĒ-mē-a) A lack of sufficient blood to a part due to obstruction or constriction of a blood vessel.

Isoantibody A specific antibody in blood plasma that reacts with specific isoantigens and causes the clumping of bacteria, red blood cells, or particles. Also called an **agglutinin.**

Isoantigen A genetically determined antigen located on the surface of red blood cells; basis for the ABO grouping and Rh system of blood classification. Also called an **agglutinogen.**

Isometric contraction A muscle contraction in which tension on the muscle increases, but there is only minimal muscle shortening so that no movement is produced.

Isotonic (ī′-sō-TON-ik) Having equal tension or tone. A solution having the same concentration of impermeable solutes as cytosol.

Isotonic (ī-sō-TON-ik) **contraction** Contraction in which the tension remains the same; occurs when a constant load is moved through the range of motions possible at a joint.

Isotopes (Ī-sō-tōps′) Chemical elements that have the same number of protons but different numbers of neutrons. Radioactive isotopes change into other elements with the emission of alpha or beta particles or gamma rays.

Isovolumetric (ī-sō-vol-yoo′-MET-rik) **contraction** The period of time, about 0.05 sec, between the start of ventricular systole and the opening of the semilunar valves; there is contraction of the ventricles, but no emptying, and there is a rapid rise in ventricular pressure.

Isovolumetric relaxation The period of time, about 0.05 sec, between the closing of the semilunar valves and the opening of the atrioventricular (AV) valves; there is a drastic decrease in ventricular pressure without a change in ventricular volume.

Isthmus (IS-mus) A narrow strip of tissue or narrow passage connecting two larger parts.

Jejunum (jē-JOO-num) The middle portion of the small intestine.

Joint kinesthetic (kin′-es-THET-ik) **receptor** A proprioceptive receptor located in a joint, stimulated by joint movement.

Juxtaglomerular (juks-ta-glō-MER-yoo-lar) **apparatus (JGA)** Consists of the macula densa (cells of the distal convoluted tubule adjacent to the afferent and efferent arteriole) and juxtaglomerular cells (modified cells of the afferent and sometimes efferent arteriole); secretes renin when blood pressure starts to fall.

Keratin (KER-a-tin) An insoluble protein found in the hair, nails, and other keratinized tissues of the epidermis.

Keratinocyte (ke-RAT-in′-ō-sīt) The most numerous of the epidermal cells; function in the production of keratin.

Ketone (KĒ-tōn) **bodies** Substances produced primarily during excessive triglyceride catabolism, such as acetone, acetoacetic acid, and b-hydroxybutyric acid.

Ketosis (kē-TŌ-sis) Abnormal condition marked by excessive production of ketone bodies.

Kidney (KID-nē) One of the paired reddish organs located in the lumbar region that regulates the composition, volume, and pressure of blood and produces urine.

Kidney stone A solid mass, usually consisting of calcium oxalate, uric acid, or calcium phosphate crystals, that may form in any portion of the urinary tract. Also called a **renal calculus.**

Kilocalorie (KIL-ō-kal′-ō-rē) **(kcal)** The amount of heat required to raise the temperature of 1000 g of water 1°C from 14°C to 15°C; the unit used to express the heating value of foods and to measure metabolic rate.

Kinesiology (ki-nē′-sē-OL-ō-jē) The study of the movement of body parts.

Kinesthesia (kin-is-THĒ-szē-a) Ability to perceive extent, direction, or weight of movement; muscle sense.

Korotkoff (kō-ROT-kof) **sounds** The various sounds that are heard while taking blood pressure.

Krebs cycle A series of biochemical reactions that occurs in the matrix of mitochondria in which electrons are transferred to coenzymes and carbon dioxide is formed. The electrons carried by the coenzymes then enter the electron transport chain, which generates a large quantity of ATP. Also called the **citric acid cycle** or **tricarboxylic acid (TCA) cycle.**

Kyphosis (kī-FŌ-sis) An exaggeration of the thoracic curve of the vertebral column, resulting in a "round-shouldered" appearance. Also called **hunchback.**

Labial frenulum (LĀ-bē-al FREN-yoo-lum) A medial fold of mucous membrane between the inner surface of the lip and the gums.

Labia majora (LĀ-bē-a ma-JŌ-ra) Two longitudinal folds of skin extending downward and backward from the mons pubis of the female.

Labia minora (min-OR-a) Two small folds of mucous membrane lying medial to the labia majora of the female.

Labium (LĀ-bē-um) A lip. A liplike structure. *Plural is* **labia** (LĀ-bē-a).

Labor The process by which a fetus is expelled from the uterus through the vagina.

Labyrinth (LAB-i-rinth) Intricate communicating passageway, especially in the internal ear.

Lacrimal canal A duct, one on each eyelid, beginning at the punctum at the medial margin of an eyelid and conveying tears medially into the nasolacrimal sac.

Lacrimal gland Secretory cells, located at the superior anterolateral portion of each orbit, that secrete tears into excretory ducts that open onto the surface of the conjunctiva.

Lacrimal sac The superior expanded portion of the nasolacrimal duct that receives the tears from a lacrimal canal.

Lactation (lak-TĀ-shun) The secretion and ejection of milk by the mammary glands.

Lacteal (LAK-tē-al) One of many lymphatic vessels in villi of the intestines that absorb triglycerides and other lipids from digested food.

Lacuna (la-KOO-na) A small, hollow space, such as that found in bones in which the osteocytes lie. *Plural is* **lacunae** (la-KOO-nē).

Lambdoid (lam-DOYD) **suture** The joint in the skull between the parietal bones and the occipital bone; sometimes contains sutural (Wormian) bones.

Lamellae (la-MEL-ē) Concentric rings of hard, calcified matrix found in compact bone.

Lamellated corpuscle Oval-shaped pressure receptor located in subcutaneous tissue and consisting of concentric layers of connective tissue wrapped around a sensory nerve fiber. Also called a **Pacinian** (pa-SIN-ē-an) **corpuscle.**

Lamina (LAM-i-na) A thin, flat layer or membrane, as the flattened part of either side of the arch of a vertebra. *Plural is* **laminae** (LAM-i-nē).

Lamina propria (PRŌ-prē-a) The connective tissue layer of a mucous membrane.

Langerhans (LANG-er-hans) **cell** Epidermal dendritic cell that functions as an antigen-presenting cell (APC) during an immune response.

Lanugo (lan-YOO-gō) Fine downy hairs that cover the fetus.

Large intestine The portion of the gastrointestinal tract extending from the ileum of the small intestine to the anus, divided structurally into the cecum, colon, rectum, and anal canal.

Laryngopharynx (la-rin′-gō-FAR-inks) The inferior portion of the pharynx, extending downward from the level of the hyoid bone that divides posteriorly into the esophagus and anteriorly into the larynx. Also called the **hypopharynx.**

Laryngotracheal (la-rin′-gō-TRĀ-kē-al) **bud** An outgrowth of endoderm of the foregut from which the respiratory system develops.

Larynx (LAR-inks) The voice box, a short passageway that connects the pharynx with the trachea.

Lateral (LAT-er-al) Farther from the midline of the body or a structure.

Lateral ventricle (VEN-tri-kul) A cavity within a cerebral hemisphere that communicates with the lateral ventricle in the other cerebral hemisphere and with the third ventricle by way of the interventricular foramen.

Leg The part of the lower limb between the knee and the ankle.

Lens A transparent organ constructed of proteins (crystallins) lying posterior to the pupil and iris of the eyeball and anterior to the vitreous body.

Lesion (LĒ-zhun) Any localized, abnormal change in a body tissue.

Lesser omentum (ō-MEN-tum) A fold of the peritoneum that extends from the liver to the lesser curvature of the stomach and the first part of the duodenum.

Lesser vestibular (ves-TIB-yoo-lar) **gland** One of the paired mucus-secreting glands with ducts that open on either side of the urethral orifice in the vestibule of the female.

Leukocyte (LOO-kō-sīt) A white blood cell.

Leukotriene (loo′-kō-TRĪ-ēn) A type of eicosanoid produced by basophils and mast cells; acts as a local hormone; produces increased vascular permeability and acts as a chemotactic agent for phagocytes in tissue inflammation.

Leydig (LĪ-dig) **cell** A type of cell that secretes testosterone; located in the connective tissue between seminiferous tubules in a mature testis. Also known as **interstitial cell of Leydig** or **interstitial endocrinocyte.**

Libido (li-BĒ-dō) Sexual desire.

Ligament (LIG-a-ment) Dense, regular, connective tissue that attaches bone to bone.

Ligand (LĪ-gand) A chemical substance that binds to a specific receptor.

Limbic system A portion of the forebrain, sometimes termed the visceral brain, concerned with various aspects of emotion and behavior; includes the limbic lobe, dentate gyrus, amygdala, septal nuclei, mammillary bodies, anterior thalamic nucleus, olfactory bulbs, and bundles of myelinated axons.

Lingual frenulum (LIN-gwal FREN-yoo-lum) A fold of mucous membrane that connects the tongue to the floor of the mouth.

Lingual lipase (LĪ-pās) Digestive enzyme secreted by glands on the dorsum of the tongue that digests triglycerides.

Lipase An enzyme that splits fatty acids from triglycerides and phospholipids.

Lipid (LIP-id) An organic compound composed of carbon, hydrogen, and oxygen that is usually insoluble in water, but soluble in alcohol, ether, and chloroform; examples include triglycerides (fats and oils), phospholipids, steroids, and eicosanoids.

Lipid bilayer Arrangement of phospholipid, glycolipid, and cholesterol molecules in two parallel sheets in which the hydrophilic "heads" face outward and the hydrophobic "tails" face inward; found in cellular membranes.

Lipogenesis (li-pō-GEN-e-sis) The synthesis of triglycerides from glucose or amino acids by hepatocytes.

Lipolysis (lip-OL-i-sis) The splitting of fatty acids from a triglyceride (fat) or phospholipid molecule.

Lipoprotein (lip′-ō-PRŌ-tēn) One of several types of particles containing lipids (cholesterol and triglycerides) and proteins that make it water-soluble for transport in the blood; high levels of **low-density lipoproteins (LDLs)** are associated with increased risk of atherosclerosis, whereas high levels of **high-density lipoproteins (HDLs)** are associated with decreased risk of atherosclerosis.

Liver Large gland under the diaphragm that occupies most of the right hypochondriac region and part of the epigastric region. Functionally, it produces bile and synthesizes most plasma proteins; converts one nutrient into another; detoxifies substances; stores glycogen, minerals, and vitamins; carries on phagocytosis of worn-out blood cells and bacteria; and helps synthesize the active form of vitamin D.

Lobe (LŌB) A curved or rounded projection.

Long-term potentiation (LTP) Prolonged, enhanced synaptic transmission that occurs at certain synapses within the hippocampus of the brain; believed to underlie some aspects of memory.

Lordosis (lor-DŌ-sis) An exaggeration of the lumbar curve of the vertebral column. Also called **swayback.**

Lower limb The appendage attached at the pelvic (hip) girdle, consisting of the thigh, knee, leg, ankle, foot, and toes. Also called **lower extremity.**

Lumbar (LUM-bar) Region of the back and side between the ribs and pelvis; loin.

Lumbar plexus (PLEK-sus) A network formed by the anterior (ventral) branches of spinal nerves L1 through L4.

Lumen (LOO-men) The space within an artery, vein, intestine, renal tubule, or other tubelike structure.

Lungs Main organs of respiration that lie on either side of the heart in the thoracic cavity.

Lunula (LOO-nyoo-la) The moon-shaped white area at the base of a nail.

Luteinizing (LOO-tē-in′-īz-ing) **hormone (LH)** A hormone secreted by the anterior pituitary gland that stimulates ovulation and progesterone secretion by the corpus luteum, and readies the mammary glands for milk secretion in females and stimulates testosterone secretion by the testes in males.

Lymph (limf) Fluid confined in lymphatic vessels and flowing through the lymphatic system until it is returned to the blood.

Lymphatic (lim-FAT-ik) **capillary** Closed-ended microscopic lymphatic vessel that begins in spaces between cells and converges with other lymphatic capillaries to form lymphatic vessels.

Lymphatic tissue A specialized form of reticular tissue that contains large numbers of lymphocytes.

Lymphatic vessel A large vessel that collects lymph from lymphatic capillaries and converges with other lymphatic vessels to form the thoracic and right lymphatic ducts.

Lymph node An oval or bean-shaped structure located along lymphatic vessels.

Lymphocyte (LIM-fō-sīt) A type of white blood cell, found in lymph nodes, associated with the immune system.

Lysosome (LĪ-sō-sōm) An organelle in the cytoplasm of a cell, enclosed by a single membrane and containing powerful digestive enzymes.

Lysozyme (LĪ-sō-zīm) A bactericidal enzyme found in tears, saliva, and perspiration.

Macrophage (MAK-rō-fāj) Phagocytic cell derived from a monocyte; may be fixed or wandering.

Macula (MAK-yoo-la) A discolored spot or a colored area. A small, thickened region on the wall of the utricle and saccule that contains receptors for static equilibrium.

Macula lutea (LOO-tē-a) The yellow spot in the center of the retina.

Major histocompatibility (MHC) antigens Surface proteins on white blood cells and other nucleated cells that are unique for each person (except for identical siblings) and are used to type tissues and help prevent rejection. Also known as **human leukocyte associated (HLA) antigens.**

Malignant (ma-LIG-nant) Referring to diseases that tend to become worse and cause death, especially the invasion and spreading of cancer.

Mammary (MAM-ar-ē) **gland** Modified sudoriferous (sweat) gland of the female that produces milk for the nourishment of the young.

Mammillary (MAM-i-ler-ē) **bodies** Two small rounded bodies posterior to the tuber cinereum that are involved in reflexes related to the sense of smell.

Marrow (MAR-ō) Soft, spongelike material in the cavities of bone. Red bone marrow produces blood cells; yellow bone marrow, formed mainly of adipose tissue, has no blood-producing function.

Mass number The total number of protons and neutrons in an atom of a chemical element.

Mast cell A cell found in areolar connective tissue along blood vessels that produces histamine, a dilator of small blood vessels during inflammation.

Mastication (mas′-ti-KĀ-shun) Chewing.

Matrix (MĀ-triks) The ground substance and fibers between cells in a connective tissue.

Matter Anything that occupies space and has mass.

Mature follicle A relatively large, fluid-filled follicle containing a secondary oocyte and surrounding granulosa cells that secrete estrogens. Also called a **Graafian** (GRAF-ē-an) **follicle.**

Maximal oxygen uptake Maximum rate of oxygen consumption during aerobic catabolism of pyruvic acid that is determined by age, sex, and body size.

Mean arterial blood pressure (MABP) The average force of blood pressure exerted against the walls of arteries; approximately equal to diastolic pressure plus one-third of pulse pressure; for example, 93 mm Hg when systolic pressure is 120 mm Hg and diastolic pressure is 80 mm Hg.

Meatus (mē-Ā-tus) A passage or opening, especially the external portion of a canal.

Mechanoreceptor (me-KAN-ō-rē-sep-tor) Receptor that detects mechanical deformation of the receptor itself or adjacent cells; stimuli so detected include those related to touch, pressure, vibration, proprioception, hearing, equilibrium, and blood pressure.

Medial (MĒ-dē-al) Nearer the midline of the body or a structure.

Medial lemniscus (lem-NIS-kus) A band of myelinated nerve fibers extending from the medulla oblongata to the thalamus on the same side. Sensory neurons in this tract transmit impulses for the sensations of proprioception, discriminative touch, hearing, equilibrium, and vibration.

Median aperture (AP-er-choor) One of the three openings in the roof of the fourth ventricle through which cerebrospinal fluid enters the subarachnoid space of the brain and cord. Also called the **foramen of Magendie.**

Median plane A vertical plane dividing the body into right and left halves. Situated in the middle.

Mediastinum (mē′-dē-as-TĪ-num) The broad, median partition between the pleurae of the lungs, that extends from the sternum to the vertebral column in the thoracic cavity.

Medulla (me-DUL-la) An inner layer of an organ, such as the medulla of the kidneys.

Medulla oblongata (ob′-long-GA-ta) The most inferior part of the brain stem.

Medullary (MED-yoo-lar′-ē) **cavity** The space within the diaphysis of a bone that contains yellow bone marrow. Also called the **marrow cavity.**

Medullary rhythmicity (rith-MIS-i-tē) **area** Portion of the respiratory center in the medulla oblongata that controls the basic rhythm of respiration.

Meiosis (mē-Ō-sis) A type of cell division that occurs during production of gametes, involving two successive nuclear divisions that result in daughter cells with the haploid (*n*) number of chromosomes.

Melanin (MEL-a-nin) A dark black, brown, or yellow pigment found in some parts of the body such as the skin and hair.

Melanocyte (MEL-a-nō-sīt′) A pigmented cell, located between or beneath cells of the deepest layer of the epidermis, that synthesizes melanin.

Melanocyte-stimulating hormone (MSH) A hormone secreted by the anterior pituitary gland that stimulates the dispersion of melanin granules in melanocytes in amphibians; continued administration produces darkening of skin in humans.

Melatonin (mel-a-TŌN-in) A hormone, secreted by the pineal gland, that helps set the timing of the body's biological clock.

Membrane A thin, flexible sheet of tissue composed of an epithelial layer and an underlying connective tissue layer, as in an epithelial membrane, or of areolar connective tissue only, as in a synovial membrane.

Membranous labyrinth (mem-BRA-nus LAB-i-rinth) The portion of the labyrinth of the inner ear that is located inside the bony labyrinth and separated from it by the perilymph; made up of the membranous semicircular canals, the saccule and utricle, and the cochlear duct.

Menarche (me-NAR-kē) The first menses (menstrual flow) and beginning of ovarian and uterine cycles.

Meninges (me-NIN-jēz) Three membranes covering the brain and spinal cord, called the dura mater, arachnoid, and pia mater. *Singular is* **meninx** (MEN-inks).

Menopause (MEN-ō-pawz) The termination of the menstrual cycles.

Menstruation (men′-stroo-Ā-shun) Periodic discharge of blood, tissue fluid, mucus, and epithelial cells that usually lasts for 5 days; caused by a sudden reduction in estrogens and progesterone. Also called the **menstrual phase** or **menses.**

Merkel (MER-kel) **cell** Type of cell in the epidermis of hairless skin that makes contact with a tactile (Merkel) disc, which functions in touch.

Merocrine (MER-ō-krin) **gland** Gland made up of secretory cells that remain intact throughout the process of formation and discharge of the secretory product, as in the salivary and pancreatic glands.

Mesenchyme (MEZ-en-kīm) An embryonic connective tissue from which all other connective tissues arise.

Mesentery (MEZ-en-ter′-ē) A fold of peritoneum attaching the small intestine to the posterior abdominal wall.

Mesocolon (mez'-ō-KŌ-lon) A fold of peritoneum attaching the colon to the posterior abdominal wall.

Mesoderm The middle primary germ layer that gives rise to connective tissues, blood and blood vessels, and muscles.

Mesothelium (mez'-ō-THĒ-lē-um) The layer of simple squamous epithelium that lines serous membranes.

Mesovarium (mez'-ō-VAR-ē-um) A short fold of peritoneum that attaches an ovary to the broad ligament of the uterus.

Metabolism (me-TAB-ō-lizm) All the biochemical reactions that occur within an organism, including the synthetic (anabolic) reactions and decomposition (catabolic) reactions.

Metacarpus (met'-a-KAR-pus) A collective term for the five bones that make up the palm.

Metaphase (MET-a-phāz) The second stage of mitosis, in which chromatid pairs line up on the metaphase plate (equatorial plane) of the cell.

Metaphysis (me-TAF-i-sis) Growing portion of a bone.

Metarteriole (met'-ar-TĒ-rē-ōl) A blood vessel that emerges from an arteriole, traverses a capillary network, and empties into a venule.

Metastasis (me-TAS-ta-sis) The spread of cancer to surrounding tissues (local) or to other body sites (distant).

Metatarsus (met'-a-TAR-sus) A collective term for the five bones located in the foot between the tarsals and the phalanges.

Micelle (mī-SEL) A spherical aggregate of bile salts that dissolves fatty acids and monoglycerides so that they can be absorbed into small intestinal epithelial cells.

Microfilament (mī-krō-FIL-a-ment) Rodlike, protein filament about 6 nm in diameter; constitutes contractile units in muscle fibers (cells) and provides support, shape, and movement in nonmuscle cells.

Microglia (mī-krō-GLĒ-a) Neuroglial cells that carry on phagocytosis. Also called **brain macrophages** (MAK-rō-fāj-ez).

Microphage (MĪK-rō-fāj) Granular leukocyte that carries on phagocytosis, especially neutrophils and eosinophils.

Microtubule (mī-krō-TOOB-yool') Cylindrical protein filament, ranging in diameter from 18 to 30 nm, consisting of the protein tubulin; provides support, structure, and transportation.

Microvilli (mī'-krō-VIL-ē) Microscopic, fingerlike projections of the plasma membranes of cells that increase surface area for absorption, especially in the small intestine and proximal convoluted tubules of the kidneys.

Micturition (mik'-too-RISH-un) The act of expelling urine from the urinary bladder. Also called **urination** (yoo-ri-NĀ-shun).

Midbrain The part of the brain between the pons and the diencephalon. Also called the **mesencephalon** (mes'-en-SEF-a-lon).

Middle ear A small, epithelial-lined cavity hollowed out of the temporal bone, separated from the external ear by the eardrum and from the internal ear by a thin bony partition containing the oval and round windows; extending across the middle ear are the three auditory ossicles. Also called the **tympanic** (tim-PAN-ik) **cavity.**

Midline An imaginary vertical line that divides the body into equal left and right sides.

Midsagittal plane A vertical plane through the midline of the body that divides the body or organs into *equal* right and left sides. Also called a **median plane.**

Milk ejection reflex Contraction of alveolar cells to force milk into ducts of mammary glands, stimulated by oxytocin (OT), which is released from the posterior pituitary gland in response to suckling action. Also called the **milk letdown reflex.**

Mineral Inorganic, homogeneous solid substance that may perform a function vital to life; examples include calcium and phosphorus.

Mineralocorticoids (min'-er-al-ō-KOR-ti-koyds) A group of hormones of the adrenal cortex that help regulate sodium and potassium balance.

Minimal volume The volume of air in the lungs even after the thoracic cavity has been opened forcing out some of the residual volume.

Minute ventilation (MV) Total volume of air inhaled and exhaled per minute; about 6000/mL.

Mitochondrion (mī'-tō-KON-drē-on) A double-membraned organelle that plays a central role in the production of ATP; known as the "powerhouse" of the cell.

Mitosis (mī-TŌ-sis) The orderly division of the nucleus of a cell that ensures that each new daughter nucleus has the same number and kind of chromosomes as the original parent nucleus. The process includes the replication of chromosomes and the distribution of the two sets of chromosomes into two separate and equal nuclei.

Mitotic spindle Collective term for a football-shaped assembly of microtubules (nonkinetochore, kinetochore, and aster) that is responsible for the movement of chromosomes during cell division.

Modality (mō-DAL-i-tē) Any of the specific sensory entities, such as vision, smell, taste, or touch.

Modiolus (mō-DĪ-ō'-lus) The central pillar or column of the cochlea.

Mole The weight, in grams, of the combined atomic weights of the atoms that make up a molecule of a substance.

Molecule (MOL-e-kyool) The chemical combination of two or more atoms covalently bonded together.

Monocyte (MON-ō-sīt') A type of white blood cell characterized by agranular cytoplasm; the largest of the leukocytes.

Monounsaturated fat A fatty acid that contains one double covalent bond between its carbon atoms; it is not completely saturated with hydrogen atoms. Plentiful in triglycerides of olive and peanut oils.

Mons pubis (monz PYOO-bis) The rounded, fatty prominence over the pubic symphysis, covered by coarse pubic hair.

Morula (MOR-yoo-la) A solid sphere of cells produced by successive cleavages of a fertilized ovum about four days after fertilization.

Motor area The region of the cerebral cortex that governs muscular movement, particularly the precentral gyrus of the frontal lobe.

Motor end plate Portion of the sarcolemma of a muscle fiber (cell) that receives neurotransmitter liberated by an axon terminal.

Motor neuron (NOO-ron) A neuron that conducts nerve impulses from the brain and spinal cord to effectors that may be either muscles or glands. Also called an **efferent neuron.**

Motor unit A motor neuron together with the muscle fibers (cells) it stimulates.

Mucin (MYOO-sin) A protein found in mucus.

Mucosa-associated lymphatic tissue (MALT) Lymphatic nodules scattered throughout the lamina propria (connective tissue) of mucous membranes lining the gastrointestinal tract, respiratory airways, urinary tract, and reproductive tract.

Mucous (MYOO-kus) **cell** A unicellular gland that secretes mucus. Two types are mucous neck cells and mucous surface cells in the stomach.

Mucous membrane A membrane that lines a body cavity that opens to the exterior. Also called the **mucosa** (myoo-KŌ-sa).

Mucus The thick fluid secretion of goblet cells, mucous cells, mucous glands, and mucous membranes.

Multiple motor unit summation Type of summation in which stimuli occur at the same time but at different locations (different motor units).

Muscarinic (mus'-ka-RIN-ik) **receptor** Receptor found on all effectors innervated by parasympathetic postganglionic axons and some effectors innervated by sympathetic postganglionic axons; so named because the actions of acetylcholine (ACh) on such receptors are similar to those produced by muscarine.

Muscle An organ composed of one of three types of muscle tissue (skeletal, cardiac, or smooth), specialized for contraction to produce voluntary or involuntary movement of parts of the body.

Muscle action potential A stimulating impulse that propagates along a sarcolemma and then into transverse tubules; in skeletal muscle, it is generated by acetylcholine, which alters permeability of the sarcolemma to cations, especially sodium ions (Na^+).

Muscle fatigue (fa-TĒG) Inability of a muscle to maintain its strength of contraction or tension; may be related to insufficient oxygen, depletion of glycogen, and/or lactic acid buildup.

Muscle spindle An encapsulated proprioceptor in a skeletal muscle, consisting of specialized intrafusal muscle fibers and nerve endings; stimulated by changes in length or tension of muscle fibers.

Muscle tissue A tissue specialized to produce motion in response to muscle action potentials by its qualities of contractility, extensibility, elasticity, and excitability; types include skeletal, cardiac, and smooth.

Muscle tone A sustained, partial contraction of portions of a skeletal or smooth muscle in response to activation of stretch receptors or a baseline level of action potentials in the innervating motor neurons.

Muscularis (MUS-kyoo-la'-ris) A muscular layer (coat or tunic) of an organ.

Muscularis mucosae (myoo-KŌ-sē) A thin layer of smooth muscle fibers (cells) located in the most superficial layer of the mucosa of the gastrointestinal tract, underlying the lamina propria of the mucosa.

Mutation (myoo-TĀ-shun) Any change in the sequence of bases in a DNA molecule resulting in a permanent alteration in some inheritable characteristic.

Myelin (MĪ-e-lin) **sheath** Multilayered lipid and protein covering, formed by Schwann cells and oligodendrocytes around axons of many peripheral and central nervous system neurons.

Myenteric plexus A network of nerve fibers from both autonomic divisions located in the muscularis coat of the small intestine. Also called the **plexus of Auerbach** (OW-er-bak).

Myocardium (mī'-ō-KAR-dē-um) The middle layer of the heart wall, made up of cardiac muscle tissue, constituting the bulk of the heart, and lying between the epicardium and the endocardium.

Myofibril (mī'-ō-FĪ-bril) A threadlike structure, running longitudinally through a muscle fiber (cell) consisting mainly of thick filaments (myosin) and thin filaments (actin, troponin, and tropomyosin).

Myoglobin (mī-ō-GLŌ-bin) The oxygen-binding, iron-containing conjugated protein complex present in the sarcoplasm of muscle fibers; contributes the red color to muscle.

Myogram (MĪ-ō-gram) The record or tracing produced by a myograph, an apparatus that measures and records the effects of muscular contractions.

Myology (mī-OL-ō-jē) The study of muscles.

Myometrium (mī'-ō-MĒ-trē-um) The smooth muscle layer of the uterus.

Myopathy (mī-OP-a-thē) Any abnormal condition or disease of muscle tissue.

Myopia (mī-Ō-pē-a) Defect in vision in which objects can be seen distinctly only when very close to the eyes; nearsightedness.

Myosin (MĪ-ō-sin) The contractile protein that makes up the thick filaments of muscle fibers (cells).

Myotome (MĪ-ō-tōm) A group of muscles innervated by the motor neurons of a single spinal segment; in embryos, that portion of a somite that develops into some skeletal muscle.

Nail A hard plate, composed largely of keratin, that develops from the epidermis of the skin to form a protective covering on the dorsal surface of the distal phalanges of the fingers and toes.

Nail matrix (MĀ-triks) The part of the nail beneath the body and root from which the nail is produced.

Nasal (NĀ-zal) **cavity** A mucosa-lined cavity on either side of the nasal septum that opens onto the face at the external nares and into the nasopharynx at the internal nares.

Nasal septum (SEP-tum) A vertical partition composed of bone (perpendicular plate of ethmoid and vomer) and cartilage, covered with a mucous membrane, separating the nasal cavity into left and right sides.

Nasolacrimal (nā'-zō-LAK-ri-mal) **duct** A canal that transports the lacrimal secretion (tears) from the nasolacrimal sac into the nose.

Nasopharynx (nā'-zō-FAR-inks) The superior portion of the pharynx, lying posterior to the nose and extending inferiorly to the soft palate.

Neck The part of the body connecting the head and the trunk. A constricted portion of an organ such as the neck of the femur or uterus.

Necrosis (ne-KRŌ-sis) A pathological type of cell death that results from disease, injury, or lack of blood supply in which many adjacent cells swell, burst, and spill their contents into the interstitial fluid, triggering an inflammatory response.

Negative feedback The principle governing most control systems; a mechanism of response in which a stimulus initiates actions that reverse or reduce the stimulus.

Neonatal (nē'-ō-NĀ-tal) Pertaining to the first four weeks after birth.

Neoplasm (NĒ-ō-plazm) A new growth that may be benign or malignant.

Nephron (NEF-ron) The functional unit of the kidney.

Nerve A cordlike bundle of nerve fibers (axons and/or dendrites) and its associated connective tissue coursing together outside the central nervous system.

Nerve fiber General term for any process (axon or dendrite) projecting from the cell body of a neuron.

Nerve impulse A wave of depolarization and repolarization that self-propagates along the plasma membrane of a neuron; also called a **nerve action potential.**

Nervous tissue Tissue that initiates and transmits nerve impulses to coordinate homeostasis.

Net filtration pressure (NFP) Net pressure that promotes fluid outflow at the arterial end of a capillary, and fluid inflow at the venous end of a capillary; net pressure that promotes glomerular filtration in the kidneys.

Neural plate A thickening of ectoderm that forms early in the third week of development and represents the beginning of the development of the nervous system.

Neuroeffector (noo-rō-e-FEK-tor) **junction** Collective term for neuromuscular and neuroglandular junctions.

Neurofibril (noo-rō-FĪ-bril) One of the delicate threads that forms a complicated network in the cytoplasm of the cell body and processes of a neuron.

Neuroglandular (noo-rō-GLAND-yoo-lar) **junction** Area of contact between a motor neuron and a gland.

Neuroglia (noo-RŌG-lē-a) Cells of the nervous system that perform various supportive functions. The neuroglia of the central nervous system are the astrocytes, oligodendrocytes, microglia, and ependymal cells; neuroglia of the peripheral nervous system include the Schwann cells and the satellite cells. Also called **glial** (GLĒ-al) **cells.**

Neurohypophyseal (noo′-rō-hī′-pō-FIZ-ē-al) **bud** An outgrowth of ectoderm located on the floor of the hypothalamus that gives rise to the posterior pituitary gland.

Neurolemma (noo-rō-LEM-ma) The peripheral, nucleated cytoplasmic layer of the Schwann cell. Also called **sheath of Schwann** (SCHVON).

Neurology (noo-ROL-ō-jē) The branch of science that deals with the normal functioning and disorders of the nervous system.

Neuromuscular (noo-rō-MUS-kyoo-lar) **junction** The area of contact between the axon terminal of a motor neuron and a portion of the sarcolemma of a muscle fiber (cell). Also called a **myoneural** (mī-ō-NOO-ral) **junction.**

Neuron (NOO-ron) A nerve cell, consisting of a cell body, dendrites, and an axon.

Neuropeptide (noo-rō-PEP-tīd) Chain of three to about 40 amino acids that occurs naturally in the nervous system, and that acts primarily to modulate the response of or to a neurotransmitter. Examples are enkephalins and endorphins.

Neurosecretory (noo-rō-SĒC-re-tō-rē) **cell** A cell in a nucleus (paraventricular and supraoptic) in the hypothalamus that produces oxytocin (OT) or antidiuretic hormone (ADH), hormones released in the posterior pituitary gland.

Neurotransmitter One of a variety of molecules within axon terminals that are released into the synaptic cleft in response to a nerve impulse and affect the membrane potential of the postsynaptic neuron. Also called a **transmitter substance.**

Neutrophil (NOO-trō-fil) A type of white blood cell characterized by granules that stain pale lilac with a combination of acidic and basic dyes.

Nicotinic (nik′-ō-TIN-ik) **receptor** Receptor found on both sympathetic and parasympathetic postganglionic neurons, so named because the actions of acetylcholine (ACh) on such receptors are similar to those produced by nicotine.

Nipple A pigmented, wrinkled projection on the surface of the breast that is the location of the openings of the lactiferous ducts for milk release.

Nociceptor (nō′-sē-SEP-tor) A free (naked) nerve ending that detects painful stimuli.

Node of Ranvier (ron-vē-Ā) A space, along a myelinated nerve fiber, between the individual Schwann cells that form the myelin sheath and the neurolemma. Also called **neurofibral node.**

Nondisjunction (non′-dis-JUNGK-shun) Failure of sister chromatids to separate properly during anaphase of mitosis (or meiosis II) or failure of homologous chromosomes to separate properly during meiosis I in which chromatids or chromosomes pass into the same daughter cell; the result is too many copies of that chromosome in the daughter cell or gamete.

Nonessential amino acid An amino acid that can be synthesized by body cells through transamination, the transfer of an amino group from an amino acid to another substance.

Norepinephrine (nor′-ep-ē-NEF-rin) **(NE)** A hormone secreted by the adrenal medulla that produces actions similar to those that result from sympathetic stimulation. Also called **noradrenaline** (nor-a-DREN-a-lin).

Notochord (NŌ-tō-cord) A flexible rod of embryonic tissue that lies where the future vertebral column will develop.

Nuclear medicine The branch of medicine concerned with the use of radioisotopes in the diagnosis of disease and therapy.

Nuclease (NOO-klē-ās) An enzyme that breaks nucleotides into pentoses and nitrogenous bases; examples are ribonuclease and deoxyribonuclease.

Nucleic (noo-KLĒ-ic) **acid** An organic compound that is a long polymer of nucleotides, with each nucleotide containing a pentose sugar, a phosphate group, and one of four possible nitrogenous bases (adenine, cytosine, guanine, and thymine or uracil).

Nucleolus (noo-KLĒ-ō-lus) Nonmembranous spherical body within the nucleus composed of protein, DNA, and RNA that functions in the synthesis and storage of ribosomal RNA.

Nucleosome (NOO-klē-ō-sōm) Elementary structural subunit of a chromosome consisting of histones and DNA.

Nucleus (NOO-klē-us) A spherical or oval organelle of a cell that contains the hereditary factors of the cell, called genes. A cluster of unmyelinated nerve cell bodies in the central nervous system. The central portion of an atom made up of protons and neutrons.

Nucleus cuneatus (kyoo-nē-Ā-tus) A group of nerve cells in the inferior portion of the medulla oblongata in which fibers of the fasciculus cuneatus terminate.

Nucleus gracilis (gras-I-lis) A group of nerve cells in the inferior portion of the medulla oblongata in which fibers of the fasciculus gracilis terminate.

Nucleus pulposus (pul-PŌ-sus) A soft, pulpy, highly elastic substance in the center of an intervertebral disc, a remnant of the notochord.

Nutrient A chemical substance in food that provides energy, forms new body components, or assists in the functioning of various body processes.

Obesity (ō-BĒS-i-tē) Body weight more than 20% above a desirable standard due to excessive accumulation of fat.

Oblique (ō-BLĒK) **plane** A plane that passes through the body or an organ at an angle between the transverse plane and either the midsagittal, parasagittal, or frontal plane.

Obstetrics (ob-STET-riks) The specialized branch of medicine that deals with pregnancy, labor, and the period of time immediately following delivery (about 42 days).

Olfactory (ōl-FAK-tō-rē) Pertaining to smell.

Olfactory bulb A mass of gray matter containing cell bodies of neurons that form synapses with neurons of the olfactory (I) nerve, lying inferior to the frontal lobe of the cerebrum on either side of the crista galli of the ethmoid bone.

Olfactory receptor A bipolar neuron with its cell body lying between supporting cells located in the mucous membrane lining the superior portion of each nasal cavity; transduces odors into neural signals.

Olfactory tract A bundle of axons that extends from the olfactory bulb posteriorly to the olfactory portion of the cerebral cortex.

Oligodendrocyte (o-lig-ō-DEN-drō-sīt) A neuroglial cell that supports neurons and produces a myelin sheath around axons of neurons of the central nervous system.

Oligospermia (ol'-i-gō-SPER-mē-a) A deficiency of sperm cells in the semen.

Olive A prominent oval mass on each lateral surface of the superior part of the medulla oblongata.

Oncogene (ONG-kō-jēn) Gene that has the ability to transform a normal cell into a cancerous cell when it is inappropriately activated.

Oncology (ong-KOL-ō-jē) The study of tumors.

Oogenesis (ō'-ō-JEN-e-sis) Formation and development of the ovum.

Oophorectomy (ō'-of-ō-REK-tō-me) The surgical removal of the ovaries.

Ophthalmic (of-THAL-mik) Pertaining to the eye.

Ophthalmologist (of'-thal-MOL-ō-jist) A physician who specializes in the diagnosis and treatment of eye disorders using drugs, surgery, and corrective lenses.

Ophthalmology (of'-thal-MOL-ō-jē) The study of the structure, function, and diseases of the eye.

Opsin (OP-sin) The glycoprotein portion of a photopigment.

Opsonization (op-sō-ni-ZĀ-shun) The action of some antibodies that renders bacteria and other foreign cells more susceptible to phagocytosis. Also called **immune adherence.**

Optic (OP-tik) Refers to the eye, vision, or properties of light.

Optic chiasm (KĪ-azm) A crossing point of the optic (II) nerves, anterior to the pituitary gland. Also called **optic chiasma.**

Optic disc A small area of the retina containing openings through which the axons of the ganglion cells emerge as the optic (II) nerve. Also called the **blind spot.**

Optician (op-TISH-an) A technician who fits, adjusts, and dispenses corrective lenses on prescription of an ophthalmologist or optometrist.

Optic tract A bundle of axons that transmits nerve impulses from the retina of the eye between the optic chiasm and the thalamus.

Optometrist (op-TOM-e-trist) Specialist with a doctorate degree in optometry who is licensed to examine and test the eyes and treat visual defects by prescribing corrective lenses.

Ora serrata (Ō-ra ser-RĀ-ta) The irregular margin of the retina lying internal and slightly posterior to the junction of the choroid and ciliary body.

Orbit (OR-bit) The bony, pyramidal-shaped cavity of the skull that holds the eyeball.

Organ A structure composed of two or more different kinds of tissues with a specific function and usually a recognizable shape.

Organelle (or-gan-EL) A permanent structure within a cell with characteristic morphology that is specialized to serve a specific function in cellular activities.

Organic (or-GAN-ik) **compound** Compound that always contains carbon in which the atoms are held together by covalent bonds. Examples include carbohydrates, lipids, proteins, and nucleic acids (DNA and RNA).

Organism (OR-ga-nizm) A total living form; one individual.

Orgasm (OR-gazm) Sensory and motor events involved in ejaculation for the male and involuntary contraction of the perineal muscles in the female at the climax of sexual intercourse.

Orifice (OR-i-fis) Any aperture or opening.

Origin (OR-i-jin) The attachment of a muscle tendon to a stationary bone or the end opposite the insertion.

Oropharynx (or'-ō-FAR-inks) The intermediate portion of the pharynx, lying posterior to the mouth and extending from the soft palate to the hyoid bone.

Orthopedics (or'-thō-PĒ-diks) The branch of medicine that deals with the preservation and restoration of the skeletal system, articulations, and associated structures.

Osmoreceptor (oz'-mō-re-CEP-tor) Receptor in the hypothalamus that is sensitive to changes in blood osmolarity and, in response to high osmolarity (low water concentration), causes synthesis and release of antidiuretic hormone (ADH).

Osmosis (os-MŌ-sis) The net movement of water molecules through a selectively permeable membrane from an area of higher water concentration to an area of lower water concentration until equilibrium is reached.

Osmotic pressure The pressure required to prevent the movement of pure water into a solution containing solutes when the solutions are separated by a selectively permeable membrane.

Osseous (OS-ē-us) Bony.

Ossicle (OS-si-kul) Small bone, as in the middle ear (malleus, incus, stapes).

Ossification (os'-i-fi-KĀ-shun) Formation of bone. Also called **osteogenesis**.

Osteoblast (OS-tē-ō-blast) Cell formed from an osteoprogenitor cell that participates in bone formation by secreting some organic components and inorganic salts.

Osteoclast (OS-tē-ō-clast') A large multinuclear cell that destroys or resorbs bone tissue.

Osteocyte (OS-tē-ō-sīt') A mature bone cell that maintains the daily activities of bone tissue.

Osteogenic (os'-tē-ō-JEN-ik) **layer** The inner layer of the periosteum that contains cells responsible for forming new bone during growth and repair.

Osteology (os'-tē-OL-ō-jē) The study of bones.

Osteon (OS-tē-on) The basic unit of structure in adult compact bone, consisting of a central (Haversian) canal with its concentrically arranged lamellae, lacunae, osteocytes, and canaliculi. Also called a **Haversian** (ha-VER-shun) **system.**

Osteoporosis (os'-tē-ō-pō-RŌ-sis) Age-related disorder characterized by decreased bone mass and increased susceptibility to fractures, often as a result of decreased levels of estrogens.

Osteoprogenitor (os'-tē-ō-prō-JEN-i-tor) **cell** Stem cell derived from mesenchyme that has mitotic potential and the ability to differentiate into an osteoblast.

Otic (Ō-tik) Pertaining to the ear.

Otolith (Ō-tō-lith) A particle of calcium carbonate embedded in the otolithic membrane that functions in maintaining static equilibrium.

Otolithic (ō-to-LITH-ik) **membrane** Thick, gelatinous, glycoprotein layer located directly over hair cells of the macula in the saccule and utricle of the inner ear.

Otorhinolaryngology (ō'-tō-rī-nō-lar'-in-GOL-ō-jē) The branch of medicine that deals with the diagnosis and treatment of diseases of the ears, nose, and throat.

Oval window A small, membrane-covered opening between the middle ear and inner ear into which the footplate of the stapes fits. Also called the **fenestra vestibuli** (fe-NES-tra ves-TIB-yoo-lē).

Ovarian (ō-VAR-ē-an) **cycle** A monthly series of events in the ovary associated with the maturation of an ovum.

Ovarian follicle (FOL-i-kul) A general name for oocytes (immature ova) in any stage of development, along with their surrounding epithelial cells.

Ovarian ligament (LIG-a-ment) A rounded cord of connective tissue that attaches the ovary to the uterus.

Ovary (Ō-var-ē) Female gonad that produces ova and the hormones estrogen, progesterone, inhibin and relaxin.

Ovulation (ō-vyo-LĀ-shun) The rupture of a mature ovarian (Graafian) follicle with discharge of a secondary oocyte into the pelvic cavity.

Ovum (Ō-vum) The female reproductive or germ cell; an egg cell.

Oxidation (ok-si-DĀ-shun) The removal of electrons from a molecule or, less commonly, the addition of oxygen to a molecule that results in a decrease in the energy content of the molecule. The oxidation of glucose in the body is called **cellular respiration.**

Oxyhemoglobin (ok′-sē-HĒ-mō-glō-bin) **(Hb·O$_2$)** Hemoglobin combined with oxygen.

Oxyphil cell A cell found in the parathyroid gland that secretes parathyroid hormone (PTH).

Oxytocin (ok′-sē-TŌ-sin) **(OT)** A hormone secreted by neurosecretory cells in the paraventricular and supraoptic nuclei of the hypothalamus that stimulates contraction of the smooth muscle fibers in the pregnant uterus and contractile cells around the ducts of mammary glands.

Palate (PAL-at) The horizontal structure separating the oral and the nasal cavities; the roof of the mouth.

Palpate (PAL-pāt) To examine by touch; to feel.

Pancreas (PAN-krē-as) A soft, oblong organ lying along the greater curvature of the stomach and connected by a duct to the duodenum. It is both exocrine (secreting pancreatic juice) and endocrine (secreting insulin, glucagon, somatostatin, and pancreatic polypeptide).

Pancreatic (pan′-krē-AT-ik) **duct** A single large tube that unites with the common bile duct from the liver and gallbladder and drains pancreatic juice into the duodenum at the hepatopancreatic ampulla (ampulla of Vater). Also called the **duct of Wirsung.**

Pancreatic islet A cluster of endocrine gland cells in the pancreas that secretes insulin, glucagon, somatostatin, and pancreatic polypeptide. Also called an **islet of Langerhans** (LANG-er-hanz).

Pancreatic polypeptide Hormone secreted by the F cells of pancreatic islets (islets of Langerhans) that regulates release of pancreatic digestive enzymes.

Papanicolaou (pap′-a-NIK-ō-la-oo) **test** A cytological staining test for the detection and diagnosis of premalignant and malignant conditions of the female genital tract. Cells scraped from the epithelium of the cervix of the uterus are examined microscopically. Also called a **Pap smear.**

Papilla (pa-PIL-a) A small nipple-shaped projection or elevation.

Paracrine (pa-RA-krin) Local hormone, such as histamine, that acts on neighboring cells without entering the bloodstream.

Paralysis (pa-RAL-a-sis) Loss or impairment of motor function due to a lesion of nervous or muscular origin.

Paranasal sinus (par′-a-NĀ-zal SĪ-nus) A mucus-lined air cavity in a skull bone that communicates with the nasal cavity. Paranasal sinuses are located in the frontal, maxillary, ethmoid, and sphenoid bones.

Paraplegia (par-a-PLĒ-jē-a) Paralysis of both lower limbs.

Parasagittal plane A vertical plane that does not pass through the midline and that divides the body or organs into *unequal* left and right portions.

Parasympathetic (par′-a-sim-pa-THET-ik) **division** One of the two subdivisions of the autonomic nervous system, having cell bodies of preganglionic neurons in nuclei in the brain stem and in the lateral gray matter of the sacral portion of the spinal cord; primarily concerned with activities that conserve and restore body energy. Also called the **craniosacral** (krā-nē-ō-SĀ-kral) **division.**

Parathyroid (par′-a-THĪ-royd) **gland** One of four (usually) small endocrine glands embedded in the posterior surfaces of the lateral lobes of the thyroid gland.

Parathyroid hormone (PTH) A hormone secreted by the parathyroid glands that increases blood calcium level and decreases blood phosphate level.

Paraurethral (par′-a-yoo-RĒ-thral) **gland** Gland embedded in the wall of the urethra whose duct opens on either side of the urethral orifice and secretes mucus. Also called **Skene's** (SKĒNZ) **gland.**

Parenchyma (par-EN-ki-ma) The functional parts of any organ, as opposed to tissue that forms its stroma or framework.

Parietal (pa-RĪ-e-tal) Pertaining to or forming the outer wall of a body cavity.

Parietal cell A type of secretory cell in gastric glands that produces hydrochloric acid and intrinsic factor. Also called an **oxyntic cell.**

Parietal pleura (PLOO-ra) The outer layer of the serous pleural membrane that encloses and protects the lungs; the layer that is attached to the wall of the pleural cavity.

Parotid (pa-ROT-id) **gland** One of the paired salivary glands located inferior and anterior to the ears and connected to the oral cavity via a duct (Stensen's) that opens into the inside of the cheek opposite the maxillary (upper) second molar tooth.

Parturition (par′-too-RISH-un) Act of giving birth to young; childbirth, delivery.

Patellar (pa-TELL-ar) **reflex** Extension of the leg by contraction of the quadriceps femoris muscle in response to tapping the patellar ligament. Also called the **knee jerk reflex.**

Patent ductus arteriosus Congenital anatomical heart defect in which the fetal connection between the aorta and pulmonary trunk remains open instead of closing completely after birth.

Pathogen (PATH-ō-jen) A disease-producing microorganism.

Pathological (path′-ō-LOJ-i-kal) **anatomy** The study of structural changes caused by disease.

Pectinate (PEK-ti-nāt) **muscles** Projecting muscle bundles of the anterior atrial walls and the lining of the auricles.

Pectoral (PEK-tō-ral) Pertaining to the chest or breast.

Pediatrician (pē′-dē-a-TRISH-un) A physician who specializes in the care and treatment of children.

Pedicel (PED-i-sel) Footlike structure, as on podocytes of a glomerulus.

Pelvic (PEL-vik) **cavity** Inferior portion of the abdominopelvic cavity that contains the urinary bladder, sigmoid colon, rectum, and internal female and male reproductive structures.

Pelvic splanchnic (PEL-vik SPLANGK-nik) **nerves** Preganglionic parasympathetic fibers from the levels of S2, S3, and S4 that supply the urinary bladder, reproductive organs, and the descending and sigmoid colon and rectum.

Pelvis The basinlike structure formed by the two hip bones, the sacrum, and the coccyx. The expanded, proximal portion of the ureter, lying within the kidney and into which the major calyces open.

Penis (PĒ-nis) The male copulatory organ, used to introduce semen into the female vagina.

Pepsin Protein-digesting enzyme secreted by chief (zymogenic) cells of the stomach in the inactive form pepsinogen, which is converted to active pepsin by hydrochloric acid.

Peptic ulcer An ulcer that develops in areas of the gastrointestinal tract exposed to hydrochloric acid; classified as a gastric ulcer if in the lesser curvature of the stomach and as a duodenal ulcer if in the first part of the duodenum.

Percussion (per-KUSH-un) The act of striking (percussing) an underlying part of the body with short, sharp blows as an aid in diagnosing the part by the quality of the sound produced.

Perforating canal A minute passageway by means of which blood vessels and nerves from the periosteum penetrate into compact bone. Also called **Volkmann's** (FŌLK-manz) **canal.**

Pericardial (per′-i-KAR-dē-al) **cavity** Small potential space between the visceral and parietal layers of the serous pericardium that contains pericardial fluid.

Pericardium (per′-i-KAR-dē-um) A loose-fitting membrane that encloses the heart, consisting of a superficial fibrous layer and a deep serous layer.

Perichondrium (per′-i-KON-drē-um) The membrane that covers cartilage.

Perilymph (PER-i-lymf) The fluid contained between the bony and membranous labyrinths of the inner ear.

Perimetrium (per-i-MĒ-trē-um) The serosa of the uterus.

Perimysium (per′-i-MĪZ-ē-um) Invagination of the epimysium that divides muscles into bundles.

Perineum (per′-i-NĒ-um) The pelvic floor; the space between the anus and the scrotum in the male and between the anus and the vulva in the female.

Perineurium (per′-i-NYOO-rē-um) Connective tissue wrapping around fascicles in a nerve.

Periodontal (per-ē⁻-o⁻-DON-tal) **disease** A collective term for conditions characterized by degeneration of gingivae, alveolar bone, periodontal ligament, and cementum.

Periodontal ligament The periosteum lining the alveoli (sockets) for the teeth in the alveolar processes of the mandible and maxillae.

Periosteum (per′-ē-OS-tē-um) The membrane that covers bone and consists of connective tissue, osteoprogenitor cells, and osteoblasts; is essential for bone growth, repair, and nutrition.

Peripheral (pe-RIF-er-al) Located on the outer part or a surface of the body.

Peripheral nervous system (PNS) The part of the nervous system that lies outside the central nervous system—nerves and ganglia.

Peristalsis (per′-i-STAL-sis) Successive muscular contractions along the wall of a hollow muscular structure.

Peritoneum (per′-i-tō-NĒ-um) The largest serous membrane of the body that lines the abdominal cavity and covers the viscera.

Peritonitis (per′-i-tō-NĪ-tis) Inflammation of the peritoneum.

Permissive (per-MIS-sive) **effect** A hormonal interaction in which the effect of one hormone on a target cell requires previous or simultaneous exposure to another hormone(s) to enhance the response of a target cell or increase the activity of another hormone.

Peroxisome (pe-ROKS-ī-sōm) Organelle similar in structure to a lysosome that contains enzymes that use molecular oxygen to oxidize various organic compounds. Such reactions produce hydrogen peroxide; abundant in liver cells.

Perspiration Sweat; produced by sudoriferous (sweat) glands and containing water, salts, urea, uric acid, amino acids, ammonia, sugar, lactic acid, and ascorbic acid; helps maintain body temperature and eliminate wastes.

pH A measure of the concentration of hydrogen ions (H^+) in a solution. The pH scale extends from 0 to 14, with a value of 7 expressing neutrality, values lower than 7 expressing increasing acidity, and values higher than 7 expressing increasing alkalinity.

Phagocytosis (fag′-ō-sī-TŌ-sis) The process by which phagocytes ingest particulate matter; especially the ingestion and destruction of microbes, cell debris, and other foreign matter.

Phalanx (FĀ-lanks) The bone of a finger or toe. *Plural is* **phalanges** (fa-LAN-jēz).

Pharmacology (far′-ma-KOL-ō-jē) The science that deals with the effects and uses of drugs in the treatment of disease.

Pharynx (FAR-inks) The throat; a tube that starts at the internal nares and runs partway down the neck, where it opens into the esophagus posteriorly and the larynx anteriorly.

Phenotype (FĒ-nō-tīp) The observable expression of genotype; physical characteristics of an organism determined by genetic makeup and influenced by interaction between genes and internal and external environmental factors.

Phlebitis (fle-BĪ-tis) Inflammation of a vein, usually in a lower limb.

Phosphorylation (fos′-for-i-LĀ-shun) The addition of a phosphate group to a chemical compound; types include substrate-level phosphorylation, oxidative phosphorylation, and photophosphorylation.

Photopigment A substance that can absorb light and undergo structural changes that can lead to the development of a receptor potential. An example is rhodopsin. Also called **visual pigment.**

Photoreceptor Receptor that detects light shining on the retina of the eye.

Physiology (fiz′-ē-OL-ō-jē) Science that deals with the functions of an organism or its parts.

Pia mater (PĪ-a MĀ-ter) The deep membrane (meninx) covering the brain and spinal cord.

Pineal (PĪN-ē-al) **gland** The cone-shaped gland located in the roof of the third ventricle. Also called the **epiphysis cerebri** (ē-PIF-i-sis se-RĒ-brē).

Pinealocyte (pin-ē-AL-ō-sīt) Secretory cell of the pineal gland that releases melatonin.

Pinna (PIN-na) The projecting part of the external ear composed of elastic cartilage and covered by skin and shaped like the flared end of a trumpet. Also called the **auricle** (OR-i-kul).

Pinocytosis (pi′-nō-sī-TŌ-sis) The process by which cells ingest liquid.

Pituicyte (pi-TOO-i-sīt) Supporting cell of the posterior pituitary gland.

Pituitary gland A small endocrine gland lying in the sella turcica of the sphenoid bone and attached to the hypothalamus by the infundibulum. Also called the **hypophysis** (hī-POF-i-sis).

Pivot joint A synovial joint in which a rounded, pointed, or conical surface of one bone articulates with a ring formed partly by another bone and partly by a ligament, as in the joint between the atlas and axis and between the proximal ends of the radius and ulna. Also called a **trochoid** (TRŌ-koid) **joint.**

Placenta (pla-SEN-ta) The special structure through which the exchange of materials between fetal and maternal circulations occurs. Also called the **afterbirth.**

Plantar flexion (PLAN-tar FLEK-shun) Bending the foot in the direction of the plantar surface (sole).

Plaque (plak) A mass of bacterial cells, dextran (polysaccharide), and other debris that adheres to teeth.

Plasma (PLAZ-ma) The extracellular fluid found in blood vessels; blood minus the formed elements.

Plasma cell Cell that produces antibodies and develops from a B cell (lymphocyte).

Plasma (cell) membrane Outer, limiting membrane that separates the cell's internal parts from extracellular fluid or the external environment.

Platelet (PLĀT-let) A fragment of cytoplasm enclosed in a cell membrane and lacking a nucleus; found in the circulating blood; plays a role in blood clotting. Also called a **thrombocyte** (THROM-bō-sīt).

Platelet plug Aggregation of thrombocytes at a damaged blood vessel to prevent blood loss.

Pleura (PLOOR-a) The serous membrane that covers the lungs and lines the walls of the chest and the diaphragm.

Pleural cavity Small potential space between the visceral and parietal pleurae.

Pleuripotent stem cell Immature stem cell in red bone marrow that gives rise to precursors of all the different mature blood cells. Previously called a **hemopoietic stem cell.**

Plexus (PLEK-sus) A network of nerves, veins, or lymphatic vessels.

Pneumotaxic (noo-mō-TAK-sik) **area** Portion of the respiratory center in the pons that continually sends inhibitory nerve impulses to the inspiratory area that limit inspiration and facilitate expiration.

Podiatry (pō-DĪ-a-trē) The diagnosis and treatment of foot disorders.

Polar body The smaller cell resulting from the unequal division of cytoplasm during the meiotic divisions of an oocyte. The polar body has no function and degenerates.

Polarized A condition in which ions having opposite charges are separated across a membrane.

Polycythemia (pol′-ē-sī-THĒ-mē-a) Disorder characterized by a hematocrit above the normal level of 55 in which hypertension, thrombosis, and hemorrhage occur.

Polysaccharide (pol′-ē-SAK-a-rīd) A carbohydrate in which three or more monosaccharides are joined chemically.

Polyunsaturated fat A fatty acid that contains more than one double covalent bond between its carbon atoms; abundant in triglycerides of corn oil, safflower oil, and cottonseed oil.

Pons (ponz) The portion of the brain stem that forms a "bridge" between the medulla oblongata and the midbrain, anterior to the cerebellum.

Positive feedback A feedback mechanism in which the response enhances the original stimulus.

Postabsorptive (fasting) state Metabolic state after absorption of food is complete and energy needs of the body must be satisfied using molecules stored previously.

Postcentral gyrus Gyrus of cerebral cortex located immediately posterior to the central sulcus; contains the primary somatosensory area.

Posterior (pos-TĒR-ē-or) Nearer to or at the back of the body. Equivalent to **dorsal** in bipeds.

Posterior column—medial lemniscus pathways Sensory pathways that carry information related to proprioception, discriminitive touch, two-point discrimination, pressure, and vibration. First-order neurons project from the spinal cord to the ipsilateral medulla in the posterior columns (fasciculus gracilis and fasciculus cuneatus). Second-order neurons project from the medulla to the contralateral thalamus in the medial lemniscus. Third-order neurons project from the thalamus to the somatosensory cortex (postcentral gyrus) on the same side.

Posterior pituitary gland Posterior lobe of the pituitary gland. Also called the **neurohypophysis** (noo-rō-hī-POF-i-sis).

Posterior root The structure composed of sensory fibers lying between a spinal nerve and the dorsolateral aspect of the spinal cord. Also called the **dorsal (sensory) root.**

Posterior root ganglion (GANG-glē-on) A group of cell bodies of sensory neurons and their supporting cells located along the posterior root of a spinal nerve. Also called a **dorsal (sensory) root ganglion.**

Postganglionic neuron (pōst′-gang-lē-ON-ik NOO-ron) The second visceral motor neuron in an autonomic pathway, having its cell body and dendrites located in an autonomic ganglion and its unmyelinated axon ending at cardiac muscle, smooth muscle, or a gland.

Postsynaptic (pōst-sin-AP-tik) **neuron** The nerve cell that is activated by the release of a neurotransmitter from another neuron and carries nerve impulses away from the synapse.

Precapillary sphincter (SFINGK-ter) A ring of smooth muscle fibers (cells) at the site of origin of true capillaries that regulate blood flow into true capillaries.

Precentral gyrus Gyrus of cerebral cortex located immediately anterior to the central sulcus; contains the primary motor area.

Preganglionic (prē′-gang-lē-ON-ik) **neuron** The first visceral motor neuron in an autonomic pathway, with its cell body and dendrites in the brain or spinal cord and its myelinated axon ending at an autonomic ganglion, where it synapses with a postganglionic neuron.

Pregnancy Sequence of events that normally includes fertilization, implantation, embryonic growth, and fetal growth and terminates in birth.

Premenstrual syndrome (PMS) Severe physical and emotional stress occurring late in the postovulatory phase of the menstrual cycle and sometimes overlapping with menstruation.

Prepuce (PRĒ-pyoos) The loose-fitting skin covering the glans of the penis and clitoris. Also called the **foreskin.**

Presbyopia (prez-bē-Ō-pē-a) A loss of elasticity of the lens of the eye due to advancing age with resulting inability to focus clearly on near objects.

Presynaptic (prē-sin-AP-tik) **inhibition** Type of inhibition in which neurotransmitter released by an inhibitory neuron depresses the release of neurotransmitter by a presynaptic neuron.

Presynaptic (prē-sin-AP-tik) **neuron** A neuron that propagates nerve impulses toward a synapse.

Prevertebral ganglion (prē-VERT-e-bral GANG-lē-on) A cluster of cell bodies of postganglionic sympathetic neurons anterior to the spinal column and close to large abdominal arteries. Also called a **collateral ganglion.**

Primary germ layer One of three layers of embryonic tissue, called ectoderm, mesoderm, and endoderm, that give rise to all tissues and organs of the organism.

Primary motor area A region of the cerebral cortex in the precentral gyrus of the frontal lobe of the cerebrum that controls specific muscles or groups of muscles.

Primary somatosensory area A region of the cerebral cortex posterior to the central sulcus in the postcentral gyrus of the parietal lobe of the cerebrum that localizes exactly the points of the body where somatic sensations originate.

Prime mover The muscle directly responsible for producing a desired motion. Also called an **agonist** (AG-ō-nist).

Primitive gut Embryonic structure composed of endoderm and mesoderm that gives rise to most of the gastrointestinal tract.

Primordial (prī-MOR-dē-al) Existing first; especially primordial follicles in the ovary.

Principal cell Cell found in the parathyroid glands that secretes parathyroid hormone (PTH); cell type in the distal convoluted tubule and collecting duct of nephron that is stimulated by aldosterone and antidiuretic hormone.

Proctology (prok-TOL-ō-jē) The branch of medicine that treats the rectum and its disorders.

Progesterone (prō-JES-te-rōn) **(PROG)** A female sex hormone produced by the ovaries that helps prepare the endometrium of the uterus for implantation of a fertilized ovum and the mammary glands for milk secretion.

Prognosis (prog-NŌ-sis) A forecast of the probable results of a disorder; the outlook for recovery.

Prolactin (prō-LAK-tin) **(PRL)** A hormone secreted by the anterior pituitary gland that initiates and maintains milk secretion by the mammary glands.

Prolapse (PRŌ-laps) A dropping or falling down of an organ, especially the uterus or rectum.

Proliferation (pro-lif′-er-Ā-shun) Rapid and repeated reproduction of new parts, especially cells.

Pronation (prō-NĀ-shun) A movement of the forearm in which the palm of the hand is turned posteriorly or inferiorly.

Prophase (PRŌ-fāz) The first stage of mitosis during which chromatid pairs are formed and aggregate around the metaphase plate of the cell.

Proprioception (prō-prē-ō-SEP-shun) The receipt of information from muscles, tendons, and the labyrinth that enables the brain to determine movements and position of the body and its parts. Also called **kinesthesia** (kin′-es-THĒ-zē-a).

Proprioceptor (prō′-prē-ō-SEP-tor) A receptor located in muscles, tendons, or joints that provides information about body position and movements.

Prostaglandin (pros′-ta-GLAN-din) **(PG)** A membrane-associated lipid; released in small quantities and acts as a local hormone.

Prostate (PROS-tāt) **gland** A doughnut-shaped gland inferior to the urinary bladder that surrounds the superior portion of the male urethra and secretes a slightly acidic solution that contributes to sperm motility and viability.

Protein An organic compound consisting of carbon, hydrogen, oxygen, nitrogen, and sometimes sulfur and phosphorus, and made up of amino acids linked by peptide bonds.

Prothrombin (prō-THROM-bin) An inactive blood-clotting factor synthesized by the liver, released into the blood, and converted to active thrombin in the process of blood clotting by the activated enzyme prothrombinase.

Proto-oncogene (prō′-tō-ONG-kō-jēn) Gene responsible for some aspect of normal growth and development; it may transform into an oncogene, a gene capable of causing cancer.

Protraction (prō-TRAK-shun) The movement of the mandible or shoulder girdle forward on a plane parallel with the ground.

Proximal (PROK-si-mal) Nearer the attachment of a limb to the trunk; nearer to the point of origin or attachment.

Pseudopods (SOO-dō-pods) Temporary, protruding projections of cytoplasm in a migrating cell.

Pterygopalatine ganglion (ter′-i-gō-PAL-a-tīn GANG-glē-on) A cluster of cell bodies of parasympathetic postganglionic neurons ending at the lacrimal and nasal glands.

Ptosis (TŌ-sis) Drooping, as of the eyelid or the kidney.

Puberty (PYOO-ber-tē) The time of life during which the secondary sex characteristics begin to appear and the capability for sexual reproduction is possible; usually occurs between the ages of 10 and 17.

Pubic symphysis A slightly movable cartilaginous joint between the anterior surfaces of the hip bones.

Puerperium (pyoo′-er-PER-ē-um) The state immediately after childbirth, usually 4–6 weeks.

Pulmonary (PUL-mo-ner′-ē) Concerning or affected by the lungs.

Pulmonary circulation The flow of deoxygenated blood from the right ventricle to the lungs and the return of oxygenated blood from the lungs to the left atrium.

Pulmonary edema (e-DĒ-ma) An abnormal accumulation of interstitial fluid in the tissue spaces and alveoli of the lungs due to increased pulmonary capillary permeability or increased pulmonary capillary pressure.

Pulmonary embolism (EM-bō-lizm) **(PE)** The presence of a blood clot or a foreign substance in a pulmonary arterial blood vessel that obstructs circulation to lung tissue.

Pulmonary ventilation The inflow (inspiration) and outflow (expiration) of air between the atmosphere and the lungs. Also called **breathing.**

Pulp cavity A cavity within the crown and neck of a tooth; is filled with pulp, a connective tissue containing blood vessels, nerves, and lymphatic vessels.

Pulse (PULS) The rhythmic expansion and elastic recoil of a systemic artery after each contraction of the left ventricle.

Pulse pressure The difference between the maximum (systolic) and minimum (diastolic) pressures; normally about 40 mm Hg.

Pupil The hole in the center of the iris, the area through which light enters the posterior cavity of the eyeball.

Pus The liquid product of inflammation containing leukocytes or their remains and debris of dead cells.

P wave The deflection wave of an electrocardiogram that reflects atrial depolarization.

Pyloric (pī-LOR-ik) **sphincter** A thickened ring of smooth muscle through which the pylorus of the stomach communicates with the duodenum. Also called the **pyloric valve.**

Pyogenesis (pī′-ō-JEN-e-sis) Formation of pus.

Pyramid (PIR-a-mid) A pointed or cone-shaped structure; one of two roughly triangular structures on the ventral side of the medulla oblongata composed of the largest motor tracts that run from the cerebral cortex to the spinal cord; a triangular-shaped structure in the renal medulla.

Pyramidal (pi-RAM-i-dal) **tracts (pathways)** *See* **Direct motor pathways.**

QRS wave The deflection waves of an electrocardiogram that represent onset of ventricular depolarization.

Quadrant (KWOD-rant) One of four parts.

Quadriplegia (kwod′-ri-PLĒ-jē-a) Paralysis of four limbs: two upper and two lower.

Radiographic (rā′-dē-ō-GRAF-ic) **anatomy** Diagnostic branch of anatomy that includes the use of x-rays.

Rami communicantes (RĀ-mē ko-myoo-ni-KAN-tēz) Branches of a spinal nerve. *Singular is* **ramus communicans** (RĀ-mus ko-MYOO-ni-kans).

Rapid eye movement (REM) sleep Stage of sleep in which dreaming occurs, lasting for 5–10 minutes several times during a sleep cycle; characterized by rapid movements of the eyes beneath the eyelids.

Reactivity (rē-ak-TI-vi-tē) Ability of an antigen to react specifically with the antibody whose formation it induced.

Receptor A specialized cell or a distal portion of a neuron that responds to a specific sensory modality, such as touch, pressure, cold, light, or sound, and converts it to an electrical signal (generator or receptor potential). A specific molecule or cluster of molecules that recognizes and binds a particular ligand.

Receptor-mediated endocytosis A highly selective process whereby cells take up specific ligands, which usually are large molecules or particles, by enveloping them within a sac of plasma membrane. Ligands are eventually broken down by enzymes in lysosomes.

Receptor potential Depolarization or hyperpolarization of the plasma membrane of a receptor that alters release of neurotransmitter from the cell; if the neuron that synapses with the receptor cell becomes depolarized to threshold, a nerve impulse is triggered.

Recessive gene A gene that is not expressed in the presence of a dominant gene on the homologous chromosome.

Reciprocal innervation (re-SIP-rō-kal in-ner-VĀ-shun) The phenomenon by which action potentials stimulate contraction of one muscle and simultaneously inhibit contraction of antagonistic muscles.

Recombinant DNA Synthetic DNA, formed by joining a fragment of DNA from one source to a portion of DNA from another.

Recovery oxygen consumption Elevated oxygen use after exercise ends due to metabolic changes that start during exercise and continue after exercise. Previously called **oxygen debt.**

Recruitment (rē-KROOT-ment) The process of increasing the number of active motor units. Also called **motor unit summation.**

Rectouterine pouch A pocket formed by the parietal peritoneum as it moves posteriorly from the surface of the uterus and is reflected onto the rectum; the most inferior point in the pelvic cavity. Also called the **pouch** or **cul de sac of Douglas.**

Rectum (REK-tum) The last 20 cm (7 in.) of the gastrointestinal tract, from the sigmoid colon to the anus.

Red nucleus A cluster of cell bodies in the midbrain, occupying a large portion of the tectum and sending fibers into the rubroreticular and rubrospinal tracts.

Red pulp That portion of the spleen that consists of venous sinuses filled with blood and thin plates of splenic tissue called splenic (Billroth's) cords.

Reduction The addition of electrons to a molecule or, less commonly, the removal of oxygen from a molecule that results in an increase in the energy content of the molecule.

Referred pain Pain that is felt at a site remote from the place of origin.

Reflex Fast response to a change (stimulus) in the internal or external environment that attempts to restore homeostasis; a reflex passes over a reflex arc.

Reflex arc The most basic conduction pathway through the nervous system, connecting a receptor and an effector and consisting of a receptor, a sensory neuron, an integrating center in the central nervous system, a motor neuron, and an effector.

Refraction (rē-FRAK-shun) The bending of light as it passes from one medium to another.

Refractory (re-FRAK-to-rē) **period** A time during which an excitable cell (neuron or muscle fiber) cannot respond to a stimulus that is usually adequate to evoke an action potential.

Regional anatomy The division of anatomy dealing with a specific region of the body, such as the head, neck, chest, or abdomen.

Regurgitation (rē-gur′-ji-TĀ-shun) Return of solids or fluids to the mouth from the stomach; backward flow of blood through incompletely closed heart valves.

Relaxin (RLX) A female hormone produced by the ovaries that increases flexibility of the pubic symphysis and helps dilate the uterine cervix to ease delivery of a baby.

Releasing hormone Hormone secreted by the hypothalamus that can stimulate secretion of hormones of the anterior pituitary gland.

Remodeling Replacement of old bone by new bone tissue.

Renal (RĒ-nal) Pertaining to the kidney.

Renal corpuscle (KOR-pus′-el) A glomerular (Bowman's) capsule and its enclosed glomerulus.

Renal pelvis A cavity in the center of the kidney formed by the expanded, proximal portion of the ureter, lying within the kidney, and into which the major calyces open.

Renal pyramid A triangular structure in the renal medulla containing the straight segments of renal tubules and the vasa recta.

Renin (RĒ-nin) An enzyme released by the kidney into the plasma, where it converts angiotensinogen into angiotensin I.

Renin–angiotensin (an′-jē-ō-TEN-sin) **pathway** A mechanism for the control of aldosterone secretion by angiotensin II, initiated by the secretion of renin by the kidney in response to low blood pressure.

Repolarization (rē-pō′-lar-i-ZĀ-shun) Restoration of a resting membrane potential following depolarization.

Reproduction (rē′-prō-DUK-shun) Either the formation of new cells for growth, repair, or replacement, or the production of a new individual.

Reproductive cell division Type of cell division in which gametes (sperm and egg cells) are produced; consists of meiosis and cytokinesis.

Residual (re-ZID-yoo-al) **volume** The volume of air still contained in the lungs after a maximal expiration; about 1200 mL.

Resistance (re-ZIS-tans) Hindrance (impedance) to blood flow as a result of higher viscosity, longer total blood vessel length, and smaller blood vessel radius. Ability to ward off disease. The hindrance encountered by an electrical charge as it moves through a substance from one point to another. The hindrance encountered by air as it moves through the respiratory passageways.

Respiration (res-pi-RĀ-shun) Overall exchange of gases between the atmosphere, blood, and body cells consisting of pulmonary ventilation, external respiration, and internal respiration.

Respiratory center Neurons in the reticular formation of the brain stem that regulate the rate and depth of respiration.

Respiratory membrane Structure in the lungs consisting of the alveolar wall and basement membrane and a capillary endothelium and basement membrane through which the diffusion of respiratory gases occurs. Also called the **alveolar–capillary** (al-VĒ-ō-lar) **membrane.**

Resting membrane potential The voltage difference between the inside and outside of a cell membrane when the cell is not responding to a stimulus; in many neurons and muscle fibers it is –70 to –90 mV, with the inside of the cell negative relative to the outside.

Retention (rē-TEN-shun) A failure to void urine due to obstruction, nervous contraction of the urethra, or absence of sensation of desire to urinate.

Rete (RĒ-tē) **testis** The network of ducts in the testes

Reticular (re-TIK-yoo-lar) **activating system (RAS)** A portion of the reticular formation that has many ascending connections with the cerebral cortex; when this area of the brain stem is active, nerve impulses pass to the thalamus and widespread areas of the cerebral cortex, resulting in generalized alertness or arousal from sleep.

Reticular formation A network of small groups of neuron cell bodies scattered among bundles of axons (mixed gray and white matter) beginning in the medulla oblongata and extending superiorly through the central part of the brain stem.

Reticulocyte (re-TIK-yoo-lō-sīt) An immature red blood cell.

Reticulum (re-TIK-yoo lum) A network.

Retina (RET-i-na) The deep coat of the posterior portion of the eyeball consisting of nervous tissue (where the process of vision begins) and a pigmented layer of epithelial cells that contact the choroid. Also called the **nervous tunic** (TOO-nik).

Retinal (RE-ti-nal) A derivative of vitamin A that functions as the light-absorbing portion of the photopigment rhodopsin.

Retraction (rē-TRAK-shun) The movement of a protracted part of the body posteriorly on a plane parallel to the ground, as in pulling the lower jaw back in line with the upper jaw.

Retroflexion (re-trō-FLEK-shun) A malposition of the uterus in which it is tilted posteriorly.

Retrograde degeneration (RE-trō-grād dē-jen-er-Ā-shun) Changes that occur in the proximal portion of a damaged axon only as far as the first neurofibral node (node of Ranvier); similar to changes that occur during Wallerian degeneration.

Retroperitoneal (re'-trō-per-i-tō-NĒ-al) External to the peritoneal lining of the abdominal cavity.

Rh factor An inherited antigen on the surface of erythrocytes (red blood cells).

Rhinology (rī-NOL-ō-jē) The study of the nose and its disorders.

Rhodopsin (rō-DOP-sin) The photopigment in rods of the retina, consisting of a glycoprotein called opsin and a derivative of vitamin A called retinal.

Ribonucleic (rī-bō-nyoo-KLĒ-ik) **acid (RNA)** A single-stranded nucleic acid constructed of nucleotides, each consisting of a nitrogenous base (adenine, cytosine, guanine, or uracil), ribose, and a phosphate group; three types are messenger RNA (mRNA), transfer RNA (tRNA), and ribosomal RNA (rRNA), each of which has a specific role during protein synthesis.

Ribosome (RĪ-bō-sōm) An organelle in the cytoplasm of cells, composed of ribosomal RNA and ribosomal proteins, that synthesizes proteins; nicknamed the "protein factory."

Right heart (atrial) reflex A reflex concerned with maintaining normal venous blood pressure.

Right lymphatic (lim-FAT-ik) **duct** A vessel of the lymphatic system that drains lymph from the upper right side of the body and empties it into the right subclavian vein.

Rigidity (ri-JID-i-tē) Hypertonia characterized by increased muscle tone, but reflexes are not affected.

Rigor mortis State of partial contraction of muscles following death due to lack of ATP; myosin heads (cross bridges) remain attached to actin, thus preventing relaxation.

Rod One of two types of photoreceptor in the retina of the eye; specialized for vision in dim light.

Root canal A narrow extension of the pulp cavity lying within the root of a tooth.

Root of penis Attached portion of penis that consists of the bulb and crura.

Rotation (rō-TĀ-shun) Moving a bone around its own axis, with no other movement.

Round ligament (LIG-a-ment) A band of fibrous connective tissue enclosed between the folds of the broad ligament of the uterus, emerging from the uterus just inferior to the uterine tube, extending laterally along the pelvic wall and through the deep inguinal ring to end in the labia majora.

Round window A small opening between the middle and inner ear, directly inferior to the oval window, covered by the secondary tympanic membrane.

Rugae (ROO-jē) Large folds in the mucosa of an empty hollow organ, such as the stomach and vagina.

Saccule (SAK-yool) The inferior and smaller of the two chambers in the membranous labyrinth inside the vestibule of the inner ear containing a receptor organ for static equilibrium.

Sacral plexus (PLEK-sus) A network formed by the ventral branches of spinal nerves L4 through S3.

Sacral promontory (PROM-on-tor'-ē) The superior surface of the body of the first sacral vertebra that projects anteriorly into the pelvic cavity; a line from the sacral promontory to the superior border of the pubic symphysis divides the abdominal and pelvic cavities.

Saddle joint A synovial joint in which the articular surface of one bone is saddle shaped and the articular surface of the other bone is shaped like the legs of the rider sitting in the saddle, as in the joint between the trapezium and the metacarpal of the thumb.

Sagittal (SAJ-i-tal) **plane** A plane that divides the body or organs into left and right portions. Such a plane may be **midsagittal (median)**, in which the divisions are equal, or **parasagittal,** in which the divisions are unequal.

Saliva (sa-LĪ-va) A clear, alkaline, somewhat viscous secretion produced mostly by the three pairs of salivary glands; contains various salts, mucin, lysozyme, salivary amylase, and lingual lipase (produced by glands in the tongue).

Salivary amylase (SAL-i-ver-ē AM-i-lās) An enzyme in saliva that initiates the chemical breakdown of starch.

Salivary gland One of three pairs of glands that lie external to the mouth and pour their secretory product (saliva) into ducts that empty into the oral cavity; the parotid, submandibular, and sublingual glands.

Salt A substance that, when dissolved in water, ionizes into cations and anions, neither of which are hydrogen ions (H^+) or hydroxide ions (OH^-).

Saltatory (sal-ta-TŌ-rē) **conduction** The propagation of an action potential (nerve impulse) along the exposed portions of a myelinated nerve fiber. The action potential appears at successive nodes of Ranvier and therefore seems to jump or leap from node to node.

Sarcolemma (sar′-kō-LEM-ma) The cell membrane of a muscle fiber (cell), especially of a skeletal muscle fiber.

Sarcomere (SAR-kō-mēr) A contractile unit in a striated muscle fiber extending from one Z disc to the next Z disc.

Sarcoplasm (SAR-kō-plazm) The cytoplasm of a muscle fiber (cell).

Sarcoplasmic reticulum (sar′-kō-PLAZ-mik re-TIK-yoo-lum) A network of saccules and tubes surrounding myofibrils of a muscle fiber (cell), comparable to endoplasmic reticulum; functions to reabsorb calcium ions during relaxation and to release them to cause contraction.

Satiety center A collection of neurons located in the ventromedial nuclei of the hypothalamus that, when stimulated, bring about the cessation of eating.

Saturated fat A fatty acid that contains only single bonds (no double bonds) between its carbon atoms; all carbon atoms are bonded to the maximum number of hydrogen atoms; prevalent in triglycerides of animal products such as meat, milk, milk products, and eggs.

Scala tympani (SKA-la TIM-pan-ē) The inferior spiral-shaped channel of the bony cochlea, filled with perilymph.

Scala vestibuli (ves-TIB-yoo-lē) The superior spiral-shaped channel of the bony cochlea, filled with perilymph.

Schwann (SCHVON) **cell** A neuroglial cell of the peripheral nervous system that forms the myelin sheath and neurolemma of a nerve fiber by wrapping around a nerve fiber in a jelly-roll fashion. Also called a **neurolemmocyte.**

Sciatica (sī-AT-i-ka) Inflammation and pain along the sciatic nerve; felt along the posterior aspect of the thigh extending down the inside of the leg.

Sclera (SKLE-ra) The white coat of fibrous tissue that forms the superficial protective covering over the eyeball except in the most anterior portion; the posterior portion of the fibrous tunic.

Scleral venous sinus A circular venous sinus located at the junction of the sclera and the cornea through which aqueous humor drains from the anterior chamber of the eyeball into the blood. Also called the **canal of Schlemm** (SHLEM).

Sclerosis (skle-RŌ-sis) A hardening with loss of elasticity of tissues.

Scoliosis (skō′-lē-Ō-sis) An abnormal lateral curvature from the normal vertical line of the backbone.

Scrotum (SKRŌ-tum) A skin-covered pouch that contains the testes and their accessory structures.

Sebaceous (se-BĀ-shus) **gland** An exocrine gland in the dermis of the skin, almost always associated with a hair follicle, that secretes sebum. Also called an **oil gland.**

Sebum (SĒ-bum) Secretion of sebaceous (oil) glands.

Secondary response Accelerated, more intense cell-mediated or antibody-mediated immune response upon a subsequent exposure to an antigen after the primary response.

Secondary sex characteristic A characteristic of the male or female body that develops at puberty under the influence of sex hormones but is not directly involved in sexual reproduction; examples are distribution of body hair, voice pitch, body shape, and muscle development.

Secretion (se-KRĒ-shun) Production and release from a gland cell of a fluid, especially a functionally useful product as opposed to a waste product.

Selective permeability (per′-mē-a-BIL-i-tē) The property of a membrane by which it permits the passage of certain substances but restricts the passage of others.

Sella turcica (SEL-a TUR-si-ka) A depression on the superior surface of the sphenoid bone that houses the pituitary gland.

Semen (SĒ-men) A fluid discharged at ejaculation by a male that consists of a mixture of sperm and the secretions of the seminiferous tubules, seminal vesicles, prostate gland, and bulbourethral (Cowper's) glands.

Semicircular canals Three bony channels (anterior, posterior, lateral), filled with perilymph, in which lie the membranous semicircular canals filled with endolymph. They contain receptors for equilibrium.

Semicircular ducts The membranous semicircular canals filled with endolymph and floating in the perilymph of the bony semicircular canals; they contain cristae that are concerned with dynamic equilibrium.

Semilunar (sem′-ē-LOO-nar) **valve** A valve between the aorta or the pulmonary trunk and a ventricle of the heart.

Seminal vesicle (SEM-i-nal VES-i-kul) One of a pair of convoluted, pouchlike structures, lying posterior and inferior to the urinary bladder and anterior to the rectum, that secrete a component of semen into the ejaculatory ducts.

Seminiferous tubule (sem′-i-NI-fer-us TOO-byool) A tightly coiled duct, located in the testis, where sperm are produced.

Senescence (se-NES-ens) The process of growing old.

Sensation A state of awareness of external or internal conditions of the body.

Sensory area A region of the cerebral cortex concerned with the interpretation of sensory impulses.

Sensory neuron (NOO-ron) A neuron that conducts nerve impulses into the central nervous system. Also called an **afferent neuron**.

Septal defect An opening in the septum (interatrial or interventricular) between the left and right sides of the heart.

Septum (SEP-tum) A wall dividing two cavities.

Serosa (ser-Ō-sa) Any serous membrane. The external layer of an organ formed by a serous membrane. The membrane that lines the pleural, pericardial, and peritoneal cavities.

Serous (SIR-us) **membrane** A membrane that lines a body cavity that does not open to the exterior. Also called the **serosa** (se-RŌ-sa).

Sertoli (ser-TŌ-lē) **cell** A supporting cell of seminiferous tubules that secretes fluid for supplying nutrients to sperm and the hormone inhibin, phagocytizes excess cytoplasm from spermatogenic cells, and mediates the effects of FSH and testosterone on spermatogenesis. Also called a **sustentacular** (sus′-ten-TAK-yoo-lar) **cell.**

Serum Blood plasma minus its clotting proteins.

Sesamoid (SES-a-moyd) **bones** Small bones usually found in tendons.

Sex chromosomes The twenty-third pair of chromosomes, designated X and Y, which determine the genetic sex of an individual; in males, the pair is XY; in females, XX.

Sexual intercourse The insertion of the erect penis of a male into the vagina of a female. Also called **coitus** (KŌ-i-tus).

Shivering Involuntary contraction of skeletal muscles that generates heat. Also called **involuntary thermogenesis.**

Shock Failure of the cardiovascular system to deliver adequate amounts of oxygen and nutrients to meet the metabolic needs of the body due to inadequate cardiac output. It is characterized by hypotension; clammy, cool, and pale skin; sweating; reduced urine formation; altered mental state; acidosis; tachycardia; weak, rapid pulse; and thirst. Types include hypovolemic, cardiogenic, vascular, and obstructive.

Shoulder joint A synovial joint where the humerus articulates with the scapula.

Sigmoid colon (SIG-moyd KŌ-lon) The S-shaped portion of the large intestine that begins at the level of the left iliac crest, projects medially, and terminates at the rectum at about the level of the third sacral vertebra.

Sign Any objective evidence of disease that can be observed or measured such as a lesion, swelling, or fever.

Simple diffusion (dif-YOO-zhun) A passive process in which there is a net or greater movement of molecules or ions from a region of high concentration to a region of low concentration until equilibrium is reached.

Sinoatrial (si-nō-Ā-trē-al) **(SA) node** A compact mass of cardiac muscle fibers (cells) specialized for conduction, located in the right atrium inferior to the opening of the superior vena cava. Also called the **pacemaker.**

Sinus (SĪ-nus) A hollow in a bone (paranasal sinus) or other tissue; a channel for blood (vascular sinus); any cavity having a narrow opening.

Sinusoid (SĪN-yoo-soyd) A microscopic space or passage for blood in certain organs such as the liver or spleen.

Skeletal muscle An organ specialized for contraction, composed of striated muscle fibers (cells), supported by connective tissue, attached to a bone by a tendon or an aponeurosis, and stimulated by somatic motor neurons.

Skin The external covering of the body that consists of a superficial, thinner epidermis (epithelial tissue) and a deep, thicker dermis (connective tissue) that is anchored to the subcutaneous layer.

Skull The skeleton of the head consisting of the cranial and facial bones.

Sleep A state of partial unconsciousness from which a person can be aroused; associated with a low level of activity in the reticular activating system.

Sliding-filament mechanism The explanation of how thick and thin filaments slide relative to one another during striated muscle contraction to decrease sarcomere length.

Small intestine A long tube of the gastrointestinal tract that begins at the pyloric sphincter of the stomach, coils through the central and inferior part of the abdominal cavity, and ends at the large intestine; divided into three segments: duodenum, jejunum, and ileum.

Smooth muscle A tissue specialized for contraction, composed of smooth muscle fibers (cells), located in the walls of hollow internal organs, and innervated by autonomic motor neurons.

Sodium–potassium ATPase An active transport pump located in the cell membrane that transports sodium ions out of the cell and potassium ions into the cell at the expense of cellular ATP. It functions to keep the ionic concentrations of these elements at physiological levels. Also called the **sodium pump.**

Soft palate (PAL-at) The posterior portion of the roof of the mouth, extending from the palatine bones to the uvula. It is a muscular partition lined with mucous membrane.

Solution A homogeneous molecular or ionic dispersion of one or more substances (solutes) in a dissolving medium (solvent) that is usually liquid.

Somatic (sō-MAT-ik) **cell division** Type of cell division in which a single starting parent cell duplicates itself to produce two identical daughter cells; consists of mitosis and cytokinesis.

Somatic nervous system (SNS) The portion of the peripheral nervous system consisting of somatic sensory (afferent) neurons and somatic motor (efferent) neurons.

Somite (SŌ-mīt) Block of mesodermal cells in a developing embryo that is distinguished into a myotome (which forms most of the skeletal muscles), dermatome (which forms connective tissues), and sclerotome (which forms the vertebrae).

Spasm (spazm) A sudden, involuntary contraction of large groups of muscles.

Spasticity (spas-TIS-i-tē) Hypertonia characterized by increased muscle tone, increased tendon reflexes, and pathological reflexes (Babinski sign).

Spermatic (sper-MAT-ik) **cord** A supporting structure of the male reproductive system, extending from a testis to the deep inguinal ring, that includes the ductus (vas) deferens, arteries, veins, lymphatic vessels, nerves, cremaster muscle, and connective tissue.

Spermatogenesis (sper′-ma-tō-JEN-e-sis) The formation and development of sperm in the seminiferous tubules of the testes.

Sperm cell A mature male gamete. Also termed **spermatozoon** (sper′-ma-tō-ZŌ-on).

Spermiogenesis (sper′-mē-ō-JEN-e-sis) The maturation of spermatids into sperm.

Sphincter (SFINGK-ter) A circular muscle that constricts an opening.

Sphincter of the hepatopancreatic ampulla A circular muscle at the opening of the common bile and main pancreatic ducts in the duodenum. Also called the **sphincter of Oddi** (OD-ē).

Sphygmomanometer (sfig′-mō-ma-NOM-e-ter) An instrument for measuring arterial blood pressure.

Spinal (SPĪ-nal) **cord** A mass of nerve tissue located in the vertebral canal from which 31 pairs of spinal nerves originate.

Spinal nerve One of the 31 pairs of nerves that originate on the spinal cord from posterior and anterior roots.

Spinal shock A period from several days to several weeks following transection of the spinal cord and characterized by the abolition of all reflex activity.

Spinothalamic (spī-nō-THAL-am-ik) **tracts** Sensory (ascending) tracts that convey information up the spinal cord to the thalamus for sensations of pain, temperature, crude touch, and deep pressure.

Spinous (SPĪ-nus) **process** A sharp or thornlike process or projection. Also called a **spine.** A sharp ridge running diagonally across the posterior surface of the scapula.

Spiral organ The organ of hearing, consisting of supporting cells and hair cells that rest on the basilar membrane and extend into the endolymph of the cochlear duct. Also called the **organ of Corti** (KOR-tē).

Spirometer (spī-ROM-e-ter) An apparatus used to measure lung volumes and capacities.

Splanchnic (SPLANK-nik) Pertaining to the viscera.

Spleen (SPLĒN) Large mass of lymphatic tissue between the fundus of the stomach and the diaphragm that functions in formation of blood cells during early fetal development, phagocytosis of worn-out blood cells, and proliferation of B cells during immune responses.

Spongy (cancellous) bone tissue Bone tissue that consists of an irregular latticework of thin plates of bone called trabeculae; spaces between trabeculae of some bones are filled with red bone marrow; found inside short, flat, and irregular bones and in the epiphyses (ends) of long bones.

Sputum (SPYOO-tum) Substance ejected from the mouth containing saliva and mucus.

Squamous (SKWĀ-mus) Flat or scalelike.

Starling's law of the capillaries The movement of fluid between plasma and interstitial fluid is in a state of near equilibrium at the arterial and venous ends of a capillary; that is, filtered fluid and absorbed fluid plus that returned to the lymphatic system are nearly equal.

Starling's law of the heart The force of muscular contraction is determined by the length of the cardiac muscle fibers just before they contract; within limits, the greater the length of stretched fibers, the stronger the contraction.

Starvation (star-VĀ-shun) The loss of energy stores in the form of glycogen, triglycerides, and proteins due to inadequate intake of nutrients or inability to digest, absorb, or metabolize ingested nutrients.

Stasis (STĀ-sis) Stagnation or halt of normal flow of fluids, as blood or urine, or of the intestinal contents.

Static equilibrium (ē-kwi-LIB-rē-um) The maintenance of posture in response to changes in the orientation of the body, mainly the head, relative to the ground.

Stellate reticuloendothelial (STEL-āt re-tik′-yoo-lō-en′-dō-THĒ-lē-al) **cell** Phagocytic cell bordering a sinusoid of the liver. Also called a **Kupffer's** (KOOP-ferz) **cell.**

Stenosis (sten-Ō-sis) An abnormal narrowing or constriction of a duct or opening.

Stereocilia (ste′-rē-ō-SIL-ē-a) Groups of extremely long, slender, nonmotile microvilli projecting from epithelial cells lining the epididymis.

Stereognosis (ste′-rē-og-NŌ-sis) The ability to recognize the size, shape, and texture of an object by touching it.

Sterile (STE-ril) Free from any living microorganisms. Unable to conceive or produce offspring.

Sterilization (ster′-i-li-ZĀ-shun) Elimination of all living microorganisms. Any procedure that renders an individual incapable of reproduction (e.g., castration, vasectomy, hysterectomy, oophorectomy).

Stimulus Any stress that changes a controlled condition; any change in the internal or external environment that excites a receptor, a neuron, or a muscle fiber.

Stomach The J-shaped enlargement of the gastrointestinal tract directly inferior to the diaphragm in the epigastric, umbilical, and left hypochondriac regions of the abdomen, between the esophagus and small intestine.

Straight tubule (TOO-byool) A duct in a testis leading from a convoluted seminiferous tubule to the rete testis.

Stratum (STRĀ-tum) A layer.

Stratum basalis (ba-SAL-is) The superficial layer of the endometrium (next to the myometrium) that is maintained during menstruation and gestation and produces a new stratum functionalis following menstruation or parturition.

Stratum functionalis (funk′-shun-AL-is) The deep layer of the endometrium (next to the uterine cavity) that is shed during menstruation and that forms the maternal portion of the placenta during gestation.

Stressor A stress that is extreme, unusual, or long-lasting and triggers the general adaptation syndrome.

Stretch receptor Receptor in the walls of blood vessels, airways, or organs that monitors the amount of stretching. Also termed **baroreceptor.**

Stretch reflex A monosynaptic reflex triggered by sudden stretching of muscle spindles within a muscle that elicits contraction of that same muscle. Also called a **tendon jerk.**

Stroke volume The volume of blood ejected by either ventricle in one systole; about 70 mL at rest.

Stroma (STRŌ-ma) The tissue that forms the ground substance, foundation, or framework of an organ, as opposed to its functional parts (parenchyma).

Subarachnoid (sub′-a-RAK-noyd) **space** A space between the arachnoid and the pia mater that surrounds the brain and spinal cord and through which cerebrospinal fluid circulates.

Subcutaneous (sub′-kyoo-TĀ-nē-us) Beneath the skin. Also called **hypodermic** (hī-pō-DER-mik).

Subcutaneous layer A continuous sheet of areolar connective tissue and adipose tissue between the dermis of the skin and the deep fascia of the muscles. Also called the **superficial fascia** (FASH-ē-a).

Subdural (sub-DOO-ral) **space** A space between the dura mater and the arachnoid of the brain and spinal cord that contains a small amount of fluid.

Sublingual (sub-LING-gwal) **gland** One of a pair of salivary glands situated in the floor of the mouth deep to the mucous membrane and to the side of the lingual frenulum, with a duct (Rivinus') that opens into the floor of the mouth.

Submandibular (sub′-man-DIB-yoo-lar) **gland** One of a pair of salivary glands found inferior to the base of the tongue under the mucous membrane in the posterior part of the floor of the mouth, posterior to the sublingual glands, with a duct (Wharton's) situated to the side of the lingual frenulum. Also called the **submaxillary** (sub′-MAK-si-ler-ē) **gland.**

Submucosa (sub-myoo-KŌ-sa) A layer of connective tissue located deep to a mucous membrane, as in the gastrointestinal tract or the urinary bladder; the submucosa connects the mucosa to the muscularis layer.

Submucosal plexus A network of autonomic nerve fibers located in the superficial portion of the submucous layer of the small intestine. Also called the **plexus of Meissner** (MĪS-ner).

Substrate A molecule upon which an enzyme acts.

Subthalamus Portion of the diencephalon lying below the thalamus and above the hypothalamus.

Subthreshold stimulus A stimulus of such weak intensity that it cannot initiate an action potential (nerve impulse).

Sudoriferous (soo′-dor-IF-er-us) **gland** An apocrine or eccrine exocrine gland in the dermis or subcutaneous layer that produces perspiration. Also called a **sweat gland.**

Sulcus (SUL-kus) A groove or depression between parts, especially between the convolutions of the brain. *Plural is* **sulci** (SUL-sī).

Summation (sum-MĀ-shun) The addition of the excitatory and inhibitory effects of many stimuli applied to a neuron. The increased strength of muscle contraction that results when stimuli follow one another in rapid succession.

Superficial (soo′-per-FISH-al) Located on or near the surface of the body or an organ.

Superficial fascia (FASH-ē-a) A continuous sheet of fibrous connective tissue between the dermis of the skin and the deep fascia of the muscles. Also called **subcutaneous** (sub′-kyoo-TĀ-nē-us) **layer.**

Superficial inguinal (IN-gwi-nal) **ring** A triangular opening in the aponeurosis of the external oblique muscle that represents the termination of the inguinal canal.

Superior (soo-PĒR-ē-or) Toward the head or upper part of a structure. Also called **cephalad** (SEF-a-lad) or **craniad.**

Superior vena cava (VĒ-na CĀ-va) **(SVC)** Large vein that collects blood from parts of the body superior to the heart and returns it to the right atrium.

Supination (soo-pi-NĀ-shun) A movement of the forearm in which the palm is turned anteriorly or superiorly.

Surface anatomy The study of the structures that can be identified from the outside of the body.

Surfactant (sur-FAK-tant) Complex mixture of phospholipids and lipoproteins, produced by type II alveolar (septal) cells in the lungs, that decreases surface tension.

Susceptibility (sus-sep′-ti-BIL-i-tē) Lack of resistance of a body to the deleterious or other effects of an agent such as a pathogen.

Suspensory ligament (sus-PEN-so-rē LIG-a-ment) A fold of peritoneum extending laterally from the surface of the ovary to the pelvic wall.

Sutural (SOO-cher-al) **bone** A small bone located within a suture between certain cranial bones. Also called **Wormian** (WER-mē-an) **bone.**

Suture (SOO-cher) An immovable fibrous joint that joins skull bones.

Sympathetic (sim′-pa-THET-ik) **division** One of the two subdivisions of the autonomic nervous system, having cell bodies of preganglionic neurons in the lateral gray columns of the thoracic segment and the first two or three lumbar segments of the spinal cord; primarily concerned with processes involving the expenditure of energy. Also called the **thoracolumbar** (thō′-ra-kō-LUM-bar) **division.**

Sympathetic trunk ganglion (GANG-glē-on) A cluster of cell bodies of postganglionic sympathetic neurons lateral to the vertebral column, close to the body of a vertebra. These ganglia extend inferiorly through the neck, thorax, and abdomen to the coccyx on both sides of the vertebral column and are connected to one another to form a chain on each side of the vertebral column. Also called **sympathetic chain** or **vertebral chain ganglia.**

Sympathomimetic (sim′-pa-thō-mi-MET-ik) Producing effects that mimic those brought about by the sympathetic division of the autonomic nervous system.

Symphysis (SIM-fi-sis) A line of union. A slightly movable cartilaginous joint such as the pubic symphysis.

Symport Process by which two substances, often Na$^+$ and another substance, move in the same direction across a cell membrane. Also called **cotransport.**

Symptom (SIMP-tum) A subjective change in body function not apparent to an observer, such as pain or nausea, that indicates the presence of a disease or disorder of the body.

Synapse (SYN-aps) The functional junction between two neurons or between a neuron and an effector, such as a muscle or gland; may be electrical or chemical.

Synapsis (sin-AP-sis) The pairing of homologous chromosomes during prophase I of meiosis.

Synaptic (sin-AP-tik) **cleft** The narrow gap that separates the axon terminal of one neuron from another neuron or muscle fiber (cell) and across which a neurotransmitter diffuses to affect the postsynaptic cell.

Synaptic delay The length of time between the arrival of the action potential at a presynaptic axon terminal and the membrane potential (IPSP or EPSP) change on the postsynaptic membrane; usually about 0.5 msec.

Synaptic end bulb Expanded distal end of an axon terminal that contains synaptic vesicles. Also called a **synaptic knob.**

Synaptic vesicle Membrane-enclosed sac in a synaptic end bulb that stores neurotransmitters.

Synarthrosis (sin′-ar-THRŌ-sis) An immovable joint; types are suture, gomphosis, and synchondrosis.

Synchondrosis (sin′-kon-DRŌ-sis) A cartilaginous joint in which the connecting material is hyaline cartilage.

Syndesmosis (sin′-dez-MŌ-sis) A slightly movable joint in which articulating bones are united by fibrous connective tissue.

Syndrome (SIN-drōm) A group of signs and symptoms that occur together in a pattern that is characteristic of a particular disease or abnormal condition.

Synergist (SIN-er-jist) A muscle that assists the prime mover by reducing undesired action or unnecessary movement.

Synergistic (syn-er-GIS-tik) **effect** A hormonal interaction in which the effects of two or more hormones acting together is greater or more extensive than the sum of each hormone acting alone.

Synostosis (sin′-os-TŌ-sis) A joint in which the dense fibrous connective tissue that unites bones at a suture has been replaced by bone, resulting in a complete fusion across the suture line.

Synovial (si-NŌ-vē-al) **cavity** The space between the articulating bones of a diarthrotic joint, filled with synovial fluid. Also called a **joint cavity.**

Synovial fluid Secretion of synovial membranes that lubricates joints and nourishes articular cartilage.

Synovial joint A fully movable or diarthrotic joint in which a synovial (joint) cavity is present between the two articulating bones.

Synovial membrane The deeper of the two layers of the articular capsule of a synovial joint, composed of areolar connective tissue that secretes synovial fluid into the synovial (joint) cavity.

System An association of organs that have a common function.

Systemic (sis-TEM-ik) Affecting the whole body; generalized.

Systemic anatomy The anatomic study of particular systems of the body, such as the skeletal, muscular, nervous, cardiovascular, or urinary systems.

Systemic circulation The routes through which oxygenated blood flows from the left ventricle through the aorta to all the organs of the body and deoxygenated blood returns to the right atrium.

Systemic vascular resistance (SVR) All the vascular resistances offered by systemic blood vessels. Also called **total peripheral resistance.**

Systole (SIS-tō-lē) In the cardiac cycle, the phase of contraction of the heart muscle, especially of the ventricles.

Systolic (sis-TO-lik) **blood pressure** The force exerted by blood on arterial walls during ventricular contraction; the highest pressure measured in the large arteries, about 120 mm Hg under normal conditions for a young adult.

Tachycardia (tak′-i-KAR-dē-a) An abnormally rapid resting heartbeat or pulse rate (over 100/min).

Tactile (TAK-tīl) Pertaining to the sense of touch.

Tactile disc Modified epidermal cell in the stratum basale of hairless skin that functions as a cutaneous receptor for discriminative touch. Also called a **Merkel** (MER-kel) **disc.**

Taenia coli (TĒ-nē-a KŌ-lī) One of three flat bands of thickened, longitudinal smooth muscle running the length of the large intestine.

Target cell A cell whose activity is affected by a particular hormone.

Tarsal gland Sebaceous (oil) gland that opens on the edge of each eyelid. Also called a **Meibomian** (mī-BŌ-mē-an) **gland.**

Tarsal plate A thin, elongated sheet of connective tissue, one in each eyelid, giving the eyelid form and support. The aponeurosis of the levator palpebrae superioris is attached to the tarsal plate of the superior eyelid.

Tarsus (TAR-sus) A collective term for the seven bones of the ankle.

T cell A lymphocyte that becomes immunocompetent in the thymus gland and can differentiate into one of several types of effector cells that function in cell-mediated immunity.

Tectorial (tek-TŌ-rē-al) **membrane** A gelatinous membrane projecting over and in contact with the hair cells of the spiral organ (organ of Corti) in the cochlear duct.

Teeth (TĒTH) Accessory structures of digestion, composed of calcified connective tissue and embedded in bony sockets of the mandible and maxilla, that cut, shred, crush, and grind food. Also called **dentes** (DEN-tēz).

Telophase (TEL-ō-fāz) The final stage of mitosis in which the daughter nuclei become established.

Tendon (TEN-don) A white fibrous cord of dense, regularly arranged connective tissue that attaches muscle to bone.

Tendon organ A proprioceptive receptor, sensitive to changes in muscle tension and force of contraction, found chiefly near the junctions of tendons and muscles. Also called a **Golgi** (GOL-jē) **tendon organ.**

Tendon reflex A polysynaptic, ipsilateral reflex that protects tendons and their associated muscles from damage that might be brought about by excessive tension. The receptors involved are called tendon organs (Golgi tendon organs).

Tentorium cerebelli (ten-TŌ-rē-um ser′-e-BEL-ē) A transverse shelf of dura mater that forms a partition between the occipital lobe of the cerebral hemispheres and the cerebellum and that covers the cerebellum.

Teratogen (TER-a-tō-jen) Any agent or factor that causes physical defects in a developing embryo.

Terminal ganglion (TER-min-al GANG-glē-on) A cluster of cell bodies of postganglionic parasympathetic neurons either lying very close to the visceral effectors or located within the walls of the visceral effectors supplied by the postganglionic fibers.

Testis (TES-tis) Male gonad that produces sperm and the hormones testosterone and inhibin. Also called a **testicle.**

Testosterone (tes-TOS-te-rōn) A male sex hormone (androgen) secreted by interstitial endocrinocytes (cells of Leydig) of a mature testis; needed for development of sperm; together with a second androgen termed **dihydrotestosterone (DHT),** controls the growth and development of male reproductive organs, secondary sex characteristics, and body growth.

Tetany (TET-a-nē) Hyperexcitability of neurons and muscle fibers caused by hypocalcemia and characterized by intermittent or continuous tonic muscular contractions; may be due to hypoparathyroidism.

Thalamus (THAL-a-mus) A large, oval structure located superior to the midbrain, consisting of two masses of gray matter covered by a thin layer of white matter.

Thermoreceptor (THER-mō-rē-sep-tor) Receptor that detects changes in temperature.

Thigh The portion of the lower limb between the hip and the knee.

Third ventricle (VEN-tri-kul) A slitlike cavity between the right and left halves of the thalamus and between the lateral ventricles of the brain.

Thirst center A cluster of neurons in the hypothalamus that is sensitive to the osmotic pressure of extracellular fluid and brings about the sensation of thirst.

Thoracic (thō-RAS-ik) **cavity** Superior portion of the ventral body cavity that contains two pleural cavities, the mediastinum, and the pericardial cavity.

Thoracic duct A lymphatic vessel that begins as a dilation called the cisterna chyli, receives lymph from the left side of the head, neck, and chest, the left arm, and the entire body below the ribs, and empties into the left subclavian vein. Also called the **left lymphatic** (lim-FAT-ik) **duct.**

Thoracolumbar (thō′-ra-kō-LUM-bar) **outflow** The fibers of the sympathetic preganglionic neurons, which have their cell bodies in the lateral gray columns of the thoracic segments and first two or three lumbar segments of the spinal cord.

Thorax (THŌ-raks) The chest.

Threshold potential The membrane voltage that must be reached to trigger an action potential.

Threshold stimulus Any stimulus strong enough to initiate an action potential or activate a sensory receptor.

Thrombin (THROM-bin) The active enzyme formed from prothrombin that acts to convert fibrinogen to fibrin during formation of a blood clot.

Thrombolytic (throm-bō-LIT-ik) **agent** Chemical substance injected into the body to dissolve blood clots and restore circulation; mechanism of action is direct or indirect activation of plasminogen; examples include tissue plasminogen activator (t-PA), streptokinase, and urokinase.

Thrombosis (throm-BŌ-sis) The formation of a clot in an unbroken blood vessel, usually a vein.

Thrombus A stationary clot formed in an unbroken blood vessel, usually a vein.

Thymus (THĪ-mus) **gland** A bilobed organ, located in the superior mediastinum posterior to the sternum and between the lungs, that plays an essential role in immune responses.

Thyroglobulin (thī-rō-GLŌ-byoo-lin) **(TGB)** A large glycoprotein molecule produced by follicle cells of the thyroid gland in which iodine is combined with tyrosine to form thyroid hormones.

Thyroid cartilage (THĪ-royd KAR-ti-lij) The largest single cartilage of the larynx, consisting of two fused plates that form the anterior wall of the larynx.

Thyroid colloid (KOL-loyd) A complex in thyroid follicles consisting of thyroglobulin and stored thyroid hormones.

Thyroid follicle (FOL-i-kul) Spherical sac that forms the parenchyma of the thyroid gland and consists of follicular cells that produce thyroxine (T_4) and triiodothyronine (T_3).

Thyroid gland An endocrine gland with right and left lateral lobes on either side of the trachea connected by an isthmus; located anterior to the trachea just inferior to the cricoid cartilage; secretes thyroxine (T_4), triiodothyronine (T_3), and calcitonin.

Thyroid-stimulating hormone (TSH) A hormone secreted by the anterior pituitary gland that stimulates the synthesis and secretion of thyroxine (T_4) and triiodothyronine (T_3).

Thyroxine (thī-ROK-sēn) (**T₄**) A hormone secreted by the thyroid gland that regulates organic metabolism, growth and development, and the activity of the nervous system.

Tic Spasmodic twitching made involuntarily by muscles that are ordinarily under voluntary control.

Tidal volume The volume of air breathed in and out in any one breath; about 500 mL in quiet, resting conditions.

Tissue A group of similar cells and their intercellular substance joined together to perform a specific function.

Tissue factor (TF) A factor, or collection of factors, whose appearance initiates the blood clotting process. Also called **thromboplastin** (throm-bō-PLAS-tin).

Tissue plasminogen activator (t-PA) An enzyme that dissolves small blood clots by initiating a process that converts plasminogen to plasmin, which degrades the fibrin of a clot.

Tongue A large skeletal muscle covered by a mucous membrane located on the floor of the oral cavity.

Tonicity (tō-NIS-i-tē) A measure of the concentration of impermeable solute particles in a solution relative to cytosol. When cells are bathed in an **isotonic solution,** they neither shrink nor swell.

Tonsil (TON-sil) An aggregation of large lymphatic nodules embedded in the mucous membrane of the throat.

Torn cartilage A tearing of an articular disc (meniscus) in the knee.

Total lung capacity The sum of tidal volume, inspiratory reserve volume, expiratory reserve volume, and residual volume; about 6000 mL in an average adult.

Trabecula (tra-BEK-yoo-la) Irregular latticework of thin plate of spongy bone. Fibrous cord of connective tissue serving as supporting fiber by forming a septum extending into an organ from its wall or capsule. *Plural is* **trabeculae** (tra-BEK-yoo-lē).

Trabeculae carneae (KAR-nē-ē) Ridges and folds of the myocardium in the ventricles.

Trachea (TRĀ-kē-a) Tubular air passageway extending from the larynx to the fifth thoracic vertebra. Also called the **windpipe.**

Tract A bundle of nerve fibers in the central nervous system.

Transcription (trans-KRIP-shun) The first step in the transfer of genetic information in which a single strand of DNA serves as a template for the formation of an RNA molecule.

Translation (trans-LĀ-shun) The synthesis of a new protein on a ribosome as dictated by the sequence of codons in messenger RNA.

Transport maximum (T_m) The maximum amount of a substance that can be reabsorbed, especially by renal tubules, under any condition.

Transverse colon (trans-VERS KŌ-lon) The portion of the large intestine extending across the abdomen from right colic (hepatic) flexure to the left colic (splenic) flexure.

Transverse fissure (FISH-er) The deep cleft that separates the cerebrum from the cerebellum.

Transverse plane A plane that divides the body or organs into superior and inferior portions. Also called a **horizontal plane.**

Transverse tubules (TOO-byools) (**T tubules**) Small, cylindrical invaginations of the sarcolemma of striated muscle fibers (cells) that conduct muscle action potentials toward the center of the muscle fiber.

Trauma (TRAW-ma) An injury, either a physical wound or psychic disorder, caused by an external agent or force, such as a physical blow or emotional shock; the agent or force that causes the injury.

Tremor (TREM-or) Rhythmic, involuntary, purposeless contraction of opposing muscle groups.

Triad (TRĪ-ad) A complex of three units in a muscle fiber composed of a transverse tubule and the sarcoplasmic reticulum terminal cisterns on both sides of it.

Tricuspid (trī-KUS-pid) **valve** Atrioventricular (AV) valve on the right side of the heart.

Triglyceride (trī-GLI-cer-īd) A lipid formed from one molecule of glycerol and three molecules of fatty acids that may be either solid (fats) or liquid (oils) at room temperature; the body's most highly concentrated source of chemical potential energy. Found mainly within adipocytes. Also called a **neutral fat.**

Trigone (TRĪ-gon) A triangular region at the base of the urinary bladder.

Triiodothyronine (trī-ī-ō-dō-THĪ-rō-nēn) (**T₃**) A hormone produced by the thyroid gland that regulates organic metabolism, growth and development, and the activity of the nervous system.

Trophoblast (TRŌF-ō-blast) The superficial covering of cells of the blastocyst.

Tropic (TRŌ-pik) **hormone** A hormone whose target is another endocrine gland.

Trunk The part of the body to which the upper and lower limbs are attached.

Tubal ligation (lī-GĀ-shun) A sterilization procedure in which the uterine (Fallopian) tubes are tied and cut.

Tubular reabsorption The movement of filtrate from renal tubules back into blood in response to the body's specific needs.

Tubular secretion The movement of substances in blood into renal tubular fluid in response to the body's specific needs.

Tumor suppressor gene A gene coding for a protein that normally inhibits cell division; loss or alteration of a tumor suppressor gene called *p53* is the most common genetic change in a wide variety of cancer cells.

Tunica albuginea (TOO-ni-ka al′-byoo-JIN-ē-a) A dense white fibrous capsule covering a testis or deep to the surface of an ovary.

Tunica externa (eks-TER-na) The superficial coat of an artery or vein, composed mostly of elastic and collagen fibers. Also called the **adventitia.**

Tunica interna (in-TER-na) The deep coat of an artery or vein, consisting of a lining of endothelium, basement membrane, and internal elastic lamina. Also called the **tunica intima** (IN-ti-ma).

Tunica media (MĒ-dē-a) The intermediate coat of an artery or vein, composed of smooth muscle and elastic fibers.

T wave The deflection wave of an electrocardiogram that represents ventricular repolarization.

Twitch contraction Brief contraction of all muscle fibers in a motor unit triggered by a single action potential in its motor neuron.

Tympanic antrum (tim-PAN-ik AN-trum) An air space in the middle ear that leads into the mastoid air cells or sinus.

Type II cutaneous mechanoreceptor A receptor embedded deeply in the dermis and deeper tissues that detects stretching of skin. Also called a **Ruffini corpuscle.**

Umbilical cord The long, ropelike structure containing the umbilical arteries and vein that connect the fetus to the placenta.

Umbilicus (um-BIL-i-kus *or* um-bil-Ī-kus) A small scar on the abdomen that marks the former attachment of the umbilical cord to the fetus. Also called the **navel.**

Upper limb The appendage attached at the shoulder girdle, consisting of the arm, forearm, wrist, hand, and fingers. Also called **upper extremity.**

Up-regulation Phenomenon in which there is an increase in the number of receptors in response to a deficiency of a hormone or neurotransmitter.

Uremia (yoo-RĒ-mē-a) Accumulation of toxic levels of urea and other nitrogenous waste products in the blood, usually resulting from severe kidney malfunction.

Ureter (YOO-re-ter) One of two tubes that connect the kidney with the urinary bladder.

Urethra (yoo-RĒ-thra) The duct from the urinary bladder to the exterior of the body that conveys urine in females and urine and semen in males.

Urinary (YOO-ri-ner-ē) **bladder** A hollow, muscular organ situated in the pelvic cavity posterior to the pubic symphysis; recieves urine via two ureters and stores urine until it is excreted through the urethra.

Urine The fluid produced by the kidneys that contains wastes or excess materials; excreted from the body through the urethra.

Urogenital (yoo′-rō-JEN-i-tal) **triangle** The region of the pelvic floor inferior to the pubic symphysis, bounded by the pubic symphysis and the ischial tuberosities, and containing the external genitalia.

Urology (yoo-ROL-ō-jē) The specialized branch of medicine that deals with the structure, function, and diseases of the male and female urinary systems and the male reproductive system.

Uterine (YOO-ter-in) **tube** Duct that transports ova from the ovary to the uterus. Also called the **Fallopian** (fal-LŌ-pē-an) **tube** or **oviduct.**

Uterosacral ligament (yoo′-ter-ō-SĀ-kral LIG-a-ment) A fibrous band of tissue extending from the cervix of the uterus laterally to the sacrum.

Uterus (YOO-te-rus) The hollow, muscular organ in females that is the site of menstruation, implantation, development of the fetus, and labor. Also called the **womb.**

Utricle (YOO-tri-kul) The larger of the two divisions of the membranous labyrinth located inside the vestibule of the inner ear, containing a receptor organ for static equilibrium.

Uvea (YOO-vē-a) The three structures that together make up the vascular tunic of the eye.

Uvula (YOO-vyoo-la) A soft, fleshy mass, especially the V-shaped pendant part, descending from the soft palate.

Vagina (va-JĪ-na) A muscular, tubular organ that leads from the uterus to the vestibule, situated between the urinary bladder and the rectum of the female.

Valence (VĀ-lens) The combining capacity of an atom; the number of deficit or extra electrons in the outermost electron shell of an atom.

Varicocele (VAR-i-kō-sēl) A twisted vein; especially, the accumulation of blood in the veins of the spermatic cord.

Varicose (VAR-i-kōs) Pertaining to an unnatural swelling, as in the case of a varicose vein.

Vasa recta (VĀ-sa REK-ta) Extensions of the efferent arteriole of a juxtamedullary nephron that run alongside the loop of the nephron (Henle) in the medullary region of the kidney.

Vasa vasorum (va-SŌ-rum) Blood vessels that supply nutrients to the larger arteries and veins.

Vascular (VAS-kyoo-lar) Pertaining to or containing many blood vessels.

Vascular spasm Contraction of the smooth muscle in the wall of a damaged blood vessel to prevent blood loss.

Vascular (venous) sinus A vein with a thin endothelial wall that lacks a tunica media and externa and is supported by surrounding tissue.

Vascular tunic (TOO-nik) The middle layer of the eyeball, composed of the choroid, ciliary body, and iris. Also called the **uvea** (YOO-vē-a).

Vasectomy (va-SEK-tō-mē) A means of sterilization of males in which a portion of each ductus (vas) deferens is removed.

Vasoconstriction (vāz-ō-kon-STRIK-shun) A decrease in the size of the lumen of a blood vessel caused by contraction of the smooth muscle in the wall of the vessel.

Vasodilation (vāz′-ō-DĪ-lā-shun) An increase in the size of the lumen of a blood vessel caused by relaxation of the smooth muscle in the wall of the vessel.

Vasomotion (vāz-ō-MŌ-shun) Intermittent contraction and relaxation of the smooth muscle of the metarterioles and precapillary sphincters that result in an intermittent blood flow.

Vein A blood vessel that conveys blood from tissues back to the heart.

Vena cava (VĒ-na KĀ-va) One of two large veins that open into the right atrium, returning to the heart all of the deoxygenated blood from the systemic circulation except from the coronary circulation.

Ventral (VEN-tral) Pertaining to the anterior or front side of the body; opposite of dorsal.

Ventral body cavity Cavity near the ventral aspect of the body that contains viscera and consists of a superior thoracic cavity and an inferior abdominopelvic cavity.

Ventral ramus (RĀ-mus) The anterior branch of a spinal nerve, containing sensory and motor fibers to the muscles and skin of the anterior surface of the head, neck, trunk, and the limbs.

Ventricle (VEN-tri-kul) A cavity in the brain or an inferior chamber of the heart.

Ventricular fibrillation (ven-TRIK-yoo-lar fib·ri-LĀ-shun) Asynchronous ventricular contractions; unless reversed by defibrillation, results in heart failure.

Venule (VEN-yool) A small vein that collects blood from capillaries and delivers it to a vein.

Vermiform appendix (VER-mi-form a-PEN-diks) A twisted, coiled tube attached to the cecum.

Vermilion (ver-MIL-yon) The area of the mouth where the skin on the outside meets the mucous membrane on the inside.

Vermis (VER-mis) The central constricted area of the cerebellum that separates the two cerebellar hemispheres.

Vertebral (VER-te-bral) **canal** A cavity within the vertebral column formed by the vertebral foramina of all the vertebrae and containing the spinal cord. Also called the **spinal canal.**

Vertebral column The 26 vertebrae of an adult and 33 vertebrae of a child; encloses and protects the spinal cord and serves as a point of attachment for the ribs and back muscles. Also called the **backbone, spine,** or **spinal column.**

Vesicle (VES-i-kul) A small bladder or sac containing liquid.

Vesicouterine (ves′-ik-ō-YOO-ter-in) **pouch** A shallow pouch formed by the reflection of the peritoneum from the anterior surface of the uterus, at the junction of the cervix and the body, to the posterior surface of the urinary bladder.

Vestibular (ves-TIB-yoo-lar) **apparatus** Collective term for the organs of equilibrium, which includes the saccule, utricle, and semicircular ducts.

Vestibular membrane The membrane that separates the cochlear duct from the scala vestibuli.

Vestibule (VES-ti-byool) A small space or cavity at the beginning of a canal, especially the inner ear, larynx, mouth, nose, and vagina.

Villus (VIL-lus) A projection of the intestinal mucosal cells containing connective tissue, blood vessels, and a lymphatic vessel; functions in the absorption of the end products of digestion. *Plural is* **villi** (VIL-ī).

Viscera (VIS-er-a) The organs inside the ventral body cavity. *Singular is* **viscus** (VIS-kus).

Visceral (VIS-er-al) Pertaining to the organs or to the covering of an organ.

Visceral effectors (e-FEK-torz) Cardiac muscle, smooth muscle, and glandular epithelium.

Visceral pleura (PLOO-ra) The deep layer of the serous membrane that covers the lungs.

Vital capacity The sum of inspiratory reserve volume, tidal volume, and expiratory reserve volume; about 4800 mL.

Vital signs Signs necessary to life that include temperature (T), pulse (P), respiratory rate (RR), and blood pressure (BP).

Vitamin An organic molecule necessary in trace amounts that acts as a catalyst in normal metabolic processes in the body.

Vitreous (VIT-rē-us) **body** A soft, jellylike substance that fills the vitreous chamber of the eyeball, lying between the lens and the retina.

Vocal folds Pair of mucous membrane folds below the ventricular folds that function in voice production. Also called **true vocal cords.**

Voltage-gated channel An ion channel in a plasma membrane composed of integral proteins that functions like a gate to permit or restrict the movement of ions across the membrane in response to changes in the voltage.

Vulva (VUL-va) Collective designation for the external genitalia of the female. Also called the **pudendum** (poo-DEN-dum).

Wallerian (wal-LE-rē-an) **degeneration** Degeneration of the portion of the axon and myelin sheath of a neuron distal to the site of injury.

Wandering macrophage (MAK-rō-fāj) Phagocytic cell that develops from a monocyte, leaves the blood, and migrates to infected tissues.

Wave summation (sum-MA-shun) The increased strength of muscle contraction that results when stimuli follow one another in rapid succession.

White matter Aggregations or bundles of myelinated axons located in the brain and spinal cord.

White pulp The portion of the spleen composed of lymphatic tissue, mostly lymphocytes (B cells).

White ramus communicans (RA-mus kō-MYOO-ni-kans) The portion of a preganglionic sympathetic nerve fiber that branches from the anterior ramus of a spinal nerve to enter the nearest sympathetic trunk ganglion.

X-chromosome inactivation The random and permanent inactivation of one X chromosome in each cell of a developing female embryo. Also called **lyonization.**

Xiphoid (ZĪ-foyd) Sword-shaped. The inferior portion of the sternum is the **xiphoid process.**

Yolk sac An extraembryonic membrane that connects with the midgut during early embryonic development but is nonfunctional in humans.

Zona fasciculata (ZŌ-na fa-sik′-yoo-LA-ta) The middle zone of the adrenal cortex consisting of cells arranged in long, straight cords that secrete glucocorticoid hormones, mainly cortisol.

Zona glomerulosa (glo-mer′-yoo-LŌ-sa) The external zone of the adrenal cortex, directly under the connective tissue covering, consisting of cells arranged in arched loops or round balls that secrete mineralocorticoid hormones, mainly aldosterone.

Zona pellucida (pe-LOO-si-da) Clear glycoprotein layer between a secondary oocyte and the surrounding granulosa cells of the corona radiata.

Zona reticularis (ret-ik′-yoo-LAR-is) The inner zone of the adrenal cortex, consisting of cords of branching cells that secrete sex hormones, chiefly androgens.

Zygote (ZĪ-gōt) The single cell resulting from the union of male and female gametes; the fertilized ovum.

CREDITS

Illustration Credits

Chapter 1: 1.1: Tomo Narashima. 1.2: Jared Schneidman Design. 1.3: Jared Schneidman Design. 1.4: Jared Schneidman Design. 1.5: Kevin Somerville. 1.6: Lynn O'Kelley. 1.8: Kevin Somerville. 1.9: Kevin Somerville. 1.10: Kevin Somerville. 1.11: Kevin Somerville. 1.12: Kevin Somerville.

Chapter 2: 2.1: Imagineering. 2.2: Jared Schneidman Design. 2.3: Jared Schneidman Design. 2.4: Imagineering. 2.5: Jared Schneidman Design. 2.6: Imagineering. 2.7: Imagineering. 2.8: Imagineering. 2.9: Imagineering. 2.10: Imagineering. 2.11: Imagineering. 2.12: Jared Schneidman Design. 2.13: Adapted from Karen Timberlake, Chemistry 6e, F8.5, p255 (Menlo Park, CA; Addison Wesley Longman, 1999). ©1999 Addison Wesley Longman, Inc. 2.14: Jared Schneidman Design. 2.15: Jared Schneidman Design. 2.16: Jared Schneidman Design. 2.17: Jared Schneidman Design. 2.18: Imagineering. 2.19: Jared Schneidman Design. 2.20: Jared Schneidman Design. 2.21: Jared Schneidman Design. 2.22: Adapted from Neil Campbell, Jane Reece, and Larry Mitchell, Biology 5e, F5.24, p75 (Menlo Park, CA; Addison Wesley Longman, 1999). ©1999 Addison Wesley Longman, Inc. 2.23: Adapted from Neil Campbell, Jane Reece, and Larry Mitchell, Biology 5e, F5.22, p74 (Menlo Park, CA; Addison Wesley Longman, 1999). ©1999 Addison Wesley Longman, Inc. 2.24: Jared Schneidman Design. 2.25: Imagineering. 2.26: Jared Schneidman Design.

Chapter 3: 3.1: Tomo Narashima. 3.2: Tomo Narashima. 3.3: Imagineering. 3.4: Imagineering. 3.5: Adapted from Bruce Alberts et al., Essential Cell Biology, F12.5, p375 and F12.12, p380 (New York: Garland Publishing Inc., 1998). ©1998 Garland Publishing Inc. 3.7: Adapted from Fundamentals of Anatomy and Physiology 4e, by Martini, Frederic H., ©1998 F3.7, p75. Reprinted by permission of Prentice-Hall, Inc., Upper Saddle River, NJ. 3.8: Jared Schneidman Design. 3.9: Imagineering. 3.10: Imagineering. 3.11: Imagineering. 3.12: Imagineering. 3.13: Imagineering. 3.14: Imagineering. 3.15: Imagineering. 3.16: Imagineering. 3.17: Tomo Narashima, Imagineering. 3.18: Tomo Narashima, Imagineering. 3.19: Tomo Narashima, Imagineering. 3.20: Tomo Narashima, Imagineering. 3.21: Tomo Narashima, Imagineering. 3.22: Tomo Narashima, Imagineering. 3.23: Tomo Narashima. 3.24: Tomo Narashima, Imagineering. 3.25: Tomo Narashima, Imagineering. 3.26: Imagineering. 3.27: Imagineering. 3.28: Imagineering. 3.29: Imagineering. 3.30: Imagineering. 3.31: Imagineering. 3.32: Imagineering. 3.33: Lauren Keswick. 3.34: Hilda Muinos.

Chapter 4: Table 4.1: Kevin Somerville, Nadine Sokol, Imagineering. Table 4.2: Kevin Somerville, Nadine Sokol. Table 4.3: Kevin Somerville, Nadine Sokol, Imagineering, Leonard Dank. Table 4.4: Kevin Somerville, Nadine Sokol. Table 4.5: Nadine Sokol. 4.1: Adapted from Lewis Kleinsmith and Valerie Kish, Principles of Cell and Molecular Biology 2e, F6.44, p237; F6.46, p238; F6.47, p239; F6.50, p241 (New York: HarperCollins, 1995). ©1995 HarperCollins College Publishers. By permission of Addison Wesley Longman. 4.2: Nadine Sokol. 4.3: Adapted from Leslie P. Gartner and James L. Hiatt, Color Textbook of Histology, F5.22, p89 (Philadelphia, PA: Saunders, 1997). ©1997 Saunders College Publishing. 4.4: Imagineering. 4.5: Imagineering.

Chapter 5: 5.1: Kevin Somerville. 5.2: Adapted from Ira Telford and Charles Bridgman, Introduction to Functional Histology 2e, p84, p261, p262 (New York: HarperCollins, 1995). ©1995 HarperCollins College Publishers. By permission of Addison Wesley Longman. 5.3: Imagineering. 5.4: Kevin Somerville. 5.5: Kevin Somerville. 5.6: Imagineering. 5.7: Lauren Keswick.

Chapter 6: 6.1: Leonard Dank. 6.2: Lauren Keswick. 6.3: Leonard Dank. 6.4: Kevin Somerville. 6.5: Kevin Somerville. 6.6: Leonard Dank. 6.8: Kevin Somerville. 6.9: Leonard Dank. 6.10: From Priscilla LeMone and Karen M. Burke, Medical-Surgical Nursing, p1560 (Menlo Park, CA: Benjamin/Cummings, 1996). ©1996 The Benjamin/Cummings Publishing Company. 6.11: Jared Schneidman Design. 6.12: Leonard Dank.

Chapter 7: Table 7.1: Nadine Sokol. Table 7.3: Leonard Dank. 7.1: Leonard Dank. 7.2: Leonard Dank. 7.3: Leonard Dank. 7.4: Leonard Dank. 7.5: Leonard Dank. 7.6: Leonard Dank. 7.7: Leonard Dank. 7.8: Leonard Dank. 7.9: Leonard Dank. 7.10: Leonard Dank. 7.11: Leonard Dank. 7.12: Leonard Dank. 7.13: Leonard Dank. 7.14: Leonard Dank. 7.15: Leonard Dank. 7.16: Leonard Dank. 7.17: Leonard Dank. 7.18: Leonard Dank. 7.19: Leonard Dank. 7.20: Leonard Dank. 7.21: Leonard Dank. 7.22: Leonard Dank. 7.23: Leonard Dank. 7.24: Leonard Dank.

Chapter 8: Table 8.1: Leonard Dank. 8.1: Leonard Dank. 8.2: Leonard Dank. 8.3: Leonard Dank. 8.4: Leonard Dank. 8.5: Leonard Dank. 8.6: Leonard Dank. 8.7: Leonard Dank. 8.8: Leonard Dank. 8.9: Leonard Dank. 8.10: Leonard Dank. 8.11a Leonard Dank. 8.12: Leonard Dank. 8.13: Leonard Dank. 8.14: Leonard Dank. 8.15: Leonard Dank. 8.16: Leonard Dank. 8.17: Leonard Dank.

Chapter 9: 9.1: Leonard Dank. 9.2: Leonard Dank. 9.3: Leonard Dank. 9.4: Leonard Dank. 9.11: Leonard Dank. 9.12: Leonard Dank. 9.13: Leonard Dank. 9.14: Leonard Dank.

Chapter 10: 10.1: Kevin Somerville. 10.3: Kevin Somerville. Adapted from Martini, Frederic H., Fundamentals of Anatomy and Physiology 4e, F10.2, p280 (Upper Saddle River, NJ: Prentice-Hall/Pearson Education, 1998). ©1998 Prentice-Hall. 10.4: Imagineering. 10.6: Imagineering. 10.7: Hilda Muinos. 10.8: Imagineering. 10.9: Imagineering. 10.10: Imagineering. 10.11: Hilda Muinos. 10.12: Imagineering. 10.13: Jared Schneidman Design. 10.14: Jared Schneidman Design. 10.15: Jared Schneidman Design. 10.16: Imagineering. 10.18: Imagineering. 10.19: Beth Willert. 10.20: Beth Willert

Chapter 11: Table 11.1: Kevin Somerville. 11.1: Leonard Dank. 11.2: Leonard Dank. 11.3: Leonard Dank. 11.4: Leonard Dank. 11.5: Leonard Dank. 11.6: Leonard Dank. 11.7: Leonard Dank. 11.8: Leonard Dank. 11.9: Leonard Dank. 11.10: Leonard Dank. 11.11: Leonard Dank. 11.12: Leonard Dank. 11.13: Leonard Dank. 11.14: Leonard Dank. 11.15: Leonard Dank. 11.16: Leonard Dank. 11.17: Leonard Dank. 11.18: Leonard Dank. 11.19: Leonard Dank. 11.20: Leonard Dank. 11.21: Leonard Dank. 11.22: Leonard Dank. 11.23: Leonard Dank.

Chapter 12: Table 12.2: Kevin Somerville. Table 12.3: Jared Schneidman Design. 12.1: Kevin Somerville, Imagineering. 12.2: Imagineering. 12.3: Sharon Ellis. 12.4: Nadine Sokol. 12.5: Nadine Sokol. 12.7: Sharon Ellis. 12.8: Imagineering. 12.9: Imagineering. 12.10: Imagineering. 12.11: Imagineering. 12.12: Imagineering. Adapted from Becker et al., The World of the Cell 3e, F22.18, p732 (Menlo Park, CA: Benjamin/Cummings, 1996) ©1996 The Benjamin/Cummings Publishing Company. 12.13: Jared Schneidman Design. 12.14: From Becker et al., The World of the Cell 3e, F22.28, p741 (Menlo Park, CA: Benjamin/Cummings, 1996) ©1996 The Benjamin/Cummings Publishing Company. 12.15: Jared Schneidman Design. 12.16: Nadine Sokol. 12.17: Nadine Sokol.

Chapter 13: 13.1: Sharon Ellis. 13.2: Sharon Ellis. 13.3: Sharon Ellis. 13.4: Sharon Ellis. 13.5: Sharon Ellis. 13.6: Leonard Dank. 13.7: Leonard Dank. 13.8: Leonard Dank. 13.9: Leonard Dank. 13.10: Sharon Ellis. 13.11: Sharon Ellis. 13.12: Imagineering. 13.13: Precision Graphics. 13.14: Precision Graphics. 13.15: Precision Graphics, Imagineering. 13.16: Imagineering. 13.17: Imagineering.

Chapter 14: Table 14.1: Nadine Sokol. Table 14.2: Hilda Muinos. 14.1: Sharon Ellis. 14.2: Sharon Ellis. 14.3: Sharon Ellis. 14.4: Sharon Ellis, Imagineering. 14.5: Sharon Ellis. 14.6: Sharon Ellis. 14.7: Sharon Ellis. 14.8: Sharon Ellis. 14.9: Sharon Ellis. 14.10: Sharon Ellis. 14.11: Sharon Ellis. 14.13: Sharon Ellis. 14.14: Sharon Ellis. 14.15: Sharon Ellis. 14.16: Hilda Muinos. 14.18: Kevin Somerville. 14.19: Kevin Somerville.

Chapter 15: Table 15.3: Kevin Somerville. Table 15.4: Kevin Somerville. 15.1: Imagineering. 15.2: Kevin Somerville. 15.3: Kevin Somerville. 15.4: Beth Willert. 15.5: Sharon Ellis. 15.6: Lynn O'Kelley. 15.7: Sharon Ellis. 15.8: Jared Schneidman Design. 15.9: Sharon Ellis. 15.10: Sharon Ellis. 15.11: Adapted from Purves et al., Neuroscience 2e, F26.1 and F26.2, p498 (Sunderland, MA: Sinauer Associates, 1997). ©1997 Sinauer Associates.

Chapter 16: Table 16.1: Kevin Somerville. Table 16.2: Kevin Somerville. 16.1: Tomo Narashima. 16.2: Lynn O'Kelley. 16.4: Sharon Ellis. 16.5: Tomo Narashima. 16.8: Lynn O'Kelley. 16.9: Tomo Narashima. 16.10: Jared Schneidman Design. 16.11: Nadine Sokol, Imagineering. 16.12: Lynn O'Kelley. 16.13: Jared Schneidman Design. 16.14: Lynn O'Kelley. 16.15: Adapted from Seeley et al., Anatomy and Physiology 4e, F15.22, p480 (New York: WCB McGraw-Hill, 1998) ©1998 The McGraw-Hill Companies. 16.16: Tomo Narashima. 16.17: Tomo Narashima. 16.18: Tomo Narashima. 16.19: Tomo Narashima. 16.20: Tomo Narashima. 16.21: Tomo Narashima, Sharon Ellis. 16.22: Tomo Narashima, Sharon Ellis.

Chapter 17: 17.1: Jared Schneidman Design. 17.2: Hilda Muinos. 17.3: Kevin Somerville. 17.4: Sharon Ellis. 17.5: Imagineering.

Chapter 18: Table 18.2: Hilda Muinos. Table 18.4: Nadine Sokol. Table 18.5: Nadine Sokol. Table 18.6: Nadine Sokol. Table 18.7: Nadine Sokol. Table 18.8: Nadine Sokol. Table 18.9: Nadine Sokol. Table 18.10: Nadine Sokol. 18.1: Lynn O'Kelley. 18.2: Jared Schneidman Design. 18.3: Jared Schneidman Design. 18.4: Jared Schneidman Design. 18.5: Lynn O'Kelley. 18.6: Imagineering. 18.7: Jared Schneidman Design. 18.8: Lynn O'Kelley. 18.9: Jared Schneidman Design. 18.10: Lynn O'Kelley. 18.11: Jared Schneidman Design. 18.12: Jared Schneidman Design. 18.13: Lynn O'Kelley. 18.14: Jared Schneidman Design. 18.15: Lynn O'Kelley. 18.16: Nadine Sokol. 18.17: Jared Schneidman Design. 18.18: Lynn O'Kelley. 18.19: Jared Schneidman Design. 18.20: Hilda Muinos. 18.21: Nadine Sokol. 18.22: Lynn O'Kelley.

Chapter 19: Table 19.3: Jared Schneidman Design. 19.1: Hilda Muinos. 19.3: Nadine Sokol. 19.4: Nadine Sokol. 19.5: Jared Schneidman Design. 19.6: Jared Schneidman Design. 19.8: Jared Schneidman Design. 19.9: Nadine Sokol. 19.11: Imagineering. 19.12: Jean Jackson. 19.13: Nadine Sokol.

Chapter 20: 20.1: Kevin Somerville. 20.2: Kevin Somerville. 20.3a, c: Hilda Muinos. 20.4: Hilda Muinos, Kevin Somerville. 20.5: Nadine Sokol. 20.6: Kevin Somerville. 20.7: Nadine Sokol, Hilda Muinos. 20.8: Hilda Muinos. 20.9: Kevin Somerville. 20.10: Kevin Somerville. 20.11: Burmar Technical Corp. 20.12: Burmar Technical Corp. 20.13: Hilda Muinos. 20.14: Kevin Somerville. 20.15: Hilda Muinos. 20.17: Hilda Muinos. 20.19: Hilda Muinos.

Chapter 21: Exhibit 21.1: Keith Ciociola. Exhibit 21.2: Keith Ciociola. Exhibit 21.3: Keith Ciociola. Exhibit 21.4: Keith Ciociola. Exhibit 21.5: Keith Ciociola. Exhibit 21.6: Keith Ciociola. Exhibit 21.7: Keith Ciociola. Exhibit 21.8: Keith Ciociola. Exhibit 21.9: Keith Ciociola. Exhibit 21.10: Keith Ciociola. Table 21.2: Imagineering. 21.1: Hilda Muinos. 21.2: Hilda Muinos. 21.3: Nadine Sokol, Imagineering. 21.4: Hilda Muinos. 21.6: Jared Schneidman Design. 21.7: Jared Schneidman Design. 21.8: Jared Schneidman Design. 21.9: Imagineering. 21.10: Imagineering. 21.11: Jared Schneidman Design. 21.12: Jared Schneidman Design. 21.13: Imagineering. 21.14: Jared Schneidman Design. 21.15: Jared Schneidman Design. 21.16: Jared Schneidman Design. 21.17: Hilda Muinos. 21.18: Hilda Muinos. 21.19: Hilda Muinos. 21.20: Kevin Somerville, Hilda Muinos. 21.21: Kevin Somerville. 21.22: Hilda Muinos. 21.23: Hilda Muinos. 21.24: Kevin Somerville. 21.25: Kevin Somerville.

21.26: Kevin Somerville. 21.27: Kevin Somerville. 21.28: Kevin Somerville. 21.29: Kevin Somerville, Nadine Sokol. 21.30: Hilda Muinos. 21.31: Hilda Muinos, Keith Ciociola. 21.32: Hilda Muinos.

Chapter 22: Table 22.3: Jean Jackson. 22.1: Sharon Ellis. 22.2: Sharon Ellis. 22.3: Sharon Ellis. 22.4: Nadine Sokol. 22.5: Steve Oh. 22.6: Sharon Ellis. 22.7: Steve Oh. 22.8: Sharon Ellis. 22.9: Nadine Sokol. 22.10: Nadine Sokol. 22.11: Jared Schneidman Design. 22.12: Jared Schneidman Design. 22.13: Jared Schneidman Design. 22.14: Jared Schneidman Design. 22.15: Jared Schneidman Design. 22.16: Jared Schneidman Design. 22.17: Jared Schneidman Design. 22.18: Jared Schneidman Design. 22.19: Jared Schneidman Design. 22.20: Jared Schneidman Design. 22.21: Nadine Sokol, Imagineering.

Chapter 23: Table 23.2: Hilda Muinos. 23.1: Lynn O'Kelley. 23.2: Kevin Somerville, Lynn O'Kelley. 23.4: Lynn O'Kelley. 23.5: Lynn O'Kelley. 23.6: Steve Oh. 23.8: Lynn O'Kelley. 23.10: Lynn O'Kelley. 23.11: Kevin Somerville. 23.12: Kevin Somerville. 23.13: Jared Schneidman Design. 23.14: Kevin Somerville. 23.15: Nadine Sokol. 23.16: Jared Schneidman Design. 23.17: Jared Schneidman Design. 23.18: Jared Schneidman Design. 23.19: Jared Schneidman Design. 23.20: Jared Schneidman Design. 23.21: Jared Schneidman Design. 23.22: Jared Schneidman Design. 23.23: Jared Schneidman Design. 23.24: Jared Schneidman Design. 23.25: Hilda Muinos. 23.26: Jared Schneidman Design. 23.27: Jared Schneidman Design. 23.28: Jared Schneidman Design. 23.29: Jared Schneidman Design.

Chapter 24: 24.1: Nadine Sokol. 24.2: Steve Oh. 24.3: Nadine Sokol. 24.4: Nadine Sokol. 24.5: Nadine Sokol. 24.6: Nadine Sokol. 24.7: Nadine Sokol. 24.8: Nadine Sokol. 24.10: Nadine Sokol. 24.11: Nadine Sokol. 24.12: Hilda Muinos. 24.13: Jared Schneidman Design. 24.14: Jared Schneidman Design. 24.15: Jared Schneidman Design. 24.16: Nadine Sokol. 24.17: Jared Schneidman Design. 24.18: Nadine Sokol. 24.19: Jared Schneidman Design. 24.20: Jared Schneidman Design. 24.21: Kevin Somerville. 24.22: Hilda Muinos. 24.24: Jared Schneidman Design. 24.25: Jared Schneidman Design. 24.26: Nadine Sokol. 24.27: Hilda Muinos. 24.28: Nadine Sokol.

Chapter 25: 25.1: Imagineering. 25.2: Imagineering. 25.3: Imagineering. 25.4: Imagineering. 25.5: Imagineering. 25.6: Imagineering. 25.7: Imagineering. 25.8: Imagineering. 25.9: Imagineering. 25.10: Imagineering. 25.11: Imagineering. 25.12: Imagineering. 25.13: Imagineering. 25.14: Imagineering. 25.15: Imagineering. 25.16: Imagineering. 25.17: Jared Schneidman Design. 25.18: Imagineering.

Chapter 26: Table 26.1: Nadine Sokol. 26.1: Kevin Somerville. 26.2: Kevin Somerville. 26.3: Steve Oh. 26.4: Nadine Sokol. 26.5: Imagineering. 26.6: Kevin Somerville. 26.7: Nadine Sokol. 26.8: Kevin Somerville. 26.9: Nadine Sokol. 26.10: Jared Schneidman Design. 26.11: Imagineering. 26.12: Jared Schneidman Design. 26.13: Jared Schneidman Design. 26.14: Imagineering. 26.15: Jared Schneidman Design. 26.16: Imagineering. 26.17: Imagineering. 26.18: Jared Schneidman Design. 26.19: Jared Schneidman Design. 26.20: Jared Schneidman Design. 26.21: Jared Schneidman Design. 26.22: Nadine Sokol. 26.23: Nadine Sokol.

Chapter 27: 27.1: Jared Schneidman Design. 27.2: Jared Schneidman Design. 27.3: Jared Schneidman Design. 27.4: Imagineering. 27.5: Imagineering. 27.6: Imagineering. 27.7: Jared Schneidman Design.

Chapter 28: 28.1: Lauren Keswick. 28.2: Imagineering. 28.3: Kevin Somerville. 28.4: Kevin Somerville. 28.5: Kevin Somerville. 28.6: Kevin Somerville. 28.7: Jared Schneidman Design. 28.8: Kevin Somerville. 28.9: Jared Schneidman Design. 28.10: Jared Schneidman Design. 28.11: Kevin Somerville. 28.12: Kevin Somerville. 28.13: Kevin Somerville. 28.14: Kevin Somerville. 28.15: Kevin Somerville. 28.17: Jared Schneidman Design. 28.18: Kevin Somerville. 28.21: Kevin Somerville. 28.22: Kevin Somerville. 28.23: Kevin Somerville. 28.24: Kevin Somerville. 28.25: Jared Schneidman Design. 28.26: Jared Schneidman Design. 28.27: Jared Schneidman Design. 28.28: Jared Schneidman Design. 28.29: Jared Schneidman Design. 28.30: Kevin Somerville. 28.31: Kevin Somerville.

Chapter 29: Table 29.2: Kevin Somerville. 29.1: Nadine Sokol. 29.2: Jared Schneidman Design. 29.3: Kevin Somerville. 29.4: Kevin Somerville. 29.5: Kevin Somerville. 29.06: Kevin Somerville. 29.07: Kevin Somerville. 29.08: Kevin Somerville. 29.09: Kevin Somerville. 29.10: Jared Schneidman Design. 29.11: Kevin Somerville. 29.12: Kevin Somerville. 29.13: Imagineering. 29.14: Jared Schneidman Design. 29.15: Jared Schneidman Design. 29.16: Jared Schneidman Design. 29.17: Jared Schneidman Design. 29.18: Jared Schneidman Design. 29.19: Jared Schneidman Design. 29.20: Jared Schneidman Design.

Photo Credits

Page 20, top left: © Biophoto / Photo Researchers; top right: The Bayer Company; bottom left: Simon Fraser / SPL / Photo Researchers; bottom right: Courtesy of Andrew Joseph Tortora and Damaris Soler; page 21, left: Technicare Corp.; right: © Howard Sochurek/Medical Images, Inc.; page 3: © John Wilson White / Addison Wesley Longman; page 13, top: Stephen A. Kieffer and E. Robert Heitzman, An Atlas of Cross-Sectional Anatomy. Harper & Row, Publishers, Inc. New York, 1979; middle: Lester V. Bergman / The Bergman Collection; bottom:1998 Martin Rotker; page 17: Mark Nielsen / Addison Wesley Longman, Inc.; page 62: Donald Fawcett / SPL / Photo Researchers; page 67: © Scott Foresman / Addsion Wesley Longman, Inc.; page 78: Kent McDonald; page 80: © D. W. Fawcett / Photo Researchers; page 81: Biophoto / Photo Researchers; page 83: Daniel S. Friend / Harvard Medical School; page 84: D. Friend / ©Don Fawcett / Photo Researchers; page 85: CNRI / SPL / Photo Researchers; page 93: CABISCO / Phototake NY; page 108, top: Biophoto / Photo Researchers; bottom: © Ed Reschke; page 109, top: © Ed Reschke; bottom: Douglas Merrill; page 110: Biophoto / Photo Researchers; page 111: Biophoto / Photo Researchers; page 112: © Ed Reschke; page 113: Bruce Iverson; page 114: Lester V. Bergman / The Bergman Collection; page 121, top: ©R. Kessel / VU; bottom: Lester V. Bergman / The Bergman Collection; page 122, top: Biophoto / Photo Researchers; bottom: © Ed Reschke; page 123, top: Biophoto / Photo Researchers; bottom: Andrew J. Kuntzman; page 124, top: © Ed Reschke; bottom: Biophoto / Photo Researchers; page 125: Biophoto / Photo Researchers; page 126, top: Biophoto / Photo Researchers; bottom: © Ed Reschke; page 127: © Ed Reschke; page 131: © Ed Reschke; page 132: © Ed Reschke; page 133: © Ed Reschke; page 143: Lester V. Bergman / The Bergman Collection; page 161: Mark Nielsen / Addison Wesley Longman, Inc.; page 170, top: The Bergman Collection; page 170, bottom:

INDEX

NOTE: Page numbers in **boldface** type indicate a major discussion. A *t* following a page number indicates a table, an *f* following a page number indicates a figure, and an *e* following a page number indicates an exhibit.

Neisseria gonorrhoeae, 1014
Neonatal period, 1022
Neonate, adjustments of at birth, 728–729, 730f, **1041–1042**
Neoplasm, 97. *See also* Cancer
Neostigmine, 283
  for myasthenia gravis, 296
Nephrogenic diabetes insipidus, 603–604
Nephrology, 914
Nephron loop (loop of Henle), 918–919, 920f, 921f
  ascending limb of, 920, 920f, 921, 921f
    histology of, 923, 923t
    $Na^+-K^+-2Cl^-$ symporters in, 934, 935f
    reabsorption in, 934, 935f
  descending limb of, 920, 920f, 921f
    histology of, 923, 923t
    reabsorption in, 934
  histology of, 923, 923t
  reabsorption in, **933–934**, 935f
Nephrons, 916, 917f, **918–924**, 920–921f
  cortical, 920, 920f
  functions of, 924–925, 924f, 942f
  histology of, **921–924**, 922f, 923t
  juxtamedullary, 920–921, 921f
  number of, **924**
  parts of, **918–921**, 920–921f
  tubular reabsorption and secretion in, **929–941**, 930t, 942f
Nephrotic syndrome, 950
Nerve block, 480
Nerve cells. *See* Neurons
Nerve endings
  encapsulated, 486, 487f, 488t
  free, 144, 486, 487f, 488t
Nerve fibers, 381. *See also specific type*
  speed of propagation and, 395–396
Nerve growth factor, 599t
Nerve impulses. *See also* Action potentials
  in homeostasis, 7
Nerve root, spinal nerve, 414
Nerves, 378. *See also specific nerve*
Nervous system, 5t, **378–381**
  aging and, **473**
  in blood pressure regulation, **683–685**, 684f, 685f
  developmental anatomy of, **472**, 473f, 474f
  endocrine system compared with, 566, 566t
  enteric, 380, 380f, 821
  metabolic rate affected by, 899
  neuronal circuits in, **403–404**, 404f
  organization of, 379–381, 380f. *See also specific division*
  structure and functions of, 378–379, 379f
Nervous tissue, 104, **132**, 133t, **378–411**. *See also specific cell type*
  disorders affecting, 406
  gray and white matter, 384–385, 388f
  histology of, **381–386**
  neuroglia, **384**, 386–387t
  neurons, **381–383**, 384f
  regeneration and repair of, **404–406**
Net diffusion, 66, 67f
Net filtration pressure
  in bulk flow, 676, 677f, 678
  in glomerular filtration, **926–927**, 927f
Neural crest, 472, 473f, 603

Neural folds, 472, 473f
Neuralgia, 480
  trigeminal (tic douloureux), 476t
Neural groove, 472, 473f
Neural plate, 472, 473f
Neural tube, 472, 473f
Neural tube defects, 473
Neuritis, **442**
Neuroendocrine system, 566, 566t. *See also* Endocrine system; Nervous system
Neurofibrillary tangles, in Alzheimer disease, 480
Neurofibrils, 381, 382f
Neurogenesis, **405**
Neurogenic shock, 687. *See also* Shock
Neuroglia, 132, 133t, **384**, 386–387t
Neurohypophyseal bud, 602, 602f
Neurohypophysis. *See* Posterior pituitary gland
Neurolemma (sheath of Schwann), 384, 385f
Neurolemmocytes. *See* Schwann cells
Neurology, 378
Neuromuscular disease, 296
Neuromuscular junction, 262, 271f, **280**, 281f
  in muscle contraction, 280, 281f, 282f
  pharmacology of, 280–283
Neuronal circuit, **403–404**, 404f
Neurons, 132, 133t, **381–383**, 383f, 384f
  adrenergic, **557–559**, 557f
  in Alzheimer disease, 479
  cholinergic, **556–557**, 557f
  damage and repair of, **404–406**, 405f
  electrical signals in, **386–396**
  first-order sensory, 486, 495, 496f
  function of, **381–383**, 382f, 401t
  motor (efferent). *See* Motor (efferent) neurons
  parts of, 381–383, 382f
  postganglionic, 550, 551f
  postsynaptic, 397, 398f
  preganglionic, 550, **550–552**, 551f
  presynaptic, 397, 398f
  second-order sensory, 495, 496f
  sensory (afferent). *See* Sensory (afferent) neurons
  structure of, **381–383**, 382f, 401t
    diversity and, 383, 383f
  third-order sensory, 495, 496f
Neuropeptides, **403**, 403t
Neurophysiology, 2t
Neurosecretory cells, 400, 573, 574f, 578–579, 579f
  negative feedback regulation and, 575, 576f
Neurosyphilis (tertiary syphilis), **497**, 1014
Neurotransmitters, **400–403**. *See also specific type*
  in autonomic responses, 550t, **556–559**
  exocytosis in secretion of, 75
  at neuromuscular junction, 280
  neuropeptides, 403, 403t
  photoreceptor release of, **527–528**, 528f
  receptors for, in signal transmission at chemical synapse, 397, 398f
  removal of from synaptic cleft, 399
  in signal transmission at chemical synapse, 397, 398f, 399
  storage of, 381
Neurotransmitter transporters, 399

Neutrons, 27, 28f
Neutrophils, 612f, 619, 619f, 623t, 749
  in connective tissue, 119, 119f
  in differential white cell count, 621t
  emigration of, 751
  formation of, 614f
  functions of, 620–621, 623t, 749
  in inflammation, 751
Nevus (mole), 156
Newborn
  adjustments of at birth, 728–729, 730f, **1041–1042**
  hemolytic disease of, **628–629**, 629f
  premature, **1042**
NFP. *See* Net filtration pressure
NGF. *See* Nerve growth factor
NH_3. *See* Ammonia
NH_4^+. *See* Ammonium ion
Niacin (nicotinamide), 907t
Nicotinamide adenine dinucleotide, 873
Nicotinic receptors, 556–557, 557f, 558t
NIDDM. *See* Non-insulin-dependent diabetes mellitus
Night blindness, 526
Nipple, 999, 1001f
Nissl bodies, 381, 382f
Nitrate ions, absorption of, 856
Nitric oxide, 402–403, 569, 569t
  in blood pressure regulation, 616–617, 686t
  hemoglobin in transportation of, 616–617
  neurotransmitter activity of, 402–403
Nitric oxide synthase, 402
Nitrogen, 26, 27t, 29f
  partial pressure of (P_{N_2}), 796
Nitrogenous bases, DNA, 53–54, 53f
NK cells (natural killer cells), 621, 623t, 749, 752t
NMJ. *See* Neuromuscular junction
NO. *See* Nitric oxide
Nociceptors, 488, 488t, 490f, 491, 494t
Nocturnal enuresis (nocturia), 950–951
Nodes of Ranvier, 384
Noise exposure, hearing loss caused by, 534
Nonciliated columnar epithelium
  pseudostratified, 113t, 115
  simple, 109t, 115
Nondisjunction, chromosomal, 1045
Nonessential amino acids, 891
Non-insulin-dependent diabetes mellitus (NIDDM), 605
Nonkeratinized stratified squamous epithelium, 115
Nonkinetochore microtubules, 93f, 94
Nonpolar covalent bond, 32f, 33
Non-rapid eye movement sleep (NREM sleep), 504–505, 505f
Nonsteroidal anti-inflammatory drugs (NSAIDs), 598
Nonvolatile acids, in acid–base balance, 967
Norepinephrine (noradrenalin), 402, 555, 591, 592t
  in blood pressure regulation, 685, 686t
  in heart rate regulation, 660
  in hypertension, 732–733
  neurons releasing, 557, 557f
  neurotransmitter activity of, 402
  receptors for, 557f, 558

EPONYMS USED IN THIS TEXT

In the life sciences, an eponym is the name of a structure, drug, or disease that is based on the name of a person. For example, you may be more familiar with the Achilles tendon than you are with its anatomically correct term, the calcaneal tendon. Because eponyms remain in frequent use, this listing correlates common eponyms with their anatomical terms.

Eponym	Anatomical Terms	Eponym	Anatomical Terms
Achilles tendon	calcaneal tendon	Kupffer's (KOOP-ferz) cell	stellate reticuloen-dothelial cell
Adam's apple	thyroid cartilage		
ampulla of Vater (VA-ter)	hepatopancreatic ampulla	Langerhans (LANG-er-hanz) cell	nonpigmented granu-lar dendrocyte
Bartholin's (BAR-tō-linz) gland	greater vestibular gland	loop of Henle (HEN-lē)	loop of the nephron
		Luschka's (LUSH-kaz) aperture	lateral aperture
Billroth's (BIL-rōtz) cord	splenic cord		
Bowman's (BŌ-manz) capsule	glomerular capsule	Magendie's (ma-JEN-dēz) aperture	median aperture
Bowman's (BŌ-manz) gland	olfactory gland	Malpighian (mal-PIG-ē-an) corpuscle	splenic nodule
Broca's (BRŌ-kaz) area	motor speech area	Meibomian (mi-BŌ-mē-an) gland	tarsal gland
Brunner's (BRUN-erz) gland	duodenal gland	Meissner (MĪS-ner) corpuscle	corpuscle of touch
bundle of His (HISS)	atrioventricular (AV) bundle	Merkel (MER-kel) disc	tactile disc
		Müllerian (mil-E rē-an) duct	paramesonephric duct
canal of Schlemm (SHLEM)	scleral venous sinus	Nissl (NIS-l) bodies	chromatophilic substances
circle of Willis (WIL-is)	cerebral arterial circle		
Cooper's (KOO-perz) ligament	suspensory ligament of the breast	node of Ranvier (ron-VĒ-ā)	neurofibral node
Cowper's (KOW-perz) gland	bulbourethral gland	organ of Corti (KOR-tē)	spiral organ
crypt of Lieberkühn (LĒ-ber-kyoon)	intestinal gland		
		Pacinian (pa-SIN-ē-an) corpuscle	lamellated corpuscle
duct of Santorini (san´-tō-RĒ-nē)	accessory duct	Peyer's (PĪ-erz) patch	aggregated lymphatic follicle
duct of Wirsung (VĒR-sung)	pancreatic duct		
		plexus of Auerbach (OW-er-bak)	myenteric plexus
end organ of Ruffini (roo-FĒ-nē)	type II cutaneous mechanoreceptor	plexus of Meissner (MĪS-ner)	submucosal plexus
		pouch of Douglas	rectouterine pouch
Eustachian (yoo-STĀ-kē-an) tube	auditory tube	Purkinje (pur-KIN-jē) fiber	conduction myofiber
Fallopian (fal-LŌ-pē-an) tube	uterine tube	Rathke's (rath-KĒZ) pouch	hypophyseal pouch
gland of Littré (LĒ-tra)	urethral gland	Schwann (SCHVON) cell	neurolemmocyte
gland of Zeis (ZĪS)	sebaceous ciliary gland	Sertoli (ser-TŌ-lē) cell	sustentacular cell
Golgi (GOL-jē) tendon organ	tendon organ	Skene's (SKĒNZ) gland	paraurethral gland
Graafian (GRAF-ē-an) follicle	vesicular ovarian follicle	sphincter of Oddi (OD-dē)	sphincter of the hepatopancreatic ampulla
Hassall's (HAS-alz) corpuscle	thymic corpuscle	Stensen's (STEN-senz) duct	parotid duct
Haversian (ha-VER-shun) canal	central canal		
Haversian (ha-VER-shun) system	osteon	Volkmann's (FŌLK-manz) canal	perforating canal
Heimlich (HĪM-lik) maneuver	abdominal thrust maneuver		
		Wernicke's (VER-ni-kēz) area	auditory association area
interstitial cell of Leydig (LĪ-dig)	interstitial endocrinocyte	Wharton's (HWAR-tunz) jelly	mucous connective tissue
islet of Langerhans (LANG-er-hanz)	pancreatic islet	Wormian (WER-mē-an) bone	sutural bone

COMBINING FORMS, WORD ROOTS, PREFIXES, AND SUFFIXES

Many terms used in anatomy and physiology are compound words; that is, they are made up of word roots and one or more prefixes or suffixes. For example, *leukocyte* is formed from the word roots *leuko-*, meaning "white," and *-cyte*, meaning "cell." Thus, a leukocyte is a white blood cell. The following list includes some of the most commonly used combining forms, word roots, prefixes, and suffixes used in the study of anatomy and physiology. Each entry includes a usage example. Learning the meanings of these fundamental word parts will help you to remember the terms that, at first glance, may seem long or complicated.

Combining Forms and Word Roots

Acr-, Acro- extremity; Acromegaly
Aden- gland; Adenoma
Angi- vessel; Angiocardiography
Arthr-, Arthro- joint; Arthritis, arthroscopy
Aut-, Auto- self; Autolysis
Audit- hearing; Auditory canal

Bio- life, living; Biopsy
Blast- germ, bud; Blastomere
Brachi- arm; Brachial plexus
Bucc- cheek; Buccal

Capit- head; Decapitate
Carcin- cancer; Carcinogen
Cardi-, Cardia-, Cardio- heart; Cardiogram
Cephal- head; Hydrocephalus
Cerebro- brain; Cerebrospinal fluid
Chondr-, Chond- cartilage; Chondrocyte
Cost- rib; Intercostal
Crani- skull; Cranial cavity

Derma-, Dermat- skin; Dermatology
Dur- hard; Dura mater

Enter-, Entero- intestine; Gastroenterology
Erythr-, Erythro- red; Erythrocyte

Gastr-, Gastro- stomach; Gastrointestinal tract
Gloss-, Glosso- tongue; Hypoglossal
Glyco- sugar; Glycogen
Gon- seed; Gonad
Gyn- female, woman; Gynecology

Hem-, Hemato- blood; Hematoma
Hepar-, Hepat- liver; Hepatic duct
Hist-, Histo- tissue; Histology
Hyster- uterus; Hysterectomy

Ischi- hip, hipjoint; Ischium

Labi- lip; Labia majora
Lingu- tongue; Sublingual glands
Lip- fat; Lipid
Lumb- lower back, loin; Lumbar

Macul- spot, blotch; Macula
Mamm- breast; Mammary gland
Meningo- membrane; Meningocele
Morpho- form, shape; Morphology
Myelo- marrow, spinal cord; Myeloblast
My-, Myo- muscle; Myocardium

Nephr-, Nephro- kidney; Nephron
Neuro- nerve; Neurotransmitter

Oculo- eye; Oculomotor nerve
Onco- mass, tumor; Oncology
Oo- egg; Oocyte
Os-, Osteo- bone; Osteocyte
Ot-, Oto- ear; Otosclerosis

Pelv- basin; Renal pelvis
Phag-, Phago- to eat; Phagocytosis
Phleb- vein; Phlebitis
Pneumo- lung, air; Pneumothorax
Pod- foot; Podiatry

Procto- anus, rectum; Proctology
Ren- kidney; Renal
Rhin- nose; Rhinology

Scler-, Sclero- hard; Atherosclerosis
Stasis- stand still; Homeostasis
Sten-, Steno- narrow; Stenosis

Therm- heat; Thermoreceptor
Thromb- clot, lump; Thrombus

Vas-, Vaso- vessel, duct; Vasoconstriction

Prefixes

A-, An- without, lack of, deficient; Atresia, anesthesia
Ab- away from, from; Abduction
Ad-, Af- to, toward; Adduction; Afferent
Andro- male, man; Androgen
Ante- before; Antepartum
Anti- against; Anticoagulant

Bas- base, foundation; Basal ganglia
Bi- two, double, twice; Bifurcate
Brady- slow; Bradychardia

Cata- down, lower, under; Catabolism
Circum- around; Circumduction
Co-, Con- with, together; Congenital
Contra- against, opposite; Contralateral
Crypt- hidden, concealed; Cryptorchidism

De- own, from; Deciduous
Di-, Diplo- two; Diploid
Dis- away from; Disarticulate
Dys- painful, difficult; Dysuria

E-, Ec-, Ef-, Ex- out from, out of; Excretion
Ecto-, Exo- outer, outside; Exocrine
Em-, En- in, on; Empyema
End-, Endo- within, inside; Endocardium
Epi- upon; Epidermis
Eu- good, easy, normal; Eupnea
Extra- outside, beyond, in addition to; Extracellular

Gen- originate, produce, form; Genetics

Hemi- half; Hemiplegia
Heter-, Hetero- other, different; Heterogeneous
Homeo-, Homo- unchanging, the same, steady; Homeostasis
Hyper- beyond, extensive; Hypersecretion
Hypo- under, below, deficient; Hypoplasia

In-, Im- in, inside, not; Incontinent, impotent
Inter- among, between; Interphase
Intra- within, inside; Intracellular
Ipsi- same; Ipsilateral

Juxta- near to; Juxtaglomerular

Later- side; Lateral

Macro- large, great; Macrophage
Mal- bad, abnormal; Malignant
Medi-, Meso- middle; Medial, mesoderm
Megalo- great, large; Megalocyte

Meta- after, between; Metacarpus
Micro- small; Microfilament
Mono- one; Monocyte

Neo- new; Neonatal

Oligo- small, deficient; Oliguria
Ortho- straight, normal; Orthopedics

Para- near, beyond, beside; Paranasal sinus
Peri- around; Pericardium
Poly- much, many, too much; Polycythemia
Post- after, beyond; Postnatal
Pre-, Pro- before, in front of; Presynaptic; pronation
Pseud-, Pseudo- false; Pseudopods

Retro- backward, located behind; Retroperitoneal

Semi- half; Semicircular
Sub- under, beneath, below; Submucosa
Super- above, beyond; Superficial
Supra- above, over; Supraventricular
Sym-, Syn- with, together; Symphysis, syndrome

Tachy- rapid; Tachycardia
Trans- across, through, beyond; Transverse colon
Tri- three; Trigone

Suffixes

-ac, -al pertaining to; Cardiac, myocardial
-algia painful condition; Myalgia
-asis, -asia, -esis, -osis condition or state of; Hemostasis

-ectomy cut out; Appendectomy

-fer carry; Efferent ducts

-gen agent that produces or originates; Pathogen
-genic producing; Carcinogenic
-gram record; Electrocardiogram

-ia state, condition; Hypertonia
-ician person associated with; Pediatrician
-itis inflammation; Neuritis

-logy, -ology the study or science of; Neurology
-lysis a loosening; Hemolysis

-malacia softening; Osteomalacia
-megaly enlarged; Cardiomegaly
-mers, -meres parts; Polymers, blastomeres

-oma tumor; Hematoma

-penia deficiency; Thrombocytopenia
-pnea breath; Eupnea
-poiesis making; Hematopoiesis

-trophy relating to nutrition or growth; Atrophy

-uria urine; Polyuria